LENGTH

scales for comparison of metric and U.S. units of measurement

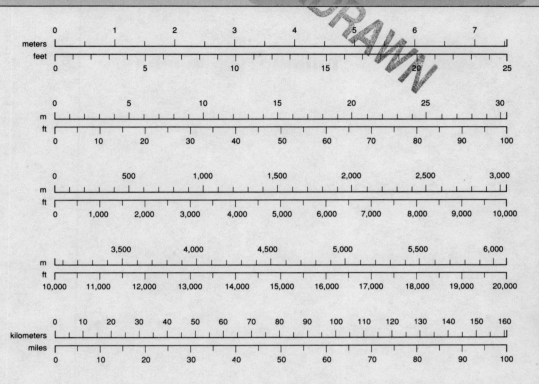

CONVERSION TABLES

	U.S. to Metric		Metric to U.S.	
	to convert	multiply by	to convert	multiply by
LENGTH	in. to mm.	25.4	mm. to in.	0.039
	in. to cm.	2.54	cm. to in.	0.394
	ft. to m.	0.305	m. to ft.	3.281
	yd. to m.	0.914	m. to yd.	1.094
	mi. to km.	1.609	km. to mi.	0.621
AREA	sq. in. to sq. cm.	6.452	sq. cm. to sq. in.	0.155
	sq. ft. to sq. m.	0.093	sq. m. to sq. ft.	10.764
	sq. yd. to sq. m.	0.836	sq. m. to sq. yd.	1.196
	sq. mi. to ha.	258.999	ha. to sq. mi.	0.004
VOLUME	cu. in. to cc.	16.387	cc. to cu. in.	0.061
	cu. ft. to cu. m.	0.028	cu. m. to cu. ft.	35.315
	cu. yd. to cu. m.	0.765	cu. m. to cu. yd.	1.308
CAPACITY (liquid)	fl. oz. to liter	0.03	liter to fl. oz.	33.815
	qt. to liter	0.946	liter to qt.	1.057
	gal. to liter	3.785	liter to gal.	0.264
MASS (weight)	oz. avdp. to g.	28.35	g. to oz. avdp.	0.035
	lb. avdp. to kg.	0.454	kg. to lb. avdp.	2.205
	ton to t.	0.907	t. to ton	1.102
	l. t. to t.	1.016	t. to l. t.	0.984

Abbreviations

avdp. avoirdupois
cc. cubic centimeter(s)
cm. centimeter(s)
cu. cubic
ft. foot, feet
g. gram(s)
gal. gallon(s)
ha. hectare(s)
in. inch(es)
kg. kilogram(s)
lb. pound(s)
l. t. long ton(s)
m. meter(s)
mi. mile(s)
mm. millimeter(s)
oz. ounce(s)
qt. quart(s)
sq. square
t. metric ton(s)
yd. yard(s)

Walker's
Mammals
of the World

Ernest P. Walker, 1891–1969

Walker's
Mammals of the World
Fifth Edition
Volume II

Ronald M. Nowak

The Johns Hopkins University Press
Baltimore and London   1991

Frontispiece photograph by Eugene Maliniak.

The Johns Hopkins University Press
2715 North Charles Street
Baltimore, Maryland 21218-4319
The Johns Hopkins Press Ltd., London

The first edition of this work appeared under the title *Mammals of the World* by Ernest P. Walker (senior author) and six coauthors: Florence Warnick, Sybil E. Hamlet, Kenneth I. Lange, Mary A. Davis, Howard E. Uible, and Patricia F. Wright. The second and third editions were revised by John L. Paradiso. The fourth edition was revised by Roland M. Nowak and John L. Paradiso.

Library of Congress Cataloging-in-Publication Data

Nowak, Ronald M.
    Walker's mammals of the world. — 5th ed. / Ronald M. Nowak.
        p.   cm.
    Rev. ed. of: Walker's mammals of the world. 4th ed. 1983.
    Includes bibliographical references (p.   ) and index.
    ISBN 0-8018-3970-X
    1. Mammals.   2. Mammals—Classification.   I. Walker, Ernest P.
(Ernest Pillsbury), 1891-1969. Walker's mammals of the world.
II. Title.   III. Title: Mammals of the world.
QL703.N69   1991
599—dc20   91-27011

A catalog record for this book is available from the British Library.

To the MAMMALS, GREAT AND SMALL,
who contribute so much to the welfare and happiness
of man, another mammal, but receive so little in return,
except blame, abuse, and extermination.

Ernest P. Walker

Contents

Volume I

Volume II

Walker's
Mammals
of the World

RODENTIA; **Family MURIDAE**

Rats, Mice, Hamsters, Voles, Lemmings, and Gerbils

This family of 267 Recent genera and 1,138 species is by far the largest mammalian family. With the exception of certain arctic islands, parts of the West Indies, New Zealand, many oceanic islands, and Antarctica, its natural range extends worldwide. Introduction through human agency has led to the establishment of several species of murids on many of the islands where the family originally was absent.

There long has been controversy regarding what groups are included in this family and how these groups are interrelated. Many authorities have tended to follow Simpson's (1945) arrangement, shown in the left column of the following table, for those groups assigned in this book to the family Muridae. To the right is the classification proposed by Chaline, Mein, and Petter (1977), which is based in part on recent paleontological evidence.

In this book the actual sequence of the right column generally has been followed. However, all of the families have been made parts of a single family, which in accordance with nomenclatural priority is called the Muridae. Although Simpson (1945) separated the Muridae and the Cricetidae, many

authorities have considered the differences between the two insufficient to warrant familial distinction. The latter position was taken by Hershkovitz (1962) and Hall (1981) and is followed here. Once the Cricetidae are made part of the Muridae, it becomes reasonable to do the same with the other families listed by Chaline, Mein, and Petter (1977). This procedure was followed by Carleton and Musser (1984) and Corbet and Hill (1986), though with some modification of sequence. Also, Carleton and Musser considered the subfamily Hydromyinae to be part of the Murinae, while, in contrast, Lidicker and Brylski (1987) included the Australian genera *Mesembriomys, Conilurus, Leporillus, Zyzomys, Pseudomys, Notomys, Leggadina,* and *Mastacomys* in the Hydromyinae. Corbet and Hill placed the subfamily Petromyscinae within the Dendromurinae. Honacki, Kinman, and Koeppl (1982) continued to recognize the families Cricetidae, Spalacidae, Rhizomyidae, Arvicolidae, and Muridae. There remains considerable support for keeping the Rhizomyidae as a separate family (Lekagul and McNeely 1977; Medway 1978). Chaline, Mein, and Petter's proposal of the subfamilies Tachyoryctinae and Taterillinae and their recognition of three microtine subfamilies were not followed by any of the other authorities cited here. The placement here of the genera *Calomyscus* and *Mystromys* in the subfamily Cricetinae follows convention but is questionable. Carleton and Musser indicated that both genera actually belong in otherwise fossil cricetid subfamilies. Corbet and Hill kept *Calomyscus* in the Cricetinae but put *Mystromys* in the Nesomyinae.

Simpson 1945

Family Cricetidae
  Subfamily Cricetinae
    Tribe Hesperomyini (New World Rats and Mice)
    Tribe Cricetini (Hamsters)
    Tribe Myospalacini (East Asian Mole-rats)
  Subfamily Nesomyinae (Malagasy Rodents)
  Subfamily Lophiomyinae (Maned Rats)
  Subfamily Microtinae
    Tribe Lemmini (Lemmings)
    Tribe Microtini (Voles, Muskrats)
    Tribe Ellobiini (Mole-voles)
  Subfamily Gerbillinae (Gerbils)

Family Spalacidae (Blind Mole-rats)

Family Rhizomyidae (African Mole-rats and Bamboo Rats)

Family Muridae
  Subfamily Murinae (Old World Rats and Mice)
  Subfamily Dendromurinae (Climbing Mice)
  Subfamily Otomyinae (African Swamp Rats)
  Subfamily Phloeomyinae (Cloud Rats)
  Subfamily Rhynchomyinae (Shrew-rats)
  Subfamily Hydromyinae (Water Rats)

Family Platacanthomyidae (Spiny Dormice and Chinese Pygmy Dormice)

Chaline, Mein, and Petter 1977

Family Cricetidae
  Subfamily Hesperomyinae (New World Rats and Mice)
  Subfamily Cricetinae (Hamsters)
  Subfamily Spalacinae (Blind Mole-rats)
  Subfamily Myospalacinae (East Asian Mole-rats)
  Subfamily Lophiomyinae (Maned Rats)
  Subfamily Platacanthomyinae (Spiny Dormice and
    Chinese Pygmy Dormice)

Family Nesomyidae
  Subfamily Nesomyinae (Malagasy Rodents)
  Subfamily Otomyinae (African Swamp Rats)

Family Rhizomyidae
  Subfamily Rhizomyinae (Bamboo Rats)
  Subfamily Tachyoryctinae (African Mole-rats)

Family Gerbillidae
  Subfamily Gerbillinae (Pygmy Gerbils)
  Subfamily Taterillinae (Gerbils)

Family Arvicolidae (Simpson's Microtinae)
  Subfamily Lemminae
  Subfamily Dicrostonychinae
  Subfamily Arvicolinae

Family Dendromuridae
  Subfamily Dendromurinae (Climbing Mice)
  Subfamily Petromyscinae (African Rock and Swamp Mice)

Family Cricetomyidae (African Pouched Rats, part of
    Simpson's Murinae)
Family Muridae
  Subfamily Murinae (Old World Rats and Mice, Including
    Simpson's Phloeomyinae)
  Subfamily Hydromyinae (Water Rats, including Simpson's
    Rhynchomyinae)

The sequence of genera used herein for the subfamily Sigmodontinae (called Hesperomyinae by Chaline, Mein, and Petter and some other authors) reflects in part an attempt to express (1) Carleton's (1980) suggestion that the genus *Nyctomys* (and presumably *Otonyctomys*) is an archaic survivor of the basal stock from which the subfamily arose, that this genus is in turn closely related to *Tylomys* and *Ototylomys*, and that the other genera of the North American branch of the subfamily (*Scotinomys–Peromyscus* in the table beginning below) form systematic groups that seem more distantly related; and (2) Reig's (1980) proposal that this North American branch is a systematic group completely distinct from the remaining, predominantly South American sigmodontines. Reig actually proposed that both groups be recognized as full subfamilies, with the name Sigmodontinae applying to the South American and the name Neotominae applying to the North American. Carleton thought it premature to formalize such a division taxonomically and indicated that the two groups might not be monophyletic assemblages. Slaughter and Ubelaker (1984), however, reported that fossil and ectoparasite data suggest a separate origin for each group.

Otherwise, the sequence of genera within subfamilies used here basically follows that of Simpson (1945) but has been adjusted in accordance with the arrangements of Cabrera (1961), Corbet (1978), Hall (1981), and Misonne (1969). Changes in the sequence also have been necessitated by the numerous systematic modifications called for by the various authorities cited in the generic accounts. This book's overall sequence of murid subfamilies and genera is as follows:

Subfamily Sigmodontinae

Nyctomys	*Abrawayaomys*	*Bibimys*
Otonyctomys	*Scolomys*	*Calomys*
Tylomys	*Nectomys*	*Zygodontomys*
Ototylomys	*Pseudoryzomys*	*Andalgalomys*
Scotinomys	*Rhagomys*	*Eligmodontia*
Baiomys	*Rhipidomys*	*Phyllotis*
Nelsonia	*Thomasomys*	*Galenomys*
Neotoma	*Aepeomys*	*Auliscomys*
Hodomys	*Megaoryzomys*	*Graomys*
Xenomys	*Phaenomys*	*Andinomys*
Ochrotomys	*Chilomys*	*Chinchillula*
Megadontomys	*Wiedomys*	*Irenomys*
Isthmomys	*Akodon*	*Punomys*
Osgoodomys	*Bolomys*	*Euneomys*
Onychomys	*Microxus*	*Neotomys*
Neotomodon	*Podoxymys*	*Reithrodon*
Podomys	*Lenoxus*	*Holochilus*
Habromys	*Juscelinomys*	*Sigmodon*
Reithrodontomys	*Oxymycterus*	*Neusticomys*
Peromyscus	*Blarinomys*	*Chibchanomys*
Oryzomys	*Notomys*	*Anotomys*
Oecomys	*Geoxus*	*Ichthyomys*
Nesoryzomys	*Chelemys*	*Rheomys*
Megalomys	*Scapteromys*	
Neacomys	*Kunsia*	

Subfamily Cricetinae

Calomyscus	*Cricetus*	*Mesocricetus*
Phodopus	*Cricetulus*	*Mystromys*

Subfamily Spalacinae

Spalax	*Nannospalax*

Subfamily Myospalacinae

Myospalax

Subfamily Lophiomyinae

Lophiomys

Subfamily Platacanthomyinae

Platacanthomys	*Typhlomys*

Subfamily Nesomyinae

Macrotarsomys	*Eliurus*	*Hypogeomys*
Nesomys	*Gymnuromys*	*Brachyuromys*
Brachytarsomys		

Subfamily Otomyinae

Parotomys	*Otomys*

Subfamily Rhizomyinae

Rhizomys	*Cannomys*	*Tachyoryctes*

Subfamily Gerbillinae

Gerbillus	*Desmodillus*	*Meriones*
Microdillus	*Desmodilliscus*	*Brachiones*
Gerbillurus	*Pachyuromys*	*Psammomys*
Tatera	*Ammodillus*	*Rhombomys*
Taterillus	*Sekeetamys*	

Subfamily Microtinae

Clethrionomys	*Arvicola*	*Lemmus*
Eothenomys	*Microtus*	*Myopus*
Alticola	*Tyrrhenicola*	*Synaptomys*
Hyperacrius	*Lemmiscus*	*Dicrostonyx*
Dinaromys	*Lagurus*	*Prometheomys*
Arborimus	*Neofiber*	*Ellobius*
Phenacomys	*Ondatra*	

Subfamily Dendromurinae

Dendromus	*Dendroprionomys*	*Deomys*
Megadendromus	*Prionomys*	*Leimacomys*
Malacothrix	*Steatomys*	

Subfamily Petromyscinae

Petromyscus	*Delanymys*

Subfamily Cricetomyinae

Beamys	*Saccostomus*	*Cricetomys*

Subfamily Murinae

Lenothrix	*Melasmothrix*	*Archboldomys*
Pithecheir	*Tateomys*	*Rhynchomys*
Hapalomys	*Palawanomys*	*Chrotomys*
Apodemus	*Eropeplus*	*Tokudaia*
Rhagamys	*Echiothrix*	*Chiruromys*
Chiropodomys	*Bunomys*	*Pogonomys*
Vernaya	*Rattus*	*Mallomys*
Vandeleuria	*Nesoromys*	*Hyomys*
Micromys	*Tryphomys*	*Anisomys*
Haeromys	*Limnomys*	*Uromys*
Anonymomys	*Tarsomys*	*Solomys*
Sundamys	*Taeromys*	*Melomys*
Kadarsanomys	*Paruromys*	*Pogonomelomys*
Abditomys	*Bullimus*	*Xenuromys*
Diplothrix	*Apomys*	*Spelaeomys*
Margaretamys	*Crateromys*	*Coryphomys*
Lenomys	*Batomys*	*Lorentzimys*
Papagomys	*Carpomys*	*Macruromys*
Paulamys	*Phloeomys*	*Mesembriomys*
Komodomys	*Crunomys*	*Conilurus*

Subfamily Murinae (cont.)

Leporillus	Maxomys	Acomys
Zyzomys	Leopoldamys	Uranomys
Pseudomys	Berylmys	Malpaisomys
Leggadina	Mastomys	Hybomys
Notomys	Myomys	Typomys
Mastacomys	Praomys	Stochomys
Golunda	Hylomyscus	Dephomys
Hadromys	Heimyscus	Rhabdomys
Bandicota	Thallomys	Lemniscomys
Nesokia	Thamnomys	Pelomys
Erythronesokia	Grammomys	Aethomys
Millardia	Stenocephalemys	Arvicanthis
Cremnomys	Oenomys	Dasymys
Srilankamys	Lamottemys	Mylomys
Dacnomys	Lophuromys	Mus
Diomys	Zelotomys	Muriculus
Chiromyscus	Colomys	
Niviventer	Malacomys	

Subfamily Hydromyinae

Celaenomys	Hydromys	Parahydromys
Leptomys	Pseudohydromys	Crossomys
Paraleptomys	Microhydromys	Mayermys
Xeromys	Neohydromys	

The various groups of murids differ substantially in form and function. Total length ranges from less than 100 mm in *Baiomys* to 800 mm or more in *Phloeomys* and *Cricetomys*. The 16 murid subfamilies can be briefly described as follows:

Sigmodontinae, mostly small animals, generalized in structure;

Cricetinae, mouselike animals with a thickset body, short tail, and cheek pouches;

Spalacinae, molelike animals with a heavyset body, short legs, and no external openings for the eyes;

Myospalacinae, stout animals resembling pocket gophers in appearance;

Lophiomyinae, arboreal, bushy-tailed rats of Africa;

Platacanthomyinae, rodents resembling dormice in external form;

Nesomyinae, a highly variable group of ratlike animals found in Madagascar;

Otomyinae, African rodents resembling the microtines and having large, high-crowned molars;

Rhizomyinae, burrowing rodents resembling pocket gophers in external appearance;

Gerbillinae, African and Asian animals often resembling jerboas, kangaroo rats, and other genera having a jumping mode of progression;

Microtinae, a Holarctic group with a heavyset body, a tail shorter than the head and body (often shorter than half the head and body length), a bluntly rounded muzzle, and short legs;

Dendromurinae, an African group, sometimes considered arboreal and usually having the hind feet naked;

Petromyscinae, small mice with a long tail, distinguished mainly by characters of the molar teeth;

Cricetomyinae, African rodents with large internal cheek pouches;

Murinae, the generalized Old World rats and mice, differing from the Sigmodontinae mainly in possessing a functional row of tubercles on the inner side of the upper molars; and

Hydromyinae, shrewlike or muskratlike rodents found in parts of the Australian-Oriental region and characterized chiefly by a reduction in the dentition.

The usual murid dental formula is: (i 1/1, c 0/0, pm 0/0, m 3/3) $\times$ 2 = 16. In certain Old World genera, and in *Daptomys* of the New World, the molars are only 2/2, and in *Mayermys*, a member of the Hydromyinae from New Guinea, the molars are 1/1. Murid molars may be rooted or rootless and either laminate, cuspidate, or prismatic.

Murids are found in a wide variety of habitats. Most species are primarily terrestrial, but some are arboreal, fossorial, or semiaquatic. They shelter in tunnels or crevices, under logs or other suitable objects, in hollow trees or logs, or in nests built on the surface or in bushes or trees. They may be diurnal or nocturnal and are usually active throughout the year. Most genera feed on plant material and invertebrates. Some take small vertebrates, including fish. Seeds and other plant matter are sometimes stored for winter use.

Some members of this family are gregarious, even highly social, while others tend to live alone or in pairs. Breeding may occur throughout the year in warm regions. Females often produce several litters annually. Most individuals probably live less than two years in the wild. Nonetheless, the high reproductive potential of certain species, especially the microtines, sometimes results in remarkable numerical increases. Peak populations are often followed by a sudden crash in numbers when food supplies in a given area are exhausted. These fluctuations may show a rather cyclic periodicity of around three or four years.

Several diseases, such as plague, tularemia, leptospirosis, and Rocky Mountain spotted fever, can be transmitted from murid rodents to people. Certain murid species are destructive to crops, young trees, and stored human food supplies.

The geological range of this family is Oligocene to Recent in North America, Europe, and Asia, Pliocene to Recent in South America, Recent in Africa, and Pleistocene to Recent in Madagascar and Australia.

RODENTIA; MURIDAE; Genus NYCTOMYS
Saussure, 1860

Vesper Rat

The single species, *N. sumichrasti*, occurs from southern Mexico to Panama (Hall 1981). Drawing on the morphology of the male reproductive tract, Voss and Linzey (1981) suggested that *Nyctomys* is not closely related to any South American murid.

Head and body length is 107–30 mm and tail length is 85–155 mm. Genoways and Jones (1972) listed weights of 39–61 grams for full-grown animals. The fine fur is short or long and has little or no gloss. The short ears are scantily covered with fine hairs. Hairs on the tail hide the scales at all points. Toward the tip of the tail the hairs are slightly longer and thicker, giving the tail a brushy appearance. The upper parts are buff, cinnamon, or tawny, with some dark hairs, especially along the middle of the back. The sides are usually paler, and the underparts are white or nearly so. The feet are usually white, but in some animals they may be dusky to brownish. The tail is usually brown. There is often a dark ring around each eye or a dark area between the eye and the base of the whiskers.

The eyes are large, and the hind feet are modified for arboreal life. The hallux (big toe) is clawed as in *Rhipidomys*, and in both *Nyctomys* and *Rhipidomys* the pad representing the pollex (thumb) may be prominent. *Nyctomys* differs from *Rhipidomys* in coloration, as well as in having a shorter, more fully haired tail. Female *Nyctomys* have four mammae.

The vesper rat is arboreal, usually being found at a considerable height in trees and rarely descending to the ground.

Vesper rat *(Nyctomys sumichrasti)*, photo from Zoological Society of Philadelphia.

It builds outside nests of twigs and fibers, like those of the red squirrel *(Tamiasciurus)*. It seems to be mainly, if not entirely, nocturnal. According to Genoways and Jones (1972), the diet of *Nyctomys* consists primarily of seeds, fruits, and other vegetable matter. It has been observed eating wild figs and avocados. Individuals occasionally arrive in the United States in bunches of bananas unloaded from ships.

Birkenholz and Wirtz (1965) maintained a captive colony of *Nyctomys*. They reported the most common vocalization to be a faint, rapid, high-pitched musical chirp or trill. The average litter size in this colony was two young. The gestation period was 30–38 days. One female gave birth to five litters over a 7-month period. Data summarized by Genoways and Jones (1972) indicated that in the wild the vesper mouse breeds during both the wet and dry seasons and probably is capable of reproduction throughout the year. Litter size evidently varies from one to four, with two probably being the mode. A captive specimen lived for 5 years and 2 months (Jones 1982).

RODENTIA; MURIDAE; **Genus OTONYCTOMYS**
Anthony, 1932

Yucatan Vesper Rat

The single species, *O. hatti*, occurs over much of the Yucatan Peninsula and in northern parts of Guatemala and Belize (Hall 1981).

Head and body length is about 104–16 mm and tail length is about 100–127 mm. Jones, Genoways, and Lawlor (1974) gave the weight of two individuals as 29.5 and 32.3 grams.

The upper parts are nearly uniformly russet to hazel, being darkest on the back. The sides are tawny to ochraceous tawny. The underparts are white with a creamy wash. The upper sides of the feet are whitish with a buffy wash, or with tawny tones, and the tail is uniformly brown and heavily haired. The external ear is similar to that of *Nyctomys*.

This rodent is much redder than *Nyctomys sumichrasti*, its pronounced russet coloration making it the showiest among the neotropical climbing rats. It resembles *Nyctomys* in structural features, but its auditory bullae are much larger, its cheek teeth are smaller, and its tarsus is narrower. Females have four mammae.

Jones, Genoways, and Lawlor (1974) collected one specimen on a rafter under the roof of a house and another in a small tree. In both instances the traps were baited with bananas. Robert T. Hatt trapped two individuals in a thatched hut, on the shelf where the rafters meet the top of the wall. The habits of *Otonyctomys* are probably much like those of *Nyctomys* and *Rhipidomys*. It is possible that *Otonyctomys* is fairly common and only seems to be rare because of its arboreal habits or because it may have a specialized diet.

RODENTIA; MURIDAE; **Genus TYLOMYS**
Peters, 1866

Climbing Rats

There are seven species (Cabrera 1961; Hall 1981):

T. bullaris, known only from the type locality in Chiapas (extreme southern Mexico);

Yucatan vesper rat *(Otonyctomys hatti)*, photo by Robert T. Hatt.

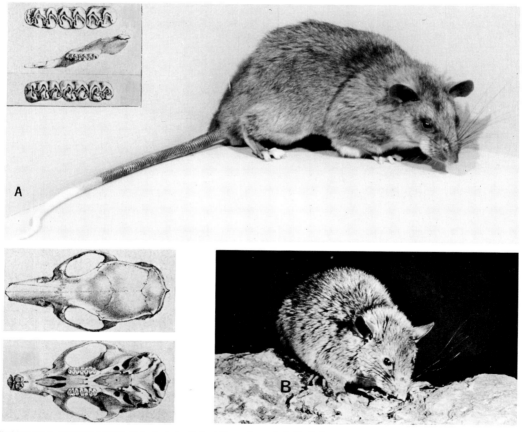

Climbing rat: A. *Tylomys nudicaudus,* photo by Robert Brown through Rollin H. Baker, Michigan State University Museum; B. *T. nudicaudus,* photo by O. J. Reichman; Skull: *T. nudicaudus,* photos from *Monatsberichte Königlich. Preuss. Akad. Wissenschaften Berlin.*

T. mirae, Colombia, northern Ecuador;
T. fulviventer, known only from the type locality in
    extreme eastern Panama;
T. nudicaudus, southern Mexico to Nicaragua;
T. panamensis, extreme eastern Panama;
T. tumbalensis, known only from the type locality in
    Chiapas (extreme southern Mexico);
T. watsoni, Costa Rica, Panama.

Cabrera (1961) suggested that *T. fulviventer* might be a sub-species of *T. mirae*.

Head and body length is 170–255 mm and tail length is usually 200–250 mm. Helm (1975) listed the average adult weight of *T. nudicaudus* as 280 grams. The upper parts are usually some shade of gray or brown; one species is cinnamon buff with an admixture of brown, and in another the middle of the back is blackish. The underparts are white or fulvous. In the species *T. fulviventer* a russet median line extends from the pectoral region to the base of the tail. The feet are brown or russet and the toes are white or brown. The tail is dark brown to blackish, the proximal part being brown, glossy black, yellowish, or whitish. In some species the fur is close, thick, and glossy and the whiskers are long, smooth, and black. The tail is slender and scantily haired. The ears are large and naked.

These rodents superficially resemble large specimens of *Rattus rattus*. The hind foot is suited to arboreal life, and all the feet are somewhat broad and short.

Heavily forested areas, often around rocky ledges, are preferred habitats. The known elevational range is 175–1,375 meters. These animals have been taken among rocks, along logs, on the banks of streams, and at the base of palm trees. Some have been shot as they climbed in palm fronds 9 meters above the ground. A captive specimen consumed bird seed, mouse breeders chow, and occasionally lettuce and apples (Baker and Petersen 1965). The gestation period of *T. nudicaudus* averages about 40 (36–51) days, and mean litter size is 2.33 young (Helm 1975). A captive individual of this species lived for 5 years and 5 months (Jones 1982).

RODENTIA; MURIDAE; **Genus OTOTYLOMYS**
Merriam, 1901

Big-eared Climbing Rat

The single species, *O. phyllotis*, is known to occur from the Yucatan Peninsula to Costa Rica. There also is a record from Guerrero, southwestern Mexico, based on remains found in an owl pellet (Hall 1981).

Head and body length is 95–190 mm and tail length is 100–190 mm. Two individuals each weighed about 120 grams, though Helm (1975) listed average adult weight as 63 grams. The fur is moderately long, soft, and full. The upper parts are some shade of brown or gray, often with black hairs intermingling, and the underparts are white or grayish. The fingers and toes are usually whitish. The insides of the hands and feet are whitish, but the central and outer parts are almost as dark as the outer sides of the arms and legs. The color of the tail ranges from dark, dull, grayish brown to blackish above and is slightly lighter below, but it may be dull yellowish in some species. The tail is naked except for a few scattered hairs, and the large ears are sparsely haired. *Ototylomys* closely resembles *Tylomys* but is smaller and has a somewhat shorter tail.

This rat occurs in a wide variety of habitats but prefers tropical forest with rocks or rocky ledges. Its altitudinal range is sea level to about 2,000 meters. It is nocturnal and mainly arboreal, foraging on creepers or branches, but also is sometimes found on the ground or among rocks (Jones, Genoways, and Lawlor 1974; Lawlor 1969). The diet apparently consists mainly of fruits and leaves (Lawlor 1982).

Available information indicates that breeding occurs throughout the year in the wild. Litter size is one to four young, averaging about two in some areas and three in others (Lawlor 1969). A report of a pregnancy lasting 174 days suggests delayed implantation (Lawlor 1982); however, in a study of captives, Helm (1975) determined gestation to average 52.8 days and to range from 49 to 69 days. A postpartum estrus was observed for all littering females. The young weighed about 10 grams at birth, and most opened their eyes at 6 days. The earliest recorded mating for a female occurred at 29 days.

RODENTIA; MURIDAE; **Genus SCOTINOMYS**
Thomas, 1913

Brown Mice

There are two species (Hall 1981):

S. xerampelinus, Costa Rica, western Panama;
S. teguina, southern Mexico to western Panama.

Head and body length is 80–85 mm and tail length is 48–70 mm. Average weight is 12 grams for *S. teguina* and 15 grams for *S. xerampelinus* (Hill and Hooper 1971). The sparsely haired tail is usually equal to about 70 percent of the head and body length. The pelage of *S. teguina* is short and somewhat harsh, while that of *S. xerampelinus* is long and soft. Colora-

Big-eared climbing rat *(Ototylomys phyllotis)*, photo by Lloyd G. Ingles.

Brown mouse *(Scotinomys xerampelinus)*, photo by Rexford Lord.

tion of the upper parts is dark yellowish brown, reddish brown, or dull dark brown, nearly black. The underparts are reddish brown, buffy brown, or grayish. The ears, hands, feet, and tail are usually blackish. Females may have either two or three pairs of mammae (Hooper 1972).

These mice seem to prefer rough, rocky terrain, but they also inhabit upland savannahs and clearings in forested areas. Their altitudinal range is 1,000–3,300 meters. They utilize runways through low-growing vegetation, are mainly or exclusively diurnal, and are insectivorous (Hooper 1972).

Hooper and Carleton (1976) provided the following information on social life and reproduction. Both species have a repertoire of vocalizations, the most elaborate of which is the "song," a sequence of pulses that in *S. teguina* lasts 7–10 seconds. Males call much more than females. Nests are built mainly or entirely by the female in *S. xerampelinus* but by both sexes in *S. teguina*. A female with young will tolerate the presence of her mate, and the male may stay with the young while the mother is out. In the laboratory, births occur throughout the year, and many females regularly produce litters at 1-month intervals. In the wild, pregnant females have been found in January, February, April, May, June, July, and August, but adequate collections are lacking for the remainder of the year. Gestation averages 30.8 days in *S. teguina* and 33.1 days in *S. xerampelinus*; litter size is one to five, usually two to three, in *S. teguina* and two to four in *S. xerampelinus*; the young of *S. teguina* are weaned at 18–21 days, those of *S. xerampelinus* at 21–24 days; females attain sexual maturity at about 33.8 days in *S. teguina* and about 51.8 days in *S. xerampelinus*, and males of both species are sexually mature at 6–8 weeks.

RODENTIA; MURIDAE; **Genus BAIOMYS**
True, 1894

Pygmy Mice

There are two living species (Cleveland 1986; Hall 1981; Stangl and Dalquest 1986; Stangl, Koop, and Hood 1983):

B. musculus, central Mexico to Nicaragua;
B. taylori, southwestern Oklahoma, Texas and southeastern Arizona to central Mexico.

A third named species, *B. hummelincki,* is actually referable to *Calomys* (Handley 1976; Hershkovitz 1962).

These are the smallest rodents in North America. Head and body length is 50–81 mm, tail length is 35–55 mm, and

Pygmy mouse *(Baiomys taylori),* photos by Ernest P. Walker.

weight is 7–8 grams. The upper parts are blackish brown to light reddish brown, and the underparts are dark gray to white or buff. This genus differs from *Peromyscus* in its smaller size, in having relatively smaller and more rounded ears, and in the shape of the coronoid process.

Pygmy mice live mainly in dry country, wherever they can find cover in dense grass, shrubs, or cultivated fields. *B. taylori* is crepuscular. It is not known to hibernate, but laboratory experiments show that it readily enters torpor at ambient temperatures near 20° C (Eshelman and Cameron 1987). It makes a network of small, narrow runways beneath a heavy canopy of matted grasses and also uses the runways of other rodents. Tiny piles of green feces are found at the intersections. According to Packard and Montgomery (1978), *B. musculus* apparently lives in underground burrows when it occurs in grasslands but under rocks when it occurs in arid rocky areas. This species shows a tendency toward diurnal and crepuscular activity. Although pygmy mice may eat some seeds and insects, the bulk of their food consists of green vegetation. Pitts (1978) found captive *B. taylori* to readily kill and eat small snakes. Schmidly (1983) stated that average home range for *B. taylori* is only 0.6 ha. in males and 0.4 ha. in females and cited population densities ranging from 1/ha. to 15/ha., the greatest being in unmowed highway rights-of-way during winter. Eshelman and Cameron (1987) listed densities of up to 84/ha. and home range estimates of 45–729 sq meters. Pygmy mice are easily kept in captivity and live together peacefully.

Breeding apparently continues throughout the year, but in *B. musculus*, at least, a decline in reproductive activity seems to occur in winter and spring (Packard and Montgomery 1978). *B. taylori* has breeding peaks in the late fall and early spring (Eshelman and Cameron 1987). Studies of captive *B. taylori* by Hudson (1974) and Quadagno et al. (1970) indicate that the estrous cycle averages 4.9 days, gestation lasts 20–25 days, litter size averages about 2.5 young and ranges from 1 to 5, newborn average 1.2 grams in weight, females may conceive young when only 28 days old, and males reach sexual maturity at about 70–80 days. One captive female had nine litters in 202 days. In the wild the young are born in a surface nest of dry grass located in a depression in the ground or under a log. Both parents share in caring for the young. The offspring seek their own food at about 18–22 days of age. According to Jones (1982), a captive *B. taylori* lived for 3 years.

In the 1950s *B. taylori* was known only as far north as southern and southeastern Texas. It subsequently extended its range across most of that state and into southwestern Oklahoma. This extension may be associated with the increasing ground cover that results from human brush eradication and the grassy rights-of-way afforded by highway construction (Austin and Kitchens 1986).

RODENTIA; MURIDAE; **Genus NELSONIA**
Merriam, 1897

Nelson's Wood Rat

The single species, *N. neotomodon*, inhabits the mountains of west-central Mexico (Hall 1981). On the basis of karyotype and other features, *Nelsonia* appears to be among the most primitive of murids (Engstrom and Bickham 1983).

Head and body length is 120–30 mm and tail length is 105–30 mm. The soft pelage of the upper parts is grayish brown, sometimes with a buffy wash; the sides are usually pale reddish; the underparts are whitish; and the feet are white to dusky. The tail is tufted terminally. Externally this

rodent looks like a large, hairy-tailed species of *Peromyscus*. In cranial and dental characters, however, *Nelsonia* most resembles *Neotoma*.

Nelson's wood rat frequents rocky areas in pine, Douglas fir, or true fir from 1,810 to 3,010 meters in elevation. Crevices in cliffs and in rimrock are used for shelters, these places apparently being cool and moist. *Nelsonia* is mainly nocturnal. It seems to feed mostly on the needles of Douglas fir and juniper. Females collected in March were not pregnant.

RODENTIA; MURIDAE; **Genus NEOTOMA**
Say and Ord, 1825

Wood Rats, Pack Rats, or Trade Rats

There are 3 subgenera and 19 species (Hall 1981):

subgenus *Neotoma* Say and Ord, 1825

N. floridana, southern South Dakota and southern New York to Texas and Florida;
N. micropus, New Mexico and southern Kansas to northeastern Mexico;
N. albigula, southwestern United States to central Mexico;
N. nelsoni, known only from the type locality in Veracruz (east-central Mexico);
N. palatina, Jalisco (west-central Mexico);
N. varia, Turner Island (Gulf of California);
N. lepida, central California and southern Idaho to southern Baja California and Arizona;
N. bryanti, Cerros Island off west-central Baja California;
N. anthonyi, Todos Santos Island off northwestern Baja California;
N. martinensis, San Martin Island off northwestern Baja California;
N. bunkeri, Coronados Island off southeastern Baja California;
N. stephensi, extreme southeastern Utah, central and eastern Arizona, western New Mexico;
N. goldmani, north-central Mexico;
N. mexicana, Colorado and southeastern Utah to western Honduras;
N. chrysomelas, central Honduras to western Nicaragua;
N. angustapalata, southern Tamaulipas (northeastern Mexico);
N. fuscipes, western Oregon to northern Baja California;

subgenus *Teonoma* Gray, 1843

N. cinerea, northwestern Canada to North Dakota and northern Arizona;

subgenus *Teanopus* Merriam, 1903

N. phenax, Sonora and Sinaloa (northwestern Mexico).

Until recently, *Hodomys*, with the single species *H. alleni*, was usually regarded as a fourth subgenus of *Neotoma*. Most authorities now recognize *Hodomys* as a distinct genus (see below). On the basis of chromosomal studies, Koop, Baker, and Mascarello (1985) suggested that *N. cinerea* is closely related to *N. fuscipes* and that both should be placed in the same subgenus, either *Neotoma* or *Teonoma*. These same authorities recognized *N. devia*, of western Arizona, as a species distinct from *N. lepida*. Such an arrangement was

Nelson's wood rat *(Nelsonia neotomodon)*, photo by Howard E. Uible of skin in U.S. National Museum of Natural History.

followed by Jones et al. (1986) but not by Corbet and Hill (1986) or Honacki, Kinman, and Koeppl (1982).

Head and body length is 150–230 mm, tail length is 75–240 mm, and weight is 199–450 grams. The fur is soft to somewhat harsh. The delicate coloration of the upper parts ranges from pale, buffy gray to darker gray and cinnamon buff. The underparts are pure white, pale grayish, or buffy. Coloration is bright rufous in *N. chrysomelas.* In some species, the so-called round-tailed wood rats, the tail is very scantily haired and resembles that of *Rattus.* In the bushy-tailed wood rat *(N. cinerea),* however, the tail is fairly well covered and somewhat resembles a short-haired squirrel tail.

Wood rats are found in a great variety of habitats, from low, hot, dry deserts, to humid jungles, to rocky slopes above the timber line. They are generally nocturnal and active throughout the year. Some species build elaborate dens or houses composed of twigs, stems, foliage, bones, rocks, or whatever material is available. These houses often rest on the ground or are placed against rocks or the base of a tree. They may be used and added to by many generations of wood rats

Top, bushy-tailed wood rat *(Neotoma cinerea),* photo by John Wanderer. Bottom, Allegheny wood rat *(N. floridana),* photo by Ernest P. Walker.

and may eventually grow to over 2 meters in both diameter and height. Bonaccorso and Brown (1972) found that a captive *N. lepida* could construct a complete house, 40 cm high and 100 cm wide, within 7–10 nights. Those species inhabiting areas where spiny cactus is available seek this material and build their houses almost entirely over this plant. The dens are so placed that it is almost impossible for an enemy to approach without being impaled, and yet the wood rats dash in and out of their homes unharmed. Several fairly well defined trails lead to these dens, but the paths are often obstructed by cactus or other material. Lowery (1974) wrote that most dens of *N. floridana* have more than one entrance leading to the actual nest. The nest lies in a cavity that measures 127–330 mm in diameter and is lined with finely shredded bark, leaves, and grass. The nest contains one alcove for sleeping and another for food storage.

Some species, at least in certain areas, do not build large houses, but utilize crevices among rocky outcrops or high on cliffs; the animals close the openings to the crevices with sticks or other material. There actually appears to be much variation in shelters, even within the same species, depending on habitat conditions and availability of materials. In the Mojave Desert of California, for example, Cameron and Rainey (1972) reported the following situations for *N. lepida*: (1) where the wood rats lived on rocky outcrops, they simply blocked crevices with debris, behind which the nest and food store were in the same chamber; (2) in desert washes, elaborate stick houses were made, averaging 52 cm high and 102 cm wide, each with five to seven entrances and three or four chambers, one for the nest and the others for food storage; and (3) in areas of cactus habitat, houses were built entirely of cholla cactus, used creosote bushes for support, averaged 55 cm high and 160 cm wide, and contained three or four chambers but no food stores or special nest structure.

Wood rats are expert climbers but do not ordinarily go far up in trees. Both *N. phenax* and *N. fuscipes,* however, regularly nest and move about in trees. Several other species also sometimes build their houses in trees. Dens at heights of about 7.5 meters have been reported for *N. lepida* (Stones and Hayward 1968) and *N. floridana* (Lowery 1974). Some populations that live near lakes and streams build their nests over water in mangrove trees, often 46–69 meters from the shore. The animals are not aquatic, however, and regularly forage on land.

Wood rats pick up material for their nests while foraging and carry it to the homesite. If they find a more attractive substance, they will drop the material they are carrying and take the new item. They often select pieces of silverware or other shining objects from camps and leave the material they had been carrying. This trait has given them the name "trade rat" or "pack rat."

The diet of wood rats consists almost entirely of such plant tissues as roots, stems, and leaves; seeds; and some invertebrates. They do not drink much water, but during dry seasons they make heavy inroads on the fleshy stems of cacti and other plants that are well filled with water.

Population density of *N. lepida* was listed as 1.2–6.1 per ha. (French, Stoddart, and Bobek 1975), and average home range was determined as 371 sq meters for males and 433 sq meters for females (Bleich and Schwartz 1975). In a radio-tracking study in California, Cranford (1977) found population density of *N. fuscipes* to vary from 14/ha. in winter to 20/ha. in late summer. There were on average 32 houses per ha. Home range in this study area varied from 590 to 4,049 sq meters and averaged 2,289 sq meters for adult males, 1,924 for adult females, and 1,719 for juveniles. Home ranges overlapped, and they expanded during the reproductive season. *Neotoma* is generally solitary. Dens are seldom occupied by more than one adult (Lowery 1974). In a group of captive *N.*

floridana a despotic social organization developed in which one individual killed or wounded all the others (Kinsey 1976). In a study of *N. cinerea,* both in the laboratory and field, Escherich (1981) found males to fight vigorously for harems of one to three females; home ranges overlapped, but most individuals occupied separate dens.

In the southern parts of the range of *Neotoma,* breeding apparently can take place at any time of the year, but the genus is not especially prolific. Lowery (1974) stated that in Louisiana only two or possibly three litters of *N. floridana* are produced annually, usual litter size is two to four young, and sexual maturity is not attained until 7 or 8 months. Overall range of litter size in this species is one to seven (Goertz 1970). Southern populations of *N. micropus* breed throughout the year, while females in the north apparently begin breeding in December or January, produce at least two and probably three litters before July, and sometimes have one or two additional litters from August to October; litter size is one to four (Birney 1973). The gestation period of *Neotoma* varies from about 30 to 40 days. According to Wiley (1980), the young of *N. floridana* weigh 11.8–14.1 grams a few hours after being born, their eyes open in 15–21 days, they are weaned in about 4 weeks, and they reach adult weight after 8 months. A captive specimen of *N. albigula* lived for 7 years and 8 months (Jones 1982).

Wood rats are neat and sanitary animals; they make pleasing pets if their extreme timidity can be overcome. In some parts of their range they live so close to farms that they are considered pests, but in general they are of little economic importance. Some populations of *N. floridana* in the eastern United States may be in jeopardy because of environmental disruption. The subspecies *N. floridana smalli,* found naturally only on Key Largo off the southeast coast of Florida, was considered endangered by Layne (1978). Its maximum population density is only about 1 per 0.8 ha., and because of human destruction of the mature tropical forest on which it depends, only about 475 ha. of suitable habitat remain. The total population on Key Largo was estimated to be 654 animals by Barbour and Humphrey (1982). *N. f. smalli* now is classified as endangered by the USDI. The IUCN classifies *N. fuscipes riparia,* found in the central San Joaquin Valley in California, as vulnerable. According to D. F. Williams (1986), the riparian habitat of this subspecies is being lost through brush removal and stream channelization.

RODENTIA; MURIDAE; Genus HODOMYS
Merriam, 1894

Allen's Wood Rat

The single species, *H. alleni,* is found in western and central Mexico, from southern Sinaloa to northern Oaxaca (Hall 1981). Until recently, most authorities, including Hall, treated *Hodomys* as a subgenus of *Neotoma.* Genoways and Birney (1974) suggested that generic rank might be warranted but did not choose that alternative, pending further investigations of the relationships of *Hodomys, Xenomys,* and *Neotoma.* Carleton's (1980) morphological studies indicated that *Hodomys* should be recognized as a separate genus, and this procedure has been followed by most subsequent authors (Carleton and Musser 1984; Corbet and Hill 1986; Eisenberg 1984; Honacki, Kinman, and Koeppl 1982).

The information for the remainder of this account was taken from Genoways and Birney (1974). Head and body length averages about 235 mm, tail length 198 mm, and weight 368 grams. The upper parts are rich reddish brown to dusky brown, the underparts are plumbeous washed with white, and the sparsely haired tail is dusky above and below in some populations, becoming whitish below in others. *Hodomys* differs from *Neotoma* in that the enamel pattern of the third lower molar is **S**-shaped, rather than having two transverse loops, and in having a greater distribution of glandular epithelium in the stomach. The glans penis of *Hodomys* resembles that of *Xenomys* in being approximately twice as long as its greatest width and having a surface indented by shallow mid-dorsal and midventral troughs.

Allen's wood rat is generally found where there is dense vegetative cover, such as mesquite, thorny scrub, and tropical deciduous forest. It shelters and nests among rocks, at the base of cliffs, and in burrows or debris beneath trees. It makes small collections of sticks, shells, and other materials but apparently does not construct the large nests characteristic of *Neotoma*. Individuals have been heard shortly after dark, calling with a long series of clear staccato "chooks." A pregnant female with a single embryo was collected in Nayarit on 3 February, a lactating female accompanied by two young was reported in Sinaloa on 3 September, and juveniles were taken in Sinaloa in February and August and in Jalisco in September.

RODENTIA; MURIDAE; **Genus XENOMYS**
Merriam, 1892

Magdalena Rat

The single species, *X. nelsoni,* has been recorded from three localities in west-central Mexico: Chamela Bay, Jalisco; Pueblo Juarez, Colima, the type locality; and Armeria, Colima (Hall 1981). There are three specimens in the U.S. National Museum of Natural History, one at the University of Michigan, and nine in the Los Angeles County Museum. Genoways and Birney (1974) suggested that both *Xenomys* and *Hodomys* might be only subgenera of *Neotoma,* but the studies of Carleton (1980) indicate that all warrant full generic rank.

Head and body length is 157–65 mm and tail length is 143–70 mm. The upper parts are deep tawny red or fulvous, and the underparts are creamy white. There is a white spot over each eye and a white spot below each ear. The upper lips and cheeks are white more than halfway to the eyes. The tail is well haired and the fur is soft. The ears, about half as long as the head, are nearly naked.

This attractive rodent resembles a small wood rat *(Neotoma)* in general external appearance and form, but the coloration is different. The skull and teeth resemble those of *Neotoma* more than they do those of any other genus. The auditory bullae of *Xenomys* are inflated, and the molar teeth are rooted.

Specimens have been collected from about sea level to elevations of approximately 450 meters. *Xenomys* is nocturnal and arboreal. It inhabits thorn forest and tropical deciduous forest in a restricted area of coastal Mexico. It is quite agile in the trees, moving about freely and without apparent hesitation. The collector of the type series stated that *Xenomys* dwells in hollow trees and that one individual was caught in a low, dense wood near a river. Pregnant females have been taken in August.

RODENTIA; MURIDAE; **Genus OCHROTOMYS**
Osgood, 1909

Golden Mouse

The single species, *O. nuttalli,* occurs in the south-central and southeastern United States (Hall 1981).

Head and body length is 51–115 mm, tail length is 50–97 mm, and weight is 15–30 grams. This mouse has a striking golden coloration and dense, soft pelage. The upper parts are rich golden brown or bright golden cinnamon, and the feet and belly are white, often washed with orange.

From *Peromyscus, Ochrotomys* differs in that the posterior palatine foramina of its skull are nearer to the interpterygoid fossa than to the posterior endings of the anterior palatine foramina; the enamel folds of its molariform teeth are compressed and thick; and its baculum is distinctly capped with a long, cartilaginous cone (Linzey and Packard 1977). Females have six mammae.

The golden mouse inhabits brushy and wooded areas and is often found in thickets of honeysuckle *(Lonicera)* and greenbrier *(Smilax).* Over most of its range the genus is mainly arboreal and builds nestlike structures in vines, bushes, and trees. At the western limits of its range, however, at least in eastern Texas, it is not arboreal. In Texas, and possibly during the summer in other parts of its range, *Ochrotomys* seems to spend more time on the ground than in trees and apparently uses nests placed under logs and stumps. Individuals have been trapped in a swamp in Florida in under-

Magdalena rat *(Xenomys nelsoni),* photo by Cornelio Sanchez Hernandez.

Golden mouse *(Ochrotomys nuttalli)*, photo by Ernest P. Walker.

ground runways. The golden mouse is an agile animal, using its tail as a balancing and prehensile organ when climbing.

Typically, two kinds of nestlike structures are used. The nest proper is usually located 0.4–4.5 meters above the ground and has the appearance of a solid mass. It has an outer covering of leaves, grass, and bark and an inner downy lining of milkweed fibers, feathers, or fur. The outer layer often envelops the abandoned nest of a bird, and the entire structure measures 100–200 mm across. The same nest may be used for several generations, and suitable nesting sites may be used year after year. The other nestlike structure, similar in appearance but less bulky and in places 15 meters above the ground, serves as a retreat for feeding. These platforms may be more numerous than the nests proper, and more than one individual may use the same feeding platform. The diet consists mainly of plant seeds.

The remainder of this account, dealing with population structure and life history, is taken largely from Linzey and Packard (1977). Reported population densities are 0.5 per ha. in Tennessee, up to 5.4 per ha. in early spring on a forested hardwood floodplain in eastern Texas, 0.47–6.89 per ha. in Great Smoky Mountains National Park, and 1.7–74.1 per ha. in Illinois. Home ranges are relatively small, recorded averages varying from about 0.053 ha. to 0.627 ha. Home ranges apparently overlap extensively, and territorial behavior is not evident. *Ochrotomys* is a fairly sociable rodent, and up to 8 individuals have been found in the same nest.

The reproductive period seems to vary considerably, having been reported as throughout the year in Louisiana, mainly from September to spring in Texas, and from March to October in Kentucky and eastern Tennessee. Captives breed throughout the year, though with marked peaks in early spring and late summer. Several litters may be produced annually in the wild, and some captive females have borne as many as 17 litters within 18 months. The gestation period is about 25–30 days. Litter size ranges from 1 to 4 young and averages about 2.7. The young have a mean weight of about 2.7 grams at birth, open their eyes after 11–14 days, are weaned at 17–21 days, and reach adult size at 8–10 weeks. Captive females have produced young when as old as 6.5 years and have lived as long as 8 years and 5 months.

RODENTIA; MURIDAE; **Genus MEGADONTOMYS**
Merriam, 1898

Thomas's Deer Mouse

The single species, *M. thomasi,* is known from several isolated localities in the mountains of the states of Guerrero, Oaxaca, Puebla, and Veracruz in southern Mexico (Hall 1981). *Megadontomys* was usually treated, as by Hall, as a subgenus of *Peromyscus* until Carleton (1980) gave it full generic rank. The latter procedure was followed by Corbet and Hill (1986), Honacki, Kinman, and Koeppl (1982), and Werbitsky and Kilpatrick (1987) but was challenged by Rogers (1983) and Rogers et al. (1984).

According to Musser (1964), total length is 300–351 mm, tail length is 150–85 mm, and weight is 61–113 grams. According to Hall (1981), the upper parts are rich tawny to darker, the underparts are creamy white, and the feet are white to grizzled. According to Carleton, *Megadontomys* is closely related to *Isthmomys,* both being characterized and distinguished from *Peromyscus* by having a relatively broad glans penis and baculum, but with the baculum also being much longer than the glans. *Megadontomys* is further distinguished by the following traits: usual occurrence of an accessory root on the first upper molar, singular articulation of the first rib, low extent of tibia-fibular fusion, moderate number of coils in the intestinal spire, complicated caecum, fluting of the glans surface, and presence of a weakly developed urethral process and crater hood.

Musser (1964) collected 15 specimens in a wet, cool oak and conifer forest at an elevation of about 2,600 meters. His records indicated that the genus is terrestrial and generally uses runways through leaf litter and humus on the forest floor. Stomach contents suggested that the diet consists in large part of berries. Heaney and Birney (1977) collected one pregnant female with three embryos on 24 April and one with two embryos on 1 June.

RODENTIA; MURIDAE; **Genus ISTHMOMYS**
Hooper and Musser, 1964

Yellow Deer Mouse and Mount Pirri Deer Mouse

There are two species (Hall 1981):

I. flavidus, western Panama;
I. pirrensis, eastern Panama.

Isthmomys was usually treated, as by Hall, as a subgenus of *Peromyscus* until Carleton (1980) gave it full generic rank. The latter procedure was followed by Corbet and Hill (1986), Eisenberg (1984), and Honacki, Kinman, and Koeppl (1982).

Measurements listed by Hall indicate that head and body length is about 150–200 mm and tail length is 155–205 mm. In the species *I. flavidus*, the yellow deer mouse, the upper parts are rich ochraceous, the underparts are yellowish white, the forefeet are white, and the hind feet are white mixed with brownish. In *I. pirrensis* the upper parts are dark brownish cinnamon to cinnamon rufous lined with black, the sides are brighter, the underparts are dull buffy white, and the feet are whitish. *Isthmomys* is distinguished from related genera mainly by unique characters of the male reproductive structures: the baculum is longer than the glans penis; the glans is cone-shaped, flared and becoming much broader distally, marked by a small groove ventrally, and armed with large, long, proximally directed spines; and the urethral opening is ventral and subterminal.

According to Goldman (1920), individuals of this genus have been collected in montane forests at altitudes of about 800–1,600 meters. They were considered common there, but none were taken at low elevations, and so the highland populations appear isolated. Two young were found in a nest about two meters from the ground behind the expanded base of a palm frond. The nest was composed of pulverized bark and plant fibers. Worn places over and under logs marked routes regularly used.

RODENTIA; MURIDAE; **Genus OSGOODOMYS**
Hooper and Musser, 1964

Michoacán Deer Mouse

The single species, *O. banderanus*, is found from Nayarit to Guerrero in southwestern Mexico (Hall 1981). *Osgoodomys* was usually treated, as by Hall, as a subgenus of *Peromyscus* until given full generic rank by Carleton (1980). The latter procedure was followed by Corbet and Hill (1986), Eisenberg (1984), and Honacki, Kinman, and Koeppl (1982). On the

basis of chromosomal data, Rogers et al. (1984) continued to recognize *Osgoodomys* as only a subgenus but noted that it has a primitive and distinctive karyotype.

Measurements listed by Hall (1981) indicate that head and body length is about 100–120 mm and tail length is 107–32 mm. Hall and Villa R. (1949) recorded weights of 48 to 68 grams. According to Hall (1981), the upper parts are ochraceous buff with a mixture of cinnamon to dusky, the underparts are creamy and there may be a broad ochraceous buff pectoral patch, the feet are white, and the tail is dark above and white to yellow below. *Osgoodomys* is distinguished by its narrow skull, with a greatly elongated posterior portion and well-developed supraoccipital beads. The baculum is only a fifth as long as the hind foot, the glans penis is small and simple, and the urethral opening is terminal.

Hall and Villa R. (1949) collected specimens among large boulders and tree roots at altitudes of 1,200–1,400 meters. Of 14 adult females taken, only one was pregnant, and it had two embryos.

RODENTIA; MURIDAE; **Genus ONYCHOMYS**
Baird, 1857

Grasshopper Mice

There are three species (Hall 1981; Hinesley 1979; Riddle and Choate 1986; Sullivan, Hafner, and Yates 1986):

O. torridus, southwestern United States, northern Mexico;
O. leucogaster, eastern Washington and southern Manitoba to extreme northern Mexico;
O. arenicola, southwestern New Mexico, extreme western Texas, and adjacent parts of north-central Mexico.

Head and body length is 90–130 mm, tail length is 30–60 mm, and weight is usually 30–60 grams. The fur is fine and dense. The upper parts of *O. leucogaster* are brownish to pinkish cinnamon, or buff, while the upper parts of *O. torridus* are grayish or pinkish cinnamon. In both species the basal two-thirds of the tail are colored like the upper parts, but the underside and terminal tip are white. The underparts of the body are white in both species. *O. arenicola* superficially resembles *O. torridus* but averages smaller in size and differs in karyotype. The genus is distinguished by the stocky body and fairly short, clublike tail. In *O. leucogaster* the tail is usually less than half the length of the head and body, but in *O. torridus* it is usually longer than half the length of the head and body. Female *Onychomys* have six mammae.

Grasshopper mice occur in shortgrass prairies and desert scrub. *O. torridus* prefers relatively xeric areas at lower elevations, while *O. leucogaster* tends to occupy more mesic areas at higher elevations (McCarty 1978). Both species are largely nocturnal and are active throughout the year. They live in practically any shelter they can find at ground level. They are good climbers but apparently do not climb regularly. The nest may be constructed in a burrow taken over from some other rodent. *O. leucogaster* reportedly excavates a U-shaped burrow with an average length of 48 cm and an average depth beneath the surface of 14 cm. Smaller burrows are constructed for such purposes as emergency retreat and seed storage (McCarty 1978). Although plant material is eaten at times, grasshopper mice are largely carnivorous. The diet includes grasshoppers, beetles, and a variety of other insects. *O. torridus* preys extensively on scorpions. Grasshopper mice also eat small vertebrates, including such rodents as *Peromyscus*, *Perognathus*, and *Microtus*. The prey is generally

Grasshopper mouse *(Onychomys leucogaster)*, photo by Ernest P. Walker.

stalked, seized with a rush, and then killed by a bite in the head. While overpowering their prey, grasshopper mice may close their eyes and lay back their ears. When confronted with relatively large prey that possess defensive mechanisms, *Onychomys* varies its strategy. For example, the jumping legs of crickets may be immobilized before the head and thorax are bitten, while the initial point of attack on a scorpion is usually the stinger-bearing tail (Langley 1981; Whitman, Blum, and Jones 1986).

Like most predators, grasshopper mice occur at relatively low densities. An average of 1.83 *O. torridus* per ha. has been reported for desert country in Nevada. Usual home range for *Onychomys* is about 2–3 ha. Adults are territorial, and captive individuals of the same sex are extremely aggressive toward one another, often fighting to the death (McCarty 1975, 1978). In the wild, according to Hafner and Hafner (1979), grasshopper mice have large, well-defined territories and may associate in male-female pairs all year. Adult vocalizations include a single sharp "chit," given during agonistic encounters; a soft chirp during mating; a quiet, pure tone lasting about 0.8 seconds, given when close to a mate; and a loud, piercing, pure tone lasting 0.7–1.2 seconds and audible to the human ear up to 100 meters away. When giving this last kind of call, the animal often stands on its hind legs with its nose pointed up. This call appears to be spontaneous but often is emitted prior to making a kill or upon detecting another *Onychomys* nearby. This shrill call, which is often repeated several times, has been compared to a miniature wolf howl in its qualities of smoothness and prolongation and in the posture of the animal when the call is given.

The following reproductive information is taken from McCarty (1975, 1978) and Pinter (1970). Grasshopper mice are capable of breeding throughout the year, but most reproductive activity in the wild occurs in the late spring and summer. The estrous cycle of *O. leucogaster* may be about 5–7 days in length. This species produces several litters per year (up to 12 in the laboratory). Female *O. torridus* apparently have a high level of sexual activity for just one breeding season; individuals born as early as April may produce two or three litters before the end of the year, while females born late in the season may give birth to as many as six litters during the following breeding season. Reported gestation periods are 26–37 days in nonlactating *O. leucogaster*, 32–47 days in

lactating *O. leucogaster*, and 26–35 days in *O. torridus*. Litter size is 1–6 young, averaging about 3.6 in *O. leucogaster* and 2.6 in *O. torridus*. The young weigh about 2.6 grams at birth, open their eyes at about 2 weeks, and are weaned at around 3 weeks. Sexual maturity is reached between 2 and 5 months in *O. leucogaster* and as early as 6 weeks in *O. torridus*. Female *O. leucogaster* remain reproductively active for up to 3 years, but female *O. torridus* seldom breed for more than 2 years.

These rodents are said to make fascinating pets and to become generally quite gentle in captivity. According to Jones (1982), a captive specimen of *O. torridus* lived for 4 years and 7 months.

RODENTIA; MURIDAE; Genus NEOTOMODON
Merriam, 1898

Mexican Volcano Mouse

The single species, *N. alstoni*, inhabits the volcanic mountains of central Mexico (Hall 1981). The original describer thought that the grinding teeth resembled those of *Neotoma*, hence the generic name. Davis and Follansbee (1945:405) stated that "on the basis of general shape of skull, dental characters, and habits, *Neotomodon* is much closer to *Peromyscus* than to any other recent genus." Nonetheless, most authorities (even as late as Corbet and Hill 1986) continued to arrange *Neotomodon* next to *Neotoma* in systematic accounts. Yates, Baker, and Barnett (1979) placed *Neotomodon* in the genus *Peromyscus* and indicated a close relationship to *P. gossypinus* and *P. floridanus* (herein referred to the separate genus *Podomys*). Carleton (1980) again elevated *Neotomodon* to generic rank though recognizing that it is closely related to *Podomys* and is in a systematic tribal group that includes *Peromyscus* but not *Neotoma*. *Neotomodon* was recognized as a separate genus by Corbet and Hill (1986), Eisenberg (1984), and Honacki, Kinman, and Koeppl (1982) but not by Rogers et al. (1984), Williams and Ramirez-Pulido (1984), or Williams, Ramirez-Pulido, and Baker (1985).

Head and body length is about 100–130 mm, tail length is about 80–105 mm, and adult weight is usually 40–60 grams. The fur is soft and dense. The upper parts are grayish to

Mexican volcano mouse *(Neotomodon alstoni)*, photo by William B. Davis.

grayish buff but occasionally, in season, become fulvous brown. The underparts are whitish, often faintly washed with buff on the chest. The relatively well-haired tail is bicolored, dusky above and white below. The ears are large and nearly naked. Females have six mammae.

The volcano mouse lives on grassy slopes in forests at an elevation of 2,600–4,300 meters. It is nocturnal and is most active shortly after dark. It does not usually make distinct runways. Hall and Dalquest (1963) excavated a burrow that measured 38 mm wide throughout its length of about 1.8 meters. It branched at a depth of about 356 mm, with one branch leading to a nest composed of a hollow ball of dry grass, 127 mm in diameter. The volcano mouse may also use the abandoned burrows of small pocket gophers *(Thomomys).*

Neotomodon lives in close association with the black-eared mouse *(Peromyscus melanotis)* and has habits much like those of this burrowing, nocturnal species of deer mouse.

Davis and Follansbee (1945) reported that the breeding season of *Neotomodon* extends at least from early June to September; that two or more litters, averaging 3.4 (2–5) young each, are born each year; and that some individuals reach sexual maturity during the same season in which they are born. Hall and Dalquest (1963) found a pregnant female with 2 embryos in November in Veracruz. In a laboratory study, Olivera, Ramirez-Pulido, and Williams (1986) found that births occurred throughout the year, with averages of 4.5 days for the estrous cycle, 27.3 days for gestation, 3.1 young for litter size, and 174.5 days for sexual maturity.

Florida mouse *(Podomys floridanus)*, photo by James N. Layne.

RODENTIA; MURIDAE; **Genus PODOMYS**
Osgood, 1909

Florida Mouse

The single species, *P. floridanus*, originally occupied most of peninsular Florida (Hall 1981). *Podomys* was usually treated, as by Hall, as a subgenus of *Peromyscus* until given full generic rank by Carleton (1980). The latter procedure was followed by Corbet and Hill (1986), Eisenberg (1984), Honacki, Kinman, and Koeppl (1982), and Jones et al. (1986) but was challenged by Rogers et al. (1984) on the basis of chromosomal data.

Except as noted, the information for the remainder of this account was taken from Layne (1978). Total length is about 179–97 mm, tail length is 79–90 mm, and adult weight is about 25–49 grams. The upper parts are brownish to brownish gray, shading to bright orange buff on the shoulders and lower sides; the underparts are white, frequently with a tawny patch on the breast; and the tail is dusky above and light below but is not sharply bicolored. The pelage is relatively long and soft, the eyes big and dark, and the ears large and sparsely haired. Unlike *Peromyscus* and the other genera separated therefrom by Carleton (1980), *Podomys* has five, rather than six, well-developed plantar tubercles on the hind foot. Hall (1981) wrote that the baculum of *Podomys* is only a quarter as long as the hind foot; the glans penis is small and simple, with the distal half and extreme basal part superficially smooth; and the urethral opening is terminal when the ventral process of the glans is folded.

The primary habitat of the Florida mouse is sand pine scrub at an early successional stage, but it also occurs commonly in pine-oak and scrubby flatwoods associations. It is a ground dweller and typically lives in burrows, favoring those of the gopher tortoise. Its diet consists of seeds, acorns, nuts, fungi, other vegetation, insects, and other small invertebrates. Breeding occurs mainly in the fall and early winter. Litter size ranges from one to six young, averaging three or four. Keim and Stout (1987) reported that a captive individual lived to an age of about 7 years and 5 months.

The Florida mouse is threatened by widespread destruction of its restricted habitat. Much of the natural sand pine scrub and pine-oak associations on which the species depends have been converted to real estate developments, citrus groves, and pine plantations. Overprotection from fire has allowed vegetation in many other areas to become too dense and shady. The species already has disappeared from much of the Florida coast and interior and is becoming increasingly rare in its remaining habitat.

RODENTIA; MURIDAE; **Genus HABROMYS**
Hooper and Musser, 1964

Crested-tailed Deer Mice and Slender-tailed Deer Mice

There are four species (Hall 1981):

H. simulatus, Veracruz (east-central Mexico);
H. chinanteco, northern Oaxaca (southern Mexico);
H. lophurus, extreme southern Mexico to northwestern El
   Salvador;
H. lepturus, northern Oaxaca (southern Mexico).

Hall treated *Habromys* as a subgenus of *Peromyscus*, but Carleton's (1980) morphological studies indicate that it warrants full generic rank, and most authorities have followed the latter arrangement (Carleton and Musser 1984; Corbet and Hill 1986; Honacki, Kinman, and Koeppl 1982).

Descriptions provided by Hall (1981) indicate that head and body length is about 80–140 mm and tail length is 87–145 mm. The upper parts are grayish brown to brownish black in color, and the underparts are white or grayish white. In *H. simulatus* and *H. lophurus* the hairs of the tail form a crest. From *Peromyscus*, *Habromys* is distinguished by having a much shorter baculum and by features of the reproductive system.

RODENTIA; MURIDAE; **Genus REITHRODONTOMYS**
Giglioli, 1874

American Harvest Mice

There are 2 subgenera and 19 species (Cabrera 1961; Hall 1981):

subgenus *Reithrodontomys* Giglioli, 1874

R. montanus, Great Plains region of the United States,
   northern Mexico;
R. burti, Sonora and Sinaloa (northwestern Mexico);
R. humulis, southeastern United States;
R. megalotis, southern British Columbia to northeastern
   Indiana and southern Mexico;
R. raviventris, San Francisco Bay area of California;
R. chrysopsis, mountains of central Mexico;
R. sumichrasti, central Mexico to western Panama;
R. fulvescens, southern Arizona and south-central United
   States to Nicaragua;
R. hirsutus, Nayarit and Jalisco (west-central Mexico);

subgenus *Aporodon* A. H. Howell, 1914

R. gracilis, southern Mexico to northwestern Costa Rica;
R. darienensis, Panama, probably extreme northwestern
   Colombia;
R. mexicanus, eastern and southwestern Mexico to western
   Panama, the Andes of Colombia and Ecuador;
R. spectabilis, Cozumel Island off northeastern Yucatan;
R. brevirostris, north-central Nicaragua to central Costa
   Rica;
R. paradoxus, southwestern Nicaragua to northwestern
   Costa Rica;
R. microdon, central Mexico to Guatemala;
R. tenuirostris, central Guatemala;
R. rodriguezi, central Costa Rica;
R. creper, Costa Rica, western Panama.

On the basis of karyological data, Carleton and Myers (1979) suggested that the above subgeneric division might not be valid, and Hood et al. (1984) and Nelson et al. (1984) suggested that *R. montanus* is closely related to *R. raviventris*, and *R. megalotis* to *R. sumichrasti*.

Head and body length is 50–145 mm, tail length is 45–115 mm, and weight is 6–20 grams. Coloration of the upper parts ranges from pale ochraceous gray and pinkish cinnamon through browns to almost black. The sides are paler and usually more ochraceous than the upper parts. The underparts are white or grayish, sometimes tinged with ochraceous buff or pinkish cinnamon. The slender, scaly, and scantily haired tail is dark above and light below, or unicolor. The juvenile pelage is more or less plumbeous, but the adult pelage is brighter. Adults apparently molt once a year.

American harvest mouse *(Reithrodontomys fulvescens)*, photo by Ernest P. Walker.

American harvest mice resemble house mice *(Mus)* but have more hair on the tail and have grooved upper incisors. The ears of harvest mice are conspicuous and sometimes large. Females have six mammae.

Habitats vary from salt marshes to tropical forests, but American harvest mice are usually associated with stands of short grass. The altitudinal range is from below sea level in places to above the tree line on some Central American mountains. The most noticeable evidence of *Reithrodontomys* is the presence of globular nests of grass about 150–75 mm in diameter. The nests are usually constructed above ground in grasses, low shrubs, or small trees. Some winter nests are located in burrows and small crevices. These mice are nocturnal and are active throughout the year. They use the ground runways of other rodents and are nimble climbers. Food consists mainly of seeds and the green shoots of vegetation. The seeds are gleaned from the ground or cut from grass stems by bending the stems to the ground. Some insects are also eaten.

Reported autumn population densities for *R. megalotis* in southern Wisconsin have varied from only about 0.75/ha. in sandy fields to 45/ha. in an abandoned field with a dense cover of low vegetation (Jackson 1961; Svendsen 1970). O'Farrell (1978) estimated an annual composite home range of 0.95 ha. for *R. megalotis* in a sagebrush desert in Nevada. Packard (1968) reported a population density of about 7.5/ha. for *R. fulvescens* in eastern Texas. He found this species to use a home range of about 0.2 ha. and observed no marked territorial behavior. There are conflicting reports regarding sociability in *R. megalotis* (Banfield 1974). A high-pitched bugling vocalization has been reported for the genus.

Reproductive activity apparently takes place throughout the year except during cold winters. In Indiana, for example, *R. megalotis* breeds from March to November (Whitaker and Mumford 1972). Banfield (1974) stated that females of this species are seasonally polyestrous and produce several litters annually but do not undergo a postpartum estrus and that the number of young averages 2.6 (1–9). In Nebraska, 75 pregnant female *R. megalotis* taken from April through November had an average of 4.3 (2–8) embryos (Jones 1964). In eastern Tennessee, Dunaway (1968) found *R. humulis* to breed mainly from late spring to late fall but occasionally during the winter and to have an average litter size of 3.4 (1–8). In Nicaragua, Jones and Genoways (1970) collected pregnant female *R. sumichrasti*, *R. fulvescens*, *R. gracilis*, *R. mexicanus*, and *R. brevirostris* from June to August; the embryos found numbered 3–5. Captive female *R. megalotis* produced up to 14 litters and 68 young per year; one gave birth when 2 years and 5 months old (Egoscue, Bittmenn, and Petrovich 1970). Gestation periods of 21–24 days have been reported for the genus. The young are born in a woven grass nest and weigh about one gram at birth. In *R. megalotis* the

young leave the nest after about 3 weeks and attain adult weight in about 5 weeks. Some females of this species breed when 17 weeks old. Very few *Reithrodontomys* live as long as 1 year, and the known record life span in the wild seems to be about 18 months (Fisler 1971).

American harvest mice are not generally considered to be detrimental to human agriculture. People, however, have had pronounced, and opposite, effects on at least two species of *Reithrodontomys*. *R. megalotis* apparently was able to extend its range eastward across Illinois and into Indiana in response to the clearing of woodlands and the destruction of tall grass prairies (Ford 1977; Whitaker and Mumford 1972). The salt marsh harvest mouse *(R. raviventris)*, one of the few mammals that can drink salt water, is classified as endangered by the IUCN, the USDI, and the California Department of Fish and Game (1978). It is dependent for survival on the marshes around San Francisco Bay, and its population has been greatly reduced and fragmented through habitat destruction caused by urban and industrial development. According to Thornback and Jenkins (1982), there are at most a few thousand individuals, and populations are now highly fragmented. Shellhammer (1989) set forth a rather gloomy future, with the species caught between increasing land development and predicted rising sea levels and the marshes also declining through loss of sediment to inland dams and water diversion. The situation is much the same for *R. megalotis limnicola*, found in salt marshes along part of the southern California coast (D. F. Williams 1986), and it now is listed as vulnerable by the IUCN.

RODENTIA; MURIDAE; **Genus PEROMYSCUS**
Gloger, 1841

White-footed Mice, or Deer Mice

There are 2 subgenera and 55 species (Carleton 1977, 1979; Carleton et al. 1982; Gunn and Greenbaum 1986; Hall 1981; Huckaby 1980; J. K. Jones and Yates 1983; Lee and Schmidly 1977; Modi and Lee 1984; Schmidly, Bradley, and Cato 1988; Schmidly et al. 1985; Walker 1980; Woodman 1988):

subgenus *Haplomylomys* Osgood, 1904

P. eremicus, southwestern United States, northern Mexico, Baja California and several nearby islands;

P. eva, southern Baja California, Carmen Island (Gulf of California);

P. merriami, southern Arizona, northwestern Mexico;

P. guardia, Angel, Granite, and Mejia islands (Gulf of California);

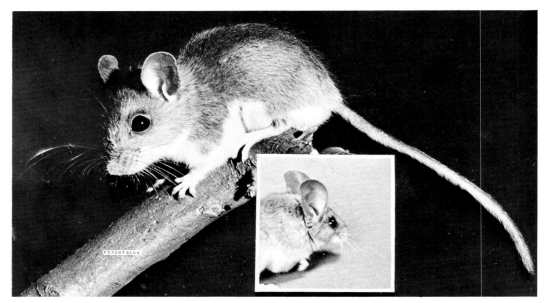

White-footed mouse *(Peromyscus leucopus)*, photo by K. H. Maslowski and W. W. Goodpaster through National Audubon Society. Inset: big-eared cliff mouse *(P. truei)*, photo by Ernest P. Walker.

P. interparietalis, Salsipuedes and San Lorenzo islands (Gulf of California);

P. collatus, Turner Island (Gulf of California);

P. dickeyi, Tortuga Island (Gulf of California);

P. pembertoni, San Pedro Nolasco Island (Gulf of California);

P. californicus, central California to northern Baja California;

subgenus *Peromyscus* Gloger, 1841

P. hooperi, Coahuila, northeastern Zacatecas (northern Mexico);

P. polionotus, southeastern United States;

P. maniculatus, southeastern Alaska, Queen Charlotte Islands, most of Canada, conterminous United States except parts of Southeast, most of Mexico, Baja California and several nearby islands;

P. sejugis, Santa Cruz and San Diego islands (Gulf of California);

P. slevini, Santa Catalina Island (Gulf of California);

P. nesodytes, formerly San Miguel Island off southern California;

P. oreas, southwestern British Columbia, northwestern Washington;

P. sitkensis, several islands off southeastern Alaska, several small islands in the Queen Charlotte group;

P. melanotis, northern and central Mexico;

P. leucopus, southeastern Alberta and Nova Scotia to Arizona and the Yucatan Peninsula;

P. gossypinus (cotton mouse), southeastern United States;

P. crinitus, central Oregon to western Colorado and northern Baja California;

P. caniceps, Monserrate Island (Gulf of California);

P. pseudocrinitus, Coronados Island (Gulf of California);

P. pectoralis, south-central Oklahoma and southeastern New Mexico to central Mexico;

P. boylii, northern California and northwestern Texas to Honduras;

P. levipes, central Mexico;

P. beatae, east-central Mexico;

P. spicilegus, western Mexico;

P. simulus, Sinaloa and Nayarit (western Mexico);

P. madrensis, Tres Marias Islands off western Mexico;

P. attwateri, southeastern Kansas to northern Arkansas and central Texas;

P. polius, Chihuahua (north-central Mexico);

P. stephani, San Esteban Island (Gulf of California);

P. aztecus, central Mexico to Honduras;

P. winkelmanni, Michoacán (western Mexico);

P. truei, southwestern Oregon and Colorado to northern Baja California and western Texas;

P. gratus, central New Mexico to southern Mexico;

P. difficilis, Colorado to south-central Mexico;

P. bullatus, Veracruz (east-central Mexico);

P. perfulvus, west-central Mexico;

P. melanophrys, Mexico;

P. mekisturus, Puebla (east-central Mexico);

P. ochraventer, southern Tamaulipas and southeastern San Luis Potosi (northeastern Mexico);

P. yucatanicus, Yucatan Peninsula;

P. mexicanus, eastern and southern Mexico to western Panama;

P. gymnotis, extreme southern Mexico, southern Guatemala, western Nicaragua;

P. mayensis, western Guatemala;

P. stirtoni, Guatemala, southern Honduras, El Salvador, west-central Nicaragua, subfossil remains from Costa Rica;

P. furvus, eastern Mexico;

P. guatemalensis, extreme southern Mexico, western Guatemala;

P. megalops, southwestern Mexico;

P. melanocarpus, north-central Oaxaca (southern Mexico);

P. melanurus, southwestern Oaxaca (southern Mexico);

P. zarhynchus, Chiapas (southern Mexico);

P. grandis, central Guatemala.

Until recently, most authorities (e.g., Hall 1981) considered *Peromyscus* to include five additional subgenera: *Habromys*,

Podomys, Osgoodomys, Isthmomys, and *Megadontomys.* Some authorities (but not Hall) also place *Neotomodon* in *Peromyscus.* The morphological studies of Carleton (1980) led to a general recognition of all of these taxa as full genera (Carleton and Musser 1984; Corbet and Hill 1986; Honacki, Kinman, and Koeppl 1982; Jones et al. 1986). New chromosomal investigations, however, suggest that these taxa should be retained within *Peromyscus* (Rogers 1983; Rogers et al. 1984; Williams, Ramirez-Pulido, and Baker 1985).

Hooper (1968) suggested that some of the named species of *Peromyscus* would eventually be shown to be synonyms or subspecies and that the actual number of biological species is about 40. On the other hand, new morphological, biochemical, amd karyological studies suggest that *P. boylii* actually comprises several distinct species, at least in the Latin American portions of its range (Bradley and Schmidly 1987; Houseal et al. 1987; Rennert and Kilpatrick 1986, 1987).

Also of interest is the status of the population in northwestern Texas originally called *P. comanche.* It long was recognized as a subspecies of *P. difficilis,* but on the basis of karyological data, Lee, Schmidly, and Huheey (1972) stated that its affinities actually are with *P. truei.* Using an electrophoretic analysis, Johnson and Packard (1974) concluded that *P. comanche* is a distinct species. Studies by Schmidly (1973) and Modi and Lee (1984), however, indicated that *P. comanche* is a subspecies of *P. truei,* and this arrangement was followed by Hall (1981) and Jones et al. (1986).

Head and body length is 70–170 mm and tail length is 40–205 mm. Klingener (1968) stated that adults range in body weight from about 15 grams in some small northern species, such as *P. crinitus* and *P. polionotus,* to over 110 grams in a few tropical forms. The pelage is usually soft and full. The coloration is variable, but the upper parts may be gray or sandy to golden or dark brown, and the underparts are white or nearly white. Some species, however, are nearly all white, and others are nearly black. In general, deer mice inhabiting the cool woods are grayish, whereas those living in open or arid country are pale. Adults molt once a year.

The ears are large in relation to the rest of the body and are covered with fine hairs. The tail is at least one-third the total length of the animal; it is fairly well haired and often tufted. In *P. boylii* the skin of the distal portion of the tail is known to break readily and freely slip off the vertebrae, apparently serving as a means of escaping predators that seize the tail (Layne 1972). Female *Peromyscus* have four or six mammae.

Deer mice are found in a great variety of habitats and are usually the most abundant mammal in the areas they occupy. *P. maniculatus* is the most widely distributed species, and, indeed, few other mammals can match its tolerance for different conditions—alpine areas, boreal forest, woodlands, grasslands, brushlands, deserts, and arid tropical areas. It does not, however, regularly occur in moist places (Baker 1968; Banfield 1974). In contrast, *P. gossypinus* is restricted to the southeastern United States, where its preferred habitat is bottomland hardwood forest and swamp (Wolfe and Linzey 1977). *P. leucopus* also occurs over a large region but keeps mainly to deep woodlands or brushy areas (Lowery 1974). *P. attwateri* seems closely associated with cliffs and other rocky areas, where it usually lives in crevices (Schmidly 1974a). Other species build nests in or under logs, in holes in trees, under brush piles, in clumps of vegetation, or in burrows dug by the mice themselves or by other animals. The nests of *P. maniculatus* are spheres of grass about 100 mm in diameter (Banfield 1974). Nests are lined with down from plants or with shredded materials. A soiled nest is abandoned, so several nests are made during the year. Deer mice are largely, though not exclusively, nocturnal and are active throughout the year. Periods of torpor, involving a reduction in body

temperature for several hours, are known to occur in *P. leucopus* and *P. maniculatus* (Lynch, Vogt, and Smith 1978). The diet includes seeds, nuts, berries, fruits, insects and other small invertebrates, and carrion. One study of *P. gossypinus* indicated that 68 percent of its diet consisted of animal matter (Wolfe and Linzey 1977). According to Banfield (1974), *P. maniculatus* is primarily a seed eater, and it may store up to three liters of food for winter use.

Normal population density in *P. maniculatus* varies from about 1/ha. to 25/ha. (Banfield 1974). In Kansas, Hansen and Fleharty (1974) found that the density of this species ranged from 4/ha. in weedy areas during the spring and summer to 19.3/ha. in the favored bluestem grass community during the autumn. In Connecticut the density of *P. leucopus* ranged from 5.1/ha. to 38.9/ha. (Miller and Getz 1977). Densities as high as 96.6/ha. have been reported for *P. gossypinus* in wet lowland forests in Tennessee (Wolfe and Linzey 1977).

According to Stickel (1968), home range in *Peromyscus* usually varies from 0.04 ha. to 4.00 ha., is generally smallest in winter and largest in summer, and is generally larger in males than in females. Banfield (1974) stated that the average home range of *P. maniculatus* is about 1.0 ha. in males and 0.6 ha. in females. In a study in Kansas, Hansen and Fleharty (1974) found the home range of this species to average 0.31 ha. in the spring and summer and 0.27 ha. in the autumn and winter. For *P. leucopus* in northern Virginia average home range was only 0.1 ha. and was about the same for both sexes, but in Quebec the home range of this species averaged 1.26 ha. in males and 0.91 ha. in females (Madison 1977; Mineau and Madison 1977). Home range data on some other species are: *P. attwateri,* average size about 0.2 ha., being about twice as large in males as in females (Schmidly 1974a); *P. gossypinus,* 0.18–0.81 ha. (Wolfe and Linzey 1977); *P. californicus,* averages about 0.15 ha. in both sexes (Merritt 1978); and *P. eremicus,* averages about 0.3 ha., there being considerable overlap in males and almost none in females (Veal and Caire 1979).

Banfield (1974) considered *P. maniculatus* to be a sociable rodent, tolerant of conspecifics regardless of age and sex, especially in winter, when up to 13 individuals may huddle together. In this species, and probably some others, the male lives with the family and helps to care for the young. Foltz (1981) concluded that populations of *P. polionotus* are organized into family groups characterized by a high degree of monogamy. Sadleir (1970) found that captive male *P. maniculatus* formed a single-line hierarchy that remained stable for a month. Studies of *P. leucopus* indicate that males and females form pairs that live together but that resident females exclude other females from their home ranges (Harland, Blancher, and Millar 1979; Madison 1977; Mineau and Madison 1977). *P. attwateri* is not aggressive, and individuals of all sex and age categories may rest together in captivity (Schmidly 1974a). In *P. californicus,* males are highly aggressive toward one another, but opposite sexes form pairs that cooperate in raising the young (Merritt 1978). Vocalizations of *Peromyscus* include thin squeaks and shrill buzzings. When excited, most species thump rapidly with the front feet, producing a drumming noise.

In some species breeding may continue throughout the year if the weather does not become too cold or too hot. The breeding season of *P. maniculatus* in the northern parts of its range is normally March through October, though some females may mate in winter under favorable conditions. This species is polyestrous and usually produces 3–4 litters annually in the wild (Banfield 1974; Godin 1977). In the laboratory, however, female *P. maniculatus* have borne up to 14 litters per year (Egoscue, Bittmenn, and Petrovich 1970). *P. gossypinus* breeds throughout the year in Texas and Florida, though there may be some decline in reproductive activity during the summer (Wolfe and Linzey 1977). In southwestern Missouri, *P. attwateri* has one breeding season in spring and another in autumn (Schmidly 1974a). Year-round reproduction, at least in some areas, has been reported in *P. pectoralis* and *P. eremicus* (Schmidly 1974b; Veal and Caire 1979). In southern Mexico both *P. mexicanus* and *P. melanocarpus* breed throughout the year, but the latter species shows a definite seasonal peak from March to July (Rickart 1977).

The estrous cycle averages 5.26 days in *P. gossypinus* and about 7 days in *P. californicus* (Merritt 1978). A postpartum estrus may occur in female *Peromyscus*. According to Layne (1968), the gestation period in this genus ranges from 21 to 27 days and averages 23.47 days in nonlactating females but may be as long as 40 days in lactating females. Somewhat longer gestations in nonlactating females have been reported for several species in southern Mexico (Lackey 1976; Rickart 1977). The overall mean litter size for the genus is 3.4 young (Layne 1968). The range for *P. maniculatus* is 1–9 in the wild (Banfield 1974) but up to 11 in the laboratory (Drickamer and Vestal 1973). Apparently, litters average larger in the north: the mean size for *P. leucopus* in central Wisconsin was 4.77 (Long 1973a), while the mean for *P. melanocarpus* in southern Mexico was about 2.0 (Rickart 1977). The overall mean weight for newborn *Peromyscus* is 2.2 grams; the eyes open at around 2 weeks. The young of most species are weaned at 3–4 weeks but may remain with the mother for another month. The average age of first estrus in females is 29.6 days in *P. polionotus*, 39.2 days in *P. eremicus*, and 48.7 days in *P. maniculatus*. These mice may begin to breed while still in subadult pelage but may continue to grow in weight for as long as 6 months (Layne 1968). Most wild individuals probably live less than 2 years, but record longevity in the laboratory for *P. maniculatus* is 8 years and 4 months (Banfield 1974).

Deer mice are used widely in physiological and genetic studies because they are clean, live well in the laboratory, can be easily fed, and have a high reproductive rate. On the other hand, Banfield (1974) considered *P. maniculatus* to be somewhat of a menace to forest regeneration through its destruction of tree seeds, especially those of conifers.

Although these mice are among the most abundant of mammals, some forms have suffered through human activity. Schmidly et al. (1985), for example, reported that the range of *P. hooperi* has become fragmented and surviving populations are seriously jeopardized by overgrazing. Perhaps the first North American mammal to disappear through human habitat modification was *P. nesodytes*, a remarkably large species known only from skeletal remains about 2,000 years old, recovered on San Miguel Island off southern California. It is possible that extinction of this mouse resulted from changes in San Miguel's flora or fauna caused by Indians, but it is more likely that it was lost when the natural vegetation of the island was destroyed by the overgrazing of sheep in the 1860s.

A number of southeastern species and subspecies of *Peromyscus* are now thought to be in danger of extinction (Bowen 1968; Dusi 1976; Holliman 1983; Humphrey and Barbour 1981; Layne 1968; Repenning and Humphrey 1986). *P. gossypinus allapaticola*, found only in mature tropical hammock forests on Key Largo, Florida, has become restricted to an area of only 120–60 ha. through the commercial and residential development of its habitat. *P. g. restrictus*, the Chadwick Beach cotton mouse of southwestern Florida, may already have vanished because of environmental disruption. Several pale-colored races of "beach mice" are in jeopardy because their small ranges are desirable to people for waterfront homes and recreational facilities. These mice include *P. polionotus ammobates*, endemic to the dunes between Mobile Bay and Perdido Bay in southern Alabama; *P. p. trissyllepsis*, found from Perdido Bay to Pensacola Bay, Florida; *P. p. allophrys* of the Gulf coastal dunes of Okaloosa and Bay counties, Florida; and *P. p. niveiventris, P. p. phasma*, and *P. p. decoloratus*, each of which is restricted to a small part of the coast of northeastern Florida. *P. p. trissyllepsis* was reduced to as few as 26 individuals after a severe hurricane in 1979 and may then have been the rarest mammal in the United States, but numbers subsequently increased to about 100 through protection and reintroduction (Fleming and Holler 1989). *P. p. decoloratus* may already be extinct. The USDI classifies *P. g. allapaticola, P. p. ammobates, P. p. trissyllepsis, P. p. allophrys*, and *P. p. phasma* as endangered and *P. p. niveiventris* as threatened.

RODENTIA; MURIDAE; Genus ORYZOMYS
Baird, 1857

Rice Rats

There seem to be 5 subgenera and 50 species (Allen 1942; Benson and Gehlbach 1979; Cabrera 1961; Contreras and Berry 1983; Contreras and Rosi 1980a; A. L. Gardner, U.S. National Museum of Natural History, pers. comm., January 1988; Gardner and Patton 1976; Goodyear and Lazell 1986; Hall 1981; Handley 1976; Hershkovitz 1960, 1966a, 1970, 1971, 1987a; Husson 1978; Massoia 1973; Massoia and Fornes 1967; Musser and Williams 1985; Myers and Carleton 1981; Myers and Wetzel 1979; Olds and Anderson 1987; Patton and Hafner 1983; Pine 1971; Spitzer and Lazell 1978):

subgenus *Oryzomys* Baird, 1857

O. palustris, southeastern United States;

O. couesi, southern Texas to northern Colombia, southern Baja California, Jamaica;

O. argentatus, lower Florida Keys;

O. nelsoni, Maria Madre Island off western Mexico;

O. fulgens, probably central Mexico;

O. dimidiatus, southeastern Nicaragua;

O. gorgasi, northwestern Colombia;

O. capito, eastern Colombia and the Guianas to northern Argentina, Trinidad;

O. talamancae, eastern Costa Rica, Panama, Colombia, northern Venezuela, Ecuador;

O. macconnelli, eastern Ecuador and Peru to the Guianas;

O. xantheolus, southwestern Ecuador, western Peru;

O. yunganus, eastern Bolivia;

O. galapagoensis, Santa Fe and San Cristobal Islands in the Galapagos;

O. melanotis, southern Sinaloa and Tamaulipas (Mexico) to Nicaragua;

O. caudatus, Oaxaca (southern Mexico);

O. alfaroi, eastern and southern Mexico to northwestern Ecuador;

O. bolivarae, southeastern Nicaragua to northern Ecuador;

O. albigularis, Costa Rica, Panama, the Andes from western Venezuela to northwestern Bolivia;

O. auriventer, the Andes of Ecuador and Peru;

O. nitidus, Ecuador, Peru, Bolivia, Brazil;

O. aphrastus, central Costa Rica, western Panama, northwestern Ecuador;

O. intectus, the Andes of central Colombia;

O. balneator, eastern and southern Ecuador;

O. polius, northern Peru;

O. melanostoma, eastern Peru;

O. subflavus, the Guianas, eastern Brazil;

Rice rat *(Oryzomys palustris)*, photo by Ernest P. Walker.

O. buccinatus, Paraguay, northern Argentina;

O. ratticeps, Paraguay, extreme southern Brazil, northeastern Argentina;

O. hammondi, the Andes of Ecuador;

subgenus *Oligoryzomys* Bangs, 1900

O. victus, St. Vincent Island in the Lesser Antilles;

O. fulvescens, northeastern and southwestern Mexico to Venezuela;

O. delicatus, Colombia to Surinam and northern Brazil;

O. munchiquensis, western Colombia;

O. microtis, Brazil, Bolivia, Paraguay, northern Argentina;

O. flavescens, northern Argentina, Uruguay;

O. andinus, northwestern Peru;

O. arenalis, northeastern Peru;

O. spodiurus, east slope of the Andes in Ecuador;

O. longicaudatus, Peru, Bolivia, Chile, southern Argentina, Tierra del Fuego;

O. delticola, northeastern Argentina, Uruguay;

O. nigripes, Paraguay, southern Brazil, Uruguay, Argentina;

O. chacoensis, central Brazil, Bolivia, Paraguay, northern Argentina;

O. utiaritensis, central Brazil;

O. mattogrossae, central Brazil;

subgenus *Microryzomys* Thomas, 1917

O. minutus, the Andes from northwestern Venezuela to Peru;

O. altissimus, the Andes of Ecuador and Peru;

subgenus *Melanomys* Thomas, 1902

O. caliginosus, Honduras to Ecuador;

O. robustulus, east slope of the Andes in Ecuador;

O. zunigae, west-central Peru;

subgenus *Sigmodontomys* J. A. Allen, 1897

O. alfari, extreme eastern Honduras to northwestern Venezuela and northern Ecuador.

Certain aspects of the systematics of this genus are indefinite, and the cited sources provide only a partial basis for the sequence of species given above. *Nesoryzomys* and *Oecomys* are sometimes considered subgenera, but Gardner and Patton (1976) recommended that both be given full generic rank; most subsequent authors have agreed, though *Oecomys* was

not recognized as a genus by Honacki, Kinman, and Koeppl (1982). Gardner and Patton also treated *O. altissimus* as a subspecies of *O. andinus*, stated that *O. albigularis* is actually a composite of several closely related species, and suggested that the name *O. rivularis* might have priority over *O. capito*. The subgenus *Sigmodontomys* once was considered part of the genus *Nectomys* but now generally is included within *Oryzomys* (Gardner and Patton 1976; Hall 1981; Hershkovitz 1970). The subgenus *Microryzomys* was listed as a full genus by Carleton and Musser (1984) but not by Corbet and Hill (1986), Honacki, Kinman, and Koeppl (1982), or Myers and Carleton (1981). The subgenus *Melanomys* also sometimes has been treated as a distinct genus. Two other former subgenera of *Oryzomys*—*Micronectomys* Hershkovitz, 1948, and *Macruroryzomys* Hershkovitz, 1948—were called *nomina nuda* by Hershkovitz (1970). The species *O. dimidiatus*, once put in the genus *Nectomys*, then made the type species of the subgenus *Micronectomys*, now is considered closely related to *O. palustris* (Hall 1981; Hershkovitz 1970). Another species of *Micronectomys*, formerly designated *Oryzomys borreoi*, actually seems referable to the genus *Zygodontomys* (Gardner and Patton 1976). The only species of *Macruroryzomys*, *O. hammondi*, is still designated a species of *Oryzomys*, but its precise systematic position is uncertain and it seems to be closely related to the extinct genus *Megalomys* (Hershkovitz 1966*b*, 1970). Hershkovitz (1966*a*, 1966*b*) suggested that most species of the subgenus *Oligoryzomys* are synonymous with *O. nigripes*, but this view was not followed by Gardner and Patton (1976), Hall (1981), Massoia and Fornes (1967), Myers and Carleton (1981), or Olds and Anderson (1987). Cabrera (1961) listed the species *O. simplex*, of eastern Brazil, but noted that its status was doubtful, and Massoia (1980) referred it to the genus *Pseudoryzomys*. Benson and Gehlbach's (1979) recognition of *O. couesi* as a species distinct from *O. palustris* was followed by Jones et al. (1986), J. K. Jones and Engstrom (1986), and most other recent authors but not by Hall (1981). Haiduk, Bickham, and Schmidly (1979) found the karyotype of *O. couesi* to be similar to that of *O. palustris texensis*. Koop, Baker, and Genoways (1983) reported greater chromosomal polymorphism in *Oryzomys* than has been found in any other higher vertebrate.

Head and body length is 93–203 mm, tail length is 75–251 mm, and weight is usually 40–80 grams. The upper parts are grayish brown to ochraceous tawny, mixed with black; the sides are paler, with less black; the underparts are white to pale buff; and the tail varies from brownish above and whitish below to uniformly dusky. The four longer hind toes in members of the subgenus *Oligoryzomys* bear tufts of silvery bristles that project beyond the ends of the claws. The form in

Oryzomys is mouselike; the pelage is coarse but not bristly or spiny; the tail is usually long, with the annulations showing through the sparse hairs; and females have eight mammae. Rice rats may be confused with cotton rats *(Sigmodon)*, but the latter have longer, grizzled fur and a shorter, stouter tail.

Rice rats live in a variety of habitats, including forests, marshes, grassy areas, and brush-covered parts of mountains. In Venezuela, Handley (1976) collected *Oryzomys* under the following conditions: *O. albigularis*, mostly on the ground in moist parts of forests; *O. capito*, mostly in houses near streams or other moist parts of forests or forest openings; *O. fulvescens*, in a variety of forested and nonforested areas, mainly in moist places; and *O. macconnelli* and *O. minutus*, mainly on the ground in montane forests. Pearson (1983) found the preferred habitat of *O. longicaudatus* in Patagonia to be blackberry and wild rose tangles.

O. palustris, the best-known species, may be active at any hour of the day throughout the year. It swims and dives readily on the surface or underwater. Esher, Wolfe, and Layne (1978) reported its swimming speed to be 0.55 meters per second. Its presence in an area can be confirmed by the feeding platforms it constructs by bending vegetation and by its woven grassy nests, about 457 mm in diameter. The nest is usually placed in a slight depression in the ground in a tangle of vegetation but may be located 1 meter above ground in areas subject to periodic inundation (Lowery 1974). In drier areas, soil burrows are constructed. The diet includes the succulent parts of grasses and sedges, seeds, fruits, insects, crustaceans, and small fish.

Benson and Gehlbach (1979) reported a population density of 1.5 per 100 sq meters for *O. couesi* in southern Texas. In a marsh on the Gulf Coast of Mississippi, Wolfe (1985) found density in *O. palustris* to fluctuate from a low of about 2/ha. in the spring to a high of about 25/ha. in the autumn and early winter. In the Panama Canal Zone, Fleming (1971) found *O. capito* to have a maximum density of 4.3/ha. and an average home range of 1.33 ha. Home ranges overlapped. Breeding in this area occurred throughout the year, with no apparent peaks, and females produced an average of 6.06 litters per year. The mean gestation period was 28.2 days, and litter size averaged 3.29 and ranged from 2 to 5. The average life span was only 4.62 months, with few individuals surviving more than a year. On the basis of collection of various species in Nicaragua, reproduction evidently also oc-

curs throughout the year there (Jones and Engstrom 1986). In Patagonia, however, *O. longicaudatus* breeds during the spring and summer (November–December); litter size averages about 5 and ranges from 2 to 11 (Pearson 1983).

In coastal Louisiana, according to Lowery (1974), *O. palustris* breeds mainly from February to November. Females are capable of producing 7 litters per year but probably average only 5–6 because of their short life span. They have a postpartum estrus and usually mate within 10 hours of giving birth. The estrous cycle is 6–9 days (Wolfe 1982). In this species gestation lasts 25 days, and litter size is 1–7, usually 3–5. Newborn *O. palustris* weigh 2–5 grams, open their eyes at 6–11 days, are usually weaned and independent at 11–13 days, and are full grown at 4 months. Females apparently are able to breed at about 7 weeks. Average longevity is probably much less than 1 year.

The northern limit of the range of *O. palustris* varies with fluctuations in population density. Occurrences in New Jersey, for example, long appeared to be sporadic, and the species actually has been considered to be threatened in that state. Arndt, Rohde, and Bosworth (1978), however, found *O. palustris* to be not uncommon along the shores of Delaware Bay in southern New Jersey. According to Richards (1980), this species apparently extended its range northward into the Ohio Valley, Illinois, and Pennsylvania during past periods of ameliorating climate and became a commensal pest in Indian settlements. *O. subflavus* of eastern Brazil also is essentially a commensal, its distribution being limited mainly to plots of sugar cane (Mares et al. 1981).

The silver rice rat *(Oryzomys argentatus)* was discovered in 1973 and now has been found on nine of the lower Florida Keys. It is rare, generally occurring at low density in salt marshes, and is in danger of extinction through human modification of its habitat (Goodyear 1987; Goodyear and Lazell 1986; Layne 1978; Spitzer and Lazell 1978). It is classified as indeterminate by the IUCN.

A number of other island species or subspecies already seem to have become extinct (Allen 1942; D. A. Clark 1984; Goodwin and Goodwin 1973; Patton and Hafner 1983). *O. palustris antillarum*, of Jamaica, apparently disappeared in the 1880s, probably because of predation by the introduced mongoose and Norway rat. *O. victus*, of St. Vincent in the Lesser Antilles, is known only by a single specimen presented to the British Museum in 1897 and probably also was de-

Silver rice rat *(Oryzomys argentatus)*, photo by Numi C. Goodyear.

Bicolored rice rat (*Oecomys* sp.), photo by Jody R. Stallings.

stroyed by the mongoose. *O. galapagoensis galapagoensis,* of San Cristobal Island in the Galapagos, has not been seen since 1835, probably having disappeared through competition with introduced Norway and black rats. In contrast, on Santa Fe Island, where there are no introduced *Rattus,* the native *O. g. bauri* seems to be thriving and numbers at least 10,000 individuals (D. B. Clark 1980). Fragmentary remains and a report from the mid-nineteenth century indicate that an unnamed species of *Oryzomys* once occurred on Barbados (Marsh 1984, 1985).

RODENTIA; MURIDAE; Genus OECOMYS
Thomas, 1906

Bicolored Rice Rats

There are three species (Alho 1982; Cabrera 1961; Hall 1981; Hershkovitz 1960; Husson 1978):

O. bicolor, tropical zone from Panama to the Guianas and Bolivia;
O. cleberi, vicinity of Brasilia in central Brazil;
O. concolor, tropical and subtropical forest zones from Costa Rica to the Guianas and Paraguay, Trinidad.

Oecomys usually has been considered a subgenus of *Oryzomys,* but on the basis of karyological features, Gardner and Patton (1976) recommended that it be given full generic rank. It was so treated by Carleton and Musser (1984) but not by

Corbet and Hill (1986) or Honacki, Kinman, and Koeppl (1982).

According to Hall (1981), head and body length of Central American specimens is about 120–50 mm and tail length is 123–62 mm. The upper parts are cinnamon brown, heavily infused with black, to bright ochraceous rufous, and the underparts are white to ochraceous. The brownish tail is not sharply bicolored, is not penicillate as in *Oryzomys,* and averages more than half of total length. The anterior margin of the maxillary plate of the skull of *Oecomys,* when viewed dorsally, does not project half so far anterior to the remainder of the maxilla as it does in *Oryzomys.*

Original descriptions of South American forms (Allen 1893; Tate 1940; Thomas 1894, 1899, 1906*b,* 1909*a,* 1910*c,* 1924) indicate that head and body length is 115–26 mm, tail length is 129–58 mm, the upper parts vary greatly from tawny through reddish fulvous to blackish brown, and the underparts are whitish or pale yellow. The tail is well haired and penciled, though not heavily so. The feet are broad and suited for climbing, and the fifth hind toe is proportionally long. The skull resembles that of *Oryzomys,* but the braincase tends to be relatively larger and more rounded. Females have eight mammae. Nitikman and Mares (1987) reported weights of 21–53 grams for *O. bicolor* and 41–72 grams for *O. concolor.*

Handley (1976) reported these rice rats to occur throughout Venezuela in a variety of habitats, including savannah, pasture, prairie, and forest. Specimens were collected both in trees and on the ground, in either dry or moist areas. Also in Venezuela, O'Connell (1982) found population densities of up to 1.25/ha., year-round reproduction, and an average litter size of 4.5 young.

RODENTIA; MURIDAE; **Genus NESORYZOMYS**
Heller, 1904

Galapagos Rice Rats

There are three species, all in Ecuador's Galapagos Islands off northwestern South America (Hutterer and Hirsch 1979; Patton and Hafner 1983):

N. darwini, Santa Cruz Island;
N. fernandinae, Fernandina Island;
N. indefessus, Fernandina, Santiago, Baltra, and Santa Cruz islands.

Nesoryzomys sometimes has been considered a subgenus of *Oryzomys,* which also occurs in the Galapagos, but was given full generic rank by the above and most other recent authors, including D. A. Clark (1984), Corbet and Hill (1986), Gardner and Patton (1976), and Honacki, Kinman, and Koeppl (1982). Usually, *N. narboroughi,* of Fernandina, and *N. swarthi,* of Santiago, have been designated separate species, but the studies of Patton and Hafner (1983) suggest that they are best treated as subspecies of *N. indefessus.*

Measurements listed by Patton and Hafner indicate that head and body length is about 100–200 mm and tail length is 79–140 mm. Descriptions cited by Allen (1942) indicate that *N. indefessus* is dark gray or blackish above and grayish below and that *N. darwini* has a general bright fulvous coloration. Both *N. darwini* and *N. fernandinae,* the latter being known only from skeletal remains found in fresh owl pellets, are much smaller than *N. indefessus.* From *Oryzomys,* *Nesoryzomys* is distinguished by its elongate, narrow snout and by its hourglass-shaped interorbital region when the skull is viewed from above.

Nesoryzomys is said to be more nocturnal than *Oryzomys* and to live in burrows or rock crevices beneath bushes (Allen 1942). On Fernandina Island, individuals have been observed at all elevations from sea level to the volcano crater rim, and reproduction there probably is limited to the rainy season of January–April (Patton and Hafner 1983). *Nesoryzomys* displays apparent fearlessness and curiosity toward people, and campers on Fernandina Island often leave their tent doors open in order to keep the rats from chewing their way in (D. A. Clark 1984). *N. indefessus narboroughi* is still common on that island, where there are no introduced Norway or black rats, but all other representatives of the genus may now be extinct through competition with *Rattus. N. i. swarthi* has not been seen alive since 1906, *N. i. indefessus* probably disappeared about 1945, and *N. darwini* was last collected in 1930 (Goodwin and Goodwin 1973; Patton and Hafner 1983).

RODENTIA; MURIDAE; **Genus MEGALOMYS**
Trouessart, 1881

West Indian Giant Rice Rats

The following three species apparently survived into historical times (Hall 1981):

M. desmarestii, known only from Martinique;
M. audreyae, known only from Barbuda;
M. luciae, known only from Santa Lucia.

According to Hershkovitz (1966b, 1970), two other fossil species have been reported: *M. curazensis,* from Curacao, and *M. curioi,* from Santa Cruz Island in the Galapagos. The latter species, however, was recently placed in a new genus, *Megaoryzomys* (Lenglet and Coppois 1979). The closest living relative to *Megalomys* appears to be *Oryzomys hammondi,* of northwestern Ecuador, and that species may represent an ancestral stock once widespread in northern South America.

Only a few complete specimens are in existence, including two mounted examples in the National Museum of Natural History in Paris. The largest species, *M. desmarestii,* had a head and body length of about 360 mm and a tail length of about 330 mm. The fur was long and harsh but not spiny. The upper parts were glossy black or dark reddish brown. The chin, throat, underparts, and base of the tail were white. The well-developed ears were nearly naked. *M. luciae* was smaller and almost all brown. *M. audreyae* is known only from a mandibular fragment and an upper incisor.

Megalomys may once have occupied most of the islands of the Lesser Antilles, but detailed information is available only for *M. desmarestii* on Martinique (Allen 1942). This species apparently occurred in large numbers among the coconut

West Indian giant rice rat (*Megalomys* sp.), photo by F. Petter.

Spiny rice rat (*Neacomys* sp.), photo by Bruce J. Hayward. Skull: *N.* sp., photos from American Museum of Natural History.

plantations. It is said to have lived in burrows and to have taken to water when driven from shelter. It was commonly eaten by the people, and the European colonists actively tried to exterminate it because of its damage to crops. Its final extinction, however, has been attributed to the great volcanic eruption of Mount Pelee in 1902. *M. luciae* probably was exterminated during the nineteenth century, though one captive individual may have lived from 1849 to 1852 at the London Zoo. *M. audreyae* apparently disappeared soon after the occupation of Barbuda by Europeans and the consequent destruction of cover.

RODENTIA; MURIDAE; Genus NEACOMYS
Thomas, 1900

Bristly Mice, or Spiny Rice Rats

There are three species (Cabrera 1961; Hall 1981; Handley 1976; Husson 1978):

N. tenuipes, eastern Panama, Colombia, Venezuela, Ecuador;

N. spinosus, Colombia, Ecuador, Peru, Mato Grosso of central Brazil;

N. guianae, southern Venezuela, Guyana, Surinam, probably northern Brazil.

Head and body length is 64–100 mm and tail length is about the same. A male *N. guianae* weighed 20 grams (Husson 1978). The hairy coat is composed of a mixture of tubular bristles interspersed with slender and fairly soft hairs. The spines are most abundant on the back, fewer on the sides, and almost absent on the belly. The presence of spines is the most conspicuous difference between this genus and *Oryzomys*.

The coloration above ranges from dark rufous, dark fulvous, or orange rufous to bright ochraceous and is finely lined and darkened with the black tips of the spines. The coloration on the sides becomes lighter and clearer as the numbers and extent of the dark tips of the spines are reduced. In some forms the sides are clear fulvous. The underparts are white, creamy, or fulvous. The tail is scantily haired, brownish above and lighter beneath.

The altitudinal range is from sea level to about 1,100 meters. These rodents have been taken under rocks and logs in dense humid forests and, in Panama, among grass and bushes along the rocky edge of a sugar cane field. Several pregnant females, with two to four embryos each, were collected in November and December. In Surinam, Genoways, Williams, and Groen (1981) took a pregnant female with three embryos on 24 September and a lactating female on 20 July.

Abrawayaomys ruschii, photo by Jody R. Stallings.

RODENTIA; MURIDAE; Genus ABRAWAYAOMYS
Cunha and Cruz, 1979

The single species, *A. ruschii*, is known only from the type locality in the state of Espirito Santo, southeastern Brazil (Cunha and Cruz 1979).

The type specimen, a young female, has a head and body length of 201 mm, a tail length of 85 mm, and a weight of 46 grams. The general coloration of the upper parts is grayish yellow, the head is darker, and the underparts are yellowish white. The pelage is composed of flattened and grooved spines mixed with finer hairs. The spines are especially numerous on the back.

Abrawayaomys resembles *Neacomys* in the texture of its pelage but is much larger and has longer dorsal hairs, a broader face, a less rounded braincase, a proportionally larger palatal foramen, a more reduced interparietal, and procumbent incisors. The molar teeth resemble those of *Oryzomys* and *Akodon*.

RODENTIA; MURIDAE; Genus SCOLOMYS
Anthony, 1924

Spiny Mouse

The single species, *S. melanops*, is known only by six specimens taken at Mera in eastern Ecuador at an elevation of about 1,150 meters by G. H. H. Tate in 1924 and now in the American Museum of Natural History and by at least six more specimens subsequently collected in the same region (A. L. Gardner, U.S. National Museum of Natural History, pers. comm., January 1988).

Head and body length is about 90 mm and tail length is about 70 mm. The pelage, except for a small patch on the throat, consists of flattened spines mixed with longer, unmodified hairs. The specimens show great differences in col-

or, but all are some shade of brown above and dull gray below. The upper parts give an impression of sooty black flecked with brown. The black is from the tips of the spines, and the brown is from the tips of the unmodified hairs. The color of the head is about the same as that of the back. The color of the sides merges from the gray of the underparts to the black and brown of the upper parts. The hands and feet are gray and the digits are whitish. The finely haired and annulated tail is the same color as the back above but somewhat lighter below.

The thumb has a broad, flat nail, and the proportions of the digits are the same as in *Neacomys*. Aside from the darker coloration, *Scolomys* closely resembles *Neacomys* in all external characteristics. In cranial characters, however, these two genera are different. The skull of *Scolomys*, for example, is short and broad, whereas that of *Neacomys* is long and slender. Female *Scolomys* have six mammae.

RODENTIA; MURIDAE; Genus NECTOMYS
Peters, 1861

Neotropical Water Rats

There are two species (Cabrera 1961; Goodwin and Greenhall 1961; Petter 1979):

N. squamipes, northern and central South America, Trinidad;
N. parvipes, French Guiana.

In *N. squamipes* head and body length is 160–255 mm and tail length is 165–250 mm. The body and hind feet of this species seem to grow in length after maturity and into old age. A series of 58 adults weighed 160–420 grams. The upper parts are buffy to tawny, with mixtures of brown, and the sides are paler. The underparts are not sharply defined; they are whitish or grayish, with a pale to pronounced wash of reddish orange, at least on the chest and belly. The pelage has

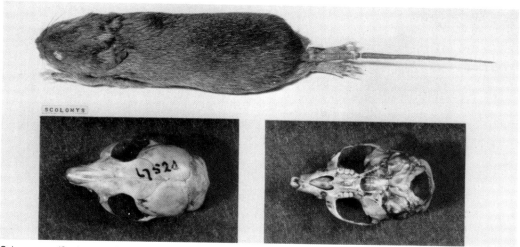

Spiny mouse *(Scolomys melanops)*, photos from American Museum of Natural History.

long, glossy guard hairs. There is a swimming fringe or keel of short, stiff hairs on the underside of the tail and a distinct fringe on the hind foot. The hind foot is large and strongly built; it is longer than wide, and the three middle digits are the longest. The webbing of the toes is well developed. Husson (1978) stated that female *N. squamipes* have two pairs of pectoral and two pairs of inguinal mammae.

These rats are usually found in forests up to about 2,200 meters in elevation. They generally occur near a swamp, lake, or stream. They are good, fast swimmers. Nests are built on the ground under old logs or brush heaps. Linares (1969) reported *N. squamipes* to occupy and nest in a cave, through which an underground stream passed.

In a 14-month study of *N. squamipes* in a gallery forest of Brazil, Ernest and Mares (1986) found individuals to occur almost exclusively near a stream or in other inundated areas. They became active at night and showed agility at both swimming and climbing. Prey was captured by rapid movements of the forepaws through water or by pouncing on land. The diet included vegetation, insects, tadpoles, and small fish.

Population density varied from 1 to 4 individuals per ha., and home range was 0.3–1.6 ha. Animals in breeding condition were collected throughout the year, though pregnancies were noted only in August, October, and November, at the start of the wet season. Records summarized by Ernest (1986) indicate that nests are usually constructed in dense undergrowth or under logs and roots and that litter size ranges from 2 to 7.

RODENTIA; MURIDAE; **Genus PSEUDORYZOMYS**
Hershkovitz, 1962

Ratos-do-Mato

There are two species (Hershkovitz 1962; Massoia 1980; Pine and Wetzel 1975; Wetzel and Lovett 1974):

P. wavrini, Bolivia, Paraguay, northern Argentina;
P. simplex, southeastern Brazil.

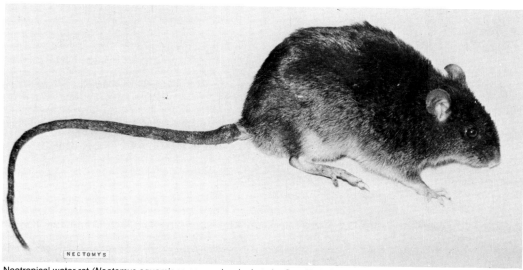

Neotropical water rat *(Nectomys squamipes amazonicus)*, photo by Cory T. de Carvalho.

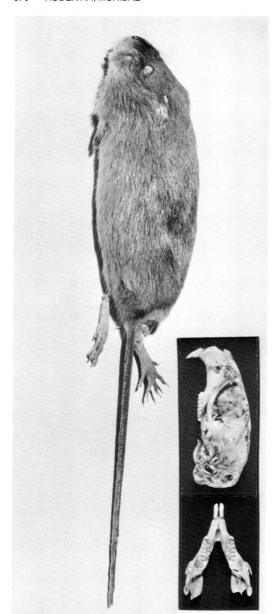

Rato-do-Mato *(Pseudoryzomys wavrini)*, photos from Field Museum of Natural History.

Head and body length is 94–140 mm, tail length is 106–40 mm, and weight is 30–56 grams. The upper parts are brown or yellowish brown, and the underparts are dull whitish with a buffy tinge. The well-haired tail is brown above and white below. This genus is distinguished from other phyllotine rodents by its moderate size, long tail, small ears, large hind feet, fully naked soles, and well-developed webbing between the hind toes.

According to Hershkovitz (1962), *Pseudoryzomys* is semiaquatic and has adaptations like those of *Oryzomys palustris*. Wetzel and Lovett (1974), however, found *Pseudoryzomys* to be pastoral at least during the dry season and not to be restricted to palustrine habitats.

RODENTIA; MURIDAE; **Genus RHAGOMYS**
Thomas, 1917

Brazilian Arboreal Mouse

The single species, *R. rufescens*, has been reported from Rio de Janeiro in eastern Brazil (Alho 1982). Specimens are in the British Museum of Natural History.

Head and body length is about 94 mm and tail length is about the same. The general color above and below is rich orange rufous. The hairs are slaty blue at the base and rufous at the tip. The underparts are slightly paler than the upper parts. The feet are yellow and the toes are whitish. The tail is scantily covered with brown hairs that form an inconspicuous tuft at the tip. The short ears, scarcely projecting above the fur, are thickly covered with rufous brown hairs.

This rodent is externally modified for an arboreal life. The fifth digit of the hind foot is long, and the hallux (big toe) in the type specimen appears to lack a claw. Females have six mammae.

RODENTIA; MURIDAE; **Genus RHIPIDOMYS**
Tschudi, 1844

Climbing Mice

There are seven species (Cabrera 1961; Guillotin and Petter 1984; Hall 1981; Handley 1976; Hershkovitz 1960):

R. latimanus, Colombia, Venezuela, Ecuador;
R. leucodactylus, southeastern Venezuela, French Guiana, the Andes from Ecuador to northwestern Argentina;
R. macconnelli, southeastern Venezuela and probably adjacent parts of Guyana and northern Brazil;
R. maculipes, eastern Brazil;
R. mastacalis, Venezuela, the Guianas, Brazil;
R. scandens, extreme eastern Panama;
R. sclateri, Venezuela, the Guianas, Trinidad.

Handley's (1976) list indicated the following differences: that *R. couesi*, of northern Venezuela and Trinidad, is a species distinct from *R. sclateri*; that *R. fulviventer*, of Colombia and Venezuela, is a species distinct from *R. latimanus*; that *R. venezuelae*, of western Venezuela, is a species distinct from *R. mastacalis*; and that *R. venustus*, of northwestern Venezuela, is a species distinct from *R. latimanus*.

Head and body length is 80–210 mm. The tail is normally longer than the head and body and may be as much as 270 mm in length. The coloration above ranges from pale grayish buffy, clay color, tawny ochraceous, and dull buffy, through fulvous and brown, to dark brown (almost black). The middle of the back is usually the darkest part of the body, because the dark tips of the hairs present in this area are fewer or completely lacking over the rest of the body. The underparts are white, creamy, buffy, fulvous, or grayish. The line of demarcation between the colors of the sides and the underparts is usually sharply defined. In some species a lateral line of buffy or yellow color separates the upper and lower surfaces. The digits are generally whitish, yellowish, or buffy, and the upper surfaces of the hands and feet are usually slightly darker than the remainder of the limbs. The tail is slightly darker than the back and is rarely bicolored.

In most species the fur is dense and soft, almost velvety. It varies from medium to long, though it may be short and dense. The tail is well haired and tufted terminally. The

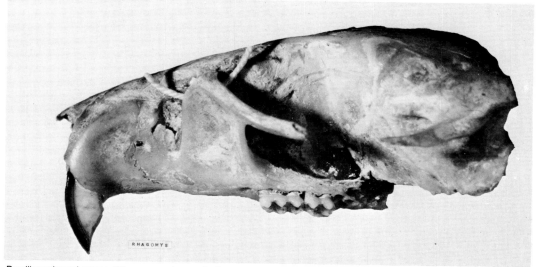

Brazilian arboreal mouse *(Rhagomys rufescens)*, photos from British Museum (Natural History).

rather large and broad feet and hands are adapted to arboreal life. The fifth toe is quite long, and the claws of both the hands and feet are large and efficiently curved for climbing. Females have six mammae.

According to Alho (1982), *Rhipidomys* is generally arboreal and nocturnal, is found in gallery forests, and may nest among passion fruit vines. He added that *R. mastacalis* lives in the trees in summer and on the ground in winter. Most specimens collected by Handley (1976) in Venezuela were taken in moist parts of forests; some were found on the

ground, some in trees, and some in houses. The only known specimen of *R. scandens* was shot in a tree 10.5 meters from the ground. This individual was active at dusk. Husson (1978) reported that all known specimens of *R. mastacalis* from Surinam were taken in buildings. Of the 12 female specimens, 3 were pregnant and 2 had newborn young; litter size was 3–5. Females apparently of this same species, with 2 or 3 young each, were observed near Rio de Janeiro, Brazil, in October and December. Dietz (1983) collected a lactating female *R. mastacalis* in central Brazil in August.

Climbing mouse *(Rhipidomys mastacalis nitela)*, photo from Zoological Society of London.

Thomas's paramo mouse *(Thomasomys hylophilus)*, photo by Norman Peterson.

RODENTIA; MURIDAE; **Genus THOMASOMYS**
Coues, 1884

Thomas's Paramo Mice

There are 25 species (Cabrera 1961; Gardner and Patton 1976; Handley 1976; Honacki, Kinman, and Koeppl 1982; Pine 1980; Voss and Linzey 1981):

T. aureus, the Andes from western Venezuela to Peru;
T. laniger, the Andes of Colombia and western Venezuela;
T. monochromos, northeastern Colombia;
T. baeops, western Ecuador;
T. kalinowskii, the Andes of central Peru;
T. vestitus, the Andes of western Venezuela;
T. hylophilus, northern Colombia, western Venezuela;
T. bombycinus, the Andes of Colombia;
T. cinereiventer, the Andes of Colombia and Ecuador;
T. rhoadsi, the Andes of Ecuador;
T. cinereus, southwestern Ecuador, northwestern Peru;
T. ischyurus, the Andes of Ecuador and northwestern Peru;
T. gracilis, the Andes of Ecuador and southeastern Peru;
T. pyrrhonotus, southern Ecuador, Peru;
T. paramorum, the Andes of Ecuador;
T. oreas, the Andes of Bolivia;
T. rosalinda, northwestern Peru;
T. taczanowskii, northwestern Peru;
T. notatus, southeastern Peru;
T. daphne, southwestern Peru, central Bolivia;
T. ladewi, the Andes of northwestern Bolivia;
T. incanus, the Andes of Peru;
T. dorsalis, eastern Brazil;
T. pictipes, extreme northeastern Argentina, southeastern Brazil;
T. oenax, extreme southern Brazil, Uruguay.

Data on the systematic affinities of the species of this genus are limited, and the above sequence is based in part on geographic distribution. Avila-Pires (1960) designated a separate genus, *Wilfredomys,* for the species *T. oenax.* Barlow (1969) did not recognize this distinction, and Pine (1980) argued that *Wilfredomys* does not warrant generic rank. Additional genera or subgenera sometimes recognized are *Erioryzomys*

Bangs, 1900, for *T. laniger* and *T. monochromos, Inomys* Thomas, 1917, for *T. incanus,* and *Delomys* Thomas, 1917, for *T. dorsalis.*

Head and body length is about 90–185 mm and tail length is 85–230 mm. Pine (1980) listed weights of about 35–61 grams for *T. oenax.* The tail in *Thomasomys* is normally longer than the head and body. The fur is usually thick and soft. The coloration above varies from olivaceous gray, dull olive fulvous, yellowish rufous, orange rufous, golden brown, reddish brown, and grayish brown to dark brown or almost black. The middorsal region is usually slightly darker than the rest of the body. The sides blend into the underparts, which are silvery grayish, soiled grayish, yellow, buffy, ochraceous buff, dark gray, or dark brownish. The underparts usually are not much paler than the upper parts. The hands and feet are about the same color as the underparts; often the central part of the upper surface is darker, and the fingers and toes lighter, sometimes whitish. The coloration of the tail varies from slightly paler than the back to slightly darker; the tail is normally moderately haired.

The hind foot usually is not modified for arboreal life, but in several species it may be very similar to the hind foot of *Rhipidomys.* It is difficult to differentiate some species of *Thomasomys* from *Rhipidomys.* The two genera can usually be distinguished, however, as follows: *Thomasomys* has an hourglass-shaped interorbital region and relatively long and robust molar rows, while *Rhipidomys* has an interorbital region with a conspicuous shelf or ridge and relatively short and narrow molar rows. Female *Thomasomys* have six or eight mammae.

Thomasomys is common in the forests on the eastern slopes of the Andes and ranges up to about 4,200 meters in such areas as the moist Urubamba Valley. It has not, however, established itself on the Altiplano. In Uruguay, Barlow (1969) found *T. oenax* in subtropical woodland. Apparently, some species are arboreal, while others are mainly terrestrial. *T. rhoadsi* is reported to dig burrows similar to those of North American microtine rodents, 25–76 mm below the surface of the ground. *T. hylophilus* has been found living among the natural galleries formed under moss-covered logs, roots, and debris. Barlow (1969) reported the stomachs of three *T. oenax* to contain plant material.

In eastern Brazil, *T. dorsalis* mates from August to at least January or February. There are two litters a year, with from two to four young in each litter.

Thomas's paramo mouse *(Thomasomys lugens)*, photo by Jaime E. Péfaur.

RODENTIA; MURIDAE; **Genus AEPEOMYS**
Thomas, 1898

There are two species (Cabrera 1961; Gardner and Patton 1976):

A. lugens, the Andes from western Venezuela to Ecuador;
A. fuscatus, the Andes of Colombia.

Aepeomys originally was placed in *Oryzomys* and long was included in *Thomasomys;* on the basis of karyological data, however, it was treated as a distinct genus by Gardner and Patton. This procedure was followed by Carleton and Musser (1984), Corbet and Hill (1986), and Honacki, Kinman, and Koeppl (1982).

In *A. fuscatus* head and body length is 100–120 mm, tail length is 85–110 mm, the upper parts are blackish with a faint wash of grayish brown, the underparts are dark gray, and the tail is light gray brown (Allen 1912). According to Thomas (1896, 1898a), two specimens of *A. lugens* have head and body lengths of 111 and 114 mm and tail lengths of 84 and 118 mm. The upper parts are olive brown and dark buffy gray, and the underparts are paler. The skull is narrow, the muzzle is long, and the thumb has a nail. The species is said to nest in trees, though it has small eyes, and most animals with small eyes are terrestrial or fossorial.

RODENTIA; MURIDAE; **Genus MEGAORYZOMYS**
Lenglet and Coppois, 1979

Galapagos Giant Rat

The single described species, *M. curioi,* is known only from bones and teeth found in cave deposits on Santa Cruz Island in the Galapagos. This species originally was placed in the genus *Megalomys,* which is otherwise known only from the West Indies. Lenglet and Coppois (1979) erected the new genus *Megaoryzomys* for this species and suggested affinity to *Oryzomys.* Steadman and Ray (1982) agreed with the need for a new genus but considered it to be most closely related to *Thomasomys.* They also noted that remains of an undesignated species of *Megaoryzomys* had been found on Isabela Island in the Galapagos.

According to Steadman and Ray, *Megaoryzomys* resembles *Thomasomys* and *Rhipidomys* but differs in having the following combination of characters: very large size (condylobasal length of skull more than 50 mm), a deep depression along the median suture of the frontals (also present in some species of *Thomasomys*), large and numerous palatal foramina, a very wide zygomatic plate, the zygomatic process of the squamosal joining the braincase at a more obtuse angle, a more rectangular braincase, a straight posterior margin of the interparietal, and planar molars.

The exact geological age of *Megaoryzomys* is not known, but the animal probably became extinct within the last two centuries. It may have survived until about 1900 and then disappeared through predation by introduced dogs, cats, pigs, and *Rattus.* Like many island animals, it had evolved in the absence of mammalian predators and would not have been wary when approached by an alien mammal (Steadman and Ray 1982).

RODENTIA; MURIDAE; **Genus PHAENOMYS**
Thomas, 1917

Rio de Janeiro Rice Rat

The single species, *P. ferrugineus*, is known from the states of Bahia and Rio de Janeiro in eastern Brazil (Alho 1982). Specimens are in the British Museum of Natural History.

Head and body length is about 150 mm and tail length is about 190 mm. The upper parts are light red or brilliant rust. The dark tips on the hairs on the top of the head and the center of the back give these areas a slightly darker appearance than the remainder of the upper parts. The ears are small, well furred, and rust-colored, with a few whitish hairs just behind them. The underparts are white with a slight yellowish tinge. The hands and feet are pale reddish brown, and the fingers and toes are partly white. The fur is thick and straight.

The skull is slender. The incisor teeth are somewhat heavy. Although the claws are slender and not especially curved, the fifth hind toe is long, and the hind foot seems to be slightly modified for arboreal life. The long tail is relatively well haired. Females have eight mammae.

RODENTIA; MURIDAE; **Genus CHILOMYS**
Thomas, 1897

Colombian Forest Mouse

The single species, *C. instans*, occurs in Colombia, western Venezuela, and Ecuador (Cabrera 1961; Carleton 1973).

Head and body length is 86–99 mm and tail length is 105–30 mm. The coloration above and below is slaty gray, though the face may be blackish from the nose to the eyes. The uniformly brown tail often has a white tip (10–20 mm). The hands and feet are brown, but sometimes the digits of all four limbs may be white. Some specimens have a white line from the throat to the middle of the belly; others have a bright buffy pectoral spot. The fur is soft and straight.

The body form is mouselike, not particularly specialized. The skull is delicate, with a large, rounded braincase and a small, slender muzzle. The upper incisor teeth project forward, and the lower incisors are long and slender. The fifth toe reaches to the base of the second phalanx of the fourth digit. This genus is similar to *Oryzomys*, being distinguished from it and other related genera mainly by characters of the skull and dentition.

The Colombian forest mouse inhabits the dark and damp forests, along with *Thomasomys hylophilus*. It is possible that *Chilomys* may be mistaken for the more common *Oryzomys*. In the cloud forests of Venezuela, Handley (1976) collected five specimens of *Chilomys*, either at the base of rotting, moss-covered logs, under moss-covered logs and fallen limbs, or under lichen- and moss-covered tree roots.

RODENTIA; MURIDAE; **Genus WIEDOMYS**
Hershkovitz, 1959

Red-nosed Mouse

The single species, *W. pyrrhorhinos*, is found in eastern Brazil, from the state of Ceara to the state of Rio Grande do Sul, and possibly in the Mato Grosso and Paraguay (Hershkovitz 1959b).

Head and body length is 100–128 mm and tail length is 160–205 mm. The coloration of the back is mixed buff and brown, and the underparts are white. In contrast, the nose,

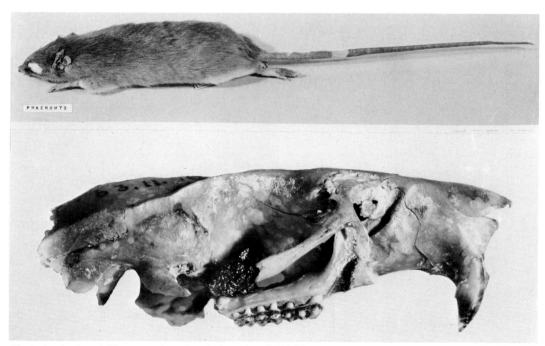

Rio de Janeiro rice rat *(Phaenomys ferrugineus)*, photos from British Museum (Natural History).

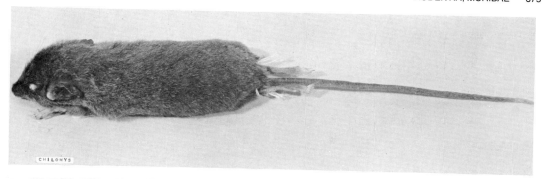

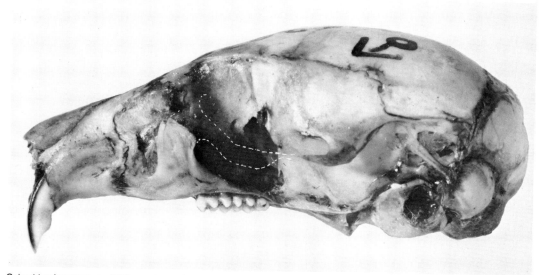

Colombian forest mouse *(Chilomys instans)*, photos from British Museum (Natural History).

Red-nosed mouse *(Wiedomys pyrrhorhinos)*, photo by Michael Mares.

eye ring, ears, outer sides of all four limbs, and rump are bright reddish orange. This rodent resembles *Thomasomys oenax* in external appearance, and *Calomys* and *Eligmodontia* in cranial and dental characters.

The red-nosed mouse inhabits scrub forests or caatingas. It is an active climber. It utilizes the abandoned nests of birds for shelter and the rearing of young. Among such nests are those of the thorn bird *(Anabates rufifrons)*. These nests are constructed of branches wound around liana vines and may be a few meters long. *Wiedomys* may also inhabit abandoned termite mud nests that were previously used by parrots. It also constructs its own nest of dry leaves and grass or cotton fibers in stone walls, hollow tree trunks, and shrub and palm thickets. Mares et al. (1981) referred to it as nocturnal and relatively rare. Streilen (1982a) reported captives to accept native seeds and to avidly chase and eat beetles and moths.

The red-nosed mouse apparently is gregarious: 8 adults and 13 young of varying size were found in one termite nest. In northeastern Brazil, Streilen (1982c) captured reproductively active females mainly at the end of the dry season, from August to October. Litter size averaged 3.8 and ranged from 1 to 6 young.

RODENTIA; MURIDAE; Genus AKODON
Meyen, 1833

South American Field Mice, or Grass Mice

There are 5 subgenera and 33 species (Cabrera 1961; Contreras 1968; Contreras and Rosi 1980b; De Santis and Justo 1980; Gardner and Patton 1976; Goodwin and Greenhall 1961; Langguth 1975a; Patterson, Gallardo, and Freas 1984; Pine 1973b, 1976a; Pine, Miller, and Schamberger 1979; Reig 1987; Voss and Linzey 1981; Ximenez and Langguth 1970):

subgenus *Akodon* Meyen, 1833

A. *albiventer*, mountains of southeastern Peru, western Bolivia, northern Chile, and northwestern Argentina;
A. *andinus*, the Andes from southern Peru to central Chile and western Argentina;
A. *boliviensis*, the Andes of southern Peru, Bolivia, and northwestern Argentina;
A. *varius*, Bolivia, Paraguay, western Argentina;
A. *puer*, mountains of central and southern Peru, Bolivia, and northwestern Argentina;
A. *olivaceus*, Chile, western Argentina;
A. *markhami*, Wellington Island off southern Chile;
A. *affinis*, western Colombia;
A. *urichi*, the Andes from western Venezuela to Bolivia, Trinidad;

A. *aerosus*, the Andes of Ecuador, Peru, and Bolivia;
A. *mollis*, the Andes from Ecuador to Bolivia;
A. *orophilus*, mountains of Peru;
A. *surdus*, mountains of southeastern Peru;
A. *azarae*, extreme southern Bolivia, Paraguay, northern Argentina, southern Brazil, Uruguay;
A. *pacificus*, mountains of western Bolivia;
A. *cursor*, southern Brazil, Paraguay, Uruguay;
A. *dolores*, central Argentina;
A. *iniscatus*, central Argentina;
A. *molinae*, Argentina;
A. *reinhardti*, eastern Brazil;
A. *serrensis*, eastern Brazil;
A. *nigrita*, eastern and southern Brazil;

subgenus *Deltamys* Thomas, 1917

A. *kempi*, islands of the Parana Delta between Argentina and Uruguay;

subgenus *Hypsimys* Thomas, 1918

A. *budini*, mountains of northwestern Argentina;

subgenus *Abrothrix* Waterhouse, 1837

A. *mansoensis*, south-central Argentina;
A. *illuteus*, mountains of northwestern Argentina;
A. *lanosus*, southern Argentina and Chile, Tierra del Fuego;
A. *longipilis*, Chile, Argentina, Tierra del Fuego;
A. *sanborni*, southern Chile and Argentina;
A. *xanthorhinus*, southern Argentina and Chile, including Tierra del Fuego and some nearby islands;
A. *hershkovitzi*, Cape Horn Islands south of Tierra del Fuego;
A. *llanoi*, Tierra del Fuego;

subgenus *Chroeomys* Thomas, 1916

A. *jelskii*, the Andes from central Peru to extreme northwestern Argentina.

The genera *Bolomys* and *Microxus* sometimes have been listed as subgenera of *Akodon*, but Reig (1987) and most other recent authorities have considered them distinct. Bianchi et al. (1971) also considered *Hypsimys, Abrothrix,* and *Chroeomys* to be separate genera. Voss and Linzey

South American field mouse (*Akodon olivaceus*), photo by Luis E. Peña.

South American field mouse *(Akodon varius)*, photo by Michael Mares.

(1981) considered such an arrangement to be consonant with data on the male reproductive organs, but it has not generally been followed by other authors. In addition, Bianchi et al. recognized the genera *Thaptomys* Thomas, 1916, and *Thalpomys* Thomas, 1916, but Reig explained that the former is part of the subgenus *Akodon* and the latter is actually a synonym of *Bolomys*, though its type species, *A. lasiotis*, has been renamed *A. reinhardti* and is also best put into the subgenus *Akodon*.

Head and body length is 75–140 mm and tail length is 50–100 mm. Dalby (1975) reported that sexually mature *A. azarae* weigh 10–45 grams. Animals of the genus *Akodon* have been described as heavy-bodied, short-limbed, short-tailed, volelike mice. The pelage is soft and full, varying above from mouse gray to dark brown. Some species have a red hue to the fur. The underside is white to dark gray, tinged with fulvous. Females have eight mammae.

South American field mice occur in a variety of habitats, including relatively arid country, grasslands, humid forests, and mountain meadows. The altitudinal range is from near sea level to about 5,000 meters. Reig (1987) noted that 82 percent of reported occurrences are from Andean environments. Some species are frequently found in human houses. Dalby (1975) reported that burrowing does not seem to occupy much of the time of *A. azarae* but that he examined one hole that went straight down for 12–15 cm, ending at a globular nest, and another that leveled off at a depth of 5–6 cm, continued for about 40 cm, and terminated at a globular nest. Various species of *Akodon* have been reported to be diurnal, nocturnal, crepuscular, or active at any time. Dalby (1975) stated that *A. azarae* is mainly herbivorous but takes a substantial amount of animal matter. Barlow (1969) found 11 stomachs of *A. azarae* to contain 20 percent plant material and 70 percent invertebrates.

On the pampas of northeastern Argentina, Dalby (1975) found that the peak density of *A. azarae* approached 200/ha.

but that density was reduced to approximately 50/ha. by late winter. At a semiarid site in north-central Chile, Fulk (1975) determined the density of *A. olivaceus* to be 30.3/ha. in August and 97.0/ha. in November. The average home range length for this species was 54.0 meters; the ranges of males overlapped considerably with one another and with the ranges of females, but there was no extensive overlap between the home ranges of females. Also in Chile, Greer (1965) found the home range of *A. olivaceus* to be 12.5 sq meters, while that of *A. longipilis* varied from 40 to 241 sq meters.

In general for the genus, the breeding season extends from August to May, there are probably two litters per year, and the number of young is usually 3 or 4, but Pine, Miller, and Schamberger (1979) reported that a female *A. xanthorhinus* taken in Chile on 2 February contained 10 embryos. Dalby (1975) reported the reproduction of *A. azarae* to be strongly seasonal, with litters born from November to April. Delayed implantation may occur in this species; gestation averaged 22.7 days, and litter size, 4.6 young. The young were weaned at 14–15 days of age and reached sexual maturity at 2 months, though young born late in the season did not become sexually mature until the following breeding season.

RODENTIA; MURIDAE; Genus BOLOMYS
Thomas, 1916

There are six described species (Macêdo and Mares 1987; Reig 1987):

B. amoenus, highlands of southeastern Peru;
B. lactens, highlands of northwestern Argentina;
B. lasiurus, eastern and southern Brazil, Bolivia, Paraguay;
B. lenguarum, Chaco of Bolivia, Paraguay, and probably Argentina, and Mato Grosso of Brazil;

Bolomys lasiurus, photo by I. Sazima.

B. obscurus, northeastern Argentina, southern Uruguay;
B. temchuki, northeastern Argentina.

In addition, Reig listed an unnamed species living in Buenos Aires Province, Argentina. The last two of the above species sometimes have been placed in the genus *Cabreramys* Massoia and Fornes, 1967, and some authorities, including Massoia (1982), continue to recognize that genus. Reig, however, considered *Cabreramys* a synonym of *Bolomys,* and most authorities now follow that arrangement. *Bolomys* itself sometimes has been listed as a subgenus of *Akodon.*

In *B. temchuki* head and body length is 85–129 mm, tail length is 60–94 mm, and weight is 22–52 grams (Massoia 1982). Original descriptions of several other species indicate measurements within this same range (Thomas 1898*b,* 1900, 1918). The typical color of the genus is agouti, which is most pronounced on the sides of the body. The back is chestnut to grayish, and the underparts are paler. The texture and length of the fur are about medium for a cricetine rodent. The tail, which is shorter than the head and body, is bicolored and tipped with fine hairs. The broad hands and feet have large nails like those of fossorial animals.

According to Reig (1987), the diagnostic characters of *Bolomys* include a broad and deep braincase, a short occipital region, a rather short and markedly tapering rostrum, the frontals always longer than the nasals, the parietals less than half the length of the frontals, the upper molars with lophs almost completely transverse, and the lower molars with lingual cusps somewhat anterior to the labial ones. Karyotypic data also support the distinction of *Bolomys* and the inclusion therein of *Cabreramys.*

Reig (1987) stated that *Bolomys* inhabits both the highest altitudes of the Altiplano and the Chacoan, pampean, and Brazilian lowlands. Barlow (1969) found *B. obscurus* to be restricted to limnic habitats and collected it in inundated places with plentiful vegetation; stomach contents indicated that the species feeds primarily on arthropods. Mares et al. (1981) reported *B. lasiurus* to inhabit cultivated and recently abandoned fields, to undergo population eruptions of great magnitude at irregular intervals, to be strictly terrestrial, and to be omnivorous in captivity. Streilen (1982*a,* 1982*b*) added that this species is primarily nocturnal and that it constructs a burrow system with several openings and one or two tunnels that lead to a spherical nest chamber about 40 cm below the surface. The nest is made of finely woven shredded grass and leaves. The fluctuating population size seems to result from the short-lived nature of the habitat and the inability of the species to withstand water deprivation very well. Streilen observed population densities as high as 187/ha. in June and July.

In a study of *B. lasiurus* on a Brazilian savannah, Alho and de Souza (1982) found an average population density of 11.8/ha. and a mean adult home range of 884 sq meters. These ranges overlapped, but those of adult males were mostly exclusive. Streilen (1982*d*) reported *B. lasiurus* to be highly intolerant of conspecifics of either sex and to readily engage in physical combat. According to Streilen (1982*a,* 1982*c*), most female *B. lasiurus* collected in April, May, and June in northeastern Brazil were pregnant. Reproductive activity in that area slowed by November, though females evidently produce more than one litter per year. There are 1–13, usually 3–6, young. Barlow (1969) took pregnant female *B. obscurus* from October to February and lactating females in December, April, and May. One female *B. obscurus* contained four embryos, another had three embryos, and a lactating female showed five placental scars.

Bolomys obscurus, photos by Abel Fornes and Elio Massoia.

RODENTIA; MURIDAE; Genus MICROXUS
Thomas, 1909

There are three species (Cabrera 1961; Reig 1987):

M. bogotensis, mountains of Colombia and northwestern Venezuela;
M. latebricola, the Andes of Ecuador;
M. mimus, the Andes of southeastern Peru.

Microxus sometimes has been designated a subgenus of *Akodon,* but Reig (1987) and most other recent authors consider it a separate genus.

Original descriptions of the type specimens are as follows: *M. bogotensis,* head and body length 91 mm, tail length 70 mm, upper parts uniformly blackish brown, underparts scarcely lighter, ears black, limbs and tail dark brown (Thomas 1895); *M. latebricola,* head and body length 83 mm, tail length 80 mm, pelage black throughout but with scattered silvery hairs on underparts (Anthony 1924); *M. mimus,* head and body length 92 mm, tail length 96 mm, fur very long and silky, upper parts dark olivaceous gray, underparts brownish gray, ears blackish brown, tail dark brown (Thomas 1901). The eyes are very small, the ears are short, and the forefeet are normal with small claws. According to Thomas (1909b), *Microxus* resembles *Akodon,* but the skull can be distinguished by its narrow zygomatic plate.

RODENTIA; MURIDAE; Genus PODOXYMYS
Anthony, 1929

Mount Roraima Mouse

The single species, *P. roraimae,* is known from five specimens taken on Mount Roraima at the point where Guyana, Venezuela, and Brazil come together (Cabrera 1961).

Head and body length of the type specimen is 101 mm and tail length is 95 mm. The original description (Anthony 1929:1) stated: "Pelage long and lax, 10–11 mm. long on back, blackish slate at base and for most of the length of the hair, only the tip being colored. Above, finely mixed clay-color and blackish, the minute specks of color at the tip of the hairs being insufficient to dominate the black and general impression resulting in a rather dark pelage; there is a tendency (shown by two out of five specimens) for the color pattern to be darkest on the rump, but otherwise the upperparts are fairly uniform; sides of head and underparts slightly lighter in tone than back; hands and feet clove-brown above; tail about half of total length, very sparsely haired, hair brown above and below; ears of fair size but partially hidden in the long pelage; eye rather small; claws of forefeet long (third claw 3 mm. beyond pad), slender, strongly compressed laterally, slightly curved, those of hind feet a trifle shorter."

Anthony also stated that this genus "appears to be somewhat intermediate in character between *Akodon* and *Oxymycterus*" and added: "External appearance that of a dark-colored, long-tailed *Akodon,* with long slender claws (which suggest the generic name); skull with long, slender rostrum, narrow zygomatic plate and general appearance of *Oxymycterus.*"

RODENTIA; MURIDAE; Genus LENOXUS
Thomas, 1909

Andean Rat

The single species, *L. apicalis,* occurs in southeastern Peru and western Bolivia (Cabrera 1961).

Head and body length is 150–70 mm and tail length is 150–90 mm. Coloration is grayish black above, paler on the sides, and grayish brown with a buffy wash, or grayish white, below. The tail is entirely brown except for the white terminal

Microxus bogotensis, photo by Norman Peterson.

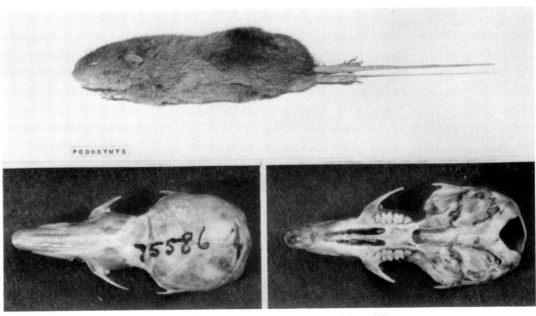

Mount Roraima mouse *(Podoxymys roraimae)*, photos from American Museum of Natural History.

part. The large size, black color, and long, white-tipped tail are distinguishing external features. The body form is heavy and ratlike, the ear is large (about 20 mm), and the claws are not enlarged. This genus could be regarded as a nonfossorial version of *Oxymycterus*. Some individuals have grooved upper incisors. *Lenoxus* lives in wooded areas between 1,850 and 2,450 meters in elevation.

RODENTIA; MURIDAE; Genus JUSCELINOMYS
Moojen, 1965

The single species, *J. candango*, is known from nine specimens taken on the grounds of the Zoobotanical Foundation in Brasilia, Federal District, Brazil.

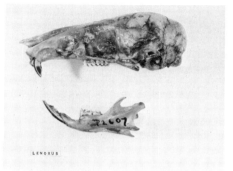
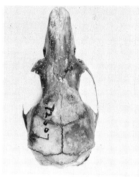
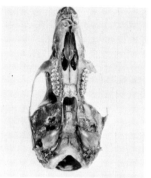

Andean rat *(Lenoxus apicalis boliviae)*, photos from American Museum of Natural History.

This genus differs from *Oxymycterus*, to which it is closely related, in having a shorter muzzle; a much thicker tail, densely covered with hair that completely covers the underlying scales; and a pelage in the ventral region composed of ferrugineous hairs with whitish bases. Cranially, *Juscelinomys* is much more strongly constructed than *Oxymycterus*, with more inflated bullae and a shorter, broader rostrum.

All specimens were taken at an elevation of 1,030 meters in a parklike area with scattered trees under which there were grasses. The entrances to two subterranean nests were recognizable by the tracks that the animals made on the ground during the excavations. The ground around the entrances to the nests was worn as smooth as pavement. The nests were poorly lined with grasses and other fine vegetable matter. Stomach contents included mostly unidentifiable vegetable material but also some large ants. Alho (1982) referred to *Juscelinomys* as terrestrial and nocturnal.

RODENTIA; MURIDAE; Genus OXYMYCTERUS
Waterhouse, 1837

Burrowing Mice

There are 13 species (Cabrera 1961; Hinojosa P., Anderson, and Patton 1987; Massoia 1963; Roguin 1986; Vitullo 1986; Ximenez, Langguth, and Praderi 1972):

O. inca, central Peru to western Bolivia;
O. hiska, southeastern Peru;
O. hucucha, central Bolivia;
O. paramensis, southeastern Peru, western and southern Bolivia, northern Argentina;
O. akodontius, northwestern Argentina;
O. rutilans, eastern and southern Brazil, Paraguay, Uruguay, northeastern Argentina;
O. nasutus, Uruguay;
O. hispidus, eastern and southern Brazil, northeastern Argentina;

O. roberti, eastern Brazil;
O. angularis, eastern Brazil;
O. delator, Paraguay;
O. iheringi, northeastern Argentina, extreme southern Brazil;
O. sanborni, central Chile and adjacent parts of Argentina.

Cabrera (1961) placed *O. iheringi* in *Microxus* and *O. sanborni* in the subgenus *Abrothrix* of *Akodon*, but both species were referred to *Oxymycterus* by Massoia (1963). Pine, Miller, and Schamberger (1979) continued to place *O. sanborni* in the genus *Akodon*. The name *O. rufus* sometimes is used in place of *O. rutilans*.

Head and body length is 93–170 mm and tail length is 70–145 mm. According to Dalby (1975), a series of sexually mature specimens of *O. rutilans* weighed 46–125 grams. The fur of *Oxymycterus* usually is not thick. The upper parts are reddish, yellowish brown, dark brown, or blackish, and the underparts are buffy, grayish brown, or grayish white. The tail, usually shorter than the head and body, is moderately or thinly haired. The snout is long and mobile, and the foreclaws are long and prominent. The outer digits of the hind feet are shorter than the central three digits. The dentition is weak in all species. Female *O. rutilans* have eight mammae, and female *O. iheringi* have six.

These rodents have been found in swamps, marshes, grasslands, brushy areas, woodlands, and forests. The following account of *O. rutilans* is based on the work of Barlow (1969) and Dalby (1975) in Uruguay and northeastern Argentina. This species seems to have an ecological role somewhat like that of *Onychomys* in North America. It apparently does not construct its own burrows or runways but may use those of *Cavia* or *Hydrochaeris* in stands of tall bunch grass. Its long foreclaws and shrewlike, pointed nose function in rooting for subsurface invertebrates. The diet is largely insectivorous, but other small invertebrates and some plant matter are eaten. *O. rutilans* seems to be almost entirely diurnal. Population density fluctuates from 5/ha. to 15/ha. Breeding occurs in all seasons of the year. Known litter size varies from one to six young, usually being two or three. The young are weaned after 14 days and reach sexual maturity near the age of 3

Burrowing mouse *(Oxymycterus rutilans)*, photo by Alfredo Langguth.

months. Alho (1982) noted that *O. roberti* is easily raised in the laboratory, that a social hierarchy is formed there, and that males mark a territory with urine and feces. A captive *O. rutilans* lived for 2 years and 7 months (Jones 1982).

RODENTIA; MURIDAE; **Genus BLARINOMYS**
Thomas, 1896

Brazilian Shrew-mouse

The single species, *B. breviceps*, has been collected in the highlands of southeastern Brazil in the states of Bahia, Espirito Santo, Rio de Janeiro, and Minas Gerais (Matson and Abravaya 1977).

According to Abravaya and Matson (1975), total length is 129–61 mm and tail length is 30–52 mm. The fur is crisp and short. Coloration is a uniform dark slaty gray throughout, with brown tips on the hairs. The back is slightly iridescent with a ruby tinge, at least when the fur is wet. The hands and feet are brown above.

The body form is modified for fossorial life. The short and conical head has extremely reduced eyes, and its short ears are hidden in the fur. The short tail is thinly haired. The hand has four functional digits and a strongly reduced fifth digit; the claws are well developed. The broad hind foot has prominent claws. No other South American mammal closely resembles *Blarinomys*. In external appearance this rodent resembles the North American short-tailed shrews of the genus *Blarina*, from which *Blarinomys* received its name.

Specimens have been taken in dense rainforests near the tops of hills at elevations of about 800 meters and at comparable localities in less humid areas. The genus is fossorial. Its burrow is made under the layer of litter on the forest floor. It is dug almost straight down for about 255 mm and then goes into a sloping tunnel that continues downward, but not at such a steep gradient. The diet is unknown but may consist primarily of insects and worms (Abravaya and Matson 1975). *Blarinomys* reportedly is docile and does not bite when cap-

tured. Pregnant females have been found in January, February, and September, and the observed number of embryos has been one or two (Matson and Abravaya 1977).

RODENTIA; MURIDAE; **Genus NOTIOMYS**
Thomas, 1890

Long-clawed Mouse

The single species, *N. edwardsii*, is known only by six specimens from southern Argentina (Pearson 1984). An additional five species have sometimes been listed for this genus, but Pearson noted that *N. angustus*, of western Argentina, is a synonym of *Akodon longipilis*; *N. macronyx* and *N. megalonyx*, of Argentina and Chile, are members of the separate genus *Chelemys*; and *N. valdivianus*, of Argentina and Chile, represents the genus *Geoxus*. *N. delfinis*, of southern Chile, was listed as belonging to *N. megalonyx* by Corbet and Hill (1986). Reig (1987) accepted Pearson's arrangement.

According to Pearson, head and body length of three specimens is 85–92 mm, tail length is 38–46 mm, and weight is 18.5–25.0 grams. The fur is not like that of a mole (as has sometimes been reported), but has bright colors (a rufous nose, bright lateral line, and whitish underparts) and a very well-developed fringe of hairs on the margins of the hind feet. The ear pinnae are small and extremely thin, especially at the margin, where they are covered with long, silky, white hairs that make it difficult to detect the margin. The nose is tipped with a dark, leathery button. The tail is relatively short, and the front claws are long, as in *Chelemys* and *Geoxus*, but these other two genera are larger and have molelike fur. The claws of *Geoxus* are stouter than those of *Notiomys* on the front feet and much longer on the hind feet. Both *Notiomys* and *Geoxus* have tiny, almost cuspless teeth, but the skull of *Notiomys* is much shorter and wider. The skull of *Chelemys* is much larger than that of *Notiomys*, and its teeth are much larger and heavier.

Pearson indicated that *Notiomys* is found on arid steppes,

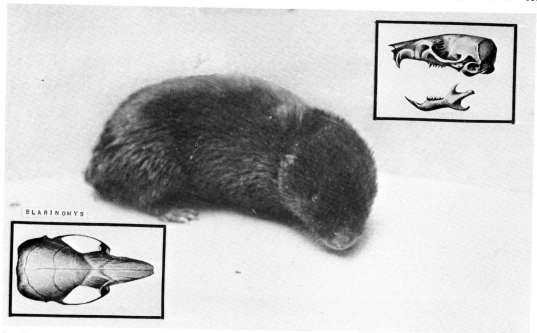

Brazilian shrew-mouse *(Blarinomys breviceps)*, photo by Joao Moojen. Insets: photos from *Bol. Mus. Paraense.*

with sandy soil and a mixture of shrubs and bunch grasses. It is largely fossorial and insectivorous but also may eat some seeds.

RODENTIA; MURIDAE; Genus GEOXUS
Thomas, 1919

Long-clawed Mole Mouse

The single species, *G. valdivianus,* is found in central and southern Chile and adjacent parts of Argentina (Cabrera 1961). This species usually has been assigned to *Notiomys* but was considered to represent a separate genus by Pearson (1984) and Reig (1987), who also suggested that there may be another species, with the name *G. michaelseni,* in extreme southern Chile.

Pearson (1983) listed overall length as 93–108 mm and adult weight as 25.5–41.0 grams. According to Reig, head and body length is less than 100 mm, tail length is less than half the head and body length, the fur is molelike and uniform in color, the front claws are stouter and longer than in *Notiomys* and *Chelemys,* the hind foot has long claws and no fringe of hairs on the margins, and the ears are small but have easily visible pinnae. The skull is slender and rather delicately built, being longer and narrower than that of *Notiomys.* According to Pearson (1984), the small and simple teeth of *Geoxus,* as well as its lack of a cecum, distinguish it from *Chelemys.*

This stout-bodied, short-tailed mouse is highly modified for a subterranean existence. The claws are powerful, sharp, curved, and long (about 7 mm). Its general appearance is shrewlike, somewhat like the North American *Blarina* or the Old World *Crocidura.* Pearson (1984) referred to it as an analogue of *Blarina.* It has a longer tail, larger ears, and larger eyes than does *Blarina,* but relative to its body length its claws

are longer than those of *Blarina, Scapanus, Thomomys,* or the other genera analyzed by Pearson for fossorial adaptation. In its search for worms and grubs it may make its own tunnels in soft soil or use those made by other species. It frequently digs down many centimeters below the surface, leaving mounds of earth up to a liter in volume.

In studies along the Andes of Patagonia, Pearson (1983) found *Geoxus* to occur in forests, marshes, and moist meadows but not to venture far out onto the steppe. Choice trapping sites were along logs in dense forest with rich humus. Activity was both diurnal and nocturnal. Stomachs of trapped individuals contained earthworms, slugs, beetle larvae, and arthropods but no spores or leafy material. Breeding took place in the spring (September–December) and probably continued through the summer. There were 3–4 embryos per pregnant female; a newly born individual weighed 3.5 grams; and males evidently attained sexual maturity at about 3 months.

RODENTIA; MURIDAE; Genus CHELEMYS
Thomas, 1903

Greater Long-clawed Mice

There are two species (Cabrera 1961; Pine, Miller, and Schamberger 1979):

C. macronyx, Chile, Argentina;
C. megalonyx, central and southern Chile.

These species usually have been assigned to *Notiomys,* but Pearson (1984) and Reig (1987) considered them to represent a distinct genus. Pearson (1983) listed overall length in *C. macronyx* as 109–46 mm and adult weight as 46–96 grams. According to Reig, head and body length of the genus is

Long-clawed mole mouse *(Geoxus valdivianus)*, photo by Richard D. Sage.

greater than 120 mm, the tail length is less than half the head and body length, and the fur is short, dense, and uniform in color (dark gray). The front claws are stout and strong but comparatively less developed than those of *Geoxus*. The hind foot has moderately long claws, but with no shaggy fringe of hairs on their margins. The small ears have visible pinnae. Unlike *Notiomys* and *Geoxus*, *Chelemys* has relatively complex teeth and a very large cecum (Pearson 1984).

Pearson's (1983, 1984) studies along the Andes of Patagonia indicate that *C. macronyx* lives primarily in forests of *Nothofagus*, from the timber line to the edge of the precordilleran steppe. It is basically a burrowing animal, a liter of earth or more sometimes being found at the mouth of its tunnels, but it evidently also moves about considerably on the surface. It is active both by day and by night. The stomachs of trapped individuals contained arthropods, green vegetation, and a substantial amount of fungi. Captives survived well on apples and rolled oats. Some animals apparently spend the winter under the snow, digging up rhizomes and amancay lilies.

Greater long-clawed mouse *(Chelemys macronyx)*, photo by Oliver P. Pearson.

When several individuals were put together in a cage, they sparred and jumped about excitedly; both males and females made grating and squeaking sounds. The breeding season in Patagonia seems to begin abruptly in the spring (about early November) and to continue for a considerable period. One pregnant female had four embryos and another had five. Young females apparently reach sexual maturity during the season of their birth.

Miller et al. (1983) noted that the species *C. megalonyx* is seldom collected and may be scarce. Its numbers might have been reduced through heavy livestock trampling of its burrow systems.

RODENTIA; MURIDAE; **Genus SCAPTEROMYS**
Waterhouse, 1837

Water Rat

The single species, *S. tumidus*, is known from southern Paraguay, extreme southern Brazil, Uruguay, and northeastern Argentina (Hershkovitz 1966a; Myers and Wetzel 1979).

Head and body length is 150–200 mm and tail length is 120–70 mm. The upper parts are grayish black and the underparts are grayish white. This rodent is thickset and heavy; it resembles the black rat *(Rattus rattus)* but has a much longer tail. The feet are long, and the middle toes of the hind feet are elongated. All the claws are long. Females have eight mammae.

Barlow (1969) stated that *Scapteromys* is semiaquatic, occupying marshy places and frequently inundated areas. Excellent swimming ability and agility in climbing onto tall plants contribute to its success in such areas. It does not seem to make runways of its own, but follows any open route through the vegetation and frequently uses the runways of *Cavia*. It digs to obtain food but apparently does not excavate burrows. It may dig shallow depressions under vegetation, in which it builds nests and bears its young. It is mainly nocturnal and crepuscular, but some specimens were collected during the day. Its diet is largely insectivorous, but other invertebrates and some plant material are also eaten. Alho (1982) stated that *Scapteromys* feeds on grasses and grass seeds.

Scapteromys has a highly developed sense of hearing and can perceive sounds of extremely high wave lengths. It is easily startled and seeks safety by plunging into the nearest body of water. Several vocalizations have been reported (Barlow 1969).

Breeding apparently takes place throughout most of the

Water rat *(Scapteromys tumidus)*, photo by Abel Fornes and Elio Massoia.

year (Barlow 1969; Hershkovitz 1966a). Males in reproductive condition have been found in every month except January, June, and July. Pregnant females have been taken in January, April, May, October, November, and December. The number of young varies from two to five and averages about four.

RODENTIA; MURIDAE; **Genus KUNSIA**
Hershkovitz, 1966

South American Giant Rats

There are two species (Avila-Pires 1972; Hershkovitz 1966a):

K. fronto, south-central Brazil, northern Argentina;
K. tomentosus, eastern Bolivia to southeastern Brazil.

Alho (1982) placed the latter species in *Scapteromys*.

These are the largest living cricetines. Head and body length is 160–287 mm and tail length is 75–160 mm. The pelage is coarse, thick on the upper parts and thinner underneath. The color of the upper parts is dark brown mixed with gray. The underparts are paler and more or less sharply defined from the upper parts. The tail is uniformly dark brown or black.

The external form is adapted for fossorial and palustrine life. All four feet are large and powerful and are provided with extremely long claws. The genus differs from *Scapteromys* in its usually larger size, much shorter tail, shorter hind feet, and four-rooted (rather than three-rooted) first lower molar.

Giant rats occur in mixed savannah-forest at elevations from sea level to 1,000 meters. *K. tomentosus* is a burrowing animal. It may live almost entirely underground during the dry season but mostly above ground during periods of heavy rains or inundations. Perhaps because of the animal's large size and powerful body, it has never been taken in ordinary snap traps. The few preserved specimens have been either seized by hand or captured by dogs. Alho (1982) reported *Kunsia* to be nocturnal.

RODENTIA; MURIDAE; **Genus BIBIMYS**
Massoia, 1979

Crimson-nosed Rats

There are three species (Contreras 1984b; Massoia 1979, 1980):

B. torresi, Parana River Delta of eastern Argentina;
B. labiosus, southeastern Brazil;
B. chacoensis, northeastern Argentina.

For three adult specimens of *B. torresi*, head and body length is 84–97 mm, tail length is 65–78 mm, and weight is 23–34 grams. The nose is bulky and of an intense crimson color. The dorsal coloration is dark chestnut with an almost black medial line; the sides of the body are yellowish chestnut; the underparts and feet are light gray; and the tail is dark gray above and light gray beneath.

This genus is about the same size as certain species of *Akodon* but has a larger skull and resembles *Scapteromys* and *Kunsia* in skeletal and dental characters. From these two genera, *Bibimys* differs in its much smaller overall size and more complex third lower molar.

Crimson-nosed rats are found in both wet and dry grassland and savannah. Alho (1982) indicated that *B. labiosus* is

South American giant rat *(Kunsia tomentosus),* photos from U.S. National Museum of Natural History.

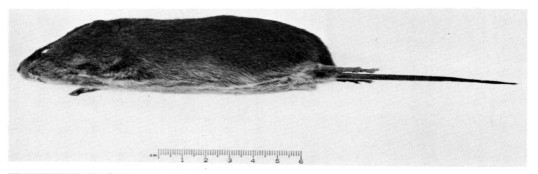

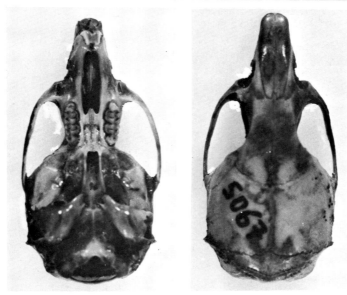

Crimson-nosed rat *(Bibimys torresi),* photos by Carlos Larrañaga.

Vesper mouse: Top, *Calomys laucha*, photo by R. B. Tesh; Bottom, *C. musculinus*, photo by Abel Fornes and Elio Massoia.

terrestrial, semifossorial, and nocturnal. It makes runways and feeds on grasses and seeds.

RODENTIA; MURIDAE; Genus CALOMYS
Waterhouse, 1837

Vesper Mice

There are seven species (Cabrera 1961; Contreras and Justo 1974; Handley 1976; Hershkovitz 1962; Pearson and Patton 1976; Petter and Baud 1981):

C. *sorellus*, the Andes of central and southern Peru;
C. *laucha*, southern Bolivia and southern Brazil to central Argentina and Uruguay;
C. *fecundus*, Bolivia;
C. *lepidus*, the Andes of southern Peru, western Bolivia, northern Chile, and northwestern Argentina;

C. *musculinus*, Argentina;
C. *callosus*, Bolivia and southern Brazil to northern Argentina;
C. *hummelincki*, coast of extreme northern Colombia and northwestern Venezuela and the nearby islands of Aruba and Curacao.

The last species was originally described from Curacao and Aruba as *Baiomys hummelincki* but was synonymized with *Calomys laucha laucha* by Hershkovitz (1962). He reported the presence of an introduced colony of *C. laucha laucha* in northeastern Venezuela and indicated that the animals on Curacao and Aruba had also been introduced. Handley (1976), however, found a wild population in northwestern Venezuela which appeared to represent the same distinct species as that on Curacao and Aruba. Corti, Merani, and Villafañe (1987) considered *C. callidus*, of southern Paraguay and northeastern Argentina, to be a species distinct from *C. laucha*.

Head and body length is about 60–125 mm and tail length

Cane mouse *(Zygodontomys brevicauda)*, photo by Marisol Aguilera M.

is about 30–90 mm. The tail is usually shorter than the head and body and is never much longer. A series of adult *C. laucha* from the Mato Grosso of Brazil weighed about 30–38 grams. The pelage usually is not very thick, but *C. lepidus* has soft fur. Coloration of the upper parts is buff, tawny, dark brown, grayish buff, or grayish. The underparts are usually grayish or white. Some species have a small white patch behind each ear. The tail is dark above and paler below in some forms and all white in others.

The body form is mouselike. The ear is prominent, the hands and feet are narrow, and the tail is moderately haired. The number of mammae apparently is unknown in some species but has been given as 8, 10, or 14; the number may vary in individual females of the same species. The low-crowned cheek teeth distinguish *Calomys* from related genera. Injuries of the tail in the form of a missing part or damaged skin are fairly common in *Calomys* and are probably caused by predators.

Vesper mice occur in a variety of habitats, including montane grasslands, brushy areas, and forest fringes. They may find shelter in bunch grass, in holes in the ground, in rotting tree stumps, or among rocks. They are active mainly at night and possibly also in the evening and early morning. Bozinovic and Rosenmann (1988) reported that captive *C. musculinus* could be induced experimentally to enter daily torpor, and they suggested that such probably occurred naturally as an adaptation to the cold, arid habitat of the species.

Barlow (1969) provided the following information on *C. laucha* in Uruguay. This species is usually found in open grasslands, on rocky hillsides, and in edge situations but occasionally is taken in the vicinity of marshes and swamps. It also frequents human dwellings and outbuildings. It is often observed at night in overgrazed rocky pasture. It does not make runways and rarely utilizes those of other species. Nests have been found under boards and rocks, in crevices in the ground, and even high above the ground in trees. The species climbs well, and on the ground it often hops on its hind legs in the manner of *Dipodomys*. The diet is predominantly vegetation, but the stomach of one individual contained beetle fragments. The breeding season apparently extends through much of the austral summer and autumn. Seven pregnant females collected in March and April contained four to eight embryos each, the average being about six. Hershkovitz (1962) added that a captive female *C. laucha* had a gestation period of about 25 days and produced two litters, the first with five young and the second with three.

In a study of captive colonies of *C. musculinus*, De Villafañe (1981) found the following averages: estrous cycle, 5.7 days; gestation, 24.5 days; litter size, 5.4 young; female sexual maturity, 72.5 days; and male sexual maturity, 82 days. Corti, Merani, and Villafañe (1987) stated that *Calomys* is a vector of the viral Argentine haemorrhagic fever.

RODENTIA; MURIDAE; **Genus ZYGODONTOMYS**
J. A. Allen, 1897

Cane Mice

There are three species (Cabrera 1961; Gardner and Patton 1976; Goodwin and Greenhall 1961; Hershkovitz 1962; Tranier 1976; Voss and Linzey 1981):

Z. brevicauda, southern Costa Rica to western Ecuador and Surinam, Trinidad;
Z. reigi, French Guiana;
Z. borreroi, northern Colombia.

Head and body length is 95–155 mm and tail length is 35–130 mm. The tail is usually shorter than the head and body. In some species the coat is full, long, and soft, while in others it is short and harsh. The upper pelage is yellowish brown, reddish, or grayish; the sides are grayish to yellowish; and the underparts are grayish white to buffy gray. The tail and ears are brown.

Zygodontomys resembles *Oryzomys* but may be distinguished by its relatively shorter tail and shorter hind feet. The anterior nails of *Zygodontomys* are smaller than those of *Akodon*. The ears of *Zygodontomys* are small. Females have eight mammae.

These animals live in open country and in areas of low bushes and thick ground cover, usually near water, from sea level to about 600 meters in elevation. They are terrestrial and their habits are much like those of meadow voles *(Microtus)*. They make runways through the dense grass and are active at night. Nests are made of grass and the down of flowers. These are built at the ends of short burrows in banks or under tree roots. The diet includes seeds, grasses (including corn and rice), and fruit. O'Connell (1982) listed population densities of up to 9.8/ha. for *Z. brevicauda* in Venezuela.

In a study of a captive colony of *Z. brevicauda* (which he referred to as *Z. microtinus*), Aguilera M. (1985) found the gestation period to be 25 days, litter size to average 4.6 young, and birth weight to be about 3.5 grams. The young opened their eyes after about 7 days and were weaned at 9–11 days. Females reached sexual maturity at around 25.6 days of age, and males at 42.3 days.

These mice have been used in the laboratory for research on yellow fever. They are easily kept in captivity and feed on a mixture of grains and fresh vegetables. In the wild they are attracted to cultivated fields and may become pests (Husson 1978).

Andalgalomys olrogi, drawing by Patricia Barquez.

RODENTIA; MURIDAE; Genus ANDALGALOMYS
Williams and Mares, 1978

There are two species (Olds, Anderson, and Yates 1987; Williams and Mares 1978):

A. olrogi, northwestern Argentina;
A. pearsoni, southeastern Bolivia, northern Paraguay.

The latter species was originally described as a member of the genus *Graomys* (Myers 1977*b*).

According to Olds, Anderson, and Yates (1987), head and body length is 86–136 mm and tail length is 97–136 mm. The upper parts are ochraceous buff or tawny, partly overlain with a suffusion of black hairs. The underparts and feet are white. The tail is brownish above and paler below and is slightly to moderately haired and penciled. The ears are moderately large and sparsely covered with fine hairs. The soles of the feet are naked.

Andalgalomys apparently is closely related to *Eligmodontia* and *Calomys*. *Andalgalomys* resembles *Eligmodontia* but can be distinguished by its longer, more penciled tail; its naked-soled, cushionless feet; its relatively longer nasals; and its more ovate molar cusps. *Andalgalomys* is most similar cranially to *Calomys* but has a more slender rostrum and much larger, more globular bullae.

A. olrogi is specialized for desert habitat, and *A. pearsoni* occupies dry grasslands. The collection of a subadult male *A. pearsoni* in July suggests that some winter breeding occurs (Myers 1977*b*).

RODENTIA; MURIDAE; Genus ELIGMODONTIA
F. Cuvier, 1837

Highland Desert Mouse

The single described species, *E. typus,* occurs from southern Peru and southern Bolivia along the Andes to the Strait of Magellan (Hershkovitz 1962; Pine, Miller, and Schamberger 1979). On the basis of size and karyotypic difference, however, Pearson, Martin, and Bellati (1987) suggested that the populations of Argentina represent a species separate from that in Peru.

Head and body length is 65–105 mm, tail length is 51–110 mm, and weight is 7–31 grams (Hershkovitz 1962; Pearson, Martin, and Bellati 1987). The pelage is long and soft. The upper parts are pale brownish yellow. In some forms the underparts are entirely white; in others the white is limited to the throat and upper breast. The tail is well haired. The young are darker and grayer than the parents.

Eligmodontia has some similarity to *Phyllotis;* both genera have a slender body, a long tail, and a silky pelage. *Eligmodontia* is distinguished by the specialization of its hind feet, inflation of the auditory bullae, and brachyodont molars. The hind feet are long, slender, and expanded toward the tips of the toes; the soles are covered with short hairs. In addition, a hairy "cushion" is located on the ball of each foot. The ears of *Eligmodontia* are long and prominent. Females have eight mammae.

This mouse has been reported from dry, gravelly plains; from areas of thornbushes or scattered grass clumps; and from bare, rocky hillsides up to about 4,575 meters in elevation. *Eligmodontia* can jump and climb with amazing agility, its period of greatest activity being during the night. It is said

Highland desert mouse *(Eligmodontia typus)*, photo by Michael Mares.

not to dig its own burrow but to inhabit the deserted dens of *Ctenomys*. It also lives under old logs and bushes. Its diet once was thought to consist largely of insects; however, Pearson, Martin, and Bellati (1987) noted that it is primarily granivorous, taking a variety of seeds, as well as some green vegetation and insects. It also has the unusual capability of drinking salt water, which suggests that it could utilize succulent parts of desert plants with a high salt content.

In Argentina, Pearson, Martin, and Bellati (1987) found population densities ranging from 0.4/ha. in spring to 3.5/ha. in autumn. The reproductive season in that area began in October (spring), continued through the summer, and ended in late April (autumn). Females could produce more than one litter during this period. The mean litter size was 5.9 young, and the range was 3–9. Both sexes attained sexual maturity after about 1.5 months of life and could breed in the season in which they were born. Few individuals lived more than 9 months.

Leaf-eared mouse *(Phyllotis darwini)*, photo by Jorge N. Artigas.

Four litters of parents caught in the wild were born in a laboratory during the autumn months of April and May and the spring months of October and November and contained 2, 2, 5, and 7 young (Mares 1988). The young weighed about 2 grams at birth and ate solid food at 17–30 days. Two females taken in March in central Chile were pregnant, one with 6 embryos and the other with 8 (Greer 1965).

Eligmodontia becomes so abundant at times that it invades human homes and becomes a pest by gnawing furniture and nibbling on unprotected food.

RODENTIA; MURIDAE; **Genus PHYLLOTIS**
Waterhouse, 1837

Leaf-eared Mice, or Pericotes

There are 11 species (Cabrera 1961; Hershkovitz 1962; Pearson 1972; Pearson and Patton 1976; Simonetti and Spotorno 1980):

P. osilae, southern Peru to northwestern Argentina;
P. andium, the Andes from central Ecuador to central Peru;
P. definitus, northwestern Peru;
P. wolffsohni, western Bolivia;
P. magister, western Peru;
P. caprinus, southern Bolivia, northwestern Argentina;
P. darwini, Peru to southern Chile and Argentina;
P. osgoodi, southern Peru, southwestern Bolivia;
P. amicus, northern and central Peru;
P. gerbillus, northwestern Peru;
P. haggardi, the Andes of Ecuador.

Cabrera (1961) placed *P. gerbillus* in a separate genus, *Paralomys* Thomas, 1926, but Pearson and Patton (1976) con-

Leaf-eared mouse *(Phyllotis magister)*, photo by Oliver P. Pearson.

sidered it merely a species of *Phyllotis*. Several additional species of *Phyllotis* were listed by Cabrera (1961) and Hershkovitz (1962), but these are now placed in the genera *Graomys* and *Auliscomys*. The subgeneric name *Loxodontomys* Osgood, 1947, used by Cabrera, is a *nomen nudum* (Pine, Miller, and Schamberger 1979).

Head and body length is 70–150 mm, tail length is 45–165 mm, and weight is 20–100 grams. The ears are so large (usually 20–30 mm long) that they give a leaflike appearance. This condition has given the scientific name to the genus and to a group of related genera, the phyllotines. The upper parts of *Phyllotis* are grayish, brownish, yellowish, orangish, buffy, or cinnamon. The underparts are grayish, whitish, or buffy. In appearance, general habits, and ecology *Phyllotis* appears to be the Andean equivalent of the North American *Peromyscus*.

Leaf-eared mice occur on savannahs, scrublands, deserts, and humid mountains. The elevational range is sea level to about 5,000 meters. Typical habitat is on bare ground, but near vegetation that will furnish food and some sort of shelter. These mice may seek rocky places, stone walls, houses, or burrows constructed by other animals. Some species are nocturnal, some are active occasionally during the day, and others are diurnal. They eat seeds, green plant material, and lichens. At a semiarid site in central Chile, Fulk (1975) found population density of *P. darwini* to be 29–46 per ha. and mean home range length to vary from 36.3 meters in February to 41.4 meters in November.

Hershkovitz (1962) stated that peak numbers of pregnant and lactating female *P. darwini* are found from November to February and in August and that embryo counts average 4.6 (2–7). For *P. osilae* the peak reproductive season is April–June and the average number of embryos is 4.4 (2–6). In central Chile, Greer (1965) collected a pregnant female and a lactating female *P. darwini* in March and juveniles in March, May, and November.

RODENTIA; MURIDAE; Genus GALENOMYS
Thomas, 1916

The single species, *G. garleppi*, is known from five specimens collected at altitudes of 3,800–4,500 meters on the Altiplano of southern Peru, western Bolivia, and northeastern Chile. Both Cabrera (1961) and Hershkovitz (1962) considered *Galenomys* to be a full genus, while Pearson and Patton (1976) considered it to be a subgenus of *Phyllotis*, pending availability of karyotypes.

Head and body lengths of two specimens are 105 and 123 mm, and corresponding tail lengths are 38 and 45 mm. The upper parts are buffy, thinly lined with brown; the head is paler than the trunk; the rump is ochraceous; and the underparts and legs are white. The body is stout, the tail is relatively short, and the ears are large. Females have eight mammae. According to Hershkovitz (1962), this genus is distinguished from *Phyllotis* by its vaulted skull; low-slung, anteriorly expanded zygomata; enlarged first upper molars; uniquely procumbent lower incisors; short, stout, thickly haired hind feet with extremely hairy soles; and very short, hairy tail.

RODENTIA; MURIDAE; Genus AULISCOMYS
Osgood, 1915

There are four species (Hershkovitz 1962; Pearson and Patton 1976; Simonetti and Spotorno 1980):

A. pictus, the Andes from central Peru to northwestern Bolivia;

A. sublimis, highlands of southern Peru, western Bolivia, northern Chile, and northern Argentina;

Auliscomys boliviensis, photo by Oliver P. Pearson.

A. boliviensis, southern Peru, western Bolivia, northern Chile;

A. micropus, southern parts of Chile and Argentina.

Cabrera (1961) treated *Auliscomys* as a subgenus of *Phyllotis*, and Hershkovitz (1962) did not use it as a genus or subgenus. Based in large part on karyological studies, however, *Auliscomys* was given generic rank by Pearson and Patton (1976) and Gardner and Patton (1976).

The following descriptions are taken from Hershkovitz (1962). In *A. pictus* head and body length is 100–147 mm; tail length is 75–118 mm; the head and shoulders are grizzled, contrasting with the brownish back and sides of the trunk; the rump, the base of the tail, and the outer thighs are more tawny or ochraceous than the back; and the underparts are gray to dirty white. In *A. sublimis* head and body length is 96–125 mm; tail length is 43–62 mm; the upper parts are buffy, finely mixed with black; the sides are paler; the cheeks are whitish; and the underparts are whitish to pale gray. In *A. boliviensis* head and body length is 105–45 mm; tail length is 78–105 mm; the upper parts are a mixture of gray, brown, and buff or ochre; the sides are paler; the muzzle and underparts are whitish; and the large ears have prominent ochraceous preauricular tufts. Pearson (1983) reported an adult weight range of 45–105 grams for *A. micropus*.

According to Hershkovitz (1962), the above species are found at high altitudes, from around 3,400 to 5,500 meters. *A. pictus* is active day and night and seems to prefer grassy places, especially near water. *A. sublimis* is nocturnal, shelters among rocks, and possibly remains below ground during much of the wet summer. *A. boliviensis* is diurnal and is found in many kinds of habitat on the Altiplano. This species is unusual in its association with the mountain viscacha *(Lagidium)*. It suns on rocks with the viscacha, scurries to shelter when *Lagidium* sounds an alarm whistle, and feeds with the larger rodent, "like an elf among grownups." Pearson (1983) reported *A. micropus* to be nocturnal, to dig burrows and climb skillfully, and to feed on green vegetation, fungi, and blossoms.

A mean population density of 0.27/ha. has been reported for *A. sublimus* in Peru (O'Connell 1982). Hershkovitz (1962) stated that *A. sublimis* seems gregarious and that nine individuals reportedly were found in a single burrow. In *A. pictus* breeding apparently begins in September or October; a pregnant female with five embryos was taken near Cuzco, Peru, on 13 April. Greer (1965) found two lactating female *A. micropus* in February–March, one with four embryos and the other with five.

Pearson (1983) reported that numerous adult *A. micropus* can live amicably in the same cage. This species evidently begins to mate by early September, as pregnant and lactating females and numerous young were taken in November. The number of embryos averaged about 4 and ranged from 1 to 7. The young become independent at an early age and may reach sexual maturity during the same season in which they are born.

RODENTIA; MURIDAE; Genus GRAOMYS
Thomas, 1916

There are three species (Cabrera 1961; Hershkovitz 1962; Pearson and Patton 1976):

G. griseoflavus, Bolivia, Paraguay, Argentina, possibly the Mato Grosso of central Brazil;

G. domorum, western and southern Bolivia, northern Argentina;

G. edithae, northwestern Argentina.

A fourth named species, *G. hypogaeus*, was recently shown to be referable to *Eligmodontia typus* (Massoia 1977; Williams and Mares 1978). Cabrera (1961) treated *Graomys* as a subgenus of *Phyllotis*. Hershkovitz (1962) did not use *Graomys* as a genus or subgenus and also considered *G. domorum* a subspecies of *G. griseoflavus*. Based in large part on karyological studies, however, Pearson and Patton (1976) and Gardner and Patton (1976) gave *Graomys* generic rank and *G. domorum* specific rank.

According to Hershkovitz (1962), head and body length is 106–65 mm; and tail length is 120–90 mm. The upper parts are buffy to tawny, more or less mixed with gray or black, and the sides and head have less black. The underparts are sharply defined white. The well-haired tail is always longer than the head and body, and its terminal part is usually bushy. The skull has a broad supraoccipital region with projecting edges.

Hershkovitz (1962) stated that *Graomys* is scansorial and adapts to most terrestrial and arboreal habitats. It has been found on stream banks, among rocks, in hollow logs, and in brush heaps. It lives in any kind of shelter, including abandoned birds' nests and holes in tree trunks. In summer it moves into human houses and does damage by eating boots and calico and gnawing on metal teapots. Its natural diet consists mainly of grasses, grains, and the fruits of mesquite. Most pregnant females have been taken in summer (December–March), and none have been found in the winter. The number of young averages about 6 and ranges from 3 to 10. Mares (1977) reported *G. griseoflavus* to be stronger and more aggressive than *Phyllotis darwini* and to be difficult to handle in the laboratory.

RODENTIA; MURIDAE; **Genus ANDINOMYS**
Thomas, 1902

Andean Mouse

The single species, *A. edax*, is known from southern Peru, Bolivia, extreme northern Chile, and northwestern Argentina (Hershkovitz 1962; Pine, Miller, and Schamberger 1979).

Head and body length is 135–65 mm and tail length is 105–50 mm. Two males weighed 80 and 91 grams (Pine, Miller, and Schamberger 1979). Coloration is dark buffy above and gray below. The body form is stocky. Females have eight mammae. This genus resembles *Punomys* but can be distinguished by its longer tail.

The Andean mouse lives on the Altiplano at elevations of 1,675–4,850 meters. It has been captured among rocks on the bank of a stream and in bushy thickets. It is nocturnal. Collectors have remarked that it lives in the branches of trees, where it makes its nest, and in round holes carpeted with very fine straw. Its diet consists of green plants.

Andinomys breeds at the end of the dry season; the young are thus born and raised when plant growth is at its peak. A female pregnant with three large embryos was taken on 18 December (Hershkovitz 1962).

RODENTIA; MURIDAE; **Genus CHINCHILLULA**
Thomas, 1898

Altiplano Chinchilla Mouse

The single species, *C. sahamae*, occurs in the highlands of southern Peru, southwestern Bolivia, northwestern Argentina, and northern Chile (Cabrera 1961).

Head and body length is 122–82 mm and tail length is 90–118 mm (Hershkovitz 1962). The color pattern is distinctive: the upper parts are buffy or grayish with black lines, and the underparts are snow white, the white hips and white rump banded with black. The tail is fully haired. The fur is thick, soft, and silky, resembling the pelage of *Chinchilla* in texture and color. The ears are large and wide. The feet are broad and the digits are normal.

This mouse inhabits the Altiplano at elevations of 3,500–5,000 meters. In its high, relatively barren habitat it is associated with such rodent genera as *Phyllotis, Akodon, Lagidium*, and *Punomys*. It lives in rocky places and is most often caught among boulders and along stone walls. It is active at night, when it feeds on plants. At times its enormous stomach is distended by more than 12 grams of green, finely ground vegetable matter (Hershkovitz 1962). The chinchilla mouse breeds in October and November, at the end of the dry season, so the young are born and raised when plant growth is at its peak.

Locally, *Chinchillula* is trapped for its warm fur. The skins

Andean mouse (*Andinomys edax*), photo by Michael Mares.

Altiplano chinchilla mice *(Chinchillula sahamae)*, photos by Hilda H. Heller.

are used as trimmings or made into robes; a robe made from the pelage of this genus may contain more than 150 skins. In some areas extensive trapping may seriously reduce the numbers of this beautiful rodent. Miller et al. (1983) trapped extensively within its former habitat in northern Chile but could not locate it.

RODENTIA; MURIDAE; **Genus IRENOMYS**
Thomas, 1919

Chilean Rat

The single species, *I. tarsalis*, occurs in southern Chile, including the islands of Chiloe and Guaitecas, and in the Andes of southern Argentina (Cabrera 1961).

Head and body length is about 100–140 mm and tail length is 134–90 mm. Pearson (1983) reported an adult weight range of 30–67 grams. The upper parts are grayish red or grayish cinnamon, and the underparts are cinnamon with a buffy or pinkish wash. The ears are a contrasting black color, the hands and feet are mainly whitish, and the tail is usually dark brown throughout. One structural feature, the grooving of the upper incisors, helps to distinguish this genus from related rodents.

The Chilean rat inhabits humid, temperate forests and is thought to be mainly arboreal. Its numbers fluctuate considerably, perhaps in response to the fruiting of various plants, such as bamboos. Meserve, Lang, and Patterson (1988) reported it to feed mainly on fruits, seeds, and green vegetation. According to Pearson, it is nocturnal and docile, and when released, it scampered across the forest floor, went down a burrow, or climbed trees or bamboo. It climbed 2.5 cm bamboo stems by "shinnying," grasping around the stem with all four limbs, then advancing both hind feet or both front feet alternately. A captive ate rolled oats, apple, and tender bamboo shoots. Mating was found to occur in the spring, and births evidently took place in summer. Two pregnant females captured in November contained 3 and 4 embryos.

RODENTIA; MURIDAE; **Genus PUNOMYS**
Osgood, 1943

Puna Mouse

The single species, *P. lemminus*, inhabits the Altiplano of Peru at elevations of 4,450–5,200 meters. Several other mammals are found at these elevations, but *Punomys* is the only one that lives entirely at such heights.

Head and body length is 130–55 mm and tail length is usually 50–70 mm. The pelage is long, soft, and loose. The upper parts are dull buffy brown or grayish brown, and the belly is whitish or grayish, occasionally with a buffy wash. The hands and feet are usually dusky above and black below. The tail is dusky above and white below, or dusky throughout.

This genus is unique in structure. The body form is stocky and volelike, and the claws are uniformly small. The dentition is peculiar: the upper incisor teeth are fairly stout and heavy, with smooth front surfaces, and the lower incisors are separated by a considerable space. The surfaces of the grinding teeth are unusually complex. The dull color and short tail help to identify this genus.

Specimens of *Punomys* were not collected until 1939, perhaps because of the extraordinary area of its occurrence. The generic name refers to the puna, or treeless zone, in the higher parts of South America. The puna mouse usually dwells in barren areas with broken rocks and where plants of the genus *Senecio* abound. *Punomys* is common within its range. It is active during the daytime, moving about from the

Chilean rat *(Irenomys tarsalis)*, photo by P. L. Meserve.

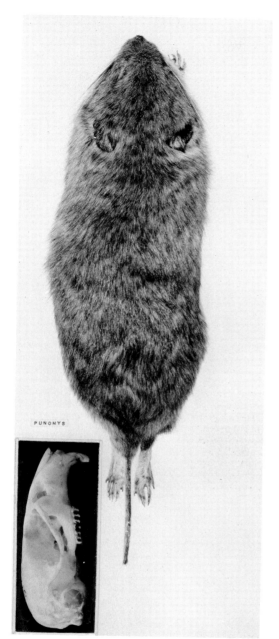

Puna mouse (Punomys lemminus), photos from Field Museum of Natural History.

after its rocky shelter has been removed and it is exposed. The young are apparently born during the warm, wet season from November to April, when plant growth is at its peak. A pregnant female contained two embryos.

RODENTIA; MURIDAE; **Genus EUNEOMYS**
Coues, 1874

Patagonian Chinchilla Mice

Yáñez et al. (1987) recognized only the single living species *E. chinchilloides*, with a range now known to include most of Chile, western and southern Argentina, and Tierra del Fuego. Earlier, Hershkovitz (1962), with considerable question, listed the following four species: *E. chinchilloides*, southern Argentina, Tierra del Fuego; *E. noei*, central Chile; *E. mordax*, west-central Argentina; and *E. fossor*, northwestern Argentina. Cabrera (1961) put *E. mordax* and *E. fossor* in a subgenus, *Chelemyscus* Thomas, 1925. Pine, Miller, and Schamberger (1979) suggested that another species of *Euneomys*, in addition to *E. noei*, occurs in central Chile.

Head and body length is 90–156 mm and tail length is 53–103 mm. The upper parts are cinnamon, reddish, reddish brown, or brownish, and the underparts are buffy or grayish. The hands and feet are usually white. The tail is usually brownish above and whitish below. A white stripe bordering the lips is present in some individuals. The body form is stocky. Females have eight mammae. *E. fossor* is especially modified externally for a burrowing life in that its fur is soft, its ear conch is greatly reduced, and its claws are elongated.

These rodents inhabit brushy and wooded areas. Individuals not quite fully grown have been seen in Patagonia in February, which suggests that breeding may take place near the end of the year.

RODENTIA; MURIDAE; **Genus NEOTOMYS**
Thomas, 1894

Andean Swamp Rat

The single species, *N. ebriosus*, is known from central and southern Peru, western and southern Bolivia, extreme northern Chile, and northwestern Argentina (Cabrera 1961; Pine, Miller, and Schamberger 1979).

Head and body length is about 120–75 mm and tail length is 62–88 mm. Three specimens from Chile weighed 63, 63, and 69 grams (Pine, Miller, and Schamberger 1979). In the subspecies *N. e. ebriosus* the upper parts are grayish brown; the chest is dirty brownish, and this color extends as a line for a short distance down the middle of the belly; the remainder of the belly is whitish; the tail is gray to buffy below and brown above; and the hands and feet are grayish to buffy, sometimes with reddish brown on the ankles. The subspecies *N. e. vulturnus* is slightly paler than *N. e. ebriosus* both above and below, and the brown sternal band is reduced or lacking.

Neotomys is easily identified by a bright rufous nose, unusually long guard hairs, and broad incisors, the upper ones having a narrow groove at the outer corner. Although it has the usual mouselike body, *Neotomys* is quite stockily built. Females have eight mammae.

Available evidence indicates that this mouse prefers grassy areas with scattered shrubs, along the banks of streams and marshes, at elevations of 3,400–4,525 meters. It is not generally present in rocky areas. It is active both day and night. Individuals living at higher elevations are possibly more ac-

shelter of one rock to another. It feeds mainly, if not entirely, on the two most offensive-smelling plants in its range: *Senecio adenophylloides*, a fleshy-leaved herb, and *Werneria digitata*, an herb. Twigs up to 510 mm long are cut from these plants to be stored under rocks in caches that may contain 30 or more cuttings. These large twigs are manipulated with considerable dexterity as the puna mouse consumes them. The plants are not chewed thoroughly, so the feces consist of large amounts of undigested material.

Caged individuals show no fear of people; in fact, in the wild the puna mouse sometimes allows itself to be picked up

Patagonian chinchilla mouse *(Euneomys chinchilloides)*, photo by Oliver P. Pearson.

tive during the warmth of the day, whereas those inhabiting somewhat lower elevations are nocturnal. Apparently, this genus is not abundant and its range is generally isolated from those of other rodents. Its shelters are usually located under isolated rocks on level ground.

RODENTIA; MURIDAE; Genus REITHRODON
Waterhouse, 1837

Coney Rat

The single species, *R. physodes,* occurs in Uruguay, Chile, and Argentina, including Tierra del Fuego (Barlow 1969; Cabrera 1961; Pine, Miller, and Schamberger 1979). According to Dalby and Mares (1974), the correct name for this species is *R. auritus.*

Head and body length is 130–200 mm and tail length is 85–105 mm. Three females from northwestern Argentina weighed 77, 94, and 95 grams (Dalby and Mares 1974). The fur is thick and soft. The upper parts usually are buffy, often with a mixture of grayish or black hairs; the underparts are whitish or grayish, often with a buffy wash. In addition to the inguinal region and the inner side of the thighs, the hands, feet, and tail are often white. The outer hind toes are reduced, and there is webbing between the middle hind toes. The soles are hairy from the back of the heel to the base of the outer digits. The ears are fairly large. Each upper incisor has a deep median groove. Females have eight mammae.

The coney rat inhabits pampas, cultivated fields, stony hills, and sandy coasts. Barlow (1969) reported that in Uruguay this animal is usually found in overgrazed pasture, among rocky outcrops, or on well-drained slopes with scanty vegetation. It either digs its own burrows or utilizes, with or without modification, the abandoned burrows of *Dasypus* and other fossorial mammals or natural crevices and holes among the rocks. The burrows constructed by *Reithrodon* were reported to have one or two entrances about 5 cm wide, to descend to about 10–25 cm below the surface and then become level, to be about 1–2 meters long, and sometimes to contain oval chambers up to 30 cm wide, in one of which was a nest platform composed of fine, dry grass. Barlow found *Reithrodon* to be strictly nocturnal in his study area but noted that this may simply have reflected a low population level at the time. *Reithrodon* is pastoral, feeding mainly on grasses,

Andean swamp rat *(Neotomys ebriosus)*, photo by B. Elizabeth Horner and Mary Taylor of specimen in Harvard Museum of Comparative Zoology.

Coney rat *(Reithrodon physodes),* photo by Abel Fornes and Elio Massoia.

other plants with tuberous rhizomes, and roots.

A mean population density of 0.69/ha. has been reported for Argentina (Dalby 1975). In Barlow's study area in Uruguay the nearness of burrows to each other suggested that *Reithrodon* may be variously gregarious or solitary. Barlow collected males in reproductive condition from January to May and two pregnant females, one with four embryos in January and one with three embryos in May. Also in Uruguay, a lactating female was taken in January and a female with four nestling young was taken in October. In Chile, Pine, Miller, and Schamberger (1979) collected a pregnant female with five embryos on 3 February. Earlier information indicates that *Reithrodon* breeds throughout the year and that the number of young varies considerably. According to Jones (1982), a captive individual lived for five years and six months.

Gudynas (1989) reported that populations of *Reithrodon* in Uruguay are declining because of grassland alterations for cattle raising and competition with introduced *Lepus.*

RODENTIA; MURIDAE; Genus HOLOCHILUS
Brandt, 1835

Web-footed Rats, or Marsh Rats

There are four species (Cabrera 1961; Handley 1976; Massoia 1980, 1981):

H. brasiliensis, central and eastern Brazil, Paraguay, Uruguay, northeastern Argentina;
H. sciureus, Venezuela, the Guianas, eastern Peru, Brazil, Bolivia;
H. chacarius, northwestern Argentina;
H. magnus, eastern Brazil, Uruguay, northeastern Argentina.

Holochilus usually is considered a close relative of *Sigmodon,* but on the basis of chromosomal data, Baker, Koop, and Haiduk (1983) suggested that it actually is systematically near *Oryzomys.* Some authorities include *H. sciureus* and *H. chacarius* in *H. brasiliensis.*

Head and body length is 130–220 mm and tail length is 130–230 mm. Twigg (1965) found *H. brasiliensis* to weigh up

to 280 grams, but *H. magnus* is the largest species. The upper parts of *Holochilus* are buffy, orangish, or tawny, usually mixed with black; the sides are paler; and the underparts vary from white to orange, except for the white and gray throat and inguinal region. The tail is thinly haired and is uniformly brown or somewhat paler beneath. Some animals have a fringe of hairs on the underside of the tail, which may serve as an aid in swimming. The toes are usually webbed. The body form is ratlike. *Holochilus* differs from the genus *Sigmodon* in the conspicuous webbing of its toes and in its soft, not harsh, pelage. According to Husson (1978), female *Holochilus* have eight mammae.

Web-footed rats are semiaquatic. They live in marshes, grassy stream banks, canebrakes, and other moist, non-wooded areas from sea level to about 2,000 meters. They construct nests on the ground or up to 3 meters above the surface in reeds or trees. The nests are made of reeds, cane, and leaves firmly interlaced and woven together. The lower half of the nest is usually packed with plant material, and the upper half contains the living quarters and the entrances. The floor has a bedding of soft plant substances. These nests are 20–40 cm in diameter and weigh as much as 100 grams. When disturbed, the occupant either leaps from the nest into the water and swims away or climbs higher in the tree. Marsh rats are active mainly at night but also may be abroad during the day. They feed primarily on marsh plants and also take mollusks.

A mean population density of 4.3/ha. has been reported for Argentina (Dalby 1975). In Uruguay, Barlow (1969) found 45 nests of *H. brasiliensis* in an area of about 50 ha. There were as many as 11 nests, some only a meter apart, in a single tree. Barlow found pregnant females in April and May and suggested that breeding activity was reduced with the onset of winter. In Guyana, Twigg (1965) found reproduction to continue throughout the year, with litter size averaging 3.12 (1–8) young and sexual maturity being attained relatively early in life. In eastern Brazil the period of pregnancy extends from February to April, and litter size is usually 5–8 but ranges up to 10.

At certain times marsh rats increase to such numbers that they destroy cultivated crops in fields and warehouses. Some investigators have connected these outbreaks with the fruiting season of the bamboo *(Merostichis).* After depleting the bamboo seeds, the rats begin to feed on cultivated crops. Eventually, however, high mortality rates of uncertain cause bring about a decline in their numbers.

Web-footed rat, or marsh rat *(Holochilus brasiliensis)*, photo by Joao Moojen.

RODENTIA; MURIDAE; Genus SIGMODON
Say and Ord, 1825

Cotton Rats

There are eight species (Cabrera 1961; Elder and Lee 1985; Hall 1981; Husson 1978; D. F. Williams 1986):

S. hispidus, southern United States to northern Venezuela and northwestern Peru;

S. alleni, western Mexico from southern Sinaloa to southern Oaxaca;

S. leucotis, central Mexico;

S. ochrognathus, north-central Mexico and adjacent parts of Arizona, New Mexico, and Texas;

S. mascotensis, southern Mexico;

S. arizonae, southeastern California, Arizona, western Mexico;

S. fulviventer, central New Mexico and southeastern Arizona to central Mexico;

S. alstoni, northern South America.

Handley (1976) placed *S. alstoni* in the genus *Sigmomys* Thomas, 1901. This procedure was followed by Williams, Genoways, and Groen (1983) but not by Carleton and Musser (1984), Corbet and Hill (1986), or Honacki, Kinman, and Koeppl (1982). Husson (1978) considered *Sigmomys* to be a subgenus.

Head and body length is 125–200 mm, tail length is 75–138 mm, and weight is 70–211 grams. The upper parts are grayish brown to blackish brown, mixed with buff, and the underparts are grayish or buff. The tail is blackish above and paler below. The fur is often harsh. The body form is stocky,

the ears are small, and the three central digits of each hind foot are larger than the other two. The generic name refers to the S-shaped pattern of the molar cusps.

Cotton rats prefer grassy and shrubby areas. Their presence is indicated by surface trails, shallow burrows beneath small plants, and small nests of grasses and sedges located in cuplike depressions in the ground. At the northern extremity of the range of *S. hispidus,* in Nebraska, nests were found 76.2 cm below the surface in the burrows of other animals (Baar, Fleharty, and Artman 1975). Cotton rats are active day and night throughout the year. They are omnivorous, feeding on vegetation, insects, and other small animals. In some areas they travel along ditches, where they feed on crayfish and fiddler crabs. They destroy both the eggs and chicks of bobwhite quail and often become pests by eating enormous quantities of sugar cane and sweet potatoes.

These animals are numerous wherever they occur, being among the most abundant rodents in the southeastern United States as well as in Mexico and Central America. They are, however, subject to population fluctuations. In the llanos of Venezuela, Vivas (1986) found *S. alstoni* to decrease during the rainy season, probably through submergence of available habitat. A peak density of 57/ha. have been reported for *S. hispidus* in southern Louisiana (Lowery 1974). In west-central Kansas the year-round average home range was about 0.4 ha. for males and 0.2 ha. for females; no indication of territoriality was found (Fleharty and Mares 1973). A male *S. ochrognathus* in the Chiricahua Mountains of Arizona had a home range of about 0.3 ha. (Hardin, Barbour, and Davis 1970). *S. hispidus* apparently is a solitary species, though close male-female pairs and dominance hierarchies are formed in captivity (Cameron and Spencer 1981).

In northeastern Kansas, near the northern extremity of

Cotton rat *(Sigmodon hispidus hispidus)*, photo by Ernest P. Walker.

the range of the genus, the breeding season is April–November (McClenaghan and Gaines 1978), but in more southerly areas *Sigmodon* breeds practically throughout the year. In Louisiana, according to Lowery (1974), reproduction occurs all year and litters are born in rapid succession. Females enter estrus every 7–9 days and also immediately after giving birth. The gestation period is 27 days, and litter size is 1–12 young, usually 5–7. The young weigh 6.5–8.0 grams and open their eyes within 24 hours. They may be weaned at 5 days but occasionally remain with the mother for 7 days or more. Sexual maturity is attained at 40–60 days, and average life span in the wild is probably less than 6 months. According to Jones (1982), a captive *S. hispidus* lived for 5 years and 2 months.

The IUCN classifies the subspecies *S. arizonae plenus* as endangered. It is known only from a few sites along the lower Colorado River in California and Arizona and may once have occurred in Nevada. Its limited marsh habitat is disappearing because of human restriction of the river's flow (D. F. Williams 1986). In contrast, *S. ochrognathus* occupies relatively dry, mountainous country and evidently is expanding its range in the southwestern United States (Davis and Dunford 1987).

RODENTIA; MURIDAE; **Genus NEUSTICOMYS**
Anthony, 1921

Fish-eating Rats, or Aquatic Rats

There are four species (Dubost and Petter 1978; Musser and Gardner 1974; Voss 1988):

N. venezuelae, known by three specimens from northeastern and southern Venezuela and one from northern Guyana;
N. peruviensis, known only by a single specimen from east-central Peru;
N. oyapocki, known only by a single specimen from southern French Guiana;
N. monticolus, the Andes of western Colombia and northern Ecuador.

The species *N. venezuelae, N. peruviensis,* and *N. oyapocki* formerly were put in the separate genus *Daptomys* Anthony, 1929, but Voss (1988) considered *Daptomys* a synonym of *Neusticomys*.

According to Voss (1988), head and body length is 98–131 mm, tail length is 82–111 mm, and average adult weight is about 40 grams. In *N. monticolus* the pelage is soft, dense, woolly, composed of underfur and fine guard hairs, and dull gray-black above when fresh. In the other species the upper parts are short, glossy, and brownish. The underparts are paler but colored like the dorsum. The tail is dark and unicolored. The external ears are conspicuous and not concealed in the fur. The hind foot is narrow and weakly fringed with hairs. As in most murids, there is a philtrum, a groove of naked skin that extends down the midline of the otherwise hairy muzzle, from the rhinarium to the margin of the upper lip. In contrast, *Chibchanomys* has pinnae hidden by the fur, no philtrum, and a broad hind foot with well-developed fringing hairs. Females of all ichthyomyine genera have six mammae. *N. oyapocki* is distinguished by the absence of the upper and lower third molars.

All specimens of *Neusticomys* have been collected in dense tropical forests along or in streams. The Venezuelan individuals were taken at elevations of 728 and 1,400 meters, and the Peruvian specimen was taken at 300 meters. Voss (1988) collected eight *N. monticolus* in Ecuador at elevations between 2,245 and 2,290 meters, along small streams descending steep, densely forested mountain slopes. Tate (1931) observed that *N. monticolus* was plentiful and could be taken along small "azequias," or irrigation brooks. He said that when trapping, the trap should be placed where the current is rapid and where the banks encroach. *Neusticomys* is frequently taken close to small waterfalls and in places exposed to spray from the water. It is assumed to feed, at least in part, on small invertebrates found in the water but is not so highly specialized for aquatic life as are the other genera of the tribe Ichthyomyini (*Chibchanomys, Anotomys, Ichthyomys,* and *Rheomys*). The hind feet of *Neusticomys* are not as broad as those of these other genera. Its incisors, however, are sharp and the outer surfaces are slightly inclined toward each other, suggesting adaptation for taking slippery prey. A lactating female *N. venezuelae* was collected in southern Venezuela on 28 July, and a pregnant female *N. monticolus* with two large embryos was collected in Ecuador on 15 May (Voss 1988).

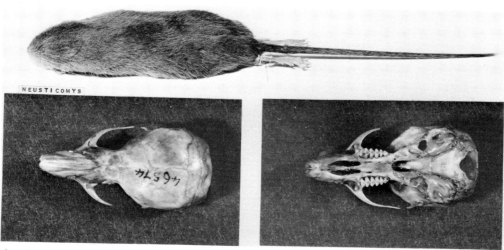

Aquatic rat *(Neusticomys monticolus)*, photos from American Museum of Natural History.

RODENTIA; MURIDAE; **Genus CHIBCHANOMYS**
Voss, 1988

Chibchan Water Mouse

The single species, *C. trichotis*, is known from the mountains of northwestern Venezuela, central and southwestern Colombia, and central Peru. This species originally was placed in *Ichthyomys*, then in *Rheomys*, and by Handley (1976) in *Anotomys*, but Voss (1988) showed that it warrants its own distinct genus.

Head and body length is 113–25 mm and tail length is 115–33 mm. The upper parts are dull gray-black, the underparts are paler, and the tail is uniformly dark. The pelage is soft, dense, woolly, and composed exclusively of underfur and long, fine guard hairs. *Chibchanomys* resembles *Anotomys* in pelage composition, postcranial proportions, and growth-invariant aspects of craniodental form but differs in its less reduced pinnae, absence of superciliary vibrissae, separate hypothenar and third interdigital pads of the manus, possession of an omohyoid muscle, and other morphological characters.

Like other rodents of the tribe Ichthyomyini, *Chibchan-*
omys preys on aquatic insects, crustaceans, and other small animals in tropical forest streams. Four specimens were taken at elevations of 2,492–2,584 meters along a swift stream descending a densely forested mountain slope. A pregnant female with a single near-term embryo was collected in northwestern Venezuela on 11 January. On 15 August 1968, Gardner (1971a) collected a subadult female (initially assigned to *Anotomys*) at an elevation of 2,400 meters on the eastern slope of the Cordillera Carpish in Peru. Subsequent study of this specimen revealed that *Chibchanomys* has a diploid chromosome number of 92, the highest reported for any mammal.

RODENTIA; MURIDAE; **Genus ANOTOMYS**
Thomas, 1906

Fish-eating Rat, or Aquatic Rat

The single species, *A. leander*, is known only from the Andes of northern Ecuador (Voss 1988).

Head and body length is 101–22 mm and tail length is 125–53 mm (Voss 1988). The upper parts are dark slaty, the

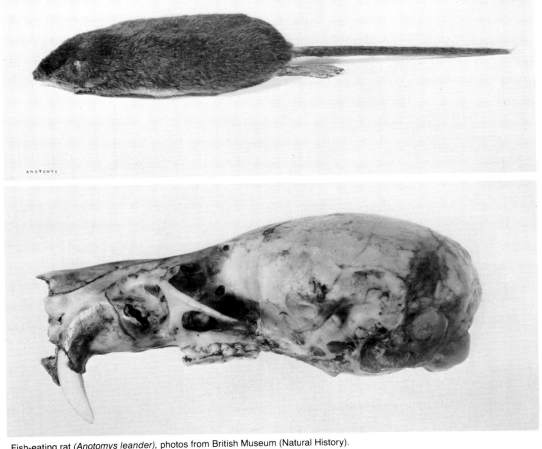

Fish-eating rat *(Anotomys leander)*, photos from British Museum (Natural History).

sides are paler, and the underparts are whitish gray. The whiskers, or vibrissae, are dark on the upper part of the face and white on the lower part. A distinct round white patch lies near each ear opening.

This genus is less highly specialized for catching and eating fish than are some of its close relatives, such as *Ichthyomys*. However, it is more specialized for an aquatic existence, as is evidenced by the following characters: the external ear is diminutive and the slitlike ear opening can be closed by muscular contractions to keep out water; the fur is velvety, like the fur of *Hydromys, Neomys, Nectogale*, and other highly specialized aquatic mammals; and the broad hind feet are lined by coarse, stiffened hairs that increase the surface area used for swimming.

Anotomys was long known only by a few specimens taken in a swift mountain stream at an elevation of 3,600 meters on the slopes of Mount Pichincha. Voss (1988) collected 12 individuals between 3,600 and 3,755 meters along two streams descending narrow ravines through dense grassland. He took a pregnant female with one large embryo on 8 March.

RODENTIA; MURIDAE; Genus ICHTHYOMYS
Thomas, 1893

Fish-eating Rats, or Aquatic Rats

There are four species (Cabrera 1961; Handley and Mondolfi 1963; Voss 1988):

I. pittieri, north-central Venezuela;
I. hydrobates, Ecuador, Colombia, northwestern Venezuela;
I. stolzmanni, eastern Ecuador, Peru;
I. tweedii, Panama, Ecuador.

Head and body length is 110–210 mm and tail length is 145–90 mm. Average weight of *I. tweedii*, the largest ichthyomyine, is 120 grams (Voss 1988). The pelage is thick and not as soft as that of allied genera. The upper parts are dark olive brown, grayish, or blackish, often with a buffy wash, and the underparts are whitish. The tail is fully haired.

These animals, about the size of common rats *(Rattus)*, are modified in body form for an aquatic and predaceous life. The head and body are flattened; the skull is fairly smooth, lacking prominent ridging; the eyes and ears are small; the whiskers are long and stout; the skin is very loose, and the fur is thick; the large hind feet are broad and have swimming fringes of hair; the hind toes are partly webbed; and the tail has a bristly undersurface. The outer corner of each incisor projects downward as a sharp point, and the cutting edges of the two upper incisors form an inverted V. Females have six mammae.

Fish-eating rats frequent streams and swampy areas from 600 to 2,800 meters in elevation. Much of their former range is now under cultivation. Söderström, who first made these animals known, stated that they usually live under rocks during the day. Observations of captives and examination of digestive tracts indicated to Voss, Silva L., and Valdes L. (1982) that *I. pittieri* consumes a wide variety of aquatic organisms, especially arthropods. While some small fish are taken, crabs and insects seem to be more important parts of the diet. Food is captured by sweeping the vibrissae through the water until contact is made, then seizing or pinning down the prey with the forepaws, and finally biting repeatedly with the sharp incisors.

RODENTIA; MURIDAE; Genus RHEOMYS
Thomas, 1906

Central American Water Mice

There are four species (Voss 1988):

R. raptor, Costa Rica, Panama;
R. thomasi, extreme southern Mexico to El Salvador;
R. underwoodi, Costa Rica, extreme western Panama;
R. mexicanus, eastern Oaxaca (southern Mexico).

The last species sometimes has been placed in a subgenus, *Neorheomys* Goodwin, 1959, but Voss (1988) considered the differences between *R. mexicanus* and *R. underwoodi* to be insufficient to justify retention of this subgenus.

Head and body length is 92–142 mm and tail length is 88–171 mm (Voss 1988). The fur is short, soft, dense, and glossy. The upper parts are dark brown or mixed black and cinnamon, and the underparts are grayish or whitish. The tail may be tipped with white. Externally, *Rheomys* closely resembles *Ichthyomys*, but the manus of *Rheomys* has four or fewer separate plantar pads, while that of *Ichthyomys* has five.

Starrett and Fisler (1970) pointed out that *R. underwoodi* is among the most highly hydrodynamically specialized mammals. Its body is streamlined, and the forelimbs do not project much, not even when it is walking. The ear pinnae are reduced and do not extend above the surface of the fur. The eyes are also small, and the nostrils open posteriorly behind flaplike valves. The hind feet are strikingly large and have laterally compressed digits interconnected by webbing. The tail flares into the body and apparently serves a stabilizing and supplementary propulsive function.

These rodents live in and near small, swift streams in tropical forests. They are known to feed on aquatic insect larvae and may also eat snails. The stomach of one of six specimens of *Rheomys* collected by Hooper (1968) contained remains of a catfish, and a captive ate pieces of sardines (Starrett and Fisler 1970). A pregnant female *R. raptor* with one large embryo was taken in Costa Rica on 20 July, and a pregnant female *R. thomasi* with a near-term embryo was taken in El Salvador on 21 February (Voss 1988).

RODENTIA; MURIDAE; Genus CALOMYSCUS
Thomas, 1905

Mouselike Hamster

The single species accepted here, *C. bailwardi*, occurs in the Caucasus, southern Turkmen S.S.R., Iran, Afghanistan, and Pakistan (Corbet 1978). This species was divided into five separate species by Vorontsov, Kartavtseva, and Potapova (1979). These species were listed, with some question, by both Corbet and Hill (1986) and Honacki, Kinman, and Koeppl (1982). They are: *C. bailwardi*, Iran, Caucasus, Turkmen S.S.R., Afghanistan, northern Pakistan; *C. baluchi*, eastern Afghanistan, western Pakistan; *C. hotsoni*, western Pakistan; *C. mystax*, Turkmen S.S.R.; and *C. urartensis*, northwestern Iran and adjacent Caucasus.

Head and body length is 61–98 mm, tail length is 72–102 mm, and weight is 15–30 grams (Hassinger 1973; Roberts 1977). The pelage is fine and soft in texture. The upper parts are pinkish buff, sandy brown, or grayish brown, and the underparts, hands, and feet are white. The tail, dark above

Fish-eating rat, or aquatic rat: Top, *Ichthyomys stolzmanni*, photo from *Proc. Zool. Soc. London;* Middle and bottom, *Ichthyomys hydrobates,* photos by Jaime E. Péfaur.

Central American water mouse *(Rheomys thomasi)*, photo from *J. Mamm.*, R. A. Sirton.

Mouselike hamster *(Calomyscus bailwardi)*, photo from *Proc. Zool. Soc. London.*

and white below, is thickly haired, tufted, and longer than the head and body. The large ears are conspicuous. There are no cheek pouches.

According to Osgood (1947), this genus is very similar to the New World *Peromyscus*, but most authorities now group it with the Old World hamsters. On the basis of a study of the structure of auditory ossicles, Pavlinov (1980b) concluded that *Calomyscus* should be placed in the subfamily Sigmodontinae (which includes *Peromyscus*) rather than in the Cricetinae. One feature distinguishing *Calomyscus* from *Peromyscus* is the number of roots the molar teeth have. Females have six mammae.

In Afghanistan, *Calomyscus* is widely distributed at elevations of between 400 and 3,500 meters and is found under evergreen oaks in monsoonal areas as well as on barren, scorched hills (Hassinger 1973). In Iran, Lay (1967) found a nest built in a narrow horizontal crevice on a rocky outcrop at 3,300 meters; it consisted of a ball of fine grass and sheep wool. *Calomyscus* is active only at night during the summer, but it is also active by day in the autumn and winter. It feeds mainly on seeds, also consumes flowers and leaves, and readily eats animal matter.

Calomyscus is not highly social, but wild individuals sometimes share favorable shelter sites, and captives sleep huddled together (Roberts 1977). There apparently is a lengthy breeding season, and some females have two litters per year. In Turkmen S.S.R., Sapargeldyev (1984) found the spring peak of reproduction to begin in late March and end in early June, with some females producing two litters during this period, each with three to seven young. In Iran, Lay (1967) found lactating females in August and December and half-grown males in early August. Hassinger (1973) collected two pregnant females in July, each with three embryos. The newborn are hairless; 13 days after being born, when their eyes open, they have a gray pelage. Adult coloration and size are attained in 6–8 months.

Osgood (1947) stated that *Calomyscus* was very limited in distribution and doubtless on the way to extinction. Hassinger (1973), however, found this genus to be abundant in Afghanistan, and Lay (1967) collected a series of 61 specimens in Iran. The Soviet Union now classifies *C. urartensis* as rare and *C. mystax* as indeterminate.

RODENTIA; MURIDAE; **Genus PHODOPUS**
Miller, 1910

Small Desert Hamsters, or Dwarf Hamsters

There are two species (Corbet 1978):

P. sungorus, Kazakh S.S.R., Mongolia and adjacent parts of Siberia, Manchuria;

P. roborovskii, western and southern Mongolia and adjacent parts of Manchuria and northern China.

There is some question regarding whether *P. campbelli*, of Mongolia and adjacent areas, is a species distinct from *P. sungorus* (Honacki, Kinman, and Koeppl 1982).

Head and body length is 53–102 mm and tail length is 7–11 mm. Tail length is generally less than one-fifth of head and body length. The upper parts are grayish to pinkish buff, and the underparts are whitish; the two colors meet to form four reentrants into the dorsal mantle, one before and one behind each leg, leaving a narrow tongue of the mantle extending down on the upper part of each leg. The ears are blackish, with white on the inside. The sides of the muzzle, upper lips, lower cheeks, lower flanks, and limbs, as well as the tail and ventral surfaces, are pure white. Sometimes there is a dark dorsal stripe. *Phodopus* has a robust body, prominent ears, and internal cheek pouches. The feet are short, broad, and densely haired throughout. Females have eight mammae.

These hamsters appear to inhabit somewhat arid areas, where there is stiff grass on plains and sand dunes. They seem to be most active in the evening and early morning but somewhat active throughout the night. Heldmaier and Steinlechner (1981) found *P. sungorus* in a laboratory colony to enter shallow daily torpor during the winter, with a maximum frequency occurring in January, when 30 percent of all individuals were torpid at a time. Loukashkin (1940) reported that he observed *P. sungorus* in association with pikas (*Ochotona*) and that the hamsters made use of their paths, tunnels, and burrows, especially in winter. Dwarf hamsters are quite clean. Undoubtedly their feeding habits are somewhat like those of other hamsters, for they eat seeds and any available plant material. Observers report that they will fill their pouches, seemingly almost to the bursting point, with millet or grain seeds, distorting the shape of the body. Then, when teased or disturbed, or upon depositing the food in their burrows, they will push the pouches with their forepaws, thus causing the grain to pour out of their mouths.

These animals are easy to tame, do well in captivity, and make good pets because of their interesting habits and amusing ways. They have a docile disposition and do not attempt to bite or run away. They are being used increasingly in behavioral and physiological research (Sawrey et al. 1984). Figala, Hoffmann, and Goldau (1973) observed reproduction in captive *P. sungorus* only from February to November. The young were born 18–19 days after breeding pairs were established. Litter size averaged five and ranged from one to nine young. According to Jones (1982), a specimen of *P. sungorus* was still living after 3 years and 2 months in captivity.

RODENTIA; MURIDAE; **Genus CRICETUS**
Leske, 1779

Common Hamster, or Black-bellied Hamster

The single species, *C. cricetus*, occurs from Belgium, across central Europe and the Soviet Union, to the Altai region of Siberia (Corbet 1978).

Head and body length is about 200–340 mm, tail length is 40–60 mm, and weight is 112–908 grams. The thick fur is usually light brown above, mostly black below, and white on the sides. Grzimek (1975) noted, however, that in addition to this normal color pattern there is much variation, from albino to melanistic animals. The small tail is almost hairless. Cheek pouches are present, and the feet are broad, usually having well-developed claws. Females have eight mammae.

The common hamster lives on steppes and cultivated land and along riverbanks. In the western parts of its range it is strictly confined to loamy and loess soils. It utilizes burrows throughout the year, hibernating there in the winter and emerging to feed during crepuscular hours at other seasons. It occasionally swims, inflating its cheek pouches with air for greater buoyancy before taking to the water. A food shortage sometimes forces mass population movements, during which large rivers may be crossed (Grzimek 1975).

Studies in Germany (cited in Grzimek 1975) indicate that burrow size is generally dependent on the hamster's age. The rooms and tunnels of the summer and autumn burrows always lie in the same plane, usually at a depth of only 50 cm, but winter burrows are often more than 2 meters below the surface. Each burrow has several oblique and vertical entrance tunnels and several compartments for the nest, food storage, and excrement. The winter burrows contain more space for food, and up to 90 kg of cereal seeds, peas, or potatoes may be stored. Winter hibernation is interrupted every five to seven days, at which time the hamster eats from its cache. The diet consists of grains, beans, lentils, roots, and green parts of plants, as well as insect larvae and frogs. In captivity the hamster does well on dog biscuits, lettuce, and corn.

The common hamster is usually a solitary animal, and each burrow has only one adult occupant. In some areas, however, burrows may be crowded together because of limited suitable construction sites, and so give the impression of a colony (Grzimek 1975). In the Kaluga District of the Soviet Union a density of about 1 burrow per 2 ha. has been reported.

The breeding season extends from early April to August. Males move onto the territory of females and are usually driven out subsequent to mating, but captive investigations have indicated that pairs may remain together and raise the young (Grzimek 1975). Females can produce a litter every month throughout the year in captivity but normally only twice a year in the wild. The gestation period is 18–20 days, and litter size is 4–12 young. The newborn weigh about 7 grams, are weaned at about 3 weeks, and reach adult size after 8 weeks (Grzimek 1975). Females attain sexual maturity at about 43 days. The life span is about 2 years.

The common hamster is trapped for its skin in some areas. It was often considered a serious pest because of its destruction of corn and other crops. The introduction of modern agricultural practices, however, resulted in drastic declines of the species, and it now is considered rare in Europe (Smit and Van Wijngaarden 1981).

Dwarf hamster *(Phodopus roborovskii):* Top, photo from East Berlin Zoo; Bottom, photo by P. Rödl.

RODENTIA; MURIDAE; **Genus CRICETULUS**
Milne-Edwards, 1867

Ratlike Hamsters

There are 12 species (Corbet 1978; Honacki, Kinman, and Koeppl 1982; Orlov and Malygin 1988):

C. migratorius, southeastern Europe to Iran and Mongolia;

C. barabensis, southern Siberia, Mongolia, northern China, Korea;

C. pseudogriseus, southern Siberia, Mongolia;

C. griseus, northeastern China;

C. obscurus, southern Mongolia, northern China;

C. sokolovi, Mongolia, northern China;

C. longicaudatus, Mongolia and adjacent parts of Siberia and China;

Black-bellied hamsters *(Cricetus cricetus)*, photo by Ernest P. Walker.

C. kamensis, Tibetan Plateau;
C. alticola, Kashmir, northern Pakistan;
C. eversmanni, northern Kazakh S.S.R.;
C. curtatus, Mongolia and adjacent parts of China;
C. triton, northeastern China, Korea, Ussuri region of Siberia.

Head and body length is 80–250 mm and tail length is 25–106 mm. *C. triton* is the largest species. The fur is quite long (about 15 mm in length on the middle of the back) and usually mouse gray in color; sometimes, however, it is reddish or buffy. The underparts are light gray or white. The feet and terminal end of the tail are white. In some species there are other white markings, and in *C. barabensis* there is a dark brown dorsal stripe. These hamsters have robust bodies, blunt muzzles, fairly short legs and tails, and huge internal cheek pouches. Females have eight mammae.

Ratlike hamsters inhabit open dry country, such as steppes and the borders of deserts. In Afghanistan *C. migratorius*

occurs at 400–3,600 meters on rocky slopes and plateaus almost devoid of vegetation (Hassinger 1973). In the spring and summer, ratlike hamsters are active both day and night, but as winter approaches they become more nocturnal. In winter they do not hibernate continuously but awaken from time to time to eat stored food. As the weather becomes colder, they sleep more heavily and the intervals between feedings are longer. Their homes are burrows in the ground, which they dig themselves. At least some species burrow almost straight down as far as 1.2 meters. Some burrows have a single entrance, but others have two or three. These burrows contain chambers that serve for food storage, nesting, living quarters, and other domestic uses.

The diet consists of young shoots and seeds. Apparently, the items most preferred are soybeans, peas, and millet seeds. The cheek pouches are so large that it does not take long to store a bushel of beans in the burrow. One author reported that he took 42 soybean seeds from the cheek pouches of an individual, and in this case the head was so large that it

Ratlike hamster *(Cricetulus triton)*, photo by Ernest P. Walker.

comprised one-third of the body. Harrison (1972) stated that the diet of *C. migratorius* is mainly vegetarian but that some insects are taken and that captives killed and ate frogs and jerboas.

Ratlike hamsters are reported to be extremely savage rodents. They make a vigorous defense by throwing themselves on their backs and opening their mouths, exposing formidable incisor teeth. In a captive study of *C. barabensis* Skirrow and Rysan (1976) found strong intolerance of conspecifics. Females were especially aggressive and were dominant over males.

Male *C. triton* reportedly leave their winter burrows in early March and enter almost all the other burrows they can find in search of females. The males and females remain together for about 10 days, and the breeding season continues to the middle of May. In Armenia *C. migratorius* breeds throughout the year if it resides in human habitations, producing 2–11 young per litter. Otherwise, according to Grzimek (1975), the length of the mating season of *Cricetulus* varies by species, sometimes beginning as early as the end of February and possibly continuing through October. Gestation lasts 17–22 days, and females produce three or four litters a year. Litter size averages about 5–6 young but occasionally is over 10. The young reach adult size in 2 months.

RODENTIA; MURIDAE; Genus MESOCRICETUS
Nehring, 1898

Golden Hamsters

There are four species (Corbet 1978; Lyman and O'Brien 1977):

M. brandti, Asia Minor, Caucasus, northern Iraq, northwestern Iran, Syria, Palestine;
M. auratus, apparently restricted to the vicinity of Aleppo in northwestern Syria;
M. raddei, north of the Caucasus between the Black and Caspian seas;
M. newtoni, eastern parts of Rumania and Bulgaria.

Many authorities include *Mesocricetus* in the genus *Cricetus*. Although *M. auratus* has the smallest natural range, it is the best-known species, because it has become common in captivity throughout the world. Corbet (1978) listed *M. brandti* as a subspecies of *M. auratus*.

Head and body length is about 170–80 mm and tail length is about 12 mm. Lyman and O'Brien (1977) gave a weight range of 137–258 grams for *M. brandti* and 97–113 grams for *M. auratus*. Ralls (1976) stated that females are larger than males. The upper parts are generally light reddish brown, and the underparts are white or creamy. One form has a distinct ashy stripe across the breast. The skin of these animals is quite loose, and the enormous cheek pouches, which open inside the lips, extend well back of the shoulders; when filled, they more than double the width of the animal's head and shoulders. In a study of 128 females, Anderson and Sinha (1972) counted 3 animals with 12 mammae each, 30 with 13, 74 with 14, 16 with 15, 4 with 16, and 1 with 17. The greater number of nipples, smaller size, and shorter tail, together with some cranial characters, distinguish *Mesocricetus* from *Cricetus*.

In the wild, golden hamsters live on dry, rocky steppes or brushy slopes. They construct burrows somewhat like those of *Cricetus*. They are essentially nocturnal but occasionally may be active by day. They appear to be almost omnivorous, eating many kinds of green vegetation, seeds, fruits, and

meat. Hibernation for uninterrupted periods of up to 28 days has been experimentally induced through exposure to cold (Lyman and O'Brien 1977). Captives that hibernated the most because of exposure to cold tended to live the longest (Lyman et al. 1981).

Adults generally live one to a burrow and will readily fight one another. Lyman and O'Brien (1977) found that litters of *M. brandti* lived amicably together until about the seventh week of life but then had to be separated. Drickamer, Vandenbergh, and Colby (1973) determined that the social rank of captive males is based on the condition of their flank glands, the hamster with the largest gland being dominant.

Litters have been born in captivity during every month of the year, but with a marked decrease in production during late fall and winter. Lyman and O'Brien (1977) reported that laboratory specimens of *M. brandti* exposed to natural day length usually were not in breeding condition between the months of November and March. Successful matings among these animals peaked in May; no female produced more than three litters per season. The reported maximum for wild populations in the Caucasus is two litters a year.

Asdell (1964) stated that in *M. auratus* the estrous cycle is 4 days long, estrus lasts 27.4 hours, and the gestation period is usually 16 and occasionally up to 19 days. Lyman and O'Brien (1977) stated that the estrous cycle of *M. brandti* lasts 4 days, gestation is 15 days, and litter size is 1–13, averaging 6. Anderson and Sinha (1972) reported the litter size of *M. auratus* to range from 2 to 16 and average about 9, but Asdell (1964) gave the range as 1–12 and the mode as 6–7. Witte (1971) observed that if the young were threatened by danger during their first 3 days of life, up to 12 of them at a time might be placed in the cheek pouches of the mother. The young are born blind and naked but grow quickly; they are weaned before 20 days and are capable of reproduction at 7–8 weeks. Peak fecundity is reached at around 1 year (Lyman and O'Brien 1977). The life span appears to be 2–3 years.

Little was known about these hamsters until 1930, when four young *M. auratus*, including a single female, were obtained in Syria and brought to Palestine. Descendants of this small group spread throughout the world as laboratory animals, zoo exhibits, and household pets. In 1938 some of these hamsters were first brought to the United States. In 1971 an additional 13 *M. auratus* were caught at Aleppo, Syria, and brought to the United States (Murphy 1985). In 1965 and 1971 specimens of *M. brandti* were captured in central Turkey, and these animals also formed the basis of a successful laboratory colony (Lyman and O'Brien 1977).

RODENTIA; MURIDAE; Genus MYSTROMYS
Wagner, 1841

White-tailed Rat

The single species, *M. albicaudatus*, occurs in South Africa and Swaziland (Meester et al. 1986).

Head and body length is 105–84 mm, tail length is 50–97 mm, and weight is 75–111 grams (Smithers 1983). The soft, long, rather woolly pelage is buffy grayish to brownish above, suffused with numerous black-tipped hairs, which become fewer on the sides, finally blending into the white underparts. The bases of the hairs on the upper surface are slate gray, whereas the hairs on the lower surface are all white. The sides of the face and the limbs are paler than the back. The hands, feet, and tail are dull white. The ears are dark brown. The tail is covered with short, stiff bristles.

This genus is thick-bodied and ratlike. It has large, broad ears and slender limbs, hands, and feet. The incisor teeth are

Golden hamsters *(Mesocricetus auratus):* A. Adults; B. A hamster with its cheek pouches filled with food; C. Dorsal view showing distended cheek pouches; D. Two-day-old hamsters; E. Eight days old; F. Thirteen days old; G. Twenty-two days old; H. Twenty-nine days old. Photos by Ernest P. Walker.

ungrooved and pale yellow. The short, slightly curved, sharp-tipped claws are almost concealed by the long white hairs of the feet. Unlike some of its relatives, *Mystromys* does not have cheek pouches. Females have two pairs of inguinal mammae.

The white-tailed rat inhabits grassy flats and dry sandy

areas. It lives in holes in the ground. It has been observed using the burrows of the carnivore *Suricata.* It also shelters in cracks in the soil. It is nocturnal and is said to be especially active and bold during rainy weather. If individuals are evident in a given area, reportedly some can be caught by placing a lighted lantern on the ground, which draws the animals to

White-tailed rat *(Mystromys albicaudatus)*, photo by H. J. Bohner, National Institutes of Health.

its light. *Mystromys* eats seeds, other vegetable matter, and insects.

Breeding apparently occurs throughout the year. In a study of captive animals, Hallett and Meester (1971) found a mean litter size of 2.9 young and a minimum interval between litters of 36 days. The young became attached to the mother's mammae immediately after being born and remained for about 3 weeks, with the female dragging them about. This process seems an excellent means of protection, and the rate of survival of the young is high for a rodent. After 3 weeks there are increasing periods of separation, and there is no suckling after about 38 days. The minimum age at which a female produced a litter was 146 days. It has been reported that the white-tailed rat does well in captivity and is tame and playful. Dean (1978) stated that maximum longevity in captivity is about 6 years. He also suggested that overgrazing of the high veld by domestic livestock might be a threat to *Mystromys*. For this reason and others involving agricultural modifications of habitat, Smithers (1986) classified the genus as vulnerable.

RODENTIA; MURIDAE; Genus SPALAX
Guldenstaedt, 1770

Ukrainian Blind Mole-rats

There are five species (Corbet 1978, 1984; Topachevskii 1976):

S. graecus, northern Rumania, Ukraine;
S. polonicus, western Ukraine;
S. arenarius, southern Ukraine;
S. microphthalmus, eastern Ukraine to the Volga River;
S. giganteus, plains around the northern Caspian Sea.

Spalax formerly was considered to comprise not more than three species, including those placed here in the genus *Nannospalax*. Topachevskii's division of *Spalax* into two genera and eight species did not meet with immediate general approval (Corbet 1978; Honacki, Kinman, and Koeppl 1982) but now seems to be gaining acceptance (Corbet 1984; Corbet and

Hill 1986). Most authorities have placed *Spalax* and *Nannospalax* in their own family, the Spalacidae, but Chaline, Mein, and Petter (1977) reduced this group to subfamilial rank, and that procedure is followed here.

According to Ognev (1963), head and body length is 170–350 mm, there is no external tail, weight is about 215–570 grams, the upper parts vary in color from wood brown to yellowish gray, the neck and chest are gray, and the belly is straw-brown. The anterior part of the head is paler than the back, sometimes has a silvery or yellowish tinge, and may have whitish or yellowish stripes on each side. The dense, soft fur is nearly reversible. A line of bristles, apparently serving as tactile aids, extends from each side of the flattened snout along a ridge to about the location of the eye.

These heavy-bodied, short-legged, powerful animals have projecting incisors and small claws. The external form is molelike. *Spalax* and *Nannospalax* can be distinguished from all other rodents by the absence of external openings for the eyes, though small eyes are present beneath the skin. From *Nannospalax, Spalax* is distinguished by having no foramina above the occipital condyles of the skull and by other cranial characters. The external ears of both genera are reduced to low ridges. Blind mole-rats recognize objects by touching them with the nose. Tactile and auditory senses are considered acute, but olfaction may be weak. The forefeet and hind feet have five digits. The incisors are broad and heavy, and the cheek teeth are rooted (not ever-growing), with Z or S enamel patterns. Females have two pairs of mammae.

Blind mole-rats seem to burrow in any area where the soil is suitable for digging, and they inhabit any type of soil that receives more than 100 mm of rainfall annually; they do not live in true desert. They are known from plains below sea level, hilly areas, upland steppes, and cultivated fields. They are extensive burrowers. Most of their digging is done with the incisors, the whole head acting as a bulldozer blade. Specialized jaws and strong muscles aid the teeth in loosening the soil. Almost all of the orbit, for example, is occupied by muscles for working the teeth. Another aid in burrowing is the broad, padded, horny snout, with which the animal packs excavated ground into the burrow walls. Although the feet are small and quite delicate, the forefeet aid to some extent in breaking the soil, and the hind feet help push the dirt out behind.

Mediterranean blind mole-rat *(Nannospalax leucodon)*, photos by P. Rödl.

Elaborate tunnel systems are constructed consisting of both upper- and lower-level passages (Ognev 1963). The upper-level passages, used mainly in the search for food, are about 10–25 cm below the surface and may be many hundreds of meters long. Their course is marked by mounds of excavated earth about 500 cm wide. Vertical shafts lead to the deeper corridors, which connect the actual living quarters. Each burrow system has one or two nesting chambers, about 20–30 cm wide and lined with bedding material, plus several storerooms and latrines. Activity goes on even during extremely hot periods or drought, though there is then perhaps less digging, and burrows are made deeper. One individual excavated 284 mounds in an area of 61 ha. from 9 May to 12 September. *Spalax* generally feeds on roots, bulbs, tubers, and various other underground parts of plants. Although they are truly fossorial, blind mole-rats are occasionally active above ground at night. During such ventures they feed on grasses, seeds, superficial roots, and perhaps insects. They, in turn, are sometimes preyed upon by owls.

Population densities in the Soviet Union are usually around 1–10 per ha. but range up to 23 per ha. (Topachevskii 1976). Blind mole-rats are solitary creatures, and except during the breeding season only a single individual is found per mound or burrow system. Vocalizations include grunting noises, quite unlike the usual rodent squeak, and a hissing threat. Breeding is thought to occur only once a year, in the spring, and litter size is about three young. Topachevskii (1976) stated that in the Soviet Union *Spalax* does considerable damage to orchards, perennial grasses, forest plantations, and cultivated crops. Some populations, however, are being adversely affected by human activity. The Soviet Union now classifies the species *S. graecus*, *S. arenarius*, and *S. giganteus* as rare.

RODENTIA; MURIDAE; **Genus NANNOSPALAX**
Palmer, 1903

Mediterranean Blind Mole-rats

There are three species (Corbet 1978, 1984; Topachevskii 1976):

N. leucodon, lower Danube Basin, Balkan Peninsula;
N. nehringi, Asia Minor, southern Caucasus;
N. ehrenbergi, Syria, Palestine, coastal Egypt and Libya.

This genus formerly was considered to be part of *Spalax* but was distinguished by Topachevskii (1976) under the name *Microspalax*. This distinction is now gaining acceptance, as is Topachevskii's division of *N. leucodon* into the three species listed above, but the more proper name *Nannospalax* is now in use (Corbet 1984; Corbet and Hill 1986). Karyological studies have revealed the existence of about 30 chromosomal forms of *N. leucodon* in the Balkans and Asia Minor, as well as 4 of *N. ehrenbergi* in Israel, and have suggested that these forms may represent additional distinct, though closely related, species of *Nannospalax* (Nevo and Bar-El 1976; Nevo, Heth, and Beiles 1982; Nevo et al. 1988; Peshev 1983; Savic and Soldatovic 1979; Soldatovic and Savic 1974). However, Giagia, Savic, and Soldatovic (1982) indicated that these forms should be treated as subspecies. Kivanc (1988) considered *N. nehringi* to be only a subspecies of *N. leucodon* and continued to place these forms in the genus *Spalax*.

Head and body length is about 150–270 mm, there is no

external tail, and weight is about 100–220 grams. The upper parts are reddish to pale brown, the underparts are grayish, and the feet are silvery. In *N. leucodon* there is a white stripe extending up the middle of the nose to the forehead. In features of the pelage, facial bristles, sensory organs, burrowing mechanisms, and general body form, *Nannospalax* resembles *Spalax* but differs from the latter genus in certain cranial characters, most notably the presence of prominent foramina above the occipital condyles of the skull (Harrison 1972; Ognev 1963; Osborn and Helmy 1980; Van Den Brink 1968).

These blind mole-rats occur in deep sandy or loamy soils in a variety of habitats, including dry brush country, fields, and woodlands, but they avoid true sand desert. They have been found in mountain clearings at an elevation of 2,600 meters in Turkey. They are both diurnal and nocturnal, but studies in Yugoslavia by Savic (1973) indicate that *S. leucodon* is more active by day than by night. The diet consists of roots, tubers, acorns, plant stems, and other vegetation that can be obtained at or below ground level. Some food is stored, and up to 18 kg of potatoes and sugar beets have been found in a burrow (Harrison 1972).

The following information on mounds and burrows was taken in part from Grzimek (1975), Harrison (1972), Nevo (1961), and Savic (1973). Except during the breeding season, blind mole-rats live mainly in an underground burrow system with many tunnels, nest rooms, storage chambers, and defecation rooms. In at least some areas, during the winter or during the hot, dry summer these burrows are placed at greater depths than at other times. Record depth is 410 cm. Generally, the living and storage rooms are about 20–50 cm deep. Tunnels to food sources, about 7 cm wide and usually 15–25 cm below the surface, radiate outward for up to 30 meters or more. The total length of tunnels in a burrow system ranges from about 65 to 195 meters. In certain areas, during the summer a series of 15–20 mounds may be built above the burrow system. The largest mound, the so-called resting mound, is surrounded by the others. It is usually about 100 cm wide and 25 cm tall. In the center is a chamber used for sleeping and resting.

In the Mediterranean region, during the wet fall and winter females build "breeding" mounds. These are elaborate structures, usually about 160 cm long, 135 cm wide, and 40 cm in height above the surface. They may even reach 250 cm in diameter and 100 cm in height. The mounds are solid domes of earth through which runs a labyrinth of permanent galleries with hard, smooth walls. The only loose, soft earth is the covering of the dome. In very wet, poorly drained areas the mounds tend to be much bigger and more elaborate, thereby protecting the occupants from flooding. In the center of each mound is a nest chamber about 20 cm wide and filled with dry grass, where the young are raised. Beneath this chamber and throughout the mound are rooms for storage and defecation. Each main breeding mound is surrounded by radial rows of smaller mounds, which are built and occupied by the males during the mating season. These smaller mounds are connected to the breeding mound by tunnels.

Nannospalax is aggressive and solitary, and generally only a single individual is found within a burrow system. Home range of *N. leucodon* averages 452 sq meters (Savic 1973). In Israel, Nevo, Heth, and Beiles (1982) found population density of *N. ehrenbergi* to average about 105/sq km and estimated that there were at least 1.6–2.0 million of the animals in the 15,500 sq km of available range in the country. During the mating season the opposite sexes evidently locate one another through a unique production of seismic signals by drumming the flattened, bony, anterodorsal surface of the head against the burrow ceiling (Rado et al. 1987). In the eastern Mediterranean region the mating season lasts from about November

to March, and the single annual litter is born from January to early April (Harrison 1972). In Yugoslavia maximum sexual activity occurs in January and February, and births usually take place in March or April (Savic 1973). The gestation period evidently lasts about a month. Litter size is one to five, usually two to four, young. The young weigh about five or six grams at birth and are naked, helpless, and pink. After about two weeks they are covered with long, gray fur, which is quite different from the short hair of the adult. At four to six weeks of age they leave the nest. The maximum life span under optimal conditions is 4.5 years (Savic 1973).

Grzimek (1975) wrote that the economic significance of these mole-rats varies from region to region. They are considered serious agricultural pests in the eastern Mediterranean region but are not regarded as being especially destructive in some other areas. In Libya some people believe that blindness will ensue if one of these mole-rats is touched, and so they will not attempt to catch the animal. Blind mole-rats have sometimes guided archeologists to digging sites by bringing to the surface objects from early civilizations.

RODENTIA; MURIDAE; Genus MYOSPALAX
Laxmann, 1769

Mole-rats, or Zokors

There are two subgenera and six species (Corbet 1978; Fan and Shi 1982):

subgenus *Eospalax* G. Allen, 1940

M. fontanieri, northern and western China;
M. cansus, north-central China;
M. baileyi, west-central China;
M. rothschildi, Gansu and Hubei (central China);
M. smithi, Gansu (central China);

subgenus *Myospalax* Laxmann, 1769

M. myospalax, southern Siberia, eastern Mongolia, northeastern China.

Corbet and Hill (1986) listed *M. aspalax*, of Transbaikalia and northern Mongolia, and *M. psilurus*, of Transbaikalia and northern China, as species separate from *M. myospalax*. However, all were treated as a single species by Mallon (1985).

Head and body length is 150–270 mm, tail length is 29–96 mm, and weight is 150–563 grams (Fan and Shi 1982; Grzimek 1975). The pelage is composed of long, soft, silky hairs and lacks coarse guard hairs; there are a few short whiskers. The coloration of the upper parts varies with the species from gray, grayish brown, or light russet to pinkish buff; usually the individual hairs are bicolored, the tips being lighter than the bases. The lower parts are generally paler than those above. In some species the tail is grayish above and white below. *M. fontanieri* has an all-white tail and a white patch on the upper lip and muzzle.

Mole-rats are stout, chunky animals with short, powerful limbs. The forefeet are armed with long, heavy claws for digging. These front claws are doubled under when the animals are walking. The middle or third claw is the largest and heaviest. The claws of the hind feet are much shorter. There are no external ears, and the eyes are so small that they are nearly hidden in the fur. Females have one pectoral and two abdominal pairs of mammae.

Mole-rats frequent cultivated and wooded areas, especially

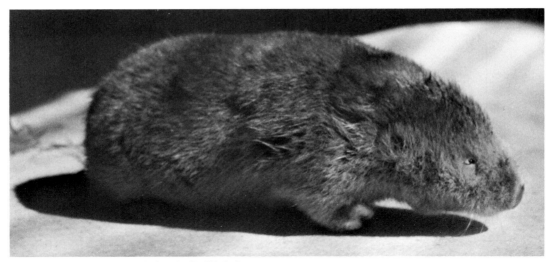

Mole-rat *(Myospalax myospalax)*, photo by Paul K. Anderson.

mountain valleys at elevations of 900–2,120 meters. They are fossorial, and their remarkable digging abilities were described by Grzimek (1975). With its powerful foreclaws one individual dug a tunnel 70 cm long in 12 minutes. It used its incisors to cut through roots and pushed loose earth aside with its head. After it had dug itself completely into the ground, the animal turned around and used its front feet and head to push the accumulated earth to the surface. After digging a tunnel 3–5 meters long, the mole-rat broke through to the surface to throw out loose soil. In the wild, in this same manner, a series of mounds develops along the route of the mole-rats, thereby allowing their presence to be detected. Their main burrows are located at a depth of about 2 meters and consist of a nest chamber, a food storage room, and a chamber for defecation. One to four food tunnels, which may reach a length of 50–100 meters, lead gradually from the nest chamber to the surface. These tunnels pass below the food plants and include storage facilities. The diet of *Myospalax* consists of roots and grains.

When frightened or angered, mole-rats utter a peculiar little squeal. Even though their eyes are small, they are sensitive to light, and if brought to the surface, they always seek a darkened place. They apparently sometimes move about on the surface at night and reportedly are active throughout the winter. In China, litters of four to six young have been found in March and April, and a pregnant female with two embryos was recorded in May (Grzimek 1975).

Mole-rats are intensively hunted in China because of their damage to crops. The Chinese farmers say that when the mole-rats leave their tunnels open, good weather will follow, but when the tunnels are closed, there will be bad weather.

RODENTIA; MURIDAE; Genus **LOPHIOMYS**
Milne-Edwards, 1867

Maned Rat, or Crested Rat

The single species, *L. imhausi,* occurs in southern Sudan, Ethiopia, Somalia, and Kenya (Misonne, *in* Meester and Setzer 1977).

Head and body length is 255–360 mm, tail length is 140–215 mm, and weight is 590–920 grams (Kingdon 1974b).

Females are usually larger than males. The coloration is usually either black and white or brown and white, in a fairly definite pattern of stripes and spots or blotches. The variation is caused mainly by the intensity or purity of the coloration and by the relative size of the differently colored areas. Variation in color is also affected by the differing extent of the light and dark parts of those hairs that are banded or tipped with light or dark pigmentation. The most common coloration gives an iron gray effect, but the color ranges from a general overall gray tone (occasioned by a preponderance of white or gray on the tips of the hairs) to almost dark brown or black (brought about by predominance of hairs largely or entirely black).

There is a prominent erectile mane from the top of the head, along the back, and along a quarter of the tail (the tip of the tail is white). On either side of the mane, along the back, the hair is shorter and lighter in color than the surrounding hair, giving the effect of a furrow in the fur and emphasizing the mane. In some animals the mane is scarcely apparent on the head and is reduced on the tail. The fur is generally dense, long, and fine to silky, except along the mane, where it is coarse. The underparts are gray to black, and the hands and feet are jet black.

The tail is densely bushy. The great toe is partially opposable, and the hands and feet have four well-developed digits and are specialized for arboreal life. The ear is small. The skull is unique among rodents in that the temporal fossae are completely roofed by bone. The surfaces of these and some of the other bones of the skull are granulated. The incisors are moderately broad.

This rodent does not resemble a rat, as the head is similar to that of a guinea pig *(Cavia)* and the body, when viewed from a distance, looks like that of a small porcupine *(Erethizon).* When the animal becomes excited or frightened, its crest erects and it exposes a glandular area along its flanks. This trait may be a protective measure to frighten its enemies into mistaking it for a porcupine.

Although the maned rat is generally believed to inhabit dense montane forests, Yalden, Largen, and Kock (1976) stated that in Ethiopia the animal has been recorded from sea level to at least 3,300 meters, is clearly tolerant of a wide range of habitats, and is by no means confined to forests. The maned rat is arboreal and quite adept in climbing, even descending headfirst, but it moves slowly. It is strictly noctur-

Maned rat, or crested rat *(Lophiomys imhausi)*, photo by Ernest P. Walker. Insets: photos by P. F. Wright of specimen in U.S. National Museum of Natural History.

nal, leaving its burrow or hole among rocks at dusk to search for food. The diet consists of leaves and tender shoots. When eating, *Lophiomys* sits on its haunches and grasps its food in its hands. The voice of this rodent is peculiar: one or two hisses or snorts, followed by a growl.

Kingdon (1974*b*) wrote that two or three young are born at once, but Delany (1975) stated that a captive female gave birth to two litters, each with a single young. These young were feeding independently of the mother by the time they were 40 days old. According to Jones (1982), one captive individual lived for seven years and six months.

RODENTIA; MURIDAE; Genus PLATACANTHOMYS
Blyth, 1859

Spiny Dormouse

The single species, *P. lasiurus,* is found in southern India (Ellerman and Morrison-Scott 1966). *Platacanthomys,* along with *Typhlomys,* has sometimes been placed in a separate family, the Platacanthomyidae, and sometimes in a sub-family, the Platacanthomyinae, of the family Gliridae. In accordance with Chaline, Mein, and Petter (1977) and Corbet (1978), the Platacanthomyinae are here considered a sub-family of the family Muridae.

Head and body length is about 130–212 mm, and tail length is 75–100 mm. The weight of one specimen, an adult female, was 75 grams. The upper parts are densely covered with sharp, flat spines intermixed with thin, delicate under-fur, but the underparts have fewer, smaller, and finer spines. The basal half of the tail is sparsely haired and scaly, and the terminal part bears long hairs that form a brush. The general coloration of the upper parts is light rufescent brown; the forehead and crown are more reddish, and the underparts are dull whitish. The tail is somewhat darker than the general body color, becoming paler at the thick, bushy tip. The feet are whitish.

This rodent is much like the Gliridae in form. The muzzle is pointed, the eyes are small, the ears are thin and naked, and the hind feet are broad and elongated. The first toe barely reaches the base of the second toe. The thumb on the forefoot, although short, is well developed. The claws of the digits are slender and compressed. Unlike the Gliridae, *Platacanthomys* has no premolar teeth, and its dental formula is the same as that of the Muridae. The incisors are smooth and com-pressed. The cheek teeth tend to be high-crowned and gener-ally have parallel oblique cross ridges of enamel on the crown. These ridges are broadened, and the depressions tend to be-come isolated, on the surface of the crown.

The spiny dormouse inhabits rocky hills and forested val-leys at elevations of 600–900 meters. It lives mainly in the cavities of trunks and branches of trees and in clefts in rocks. Its nest is constructed mostly of leaves and moss. The long, tufted tail is undoubtedly helpful as a balancing organ as the animal moves about and leaps in trees. The diet consists of fruits, seeds, grains, and roots. A captive specimen was slug-gish during the day, and although it allowed itself to be han-dled without showing fear or attempting to escape, it would inquisitively bite at a finger. Rajagopalan (1968) reported that a female caught in the wild was still living after 20 months of captivity.

In some areas the spiny dormouse is so plentiful that it is considered a pest. The native people call it the "pepper rat," because it destroys large quantities of ripe peppers. It also is said frequently to get into "toddy-pots," containers in which palm juice is collected.

PLATACANTHOMYS

Spiny dormouse *(Platacanthomys lasiurus)*, photo from *Proc. Zool. Soc. London*, 1865.

RODENTIA; MURIDAE; **Genus TYPHLOMYS**
Milne-Edwards, 1877

Chinese Pygmy Dormouse

The single species, *T. cinereus*, occurs in southern China and northern Viet Nam (Ellerman and Morrison-Scott 1966; Wu and Wang 1984). *Typhlomys*, along with *Platacanthomys*, has sometimes been placed in a separate family, but the two genera are considered here to represent a subfamily of the Muridae (see the account of *Platacanthomys*).

Head and body length is 70–98 mm and tail length is 95–135 mm. The pelage is short, dense, soft, and spineless. The upper parts are uniformly deep mouse gray; the underparts and insides of the limbs are pale grayish, the individual hairs having a gray base and a white tip. The hands are white and the feet are dusky. The long gray tail is sparsely haired and scaly on the basal half but more heavily covered toward the tip, with longer hairs that form a distinct terminal brush. The tip is usually white.

Typhlomys is mouselike in external appearance. It has prominent, scantily haired ears; small eyes; and long, slender hind feet. The claws of all the digits are slender and compressed. The dental formula and structure of the molar teeth are the same as those of *Platacanthomys*.

This rodent is found at elevations of 1,200–2,100 meters in mountains abundantly covered with dwarfed, moss-laden deciduous trees and an undergrowth of small bamboos. Practically nothing of its natural history has been recorded, but the native people seem to understand its habits and trap it quite readily. The natives claim that cats will not eat this rodent.

RODENTIA; MURIDAE; **Genus MACROTARSOMYS**
Milne-Edwards and Grandidier, 1898

There are two species (Misonne, *in* Meester and Setzer 1977):

M. bastardi, western Madagascar;
M. ingens, northwestern Madagascar.

In *M. bastardi* head and body length is 80–100 mm and tail length is 100–145 mm. This is the smallest of the Malagasy

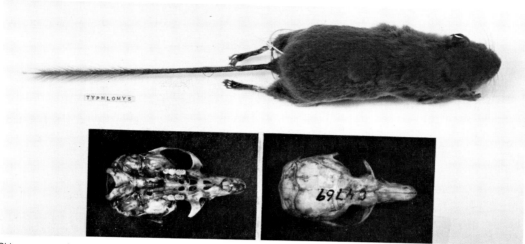

Chinese pygmy dormouse *(Typhlomys cinereus):* Skin: photo by P. F. Wright of specimen in U.S. National Museum of Natural History; Skull: photos from American Museum of Natural History.

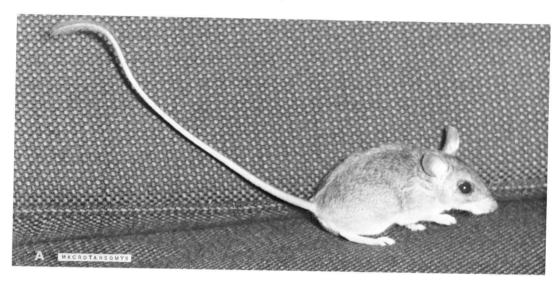

Macrotarsomys ingens: A. Photo by F. Petter; B. Photo by Howard E. Uible.

mice. The only other species on the island sometimes under 100 mm in head and body length is *Eliurus minor*. In *M. ingens* the head and body length is about 120 mm and tail length is about 210 mm (Petter 1972). Both species of *Macrotarsomys* have brownish fawn upper parts, whitish underparts, and a long, stiff tail with a terminal tuft.

The body form of *Macrotarsomys* is like that of some Gerbillinae. The ear is large (20–25 mm long in *M. bastardi*), the hind foot is long (22–28 mm in *M. bastardi*), and the tail is elongated. The three central toes are longer than the first and fifth digits, and the fifth toe is longer than the reduced first toe. The first toe bears a claw that is located fairly low on the foot. The moderate inflation of the bullae gives the skull a unique appearance among the genera of mice from Madagascar.

M. bastardi inhabits dry wooded areas, scrublands, and grassy plains. It lives in burrows often concealed under a rock or small bush. The burrows may be 1.5 meters or more in length but not very deep, and the outlets are usually kept closed. This species is strictly terrestrial and nocturnal. Its diet consists of berries, fruits, seeds, roots, and plant stems. *M. ingens* digs burrows, which can be located by the soil thrown up, but its nocturnal activity seems largely arboreal (Petter 1972). *M. bastardi* lives in pairs (Petter 1972). Litters usually consist of two or three young.

RODENTIA; MURIDAE; Genus NESOMYS
Peters, 1870

The single species, *N. rufus*, is found in the eastern forests and on the western coast of Madagascar (Misonne, *in* Meester and Setzer 1977).

Head and body length is 186–230 mm and tail length is 160–90 mm. The pelage is fairly long, soft, smooth, and shiny. The fine-textured black whiskers on the muzzle extend beyond the tips of the blackish brown ears. The coloration of the upper parts is dark rust brown mixed with brownish yellow, which gives the animal a fawn color. The hairs are usually slate-colored at the base. The sides of the head, body, and limbs are reddish, whereas the lips, throat, breast, middle of the belly, and underside of the tail are white; the upper part of the tail is rust red. A white tuft occurs on the end of the tail in the subspecies *N. r. auderbeti*, and the subspecies *N. r. lambertoni* is remarkable for its hairy tail.

The incisors are smooth and whitish toward the points; the upper teeth are brownish orange, and the lower ones are yellowish. The long, well-developed feet help *Nesomys* to leap about freely. The three middle toes are elongated (especially the center one), whereas the two outer toes are much shorter. The claws of the fingers are slightly more than half as big as those of the toes, the latter being very strong. The almost straight, brownish white claws are partially concealed by long, rigid hairs growing from the roots of the nails.

According to Petter (1972), this mouse inhabits wet forests and is terrestrial and nocturnal.

RODENTIA; MURIDAE; Genus BRACHYTARSOMYS
Günther, 1875

The single species, *B. albicauda*, is known from the eastern forests of Madagascar (Petter 1972).

Head and body length is 200–250 mm and tail length is about the same (Petter 1972). The tail is prehensile (Misonne, *in* Meester and Setzer 1977). The soft, dense, woolly pelage is grayish brown on the upper parts, rufous on the sides, and white on the underparts. The head is reddish brown, and the nose and lips are blackish. The anterior half of the sparsely haired tail is black, but the posterior half is white.

The appearance of this genus is similar to that of *Nesomys*, but *Brachytarsomys* differs in its cranial structure and in having shorter, broader hind feet. *Brachytarsomys* has a short snout, small eyes and ears, five rows of mustache hairs, and prominent claws. The hind foot has a long fifth digit. The incisors are not grooved.

Petter (1972) wrote that *Brachytarsomys* is strictly arboreal, lives in holes of trees, and feeds mainly on fruit.

RODENTIA; MURIDAE; Genus ELIURUS
Milne-Edwards, 1885

There are two species (Misonne, *in* Meester and Setzer 1977):

E. myoxinus, Madagascar;
E. minor, eastern Madagascar.

Head and body length is 80–175 mm, tail length is 110–200 mm, and weight is 35–103 grams. The pelage is fairly soft. The upper parts are uniformly brownish gray or yellowish gray, the latter effect being caused by the gray base and fawn tip of the individual hairs. The feet and lower surfaces are generally light gray. In the subspecies *E. myoxinus myoxinus* the tail is so well clothed with moderately stiff, deep brown hair that it looks almost bushy. In other subspecies the basal third is almost naked, and the rest of the tail is slightly bushy

Nesomys rufus, photo by Constance P. Warner.

Brachytarsomys albicauda, photo by F. Petter.

or penciled. The pencil on the tail of *E. myoxinus tanala* and *E. m. pencillatus* is white. *E. m. tanala* has a dark gray spot in the middle of the back and yellowish white underparts. In *E. myoxinus majori* there is an indistinct ring around the eye. The ears are dark and almost naked, the palms are pink, the mustache is black with long whiskers, and the eyes are large and conspicuous. Females have two pairs of inguinal mammae.

Eliurus minor, photo by Howard E. Uible. Insets: *E. myoxinus*, photos by P. F. Wright of specimen in U.S. National Museum of Natural History.

These mice apparently prefer heavily forested areas. *E. myoxinus* is mainly nocturnal and arboreal (Petter 1972), but *E. minor* burrows in the ground (Grzimek 1975). A female *Eliurus* was observed nursing in early November.

RODENTIA; MURIDAE; **Genus GYMNUROMYS**
Forsyth Major, 1896

Voalavoanala

The single species, *G. roberti,* is known from eastern Madagascar (Misonne, *in* Meester and Setzer 1977).

Head and body length is 125–60 mm and tail length is 152–75 mm. The upper parts are blackish gray or slaty, and the underparts are white or yellowish white. Long whiskers (50–60 mm) are present, and the bicolored, scaly tail is scantily haired. The body form is ratlike, and the feet are quite broad, with fairly long fifth digits. This genus is best characterized by the pattern of the cheek teeth, which are completely flat-crowned, laminate, and tightly compressed.

This genus inhabits forests. Pregnant females, each bearing two embryos, have been collected in June and July. Schlitter (1989) classified *Gymnuromys* as rare, noting that fewer than a dozen specimens have been collected.

RODENTIA; MURIDAE; **Genus HYPOGEOMYS**
Grandidier, 1869

Malagasy Giant Rat

The single species, *H. antimena,* occurs around Morondava on the western coast of Madagascar (Misonne, *in* Meester and Setzer 1977).

This is the largest rodent in Madagascar. Head and body length is 300–350 mm and tail length is 210–50 mm. The ears are large (50–60 mm). The pelage is harsh. The upper parts are gray, grayish brown, or reddish; the head is darkest.

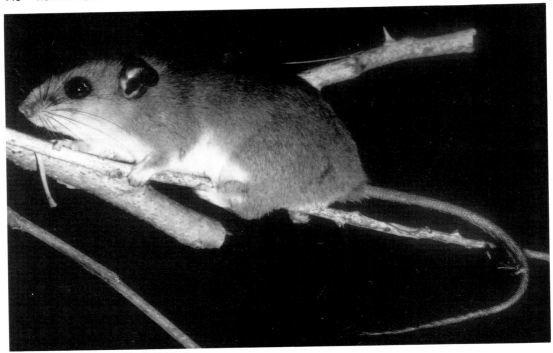

Eliurus myoxinus, photo by Martin E. Nicoll.

The limbs, hands, feet, and underparts are white. The dark tail is covered with stiff, short hairs. The hind foot is quite long and has relatively well developed claws.

This giant rat is known only from sandy coastal forests. It is a jumping and running rodent, strictly nocturnal in habit (Petter 1972). It apparently has about the same ecological role as a rabbit. It builds long, deep burrows and feeds mainly on fallen fruit. The usual number of offspring seems to be one.

Unfortunately, the numbers and distribution of this unique mammal have declined sharply, because suitable habitat has been reduced through the cutting and burning of virgin forests (Grzimek 1975).

RODENTIA; MURIDAE; Genus BRACHYUROMYS
Forsyth Major, 1896

There are two species (Misonne, *in* Meester and Setzer 1977):

B. ramirohitra, eastern Madagascar;
B. betsileonensis, southeastern Madagascar.

Head and body length is 145–80 mm, tail length is 60–105 mm, and weight is 85–105 grams. The upper parts are brown mixed with red and black, and the underparts are reddish. The fur is thick and soft; the individual hairs have a gray base and a brown apex. The external form is somewhat volelike. The tail is short, dark, and somewhat hairy. The skull is broad and massive, and the head is broad and rounded. The ears are well haired, and the palms are dark. Females have three pairs of mammae.

Brachyuromys has been trapped in a large, moist meadow on the central plateau of Madagascar. The animals were living in dense, matted grass and reeds, in association with *Microgale*, *Eliurus*, and *Rattus*. The grass and reed stems were so matted down and so densely tangled that no sunlight reached the runways where the mice were traveling. They were active at all hours during both the night and the day.

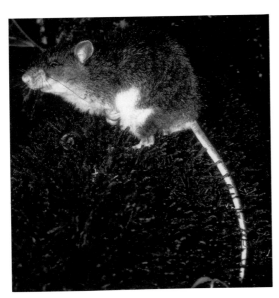

Voalavoanala *(Gymnuromys roberti)*, photo by Martin E. Nicoll.

Malagasy giant rat *(Hypogeomys antimena)*, photo by F. Petter.

RODENTIA; MURIDAE; **Genus PAROTOMYS**
Thomas, 1918

Karroo Rats

There are two species (Misonne, *in* Meester and Setzer 1977; Smithers 1971):

P. brantsii, southern Namibia, southern Botswana, Cape Province of South Africa;
P. littledalei, Namibia, Cape Province of South Africa.

Head and body length is 135–70 mm and tail length is 75–120 mm. Smithers (1971) listed weights of 89–155 grams for *P. brantsii.* In *P. brantsii* the general coloration of the upper parts is rusty yellowish, variegated with blackish or brownish to form delicate streaks. The sides of the head, neck, and body, as well as the underparts, are grayish white. In one form the basal half of the tail is reddish orange and the terminal half is brownish red, but in another form the tail is black except for a

tawny buffish base. In *P. littledalei* the upper parts vary from tawny to cinnamon buff, and the sides and underparts are buff to whitish buff. The upper portion of the tail is colored like the back, and the underside is paler.

Parotomys has small external ears, but the auditory bullae are much enlarged and spherical. The bullae of *Otomys* are relatively much smaller. *P. brantsii* has definite grooves on the upper incisor teeth, but the incisors of *P. littledalei* lack grooves. Females have four mammae.

Karroo rats inhabit sandy velds and high and low flats, generally near karroo or saltbushes. They are diurnal and live in burrows with many entrances dug in hard sandy ground. *P. brantsii* builds its nests within chambers in the burrows (Smithers 1983). *P. littledalei,* however, constructs conspicuous nests of interwoven sticks and grass up to about 600 cm in height among the tangled roots and branches of shrubs above the burrows. Karroo rats seldom venture far from their shelters. They are quite wary, and at the slightest indication of an intruder they quickly seek the safety of their homes. Before diving for shelter, however, they give a loud, sharp, piercing whistle and stamp their back feet (Smithers 1983). They are

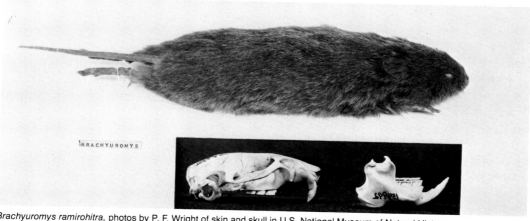

Brachyuromys ramirohitra, photos by P. F. Wright of skin and skull in U.S. National Museum of Natural History.

Karroo rat *(Parotomys littledalei)*, photo by John Visser.

vegetarian, feeding on grass, grass seeds, and fresh green shoots of low-growing vegetation (Smithers 1971).

While *P. littledalei* seems to be solitary, *P. brantsii* is gregarious. It is said that breeding takes place as often as four times per year, that litters contain up to four young, and that the young become sexually mature at three months. Of nine pregnant females taken in Botswana in February 1967, one carried a single embryo, six had two embryos each, and two had three embryos. A female has been observed dragging the young attached to the nipples (Smithers 1971).

RODENTIA; MURIDAE; Genus OTOMYS
F. Cuvier, 1823

African Swamp Rats, or Groove-toothed Rats

There are nine species (Kingdon 1974*b*; Misonne, *in* Meester and Setzer 1977):

O. typus, mountainous areas from southern Sudan and Ethiopia to northeastern Zambia;

O. laminatus, South Africa;

O. anchietae, Angola, southern Tanzania;

O. angoniensis, savannahs from Kenya to Angola and South Africa;

O. saundersiae, Cape Province of South Africa;

O. irroratus, Cameroon, eastern Zaire, Uganda, Kenya, Zimbabwe, South Africa;

O. denti, mountainous areas from Uganda to Malawi;

O. sloggetti, eastern South Africa, Lesotho;

O. unisulcatus, Cape Province of South Africa.

The species *O. unisulcatus* has sometimes been placed in a separate genus, *Myotomys* Thomas, 1918. Misonne (*in* Meester and Setzer 1977) distinguished *O. tropicalis,* found in tropical Africa, from *O. irroratus* but noted that the two might be conspecific, as they were treated by Delany (1975), Kingdon (1974*b*), and Rosevear (1969). F. Petter (1982) suggested that *O. typus, O. laminatus,* and *O. anchietae* might be part of the species *O. irroratus. O. maximus,* found in a small region where Angola, Zambia, Zimbabwe, Botswana,

African swamp rat *(Otomys irroratus)*, photo from South African Union Health Department through D. H. S. Davis.

African swamp rats (Otomys unisulcatus), photos by John Visser.

and Namibia come together, was treated as a species distinct from O. angoniensis by Swanepoel, Smithers, and Rautenbach (1980). This procedure was followed by Corbet and Hill (1986), Honacki, Kinman, and Koeppl (1982), and Smithers (1983) but not by Meester et al. (1986).

Head and body length is 124–217 mm, tail length is 55–150 mm, and weight is 60–255 grams (Kingdon 1974b; Smithers 1971). The pelage varies in density, texture, and length but is usually long, thick, and shaggy. The general coloration of the upper parts ranges from pale buffy through shades of brownish to dark brown or bright rusty. The underparts are white, creamy, buffy, brownish, or dull dark grayish, generally paler than the upper parts. The relatively short tail is fairly well haired, but it is not penciled and is usually darker above than below.

African swamp rats are characterized by having at least one conspicuous groove on all four incisor teeth. The upper incisors are strongly curved backward into the mouth. The head of Otomys is rounded and rather volelike, and the ears are relatively small. Dasymys closely resembles Otomys in external appearance, but that genus lacks the grooved incisors and has a longer tail. Otomys has a stocky body form. Females have four mammae.

The subfamily Otomyinae, of which Otomys is the most widespread living genus, has been called the ecological equivalent of the Microtinae, the voles and lemmings of the Holarctic (Kingdon 1974b). Both groups comprise grass and herb eaters that live mainly in moist, grassy areas and make runways under dense cover. The general habitats of Otomys include swamps, grasslands, savannahs, brush country, and alpine meadows. In Ethiopia, O. typus is found at elevations of 1,800–4,000 meters, where it frequents the edges of streams and marshes (Yalden, Largen, and Kock 1976). In Botswana O. angoniensis usually inhabits the heavily vegetated fringes of rivers and swamps, but during the wet season it may move some distance from waterways (Smithers 1971). O. sloggetti seems to prefer rocky areas, and O. unisulcatus utilizes dry habitat (Misonne, in Meester and Setzer 1977).

Most species usually shelter above ground, where they construct a nest of shredded plant material and twigs, often at the base of a small shrub or clump of grass. Others build a nest of fine grass at the bottom of a simple burrow, and still others appear not to make a nest at all. A conspicuous system of runways or tunnels is made through the grass to favorite feeding areas. The animals may be either diurnal or nocturnal, but Kingdon (1974b) found them to be most active around dawn and dusk. They are not primarily aquatic, but they enter water freely to swim from one reed patch to another, and they can dive and stay underwater for a short time to escape their enemies.

In contrast to most other species of Otomys, O. uni-

sulcatus frequents dry, sandy areas in shrub growth and piles of rock. It is active mainly during the day. Its shelter is usually constructed above ground in shrubs and less commonly in trees or around matted clumps of grass. This species sometimes excavates extensive burrow systems in sandy areas, often under sandstone rocks. In such locations it may not build a home above ground. The tunneled shelters are made of sticks or of weeds and grasses, the burrows extending below the ground. The female makes a soft nest of dry grass and twigs in a subterranean burrow. Surface paths or runs connect one refuge with another, and when the animals are disturbed, they use these paths to reach another shelter.

Otomys is almost exclusively herbivorous. It feeds mainly on green grasses, semiaquatic plants, and tender shoots. It also takes a variety of grains, seeds, berries, roots, and bark. It apparently does not harm cultivated crops but has done extensive damage to forest nurseries by gnawing the bark and cambium of young trees (Kingdon 1974b).

Population densities of around 16–42 per ha. have been reported for *O. irroratus*. Marked individuals of this species have been found to stay within an area of about 20 ha. (Kingdon 1974b). Although *Otomys* is sometimes said to occur in pairs, family groups, or colonies, a study of captive *O. irroratus* showed adults to be highly unsocial and aggressive toward one another (Davis 1972). In *O. unisulcatus* a pair of animals usually occupies a single shelter.

According to Kingdon (1974b), breeding in East African *Otomys* apparently is continuous, there may be up to five litters annually, and litter size is only 1–2 young. In southern Africa pregnant and lactating female *O. angoniensis* have been captured from August to May; breeding evidently peaks in the wet summer months and declines in the dry winter, when food availability decreases. Females of this species produce 1–5, usually 2–3, young (Bronner and Meester 1988). In a study in southern South Africa, Swanepoel (1975) found reproduction in August 1973, January 1974, and June 1974; 10 pregnant female *O. irroratus* contained an average of 2.1 (1–3) embryos, and 4 pregnant female *O. unisulcatus* had 1–3 embryos each. In a study of *O. irroratus* captured near Pretoria, and their offspring, Davis and Meester (1981) found little or no breeding from May to July, but litters were produced throughout the remainder of the year. Gestation was estimated at 40 days, with litter size averaging 2.33 and ranging from 1 to 4. The precocial young weighed about 12.5 grams each at birth, and although they were well-furred and able to move about at birth, they clung firmly to the nipples of the mother for 7 days. They opened their eyes in their second day of life and were fully weaned after 13 days. Females attained sexual maturity at 9–10 weeks, and males at 13 weeks.

Schlitter (1989) classified *O. denti* as vulnerable because of its patchy distribution in limited suitable areas of mountain forest habitat.

RODENTIA; MURIDAE; **Genus RHIZOMYS**
Gray, 1831

Bamboo Rats

There are two subgenera and three species (Ellerman and Morrison-Scott 1966; Sung 1984):

subgenus *Rhizomys* Gray, 1831

R. sinensis, central and southern China, northern Burma, Viet Nam;

R. pruinosus, Assam to southeastern China and Malay Peninsula;

subgenus *Nyctocleptes* Temminck, 1832

R. sumatrensis, Burma, Thailand, Indochina, Malay Peninsula, Sumatra.

Rhizomys, along with *Cannomys* and *Tachyoryctes*, usually has been placed in a distinct family, the Rhizomyidae, but the three genera are considered here to represent only a subfamily of the Muridae, for reasons discussed in the above account of that family.

Head and body length is 230–480 mm, tail length is 50–200 mm, and weight is 1–4 kg. *R. sumatrensis* is the largest species. In northern parts of the range of the genus the pelage is soft, thick, and silky, but in the tropics the fur becomes harsh and scanty. The coloration of the upper parts ranges from slate through pinkish gray to brownish gray; the underparts are generally paler. Some hairs may be tipped with white.

Bamboo rats resemble American pocket gophers (family Geomyidae) but lack external cheek pouches. They have stout, heavy bodies; short legs; short, naked, scaleless tails; small eyes and ears; and strong digging claws, the third digit of which has the longest nail. The pads of the feet are granulated; the two posterior sole pads are joined in the subgenus *Nyctocleptes* and distinct in the subgenus *Rhizomys*. The zygomatic arches of the skull are extremely wide and strong. The stout incisor teeth are orange, nearly vertically directed, and not covered by the lips. The dental formula is the same as that of most other Muridae, but there have been suggestions that the first of the three molars, both upper and lower, is actually a premolar. Female *R. sumatrensis* have two pectoral and three abdominal pairs of mammae, and female *R. pruinosus* have one or two pectoral and three abdominal pairs (Lekagul and McNeely 1977).

These rodents generally inhabit bamboo thickets at elevations of 1,200–4,000 meters. They spend much of their life underground among the roots of dense stands of bamboo. They dig extensive burrows, using both their teeth and their claws. An individual bamboo rat usually has several burrows, only one or two of which may be in active use. These animals commonly remain underground to cut and eat the bamboo roots that form their staple diet. According to Medway (1978), they sometimes come up at night and roam widely and may climb bamboo. They straddle the stem, which is gripped between the legs, and with their teeth they cut out sections of stem wall, which are then taken down to the burrow. Other grasses, seeds, and fruits are also eaten. If enough fruit is available, free water will not be drunk.

Although powerfully built, these animals move slowly; their gait is a cumbersome waddle. If they sense that they are going to be overtaken, or if they are cornered, bamboo rats become quite fierce, making short rushes at anything put in front of them and biting savagely. At the same time, a grunting noise is emitted, coupled with a peculiar grinding action of the teeth. Even captives are said to be vicious; if taken young, however, bamboo rats become very tame and tractable in captivity (Medway 1978).

In northern Viet Nam, *R. pruinosus* and *R. sumatrensis* breed from February to April and from August to October and have litters of one to four young (Tien and Sung 1971). According to Lekagul and McNeely (1977), the gestation period is at least 22 days, and three to five young are born, naked and blind, in an underground nest. Hair begins to grow at around 10–13 days of age, and the eyes open at about 24 days. The young first take solid food at around 1 month but

Bamboo rat *(Rhizomys pruinosus)*, photo by Wang Sung.

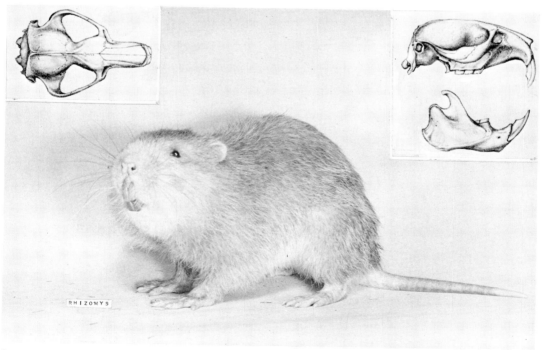

Sumatran bamboo rat *(Rhizomys sumatrensis)*, photo by Ernest P. Walker. Insets: *R. pruinosus*, photos from *Anatomical and Zoological Researches: Zoological Results of the Two Expeditions to Western Yunnan in 1868 and 1875*, John Anderson.

still suckle occasionally at 3 months. Longevity is about 4 years.

Wild bamboo rats raid plantations for tapioca and sugar cane roots (Medway 1978). The animals, in turn, are trapped or dug out by native people for use as food.

RODENTIA; MURIDAE; Genus CANNOMYS
Thomas, 1915

Lesser Bamboo Rat

The single species, *C. badius*, is found in Nepal, Assam, northern Bangladesh, Burma, Thailand, Laos, and northern Viet Nam (Ellerman and Morrison-Scott 1966; Kock and Posamentier 1983; Sung 1976). *Cannomys*, along with *Rhizomys* and *Tachyoryctes*, has usually been placed in a distinct family, the Rhizomyidae, but the three genera are here considered to represent only a subfamily of the Muridae, for reasons discussed in the above account of that family.

Head and body length is 147–265 mm, tail length is 60–75 mm, and weight is 500–800 grams (Lekagul and McNeely 1977). The fur is fairly thick on the head and body and very thin on the tail. The color ranges from reddish cinnamon and chestnut brown to ashy gray and plumbeous. Individuals are almost uniformly colored throughout; there may be a longitudinal white band on the top of the head and a narrow white band from the chin to the throat.

Like *Rhizomys*, *Cannomys* resembles the American pocket gophers but lacks cheek pouches. *Cannomys* has a thick, heavy body; very wide zygomatic arches in the skull; small eyes and ears; and short legs with long but powerful digging claws. *Cannomys* differs from *Rhizomys* in being smaller, having incisor teeth that protrude forward, and having smooth

sole pads. Females have two pectoral and two abdominal pairs of mammae (Lekagul and McNeely 1977).

The lesser bamboo rat constructs burrows in grassy areas, forests, and sometimes gardens. It digs rapidly, using its powerful teeth as well as its claws. The tunnels often are very deep and located in hard, stony ground. Above ground, the lesser bamboo rat moves slowly, though it is said to be fearless when surprised by an enemy. It leaves its burrow in the evening to feed on various plant materials, including shrubs, the young shoots of grasses and cereals, and roots. "Bamboo" rat is something of a misnomer, as the animal consumes all kinds of vegetation.

Captives maintained by Eisenberg and Maliniak (1973) showed peaks of activity in the morning and evening but slept a great deal during the day. The gestation period in three pregnancies was 40–43 days. Litter size in five deliveries was one or two young. The young developed relatively slowly. One captive lived for 3 years and 3 months (Jones 1982).

The lesser bamboo rat is sometimes common in tea gardens and has been reported to damage tea plants, but the amount of damage may be exaggerated. This animal is eaten by many of the Burmese hill tribes.

RODENTIA; MURIDAE; Genus TACHYORYCTES
Rüppell, 1835

African Mole-rats

There are two species (Misonne, *in* Meester and Setzer 1977; Yalden 1975):

T. macrocephalus, highlands of Ethiopia;
T. splendens, Ethiopia, Somalia, Uganda, Rwanda, Burundi, Kenya, northern Tanzania.

Lesser bamboo rat *(Cannomys badius)*, photo by Constance P. Warner. Insets: photos from *Anatomical and Zoological Researches: Zoological Results of the Two Expeditions to Western Yunnan in 1868 and 1875*, John Anderson.

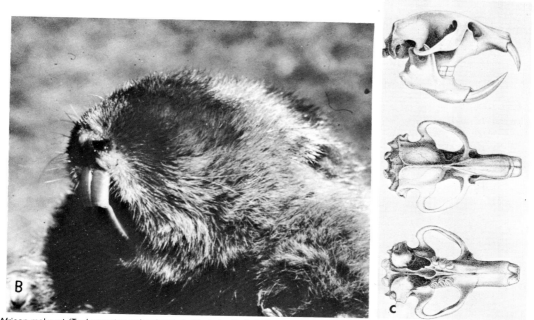

African mole-rat *(Tachyoryctes splendens):* A. Photo by C. A. Spinage; B. Photo by P. Morris; C. Photo from Museum Senckenbergianum.

This fossorial genus, along with *Rhizomys* and *Cannomys,* has usually been placed in a distinct family, the Rhizomyidae, but is here considered part of the Muridae, for reasons discussed in the above account of that family.

In *T. splendens* head and body length is 160–260 mm, tail length is 50–95 mm, and weight is 160–280 grams (Kingdon 1974b). *T. macrocephalus* is a larger species, its head and body length reaching 313 mm and its normal weight range being 330–930 grams (Yalden 1975). The short tail of *Tachyoryctes* is about twice the length of the hind foot and is usually well haired. The fur is thick and soft. The coloration is variable: some animals are shining black throughout, while others are brownish, reddish brown, pale gray, or cinnamon buff. Most young are black. Albinos and partial albinos are as common as black adults. The underparts are usually slightly paler than the upper parts and often have a silvery effect.

The body form is stocky and molelike. The eyes are small but plainly visible and functional, and the ears are small. The stiff hairs on the face are undoubtedly tactile. The thick, projecting incisors are not grooved and are deep orange in color. Although the claws are not particularly large, the hands and feet are well developed, and the short, powerful legs are suited for digging. African mole-rats resemble American pocket gophers (Geomyidae) but lack external cheek pouches and have somewhat longer and more fluffy fur.

African mole-rats usually are found in areas with more

than 500 mm of annual rainfall and flourish best in wet uplands. They favor open grassland, thinly treed savannah, moorland, and cultivated areas. They can withstand intense cold and are found at elevations of up to 4,150 meters (Kingdon 1974b; Yalden 1975). These animals construct a burrow system consisting of a nest chamber, a nearby bolt hole, and a series of foraging tunnels. The nest chamber contains a nest lined with grasses and herbs, a food store, and a sanitary area. The foraging tunnels may be up to 52 meters long; they usually run just below the level of the grass roots on which the animals feed, about 15–30 cm deep, but in the dry season *Tachyoryctes* may dig to a depth of 1 meter and become less active. Mounds of earth about 15–40 cm wide and 7–15 cm high develop as the burrow is excavated (Delany 1975; Jarvis and Sale 1971; Kingdon 1974b).

Based on observations of captives, Jarvis and Sale (1971: 454–55) described the burrowing procedure of *T. splendens* as follows: "Except where the soil is very soft, all burrowing is done with the incisors. Strong forward and upward sweeps of the lower incisors cut away at the soil face. While biting in this way the mole-rat braces itself by gripping the burrow sides with laterally rotated hind feet. The fore feet scratch up the loosened soil and push it underneath the animal where a pile gradually accumulates. Periodically, when the pile under the abdomen is large enough, the hind feet relax their grip, are rapidly brought forward, collect the soil and kick it vigorously behind the animal. . . . Excavation continues and soil collects until the burrow behind the mole-rat is completely blocked with loose earth. The next phase in burrowing, namely the transportation of excavated soil, follows. The mole-rat first turns round by a lateral 'somersault' in which it curls up and pushes itself round with its fore feet. Then, using one side of its face and one fore foot held close to the chin, it consolidates the earth and pushes it along the burrow."

Kingdon (1974b) reported a captive *T. splendens* to be nocturnal. Through the use of radioactive tagging, however, Jarvis (1973a) found wild individuals to be active mainly from 1000 to 1900 hours and to remain in the nest at other times. *T. splendens* sometimes comes to the surface to forage for nesting materials and food, such as grasses and cultivated legumes; however, its diet consists mainly of the under-

ground parts of plants—roots, rhizomes, tubers, bulbs, and corms. Some food is stored. Yalden (1975) found *T. macrocephalus*, in contrast to *T. splendens*, to depend largely on aboveground vegetation, mainly grass. Individuals of this species were observed extensively during daylight. They generally foraged by bringing just the head and shoulders out of the burrow and gathering whatever plants could be reached.

Populations of *T. splendens* sometimes attain high densities—about one individual per 140 sq meters was calculated in one area—but each adult lives alone in its own burrow system. Captives fight savagely with each other for food (Kingdon 1974b). They have a characteristic attitude of defense when cornered, holding the head erect with the mouth wide open, and can give a vicious bite.

Yalden (1975) reported that six adult *T. macrocephalus* occupied an area of approximately 1,100 sq meters on a moor at an elevation of 3,900 meters. No social structure was evident, and each animal appeared to be solitary. A single immature individual was also present, which suggests that litter size in the species is small.

A considerable amount of reproductive data on *T. splendens* has now accumulated (Delany 1975; Jarvis 1969, 1973b; Kingdon 1974b; Rahm 1969b). The capability for breeding continues throughout the year, though activity is highest during the rains and lowest in the dry season. Females are polyestrous and generally have 2 litters in fairly rapid succession, the second probably being conceived during a postpartum estrus. In a study in Kenya, Jarvis (1973b) found the average annual number of litters per female to be 2.1 and the gestation period to be 37–40 days. In eastern Zaire, however, Rahm (1969b) determined gestation to last 46–49 days. There may be up to 4 young per litter, but usually there are only 1 or 2. The young are weaned at 4–6 weeks, leave the mother's burrow about 1 month later, and attain sexual maturity at 6 months. Average life expectancy is around 1 year, but maximum known longevity in the field is 3.1 years.

African mole-rats are sometimes agricultural pests, because of their raids on sweet potatoes, corn, beans, peas, groundnuts, and root crops. They were formerly regularly used for food by some native tribes. Water was poured into their burrows, which brought them to the surface and made

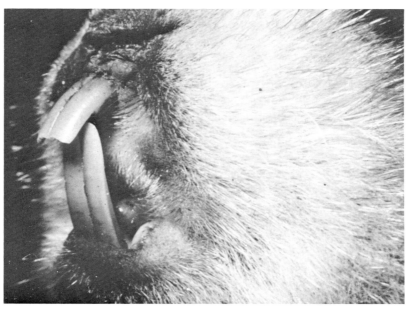

African mole-rat *(Tachyoryctes macrocephalus)*, photo by D. W. Yalden.

them easy to catch. They have become rare in the Buganda area of Uganda, probably because their skins have value as charms (Kingdon 1974b).

RODENTIA; MURIDAE; **Genus GERBILLUS**
Desmarest, 1804

Northern Pygmy Gerbils

The following 3 subgenera and 38 species are tentatively recognized:

subgenus *Handecapleura* Lataste, 1882

G. *campestris*, Atlantic coast of the Sahara to Egypt and Somalia;
G. *dasyurus*, Nile Delta (Corbet 1984), Sinai, Syria, Iraq, Arabian Peninsula;
G. *mackilligini*, southeastern Egypt and probably adjacent Sudan;
G. *jamesi*, Tunisia;
G. *maghrebi*, known only from the type locality in northern Morocco;
G. *stigmonyx*, Sudan;
G. *amoenus*, Libya, Egypt, possibly Mauritania to Tunisia;
G. *poecilops*, southwestern Arabian Peninsula;
G. *famulus*, southwestern Arabian Peninsula;
G. *nanus*, Morocco to Somalia and western India;
G. *watersi*, Sudan, Somalia, Djibouti;
G. *mesopotamiae*, Tigris-Euphrates Valley in Iraq and western Iran;
G. *henleyi*, Algeria to Israel and Arabian Peninsula;
G. *syrticus*, northeastern Libya;
G. *pusillus*, southern Sudan (Dieterlen and Nikolaus 1985), southwestern Ethiopia, Kenya;
G. *ruberrimus*, Kenya, Somalia;
G. *bottai*, Sudan, Kenya;
G. *muriculus*, western Sudan;
G. *mauritaniae*, known only by a single specimen from central Mauritania;

subgenus *Dipodillus* Lataste, 1881

G. *simoni*, eastern Morocco to Egypt;
G. *zakariai*, Kerkennah Islands off eastern Tunisia;

subgenus *Gerbillus* Desmarest, 1804

G. *gerbillus*, Morocco and northern Nigeria to Jordan and Kenya;
G. *cheesmani*, Arabian Peninsula to southwestern Iran;
G. *aquilus*, eastern Iran, southern Afghanistan, western Pakistan;
G. *gleadowi*, Pakistan, northwestern India;
G. *andersoni*, coastal areas from Tunisia to Israel;
G. *pyramidum*, throughout northern Africa, Israel;
G. *allenbyi*, coastal Israel;
G. *hesperinus*, northern Morocco;
G. *pulvinatus*, Somalia, Ethiopia, Djibouti, Kenya;
G. *perpallidus*, northwestern Egypt;
G. *riggenbachi*, Morocco, Western Sahara;
G. *latastei*, Tunisia, Libya, possibly Algeria;
G. *hoogstraali*, southwestern Morocco (Zyadi 1988);
G. *occiduus*, known only from the type locality in southwestern Morocco;
G. *rosalinda*, central Sudan;
G. *agag*, southern Mauritania to northern Nigeria and Sudan;
G. *nancillus*, central Sudan.

The above list is probably far from definitive, but it does represent an attempt at resolving differences among numerous recent revisors (Corbet 1978; Cockrum 1977; Cockrum, Vaughan, and Vaughan 1976b; Harrison 1972; Hubert 1978b; Kock 1978b; Lay 1975; Lay and Nadler 1975; Misonne 1974; Nevo 1982; Osborn and Helmy 1980; F. Petter, *in* Meester and Setzer 1977; Schlitter 1976) and between the work of some of these authorities and Lay's (1983) conclusion that this work involved placing many specific names in the synonymy of other species, without sufficient data. Lay himself recognized 62 full species. Corbet and Hill (1986), utilizing Lay's paper as a source, listed 49. Prior to publication of Lay's paper, Honacki, Kinman, and Koeppl (1982) had recognized 34 species. In formulating the above list, the general rule was to accept the synonymy of the earlier authors except in those instances for which Lay (1983) presented substantial evidence indicating that the involved species are valid. G. *dunni*, of Ethiopia, Somalia, and Djibouti, was treated as a full species by Lay and by F. Petter (*in* Meester and Setzer 1977) but was included in G. *pulvinatus* by Schlitter (1976). Cockrum (1977) suggested that G. *latastei* might include G. *dunni* as well as G. *perpallidus*, G. *riggenbachi*, and G. *rosalinda*. Yalden, Largen, and Kock (1976) thought that G. *dunni* and G. *pul-*

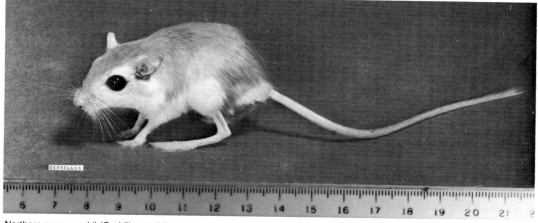

Northern pygmy gerbil *(Gerbillus gerbillus)*, photo by Ernest P. Walker.

vinatus were inseparable from *G. pyramidum*.

The genera *Microdillus* and *Gerbillurus* sometimes have been considered only subgenera of *Gerbillus*, but they are treated separately here in accordance with most recent authorities. *Dipodillus* generally has been given full generic rank (Corbet 1978; Cockrum, Vaughan, and Vaughan 1976a; Honacki, Kinman, and Koeppl 1982; F. Petter, *in* Meester and Setzer 1977; Schlitter and Setzer 1972) but was considered neither a genus nor a subgenus by Lay (1983), and the latter procedure was followed by Corbet and Hill (1986). In contrast, Osborn and Helmy (1980) not only maintained *Dipodillus* as a genus but included within it species assigned here to the subgenus *Handecapleura*. The species *G. mauritaniae* had been placed in its own genus, *Monodia* Heim de Balsac, 1943, but was based on a single abnormal specimen that actually is referable to the subgenus *Handecapleura* (F. Petter, *in* Meester and Setzer 1977; Schlitter 1976). Pavlinov (1982a) continued to recognize both *Monodia* and *Dipodillus* as full genera and also described a new subgenus, *Petteromys*, to include an undetermined number of the species of the genus *Gerbillus*.

Head and body length is 50–130 mm, tail length is 69–180 mm, and weight is 10–63 grams. The upper parts range from pale yellowish gray, clay color, or sandy buff to sandy red, mouse gray, dark fulvous, or brilliant reddish brown. The median line of the back is usually slightly darker. The sides and flanks are paler, blending into the white, creamy, or pale gray of the underparts. The tail is moderately to well furred, generally with a slight tuft near the tip. In some species there is a slight crest on the upper side of the distal half. The underside of the tail is usually light, the upper part near the body is darker, and the terminal portion is darkest (sometimes being dark brown or black).

Gerbillus is characterized by a slender form and fairly long ears and claws. The slender upper incisors are grooved, though sometimes indistinctly. In the subgenera *Gerbillus* and *Handecapleura* the tail is longer than the head and body, the cheek teeth are not hypsodont (Corbet 1978), and the hind feet are elongated, usually being over 25 percent of the head and body length. In the subgenus *Handecapleura* the soles of the hind feet are naked, while in the subgenus *Gerbillus* the soles are hairy. *Dipodillus* differs from the other subgenera in having a tail that is usually shorter than the head and body, no tuft on the tail, hind feet that are usually less than 25 percent of the head and body length, and slightly hypsodont teeth; the soles of its hind feet are naked. Lay (1983) questioned the reliability of some of the characters commonly used to distinguish these subgenera.

The remainder of this account is based in large part on information compiled by Harrison (1972). Northern pygmy gerbils are found in dry country, sandy or rocky, sometimes with only scarce and coarse vegetation. Some species, however, tend to concentrate in moist, well-vegetated places, and *G. poecilops* seems to prefer cultivated areas and has a tendency to invade farm buildings. All species are primarily nocturnal, and some are also crepuscular. They construct burrows, which vary from small, simple holes to fairly elaborate excavations. The burrows of certain species, such as *G. mesopotamiae*, may be two to three meters long and have several entrances, food storage tunnels, and an enlarged chamber with a nest of dry vegetation about a meter below the surface. The burrow entrances are usually kept closed with sand. The diet consists of seeds, roots, nuts, grasses, and insects.

The individuals of most species tend to build their burrows close to one another and thereby give the impression of an extensive colony. Soliman (1983) reported population densities of 2.8–7.0 per ha. for *G. andersoni* and up to 2.6 per ha. for *G. gerbillus* and home ranges of around 2,000 sq meters. The females of most species are polyestrous, and breeding has been reported at all seasons of the year. The gestation period is 20–22 days, and litter size is one to eight young, usually about four or five. The offspring are born naked and helpless, open their eyes at around 16–20 days, and are weaned at about 1 month. According to Jones (1982), a captive specimen of *G. pyramidum* lived 8 years and 2 months.

RODENTIA; MURIDAE; Genus MICRODILLUS
Thomas, 1910

The single species, which occurs in Somalia, was described as *Gerbillus peeli* by De Winton (1898). Later, Thomas (1910a) stated that the originally described skin was actually referable to the genus *Ammodillus* but that the original skull, along with subsequently collected material, represented an entirely new genus, which he called *Microdillus*. Although long considered to be only a subgenus of *Gerbillus*, *Microdillus* was listed as a distinct genus by F. Petter (*in* Meester and Setzer 1977).

According to Funaioli (1971), head and body length is 60–80 mm and tail length is 56–62 mm. The general color is pale yellowish brown; the underparts, hands, and feet are whitish. There is a round, white spot behind each ear. The soles are naked.

The skull, unlike that of other gerbils, is peculiarly square and short. It is abnormally bowed, with a strongly convex cranial profile. Like that of *Dipodillus*, the tail is shorter than the head and body and lacks a terminal tuft. Whereas the upper third molar of *Dipodillus* lacks cusps, this tooth in *Microdillus* contains three or four cusps (F. Petter, *in* Meester and Setzer 1977; Schlitter and Setzer 1972).

Funaioli (1971) stated that the few known specimens have been collected on the predesert steppes of north-central Somalia. *Microdillus* is nocturnal. During the day it apparently takes refuge in an underground burrow.

RODENTIA; MURIDAE; Genus GERBILLURUS
Shortridge, 1942

Southern Pygmy Gerbils

There are four species (Davis, *in* Meester and Setzer 1977; Schlitter 1973; Schlitter, Rautenbach, and Coetzee 1984):

G. paeba, southwestern Angola, Namibia, Botswana, South Africa;

G. tytonis, southern Namibia;

G. vallinus, southwestern Angola, Namibia;

G. setzeri, Namibia.

Gerbillurus long was considered a subgenus of *Gerbillus*. Davis (*in* Meester and Setzer 1977) listed *Gerbillurus* as a separate genus, but with some question as to whether *G. paeba* and *G. vallinus* belonged to the same genus. Schlitter (1976) thought *Gerbillurus* to be more closely related to *Tatera* than to *Gerbillurus*, and this view was supported by the chromosomal studies of Qumsiyeh (1986). Pavlinov (1982a) designated a new subgenus, *Progerbillurus*, to include *G. paeba*, and F. Petter (1983a) described the subgenus *Paratatera* for the species *G. tytonis*. However, Schlitter, Rautenbach, and Coetzee (1984) considered *G. paeba* and *G. tytonis* to be closely related; they agreed with designation of the subgenus *Progerbillurus* but doubted the need for *Paratatera* (if *G. paeba* and *G. tytonis* are put into one subgenus, *G.*

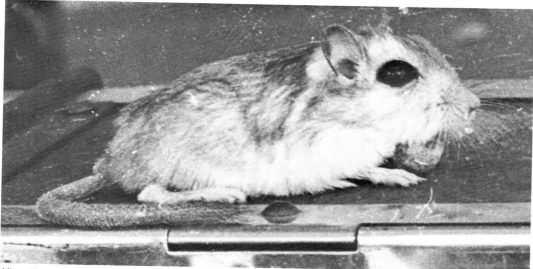

Microdillus peeli, photo by Alberto Simonetta.

vallinus and *G. setzeri* would fall into another subgenus, which would have the name *Gerbillurus*).

Head and body length is about 90–120 mm and tail length is 95–156 mm. The weight of *G. paeba* is 20–37 grams (Smithers 1971). The upper parts are brownish or gray in *G. paeba,* sandy buff in *G. vallinus,* hazel in *G. tytonis,* and light pinkish cinnamon in *G. setzeri.* The underparts, feet, and ventral surface of the tail are generally white.

Externally, *Gerbillurus* resembles *Gerbillus.* The tail is longer than the head and body and has a tufted tip. Unlike in

Gerbillus, the molar laminae in the upper and lower jaws have no sign of median crest connections. Unlike in *Tatera,* the soles of the hind feet of *Gerbillurus* are well haired, and the zygomatic plate of the skull does not project far forward (Davis, in Meester and Setzer 1977).

All species inhabit desert or semidesert regions. Smithers (1971) stated that *G. paeba* is confined to sandy ground or sandy alluvium with a grass, scrub, or light woodland cover. This species is nocturnal and terrestrial and lives in small warrens with many entrances. The entrances are often con-

Southern pygmy gerbil *(Gerbillurus tytonis),* photos by Mary Seely.

Southern pygmy gerbil *(Gerbillurus paeba)*, photo by John Visser.

cealed under a tuft of grass or at the base of bushes. The diet consists of the seeds of grasses, bushes, and trees. *G. paeba* apparently breeds throughout the year. The number of embryos per pregnant female averages 3.7 and ranges from 2 to 5.

RODENTIA; MURIDAE; **Genus TATERA**
Lataste, 1882

Large Naked-soled Gerbils

There are 2 subgenera and 13 species (Bates 1985, 1988; Corbet 1978; Davis, *in* Meester and Setzer 1977; Harrison 1972; Hubbard 1970; Hubert, Adam, and Poulet 1973; Yalden, Largen, and Kock 1976):

subgenus *Gerbilliscus* Thomas, 1897

T. boehmi, southwestern Kenya to Angola and Zambia;

subgenus *Tatera* Lataste, 1882

T. indica, Syria to India, Sri Lanka;
T. leucogaster, Angola and southwestern Tanzania to South Africa;
T. nigricauda, southern Ethiopia to northern Tanzania;
T. robusta, Guinea-Bissau to Somalia and central Tanzania;
T. guineae, Senegal to Ghana;
T. phillipsi, Ethiopia, Somalia, Kenya;
T. afra, southwestern South Africa;
T. valida, savannah zones from Senegal to western Ethiopia, and south to Angola and Zambia;
T. gambiana, Senegal;
T. brantsi, southern Africa;
T. inclusa, Tanzania to Zimbabwe and Mozambique;
T. pringlei, northeastern Tanzania.

Pavlinov (1982*a*) considered *Tatera* to include only a single species and *Gerbilliscus* to be a separate genus with most of the above species.

Head and body length is 90–200 mm, tail length is 115–245 mm, and weight is 30–227 grams. The pelage is soft or medium. The coloration above ranges from pale sandy gray or pale buffy gray to dark buffy, blackish cinnamon, sandy brown, or dark gray; the underparts are white or whitish. The hands and feet are light in color. In most forms the tail is slightly darker, both above and below, than the back. The sides of the tail are light, and in most forms the terminal part is white or tufted with long dark hairs. There are usually light-colored or white spots above and behind the eyes and ears and on the sides of the nose.

The body form is heavy and ratlike. The zygomatic plate of the skull is always projected very strongly forward. In the subgenus *Gerbilliscus* the upper incisors usually have two shallow grooves. In the subgenus *Tatera* the upper incisors are single-grooved or plain (Davis, *in* Meester and Setzer 1977). The soles of the hind feet are naked. Females have six or eight mammae.

These gerbils inhabit sandy plains, grasslands, savannahs, woodlands, and cultivated areas. They are nocturnal and terrestrial. They usually walk on all four limbs, but when alarmed they flee by means of running bounds of up to 1.5 meters in height. *T. indica* is said to be able to cover 3.5 meters in one leap. Short burrows, the so-called bolt holes, are excavated for sudden retreats, but deeper burrows are used as the living quarters. The animals burrow in sandy soil to a depth of 1 meter or more. Their tunnel systems are practically underground labyrinths in some areas, with numerous entrances. *T. indica* often has the enlarged nest chamber located in the center of the tunnel system. The entrances to the burrows are usually blocked with loose earth. The diet con-

Large naked-soled gerbils *(Tatera leucogaster)*, photo by Ernest P. Walker.

Large naked-soled gerbil *(Tatera afra)*, photo by John Visser.

sists of roots, bulbs, seeds, green plant material, and insects. *T. indica* also eats eggs and young birds.

Prakash, Jain, and Rana (1975) found that *T. indica* resided in shallow burrows on the fringes of main streets in an Indian town. In this area the density of the gerbils was 175–460 per ha., but in rural grasslands density was only 10 per ha. Prakash (1975) reported the home range of *T. indica* in the Rajasthan Desert to be 1,875 sq meters for males and 1,912 sq meters for females. *Tatera* appears to be a gregarious genus, as a number of interconnecting burrows are usually located in the same area, but burrows housing only 1 or 2 adults are sometimes found. Cannibalism is apparently frequent in captive *T. indica*, females consuming their own newborn offspring. Wild individuals of this species have been reported to eat smaller or trapped individuals of their own kind.

Smithers (1971) stated that *T. leucogaster* breeds all year in Botswana, but Sheppe (1973) found reproduction in this species to be heavily concentrated in the rainy season, January–February, in Zambia. Prakash (1975) reported that *T. indica* breeds all year in the Rajasthan Desert of India, with a major peak during the monsoon of July–August and minor peaks in February and November. Females of this species may have three or four litters annually, the estrous cycle averages 4.82 days, and gestation lasts 26–30 days. An average gestation period of 22.5 days has been reported for *T. afra*. Overall litter

size in the genus is 1–13 young, with averages of 3.3 in *T. brantsii* and 4.5 in *T. leucogaster* (Smithers 1971). In *T. indica* the young usually number 4–6; they open their eyes at 14 days, are weaned and independent at 21–30 days, and reach sexual maturity at 10–16 weeks (Prakash 1975; Roberts 1977). A captive *T. indica* lived for 7 years (Jones 1982).

Gerbils of this genus are suitable hosts for bubonic plague and are said to be important agents in spreading the disease in southern Africa and the Middle East, especially because of their tendency to enter human habitations. *T. indica* also is considered a serious agricultural pest in some areas because of its destruction of grains and legumes (Harrison 1972; Roberts 1977).

RODENTIA; MURIDAE; Genus TATERILLUS
Thomas, 1910

Small Naked-soled Gerbils

There are apparently eight species (Gautun, Tranier, and Sicard 1985; F. Petter, *in* Meester and Setzer 1977; Robbins 1974, 1977; Sicard, Tranier, and Gautun 1988):

T. gracilis, Senegal to Nigeria, probably Cameroon;

T. petteri, Burkina Faso;

T. pygargus, southern Mauritania, Senegal, Gambia, southwestern Mali;

T. arenarius, southern parts of Mauritania, Mali, and Chad;

T. lacustris, northeastern Nigeria, northern Cameroon, possibly Niger and Chad;

T. congicus, eastern Cameroon to Sudan;

T. harringtoni, central Sudan and eastern Central African Republic to southern Somalia and northeastern Tanzania;

T. emini, Chad and Central African Republic to northwestern Kenya.

Head and body length is 100–140 mm and tail length is 140–76 mm. *T. arenarius* weighs 42–52 grams (Robbins 1974). The coloration above ranges from pale yellow, through light buffy, clay color, light reddish brown, and tawny olive, to dark tawny brown. The sides are paler, and the underparts, including the hands and feet, are white or almost white. Most forms have light areas or spots on the sides of the face, above the eyes, and in back of the ears. Some forms have faint dark markings on the face. The tail is usually well tufted and darkest toward the tip. Externally, this genus is similar to *Tatera*, but it is usually smaller and the soles are sometimes

Small naked-soled gerbil *(Taterillus gracilis)*, photo by Jane Burton.

partially haired, not completely naked. As in *Tatera*, the upper incisors are grooved.

These gerbils usually inhabit treeless plains or thorny scrub savannahs. Robbins (1973), however, observed that specimens of *T. gracilis* had been taken in a wide spectrum of habitats ranging from moist woodland to very arid Sahel woodland. These gerbils live in underground burrows and excavate the earth in the form of surface mounds.

In studies on a dry thornbush savannah in northern Senegal, Poulet (1972, 1978) found *T. pygargus* to be nocturnal, granivorous, and insectivorous. Population density in this area was calculated to be about 2–6 per ha.; following heavy rainfall, however, density increased to as high as 180 per ha. Adult home range averaged 1,100 sq meters for males and 300 sq meters for females. The home range of an adult male overlapped those of several females. Births occurred only after rains, in the period from September to March. A female could have a litter every six weeks and would change her burrow each time. The gestation period was three weeks, and there usually were four young per litter. Juveniles were nomadic for a brief period and then settled onto permanent home ranges by the time they were three to five months old.

RODENTIA; MURIDAE; **Genus DESMODILLUS**
Thomas and Schwann, 1904

Cape Short-eared Gerbil

The single species, *D. auricularis*, occurs in Namibia, Botswana, and South Africa (F. Petter, *in* Meester and Setzer 1977).

Head and body length is 90–125 mm, tail length is 84–99 mm, and weight is 39–70 grams. The upper parts vary in color from uniform orangish to tawny brown. There is a white spot behind the ears on the back of the neck. The underparts, hands, and feet are white.

Desmodillus is characterized by short ears with a white spot behind them, enlarged auditory bullae, hind feet with hairy soles, laminate molars, and a tail that is haired but not tufted. The tail is about four-fifths as long as the head and body, and the hind foot is about one-fifth as long. The hind feet are not especially modified for jumping.

This gerbil inhabits open sandy plains, dry tablelands, and cultivated areas. It runs like an ordinary rat and is not noticeably saltatorial. Its burrows are recognizable by the small surface heaps of earth thrown up but are not as complicated as those built by some other genera of gerbils. The tunnels usually do not penetrate more than two meters, and they have blind passages or vertical escape holes. A nest of dried straw or sticks may be located in a chamber at the end of the tunnel. When the burrow is occupied, a heap of prickly seeds of a terrestrial creeping plant lies in front of the entrance on the mound of earth. *Desmodillus* is nocturnal. Its diet consists of seeds, grain, and insects such as locusts and grasshoppers. It usually takes the food to the entrance of the burrow, where it sits down and eats, leaving husks and wings around the entrance. During a drought this rodent may move to a different area. Laboratory investigations by Pettifer and Nel (1977) have shown *Desmodillus* to store a larder of food in its nest box.

This gerbil is not sociable. Although there may be many burrows within a short distance of each other, they do not form interconnected warrens (Smithers 1971). In a study of captives, Nel and Stutterheim (1973) found adults to be apparently solitary except briefly during mating. Two weeks after two males and two females had been released in an outdoor enclosure, the males had been killed and eaten by the females. Three weeks later one female was killed and eaten by the other.

Available evidence (Keogh 1973; Nel and Stutterheim 1973; Smithers 1971) indicates that females are polyestrous and may bear young at any time of the year. The gestation period is normally 21 days, but a postpartum pregnancy lasted 35 days. Litter size ranges from one to seven young and usually is about two to four. The young develop slowly, opening their eyes at 22 days and being weaned at 33 days.

RODENTIA; MURIDAE; **Genus DESMODILLISCUS**
Wettstein, 1916

The single species, *D. braueri*, is known from northern Senegal, northern Nigeria, and Sudan (F. Petter, *in* Meester and Setzer 1977; Poulet 1972).

Head and body length is 50–70 mm, tail length is about three-fourths that of the head and body, and Poulet (1972) stated that adults scarcely exceed 10 grams in weight. The pelage is quite soft, sandy fawn on the upper parts, somewhat paler on the sides, and white on the undersurface. Postauricular white spots are evident. The tail is haired but not tufted.

Desmodilliscus is similar to *Desmodillus* and *Pachyuromys*. The auditory bullae are enlarged. The presence of 3 upper and 2 lower cheek teeth (making a total of 10 cheek teeth) is unique in the family Muridae. The incisors are narrow and furrowed. The skull resembles that of a *Desmodillus* in miniature.

According to Poulet (1972), this tiny gerbil occupies the same kind of habitat as *Taterillus*—dry thornbush savannah. It is nocturnal and granivorous and constructs more complex burrows than those of *Taterillus*, with many entrances. Poulet (1978) stated that the abundance of *Desmodilliscus* apparently varies inversely with that of all other rodents in its habitat; it is rare during periods of heavy rainfall and becomes more numerous during droughts.

Cape short-eared gerbil *(Desmodillus auricularis)*, photo from South African Union Health Department through D. H. S. Davis.

Desmodilliscus braueri, photo by Francois M. Catzeflis.

RODENTIA; MURIDAE; **Genus PACHYUROMYS**
Lataste, 1880

Fat-tailed Gerbil

The single species, *P. duprasi*, inhabits the northern part of the Sahara Desert region from western Morocco to Egypt (Corbet 1978).

Head and body length is 105–35 mm and tail length is 45–60 mm. The soft pelage of the upper parts ranges from light yellowish gray to buffy brown; the underparts, hands, and feet are white. There is a white spot behind each ear. The tail is bicolored: coloration above is like that of the back, and below it is whitish. Some forms have pinkish buff along the sides and on the rump. The ears are short and buffy white, but the extreme edges are brown.

Pachyuromys is stockily built. It has well-developed claws on the fingers and slightly grooved upper incisors. As in the case of many desert mammals, the auditory bullae and mastoids are so inflated that they extend beyond the foramen magnum. The common name of this genus is derived from the peculiar shape of the tail, which is short, noticeably thickened, and club-shaped.

Daly (1979) wrote that *Pachyuromys* is insectivorous and confined to the type of desert habitat known as the hamada—gravelly plains with small patches of bushy perennial vegetation. The gestation period is 19–22 days, and litter size is three to six young. Captive specimens are used in laboratory studies. One such individual lived for 4 years and 5 months (Jones 1982).

RODENTIA; MURIDAE; **Genus AMMODILLUS**
Thomas, 1904

Walo

The single species, *A. imbellis*, occurs in Somalia and southwestern Ethiopia (F. Petter, *in* Meester and Setzer 1977; Yalden, Largen, and Kock 1976).

Head and body length is 85–106 mm and tail length is 134–60 mm (Funaioli 1971). The coloration of the upper parts is reddish fawn; the hairs of the back are tipped with black, as are the hairs of the sides, but to a lesser degree, thus being a clearer, lighter fawn. On the eyebrows, cheeks, and fronts of the arms the hairs are tipped with fawn. Distinct white spots are visible at the bases of the ears and above the eyes. The hands, feet, chin, cheeks, and underparts are white. The tail is darker above than below and scantily haired on the

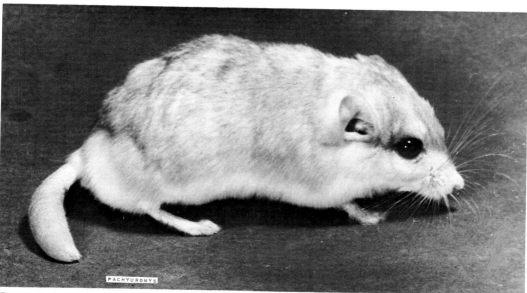

Fat-tailed gerbil *(Pachyuromys duprasi)*, photo by Ernest P. Walker.

Walo (*Ammodillus imbellis*), adapted from *Proc. Zool. Soc. London*, St. Leger.

basal portion but terminally tufted with brownish hairs 8–10 mm in length. The feet and hands are scantily covered. The hairless pads and soles are covered with conspicuous scalelike granulations.

Ammodillus is a desert rodent. Although individuals are noted for fighting among themselves, they have an extraordinary weakness in the lower jaw, as the coronoid process of the mandible is absent. This structural feature led to the choosing of the specific name, *imbellis*, which means "feeble."

RODENTIA; MURIDAE; Genus SEKEETAMYS
Ellerman, 1947

Bushy-tailed Jird

The single species, *S. calurus*, is known from eastern Egypt, Sinai, southern Israel and Jordan, and near Riyadh in central Saudi Arabia (Harrison 1972; Nader 1974). This species originally was described as *Gerbillus calurus* and later was regarded as *Meriones calurus*. In 1947 a new subgenus, *Sekeetamys*, was set up for the retention of this species in *Meriones*. Chromosomal studies, however, have not supported its inclusion in *Meriones* or *Gerbillus*. Since 1956 *Sekeetamys* has generally been considered a full genus (Corbet 1978).

Head and body length is 100–125 mm and tail length is 110–60 mm. The upper parts are yellowish or reddish, washed with black; the underparts, hands, and feet are whitish. The brownish, bushy tail usually bears a white tuft on the tip and sometimes has a central black terminal tuft with white tufts on either side. The soles of the hind feet are naked.

The bushy-tailed jird inhabits deserts and rocky slopes that are characterized by broken ground and hard surfaces. It bur-

rows under boulders and rocky ledges. It shows considerable climbing ability in its rocky habitat. The average number of young for 47 litters born in the Giza Zoological Gardens in Cairo was 2.9, and the largest number in a single litter was 6; births occurred throughout the year. A captive specimen lived for five years and five months (Jones 1982).

RODENTIA; MURIDAE; Genus MERIONES
Illiger, 1811

Jirds

There are 4 subgenera and 14 species (Corbet 1978, 1984; Ellerman and Morrison-Scott 1966; F. Petter, *in* Meester and Setzer 1977):

subgenus *Parameriones* Heptner, 1937

M. persicus, eastern Turkey to Pakistan;
M. rex, southwestern Arabian Peninsula;

subgenus *Cheliones* Thomas, 1919

M. hurrianae, southern Afghanistan, southeastern Iran, Pakistan, western India;

subgenus *Meriones* Illiger, 1811

M. vinogradovi, eastern Asia Minor, northwestern Iran, Syria;
M. tamariscinus, lower Volga region of Soviet Union to northwestern China;
M. tristrami, Turkey to northwestern Iran and Sinai;

Bushy-tailed jird *(Sekeetamys calurus)*, photo by Ernest P. Walker.

subgenus *Pallasiomys* Heptner, 1933

M. unguiculatus, Mongolia and adjacent parts of southern Siberia and northern China, Sinkiang, Manchuria;

M. meridianus, Caspian Sea region to Mongolia and northern China;

M. chengi, Sinkiang;

M. shawi, Morocco to Egypt;

M. libycus, Western Sahara to western Sinkiang;

M. crassus, northern Africa, southwestern Asia;

M. sacramenti, southern Israel;

M. zarudnyi, southern Turkmen S.S.R., northeastern Iran, northern Afghanistan.

M. caudatus, of Libya, sometimes has been regarded as a full species, but has been considered to be part of *M. libycus* by most recent authors (Corbet and Hill 1986; Honacki, Kinman, and Koeppl 1982; Pavlinov 1982b; F. Petter, *in* Meester and Setzer 1977).

Head and body length is 95–180 mm and tail length is 100–193 mm. In examining captive individuals of *M. unguiculatus,* Norris and Adams (1972) found weight to average 50–55 grams in females and 60 grams in males. Roberts (1977) reported the weight of *M. crassus* to be 29–40 grams and that a male *M. persicus* weighed 100 grams. The dense pelage is fairly long and soft in most forms but is short and harsh in others. The covering on the tail is short near the base and progressively longer toward the tip, so that the tail is slightly bushy in most of the forms and has a slight crest in some of the others. Coloration of the upper parts varies from pale, clear yellowish, through sandy and grayish, to brownish. The sides of the body are generally lighter than the back because of the absence of black-tipped hairs. The underparts, including the hands and feet, are white, pale yellowish, buffy, or pale gray. There are frequently light areas about the face. Heptner (1975) found the different colors of the subspecies of *M. meridianus* to harmonize with the sand shades found in the habitats of the animals.

Externally, *Meriones* is quite ratlike. It has narrow, well-developed ears; a tail length approximating that of the head and body; upper incisors with a narrow groove on the anterior surface; slightly elongated hind legs for leaping; and strong claws. Harrison (1972) listed the number of mammae as eight in females of *M. persicus, M. tristrami,* and *M. crassus.*

Jirds inhabit clay and sandy deserts, bush country, arid steppes, low plains, cultivated fields, grasslands, and mountain valleys. They are terrestrial and construct burrows in soft soil, where they spend much of their time. The complexity of the burrows varies both within and between species. Young individuals of *M. crassus* may excavate a burrow that descends at an angle of 15°–30° to a depth of half a meter and has only one entrance. The most complex galleries of this same species often attain a combined shaft length of 30–40 meters and have as many as 18 entrances (Koffler 1972). The tunnels of *M. unguiculatus* extend about 0.45–0.60 meters underground and are about 4 cm in diameter; a nest and one or two storerooms are in the central part of the system (Gulotta 1971). The burrows of *M. libycus* may be over 1.5 meters deep and radiate outward 3–4 meters in a tortuous tunnel system (Harrison 1972). Naumov and Lobachev (1975) classified the burrows of *M. persicus, M. tamariscinus,* and *M. unguiculatus* as simple and referred to those of *M. vinogradovi, M. meridianus,* and *M. libycus* as complex. The

Jird *(Meriones unguiculatus),* photo by Ernest P. Walker.

burrows usually include several food storage chambers near the surface and one or more nest chambers at a greater depth. Nests are usually composed of dried vegetation.

Roberts (1977) wrote that *M. persicus* and *M. crassus* are nocturnal, while *M. hurrianae* and *M. libycus* are diurnal. Gulotta (1971) stated that *M. unguiculatus* is active both day and night throughout the year. She reported that this species adapts to a wide range of temperature and humidity, is active on the surface both in sub-zero winters and on summer days of more than 38° C, and neither hibernates nor estivates. Reports of true hibernation in *Meriones* may have resulted from the tendency of some species to remain underground for long periods in the winter and depend on hoarded food. Roberts (1977) stated that *M. persicus* may undergo intervals of torpidity at such times. Naumov and Lobachev (1975) stated that *M. tristrami* may remain in its burrow for two months during the winter and live entirely off of stored food. The diet of *Meriones* consists of green vegetation, roots, bulbs, seeds, cereals, fruits, and insects. *M. libycus* stores up to 10 kg of seeds in the northern part of its range (Naumov and Lobachev 1975).

Maximum recorded population densities under favorable conditions are: 5–10 per ha. for *M. meridianus*; 20–30 per ha. for *M. tamariscinus* (Naumov and Lobachev 1975); 11 per ha. for *M. crassus* (Koffler 1972); and 477 per ha. for *M. hurrianae* (Roberts 1977). The density of *M. vinogradovi* in an area of 401,000 ha. was reported as 37.4/ha. before a poisoning control program and 0.8/ha. afterward (Klimchenko et al. 1975). *M. crassus* has been reported to use a home range of 1,200–10,000 sq meters (Koffler 1972), and the individual home range of *M. libycus* varies from 50 to 120 meters in diameter (Naumov and Lobachev 1975). In the Rajasthan Desert of India, Prakash (1975) determined the home range of *M. hurrianae* to average 88.7 sq meters for males and 154.7 sq meters for females. This species usually forages within 20 meters of its burrow (Roberts 1977). In contrast, the daily summer movements of *M. unguiculatus* may cover 1.2–1.8 km, and a marked individual moved 50 km (Naumov and Lobachev 1975).

Investigations of *M. libycus* in Algeria (Daly 1979; Daly and Daly 1975a) have revealed that females remain for many months in a single *daya* (a vegetated depression surrounded by desert). In one such *daya* five adult females occupied home ranges of about 0.6–4.0 ha. each. There was little or no overlap between the home ranges of females. Individual adult males, however, occupied relatively large home ranges covering several *dayas*. Male home ranges overlapped extensively both with each other and with those of females, but interaction was rare, and all adults occupied separate burrows.

Social behavior seems to vary considerably both within and between species of *Meriones*. In contrast to the above information on *M. libycus*, both Roberts (1977) and Naumov and Lobachev (1975) referred to this species as colonial and noted that many individuals burrowed in the same vicinity. Roberts added that *M. crassus* occurred in smaller colonies than did *M. libycus* and that *M. persicus* was not gregarious. He also wrote that adults of *M. hurrianae* were not aggressive and that several sometimes shared a burrow, though probably not the same nest chamber. Naumov and Lobachev stated that two or more families of *M. vinogradovi* were sometimes found in the same burrow and that a burrow of *M. unguiculatus* might contain 3–14 animals, probably a mated pair and their most recent litter. *M. unguiculatus* has been studied extensively in the laboratory, resulting in the consensus that adults can be kept together but that introduction of a stranger may result in a fight to the death (Gulotta 1971). Males scent-mark a territory with a ventral sebaceous gland (Yahr 1977). Females are at least as territorial and aggressive as males (Swanson 1974). Monogamous pairs seem to do well

in captivity (Norris and Adams 1972), but there is some question about the role of the male. Ahroon and Fidura (1976) suggested that the male disrupts maternal behavior to the extent that many young are lost. Elwood (1975), however, reported that the male shares in caring for the litter, spending considerable time cleaning, grooming, and warming the newborn. Ostermeyer and Elwood (1984) found that fathers, and sometimes juveniles, assisted in the rearing of younger animals. Jirds have several vocalizations and also apparently communicate by thumping the hind feet.

Captive *M. unguiculatus* are capable of breeding throughout the year; in the wild, however, the reproductive season extends from about February to October, and up to three litters are produced (Gulotta 1971; Naumov and Lobachev 1975). The estrous cycle lasts 4–6 days, and a postpartum estrus may occur. Gestation periods of 24–30 days have been reported in this species, but gestation may actually last only 19–21 days. Litter size is 1–12, usually 4–7, young. Newborn weigh about 2.5 grams each, open their eyes after 16–20 days, and are weaned at 20–30 days. Sexual maturity is attained at 65–85 days. Females are capable of reproduction until they are as much as 20 months old, but average longevity in the wild is only 3–4 months.

Reproduction continues all year in *M. hurrianae*, but there are peaks of breeding activity in the late winter, midsummer, and, in the Rajasthan Desert of India, autumn (Prakash 1975; Roberts 1977). Females of this species have an estrous cycle averaging 6.22 days, may experience a postpartum estrus, and give birth to three or four litters per year. The reported gestation period is 28–30 days. Litter size averages 4.4 and ranges from 1 to 9. The young open their eyes at 15–16 days, are weaned at 3 weeks, and reach sexual maturity at 15 weeks.

Data compiled on other species of the Soviet Union and Middle East (Harrison 1972; Koffler 1972; Naumov and Lobachev 1975; Roberts 1977) indicate that reproduction may occur throughout all or most of the year but usually takes place from late winter to early autumn and that two or three litters are produced by each female. Gestation periods of 20–31 days and litter sizes of 1–12 young have been reported. According to M. L. Jones (1982), a captive specimen of *M. crassus* lived for 5 years and 7 months.

Several species of jirds are used in laboratory research. *M. unguiculatus* was brought to the United States for such purposes in the early 1950s and has been found to be highly adaptable, clean, and in need of little care (Gulotta 1971; Kaplan and Hyland 1972). It has become increasingly popular in medical, physiological, and psychological research and also is kept now as a pet by many persons (usually being referred to as a "gerbil"). There is concern, however, that feral populations could become established in North America and that these might destroy crops and embankments and even displace native rodents (Fisler 1977). Several species of *Meriones* are sometimes considered pests in the Old World, because they eat cultivated plants, damage irrigation structures by burrowing, and spread disease (Klimchenko et al. 1975; Naumov and Lobachev 1975).

RODENTIA; MURIDAE; Genus **BRACHIONES**
Thomas, 1925

Przewalski's Gerbil

The single species, *B. przewalskii*, occurs in northwestern China from western Sinkiang to northern Gansu; it is not found in Mongolia (Corbet 1978).

Head and body length is 80–95 mm and tail length is 70–80 mm. The upper parts are pale grayish yellow, pale yellow-

Przewalski's gerbils *(Brachiones przewalskii):* A. Photo from *Mammalia of Central Asia,* E. Büchner; B. Photo by P. F. Wright of specimen in U.S. National Museum of Natural History.

ish buff, or light sandy gray. The underparts, hands, and feet are white. The tail is buffy or whitish throughout; it is slender and tapering and is not tufted. The soles are densely haired, whereas the palms are naked.

The thickset body, reduced ears, shortened tail, long foreclaws, and hairy soles indicate that this rodent has developed burrowing habits. It is found in desert areas.

RODENTIA; MURIDAE; Genus PSAMMOMYS
Cretzschmar, 1828

Fat Sand Rats

There are two species (Cockrum, Vaughan, and Vaughan 1977; Corbet 1978; F. Petter, *in* Meester and Setzer 1977):

P. obesus, Libya, Egypt, coastal Sudan, Palestine, Saudi Arabia;

P. vexillaris, Algeria, Libya.

Head and body length is 130–85 mm and tail length is 110–50 mm. The upper parts are reddish brown, reddish, yellowish, or sandy buff, and the underparts are yellowish, buffy, or whitish. The tail is fully haired and has a terminal tuft. The external form is stocky. From *Meriones,* this genus is distinguished by its nongrooved incisors and short (10–15 mm), thick, rounded ears.

Fat sand rats are mainly diurnal and live in sandy areas with scant vegetation. They construct complex burrows with several entrances and food storage chambers and a chamber containing a nest of finely cut vegetation. They sit up on their hind legs and tail but quickly retreat into their burrows when alarmed. Studies in the Sahara Desert (Daly 1979; Daly and

Fat sand rat *(Psammomys obesus)*, photo by Robert E. Kuntz.

Daly 1973) have shown that the preferred food of fat sand rats is the leaves and stems of succulent plants of the family Chenopodiaceae, which contain much water but also a high proportion of salt. Most mammals could not survive by eating these plants without also having an abundant source of fresh water, but fat sand rats thrive on this diet through the aid of their extremely powerful kidneys, which produce a urine with a high salt concentration. *Psammomys* also feeds on other kinds of plants and reportedly can be very destructive to grain; 500 heads of barley were once found stored in a burrow (Harrison 1972).

During a five-month study of fat sand rats in the A!gerian Sahara, Daly and Daly (1975b) calculated average home range length to be 189.6 meters in males and 75.8 meters in females. Mean weekly range lengths were 67.7 meters in males and 11.7 meters in females. Females tended to remain in small areas around the bushes they used for food and to move only when edible vegetation was exhausted. Subordinate males also used a small home range but moved more frequently. Dominant males ranged over relatively large areas encompassing the ranges of several females and subordinate males. A conspicuous form of communication observed in this study was audible footdrumming, sometimes accompanied by a high-pitched squeak. Throughout this study, which lasted from December to April, females were pregnant, lactating, or both; and local people said that young could be found in any month of the year. Associated laboratory work indicated that the gestation period normally is about 25 days but is extended to about 36 days after postpartum mating. Litter size in captivity was 2–5 young, weaning occurred at around 3 weeks, and females first conceived at 3.0–3.5 months.

RODENTIA; MURIDAE; Genus RHOMBOMYS
Wagner, 1841

Great Gerbil

The single species, *R. opimus*, is found in Soviet Central Asia, Iran, Afghanistan, western Pakistan, Sinkiang, and southern Mongolia (Corbet 1978).

Head and body length is 150–200 mm and tail length is 130–60 mm. The upper parts are sandy yellow, orangish buff, or dark grayish yellow, and the underparts are whitish. The fur is thick and soft, and the tail is hairy, almost bushy. The external form is stocky, the claws are large, and the soles are hairy. Two longitudinal grooves are present on each upper incisor.

The great gerbil usually inhabits sandy deserts but also occurs in clay deserts and in the foothills of the mountains of Central Asia. According to Naumov and Lobachev (1975), the most important factor in its distribution is proper soil conditions for excavation of its deep and complex burrow. Subsandy soils are best. The burrows are variable in structure but may contain several levels with nest chambers, lengthy tunnels, and food storage chambers. The winter nest chambers are 1.5–2.5 meters below the surface, where the microclimate is stable all year. The great gerbil is mainly diurnal, being most active at dawn but sometimes remaining on the surface until dusk. It does not hibernate, though its activity apparently is greatly reduced during the winter in many areas.

The winter activity of *Rhombomys* is in direct proportion to the air temperature and in inverse proportion to the depth of the snow. In those parts of the range with a snow cover, only a few of the burrow exits are open by midwinter, and air holes are predominant. Where the burrows are completely covered by snow for a long time, in many colonies the animals come to the surface only rarely.

Little contact is made between individual *Rhombomys* and those living in neighboring colonies, at least in winter, when many areas are covered with snow. In a study conducted during the winter in the northern Lake Aral region, the maximum length of a trail in the snow was 20 meters, and 83 percent of the tracks examined were in the immediate vicinity of the burrow exit. Following a harsh winter in Turkmen S.S.R., Marinina (1971) found a population density of only 1 per every 5–10 ha. in areas of open sand ridges and a density of 2 per ha. in areas of high ridges and hills.

The diet of *Rhombomys* consists of a variety of desert plants. In some areas the food is stored for winter use (up to 60 kg or more per burrow), but in other regions the animals feed mainly on the surface. The food reserves for winter are located in compartments in the burrow, but in some places the plants are stored in heaps on the surface.

The following information on social life and reproduction is taken mainly from Naumov and Lobachev (1975) and

Great gerbil *(Rhombomys opimus)*, photo by N. Mocroucov, through D. Bibikov.

Roberts (1977). The great gerbil is gregarious, and many individuals may burrow in the same vicinity to form a large colony. Several adults may share a burrow and work together in its construction. Captives can be kept together without fighting. Females are polyestrous and in captivity can produce six litters in 6 months. In the wild the breeding season lasts from about April to September, and two or three litters are born to each female. The gestation period is 23–32 days. Litter size is 1–14 young, usually 4–7. Some females reach sexual maturity at 3–4 months. Maximum longevity is 3–4 years for females and 2–3 years for males.

This rodent is considered a pest in the Soviet Union. It is a reservoir for plague, and it damages crops, railway embankments, and the sides of irrigation channels. In some areas it is trapped for its skin.

RODENTIA; MURIDAE; Genus CLETHRIONOMYS
Tilesius, 1850

Red-backed Mice, or Bank Voles

There are six species (Aimi 1980; Corbet 1978; Hall 1981; Stenseth 1985):

C. rufocanus, Scandinavia to northeastern Siberia and Korea, Hokkaido, Kuril Islands;

C. rutilus, northern Scandinavia to the Bering Strait and Manchuria, Sakhalin, Hokkaido and nearby Rishiri Island, Alaska to Keewatin;

C. gapperi, British Columbia to mainland Newfoundland, northern conterminous United States, Rocky Mountains, Appalachians;

C. californicus, western Oregon, northern California;

C. glareolus, Europe, parts of Asia Minor and Central Asia;

C. centralis, Altai and Tien Shan mountains of Siberia and Sinkiang.

Two additional species, *C. andersoni* and *C. rex,* are often recognized, but Aimi (1980) assigned the former to the genus

Eothenomys and the latter (found on Rishiri Island) to *C. rutilus.* According to Stenseth (1985), *C. centralis* sometimes is considered a synonym of *C. glareolus,* and *C. rutilus* and *C. gapperi* sometimes are considered conspecific. On the basis of myological and osteological analysis, Stein (1987) reported *C. rutilus, C. gapperi,* and *C. glareolus* to be closely related, *C. rufocanus* to be more distinctive and primitive within the genus, and *C. rex* to not warrant specific status.

Head and body length is 70–112 mm, tail length is 25–60 mm, and weight is about 15–40 grams. The fur is dense, long, and soft in winter but shorter and harsher in summer. The general coloration above is dark lead gray with a pronounced reddish wash. The wash becomes less prominent on the sides, which are grayish. The underparts are dark slate gray to almost white. The tail has a slight terminal pencil of hairs. The short thumb is provided with a flat nail. The ears are slightly more conspicuous through the fur in *Clethrionomys* than in *Microtus,* and the eyes are also more prominent. Females have eight mammae.

These rodents inhabit cold, mossy, rocky forests and woodlands in both dry and moist areas. They also inhabit tundra and bogs. They are active night and day, summer and winter, scurrying and climbing about stumps, fallen logs, and rough-barked trees. According to Banfield (1974), *C. gapperi* does not make runways of its own but often uses those of other small mammals. It does construct spherical nests of grasses, mosses, lichens, or shredded leaves. These nests are usually hidden under the roots of stumps, logs, or brush piles but may be located in holes or branches of trees high above the ground. During the winter, globular nests of grass may be placed directly on the ground under the snow, with tunnels radiating from the nest under the snow. The diet of *Clethrionomys* consists of tender vegetation, nuts, seeds, bark, lichens, fungus, and insects. Food is often stored in the nest for use when the supply is short.

French, Stoddart, and Bobek (1975) listed population densities for various species in Eurasia ranging from fewer than 1 to nearly 100 individuals per ha. The average home range of *C. glareolus* in Czechoslovakia has been calculated to be 0.82 ha. for males and 0.71 ha. for females. Banfield (1974) wrote that the home range of *C. gapperi* may be as large as 1.4 ha. in the summer and as small as 0.14 ha. in the winter, when

Red-backed mouse *(Clethrionomys glareolus)*, photo by P. Morris.

foraging is restricted by a blanket of snow. He noted that populations fluctuate widely from year to year, but with no apparent periodicity. Merritt and Merritt (1978) cited population densities of 2.0–74.1 per ha. for *C. gapperi*. In their own study in Colorado, the vole population stabilized at about 18/ha. over the winter, declined to 10/ha. during the spring thaw, and then increased to about 42/ha. in November, following the reproductive season.

When disturbed, red-backed mice may utter a chirplike bark that can be heard from one to two meters away, and they flee or freeze in position, depending on their location and preceding activity. They also gnash or chatter their teeth. Mihok (1976) reported that encounters between captive male and female *C. gapperi* were generally amicable and resulted in female dominance if the female was in breeding condition but resulted in male dominance and pursuit of the female if she was not in breeding condition. Encounters between animals of the same sex resulted in avoidance behavior and aggression. West (1977) found a lessening of aggressive behavior in *C. rutilus* in Alaska during midwinter, which apparently al-

Père David's vole *(Eothenomys smithi)*, photo by K. Tsuchiya.

lowed the animals to huddle and thereby conserve heat.

Breeding may begin as early as late winter and continue to late fall; the gestation period is 17–20 days; and overall litter size is 1–11 young. Data compiled by Innes (1978) indicate that litter size in *C. gapperi* increases with latitude and elevation. It was found to average 3.33 young in one study in southeastern Kentucky and 7.25 in an investigation in central Alberta. Innes stated that the modal number of offspring is 5 in *C. gapperi* and 7 in *C. rutilus*. Banfield (1974) stated that the reproductive period of *C. gapperi* in Canada is April to early October and that three or four litters are borne by each female per season. The newborn weigh about 1.9 grams, open their eyes at 9–15 days, and are weaned and independent at 17–21 days. The young of the spring litter bear their own first young at about 4 months. According to Jones (1982), a specimen of *C. glareolus* lived in captivity for 4 years and 11 months.

Red-backed mice destroy quantities of insect larvae and also form a major food source of fur-bearing animals. Mice of this genus have been found to be important in Great Britain, and probably elsewhere, as agents in transporting, burying, and eating tree seeds and in damaging or killing seedlings (Ashby 1967). The species *C. glareolus* was first recorded in Ireland only in 1964 and was probably introduced there through human agency (Corbet 1978; Fairley 1971).

RODENTIA; MURIDAE; **Genus EOTHENOMYS**
Miller, 1896

Père David's Voles, or Pratt's Voles

There are 12 species (Aimi 1980; Corbet 1978; Lekagul and McNeely 1977):

E. melanogaster, central and southeastern China, northern Burma, northern Thailand;
E. olitor, Yunnan (southern China);
E. proditor, Yunnan and Sichuan (southern China);
E. chinensis, Yunnan and Sichuan (southern China);
E. custos, Yunnan and Sichuan (southern China);
E. smithi, Japan;
E. regulus, Hebei (northeastern China), Korea;
E. shanseius, Shanxi and possibly Hebei (northeastern China);
E. inez, northeastern China;
E. eva, central China;
E. lemminus, northeastern Siberia;
E. andersoni, Honshu (Japan).

The last two species sometimes have been assigned to a separate genus or subgenus, *Aschizomys* Miller, 1898, and sometimes *E. lemminus* has been placed in *Alticola*, and *E. andersoni* in *Clethrionomys*. A detailed study led Aimi (1980) to assign *E. andersoni* to *Eothenomys*, and this procedure was followed by Honacki, Kinman, and Koeppl (1982), though not by Corbet and Hill (1986) or Stenseth (1985).

Head and body length is 88–126 mm and tail length is 30–55 mm. Lekagul and McNeely (1977) gave the weight of *E. melanogaster* as 27 grams. Coloration above is usually dark reddish brown with a peculiar metallic reflection caused by the burnished tips of a portion of the longer hairs, an effect referred to by some writers as a brassy shine; the underparts are bluish gray. The young are blackish. The pelage is fairly short and smooth. The soles of the feet are hairy behind the pads. The tail is covered with stiff hairs that form a short, thin

terminal pencil. Females of most species have four mammae, but those of *E. andersoni* have eight (Aimi 1980). *Eothenomys* is perhaps phylogenetically intermediate to *Clethrionomys* and *Microtus*, as it possesses some characteristics common to both genera.

These animals frequent easily tunneled banks and slopes in both wooded growth and mountain meadows. Their elevational range is from about 1,800 to at least 4,400 meters. They tend to be active both day and night. Some forms seem to spend more time underground than others. Burrows open to the surface at frequent intervals, and numerous surface runways may be evident. Surface runways of other animals, such as moles, may also be used. The habits of *Eothenomys* are presumably like those of *Clethrionomys* and *Microtus*.

Breeding apparently occurs throughout the year, or nearly so, as pregnant females have been found in the wild from February to October. In studies of a laboratory colony of *E. smithi*, Ando, Shiraishi, and Uchida (1987, 1988) found the gestation period to be about 19 days, postpartum mating to be normal, litter size to average 4.45 and range from 1 to 9 young, birth weight to be about 2.13 grams, the eyes to open at 13.8 days, weaning to occur at around 25 days, and maximum longevity to be 3.5 years.

RODENTIA; MURIDAE; **Genus ALTICOLA**
Blanford, 1881

High Mountain Voles

There are four species (Corbet 1978):

A. roylei, mountains of Central Asia from Mongolia to the western Himalayas;
A. stoliczkanus, Altai Mountains of Mongolia, Nan Shan Mountains of north-central China, Tibet, Himalayas;
A. strelzowi, northeastern Kazakh S.S.R., Altai Mountains of Mongolia;
A. macrotis, mountains of Mongolia and adjacent parts of Siberia.

Head and body length is 80–140 mm and tail length is 15–54 mm. Ten specimens, with an average length of 108 mm, weighed 35–49 grams. The coloration above is usually some shade of gray or brown; the gray is mixed with ash, silver, buff, or yellow, and the brown, when present, with grays or reds. The underparts are usually white or buffy white. The feet are generally white or grayish white. In most forms the well-haired tail is white or buffy white throughout. The soles are usually haired posteriorly. The fur is thick, soft, and long. It is possible that the winter coat is longer and paler in color than the summer coat.

The external form is not modified for burrowing. The eyes are moderately large, and the ears are of moderate size, though they are often concealed in the long fur. The flattened skull of *A. strelzowi* is probably an adaptation for living under rocks. Female high mountain voles have eight mammae.

Alticola seems to be the Asian equivalent of the snow vole, *Microtus nivalis*, of Europe, which has a white tail and lives at high elevations in the mountains. The elevational range of *Alticola* is about 900 meters to at least 5,700 meters. *A. strelzowi* is mainly diurnal, whereas *A. roylei* is partly diurnal but mainly nocturnal. Most, if not all, forms store food for the winter. The stems and leaves of herbaceous plants and shrubs may be dried first in the sun and then piled into heaps or stuffed into stone crevices. *A. roylei* deposits its banana-shaped droppings at the entrances of the burrow, a definite sign of its presence.

High mountain vole *(Alticola roylei)*, photo by V. Masin through D. Bibikov.

A. roylei is said to be fearless, sometimes coming within 4 cm of a person's hand. *A. strelzowi* is known to live in colonies. In the Altai Mountains the breeding season of this species begins in late April or early May and lasts about 2.5 months; there are 1–2 litters per season, with litter size ranging from 3 to 13 and averaging 7.1 young (Eshelkin 1976). In the mountains of eastern Kazakh S.S.R., *A. roylei* generally breeds from March or April to September or October, with a peak from April to June, but winter reproduction also has been observed. There are 2–3 litters per year, and litter size ranges from 2 to 6 and averages 3.6 (Obidina 1972).

RODENTIA; MURIDAE; Genus HYPERACRIUS
Miller, 1896

Kashmir Voles and Punjab Voles

There are two species (Corbet 1978):

H. wynnei, extreme northern Pakistan;
H. fertilis, Kashmir, northern Pakistan.

Head and body length is 96–138 mm, tail length is 24–40 mm, and weight is 21.5–60 grams (Roberts 1977). The pelage is shorter and more dense than that of *Alticola*. Coloration of *H. fertilis* is deep reddish brown on the upper parts, gradually becoming dull ochraceous below. The feet are dusky. The tail is slightly bicolored, sepia above and dirty white below. *H. wynnei* apparently occurs in two distinct color phases. In the light phase the upper parts are yellowish brown and the underparts are grayish tinged with wood brown; the feet and hands are grayish brown on the upper surface; and the tail is slightly bicolored, dark grayish brown above and paler gray below. In the dark phase the upper parts vary from a lustrous seal brown to a glossy blackish brown. The underparts are duller and somewhat paler and are lightened by short whitish-

or yellowish-tipped hairs of the ventral surface. The hands and feet are dark gray above, and the tail is dusky gray and only slightly bicolored. The whiskers are short, scarcely reaching to the ears, and the feet are well haired. The dark phase of *H. wynnei*, unusual among animals of this kind, may be a natural result of retaining the darker coat of the immature form, owing to the fact that the animal has adopted a subterranean life away from light. *H. wynnei* differs from *H. fertilis* in being larger in size and in having longer, more lax, more luxuriant pelage.

The general appearance of *Hyperacrius* is similar to that of *Microtus* and *Alticola*, the principal means of differentiation being in cranial and dental structure. The well-developed claws on all four limbs are long and slender; the pollex is provided with a flattened, fairly large nail. Females have four inguinal mammae.

According to Roberts (1977), *H. wynnei* occurs in moist temperate forests and meadows at elevations of 1,850–3,050 meters, while *H. fertilis* is found in the subalpine scrub zone and alpine meadows at 2,450–3,600 meters. *H. wynnei* is highly fossorial, excavating extensive networks of shallow feeding tunnels, about 32–75 mm deep, as well as much deeper burrows descending almost vertically into the ground that are used for sleeping and breeding. During the winter this species constructs tunnels near or on the surface, through the snow, and lines them with soil and vegetation. It does not hibernate and is active at any time of the day or night. *H. fertilis* is not so fossorial and will forage above the ground. Both species are completely herbivorous, eating grass, stems, and roots. *H. wynnei* reportedly does considerable damage to potato crops.

Roberts (1977) stated that *H. wynnei* is loosely colonial. The female constructs a grass-lined nest at the end of a deep burrow and probably produces two to three litters in a breeding season, with two or three offspring each. Young animals approximately four to six weeks old have been found in early June and early October. *H. fertilis* also produces two or three litters, during the spring and summer.

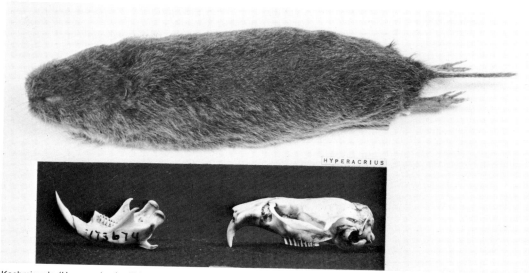

Kashmir vole *(Hyperacrius fertilis)*, photos by P. F. Wright of specimen in U.S. National Museum of Natural History.

RODENTIA; MURIDAE; **Genus DINAROMYS**
Kretzoi, 1955

Martino's Snow Vole

The single species, *D. bogdanovi*, occurs in the mountains of Yugoslavia and possibly Albania and northwestern Greece and also is known by fossil remains from northern Italy and eastern Greece (Petrov and Todorovíc 1982). This species sometimes has been referred to the genus *Dolomys* Nehring, 1898, which is based on a Pleistocene fossil from Hungary, but evidence now indicates that *D. bogdanovi* represents a separate, Recent genus (Corbet 1978).

Head and body length is 99–152 mm, tail length is 74–119 mm, and weight is 30–82 grams (Petrov and Todorovíc 1982; Van Den Brink 1968). The tail is usually equal to more than half the head and body length. The fur is dense, soft, and moderately long. The upper parts are grayish brown, buffy grayish, or bluish gray, and the underparts are usually paler, often grayish white. The feet are white, and the tail is usually dark brown above and white on the sides and below. The thinly haired tail has a short, stiff, thin pencil line about 5–6 mm in length. The palms and soles are naked except near the heels. The large ears are densely haired.

The small thumb bears a small, flattened nail. The other digits have short, sharp claws, about equally long on the hands and feet. Notable cranial characters include the broad and faintly grooved upper incisors and the large, inflated bullae, though the mastoids are comparatively small.

This vole is found in the mountains, generally at elevations of 1,300–2,200 meters and rarely below 680 meters. It is nocturnal and active throughout the year. On Mount Trebevic (south of Sarajevo) it finds shelter and builds nests under rocks and isolated stones. Thawing snows and spring rains force it to move to drier ground. It feeds on grasses and stores food for the winter.

Breeding usually occurs twice a year, in March and again in June. The time of the second breeding season varies more than that of the first. A study on Mount Trebevic showed that only one litter was produced during the dry year of 1946. The gestation period is thought to be at least one month. There are usually two or three good-sized young per litter. The young are reported to come out of the burrows in July.

Martino's snow vole *(Dinaromys bogdanovi)*, photos by Ivo R. Savi.

RODENTIA; MURIDAE; **Genus ARBORIMUS**
Taylor, 1915

Tree Voles

There are two species (Hall 1981):

A. albipes, western Oregon, northwestern California;
A. longicaudus, western Oregon, northwestern California.

Arborimus originally was described as a subgenus and for many years was generally regarded as part of the genus *Phenacomys*. Johnson (1973) and Johnson and Maser (1982), however, listed numerous characters indicating that *Arborimus* should be given full generic rank. Such rank was recognized by Corbet and Hill (1986) and Honacki, Kinman, and Koeppl (1982), though not by Carleton and Musser (1984), Hall (1981), or Jones et al. (1986).

Head and body length is about 95–110 mm, tail length is 60–87 mm, and weight is 25–56 grams. The pelage is fairly long and finely textured. The upper parts are generally dark brown in *A. albipes* and reddish brown or cinnamon in *A. longicaudus*. The underparts are whitish, gray, or pinkish buff. *Arborimus* shares with *Phenacomys* several dental characters that distinguish both genera from *Microtus*. From *Phenacomys*, *Arborimus* differs in such characters as its much longer tail, smaller ears, longer feet, and wider and shorter molar teeth. Female *A. longicaudus* have only 4 mammae (compared with 8 in *Phenacomys*), while 4, 5, and 6 mammae have been recorded in *A. albipes* (Johnson 1973; Johnson and Maser 1982).

Both species of *Arborimus* are restricted largely to lower elevations in humid, temperate, coastal forests (Johnson 1973). *A. longicaudus* is arboreal and has a highly specialized way of life. Its movements, unlike those of most mice, are slow and cautious, especially along branches, where care is taken to obtain a firm grasp on a new support before leaving a safe foothold. This species is active throughout the year. The males live in burrows on the ground or in piles of debris, but the females live in trees. The males climb the trees to mate and to build small temporary nests. The females also build nests, 228–305 mm in diameter and 4.5–15.0 meters above the ground. The nests are made of accumulations of twigs and discarded spines of fir and spruce leaves; they are generally placed at a forked branch or over a nest that has been abandoned by a bird or squirrel. The tender terminal twigs of coniferous trees are cut for food. They are carried to the nest, where the needles are eaten and the spines are used to enlarge the nest. Comparatively little is known about the ecology of *A. albipes*, but data collected by Voth, Maser, and Johnson (1983) indicate that this species occupies a spatial niche extending from ground level into the tree canopy and that its diet consists entirely of leaves from vascular plants.

Reproductive aspects of *A. longicaudus* differ markedly from those of *Phenacomys* (Hamilton 1962). The breeding season lasts from February to September, the estrous cycle

A. Tree vole *(Arborimus longicaudus)*, photo by Alex Walker. B. Heather voles *(Phenacomys intermedius)*, mother and ten-day-old young, photo by J. B. Foster.

apparently averages 5.9 days, and the gestation period is normally 27.0–28.5 days but may possibly be extended to as long as 48 days if the female is nursing a previous litter. Litter size is only 1–3, and the young are not weaned until 30–35 days. The lengthy time required for gestation and weaning are thought to be physiological adjustments to the difficulties of producing energy from a diet of coniferous needles. Litter size in *A. albipes* reportedly ranges from 2 to 4 and is most commonly 3 (Johnson and Maser 1982).

Arborimus long was considered difficult to find and was unusual in museum collections. *A. longicaudus* now is known from a large number of specimens, though it seems to have declined in response to logging of old-growth forests (Corn and Bury 1988; D. F. Williams 1986). *A. albipes* apparently is extremely rare throughout its entire range and may be highly sensitive to logging and other modifications of its habitat (Olterman and Verts 1972; D. F. Williams 1986).

RODENTIA; MURIDAE; Genus PHENACOMYS
Merriam, 1889

Heather Vole

The single species, *P. intermedius*, occupies most of subarctic Canada, northern Minnesota, and the mountains of the western conterminous United States (Hall 1981). This species occurred across much of the central and eastern United States in the late Pleistocene and inhabited the Great Basin region until at least 5,300 years ago (Grayson 1981, 1987; McAllister and Hoffmann 1988). Until recently, most authorities, including Hall, considered the genus *Arborimus* (see account thereof) to be part of *Phenacomys*. Hall also noted that *P. intermedius* may actually be a composite of two or three allopatric species.

Head and body length is about 90–117 mm, tail length is 26–41 mm, and weight is 25–40 grams. Coloration varies, but generally the upper parts are grayish brown and the underparts and feet are silvery gray. The pelage is long and silky. *Phenacomys* so closely resembles the genera *Microtus* and *Clethrionomys* that positive identification may not be possible until the skull and dentition are carefully examined. Unlike *Microtus*, adult *Phenacomys* have rooted cheek teeth. Unlike in *Clethrionomys*, in *Phenacomys* the inner reentrant angles of the lower molar teeth are much deeper than the outer. Females have eight mammae.

As pointed out by Johnson (1973), the ecology of *Phenacomys* differs markedly from that of *Arborimus,* and these differences constitute part of the case for allotting the two species to separate genera. According to Banfield (1974), the heather vole occurs in a wide range of habitats from sea level to the tree line in the Rocky Mountains. It seems to prefer open coniferous forests with an understory of heaths or areas of shrubby vegetation on forest borders or in meadows. It is terrestrial and largely nocturnal or crepuscular. It is active throughout the winter, at which time it builds a spherical nest about 150 mm in diameter composed of twigs, lichens, and grass. The nest is placed on the ground between rocks or under a shrub, and from it radiate tunnels out under the snow. The summer nest is built about 100–250 mm underground, and tunnels radiate outward and upward to the surface. The heather vole feeds on a variety of plant materials including bark, buds, heaths, forbs, berries, and seeds. Nagorsen (1987) reported that food caches are made in both summer and winter, those of the latter season consisting of twigs piled on the forest floor.

McAllister and Hoffmann (1988) cited population densities of 10 per ha. in alpine communities and 0.5–4.3 per ha. in montane forest. They noted also that captives are mild-tempered and docile but that males are aggressive toward one another during the breeding season and that females with young are aggressive toward any intruders. Banfield (1974) stated that females are seasonally polyestrous and that the breeding season extends from May to August. The gestation period is 19–24 days, and litter size ranges from 2 to 8 young, averaging about 5. The newborn weigh approximately 2.4 grams, open their eyes at 14 days, and are weaned at 17–21 days. Females reach sexual maturity within 4–6 weeks, but males do not breed until the following spring.

Phenacomys long was considered one of the rarest genera of North American mammals. In the mid-1950s, however, many specimens were collected in Canada (Banfield 1974). It also seems to have extended its range in the Cascade Mountains of Oregon, perhaps in response to logging, which provides the more open habitat preferred by the genus (Corn and Bury 1988).

RODENTIA; MURIDAE; Genus ARVICOLA
Lacépède, 1799

Water Voles, or Bank Voles

There are two species (Corbet 1978; Mallon 1985):

A. terrestris, most of Europe and Siberia, Mongolia, parts of southwestern Asia;

A. sapidus, France, Spain, Portugal.

Hall (1981) considered *Arvicola* to be a subgenus of *Microtus* and to include the species *M. richardsoni*. Some other authorities, including Banfield (1974), have recognized *Arvicola* as a separate genus that includes *M. richardsoni*. Nadler et al. (1978) employed the name *Arvicola richardsoni* and pointed out that in certain biochemical features this species differs strikingly from *Microtus* but observed also that there are some resemblances and that further study was needed to determine whether the species should be allocated to *Arvicola*. Of late, the usual practice has been to regard *Arvicola* as a full genus, but to assign *M. richardsoni* to *Microtus* (Carleton and Musser 1984; Chaline and Graf 1988; Corbet and Hill 1986; Honacki, Kinman, and Koeppl 1982; Jones et al. 1986; Ludwig 1984).

Head and body length is 120–220 mm, tail length is 65–125 mm, and weight is 70–250 grams. *A. terrestris* is the largest microtine rodent native to the Old World. The fur is thick. The general coloration ranges from moderately light brown to dark brown above and from buffy to slate gray below. Black individuals sometimes occur locally. Slight aquatic modifications are developed in some species, such as the small, hairy swimming fringe on the foot. The tail is well haired, and the sole of the hind foot is generally haired. The claws on all four limbs are well developed. Both sexes possess flank glands, and females have eight mammae. Individuals seem to grow throughout life. The cheek teeth are ever-growing.

There apparently are both aquatic and nonaquatic populations of these voles. The aquatic populations, which include most *A. terrestris* in the western parts of Europe, live along the banks of streams, ponds, and canals, where the water level is constant. They dive freely and swim either on the surface or underwater, sometimes near the bottom of the pool or stream, where they enter burrows that terminate below water level. In parts of Europe, fossorial, nonaquatic populations of *A. terrestris* predominate (Meylan 1977). These populations occupy grasslands and cultivated areas, construct extensive burrow systems, and leave conspicuous mounds on the surface. The burrows average 40.2 meters in length, contain one

Water vole *(Arvicola terrestris)*, photo from A. Van Wijngaarden.

or two nests, and in autumn and winter contain storage chambers. Grzimek (1975) noted that *A. terrestris* also freely uses the tunnels of moles.

Bank voles are either nocturnal or diurnal and are active throughout the year. They feed on aquatic plants, herbs, grasses, twigs, buds, roots, bulbs, and fallen fruit. Some food is usually stored for the winter, though the animals continue to forage at this time.

Corbet and Southern (1977) gave the average home range length of *A. terrestris* as 130 meters for males and 77 meters for females. Pelikan and Holisova (1969) reported the average number of *A. terrestris* per 100 meters of stream to vary from 4.0 to 9.3 individuals. Meylan (1977) stated that densities of *A. terrestris* sometimes exceed 1,000 per ha. during peaks of the population cycle, which seem to occur every four to six years. Although water voles are sometimes found under crowded conditions, they do not form large social groups. Males mark territories with secretions from flank glands and may fight fiercely when populations are dense, making high, shrill squeaks (Grzimek 1975). The process of scent marking, which involves raking the hind feet over the flank glands, is peculiar to both *A. terrestris* and *Microtus richardsoni* but not to other *Microtus* (Jannett and Jannett 1974). Females also fight each other, but not to the same extent as males. Meylan (1977) reported that each burrow of fossorial populations of *A. terrestris* is inhabited by a family group, which includes an adult pair plus two generations of young. These populations

usually breed from March to October, but winter reproduction is not unusual.

Other populations of *A. terrestris* also breed from the spring to the early fall (Corbet and Southern 1977; Grzimek 1975). The gestation period is 20–22 days. There are one to four litters per female per year, and litter size is 2–8 young, usually 4–6. The newborn weigh about 5 grams, open their eyes at 8 days, and are weaned at 14 days. Corbet and Southern (1977) stated that mean longevity of *A. terrestris* is 5.4 months under natural conditions but that captive specimens have lived up to 5 years.

In the Soviet Union *A. terrestris* is hunted for its fur, but it may transmit tularemia to persons handling it. The fossorial populations of *A. terrestris* reportedly sometimes cause extensive damage to pastures, the root systems of orchard trees, grapevines, tulip bulbs, and vegetable crops (Meylan 1977).

RODENTIA; MURIDAE; Genus MICROTUS
Schrank, 1798

Voles, or Meadow Mice

There are 8 subgenera and 67 species (Bolshakov and Shubnikova 1988; Carleton and Musser 1984; Corbet 1978, 1984; Ellerman and Morrison-Scott 1966; Hall 1981; Hassinger

Vole *(Microtus montebelli)*, photo by Ernest P. Walker.

1973; Kivanc 1986; Kovalskaya and Sokolov 1980; Lawrence 1982; Madureira 1981; Mallon 1985; Malygin and Yatsenko 1986; Meier and Yatsenko 1980; Morlok 1978; Moyer, Adler, and Tamarin 1988; Orlov and Kovalskaya 1978; Roguin 1988; Woods, Post, and Kilpatrick 1982):

subgenus *Aulacomys* Rhoads, 1894

M. *richardsoni* (American water vole), mountains of southwestern Canada and northwestern conterminous United States;

subgenus *Chionomys* Miller, 1908

M. *nivalis* (snow vole), mountainous areas of southern Europe and southwestern Asia;
M. *roberti*, Caucasus, northeastern Asia Minor;
M. *gud*, Caucasus, northeastern Asia Minor;

subgenus *Microtus* Schrank, 1798

M. *socialis*, southeastern Europe to eastern Kazakh S.S.R. and Iran, northeastern Libya;
M. *guentheri*, Greece, Asia Minor, Syria, Palestine;
M. *irani*, Asia Minor, Iraq, Iran;
M. *cabrerae*, southern France, Spain;
M. *oeconomus*, Scandinavia and the Netherlands to eastern Siberia and north-central China, Alaska, northwestern Canada, many northern Pacific continental islands;
M. *middendorffi*, northern Siberia;
M. *arvalis*, Denmark and northern Spain to central Siberia and Asia Minor, and (probably through introduction) Spitsbergen;
M. *rossiaemeridionalis*, Finland to Greece and the southern Urals;
M. *transcaspicus*, southern Turkmen S.S.R., northern Afghanistan, probably northern Iran;
M. *kermanensis*, south-central Iran;
M. *kirgisorum*, Tien Shan Mountains of Kirghiz S.S.R.;
M. *mongolicus*, Mongolia and adjacent parts of Siberia;
M. *maximowiczii*, Lake Baikal to upper Amur Basin in southern Siberia;
M. *mujanensis*, Muja River Valley in south-central Siberia;
M. *evoronensis*, coast of Lake Evoron in southeastern Siberia;
M. *fortis*, southeastern Siberia, Mongolia, Manchuria, Korea, eastern China, possibly Sakhalin;
M. *sachalinensis*, Sakhalin;
M. *kikuchii*, Taiwan;
M. *clarkei*, Yunnan (southern China), northern Burma;
M. *millicens*, Sichuan (central China);
M. *musseri*, Sichuan (central China);
M. *montebelli*, Japan;
M. *agrestis* (field vole), Britain and northern Spain to east-central Siberia;
M. *pennsylvanicus* (meadow vole), Alaska, Canada, the conterminous United States as far south as Georgia and New Mexico, west-central coast of Florida, north-central Mexico;
M. *breweri*, Muskeget Island off Massachusetts;
M. *nesophilus*, Great Gull and Little Gull islands off eastern Long Island (New York);
M. *montanus*, south-central British Columbia to east-central California and northern New Mexico;
M. *canicaudus*, northwestern Oregon;
M. *oaxacensis*, north-central Oaxaca (southern Mexico);
M. *californicus*, western Oregon, most of California, northern Baja California;

M. *guatemalensis*, extreme southern Mexico, southern Guatemala;
M. *townsendii*, Vancouver and nearby islands, extreme southwestern mainland of British Columbia to northwestern California;
M. *longicaudus*, eastern Alaska to California and New Mexico;
M. *coronarius*, Coronation and nearby islands off southeastern Alaska;
M. *oregoni*, southwestern British Columbia to northwestern California;
M. *chrotorrhinus*, southeastern Canada, northeastern Minnesota, northern New York and New England, Appalachians;
M. *xanthognathus*, central Alaska to northern Manitoba and west-central Alberta;
M. *ochrogaster* (prairie vole), central Alberta to western West Virginia and northeastern New Mexico, also originally in southeastern Texas and southwestern Louisiana;
M. *mexicanus*, southwestern United States to southern Mexico;
M. *umbrosus*, Oaxaca (southern Mexico);

subgenus *Proedromys* Thomas, 1911

M. *bedfordi*, known only from the type locality in Gansu (central China);

subgenus *Lasiopodomys* Lataste, 1887

M. *brandti*, Mongolia and adjacent parts of Siberia;
M. *mandarinus*, north-central Mongolia and adjacent part of Siberia south of Lake Baikal, parts of north-central and northeastern China;

subgenus *Neodon* Hodgson, 1849

M. *blythi*, Tibetan Plateau and Himalayas;
M. *sikimensis*, western Nepal to northern Burma and southern China;
M. *juldaschi*, Tien Shan and Pamir mountains of Central Asia;
M. *afghanus*, mountains of southern Soviet Central Asia and Afghanistan;

subgenus *Pitymys* McMurtrie, 1831

M. *subterraneus*, France to western Soviet Union;
M. *multiplex*, Appenines and southern side of Alps in Switzerland, Austria, France, and Italy;
M. *tatricus*, Tatra Mountains between Poland and Czechoslovakia;
M. *bavaricus*, Bavaria (southern Germany);
M. *liechtensteini*, northwestern Yugoslavia;
M. *schelkovnikovi*, Talysh and Elbruz mountains south of Caspian Sea;
M. *majori*, Thrace, Caucasus, Asia Minor;
M. *savii*, France, northern Spain, Italy, Sicily, possibly Macedonia;
M. *lusitanicus*, southwestern France, northwestern Spain, Portugal;
M. *duodecimcostatus*, southeastern France, Spain, Portugal;
M. *thomasi*, southern coastal Yugoslavia, Greece, probably Albania;
M. *pinetorum* (pine vole), extreme southern Ontario, eastern United States;
M. *quasiater*, east-central Mexico;

subgenus *Stenocranius* Kastschenko, 1901

M. gregalis, the Eurasian tundra zone from the White Sea
to Bering Strait, southern Siberia, Mongolia, parts of
Soviet Central Asia and northern and western China;
M. miurus (singing vole), Alaska, Yukon, western
Mackenzie;
M. abbreviatus, Hall and St. Matthew islands (Bering Sea).

The above employment of subgenera basically follows that of
Hall (1981) and Ellerman and Morrison-Scott (1966). Several
additional subgenera are sometimes utilized, as by S. Anderson (1985) and Honacki, Kinman, and Koeppl (1982). As
explained in the account of *Arvicola,* the species *M.
richardsoni* sometimes has been assigned to that genus (or
subgenus); in accordance with Carleton and Musser (1984),
however, this species is here placed in the subgenus *Aulacomys* of *Microtus.* European authorities, such as, for example, Corbet (1978), Corbet and Hill (1986), and Ellerman and
Morrison-Scott (1966), have tended to maintain *Pitymys* as a
full genus, sometimes including *Neodon* as a subgenus. Most
American authorities, including Carleton and Musser (1984),
Hall (1981), and Jones et al. (1986), have considered *Pitymys*
to be part of *Microtus.* Honacki, Kinman, and Koeppl (1982)
did list *Pitymys* as a full genus, with some question, and S.
Anderson (1985) left the question open. In contrast, Chaline
and Graf (1988) regarded *Pitymys* as a subgenus but noted
that the species *M. pinetorum,* on which *Pitymys* was originally based, does not belong in the same systematic group as
the Eurasian species that have been assigned to *Pitymys.*
These authorities also believed that biochemical data supported designation of *Chionomys* as a distinct genus. Ellerman and Morrison-Scott (1966) placed *M. afghanus* in the
separate genus *Blanfordimys* Argropulo, 1933, and *M. blythi*
(= *Pitymys leucurus*) in the subgenus *Phaiomys.* Corbet
(1978), however, indicated that both these species should be
joined with *M. sikimensis* and *M. juldaschi* in the subgenus
Neodon. Honacki, Kinman, and Koeppl (1982) recognized
Proedromys as a distinct genus, but this procedure was not
followed by Carleton and Musser (1984) or Corbet and Hill
(1986). Repenning (1983) considered *M. nemoralis,* found
from the Great Lakes region to central Texas, to be a species
distinct from *M. pinetorum;* he recognized *Pitymys* as a
full genus and placed both species therein, along with *M.
ochrogaster.*

Head and body length is 83–178 mm and tail length is 15–
98 mm. The tail is always shorter than the head and body,
usually being less than half as long. Weight is 17–20 grams in
M. oregoni, a small species; 28–70 grams in *M. pennsylvanicus;* and 113–70 grams in *M. xanthognathus,* one of the
largest species (Burt and Grossenheider 1976). The pelage is
usually quite long and loose. The general coloration of the
upper parts is grayish brown, the darker forms approaching a
sooty black and the paler forms being tawny, reddish, or
yellowish. The underparts range from grayish through pale
brown to white.

These voles have short, rounded ears that are nearly concealed by the pelage. The subgenus *Pitymys* has become
modified for a semifossorial existence through the reduction
of the eyes, external ears, and tail and in having a close,
velvety pelage. The foreclaws are somewhat enlarged for digging. The species *M. afghanus* has enormous bullae, which
distinguish it from all other microtines. Female *Microtus*
usually have eight mammae, but the species *M. ochrogaster*
has only six.

As indicated by the following examples from Banfield
(1974), many kinds of habitat are occupied: *M. pinetorum,*
dry deciduous forest; *M. ochrogaster,* dry grasslands; *M.
miurus,* alpine tundra; *M. pennsylvanicus,* wet meadows;
M. montanus, arid grasslands; *M. townsendii,* salt marshes;
M. oeconomus, damp tundra; *M. longicaudus,* grassy areas
within forests; *M. chrotorrhinus,* damp talus slopes and
rocky outcrops; *M. xanthognathus,* forests and bogs; and *M.
oregoni,* humid coniferous forests. *M. pinetorum* is mostly
fossorial and seldom comes above ground. Its burrows
through thick leafmold and loose soil, however, are normally
not more than 100 mm below the surface. It builds a globular
nest of shredded leaves about 150 mm in diameter under

Pine vole *(Microtus pinetorum),* photo by Ernest P. Walker.

either a log or a rock or in a passage of its burrow. Many other species make both burrows and runways. *M. ochrogaster*, for example, excavates an elaborate system of feeding tunnels about 50 mm in diameter, 50–100 mm deep, and up to 110 meters long (Banfield 1974). Its winter nest is located in a globular chamber 200 mm wide and 150–450 mm below the surface. The burrow also contains a food storage chamber which may be nearly 1 meter in length. Surface runways are made by cutting and trampling vegetation and scattering dirt from the connecting underground burrow to make paths 25–50 mm wide (Schwartz and Schwartz 1959). The summer nest may be above ground in a clump of vegetation.

M. richardsoni occurs primarily in alpine meadows and along mountain streams at elevations of about 1,500–2,000 meters (Banfield 1974). It constructs large burrow systems in stream banks and beneath the snow in winter. Nests of grass or other plant tissues are built in the burrows, slightly above the water level, or are placed under logs, driftwood, or dense vegetative cover on the surface. According to Ludwig (1984), the subterranean passages and nest chambers are excavated from June to late September. Tunnels are located immediately below the thick (4–6 cm) network of roots and mosses that covers the soil surface. Subterranean and surface runways form branching systems that lead to nest chambers, feeding areas, and stream edges. Average burrow diameter is 8 cm, and average nest chamber diameter is 9.7 cm. There are numerous entrances, some directly from water level.

Voles are active throughout the year; they do not hibernate. Although they may move about at any time, most species seem to concentrate their activity at night or in the early morning and late afternoon. *M. chrotorrhinus* has sometimes been called a diurnal species, but Timm, Heaney, and Baird (1977) found it to be largely nocturnal. Voles are terrestrial, but many species swim and dive well. They are strictly vegetarian and within a 24-hour period may consume their own weight in grasses, leaves, twigs, bulbs, tubers, seeds, nuts, or other plant matter. Food may be stored for the winter.

Home range of *Microtus* generally seems to be about 1,000 sq meters or less (Banfield 1974), though in the Northwest Territories the home range of *M. xanthognathus* was found to vary from 116 to 27,500 sq meters (Douglass 1977). In a study of *M. pennsylvanicus* in Kentucky, utilizing radioisotopes for tracking, Ambrose (1969) determined home range of males to average only about 300 sq meters and that of females, 140 sq meters.

Population densities of *Microtus* vary considerably and seem to run in cycles, at least in some areas. Gaines and Rose (1976) found a two-year cycle for *M. ochrogaster* in eastern Kansas; in one study area the density fluctuated from about 4/ha. to 115/ha. Maximum reported density in this species is 358/ha. (Rose and Gaines 1978). Dramatic fluctuations, with peaks every three or four years, have been reported for *M. pennsylvanicus* (Banfield 1974). Normal densities in that species are about 37–117 per ha. in old fields and 117–370 per ha. in marshes, but peak populations may reach 1,000 per ha. In the Humboldt Valley of northwestern Nevada the density of *M. montanus* reportedly reached about 20,000–30,000 per ha. during a "mouse plague" in 1907–8 (Hall 1946). Possibly the highest scientifically documented density of *Microtus* is 1,300/ha., which was attained by *M. oregoni* near Vancouver, British Columbia (Boonstra and Krebs 1978). Some other recorded densities are: *M. pinetorum*, up to 500/ha. (Banfield 1974); *M. ochrogaster*, up to 35/ha. on the xeric prairie of western Nebraska but up to 250/ha. on the eastern tall-grass prairie (Meserve 1971); *M. oregoni*, an average of 2/ha. in spring and 4/ha. in autumn in Oregon forests (Gashwiler 1972); and *M. agrestis*, 5–60 per ha. over a four-year period in Sweden (Hansson 1968). Banfield (1974) noted

that populations of *M. richardsoni* fluctuate dramatically from abundance to scarcity. Lidicker (1988) suggested that vole cycles depend on a mosaic of factors, including fecundity, social spacing, vegetation production, and predation.

Myllymäki (1977*a*, 1977*b*) reported that Scandinavian populations of *M. agrestis* generally have cycles with peaks every three to four years. Populations are divided into reproductive males, which have fixed home ranges and strict territoriality; reproductive females, which have fixed home ranges but little or no territoriality; nomadic subadults; and juveniles (under 35 days old). Individual home range is 400–800 sq meters in males and 200–400 sq meters in females. The home ranges of males show little or no overlap, while females share even areas of intensive use. The only social bond is that between the female and her nursing offspring. Resident males force young males to disperse, but young females may remain in the vicinity of their mother's home range.

In general the social life of *Microtus* is something of an enigma. A number of species are known to live in what appear to be colonies of hundreds of individuals. The animals therein, however, may be totally uncooperative and extremely aggressive toward one another. Banfield (1974) stated that *M. ochrogaster* is a socially tolerant rodent, with up to nine individuals (probably a mated pair plus a late-season litter) sharing the same winter nest and food cache. That this species is highly aggressive, however, is indicated by the presence of wounds in 44 percent of the specimens examined by Rose and Gaines (1976). *M. pennsylvanicus* is also socially aggressive; the males fight viciously among themselves, and the females tend to dominate the males (Banfield 1974). There is no evidence of lasting pairs or formal social structure (Getz 1972). Perhaps social tolerance in *Microtus* exists up to a certain point, but aggressiveness increases with density and serves to limit and disperse populations in accordance with available resources.

Compared with *Clethrionomys*, *Microtus* appears to be slow-moving, docile, and easily trapped and tamed. When upset, these voles may emit a high-pitched squeak, gnash their teeth, and either flee or freeze, depending upon their location and previous activity. Several species are known to have a variety of well-defined sounds. The most vocal vole is *M. miurus*, which produces a series of thin, high, pulsating chirps whenever danger threatens (Banfield 1974).

Female voles are polyestrous and usually give birth several times per year. A postpartum estrus may occur. Breeding generally takes place throughout the year in southern parts of the range of *Microtus* and from spring to early autumn in the north; under good conditions, however, some winter reproduction may occur even in the north. Jannett (1984) reported *M. montanus* to breed under a deep layer of snow in the mountains of northwestern Wyoming and listed records of midwinter breeding in several other species of *Microtus*, *Dicrostonyx*, *Lemmus*, and *Synaptomys*. Captive female *M. pennsylvanicus* have produced up to 17 litters in one year, but for wild individuals in Canada the annual average is 3.5 (Banfield 1974). *M. pinetorum* gives birth to 4–6 litters between January and October in Canada but breeds all year in Louisiana (Lowery 1974) and Oklahoma (Goertz 1971). In the latter area, peak reproduction occurs between October and May, and from 1 to 4 litters are produced. In eastern Kansas there is no distinct breeding season for *M. ochrogaster*, though there seems to be less reproduction in summer and winter, as well as during periods of sharp population decline (Gaines and Rose 1976; Rose and Gaines 1978). In a study of *M. richardsoni* in Montana, Brown (1977) determined that one litter was produced in late June or early July and a second in August. Banfield (1974) indicated that the females of this species born in the season's first litter could mature in about 5 weeks and produce small litters (with as few as 2 young) of their own during the summer.

Reported gestation periods in *Microtus* range from 19 to 25 days (Hasler 1975). Gestation was measured precisely as 24 days in *M. pinetorum* by Kirkpatrick and Valentine (1970). A wild female *M. guentheri* gave birth to a record 17 young (Hasler 1975). Usual litter size for *Microtus* is 1–12, and the number of young seems to tend to be larger in the north. Some average (and extreme) litter sizes are: *M. mexicanus* in Mexico, 2.3 (1–4) (Choate and Jones 1970); *M. pinetorum* in Oklahoma, 2.6 (1–5) (Goertz 1971); *M. richardsoni* in Montana, 7.8 (6–10) (Brown 1977); *M. ochrogaster* in Canada, 3.4 (1–7) (Banfield 1974); and *M. miurus* in northern Alaska, 8.2 (4–12) (Bee and Hall 1956). Innes (1978) found litter size in the genus *Microtus* to be significantly correlated with latitude and elevation, though within particular species litter size was not significantly correlated with either variable. The newborn of *M. pennsylvanicus*, which are typical for the genus, weigh about 2.1 grams, open their eyes at 9 days, and are weaned at 12 days. Sexual maturity is reached at 25 days by females and at 45 days by males. One study indicated an average longevity of less than 1 month in this species, though some wild individuals live for up to 1 year (Banfield 1974). A captive specimen of *M. guentheri* lived for 3 years and 11 months (Jones 1982).

When their populations are high, voles may become serious pests because of their destruction of growing grain, forest plantings, and hay (Jackson 1961). During severe winters, *M. pinetorum* may cause damage to fruit orchards by eating the bark from the roots and lower trunks of trees. Such losses are very heavy when the snow and fallen leaves are mounded around the base of the tree. *M. richardsoni* usually is not an economic problem, but during population peaks it may destroy potatoes, alfalfa, and turnips (Banfield 1974). Moles living within the range of the pine vole are often unjustly accused of damaging bulbs, seeds, and other crops actually taken by *M. pinetorum*. Despite the general high numbers of voles, they are declining in some areas. *M. oeconomus*, for example, is abundant over a vast part of Eurasia and North America, but many populations in central and western Europe are now isolated and in jeopardy because of human habitat modification (Smit and Van Wijngaarden 1981). The species *M. bavaricus*, of Germany, is now listed as extinct by the IUCN. Pucek (1989) noted that its only known habitat had been destroyed by construction, but suggested the possibility that it survives elsewhere.

The species *M. nesophilus*, found only on the Gull Islands off the east coast of Long Island, New York, apparently became extinct when its habitat was destroyed by the construction of fortifications during the Spanish-American War in 1898 (Allen 1942). *M. breweri*, which occurs only on tiny Muskeget Island off southeastern Massachusetts, also occasionally has been thought to be in jeopardy and now is listed as rare by the IUCN. Both species sometimes have been considered to be only subspecies of *M. pennsylvanicus* of the neighboring mainland, though Moyer, Adler, and Tamarin (1988) recently confirmed the specific status of *M. breweri*.

Numerous populations of *Microtus* in the southwestern United States may be near extinction, because the highly limited moist habitat on which they depend is also attractive to people and livestock. *M. californicus scirpensis*, restricted to an isolated area of marshes near Death Valley in southeastern California, long was thought to be extinct but recently was rediscovered (Bleich 1979). Human development in the area now jeopardizes the few surviving voles. Even more critical is the situation of *M. mexicanus hualpaiensis*, of northwestern Arizona. Recent surveys have found it in only three spots, with a total suitable habitat of 0.31 ha., which are being adversely affected by cattle grazing. Both subspecies

now are classified as endangered by the USDI, and *M. c. scirpensis* is so listed by the IUCN.

Problems also center on subspecies in the Southeast, which are isolated by hundreds of kilometers from the rest of their species. *M. ochrogaster ludovicianus* is known by 26 specimens collected on the prairies of southwestern Louisiana in 1899 and by a single specimen taken about the same time in southeastern Texas. This subspecies has not since been collected, despite persistent efforts, and may be extinct (Lowery 1974). *M. pennsylvanicus dukecampbelli* was discovered in 1980 in a tiny salt marsh on Waccasassa Bay on the Gulf coast of Florida. It evidently has survived there in isolation for thousands of years, perhaps climbing up into vegetation during times of high water, but now is constantly vulnerable to human and natural disruption of its habitat (Woods, Post, and Kilpatrick 1982).

RODENTIA; MURIDAE; **Genus TYRRHENICOLA**
Forsyth Major, 1905

Tyrrhenian Vole

The single species, *T. henseli*, formerly occurred on Corsica and Sardinia (E. Anderson 1984; Kurten 1968). It is thought to be closely related to *Microtus* and may be only a subgenus thereof. Its remains have been recovered in great numbers from cave breccias dating from the middle Pleistocene to the postglacial period. It evidently evolved greater size during this time, as postglacial forms average about 10 percent larger than those of the middle Pleistocene. Extinction occurred about 2,000 years ago and may be associated with ecological changes wrought by modern people and their domestic animals (Vigne 1982, 1987).

RODENTIA; MURIDAE; **Genus LEMMISCUS**
Thomas, 1912

Sagebrush Vole

The single species, *L. curtatus*, is found in the Snake River Valley and Great Basin region of the western United States and in the Great Plains region from southern Alberta and Saskatchewan to northern Colorado (Hall 1981). *Lemmiscus* originally was named as a subgenus of *Lagurus* and still is regarded as such by some authorities, including Hall. It was recognized as a distinct genus, more closely related to *Microtus* than to *Lagurus*, by Carleton and Musser (1984). This procedure was followed by Jones et al. 1986 but not by Corbet and Hill (1986) or Honacki, Kinman, and Koeppl (1982).

Head and body length is 90–130 mm, tail length is 16–30 mm, and weight is 17–38 grams. The upper parts are pale buffy gray to ashy gray, the sides are paler, and the underparts are silvery or soiled whitish to buffy. The dense, lax pelage is usually longer and softer than that of *Microtus*. The body is somewhat stocky and lemminglike, and the tail is about the length of the hind foot. Some modifications for burrowing in sandy soil are the small ears, short tail, stout claws, and haired palms and soles. Females have eight mammae.

From Old World *Lagurus*, *Lemmiscus* is distinguished by a relatively longer tail, the ears being more than half as long as the hind foot, the presence of cement in the reentrant angles of the molar teeth, the presence of four (rather than five)

Sagebrush vole *(Lemmiscus curtatus)*, photo by Murray L. Johnson.

closed triangles in the third lower molar, the absence of a conspicuous dorsal stripe, and various morphological features of the stomach. From other North American arvicoline rodents, *Lemmiscus* is distinguished by its dental characters, large bullae, short tail, and pale coloration.

The ecology of *Lemmiscus* was covered in detail by Carroll and Genoways (1980) and Maser (1974). This genus is generally restricted to semiarid prairies, rolling hills, or brushy canyons, which are usually dominated by sagebrush and bunch grasses. Its burrows usually occur in clusters having 8–30 entrances each. The entrances are usually in cover but sometimes are in the open. There are numerous short tunnels at depths of 80–460 mm and a nest chamber up to 250 mm in diameter. The nest is constructed mainly of leaves and stems. Abandoned tunnels of *Thomomys* are often incorporated into the burrow system. *Lemmiscus* also utilizes surface runways, which are usually indistinct, and short adjacent tunnels, which may serve as escape holes. As the food supplies around a given burrow system become exhausted, the animals move to another system, and then to still another or back to the original system. The sagebrush vole is active throughout the year and 24 hours a day, but its main periods of activity are from 2–3 hours before sunset to 2–3 hours after full darkness and from 1–2 hours before sunrise to 1–2 hours afterward. The diet consists mainly of the green parts of plants, rather than seeds. Freshly cut vegetation may be heaped before use and often pulled into burrows, but no long-term storage has been noted.

Population densities of 4–16 individuals per ha. were found on the sagebrush steppe of southeastern Idaho (Mullican and Keller 1986). There is no evidence of overall social organization in colonies of *Lemmiscus*, but captive females sometimes share nests and suckle each other's offspring. It also has been reported that wild females often live together while raising the young, but Mullican and Keller (1987) found no evidence of communal nesting. Aggressive behavior sometimes occurs in this genus, especially in breeding males (Carroll and Genoways 1980).

The sagebrush vole appears to breed throughout the year, though the main season may extend only from March to early December in the north. The estrous cycle of this species seems to be about 20 days, and a postpartum estrus occurs within 24 hours of birth (Carroll and Genoways 1980). A captive female produced 14 litters and 80 young in 1 year (Egoscue, Bittmenn, and Petrovich 1970). According to information compiled by Carroll and Genoways, the gestation period is 24–25 days, and the number of embryos averages about 5 and ranges from 1 to 13. The newborn are naked and blind, weigh about 1.5 grams, open their eyes at 11 days, and are weaned and independent at 21 days. Females usually reach sexual maturity at about 60 days, and males at 60–75 days.

RODENTIA; MURIDAE; Genus LAGURUS
Gloger, 1841

Steppe Lemmings

There are three species (Corbet 1978; Honacki, Kinman, and Koeppl 1982):

L. luteus, Kazakh S.S.R. to western Mongolia;

L. przewalskii, Sinkiang, southern Mongolia, northern Tibet;

L. lagurus, steppe region from Ukraine to western Mongolia and Sinkiang.

Lemmiscus sometimes is considered part of *Lagurus*. The species *L. luteus* and *L. przewalski* sometimes have been put in their own subgenus, *Eolagurus* Argyropulo, 1946. *Eolagurus* was recognized as a full genus by Corbet and Hill (1986) and Honacki, Kinman, and Koeppl (1982) but not by Carleton and Musser (1984).

According to Ognev (1964), head and body length is 80–195 mm and tail length is 7–22 mm. In the relatively small species *L. lagurus* the back is light gray or cinnamon gray with a black stripe along the spine. *L. luteus* has a sand yellow back. The underparts are whitish. *Lagurus* resembles *Lemmiscus* but is distinguished by a tail that is usually shorter than the hind foot, extremely small ears, and various internal characters, as noted in the account of the latter genus. Females have eight mammae.

These animals inhabit steppes and semidesert areas and occasionally pastures and cultivated fields. They construct temporary burrows to shelter from danger and larger burrows for residence. The latter generally have only two or three openings and may extend to 90 cm below the surface. About halfway down is a spherical, grass-lined nest chamber with an average diameter of 10 cm. Activity is mainly nocturnal. The diet consists of a great variety of green vegetation, tubers, and bulbs. Some food is dried and stored for winter use.

Populations of *Lagurus* may fluctuate substantially. *L. lagurus* is said to increase dramatically in some years and undergo mass migrations. Shubin (1972) called this species the most abundant rodent in the Kazakh steppe zone and reported population densities of 30–50 per ha. *L. lagurus* produces up to 5 litters a year, with reproduction occurring mainly from April to October but sometimes also in winter; the average number of embryos per female is 8.1, and the maximum is 12 (Shubin 1972). *L. luteus* breeds during the summer in the wild, with females giving birth to at least 3 litters of 4–10 young each (Shubin 1974). Sexual maturity in this species is attained at only 3–4 weeks of age.

L. lagurus is considered a serious pest because of damage to crops and pastures (Shubin 1972). This species was once found in the vicinity of Kiev, about 300 km west of its present range, but died out in this area around 1900 because of human disturbance of its habitat (Pidoplichko 1973). *L. luteus* also has declined in range, having disappeared in Kazakh S.S.R., where it was formerly widespread (Corbet 1978).

RODENTIA; MURIDAE; Genus NEOFIBER
True, 1884

Round-tailed Muskrat, or Florida Water Rat

The single species, *N. alleni,* inhabits most of Florida and the Okefenokee Swamp area of southeastern Georgia (Hall 1981; Wassmer and Wolfe 1983).

Head and body length is 163–227 mm, tail length is 95–170 mm, and weight is 187–357 grams. The fairly long pelage is composed of glistening guard hairs and short, soft underfur. The general coloration of the upper parts is brown to blackish brown, whereas the underparts are whitish or tinged with buff. The feet are not modified for aquatic life to the pronounced degree found in *Ondatra,* but they do have small swimming fringes. The scaly, scantily haired tail is round. The ears are small and nearly concealed in the fur. Females have six mammae.

The round-tailed muskrat inhabits fresh-water bogs, swamps, lake margins, stream banks, and brackish waters of river deltas. It seems to choose bogs rather than the aquatic habitat preferred by *Ondatra.* It apparently likes thickly vegetated areas. Although more terrestrial then *Ondatra, Neofiber* is a good swimmer. It is active throughout the year and is primarily nocturnal. Information for this account was taken in large part from Birkenholz (1963, 1972) and Layne (1978).

This rodent constructs a nearly spherical to dome-shaped house about 305–457 mm high and 180–610 mm in diameter. The house rests on a base of decaying vegetation and is composed of tightly woven grass or emergent aquatic plants. In the middle of the house is a nest chamber about 100 mm in diameter, and there are usually two exits leading downward in opposite directions from the chamber. These passages are dug through the soft, moisture-soaked ground, so they actually become water-filled tunnels. A network of surface trails about 75 mm wide entwine the vegetation near the house and lead into the water. Before giving birth, a female will thicken the walls and raise the floor of her house. Some individuals dig burrows near a marsh and also burrow into the marsh bottom during a drought. From one to six floating feeding platforms 100–150 mm in diameter and containing one or two plunge holes are located above the water near the house. Sometimes a protective cover is added to a platform.

The round-tailed muskrat is entirely herbivorous. Its main food is maiden cane *(Panicum hemitomon).* It also eats rushes, sedges, sawgrass, and mangrove bark. Food may be consumed where it is found, carried to a feeding platform, or brought back to the nest.

Population densities of 100–300 per ha. have been found in favorable areas. The average space used by one individual is only about 50 sq meters. There is little evidence of territoriality, and individuals appear to be socially tolerant, but there is only one adult occupant per house. Breeding seems to peak in the late fall and early winter but to continue to some

Round-tailed muskrat *(Neofiber alleni),* photo by James N. Layne.

extent throughout the year. Females normally produce four or five litters annually. The estrous cycle lasts 15 days or less, the gestation period is 26–29 days, and litter size averages 2.2 young and ranges from 1 to 4. The young weigh about 12 grams at birth, open their eyes by 14–18 days, and reach sexual maturity at 90–100 days.

The round-tailed muskrat is not usually hunted for its pelt. It may occasionally damage crops, but it is generally of little economic significance. Some populations have declined because canals have drained marshes and allowed salt-water intrusion.

RODENTIA; MURIDAE; Genus ONDATRA
Link, 1795

Muskrat

The single species, *O. zibethicus*, occurs naturally in northern Baja California and in most of Canada and the United States; notable areas of absence include most northern tundra regions, much of California and Texas, the coastal parts of Georgia and South Carolina, and all of Florida (Hall 1981).

Head and body length is 229–325 mm, tail length is 180–295 mm, and weight is 681–1,816 grams. The pelage is composed of a short, soft, dense, fine underfur interspersed with a thick protective coat of long, coarse, dark, shining guard hairs that produce the dominant color of the upper parts. The general coloration ranges from medium silvery brown to dark brown or almost black. The lower surface is generally somewhat paler than the upper parts. The forefeet, hind feet, and tail are dark brown to black.

The muskrat is morphologically adapted to swimming. The hind foot is partially webbed, and along its edge is a row of closely set, short, stiff hairs commonly called the swimming fringe. The scaly, almost hairless tail is flattened laterally and is used as a rudder. The muskrat gets its name from the pronounced musky odor that comes from the secretions of glands in the perineal area. These glands are especially large in males and open within the foreskin of the penis, where the secretions mix with the urine and are then deposited along the route of the animal. Females usually have two pairs of inguinal and two pairs of pectoral mammae (Lowery 1974).

Except as noted, the remainder of this account is based primarily on information provided by Lowery (1974) and Banfield (1974). The muskrat is found in a wide range of aquatic habitats—fresh- and salt-water marshes, lakes, ponds, rivers, and sloughs. The water should be deep enough that it

does not freeze all the way to the bottom but not so deep that there is no submerged vegetation. The muskrat spends much of its time in the water and can swim nearly 100 meters underwater. In a stressful situation it can remain submerged for up to 17 minutes, but the usual underwater interval is about 2–3 minutes. The muskrat is largely nocturnal and crepuscular but sometimes is seen abroad by day, especially in the winter.

Two main kinds of shelters are constructed. When living along a flowing stream or canal, the muskrat makes a burrow in the bank or dike. From an entrance located below the lowest depth to which the water freezes, the tunnel extends for up to 10 meters to a dry chamber located above the high water level. Several burrows may be interconnected.

When living in a flat marsh or swamp, the muskrat builds a mound-shaped house of grass, cattails, or other vegetation and keeps the outside plastered with mud. The house is situated on a stump or clump of vegetation or directly on the ground. It is usually about 1–2 meters wide and 1 meter tall. A dry nest is located inside. One or more tunnels lead from the bottom of the house, pass under the substrate, and sometimes exit underwater.

Houses may be destroyed by spring flooding, and in some areas the muskrat uses a burrow during the spring and builds a new house in the late summer or early fall. The muskrat also may construct fairly well-defined channels through vegetation, slides along stream banks, and feeding platforms above the water of a marsh or pond. In the north a characteristic winter structure is the "push-up," a dome of frozen vegetation over a hole in the ice, where the muskrat can come up for air and to eat what has been found underwater. The most important foods are cattails *(Typha)* and bulrushes *(Scirpus)* in North America and the water lily *(Nymphaea alba)* in Europe (Willner, Chapman, and Goldsberry 1975). The muskrat also takes many other kinds of vegetation and a substantial amount of animal matter. Studies indicate that in the delta marshes of Louisiana 70 percent of the diet is cattails, 15 percent is grass, 10 percent is other vegetation, and 5 percent is crabs, crayfish, mussels, and small fish.

Population density varies from about 7/ha. in open ponds to 87/ha. in cattail marshes. Population levels fluctuate periodically, and excessive numbers may denude marshes of their vegetation. Shifts of populations sometimes occur, especially in the spring, when there is competition for denning sites and mates. Individuals have been known to travel nearly 13 km; however, MacArthur (1978) found few winter movements to extend beyond 150 meters from the den.

The muskrat is a vicious fighter and utters a whinish growl when disturbed. It appears to be monogamous, but especially

Muskrat *(Ondatra zibethicus)*, photo by Ernest P. Walker.

quarrelsome, during the breeding season. When the female is suckling young, the male lives in a separate nest on the outside of the house. Later the mated pair and one litter of young work together to maintain the house and to defend a feeding territory about 60 meters in diameter. When the young reach sexual maturity, they are evicted, but they may move only about 8 meters away to build houses of their own.

Females are polyestrous, with the estrous cycle lasting an average of 6.1 days; a postpartum estrus may occur. In Canada the breeding season extends from about March to September, and 2 litters are normally produced. In the southern United States reproduction is continuous, with peaks in November and March, and there are usually 5–6 litters per year. The gestation period is 25–30 days. Litter size varies from 1 to 11 young but averages as low as 2.4 in some parts of the south and as high as 7.1 in the north. The newborn are naked and blind and weigh about 22 grams. Their eyes open at 14–16 days, they are weaned at 21–28 days, and they are independent at 1 month. In the south some females may reach sexual maturity as early as 6–8 weeks, but in Canada the young are not able to breed until the year following their birth. Longevity is thought to be about 3 years in the wild and up to 10 years in captivity.

The muskrat yields a thick, glossy, durable fur, but for a long time it was not fully appreciated by commercial interests. In the early 1900s muskrat skins sold for only $0.05 or $0.10 each. Subsequent large-scale exploitation of fur resources in the marshes of Louisiana led to that state's becoming the foremost fur-producing area in North America and to the muskrat's often surpassing all other fur bearers of the continent in total annual market value. In recent years the muskrat has lost this position to the raccoon, but it continues to be of major importance in the fur trade. For the 1976/77 season, 45 states of the United States reported a total harvest of 7,148,370 muskrat skins, with an average value of $4.75. In Canada the take was 2,554,879 skins, with an average value of $4.20 (Deems and Pursley 1978). In 1983/84 the take in the United States was 5,453,925, while in Canada 1,506,027 skins were sold at an average price of only $1.52 (Novak, Obbard, et al. 1987). The muskrat also is sought by some persons for use as food.

In 1905 a few muskrats escaped from captivity in central Europe, and their descendants eventually spread over a wide area. Later, when the value of muskrat fur had increased, deliberate introductions were made in several parts of Europe and Asia. The species now ranges from Sweden and France to eastern Siberia and central Honshu. It is considered a serious pest, especially in the Netherlands, because of its undermining of dikes and dams (Corbet 1978; Danell 1977; Grzimek 1975). Introductions also have been made to California, islands off the coast of British Columbia, and southern South America (Willner et al. 1980).

RODENTIA; MURIDAE; Genus LEMMUS
Link, 1795

True Lemmings

There are three species (Burns 1980; Corbet 1978; Hall 1981; Smith and Edmonds 1985):

L. lemmus, Norway, Sweden, Finland, extreme northwestern Soviet Union;
L. sibiricus, arctic parts of Soviet Union from the White Sea to Bering Strait, Alaska to Baffin Island and central British Columbia, also in mountains of southeastern Alberta;
L. amurensis, parts of eastern Siberia.

Soviet karyological studies suggest that *L. chrysogaster*, of northeastern Siberia, is specifically distinct from *L. sibiricus* but possibly conspecific with North American populations, which then would take the species name *L. trimucronatus* (Corbet 1984; Corbet and Hill 1986). *L. nigripes*, of the Pribilof Islands, was listed as a separate species by Corbet and Hill (1986). Jones et al. (1986) continued to recognize *L. sibiricus* as the only North American species.

Head and body length is 100–135 mm, tail length is 18–26 mm, and weight is 40–112 grams. The coloration remains the same throughout the year. The upper parts are grayish or

True lemming *(Lemmus sibiricus)*, photo by Ernest P. Walker.

brownish, with a heavy tinge of buffy and dark brown in some species; the underparts are light gray, buffy, or brownish. The subspecies *L. sibiricus nigripes*, of St. George Island in the Bering Sea, has black feet throughout the year. True lemmings are heavily furred, stockily built, and well adapted to the rigorous conditions of their environment. They have a short tail and ears so small that they are almost concealed by the fur. A claw on the thumb, which is most conspicuous in winter, is long and flat.

True lemmings are found mainly in alpine and tundra areas. They are active both day and night throughout the year. Apparently, in some areas there are regular movements in the spring and autumn to find suitable nesting sites and materials (Grzimek 1975). During the summer the lemmings construct burrows up to about 1 meter in length and about 50–300 mm below the surface. A nest about 150–80 mm in diameter and lined with grass and lemming fur may be located in one chamber of the burrow, and there are other chambers for retreat and defecation. Winter nests are hollow cones or balls of dry vegetation, usually placed on the surface of the ground under the snow, from which subniveal trails radiate to feeding areas (Banfield 1974; Grzimek 1975). All lemmings feed on vegetable matter—sedges, grasses, bark, leaves, berries, lichens, and roots.

In a study of *L. sibiricus* in northern Alaska, Banks, Brooks, and Schnell (1975) found home range to vary from 0.0004 ha. to 2.85 ha. and to average about 1.0 ha. for males and 0.5 ha. for females. Banfield (1974) wrote that densities vary from about 50 per ha. to 325 per ha. in an expanding lemming population. There are regular fluctuations, with peak numbers reached every two to five years. Eventually an excess population may overutilize the food supply and force the lemmings to begin a large-scale movement. At such times the animals may swim across rivers and lakes and appear in human settlements. Many of them die because they are unable to find enough suitable habitat. There have been reliable accounts, especially in northern Europe, of great swarms of lemmings moving over wide areas and eventually plunging into the sea to drown. Such phenomena appear to be unusual and to represent a modification of the typical microtine population cycle rather than any special behavioral aspect of *Lemmus* (Grzimek 1975).

True lemmings are prolific animals. In a captive colony,

Rausch and Rausch (1975) found litters to be delivered regularly throughout the year. Some females became pregnant when they were only 14 days old. One pair produced eight litters in 167 days, after which the male died. The breeding season in the wild usually lasts from spring to autumn, with each female having one to three litters, but under ideal conditions reproduction may continue through the winter. A postpartum estrus may take place. The gestation period is 16–23 days, and litter size is 1–13, averaging 7.3 young in Canada. The newborn weigh about 3.3 grams, open their eyes at 11 days, and are weaned at 14–16 days. Life expectancy is around 1–2 years (Banfield 1974; Grzimek 1975; Hasler 1975).

RODENTIA; MURIDAE; **Genus MYOPUS**
Miller, 1910

Wood Lemming

The single species, *M. schisticolor*, occurs in the coniferous forest zone from Norway to eastern Siberia and northern Mongolia (Corbet 1978, 1984). *Myopus* was regarded as a subgenus of *Lemmus* by Koenigswald and Martin (1984).

Head and body length is 75–110 mm and tail length is 14–18 mm. Weight is 20–30 grams (Grzimek 1975). The pelage is soft, dense, and slaty black above, with a definite reddish brown area on the center of the back extending from the shoulders to within about 15 mm of the base of the tail. The remainder of the coat is uniformly a dark slaty gray, slightly paler on the ventral surface. The upper parts have a peculiar metallic luster that is produced by silvery tips on the shorter hairs, with an indistinct showing of black guard hairs. The feet and tail are black, but the hairs on the undersurface of the tail have a silvery luster. The tail is heavily furred. The palms are naked. The hind feet are densely haired behind the pads but are naked in front of the pads, like the palms of the forefeet. There is no substantial difference in the coloration of the sexes, nor is there much seasonal variation in color. The winter coat is slightly longer than the summer pelage.

This tiny, thickset rodent has ears that are small and project little beyond the fur but that are well developed,

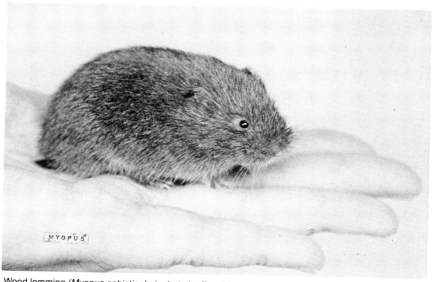

Wood lemming *(Myopus schisticolor)*, photo by Erna Mohr.

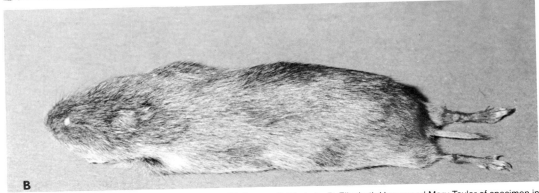

Bog lemming *(Synaptomys cooperi):* A. Photo by Paul F. Connor; B. Photo by B. Elizabeth Horner and Mary Taylor of specimen in Harvard Museum of Comparative Zoology.

rounded, and well haired. The ears have valves that can regulate the size of the ear openings. The thumb of the hand is small but bears a large, flattened nail with parallel sides and a notch at the end, somewhat resembling the thumbnail of *Lemmus* but smaller. Female *Myopus* have eight mammae.

The wood lemming lives in mossy bogs and coniferous forests at elevations of 600–2,450 meters. Its runways have been found in moss, under the roots of trees, and beneath fallen tree trunks. It is known to eat certain mosses *(Dicranum),* stems of red wortleberry, and the bark of juniper. It is apparently rare and is taken only by accident or in years when its populations become unusually large. Its numbers fluctuate, and it makes local migrations. Following a study of captives, Ilmen and Lahti (1968) reported that one female delivered 11 litters in 299 days. Average time between litters was 25 days, and litter size was 1–6 young. Females could successfully mate at about 1 month. Mysterud, Viitala, and Lahti (1972) suggested that breeding during the winter takes place in Norway.

RODENTIA; MURIDAE; Genus SYNAPTOMYS
Baird, 1857

Bog Lemmings

There are two subgenera and two species (Hall 1981; Linzey 1983):

subgenus *Synaptomys* Baird, 1857

S. *cooperi,* southeastern Canada, central and northeastern United States;

subgenus *Mictomys* True, 1894

S. *borealis,* Alaska and Canada south of the tundra, parts of the extreme northern conterminous United States.

Koenigswald and Martin (1984) considered *Mictomys* a full genus, but this procedure was not followed by Jones et al. (1986).

Head and body length is 100–130 mm, tail length is 16–27 mm, and weight is 21–50 grams. The general coloration above is a grizzled, grayish cinnamon brown composed of a mixture of gray, yellowish brown, and black in varying degrees. The underparts are soiled whitish over the slate-colored bases of the hairs. The tail is brownish above and whitish below. The body form is thickset, with long, loose, and coarse pelage. The incisors are orange. The upper incisors have a shallow longitudinal groove at the outer edge. The nail of the first finger is flat and strap-shaped. There are three pairs of mammae in female S. *cooperi* and four pairs in female S. *borealis.*

These animals seem to occupy small, local, isolated areas; their spotty occurrence makes a general map of their range misleading. In the southern parts of their range they are limited mainly to scattered cold bog and cold spring areas.

Farther north they have a more general distribution, usually living in moist areas, though in the humid Pacific Northwest they have been found near rocky cliffs. They do not hibernate and are active at any hour of the day or night. They make surface runways and subsurface tunnels, as do voles, though they occasionally occupy the tunnels and use the runways made by other small mammals. Nests of dry, shredded grasses and sedges, about 100–150 mm in diameter, are concealed in hollows under sphagnum mounds, logs, or stumps or in grass tussocks (Banfield 1974). The powerful jaws of bog lemmings suggest that they gnaw through tangles of roots, moss, and soil. Apparently they feed largely on plant tissues, mainly green parts of low vegetation, and to a lesser extent on slugs, snails, and other invertebrates.

Population density varies from about 6 per ha. to 35 per ha., with regular cycles from a low in the spring to an autumn peak. Individual home ranges are about 0.08–0.20 ha., with males usually traveling more extensively than females. The area around the nest is defended (Banfield 1974). It is believed that each sex makes its presence known by secretions from glands in the groin. The voice has been described as a sharp, short squeak.

According to Banfield (1974), the breeding season of *S. cooperi* usually lasts from April to September but may extend through the winter under ideal conditions of food and cover. The breeding season of *S. borealis* lasts from May to August. Females are seasonally polyestrous and may have a postpartum estrus. A captive female bore six litters in 22 weeks, but wild females probably have litters every 67 days during the spring and summer. The gestation period is 23 days. Litter size ranges from one to eight young, averaging three in *S. cooperi* and four in *S. borealis*. The newborn weigh about 3.7 grams, open their eyes after 10–12 days, and are weaned at 16–21 days. Males reach sexual maturity at 5 weeks. A female captured as an adult by Linzey (1983) lived for another 2 years and 5 months.

Handley (1980b) stated that the subspecies *S. cooperi heletes*, restricted to the Dismal Swamp of Virginia and North Carolina, is at least endangered and possibly extinct. Although seemingly common in the 1890s, it subsequently declined, probably because of habitat destruction and the drying up and overgrowth of sphagnum bogs. Recently, Rose (1981) discovered a living population of this subspecies. The subspecies *S. cooperi paludis*, of southwestern Kansas, and *S. c. relictus*, of southwestern Nebraska, are known only from local, isolated sites and may be in imminent danger of extinction through loss of habitat. The subspecies *S. borealis sphagnicola*, which occupies a small part of New England and adjacent Canada, evidently is very rare, perhaps because of environmental disruption, though Clough and Albright (1987) recently collected several specimens in Maine.

RODENTIA; MURIDAE; Genus DICROSTONYX
Gloger, 1841

Collared Lemmings, or Varying Lemmings

There are five species (Corbet 1978, 1984; Hall 1981):

D. torquatus, tundra region of northeastern Europe and Siberia;
D. vinogradovi, Wrangel Island off northeastern Siberia;
D. groenlandicus, tundra region of Alaska, northwestern and north-central Canada, and Greenland;
D. exsul, St. Lawrence Island (Bering Sea);
D. hudsonius, northern Quebec, mainland Newfoundland.

Corbet and Hill (1986) considered *D. groenlandicus* conspecific with *D. torquatus* but otherwise followed the above systematic arrangement. Based on studies of chromosomal variation, however, Honacki, Kinman, and Koeppl (1982) divided what is listed above as *D. groenlandicus* into the following six species: *D. groenlandicus*, Greenland and arctic islands of Canada; *D. kilangmiutak*, Victoria and Banks islands and adjacent mainland of Canada; *D. nelsoni*, western Alaska; *D. richardsoni*, Canada west of Hudson Bay; *D. rubricatus*, northern Alaska; and *D. unalascensis*, Umnak and Unalaska islands in the Aleutians. Jones et al. (1986) listed *D. nelsoni*,

Collared lemming *(Dicrostonyx hudsonius)*. Inset: left foreclaws in winter. Photos by Ernest P. Walker.

D. richardsoni, D. rubricatus, and *D. stevensoni* (Umnak Island) as full species.

Head and body length is 100–157 mm, tail length is 10–20 mm, and weight is 30–112 grams. In summer the general coloration above varies from light grayish buff to dark gray, strongly tinged with buffy to reddish brown. Some *Dicrostonyx* are colored a deep, rich buff, and others are merely grayish tinged with reddish brown. The underparts are whitish, grayish, or buffy. There are definite dark lines from the top of the head along the middle of the back and on the sides of the head. In winter all species are entirely white. They are the only rodents that turn white in winter. Both the winter and summer coats are thick and heavy.

These lemmings are short, stockily built creatures. The third and fourth claws of their forefeet are unusually large and well adapted to digging. In summer these claws are about normal in appearance, but with the approach of winter they develop a peculiar double effect in a vertical plane, which makes them very strong and well suited for burrowing in the frozen earth or in snow and ice. This remarkable development of the third and fourth claws in winter distinguishes *Dicrostonyx* from all other small rodents. The ears of this genus are so small that they are little more than rims, completely hidden in the fur. Females have eight mammae.

These lemmings occur on the treeless arctic tundra, mainly in dry, sandy, and gravelly areas with some plant cover. They are active throughout the year. During the summer they live in shallow underground burrows or under rocks. The burrows are simple structures up to 6 meters long but usually less than 0.5 meters (Brooks and Banks 1973). The tunnels are about 100–200 mm deep and 50–75 mm wide. They lead to nesting chambers about 100–150 mm in diameter and lined with dry grasses or other soft material; there are also small, bare resting chambers and toilet areas (Banfield 1974). An individual male may use up to 50 burrows, many only for retreat but several with nests (Brooks and Banks 1973). Faint runways on the surface of the ground may also be made, but these paths are not as pronounced as those of some other microtines. In the winter collared lemmings place their grass nests on the surface of the earth under a snowbank or even in the middle of a snowbank. They dig their tunnels through the snow with their special winter claws (Banfield 1974). They swim readily. In the winter they eat buds, twigs, and bark, and their summer diet consists of fruits, flowers, grasses, and sedges.

Population densities fluctuate dramatically, from 0.6 per ha. to 400 per ha. (Banfield 1974), but normal peak densities are 15–40 per ha. (Brooks and Banks 1973). At times these lemmings become so plentiful that they are forced to make short migrations in search of food, but they rarely make the lengthy mass movements sometimes observed in *Lemmus.* In a radiotracking study of *D. groenlandicus* in Manitoba, Brooks and Banks (1971) found home range to average 2.03 ha. for males and only 0.16 ha. for females. Although nesting burrows are defended (Banfield 1974) and captive males fight and form dominance relationships, there is no clear evidence of territoriality (Bowen and Brooks 1978). Several distinctive sounds have been identified, including a "squeal-squawk-grind" complex associated with agonistic behavior (Brooks and Banks 1973).

The breeding season usually extends from early March to early September but under ideal winter conditions may begin in January (Banfield 1974). Females are seasonally polyestrous and have a postpartum estrus. An average estrous cycle of 9.6 days has been reported for *D. groenlandicus* (Hasler 1975). In captivity, females may give birth up to five times from April to September, but in the wild they probably produce only two or three litters per year (Banfield 1974). They are very aggressive after giving birth, often attacking

and occasionally killing males that attempt to mate with them. Without such aggressiveness, however, the males may kill the young already born. The normal gestation period is 19–21 days, but gestation can be extended to as long as 31 days if the female is nursing a previous litter and is able to protect it from the male (Mallory and Brooks 1978). Litter size is 1–11 (Brooks and Banks 1973) and averages 3.3 in captivity (Hasler and Banks 1975) and 4.5 in the wild (Banfield 1974). The young weigh an average of 3.8 grams at birth, open their eyes after 12–14 days, and are weaned and disperse at 15–20 days (Banfield 1974; Brooks and Banks 1973). In a study of captives, Hasler and Banks (1975) found the youngest female to give birth at 58 days and the oldest at 580 days; the respective figures for males were 85 and 683 days. Captives have lived for as long as 3 years and 3.5 months (Banfield 1974).

The beautiful white winter coat of *Dicrostonyx* is used by the Eskimos for garment trimming. The Eskimo children use the little skins to make doll clothes.

RODENTIA; MURIDAE; **Genus PROMETHEOMYS**
Satunin, 1901

Long-clawed Mole-vole

The single species, *P. schaposchnikowi,* occurs in the Caucasus and extreme northeastern Asia Minor (Corbet 1978).

Head and body length is 125–60 mm and tail length is 45–

Long-clawed mole-vole (*Prometheomys schaposchnikowi*), photos by R. Zimina.

Mole-vole, or mole-lemming *(Ellobius talpinus)*, photo by Olga Ilchenko through Sergei Popov, Moscow Zoo.

65 mm. Grzimek (1975) gave the weight as 70 grams. The upper parts are dull grayish brown suffused with cinnamon, while the underparts are pale plumbeous gray with a cinnamon tint. The fully haired tail is unicolored brownish gray. This stocky rodent has a round head and large claws on the forefeet. The claw of the middle digit of the hand is the longest, being over 5 mm in length. The eyes and ears are reduced. The upper incisors are grooved, and the molar teeth are rooted. The tail is swollen at the tip. Females have eight mammae.

Prometheomys is found in alpine and subalpine meadows and open places at elevations of 1,500–2,800 meters (Grzimek 1975). It builds complex burrows and throws the excavated dirt on top of the ground in the form of heaps. The claws, rather than the teeth, are used in digging. A grass-lined nest appears to be the center of the burrow system, as many tunnels radiate from there, gradually approaching the surface. This rodent is active throughout the year. Zimina and Yasny (1977) found maximum activity to be from 1100 to 1300 hours and the diet to consist entirely of the green parts of plants found above the ground. *Prometheomys* also is said to feed on subterranean roots.

Dzneladze (1974) reported population densities of 3–32 per ha. Grzimek (1975) stated that several older males live together with the females in a single burrow and that litter size is 2–6 young. Zimina and Yasny (1977) reported that females produce two litters during the summer, with an average of 3 young each.

RODENTIA; MURIDAE; **Genus ELLOBIUS**
Fischer, 1814

Mole-voles, or Mole-lemmings

There are four species (Corbet 1978, 1984):

E. talpinus, steppe region from the Ukraine to Mongolia and northern Afghanistan;
E. alaicus, Kirghiz S.S.R.;
E. fuscocapillus, southern Turkmen S.S.R., Afghanistan, Pakistan;
E. lutescens, Caucasus, eastern Turkey, Iraq, Iran.

Head and body length is 100–150 mm and tail length is 5–22 mm. The upper parts are dull or bright cinnamon, pinkish buff, buff, or brownish, and the underparts are grayish, brownish, or whitish. A dark face patch may be present. The fur is velvety or plushlike, resembling the pelage of other mammals that spend most of their lives underground.

These burrowing rodents have a round head and reduced eyes and ears. The claws are small. The large incisors, those of the upper jaw protruding forward and downward, are used to loosen earth and cut roots. *Ellobius* is probably the most specialized fossorial microtine genus.

According to Grzimek (1975), mole-voles live mainly in steppe and semidesert areas, only rarely moving into woodlands. In Mongolia they prefer the deep, moist soil on the banks of lakes and streams. They construct a system of branching tunnels, including food storage rooms, at a depth of 20–30 cm and a nest chamber at a depth of at least 50 cm. They feed on the underground parts of plants; tulip bulbs and pulpy tubers are especially favored. The gestation period is 26 days. Females produce six or seven litters annually, each with three to five young. The offspring remain in the parents' nest for 2 months before becoming independent and attain sexual maturity by 96 days.

Mole-voles damage crops in some areas. They have some value in the fur trade.

RODENTIA; MURIDAE; **Genus DENDROMUS**
A. Smith, 1829

African Climbing Mice, or Tree Mice

There are six species (Dieterlen 1969; Kingdon 1974*b*; Meester 1973; F. Petter, *in* Meester and Setzer 1977):

D. lovati, mountains of Ethiopia;
D. nyikae, central Angola to Malawi, eastern escarpment of Zimbabwe and Transvaal;
D. melanotis, Guinea to Ethiopia, and south to South Africa;
D. mesomelas, Cameroon to Ethiopia, and south to South Africa;

African climbing mouse (*Dendromus mesomelas*), photos by John Visser.

D. *mystacalis*, southeastern Nigeria to Ethiopia, and south to South Africa;

D. *kahuziensis*, known only by a single specimen from Mount Kahuzi in eastern Zaire.

The following subgeneric terms sometimes are applied: *Chortomys* Thomas, 1916, for *D. lovati*; and *Poemys* Thomas, 1916, for *D. nyikae* and *D. melanotis* (see Rosevear 1969). Corbet and Hill (1986) listed *D. messorius*, found from Nigeria to Zaire, as a species distinct from *D. mystacalis*.

Head and body length is 50–100 mm, tail length is 65–132 mm, and weight is 5–21 grams. The pelage is soft and woolly. The upper parts are grayish or brownish; there is usually one black dorsal stripe, but *D. lovati* has three stripes on the back. The underparts are whitish or yellowish. The finely scaled and scantily haired tail is either entirely dull brownish or slightly bicolored, in which case the underside is paler. Like many nocturnal tree mammals, these mice usually have well-defined, dusky "spectacles" encircling the eyes. The remarkably long tail is semiprehensile. The forefeet have only three well-developed digits, but the hind feet are normal. The upper

incisors are grooved. Females have eight mammae.

African climbing mice are found in a variety of habitats at elevations ranging from sea level to more than 4,300 meters. Although these animals are sometimes said to occur mainly in trees, Rosevear (1969) pointed out that positive and reliable recorded evidence that they are arboreal is almost nonexistent. Some populations are known to spend most of their time on the ground or even in subterranean burrows. Kingdon (1974b) listed the following ecological conditions: *D. mesomelas*, the driest and most terrestrial habitats whenever other species of *Dendromus* are present but very wet areas of grass and herbaceous growth in alpine and subalpine zones and in other places where other species of *Dendromus* are absent; *D. nyikae*, savannahs and woodlands; *D. melanotis*, usually relatively open, dry savannahs, often in short grass and scrub growing on sandy plains; and *D. mystacalis*, forests and other densely vegetated areas. In Botswana the species *D. melanotis*, *D. mesomelas*, and *D. nyikae* all have been found in swampy lowlands (Smithers 1971).

Most species climb with great agility, though Yalden, Largen, and Kock (1976) observed that *D. lovati* lives in grasslands, nests beneath boulders, and shows no inclination to climb. All species seem to be primarily nocturnal and to spend daylight hours in globular nests of shredded vegetation (Kingdon 1974b; Rosevear 1969). *D. mystacalis*, evidently the species best adapted to climbing, may build its nest 3 meters or more above the ground in banana trees or thatched roofs, though more often nests are placed somewhat lower in thick vegetation. *D. melanotis* and *D. nyikae* may build their nests in underground burrows, apparently to avoid annual fires. The burrows of *D. melanotis* are simple, being about 30–60 cm deep and consisting of an open entrance tunnel, a small grass-lined nest chamber, and an emergency exit hole opposite the entrance. African climbing mice also have been known to reline and occupy the nests originally built by sunbirds and weaverbirds. Their diet consists of seeds, berries, insects, small lizards, birds' eggs, and nestlings.

Recorded population densities have varied as follows: *D. mystacalis*, from 0.6/ha. in savannah to 19.8/ha. in herbaceous growth bordering marsh; and *D. mesomelas*, from 0.5/ha. in herbaceous growth bordering marsh to 6.6/ha. in savannah (Kingdon 1974b). In a laboratory study, Choate (1972) found *D. mesomelas* to be a socially tolerant species, with groups containing individuals of all ages and sexes, including several mature males, living together without antagonism. Rosevear (1969) noted, however, that male and female *D. mystacalis* are solitary and live in separate nests, and Smithers (1971) stated that individuals of *D. melanotis* are territorial and may fight to the death.

Evidence of reproductive activity in *Dendromus* has been collected throughout the year, though breeding apparently is restricted to certain periods in many populations. The following information on seasonality is taken from Kingdon (1974b), Delany (1975), and Smithers (1971): *D. nyikae*, a lactating female collected in August and a litter of young found in June in Tanzania; *D. melanotis*, probably a strictly seasonal breeder where subject to climatic fluctuation and fires, a pregnant female taken in Uganda in July, young observed from December to March in Zimbabwe; *D. mesomelas*, probably a seasonal breeder, many young found in February in Tanzania; *D. mystacalis*, near the equator a birth peak from November to January, but scattered records of breeding in other months in eastern Zaire and Uganda. Overall litter size in the genus is two to eight young. In *D. melanotis* the gestation period is 23–27 days, and the young leave the nest at 30–35 days (Delany 1975). A captive *D. mesomelas* lived for 3 years and 3 months (Jones 1982).

RODENTIA; MURIDAE; Genus MEGADENDROMUS
Dieterlen and Rupp, 1978

The single species, *M. nikolausi*, is known from the highlands of eastern Ethiopia (Dieterlen and Rupp 1978).

In three specimens, head and body length is 117–29 mm, tail length is 92–106 mm, and weight is 49–66 grams. Respective figures for a fourth, reported by Demeter and Topál (1982), are 97.6 mm, 86.4 mm, and 40 grams. The upper parts are brown, except for a black dorsal stripe extending from behind the ears to the rump, and the underparts are light brown with some gray showing through. The sides of the head and the shoulders are reddish brown, and there are dark "spectacles" around the eyes. The hands and feet are silvery gray. The tail is whitish gray except for a thin black stripe on top.

Megadendromus has a striking resemblance to *Dendromus* in color pattern but differs in being much larger and in having a tail length that is only about 80 percent of the head and body length. There also are important cranial and dental distinctions, including the presence of two lingual cusps on the first upper molar of *Megadendromus*.

This genus inhabits mountainous areas thickly vegetated with grasses and scrub at elevations of 3,000–4,000 meters. All specimens were collected at night. Schlitter (1989) classified *Megadendromus* as rare and noted that it probably is less arboreal than is *Dendromus*.

RODENTIA; MURIDAE; Genus MALACOTHRIX
Wagner, 1843

Gerbil Mouse, or Long-eared Mouse

The single species, *M. typica*, occurs in southern Angola, Namibia, southern Botswana, and South Africa (F. Petter, *in* Meester and Setzer 1977; Smithers 1971).

Head and body length is 65–95 mm and tail length is 28–42 mm. Smithers (1983) gave the weight as 7–20 grams. The pelage is long, dense, and silky. The upper parts range from pale brownish or buffish to reddish brown and have a more or less distinct dark dorsal stripe and a dark crown spot. There is a sprinkling of black or brown hairs on the back and sides. The underparts, feet, and tail are white, the latter being scantily haired. The hind feet have only four toes each, the soles are hairy, the limbs are slender, and the ears are large. The teeth are similar to those of *Dendromus* and *Steatomys*. Females have eight mammae.

Long-eared mouse, or gerbil mouse *(Malacothrix typica)*, photo by John Visser.

The gerbil mouse is associated with sandy plains and inland grassy velds. It is nocturnal and terrestrial. It constructs a burrow by digging a passageway that slants downward to a depth of 60–120 cm, where a chamber is excavated and the soil is used to fill up the original tunnel. A nest is made of grass and often of feathers. From the nest chamber a new tunnel is made to the surface. The soil from this tunnel is also transported to the original passageway beyond the chamber, so that a mound of dirt is not formed near the tunnel exit. The deceiving pile of freshly excavated soil is some distance away. The gerbil mouse may travel up to 4 km from its burrow. It generally follows cattle tracks, footpaths, or sandy roads, and its presence is frequently noted by tiny footprints along these routes. It returns to its burrow before dawn. According to Smithers (1971), the natural diet appears to consist entirely of green vegetable matter.

Smithers (1971) stated that the breeding season in Botswana appears to fall during the warmer, wetter months of the year, from about August to March. More than one litter may be produced during this season. The gestation period evidently can vary from 22 to 35 days but usually is around 22–26 days. Litters consist of two to seven young, which are born naked and weigh about 1.1 gram. They may reach sexual maturity at 51 days.

RODENTIA; MURIDAE; Genus DENDROPRIONOMYS
F. Petter, 1966

The single species, *D. rousseloti*, is known only by four specimens from the city of Brazzaville, southern Congo (F. Petter, *in* Meester and Setzer 1977).

Dendroprionomys, as implied by its generic name, exhibits certain characters of *Dendromus* and others of *Prionomys* and appears to be intermediate to these other genera.

The pelage is velvety and molelike, as in *Prionomys*, and the dorsal coloration is brownish, like that of *Dendromus*. The flanks are paler, and the line of demarcation between the dorsal and ventral surfaces is tawny. The venter is entirely white, but the bases of the hairs are gray for half their length. A black area underlines the eyes.

The ears appear to be relatively well developed and are

Dendroprionomys rousseloti, photo by F. Petter.

pigmented gray. The tail is scaled and is longer than the head and body. The forefeet are provided with four well-developed toes, as in *Prionomys*. The hind feet have five digits, as in both *Prionomys* and *Dendromus*. The dentition is such as to imply a diet that is at least partially insectivorous. The upper incisors are furrowed longitudinally.

RODENTIA; MURIDAE; **Genus PRIONOMYS**
Dollman, 1910

Dollman's Tree Mouse

The single species, *P. batesi*, is known only from Bitye in southern Cameroon and from around Bangui in the Central African Republic (F. Petter, *in* Meester and Setzer 1977).

Head and body length is about 60 mm and tail length is about 100 mm. The short, velvety fur resembles that of shrews. The coloration of the upper parts is pinkish chocolate. The sides are paler and blend almost imperceptibly into the grayish pinkish buff underparts. The face is slightly lighter than the remainder of the upper parts. The sides of the face below the eyes are washed with pinkish buff, and the eyes are encircled with narrow black rings. The scantily haired,

finely scaled tail is dark brown, almost black. There appears to be no hair on the dorsal tip of the tail, which suggests that it may be prehensile.

The size of *Prionomys* is comparable to that of a large *Dendromus*. The ears are rather small and rounded. The forefeet have four well-developed fingers, though the thumb is absent. The second and fourth fingers are moderately elongated and have short claws. The outer finger is only half the length of the two middle fingers and bears a small nail. The hind foot has five toes, all armed with claws. The inner toe is about half as long as the middle digit, and its claw is smaller and more blunt than those of the other toes. The upper incisor teeth are short, slender, and ungrooved and project forward. The slender lower incisors are sharply pointed.

The type specimen was taken at Bitye, Ja River, at an elevation of about 600 meters. F. Petter (*in* Meester and Setzer 1977) referred to *Prionomys* as a burrowing animal. Genest-Villard (1980) reported it to be strictly insectivorous.

RODENTIA; MURIDAE; **Genus STEATOMYS**
Peters, 1846

Fat Mice

There are six species (Anadu 1979*a*; Coetzee, *in* Meester and Setzer 1977; Swanepoel and Schlitter 1978):

S. pratensis, Cameroon to southern Sudan, and south to northern Namibia and eastern South Africa;
S. caurinus, Senegal to Nigeria;
S. jacksoni, western Ghana, southwestern Nigeria;
S. parvus, southern Sudan and Somalia to northern Namibia and eastern South Africa;
S. cuppedius, Senegal, southern Niger, northern Nigeria;
S. krebsii, Angola, western Zambia, northern Botswana, South Africa.

Genest-Villard (1979) recognized *S. opimus*, of Cameroon, the Central African Republic, and Zaire, as a species distinct from *S. pratensis*.

Dollman's tree mice *(Prionomys batesi)*, photo by F. Petter.

Fat mice *(Steatomys pratensis)*, photo by Ernest P. Walker.

Head and body length is 65–145 mm, tail length is 34–59 mm, and weight is 5–70 grams (Coetzee, *in* Meester and Setzer 1977; Swanepoel and Schlitter 1978). The pelage is thick, short, soft, and silky. The general coloration of the upper parts ranges through various shades of brown and buff, frequently mixed with black; the underparts are white. The dorsal surface of the scantily bristled tail is generally colored like the back, and the ventral surface is white, but in at least one form the tail is all white.

These animals have a thick body, rather large and rounded ears, and a thick tail that tapers to a point. The four fingers have slightly curved claws that are suited for burrowing, and there are five toes. The upper incisor teeth are grooved. The females of most species have 8 mammae, but female *S. pratensis* and *S. caurinus* usually have 12, sometimes up to 16 (Coetzee, *in* Meester and Setzer 1977).

Fat mice are found in forest clearings, cultivated fields, savannahs, and semidesert areas (Coetzee, *in* Meester and Setzer 1977). They are nocturnal and terrestrial and construct burrows in sandy soil. The excavated ground is deposited in mounds on the surface. The burrows of fat mice may reach a depth of 1.2 meters and a length of 1.7 meters. A nest constructed of shredded vegetation is concealed in a large chamber at the end of the tunnel. According to Coetzee (*in* Meester and Setzer 1977), the burrows of *S. pratensis* are simple, with a main entrance dug at an angle of 30° to the surface, a nest chamber about 30 cm below the ground, and two or more side passages that are normally sealed at ground level and used as escape routes. The diet of fat mice consists of seeds, grass bulbs, and insects.

Rosevear (1969) declared that the most noted fact about fat mice is their ability to accumulate thick layers of fat, enabling them to sleep, or at least to remain highly inactive, in their underground nests throughout the dead season when food is scarce. In the tropics this season is not associated with wintry conditions, as there is still sunshine and heat, and the process of dormancy is called estivation, not hibernation. In West African tropics the fat mice estivate during the dry season, from November to March. In temperate zones farther south, estivation corresponds more to the cooler season, approximately April–October.

Fat mice evidently live alone in their burrows, except for females with young (Coetzee, *in* Meester and Setzer 1977; Genest-Villard 1979). Numerous burrows, however, are often present in a small area. Although *Steatomys* is well known and frequently collected, reproductive data are limited. Young of *S. pratensis*, *S. caurinus*, and *S. parvus* have been recorded from early summer to late autumn in temperate areas (Coetzee, *in* Meester and Setzer 1977). *S. cuppedius*

may breed during the dry season and produce two litters (Rosevear 1969). Pregnant females, one with two embryos and the other with seven, were reported by Smithers (1971), and other available records of litter size fall between these extremes.

Because of the high fat content of its body, *Steatomys* is regarded as a special delicacy by native peoples throughout its range. It is easily dug up, particularly during estivation (Coetzee, *in* Meester and Setzer 1977).

RODENTIA; MURIDAE; Genus DEOMYS
Thomas, 1888

Congo Forest Mouse

The single species, *D. ferrugineus*, is found from southern Cameroon to Uganda and southwestern Congo (F. Petter, *in* Meester and Setzer 1977).

Head and body length is 120–44 mm, tail length is 150–215 mm, and weight is 40–70 grams (Kingdon 1974*b*). The coloration of the upper parts is pale red or reddish fawn, mixed with black along the middle of the back. The individual hairs on the upper parts have a white base, a slate-colored middle portion, and a reddish tip. Some of the hairs along the back are completely black. The face and sides of the head are paler and duller in color than the back, except for a poorly defined blackish ring around each eye. The underparts, the insides of the arms and legs, and the hands are white. The slender, coarsely haired tail is distinctly bicolored in one form, slate gray above and white below, but in another form the terminal 25–76 mm of the tail is entirely white.

Deomys has large, rounded ears; long, narrow hind feet; orange upper incisor teeth, each with two minute grooves; and smooth yellow lower incisors. Females have four mammae.

According to Kingdon (1974*b*), *Deomys* occurs in tropical forests, especially in or near seasonally flooded swamps. It constructs leaf and fiber nests in holes or crevices at the base of trees. When frightened it may leap up to 50 cm off the ground using its powerful hind legs, but it normally progresses on all fours across the forest floor in search of prey. It also climbs well. It apparently is crepuscular but may also be active at night or in the late afternoon. During the rainy season in Uganda, *Deomys* may be found in well-drained areas, but in dry periods the animals concentrate in moist swamps. The diet consists of insects, other small invertebrates, and occasionally vegetable matter.

Kingdon (1974*b*) stated that the main social units are

Congo forest mouse *(Deomys ferrugineus)*, adapted from *Proc. Zool. Soc. London*, St. Leger.

male-female pairs and mothers with young. There seem to be extensive reproductive peaks, and *Deomys* may produce closely spaced litters throughout the year. Litter size is one to three, usually two young.

RODENTIA; MURIDAE; **Genus LEIMACOMYS**
Matschie, 1893

Groove-toothed Forest Mouse

The single species, *L. buettneri,* is known only by two specimens collected in 1890 near Yege in central Togo. The specimens are now in the Zoologisches Museum of Humboldt University in Berlin. Although sometimes considered to be-

long to the subfamily Murinae, *Leimacomys* now is thought to be a dendromurine (Rosevear 1969).

Head and body length of the type specimen is 118 mm and tail length is 37 mm. The upper parts are dark brown to grayish brown, the shoulders and flanks are light brown, and the underparts are pale grayish brown. The feet are covered by short brown hairs. The small ears are well haired. The claws on the forefeet are fairly long but somewhat shorter than those of the hind feet. The tail is extremely short and apparently naked. *Leimacomys* differs from the other dendromurines, *Dendromus* and *Steatomys,* in that the third upper and lower molars are not so reduced, the incisors are only weakly grooved, and the zygomatic plate is broader and well forward of the upper branch, its messeteric knob being relatively poorly developed (Rosevear 1969).

Schlitter (1989) classified *Leimacomys* as extinct, noting that it had not been found by two surveys in Togo during the

1960s. He added that it probably had been terrestrial and could have been at least partially insectivorous.

RODENTIA; MURIDAE; Genus PETROMYSCUS
Thomas, 1926

Rock Mice

There are two species (F. Petter, *in* Meester and Setzer 1977):

P. monticularis, southern Namibia;
P. collinus, southern Angola, Namibia, western South
    Africa.

Head and body length is about 70–90 mm and tail length is 80–100 mm. Withers (1983) reported weight to be 20 grams. The pelage is fine, straight, soft, silky, and rather thin and lacks guard hairs. The coloration of the upper parts ranges from various shades of buffy to brownish or drab gray. The individual hairs of the underparts have a slaty base with a white tip, giving a grayish appearance. The hands, feet, and tail are uniformly white to buffy in all forms except one, in which the bicolored tail is drab gray above and white below. The tail is moderately well haired, but the hairs are too scanty to hide the coarse scales. In *P. collinus* the tail is longer than the head and body, but in *P. monticularis* the tail is shorter than the head and body. The ears are rather long, especially in *P. collinus,* and the legs are short. The short, broad feet have the normal number of digits. The upper incisor teeth are not grooved. Females usually have six mammae but occasionally have only four.

Rock mice inhabit dry, barren mountains; they prefer areas where large quantities of loose boulders and rocky outcrops predominate. They are nocturnal, hiding by day under rocks or in crevices and stealthily creeping between the boulders at night. Reports indicate that rock mice are omnivorous in diet.

In a study of *P. collinus* in the Namib Desert, Withers (1983) found a 6 ha. site supporting 30–40 individuals in June and 22 in February. Reproduction was highly seasonal, with pregnancies peaking in summer (January), prior to extensive rainfall. Females probably produced one litter annually, with

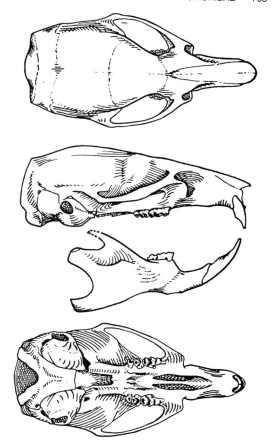

Skull of groove-toothed forest mouse *(Leimacomys buettneri)*, from *The Rodents of West Africa*, Rosevear, British Museum (Natural History), 1969.

Rock mouse *(Petromyscus collinus)*, photo by F. Petter.

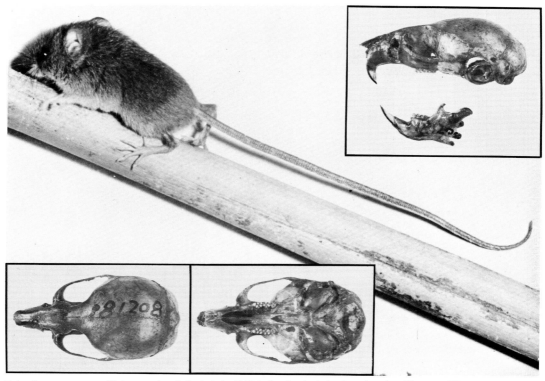

Delany's swamp mouse *(Delanymys brooksi)*, photo by F. Dieterlen. Inset: photos from American Museum of Natural History.

two or three young. A female *P. collinus* collected in September had two embryos.

RODENTIA; MURIDAE; **Genus DELANYMYS**
Hayman, 1962

Delany's Swamp Mouse

The single species, *D. brooksi*, occurs in extreme southwestern Uganda, in Rwanda, and in the mountainous area near Lake Kivu in eastern Zaire (F. Petter, *in* Meester and Setzer 1977; Van der Straeten and Verheyen 1983).

Head and body length is 50–63 mm, tail length is 87–111 mm, and weight is 5.2–6.5 grams (Delany 1975). The upper parts are a warm russet, the chin is whitish, and the remaining underparts are a warm buff. The bases of the hairs, both dorsally and ventrally, are grayish. *Delanymys* and *Petromyscus* are placed together in the subfamily Petromyscinae because of the similar structure of their molar teeth, but the two genera are very different in external appearance and habitat (Kingdon 1974*b*).

Delanymys has been taken at elevations of 1,700–2,625 meters, usually in vegetation associated with marshes within bamboo and montane forests (Delany 1975; Kingdon 1974*b*). It is nocturnal and a very good climber. It constructs a small, round, grass nest with two entrances. One such nest was located 50 cm above the ground in a bush. The diet consists mainly of seeds. Two pregnant females had three embryos each, and a nest found in June contained four blind young.

RODENTIA; MURIDAE; **Genus BEAMYS**
Thomas, 1909

Long-tailed Pouched Rat

The single species, *B. hindei*, is known from the coast of Kenya, Tanzania, Malawi, and northeastern Zambia (Ansell and Ansell 1973; Kingdon 1974*b*). *B. major* of southwestern Tanzania, Malawi, and Zambia was considered by Schlitter (1989) to be a distinct species.

Head and body length is 130–87 mm, tail length is 100–155 mm, and weight is 55–150 grams (Kingdon 1974*b*). Coloration is grayish or grayish brown above and white below. *Beamys* is distinguished externally by the naked, white-tipped tail of medium length. The underside of the tail is dark on the proximal quarter and whitish for the rest of its length. The tail is flattened, with sharp edges, and the lower side is wider than the upper, so that in cross section its shape resembles that of a truncated pyramid. The tail is not noticeably scaly, but only the basal centimeter is hairy. Like the other two genera of the subfamily Cricetomyinae, *Cricetomys* and *Saccostomus*, *Beamys* has cheek pouches.

According to Kingdon (1974*b*), *Beamys* lives in forest and moist woodland, from sea level to 2,100 meters. It must live near water on soft and sandy soil. It constructs burrows up to 9 meters in length and about 60 cm deep. The burrows consist of a vertical entrance tunnel, passages with enlarged chambers for nesting and food storage, and at least one cul-de-sac, the end of which is used as a latrine. The nest is made of dry vegetation. *Beamys* is completely nocturnal and mainly ter-

Long-tailed pouched rat *(Beamys hindei)*, photo from Nyasaland Museum through P. Hanney.

restrial, but it is a confident climber and harvests some food that is still attached to plants. Intensive food collection and storage activity continues throughout the year. The diet consists mainly of seeds and fruits.

Kingdon (1974b) added that burrows are occupied by a single adult but that burrows of a male and female may be only a few meters apart and that usually a number of burrows occur in the same vicinity. In Malawi, breeding has been found to coincide with the rainy season. Litter size in the wild is reportedly 4–7 young. In a study of captives, Egoscue (1972) found females to be reproductively active throughout the year and to produce up to five litters annually. The gestation period was 22–23 days, and litter size averaged 2.8 (1–5) young. The young weighed an average of 3.2 grams at birth, opened their eyes at about 3 weeks, and were completely weaned at 5–6 weeks. The youngest breeding female gave birth to her first litter at 5.5 months, but most females and males did not reach sexual maturity until another 1.5–3.0 months had passed. One female produced two litters when she was at least 3 years old.

Schlitter (1989) classified *B. hindei* (and *B. major*) as vulnerable. These animals are dependent on dwindling forests and now can be found only in scattered areas of remnant habitat.

RODENTIA; MURIDAE; Genus SACCOSTOMUS
Peters, 1846

African Pouched Rats

Two species now are recognized (Hubert 1978a):

S. campestris, Angola, Zambia, Malawi, Mozambique, Zimbabwe, Botswana, South Africa;
S. mearnsi, Ethiopia, Somalia, Uganda, Kenya, Tanzania.

On the basis of preliminary electrophoretic data and the distribution of chromosomal forms, Gordon (1986) reported that *S. campestris* probably comprises at least two species.

Head and body length is 94–188 mm, tail length is 30–81 mm, and weight is 40–85 grams (Delany 1975; Hubert 1978a; Kingdon 1974b). The pelage is quite long, dense, and fine in texture. The upper parts are gray or brownish gray, being darker on the back than on the sides. The underparts are white in *S. campestris* and gray in *S. mearnsi*. The tail, lightly covered with short hairs, is usually bicolored (dark on top and pale below).

African pouched rats have a robust body; a broad, thick head; and short, strong legs and toes. The ears are short and rounded, and the eyes are small. The tail is short and thick at the base. *Saccostomus* derives its common name from the cheek pouch that opens inside the lips at each side of the head and extends back to about the shoulders. The incisor teeth are not grooved. Females have 10 or 12 mammae.

African pouched rats inhabit savannahs, scrubby areas, grassy places in open forests, cultivated fields, and sandy plains. They are nocturnal, terrestrial, and rather slow-moving. They may utilize holes dug by other animals or dig their own simple burrows, which often have two entrances and a single enlarged sleeping and storage chamber (Kingdon 1974b). At least in the southern portion of their range, they accumulate seeds during the summer for use in the winter. This food is transported from the field to the burrow in the cheek pouches. The diet consists of seeds, berries, grains, acacia nuts, and occasionally insects.

According to Kingdon (1974b), burrows are often in close proximity, but each contains only a single adult. Breeding evidently continues throughout the year in East Africa. Smithers (1971) stated that in Zambia pregnant females were taken from January to April, and juveniles in September, October, and February; the number of embryos ranged from 5 to 10. According to Delany (1975), there may be as few as 2 young per litter. A captive *S. campestris* lived for 2 years and 9 months (Jones 1982).

African pouched rats *(Saccostomus campestris)*, photo by Ernest P. Walker.

RODENTIA; MURIDAE; **Genus CRICETOMYS**
Waterhouse, 1840

African Giant Pouched Rats

There are two species (F. Petter, *in* Meester and Setzer 1977):

C. gambianus, Senegal to central Sudan, and south to
  South Africa;
C. emini, Sierra Leone to Lake Tanganyika, island of Bioko
  (Fernando Poo).

Head and body length is 240–450 mm and tail length is 365–460 mm. Smithers (1983) listed weights of 1.0–2.8 kg for males and 0.96–1.39 kg for females. The fur is short and thin; in some forms it is coarse and harsh, but in others it is relatively fine and sleek. The coloration of the middorsal region ranges from dark grayish brown to medium grayish with a tinge of brown or clear reddish brown. The general coloration becomes paler on the sides, sides of the face, and flanks, ranging from soft gray with a brownish tinge to reddish brown, vinaceous, or buffy. The underparts are soiled white, white, or creamy. The conspicuous ears are practically naked. The tail is almost naked; the dorsal two-thirds are dark

African giant pouched rat *(Cricetomys gambianus).* Inset: animal with its cheek pouches well filled. Photos from New York Zoological Society.

grayish, and the remainder is soiled white or soiled creamy. Many individuals are mottled with gray, or almost spotted, mainly on the anterior half of the upper parts. This mottling may be inconspicuous, may form irregular small spots or large blotches, or may cover almost all of the upper parts. In some forms there is a fairly definite, almost white stripe across the back just behind the shoulders.

Cricetomys has a rather long, narrow head; ungrooved incisor teeth; cheek pouches; and a scaly tail. The eyes are very small, which is unusual in a nocturnal animal, and this condition suggests that *Cricetomys* relies more on its senses of smell and hearing than on vision (Smithers 1983). Females have eight mammae.

These rats dwell in forests and thickets. They are nocturnal but have been seen foraging during the day, at which time they behave as if they were almost blind, sitting on their haunches and sniffing in all directions. They can climb well, using the long tail for balancing, and can swim. For shelter they often use natural crevices and holes, termite mounds, or hollow trees, but they can dig their own burrows, which consist of a long passage with side alleys and several chambers, one for sleeping and the others for storage (Kingdon 1974b). The burrows have two to six openings and are frequently located at the base of a tree or among dense vegetation; the entrances are often closed from the inside with leaves. Smithers (1983) reported that in Zimbabwe many less complex burrows have been found, consisting of a straight tunnel, some not more than 1.5 meters long, with a single nesting chamber at the end.

Ajayi (1975) stated that these rats are completely omnivorous, feeding on vegetables, insects, crabs, snails, and other items but apparently preferring palm fruits and palm kernels. Ewer (1967) stated that the cheek pouches are used to carry food and bedding material and that there is regular coprophagy. According to Kingdon (1974b), these rats store a considerable amount of food, as well as many nonedible items, such as coins, metal, and bits of cloth.

Cricetomys is generally solitary; a captive male and female may be kept together, but two mature males may fight to the death (Ewer 1967). Reproduction takes place at various times of the year. In a study of captives, Ajayi (1975) observed one female to give birth 5 times in 9 months, and he thought that probably females could produce 10 litters annually. The gestation period was 27–36, usually 30–32, days. Litters numbered 1–5, most commonly 4, young. Sexual maturity was attained at about 20 weeks. According to Jones (1982), a captive specimen lived for 7 years and 10 months.

These animals are timid but soon become tame in captivity and make delightful pets. They are in great demand as food by the native tribes. Studies have been made of their potential for domestic production of food. In some West African towns, *Cricetomys* has become a sewer rat and is killed along with *Rattus* by the rat catchers. In contrast, *C. gambianus* is classified as rare in South Africa (Smithers 1986).

RODENTIA; MURIDAE; **Genus LENOTHRIX**
Miller, 1903

Gray Tree Rat

The single species, *L. canus*, is known from the Malay Peninsula, Sarawak, and Tuangku Island off northwestern Sumatra (Musser 1981a). *Lenothrix* long was considered to be only a subgenus of *Rattus* and to comprise a number of species in addition to *L. canus* (Laurie and Hill 1954). Misonne (1969), however, explained that *Lenothrix* has no close relationship to *Rattus*, is one of the most primitive genera in the sub-

family Murinae, and includes only the species *L. canus*. This view was supported by Medway (1977, 1978), Medway and Yong (1976), Musser and Boeadi (1980), and Musser and Newcomb (1983). Nonetheless, on the basis of an electrophoretic analysis, Chan, Dhaliwal, and Yong (1978) supported retention of *L. canus*, as well as other species formerly assigned to the subgenus *Lenothrix*, within *Rattus*.

Head and body length is 165–220 mm, tail length is 190–265 mm, and weight is 81–273 grams. The soft, rather woolly fur lacks spines. The upper parts are gray or brownish gray, and the underparts are white. The tail is pigmented above and below at the base and is distally unpigmented. The only other rat having a tail with this same color pattern is *Berylmys bowersi*, and that species is larger than *L. canus. Lenothrix* is distinguished from *Rattus* by dental characters, most notably the presence of a rudimentary extra cusp or posteriorly directed fold of enamel in the midline of the posterior face of the transverse ridges of the first and second upper molars. Female *Lenothrix* have 10 pairs of mammae (Musser 1981a).

According to Medway (1978), the gray tree rat is locally common on the Malay Peninsula in plantations and primary and secondary forests. It is arboreal. Litter size averages three and ranges from two to six young. The estimated mean life span in the wild is five months, but record longevity in captivity is three years and nine months.

RODENTIA; MURIDAE; **Genus PITHECHEIR**
F. Cuvier, 1842

Monkey-footed Rats, or Malay Tree Rats

There are two species (Musser and Newcomb 1983):

P. parvus, Malay Peninsula;
P. melanurus, Java.

Chasen (1940) indicated that *P. melanurus* also occurred in Sumatra, but Musser (1982c) regarded the single record from that island to be incorrect. Misonne (1969) considered *Pithecheir* a close relative of *Lenothrix*, and Musser and Newcomb (1983) also suggested affinity.

Head and body length is 122–80 mm, tail length is 157–215 mm, and weight is 58–146 grams (Lim and Muul 1975; Muul and Lim 1971). The long, soft, dense pelage is reddish to tawny on the upper parts but becomes buffy on the sides of the abdomen and rump and blends into pure white on the underparts. The bases of the individual hairs on the back are mouse gray. The scantily haired tail, ears, hands, and feet are usually reddish, though in some animals the feet are white. The nails are horn brown, and the whiskers are long and numerous.

The hind foot is modified for arboreal life. The first toe on the hind foot is thumblike and opposable, hence the origin of the scientific name, which means "ape hand." The pollex is very small, the tail is prehensile, and the incisor teeth are moderate. Females have four mammae.

Monkey-footed rats are found in dense forests up to about 1,600 meters in elevation. They build globular nests of leaves or moss in the branches or hollows of trees. Two nests found in Java were about 150 mm in diameter and located about 2 and 3.5 meters above the ground. Lim and Muul (1975) reported that captives were active only at night and spent most of their time climbing. The movements of these rodents are quite slow and cautious. The tail is used for grasping as the animals climb and scamper about limbs and branches. The

Malay tree rat *(Pithecheir melanurus)*, photo by Lim Boo Liat. Inset: photo from Museum Zoologicum Bogoriense.

diet may be mostly green plant material, though a captive lived on plantains and crickets.

Medway (1978) stated that pregnancies have been recorded at all times of the year on the Malay Peninsula and that normally two young are born. In Java, single young have been noted from April to September. Lim and Muul (1975) reported that some captives bred four times per year, that the young are carried about attached to the mother's nipples until they are weaned, and that captive specimens have lived up to three years.

RODENTIA; MURIDAE; **Genus HAPALOMYS**
Blyth, 1859

Asiatic Climbing Rats, or Marmoset Rats

There are two species (Musser 1972*a*):

H. longicaudatus, southeastern Burma, southern Thailand, Malay Peninsula;
H. delacouri, Indochina, Hainan.

Head and body length is 121–68 mm and tail length is 140–203 mm. In some forms the tail is considerably longer than the head and body, but in others it is only slightly longer. The fur is thick and soft. The coloration above is buffy, dull reddish gray, or grayish brown; the sides are generally paler than the back; and the underparts, including all four legs to below the knee, are white. The limbs are usually buffy gray, and the hands and feet are buffy gray or brownish white. The tail is thinly haired and sometimes penciled terminally. The vibrissae are prominent and slightly longer than the head.

The hind feet are specialized for arboreal life in that they have enlarged toe pads consisting of two flat plates with a groove between them. The toes are long and slender, the middle three being almost equal in length; the large toe is wide, without a claw, and opposable. There are four fingers; the second, third, and fourth each have a small claw nearly embedded in the pad, and the first finger is said to be little more than a slight projection on the inner side of the hand and to have no trace of a nail. The general body form of *Hapalomys* is ratlike. Females have eight mammae.

The incisor teeth are broadened and powerful. *Hapalomys* is the only murid genus with three rows of approximately equally developed cusps on the lower cheek teeth, a character that some authorities consider to be very primitive.

These mice are arboreal and inhabit tropical forests. According to Medway (1964, 1978), *H. longicaudus* is a skillful climber and appears to be associated strictly with bamboo. By day it retires to a nest in the internodal cavities of dead or living bamboo stems. It gains access to the internodes by gnawing a circular hole about 35 mm in diameter in the outer

Asiatic climbing rat *(Hapalomys longicaudatus)*, photos by Jane Burton. Inset: skull, photo by P. F. Wright of specimen in U.S. National Museum of Natural History.

wall. All nests examined have been lined exclusively with bamboo leaves. The natural diet appears to consist of the shoots, flowers, and fruit of bamboo.

Muul (1989) indicated that the survival of *H. longicaudus* is threatened through human destruction of stands of bamboo. Wang, Zheng, and Kobayashi (1989) designated *H. delacouri* as endangered, also because of habitat loss.

RODENTIA; MURIDAE; Genus APODEMUS
Kaup, 1829

Old World Wood and Field Mice

There are 2 subgenera and 14 species (Corbet 1978, 1984; Dolan and Yates 1981; Xia 1985):

subgenus *Sylvaemus* Ognev, 1924

A. mystacinus, Balkan Peninsula, Asia Minor and adjacent parts of Georgia and Iraq, Palestine, Crete, Rhodes, several Aegean islands;
A. flavicollis, most of Europe, Asia Minor, Palestine;
A. sylvaticus, Europe, parts of central and southwestern Asia, Himalayas, northwestern Africa, British Isles and many nearby islands, Iceland;
A. krkensis, Krk Island off northwestern Yugoslavia;
A. peninsulae, eastern and southern Siberia, Manchuria,

Korea, northeastern and central China, Sakhalin, Hokkaido;
A. draco, southern and eastern China, Assam, Burma, Taiwan;
A. speciosus, Japan;
A. navigator, Honshu, Dogo, and Kunashir islands;
A. argenteus, Japan;
A. latronum, Sichuan and Yunnan (southern China), northern Burma;
A. microps, east-central Europe;
A. gurkha, Nepal;

subgenus *Apodemus* Kaup, 1829

A. agrarius, central and eastern Europe, parts of Central Asia and southern Siberia, Manchuria, Korea, eastern and southern China, Taiwan;
A. chevrieri, central China.

Head and body length is 60–150 mm, tail length is 70–145 mm, and weight is 15–50 grams. The fur is usually soft, though it may be bristly in *A. speciosus,* and the tail is moderately haired. The general coloration above is grayish buff, grayish brown, brown mixed with yellow or red, light brown, or pale sand color. The underparts are white or grayish, often suffused with yellow, and the hands and feet are usually white. Some forms have a reddish yellow chest patch, and *A. agrarius* has a black middorsal stripe. The tail may be longer or shorter than the head and body. The tail is not prehensile

Yellow-necked mouse *(Apodemus flavicollis)*, photo by K. Rudloff through East Berlin Zoo.

like that of *Micromys,* a genus that resembles *Apodemus* in general appearance. Female *Apodemus* have six or eight mammae.

These mice inhabit grassy fields, cultivated areas, woodlands, and forests. They climb well, though not as well as *Micromys,* and are more active jumpers. They are also good swimmers. Grzimek (1975) wrote that while *A. agrarius* is diurnal, *A. flavicollis* and *A. sylvaticus* are nocturnal or crepuscular. They may move into human habitations in the fall and winter but generally dig deep burrows, usually with a nest of shredded grass and leaves at the end of the tunnel. *A. flavicollis* sometimes nests between roots, in rocky crevices, or in hollow trees. Jennings (1975) stated that burrows of *A. sylvaticus* are usually about 3 cm wide and 8–18 cm below the surface; tunnels of other animals are sometimes incorporated. All burrow systems that he examined consisted of a circular tunnel around the roots of a hazel tree, another tunnel penetrating below the tree to a nest chamber, and other tunnels leading out to entrances. As has been observed in some other species, the burrows contained stores of food. The diet of *Apodemus* includes roots, grains, seeds, berries, nuts, and insects.

These rodents normally spend their life within an area about 180 meters in diameter. In a study of *A. sylvaticus* in Ireland, Fairley and Jones (1976) found average range lengths of about 109 meters for males and 64 meters for females. For the same species in England, Crawley (1969) calculated average home ranges of 2,250 sq meters for males and 1,817 sq meters for females. Over a seven-year period in England, Montgomery (1976) found population density to be 8.7–40.1 per ha. in *A. flavicollis* and 2.9–18.2 per ha. in *A. sylvaticus.* The latter species may undergo periodic fluctuations in population. Babinska-Werka, Gliwicz, and Goszczynski (1981) re-

ported an average density of 54/ha. for a population of *A. agrarius* in an old cemetery in Warsaw.

Jennings (1975) observed possible cooperative burrowing in *A. sylvaticus.* In this species, at least, several adults may live in the same nest, though females apparently do not permit the males to enter when young are present. The breeding season of *Apodemus* may vary from year to year, and females produce up to six litters annually. While Larsson, Hansson, and Nyholm (1973) reported winter reproduction for *A. sylvaticus* in north Sweden, Grzimek (1975) stated that both this species and *A. flavicollis* generally breed from March to September. They usually produce four litters annually, averaging 5 young each, after a gestation period of 23 days. The young of these species weigh about 2.5 grams at birth, open their eyes 13 days later, are weaned at 3 weeks, and reach sexual maturity at 2 months. In *A. agrarius* the gestation period is 21–23 days and litter size averages 6 young. In a study of *A. microps* in Czechoslovakia, Holisova, Pelikan, and Zejoa (1962) found the reproductive season to be late February to early October, most pregnancies to occur in March–August, and litter size to average 6.1 and range from 2 to 10 young. Lay (1967) reported a range of 2–10 embryos in pregnant female *A. sylvaticus* in Iran. A captive individual of this species lived for 4 years and 5 months (Jones 1982), but average life span in wild members of the genus is probably 1 year or less.

In woodland ecosystems of Great Britain, these mice have been found to be important as agents for the transportation and burying of tree seeds, as well as for damage and destruction of seedlings. Their influence on regeneration of forests is complex and needs further detailed study (Ashby 1967). *A. speciosus,* of Japan, is a carrier of scrub typhus, and *A. agrarius* is a possible carrier of hemorrhagic fever.

Old World field mouse *(Apodemus agrarius),* photos by Ernest P. Walker.

Pencil-tailed tree mouse *(Chiropodomys gliroides)*, photo by Ernest P. Walker.

RODENTIA; MURIDAE; **Genus RHAGAMYS**
Major, 1905

Hensel's Field Mouse

The single species, *R. orthodon*, formerly occurred on Corsica and Sardinia (E. Anderson 1984; Kurten 1968). It is thought to be related to *Apodemus* but was larger. Remains of the species have been found in caves, and it appears to have been common from the middle Pleistocene until well into the postglacial period. According to Vigne (1987), *Rhagamys* appears to have been evolving larger size through the end of the Pleistocene but then to have undergone a sharp reduction in size following the arrival of modern humans and their domestic animals. The ecological disturbances and predation that resulted from this invasion led to the extinction of *Rhagamys* about 2,000 years ago.

RODENTIA; MURIDAE; **Genus CHIROPODOMYS**
Peters, 1868

Pencil-tailed Tree Mice

There are six species (Jenkins and Hill 1982; Musser 1979; Wu and Deng 1984):

C. karlkoopmani, Siberut and North Pagi islands off
    Sumatra;
C. major, northern Borneo;
C. calamianensis, Busuanga, Palawan, and Balabac islands
    (Philippines);
C. muroides, northeastern Borneo;
C. gliroides, Assam to southeastern China and Malay
    Peninsula, Sumatra, Java, Bali, Borneo, Natuna Islands;
C. jingdongensis, known only from the type locality at
    Jingdong in Yunnan.

Musser and Newcomb (1983) suggested that *Chiropodomys* is closely related to *Hapalomys* and perhaps to *Haeromys*.
Head and body length is 66–122 mm, tail length is 85–171 mm, and weight is 15–43 grams (Musser 1979). The pelage is soft, dense, and uniform in length, without conspicuous guard hairs or spines. The general coloration of the upper parts is dull grayish brown, buffy brown, or chestnut, and that of the underparts is white to gray or orange red. The tail is brown, except in *C. karlkoopmani,* in which the basal third

of the tail is brown and the remainder is white. The tail is rather thinly covered with short hairs near the base, becoming more or less penciled terminally.
These rodents have short, broad feet and a long tail. The first digit of both the hands and the feet is short and stumpy and has a flat nail. The other digits have short, slightly curved claws. The digit pads are large, and the sole of the hind foot is naked. The ears are moderately large, thin, and nearly naked. Females usually have four mammae.
Pencil-tailed tree mice occur in forests and are primarily arboreal. The best-known species is *C. gliroides* (Medway 1978; Musser 1979). It inhabits a variety of forest types and is especially common where there is bamboo. It is most active at night and spends the day in nests in hollow trees or the internodes of bamboo. To reach the internodal space of a standing bamboo stem, the mouse gnaws a circular hole 25 mm in diameter in the side of the internode. The opening is then lined with leaves. *C. gliroides* is apparently herbivorous, but little is known about its diet in the wild. Captives have thrived on mixed roots, sweet potatoes, grains, and fruits, occasionally supplemented with fresh bones and raw meat. Individuals are very fierce and are not easy to tame even when born and reared in captivity. Females are polyestrous, with estrous periods of 1 day recurring at intervals of not less than 7 days. One female produced four litters in a 10-month period. Breeding occurs throughout the year in the Malay Peninsula but apparently peaks in September–March. A gestation period lasted 19–21 days. Litter size is 1–4, averaging 2.4, young. The newborn cling persistently to their mother's nipples and will be dragged about if she is disturbed. They are partly independent at 17 days, completely weaned at 1 month, and fully mature at 100 days. Mean longevity in the wild has been calculated as 23.8 months, but captives have lived as long as 43 months.

RODENTIA; MURIDAE; **Genus VERNAYA**
Anthony, 1941

Vernay's Climbing Mice

There are two species (Musser 1979; Wang, Hu, and Chen 1980):

V. fulva, known only by two specimens from northern
    Burma and one from Yunnan (southern China);
V. foramena, Sichuan (central China).

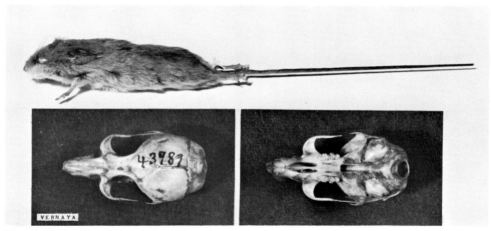

Vernay's climbing mouse *(Vernaya fulva)*, photos from American Museum of Natural History.

It is possible that other specimens of *Vernaya* are in collections under the name *Chiropodomys, Vandeleuria,* or *Micromys,* as these genera are so similar in appearance that they can be distinguished only by careful examination.

In *V. fulva* head and body length is about 90 mm and tail length is about 115 mm. This species resembles *Vandeleuria* in the nature of the pelage, the reddish color of the coat, and the long, nontufted tail. In *V. foramena* the tail is also longer than the head and body, the general color is cinnamon brown, and the dorsal pelage is long, thick, and velvety.

Vernaya differs from *Vandeleuria* and *Chiropodomys* in the development of the digits of the hands and feet. In *Vernaya* all digits except the pollex, which is vestigial and has an extremely small, flat nail, have characteristic pointed claws. The incisor teeth are not grooved. Female *V. foramena* have six mammae.

The specimens of *V. fulva* were collected at elevations of around 2,100–2,700 meters; the Burmese material was taken in areas of low, dense vegetation among low cliffs and rocky outcrops (Musser 1979). The specimens of *V. foramena* were collected in subalpine coniferous forest at elevations of 2,460–2,500 meters (Wang, Hu, and Chen 1980).

Long-tailed climbing mouse *(Vandeleuria oleracea)*, photo by Rita Maser.

RODENTIA; MURIDAE; **Genus VANDELEURIA**
Gray, 1842

Long-tailed Climbing Mice

There are apparently two species (Lekagul and McNeely 1977; Musser 1979):

V. oleracea, western India to southern China and Viet Nam, lowlands of Sri Lanka;
V. nolthenii, highlands of Sri Lanka.

Head and body length is about 55–85 mm and tail length is usually 90–130 mm. Weight is approximately 10 grams (Lekagul and McNeely 1977). The upper parts vary from pale buffy through dull brown to dark reddish brown, and the underparts are white or cream color. The fur is full, soft, and almost silky. The tail is fairly well haired but is not tufted at the tip as in *Chiropodomys*.

This genus of small tree mice is distinguished by the structure of the hands and feet and by features of the teeth and skull. The first and fifth digits on both the hand and the foot have a flat nail instead of a claw; the claws on the remaining digits are small. The limbs are adapted to grasping and climbing by the opposition of the first digit on both the hand and the foot, the shortness of the foot, and the development of the terminal pads of the fingers and toes. This genus may be distinguished from *Chiropodomys*, which has similar hands and feet, by its mouselike size and nontufted tail. Females of *Vandeleuria* have eight mammae.

These mice are nocturnal and essentially arboreal. They usually spend the day in a nest in the branches of a tree, in a hole in a tree trunk, or, occasionally, on rooftops. The nest is made of grass, sometimes mixed with dried leaves. In Thailand they may live in tall cane, where they build a globular nest of grass a couple of meters above the ground (Lekagul and McNeely 1977). These mice are quite active as they run among branches and twigs, using the tail as a balancing organ while they run up or down vertical shoots and out to the tips of slender twigs. Sometimes the long, almost prehensile tail is wound loosely around a twig to steady the animal. Long-tailed climbing mice are much less active and are slower on the ground, where they come only to collect leaves and grasses for their nests.

The diet in the wild is probably made up of fruits and the buds and shoots of trees and shrubs. An individual kept in captivity thrived on bread, milk, and such fruits as plantains and pawpaws. This captive specimen became fairly tame after a few weeks and spent most of the day asleep with its tail curled around its body.

Apparently, the older males live alone. The young are born and raised in a nest. Usually three or four are born at one time, but litters of six young have been seen.

RODENTIA; MURIDAE; **Genus MICROMYS**
Dehne, 1841

Old World Harvest Mouse

The single species, *M. minutus*, occurs from western Europe to east-central Siberia and Korea, in parts of southern China and Assam, in Great Britain and Japan, and on Taiwan (Corbet 1978, 1984).

This is among the smallest of rodents. Head and body length is 55–75 mm, tail length is 50–75 mm, and weight is usually 5–7 grams. The pelage is brownish with a yellowish

Old World harvest mice *(Micromys minutus)*, photo by Eric J. Hosking.

or russet tinge on the upper parts and white to buffy beneath. The tail is bicolored. The fur is somewhat longer in winter than in summer. Typical features include the short, rounded head; small, rounded ears; fairly broad feet; and the hairless condition of the upper portion of the tip of the tail. The foot structure facilitates scampering up stems, and the nakedness of the tail suggests that it is semiprehensile.

The remainder of this account is based in large part on the review papers by Trout (1978*a*, 1978*b*). The Old World harvest mouse usually is found in tall vegetation, such as hedgerows, weed beds, tall grass, reeds, grain or rice fields, and bamboo thickets. During the spring and summer breeding season, special nests are constructed, one for each litter of young. These globe-shaped nests, about 60–130 mm in diameter, usually are suspended between vertical grass stems about 100–130 cm above the ground. They are composed of three layers of grass leaves compactly woven together. The leaves of the inner layer are finely shredded to form a soft lining for the young. There are one or more entrance holes, but these are kept closed by the female during the first week following birth. Such a nest requires 2–10 days to build. Nonbreeding animals may construct similar nests, but these are more flimsy and lack the inner lining. *Micromys* sometimes builds its nest on the ground or in a hole, especially during cooler months, and at such times it also may live under haystacks or straw or in structures made by people. It does not hibernate and may be active by day or night; however, most investigations have found it to be primarily nocturnal, with peaks of activity after dark and before dawn. Its diet consists mainly of seeds, green vegetation, and insects, but it is known sometimes to take the eggs of small birds and to eat meat in captivity. Storage of food has been noted in captivity but has not been confirmed in the wild.

Micromys normally lives within small, overlapping home ranges. One study found the individual range size to average 400 sq meters for males and 350 sq meters for females. Population levels fluctuate over the years and seasonally, generally reaching a peak in the autumn. Maximum annual densities during a three-year period at three different sites in England varied from 17/ha. to 207/ha. During the winter, however, large numbers of mice may congregate in a barn or grain storage facility. As many as 5,000 individuals have been found in such structures. Studies have shown that the animals tolerate one another under such conditions but that aggression increases as density declines and the breeding season approaches. Two captive males placed together will fight savagely. Adults of opposite sexes come together only to construct a breeding nest and to mate; the female then drives the male away. Vocalizations include a chatter during courtship and an aggressive squeal.

Reproduction is generally concentrated during the warmer, drier months, starting around April or May and peaking from July to September but sometimes lasting until late autumn. Females are polyestrous, undergo a postpartum estrus, and under favorable conditions can give birth several times in rapid succession. Because of the short natural longevity, a female normally produces only one or two litters in its lifetime, but captives have produced up to nine. The gestation period and minimum interval between litters are about the same, 17–18 days. The number of young per litter ranges from 1 to 13, usually is around 3–8, and has averaged about 5 in England and 7 in Bulgaria. The young weigh about a gram at birth, open their eyes and are fully furred at 8–10 days, are weaned and leave the nest at 15–16 days, and reach sexual maturity at as early as 35 days. As many as four generations of mice may breed during one reproductive season. Few individuals live over 6 months; maximum known longevity is 16–18 months in the wild and just under 5 years in captivity.

Although *Micromys* is widely distributed, in areas where modern farm machinery is used this little mouse appears to be on the decline. Its food and shelter may be destroyed by reaping machines, which leave a much shorter stubble than does a manual scythe. Pucek (1989) indicated that it now is considered rare in Austria, Czechoslovakia, and Switzerland.

RODENTIA; MURIDAE; **Genus HAEROMYS**
Thomas, 1911

Pygmy Tree Mice

There are three named species (Laurie and Hill 1954; Medway 1977):

H. margarettae, Borneo;
H. pusillus, northern and eastern Borneo;
H. minahassae, northern Sulawesi.

A fourth, unnamed species occurs on Palawan Island in the Philippines (Musser 1986). *Haeromys* is a systematically primitive murid, perhaps with some affinity to *Anonymomys* (Musser and Newcomb 1983).

Haeromys is among the smallest genera of rodents. In *H. pusillus* head and body length is 56–74 mm and tail length is 108–22 mm (Musser and Newcomb 1983); head and body length of the type specimen of *H. margarettae* is 76 mm and tail length is 144 mm; and in *H. minahassae* head and body length is 72 mm and tail length is about 110 mm. The pelage is long and soft. *H. minahassae* is rufous on the upper parts, with a dull back and bright sides. In *H. margarettae* and *H. pusillus* the general coloration of the upper parts is deep chestnut rufous, with a grayish cast resulting from gray hair bases, but the coloration is clearer on the sides, where a rufous

Pygmy tree mouse *(Haeromys minahassae)*, photo by Margareta Becker through Guy G. Musser.

lateral band is formed. In *H. minahassae* the tail is brown, and in *H. margarettae* and *H. pusillus* it is greenish gray. In all species the underparts are white, the whiskers are black, and the tail is unicolored and covered with short hairs.

This genus resembles *Chiropodomys* in having a muzzle that is short and slender and ears that are relatively small, oval, and sparsely haired. The hands and feet are adapted for arboreal life; the great toe is opposable. The thumb has a large nail, whereas the other digits have short, sharp, curved claws. Females of all species have six mammae (Musser and Newcomb 1983).

Musser (1979:438) wrote: "Species of *Haeromys* inhabit tropical evergreen forests in both lowlands and mountains. They build globular nests in cavities in trees. The animals I observed in Celebes ate only small seeds, mostly from figs." Muul (1989) indicated that *H. margarettae* and *H. pusillus* are threatened by human habitat destruction.

RODENTIA; MURIDAE; Genus ANONYMOMYS
Musser, 1981

Mindoro Rat

The single species, *A. mindorensis,* is known only by three specimens from Ilong Peak, northeastern Mindoro Island, Philippines (Musser 1981*a*).

In an adult male, head and body length is 125 mm and tail length is 206 mm. The upper parts are bright buffy or tawny brown, and the underparts are cream. The tail is monocolored and tipped by a tuft. The long, dense pelage has numerous spines. The body is stocky, the ears are small and sparsely haired, and the hind feet are short and wide.

From *Rattus* and *Limnomys, Anonymomys* is distinguished in having smaller bullae, the squamosal roots of the zygomatic arches originating higher on the sides of the braincase, the posterior margin of the palatal bridge not extending far past the molar toothrow, the pterygoid fossa nearly flat and not perforated by a large foramen, and the mesopterygoid fossa at least as wide as the palatal bridge. *Anonymomys* resembles *Niviventer cremoriventer* externally but differs in having paler and less spiny pelage, a shorter and broader rostrum, a more domed braincase, much larger bullae, and relatively larger teeth.

The Mindoro rat is considered to be arboreal. All specimens were taken in a mountainous area at an elevation of about 1,350 meters.

RODENTIA; MURIDAE; Genus SUNDAMYS
Musser and Newcomb, 1983

Giant Sunda Rats

There are three species (Medway 1984; Musser and Newcomb 1983):

S. muelleri, Malay Peninsula, Sumatra, Borneo, Palawan (Philippines), and many small nearby islands;
S. maxi, western Java;
S. infraluteus ("giant rat of Sumatra"), Sumatra, Sarawak and Sabah (northern Borneo).

These species generally were placed in *Rattus,* and sometimes in the subgenus or separate genus *Bullimus,* until Musser and Newcomb demonstrated that they belong in a

Giant Sunda rat *(Sundamys muelleri),* drawing by Fran Stiles through Guy G. Musser.

separate, more primitive genus, perhaps with affinity to *Kadarsanomys.*

Head and body length is 179–299 mm, tail length is 175–370 mm, and weight is 206–643 grams. The fur is thick and slightly harsh to the touch above. The upper parts are dark brown or brownish gray, the sides are paler, and the usually well-demarcated underparts are white, cream, buff, or grayish. The ears are small, round, inconspicuously haired, and dark brown. The feet are brown, and the tail is dark brown. The body is chunky, the tail generally longer than the head and body, the cranium large, the braincase long, the rostrum long and wide, the auditory bullae relatively very small, and the molar teeth large. There are three pairs of mammae in female *S. infraluteus* and four pairs in the other species.

From *Rattus, Sundamys* is distinguished by a combination of characters, including larger body size, slightly swollen rostrum, alisphenoid canal concealed behind a lateral strut of alisphenoid bone (rather than being an open channel as in *Rattus*), very small bullae, and large and stocky molars, the third upper and lower ones being relatively much larger when compared with the other teeth. From *Bullimus, Sundamys* differs in having a much wider rostrum, a less constricted interorbital region, a narrower braincase, and less chunkier and lower-crowned molars.

Giant Sunda rats are found in tropical forests from coastal lowlands to elevations of about 2,700 meters. They swim readily and seem to prefer marshy areas, valley bottoms, and stream sides but also occur in cultivated fields, where they reportedly become serious pests. Activity is largely nocturnal and terrestrial, with daylight hours spent in holes or under logs, but some individuals have been found on tree branches.

The diet includes fruits, leaves, shoots, insects, crabs, land snails, and small lizards.

Individuals may maintain a permanent den within an elongated territory beside a stream. Four such areas were 76, 91, 137, and 183 meters long. According to Medway (1978), the average lifetime home range of *S. muelleri* is 409 meters. The young of that species are born in a nest built in a sheltered position at or above ground level. Pregnant females were found in all months on the Malay Peninsula, most frequently from July to September and least from January to March. Litter size ranged from 1 to 9 young and averaged 3.8. Longevity averaged only 5.6 months in the wild but has exceeded 2 years in captivity.

RODENTIA; MURIDAE; Genus KADARSANOMYS
Musser, 1981

The single species, *K. sodyi*, is known by 17 specimens from the southwestern slopes of Gunang Pangrango-Gede, a volcanic massif in western Java. Fossil fragments have been found in central Java (Musser 1982d). Although *sodyi* was long considered only a subspecies of *Lenothrix canus*, Musser (1981c) made it the basis of an entirely new genus which he considered not to be closely related to any other murid genus. Musser and Newcomb (1983) indicated distant affinity to *Rattus*.

Head and body length is 182–210 mm and tail length is 263–305 mm. The middle of the head and back is dark brown, and this color pales to grayish brown on the sides and cheeks. The underparts are white. The tail, which is longer than the head and body, is pale brown and covered with scales. On each scale are three hairs, which are short near the base of the tail but increase in length toward the tip. The front and hind feet are long and broad. Females have four pairs of mammae.

Kadarsanomys resembles *Lenothrix* in body size and proportion but has smaller ears, shorter hind feet, a different color pattern, and different proportions of the feet and skull. *Kadarsanomys* has a thicker cranium, a shorter rostrum, a higher braincase, narrower zygomatic plates, longer incisive foramina, a narrower palatal bridge, and larger bullae.

Kadarsanomys is arboreal. All specimens were taken in forest at an elevation of about 1,000 meters. Most were caught inside bamboo stems, in which they had gnawed holes 3–4 cm in diameter. A nest of dry leaves containing four young was found in such a stem on 24 January. In addition, juveniles were caught in June.

RODENTIA; MURIDAE; Genus ABDITOMYS
Musser, 1982

The single species, *A. latidens*, is known only by one specimen from Mount Data in northern Luzon, Philippines, and one from Los Banos in the southern part of the same island.

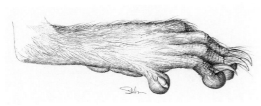

Abditomys latidens, drawing of foot by Fran Stiles through Guy G. Musser.

This species originally was assigned to *Rattus* but is now considered to represent a distinct genus, not known to be closely related to any other, though with some resemblance to *Kadarsanomys* (Musser 1982c).

Head and body lengths of the two specimens are 232 and 216 mm, and respective tail lengths are 242 and 271 mm; weight of one is 269 grams. The pelage is long and coarse but not spinous. The upper parts are brown with tawny highlights, the underparts are buffy yellow with gray suffused through the base of the coat, the ears are large and brown, the feet are brown, and the tail is completely brown and covered with inconspicuous hairs. One of the specimens is a female and has four pairs of mammae.

The skull of *Abditomys* is relatively deeper than that of *Rattus* and has much more inflated bullae and larger molar teeth. *Abditomys* also differs from *Rattus*, and from almost all other rodents of the subfamily Murinae, in that there is a prominent nail, rather than a claw, on each hallux (big toe); the other digits of the long and wide hind feet terminate in elongate, sharp claws. The following murine genera also have species with a nail on the hallux: *Hapalomys*, *Chiropodomys*, *Vandeleuria*, *Pithecheir*, *Chiromyscus*, and *Kadarsanomys*. Of these genera, *Kadarsanomys* has the greatest resemblance to *Abditomys* in cranial and dental features. Both genera have a high cranium, a small interparietal, long incisive foramina, large bullae, and large molars. *Kadarsanomys* differs, however, in having a shorter and relatively broader rostrum and a more expansive braincase with straighter sides. In addition, the nail on the hallux of *Abditomys* is long and pointed, while that of *Kadarsanomys* is small and truncate.

The specimen of *Abditomys* from Los Banos was taken in an area of intensive rice cultivation at an elevation of not more than 60 meters. The specimen from Mount Data, however, was collected in a dense, mossy forest at about 2,000 meters. The long tail, large hind feet, and sharp claws and nails of the genus are among a number of features suggesting that it is highly arboreal and, like *Kadarsanomys*, associated with bamboo. The massive structure of the skull and teeth indicates that the animal may bite into hard substances, such as the internodes of large bamboo stems, to seek food and nesting sites. The diet may be predominately herbaceous.

RODENTIA; MURIDAE; Genus DIPLOTHRIX
Thomas, 1916

The single species, *D. legata*, is found on the Ryukyu Islands south of Japan. Thomas (1906a) originally placed this species in the genus *Lenothrix*, along with *L. canus*, but subsequently (1916) erected for it the separate genus *Diplothrix*. Ellerman and Morrison-Scott (1966) questionably arranged *Diplothrix* as a subgenus of *Rattus*, but Misonne (1969) indicated that *Diplothrix* is a distinct genus related to *Lenothrix*. Musser and Boeadi (1980) stated that *Diplothrix* is a good genus and is not closely related to *Rattus*.

Head and body length of the type specimen is 230 mm and tail length is 246 mm. The fur is very long and thick. The general color above approaches clay color but is more grayish. The underparts are dirty grayish. The upper surfaces of the hands and feet are dark brown. The tail is evenly well haired throughout its length; its color is dark brown on the basal three-fifths and white beyond. The ears are short and thinly haired, with a buffy patch behind their posterior base. From that of *Lenothrix*, the skull of *Diplothrix* differs mainly in being much broader between the parietal ridges. In addition, the posterior lamina of the third upper molar of *Diplothrix* consist of two elements, an internal and a median cusp, not of a single cusp as is the case in *Lenothrix canus*.

Diplothrix legata, photo by Shogo Asao / Nature Production, Tokyo.

RODENTIA; MURIDAE; **Genus MARGARETAMYS**
Musser, 1981

Margareta's Rats

There are three species (Musser 1981*a*):

M. beccarii, northeastern and central Sulawesi;
M. elegans, central Sulawesi;
M. parvus, central Sulawesi.

Musser (1977*b*) suggested affinity between *M. beccarii* (then known as *Rattus beccarii*) and the genus *Limnomys*, of Mindanao, but later (1981*a*) explained that the resemblance is superficial, that *Margaretamys* is not closely related to *Lim-nomys* or *Rattus*, and that it may actually have phylogenetic ties to *Lenothrix* or *Lenomys*.

In *M. beccarii* head and body length is 117–52 mm and tail length is 150–200 mm. The pelage is thick, short, and semi-spinous. The upper parts are generally grayish brown; the underparts vary from cream through yellow to dark ochrace-ous buff; there are wide, dark brown rings around the eyes; and the tail is usually brown above and paler below. In *M. elegans* head and body length is 183–97 mm and tail length is 248–86 mm. The fur is thick, long, and very soft. The upper parts are brown; the underparts are whitish gray; and the tail is brown on the proximal half to two-thirds and white on the remainder. In *M. parvus* head and body length is 96–114 mm and tail length is 154–84 mm. The pelage is dense, short, and soft. The upper parts are reddish brown; the underparts are dark gray; and the tail is monocolored (Musser 1981*a*).

From *Rattus* and *Limnomys*, *Margaretamys* is distin-guished in having a prominently penicillate tail, smaller bul-lae, the squamosal roots of the zygomatic arches originating higher on the sides of the braincase, the posterior margin of the palatal bridge not extending past the molar toothrow, the pterygoid fossa nearly flat and not perforated by a large fora-men, and the mesopterygoid fossa at least as wide as the palatal bridge. From *Niviventer*, *Margaretamys* is distin-guished in having a more penicillate tail, larger bullae, a more rounded cranium, and a much larger coronoid process relative to the size of the dentary.

Margareta's rats are arboreal forest dwellers. *M. beccarii* occurs from sea level to the upper limits of lowland tropical rainforest. *M. elegans* and *M. parvus* are found in montane forest at elevations of up to 2,300 meters.

RODENTIA; MURIDAE; **Genus LENOMYS**
Thomas, 1898

Trefoil-toothed Giant Rat

The single species, *L. meyeri*, is endemic to Sulawesi (Groves 1976; Musser 1970*c*:18).

Head and body length is 235–90 mm and tail length is 210–85 mm. The pelage is thick, soft, and rather woolly, with elongated guard hairs. The general coloration of the upper parts is fuscous to drab olive black, with a sprinkling of whitish; a dorsal line is present and is somewhat darker than the sides. The head is the same color as the body. The under-parts are buffy white, blending gradually into the color of the sides and leaving no conspicuous line of demarcation. The individual hairs of the underparts have a gray base and a white tip. The feet and hands are brownish gray. The terminal section (one-half to three-fifths) of the quite scantily haired tail is flesh-colored.

According to Musser (1970*c*), *Lenomys* differs from *Rat-tus callitrichus*, of Sulawesi, in having a much larger and more massive skull, shallower zygomatic arches, an hourglass-shaped interorbital region, shorter incisive foramina, a longer and arched bony palate, and wider incisor teeth. The mo-lariform teeth of *Lenomys* have three distinct rows of cusps; the lateral and lingual cusps are distinct and well set off from the middle row.

RODENTIA; MURIDAE; **Genus PAPAGOMYS**
Sody, 1941

Flores Giant Rat

The single living species, *P. armandvillei*, occurs only on Flores Island in the East Indies. Another species, *P. theodor-*

Margareta's rats: Left, *Margaretamys elegans;* Upper right, *M. parvus;* Lower right, *M. beccarii.* Drawings by Fran Stiles through Guy G. Musser.

Trefoil-toothed giant rat *(Lenomys meyeri)*, photo by Margareta Becker through Guy G. Musser.

verhoeveni, is known from subfossil material collected on Flores estimated to be 3,550 years old. The same deposits contained the remains of the extinct genus *Paulamys,* which seems closely related to *Papagomys* (Musser 1981*b*).

Head and body length is 410–50 mm and tail length is 330–70 mm. According to Musser (1981*b*), the upper parts are dark brown or tan, the middle of the head and body is darker than the sides, and the underparts are pale gray with a slight tan suffusion. The pelage is dense and harsh, especially on the upper parts, where it consists of flattened, flexible spines mixed with regular hairs. The stout tail, shorter than the head and body, appears naked but is covered with large

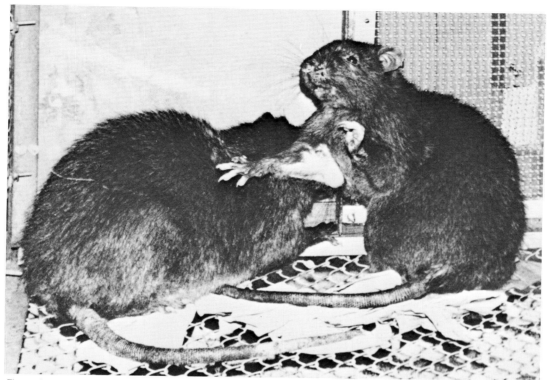

Flores giant rats *(Papagomys armandvillei)*, photo by M. Sukaeri through Guy G. Musser.

scales, each with three short, stiff hairs emerging from the base. The basal two-thirds of the tail is blackish brown, and the distal third may be either the same color, pale brown, or white. The ears are small, rounded, and covered with fine hairs. The front feet are short and wide, and the hind feet are long and wide.

Papagomys has sometimes been considered to have phylogenetic affinity to *Mallomys*, but the two are not closely related. *Papagomys* is larger, has a tail that is shorter (rather than longer) than the head and body, has short and harsh (rather than long, soft, and woolly) pelage, has a long and slender (rather than short and wide) rostrum, has relatively much larger bullae, and differs in numerous other cranial and dental characters (Musser 1981*b*).

The body structure of *Papagomys* is like that of a rat adapted for life on the ground and refuge in burrows. The diet may include leaves, buds, fruits, and certain kinds of insects (Musser 1981*b*).

RODENTIA; MURIDAE; **Genus PAULAMYS**
Musser, 1986

The single species, *P. naso*, is known only by cranial and dental fragments from several cave deposits, one of which was determined by radiocarbon dating to be about 3,550 years old, on western Flores in the East Indies. The genus originally was given the preoccupied name *Floresomys*. It apparently is a close relative of the living genus *Papagomys*, also found on Flores (Musser 1981*b*; Musser, Van de Weerd, and Strasser 1986).

Paulamys was probably about the same size as *Rattus norvegicus*. The available specimens are characterized by slim

incisor teeth and small molars set in a large mandible, in which the dentary anterior to each molar row is elongate. Such a configuration suggests that *Paulamys* was a long-nosed rat.

The molars of *Paulamys* are superficially similar to those of *Bunomys*, a genus of Sulawesi, but the dentaries of the latter genus are slender and have only a short segment in front of the toothrow. *Paulamys* generally resembles *Papagomys*, but its dentary is comparatively low and long. The morphology of *Paulamys* suggests that it was a terrestrial forest rat that ate insects, snails, earthworms, fungi, and fruits.

RODENTIA; MURIDAE; **Genus KOMODOMYS**
Musser and Boeadi, 1980

The single species, *K. rintjanus*, is currently found on the islands of Rintja and Padar, between Sumbawa and Flores in the East Indies (Musser and Boeadi 1980). The species is also known from subfossil material collected on Flores Island estimated to be 3,550 years old (Musser 1981*b*). *K. rintjanus* originally was described as a species of *Rattus* but is now known to represent a distinct genus not closely related to *Rattus* and apparently with affinity to *Papagomys* and other relictual murid genera of the eastern Sunda Islands.

Head and body length is 125–200 mm and tail length is 112–63 mm. The dorsal pelage is thick, coarse, and spinous. The upper parts are sandy-colored, with the middle of the head and body from the nose to the rump being darker and suffused with brown and the cheeks and sides of the head and body being paler and suffused with gray. The feet are hairy and white. The ears are tan and covered with fine short hairs.

Lesser Sulawesian shrew rat *(Melasmothrix naso),* photo by Guy G. Musser.

Top right, greater Sulawesian shrew rat *(Tateomys rhinogradoides).* Left, greater Sulawesian shrew rat *(Tateomys macrocercus).* Bottom, lesser Sulawesian shrew rat *(Melasmothrix naso).* Drawings by Fran Stiles through Guy G. Musser.

The tail is coarsely scaled and covered with long silver hairs that give a hairier appearance than that of *Rattus*. The tail is brown on the top and sides and paler below.

Komodomys is a medium-sized rat with a tail much shorter than the head and body. The hind feet are long and narrow, and each has six plantar tubercles. The digits and nails are long. The dorsal profile of the cranium is strongly arched, the rostrum is long and narrow, and the incisive foramina are very long and narrow. The bullae are relatively larger than those of *Rattus*. Also unlike in *Rattus*, but like in *Papagomys*, the molar teeth of *Komodomys* have very high cusps. *Papagomys* is a much larger rat than *Komodomys*. Female *Komodomys* have 10 mammae.

This rodent may be associated with monsoon forest. Its sandy-colored upper parts, densely haired white feet, moderately large ears, and short, hairy tail suggest that it is a ground dweller in dry scrub or forest. Its cranial and dental specializations may reflect adaptation to a dry, or seasonally dry, tropical forest habitat, where the tall scrub and short, partly deciduous trees provide dense cover above sparse undergrowth at ground level.

RODENTIA; MURIDAE; **Genus MELASMOTHRIX**
Miller and Hollister, 1921

Lesser Sulawesian Shrew Rat

The single species, *M. naso*, is found in central Sulawesi. The information for this account was taken from Musser (1982*b*).

Head and body length is 111–26 mm, tail length is 81–94 mm, and weight is 40–58 grams. The pelage is short, very dense, and velvety. Both the upper parts and underparts are a rich, dark chestnut color, so dark that the animals appear black in most lights. The ears are black, and each eye is surrounded by a dark gray ring. The tail is usually black above and gray below, but sometimes it is entirely black. All four feet are usually black over all surfaces. Females have three pairs of mammae.

Melasmothrix is characterized by a small and chunky volelike body, an elongate shrewlike head and muzzle, small eyes, small and round ears, a tail much shorter than the head and body, wide front feet with long and stout claws, and narrow hind feet with large claws. From *Tateomys*, evidently its closest relative, *Melasmothrix* is distinguished by its coloration, short tail, number of mammae, and various cranial and dental characters. In *Melasmothrix* there is a spacious postglenoid vacuity extending up under the squamosal zygomatic root, large inflated auditory bullae and hardly any eustachian tube, relatively long incisive foramina, and a discrete and usually large posterior cingulum on each first and second upper molar.

Melasmothrix has a striking superficial resemblance to *Archboldomys*, but the two genera are not closely related. In *Melasmothrix*, for example, the cusps of the occlusal surfaces of the molar teeth are discrete high cones; in *Archboldomys* these cusps are broadly connected to form a tubercularlike laminar hypsodonty.

Melasmothrix long was known only by a single specimen collected in 1918 at Rano Rano at an elevation of 1,800 meters. In 1973 and 1975, however, 35 new specimens were taken on a nearby mountain at about 1,900–2,300 meters. The genus is considered common in the cool and wet moss forest at these altitudes. It uses narrow runways along and under fallen tree limbs, logs, and boulders. It is diurnal and terrestrial and was often seen darting out from under moss-covered roots or decaying tree trunks and scooting across small clearings. Its wide front feet and robust claws form a

superb structure for digging into soft moist earth. Its diet consists primarily of earthworms and insect larvae, which are dug out of the soil, wet moss, and leaf litter of the forest floor.

RODENTIA; MURIDAE; **Genus TATEOMYS**
Musser, 1969

Greater Sulawesian Shrew Rats

Musser (1969*a*, 1982*b*), from whom the information for this account was taken, described two species:

T. rhinogradoides, known by nine specimens from central Sulawesi;
T. macrocercus, known by eight specimens from central Sulawesi.

Tateomys is thought to be closely related to *Melasmothrix* but not to the larger Sulawesian shrew rat, *Echiothrix*.

In *T. rhinogradoides* head and body length is 137–56 mm, tail length is 152–68 mm, and weight is 70–98 grams. The pelage is thick, short, and velvety. The upper parts are dark brownish gray with burnished highlights, the underparts are gray overlain with a pale buff wash, the ears and all four feet are dark brownish gray, and the tail is dark grayish brown above and white below. The tail of this species is thick, sturdy, and only slightly longer than the head and body. The broad, stout front feet have heavy, robust claws, while the long, narrow hind feet have long claws. In *T. macrocercus* head and body length is 110–20 mm, tail length is 160–75 mm, and weight is 35–55 grams. The fur resembles that of *T. rhinogradoides* but is more woolly and is grayer, with no burnished highlights. The front feet are white, the tail is dark blue-gray above and pale gray below, and seven of the specimens have a white tail tip. In this species the tail is much longer than the head and body, and the small front feet have long, narrow claws.

The external form of *Tateomys* resembles that of a terrestrial species of *Rattus*, but the genus is shrewlike in its elongate muzzle, tiny eyes, and velvety pelage. The incisors are short, weak, thin, and orthodont. The maxillary teeth are small and low-crowned, and their occlusal surface is basined and simple in structure. *Tateomys* resembles *Melasmothrix* but can be distinguished by its larger head and body, relatively smaller eyes, more elongate and delicately built skull, relatively shorter rostrum, and weaker incisors. There is a postglenoid vacuity that does not extend up under the squamosal zygomatic root, the bullae are small with long eustachian tubes, the incisive foramina are relatively short, and the cingulum on the first and second upper molar is absent or indistinct. Female *Tateomys* have only two pairs of mammae.

Like *Melasmothrix*, *Tateomys* dwells in cool mountain forests, where humidity is very high and nearly all surfaces are covered by moss. Specimens have been taken at elevations from about 2,000 to 2,300 meters, in the same kind of habitat occupied by *Melasmothrix*. However, *Tateomys* is nocturnal, and both species apparently eat only earthworms; captives refused other food. The morphology of *T. rhinogradoides* suggests that it is terrestrial. Its wide, spatulatelike front feet and large foreclaws suggest that it can dig deep into moss, soil, and rotten wood to extract earthworms. In contrast, the smaller body, very long tail, slender feet, and smaller foreclaws of *T. macrocercus* suggest that the species is scansorial. It probably works over the forest floor but also scampers up onto trunks, boulders, and other surfaces above ground level. Its delicate claws suggest that it may be restricted to extracting earthworms from moss. In its habitat earthworms are

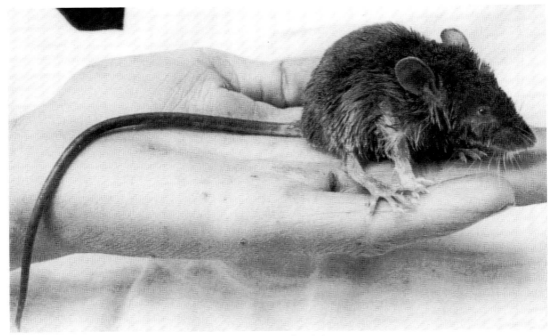

Greater Sulawesian shrew rat *(Tateomys macrocercus)*, photo from Guy G. Musser.

found on moss-covered surfaces up to about 1.5 meters above the ground.

RODENTIA; MURIDAE; **Genus PALAWANOMYS**
Musser and Newcomb, 1983

Palawan Rat

The single species, *P. furvus*, is known only by four specimens collected in 1962 at elevations of around 1,100–1,360 meters on Mount Mantalingajan on the island of Palawan in the southwestern Philippines (Musser and Newcomb 1983). *Palawanomys* may be a close relative of *Rattus*.

Head and body lengths of the four specimens are 125, 135, 160, and 160 mm, and respective tail lengths are 126, 140, 130, and 145 mm. The fur of the head and body is long, thick, and glossy. The upper parts are chocolate brown, and the underparts are paler, dark grayish brown. The ears are blackish brown and covered with short hairs. The tail is dark chocolate brown and covered with short, stiff hairs. The feet are brown, the hind ones being long and narrow. The cranium is small, the rostrum moderately long and wide, the braincase elongate, and the molar teeth large and chunky. The holotype, a female, has four pairs of mammae.

Palawanomys resembles some species of *Rattus* in external and certain cranial features but is considered to be a separate and more primitive genus. From *Rattus*, and all other murids, *Palawanomys* is distinguished by a combination of characters, including small body size; dark pelage; relatively short tail; low and weak interorbital, postorbital, and temporal ridges; short incisive foramina ending just before, at, or barely past the front faces of the first upper molars; long palatal shelf; wide, nearly flat pterygoid fossae; each first upper molar with one lingual root instead of two; and little overlap among the upper molars and among the lower molars.

The genus is considered terrestrial.

RODENTIA; MURIDAE; **Genus EROPEPLUS**
Miller and Hollister, 1921

Sulawesian Soft-furred Rat

The single species, *E. canus*, is known by five specimens, all obtained from elevations above 1,500 meters in middle Sulawesi (Musser 1970c:21).

Head and body length is 195–240 mm and tail length is 210–315 mm. The soft pelage is made up of a coat of underfur, about 25–28 mm in length, and longer guard hairs, 35–45 mm long on the back and flanks. The general coloration of the upper parts is brownish gray. Each hair has a pale slate base that terminates in a 3–5 mm tip of pale buff color. The underparts are light gray, but there is no sharp line of demarcation. The feet are scantily covered with short black hairs, and the whiskers are also black. The moderately haired tail is long, and the terminal third or half is white.

Eropeplus resembles *Rattus callitrichus*, of Sulawesi, in external features but differs in cranial and dental characters (Musser 1970c). In *Eropeplus*, for example, the skull is smaller, the rostrum is shorter and narrower, the interorbital region is shaped like an hourglass, the incisive foramina are short and do not extend to or beyond the front margins of the toothrows as in *Rattus callitrichus*, and the third upper molars are conspicuously longer than wide.

RODENTIA; MURIDAE; **Genus ECHIOTHRIX**
Gray, 1867

Sulawesian Spiny Rat, or Shrew Rat

The single species, *E. leucura*, is known from northern and central Sulawesi (Laurie and Hill 1954). Misonne (1969) suggested a distant relationship between *Echiothrix* and the sub-family Hydromyinae. Musser (1969*a*) summarized other views on the affinities of *Echiothrix* and noted that specimens had been taken only on coastal plains and adjoining foothills.

Head and body length is 200–250 mm and the tail is always shorter. The upper parts are grayish or dark gray brown, with black-tipped hairs on the back and sides. The underparts are whitish, yellowish, creamy buff, or reddish buff. The pelage contains both soft hairs and bristles. The nearly naked, cylindrical tail has rings of square scales. The head and nose are elongated, giving *Echiothrix* a somewhat shrewlike appearance.

The upper incisor teeth are short, and each has two faint longitudinal grooves. The lower incisors are elongated, arched, and widely divergent from each other, and "it is difficult to see how they can function, as they close on either side of the premaxillae, more or less, and evidently do not touch the upper incisors at all" (Ellerman 1941:269).

RODENTIA; MURIDAE; Genus BUNOMYS
Thomas, 1910

Four species have been described (Chasen 1940; Musser 1973*a*, 1981*b*):

B. andrewsi, Sulawesi and Butung Island to southeast;
B. penitus, central and southwestern Sulawesi;
B. fratorum, northeastern Sulawesi;
B. chrysocomus, Sulawesi.

These species generally were included in *Rattus* until Musser (1981*b*) stated that they were better placed in a separate genus. Musser and Newcomb (1983), from whom the information for the remainder of this account was taken, indicated that two additional species have been discovered and would soon be described.

Head and body length is 147–97 mm, tail length is 102–78 mm, and weight is 60–150 grams. All species have soft, thick pelage that is either silky or woolly to the touch. The upper parts are blue-gray to dark brown, and the underparts are blue-gray, buffy gray, brownish gray, or whitish gray. There is usually no contrast in color between the upper and lower parts. Females have only four mammae. *Bunomys* resembles *Berylmys*, and differs from *Rattus*, in that the interorbital and postorbital regions of the skull, as well as the dorsolateral sides of the braincase, are outlined only by low and inconspicuous ridges. From *Berylmys*, *Bunomys* differs in having a nontriangular cranial outline and orange incisor teeth. The mandibles of both genera are similar, but the coronoid process of *Bunomys* is larger relative to size of ramus. The two genera are not considered to be closely related.

All species of *Bunomys* are terrestrial and live in tropical forest and scrub. *B. chrysocomus* occurs at middle and high elevations, *B. fratorum* in both lowlands and mountains, *B. andrewsi* in lowlands, and *B. penitus* in mountains. The diet consists of fruits, insects, snails, and earthworms.

RODENTIA; MURIDAE; Genus RATTUS
Fischer, 1803

Rats

The following 2 subgenera and 51 species are recognized here (Agrawal and Ghosal 1969; Barbehenn, Sumangil, and Libay

Sulawesian soft-furred rat *(Eropeplus canus)*, photo by Howard E. Uible of specimen in U.S. National Museum of Natural History. Inset of skull: photo by P. F. Wright of specimen in U.S. National Museum of Natural History.

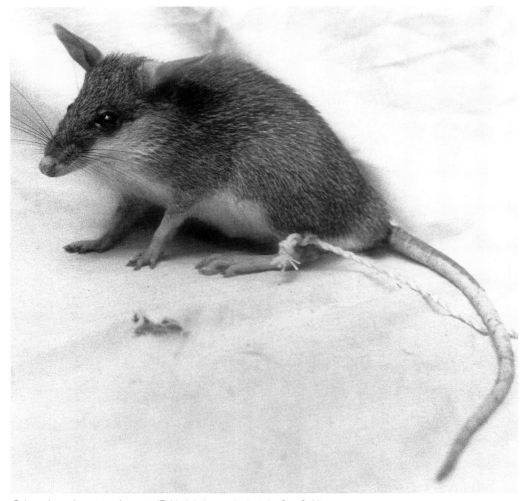

Sulawesian spiny rat, or shrew rat *(Echiothrix leucura)*, photo by Guy G. Musser.

1972–73; Calaby and Taylor 1980; Chasen 1940; Corbet 1978; Dennis and Menzies 1978; Ellerman and Morrison-Scott 1966; Laurie and Hill 1954; J. T. Marshall, *in* Lekagul and McNeely 1977; Medway 1977, 1978; Medway and Yong 1976; Menzies 1978; Menzies and Dennis 1979; Misonne 1969, 1979; Musser 1970a, 1970b, 1970c, 1971a, 1971b, 1971c, 1971d, 1971e, 1972a, 1972b, 1973a, 1973b, 1973c, 1977a, 1977b, 1979, 1984, 1986; Musser and Boeadi 1980; Musser and Califia 1982; Musser and Chiu 1979; Musser and Heaney 1985; Musser, Marshall, and Boeadi 1979; Musser and Newcomb 1983, 1985; Sanborn 1952; Taylor 1934; Taylor, Calaby, and Smith 1983; Taylor, Calaby, and Van Deusen 1982; Taylor and Horner 1973a; Tien 1960, 1966):

subgenus *Rattus* Fischer, 1803

R. rattus (black rat, roof rat, house rat), probably originated in Malaysian region, now found throughout the world as a human commensal;

R. ranjiniae, extreme southern India;

R. nitidus, apparently occurs naturally from Nepal to southeastern China and Viet Nam, now also found as a human commensal in highland areas of Luzon,

Sulawesi, New Guinea, and other islands of the East Indies;

R. turkestanicus, Soviet Central Asia and northeastern Iran to northeastern India, southern China;

R. tiomanicus, found partly as a human commensal on Malay Peninsula, Nicobar Islands, Sumatra, Java, Borneo, Palawan, Mindoro, and many small nearby islands;

R. baluensis, Mount Kinabalu in northern Borneo;

R. palmarum, Nicobar and Andaman Islands (Bay of Bengal);

R. lugens, Siberut, Sipura, and Pagi Islands off western Sumatra;

R. adustus, Enggano Island off southwestern Sumatra;

R. argentiventer (rice field rat), found largely as a human commensal and partly through introduction from southern Thailand and Viet Nam to the Philippines and New Guinea;

R. losea, found partly as a human commensal from southeastern China to peninsular Thailand, and on Taiwan and Hainan;

R. osgoodi, highlands of southern Viet Nam;

R. hoxaensis, central Viet Nam;

R. sikkimensis, Nepal to southeastern China and Viet

Bunomys chrysocomus, photo by Margareta Becker through Guy G. Musser.

Nam, Hainan, islands off west coast of peninsular Thailand;

R. brunneus, Nepal;

R. exulans (Polynesian rat), found as a human commensal and largely through introduction from Bangladesh to Viet Nam, throughout the East Indies, and on the Hawaiian and many other Pacific islands;

R. norvegicus (Norway rat, brown rat), may have originated in northern China, now found throughout the world as a human commensal;

R. stoicus, Andaman Islands (Bay of Bengal);

R. hoogerwerfi, Sumatra;

R. enganus, Enggano Island off southwestern Sumatra;

R. macleari, Christmas Island (Indian Ocean);

R. nativitatis, Christmas Island (Indian Ocean);

R. everetti, Philippines;

R. tyrannus, known only by the type specimen from Ticao Island (Philippines);

R. gala, Mindoro Island (Philippines);

R. annandalei, Malay Peninsula, Sumatra, and nearby islands;

R. montanus, Sri Lanka;

R. korinchi, central Sumatra;

R. tawitawiensis, Tawitawi Island off northeastern Borneo;

R. hoffmanni, Sulawesi;

R. xanthurus, Sulawesi and nearby Peleng Island;

R. marmosurus, northern and central Sulawesi;

R. jobiensis, islands in Geelvink Bay off northwestern New Guinea;

R. lutreolus, southeastern Australia, Tasmania;

R. fuscipes, eastern and southern coasts of Australia;

R. leucopus, southern New Guinea, northern Queensland;

R. novaeguineae, mountains of eastern New Guinea;

R. praetor, New Guinea and nearby islands from the Moluccas to the Solomons;

R. doboensis, Aru Islands southwest of New Guinea;

R. rennelli, Rennell Island (Solomons);

R. elaphinus, Sula Islands east of Sulawesi;

R. feliceus, Ceram Island between Sulawesi and New Guinea;

R. morotaiensis, Morotai Island in the northern Moluccas.

R. steini, mountains of New Guinea;

R. mordax, eastern New Guinea and islands to southeast;

R. giluwensis, mountains of east-central New Guinea;

R. tunneyi, northern, central, and southwestern Australia;

R. sordidus, southern New Guinea, central and eastern Australia;

subgenus *Stenomys* Thomas, 1910

R. niobe, mountains of New Guinea;

R. richardsoni, mountains of western New Guinea;

R. verecundus, New Guinea.

In number of species *Rattus* is among the largest of mammalian genera. *Rattus* once was considered to be even more extensive, but many of its former species are now generally assigned to other genera (see accounts of *Lenothrix*, *Anonymomys*, *Sundamys*, *Kadarsanomys*, *Abditomys*, *Diplothrix*, *Margaretamys*, *Lenomys*, *Komodomys*, *Palawanomys*, *Bunomys*, *Nesoromys*, *Tryphomys*, *Limnomys*, *Tarsomys*, *Taeromys*, *Paruromys*, *Bullimus*, *Apomys*, *Millardia*, *Srilankamys*, *Niviventer*, *Maxomys*, *Leopoldamys*, *Berylmys*, *Mastomys*, *Myomys*, *Praomys*, *Hylomyscus*, *Heimyscus*, *Stochomys*, *Dephomys*, and *Aethomys*). A major turning point in our understanding came with Misonne's (1969) arrangement of the subfamily Murinae, especially in regard to his separation of several African lineages from *Rattus*. The work of Musser and colleagues (as cited above) subsequently showed that the content of *Rattus* in Southeast Asia is also much more restricted than once thought. Misonne listed *Bullimus* and *Leopoldamys* (in addition to *Rattus* and *Stenomys*) as subgenera of *Rattus*, and J. T. Marshall (*in* Lekagul and McNeely 1977) recognized still another, *Berylmys*; all three are now considered distinct genera. Musser and Newcomb (1983) indicated that *Stenomys* is still a valid subgenus (if not a full genus), though not with exactly the same content set forth by Misonne.

Some additional questions and disagreements regarding the classification of this complex genus are as follows. Tiwari, Ghose, and Chakraborty (1971) suggested that *R. rufescens*, of India, is a species distinct from *R. rattus*. The name *R. rattoides* sometimes is used for the species listed above as *R. turkestanicus*, though Schlitter and Thonglongya (1971) showed the latter to be the proper designation. The name *R. koratensis* sometimes is used for the species listed above as *R.*

sikkimensis, though Musser and Heaney (1985) used the latter. Medway and Yong (1976) indicated that *R. diardii*, of the Malay Peninsula and East Indies, is a species distinct from *R. rattus*, rather than only an ecological subspecies as it is generally considered to be. J. T. Marshall (*in* Lekagul and McNeely 1977) suggested that *R. tiomanicus* may not be a species distinct from *R. rattus*. Chromosomal data gathered by Baverstock et al. (1977) may indicate that the three Australian subspecies of *R. sordidus* recognized by Taylor and Horner (1973a)—*R. s. sordidus* in the northeast, *R. s. villosissimus* of the central and southeastern regions, and *R. s. colletti* of Arnhem Land—are actually separate species. Also on the basis of chromosomal data, Menzies and Dennis (1984) disagreed with Taylor, Calaby, and Van Deusen's (1982) conclusion that *R. gestroi*, of southeastern Papua New Guinea, is not a species distinct from *R. sordidus*. Musser (1982b) suggested that both *R. gala* and *R. tyrannus* may be subspecies of *R. everetti*. Laurie and Hill (1954) stated that *R. doboensis* is possibly a race of *R. praetor*. Musser (1971e) suggested that *R. germaini*, described from the Condor Islands off southern Viet Nam, might be specifically distinct from *R. rattus*, but it was not listed as such by Musser and Newcomb (1983) or any other recent authorities.

Head and body length is 80–300 mm. The tail varies in length by species, ranging from substantially shorter to much longer than head and body length. The best-known species, *R. norvegicus*, usually weighs 200–400 grams, with a few individuals reaching 500 grams (Grzimek 1975). Most other species are considerably smaller. In Hawaii, average weights for *R. exulans* were only 37 grams in males and 39 grams in females, while respective averages for *R. rattus* were 108 and 77 grams (Tamarin and Malecha 1972). In some species the pelage is soft, in others it is coarse, and in still others the hairs are enlarged and stiffened into bristles or spines. The coloration of the upper parts ranges through a great variety of black, brown, gray, yellow, orange, and red shades. The underparts are usually white or grayish. The body form may be stocky or somewhat slender. Certain digits are reduced in some forms but are long in others. In some species the feet are modified for a terrestrial life, while in others the feet are adapted to an arboreal existence. The number of mammae in females is 4, 6, 8, 10, or 12.

The lifestyle of the several well-known commensal species is not representative of the genus as a whole. Most species of *Rattus* live in natural forest, ranging from tropical lowland to montane type, and tend to avoid human settlements. In Australia *Rattus* is found in grasslands, heaths, savannahs, and sandy flats, as well as forests (Ride 1970). Of the 12 species in Thailand discussed by Lekagul and McNeely (1977), 6 are fully wild and occur primarily in forests. The remaining 6 Thai species are commensals: *R. rattus*, the most abundant mammal in the nation, occupies cities, villages, cultivated fields, and some natural habitat; *R. nitidus* is found exclusively inside the houses of hill villages; *R. norvegicus* usually occurs in urban areas; *R. argentiventer* is entirely dependent on human rice fields and plantations; *R. losea* is found mainly in gardens and rice fields but also is known to occur in the wild state in forests; and *R. exulans* inhabits houses and rice fields.

According to Musser (1973a, 1977a), three related commensal rats of Southeast Asia have divided the available habitat among them. *R. rattus diardii* lives in and around human dwellings; *R. tiomanicus* is found in gardens, plantations, scrub country, and secondary forests; and *R. argentiventer* occurs primarily in rice fields and grasslands. Musser's studies in the Philippines and Sulawesi indicate that these commensals, along with *R. nitidus*, *R. norvegicus*, and *R. exulans*, do not occur in primary forest but are widely distributed in close association with people. In contrast, the many indigenous species of *Rattus* in the Philippines and Sulawesi

have very restricted ranges and occur mainly in primary forest.

Rats shelter in such places as burrows, under rocks, in logs, and in piles of rubbish. Species that are good climbers sometimes build nests in trees or other elevated positions. *R. rattus* is an extremely agile climber and can run along a wire only 1.6 mm in diameter (Ewer 1971). It constructs a loose, spherical nest of shredded vegetation, cloth, or other suitable material (Medway 1978). When occupying buildings, this species is usually found in the dry upper levels. *R. norvegicus* is a ground dweller. It may originally have lived along stream banks in Asia and then spread rapidly as canals and rice fields were developed. Its underground burrows contain long, branching tunnels, one or more exits, and rooms for nests and food storage. When using buildings, it generally occurs in cellars, basements, and lower floors. It also occupies sewers and rubbish heaps. It swims and dives well (Grzimek 1975). *R. argentiventer* also burrows in the soil, while *R. tiomanicus* constructs a loose, spherical nest of vegetation on the surface of the ground under some shelter (Medway 1978).

Rats are omnivorous, eating a wide variety of plant and animal matter. Seeds, grains, nuts, vegetables, and fruits may be preferred by most species, but insects and other invertebrates sometimes predominate in the diet. Kingdon (1974b) stated that *R. rattus* and *R. norvegicus* eat everything that people eat and much else, such as soap, hides, paper, and beeswax. Food may be carried back to the nest and stored. Grzimek (1975) noted that *R. norvegicus* seems to prefer animal matter, including birds and eggs, and is excellent at catching fish. This species also feeds on mice, poultry, and young lambs and pigs; it may attack larger animals, even humans.

The reported density of rat populations varies widely: for example, *R. fuscipes* in Australia, 0.38–3.04 per ha. (Barnett, How, and Humphreys 1977); *R. exulans* in Hawaii, 70–188 per ha. (Wirtz 1972); and *R. tiomanicus* on a Malay oil palm plantation, up to 1,200 per ha. (Medway 1978). In the United States there are an estimated 100–175 million rats, mostly *R. norvegicus* (Banfield 1974; Jackson 1961; Pratt, Bjornson, and Littig 1977; Schwartz and Schwartz 1959). Densities of this species range from about 25 to 150 per block in some U.S. cities and from 50 to 300 per country farm. As in some other species of *Rattus*, the population density of *R. norvegicus* appears to be cyclic and sometimes increases sharply. Large-scale movements from an area may then seem to occur. Favorable conditions of food and shelter have resulted in the taking of over 200 *R. norvegicus* during a five-day period in a town area of less than 0.2 ha., and of 4,000 rats on about 2.5 ha. of a midwestern farm. The normal home range of *R. norvegicus* is only about 25–150 meters in diameter; according to Grzimek (1975), however, individuals have been known to move 3 km from their nests and return in a single night. Calculated home range diameters of some other species are: *R. tiomanicus*, 315–57 meters; *R. argentiventer*, 273 meters; and *R. exulans*, 280 meters (Medway 1978). A radiotracking study of *R. exulans* in Hawaiian sugar cane fields showed an average home range of 1,845 sq meters for males and 607 sq meters for females (Nass 1977).

Several forms of social structure have been reported for *R. norvegicus*, but the differences were basically resolved by Calhoun (1963). In a study of rats caught in the wild and their descendants in a 0.10 ha. enclosure, he demonstrated the functioning of parallel social systems. In one part of the enclosure higher-ranking males established individual territories around burrows containing several females. Each male excluded other males, and only he mated with the resident females. The females of a colony collectively nurtured their young, and when they were reproductively active, they excluded other rats from the burrow. These territorial colonies kept well-organized burrows and carefully maintained

A. Norway rat *(Rattus norvegicus)*, photo from U.S. Fish and Wildlife Service. B. Black rat *(R. rattus)*, photo by Ernest P. Walker. C. Black rat *(R. rattus)*, photo from U.S. Fish and Wildlife Service.

nests. Reproduction occurred regularly and successfully. As the young matured, subordinate individuals among them were forced into another part of the enclosure. This downward change in status, involving mostly males, was not reversible; there was no permanent movement of lower-ranking individuals into the area occupied by the dominant rats. No territories were established in the area used by the lower-ranking animals. Large packs were formed, and when a female entered estrus, she would be followed by numerous males and mounted several hundred times in a night. Such stressful conditions resulted in poor reproduction; burrows

were not well organized, and nests were poorly maintained. This process was considered to be responsible for limiting the size of the overall rat population, so that there were never more than about 200 animals in the pen. Grzimek (1975), however, noted that size of wild packs may exceed 200 individuals. Watts (1980) reported that adult *R. rattus, R. norvegicus,* and seven species of native Australian *Rattus* all have the same basic vocal repertoire, consisting mainly of squeals and whistles used in aggressive encounters.

In a study of *R. rattus* in Accra, Ghana, Ewer (1971) found social groups to contain a single dominant male and some-

Wild Sulawesian rat *(Rattus hoffmanni)*, photo by Guy G. Musser.

times a linear male hierarchy. There also were two or three equally ranking top females, which were subordinate to the dominant male but dominant to all other members of the group. Females were much more aggressive than males. A group territory was formed around a feeding place and defended against outsiders, but only the females drove away other females. Attacks were frequent and directed downward, but serious fights were usually avoided through appeasement by the subordinate. Fighting took several forms, including biting, jumping, standing on hind legs and hitting with the front paws, and striking with the rump from the side. Infants had almost complete immunity from aggression and could take food literally from under the nose of dominant adults.

During Ewer's study, young appeared throughout the year. The gestation period was found to be 21–22 days in nonlactating females and 23–29 days in lactating females. Litters most commonly numbered 8 young, and females could give birth at 3–5 months. Medway (1978) reported that in the Malay Peninsula *R. rattus* also breeds throughout the year, has a litter size of 1–11 young, and reaches sexual maturity at about 80 days. Captive individuals of this species have lived as long as 4 years and 2 months.

In *R. norvegicus* reproduction may occur all year in captive populations and some wild populations, but there are usually spring and autumn peaks. Females are polyestrous, and each may have 1–12 litters per year. There is a postpartum estrus within 18 hours of birth. The estrous cycle lasts 4–6 days, estrus 20 hours, and the gestation period 21–26 days. Litter size is 2–22 young, averaging about 8–9. The young are born naked and blind, but they are fully furred and their eyes are open at about 15 days. The are weaned and leave the nest at around 22 days and attain sexual maturity at 2–3 months (Asdell 1964; Banfield 1974; Grzimek 1975; Lowery 1974).

Most species of *Rattus* that have been investigated seem less prolific than *R. rattus* and *R. norvegicus*. Medway (1978) indicated that the other species of the Malay Peninsula breed all year but that in most litter size averages only around 3–6 young. In New Guinea, reproduction of several native species apparently is reduced, or ceases altogether, during the dry season, about June–October, and litters commonly number only about 1–3 young (Dwyer 1975b; Menzies and Dennis 1979). All native *Rattus* of Australia appear to be seasonal breeders in the wild (Taylor and Horner 1973b). The more northerly populations of that continent have only a brief

winter lull in reproduction and a peak during the wet summer. More southerly populations undergo a lull of up to 7 months and have their breeding peaks in spring and early summer. Laboratory colonies are capable of reproduction throughout the year. Females of most Australian species are known to be polyestrous and to undergo a postpartum estrus. Their gestation periods vary from 21 to 30 days, and litter sizes from 3 to 14 young.

The commensal species *R. exulans* has been studied extensively on the Pacific islands (Egoscue 1970; Tamarin and Malecha 1972; Wirtz 1972, 1973). Females are polyestrous and capable of breeding all year in the laboratory, with up to 13 litters annually, but wild populations have reproductive seasons centered in the summer months and usually produce only 1–3 litters per year. The gestation period is 19–30 days, litter size averages about 4 young, and weaning takes place at 2–3 weeks. Some young females do not reach sexual maturity until the season following their first winter.

Most species of *Rattus* are not known to be of any direct importance to people. They live in natural habitat and do not enter human habitations or intensively cultivated areas. Many of these species have a relatively restricted range and habitat tolerance. Some species and subspecies, such as *R. sikkimensis remotus* of Thailand, may be threatened with extinction. Taylor and Horner (1973a) observed that *R. tunneyi* had disappeared in southern and southwestern Australia, probably because of excessive grazing by domestic livestock. Both *R. macleari* and *R. nativitatis*, found only on Christmas Island in the Indian Ocean, apparently became extinct by 1908, perhaps because of disease transmitted by introduced *R. rattus* (Harper 1945).

The activities of the few commensal and pest species of *Rattus* have adversely affected the reputation of the entire genus and, indeed, of all mammals that contain the term *rat* in their vernacular names. Most economic and health problems are actually associated with only seven species: *R. rattus* (black rat), found throughout the world and responsible for the spread of disease and immense agricultural losses; *R. nitidus* and *R. tiomanicus*, agricultural pests and residents of human homes in Southeast Asia; *R. argentiventer* and *R. losea*, known mainly for their depredations on rice fields and gardens in Southeast Asia; *R. exulans*, apparently carried over a vast part of Southeast Asia and the Pacific in association with early human migrations and now sometimes considered a menace to health and agriculture; and *R. norvegicus* (Norway rat), known over much of the world for its destruction of property and stored food and its threat to the health and safety of people, domestic animals, and wildlife. Under certain conditions, several other species of *Rattus* have been known to act as human commensals and/or become agricultural pests in local areas (Braithwaite 1980; Menzies and Dennis 1979).

The numbers and distributions of all seven major commensal species of *Rattus* have apparently been increased through human agency, either directly by transportation in boats and caravans or indirectly by habitat modification, poor sanitary practices, and elimination of predators. The most spectacular range expansions have been those of the black and Norway rats. There have been reports that *R. rattus* was present in Europe as early as the Pleistocene (Grzimek 1975), but Kurten (1968) stated that both the black and Norway rats probably entered Europe in postglacial times as human commensals. *R. rattus* may actually have been brought to Europe at the time of the Crusades and then to the Western Hemisphere during the explorations of the sixteenth century. *R. norvegicus* was not known in Europe until about 1553 and did not reach North America until 1775. Although *R. rattus* is by far the more common and widespread of the two species in the tropics, *R. norvegicus* has proved the more adaptable in temperate zones, especially in urban areas. As the latter species spread through North America and Europe, it excluded *R. rattus* from favorable habitat on the ground and in the lower levels of buildings. The black rat thus has become rare in many areas and has even been designated as endangered in the state of Virginia (Handley 1980b). Remarkably, Pucek (1989) reported that it has disappeared in much of Europe and referred to it as one of the "top ten endangered rodents" on the continent. Small populations continue to maintain themselves in certain temperate areas such as Maryland (Feldhamer and Gates 1980), Hungary (Jabir, Bajomi, and Demeter 1985), and Chile (Simonetti 1983).

Both *R. rattus* and *R. norvegicus* have occupied many oceanic islands, where they have had a devastating effect on the native birds, reptiles, and vegetation (Wace 1986). On South Georgia in the South Atlantic, *R. norvegicus* has returned to a fully feral life (Pye and Bonner 1980). It digs burrows in stools of tussock grass and feeds on the grass, as well as insects, carrion, and ground-nesting birds.

Norway and black rats consume vast quantities of food stored for people and their livestock and contaminate much more than they eat. They also gnaw the insulation from electrical wires, sometimes causing fires, and occasionally cut through lead pipe and concrete dams. On a world basis, the direct damage caused each year by these two species amounts to billions of dollars. In the United States alone, direct economic losses to rats have been estimated at $500 million to $1 billion annually (Pratt, Bjornson, and Littig 1977). Further expenses are incurred in the process of poisoning campaigns and other efforts at control, but rats can be permanently eradicated only through elimination of garbage and carelessly stored food supplies and of environmental conditions that provide suitable shelter for the animals.

Among the many diseases spread by Norway and black rats are bubonic (black) plague, murine typhus, food poisoning *(Salmonella)*, leptospirosis, trichinosis, tularemia, and rat-bite fever. Gratz (1984) listed about 40 diseases spread by rats, including schistosomiasis, which now infects as many as 200 million people worldwide. Rat-borne diseases are thought to have taken more human lives in the last 10 centuries than all the wars and revolutions ever fought. Bubonic plague, which is transmitted to humans by rat fleas, killed more than one-quarter of the population of Europe from 1347 to 1352, about 11 million people in India from 1892 to 1918,

and 60,000 people in Uganda from 1917 to 1942 (Grzimek 1975; Kingdon 1974b). Serious outbreaks of plague recurred at San Francisco (1902–41), Galveston (1920–22), and New Orleans (1912–26) (Lowery 1974). Rats make direct attacks on about 14,000 persons annually in the United States, occasionally inflicting mortal wounds (Pratt, Bjornson, and Littig 1977). Rats also kill poultry, domestic livestock, and game birds. Rat predation and competition has contributed to the endangerment or extinction of many species of wildlife.

Lowery (1974) stated that *R. norvegicus*, especially in its white mutant form, has partially redeemed itself through its role in medical and basic research. The species is used by biological laboratories throughout the world and has led to important discoveries in immunology, pathology, epidemiology, genetics, and physiology.

RODENTIA; MURIDAE; **Genus NESOROMYS**
Thomas, 1922

Ceram Island Rat

The single species, *N. ceramicus*, is known only from the island of Ceram, between Sulawesi and New Guinea (Laurie and Hill 1954). The type specimen was trapped in heavy jungle on Mount Manusela at an elevation of about 1,800 meters. This species originally was placed in the genus *Stenomys*, which now is considered a subgenus of *Rattus*. Musser (1981b) and Musser and Newcomb (1983) continued to follow this procedure, but Misonne (1969) observed that while *Nesoromys* seems to be related to *Stenomys*, it is differentiated enough to be a separate genus.

Head and body length of the type specimen is 135 mm and tail length is 140 mm. The pelage is fine, soft, and thick. The general coloration of the upper parts is a fine speckling of olive brown. The underparts are somewhat lighter because of the dull, drab tips of the hairs. The short ears are almost black; the feet are dark brown; and the tail, which is almost naked, is also dark brown. *Nesoromys* is distinguished from *Rattus* by its long, narrow muzzle. The hind feet are also narrow.

RODENTIA; MURIDAE; **Genus TRYPHOMYS**
Miller, 1910

Mearns's Luzon Rat

The single species, *T. adustus*, is now known by at least 10 specimens from lowland and highland parts of Luzon Island in the Philippines. *Tryphomys* was considered a synonym of *Rattus* by Sanborn (1952) and Misonne (1969) but was treated as a full genus by Taylor (1934), Musser (1981b), and Musser and Newcomb (1983).

Head and body length is 131–74 mm and tail length is 130–64 mm. The fur is thick, shaggy, harsh, and of a grizzled appearance. The upper parts are dark brown flecked with buff, the underparts are whitish gray or deep buffy gray, and the ears are brown. The dorsal surface of the feet is whitish gray or whitish brown. The soles are naked and and have five well-developed tubercles. The claws also are well developed, but the nail of the thumb is small and appressed. The long and narrow hind feet have especially large claws. The body is stocky. The tail is shorter than the head and body, coarsely and conspicuously scaled, and uniformly brown. Females have five pairs of mammae.

The skull is large and heavily built, the rostrum short and wide, the zygomatic plates very deep, the interorbital region

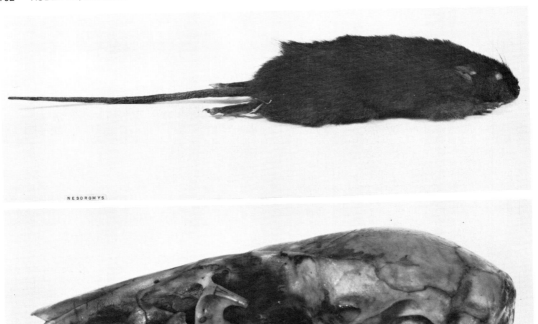

Ceram Island rat *(Nesoromys ceramicus)*, photos from British Museum (Natural History).

narrow, the incisive foramina long and narrow, and the auditory bullae large and inflated. Based on molar occlusal pattern, Misonne considered *Tryphomys* to be closely related to *Sundamys*, but Musser and Newcomb disagreed. In *Tryphomys* the first and second laminae of each first upper molar and the first lamina of each second upper molar consist of discrete cusps that are narrowly joined to one another in each lamina; this configuration is cuspidate, rather than noncuspidate as in *Sundamys*. Also, in young adult *Tryphomys* the three laminae on each first lower molar consist of large, discrete cusps, whereas in *Sundamys* these laminae are noncuspidate.

RODENTIA; MURIDAE; Genus LIMNOMYS
Mearns, 1905

The single species, *L. sibuanus,* is known by four specimens collected in 1904 and 1906 on the island of Mindanao in the Philippines. The complex history of these specimens, during which they were sometimes thought to represent three separate species and sometimes placed in the genus *Rattus,* was explained by Musser (1977*b*).

The single adult specimen is a female with a head and body length of 125 mm and a tail length of 150 mm. The top and sides of the head and body are covered with tawny, dense, long pelage. Long, black guard hairs are scattered over the back and rump. The underparts are cream, the ears are dark brown, and the densely haired tail is dark brown on all sur-

faces. The front and hind feet are brownish white, and the hind feet have a dark brown stripe from the ankle to the bases of the digits. The hind feet are short and broad. There are six mammae.

Musser considered *Limnomys* a genus distinct from *Rattus* because of a combination of characters. In *Limnomys* these are: short body, long and hairy tail, short and broad hind feet, six mammae, short rostrum, narrow zygomatic plates, short bony palatal bridge, large bullae, and details of molar topography.

The adult female specimen of *Limnomys* was collected on 30 June 1904 at an elevation of about 2,000 meters on a wet, mossy growth of vegetation beside a small stream. The mammae were probably then functional. The three other specimens were taken on 6 June 1906 in a montane forest at an elevation of about 2,800 meters. These three specimens represent immature animals, all of about the same age, perhaps from the same litter.

RODENTIA; MURIDAE; Genus TARSOMYS
Mearns, 1905

The single species, *T. apoensis,* is known by five specimens collected in montane forests on the island of Mindanao in the Philippines (Taylor 1934). Some authorities, such as, for example, Ellerman and Morrison-Scott (1966) and Misonne (1969), have considered *Tarsomys* a synonym of *Rattus,* but Musser (1977*a*, 1977*b*, 1981*b*) treated it as a distinct genus.

inflated. The rostrum is elongated, similar to that of *Apomys*, but the auditory bullae and anterior palatal foramen are different. The bullae resemble those of *Rattus* but are more flattened and compressed externally.

RODENTIA; MURIDAE; **Genus TAEROMYS**
Sody, 1941

There are six species (Musser 1982*b*, 1984; Musser and Newcomb 1983):

T. celebensis, Sulawesi;
T. callitrichus, Sulawesi;
T. arcuatus, southeastern Sulawesi;
T. taerae, northeastern peninsula of Sulawesi;
T. hamatus, central Sulawesi;
T. punicans, central and southwestern Sulawesi.

These species usually were placed in the genus *Rattus* until the authorities cited above, from whom the data for this account were taken, restored *Taeromys* to generic rank.

Head and body length is about 185–249 mm and tail length is about 156–306 mm. The upper parts are generally brown, or bluish or brownish gray, and the underparts are gray or buffy. The tail is dark basally and white distally in most species but completely blackish brown in *T. punicans*. The pelage is commonly short, soft, dense, and silky. The body is stout, the tail usually longer than the head and body, and the hind feet long and slender. The braincase, bullae, and molars are relatively large, while the rostrum is rather slender and delicate. In contrast to that of *Rattus*, the skull of *Taeromys* is delicate and smooth; the ridges bounding the dorsolateral margins of the interorbital and postorbital regions are low and inconspicuous. Females have six mammae.

All species live in forests, though *T. celebensis* and *T. punicans* are found mainly at lower elevations, and *T. taerae* and *T. hamatus* in the mountains. Most species are terrestrial, but *T. celebensis* also occurs in trees and vines. *T. hamatus* is known to feed on fruits, and *T. callitrichus* on fruits, leaves, and insects. Neither *T. celebensis* nor *T. punicans* is common, and they may have been eliminated in some areas through hunting by people for use as food.

RODENTIA; MURIDAE; **Genus PARUROMYS**
Ellerman, 1954

Sulawesian Giant Rat

The single species, *P. dominator*, occurs on Sulawesi and originally was described as a species of *Rattus*. *Paruromys* originally was proposed as a subgenus for this species, but Musser (1984) and Musser and Newcomb (1983) considered it a full genus with possible close affinity to *Taeromys*.

This is the largest murid on Sulawesi. Head and body length is 200–257 mm, tail length is 237–310 mm, and weight is 350–500 grams. The fur is soft and thick but not long. The upper parts are grayish brown, the underparts white or cream, and the ears and feet dark brown. The basal portion of the tail is blackish brown, and the distal half to two-thirds is white. The body is stocky, the face elongate, and the tail much longer than the head and body. *Paruromys* is characterized by an elongate cranium with very wide zygomatic plates, a low braincase, small bullae, short incisive foramina, large and extremely ophisthodont upper incisor teeth,

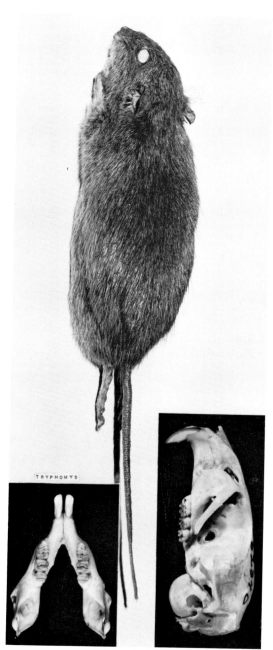

Mearns's Luzon rat *(Tryphomys adustus)*, photos from Field Museum of Natural History.

According to Taylor's (1934) description, head and body length of the type specimen is 135 mm and tail length is 120 mm. The pelage is long and rather coarse but not spiny. The upper parts are brownish slate, the underparts are grizzled yellow brown, and the ears are dark purplish brown. The hairy tail is dark purplish throughout. The feet are naked below but covered with slightly grizzled, drab brown hairs above.

The skull and teeth of *Tarsomys* resemble those of *Batomys* but are relatively broader, and the braincase is more

Sulawesian giant rat *(Paruromys dominator)*, photo by Guy G. Musser.

and molars with simple cusp patterns. Females have three pairs of mammae.

This giant rat is common in forests from coastal lowlands to the highest mountains. It is easily trapped on the ground, on top of fallen tree trunks, or in tangles of woody vines and crowns of short trees. Nests are found on the ground between or inside rotting tree trunks, under boulders, in the interstices of strangler fig roots, and in underground burrows. The diet consists exclusively of fruits. *Paruromys* is a favorite food of the mountain people of central Sulawesi.

RODENTIA; MURIDAE; Genus BULLIMUS
Mearns, 1905

Three species now are recognized (Heaney and Rabor 1982; Musser 1982*b*):

B. luzonicus, Luzon;
B. bagobus, Samar, Dinagat, Siargao, Calicoan, and
    Mindanao islands in the Philippines;
B. rabori, Mindanao.

Misonne (1969) considered *Bullimus* a subgenus of *Rattus*, and he included therein numerous other species from the Philippines, Sulawesi, and other parts of Southeast Asia, including *Sundamys muelleri*. Musser (1982*b*) indicated that *Bullimus* is a distinct genus known to comprise only the above three species but that there may be others. He also reported the overall range of the genus to include the Philippine islands of Samar, Calicoan, Leyte, Dinagat, and Bohol.

In the type specimen of *B. luzonicus* head and body length is 240 mm and tail length is 200 mm; in the type of *B. bagobus* head and body length is 275 mm (Taylor 1934); and in the type of *B. rabori* head and body length is 305 mm and tail length is 243 mm (Sanborn 1952). Total lengths of 404–46 mm and tail lengths of 155–87 mm were recorded for *B. bagobus* (Heaney and Rabor 1982). According to Musser (1982*b*), *Bullimus* is characterized by brown upper parts, grayish white or white underparts, coarse and short pelage, a long nose, a large and bulky body, a stubby and short tail, long and narrow hind feet, a large, conspicuous midventral cutaneous glandular area in males, and four pairs of mammae in females.

According to Musser and Newcomb (1983), the skull of *Bullimus* is distinguished by being very deep and having a long and slender rostrum, a constricted interorbital region, and a wide braincase. Compared with those of *Rattus*, the molar teeth of *Bullimus* are higher-crowned and have wider and straighter lamina. The first upper molar usually lacks a posterior cingulum. These rats have been found in forests, scrub, and cultivated fields, both in lowlands and mountains. They are terrestrial and apparently utilize burrows under thick vegetation.

Bullimus sp., photo by Paul D. Heideman.

RODENTIA; MURIDAE; **Genus APOMYS**
Mearns, 1905

There are eight species, all in the Philippines (Heaney and Rabor 1982; Musser 1982*a*):

A. *datae*, northern Luzon;
A. *abrae*, northern Luzon;
A. *sacobianus*, known by a single specimen from
    southwestern Luzon;
A. *musculus*, northern Luzon, Mindoro;
A. *microdon*, Catanduanes, Leyte, Dinagat;
A. *insignis*, Mindanao, Dinagat;
A. *hylocoetes*, southern Mindanao;
A. *littoralis*, known by two specimens from southern
    Mindanao and, provisionally, by four from Negros.

Apomys was included within *Rattus* by Sanborn (1952) and Ellerman and Morrison-Scott (1966), but more recent authorities have indicated that it is a distinct genus and not closely related to *Rattus* (Misonne 1969; Musser 1977*a*, 1977*b*, 1982*a*).

The information for the remainder of this account was taken primarily from Musser (1977*a*, 1982*a*), with some aug-mentation from Johnson (1962), Sanborn (1952), and Taylor (1934). Head and body length is 76–143 mm, tail length is 82–176 mm, and weight is about 20–50 grams. The tail ranges from slightly shorter than the head and body to considerably longer. The pelage is long, soft, and thick. The upper parts are some shade of brown or buff, the sides are generally paler, and the underparts are grayish, buff, or fawn. In most species the scantily haired tail is darker above than below. The skull is elongate, the feet are long and narrow, and there are six plantar pads. Females have two pairs of mammae.

Apomys is distinguished from *Rattus* and all other murid genera by a combination of features. The interorbital region of the skull is wide, and the margins of the interorbital and postorbital areas are rounded, without ridges or shelves. The braincase is smooth and globular, without the supraorbital ridges that extend back along the margins of the braincase, a characteristic of most species of *Rattus*.

The squamosomastoid foramen is generally much more prominent in *Apomys* than in *Rattus*. In *Apomys* the bullae are small and loosely attached to the braincase. The occlusal surfaces of its first and second upper molars are simple, and the third upper molar is tiny relative to the first and second. The configurations of the occlusal surfaces and other dental characters of *Apomys* resemble those of *Melomys* to some extent, and the two genera might be closely related. A unique

Apomys littoralis, photo by Paul D. Heideman.

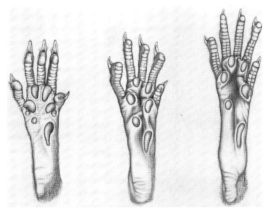

Left, *Rattus exulans*. Center, *Apomys hylocetes*. Right, *Apomys insignis*. Drawings of hind feet by Patricia Wynne through Guy G. Musser.

character of *Apomys,* however, is the combination of a pitted and perforated bony palate with a truncate margin ending just posterior to the molar rows.

Most specimens of *Apomys* have been collected in montane forests at elevations of 300–2,800 meters. One specimen of *A. littoralis* was found at only 15 meters, but the others were taken at around 1,000–1,500 meters. *A. sacobianus* has been termed a lowland species. All species dwell in forests, and individuals have been found mainly in dense vegetative cover or among rocks. They evidently are terrestrial, and some species may also be scansorial.

RODENTIA; MURIDAE; Genus CRATEROMYS
Thomas, 1895

Bushy-tailed, Ilin Island, and Dinagat Cloud Rats

There are three species (Musser and Gordon 1981; Musser, Heaney, and Rabor 1985; Sanborn 1952; Taylor 1934):

Bushy-tailed cloud rats *(Crateromys schadenbergi)*, photo by Ernest P. Walker. Insets: A. Upper molars; B. Lower molars; photos from *Trans. Zool. Soc. London*, "Mammals from the Philippines," Oldfield Thomas.

C. australis, known only by the holotype from Dinagat Island (Philippines);

C. paulus, known only by the holotype from Ilin Island (Philippines);

C. schadenbergi (bushy-tailed cloud rat), northern Luzon.

In the single specimen of *C. paulus* head and body length is 255 mm, tail length is 215 mm, the pelage is short and coarse, the upper parts are generally dark grizzled brown, and the underparts are cream-colored. The tail is densely furred but not bushy and is tricolored—the same color as the rump on the basal 45 mm, blackish brown on the next 150 mm, and cream-colored on the distal 20 mm (Musser and Gordon 1981). In the single specimen of *C. australis* head and body length is 265 mm, tail length is 281 mm, the fur is rough, the upper parts are tawny peppered with black, and the underparts are bright orange-brown. The relatively long tail is tricolored. The basal 30 mm is about the same color and texture as that of the upper parts of the body, and the rest of the tail is covered with short, bristly hairs; the proximal half is black, and the distal half is white. In coloration, body proportions, and some cranial and dental characters, *C. australis* appears to be the most primitive of the three species (Musser, Heaney, and Rabor 1985).

Except as noted, the remainder of this account applies only to *C. schadenbergi*. Head and body length is 325–94 mm and tail length is 355–475 mm. The pelage is quite dense, consisting of woolly underfur and long, straight or wavy guard hairs. The general coloration is highly variable, but the animal is usually dark brown to black on the upper parts, dark grayish on the sides, and iron gray on the underparts; some individuals, however, are white or brownish on the anterior part of the body, and occasionally the underparts are irregularly white. The long tail is exceptionally heavily haired and is thickly bushy, a unique characteristic in the family Muridae.

This large rodent has an elongated body, a slender muzzle, and small eyes and ears. The hands and feet each have five digits; the thumb has a well-developed flattened nail, and the remaining fingers and toes have powerful, slender claws. There are strong tufts of hair at the base of each claw. The hind foot is broad.

The bushy-tailed cloud rat appears to be quite common among the high mountains and plateaus of northern Luzon. It prefers a forested habitat, is arboreal, and is most active after sunset. During the day it sleeps in tree cavities or in holes among the roots of trees. It has a strange cry, so shrill that it seems like that of some insects. It is said to feed upon the buds and bark of young pine tree sprouts and on fruits, the latter being eaten while on the tree rather than after they have fallen to the ground.

Igorote natives from Mount Data trap this animal and sell its pelt in the market at Baguio. The wool-like pelt is attractive and serviceable. Several bushy-tailed cloud rats have been kept as pets, but no notes were made on their temperament. One captive lived for four years and three months (Jones 1982). On the basis of a visit to Ilin Island in November 1988, during which he found no evidence of *C. paulus* and observed that its forest habitat had been destroyed by human activity, Pritchard (1989) concluded that the species is extinct.

RODENTIA; MURIDAE; Genus BATOMYS
Thomas, 1895

Luzon and Mindanao Forest Rats

There are three species (Sanborn 1952, 1953; Taylor 1934):

B. granti, known only by five specimens from Mount Data in northern Luzon;

B. dentatus, known only by the type specimen collected in Benguet Province in northern Luzon;

B. salomonseni, known only by three specimens from Bukidnon Province in Mindanao.

The specimens of *B. salomonseni* were collected in 1951 and described by Sanborn (1953) as a new genus, *Mindanaomys*. Misonne (1969), however, listed *Mindanaomys* as a synonym of *Batomys*. Heaney and Rabor (1982) agreed and also reported specimens from Dinagat Island, which evidently represent a new species.

In *B. granti* the head and body length of the type specimen is 204 mm and the tail length (probably not complete) is 121 mm. The upper parts are fulvous and black, becoming rufous toward the rump; the underparts are slaty buff; the hands and feet are brown with whitish digits; the tail is dark brown to black and covered with thick hair; and the eyes are surrounded by a seminaked or finely haired ring. In *B. dentatus* the head and body length is 195 mm and the tail length is 185 mm. The upper parts are uniformly light brown; the underparts are ochraceous buff, more buffy than in *B. granti*; the hands and feet are dull buffy gray; the tail is uniformly blackish brown basally, white toward the tip, and thinly covered with hair; and the area around the eyes is furred normally. In *B. salomonseni* the head and body length of the type specimen is 175 mm and the tail length is 140 mm. The upper parts are generally dark brown, becoming lighter and more buffy on the sides; the hairs of the underparts have a gray base and a long buffy yellow tip; one of the hands, both of the feet, and all of the digits of the type specimen are white, except for a dark line from the wrist and the ankle to the digits (the other hand is wholly white); and the tail is blackish brown above and below and fairly well haired. The fur of *Batomys* is thick and soft.

These rats resemble *Carpomys* externally but can be distinguished by their relatively shorter tail and by cranial and dental characters. In *Batomys*, for example, the skull is more elongate and the bullae are smaller than in *Carpomys*.

Specimens of *Batomys* have been taken in areas of dense vegetation at elevations of about 1,600–2,400 meters. Native people, with the aid of their small terriers, captured the specimens of *B. granti*. The IUCN now classifies this species as indeterminate.

RODENTIA; MURIDAE; Genus CARPOMYS
Thomas, 1895

Luzon Rats

There are two species (Sanborn 1952; Taylor 1934):

C. melanurus, known only by four specimens from Mount Data in northern Luzon;

C. phaeurus, known only by three specimens from Mount Data and one from Mount Kapilingan in northern Luzon.

In *C. melanurus* head and body length is about 200 mm and tail length is about 210 mm. The tail is a deep, shining black and is thickly furred for 25–50 mm next to the body. The incisor teeth are quite large. In *C. phaeurus* head and body length is 175–95 mm and tail length is 160–80 mm. The tail is dark brown or blackish, but never shining black, and is thickly furred for only a short distance. The incisors are much smaller than in *C. melanurus*.

Batomys sp., photos by Paul D. Heideman.

Otherwise, both species look much alike. The soft fur is fluffy and thick. The upper parts are deep fulvous, coarsely lined with black, and the underparts are yellowish white. The bases of the hairs are slate or buffy white in *C. melanurus*, but they are not slaty in *C. phaeurus*. The body form in both species is somewhat heavy. The broad hind foot is adapted to arboreal life, but the large toe is clawed and not opposable. The thumb is the only digit with a nail instead of a claw. Females have four mammae.

The known specimens were collected in areas of dense vegetation at elevations of about 2,100–2,400 meters. Taylor (1934) stated that both species are arboreal, but the specimens of *C. phaeurus* from Mount Data were dug out from among the roots of trees.

RODENTIA; MURIDAE; **Genus PHLOEOMYS**
Waterhouse, 1839

Slender-tailed Cloud Rats

There are two species (Taylor 1934):

P. cumingi, southern Luzon, Mindoro, and Marinduque islands (Philippines);
P. pallidus, northern Luzon.

Schauenberg (1978) suggested that *P. pallidus* is only an individual or seasonal variant of *P. cumingi*. He also recog-

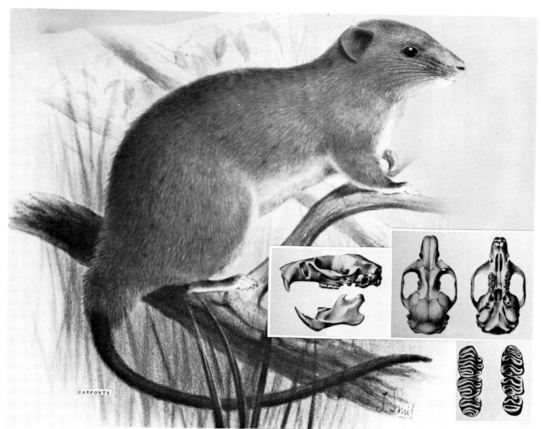

Luzon rat *(Carpomys melanurus)*, photos from *Trans. Zool. Soc. London*.

Slender-tailed cloud rat *(Phloeomys cumingi)*, photo by Ernest P. Walker.

nized the establishment of a separate family, the Phloeomyidae, for *Phloeomys*. Taylor (1934) listed another species, *P. elegans*, known only by a single specimen from an unknown locality in the Philippines, but he suspected that it actually belonged to one of the other two named species. Honacki, Kinman, and Koeppl (1982) included *P. elegans* in *P. cumingi*.

These rats are the largest members of the subfamily Murinae. Head and body length is 280–485 mm, tail length is 200–350 mm, and weight is about 1.5–2.0 kg. In *P. cumingi* the pelage is rather rough, suberect, and intermixed with long hairs. The general coloration of the upper parts is blackish brown washed with dirty yellowish or reddish yellow, and there may be an irregular reddish brown blotch on the dorsal surface; the underparts are paler. In *P. pallidus* the pelage is long, dense, and soft in comparison with that of *P. cumingi*. The general coloration of the upper parts results from the longer individual hairs, which have a brown or reddish brown base and a white tip, and the shorter fur is uniformly brownish; the hairs on the anterior part of the body have a gray to brown base and a black tip; the ears are black; the cheeks are gray; and the tail is black to brownish black. In both species the well-haired tail is not bushy.

Phloeomys has broad incisor teeth, a blunt muzzle, and small ears. The feet are large and wide, adapted to arboreal life, and the foreclaws are large. Females have four mammae.

Slender-tailed cloud rats live in forested areas from sea level to high mountains. Information compiled by Schauenberg (1978) suggests that they may shelter in either burrows or hollow tree trunks. They apparently are mainly arboreal and nocturnal. Their movements seem relatively sluggish. The natural diet is not known, but captives have taken a wide variety of vegetable matter.

According to Schauenberg (1978), births in captivity have been recorded in all months except January, March, and May, and a wild pregnant female was taken on 2 August. There is only a single young at a time, which the mother carries firmly attached to a nipple. *Phloeomys* thrives in captivity; one specimen lived for 13 years and 7 months.

RODENTIA; MURIDAE; Genus CRUNOMYS
Thomas, 1897

Philippine and Sulawesian Shrew Rats

Four species have been described (Musser 1982*b*; Taylor 1934):

C. fallax, known by a single specimen from northeastern Luzon;

C. rabori, known by a single specimen from north-central Leyte;

C. melanius, known by two specimens from Mindanao;

C. celebensis, known by three specimens from central Sulawesi.

Misonne (1969) stated that *Crunomys* showed evolutionary trends that associated it with, though did not place it within, the subfamily Hydromyinae. Musser (1982*b*) suggested that *Crunomys* might be related to *Rhynchomys*, *Chrotomys*, and *Celaenomys*, the last of which is here placed in the Hydromyinae, but also to other Philippine species of the Murinae. He noted that *Crunomys* is structurally a primitive rat.

The type specimen of *C. fallax* has a head and body length

of 105 mm and a tail length of 79 mm. The pelage consists of close, short fur, which is profusely mixed with flattened spines. The whiskers are long, and the tail is uniformly covered with short hair. The upper parts are generally grayish but are rather yellowish on the back; the underparts are grayish white. The dorsal spines have a white base and a black tip. The sides of the nose and the ears are brown, the hands and feet are grayish brown, and the digits are white. The tail is black above and paler on the undersurface.

The two specimens of *C. melanius* have head and body lengths of 98 and 122 mm and tail lengths of 68 and 79 mm. The fur, which is close and fine, is intermixed with flattened spines. The general coloration of the upper parts is blackish brown, and the limbs, hands, and feet are colored much the same. The tail is uniformly blackish. The underparts are only slightly paler than the upper parts. One specimen, a female, has six mammae.

The type specimen of *C. rabori* has a tail length of 90 mm. The head and body length can not be precisely determined, because of damage, but is greater than the tail length. The fur consists mostly of flattened spinelike hairs. The upper parts are very dark brown, the underparts are pale brownish gray, the ears are blackish brown, the feet are brown, and the tail is blackish brown above and mottled below (Musser 1982*b*).

The three specimens of *C. celebensis* have head and body lengths of 115, 118, and 127 mm, tail lengths of 80, 82, and 84 mm, and weights of 55, 55, and 35 grams. The fur is short, thick, and soft and not spinous as in the other species. The upper parts are dark chestnut, the underparts are paler and partly suffused with gray, and there are dark markings on the flanks and limbs. The head is short and broad, the ears small and round, the body stocky, the legs short, the hind feet narrow, and the claws moderately narrow and not recurved. One specimen, a female, has eight mammae (Musser 1982*b*).

Musser (1982*b*) wrote that *Crunomys* is characterized by small size, a tail much shorter than the head and body, a moderately long rostrum relative to cranial length, smooth dorsolateral margins of interorbital and postorbital regions of the skull, a round and smooth braincase, zygomatic plates set anterior to the molar rows, upper molars anchored by three roots, and other cranial and dental features. *Crunomys* differs from most murids in the configuration of its pterygoid region and carotid arterial pattern. In most murids, including the new genus *Archboldomys*, a large stapedial artery branches off from the common carotid, goes through the otic region, emerges as the internal maxillary artery onto the posterior pterygoid plate, and then passes beneath the plate into the alisphenoid canal. In *Crunomys* there is only a small stapedial artery that does not continue past the otic region; the internal maxillary artery branches off more anteriorly and then courses across the alisphenoid wing of the pterygoid plate.

The type specimen of *C. fallax* was collected in 1894 in a forest at an elevation of approximately 307 meters. It was shot beside a stream, where it was foraging for food. One specimen of *C. melanius* was collected in 1906 at an elevation of about 924 meters on Mount Apo. The other specimen was obtained in 1923 in a forest at sea level (Taylor 1934). The type of *C. rabori* was taken in a mountainous area; its body proportions are those of a terrestrial animal. *C. celebensis* was found in tropical rainforest at elevations around 1,000 meters. All three specimens were caught near or on forested terraces next to wet ravines or small streams. This species is considered terrestrial and probably diurnal (Musser 1982*b*).

Crunomys celebensis, drawing by Fran Stiles through Guy G. Musser.

RODENTIA; MURIDAE; **Genus ARCHBOLDOMYS**
Musser, 1982

Mount Isarog Shrew Rat

The single species, *A. luzonensis*, is known only by the holotype, a young adult male from Mount Isarog, a volcanic peak on the southeastern peninsula of Luzon Island in the Philippines (Musser 1982*b*).

Head and body length of the holotype is 70 mm and tail length (with tip missing) is 70 mm. The pelage is dense, soft, and slightly woolly. The upper parts are dark chestnut, the underparts are mostly dark gray overlaid with pale buff and silver flecking, and there are no patterns or sharp demarcations in color. The ears are small and dark chestnut, the feet are small and brown, and the relatively short tail is brown above and slightly paler below.

Archboldomys is thought to be closely related to *Crunomys*, and resembles the latter in size of head and body, tail, feet, and ears. *Archboldomys* differs, however, in having such features as long and thick fur, elongate claws on the front feet, a slender tapered rostrum, inflated bullae that are longer relative to length of skull, and a configuration of the pterygoid region and carotid arterial pattern as described in the account of *Crunomys*.

The holotype was collected at an elevation of about 2,000 meters, probably in montane forest. The species is thought to be terrestrial and diurnal and to feed on earthworms and other soft-bodied invertebrates dug out of the forest floor.

RODENTIA; MURIDAE; **Genus RHYNCHOMYS**
Thomas, 1895

Shrewlike Rats

There are two species (Musser and Freeman 1981; Taylor 1934):

R. soricoides, known by seven specimens from Mount Data in northern Luzon;

R. isarogensis, known by a single specimen from Mount Isarog in southeastern Luzon.

In *R. soricoides* head and body length is 188–215 mm and tail length is 132–46 mm. In the type specimen of *R. isarogensis* head and body length is 187 mm and tail length is only 105 mm. The pelage is thick, short, and velvety. The upper parts are uniformly dark olivaceous gray. The underparts are dirty gray and not sharply defined from the back in *R. soricoides* but appear whitish gray in *R. isarogensis*. The dorsal surfaces of the feet are dark brown in *R. soricoides* but white in *R. isarogensis*. The tail is fairly well covered with hair and is not tufted; in *R. soricoides* it is blackish above and slightly paler

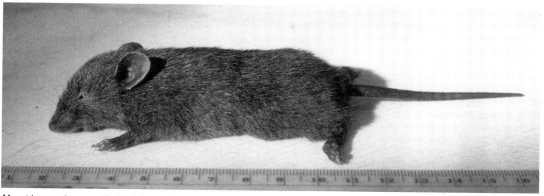

Mount Isarog shrew rat *(Archboldomys luzonensis)*, photo of dead specimen by Paul D. Heideman.

Shrewlike rat *(Rhynchomys isarogensis)*, photos by Paul D. Heideman.

below, but in *R. isarogensis* it is dark brown above and mostly unpigmented below.

These rodents resemble shrews in having an elongated muzzle and relatively small eyes. The rostrum of the skull is greatly elongated, and the lower jaw is long, slender, and straight. There are no third molar teeth, and the remaining molars are extraordinarily reduced in size and complexity. The upper incisors are white, the lower ones pale yellow. The feet are similar to those of *Rattus*. The great toe has a small nail.

Shrewlike rats are apparently rare, and little is known about their biology. The specimens of *R. soricoides* were taken in thick bushes and mossy forests at elevations of 2,286–2,460 meters. *R. isarogensis* was collected in a montane forest at 1,660 meters. Because of the small cheek teeth

of these animals, it has been suggested that their diet consists of soft food, mainly insects and worms, rather than plant material.

RODENTIA; MURIDAE; Genus CHROTOMYS
Thomas, 1895

Luzon Striped Rat

There are two species (Musser, Gordon, and Sommer 1982):

C. whiteheadi, mountains of northern Luzon;
C. mindorensis, lowlands of Luzon and Mindoro.

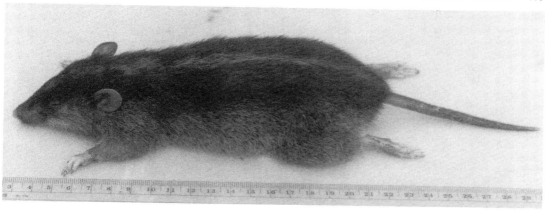

Luzon striped rat (*Chrotomys* sp.), photos by Paul D. Heideman.

side by broad, shining black bands, extends from the middle of the face along the middle of the back almost to the tail. The ground color of the back gradually shades lighter on the sides to slaty gray on the underparts. The fingers and toes are white, and the remaining portions of the hands and feet are shiny gray. The tail is thinly haired, black above and paler beneath, with the tip sometimes white. The eyes are quite small, and the ears are fairly large and covered with short, fine hairs. The pollex has a rounded nail, whereas the other digits have well-developed, slightly curved claws. The incisor teeth are pale yellow and project somewhat forward. Females have two pairs of mammae.

Specimens have been taken in montane forests at elevations of up to 2,600 meters but also in lowland rice fields and grasslands. The diet is said to include sweet potatoes, grass, and earthworms.

RODENTIA; MURIDAE; Genus TOKUDAIA
Kuroda, 1943

Ryukyu Spiny Rat

The single species, *T. osimensis,* is known from Amami-Oshima, Tokuno-Shima, and Okinawa in the Ryukyu Islands

Head and body length is 150–96 mm, tail length is 90–120 mm, and weight is 115–60 grams (Temme 1974). The fur is short, soft, and straight. The general coloration above is grayish brown, and some individuals have a rufous tinge. A well-defined bright buff or orange line, bordered on either

Ryukyu spiny rat *(Tokudaia osimensis),* photo by K. Tsuchiya.

south of Japan (Ellerman and Morrison-Scott 1966; Wang, Zheng, and Kobayashi 1989).

Head and body length is 125–75 mm and tail length is 100–125 mm. The upper parts are mixed black and orange tawny, and the underparts are grayish white with a faint orange wash. The tail is bicolored throughout. Externally, this rodent looks like a large vole. The body is short and thick, and the dense pelage consists of fine hairs and coarse, grooved spines. The latter are present on all parts of the body except the region around the mouth and ears, the feet, and the tail. The dorsal spines are black throughout, and the ventral spines are usually white with a rufous tip. Females have four mammae.

In the U.S. National Museum of Natural History are 13 specimens that were collected on northern Okinawa in a thick, shrubby forest about three meters high with an undergrowth of coarse grasses and brake ferns. The IUCN (1972) classifies *T. osimensis* as indeterminate and believes the species to be seriously endangered by the general destruction of natural areas on the Ryukyu Islands. Wang, Zheng, and Kobayashi (1989) indicated that the species is very low in numbers but is protected.

RODENTIA; MURIDAE; Genus CHIRUROMYS
Thomas, 1888

There are four species (Dennis and Menzies 1979):

C. forbesi, southeastern New Guinea, D'Entrecasteaux and Louisiade archipelagoes;
C. kagi, southeastern New Guinea;
C. lamia, southeastern New Guinea;
C. vates, eastern New Guinea.

Chiruromys was long considered to be only a subgenus of *Pogonomys*, but following a study of morphology and ka-

ryology, Dennis and Menzies (1979) concluded that the two taxa are not closely related and should be designated separate genera. Corbet and Hill (1986) included *C. kagi* within *C. lamia*.

Head and body length is 115–75 mm and tail length is 160–245 mm (Tate 1951b). *Chiruromys* resembles *Pogonomys* in external appearance. The pelage is soft and dense. The upper parts vary in color from grayish through fawn and brown to rich reddish brown, and the underparts are usually white. Females have six mammae. The tail characteristics are generally the same as in *Pogonomys*, but *Chiruromys* can be distinguished by the rough appearance of its tail: the scales stand out like teeth on a rasp, whereas the tail of *Pogonomys* has a smoother appearance.

These rodents are found in forested areas from sea level to 2,000 meters. Unlike *Pogonomys*, they are completely arboreal, nest in hollow trees, and probably do all their foraging in the forest canopy. The diet and colonial habits are the same as in *Pogonomys*. A pregnant female and a newborn individual were found in December. Litter size is one to three young (Menzies and Dennis 1979).

RODENTIA; MURIDAE; Genus POGONOMYS
Milne-Edwards, 1877

Prehensile-tailed Rats

There are four species (Dennis and Menzies 1979; Flannery 1988):

P. macrourus, New Guinea;
P. championi, highlands of western Papua New Guinea;
P. sylvestris, highlands of New Guinea;
P. loriae, highlands of New Guinea, D'Entrecasteaux Archipelago.

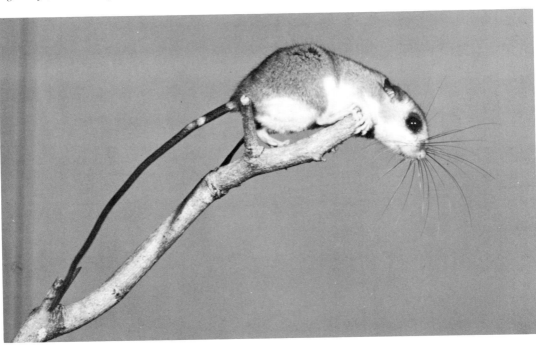

Chiruromys vates, photo by J. I. Menzies.

Prehensile-tailed rat *(Pogonomys sylvestris)*, photo by P. A. Woolley and D. Walsh.

Pogonomys also recently was discovered on the Cape York Peninsula of northern Queensland. The specimens were referred to *P. loriae* by Corbet and Hill (1986), *P. macrourus* by Honacki, Kinman, and Koeppl (1982), and *P. mollipilosus* (probably a synonym of one of the other species) by J. W. Winter (*in* Strahan 1983). *Chiruromys*, long considered a subgenus of *Pogonomys*, was regarded as a separate genus by Dennis and Menzies (1979), and that position is followed here.

Head and body length is 95–148 mm and tail length is 150–214 mm (Tate 1951*b*). Mean adult weight is about 40 grams in *P. sylvestris*, 74 grams in *P. loriae* (Dwyer 1975*b*), and 55 grams in *P. macrourus* (McPhee 1988). The pelage is dense, woolly, and quite soft, with few guard hairs. The numerous whiskers are long, dark, and conspicuous. The general coloration of the upper parts ranges from grayish through rufous and reddish brown to dark brown or almost black; the paler-colored forms have a sprinkling of silver hairs on the upper parts. The underparts are white, buffy, or grayish. The long, scantily haired, prehensile tail is coarsely scaled and mostly brown in color. In most species, however, the terminal 10–20 mm of the upper surface of the tail is flesh-colored, naked, and smooth. *P. championi* is unique in that its tail lightens with age, beginning from the distal end. Unlike most prehensile tails, the tail of *Pogonomys* curls backward or up at the tip, rather than downward.

Pogonomys has a short head and large eyes. The hands and feet are short, broad, and modified for climbing. There is a slightly opposable hallux with a fully developed claw. The molar teeth have a complex folded pattern. Females have six mammae.

Prehensile-tailed rats are found in forests from sea level to about 2,700 meters in elevation. Menzies and Dennis (1979) reported that they are nocturnal and only partly arboreal, foraging in low vegetation or on the ground and nesting in underground burrows. The nests of *P. loriae* are said to have several entrance tunnels that may be many meters in length. The diet consists largely of young leaves and shoots of grass and bamboo.

Menzies and Dennis (1979) stated that these rats are colonial. A single nest may contain a group of up to 15 individuals of all ages and sexes, though usually there are only 3–5 animals per colony. In a study in eastern Papua New Guinea, Dwyer (1975*b*) found that breeding apparently ceases during the dry season from July to September but is otherwise continuous. In both *P. loriae* and *P. sylvestris* the usual litter size is 2–3 young. There are 3–4 young in *P. macrourus* (McPhee 1988). According to Jones (1982), a captive *P. macrourus* lived for two years and five months.

RODENTIA; MURIDAE; **Genus MALLOMYS**
Thomas, 1898

Giant Tree Rats

There are four species (Flannery et al. 1989):

M. rothschildi, mountains of New Guinea;
M. aroaensis, highlands of Papua New Guinea;
M. istapantap, mountains of east-central New Guinea;
M. gunung, Snow Mountains of west-central New Guinea.

Prehensile-tailed rat *(Pogonomys loriae)*, photo by P. A. Woolley and D. Walsh.

Giant tree rat *(Mallomys rothschildi)*, photo by J. I. Menzies.

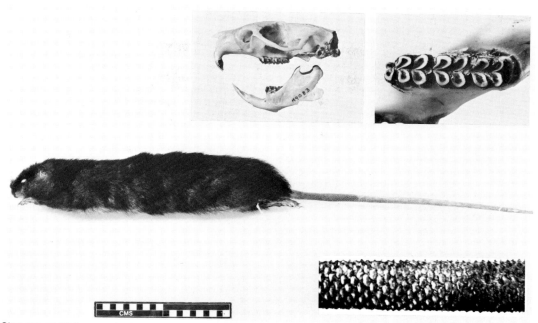

Giant tree rat *(Mallomys rothschildi)*, photo by C. W. Turner through Basil Marlow. Insets: section of tail from *Zeitschr. Saugetierk.*; photos of skull and right lower molar row by C. W. Turner through Basil Marlow.

Head and body length is 292–477 mm, tail length is 280–440 mm, and weight is 925–1,950 grams (Flannery et al. 1989). There appears to be considerable variation in both the texture and the color of the pelage. The fur is quite long and thick and has a tendency to be somewhat woolly in most specimens. The general coloration of the upper parts is dark brown to fuscous brown or grayish, with the guard hairs being all black or black with a white tip; the shades of coloration are dependent upon the abundance and color of the guard hairs. In at least one form the crown is light brown. The underparts are creamy or dull white, and the individual hairs are unicolored. In a few specimens a white band is present, extending across the middle of the underparts and well up onto the sides. The hands, feet, and whiskers are black. The tail is scantily haired and coarsely scaled; the basal part is brownish, and approximately the terminal half is white.

The giant tree rat has the usual murid body form and large, broad, heavy feet. The pollex has a short nail, and the other digits have considerably enlarged and slightly curved claws. The muzzle is short, and the skull is heavy and thick. The moderately broad incisors are ungrooved and project slightly forward. Menzies and Dennis (1979) stated that the molar teeth are very distinct, their crowns being divided into an intricate triple series of cusps with a regularity not seen in any other genus. Females have six mammae.

According to Menzies and Dennis (1979), *Mallomys* is evidently common in montane forests throughout New Guinea. It seems to be mainly arboreal. Its lair is usually in a hollow tree, often at a considerable height, but is sometimes on the ground. It is entirely vegetarian, feeding largely on shoots such as those of climbing bamboo. In some areas the local people said that there is a second kind of *Mallomys* that is more terrestrial, is found in open country as well as forest, and lives in holes in the ground. This claim seems to have been borne out by the report of Flannery et al. (1989) that while *M. rothschildi* almost always nests in tree hollows, the newly described species *M. aroaensis* usually nests in bur-

rows. They also reported the collection of pregnant females, each with a single embryo, in July, October, and December. The large size of *Mallomys* makes it a desirable game animal. Its teeth are sometimes extracted and used for engraving.

RODENTIA; MURIDAE; Genus HYOMYS
Thomas, 1903

White-eared Giant Rat

The single species, *H. goliath*, is found in New Guinea (Laurie and Hill 1954).

Head and body length is 295–390 mm and tail length is 255–380 mm. The weight is around 1 kg (Menzies and Dennis 1979). The fur is coarse and harsh. The color of the upper parts is usually mixed gray and fuscous or dark slaty gray. In some individuals the guard hairs are gray with a white subterminal band, and in others they are black with a white tip or white throughout. The underparts are grayish, buffy gray, or dull white. The tail in *Hyomys* is coarsely scaled and almost naked, and the scales appear even rougher and more naked than those of *Mallomys*.

Hyomys is a large rat with a somewhat bulky form. Growth apparently continues for a period after maturity is reached. The large toe has a broad nail. *Hyomys* has small ears, well-developed claws, and a tail with large, overlapping scales. The scales are subject to considerable wear. The skull is massive, and the molar teeth are large. Menzies and Dennis (1979) stated that *Hyomys* can be distinguished from *Mallomys* in that the three cusps of each loph of its molar crowns are merged and form transverse ovals, while in *Mallomys* the separate cusps are distinct. Female *Hyomys* have four mammae.

According to Menzies and Dennis (1979), *Hyomys* is fairly common in hill forest from 1,200 to 3,000 meters throughout

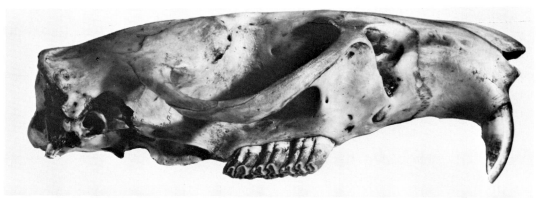

White-eared giant rat *(Hyomys goliath),* photos from British Museum (Natural History). Inset of tail from *Zeitschr. Saugetierk.*

New Guinea. Although it climbs readily, it seems more terrestrial than *Mallomys* and is said to make a large nest of leaves between tree roots, between rocks, or in a hollow log. The diet consists mainly of bamboo and other such shoots, but people occasionally complain that it raids gardens. Litters may ordinarily contain only a single young.

RODENTIA; MURIDAE; Genus ANISOMYS
Thomas, 1903

Powerful-toothed Rat

The single species, *A. imitator,* occurs in New Guinea (Laurie and Hill 1954).

Head and body length is 244–300 mm and tail length is 285–330 mm. Weight is around 500–600 grams (Menzies and Dennis 1979). The coat is composed of short, coarse hairs. The upper parts are blackish fawn, and the underparts are dull buffy white. The head is nearly black. The arms and legs are grayish, and the hands and feet are brown, becoming white on the digits. Hexagonal scales show through the scantily haired tail. *Anisomys* has small ears, strong scansorial feet, and a long tail. The great toe has a broad nail. Females have six mammae.

Anisomys resembles *Uromys* in external appearance but has highly distinctive dental characters (Menzies and Dennis 1979). The lower incisor teeth are laterally compressed and bladelike and are less than half the width of the upper incisors. The molar teeth are remarkably small, the length of the toothrow in a given specimen being less than half that of a specimen of *Uromys* with the same external measurements.

Anisomys is apparently distributed throughout the rainforests of New Guinea between elevations of 900 and 2,700 meters (Laurie and Hill 1954). Most specimens have been trapped on the ground, but the broad scansorial feet indicate an ability to climb easily. The small size of the molars may be associated wth a diet that includes nuts with hard shells but soft, pulpy interiors. *Anisomys* also eats other vegetable matter and is said to raid gardens (Menzies and Dennis 1979).

RODENTIA; MURIDAE; Genus UROMYS
Peters, 1867

Giant Naked-tailed Rats

There are six species (Laurie and Hill 1954; Menzies and Dennis 1979):

U. caudimaculatus, New Guinea, Kei and Aru islands, northern Queensland;
U. anak, New Guinea;
U. neobritannicus, New Britain Island (Solomons);
U. rex, Guadalcanal Island (Solomons);
U. imperator, Guadalcanal Island (Solomons);
U. salamonis, Florida Island (Solomons).

Head and body length is about 200–400 mm and tail length is about 200–350 mm. The weight of *U. caudimaculatus* is around 600–700 grams and that of *U. anak* is over 1,000 grams (Menzies and Dennis 1979). The pelage is fairly short, dense, and relatively coarse, except in *U. caudimaculatus,* in which it is rather scant. The coloration of the upper parts ranges from grayish through various shades of brown to gray black; the underparts are white or grayish. In some species, such as *U. anak,* the tail is entirely dark, but in others, such as *U. caudimaculatus,* the tail is white or yellowish toward the tip. The hairs of the tail number only one per scale. The scales of the tail form a mosaic pattern, whereas in most other

Giant naked-tailed rat *(Uromys caudimaculatus)*, photo by Stanley Breeden.

genera of the subfamily Murinae the scales of the tail are flat and overlapping.

In *U. neobritannicus,* the largest species, the squamosal crest of the skull is well developed and the ears are relatively large. The other species lack a well-developed squamosal crest. The ears are relatively long in *U. caudimaculatus* and *U. anak* but short in the remaining three species. The teeth of *Uromys* are similar to those of *Melomys* but much longer. The incisors are large and powerful, with uppers and lowers being about the same size. The hands and feet are broad and well clawed. Females have two pairs of mammae.

Giant naked-tailed rats are arboreal and are excellent climbers. Their tail is not truly prehensile but does facilitate climbing by curling over limbs and gripping with the rasplike scales. According to Menzies and Dennis (1979), *U. anak* inhabits forests at elevations of 1,000–2,500 meters, and *U. caudimaculatus* occurs in both forests and grasslands in the lowlands. Nests are usually made in hollow trees, though *U. caudimaculatus* sometimes seems to raise its young on bare rock in caves or old mine tunnels. The diet consists of coconuts, other nuts, fruits, and flowers. Litters of *U. caudimaculatus* have been recorded in New Guinea during November and December and may contain one to three young. The young remain with the mother until they are at least half-grown. *U. anak* is often hunted for food by the highland natives of New Guinea.

Referring to Australia, Wellesley-Whitehouse (*in* Strahan 1983) stated that *U. caudimaculatus* sometimes becomes a pest through its raids on coconut plantations and kitchens. With its formidable incisors, it is even able to open cans of food. Population density varies greatly, and home range appears to be large. Individuals are solitary and probably territorial. Males become highly aggressive during the breeding season, which begins in October–November with the onset of the summer rains. The gestation period is 36 days, and litters usually contain two or three young.

RODENTIA; MURIDAE; Genus SOLOMYS
Thomas, 1922

Naked-tailed Rats

There are two subgenera and three species (Laurie and Hill 1954):

subgenus *Solomys* Thomas, 1922

S. sapientis, Santa Ysabel Island (Solomons);
S. salebrosus, Bougainville Island (Solomons);

subgenus *Unicomys* Troughton, 1935

S. ponceleti, Bougainville Island (Solomons).

Some authorities have included *Solomys* in the genus *Melomys,* but Laurie and Hill (1954) did not agree, noting that *Solomys* is much larger and has a larger and much heavier skull with large bullae. Flannery et al. (1988) treated *Unicomys* as a full genus and reported it from an archaeological site about 1,860 years old on Buka Island in the Solomons.

The species *S. sapientis* has a head and body length of about 250 mm and a tail length of about 250 mm; coloration is cinnamon brown above and on the sides and pinkish buff below. The species *S. salebrosus* has a head and body length of about 230 mm and a tail length of about 215 mm; coloration is yellowish brown above, cinnamon buff on the sides, and pinkish buff below. *S. ponceleti* is the largest species; a young adult female measured 330 mm in head and body length and 340 mm in tail length. The color of this species is brownish black above and below. The hair of *S. ponceleti* is long and fine, without woolly underfur, whereas it is coarser in the other two species.

The tail of *Solomys* appears to be prehensile and lacks hairs

Powerful-toothed rat *(Anisomys imitator)*, photos from Museum Zoologicum Bogoriense.

Naked-tailed rat *(Solomys salebrosus)*, photo from Australian Museum, Sydney.

for most of its length. The well-padded feet are supplied with mobile and strongly clawed digits. The incisor teeth are broad and stout. Females have four mammae.

Naked-tailed rats inhabit thick woods and are apparently arboreal. According to Troughton (1936:347), *S. sapientis* "cracks the ngali *(Canarium)* nuts and gnaws coconuts, and is found in trees felled by the natives." The natives of the Solomon Islands occasionally eat these rats.

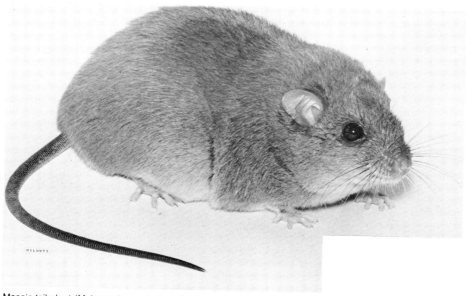

Mosaic-tailed rat *(Melomys burtoni)*, photo by Stanley Breeden.

RODENTIA; MURIDAE; **Genus MELOMYS**
Thomas, 1922

Mosaic-tailed Rats, or Banana Rats

There are 21 species (Baverstock et al. 1980; Knox 1978; Laurie and Hill 1954; Menzies and Dennis 1979; Ride 1970; Tate 1951*b*; Watts 1979; Winter 1984*a*):

M. albidens, known only from the type locality at Lake Habbema in western New Guinea;

M. fellowsi, mountains of eastern New Guinea;

M. levipes, New Guinea;

M. hadrourus, known by six specimens from northeastern Queensland;

M. lorentzi, New Guinea;

M. aerosus, Ceram Island between Sulawesi and New Guinea;

M. cervinipes, coastal Queensland, northeastern New South Wales;

M. capensis, rainforests of northeastern Queensland;

M. rubicola, Bramble Cay off south-central New Guinea;

M. moncktoni, New Guinea;

M. obiensis, Obi Island between Sulawesi and New Guinea;

M. platyops, New Guinea;

M. rubex, New Guinea;

M. lutillus, New Guinea, Cape York Peninsula of northern Queensland;

M. burtoni, coastal areas from northeastern Western Australia to northeastern New South Wales, Melville and Groote Eylandt islands;

M. rufescens, New Guinea, Bismarck Archipelago, Solomon Islands;

M. leucogaster, southern New Guinea;

M. fulgens, Ceram and Talaud islands between Sulawesi and New Guinea;

M. arcium, Rossel Island off eastern New Guinea;

M. porculus, Guadalcanal Island (Solomons);

M. fraterculus, Ceram Island between Sulawesi and New Guinea.

Winter (1984*a*) suggested that *M. hadrourus* might actually represent a new genus, systematically intermediate to *Uromys* and *Melomys*. Baverstock et al. (1980) and Knox (1978) indicated that the Australian populations, at least, of *M. lutillus* belong in the species *M. burtoni,* and this arrangement was followed by Redhead (*in* Strahan 1983). However, Corbet and Hill (1986) and Honacki, Kinman, and Koeppl (1982) continued to list *M. lutillus* as a full species.

Head and body length is 90–180 mm, tail length is 110–97 mm, and weight is 30–200 grams. The pelage is usually soft, thick, and woolly and sometimes is fairly long. The coloration of the upper parts ranges from tawny through several shades of brown to russet. The underparts are white, creamy, or pale brownish or gray. Like *Uromys,* these rodents are easily recognized by their almost naked, filelike tail. The scales of the tail form somewhat of a mosaic pattern, which differs strikingly from the evenly ringed arrangement seen on the tails of the typical rats. In some species the tail is slightly prehensile (Menzies and Dennis 1979). The feet are broad and have well-developed pads. From *Uromys, Melomys* is distinguished externally by its smaller size. Females of most species of *Melomys* have two pairs of mammae, but female *M. lorentzi* have only a single pair (Menzies and Dennis 1979).

Most species of *Melomys* are forest dwellers, but *M. lutillus, M. rufescens,* and *M. burtoni* are primarily inhabitants of grassland and open scrub country; the elevational range of the genus is sea level to over 3,000 meters. Most species are basically terrestrial but capable of climbing. *M. lorentzi* and *M. rubex* live in burrows in the forest floor, while *M. lutillus, M. rufescens,* and *M. cervinipes* may nest in vegetation up to 3 meters above the ground (Dwyer 1975*b*; Menzies and Dennis 1979; Ride 1970; Taylor and Horner 1970*b*). Nests are also located among roots, in hollow timber, and under rock ledges. Aboveground nests, made of grass or leaves, are spherical and usually 125–200 mm in diameter. The diet of *Melomys* is thought to consist of fruits, berries, and other vegetable matter. Wood (1971) found *M. cervinipes* to be extensively arboreal and strictly nocturnal. Barnett, How, and Humphreys (1977) reported population densities of 0.37–0.97 per ha. for this species. In an investigation of *M. burtoni* in the Northern Territory, Begg et al. (1983) found males to reach a density of 20.8/ha. and to occupy overlapping home ranges of about 0.43 ha.; respective figures for females were 17.7/ha. and 0.23 ha. Breeding occurred throughout the year.

Studies in eastern Australia (Taylor and Horner 1970b; Wood 1971) indicate that *M. cervinipes* and *M. burtoni* breed throughout the year but have seasonal peaks that differ by locality. Female *M. cervinipes* are known to be polyestrous. In more southerly areas, reproduction occurs at least in late spring and summer, and farther north it extends into the autumn and winter. In New Guinea some species apparently breed all year, but data are limited (Dwyer 1975b; Menzies and Dennis 1979). Overall litter size for the genus is one to five, with two young being most common. The young cling tenaciously to the nipples of the mother and are dragged about almost continuously during their first two weeks of life. Subsequently the young become more independent but still hold onto the mammae when disturbed. Their eyes open at about 10 days, and weaning occurs after 20 days (Redhead, *in* Strahan 1983). Captives usually produce their first litter at 7 months.

Watts (1979) considered *M. hadrourus*, which was initially thought possibly to be a newly discovered population of *M. levipes*, to be at risk but noted that the latter species is common in New Guinea. Limpus (1983) stated that *M. rubicola* numbers only a few hundred individuals and is in danger of extinction because of the natural erosion of the small coral cay on which it lives but noted also that this species may be only an island race of *M. capensis*. *Melomys* is not known to occur in plague proportions in Australia, but *M. burtoni* becomes locally abundant in cane plantations and is regarded there as a major pest (Taylor and Horner 1970b).

RODENTIA; MURIDAE; Genus POGONOMELOMYS
Rümmler, 1936

Rümmler's Mosaic-tailed Rats

There are four species (Laurie and Hill 1954; Menzies and Dennis 1979):

P. mayeri, mountains of western New Guinea;
P. bruijni, extreme western and south-central New Guinea;
P. sevia, mountains of northeastern New Guinea;
P. ruemmleri, mountains of New Guinea.

The name *Rattus shawmayeri*, which was associated with *Niviventer eha* by Laurie and Hill (1954) and with the subgenus *Stenomys* of *Rattus* by Misonne (1969), is probably a synonym of *Pogonomelomys ruemmleri* (G. G. George 1979).

Head and body length is 130–90 mm and tail length is about 140–210 mm. Two adult male *P. sevia* weighed 57.1 and 64.7 grams (Dwyer 1975b). The dorsal pelage of *P. bruijni* is slightly crisp, and that of *P. sevia* and *P. ruemmleri* is long and dense. The upper parts are brownish, usually reddish brown or dark brown, and the underparts are whitish, grayish, or buffy. The hands and feet are yellowish brown to white. The tail is prehensile. Each of the six-sided scales on the tail has one, three, or many hairs. The feet have large claws. Female *P. mayeri*, *P. bruijni*, and *P. sevia* have four mammae, and female *P. ruemmleri* have six.

One species, *P. bruijni*, occurs in lowland rainforest, and the other three inhabit montane forest or alpine grassland at elevations of up to 3,300 meters. Both *P. bruijni* and *P. sevia* are evidently arboreal, but *P. ruemmleri* is at least partly terrestrial. Limited evidence indicates that the diet of *P. ruemmleri* consists only of vegetable matter, including leaves (Menzies and Dennis 1979). *P. sevia* appears to live off the ground beneath the dead fronds of pandanus palms and to be solitary. Observations suggest that *P. sevia* begins to breed in early October and that pregnancy may be concurrent with lactation. Two females taken in October were pregnant, each with a single embryo (Dwyer 1975b).

RODENTIA; MURIDAE; Genus XENUROMYS
Tate and Archbold, 1941

White-tailed New Guinea Rat

The single species, *X. barbatus*, is known only by 13 specimens from New Guinea (Flannery 1985; Menzies and Dennis 1979). One was collected in about 1900 in the southeastern part of the island; one was taken in 1939 in a heavy forest at an elevation of 75 meters near the Idenburg River in the northwestern part; one was taken in the mountains along the upper Sepik River in the north-central part; three are from Mount Dayman in the extreme southeastern part; and seven were collected in 1984 in the Yapsiei area of extreme western Papua.

Three male specimens measure 275, 310, and 430 mm in head and body length, two measure 220 and 281 mm in tail length, and one had a weight of 1,000 grams. One male is reddish brown above and buffy below, while another is gray above and white below. The hands and feet are so scantily haired that they appear naked. The tail is white for at least half its length. The tail is coarsely scaled, and the unkeeled scales are arranged annularly instead of spirally.

According to Tate (1951b:284–85), *Xenuromys* comprises "large rats having the general appearance of *Uromys*, but with the normal overlapping scales of *Rattus*. . . . Unique or unusual habits may be implied by its rarity, but its feet and tail show no structural feature to indicate whether it is particularly arboreal or fossorial, nor does the pelage suggest aquatic habits."

According to Flannery (1985), *Xenuromys* seems to be widely distributed at elevations from 75 to 1,200 meters. The single live specimen that he collected was located during the afternoon in a lair in a cleft stone in a dense rainforest. This individual attempted to escape by running along the ground rather than by climbing a tree. *Xenuromys* apparently feeds on soft fruits, seeds, and insects. It, along with other large murids of New Guinea, is killed by local people who value its dentary for use as an engraving tool.

RODENTIA; MURIDAE; Genus SPELAEOMYS
Hooijer, 1957

The single species, *S. florensis*, is known only by cranial and dental fragments from sediments, determined by radiocarbon dating to be about 3,550 years old, in Liang Toge, a cave on western Flores in the East Indies. The same deposits also contained subfossil remains of *Paulamys*, *Komodomys*, and *Papagomys*, but *Spelaeomys* is not considered to be closely related to any of those genera; instead, its systematic affinities appear to be with the endemic murid genera of New Guinea (Musser 1981b).

The specimens of *Spelaeomys* consist of 2 maxillary fragments and 30 pieces of mandibles and represent 32 individuals. The genus probably was similar to *Papagomys* in body size and superficial cranial features, but the cusp patterns on

Rümmler's mosaic-tailed rat (*Pogonomelomys* sp.), photo by J. I. Menzies.

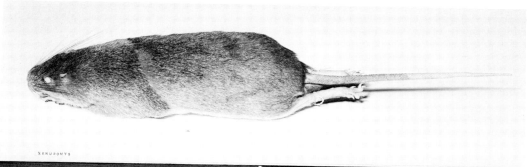

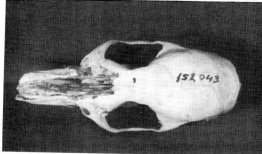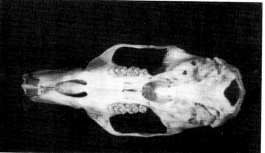

White-tailed New Guinea rat *(Xenuromys barbatus)*, photos from American Museum of Natural History.

the molar teeth of *Spelaeomys* are more complex than those found in *Papagomys* or any other native murid of Flores. The upper molars of *Spelaeomys* have many large, tall, erect, and discrete cusps, which remain separate even after appreciable wear, and there is a prominent posterior cingulum at the back of each first and second molar. The lower molars also have large, high, and discrete cusps and large and numerous cusplets along the labial margin of each tooth. The cusps resemble those of the New Guinean genus *Mallomys* in relative position, shape, and quality of being nearly erect and separate, though the overall occlusal pattern of *Mallomys* is much simpler. The hypsodont molar teeth and extensive occlusal surfaces of *Spelaeomys* suggest that the genus may have been partially arboreal and fed on leaves, buds, flowers, and insects such as moths and katydids.

RODENTIA; MURIDAE; Genus CORYPHOMYS
Schaub, 1937

The single species, *C. buehleri*, is known only by subfossil fragments from prehistoric deposits in limestone caves on the

New Guinea jumping mouse *(Lorentzimys nouhuysi)*, photo by P. A. Woolley and D. Walsh.

island of Timor in the East Indies. It was a giant rat with a complex occlusal configuration on its molar teeth closely resembling the pattern in *Spelaeomys* (Musser 1981*b*). According to Hooijer (1965), the length of the three lower molar teeth is 17.6–20.1 mm. There are no accessory buccal tubercles on these teeth, and the posterior loph of the third lower molar is straight and undivided. This combination of characters is found in no other genus.

RODENTIA; MURIDAE; Genus **LORENTZIMYS**
Jentink, 1911

New Guinea Jumping Mouse

The single species, *L. nouhuysi,* is found throughout forested parts of New Guinea (Menzies and Dennis 1979).

Head and body length is 75–85 mm and tail length is about 103–10 mm. Two males weighed 12.6 and 12.9 grams, and two females weighed 15.1 and 16.8 grams (Dwyer 1983). In the subspecies *L. n. nouhuysi,* which occurs up to an elevation of about 900 meters, the pelage is relatively short and crisp. In

the subspecies *L. n. alticola,* which occurs on mountain slopes from about 600 to 3,000 meters, the pelage is long, soft, and dense. The upper parts of *Lorentzimys* are reddish brown or grayish, and the underparts are buff or grayish white. The feet are white. The head is slaty around the eyes and ears, and the underside of the neck is whitish. In one form the hairs of the tail are white, and in the other they are black, but in both forms a pencil of hairs appears at the tip of the tail. The whiskers are numerous and fairly long. Females have six mammae.

This genus is distinguished by the cranial structure. The head is short and broad, and the upper incisor teeth are inclined slightly forward. The feet are slender, and the narrow ears taper to a rounded point. There are five fingers; four are armed with sharp, arched claws, but one is so small that only a flat nail is visible. Females have six mammae.

Dwyer (1983) reported this jumping mouse to occur in forest and to sometimes be active during daylight. According to Menzies and Dennis (1979), *Lorentzimys* seems to be completely arboreal but may prefer open or disturbed forest with abundant low shrubbery rather than mature forest. It nests in *Pandanus* and similar trees. Its morphology suggests that it runs and jumps through the branches rather than climbs. Several individuals may live together in one nest, and one pregnant female carried two embryos.

RODENTIA; MURIDAE; Genus **MACRUROMYS**
Stein, 1933

New Guinean Rats

There are two species (Laurie and Hill 1954; Menzies and Dennis 1979):

M. elegans, Weyland Mountains of western New Guinea;
M. major, mountains from west-central to northeastern New Guinea.

In *M. elegans* head and body length is 150–60 mm and tail length is 205–20 mm. Coloration is grayish brown above and whitish below, and the tail is dark above and white below. In *M. major* head and body length is 225–50 mm and tail length is 315–40 mm. Coloration is mottled yellowish black above and grayish below, and the terminal two-thirds of the tail is white. *M. major* has coarser pelage than does *M. elegans.* Females have two pairs of mammae (Menzies and Dennis 1979).

Macruromys superficially resembles *Uromys caudimaculatus,* but its tail has overlapping scales like those of *Rattus,* with each scale carrying three hairs. *Macruromys* is also distinguished by its remarkably small and simple molar teeth, a character partly shared with *Anisomys.* The length of the molar toothrow in *Macruromys major* is only about half that of a specimen of *Uromys* of comparable size (Menzies and Dennis 1979).

Both species of *Macruromys* occur in mountains. The elevational range of *M. elegans* is 1,400–1,800 meters, and that of *M. major* is 1,200–1,500 meters. Ziegler (1982*b*) stated that the two species are very scarce, terrestrial herbivores probably subsisting primarily on very soft vegetation.

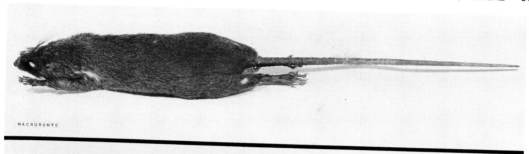

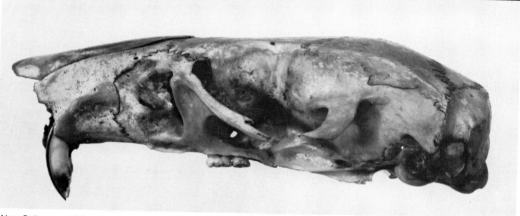

New Guinean rat *(Macruromys major)*, photos from British Museum (Natural History).

RODENTIA; MURIDAE; **Genus MESEMBRIOMYS** *Palmer, 1906*

Tree Rats

There are two species (Ride 1970):

M. macrurus, northern parts of Western Australia and Northern Territory;

M. gouldii, northeastern Western Australia to Cape York Peninsula of northern Queensland, Melville and Bathurst islands.

In *M. gouldii* head and body length is 240–350 mm and tail length is 270–390 mm. Calaby (*in* Strahan 1983) listed weight of this species as 430–870 grams. In *M. macrurus* head and body length is 188–245 mm, tail length is 291–360 mm, and weight is 207–330 grams (McKenzie, *in* Strahan 1983). The fur in both species is rather rough. The upper parts of *M. gouldii* are yellowish gray or yellowish brown, with black guard hairs, and the underparts are whitish, creamy, or slaty gray. The hands and feet are black, though the feet are irregularly blotched with cream and brown. The upper parts of *M. macrurus* are usually bright rufous or buffy brown, the sides are buffy gray, and the underparts are creamy white. The hands and feet are white. In both *M. gouldii* and *M. macrurus* the tip of the white tail is tufted, but the tail is not so well haired near its base, where the scales can be traced.

The ears are moderate in size, and their outer surfaces are quite densely haired. The foreclaws are fairly large, and the hind foot is somewhat broad, with large claws. The skull is heavily built. Females have four mammae.

Tree rats are found in woodland savannah (Ride 1970).

They are nocturnal and arboreal but may descend to the ground to feed. By day they shelter in hollow trees. At times they live near wooden buildings and climb about the rafters at night. A nest, mainly of leaves and bark, may be located in a tree. One nest covering part of the main cavity of the trunk was found in a tree about 3.5 meters above the ground. Studies of fecal samples showed that in Queensland large seeds or nuts are a major food of *M. gouldii* and that some insects also are taken (Watts 1977). According to Ride (1970), all naturalists who have handled *M. gouldii* have noted its savage temper and the severity of its bite. When enraged, it is loud-voiced, and it is said to raise its voice progressively to a sort of whirring, machinelike crescendo.

In Western Australia *M. macrurus* has been reported to breed throughout the year; a pregnant female *M. gouldii* has been recorded in mid-August (Taylor and Horner 1971). In a laboratory study of *M. gouldii,* Crichton (1969) found females to be polyestrous, to have a mean estrous cycle of 26 days, and to undergo postpartum estrus and mating. The gestation period was 43–44 days, and litters contained one to three young. The young weighed about 35 grams at birth, opened their eyes after 11 days, clung tenaciously to the teats for the first few weeks of life, and were weaned at 42 days. Females attained sexual maturity at about 2–3 months. According to Jones (1982), a captive *M. gouldii* lived for 3 years and 11 months.

Crichton (1969) indicated that *M. gouldii* has declined considerably in numbers. Watts (1979) listed both *M. gouldii* and *M. macrurus* as rare and thought the latter species to be "at risk."

Tree rat *(Mesembriomys gouldii)*, photo by A. C. Robinson.

RODENTIA; MURIDAE; Genus CONILURUS
Ogilby, 1838

Tree Rats, or Rabbit Rats

There are two species (Menzies and Dennis 1979; Ride 1970):

C. penicillatus, western Papua New Guinea, northeastern Western Australia, northern Northern Territory and nearby Melville Island;
C. albipes, southern Queensland, New South Wales, Victoria, South Australia.

Head and body length is 165–200 mm and tail length is 180–215 mm. The pelage of *C. albipes* is close and soft, while that of *C. penicillatus* is rigid and almost spiny. Coloration of the upper parts ranges from blackish brown to grayish and sandy, with tinges of buff. The underparts and the feet are white or buffy, and occasionally there is a rusty tinge on the nape and crown. The tail of *C. albipes* is distinctly bicolored, dark brown above and white below for its entire length, whereas the tail of *C. penicillatus* is wholly black or has a terminal black tip. The tail is usually uniformly haired throughout and tufted at the tip. *Conilurus* has fairly large ears and a moderately long hind foot. Females have four mammae.

Rabbit rats are found in a variety of habitats—beaches along salt water, swamps, grassy plains, and well-timbered areas. *C. penicillatus* is nocturnal in habit, sleeping by day in hollow trees and limbs, where it constructs a comfortable, well-lined nest of pieces of vegetation. In the evenings it may run along the edge of the surf, apparently feeding on matter that is thrown up on the beach by the waves.

Taylor and Horner (1971) studied a population of *C. penicillatus* on the Cobourg Peninsula of the Northern Territory. Favored habitats were near the beach in either grassy or sand dune areas. The animals were nocturnal, and both arboreal and terrestrial activity was noted. Cited observations indicate that the young cling tenaciously to the nipples, even when the female scampers about away from the nest. According to

Baverstock (*in* Strahan 1983), there is no evidence of a regular breeding season in *C. penicillatus*, females are polyestrous and probably mate soon after giving birth, the gestation period is 33–35 days, and the young become independent at about 20 days.

Although *C. penicillatus* is still common in the Northern Territory, *C. albipes* has not been seen alive in the twentieth century (Ride 1970). Watts (1979) presumed the latter species to be extinct (as it now is listed by the IUCN) and suggested that possible causes were habitat disruption by introduced rabbits and livestock and predation by introduced foxes.

RODENTIA; MURIDAE; Genus LEPORILLUS
Thomas, 1906

Australian Stick-nest Rats

There are two species (Ride 1970):

L. conditor, South Australia and nearby Franklin Island, western New South Wales, possibly northwestern Victoria;
L. apicalis, central and southeastern Australia.

Head and body length is about 140–200 mm. In *L. conditor* the tail is shorter than the head and body, while in *L. apicalis* it is longer and may be as much as 250 mm. The tail is fairly well haired, with somewhat longer hairs toward the tip. The pelage is thick and soft. The upper parts are light yellowish brown, light brown, dull brown, or pale grayish brown. The underparts in *L. conditor* are usually grayish, whereas in *L. apicalis* they are white. When resting, these fluffy-haired animals, with their blunt nose and large ears, look like small rabbits with ratlike tails. Females have four mammae.

According to Ride (1970), stick-nest rats inhabit sclerophyll woodland, shrub and tree heaths, and saltbush plain. They are nocturnal and build houses of sticks. These shelters presumably enable them to withstand the desiccating heat of

Rabbit rat *(Conilurus penicillatus)*, photo by H. Beste and J. Beste.

the desert days. Robinson (1975) found that these houses range from just a few twigs against a bush to elaborate structures up to 1.5 meters high. Size and construction method depend on local conditions. Houses are often built around a small bush or placed against a rock, but in areas where there is little woody growth *Leporillus* may live in loose heaps of sticks placed over rabbit warrens, which allow instant escape.

The latter type of home often has small stones placed on top and among the sticks to anchor the structure against strong winds. The shelters, or "wurlies," contain numerous passageways with several grass nests. They are shared at times with bandicoots, penguins, and even snakes. Along the beaches the shelter may consist of some debris or seaweed placed among the boulders. The species *L. apicalis* is not well

Australian stick-nest rat *(Leporillus conditor)*, photo by H. J. Aslin / National Photographic Index of Australian Wildlife.

known; it was said to shelter in hollow trees and to occupy the abandoned homes only of *L. conditor,* but it probably did build its own nest (Ride 1970).

The population of *L. conditor* on Franklin Island is independent of fresh water, and the staple diet there was once reported to consist of the succulent *Tetragona implexicana* (Ride 1970). More recently, Robinson (1975) found no remains of this plant in stomachs of *L. conditor* collected on Franklin Island but did confirm that the diet is entirely vegetarian.

Stick-nest rats are usually gentle and tame in captivity. They are gregarious, and some large houses may shelter a colony of animals, each family occupying one of a number of interconnected cavities (Australian National Parks and Wildlife Service 1978). Robinson (1975) found *L. conditor* to be an opportunistic breeder on Franklin Island, with reproduction apparently possible at any time of the year when conditions are proper. Gestation lasts about 44 days, and litters contain one or two young. Ride (1970) stated that females drag their young around attached to their nipples.

Remains from aboriginal sites show that less than 1,000 years ago both species of *Leporillus* ranged across the southern arid zone of Australia to the west coast (Baynes 1984). However, both have declined drastically, and *L. apicalis* already may be extinct (Australian National Parks and Wildlife Service 1978; Robinson 1975; Thornback and Jenkins 1982; Watts 1979). Suggested factors in the decline include competition from introduced sheep and rabbits, habitat destruction by people and livestock, and killing by introduced predators. Robinson (*in* Strahan 1983) noted that some reports suggest that *L. apicalis* may yet survive in Western Australia. *L. conditor* may survive in some places on the mainland, but specimens were last collected there in 1921. The only currently known population is that on the 500 ha. Franklin Island, and estimates of its numbers have ranged from 1,500 to 5,000. *L. conditor* is listed as endangered by the USDI and as rare by the IUCN and is on appendix 1 of the CITES.

RODENTIA; MURIDAE; Genus ZYZOMYS
Thomas, 1909

Thick-tailed Rats, or Rock Rats

There are three species (Ride 1970):

Z. argurus, northern Western Australia to northern Queensland;

Z. woodwardi, northeastern Western Australia, northern Northern Territory;

Z. pedunculatus, southern Northern Territory.

Head and body length is 85–177 mm and tail length is 91–135 mm. Gordon and Johnson (1973) listed weights of 35–65 grams for eight specimens of *Z. argurus* from Queensland. Begg (1981) reported maximum weight for *Z. woodwardi* to be 130 grams in a male and 143 grams in a probably pregnant female. The pelage is crisp, harsh, and almost spinous. It is brownish, grayish, or buffy to reddish sandy on the dorsal surface and shades into white on the ventral surface. On some animals all ventral hairs are white, whereas on others the white-tipped hairs have a grayish base. The hands and feet are white. The tail is bicolored in some forms and entirely white in others.

These delicately built animals have a characteristic thickening of the tail caused by a proliferation of the skin; it is not primarily a reservoir of fat, as is the case in some mammals.

The oldest individuals have the thickest tails, while the youngest specimens show only a slight enlargement. The tail is very fragile. Although some other rats and mice lose their tail with ease, the hair and flesh strip from the vertebrae of rock rats especially easily; the animals soon amputate the remaining skeletal section to leave a shortened stump (Ride 1970). The feet of *Zyzomys* are short and broad, the ears are fairly small and rounded, and females have four mammae. The skin is quite thin and easily torn.

According to Begg (1981), all three species are restricted to suitable rock outcrops. *Z. woodwardi* is found mainly in places with substantial rainfall and dense vegetation, such as closed forest and vine thickets. *Z. argurus* is found in a wide variety of drier habitats, including open forest, woodland, and scrub. Activity is nocturnal. Watts (1977) found the feces of wild *Z. woodwardi* to contain the remains of the fruit and seeds of a dicotyledon, and the feces of *Z. argurus* to contain mostly plant remains but also some insect remains. Begg and Dunlop (1985) identified 43 species of plants in fecal samples from *Z. argurus* and *Z. woodwardi.*

There have been several recent studies of population ecology in the tropical part of northern Australia. Bradley et al. (1988) reported home range of *Z. argurus* there to average 686 sq meters for males and 1,168 sq meters for females. Population densities of this species in different areas varied from 2.3/ha. to 12.8/ha. Numbers were highest from September (late in the dry season) to January (in the middle of the wet season) and then dropped sharply by April (end of the wet season). Reproduction apparently took place all year, but pregnant females were most abundant in April. Litter size averaged 2.7 and ranged from 1 to 6 young. In the same region, Begg (1981) found that *Z. argurus* and *Z. woodwardi* did not occur in the same locality and that males wandered over much larger areas than did females. Pregnant and lactating females of both species were found throughout the year, with a breeding peak evident from March to May, early in the dry season. Known individuals gave birth at intervals of 2–14 months. Sexual maturity was attained at 4–5 months, and some individuals lived at least 2 years. Calaby and Taylor (1983) also found the two species to breed all year, with no evident peaks. Females were polyestrous, experienced a postpartum estrus, and produced several litters annually, each with 1–3 young. The gestation period of *Z. argurus* has been recorded as 32–34 days.

Watts (1979) considered *Z. argurus* to be generally common but *Z. woodwardi* and *Z. pedunculatus* to be rare. *Z. pedunculatus* was last collected in 1960; the species is listed as endangered by the USDI and indeterminate by the IUCN and is on appendix 1 of the CITES.

RODENTIA; MURIDAE; Genus PSEUDOMYS
Gray, 1832

Australian Native Mice

There are 23 species (Baverstock, Watts, and Cole 1977; Borsboom 1975; Fox and Briscoe 1980; Hocking 1980; Kitchener 1980*b,* 1985; Kitchener, Adams, and Baverstock 1984; Kitchener and Humphreys 1986, 1987; Mahoney and Posamentier 1975; Menkhorst and Seebeck 1981; Posamentier and Recher 1974; Ride 1970; Watts 1979):

P. delicatulus, northern Australia;

P. pilligaensis, northern New South Wales;

P. hermannsburgensis, arid parts of Australia;

P. bolami, Western Australia;

Thick-tailed rat: Top, *Zyzomys argurus*, photo from Queensland Museum; Bottom, *Z. woodwardi*, photo by H. Beste and J. Beste.

Australian native mouse *(Pseudomys nanus)*, photo by W. D. L. Ride.

P. chapmani, Hammersley Range in northwestern Western Australia;

P. johnsoni, central Northern Territory;

P. laborifex, Kimberley region of Western Australia;

P. novaehollandiae, coastal New South Wales, Mornington Peninsula of Victoria, northeastern Tasmania;

P. shortridgei, southwestern Western Australia, western Victoria;

P. albocinereus, southwestern Western Australia and islands of adjacent Shark Bay;

P. apodemoides, southeastern Australia;

P. fumeus, Victoria, subfossil remains from southeastern New South Wales;

P. glaucus, southern Queensland;

P. occidentalis, southwestern Western Australia;

P. praeconis, Peron Peninsula and islands of Shark Bay in Western Australia;

P. gouldii, southern Western Australia, South Australia, western New South Wales;

P. australis, central and southern Queensland, inland New South Wales, South Australia;

P. fieldi, vicinity of Alice Springs in southern Northern Territory;

P. higginsi, Tasmania;

P. oralis, southeastern Queensland, northeastern New South Wales;

P. desertor, western and central Australia;

P. gracilicaudatus, northeastern Queensland to coastal New South Wales;

P. nanus, Western Australia, Northern Territory.

Ride (1970) regarded *Leggadina* Thomas, 1910, as part of *Pseudomys*, and thus included the species *L. forresti* within the latter genus. Subsequent authorities have tended to maintain *Leggadina* as a separate genus with *L. forresti* as one of its species (Fox and Briscoe 1980; Mahoney and Posamentier 1975; Watts 1979). *Leggadina* (see account thereof) also sometimes is considered to include certain of the first eight species in the above list. *Gyomys* Thomas, 1910, and *Thetomys* Thomas, 1910, are sometimes employed as genera or subgenera for a number of the species listed above. According to Mahoney and Posamentier (1975), however, the charac-

ters of these two taxa do not hold up and it now seems best not to use subgenera within *Pseudomys*. Lidicker and Brylski (1987) recognized *Gyomys* as a distinct genus with the species *P. shortridgei*, *P. albocinereus*, *P. fumeus*, and *P. apodemoides*.

Head and body length is 60–160 mm, tail length is 60–180 mm, and weight is 12–90 grams. The pelage is soft in most species but harsh in *P. desertor*. The upper parts of the different species are various shades of gray, brown, or yellow; the underparts are usually white, grayish white, or yellowish white. The head and ears are usually paler than the remainder of the upper parts. The tail is moderately haired and in most species is bicolored, being dark above and white to buffy below. Ride (1970) observed that these mice are rather unspecialized in appearance, some species resembling big fluffy rats and others being as small as very tiny mice. The ears are relatively small in some species but large in others. Depending on the species, tail length is either less than, equal to, or more than head and body length. Females have two inguinal pairs of mammae.

Habitats of the different species include sandy plains, stony ridges, coastal dunes, grasslands, shrub and tree heaths, swamps, timbered areas, and rainforests. Ecology and behavior are not well known, but evidently these mice are nocturnal and usually spend the day in burrows. *P. australis* sometimes burrows in clay flats and riverbanks. On the flats its burrows are not more than 150–200 mm deep, usually have only one opening, and lack side passages. The burrows of *P. delicatulus* are located some 60 mm below the surface, and after about 1.5 meters they terminate in a circular space containing a nest of dried grass. The burrows of *P. hermannsburgensis* are about 200 mm in depth and 1 meter in length.

The species *P. chapmani* builds low mounds of pebbles over its burrow systems, and *P. hermannsburgensis* may use these mounds after they are constructed (Dunlop and Pound 1981). The pebbles are of a uniform size and cover a large area, often 1 meter in diameter. The pebbles are probably collected both by excavation and from the surface. Some local mammalogists believe these are used as dew traps. Since the air around the pebbles warms more rapidly as the sun rises than do the pebbles themselves, dew forms on the pebbles by condensation. As the areas in which these mounds are found

Australian native mouse *(Pseudomys delicatulus),* photo by Basil Marlow. Inset: pile of pebbles assembled by *P. chapmani,* photo by Stephen Davis.

are quite dry, except after heavy rain, these dew traps solve the problem of water shortage. Local farmers use the many pebble mounds for mixing concrete. It is believed that the ancient people of the Mediterranean region used a dew trap method comparable to that of *P. chapmani.*

The species *P. albocinereus* tunnels in the sand to about one meter below the surface. The burrows have several openings, at least one of which is used exclusively for throwing sand to the surface from the tunnel. Many of the entrances have surplus sand piled around the opening. Most of the occupied burrows are difficult to locate, as the mice often close the entrances from the inside. One observer on Bernier Island noted that the mice surrounded the open "escape exits" with a delicate network of small twigs and shredded vegetation. It is believed that this barricade prevents loose sand from filling the openings.

The diet of Australian native mice is thought to consist of a variety of seeds, roots, other vegetable matter, and sand-dwelling insects. Watts (1977) suggested that several species of *Pseudomys* are primarily grass eaters, while other species eat very little grass. In captivity some of these mice readily take mealworms, sunflower seeds, and fruits.

Australian native mice reportedly have a relatively mild and gentle disposition. Some species appear to be sociable; the highest number of *P. hermannsburgensis* to be taken from a single burrow is five. Happold (1976a) found that captive groups of *P. albocinereus* lived amicably together but that intruders might be attacked and killed. In the wild, several individuals of this species are found in the same burrow, generally an adult male, one or two adult females, and the young of one or more generations. In contrast, individuals of *P. desertor* were found to exhibit strong mutual repulsion and to be solitary except when pairs associate briefly for mating and when a female has dependent young. Watts (1976a) identified 11 basic calls in *P. australis,* including a defensive screech heard in agonistic encounters between adults.

Taylor and Horner (1972) observed that *Pseudomys* appears to have a regular breeding season where it occurs in areas with a regular climate and assured rainfall but that in locations with erratic precipitation the genus may be opportunistic and breed following adequate rains at a time coincident with the proliferation of grass and herbs. In a summary of reproduction in the genus, Watts (1974) listed gestation periods of 28–40 days and litter sizes of 3–5 young. Happold (1976b) provided the following data: *P. albocinereus,* reproduction all year in Victoria but probably a peak season in August, after winter rainfall, a normal estrous cycle of 7–10 days, postpartum estrus and mating within 12 hours of parturition, intervals of 38 and 39 days between copulation and birth, and an average litter size of 3.8 (2–5) young; *P. shortridgei,* pregnant females taken only from October to December in Victoria, no postpartum estrus and only a single litter per year, and 3 young born in each of six litters; *P. desertor,* probably breeds throughout the year, estrous cycles of 7–9 days, postpartum estrus and mating within 7 hours of parturition, intervals of 28–34 days between successive births, and a mean litter size of 3 (1–4).

In a laboratory study of *P. novaehollandiae,* Kemper (1976a, 1976b) found a mean estrous cycle of 6 days and a postpartum estrus and mating. The gestation period averaged 31.5 (29–33) days in nonlactating females and 33.2 (32–37) days in lactating females. Litter size averaged 4 and ranged from 1 to 6 young. The young weighed about 1–2 grams at birth, opened their eyes after about 15 days, and were weaned at 3–4 weeks. Females reached sexual maturity at about 13 weeks, and males at 20 weeks.

Smith, Watts, and Crichton (1972) reported the mean length of 52 estrous cycles in captive *P. australis* to be 8.5 days. The gestation period was 30–31 days in five cases, and the interval between litters was 30–100 days. Mean size of 140 litters was 3.6 young, with the range being 1–7. The young clung to the nipples of the mother and were dragged about wherever she went. Weaning occurred at 22–30 days, and females were capable of giving birth at 12 weeks. According to Jones (1982), a captive specimen of *P. australis* lived for 5 years and 7 months.

Several species of *Pseudomys* have been adversely affected by the European colonization of Australia. Ride (1970) noted that the hardest-hit species were those that inhabited areas taken over by people for settlement or agriculture but that good populations of certain species still occupy offshore islands, inland areas, undeveloped coastal sand heaths, and rainforests. Cockburn (1978) suggested that the decline of some species was associated with a reduction in summer burning of heath lands by aboriginal people and the resulting increased destructiveness of subsequent fires.

The IUCN now classifies *P. fieldi* as extinct, *P. oralis* as indeterminate, and *P. praeconis* as rare. The USDI lists the following species as endangered: *P. fieldi, P. gouldii, P. novaehollandiae, P. praeconis, P. shortridgei, P. fumeus,* and *P. occidentalis*. The Australian National Parks and Wildlife Service (1977, 1978) covered the following species in the series *Australian Endangered Species: P. fumeus, P. occidentalis, P. praeconis,* and *P. shortridgei*. The CITES covers *P. praeconis* on appendix 1. Watts (1979) listed *P. fieldi* and *P. gouldii* as presumed extinct; *P. australis, P. desertor,* and *P. gracilicaudatus* as rare; and *P. occidentalis, P. oralis, P. praeconis,* and *P. shortridgei* as rare and at risk. The range of *P. occidentalis* apparently has declined drastically because of clearing of its habitat for cereal growing, but some remaining populations are secure in national parks. *P. oralis* had been known only from two specimens collected in the first half of the nineteenth century, but a few more specimens were recently taken in southern Queensland. Subfossils of *P. prae-*

conis have been found in southwestern Australia, and a specimen was collected in 1858 on the Peron Peninsula, but the species now seems to survive only on Bernier Island in Shark Bay. *P. shortridgei* was thought to have disappeared from Western Australia, where it was discovered in 1906, but a living population was discovered in 1983 (Baynes, Chapman, and Lynam 1987). Good populations also were found in Victoria in 1961.

The New Holland mouse *(P. novaehollandiae)* long was known with certainty only from specimens collected in the early nineteenth century in New South Wales. In 1967, however, the species was rediscovered near Sydney (Mahoney and Marlow 1968). Subsequent investigation indicated a wide if somewhat patchy distribution from the northern coast of New South Wales to the Mornington Peninsula of southern Victoria and that the species is not in immediate danger of extinction (Keith and Calaby 1968; Posamentier and Recher 1974). Its optimum habitat is dry heath that has been disturbed by fire and is actively regenerating. *P. novaehollandiae* is known from subfossil remains found on Tasmania and Flinders Island, in addition to its mainland range, and in 1976 a living population was located along the northeastern coast of Tasmania (Hocking 1980).

RODENTIA; MURIDAE; Genus LEGGADINA
Thomas, 1910

There are two species (Morton 1974; Ride 1970; Watts 1976*b*):

Leggadina forresti, photo by B. G. Thomson.

L. forresti, arid areas from west-central Western Australia to central Queensland and western New South Wales; *L. lakedownensis,* northeastern Queensland.

Ride (1970) included *Leggadina* within the genus *Pseudomys,* but Misonne (1969) stated that the two taxa have little to do with one another. The species *Pseudomys delicatulus, P. hermannsburgensis,* and *P. novaehollandiae* sometimes have been placed in the genus *Leggadina.* Mahoney and Posamentier (1975) considered the latter two species to be part of *Pseudomys* but stated, with some doubt, that *P. delicatulus* belongs in *Leggadina.* Lidicker and Brylski (1987) included *P. delicatulus,* as well as *P. hermannsburgensis,* in *Leggadina.* Fox and Briscoe (1980) placed all three species in *Pseudomys,* thereby leaving *Leggadina* with only *L. forresti* and *L. lakedownensis.*

Head and body length is 60–100 mm and tail length is 40–69 mm (Watts 1976*b*). Weight is 15–25 grams in *L. forresti* (Morton, *in* Strahan 1983) and 15–20 grams in *L. lakedownensis* (Covacevich, *in* Strahan 1983). The tail is noticeably shorter than the head and body. The pelage is crisp, with a considerable portion of stout guard hairs. *L. forresti* is yellowish olive brown above and white below. In *L. lakedownensis* the hairs on the back have a buffy brown tip and grade to pale olive buff on the sides; the underparts are white. The head is broad, the muzzle blunt, and the ears short and broad. Females have four mammae.

Watts (1979) stated that *L. lakedownensis* is found in well-grassed areas within or near tropical open woodland. He noted that this species normally occurs in low numbers, and he considered it rare and at risk. Covacevich (*in* Strahan 1983) reported that this species breeds readily in captivity, with a gestation period of about 30 days and a litter size of 2–4 young.

According to Morton (*in* Strahan 1983), *L. forresti* is re-stricted largely to arid country, usually in tussock grasslands and low shrublands. It is nocturnal and known to eat seeds and green vegetation and may be able to extract sufficient moisture from its food to survive without drinking. Individuals have been dug from shallow burrows about 15 cm deep containing nests of grass and one or two blind tunnels. The burrows are inhabited by single animals or females with young. Females have been found breeding after winter rains. Laboratory studies show that the estrous cycle covers 7 days. Litter size is usually 3–4 young.

RODENTIA; MURIDAE; **Genus NOTOMYS**
Lesson, 1842

Australian Hopping Mice, or Jerboa Mice

There are nine species (Mahoney 1975, 1977; Ride 1970; Watts 1979):

N. mitchelli, southern Australia;
N. mordax, known only by a single specimen from southeastern Queensland;
N. alexis, central Australia;
N. fuscus, southeastern Western Australia to southwestern Queensland;
N. cervinus, south-central Australia;
N. macrotis, known only by two specimens from southwestern Western Australia;
N. longicaudatus, southwestern Western Australia, southern Northern Territory, northwestern New South Wales;

Australian hopping mouse *(Notomys fuscus),* photo from Queensland Museum. Inset: *N.* sp., photo by Shelley Barker.

N. *amplus*, known only by two specimens from the extreme southern part of the Northern Territory;

N. *aquilo*, northeastern Northern Territory and the nearby island of Groote Eylandt, western Cape York Peninsula of northern Queensland.

Head and body length is 91–177 mm and tail length is 125–225 mm. Weight is 20–50 grams (Watts 1975a). The general coloration of the upper parts varies from pale sandy brown through yellowish brown to ashy brown or grayish. The underparts are white in all species except N. *mitchelli*, which is whitish gray below. The body covering is fine, close, and soft in most forms. The long hairs near the tip of the tail give the effect of a brush. These rodents are characterized by their strong incisor teeth, long tail, large ears, and extremely lengthened and narrow hind feet, which have only four sole pads. Females have four mammae.

All species that have been examined have a well-developed sebaceous glandular area on the underside of the neck or chest. Watts (1975b) determined this area to be active in all adult males but to be active in females only during pregnancy and lactation. He suggested that the glands are used for territorial marking and marking of group members, including newborn young.

Australian hopping mice inhabit sand dunes, grasslands, tree and shrub heaths, and lightly wooded areas. They are nocturnal and saltatorial and are often considered the ecological equivalent of the gerbils *(Meriones, Gerbillus)*, jerboas *(Jaculus)*, and kangaroo rats *(Dipodomys)* found in the deserts of the Northern Hemisphere (Watts 1975a). Normally, they move about awkwardly on all fours or make short hops, but if startled they bound rapidly and gracefully using only the large hind feet. They dig their own burrows, some being of simple construction but others being rather complex. In sandy soil the burrow of N. *mitchelli* leads downward at an angle of about 40° for 1.2–1.5 meters and then levels off and leads into a nest chamber, from which it leads almost straight upward to the exit. There often is no mound of excavated soil around the opening. *Notomys* is commonly found in association with *Antechinomys*, its marsupial counterpart, and sometimes both genera inhabit the same tunnel system.

The diet of *Notomys* consists of berries, leaves, seeds, and other available vegetation. Ride (1970) stated that although hopping mice will drink water readily when it is available, at least some species do not require water at all. They survive through their ability to excrete waste nitrogen in the form of one of the most highly concentrated urines yet known in a rodent. They also avoid extreme heat and desiccation by remaining underground during the day. They are subject to great fluctuations in numbers; following seasons of good rainfall and plant growth some localities may be swarming with them.

The species N. *alexis* and N. *cervinus* apparently live and move about in groups. In a study of captive N. *alexis* by Happold (1976a), animals placed together displayed little aggression and soon became amicable. Tolerance was shown toward newborn and juveniles, and females suckled young other than their own. Watts (1975a) identified about eight vocalizations in four species of *Notomys*. The calls included twittering during aggressive chases, but none of the sounds was associated with threats or fighting.

The following reproductive information was taken from Aslin and Watts (1980); Crichton (1974); Smith, Watts, and Crichton (1972); and Watts (1974). N. *alexis*, N. *fuscus*, and N. *cervinus* apparently are opportunistic breeders in the wild, giving birth after suitable rainfall. N. *mitchelli*, however, inhabits areas of regular winter rainfall and breeds during that season. In captivity the females of all four species are polyestrous, with no evidence of seasonality. The estrus cy-

cle lasts about 7–8 days in N. *alexis*, N. *mitchelli*, and N. *fuscus* but varies from 8 to 38 days in N. *cervinus*. The gestation period for nonlactating females is 32 days in N. *alexis*, 34–37 days in N. *mitchelli*, and 38–43 days in N. *cervinus*. All three of these species have a postpartum estrus and mating. Lactation seems to delay implantation up to 11 days in N. *cervinus* and probably in N. *mitchelli* but not in N. *alexis*. In N. *fuscus* gestation lasts 32–38 days; a postpartum estrus is not common in this species, but some females entered estrus 14–22 days after giving birth. Usual litter size is 2–4 young in N. *fuscus*, N. *cervinus*, and N. *mitchelli*. In 176 litters of N. *alexis* mean litter size was 4 young, with a range of 1–9. The young weigh about 2–4 grams at birth and open their eyes at 18–28 days. They cling tenaciously to the nipples of the mother and are dragged about wherever she goes; weaning occurs at around 1 month. Female N. *alexis* and N. *mitchelli* can give birth at 3 months, but female N. *cervinus* do not bear young until at least 6 months. Both sexes of N. *fuscus* reach reproductive maturity at 70 days; one female of this species produced 9 litters in its lifetime and died at 26 months old; males of this species are capable of breeding up to the age of 38 months. According to Jones (1982), a captive N. *alexis* lived for 5 years and 2 months.

The species N. *macrotis*, N. *mordax*, N. *longicaudatus*, and N. *amplus* are known only by specimens collected between 1843 and 1901 (Ride 1970). The reasons for their disappearance are unknown but may be associated with predation by introduced foxes and cats (Thornback and Jenkins 1982). Watts (1979) presumed all four to be extinct and also listed N. *aquilo* and N. *fuscus* as rare. The IUCN now lists the same four species as extinct and designates N. *fuscus* as indeterminate. N. *aquilo* is classified as endangered by the USDI. It was thought to have disappeared from the mainland, but a specimen was taken in Arnhem Land in 1979, and there are relatively large populations on Groote Eylandt (Thornback and Jenkins 1982).

RODENTIA; MURIDAE; **Genus MASTACOMYS**
Thomas, 1882

Broad-toothed Rat

The single species, M. *fuscus*, occurs in southeastern New South Wales, Victoria, and Tasmania. Certain late Pleistocene and subfossil remains, found over a somewhat larger region of southeastern Australia, formerly were considered to represent other species, but all have now been assigned to M. *fuscus* (Wakefield 1972).

Head and body length is 145–95 mm and tail length is 95–135 mm. Green (1968) listed weights of 100–196 grams. The dense hair is long and silky, and the coloration throughout is sooty brown to grayish brown. The skin of the tail and feet is dark-colored. The broadened molar teeth and narrowed palate are characteristic of the genus. It is volelike in appearance and habits. Females have four mammae.

The remainder of this account is based largely on Ride (1970). The broad-toothed rat appears to be a relict genus, surviving only in isolated colonies in places that provide it with the cold, humid, or alpine conditions that it requires. In Kosciusko State Park, of southeastern New South Wales, this rat lives along small creeks among shrubs and long grasses. The climate is rigorous, with a heavy winter snowfall. Conditions beneath the snow are relatively mild, and *Mastacomys* dwells in tunnels among the shrubs. By contrast, Victorian colonies live at low altitudes, which are wet but less cold. In Tasmania *Mastacomys* also lives in areas with a severe winter climate, but it is very localized in distribution and is found in

Broad-toothed rat *(Mastacomys fuscus)*, photo by Ederic Slater.

only one type of habitat, consisting of vegetation on very boggy ground, such as the drainage systems of wet sedge-lands in openings in rainforest, or button grass or tussock grass areas. In Tasmania *Mastacomys* is invariably found in association with *Rattus lutreolus* and *Antechinus minimus*.

In Kosciusko State Park breeding probably begins in the spring, and litters are produced in the summer. In Tasmania the breeding season is from October to March, and during this period each female produces more than one litter. Captive observations indicate that the female enters estrus soon after one litter is born and then subsequent to mating, aggressively keeps the male away until the young are independent. Gestation lasts about 5 weeks, and litters of up to three young have been recorded. The young are large and well furred at birth; they attach themselves firmly to the nipples of the mother and are dragged about on their back. They apparently attain independence at 30 days and adult weight at 16 weeks.

Green (1968) observed that while the range of *Mastacomys* covers an extensive part of Tasmania, its habitat is specialized and subject to human environmental alterations. He cautioned that the survival of the genus is threatened by hydroelectric development, road construction, and cattle grazing.

RODENTIA; MURIDAE; Genus GOLUNDA
Gray, 1837

Indian Bush Rat, or Coffee Rat

The single species, *G. ellioti*, occurs in Pakistan, India, Nepal, Bhutan, and Sri Lanka (Ellerman and Morrison-Scott 1966).

Head and body length is usually 110–55 mm and tail length is usually 90–130 mm. Four specimens weighed 50–80 grams (Roberts 1977). The texture of the pelage varies considerably. In some animals the covering is fairly soft with only a few harsh hairs; in others it is coarse, and the longer hairs, some of which are spiny, are flattened and grooved. The coat is generally thin, but the hairs are rather long. The coloration above is grayish, yellowish brown, reddish brown, or fairly dark brown; the darker colors are produced by a fine speckling of fulvous and black hairs. The coloration below is light gray, bluish gray, or white.

Golunda is a thickset, heavy rodent, rather volelike in appearance. The head is short and rounded, and the ears are rounded and hairy. The tail, which is shorter than the head and body, is stout at the base, tapers toward the tip, and is covered with coarse, short hairs. The upper incisor teeth are grooved, and the pattern of the molars is somewhat unusual. Females have eight mammae.

This rodent usually inhabits swamps, grasslands, the edge of cultivated areas, and (at least in Sri Lanka) jungles. Prakash and Rana (1972) observed that in a grassy area of the Indian Desert the burrows of *Golunda* are invariably under bushes. Otherwise, this rodent has been reported to burrow only rarely and to locate its nest on the ground or just above the ground, usually in dense grass. The globular nest is made of plant fibers and is 150–230 mm i n diameter. *Golunda* is a good climber but usually moves about on the ground in definite runways. Some observers have said that *Golunda* is slow in its movements; others have considered it to be quick. It is most active in the early morning and evening. Its food is almost exclusively plant material, normally the roots and stems of such plants as the "dub" or "nariyali grass" (Cynodon).

Individuals generally live alone or in family groups. In studies in the Indian Desert, Prakash (1975) found pregnant females from March to August. The number of embryos averaged 6.6 and ranged from 5 to 10. Earlier reports indicated the usual number of young to be 3–4.

When coffee was grown in Sri Lanka, the bush rat became extremely plentiful and did great damage to the crops by eating the buds and blossoms of the coffee plant. It has been suggested that the animals migrated to the plantations, but it is probable that they simply increased in numbers in response to the abundant food supply. Coffee is no longer planted in Sri Lanka, and bush rat numbers have decreased substantially.

Indian bush rat *(Golunda ellioti gujerati)*, photo by P. J. Deoras.

RODENTIA; MURIDAE; **Genus HADROMYS**
Thomas, 1911

Manipur Bush Rat

The single species, *H. humei*, occurs in Assam at the eastern edge of India and in Yunnan in southwestern China (Ellerman and Morrison-Scott 1966; Yang and Wang 1987).

Head and body length is about 95–140 mm, tail length is about 105–40 mm, and weight is 41–77 grams. The fur is thick and soft. The upper parts are blackish gray. The middle area of the back is sometimes darker than the head and shoulders, and the rump has a reddish wash. The belly is yellow or orange, with a slaty tinge, or is white. The inner sides of the thighs are rufous, and the hands and feet are yellowish white. The fairly well haired tail is distinctly bicolored—black above and whitish below.

The body has a fairly stout form. The fifth finger and the fifth toe are short. The teeth are large and powerful; the incisors are not grooved, and the molars are broad and heavy. *Hadromys* resembles *Golunda* in general external appearance, but its tail is slightly longer, and its fur is not spiny or less spiny. The upper incisors of *Golunda* are grooved. Females have eight mammae.

Seven specimens were collected for the Zoological Survey of India in connection with a scrub typhus outbreak in 1945. These specimens came mainly from an oak parkland habitat described by Roonwal (1949) as follows: "Fairly thickly covered semi-scrub of moderately tall oak trees (up to 9–12 meters high) . . . on hill-sides at 1,200 meters and probably higher altitudes. . . . Soil loamy with rock beneath and with practically no humus. Usually not associated with streams. . . . Tree canopy very open. Very tall grass, over 3–4 meters high. Hardly any shrubs. Ground canopy also very open."

Leaves of grass that had been cut and eaten were found in the stomach of an adult male. A pregnant female was collected in September.

HADROMYS

Manipur bush rat *(Hadromys humei)*, photo from *Proc. Zool. Soc. London.*

RODENTIA; MURIDAE; **Genus BANDICOTA**
Gray, 1873

Bandicoot Rats

There are three species (Ellerman and Morrison-Scott 1966; Lekagul and McNeely 1977; Medway 1978):

B. bengalensis, Pakistan to Burma, Sri Lanka, Penang Island off west coast of Malay Peninsula, Sumatra, Java;

B. indica, western India to southeastern China and Indochina, northern Malay Peninsula, Sri Lanka, Taiwan, Sumatra, Java;

B. savilei, central Burma to southern Viet Nam.

Musser (1982c) and Musser and Newcomb (1983) doubted the occurrence of *B. indica* on Sumatra and considered the presence of both that species and *B. bengalensis* on the Malay Peninsula, Penang, Sumatra, and Java to be the result of unintentional introduction by people. *Gunomys* Thomas, 1907, is a synonym of *Bandicota*.

Head and body length is 160 to at least 360 mm and tail length is 130–258 mm. The tail is usually somewhat shorter than the head and body but may be slightly longer. Lekagul and McNeely (1977) listed average weights of 199 grams for *B. savilei* and 545 grams for *B. indica,* but Grzimek (1975) noted that weight could reach 1,500 grams. The texture of the pelage varies from soft and fairly dense to thin and coarse, and there is much variation in the length of the guard hairs.

The general coloration of the upper parts ranges from light grayish to various shades of brown or almost black. The lower parts are a dirty white. The tail is scantily haired. The front claws are large, and the muzzle is short and broad. The incisor teeth are yellow or orange in color. Females have 12–18 mammae.

The natural habitat of *B. bengalensis* has been reported to be evergreen jungle and oak scrub, but this species has become an urban commensal of humans. *B. indica* is also a commensal in some areas and is now especially common in lowland rice fields. Lekagul and McNeely (1977) did not locate *B. indica* in purely natural habitat in Thailand. Bandicoot rats are good diggers and construct elaborate burrows at the edges of fields, in dikes and stream banks, and even in city streets. There are two to six entrances, each marked by a small mound of earth (Medway 1978). The burrow may extend to a depth of 60 cm and contains enlarged chambers used for nesting and food storage. Ordinarily these rodents leave their retreats only under cover of darkness. They are good swimmers and divers. They are omnivorous in diet but in some areas feed largely on products of cultivation, such as rice, grains, sugar cane, fruits, nuts, and potatoes.

There is usually only a single adult per burrow, but high population densities can occur in favorable areas (Grzimek 1975). A two-month radiotracking study in India found 10 individual *B. bengalensis* to confine their movements to part of a 0.4 ha. field (Fulk, Smiet, and Khokhar 1979). Female *B. bengalensis* have an estrous cycle of about 5 days (Sahu and Maiti 1978). This species breeds throughout the year in Pakistan and India (Prakash 1975; Roberts 1977). In the city

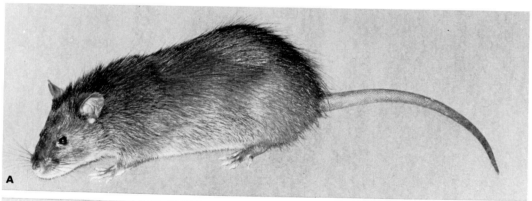

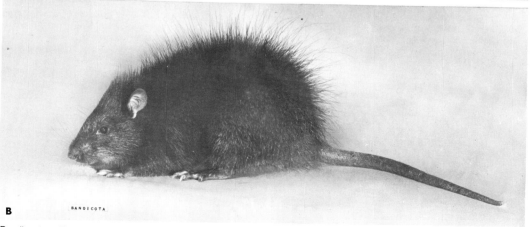

Bandicoot rat: Top, *Bandicota bengalensis*, photo by Ernest P. Walker; Bottom, *B. indica*, photo by Robert E. Kuntz.

of Rangoon, Burma, some reproduction occurs at all times of the year, but there are peaks during the dry season from November to April (Walton et al. 1978). Studies in this city indicate that females have an average of about 62 days between litters, whereas earlier investigations in Calcutta determined a wait of only 31–34 days and thus a probable postpartum estrus and mating. In Rangoon litter size averaged 7.4 young and ranged from 1 to 14. In Viet Nam litter size of *B. indica* ranged from 2 to 12 young (Medway 1978). In southern Pakistan the largest litters of *B. bengalensis,* consisting of 14–18 young, are born from September to November, while during the remainder of the year the usual litter size is 5–10. The young weigh an average of 3.5 grams, open their eyes at 14 days, are weaned at 28 days, and reach sexual maturity at around 3 months (Roberts 1977).

Bandicoot rats are characterized by their ferocious nature. When disturbed, they emit a harsh, nasal barking. Lekagul and McNeely (1977) reported that an effort to breed *B. indica* for laboratory use was not successful, because the animals remained too savage. Lekagul and McNeely also noted that since this species is large, delicious to eat, and generally found as a commensal, it may have been geographically spread through human agency. In 1946 an introduction was made into the states of Kedah and Perlis in mainland Malaysia. The population of *B. indica* on Taiwan reportedly originated from releases made by the Dutch. Several other peculiarities in the range of both *B. indica* and *B. bengalensis* might be interpreted as having come about through introduction. Whatever the case, these animals are now human commensals over large areas and are often considered serious pests in both agricultural and urban settings (Grzimek 1975; Walton et al. 1978). They raid crops and eat and contaminate stored foods. Their tunnels and mounds damage irrigation structures, block gutters, and cause the collapse of streets and buildings. They are involved in the spread of plague and other diseases. In return, however, many people hunt bandicoot rats for food and dig up their burrows to obtain the food stored by the animals.

RODENTIA; MURIDAE; Genus NESOKIA
Gray, 1842

Pest Rat, or Short-tailed Bandicoot Rat

The single species, *N. indica,* occurs from Egypt to Sinkiang and northern India (Corbet 1978). A specimen was first taken in Bangladesh in 1978, and the species now has spread throughout that country (Poché et al. 1982).

Head and body length is 140 to about 215 mm, tail length is 88–129 mm, and adult weight is approximately 112–75 grams. The texture of the pelage varies from coarse, short, harsh, and semispinous to long, fine, and silky. In some forms the coat is dense, and in others it is thin, but in all forms the tail is practically naked and the hands and feet are scantily haired. The general coloration of the upper parts ranges from fawn or yellowish brown to grayish brown, mixed in places with red, and the underparts are grayish to whitish. The color of the back gradually merges into that of the sides and flanks without a distinctive line of demarcation.

Nesokia is characterized by a robust form; a short, rounded head; a short, broad muzzle; rounded ears; the relatively short, almost naked tail; and broad feet. All the claws except the rudimentary thumb are armed with strong, nearly straight nails. The incisor teeth are stout and broad. Females have eight mammae.

The short-tailed bandicoot rat usually lives in moist areas or along streams and canals within a variety of general habitats, including deserts, steppes, cultivated areas, and forests. The elevational range is from about 26 meters below to 1,500 meters above sea level. This animal burrows extensively, usually at depths of 15–60 cm, and throws the loose soil onto the surface of the ground around the tunnel entrance. All individuals observed by Lay (1967) seemed to confine their activities to their burrow systems, each of which consisted of several to many tunnels, a single enlarged nest chamber lined with finely chewed vegetation, and several enlarged, unlined chambers. In Iraq the holes and prodigious diggings of *Nesokia* are seen frequently along the banks of irrigation canals (Harrison 1972). One burrow system excavated in this area consisted of a nest chamber about 30 cm in diameter and a connecting main tunnel that ran for about 4.5 meters, with periodic side tunnels and ventilation holes leading to the surface. Taber, Sheri, and Ahmad (1967) reported that in Pakistan the burrows are as long as 23 meters. In some areas at least, *Nesokia* also makes well-defined runways along the surface. The diet consists of grass, grains, roots, and cultivated fruits and vegetables. Sometimes food is stored in the burrows. Up to 454 grams of grain have been found in the storage chambers of a single burrow.

Lay (1967) reported that usually only a single rat occupies each burrow and that all individuals captured alive fought fiercely. A captive colony in Teheran bred throughout the year. Al-Robaae (1977) indicated that wild populations in Iraq probably breed all year and that litter size there is 1–8 young. Litter sizes listed by Prakash (1971) ranged from 2 to 10 young. Beg, Adeeb, and Rana (1981) suggested a 17-day gestation period.

In Iraq this rat does considerable damage to watermelons, sweet melons, and tomatoes (Al-Robaae 1977). It also causes extensive damage by raiding grain fields and tunneling through irrigation walls. Native people eat *Nesokia*, as well as the grain stored in its burrows. They smoke the tunnels, or

Pest rat *(Nesokia indica),* photo by Harry Hoogstraal.

pour water into them, until the animal comes to the surface, where it is captured, sometimes with the aid of a dog.

RODENTIA; MURIDAE; Genus ERYTHRONESOKIA
Khajuria, 1981

Red Pest Rat

Khajuria (1981) described this genus and the single species, *E. bunnii*, on the basis of two specimens collected in 1974 in Basra Province, southern Iraq. Carleton and Musser (1984) included *Erythronesokia* in *Nesokia*, without comment, but Corbet and Hill (1986) listed it as a full genus.

In the holotype, an adult male taken on 23 March, head and body length is 250 mm and tail length is 205 mm. The upper parts are bright red, fading to orange or buff on the flanks, and the underparts and legs are white. The feet are darker, and the tail is generally black but covered with short whitish hairs. The pelage is coarse. From *Nesokia*, considered to be its closest relative, *Erythronesokia* differs by its larger size, relatively longer and nonnaked tail, reddish dorsal coloration, and tendency to develop postorbital processes in the skull.

Although it is the largest murid in Iraq, the red pest rat was not known to science until recently, and several attempts to trap additional specimens were unsuccessful. It is, however, well known to the local people in its habitat because of its size and distinctive coloration. Farmers have reported that it is very rare and is seen only during floods near the marshes where it lives. It then swims to shore and may climb date palm trees, which it is said to damage by eating the soft inner parts. The second specimen, a juvenile female, was taken on 16 November. Nader (1989) regarded *Erythronesokia* as endangered. Its limited habitat has been the scene of intensive military activity in the last decade.

RODENTIA; MURIDAE; Genus MILLARDIA
Thomas, 1911

Asian Soft-furred Rats, or Cutch Rats

There are four species (Ellerman and Morrison-Scott 1966; Mishra and Dhanda 1975; Misonne 1969):

M. meltada, Pakistan, India, Nepal, Sri Lanka;
M. kondana, southwestern India;
M. kathleenae, central Burma;
M. gleadowi, Pakistan, western India.

Millardia sometimes is considered to be only a subgenus of *Rattus*, or, if regarded as a full genus, it sometimes is considered to include *Cremnomys* (see account thereof).

The following descriptive data were taken in part from Mishra and Dhanda (1975). Head and body length is 80–200 mm and tail length is 68–186 mm. A specimen of *M. meltada* weighed 70 grams. The fur is soft. In *M. meltada* the upper parts are sandy gray, grayish brown, or whitish buff lined with brown; the underparts are whitish or grayish; and the tail is dark above and light below. In *M. kondana* the upper parts are dark brown, the underparts are grayish white, and the tail is dark above and grayish below. In *M. gleadowi* the upper parts are pale grayish brown or sometimes fawn-colored, the underparts are white, and the tail is light brown

above and white below. *M. kathleenae* is a pale-colored species with a white-tipped tail.

Millardia resembles *Rattus* in external appearance but has been characterized as follows: total or partial suppression of the pads of the soles of the feet; toes shortened, the fifth scarcely reaching the base of the fourth; tail usually subequal in length to the head and body; and palate long, exceeding half the occipitonasal length. In most species the ears are rather large and the tail is well haired. *M. kondana*, has relatively small ears and a poorly haired tail and also comparatively well developed pads on the feet. Female *M. meltada* and *M. kondana* have eight mammae; female *M. gleadowi*, six; and female *M. kathleenae*, four.

The species *M. meltada* has been found in grasslands, swamps, broken rocky ground, and cultivated fields. It shelters in short burrows, stone heaps, and the abandoned burrows of other rodents. During the hot months it often seeks the cracks formed in drying soil. The diet consists of grain, seeds, and swamp vegetation. At times *M. meltada* may increase in numbers to plague proportion and cause considerable damage to crops. Roberts (1977) stated that both *M. meltada* and *M. gleadowi* are nocturnal and spend the day in shelters. The latter species makes short burrows, usually under the roots of a bush, extending at an angle of about 45° to a depth of not more than 61 cm.

Prakash (1975) reported the average home range size for *M. meltada* in the Rajasthan Desert to be 1,217 sq meters. In at least some parts of India this species is said to live in groups of 2–6 individuals. In the Rajasthan Desert it apparently is capable of breeding throughout the year, with pregnancies reaching a peak in October and litter size ranging from 4 to 10 and averaging 5.9 young (Prakash 1971). Farther north, in the Punjab, Bindra and Sagar (1968) found captive *M. meltada* to breed during the hot months from March to May and August to October. Females produced 2–7 litters annually, gestation lasted about 20 days, and litter size ranged from 1 to 8 young, with averages of 3.4 in the laboratory and 6 in the field. Females attained sexual maturity at 3–4.5 months.

Prakash (1971) observed breeding in *M. gleadowi* only from August to October in the Indian Desert. Litters consisted of two or three young. According to M. L. Jones (Zoological Society of San Diego, pers. comm., 1978), a captive *M. gleadowi* lived for four years.

RODENTIA; MURIDAE; Genus CREMNOMYS
Wroughton, 1912

There are three species (Ellerman and Morrison-Scott 1966):

C. blanfordi, India, Sri Lanka;
C. cutchicus, India;
C. elvira, southern India.

Both *Millardia* and *Cremnomys* sometimes are considered only subgenera of *Rattus*. Misonne (1969) suggested that *Millardia* is distinct and that *Cremnomys* is a subgenus thereof. Mishra and Dhanda (1975), while recognizing *Millardia* as a full genus, considered the species of *Cremnomys* to be representative of a separate genus. *Cremnomys* was listed as a full genus by Carleton and Musser (1984), Corbet and Hill (1986), and Honacki, Kinman, and Koeppl (1982).

Original descriptions of type specimens are as follows: *M. blanfordi*, head and body length 105 mm, tail length 155 mm, upper parts slate-colored tipped with fawn, underparts white, tail dark basally and white distally (Thomas 1881); *M. cutchicus*, head and body length 117 mm, tail length 141 mm, pelage long and soft, upper parts drab gray, underparts white,

Asian soft-furred rat: Top, *Millardia meltada*, photo by K. Tsuchiya; Bottom, *M. kondana*, photo by A. C. Mishra.

tail grayish (Wroughton 1912); *M. elvira*, head and body length 149 mm, tail length 196 mm, upper parts brownish gray, underparts grayish white, tail bicolored (Ellerman 1946).

According to Wroughton (1912), *Cremnomys* resembles *Millardia*, but the former can be distinguished by its relatively much longer tail, a longer fifth toe on the hind foot, six (rather than five) pads on the hind foot, a flatter skull profile, and a shorter bony palate. Females have six mammae. The length of the tail is thought to be an adaptation to movement among rocks.

Prakash and Rana (1972) found *C. cutchicus* to be common in rocky parts of the Indian Desert and to shelter in rock crevices. Reproduction there evidently is affected by rainfall (Prakash 1971). During 1968 there was a drought in this area, and only a single pregnant female was found from November 1968 to April 1969. In 1970, however, the rains exceeded the normal level, and pregnant females were taken from March to October, with a peak in August. Litter size averaged four and ranged from two to eight young.

RODENTIA; MURIDAE; **Genus SRILANKAMYS**
Musser, 1981

Ceylonese Rat

The single species, *S. ohiensis*, occurs in central Sri Lanka. This species was long included in the genus *Rattus*, but Musser (1981*a*) showed that it belongs in its own genus.

In an adult female, head and body length is 145 mm and tail length is 173 mm. The upper parts are glossy dark gray, slightly suffused with brown. The underparts are pale cream and are sharply demarcated from the dark dorsum. The tail is sharply bicolored, being dark grayish brown above and white below. The upper surfaces of the feet are mostly white. The ears are small and scantily haired. Females have eight mammae.

Srilankamys is distinguished from *Rattus* in having shorter, softer, and finer pelage; smaller bullae; the alisphenoid canal of the skull hidden by a lateral strutlike wall; the squamosal roots of the zygomatic arches originating higher on the sides of the braincase; the posterior margin of the palatal bridge not extending far past the molar toothrow; the pterygoid fossa nearly flat and not perforated by a large foramen; the mesopterygoid fossa at least as wide as the palatal bridge; thinner, more delicately built dentaries, with smaller coronoid processes; relatively smaller teeth; and ivory-colored, rather than bright orange, incisors.

Srilankamys occurs in mountain forests and is considered to be terrestrial. The type specimen was collected at an elevation of about 1,800 meters.

RODENTIA; MURIDAE; **Genus DACNOMYS**
Thomas, 1916

Large-toothed Giant Rat, or Millard's Rat

The single species, *D. millardi*, is known from the Darjeeling district of northeastern India, Assam, and Laos (Ellerman and Morrison-Scott 1966). The elevational range is about 1,000–1,800 meters.

Head and body length is 228–90 mm and tail length is 308–35 mm. The somewhat coarse pelage is short and thin, and the individual hairs on the back are 15–16 mm in length. The general coloration of the upper parts is olive brown to grayish or smoky brown, becoming lighter on the sides as it blends into the cinnamon drab or pale brownish of the underparts. The gular, inguinal, and axillary areas are creamy or white. The ears and hands are brown, and the fingers are white. The finely ringed tail (about 10 scales per cm) is scantily sprinkled with fine, short hairs.

Dacnomys is a large, ordinary-looking rat with no special external distinguishing characters. It derives one of its common names from the relatively large size of the molar teeth. The sole pads are large and rounded, and the fifth toe of the hind foot lacks a claw. Females have eight mammae.

RODENTIA; MURIDAE; **Genus DIOMYS**
Thomas, 1917

Manipur Mouse, or Crump's Mouse

The single species, *D. crumpi*, was long known only by a single specimen, a skull collected at an elevation of 1,300 meters on Mount Paresnath in east-central India, among rocks. In 1942, 25 years after publication of the description, a series of 30 specimens was secured at Bishenpur, Manipur, Assam, in extreme eastern India. In 1978 two skins and four skulls were collected in south-central Nepal at an elevation of 200 meters in moist deciduous forest (Ingles et al. 1980).

Head and body length is 100–145 mm and tail length is 105–35 mm. The back is blackish gray, the rump is usually black, and the feet are white. The underparts are whitish gray, and the tail is black above and white below. The fur is thick and soft, but the tail is scantily haired.

Ellerman (1946:205) wrote: "The genus *Diomys* stands on the important character that the incisors are so proodont, that the condylobasal length normally is at least as long as the

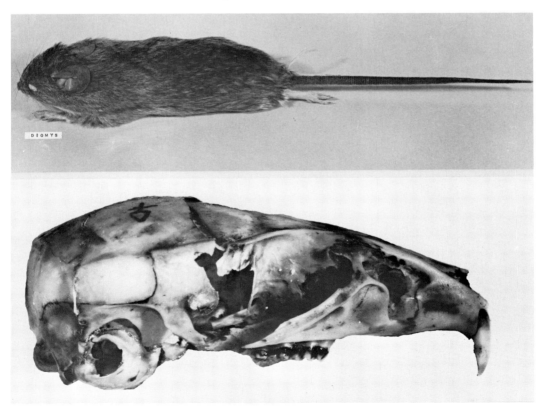

Manipur mouse *(Diomys crumpi)*, photos from British Museum (Natural History).

occipitonasal length, and often exceeds it (three exceptions in twenty measurable skulls). This is very rare in Asiatic Muridae; the only other murid genera from Asia with this character normally present are the very different Bandicoot Rats, *Nesokia* and *Bandicota*."

Musser and Newcomb (1983) made an extensive comparison of *Diomys* and *Berylmys* and found some superficial resemblance but concluded that the two genera are not closely related. In its short and narrow rostrum, the vertical sides and dorsolateral ridges of its braincase, its very long and slender incisive foramina, its high-crowned molar teeth, its silky pelage, and various other characters, *Diomys* seems to have affinity to *Millardia* and *Cremnomys*.

RODENTIA; MURIDAE; Genus CHIROMYSCUS
Thomas, 1925

Fea's Tree Rat

The single species, *C. chiropus,* occurs in eastern Burma, northern Thailand, Laos, and Viet Nam (Marshall, *in* Lekagul and McNeely 1977). The species was first described as *Mus chiropus* by O. Thomas in 1891 from a specimen that had been collected in the Karin Hills of Burma and preserved in alcohol. It did not again come to attention until 1924, when H. Stevens, on the Sladen-Godman Expedition to northern Viet Nam, collected a specimen at Bao-Ha at an elevation of 90 meters in the forest country of the Red River. This specimen became the type of the genus established by Thomas.

Head and body length is 127–60 mm. Van Peenen, Ryan, and Light (1969) listed tail length as 198–252 mm. The pelage is rather bristly. Coloration of the upper parts is "warm-lined buffy" suffused with black. The sides of the face are ochraceous, the color extending up behind the ears, and there is a dark area surrounding each eye. The rump is bright ochraceous, the throat is rusty, and the underparts are creamy. The flesh-colored ears are covered with short hairs. The long, slender tail is also covered with fine, short, light-colored hairs and is slightly penciled.

This genus is similar to *Niviventer* but differs in having a nail, rather than a claw, on the hallux (Musser 1981*a*). The hind foot of *Chiromyscus* is specialized for climbing; the great toe is opposable, and large digital pads are present. The claws on all the digits are short. Females have eight mammae.

Chiromyscus lives in forests and is probably arboreal. No other information on its natural history has been recorded.

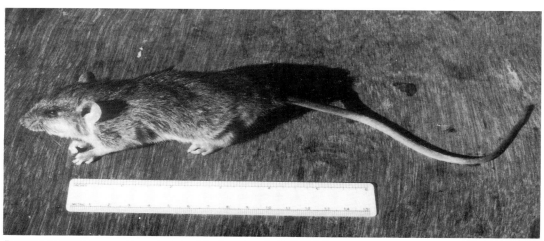

Fea's tree rat *(Chiromyscus chiropus),* photo of dead specimen by D. S. Melville.

White-bellied rat *(Niviventer cremoriventer),* photo by Louise H. Emmons.

RODENTIA; MURIDAE; **Genus NIVIVENTER**
Marshall, 1976

White-bellied Rats

There are 15 species (Abe 1983; Marshall, *in* Lekagul and McNeely 1977; Musser 1981*a*):

N. andersoni, mountains of central and southern China;
N. excelsior, Sichuan (central China);
N. brahma, Assam, northern Burma;
N. eha, Nepal, Sikkim, northeastern India, northern
    Burma, Yunnan (south-central China);
N. langbianis, Assam, Burma, northeastern Thailand,
    Laos, Viet Nam;
N. cremoriventer, Malay Peninsula, Sumatra, Java, Borneo,
    and many small nearby islands;
N. niviventer, China, extreme northeastern Pakistan,
    northern India, Nepal, Sikkim;
N. confucianus, much of China, northern Burma, northern
    Thailand, northern Indochina, Taiwan, Hainan;
N. tenaster, mountains of Assam, southern Burma, and
    Viet Nam;
N. fulvescens, southeastern Tibet, Yunnan (south-central
    China), northern India, Nepal, Sikkim, Bangladesh,
    Assam, Burma, Thailand, Indochina, Hainan;
N. bukit, southern Burma, southern Thailand, Malay
    Peninsula, Sumatra, Java, Bali;
N. hinpoon, central Thailand;
N. coxingi, northern Burma, Taiwan;
N. rapit, known from scattered localities in Malay
    Peninsula, Sumatra, and Borneo;
N. lepturus, western and central Java.

The above species were long considered part of *Rattus*. Misonne (1969) placed them in the genus *Maxomys*. Musser, Marshall, and Boeadi (1979), however, showed that these species did not belong in *Maxomys*. Marshall (*in* Lekagul and McNeely 1977) treated *Niviventer* as a subgenus of *Rattus*. Finally, Musser (1981*a*) elevated *Niviventer* to generic rank, established its contents as given in the above list, and determined that it was most closely related to the genus *Chiromyscus*. Marshall (*in* Lekagul and McNeely 1977) considered *N. langbianis* to be a subspecies of *N. cremoriventer*, but Musser (1973*c*, 1981*a*) treated the two as separate species. On the basis of specimens from Thailand, *N. bukit* was regarded as a subspecies of *N. fulvescens* by Abe (1983), but it was retained as a full species by Corbet and Hill (1986).

Head and body length is 110–208 mm and tail length is 120–270 mm (Medway 1978; Musser 1970*a*, 1973*c*; Musser and Chiu 1979). Some average weights are: *N. confucianus*, 65 grams; *N. rapit*, 80 grams; *N. bukit*, 73 grams; and *N. hinpoon*, 61 grams (Lekagul and McNeely 1977). The upper parts of the head and body are grayish brown in some species, reddish brown in a few, and either gray, buffy gray, or reddish brown in one. The underparts are white or cream in most species but dark gray in a few. The tail is usually dark above and light below, but it is monocolored in some species. The pelage is dense and either soft and woolly or semispinous. Females have six or eight mammae (Musser 1981*a*).

From *Rattus*, both *Niviventer* and *Maxomys* are distinguished in having much smaller bullae, the squamosal roots of the zygomatic arches originating higher on the sides of the braincase, the posterior margin of the palatal bridge not extending far past the molar toothrow, the pterygoid fossa nearly flat and not perforated by a large foramen, and the mesopterygoid fossa as wide as or wider than the palatal bridge. From *Maxomys*, *Niviventer* is distinguished in having the tail usually much longer than the head and body, a relatively shorter rostrum, a less inflated braincase, smaller lacrimals, and relatively narrower incisive foramina.

These rats occur in various kinds of forest in both lowlands and mountains. They are cursorial, scansorial, or arboreal (Marshall, *in* Lekagul and McNeely 1977; Musser 1981*a*; Musser and Chiu 1979). Litters of two to five young have been recorded for *N. cremoriventer* (Medway 1978). *N. hinpoon*, which is restricted to a remote area of wooded limestone cliffs, may be threatened with extinction.

RODENTIA; MURIDAE; **Genus MAXOMYS**
Sody, 1936

Rajah Rats, or Spiny Rats

There are 16 species (Chasen 1940; Medway 1977, 1978; Musser 1986; Musser, Marshall, and Boeadi 1979):

M. surifer, southern Burma, Thailand, Indochina, Malay
    Peninsula, Sumatra, Java, Borneo, many small islands of
    East Indies;
M. moi, Laos, Viet Nam;
M. pagensis, Mentawai Islands off western Sumatra;
M. panglima, Balabac, Palawan, Busuanga, and Culion
    islands (Philippines);

Rajah rat *(Maxomys surifer)*, photo by Lim Boo Liat.

M. rajah, Malay Peninsula, Rhio Archipelago, Sumatra, Borneo;

M. hellwaldi, Sulawesi;

M. dollmani, Sulawesi;

M. bartelsii, Java;

M. alticola, mountains of northern Borneo;

M. ochraceiventer, Borneo;

M. inas, mountains of Malay Peninsula;

M. hylomyoides, Sumatra;

M. musschenbroekii, Sulawesi;

M. baeodon, northern Borneo;

M. whiteheadi, Malay Peninsula, Sumatra, Borneo, many small islands of East Indies;

M. inflatus, mountains of Sumatra.

Maxomys was originally proposed as a genus for the species *M. bartelsii*. Subsequently it came to be generally considered a subgenus of *Rattus* and to include a number of species, particularly those now placed in the genus *Niviventer*. Most of the species in the above list were actually put in *Lenothrix*, which also was treated as a subgenus of *Rattus* (Ellerman and Morrison-Scott 1966; Laurie and Hill 1954; Musser, Marshall, and Boeadi 1979). Misonne (1969) reinstated *Maxomys* as a full genus including *M. bartelsii*, as well as species now placed in the genus *Niviventer*. Misonne also, however, put the species listed above as *M. baeodon, M. hellwaldi, M. musschenbroekii, M. whiteheadi, M. rajah*, and *M. surifer* into the subgenus *Leopoldamys* of *Rattus*. Later authorities (Marshall, *in* Lekagul and McNeely 1977; Medway and Yong 1976; Van Peenen et al. 1974) indicated that *M. bartelsii* is not closely related to the other species placed in *Maxomys* by Misonne but is actually associated with most of the species that Misonne put into *Leopoldamys*. Finally, Musser, Marshall, and Boeadi (1979) removed *Rattus niviventer* and its relatives from *Maxomys* and established the contents of the genus as given in the above list. Musser and Newcomb (1983) confirmed that *Maxomys* is not closely related to *Rattus* and indicated that three additional species of the former would soon be described.

The following descriptive data were taken from Marshall (*in* Lekagul and McNeely 1977), Medway (1977, 1978), Musser (1969b), and Musser, Marshall, and Boeadi (1979). Head and body length is 100–235 mm, tail length is 88–226 mm, and weight is 35–284 grams. The dorsal pelage is short, soft, and dense or short and spiny. The upper parts range in color from dark brown, bright reddish brown, and iron gray to bright yellowish mixed with black. The underparts are white, grayish, or chestnut. The finely scaled tail is usually dark above and white below.

Maxomys comprises small to medium-sized rats of a generalized appearance. Key characters of the genus include a long and slender hind foot with smooth and naked plantar surfaces, a tail usually shorter or only slightly longer than the head and body, a broad and inflated braincase, broad incisive foramina, and small bullae. Species of *Rattus* generally have a wider hind foot with well-developed plantar pads, a less inflated braincase, narrower incisive foramina, and larger bullae. Female *Maxomys* have six or eight mammae.

Musser, Marshall, and Boeadi (1979) stated that these rats are cursorial animals, adapted for life on the floor of evergreen or semievergreen forests in tropical lowlands and mountains. Marshall (*in* Lekagul and McNeely 1977) wrote that *Maxomys* is never found in trees. He noted that *M. surifer* occurs in all types of forest and is almost as ubiquitous as *Rattus rattus* but is not found in association with people. Also, in contrast to *R. rattus*, *M. surifer* was described as being gentle and naturally tame and friendly.

According to Medway (1978), *M. surifer* lives in burrows in the forest floor, each with one or two entrances and a nest chamber lined with leaves. Its diet consists mainly of vegetation, including roots and fallen fruits, but also includes some invertebrates and small vertebrates; some food, such as nuts, may be stored in the burrow. Medway listed the following average diameters of effective lifetime ranges: *M. surifer*, 280 meters; *M. rajah*, 263 meters; and *M. whiteheadi*, 476 meters. Medway wrote that *M. whiteheadi* breeds throughout the year in the Malay Peninsula and has an average of 3 (1–6) young per litter. For *M. surifer* and *M. rajah* in the same area litter size averaged 3.3 and ranged from 2 to 5.

RODENTIA; MURIDAE; Genus LEOPOLDAMYS
Ellerman, 1947–1948

Long-tailed Giant Rats

There are four species (Chasen 1940; Marshall, *in* Lekagul and McNeely 1977; Musser 1981a):

L. neilli, southern Thailand;

L. edwardsi, northeastern India to southeastern China and central Viet Nam, mainland Malaysia, Sumatra;

L. sabanus, Bangladesh, Thailand, Indochina, Malay Peninsula, Sumatra, Java, Borneo, and many small nearby islands;

L. siporanus, Mentawai Islands off western Sumatra.

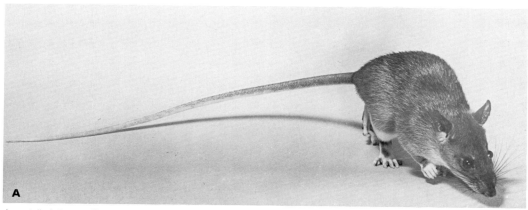

Long-tailed giant rat *(Leopoldamys sabanus)*, photo by Ernest P. Walker.

Leopoldamys originally was described as a subgenus of *Rattus*. Misonne (1969) included in this subgenus the species now known as *Maxomys surifer*, *M. rajah*, *M. musschenbroekii*, *M. whiteheadi*, *Srilankamys ohiensis*, and *Rattus nativitatis*, as well as *Leopoldamys edwardsi*, *L. sabanus*, and *L. siporanus*. Subsequent work (Marshall, *in* Lekagul and McNeely 1977; Musser, Marshall, and Boeadi 1979; Musser 1981*a*) established *Leopoldamys* as a full genus, with contents as given in the above list. Musser and Newcomb (1983) concluded that this genus is closely related to *Niviventer*, *Maxomys*, and *Berylmys* but not to *Rattus*.

Head and body length is 180–275 mm, tail length is 300–420 mm, and weight is 200–495 grams (Medway 1978; Yong 1970). The pelage is short and sleek. The upper parts, which are generally brown or buffy, are sharply demarcated from the white underparts. The body is large, and the length of the tail is much greater than that of the head and body. The hind feet are long and slender, and the ears are small and scantily haired. Females have four pairs of mammae (Musser 1981*a*).

Leopoldamys is distinguished from *Rattus* in having a usually larger body and relatively much longer tail, shorter and sleeker pelage, much smaller bullae, the alisphenoid canal of the skull hidden by a lateral strutlike wall, the squamosal roots of the zygomatic arches originating higher on the sides of the braincase, the posterior margin of the palatal bridge not extending past the molar toothrow, the pterygoid fossa nearly flat and not perforated by a large foramen, the mesopterygoid fossa at least as wide as the palatal bridge, and smaller incisor teeth. *Leopoldamys* superficially resembles *Niviventer* but is much larger and has relatively smaller bullae and larger incisors (Musser 1981*a*).

According to Medway (1978), *L. edwardsi* is found in montane forest above 750 meters, and *L. sabanus* occurs in evergreen forest, generally below 750 meters. The latter species is usually found at ground level but is able to climb freely. The diet includes insects, other invertebrates, and a great variety of vegetable matter. The young are born in short burrows dug into sloping hillsides and provided with one or two entrances and a nest chamber lined with leaves. Pregnant females have been taken in all months of the year but most frequently from July to September and least frequently from January to March. The number of young per litter averages 3.1 and ranges from 1 to 7. Mean longevity in the wild is only 4.1 months, but captives have lived up to 2 years. *L. neilli*, only recently discovered in an area of remote limestone cliffs, may be threatened with extinction.

RODENTIA; MURIDAE; **Genus BERYLMYS**
Ellerman, 1947

White-toothed Rats

There are four species (Musser and Newcomb 1983):

B. manipulus, Assam, northern and central Burma;
B. berdmorei, southern Burma to Viet Nam, Con Son Island;
B. mackenziei, Assam, central and southern Burma, Sichuan (south-central China), southern Viet Nam;
B. bowersi, Assam to southeastern China and Malay Peninsula, northwestern Sumatra.

Berylmys originally was described as a subgenus of *Rattus*. Misonne (1969), however, assigned its species to *Bullimus*, which he regarded as another subgenus of *Rattus*. Marshall (*in* Lekagul and McNeely 1977) again recognized *Berylmys* itself as a subgenus of *Rattus*. Musser and New-

comb (1983) considered it a full genus with affinity to perhaps *Sundamys* on the one hand and to *Maxomys*, *Niviventer*, and *Leopoldamys* on the other. Except as noted, the information for the remainder of this account was taken from Musser and Newcomb.

Head and body length is 135–300 mm, tail length is 140–310 mm, and weight is 177–650 grams. The pelage is dense and crisp. The upper parts of living animals are usually iron gray and sharply demarcated from the white underparts. The tail is monocolored in some populations and bicolored in others; in the latter case the basal portion is dark and the distal section, or just the tip, is white. The ears are large and dark brown. The upper surfaces of the feet are white, gray, or brown. The hind feet are long and narrow. Females of *B. bowersi* have 8 mammae, and females of the other species have 10.

From *Rattus*, *Berylmys* can be distinguished by the texture and coloration of the fur and by the sharp demarcation between dorsum and venter. In *Berylmys* the braincase is triangular in dorsal view, and its dorsolateral margins, as well as those of the postorbital and interorbital regions, are outlined by weak and low ridges. In *Rattus* the cranium is rectangular in dorsal outline, with the most conspicuous feature being the high and thick ridges that define the dorsolateral margins from the interorbital region to the occiput. Also in *Berylmys* the incisor teeth are usually white to pale orange in color, while those of *Rattus* are always bright orange.

All the species of *Berylmys* live in tropical forest or scrub, sometimes in marshy areas and sometimes around the margins of cultivated fields. They are primarily terrestrial and dwell in burrows in the ground. Specimens have been collected at elevations from about 80 to 2,800 feet. The diet includes leaves, grass, seeds, fruits, insects, molluscs, and earthworms. According to Medway (1978), *B. bowersi* is remarkably tame; even adults caught in the wild are immediately docile. Litter size in this species is 2–5 young, and some individuals caught as adults have survived 35 months in captivity.

RODENTIA; MURIDAE; **Genus MASTOMYS**
Thomas, 1915

Multimammate Rats

There are seven species (Gordon 1978; Green et al. 1980; Meester et al. 1986; Misonne, *in* Meester and Setzer 1977; F. Petter 1975, 1977):

M. pernanus, southwestern Kenya, northern Tanzania, Rwanda;
M. natalensis, Africa (except West) south of the Sahara;
M. huberti, West Africa south of the Sahara;
M. coucha, southern Africa;
M. erythroleucus, southern Morocco, Sahel and Sudan zones of West and Central Africa;
M. angolensis, western Zaire, Angola;
M. shortridgei, northeastern Namibia, northwestern Botswana.

Mastomys often was included within *Rattus* until Misonne (1969) showed that the two are not closely related. He and various other authorities considered *Mastomys* to be a subgenus of *Praomys* (see account thereof), but most recent writers, such as Carleton and Musser (1984), Happold (1987), and Meester et al. (1986), regard the two as distinct genera. There are systematic difficulties involving *M. natalensis*, *M. huberti*, *M. coucha*, and *M. erythroleucus*. Many authors, such

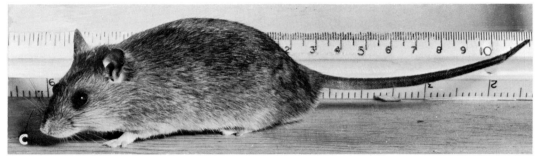

Multimammate rat *(Mastomys natalensis)*, photo by H. S. Davis.

as, for example, Rosevear (1969) and Kingdon (1974b), have dealt with this group under a single specific name, M. natalensis. Misonne (in Meester and Setzer 1977) stated that the group included two species, M. natalensis and M. erythroleucus, but he did not distinguish the two by distribution or morphology. Gordon (1978) showed that in Zimbabwe the populations formerly called M. natalensis actually comprise two species, one with a diploid chromosome number of 32, and the other with 36. Meanwhile, F. Petter (1977) pointed out that in West Africa there are also two species—M. huberti, with 32 chromosomes, and M. erythroleucus, with 38. The latter species was assigned a range in the Sahel and Sudan zones of West and Central Africa. Robbins, Krebs, and Johnson (1983) found biochemical distinctions between the species with 32 and 38 chromosomes in Sierra Leone but suggested that there was some doubt as to the correct names to be applied. In addition, Corbet (1978) reported M. erythroleucus to be the only multimammate rat in Morocco. Green et al. (1980) showed that the species with 36 chromosomes is widespread in southern Africa and that the correct name is actually M. coucha. The species with 32 chromosomes also was found to occur in much of this region; on the basis of the karyotype of specimens from the type locality, the name M. natalensis was thought to apply to this species.

Descriptive information from Kingdon (1974b), Rosevear (1969), and Smithers (1983) indicates that head and body length is about 60–170 mm, tail length is 60–150 mm, and weight is 20–80 grams. Coloration of the upper parts ranges through a considerable variety of brownish, reddish, yellowish, and grayish shades. The underparts are white or grayish. The fur is usually long, soft, and springy but sometimes is short and harsh. The tail, which is approximately equal in length to the head and body, appears to be naked but actually is finely haired. The ears are moderately sized and rounded, and the feet are narrow.

Mastomys superficially resembles *Rattus* but differs in the same characters that separate *Praomys*. From *Praomys*, *Myomys*, and *Hylomys*, *Mastomys* is distinguished by its relatively shorter tail and by a combination of cranial characters, particularly that the palatine bone extends no more forward than the middle of the second upper molar and that the anterior palatal foramina always reach to well between the molar rows. Females of the species M. natalensis, M. huberti, M. erythroleucus, and M. coucha have 16–24 mammae, more than any other rodent. M. shortridgei has only 10 mammae.

Multimammate rats occur in many types of habitat (Kingdon 1974b; Rosevear 1969; Smithers 1983). The species M. natalensis may once have been restricted mainly to savannahs but now can be found throughout much of sub-Saharan Africa, chiefly in association with people. It occupies cultivated and abandoned fields, buildings, and villages but not large towns, probably because of competition with *Rattus*

rattus. M. erythroleucus inhabits relatively dry but well-vegetated areas around the edges of the Sahara. M. shortridgei occupies a much more restricted wetland area along the Okavango River. Smithers (1983) referred to M. natalensis as nocturnal and terrestrial, but Delany (1975) noted that it climbs and swims well. This species uses a crevice or digs its own burrow, which usually consists of a network of galleries without a central chamber. Smithers (1971) stated that under field conditions M. natalensis eats mostly grass and other seeds. Kingdon (1974b) observed, however, that in suitable habitat this species may take as much animal matter—mainly insects—as vegetation. In the vicinity of human settlements M. natalensis eats nearly everything that people do.

Periodic population outbreaks have been reported for M. natalensis, and at times hundreds of individuals may be seen at once. This species seems to have a relatively mild disposition and lacks the aggressiveness of *Rattus rattus*. Several families may live together in one burrow, and strangers are accepted into the group (Choate 1972; Kingdon 1974b).

The reproductive features of M. natalensis are well known (Delany 1975; Kingdon 1974b; Neal 1977b; Rosevear 1969; Sheppe 1972, 1973; Smithers 1971). There is general agreement that this species is capable of breeding throughout the year but that peak reproduction comes toward the end of the rains and in the early part of the dry season. In Uganda these peak periods are centered around May and November. Postpartum estrus and mating have been reported, and females may have two litters per season. The gestation period is 23 days, and litter size is 1–22, usually about 10–12. The young weigh about 1.8 grams at birth, open their eyes at around 16 days, are weaned and independent at 3 weeks, and are usually able to breed at 3.5 months.

Kingdon (1974b) thought it likely that M. natalensis was originally restricted to the savannahs of southern Africa but spread northward as a commensal of people. It is now declining in many areas through competition with *Rattus rattus* but still is the dominant commensal in more remote villages. It is sometimes a serious pest in human habitations and to agriculture, and it is also the principal African indigenous host of human plague. Smithers (1983) explained that plague originally was introduced to southern Africa in 1899 by shipborne *Rattus*. Infected fleas from that genus then passed to *Mastomys*, which helped spread the disease through much of the region.

RODENTIA; MURIDAE; Genus MYOMYS
Thomas, 1915

There are seven species (Misonne, in Meester and Setzer 1977; Nader, Kock, and Al-Khalili 1983; Van der Straeten and Dieterlen 1983; Van der Straeten and Verheyen 1978b):

M. albipes, central Sudan, highlands of Ethiopia;

M. ruppi, southwestern Ethiopia;

M. fumatus, southern Sudan, Ethiopia, Somalia, Uganda, Kenya, southwestern Arabian Peninsula;

M. verreauxi, southwestern South Africa;

M. butleri, southwestern Sudan;

M. daltoni, Senegal to Nigeria;

M. derooi, Ghana to Nigeria.

Like *Mastomys* (see account thereof), *Myomys* once was considered part of *Rattus* and also has sometimes been regarded as a subgenus of *Praomys.* To add to the confusion, some authorities have used the generic or subgeneric name *Myomyscus* Shortridge, 1942, in place of *Myomys* (see Meester et al. 1986:287). Honacki, Kinman, and Koeppl (1982) listed *Myomys* as a subgenus of *Praomys* including only the species *M. derooi* and listed *Myomyscus* as a subgenus for the other five species named above. Carleton and Musser (1984) recognized *Myomys* as a full genus that includes *Myomyscus,* and that procedure has been followed here.

The following descriptive information was taken from Happold (1987), Kingdon (1974b), Rosevear (1969), and Smithers (1983). Head and body length is 75–124 mm, tail length is 84–155 mm, and weight is 20–45 grams. The upper parts range in color from blackish gray through various shades of brown to yellowish gray, and the underparts are white or gray. The head is narrow and pointed, the ears are fairly large and rounded, and the tail is substantially longer than the head and body. From *Rattus, Myomys* differs in the same characters that distinguish *Praomys.* The tail of *Myomys* is relatively longer than that of *Mastomys,* and the palatine bone of the skull of *Myomys* does not extend as far forward as that of *Praomys,* reaching only to the level of the junction of the second and first upper molars. Female *Myomys* have 10 mammae.

Myomys occurs in a variety of habitats, from dry woodland to lush savannahs and forests. It is especially common in rocky areas and is sometimes found in cultivated fields and human habitations. Smithers (1983) indicated that *M. verreauxi* is especially common in riverine forest, is nocturnal and terrestrial, and evidently feeds mainly on insects. In contrast, *M. fumatus* in Arabia lives in rocky grassland and feeds largely on leaves and shoots. Population densities of around 1–4 per ha. and home ranges of about 1,500 sq meters were found in this last area. Pregnant females were found there from December to March and in May and October (Al-Khalili, Delany, and Nader 1988).

According to Happold (1987), *M. daltoni* lives in burrows and, in captivity at least, makes nests from shredded grass. It breeds throughout the year, with peaks in the dry season and early wet season (February–April) and in the late wet season (October–November). Pregnant female *M. derooi* have been taken in Togo during all months in which collections have been made (July–December). Reported litter size is 2–4 in *M. fumatus* (Delany 1975), 3–10 in *M. daltoni* (Anadu 1979b), and 2–5 in *M. derooi* (Van der Straeten and Verheyen 1978b). Sexual maturity is attained at 4.5–5.5 months by *M. daltoni* (Anadu 1979b).

RODENTIA; MURIDAE; Genus PRAOMYS
Thomas, 1915

African Soft-furred Rats

There are eight species (Ansell and Ansell 1973; Hubert, Adam, and Poulet 1973; Misonne, *in* Meester and Setzer

1977; Van der Straeten and Dieterlen 1987; Van der Straeten and Verheyen 1981):

P. tullbergi, Senegal to Central African Republic;

P. misonnei, Irangi in east-central Zaire;

P. rostratus, Liberia, Ivory Coast;

P. hartwigi, mountains of Cameroon;

P. morio, Cameroon;

P. jacksoni, Cameroon to southern Sudan and Zambia;

P. lukolelae, Cameroon, Central African Republic, Congo;

P. delectorum, Kenya, Tanzania, northeastern Zambia, Malawi.

Praomys sometimes has been included within the genus *Rattus;* in turn, the genera *Mastomys, Myomys, Hylomyscus,* and *Heimyscus* sometimes have been included in *Praomys.* Misonne (1969) showed *Praomys* to be a separate genus, not closely related to *Rattus,* and he considered *Mastomys, Myomys,* and *Hylomyscus* to be subgenera of *Praomys.* This procedure was followed by some authorities, such as Kingdon (1974b), while others, such as Rosevear (1969) and Robbins, Choate, and Robbins (1980), treated these subgenera as full genera. Misonne also named *Heimyscus* as a distinct genus. Carleton and Musser (1984) listed all of these taxa as full genera, while Corbet and Hill (1986) and Honacki, Kinman, and Koeppl (1982) continued to include *Mastomys* and *Myomys* within *Praomys* and recognized only *Hylomyscus* as a separate genus (which included *Heimyscus*).

The following descriptive data were taken primarily from Rosevear (1969) and Kingdon (1974b). Head and body length is 89–140 mm, tail length is 110–67 mm, and weight is 21–57 grams. All species have soft, gray-based fur which is either long or short. The coloration of the upper parts ranges through a great variety of red, brown, gray, and yellow shades; the underparts are white, gray, or yellowish. Both the tail and the ears appear to be naked but actually have a light covering of hairs.

The body is slender, the head is pointed, and the ears are large and rounded. The tail ranges up to about one-third longer than the head and body. *Praomys* is usually considerably smaller than those species of *Rattus* that have been introduced to Africa. Whereas those species have a skull length of over 36 mm, a first upper molar with five roots, and harsh fur, *Praomys* has a skull length of below 36 mm, a first upper molar with three or four roots, and soft fur. From *Mastomys, Myomys,* and *Hylomyscus, Praomys* differs in that the palatine bone of its skull extends forward all the way to the level of the first upper molar tooth. Females have 6–8 mammae.

African soft-furred rats occur primarily in humid forests and are largely nocturnal; apparently all species are at least partly arboreal, though some may nest underground (Kingdon 1974b; Rosevear 1969). *P. jacksoni* and *P. delectorum* nest in short burrows and also make concealed runways. Other species nest in hollow logs or trees. The diet consists of fruits, seeds, green vegetation, and insects (Happold 1987; Rosevear 1969). Average home ranges of 0.23 ha. for males and 0.22 ha. for females have been reported in a population of *P. tullbergi* (Happold 1977).

According to Happold (1987), *P. tullbergi* breeds throughout the year, with one peak at the end of the main dry season and beginning of the wet season (February–April) and another after the secondary dry season (October–November). Litter size usually is 3–4 young, though litters of up to 6 have been recorded. Growth is rapid, and adult size is attained after 10–12 weeks. An adult female may have several litters in rapid succession. *P. jacksoni* is thought to breed throughout the year in Uganda (Delany 1975). Studies of captive *Praomys* indicate that gestation lasts 26–27 days, young usually number 4–5, their eyes open after 17–19 days, weaning occurs by the 29th day, and sexual maturity is attained at 2.5 months (Rosevear 1969). A litter size of 1–8 young has been reported for *P. jacksoni* (Delany 1975). In a wild population of *P. tullbergi* in Nigeria, most individuals were replaced after only 6 months, but some lived to at least 20 months (Happold 1977). A captive individual of this species lived for 5 years and 2 months (Jones 1982).

RODENTIA; MURIDAE; Genus HYLOMYSCUS
Thomas, 1926

African Wood Mice

There are seven species (Ansell and Ansell 1973; Bishop 1979; Misonne, *in* Meester and Setzer 1977; Robbins, Choate, and Robbins 1980; Robbins and Setzer 1979):

H. baeri, Ivory Coast, Ghana;
H. aeta, Cameroon and Gabon to Uganda, island of Bioko (Fernando Poo);
H. carillus, Angola;
H. denniae, mountains from northeastern Zaire and Uganda to Zambia;
H. parvus, Cameroon, Congo, Gabon;
H. alleni, Guinea to Central African Republic and Gabon, Bioko;
H. stella, Nigeria to Zaire and Kenya.

Hylomyscus often was included within *Rattus* until Misonne (1969) showed that the two are not closely related. He and certain other authorities considered *Hylomyscus* to be a subgenus of *Praomys* (see account thereof), but most recent writers, such as Carleton and Musser (1984), Corbet and Hill (1986), Happold (1987), and Honacki, Kinman, and Koeppl (1982), regard the two as distinct genera. Based on electrophoretic and karyological data, Iskandar et al. (1988) suggested that *H. stella* comprises two species in Gabon.

The information for the remainder of this account was taken from Delany (1975), Happold (1987), Kingdon (1974b), and Rosevear (1969). Head and body length is 69–120 mm, tail length is 84–172 mm, and weight is 8–42 grams. The upper parts are reddish brown to dull gray brown, and the underparts are whitish gray. The fur is soft, the head small, the muzzle pointed, the ears relatively large, and the tail very long. Both the ears and the tail appear to be naked but are actually covered with fine hairs. From related genera, *Hylomyscus* can be distinguished by its small size, the narrow zygomatic plate of its skull, its hind foot measuring 20 mm or less, and its tail being at least 25 percent longer than its head and body. The palatine bone of the skull extends forward to the junction between the first and second upper molar teeth. Females have six or eight mammae.

These wood mice live in forests and dense vegetation. *H. denniae* occupies montane forests at elevations of up to 3,500 meters. They are strictly nocturnal and highly arboreal, running and climbing among branches and vines. They build nests of leaves or other vegetation in tree holes, forks of branches, and palm or banana leaf axils. They evidently feed mainly on fruits and other vegetation but also eat some animal matter.

Breeding probably occurs throughout the year in West Africa, with peaks during the wet seasons. In the very wet mountains of Uganda, however, reproductive peaks for *H. denniae* occur in the relatively dry periods of July–August and December; litter size there is 3–6 young. Studies of *H.*

stella indicate that 71 percent of pregnancies in Uganda occur in June, gestation lasts 25–33 days, litter size is 1–5 young, birth weight is 1–2 grams, the eyes open after 15–21 days, and the young leave the mother at about 30 days.

RODENTIA; MURIDAE; Genus HEIMYSCUS
Misonne, 1969

The single species, *H. fumosus*, occurs in Cameroon, Gabon, and possibly the Central African Republic. It was named as a species of *Hylomyscus* by Brosset, Dubost, and Heim De Balsac (1965). Misonne (1969) thought that the dentition of this species differed widely from that of *Hylomyscus* but was close to that of *Dephomys*; he suggested a new generic name, *Heimyscus*, for the species. Robbins, Choate, and Robbins (1980) indicated that they planned a review of the status of the species. *Heimyscus* was listed as a full genus by Carleton and Musser (1984) but not by Corbet and Hill (1986) or Honacki, Kinman, and Koeppl (1982).

Based on three specimens, head and body length is 69–73 mm, tail length is 98–117 mm, and weight is 13.5–15.0 grams. The upper parts are generally dark smoky brown with reddish highlights, and the underparts are whitish. From *Hylomyscus*, *Heimyscus* is distinguished by its relatively shorter tail but relatively longer hind feet. The first and fifth digits of these feet, however, are shorter than those of *Hylomyscus* and have smaller plantar tubercles. The molar teeth of *Heimyscus* are more deeply sculptured and more separated from one another than are those of *Hylomyscus*. The specimens were collected in agricultural areas or secondary forest bordering farms.

RODENTIA; MURIDAE; Genus THALLOMYS
Thomas, 1920

Acacia Rat

The single generally accepted species, *T. paedulcus*, occurs from southern Ethiopia and western Somalia to Angola and South Africa (Misonne, *in* Meester and Setzer 1977; Yalden, Largen, and Kock 1976). Recently, Gordon (1987) tentatively recognized a second species, *T. nigricauda*, in the arid country of Namibia and Botswana.

Head and body length is 120–62 mm, tail length is 130–210 mm, and weight is 63–100 grams (Kingdon 1974b; Smithers 1971). The fur is generally quite soft. In some animals it is soft and fine, while in others it is rather coarse. The coloration of the upper parts ranges from pale buff through yellow fawn to brownish gray. The dorsal surface is usually darker than the sides of the body, and the sides of the face are generally grayish. The underparts are white. The tail is generally brown, the hands and feet are pale grayish or white, and the scantily haired ears have a reddish tinge. *Thallomys* is characterized by its long tail and short feet. Females have three pairs of mammae.

As the common name suggests, this rodent is usually associated with acacia trees, where it nests in forks or branches, in crevices in the trunk, in hollow limbs, under loose bark, or in holes in the ground at the base of the tree. The nests are built of twigs and other vegetable matter. *Thallomys* is shy but quite active; it peers from its nest at the slightest strange noise and runs to the dense, thorny foliage to hide from predators. It is an expert climber but occasionally will drop to the ground in an effort to escape danger. It normally emerges

Acacia rat *(Thallomys paedulcus)*, photo from Nyasaland Museum through P. Hanney.

from its retreat in the early evening to search for food. The diet is based principally on acacia trees (Kingdon 1974*b*). Buds and leaves are staples and are eaten in the canopy, gum is gnawed directly off the bark, and acacia and grass seeds are foraged for on the ground beneath the trees. Berries, roots, and an occasional insect are also taken.

A large tree occasionally harbors a small colony of *Thallomys*. The group probably represents a family, since not more than two pairs of adults and their young have been found together in a single tree. In a study of captives from the Transvaal, Meester and Hallett (1970) learned that no litters were produced in the winter period from mid-April to mid-July, that the minimum interval between litters was 26 days, that mean litter size was 2.7 young, and that the minimum age at which a female produced a litter was 107 days. For Botswana, Smithers (1971) reported that young apparently are born from October through May and litter size ranges from 2 to 5 young. In Zambia, Sheppe (1973) found a pregnant female with 7 embryos on 26 March. Kingdon (1974*b*) stated that in East Africa the breeding season centers on the rains, with pregnant females recorded in April and young animals in July and August. According to Jones (1982), a captive specimen lived for 3 years and 6 months.

RODENTIA; MURIDAE; Genus THAMNOMYS
Thomas, 1907

Thicket Rats

There are two species (Hutterer and Dieterlen 1984; Misonne, *in* Meester and Setzer 1977):

T. venustus, eastern Zaire and adjacent parts of Uganda and Rwanda;
T. rutilans, Guinea to Uganda and northwestern Angola.

There has been disagreement regarding the systematics of this genus. *Grammomys* (see account thereof) often has been treated as a subgenus, and the species *T. rutilans* sometimes has been placed therein.

Head and body length is 121–61 mm, tail length is 160–222 mm, and weight is 50–100 grams (Kingdon 1974*b*). The pelage is fairly long and soft in some animals but crisp in others. At higher elevations the fur on the back may be 20 mm long. The coloration of the back ranges from dark brown or reddish brown to orange rufous and clay color. The flanks are usually paler, and the underparts are white, buffy, or gray. In *T. venustus* the tail is fairly well haired and slightly tufted at the tip. In *T. rutilans* the tail is scantily haired, except for the terminal third, which is somewhat bushy. *Thamnomys* is characterized by a long tail and short, curved claws. Females usually have four inguinal mammae but sometimes have two additional abdominal mammae.

According to Kingdon (1974*b*), *T. venustus* inhabits montane forests and *T. rutilans* prefers farms or areas of rank secondary growth within forests. Both species are arboreal and nocturnal. They construct formless nests of leaves and dried grass in hollow trees, forks, tangles of vegetation, or human buildings. Rosevear (1969) stated that *T. rutilans* normally lives at heights of about 1–2 meters above the ground and makes a nest of very fine, threadlike strips of grass gathered into a kind of bag about 150 mm long by 100 mm wide. The diet includes leaves and seeds and probably resembles that recorded for *Grammomys* (Kingdon 1974*b*).

Thamnomys seems usually to be solitary, though one observer reported seeing the parents and young of two litters in the same nest. Other reproductive data, reported by Delany (1975), are that pregnant female *T. venustus* collected in Zaire contained one or two embryos each, that pregnant female *T. rutilans* taken in Uganda in March had one to three embryos, and that the gestation period of the latter species is about 25 days. Kingdon (1974*b*) observed that the young of *Thamnomys* remain firmly attached to the nipples of the mother for a period of about 2 weeks.

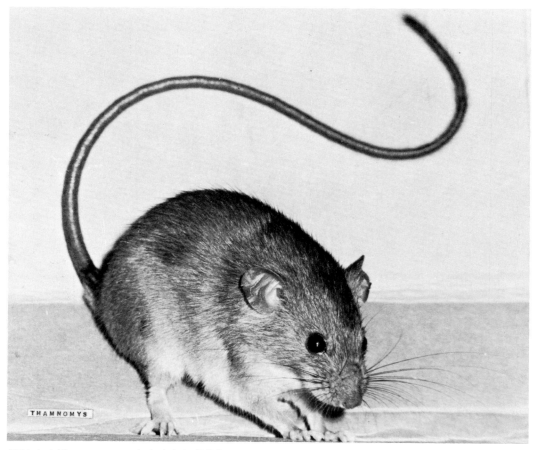

Thicket rat *(Thamnomys venustus)*, photo by F. Petter.

RODENTIA; MURIDAE; Genus GRAMMOMYS
Thomas, 1915

There are eight species (Dieterlen and Nikolaus 1985; Hubert, Adam, and Poulet 1973; Hutterer and Dieterlen 1984; Olert, Dieterlen, and Rupp 1978):

G. gigas, Kenya;
G. cometes, southern Sudan to Zimbabwe;
G. dolichurus, Central African Republic to Angola and southeastern South Africa;
G. buntingi, Guinea, Sierra Leone, Liberia, Ivory Coast;
G. macmillani, Sudan, Ethiopia, Uganda, Zaire, and possibly farther west;
G. aridulus, Sudan;
G. minnae, Ethiopia;
G. caniceps, Kenya.

Primarily on the basis of dental characters, Rosevear (1969) considered *Grammomys* to be a genus separate from *Thamnomys* but stated that it was very possible that *T. rutilans* was more closely related to *Grammomys* than to *T. venustus*. Some subsequent authorities (Ansell and Ansell 1973; Delany 1975; Misonne, *in* Meester and Setzer 1977; Olert, Dieterlen, and Rupp 1978) have treated *Grammomys* as a

subgenus that includes *T. rutilans.* Honacki, Kinman, and Koeppl (1982) listed *Grammomys* as a full genus and *T. rutilans* as a species thereof. Corbet and Hill (1986) included *Grammomys* within *Thamnomys,* while Carleton and Musser (1984) and Meester et al. (1986) considered the two distinct. Studies by Hutterer and Dieterlen (1984) suggest that, after all, *Grammomys* is a separate genus and that *T. rutilans* should remain in *Thamnomys.* Based on karyology, Roche et al. (1984) reported an evidently undescribed species of *Grammomys* from Somalia.

Head and body length is 97–140 mm, tail length is 144–202 mm, and weight is 18–65 grams (Hutterer and Dieterlen 1984). The upper parts vary widely in color, from reddish brown to yellowish, but often are paler than those of *Thamnomys;* the white underparts usually are clearly demarcated from the flanks. The body is slender, and the tail is long and well-haired at the tip. Externally, from *Thamnomys, Grammomys* can be distinguished by its relatively longer tail, narrower hind foot, and softer pelage. Although there is some resemblance between the molar teeth of *Grammomys* and those of the species *Thamnomys rutilans, Grammomys* lacks a postero-internal cusp on the first upper molar, which is a character of all species of *Thamnomys* (Rosevear 1969).

According to Kingdon (1974*b*), *G. dolichurus* is adapted primarily to tall grass and secondary scrub, and *G. cometes* lives in wet forests. Both species are arboreal and entirely nocturnal. Their nests may be located in tree hollows, human habitations, or tangles of vegetation and may take the form of a woven globe of rough grasses with a finer inner lining of chewed grass. The nests of *G. dolichurus* are placed at heights of 0.5–4.0 meters (Delany 1975). The diet of *G. dolichurus* consists mainly of green stems, fruits, nuts, flowers, and other vegetable matter but also includes insects (Kingdon 1974b).

Home ranges of 600–650 sq meters and a population density of 2.5/ha. have been reported for *G. dolichurus*, and the species is solitary (Kingdon 1974b). Delany (1975) stated that in Uganda female *G. dolichurus* give birth from May to July and in November, apparently at intervals of 5–6 weeks. In southern Africa, breeding occurs during the warm, wet months from about October to February (Smithers 1983). In a study of captive *G. dolichurus*, however, Bland (1973) determined that pregnancy could occur every 28–29 days. It was also learned that the gestation period of six litters was 24 days, that litter size ranged from one to seven (usually being two to four) young, and that sexual maturity was attained at 50–70 days. A captive specimen of *G. dolichurus* was still living after 4 years and 5 months (Jones 1982). Kingdon (1974 b) stated that *G. dolichurus* has become an important experimental animal in malarial research.

RODENTIA; MURIDAE; **Genus STENOCEPHALEMYS** *Frick, 1914*

Ethiopian Narrow-headed Rats

There are two species (Yalden, Largen, and Kock 1976):

S. albocaudata, northern and central Ethiopia;
S. griseicauda, south-central Ethiopia.

Head and body length is 120–95 mm and tail length is 114–65 mm (Van der Straeten 1981). The thick fur is soft and long, and the tail is fairly well haired. The coloration of the back is mottled umber, the mottling caused by intermixed black hairs; the sides are pinkish buff; and the underparts are gray. The outer parts of the upper arms and thighs are pinkish buff mixed with gray, and the inner parts are gray. The lower arms and legs, and the hands and feet, are generally white. In *S. albocaudata* the tail is mostly white, while in *S. griseicauda* the tail is usually gray on the dorsal side.

Stenocephalemys has large ears and a small first digit on the hand, which has a nail instead of a claw. *Stenocephalemys* is distinguished from *Rattus* by the extremely constricted

Ethiopian narrow-headed rat *(Stenocephalemys albocaudata)*, photos by D. W. Yalden.

frontal bones of its skull. *Dasymys* also has narrow frontals but differs from *Stenocephalemys* in molar pattern and other cranial features. Misonne (1969) considered *Stenocephalemys* to be an advanced version of *Thamnomys*. Yalden, Largen, and Kock (1976) stated that *Stenocephalemys* is closely related to *Praomys* but is much larger and has a thicker coat.

According to Yalden, Largen, and Kock (1976), *Stenocephalemys* occurs in scrub and grassland at elevations of 3,000–4,000 meters. Yalden (University of Manchester, pers. comm.) reported that *S. albocaudata* is strictly nocturnal and often uses burrows and runways that are utilized by *Arvicanthis* during the daytime.

RODENTIA; MURIDAE; Genus OENOMYS
Thomas, 1904

Rufous-nosed Rat

The single species, *O. hypoxanthus*, occurs from Sierra Leone to central Ethiopia, and south to northern Angola and southwestern Tanzania (Dieterlen and Rupp 1976; Misonne, *in* Meester and Setzer 1977).

Head and body length is 105–220 mm, tail length is 135–205 mm, and weight is 50–121 grams (Delany 1975; Kingdon 1974b; Rosevear 1969). The pelage consists of soft, woolly underfur about 15–17 mm in length and long, fine guard hairs, which on the back reach a length of 25–30 mm. The general coloration of the upper parts ranges from sepia, rufous, and olive to slate, occasionally brushed with black. The rump is generally somewhat reddish, and the flanks are somewhat paler than the rump. The nose, or at least the side of the face, is reddish, thereby providing the common name. The underparts are white, often tinged with buff, and some forms have a definite buffy line where the pale color of the underparts joins the darker coloration of the sides. The feet and hands are white, brownish gray, or rufous.

The tail is practically naked, so the large scales are exposed (there are about 12 scales per cm). The ears are fairly large, rounded, and hairy. The hind feet have naked soles and large plantar pads, and the fifth digit is slightly longer than the first. Females have four, six, or eight mammae (Kingdon 1974b).

The rufous-nosed rat is plentiful in forest clearings at elevations of about 300–3,000 meters. It favors moist areas with thick vegetation. Available information (Delany 1975; Kingdon 1974b; Rosevear 1969) indicates that *Oenomys* is semi-arboreal and a good climber, often being found in vegetation a few meters above the ground. Its nests may be built underground in tunnels, directly on the surface among grass roots or under piles of debris, or in shrubs or the fork of a tree at a height of 1–3 meters. The nests, which are used for both sleeping and breeding, are woven of dry grass and other plant material and measure about 150–200 mm in diameter. Both nocturnal and diurnal activity has been reported. The diet consists mainly of fresh green vegetation, but insects also are taken. *Oenomys* reportedly damages growing rice, millet, and other crops. Recorded population densities are 17.3/ha.

Rufous-nosed rat *(Oenomys hypoxanthus)*, photo by F. Petter.

along the borders of a marsh and 12.5/ha. in marshland. Adult individuals are usually found alone. Breeding probably occurs throughout the year, at least in Zaire and western Uganda. Litter size is one to six, usually two to four, young.

RODENTIA; MURIDAE; Genus LAMOTTEMYS
Petter, 1986

The single species, *L. okuensis,* is known by four specimens, two males and two females, from elevations of 2,100–2,900 meters on Mount Oku in southwestern Cameroon (Dieterlen and Van der Straeten 1988; Petter 1986).

Head and body length is 108–35 mm, tail length is 107–32 mm, and weights of two specimens are 43 and 69 grams. The pelage is generally brown, somewhat paler below, and composed of fine hairs 10–12 mm long. The grayish tail is covered with fine scales and short hairs. The digits resemble those of the subgenus *Desmomys* of *Pelomys,* the short fifth finger bearing a nail instead of a claw. The incisors have a weak groove, also as in *Desmomys;* however, the structure of the molar teeth is more like that of *Oenomys.* In both *Lamottemys* and *Oenomys* the molars are stephanodont, and the tubercles of the upper molars are contracted as a result of the widening and deepening of the intervening valleys.

RODENTIA; MURIDAE; Genus LOPHUROMYS
Peters, 1874

Brush-furred Mice

Dieterlen (1976) recognized two species groups and nine species:

woosnami group

L. woosnami, mountains of northeastern Zaire, southwestern Uganda, and western Rwanda;
L. luteogaster, northeastern Zaire;
L. medicaudatus, mountains around Lake Kivu in northeastern Zaire and western Rwanda;

sikapusi group

L. sikapusi, Sierra Leone to western Kenya and northern Angola;

L. flavopunctatus, Ethiopia to northeastern Angola and northern Mozambique;
L. rahmi, mountains around Lake Kivu in northeastern Zaire and western Rwanda;
L. nudicaudus, southern Cameroon, Equatorial Guinea including island of Bioko (Fernando Poo);
L. cinereus, mountains west of Lake Kivu in northeastern Zaire;
L. melanonyx, mountains of south-central Ethiopia.

Head and body length is 84–160 mm, tail length is 46–148 mm, and weight is 23–111 grams (Delany 1975; Dieterlen 1976; Kingdon 1974b). In the *woosnami* group of species the tail length is about equal to the head and body length, while in the *sikapusi* group the tail length averages only about half the head and body length. The pelage in most forms is fairly long, sleek, and thick and is unique in structure. The slightly coarse hairs are flattened and tapered at either end. The coloration is unusual for a murid. The upper parts range from sandy buff through olive gray to dark brown. In some species there is fine streaking or various degrees of speckling from whitish to orangish. The underparts range from cream through cinnamon to dark orange or wood brown. The upper surfaces of the hands and feet are usually somewhat lighter than the back. The tail is well furred in some forms, but in others it is scantily covered with fine, short hairs. The upper surface of the tail is almost as dark as the back, whereas the lower surface is usually light-colored.

In addition to their coloration and peculiarly textured fur, these animals are distinguished by their short legs and chunky shape (Kingdon 1974b). They have five fingers and five toes, all provided with claws. According to Dieterlen (1976), the *woosnami* group is characterized by its long tail, long feet, big ears, short claws, and sausagelike stomach glands, whereas the *sikapusi* group has a short tail, relatively short hind feet and ears, relatively long claws, and packet-shaped stomach glands. The thin incisor teeth of *Lophuromys* occasionally are inclined forward. The skin is delicate and easily torn, and it is not unusual for the tip of the tail to be missing.

Brush-furred mice occur in a variety of habitats, from grassy plains and swampy areas to montane forests at elevations above 4,000 meters. They are exclusively terrestrial (Rosevear 1969). According to Kingdon (1974b), moisture and grass are probably essential for these animals, as they are not found in closed canopy forest or in areas subject to periodic drought. They construct nests of dry grass and leaves under rocks, roots, or fallen timber or in crevices or short, straight burrows that they dig themselves. Activity seems to

Brush-furred mouse (*Lophuromys sikapusi*), photo by D. C. D. Happold.

vary by species, with *L. flavopunctatus* tending to be more diurnal, and *L. woosnami* more nocturnal. Kingdon noted that the success of these mice may stem from their specializing in eating insects, especially ants. They also feed on other invertebrates, frogs and other small vertebrates, and vegetation. Dieterlen (1976) stated that the amount of animal food taken by each species of *Lophuromys* varied from 40 percent to 100 percent.

Happold (1977) reported a mean home range of 2,100 sq meters for *L. sikapusi* in western Nigeria. For *L. flavopunctatus*, Kingdon (1974b) cited a home range of 350 sq meters in Uganda and a density of 14–19 per ha. in the Ituri Forest; he thought that densities would be considerably higher in arable areas of Uganda. Rosevear (1969) stated that brush-furred mice are generally solitary and fight extensively among themselves. Collectors have long been impressed by the large proportion of specimens with body wounds, torn ears, or partial or complete loss of the tail.

Breeding cycles have been shown to be highly plastic and responsive to ecological conditions. In tropical areas reproductive activity tends to be continuous, and in southern parts of the range of *Lophuromys* there is a single prolonged breeding season during the rains from October to May (Kingdon 1974 b). In studies in Uganda, Delany (1971, 1975) found pregnant female *L. flavopunctatus* in every month of the year, but he indicated that breeding peaks occurred during the rainy seasons, October–December and March–June. The reproductive cycle of *L. sikapusi* in Uganda was considered to be much the same. Happold (1977) found pregnant female *L. sikapusi* in western Nigeria in February, March, April, May, July, and October and noted that one female had consecutive litters in October, February, and March. Reported gestation periods are 30.5 days for *L. flavopunctatus* (Delany 1971) and 32 days for *L. woosnami* (Dieterlen 1976). Litter sizes are one to four young, averaging about two in *L. flavopunctatus*, two to five in *L. sikapusi*, and one to two in *L. woosnami* (Delany 1975). Young *L. flavopunctatus* weigh 5–8 grams at birth and open their eyes after 4–7 days (Kingdon 1974b).

Live animals have been reported to be very sensitive and hard to handle without causing injury or death. Nonetheless, Kingdon (1974b) stated that they have been used in the laboratory and are clean and easy to keep and breed. Captives may survive up to 2 years. Schlitter (1989) expressed concern for several species restricted to highland forests in eastern Africa, and classified *L. medicaudatus*, *L. rahmi*, and *L. melanonyx* as rare.

RODENTIA; MURIDAE; **Genus ZELOTOMYS**
Osgood, 1910

Broad-headed Mice

There are two species (Germain and Petter 1973; Meester 1976; Misonne, *in* Meester and Setzer 1977; Smithers 1971):

Z. hildegardeae, Central African Republic and Kenya to Angola and Zambia;
Z. woosnami, northern and eastern Namibia, Botswana, northwestern South Africa.

Germain and Petter (1973) recognized *Z. instans* of northern Zaire and the Central African Republic as a species separate from *Z. hildegardeae*.

Head and body length is 104–37 mm, tail length is 80–118 mm, and weight is 38–85 grams (Delany 1975; Kingdon 1974 b; Rautenbach and Nel 1975; Smithers 1971). The upper parts range from pinkish buff to dark gray, slate gray, and buffy brown. The flanks are grayish to yellowish; the underparts are light gray, pinkish buff, or whitish; and the hands and feet are pinkish buff to white. The short, stiff hairs that form the sparse tail covering are white to grayish brown.

There are no outstanding external peculiarities except the slightly broadened head, a feature reflected in the common name. The generic name, meaning "imitator," refers to the external resemblance of *Zelotomys* to the genus *Praomys*, subgenus *Mastomys*. The incisor teeth of *Zelotomys* are rather slender and slightly protruding, whereas in most other rodents the incisors curve back toward the mouth. The ears are of moderate size and rounded. Females have 10 mammae.

According to Kingdon (1974b) and Delany (1975), *Z. hil-*

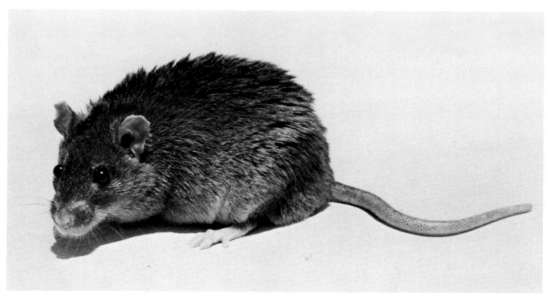

Broad-nosed mouse (*Zelotomys* sp.), photo by C. L. Cheeseman.

degardeae is associated mainly with moist, grassy savannahs and scrub. It is insectivorous and probably forages for its food under the grass cover. Its density apparently is kept low by fires, predators, and the scarcity of insects during the dry season. Pregnant females have been taken in November in Kenya and in February, March, May, June, and July in Uganda. Litter size has ranged from three to seven young.

In Botswana, Smithers (1971) collected *Z. woosnami* mostly on sandy ground with sparse grass and a thin, open scrub cover. He observed that this species is nocturnal, terrestrial, and graminivorous but that it may eat some meat. Birkenstock and Nel (1977) reported that *Z. woosnami* lives in burrows probably dug by other animals but that it is capable of digging and making its own nest. Individuals were found to be widely spaced and asocial. When two captives were put together, one was usually killed, even if it was of a different sex and age. Birkenstock and Nel also stated that the breeding season of *Z. woosnami* corresponded with the warm, wet months from December to March, that the minimum period between births was 31 days, and that litters numbered 4 or 5 young. Meester (1976) reported that three litters of *Z. woosnami*, numbering 4, 5, and 5, arrived between March and the end of July. He considered the species to be rare but not endangered. Smithers (1986) noted that females with up to 11 embryos have been recorded, and he also classified *Z. woosnami* as rare.

RODENTIA; MURIDAE; Genus COLOMYS
Thomas and Wroughton, 1907

African Water Rat

The single species, *C. goslingi*, has been recorded at scattered sites from western Cameroon to Ethiopia, Kenya, and northern Zambia and Angola (Dieterlen 1983). The generic name *Nilopegamys* Osgood, 1928, is a synonym of *Colomys*, and the name *Nilopegamys plumbeus* probably represents a subspecies of *Colomys goslingi* (Misonne, *in* Meester and Setzer 1977).

Head and body length is 117–40 mm, tail length is 145–80 mm, and weight is 50–75 grams (Kingdon 1974*b*). The pelage is thick, short, and soft, with a velvety texture. The upper parts range in color from cinnamon to wood brown and dark brown, the darkest coloration being just behind the middle of the back. The underparts, including the hands, feet, forearms, and lower legs, are generally white. Occasionally the coloration of the underparts extends in a line onto the upper arms, sometimes appearing as a ring about the ankles. The dark color of the upper parts and the light color of the underparts meet in a distinct line of demarcation. There is generally a white or pale-colored spot just below the ear. The large grayish ears are practically naked. The tail varies from light brown to almost black on the dorsal surface and is slightly paler beneath; the hairs on the distal part of the tail are white or pale in color and slightly longer than those of the basal part. The tail is, however, scaly and poorly haired.

Colomys has a light and slender body form. The feet are elongate, and the four fingers are long. Females have eight mammae. The incisor teeth are not grooved and are flat or slightly concave on the front surface. Kingdon (1974*b*) pointed out that in the related genus *Malacomys* the upper lip slopes sharply away from the nose to allow maximum exposure of the incisors and that the animal is incapable of closing its mouth. In *Colomys*, however, both the upper lip and the chin are swollen and cover the incisors. This modification may be associated with the semiaquatic lifestyle of *Colomys*.

According to Delany (1975), Dieterlen and Statzner (1981), and Kingdon (1974*b*), *Colomys* is found close to running water in swamps and along rivers in forested country, at both high and low elevations. It is nocturnal and constructs short tunnels in stream banks. It swims rapidly with powerful thrusts of its long hind legs. It searches for food by wading through watery mud, with its nose held near the water. It evidently obtains prey both by passive filtration through its vibrissae and by actively pouncing on worms, slugs, crustaceans, and aquatic insects. It also eats small vertebrates and some vegetable matter. Individuals are probably solitary. Records suggest a breeding season of March to July in western Uganda and eastern Zaire. Pregnant females have contained one to three embryos.

African water rat *(Colomys goslingi)*, photo by F. Dieterlen.

Big-eared swamp rat *(Malacomys edwardsi)*, photo by D. C. D. Happold.

RODENTIA; MURIDAE; **Genus MALACOMYS**
Milne-Edwards, 1877

Big-eared Swamp Rats, or Long-eared Marsh Rats

There are four species (Ansell 1974; Dieterlen and Van der Straeten 1984; Gautun, Sankhon, and Tranier 1986; Rautenbach and Schlitter 1978; Van der Straeten and Verheyen 1979 *a*; Verheyen and Van der Straeten 1977):

M. edwardsi, Guinea to southern Nigeria;
M. cansdalei, eastern Liberia to Ghana;
M. longipes, Guinea, southeastern Nigeria to Uganda and
    northwestern Zambia;
M. verschureni, known only by four specimens from
    northeastern Zaire.

Rautenbach and Schlitter (1978), as well as Cole (1972), classified the forms *cansdalei* and *giganteus,* which occur from eastern Liberia to Ghana, as subspecies of *M. longipes*. Van der Straeten and Verheyen (1979a), however, considered *M. cansdalei* to be a full species, more closely related to *M. edwardsi* than to *M. longipes,* and considered *M. l. giganteus* to be a synonym of *M. cansdalei*.

Head and body length is 117–90 mm, tail length is 117–222 mm, and weight is 50–145 grams (Cole 1972; Kingdon 1974b; Van der Straeten and Verheyen 1979a). The pelage is thick, soft, and velvety, much like that of certain shrews. The texture of the fur is so fine that it has been compared to the bloom of a peach. General coloration of the upper parts is dark brown with a sprinkling of grayish, the sides are ochraceous, and the underparts are gray. *M. edwardsi* has an oval, sooty-colored patch around each eye. The hands and feet are plentifully sprinkled with fine-textured short hairs. The whiskers are fine, long, and dark. The tail is so poorly haired that in some specimens it appears to be almost naked.

These medium-sized, slender-bodied rats have large, naked ears; a long, narrow skull; long, slender hind feet; and a long tail. It has been suggested that the structure of the feet permits the paws to spread more than those of most terrestrial rats, as an adaptation for traveling over swampy ground. Females usually have four or six mammae.

These rodents seem to be mainly nocturnal and usually inhabit areas of dense vegetation near water. Kingdon (1974b) stated that *M. longipes* is a capable climber but makes its grass or leaf nests in crevices or among tree roots. It apparently eats animal and vegetable matter in nearly equal proportions, feeding on fallen fruits, seeds, nuts, roots, insects, slugs, snails, and crabs. Misonne and Verschuren (1976) reported that eight stomachs of *M. edwardsi* from Liberia contained fruit and that one had a small proportion of insect remains.

Kingdon (1974b) wrote that *Malacomys* is generally solitary, has a mild disposition, and has no clearly defined breeding season. In a study of *M. edwardsi* in the rainforests of western Nigeria, Happold (1977) found home range to average 0.42 ha. for males and 0.37 ha. for females. Pregnant females were taken from November to July, which is mainly a dry period, and peak reproduction occurred in January. Females could produce a litter every one or two months during this season. Misonne and Verschuren (1976) collected four pregnant female *M. edwardsi* in Liberia in December and January, each with two embryos. Delany (1975) stated that pregnant female *M. longipes,* with three or four embryos each, had been taken in Uganda in May, July, and September and that the number of embryos reported in specimens from Zaire was one to five. In Zambia two pregnant females, each with three embryos, were taken in August (Ansell 1974).

Egyptian spiny mouse *(Acomys cahirinus)*, light and dark subspecies, photos by Ernest P. Walker. The animal in the lower picture has lost the tip of its tail.

RODENTIA; MURIDAE; **Genus ACOMYS**
I. Geoffroy St.-Hilaire, 1838

Spiny Mice

There are eight species (Corbet 1978, 1984; Dippenaar and Rautenbach 1986; Kingdon 1974*b*; F. Petter 1983*b*; Petter and Roche 1981; Setzer, *in* Meester and Setzer 1977; Spitzenberger 1978; Yalden, Largen, and Kock 1976):

A. cahirinus, Mauritania to southern Pakistan and
    Tanzania, Crete, Cyprus;
A. cilicius, Asia Minor;
A. russatus, northeastern Egypt, Palestine, Jordan, Arabian
    Peninsula;
A. whitei, Oman;
A. spinosissimus, Tanzania, Zambia, Zimbabwe,
    Mozambique, eastern Botswana, northeastern South
    Africa;
A. subspinosus, southwestern South Africa;
A. wilsoni, southeastern Sudan, southern Ethiopia, Kenya;
A. louisae, Ethiopia, Somalia.

The last species was referred to a newly erected subgenus, *Peracomys,* by F. Petter (1983*b*) and Petter and Roche (1981).

Some authorities (such as Harrison 1972; Kingdon 1974*b*; and Setzer, *in* Meester and Setzer 1977) recognize *A. dimidiatus,* of southwestern Asia and northeastern Africa, as a full species distinct from *A. cahirinus.* Ellerman and Morrison-Scott (1966) considered *A. cahirinus* to be a small commensal form of *A. dimidiatus,* but since the name *A. cahirinus* had priority, it was applied to the entire species. Neither Rosevear (1969), Corbet (1978), nor Osborn and Helmy (1980) recognized *A. dimidiatus* as a separate species, though Corbet suggested that *A. cahirinus* might actually comprise several closely related species. Subsequently, Corbet and Hill (1986) recognized *A. minous* of Crete as a species separate from *A. cahirinus.* Pucek (1989), however, stated that *Acomys* probably had been introduced to Crete by people. Until Dippenaar and Rautenbach (1986) demonstrated that *A. subspinosus* is endemic to South Africa, most authorities considered its range to extend as far north as Ethiopia.

Head and body length is 70–175 mm, tail length is 42–125 mm, and weight is 11–90 grams. The brittle tail is easily separated, partly or entirely, from the rest of the animal, and therefore measurements of tail length are often misleading. Coloration above is pale yellowish or reddish brown, reddish, or dark grayish. The underparts are white. *A. cahirinus* and *A. russatus* are sometimes melanistic. The back and tail are covered with coarse, inflexible spines. The tail appears to be practically naked, and the scales are conspicuous. It may be whitish above and below, bicolored, or uniformly dark. The ears are large and erect. Females have four or six mammae.

Spiny mice are found in arid regions. The African forms inhabit rocky country, semidesert areas, and dry woodland or savannah (Kingdon 1974b). They shelter in rocky crevices, cracked soil, gerbil or other rodent burrows, and termite mounds. In some areas, most notably Egypt, A. cahirinus has become a human commensal and lives in and around buildings. Most species are primarily nocturnal but are sometimes active in the early morning and late afternoon. In the Arabian Peninsula, A. russatus is largely diurnal in its extremely rocky habitat and seems to be much better adapted to high temperatures than the sympatric A. cahirinus (Harrison 1972). Spiny mice are omnivorous but feed mainly on plant material, especially grains and grasses. Osborn and Helmy (1980:299) wrote that dates are a staple in some areas and that "the dried flesh and bone marrow of mummified humans is a source of food for A. cahirinus in the tombs of Gebel Drunka southwest of Asyut."

Most knowledge of the reproduction of Acomys has been derived from captive studies, mainly of Middle Eastern A. cahirinus (Grzimek 1975; Harrison 1972; Kingdon 1974b; Rosevear 1969). Spiny mice are gregarious; females that already have given birth will help to clean new mothers and bite their umbilical cords. Reproduction is continuous, there being a postpartum estrus and mating and sometimes a series of 12 or more litters without cessation. The gestation period is five to six weeks, which is quite long for a murid. There are one to five young, with the larger litters generally being born to older mothers. Nest construction is rudimentary, but the young are remarkably well developed. They weigh up to seven grams when born, their eyes are open either at birth or two or three days later, and they are weaned after about two weeks. Sexual maturity is attained at two to three months. Average longevity is around three years, but some individuals have lived up to five years.

Under good conditions, wild Acomys probably breed over the greater part of the year (Kingdon 1974b). In Kenya, Neal (1983) found no seasonal changes in breeding activity despite pronounced seasonal changes in climate. Collections from the Arabian Peninsula indicate that reproduction in A. cahirinus

there continues at least throughout the spring and summer months (Harrison 1972). In Tanzania, Hubbard (1972) collected pregnant female A. spinosissimus, each with one to four embryos, from October to February. The young of this species were found to travel with the mother during their first night of life but not to hang from the mammae. In contrast, the young of A. cahirinus hystrella in this area were born naked and blind and did attach themselves firmly to the mammae. Hubbard noted that neither species usually built a nest. In Botswana, however, Smithers (1971) observed A. spinosissimus to construct a rough nest of grass and leaves, in which the young were born. Pregnant females, with two to five embryos each, were taken in Botswana in the warm, wet months of December, January, March, and April.

Spiny mice have become popular as pets and are easy to keep and breed in the laboratory. In Egypt, however, they may be a public health menace, as a significant proportion have been found to carry the causal organism of typhus (Harrison 1972).

RODENTIA; MURIDAE; Genus URANOMYS
Dollman, 1909

White-bellied Brush-furred Rat

The single species, *U. ruddi*, has been recorded from Senegal to Nigeria and in northeastern Zaire, Uganda, Kenya, Malawi, and Mozambique (Kingdon 1974b; Misonne, *in* Meester and Setzer 1977).

Head and body length is 84–134 mm, tail length is 53–79 mm, and weight is 41–53 grams (Delany 1975; Kingdon 1974b). The hairs on the back are fairly long, especially on the rump, and are slightly stiffened, thereby giving the coat a crisp or brittle texture. The upper parts vary in color from dark brown to reddish and gray. The underparts vary from grayish to white, with a tinge of buff in some individuals. The tail, which is scantily covered with short hairs, is dark above and slightly paler below.

In external appearance *Uranomys* resembles *Lophuromys*, except that the backs of its hands and feet are covered with fine white hairs. Also, the underparts of *Lophuromys* are never pure white. The skull of *Uranomys* resembles that

White-bellied brush-furred rat *(Uranomys ruddi)*, photo by Jane Burton.

of *Acomys*, most notably in having the palate extended well behind the molars by a roof to the mesopterygoid fossa (Rosevear 1969). The cheek teeth also closely resemble those of *Acomys* but are unique in that the anterior cusps of the front lower molars are flattened (Kingdon 1974b). The incisors usually project forward. The hind foot is broad, and the middle three digits are relatively long. Females have 12 mammae.

This genus is generally considered rare but has been taken in large numbers in the Ivory Coast. It lives mainly on savannahs and is reportedly nocturnal and insectivorous (Kingdon 1974b). It constructs a burrow about 15 cm deep with two exits, each topped by a mound of soil, and a blind tunnel up to 40 cm long. A grass-lined nest is built within the burrow, and there is no food storage. Each burrow system is occupied by two adults. In the Ivory Coast, breeding occurs throughout the year, but litters are larger from September to December, when they average 4.0–5.7 young, than at other times of the year, when they average 2.6–3.7 young (Delany 1975).

RODENTIA; MURIDAE; Genus MALPAISOMYS
Hutterer, Lopez-Martínez, and Michaux, 1988

Canarian Lava Fields Rat

The single species, *M. insularis*, is known only from sub-fossil skeletal material found in deposits about 1,000–5,000 years old on Fuerteventura, Lanzarote, and Graciosa in the eastern Canary Islands (Hutterer, Lopez-Martínez, and Michaux 1988). *Malpaisomys* is not thought to be closely related to *Canariomys* Crusafont-Pairo and Petter, 1964, a large fossil rat found in volcanic deposits of early Quaternary age, also in the Canary Islands.

Malpaisomys is described as a small-bodied terrestrial rat with a stout skull, wide zygomatic arches, an extremely narrow interorbital constriction, long incisive foramina, a palate extending posteriorly beyond the third upper molars and covering most of the mesopterygoid fossa with bone, small bullae, a mandible having a prominent mandibular crest passing almost to the distal part of the condyle, the condyle sigmoid-shaped in lingual view, and brachyodont teeth. Superficially, the skull of *Malpaisomys* resembles that of *Nesoryzomys*, of the Galapagos, and both genera are endemic to volcanic islands with rough lava fields. However, the characters of the molar teeth and the remarkable posterior extension of the palatal bridge indicate affinity to the African genera *Acomys* and *Uranomys*.

RODENTIA; MURIDAE; Genus HYBOMYS
Thomas, 1910

Eastern Back-striped Mice

Four species are currently recognized (Carleton and Robbins 1985; Van der Straeten 1985; Van der Straeten and Hutterer 1986; Van der Straeten, Verheyen, and Harrie 1986; Verheyen and Van der Straeten 1985):

H. univittatus, southeastern Nigeria to Uganda, extreme northeastern Zambia;
H. lunaris, eastern Zaire, Uganda, Rwanda;
H. eisentrauti, Cameroon;
H. basilii, Bioko.

The authorities listed above are not consistent with respect to whether *Hybomys* includes *Typomys* as a subgenus (see ac-

count of the latter) or with respect to the distributions of *H. univittatus* and *H. lunaris*.

Head and body length is 100–150 mm, tail length is 77–134 mm, and weight is 30–70 grams (Genest-Villard 1978a; Kingdon 1974b; Rosevear 1969). The fur is usually soft, but in some individuals it is inclined to be coarse. The upper parts range from brown to almost black. There is usually a dark brown or black middorsal stripe that extends from the nape to the base of the tail, though it may be almost undetectable in some individuals. The underparts range from tawny or ochraceous to grayish white. The upper surfaces of the hands and feet are usually the same color as the back. The thinly haired tail is about the same color, though in some animals it is entirely black.

Although *Hybomys* resembles *Arvicanthis* externally in such features as the dark dorsal stripe, it differs from the latter genus by its more slender form and almost naked tail. The soles of the feet have five well-developed tubercles and a rudimentary sixth, whereas the palms have six tubercles, all well developed. The thumb is small and has a blunt nail. The hind foot is narrow. Females have two or three pairs of mammae.

These mice inhabit forests, thick brush, and the edges of cultivated areas. They are strictly terrestrial but seem to require moist places and reportedly are good swimmers (Rosevear 1969). *H. univittatus* is an animal of the forest floor, preferring areas near rivers or swamps, with abundant leaf litter and shade (Kingdon 1974b). Genest-Villard (1978a) found *H. univittatus* to be entirely diurnal. Individuals spent the night, and resting periods during the day, in burrows, within which they built nests of twigs. Each animal used several burrows, but males, at least, spent most inactive time in one favored lair. *Hybomys* also may shelter in rotting logs, piles of twigs, or termite mounds. *H. univittatus* has been reported to be mainly frugivorous and to be destructive to cassava roots, but Genest-Villard (1980) found it to feed mostly on insects.

In a radiotracking study of *H. univittatus* in a dense equatorial forest, Genest-Villard (1978a) determined home range to be 4,500–6,100 sq meters for males and 1,400–1,800 sq meters for females. A male spent its entire life (about 12 months) in one area and after death was replaced by a younger animal. Home ranges of males did not overlap, and intruders were chased and viciously bitten. The home range of one male could overlap that of one or two females, but individuals were solitary and did not regularly share burrows. Males did occasionally visit the burrows of females.

In captivity, and in cultivated areas where food is abundant, *H. univittatus* is capable of reproduction throughout the year. In the forests of Zaire, however, the main breeding season extends from February to May, the period corresponding with the rains and maximum fruit production (Kingdon 1974b). In Uganda, pregnant females of this species have been collected in April, July, and October (Delany 1975). Laboratory studies of *H. univittatus* show that gestation lasts 29–31 days, litter size averages about 3 (1–4) young, the eyes of the young open at 7–10 days, and females attain sexual maturity by 3 months (Rosevear 1969).

RODENTIA; MURIDAE; Genus TYPOMYS
Thomas, 1911

Western Back-striped Mice

There are two species (Carleton and Robbins 1985):

T. trivirgatus, Sierra Leone to southwestern Nigeria;
T. planifrons, Guinea, Sierra Leone, Liberia, Ivory Coast.

Western back-striped mouse *(Typomys trivirgatus)*, photo by Jane Burton.

Until distinguished at the generic level by Van der Straeten (1984), *Typomys* usually was considered part of *Hybomys* (e.g., by Rosevear 1969). Van der Straeten's subsequent papers (Van der Straeten 1985; Van der Straeten and Hutterer 1986; Van der Straeten, Verheyen, and Harrie 1986; Verheyen and Van der Straeten 1985) have not been consistent concerning whether *Typomys* should be treated as a genus or subgenus. Carleton and Robbins accepted the latter status and indicated that further study would be needed to evaluate the question of generic distinction. Corbet and Hill (1986) listed *Typomys* as a full genus.

Head and body length is 100–160 mm, tail length is 77–120 mm, and weight is 54–68 grams. The pelage is short and dense. In *T. trivirgatus* the upper parts are brownish gray or yellowish brown, and a black dorsal stripe extends from the forehead to the base of the tail, flanked to each side by a less distinct stripe. In *T. planifrons* the upper parts are a more uniform dark brown, and the single dorsal stripe begins on the back of the head. The underparts of both species are more variable in color, but *T. planifrons* typically has a strong suffusion of reddish overlaying a dark gray base, while *T. trivirgatus* has a more yellowish wash with a base of pale gray (Carleton and Robbins 1985; Rosevear 1969).

Typomys resembles *Hybomys* in general size and postcranial structure but differs markedly in characters of the skull and teeth (Carleton and Robbins 1985; Rosevear 1969). For example, the profile of the skull of *Typomys* is practically flat for the anterior half; its rostrum is shallow, narrow, and sharp; the anterior palatal foramina are relatively broad and open; and the first lower molar has little or no sign of an anterior supplementary cusp. *Hybomys*, on the other hand, has a strongly arched skull; a deep, broad, and blunt rostrum; relatively narrow anterior palatal foramina; and nearly always a well-developed anterior supplementary cusp on the first lower molar. As far as is known, female *Typomys* have only two pairs of mammae.

According to Happold (1987), these mice occur mostly in undisturbed rainforest but may also be found in cocoa plantations adjacent to natural forest. They are both nocturnal and diurnal and are mainly terrestrial, though there is one record of an individual climbing trees. Captives have made nests of leaves. The diet is largely insectivorous and includes termites, beetles, grasshoppers, slugs, and some vegetation. They may live in small groups that reside in a particular place for only a short period of time before moving on to another locality. Pregnant females carrying two or three young each have been recorded in Nigeria during March, July, and September. In addition, Happold (1977) found young in Nigeria during November and December, which is the dry season, and Misonne and Verschuren (1976) collected pregnant females, each with 2 embryos, in Liberia in January and February.

Western back-striped mice have a patchy and irregular distribution and are rare even in their preferred habitat. They seem to require undisturbed areas and have not been recorded in secondary forest or farmlands (Happold 1987). They apparently have been eliminated through human environmental disruption in Benin, Togo, and southeastern Ghana (Carleton and Robbins 1985).

RODENTIA; MURIDAE; **Genus STOCHOMYS**
Thomas, 1926

Target Rat

The single species, *S. longicaudatus*, is found from Nigeria to western Uganda (Kingdon 1974b; Misonne, *in* Meester and Setzer 1977). This species, along with *Dephomys defua*, sometimes has been placed in *Rattus* or *Aethomys*. Both Misonne (1969) and Rosevear (1969), however, used the generic names *Stochomys* for *S. longicaudatus* and *Dephomys* for *D. defua*. Some subsequent authorities (Cole 1975; Honacki, Kinman, and Koeppl 1982; Misonne, *in* Meester and Setzer 1977; Misonne and Verschuren 1976) recognized *Stochomys* as a genus that includes *Dephomys*. Then Van der Straeten (1984) again concluded that *Dephomys* warrants generic rank, and this procedure was followed by Carleton and Musser (1984) and Corbet and Hill (1986).

Except as noted, the remainder of this account is based on Happold (1987), Kingdon (1974b), and Rosevear (1969). Head and body length is 120–72 mm, tail length is 177–250 mm, and weight is 45–104 grams. The pelage is long and rather coarse. Remarkably long guard hairs project above the general coat contour. The common name of the genus is derived from the fancied resemblance of these bristles to arrows sticking in a target. The upper parts, face, and neck are reddish brown, becoming more gray on the flanks, and the underparts are white. Both the long tail and the small to moderate-sized ears appear to be naked but are sparsely covered with short hairs.

Stochomys is readily distinguished from related murids by

Target rat *(Stochomys longicaudatus)*, photo by D. C. D. Happold.

its large size, its relatively long and seemingly naked tail, and the peculiarities of its pelage. The hands and feet each have five digits. The thumb is greatly reduced and bears a flat nail. Females have six mammae.

The target rat is found principally in rainforest, where it has been collected in swampy and densely vegetated areas, but some individuals occur in secondary forest and oil palm plantations. Activity is mainly nocturnal and terrestrial. Captives have constructed spherical nests of shredded grass. Examination of stomach contents indicates that *Stochomys* feeds largely on fruit but also takes some green vegetation and insects (Misonne and Verschuren 1976; Rahm 1972). A female used a home range of only 0.15 ha. in a two-year period (Happold 1977).

A pregnant female carrying two embryos was taken on 20 July in Cameroon, and young individuals were found there from 24 July to 3 August. In Nigeria, pregnancies in this species were recorded in March and June, and young were found in January, June, and December; two litters each contained three young. Breeding appears to go on throughout the year in East Africa.

mainder of this account was taken, the upper parts are warm brown with a hint of red, the flanks are only slightly paler, and the underparts are white. The guard hairs are limited mostly to the hind part of the back and are neither so long nor so strongly built as those of *Stochomys*. The face is long and pointed, the ears relatively small and practically naked, and the tail extremely long. The hind foot is short and broad compared with that of *Stochomys*. The cusps of the first upper molar tooth are rather well defined in *Dephomys* but not in *Stochomys* (Misonne, *in* Meester and Setzer 1977). Females have four mammae.

Defua rats apparently are rare inhabitants of tropical forest and may prefer swampy, densely vegetated areas. The broad hind foot and long tail suggest that these animals spend more time in the trees than on the ground. Activity is nocturnal. Examination of stomach contents indicates that the diet consists largely of fruit but also includes some insects (Misonne and Verschuren 1976; Rahm 1972). Juvenile *S. defua* have been taken in early October and mid-December, and Misonne and Verschuren collected four pregnant females, each with two or three embryos, in Liberia in January and February.

RODENTIA; MURIDAE; **Genus DEPHOMYS**
Thomas, 1926

Defua Rats

There are two species (Misonne, *in* Meester and Setzer 1977; Van der Straeten 1984):

D. defua, Guinea to Ghana;
D. eburnea, Ivory Coast.

Dephomys sometimes has been considered part of *Rattus* or *Stochomys* (see account of latter).

Head and body length is 109–50 mm and tail length is 176–223 mm (Van der Straeten 1984). Reported weight is 30–60 grams for *S. defua* (Misonne and Verschuren 1976). According to Rosevear (1969), from whom most of the re-

RODENTIA; MURIDAE; **Genus RHABDOMYS**
Thomas, 1916

Four-striped Grass Mouse

The single species, *R. pumilio,* is found from Uganda and Kenya to Angola and South Africa (Misonne, *in* Meester and Setzer 1977).

Head and body length is 90–137 mm, tail length is 80–135 mm, and weight is 30–73 grams (Kingdon 1974*b*; Swanepoel 1975). The fur is coarse. The striping on the dorsal surface of most forms is constant, but the tones of the color vary considerably. The striping consists of a pale middorsal streak, ranging from pale yellowish gray to buff, that extends from the back of the neck to the base of the tail. This streak is usually bordered on each side by two dark lines that range from light

Four-striped grass mouse *(Rhabdomys pumilio)*, photo by Don Davis.

golden brownish or clay color to dark chocolate brown or almost black. The remainder of the upper parts vary in color from yellowish gray or pale yellowish brown to light grayish brown; the underparts are paler. The tail is scaly and thickly covered with short hairs.

These medium-sized mice have a tail that is generally shorter than the head and body. From *Lemniscomys, Rhabdomys* is distinguished by its functional fifth digit. The skull is similar to that of *Arvicanthis*. Females have four pairs of mammae.

In East Africa the four-striped grass mouse has a discontinuous range, being strictly limited to grassy uplands, moorlands, and the subalpine zone at elevations of 1,700–3,500 meters (Kingdon 1974b). Farther to the south it is more widely distributed, occurs at lower elevations, and has been reported in a variety of habitats, including dry riverbeds, bush and scrub country, forest edge, and cultivated areas. It is sometimes found along stone walls and in outbuildings. It is rather common and is primarily diurnal (Christian 1977; Smithers 1971). It seems to enjoy basking in the sun on cool days. Although not classified as arboreal, it has been seen climbing about the low branches of shrubs. It has sometimes been reported to live in burrows with a single slanting entrance that usually is located near a shrub. Its nests, however, are more often located above the ground in a dense shrub, in a bunch of grass, or among tree roots. In the wild the nests are made of grass, leaves, fibers, and moss. In a captive study comparing *Rhabdomys* with *Aethomys* and *Praomys, Rhabdomys* built the best nests, these being spherical or cuplike structures of cotton and paper strips (Stiemie and Nel 1973). The diet of *Rhabdomys* is primarily vegetarian, consisting of roots, seeds, berries, and cultivated grains but also including snails, insects, and eggs. A home range of about 20 ha. has been calculated (Kingdon 1974b).

Males in the wild are well spaced, and studies by Choate (1972) indicate that they do not tolerate one another in captivity. A mated pair, however, jointly constructs the nest and remains together until the female gives birth, at which time she sometimes excludes the male. The weaned young of one litter may be allowed to remain when a second litter is produced three to four weeks after a postpartum estrus. This process may be responsible for reports of groups of 12–30 individuals living in the same nest. In the wild in Botswana, breeding apparently continues throughout the year, with embryo counts averaging 5 and ranging from 3 to 9 (Smithers 1971).

Laboratory investigations have shown that females can produce four litters per year, containing 3–12 young each

(averaging about 6), with a minimum interval between litters of 23 days. There is a postpartum estrus, the estrous cycle averages 7.4 days, and gestation averages 26.4 days. The young weigh about 2.5 grams each at birth, open their eyes after 7 days, leave the nest at 14 days, and are nearly full-grown at 24 weeks. Females first give birth at about 90 days (Brooks 1982; Dewsbury, Ferguson, and Webster 1984; Kingdon 1974b). A captive lived for 2 years and 11 months (Jones 1982).

RODENTIA; MURIDAE; Genus LEMNISCOMYS
Trouessart, 1881

Striped Grass Mice

There are eight species (Kingdon 1974b; Misonne, *in* Meester and Setzer 1977; Van der Straeten 1975, 1976, 1980a, 1980b; Van der Straeten and Verheyen 1978a, 1979b, 1980):

L. griselda, Angola;
L. linulus, Senegal to Ivory Coast;
L. rosalia, Kenya to eastern South Africa;
L. roseveari, known from two localities in Zambia;
L. striatus, savannah zones from Sierra Leone to Ethiopia, and south to northern Angola and Malawi;
L. macculus, northern and eastern Zaire, southern Sudan, Uganda, Kenya;
L. barbarus, Morocco to Tunisia, Senegal to Sudan and Tanzania;
L. bellieri, Ivory Coast.

Van der Straeten and Verheyen (1980) indicated that the form *mittendorfi*, from Cameroon, which was included in *L. striatus* by Misonne (*in* Meester and Setzer 1977), might actually be a separate species with affinity to *L. macculus* and *L. bellieri*. Corbet and Hill (1986) listed *mittendorfi* as a full species.

Head and body length is 83–140 mm, tail length is 75–159 mm, and weight is 18–68 grams (Kingdon 1974b; Rosevear 1969). The thin fur is unusually rough and coarse. This genus shows three main kinds of color pattern. In *L. rosalia* there is a single dark brown or black stripe down the middle of the back, sometimes extending between the eyes and onto the tail. In *L. striatus, L. macculus,* and *L. bellieri* there is again a dark middorsal stripe, but on both sides of it there are about five or six longitudinal rows of pale spots extending from the

Striped grass mouse *(Lemniscomys striatus striatus)*, photo from New York Zoological Society.

shoulders to the rump; in some forms the spots are definitely separated, while in others they tend to run together on their long axes. In *L. barbarus* the rows of spots are replaced by solid pale stripes, extending from shoulders to rump, on either side of the dark middorsal line. The general color of the upper parts of *Lemniscomys*, other than the spots and stripes, ranges from medium clay to dull dark brown. The pale spots and stripes range from light buffy to a dull dark clay color. The flanks range from clear buffy through light clay to fairly dark gray tinged with buff. The underparts vary from white or buffy white to medium buffy gray. The tail above is almost as dark as the middorsal stripe, and the remainder of the tail is almost as pale as the pale spots or stripes on the body. Females have two pairs of pectoral and two pairs of inguinal mammae.

According to Kingdon (1974*b*), *L. rosalia* is the most primitive species and is declining but remains common along the edge of grassy pans within the woodland zone. *L. striatus* is a more successful species and occupies a wide variety of grassy habitats from sea level to 3,500 meters. It is especially common in forest clearings and also is found along streams, in swampy places, and in cultivated areas. The considerably smaller species *L. macculus* and *L. barbarus* are partly sympatric with *L. striatus* but generally live in more arid country. *L. barbarus* is typical of the drier savannahs and steppes that border the Sahara. All species construct small, spherical nests of grass and leaves near the ground and make well-defined surface runways around the nests for foraging. They may sometimes seek refuge in burrows dug by other animals. *Lemniscomys* is terrestrial and has been reported to be active by day, but studies have indicated that *L. striatus* and *L. barbarus* are sometimes crepuscular and nocturnal (Rosevear 1969). The diet includes grass, soft seeds, cultivated crops, and occasionally insects.

Relatively high population densities have been recorded, but these mice do not seem to be gregarious (Kingdon 1974*b*). Neal (1977*a*) stated that mean life expectancy is very short in the wild and nearly all adults die after each breeding season, resulting in a nearly total population turnover twice a year. A captive *L. striatus*, however, lived for 4 years and 10 months (Jones 1982).

There is a general consensus that breeding in the wild corresponds mainly with the two annual rainy seasons (Chidumayo 1977; Delany 1975; Gautun 1975; Kingdon 1974*b*; Neal 1977*a*; Rosevear 1969). In Uganda these periods extend from April to June and from September to December. The reported number of young per litter is 1–12, most commonly being 4–5. Most wild females apparently give birth only once during a wet season, but a captive pair produced 4 litters in 15 weeks. The young weigh about 3 grams at birth, open their eyes at 6–8 days, and attain full size by 6 months. Recent investigations of captive *L. striatus* (D. C. Wharton, New York Zoological Society, pers. comm.) have shown that the estrous cycle lasts 3–5 days, the precise gestation period is 21 days (some older reports gave it as 28 days), consecutive litters may be produced 21–25 days apart when a male and female are kept together, young begin eating solid food at 14 days, males attain sexual maturity by 2.5 months, females do not give birth until about 5 months and produce their last litter at 14 months, and longevity seldom exceeds 2.5 years.

RODENTIA; MURIDAE; Genus PELOMYS
Peters, 1852

Groove-toothed Creek Rats

There are three subgenera and seven species (Dieterlen 1974; Misonne, *in* Meester and Setzer 1977; Yalden, Largen, and Kock 1976):

subgenus *Komemys* de Beaux, 1924

P. isseli, islands in northwestern Lake Victoria (Uganda);
P. hopkinsi, southwestern Uganda, Rwanda;

subgenus *Desmomys* Thomas, 1910

P. harringtoni, highlands of Ethiopia;
P. rex, known only by the type skin from southwestern Ethiopia;

subgenus *Pelomys* Peters, 1852

P. minor, northeastern Angola, southern Zaire, northwestern Zambia;
P. fallax, eastern Zaire and southern Kenya to Botswana and northern Mozambique;
P. campanae, southwestern Zaire, western Angola.

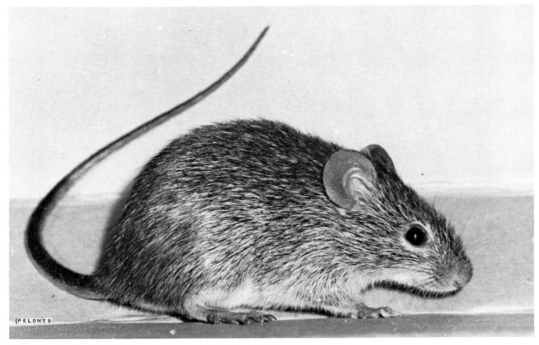

Groove-toothed creek rat *(Pelomys campanae)*, photo by F. Petter.

Misonne (1969) thought there was no justification for the use of subgenera in *Pelomys*, but, on the other hand, Petter (1986) suggested that *Desmomys* warrants full generic rank. Dieterlen (1974) doubted the validity of *P. rex*, but Yalden, Largen, and Kock (1976) accepted it as a species. An additional species listed by Misonne *(in* Meester and Setzer 1977), *P. dembeensis,* is now referred to the genus *Arvicanthis.*

Head and body length is 100 to about 215 mm and tail length is usually 100–180 mm. The tail may be shorter than, the same length as, or longer than the head and body. Reported weights are 46–170 grams (Delany 1975; Kingdon 1974b; Smithers 1971). The hairs are usually coarse, slightly stiff, and shiny, sometimes with a greenish brown or olive iridescence. The upper parts are tawny yellowish, yellow and black, dusky and tawny, or buffy clay and blend into the gray, buff, or whitish of the underparts. Some forms have a single dark middorsal stripe, which is often indistinctly defined from the color of the upper parts. The moderately haired tail is frequently dark above and pale below. The nose may be more brightly colored than the face, and the ears are usually thinly covered with reddish hairs.

The short fifth finger is clawed in the subgenus *Komemys* but has a nail in the subgenera *Pelomys* and *Desmomys.* The outer toes are short. The genus *Pelomys* has grooved upper incisor teeth, and this character distinguishes it from the closely related genera *Lemniscomys, Arvicanthis,* and *Dasymys.* It is difficult, however, to distinguish *Pelomys minor* from *Lemniscomys griselda,* as both species have a narrow dark stripe along the middle of the back and other common external features. Female *Pelomys* have eight mammae.

In Ethiopia, *P. harringtoni* lives on plateaus at elevations of 1,800–2,800 meters and is semiarboreal (Yalden, Largen, and Kock 1976). Otherwise, the genus is found mainly in marshes and wet grasslands and along streams. *P. fallax* is a good swimmer and in some places has become semiaquatic. It usually makes a nest of grass or leaves at ground level but has been reported to burrow in some areas. This species is active both by day and by night. Its diet consists of grasses, swamp vegetation, and cultivated crops (Delany 1975; Kingdon 1974b).

In Zambia, according to Sheppe (1972), *P. fallax* apparently has no fixed breeding season, and seven pregnant females contained seven to nine embryos each. Earlier reports indicated that litter size in this species is usually only two or three young, though Delany (1975) referred to overall embryo counts of one to nine. Delany (1969) also reported that of four pregnant female *P. isseli* collected on Bugala Island in Lake Victoria in April 1966, two were lactating and one was in estrus.

Schlitter (1989) designated both *P. isseli* and *P. hopkinsi* as vulnerable. The former has been located only on the three small islands of Bugala, Bunyama, and Kome, despite searches elsewhere in Lake Victoria. The latter is restricted to small, tenuous strips of habitat in the edge vegetation of papyrus swamps.

RODENTIA; MURIDAE; **Genus AETHOMYS**
Thomas, 1915

Bush Rats, or Rock Rats

There are two subgenera and nine species (Ansell 1978; Davis, *in* Meester and Setzer 1977; Visser and Robinson 1986):

subgenus *Aethomys* Thomas, 1915

A. kaiseri, eastern Angola to southern Kenya and Malawi;
A. thomasi, western and central Angola;
A. hindei, northern Nigeria to southern Sudan and
    northeastern Tanzania;

Bush rat, or rock rat (*Aethomys kaiseri*), photo by Eugene Maliniak.

A. bocagei, western and central Angola, possibly Zaire and southern Congo;

A. silindensis, known only by two specimens from Mount Chirinda in eastern Zimbabwe;

A. chrysophilus, southeastern Kenya to Angola and eastern South Africa;

A. nyikae, southern Zaire, northeastern Angola, Zambia, Malawi, eastern Zimbabwe;

subgenus *Micaelamys* Ellerman, 1941

A. namaquensis, Angola and Zambia to South Africa;
A. granti, the Central Karroo of South Africa.

Aethomys sometimes has been regarded as only a subgenus of *Rattus,* but Misonne (1969) considered it a separate genus, and all authorities cited in this account have taken the same position. *Stochomys* sometimes has been included as a subgenus of *Aethomys* but was distinguished as a full genus by Davis (*in* Meester and Setzer 1977).

The following descriptive data were compiled from Ansell (1960, 1974), Ansell and Ansell (1973), Davis (*in* Meester and Setzer 1977), Kingdon (1974b), and Smithers (1971). Head and body length is about 100–184 mm, tail length is 120–202 mm, and weight is 63–150 grams. The tail varies in length, according to the species, from about 70 percent to 150 percent of the head and body length. The fur is short and may be sleek or rough. The upper parts range in color through a variety of brown, gray, red, and yellow shades. The hairs of the upper parts are basally gray in all species. The hairs of the underparts may be either gray or white at the base and either gray or white at the tip.

Aethomys has a ratlike shape, a pointed head, and limbs of equal size. Unlike in species of *Rattus* that may occur in the same region, in *Aethomys* the feet are usually white above. The tail may appear to be naked, but it has a light to moderate covering of hairs. Unlike those of *Pelomys,* the upper incisor teeth of *Aethomys* are not grooved. Females have four to six mammae.

The following ecological data were compiled from Ansell (1960), Ansell and Ansell (1973), Hubbard (1972), Kingdon (1974b), Okia (1976), and Smithers (1971). These rats inhabit grassland with open scrub, cultivated areas, savannahs, and forest edge. Several species, especially *A. namaquensis,* favor localities with rocky outcrops or kopjes. *A. nyikae* has been collected in montane country at an elevation of 2,120 meters. All species are mainly nocturnal and largely terrestrial, though *A. chrysophilus* and *A. namaquensis,* at least, are partly arboreal. Shelters include burrows excavated under bushes or rocks, crevices among rocks, holes in termite mounds, and hollow trees and logs. Nests may be large accumulations of grass, twigs, and debris. *A. namaquensis* sometimes constructs its nest in the fork or hole of a tree, though usually not more than two meters above the ground. The diet is almost entirely vegetarian and includes grains, grass seeds, roots, nuts, fallen fruits, and cultivated crops.

In Uganda, Okia (1976) found *A. hindei* to use individual home ranges of 30–1,600 sq meters. Kingdon (1974b) stated that *A. chrysophilus* lives in small social groups but that *A. kaiseri* appears to be less social and therefore may achieve a more scattered distribution. Hubbard (1972), however, located a large group of *A. kaiseri* concentrated in one area, apparently because of dry surrounding conditions. Ansell (1960) noted that *A. nyikae* is thought to be communal or semicommunal, and Smithers (1971) referred to *A. namaquensis* as colonial. In studies of *A. namaquensis* in Zimbabwe, Choate (1972) found each family group to have a minimum exclusive space of 100 sq meters centered on the nest. Captives showed territorial behavior and marked with urine. When placed together in small cages, two males would fight to the death. Individuals of opposite sexes usually lived peacefully together and cooperated in building the nest. Choate reported wild populations of *A. chrysophilus* to consist of widely spaced, apparently territorial pairs. Captives, however, showed less aggression than *A. namaquensis,* and small groups of males could be kept together. Hubbard (1972) described the call of *A. kaiseri* as a medium-pitched, rhythmical "chit-chit-chit."

The following reproductive data were compiled from Ansell (1960), Brooks (1972), Choate (1972), Hubbard (1972), Kingdon (1974b), Okia (1976), and Smithers (1971). In captivity, and in the wild in East Africa, reproduction seems to continue throughout the year, sometimes with summer peaks. In Botswana, however, *A. namaquensis* evidently does not breed during the cool, dry period from June to September. There may be a postpartum estrus. The average time between successive litters of captive *A. chrysophilus* was about 29 days. Gestation periods of 21–25 days have been reported. Most litters of *A. hindei* consist of only one or two young; in other species the range is one to seven, and the average is about three or four. The young attach themselves tenaciously to the nipples of the mother and may be dragged about for 3 weeks. They open their eyes at around 14 days. Female *A. chrysophilus* produce their first litter at an average age of 138 days.

RODENTIA; MURIDAE; Genus **ARVICANTHIS**
Lesson, 1842

Unstriped Grass Mice, or Kusu Rats

The following five species now appear to be recognized (Allen 1939b; Harrison 1972; Kingdon 1974b; Misonne, *in* Meester and Setzer 1977; Rosevear 1969; Yalden, Largen, and Kock 1976):

A. niloticus, Nile Delta of Egypt, southwestern Arabian Peninsula;
A. abyssinicus, Senegal to Ethiopia and Zambia;
A. blicki, central Ethiopia;
A. dembeensis, southern Sudan and Ethiopia to Tanzania;
A. somalicus, east-central Ethiopia, Somalia, Kenya.

Misonne (*in* Meester and Setzer 1977) listed only one species, *A. niloticus*, but acknowledged that others might be distinct. Rosevear (1969) suggested agreement that *A. niloticus* is the

only species. Yalden, Largen, and Kock (1976), however, explained that studies in progress indicated that populations in Egypt, the type locality of *A. niloticus*, are distinct and isolated from those farther to the south in Africa. Therefore, the proper name for the most widespread African species would be *A. abyssinicus*. Yalden, Largen, and Kock also considered *A. blicki* and *A. somalicus* to be separate species. Various authorities, such as Kingdon (1974b), used the name *A. lacernatus* for the population here called *A. dembeensis*. The type of *A. lacernatus*, however, is now known to represent the genus *Meriones* (Yalden, Largen, and Kock 1976). *A. dembeensis* was listed as a species of *Pelomys* by Misonne (*in* Meester and Setzer 1977), but Dieterlen (1974) showed this species to be referable to *Arvicanthis*, and Yalden, Largen, and Kock used its name for the population formerly called *A. lacernatus*. Corbet and Hill (1986) questionably referred the Arabian population to *A. dembeensis*, not *A. niloticus*.

Head and body length is 106–204 mm, tail length is 90–160 mm, and weight is 50–183 grams (Delany 1975; Kingdon 1974b; Rosevear 1969). The tail is usually shorter than the head and body. The hairs are generally coarse and stiff, so the coat is slightly harsh and somewhat spiny. The upper parts vary from blackish chestnut to tawny olive, grayish olive, and light gray. A definite buffy tinge is present in some animals, but others are quite pale in color. Some forms have a slight dorsal stripe. The hairs usually have a dark or black tip, but a subterminal band of gray or buffy hair produces a lighter overall tone in some individuals. The underparts are slightly paler than the upper parts, buffy white or grayish in the pale-colored forms and lead color or brownish dark gray in the darker forms. There usually is no sharp line of demarcation between the upper parts and underparts. The ears are usually reddish (occasionally brick red); the hands and feet are gray, yellowish, or almost as dark as the body; and the fairly well haired tail is definitely bicolored.

The fifth finger is considerably reduced but is functional

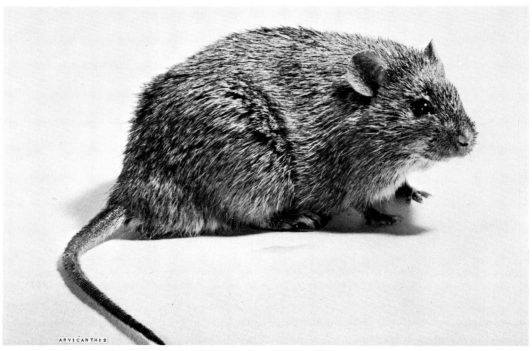

Kusu rat *(Arvicanthis abyssinicus)*, photo by Ernest P. Walker.

and bears a claw. The second, third, and fourth toes are rather long, though the first and fifth are short. The ears are fairly large and rounded. The incisor teeth are not grooved. Females have six mammae (Rosevear 1969).

These mice are most common in grasslands and savannahs but also have been reported from forests and scrubby thickets. In Ethiopia they are found at elevations of up to 3,700 meters (Müller 1977). In Uganda they tend to live under dense herb mats, where they construct long tunnels radiating out from their holes. The latter may be dug in soft earth but are usually in crevices, under rocks or fallen trees, or in termite mounds (Kingdon 1974b). Nests of fine grass are built in burrows or on the surface. Both diurnal and nocturnal activity have been recorded (Harrison 1972; Rosevear 1969). The diet is primarily vegetarian and includes seeds, leaves, grass, and cultivated crops.

Unstriped grass mice are gregarious and often live in colonies. Such groups contain several adult males and females, which are aggressive toward unrelated individuals of the same sex (Delany 1986). Population density varies and may increase greatly under proper conditions. Hubbard (1972:425) described an October population explosion in the Serengeti of Tanzania, during which the mice became "so numerous, one could hardly avoid stepping on them and many were killed by passing trucks. Survivors were feeding on the dead." Poulet and Poupon (1978) reported that density reached about 100/ha. in the Sahel of Senegal in February 1976 but collapsed in 1977. In a study in the Semien Mountains National Park in Ethiopia, Müller (1977) found densities of 65–250 per ha. Average individual home range during the rainy season in this area was 2,750 sq meters for males and 950 sq meters for females and juveniles. Respective figures for the dry season were 1,400 and 600 sq meters. Adult home ranges overlapped considerably.

In Müller's study area breeding occurred in the first half of the dry season. However, Kingdon (1974b) indicated that several litters might be born to female *Arvicanthis* during the wet season. Rosevear (1969) thought it likely that this genus breeds throughout the year. Delany (1975) stated that breeding does occur all year in Uganda; he cited a laboratory investigation that showed that the gestation period is approximately 18 days and litter size averages 5.3 young and ranges from 1 to 11. Ghobrial and Hodieb (1982) reported that females mated within 1–2 days of giving birth, that gestation actually lasts 21–23 days, that there may be up to 12 young, and that sexual maturity is reached at 3–4 months. According to Jones (1982), a captive individual lived for 6 years and 8 months.

Shaggy Swamp Rat

The single species, *D. incomtus*, occurs from Senegal to Ethiopia and south to South Africa (Hubert, Adam, and Poulet 1973; Misonne, *in* Meester and Setzer 1977; Yalden, Largen, and Kock 1976).

Head and body length is 113–90 mm, tail length is 97–185 mm, and weight is 48–150 grams (Kingdon 1974b; Sheppe 1973; Smithers 1971). The tail is usually somewhat shorter than the head and body. The texture of the fur varies considerably. In some animals it is long, soft, straight, and silky; in others it is loose and coarse; in still others it is shaggy. The upper parts are olive brown, yellowish brown, brownish mixed with black, or slaty black. The underparts are grayish white, whitish, pale buff, or olive buff. The hands and feet are usually light brownish and only scantily haired. The tail, which is almost naked, is dark brown above and below or slightly paler below.

Dasymys is a thickset, heavy rodent. Both the hands and the feet have five digits. The toes are fairly short, and the claws are not suited to climbing. The ears are small, rounded, and evenly fringed with hair. The incisor teeth are not grooved. Females have six mammae.

Although it is reported to occur in forests and savannahs, *Dasymys* is usually found in wet, grassy areas, such as marshes and reed beds. Kingdon (1974b) wrote that this rat is very successful in highland bogs and marshes, occurring at elevations of up to 4,000 meters in the Ruwenzori Mountains. It is primarily nocturnal but is sometimes active by day. It is basically terrestrial but swims and dives readily. Smithers (1971) stated that it builds domed nests of cut grass and other vegetation at ground level, with a nearby short refuge tunnel running into the ground. It also establishes distinct runways that radiate from the nest to feeding areas. Delany (1975) added that the burrow may be up to 2 meters long and 30 cm deep. *Dasymys* also has been reported to find shelter in holes along the banks of streams. Its diet apparently consists largely of plants growing in or near the water, but traces of insects have been found in some stomachs.

Reported population densities are 34/ha. in marshland and 2.6/ha. in savannah (Kingdon 1974b). Breeding records are scattered but extend over most of the year in eastern and southern Africa (Ansell 1960; Delany 1975; Sheppe 1973; Smithers 1971). Pregnant females were taken in Liberia in

Shaggy swamp rat *(Dasymys incomtus rufulus)*, photo by Jane Burton.

African groove-toothed rat *(Mylomys dybowskyi)*, photo by C. L. Cheeseman.

December and February (Misonne and Verschuren 1976). Reported litter size in *Dasymys* ranges from one to nine young but seems usually to be about two to five.

RODENTIA; MURIDAE; **Genus MYLOMYS**
Thomas, 1906

African Groove-toothed Rat

The single species, *M. dybowskyi*, is found from Ivory Coast to Kenya and the lower Congo River (Misonne, *in* Meester and Setzer 1977).

Head and body length is 120–94 mm, tail length is 104–80 mm, and weight is 46–190 grams (Delany 1975; Kingdon 1974b). The upper parts have a grizzled appearance much like that of *Arvicanthis*, but the hairs are longer and glossier. The coloration above is yellowish buff lined with black, brown or dull buff, or mixed black and buff. The type specimen of one subspecies (*M. d. richardi* from Zaire) has a pronounced bluish green iridescence on the upper parts. The rump may be ochraceous or rufescent. The underparts are usually white (buff in the subspecies *M. d. richardi*). The upper surfaces of the hands and feet are pale buff, tawny, or black, and the moderately haired tail is dark above and lighter below.

The body form is stocky. Only three digits are well developed on the hands and feet. Each of the upper incisors has a single groove, but the lower incisors are not grooved. *Pelomys* is similar to *Mylomys* in external appearance, but the grooving of its incisor teeth is much less pronounced, and the surface structure of its molars is different.

The following data were taken from Delany (1975) and Kingdon (1974b). *Mylomys* has been found in a variety of moist grassland habitats, often at high elevations of up to 2,400 meters. It is mainly diurnal but may also be active by night. It nests on the surface and does not burrow. The diet seems to be entirely herbivorous and consists of grass stems and leaves. In Uganda, pregnant females have been taken during all months of the year except January and February,

and they have been taken in higher proportions during the wet months.

RODENTIA; MURIDAE; **Genus MUS**
Linnaeus, 1758

Mice

There are 4 subgenera and 40 species (Ansell 1978; Ellerman and Morrison-Scott 1966; Grzimek 1975; Hubert, Adam, and Poulet 1973; Kingdon 1974b; J. T. Marshall 1977, 1979, 1981b, 1986, and *in* Lekagul and McNeely 1977; Marshall and Sage 1981; Musser and Newcomb 1983; F. Petter 1978a, 1981, and *in* Meester and Setzer 1977; Vermeiren and Verheyen 1980; Yalden, Largen, and Kock 1976):

subgenus *Pyromys* Thomas, 1911 (spiny mice)

M. shortridgei, central Burma to Viet Nam;
M. saxicola, Pakistan, India, Nepal;
M. platythrix, India;
M. phillipsi, India;
M. fernandoni, Sri Lanka;

subgenus *Coelomys* Thomas, 1915 (shrew mice)

M. mayori, Sri Lanka;
M. pahari, Sikkim to Indochina;
M. criciduroides, mountains of Sumatra;
M. vulcani, mountains of Java;
M. famulus, southern India;

subgenus *Mus* Linnaeus, 1758 (house and rice field mice)

M. caroli, found partly as a human commensal and perhaps partly through introduction in southern China, Thailand, Indochina, Malay Peninsula, Ryukyu Islands, Taiwan, Hainan, Sumatra, Java, Madura, and Flores;

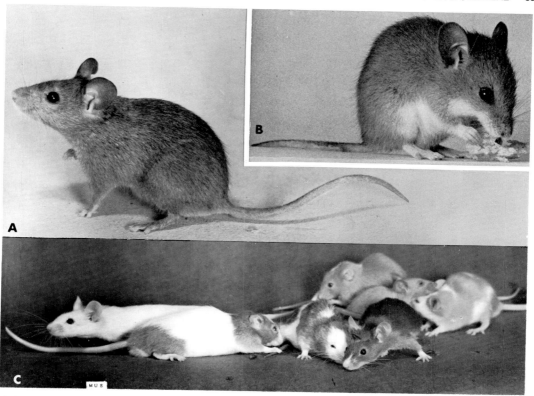

A. House mouse (*Mus musculus*), photo by Ernest P. Walker. B. African pygmy mouse (*M. minutoides*), photo by K. B. Newman. C. Laboratory mice (*M. musculus*), photo by Ernest P. Walker.

M. cervicolor, found partly as a human commensal and perhaps partly through introduction from Nepal to Indochina and on Sumatra and Java;

M. cookii, Nepal to Indochina;

M. booduga, India, Burma, Sri Lanka;

M. dunni, India and, probably through introduction, Sumatra;

M. terricolor, Pakistan, India, Nepal;

M. musculus (house mouse), perhaps naturally distributed from Sweden and the Mediterranean region to Japan and Nepal, now found partly as a human commensal throughout the world;

M. spicilegus, Mediterranean region, steppe country from Austria to western Iran;

subgenus *Nannomys* Peters, 1876 (African pygmy mice)

M. callewaerti, southern Zaire, Angola;

M. setulosus, Guinea to Congo;

M. baoulei, Ivory Coast;

M. pasha, northern Zaire;

M. triton, northwestern Zaire, Uganda, Tanzania, Zambia;

M. bufo, eastern Zaire, western Uganda;

M. tenellus, Sudan and Somalia to Tanzania;

M. haussa, the Sahel from Senegal to Nigeria;

M. mattheyi, savannah zone from Senegal to Ghana;

M. indutus, Botswana, South Africa;

M. setzeri, Zambia, Botswana;

M. gratus, western Uganda;

M. minutoides, throughout Africa south of the Sahara;

M. proconodon, Ethiopia;

M. mahomet, Ethiopia;

M. sorella, western Kenya;

M. neavei, Zambia, Zimbabwe;

M. wamae, Kenya;

M. acoholi, Kenya;

M. kasaicus, central Zaire;

M. oubanguii, Central African Republic;

M. goundae, northern Central African Republic.

The systematics of the subgenus *Mus,* particularly with respect to the species listed above as *M. musculus* and *M. spicilegus,* have fluctuated remarkably within the last two decades. Until recently most authorities considered *M. musculus* to include *M. spicilegus,* though it long was recognized that the resulting entity consisted of several fairly distinct kinds of mice, some of which might not regularly interbreed with one another. Certain of these kinds appear to be wild-living animals, native to the areas they now inhabit. Others are usually wild but apparently spread over much of their range through human agency. Still others are primarily commensals of people.

Subsequently, Marshall and Sage (1981) pointed out that biochemical and morphological analyses indicated that what are listed above as *M. musculus* and *M. spicilegus* actually together comprise eight separate species. That report, as augmented by earlier information from Ellerman and Morrison-Scott (1966) and Grzimek (1975), suggested that there were three primarily commensal species, with long tails, and five basically wild-living species, with short tails. Both wild and commensal species were thought to sometimes occupy the same area. The commensal species were: *M. domesticus,* found in Europe west of the Elbe River and from the Mediterranean region to Nepal; *M. poschiavinus,* restricted to

Switzerland and northern Italy; and *M. castaneus,* found in the cities of India, China, and Southeast Asia. The wild species were *M. musculus,* Sweden, Europe east of the Elbe River, Siberia, Mongolia, and China; *M. spretus,* western Mediterranean region; *M. abbotti,* eastern Mediterranean region; *M. hortulanus,* steppe country from Austria to western Iran; and *M. molossinus,* Manchuria, Korea, central China, and Japan. All eight of these species were recognized by Corbet and Hill (1986) and Honacki, Kinman, and Koeppl (1982).

Some recent biochemical and karyological work, however, now indicates that there are fewer full species. There again is general recognition that *M. domesticus* is a subspecies of *M. musculus,* though the two are still held to be usually distinguishable where their ranges meet in Europe (Bonhomme et al. 1984; Corbet 1988*b*; Orsini et al. 1983). Morphological studies by Marshall (1986) show that these two intergrade in parts of their Asian range and that the resulting species also includes *castaneus, molossinus,* and *poschiavinus.* In addition, Marshall concluded that *spretus, abbotti,* and *hortulanus* represent a single intergrading species, the proper name of which is *M. spicilegus.* This classification has been used in the above list, though it should be emphasized that *M. musculus* and *M. spicilegus* both comprise several forms with morphological, biochemical, and behavioral distinctions.

Misonne's (1969) classification of *Mus* is substantially different from that given in the above list. He did not recognize subgenera but divided the genus into three sections of related species: a *booduga* section, with the species *M. fernandoni, M. booduga, M. musculus, M. cervicolor, M. triton,* and *M. pasha;* a *minutoides* section, with the species *M. minutoides, M. musculoides* (found in West Africa and considered a subspecies of *M. minutoides* by F. Petter [*in* Meester and Setzer 1977] but also treated as a separate species by Rosevear [1969] and Hubert, Adam, and Poulet [1973]), *M. setulosus, M. gratus, M. birungensis* (found in eastern Zaire and not recognized as a species by F. Petter [*in* Meester and Setzer 1977]), *M. neavei, M. sorellus, M. tenellus, M. wamae* (found in Kenya and not recognized as a species by F. Petter [*in* Meester and Setzer 1977]), and *M. platythrix;* and a *pahari* section, with the species *M. pahari, M. shortridgei, M. crociduroides, M. mayori,* and probably *M. bufo* and *M. callewaerti.* Misonne also did not consider *Muriculus* to be a genus distinct from *Mus* and placed its one species, *imberbis,* in his *booduga* section.

At one time or another certain of the species in the above list have been put into various additional genera or subgenera. Perhaps most important is *Leggada* Gray, 1837, the type species of which is *M. booduga* of India but which eventually came to be used as the genus for all native African species of *Mus* (Allen 1939*b*). *Leggada* was considered a synonym of *Mus* by Ellerman and Morrison-Scott (1966) and Misonne (1969) but was used as a subgenus for African species by Ansell (1978), Rosevear (1969), and Yalden, Largen, and Kock (1976). J. T. Marshall (1977, 1979) placed *M. booduga* in the subgenus *Mus,* while for the native African species he used the subgenus *Nannomys* Peters, 1876, the type species of which is *Mus setulosus.* Marshall's arrangement has been followed in the above list.

The generic name *Gatamiya* Deraniyagala, 1965, was established for a single specimen that had been run over by a car in Sri Lanka. The animal appeared to closely resemble *Mus* but reportedly differed in not possessing a bristly or spiny coat, in having a completely hairy tail, and in being smaller. Eisenberg and McKay (1970:91) did not recognize *Gatamiya* as a genus and noted: "An examination of the single type of *Gatamiya* indicates no departure in the hairiness of the tail than is found in many immature specimens of the Muridae."

In *Mus musculus* head and body length is generally 65–95 mm, tail length is about 60–105 mm, and weight is usually 12–30 grams. In *M. minutoides,* one of the smallest of living mammals, head and body length is 45–82 mm, tail length is 28–63 mm, and weight is 2.5–12.0 grams (Delany 1975). For the genus as a whole, head and body length is up to 125 mm. The tail ranges from substantially shorter to slightly longer than the head and body. Lekagul and McNeely (1977) listed the following ranges of average measurements for six species in Thailand: head and body length, 74–110 mm; tail length, 58–86 mm; and weight, 12–34 grams.

The fur of *Mus* may be soft, harsh, or spiny. The tail appears to be naked but has a covering of fine hairs. The coloration above ranges from pale buff or pale gray through dull grayish browns and grays to dark lead color or dull brownish gray. The sides may be slightly lighter, and the underparts are usually lighter than the upper parts. *Mus musculus* is commonly light brown to black above, and white below, often with a buffy wash, and the tail is lighter below than above. Commensal forms of mice tend to have a longer tail and a darker coat than the wild forms. Domesticated strains of *M. musculus* have been developed, the most common being the albinos; other such strains have a black and white piebald pattern, and some carry various shades of black or gray.

The "singing," "waltzing," and "shaker" mice are common house mice. Faint but audible twittering sounds, emitted by house mice when in their shelters, have been reported from various parts of the world. The other types mentioned are defective in their balancing apparatus; they "waltz" or "shake" instead of moving about as normal mice do.

Some of the larger species of *Mus* are larger than some of the smaller species of *Rattus,* and the two genera are often confused, though they are not closely related. With respect to those kinds found in the Americas, the two genera can be distinguished as follows. In *Mus* total length is less than 250 mm, tail length is less than 110 mm, occipitonasal length of the skull is less than 35 mm, and the length of the first upper molar is greater than the combined length of the second and third upper molars. In *Rattus* total length is more than 250 mm, tail length is more than 110 mm, occipitonasal length is more than 35 mm, and the length of the first upper molar is less than the combined length of the second and third upper molars (Hall 1981). The upper incisors of *Mus* are notched, but those of *Rattus* are not. The skull of *Mus* is light and usually rather flat. The hind foot is usually narrow, and the outer digits tend to be shortened. Most female *Mus* have 10 or 12 mammae.

The natural habitat of *Mus* includes forests, savannahs, grasslands, and rocky areas. Some species are dependent on human habitations or cultivated fields. As noted in the above systematic discussion, a number of wild and commensal kinds of mice are grouped under the names *Mus musculus* and *M. spicilegus.* Wild forms have been found from the tropics to the Faeroes (62° N) and Macquarie (54° S) islands, from the swamps of Georgia to the deserts of Peru and central Australia, from sea level to high mountains, and even in coal mines at depths of 550 meters (Bronson 1979). Some wild populations exist for only part of the year, disappearing during the winter and being replenished in the spring by stock that had sheltered in buildings. In Great Britain, three categories of populations exist: those occupying buildings, those inhabiting stacks of grain, and those living free in fields (Berry 1970). There is a net movement into the grain stacks after they are erected in the autumn, and a net movement out into the fields in the spring. Colonies in British buildings, however, tend to be isolated, and there is little interchange with field populations. Some other species of *Mus,* including most of those in the subgenus *Mus,* as well as *M. minutoides,* of Africa, also sometimes act as human commensals.

African pygmy mouse *(Mus minutoides)*, photo by John Visser.

Wild-living populations of *M. musculus* dwell in cracks in rocks or walls or make underground burrows, which usually consist of a complex network of tunnels, several chambers for nesting and storage, and three or four exits (Berry 1970). When occupying human habitations, these species nest behind rafters, in woodpiles, in storage areas, or in other hidden spots near a source of food. The nest is a loose structure of rags, paper, or other soft substances lined with finer, shredded material (Jackson 1961). *M. minutoides* may shelter in any sort of cover during the day and also digs shallow burrows in soft soil. Its natal nest consists of a ball of soft grass or other fibers (Smithers 1971). *M. caroli* and *M. cervicolor* are burrowing species, but *M. shortridgei* and *M. pahari* build grass nests and do not burrow (Lekagul and McNeely 1977).

The commensal forms of *Mus* are active at any hour, but the wild forms seem to be mainly nocturnal. Most species are basically terrestrial but are good climbers, and *M. musculus* also swims well. The daily movements of a commensal mouse usually cover only a few square or cubic meters, but dispersing feral individuals have been known to wander up to 2 km.

Wild mice eat many kinds of vegetable matter, such as seeds, fleshy roots, leaves, and stems. Insects and some meat may be taken when available. Delany (1975) wrote that *M. triton* is mainly insectivorous. Commensal mice feed on any human food that is accessible, as well as on paste, glue, soap, and other household materials. Some mice store food or live within a human storage facility.

Bronson (1979) explained that commensal populations of *M. musculus* are characterized by relative stability, high densities of up to 10 per sq meter, individual home ranges of under 10 sq meters, and little potential for sharp increase; and that field populations are characterized by instability, densities of up to 1 per 100 sq meters, individual home ranges of a few hundred to a few thousand sq meters, and potential for irruption. Such irruptions have sometimes attained plague proportions. Lowery (1974) reported that densities have reached nearly 1,250/ha. in Louisiana. During 1926 and 1927 in the dry bed of Buena Vista Lake in Kern County, California, the number of mice in places was estimated at over 205,000/ha.

There now seems to be a general consensus that *M. musculus* is both territorial and colonial when living under commensal or laboratory conditions (Berry 1970; Bronson 1979; Lidicker 1976; J. A. Lloyd 1975; Mackintosh 1973; Poole and Morgan 1973, 1976). Territoriality, however, evidently is not pronounced among wild populations and may break down in overcrowded laboratory colonies. Studies show that when a number of unrelated mice are placed together in an enclosure, considerable fighting will result. A male and a female, or several nonpregnant females, will soon establish an amicable relationship, but two or more males may fight savagely until one establishes dominance. A dominant male sets up a territory with definite boundaries. This territory eventually will include a family group of several females and their young. There also may be one or more subordinate males, though there is some evidence that several males occasionally share a territory on an equal basis. The female members of a family may establish a loose hierarchy among themselves, but they are far less aggressive than the males. Normally, aggression within a family is rare, but the members join in strict defense of the territory against outsiders. Territories are cohesive and long-lasting; some have been observed to remain stable for 11 months. As young mice mature they are generally made to disperse through adult aggression, though some, especially females, may remain in the vicinity of the parents. The forced dispersal of young is known to take place in nonterritorial, wild-living populations, as well as in territorial populations. This kind of social arrangement tends to resist genetic modification of established colonies but to encourage strongly the formation of new groups. A system of pheromonal cues seems to promote successful colonization by allowing females to avoid pregnancy before dispersal but then to ovulate rapidly once a new home is established. Such a situation, along with the remarkable adaptiveness and reproductive potential of the species, is considered to be responsible for the success of house mice in spreading over much of the world.

The following reproductive information on *M. musculus* was taken largely from Berry (1970), Bronson (1979), and Lowery (1974). Breeding continues throughout the year in laboratory, most commensal, and some wild populations. In

Great Britain, however, free-living mice have a reproductive season extending from April to September. The estrous cycle is 4–6 days, with estrus lasting less than a day. Females may experience a postpartum estrus 12–18 hours after giving birth. There are usually 5–10 litters per year if conditions are suitable, but there may be as many as 14. The gestation period is 19–21 days but may be extended by several days if the female is lactating. Litters consist of 3–12, usually 5–6, young. They weigh about 1 gram at birth and are naked and blind. They are fully furred after 10 days, open their eyes at 14 days, are weaned at about 3 weeks, and reach sexual maturity at 5–7 weeks. There is usually 60–70 percent mortality before independence. Average life span is 2 years in the laboratory, but some captive individuals have lived for 6 years.

Considerably less is known about the reproduction of other species. In the Rajasthan Desert of India, *M. booduga* breeds over much of the year and produces litters of 1–13 young, *M. cervicolor* has been known to give birth to 2–6 young in July and December, and *M. platythrix* has a litter size of 3–10 (Prakash 1975). *M. minutoides* is capable of reproduction throughout the year in the laboratory but breeds mainly during the wet seasons, April–June and September–December, in the wild in East Africa; its recorded litter size is 2–8 young (Delany 1975; Kingdon 1974b). A captive specimen of *M. minutoides* lived for 3 years and 1 month (Jones 1982).

Mus musculus is known from Pleistocene fossils in Europe (Kurten 1968) and is evidently a natural inhabitant of parts of that continent, as well as much of Asia. The earliest known association of the house mouse with an urban community was at a neolithic site in Turkey about 8,000 years ago (Brothwell 1981). The species was familiar to the ancient Egyptians and Greeks and may have reached Britain during Roman times. Its subsequent spread over most of the world was facilitated by human construction of houses and barns, which provided shelter to the mouse, and the development of agriculture, which provided food. It was also inadvertently carried about in ships and caravans. Several other species of the subgenus *Mus* expanded their range in Asia and now live partly as human commensals. *M. musculus* and its relatives do not cause such serious health and economic problems as *Rattus norvegicus* and *R. rattus*. Mice are, however, agricultural pests in some areas, and they do consume some stored human food and contaminate far more. They also destroy woodwork, furniture, upholstery, and clothing. They contribute to the spread of such diseases as murine typhus, rickettsial pox, tularemia, food poisoning *(Salmonella)*, and bubonic plague. On the other hand, the albino strains of *M. musculus* are used extensively in laboratory research and have added immeasurably to our knowledge of medicine and genetics. In 1965 alone, some 800,000 mice were used at the National Institutes of Health in Bethesda, Maryland (Berry 1970; Grzimek 1975; Lowery 1974).

RODENTIA; MURIDAE; **Genus MURICULUS**
Thomas, 1902

Stripe-backed Mouse

The single species, *M. imberbis*, is known only from the highlands of Ethiopia. Misonne (1969) considered *Muriculus* to be a synonym of *Mus*, F. Petter (*in* Meester and Setzer 1977) wrote that it probably should be included in *Mus*, and Yalden, Largen, and Kock (1976) stated that it is at least debatable whether *Muriculus* is distinct from *Mus*. Nonetheless, *Muriculus* was listed as a full genus by Corbet and Hill (1986).

Head and body length is 70–95 mm and tail length is about 45–60 mm. The thick pelage is rather crisp, or it may be soft with no suggestion of spines. The hairs of the upper parts, at least in some individuals, have a dark blue gray base and a dark yellowish brown tip. A middorsal stripe is usually present from about the middle of the back to the base of the tail, though it may extend forward as a faint line almost to the shoulders. The underparts are tawny ochraceous, creamy, or almost white. Sometimes there is a tawny ochraceous tuft of hairs at the front base of the ears. The closely and finely haired tail is usually dark brown above and yellowish gray below.

Muriculus resembles *Mus* in external appearance but is easily distinguished by the dark dorsal stripe; this stripe, however, may be poorly defined in some individuals. The head is relatively short, and the ears are of medium size and rounded. The claws are short. The incisor teeth are narrow and have a smooth front surface; they are inclined so far forward that the tips of the upper incisors do not curve backward toward the body.

The stripe-backed mouse inhabits open slopes in grassy or rocky areas extending to the upper limit of forests. It shelters in holes in the ground and is associated with *Arvicanthis abyssinicus*, at least in some areas. Of *Muriculus*, Yalden, Largen, and Kock (1976:32) wrote: "This little-known endemic mouse has been recorded from altitudes between 1900–3400 m on both sides of the Rift Valley. It has not been found by the most recent collectors and it seems possible that the species has become rarer since 1940, perhaps as a result of the conversion of its habitat to agricultural land." Schlitter (1989) reported the collection of a specimen near Addis Ababa in 1976 and designated *M. imberbis* as vulnerable.

Stripe-backed mouse *(Muriculus imberbis)*, photo from Museum Senckenbergianum.

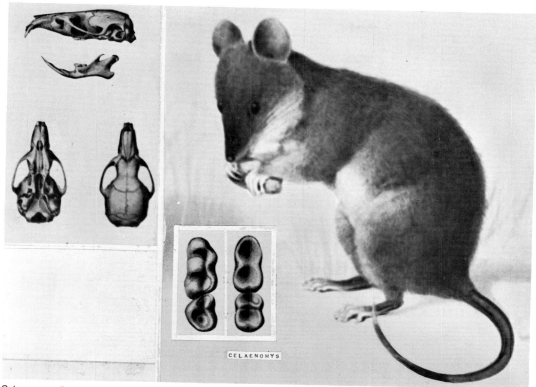

Celaenomys silaceus, photos from *Trans. Zool. Soc. London*, "Mammals from the Philippines," Oldfield Thomas.

RODENTIA; MURIDAE; Genus CELAENOMYS
Thomas, 1898

The single species, *C. silaceus*, is known by six specimens from northern Luzon (Sanborn 1952; Taylor 1934).

Head and body length of the type specimen is 195 mm, and tale length is 110 mm. The pelage is soft, close, and velvety. The upper parts are uniformly gray, and the underparts are paler, but there is no sharply defined line of demarcation. Most individual hairs are slaty gray on the basal part and washed with buffy white on the tip. The sides of the muzzle are almost black, the ears are grayish, and the feet are dark gray as far as the digits, which are whitish or flesh-colored. The thinly haired tail is white except for the upper basal part, which is brownish.

Celaenomys has a shrewlike body form and at a glance could be mistaken for *Rhynchomys*, but it is distinguished from that genus by its larger teeth and shorter muzzle. The incisors, like those of *Rhynchomys*, project forward, but they are larger and more powerful. The eyes are small, and the ears are short. The sharply pointed muzzle is produced by a wedge-shaped skull.

Specimens have been obtained in mountainous areas at elevations of 2,100–2,460 meters. Some have been collected in densely vegetated gullies and mossy forests.

The elevational range is sea level to about 3,000 meters.

Head and body length is 144–62 mm and tail length is 150–60 mm. The fur is close, soft, and velvety. The general coloration of the upper parts is rufous fawn to brownish. The shoulders, flanks, and hips are bright rufous. The black on the upper side of the muzzle extends backward to form an indistinct ring around the eyes. The cheeks, the inner side of the arms, and the underparts from the chin to the arms are creamy white, the hairs evenly colored to the base; one form, however, has a grayish brown abdomen owing to the gray base of the individual hairs. The upper surfaces of the thinly covered hands and feet are white. The finely scaled tail is brown on the basal dorsal surface and has a white tip and white undersurface.

This genus does not show any particular modification for aquatic life. The third molar teeth are retained, though they are much reduced in size (Menzies and Dennis 1979). The incisor teeth are broad, flattened in front, and pale yellow with a white tip. The eyes are small, and the ears are rather small and naked. The forefeet are normal for a murid, but the hind feet are elongated, suggesting the possibility of leaping habits. The three center toes of the hind feet are considerably larger than the others. There are five sole pads on the front feet and six on the hind feet. Females have four mammae.

RODENTIA; MURIDAE; Genus LEPTOMYS
Thomas, 1897

The single species, *L. elegans*, occurs over much of New Guinea (Laurie and Hill 1954; Menzies and Dennis 1979).

RODENTIA; MURIDAE; Genus PARALEPTOMYS
Tate and Archbold, 1941

There are two species (Laurie and Hill 1954; Menzies and Dennis 1979):

Leptomys elegans, photo by P. A. Woolley and D. Walsh.

P. wilhelmina, mountains of western and central New
    Guinea;
P. rufilatus, Cyclops Mountains of north-central New
    Guinea.

Head and body length is 120–40 mm and tail length is 130–
40 mm. *P. wilhelmina* is grayish brown above and dusky
white below. Its tail is gray above and white below and has a
white tip. *P. rufilatus* is distinguished from *P. wilhelmina* by
its white throat, broad reddish brown lateral line, and more
reddish head and hind legs. *Paraleptomys* is similar to *Leptomys* but differs from that genus in having neither an elongated foot nor the third upper and lower molars.

Although *P. wilhelmina* is common at elevations of 1,800–
2,700 meters, little is known of its biology. *P. rufilatus* apparently is known only by two specimens taken at 1,450 meters
on Mount Dafonsero.

RODENTIA; MURIDAE; Genus XEROMYS
Thomas, 1889

False Water Rat

The single species, *X. myoides*, is known by six specimens
from the Mackay area of southeastern Queensland, five from
the coast of the Northern Territory, and three from nearby
Melville Island (Watts 1979).

Head and body length is 110–14 mm and tail length is
about 88 mm. The general color of the upper parts is dark
slaty gray, which gradually blends into the lighter underparts, with no distinctive line of demarcation. The limbs are
like the back in color, and the hands, feet, and tail are scantily
covered with fine white hairs. Fine scales are conspicuous on
the tail.

The body is streamlined, the head is long and like that of
Hydromys, the ears are short and round, the eyes are small,
the tail is much shorter than the head and body, and the coat
is water-resistant. Nonetheless, the feet are not webbed, and
there is some question whether *Xeromys* is aquatic. There are
only two molar teeth on each side of the skull. The ungrooved
incisors project slightly outward; the upper incisors are
yellow or orange, and the lower ones are white. Females have
four mammae.

According to the Australian National Parks and Wildlife
Service (1978), the false water rat has been recorded in coastal
swamps with a variable cover of mangrove forest, in swamps
lined with mangroves and paperbark trees, and in reed
swamps with tall grass, shrubs, and pandanus palms. It is an
agile climber and is well adapted to tidal areas. It builds its
nest among the roots of mangrove trees. The nest is a mound
of leaves and mud about 60 cm high with a single opening
near the apex. A narrow tunnel from the opening leads to a
nest chamber, from which radiate numerous interconnecting
tunnels, some going well below ground level.

Redhead and McKean (1975) found a captive animal to be
nocturnal, spending the day in a burrow dug into the bottom
of its cage. Although those authors suggested that *Xeromys* is
a capable swimmer and adapted for aquatic life, Magnusson,
Webb, and Taylor (1976) spent thousands of hours observing
the rivers of Arnhem Land and never saw this genus in the
water, day or night. It was seen only in small trees lining the
rivers, climbing agilely among the branches. A captive took
animal food in preference to vegetable matter; it ate insects,
fish, and lizards and was able to attack and kill crabs larger
than itself.

The false water rat long was known only from southeastern Queensland and by a single specimen taken in 1903 in
the Northern Territory. Since 1970 additional animals have
been found on the coast of the Northern Territory and on
Melville Island. Nonetheless, *X. myoides* is considered rare,
and its habitat is thought to be jeopardized by agriculture,
livestock grazing, swamp drainage, and urban and recreational development (Australian National Parks and Wildlife Service 1978; Thornback and Jenkins 1982; Watts 1979). The
species is listed as endangered by the USDI and is on appendix
1 of the CITES.

RODENTIA; MURIDAE; Genus HYDROMYS
E. Geoffroy St.-Hilaire, 1805

Water Rats, or Beaver Rats

There are four species (Laurie and Hill 1954; Menzies and
Dennis 1979; Musser and Piik 1982; Ride 1970; Ziegler
1984):

H. chrysogaster, New Guinea, Aru and Kei islands,
    Australia, Tasmania;

H. hussoni, western and central New Guinea;
H. neobritannicus, New Britain Island east of New Guinea;
H. habbema, mountains of New Guinea.

The generic name *Baiyankamys* Hinton, 1943, was proposed for the species *B. shawmayeri.* Mahoney (1968), however, showed *B. shawmayeri* to be a junior subjective synonym of *Hydromys habbema.* It was based on a composite specimen, the skin and skull of which represented *Hydromys habbema,* while the mandible belonged to a *Rattus niobe.*

Head and body length is 136–350 mm, tail length is 130–350 mm, and weight is 400 to approximately 1,300 grams. The pelage is composed of shiny guard hairs and dense, soft underfur. Coloration of the upper parts ranges from dark brown, almost black, to golden brown or dark gray. The underparts range from brownish to yellowish white or bright orange. The well-haired tail is dark except for a white tip.

These large water rats have a sleek, streamlined appearance. Aquatic adaptations include the long and flattened head, forward-thrust nostrils, high-set eyes, small ears, seal-like fur, and broad, partially webbed feet. Females have four mammae.

Beaver rats are aquatic and occur wherever suitable habitat is provided by rivers, swamps, marshes, backwaters, or estuaries. *H. habbema,* of New Guinea, is found along mountain streams at elevations up to and above 3,000 meters (Menzies and Dennis 1979). Beaver rats live in a nest of shredded weeds in a hollow log or a burrow in a riverbank. They hunt along the bottom of streams but generally carry their prey to a favorite log, stone, or island before eating.

In a study in New South Wales, Woolard, Vestjens, and MacLean (1978) observed *H. chrysogaster* to swim at any time but especially in the evening. Its eyesight was good, and it kept its eyes open underwater. All prey was caught and carried by mouth. Fish (some up to 30 cm long) and large aquatic insects were the most important foods. Other prey included spiders, crustaceans, mussels, frogs, turtles, birds, and bats.

The main reproductive season in Australia is spring and summer (September–March), when each female usually produces 2–3 litters at 2-month intervals. Estrus lasts about 10 days, and gestation is about 34 days. Litter size is 1–7, usually 3–4, young. The young are born naked and blind, open their eyes after 14 days, are weaned after 29 days, and reach sexual maturity by about 4 months. Captive females have bred until 3.5 years old (Olsen 1982; Ride 1970; Watts 1974).

Beginning about 1937 in Australia, *H. chrysogaster* was trapped extensively for its beautiful pelt. At that time there was a shortage of muskrat fur, and a single skin of *Hydromys* would bring up to 65 cents. Trapping regulations have since been established, at least in some areas. This species may have increased in response to human irrigation projects and is sometimes considered a pest in the vicinity of fish farms (Smales 1984).

RODENTIA; MURIDAE; Genus PSEUDOHYDROMYS
Rümmler, 1934

New Guinea False Water Rats

There are two species (Laurie and Hill 1954; Lidicker and Ziegler 1968; Menzies and Dennis 1979):

P. murinus, known by 10 specimens from the mountains of Papua New Guinea;
P. occidentalis, known by 5 specimens from the mountains of western and central New Guinea.

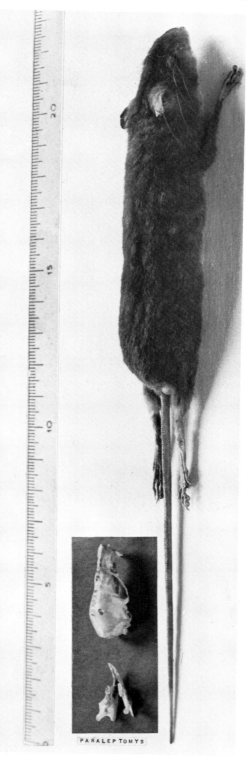

Paraleptomys wilhelmina, photos from Museum Zoologicum Bogoriense.

False water rat *(Xeromys myoides)*, photo by T. Redhead.

Head and body length is 85–115 mm and tail length is 90–115 mm. A female taken by Lidicker and Ziegler (1968) weighed 19.9 grams. The pelage is dense, short, and soft. The upper parts are dark gray or gray brown, the underparts are paler, and the tail is brown. The skull is elongate and rather flat, contributing to an overall shrewlike appearance. Females have four mammae (Menzies and Dennis 1979).

These animals appear to be nonaquatic; their feet have pads and soles of a type associated with terrestrial life. They evidently live on the ground in montane forests and feed on insects (Menzies and Dennis 1979). Specimens have been taken at elevations of 2,100–3,600 meters. A female *P. murinus* trapped on 23 October by Lidicker and Ziegler (1968) was pregnant with a single embryo.

Water rat *(Hydromys chrysogaster)*, photo by Stanley Breeden.

New Guinea false water rat (*Pseudohydromys* sp.), photo from American Museum of Natural History, Archbold Expeditions Collection.

RODENTIA; MURIDAE; Genus MICROHYDROMYS
Tate and Archbold, 1941

The single species, *M. richardsoni,* is known by a very few records from west-central to southeastern New Guinea (Menzies and Dennis 1979).

The type specimen has a head and body length of 80 mm and a tail length of 92 mm. The upper parts are grayish black, the underparts are slightly paler, and the terminal 10 mm of the tail is white. The genus is distinguished from other members of the subfamily Hydromyinae by its small size, short muzzle, and grooved upper incisors.

Microhydromys evidently lives in hill forests at somewhat lower elevations than related genera. The type specimen was taken in 1939 at 850 meters, and Tate (1951*b*) considered the likely elevational range to be 600–900 meters. *Microhydromys* apparently is not aquatic and probably lives on the ground and feeds on insects (Menzies and Dennis 1979).

RODENTIA; MURIDAE; Genus NEOHYDROMYS
Laurie, 1952

The single species, *N. fuscus,* is known from scattered localities in the montane rainforests of central and eastern New Guinea at elevations of 2,500–3,000 meters (Laurie and Hill 1954; Menzies and Dennis 1979).

Head and body length of the type specimen, an adult female, is 92 mm and tail length is 78 mm. An immature female taken by Dwyer (1983) on 26 July had a head and body length of 90 mm, a tail length of 82 mm, and a weight of 18 grams. The upper parts are smoky gray, and the underparts are slightly paler. The tail is brownish above and below, except for a whitish terminal area of some 15 mm. This small, mouselike rodent is not modified for aquatic life. It is distinguished from other members of the subfamily Hydromyinae by its small molar teeth and fairly long muzzle, as well as other dental and skeletal features. Females have four mammae.

RODENTIA; MURIDAE; Genus PARAHYDROMYS
Poche, 1906

Mountain Water Rat

The single species, *P. asper,* is widely distributed in the mountains of New Guinea (Laurie and Hill 1954; Menzies and Dennis 1979).

Head and body length is about 240 mm and tail length is about 260 mm. The pelage is firm to bristly, short, and not as

New Guinea false water rat (*Pseudohydromys* sp.), photo by J. I. Menzies.

Microhydromys richardsoni, photo from American Museum of Natural History.

dense or sleek as the pelage of *Hydromys*. The general coloration of the upper parts is brownish gray; the ends of the
longest hairs are black, and those of the shorter ones are dull
creamy whitish. The underparts are dull white with a buffy
wash, and a sharp demarcation from the color of the sides is
lacking. The numerous whiskers, the upper ones black and
the lower ones white, are stiff. The finely haired ears are
grayish brown; the upper surfaces of the hands and feet are
pale brownish; and the well-haired tail is basally brownish
black, with a distinctive white, brushed tip.

The muzzle in this genus is exceptionally wide and bears
highly developed vibrissae, which may be associated with the
detection of food. The hind feet are partially webbed; the
soles are smooth and slightly granulated, and the pads are
distinct. The incisor teeth grow with their roots wide apart
and their tips converging. Females have four mammae.

Parahydromys occurs both in forest and in vegetation
along mountain streams, even when these run through cultivated areas. The elevational range is about 600–2,700 meters.
This rodent does not have marked aquatic adaptations but
may be associated with waterways. It is said to feed on insects

and other invertebrates, which it digs up (Menzies and Dennis 1979).

RODENTIA; MURIDAE; **Genus CROSSOMYS**
Thomas, 1907

Earless Water Rat

The single species, *C. moncktoni*, is known from the highlands of eastern New Guinea and probably occurs throughout
the island (Laurie and Hill 1954; Menzies and Dennis 1979).

Head and body length is about 205 mm and tail length is
about 220 mm. The dorsal pelage is long and soft, with dense,
woolly, glossy underfur. The general coloration of the upper
parts is mottled brownish gray with a pale yellowish olivaceous wash. The few scattered guard hairs have a black tip and
a subterminal brownish ring. The soft, cottony hairs of the
underparts are pure white to the base. The well-defined line
where the two colors meet occurs quite high on the sides of

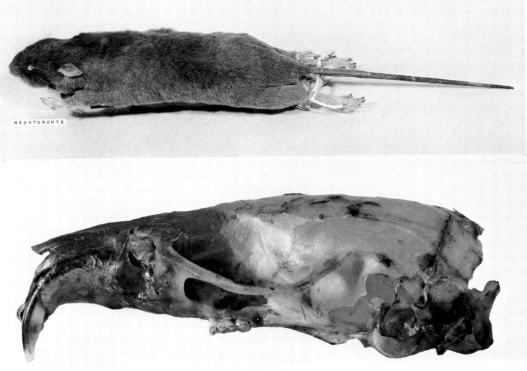

Neohydromys fuscus, photos from British Museum (Natural History).

Mountain water rat *(Parahydromys asper)*, photo of head of dead specimen by J. I. Menzies.

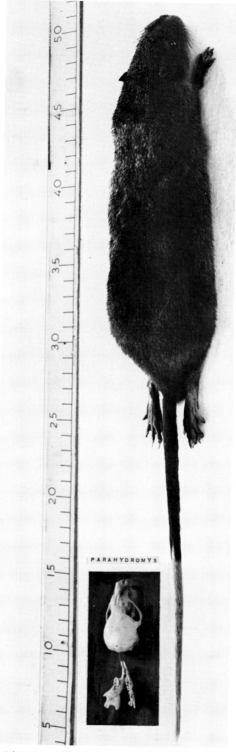

the animal. The thick tail is light gray above and white below. Two rows of long white hairs begin on the sides of the body and gradually converge into a single raised crest along the underside of the tail. This development is similar to the swimming fringe in the European water shrew *(Neomys)* and may assist the tail in serving as a rudder.

In addition to this tail arrangement, *Crossomys* has waterproof fur, greatly reduced ears, small eyes, a smoothly rounded head, and very large, webbed hind feet. These adaptations make *Crossomys* even more specialized for aquatic life than *Hydromys*. The hands of *Crossomys* are relatively small, the wrists are slender, and the claws are small, delicate, and strongly curved. The incisor teeth are narrow and beveled on the sides. Females have four mammae.

The elevational range of *Crossomys* is about 600–3,000 meters, and this animal apparently occurs only along waterways. The typical habitat is a small, swift mountain stream, but when frogs are spawning, *Crossomys* may move away from the rivers to hunt for tadpoles. It lives in holes in stream banks and, in addition to tadpoles, feeds on insects, mollusks, and small aquatic vertebrates (Menzies and Dennis 1979).

RODENTIA; MURIDAE; **Genus MAYERMYS**
Laurie and Hill, 1954

Shaw-Mayer's Mouse

The single species, *M. ellermani*, is known by eight specimens from the mountains of northeastern New Guinea (Laurie and Hill 1954; Lidicker and Ziegler 1968).

Head and body length is 92–103 mm, tail length is 97–107 mm, and weight is about 17–21 grams. The upper parts are predominantly smoky gray; the underparts are pale gray, and some specimens have a small white spot on the middle of the chest. The tail is brownish, both above and below, and may have patches of white or pale gray brown.

Mayermys is distinguished from all other members of the family Muridae, and indeed from all other rodents, in having only one molar tooth on each side of the upper and lower jaws. The incisor teeth are well developed, the uppers slightly curved forward and not grooved. The feet seem to be adapted for terrestrial life, but as in *Pseudohydromys* and *Neohydromys*, there is a slight membrane between the fingers and toes.

Specimens have been collected at elevations of 2,100–2,700 meters in montane forests. Four were trapped under logs. All reported specimens are males.

Mountain water rat *(Parahydromys asper)*, photo from Museum Zoologicum Bogoriense.

Earless water rat (Crossomys moncktoni), photo from Basil Marlow, Australian Museum, Sydney.

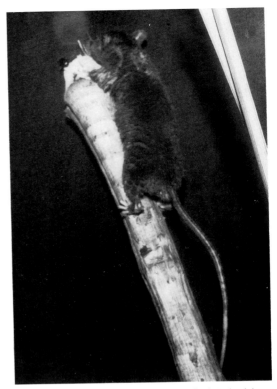

Shaw-Mayer's mouse (Mayermys ellermani), photo by J. I. Menzies.

RODENTIA; **Family GLIRIDAE**

Dormice

This family consists of 9 Recent genera and 20 species. One genus, *Graphiurus*, the sole representative of the subfamily Graphiurinae, inhabits Africa south of the Sahara. The other genera, which compose the subfamily Glirinae, occur in the Palearctic region—Europe, northern Africa, central and southwestern Asia, and Japan. Ellerman and Morrison-Scott (1966), among others, recognized two additional glirid subfamilies: the Platacanthomyinae, considered here to be a subfamily of the Muridae; and the Seleviniinae, treated here as a full family. Corbet's (1978) use of the name Gliridae is followed here, but the name of this family is sometimes given as Muscardinidae or Myoxidae. The sequence of genera presented here follows that of Simpson (1945).

Dormice look like squirrels but are smaller, and one genus (*Glirulus*) resembles chipmunks. Head and body length is 60–190 mm and tail length is 40–165 mm. The pelage is soft, and the tail (except in *Myomimus*) is bushy. The eyes are prominent and the ears are rounded. The legs and toes are short, and the short, curved claws are adapted to climbing. The forefeet have four digits and the hind feet have five. The underside of the feet and digits is naked. Female dormice have 6–12 mammae.

The skull has well-developed zygomatic bones, no postorbital processes, and relatively large bullae (Arata 1967; Ognev 1963). The dental formula is: (i 1/1, c 0/0, pm 1/1, m 3/3) × 2 = 20. The cheek teeth are low-crowned and have a series of parallel ridges of enamel across the crown.

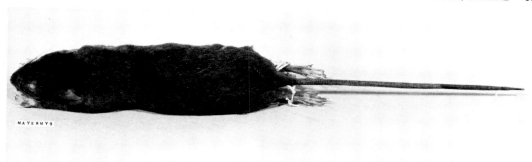

Shaw-Mayer's mouse *(Mayermys ellermani)*, photos from British Museum (Natural History).

African dormouse *(Graphiurus murinus)*, photo by John Visser.

Fat dormouse *(Glis glis)*, photo by R. Pucholt.

Dormice live in wooded areas, hedgerows, gardens, and rocky places. They are generally nocturnal and scansorial and are squirrel-like in some habits. They shelter in hollow trees, on the branches of trees or shrubs, in rocky crevices, in the deserted burrows of other animals, and in the attics of buildings, often in a nest of plant material. During the late summer and early autumn they generally become quite fat, and in the Palearctic from October to April they hibernate in a curled-up, circular position; they may awake from time to time to eat food that they have stored. The diet includes fruits, nuts, insects, eggs, and small vertebrates.

There are usually one or two litters per year, each consisting of 2–10 young. Gestation periods of 21–30 days have been reported. The young are born in a nest, often lined with moss, in a tree hollow, on a branch, or sometimes in a ground shelter. Longevity is up to about 5.5 years in the wild.

The known geological range of the Gliridae is middle Eocene to Recent in Europe, upper Miocene to Recent in Africa, and Recent in Asia (Klingener 1984). One of the Recent genera is extinct and most others have become rare because of human habitat disruption.

RODENTIA; GLIRIDAE; Genus GLIS
Brisson, 1762

Fat Dormouse, or Edible Dormouse

The single species, *Glis glis*, occurs from France and northern Spain to the Volga River and northern Iran and on the islands of Sardinia, Corsica, Sicily, Crete, and Corfu (Corbet 1978); it was introduced to England in 1902 and has since become common in parts of that country. There is some support for use of the name *Myoxus* Zimmermann, 1780, in place of *Glis*, but Corbet (1978) considered such a change to be neither necessary nor desirable.

Head and body length is 130–90 mm, tail length is 110–50 mm, and weight is 70–180 grams (Van Den Brink 1968). The short, soft, thick pelage is silvery gray to brownish gray on the upper parts, lighter on the flanks, and white or yellowish on the underparts. This squirrel-like animal has large and rounded ears, small eyes, and a long, densely bushy tail. The hands and feet, with their rough pads, are adapted to climbing. Females have 10 or 12 mammae (Ognev 1963).

Except as noted, the information for the remainder of this account was taken from Grzimek (1975) and Ognev (1963). The edible dormouse inhabits deciduous or mixed forests and fruit orchards in both lowlands and mountains. Its populations in any given area are partly dependent on the existence of a suitable number of hollow trees, as these are the most common sites for daily shelter, hibernation, and natal nests. The hollows may be heavily lined with grass or other vegetation, especially if being used for hibernation or rearing of young. *Glis* also shelters in crevices between rocks, burrows among tree roots, woodpecker holes, piles of mulch, attics, barns, and artificial nest boxes. Hibernation burrows sometimes are 50–100 cm below the surface of the ground.

The edible dormouse is primarily nocturnal and crepuscular, though occasionally it is active by day. It is highly arboreal, and its agility in the trees may exceed that of squirrels. Some of its leaps have been reported to cover 7–10 meters. It has exceptionally good senses of vision, hearing, smell, and touch (through its vibrissae). Individuals visit many trees each night in the search for food. If a food shortage occurs, there may be movement to a different area. During winter hibernation *Glis* sleeps on its side or back and curls its body so that the feet touch the muzzle. The hibernation period is very long, with the animals entering torpor from September to November and emerging from early May to early June. The diet consists mainly of seeds, nuts, acorns, berries, and soft fruits. Insects may be important at certain times, and small birds are taken on rare occasions.

The population density in the northern Caucasus was calculated at 30/ha. In Moravia, however, Gaisler, Holas, and Homolka (1976) found a minimum density of only 1/ha. and an individual home range diameter of about 200 meters. The edible dormouse is apparently territorial and marks its space with glandular secretions. It is highly vocal; a variety of chirps, whistles, squeaks, and squeals have been noted. Individuals are quarrelsome, and males are reported to fight savagely during the breeding season. Nonetheless, small groups may hibernate together, and as many as eight individuals have been found in a single tree hollow. In studies of captives, males have been observed to remain in the vicinity of females after mating and to be allowed back into the nest about 16 days following birth, where they help to clean and protect the young. Families may stay together through winter hibernation; a wild male, however, probably leaves a female after mating, in order to pursue other estrous females.

The mating season extends from June to early August. In Moravia the young are born in August and early September (Gaisler, Holas, and Homolka 1976). Although two litters per year have been reported in some areas, the very short active season of *Glis* suggests that a single annual litter per female is the usual case. Gestation periods of 20–30 days have been reported. Litter size is 2–10 young, and the average in Gaisler, Holas, and Homolka's (1976) study was 4.5. The young are born naked and blind, open their eyes and are weaned after about 4 weeks, and do not begin to mate until after their first hibernation. Wild individuals have been known to live for more than 4 years, and a captive survived for 8 years and 8 months (Jones 1982).

In some areas *Glis* is considered extremely harmful to the production of fruit and wine. It consumes large amounts of apples, pears, plums, and grapes and has been reported to destroy one-third of the grape crop in the northern Caucasus. However, it is easily trapped, there is some demand for its luxuriant fur, and it is hunted for use as food and a source of fat. In ancient Rome, *Glis* was considered a delicacy, and colonies were kept in large enclosures planted with nut-bearing bushes and provided with nesting sites. Prior to a feast, individual animals would be confined to earthen urns and fattened on acorns and chestnuts. The meat of *Glis* is still a gourmet dish in some parts of Europe. Pucek (1989) indicated that *Glis* has become rare in much of Europe through destruction of forest habitat.

RODENTIA; GLIRIDAE; **Genus MUSCARDINUS**
Kaup, 1829

Common Dormouse, or Hazel Mouse

The single species, *M. avellanarius*, occurs from France and southern Sweden to the European part of the Soviet Union and northern Asia Minor, as well as in southern Great Britain, Sicily, and Corfu (Corbet 1978).

Head and body length is 60–90 mm, tail length is 55–75 mm, and weight is 15–40 grams (Van Den Brink 1968). The general coloration of the upper parts is rich yellowish brown or yellowish red. The throat and chest are creamy white, and the belly is pinkish buff. The well-haired tail is brownish above and paler below. The snout is blunt, the eyes are large, and the ears are relatively small. Females have eight mammae (Ognev 1963).

The common dormouse dwells in thickets and in forests with an abundance of secondary growth. It is arboreal and scrambles about with great agility but does not usually ascend as high into the trees as does *Glis*. The common dormouse is

Hazel mouse *(Muscardinus avellanarius)*, photo from Zoological Society of London.

nocturnal and crepuscular, sleeping by day in a globular nest constructed of shredded bark, leaves, grass, or moss and located in a bush or the lower branches of a tree. The nest is usually about 1–2 meters above the ground (Grzimek 1975). The nest of an individual animal is 60–80 mm in diameter; the nest of a female with young is more extensively lined and measures about 120 mm in diameter.

Winter hibernation is spent in another nest, located in a burrow, in a stump, or on the ground beneath moss, leaves, and other debris. This nest is constructed of vegetation and is bound together with a sticky secretion from the salivary glands. The period of dormancy may begin as early as August and extend to May in the Soviet Union (Ognev 1963). In England this period lasts from late October to early April. Apparently, hibernation is triggered when the external temperature falls below 16° C. During deepest dormancy the blood temperature of *Muscardinus* falls to 0.25°–0.50° C, whereas its normal temperature is 34°–36° C.

The common dormouse appears to be associated with nut-bearing trees, and hazelnuts are known to be an important part of its diet (Grzimek 1975; Ognev 1963). Nonetheless, one observer considered *Muscardinus* to be just as incapable as other small rodents of opening hard hazelnuts. *Muscardinus* is able to gnaw a hole in fresh nuts, thereby getting the seed out in little pieces. Its diet also includes other seeds, buds, berries, insects, and possibly birds' eggs and nestlings. Some food may be stored during the summer and utilized during brief periods of winter arousal.

Van Den Brink (1968) wrote that *Muscardinus* sometimes lives in small colonies. It is not as vocal as *Glis* but does produce soft chirping and whistling sounds. Ognev (1963) stated that young had been found from June to mid-October, that there are probably at least two litters per year, and that litter size is 3–5 young. In a study in Moravia, Gaisler, Holas, and Homolka (1976) learned that there was usually only one annual litter, occasionally two; that average litter size was 4.7 young; and that the young normally bred after their first hibernation but rarely bred in their first summer of life at an age of only 2.5 months. According to Corbet and Southern

Garden dormouse *(Eliomys quercinus):* A. Photo by Eviatar Nevo; B. Individual with extra fatty tissue in preparation for hibernation, photo from Archives of Zoological Garden Berlin-West.

(1977), birth peaks occur in Europe from June to early July and from late July to early August, the gestation period is 22–24 days, litters contain 2–7 young, the eyes open after 18 days, independence is attained at 40 days, and maximum known longevity is 4 years in the wild and 6 years in captivity.

Pucek (1989) indicated that *Muscardinus* has become rare in much of its range because of destruction of forest habitat. It now receives full legal protection in most countries where it occurs.

RODENTIA; GLIRIDAE; **Genus ELIOMYS**
Wagner, 1843

Garden Dormouse

The single species, *E. quercinus,* occurs from France and Spain to the Ural Mountains, in northern Africa and the lands at the eastern end of the Mediterranean Sea, on Sicily and most other islands of the western Mediterranean, and on the Arabian Peninsula (Corbet 1978; Nader, Kock, and Al-Khalili 1983). Harrison (1972) considered the populations of *Eliomys* in Asia and Africa to represent a distinct species with the name *E. melanurus*. Skeletal remains from archeological sites dating to Roman times in Britain probably represent individuals imported for culinary purposes (O'Connor 1986).

Head and body length is 100–175 mm, tail length is 90–135 mm, and weight is 45–120 grams (Grzimek 1975; Van Den Brink 1968). The pelage is short except at the tip of the tail, where it is long and forms a tuft. The general coloration of the upper parts ranges through several gray and brown shades. The underparts are creamy or white. There are usually some black markings on the face. In European populations of *Eliomys* the tail is distinctly tricolored dorsally, having a cinnamon brown proximal half, a broad black preterminal band, and a white tip (Ognev 1963). In Asian and African populations the proximal third of the tail is brownish white, and the remainder of the tail is usually black (Harrison 1972). Females have eight mammae.

The common name garden dormouse is not fully appropriate, as *Eliomys* is generally found in extensive forests in Europe (Grzimek 1975; Ognev 1963). It also occurs in a variety of other habitats, including swamps, rocky areas, and cultivated fields. In the Middle East it has been found from the steppe desert of Hejaz to the high mountains of Lebanon (Harrison 1972). It shelters in such diverse places as hollow trees, branches of shrubs, crevices among rocks, and foresters' huts. A bird or squirrel nest may be used as a foundation for a shelter, but when *Eliomys* builds a complete nest, it is made of leaves and grass, is compact and globular in shape, and usually is located 0.8–3.0 meters above the ground. In some areas *Eliomys* is highly arboreal and is reported to be extremely graceful and agile (Ognev 1963), but in other places it occurs where there are no trees at all (Harrison 1972). It is apparently most active by night. In some areas it is known to gain weight in the fall and to become dormant during the coldest part of winter. The diet consists of acorns, nuts, fruits, insects, small rodents, and young birds. Ognev (1963) suggested that *Eliomys* is primarily a predator.

According to Grzimek (1975), large numbers of individuals may live in close proximity and share sleeping and feeding sites. Except during the mating season there is no fighting, not even when two unrelated groups come together. *Eliomys* is very noisy and has a variety of calls (Ognev 1963). Females are thought to be polyestrous and to breed from May to October, at least in Europe. Births evidently occur during the same period in Morocco (Moreno and Delibes 1982). In southwestern Spain there are birth peaks in March–May and August–October (Moreno 1988). The gestation period is about 22–28 days, and litter size is 2–8 young. The eyes of the newborn open after about 21 days. A captive specimen lived for 5 years and 6 months (Jones 1982).

Pucek (1989) indicated that *Eliomys* has become rare in much of its range because of destruction of forest habitat. It now is regarded as endangered in Czechoslovakia, Poland, and Finland.

RODENTIA; GLIRIDAE; **Genus DRYOMYS**
Thomas, 1906

Forest Dormice

There are two species (Corbet 1978; Mallon 1985; Nader 1989):

D. nitedula, central Europe to Mongolia and Iran;
D. laniger, known only from a few localities in
    southwestern Turkey.

Forest dormouse *(Dryomys nitedula),* photo by R. Pucholt.

Head and body length is 80–130 mm, tail length is 60–113 mm, and weight is 18–34 grams (Gaisler, Holas, and Homolka 1976; Ognev 1963; Van Den Brink 1968). The general coloration is grayish brown to yellowish brown on the upper parts and buffy white on the underparts. *Dryomys* is similar to *Eliomys* but is smaller, has a skull with much smaller tympanic bullae and a more rounded braincase, and has a more uniform tail, which is flattened and moderately bushy. Females have eight mammae.

Forest dormice inhabit dense forests and thickets at elevations of up to 3,500 meters; they sometimes utilize cultivated areas, gardens, and rocky meadows (Grzimek 1975). Their nests are usually located in dense shrubbery or the lower branches of trees. Hibernation sites may be in hollow trees, among tree roots, or in underground burrows. The temporary nest of an individual animal is rather flimsy, but natal nests are solidly constructed (Ognev 1963). Studies in Israel showed that nests are located 1–7, usually about 3, meters above the ground in the branches of trees. They are globular, measure 150–250 mm in diameter, have an outer layer of twigs and leaves and an inner lining of bark or moss fragments, and have an entrance hole on the side or top (Harrison 1972).

Forest dormice are nocturnal and arboreal. They climb with great agility, and their leaps from branch to branch cover up to 2 meters. In Israel they are active all year, even in the high mountains, though they may undergo torpor for some hours during winter days (Harrison 1972). Farther north the data are conflicting. Evidently, hibernation does sometimes occur from October to April in Europe. During this period the animals curl up like a ball while sitting on the hind legs; the tail is wrapped around the body, and the hands are pressed onto the cheeks. There may be occasional emergence to eat from stores of food. Observations in the Soviet Union, however, suggest that extensive, deep hibernation does not necessarily occur and that forest dormice may be active throughout most or all of the winter (Ognev 1963). The diet of *Dryomys* consists of seeds, acorns, buds, fruits, arthropods, eggs, and young birds; animal matter seems to be preferred during the summer (Grzimek 1975).

Nests tend to be clustered in small groups in the same tree or adjacent trees (Harrison 1972). In one area of about 8,000 sq meters, 11 inhabited nests were found (Ognev 1963). *Dryomys* has a variety of vocalizations, most notably a delicate, melodious squeak that serves as an alarm call. The breeding season extends from March to December in Israel, where each female gives birth two or three times annually, and from May to August in Europe, where there usually seems to be just a single litter per year (Gaisler, Holas, and Homolka 1976; Grzimek 1975; Harrison 1972; Ognev 1963). Gestation lasts at least 1 month. There are usually two to five, occasionally up to seven, young. The young weigh about 2 grams at birth, open their eyes at 16 days, and attain independence after 4–5 weeks. In Europe they do not mate until after their first winter.

Forest dormice are quite aggressive and never really become tame in captivity. They may be brought to a point at which they will allow themselves to be petted, but they usually bite with their sharp incisors when an attempt is made to hold them. If disturbed while resting, they often lie on their back or side and scratch with their hind legs. If disturbed further, they may suddenly leap high into the air and spit and hiss. Wild populations sometimes cause local damage by raiding fruit orchards and gnawing the bark of coniferous trees. Pucek (1989) reported that, because of the destruction of forest habitat, *D. nitedula* is regarded as endangered in Czechoslovakia and as rare in most other European countries. Nader (1989) designated *D. laniger* as rare.

RODENTIA; GLIRIDAE; Genus HYPNOMYS
Bate, 1919

Balearic Dormice

There are two species, both now extinct (E. Anderson 1984; Kurten 1968):

H. morphaeus, Mallorca in the Balearic Islands (western Mediterranean);
H. mahonensis, Menorca in the Balearic Islands.

Japanese dormouse *(Glirulus japonicus)*, photo by Michi Nomura.

According to Kurten (1968), *H. morphaeus* was about as large as *Glis*, while *H. mahonensis* was intermediate in size to *Glis* and *Leitha* Lydekker, 1896. *Leitha* was a giant dormouse about the size of a hamster, known from the middle and late Pleistocene of Malta and Sicily. Both species of *Hypnomys* survived the Pleistocene but subsequently disappeared. Extinction apparently came about 4,000–5,000 years ago and was associated with ecological disturbances caused by the spread of people and their domestic livestock across the Balearic Islands (P. S. Martin 1984; Vigne 1987).

RODENTIA; GLIRIDAE; Genus GLIRULUS
Thomas, 1906

Japanese Dormouse

The single species, *G. japonicus*, is found on the Japanese islands of Honshu, Shikoku, and Kyushu (Corbet 1978).

Head and body length is 65–80 mm and tail length is 40–55 mm. The coloration is pale olive brown with a dark brown to black dorsal stripe. This stripe varies in width and sometimes is very obscure. The fur is soft and thick. A tuft of long hairs lies in front of the ear, and the tail is flattened from top to bottom.

This dormouse inhabits mountain forests, usually from about 400 to 1,800 meters in elevation; one individual, however, was captured in a cottage at 2,900 meters. *Glirulus* is arboreal and nocturnal. It shelters in a tree hollow or in a nest among branches. The round nest is covered on the outside with lichens and lined on the inside with bark. Winter hibernation is spent in hollow trees, cottages, or birdhouses. A semidormant individual was once captured in July in a snow depression in a ravine of the Japanese Alps, but after several minutes it awakened and escaped from the container where it had been placed. The diet of *Glirulus* consists of seeds, fruits, insects, and birds' eggs. In captivity it does well on rice, peanuts, sweet potatoes, fruits, and insects.

The young are usually born in June or July. Occasional births in October probably represent the second litters of certain females. There may rarely be up to seven young, but observations of captives by Michi Nomura (Oiso-Machi, Kanagawa Pref., Japan, pers. comm.) indicate the usual range to be three to five and the average to be four; she also determined the gestation period to be about 1 month.

Wang, Zheng, and Kobayashi (1989) indicated that *Glirulus* is very low in numbers and is legally protected. It evidently has declined mainly through loss of forest habitat.

RODENTIA; GLIRIDAE; Genus CHAETOCAUDA
Wang, 1985

Chinese Dormouse

The single species, *C. sichuanensis*, is known by two female specimens taken in northern Sichuan Province of China (Wang Youzhi 1985).

Head and body lengths are 90 and 91 mm, tail lengths are 92 and 102 mm, and weights are 24.5 and 36 grams. *Chaetocauda* resembles *Myomimus* but is distinguished by a dark chestnut color around the eyes, larger ears and bullae, and a tail that is terminally club-shaped and covered with dense hairs, rather than pointed and thinly haired as in *Myomimus*. *Chaetocauda* also has a more squarish palate than does *Myomimus*, a thinner mandible, a relatively greater interorbit-

al width, a much less complex molar structure, and deeply grooved, rather than smooth, incisor teeth.

Chaetocauda has been taken at elevations of about 2,500 meters in subalpine, mixed forest. It is nocturnal and nests in small trees, 3.0–3.5 meters above the ground. The stomachs contained a mixture of green vegetation and starch.

RODENTIA; GLIRIDAE; Genus MYOMIMUS
Ognev, 1924

Mouselike Dormouse

Corbet (1978) recognized only a single species, *M. personatus*, which occurs in southeastern Bulgaria, Thrace, western Turkey, and extreme southwestern Turkmen S.S.R. near the shores of the Caspian Sea and also is known from subfossil material in southern Asia Minor and Palestine. Rossolimo (1976a, 1976b), however, considered the Bulgarian populations to represent a distinct species, *M. bulgaricus*, and described a new species, *M. setzeri*, from northwestern Iran. Honacki, Kinman, and Koeppl (1982) accepted three species, but indicated that the proper name for the one in Bulgaria, Turkey, and Palestine is *M. roachi*. Ognev (1963) suggested that the range of *M. personatus* may extend as far east as Afghanistan. Subfossil material of *Myomimus* now has been found in Afghanistan (Corbet 1984).

Head and body length is 61–112 mm and tail length is 53–94 mm (Rossolimo 1976a, 1976b; Van Den Brink 1968). The general coloration of the upper parts is a closely mixed combination of ochraceous and gray. The underparts, insides of the limbs, and the feet are white. There is a sharply defined line of demarcation between the upper and lower parts. Unlike other dormice, which have rather bushy tails, *Myomimus* has a thinly haired, mouselike tail covered with short, white hairs.

Ognev (1963) indicated that *Myomimus* is the only glirid that is not specialized for arboreal life and that one specimen was caught among stones scattered amidst small bushes. Van Den Brink (1968) stated that *Myomimus* seems to live on and under the ground. The Soviet Union classifies *M. personatus* as indeterminate. Pucek (1989) designated *Myomimus* as one of the "top ten endangered rodents in Europe."

RODENTIA; GLIRIDAE; Genus GRAPHIURUS
Smuts, 1832

African Dormice

There are ten species (Genest-Villard 1978b; Robbins and Schlitter 1981; Schlitter, Robbins, and Williams 1985; Smithers 1971):

G. parvus, Mali and Sierra Leone to Nigeria, southern Ethiopia and Somalia to Zimbabwe and probably Angola;

G. murinus, throughout Africa south of the Sahara;

G. lorraineus, Sierra Leone to Zaire;

G. olga, Niger, northern Cameroon;

G. christyi, southern Cameroon, northern Zaire;

G. surdus, southern Cameroon, Equatorial Guinea;

G. crassicaudatus, Liberia to Cameroon, island of Bioko (Fernando Poo);

G. platyops, southern Zaire, Zambia, Zimbabwe, Namibia, Botswana, probably eastern South Africa;

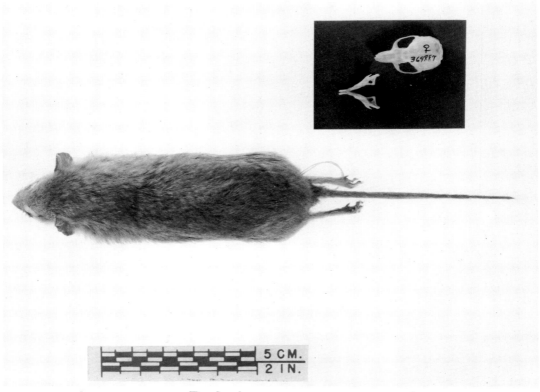

Mouselike dormouse *(Myomimus personatus)*, photo from U.S. National Museum of Natural History.

G. hueti, Senegal to Central African Republic and Angola; *G. ocularis,* western South Africa.

Head and body length is 70–165 mm and tail length is 50–135 mm. The weight of *G. murinus* is 18–30 grams (Kingdon 1974*b*). The fur varries in texture: it is rather soft and dense in most forms, quite soft in a few, and slightly coarse in others. The general coloration of the upper parts ranges from pale ashy gray to dark slaty gray, and from buffy to reddish brown, tinged with grayish. The underparts are white to grayish, often tinged with buff or reddish brown. Occasionally the fur on the throat and chest is stained by plant or fruit juices. There are variously arranged black and white markings on the face. The top of the tail is usually black or dark brown, and the bottom is whitish. In most forms the tail is well furred. As is the case with many crevice-dwelling mammals, the skull is flattened. Females have six or eight mammae.

African dormice inhabit forests and, in southern and eastern Africa, rocky areas in the dry tablelands, generally along waterways. They are arboreal but may frequently be found on the ground. They often shelter in trees and shrubs, making their globular nests in cavities and among the branches. Some animals use crevices in cliffs or stone fences, and others live in thatched roofs. They are sometimes found in boxes of rubble or in the upholstery of old furniture in houses and storerooms. Their occurrence in human habitations has declined through competition with the introduced *Rattus rattus*

African dormouse *(Graphiurus murinus)*, photo by Ernest P. Walker.

(Kingdon 1974*b*). They are principally nocturnal, but in dense, dark forests they are occasionally active by day. There is little precise information on hibernation, but apparently in certain areas *Graphiurus* becomes fat by the end of autumn and retires to a shelter, where it may remain dormant through the winter. The diet includes grains, seeds, nuts, fruits, insects, eggs, and small vertebrates. African dormice occasionally become a nuisance by raiding poultry yards.

Channing (1984) found a mean population size of 7 adult *G. ocularis* in a 7.75 ha. study area of South Africa. Breeding in that area occurred in spring and summer, with litter sizes of 4–6 young. According to Kingdon (1974*b*), *G. murinus* shows strong signs of territoriality in captivity, and males have been known to kill and eat one another, but in the wild as many as 11 adults of both sexes have been found in a single nest. Vocalizations include various twittering sounds and a surprisingly loud shriek. There is no conclusive evidence of reproductive seasonality, but there do appear to be breeding peaks. In Uganda many pregnant and lactating females were recorded in March, and 1 pregnant female was found in October. In Kenya 5 pregnant females were found in November, and in southwestern Tanzania a pregnant female was taken in January and a juvenile in March. Litters contain 1–5 young, each weighing about 3.5 grams at birth. A captive specimen of *G. murinus* lived for 5 years and 9 months (Jones 1982).

Smithers (1986) classified *G. ocularis* as rare in South Africa.

RODENTIA; **Family SELEVINIIDAE**; **Genus SELEVINIA**
Belosludov and Bashanov, 1939

Desert Dormouse

The single known genus and species, *Selevinia betpakdalaensis*, occurs in deserts to the west and north of Lake Balkhash in eastern Kazakh S.S.R. (Corbet 1978, 1984). The family Seleviniidae has a short but confusing taxonomic and nomenclatural history (see Arata 1967).

Head and body length is 75–95 mm and tail length is 58–77 mm (Ognev 1963). Two pregnant females weighed 21.4 and 24 grams, and a small male weighed 18 grams. The dense fur is grayish above and whitish below. The method of molt is peculiar: instead of a sloughing off of individual hairs, the epidermis becomes detached along with the hairs growing on it. A dense growth of new hair is found already in place in areas where patches of skin have fallen off. The molt begins at the back of the neck (between the ears) and then proceeds along the back and sides; the entire process takes about 1 month. The new hairs come in quickly, growing up to 1 mm in a 24-hour period. In winter the individual hairs attain a length of 10 mm. The tail is covered with short hairs, and the scales are not visible. The palms and soles are naked.

It has been suggested that *Selevinia* is a highly modified and aberrant dormouse. It has a round, stocky body and a long tail. The hand has four digits and the foot has five. The external ears extend beyond the fur. The skull has enormous tympanic bullae and relatively weak zygomatic arches. The dental formula is usually given as: (i 1/1, c 0/0, pm 0/0, m 3/3) × 2 = 16; however, there are also two upper premolars that are lost early in life (Klingener 1984). The upper incisor teeth are massive and have a deep groove on their front surface. The small, single-rooted cheek teeth have a relatively simple enamel pattern.

Selevinia has a sporadic distribution in clay and sandy deserts. It occurs among thickets of spirianthus, in growths of wormwood, and among boyalych *(Salsola laricifolia)*. It may live in burrows under the bushes, but a captive dug a burrow (28 cm long) only when the temperature was low; at other times this individual sheltered under a leaf or small stone. Some observers have reported *Selevinia* to be diurnal, but most individuals come out at twilight and remain active throughout the night, thus avoiding the heat. When the cap-

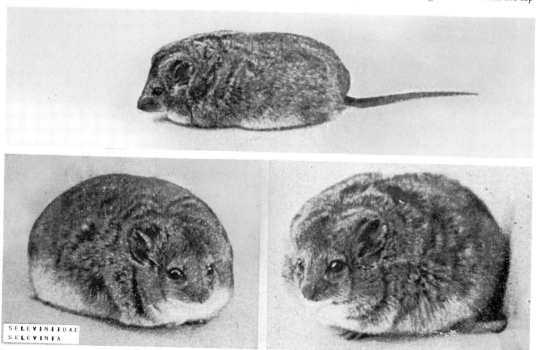

Desert dormouse *(Selevinia betpakdalaensis)*, photos from *Zeitschr. Saugetierk.*

Desert dormouse (Selevinia betpakdalaensis), photo by A.
Pechenyuk through D. Bibikov.

tive mentioned above was brought out into the sunlight for
five minutes at midday in March, its ear muscles were so
badly burned that it almost died.

When undisturbed, the desert dormouse travels at a lei-
surely, ambling gait, but when alarmed, it progresses by small
leaps. It does not jump higher than about 20 cm but is a good
climber. It is quite active at moderate temperatures, when it
may venture far from its shelter. Signs of its presence are not
extensive in a given area, and there have been suggestions
that it leads a nomadic life in the summer. When temperatures fall below about 5° C, however, Selevinia goes into a
state of dormancy, during which its breathing rate may decrease from 108 to 25 per minute. Although such behavior
has been observed only in captivity, it is reasonable to assume
that it takes place in nature. Such a period of lethargy might
explain the apparent scarcity of Selevinia during the colder
parts of the year.

The diet seems to consist solely of invertebrates such as
insects and spiders. The intestinal tracts of insects, at least of
mealworms, are not eaten. Selevinia eats up to three-fourths
of its weight in a 24-hour period. About one-fourth of the
insect food eaten by a captive was discharged as feces, mostly
in the form of indigestible chitinous material. Selevinia
readily drinks water in captivity.

The desert dormouse is quite gentle in captivity. It does not
attempt to bite when caught and held in the hand. Peculiar
chirping sounds are emitted when it is feeding or disturbed.

Mating takes place in May and possibly again in July. Pregnant females with six or eight embryos have been reported.

The Soviet Union designates S. betpakdalaensis as rare.
The only known fossils referable to this family are two mandibles from the Pliocene of Poland (Klingener 1984).

RODENTIA; Family ZAPODIDAE

Birch Mice and Jumping Mice

This family of 4 living genera and 17 species occurs from
central Europe to China and in much of Canada and the
United States. There are two subfamilies: Sicistinae, with the
single genus Sicista; and Zapodinae, with Zapus, Eozapus,
and Napaeozapus. Both were placed in the family Dipodidae
by Klingener (1984).

The Zapodidae are small, mouselike animals modified for
jumping. In the subfamily Zapodinae the hind limbs and feet
are especially long, while in the Sicistinae the hindquarters
are only slightly modified for jumping. Head and body length
is 50–110 mm, tail length is 65–165 mm, and weight is
normally 6–28 grams. Females have four pairs of mammae.

In the Zapodidae the neck vertebrae are not fused, the
auditory bullae are relatively small, and the three central
bones of the hind feet are not united as they are in most
members of the family Dipodidae. The dental formula is:
(i 1/1, c 0/0, pm 0–1/0, m 3/3) × 2 = 16 or 18. The upper
incisors are grooved in the Zapodinae but smooth in the Recent Sicistinae. The cheek teeth are low-crowned in Sicista
but tend to be high-crowned and cuspidate in the Zapodinae.

These scampering and jumping rodents inhabit thickets,
forests, meadows, swamps, bogs, and sagebrush flats. When
startled, members of the Zapodinae can jump up to 2 meters.
The long tail is used as a balancing organ during leaps. These
rodents shelter under logs or in underground burrows either
dug by themselves or abandoned by other animals. Their
rather inconspicuous burrows are not marked by a mound of
earth. Runways are not generally made. Activity is mainly
nocturnal. Zapodids gain weight in the fall and then become
dormant for six to eight months. Their diet consists of berries, seeds, fungi, and small invertebrates. Food is not stored.
They drink free water.

The geological range of this family is Oligocene to Recent
in Europe, lower Pliocene to Recent in Asia, and early Miocene to Recent in North America.

RODENTIA; ZAPODIDAE; Genus SICISTA
Gray, 1827

Birch Mice

There are 12 species (Corbet 1978; Sokolov and Baskevich
1988; Sokolov, Baskevich, and Kovalskaya 1981, 1986a,
1986b; Sokolov, Kovalskaya, and Baskevich 1982):

S. subtilis, steppes from eastern Austria to Lake Baikal in
south-central Siberia;
S. severtzovi, Voronezh region of southern Soviet Europe;
S. betulina, forests from Norway and Austria to the Ussuri
region of southeastern Siberia;
S. napaea, northwestern Altai Mountains in south-central
Siberia;
S. pseudonapaea, southern Altai Mountains;
S. concolor, Kashmir, mountains of Gansu and Sichuan
(central China);

S. caucasica, northern Caucasus;
S. kluchorica, northern Caucasus;
S. kazbegica, Georgia;
S. armenica, northwestern Armenia;
S. tianshanica, Tien Shan Mountains of Central Asia;
S. caudata, Sakhalin Island.

Head and body length is 50–90 mm, tail length is 65–110 mm, hind foot length is 14–18 mm, and weight is 6–14 grams. The upper parts are light or dark brown to brownish yellow, and the underparts are paler, usually a lighter brown. *S. betulina* and *S. subtilis* differ from the other species in having a sharply defined black stripe down the middle of the back. *Sicista* is a small, mouselike genus with a fairly long and semiprehensile tail but without special elongation of the legs or feet.

Birch mice inhabit forests, thickets, moors, subalpine meadows, and steppes. They excavate shallow burrows, in which they build oval-shaped nests of dry grass and cut plant stems. They travel on the ground by leaping and readily climb bushes and shrubs—the outer toes hold on to twigs, and the tail curls around other branches for additional support. They are generally active at night. It has been suggested that *S. betulina* spends the summer in wet meadows and migrates to forests in the winter. For at least six months of the year *Sicista* hibernates in an underground burrow. The species *S. napaea,* of the Altai Mountains, is active only from mid-May to early September (Shubin and Suchkova 1975). There is a considerable weight loss during hibernation; an *S. betulina* weighing about 12 grams in the fall may weigh only about 6 grams in the spring. The diet includes seeds, berries, and insects.

The reproductive season of *S. napaea* in the Altai Mountains is late June to mid-August, gestation lasts about 20 days, and most females produce litters of 1–6 young (Shubin and Suchkova 1975). In the Karelian A.S.S.R. of the northwestern Soviet Union, Ivanter (1972) learned that during the summer each adult female *S. betulina* raised a single litter containing 3–11 young. Sexual maturity was attained after almost 1 year of life, and a few individuals lived for 3–4 years. In another study of *S. betulina,* in the Bialowieza National Park of Poland, females were found to be in estrus from May to about the end of June. There was a single litter per year, and each female evidently gave birth only twice in her life-

Birch mice *(Sicista subtilis),* photos by K. A. Rogovin.

time. The gestation period was estimated to be 4–5 weeks, and the duration of parental care was 4 weeks. Sexual maturity was attained following the first winter hibernation, and maximum longevity was estimated at 40 months.

RODENTIA; ZAPODIDAE; **Genus ZAPUS**
Coues, 1876

Jumping Mice

There are three species (Hafner, Petersen, and Yates 1981; Hall 1981):

Birch mouse *(Sicista betulina)*, photo by Liselotte Dorfmüller.

Z. hudsonius, southern Alaska to Labrador and northern Georgia, isolated populations in mountains of Arizona and New Mexico;

Z. princeps, southern Yukon to northeastern South Dakota and the southwestern United States;

Z. trinotatus, southwestern British Columbia to northwest coast of California.

Head and body length is 75–110 mm, tail length is 108–65 mm, hind foot length is 28–35 mm, and summer weight is about 13–20 grams. The pelage is coarse. The back is grayish brown, the sides are yellowish brown, and the underparts are white. The posterior part of the body is much heavier than the forepart, and the hind limbs are much larger and more powerful than the forelimbs. Although cheek pouches have sometimes been reported in this genus, none are present. The upper incisor teeth are narrow and grooved in front.

Except as noted, information for the remainder of this account was taken from Banfield (1974) and Whitaker (1972). Jumping mice live in wooded areas, grassy fields, and alpine meadows. They are especially common in the thick vegetation bordering streams, ponds, and marshes. During the summer they build spherical nests of woven grass about 100 mm in diameter. These are usually placed under a log, board, root, or clump of vegetation. The winter hibernation nest is made of grass and leaves and is usually located in a burrow about 30–90 cm below the ground. These rodents do not normally progress by jumping but crawl slowly on all fours or make short hops. When startled, however, they often take several leaps of up to 1 meter in length, using their powerful hind legs for propulsion and their long tail for balance. They are capable of climbing in bushes and are excellent swimmers, sometimes diving to depths of over 1 meter.

Jumping mice are primarily nocturnal but occasionally are seen by day. They are evidently more nomadic than most small rodents, and during the summer they may wander over nearly 1 km in search of moist habitat. They are among the most profound of mammalian hibernators, generally entering dormancy by late September or October and emerging in late April or early May. When in torpor, they are rolled up in a ball, the head tucked between the hind feet and the tail curled around the body. The normal body temperature is about 37° C but drops to as low as 2° C during torpor. Over a period of about two weeks in the fall, prior to hibernation, *Zapus* accumulates fat, and its body weight may double, reaching as much as 37 grams. Seeds are generally the most significant component of the diet, but *Zapus* also eats fungus, nuts, berries, fruits, and insects, the last item being especially important in the spring. There is evidently no storage of food. Captives have been observed to drink water regularly.

Populations of *Z. hudsonius* fluctuate from year to year; there are usually about 5–20 per ha., but density has been known to reach 48 per ha. Average individual home range in this species varies from about 0.15 ha. to 1.10 ha. and is usually slightly larger in males. Mean home ranges of *Z. princeps* in the mountains of Colorado have been reported as follows: 0.31 ha. for males and 0.24 ha. for females (Myers 1969), and 0.17 ha. for males and 0.10 ha. for females (Stinson 1977). The latter authority found males to have overlapping home ranges and to be tolerant of one another but found females to have more exclusive territories. In general, jumping mice are considered to be solitary creatures but are not

Jumping mouse *(Zapus hudsonius)*, photo by Karl H. Maslowski.

antagonistic even when strangers are placed together. Captives have been reported to be docile, but Gannon (1988) noted that *Z. trinotatus* is high-strung and may attempt to bite the handler. It produces a squeaking sound when fighting

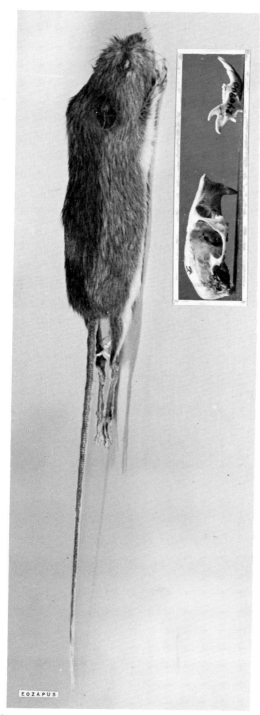

Chinese jumping mouse *(Eozapus setchuanus)*, photos by Howard E. Uible (skin) and P. F. Wright (skull) of specimens in U.S. National Museum of Natural History.

and vibrates its tail rapidly against the substrate to produce a drumming noise.

The reproductive season extends from about May to September, with most young being born from June to August. In some areas *Z. hudsonius* is known to produce at least two, and occasionally three, litters per season, but *Z. princeps* and *Z. trinotatus* evidently have only a single annual litter. The gestation period is 17–23 days, and litter size is 2–9, usually 4–6, young. The newborn weigh about 0.8 grams each and are naked and blind. During the fourth week of life their eyes open and they become fully furred, and by the end of that week they are weaned. Female *Z. hudsonius* born early in the breeding season may be able to produce their own litters in the same season, but most individuals do not mate until after their first hibernation. *Z. princeps* has a relatively long life span for a rodent, with some individuals surviving up to 4 years in the wild (Brown 1970). A captive *Z. hudsonius* lived for 5 years (Jones 1982).

The IUCN classifies *Z. trinotatus orarius* as indeterminate. According to D. F. Williams (1986), this subspecies is restricted to a small area on the Point Reyes Peninsula on the northern California coast. It appears to be scarce, and its limited habitat may be decreasing because of grazing by livestock and introduced Eurasian deer. The subspecies *Z. hudsonius preblei*, found in eastern Wyoming and Colorado, may have declined to the point of extinction because of agricultural modification of its grassland habitat.

RODENTIA; ZAPODIDAE; Genus EOZAPUS
Preble, 1899

Chinese Jumping Mouse

The single species, *E. setchuanus*, occurs in central China, in southeastern Qinghai, southern Gansu, and western Sichuan (Corbet 1978). *Eozapus* is sometimes considered to be only a subgenus of *Zapus*.

Head and body length is 80–100 mm, tail length is 100–150 mm, and hind foot length is 25–33 mm. The upper parts are tawny orangish, the underparts are white, with or without a median tawny or buffy stripe, and the tail is dark above and white below. As in *Zapus* and *Napaeozapus*, the hind legs, hind feet, and tail of *Eozapus* are very long. *Eozapus* is distinguished from the American jumping mice by the pattern of its molar teeth, the dark streak down the middle of its breast and belly, and the white tip on its tail.

Only about a dozen specimens are known to have been collected. These are in the Paris Museum of Natural History, the Museum of the Academy of Sciences at Leningrad, the British Museum of Natural History, the U.S. National Museum of Natural History, the American Museum of Natural History, and the Academy of Sciences in Philadelphia. The specimens have come from high mountain elevations, and there is some indication that the preferred habitat of *Eozapus* is beside streams in cool forests. Mainly because of the destruction of forest habitat, Wang, Zheng, and Kobayashi (1989) designated *Eozapus* as endangered.

RODENTIA; ZAPODIDAE; Genus NAPAEOZAPUS
Preble, 1899

Woodland Jumping Mouse

The single species, *N. insignis*, ranges from southeastern Manitoba to Labrador and Pennsylvania and south along the

Woodland jumping mouse *(Napaeozapus insignis)*, photo by Ernest P. Walker.

Appalachian Mountains to northern Georgia. The information in this account was taken entirely from Whitaker and Wrigley (1972) and Wrigley (1972).

Head and body length is 80–100 mm, tail length is 115–60 mm, hind foot length is 28–34 mm, and weight (without embryos or fat accumulated prior to hibernation) is 17–26 grams. The pelage is coarse because of the stiff guard hairs and has a tricolor pattern. The back is brown to black, the sides are orange with a yellow or red tint, and the underparts are white. This pattern probably affords camouflage protection by simulating dead vegetation. The tail is distinctly bicolored, grayish brown above and white below, and nearly always has a white tip. *Napaeozapus* differs from *Zapus* in having this white tail tip, as well as in having three, rather than four, molariform teeth. As in *Zapus*, the upper incisors are grooved and females have four pairs of mammae.

Napaeozapus inhabits spruce-fir and hemlock-hardwood forests, within which it selects cool, moist places with dense vegetation. It is often found in bogs and swamps or along streams but also may occur far from free surface water. It shelters in piles of brush or in underground burrows that it either digs itself or takes over from other small animals. The entrance is kept concealed during the day. A globular nest is constructed from dry leaves and grass. *Napaeozapus* uses a quadrupedal walk when moving slowly and a quadrupedal hop for greater speed. Such a hop can cover up to 1.8 meters but normally is about 0.6–0.9 meters long and 0.3–0.6 meters high. *Napaeozapus* climbs well in bushes but does not ascend trees. It can swim rapidly for short distances. It is mainly nocturnal but may be active in the late morning or early evening. Most mature individuals enter a state of hibernation by late September, but most young begin in October. Emergence is usually in May, with males appearing about two weeks before females. There is a rapid accumulation of body fat in the several weeks before hibernation, and an adult that weighed 20 grams in the spring may exceed 30 grams by autumn. There is a weight loss of 30–35 percent during hibernation. Seeds, fungi, and insects are important components of the diet, and fruits, nuts, and other kinds of vegetation are also eaten. There is evidently little or no food storage in the wild.

Population densities of 0.64–59.0 per ha. have been reported, but the average in good habitat is about 7.5 per ha. Home ranges vary from 0.4 to 3.6 ha., and those of both sexes overlap. Individuals are highly tolerant of one another, and groups can be kept together in captivity without displays of aggression. *Napaeozapus* is usually silent but may utter a soft clucking sound, or squeal if disturbed.

The reproductive season extends from early May to early September, with pronounced peaks in June and August.

Many females give birth twice during this period, especially in more southerly populations, but many others have only a single annual litter. The gestation period is about 23 days, and litter size averages approximately 4.5 young, ranging from 2 to 7. The newborn are naked and blind and weigh about 0.9 grams each. They are fully furred after 24 days of life, their eyes open at 26 days, and they are weaned at 34 days. Neither sex is able to breed until after the first hibernation. By autumn about 70 percent of the animals in a given population of *Napaeozapus* are young of the year; however, many wild individuals live for 2, 3, or even 4 years.

RODENTIA; Family DIPODIDAE

Jerboas

This family of 11 living genera and 32 species occurs throughout the arid part of the southern Palearctic region, from the Sahara, across southwestern and Central Asia, to the Gobi. There are four subfamilies: Dipodinae, with the genera *Dipus*, *Paradipus*, *Jaculus*, and *Stylodipus*; Allactaginae, with the genera *Allactaga*, *Alactagulus*, and *Pygeretmus*; Cardiocraniinae, with the genera *Cardiocranius*, *Salpingotus*, and *Salpingotulus*; and Euchoreutinae, with the single genus *Euchoreutes* (Corbet 1978; Klingener 1984; Pavlinov 1980*a*).

Head and body length is 36–263 mm, tail length is 70–308 mm, and the extraordinarily great hind foot length is 18–98 mm. The pelage is generally satiny or velvety. The coloration is usually sandy or buffy, resembling the color of the ground in which these animals burrow. The tail is longer than the head and body and usually has a terminal tuft. The eyes, like those of most nocturnal animals, are relatively large. Either the external ears or the tympanic bullae, or both, are enlarged. The dental formula is: (i 1/1, c 0/0, pm 0–1/0, m 3/3) × 2 = 16 or 18. The cheek teeth are rooted, high-crowned, and cuspidate.

The Dipodidae are characterized by their remarkable adaptations for jumping. The hind legs are at least four times longer than the front legs. Except in the subfamily Cardiocraniinae, the three central foot bones are fused to form a single cannon bone that gives great strength and support. The number of hind toes is sometimes reduced to three. If five hind digits are present, the first and fifth (the lateral digits) are so small that only the three central digits actually support the foot. In the species *Allactaga tetradactyla* there is only one lateral digit.

Jumping is probably an adaptation for escape from predators in open country. When moving rapidly, jerboas leap and

spring with their hind legs, covering up to three meters in a single bound. The long tail serves as a balancing organ. Even when moving slowly, jerboas generally use only their hind legs, bringing them forward alternately in a bipedal walk. Jerboas sometimes, however, hop slowly, rabbitlike, on all four legs.

The Dipodidae are also characterized by their adaptations for burrowing in the arid desert, semidesert, and steppe regions where they live. Jerboas living in sandy soils have tufts of bristly hairs under the digits and soles of the hind feet. These tufts act as friction pads and tend to support the animal on loose sand; they also aid in kicking the soil backward after it has been accumulated by the forelimbs in burrowing. Another adaptation of jerboas that live in sandy areas is the tuft of bristly hairs around the opening of the external ear, which guards against the entrance of wind-blown sand. These ear and foot tufts are lacking, or are not so well developed, in jerboas living in loamy soils. At least in the genus *Jaculus*, a thickened fold of skin over the nose can be drawn forward to protect the nostrils when the animal is pushing earth with its snout. A bony mechanism in front of the orbital cavity on the skull of some jerboas protects the eye when the head is used in digging.

In the late spring and summer, jerboas often plug their burrows during the day and sometimes after leaving at night. Thus they keep the heat out and the moisture in, maintaining a suitable microclimate within the burrow during the hottest parts of the year. This is characteristic of many burrowing mammals and is an essential survival mechanism in arid regions. The permanent burrows may have emergency exits—side tunnels ending at or near the surface—through which the jerboa "bursts" when threatened by a predator. Jerboas often lie on their side when sleeping in the burrow so as to better accommodate their long legs. Most species become dormant during the winter. Dipodids feed on seeds, the succulent parts of plants, and insects. They apparently do not require free water in nature, but they drink readily in captivity.

When handled or disturbed, these animals sometimes emit grunting noises or shrill shrieks, but they are usually silent. Tapping sounds, made within the burrow by the hind legs, have been reported for some species. Except for females with dependent young, generally only a single animal inhabits a burrow. Females often give birth to more than one litter per year.

The known geological range of the Dipodidae is Oligocene to Recent in Europe, Asia, and North America and Pleistocene to Recent in North Africa (Klingener 1984).

Lesser five-toed jerboa *(Alactagulus pumilio)*, photos by K. A. Rogovin.

RODENTIA; DIPODIDAE; **Genus DIPUS**
Zimmermann, 1780

Rough-legged Jerboa, or Northern Three-toed Jerboa

The single species, *D. sagitta*, is found from the northern Caucasus and northeastern Iran to Manchuria (Brown 1978; Corbet 1978).

Head and body length is 105–57 mm, tail length is 140–90 mm, and hind foot length is 60–70 mm. Weight is about 70–85 grams during the spring and summer and about 90–110 grams in the fall, just before dormancy. The upper parts are apparently bright orangish mixed with black during the winter and pale sandy buff during the summer. The hip stripe, underparts, and tail tuft are white. There are three digits on each hind foot and a stiff brush of hairs on each toe. The upper incisors are grooved and yellow. Females have eight mammae (Ognev 1963).

The rough-legged jerboa inhabits sandy areas with a covering of shrubs, pines, or other vegetation. It lives in burrows,

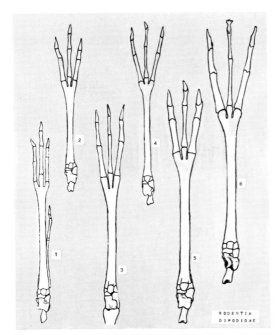

The bones of the specialized feet of members of the Dipodidae: 1. *Allactaga tetradactyla*; 2. *Stylodipus telum*; 3. *Dipus sagitta*; 4. *Jaculus lichtensteini*; 5. *J. orientalis*; 6. *Paradipus ctenodactylus*. Photos from *Faune de l'URSS, Mammifères*, B. S. Vinogradov.

Rough-legged jerboa, or northern three-toed jerboa *(Dipus sagitta)*, photo by K. A. Rogovin.

usually located in the larger sand hillocks. Three main kinds of burrows have been reported (Ognev 1963). In the spring, following hibernation, *Dipus* excavates a permanent burrow about 40–100 cm below the surface. The burrow includes a nest chamber in a side passage, a refuge tunnel extending below the nest, a regular entrance passage, and one or more emergency exits, which extend to just below the surface. The entrance is kept sealed during the summer to maintain a suitably cool and moist microclimate but is left open in the spring and fall. There is a characteristic mound of excavated soil at the entrance. The nest is a loose structure made of dry vegetation. The recorded average length of burrow passageways is 319 cm for males and 480 cm for females. The females use such burrows to rear their young. In addition to this permanent home, individuals dig holes for temporary shelter during surface movements. These tunnels are nearly straight and extend only 50–60 cm into the ground at an angle of 10°–30°. The third kind of burrow, used for winter hibernation, consists of a sealed tunnel leading to a nest chamber about 80–100 cm below the surface. Instead of leaving a mound at the entrance of its hibernation burrow, *Dipus* scatters and tramples the excavated soil. The rough-legged jerboa is a rapid digger, tunneling through as much as 125 cm of soil in 10–15 minutes (Ognev 1963). Naumov and Lobachev (1975) noted that the burrows of *Dipus* may be up to 150 cm deep and contain three to five separate chambers.

Dipus is nocturnal and terrestrial but able to climb bushes. Its bipedal hops normally cover 10–15 cm but extend for 120–40 cm if the animal is disturbed (Ognev 1963). All individuals in a given area tend to leave their burrow at about the same time each evening and travel to the feeding areas, which may be several hundred meters away. On colder nights, activity may last less than an hour. *Dipus* generally hibernates from

November to March, but in the Kyzyl Kum Desert of Uzbek S.S.R. the period of dormancy lasts only 36–60 days (Naumov and Lobachev 1975); in certain areas there may be no hibernation at all (Ognev 1963). The diet of *Dipus* includes all parts of vegetation (roots, stems, seeds, flowers, and fruits); insects may be important in some areas. *Dipus* relishes plants with milky juices; it uses its sense of smell in digging for subsurface sprouts and the insect grubs in gall nuts on the underground parts of certain plants.

Population density generally peaks in July or August (Naumov and Lobachev 1975). An average of 7–9 per ha. and a maximum of 21 per ha. have been reported at that time. Most of these animals die by autumn, however, and spring density is only 2–3 per ha. Individuals spend most of their time in a home range of two to three hectares but may shift their range several times during a summer. According to Ognev (1963), there is only a single adult per burrow, and males fight savagely among themselves during the breeding season.

In some areas, reproduction continues from February to October, and up to three litters may be produced by each female, but mating usually does not begin until March. There is generally an intensive breeding peak from March to May and a lesser peak from July to August. The gestation period is 25–30 days. Litters contain one to eight young, the average being about three or four. A female born during the first breeding peak may give birth to its own litter during the second period (Naumov and Lobachev 1975).

RODENTIA; DIPODIDAE; **Genus PARADIPUS**
Vinogradov, 1930

Comb-toed Jerboa

The single species, *P. ctenodactylus,* occurs from the Syr Darya River in south-central Kazakh S.S.R. to the shores of the Caspian Sea in southwestern Turkmen S.S.R. (Corbet 1978). The remainder of this account is based largely on works by Ognev (1963) and Naumov and Lobachev (1975).

Head and body length is 110–55 mm, tail length is 206–21 mm, and hind foot length is 73–82 mm. The upper parts are hazel to pinkish cinnamon, and the underparts are white. There is a broad white zone around each eye, and there are rusty yellow patches on each cheek and the chest. The long tail has a large, white terminal tuft. The length of the ears, about 33–38 mm, exceeds that of most other jerboas. The upper incisors are white and ungrooved. There are no premolar teeth.

The hind feet of *Paradipus* have three digits, the medial one being noticeably the longest. Along each side of each hind digit is a comb of stiff hairs. These bristles are longest on the middle toe and on the internal edges of the two lateral toes. This arrangement is thought to assist *Paradipus* in moving about on sandy surfaces. The forelimbs lack such hairy digits but have prominent claws, up to 7 mm long, which are employed in burrowing.

The comb-toed jerboa inhabits sandy deserts with a sparse covering of shrubby vegetation. Its relatively simple burrow extends down at nearly a right angle to the slope of a sand dune. The entrance tunnel, about 10 cm in diameter, is not kept sealed. There are usually no side passages or emergency exits. The nest chambers of summer burrows are located about 1.5–2.5 meters below the surface. Some summer burrows are occupied for only a day, but others are used for up to 30 days. During winter hibernation the burrow is located on the protected side of a dune, and the nest chamber, often 4–5 meters deep, is lined with vegetation.

Paradipus is nocturnal and basically terrestrial, but it can climb bushes and small trees. It is reportedly among the fastest of jerboas, being able to cover up to 3 meters in a single leap and 180 meters per minute. Individuals travel 7–11 km per night, depending on the distance to feeding areas. Hibernation occurs from early December to mid-February, a period when food supplies are low. The diet consists entirely of desert vegetation, and all parts of plants are utilized—shoots, stems, buds, seeds, flowers, and fruits.

Each individual occupies a home range with a radius of 1.5–2.0 km, but there is considerable overlap of ranges. The observed number of burrows and tracks indicates that *Paradipus* is common in the area it inhabits. A characteristic tapping sound of two quick beats and then a long pause has been reported to occur when an animal is in its burrow. There are apparently two periods of reproduction, April–May and July. The recorded number of embryos per pregnant female is two to six.

RODENTIA; DIPODIDAE; **Genus JACULUS**
Erxleben, 1777

Desert Jerboas

There are two subgenera and five species (Corbet 1978; Ellerman and Morrison-Scott 1966; Harrison 1972; Hassinger 1973):

subgenus *Eremodipus* Vinogradov, 1930

J. lichtensteini, southern Soviet Central Asia from the southeastern shore of the Caspian Sea to the area south of Lake Balkhash;

subgenus *Jaculus* Erxleben, 1777

J. jaculus, desert and semidesert areas from Morocco and Mauritania to southwestern Iran and Somalia;
J. blanfordi, southern and eastern Iran, southern and western Afghanistan, southwestern Pakistan;
J. orientalis, Morocco to southern Israel;
J. turcmenicus, southern Soviet Central Asia from the southeastern shore of the Caspian Sea to the Kyzyl Kum Desert.

Eremodipus was treated as a distinct genus by Naumov and Lobachev (1975). Ranck (1968) considered *J. jaculus* to comprise two sibling species, *J. jaculus* and *J. deserti*, but Harrison (1978) rejected this view.

Head and body length is 95–160 mm, tail length is 128–250 mm, and hind foot length is 50–75 mm. Osborn and Helmy (1980) listed average weights of 55 grams for *J. jaculus*, the smallest species in the genus, and 134 grams for *J. orientalis*, the largest species. The upper parts are pale to dark sandy or buffy, the underparts are whitish, and there is a whitish stripe on the hip. Each hind foot has three toes, and there is a cushion or fringe of hairs under the toes. The eyes and ears are relatively large, the upper incisor teeth are

Comb-toed jerboa *(Paradipus ctenodactylus)*, photo from *Zeitschr. Saugetierk.*

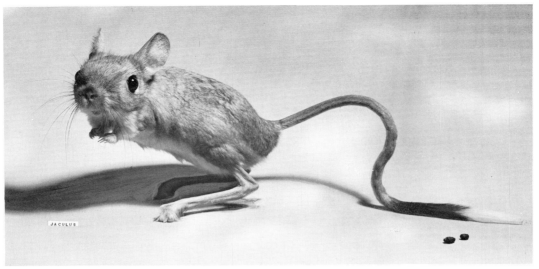

Desert jerboa *(Jaculus jaculus)*, photo by Ernest P. Walker.

grooved, and there are no premolars. Females have eight mammae (Harrison 1972).

These jerboas are found in a variety of habitats, including both rolling and relatively flat sandy deserts, saline deserts, rocky valleys, and meadows. Osborn and Helmy (1980) stated that the burrows of *J. jaculus* and *J. orientalis* are generally excavated in hard ground, are kept sealed during summer and left open in winter, slant to a depth of about 1–2 meters, and include a sleeping chamber with a nest of camel's hair or shredded vegetation, one or two escape tunnels, and in the case of *J. orientalis* a food storage chamber. According to Harrison (1972), the burrow of *J. jaculus* descends in a counterclockwise spiral to a depth of about 1.2 meters, where the nest is placed; several additional entrances radiate from the main burrow.

These jerboas progress along the ground with leaps that may measure several meters in length. Captives were observed to jump from a standing position to a height of almost a meter. When they are standing, the long tail is held curved and the tuft touches the ground, thus providing support. *Jaculus* is exclusively nocturnal and may move over a considerable distance in a single night; the track of one individual was reportedly followed for 14 km (Osborn and Helmy 1980). Wet weather is thought to cause reduced activity, but there is conflicting information on winter dormancy. According to Harrison (1972), *J. jaculus* remains active in Iraq on even the coldest nights. Nonetheless, hibernation has been found for some individual *J. jaculus* in Iraq (Hatt 1959) and for *J. orientalis* in Morocco (El Hilali and Veillat 1975). It is also apparent that *Jaculus* retires to its burrow and estivates during long, hot, dry periods. The diet of *Jaculus* includes roots, sprouts, seeds, grains, and cultivated vegetables (Osborn and Helmy 1980). Sufficient metabolic water is probably derived from vegetation, as captive animals have refused to drink for long periods, even though water was placed in their cages daily.

From 1 to 50 or more *J. orientalis* have been counted over a distance of 0.8 km. This species is evidently sociable, nearly always being found in small groups, but *J. jaculus* is strictly solitary (Osborn and Helmy 1980). Two or three desert jerboas have often been found sleeping together in a single nest. Five captive adults were quite gentle and slept together in huddled masses.

Captive female *J. jaculus* have given birth to litters of 3–4 young every 3 months and have a gestation period of 25 days or less (Harrison 1972). Wild young of this species appear in May in northern Saudi Arabia, and a litter of 4 was found in Iraq in December. Osborn and Helmy (1980) wrote that reproductive periods in *J. jaculus* have lasted from February to September in Egypt and from October to February in Sudan, that litter size in this species is 2–10 young, and that *J. orientalis* evidently has a 5- to 6-month breeding season beginning in February and a mean litter size of 3. A gestation period of about 40 days has been reported for *J. orientalis*. Naumov and Lobachev (1975) reported that *J. turcmenicus* may have three annual litters of 3–6 young each, in the spring, summer, and autumn, but that *J. lichtensteini* apparently has only one litter per year, the size of which is 2–8. A captive *J. jaculus* lived for 6 years and 3 months (Jones 1982).

J. orientalis sometimes raids sprouting barley and ripe grain in the fields of Bedouins and in turn is captured and eaten by these people (Osborn and Helmy 1980). The Soviet Union classifies *J. turcmenicus* as rare, and Nader (1989) regarded *J. orientalis orientalis* of Palestine and Sinai as rare.

RODENTIA; DIPODIDAE; **Genus STYLODIPUS**
G. M. Allen, 1925

Thick-tailed Three-toed Jerboas

There are three species (Corbet 1978; Sokolov and Shenbrot 1987):

S. telum, Ukraine to Mongolia;
S. andrewsi, Mongolia;
S. sungorus, southwestern Mongolia.

The generic name *Scirtopoda* Brandt, 1843, often is used for this species.

Head and body length is 100–130 mm, tail length is 132–63 mm, and hind foot length is 45–60 mm. The upper parts are sandy or buffy, being darkened somewhat by a sprinkling of black-tipped and completely black hairs. The hairs along the sides of the body have a white base and a bright buffy tip. The underparts, the backs of the feet, and the hip stripe are

Thick-tailed three-toed jerboa *(Stylodipus telum)*, photo by K. A. Rogovin.

white. The tail is about the same color as the back, except that the base may be encircled by white; there is no distinct terminal tuft or white tip.

Each hind foot has three digits, the middle one being the longest. Each of these toes has a stout claw concealed by stiff hairs; the soles of the hind feet also are haired. The ears are relatively short. The incisor teeth are white and grooved. Females have eight mammae (Ognev 1963).

The remainder of this account is based largely on Naumov and Lobachev (1975). *Stylodipus* inhabits deserts and steppes and occasionally has been reported in cultivated fields and pine forests. It generally requires compact soils. It excavates two kinds of burrows for summer use. Simple temporary holes are dug for one day's rest or for escape during the night. The permanent burrows are more complex, usually having a main entrance, emergency exits, and one or more chambers. Overall length of the passageways is 65–270 cm, and depth is 20–70 cm below the surface. The entrance is kept sealed by day. *Stylodipus* is generally nocturnal, with peaks of activity from about 2200 to 2400 hours and at around 0300 hours. It hibernates from September or October to mid-March. The diet consists of lichens, rhizomes, bulbs, seeds, and wheat.

Population density sometimes reaches 12–20 per ha. Individual home ranges are only 20–45 meters in diameter during the summer and do not overlap. Following its participation in reproductive activity, however, an individual may shift its range once or twice a month. The overall breeding season lasts from March to August, but it is not known whether females give birth more than once. The number of young per litter is 2–8, usually 3–5.

RODENTIA; DIPODIDAE; Genus ALLACTAGA
F. Cuvier, 1836

Four- and Five-toed Jerboas

There are 3 subgenera and 11 species (Corbet 1978, 1984; Ellerman and Morrison-Scott 1966; Hassinger 1973; Roberts 1977; Sokolov 1981; Womochel 1978):

subgenus *Allactaga* F. Cuvier, 1836

A. elater, lower Volga River and eastern Asia Minor to Sinkiang and western Pakistan;
A. euphratica, Asia Minor and Jordan to Afghanistan;
A. hotsoni, southern Afghanistan, southeastern Iran, southwestern Pakistan;
A. firouzi, southwestern Iran;
A. bobrinskii, deserts of Uzbek and Turkmen S.S.R.;
A. severtzovi, Soviet Central Asia;
A. major, Moscow and Ukraine to southwestern Siberia and Tien Shan Mountains;

subgenus *Orientallactaga* Shenbrot, 1984

A. sibirica, Caspian Sea to Manchuria;
A. nataliae, Mongolia;
A. bullata, western and southern Mongolia and adjacent parts of China;

subgenus *Scarturus* Gloger, 1841

A. tetradactyla, coastal plains of Libya and Egypt.

Head and body length is 90–263 mm, tail length is 142–308 mm, and hind foot length is 46–98 mm. Ognev (1963) listed weights of 44–73 grams for *A. elater*, the smallest species in the genus; 95–140 grams for *A. sibirica*; and 280–420 grams for *A. major*, the largest of all jerboas. Osborn and Helmy (1980) gave the average weight of *A. tetradactyla* as 52 grams. The coloration of the upper parts is mixed russet and black to sandy and grayish buff. The underparts are whitish, and there is a white stripe on the hip.

In all species except *A. tetradactyla* there are five digits on each hind foot. The two lateral toes are much smaller than the three central digits and do not contribute substantially to the support of the foot. *A. tetradactyla*, the four-toed jerboa, has only one small lateral digit. *A. bobrinskii* differs from other species in the Soviet Union in that the hind digits are covered beneath by a brush of hairs. The sole of the hind foot of *Allactaga* has a tuft of stiff hairs. The eyes are large, and the

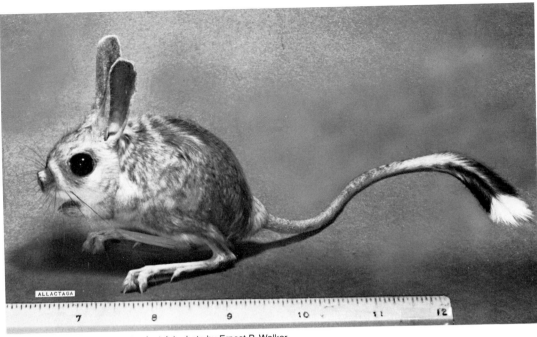

Four-toed jerboa (Allactaga tetradactyla), photo by Ernest P. Walker.

ears are long and slender, being about the same length as the head. There is one small premolar tooth on each side of the upper jaw. Females have eight mammae.

These jerboas inhabit deserts, semideserts, and steppes. At least some species prefer areas of compact soil with a thick cover of vegetation. A. tetradactyla occurs in coastal salt marshes and clay deserts; its burrows are simple, about 60–150 cm deep, and occupied only briefly (Osborn and Helmy 1980). A. major has been reported to construct four different kinds of burrows (Ognev 1963). Its permanent summer home consists of a main passage up to 3 meters long that leads to a chamber 65–150 cm below the surface. An extra escape hole is sometimes added. Only females actually construct a nest in the summer home. For temporary shelter on summer days, A. major may dig a long but simple tunnel up to 13 meters in length and 20–30 cm deep. For emergency refuge at night it digs a blind tunnel, 80–150 cm long, straight into the ground. Finally, for winter hibernation, A. major establishes a nest chamber at a depth of 100–250 cm. The other species of Allactaga also excavate simple temporary shelters, as well as permanent homes that range from the simple to the complex (Naumov and Lobachev 1975). The burrow of A. euphratica is a relatively straight tunnel about 120–70 cm long and extending 45–70 cm below the surface. The burrow of A. severtzovi is much longer and is especially deep and complex. Females of the latter species dig tunnels up to 5.3 meters long, and there may be several chambers, escape exits, and side branches. The small species A. elater burrows in hard soil, sometimes making several chambers at considerable depths. Nonetheless, its heaps of excavated soil are not much larger than those left by certain ground beetles. A. elater often digs in wagon tracks or other depressions in the ground, and the excavated soil may fill the depression and make it difficult to find the burrow. Unlike those of some other species of Allactaga, the occupied burrows of A. elater are often left unsealed (Ognev 1963).

Allactaga is primarily nocturnal, but some species are known to emerge from their burrows before sunset to begin foraging (Naumov and Lobachev 1975). When undisturbed, these jerboas move with a slow bipedal walk, but if startled, they begin leaping in a zigzag pattern, covering a meter or more per jump (Ognev 1963). A. elater has been timed at a speed of 48 km/hr (Roberts 1977). In A. euphratica, one of the more southerly species, winter dormancy lasts only from late October to February, and some populations do not hibernate at all. In the other species that occur in the Soviet Union, hibernation generally extends from September or October to about March or April (Naumov and Lobachev 1975). Most species of Allactaga are primarily vegetarian. A. major is almost exclusively a seed eater but also takes bulbs and some other plant matter. In contrast, A. sibirica is mainly insectivorous (Naumov and Lobachev 1975). Observations in Uzbek S.S.R. indicate that when A. elater initially emerges from hibernation, it feeds mostly on vegetation, but that in May it begins to concentrate on insects (Pavlenko and Daveletshina 1971).

Naumov and Lobachev (1975) reported population densities of 1/ha. for A. euphratica in the Talish Mountains and up to 7/ha. for A. elater in favorable habitat. Pavlenko and Daveletshina (1971) stated that the mean density of A. elater was not over 3/ha. in Uzbek S.S.R., but Mazin (1975) gave the density of this species as 12–15 per ha. in southeastern Kazakh S.S.R. According to Ognev (1963), there is usually only a single adult A. elater per burrow.

Extensive data compiled by Bekenov and Mirzabekov (1977), Lay (1967), Mazin (1975), and Naumov and Lobachev (1975) indicate that the breeding season of A. elater extends from March to as late as November but that there are peak periods of reproductive activity that vary from place to place. There are generally two or three such peaks, corresponding with spring, summer, and fall. Many females evidently have three annual litters. Most other Soviet species also are thought to have lengthy breeding seasons, with several peak periods, and possibly two or three litters per year. There are usually about three to five young at a time, but overall recorded litter size in the Soviet Union is one to eight. The young of

A. *elater* remain with the mother for 30–35 days and attain sexual maturity by 3.5 months. Therefore, it is possible for a female of this species born early in the reproductive season to give birth to its own litter before the season ends. The young of A. *major* remain with the mother for at least 1.5 months and apparently do not mate until the year after they are born.

According to Harrison (1972), a pregnant female A. *euphratica* with nine embryos was taken in Syria on 23 April. Several litters per year are apparently the rule in this area. Studies of captives indicate that the young of this species open their eyes after about two weeks and that adult females will care for young other than their own. One captive A. *euphratica* lived for four years and two months (Jones 1982).

Ognev (1963) wrote that A. *major* sometimes destroys melons and pumpkins, as it tries to get at the seeds. It also reportedly does damage to wheat, other grains, and rubber plants. Its skin is sometimes utilized by industry but is too delicate to be of much value. The unique four-toed African species, A. *tetradactyla*, has a very restricted range and ecological tolerance and may be threatened with extinction by desert reclamation projects (Osborn and Helmy 1980). A. *bullata* was regarded as rare by Wang, Zheng, and Kobayashi (1989), and the subspecies A. *euphratica euphratica*, known only from two localities in Saudi Arabia, was designated as rare by Nader (1989).

RODENTIA; DIPODIDAE; Genus ALACTAGULUS
Nehring, 1897

Lesser Five-toed Jerboa, or Little Earth Hare

The single species, A. *pumilio*, is found from the Don River and northeastern Iran to Inner Mongolia (Corbet 1978). *Alactagulus* sometimes is treated as a subgenus of *Pygeretmus*, and A. *pumilio* sometimes is called A. *pygmaeus* (Corbet 1984).

Head and body length is 88–125 mm, tail length is 107–85 mm, and hind foot length is 34–52 mm. Ognev (1963) listed weight as 44–61 grams. The back and proximal part of the tail are generally dull buffy mixed with black, the flanks and face are paler, and the underparts are white. The tail has a pronounced tuft that is mostly dark but white at the tip. Each hind foot has five toes; the first and fifth are extremely short, but the three central digits are long. The toes are fringed with hair on the sides, but the soles are naked. *Alactagulus* differs from *Allactaga* in having shorter ears, no premolar teeth, and a simpler enamel pattern on the molars. Female *Alactagulus* have eight mammae.

The lesser five-toed jerboa inhabits steppes and clay and saline deserts. It is nocturnal and terrestrial. According to Naumov and Lobachev (1975), it builds both permanent and temporary burrows. The latter are used for emergency refuge and are excavated in soft soil with a cover of vegetation. The permanent burrows are usually located in hard, bare soil. The excavated earth is deposited in a characteristic, flat mound above the burrow. The initial passage of the permanent burrow may be up to 6 meters long, but it is sealed and not used after the entire burrow is complete. The initial tunnel usually divides into two branches, one leading downward to a nest chamber and the other leading first to the side and then up to a regular entrance. There may also be a narrow airhole or vent leading to the surface. The main entrance is kept plugged with earth during the day. The chamber is lined with soft nesting material. The nests of breeding females are about 30–70 cm beneath the surface of the ground. Hibernation chambers contain no bedding material and may be more than 60 cm deep. In the region west of the Aral Sea hibernation lasts from November to February or March. The diet includes succulent vegetation, bulbs, roots, and rhizomes. Some food may be stored in the burrow.

Naumov and Lobachev (1975) reported a population density of about 10/ha. in the clay desert near the Aral Sea. The reproductive season extends from March to September. There are two peak breeding periods, one in the spring or early summer and one in the late summer or early fall. Some females have more than one annual litter. The number of young ranges from 1 to 7 and usually is 2–4. A female born in the spring may produce its own litter in the fall. According to Rogovin (1981), young females settle on the periphery of the mother's range and may occupy it if she dies, while young males migrate to other areas.

Lesser five-toed jerboa (*Alactagulus pumilio*), photo by F. Petter.

Fat-tailed jerboa *(Pygeretmus platyurus)*, photo by V. Masin through D. Bibikov.

RODENTIA; DIPODIDAE; **Genus PYGERETMUS**
Gloger, 1841

Fat-tailed Jerboas

There are two species (Corbet 1978):

P. platyurus, western and extreme eastern Kazakh S.S.R.;
P. shitkovi, eastern Kazakh S.S.R. in the area of Lake
   Balkhash.

In *P. platyurus* head and body length is 75–95 mm, tail length is 78–90 mm, and hind foot length is 30–35 mm. In *P. shitkovi* head and body length is 97–122 mm, tail length is 94–128 mm, and hind foot length is 38–45 mm. Coloration is sandy brown above and white below. The tail has no terminal tuft. Considerable subcutaneous fat deposits accumulate in the tail, often causing it to become nearly cylindrical and to be

as much as 12 mm in diameter (Ognev 1963). Each hind foot has five digits, the first and fifth of which are extremely short and do not contribute substantially to the support of the foot. The ears, when pressed forward, do not reach the nose. There are no premolar teeth.

Fat-tailed jerboas inhabit clay and saline deserts and semideserts. They are nocturnal and terrestrial and have been reported from areas of both scant and thick vegetation. Compared with most other jerboas, they seem to be poor jumpers and diggers (Ognev 1963). According to Naumov and Lobachev (1975), their burrows are simple, consisting of a passage about 20 cm long and 10–15 cm deep and a chamber for both daily use and hibernation. Burrows seem to be changed frequently. Hibernation lasts from October to April. The diet includes green vegetation, bulbs, spiders, and insects. There is apparently only one reproductive season, with young having been collected in May and June. Litters of five and six young have been recorded.

Five-toed dwarf jerboa *(Cardiocranius paradoxus)*, photo by V. Masin through D. Bibikov.

RODENTIA; DIPODIDAE; **Genus CARDIOCRANIUS**
Satunin, 1903

Five-toed Dwarf Jerboa

The single species, *C. paradoxus*, has been recorded from the area north of Lake Balkhash in Kazakh S.S.R., the Tuva Autonomous Region of south-central Siberia, various parts of Mongolia, and the Nan Shan Mountains of northern China (Corbet 1978).

Head and body length is 50–75 mm, tail length is 70–78 mm, and hind foot length is 25–30 mm. The upper parts are grayish buff, the underparts are white, and the scantily haired tail is light brown above and white below. The tail is constricted at the base but suddenly expands and then tapers toward the tip, where there is a small tuft of hairs.

In addition to its small size and unusual tail, *Cardiocranius* differs from most jerboas in several characters. The three central metatarsals of the hind foot are not fused into a single cannon bone. The hind foot has five separate toes. The outermost toe (digit 5) is only about 4 mm shorter than the fourth digit, but the innermost toe (digit 1) is very small and extends only about 8 mm beyond the base of the second digit. The sole of the hind foot has a tuft of bristly hairs. The external ears are remarkably small, but the tympanic bullae are enormously inflated. The skull has a heart-shaped appearance when viewed from above, hence the origin of the generic name. The upper incisor teeth are deeply grooved on the front surface. There is a small upper premolar on each side of the jaw. Females have eight mammae (Ognev 1963).

According to Naumov and Lobachev (1975), *Cardiocranius* is a specialized genus, adapted to rocky deserts. It is nocturnal, foraging mainly from 2200 to 0400 hours. It is a seed eater and accumulates considerable body fat toward midsummer. Specimens in reproductive condition have been found in Kazakh S.S.R. in July.

Two specimens taken in the northern Gobi Desert "were dug out of the burrows which were found among sandhills covered with *Nitaria schoberi*. Both these specimens lived some time in captivity; they refused whatever vegetable food was offered to them. When they were put together into a box they fought very fiercely, this made it necessary to separate them" (Allen 1940:1084).

The Soviet Union classifies *C. paradoxus* as rare. The species also is regarded as rare in China (Wang, Zheng, and Kobayashi 1989).

RODENTIA; DIPODIDAE; **Genus SALPINGOTUS**
Vinogradov, 1922

Three-toed Dwarf Jerboas

There are five species (Corbet 1978; Hassinger 1973; Wang, Zheng, and Kobayashi 1989; Vorontsov and Shenbrot 1984):

S. kozlovi, Gobi Desert of southern Mongolia and northern China;

S. crassicauda, desert areas from Aral Sea to Mongolia;

S. thomasi, known only by the type specimen from either Afghanistan or Tibet;

S. heptneri, known only from the type locality south of the Aral Sea;

S. pallidus, Kazakh S.S.R.

Vorontsov and Shenbrot (1984) placed *S. thomasi*, *S. heptneri*, and *S. pallidus* in the new subgenus *Prosalpingotus*, and *S. crassicauda* in the new subgenus *Anguistodontus*. They also listed *Salpingotulus* as a subgenus of *Salpingotus*.

Head and body length is about 41–57 mm, tail length is 93–126 mm, and hind foot length is 21–25 mm (Ognev 1963). The upper parts are sandy or buffy, and the underparts are pale yellowish. There is no terminal tuft on the tail of *S. crassicauda*; a small, pale-colored tuft on the tail of *S. kozlovi*; and a dense, black terminal tuft on the tail of *S. heptneri*. In *S. crassicauda* and *S. thomasi* the proximal third or half of the tail is swollen from fat accumulation under the skin, but the tail of *S. kozlovi* is not expanded.

Each hind foot has only three toes, but the metatarsals are not fused to form a cannon bone. There are tufts of hair beneath the toes. The external ears are short, but the auditory bullae are greatly inflated. The upper incisor teeth are not grooved. There is a small premolar on each side of the upper jaw. Females have eight mammae.

In Mongolia and the Soviet Union *S. crassicauda* inhabits sand dunes overgrown with tamarisk, saxaul, and saltwort

Three-toed dwarf jerboa *(Salpingotus crassicauda)*, photo by V. Masin through D. Bibikov.

(Ognev 1963). Its burrows are up to three meters in length. It eats either animal (insects and arachnids) or vegetable food. In the Zaysan Depression in extreme eastern Kazakh S.S.R. there are two consecutive breeding peaks in the period from May to July. Average litter size is about 2.7 young. Sexual maturity is not attained within the year of birth (Naumov and Lobachev 1975).

The Soviet Union classifies *S. crassicauda* and *S. heptneri* as rare. *S. crassicauda* and *S. kozlovi* are regarded as rare in China (Wang, Zheng, and Kobayashi 1989).

RODENTIA; DIPODIDAE; Genus SALPINGOTULUS
Pavlinov, 1980

Baluchistan Pygmy Jerboa

The single species, *S. michaelis*, occurs in southwestern Pakistan and possibly the adjacent part of Afghanistan (Hassinger 1973; Roberts 1977). This species was not discovered until 1966 and was placed in the genus *Salpingotus* until Pavlinov (1980a) erected for it the new genus *Salpingotulus*. Vorontsov and Shenbrot (1984) treated *Salpingotulus* as a subgenus of *Salpingotus*, and this procedure was followed by Corbet and Hill (1986).

Head and body length is 36–47 mm, tail length is 72–94 mm, and hind foot length is 18–19 mm. The upper parts are pale yellow ochre or sandy buff, the hind legs are pinkish, and the underparts are pure white. The body fur is soft and silky. The well-haired, long tail has a terminal tuft of dark hairs (Roberts 1977).

Salpingotulus resembles *Salpingotus* but is generally smaller. Each hind foot has three toes and a brush of stiff hairs. The forelimbs have four digits each. The eyes are comparatively large. Females have only six mammae (Roberts 1977). The tail is not generally swollen from fat storage (Corbet 1978); only 2 of 15 specimens had fat accumulation in the tail.

According to Roberts (1977), *Salpingotulus* inhabits sand dunes, barren gravel flats, and sandy plains. It is exclusively nocturnal and resides in extensive, self-excavated burrow systems. Locomotion is by small, rapid hops on the hind limbs, and digging is accomplished with both the fore- and hind limbs. There evidently is no winter hibernation, but captives became torpid each day during the colder months and could then be picked up and handled. The diet consists of grass seeds and stems and other vegetation. The genus is loosely colonial, there being a number of burrows in the same locality. Up to six adults will sleep huddled together. There seem to be two breeding periods: the first litter is born at the end of June, and the second is produced in August, almost

Baluchistan pygmy jerboa *(Salpingotulus michaelis)*, photo by J. FitzGibbon.

Long-eared jerboa *(Euchoreutes naso)*, photo by K. A. Rogovin.

immediately after the first is weaned. The usual litter size is two to four young.

RODENTIA; DIPODIDAE; Genus EUCHOREUTES
Sclater, 1891

Long-eared Jerboa

The single species, *E. naso*, is known from western Sinkiang, north-central China, and extreme southern Mongolia (Corbet 1978).

Head and body length is 70–90 mm, tail length is 150–62 mm, and hind foot length is 40–46 mm. The upper parts are reddish yellow to pale russet, rather than sandy or buffy as in most jerboas; the underparts are white. The tail is mostly covered with short hairs but has a terminal tuft. For most of its length the tail is colored like the body, but it becomes white at the beginning of the tuft, then black in the middle of the tuft, and then white again at the tip.

The large ears are about a third longer than the head. Each hind foot has five toes, the two lateral digits being much shorter than the three central digits. The central metatarsals are fused for only a part of their length. The toes are tufted beneath with bristly hairs. The incisor teeth are narrow and white. There is a small premolar on each side of the upper jaw. Females have eight mammae.

The habits and biology of *Euchoreutes* are not known, and there are only a few specimens in study collections. They have been taken in sandy valleys covered with low bushes of *Haloxylon ammodendron* and sometimes in the huts of nomads. One long-eared jerboa was kept alive in captivity for four days; it had a strong tendency to bite. Mainly on the basis of habitat disturbance by people, Wang, Zheng, and Kobayashi (1989) designated *Euchoreutes* as endangered.

RODENTIA; Family HYSTRICIDAE

Old World Porcupines

This family of 3 living genera and 11 species occurs in parts of southern Europe (possibly through introduction), all across southern Asia, on many islands of the East Indies, and throughout Africa. The sequence of genera presented here follows that of Van Weers (1979).

Head and body length is 350–930 mm, tail length is 25–260 mm, and weight is 1.5–30.0 kg. The head and body, and in some species the tail, are covered by spines or quills—thick, stiff, sharp hairs as much as 350 mm in length. The coloration is brownish or blackish, often with white bands on the hairs and spines.

These large rodents have a heavyset body, a tail not more than half as long as the head and body, and short limbs. There are five digits on the forefoot, the thumb is reduced, and there are five digits on the hind foot. The soles of the feet are smooth. The intestinal tract is approximately 790 mm in length, not including the stomach or cecum. The facial part of the skull is inflated by pneumatic cavities. The dental formula is: (i 1/1, c 0/0, pm 1/1, m 3/3) × 2 = 20. The four cheek teeth are flat and have a wavy enamel pattern. They are incompletely rooted and moderately to strongly high-crowned. Females have two or three pairs of mammae, which are located on the sides of the body.

Old World porcupines inhabit deserts, savannahs, and forests. They shelter in caves, crevices, holes dug by other animals, or burrows they have excavated themselves and construct a nest of plant material within the den. They are mainly nocturnal and terrestrial, walking ponderously on the sole of the foot with the heel touching the ground. They run with a shuffling gait or gallop clumsily when pursued. Except for *Trichys*, they rattle their spines when moving about and may also stamp their feet when alarmed. The diet consists mainly of vegetation, but carrion feeding and bone gnawing have been reported. Temperament, at least in captivity, differs by species and individuals, ranging from shy and nervous to docile. At least some species are moderately gregarious, and up to 10 individuals have been found in a single burrow. Litters usually contain 1–2 young, born with their eyes open and with soft, short quills. After a week or so, when the quills have hardened, the young may leave the nest with the mother. Old World porcupines have a relatively long life span and do well in zoos; captive members of all genera are known to have lived at least 10 years.

The geological range of the Hystricidae is Miocene to Recent in Asia, Pliocene to Recent in Europe, and Pleistocene to Recent in Africa (Woods 1984).

RODENTIA; HYSTRICIDAE; Genus TRICHYS
Günther, 1876

Long-tailed Porcupine

The single species, *T. fasciculata*, is found on the Malay Peninsula, Sumatra, and Borneo (Van Weers 1976).

Head and body length is 350–480 mm, tail length is 175–230 mm, and weight is 1,750–2,250 grams. The body is covered with flattened, flexible, grooved spines of moderate length. Interspersed among these spines are a few longer, cylindrical, spinelike hairs. The head and underparts are hairy, and the underfur is rather woolly. The tail is scaly for most of its length, but its base is clothed like the back, and its tip has a tuft of flat, stiff bristles. The general coloration of the upper parts is brownish; the individual hairs have a white base and a brown tip. The coloration becomes paler on the sides and finally blends into the white underparts.

Trichys is the most primitive of the Old World porcupines. With its relatively long, scaly tail and short spines, it bears a superficial resemblance to *Rattus* (Grzimek 1975). Many specimens of *Trichys*, however, lack any appreciable tail, thus

Long-tailed porcupine *(Trichys fasciculata)*, photos by Ernest P. Walker.

suggesting that this organ breaks easily from the body. *Trichys* cannot bristle or rattle its quills as most other porcupines do. Its broad forefoot has four well-developed digits, each armed with a thick claw. The skull has well-marked postorbital processes (Van Weers 1979).

The long-tailed porcupine is a forest dweller. It is an agile climber and gets its food from the tops of trees and bushes (Grzimek 1975). It is reported to destroy pineapples in some areas. The tail seems to have value to some native people, who therefore sever it from the hide. A captive specimen was still alive after 10 years and 1 month (Jones 1982).

RODENTIA; HYSTRICIDAE; Genus ATHERURUS
F. Cuvier, 1829

Brush-tailed Porcupines

There are two species (Chasen 1940; Lekagul and McNeely 1977; Misonne, *in* Meester and Setzer 1977; Van Weers 1977):

A. macrourus, south-central China, Assam, Burma, Thailand, Indochina, Malay Peninsula and several small nearby islands, Hainan;
A. africanus, Gambia to western Kenya and southern Zaire.

Head and body length is 365–570 mm, tail length is 102–260 mm, and weight is 1.5–4.0 kg (Kingdon 1974*b*; Medway 1978; Van Weers 1977). The pelage is almost entirely spiny, though the spines are softer on the head, legs, and underparts. The middorsal spines are the longest. Most of the spines are flattened and have a groove on their dorsal surface; interspersed among the flattened, grooved spines on the lower back are a few round, thick bristles. The basal part of the tail is clothed like the back, the middle part is scaly, and the terminal part has a thick tuft of bristles. The general coloration of the upper parts is grayish brown to blackish brown, and the individual hairs have a whitish tip. The sides are paler than the back, and the underparts are nearly white. The tail tuft is whitish to creamy buff.

Atherurus has a relatively long body; short, stout limbs;
and short, rounded ears. The partially webbed feet are armed with blunt, straight claws. Postorbital processes are either lacking or very weak in the skull (Van Weers 1979). The tail is about one-fourth to one-half as long as the head and body; it breaks easily and is often lost. Females have two pairs of lateral thoracic mammae (Weir 1974*b*).

These porcupines live in forests at elevations of up to 3,000 meters (Grzimek 1975). They are strictly nocturnal, sheltering by day in a hole among tree roots, a rocky crevice, a termite mound, a cave, or an eroded cavity along a stream bank. Evidently they do not usually dig their own burrows but do line the natal chamber with vegetation. They run swiftly and are basically terrestrial but are capable of climbing, can jump to a height of over 1 meter, and can swim well (Kingdon 1974*b*; Lekagul and McNeely 1977). The diet consists mainly of green vegetation, bark, roots, tubers, and fruit and may sometimes include cultivated crops, insects, and carrion. Emmons (1983) found home range of *A. africanus* to average 13.3 ha. during a month and a half; males foraged throughout this area each night, while females had no consistent movement pattern.

Except as noted, the information for the remainder of this account was taken from Kingdon (1974*b*) and applies to *A. africanus.* Small colonies of 6–8 individuals may share a burrow and a feeding territory. Group territories have been found to measure 0.18–5.00 ha. and to be regularly marked in certain spots with dung deposits. Evidently the groups represent a mated pair and their offspring of several litters. A number of such families, comprising up to 20 porcupines, may live in close proximity in a favorable area. Lekagul and McNeely (1977) indicated that *A. macrourus* also may live and move about in small groups.

There seems to be no definite breeding season, up to three litters may be born per year, and the gestation period is 100–110 days. Overall litter size is one to four young, but data compiled by several authorities (Asdell 1964; Lekagul and McNeely 1977; Rahm 1962) suggest that a single offspring is the usual case in both Africa and Asia. The newborn are well developed, weigh 150 grams each, and open their eyes within hours of birth. They are able to eat solid food after 2–3 weeks, but lactation continues until they are 2 months old. Adult weight is attained at 2 years. A captive specimen of *A. africanus* lived for 22 years and 11 months (Snyder and Moore 1968).

A. West African brush-tailed porcupine *(Atherurus africanus)*, photos by Ernest P. Walker. B. Asiatic brush-tailed porcupine *(A. macrourus)*, photo by Lim Boo Liat.

RODENTIA; HYSTRICIDAE; Genus HYSTRIX
Linnaeus, 1758

Old World Porcupines

There are three subgenera and eight species (Corbet 1978; Kingdon 1974*b*; Misonne, *in* Meester and Setzer 1977; Van Weers 1978, 1979):

subgenus *Thecurus* Lyon, 1907

H. pumila, Palawan, Busuanga, and Balabac islands (Philippines);
H. sumatrae, Sumatra;
H. crassispinis, Borneo;

subgenus *Acanthion* Cuvier, 1822

H. brachyura, Nepal to eastern China and Malay Peninsula, Hainan, Sumatra, Borneo;
H. javanica, islands from Java to Flores;

subgenus *Hystrix* Linnaeus, 1758 (crested porcupines)

H. indica, Asia Minor and Arabian Peninsula to Soviet Central Asia and India, Sri Lanka;

H. cristata, Italy, Sicily, possibly the Balkans, Mediterranean coast of Africa to northern Zaire and Tanzania;
H. africaeaustralis, southern half of Africa.

Thecurus often has been treated as a distinct genus. Its head and body length is 420–665 mm, its tail length is 25–190 mm, and its weight is 3.8–5.4 kg (Grzimek 1975; Van Weers 1978). Its body is densely covered with flattened spines, each of which is deeply grooved longitudinally and increases in rigidity toward the tip. The spines are smaller along the tail and are more flexible on the underparts. Coarse, bristlelike hairs cover the feet. The back is dark brown to black, the sides are paler, and the underparts are buffy white. *Thecurus* resembles the other subgenera of *Hystrix* in having a relatively short tail with rattle quills. It resembles the subgenus *Acanthion* in that it lacks a well-developed crest and its quills have only one black band.

In the subgenus *Acanthion* head and body length is 455–735 mm, tail length is 60–115 mm, and weight is about 8 kg (Lekagul and McNeely 1977; Medway 1978; Van Weers 1979). Quill structure and arrangement generally resemble those of the subgenus *Thecurus*. In the species *H. brachyura*, according to Lekagul and McNeely (1977), the front half of the body is covered with short, dark brown spines, while the hindquarters have long, pointed, whitish quills, usually with

Sumatran porcupine *(Hystrix sumatrae)*, photo by Ernest P. Walker.

one distinct blackish ring. There is a short whitish crest on the neck and upper back. The short tail has both long, pointed quills and rattle quills.

In the subgenus *Hystrix* head and body length is 600–930 mm, tail length is 80–170 mm, and weight is about 10–30 kg. The head, neck, shoulders, limbs, and underside of the body are covered with coarse, dark brown or black bristles. There are long quills along the head, nape, and back, and these quills can be raised into a crest. The sides and back half of the body are covered with stout, cylindrical quills up to 350 mm long and mostly marked with alternating light and dark bands. There may be some longer, more slender, and more flexible quills, usually all white. The rattle quills of the short tail are better developed than those in other subgenera.

Hystrix is distinguished externally from the other genera of Old World porcupines by its relatively shorter tail and the presence of rattle quills. These quills are located at the end of the tail. They are slender for most of their length but are of much greater diameter for about the terminal fifth. The expanded portion is hollow and thin-walled, so several quills vibrating together produce a hisslike rattle.

The broad forefoot of *Hystrix* has four well-developed digits, each armed with a thick claw, and the hind foot has five digits. The eyes and external ears are very small. The facial region of the skull is inflated by pneumatic cavities, and the nasal bones are enlarged; these characters are especially pronounced in the subgenus *Hystrix*. Female *H. cristata* have two or three pairs of lateral thoracic mammae (Weir 1974b).

These porcupines are highly adaptable, being found in all types of forests, plantations, rocky areas, mountain steppes, and sandhill deserts; the elevational range is sea level to 3,500 meters (Kingdon 1974b; Medway 1978; Roberts 1977). They shelter in caves, rock crevices, aardvark holes, or burrows they dig themselves. These burrows often have several entrances, are sometimes used for many years, and can become quite extensive. One burrow was 18 meters in length, terminated in a chamber 1.5 meters below the surface, and had three escape holes. *Hystrix* is nocturnal and terrestrial. It does not usually climb trees but does swim well. Its movement was described in Grzimek (1975) as "easy and grace-

ful." Kingdon (1974b), however, wrote that the normal gait is a ponderous, plantigrade walk, while it trots or gallops when alarmed. It tends to follow paths and may cover up to 15 km per night in search of food. According to Grzimek (1975), European animals sometimes remain in their holes through the winter but do not truly hibernate. The diet of *Hystrix* includes bark, roots, tubers, rhizomes, bulbs, fallen fruits, and cultivated crops. Insects and small vertebrates are occasionally taken. Carrion feeding has been reported but is not common. Bones are frequently found in and around burrows, probably having been carried there and gnawed to obtain calcium and to hone the incisor teeth.

Piping calls and a piglike grunt have been reported for *Hystrix*. Kingdon (1974b) stated that there is considerable grunting and quill rattling as the porcupines shuffle around by night. At the least encounter with another animal, they raise and fan their quills, thereby more than doubling their apparent size. If still bothered, they stamp their feet, whirr their quills, and finally charge backward, attempting to drive the thicker, shorter quills of the rump into the enemy. Lions, leopards, hyenas, and even humans are sometimes injured or killed in this manner.

Small family groups commonly share a burrow, but the female may establish a separate den in which to bear its young, and foraging is generally done alone except when parents are accompanying their young (Kingdon 1974b; Roberts 1977; Van Aarde 1987b). Breeding has been reported throughout the year at the London Zoo (Asdell 1964), from March to December in Indian zoos (Prakash 1975), and from July to December in Central Africa (Kingdon 1974b). In South Africa, captive females produced litters throughout the year, mainly from August to March, with a peak during the summer month of January (Van Aarde 1985). In this last region there is only one litter per year, though two have been reported on occasion elsewhere. Weir (1974b) listed an estrous cycle of about 35 days and a gestation period of 112 days in Africa, but Van Aarde (1985) found South African female *H. africaeaustralis* to have a mean cycle of 31 days and a mean gestation of 93.5 days. Overall litter size in the genus is one to four, but there seem usually to be one or two offspring.

A. African porcupines *(Hystrix cristata),* mother and 51-day-old young, photo by Ernest P. Walker. B. Malayan porcupine *(H. brachyura),* photo by Lim Boo Liat.

They are born in a grass-lined chamber within the burrow and are well developed at birth. Young *H. africaeaustralis* weigh a mean 351 grams at birth, are suckled for a mean 101 days, and attain full size at about 1 year (Van Aarde 1985, 1987*a*). Females of this species reach sexual maturity at 9–16 months (Van Aarde 1985), males at 8–18 months (Van Aarde and Skinner 1986*a*). Old World porcupines have a remarkably long life span; some individuals probably survive 12–15 years in the wild. A captive *H. pumila* lived 9 years and 6 months, and a captive *H. brachyura* lived 27 years and 3 months (Jones 1982). The latter specimen apparently holds the longevity record for the order Rodentia.

There is controversy regarding the status of *H. cristata* in Europe. Corbet (1978) gave the range there as Italy, Sicily, Albania, and northern Greece and noted that the species had perhaps been introduced to Europe. Smit and Van Wijngaarden (1981) stated that most authorities think that *H. cristata* was introduced to Italy and Sicily by the Romans, but they concluded that it probably does not occur in the Balkans. They noted that the species is rare and decreasing in numbers in the Mediterranean region, because it is killed by persons who consider it to be an agricultural pest or who use it for food. Pucek (1989), however, reported that it increased in Italy following protection in 1974. Osborn and Helmy (1980) wrote that in Egypt *H. cristata* is probably now found only around the cliffs north of Salum, if it has not already disappeared. Farther south in Africa, and in parts of Asia, *Hystrix* reportedly does much damage by gnawing the bark of trees on rubber plantations and by eating corn, pumpkins, sweet potatoes, cassava, and young cotton plants. Because of this factor, and because its quills are used as ornaments and talismans, *Hystrix* has been exterminated in heavily settled parts of Uganda (Kingdon 1974*b*). It may also have disappeared on Singapore (Medway 1978). Nader (1989) stated that the subspecies *H. indica indica* of Arabia is rare and is killed on sight by farmers. Wang, Zheng, and Kobayashi (1989) regarded *H. brachyura yunnanensis* of Yunnan as vulnerable.

RODENTIA; Family ERETHIZONTIDAE

New World Porcupines

This family of 4 living genera and 10 species is found from the arctic coast of North America to northern Mexico and the Appalachian Mountains and from southern Mexico to Ecuador and northern Argentina. The sequence of genera presented here is based on that of Cabrera (1961) and on data cited by Woods (1973) indicating that *Coendou* is ancestral to *Erethizon*. The genus *Chaetomys*, formerly assigned to this family, recently was transferred to the Echimyidae (Woods 1984).

Head and body length is 300–860 mm and tail length is 75–450 mm. Some of the hairs are modified into short, sharp spines with overlapping barbs. The quills are embedded singly in the skin musculature, rather than laterally grouped into clusters of four to six as in the Old World porcupines (Hystricidae) (Woods 1984). The pelage is marked with blackish to brownish, yellowish, or whitish bands.

Like the Old World porcupines, the New World porcupines are heavyset, relatively large rodents. In the Erethizontidae, however, the foot is modified for arboreal life: the sole is widened, and in some forms the first digit on the hind foot is replaced by a broad, movable band. The functional digits have strong, curved claws. The limbs are fairly short. The tail is relatively short in *Echinoprocta* and *Erethizon* but long and prehensile in *Coendou*. In *Erethizon* and *Coendou* the axis

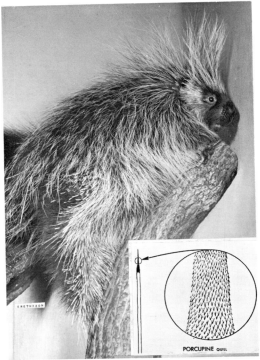

North American porcupine *(Erethizon dorsatum)*, photo by Ernest P. Walker. Inset from *The Mammal Guide*, Ralph S. Palmer.

and third cervical vertebra are fused, as in the genus *Dinomys*. The dental formula is: (i 1/1, c 0/0, pm 1/1, m 3/3) × 2 = 20. The cheek teeth are rooted and have reentrant folds.

The geological range of the Erethizontidae is Oligocene to Recent in South America and late Pliocene to Recent in North America.

RODENTIA; ERETHIZONTIDAE; Genus ECHINOPROCTA
Gray, 1865

Upper Amazonian Porcupine

The single species, *E. rufescens*, is known from Colombia (Cabrera 1961).

The length of this animal, from the tip of its snout to the end of its tail, is about 500 mm. The tail is short, being about as long as the hind foot. The coloration of the back and sides is pale brown to blackish. The chin, throat, and undersides are pale brown, and the feet and tail are dark gray to black. There is a short white streak in the center of the nose, and there is a slight crest on the nape formed by a few white spines. The spines gradually become thicker, stronger, and shorter from the head to the rump. On the posterior part of the back, above the tail, the spines are well developed, short, and thick. The tail is hairy and not prehensile. The whiskers are black, and the incisor teeth are slender and yellow.

The habits and biology of this porcupine are not well known. It is arboreal and appears to be fairly common around Bogota at elevations of 800–1,200 meters. Specimens are in

Upper Amazonian porcupine *(Echinoprocta rufescens)*, photo by Eugene Maliniak.

the Paris Natural History Museum, the British Museum of Natural History, the American Museum of Natural History, and the La Salle Institute in Bogota.

RODENTIA; ERETHIZONTIDAE; Genus COENDOU
Lacépède, 1799

Prehensile-tailed Porcupines, or Coendous

There are two species (Cabrera 1961; Goodwin and Greenhall 1961; Hall 1981; Handley 1976; Husson 1978):

C. bicolor, Panama to Bolivia;
C. prehensilis, Venezuela, the Guianas, Brazil, Bolivia, Trinidad.

Sphiggurus (see account thereof) sometimes is considered a subgenus of *Coendou*, and certain of its species sometimes are referred to *Coendou*.

Head and body length is about 300–600 mm, tail length is 330–485 mm, and weight is about 0.9–5.0 kg. The body is clothed with short, thick spines. In some forms there is a woolly underfur. The general coloration of the upper parts varies considerably from light yellowish through shades of brown to almost black, and some forms are speckled; the underparts are usually grayish.

The long tail is prehensile and lacks spines, in contrast to the short, spine-clad tail of *Erethizon*. The upper surface of the terminal part of the tail in *Coendou* is naked and modified for direct contact in coiling about branches. The tip coils upward and has a callus pad on the hairless upper side near the tip. The hands and feet are highly specialized for climbing. There are four digits on each limb, each of which is armed with a long, curved claw.

Prehensile-tailed porcupines live mainly in forests, though sometimes they enter cultivated areas. In Peru they occur in both coastal and Amazonian areas, at elevations of 150–2,500 meters (Grimwood 1969). They are principally nocturnal and arboreal. Handley (1976) collected 91 percent of his specimens of *C. prehensilis* in trees and the remainder on the ground. These porcupines are slow in their movements but are sure-footed climbers, using their tails in conjunction with their hands and feet. They prefer to sleep in tangled vegetation among the treetops but also shelter in hollow limbs, tree trunks, and shallow burrows. The diet includes leaves, tender stems, fruits, blossoms, and roots. The bark of certain trees

Prehensile-tailed porcupine *(Coendou prehensilis)*, photo by Bernhard Grzimek.

Prehensile-tailed porcupine (*Coendou* sp.), baby, photo by R. B. Tesh.

may be peeled away to reach the cambium layer (Starrett 1967).

In a study of *C. prehensilis* on the llanos of Venezuela, Montgomery and Lubin (1978) found individuals to move up to 700 meters per night and to rest by day in trees, usually at heights of 6–10 meters. The home ranges of three radiotracked animals were 8, 10, and 38 ha.

These porcupines are said to be pugnacious and to show no fear upon capture; they bite and try to hit an adversary with their spines. They also have been observed to stamp their hind feet when excited and to roll up in a ball if caught in the open (Starrett 1967). They frequently sit on their haunches and shake their spines by moving the skin. Both deep growls and plaintive cries have been reported. Births and pregnant females have been reported in the wild in January, February, March, May, July, August, and September. Observations of captive colonies (Roberts, Brand, and Maliniak 1985; Roberts et al. 1987) indicated no reproductive seasonality. Females usually mated immediately postpartum, had a gestation period of about 203 days, and bore a single large, precocial offspring. At birth the young weighed about 415 grams, had their eyes open, and were able to climb. Weaning occurred as early as 10 weeks, nutritional independence at 10 weeks, adult size at 48 weeks, and female sexual maturity at about 19 months. One female produced 10 litters in 8.5 years and still was reproductively active at an estimated age of 11.5 years. A captive *C. prehensilis* lived for 17 years and 4 months (Jones 1982).

Coendous occasionally raid plantations to feed on guavas, bananas, and corn (Starrett 1967). They are used as food in some parts of Peru but are not regarded as endangered (Grimwood 1969).

RODENTIA; ERETHIZONTIDAE; Genus SPHIGGURUS
F. Cuvier, 1825

There are six species (Cabrera 1961; Hall 1981; Handley 1976; Husson 1978; Woods 1984):

S. mexicanus, east-central Mexico to western Panama;
S. pallidus, known only by two specimens purported to be from somewhere in the West Indies;
S. insidiosus, Surinam, Brazil;
S. villosus, southeastern Brazil;

S. spinosus, southeastern Brazil, Paraguay, Uruguay, northeastern Argentina;
S. vestitus, Colombia, western Venezuela.

Sphiggurus often is treated as a subgenus of *Coendou* but was considered to be a separate genus by Honacki, Kinman, and Koeppl (1982), Husson (1978), and Woods (1984). The species *C. mexicanus* and *C. pallidus*, usually placed in *Coendou*, were transferred to *Sphiggurus* by Woods (1984).

According to Husson (1978), two specimens of *S. insidiosus* have head and body lengths of 363 and 380 mm, tail lengths of 375 and 370 mm, and weights of 1,340 and 1,150 grams. *Sphiggurus* resembles *Coendou* in general structure and coloration, but the spiny covering of the back is mixed with or covered by long, thick fur, and the covering of the chest and underparts is considerably softer than in *Coendou*. Santos, Oliver, and Rylands (1987) reported a white form of *S. insidiosus* to be common in some areas. Husson noted that the skull of *Sphiggurus* is more slender and flattened than that of *Coendou*, the rostrum is relatively long, in anterior view the nasal opening is more triangular than heart-shaped, and the diastema is relatively longer than the row of cheek teeth. He noted that female *S. insidiosus* have two or three pairs of abdominal mammae, but Alvarez del Toro (1967) reported that female *S. mexicanus* have 12 mammae.

The ecology of *Sphiggurus* is thought to be much like that of *Coendou*. It is nocturnal and arboreal, usually living high in trees. All but one of the specimens taken in Veracruz by Hall and Dalquest (1963) were found in tall trees, 18–30 meters above the ground, and one was found at a height of 6 meters. However, several specimens of *S. mexicanus* reported by Jones, Genoways, and Lawlor (1974) were dug from retreats in a rocky cave.

In the Tuxtla Gutierrez Zoo in southern Mexico, Alvarez del Toro (1967) found that male *S. mexicanus* could not be kept together, because of fighting, but that a male could live peacefully with two or three females. One pair bred three times in six years. The female never made a nest, and the young were born in branches provided in the cage. A single young is probably normal, and it is relatively large, about 250 mm long. It resembles the parents in shape and color but is hairier and has soft, short quills; the quills harden in about a week. Pregnant female *S. mexicanus* with a single embryo have been reported by Hall and Dalquest (1963) and Jones, Genoways, and Lawlor (1974), and a single young also is on record for *S. insidiosus* (Asdell 1964).

Sphiggurus vestitus, photo by E. La Marca and J. Molinari.

RODENTIA; ERETHIZONTIDAE; **Genus ERETHIZON**
F. Cuvier, 1822

North American Porcupine

The single species, *E. dorsatum*, occurs from northern Alaska to Newfoundland and south to northern Mexico and Tennessee (Hall 1981).

Head and body length is 645–80 mm, tail length is 145–300 mm, and weight is usually 3.5–7.0 kg, but large males occasionally weigh as much as 18 kg. The upper parts of the body are covered with thick, sharp, barbed quills distributed among longer, stiff guard hairs; the underfur is woolly. The quills are about 2 mm in diameter, up to 75 mm long, and may exceed 30,000 in number (Woods 1973). The general coloration of the upper parts is usually dark brown or blackish; the individual quills have a yellowish white base and a dark tip. The underparts lack the sharp quills and are covered only with stiff, dark hairs.

Erethizon has a robust body, a small head, moderately small ears, short legs, and a short, thick tail that is armed with quills above and stiff bristles below. The heavy feet have naked soles and are modified for arboreal life. There are four toes on the forefoot and five on the hind foot, all with strong, curved claws. The skull is massive and somewhat swollen

North American porcupine *(Erethizon dorsatum)*, photos by Uldis Roze.

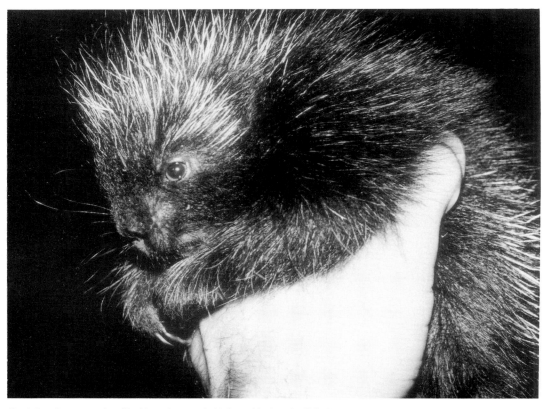

North American porcupine *(Erethizon dorsatum)*, 10 days old, photo by Uldis Roze.

between the orbits. The incisor teeth are heavy, project ante-riorly, and are deep orange or almost ochraceous in color. Females have one pectoral and one abdominal pair of mam-mae (Jackson 1961).

The preferred habitat of *Erethizon* is forest with mixed hardwood and softwood trees. The genus is highly adaptable and may even be found in open tundra, rangeland, and desert, but it usually stays in vegetated riparian areas when it is away from forest (Woods 1973). It is primarily terrestrial but fre-quently ascends trees to heights of 18 meters in order to obtain food. It climbs surely, though slowly, and has excellent balance. It does not jump in climbing or descending. In a deciduous forest or mixed hemlock-hardwood forest the por-cupine spends most of its time in trees, as that is where it finds food. In coniferous forests most of the spring and summer is spent on the ground (U. Roze, Queens College, pers. comm., 1989). *Erethizon* swims well, the hollow quills furnishing considerable buoyancy. It dens in a cave, crevice, hollow log or tree, burrow, snowbank, or crude nest in a tree. More than one den is used, and temporary shelters often supplement the main habitation. Brander (1973) found that dens in northern Michigan were in trees, logs, and stumps and that entrances were usually at ground level but sometimes over 6 meters high.

This porcupine is thought to have relatively poor vision but good senses of hearing and smell; it sniffs the air almost continuously. It is mainly nocturnal but sometimes forages by day. It does not hibernate and is active throughout the winter but may remain in its den when the weather is very cold or stormy. Seasonal migrations occur in some areas, depending on the availability of food. One individual recap-tured six times was found to have moved an average of 1,500

meters from its winter den to the point of capture in the summer (Woods 1973). Brander (1973) observed an autumn dispersal in northern Michigan, during which individuals moved an average of 480 meters. In contrast, the mean daily distance between the den and feeding site was found to be only 8 meters in winter and 150 meters in summer. Other reports of such daily movements have ranged up to an aver-age of 130 meters in winter and 1,200 meters for one animal in May. *Erethizon* uses regular runways through the vegeta-tion or snow and may wear a distinct path from the den to the feeding area.

The winter diet consists mainly of evergreen needles and the cambium layer and inner bark of trees. During the spring and summer the porcupine eats buds, tender twigs, roots, stems, leaves, flowers, seeds, berries, nuts, and other vegeta-tion. It is fond of salt and avidly gnaws bones and antlers found on the ground because of their high mineral content (Banfield 1974). Food may be held in the hands. The digestive tract is about 850 cm long, 46 percent of which is small intestine containing bacteria that can decompose cellulose (Woods 1973). Sometimes, 450 grams of food is consumed in a day, and about one-fifth of this weight is excreted in the form of brown, crescent-shaped droppings.

There may be population cycles with peaks 12–20 years apart. Reported population densities have varied from about 1 to 37 individuals per sq km but usually have been about 5–10 per sq km (G. W. Smith 1977; Woods 1973). A winter feed-ing range was calculated to cover 5.4 ha., and summer ranges of 13.0 and 14.6 ha. were determined during a 30-day radi-otracking study. *Erethizon* evidently does not defend a terri-tory and shows little aggression but may defend a specific feeding tree (Woods 1973). This genus is generally solitary, with only one adult per den, but pairs may share a single feeding tree, several animals sometimes share a den on a rotating basis, and during the winter a number of individuals may occupy a den simultaneously. Brander (1973) found that during the summer and early autumn about a dozen porcupines might gather at certain nocturnal feeding sites and spend much time vocalizing. A great variety of sounds has been reported for *Erethizon*, including moans, grunts, coughs, wails, whines, and tooth chatters.

This slow-moving, seemingly clumsy creature is said never to attack. It may protect itself by climbing or fleeing, but if cornered, it erects the quills, turns the rump toward the source of danger, and rapidly lashes out with the barbed tail. The quills are not thrown or shot, but they are so lightly attached that when they enter the skin of an enemy, they become detached from the porcupine. The quills then continue to work their way deeper at a rate of up to 1 mm or more per hour and can cause death if they puncture vital organs. Despite this defense mechanism, the porcupine is preyed upon by the great horned owl and many carnivorous mammals. The bobcat, wolverine, and fisher are especially adept at flipping the porcupine on its back to expose the unprotected underparts (Banfield 1974). The fisher *(Martes pennanti)* has even been reintroduced in some areas in an effort to control porcupine numbers. Powell and Brander (1977) noted a 76 percent decline in the porcupine population of a study area in northern Michigan during a 13-year period subsequent to the establishment of the fisher.

In the fall or early winter, males seek out females for mating. There is an elaborate courtship, involving extensive vocalization and a comical kind of dance (Banfield 1974). The male usually showers the female with urine before mating, and afterward the female repels the male. Females may be polyestrous and will recycle in 25–30 days if fertilization does not occur at the time of first ovulation; heat lasts 8–12 hours (Woods 1973). The gestation period is 205–17 days, and the young are born from April to June. Litters generally contain a single offspring, though there are rare records of twins. The newborn is well developed; it weighs 340–640 grams, its eyes are open, it can walk about unsteadily, and it is covered with long, black hair and short, soft quills. It exhibits the typical defense reaction of turning the rump toward danger, and within days it can climb trees. Although nursing has lasted several months in the laboratory, weaning apparently occurs

much sooner in the wild, as the young can survive on a diet of vegetation within 2 weeks of birth. It gains weight at a rate of about 450 grams per day. Some males are capable of mating at 16 months, though Banfield (1974) stated that sexual maturity in *Erethizon* is reached at approximately 2.5 years. The natural life span is evidently long. Wild individuals are known to have lived up to 18 years (Earle and Kramm 1980).

The porcupine sometimes causes the death of timber and ornamental trees by girdling the bark of the trunk or stripping off all the bark above the snow line in winter. There have been extensive debates and analyses regarding the seriousness of such activity (Woods 1973). While Banfield (1974) indicated that the porcupine is one of the most important mammalian forestry pests, Jackson (1961) suggested that damage has been greatly exaggerated. There is general agreement that individual porcupines sometimes become a nuisance by gnawing woodwork, furniture, tools, cooking utensils, saddles, canoe paddles, and other objects that have received salt deposits through contact with human perspiration. The porcupine is considered tasty food by some persons, and its quills have been used in decorative work by Indian tribes. Perhaps in part through human agency, *Erethizon* has vanished, except for occasional wandering animals, from that section of the eastern United States to the south of Minnesota, northern Wisconsin, Michigan, and Pennsylvania (Handley 1980*b*; Jackson 1961). The genus may, however, already have been rare in the southern Appalachians during colonial times, and its disappearance there could be associated with climatic changes.

RODENTIA; Family CAVIIDAE

Cavies and Patagonian "Hares"

This family of 5 Recent genera and 17 species occurs over most of South America. The sequence of genera presented here follows that of Cabrera (1961), who recognized two subfamilies: Caviinae, with *Microcavia*, *Galea*, *Cavia*, and *Kerodon*; and Dolichotinae, with *Dolichotis*. Based on serological, behavioral, and reproductive data, Woods (1984) placed *Kerodon* in the Dolichotinae.

Wild cavies *(Cavia aperea)*, photo by J. P. Rood.

Head and body length is 200–750 mm and the tail is vestigial. The pelage in wild forms is fairly coarse or crisp. In the subfamily Caviinae (cavies) the body form is robust, the head is large, and the ears and limbs are short. In the Dolichotinae (Patagonian "hares") the proportions are rabbitlike, the ears are long, and the limbs are long and thin. In both subfamilies there are only four digits on the forefoot and three on the hind foot. The nails are short and sharp in *Cavia* and *Microcavia*, blunt but well developed in *Kerodon*, and hooflike on the hind foot and clawlike on the forefoot in *Dolichotis*. The soles of the feet are naked in the Caviinae but mostly haired in the Dolichotinae.

The dental formula is: (i 1/1, c 0/0, pm 1/1, m 3/3) × 2 = 20. The incisors are short. The toothrows tend to converge anteriorly, and the cheek teeth are rootless (ever-growing), with a rather simple pattern of two prisms having sharp folds and angular projections.

Caviids occur in habitats ranging from marshy, tropical floodplains to dry, rocky meadows at elevations of up to 4,000 meters. They live in a wide variety of habitats—open grasslands, dry steppes, forest edges, swamps, and rocky mountainous areas (Woods 1984). None of these rodents hibernates, even when living at high altitudes and/or when temperatures are very low. The diet consists of many kinds of plant material. Breeding may continue throughout the year if climatic conditions are favorable. The gestation period, ranging from about 50 to 70 days, is relatively short for the suborder Hystricomorpha. Nonetheless, the young are well developed at birth and reach sexual maturity rather early. The potential life span is relatively long.

The geological range of the Caviidae is middle Miocene to Recent in South America.

RODENTIA; CAVIIDAE; Genus MICROCAVIA
Gervais and Ameghino, 1880

Mountain Cavies

There are two subgenera and three species (Cabrera 1961; Mares and Ojeda 1982; Pine, Miller, and Schamberger 1979):

subgenus *Microcavia* Gervais and Ameghino, 1880

M. australis, Argentina, southern Chile, possibly southern Bolivia;
M. shiptoni, mountains of northwestern Argentina;

subgenus *Monticavia* Thomas, 1916

M. niata, mountains of Bolivia, possibly southeastern Peru and northern Chile.

The information for the remainder of this account was taken entirely from Rood's (1970*b*, 1972) papers, which dealt primarily with studies of *M. australis*. Superficially, this species resembles a small, tailless ground squirrel. Head and body length is about 200–220 mm, there is no external tail, and weight is about 200–500 grams. The upper parts are olive gray agouti, and the underparts are pale gray. There is a prominent white ring around the eye.

Microcavia closely resembles *Galea*, but the latter genus has a less prominent eye ring, darker pelage, yellow incisors (white in *Microcavia*), a smaller eye, and a prominent sub-mandibular gland (absent in *Microcavia*). Females of both genera have four mammae.

The species *M. shiptoni* and *M. niata* live in high mountains, but *M. australis* occurs throughout Argentina except in the humid northeastern provinces. It is the most abundant member of the Caviinae in the semiarid thornbush of central Argentina. In such habitat it uses clumps of thornbush for cover and makes runways through the open areas between clumps of bushes. Under the bushes it digs a shallow depression, in which it rests and sleeps, and it keeps the surrounding area clear. It also may dig a burrow; the main tunnel of one was 134 cm long, 5 cm wide, and 28 cm below the surface at the deepest point. Most activity is diurnal. *Microcavia* sometimes climbs bushes and low trees to feed. The diet consists mostly of leaves and also includes fruits. Animals were never seen to drink and presumably got sufficient water from succulent vegetation.

Population size in a given area varies substantially, depending on reproductive activity and movements. Average density during one year of study was calculated at 24/ha. Mean home range was 7,720 sq meters for males and 3,525 sq meters for females. The ranges of males overlap those of both other males and females. A male regularly wanders about visiting several females during a day. He sometimes forms a temporary association with a female and her young and will follow her about closely, nose to rump, especially if she is near estrus. Encounters between males, however, almost always involve aggression. The males in an area establish a linear dominance hierarchy, and subordinates are chased. Sometimes there are fights to the death. Females tend to stay within a small area of bushes and do not often meet members of their own sex. If they do, the reaction can be friendly or hostile. Two females, especially a mother and daughter, may kiss upon meeting and then sit or forage together. Young animals up to about 1 month old are tolerated by adults of both sexes. *Microcavia* apparently has an amicable relationship with the sympatric *Galea*, and cavies of both genera may be seen together in groups. The vocalizations of *Microcavia* include an alarm "tsit," a "twitter" of annoyance, and a "shriek" of fear.

The breeding season in east-central Argentina lasts from August to April, with most births occurring in the spring (September–December). Females are polyestrous, have a postpartum estrus immediately after giving birth, and may be able to mate again within 15 days if fertilization does not occur at the postpartum estrus. When a pregnant female is near term, males gather about it and display much aggression toward one another. As soon as the offspring are delivered, as many as six males pursue the female. A female can theoretically produce five litters during a season, and one individual wild female is known to have given birth to four. The gestation period averages about 54 days. Litters of one to three young were observed in the wild, but captive females sometimes had litters of four or five. The young weigh around 30 grams at birth, are able to run about and nibble solid food during their first day of life, and are ordinarily weaned at 3 weeks. A young animal sometimes nurses from a female other than its own mother. Young females usually come into estrus at 40–50 days but do not necessarily conceive at that time. One female is known to have become pregnant at 82 days. Life expectancy in the wild is apparently short, mainly because of predation by the grison *(Galictis cuja)*. Of the 44 cavies marked on Rood's study area in April 1966, only 15 were left in February 1967.

Microcavia is generally considered a pest by the people in the area in which it lives. It supposedly destroys crops, and its holes are a hazard to horses. It also is sometimes hunted by people for use as food. Nonetheless, it seems to adapt well to civilization and continues to exist in large numbers.

Mountain cavy *(Microcavia australis)*, photo by J. P. Rood.

Cui *(Galea musteloides)*, photo by T. D. Dennett. Inset: chin gland and view of the yellow-colored incisors, photo by Barbara J. Weir.

RODENTIA; CAVIIDAE; **Genus GALEA**
Meyen, 1833

Yellow-toothed Cavies, or Cuis

There are three species (Cabrera 1961; Wetzel and Lovett 1974):

G. flavidens, Brazil;
G. musteloides, southern Peru, Bolivia, Argentina;
G. spixii, Brazil, eastern Bolivia, Paraguay.

The following descriptive data were provided mostly by Barbara J. Weir (Wellcome Institute, Zoological Society of London, pers. comm., c. 1970). Head and body length is 150–250 mm, there is no external tail, and weight is 300–600 grams.

The upper parts are agouti-colored, and the underparts are grayish white. The pelage is paler than that of *Cavia* and does not molt so readily, and it is not as long or coarse as that of *Microcavia*. The body form is more compact and stocky than that of *Cavia*. The incisor teeth are yellowish or light orange, rather than white as in *Cavia* and *Microcavia*. There is a prominent submandibular gland in *Galea* but not in the other two genera. The hind foot has three toes, the forefoot has four, and all digits have strong, sharp claws. Females have one pair of inguinal and one pair of lateral thoracic mammae.

Yellow-toothed cavies inhabit grasslands at high or low elevations, as well as rocky and brushy areas. They sometimes excavate their own dens, and extensive colonies reportedly honeycomb the ground with their burrows in the Bolivian highlands (Starrett 1967). They also use the abandoned holes of larger mammals, such as armadillos (*Chaetophractus*), viscachas (*Lagostomus*), and tuco-tucos (*Ctenomys*), as well as crevices in stone walls (Rood 1972). They are primarily terrestrial and diurnal. The diet consists of grasses, forbs, and other kinds of vegetation.

Rood (1972) studied *G. musteloides* both in captivity and in the wild. One female had a home range of 4,275 sq meters. Animals lived in groups, but only a single adult male could be kept in an enclosure. Free-living males established a linear dominance hierarchy that was maintained by aggression. The female hierarchy was less stable than that of males. Generally, however, females were dominant to males, and older animals were dominant to younger ones. Subordinates usually retreated, but aggressive displays, fighting, and serious wounds were common. Immature individuals were generally tolerated until they were about one month old, when adults would begin to chase them. Vocalizations included a "churr," indicating sexual arousal; a tooth chatter, signifying a threat; and a "rusty gate" screech during an attack.

Observations by Lacher (1981), Rood (1972), and Weir (1974b) indicate that breeding continues throughout the year but that most litters are born during spring and summer in the wild. Females are polyestrous and can produce 7 litters per year; one had 12 in 21 months. The estrous cycle averages about 22 days, receptivity lasts 2–3 hours, and there is a postpartum estrus. Females, however, are apparently induced, rather than spontaneous, ovulators, and the presence of a male is needed to stimulate estrus. The submandibular gland, which is more developed in the male, and a behavioral pattern in which the male closely follows the female, chin to rump, seem to be involved in inducing estrus. As soon as females give birth, they are pursued by males. The average gestation period is 53 days, and the range is 49–60 days. The mean number of young per litter is about two in *G. spixii* and three in *G. musteloides*, and the range is one to seven. The young weigh about 40 grams each at birth, are well developed, and can move about. They normally nurse for 3 weeks. Sexual maturity is generally attained at 3 months in males and 2 months in females, but some females reach puberty much earlier; one conceived when only 9 days old, a record for mammals. *G. musteloides* is known to have lived over 15 months in the wild and over 22 months in captivity. According to Jones (1982), a captive *G. spixii* survived for 4 years and 7 months.

RODENTIA; CAVIIDAE; **Genus CAVIA**
Pallas, 1766

Cavies, or Guinea Pigs

There are eight species (Cabrera 1961; Gade 1967; Handley 1976; Husson 1978; Pine, Miller, and Schamberger 1979; Rood 1972; Schlieman 1982; Weir 1974c; Ximenez 1980):

C. aperea, central Ecuador, southern Surinam, eastern and southern Brazil, Paraguay, Uruguay, northern Argentina;

C. tschudii, Peru, southern Bolivia, northern Chile, northwestern Argentina;

C. porcellus, about 500 years ago found in domestication from northwestern Venezuela to central Chile;

C. guianae, southern Venezuela, Guyana, and probably adjacent parts of northern Brazil;

C. anolaimae, vicinity of Bogota in Colombia;

C. nana, western Bolivia;

C. fulgida, southeastern Brazil;

C. magna, extreme southern Brazil, Uruguay.

Cabrera (1961) listed the wild species *C. anolaimae* and *C. guianae* as subspecies of the domestic guinea pig, *C. porcellus*, because the original describers of these wild species had suggested strong affinity to the domestic species. Weir (1974c), however, stated that there is no wild form of *C. porcellus* and that this species probably originated within *C. aperea*, *C. fulgida*, or *C. tschudii*. It therefore becomes necessary to treat *C. anolaimae* and *C. guianae* as separate species again.

The following descriptive information was provided in part by B. J. Weir (Wellcome Institute, Zoological Society of London, pers. comm.). Head and body length is 200–400 mm, there is no external tail, and weight is 500–1,500 grams. Wild species have fairly coarse and long pelage, a crest of hairs at the neck, and agouti-banded hairs that produce an overall grayish or brownish color. In the domestic guinea pig the pelage may be smooth and short, smooth and long, or coarse and short, and in some breeds rosettes are formed; there is a wide variety of colors and patterns. In both wild and domestic forms the body hairs are easily shed.

Cavies have a stocky build, fairly short legs, and short, unfurred ears. There are three digits on the hind foot and four on the forefoot, all armed with sharp claws. The incisor teeth are white. Females have a single pair of inguinal mammae (Weir 1974b).

Cavies occur in a wide variety of habitats, including open grasslands, forest edge, swamps, and rocky areas. They are sometimes found at elevations of up to 4,200 meters (Grzimek 1975). They have been reported either to dig their own burrows or to take over the abandoned holes of other mammals. In a study in eastern Argentina, however, Rood (1972) found *C. aperea* not to burrow but to shelter in brush piles and clumps of vegetation. Feeding activity was mainly crepuscular in this area but occasionally extended throughout the day during cold weather. *Cavia* is terrestrial and seems to follow well-defined tracks from den to feeding area but has also been reported to swim several kilometers during floods (Rood 1972). The diet consists of many kinds of vegetation.

Cavies generally associate in small groups usually comprising 5–10 individuals, but in favorable habitat a number of such groups may converge and give the impression of a large colony (B. J. Weir, Wellcome Institute, Zoological Society of London, pers. comm.). In his study area, Rood (1972) found *C. aperea* to have a maximum population density of about 38/ha. Home ranges there averaged 1,387 sq meters for males and 1,173 sq meters for females. The ranges were stable and centered around a clump of vegetation that was used for food and shelter. Neither sex maintained an exclusive territory, and ranges frequently coincided. Each sex had its own linear dominance hierarchy, and subordinate animals either retreated or were attacked. If an alpha male eventually lost a fight, it became passive and withdrawn and sometimes died. Vocalizations included bubbly squeaks of excitement, "chirps" of anxiety, and a tooth-chatter threat.

Data compiled by Rood (1972) show that in captivity both

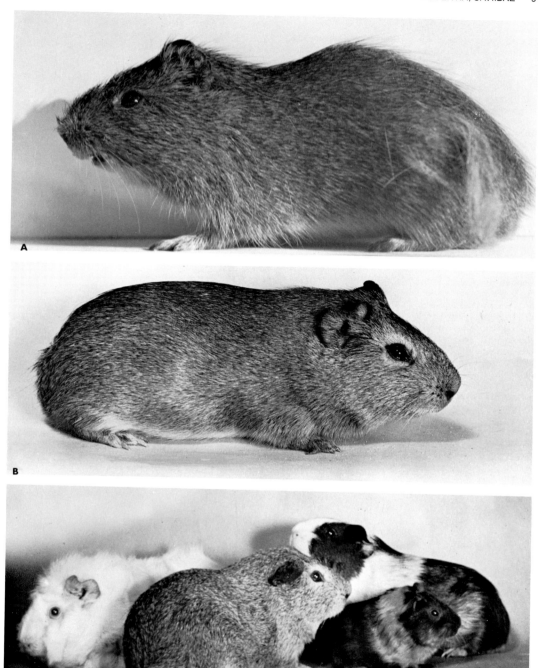

A. Wild cavy *(Cavia tschudii),* photo by Hilda H. Heller. B & C. Guinea pigs, domestic strain *(C. porcellus),* photos by Ernest P. Walker.

C. aperea and *C. porcellus* breed throughout the year, with birth peaks in the spring. Females are polyestrous and produce a maximum of five litters annually. Wild *C. aperea* may also reproduce throughout the year under favorable conditions but did not breed during the severe winter in Rood's study area. The estrous cycle averages about 16.5 days in *C. porcellus* (Asdell 1964) and 20.5 days in *C. aperea* (Weir 1974*b*). Females of both the wild and domestic species experience a postpartum estrus immediately after giving birth, and receptivity lasts less than half a day. When a pregnant female is about to give birth, males gather about her. The most dominant male aggressively guards the female from the others, and he usually mates first. If he subsequently can no longer protect the female, several subordinates pursue and mate with her (Rood 1972).

The overall gestation range in the genus is 56–74 days, with averages of about 62 days in *C. aperea*, 63 in *C. tschudii*, and 68 in *C. porcellus*. The mean (and extreme) numbers of young per litter are 4.0 (1–13) in *C. porcellus*, 2.3 (1–5) in *C. aperea*, 1.35 in *C. fulgida*, and 1.9 (1–4) in *C. tschudii*. A newborn *C. aperea* weighs about 60 grams, and one of *C. porcellus* weighs about 100 grams (Weir 1974*b*). The young are well developed at birth and can run about and eat solid food on their first day of life (Rood 1972). They can survive without further nursing after 5 days but normally suckle for 3 weeks. Sexual maturity is usually reached after 3 months in males and 2 months in females, but the minimum recorded age of conception in a female is only 21 days (Weir 1974*b*). Longevity may be as much as 8 years, at least in captivity.

The domestic guinea pig *(C. porcellus)* apparently has been bred for meat production in South America for at least 3,000 years (Gade 1967; Weir 1974*c*). Its precise origin is unknown, but domestication probably occurred just once. Crossbreeding and biochemical studies indicate that *C. porcellus* probably was derived from *C. aperea*, *C. tschudii*, or *C. fulgida* but that it is distinct from each of these other species. During the period of the Inca Empire, from about 1200 to 1532, highly selective breeding produced a variety of strains, differentiated by color pattern and flavor. At that time the range of domestication extended from northwestern Venezuela to central Chile. Subsequent to the Spanish conquest there was a disruption of breeding and a contraction of range. The guinea pig is still widely kept as a source of food by the native people of Ecuador, Peru, and Bolivia. In some areas the animals are restricted to huts and are treated well. In other areas they are allowed to run free and scavenge, and under such conditions some feral populations have become established. The guinea pig was brought to Europe in the sixteenth century, and since the mid-1800s it has been used by laboratories around the world for research on pathology, nutrition, genetics, toxicology, and development of serums. It also makes an ideal pet.

RODENTIA; CAVIIDAE; Genus KERODON
F. Cuvier, 1825

Rock Cavy, or Moco

The single species, *K. rupestris*, is found in eastern Brazil from the state of Piaui to the northern part of Minas Gerais (Cabrera 1961).

Kerodon is about the same size as *Cavia* or somewhat larger. The tail is either absent or only a vestigial projection. Adult weight is 900–1,000 grams. The upper parts are generally grayish with white and black mottling, the throat is whitish, and the underparts are yellowish brown. Unlike other genera in the subfamily Caviinae, the well-developed nails of the digits are blunt rather than sharp.

The moco inhabits arid and pebbly areas near stony mountains or hills. It seeks shelter under rocks or in the fissures between stones, sometimes making a burrow under the stones. It leaves its shelter late in the afternoon or evening and runs on the ground or climbs trees looking for food, mainly tender leaves. It descends by leaps or by a single jump at the slightest sign of danger. The feces are small, green, and capsule-shaped.

Lacher (1981) found *Kerodon* to be an extreme habitat specialist and to exhibit resource defense polygyny. Individuals maintain fidelity to specific rock piles; males actively defend these sites and may accumulate a harem of females indirectly as the latter seek shelter there. Both sexes have a linear dominance hierarchy, but that of the females appears much more pronounced. There are several vocalizations, including a slow whistle emitted when animals leave the rock

Moco *(Kerodon rupestris)*, photo by Luiz Claudio Marigo.

piles to forage and an alarm whistle that is recognized by all the animals in a given area.

Lacher (1981) reported males to follow estrous females chin to rump, as in *Galea*, but also to circle the females to control their movements. Births apparently occur throughout the year except from April to June. Females exhibit a postpartum estrus and mate within a few hours of giving birth; they may produce several litters annually, each with one or two young. The gestation period averages about 75 days. The precocial young weigh about 76 grams each at birth and are weaned after about 35 days. According to Roberts, Maliniak, and Deal (1984), the young reach adult size at 200 days, and the mean age of females at first conception is 151 days. One individual reportedly lived in captivity for 11 years.

Kerodon is easily domesticated and is suitable as a pet. Persons native to its range are very fond of its flesh and prepare it as food or medicine. Hunters attempt to attract it by imitating its shrill whistle. Lacher (1979) suggested that *Kerodon*, reared in captivity, could be of potential value as a food source but also warned that the genus was threatened in the wild through extensive destruction of its limited habitat.

RODENTIA; CAVIIDAE; **Genus DOLICHOTIS**
Desmarest, 1819

Patagonian Cavies, Patagonian "Hares," or Maras

There are two species (Cabrera 1961):

D. salinicola, southern Bolivia, Paraguay, northern Argentina;

D. patagonum, central and southern Argentina.

Cabrera placed *D. salinicola* in the separate genus *Pediolagus* Marelli, 1927. This procedure was not followed by Wetzel and Lovett (1974) or Woods (1984).

Head and body length is 690–750 mm in *D. patagonum* but only about 450 mm in *D. salinicola*. In both species the maximum length of the tail is about 45 mm. Large individuals of *D. patagonum* weigh 9–16 kg. The general coloration of the upper parts is grayish, and that of the underparts is whitish. In *D. salinicola* there is a horizontal band of white or yellowish fur running from the tail around the flank to the belly. The pelage is dense. Although the hairs are fine, they have a crisp texture and stand at nearly a right angle to the skin.

Dolichotis is modified for a cursorial life. The body form is similar to that of long-legged rabbits and hares. The hind limbs are long, the hind foot has three digits, and each digit has a hooflike claw. The forefoot has four digits, each bearing a sharp claw. Female *D. salinicola* have two pairs of mammae, and female *D. patagonum* have four pairs (Weir 1974b).

Maras inhabit arid areas with coarse grass or scattered shrubs. They shelter in burrows of their own construction or in the abandoned holes of other animals, such as *Lagostomus*. They are terrestrial and diurnal. In a study of an introduced colony of *D. patagonum* in Brittany, Dubost and Genest (1974) found that individuals rested at night in wooded areas and then traveled to open grassland to feed during the day. *Dolichotis* uses a variety of locomotions: walking when undisturbed, hopping like a rabbit or hare, galloping, and stotting—a sort of bounce on all four limbs at once—for

Patagonian cavy *(Dolichotis patagonum)*, photo from East Berlin Zoo.

Dwarf Patagonian cavy (*Dolichotis salanicola*), photo by Eugene Maliniak.

covering long distances at high speed. Rood (1972) clocked a mara running beside his car at 45 km/hr over a distance exceeding 1 km. When not moving about, *Dolichotis* stands with its legs straight, sits on its haunches with the forepart of the body resting on fully extended front legs, or reclines like a cat, with the front limbs turned under the chest—an unusual position for a rodent. It spends considerable time basking in the sun but is ever on the alert for danger. The diet consists of any available vegetation.

Three or four wild individuals are often seen traveling together in single file, but occasionally a group of up to 40 may be observed. Studies of an introduced colony of about 70 *D. patagonum* in Brittany (Dubost and Genest 1974; Genest and Dubost 1974) showed that the group dispersed by night and concentrated in feeding areas by day. There was no territoriality, and the entire colony shared available space, but males had a dominance hierarchy maintained by aggression. The basic social unit was the mated pair, and strict monogamy was practiced throughout the year. The female initiated most movements, and the male always remained close and defended her against other males. Eisenberg (1974) listed a number of vocalizations, including a "wheet" for seeking contact and a grunt used as a threat.

According to Dubost and Genest (1974), the introduced colony of *D. patagonum* in Brittany breeds throughout the year but most intensively during the European winter, a period corresponding with summer in the natural range of the species. Females are polyestrous, experience a postpartum estrus, and produce three or four litters annually. The interval between litters is about 3 months during the European winter and 3.5 months during summer. Litters usually contain two young but sometimes have one or three. The young

are born outside of the burrow but move inside soon afterward. They are well developed, their eyes are open, and they can move about freely. Weaning occurs by the age of 11 weeks, and 8-month-old females may give birth to their own young. Asdell (1964) stated that a pregnant female *D. salinicola* with two embryos was taken in Argentina in August. Weir (1974*b*) listed the gestation period of *D. salinicola* as approximately 77 days. Grzimek (1975) reported that a captive *D. patagonum* lived for almost 14 years but that most individuals do not survive beyond 10 years.

Rood (1972) noted that *D. patagonum* appears to be declining in numbers in the wild and now is rare in the province of Buenos Aires, Argentina, where it was formerly abundant. The decline is attributable to habitat destruction by people and competition with the introduced and more adaptable European hare *(Lepus capensis)*.

RODENTIA; **Family HYDROCHAERIDAE; Genus HYDROCHAERIS**
Brünnich, 1771

Capybara

The single living genus and species, *Hydrochaeris hydrochaeris*, occurs in Panama, to the east of the Canal Zone, and on the east side of the Andes in South America, from Colombia and the Guianas to Uruguay and northeastern Argentina (Cabrera 1961; Hall 1981). The use of the above scientific names, rather than Hydrochoeridae, *Hydrochoeris*, and *hydrochoeris*, is in keeping with the explanation of Husson

(1978). Some authorities, including Mones and Ojasti (1986), regard *H. isthmius*, of Panama, western Colombia, and northwestern Venezuela, as a separate species.

This is the largest living rodent. Head and body length is 100–130 cm, the tail is vestigial, shoulder height is up to 50 cm, and weight is 27–79 kg. Zara (1973) reported the average weight of captive adults to be 50 kg in males and 61 kg in females. The long, coarse pelage is so sparse that the skin is visible. The coloration is generally reddish brown to grayish on the upper parts and yellowish brown on the underparts. Occasionally there is some black on the face, the outer surface of the limbs, and the rump. In the mature male a bare, raised area on the top of the snout contains greatly enlarged sebaceous glands.

Hydrochaeris resembles *Cavia* but is much larger and has a proportionally shorter body. The limbs are short, the head is relatively large and broad, the ears are short and rounded, and the small eyes are placed dorsally and relatively far back on the head. The muzzle is heavy and truncate, and the upper lip is enlarged. The forefoot has four digits and the hind foot, three. The digits are arranged in a radial pattern, and all are partially webbed and armed with short, strong claws. Females have five pairs of ventral mammae (Weir 1974b).

The dental formula is: (i 1/1, c 0/0, pm 1/1, m 3/3) × 2 = 20. The toothrows tend to converge anteriorly. The incisor teeth are white and shallowly grooved. The cheek teeth are rootless (ever-growing), have an unusually large amount of cement, and have a surface pattern that is similar to, but more complex than, that of the Caviidae.

The capybara inhabits densely vegetated areas around ponds, lakes, rivers, streams, marshes, and swamps. It does not use a den but shelters in thickets (Schaller 1976). In areas where it is not disturbed it is active in the morning and evening, resting during the heat of the day in a shallow bed in the ground. Like some other mammals, however, it apparently has become nocturnal in areas where molested by people. When alarmed on land, it runs like a horse, and when closely

pursued, it enters the water, where it swims and dives with ease. While it is swimming, only the nostrils, eyes, and ears project above the water. The capybara also swims completely underwater, often for a considerable distance, or it may hide in floating vegetation, exposing only its nostrils. Although it thus seems to be an aquatic mammal, water is used primarily as a place of refuge, and most normal activity is on land (Schaller 1976). The diet consists mainly of grasses, and the capybara is sometimes seen grazing with cattle. It also eats aquatic plants, grains, melons, and squashes. Characteristic mounds of elongate fecal pellets are deposited.

Schaller and Cranshaw (1981) reported a population density of about 12.5/sq km in one part of Venezuela. *Hydrochaeris* is commonly found in groups of about 20. Schaller (1976) once saw 64 together, and old reports told of colonies of 100 or more, but such large congregations are apparently not stable. Mones and Ojasti (1986) stated that the capybara lives in herds ranging from a pair or family to complex groups of several adults of both sexes and their offspring; a typical herd is controlled by a dominant male. Observations in captivity (Donaldson, Wirtz, and Hite 1975) indicate that the basic social unit is a family group and that outsiders are not readily accepted. There is a stable and viciously enforced social hierarchy, with much aggression and fighting, and it is inadvisable to place two adults of the same sex in a small area. Reported vocalizations include low, clicking sounds of contentment; sharp, prolonged whistles; abrupt grunts; and weak barks.

Breeding occurs throughout the year in Venezuela, but mating reaches a peak in April and May, shortly before the rains begin. In Argentina, mating was observed in October, also just before the rainy season (Schaller 1976). There usually is one litter per year, though some females may have two if conditions remain favorable (Mones and Ojasti 1986). Two definite gestation periods, lasting 149 and 156 days, were recorded by Zara (1973). Weir (1974b) listed a single gestation of 104–11 days in the small northern subspecies *H. h. isth-*

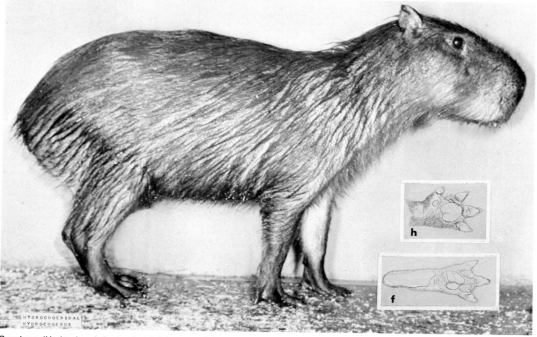

Capybara *(Hydrochaeris hydrochaeris)*, just out of the water, photo by Ernest P. Walker. Insets: bottoms of hand and foot, photos from *Proc. Zool. Soc. London.*

Capybara *(Hydrochaeris hydrochaeris)*, photo from Riverbanks Zoo, Colombia, South Carolina.

mius. She also reported litter size in the genus to average five (one to eight) young, weight of each newborn to be about 1,500 grams, and normal lactation to last 16 weeks. The young are precocial and able to follow their mother and eat grass shortly after being born (Schaller 1976). Both sexes reach puberty at 15 months (Zara 1973). Life expectancy in the wild is apparently 8–10 years, but some captives have survived over 12 years (Grzimek 1975).

The capybara is sometimes killed by persons who consider it to be an agricultural pest, and it is also intensively hunted for its meat and hide. Its thick, fatty skin provides a grease used in the pharmaceutical trade, and its incisors are used as ornaments by native people. In Venezuela, capybara populations have declined in the face of heavy market hunting, but efforts have been made to regulate the kill (Ojasti and Padilla 1972). In Peru this genus has disappeared from many waterways where it was formerly common (Grimwood 1969). Nonetheless, the capybara remains widespread and common in much of South America and has even been raised commercially on some ranches (Schaller 1976). Such ranching is thought to have the advantage of maintaining natural wetlands, rather than draining them as might be required in the case of intensive cattle raising (Vasey 1979).

The geological range of the Hydrochaeridae is early Pliocene to Recent in South America, Recent in Central America, and late Pliocene to late Pleistocene in the southeastern corner of North America (Mones 1973, 1975; Woods 1984).

RODENTIA; **Family DINOMYIDAE; Genus DINOMYS**
Peters, 1873

Pacarana

The single living genus and species, *Dinomys branickii,* occurs in the highlands from northwestern Venezuela and Colombia to western Bolivia (Boher B., Naveua S., and Escobar M. 1988; Cabrera 1961).

Head and body length is 730–90 mm, tail length is about 200 mm, and weight is 10–15 kg. The upper parts are black or brown, with two more or less continuous broad stripes on each side of the midline of the back and two shorter rows of white spots on the sides. In older individuals the stripes seem to be broader and more conspicuously white. The underparts are paler than the upper parts and are not marked. The pelage is rather coarse, scant, and of varied length. The tail is fully haired. The whiskers are about the same length as the head.

"Pacarana," a Tupí Indian term meaning "false paca," signifies the resemblance, in size and color pattern, between *Dinomys* and the pacas *(Agouti). Dinomys,* however, has a more thickset body than *Agouti* and also has a stout tail approximately one-fourth the length of the head and body. The overall appearance of *Dinomys* is reminiscent of an immense guinea pig *(Cavia).* Its head is broad, its ears are short and rounded, and its limbs are short. Each of the broad feet bears four digits, all armed with a long, powerful claw. Females have two lateral thoracic and two lateral abdominal pairs of mammae (Weir 1974*b*).

The dental formula is: (i 1/1, c 0/0, pm 1/1, m 3/3) × 2 = 20. The incisors are broad and heavy. The cheek teeth are probably rootless (ever-growing) and are extremely high-crowned, each tooth consisting of a series of transverse plates.

The pacarana occurs in valleys and on lower mountain slopes. Grimwood (1969) described it as an animal of the high selva (rainforest) zone and of the upper, or better-drained, parts of the low selva zone. He gave the elevational range as 240–2,000 meters. Wild individuals are said to shelter in natural crevices, which they enlarge by digging with their strong claws, but captives have not been seen to dig. Captives have used their claws to climb trees ably and have preferred to sleep in elevated places. Captives are active mainly after dark, and *Dinomys* is thought to be naturally nocturnal (Collins and Eisenberg 1972). When on the ground, the pacarana normally progresses by a waddling gait. When threatened by an agile predator, such as an ocelot or coati, *Dinomys* backs against a cliff or into a hole to protect its vulnerable hind parts. When eating, *Dinomys* sits upon its haunches and, in a dexterous manner, carefully examines its food before consuming it. The diet seems to consist mainly of fruits, leaves, and tender stems.

Although its generic name means "terrible mouse," the pacarana fights only as a last resort. Captives are generally slow-moving, good-natured, and peaceful and do not attempt to bite or escape. They allow themselves to be scratched, occasionally indicating displeasure by a low, guttural growl. Collins and Eisenberg (1971) reported the most common vocalizations, heard in encounters between a male and female pacarana, to be a hiss, a growl, and a staccato whimper.

Pregnant females with two embryos each were taken in the wild in February and May. Another female, which was pregnant when captured alive, gave birth to two young on 8 January 1970. Still another female produced a litter of two young, both dead, on 1 February 1971, following a gestation period of 223–83 days (Collins and Eisenberg 1972). According to Grzimek (1975), of those litters born in captivity in 1974, half contained a single offspring and half contained two.

Pacarana *(Dinomys branickii)*, photo from New York Zoological Society.

Weir (1974*b*) listed newborn weight as 900 grams. Although the pacarana tames readily and makes an interesting pet, most captives do not survive for very long; one, however, lived for 9 years and 5 months (Jones 1982).

Dinomys was not discovered by science until 1873, has always seemed rare, and at several times was feared to be extinct (Allen 1942; Grzimek 1975). Although the number of captive specimens has increased since the 1940s, wild populations are thought to be declining through habitat destruction and excessive hunting by people for use as food (Grimwood 1969).

On the basis of paleontological, anatomical, parasitological, and behavioral evidence, Grand and Eisenberg (1982) suggested a close relationship between the Dinomyidae and the Erethizontidae. The geological range of the Dinomyidae is early Miocene to Recent in South America. One fossil member of the family, *Telicomys*, was the largest of all rodents, reaching the size of a small rhinoceros.

RODENTIA; Family HEPTAXODONTIDAE

This apparently extinct family is known only from skeletal remains found in cave deposits in the West Indies. Hall (1981) listed two genera—*Elasmodontomys* and *Quemisia*—that evidently still lived when humans occupied the islands and four other genera—*Clidomys* Anthony, 1920, *Spirodontomys* Anthony, 1920, *Speoxenus* Anthony, 1920, and *Amblyrhiza*—that are thought to have disappeared before the arrival of people. He also referred to an additional genus—*Alterodon* Anthony, 1920—that is related to the others but is not definitely a member of the Heptaxodontidae. MacPhee, Woods, and Morgan (1983) concluded that *Alterodon* is a heptaxodontid (not an octodontid, as it sometimes is designated) and that it may be no more than a variant of *Clidomys*. MacPhee (1984) referred *Alterodon*, *Spirodontomys*, and *Speoxenus*, each known only from Jamaica, to *Clidomys* and noted that the latter is not known to have survived into the Holocene. According to Allen (1942), it is probable that *Amblyrhiza* did live well after Pleistocene times, and therefore an account of it appears below, even though it may not have been a contemporary of people.

The skulls of this family are massive, and remains suggest

that the bodies were stout. Each of the cheek teeth has four to seven laminae, the crests of which are nearly parallel and are arranged obliquely to the long axis of the skull. All genera were probably terrestrial. The presence of their remains with human refuse and artifacts suggests that some of these animals, at least, were used as food.

RODENTIA; HEPTAXODONTIDAE; Genus ELASMODONTOMYS
Anthony, 1916

The single species, *E. obliquus*, is known only by osseous material from cave deposits at Utado, Morovis, and Ciales in Puerto Rico (Hall 1981). The genus *Heptaxodon* Anthony, 1917, with the single species *H. bidens*, evidently was based on young individuals of *E. obliquus* and now is considered a synonym of the latter (Ray 1964).

Elasmodontomys appears to have been about the size of a paca *(Agouti)* and to have had a heavy body. The skull resembles that of a nutria *(Myocastor)*; it is flat-topped and has rather large, laterally compressed bullae. The short bones of the digits indicate that *Elasmodontomys* was terrestrial rather than arboreal. Available evidence suggests that this genus was distributed throughout the forested parts of Puerto Rico and apparently survived until about the time that European explorers arrived.

RODENTIA; HEPTAXODONTIDAE; Genus QUEMISIA
Miller, 1929

The single species, *Q. gravis*, is known only by osseous material from cave deposits near St. Michel in Haiti and Samana Bay in the Dominican Republic (Hall 1981).

Quemisia was apparently about the same size as *Elasmodontomys*. Its cranial peculiarities include a long union of the lower jaws, short lower incisors, and an unusual twisting of the enamel pattern of the cheek teeth.

Allen (1942) summarized the available data on *Quemisia*. The original describer of the genus, G. S. Miller, considered it to be identical with an animal called the "quemi" by Oviedo

Elasmodontomys obliquus, photos from American Museum of Natural History.

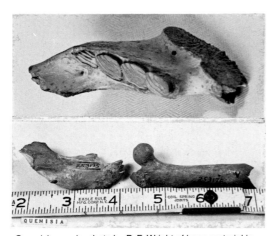

Quemisia gravis, photo by P. F. Wright of bone material in U.S. National Museum of Natural History.

in his account of the animals of Hispaniola, published slightly more than 25 years after the Spanish discovery of this island. Oviedo's brief description indicated that the "quemi" resembled the hutias in color but was larger. That *Quemisia* was utilized as food by the native Indians of Hispaniola is suggested by the presence of its limb bones at a depth of about 1.2 meters in a kitchen midden near the entrance of a cave. The genus probably became extinct during the first half of the sixteenth century. Recently, Woods (1989) suggested that Oviedo's description of the quemi actually refers to *Plagiodontia velozi*.

RODENTIA; HEPTAXODONTIDAE; Genus AMBLYRHIZA
Cope, 1868

The single species, *A. inundata*, is known only by cranial fragments and teeth from cave deposits on the islands of Anguilla and St. Martin at the extreme northern point of the Lesser Antilles (Allen 1942).

Amblyrhiza was related to *Quemisia* and *Elasmodontomys* but was much larger—nearly the size of an American black bear. Its skull is estimated to have been about 400 mm long. Although *Amblyrhiza* probably lived well after Pleistocene times, there is no evidence to indicate how it reached the tiny islands where its remains were found or when it became extinct.

RODENTIA; Family DASYPROCTIDAE

Agoutis and Pacas

This family of 3 living genera and 15 species occurs from east-central Mexico to southern Brazil and on the Lesser Antilles. There are two subfamilies: Dasyproctinae (agoutis), with the genera *Dasyprocta* and *Myoprocta*; and Agoutinae (pacas), with the single genus *Agouti*. Cabrera (1961) considered the Agoutinae to be a separate family and *Stictomys* to be a genus distinct from *Agouti*, but otherwise his nomenclature and sequence are followed here. The Agoutinae also were given familial rank by Honacki, Kinman, and Koeppl (1982) and Woods (1984) but not by Corbet and Hill (1986). Some authorities have used the name Cuniculinae in place of Agoutinae. It is perhaps unfortunate, and should be remembered, that the animals with the common name agouti are not the same as those with the scientific name *Agouti*.

Head and body length is 320–795 mm, tail length is 10–70

Amblyrhiza inundata, the last two teeth of the upper molar series, photo from *Bull. Amer. Mus. Nat. Hist.*, "New Fossil Rodents from Porto Rico," H. E. Anthony.

mm, and weight is approximately 1–10 kg. The pelage is coarse and thick. Pacas display a spotted pattern on a brownish or blackish ground color, but agoutis are usually uniformly colored above and pale below. Both groups have a piglike body and a rabbitlike head. The external form is modified for running in that the limbs, especially the hind legs, are lengthened and the lateral toes are reduced in size. The thumb is vestigial. In the pacas the forefoot has four functional digits and the hind foot has five; in the agoutis the forefoot has four functional digits and the hind foot has three. The claws are thick and hooflike. Females have four pairs of ventral mammae.

The dental formula of the Dasyproctidae is: (i 1/1, c 0/0, pm 1/1, m 3/3) × 2 = 20. The incisors are fairly thin, and the cheek teeth are high-crowned and semirooted. Part of the zygomatic arch of the skull in *Agouti* is greatly enlarged and contains a large sinus, a condition not found in any other mammal. The surface of this enlargement becomes extremely rugose in the adult. This character is the principal basis on which some authorities place *Agouti* in its own family.

The geological range of the Dasyproctidae is Oligocene to Recent in South America (Woods 1984).

RODENTIA; DASYPROCTIDAE; Genus DASYPROCTA
Illiger, 1811

Agoutis

There are 11 species (Cabrera 1961; Hall 1981; Husson 1978; Ojasti 1972):

D. mexicana, southern Mexico, introduced in Cuba;

D. ruatanica, Ruatan Island off northern Honduras;

D. coibae, Coiba Island off southwestern Panama;

D. punctata, southern Mexico to northern Argentina, introduced in Cayman Islands;

D. fuliginosa, Colombia, Venezuela, Surinam, Amazon Basin of Peru and northern Brazil;

D. kalinowskii, southeastern Peru;

D. leporina, Venezuela, Guianas, Brazil, introduced in the Lesser Antilles;

D. azarae, central and southern Brazil, Paraguay, northeastern Argentina;

D. guamara, Orinoco Delta in northeastern Venezuela;

D. cristata, the Guianas;

D. prymnolopha, eastern Brazil.

The use of the name *D. leporina,* instead of *D. aguti,* follows Husson (1978).

Head and body length is 415–620 mm and tail length is 10–35 mm. Grzimek (1975) listed weight as 1.3–4.0 kg. The pelage is usually quite coarse and glossy, with the longest and thickest hairs on the posterior part of the back. In most forms the general coloration of the upper parts ranges from pale orange through several shades of brown to almost black. The underparts are generally whitish, yellowish, or buffy. In some forms slight stripes may be present, and in others the rump color contrasts with the color of the remainder of the back. The body form is slender, the ears are short, and the hind foot has three toes with hooflike claws. Females have four pairs of ventral mammae.

Agoutis are found in forests, thick brush, savannahs, and cultivated areas. In Peru, according to Grimwood (1969:34–35), *Dasyprocta* "is confined to the Amazonian region, where it is found in all parts of the low selva rainforest zone and

Agouti (*Dasyprocta* sp.), photo from San Diego Zoological Garden. Inset: right forepaw and hind foot, photo from *Proc. Zool. Soc. London.*

many parts of the high selva zone, in which it appears to extend to greater altitudes than *Cuniculus* [= *Agouti*] being regularly reported up to 2,000 m. or more. Agoutis live closely associated with water, but they are to be found on the banks of all types of streams, down to the merest runnel, and are therefore more widespread and common than pacas."

In some areas *Dasyprocta* constructs burrows among limestone boulders, along riverbanks, or under the roots of trees. In a study of *D. punctata* on Barro Colorado Island, Panama, however, Smythe (1978) found that each animal had several sleeping spots, usually located in hollow logs, among tree roots, or under tangled vegetation. Well-defined paths radiate out from the shelters of *Dasyprocta*.

Agoutis are basically diurnal, but in areas where they have been molested by people they may not leave their shelters until dusk. They are terrestrial and are adapted for a cursorial life. They walk, trot, or gallop on their digits and also can jump vertically at least two meters from a standing start. They often sit with the body erect and the ankles flat on the ground, a position from which they can dart off at full speed. If danger threatens, they may pause, motionless, with one forefoot raised, but if discovered, they can travel with remarkable speed and agility. They usually sit erect to eat, holding the food in their hands. The diet consists of fruits, vegetables, and various succulent plants. On Barro Colorado Island, Smythe (1978) learned that *D. punctata* lived mainly on fruit when it was available, then buried the fruit seeds at scattered sites, to be eaten when fruit was not in season. Individuals often followed bands of monkeys and picked up the fruit that was dropped from trees. *D. punctata* also sometimes browsed and ate crabs.

Reported population densities of *D. punctata* are about 0.1/ha. in Tikal National Park, Guatemala, and 1.0/ha. on Barro Colorado Island (Cant 1977). In the latter area, Smythe (1978) found the basic social unit to be the mated pair, that bond apparently lasting until death. Each pair occupied a territory of about 1–2 ha., which included several sleeping spots and food trees, as well as a length of creek bed, where the natal nests were located. The pair, especially the male, aggressively drove off intruding agoutis. Territorial defense sometimes resulted in vicious fighting and severe wounds. When behaving aggressively, *D. punctata* sometimes erected the long hairs of its rump. When disturbed, this species may thump the ground with its hind feet. It also has a number of vocalizations, most notably an alarm bark like that of a small dog, which it makes as it runs away from danger.

In Venezuela, wild *Dasyprocta* apparently breed throughout the year (Ojasti 1972). Both seasonal and continuous reproduction have been reported for captive agoutis (Asdell 1964; Weir 1974*b*). On Barro Colorado Island, Smythe (1978) observed young almost every month, with a maximum from March to July, when fruit was abundant. Some populations of *Dasyprocta* evidently mate twice a year. During the courtship the male sprays the female with urine, causing her to go into a "frenzy dance"; after several sprays she allows the male to approach (Smythe 1978). Females have an average estrous cycle of 34 days, experience a postpartum estrus, and have a gestation period of about 104–20 days (Weir 1974*b*). Litters usually contain one or two young, there are sometimes three, and there is one record of four (Asdell 1964; Ojasti 1972; Smythe 1978; Weir 1974*b*). The newborn are fully furred, have their eyes open, and are able to run in their first hour of life (Smythe 1978). Lactation normally lasts about 20 weeks (Weir 1974*b*). The breakup between parents and offspring is associated with the coming of a new litter, increasing aggression by the adults, and a decline in food supplies (Smythe 1978). On Barro Colorado Island it was found that young not born during the fruiting season did not survive, and even those born in that favorable period suffered

70 percent mortality in the following nonfruiting season. Deaths were caused both by lack of food and by predation from the coati *(Nasua)*, males of which become carnivorous in the nonfruiting season. Subsequently the chances for survival greatly increased, and *Dasyprocta* has a potentially long life span. One captive *D. leporina* survived for 17 years and 9 months (Jones 1982).

Agoutis tame easily and make affectionate pets, but they are intensively hunted throughout their range by people for use as food. In some areas their numbers have declined seriously because of both hunting and habitat destruction (Grimwood 1969; Mares et al. 1981; Smythe 1978). In parts of the Amazon Basin, however, the breaking up of virgin forests seems to have created more favorable habitat for *D. leporina*. There have been suggestions that in this area the species would be superior to beef cattle for commercial food production (Smith 1974).

As noted in the above systematic list, several species of *Dasyprocta* were introduced in the West Indies. These introductions apparently were carried out by native Caribbean peoples before the coming of European explorers, for the purpose of supplying food. Populations of *D. leporina* were established as far north as the Virgin Islands and were present long enough to develop into several named subspecies (Hall 1981). With the increase of humans on the islands, the clearing of the natural forests, and the introduction of predators, such as the mongoose, agoutis became very rare. They survived on some islands, however, through the early twentieth century (Allen 1942).

RODENTIA; DASYPROCTIDAE; **Genus MYOPROCTA**
Thomas, 1903

Acouchis

There are two species (Cabrera 1961; Husson 1978):

M. exilis (red acouchi), to the east of the Andes in Colombia, southern Venezuela, the Guianas, Ecuador, northern Peru, and the Amazon Basin of Brazil;
M. acouchy (green acouchi), to the east of the Andes in southern Colombia, eastern Ecuador, northern Peru, and the Amazon Basin of Brazil.

Cabrera used the name *M. acouchy* for the red acouchi and *M. pratti* for the green, but Husson pointed out that *M. exilis* is the proper name for the red and that *M. acouchy* applies to the green. Husson also indicated that both forms might be merely color variations of the same species. The distributions mapped for each species by Mares and Ojeda (1982) follow the nomenclature of Cabrera.

The following descriptive data were provided in part by B. J. Weir (Wellcome Institute, Zoological Society of London, pers. comm.). Head and body length is 320–80 mm, tail length is 45–70 mm, and weight is 600–1,300 grams. The pelage is coarse. The upper parts are generally reddish to blackish or greenish. The muzzle and sides of the head are often brightly colored with yellow, orange, or red. The underparts are brownish, orangish, yellowish, or whitish. The short-haired tail, which is white underneath, perhaps is used as an intraspecific signaling device.

Myoprocta closely resembles *Dasyprocta* but is usually much smaller and has a more prominent tail. The incisor teeth are pale orange in color. Females have four pairs of ventral mammae.

Myoprocta seems restricted to the tropical forests of the Amazon Basin. Handley (1976) collected all specimens in

Acouchi *(Myoprocta acouchy)*, photo by Barbara J. Weir.

moist areas of evergreen forest. Grimwood (1969) stated that *Myoprocta* generally occurs in the same areas as *Dasyprocta* in the low selva zone; it is reportedly less dependent on water than are other members of the Dasyproctidae but is never found far from it. It is basically diurnal and terrestrial and sometimes lives in holes in riverbanks. Its diet is thought to be about the same as that of *Dasyprocta,* and it also sits erect to eat and buries seeds in scattered places for future use (Smythe 1978).

In French Guiana, Dubost (1988) found individual *M. exilis* to occupy home ranges of 9,600–12,000 sq meters during the wet season and 6,500–7,300 sq meters in dry months. Each range contained several areas of high use, with males tending to stay in open forest and females in dense vegetation. Individuals did not dig their own burrows but made leaf nests in hollow logs or old armadillo holes. Social units comprised 1–7 individuals, frequently an adult male and female with juveniles. However, each individual generally used a separate home range.

Observations of captives indicate that females tolerate and initiate contact with males to a much greater extent than do female *Dasyprocta.* As in the latter genus, urine spraying is a prominent feature of courtship (Kleiman 1971). Eisenberg (1974) listed about a dozen different vocalizations, including a sharp twit when startled; a tooth-chatter threat; peeping when seeking contact; and a snort before an attack.

In captivity in London, breeding occurs at any time of the year, but there is a birth peak during the summer. Males are fertile throughout the year, but females may go into anestrus during the summer. The estrous cycle averages about 42 days. A postpartum estrus may occur, but females usually mate after lactation has ceased, unless the young are lost before that time. The mean gestation period is about 99 days. The average number of young is two, and the range is one to three. The newborn weigh about 100 grams each, are fully furred, and have their eyes open. The minimum period of lactation needed for survival is 14 days. Both sexes reach puberty at 8–12 months (Kleiman 1970; Weir 1974*b*). Ac-

cording to Jones (1982), a captive specimen lived for 10 years and 2 months.

Grimwood (1969) stated that *Myoprocta* is less common in Peru than is *Dasyprocta* but is an important source of food for the native people. In French Guiana, Dubost (1988) found that unlike *Dasyprocta, Myoprocta* inhabits only undisturbed areas and is generally absent from areas modified by people, such as plantations and secondary forests.

RODENTIA; DASYPROCTIDAE; Genus AGOUTI
Lacépède, 1799

Pacas

There are two species (Cabrera 1961; Grimwood 1969; Hall 1981):

A. paca, east-central Mexico to Paraguay;

A. taczanowskii, the Andes of northwestern Venezuela, Colombia, Ecuador, and possibly Peru.

The name *Cuniculus* Brisson, 1762, often is used in place of *Agouti.* Cabrera (1961), along with various other authorities, placed *A. taczanowskii* in a separate genus, *Stictomys* Thomas, 1924. Handley (1976), however, put this species in *Agouti.*

Head and body length is 600–795 mm, tail length is 20–30 mm, and weight is about 6.3–12.0 kg. The pelage of *A. paca* is rather coarse and lacks fine underfur. *A. taczanowskii* has a thicker and softer pelage. The upper parts of *Agouti* are brownish to black and usually have four longitudinal rows of white spots on each side. The underparts are white or buffy.

The body is robust, the ears are medium-sized, the forefoot has four digits, and the hind foot has five digits. Part of the zygomatic arch of the skull in *A. paca* is specialized as a resonating chamber, a feature that is absent in all other mam-

Paca *(Agouti paca)*, photo by Ernest P. Walker. Insets: photos by Howard E. Uible.

mals. Females have four pairs of mammae (Weir 1974*b*).

Pacas live in a variety of habitats but usually seek forested areas near water. *A. taczanowskii* lives on the high Andean paramo. Grimwood (1969) reported that *A. paca* is confined to the Amazonian part of Peru, where it is found throughout the low selva (rainforest) zone but may occur locally up to elevations of 3,000 meters. It prefers small, swift, tree-shaded streams to open riverbanks.

Pacas are nocturnal, often spending the day in burrows that they excavate themselves or take over from some other animal. These burrows may be located in banks, on slopes, among tree roots, or under rocks and usually have one or more escape exits. Leopold (1959) stated that an individual generally has several burrows or uses such alternative shelters as hollow logs, stumps, and rock piles. In the limestone country of the Yucatan Peninsula, pacas frequent caves or sinkholes and do not dig burrows. Hall and Dalquest (1963) reported that all the burrows they examined were simple tubes, mostly about two meters below the surface of the ground; one was excavated for a distance of over six meters, but the end was not reached.

After dusk, pacas emerge to follow well-defined pathways that lead to feeding grounds and water. Although pacas are terrestrial, they enter water freely and swim well. If alarmed, they generally attempt to make their escape in water. Their diet consists of leaves, stems, roots, seeds, and fallen fruit; their favorite foods appear to be avocados and mangoes.

Collett (1981) estimated population density to be 84–93 (including 38–54 adults) per sq km and noted little interaction between individuals. Leopold (1959) stated that pacas are not social animals, but live alone, each with its own holes and runways. Only about half as many vocalizations have been noted for *Agouti* as for the more sociable *Myoprocta* (Eisenberg 1974). In the Yucatan, pacas reportedly mate in early winter, and the females produce litters of two young in

the dry season (winter to early spring). Other reports indicate that a single young is usually born, that twins are rare, and that there are probably two litters per year. In Rio de Janeiro, one litter was noted in February and another in July. According to Kleiman, Eisenberg, and Maliniak (1979), *A. paca* has a gestation period of 118 days and produces a single young that weighs 710 grams. In a study of *A. paca* in Colombia, Collett (1981) found no indication of seasonal reproduction, a postpartum estrus, a mean interbirth interval of 191 days, all pregnancies to involve a single embryo, females to first give birth at about 1 year, and the age of the oldest individuals in the population to be 12–13 years. A captive specimen in the National Zoological Park in Washington, D.C., lived slightly more than 16 years.

Pacas are sometimes considered agricultural pests, as they may destroy yams, cassava, sugar cane, corn, and other crops. They are killed by people for this reason and are also hunted intensively throughout their range for their flesh, which is said to be unusually delicious and to have a far higher price than any other meat, domestic or wild. Such hunting, in combination with habitat destruction, has resulted in pacas' being exterminated or becoming rare over large areas (Baker 1974; Grimwood 1969; Hall and Dalquest 1963; Leopold 1959). Smythe (1987) reported that pacas have potential for domestication as a food source.

RODENTIA; Family CHINCHILLIDAE

Viscachas and Chinchillas

This family of three Recent genera and six species occurs in western and southern South America. The sequence of genera presented here follows that of Cabrera (1961).

Head and body length is 225–660 mm and tail length is 75–400 mm. All genera have thick fur, that of the montane genera, *Lagidium* and *Chinchilla*, being finer than that of *Lagostomus*, which occurs on the pampas. The pelage is coarser on the tail than on the body. The palms and soles of most forms are naked. Coloration varies considerably among the genera.

The body form is slender. The forelimbs are relatively short, but the hind limbs are long and muscular. The forefoot has four long, flexible digits. The hind foot is elongate. *Lagidium* and *Chinchilla* have four digits on the hind foot, with relatively weak claws. *Lagostomus* has only three hind digits, but they are armed with strong claws. Fleshy pads or pallipes are present on the feet. *Chinchilla* and *Lagostomus* have stiff bristles on the hind digits, which may be used for grooming.

The head, eyes, and ears are relatively large. The bullae are small in *Lagostomus*, intermediate in *Lagidium*, and enormously expanded in *Chinchilla*. The dental formula is: (i 1/1, c 0/0, pm 1/1, m 3/3) × 2 = 20. The incisors are fairly narrow. The cheek teeth are ever-growing and have a pattern of tightly pressed transverse laminae without cement.

Chinchillids live in burrows or rocky crevices. They are basically cursorial but often jump bipedally and sit erect while eating, sunbathing, or grooming. All three genera have been observed to dustbathe. They are active throughout the year and are mainly vegetarian. They tend to be gregarious and are seen in groups ranging from families to colonies of several hundred animals. There is a long gestation period, and the young are born fully furred and with their eyes open. All three genera are intensively hunted by people.

The geological range of the Chinchillidae is early Oligocene to Recent in South America (Woods 1984).

RODENTIA; CHINCHILLIDAE; **Genus LAGOSTOMUS**
Brookes, 1828

Plains Viscacha

The single living species, *L. maximus*, occurs in extreme southern Paraguay and in northern and central Argentina (Cabrera 1961). Another species, *L. crassus*, is known only by a single skull found in southern Peru. Since the specimen was not fossilized, it may have represented an animal that lived in Recent times, but any associated population evidently disappeared long ago.

Head and body length is 470–660 mm and tail length is 150–200 mm. Weight is about 2.0–4.5 kg in females and 5.0–8.0 kg in males (Weir 1974a). The guard hairs of the pelage are dark and coarse, but the underfur is very soft. The general coloration of the upper parts varies with the habitat, ranging from light brown in sandy areas, such as southern Buenos Aires province, to dark gray in Entre Rios. The underparts are white. The face is strikingly marked in black and white. The mustache is much thicker and more pronounced in males than in females. The tail is fully furred, the hairs on the dorsal surface being the longest. The base of the underside, which is usually bare and horny, is used as a third leg when the animal sits erect on its haunches (Weir 1974a).

The head is large and blunt. The forefoot has four well-developed digits, and the hind foot has three digits armed with stout, sharp claws. The middle hind digit has the largest claw and also has a pad of stiff bristles on the inner side. The forefeet are used mainly for digging, and soil is pushed out with the nose. The rhinarium is furred and intricately folded,

Plains viscacha *(Lagostomus maximus)*, photo by Ernest P. Walker. Undersurfaces of right front and hind feet, photos from *Proc. Zool. Soc. London*.

thus preventing earth from entering the nostrils. Females have two pairs of mammae, both placed laterally on the thorax (Weir 1974b).

The following ecological information was taken largely from Weir (1974a) and Woods (1984). *Lagostomus* inhabits relatively barren parts of the pampas at elevations of up to 2,680 meters. It constructs extensive burrow systems, called viscacheras, some of which are used for centuries. Such a system may contain 4–30 entrances and cover up to 600 sq meters. Some entrances are so large that a person could stand waist-deep in one. A round chamber is generally located about 1.5 meters from the entrance, and tunnels lead off from it in different directions. During construction of a burrow system, 80 cubic meters of soil may be moved, and much of it deposited in flattened mounds about 50 cm high. The animals carry many inedible items—sticks, stones, bones, dung, and objects dropped by people—from the surrounding area and place them on the ground above the burrows. Various kinds of animals, such as insects, toads, lizards, snakes, burrowing owls, other rodents, and skunks, sometimes share the burrows with *Lagostomus*. Even foxes and boa constrictors may move in and prey on the viscacha.

Lagostomus is crepuscular, emerging from its burrow at dawn and dusk to feed. The surface areas in the vicinity of long-established viscacheras become denuded of food supplies, and the animals travel some distance over well-worn pathways to reach new feeding grounds. If a viscacha is pursued, it can travel at 40 km/hr, alternating running with 3-meter leaps and sharp turns. The diet consists mostly of grass and seeds, but nearly any kind of vegetation may be eaten.

Colonies contain 15–30, sometimes up to 50, individuals (Grzimek 1975). The exact composition probably varies from place to place; in some areas the adult males are thought to live apart from the main colony at certain times of the year, especially when the young are born (B. J. Weir, Wellcome Institute, Zoological Society of London, pers. comm.). Although individuals tend to be concentrated in groups, overall population density in eastern Argentina has been estimated at 2–5 per ha. At the onset of the breeding season, adult males become more aggressive and engage in frequent fights with one another (Woods 1984). Nearly all males bear scars from such conflicts, and deaths sometimes occur. Males express displeasure or aggression by a fearsome crescendo of notes culminating in a high whine and accompanied by foot stomping and tail wagging. The most common vocalization of *Lagostomus*, however, is a two-syllable "uh-huh?" made when the animal is investigating (Weir 1974a).

Lagostomus is capable of breeding throughout the year in captivity. In some northern parts of the range, where the climate is favorable, wild females produce two litters annually. In east-central Argentina, however, there is a single breeding season, with mating in March or April and births in July or August. Females have an estrous cycle of 45 days and rarely experience a postpartum estrus, but there is generally a fertile postlactation estrus. The gestation period averages 154 (145–66) days. The mean number of young per litter is two, and the range is one to four. The precocial offspring are fully furred and weigh about 200 grams each. The minimum period of lactation is 21 days, but weaning generally occurs by 8 weeks. Both sexes may attain sexual maturity as early as 5.5 months, but the average is 8.5 months for females and 15 months for males (Asdell 1964; Weir 1974b; Woods 1984). A captive specimen lived for 9 years and 5 months (B. J. Weir, Wellcome Institute, Zoological Society of London, pers. comm.).

A hand-reared viscacha makes a delightful pet, but in the wild *Lagostomus* is generally considered to be a pest. It competes with domestic animals for forage, it sometimes raids cultivated crops, its huge burrows may cause livestock and even humans to stumble and break limbs, and its probably acidic urine devalues the soil for many years. For these reasons, and also because of its valuable flesh and fur, *Lagostomus* has been intensively hunted and poisoned by people. It once occurred in large numbers on the pampas, and it is said to have been possible for a person to ride 800 km without losing sight of the viscacha. In recent decades the range and numbers of *Lagostomus* have been greatly reduced by systematic extermination campaigns, and Weir (1974a) stated that the genus probably would be nearly extinct in about 20 years. Barlow (1969) indicated that the viscacha had once been introduced into Uruguay but had subsequently been extirpated there.

RODENTIA; CHINCHILLIDAE; Genus LAGIDIUM
Meyen, 1833

Mountain Viscachas

There are three species (Cabrera 1961; Grimwood 1969; Rowlands 1974):

L. peruanum, Peru;
L. viscacia, extreme southern Peru, western and southern Bolivia, northern and central Chile, western Argentina;
L. wolffsohni, southern Chile, southwestern Argentina.

Some authorities, including Rowlands (1974) and B. J. Weir, recognize L. boxi, of northwestern Patagonia, as a species distinct from L. viscacia.

The remainder of this account is based in part on data provided by B. J. Weir (Wellcome Institute, Zoological Society of London, pers. comm.). Head and body length is 300–450 mm and tail length is 200–400 mm. The smallest species, L. peruanum, weighs 900–1,600 grams, but the other species may weigh up to 3,000 grams. The pelage is thick and soft except on the upper part of the tail, where it is coarse. The coloration of the upper parts ranges, as elevation increases, from dark gray to almost chocolate brown. There is a black middorsal stripe, which is more prominent in southerly populations. The underparts are usually white, yellow, or pale gray. The tip of the tail is black to reddish brown.

Mountain viscachas look like long-tailed rabbits, and in Patagonia they are called rock squirrels. The ears are long and covered with hair and have a fringe of white hairs along the edge. Both the hind foot and the forefoot have four digits. The enamel of the incisor teeth is not colored. Females have a single pair of lateral thoracic mammae.

These viscachas inhabit dry, rocky, rugged, mountainous country with sparse vegetation. In Peru they are most commonly found at elevations of 3,000–5,000 meters (Grimwood 1969). They are diurnal and do not hibernate. They live in rocky clefts and crevices and spend much of the day on suitable perches, sunning themselves and dressing their fur. They are poor diggers and are rarely found in earth burrows. They run and leap among the rocks with great agility. Pearson (1948) found that L. peruanum never traveled more than about 70 meters from a rocky shelter and always lived near water, possibly because of the succulent vegetation in the vicinity. *Lagidium* eats almost any kind of plant, including lichens, moss, and grass.

In Peru, L. peruanum has a remarkably sparse distribution, with colonies often being separated by 10–20 km of seemingly suitable habitat (Grimwood 1969). Pearson (1948) referred to this species as being among the most abundant mammals in the highlands of southern Peru but gave its

Mountain viscachas *(Lagidium viscacia)*, photo by Ernest P. Walker. Insets: hand and foot, photos from *Proc. Zool. Soc. London.*

overall population density as only about 11/sq km. He found colonies of 4–75 individuals, with the larger assemblies divided into probable families of 2–5 animals each. There also have been reports of colonies containing several hundred individuals. Although each family maintained its own burrow and sunning rocks, Pearson found intergroup relations to be generally peaceful, and he observed no serious fighting. At the onset of the breeding season, however, females drove males from the burrows, and the males then moved about considerably and became more promiscuous. A variety of vocalizations were heard, most notably a high-pitched, mournful whistle that evidently serves to warn others of danger.

In Peru, mating begins in October, and all adult females are pregnant by December (Pearson 1948). In Patagonia, mating occurs in May or June. Females have a mean estrous cycle of 56.7 days (Weir 1974b). Pearson (1948) indicated that there is sometimes a postpartum estrus in *L. peruanum* and that it might be possible for females to produce two or even three litters per year; this is unlikely, however, as the gestation period of *L. peruanum* is now known to be 140 days, and that of *L. viscacia* has been estimated at 120–40 days. Only a single young is born at a time. It is fully haired, its eyes are open, and it can eat solid food on its first day of life. Newborn weight is about 180 grams in *L. peruanum* and 260 grams in *L. viscacia.* The young are normally weaned after about 8 weeks, and both sexes reach puberty at 1 year (Weir 1974b). Although few wild viscachas live more than 3 years (Pearson 1948), one captive *L. peruanum* survived for 19 years and 6 months (Jones 1982).

People hunt mountain viscachas for both meat and fur, and their numbers seem to have declined in some areas. In Chile they are protected by law, except in the extreme north, where they are still abundant, but illegal hunting is resulting in their becoming increasingly rare in the central and southern parts of the country (Miller at al. 1983; Pine, Miller, and Schamberger 1979). About a million viscacha pelts (*Lagidium* and *Lagostomus*) were exported from Argentina from 1972 to 1979; the value was U.S. $3.00–5.00 per pelt (Gudynas 1989).

RODENTIA; CHINCHILLIDAE; **Genus CHINCHILLA**
Bennett, 1829

Chinchillas

Cabrera (1961), as well as Thornback and Jenkins (1982), recognized two species:

C. brevicaudata, the Andes of Peru, Bolivia, northern Chile, and northwestern Argentina;
C. laniger, the Andes of northern Chile.

Some authorities, such as Pine, Miller, and Schamberger (1979), treat *C. brevicaudata* as a subspecies of *C. laniger.*

The remainder of this account is based in large part on data provided by B. J. Weir (Wellcome Institute, Zoological Society of London, pers. comm.). Head and body length is 225–380 mm and tail length is 75–150 mm. The female is the larger sex and weighs up to 800 grams, while males rarely weigh over 500 grams. The silky pelage is extremely dense and soft, with as many as 60 hairs growing from each hair follicle. The general coloration of the upper parts is bluish, pearl, or brownish gray, and usually each hair has a black tip. The underparts are yellowish white. The furry tail is covered with coarse hairs on the dorsal surface. Chinchillas of many

Chinchilla *(Chinchilla laniger)*, with the well-formed feces clearly shown, photo by Ernest P. Walker. Insets: undersurfaces of right front and right hind foot, photos from *Proc. Zool. Soc. London.*

colors, ranging from white to black, have been bred in captivity.

The head is broad, the external ears are large, and the auditory bullae are relatively enormous. The large, black eyes have a vertical slit pupil. Both the short forefoot and the narrow hind foot have four digits, with stiff bristles surrounding the weak claws. There are vestigial cheek pouches. The enamel of the incisor teeth is usually colored. Females have one pair of inguinal and two pairs of lateral thoracic mammae.

The natural habitat is the relatively barren areas of the Andes Mountains at elevations of 3,000–5,000 meters. Chinchillas shelter in crevices and holes among the rocks. They are basically crepuscular and nocturnal, though they have been observed on bright days, sitting in front of their holes and climbing and jumping about the rocks with remarkable agility (Grzimek 1975). To eat, they sit erect and hold the food in their forepaws. The diet consists of any available vegetation.

In former times, colonies of about 100 individuals were frequently seen. Although chinchillas are sometimes referred to as monogamous, there is no substantial evidence for this view. Females are very aggressive toward each other and toward males, even when in estrus, but serious fighting probably is rare in the wild. Threats are expressed by growling, chattering the teeth, and urinating.

The breeding season of *C. laniger* extends from November to May in captive populations in the Northern Hemisphere and from May to November in the Southern Hemisphere. There are usually two litters during this season. *C. brevicaudata* occasionally produces three litters annually (Grzimek 1975). Females have an estrous cycle averaging 38 days and a postpartum estrus (Weir 1974b). The mean gestation period is 111 days. Litters contain one to six, usually two to three, young. The newborn weigh about 35 grams each, are fully furred, and have their eyes open. Lactation normally lasts 6–8 weeks. Sexual maturity is attained at an average age of 8 months by both males and females but may occur as early as 5.5 months. The life span is probably about 10 years in the wild, but captives have lived for over 20 years, and some have bred at 15 years.

Since the time of the ancient Incas, chinchilla fur has been highly prized for human apparel (Allen 1942). The coming of Europeans to the Western Hemisphere resulted in a greatly increased demand for this fur and a corresponding decline in the number and distribution of wild chinchillas. It is said that it was once possible for a person to see thousands of chinchillas in the course of a day's journey. As late as 1900 an estimated 500,000 skins were being exported annually from Chile (IUCN 1972). Shortly thereafter, however, the genus became rare and the price of its skin rose. Indeed, considering its size and weight, a chinchilla pelt is the most valuable pelt in the world. Coats made of wild chinchilla fur have sold for as much as $100,000, and that was years ago. In Japan in 1981, coats were being sold for about $49,000. The pelt of *C. brevicaudata* is more valuable than that of *C. laniger* (Thornback and Jenkins 1982).

Although chinchillas are now protected by law in their natural habitat, enforcement is difficult in the remote areas involved, and hunting continues. Populations also have declined through burning and harvesting of the algarrobilla shrub *(Balsamocarpon brevifolium)* and possibly through competition with the genera *Octodon* and *Abrocoma* (Miller et al. 1983). Fewer than 10,000 *C. laniger* are thought to survive (E. K. Rice 1988), and *C. brevicaudata* is presumed to be very rare (Thornback and Jenkins 1982). All wild populations are on appendix 1 of the CITES, both species are classified as indeterminate by the IUCN, and the subspecies *C. brevicaudata boliviana* is listed as endangered by the USDI. Some chinchillas were brought into captivity in the late nineteenth and early twentieth centuries, and today probably hundreds of thousands, if not millions, of their descendants are bred commercially throughout the world (Grzimek 1975). Attempts have been made to introduce chinchillas into the wild in various places, including back into the Andes, but no successes have yet been reported.

RODENTIA; **Family CAPROMYIDAE**

Hutias and Nutria

This family contains 8 described Recent genera, only 4 of which still survive, and 33 species. Remains of another Recent, but extinct and unnamed, genus have been discovered in southwestern Haiti (Morgan and Woods 1986). One of the living genera, *Myocastor* (nutria), is native to southern South America. All other Recent genera (hutias) occur only in the West Indies. The sequence of genera presented here follows that of Hall (1981), who recognized three subfamilies: Capromyinae, with *Capromys, Geocapromys, Macrocapromys,* and *Hexolobodon;* Plagiodontinae, with *Plagiodontia, Isolobodon,* and *Hyperplagiodontia;* and Myocastorinae, with *Myocastor.* Some authorities, such as Woods (1984), consider the Myocastorinae to be a separate family, and others have classed it as a subfamily of the Echimyidae. Cabrera (1961) listed one additional genus, *Procapromys* Chapman, 1901, with the species *P. geayi,* supposedly from northern Venezuela. The place of origin of the single known specimen is questionable, however, and it may actually represent a young individual of the species *Capromys pilorides* (Packard 1967).

Head and body length is 200–635 mm, tail length is 35–425 mm, and weight is up to 9 kg. The pelage is harsh in hutias but soft and thick in the nutria. Coloration is generally brownish or grayish. The head is broad in hutias and somewhat triangular in the nutria. All genera have a robust body form, small eyes and ears, short limbs, a reduced thumb, and prominent claws. There are two pairs of mammae in female hutias and four pairs in the female nutria.

The dental formula is: (i 1/1, c 0/0, pm 1/1, m 3/3) × 2 = 20. In hutias the incisors are narrow, and the rootless cheek teeth are high-crowned and grow throughout life. In the nutria the incisors are broad and strong, and the extremely high-crowned cheek teeth are semirooted and decrease in size as they converge anteriorly.

Hutias are terrestrial and live in forests and plantations. The nutria is modified for an aquatic life and inhabits marshes, lakes, and sluggish streams. The diet consists of vegetation and small animals. The young are born after a relatively long gestation period and are highly precocious. All genera have been either completely exterminated or reduced in numbers in their native habitat, through human agency. The nutria, however, also has been introduced in various parts of the Northern Hemisphere, where it has become abundant.

The known geological range of the subfamilies Capromyinae and Plagiodontinae is Pleistocene to Recent in the West Indies, though they probably were present there from the Oligocene. The geological range of the Myocastorinae is late Oligocene to Recent in South America (Woods 1984).

RODENTIA; CAPROMYIDAE; **Genus CAPROMYS**
Desmarest, 1822

Cuban Hutias

The following 5 subgenera and 13 species apparently lived during Recent times (Hall 1981; Kratochvil, Rodriguez, and Barus 1978; Varona 1979; Varona and Arredondo 1979):

subgenus *Capromys* Desmarest, 1822

C. garridoi, Cayos Maja off south-central Cuba;
C. arboricolus, eastern Cuba;
C. pilorides, Cuba and several small nearby islands;

subgenus *Mysateles* Lesson, 1842

C. melanurus, eastern Cuba;
C. prehensilis, Cuba and Isle of Pines;

subgenus *Mesocapromys* Varona, 1970

C. auritus, Cayo Fragoso off north-central Cuba;
C. sanfelipensis, Cayo Juan Garcia off southwestern Cuba;

subgenus *Stenocapromys* Varona, 1979

C. gracilis, known only by skeletal remains from cave deposits in western Cuba;

subgenus *Pygmaeocapromys* Varona, 1979

C. minimus, known only by skeletal remains from eastern Cuba;
C. beatrizae, known only by skeletal remains from western Cuba;
C. silvai, known only by skeletal remains from south-central Cuba;
C. nanus, Cuba;
C. angelcabrerai, Cayos de Ana Maria off south-central Cuba.

According to Kratochvil, Rodriguez, and Barus (1978), the living hutias of Cuba should be arranged as follows: (1) the genus *Capromys* should contain only the single species *C. pilorides;* (2) *Mysateles* should be a full genus containing the subgenus *Mysateles,* with the species *C. melanurus* and *C. prehensilis,* and the new subgenus *Leptocapromys,* with the species *C. garridoi* and *C. arboricolus;* and (3) *Mesocapromys* should be a full genus containing the subgenus *Mesocapromys,* with the species *C. auritus,* and the new subgenus *Paracapromys,* with the species *C. nanus* and *C. sanfelipensis.*

Head and body length is about 220–500 mm, tail length is 150–300 mm, and weight is about 0.5–7.0 kg. The furry pelage consists of long, coarse guard hairs and moderately dense, softer underfur. The coloration of the upper parts ranges through a variety of buff, yellow, red, gray, brown, and black shades. The underparts are usually paler. The tail is well haired and unicolored. In the subgenus *Mysateles* the tail is known to be prehensile. The feet are broad and have prominent claws. The stomach, at least in the subgenus *Capromys,* has two constrictions that divide it into three compartments, and it is thus one of the most complex stomachs in the Rodentia. Female *C. pilorides* and *C. nanus* have two lateral thoracic pairs of mammae (Weir 1974b).

These hutias inhabit forests or rocky areas and seem to be basically arboreal. Some species are diurnal and sun themselves during the morning on leafy boughs of tall trees, curling up in such a way that from the ground they look like clumps of foliage. *C. melanurus* has been reported to be nocturnal and to run along branches and leap from tree to tree like a squirrel (Allen 1942). Although some species are said to seek refuge in holes in the ground, Varona (1979) reported *C. angelcabrerai* and *C. auritus* to construct communal nests in mangrove trees. These nests are circular, about 1 meter in diameter, and are composed of mangrove branches and leaves. The diet of Cuban hutias includes leaves, bark, fruits, lizards, and other small animals.

In captivity, *C. pilorides* breeds throughout the year but has a birth peak in June. Scattered records of gestation indi-

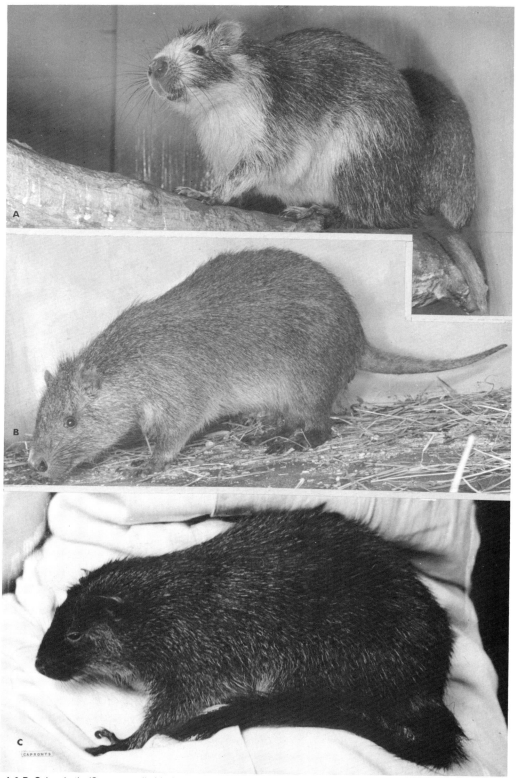

A & B. Cuban hutia *(Capromys pilorides)*, photos by Ernest P. Walker. C. *C. melanurus*, photo by Erna Mohr.

cate a usual period of around 110–40 days. There is apparently only a single offspring per birth in *C. nanus*, but one to six, usually two, in *C. pilorides*. The newborn are fully furred, have their eyes open, and are capable of moving about freely. In *C. pilorides* nursing continues for 153 days, and sexual maturity is attained by 10 months (Smith Canet and Berovides Alvarez 1984; Johnson, Taylor, and Winnick 1975; Oliver 1977; Weir 1974*b*). A captive *C. pilorides* lived for 11 years and 4 months (Jones 1982).

As indicated in the above systematic list, four species of *Capromys* are known only by skeletal remains, and these species are presumed to be extinct. The remains were often found in association with Indian artifacts and kitchen middens. It is thus evident that hutias have long been used as food by people, and some species may have disappeared before the time of Columbus. The arrival of European colonists accelerated the decline of hutias through increased demand for meat, clearing of forests, and introduction of predators (Allen 1942; Oliver 1977).

The only relatively common species of *Capromys* is *C. pilorides*, but even it has declined in numbers. It can still be hunted legally for two or three months of the year. All other species are now fully protected by law but illegal hunting is jeopardizing most species. The species *C. garridoi*, *C. auritus*, *C. sanfelipensis*, and *C. angelcabrerai* occur only on small islands, where they are especially vulnerable to hunting by visiting fishermen (Thornback and Jenkins 1982).

There are serious problems in addition to hunting. *C. nanus* was first known only by skeletal fragments from cave deposits but was found alive in 1917. It formerly occurred throughout Cuba but now is restricted to the Zapata Peninsula in the south-central part of the island, where it is highly endangered through agricultural development and predation by the introduced mongoose. *C. melanurus*, of eastern Cuba, is jeopardized mainly by deforestation. Both the IUCN and

USDI now classify *C. auritus*, *C. sanfelipensis*, *C. nanus*, and *C. angelcabrerai* as endangered, and the IUCN also designates *C. garridoi* and *C. melanurus* as indeterminate. All surviving species, however, are probably in serious danger of extinction, except *C. pilorides* (Oliver 1977).

RODENTIA; CAPROMYIDAE; **Genus GEOCAPROMYS**
Chapman, 1901

Bahaman and Jamaican Hutias

There are five species (Anderson et al. 1983; Hall 1981):

G. brownii, Jamaica;
G. thoracatus, Little Swan Island off northeastern Honduras;
G. columbianus, known only by skeletal remains from Cuba;
G. ingrahami, Bahamas;
G. pleistocenicus, known only by skeletal remains from Cuba.

Geocapromys was treated as a subgenus of *Capromys* by Hall (1981) and Corbet and Hill (1986), but Anderson et al. (1983), Honacki, Kinman, and Koeppl (1982), and Woods (1985) considered it a full genus. Unidentified and extinct species of *Geocapromys* are known from Atwood Cay in the Bahamas and Cayman Brac in the Grand Cayman Islands (Hall 1981; Oliver 1977).

In *G. brownii* head and body length is 330–455 mm, tail length is 35–64 mm, and weight is 1–2 kg. The short, dense pelage is yellowish gray, dark brown, or blackish above and buffy gray or dusky brown below. The short, rounded ears are

Bahaman hutia *(Geocapromys ingrahami)*, photo by Garrett C. Clough.

covered with minute hairs. *Geocapromys* is similar to *Capromys* but is distinguished externally by its shorter tail, less prominent claws, and shorter and finer fur. The skull of *Geocapromys* is characterized by a tendency toward anterior convergence of the toothrows, resulting in contact between the bases of the alveoli of the right and left fourth upper premolars; a broad, vertically or posteriorly oriented superior zygomatic root of the maxilla, and the presence of a small third reentrant angle on the anterolingual surface of the fourth upper premolar (Anderson et al. 1983).

Clough (1972, 1973, 1974) made a detailed field study of *G. ingrahami* on East Plana Cay, a small semiarid island in the Bahamas. The animals were found to be almost completely nocturnal. By day they stayed underground in caves, crevices, or holes in the limestone. They showed no tendency to dig their own burrows and probably did not make nests. They normally moved with a slow waddle but were quicker when alarmed. They were facile climbers, ascended trees to look for food, and descended headfirst. They ran and climbed up steep, rocky cliffs with agility and speed. They stood on their hind limbs to reach into shrubs and used their forepaws to grasp and manipulate food. Most of the night was spent eating, and the diet consisted mainly of bark, small twigs, and leaves. A supplementary laboratory investigation showed that individuals could survive well without water and could tolerate water with half the salt concentration of sea water for over a week.

There were calculated to be 30 individual *G. ingrahami* per ha. on East Plana Cay, and this density was thought to have been maintained for many years. There were no natural predators on the island, but the hutias appeared to be in balance with available food supplies. Despite the crowded conditions, there was little social antagonism and no serious fighting. *G. ingrahami* was basically solitary, but there was no territoriality and considerable intraspecific tolerance. Two or more adults were often seen together, and 15 captives were kept in a small enclosure without trouble. In a laboratory investigation of *G. ingrahami*, R. J. Howe (1976) also observed a minimum of agonistic behavior but found that when strangers were placed together there would be some strife and eventually establishment of an amicable dominance hierarchy. Threats between males increased when a female was in estrus. A possibly unique kind of nonagonistic wrestling was observed, generally in combination with mutual grooming. Howe (1974) also learned that individuals scent-marked with urine, not as a territorial measure but as a means of communication, especially of sexual information.

In the wild, breeding probably continues throughout the year. In captivity, *G. brownii* may produce two or more litters annually. The gestation period of this species is about 123 days, with an average interbirth interval of 168 days (Anderson et al. 1983). Apparently, there is only a single offspring per birth in *G. ingrahami* but one to three in *G. brownii.* Newborn *G. ingrahami* weigh about 80 grams. Those of *G. brownii* are extremely precocial and are capable of most adult movements and eat solid food within 30 hours of birth (Anderson et al. 1983; Clough 1972, 1974; Johnson, Taylor, and Winnick 1975; Oliver 1977; Weir 1974*b*). A wild female *G. ingrahami* was recaptured over 5 years after being tagged, and within 20 meters of the original site of capture and release (Clough 1974). A captive *G. brownii* lived for 8 years and 4 months (Jones 1982).

The extinct species probably disappeared for the same reasons and at about the same time as given above in the account of *Capromys.* The subspecies *G. ingrahami abaconis* and *G. i. irrectus,* which once occurred throughout the Bahamas, were probably largely exterminated in early colonial times. A population was reported living on Samana Cay as recently as 1931 (Buden 1986). The well-studied subspecies *G. i. ingrahami* was not discovered until 1891; it probably survived because East Plana Cay was never inhabited by people. There are an estimated 6,000–12,000 of the animals on the 456 ha. island, and some have also been introduced to a small cay in Exuma National Sea and Land Park. Clough (1985) reported that the population on the latter island had grown from 11 to at least 750 individuals.

The species *G. brownii* of Jamaica has declined drastically in the last 30 years because of the clearing of forests for agriculture and excessive hunting, though Oliver (1982) found it to be more widespread than once feared. A program of captive breeding and reintroduction is being carried out (Oliver et al. 1986). *G. thoracatus,* of Little Swan Island, seemed common in the early twentieth century but apparently disappeared following a severe hurricane in 1955 and subsequent predation by introduced house cats. The IUCN classifies *G. ingrahami* as rare and *G. brownii* as vulnerable.

RODENTIA; CAPROMYIDAE; Genus MACROCAPROMYS
Arredondo, 1958

The single species, *M. acevedoi,* is known only by skeletal remains from cave deposits in western Cuba and is apparently extinct. The humerus is one-fifth longer than that in the largest species of *Capromys* (Hall 1981).

RODENTIA; CAPROMYIDAE; Genus HEXOLOBODON
Miller, 1929

There are two species (Hall 1981):

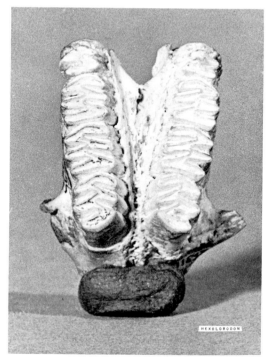

Hexolobodon phenax, upper molars, photo by Howard E. Uible of specimen in U.S. National Museum of Natural History.

H. *poolei*, known only by skeletal remains from the type locality in north-central Haiti;

H. *phenax*, known only by skeletal remains from north-central Haiti.

Hexolobodon was about the size of *Capromys pilorides*, but the skull differs from that of *Capromys* in having a shorter rostrum and a generally more robust form. *Hexolobodon* resembles the subgenus *Geocapromys* but differs in that the root of the premolar is thrown conspicuously forward and away from that of the first molar; the roots of the cheek teeth are less specialized and appear to close with age, instead of continually growing; and the lower incisor root is extended to the outer side of the mandibular toothrow. The remains of *Hexolobodon* have been found in cave deposits. The genus presumably was exterminated soon after the coming of European colonists.

RODENTIA; CAPROMYIDAE; Genus PLAGIODONTIA
F. Cuvier, 1836

Hispaniolan Hutias

The following seven species apparently lived during Recent times (Hall 1981):

P. *aedium*, Hispaniola;

P. *hylaeum*, Hispaniola;

P. *ipnaeum*, known only by skeletal remains from Hispaniola;

P. *spelaeum*, known only by skeletal remains from the type locality in Haiti;

P. *araeum*, known only by a single tooth from the type locality in Haiti;

P. *caletensis*, known only by skeletal remains from the type locality in the Dominican Republic;

P. *velozi*, known only by skeletal remains from the type locality in Haiti.

In P. *aedium* head and body length is about 312 mm and tail length is about 153 mm. Kleiman, Eisenberg, and Maliniak (1979) listed adult weight as 1,267 grams. In P. *hylaeum* head and body length is 348–405 mm and tail length is 125–45 mm. Judging from the skeletal remains, the largest species in the genus is P. *ipnaeum* and the smallest is P. *spelaeum*. In the living species, the short, dense pelage is brownish or grayish on the upper parts and buffy on the underparts. The tail is scaly and practically naked. Both the forefoot and the hind foot have five digits, all armed with claws except the thumb, which has a short, blunt nail. Females have three pairs of lateral thoracic mammae (Weir 1974b).

Hispaniolan hutias inhabit forests. Woods (1981) reported that they occupy rough hillsides and ravines from sea level to 2,000 meters in elevation, that some populations use burrows and feed near the ground, and that other populations may den in tree cavities and move through the trees, rather than come to the ground. Captives have been observed to be nocturnal and arboreal and to use nest boxes placed high off the ground (Oliver 1977). Wild P. *aedium* were reported to be active only at night, to hide during the day, to feed principally on roots and fruits, and to live in male-female pairs. Woods (1981) stated that three or four individuals commonly occupy the same burrow system. Specimens of P. *hylaeum* were caught in December in hollow trees near a lagoon; four pregnant females each contained a single embryo (Allen 1942). According to Johnson, Taylor, and Winnick (1975), captive female P. *aedium* have an estrous cycle of 10 days, a gestation period of 119 days, and apparently a single offspring. Weir (1974b) listed gestation as 123–50 days and litter size as one to two young in this species. A captive P. *aedium* was still alive after 9 years and 11 months (Jones 1982).

Five of the seven species in this genus are known only by skeletal remains, often found in association with human kitchen middens. These five species probably disappeared by

Hispaniolan hutia *(Plagiodontia hylaeum)*, photo from New York Zoological Society.

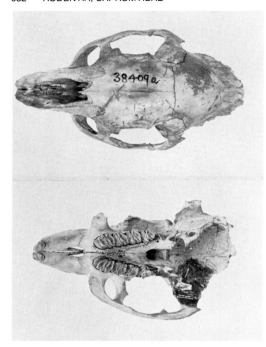

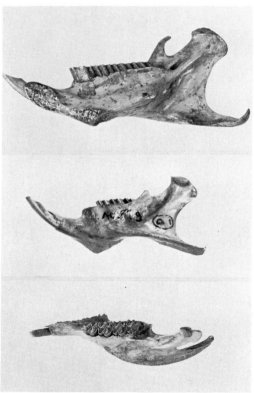

Isolobodon portoricensis, photos from American Museum of Natural History.

the seventeenth century because of excessive hunting by people (Allen 1942; Goodwin and Goodwin 1973; Oliver 1977). *P. aedium* and *P. hylaeum* have been greatly reduced in range and numbers and are threatened by deforestation, hunting, and predation by the introduced mongoose. The human population of Hispaniola is increasing, most of the island's forest cover is being cleared for agriculture, and hutias are killed whenever encountered (Thornback and Jenkins 1982). The IUCN regards *P. hylaeum* as a subspecies of *P. aedium* and classifies the latter as indeterminate. Recent surveys in Haiti have found *P. aedium* to be somewhat more common than once feared, though still in jeopardy, and also have received reports of the possible survival of *P. velozi* (Woods 1981; Woods, Ottenwalder, and Oliver 1985).

RODENTIA; CAPROMYIDAE; Genus ISOLOBODON
J. A. Allen, 1916

There are two subgenera and three species (Hall 1981):

subgenus *Isolobodon* J. A. Allen, 1916

I. portoricensis, known only by skeletal remains from the Dominican Republic, Puerto Rico, and the Virgin Islands;
I. levir, known only by skeletal remains from Hispaniola;

subgenus *Aphaetreus* Miller, 1922

I. montanus, known only by skeletal remains from Hispaniola.

These extinct hutias were about the size of *Plagiodontia* (Allen 1942). They are distinguished mainly by dental characters. According to Hall's (1981) key, in *Plagiodontia* the root of the upper incisor occurs at the anterior margin of the zygomatic process of the maxillary, and the upper premolar has one external reentrant angle. In *Isolobodon* the root of the upper incisor occurs in the infraorbital foramen, and the upper premolar has two external reentrant angles.

Available evidence suggests that *I. portoricensis* was native to Puerto Rico but was transported to other islands by the early Indians. It may have been domesticated, and its abundant remains in kitchen middens indicate that it formed a staple article of diet for the aborigines. *I. levir* was evidently native to Hispaniola and seems not to have been so heavily utilized as human food. Both species may have become extinct shortly after the coming of European explorers. *I. montanus* inhabited mountainous areas. Its remains have not been found in association with people, and its date of extinction is not known. Skeletal fragments of all three species are present in cave deposits made by the extinct giant barn owl (*Tyto ostolaga*). This owl preyed on the hutias and perhaps disappeared when their numbers were reduced by excessive human hunting (Allen 1942).

Woods (1981) stated that *I. portoricensis* was a hundred times more common on Hispaniola than was *Plagiodontia* but disappeared because of its less secretive habits. In a survey in 1984–85, Woods, Ottenwalder, and Oliver (1985) received reports of living animals in Haiti and Puerto Rico that resemble *Isolobodon*. Nonetheless, these authorities concluded that the genus is probably extinct, though it evidently persisted in small numbers until wiped out by the introduced mongoose early in the twentieth century. Apparently fresh remains of *Isolobodon* in owl pellets were still being found at that time.

RODENTIA; CAPROMYIDAE; Genus HYPERPLAGIODONTIA
Rímoli, 1977

The single species, *H. stenocoronalis*, is known only by skeletal remains from the type locality in north-central Haiti. This evidently extinct hutia was larger than *Plagiodontia*, its nearest generic relative. Its molariform teeth are slightly highcrowned and have short roots. The folds in the molariform teeth are more compressed anteriorly and are sharper and more conspicuous than in *Plagiodontia* (Hall 1981).

RODENTIA; CAPROMYIDAE; Genus MYOCASTOR
Kerr, 1792

Nutria, or Coypu

The single species, *M. coypus*, is native to southern Brazil, Bolivia, Paraguay, Uruguay, Argentina, and Chile (Cabrera 1961).

Head and body length is 430–635 mm and tail length is 255–425 mm. Weight is usually 5–10 kg but occasionally reaches 17 kg (Grzimek 1975). The pelage of the upper parts contains many long, coarse guard hairs with a general coloration of yellowish brown or reddish brown. These hairs nearly conceal the soft, thick, velvety, dark gray underfur. The belly fur is pale yellow and is not as coarse as that of the upper parts. The tail is scaly and thinly haired except at the base.

Myocastor looks like a large, robust rat. The eyes and ears are small. The tail is cylindrical, not laterally compressed as in the muskrat *(Ondatra)*. The webbed hind foot is much longer than the forefoot and contains five digits; the first four are connected by skin, and the fifth is free and used for grooming the fur. The forefoot has four long, flexible, unwebbed digits and a vestigial thumb. The claws are sharp and strong. The

incisor teeth are large and colored a brilliant orange. Females have four pairs of thoracic mammae situated well up on the sides of the body.

The nutria is semiaquatic, inhabiting marshes, lake edges, and sluggish streams. Although it generally prefers fresh water, the population of the Chonos Archipelago in Chile occurs in brackish and salt water. According to Lowery (1974), *Myocastor* is adept on land but at home in the water and is often seen by day but is most active at night. It commonly makes platforms of vegetation, where it feeds and grooms itself. For shelter the nutria takes over the hole of another animal or constructs its own burrow. The latter may be a simple tunnel or a complex system containing passages that extend 15 meters or more and chambers that hold crude nests of vegetation. The nutria also makes runways through the grass and wanders within a radius of about 180 meters of its den. Much time is spent swimming and browsing on aquatic plants. The diet is strictly vegetarian (Lowery 1974). A study in Maryland determined roots to be the most important food and population density to range from 2.7/ha. to 16.0/ha. (Willner, Chapman, and Pursley 1979).

Myocastor lives in pairs, but the presence of many animals in favorable habitat may give the impression of a large colony. Breeding occurs throughout the year in captivity and in at least some wild populations. Females are polyestrous and may produce two or three litters annually (Brown 1975; Grzimek 1975; Willner, Chapman, and Pursley 1979). Females experience a postpartum estrus within 1–2 days of giving birth, have an estrous cycle of about 24–26 days, and have a period of receptivity lasting 1–4 days (Lowery 1974). Ovulation may be induced. The gestation period is usually 128–30 days. Litter size averages about 5 and ranges from 1 to 13 young. The newborn weigh approximately 225 grams each, are fully furred, and have their eyes open. They are capable of surviving away from the mother after only 5 days of nursing, but they usually remain for 6–10 weeks. Young born in the summer may attain sexual maturity at 3–4 months, and those born in the fall, at 6–7 months (Weir 1974b). In the wild there is 80 percent mortality in the first

Nutria, or coypu *(Myocastor coypus)*, photo by P. Rödl.

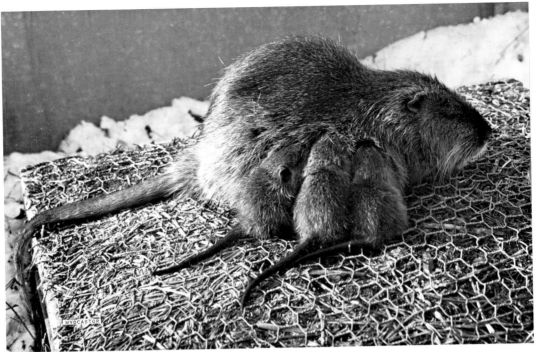

Nutria, or coypu *(Myocastor coypus)*, with nursing young, photo by H. L. Dozier from U.S. Fish and Wildlife Service.

year of life, and few animals live more than 2 or 3 years (Willner, Dixon, and Chapman 1983). Captives maintained by Gosling and Baker (1981) lived up to 6 years and 7 months, and there are reports of individuals surviving for about 10 years.

A demand for the velvety, plushlike underfur of the nutria developed in the early nineteenth century and has continued to the present. *Myocastor* also has been hunted by people for use as food. The genus became rare in much of its natural habitat, but by the early 1900s efforts were being made to regulate the harvest and to establish captive breeding farms (Allen 1942). Such farms were begun both in the original range of the nutria and in many other parts of the world. Some animals escaped from captivity or were deliberately introduced, and feral populations developed in the United States, Canada, England, France, Holland, Scandinavia, Germany, Asia Minor, the Caucasus, Soviet Central Asia, and Japan (Corbet 1978; Hall 1981; Van Den Brink 1968). Although intensively trapped for its fur, the nutria is considered a pest in some of these places because its burrows damage dikes and irrigation facilities, it raids rice and other cultivated crops, and it competes with native fur-bearing animals.

The most extensive populations in the United States are in the south-central part of the country, especially in the marshes of Louisiana. Most of these animals are apparently descended from 20 individuals brought to Louisiana in 1938. By the late 1950s there were an estimated 20 million feral nutria in Louisiana, and by 1962 the nutria had replaced the muskrat as the leading fur animal in the state (Lowery 1974). In the 1976/77 season, 1,890,853 nutria pelts from Louisiana were marketed. The total number for the United States was 2,018,815, and the value was $10,598,779 (Deems and Pursley 1978). The U.S. take in 1983/84 fell to 917,311 (Novak et al. 1987). Several million pelts also are exported annually from South America, though there seems to have been a decline in recent years through overexploitation; the average

value per pelt approached U.S. $20.00 in 1983 (Gudynas 1989).

RODENTIA; **Family OCTODONTIDAE**

Octodonts

This family of five Recent genera and nine species occurs in Peru, Bolivia, Argentina, and Chile. The sequence of genera presented here follows that of Cabrera (1961).

Head and body length is 125–310 mm and tail length is 40–180 mm. In the genera *Octodon, Octodontomys,* and *Octomys* the tail is about as long as the head and body, while in *Aconaemys* and *Spalacopus* it is much shorter than the head and body. The fur on the body is usually long, thick, and silky. The coarse hair of the tail is short at the base and becomes longer toward the tip, where it forms a tuft.

The head is large, and the nose is pointed. The moderate-sized ears are rounded and covered with short hair. The whiskers are long. The hind foot has four well-developed toes, each with a sharp, curved claw and a row of stiff bristles extending beyond the claw. The thumb usually is reduced (Woods 1984). The dental formula is: (i 1/1, c 0/0, pm 1/1, m 3/3) × 2 = 20. The grinding surfaces of the cheek teeth are shaped like a figure eight, or sometimes like a kidney. Females have three or four pairs of mammae.

Octodonts inhabit coastal areas, foothills, and the Andes Mountains to elevations of 3,500 meters or more. They live both in cultivated areas and on barren, rocky slopes. They are good diggers, especially *Aconaemys* and *Spalacopus*. They either construct their own burrows or dwell in crevices, rock piles, fences, hedgerows, or the holes of other animals. These agile rodents carry their tail erect while running and sometimes climb shrubs or trees. They often sit on their haunches.

The diet consists of plant matter, such as bulbs, tubers, bark, and cactus; occasionally growing cornstalks are eaten. Females may give birth to two litters annually, each with 1–10 young. The geological range of this family is early Oligocene to Recent in South America (Woods 1984).

RODENTIA; OCTODONTIDAE; Genus OCTODON
Bennett, 1832

Degus

There are three species (Cabrera 1961; Pine, Miller, and Schamberger 1979):

O. *degus*, Atacama to Curico Province in northern and central Chile;

O. *bridgesi*, base of the Andes in the provinces of O'Higgins, Colchagua, Curico, and Concepción in central Chile;

O. *lunatus*, coastal parts of Coquimbo, Aconcagua, and Valpariso provinces in central Chile.

A report of O. *degus* from Peru may have been based on an escaped pet (Woods and Boraker 1975). However, based on the recent discovery of nonfossilized remains, Mares and Ojeda (1982) indicated that O. *bridgesi* may possibly occur in southern Argentina.

Head and body length is 125–95 mm and tail length is 105–65 mm. Woods and Boraker (1975) gave the weight of O. *degus* as 170–300 grams and noted that the other species are larger. The upper parts are grayish to brownish, often with an orange cast, and the underparts are creamy yellow. A black brush at the tip of the tail is most prominent in O. *degus*.

When holding a degu by the tail, care must be taken or the animal may spin like a top and leave the person holding only the skin (Woods and Boraker 1975). The degu then bites off the naked vertebrae and tendons, there is little bleeding, and the wound heals. The lost section is not replaced, and many animals are found with a portion of the tail missing. When the degu is familiar with the handler, tail shedding rarely occurs.

On all four limbs the first four digits are well developed and

have sharp claws, but the fifth is reduced. Long, comblike bristles extend over the claws of the hind feet. The auditory bullae are moderate in size and smaller than those of *Octodontomys* and *Octomys*. The anterior upper cheek teeth of O. *degus* are kidney-shaped, with only moderate internal indentations. In the other two species these teeth have deep indentations that nearly touch the opposite side of the teeth (Woods and Boraker 1975). The incisor enamel is pale orange. Females have eight mammae.

The only species for which there is substantial natural history information is O. *degus*, and it has been investigated both in the wild and in the laboratory (Boulenge and Fuentes 1978; Fulk 1975, 1976; Meserve, Rodriguez-M., and Martin 1978; Weir 1974b; Woods and Boraker 1975). This species inhabits the west slope of the Andes at elevations of up to 1,200 meters. It is found in relatively open areas near thickets, rocks, or stone walls. It is active throughout the year and is diurnal, with peaks of activity in the morning and late afternoon. It feeds on the ground and will also climb into the branches of small trees and bushes. An elaborate communal burrow system is constructed, with the main section under rocks or shrubs. A complex network of tunnels and surface paths leads out to feeding sites. Piles of sticks, stones, and dung accumulate on the ground above the tunnels and apparently serve as territorial markers. The diet includes grass, leaves, bark, herbs, seeds, and fruits and during the dry season may be supplemented with fresh droppings of cattle and horses. The degu sometimes becomes an agricultural pest by raiding orchards, vineyards, and wheat fields. It stores food for winter use.

This species is the most common mammal of central Chile. Reported population densities have ranged from 10 per ha. to 259 per ha., with 40–80 per ha. possibly being near average. Densities are highest in the spring (September–November), when juveniles are entering the population. O. *degus* lives in small colonies with a strong social organization based on group territoriality. The burrow is the center of the defended territory. Females of the same social group may rear their young in a common burrow. Captive animals exhibit considerable social tolerance, and a group of strangers may be placed together without fighting. Eisenberg (1974) reported a variety of vocalizations.

Captive colonies in the Northern Hemisphere breed throughout the year, and females may have more than one annual litter. In the natural range in central Chile, however,

Degu *(Octodon degus)*, photo by Luis E. Peña. Inset from *Trans. Zool. Soc. London.*

pregnant and lactating females have been found only in September. In northern Chile pregnant females or juveniles have been taken in February, April, and November. There seems to be no regular estrous cycle, and female O. degus may require the presence of a male to induce ovulation. They experience a postpartum estrus but do not usually mate at this time. The gestation period averages about 90 days. Litters contain 1–10 young, with an average of 6.8 in a laboratory colony in Vermont. In this colony the young have been born fully furred and with their eyes open. In another group in London, however, the newborn have been sparsely haired, and their eyes have not opened for 2–3 days. Each newborn weighs about 14 grams. The offspring must nurse for at least 14 days and normally do so for 4 weeks. Wild parents have been observed to cut fresh grass and bring it into the burrow for the young to eat. The age of sexual maturity has been reported to be as early as 45 days and as late as 20 months, but the average is about 6 months. According to Jones (1982), a specimen of O. degus was still alive after 7 years and 1 month in captivity.

RODENTIA; OCTODONTIDAE; **Genus OCTODONTOMYS**
Palmer, 1903

Chozchoz (plural = chozchoris)

The single species, O. gliroides, occurs in the highlands of southwestern Bolivia, northern Chile, and northwestern Argentina (Cabrera 1961; Pine, Miller, and Schamberger 1979). Except as noted, the information for this account was provided by B. J. Weir (Wellcome Institute, Zoological Society of London, pers. comm.).

Head and body length is about 180 mm, tail length is about the same, and weight is 100–200 grams. An adult female from Chile had a head and body length of 170 mm, a tail length of 132 mm, and a weight of 158 grams (Pine, Miller, and Schamberger 1979). The fur resembles that of *Chinchilla* in being very dense and soft. Coloration is pearly gray above and white below. The eyes and external ears are relatively large. The auditory bullae and the tail brush are larger than those of *Octodon*. There are no apparent modifications for fossorial life. Females have one inguinal and two lateral pairs of mammae (Weir 1974b).

The chozchoz is found in mountains and on the Altiplano. It lives in burrows at the base of cacti and scrubby acacias, in rock crevices, in caves, and in old Indian tombs. In northern

Argentina, at least, this rodent is active only after dusk. It eats acacia pods during the winter and fruits of the cactus during summer. Burrows can be detected by the piles of empty yellow seed pods outside. Eisenberg (1974) stated that the threat vocalization of *Octodontomys* is a twitter.

On 17 November, Pine, Miller, and Schamberger (1979) collected both a lactating female and a juvenile in northern Chile. Weir (1974b) reported that females have a postpartum estrus, gestation lasts 100–109 days, and litters contain one to three young, each weighing 15 grams at birth. The young are born fully furred and with their eyes open. One specimen was still living after 7 years and 4 months in captivity (Jones 1982).

RODENTIA; OCTODONTIDAE; **Genus OCTOMYS**
Thomas, 1920

Viscacha Rats

There are two species (Cabrera 1961):

O. mimax, mountains of northwestern Argentina;
O. barrerae, plains of eastern Mendoza Province in west-central Argentina.

Cabrera put O. barrerae in the separate genus *Tympanoctomys* Yepes, 1940, and this procedure was followed by Torres-Mura, Lemus, and Contreras (1989), but Woods (1984) and most other recent authorities have included *Tympanoctomys* in *Octomys*.

Head and body length is 160–70 mm and tail length is 170–80 mm. A specimen of O. (Tympanoctomys) barrerae weighed 82 grams (Ojeda et al. 1989). The upper parts are pale buff; the underparts, hands, and feet are white. The body form is not modified for fossorial life, and the tail is bushy. This genus resembles *Octodontomys* externally, but the re-entrant folds of the upper molars in *Octomys* meet in the middle of the teeth. The auditory bullae of *Octomys* are the largest in the family Octodontidae (Packard 1967).

Viscacha rats appear to be rare and are difficult to trap. There is no evidence that the scarcity has been caused by human agency. According to Contreras, Torres-Mura, and Yáñez (1987), viscacha rats live in desert scrub habitats in mountainous regions and are nocturnal burrowers and herbivores. These rodents spend the day sleeping in their burrows. Females produce several litters per year.

The species O. (Tympanoctomys) barrerae had not been

Chozchoz (*Octodontomys gliroides*), photo by Barbara J. Weir.

Viscacha rat *(Octomys mimax)*, photo by Michael Mares.

collected since 1940 but was taken recently in Mendoza Province by Torres-Mura, Lemus, and Contreras (1989) and Ojeda et al. (1989). The new specimens show that the species is found only in places with halophytic (salt-rich) vegetation and is a strict herbivore, feeding mainly on leaves and stems. A pregnant female was captured on 12 December 1987.

RODENTIA; OCTODONTIDAE; Genus SPALACOPUS
Wagler, 1832

Coruro

The single species, *S. cyanus*, occurs in central Chile, from the coast to the Andean slopes and from Coquimbo to Maule Province (Cabrera 1961; Pine, Miller, and Schamberger 1979; Reig 1970).

Head and body length is 140–60 mm, tail length is 40–50 mm, and weight is about 60–110 grams. The pelage is soft and glossy. The coloration is brownish black or black throughout, though one specimen had large, irregular white pectoral and pelvic patches. The feet are dark gray.

The body form is stocky. The tail is cylindrical, scaly, almost hairless, and only about as long as the hind foot. The claws are not greatly enlarged, but the forefeet are specialized for digging. The eyes and external ears are small. The cheek teeth are ever-growing and have crowns with an **8**-shaped pattern, but the reentrant folds do not meet in the middle of the teeth. The incisors are long, broad, and strongly curved forward, being even more procumbent than in *Ctenomys* (Reig 1970).

The coruro is fossorial. It inhabits both coastal areas and mountain slopes and has an elevational range of sea level to about 3,000 meters. Pine, Miller, and Schamberger (1979) found the genus to be very abundant on wet flats along streams. Reig (1970) studied colonies in a semiarid area with scattered clumps of vegetation. Except as noted, the remainder of this account is based on that study.

Spalacopus maintains a complex communal burrow system. There are long, interconnecting tunnels about 5–7 cm in diameter and generally 10–12 cm below the surface. Leading from the main tunnels are short passages, some of which are used to push excavated earth to surface mounds and some of which terminate under clusters of food plants. Entrances are not kept plugged. In one colony there were 350 recently made openings, of which 314 were in an area of 400 sq meters. *Spalacopus* uses its incisors and forefeet to dig in a manner much like that of *Ctenomys*. It is evidently active only during daylight and is rarely seen above ground, except when excavating soil. Colonies are nomadic, using up the food resources of one area in a few days and then tunneling on to new sites. The diet consists primarily, if not exclusively, of the tubers and underground stems of the lily *Leucoryne ixiodes*. Free water evidently is not required. According to reports made prior to Reig's investigation, *Spalacopus* hoards bulbs and tubers in its burrows for the winter, and these stores are raided by people for use as food.

The colonies are small but usually adjoin one another, so *Spalacopus* can sometimes be found almost continuously over an extensive area. One colony consisted of 15 animals, including at least 3 adult males, 4 nonpregnant adult females, 2 pregnant females, 2 juvenile males, and 2 juvenile females. The pregnant females, which Reig caught in late February,

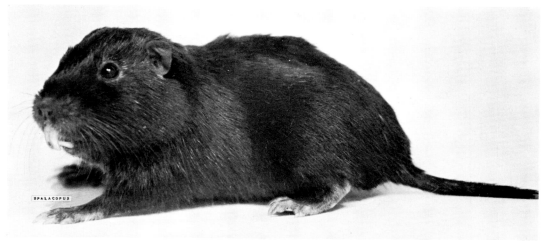

Coruro *(Spalacopus cyanus)*, photo by Luis E. Peña.

each contained 3 small fetuses. Another source indicated that females have two litters per year.

The call of the coruro as heard when wild individuals push soil to the surface is a sequence of three or four trills, each lasting about 5 seconds and separated by about 3 seconds of silence. Captives were active and noisy. Eisenberg (1974) listed several other sounds, including a tooth chatter, used as a threat. According to Jones (1982), one captive was still living after 5 years and 10 months.

RODENTIA; OCTODONTIDAE; Genus ACONAEMYS
Ameghino, 1891

Rock Rats

There are two species (Cabrera 1961; Pearson 1984; Pine, Miller, and Schamberger 1979):

A. *fuscus*, highlands of west-central Argentina and that part of Chile from Curico to Osorno Province;

A. *sagei*, southern Argentina.

Head and body length is 140–87 mm and tail length is 55–80 mm. Greer (1965) listed weights of 118.9 grams for a male, 135.3 grams for a nonpregnant female, and 152.2 grams for a pregnant female. Pearson (1984) reported weights of 121–43 grams for A. *fuscus* and 83–110 grams for A. *sagei*. The coloration of the subspecies A. *fuscus fuscus* is usually dark brown throughout. Some specimens have a bicolored tail and a suffusion of white or cream-colored hairs on the dorsal surface of the feet (Pine, Miller, and Schamberger 1979). The southerly subspecies, A. *f. porteri*, has a more woolly pelage and a bicolored tail, black above and white below. A. *sagei* also is brown and has a slightly bicolored tail.

This stocky rodent is modified for fossorial life. The tail is much shorter than the head and body. *Aconaemys* differs from *Spalacopus* in that the reentrant folds of the upper

Rock rat *(Aconaemys fuscus)*, photo by Richard D. Sage.

molars meet in the middle of the teeth and the upper incisors do not curve forward so markedly.

Aconaemys is found in the high Andes and in coastal mountains. Most of what is known about this genus was recorded by Osgood (1943), who, however, thought that it was restricted to coniferous forests and that it would disappear as the large trees *(Araucaria)* under which it burrowed were cut down by people. Greer (1965) found *Aconaemys* to inhabit open, cleared areas as well as forest and to burrow under bushes as well as trees. Both authorities agreed that *Aconaemys* makes a complex, though shallow, burrow system and prefers well-drained ground near ledges and boulders. A series of entrances open flush with the ground and are connected by numerous branching tunnels. The entrances are also connected by surface runways, which are either open or partly concealed by vegetation. In the mid-nineteenth century *Aconaemys* was reported to be very common on the east side of the Andes and to honeycomb the ground with its burrows, thus posing a danger to horses and riders. The genus is mainly nocturnal but occasionally is seen by day. It is active throughout the winter, under the snow. According to Pearson (1983, 1984), captives ate a variety of green vegetation, roots, fruits, and bread. He stated also that *A. fuscus* uttered high-pitched squeaks and that *A. sagei* made chuckling sounds and squeaks like those of guinea pigs. Newborn young and pregnant female *A. fuscus* were collected in October and November. Three pregnant females contained 2, 3, and 5 embryos.

RODENTIA; Family CTENOMYIDAE; Genus CTENOMYS
De Blainville, 1826

Tuco-tucos

The single Recent genus, *Ctenomys*, contains 2 subgenera and 44 species (Altuna and Lessa 1985; Anderson, Yates, and Cook 1987; Cabrera 1961; Contreras 1984*a*; Contreras and Berry 1982*a*, 1982*b*; Contreras and Contreras 1984; Contreras, Roig, and Suzarte 1977; Freitas and Lessa 1984; Honacki, Kinman, and Koeppl 1982; Pine, Miller, and Schamberger 1979; Roig and Reig 1969; Travi 1981; Vitullo, Roldan, and Merani 1988; Ximenez, Langguth, and Praderi 1972):

subgenus *Ctenomys* De Blainville, 1826

C. yolandae, northeastern Argentina;
C. brasiliensis, eastern Brazil;
C. flamarioni, southern Brazil;
C. torquatus, southern Brazil, Uruguay, northeastern Argentina;
C. pearsoni, Uruguay;
C. rionegrensis, Uruguay;
C. frater, southern Bolivia, northwestern Argentina;
C. knighti, northwestern Argentina;
C. tuconax, northwestern Argentina;
C. emilianus, west-central Argentina;
C. pontifex, west-central Argentina;
C. mendocinus, Argentina;
C. occultus, northwestern Argentina;
C. tucumanus, northwestern Argentina;
C. latro, northwestern Argentina;
C. saltarius, northwestern Argentina;
C. perrensis, northeastern Argentina;
C. talarum, east-central Argentina;
C. australis, east-central coast of Argentina;
C. porteousi, east-central Argentina;
C. azarae, central Argentina;
C. argentinus, northern Argentina;
C. dorbignyi, northeastern Argentina;
C. bonettoi, northern Argentina;
C. boliviensis, Mato Grosso of central Brazil, east-central Bolivia;
C. steinbachi, eastern Bolivia;
C. dorsalis, Paraguay;
C. minutus, southern Brazil, eastern Bolivia, Uruguay;
C. leucodon, region of Lake Titicaca in southern Peru and western Bolivia;
C. lewisi, southern Bolivia;
C. peruanus, extreme southern Peru;
C. robustus, northern Chile;
C. maulinus, central Chile;
C. magellanicus, extreme southern Chile and Argentina, including Tierra del Fuego;
C. opimus, southern Peru, southwestern Bolivia, northern Chile, northwestern Argentina;
C. fulvus, northern Chile;
C. coludo, northwestern Argentina;
C. famosus, west-central Argentina;
C. johanis, west-central Argentina;

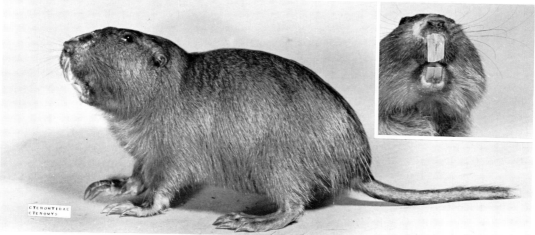

Tuco-tuco (*Ctenomys* sp.), photos by Ernest P. Walker.

Tuco-tuco *(Ctenomys knighti)*, photo by Michael Mares.

C. tulduco, west-central Argentina;
C. validus, west-central Argentina;
C. colburni, extreme southwestern Argentina;
C. sericeus, extreme southwestern Argentina;

subgenus *Chacomys* Osgood, 1946

C. conoveri, Paraguay.

Systematic understanding of this complex genus is in a state of flux, and it is likely that there will be much change in the number and relationships of species indicated above. Recent studies of genetic variation (Sage et al. 1986) and sperm morphology (Vitullo, Roldan, and Merani 1988) suggest taxonomic arrangements somewhat different from the above and from each other.

Part of the information used in the remainder of this account was provided by B. J. Weir (Wellcome Institute, Zoological Society of London, pers. comm.). Head and body length is 150–250 mm and tail length is 60–110 mm. Weight ranges from 100 grams in *C. talarum* to 700 grams in *C. tucumanus.* The skin on the body is loose. The pelage ranges from short to long and is usually thick. The coloration of the upper parts ranges from gray or creamy buff through various shades of brown to almost black, and the underparts are usually paler.

In outward appearance these South American burrowing rodents strongly resemble the North American pocket gophers (Geomyidae), and their habits are much the same. Pocket gophers have external cheek pouches, which tuco-tucos lack, and have better-developed fringes of hair on the forefeet. In *Ctenomys,* however, the hair fringes on the hind feet are stiffer and form comblike bristles. These "combs," which are the basis of the familial and generic names, are used to groom dirt from the fur.

The body is robust and cylindrical, the head is large, the tail is short and sparsely haired, and the neck and limbs are short and muscular. The forelimbs are somewhat shorter than the hind limbs. All the digits have strong claws, those on the forefeet being the longest. The eyes are small, though perhaps somewhat larger than would be expected for fossorial rodents, and the external ears are reduced. The auditory bullae are enlarged. The dental formula is: (i 1/1, c 0/0, pm 1/1, m 3/3) × 2 = 20. The prominent incisor teeth have a bright orange enamel and are large and thick, but they are not as procumbent as those of *Spalacopus.* The crown pattern of the molars is kidney-shaped, and the last molar is reduced. Females, at least those of *C. opimus, C. peruanus,* and *C. talarum,* have one inguinal and two lateral thoracic pairs of mammae.

Tuco-tucos are found from the tropics to the sub-Antarctic but seem to prefer coastal areas, grassy plains, forests, and the Altiplano. The elevational range is sea level to 4,000 meters (Weir 1974a). *C. torquatus* has been found in cultivated fields, pastures, and vacant lots in the city of Montevideo (Barlow 1969). The main requirement of this fossorial genus seems to be sandy, somewhat dry soil. One species, *C. lewisi,* tunnels in stream banks and may be semiaquatic.

Burrows usually consist of a long main tunnel, either sinuous or nearly straight, and several short passages that either have a blind ending or lead to feeding sites at the surface. The main tunnel may be 14 meters long and is about 5–7 cm wide and 30 cm below the surface (Barlow 1969; Packard 1967). There is a grass-lined nest chamber, usually below the level of the main tunnel, as well as several chambers for food storage. The surface entrances may be marked by heaps of excavated soil and are often plugged if the burrow is occupied. Short lengths of chopped grass are commonly found in the burrows of *C. talarum* and may serve as air filters. Both this species and *C. torquatus* keep burrow temperature at an average of 20°–22° C by opening and closing the surface entrances according to the direction of wind and sun. A burrow is sometimes shared with other animals, such as lizards, mice, and cuis *(Galea).*

Most digging takes place during the daylight hours, mainly during early morning and late afternoon. The earth is loosened with the forefeet and then swept away with the hind feet. The incisors are used mainly to cut through protruding roots (Weir 1974*a*). *C. talarum* is known to pack the earth of its tunnel walls firmly by standing on its forefeet, walking the hind feet up the wall, urinating, and then stamping the moistened soil into position as it lowers the hind feet. Tuco-tucos are rarely seen completely out of their burrows, but short surface forays may occur on sunny days. The position of the eyes, almost level with the top of the head, enables an animal to look out of its shelter without appreciably exposing itself. If danger threatens, the animal rapidly retreats backward, its nearly naked tail apparently acting as a sensory organ. *C. peruanus* and *C. opimus* can distinguish a moving human at a distance of about 50 meters. Some food, in the form of roots, stems, and grass, is probably pulled into the burrows from below. *Ctenomys* also obtains the aboveground parts of plants by reaching out from the surface entrance of the burrow (Barlow 1969). The diet is thought to be entirely vegetarian.

Population densities as high as 207/ha. have been recorded (Pearson et al. 1968), but the figure is usually much lower. Barlow (1969) noted 80 burrow systems in an area of 7 ha. In most species a burrow is occupied by a single individual, but several female *C. peruanus* may share the same system.

The common name, "tuco-tuco," is an attempt to express in words the vocalization of some species, such as *C. talarum.* This call is made mainly by the males, probably to show territoriality or fear. The actual sound is something like "tloc-tloc-tloc," and it seems to come bubbling up from the ground. A single vocalization lasts from 10 to 20 seconds, starting slowly and then becoming more rapid.

Females are generally monestrous and produce only a single litter per year. They may experience a postpartum estrus but do not usually mate at that time (Weir 1974*b*). In Peru *C. opimus* mates in the dry season, and most births occur in the wet season, when plant growth is at its peak. In Uruguay, however, *C. torquatus* mates in the wet season, and pregnant females are found from September to January (Barlow 1969). In Argentina *C. talarum* mates from May to July. In Chile, Pine, Miller, and Schamberger (1979) found a lactating female *C. opimus* on 15 September and one of *C. robustus* on 26 April. Weir (1974*b*) listed average gestation periods of 120 days for *C. peruanus*, 102 days for *C. talarum*, and 107 days for *C. torquatus*; and litter sizes of one to three young in *C. opimus*, one to five in *C. peruanus*, and one to seven in *C. talarum.* Newborn *C. peruanus* are well developed and are able to give the same call as adults, to leave the nest, and to feed on green vegetation almost immediately. The young of captive *C. talarum*, however, do not have their eyes open at birth and have only the guard hairs projecting through the skin. They weigh about eight grams each. They grow quickly and can usually fend for themselves at 10 days, but they normally nurse for 5 weeks. Both sexes of *C. talarum* reach puberty at an average age of 8 months, in the breeding season following their birth. Mean longevity in the wild is probably less than 3 years.

Horseback riding in country frequented by tuco-tucos is sometimes dangerous, as the burrows cave in and cause the horses to break their legs. Tuco-tucos also are sometimes considered pests, because they damage cultivated crops and compete with livestock for forage, and thus have been intensively persecuted by people. They are now absent or greatly reduced in numbers in some areas in which they were formerly abundant, for example, in southern Patagonia to the east of the Andes, an area now largely fenced and devoted to sheep raising. Miller et al. (1983) reported that intensive grazing of sheep and other agricultural activity also has contributed to the decline and has resulted in *C. magellanicus* becoming rare and endangered.

The geological range of the family Ctenomyidae is Pliocene to Recent in South America. This family is thought to be a fossorial offshoot from the Octodontidae. Indeed, some authorities consider it only a subfamily of the latter (Mares and Ojeda 1982).

RODENTIA; **Family ABROCOMIDAE; Genus ABROCOMA**
Waterhouse, 1837

Chinchilla Rats, or Chinchillones

The single Recent genus, *Abrocoma*, contains two living species (Cabrera 1961; Pine, Miller, and Schamberger 1979):

A. bennetti, central Chile;
A. cinerea, southeastern Peru, southern Bolivia, northern Chile, northwestern Argentina.

Remains of an extinct species, *A. oblativa*, have been recovered from burial sites in Peru.

In *A. cinerea* head and body length is 150–201 mm and tail length is 59–144 mm. In the somewhat larger *A. bennetti* head and body length is 195–250 mm and tail length is 130–80 mm. An adult male *A. bennetti* weighed 224 grams, and an adult female weighed 307 grams (Pine, Miller, and Schamberger 1979). The pelage consists of long, soft, dense underfur and fine, long guard hairs; it resembles that of *Chinchilla* but is not so woolly. Coloration is silvery gray above and white or yellowish below in *A. cinerea* and brownish gray above and brownish below in *A. bennetti.* There may be a white or yellowish patch on the chest, marking a glandular region.

Abrocoma bears some resemblance to *Chinchilla* but the head and ears are proportionally longer, giving it a ratlike appearance—hence the common name, "chinchilla rats." The nose is pointed, the eyes are large, and the ears are large and rounded. The tail is cylindrical, shorter than the head and body, and covered with fine, short hairs. The limbs are short. The forefoot has four digits and the hind foot, five. The soles of the feet are naked and covered with small tubercles. The small, weak claws are hollow on the underside. Stiff hairs project over the claws of the three middle toes of the hind foot (as in the families Chinchillidae, Octodontidae, and Ctenomyidae), forming combs that are probably used to remove parasites and groom the pelage, or perhaps to aid in removing dirt loosened by digging.

The intestinal tract of *Abrocoma* is long; the small intestine measures 1.5 meters and the large intestine, 1 meter. The voluminous cecum is 205 mm in length. *A. bennetti* has more ribs than any other rodent, 17 pairs. Female *Abrocoma* have four mammae.

The rostrum of the skull is long and narrow, the braincase is rounded, and the bullae are enlarged. The dental formula is: (i 1/1, c 0/0, pm 1/1, m 3/3) × 2 = 20. The incisors are

Chinchilla rat (*Abrocoma bennetti*), photo by Ernest P. Walker.

narrow, and the cheek teeth continue to grow throughout life. The lower cheek teeth differ from the uppers in having two deep, sharply angular, inner folds (Packard 1967).

The species *A. cinerea* is known only from the Altiplano, generally occurring in rocky areas at elevations of 3,700–5,000 meters. *A. bennetti* is known from the Andes and coastal hills of Chile up to elevations of 1,200 meters. Chinchillones live in tunnels in the ground and among the crevices of rocks. The entrances to these tunnels are usually located at the base of a bush or under rocks. Fulk (1976) found that *A. bennetti* sometimes shared a burrow with *Octodon degus*, a rodent of comparable size; on two occasions nest burrows were determined to be occupied by mothers and young of both species. *A. bennetti* has been reported to climb bushes and trees. The diet of *Abrocoma* includes many kinds of plant material.

Chinchillones appear to be colonial; about half a dozen *A. cinerea* were noted living within 18 meters of each other. This species is reported to grunt before or after an attack, to squeak while avoiding intruders, and to gurgle when being groomed (Eisenberg 1974). A pregnant female *A. cinerea* with two embryos was taken in December in Peru, and a mother with two young about 1 week old was collected in April in northern Chile. Weir (1974b) listed a single gestation period of 115–18 days for this species and gave litter size as one to two young. Fulk (1976) collected a pregnant female *A. bennetti* near Santiago, Chile, on 25 July, took another 300 km to the north on 1 June, and observed a female carrying two newborn at Santiago on 30 August. He gave the embryo counts for three pregnant females of this species as four, five, and six. According to Jones (1982), a captive *A. bennetti* lived 2 years and 4 months.

Miller et al. (1983) reported that chinchillones have become rare in Chile through hunting and possibly habitat destruction. In some areas the native people sell the pelts of these animals to gullible travelers as *Chinchilla*. In parts of Chile the skins are taken to local fur markets, but they do not bring high prices.

The geological range of the family Abrocomidae is late Miocene to Recent in South America.

RODENTIA; Family ECHIMYIDAE

Spiny Rats

This family contains 17 Recent genera and 69 species. One genus, *Heteropsomys*, occurred in the West Indies but appar-

ently became extinct in the nineteenth century. The living genera are found from southern Honduras to Peru, Paraguay, and southeastern Brazil. The sequence of genera presented here is based on those of Cabrera (1961) and Hall (1981), who recognized two subfamilies: Echimyinae, for most genera; and Dactylomyinae, for *Dactylomys*, *Kannabateomys*, and *Olallamys*. However, two additional subfamilies are recognized: Chaetomyinae, based on the transfer of *Chaetomys* from the Erethizontidae; and Heteropsomyinae, for *Heteropsomys*, though the latter may actually belong in the family Capromyidae (Woods 1984).

Head and body length is 105–480 mm and tail length is 50–430 mm. The tail ranges from less than one-fourth the head and body length (in some *Euryzygomatomys*) to more than the head and body length (Packard 1967). All living genera except the dactylomyines and *Thrichomys* and *Isothrix* have a spiny or bristly pelage consisting of flattened, stiff, sharp-pointed hairs attached to the flesh by a narrow basal stalk. The tail varies from well haired to scantily haired. The tail is often lost, because it fractures easily, which may help it to escape from predators.

Echimyids are ratlike in general appearance, with pointed or somewhat truncate noses and moderate-sized eyes and ears. The rounded or bluntly pointed ears extend beyond the pelage. The limbs are of moderate length. The first digit of the forefoot (pollex) is vestigial. The other digits are short in most genera; moderately elongate in the semiarboreal *Isothrix*, *Diplomys*, *Echimys*, and *Makalata*; and greatly elongate, somewhat united, and modified in the Dactylomyinae, so the first and second toes of the hind foot, and the second and third of the forefoot, are on one side of a limb, while the more lateral toes are on the other side (Packard 1967). The dental formula is: (i 1/1, c 0/0, pm 1/1, m 3/3) × 2 = 20. The cheek teeth are flat-crowned and rooted.

Spiny rats inhabit forests or clearings, often near water. They are often common and the most abundant mammal in parts of their range, but most genera are little known. They may dig their own burrows or seek shelter beneath stumps, logs, or rocks. Several genera live in the hollows of trees, and *Euryzygomatomys* is apparently fossorial. *Thrichomys* has been noted in captivity leaping on its hind legs, but the other genera are climbing or scampering animals. Echimyids are reported to be good swimmers. They are usually active in the evening or at night. Most spiny rats die when exposed to heat and dryness. The herbivorous diet includes grass, sugar cane, bananas, other fruits, and nuts. Some echimyids consume large amounts of water.

These rodents sometimes live in small groups. Various scolding sounds have been noted, but captives are often peaceful. If handled carefully, they do not attempt to bite. Breeding may occur throughout the year in much of the range of the Echimyidae. Litter size is one to seven young. The newborn are well furred, and their eyelids are formed. Within a few hours they are active and nimble and can emit soft whistling sounds. They begin to eat solid food at about 11 days and leave the mother at about 2 months.

The geological range of this family is late Oligocene to Miocene, and Pleistocene to Recent, in South America, and Pleistocene to Recent in Central America and the West Indies.

RODENTIA; ECHIMYIDAE; Genus PROECHIMYS
J. A. Allen, 1899

Spiny Rats, or Casiragua

There are 2 subgenera and 28 species (Cabrera 1961; Gardner and Emmons 1984; Goodwin and Greenhall 1961; Hall 1981; Handley 1959, 1976; Husson 1978; Martin 1970; Patton 1987; Patton and Gardner 1972; F. Petter 1978*b*):

subgenus *Proechimys* J. A. Allen, 1899

P. mincae, northern Colombia;
P. poliopus, northwestern Venezuela;
P. guairae, northeastern Colombia, northwestern Venezuela;
P. hoplomyoides, southeastern Venezuela, and adjacent Guyana and Brazil;
P. urichi, northern Venezuela;
P. trinitatis, Trinidad and nearby mainland of Venezuela;
P. magdalenae, north-central Colombia;
P. chrysaeolus, north-central Colombia, western Venezuela;
P. canicollis, northern Colombia, northwestern Venezuela;
P. cuvieri, eastern Venezuela, the Guianas, northern Brazil, northeastern Peru;
P. decumanus, southwestern Ecuador, northwestern Peru;
P. longicaudatus, Peru, western Brazil, Bolivia, northern Paraguay;
P. brevicauda, southern Colombia, eastern Peru, northwestern Brazil;
P. gularis, eastern Ecuador;
P. amphichoricus, southern Venezuela and adjacent Brazil;

P. quadruplicatus, eastern Ecuador, northeastern Peru;
P. steerei, eastern Peru, western Brazil;
P. goeldii, Amazonian Brazil;
P. guyannensis, Colombia, Venezuela, Trinidad, the Guianas, Ecuador, northeastern Peru, Brazil, Bolivia;
P. oris, northeastern Brazil;
P. simonsi, eastern Ecuador, northeastern Peru, southern Colombia;
P. semispinosus, southern Honduras to coastal Colombia and Ecuador;
P. oconnelli, central Colombia;

subgenus *Trinomys* Thomas, 1921

P. albispinus, eastern Brazil;
P. myosuros, eastern Brazil;
P. setosus, eastern Brazil;
P. dimidiatus, southeastern Brazil;
P. iheringi, southeastern Brazil.

Petter (1973) suggested that the generic name *Cercomys* Cuvier, 1829, should replace *Proechimys*. As explained by Patton and Gardner (1972), the systematics of *Proechimys* are highly unstable, and the above list could easily be altered through further investigation or different taxonomic interpretation. Recent studies by Patton (1987) have clarified the situation to some extent, but he indicated that many questions remain. Hall (1981) listed *P. corozalus*, known only by a fossilized jawbone of uncertain geologic age from Puerto Rico, but noted that it may not actually represent *Proechimys*, and Woods (1985) indicated that the specimen is referable to the family Capromyidae.

Head and body length is 160–300 mm and tail length is 109–320 mm. *P. trinitatis* weighs 300–380 grams, and *P. semispinosus* weighs up to 500 grams (Gliwicz 1973). The fur is spiny but not as much so as in other genera of the Echimyidae. The general coloration of the upper parts is orangish brown, reddish brown, or light tawny mixed with black; the underparts are usually whitish. The thinly haired tail is brown above and white below. The tail is frequently absent through autotomy—it readily fractures at the fifth caudal vertebra and does not regenerate; this process probably provides a means of escape from predators that seize the tail. Females have one inguinal and two lateral thoracic pairs of mammae (Weir 1974*b*).

Casiragua usually inhabit forests, often near coasts and

Spiny rat *(Proechimys setosus)*, photo by Joao Moojen.

waterways. They occasionally move into human habitations but generally live in burrows in the ground, beneath tree roots, and among rocks. Except for the members of the subgenus *Trinomys*, they are not thought to dig their own burrows (Emmons 1982; Gardner and Emmons 1984). They are capable of climbing but rarely do so. They are nocturnal, leaving their dens in the evening to forage on the forest floor. Many kinds of plant material are eaten. In captivity they accept such items as fruit, corn, coconuts, cereals, potatoes, and cabbages. *P. semispinosus* is known to cache food in its burrow (Maliniak and Eisenberg 1971). Home ranges are generally small, about 0.1–1.5 ha., with those of males averaging larger than those of females (Emmons 1982).

In the Panama Canal Zone, Fleming (1971) determined that population density of *P. semispinosus* increased from 1.1/ha. in June to 3.8/ha. in October (the wet season) and then declined to 0.6/ha. in the following June. Gliwicz (1973) found an average density of 8.5/ha. in the same area. *P. semispinosus* defends its burrow against adults of the same sex, but there is considerable overlap in individual home ranges (Maliniak and Eisenberg 1971). In a study of *P. trinitatis* on Trinidad, Everard and Tikasingh (1973) found a maximum population density of about 14/ha. and an average home range of 0.1 ha. *Proechimys* is generally docile in captivity, but adults may be aggressive toward one another and inflict serious wounds. *P. semispinosus* growls and chatters its teeth when threatening a rival and has several other vocalizations (Eisenberg 1974).

There is breeding throughout the year by *P. semispinous* in the Panama Canal Zone and by *P. guyannensis* on Trinidad (Everard and Tikasingh 1973; Fleming 1971), and the same is probably true for most other populations of *Proechimys*. Litters may follow in rapid succession, the average annual number per female *P. semispinosus* in Panama being 4.68. Female *P. guairae* have a mean estrous cycle of about 23 days, 50 percent of them experience a fertile postpartum estrus, and 50 percent have a fertile prepartum estrus 1–2 days before giving birth (Weir 1974b). Patton and Gardner (1972) reported collection of pregnant female *P. guyannensis*, *P. simonsi*, and *P. brevicauda*, each with 1–3 embryos, from June to August in northeastern Peru. Everard and Tikasingh (1973) reported a gestation period of 62–64 days and a mean

litter size of 2.4 young for *P. guyannensis*. Weir (1974b) listed gestations of 61–64 days for *P. guairae* and 63–66 days for *P. semispinosus* and the following average (and extreme) numbers of young per litter: *P. dimidiatus*, 3 (1–5); *P. guairae*, 3 (1–7); and *P. semispinosus*, 3 (1–5). She also reported that the newborn are highly precocious and that those of *P. guairae* weigh 22 grams each. The normal period of nursing in this species is 3 weeks; puberty may be reached at as early as 35 days but usually occurs at 2 months in females and 3 months in males. Everard and Tikasingh (1973) reported that the young of *P. guyannensis* are weaned at 21–35 days and become sexually mature at 146 days, that wild individuals have been recaptured after 20 months, and that captives have been kept for 3.5 years. Fleming (1971) stated that *P. semispinosus* attains sexual maturity at 6–7 months and that wild individuals over 2 years old are common. According to Jones (1982), a captive *P. semispinosus* lived 4 years and 10 months.

RODENTIA; ECHIMYIDAE; Genus **HOPLOMYS**
J. A. Allen, 1908

Armored Rat, or Thick-spined Rat

The single species, *H. gymnurus*, is found from southern Honduras to the western side of the Andes in Colombia and northwestern Ecuador (Hall 1981; Handley 1959).

Head and body length is 220–320 mm and tail length is 150–255 mm. Weight is about 450 grams (Kleiman, Eisenberg, and Maliniak 1979), and males are heavier than females. The pelage is more spiny than in other echimyid genera. The spines are best developed in the middle of the anterior part of the back, where they are up to 33 mm long and up to 2 mm in diameter. They are white basally and colored toward the tip. The spines on the upper parts are tipped with black; those on the sides are usually tipped with orange or banded with orange and black. The spines come out easily when brushed the wrong way. Beneath the spines are soft hairs, usually orange or yellow in color. In one form

Thick-spined rat *(Hoplomys gymnurus)*, photo by R. B. Tesh. Skull, photo by P. F. Wright of specimen in U.S. National Museum of Natural History.

these soft hairs are blackened and form a distinct stripe along the middle of the back from the snout to the base of the tail. The underparts are whitish. The scantily haired, scaly tail is usually brownish above and whitish below. The sparsely furred ears are dark brown to black. *Hoplomys* is similar in body form to *Proechimys*. The hind feet are long and narrow, and the tail is shorter than the head and body. The tail is easily lost through autotomy.

The thick-spined rat has been found in fairly open rainforests, thickets, and grassy clearings, usually near large decaying logs. The maximum recorded elevation is about 780 meters. This rodent is terrestrial and burrows in creek banks. It digs a short, simple horizontal shaft with an enlarged chamber near the entrance and a nest chamber at the end. The nests are dry and kept free from feces. Runways are made in the vegetation above the ground. A wide variety of plant food is eaten, but insects are also taken. One to three precocious young are produced, probably throughout the year. In Panama, Tesh (1970b) took pregnant females from February to July but did not collect any females from October to January. He found a mean litter size of 2.1. The newborn were furred, open-eyed, and ambulatory.

RODENTIA; ECHIMYIDAE; Genus EURYZYGOMATOMYS
Goeldi, 1901

Guiara

The single species, *E. spinosus*, occurs in eastern and southern Brazil, Paraguay, and northeastern Argentina (Cabrera 1961).

Head and body length is 170–270 mm and tail length is 50–55 mm. Males may be somewhat larger than females. As in the other Echimyidae, spines are present on the back. The coloration is drab brown above and whitish below. The hands and feet are brown. Females have one inguinal and two lateral thoracic pairs of mammae (Weir 1974b).

The guiara lives in areas covered with grass or bushes, preferably near water. The well-developed claws and short tail indicate fossorial habits, but very little information on natural history is available. Alho (1982) referred to *Euryzygomatomys* as terrestrial, fossorial, and nocturnal; he added that it lives in low bushes or on the edges of clearings and that some individuals were observed to dig tunnels. Mares and Ojeda (1982) indicated that the diet may be herbivorous or frugivorous. A pregnant female with three embryos was taken near Rio de Janeiro in November. Two pregnant females, each with a single embryo, were collected in the Brazilian state of Minas Gerais.

RODENTIA; ECHIMYIDAE; Genus CLYOMYS
Thomas, 1916

There are two species (Avila-Pires and Caldatto Wutke 1981):

C. laticeps, central and southern Brazil, Paraguay;
C. bishopi, southeastern Brazil.

Head and body length is 105–230 mm, tail length is 55–87 mm, and weight is 21–39 grams. The coloration above is grayish brown and black, mixed with rufous; the underparts are whitish or buffy. Grayish patches may be present in the throat region and in the middle of the chest and belly. The fur is bristly. *Clyomys* is distinguished from other genera of the family Echimyidae by the well-developed fossorial claws on the forefeet and the enlarged bullae of the skull.

Burrows of *Clyomys* have been found in a natural clearing, and individuals of this genus appear to live in colonies. Joao Moojen (pers. comm. to Ernest P. Walker) described the collection of two specimens as follows: "The traps that caught both specimens alive were set at the mouth of the most elaborate rodent nest I ever saw. A very regular three-spiral helix wound around a large vertical root. The sloping passageway was 80 mm. wide and 90 mm. high and went down 850 mm. vertically to the main nest that was 250 mm. in diameter.

Guiara (*Euryzygomatomys spinosus*), photo by Ernest P. Walker.

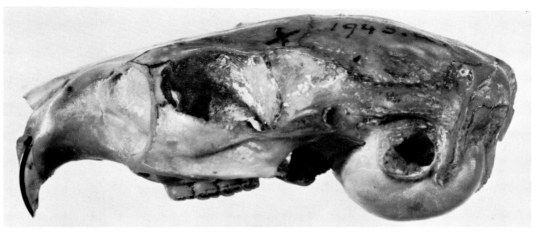

Clyomys laticeps, photos by Joao Moojen. Skull: photo from British Museum (Natural History).

Another nest was midway in the spiral and was only 170 mm. in diameter. The lower nest was covered with scant grassy materials, quite worn and old; the midway nest was clean with a few bits of a meaty sweet root (I tasted it) which I was not able to identify."

Lacher and Alho (1989) found the characteristic burrows of *C. laticeps* throughout a study area in the Pantanal, a large region of wet grassland and subtropical forest in southwestern Brazil. Mean population density of the species was 3.51/ha. Avila-Pires and Caldatto Wutke (1981) indicated that *C. bishopi* is extending its range as a result of the establishment of pine plantations.

RODENTIA; ECHIMYIDAE; Genus CARTERODON
Waterhouse, 1848

The single species, *C. sulcidens*, occurs in eastern Brazil (Cabrera 1961). The species was first described in 1841 from fossil remains collected at Lagoa Santa in the state of Minas Gerais. In 1851 living animals were found on the open pampas in Lagoa Santa. Another specimen, collected by Miranda Ribeiro in Campos Novos, Mato Grosso, is now in the Museu Nacional, Rio de Janeiro, Brazil. Two more specimens were taken by Joao Moojen in Brasilia.

Head and body length is about 155–200 mm and tail length is about 68–80 mm. The upper parts are yellow brown, shaded with black, the sides are grayish, the lower parts of the

neck and throat are reddish, the sides of the belly are yellowish red, and the middle of the belly is white; the well-haired tail is black above and pale yellow below. The back of *Carterodon* is supplied with bristles and spines; the relatively soft spines end in a hairlike point and are long and flexible. The upper incisor teeth are grooved.

Carterodon inhabits the large mesa, savannah-like cerrado region about 400 meters above sea level, where there are two definite seasons, one from October to March, when rainfall measures about 2,030 mm annually, and the other for the remaining months of the year, when the rains stop completely. The vegetation is grass with scattered trees.

A few notes on the natural history of *Carterodon* were made many years ago by John Reinhardt, who stated: "It inhabits the open *Campos*, overgrown with shrubs and trees, where it digs its residence, consisting of a rather long tube 3 to 4 inches [76–102 mm] in diameter, and leading in a slanting direction into a chamber, scarcely beyond a foot [300 mm] from the surface of the ground, which the animal lines with grass and leaves. The stomach of the two specimens which I examined was entirely filled with a yellow pasty substance, evidently of vegetable origin." The specimens analyzed by Reinhardt (1852), apparently taken in the Southern Hemisphere in late fall or early winter, were an immature male and a pregnant female bearing a 25–37 mm embryo.

The two specimens collected by Joao Moojen (pers. comm. to Ernest P. Walker) were a young male with only two upper molariform teeth and a female in puberty, about adult size, with three molariform teeth, the third molar just breaking

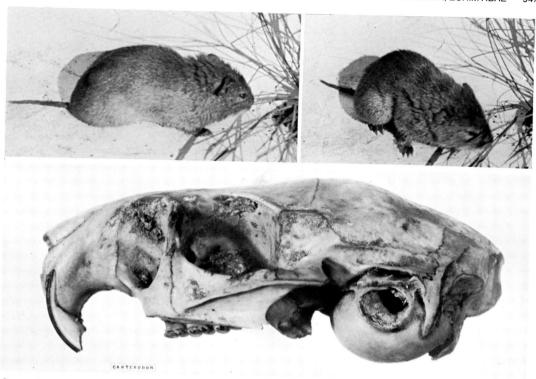

Carterodon sulcidens, photos by Joao Moojen. Skull: photo from British Museum (Natural History).

through. These were found in a nest about 200 mm in diameter filled with dry grass and chewed roots at the end of a tube 800 mm in length and 70 mm in diameter. Moojen cited Burmeister, who noted that *Carterodon* remains in its underground galleries during the day but emerges in the late afternoon or night, when it is easily captured by owls *(Tyto alba)*. For this reason, the genus was first known from balls vomited by this owl, particularly inside the caves of the Lagoa Santa area.

RODENTIA; ECHIMYIDAE; **Genus THRICHOMYS** Trouessart, 1880

Punaré

The single species, *T. apereoides*, occurs in eastern Brazil and Paraguay (Cabrera 1961). The generic name *Cercomys* Cuvier, 1829, and the specific name *Cercomys cunicularis* generally have been applied to this animal. Petter (1973), however, found that the type specimen of *Cercomys* and *C. cunicularis* is actually a composite consisting of the skin of a cricetid rodent, the skull of a *Proechimys*, and the mandible of an *Echimys*. He thus concluded that the name *Cercomys cunicularis* is invalid and that the animals that have been referred to *Cercomys* should be given the later name, *Thrichomys apereoides*. He added, however, that the name *Cercomys* is applicable to the genus currently called *Proechimys*.

Head and body length is 200–290 mm and tail length is 180–220 mm. Weight is 300–450 grams (Roberts, Thompson, and Cranford 1988). The upper parts are dull brown, and the underparts are grayish to white. The fur is soft, the tail is

hairy, and, unlike in most echimyids, there are no spines or bristles.

Thrichomys usually inhabits rocky and thickly vegetated areas (Streilen 1982a) but in some places is common even in swamps. It lives in rock crevices and under such plants as cactus. Its straw-lined nest is built in a hollow tree or log, a stone fence, or under a rock. Mares et al. (1981) reported it to be generally crepuscular and scansorial. In captivity the punaré has been seen to leap only on its hind legs. It eats cotton seeds or cotton fruits, and during the dry season it feeds on coconuts or the tender parts of some cacti. Streilen (1982a) stated that laboratory specimens readily accepted all types of native seeds and also quickly ingested insects. Lacher and Alho (1989) reported an average population density of 2.61/ha. in a Brazilian wetland.

In northeastern Brazil, Streilen (1982c, 1982d) found a somewhat higher density during certain months and an average home range size of 1,906 sq meters. There was little overlap between ranges of individuals of the same sex, but male ranges overlapped those of females. There was relatively little agonistic interaction, though physical confrontation was observed in the laboratory. Reproductive activity occurred throughout the year, being lowest in December and January, a period just after the peak of seed maturation and prior to the start of fruit maturation. Adult females produced 2 or 3 litters annually, with a 4–6 month interval between births and a litter size averaging 3.1 and ranging from 1 to 6 young. The young reached sexual maturity at 7–9 months.

Roberts, Thompson, and Cranford (1988) observed reproduction throughout the year in captive colonies in the northeastern United States. The gestation period was 95–98 days, and some females experienced a postpartum estrus. There were 1–7, usually 2–4, young per litter. At birth the precocial

Punaré *(Thrichomys apereoides)*, photo by Joao Moojen.

young averaged 21.1 grams each, were fully furred, and had their eyes open. They were weaned at about 6 weeks and reached adult weight at 200 days. Females were sexually mature at about 90 days. One female captured as an adult subsequently produced 14 litters, with 56 young, over a period of 47 months. The unusually long gestation and weaning periods are associated with a low metabolic rate and are thought to be adaptations to minimize maternal reproductive expenditure in a harsh and unpredictable environment.

RODENTIA; ECHIMYIDAE; Genus MESOMYS
Wagner, 1845

There are four species (Alho 1982; Cabrera 1961; Husson 1978; Lemke et al. 1982):

M. hispidus, southern Colombia and Venezuela, eastern
    Ecuador, northern Peru, Amazonian Brazil;
M. stimulax, Surinam, northern Brazil;
M. obscurus, known only by description of a specimen
    from an indefinite locality in Brazil, and possibly a
    subspecies or synonym of *M. hispidus*;
M. didelphoides, probably Brazil.

Head and body length is 150–200 mm and tail length is 113–220 mm. The heavily spined pelage resembles that of *Hoplomys*. The coloration of the upper parts ranges through various shades of brown; dark brown and pale buffy bands on the spine of some species produce a speckled effect. The underparts are orangish, pale buffy, or whitish. The tail is usually brown throughout and tufted at the tip.

These rodents live mainly in forests at elevations of 90–1,950 meters. The short, rather broad feet suggest that *Mesomys* is arboreal. In the lowland forests of southern Venezuela, Handley (1976) collected most specimens in trees or houses

and near streams or in other moist places. A female *M. hispidus* with a single embryo was taken in March near the Madeira River in the center of Amazonian Brazil.

RODENTIA; ECHIMYIDAE; Genus LONCHOTHRIX
Thomas, 1920

The single species, *L. emiliae*, is found in Brazil to the south of the Amazon River (Cabrera 1961). Specimens have been taken along the Madeira and Tapajos rivers. They are in the British Museum of Natural History, the American Museum of Natural History, and the Departamento de Biologia, Sao Paulo, Brazil.

Head and body length is about 200 mm and tail length is about 190 mm. The upper parts are dark brown, with buffy spots in the shoulder region; the sides are tawny; and the underparts are pale tawny. The spines on the body are well developed, and even the belly is somewhat spiny. The tail is mostly furless, scaly above and covered with short spines, but tufted at the end with hairs that may be 70 mm long. *Lonchothrix* lives in forests, and the broadened hind feet suggest that it is arboreal in habit.

RODENTIA; ECHIMYIDAE; Genus ISOTHRIX
Wagner, 1845

Toros

There are three species (Patton and Emmons 1985; Petter and Cuenca Aguirre 1982):

I. bistriatus, Amazon and Orinoco basins of southern
    Colombia and Venezuela, eastern Peru and Bolivia, and
    western and central Brazil;

Lonchothrix emiliae, photo from British Museum (Natural History).

Mesomys hispidus, photos by Louise Emmons.

I. pagurus, Amazon Basin of central Brazil;
I. pictus, coastal forests of eastern Brazil.

Head and body length is 220–410 mm and tail length is about 220–50 mm. *I. pictus* has a black and white, or a dark brown and white, color pattern, but the other species are dull brown. The soft fur lacks spines and bristles, and the long bushy tail is almost like that of a squirrel. The hind feet are of the arboreal type.

Isothrix is said to live in holes about 10 mm from the base of large trees along riverbanks and, during the afternoon, to sit in the entrance with only its head exposed. Handley (1976) collected all his specimens of *I. bistriatus* near streams in evergreen forests, mostly in trees. A female *I. bistriatus* with a single embryo was taken in May near the Cunuconuma River, Amazonia, Brazil.

RODENTIA; ECHIMYIDAE; Genus DIPLOMYS
Thomas, 1916

There are three species (Cabrera 1961; Hall 1981):

D. labilis, Panama and nearby San Miguel Island, and
  probably the adjacent part of northwestern Colombia;
D. caniceps, western Colombia, northern Ecuador;
D. rufodorsalis, extreme northern Colombia.

According to L. H. Emmons (U.S. National Museum of Natural History, pers. comm., 1989), *D. rufodorsalis* does not belong in *Diplomys.*

Head and body length is 250–480 mm and tail length is 200–280 mm. Specimens of *D. labilis* collected by Tesh (1970a) weighed up to 492 grams, all those over 400 grams being pregnant females. The fur is harsh but lacks spines. The upper parts are rusty brown, reddish, or orange buffy mixed with black; the underparts are buffy, reddish, or reddish white. The tail is mostly brownish, thinly haired, and slightly tufted with black or white. In *D. labilis* the ears are short and conspicuously tufted. The feet are of the arboreal type—short and broad with strongly curved claws.

In central Panama, Tesh (1970a) collected 74 specimens of *D. labilis* in a hilly area covered with patches of moist tropical forest and scrubby secondary growth. All specimens were taken either from tree holes, usually some distance above the ground, or in trees near streams. Observations of both wild and captive animals indicated that activity is strictly nocturnal. They climbed easily and could jump about a meter through the air. Captives ate bananas, apples, mangoes, sunflower seeds, almonds, and coconuts. They were difficult to maintain and much more hostile than *Proechimys* and *Hoplomys,* readily attacking objects placed in their cage. Pregnant females were taken in January, February, March, April, May,

Toro *(Isothrix bistriatus)*, photo by James L. Patton.

August, September, and November. Most of these females contained one embryo, and some had two.

RODENTIA; ECHIMYIDAE; **Genus ECHIMYS**
G. Cuvier, 1809

Arboreal Spiny Rats

There are 10 species (Cabrera 1961; L. H. Emmons, U.S. National Museum of Natural History, pers. comm., 1989):

E. grandis, northeastern Peru, Amazonian Brazil;
E. chrysurus, the Guianas, northeastern Brazil;
E. saturnus, eastern Ecuador;
E. semivillosus, northern Colombia, Venezuela, Margarita Island;
E. macrurus, south of the Amazon in Brazil;
E. blainvillei, southeastern Brazil;
E. dasythrix, southeastern Brazil;
E. nigrispinus, southeastern Brazil;
E. braziliensis, southeastern Brazil;
E. unicolor, an undetermined part of Brazil.

An additional and widespread species, *E. armatus,* was assigned to a distinct genus, *Makalata*, by Husson (1978), though some authorities, including Corbet and Hill (1986), have returned it to *Echimys.*

Head and body length is 170–350 mm and tail length is

Diplomys sp., photo by R. B. Tesh.

150–320 mm. An adult female *E. chrysurus* weighed 640 grams (Husson 1978). The fur is bristly or spiny. The upper parts are usually some shade of brown or red, and the underparts are usually white or buffy. In *E. saturnus* the upper parts are dark, the middle of the back being black, and the underparts are whitish or coppery brown. Some species have a hairy tail, and in a few the tail is slightly tufted or bushy, but in others the tail is scaly. The broad feet have prominent claws and are adapted to arboreal life. Female *E. chrysurus* have one

Arboreal white-faced spiny rat *(Echimys chrysurus)*, photo of dead specimen by Louise Emmons.

inguinal and three abdominal pairs of mammae (Husson 1978).

Echimys lives in trees, often on riverbanks or in flooded areas. It is reported to slink along the branches rather than to scamper like a squirrel. In Venezuela, Handley (1976) collected nearly all specimens of *E. semivillosus* in trees, most near streams or in other moist places within thorn forest. The nests of *Echimys* are built of dry leaves in hollow tree trunks or in holes in trees. The animals remain in their upper hole (usually they have two) during the day and come out on the branches at night. Small groups may live together. At night they give loud calls that sound like "cró, cró" or "tró, tró." The usual number of young appears to be one or two. A pregnant female *E. blainvillei* with two embryos was taken in September, and Husson (1978) collected a pregnant female *E. chrysurus* with two embryos on 2 September in Surinam.

RODENTIA; ECHIMYIDAE; Genus MAKALATA
Husson, 1978

Husson (1978) erected this new genus for the species previously known as *Echimys armatus* and now designated as *Makalata armata*. It is known to occur in Ecuador, Colombia, Venezuela, Trinidad, the Guianas, and northeastern Brazil. This species also has been reported to be present, through introduction, on the island of Martinique in the Lesser Antilles, but Hall (1981) indicated that the basis of the report is a single specimen that was brought in by a vessel and died without reproducing.

Head and body length is 170–202 mm, tail length is 182 mm in each of two specimens, and reported weight is 147–317 grams. The upper parts are dark yellowish brown, heavily lined with black. The posterior third of the back is conspicuously speckled with yellow. The spines are pale gray at the base, becoming darker distally, and sometimes, in the speckled area of the body, have a distinct pale yellowish termi-

Arboreal spiny rat *(Echimys nigrispinus)*, photo by Jody R. Stallings.

Makalata armata, painting by R. van Assen, Rijksmuseum van Natuurlijke Historie, Leiden.

nal band. The sides of the body are lighter than the back and have fewer spines. The ventral surface is pale yellowish or grayish brown and has no spines. The tail is relatively short, has no tuft of hairs on the tip, and frequently is lost by autotomy. Compared with *Echimys, Makalata* has a more uniform body color and smaller spines, and the folds of the cheek teeth open lingually instead of bucally. One female specimen had two well-developed abdominal mammae, and some other specimens also had a pair of inguinal mammae, perhaps not functional.

Available information, as summarized by Husson (1978), suggests that in Surinam this spiny rat lives in trees, preferably along rivers, and nests between the roots of trees. A female with one embryo was taken in November, and newborn were collected in October. According to Jones (1982), a captive specimen lived for three years and one month. *Makalata* is reportedly harmful to the cultivation of bananas in Surinam, climbing the trees at night and eating the green fruit.

RODENTIA; ECHIMYIDAE; Genus DACTYLOMYS
I. Geoffroy St.-Hilaire, 1838

Coro-coros

There are three species (Cabrera 1961):

D. dactylinus, eastern Ecuador, Amazonian Brazil,
    probably Colombia;
D. boliviensis, southeastern Peru, central Bolivia;
D. peruanus, the Andes of southeastern Peru.

The last species has sometimes been placed in a separate genus, *Lachnomys* Thomas, 1916.

In *D. boliviensis* and *D. dactylinus* head and body length is about 300 mm, tail length is 400–430 mm, the upper parts are olivaceous gray or grayish, often with a rusty suffusion, and the underparts are white. Emmons (1981) stated that *D. dactylinus* weighs 600–700 grams. In *D. peruanus* head and body length is about 250 mm, tail length is about 320 mm, the back is yellowish brown, the sides are lighter yellowish brown, most of the head and the basal part of the tail are gray, and the underparts are white.

The soft fur of *Dactylomys* lacks spines and bristles. In *D. boliviensis* and *D. dactylinus* the tail is naked, except for the well-haired basal part. In *D. peruanus* the pelage of the body is thicker and the tail is hairy, being bushy on the basal half and more or less tufted at the end. In each species the third and fourth digits of all four feet are elongated and broadened, the palate is constricted anteriorly, and the main lobes of the upper cheek teeth are not united by enamel bridges.

These rodents have been found in dense vegetation near water. The type specimen of *D. peruanus* was taken at an elevation of 1,800 meters. Coro-coros are apparently arboreal and nocturnal. They grasp branches between the third and fourth digits and can run swiftly. Emmons (1981) found *D. dactylinus* to be most common in forested areas disturbed by selective logging. By day it rested in dense vegetation within trees and vines. Individuals were observed to strip bark from green bamboo stems and eat the inner surface. One had a home range of 0.23 ha. within a bamboo patch on the edge of a lake and covered 45–68 meters a night. Movements were slow and cautious. The animals often were associated in pairs, probably a male and a female. Several vocalizations were noted. LaVal (1976:403) described the most prominent call of *D. dactylinus* as follows: "The rather low-pitched notes, all having approximately the same frequency range, might be described as a long succession of 'boop . . . boop . . . boop.'" He reported that the call was restricted to the night, mainly from 1800 to 0100 hours, and could be heard all year. He

Coro-coro (*Dactylomys* sp.), photo by Louise Emmons.

suggested that the call is territorial in function. The call was heard mostly in disturbed forest, where bamboo was common, but seemed to originate from broad-leaved evergreens in the canopy.

RODENTIA; ECHIMYIDAE; Genus **KANNABATEOMYS**
Jentink, 1891

Rato de Taquara

The single species, *K. amblyonyx*, occurs in southeastern Brazil, Paraguay, and northeastern Argentina (Cabrera

1961). There are specimens in the British Museum of Natural History and the Departamento de Biologia, Sao Paulo, Brazil.

Head and body length is about 250 mm and tail length is about 320 mm. The upper parts are dull buffy yellowish, sometimes with an orangish cast, and the underparts are paler, being buffy or whitish. The thick, soft fur lacks spines and bristles, and the tail is haired. The third and fourth digits of all four limbs are elongated and broadened. *Kannabateomys* differs structurally from *Dactylomys* in that the palate is not appreciably constricted and the main lobes of the upper cheek teeth are united by narrow enamel ridges.

In Brazil this rodent usually lives in bamboo thickets along stream banks. By grasping the bamboo shoots and stalks between its third and fourth digits, it climbs the plants and

Rato de Taquara *(Kannabateomys amblyonyx)*, photo by Luis Fernando B. M. Silva through Jody R. Stallings.

eats the succulent parts of the bamboo during the night. A pregnant female was noted in November. The usual number of young per birth appears to be one. A captive specimen lived for a year and seven months (Jones 1982).

RODENTIA; ECHIMYIDAE; Genus OLALLAMYS
Emmons, 1988

There are two species (Cabrera 1961):

O. albicauda, northwestern and central Colombia;
O. edax, northwestern Venezuela and probably adjacent parts of northern Colombia.

Emmons (1988a) proposed the generic name *Olallamys* to replace *Thrinacodus* Günther, 1879, which is preoccupied by the name of a fossil shark.

Head and body length is 180–240 mm and tail length is 250–350 mm. The upper parts are reddish brown or yellowish brown, and the underparts are yellowish white or white.

A yellow line may occur along the sides. The fur is thick and soft, there are no spines or bristles, and the tail is haired. The third and fourth digits of the limbs are elongated and broadened but not to a marked degree. In *O. edax* the tail is completely white below and on the terminal half of the dorsal surface.

Specimens are in the American Museum of Natural History and the British Museum of Natural History. Specimens have been collected at elevations of 2,000 and 2,800 meters. In the mountains of Colombia these rodents live in thickets of the bamboo *Chusquea*. They emit a whistling cry.

RODENTIA; ECHIMYIDAE; Genus CHAETOMYS
Gray, 1843

Thin-spined "Porcupine"

The single species, *C. subspinosus*, is apparently restricted to a small part of southeastern Brazil, in the states of Sergipe, Bahia, Espirito Santo, and Rio de Janeiro (Avila-Pires 1967;

Olallamys albicauda, photos by R. B. Mackenzie.

Thin-spined "porcupine" (Chaetomys subspinosus), photo by Joao Moojen.

Coimbra-Filho 1972; Santos, Oliver, and Rylands 1987). Although originally classified as a porcupine in the family Erethizontidae, *Chaetomys* recently was transferred to the Echimyidae (Woods 1984).

Head and body length is about 430–57 mm and tail length is 255–80 mm. The pelage of the back is peculiar, as the hairs are more like bristles than spines. The hairs of the head, neck, and forelimbs, however, are spinelike and less flexible than those on other parts of the body. All the large, coarse hairs are cylindrical and wavy. The general coloration of the upper parts is dull brownish, or sometimes grayish white, and the underparts are slightly rufous brown. The feet and tail are brownish black. The fairly long tail is scaly throughout and only moderately enlarged in the basal part. The underside of the tail is covered with short, stiff hairs.

Although the tip of the tail is naked and resembles that of *Coendou*, most authorities agree that it is not prehensile. The hands and feet of *Chaetomys* each have four digits, all armed with long, curved claws. The skull is unique among rodents, as the orbit is almost ringed by bone, there being a broadened jugal and pronounced postorbital process of the frontal. The incisor teeth are narrow.

The thin-spined "porcupine" inhabits densely vegetated brush country and woodland around open savannahs and cultivated areas. It is known to live in the vicinity of cocoa trees, the nuts of which it eats. It generally moves slowly but is a quick jumper and climber. It is nocturnal (Coimbra-Filho 1972). It utters hoarse sounds and is docile in captivity. It occurs in a limited area, where its natural habitat is being destroyed or modified by deforestation, agriculture, and industrial development. It is thought to be declining, though because of its ability to survive on forest edge and cocoa plantations, its situation may not be as critical as that of certain other mammals of the same region, such as the primates *Leontopithecus* and *Brachyteles* (Thornback and Jenkins 1982). *Chaetomys* was not reliably reported from 1952 until Santos, Oliver, and Rylands (1987) found it at several places in the Atlantic coastal forests of southeastern Brazil. It is classified as indeterminate by the IUCN and as endangered by the USDI.

RODENTIA; ECHIMYIDAE; Genus HETEROPSOMYS
Anthony, 1916

There are three subgenera and six species, all probably extinct (Hall 1981):

subgenus *Heteropsomys* Anthony, 1916

H. insulans, known only by skeletal remains from the type locality near Utuado in Puerto Rico;
H. antillensis, known only by skeletal remains from Puerto Rico;

subgenus *Brotomys* Anthony, 1916

H. voratus, known only by skeletal remains from the Dominican Republic;
H. contractus, known only by skeletal remains from the type locality in north-central Haiti;

subgenus *Boromys* Anthony, 1916

H. offella, known only by skeletal remains from Cuba and the Isle of Pines;
H. torrei, known only by skeletal remains from Cuba and the Isle of Pines.

On the basis of the morphology of the molariform teeth, the position of the root of the incisor, and the nature of the mandibular foramen (see below), Woods (1984) recommended tentative generic rank for *Heteropsomys, Brotomys,* and *Boromys*. He also indicated that they might best be transferred to the family Capromyidae.

These rodents can be described only on the basis of their skulls and teeth. In the subgenus *Heteropsomys* the skull is somewhat smaller than that of an agouti (*Dasyprocta*), having a total length of about 70 mm. The upper incisors are long but weak. The upper cheek teeth have a single conspicuous internal fold with three separate, transverse, enamel-surrounded lakes. In the subgenus *Brotomys* the skull is about the same size as that of *Proechimys* but has a shorter and broader rostrum. The infraorbital canal is very large and lacks an accessory canal for the passage of a nerve. The teeth are weakly developed. Each of the upper cheek teeth has three short roots and a deep enamel infolding from about the middle of each side. The subgenus *Boromys* closely resembles *Brotomys* but differs in having a swelling on the bone over the end of the root of the upper incisors, a channel for the passage of a nerve on the floor of the antorbital foramen, and two inner and two outer reentrant folds on each of the cheek teeth. The incisors of *Boromys* are orange yellow in color (Allen 1942; Hall 1981).

The subgenus *Heteropsomys*, of Puerto Rico, is thought to have survived until relatively recent times but has not been

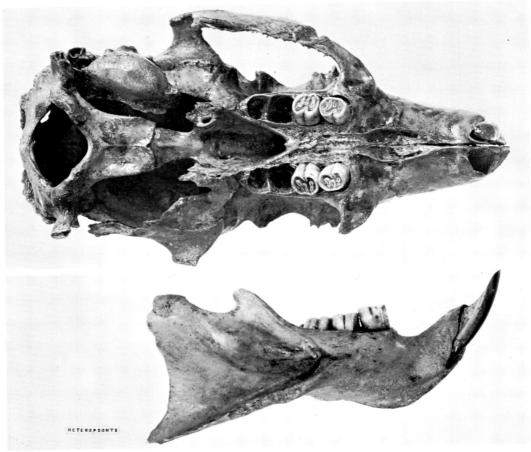

Heteropsomys insulans, photos from American Museum of Natural History.

found in direct association with human artifacts. The sub-genus *Brotomys* was widely distributed on Hispaniola, and its remains have been found both in cave deposits, probably made by owls, and in abundance in aboriginal kitchen middens. Early Spanish writings suggest that it was a prized food source of the natives. Its extinction, at some point after the arrival of Europeans, may have been associated with intro-duction of *Rattus*. The subgenus *Boromys* evidently occurred over much of Cuba and the Isle of Pines. The species *H. offella* was probably used as food by people, as its bones were first discovered during excavation of an old Indian village. The remains of both this species and *H. torrei* also have been found in abundance in cave deposits. Some of the bones are remarkably fresh, and *Boromys* probably survived until the latter half of the nineteenth century (Allen 1942; Goodwin and Goodwin 1973).

RODENTIA; Family THRYONOMYIDAE; Genus THRYONOMYS
Fitzinger, 1867

Cane Rats

The single Recent genus, *Thryonomys*, contains two species (Misonne, *in* Meester and Setzer 1977; Yalden, Largen, and Kock 1976):

T. swinderianus, Gambia to southern Sudan, and south to northern Namibia and eastern South Africa;

T. gregorianus, Cameroon and western Ethiopia to Zimbabwe.

Head and body length is about 350–610 mm and tail length is 65–260 mm. Adults generally weigh 4–7 kg, but a few indi-viduals approach 9 kg. The pelage is coarse. The bristlelike hairs are flattened, grooved longitudinally along their upper surface, and usually grow in groups of five or six. Underfur is lacking. The general coloration of the upper parts is speckled yellowish brown or grayish brown, and that of the underparts is grayish or whitish. The tail is scantily covered by short, bristly hairs, with scales between the hairs; its color is brownish above and buffy white below.

The body is large and heavyset. The ears are short and rounded and barely extend above the spiny pelage. The fore-foot has three well-developed central digits, but the first digit (thumb) is reduced and the fifth digit is small and nearly functionless. The digits of the hind foot are somewhat larger, though the first digit is absent. The palms and soles are naked, and the claws are thick and heavy. The massive skull is strongly ridged. The dental formula is: (i 1/1, c 0/0, pm 1/1, m 3/3) × 2 = 20. The orange-colored incisors are broad and powerful; the upper incisors have three deep grooves. The cheek teeth are rooted and moderately high-crowned. Females have two or three pairs of lateral mammae (Weir 1974b).

Cane rat *(Thryonomys swinderianus)*, photo by Ernest P. Walker.

While both species of *Thryonomys* may occur in the same general area, Kingdon (1974*b*) explained that there are distinct primary ecological niches. *T. swinderianus* is a semiaquatic inhabitant of marshes and reedbeds, and *T. gregorianus* lives mainly on dry ground in moist savannahs. Both species depend on grass for cover and food. They may shelter only in tall grass but sometimes use rock crevices, termite mounds, or the abandoned holes of the aardvark *(Orycteropus)* and porcupine *(Hystrix)*. If none of these is available, *Thryonomys* sometimes excavates a shallow burrow. Well-defined paths lead from the shelters to the feeding sites, which may be 50 meters away. Cane rats are generally nocturnal but are sometimes active by day. They apparently cannot see well but have good senses of smell and hearing. Despite their appearance, they are fast and agile. If startled while above ground, they bolt through the matted vegetation with sudden and amazing speed, often toward water. They swim and dive with ease. The diet consists mainly of grass and cane and also includes bark, nuts, fallen fruits, and cultivated crops. Rocks, bones, and ivory are sometimes gnawed, apparently to hone the incisors rather than as a source of nutrients.

Kingdon (1974*b*) reported that during a brush-clearing operation in Uganda a count of cane rats indicated an overall population density of nearly 1/ha. During part of the year *T. swinderianus* seems to live in groups of mixed ages and sexes, but in the dry season there is reportedly a division into solitary males and groups of females. *T. gregorianus* lives in small family groups, each of which may have a territory of 3,000–4,000 sq meters. Captive males live peacefully with females and young but fight other adult males. Ewer (1971) described the highly specialized fighting technique of *Thryonomys*: the two contestants have a nose-to-nose pushing duel; if one relaxes pressure for a moment, the other may whip the rump around to try to knock him off balance.

In East Africa both species of *Thryonomys* apparently have two breeding seasons during the wetter months from March to November (Kingdon 1974*b*); under favorable conditions a female can produce two litters in this period. In South Africa the young are born from June to August. In Ghana, Asibey (1974*b*) determined that *T. swinderianus* breeds throughout the year, with peaks in January–March and July–September. In that country, females probably have two litters annually. The gestation period was calculated to vary from 137 to 172

days and to average 155 days. Embryo counts ranged from one to eight, and mean litter size was four young. *Thryonomys* makes a special natal nest, usually scooping out a hollow depression in a sheltered area and lining it with grass and leaves. The newborn weigh about 129 grams each, are covered with hair, have their eyes open, and are soon able to run about. The young begin to breed at about 1 year (Kingdon 1974*b*). A captive *T. swinderianus* lived for 4 years and 4 months (Jones 1982).

Cane rats may do considerable damage in sugar cane fields, and in an effort to reduce their numbers on plantations, pythons are often protected there. Cane rats also become pests by eating maize, millet, cassava, groundnuts, sweet potatoes, and pumpkins (Kingdon 1974*b*). The flesh of *Thryonomys* is an important source of protein for many native people. There are intensive organized hunts with dogs and spears, and investigations into domestication have been made. In Ghana, where *Thryonomys* is known as the "grasscutter," its meat sells for nearly twice the price of beef, mutton, and pork (Asibey 1974*a*). From July 1970 to June 1971 the markets of Accra are estimated to have sold nearly 200,000 kg of cane rat meat, with a value of about U.S. $220,000.

The geological range of the family Thryonomyidae is Oligocene to Recent in Africa and Miocene in Europe and Asia (Woods 1984).

RODENTIA; Family PETROMURIDAE; Genus PETROMUS
A. Smith, 1831

Dassie Rat

The single known genus and species, *Petromus typicus*, is found only in southern Angola, Namibia, and northwestern South Africa (Misonne, *in* Meester and Setzer 1977). The names Petromyidae and *Petromys*, respectively, sometimes have been applied to the family and genus.

Head and body length is 140–200 mm and tail length is 130–80 mm. The pelage is soft and silky, but the underfur is absent, so the hairs stand out separately and have a wiry appearance on the living animal. The hairs grow in clusters of

Dassie rat *(Petromus typicus)*, photo by Chris Stuart.

three to five, and the facial whiskers are long and black. When this rodent lies flattened out on a rock, its only distinguishing mark is the yellowish nose, as the coloration of *Petromus* blends with that of the rocks on which it lives. Above, the foreparts are dark gray, light gray, yellowish, or pale buffy. The lower back is tawny, orangish tawny, or yellowish. Occasionally the upper parts are grayish throughout. The underparts are grayish or yellowish. The skin is very soft and tears easily when a specimen is being prepared. The tail is covered with scattered long hairs; it has soft joints and breaks readily, generally at the base.

This rodent is somewhat squirrel-like in external appearance, but the tail is not bushy. The limp body can be greatly compressed, so that with its flat skull this rodent can squeeze into narrow crevices. The ribs are so flexible that *Petromus* can be pressed almost flat without injury to the body. The ears are short. The feet are narrow. The forefoot has four digits and the hind foot has five. The claws are short, and some stiff, bristlelike hairs are associated with the claws of the hind foot. The dental formula is: (i 1/1, c 0/0, pm 1/1, m 3/3) × 2 = 20. The incisors are fairly narrow, and the cheek teeth are high-crowned and rooted. Females usually have three pairs of mammae, but sometimes there are only two pairs. The mammae are lateral, placed high up on a level with the shoulder blades, enabling the young to suckle from the side when hiding in narrow rock crevices.

The dassie rat is associated with rocky areas on hills and mountains. It generally emerges from its rock shelter during the day, especially in the early morning and late afternoon, but occasionally forages after sunset. While resting or sunning itself, it generally selects a spot under a projecting rock that provides protection from attack by birds of prey. *Petromus* usually moves on the rocks by running, rather than jumping, but does occasionally spring from one rock to another. When jumping, it spreads its flattened body, somewhat in the manner of flying squirrels during a glide. The dassie rat seeks food on the ground or in bushes. It is a vegetarian, eating a variety of green plant material, seeds, and berries.

Petromus usually travels alone or in pairs. When undisturbed, it is playful, often whisking about and playing with plant stems, but when alarmed, it darts quickly to shelter, often giving a warning note (a whistling call) upon reaching safety. In a study in the Namib Desert, Withers (1983) re-

corded about 15 individuals on a 6 ha. study area. Pregnant females were found in January (summer), before the rainy season. There was a single annual litter containing 1 or 2 young. Mating evidently occurs during the early summer months (November–December). The young are born in an advanced stage, being quite large and covered with hair.

The geological range of the Petromuridae is Oligocene to Miocene, and Recent in Africa (Woods 1984).

RODENTIA; **Family BATHYERGIDAE**

African Mole-rats, or Blesmols

This family of five Recent genera and eight species occurs in Africa to the south of the Sahara Desert. The sequence of genera presented here follows that of Simpson (1945).

Head and body length is 80–330 mm and tail length is 10–70 mm. The genus *Heterocephalus* is practically hairless, but the other genera have a thick, soft, and woolly or velvety pelage. The external form resembles that of other fossorial mammals. The body is stocky, the tail and limbs are short, and the eyes and ears are minute. Blesmols apparently see indistinctly for a very short distance and generally close the eyelids when the head is touched. External ears are represented by a small circle of bare skin around the ear opening. The hands and feet are large, and the palms and soles are naked. The five fingers and five toes have either long or short claws, depending on the genus. The hind claws are usually hollow on the underside and shorter than the foreclaws.

The skull is stoutly built and modified to support the large incisor teeth. In all genera except *Bathyergus* the lower jaws are not tightly joined in front. The upper incisors are grooved in some forms, and the cheek teeth are high-crowned and rooted. The dental formula is: (i 1/1, c 0/0, pm 2–3/2–3, m 0–3/0–3) × 2 = 12–28. The cheek teeth (premolars plus molars) number 4/4 in *Georychus, Cryptomys,* and *Bathyergus;* 2/2 or 3/3 in *Heterocephalus;* and 6/6 in *Heliophobius,* though all are not usually in place simultaneously (Meester and Setzer 1977). The increase over the usual rodent number of cheek teeth in *Heliophobius* may result from retention of the milk dentition (Packard 1967).

The usual habitat of African mole-rats is an area of loose,

sandy soil. All species feed primarily on subterranean bulbs and roots found by tunneling 10–300 mm below the surface. The depth of the tunnel is apparently directly related to the looseness of the soil. The burrow systems include large chambers for sleeping and storing food. The excavated soil is thrown on the surface at intervals in the form of mounds. In some areas these rodents may so honeycomb the ground that a person sinks deep into the sand at almost every step. It is sometimes difficult to distinguish between the mounds of the different genera. *Cryptomys, Georychus,* and *Bathyergus* reportedly occur together in parts of the Cape Province of South Africa. This close association is unusual, as only one genus of a particular family of strictly burrowing mammals generally occupies a single habitat. Presumably, the different sizes of the tunnels eliminate actual contact between the genera. Bathyergids can orientate well; when their burrows are destroyed, they dig new tunnels directly to their exits or special chambers. They seldom emerge above ground, except when flooded out or when seeking new quarters.

In *Bathyergus* the foreclaws are the principal tools for digging. In the other four genera the incisor teeth are used to a greater extent than the claws. Blesmols use their hind limbs and the flattened rows of bristles on each side of the tail to remove the soil excavated by the foreclaws or incisors. Soil is prevented from entering the mouth during digging by fusion and apposition of lateral folds of the lips.

These rodents are reported to make interesting pets, but some species bite unexpectedly; even the newborn can inflict severe bites. Blesmols often destroy the tuber crops of native people, but like all burrowing animals, they help to furrow and aerate the soil.

The geological range of this family is Miocene, and Pleistocene to Recent, in Africa and Oligocene in Mongolia (Kingdon 1974b; Packard 1967). *Gypsorhychus,* an extinct giant mole-rat allied with the Recent genus *Georychus,* is represented by skull and dental remains from Taungs, in Mongolia, and from Namibia.

RODENTIA; BATHYERGIDAE; Genus GEORYCHUS
Illiger, 1811

Cape Mole-rat

The single species, *G. capensis,* occurs in the southwestern and southern parts of Cape of Good Hope Province in South Africa. There are also relictual populations in Natal and southeastern Transvaal (Meester et al. 1986).

Head and body length is 150–205 mm and tail length is 15–40 mm. Mean body mass is 181 grams (Bennett and Jarvis 1988a). The soft pelage is so thick, fluffy, and long that it practically conceals the tail in many individuals. The tail has a flattened appearance, because it is haired mainly on the sides. The general coloration of the upper parts is buffy to buffy orange, frequently with a brownish tinge; the underparts are somewhat paler. The head has various black or dark brown markings, with white spots. The hands, feet, and tail are white. Partial albinism is not uncommon.

The general form of *Georychus* is similar to that of North American pocket gophers (Geomyidae), but the Cape mole-rat lacks cheek pouches and is smaller. The external ears of the Cape mole-rat are represented by round openings surrounded with thickened skin around the edges. The claws are only moderately developed and are relatively weak. The incisors are white and ungrooved, and project forward to an unusual degree. Normally, there are three pairs of mammae, but four pairs are not unusual.

Georychus is strictly a burrower and spends almost its entire life underground. Apparently the incisors are used more than the claws for digging, especially in hard soil. The earth from the tunnels is thrown out at intervals, thereby plainly marking the line of excavation. The burrows are occasionally near the surface and branch into blind tunnels. The main burrow eventually leads into a smooth-walled, somewhat globular chamber, in which the mole-rat stores tubers, roots, and bulbs. It is said that *Georychus* bites off the buds ("eyes") of bulbs and tubers to prevent them from sprouting. *Georychus* is destructive to tuber crops. It makes an interesting pet but has the habit of biting unexpectedly.

In studies of wild and captive animals, Bennett and Jarvis (1988a) found adults to live alone. Their burrow systems sometimes came to within a meter of one another but did not interconnect. If two strange adults were put together, fighting would result, and if it were allowed to continue, it would result in death. In June (winter in South Africa), however, males began to drum with their hind feet, evidently signaling the start of the reproductive season, and females eventually responded. Actual breeding occurred from August to December. Females experienced a postpartum estrus, mated within 10 days of giving birth, and could produce two litters during the season. Gestation lasted 44 days. The number of young averaged 5.9 and ranged from 3 to 10. At birth they were 30–40 mm long and naked. After 9 days of life they were fully

Cape mole-rat *(Georychus capensis),* photo by G. B. Rabb, Chicago Zoological Park, Brookfield, Illinois.

furred and their eyes were open, and by 17 days they were eating solid food. Intersibling aggression began at 35 days, dispersion occurred at 60 days, and adult size was attained after about 260 days. Taylor et al. (1985) indicated that sexual maturity came at around 10 months of age and that some wild individuals lived at least 3 years.

RODENTIA; BATHYERGIDAE; Genus CRYPTOMYS
Gray, 1864

Common Mole-rats

There are three species (Ansell 1978; De Graaff, *in* Meester and Setzer 1977; Kingdon 1974*b*):

C. ochraceocinereus, Ghana to Sudan and northern Uganda;
C. mechowi, Angola, southern Zaire, Zambia, Malawi, southern Tanzania;
C. hottentotus, southern Zaire and Tanzania to Namibia and South Africa.

Bennett and Jarvis (1988*b*) regarded *C. damarensis,* found from western Zambia to northwestern South Africa, as a species distinct from *C. hottentotus.* Nevo et al. (1987) thought that *C. damarensis* might even warrant generic separation and also considered *C. natalensis,* of eastern South Africa, to be a species distinct from *C. hottentotus.*

Head and body length is 100–215 mm and tail length is 10–30 mm. Reported weights are 46–221 grams for *C. hottentotus* (Bennett and Jarvis 1988*b*; Smithers 1971) and 200 grams for *C. ochraceocinereus* (Kingdon 1974*b*). The coloration is variable—whitish, yellowish, clay, fawn, grayish, brown, reddish brown, cinnamon buff, and blackish. There may or may not be a white spot on the head. The fur is thick and often velvety. The eyes and external ears are very small,

but the cornea of the eye is sensitive to air currents and the animals are remarkably receptive of sounds and vibrations (Kingdon 1974*b*). The large lower incisors of *Cryptomys* are separately movable. Female *C. ochraceocinereus* were reported to have two pairs of mammae by Weir (1974*b*) and three or four pairs by Kingdon (1974*b*).

According to Kingdon (1974*b*), *Cryptomys* occupies a wide variety of soils in woodlands, savannahs, and secondary forests. *C. hottentotus* occurs at elevations of up to 2,200 meters. Tunnel depth is adapted to soil consistency; the looser the soil, the deeper the burrow. In seasonally flooded areas, *Cryptomys* constructs an extensive mound in which living chambers and food stores are above the high-water mark. Even in other areas, the main living and storage rooms are located in higher ground, and from there tunnels extend in several directions. Fresh tunnels are characterized by mounds of excavated earth at intervals of about 1.25–6.25 meters. In one chamber is a large communal nest made of vegetation and used for sleeping. Hickman (1979) excavated four burrow systems of *C. hottentotus.* Each contained a single functional nest at a depth of 17 cm. The total length of the systems varied from 58 to 340 meters, which indicates that *Cryptomys* may have the longest constructed and maintained burrow of any animal.

Activity may occur throughout the day and night and alternates with about five major rest periods during a 24-hour period (Hickman 1980). *Cryptomys* digs with the protuberant lower incisors and the forefeet, unlike *Bathyergus,* which uses mainly the forefeet. The ground excavated from the main tunnel is pushed with the hind feet and tail up a side tunnel, until the earth comes up in a cone and topples over to form a mound. Most burrowing activity occurs during the wet season (Kingdon 1974*b*; Smithers 1971); at this time the mole-rats extend new tunnels to feeding areas and carry back food for storage. In the dry season new mounds are not made, but soil is redistributed in old, unused tunnels. Surface activity is rare, but the animals evidently sometimes go above ground to gather nesting materials and dig up seeds. The diet

Common mole-rat (*Cryptomys* sp.), photo by Ernest P. Walker.

consists mainly of roots, bulbs, tubers, and aloe leaves. Invertebrates, such as earthworms, cockchafer larvae, and white ants, are occasionally eaten.

Cryptomys lives in small colonies that share a sleeping nest and burrow system. Captives from the same system live peacefully together but are extremely aggressive toward individuals originating in other colonies. *Cryptomys* can bite savagely and has a variety of squeaks, grunts, and growls.

In studies of *C. hottentotus*, Bennett and Jarvis (1988*b*) found colonies to consist of up to 25 individuals. These groups were thought to be extended families comprising several generations. There was some division of labor in the colony, in which reproduction was limited to the largest female and to one or two large males, and most of the work was done by smaller animals. The social system thus resembles that of *Heterocephalus*, though in *C. hottentotus* the workers give less direct care to the pups. There was no evidence of seasonal breeding in captivity, gestation lasted 78–92 days, and litters averaged 2.8 young. No female had more than two litters per year in a captive colony. The young were highly precocial, wandering out of the nest after 24 hours and being fully weaned by 3 weeks.

Other reports indicate that there is probably a single annual litter per female in the wild and that its timing probably varies by location. In Botswana, pregnant females were collected in February and July, and very young individuals were found in January, February, July, August, and November; in eastern Uganda a pregnant female was taken in December; in the Kalahari Desert two pregnant females were taken in Ap-

ril; and in Angola young were produced in January and February. Litter size is 1–5 young (De Graaff 1972; Hickman 1979; Kingdon 1974*b*; Smithers 1971). A captive specimen lived for 2 years and 5 months (Jones 1982).

Common mole-rats often raid the tuber crops of native people and become pests in vegetable and flower gardens by destroying bulbs. These animals are sometimes hunted for use as human food (Kingdon 1974*b*).

RODENTIA; BATHYERGIDAE; Genus HELIOPHOBIUS
Peters, 1846

Silvery Mole-rat, or Sand Rat

The single species, *H. argenteocinereus*, occurs in southern Kenya, Tanzania, southeastern Zaire, eastern Zambia, Malawi, and Mozambique (Ansell 1978; De Graaff, *in* Meester and Setzer 1977; Kingdon 1974*b*). Some authorities recognize *H. spalax*, known only from the slopes of Mount Kilimanjaro, as a separate species (Corbet and Hill 1986; Honacki, Kinman, and Koeppl 1982).

Head and body length is 100–200 mm and tail length is about 15–40 mm. Average weight is 160 grams (Kingdon 1974*b*). The pelage is short and dense. The upper parts are pale sandy, reddish, or grayish, and the underparts are usually paler. Some individuals have a white frontal spot and white spots on the belly. Modifications for fossorial life in-

Sand rat *(Heliophobius argenteocinereus)*, photo from Nyasaland Museum through P. Hanney. Insets: photos from *Naturw. Reise nach Mossambique*, W. Peters.

clude extreme reduction of the eyes and external ears; a short, stubby tail bearing a stiff fringe of hairs; a cylindrical body; and short limbs with large, broad hands and feet, the latter also fringed with stiff hairs (Jarvis and Sale 1971). The incisors are prominent, and folds of the lips extend downward and prevent the entrance of soil into the mouth during digging.

The following ecological information was taken largely from Jarvis (1973a) and Jarvis and Sale (1971). *Heliophobius* is usually found in sandy areas with low rainfall (250–630 mm annually). It constructs burrows consisting of a single main tunnel about 47 meters long, 15–25 cm deep, and 5 cm wide and numerous side branches. Mounds of excavated soil mark the course of the main tunnel. The nest is located in a small circular chamber 8–10 cm in diameter and 30 cm below the ground. The nest chamber is not used for food storage or defecation. Bolt holes for emergency refuge may extend from the main tunnel to depths of about 50 cm.

In burrowing, *Heliophobius* uses its powerful incisors to loosen the soil, which is then pushed under the abdomen with the forefeet and kicked behind the animal with the hind feet. When enough soil has accumulated to the rear of the mole-rat, the animal backs up, thrusts its nose and upper incisors into the top of the burrow for support, and uses its hind feet to force the plug of earth back through the tunnel and out onto the surface.

Heliophobius appears to be unreceptive to light stimuli but is said to have acute senses of smell and hearing. It spends about half of the day outside of its nest chamber. It may be active at any time, the peak period being 0900–2200 hours. It rarely forages above ground but apparently sometimes moves about on the surface, as its skulls have been found in owl pellets and one individual was captured when seemingly attracted to a campfire at night. Food storage might occur but was not evident during the studies of Jarvis and Sale. Instead, *Heliophobius* fed on underground portions of plants along the foraging tunnels and allowed them to continue growing in place. The diet consists mainly of tubers and bulbs.

This mole-rat is solitary and extremely aggressive toward others of its kind. The burrows excavated by Jarvis and Sale (1971) contained a single individual each. A female captured on 24 May 1967 and subsequently held in isolation gave birth to two blind and hairless young on 19 August 1967 (Jarvis 1969). According to Kingdon (1974b), the one annual litter is reportedly produced at the beginning of the long rains, pregnant females containing two or three embryos were collected in Kenya in May, and overall litter size is one to four. An adult female held captive for two years by Jarvis (1973a) was released for a month and then recaptured and was still alive and healthy a year later.

Heliophobius reportedly sometimes damages cultivated crops, such as potatoes. In many areas this animal is eaten by the native people.

RODENTIA; BATHYERGIDAE; **Genus BATHYERGUS**
Illiger, 1811

Dune Mole-rats

There are two species (De Graaff, *in* Meester and Setzer 1977):

B. janetta, extreme southwestern Namibia, coastal part of northwestern South Africa;

B. suillus, coastal parts of southwestern and southern South Africa.

Dune mole-rat *(Bathyergus suillus)*, photo by John Visser.

Head and body length is 175–330 mm and tail length is 40–70 mm. Adult males weigh about 1,500 grams and adult females, about 850 grams (Jarvis 1969). The pelage is thick and rather woolly. The upper parts are cinnamon, drab gray, or silvery buff. In *B. janetta* there is a broad, dark dorsal band. In most forms the underparts are colored like the upper parts, but in one form the ventral surface is black. White and piebald varieties of *Bathyergus* are common.

These mole-rats have short legs and five well-developed foreclaws, that of the second digit being the longest. The third toe of the hind foot also has an especially long claw. The pale brown tail has a flat, featherlike appearance resulting from the fringe of long, pale slate hairs growing out from either side. The eyes are very small, and there are no external ears. The incisor teeth are white; the upper ones are heavily grooved and the lower ones are not grooved. Both sets of incisors project forward to a remarkable degree, but unlike those of all other bathyergid genera, the two lower ones cannot be moved independently, as the lower jaws are ossified at the mandibular symphysis. Females have six mammae.

The habitat of *Bathyergus* appears to be restricted to sand dunes and sandy flats, especially near the coast but up to elevations of about 1,500 meters. Dune mole-rats spend their life in extensive burrow systems that they themselves have excavated. They are the only bathyergids that dig mainly with the forefeet rather than the incisor teeth (Jarvis 1969). They push soil in front of them through the burrows and then throw it out through short side tunnels leading to the surface. A captive individual made a chattering noise while burrowing and would turn and bite if interrupted. The diet consists principally of bulbs and fleshy roots. In agricultural areas, *Bathyergus* may collect a great surplus of food and store it within the burrow.

Small groups of dune mole-rats probably occupy the same burrow system and form a close-knit colony. Females are probably monestrous, and mating occurs toward the end of the winter rains (winter extends from May to August in South Africa). Pregnant females have been found from midwinter to early spring (July–October), with a peak in August. The gestation period apparently lasts about 2.0–2.5 months. Embryo counts in 34 pregnant females averaged 2.4 and ranged from 1 to 4. Births seem to be timed so that the young are weaned, and establish their own burrows, mainly during the early spring, when the rainfall is slight but the soil is still easily worked and the vegetation is lush (Jarvis 1969; Van der Horst 1972).

Dune mole-rats are sometimes considered pests, because they raid cultivated crops, especially potatoes. Their burrows are a serious menace to persons riding horseback, as the ground is weakened to such an extent that it collapses under the weight. Railroad ties may also be undermined, causing the rails to drop under the weight of a passing train. The species B. janetta is considered rare in South Africa and is losing part of its limited habitat when sand dunes are removed in the course of diamond mining (Smithers 1986).

RODENTIA; BATHYERGIDAE; Genus HETEROCEPHALUS
Rüppell, 1842

Naked Mole-rat, or Sand Puppy

The single species, H. glaber, occurs in central and eastern Ethiopia, central Somalia, and Kenya (De Graaff, in Meester and Setzer 1977).

Head and body length is 80–92 mm, tail length is 28–44 mm, and weight is 30–80 grams (Kingdon 1974b). This is the smallest bathyergid, the average weight being only about 35 grams (Jarvis 1978). The animal appears to be naked at first glance, but a few pale-colored hairs are scattered about the body and tail, there are prominent vibrissae on the lips, and a fringe of fine hairs is present along the edges of the feet. This fringe helps to collect and sweep back loose soil during digging. The wrinkled skin is pinkish or yellowish in color.

Heterocephalus has a short and chunky head, minute eyes and external ears, and very prominent incisor teeth. The front feet are large and have five broad, flattened digits, each of which has a small, conical claw. The hind feet also have five clawed digits. The body lacks sweat glands and the normal mammalian layer of subcutaneous fat (Jarvis 1978). One female was found to have seven functional pairs of mammae (Jarvis 1969).

The naked mole-rat is completely fossorial, is adapted exclusively to arid conditions, and is found in a variety of soil types (Kingdon 1974b). The elevational range in Ethiopia is 400–1,500 meters (Yalden, Largen, and Kock 1976). The investigations of Jarvis (1969, 1978, 1985) and Jarvis and Sale (1971) have added substantially to our knowledge of this remarkable genus, and the remainder of this account is based largely on those studies.

Heterocephalus constructs an extensive and complex burrow system. A large communal nest chamber is either unlined or partly floored with dry vegetation. Radiating outward are a series of foraging tunnels usually about 3 cm wide and 15–40 cm below the surface. One system was estimated to have more than 300 meters of such passages. There is much branching of the tunnels in the vicinity of food supplies. Bolt holes, reaching depths of about 70 cm, are sometimes present and may serve as emergency refuges. Above the burrow are the mounds of excavated earth, which are uniquely volcanolike in shape and in the presence of a central, unplugged hole.

Digging is done primarily with the incisor teeth, and the feet are used to kick and push the loosened earth. An individual near the surface may work alone in much the same manner as Heliophobius, but deeper burrowing is done by relay. One animal remains for a while at the earth face, loosening the soil and kicking it backward. Behind this animal is a chain of several others. When enough earth has accumulated, the first animal of the chain pushes the soil backward with its hind feet through the tunnel to a point near the surface and then takes a place at the end of the chain. Meanwhile, the other animals of the chain have advanced in position by straddling the one moving to the rear. Near the surface another individual remains for a while to collect the soil brought up by the others and kick it out through the central hole of the forming mound. At irregular intervals, this individual and the one doing the drilling are relieved by other members of the team.

Heterocephalus has the poorest capacity for thermoregulation of any known mammal. Its body temperature is relatively low, only 32° C. Nonetheless, the temperature within its burrow is maintained at 30°–32° C, and the humidity at 90 percent, regardless of external conditions. The animals avoid extreme temperatures near the surface by restricting burrowing activity to early morning and late afternoon. Most burrowing seems to be done following the rains, when the soil is more easily worked. Therefore, the important task of locating new food supplies is accomplished mainly during energetically favorable periods. The tubers, roots, and corms, which compose most of the diet, are generally left growing in place after discovery and are visited by individuals as the need arises.

Heterocephalus lives in well-organized colonies usually comprising about 20–30 individuals, though there have been reports of up to 100 members. As colonies may be located less than 100 meters apart in suitable habitat, it is possible that the reports of large groups have resulted from simultaneous observation of several colonies. In any event, detailed studies of captive colonies indicate that they are led by a single large female and that there are both working and nonworking classes. In one group the dominant female weighed 53 grams, the 3 nonworkers (2 males and 1 female) had an average weight of 38 grams, and the 10 workers (mixed sexes) averaged 32 grams. Although the working class comprised mostly younger animals, the weight difference did not appear to be associated mainly with age variation, as all animals were fully adult and over 3.5 years of age. The dominant female and the nonworkers spent most of their time in the nest chamber, coming out to urinate or defecate. The workers cooperated in burrowing, gathered nest materials, and brought food to the nest for the dominant female and nonworkers. All animals huddled together to sleep. Before full development of the social hierarchy, females coming into estrus fought and frequently killed one another. Once the dominant female established her position, she suppressed breeding by the other females. Only she mated, and she initiated courtship, generally with a nonworking male.

There is evidently a well-defined breeding season, with births occurring mainly from February to April, during the long rains. If the first litter survives, the female apparently breeds only once a year, but if not, she mates again. One female gave birth to three litters in a period of only six months. Recorded litter size is 3–12 young. The newborn are blind and, of course, naked. Their average weight is 1.9 grams. Members of the working class bring food to the young, keep them warm, and move them in case of danger. The young mature slowly, their eyes do not open for several weeks, and they do not reach adult size for at least a year. Nonetheless, they enter the working class when fully

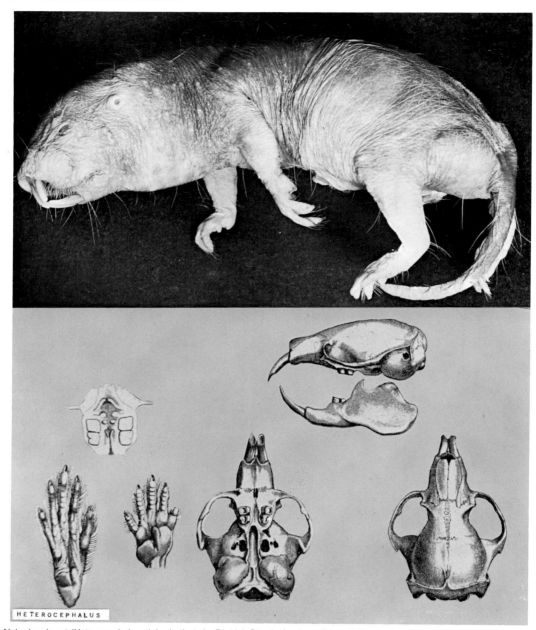

HETEROCEPHALUS

Naked mole-rat *(Heterocephalus glaber)*, photo by Dietrich Starck. Anatomy *(H. g. phillipsi)*, photo from *Proc. Zool. Soc. London*.

weaned, at 1–2 months. Many captives have lived for at least 5 years.

RODENTIA; Family CTENODACTYLIDAE

Gundis

This family of four Recent genera and five species occurs in northern Africa. The sequence of genera presented here follows that of George (1985).

Head and body length is 160–240 mm and tail length is 10–50 mm. The fur is soft and the tail is fully haired. Gundis resemble guinea pigs *(Cavia)* in external appearance, having a thickset and compact body. There are four digits on each foot; the two inner digits of the hind foot have comblike bristles. The claws are not enlarged but are very sharp. The skull is flattened and broad posteriorly. The bullae are inflated. The dental formula is: (i 1/1, c 0/0, pm 1–2/1–2, m 3/3) × 2 = 20 or 24. The upper incisors are slightly grooved *(Felovia)* or are ungrooved, and the cheek teeth are rootless (ever-growing). This family is unusual in that females have a cervical pair of mammae in addition to a single pair placed laterally on the anterior thorax (B. J. Weir, Wellcome Institute, Zoological Society of London, pers. comm.).

According to George (1974), all four genera are found in desert or semidesert habitats at elevations of sea level to 2,400 meters. They never excavate or occupy burrows but dwell in caves or rocky crevices. Ideal living sites have permanent shelters, temporary shelters, and ledges for sunbathing. Nests are not constructed. All genera are diurnal and are not known to hibernate or estivate. When the weather is cold, windy, or wet, activity is restricted, and the animals may not emerge at all. When the temperature is high, however, they come out at first light and remain active for five hours. Activity declines during the hot, middle part of the day but increases in the two to four hours before dusk. There are three main kinds of activity: foraging, sunbathing, and general (play, chasing, and exploring). The diet includes leaves, stalks, seeds, and flowers; animal food is not known to be eaten. Food is not stored, and fat reserves are not accumulated by the body. After a gestation period of about 55 days, the young are born fully furred with their eyes open.

The systematic affinities of the Ctenodactylidae are uncertain, and the family perhaps should be placed within the suborder Sciuromorpha rather than the Hystricomorpha. The earliest fossils assigned to the family date from the middle Eocene of Asia. There are also Oligocene, Miocene, and Pliocene remains from Asia; Miocene and Pleistocene remains from Sardinia and Sicily; and Miocene and Pleistocene remains from North Africa. The living genera evidently originated in Africa and are known only from the Recent of that continent (George 1979; Wood 1977).

RODENTIA; CTENODACTYLIDAE; Genus PECTINATOR
Blyth, 1856

Speke's Pectinator

The single species, *P. spekei*, occurs in coastal and eastern Ethiopia, Djibouti, and Somalia (De Graaff, *in* Meester and Setzer 1977).

Head and body length is 140–90 mm and tail length is 40–60 mm (Funaioli 1971). Adult females weigh about 178 grams (George 1978). The pelage is quite soft. The upper parts are ashy gray, suffused with black or brown, and the sides are grayish. The hands, feet, and underparts are grayish white.

Individuals taken from near the coast and from an elevation of 1,800 meters seem to differ only slightly in color and thickness of fur. The whiskers are fairly long, and the tail is bushy. The skin is said to be extremely thin and easily torn. The digits of the hind feet have brushes of comblike bristles. Females have two pairs of mammae.

The basic ecology is as given above in the family account. In Ethiopia, according to Yalden, Largen, and Kock (1976), *Pectinator* is a denizen of rocky cliffs in desert or semidesert areas and is often found together with the hyrax *(Procavia)*. The genus has been recorded reliably in Ethiopia at elevations from sea level to 1,200 meters. *Pectinator* shelters in rock crevices, emerges shortly after sunset, often basks in the sun, and feeds only on vegetation. In a study in Ethiopia, George (1974) found three colonies to live within well-defined boundaries encompassing areas of 1,500–2,000 sq meters. A whistling call was heard during the approach of a goshawk. The call has been described as a long, drawn-out "whee, whee."

George (1978) observed many young in the Danakil Desert of Ethiopia probably born from late August to mid-September. Captive females were in anestrus only in July; perhaps *Pectinator* is a more opportunistic breeder than *Ctenodactylus* and *Massoutiera*. The estrous cycle averages 22.7 days. Five captive-born litters contained a single young each, and one litter contained two young. A female was held in captivity for 4 years.

RODENTIA; CTENODACTYLIDAE; Genus FELOVIA
Lataste, 1886

The single species, *F. vae*, is known from Mauritania, Senegal, and Mali (De Graaff, *in* Meester and Setzer 1977).

Head and body length is about 170–230 mm and tail length is about 20–30 mm. The mean weight of 10 captive adult males was about 186 grams (George 1978). The upper parts are dark yellowish red, and the underparts are paler red. *Felovia* resembles *Massoutiera* externally but can be distinguished by its slightly grooved upper incisors and less prominent hair fringe on the inner edge of the ear. Females have two pairs of mammae.

The basic ecology is as given above in the family account. George (1974) stated that *Felovia* lives on the edge of very dry

Speke's pectinator *(Pectinator spekei)*, photo by Wilma George.

Felovia vae, photo by Wilma George.

tropical forest but was never seen to climb a tree. She studied a colony in western Mali in 1972 that was present on the same site where it had been discovered in 1885. In this area, George (1978) found young in mid-March that probably had been born from mid-December to January, 2–3 months after the rains. A captive female had an estrous cycle of 23 days. Four litters contained a single young each. According to B. J. Weir (Wellcome Institute, Zoological Society of London, pers. comm.), George's captives appeared to be very nervous and would vocalize with a birdlike trill when disturbed. Because of the ongoing deforestation and desertificaton of the Sahel region, Schlitter (1989) classified *Felovia* as vulnerable.

RODENTIA; CTENODACTYLIDAE; Genus MASSOUTIERA
Lataste, 1885

The single species, *M. mzabi*, occupies the central Sahara Desert in Algeria, northern Niger, northwestern Chad, and probably southwestern Libya (De Graaff, *in* Meester and Setzer 1977; Gouat, Gouat, and Coulon 1984).

Head and body length is 170–240 mm and tail length is about 35 mm. The weight of adults held in captivity by George (1978) averaged about 172 grams for males and 194

Massoutiera mzabi, photo by Jean-Marie Baufle through F. Petter.

Gundi *(Ctenodactylus gundi)*: A. Photo by Schomber-Kock; B & C. Right hind and front foot, photos by P. F. Wright of specimen in U.S. National Museum of Natural History; D. Photo from Archives of Zoological Garden Berlin-West.

grams for females. The coloration ranges through various yellow and brown shades. The ears are extraordinary in being flat against the head and immovable. Like the eyes, they are rounder in *Massoutiera* than in *Ctenodactylus*. A fringe of hairs around the inner margin of the ear, like that in jerboas and other desert rodents, protects the meatus from wind-blown sand. Females have two pairs of mammae.

The basic ecology is as given above in the family account. *Massoutiera* is found mainly in mountainous areas that rise above the Sahara Desert. It lives in rock crevices, is diurnal, does not emerge in cold or stormy weather, often sunbathes, and feeds on plant material. Despite the harsh, dry conditions under which it usually lives, *Massoutiera* has very little physiological adaptation for water conservation or tempera-ture control (George 1988). Its survival seems to depend in-stead on taking advantage of available shade and wind, as well as on being able to feed on a variety of plants with a high water content. In studies in Algeria, George (1974) found that colonies used many temporary shelters. *Massoutiera* thumps with its hind feet when excited or alarmed.

George (1978) found young in the wild from March to June. In April and May two females that were collected were pregnant, but four were not. Captive females had an estrous cycle averaging 24.9 days and were in anestrus from October to March. Four litters of two young, and one of three young, were produced. Two newborn weighed 20 and 21 grams. They were fully furred and had their eyes open. Of seven

captive females held by George, three remained with a male for approximately 2 years, two were paired each with a single male for varying periods, and two lived together.

RODENTIA; CTENODACTYLIDAE; **Genus CTENODACTYLUS**
Gray, 1828

Gundis

There are two species (De Graaff, *in* Meester and Setzer 1977; George 1982):

C. gundi, Morocco to northwestern Libya;
C. vali, southern Morocco to northwestern Libya.

Head and body length is about 160–208 mm and tail length is 10–25 mm. The average weight of captive adults is 174.5 grams in *C. vali* and 288.8 grams in *C. gundi* (George 1978, 1982). The upper parts are buffy in color, occasionally pinkish buff, and the underparts are paler, usually whitish or slaty. The generic name, like that of the tuco-tucos *(Ctenomys),* is derived from the comblike bristles on the toes of the hind foot. *Ctenodactylus* resembles *Pectinator* externally but differs in having a tail that is shorter than the hind foot. Females have two pairs of mammae.

The basic ecology is as given above in the family account. Gundis live on rocky slopes within warm temperate and subtropical deserts. The normal movement on a level surface is a quick run, the belly almost touching the ground. On sloping surfaces, however, C. gundi presses its body against the wall and uses the slightest irregularities in the obstruction to ascend almost perpendicularly. Flat stones or logs are preferred for sunning and resting. It has been reported that Ctenodactylus will "play possum" when threatened and may show "fear paralysis" for up to 12 hours. Such behavior, however, has never been observed by George, who has studied gundis in the wild and in captivity for many years. Gundis rely mainly on speed and agility to squeeze into very small cracks to escape from predators. These rodents seem to feed solely on plant material. They do not drink, but obtain sufficient water from the plants that they eat. The main call is a birdlike whistle, and they also thump with their hind feet when excited or alarmed.

In Algeria, according to George (1978), semicaptive C. vali produced litters mainly in March and April, but local nomads reported the overall breeding season to extend from February to June. In Tunisia a female C. gundi was found with a juvenile probably born in January. Females evidently produce only one litter per year and do not experience a postpartum estrus. The estrous cycle averages 28.7 days in C. gundi and 23.4 days in C. vali. Anestrus extends from September to May in C. gundi and May to December in C. vali. A single gestation period in C. vali was determined to be 56 days. Litter size is one to three young. The newborn weigh about 20 grams each, are fully furred, have their eyes open, and are able to run. They are capable of feeding by themselves after a few days but usually nurse for several weeks. They may require 9–12 months to reach full adult size and sexual maturity. One of the females kept by George lived for 5 years.

Order Cetacea

Whales, Dolphins, and Porpoises

This order of wholly aquatic mammals occurs in all the oceans and adjoining seas of the world, as well as in certain lakes and river systems. Living cetaceans are traditionally divided into two suborders: the Odontoceti, individuals of which have teeth (generally all of one kind) and an asymmetrical skull and which comprise the families Iniidae, Lipotidae, Pontoporiidae, Platanistidae, Delphinidae, Phocoenidae, Monodontidae, Ziphiidae, and Physeteridae; and the Mysticeti, individuals of which have plates of baleen, instead of teeth, and a symmetrical skull and which comprise the families Eschrichtidae, Balaenopteridae, Balaenidae, and Neobalaenidae. Rice (1977, 1984) considered the Odontoceti and the Mysticeti to be distinct orders, because it is questionable whether the two groups had a common origin and because the differences between them are as great as those between some of the universally recognized orders of mammals. Certain authorities, such as Ellis (1980), have followed Rice's procedure. Gaskin (1976), however, summarized data supporting a monophyletic origin for cetaceans and treated the Odontoceti and Mysticeti as suborders. The latter procedure also is supported by new cytological data (Arnason, Höglund, and Widegren 1984) and now has returned to general acceptance (Barnes, Domning, and Ray 1985; Corbet and Hill 1986; Honacki, Kinman, and Koeppl 1982; Jones et al. 1986).

There are several problems regarding the division of the Cetacea at the family level, perhaps the most extensive involving the fresh water or river dolphins of the genera *Inia*, *Lipotes*, *Pontoporia*, and *Platanista*. These genera resemble one another superficially, as well as in some internal characters, and often have been united in a single family, the Platanistidae. However, there also long has been recognition by some authorities that the resemblances might result from convergent evolution and that at least some of the genera belonged in different families. Rice (1977) placed all of these genera in the family Platanistidae but recognized three subfamilies: Iniinae, with the genera *Inia* and *Lipotes*; Pontoporiinae, with *Pontoporia*; and Platanistinae, with *Platanista*. Kasuya (1973) gave familial rank to all three groups, and Zhou, Qian, and Li (1979) thought that *Lipotes* should also be in a monotypic family, the Lipotidae. Although the new general mammal lists by Corbet and Hill (1986) and Honacki, Kinman, and Koeppl (1982) continued to place all four genera in the Platanistidae, specialists in cetacean systematics now generally have accepted the Iniidae, Lipotidae, Pontoporiidae, and Platanistidae as full families but have included all within the superfamily Platanistoidea (Barnes, Domning, and Ray 1985; Pilleri and Gihr 1981b; Rice 1984; Zhou 1982). Pilleri and Gihr used the name Stenodelphidae in place of Pontoporiidae.

Considering the above, as well as the increasing recognition of the Phocoenidae as a family distinct from the Delphinidae and of the Neobalaenidae as distinct from the Balaenidae, 13 Recent cetacean families are accepted here, along with 41 genera and 79 species. Rice's (1977, 1984) sequence of these taxa generally is followed here, except that the odontocete families are here placed before the mysticete families.

The length of the cetacean head and body is taken in a straight line from the tip of the snout to the notch between the tail flukes. Head and body length varies from about 1.2 to 31.0 meters, and weight ranges from 23 to 160,000 kg. The tail flukes are set in a horizontal plane, thereby immediately distinguishing cetaceans from fish, the tail fins of which are in a vertical position. Other external features conspicuous in cetaceans are the torpedo-shaped body, front limbs that are modified into flippers (pectoral fins) and ensheathed in a covering, the absence of hind limbs, and the usual presence of a dorsal fin. There is no covering of fur, but all whales have some hairs in the embryonic stage, and in adult mysticetes a few bristles persist around the mouth. Cetaceans lack sweat glands and sebaceous glands. They have a fibrous layer (blubber) filled with fat and oil just beneath the skin, which assists in heat regulation. There are no external ears or ear muscles and no scales or gills.

The nostrils open externally, usually at the highest point of the head. The odontocetes have a single blowhole, and the mysticetes have a double one. The asymmetry of the skull of the toothed whales is in correlation with the reduction of one of the nasal passages; both features seem to have evolved in connection with modifications needed for sound production (Yurick and Gaskin 1988). There is a direct connection between the blowhole and the lungs, so that a suckling calf cannot get milk into its lungs. The blowhole is closed by valves when the animals are submerged. Cetaceans do not blow liquid water out of the lungs. When the animals exhale, the visible spout is from the condensation of water vapor entering the air from the lungs, and possibly from the discharge of the mucous oil foam that fills the air sinuses.

The bones are spongy in texture, and the cavities are filled with oil. In some genera the vertebrae in the neck are fused, but all lack the complex articulation of vertebrae of land mammals and present a graded series from head to tail. There are no bony supports for the dorsal fin and the tail flukes. The pelvic girdle, represented by two small bones embedded in the body wall and free from the backbone, serves only as the attachment for the muscles of the external reproductive organs.

Propulsion is obtained by means of up-and-down movements of the tail such that the flukes present an inclined surface to the water at all times. The force generated at right angles to the surface of the flukes is resolvable into two com-

The back of a humpback whale *(Megaptera novaeangliae)* that has surfaced to breathe. The two openings on top of the head are the "blowholes," which the whale closes before submerging. These are comparable to the nostrils of other mammals. Photo by Vincent Serventy.

ponents, one raising and lowering the body and the other driving the animal forward. The fins serve as balancing and steering organs. Cetaceans are the swiftest animals in the sea; some dolphins can maintain a sustained speed of 26–33 km/hr (Rice 1967).

Certain questions regarding the physiological adaptations of cetaceans in diving and temperature tolerance remain unanswered, but some of the general mechanisms are known. Before diving, a cetacean expels the air from its lungs. Some of the adaptations that make long dives possible are as follows: (1) the oxygen combined with the hemoglobin of the blood and with the myoglobin of the muscles accounts for 80–90 percent of the oxygen supply utilized during prolonged diving; (2) arterial networks seem to act as shunts, maintaining the normal blood supply to the brain but effecting a reduced supply to the muscles and an oxygen debt that the animal can repay when it again surfaces; (3) a decreased heartbeat further economizes the available oxygen; and (4) the respiratory center in the brain is relatively insensitive to an accumulation of carbon dioxide in the blood and tissues. The hydrostatic pressures encountered at great depths are alleviated by not breathing air under pressure and by the permeation of the body tissues with noncompressible fluids. The only substances in the body of a cetacean that can be compressed appreciably by the pressure of great depths are the free gases, found mainly in the lungs. The collapse of these gases drives them into the more rigid, thick-walled parts of the respiratory system. The body temperature is regulated by the insulation of the blubber, which retains body heat when the animals are in cold water, and by the thin-walled veins associated with arteries in the fins and flukes.

Some cetaceans are the only animals other than elephants that have a brain larger than that of humans; brain weight varies from about 0.2 kg in *Pontoporia blainvillei* to 9.2 kg in *Physeter catodon* (Kamiya and Yamasaki 1974; Rice 1967). In adult *Homo sapiens* the weight is about 1.3 kg. Like that of humans, the brain of cetaceans is highly convoluted. Moreover, cetaceans evolved a brain the size of humans' 30 million years ago, whereas the human brain has been its present size for only about 100,000 years (Lilly 1977). These factors, along with the ability of captive odontocetes to learn rapidly and to

form social bonds with humans, suggest the existence of a high level of intelligence in the Cetacea.

Most cetaceans have eyes well adapted for underwater vision and can also see well above water. A greasy secretion of the tear glands protects the eyes against the irritation of salt water. The sense of smell is vestigial in mysticetes and absent in odontocetes. Cetaceans have good directional hearing underwater (Rice 1967).

Hydrophonic studies have shown that cetaceans produce numerous underwater sounds and that some genera depend to a large extent on echolocation for orientation and securing food (Caldwell and Caldwell 1977, 1979; McNally 1977; Rice 1984; Thompson, Winn, and Perkins 1979). Odontocetes commonly produce two kinds of sounds: clicklike pulses, which last 0.001–0.010 second, have a frequency band of 100 to over 200,000 cycles per second, and are uttered in series of five to several hundred per second; and whistles, which last about 0.5 second and have a frequency band of 4,000–20,000 cps. Clicks are used in echolocation and sometimes in communication, while whistles may be highly individualized and are used primarily for communication. There is still some question where sounds are produced, but there is increasing evidence that both clicks and whistles originate in a series of air sacs that lie above the bony cranium in the soft tissues around the blowhole on top of the head. Sounds are reflected off the concave dorsal surface of the skull and are then focused and directed by the melon, a large pocket of fat on the forehead of most odontocetes. This pocket is especially well developed in genera that regularly feed in the lightless bathypelagic zone (such as *Hyperoodon*) and those that live in turbid rivers (such as *Platanista*). The enormous square head of the sperm whale *(Physeter)* is filled in large part by the spermaceti organ, which probably functions much like the melon in smaller odontocetes. Mysticetes seem to employ only a crude form of echolocation but do produce a variety of moans and squeals, which are sometimes combined into elaborate "songs" and are primarily for communication. It is probable that the large size of the cetacean brain is associated with the development of a precise sense of hearing and the analysis of complex echolocative and communicative signals.

Odontocetes generally feed on fish, cephalopods (such as squid and octopus), and crustaceans. The conical teeth seize slippery prey but are not adapted for chewing. The killer whale *(Orcinus)* regularly takes warm-blooded prey, such as penguins, pinnipeds, and other cetaceans. Mysticetes feed on many different kinds of small animals, mostly crustaceans, which are collectively referred to as zooplankton. Mysticetes are often called filter feeders or grazers. Their baleen is composed of modified mucous membrane and arranged in thin plates, one behind the other and suspended from the palate. There are two rows of plates, one on each side of the mouth. The plates hang at right angles to the longitudinal axis of the head. The outer borders of the plates are smooth, whereas the inner parts are frayed into brushlike fibers. These plates act as sieves or strainers. Most mysticetes feed by swimming with their mouth open through swarms of plankton. They then use their huge tongue to force water out through the lowered baleen, within which food organisms are trapped.

Most cetaceans are gregarious to some extent, and most have a relatively long period of parental care and maturation. The gestation period ranges from 9.5 to 17 months. There is almost always a single offspring, which at birth is usually one-fourth to one-third the length of the mother. Immediately after being born in the water, baby cetaceans must reach the surface for a supply of air. Most, if not all, cetacean mothers probably push their offspring to the surface. When nursing, the mother floats on her side so that the calf can breathe. Later the calf can suckle underwater. The teats of the mammary glands lie within paired slits on either side of the

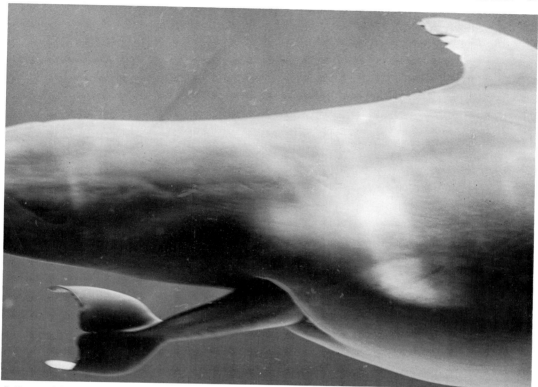

Bottle-nosed dolphin *(Tursiops truncatus)*, showing a baby being born tail first, photo from Miami Seaquarium.

reproductive opening. The mammary glands have large reservoirs in which the milk collects, and the contraction of the body muscles forces the milk, by way of the teats, into the mouth of the young. The rapid growth rate of most cetaceans is at least partly related to the high calcium and phosphorus content of the milk. Because cetaceans live in an aquatic environment, and need not support their own weight, they can attain great size. Maximum size sometimes is not reached until many years after sexual maturity. Most species have a potentially long life span, and some individuals are thought to have lived over 100 years.

Cetaceans were spared from the wave of extinction of large mammals that swept the earth toward the close of the Pleistocene (about 10,000 years ago) and that was probably associated with the spread of advanced human hunters (Martin and Wright 1967). Some whales and dolphins were killed by people in ancient times, but generally in small numbers and near the shore. By about the year A. D. 1000 the Basques of coastal France and Spain had developed a fishery involving the pursuit and harpooning of the right whale *(Eubalaena glacialis)* from small boats. As the whales became fewer, the Basques went farther and farther out to sea, perhaps reaching waters off Newfoundland just before Columbus's arrival in the West Indies. Over the next 500 years ships became bigger and faster, hunting and processing techniques steadily improved, human population and demand for cetacean products increased, and whale populations declined (Allen 1942; Brownell, Ralls, and Perrin 1989; Committee for Whaling Statistics 1980, 1984; Ellis 1980; Grzimek 1975; D. O. Hill 1975; McHugh 1974; Small 1971; U.S. National Marine Fisheries Service 1981, 1984, 1987, 1989).

For centuries, one of the most valuable cetacean products was baleen (sometimes erroneously called bone), thin strips of which were used to stiffen various articles of clothing and for other purposes requiring a combination of strength and flexibility. Changes in fashion and the development of spring steel and plastics greatly reduced the demand for baleen. The most consistently important objective of whaling was the oil derived from blubber, which was a major fuel for lamps. Some approximate average yields of barrels of oil (1 barrel = about 105 liters) for individuals of certain species of whales are: sperm, 35; fin, 40; blue, 75; humpback, 30; and bowhead, 90. The invention of kerosene and eventually of the electric light helped to alleviate the pressure on whale populations in the latter half of the nineteenth century. In 1905, however, a new process was developed for hardening fat, and this led to the use of baleen whale oil in the production of margarine. Whale oil also came to be used in the manufacture of soaps, lubricants, waxes, explosives, and numerous other products. The true bones of whales were ground up and used to make glue, gelatin, and fertilizer. Whale meat was eaten by the people of some nations and was used as dog food and (when ground up) as cattle feed.

For many years the right whale, rich in oil and baleen, remained the prime target of the industry. It was hunted from bases on both sides of the Atlantic and in Iceland. As it became rare, the fishery slackened, finally terminating around 1700 in Europe and 1800 in America. Meanwhile, however, whalers had been turning increasingly toward arctic waters in search of the bowhead *(Balaena mysticetus)*. From the early seventeenth century to the late nineteenth century, first at Spitsbergen and then in the waters off northern North America and Russia, one population after another was devastated. Both the right whale and the bowhead, as well as the gray *(Eschrichtius robustus)* and the humpback *(Megaptera novaeangliae)*, which also were sometimes taken in the early days of whaling, tended to remain near the coast and so could be easily killed and processed. With the decline of these species in the Northern Hemisphere, deep-sea whaling intensified, lengthy voyages became the rule, and the sperm whale *(Physeter catodon)* became the main target. The pursuit of this species was led by the United States from the early eighteenth century to the middle of the nineteenth. Populations of the right whale in the North Pacific and the Southern Hemisphere also came under intensive exploitation, starting about 1800. When the industry reached its peak in 1847, nearly 700 American whaling ships were operating throughout the world. Over the next 20 years the industry declined, with some suggested causes being a reduction in whale numbers, increasing expenses, destruction of much of the U.S. whaling fleet in the Civil War, and the development of kerosene, which could be used in place of whale oil for illumination.

American whaling never again attained the importance of the early 1800s, but developments elsewhere allowed the industry to recover and even expand. During the 1860s in Norway a gun was perfected that fired a harpoon with an explosive head. Subsequently, people learned how to inflate the carcass of a dead whale with air so that it would not sink. These techniques encouraged large-scale exploitation of the blue *(Balaenoptera musculus)*, fin *(B. physalus)*, and sei *(B. borealis)* whales, which previously had been too big or too fast to approach, harpoon by hand, and keep afloat. In the early 1900s, enormous stocks of these species, as well as the humpback, were discovered in the waters around Antarctica. Starting in 1904, whaling stations were established on islands in this region. The humpback, often found in the vicinity of land, was quickly and drastically reduced. In the 1920s came the first floating factories, huge ships with slip sternways for drawing up dead whales, no matter how large, and with all the facilities for processing them. Accompanied by a fleet of swift, diesel-powered catcher boats, the factories could remain for lengthy periods in the midst of whale populations in remote waters. Eventually the whalers began to use such sophisticated methods as sonar and aircraft observation to locate their objective.

The value of cetacean products, together with improvements in human technology, resulted in the twentieth century's becoming the most destructive period in whaling history. From 1904 to 1939, 580,931 blue, fin, and humpback whales were recorded killed in the Southern Hemisphere. The peak annual tonnage taken was in the 1930/31 season (a season refers to the austral summer, December–March), when 29,410 blue, 10,017 fin, and 576 humpback whales were killed in antarctic waters. The rate of kill fell sharply during World War II but rapidly built up by 1947. With the decline of the blue whale, pressure increased on the fin and sei whales and, once again, on the sperm whale. There also was renewed interest in whaling in the Northern Hemisphere, especially in the Pacific Ocean. The peak annual catch of individual whales was reached during the 1961/62 season, when the worldwide kill was 65,966 blue, fin, sei, humpback, and sperm whales. During that season, 21 factory ships and 269 catcher boats operated in the Antarctic. The major whaling nations were Japan, Norway, the Soviet Union, Great Britain, and the Netherlands. As the larger species became rare, and as international regulations became more effective, the harvest fell. The blue and humpback whales received complete protection in 1966. The recorded kill of fin, sei, and sperm whales was 41,640 in the 1968/69 season and 9,429 in 1978/79. There was, however, a corresponding rise in the take of the smaller minke *(Balaenoptera acutorostrata)* and Bryde's *(B. edeni)* whales, the total catch of which was 4,238 in 1968/69 and 10,777 in 1978/79.

International control of whaling was first seriously dis-

cussed in 1927 at a meeting of the League of Nations. In 1935 the United States, Norway, Great Britain, and several other nations entered into an agreement under which the right and bowhead whales were protected (effective in 1937) and certain other whaling regulations were set forth. In 1946 the International Whaling Commission, a regulatory body now consisting of 35 member nations, was formed. Although the commission established various limits and refuge areas, its objectives were long thwarted by lack of cooperation by those members most involved in the whaling industry. Finally, as all the large species of whales approached or passed commercial extinction, regulations became more meaningful. The total kill quota fell from 45,673 whales in the 1973/74 season to 14,523 in 1980/81, only 2,121 of which were from the larger species (fin, sei, sperm). By the latter season only Japan and the Soviet Union still conducted major whaling operations, each employing one or two floating factories. Several other countries had small, land-based whale fisheries, and illegal "pirate" whaling ships were sometimes active.

At its 1982 meeting, the International Whaling Commission voted to end commercial whaling altogether by the end of the 1984/85 season; the last annual quota was 6,623, of which none were sperm whales. The U.S. government, reacting to immense public concern, long had advocated a total ban on commercial whaling. In 1972 the U.S. Congress passed the Marine Mammal Protection Act, which essentially ended the taking and importation of cetaceans and their products by persons subject to U.S. jurisdiction. Even after 1985 several nations continued to conduct whaling operations, first under a formal "objection," which legally removed their obligation to comply with the International Whaling Commission's moratorium, and later on the grounds that the whales were being taken for legal research purposes. Prospects for enforcement of U.S. laws providing for trade restrictions against these countries, especially Japan, helped to end the strictly commercial operations by 1987, but "research" whaling has continued, as has limited subsistence hunting by native peoples. The latter two activities were responsible for a kill of fewer than 700 whales during 1988. All cetaceans are on either appendix 1 or appendix 2 of the CITES.

One of the immediate goals of the U.S. Marine Mammal Protection Act was to reduce the number of small cetaceans (notably *Stenella* and *Delphinus*) being killed or injured by commercial fishing operations. The most critical problem was the incidental catch of these mammals in seines intended for tuna, especially in the eastern Pacific. Although many kinds of small cetaceans had been intentionally hunted by people, and while some populations had declined, there had been no worldwide pursuit and devastation comparable to what befell the larger whales (E. D. Mitchell 1975a, 1975b). In the 1960s, however, improved types of tuna seines began trapping large numbers of dolphins commonly found in association with the fish. The total kill from 1959 to 1972 has been estimated at 4.8 million (Lo and Smith 1986). The number of dolphins killed or seriously injured in 1972 was estimated at 368,600 for U.S. fishing vessels and 55,078 for vessels of other countries. Subsequent regulations, modifications of fishing gear, and cooperation by fishermen reportedly reduced the respective figures for 1979 to 17,938 and 6,837. Unfortunately, the kill by foreign fishing fleets rose drastically in the 1980s and now sometimes exceeds 100,000 per year.

The known geological range of the Cetacea is middle Eocene to Recent. Possibly, however, the order diverged from an extinct order of terrestrial carnivores, the Creodonta, at the end of the Cretaceous and entered the sea in the Paleocene (Gaskin 1976). The earliest known fossil cetaceans are assigned to an extinct suborder, the Archaeoceti. In this group the skull is symmetrical and the teeth are clearly differentiated into incisors, canines, premolars, and molars. The Odontoceti first appear in upper Eocene strata, and the Mysticeti first appear in the lower Oligocene.

CETACEA; **Family INIIDAE; Genus INIA**
D'Orbigny, 1834

Boutos, or Amazon Dolphins

The single Recent genus, *Inia*, contains two species (Pilleri and Gihr 1977a, 1981a; Rice 1977):

I. boliviensis, upper Madeira River system of Bolivia;
I. geoffrensis, Amazon and Orinoco basins of South
    America.

Some authorities, including Corbet and Hill (1986), regard *I. boliviensis* as only a subspecies of *I. geoffrensis.*

Head and body length is about 170–300 cm and expanse of the tail flukes is about 51 cm. Two adult males weighed 59 and 122 kg, and an adult female weighed 71 kg (Harrison and Brownell 1971). The variable coloration of this genus is apparently associated with age. All of the younger animals kept by Trebbau (1975) had dark bluish metallic gray upper parts and paled to silvery gray on the lateral and ventral parts.

Older and larger individuals were much lighter and generally pink in color. The lateral and ventral parts were clear pinkish gray, the flukes and most of the snout were pinkish, and the melon (forehead) was bluish and pink. The darkest parts of the older animals were the regions around the blowhole, eyes, neck, and middorsum. There were no distinct borders to the colors.

The skull is nearly symmetrical. The rounded head bears the blowhole on its summit. The eyes are small. The forehead is blunt, and there is a distinct external neck which has all its vertebrae free. The beak is long, slender, slightly down-curved, and clearly distinguishable from the rest of the head. For most of their length the lower jaws are fused or at least closely appressed. The teeth number about 33–34 on each side of each jaw, making a total of 132–36; the back 8–9 teeth have a distinct keel. The dorsal fin is low, long, and ridgelike, and the pectoral fin is short and broad. The upper arm bone is longer than the lower ones.

Inia has a sparse covering of short, stiff bristles on its snout, whereas *Lipotes*, *Pontoporia*, and *Platanista* lack such hairs. Other distinguishing characters, as listed by Zhou (1982), include a transverse and crescentic blowhole, the presence of a fore-stomach and a single main stomach, the absence of a caecum, the absence of a maxillary crest, the separation of the palatal portion of the maxillas by a vomer, and teeth that have crowns with nodular enamel rugosity.

These dolphins are restricted to fresh water. According to

Amazon dolphins *(Inia geoffrensis):* A. Photo from Fort Worth Zoological Park through Lawrence Curtis; B. Photo by James N. Layne.

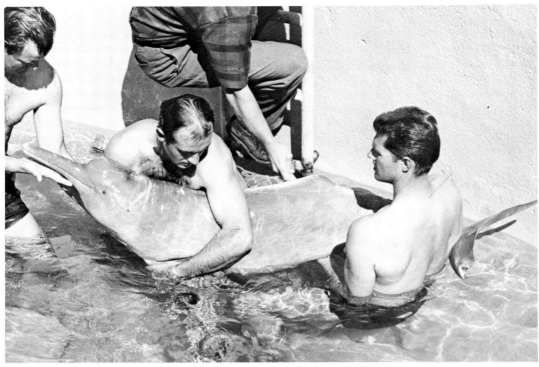

Amazon dolphin *(Inia geoffrensis)*, photo by Bob Noble through Marineland of the Pacific.

Trebbau and Van Bree (1974), they generally are found in brownish, turbid, and slow-moving or temporarily stagnant streams. They may migrate into flooded forests, as well as small streams and lakes, during periods of high water. They then sometimes become trapped in lakes during the dry season but are able to survive by preying on fish that also are trapped. If fish are abundant in an area, the dolphins can be seen there for weeks at a time. They surface to breathe every 30–60 seconds. Sometimes only the blowhole and top of the head emerge, but frequently the dorsal fin and ridge of the back are also exposed. When swimming rapidly, and apparently when feeding, *Inia* rolls to breathe. This genus seems to be less active than the Delphinidae but will occasionally leap out of the water to heights of 125 cm.

The senses of touch and hearing are probably acute; *Inia* probes for food on the bottom with its sensitive snout and may use echolocation to detect underwater obstacles and prey. The eyes are small, but Caldwell, Caldwell, and Evans (1966) found vision to be acute and apparently to be the preferred method of environmental investigation. They also recorded 12 types of sounds, which generally were less varied, lower in intensity, and of slightly lower frequencies than those heard in most other odontocetes. The diet consists of fish, usually less than 30 cm long. Captive adults ate 4–5 kg of fish per day (Trebbau and Van Bree 1974). Observations by Defler (1983) suggest that *Inia* may sometimes associate with the giant otter *(Pteronura)* in the hope of securing fish driven by the latter from shallow water.

Trebbau (1975) found *I. geoffrensis* mostly in small groups that had a tendency to occupy a defined territory. When an individual was taken captive, others would come to its assistance. Up to 8 captives were held together, evidently without strife. Johnson (1982) observed a large captive male to share its food with a female and a smaller male held in the same tank. Caldwell and Caldwell (1969) observed wild *I. geoffrensis* in schools of 12–15, occasionally up to 20, individuals and noted that captives tended to cluster together. Pilleri and Gihr (1977a) reported *I. boliviensis* usually to occur alone or in pairs. Harrison and Brownell (1971) collected a pregnant female *I. geoffrensis* in February and pregnant and lactating females and a calf in April. They suggested that implantation can occur in October and November, and probably earlier, and that births take place from July to September in the upper Amazon. Length at birth is 76–80 cm. According to Brownell (1984), one specimen was still alive after 18 years in captivity, and longevity for the genus probably is about 30 years.

These curious and inoffensive dolphins often swim around fishing boats. Pilleri (1979) stated that *I. geoffrensis* seems common in the Orinoco Basin and is not hunted there but is threatened by motorboats, pollution, and dam construction. Pilleri and Gihr (1977a) reported that *I. boliviensis*, which is isolated from *I. geoffrensis* by a 400 km stretch of rapids, has been severely reduced in numbers, mainly through hunting by people for its hide and fat. *I. geoffrensis* (including *I. boliviensis*) is classified as vulnerable by the IUCN.

The geological range of the Iniidae is early Miocene to Recent in South America, Miocene in Florida, and possibly late Miocene in western North America (Rice 1984). Grabert (1984) theorized that during the Miocene the Iniidae crossed from the Pacific through former gaps in the Andes into central South America. There *I. boliviensis* evolved, and in the early Pleistocene some populations expanded into the Amazon Basin, where they gave rise to *I. geoffrensis*. About 10,000 years ago a barrier of nonturbid, "blackwater" formed, which separated the subspecies *I. g. geoffrensis* in the Amazon and *I. g. humboldtiana* in the Orinoco system.

CETACEA; **Family LIPOTIDAE; Genus LIPOTES**
Miller, 1918

Baiji, or Whitefin Dolphin

The single known genus and species, *Lipotes vexillifer*, has been recorded from the mouth of the Chang Jiang (Yangtze) to a point about 1,900 km up that river, from Dongting and Poyang lakes, and from the Quiantang River just south of the mouth of the Chang Jiang (Zhou, Qian, and Li 1977). Early reports indicating that the species was restricted to Dongting Lake and some nearby waters were incorrect. As discussed above in the account of the order Cetacea, *Lipotes* long was placed in the family Platanistidae or in the Iniidae but now generally is considered to represent a distinct family.

Head and body length is 200–250 cm, and one specimen weighed 160 kg. Coloration is pale blue gray above and whitish below. The long, beaklike snout is curved upward, and there are 32–36 teeth in each side of each jaw. The eye is greatly reduced but is functional. Externally, *Lipotes* resembles *Inia* but differs in having an up-curved rostrum, a relatively smaller and more rounded pectoral fin, and a larger and more triangular dorsal fin (Brownell and Herald 1972). Other distinguishing characters of *Lipotes*, as listed by Zhou (1982), include the lack of hairs on the snout, a blowhole that is longitudinal and elliptic, no fore-stomach but a three-compartment main stomach, the absence of a caecum, the absence of a maxillary crest, contact between the palatal portion of the maxillas, and teeth that have crowns with reticulate enamel rugosity.

Zhou, Qian, and Li (1977) considered *Lipotes* to be a fluvi-atile and estuarine odontocete, contrary to early reports that the genus was primarily an inhabitant of shallow lakes. During periods of high water, as in the late spring and summer, *Lipotes* also goes upstream into lakes and small rivers. The long beak is used to probe muddy bottoms for food. Dives are short, usually lasting only 10–20 seconds (Zhou, Pilleri, and Li 1979). Several underwater acoustic signals are emitted, including a whistle, which apparently function in communication (Jing, Xiao, and Jing 1981). The diet consists of fish, and the stomach of the type specimen contained 1.9 liters of an eel-like catfish (Brownell and Herald 1972).

Zhou, Pilleri, and Li (1979, 1980) found an overall average of one *Lipotes* every four km along part of the Chang Jiang. The genus usually occurred in pairs, which in turn made up a larger social unit of about 10 individuals. Births were thought to take place in March and April; Brownell and Herald (1972), however, stated that a lactating female was captured on 21 December.

Lipotes long was protected by custom, though if an individual was killed accidentally, its meat would be eaten and its fat used for medicinal purposes. In recent decades, numbers and distribution have declined seriously through hunting, accidental catching by fishermen, collision with motorboats, and development of irrigation facilities (Chen et al. 1979; Zhou, Pilleri, and Li 1979; Zhou, Qian, and Li 1977). The genus no longer occurs on Dongting Lake, because of sedimentation. Legal protection has been provided in China since 1975. Brownell, Ralls, and Perrin (1989) stated that the baiji probably is the most endangered of all cetaceans, with only a few hundred surviving. It now is classified formally as endangered by the IUCN and USDI and is on appendix 1 of the CITES.

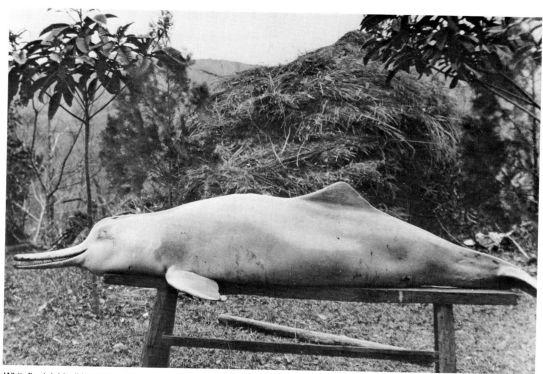

Whitefin dolphin *(Lipotes vexillifer)*, photo by Clifford Pope through Robert L. Brownell, Jr.

La Plata dolphin *(Pontoporia blainvillei)*, photos by Cory T. de Carvalho.

CETACEA; **Family PONTOPORIIDAE; Genus PONTOPORIA**
Gray, 1846

Franciscana, or La Plata Dolphin

The single Recent genus and species, *Pontoporia blainvillei*, occurs in the coastal waters and estuaries off southeastern South America, from the Tropic of Capricorn near Ubatuba, Brazil, south to the Valdez Peninsula, Argentina (Brownell 1975a). As discussed above in the account of the order Cetacea, Pontoporiidae sometimes has been considered to be a subfamily of Platanistidae but now generally is recognized as a distinct family. In addition, as noted by Rice (1984), some earlier workers had regarded the Pontoporiidae as a subfamily of the Delphinidae. Some authorities, such as Pilleri and Gihr (1981b), use the name Stenodelphidae in place of Pontoporiidae.

Head and body length is 130–75 cm and weight is 20–61 kg. Females are generally larger than males. The color is grayish above and paler below. The young are usually brownish. The beak is extremely long, slender, slightly downcurved, and moderately demarcated from the bulging forehead (Rice 1984). There are 48–61 teeth on each side of each jaw (Kasuya and Brownell 1979), and the total number of teeth per individual has been counted at 210–42.

Pontoporia bears some resemblance to *Inia* in external features of the head, neck, and pectoral fin but differs in that the beak is more slender and the dorsal fin is prominent and triangular in shape. The neck, while distinct and having all the vertebrae free, is not so obvious as in *Inia*. Other distinguishing characters, as listed by Zhou (1982), include the lack of hairs on the snout, a blowhole that is transverse and crescentic, no fore-stomach and a single main stomach, the absence of a caecum, the absence of a maxillary crest, contact between the palatal portion of the maxillas, and teeth that have a simple crown.

This is the only member of the superfamily Platanistoidea that occurs in salt water. It is found in the estuary of the Rio de la Plata but has not been recorded from the adjoining Parana and Uruguay rivers (Brownell 1975a). According to Kellogg (1940), *Pontoporia* is rarely seen in the Rio de la Plata during winter, perhaps because most schools then migrate out to sea or northward along the Brazilian coast. *Pontoporia* presumably locates its prey by echolocation and by probing the bottom with its long snout. Examination of stomach contents suggests that most food is taken at or near the bottom. The diet includes fish, squid, and shrimp (Fitch and Brownell 1971).

Examination of specimens taken off the coast of Uruguay (Brownell 1975a; Harrison, Bryden, and McBrearty 1981;

Susus (*Platanista* sp.), photos by G. Pilleri.

or early Pliocene and Pleistocene in western North America and Pliocene and Recent in South America (Rice 1984).

CETACEA; **Family PLATANISTIDAE; Genus PLATANISTA**
Wagler, 1830

Ganges and Indus Dolphins, or Susus

The single Recent genus, *Platanista,* contains two species (Pilleri and Gihr 1971; Rice 1977):

P. gangetica, Ganges-Brahmaputra-Meghna river system and the Karnaphuli River of India, Bangladesh, and Nepal;

P. indi, Indus River system of Pakistan and India.

As discussed above in the account of the order Cetacea, the family Platanistidae sometimes has been considered to include the families Iniidae, Lipotidae, and Pontoporiidae, but now each group generally is given familial rank. Van Bree (1976) argued that the name *P. minor* has priority over *P. indi,* but Pilleri and Gihr (1977b) rejected this view.

Head and body length is variable, but adults are usually 200–300 cm long. The female is the larger sex, there being a questionable report of one that measured about 400 cm. The expanse of the tail flukes is about 46 cm. The back is dark lead gray to lead black, and the belly is somewhat lighter. The snout is slender, slightly up-curved, well differentiated from the steeply rising forehead, and about 18–21 cm long. The neck is distinctly constricted, the dorsal fin is low and ridgelike, and the pectoral fin is cut off squarely at the end (Kellogg 1940). There are 26–37 teeth on each side of each

Harrison and Brownell 1971; Kasuya and Brownell 1979) indicates that most females have a 2-year reproductive cycle, with mating from December to February, births from September to December, lactation until the following August or September, and then a rest of several months. However, some females may undergo a postpartum estrus and breed once a year. The gestation period is about 10.5 months. The single young is generally 70–75 cm long at birth. It starts to take solid food at 3 months, but lactation may last at least 9 months. Both males and females become sexually mature at 2–3 years and physically mature 1–2 years later. One male specimen was 16 years old at the time of death.

The Franciscana is taken regularly off the coast of Uruguay in the nets of fishermen who are primarily seeking sharks (E. D. Mitchell 1975a; Kasuya and Brownell 1979; Pilleri 1971). About 1,500–2,000 individual *Pontoporia* are killed annually in this manner. They are used for pig feed and as a source of oil. This kill may be adversely affecting overall population size (Brownell, Ralls, and Perrin 1989).

The geological range of the Pontoporiidae is late Miocene

jaw. The upper toothrows are merged together and almost in contact.

Platanista resembles *Inia* in external features of the head, beak, neck, and dorsal fin but differs in the truncate pectoral fin. Other distinguishing characters, as listed by Zhou (1982), include the lack of hairs on the snout, a longitudinal and slitlike blowhole, the presence of a fore-stomach and a two-compartment main stomach, the presence of a caecum, the presence of a maxillary crest, contact between the palatal portions of the maxillas, and teeth with a simple crown. The maxillary crest is a unique feature and involves the development of thin plates of bone, one on each side of the skull, that project outward from the base of the rostrum and nearly meet in front of the blowhole.

Susus occur only in fresh water but are found from the tidal limits of rivers to the foothills of the Himalayas. They may undergo local migrations, moving into small tributaries during the monsoons but returning to large rivers in the dry winter (Kasuya and Haque 1972). *Platanista* rises to breathe about every 30–120 seconds, usually plunging out of the water in an upward or forward direction but occasionally exposing only the blowhole. Captives normally swim on their side, apparently because the tiny, deeply set eye can receive light only at such an angle (Herald et al. 1969; Purves and Pilleri 1974–75). Although the eye of *Platanista* lacks a lens and usually is not visible externally, and the genus is sometimes referred to as being blind, the eye does seem to function as a direction-finding device. Waller (1983) suggested that the eye may form a visual image in air when the dolphin lifts its head above water. Captives swim about continuously over a 24-hour period, perhaps because the streams they naturally inhabit flow swiftly during the monsoons and constant activity is necessary to avoid injury (Pilleri et al. 1976). There is also a continuous transmission of sound over a 24-hour period. Mizue, Nishiwaki, and Takemura (1971) found 87 percent of the sounds to be clicks for echolocation and 5 percent to be communicative. To find food, *Platanista* probably uses echolocation, and also probes with its sensitive snout for fish, shrimp, and other organisms in the bottom mud.

There are reports that *Platanista* travels and feeds in schools of 3–10 or more individuals. In 90 percent of the sightings of *P. gangetica* made by Kasuya and Haque (1972), however, only a single animal was seen. Births of *P. gangetica* may occur at any time of the year but are mostly from October to March, with a peak in December and January, at the beginning of the dry season (Kasuya 1972b). The single young is 70 cm long at birth and is weaned within a year. Sexual maturity is attained at about 10 years. Two males were reported to be still growing at 16 and 28 years of age.

Both species of *Platanista* are on appendix 1 of the CITES. According to the U.S. National Marine Fisheries Service (1978), *P. gangetica* is still fairly numerous and is not endangered, but Brownell, Ralls, and Perrin (1989) reported that it seems to be declining, and the IUCN now classifies it as vulnerable. A few are captured incidentally in the nets of fishermen, who attempt to release the dolphins alive. The species has apparently disappeared in the Karnaphuli River, which empties into the Bay of Bengal to the east of the Ganges. *P. indi* has declined drastically in range and numbers and is classified as endangered by the IUCN (1976). The main threat seems to be construction of numerous barrages—barriers to impound water for subsequent use in irrigation. Therefore, dolphin populations are split up, seasonal movements are restricted, and water quality deteriorates. The animals have also been hunted for their meat and oil. In 1974 the total population was estimated to contain only 450–600 individuals, most of which were in a 130 km stretch of the Indus

between Sukkur and Guddu barrages (Kasuya and Nishiwaki 1975). Although protected by law, *P. indi* is still regularly taken by fishermen (Pilleri and Pilleri 1979a).

The geological range of the Platanistidae is middle Miocene in eastern North America, early and middle Miocene in western North America, and Recent in south-central Asia (Rice 1984).

CETACEA; **Family DELPHINIDAE**

Dolphins

This family of 17 Recent genera and 34 species inhabits all the oceans and adjoining seas of the world, as well as the estuaries of many large rivers. Some species occasionally ascend rivers. The porpoises of the family Phocoenidae (see account thereof) sometimes have been included in the Delphinidae. Rice (1984) divided the Delphinidae into five subfamilies: Stenoninae, with the genera *Steno, Sousa,* and *Sotalia;* Delphininae, with *Tursiops, Stenella, Delphinus, Lagenodelphis, Lagenorhynchus,* and *Grampus;* Globicephalinae, with *Peponocephala, Feresa, Pseudorca, Globicephala, Orcinus,* and *Orcaella;* Lissodelphinae, with *Lissodelphis;* and Cephalorhynchinae, with *Cephalorhynchus.* Barnes, Domning, and Ray (1985) accepted the same subfamilies but placed *Orcaella* in its own subfamily, Orcaellinae, and transferred it to the family Monodontidae. Pilleri and Gihr (1981b) considered the Stenoninae to be a full family with the name Stenidae.

Head and body length ranges from as little as 139 cm in some specimens of *Sotalia* to as much as 980 cm in *Orcinus,* and weight in full-grown individuals ranges from about 50 to 9,000 kg. The common name "dolphin" is generally applied to small cetaceans having a beaklike snout and a slender, streamlined body form, whereas the term "porpoise" refers to those small cetaceans with a blunt snout and a rather stocky body form. The blowhole is located well back from the tip of the beak or the front of the head. The pectoral and dorsal fins are sickle-shaped, triangular, or broadly rounded, and the dorsal fin is located near the middle of the back. The genus *Lissodelphis* lacks a dorsal fin.

In the Delphinidae the vertebrae number 50–100 and the first 2 neck vertebrae are fused, whereas in the Iniidae, Lipotidae, Pontoporiidae, and Platanistidae all the neck vertebrae are free. Delphinids also have a lesser number of double-headed ribs than do members of these other four families. The fusion of the lower jaws in delphinids does not exceed one-third of their length, but this fusion is more than half the length in the other four families. In the Delphinidae, unlike in the Platanistidae, the skull lacks a crest. There are usually many functional teeth in both the upper and lower jaws of delphinids, the maximum number being about 260; in the genus *Grampus,* however, there are only 4–14 teeth, and these are confined to the lower jaw.

The spout of dolphins usually is not well defined, but that of *Globicephala* extends about 150 cm. Breathing is often accompanied by a low hissing or puffing noise. The respiration rate of bottle-nosed dolphins *(Tursiops)* taken in an aquarium where they could swim freely was 1.5–4.0 respirations per minute. This genus has a heart rate of 81–137 beats per minute (the average is 100 per minute). Bottle-nosed dolphins have been observed sleeping in calm water about 30 cm below the surface; slight movements of the tail brought the head above water so that the animals could breathe.

The Delphinidae include the most agile and some of the most speedy cetaceans. They commonly surface several times a minute and frequently leap clear of the water. Many species

Pacific white-sided dolphins *(Lagenorhynchus obliquidens)* trained to leap unusually high. Many of the smaller cetaceans regularly leap out of the water but rarely go more than a foot or two above the surface and reenter the water at a distance of only a few feet. Photo from Marineland of the Pacific.

follow ships and seem to frolic about the bow. They have remarkable group precision and regularity of movement. Migration is known to occur in some species.

The ability to perceive objects by means of reflected sound (echolocation) was demonstrated in *Tursiops* in 1958, the first time for a cetacean. The sense of vision is not as well developed as that of hearing, but *Tursiops* can see moving objects in the air at least 15 meters away. Observations of captive dolphins indicate that these animals are very intelligent. Adult delphinids will engage in complex play with various objects and can be trained to perform tricks.

Dolphins and porpoises usually associate in schools of five to several hundred individuals, though they are sometimes seen alone or in pairs. Groups may assemble when frightened and will attack intruders. They sometimes kill large sharks by ramming them. They utter a wide variety of underwater sounds that appear to function in communication. Cooperative behavior has often been observed; one or more individuals will come to the aid of another that is injured, sick, or giving birth, pushing it to the surface so that it can breathe.

The oldest documented delphinid is from the late Miocene (Barnes, Domning, and Ray 1985). The known geological range of the family extends from then to Recent in North America and Europe, upper Pliocene to Recent in Japan, Pleistocene to Recent in New Zealand, and Recent in all oceans.

CETACEA; DELPHINIDAE; **Genus STENO**
Gray, 1846

Rough-toothed Dolphin

The single species, *S. bredanensis*, occurs in tropical and warm temperate waters of all oceans and adjoining seas (Rice 1977). On the basis of reported hybridization between *Steno* and *Tursiops* and between *Tursiops* and *Grampus*, Van Gelder (1977b) recommended making *Steno* a synonym of *Grampus* (the earliest available name of the three).

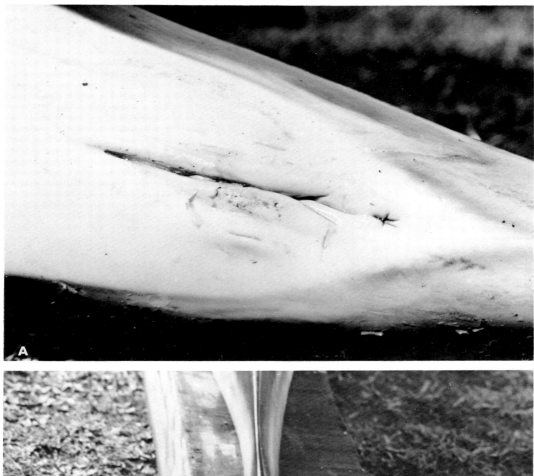

Pacific white-sided dolphin *(Lagenorhynchus obliquidens):* A. Urogenital region of a young female, showing the mammary slits on each side of the urogenital opening, the anus to the right; B. Dorsal view of caudal flukes, showing the deep tail notch typical of the Delphinidae and the prominent ridge on the top of the tail. Photos by Warren J. Houck.

Rough-toothed dolphin *(Steno bredanensis):* Top, photo by Gary L. Friedrichsen through National Marine Fisheries Service; Bottom, photos by Michael S. Sinclair.

with scattered spots and markings. The belly is pinkish white or rose-colored, with slaty spots. The pectoral fin, dorsal fin, and tail flukes are dark. The beak is white, slender, and compressed from side to side. There are 20–27 teeth on each side of each jaw. The surfaces of the teeth are roughened and furrowed by vertical ridges and wrinkles.

Steno is found mainly in tropical and subtropical waters, but a few individuals have stranded in colder areas outside the normal range. Stomach contents have included fish and octopus. Groups usually are made up of 50 individuals or fewer, but one school of over 100 animals reportedly stranded. *Steno* has been found in schools together with *Tursiops* and *Stenella*. Captives are easily trained and are reportedly even more intelligent than *Tursiops*. Four females taken off West Africa in May contained fetuses 60–87 cm long (Leatherwood, Caldwell, and Winn 1976; E. D. Mitchell 1975a; Perrin and Walker 1975; Rice 1967). Individuals estimated to be up to 32 years old have been found (Perrin and Reilly 1984).

In the Mediterranean Sea near Sicily, Watkins et al. (1987) observed an aggregation of approximately 160 rough-toothed

Head and body length is 175–275 cm, pectoral fin length is about 30 cm, and dorsal fin height is about 15 cm. A male specimen 216 cm long weighed 114 kg (Perrin and Walker 1975). The upper parts are slate-colored or purplish black,

dolphins, divided into eight groups of about 20 individuals each. The animals varied in size and included many cow-calf pairs. A few broke away from their groups and rode the bow wave of the ship, traveling at 16 km/hr. Some leaped clear of the water, and some dived to a depth of 70 meters, where they vocalized into a hydrophone. Both broad-spectrum clicks and whistles were recorded.

There is little human effort to hunt this genus, but some individuals are caught and used for food in Japan, Africa, and the Caribbean, and some are captured accidentally in the tropical Pacific tuna fishery (E. D. Mitchell 1975*b*).

CETACEA; DELPHINIDAE; **Genus SOUSA**
Gray, 1866

Humpback Dolphins

E. D. Mitchell (1975*a*) and Rice (1977) recognized two species:

S. chinensis, coastal waters of eastern and southern Africa, southern Asia, the East Indies, and Australia;
S. teuszii, coastal waters from Mauritania to Angola.

The above arrangement was accepted by Corbet and Hill (1986) and Honacki, Kinman, and Koeppl (1982). However, some authorities, including Lekagul and McNeely (1977) and Medway (1977), consider *Sousa* to be part of *Sotalia*. Pilleri and Gihr (1981*b*) accepted *Sousa* as a full genus and divided it into the following distinct species: *S. plumbea*, coastal waters from eastern Africa to Thailand; *S. lentiginosa*, coastal waters from eastern Africa to Thailand; *S. chinensis*, coastal waters of southern China; *S. teuszi*, coastal waters of western Africa; and *S. borneensis*, coastal waters of Borneo and Australia. Meester et al. (1986) and Ross (1984) regarded *S. plumbea*, but not *S. lentiginosa*, as a species distinct from *S. chinensis*. Rice (1977) and Ross (1984) suggested that *S. teuszii* might be only a subspecies of *S. chinensis* (including *S. plumbea*).

Head and body length is 120–250 cm, pectoral fin length is 30 cm or less, dorsal fin height is about 15 cm, and width of the tail flukes is about 45 cm. A specimen of *S. teuszii* 230 cm long weighed 139 kg. Coloration is variable; most forms are brown, gray, or black above and paler beneath, but some populations are whitish, speckled with gray, or freckled with brown spots.

The skull of *Sousa* differs from that of other dolphins in the rather long symphysis of the jaws and in the widely separated pterygoid bones that do not close together behind the palate. There are 23–50 teeth on each side of each jaw. *Sousa* resembles *Tursiops* and *Steno* but is distinguished by having more teeth and 10–15 lower vertebrae. The rostrum of *Sousa* is always narrower than that of *Tursiops*.

These dolphins are found in both salt and fresh water, inhabiting seas, estuaries, and the mouths of rivers. They sometimes ascend rivers and have been reported 1,200 km up the Chang Jiang (Yangtze) in China (Lekagul and McNeely 1977). One population of about 500 individuals seems to remain all year in muddy mangrove creeks in the delta of the Indus River (Pilleri and Pilleri 1979*b*). A population off South Africa also does not appear to migrate, but its habitat consists of deep waters over sand and reefs (Saayman and Tayler 1979).

Sousa sometimes leaps out of the water to a height of 120 cm but seems to be slower than most dolphins. It rolls to breathe and appears to roll faster when feeding. Saayman and Tayler (1979) found *Sousa* to ride waves rarely, to remain submerged for up to three minutes, and to hunt individually, generally in the vicinity of reefs. The diet evidently consists of fish. Purves and Pilleri (1983) reported the animals in the Indus Delta to produce clicks, whistles, and screams.

Sousa often is seen alone or in pairs, but Ross (1984) and Saayman and Tayler (1979) reported group size off South Africa to average about 7 and range up to about 30 individuals. In that area births occurred all year, with a peak in summer (December–February). The number of offspring per birth is 1.

Small numbers of humpback dolphins are caught in the Arabian Sea, the Red Sea, and the Persian Gulf and are used for human consumption (E. D. Mitchell 1975*a*). The population in the Indus Delta is not deliberately fished but is threatened by accidental catching and industrial pollution (Pilleri and Pilleri 1979*b*). *Sousa* is on appendix 1 of the CITES.

CETACEA; DELPHINIDAE; **Genus SOTALIA**
Gray, 1866

Tucuxi, or River Dolphin

Rice (1977) recognized a single species, *S. fluviatilis*, with two subspecies: *S. f. fluviatilis*, in the Amazon River and its tributaries; and *S. f. guianensis*, in coastal waters and the lower reaches of rivers from northwestern Venezuela to southern Brazil. Husson (1978) treated *guianensis* as a full species. *Sotalia* may also occur along the Atlantic coast of Central America (E. D. Mitchell 1975*a*).

Head and body length is 139–87 cm, (Perrin and Reilly 1984), dorsal fin height is 11–13 cm, and greatest girth is 70–98 cm. An adult male 160 cm long weighed about 47 kg. Best and Da Silva (1984) listed weights of 40–53 kg for sexually mature animals. In *S. f. guianensis* the upper parts range from pale bluish gray to brown or blackish. In at least some populations the color of the back extends to a circle around the eye, onto the pectoral fin, and to the sides of the tail. In some populations the sides are yellowish orange and there is a bright yellow patch on each side of the dorsal fin near the top. The underparts of *S. f. guianensis* are white, pinkish, or grayish. In *S. f. fluviatilis* the upper parts are bluish or pearl gray, the color being darker anteriorly. Larger animals are noticeably paler above. The pectoral fin, both above and below, is the same color as the back. The underparts are pinkish white to white. A prominent band of the ventral coloration extends upward on the sides of the body to slightly above the level of the eye. The dorsal color, however, extends down to a distinct line from the corner of the mouth to the base of the pectoral fin and includes the eye. The tip of the beak and the apex of the dorsal fin are conspicuously white.

Sotalia is often found in the same area as *Inia*, but the latter has a more prominent dorsal fin, a longer beak, and a more prominently bulging forehead. From *Steno* and *Tursiops*, *Sotalia* is distinguished by the separation of its pterygoid bones and by having fewer caudal vertebrae and more teeth. There are 26–35 teeth on each side of each jaw.

Sotalia is found in both salt and fresh water. *S. f. guianensis* occurs along the coast but often enters large rivers. *S. f. fluviatilis* occurs in much of the Amazon Basin, as far west as the Andes (E. D. Mitchell 1975*a*). The latter subspecies has been reported to be more active but less inquisitive than *Inia*. *Sotalia* swims rather slowly and rarely leaps clear of the water but has been seen to jump 120 cm above the surface. When breathing, it always rolls out of the water, the head and trunk usually appearing in smooth sequence and a short puff being the only sound heard. The reported interval between breaths is 5–85 seconds, with a mean of 33 seconds. *Sotalia* is most

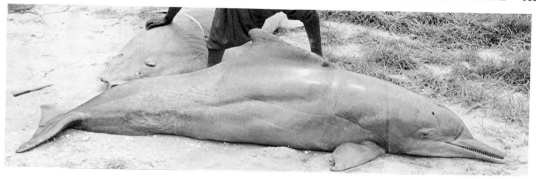

Humpback dolphin: Top, *Sousa teuszii*, photo by J. Cadenat; Middle and bottom, *S. chinensis*, photos by Michael M. Bryden.

River dolphins *(Sotalia fluviatilis guianensis)*, photos by W. Gewalt.

active in the early morning and late afternoon. Captives have been found to emit echolocation-type clicks (Caldwell and Caldwell 1970) and have been trained to perform various acrobatics (Gewalt 1979a). They have been reported to be more difficult to train than is *Tursiops* but to be more reliable in day-to-day performance (Terry 1986). The diet consists mostly of fish, but shrimp are also eaten.

The tucuxi has regular movements within a home range and is gregarious. Groups often swim and roll in tight formation and in nearly perfect synchrony, the individuals almost touching sides when they appear at the surface. No social interaction between *Sotalia* and *Inia* has been observed, though the two genera are often in close proximity. Husson (1978) reported *Sotalia* to be rather common in the mouths of large rivers in Surinam, where it occurs in groups of about 10 individuals. A female taken in this area between 15 February and 13 April 1971 had a fetus about 60 cm long. Best and Da Silva (1984) found that a birth peak apparently occurs in October and November, during the low-water period of the Amazon River system. The gestation period there is estimated to be 10.0–10.3 months, and the newborn is 71–83 cm long.

Some South American natives regard *Sotalia* as a sacred animal, consider it a good friend and protector, and believe that it will bring the bodies of drowned persons to shore. *Sotalia*, however, often is taken accidentally in the nets of fishermen (E. D. Mitchell 1975a) and is losing its habitat to rapid human development in the Amazon (Brownell, Ralls, and Perrin 1989). The genus is on appendix 1 of the CITES.

CETACEA; DELPHINIDAE; **Genus TURSIOPS**
Gervais, 1855

Bottle-nosed Dolphins

There are apparently three species (Ellerman and Morrison-Scott 1966; Gaskin 1968; Hall 1981; Lekagul and McNeely 1977; Rice 1977; Ross 1977):

T. truncatus, primarily in temperate and tropical waters of the Atlantic Ocean and adjoining seas;

T. aduncus, primarily in temperate and tropical waters of the Indian, South Pacific, and western and southern North Pacific oceans and adjoining seas;

T. gillii, primarily in temperate waters of the eastern North Pacific Ocean and Gulf of California.

Many authorities, including Rice (1977), have referred all *Tursiops* to the single species *T. truncatus*. Hall (1981) used the name *T. nesarnack* in place of *T. truncatus* and treated *T. aduncus* as a subspecies thereof. Zhou and Qian (1985) reported the presence of both *T. truncatus* and *T. aduncus* off the coast of China. Van Gelder (1977b) recommended making *Tursiops* a synonym of *Grampus*, on the basis of reported hybridization between the two genera. Such hybridization now is known to have occurred both in the wild and in captivity, and hybrids also have been produced in captivity through interbreeding of *Tursiops* with the genera *Pseudorca, Steno,* and *Globicephala* (Sylvestre and Tasaka 1985).

Head and body length is 175–400 cm, pectoral fin length is 30–50 cm, dorsal fin height is about 23 cm, and width of the tail flukes is about 60 cm. Adults usually weigh 150–200 kg, but weights in excess of 650 kg have been reported (Leather-

Bottle-nosed dolphin *(Tursiops truncatus)*, photo from Marineland of the Pacific. Inset: *T. gillii*, photo by Warren J. Houck.

wood, Caldwell, and Winn 1976). The upper parts are usually dark gray or slaty blue, the flippers and flukes are darker, and the underparts are paler. The genus is distinguished by the short, well-defined snout or beak, about 8 cm long and supposedly resembling the top of an old-fashioned gin bottle. There are 20–28 teeth on each side of each jaw. Each tooth is about 1 cm in diameter.

Tursiops is found mainly in coastal waters, often in bays and lagoons, and sometimes ascends large rivers. In certain areas, however, it ranges as far offshore as the edge of the continental shelf. Many researchers recognize two forms of *Tursiops*, the smaller staying in shallow waters near the mainland and the larger occurring farther offshore. The genus appears to be migratory in some areas, and movements may relate to seasonal changes in temperature or food distribution. *Tursiops* sometimes rides the surf or the bow wave of vessels and may jump to heights of six meters above the surface (Leatherwood, Caldwell, and Winn 1976; Leatherwood and Reeves 1978; E. D. Mitchell 1975b). In 26 separate observations along a tidal creek on the Georgia coast, Hoese (1971) saw *Tursiops* rush its entire body, except the tail, up onto a mudbank to capture small fish that had been driven in front of it.

Tursiops is commonly seen along the coasts of the United States and is also familiar to people through its displays of agility and sagacity under captive conditions. Compared with most wild animals, it is easily trained to perform acrobatics, locate hidden objects, and play with balls. It is also used widely in research work involving cetacean physiology, psychology, and sociology. Such studies have demonstrated that *Tursiops* has a high degree of what humans commonly refer to as intelligence. Some of these dolphins were even able to learn complex procedures quickly merely by watching other individuals perform (Adler and Adler 1978).

Tursiops is also noted for having a brain that is larger than that of *Homo* and a variety of vocalizations. Certain authorities have argued that these dolphins, and probably other cetaceans as well, have a complex language and may even-

tually be able to communicate meaningfully with people. The preponderance of current evidence, however, suggests that the large cetacean brain is associated mainly with development of the sense of hearing and the processing of echolocation and communicative data. Each individual *Tursiops* appears to have its own "signature" whistle, by which it can communicate a limited amount of information on its identity, location, and condition to others of its kind (Caldwell and Caldwell 1977, 1979; Hickman and Grigsby 1978; Jerison 1986; Ridgway 1986a). Other sounds and visual signals are also used in communication. Both the whistles and the high-frequency, clicklike pulses used for echolocation are thought to be produced not in the larynx but in the nasal sacs within the forehead (Hollien et al. 1976). Echolocation—the detection of objects by bouncing sound waves off of them—is utilized by *Tursiops* to find obstacles and food underwater. Hult (1982) found that series of high-intensity clicks are used also to herd schools of fish. The diet consists mainly of bottom-dwelling fish. Captive dolphins eat 6–7 kg of food per day (E. D. Mitchell 1975a).

According to Shane, Wells, and Würsig (1986), feeding peaks in the morning and afternoon, but activity may occur throughout the day and night. Individuals seem to make regular use of a particular area, but some animals move to different areas seasonally. Off the west coast of Florida, a population of about 105 dolphins appears to maintain an overall home range of approximately 85 sq km. Within this group, female-calf pairs and subadult males have an average range of 40 sq km, and other adult females, subadult females, and adult males have a range of 15–20 sq km. The density in this region is about average for the species, but other reported densities range from 0.06 to 4.80 per sq km. Territorial defense has not been observed in *Tursiops*, though individuals sometimes appear to recognize range boundaries. At least some groups in the eastern Pacific appear to have a limited home range associated with particular islands (Leatherwood and Reeves 1978).

Tursiops may form large aggregations, which usually

Bottle-nosed dolphins (*Tursiops truncatus*), photo by Bob Noble through Marineland of the Pacific.

comprise smaller groups of 2–15 individuals (Leatherwood, Caldwell, and Winn 1976; Shane, Wells, and Würsig 1986). The average size of 688 observed subunits within a population off the west coast of Florida was 4.8 (Irvine et al. 1981). In the eastern Pacific, group size is usually 15 or less near the shore but 25–50 far offshore (Leatherwood and Reeves 1978). Schools of up to 600 were reported off North Carolina in the late nineteenth century, and sexes were about equally represented there (Mead 1975). In 93 sightings of *T. aduncus* in the Indian Ocean, Saayman and Tayler (1973) found the number of individuals per group to range from 3 to 1,000 and to average about 140. The members of a group may cooperate in hunting fish, with some individuals crowding the fish toward shore and others patrolling offshore to prevent the fish from escaping, or with different units carrying out a synchronized attack at opposite ends of a school of fish (Würsig 1986). Social hierarchies are evident in captivity, with the largest adult male in a tank being dominant to all other individuals and the larger females ranking above smaller females. Dominance is expressed by biting, ramming, and tail slapping. In the wild, however, some groups may consist solely of females and young, with these units being joined temporarily by small bands of adult males (Shane, Wells, and Würsig 1986). Subadult males seem to form their own groups in the wild (Irvine et al. 1981).

In European waters the young of *T. truncatus* are born in midsummer, and most births off Florida occur from February to May (E. D. Mitchell 1975a). *T. aduncus* has a prolonged reproductive season off South Africa, with most births occurring in late spring and summer (Ross 1977). The normal interval between calves is two years, but if the young dies at birth, another offspring may be produced a year later. The gestation period is 12 months. The single newborn weighs about 9–12 kg and is 100–125 cm long. Lactation usually lasts 12–18 months, but the young may begin to take some solid food at less than 6 months. Female-calf bonds usually last 3–6 years. Sexual maturity is reportedly attained at some point between 5 and 12 years in females and between 9 and 13 years in males. Specimens taken from Florida waters have lived at least 25 years, while some taken off Japan have lived 30–35

years (E. D. Mitchell 1975a; Odell 1975; Ross 1977; Sergeant, Caldwell, and Caldwell 1973; Shane, Wells, and Würsig 1986).

Bottle-nosed dolphins have been deliberately hunted by people in many parts of the world (Mead 1975; E. D. Mitchell 1975a, 1975b). Products include meat for human consumption, fertilizer, body oil for cooking and illumination, and jaw oil for use as a lubricant in watches and precision instruments. An intensive fishery in the Black and Azov seas led to a sharp reduction in the number of *Tursiops*, and finally to a ban on dolphin catching by the Soviet Union in 1966. Several commercial fisheries existed in the eastern United States until the early twentieth century. The largest was that on Cape Hatteras, North Carolina, where 2,000 or more dolphins were taken annually during the peak years 1885–90. There are still small directed fisheries in West Africa, Sri Lanka, Indonesia, and Japan. *Tursiops* also is occasionally taken accidentally in nets set for tuna and is sometimes shot by fishermen, who consider it a competitor. Many bottle-nosed dolphins have been taken alive in waters off the United States, Japan, and Italy for purposes of display and human entertainment. Under the U.S. Marine Mammal Protection Act of 1972 such collecting requires a permit, and all other taking and importation of cetaceans and their products is prohibited. The U.S. National Marine Fisheries Service (1989) estimated that there were 14,000–23,000 bottle-nosed dolphins off the Gulf and east coasts of the United States.

CETACEA; DELPHINIDAE; **Genus STENELLA**
Gray, 1866

Spinner, Spotted, and Striped Dolphins

There seem to be five species (Brownell and Praderi 1976; Hall 1981; Hubbs, Perrin, and Balcomb 1973; Jones et al.

Spotted dolphin *(Stenella attenuata)*, photo by R. L. Pitman through National Marine Fisheries Service.

1986; Mullen 1977; Perrin 1975a, 1975b; Perrin et al. 1981, 1987; Rice 1977):

S. longirostris (pantropical spinner dolphin), tropical and warm temperate waters of the Atlantic, Indian, and Pacific oceans and adjoining seas;

S. clymene (Atlantic spinner dolphin), tropical and subtropical waters of the Atlantic Ocean and adjoining seas;

S. coeruleoalba (striped dolphin), tropical and temperate waters of the Atlantic, Indian, and Pacific oceans and adjoining seas;

S. attenuata (bridled or pantropical spotted dolphin), tropical and warm temperate waters of the Atlantic, Indian, and Pacific oceans and adjoining seas;

S. frontalis (Atlantic spotted dolphin), tropical and warm temperate waters of the Atlantic Ocean and adjoining seas.

Although Hall (1981) suspected that there are only two species of spotted dolphins, he, along with Schmidly, Beleau, and Hildebran (1972) and Leatherwood, Caldwell, and Winn (1976), recognized a third species to take the place of *S. attenuata* in the Atlantic and adjoining seas. Hall used the name *S. frontalis* for that species and *S. pernettensis* in place of what is listed above as *S. frontalis*. Hall also listed a fourth specific name, *S. dubia*, that might apply to certain spotted dolphin populations. In a detailed revision of the spotted dolphins, Perrin et al. (1987) recognized only two species, found *S. frontalis* to be the correct name of the Atlantic spotted dolphin, placed *S. plagiodon* and *S. pernettensis* in the synonymy thereof, and considered *S. dubia* a *nomen nudum*.

Head and body length is 150–350 cm, pectoral fin length is 15–20 cm, dorsal fin height is 15–30 cm, width of the tail flukes is about 35–50 cm, and weight is about 60–165 kg. Males average larger than females, but the latter have relatively longer rostra (Perrin 1975b; Schnell, Douglas, and Hough 1985). There is much variation in color pattern. In general, spinner dolphins are dark gray to black above, light gray or tan on the sides, and paler below; mature spotted dolphins have a sprinkling of pale spots on a dark background above and a sprinkling of dark spots on a pale background below; and the striped dolphin is grayish above, has mostly pale underparts and sides, and has one dark stripe extending

from each eye to the anus and another stripe extending from the eye to the pectoral fin.

There is a distinct beak, the rostrum is long and narrow, the palate is not grooved on the inner side of the toothrow, and the union of the two branches of the lower jaw is relatively short. There are 29–65 teeth on each side of each jaw and 68–81 vertebrae (Hall 1981; Leatherwood, Caldwell, and Winn 1976).

These dolphins seem to prefer the deeper, clearer offshore waters, mainly in tropical and subtropical parts of the world. *S. frontalis* is generally found over 8 km from the shore but in the Gulf of Mexico may move closer during spring and summer (Leatherwood, Caldwell, and Winn 1976). There may be two forms of *S. frontalis*, the larger inhabiting the continental shelf and the smaller being oceanic, though the size difference could reflect differential feeding habits (Perrin et al. 1987). There also are several morphologically separate stocks of *S. attenuata, S. coeruleoalba,* and *S. longirostris* in the eastern Pacific (Perrin et al. 1985; Schnell, Douglas, and Hough 1986). With respect to *S. attenuata,* analysis of skulls indicates a sharp demarcation between a smaller pelagic and a larger coastal form; masticatory structure of the latter suggests adaptation for feeding on larger prey (Douglas, Schnell, and Hough 1984). In this same region, *S. attenuata* and *S. longirostris* have been found to occur primarily in tropical waters north of the equator and south of the Galapagos, where there is relatively small annual variation in surface temperatures; *S. coeruleoalba* appears to prefer equatorial and subtropical waters with relatively large seasonal changes in surface temperature (Au and Perryman 1985). *S. longirostris* is often found far out to sea, where its distribution fluctuates in accordance with changing oceanographic conditions (E. D. Mitchell 1975a). *S. coeruleoalba* apparently is migratory in the western Pacific, with some populations approaching Japan in the autumn, swimming along the coast, and then moving back out to sea the following spring (Nishiwaki 1975).

These dolphins can swim at speeds of 22–28 km/hr. They often seem to frolic about ships and ride the bow waves. They sometimes jump clear of the water, and the spinner dolphins derive their name from their habit of leaping above the surface and turning two or more times on their longitudinal axis (Leatherwood, Caldwell, and Winn 1976). *S. frontalis* has been found to produce two general categories of sounds: whistles, probably highly individualized, for communica-

Spinner dolphins *(Stenella longirostris)*, photo by Gary L. Friedrichsen through National Marine Fisheries Service.

tion; and regular trains of clicks for exploring the environment and finding food through echolocation (Caldwell and Caldwell 1971a; Caldwell, Caldwell, and Miller 1973). *S. attenuata* feeds near the surface, but *S. longirostris* may sometimes feed at depths of 250 meters or more (Fitch and Brownell 1968; E. D. Mitchell 1975a). The diet of *Stenella* consists mainly of squid and small fish.

According to Perrin et al. (1987), schools of *S. attenuata* may number up to several thousand dolphins, and oceanic populations may have home ranges several hundred kilometers or more in diameter. Earlier reports (perhaps referring to permanent social units) indicated that group size is 5–30 in *S. attenuata*, up to several hundred in *S. longirostris*, and up to 100, but more commonly 6–10, in *S. frontalis* (Leatherwood, Caldwell, and Winn 1976). Schools of up to 3,000 *S. coeruleoalba* are found off Japan. In this area the young of *S. coeruleoalba* evidently leave their mother at about 2–3 years and join in their own school until they reach sexual maturity, when they return to the main group (Nishiwaki 1975). Estrous females also sometimes form a separate school (Kasuya 1972a). Much the same pattern has been reported in Japanese populations of *S. attenuata* (Kasuya, Miyazaki, and Dawbin 1974), but no evidence of age or sex segregation has been found in schools of this species in the eastern tropical Pacific (Perrin, Coe, and Zweifel 1976).

Strandings of *S. longirostris* in Florida indicate a calving season from May to July, and newborn *S. frontalis* are 76–120 cm long; otherwise little is known about the biology of *Stenella* in Atlantic waters (Mead et al. 1980; Perrin et al. 1987). In the last two decades, however, studies of many individuals killed through direct exploitation off Japan and incidental catching by the tuna fishery of the eastern Pacific have greatly increased our knowledge of the reproduction and life history of *Stenella* (Kasuya 1972a, 1976, 1985; Kasuya, Miyazaki, and Dawbin 1974; Miyazaki 1984; Myrick et al. 1986; Nishiwaki 1975; Perrin, Coe, and Zweifel 1976; Perrin and Henderson 1984; Perrin, Holts, and Miller 1977; Perrin, Miller, and Sloan 1977; Perrin et al. 1987). Births may occur at any time, but there are pronounced spring and fall peaks on both sides of the Pacific. The interval between births is about 26 months in *S. longirostris* in the eastern Pacific. Off Japan the birth interval in an intensely exploited population of *S. coeruleoalba* declined from about 4 years in 1955 to 2.8 years in 1977, and the age of female sexual maturity declined from 9.7 to 7.2 years. The birth interval of *S. attenuata* is about 48 months off Japan, where there is little disturbance, but only 26–36 months in the eastern Pacific, where there have been heavy losses to the tuna fishery. Such developments apparently reflect a greater compensatory reproductive effort by the exploited populations. However, such a compensation factor has not been determined to differentiate between the

especially hard-hit population of *S. longirostris* found within several hundred kilometers of the Pacific coast of Mexico and Central America and the less heavily exploited population of that species found farther out to sea.

Reported gestation periods are about 10.6 months in *S. longirostris*, 11.2–11.5 months in *S. attenuata*, and 11–12 months in *S. coeruleoalba*. There is normally a single young, but twins have been reported to occur on rare occasion. The mean length of the newborn is about 77 cm in *S. longirostris*, 83 cm in *S. attenuata*, and 100 cm in *S. coeruleoalba*. Lactation continues for about 10.1 months in *S. longirostris*, 11.2–19.9 months in *S. attenuata*, and 18 months in *S. coeruleoalba*. A small percentage of females have been found to be simultaneously lactating and pregnant. In *S. longirostris* males reach sexual maturity at 6.0–11.5 years, and females probably at 5 years. In *S. attenuata* average age at sexual maturity is 14.7 years for males and 10–12 years for females. In *S. coeruleoalba* both sexes become sexually mature at 5–9 years and physically mature at 14–17 years. Pregnant female *S. attenuata* up to 35 years old have been found. Natural annual mortality is less than 14 percent for both adults and juveniles, and life span is relatively long. Maximum age has been estimated at 46 years for *S. attenuata* and 50 years for *S. coeruleoalba*.

There are fisheries directed against *Stenella* in various parts of the world, the most important being that of Japan (Kasuya 1976; E. D. Mitchell 1975a, 1975b; Nishiwaki 1975). Fishermen of that country have annually taken about 20,000 *S. coeruleoalba* and 500–2,000 *S. attenuata* by driving and harpooning the animals. The dolphins are used locally for human consumption. The population of *S. coeruleoalba* off Japan is estimated to have contained 400,000–600,000 individuals originally, but excessive exploitation has reduced this number by about half.

Stenella has been the genus most seriously affected through the incidental killing of dolphins by tuna fisheries (E. D. Mitchell 1975a, 1975b; U.S. National Marine Fisheries Service 1978, 1981, 1987, 1989). Fishermen know that *Stenella* often associates with schools of tuna, and they therefore set their nets around groups of dolphins. Deployment of a new type of purse seine in the 1960s led to the accidental death of great numbers of dolphins, which became entangled as the nets were closed. The most serious losses were caused by American vessels, to the species *S. attenuata* and *S. longirostris* in the eastern tropical Pacific. In 1972, one of the worst years on record, the estimated number of dolphins taken by U.S. fishermen was 368,600, of which 178,000 were *S. attenuata* and 42,000 were *S. longirostris*. That same year, however, the U.S. Marine Mammal Protection Act was passed, prohibiting the killing and importation of cetaceans by persons subject to U.S. jurisdiction, except under permit.

To avoid economic hardship, the U.S. government issued permits allowing a certain number of dolphins to be taken accidentally, but the set quota was reduced each year, dropping from 78,000 in 1976 to 31,150 in 1980. Regular inspection, development of improved fishing gear, and cooperation by fishermen has led to a marked reduction in mortality. In 1980 the U.S. fishing fleet was estimated to have accidentally taken only 15,000 dolphins of all kinds. In subsequent years the kill by U.S. tuna fishermen has remained about the same or increased slightly, but there has been an enormous rise in the incidental take by foreign fishing fleets in the eastern Pacific. The total annual kill now is around 100,000, and there again is concern for the survival of the species. Current population estimates for the eastern tropical Pacific are: *S. longirostris*, 900,000; *S. attenuata*, 2.2 million; and *S. coeruleoalba*, 2.3 million. As early as 1979 the morphologically distinct population of *S. longirostris* found within several hundred kilometers of the coast of Mexico and Central America was calculated to be only about 20 percent of its original size (Perrin and Henderson 1984), and overall numerical estimates for the species are now less than half what they were then.

CETACEA; DELPHINIDAE; Genus DELPHINUS
Linnaeus, 1758

Common Dolphin, or Saddleback Dolphin

Rice (1977) recognized only a single species, *D. delphis*, occurring throughout the tropical and warm temperate oceans and adjoining seas of the world. Banks and Brownell (1969) recognized an additional species, *D. bairdii*, based on specimens from waters off California and Baja California. Neither Van Bree and Purves (1972) nor Hall (1981) accepted *D. bairdii* as a distinct species. Van Bree and Gallagher (1978), however, suggested that *D. tropicalis*, of the northern Indian Ocean and adjoining waters, is a species distinct from *D. delphis*, which occurs in the same region.

Head and body length is 150–250 (rarely 260) cm, pectoral fin length is about 30 cm, dorsal fin height is about 40 cm, and width of the tail flukes is about 50 cm. Lekagul and McNeely (1977) listed weight as 60–75 kg. *Delphinus* is among the most colorful of cetaceans. Over most of its range, the back is brown or black, the belly is whitish, and the sides have bands and stripes of gray, yellow, and white. Some populations, however, lack the side markings. The well-defined beak is narrow and sharply set off from the forehead by a deep V-shaped groove. There are 40–50 teeth on each side of each jaw.

Delphinus is a pelagic genus, generally occurring well out to sea, but it has been reported to enter fresh water occasionally. Populations off southern California evidently shift northward and farther offshore during the spring and summer (U.S. National Marine Fisheries Service 1978). Intensive radiotracking studies have shown that the common dolphin tends to occur in places with significant bottom relief—canyons, escarpments, sea mounts—where currents are interrupted and the resultant upswell supports high levels of oceanic productivity. Schools tend to divide into small parties to feed in the afternoon and night and then regroup at dawn. Dives generally last two to three minutes and reach depths of up to 280 meters (Leatherwood and Reeves 1978; Leatherwood et al. 1982). *Delphinus* is among the swiftest of cetaceans; although it usually travels at about 10 km/hr, it is capable of reaching over four times that speed. It often rides the bow waves of ships and seems to frolic about them. It sometimes feeds in the company of *Lagenorhynchus*. The diet includes cephalopods and small fish, including flying fish.

Delphinus is among the most gregarious of mammals, often being seen in groups of 1,000 or more (Leatherwood and Reeves 1978). Schools of up to 300,000 individuals sometimes formed over concentrations of fish in the Black Sea (E. D. Mitchell 1975b). During the spring and summer the large schools may break up into groups of 50–200 animals, and the sexes may segregate between mating seasons (U.S. National Marine Fisheries Service 1978). Individuals have been seen aiding wounded members of their group, supporting them in the water and pushing them to the surface to breathe (Lekagul and McNeely 1977).

In the northeastern Pacific there are apparently two periods of mating, January–April and August–November, and two periods of calving, March–May and August–October.

Common dolphin *(Delphinus delphis)*, photo by M. Scott Sinclair through National Marine Fisheries Service.

Each adult female gives birth every 2 or 3 years. Gestation lasts 10–11 months. The offspring is 79–105 cm long at birth and is weaned after about 19 months. The range of estimates of age at sexual maturity in different regions is 2–7 years in males and 3–12 years in females. Potential longevity is over 20 years (Leatherwood and Reeves 1978; Perrin and Reilly 1984; U.S. National Marine Fisheries Service 1978).

Delphinus has long been hunted by people in various parts of the world. Perhaps the largest fishery was in the Black Sea, where the population was estimated to have been 1 million individuals and where up to 200,000 were taken annually in the 1930s. This population subsequently underwent a precipitous decline, and its hunting was banned by the Soviet government in 1966 (E. D. Mitchell 1975*b*). Turkish fishermen, however, continued to catch large numbers, perhaps 88,000 in 1971 alone (Berkes 1977). In the eastern Pacific, *Delphinus* has been accidentally taken in the tuna fishery, in the same manner described in the account of *Stenella*. The kill was around 22,000 individuals in 1973 but has subsequently declined in response to regulation. The total current population of the common dolphin in the eastern tropical Pacific is estimated at 900,000 (U.S. National Marine Fisheries Service 1978, 1981, 1987, 1989).

CETACEA; DELPHINIDAE; **Genus LAGENODELPHIS**
Fraser, 1956

Short-snouted White-belly Dolphin, or Fraser's Dolphin

The single species, *L. hosei*, occurs in the tropical and warm temperate waters of the Atlantic, Indian, and Pacific oceans and adjoining seas (Caldwell, Caldwell, and Walker 1976; Hersh and Odell 1986; Rice 1977). Until about 20 years ago, *Lagenodelphis* was known to science only by a single skeleton collected before 1895 at the mouth of the Lutong River in Borneo.

An adult male and an adult female taken off South Africa had the following measurements: head and body length, 264 and 236 cm; pectoral fin length, 29 and 27 cm; dorsal fin height, 22 and 17 cm; width of the tail flukes, 59 and 57 cm; and weight, 209 and 164 kg (Perrin et al. 1973). The back is gray and the belly is white. Along each side, extending from the region of the eye to the anus, are three stripes, a creamy white one above, a blackish one in the middle, and another

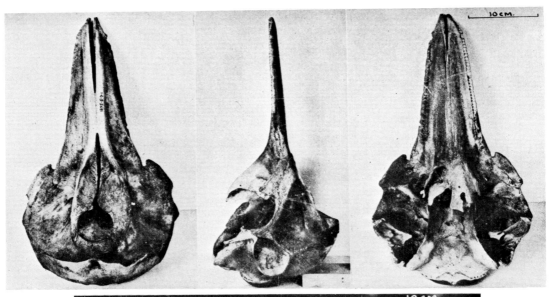

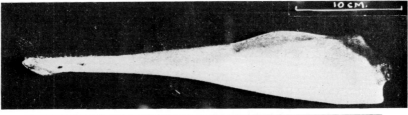

Fraser's dolphin *(Lagenodelphis hosei)*, photos from *Sarawak Museum Journal*, F. C. Fraser.

white one below. The body is robust, the pectoral and dorsal fins are relatively small, and the beak is short and indistinct. There are 39–44 teeth on each side of each jaw (Leatherwood, Caldwell, and Winn 1976). The skull is similar to that of *Lagenorhynchus*, but the facial part has a pair of deep palatal grooves and the premaxillary bones are fused dorsally in the midline; this fusion resembles that seen in *Delphinus* but is not so extensive.

Most records are from well offshore. *Lagenodelphis* appears to be a deep diver, and when it surfaces to breathe, it often charges up, creating a spray from its head. It has been reported to leap clear of the water. *Lagenodelphis* occurs in groups of up to 1,000 and is occasionally seen in the company of *Stenella attenuata* (Leatherwood, Caldwell, and Winn 1976; Leatherwood et al. 1982). Recorded stomach contents include deep-sea fish, squid, and shrimp (Caldwell, Caldwell, and Walker 1976; E. D. Mitchell 1975a). A pregnant female was collected off South Africa on 17 February 1971 (Perrin et al. 1973). Length of the single newborn is approximately 100 cm. Many of the known specimens have been taken accidentally in the tuna fishery, in the same manner described in the account of *Stenella* (U.S. National Marine Fisheries Service 1978).

CETACEA; DELPHINIDAE; Genus **LAGENORHYNCHUS**
Gray, 1846

White-sided and White-beaked Dolphins

There are six species (Rice 1977; Testaverde and Mead 1980):

L. albirostris (white-beaked dolphin), North Atlantic from Davis Strait and Newfoundland to Barents Sea and North Sea;

L. acutus (Atlantic white-sided dolphin), North Atlantic from southern Greenland and Virginia to western Norway and the British Isles;

L. obliquidens (Pacific white-sided dolphin), waters from southeastern Alaska to Baja California and off Japan and the Kuril Islands;

L. obscurus (dusky dolphin), temperate waters near the coasts of South America, South Africa, Kerguelen Island, southern Australia, and New Zealand;

L. australis (Peale's dolphin, or blackchin dolphin),

Pacific white-sided dolphin *(Lagenorhynchus obliquidens)*, photo from New York Zoological Society.

Lagenorhynchus sp., photo by R. L. Pitman.

temperate waters off southern South America and the Falkland Islands;

L. cruciger (hourglass dolphin), temperate waters of the Southern Hemisphere.

Head and body length is 150–310 cm, pectoral fin length is about 30 cm, dorsal fin height is up to 50 cm, and the width of the tail flukes is 30–60 cm. Sergeant, St. Aubin, and Geraci (1980) stated that adult weight of *L. acutus* is up to 234 kg in males and 182 kg in females. Banfield (1974) gave the weight of *L. obliquidens* as 82–124 kg. Most species have a gray or black back, a paler belly, and various bands and stripes on the sides. In *L. albirostris* the beak is white, and in *L. obliquidens* there are conspicuous yellowish brown and grayish streaks on the sides. The dorsal and pectoral fins are pointed. The beak is poorly defined and only about 5 cm long. There are 73–92 vertebrae, and there are 22–45 teeth on each side of each jaw (Hall 1981).

These dolphins generally occur in cool waters. *L. acutus* and *L. cruciger* are primarily pelagic species, while *L. obscurus* and *L. australis* are usually found near coasts (E. D. Mitchell 1975a; Rice 1977). In the western part of its range, *L. albirostris* moves north into Davis Strait during the spring and summer, then turns back in the autumn and spends the winter as far south as Cape Cod (Leatherwood, Caldwell, and Winn 1976). *L. obliquidens* may move close to the southern California shore in the winter and spring but then moves north and farther out to sea in the summer and fall (Leatherwood and Reeves 1978). Such movements are probably related to the availability of food.

Neither *L. acutus* nor *L. albirostris* commonly rides the bow waves of ships (Leatherwood, Caldwell, and Winn 1976). *L. obliquidens* often leaps clear of the water and is the only dolphin of the eastern Pacific known to turn complete somersaults under natural conditions (Leatherwood and Reeves 1978). *L. obscurus* also is highly acrobatic, its displays apparently serving to communicate social information (Würsig and Würsig 1980). In radiotracking studies off the coast of southern Argentina, Würsig (1982) found the minimum mean daily movement of *L. obscurus* to be 19.2 km, and Würsig

and Bastida (1986) determined several individuals to have ranges extending about 50–160 km from the site of capture. The diet of *Lagenorhynchus* includes herring, mackerel, capelin, anchovies, hake, squid, crustaceans, and whelks (gastropods).

Dusky dolphin *(Lagenorhynchus obscurus)*, photo by Stephen Leatherwood.

These dolphins are gregarious. Group size may be up to 1,000 in *L. acutus* and 1,500 in *L. albirostris*. Groups are usually much smaller, however, especially in the western Atlantic (Leatherwood, Caldwell, and Winn 1976; E. D. Mitchell 1975a); off Canada, *L. acutus* generally is seen in groups of only 6–8 (Banfield 1974). Investigations based on strandings (Sergeant, St. Aubin, and Geraci 1980) determined that large groups contained a great excess of adult females over adult males and no immature animals (2.5–6.0 years old) but that immature animals predominated in individual strandings and occasionally formed schools of their own. *L. obliquidens* is found in groups of up to several thousand individuals (Leatherwood and Reeves 1978) but segregates by age and sex during the breeding season (Banfield 1974). In a study of *L. obscurus* off Argentina, Würsig and Würsig (1980) observed group size to range from 6 to 300 but noted that the larger schools were feeding aggregations, that nonfeeding groups usually contained only 6–15 animals, and that the latter individuals included both sexes and sometimes young. Aggregation apparently is facilitated through signals involving leaping and vocalization, and the different groups seem to cooperate in herding schools of fish and keeping them from escaping. After coming together and feeding in such a manner, there may be considerable social interaction and sexual activity (Würsig 1986). Two individuals marked off Argentina on 3 January 1975 were observed to be still together on 1 December 1982 (Würsig and Bastida 1986).

All species that have been investigated seem to give birth in the late spring and summer. Gestation periods of 10–12 months have been estimated. There is normally a single young per birth, but there is at least one record of a pregnant female with two large embryos. Birth size is about 76 cm and 14 kg in *L. obliquidens*, 100 cm in *L. acutus*, and 122 cm in *L. albirostris*. In *L. acutus* nursing lasts about 18 months, the young leave the mother at an age of 2 years, and maximum longevity is estimated to be 22 years for males and 27 years for females. Captive *L. obliquidens* have been maintained for approximately 20 years (Banfield 1974; E. D. Mitchell 1975a; Sergeant, St. Aubin, and Geraci 1980; U.S. National Marine Fisheries Service 1978; Würsig and Würsig 1980).

There is a large directed fishery for *L. obscurus* along the coast of Peru, with the fresh meat being sold in the markets of Lima for U.S. $1.00–$1.25 per kg; nearly 10,000 individuals are killed there each year (Read et al. 1988). Fishermen also deliberately have taken small numbers of *L. acutus* and *L. albirostris* off Newfoundland, Norway, and the British Isles. *L. obliquidens* has been taken both intentionally and incidental to other fisheries off Japan (E. D. Mitchell 1975a, 1975b). The population of the latter species in Japanese waters has been estimated at 30,000–50,000 individuals, but this figure may not be reliable (U.S. National Marine Fisheries Service 1981).

CETACEA; DELPHINIDAE; Genus **GRAMPUS**
Gray, 1828

Risso's Dolphin, or Gray Grampus

The single species, *G. griseus*, occurs in all temperate and tropical oceans and adjoining seas (Rice 1977).

Head and body length is 360–400 cm, pectoral fin length is about 60 cm, dorsal fin height is about 40 cm, and width of the tail flukes is about 76 cm. An adult male stranded in Florida was estimated to weigh 400–450 kg (Paul 1968). Coloration

varies with age, but adults are usually slaty or black above, tinged with blue or purple, and paler beneath. The fins and tail are black. There are usually whitish streaks on the body, probably healed scars from attacks by other *Grampus* and perhaps by squid.

Grampus is distinguished from other dolphins in having only three to seven pairs of teeth. These pairs are in the front end of the lower jaw, but occasionally one or two vestigial teeth are present in the upper jaw. A beak is lacking, and the front of the head rises almost vertically from the tip of the upper jaw. Field marks include the blunt snout and the pale gleam of the back in front of the high, pointed, recurved dorsal fin.

Grampus generally occurs well out to sea in waters deeper than 180 meters (E. D. Mitchell 1975a). It apparently migrates northward during the warmer months (U.S. National Marine Fisheries Service 1978), and it has been observed as far north as the Gulf of Alaska (Braham 1983). It is frolicsome, sometimes riding the bow waves of vessels and leaping clear of the water. It may fall back into the water head up and occasionally waves the back third of its body in the air. The diet includes cephalopods and fish. *Grampus* usually is seen in groups of fewer than 12, but several such schools may associate in one area (Leatherwood, Caldwell, and Winn 1976). An aggregation of over 2,000 was observed in an area 2–4 km long and 4 km wide off the coast of Washington in August (Braham 1983). A pregnant female with a near-term embryo was taken in December. The newborn is approximately 150 cm long (E. D. Mitchell 1975a).

The gray grampus has been taken sporadically by small whale fisheries in Newfoundland, the Lesser Antilles, Japan, and Indonesia (E. D. Mitchell 1975a, 1975b). The famous dolphin "Pelorus Jack," which received lifelong protection from the government of New Zealand, is believed to have been a *Grampus*. This individual had the habit of playing about ships and seemed to guide them into Pelorus Sound. Observations of Pelorus Jack extended from about 1896 to 1916, thus giving some idea of the longevity of *Grampus*.

CETACEA; DELPHINIDAE; Genus **PEPONOCEPHALA**
Nishiwaki and Norris, 1966

Many-toothed Blackfish, or Melon-headed Whale

The single species, *P. electra*, occurs in tropical waters of the Atlantic, Indian, and Pacific oceans, and also is known from Japan, southeastern Australia, and South Africa (Best and Shaughnessy 1981; Caldwell, Caldwell, and Walker 1976; Perrin 1976; Rice 1977). This species long was placed in the genus *Lagenorhynchus*, but Nishiwaki and Norris (1966) considered it to resemble *Feresa* and *Pseudorca* more closely and placed it in a distinct genus.

Head and body length at physical maturity is about 250–80 cm, pectoral fin length is about 50 cm, dorsal fin height is about 25 cm, and width of the tail flukes is about 60 cm. Best and Shaughnessy (1981) reported an adult male to weigh 206 kg. The upper parts are black, and the belly is slightly paler. The lips and an oval patch on the abdomen are often white or unpigmented. The body is rather long and slim, and the tail stock is long. Unlike in *Lagenorhynchus*, there is no beak, the forehead being rounded, curving smoothly from the anterior tip of the rostrum to the blowhole, and overhanging the lower jaw to some extent. *Peponocephala* resembles *Pseudorca* but

Gray grampus *(Grampus griseus)*, photos by Warren J. Houck.

is smaller, has a sharper appearance to the snout, and has more teeth. There are 20–26 teeth on each side of each jaw. The anterior three cervical vertebrae are fused (Bryden, Harrison, and Lear 1977; Caldwell, Caldwell, and Walker 1976; Nishiwaki and Norris 1966).

Most of what is known about the natural history of this genus was summarized by Bryden, Harrison, and Lear (1977) and Perrin and Reilly (1984). *Peponocephala* seems normally to stay well out to sea but may be seen near land at any time of the year. One group was observed to swim at 20 km/hr, and many individuals leaped almost clear of the water. The diet consists of fish, squid, and possibly pelagic mollusks. *Peponocephala* is probably a social animal. Schools containing 15–500 individuals have been observed. One stranding of 53 specimens in Australia included 16 males, 35 females, and 2 individuals of undetermined sex. Limited data suggest that in the Southern Hemisphere births occur from August to December and that gestation lasts 12 months. A newborn individual was found in Hawaii in June. Length at birth is about 94 cm. Sexual maturity seems to come before 7 years in males and 12 years in females. Maximum longevity has been estimated to be 47 years.

CETACEA; DELPHINIDAE; **Genus FERESA**
Gray, 1871

Pygmy Killer Whale

The single species, *F. attenuata*, occurs in tropical and warm temperate waters of the Atlantic, Indian, and Pacific oceans and adjoining seas (Leatherwood, Caldwell, and Winn 1976; Rice 1977).

Head and body length is usually 240–70 cm, and dorsal fin height is usually about 20–30 cm (Leatherwood, Caldwell, and Winn 1976). A female collected in Japan had a head and body length of 235 cm, a pectoral fin length of 44 cm, a dorsal fin height of 24 cm, and a tail fluke width of 36 cm. The back is black or dark gray, the sides are often paler, and the chin is often white. There also are commonly white areas around the lips and on the belly. The forehead is rounded. There are 10–13 teeth on each side of each jaw. There is some resemblance to *Orcinus*, but that genus is far larger.

Until 1952 *Feresa* was known only by two skulls, but subsequently the genus was found to be widely distributed,

Melon-headed whale *(Peponocephala electra)*, photo by Michael M. Bryden.

though perhaps rare. Most records are from places with little or no continental shelf. *Feresa* apparently is a year-round resident off Japan, Hawaii, and South Africa, but short, seasonal migrations may occur (Caldwell and Caldwell 1971*b*). The diet is not well known, but a captive in Japan ate sardines, horse mackerel, sauries, and squid (Nishiwaki 1966). According to the U.S. National Marine Fisheries Service (1978, 1981), *Feresa* feeds on other kinds of dolphins, especially the young. Captives are aggressive, and other kinds of dolphins kept with them show fright reactions. *Feresa* usually travels in groups of 5–10, but aggregations of several hundred have been seen.

CETACEA; DELPHINIDAE; **Genus PSEUDORCA**
Reinhardt, 1862

False Killer Whale

The single species, *P. crassidens*, occurs in all the temperate and tropical oceans and seas of the world (Rice 1977). The species was described in 1846 on the basis of a fossilized skull found in the Lincolnshire Fens in England. Stranded animals collected in 1862 in the Bay of Kiel allowed assessment of the external morphology (Allen 1942).

Head and body length is up to 610 cm in males and 490 cm in females. Dorsal fin height is about 40 cm, and weight reaches 1,360 kg (Scheffer 1978*b*; U.S. National Marine Fisheries Service 1978). The coloration is black throughout. *Pseudorca* bears some resemblance to *Orcinus* but can be distinguished by its uniformly dark color, more slender build, more tapering head, smaller and more backwardly curving dorsal fin, and tapering flippers, which average about one-tenth of the head and body length. There are usually 8–11 teeth on each side of each jaw.

Pseudorca usually is found well out to sea. It often rides the bow waves of ships and jumps completely out of the water (Leatherwood, Caldwell, and Winn 1976). The diet is known to include squid, medium to large fish, and young dolphins (U.S. National Marine Fisheries Service 1981). A captive, however, was not aggressive toward other kinds of dolphins in the same pool and formed a close social relationship with them. This captive also was friendly to people and learned tricks rapidly. The wild school from which it was taken contained about 300 individuals that comprised small subgroups of 2–6 animals each. They emitted a variety of vocalizations, including a piercing whistle (Brown, Caldwell, and Caldwell 1966). Schools of up to 835 individuals have stranded, and these have included animals of all ages and sexes (U.S. National Marine Fisheries Service 1978). There does not appear to be a restricted breeding season (Scheffer 1978*b*). The young are usually 160–200 cm long at birth, gestation lasts 15.5 months, lactation is estimated to continue for about 18

Pygmy killer whales *(Feresa attenuata)*, photo by Kenneth S. Norris (top) and Gary L. Friedrichsen (bottom) through National Marine Fisheries Service.

months, and sexual maturity is attained at 8–14 years. Some individuals have lived up to 22 years (Perrin and Reilly 1984; Purves and Pilleri 1978).

Pseudorca is notorious to fishermen in various parts of the world because of its habit of stealing fish from lines (Leatherwood, Caldwell, and Winn 1976; Scheffer 1978b). *Pseudorca* is taken in small numbers for human consumption and incidental to the tuna fishery of the eastern tropical Pacific (E. D. Mitchell 1975a, 1975b).

CETACEA; DELPHINIDAE; Genus GLOBICEPHALA
Lesson, 1828

Pilot Whales, or Blackfish

There appear to be two species (Hall 1981; Kasuya 1975; E. D. Mitchell 1975a; Reilly 1978; Rice 1977; Van Bree 1971):

G. melaena, cool temperate waters of the Southern Hemisphere and the North Atlantic, also occurred in the North Pacific off Japan until at least the tenth century A.D.;

G. sieboldii, tropical and temperate waters of the Atlantic, Indian, and Pacific oceans and adjoining seas.

The latter species often has been referred to as *G. macrorhynchus.*

Head and body length is 360–850 cm, pectoral fin length is about one-fifth the head and body length, dorsal fin height is about 30 cm, and width of the tail flukes is about 130 cm. Males are usually larger than females. Banfield (1974) stated that the average weight of *G. melaena* is 800 kg and that the largest recorded male weighed 2,750 kg. The coloration is black throughout, except for a white area often present below the chin. The head is swollen, so the forehead bulges above the upper jaw. The pectoral fin is narrow and tapering, and the dorsal fin is located just in front of the middle of the

False killer whales *(Pseudorca crassidens):* A. Photo by K. C. Balcomb; B. Photo from Marineland of the Pacific.

back. There are 7–11 teeth on each side of each jaw.

According to Banfield (1974), *G. melaena* may occur either near the shore or well out to sea. During the colder months a population of this species remains in the Gulf Stream to the south of the Grand Banks of Newfoundland. In the summer this population shifts shoreward and northward to the Gulf of St. Lawrence, the Labrador Sea, and Greenland. Such movements are evidently in response to the migrations of the squid *Illex*, the major source of food. There is an apparently separate population of *G. melaena* off northwestern Europe (E. D. Mitchell 1975*b*). In the eastern Pacific, *G. sieboldii* seems to become more abundant near the shore during the winter (Reilly 1978).

Pilot whales sleep by day and feed by night. They normally swim at about 8 km/hr but can attain speeds of 40 km/hr (Banfield 1974; U.S. National Marine Fisheries Service 1978). A radio-tagged individual was tracked to a depth of 610 meters, and a captive was used in a U.S. Navy experiment to recover objects about 500 meters below the surface. Pilot whales are among the most intelligent and affable of cetaceans, adapting well to captivity and being easily trained. They have been demonstrated to use echolocation efficiently and to have a variety of sounds for communication, including an individualized "signature" whistle (Banfield 1974; Reilly 1978; Taruski 1979). The preferred diet is squid, up to 27 kg of

which may be consumed daily. Other cephalopods and small fish are also eaten.

Group size varies. The U.S. National Marine Fisheries Service (1978) stated that *G. melaena* occurs in schools of hundreds and thousands, but Banfield (1974) gave average group size as only 20 individuals. Each of five stranded groups of *G. melaena* in Great Britain contained 23–40 animals and consisted mostly of females but also had more than one mature male (Martín, Reynolds, and Richardson 1987). In the eastern Pacific, three kinds of herds of *G. sieboldii* have been identified (Reilly 1978): the traveling-hunting group, a large and well-organized unit that may be divided into harems of females and young led by one or a few large males, and other subgroups based on age and sex; the feeding group, a loose aggregation of animals that come together only to exploit a source of food; and the loafing group, a cluster of 12–30 animals that float in one area for such purposes as mating and nursing young. Polygamy seems to be the rule in both species of *Globicephala*. Adult males are often scarred, presumably from battles over control of a harem. Fighting involves biting, butting with the large head, and slapping with the tail.

Social organization seems to be highly developed in *Globicephala*, and this factor sometimes works to the disadvantage of the animals. Individuals harpooned by people usually rush forward in panic and can be driven toward the shore.

Pacific pilot whale *(Globicephala sieboldii)*, photos from Marineland of the Pacific.

The rest of the group seems to respond to the movements and cries of the wounded animals and to follow them into shallow water, where all may be easily killed. The same tendency may be associated with mass stranding, an especially common occurrence in *Globicephala*. It is possible that if a leader's echolocation mechanism fails to function properly, because of pathology or unusual environmental conditions, it will guide the entire group onto the shore. The cries of the first animals to become stranded attract others to the same fate (Banfield 1974; Reilly 1978).

Analysis of specimens of *G. sieboldii* taken in fisheries off the Pacific coast of Japan (Kasuya and Marsh 1984) show that there is a diffusely seasonal reproductive season, with matings peaking in April–May and births in July–August. The gestation period averages 452 days, and mean length of the single newborn is 140 cm. The young begin to take some solid food at 6–12 months, but lactation continues for at least 2 years. Many calves suckle until they are 6 years old, and a few, particularly those of older mothers, do so until past 10 years. Young males may move to another school after weaning, but females remain in their maternal group for life, thus creating a matrilineal society. Males reach sexual maturity at an average age of 14.6 years and social maturity at 17 years and have a maximum known longevity of 46 years. Females attain sexual maturity at 7–12 years and subsequently produce up to 4 or 5 calves. Their reproductive activity slows after age 28 and stops altogether by age 40, but they may live up to 63 years.

Strandings of *G. melaena* on the British coast indicate no evident reproductive seasonality, that lactation lasts 20.6 months and interbirth interval is 3.5 years, and that sexual

maturity comes at 7 years in females and 9–14 years in males (Martín, Reynolds, and Richardson 1987). In the western Atlantic, mating takes place mainly in April and May, before departure from the wintering area. The gestation period is 15–16 months, and the birth peak is in August. The single newborn is 150–210 cm long. Lactation lasts 21–22 months. Calving interval is 3.3 years. Sexual maturity is attained at 6–7 years in females and 12 years in males. Longevity is up to 40 years in females and 50 years in males (Banfield 1974; Leatherwood, Caldwell, and Winn 1976; Perrin and Reilly 1984; Reilly 1978).

Pilot whales have been among the more heavily exploited of the small cetaceans (Banfield 1974; Kasuya and Marsh 1984; E. D. Mitchell 1975a, 1975b; Reilly 1978; U.S. National Marine Fisheries Service 1978). There have been active fisheries in Japan, the Caribbean, the northeastern United States, Newfoundland, Ireland, Norway, and several islands north of Great Britain. The main products are meat for human and domestic animal consumption, blubber oil, and oil from the bulbous head. Records of the Faeroe Islands fishery extend back nearly 400 years. The cumulative kill there was 117,546 animals from 1584 to 1883. Despite this large take, the annual average has not declined, and the population of *Globicephala* in the northeastern Atlantic appears to be in balance with the harvest. In contrast, the western Atlantic population, estimated to have once contained 60,000 animals, evidently declined sharply following commercial organization of the Newfoundland fishery in 1947. About 40,000 pilot whales were killed there from 1951 to 1959, but only 6,902 were taken from 1962 to 1973. The catch on the Pacific coast of Japan was about 1,000 annually in the late 1940s and early 1950s and has been 400–700 per year more recently.

CETACEA; DELPHINIDAE; **Genus ORCINUS**
Fitzinger, 1860

Killer Whale

Rice (1977) and most other authorities have recognized a single species, *O. orca*, which occurs in all the oceans and adjoining seas of the world, chiefly in coastal waters and cooler regions. The species *O. nanus* and *O. glacialis*, described from antarctic waters by Soviet authorities, were not accepted by Heyning and Dahlheim (1988).

Killer whale *(Orcinus orca)*, photo by Scanlan. Skull: photo by Warren J. Houck.

Orcinus is the largest member of the dolphin family. Sexual maturity is attained at a length of about 550 cm in males and 490 cm in females (E. D. Mitchell 1975a). Subsequently, as in all cetaceans, growth continues for some years. Maximum head and body length is about 980 cm in males and 850 cm in females. Maximum weight is about 9,000 kg in males and 5,500 kg in females (Scheffer 1978a). In males the body becomes stocky with age, and there is a disproportionate increase in the size of certain organs: the broad, round pectoral fin reaches a length of 200 cm; the erect, triangular dorsal fin reaches a height of 180 cm; and the tail flukes reach an expanse of 275 cm. In females the body remains less stocky, and the dorsal fin is smaller and backwardly curved. There is a sharply contrasting color pattern. The upper parts are black except, usually, for a light gray area behind the dorsal fin; the underparts are white; and there are white patches on the sides of the head and on the posterior flanks. The snout is bluntly rounded. There are 10–14 large teeth on each side of each jaw.

The killer whale seems to prefer coastal waters and often enters shallow bays, estuaries, and the mouths of rivers. Some populations appear to reside permanently in relatively small areas, but others migrate, probably in association with food supplies. In the North Atlantic during the spring and summer there appear to be movements both closer to the southeastern Canadian coast and northward into arctic waters. *Orcinus* usually swims at about 10–13 km/hr but can exceed 45 km/hr. Its dives usually last 1–4 minutes, but one individual is known to have remained underwater for 21 minutes; another was found entangled in a submarine cable at a depth of about 1,000 meters.

Orcinus produces clicklike sounds such as those known to be used by other delphinids for echolocation, and its ability to echolocate has been demonstrated experimentally. Also emitted are a variety of other sounds apparently used in communication, including screams, distinct tonal whistles, and pulsed calls. Whistles usually are associated with low levels of whale activity. Most sounds heard in stable social groups engaged in active behavior are repetitious pulsed calls. Each such group has a repertoire of discrete pulsed calls that is unique to the group and that may persist for at least 10 years (Dahlheim and Awbrey 1982; Ford and Fisher 1983; Heyning and Dahlheim 1988).

Orcinus adjusts to captivity and learns quickly. Within two months of its first training session, one individual was performing before the public, and after five months its trainer could safely place his head into its mouth (Banfield 1974; Leatherwood, Caldwell, and Winn 1976; Scheffer 1978a).

Orcinus received its common name because it is the world's largest predator of warm-blooded animals, though in relative terms it is no more of a "killer" than an insectivorous frog or a beef-eating human. And while marine mammals are evidently the preferred prey in certain areas at certain times, an examination of the stomach contents of 364 *Orcinus* taken off Japan showed that fish and cephalopods predominated over cetaceans and pinnipeds (Scheffer 1978a). *Orcinus* does have a throat large enough to swallow seals, young walrus, and the smaller cetaceans. Steltner, Steltner, and Sergeant (1984) described an attack by 30–40 killer whales on a herd of several hundred narwhals *(Monodon)*. A group of killer whales sometimes joins in an attack on a large baleen whale, literally tearing it to pieces. Penguins and other aquatic birds are also seized. When *Orcinus* detects a seal or bird near the edge of the ice in arctic waters, it may dive deeply and then rush to the surface, breaking ice up to a meter thick and dislodging its prey into the water. The approach of a group of killer whales usually seems to panic other marine vertebrates. The U.S. National Marine Fisheries Service (1981) stated that in Argentina *Orcinus* has been seen flopping well up onto the land to drive pinnipeds into the water. Of 568 observations made by Lopez and Lopez (1985) of *Orcinus* hunting pinnipeds along the Argentine coast, 203 involved two or more whales corralling prey in the water, and 365 involved one or more whales rushing up onto the surf zone of the beach and then moving back to deep water; 122 of these beachings ended in the capture of a sea lion or elephant seal.

Schools of up to 250 individuals have been reported (Hoyt 1977). According to Scheffer (1978a), however, groups usually consist of 2–40 animals. Heyning and Dahlheim (1988) noted that average pod (family group) size is about 5 or 6 and that reports of very large groups may refer to temporary joining of a number of pods, possibly in relation to seasonal peaks in prey availability or to social activity. In the enclosed waters from northeast of Vancouver Island to Puget Sound, surveys have found a permanent population of about 260 killer whales divided into pods of about 10 members each. Pods generally are well organized and led by a large male, though females and young may sometimes separate from the older males. A pod observed by Cousteau and Diole (1972)

Killer whale *(Orcinus orca)*, photo by Dr. Bubenik through Bernhard Grzimek.

7–10 adult males, 22 adult females, and 17–22 immatures (Balcomb and Bigg 1986). However, the large pods seem to be divided into cohesive subgroups, often consisting of a mated pair, a barren female, and a calf (Heimlich-Boran 1986). Individuals have been seen to go to the assistance of others that were trapped or injured.

Breeding can occur at any time of the year, though in the Northern Hemisphere mating may peak from May to July, and births take place mainly in the autumn. The single newborn weighs about 180 kg and is about 240 cm long (U.S. National Marine Fisheries Service 1978, 1981). Observations in captivity indicate a gestation period of 517 days and a weaning age of 14–18 months (Asper, Young, and Walsh 1988). Various reported calculations of calving interval range from 3.0 to 8.3 years (Perrin and Reilly 1984). Maximum longevity is at least 50 years (Haley 1978).

There long has been controversy about whether the killer whale is a threat to human swimmers and boaters. Documented records of attacks are rare and usually involve provocation by the persons involved (Leatherwood, Caldwell, and Winn 1976; Scheffer 1978a). In contrast, people often kill or harass *Orcinus* because it is considered to be a competitor to fisheries. In the 1950s the U.S. Navy reportedly machine-gunned hundreds of killer whales at the request of the government of Iceland (E. D. Mitchell 1975b). *Orcinus* also has been killed by people in such places as Norway, Greenland, and Japan to obtain meat and oil. During the 1979/80 season the whaling fleet of the Soviet Union took 916 killer whales (Committee for Whaling Statistics 1980). Only 16 were reported taken in the 1982/83 season, however, and subsequently during the 1980s all commercial hunting ceased. *Orcinus* has been live-captured regularly off the coasts of British Columbia and Washington, mainly for eventual use in human entertainment (Bigg and Wolman 1975).

CETACEA; DELPHINIDAE; **Genus ORCAELLA** *Gray, 1866*

Irrawaddy River Dolphin

The single species, *O. brevirostris*, occurs in coastal waters from the Bay of Bengal to New Guinea and northern Aus-

contained 1 large alpha male, 1 large female, 7 or 8 medium-sized females, and 6–8 calves. Some of the larger pods in the vicinity of Vancouver Island contain several adult males; from 1974 to 1981 the largest pod there was observed to have

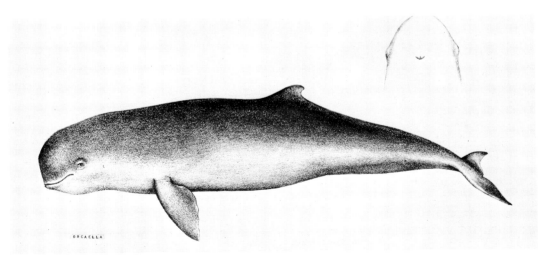

Irrawaddy River dolphin *(Orcaella brevirostris)*, photo from *Anatomical and Zoological Researches: Zoological Results of the Two Expeditions to Western Yunnan in 1868 and 1875*, John Anderson.

tralia and ascends far up the Ganges, Irrawaddy, Mekong, and other rivers (Rice 1977).

Head and body length is 180–275 cm (Lekagul and Mc-Neely 1977). The coloration is slaty blue or slaty gray throughout, or the underparts may be slightly paler. There is a bulging forehead, a short and shelflike beak, and 12–19 teeth on each side of each jaw. The pectoral fin is broadly triangular. The dorsal fin is small, sickle-shaped, and located on the posterior half of the back. There is great flexibility of the neck and tail (E. D. Mitchell 1975a).

Orcaella lives in warm, tropical, and often silty waters. It regularly enters rivers, has been recorded from 1,440 km up the Irrawaddy, and can live permanently in fresh water. It often accompanied river steamboats. It breathes at intervals of 70–150 seconds; the head appears first and then disappears, and then the back emerges, but the tail is rarely seen. The diet consists of fish and crustaceans, often found on the bottom (Lekagul and McNeely 1977; E. D. Mitchell 1975a). Although groups usually consist of no more than 10 animals, solitary individuals are rarely seen (Leatherwood and Reeves 1983). A pregnant female about 210 cm in length contained an embryo 90 cm long.

This dolphin occasionally is taken accidentally, and its oil reportedly has been used as a remedy for rheumatism in parts of India, but it is generally unexploited. According to Thein (1977), the fishermen of Burma attract *Orcaella* by tapping the sides of their boats with oars. The dolphin swims around the boat in ever-diminishing circles, thereby forcing the fish into nets. The fishermen share their catch with *Orcaella* and consider it a friend that is not to be harmed.

CETACEA; DELPHINIDAE; **Genus LISSODELPHIS**
Gloger, 1841

Right Whale Dolphins

There are two species (Rice 1977):

L. borealis, temperate waters of the North Pacific, from Japan and the Kuril Islands to British Columbia and California;
L. peronii, temperate waters of the oceans of the Southern Hemisphere.

In *L. borealis* head and body length is up to 307 cm in males and 230 cm in females, pectoral fin length is about 28 cm, and width of the tail flukes is about 35 cm. A mature male 282 cm long weighed 113 kg. A mature female 226 cm long weighed 81 kg (Leatherwood and Walker 1979). *L. peronii* is smaller, reaching a length of about 210–40 cm at maturity (Gaskin 1968). *L. borealis* is mostly black or dark brown, but there is a broad white patch on the chest and a narrow midventral white band from the chest to the tail. In *L. peronii* the top of the head and back are black or bluish black, and the beak, flippers, tail, sides, and underparts are white. The beak is short but distinct. There are 36–47 sharp, pointed teeth on each side of each jaw. The vernacular name refers to the lack of a dorsal fin in both *Lissodelphis* and the right whales *(Balaena)*.

These dolphins are primarily oceanic, usually remaining well offshore. Their numbers apparently increase over the continental shelf of California during the fall and winter, a period corresponding with the maximum abundance of squid in the region. In the late spring and summer the dolphins move farther north and out to sea. *Lissodelphis* is quick and active and sometimes travels by smooth, low leaps above the surface, each of which may cover up to 7 meters. Speeds as great as 45 km/hr may be briefly attained; a boat traveling at

33 km/hr was easily outdistanced during a 10-minute chase. Bow waves of vessels are usually avoided. Entire schools have been observed to dive for as long as 6 minutes and 15 seconds. Otoliths found in the stomach of an adult *L. borealis* indicated that the animal had been feeding on fish that dwell at depths of at least 200 meters. The diet consists of various kinds of fish, especially lanternfish, and cephalopods, such as squid. *L. borealis* produces series of clicks much like those used by other genera for echolocation. Whistles, apparently for communication, are also emitted but are not common. Groups contain up to 2,000 individuals but usually number 10–50. Newborn *L. borealis* are seen most frequently in early spring and are about 60 cm long. Fishermen take small numbers of *L. peronii* off Chile and of *L. borealis* off Japan. The eastern Pacific population of *L. borealis* has been estimated to contain 17,800 animals (Fitch and Brownell 1968; Leatherwood et al. 1982; Leatherwood and Walker 1979; E. D. Mitchell 1975a; U.S. National Marine Fisheries Service 1978).

CETACEA; DELPHINIDAE; **Genus
CEPHALORHYNCHUS**
Gray, 1846

Southern Dolphins, or Piebald Dolphins

There are four species (Brownell and Praderi 1985; Goodall, Galeazzi, et al. 1988; Goodall, Norris, et al. 1988; Rice 1977):

C. commersonii, coastal waters of Argentina, Tierra del Fuego, Falkland Islands, Kerguelen Islands, and possibly South Georgia and South Shetland Islands;
C. eutropia (black dolphin), coastal waters of Chile and Tierra del Fuego;
C. heavisidii, coastal waters of southwestern Africa;
C. hectori, coastal waters of New Zealand.

Head and body length is 110–80 cm, pectoral fin length is 15–30 cm, dorsal fin height is about 10–15 cm, and width of the tail flukes is about 30–40 cm. The average size of females is slightly larger than that of males. The various species do not differ greatly in weight, and the overall range is about 26–74 kg (P. B. Best 1988; Goodall, Galeazzi, et al. 1988; Goodall, Norris et al. 1988; Slooten and Dawson 1988). The color pattern consists of various amounts of sharply contrasting black and white. Generally the head, pectoral and dorsal fins, tail region, and posterior part of the back are black, and the chin, sides, belly, and anterior part of the back are white; sometimes the upper parts are entirely black. There is no beak, and there are 24–33 teeth on each side of each jaw.

These dolphins are found in coastal waters (Rice 1977). According to Baker (1978), *C. hectori* usually occurs within 8 km of the coast of New Zealand, often in shallow, muddy waters. It often is attracted to boats and briefly rides the bow waves and follows in the wake. It rarely jumps clear of the water. It feeds on the sea floor, taking various kinds of fish and perhaps such invertebrates as squid and crustaceans. Groups usually contain 2–8 individuals, but occasionally over 20 are seen together, and there is one record of an aggregation of 200–300. Slooten and Dawson (1988) indicated that social interaction takes place within such aggregations. They reported also that mating peaks early in the winter, that *C. hectori* gives birth during the austral spring and early summer (November–February), that the calving interval probably is at least 2 years, and that the young evidently begin taking solid food by the age of 6 months.

Right whale dolphins (Lissodelphis borealis): Top and middle, photos by M. Scott Sinclair through National Marine Fisheries Service; Bottom, photo by K. C. Balcomb.

The behavior and ecology of C. commersonii and C. eutropia is basically similar to that of C. hectori (Goodall, Galeazzi, et al. 1988; Goodall, Norris, et al. 1988). Both are strong swimmers and often ride bow waves. C. eutropia is found in areas of strong tidal flow above a steeply dropping continental shelf and is known to ascend rivers. C. commersonii occurs along the coast, where the continental shelf is wide and flat; its diet consists largely of bottom-dwelling invertebrates. Such habitat, which is characterized by turbid water, probably makes echolocation important for navigation and locating prey. The narrow-band, high-frequency pulses emitted by Cephalorhynchus are very much like those of Phocoena, which occupies the same kind of habitat. Watkins, Schevill, and Best (1977) recorded three kinds of sounds made by all four species of Cephalorhynchus: clicks at slow rates, short bursts of clicks, and a cry.

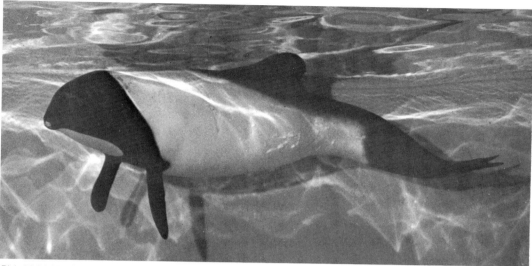

Piebald dolphin *(Cephalorhynchus commersonii)*, photo by Stephen Spotte, Mystic Marinelife Aquarium.

Groups of *C. eutropia* usually consist of 2–5 individuals, but aggregations of 20–50 frequently are seen (Goodall, Norris, et al. 1988). For *C. commersonii,* groups of over 110 individuals have been reported, but the usual number is about 7–9 (Goodall, Galeazzi, et al. 1988; Leatherwood, Kastelein, and Miller 1988). A captive group of 6 *C. commersonii* was kept together without strife, each animal consuming 3–4 kg of food per day (Gewalt 1979*b*; Spotte, Radcliffe, and Dunn 1979). In the wild, *C. commersonii* gives birth during the summer (December–March). The gestation period is 11–12 months, the newborn is about 75 cm long and weighs 7.5 kg, sexual maturity is reached at 5–6 years, and individuals up to 18 years old have been found (Goodall, Galeazzi, et al. 1988; Lockyer, Goodall, and Galeazzi 1988).

Cephalorhynchus is taken deliberately off the coast of South America and is subject to accidental killing by fishermen and crabbers throughout its range; the genus is considered to be especially vulnerable to coastal purse seining (Goodall, Galeazzi, et al. 1988; Goodall, Norris, et al. 1988;

E. D. Mitchell 1975*a,* 1975*b*). *C. hectori,* of New Zealand, appears particularly susceptible because of the trawl fishing and pollution in its limited habitat, but Baker (1978) did not consider reports of a decline to be substantiated. Cawthorn (1988) estimated this species to number 5,000–6,000 and indicated that entanglement in gill nets posed a threat. Dawson and Slooten (1988) concluded that the species was declining because of this problem and estimated that only 3,000–4,000 survived.

CETACEA; Family PHOCOENIDAE

Porpoises

This family of four Recent genera and six species is found in the North Pacific Ocean and adjoining seas, in coastal waters on both sides of the North Atlantic and around South Amer-

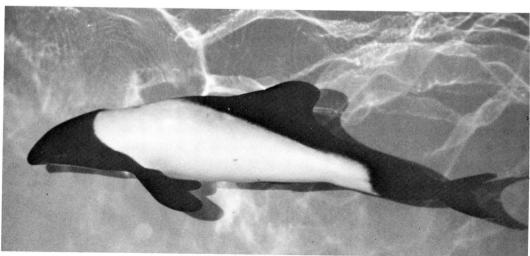

Piebald dolphin *(Cephalorhynchus commersonii)*, photo by Stephen Spotte, Mystic Marinelife Aquarium.

ica, in the Black Sea, off southern Asia and the East Indies, and around certain sub-antarctic islands. The Phocoenidae sometimes are included in the Delphinidae but were given familial rank by Barnes, Domning, and Ray (1985), Pilleri and Gihr (1981b), and Rice (1984). Barnes (1985) recognized two phocoenid subfamilies: Phocoeninae, with the genera *Phocoena* and *Neophocaena;* and Phocoenoidinae, with *Australophocaena* and *Phocoenoides.*

These are relatively small cetaceans, with a head and body length of about 120–220 cm and a normal weight range of 25–125 kg. They share many characters with the Delphinidae, but as noted in the account of that group, porpoises have a blunt snout and a rather stocky body form. The genus *Neophocaena* lacks a dorsal fin. The Phocoenidae have 15–30 teeth on each side of each jaw. These teeth are spade-shaped (though greatly reduced in *Phocoenoides*), with laterally compressed, weakly two- or three-lobed crowns, whereas the teeth of the Delphinidae are unlobed (Rice 1984).

According to Barnes (1985), all phocoenids share a unique suite of cranial characters: an eminence or boss on each premaxilla anterior to the narial opening, a small (atrophied) posterior termination of the premaxilla that projects posteriorly adjacent to each naris and does not reach the nasal bone, palatine bones that are relatively widely exposed on the palate and thereby separate the hamular processes of the pterygoids, a branch of the preorbital lobe of the air sinus system that extends from the orbit dorsally into a recess that lies between the frontal bone and the facial portion of the maxillary bone on either side of the skull, and an asymmetrical cranial vertex that is only slightly offset to the left side. Each of these characters is found also in at least one species of some other odontocete family, but the combination is found only in the Phocoenidae.

According to Rice (1984), porpoises of the genera *Phocoena* and *Neophocaena* inhabit coastal waters, bays, estuaries, swamps, and large rivers. They normally are slow swimmers and almost never accompany vessels. *Phocoenoides* usually inhabits oceanic waters, though it also enters fjords and straits with deep, clear water. It is among the swiftest of cetaceans and habitually races ahead of vessels. It feeds on deep-dwelling squid and fish. The biology of *Australophocaena* is not well known, but Barnes (1985) noted that it also appears to be an offshore animal. Porpoises usually travel in small groups. Females have a gestation period of

about 11 months and give birth during the warm months.

The geological range of the Phocoenidae is late Miocene to early Pliocene in western North America, early Pliocene in western South America, Pleistocene in eastern Asia, and Recent in the present range (Barnes 1985; Rice 1984).

CETACEA; PHOCOENIDAE; **Genus PHOCOENA**
G. Cuvier, 1817

Common Porpoises, or Harbor Porpoises

There are three species (Brownell and Praderi 1984; Gaskin 1984; Rice 1977; Villa 1976):

P. phocoena, coastal waters from Davis Strait to North Carolina, from Iceland and Novaya Zemlya to the Cape Verde Islands and Gambia, from northern Alaska to Japan and Baja California, and in the Mediterranean and Black seas;

P. sinus, Gulf of California;

P. spinipinnis, coastal waters from northern Peru and Uruguay to Tierra del Fuego.

Guiler, Burton, and Gales (1987) identified a skull from Heard Island in the southern Indian Ocean as *P. spinipinnis,* but Brownell, Heyning, and Perrin (1989) referred the specimen to *Australophocaena.*

Head and body length is about 120–200 cm, pectoral fin length is 15–30 cm, dorsal fin height is 15–20 cm, and width of the tail flukes is 30–65 cm. In *P. phocoena* the mean head and body length is about 150–60 cm and the usual weight is 45–60 kg. The respective maximums are 186 cm and 90 kg (Gaskin, Arnold, and Blair 1974). Four physically mature specimens of *P. sinus* were about 135–44 cm long and weighed 43–47 kg (Brownell et al. 1987). The sexes are approximately equal in size. Coloration is variable but usually is dark gray or black above, grading to whitish below, or is entirely dark. The conical head is not beaked, and the dorsal fin is usually triangular and located just behind the middle of the back. There are 16–28 teeth on each side of each jaw. The

Common porpoise *(Phocoena phocoena)*, photo by Stephen Spotte, Mystic Marinelife Aquarium.

spade-shaped teeth are entirely crowned or bear crowns with two or three lobes.

Common porpoises frequent coastal waters, bays, estuaries, and the mouths of large rivers. They sometimes ascend the rivers. In the western Atlantic, at least, *P. phocoena* apparently moves well out to sea at the end of summer and reappears in the spring. In other regions *P. phocoena* also moves away from shore and/or toward the south during the fall and winter in order to avoid the build-up of ice (Gaskin 1984). This species generally swims quietly near the surface, rising about four times per minute to breathe. It seems to be less frolicsome than are most members of the family Delphinidae; seldom jumps out of the water, ignores boats, and rarely rides bow waves. When pursued, it can attain a speed of about 22 km/hr. When diving for food it stays underwater for an average of about four minutes. One individual was caught in a fish trap at a depth of 79 meters. *P. phocoena* produces clicklike sounds such as those known to be used by some members of the Delphinidae for echolocation. Its diet consists mainly of smooth, nonspiny fish about 10–25 cm long, such as herring, pollack, mackerel, sardines, and cod (Gaskin, Arnold, and Blair 1974). Captives do not seem to differ much from *Tursiops* in trainability (Andersen 1976).

In a study of *P. phocoena* off New Brunswick, Gaskin and Watson (1984) learned that when known individuals returned each spring and summer, they reestablished specific ranges of about 1.0–1.5 sq km and patrolled these areas in a territorial fashion. Although *P. phocoena* may sometimes travel in schools of nearly 100 individuals, it is usually seen in pairs or in groups of 5–10 (Leatherwood, Caldwell, and Winn 1976). The reproductive season of this species could be extensive, but mating seems to occur mainly from June to September and calving from May to August. Gestation probably lasts 10–11 months. The single newborn weighs 6–8 kg and is 70–100 cm long. The young is brought to a sheltered cove by the mother and nurses for about 8 months. Sexual maturity is attained at 3–4 years. The life span is short for a cetacean, usually about 6–10 years and rarely exceeding 13 years (Gaskin 1977; Gaskin, Arnold, and Blair 1974; Gaskin and Blair 1977; Gaskin et al. 1984; Simons 1984).

Little is known about the life history of the other species of *Phocoena*. A pregnant female *P. spinipinnis* with a near-term fetus was collected off Uruguay in late February or early March (E. D. Mitchell 1975a).

Phocoena has been exploited heavily in various areas, its meat used for human and domestic animal consumption, and its oil used for lamps and lubricants. Indian tribes in eastern Canada formerly caught several thousand *P. phocoena* yearly and sold the oil. This fishery ended by the late nineteenth century, but the population of the region may still be jeopardized by accidental catches in nets set for fish and is known to be declining in some areas. Other populations also are being reduced by entanglement in modern synthetic fishing nets, which are visually and acoustically almost undetectable to porpoises. In the Baltic Sea, *P. phocoena* now has declined to a critically low level following centuries of exploitation, during which catches reached up to 3,000 per year off Denmark alone. Numbers also have fallen sharply in the Netherlands, probably mainly because of water pollution, as well as entanglement and excessive harvest of food fishes. The isolated population of *P. phocoena* in the Black Sea and adjoining Sea of Azov, once estimated to contain 25,000–30,000 individuals, sustained an annual harvest of up to 2,500, but this fishery was terminated by order of the Soviet government in 1966 and by the Turkish government in 1983. Very few porpoises survive there, and the population of the Mediterranean Sea, never very numerous, may now be extinct. About 700–1,500 *P. phocoena* are still deliberately or accidentally taken off western Greenland each year, and this kill may be

adversely affecting the overall resident population of 10,000–15,000 in that region. *P. spinipinnis* is taken commercially off the coasts of Chile and Peru; in Peru about 113,500 kg of meat are marketed annually (Gaskin 1977, 1984; E. D. Mitchell 1975a, 1975b; Read and Gaskin 1988; Smeenk 1987; U.S. National Marine Fisheries Service 1978).

The species *P. sinus* is classified as vulnerable by the IUCN and as endangered by the USDI and is on appendix 1 of the CITES. Villa (1976) considered it to be nearly extinct. It formerly may have occurred throughout the Gulf of California, but Brownell (1986) reported that all precise and all recent records are from the upper part of the gulf and stated that the species has the most limited natural distribution of any marine cetacean. *P. sinus* was considered abundant in the early twentieth century but declined in conjunction with the intensification and modernization of commercial fisheries, starting in the 1940s. A main problem may be accidental catching in nets set for fish and shrimp. Several hundred may have died annually in this manner during the early 1970s (Brownell 1983). In addition, fishing might be reducing food supplies, and dams in the southwestern United States and western Mexico may be cutting off the flow of nutrients to the Gulf of California. Silber (1988) reported several new sightings but stated that the species is exceedingly rare.

CETACEA; PHOCOENIDAE; Genus NEOPHOCAENA
Palmer, 1899

Finless Porpoise

Rice (1977) recognized a single species, *N. phocaenoides*, occurring in warm coastal waters and certain rivers from Pakistan to Korea, Japan, Borneo, and Java. Pilleri and Gihr (1975, 1981b) thought that there were actually three species: *N. phocaenoides*, of southern Asia (as far west as the Persian Gulf) and the East Indies; *N. asiaeorientalis*, of the Chang Jiang (Yangtze) River Valley in China; and *N. sunameri*, of Japan and Korea.

This is the smallest cetacean. Head and body length is 120–60 cm, pectoral fin length is about 28 cm, and expanse of the tail flukes is about 55 cm. Weight is 25–40 kg (Lekagul and McNeely 1977). Although there are numerous literary references indicating that the color is black or dark plumbeous gray, observations of live animals by Pilleri, Zbinden, and Gihr (1976) show that the upper parts are actually pale gray, with a bluish tinge on the back and sides, and that the underparts are whitish. Apparently, however, the skin quickly darkens after death. Distinctive features are the small size, the abruptly rising forehead, and the absence of a dorsal fin. There are 15–21 spade-shaped teeth on each side of each jaw.

In southern Asia and the East Indies *Neophocaena* seems to be closely associated with mangroves and salt marshes (Pilleri and Gihr 1975). In China the genus is found in shallow coastal waters, in the middle and lower Chang Jiang (Yangtze), and in Dongting Lake, often in association with *Lipotes* (Chen et al. 1979; Zhou, Pilleri, and Li 1979). Part of the Japanese population is migratory, concentrating in the Inland Sea in the spring and moving out to the Pacific coast from late summer to midwinter (Kasuya and Kureha 1979). *Neophocaena* moves rather slowly, rolls to the surface to breathe, and does not jump out of the water. The diet includes fish, shrimp, and small squid.

About half of all sightings made in Japan by Kasuya and Kureha (1979) involved solitary animals, and most of the others were of two animals, generally a cow and calf or a mated pair. There is usually a 2-year breeding cycle in Japan, with gestation lasting approximately 11 months, births oc-

Finless porpoise (Neophocaena phocaenoides), photos by T. Kasuya.

curring from April to August, and weaning taking place from September to June. Sexual maturity may possibly be attained by the age of 2 years. Some individuals are estimated to have lived up to 23 years.

In China *Neophocaena* is usually seen in groups of 3–5 individuals and occasionally in schools of up to 20. Births occur from February to May, gestation may be about 10 months, and the newborn is about 68–74 cm long. The calf clings with its flippers and is carried on the back of the female in a spot where her skin is roughened. The extent of this roughened area, however, is considerably less in the Chinese population than in that of southern Asia. Sometimes a group of as many as 6 females is seen, each carrying its calf in this manner (Chen et al. 1979; Chen, Liu, and Harrison 1982; Pilleri and Chen 1979).

Neophocaena has been taken regularly in nets set for shrimp and fish and is also deliberately hunted for its meat, skin, and oil (E. D. Mitchell 1975*b*). For these reasons, and also perhaps because of pollution and collision with motor-boats, its numbers have declined in some areas. Modification of fishing methods in Japan has reduced the accidental catch there. The population using the Inland Sea, perhaps the largest in Japanese waters, contains about 4,900 individuals (Kasuya and Kureha 1979). *Neophocaena* is on appendix 1 of the CITES.

CETACEA: PHOCOENIDAE; **Genus**
AUSTRALOPHOCAENA
Barnes, 1985

Spectacled Porpoise

The single species, *A. dioptrica*, is known from the coastal waters of Uruguay, Argentina, the Falkland Islands, South Georgia, Heard Island, the Kerguelen Islands, New Zealand, the Auckland Islands, and Macquarie Island (Baker 1977; Brownell 1975*b*; Brownell, Heyning, and Perrin 1989; Fordyce, Mattlin, and Dixon 1984). This species long was assigned to *Phocoena* but was shown by Barnes (1985) to represent a separate genus, more closely related to *Phocoenoides* and perhaps the Southern Hemisphere counterpart of the latter.

Pagnoni and Saba (1989) reported that an adult male from Argentina was 202 cm long and weighed 115 kg. According to Brownell (1975*b*), an adult male and an adult female measured 204 and 186 cm in length, and 255 and 150 cm in dorsal fin height. The dorsal and upper lateral surfaces to just above the eyes are black, the sharply demarcated lower lateral and ventral surfaces are white, the pectoral flippers are white, and the eyes are surrounded by a black patch. There is some external resemblance to *Phocoena*, but *Australophocaena* is distinguished by its large, triangular dorsal fin and its relatively smaller and more rounded pectoral fins. There are 18–23 upper and 16–19 lower teeth on each side.

Barnes (1985) observed that the skull of *Australophocaena* resembles that of *Phocoenoides* in having a high, antero-posteriorly elongated and broad cranial vertex; a relatively small temporal fossa; an inflated zygomatic process of the squamosal; a short and anteroposteriorly expanded postorbital process of the frontal; and a flat rostrum that does not curve downward distally. In *Phocoena* the skull has a lower and generally more transversely compressed and knoblike cranial vertex, a larger and more circular temporal fossa, a more slender zygomatic process, a longer and pointed postorbital process, and a rostrum on which the premaxillary bones arch upward and the anterior tip and lateral margins curve more ventrally. *Australophocaena* has more vertebrae than does *Phocoena* but fewer than *Phocoenoides*. The teeth of *Australophocaena* have spade-shaped crowns, as do those of *Phocoena*, and are not so reduced as those of *Phocoenoides*. From *Phocoenoides*, *Australophocaena* also differs in having larger nasal bones, a much shorter rostrum, and a higher and more compressed braincase.

The biology of the spectacled porpoise is not well known, but it appears to be an offshore animal, like *Phocoenoides* but in contrast to *Phocoena* and *Neophocaena* (Barnes 1985). Pregnant females were collected in July and August (Brownell 1975b).

CETACEA; PHOCOENIDAE; **Genus PHOCOENOIDES**
Andrews, 1911

Dall Porpoise

The single species, *P. dalli*, occurs from Japan, around the rim of the North Pacific, to Baja California (Kasuya 1978; E. D. Mitchell 1975a; Morejohn 1979; Rice 1977). The name *truei* has sometimes been used in a specific, subspecific, or other categorical sense for most of the animals found off the Pacific coast of Japan. This name, however, seems to be based on a color variant of *P. dalli*. Miyazaki, Jones, and Beach (1984) observed both forms in the same school during the breeding season and also cited records of females of one form giving birth to young of the other form.

Head and body length is about 170–236 cm, pectoral fin length is 18–25 cm, dorsal fin height is 13–20 cm, and expanse of the tail flukes is 40–56 cm. Weight is usually 80–125 kg, but the U.S. National Marine Fisheries Service (1978) reported a maximum of 218 kg. Both sexes are about the same size (Morejohn 1979). The upper parts of *Phocoenoides* are black. In most populations there is a large white area on each side of the body, but it does not extend far in front of the anterior margin of the dorsal fin. In the population off the Pacific coast of Japan, to which the name *truei* is often applied, the white area usually extends laterally to above the flipper, and there also may be a white throat patch. Some individuals have intermediate color patterns, a few are entirely black, and a few are mostly brownish or gray.

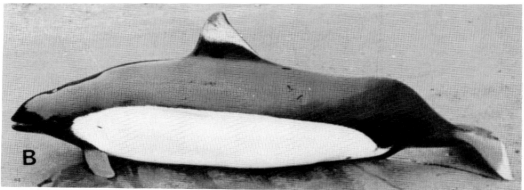

Dall porpoise *(Phocoenoides dalli)*: A. Color pattern of most populations; B. Color pattern commonly found off Pacific coast of Japan. Photos by Warren J. Houck.

The head of *Phocoenoides* is sloping, and the lower jaw projects slightly beyond the upper. The dorsal fin is low and triangular. There are 97–98 vertebrae, compared with only 64–65 in *Phocoena* (Hall 1981). The spade-shaped teeth, the smallest of any phocoenid or delphinid (Morejohn 1979), number 19–27 on each side of each jaw. An unusual feature is the presence of horny protuberances of the gums between the teeth. These "gum teeth" function as gripping organs and probably wear down with use to expose the teeth; the true teeth thus seem to function mainly in older animals.

Phocoenoides occurs mainly in cool waters, sometimes near land but generally well offshore in seas more than 180 meters deep. In the western Pacific there is a well-defined annual migration in which most Japanese animals shift northward for the summer to the Sea of Okhotsk and the Kuril Islands. In the eastern Pacific there are apparently no large-scale migrations, and large numbers of *Phocoenoides* remain throughout the year from Alaska to California. There is, however, a tendency for concentration near the shore and to the south during the fall and winter, and offshore and to the north in the spring and summer. Such seasonal movements are probably related to distributional changes in prey organisms (Kasuya 1978; Leatherwood and Reeves 1978; E. D. Mitchell 1975a; Morejohn 1979; U.S. National Marine Fisheries Service 1978).

Phocoenoides, unlike *Phocoena*, often plays about ships and leaps out of the water and also emerges farther from the water when rolling. It is reputedly the fastest cetacean, sometimes briefly attaining speeds of 55 km/hr. It does not bother to ride bow waves unless the vessel is moving at least 26 km/hr. It commonly preys on animals that live at depths in excess of 180 meters. It produces clicklike sounds that presumably are used for echolocation. The diet includes fish, mainly small unarmed fish such as herring and anchovies, and squid (Jefferson 1988; Leatherwood and Reeves 1978; Morejohn 1979).

Groups of 2–12 individuals are usual, but aggregations of several thousand have been seen (Jefferson 1988). Available evidence indicates that births occur between early spring and early fall in the eastern Pacific, with a strong peak in June, July, and August (Jefferson 1989). More detailed studies in Japan (Jefferson 1988; Kasuya 1978) have shown that births in the western Pacific take place in August and September and

that females apparently give birth at intervals of 3 years. The gestation period averages 11.4 months. Estimates of the length of lactation vary widely, from as short as 1.6–4.0 months to 2 years. The single newborn is about 100 cm long. The age of sexual maturity is 5–8 years in males and 3–7 years in females. Maximum known age attained in the western Pacific is 22 years.

The meat of *Phocoenoides* has been used for human consumption, its blubber is a source of oil, and its bones have been ground up and used for fertilizer (Banfield 1974). Direct exploitation has been especially heavy in Japan, and the population there appears to have declined as a result. An even greater problem in this area, however, has been the accidental kill by fishermen using gill nets to catch salmon. This kill has approached 20,000 porpoises annually and seems to involve mainly immature animals (Kasuya 1978; E. D. Mitchell 1975a, 1975b; U.S. National Marine Fisheries Service 1978). The current estimate of the total number of *Phocoenoides* in the Pacific is 2,150,000 (U.S. National Marine Fisheries Service 1989).

CETACEA; Family MONODONTIDAE

Beluga and Narwhal

This family of two known genera, each with one species, occurs in the Arctic Ocean and nearby seas. The family has sometimes been considered part of the Delphinidae. In contrast, Barnes, Domning, and Ray (1985) transferred the genus *Orcaella* from the Delphinidae to the Monodontidae.

Head and body length (not including the tusk of the narwhal) is usually 280–490 cm. The body form resembles that of the Delphinidae. The forehead is high and globose, the snout is blunt, and there is no beak. The blowhole is located well back from the tip of the snout. There are no external grooves on the throat. The pectoral fin is short and rounded, and there is no dorsal fin. The skull lacks crests. There are 50–51 vertebrae. The cervical vertebrae are not fused, allowing the neck to be more flexible than in most cetaceans.

These animals live in cold waters. They migrate in response to shifting pack ice and hard winters. Individuals that

Beluga whale *(Delphinapterus leucas)*, photos by Warren J. Houck.

become trapped in ice fields can sometimes break through the ice by ramming it from the underside; the cushion on top of the head lessens the shock to the animal. The small spout is not well defined. Both genera sometimes ascend rivers. They emit various sounds and are sensitive to sounds in the water but apparently disregard noises originating on land. They seem to feed mainly on the bottom. The diet includes fish, cephalopods, and crustaceans. Groups sometimes comprise more than 100 individuals. Both genera are extensively hunted by natives of the Arctic.

Geologically, the Monodontidae now are known to have occupied temperate waters as far south as Baja California during the late Miocene and Pliocene (Barnes, Domning, and Ray 1985).

CETACEA; MONODONTIDAE; **Genus**
DELPHINAPTERUS
Lacépède 1804

Beluga, or White Whale

The single species, *D. leucas*, occurs primarily in the Arctic Ocean and adjoining seas, the Sea of Okhotsk, the Bering Sea, the Gulf of Alaska, Hudson Bay, and the Gulf of St. Lawrence (Rice 1977). It also ascends large rivers, such as the Amur, Anadyr, Ob, Yenesei, Yukon, Churchill, and St. Lawrence. Individuals have appeared as far south as Japan, Washington, Connecticut, New Jersey, Ireland, Scotland, the Rhine River, and the Baltic Sea (Banfield 1974; Ellerman and Morrison-Scott 1966; Ellis 1980; Hall 1981). The use of the name beluga has sometimes caused confusion, since the name is also applied to the great white sturgeon, one of the principal sources of caviar.

According to Fay (1978), head and body length is usually about 340–460 cm in males and 300–400 cm in females, and average weight is about 1,500 kg in males and 1,360 kg in females. Kleinenberg et al. (1969) listed maximum size as 700 cm and 2,000 kg. Pectoral fin length is about 20–45 cm. The general adult color is creamy white. The young are dark gray, black, or bluish in the first year of life; then become yellowish, mottled brown, or pale gray; and finally attain the white coloration at around five years. The lightening in color is caused by a reduction in the melanin of the skin.

The body is fusiform, the tail is strongly forked, there is a constriction at the neck, and the snout is blunt. The upper jaw protrudes slightly ahead of the bulbous melon. There is no dorsal fin, but a low dorsal ridge is present on the back. There are 8–10 teeth on each side of each jaw.

The beluga occurs both along coasts and in deep offshore waters. Some populations, such as that of the Gulf of Alaska, seem to reside permanently in relatively small areas, while others undertake extensive migrations. Wintering sites may be either north or south of summering sites and may be either nearer to or farther from shore. The environmental conditions that prevail in the range of a particular population, such as prey distribution and extent of the pack ice, apparently affect the choice of seasonal movements. One population of about 11,500 whales concentrates in the Mackenzie River estuary from late June to early August, spreads through the deep waters of the Beaufort Sea during the late summer, and then migrates about 5,000 km to the southwest and is distributed mainly in shallow coastal areas, bays, and river mouths around the Bering Sea in winter. Another population concentrates in river estuaries adjoining southern Hudson Bay in the summer but moves northward and into the open waters of the bay in winter. In the waters off northern Eurasia there seems to be a general tendency to move away from

shore in the autumn and toward shore in the spring, but there are exceptions, and times of migration vary from year to year. An advantage of being close to shore in the summer is that rivers are unfrozen and may be entered for purposes of finding food and rearing young. Some individuals have ascended rivers in Siberia for distances of up to 2,000 km (Banfield 1974; Fay 1978; Finley, Hickie, and Davis 1987; Harrison and Hall 1978; Kleinenberg et al. 1969; Sergeant 1973).

Delphinapterus, a very supple animal, can scull with its tail and thus swim backward. When swimming normally it moves at about 3–9 km/hr, and when pursued, it can attain a speed of 22 km/hr. It usually surfaces to breathe every 30–40 seconds, but feeding dives usually last 3–5 minutes. One individual is known to have remained underwater for 20 minutes on a dive to a depth of 647 meters (Ridgway 1986*b*). The beluga is capable of covering a distance of 2–3 km underwater, and therefore the presence of surface ice is not necessarily a problem (Kleinenberg et al. 1969). The beluga cannot remain in waters extensively covered by thick ice, however, because it is unable to break through more than a very thin layer to make breathing holes (Harrison and Hall 1978).

Delphinapterus is known to produce a variety of sounds. One, "a low liquid trill, like the cries of curlews in the spring," has given rise to the vernacular name "sea canary." Many other animals produce underwater noises, but few such sounds can be heard so readily above water as those of the beluga. Also emitted are whistles and clicks comparable to those known to be used by certain other cetaceans for communication and echolocation, respectively (Fay 1978; Kleinenberg et al. 1969). That *Delphinapterus* employs echolocation extensively is also suggested by its huge frontal melon, an organ thought to be used in the directing of sonar signals, and by the fact that this genus lives in regions that lack sunlight for months at a time, and in waters that may be partly covered by ice. Echolocation presumably is used to avoid obstacles and to search for prey on the bottom. The diet includes a great variety of fish, such as cod and herring, as well as octopus, squid, crabs, and snails (Fay 1978; Kleinenberg et al. 1969).

Schools of as many as 10,000 individuals have been reported, but such aggregations are formed only temporarily for migration or in the presence of abundant food sources. Permanent social units consist of only about 10 animals and seem to be led by a large male (Kleinenberg et al. 1969). There is conflicting information on reproduction, but studies in arctic Canada (Braham 1984*b*; Brodie 1971; Sergeant 1973) indicate that the calving season lasts from April to September and peaks in late June and July, that each adult female gives birth every 2 or 3 years, and that the gestation period is about 14–15 months. The newborn is about 160 cm long and weighs about 80 kg. Lactation lasts for 20–24 months. Sexual maturity is attained at 4–7 years in females and 8–9 years in males. Potential longevity is 25–30 years.

The beluga has been exploited by the native people of the Arctic since ancient times and has been subject to commercial fishing operations in the twentieth century. It is often caught in nets set across migratory routes. The meat is used for human and domestic animal consumption. The 10- to 25-cm-thick layer of blubber yields 100–300 liters of oil, which is used in the production of soap, lubricants, and margarine. The fat of the head is rendered into a high-quality lubricant. The bones are ground up for fertilizer, and the skin is used to make boots and laces. The tanned hide of the beluga is sometimes called "porpoise leather." In recent decades the demand for these products has declined. From the 1950s to the 1980s, annual catches fell from about 4,000 to 1,200 in the Soviet Union. There is also an annual kill of about 1,000 in Canada and 200 in Alaska. The current number of *Delphinapterus* in the world is estimated at 62,000–88,000, of which at least

Beluga whales *(Delphinapterus leucas)*, photo by Warren J. Houck.

30,000 are in North American waters (Braham 1984*b*; Kleinenberg et al. 1969; E. D. Mitchell 1975*a*, 1975*b*; Sergeant and Brodie 1975; U.S. National Marine Fisheries Service 1989).

Some North American populations, such as that in Ungava Bay, have been almost eliminated (Reeves and Mitchell 1987*a*), and others are being overexploited. Originally, about 7,000 white whales summered in Hudson and James bays and associated rivers, but that number was reduced by about 75 percent through hunting, and recovery may now be impossible because of river diversion, harbor construction, and other human activities in the habitat (Reeves and Mitchell 1987*b*; Smith and Hammill 1986). The population off western Greenland and in Lancaster Sound has been reduced substantially but is still large (Reeves and Mitchell 1987*c*), and the population that summers in the Beaufort Sea is considered to be relatively safe, though there is concern about the potential effects of human development of the Mackenzie River estuary (Finley, Hickie, and Davis 1987). The isolated population in the St. Lawrence estuary numbered at least 5,000 individuals in 1885 but declined drastically through commercial, sport, and bounty hunting. It now contains about 500 animals and is legally protected, but its reproductive rate is low, evidently because of chemical pollution of its habitat (Reeves and Mitchell 1984; Sergeant and Hoek 1988).

CETACEA; MONODONTIDAE; Genus MONODON
Linnaeus, 1758

Narwhal

The single species, *M. monoceros*, occurs in the Arctic Ocean and nearby seas, from 65° to 85° N, mainly between 70° and 80° N. There have been occasional records as far south as the Alaska Peninsula, Newfoundland, Great Britain, and Germany (Reeves and Tracey 1980).

Head and body length, exclusive of the tusk, is 360–620 cm, pectoral fin length is 30–40 cm, and expanse of the tail flukes is 100–120 cm. According to Reeves and Tracey (1980), average head and body length is about 470 cm in males and

400 cm in females, and average weight is 1,600 kg in males and 900 kg in females. About one-third of the weight is blubber. Coloration becomes paler with age. Adults have brownish or dark grayish upper parts and whitish underparts, with a mottled pattern of spots throughout. The head is relatively small, the snout is blunt, and the flipper is short and rounded. There is no dorsal fin, but there is an irregular ridge about 5 cm high and 60–90 cm long on the posterior half of the back. The posterior margins of the tail flukes are strongly convex, rather than concave or straight as in most cetaceans.

There are only two teeth, both in the upper jaw. In females the teeth usually are not functional and remain embedded in the bone. In males the right tooth remains embedded, but the left tooth erupts, protrudes through the upper lip, and grows forward in a spiral pattern to form a straight tusk. The tusk is about one-third to one-half as long as the head and body and sometimes reaches a length of 300 cm and a weight of 10 kg. In rare cases the right tooth also forms a tusk, but both tusks are always twisted in the same direction. Occasionally one or even two tusks develop in a female. The distal end of the tusk has a polished appearance, and the remainder is usually covered by a reddish or greenish growth of algae. There is an outer layer of cement, an inner layer of dentine, and a pulp cavity that is rich in blood. Broken tusks are common, but the damaged end is filled by a growth of reparative dentine. The form of this plug sometimes gives the erroneous impression that another narwhal has jammed its own tusk directly into the damaged one (Newman 1978; Reeves and Tracey 1980).

It traditionally has been suggested that the tusk of *Monodon* is used to break through surface ice, to probe the bottom for food, or to skewer prey (Ellis 1980). More recent hypotheses are that the tusk serves to radiate heat or to enhance echolocation (Reeves and Mitchell 1981). New evidence, however, supports an old view that the tusk is exactly what it appears to be, namely, a weapon. According to Silverman and Dunbar (1980), the presence of many scars on the heads of adult males, the large proportion of broken tusks seen, the discovery of a tusk tip embedded in the jaw of a male, and actual observations of narwhals crossing and striking their tusks strongly indicate that the tusk is used in intraspecific aggression, most probably during the mating season. It also

Narwhals *(Monodon monoceros)*, painting by Richard Ellis.

may serve to communicate dominance rank, or in ritualized displays, much like the horns and antlers of some ungulates. Best (1981) emphasized the role of tusk length as a means of nonviolent assessment of hierarchical status and doubted that the tusk actually was used to physically damage an opponent, but Gerson and Hickie (1985) rejected this view in favor of the weapon hypothesis. They found the degree of head scarring in adult males to be correlated positively with body size and tusk girth and weight (but not tusk length). Based on knowledge of other mammals, the amount of scarring resulting from tusk-inflicted injuries would be expected to be greatest in dominant males, who must fight more often and more intensively in order to defend a harem. Females and immature males were found to have significantly fewer head scars.

Monodon has the most northerly distribution of any mammal; its normal range is almost entirely above the Arctic Circle. It is generally found in deeper waters than those used by *Delphinapterus* and tends to avoid shallow seas like those around Alaska. It seems to be most common in the eastern Canadian Arctic and in waters off Greenland. One or more major populations winter in the open waters of Baffin Bay and Davis Strait. When the ice breaks up in the summer, there are large-scale migrations to the north, east, and west, into the fjords and inlets of Greenland and the islands of the eastern Canadian Arctic. Another group winters in the Greenland Sea and summers along the east coast of Greenland and around Svalbard (Newman 1978; Reeves and Tracey 1980).

The narwhal is a rapid swimmer. It has been reported to dive to depths as great as 370 meters and has been timed to stay underwater for 15 minutes. It maintains breathing holes in pack ice by upward thrusts of the large melon on the forehead (not the tusk) and can break through a layer of ice 18 cm thick. Sometimes several individuals cooperate to break the ice. *Monodon* emits a wide variety of sounds. The deep groans, roars, and gurgles that have been reported are probably associated with respiration. There are also screeches and whistles, probably for communication, and series of clicks, much like those known to be used by some other cetaceans for echolocation (Ford and Fisher 1978; Reeves and Tracey 1980). There has been no conclusive demonstration that *Monodon* actually does echolocate, but such an ability would seem useful in finding obstacles and prey in ice-covered waters or on long arctic nights. The diet consists of fish, cephalopods, and crustaceans.

Up to 2,000 individuals may join in a migratory school, but permanent groups contain 3–20. Most of the smaller groups are apparently families, and each seems to have only 1 large male. There are also groups consisting entirely of males or of females and young. Births occur from June to August, the period when the breakup of pack ice allows entrance to protected bays and fjords. Each female apparently produces a single calf every 3 years. Gestation is thought to last about 15.3 months. The newborn is around 160 cm long and weighs just over 80 kg. The period of lactation is unknown but is presumed to be about the same as in *Delphinapterus*. Sexual maturity may come at about 8 years of age in males and 6 years in females. Maximum longevity evidently approaches 40 years (Best and Fisher 1974; Ellis 1980; Mansfield, Smith, and Beck 1975; Reeves and Tracey 1980; Strong 1988).

The narwhal long has been hunted for subsistence by the native people of the Arctic. The animals are traditionally harpooned from kayaks when they enter inlets during the summer. Occasionally a large number can be easily killed when they become trapped by the rapid formation of pack ice at the end of summer. The sinews are used for thread, the meat is sometimes eaten by humans and sled dogs, and the skin, high in vitamin C, is probably what prevented scurvy in the arctic peoples (Reeves 1977).

Commercial hunting of *Monodon* seems to have begun in the tenth century A.D., soon after the establishment of Viking colonies in Greenland. The tusk was long sold at high prices, as it was said to be the horn of the mythical unicorn and to have magical properties, such as rendering poison harmless. By the seventeenth century the true origin and general availability of the tusk were recognized, and prices fell accordingly. Some demand continued, especially in the Orient, where the tusk was used in decorative work and medicinal preparations. Skins were imported to Europe to make boots, laces, and gloves. In the 1960s, as in the case of other kinds of ivory, the value of the narwhal's tusk again began to increase. The price went from about U.S. $2.75 per kg in 1965 to nearly $100 per kg in 1979. Most tusks, however, go whole to collectors and decorators, who have paid up to $4,500 for a choice specimen. Such a market caused an intensification of hunting, which is now often done with motorboats, explosive harpoons, and high-powered rifles. In 1976 the Canadian government established regulations that set quotas for various areas and required full utilization of the carcass, as well as the tusk, but these rules are nearly impossible to enforce. The annual kill of *Monodon* in Canada and Greenland is now thought to be just over 1,000. There are probably about 29,000 narwhals to the west of Greenland and several thousand more to the east. It is not known whether overall populations in Canada and Greenland are being reduced by human hunting, but there apparently have been severe declines in Soviet waters (Braham 1984*b*; Broad, Luxmoore, and Jenkins 1988; Davis et al. 1978; Ellis 1980; Mansfield, Smith, and Beck 1975; Newman 1978; Reeves 1977; Reeves and Mitchell 1981; Reeves and Tracey 1980; Smith et al. 1985; Strong 1988).

CETACEA; Family ZIPHIIDAE

Beaked Whales

This family of 6 Recent genera and 18 species occurs in all the oceans and adjoining seas of the world.

Head and body length is about 4–13 meters and weight ranges from 1,000 to over 11,000 kg. The vernacular name is derived from the long, narrow snout, which is sharply demarked from the high, bulging forehead in *Berardius* and *Hyperoodon* and which forms a continuous smooth profile with the head in *Ziphius, Tasmacetus,* and *Mesoplodon* (the external appearance of the sixth genus, *Indopacetus,* is not known). The pectoral fin is rather small and ovate. The dorsal fin is small, usually sickle-shaped, and located on the posterior half of the back. The genus *Ziphius* has a low median keel from the dorsal fin to the tail. The tail flukes of beaked whales usually are not so strongly notched in the center as in other cetaceans. There is a pair of grooves on the throat that converge anteriorly to form a V pattern at the chin. The stomach has 4–14 chambers, but no esophageal chamber (Rice 1967).

The genus *Tasmacetus* has one pair of large functional teeth in the lower jaw and 17–28 small functional teeth on each side of both the upper and lower jaws. In the other genera there are only one or two pairs of large functional teeth, and these are in the lower jaw. These teeth push through the gums sooner in males than in females and often never erupt in the latter sex. There also are frequently series of small nonfunctional teeth in the upper and lower jaws. Except in *Berardius,* the bones of the skull are asymmetrical. Certain cranial bones are crested or elevated, forming large

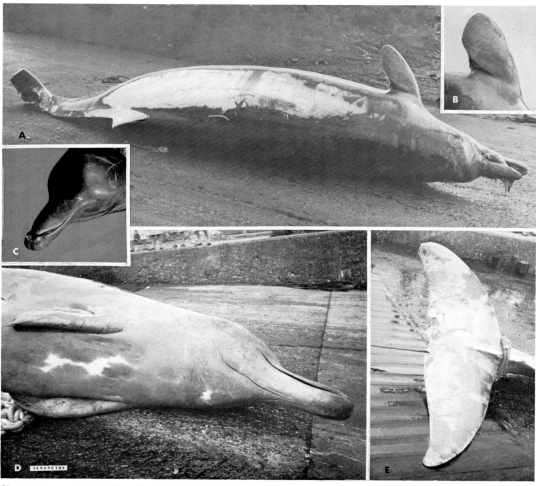

Giant bottle-nosed whale *(Berardius bairdii):* A & B. Photos from Fisheries Research Board of Canada through I. B. MacAskie; C. Head showing teeth in lower jaw, photo from Tokyo Whales Research Institute through Hideo Omura; D & E. Photos by Warren J. Houck.

ridges in *Hyperoodon* and lesser ridges in some of the other genera. There are 43–49 vertebrae, and those of the neck tend to fuse.

Beaked whales are the least known of cetacean families. They usually remain well out to sea, avoid ships, and dive to great depths to secure cephalopods and fish. The presence of a well-developed melon on the forehead suggests that ziphiids are echolocators (U.S. National Marine Fisheries Service 1981). Some appear to be solitary, some travel in groups of 2–12, and some, particularly *Ziphius*, associate in schools of 40 or more. They generally swim and dive in unison. The hides of many individuals are scratched from intraspecific fighting. The newborn are about one-third as long as the mother.

The geological range of this family is early Miocene to late Pliocene in Europe, early Miocene in Africa, middle to late Miocene and Pliocene in North America, early Miocene in South America, early Miocene and Pliocene in Australia, Miocene in New Zealand, and Recent in all oceans (Rice 1984).

CETACEA; ZIPHIIDAE; **Genus BERARDIUS**
Duvernoy, 1851

Giant Bottle-nosed Whales

There are two species (Goodall 1978; Rice 1977):

B. arnuxii, known from waters off South Africa, Australia, New Zealand, Argentina, Tierra del Fuego, the Falkland Islands, South Georgia, the South Shetlands, and the Antarctic Peninsula;

B. bairdii, the North Pacific from the Bering Sea to Japan and California.

These are the largest ziphiids. *B. bairdii* attains sexual maturity at a head and body length of about 10 meters, and growth continues to a maximum of 12.8 meters in females and 12.0 meters in the slightly smaller males. A female *B. bairdii* 11.1 meters long weighed about 11,380 kg (Rice 1984). Pectoral fin length is about 100 cm, height of the dorsal fin is about 30 cm, and expanse of the tail flukes is 250–300 cm. *B. arnuxii* is smaller than *B. bairdii*, not being known to exceed 9.9 meters in head and body length (McCann 1975). Coloration is uniformly blackish brown or dark gray, sometimes with white blotches on the underparts. The skin is always extensively covered with pairs of parallel scratches, probably made by the teeth of conspecifics.

The skull is more nearly symmetrical than in the other beaked whales. The snout is tapered, and the forehead is well defined. Both sexes have two pairs of large teeth in the lower jaw. The anterior pair is completely visible, even when the mouth is closed, because the lower jaw protrudes well beyond the upper.

Giant bottle-nosed whales usually stay well offshore in waters over 1,000 meters deep. Substantive natural history data are available only for *B. bairdii*. This species is alleged to have dived to depths as great as 2,400 meters, after being harpooned. It raises its flukes in the air before diving. It is said to be alert and hard to capture. It normally stays underwater for 15–20 minutes at a time but has remained submerged for just over an hour. Its diet consists mostly of squid and also includes octopus, crustaceans, and deepwater fish (Ellis 1980; Rice 1978*d*, 1984). One population appears off the Pacific coast of southern Japan in the summer and moves northward to the waters east of Hokkaido in the autumn. During this period it concentrates over the continental shelf in waters 1,000–3,000 meters deep and feeds on organisms living on the ocean floor. Schools spend four or five times longer in dives than they do at the surface (Kasuya 1986).

Groups of *B. bairdii* are tightly organized and contain 3–30 individuals of all ages and sexes (Ellis 1980; Rice 1978*d*). Studies of this species in Japan (Kasuya 1977) indicate that mating peaks in October and November and births occur mainly in March and April. The gestation period is about 17 months, apparently the longest of any cetacean. The newborn is around 460 cm long. Sexual maturity is attained at 8–10 years, and physical maturity at over 20 years. Mead (1984) listed maximum known longevity as 71 years.

Hunting of *B. bairdii* has been undertaken in Japan since at least 1612. The dried meat is considered a delicacy in some parts of the country. For many years the annual kill amounted to only a few individuals, but following World War II modern fishing methods resulted in a sharply increased harvest. The peak kill was 382 whales in 1952. A subsequent decline in the take, which averaged 69 per year from 1969 to 1977, may be partly associated with a reduction in the whale population (Ellis 1980; E. D. Mitchell 1975*a*, 1975*b*; Nishiwaki and Oguro 1971; Rice 1978*d*). *Berardius* is on appendix 1 of the CITES.

CETACEA; ZIPHIIDAE; **Genus ZIPHIUS**
G. Cuvier, 1823

Goose-beaked Whale

The single species, *Z. cavirostris*, occurs in the temperate and tropical waters of all oceans and adjoining seas (Rice 1977).

Females reach sexual maturity at a head and body length of 6.1 meters and then continue to grow to a maximum of 7 meters. Males reach sexual maturity at 5.4 meters and then continue to grow to a maximum of 6.7 meters. A frequently cited report of an individual 8.5 meters long is erroneous (E. D. Mitchell 1975*a*). Pectoral fin length is about 50 cm, dorsal fin height is about 40 cm, and expanse of the tail flukes is about 150 cm. An adult male 5.7 meters long weighed 2,450 kg (Ross 1984), and an adult female 6.6 meters long weighed 2,952.5 kg. Coloration is variable, but two frequently observed patterns are: face and upper back creamy white, remainder of body black; and entire body grayish fawn, with some small blotches of slightly darker gray below.

The beak is short and blends into the sloping forehead. The opening of the mouth is relatively small. Males have two functional teeth, which protrude from the tip of the lower jaw; these teeth usually are not visible in females. Rows of small rudimentary teeth usually are present in both jaws. As in other ziphiids, the tail flukes generally lack a median notch, but some specimens of *Ziphius* do have such a notch.

According to Leatherwood, Caldwell, and Winn (1976), *Ziphius* seems to be primarily tropical in distribution but moves northward into temperate waters during the summer. There are also records from as far north as the Aleutian Islands and the Gulf of Alaska (C. S. Harrison 1979; Rice 1978*d*). *Ziphius* is generally found well offshore and often dives to great depths, remaining underwater for 30 minutes or more. Before diving, it raises its tail flukes straight above the surface. It has been observed to leap clear of the water. Its diet consists primarily of squid and also includes deepwater fish.

Groups of up to 40 individuals have been reported, but schools usually contain fewer than half that number. Some white-headed adults, possibly old males, are solitary (Rice 1978*d*). The members of a group often travel, dive, and feed together in fairly close association. Banfield (1974) stated that births occur in late summer or early autumn. A gestation

Goose-beaked whales *(Ziphius cavirostris):* A. Photo from National Geographic Society of painting by Else Bostelmann; Inset: lower jaw showing the two teeth, photo by P. F. Wright of specimen in U.S. National Museum of Natural History; B. Photo by Warren J. Houck.

period of about 1 year has been reported. The newborn is about 200–300 cm long (E. D. Mitchell 1975a). According to Mead (1984), newborn average 270 cm in length, and maximum known longevity is 36 years.

CETACEA; ZIPHIIDAE; **Genus TASMACETUS**
Oliver, 1937

Shepherd's Beaked Whale

The single species, *T. shepherdi,* is known by 13 specimens stranded in Australia, New Zealand, Stewart and Chatham islands near New Zealand, the Juan Fernandez Islands off central Chile, the Valdez Peninsula of east-central Argentina, Tierra del Fuego, and Isla Gable east of Tierra del Fuego; and by a probable sighting of a live individual off New Zealand

(Brownell, Aguayo L., and Torres N. 1976; Goodall 1978; Lichter 1986; Mead and Payne 1975; Watkins 1976).

On the basis of four specimens, Mead and Payne (1975) provided the following data. Head and body length is 6.1–7.0 meters, pectoral fin length is 69 cm, dorsal fin height is 34 cm, and expanse of the tail flukes is 135–52 cm. The general coloration is thought to be dark, but the ventral surface is pale, there is a light area anterior to the pectoral fin, and there are two light stripes along part of the side. In males there are two large teeth at the tip of the lower jaw, but in the one known female specimen these teeth did not erupt. In contrast to all other ziphiids, these ziphiids also have numerous smaller functional teeth. The total tooth count is 17–21 on each side of the upper jaw and 18–28 on each side of the lower jaw. The stomach of one specimen contained a number of fish, as well as a small crab and a small squid beak that may have been eaten by the fish. The stomach contents suggest that *Tasmacetus* had been feeding on the bottom in fairly deep water.

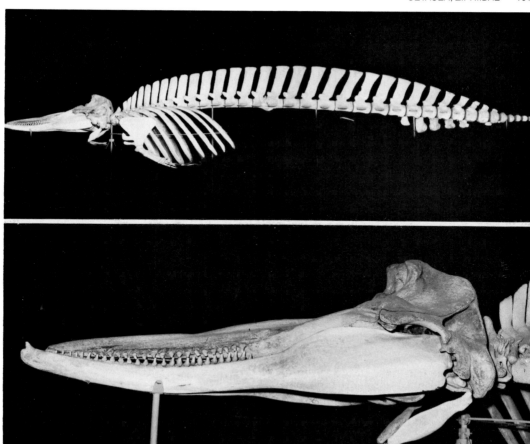

Shepherd's beaked whale *(Tasmacetus shepherdi)*, photos by Eldon V. Burkett through Wanganui Public Museum, New Zealand.

CETACEA; ZIPHIIDAE; **Genus HYPEROODON**
Lacépède, 1804

Bottle-nosed Whales

There are two species (Ellis 1980; Rice 1977):

H. ampullatus, the North Atlantic from Davis Strait and
    Novaya Zemlya to Rhode Island and the English
    Channel;
H. planifrons, known from waters off Australia, New
    Zealand, Brazil, Argentina, Tierra del Fuego, the
    Falkland Islands, South Georgia, the South Orkney
    Islands, South Africa, and the Pacific and Indian ocean
    sectors of Antarctica.

Leatherwood et al. (1982) reported a sighting by K. C. Balcomb of a group of *Hyperoodon,* unidentified as to species,
near the equator in the central Pacific. Moore (1968) placed *H.
planifrons* in its own subgenus, *Frasercetus.*

Maximum head and body length is 9.1 meters in males
and 7.5 meters in females. The dorsal fin is at least 30 cm high
and is distinctly hooked (Leatherwood, Caldwell, and Winn
1976). A female 6 meters long weighed 2,500 kg. Coloration
becomes lighter with age. Calves are grayish brown to black,
immature animals are often spotted yellowish brown and
white, and old individuals may be completely yellowish
white. There are usually two large teeth in the lower jaw of
young males, but in older males one or both may be lost. In
females the functional teeth are smaller or do not emerge at
all. Rows of vestigial teeth are often present in the lower and
upper jaws. In females and young males the forehead slopes
rather smoothly into the beak (Ellis 1980). In older males the
forehead rises abruptly from the beak and may become so
bulbous that when viewed in profile, it appears to protrude
forward at the top. This development results in part from the
enlargement of crests on the maxillary bones. The forehead
encloses a huge melon filled with oil that is much the same in
appearance, and probably in function, as that of the spermaceti of *Physeter.*

Bottle-nosed whales generally stay well out to sea. They
seem to occur mainly in cooler waters and may approach the
polar ice packs during the summer. Limited observations
suggest that *H. planifrons* appears off southern Africa during
the summer months of November–January (G. J. B. Ross

Bottle-nosed whale *(Hyperoodon ampullatus)*, photo from *Endeavour*.

1984). Substantive natural history data are available only for
H. ampullatus. This species is usually found in waters over
1,000 meters deep. During the spring and early summer it
migrates to the northern parts of its geographic range, and in
the late summer and autumn it returns south. It dives suddenly and with great speed. Although dives usually last about
30–45 minutes, there are reports that this whale has remained underwater for 2 hours, which, if accurate, would be a
record for the Cetacea. One harpooned individual is said to
have dived vertically and pulled out 1,000 meters of line.
Feeding probably takes place at great depths. The diet consists
mainly of squid and also includes fish and bottom-dwelling
echinoderms (Benjaminsen and Christensen 1979; Leatherwood, Caldwell, and Winn 1976; Rice 1967).

A variety of sounds has been reported for *H. ampullatus*,
most notably series of clicks and whistles such as are known to
be used by certain other odontocetes for, respectively, echolocation and communication (Winn, Perkins, and Winn
1970). Groups usually comprise 2–4 individuals, generally of
the same age and sex. Larger groups are sometimes seen and
usually contain mature animals of both sexes. Solitary
whales are common and are often young individuals. Groups
are said not to abandon an injured member (Benjaminsen and
Christensen 1979). The group reported by Leatherwood et al.
(1982) comprised about 25 individuals.

Based on observations off South Africa, the calving season
of *H. planifrons* appears to be in spring or summer (G. J. B.
Ross 1984). Both the mating and calving seasons of *H. ampullatus* peak in April. Females probably give birth every 2
years. The gestation period is thought to last 12 months. The
single newborn is about 300–330 cm long, and lactation may
last about a year. Sexual maturity is attained at 8–12 years in
females and 7–11 years in males. Maximum longevity is at
least 37 years (Benjaminsen and Christensen 1979).

A large bottle-nosed whale can yield up to 200 kg of sper-
maceti oil, the uses of which are given in the account of
Physeter, as well as about 2,000 kg of blubber oil (Ellis 1980).
Following the decline of the bowhead whale *(Balaena mys-*

ticetus) in the late nineteenth century, there was increasing
commercial interest in *H. ampullatus*, especially in Norway.
From 1882 to 1920 approximately 50,000 individuals were
taken off northwestern Europe. The annual kill fell from
2,000–3,000 in the 1890s to 20–100 in the 1920s. After some
years of reduced hunting, and perhaps of recovery by the
whale population, the fishery again intensified: the annual
kill peaked at around 700 individuals in 1965 but then declined to near zero. The population in the eastern Atlantic is
estimated to have originally contained 40,000–100,000
whales (Benjaminsen and Christensen 1979; Ellis 1980; U.S.
National Marine Fisheries Service 1978). The current number
is much lower, and the western Atlantic population is also
small, but exact status is unknown. *H. ampullatus* is classified
as vulnerable by the IUCN and as endangered by the Soviet
Union, and both species are on appendix 1 of the CITES.

CETACEA; ZIPHIIDAE; **Genus INDOPACETUS**
Moore, 1968

Indo-Pacific Beaked Whale

The single species, *I. pacificus*, is known with certainty only
by two skulls, one found at Mackay on the east coast of
Queensland and the other at Danane on the east coast of
Somalia (Moore 1968).

Some authorities include *Indopacetus* within *Mesoplodon*.
Among them are Pitman, Aguayo L., and Urbán (1987), who
also reported that an unidentified beaked whale observed
many times in the eastern Pacific off Mexico, Central America, and northern South America might represent *I. pacificus*.
The unidentified whale has an estimated length of 5.0–5.5
meters, a relatively flat head, a moderately long beak, and a
distinctive dorsal fin which is low, wide-based, and triangular.
There are two morphs: the larger, evidently representing
adult males, is black with a broad white or cream-colored

swathe originating immediately posterior to the dorsal surface of the head and running posteroventrally on either side of the animal and is marked with scratches and scars; and the smaller, probably being females and young, is uniform gray-brown or bronze-colored and is unscarred. The two kinds have been seen together in groups of up to eight individuals.

The first skull to be found was nearly 122 cm long and was thought to have come from a fully mature animal about 7.6 meters in length (Ellis 1980). Moore (1968, 1972) listed the following cranial characters as distinguishing the genus: (1) the alveoli of the developed teeth are a single pair, apical on the mandible, and in an old adult male become progressively at least as shallow as 30 mm; (2) the frontal bones occupy an area of the synvertex of the skull approximating or exceeding that occupied by the nasal bones; (3) there is almost no posterior process of the premaxillary crest extending posteriorly on the synvertex between the nasal and maxillary bones or between the frontal and maxillary bones; (4) in the lateral extension of the maxillary bone over the orbit there is a deep groove about half as long as the orbit; (5) at about the mid-length of the beak there is a swelling caused by the lateral margins proceeding forward a short distance without convergence, or even with a little divergence and then convergence again; (6) there is fusion of a considerable length of the mesethmoid bone to both premaxillary rims of the mesorostral canal; and (7) proliferation of bone from the vomer (and distally the premaxillae) into the mesorostral canal at the onset of adulthood is absent or minimal in the one adult specimen.

CETACEA; ZIPHIIDAE; **Genus MESOPLODON**
Gervais, 1850

Beaked Whales

There are 3 subgenera and 11 species (Ellis 1980; Goodall 1978; Lichter 1986; Loughlin and Perez 1985; Mead 1981;

Mead and Baker 1987; Moore 1968; Reiner 1986; Rice 1977, 1978*d*):

subgenus *Mesoplodon* Gervais, 1850

M. hectori, known by 20 specimens from southern California, Argentina, Tierra del Fuego, the Falkland Islands, South Africa, Tasmania, and New Zealand;

M. mirus, temperate waters from Nova Scotia and Florida to the British Isles, and off South Africa;

M. europaeus, western North Atlantic from New York to the West Indies, one record from the English Channel;

M. ginkgodens, known by 10 specimens from Sri Lanka, Taiwan, Japan, and southern California;

M. grayi, known from the Netherlands, South Africa, southern Australia, New Zealand, the Chatham Islands, Chile, Tierra del Fuego, Argentina, and the Falkland Islands;

M. carlhubbsi, temperate waters of the North Pacific from Japan to British Columbia and California;

M. bowdoini, known from Western Australia, Victoria, Tasmania, New Zealand, Campbell Island, and Kerguelen Island;

M. stejnegeri, cool temperate waters from the Bering Sea to Japan and California;

M. bidens, cool temperate waters of the North Atlantic from Newfoundland and Massachusetts to southern Norway and the Azores;

subgenus *Dolichodon* Gray, 1866

M. layardii, known from South Africa, Australia, Tasmania, New Zealand, Tierra del Fuego, Argentina, Uruguay, and the Falkland Islands;

subgenus *Dioplodon* Gervais, 1850

M. densirostris, tropical and warm temperate waters of all oceans.

True's beaked whales *(Mesoplodon mirus),* photo from National Geographic Society of painting by Else Bostelmann. Inset: *M. sp.,* photo from American Museum of Natural History.

Hall (1981) used the generic name *Micropteron* Eschricht, 1849, in place of *Mesoplodon*.

Head and body length is 3–7 meters, pectoral fin length is 20–70 cm, dorsal fin height is about 15–20 cm, and expanse of the tail flukes is about 100 cm. *M. carlhubbsi* reaches a maximum length of about 530 cm and a weight of about 1,500 kg, with no discernible size difference between the sexes (Mead, Walker, and Houck 1982). Color is variable but is usually slaty black to bluish black above and paler below. Only two teeth become well developed, one on each side of the lower jaw, and these may be lost in old age. In most species they are exposed when the mouth is closed and are used in intraspecific combat. In females the functional teeth are much smaller than those of males and often do not erupt above the gums. There may be small nonfunctional teeth in both jaws. A row of such vestigial teeth on each side of the upper jaw seems to be characteristic of *M. grayi*.

The structure of the pair of functional teeth in males varies remarkably (Ellis 1980; Rice 1978d). They can be briefly described as follows: *M. hectori*, shaped like laterally flattened triangles, located at tip of jaw; *M. mirus*, small and angled slightly forward, located at tip of jaw; *M. europaeus*, shaped like laterally flattened triangles, located one-third of the way from tip to apex of mouth; *M. ginkgodens*, shaped in profile like the leaf of the ginkgo tree, located one-third of the way from tip to apex of mouth; *M. grayi*, shaped in profile like a flattened onion, located in middle of mouth; *M. carlhubbsi*, large and straight-sided, located near middle of mouth and exposed when mouth closed; *M. bowdoini*, large and flattened, located in raised sockets near front of mouth; *M. stejnegeri*, large and tusklike, exposed, located in back of mouth; *M. bidens*, small and pointed slightly to rear, located near middle of mouth; *M. layardii*, shaped somewhat like the tusks of a boar (Suidae), come out of mouth and curve over upper jaw; and *M. densirostris*, set at a point where the lower jaw becomes extremely high and broad, so the teeth are well exposed and extend above the upper jaw.

Mesoplodon is apparently a pelagic genus, staying mainly in deep waters far from shore. It is relatively sluggish when at the surface (Leatherwood et al. 1982). The diet includes squid, other cephalopods, and fish. Rice (1978d) observed a group of 10–12 *M. densirostris* dive; after 45 minutes they still had not resurfaced. Most of his other sightings of the genus have been of schools of 2–6 individuals. There is one record of a stranding of 28 *M. grayi* (Ellis 1980). Usually, however, *Mesoplodon* is encountered alone or in small groups (Leatherwood, Caldwell, and Winn 1976). *M. stejnegeri* has been observed in tight groups of 5–15 individuals of various sizes swimming and diving in unison (Loughlin et al. 1982). Adult male *Mesoplodon* often are heavily scarred from fighting one another. The scars evidently are inflicted by the two functional teeth. Heyning (1984) showed that in *M. carlhubbsi* the scars are often parallel and probably result from attacks delivered with the mouth closed.

Reproductive data are very limited (Ellis 1980; Mead, Walker, and Houck 1982; Rice 1978d). In *M. bidens* mating usually takes place in late winter and spring, gestation lasts about 1 year, the newborn is about 213 cm long, and weaning occurs after about 1 year. *M. carlhubbsi* evidently gives birth during the summer after a gestation period of 12 months, the newborn averaging 250 cm. A female *M. hectori* and a young calf stranded near San Diego in May. A female *M. europaeus* and a young calf stranded in Florida in October. Female *M. densirostris* attain sexual maturity at about 9 years and are known to have lived for 27 years (Mead 1984). Very small numbers of *Mesoplodon* are taken by commercial fisheries (E. D. Mitchell 1975b).

CETACEA; Family PHYSETERIDAE

Sperm Whales

This family of two Recent genera and three species occurs in all the oceans and adjoining seas of the world. The genus *Kogia* sometimes is placed in a subfamily or family, the Kogiidae, distinct from that containing the genus *Physeter*. The sequence of genera presented here follows that of Hall (1981) and not that of Rice (1977).

The characters common to both genera include: a broad, flat rostrum; a great depression in the facial part of the skull to accommodate a highly developed spermaceti organ, which is bounded posteriorly by a high occipital crest; an S-shaped blowhole located on the left side of the snout; no preorbital or postorbital processes on the skull; a simple air sinus system; a left nasal passage for respiration and a right one apparently modified for sound production; numerous short grooves on the throat; a small, narrow lower jaw that ends well short of the anterior end of the snout; functional teeth only in the lower jaw; and relatively small pectoral appendages (Rice 1967; Hall 1981). The two genera differ markedly in size and certain other morphological characters. There are also major differences in known aspects of natural history.

The geological range of the Physeteridae is middle Miocene to Pleistocene in North America, early Miocene in South America, early Miocene to Pleistocene in Europe, late Pliocene in Japan, early Pliocene in Australia, and Recent in all oceans (Rice 1984).

CETACEA; PHYSETERIDAE; Genus KOGIA
Gray, 1846

Pygmy and Dwarf Sperm Whales

There are two species (Hall 1981; Nagorsen 1985; Pinedo 1987; Rice 1977):

K. breviceps (pygmy sperm whale), worldwide in tropical and temperate waters;
K. simus (dwarf sperm whale), worldwide in tropical and temperate waters.

Head and body length is 210–70 cm in *K. simus* and 270–340 cm in the larger *K. breviceps*. Pectoral fin length is about 40 cm, and expanse of the flukes is about 61 cm. The usual weight has been estimated at 136–272 kg for *K. simus* and 318–408 kg for *K. breviceps*, but the U.S. National Marine Fisheries Service (1981) indicated that the latter species sometimes exceeds 772 kg. Males are generally larger and heavier than females. Both species are dark steel gray on the back, shade to a lighter gray on the sides, and gradually fade to a dull white on the belly (Leatherwood, Caldwell, and Winn 1976).

The head is only about one-sixth the total length and resembles that of some delphinid genera in external outline, but the blunt snout projects beyond the lower jaw. Among cetaceans, the facial part of the skull of *Kogia* is among the shortest. The blowhole of *Kogia*, like that of *Physeter*, is on the left side of the head, but it is located on the forehead rather than toward the tip of the snout. *Kogia* resembles *Physeter* in having a spermaceti organ in the head and in having the functional teeth confined to the lower jaw. The teeth are sharp and curved; the number on each side of the lower jaw is 8–11 in *K. simus* and 12–16 in *K. breviceps* (Hall 1981). A few smaller, nonfunctional teeth may be present in the upper jaw

of *K. simus*. The dorsal fin is hooked and curved like a sickle; it is relatively high and located near the middle of the back in *K. simus*, while it is much lower and located posterior to the middle of the back in *K. breviceps*. As for *Physeter* (see account thereof), there has been debate regarding the function of the spermaceti organ, with the most convincing evidence in the case of *Kogia* being that it serves as an acoustical transducer for the production of sound pulses for echolocation (Nagorsen 1985).

A few observations of live animals indicate that *Kogia* is rather sluggish and sometimes floats motionless on the surface. Otherwise, the genus is known largely on the basis of strandings. Some of these records suggest that *K. breviceps* moves away from the shores of the eastern United States and South Australia during the summer and that *K. simus* migrates toward Japan in the summer. South African records, however, indicate that both species are present throughout the year and that neither makes major seasonal movements (Fitch and Brownell 1968; G. J. B. Ross 1979). The diet includes mostly cephalopods, as well as a variety of fish and some crustaceans. The stomachs of two specimens of *K. simus* from the northeastern Atlantic contained crustaceans known to inhabit depths of 500–1,300 meters (Nagorsen 1985). Analysis of diet by G. J. B. Ross (1984) showed that 93 percent of the cephalopods consumed by immature *K. simus*, as well as by calves and accompanying females, consisted of species that dwell over the continental shelf, while 71 percent of the food of adults consisted of oceanic species. Respective figures for *K. breviceps* were 45 percent and 85 percent. These figures suggest that juvenile and immature *Kogia*, especially those of *K. simus*, live closer inshore than do adults.

According to Rice (1978c), *K. breviceps* usually is seen in pairs, perhaps a female and a calf, or in groups of 3–5. Data summarized by G. J. B. Ross (1979) suggest that the school size of *K. simus* is 10 animals or fewer, that females with calves may group together, that immature animals may form their own groups, and that sexually mature males and females can be found in the same school. Intraspecific fighting has been witnessed. Ross also analyzed records of small and

Pygmy sperm whale *(Kogia breviceps)*, photos from Marineland of Florida.

near-term fetuses in *K. breviceps*, which indicate that in both the Northern and Southern hemispheres the mating and calving seasons extend over a period of approximately 7 months, from autumn to spring. There were too few records of *K. simus* to determine whether there is a distinct breeding season in that species, though there does appear to be an extended calving season of at least 5–6 months. In *K. breviceps* the gestation period may last over 11 months. A number of females have been found to be pregnant and lactating at the same time, and it seems likely that many females give birth every year. The single newborn is about 100 cm long in *K. simus* and 120 cm long in *K. breviceps*.

Pygmy and dwarf sperm whales seem to be rare, but occasionally they are taken and utilized by fishermen, and Sylvestre (1983) reported data on 33 stranded specimens that had been maintained alive for brief periods in captivity. The rarity may reflect an early, unrecorded exploitation. *Kogia* is lethargic and easy to approach, and a few records indicate that this genus was sometimes harpooned by whalers (Ellis 1980; E. D. Mitchell 1975*b*).

CETACEA; PHYSETERIDAE; **Genus PHYSETER**
Linnaeus, 1758

Sperm Whale

The single species, *P. catodon*, occurs in all oceans and adjoining seas of the world, except in polar ice fields (Rice 1977). Many authorities have followed Husson and Holthuis (1974) in using the name *P. macrocephalus* for this species, but Hall (1981) considered *P. catodon* to be the proper designation. In recent debate, detailed arguments have been presented by Schevill (1986) in support of *P. catodon* and by Holthuis (1987) in support of *P. macrocephalus*.

This is the largest toothed mammal in the world and is the most sexually dimorphic of all cetaceans. Sexual maturity is attained at a length of about 12.2 meters in males and 8.5 meters in females; growth subsequently continues for a number of years. Maximum head and body length is nearly 20 meters in males, though individuals over 15.2 meters are rare. Head and body length in females seldom exceeds 12 meters. Physically mature males usually weigh 35,000–50,000 kg, while females weigh only about one-third as much. The pectoral fin is about 200 cm long, and the expanse of the tail flukes is usually 400–450 cm. The color is gray to dark bluish gray or black. With increasing age, males may become paler and sometimes piebald. Most specimens have

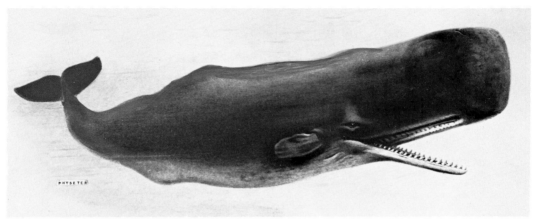

Sperm whale *(Physeter catodon)*, photos from *British Mammals*, Archibald Thorburn (top) and R. L. Pitman (bottom).

some white in the genital and anal regions and on the lower jaw (Best 1974, 1979; Ellis 1980; Leatherwood, Caldwell, and Winn 1976). Several completely white individuals are on record (Hain and Leatherwood 1982).

The head is enormous and squarish, especially in males. The skull is the most asymmetrical of any mammal's. The blowhole is located toward the tip of the snout and on the left side. The lower jaw is extraordinarily slender and has 16–30 strong, conical teeth on each side. These teeth, which may be more than 20 cm long, fit into sockets in the palate when the mouth is closed. Smaller, nonfunctional teeth are present in the upper jaw. *Physeter* is the only cetacean with a gullet large enough to swallow a man. There is no dorsal fin, but there is a longitudinal row of bumps on the posterior half of the back. The vertebrae number 50–51, and most of the neck vertebrae are fused. The blubber is up to 35 cm thick. A specimen of *Physeter* 13 meters long and weighing 20,000 kg had a heart that weighed 116 kg. The brain of *Physeter*, the largest of any mammal's, weighs about 9.2 kg (Rice 1967).

The most striking morphological feature, and the one that gives *Physeter* its vernacular name, is the huge spermaceti organ in the head, filled with up to 1,900 liters of waxy oil. At the anterior and posterior ends of the organ are air sacs. The left nasal passage goes along the left side of the spermaceti organ, leads directly to the blowhole, and also connects to the anterior air sac. The much narrower right nasal passage goes through the lower part of the spermaceti organ, opens into the anterior air sac through large internal lips, and also connects to the posterior air sac. This general structure has been known for centuries, but its exact functions are still being debated (Best 1979). It may either assist in evacuating the lungs prior to a dive and in absorbing nitrogen at extreme pressures, regulate buoyancy during deep diving, or reverberate and focus sounds. Clarke (1979), noting that *Physeter* apparently lies motionless at great depths to await and snap up squid, suggested that buoyancy is then controlled by varying the temperature and thus the density of the spermaceti oil. Cooling might be accomplished by drawing water into the nasal passages and around the oil. Norris and Harvey (1972) opposed this theory and suggested that the spermaceti organ is involved mainly in the production of burst pulses for long-range echolocation. According to this view, air goes through the right nasal passage, produces sounds at the lips leading into the anterior air sac, goes back up the left nasal passage, and is then pumped around again. The two air sacs may serve as sound mirrors.

Physeter produces a variety of sounds, including groans, whistles, chirps, pings, squeaks, yelps, and wheezes (Ellis 1980). The voice is very loud and can be heard many kilometers away by persons with proper underwater listening equipment. The most common sounds are series of clicklike pulses, which are generally much lower in frequency than those of the Delphinidae. The spermaceti organ may be used in the directional beaming as well as the production of these pulses, which are apparently involved in both echolocation and communication. In addition to other sounds, each whale has its own individualized "coda," a stereotyped, repetitive sequence of 3–40 or more clicks that is heard only when one animal meets another (Rice 1978c; Norris and Harvey 1972).

The sperm whale generally is found in waters conducive to the production of squid—at least 1,000 meters deep and with cold-water upwellings (Ellis 1980). The best areas are off the coasts of South America and Africa, in the North Atlantic and the Arabian Sea, between Australia and New Zealand, in the western North Pacific, and all along the Equator. Most animals stay between 40° N and 40° S, but during the summer the bachelor males of medium size move to between 40° and 50°, and at least some of the older males venture beyond 50° into or near arctic and antarctic waters. Although the other animals do not wander as far, there are northward shifts of concentration in the boreal summer and southward shifts in the austral summer as the movements of squid are followed (Berzin 1972; Best 1974, 1979; Rice 1978c). Certain major populations or stocks seem to move together on a seasonal basis, but migrations are not as predictable as those of the baleen whales. *Physeter* normally swims at a maximum of around 10 km/hr, but when pursued it can attain speeds of up to 30 km/hr (Berzin 1972). It sometimes lifts its head vertically out of the water, apparently to look or listen. Before a dive it gives a spout, characteristic in being directed obliquely forward to the left. It then lifts the tail flukes high in the air and descends almost vertically. There is commonly a prolonged initial submergence at around 360 meters for 20–75 minutes, followed by a series of shallow dives. Females do not generally go as deep as the larger males (Ellis 1980). There are at least 14 instances on record of sperm whales becoming

Sperm whale *(Physeter catodon)*, photo from American Museum of Natural History.

entangled in underwater communication cables, one at a depth of 1,135 meters. *Physeter* has been tracked by sonar to a depth of 2,500 meters. An individual killed after an 82-minute dive in waters 3,200 meters deep was found to have just consumed a kind of shark known to dwell on the bottom (Rice 1978c). Several species of sharks and other fish are included in the diet, but the predominant food is squid. Most of the squid taken are less than 1 meter long, but the stomach of one sperm whale contained a squid about 10.5 meters long. It has been estimated that each whale eats about 3 percent of its weight in squid per day (Ellis 1980). *Physeter* often bears the scars of combat with large squid.

Hundreds of individuals sometimes join in migratory schools, and there is one record of 3,000–4,000 being seen together off Patagonia. The basic social unit, however, is the mixed school, which has a year-round membership of adult females, calves, and some juveniles, usually 20–40 individuals in all. These units are stable, and there is evidence that the bonds between females last many years. If nursing calves are present, the group may be referred to as a nursery school. Shortly after weaning, young males and some young females begin to leave and to join in juvenile schools, usually with 6–10 individuals. The females eventually return to a mixed school before reaching puberty. Males from the juvenile schools and some coming directly from mixed schools form bachelor schools. When the bachelors are small, these schools contain 12–15 whales, though several groups may aggregate. As the bachelors grow larger, they divide into smaller groups, and then often into pairs and lone animals. Even if an association is maintained by older males, the individuals may stay some distance from each other. During the breeding season, each mixed school is commonly joined by 1–5 large males. The groups then become known as breeding or harem schools, and there usually is about 1 adult male for every 10 adult females. New evidence suggests that the males may remain only a few hours before moving off in search of another group with estrous females and that there is no long-term control of harems. The exact relationship between the adult males is not well understood. There is apparently some fighting for the right to join a breeding school, and the scars from such battles frequently cover the head of males. Nonetheless, it may be that several males establish a dominance hierarchy and that they share the females of one or more mixed schools. Such males may even have composed an organized bachelor school, prior to the breeding season. In any event, only 10–25 percent of the fully adult males in a population are able to get into a breeding school. The others move toward the Arctic and Antarctic. It is not known whether some or all of the adult males in a breeding school also eventually spend some time in polar waters (Best 1979; Ellis 1980; Gaskin 1970; Ohsumi 1971; Whitehead and Arnbom 1987).

The breeding season varies, but mating generally peaks in the spring, and calving in the fall, in both the Northern and Southern hemispheres. Females generally give birth at intervals of 5–7 years. Gestation has been variously calculated at 14–19 months. The single newborn is about 400 cm long and weighs nearly 1,000 kg. Some solid food is taken by the age of 1 year, but nursing usually lasts just over 2 years and seems to continue in some instances for up to 13 years. Sexual maturity is attained by females at 7–9 years. Males may have the physiological ability to produce offspring at less than 19 years, but they do not reach social maturity—the ability to enter a breeding school—until they reach 25–27 years. Full physical maturity comes at 35 years in males and 28 years in females. Some individuals are estimated to have lived up to 77 years (Berzin 1972; Best 1968, 1970a, 1974, 1979; Best, Canham, and Macleod 1984; Caldwell, Caldwell, and Rice 1966; Frazer 1973; Haley 1978; Ohsumi 1965, 1966; Rice 1978c).

The sperm whale has been hunted regularly by people since 1712. The meat is not generally used for human consumption, and of course there is no yield of baleen. The large teeth of *Physeter* were valued, however, as a medium for the artistic form of engraving and carving known as scrimshaw. A product unique to the sperm whale is ambergris, a waxy substance probably formed in the intestines from solid wastes coalescing around a matrix of indigestible material. It is used as a fixative and has the property of retaining the fragrance of perfumes. Apparently the heaviest mass of ambergris from a single animal weighed about 450 grams. Such an amount of the substance would have sold for about \$10–\$50 in the early 1960s, depending on color, but prices have fallen in recent years because of the development of synthetic fixatives. The most important product of the sperm whale is oil, though unlike that of baleen whales, it cannot be used in the manufacture of margarine. Sperm whale oil was once the major source of fuel for lamps and recently has served as a lubricant and as the base for skin creams and cosmetics. The oil of the spermaceti organ solidifies into a white wax upon exposure to air and is used in making ointments and fine, smokeless candles. It also has served as a high-quality lubricant for precision instruments and machinery and as a component of automatic transmission fluid.

An intensification of sperm whale hunting came around 1750 with the invention of the spermaceti candle and the development of onboard processing facilities, which allowed whaling vessels to remain at sea until their holds were filled with oil (Ellis 1980). For the next 100 years Americans dominated the industry, probably taking up to 5,000 sperm whales annually. This kill may not have been enough to reduce overall populations, but in the second half of the nineteenth century both sperm whale hunting and American whaling declined. Some of the suggested reasons are increasing costs, replacement of whale oil by kerosene, destruction of much of the U.S. fleet in the Civil War, and opening of the western Arctic and North Pacific to hunting of the bowhead whale (*Balaena mysticetus*). The development of the harpoon gun in the 1860s brought more hunting of the large baleen whales and less emphasis on *Physeter*.

By the 1930s a decline in some of the baleen species had resulted in renewed interest in sperm whale hunting. Floating factory ships could remain for lengthy periods within the prime habitat of *Physeter*, especially the North Pacific. In the 1936/37 season, for the first time in many years, the annual kill rose above 5,000. Subsequent efforts at international regulation were largely unsuccessful. In the 1950/51 season the take was 18,264 sperm whales, in 1963/64 it peaked at 29,255, and it remained above 20,000 in all but one season until 1975/76. About one-fourth of the total catch was made by shore-based operations, and the remainder by pelagic expeditions. Finally, in response to immense scientific and public concern, the International Whaling Commission began to reduce quotas substantially. The kill in the 1978/79 season was 8,536, the quota set for 1980/81 was only 1,849, and no kill was authorized for 1981/82. Moreover, the use of floating factories was banned in the hunting of *Physeter*, and most shore-based sperm whale fisheries closed down. All commercial whaling was halted by 1985 (Committee for Whaling Statistics 1980; Gulland 1974; Ellis 1980; McHugh 1974; Rice 1978c; U.S. National Marine Fisheries Service 1978, 1981, 1984, 1987).

There had been an authoritative consensus that protective measures had come in time to maintain large stocks of *Physeter* and that it remained by far the most numerous of the great whales. In 1946 the estimated number of males over 9.2 meters long (the minimum legal limit for catching) and sexually mature females was about 1.1 million. Even into the 1980s the estimate for these categories was just over 700,000,

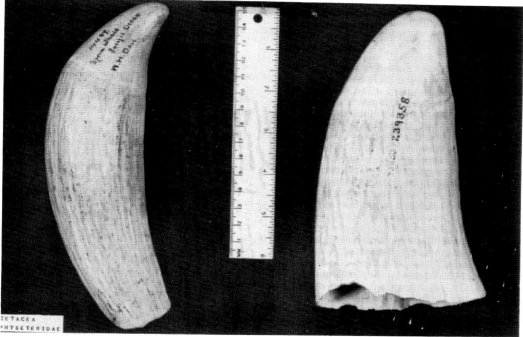

Top, dead sperm whale *(Physeter catodon)*, photo by Warren J. Houck. Bottom, teeth of *P. catodon*, photo by P. F. Wright of specimen in U.S. National Museum.

and the total population was thought to be nearly twice as great (U.S. National Marine Fisheries Service 1978, 1981, 1987). Then, at the 1989 meeting of the International Whaling Commission came the most shocking mammalian conservation news in many years. It was reported that direct counts by whaling cruises over the previous eight years had led to new and drastically lower estimates for most of the large whales. *Physeter* now is thought to number only 5,000–10,000 individuals in the Southern Hemisphere, having been nearly exterminated by the intensive commercial kills that continued through the 1970s (*Marine Mammal News* 15[1989]:5). There doubtless are fewer in the Northern Hemisphere. The official USDI classification of *Physeter* as endangered, which had been criticized by some parties as

being alarmist, now can be recognized as fully accurate. *Physeter* also is on appendix 1 of the CITES.

CETACEA; Family ESCHRICHTIDAE; Genus ESCHRICHTIUS
Gray, 1864

Gray Whale

The single known genus and species, *Eschrichtius robustus*, occurs in coastal waters from the Sea of Okhotsk to southern Korea and Japan and from the Chukchi and Beaufort seas to

Gray whales *(Eschrichtius robustus)*, painting by Richard Ellis.

the Gulf of California (Rice 1977). Other populations apparently once existed in the North Atlantic, there being records from the east coast of the United States, western Europe, the Baltic Sea, and Iceland (Fraser 1970; Mead and Mitchell 1984; Mitchell and Mead 1977). Some of the Atlantic records are based on material that is subfossil but from historical times. There also are a few old specimens indicating that the western Pacific population once extended as far south as Hainan (Wang 1984). Geologically, the family Eschrichtidae is known only from the Recent and from a single Pleistocene specimen of *E. robustus* about 50,000–120,000 years old found in southern California (Barnes and McLeod 1984). Hall (1981) used the name *E. gibbosus* in place of *E. robustus.*

At sexual maturity head and body length is about 11.1 meters, at physical maturity the length averages 13 meters in males and 14.1 meters in females, and the maximum reliably recorded length is 14.3 meters in males and 15 meters in females (Rice and Wolman 1971). Pectoral fin length is approximately 200 cm, and expanse of the tail flukes is about 300 cm. There is no dorsal fin, but there are 8–14 low humps along the midline of the lower back. Adults weigh 20,000–37,000 kg. Coloration is black or slaty gray, with many white spots and skin blotches, some being discolored patches of skin and others being areas of white barnacles.

The snout is high and rigid, and the throat has two or three (rarely four) short, shallow, curved furrows. *Eschrichtius* is a baleen whale and has no teeth. The baleen is yellowish white in color and is arranged in series of 130–80 plates on each side of the mouth. The largest plates are 34–45 cm long. The two rows of plates do not meet in front, as they usually do in the

family Balaenopteridae. There are 56 vertebrae in *Eschrichtius*, and the neck vertebrae are separate.

The gray whale is found primarily in shallow water and generally remains closer to shore than any other large cetacean. The eastern Pacific population makes an annual migration of over 18,000 km. From late May to early October this population concentrates in the shallow waters of the northern and western Bering Sea, the Chukchi Sea (between northern Alaska and Siberia), and the Beaufort Sea (off northeastern Alaska). Some individuals, however, spend the summer farther south, scattered along the coast of Oregon and northern California. From October to January the main part of the population moves down the east side of the Bering Sea, goes through Unimak Pass in the Aleutians, and proceeds down the west coast of North America. In January and February most whales are along the west coast of Baja California and on the eastern side of the Gulf of California off Sonora and Sinaloa. In these areas are five bays and lagoons, within which the young are born. From late February to June the population migrates northward to its summering sites in the Arctic (Rice and Wolman 1971; Wolman and Rice 1979).

During the southward phase of the migration of the eastern Pacific population the animals are more concentrated and may move somewhat nearer to the shore than during the northward phase. For example, most individuals pass San Diego during a six-week period from late December to early February, and in certain past years 95 percent of the population traveled 3–5 km offshore. At other points the whales sometimes move within 1 km of land. The migration is therefore witnessed by large numbers of people. Unfortunately, perhaps, increasing boat traffic seems to have forced the whales to stay farther from shore (Wolman and Rice 1979). Observations by Leatherwood (1974) indicated that half the population passed more than 64 km off San Diego.

Considerably less is known about the western Pacific population, which has been reduced to very small numbers. Apparently it spent the summer in the Sea of Okhotsk and then migrated through the Sea of Japan and along the Pacific coast of Japan to wintering sites off the southern coast of Korea and in the Inland Sea of Japan. Some animals may once have migrated southward along the coast of China to breeding waters around Hainan. The extinct population of the eastern Atlantic probably summered in the Baltic Sea and wintered along the Atlantic and Mediterranean coasts of southern Europe and North Africa. The western Atlantic population may have bred and given birth in the shallow lagoons and bays of south-central and southeastern Florida before migrating northward (Omura 1974, 1984; Reeves and Mitchell 1988; Rice and Wolman 1971; Wang 1984).

The normal swimming speed of the gray whale is 7–9 km/hr (Rice and Wolman 1971), but when pursued it may reach about 13 km/hr. Migrating individuals usually submerge for four or five minutes and then surface to blow three to five times. The spout rises to a height of about 300 cm. The tail flukes usually appear above the water just before a deep dive but not before a shallow dive. The back is not arched before diving, as in the humpback whale (Megaptera). The gray whale often seems to play in heavy surf and in shallow water along the shore, sometimes throwing itself clear of the water and falling back on its back or side. It occasionally becomes stranded in less than 1 meter of water but refloats on the next tide, without injury. Visual stimuli appear to be involved in orientation during migration. The habit of spyhopping—lifting the head vertically out of the water and appearing to look around—is commonly observed in Eschrichtius and may be involved in social interaction (Samaras 1974). A variety of sounds has been reported, including grunts, groans, rumbles, whistles, and clicks. The clicks seem to be of a frequency too low for use in precise echolocation but may possibly assist in sonar navigation or in the detection of dense concentrations of food organisms (Fish, Sumich, and Lingle 1974).

Feeding takes place primarily on the bottom, with individuals apparently sometimes plowing their heads sideways through the mud or sand to stir up prey (Ellis 1980). Water and organisms are sucked into the mouth, and then the water is forced out, leaving the food trapped within the baleen. The diet consists mainly of gammaridean amphipods (small crustaceans), especially Ampelisca macrocephala, a creature about 25 mm long and found on sandy bottoms at depths of 5–300 meters (Rice and Wolman 1971). Other crustaceans, certain mollusks and worms, and small fish are also eaten. A captive yearling female consumed about 900 kg of squid per day (Ray and Schevill 1974). Available evidence suggests that there is little or no feeding, except when the animals are in the northern part of their annual migratory route, and that fasting lasts up to six months (Rice 1978b). Individuals moving south past San Francisco lost 11–39 percent of their body weight by the time they moved north past the city (Rice and Wolman 1971).

Aggregations of up to 150 individuals have been seen in the arctic feeding waters (Rice 1967), but Eschrichtius is not a particularly social cetacean. It usually migrates alone or in groups of 2–3, though sometimes up to 16 travel together. Moreover, there is segregation by age and sex during migrations. Generally females precede males and adults precede immature animals. Newly pregnant females are the first to leave on the northward migration (Rice 1978b; Rice and Wol-

man 1971). Segregation also has been reported at the wintering sites, with calving and nursing females remaining well within protected lagoons and males positioning themselves at the entrances (Ellis 1980). Individuals have been observed to aid others that are injured or giving birth by pushing them to the surface so they can breathe. Females are reportedly very protective of their young, even to the point of attacking whaling boats.

Females have a 2-year reproductive cycle. Most enter estrus and mate during a 3-week period in late November and early December, while still migrating south, but some do not mate until in the wintering lagoons or even on the northward migration. Births occur mainly from late December to early February. Gestation lasts about 13 months, lactation lasts about 7 months, and there are then 3–4 months of anestrus. The single newborn averages about 490 cm in length and 500 kg in weight. It is weaned around August in the summer feeding area, after it has grown to a length of 850 cm. The mean age of puberty is about 8 years in both sexes, and full physical maturity comes at about 19 years in males and 17 years in females (Rice 1978b; Rice and Wolman 1971; Yablokov and Bogoslovskaya 1984). Maximum estimated longevity is 70 years (Haley 1978).

Like various other large cetaceans, the gray whale has been killed by people for its oil, meat, hide, and baleen. It was hunted in ancient times by the native people of northwestern North America and eastern Siberia and probably also by Europeans and Japanese. Because of its migrations close to shore, it was comparatively simple to locate and secure. According to Krupnik (1984), aboriginal whaling began to develop in the northern Pacific nearly 2,000 years ago and eventually became the major economic activity along most coasts of the Chukchi, Bering, and Okhotsk seas. The activity declined by the mid-nineteenth century in connection with a reduction in the number of native people, cultural changes, and usurpation of whaling by Americans. The European gray whale population may have disappeared around A.D. 500, though there is a likely record for Iceland in the early seventeenth century. The western Atlantic population apparently survived until about 1700 (Fraser 1970; Mead and Mitchell 1984; Mitchell and Mead 1977; Rice and Wolman 1971).

The western Pacific population seems to have always been rather small, with about 50 individuals being taken annually in the eighteenth and nineteenth centuries by Japanese whalers. The branch that bred in the Inland Sea of Japan probably contained fewer than 1,000 whales. It was gone by 1900, the last survivors perhaps being driven away by increasing boat traffic and industrialization (Omura 1974, 1984). The branch that bred off southern Korea may have numbered 1,000–1,500 when modern whaling began in that area around the turn of the century. By 1933 the Korean population appeared to be extinct, but there were 67 known kills from 1948 to 1966 and sightings in Korean waters as late as 1974 (Brownell and Chun 1977; Rice and Wolman 1971; Wolman and Rice 1979). An individual was seen off the Pacific coast of Japan in 1982 (Furuta 1984), and more recently others have been observed in the Sea of Okhotsk and near the Kuril Islands. Some authorities have considered these reports to refer to stray animals from the eastern Pacific population, but Reeves and Mitchell (1988) suggested that they represent survival of the original eastern stock and estimated that this population numbers in the tens or low hundreds.

The wintering lagoons of the eastern Pacific population were discovered by American whalers in 1846. Shore-whaling stations were established in the area, and from 1846 to 1874 the known kill was 10,800 animals. By about the turn of the century regular shore whaling stopped, and the population seemed extinct, but there may still have been several

Gray whale *(Eschrichtius robustus):* Top, photo by David Withrow; Bottom, photo by David Rugh.

thousand individuals left. Shortly thereafter, factory ships came into use for the hunting of the gray whale. From 1921 to 1947, 1,153 individuals are known to have been killed, mainly by Norwegian, Japanese, and Russian vessels. Since 1946 the species has received protection under the International Whaling Convention, except that about 160 individuals have been taken legally each year by Siberian Eskimos, and also a few by Alaskan natives. Several more perish accidentally after becoming entangled in fishing gear. The eastern Pacific population evidently increased steadily in response to protection and currently is estimated to contain 18,000–21,000 whales, perhaps not far below the number present before exploitation began. The main current problem is disturbance of the animals and their habitat, especially in the calving lagoons, by industrialization, shipping, and even well-meaning tourists, who follow the whales in motorboats (Reeves and Mitchell 1988; Rice 1978b; Rice and Wolman 1971; Rice, Wolman, and Braham 1984; Storro-Patterson 1977; U.S. National Marine Fisheries Service 1987, 1989; Wolman and Rice 1979). The gray whale is listed as endangered by the USDI and is on appendix 1 of the CITES.

CETACEA; **Family BALAENOPTERIDAE**

Rorquals

This family of two Recent genera and six species occurs in all the oceans and adjoining seas of the world.

Baleen from a whale (*Balaenoptera* sp.), photo by P. F. Wright of specimen in U.S. National Museum of Natural History.

Head and body length of sexually mature adults is 6.7–31.0 meters, and weight is estimated to range up to 160,000 kg (Rice 1967). In each species the females are larger than the males. The pectoral fin is tapering, and the dorsal fin is located on the posterior part of the back. The rostrum is broad and flat, and the sides of the lower jaw bow outward. The Balaenopteridae are baleen whales and have no teeth past the embryonic stage. There is one row of baleen plates on each side of the top of the mouth, the two rows usually joining anteriorly. Each row may consist of more than 300 plates which range in vertical length from 20 to 101 cm. The vertebrae number 42–65; the neck vertebrae are generally separate.

The term "rorqual" is Norwegian and means "tube whale" or "furrow whale" (Ellis 1980). It refers to the longitudinal folds or pleats, 10–100 in number and 25–50 mm deep, that are present on the throat, chest, and, in some species, belly. These furrows allow the throat to expand enormously and so greatly increase the amount of material that the whale can take in when feeding.

The Balaenopteridae are distinguished from the Balaenidae by the presence of the throat and chest furrows, the more elongate and streamlined body form, the relatively smaller head, the more tapering pectoral fin, the softer and less massive tongue, and the usually shorter and less flexible baleen. Unlike the genera *Balaena* and *Eschrichtius*, the Balaenopteridae have a dorsal fin.

The feeding habits and dietary preferences of the Balaenopteridae are associated with the characteristics of the head, mouth, tongue, and baleen plates. The sei whale (*Balaenoptera borealis*) has finely meshed baleen fringes and sometimes uses the feeding method known as skimming, as do the Balaenidae (Ellis 1980). In skimming the animal swims slowly through swarms of tiny food organisms with its mouth open and its head above water to just behind the nostrils. When a mouthful of organisms has been filtered from the water by the baleen plates, the whale dives, closes its mouth, and swallows the food. The other rorquals use the feeding method known as swallowing: the animal may turn on its side and often has part of its head above the water; after taking in a great amount of food organisms and water, the whale forces the water out with its tongue, leaving the food trapped in the baleen.

All species are migratory, most giving birth in warm areas and feeding predominantly in colder waters. The diet depends partly on location. The most important foods are shrimplike

creatures of the family Euphausiidae (mainly the genera *Euphausia*, *Thysanoessa*, and *Meganyctiphanes*) and copepods (*Calanus* and *Metrida*). These crustaceans form part of the zooplankton, the mass of minute floating and weak-swimming animals found near the surface of the ocean. In colder areas there are fewer kinds of plankton, but the numbers of each kind are greater. During the summer, shoals of these organisms concentrate in the upper layers of polar waters, and the whales are attracted to them. In the Antarctic the diet of the minke (*B. acutorostrata*), fin (*B. physalus*), blue (*B. musculus*), and humpback (*Megaptera*) whales consists mainly of the shrimp *Euphausia superba* (commonly called krill). Outside of the Antarctic the blue whale feeds predominantly on other euphausiids, but the fin, minke, and humpback take substantial amounts of copepods and fish. The fine, fleecy baleen of the sei whale is suited for trapping smaller organisms than those eaten by other rorquals. The sei feeds on *Euphausia superba* in the Antarctic, but also on the amphipod *Parathemisto gandichaudi*, and in other areas it seems to prefer copepods. In contrast, the Bryde's whale (*B. edeni*) has coarse baleen, does not enter the Antarctic, and feeds mainly on fish (Ellis 1980).

All rorquals have been hunted by people for oil, meat, baleen, and other products. The most intensive exploitation has taken place in antarctic waters, where large numbers of most species gather for part of the year to feed on plankton. The whaling activity in this region followed a systematic pattern (D. G. Chapman 1974; Gulland 1974, 1976; McHugh 1974). The first species to be depleted was the humpback, which has a localized, coastal distribution and could be easily taken from the island whaling stations established in the first years of the twentieth century. The introduction in the 1920s of factory ships with slip sternways, which allowed whalers to operate independently of land facilities, led to increased killing of other species.

The blue, fin, and sei whales, in that order, become smaller in size and concentrate farther from the antarctic ice pack. Exploitation proceeded in this same order. The blue whale, the most valuable species in relation to individual hunting effort, received initial maximum attention and was evidently in decline by the late 1930s. Whalers then centered their activity somewhat farther north, and the fin whale became the mainstay of the industry for the next 30 years. Meanwhile, efforts at protection were largely ineffective. The International Whaling Commission established quotas on the basis of "blue whale units." Each unit equaled 1 blue whale, 2 fin whales, 2.5 humpbacks, or 6 sei whales. Since quotas were assigned by units rather than by species, it remained more profitable to hunt the larger species. By the 1960s, however, the blue whale seemed to be nearly extinct, and the fin whale greatly reduced in numbers, and attention was then directed still farther north, to the smaller sei whale. Another factor in the switch to the sei was a lower demand for whale oil, along with an increase in the value of the meat. The sei has less blubber and relatively more and better meat than the larger species. Concern for the future of the sei, and the setting of lower quotas specifically for it and the larger whales, contributed to an increased kill of the minke whale, smallest of the rorquals, in the 1970s. There also was a rise in the take of Bryde's whale, a species considerably larger than the minke but one that does not migrate to the Antarctic.

All large-scale commercial whaling ended by the late 1980s, and there was hope that major stocks of most of the rorquals had survived and would form the basis for recovery of the once great Antarctic herds. However, at the 1989 meeting of the International Whaling Commission came the shocking new estimates that only 500 blue and 2,000–3,000 fin whales survived. These low figures added weight to earlier studies showing that there had been major changes in the

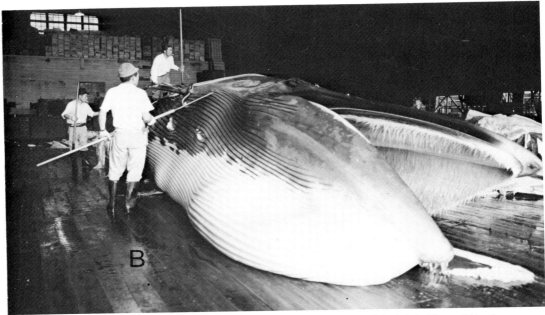

A. Blue whale *(Balaenoptera musculus)*, photo by K. C. Balcomb. B. Sei whale *(B. borealis)*, photo by Warren J. Houck.

Antarctic ecosystem as a result of the decline of the great whales. This reduction meant that more than 100 million metric tons of krill, formerly eaten by the large whales, had become available to other animals (Laws 1985). As a result, there have been remarkable increases in such species as squid, penguins, the crabeater seal *(Lobodon carcinophagus)*, and the Antarctic fur seal *(Arctocephalus gazella)*. As these other species became more abundant, the krill surplus was taken up and the recovery potential for the whales further reduced. There now also is a growing human harvest of krill and thus a threat to the entire ecosystem.

The geological range of the Balaenopteridae is middle Miocene to Pleistocene in North America and Europe, Miocene in eastern Asia, early Pliocene and Pleistocene in South America, and Recent in all oceans (Rice 1984).

CETACEA; BALAENOPTERIDAE; **Genus**
BALAENOPTERA
Lacépède, 1804

Minke, Bryde's, Sei, Fin, and Blue Whales

There are five species (Rice 1977):

B. acutorostrata (minke whale), all oceans and adjoining seas;

B. edeni (Bryde's whale), tropical and warm temperate waters of the Atlantic, Indian, and Pacific oceans and adjoining seas;

B. borealis (sei whale), all oceans and adjoining seas except in tropical and polar regions;

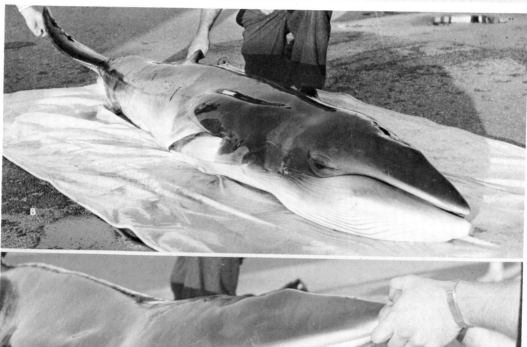

A. Fin whale *(Balaenoptera physalus)*, photo from Field Museum of Natural History. B & C. Minke whale *(B. acutorostrata)*, photos from Marineland of the Pacific.

B. *physalus* (fin whale), all oceans and adjoining seas, but rare in tropical waters and among pack ice;

B. *musculus* (blue whale), all oceans and adjoining seas.

In addition to the characters set forth in the familial account, *Balaenoptera* is distinguished by a relatively slender body, a compressed tail stock that abruptly joins the flukes, a pointed snout, a relatively short pectoral fin (usually 8–11 percent of head and body length), and a usually sickle-shaped dorsal fin. Additional information is provided separately for each species.

Balaenoptera acutorostrata (minke whale).

In the more common form, head and body length at sexual maturity averages about 7.3 meters in males and 7.9 meters in females. At physical maturity the respective averages are about 8.3 and 8.8 meters (E. D. Mitchell 1975b; Ohsumi and Masaki 1975; Ohsumi, Masaki, and Kawamura 1970). Maximum size is around 10.2 meters in length and 10,000 kg in weight. As in all rorquals, females average slightly larger than males. The upper parts are dark gray to black, and the underparts are white. There may also be a white patch on the pectoral fin. The rostrum is markedly triangular in shape and is shorter than that of any other rorqual. The dorsal fin is high and curved strongly backward. There are 50–70 ventral grooves, and these do not extend as far as the navel. On each side of the upper part of the mouth is a row of about 300 baleen plates mostly yellowish white in color but sometimes asymmetrical in pattern (Ellis 1980; Leatherwood, Caldwell, and Winn 1976).

There recently has been recognition of a second form of minke whale in waters near Australia, New Zealand, New Caledonia, and South Africa (Arnold, Marsh, and Heinsohn 1987; Best 1985). Referred to as the dwarf or diminutive form, this whale may reach sexual maturity when 6–7 meters long and full size at 7–8 meters. There is a large white area on the base of the pectoral fin and the adjacent thorax. Coloration of the upper surface of the head and rostrum is paler than in the other form. The baleen plates are predominantly light and do show an asymmetrical pattern. There is uncertainty whether the dwarf minke whale may represent a subspecies or even a separate species.

The minke whale is not as coastal an animal as *Eschrichtius* but seldom goes farther than 160 km from land and often enters bays and estuaries. It also moves farther into polar ice fields than any other rorqual. There are a number of discrete populations, at least some of which make extensive migrations. The populations of the Northern Hemisphere winter in tropical waters, but during the spring and summer they spread northward, often beyond the Arctic Circle. While these populations are in the north, those of the Southern Hemisphere are in the tropics for the austral winter. When the northern populations return to the tropics in the boreal autumn, the southern populations set out in the direction of Antarctica (Ellis 1980; Mitchell 1978; U.S. National Marine Fisheries Service 1978).

The minke whale is a fast swimmer and is the most acrobatic member of its genus. It has a tendency to approach ships and often leaps completely out of the water (Leatherwood, Caldwell, and Winn 1976; Mitchell 1978). Its spout, however, is invisible. It has a wide variety of vocalizations, including a series of thumps that may be distinctive for each individual (Thompson, Winn, and Perkins 1979). In the Antarctic the diet consists almost exclusively of plankton (see familial account), but in the Northern Hemisphere *B. acutorostrata* also eats squid, herring, cod, sardines, and various other kinds of small fish (Ellis 1980).

This rorqual is generally seen alone or in groups of two to four individuals, but several hundred sometimes gather in the vicinity of abundant food (Leatherwood, Caldwell, and Winn 1976; E. D. Mitchell 1975a). There is some segregation by age and sex off Newfoundland, with adult males evidently remaining farther from shore than females and young (Mitchell and Kozicki 1975). Observations of a population along the coast of Washington revealed the presence of three exclusive adjoining ranges in about 600 sq km, each shared by at least seven whales, though there was no indication of group relationships or territoriality (Dorsey 1983).

There is some question about the reproductive cycle. Mitchell (1978) gave the calving interval as only 1 year, and Mitchell and Kozicki (1975) found over 85 percent of mature females off Newfoundland to be pregnant. The U.S. National Marine Fisheries Service (1978), however, stated that females are now known to give birth once every 2 years. The breeding period is extensive (E. D. Mitchell 1975a). In the North Atlantic mating occurs from December to May, and calving from October to March. In the North Pacific reproduction occurs throughout the year, with mating peaks in January and June and birth peaks in December and June. The gestation period is 10.0–10.5 months. There is normally a single offspring about 280 cm long. Lactation is thought to last less than 6 months. Sexual maturity is attained at 7–8 years, and physical maturity when males are 18–20 years and females 20–22 years (Ohsumi and Masaki 1975; Ohsumi, Masaki, and Kawamura 1970). Maximum longevity is estimated at 47 years (Haley 1978).

The minke whale has been hunted in Norway since the Middle Ages and was regularly taken by local fisheries there and in some other countries from the 1920s to 1970s but long was not considered of major commercial importance (E. D. Mitchell 1975b). With the decline of the larger rorquals in the 1960s, the minke became subject to more intensive hunting, especially in the Antarctic. In the 1976/77 season the worldwide kill peaked at 12,398 individuals, surpassing the take of any other species (Committee for Whaling Statistics 1980). With growing public demands for an end to commercial whaling, and under reduced quotas set by the International Whaling Commission, the kill dropped to 6,055 during the 1983/84 season (Committee for Whaling Statistics 1984). Kills continued at this level for several more years, despite the commission's official moratorium on commercial whaling starting in 1985, because several nations chose to issue a formal objection to the ban. Such measures were dropped by 1988, however, and the annual kill is now fewer than 1,000 minke whales that are taken either for aboriginal subsistence or for alleged research purposes.

The U.S. National Marine Fisheries Service (1989) reported the current population estimate of minke whales to be 258,300 in the Southern Hemisphere, 13,500 in the North Pacific, and 44,000–60,000 in the North Atlantic. Ohsumi (1979), however, gave an estimate of 416,700 for the Southern Hemisphere and suggested that this figure represented an eightfold increase since the 1930s, because the minke whale had multiplied in the absence of competition from the larger rorquals. All populations of *B. acutorostrata* except that off western Greenland are on appendix 1 of the CITES.

Balaenoptera edeni (Bryde's whale).

Head and body length at sexual maturity averages 12.2 meters in males and 12.5 meters in females (U.S. National Marine Fisheries Service 1978). Maximum head and body length is 14 meters, and dorsal fin height is up to 46 cm (Leatherwood, Caldwell, and Winn 1976). The coloration is dark blue gray above and paler below. *B. edeni* resembles *B. borealis* but can be distinguished by the presence of two dorsal ridges running from the tip of the snout to beside the blowhole, in addition to another ridge down the middle of the snout. The dorsal fin is strongly sickle-shaped. There are approximately

45 ventral grooves, extending to the navel. On each side of the mouth is a row of about 300 baleen plates, and often the rows do not join anteriorly. The plates are relatively short; the front ones are whitish and the rear ones are dark (Ellis 1980).

Certain populations inhabit waters of high productivity near the shore, while others are mainly pelagic. Unlike other rorquals, *B. edeni* makes no large-scale movements toward the poles in the spring. Tropical populations may be sedentary, but those in temperate waters seem to make limited migrations in response to shifting food sources. *B. edeni* has a tendency to approach ships and is a relatively deep diver. Its diet consists mainly of small schooling fish, such as anchovies and sardines, but in some areas it also feeds extensively on shrimplike crustaceans of the family Euphausiidae (Ellis 1980; Leatherwood, Caldwell, and Winn 1976; U.S. National Marine Fisheries Service 1978).

Bryde's whale is often seen in groups, usually of 2–10 individuals but sometimes of over 100. The reproductive season is usually winter, but in some regions, as in the waters off South Africa, it extends throughout the year. Females generally give birth every other year to a single offspring about 335 cm long. The gestation period is 12 months, and lactation has been estimated to last 6 months. The age of sexual maturity is 7–10 years and that of physical maturity is 15–18 years. Some individuals are estimated to have lived up to 72 years (Ellis 1980; Haley 1978; Lockyer 1984; Mitchell 1978; U.S. National Marine Fisheries Service 1978, 1981).

Bryde's whale long has been hunted along the coasts of Japan, South Africa, and Baja California but did not receive major commercial attention until the 1970s, subsequent to the decline of most other rorquals. The maximum recorded kill was 1,882 in the 1973/74 season (Committee for Whaling Statistics 1980). The 1981/82 quota set by the International Whaling Commission was 1,392. Quotas subsequently dropped and were halted altogether with the formal moratorium on commercial whaling after 1985. The U.S. National Marine Fisheries Service (1989) estimated the Pacific Ocean to have 30,200–55,500 *B. edeni.* Estimates for other regions are not available. The species is on appendix 1 of the CITES.

Balaenoptera borealis (sei whale).

Head and body length at sexual maturity averages 13.1 meters in males and 13.7 meters in females (U.S. National Marine Fisheries Service 1978). The biggest specimen on record, a female, was 20 meters long, but most individuals now measure 12.2–15.2 meters. Adults typically reach a weight of about 20,000 kg and rarely up to 40,000 kg (Horwood 1987). The general coloration is dark steel gray, with irregular white markings ventrally. The dorsal fin, set farther back than on *B. acutorostrata*, is 25–61 cm in height. There are 38–56 ventral grooves, and these end well before the navel. On each side of the upper part of the mouth is a row of 300–380 baleen plates. The color of the plates is ashy black, but the fine inner bristles are whitish (Ellis 1980; Leatherwood, Caldwell, and Winn 1976).

The sei whale is a pelagic species, normally occurring far from shore. Like most rorquals, it feeds in temperate and subpolar regions in summer and migrates to subtropical waters for the winter. It does not, however, go as far into the Antarctic as do the blue and fin whales. It is one of the fastest cetaceans, reportedly being able to reach speeds of up to 50 km/hr. It is not usually a deep diver. Periods of submergence generally last 5–10 minutes. When feeding, the sei whale remains near the surface, often twisting on its side as it swims through swarms of prey. Unlike other rorquals, it sometimes obtains food by the skimming method (see familial account). It generally eats smaller organisms than do the other rorquals, mostly copepods and amphipods, but also takes euphausiids and small fish. Estimates of daily consumption per whale are about 200–900 kg (Banfield 1974; Ellis 1980; Horwood 1987; Kawamura 1973, 1974; Mitchell 1978). *B. borealis* is not known to use echolocation to search for prey but has been heard to emit a "sonic burst of 7–10 metallic pulses" (Thompson, Winn, and Perkins 1979).

Groups usually consist of two to five individuals, but

Sei whale *(Balaenoptera borealis)*, painting by Richard Ellis.

sometimes thousands may gather in the vicinity of abundant food (Ellis 1980; Leatherwood, Caldwell, and Winn 1976). According to Banfield (1974), there is a protracted mating season that extends from November to February in the Northern Hemisphere and from May to July in the Southern Hemisphere. Females generally give birth every other year, but Masaki (1978) reported that the percentage of pregnant females being taken has recently increased, perhaps as a natural compensatory response to human exploitation. The gestation period was reported as 10.5 months by Masaki (1976), 11.5 months by Frazer (1973), and 12 months by Ellis (1980). There are rare cases of multiple fetuses, but normally a single offspring about 450 cm long is produced. Lactation lasts 6–7 months. Sexual maturity is attained at around 10 years, and physical maturity at 25 years (Mitchell 1978). Some individuals are estimated to have lived up to 74 years (Haley 1978).

The sei whale has been taken regularly by people since the 1860s but did not achieve major commercial importance for another 100 years. The annual kill in antarctic waters did not exceed 1,000 individuals until 1950. There was a pronounced increase in the kill during the 1950s and 1960s in conjunction with a declining harvest of blue and fin whales (see familial account). The worldwide take of the sei whale peaked at 25,454 in the 1964/65 season (McHugh 1974); the take then steadily dropped to only 150 in 1978/79 (Committee for Whaling Statistics 1980). The annual quota set by the International Whaling Commission for most subsequent seasons was 100, and after 1985 all commercial whaling was officially halted. The original number of sexually mature sei whales, exclusive of those in the North Atlantic, is estimated at 200,000; the total population, including immature animals, would be about 50 percent larger (U.S. National Marine Fisheries Service 1978). Recent estimates given by the U.S. National Marine Fisheries Service (1989) are 24,000 in southern waters, 4,600 in the North Atlantic, and 22,000–37,000 in the Pacific. However, reports at the 1989 meeting of the International Whaling Commission suggest that far fewer now survive. *B. borealis* is listed as endangered by the USDI and is on appendix 1 of the CITES.

Balaenoptera physalus (fin whale).

Head and body length at sexual maturity averages about 17.7 meters in males and 18.3 meters in females. At physical maturity the respective averages are about 19 and 20 meters. Maximum known length is 25 meters in males and 27 meters in females. Adults have not been weighed, but calculations suggest that a 25-meter animal would weigh about 70,000 kg (Gambell 1985). The strongly curved dorsal fin is up to 61 cm high, and the expanse of the tail flukes is about 25 percent of the head and body length. The general coloration is brownish gray above and white below; the pattern, however, is asymmetrical, especially in that the lower jaw is white on the right and dark on the left. *B. physalus* is slimmer than *B. musculus*; if two individuals, one of each species, are the same length, the *B. musculus* will be much heavier. Indeed, even though *B. physalus* is the second longest cetacean, it often weighs less than the thickset *Balaena*, *Eubalaena*, and *Physeter*. The rostrum of *B. physalus* is sharply pointed. There are on average 85 ventral grooves, and these terminate at the level of the navel. On each side of the upper part of the mouth is a row of 350–400 baleen plates. Those plates in the forward third of the right row are whitish, but those in the rear two-thirds of the right row, and all of those in the left row, are dark grayish blue (Ellis 1980; Leatherwood, Caldwell, and Winn 1976; Mitchell 1978; U.S. National Marine Fisheries Service 1978).

The fin whale is a pelagic species, seldom seen in water less than 200 meters deep. Numerous discrete populations have been identified, most of which are known to be highly migratory. In the spring and early summer, populations generally move into cold temperate and polar waters to feed, and in the autumn they return to warm temperate and tropical regions. Since the seasons in the northern half of the world are opposite those of the southern half, the fin whale populations of one hemisphere do not meet those of the other in equatorial waters (Banfield 1974; D. G. Chapman 1974; Ellis 1980).

The fin whale is among the fastest of cetaceans, being able to sustain a speed of around 37 km/hr. It is probably a deeper diver than the blue and sei whales, sometimes reaching depths of at least 230 meters and remaining underwater for up to 15 minutes. It occasionally leaps completely out of the water. A swallowing method of feeding is employed, during which the throat becomes distended to nearly double normal diameter (see familial account). There is very little, if any, feeding during the fall and winter, when the whales are in lower latitudes. In the Antarctic the diet consists almost entirely of small, shrimplike crustaceans of the family Euphausiidae. In northern waters *B. physalus* also eats other crustaceans and various kinds of small fish (Ellis 1980; Leatherwood, Caldwell, and Winn 1976). Banfield (1974) indicated that the fin whale uses echolocation to find its food, but this activity has not actually been demonstrated. The species is known to produce a great variety of low-frequency sounds and probably also some high-frequency pulses (Thompson, Winn, and Perkins 1979).

The fin whale may be monogamous and is regularly seen in pairs. Usual group size is 6–7 individuals, there are often up to 50, and occasionally as many as 300 travel together on migrations. Both mating and calving occur during the late fall or winter, when the whales are in warm waters. Each mature female gives birth every 2–3 years. The gestation period is 11.0–11.5 months. There are rare records of as many as six fetuses, but normally a single offspring is produced. The newborn is about 650 cm long and weighs 1,800 kg. It nurses for about 6–7 months, until it is around 11.5–12.2 meters long, and then travels with the female to the polar feeding areas. Sexual maturity is attained at 6–11 years (Banfield 1974; Ellis 1980; Frazer 1973; Lockyer 1984; Mitchell 1978). Some individuals are estimated to have lived up to 114 years (Haley 1978).

Following development of the harpoon gun in the 1860s, *B. physalus* was regularly hunted by Norwegian whalers in the North Atlantic. Exploitation of the vast antarctic populations began after establishment of the first island whaling station in that region in 1904 and intensified after introduction of floating factories there in the 1920s. As the blue whale declined, the fin received increased hunting emphasis. In the 1937/38 season the take of *B. physalus* in the Antarctic was 28,009 individuals, nearly twice as great as that of *B. musculus*. After a lull during World War II, large-scale pelagic whaling resumed in the Antarctic and increased in other areas, especially the North Pacific. The worldwide kill of the fin whale exceeded 10,000 animals in every season from 1946/47 to 1964/65 and averaged around 30,000 annually from 1952 to 1962. Such a harvest was far in excess of sustainable yield, and by 1960 there was evidence of seriously reduced populations. Initial conservation efforts were largely unsuccessful, but by the mid-1970s the International Whaling Commission had drastically lowered annual quotas and had completely banned hunting in some regions. The kill dropped to 5,320 in 1968/69 and 743 in 1978/79 (Committee for Whaling Statistics 1980; Ellis 1980; Gulland 1974; McHugh 1974). Legal quotas fell subsequently and ceased altogether after 1985, except that a small aboriginal subsistence catch (10 in 1987) was allowed off western Greenland

(U.S. National Marine Fisheries Service 1987). Some countries, notably Iceland, also continued to hunt *B. physalus* for alleged research purposes.

The number of mature *B. physalus* in the world prior to exploitation is estimated at 470,000, of which 400,000 were in the Southern Hemisphere. Inclusion of immature individuals would increase these figures by roughly 50 percent (U.S. National Marine Fisheries Service 1978). A more recent numerical estimate by the U.S. National Marine Fisheries Service (1989) was 105,200–121,900, including 85,000 in southern waters. However, at the 1989 meeting of the International Whaling Commission it was reported that direct counts made by recent whaling cruises had led to a new estimate that probably only 2,000–3,000, and certainly not more than 5,000, fin whales were left in the Southern Hemisphere. These sad figures clearly reflect the commercial slaughter that persisted unchecked through the 1960s and continued even in the 1980s as a supposed research activity. *B. physalus* is classified as vulnerable by the IUCN and, more appropriately, as endangered by the USDI and is on appendix 1 of the CITES.

Balaenoptera musculus (blue whale).

This is the largest animal ever known to have existed. There is, however, a pygmy subspecies *(B. m. brevicauda)*, described by Ichihara (1966) and reported to inhabit a restricted zone of the Southern Hemisphere to the north of 54° S and between 0° and 80° E. Most authorities now seem to accept this subspecies as valid (Ellis 1980; Hall 1981; Rice 1977), but others, such as Small (1971), thought it to represent only young individuals of another subspecies. That part of the body of the pygmy blue whale posterior to the dorsal fin is described as being relatively shorter than that of other populations, and hence overall head and body length is reduced. Sexual maturity in this subspecies is attained at an average length of 19.2 meters, and maximum length is about 24.4 meters (Ellis 1980).

Disregarding the pygmy subspecies, head and body length of the blue whale at sexual maturity averages 22.5 meters in males and 24.0 meters in females (U.S. National Marine Fisheries Service 1978). Average length at physical maturity for antarctic specimens is 25 meters in males and 27 meters in females (Banfield 1974). Animals from the Northern Hemisphere are somewhat smaller. There is some question about maximum size, and it is possible that individuals grew larger prior to the period of intensive exploitation that began in the 1920s. According to Yochem and Leatherwood (1985), the

longest specimen measured in the scientifically correct manner (in a straight line from the point of the upper jaw to the notch in the tail) was a 33.58-meter female taken in the Antarctic sometime between 1904 and 1920, and the heaviest was a 27.6-meter female that weighed 190,000 kg, taken in the Antarctic in 1947.

The general coloration is slate or grayish blue, mottled with light spots, especially on the back and shoulders. Although the underparts have about the same basic color as the upper parts, they sometimes acquire a yellowish coating of microorganisms. Thus, the vernacular name "sulphurbottom" is often applied to *B. musculus*. The dorsal fin is relatively small, usually reaching a height of only 33 cm. The rostrum is less sharply pointed than that of any other member of the genus. There are approximately 90 ventral grooves, and these extend to the navel. On each side of the upper part of the mouth is a row of 300–400 baleen plates. They are black in color and range in length from 50 cm in front to 100 cm in back (Ellis 1980; Leatherwood, Caldwell, and Winn 1976).

Prior to exploitation, at least 90 percent of all blue whales lived in the Southern Hemisphere, but there are populations in the North Atlantic and North Pacific. Banfield (1974) described the species as a pelagic denizen of polar and temperate seas. Mitchell (1978) noted that whereas the blue whale goes farther into the Antarctic than do the other large rorquals, it does not move as close to the northern polar ice fields as does the fin whale. Populations generally spend the winter in temperate and subtropical zones, migrate toward the poles in the spring, feed in high latitudes during the summer, and move back toward the equator in the fall. Because of the difference in seasons between the northern and southern parts of the earth, all migrating populations move in roughly the same direction at the same time, and thus those of the Northern Hemisphere do not meet those of the Southern Hemisphere. There seem to be a number of distinct stocks, and though these may overlap to some extent in the summer feeding areas, they separate and return to their own discrete breeding sites each year (Gulland 1974; Mackintosh 1966). The stock of the pygmy subspecies around Kerguelen, Crozet, and Heard islands does not appear to migrate to the same extent as other populations (Ellis 1980).

The blue whale normally swims at a speed of around 22 km/hr but may swim as fast as 48 km/hr if alarmed. It usually feeds at depths of less than 100 meters, but harpooned individuals have gone deeper than 500 meters below the surface. Dives normally last 10–20 minutes and are followed by

Blue whale *(Balaenoptera musculus)*, painting by Richard Ellis.

a series of 8–15 blows. The spout reaches a height of 9.1 meters. The blue whale is the only member of its genus that commonly lifts its tail flukes out of the water before a dive. To feed, *B. musculus* takes in large amounts of water and organisms, greatly distending its throat, and then forces the water out, leaving the food trapped in the inner fibers of the baleen (see familial account). The highly restricted diet consists almost entirely of shrimplike crustaceans of the family Euphausiidae. These organisms, some of which are known as krill, are generally less than 5 cm long. When in the summer feeding areas, these whales each have a daily consumption of probably about 40 million individual euphausiids, with a total weight of 3,600 kg. During the rest of the year—a period of up to eight months—the blue whale apparently does not eat at all and lives off of stored fat (Ellis 1980; Leatherwood, Caldwell, and Winn 1976; Mitchell 1978; Rice 1978*a*). *B. musculus* emits deep, low-frequency sounds, as well as series of clicks that may possibly be used in the echolocation of swarms of krill (Thompson, Winn, and Perkins 1979).

Off the coast of California, aggregations of up to 60 blue whales are common (Rice 1978*a*). Usually, however, the species is seen alone or in groups of two or three individuals (Banfield 1974). In the Southern Hemisphere, mating peaks in the summer (June–July), and births peak in the spring (Lockyer 1984). Females give birth every 2–3 years (U.S. National Marine Fisheries Service 1978). According to Mizroch, Rice, and Breiwick (1984), females appear to be seasonally monestrous, though if they fail to conceive, they may ovulate two or three times during one estrous cycle. The gestation period, usually reported at 10–12 months, is surprisingly short for so large an animal. Frazer (1973) listed gestation as only about 9.6 months, explaining that if the period lasted much longer, the young would be born at a disadvantageous time—just before or during the season spent in cold waters, and before it had accumulated much protective blubber. Twins have been reported on rare occasion, but there is normally a single offspring. At birth it is about 7 meters long and weighs 2,000 kg. It gains 90 kg per day and is weaned after 7–8 months, when it is about 15 meters long. Sexual maturity is attained at 5–10 years (Banfield 1974; Ellis 1980; Yochem and Leatherwood 1985). Some individuals are estimated to have lived for up to 110 years (Haley 1978).

Because of its size, speed, strength, and remote habitat, the blue whale was generally considered too difficult a target in the early days of whaling. This situation was changed by a series of developments from the 1860s to 1920s, as described in the account of the order Cetacea. Regular hunting of *B. musculus* began off Norway, steadily spread across the Northern Hemisphere as one population after another was depleted, and finally came to center in the vast feeding waters of the Antarctic. The annual kill went up dramatically following the introduction of factory ships with slip sternways. The total recorded kill in the twentieth century is approximately 350,000 individuals, of which over 90 percent were taken in the Antarctic. The peak season was 1930/31, when the kill was 29,410 in the Antarctic and 239 in other parts of the world. There subsequently was a general decline in the worldwide seasonal harvest, to 12,559 in 1939/40, 6,313 in 1949/50, and 1,465 in 1959/60. Various international efforts to set size limits, sanctuary areas, and quotas had been attempted since the 1930s, but they were inadequate and did not receive full compliance. By the early 1960s it was clear to almost all concerned persons that the blue whale was nearing extinction, and groups of scientists were recommending that the International Whaling Commission establish total protection for the species. Because of a continued lack of cooperation from whaling interests, such protection did not come until the year after the 1965/66 season, when 613 blue whales were killed, only 20 of them in the Antarctic (Allen 1942; Ellis 1980; Gulland 1974; McHugh 1974; Small 1971).

The estimated numbers of blue whales prior to exploitation are: Southern Hemisphere, 150,000–210,000; North Pacific, 4,500–6,000; western North Atlantic, 1,100; and eastern North Atlantic, 6,000 (Yochem and Leatherwood 1985). Although Small (1971), among others, warned that these numbers had been so drastically reduced that recovery might be impossible, there was a general authoritative view through the early 1980s that substantial and viable stocks remained. Numbers estimated by the U.S. National Marine Fisheries Service (1978, 1981, 1987, 1989) were: Southern Hemisphere, 10,000 (about half of which are the pygmy subspecies); North Pacific, 1,600; and North Atlantic, a few hundred. There even had been indications of slowly increasing numbers, and populations off the east and west coasts of North America appeared to be larger than once feared (Berzin 1978; Leatherwood, Caldwell, and Winn 1976). International protection against direct killing seemed to be working.

Then, new data from systematic survey cruises became available. Horwood (1986) provided an estimate of only about 1,300–2,000 blue whales left in the Southern Hemisphere. Subsequently, at the 1989 meeting of the International Whaling Commission, it was reported that only 500 blue whales survived there. The earlier dire predictions now seem to have been justified, and there is fear that remaining populations of the world's largest animal may not be viable. There are other threats to any potential recovery, especially regarding the increasing harvest of Antarctic krill for use as human food (Gulland 1974; Laws 1985; McWhinnie and Denys 1980). The blue whale is classified as endangered by the IUCN and the USDI and is on appendix 1 of the CITES.

CETACEA; BALAENOPTERIDAE; **Genus MEGAPTERA**
Gray, 1846

Humpback Whale

The single species, *M. novaeangliae*, occurs in all oceans and adjoining seas of the world (Rice 1977).

Head and body length at sexual maturity averages 11.6 meters in males and 11.9 meters in females (U.S. National Marine Fisheries Service 1978). Average length at physical maturity for North Pacific specimens is 12.5 meters in males and 13.0 meters in females (Banfield 1974). Individuals over 15 meters long are very rare. Average weight is around 30,000 kg. The pectoral fin is about one-third as long as the head and body and is the largest belonging to any cetacean, both absolutely and relatively. The dorsal fin varies from bumplike to strongly curved in shape and from 15 to 60 cm in height. The expanse of the tail flukes is also about one-third the length of the head and body. The general coloration is usually black above and white below, but there is much variation. Clusters of white barnacles are often present. The body is stocky compared to that of *Balaenoptera*. There are 10–36 grooves extending from beneath the tip of the snout to the area of the navel. Irregular knobs and protuberances occur along the snout and lower lip. On each side of the upper part of the mouth is a row of about 340 baleen plates, gray to black in color.

The humpback whale is basically oceanic but enters shallow, tropical waters for the winter breeding season. At this time, each of the discrete populations utilizes a certain archipelago or stretch of continental coastline. Johnson and Wolman (1984) identified the major wintering areas in the North Pacific as follows: (1) west coast of Baja California, Gulf of California, mainland Mexican coast from southern Sonora to

Humpback whales *(Megaptera novaeangliae)*, painting by Richard Ellis.

Jalisco, and around Islas Revillagigedo; (2) Hawaiian Islands from Kauai to Hawaii; and (3) around Mariana, Bonin, and Ryukyu islands and Taiwan. During the spring there are movements along well-defined migratory routes toward the high-latitude feeding areas. The Northern Hemisphere migration extends as far as the Chukchi Sea and Spitsbergen. Some of the whales wintering off of Hawaii and Mexico form several summer feeding herds in the waters south of Alaska. There evidently is limited exchange of individuals between these two winter groups, and it may be that all *Megaptera* in the central and eastern North Pacific are part of a single structured stock (Baker et al. 1986; Darling and Jurasz 1983; Payne and Guinee 1983). In the Southern Hemisphere there are summer concentrations off the southern ends of continents and around such subantarctic islands as South Georgia. In the autumn there are return migrations toward the equator. Because of the reversal of seasons, the populations of the Northern Hemisphere are not in equatorial waters at the same time as those of the Southern Hemisphere (Dawbin 1966; Wolman 1978).

Despite its common name, *Megaptera* is a graceful swimmer and among the most acrobatic of large cetaceans. It often makes a somersault by leaping completely out of the water with its belly up, plunging back headfirst, and then circling underwater to the original position. It normally travels at about 7 km/hr, but maximum speed has been estimated at 27 km/hr. Just before a dive it emits a series of three to six short, broad spouts and raises its tail flukes high above the water. Feeding is usually accomplished by taking in a large amount of water and organisms and then forcing out the water to strain the food in the baleen (see familial account). There have been observations of one or two whales diving beneath a school of fish and then spiraling upward while emitting bubbles. The bubbles rose in a cylindrical pattern around the fish and seemed to form a barrier, through which they would not pass. The whales then rushed into the school of fish to feed. Small fish evidently form a major part of the diet, but shrimplike crustaceans, especially of the family Euphausiidae, are the most important foods in the Antarctic. As is the general case with rorquals, most, if not all, feeding is done during the summer in high latitudes (Ellis 1980; Wolman 1978).

Experiments with a captive humpback suggested to

Beamish (1978) that the genus does not use echolocation to find food. *Megaptera* is, however, among the most loquacious of cetaceans. It emits a great variety of sounds, certain of which are sometimes combined into an elaborate song. Winn and Winn (1978) reported the song to evolve through "moans and cries," to "yups or ups and snores," to "whos or wos and yups," to "ees and oos," to "cries and groans," and finally to various "snores and cries." These authorities considered the song to be one of the longest, and possibly the most patterned, in the animal kingdom. It lasts 6–35 minutes, but the highly rhythmic pattern is repeated over and over for hours and possibly months at a time. It is heard only in the tropics during the winter breeding season. It is produced only by single individuals, which are thought to be young but sexually mature males. It may serve to space the males, to attract females, or to locate group members. There apparently are dialects, in that the song of the whales in one area differs to some extent from that heard in other areas.

Winn et al. (1981) determined the dialects of the Hawaiian and Mexican breeding groups to be essentially identical, thereby supporting the view that both are part of the same overall population. This dialect differs from one shared by whales of the Cape Verde Islands and West Indies, while the population at Tonga in the South Pacific has still another distinct dialect. Perhaps the most remarkable feature of the song is that in each area it seems to change from year to year and that the new versions are effectively spread throughout the singing population. Payne and Payne (1985) found that while songs always overlap in part during successive years, they become increasingly different as time elapses, and after three or four years nearly all elements of the song have been modified. Payne, Tyack, and Payne (1983) concluded that the changes are transmitted by listening and learning and that humpback singing seems to fill an evolutionary gap between birds, which increase their repertoire through mimicry, and people, who create new songs. Baker and Herman (1984) showed that some of the same animals known to be singers are also those that escort females and young (see below).

There is segregation by age and sex during migrations (Dawbin 1966). In autumn the order of progression is: females with their recently weaned calves, then independent juveniles, then mature males and females that are not reproductively active, and finally females in late pregnancy.

During the spring the order is: females in early pregnancy, then independent juveniles, then mature males and females that are not reproductively active, and finally females in the early stage of lactation. This arrangement ensures that pregnant females spend maximum time in the feeding waters and that young calves spend maximum time in warm regions. In feeding areas *Megaptera* sometimes is seen in aggregations of up to 150; in the breeding season, however, animals are usually found alone or in groups of 2–9. The groups seem often to represent a number of males competing for proximity to a female (Tyack and Whitehead 1983). A female and young calf commonly are accompanied by another adult, evidently a male (Ellis 1980; Herman and Antinoja 1977). This escort male is aggressive toward other males that approach, often inflating its throat with water or air and lunging at its rivals, or sometimes striking the latter with its tail. The pair bonds are not stable, however, and both males and females associate with a number of animals of the opposite sex during the breeding season (Baker and Herman 1984).

The mating and calving season is October–March in the Northern Hemisphere and April–September in the Southern Hemisphere. Females are seasonally polyestrous and usually give birth every 2 years to a single young about 400–500 cm in length and 1,350 kg in weight. However, several females have been observed to produce a calf in successive years. The gestation period is 11.0–11.5 months, and lactation lasts about 11 months. Sexual maturity is attained at 4 or 5 years (Banfield 1974; Glockner-Ferrari and Ferrari 1984; U.S. National Marine Fisheries Service 1978; Winn and Reichley 1985). Some individuals are estimated to have lived up to 77 years (Haley 1978).

The humpback has been regularly hunted for a much longer period than other large rorquals. It is relatively easy to secure, because it tends to stay near the shore in the breeding season, is a slow swimmer, often approaches vessels, and is highly visible. It has been considered a valuable source of oil, meat, and baleen. It was taken by aboriginal peoples in northwestern North America in ancient times and has recently been hunted for subsistence on certain West Indian and Pacific islands and in Greenland. Coastal fisheries began in Japan and Bermuda in the 1600s and in eastern North America by the 1700s. Major commercial exploitation of the genus spread over the North Atlantic and North Pacific in the nineteenth and early twentieth centuries. The most intensive period of hunting, however, began with the establishment of whaling stations on the islands around Antarctica in the early 1900s. The total kill in the Southern Hemisphere from 1904 to 1939 was 102,298 individuals. Populations were quickly and drastically reduced. Humpbacks composed 96.8 percent of the catch of whales at South Georgia in the 1910/11 season but only 9.3 percent in 1916/17. Nonetheless, large-scale exploitation continued for many years. The annual worldwide kill was above 2,400 in every season from 1948/49 to 1963/64. By this time there was general realization that the genus was nearing extinction. From 1964 to 1966 the International Whaling Commission extended protection to all populations, except that a small aboriginal quota continued (Allen 1942; Ellis 1980; Gulland 1974; McHugh 1974; Mitchell and Reeves 1983). The quota was 10 whales for the 1981/82 season (*Marine Mammal News* 7[1981]:1–3) but zero in 1986/87 (U.S. National Marine Fisheries Service 1987).

Original populations are estimated at 100,000 animals in the Southern Hemisphere, 15,000 in the North Pacific, and 10,000 in the northwestern Atlantic. The current worldwide estimate is 9,500–10,000, about 5,800 of which are in the North American Atlantic. The largest group in the North Pacific, about 700 individuals, winters in the waters around Hawaii, but, interestingly, this population seems to have become established only in the last 200 years. It now may be

jeopardized by increasing boat traffic and disturbance from tourists. There are similar problems in the summer waters of Glacier Bay, Alaska. There is evidence that the North Atlantic population has increased since 1915, but individuals sometimes become entangled in fishing nets off eastern Canada (Herman 1980; Johnson and Wolman 1984; U.S. National Marine Fisheries Service 1978, 1981, 1987, 1989; Winn, Edel, and Taruski 1975; Winn and Reichley 1985; Wolman 1978). The humpback is classified as endangered by the IUCN and USDI and is on appendix 1 of the CITES.

CETACEA; Family BALAENIDAE

Right Whales

This family of two Recent genera and species, *Eubalaena glacialis* and *Balaena mysticetus,* is found in all oceans and adjoining seas of the world except in tropical and south polar regions (Rice 1984). A third genus, *Caperea* (see account thereof), often has been placed in the Balaenidae.

The two genera are among the largest of baleen whales, and females average larger than males. Both genera have a massive body and a relatively large head, accounting for one-fourth to one-third of the total length. The rostrum is narrow and highly arched, resulting in the cleft of the mouth being curved. The lower jaw is enormous and has a large, fleshy lip. There are no throat and chest grooves, as in the Balaenopteridae, and no dorsal fin. The pectoral fin is short, broad, and rounded. The tongue is heavy and muscular. This family, like the Balaenopteridae and Eschrichtiidae, has baleen instead of teeth. The baleen plates of the Balaenidae are the longest in the Cetacea and are usually narrower than in the other two families. There is a row of plates on each side of the upper part of the mouth, and the two rows do not join anteriorly. The plates fold in the floor of the closed mouth and straighten when the mouth opens. The vertebrae number 54–57; the 7 neck vertebrae are fused into a single unit. The manus has five digits. The tail stock is constricted and tapers into the flukes.

The diet consists mainly of the small forms of animal life known collectively as zooplankton (and sometimes called krill). These organisms include crustaceans, such as the shrimplike euphausiids and copepods, and the free-swimming mollusks known as pteropods. Unlike most species of the Balaenopteridae, the Balaenidae use the feeding method called skimming. The whales swim through swarms of prey, with the mouth open and the head above water to just behind the nostrils. When a sufficient mouthful of organisms has been filtered from the water by the inner bristles of the baleen plates, the whales force out the water, dive, and swallow the food.

The common name "right" whale refers to the consideration of *Eubalaena* and *Balaena* in the early days of commercial whaling as the proper or best kind of whales to hunt. They were slower and less active than the sperm whale and most other baleen whales, and they often came closer to land. Unlike the others, they had greater buoyancy and were less likely to sink after being killed. Moreover, they yielded the most valuable products. The amount of oil usually derived from an adult was 80–100 barrels (1 barrel = about 105 liters), more even than from a blue whale *(Balaenoptera musculus)*. Some large individual *Eubalaena* and *Balaena* were said to have produced several hundred barrels each. The oil was used mainly as a fuel for lamps and in cooking but also served as a lubricant, in leather tanning, and in the manufacture of soap and paint.

The baleen of *Eubalaena* and *Balaena* was the finest avail-

Dr. Roy Chapman Andrews, 6 ft. 1 in. in height, standing beside the skull of a right whale *(Balaena glacialis).* The plates of baleen are smooth on the outside, but on the inside they are fringed to form an effective strainer. Photo from American Museum of Natural History.

able, and the plates were far longer than those of other species. The usual yield from an adult *Balaena* was 680–760 kg. The hard outer portion of the plates was split into strips that could be used in the manufacture of umbrella ribs, fishing rods, carriage springs, whips, and numerous other items that required a combination of strength and elasticity. Baleen strips were especially useful as a stiffening in fashionable garments, such as the farthingale (a framework for supporting the elaborate dresses of the Elizabethan era) and the hoop skirt of the nineteenth century. The fine inner fibers of the baleen plates were also important in fashion. According to Gilmore (1978), they were woven into fabrics, giving a stiffness and rustle to taffeta and crinoline, and could be washed without softening or loss of elasticity.

The value of whale products fluctuated over the years. Typical nineteenth-century prices in the United States, however, were around $30 per barrel for oil and $10 per kg for baleen. At such rates the average *Balaena* could bring over $10,000, at a time when the cost of living was about 5 percent of what it is today. By the early twentieth century the last great concentrations of *Eubalaena* and *Balaena* had been eliminated, whale oil was no longer being used for illumination, and spring steel was replacing baleen. Although whaling was about to enter its most intensive era, right whales were not to be of major commercial importance.

The geological range of the family Balaenidae is early Miocene to Pleistocene in South America, middle Miocene to Pleistocene in western North America, late Miocene in Australia, early Pliocene to Pleistocene in Europe, and Recent in all oceans (Rice 1984).

CETACEA; BALAENIDAE; Genus EUBALAENA
Gray, 1864

Right Whale

A single species, *E. glacialis,* occurring mainly in temperate parts of the Atlantic, Indian, and Pacific oceans and adjoining seas, was recognized by Banfield (1974), Gaskin (1982, 1987), Meester et al. (1986), and Rice (1977, 1984). The populations of the Northern and Southern Hemispheres generally are separated from one another by several thousand kilometers,

and some authorities, such as Cummings (1985) and Payne and Dorsey (1983), regard the southern populations as a full species, *E. australis.* Brownell, Best, and Prescott (1986) indicated that the question was unsettled and recommended further systematic work. *Eubalaena* often is included within the genus *Balaena,* but there has been a recent trend by authorities to treat the two as distinct genera (Barnes and McLeod 1984; Brownell, Best, and Prescott 1986; Cummings 1985; Gaskin 1982, 1987; Payne and Dorsey 1983).

Head and body length at physical maturity is usually 13.6–16.6 meters. Banfield (1974) listed average size as 13.7 meters and 22,000 kg for males, and 14 meters and 23,000 kg for females. Rice (1984) stated that maximum length is 18 meters and that one female 17.4 meters long weighed 106,500 kg. Pectoral fin length is 180–210 cm. There is no dorsal fin. The coloration is usually black throughout, but there are sometimes large white patches, especially on the belly. Entirely white calves have been observed (Best 1970b). Around the head are series of horny protuberances representing accumulations of cornified layers of skin and commonly infested with barnacles and parasitic crustaceans. The most conspicuous of these callosities is located on the tip of the upper jaw and is known as the bonnet. The callosities are present from the time of birth, but their exact function is unknown. Payne and Dorsey (1983) found that males have more and larger callosities than do females and suggested that they are used as weapons for intraspecific aggression.

The head of *Eubalaena* is smaller than that of *Balaena,* about one-fourth of the total length, and the upper jaw is not so strongly arched. The narrow upper jaw is practically concealed by the high, massive lower jaw when the mouth is closed. On each side of the upper part of the mouth is a row of 225–50 baleen plates. These plates are usually 180–220 cm in vertical length and range in color from dark gray or brown to black. The two blowholes are set well apart, and thus there are two spouts, which form a V pattern (Gilmore 1978; Leatherwood, Caldwell, and Winn 1976).

The right whale is primarily a species of temperate waters, though some individuals apparently move just north of the Arctic Circle and just south of the Tropic of Cancer. It is usually found closer to land than are most large whales, especially in the breeding season. Although it was largely eliminated by whalers before much scientific information could be gathered, available evidence indicates that there were well-

Right whale *(Eubalaena glacialis)*, photo by Roger Payne.

defined migrations to higher latitudes for summer feeding and back to lower latitudes for winter breeding. Populations of the Southern Hemisphere wintered mainly between 30° and 50° S, off the coasts of South America, southern Africa, Australia, and New Zealand, and then moved toward the Antarctic for the summer. The population of the eastern North Atlantic wintered in the Bay of Biscay and summered between Great Britain and Iceland. In the western North Atlantic, from about November to March the whales remained from Cape Cod to Bermuda and the Gulf of Mexico. In late winter they began to move north, and by spring they were concentrated off the northeastern United States. Some remained in the latter region, but the major summer feeding waters were around Newfoundland and in the Labrador Sea. Part of the remnant population in the western North Atlantic calves during winter off Georgia and Florida, feeds off New England in the spring, feeds in the Bay of Fundy and off Nova Scotia in the summer and early autumn, and then migrates southward in late October and November. Pacific populations seem to have passed Japan in April and fed off southern Alaska from May to September; wintering waters are unknown but may possibly have included those around Hawaii and the Ryukyu Islands (Allen 1942; Banfield 1974; Gaskin 1968; Gilmore 1978; Herman et al. 1980; Omura 1986; Reeves, Mead, and Katona 1978; Winn, Price, and Sorensen 1986).

The right whale is a relatively slow swimmer, averaging about 8 km/hr, but frequently leaps clear of the water and engages in other acrobatics. It commonly makes a series of five or six shallow dives and then lifts its tail flukes above the water and submerges for about 20 minutes. It is usually not wary of boats and can be easily approached (Leatherwood, Caldwell, and Winn 1976). It appears to frolic in stormy seas and even to use its tail to sail in the wind (Payne 1976). To feed it employs the skimming method, generally swimming near the surface with its mouth open to strain out organisms from the water with its baleen (see familial account). Its diet consists mainly of copepods—crustaceans only a few millimeters in diameter—and also includes euphausiids, pteropods, and small fish (Gilmore 1978).

According to Banfield (1974), *Eubalaena* sometimes was found in aggregations of 100 or more. Social activity, however, is generally limited to mating concentrations and the relationship between mother and young. *Eubalaena* emits a number of low-frequency bellows, moans, pulses, and other sounds, mostly during courtship (Cummings 1985; Gilmore 1978). The primary contact call is a low, tonal upsweep that intensifies toward the end (Clark 1982). A male may attempt to mate with several females, and a female may sometimes be courted by several males at once (Ellis 1980; Payne 1976). There is some question whether physical confrontation occurs between males during the breeding season, though sometimes one male supports a female while another male mates with her (Gaskin 1987). Females give birth to a single offspring about every 2–5 years; at that time they separate from the rest of the population and move closer to shore (Kraus et al. 1986). Mothers with young calves in the population that winters off Peninsula Valdes in Argentina concentrate along the coast in water only about 5 meters deep (R. Payne 1986). In the Southern Hemisphere, mating peaks from August to October, calving peaks from May to August, and gestation is thought to last 10 months (Lockyer 1984). The young is about 350–550 cm long at birth and nurses for about 7 months. The age of sexual maturity has been estimated to be about 6 years (Gaskin 1987) and 10 years (Cummings 1985).

Because of its coastal habitat and economic value (see fa-

Right whales *(Eubalaena glacialis)*, photo by Roger Payne.

milial account), *Eubalaena* was among the first of the large whales to be extensively exploited. It may have been hunted as early as the ninth century A.D. off Norway and was regularly taken in the Bay of Biscay in the tenth century. By the late fifteenth century it had become rare in the latter area, and the Basque whalers shifted their emphasis to other waters, eventually reaching Newfoundland. An estimated 25,000–40,000 whales were killed around Newfoundland from 1530 to 1610, after which the Basques began to journey northward in pursuit of the bowhead. The right whale, however, became the basis of a major industry in the American colonies during the seventeenth century, with hunting initially being done from small boats in such places as Delaware Bay and Cape Cod Bay and later involving extended voyages in large ships. By 1700 in Europe and 1800 in the United States, the right whale had become too rare to be of commercial interest. Toward the end of the eighteenth century, however, large stocks of the species were discovered in the wintering waters of the Southern Hemisphere and in the summer feeding areas of the North Pacific. By this time the industry was dominated by American companies. On the basis of catch records and oil and baleen imports, it has been estimated that 70,343–74,693 right whales were taken by American vessels from 1805 to 1914 (see Best 1970*b* and 1987 for an explanation of these figures). By the latter year the right whale was very rare throughout its range. Some hunting continued in the early twentieth century, but takes were usually considerably fewer than 100 per year. In 1937, in accordance with the International Agreement for the Regulation of Whaling, *Eubalaena* received complete protection (Aguilar 1986; Allen 1942; Banfield 1974; Reeves, Mead, and Katona 1978).

Braham and Rice (1984) estimated the original number of right whales in the world as 100,000–300,000, about two-thirds of them in the Southern Hemisphere. The species now has become one of the rarest of large mammals. According to the U.S. National Marine Fisheries Service (1989), current estimated numbers are 3,000 in the Southern Hemisphere, 100–200 in the North Pacific, and a few hundred in the North Atlantic (almost all on the American side). There is evidence that the populations of the Southern Hemisphere have grown since the granting of protection. The group wintering off Peninsula Valdes, Argentina, has been increasing at 6.8 percent annually and now contains 450–600 whales (Whitehead, Payne, and Payne 1986). There have been suggestions that an increase also has occurred in the North Atlantic, but Reeves, Mead, and Katona (1978) cautioned that recovery has been modest at best and that the species may be jeopardized by such factors as competition for food with the sei whale, accidental trapping in fishnets, pollution, and boat collisions. Braham and Rice (1984) added that the right whale might be the most vulnerable of all great whales, especially because females move into shallow coastal bays to give birth. Gaskin (1987) doubted that there had been any recovery of the western North Atlantic stock and stated that the eastern North Atlantic population is virtually extinct. Brown (1986) listed recent records indicating that there still are a few right whales in the eastern North Atlantic but noted that they might be stragglers from the west. Scarff (1986) reported no evidence of recovery in the eastern Pacific. *Eubalaena* is classified as endangered by the USDI and is on appendix 1 of the CITES. The IUCN now recognizes two species, designating *E. glacialis* as endangered and *E. australis* as vulnerable.

CETACEA; BALAENIDAE; **Genus BALAENA**
Linnaeus, 1758

Bowhead Whale, or Greenland Right Whale

The single species, *B. mysticetus*, is found in the Arctic Ocean and adjoining seas, the Sea of Okhotsk, Hudson Bay, and the Gulf of St. Lawrence (Banfield 1974; Rice 1977). The genus *Eubalaena* (see account thereof) often is included within *Balaena*.

Head and body length at sexual maturity averages 11.6 meters in males and 12.2 meters in females (U.S. National Marine Fisheries Service 1978). At physical maturity the usual length is 15–18 meters. The largest specimens reach about 19.8 meters in length and over 100,000 kg in weight (Ellis 1980; Reeves and Leatherwood 1985). Pectoral fin length is about 200 cm, and expanse of the tail flukes is 550–790 cm. There is no dorsal fin. The adult coloration is black, except that the anterior part of the lower jaw is cream-colored, the belly occasionally has white patches, and the junction of the body and tail is sometimes gray.

Right whale *(Eubalaena glacialis):* Top photo shows the baleen in place and the very large tongue. Inset: a smaller piece of baleen of lighter color. Bottom photo shows the tongue and the rough surface of the top of the front portion of the head (the baleen has been removed). Photos by G. C. Pike through I. B. MacAskie.

The skull is about 40 percent as long as the entire animal (Hall 1981). Seen from the front, the enormous lower jaw forms a U around the relatively narrow upper jaw. Seen from the side, both jaws are highly arched, and when the mouth is closed the lower jaw may partly conceal the upper. On each side of the mouth is a row of about 300 baleen plates. These plates, the largest of any whale's, are 300–450 cm in vertical length and are black in color. As in *Eubalaena*, there are two separated blowholes that produce a V-shaped spout. According to Ellis (1980), the blubber of *Balaena* is 25–50 cm thick.

Some early commercial and modern subsistence whalers have recognized a relatively short, stocky form of bowhead with shorter, thinner baleen. This form, referred to as the "ingutuk," sometimes even has been considered to be a separate species, but Braham et al. (1980) concluded that it was a morphological variant of *B. mysticetus.*

The bowhead is primarily an arctic species, seldom occurring south of about 45° N, but in 1969 a specimen was discovered off Japan at 33°28′ N (Nishiwaki and Kasuya 1970). Normally the bowhead is found in association with ice floes, appearing to move seasonally in response to the melting and freezing of the ice. During the summer it frequents bays, straits, and estuaries. Prior to exploitation there seem to have been four major populations. One wintered off the Kuril Islands and summered in the Sea of Okhotsk. Another, the western Arctic stock, wintered in the Bering Sea, migrated north through the Bering Strait in the spring and early summer, remained in the Chukchi and Beaufort seas from late summer to early autumn, and then returned south. A third group apparently wintered along the east coast of Canada as far south as the Gulf of St. Lawrence. In the spring and summer this group moved northward and westward into Davis Strait, Hudson Bay, Baffin Bay, and adjoining waters. The fourth population apparently wintered off southeastern Greenland and migrated to and beyond the Spitsbergen region during the spring and summer (Banfield 1974; Bockstoce and Botkin 1983; Marquette 1978; Rice 1977; W. G. Ross 1979). There have been some suggestions that the bowheads of Hudson Bay form a fifth discrete population (Braham 1984).

During migration the bowhead swims at about 6 km/hr; its maximum speed is around 15 km/hr. It normally surfaces

Bowhead whale *(Balaena mysticetus)*, accompanied by belugas, painting by Richard Ellis.

for up to 2 minutes, blows four to nine times, and then submerges for 5–10 minutes. Harpooned individuals are said to have stayed underwater for over an hour. Before a dive the bowhead lifts its tail flukes above the surface. It sometimes leaps almost entirely out of the water. To feed, it employs the skimming method (see familial account). The diet consists mainly of zooplankton—copepods, amphipods, euphausiids, and pteropods. The estimated daily intake during the summer feeding season is 1,800 kg. At other times of the year, like most baleen whales, the bowhead lives off of stored fat reserves and eats little or nothing (Banfield 1974; Marquette 1978).

Würsig et al. (1985) observed bowheads to sometimes feed in a V-shaped formation, with up to 14 animals, mouths wide open, moving in the same direction at the same speed. Groups consisting of two adults and a calf also were seen to persist for at least a few weeks. Previous reports (Banfield 1974; Ellis 1980; Marquette 1978) indicate that *Balaena* usually is found alone or in groups of two or three individuals but that larger groups may form during migration, and in the past such schools often included several hundred whales. There is sometimes segregation by age and sex. During the spring migration of the Bering Sea population, young animals move north first, followed by large males and females with calves. Clark and Johnson (1984) reported bowhead sounds to vary greatly but to consist mostly of low tones and to include moans, rich purrs, roars, and complex pulsive calls. The voice of *Balaena* has been described as a drawn-out hooting or humming sound.

Available data (Banfield 1974; Breiwick, Eberhardt, and Braham 1984; Ellis 1980; Marquette 1978; Nerini et al.

1984) indicate that mating occurs mainly in the late winter, and births in the spring and summer, peaking in the western Arctic in May. The female gives birth every 3–6 years, normally to a single calf. The gestation period is thought to last about 13 months. The young is 400–450 cm long at birth and apparently is weaned sometime from 6 to 12 months later. The age of sexual maturity is unknown, but there have been various estimates of 4–9 years. There are questionable reports, based on the finding of old harpoons in newly taken whales, of individuals having lived about 40 years and also of individuals having made a transpolar passage from the Atlantic to the Pacific side of the Arctic.

Like *Eubalaena*, the bowhead was one of the "right" whales of commerce. Indeed, because of the length of its baleen and the thickness of its blubber, it was the most economically valuable of all cetaceans (see also familial account). Regular hunting by Europeans began in the sixteenth century off Greenland. In the early seventeenth century Spitsbergen became the center of the industry. Whaling bases were established there by expeditions from the Netherlands, England, France, Spain, Denmark, and Germany. By the early 1700s the bowhead had become so rare in this region that hunting was no longer profitable. The focus of exploitation then shifted to the west of Greenland. Intensive whaling began in Davis Strait in 1719, Baffin Bay and adjacent sounds in 1818, and Hudson Bay in 1860. In these regions whaling was dominated by the Dutch for most of the eighteenth century and by the British and Americans in the nineteenth century. In some years several hundred ships were involved. By the mid-nineteenth century the bowhead was becoming rare in most areas from Greenland to eastern Canada, and after 1887 not more than 10 ships per year hunted in these waters. The last 2 vessels to try, in 1912 and 1913, did not take a single whale. Meanwhile, however, starting in 1848, American whalers had sailed north through the Bering Strait to exploit the last major summer concentration of bowheads, that of the Beaufort and Chukchi seas. These whales also were exploited in the wintering waters in the Bering Sea and were eliminated there progressively from south to north. Following a relaxation of the demand for whale oil in the mid-nineteenth century, the intensity of hunting corresponded with the price being paid for baleen for use in fashion. The western Arctic bowhead population probably was approaching extinction by the turn of the century, but the price of baleen per pound fell from $5.00 in 1907 to $0.075 in 1912, and hunting ceased. There is some question whether the present number of bowheads in the western Arctic represents a subsequent moderate increase. The total kill of bowheads had been about 19,000 in the western Arctic since 1848 and about 37,000 in the waters from northeastern Canada to Greenland since 1719 (Allen 1942; Banfield 1974; Bockstoce 1980a, 1980b, 1986; Bockstoce and Botkin 1983; Breiwick, Eberhardt, and Braham 1984; Ellis 1980; W. G. Ross 1974, 1979).

The bowhead whale received protection under the International Agreement for the Regulation of Whaling in 1937, the International Whaling Convention of 1946, the United States Marine Mammal Protection Act of 1972, and the U.S. Endangered Species Act of 1973. In each case, however, an exception was made for subsistence hunting by aboriginal peoples. Certain groups of Eskimos in northwestern Alaska had been whaling since ancient times. Their cultural and economic status is still associated with the pursuit and kill of the bowhead during its annual migrations. They once utilized the entire animal, including the blubber for fuel and the baleen for making tools, but now the meat (around 10,000 kg per whale) is the main economic objective. A few bowheads also are taken by natives of northeastern Siberia.

Until about 1970 the usual take by the Alaskan Eskimos was 10–15 whales per year. Subsequently the kill rose, apparently because the petroleum industry had made money available for native Alaskans to purchase the boats, guns, and other equipment necessary for the culturally prestigious venture of whaling. In 1976, 48 bowheads were killed and another 46 were known to have been wounded; in 1977 the respective figures were 29 and 82. Because of growing concern that such exploitation would jeopardize the survival of the remaining bowhead population, the International Whaling Commission decided in 1977 to rescind the exemption for Eskimo whaling. The United States government, in a reversal of its general stand against whaling, then urged that some hunting be allowed. The commission thus began to set quotas for the bowhead, just as for commercially harvested species. The quota for 1980 was 18 whales landed or 26 struck, whichever came first ("struck" means either having been landed or hit by a harpoon but then lost). Some Eskimos argued for a higher kill and announced that they would not abide by the commission's quota. Indeed, the quota was exceeded during the 1980 hunts, 16 whales having been landed and 18 others struck and lost. Subsequently, however, the take has been kept within the quota. Starting in 1984, the quota has been set by strikes only, with the provision that strikes not used may be added to the limit, so long as the total number of strikes does not exceed 32. In 1986 the full 32 strikes were available, but only 28 were used (Bockstoce 1980a; Ellis 1980; Evans and Underwood 1978; Marquette 1978, 1979; R. Rau 1978; U.S. National Marine Fisheries Service 1978, 1981, 1987). Breiwick, Eberhardt, and Braham (1984) cautioned that any regular kill of more than 22 whales per year might eliminate the hope of a future population increase.

There is disagreement regarding how many bowheads there are and were. Reported estimates of the western Arctic population in the mid-nineteenth century range from 4,000 to 40,000 individuals. The most commonly mentioned range is 10,000–30,000 (Breiwick, Eberhardt, and Braham 1984; Breiwick and Mitchell 1983; Breiwick, Mitchell, and Chapman 1981; Ellis 1980; Evans and Underwood 1978; Marquette 1978). A "best estimate" of 18,000 for the original western Arctic stock recently has been made by the International Whaling Commission, and there also are estimates of 25,000 originally in the eastern Atlantic, about 6,000–12,000 for the western Atlantic, and 6,500 for the Sea of Okhotsk (Braham 1984a; Braham, Krogman, and Carroll 1984; Reeves and Leatherwood 1985). However, a detailed analysis of logbooks and journals of early whaling cruises led to an estimate for the western Arctic in 1847 of no fewer than 20,000 and no more than 40,000 (Bockstoce and Botkin 1983). The U.S. National Marine Fisheries Service (1987) gave the estimate of the current population in the western Arctic as 4,417 whales but later (1989) listed estimates of 4,855 and 7,800. According to sources cited by Braham (1983), there may still be a few hundred individuals in both the western Atlantic and Okhotsk populations, but there have been only a few recent reports of the bowhead from the Spitsbergen-Greenland region. Reeves et al. (1983) concluded that while the western Atlantic population has been severely reduced, it continues to occupy much of its former range and to follow the original migratory schedule. Mitchell and Reeves (1982) warned that low-level but persistent hunting by native peoples and predation by killer whales might be preventing recovery of this population. McQuaid (1986) suggested that recent sightings indicate the survival of a small relict population in the eastern North Atlantic. The species is classified as endangered by the IUCN and USDI and is on appendix 1 of the CITES.

CETACEA; Family NEOBALAENIDAE; Genus CAPEREA
Gray, 1864

Pygmy Right Whale

The single known genus and species, *Caperea marginata*, is known from strandings or sightings in western and southern Australia, Tasmania, New Zealand, the Falkland Islands, the South Atlantic Ocean, South Africa, the Crozet Islands, and possibly Argentina and Chile (Ross, Best, and Donnelly 1975). Many authorities, including Rice (1984), treat the Neobalaenidae as a subfamily of the Balaenidae, but Barnes, Domning, and Ray (1985) and Barnes and McLeod (1984) regarded the two groups as separate families. There are no known fossils.

Caperea is the smallest baleen whale. Females apparently are larger than males. According to Baker (1985), the largest female on record was 645 cm long, while the largest male measured 609 cm. A female 621 cm long weighed 3,200 kg, and a male 547 cm long weighed 2,850 kg. Another adult female specimen measured as follows: pectoral fin length, 66 cm; dorsal fin height, 25 cm; and expanse of the tail flukes, 181 cm. The coloration is black or dark gray above and paler below. Young animals may be lighter than adults, and one juvenile appears to have been albinistic (Ross, Best, and Donnelly 1975). The tongue and interior part of the mouth are pure white. In profile, the ventral outline of the anterior part of the throat is concave (Ross, Best, and Donnelly 1975). There are 213–30 baleen plates on each side of the mouth. They are narrow, have a fringe of fine soft hair, and are colored whitish yellow with a narrow, dark brown marginal

Pygmy right whale *(Caperea marginata)*, photos by Mr. T. Dicks through Peter B. Best.

band on the external edge (Baker 1985). The plates are up to 70 cm long and are ivory white in color but have a dark outer margin. The sickle-shaped dorsal fin is located on the posterior part of the back.

Caperea resembles *Eubalaena* and *Balaena* in having a relatively large head, about one-fourth of the total length, and a strongly arched lower jaw. Aside from its much smaller size, it differs from these other genera in having a dorsal fin, incipient throat grooves, a different type of baleen, a smaller

head size in proportion to the body, a proportionally shorter humerus, and four instead of five digits in the manus. Its skull is very different, having a larger, more anteriorly thrust occipital shield; a shorter, wider, and less arched rostrum; smaller nasal bones; shorter supraorbital processes of the frontals; and orbits and glenoid fossae that are not located so far ventrally (Barnes and McLeod 1984). There are 40–44 vertebrae, fewer than in the other genera (Baker 1985).

Caperea has 34 ribs, more than any other cetacean genus.

These ribs also extend farther posteriorly than those of any other cetacean, so only two vertebrae without ribs intervene between those with ribs and the tail. The ribs become increasingly flattened and widened toward the tail and thereby probably provide additional protection to the internal organs.

The pygmy right whale is known only by 71 specimens and a small number of sightings. Available information on natural history was summarized by Baker (1985), E. D. Mitchell (1975a), and Ross, Best, and Donnelly (1975). The genus appears to be restricted to temperate waters of 5°–20° C in the Southern Hemisphere. Although it is primarily pelagic, stranding records suggest a movement, mainly by juveniles, toward the shore in spring and summer. At such times *Caperea* is frequently found in sheltered, shallow bays. Records indicate that this genus occurs throughout the year around Tasmania but only from August to December off Australia and from December to February off South Africa.

Old reports suggested that *Caperea* dives deeply and may remain for long periods near the bottom. More recent observations indicate that it stays underwater only for three or four minutes at a time, at a depth of two or three meters. It is a relatively slow swimmer, averaging around 7 km/hr, and proceeds with a flexing of the entire body. The diet evidently consists mainly of the minute crustaceans known as copepods.

Schools of up to eight individuals have been observed. Records of pregnant females suggest a mating and calving season of several months. At least some young are thought to be born during the autumn or winter, apparently well away from land. Theoretical length at birth is 190 cm. Nursing is presumed to last about five or six months and to be followed by a movement of the young toward shore in the spring and summer.

Caperea now is on appendix 1 of the CITES.

Order Carnivora

Dogs, Bears, Raccoons, Weasels, Mongooses, Hyenas, and Cats

This order of 7 families, 92 genera, and 240 species occurs naturally throughout the world, except for Australia, New Guinea, New Zealand, Antarctica, and many oceanic islands. One species, *Canis familiaris*, apparently was introduced into Australia by human agency in prehistoric time and subsequently established wild populations on that continent. Two living superfamilies of carnivores are usually recognized (Hall 1981; Stains 1984): the Arctoidea (or Canoidea), with the families Canidae, Ursidae, Procyonidae, and Mustelidae; and the Aeluroidea (or Feloidea), with the families Viverridae, Hyaenidae, and Felidae. Some authorities, such as Simpson (1945), have considered these 7 families to be comprised by 1 suborder, the Fissipedia, and that there is a second living carnivore suborder, the Pinnipedia. Other authorities, such as Tedford (1976) and Rice (1977), place the pinnipeds in the arctoid group of the Carnivora. The Pinnipedia are here treated as a full order, the account of which should be seen for further information on questions regarding its classification.

The smallest living carnivore is the least weasel *(Mustela nivalis)*, which has a head and body length of 135–85 mm, a tail length of 30–40 mm, and a weight of 35–70 grams. The largest is the grizzly or brown bear, some individuals of which, particularly along the coast of southern Alaska, attain a head and body length of 2,800 mm and a weight of 780 kg.

Carnivores have four or five clawed digits on each limb. The first digit (pollex and hallux) is not opposable and sometimes is reduced or absent. Some carnivores, including canids and felids, are digitigrade, walking only on their toes. Others, such as ursids, are plantigrade, walking on their soles with the heels touching the ground. The brain has well-developed cerebral hemispheres, and the skull is heavy, with strong facial musculature. The articulation of the lower jaw is such as to permit only open-and-shut (not side-to-side) movements. The stomach is simple. Males have a baculum. The number of mammae in females is variable; they are located on the abdomen, except that in the Ursidae some are pectoral.

The teeth are rooted. The small, pointed incisors number 3/3 in all species except *Ursus ursinus*, which has 2/3, and *Enhydra lutris*, which has 3/2. The first incisor is the smallest, and the third is the largest, the difference in size being most marked in the upper jaw. The canine teeth are strong, recurved, pointed, elongate, and round to oval in section. The premolars are usually adapted for cutting, and the molars usually have four or more sharp, pointed cusps. The last upper premolar and the first lower molar, called the car-nassials, often work together as a specialized shearing mechanism. The carnassials are most highly developed in the Felidae, which have a diet consisting almost entirely of meat, and are least developed in the omnivorous Ursidae and Procyonidae.

Most carnivores are terrestrial or climbing animals. Two genera, *Potos* and *Arctictis*, have prehensile tails. Apparently, all carnivores can swim if necessary, but the polar bear *(Ursus maritimus)* and the river otters *(Lutra, Aonyx, Pteronura)* are semiaquatic, and the sea otter *(Enhydra)* spends practically its entire life in the water. Land-dwelling carnivores shelter in caves, crevices, burrows, and trees. They may be either diurnal or nocturnal.

Most species of the Canidae, Mustelidae, Viverridae, and Felidae live solely or mainly on freshly killed prey. Their whole body organization and manner of living are adapted for predation. The diet may vary by season and locality. Hunting is done by scent and sight, and the prey is captured by a surprise pounce from concealment *(Panthera pardus)*, a stalk followed by a swift rush *(Mustela frenata)*, or a lengthy chase *(Canis lupus)*. Some species regularly eat carrion. *Arctictis* is largely frugivorous, and the diet of certain other viverrids consists partly of fruit. The Hyaenidae include two genera *(Crocuta* and *Hyaena)* that both hunt large animals and feed on carrion and one genus *(Proteles)* that is largely insectivorous. The Procyonidae and Ursidae, except for the carnivorous polar bear, are omnivorous, eating a wide variety of plant and animal life.

Carnivores are solitary or associate in pairs or small groups. Females commonly produce a single litter each year, but those of a few species may give birth two or three times annually, and those of some large species usually mate at intervals of several years. Most species have gestation periods of about 49–113 days. Delayed implantation of the fertilized egg occurs in ursids and some mustelids, so the period from mating to birth is considerably longer than average. Litter size commonly ranges from 1 to 13. The offspring usually are born blind and helpless, but with a covering of hair. They are cared for solicitously by the mother and, in some species, by the father. There is often a lengthy period of parental care and instruction.

The Carnivora were once considered to include the Creodonta, an extinct group dating back to the late Cretaceous, as a suborder. It now appears, however, that the Carnivora arose independently from ancestral insectivore stock and were initially represented by the family Miacidae of the late Paleocene and Eocene (Hall 1981; Romer 1968).

A. Maned wolf *(Chrysocyon brachyurus)*. B. Linsang *(Prionodon linsang)*. C. Leopard *(Panthera pardus)*. D. Giant panda *(Ailuropoda melanoleuca)*. Photos by Bernhard Grzimek.

CARNIVORA; **Family CANIDAE**

Dogs, Wolves, Coyotes, Jackals, and Foxes

This family of 16 Recent genera and 36 species has a natural distribution that includes all land areas of the world except the West Indies, Madagascar, Taiwan, the Philippines, Borneo and islands to the east, New Guinea, Australia, New Zealand, Antarctica, and most oceanic islands. There are wild populations of the species *Canis familiaris* in Australia and New Guinea, but these apparently originated through introduction by human agency. The living Canidae traditionally have been divided, mainly on the basis of dentition, into three subfamilies: Caninae, with the genera *Canis, Alopex, Vulpes, Fennecus, Urocyon, Nyctereutes, Dusicyon, Cerdocyon, Atelocynus,* and *Chrysocyon;* Simocyoninae, with *Speothos, Cuon,* and *Lycaon;* and Otocyoninae, with *Otocyon.*

Recent studies have indicated that subfamilial distinction for the Simocyoninae and Otocyoninae is not warranted and also have revealed considerable controversy regarding the systematics of the Caninae. Langguth (1975*b*) referred most species of the South American *Dusicyon* to a subgenus

A. Maned wolf pup *(Chrysocyon brachyurus)*, photo from Los Angeles Zoological Society. B. Red fox pup *(Vulpes vulpes)*, photo by Leonard Lee Rue III.

Bush dogs *(Speothos venaticus)*, photo by Bernhard Grzimek.

(Pseudalopex) of *Canis,* referred one species *(D. vetulus)* to the genus *Lycalopex,* and retained only one species *(D. australis)* in *Dusicyon.* Clutton-Brock, Corbet, and Hills (1976) considered the genus *Vulpes* to include *Fennecus* and *Urocyon,* and the genus *Dusicyon* to include *D. australis,* the species referred by Langguth to *Pseudalopex* and *Lycalopex,* and also *Cerdocyon* and *Atelocynus.* Van Gelder (1978) expanded the genus *Canis* to contain the following as subgenera: *Dusicyon,* with the single species *D. australis; Pseudalopex,* with most species traditionally assigned to *Dusicyon; Lycalopex,* with the species formerly called *Dusicyon vetulus; Cerdocyon; Atelocynus; Vulpes,* including *Fennecus* and *Urocyon;* and *Alopex.* Berta (1987) considered *Otocyon* and *Urocyon* to be closely related to *Vulpes, Lycalopex* to be part of *Pseudalopex, D. australis* to be the only modern species of *Dusicyon, Nyctereutes* to be the closest living relative of *Cerdocyon,* and *Speothos* to be the closest living relative of *Atelocynus.*

Because of the controversy, a conservative position has been taken here, and all traditionally recognized genera have been maintained. In addition, *Pseudalopex* and *Lycalopex* have been given generic rank. The sequence of genera presented here is based partly on information given by Langguth (1975b) and Nowak (1978, 1979) indicating that *Vulpes* and the other foxes are more primitive than *Canis.* Further systematic comments occur in the generic accounts.

In wild species, head and body length is 357–1,600 mm, tail length is 125–560 mm, and weight is 1–80 kg. *Fennecus zerda* is the smallest species, and *Canis lupus* is the largest. In a given population, males generally average larger than females. Most species are uniformly colored or speckled, but one species of jackals *(Canis adustus)* has stripes on the sides of its body, and *Lycaon* is covered with blotches.

Canids have a lithe, muscular, deep-chested body; usually long, slender limbs; a bushy tail; a long, slender muzzle; and large, erect ears. There are four digits on the hind foot and five on the forefoot, except in *Lycaon,* which has four on both the front and the back foot. The claws are blunt. Males have a well-developed baculum, and females generally have three to seven pairs of mammae.

The skull is elongate. The bullae are prominent but usually are not highly inflated. The dental formula in all but three species is: (i 3/3, c 1/1, pm 4/4, m 2/3) × 2 = 42. The molars are 1/2 in *Speothos,* 2/2 in *Cuon,* and 3/4 or 4/4 in *Otocyon.*

Canids occur from hot deserts *(Fennecus)* to arctic ice fields *(Alopex).* For dens they may use burrows, caves, crevices, or hollow trees. These alert, cunning animals may be diurnal, nocturnal, or crepuscular. They are generally active throughout the year. They walk, trot tirelessly, amble, or canter, either entirely on their digits or partly on more of the foot. At full speed they gallop. The gray foxes *(Urocyon)* often climb trees, an unusual habit for canids. The senses of smell, hearing, and sight are acute. Prey is captured by an open chase or by stalking and pouncing. The diet may vary by season, and vegetable matter is important to some species at certain times.

Some canids, especially the larger species, occur in packs of up to 30 members and seek prey animals that are larger than themselves. Most smaller canids hunt alone or in pairs, preying on rodents and birds. There is usually a regular home range, part or all of which may be an exclusive territory. Females generally give birth once a year. Litters usually contain 2–13 young. Gestation averages around 63 days. The offspring are blind and helpless at birth but are covered with hair. They are cared for solicitously by the mother and often by the father and other group members as well. Sexual maturity comes after 1 or 2 years. Potential longevity is probably at least 10 years in all species.

The geological range of this family is late Eocene to Recent in North America and Europe, early Oligocene to Recent in Asia, late Pliocene to Recent in South America, Pliocene to Recent in Africa, and late Pleistocene to Recent in Australia (Berta 1987; Langguth 1975b; Macintosh 1975).

CARNIVORA; CANIDAE; Genus VULPES
Bowdich, 1821

Foxes

There are 10 species (Clutton-Brock, Corbet, and Hills 1976; Coetzee, *in* Meester and Setzer 1977; Corbet 1978; Ellerman and Morrison-Scott 1966; Hall 1981; Mendelssohn et al. 1987; Roberts 1977):

V. vulpes (red fox), Eurasia except the southeastern tropical zone, northern Africa, most of Canada and the United States;

V. corsac (corsac fox), dry steppe and subdesert zone from the lower Volga River to Manchuria and Tibet;

V. ferrilata (Tibetan sand fox), high plateau country of Tibet, Nepal, and north-central China;

V. cana (Blanford's fox), mountain steppe zone of southern Turkmen S.S.R., Iran, Pakistan, and Afghanistan and an apparently isolated population in Israel and Sinai;

V. velox (swift fox), southern Alberta and North Dakota to northwestern Texas;

V. macrotis (kit fox), southern Oregon to Baja California and north-central Mexico;

V. bengalensis (Bengal fox), Pakistan, India, Nepal;

V. rueppellii (sand fox), desert zone from Morocco and Niger to Afghanistan and Somalia;

V. pallida (pale fox), savannah zone from Senegal to northern Sudan and Somalia;

V. chama (Cape fox), dry areas of southern Angola, Namibia, Botswana, western Zimbabwe, and South Africa.

Treatment of *Vulpes* as a distinct genus that does not include *Fennecus, Urocyon,* or *Alopex* is in keeping with the arrangements of such authorities as Coetzee (*in* Meester and Setzer 1977), Hall (1981), Jones et al. (1986), and Rosevear (1974). Other views have been to consider *Vulpes* a full genus that includes *Alopex* (Youngman 1975); a full genus that includes *Fennecus* and *Urocyon* but not *Alopex* (Clutton-Brock, Corbet, and Hills 1976); and a subgenus of *Canis* that includes *Fennecus* and *Urocyon* but not *Alopex* (Van Gelder 1978). The North American red fox has sometimes been designated a separate species, *V. fulva,* but most authorities now consider it to be conspecific with the Palearctic *V. vulpes.* Hall (1981) treated *V. macrotis* as being conspecific with *V. velox,* because Rohwer and Kilgore (1973) had reported interbreeding between these two kinds of fox where their ranges meet in eastern New Mexico and western Texas. Thornton and Creel (1975) and Stromberg and Boyce (1986), however, argued for continued recognition of the two as separate species.

Vulpes is characterized by a rather long, low body; relatively short legs; a long, narrow muzzle; large, pointed ears; and a bushy, rounded tail that is at least half as long, and often fully as long, as the head and body. The pupils of the eyes generally appear elliptical in strong light. Some species have a pungent "foxy" odor, arising mainly from a gland located on the dorsal surface of the tail, not far from its base. Females usually have six or eight mammae. Additional information is provided separately for each species.

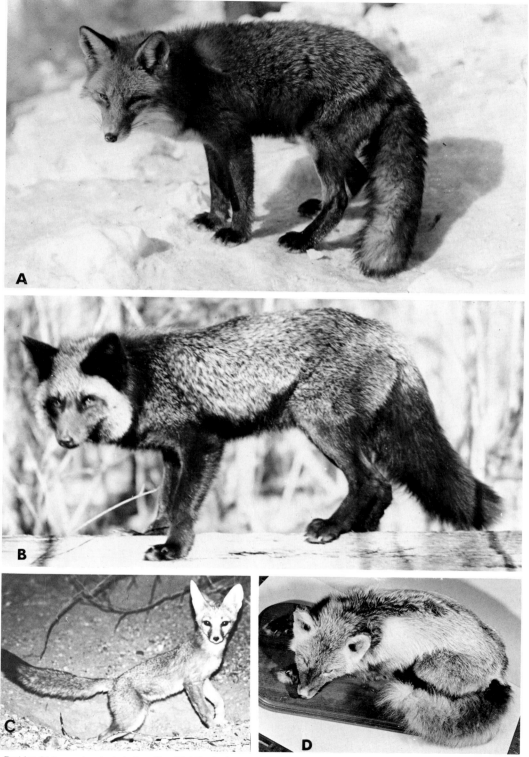

A. Red fox *(Vulpes vulpes)*, photo from New York Zoological Society. B. Silver fox *(V. vulpes)*, photo from Fromm Brothers, Inc. C. Kit fox *(V. macrotis)*, photo by O. J. Reichman. D. Cross fox *(V. vulpes)*, photo by Howard E. Uible of mounted specimen in U.S. National Museum of Natural History.

Vulpes vulpes (red fox).

Head and body length is 455–900 mm, tail length is 300–555 mm, and weight is 3–14 kg. Average weights in North America are 4.1–4.5 kg for females and 4.5–5.4 kg for males (Ables 1975). The usual weight in central Europe is 8–10 kg (Haltenorth and Roth 1968). The typical coloration ranges from pale yellowish red to deep reddish brown on the upper parts and is white, ashy, or slaty on the underparts. The lower part of the legs is usually black, and the tail is generally tipped with white or black. Color variants, known as the cross fox and silver fox, represent, respectively, about 25 percent and 10 percent of the species. The cross fox is reddish brown in color and gets its name from the cross formed by one black line down the middle of the back and another across the shoulders. The color of the silver fox, which has the most prized fur of any fox, ranges from strong silver to nearly black. The general color effect depends on the proportion of white or white-tipped to black hairs. An individual with only a few white hairs is sometimes called a black fox.

The red fox rivals the gray wolf *(Canis lupus)* for having the greatest natural distribution of any living terrestrial mammal besides *Homo sapiens*. Habitats range from deep forest to arctic tundra, open prairie, and farmland, but the red fox prefers areas of highly diverse vegetation and avoids large homogeneous tracts (Ables 1975). Elevational range is sea level to 4,500 meters (Haltenorth and Roth 1968). Daily rest may be taken in a thicket or any other protected spot, but each individual or family group usually has a main earthen den and one or more emergency burrows within the home range. An especially large den may be constructed during the late winter and subsequently used to give birth and rear the young. Some dens are used for many years by one generation of foxes after another. The preferred site is a sheltered, well-drained slope with loose soil. Often a marmot burrow is taken over and modified. Tunnels are up to 10 meters long and lead to a chamber 1–3 meters below the surface. There is sometimes only a single entrance, but there may be as many as 19. A system of pathways connects the dens, other resting sites, favored hunting areas, and food storage holes (Ables 1975; Banfield 1974; Haltenorth and Roth 1968; Stanley 1963).

The red fox is terrestrial, normally moving by a walk or trot. It has great endurance and can gallop for many kilometers if pursued. It can run at speeds of up to 48 km/hr, can leap fences 2 meters high, and can swim well (Haltenorth and Roth 1968). It has keen senses of sight, smell, and hearing. Its ability to survive in the close proximity of people, and often to elude human hunters and their dogs, has given it a reputation for cunning and intelligence. Most activity is nocturnal and crepuscular. Individuals cover up to 8 km per night as they move on circuitous routes through the home range (Banfield 1974). During the autumn, the young born the previous spring disperse from the parental home range. The usual distance traveled at this time is about 40 km and 10 km for females; maximum known is 394 km (Ables 1975; Storm et al. 1976). Once the young animals establish themselves in a new area, they generally remain there for life.

The diet is omnivorous, consisting mostly of rodents, lagomorphs, insects, and fruit. To hunt mice, the red fox stands motionless, listens and watches intently, and then leaps suddenly, bringing its forelegs straight down to pin the prey. Rabbits are stalked and then captured with a rapid dash (Ables 1975). Daily consumption is around 0.5–1.0 kg. Sometimes a hole is dug and excess prey placed therein and covered over, to be eaten at a later time (Haltenorth and Roth 1968).

The most favorable areas usually support an average of one or two adults per sq km (Ables 1975; Haltenorth and Roth 1968; Insley 1977). Home range size varies with habitat conditions and food availability and becomes larger in winter and smallest around the time of the arrival of newborn (Ables 1975). According to Grzimek (1975), the home range is 5–12 sq km in good habitat and 20–50 sq km in poor habitat. Jones and Theberge (1982) reported home range to average about 16 sq km in tundra habitat of northwestern British Columbia, larger than in temperate environments. *V. vulpes* is apparently territorial. There is little overlap of home ranges, and individuals on different ranges avoid one another (Storm and Montgomery 1975). Captive males were found continually to harass and chase foxes newly introduced to the enclosure, while females seldom became involved in such interaction (Preston 1975). In the breeding season, however, females do exhibit territorial behavior (Haltenorth and Roth 1968).

A home range is typically occupied by an adult male, one or two adult females, and their young (Storm and Montgomery 1975). Occasionally, two females have litters in the same den (Pils and Martin 1978). Males may fight one another during the breeding season. A vixen sometimes mates with several males, but she later establishes a partnership with just one of them (Haltenorth and Roth 1968). For a period extending from shortly before birth to several weeks thereafter, the female remains in or very near the den. The male then brings her food but does not actually enter the maternal den.

The mating season varies with latitude. In Europe, it is December–January in the south, January–February in central regions, and February–April in the north (Haltenorth and Roth 1968). In North America, mating occurs over about the same period (Ables 1975; Storm et al. 1976). Females are monestrous, have an estrus of 1–6 days, and have a gestation period of 49–56, usually 51–53, days. Litter size is 1–13 young but averages about 5 throughout the range of the fox (Ables 1975; H. G. Lloyd 1975). The young weigh 50–150 grams each at birth, open their eyes after 9–14 days, emerge from the den at 4–5 weeks, and are weaned at 8–10 weeks. They may be moved to a new den at least once. The family remains together until the autumn. Sexual maturity is reached at about 10 months. Potential natural longevity is around 12 years, though few individuals live more than 3–4 years, at least where the species is heavily hunted and trapped (Ables 1975).

The red fox is killed by people for sport, to protect domestic animals and game, to prevent the spread of rabies, and to obtain the valuable pelt. Sport hunting may involve an elaborate daytime chase by large numbers of riders and dogs or a nocturnal effort by one person to lure the fox with a call imitating that of a wounded rabbit. In Great Britain, *V. vulpes* is traditionally valued as a game animal, but H. G. Lloyd (1975) noted that it is also the only mammal in the country subject to a government-approved bounty. It has become common in parts of London and other cities, and control efforts there have not substantially reduced its numbers (Harris and Smith 1987). The red fox is often considered to be a threat to poultry, but depredations are generally localized, and many of the birds eaten are taken in the form of carrion. Studies have indicated that the red fox has little effect on wild pheasant populations (Ables 1975). Rabid foxes are said to be a serious menace in some areas, especially Europe, and intensive persecution there may be threatening the species in certain parts of its range; 180,000 individuals are taken every year in Germany alone (Grzimek 1975). A rabies epizootic, the main vector of which is the fox, spread from Poland across much of Europe from the 1940s to 1970s; however, only 5–10 percent of reported cases of rabies in domestic animals in the involved region have resulted from this epizootic (Steck and Wandeler 1980). Most cases of rabies in Canada from 1958 to 1986 were reported from Ontario, and most of those occurrences (17,982) were in the red fox (Rosatte 1988). Problems caused by the fox are perhaps more than balanced by its control of rodent populations, which might otherwise multiply and damage human interests.

From 1900 to 1920 in North America, and to some extent in other parts of the world, catching wild foxes and raising them in captivity developed into an important industry. In the early stages of the breeding effort, choice animals often sold for more than $1,000 each. Through selective breeding, strains were developed that nearly always produce silver-colored offspring. The number of foxes being raised for their fur now exceeds that of all other normally wild animals, except possibly the mink *(Mustela vison)*. One fox farm permanently employed about 400 people and sold pelts worth more than $18 million. The fur is used in coats, stoles, scarves, and trimming. The value of fox pelts has varied widely, depending on fashions, availability, and economic conditions. According to Banfield (1974), the average price of a silver fox skin was $246.46 in 1919/20 but only $17.94 in 1971/72. The average price of a wild-caught U.S. red fox skin rose from $12.00 in the 1970/71 season to about $48.00 in 1976/77. The reported number of red foxes trapped for their fur during the latter season in the United States and Canada was 421,705 (Deems and Pursley 1978). The number of pelts taken annually rose above 500,000 during the early 1980s, with average prices peaking at over $60.00, but there was a decline to under $20.00 by 1984 (Voigt 1987).

Despite human persecution, *V. vulpes* has maintained or even increased its numbers in many parts of its range. There are now probably more in Great Britain than there were in medieval times, because of improved habitat conditions resulting from the establishment of hedgerows and crop rotation (H. G. Lloyd 1975). This species is able to carry on its mode of life in intensively farmed areas and sometimes even in large cities (Ables 1975; Grzimek 1975). It also has been successfully introduced in some areas, especially by persons of English background who desired to continue traditional fox hunting. The species was brought to Australia in 1868 and subsequently spread over much of that continent, to the lasting detriment of the native fauna (Clutton-Brock, Corbet, and Hills 1976; Ride 1970).

Introductions from England also were made in eastern North America in colonial times. The species was naturally present in this region but apparently was not abundant. It subsequently increased in numbers and became established in areas not previously occupied, mainly because of the breaking up of the homogeneous forests by people and continuous introduction by hunting clubs. In the twentieth century the red fox has greatly extended its range in the southeastern United States and has occupied Baffin Island and moved as far north as the southern coast of Ellesmere Island. It has spread westward across the Great Plains, possibly in response to a human-caused reduction of coyote *(Canis latrans)* numbers (Banfield 1974; Hall 1981; Hatcher 1982; Lowery 1974). The only major North American population that may be in trouble from a conservation viewpoint is that of the Sierra Nevada of California, where surveys indicate that the native subspecies *(V. v. necator)* is very rare and evidently declining (Schempf and White 1977).

Vulpes corsac (corsac fox).

Head and body length is 500–600 mm, and tail length is 250–350 mm. The fur is thick and soft. The general coloration of the upper parts is pale reddish gray, or reddish brown with silvery overtones. The underparts are white or yellow. *V. corsac* is externally similar to *V. vulpes* but has relatively longer legs. Its ears are large, pointed, and very broad at the base (Novikov 1962).

The corsac fox is a typical inhabitant of steppes and semidesert. It avoids forests, thickets, plowed fields, and settled areas. It lives in a burrow, often taken over from another mammal, such as a marmot or badger. Self-excavated bur-

Corsac foxes *(Vulpes corsac),* photo from Amsterdam Zoo.

rows are simple, usually very shallow, and sometimes are found in groups (Novikov 1962). Although usually reported to be nocturnal in the wild, *V. corsac* is active by day in captivity; it is said to be an excellent climber (Grzimek 1975). It runs with only moderate speed and can be caught by a slow dog, but it has excellent senses of vision, hearing, and smell (Stroganov 1969). Most reports indicate that it is nomadic and does not keep to a fixed home range (Ognev 1962). It may migrate southward when deep snow and ice make hunting difficult (Stroganov 1969). The diet consists mostly of small rodents but also includes pikas, birds, insects, and plant material.

This species is more social than other foxes, with several individuals sometimes living together in the same burrow (Ognev 1962). Small hunting packs are said to form in the winter (Stroganov 1969), though perhaps these represent mated pairs and their grown young of the previous spring. Males fight one another during the breeding season but then remain with the family (Grzimek 1975; Novikov 1962). Mating occurs from January to March, gestation lasts 50–60 days, and litters usually contain 2–11 young (Stroganov 1969). Females in the Berlin Zoo did not reach sexual maturity until their third year of life (Grzimek 1975).

The corsac fox lacks the penetrating odor of most *Vulpes* and was frequently kept as a pet in eighteenth-century Russia (Grzimek 1975). Its warm and beautiful fur led to large-scale commercial trapping; up to 10,000 pelts were sold annually in the western Siberian city of Irbit in the late nineteenth century. For this reason, and also because of the settlement and plowing of the steppes, the corsac fox has disappeared in much of its range (Ognev 1962; Stroganov 1969).

Vulpes ferrilata (Tibetan sand fox).

Head and body length is 575–700 mm, tail length is 400–475 mm, and males weigh up to 7 kg (Mitchell 1977). The fur is soft and thick and the tail is bushy. The upper parts are pale gray agouti or sandy, with a tawny band along the dorsal region. The underparts are pale, the front of the leg is tawny, and the tip of the tail is white. The skull is peculiarly elongated and has a very narrow maxillary region (Clutton-Brock, Corbet, and Hills 1976).

Mitchell (1977) found this fox on barren slopes and in stream beds at 3,000–4,000 meters in the Mustang district of Nepal. In this area, dens are made in boulder piles or in burrows under large rocks. The diet consists of rodents, lagomorphs, and ground birds. Mitchell observed pairs hunting along stream beds, on boulder heaps, and in wheat fields. Mating occurs in late February, and two to five young are born in April or May.

Vulpes cana (Blanford's fox).

Novikov (1962) reported head and body length to be less than 500 mm and tail length to be 330–410 mm. Three specimens listed by Mendelssohn et al. (1987) had head and body lengths of 406–38 mm, tail lengths of 324–28 mm, and weights of 710–956 grams. Clutton-Brock, Corbet, and Hills (1976:155) described *V. cana* as "a small fox with extremely soft fur and a long very bushy tail. The colouring is blotchy black, grey and white with a dark tip to the tail and a dark patch over the tail gland. There is an almost black mid-dorsal line and the hind legs may be dark. . . . The underparts are white, the ears are grey, and there is a small dark patch between the eyes and nose."

According to Roberts (1977), the habitat of *V. cana* is mountain steppe. It is reportedly more frugivorous than the other foxes of Pakistan, being fond of ripe melons and seedless grapes, and sometimes damaging crops. Mendelssohn et al. (1987) found it not to be rare in the rocky habitats of the Negev and Judean deserts. They added that it has an aston-

ishing jumping ability and can move upward among cliffs by pushing itself from one vertical wall to another. Its diet in that area evidently consists mainly of arthropods, and females captured there gave birth to litters of 1 and 3 young in February and April. The skin of *V. cana* is valued in commerce and is heavily hunted. Novikov (1962) called this species one of the rarest predators of the Soviet Union. It is on appendix 2 of the CITES.

Vulpes velox (swift fox).

Head and body length is 375–525 mm, tail length is 225–350 mm, and weight is 1.8–3.0 kg. Males average larger than females. The winter coat is long and dense; the upper parts are dark buffy gray; the sides, legs, and lower surface of the tail are orange tan; and the underparts are buff to pure white. In summer the coat is shorter, harsher, and more reddish. *V. velox* differs from the closely related *V. macrotis* in having smaller ears, a broader snout, and a shorter tail (Egoscue 1979).

The swift fox inhabits prairies, especially those with grasses of short and medium height. For shelter it depends on burrows, which are either self-excavated or taken over from another mammal. The burrows are usually simple and located on high, well-drained ground. The tunnels may be 350 cm long and lead to a chamber as much as 150 cm below the surface. There are one to seven or more entrances. The swift fox is primarily nocturnal but sometimes suns itself near the den. Its diet consists mostly of lagomorphs and also includes rodents, birds, lizards, and insects (Egoscue 1979; Kilgore 1969).

The usual social unit is a mated pair and their young, but occasionally a male will live with two adult females. The mating season in Oklahoma is late December to early January, and most young are born in March or early April. Females are monestrous. Litters consist of three to six young. Their eyes open after 10–15 days, weaning occurs after 6–7 weeks, and they probably remain with the parents until August or early September. A captive lived for 12 years and 9 months (Egoscue 1979; Kilgore 1969).

The swift fox is not as cautious as *V. vulpes* and seems to take poison baits readily. In the mid- and late nineteenth century, intensive poisoning was carried out on the Great Plains, mainly to eliminate wolves, coyotes, and other predators, and many swift foxes were accidentally killed. Subsequently, much habitat was lost as the prairies were converted to agriculture. The swift fox was also taken for its fur, though it has never been of major commercial value. By the 1920s, the northern subspecies, *V. velox hebes*, apparently had disappeared, though occasional reports continued in Canada, and the southern subspecies, *V. velox velox*, survived only in Colorado, New Mexico, western Texas, and possibly western Kansas. For reasons not fully understood, a moderate comeback seems to have occurred in the last few decades. The species reappeared in Oklahoma, much of Kansas, Nebraska, and Wyoming in the 1950s and in South Dakota, North Dakota, and Montana in the 1960s and 1970s (Egoscue 1979; Floyd and Stromberg 1981; Kilgore 1969; Moore and Martin 1980; Zumbaugh and Choate 1985).

V. velox hebes is on appendix 1 of the CITES. The USDI lists the subspecies as endangered, but this designation officially applies only in Canada. The swift fox populations now on the northern plains of the United States may be descended from animals that moved north from the range of *V. velox velox*. There has been considerable controversy within the last decade regarding both the systematic status of the swift fox subspecies and a reintroduction program in Canada. The latter project, which began in 1983, involves the capture of foxes in Colorado, Wyoming, and South Dakota and the release of them or their offspring in Alberta and Saskatchewan.

Stromberg and Boyce (1986) argued on the one hand that *hebes* probably is not a valid subspecies but on the other that there is significant geographic variation in *V. velox* and that gene flow from the transplanted animals might adversely affect the viability of natural populations on the U.S. side of the border. Herrero, Schroeder, and Scott-Brown (1986) replied that the transplanted stock was taken from the northernmost populations that were readily available and that more northerly animals in the United States may be recently descended from the same stock. Despite the USDI listing, the original swift fox population of Canada evidently disappeared by the 1930s. Carbyn (1989) reported that the program had released about 250 animals and that many are known to have survived.

Vulpes macrotis (kit fox).

The account of this species is based in large part on McGrew (1979). Head and body length is 375–500 mm and tail length is 225–323 mm. Average weight is 2.2 kg for males and 1.9 kg for females. The back is generally light grizzled or yellowish gray, the shoulders and sides are buffy to orange, and the underparts are white. The ears, which are proportionally the largest of any North American canid, are set closely together.

The kit fox is closely associated with steppe and desert habitat, generally with a covering of shrubs or grasses. Dens usually have multiple entrances, the number varying from 2 to 24. There are groups of dens in favorable areas, and a fox family may move from one to another during the year, leaving most vacant at any given time. *V. macrotis* is nocturnal and may travel several kilometers per night during hunts. Its diet consists largely of rodents, such as kangaroo rats, and lagomorphs.

Optimal habitat in Utah was found to support two adults per 259 ha. Other areas had densities of one fox per 471–1,036 ha. In the San Joaquin Valley, Morrell (1972) found that each fox apparently spent its entire life in an area of 260–520 ha. Home ranges overlap extensively, and there apparently is no definite territory. Usually an adult male and female live together, though not necessarily permanently, and a second female is sometimes present. When a female is nursing young, it rarely leaves the den, and the male supplies it with food. Several vocalizations are known, including a bark by mothers to recall the young. Females are monestrous. Mating occurs from December to February, and the young are born in February and March. There are usually four or five offspring, each of which weighs about 40 grams. The young emerge from the den at 1 month and begin to accompany the parents at 3–4 months. The family splits up in the fall, with the young dispersing beyond the parental home range. According to Jones (1982), one specimen was still living after 20 years in captivity.

The kit fox is not a particularly cautious animal, and its numbers have been greatly reduced in some areas by poisoning, trapping, and shooting. Habitat disruption also has led to declines, especially in California. The subspecies *V. m. macrotis*, of the southwestern corner of the state, disappeared by 1910. The San Joaquin Valley subspecies, *V. m. mutica*, is classified as endangered by the USDI and as rare by the California Department of Fish and Game (1978). It has declined because of conversion of areas of natural vegetation to irrigated agriculture. Most states classify *V. macrotis* as a fur bearer and allow trapping; in recent years the annual legal kill has been 5,400–7,800, and pelts have sold for $12–$23 (O'Farrell 1987).

Vulpes bengalensis (Bengal fox).

The account of this species is based largely on Roberts (1977). Head and body length is 450–600 mm and tail length is 250–350 mm. Males weigh 2.7–3.2 kg and females weigh less than 1.8 kg. The upper parts are yellowish gray or silvery gray, the underparts are paler, and the backs of the ears and the tip of the tail are dark.

The Bengal fox is generally found in open country with a scattering of trees. It avoids deserts and mountains. It digs its own burrow and hunts mainly by night. A captive could climb low trees and a vertical wire net. The diet is omnivorous and includes small vertebrates, insects, and fruit. *V. bengalensis* is adept at catching frogs and digging lizards out of their burrows. The young, usually four to a litter, are born from February to April.

Vulpes rueppellii (sand fox).

Head and body length is 400–520 mm and tail length is 250–350 mm (Roberts 1977). Weight is about 1.5–3.0 kg. The coat

Kit fox *(Vulpes macrotis)*, photo from San Diego Zoological Society.

Sand fox *(Vulpes rueppelli)*, photo from Antwerp Zoo.

is very soft and dense. The upper parts are silvery gray, the sides are grayish buff, and the underparts are whitish. *V. rueppellii* is much more lightly built than *V. vulpes;* it has rather short legs and broad ears (Dorst and Dandelot 1969).

The usual habitat is stony or sandy desert. Activity is mainly nocturnal. In a study in Oman, Lindsay and Macdonald (1986) found individuals to spend the day in underground dens and to change their den site on an average of once every 4.7 days. By night these animals hunted over a large home range calculated to average 30.4 sq km by one method and 69.1 sq km by another. The diet in that area consisted mostly of small mammals and also included lizards, insects, and grass. There were three monogamous pairs, the members of each of which shared a den site and a portion of hunting range, but there also was some overlap in the ranges of different pairs. Two of the pairs had cubs when the study was conducted, during January–April. Other reported information (Dorst and Dandelot 1969; Roberts 1977) is that a captive female had a litter of three young, and a litter of two young evidently was born in March.

Vulpes pallida (pale fox).

Head and body length is 406–55 mm, tail length is 270–86 mm, and weight is 1.5–3.6 kg. The upper parts are pale sandy fawn, variably suffused with blackish; the flanks are paler, and the underparts are buffy white. The tail is long and bushy (Dorst and Dandelot 1969; Rosevear 1974).

The habitat is savannah. Burrows are large, with tunnels extending 10–15 meters and opening into chambers lined with dry vegetation. Activity is mainly nocturnal. The diet includes rodents, small reptiles, birds, eggs, and vegetable matter. The pale fox is gregarious (Coetzee, in Meester and Setzer 1977; Dorst and Dandelot 1969). Three captive adults, a female and two males, seemed to get along amicably. The female gave birth to a litter of four young in June 1965 (Bueler 1973).

Vulpes chama (Cape fox).

Head and body length is about 560 mm, tail length is about 330 mm, and weight is about 4 kg. The upper parts are silvery gray, and the underparts are pale buff. The tail is very bushy and has a black tip. The ears are pointed. The muzzle is short but pointed (Dorst and Dandelot 1969).

The Cape fox inhabits dry country, mainly open plains and karroo. It is nocturnal, hiding by day under rocks or in burrows in sandy soil. The diet consists mainly of small vertebrates and insects. This species lives alone or in pairs. Its call is a yell followed by several yaps. The breeding season is September–October, gestation lasts 51–52 days, and litters contain three to five young. Human persecution has been responsible for numerical and distributional declines, though studies have shown that *V. chama* is not a harmful predator (Bekoff 1975; Bothma 1966; Dorst and Dandelot 1969).

CARNIVORA; CANIDAE; **Genus FENNECUS**
Desmarest, 1804

Fennec Fox

The single species, *F. zerda,* occurs in the desert zone from southern Morocco and Niger to Egypt and Sudan (Coetzee, *in* Meester and Setzer 1977). There also are records from Sinai, southern Iraq, Kuwait, and the southeastern Arabian Peninsula (Gasperetti, Harrison, and Büttiker 1985). *Fennecus* was included in the genus *Vulpes* by Clutton-Brock, Corbet, and Hills (1976) and Corbet (1978) and in the subgenus *Vulpes* of the genus *Canis* by Van Gelder (1978).

This is the smallest canid, but it has the proportionally largest ears in the family. Head and body length is 357–407 mm, tail length is 178–305 mm, and weight is about 1.0–1.5 kg. The ears are 100–150 mm long. The coloration of the upper parts, the palest of any fox's, is reddish cream, light fawn, or almost white. The underparts are white, and the tip of the tail is black. The coat is thick, soft, and long. The tail is heavily furred. The bullae are exceedingly large, and the dentition is weak (Clutton-Brock, Corbet, and Hills 1976). The feet have hairy soles, enabling the animal to run in loose sand (Bekoff 1975).

Cape fox *(Vulpes chama)*, photo from San Diego Zoological Society.

Fennec foxes *(Fennecus zerda)*, photo from New York Zoological Society.

Fennecus occurs in arid regions and usually lives in burrows several meters long in the sand. It digs so rapidly that it has gained the reputation of being able to sink into the ground. It is nocturnal and quite agile; a captive could spring 60–70 cm upward from a standing position and could jump about 120 cm horizontally. Some food apparently is obtained by digging, as evidenced by the pronounced scratching or raking habit of captives. The diet includes plant material, small rodents, birds and their eggs, lizards, and insects, such as the noxious migratory locusts. Although an abundance of fennec tracks around some water holes indicates that this fox drinks freely when the opportunity arises, travelers also report tracks in the desert far from oases. Laboratory studies suggest that *Fennecus* can survive without free water for an indefinite period (Banholzer 1976).

The fennec lives in groups of up to 10 individuals. Males mark their territory with urine and become aggressive during the breeding season. One captive male became dominant over

a group at the age of 4 years and then killed its 8-year-old father. Females are aggressive and defend the nest site when they have newborn offspring. Males remain with their mate after the young are born and defend them but do not enter the maternal den. Mating occurs in January and February in captivity, and the young are born in late winter and early spring. Females normally give birth once a year, but if the first litter is lost, another may be produced 2.5–3 months later. The gestation period is about 50–52 days, and there are 2–5 young per litter. The young are weaned at 61–70 days and become sexually mature at 11 months (Bekoff 1975; Koenig 1970). One captive lived for 14 years and 7 months (Jones 1982).

Although it does no harm to human interests, the fennec is intensively hunted by the native people of the Sahara. It has become rare in some parts of northwestern Africa (Grzimek 1975). It is on appendix 2 of the CITES.

Gray fox *(Urocyon cinereoargenteus)*, photo from San Diego Zoological Society.

CARNIVORA; CANIDAE; **Genus UROCYON**
Baird, 1858

Gray Foxes

There are two species (Cabrera 1957; Hall 1981):

U. cinereoargenteus, Oregon and southeastern Canada to
western Venezuela;
U. littoralis, San Miguel, Santa Rosa, Santa Cruz, Santa
Catalina, San Nicolas, and San Clemente islands off
southwestern California.

Urocyon was included in the genus *Vulpes* by Clutton-Brock,
Corbet, and Hills (1976) and in the subgenus *Vulpes* of the
genus *Canis* by Van Gelder (1978). All other authorities cited
in this account treated *Urocyon* as a full genus.

In *U. cinereoargenteus* head and body length is 483–685
mm and tail length is 275–445 mm. In *U. littoralis* head and
body length is 480–500 mm and tail length is 110–290 mm.
Usual weight in the genus is 2.5–7.0 kg. The face, upper part
of the head, back, sides, and most of the tail are gray. The
throat, insides of the legs, and underparts are white. The sides
of the neck, lower flanks, and ventral part of the tail are rusty.
The hairs along the middle of the back and top of the tail are
heavily tipped with black, which gives the effect of a black
mane. Black lines also occur on the legs and face of most
individuals. A concealed mane of stiff hairs occurs on top of
the tail. The pelage is coarse. The skull of *Urocyon* is distinguished from that of *Vulpes* in having a deeper depression
above the postorbital process, much more pronounced and
more widely separated ridges extending from the postorbital
processes to the posterior edges of the parietals, and a conspicuous notch toward the rear of the lower edge of the mandible (Lowery 1974).

Gray foxes frequent wooded and brushy country, often in
rocky or broken terrain, and are possibly most common in the

arid regions of the southwestern United States and Mexico.
The preferred habitat in Louisiana is mixed pine-oak woodland bordering pastures and fields with patches of weeds
(Lowery 1974). Several sheltered resting sites may be used on
different days. The main den is in a pile of brush or rocks, a
crevice, a hollow tree, or a burrow that is either self-excavated
or taken over from another animal. Dens in hollow trees have
been found up to 9.1 meters above the ground. Dens used for
giving birth may be lined with vegetation (Trapp and Hallberg 1975).

Urocyon is sometimes called a tree fox, because it frequently climbs trees, a rather unusual habit for a canid. It
lacks the endurance of the red fox, and when pursued, it often
will seek refuge in a tree. It also climbs without provocation,
shinnying up the trunk and then leaping from branch to
branch. Most activity is nocturnal and crepuscular. Nightly
movements cover about 200–700 meters (Trapp and Hallberg
1975). The diet includes many kinds of small vertebrates as
well as insects and vegetable matter. *Urocyon* seems to take
plant food more than do other foxes, and its diet may consist
mostly of fruits and grains at certain seasons and places.

Reported population densities are about 0.4–10.0 foxes per
sq km (Trapp and Hallberg 1975). Reported home range size
varies from 0.13 to 7.7 sq km. Four females in central California followed by radiotracking for various periods from January to July used ranges of 0.3–1.85 sq km (Fuller 1978).
Apparently, each family group uses a separate area. The normal social unit is an adult pair and their young. There is one
litter per year. Mating occurs from late December to March in
the southeastern United States and from mid-January to late
May in New York. The gestation period has never been precisely determined but has been variously estimated at 51–63
days. Litters usually contain about 4 young but may have
from 1 to 10. The young weigh around 100 grams each at
birth, are blackish in color, and open their eyes after 9–12
days. They can climb vertical tree trunks after 1 month of life
and begin to take solid food in 6 weeks. They forage independently by late summer or early fall but apparently remain in

the parental home range until January or February. Most females breed in their first year of life (Trapp and Hallberg 1975). A captive lived for 13 years and 8 months (Jones 1982).

If captured when small, gray foxes tame readily, are as affectionate and playful as domestic dogs, and make more satisfactory pets than do red foxes. The attractive skins of *Urocyon* are used commercially but are not classed as fine furs. For much of the twentieth century the price per pelt averaged around $0.50, but it began to rise in the 1960s (Lowery 1974). For the 1970/71 season the reported harvest in the United States was 26,109 skins, with an average value of $3.50. By the 1976/77 season the reported take had grown to 225,277 pelts, and the average price to $34.00 (Deems and Pursley 1978). The island fox, *U. littoralis,* is classified as rare by the California Department of Fish and Game (1978) and is fully protected by California state law.

CARNIVORA; CANIDAE; **Genus ALOPEX**
Kaup, 1829

Arctic Fox

The single species, *A. lagopus,* occurs on the tundra and adjacent lands and ice-covered waters of northern Eurasia, North America, Greenland, and Iceland (Banfield 1974; Chesemore 1975; Corbet 1978; Hall 1981). *Alopex* was included within the genus *Vulpes* by Youngman (1975) and was considered a subgenus of *Canis* by Van Gelder (1978) but was treated as a full genus by Clutton-Brock, Corbet, and Hills (1976), as well as by all other authorities cited in this account.

Head and body length is 458–675 mm, tail length is 255–425 mm, shoulder height is about 280 mm, and weight is 1.4–

Arctic foxes *(Alopex lagopus):* A. White and blue phases; B. Blue phase; C. Summer coat; photos by A. Pedersen. D. Winter coat, photo by Ernest P. Walker.

9.0 kg. The dense, woolly coat gives *Alopex* a heavy appearance. There are two color phases. Individuals of the "white" phase are generally white in winter and brown in summer but may remain fairly dark throughout the year in areas of less severe climate. Individuals of the "blue" phase are pale bluish gray in winter and dark bluish gray in summer. The blue phase constitutes less than 1 percent of the arctic fox population of most of mainland Canada and less than 5 percent of that of Baffin Island but makes up 50 percent of the population of Greenland. The winter pelage develops in October and is shed in April. In addition to its coloration, *Alopex* differs from *Vulpes* in having short, rounded ears and long hairs on the soles of its feet.

Alopex is found primarily in arctic and alpine tundra, usually in coastal areas. It generally makes its den in a low mound, 1–4 meters high, on the open tundra. Dens usually have 4–12 entrances and a network of tunnels covering about 30 sq meters. Some dens may be used for centuries, by many generations of foxes, and eventually become very large, with up to 100 entrances. The organic matter that accumulates in and around a den stimulates a much more extensive growth of vegetation than is found on most of the tundra. *Alopex* sometimes dens in a pile of rocks at the base of a cliff. It may use its den throughout the year but often seems not to have a fixed home site except when rearing young. It is active at any time of the day and throughout the year. It moves easily over snow and ice and swims readily (Banfield 1974; Chesemore 1975; Macpherson 1969; Stroganov 1969). During blizzards it may shelter in burrows dug in the snow. Several individuals were noted in winter on the Greenland icecap, more than 450 km from the nearest ice-free land and with the temperature below −50° C. Captives have survived experimental temperatures as low as −80° C.

The arctic fox makes the most extensive movements of any terrestrial mammal other than *Homo sapiens*. Over much of its range, including Alaska, there is a basic seasonal pattern, probably associated with food availability (Chesemore 1975). In the fall and early winter the animals shift toward the shore and out onto the pack ice. They are capable of remarkably long travels across the sea ice, having been sighted 640 km north of the coast of Alaska. One individual reportedly reached a latitude of 88° N, at a point 800 km from the nearest land in the Soviet Union. Another fox, tagged 8 August 1974 on Banks Island, was trapped on 15 April 1975 on the northeastern mainland of the Northwest Territories, having covered a straight-line distance of 1,530 km (Wrigley and Hatch 1976). Still another animal, tagged in northeastern Alaska, reached a point on Banks Island, 945 km away (Eberhardt and Hanson 1978). Some foxes have been carried by ice floes as far as Cape Breton and Anticosti islands in the Gulf of St. Lawrence (Banfield 1974).

In some areas, though apparently not in Alaska, there are inland migrations to the forest zone during the winter, in addition to or instead of a shift onto the sea ice. These movements sometimes involve large numbers of foxes and seem especially extensive following a crash in lemming populations, a major food source (Banfield 1974; Chesemore 1975; Pulliainen 1965). *Alopex* has occasionally traveled overland to south-central Ontario and the St. Lawrence River. The deepest known penetration in North America was made by an individual taken in December 1974 in southern Manitoba, nearly 1,000 km south of the tundra. In the Soviet Union, however, inland movements have extended as far as 2,000 km (Wrigley and Hatch 1976).

Alopex takes any available animal food, alive or dead, and frequently stores food for later use. It may follow polar bears on the pack ice, and wolves on land, in order to obtain carrion (Banfield 1974). Its winter diet includes the remains of marine mammals, invertebrates, sea birds, and fish (Chesemore

1975). It also is known to be an important predator of the ringed seal *(Phoca hispida)*, digging through the snow to reach the pups in their subnivean lairs (Smith 1976).

In winter, for those populations that move inland, and in summer, throughout most of the range of *Alopex* the diet consists mainly of lemmings. Other small mammals, ground-nesting birds, and stranded marine mammals are also utilized, but arctic fox populations seem to be associated with those of lemmings in a three- to five-year cycle (Chesemore 1975; Wrigley and Hatch 1976). Following a lemming crash, population density has been estimated at only 0.086 foxes per sq km (Banfield 1974).

Although a number of dens may be found in a favorable area, most are not utilized at any one time. Occupied dens are at least 1.6 km apart and are usually distributed such that there is only one family per every 32–70 sq km (Macpherson 1969). In a nonmigratory population on the coast of Iceland, home ranges were found to vary from 8.6 to 18.5 sq km and evidently to represent defended territories. The foxes there lived in social groups comprising one adult male, two vixens, and young of the year. In such situations, one of the vixens apparently is a nonbreeding animal, born the previous year, that stays on to help care for the next litter before dispersing (Hersteinsson and Macdonald 1982). *Alopex* is monogamous and may mate for life (Chesemore 1975). A number of adults sometimes gather temporarily around a food source, such as a stranded whale, but they then may fight one another. Vocalizations include barks, screams, and hisses (Banfield 1974).

Mating occurs from February to May, and births take place from April to July. Females are monestrous and have an estrus of 12–14 days and an average gestation period of 52 (49–57) days. The number of young per litter varies, depending on environmental conditions, but ranges from 2 to as many as 25. The usual number seems to be about 6–12. The young weigh an average of 57 grams each at birth. They emerge from the den, and are weaned, at around 2–4 weeks. Subsequently, for a brief period, they are brought food by both parents, but by autumn they disperse. They are capable of breeding at 10 months. Most young do not survive their first 6 months of life, and few animals live more than several years in the wild (Banfield 1974; Chesemore 1975; Macpherson 1969; Stroganov 1969). A captive, however, lived for 15 years (Jones 1982).

The raising of blue-phase foxes has been an important industry, as the undressed skins have sold for up to $300 each. However, prices have sometimes fallen below $10, thereby forcing many fox farmers out of business. Some operations have pens, where selective breeding is practiced, but others are on small islands, where the animals run at liberty. Fox farming began in Alaska in 1865, and blue-phase animals were extensively introduced on the Aleutian Islands (Chesemore 1975). These animals increased in numbers and jeopardized or exterminated some populations of native birds. *Alopex* apparently did not originally occur in the Aleutians, except for the islands at the extreme western end of the chain (Murie 1959).

White-phase foxes also have desirable furs and have been subject to intensive trapping. The skins may be left the natural white or dyed one of many colors, especially "platinum" or "blue" imitation. The arctic fox has long been important to the economy of the native people living within its range. Fluctuations in fox numbers and fur prices have frequently combined to cause hardship for native trappers. Trade in white fox skins developed in northern Alaska during the nineteenth century, in conjunction with the whaling industry. In the 1920s, fox trapping was the most important source of income in the area, and the price per pelt averaged $50.00. The Depression destroyed the market, and by 1931 skins sold for $5.00 or less (Chesemore 1972). Recently, prices began to rise again. During the 1976/77 season 4,261 skins were marketed in Alaska at an average price per pelt of $36.00; the figures for Canada were 36,482 and $54.20 (Deems and Pursley 1978). The kill fell to 720 in Alaska and 16,405 in Canada during the 1983/84 season, with the average price in the latter country being only $6.00 (Novak, Obbard, et al. 1987).

In Iceland the arctic fox is persecuted because of its reputed depredations on sheep and lambs. Legislation there has promoted its destruction since at least as early as A.D. 1295. State-subsidized hunting has been so efficient that large parts of the island are now devoid of the species (Hersteinsson and Macdonald 1982). In northern Alaska there is concern about the effects of petroleum exploitation and other human development on the limited number of long-term favorable den sites (Garrott, Eberhardt, and Hanson 1983).

CARNIVORA; CANIDAE; **Genus LYCALOPEX**
Burmeister, 1854

Hoary Fox

Langguth (1975b) considered *Lycalopex* to be a full genus with a single species, *L. vetulus*, occurring in the states of Mato Grosso, Goias, Minas Gerais, and Sao Paulo in south-central Brazil. *Lycalopex* was treated as a subgenus of *Dusicyon* by Cabrera (1957) and as a subgenus of *Canis* by Van Gelder (1978). *L. vetulus* was included within *Dusicyon* by Clutton-Brock, Corbet, and Hills (1976), who noted, however, that it was the most foxlike member of that genus. Berta (1987) included *Lycalopex* within *Pseudalopex* and suggested a relationship to *P. sechurae*.

Head and body length is about 585–640 mm, tail length is 280–320 mm, and weight is about 4 kg. The coat is short. The upper parts are a mixture of yellow and black, giving an overall gray tone. The ears and the outsides of the legs are reddish or tawny. The tail has a black tip and a marked dark stripe along the dorsal line. The underparts are cream to fawn. Compared with *Dusicyon*, *Lycalopex* has a short muzzle, small skull and teeth, reduced carnassials, and broad molars (Bueler 1973; Clutton-Brock, Corbet, and Hills 1976; Grzimek 1975).

The habitat is grassy savannah on smooth uplands, or savannahs with scattered trees. A deserted armadillo burrow may be used for shelter and for the natal nest. The diet consists of small rodents, birds, and insects, especially grasshoppers. The dentition suggests substantial dependence on insects, but *Lycalopex* is persecuted by people because of

Hoary fox *(Lycalopex vetulus)*, photo by Luiz Claudio Marigo.

presumed predation on domestic fowl (Langguth 1975b). It is usually timid but courageously defends itself and its young. Births occur in the spring, there usually being two to four young per litter (Bueler 1973; Grzimek 1975).

CARNIVORA; CANIDAE; **Genus PSEUDALOPEX**
Burmeister, 1856

South American Foxes

There are four species (Berta 1987; Clutton-Brock, Corbet, and Hills 1976; Langguth 1975b; Van Gelder 1978):

P. gymnocercus, humid grasslands of southern Brazil, Paraguay, northern Argentina, and Uruguay;
P. culpaeus, the Andes region from Ecuador to Patagonia;
P. griseus, plains and low mountains of Chile, western Argentina, and Patagonia;
P. sechurae, arid coastal zone of southwestern Ecuador and northwestern Peru.

Pseudalopex was considered a subgenus of *Canis* by Langguth (1975b) and Van Gelder (1978). The four species listed above were included in the genus *Dusicyon* by Clutton-Brock, Corbet, and Hills (1976), along with the type species of that genus, *D. australis*. Berta (1987) recognized *Pseudalopex* as a full genus comprising the above four species and considered *D. australis* to be the only living species of *Dusicyon*. Information provided by these various authorities suggests that *D. australis* is at least as different from the species of *Pseudalopex* as it is from *Canis* and that *Pseudalopex* is at least as different from *Canis* as it is from *Vulpes*. Therefore, since *Dusicyon* (for the species *D. australis*) and *Vulpes* are here being maintained as separate genera, it is also advisable to give generic rank to *Pseudalopex*. Miller et al. (1983) treated *P. fulvipes*, of Chiloe Island, as a species distinct from *P. griseus*.

Head and body length is 600–1,200 mm, tail length is 300–500 mm, and weight is 4–13 kg. *P. culpaeus* is the largest species, and *P. sechurae* the smallest. The coat is usually heavy, with a dense underfur and long guard hairs. The upper parts are generally gray agouti with some ochraceous or tawny coloring (Clutton-Brock, Corbet, and Hills 1976). The head, ears, and neck are often reddish. The underparts are usually pale. The tail is long, bushy, and black-tipped. There is some resemblance to a small coyote *(Canis latrans)*. The dentition, however, is more foxlike than doglike; the molars are well developed, and the carnassials are relatively short (Clutton-Brock, Corbet, and Hills 1976).

Habitats include sandy deserts for *P. sechurae*; low, open grasslands and forest edge for *P. griseus*; pampas, hills, deserts, and open forests for *P. gymnocercus*; and dry rough country and mountainous areas up to 4,500 meters in elevation for *P. culpaeus* (Crespo 1975; Grzimek 1975; Langguth 1975b). Dens are usually among rocks, under bases of trees and low shrubs, or in burrows made by other animals, such as viscachas and armadillos. Most activity is nocturnal, but some individuals are occasionally active during the day. *P. gymnocercus* sometimes collects and stores objects, such as strips of leather and cloth. This species may freeze and remain motionless upon the appearance of a human being; in one case it reportedly did not move even when approached and struck with a whip handle. The omnivorous diet includes rodents, lagomorphs, birds, lizards, frogs, insects, fruit, and sugar cane.

Studies of stomach contents indicate that *P. gymnocercus* takes an equal amount of plant and animal food. *P. culpaeus*

South American fox *(Pseudalopex culpaeus)*, photo by Ernest P. Walker.

seems to be more carnivorous than other species and reportedly sometimes preys heavily on introduced sheep and European hares. In western Argentina, during the spring part of the population of *P. culpaeus* shifts 15–20 km into the higher mountains, in response to the seasonal movements of the sheep and hares. Its normal home range in that area is 4 km in diameter (Crespo 1975). *P. griseus* has been found to occur at an average density of 1 individual per 43 ha. in southern Chile (Durran, Cattan, and Yáñez 1985).

The voice of *Pseudalopex* has been described as a howl or a series of barks and yaps. It is heard mainly at night, especially during the breeding season. According to Crespo (1975), *P. culpaeus* and *P. gymnocercus* mate from August to October and give birth from October to December (the austral spring). Females are monestrous. The gestation period is 55–60 days. Embryo counts range from one to eight, with averages of about four in *P. gymnocercus* and five in *P. culpaeus*. The male helps to provide food to the family. At 2–3 months, the young begin to hunt with the parents. Few individuals live more than several years in the wild, but a captive *P. gymnocercus* survived 13 years and 8 months (Jones 1982).

These canids are killed by people because they are alleged to prey on domestic fowl and sheep and because their fur is desirable. Populations of *Pseudalopex* have thus declined in some areas, such as in Buenos Aires Province, Argentina. In southern Chile, *P. griseus* is legally protected but is threatened by persecution for alleged depredations and environmental disturbance (Durran, Cattan, and Yáñez 1985; Miller et al. 1983). Crespo (1975) noted, however, that *P. culpaeus* occurred in relatively low numbers in Neuquen Province, western Argentina, until the early twentieth century and then greatly increased in response to the introduction of sheep and European hares. *P. griseus*, *P. gymnocercus*, and *P. culpaeus* are on appendix 2 of the CITES. Records of CITES indicate an annual export during the 1980s of about 100,000

skins of *P. griseus* and several thousand of *P. culpaeus*, virtually all from Argentina (Broad, Luxmoore, and Jenkins 1988).

CARNIVORA; CANIDAE; **Genus DUSICYON**
Hamilton-Smith, 1839

Falkland Island Wolf

Berta (1987) considered this genus to include a single modern species, *D. australis*, which formerly occurred on West and East Falkland Islands, off the southeastern coast of Argentina. She added that another species, *D. avus*, occurred in the late Pleistocene of southern Chile and survived into the Recent of southern Argentina. Additional species usually have been assigned to *Dusicyon*. Cabrera (1957) recognized two subgenera: *Dusicyon*, with the species here placed in the genus *Pseudalopex*; and *Lycalopex*, which is here treated as a full genus. Clutton-Brock, Corbet, and Hills (1976) included within *Dusicyon* all of those species that are here placed in the genera *Lycalopex*, *Pseudalopex*, *Dusicyon*, *Cerdocyon*, and *Atelocynus*. These authorities indicated, however, that *D. australis* is at least as close, systematically, to some species of *Canis* as it is to the species here assigned to *Pseudalopex*. They even discussed the possibility that *D. australis* is a form of *Canis familiaris*. Both Langguth (1975b) and Van Gelder (1978) considered *Dusicyon* to be a subgenus of *Canis* and to include only *D. australis*.

Only 11 specimens of *D. australis* are known, and not all include skins. In one specimen, head and body length was 970 mm and tail length was 285 mm. The upper parts are brown with some rufous and a speckling of white, and the underparts are pale brown. The coat is soft and thick. The tail is short, bushy, and tipped with white. The face and ears are short, and the muzzle is broad. The skull is large and has inflated frontal sinuses, more like the situation of *Canis* than that of *Pseudalopex* (Allen 1942; Clutton-Brock, Corbet, and Hills 1976).

Dusicyon was the only terrestrial mammal found on the Falkland Islands by the early explorers. Its natural diet consisted mainly of birds, especially geese and penguins, and also included pinnipeds. Its presence on the islands, about 400 km from the mainland, is something of a mystery.

Clutton-Brock, Corbet, and Hills (1976) thought it most likely that *Dusicyon* had been taken to the Falklands as a domestic animal by prehistoric Indians. Possible descent from either *Pseudalopex* or *Canis* was suggested. Berta (1987), however, pointed out that lowered sea level during the Pleistocene would have facilitated natural movement and that the distinguishing characters of *Dusicyon* probably result from subsequent isolation rather than from domestication.

Dusicyon demonstrated remarkable tameness toward people. Individuals waded out to meet landing parties. Later they came into the camps in groups, carried away articles, pulled meat from under the heads of sleeping men, and stood about while their fellow animals were being killed. The dogs were often killed by a man holding a piece of meat as bait in one hand and stabbing the dog with a knife in the other hand.

Although *Dusicyon* was discovered in 1690, it was still common, and still behaving in a very tame manner, when Darwin visited the Falklands in 1833. In 1839, however, large numbers were killed by fur traders from the United States. In the 1860s Scottish settlers began raising sheep on the islands. *Dusicyon* preyed on the sheep and was therefore intensively poisoned. The genus was very rare by 1870, and the last individual is said to have been killed in 1876 (Allen 1942).

CARNIVORA; CANIDAE; **Genus CERDOCYON**
Hamilton-Smith, 1839

Crab-eating Fox

The single species, *C. thous*, has been recorded from Colombia, Venezuela, Guyana, Surinam, eastern Peru, Bolivia, Paraguay, Uruguay, northern Argentina, and most of Brazil outside of the lowlands of the Amazon Basin (Grimwood 1969; Husson 1978; Langguth 1975b). *Cerdocyon* was recognized as a distinct genus by Berta (1982, 1987), Cabrera (1957), and Langguth (1975b), as a part of *Dusicyon* by Clutton-Brock, Corbet, and Hills (1976), and as a subgenus of *Canis* by Van Gelder (1978).

Head and body length is 600–700 mm, tail length is about 300 mm, and weight is 6–7 kg. The coloration is variable, but the upper parts are usually grizzled brown to gray, often with a yellowish tint, and the underparts are brownish white. The short ears are ochraceous or rufous. The tail is fairly long, bushy, and either totally dark or black-tipped (Clutton-Brock, Corbet, and Hills 1976). The relatively short and robust legs may be tawny (Brady 1979).

Except as noted, the information for the remainder of this account was taken from Brady (1978, 1979). *Cerdocyon* inhabits woodlands and savannahs and is mainly nocturnal. On the llanos of Venezuela, during the wet season it uses high ground and shelters by day under brush. During the dry season it occupies lowlands and spends the day in clumps of matted grass. These grass shelters have several entrances, are used repeatedly, and may serve as natal dens. *Cerdocyon* forages from about 1800 hours to 2400 hours. It stalks and pounces on small vertebrates and apparently listens for crabs

Crab-eating fox *(Cerdocyon thous)*, photo from San Diego Zoological Garden.

in tussocks of grass. In the dry season the percentage composition of its diet is: vertebrates, 48; crabs, 31; insects, 16; carrion, 3; and fruit, 2. In the wet season the percentage breakdown is: insects, 54; vertebrates, 20; fruit, 18; and carrion, 7.

Three pairs occupied home ranges of approximately 54, 60, and 96 ha. The ranges overlapped to some extent but were regularly marked with urine. Tolerance of neighbors was shown in the wet season, but aggression increased in the dry season. Mated pairs form a lasting bond and commonly travel together but do not usually hunt cooperatively. There is a variety of vocalizations, including a siren howl for long-distance communication between separated family members (Brady 1981). Breeding may take place throughout the year, but births peak in January and February on the llanos of Venezuela. Captive females produced two litters annually at intervals of about 8 months. The gestation period is 52–59 days, and litter size is three to six young. The young weigh 120–60 grams at birth, open their eyes at 14 days, begin to take some solid food at 30 days, and are completely weaned by about 90 days. Both parents guard and bring food to the young. Independence comes at 5–6 months, and sexual maturity at about 9 months. According to Smielowski (1985), one specimen lived in captivity for 11 years and 6 months.

CARNIVORA; CANIDAE; Genus NYCTEREUTES
Temminck, 1839

Raccoon Dog

The single species, *N. procyonoides,* originally occurred in the woodland zone from southeastern Siberia to northern Viet Nam, as well as on all the main islands of Japan (Corbet 1978). Although their present distributions may suggest otherwise,

Nyctereutes and *Cerdocyon* evidently are closely related. This view is supported by the fossil record, which suggests the presence of a common ancestor that ranged over Eurasia and North America 4–10 million years ago (Berta 1987).

Head and body length is 500–680 mm and tail length is 130–250 mm. Weight is 4–6 kg in the summer but 6–10 kg prior to winter hibernation (Novikov 1962). The pelage is long, especially in winter. The general color is yellowish brown. The hairs of the shoulders, back, and tail are tipped with black. The limbs are blackish brown. The facial markings resemble those of raccoons *(Procyon).* There is a large dark spot on each side of the face, beneath and behind the eye. *Nyctereutes* is somewhat like a fox in external appearance but has proportionately shorter legs and tail.

According to Novikov (1962), the raccoon dog occurs mainly in forests and in thick vegetation bordering lakes and streams. It usually dens in a hole initially made by a fox or badger or in a rocky crevice but sometimes digs its own burrow. It is primarily nocturnal but may wander about by day if pressed by hunger. *Nyctereutes* is the only canid that hibernates, though the process is neither profound for individuals nor universal for the species. In northern parts of the range, animals that are well nourished hibernate from as early as November to as late as March but may awaken occasionally to forage on warm days. In the southern parts of the range there is no winter sleep. Poorly nourished individuals do not hibernate, even in the north. Successful hibernation may be preceded by a period of intensive eating, which increases weight by nearly 50 percent. The summer diet may consist in large part of frogs and also includes rodents, reptiles, fish, insects, mollusks, and fruit. In the fall, vegetable matter, such as berries, seeds, and rhizomes, becomes important. Northern individuals that do not hibernate have a difficult time in the winter but may subsist on small mammals, carrion, and human refuse.

In the Ussuri region of southeastern Siberia, Kucherenko

Raccoon dog *(Nyctereutes procyonoides)*, photo by Ernest P. Walker.

and Yudin (1973) found population density to average 1–3 individuals per 1,000 ha. and to reach 20 per 1,000 ha. in the best habitat. Home ranges of 100–200 ha. have been reported for the introduced European populations but were found to be only 8–48 ha. in studies in Kyushu (Ikeda 1986). In the latter area, ranges overlapped extensively, and *Nyctereutes* was observed to undergo much amicable social activity. At major food sources, such as fruiting trees and garbage dumps, up to 20 individuals were seen to intermingle without belligerence. There also was communal use of latrines, which may serve as sites to pass on information. Antagonism increased during the breeding season. Male-female pairs formed in late winter, and both parents cared for the young, though each sex foraged separately. The family dissolved at the end of summer, but it was suspected that pair bonds persisted from year to year. Vocalizations include growls and whines but not barks.

Mating occurs from January to March, estrus lasts about 4 days, and the gestation period is usually 59–64 days. There are commonly 4–8 young, but as many as 19 have been reported. They weigh 60–90 grams each at birth, open their eyes after 9–10 days, and nurse for up to 2 months. Both parents, however, begin to bring them solid food after only 25–30 days. They are capable of an independent existence by 4–5 months and attain sexual maturity at 9–11 months (Novikov 1962; Valtonen, Rajakoski, and Mäkelä 1977). A captive specimen lived for 10 years and 8 months (Jones 1982).

In Japan the flesh of the raccoon dog has been used for human consumption, and the bones for medicinal preparations. The skin, known commercially as "Ussuri raccoon," is used widely in the manufacture of such items as parkas, bellows, and decorations on drums. Although about 70,000 are taken every year in Japan for the fur trade, *Nyctereutes* is still common there and sometimes is found within large cities (Ikeda 1986). Populations have declined in southeastern Siberia through overhunting and habitat disturbances (Kucherenko and Yudin 1973).

From 1927 to 1957 over 9,000 raccoon dogs were released by people in regions to the west of the natural range, especially in European Russia. The hope was to create new and valuable fur-producing populations. *Nyctereutes* did become established, eventually spreading almost throughout European Russia, and did increase in importance in the Soviet fur trade. It also, however, extended its range westward, reaching Finland in 1935, Sweden in 1945, Romania in 1951, Poland in 1955, Czechoslovakia in 1959, Germany and Hungary in 1962, and France in 1979 (Artois and Duchêne 1982; Kubiak 1965; Mikkola 1974; Novikov 1962). It now occurs throughout West Germany (Roben 1975), and an individual was recently caught in England (*Oryx* 13[1977]:434). The rate of expansion has slowed in recent years, though the prognosis is that the Balkans, Scandinavia, and France will be colonized (Nowak 1984). *Nyctereutes* is generally considered a nuisance to the west of the Soviet Union. It destroys small game animals and fish, and its fur, which does not become as long as in the native habitat, is almost worthless.

CARNIVORA; CANIDAE; Genus ATELOCYNUS
Cabrera, 1940

Small-eared Dog

The single species, *A. microtis*, inhabits the Amazon, upper Orinoco, and upper Parana basins in Brazil, Peru, Ecuador, Colombia, and probably Venezuela. *Atelocynus* was recognized as a distinct genus by Berta (1986, 1987), Cabrera (1957), and Langguth (1975b), as a part of *Dusicyon* by Clutton-Brock, Corbet, and Hills (1976), and as a subgenus of *Canis* by Van Gelder (1978).

Head and body length is 720–1,000 mm, tail length is 250–350 mm, shoulder height is about 356 mm, and weight is around 9 or 10 kg. The ears, rounded and relatively shorter than those of any other canid, are only 34–52 mm in length. The upper parts are dark gray to black, and the underparts are rufous mixed with gray and black. The thickly haired tail is black, except for the paler basal part on the underside. It has

Small-eared dog *(Atelocynus microtis)*, photo from Field Museum of Natural History.

been reported to sweep the ground when hanging perpendicularly; however, when captives at the Brookfield Zoo, in Chicago, were standing, they curved the tail forward and upward against the outer side of a hind leg, so that the terminal hairs did not drag on the ground. The temporal ridges of the skull are strongly developed, and the frontal sinuses and cheek teeth are relatively large (Clutton-Brock, Corbet, and Hills 1976).

Atelocynus inhabits tropical forests from sea level to about 1,000 meters. It moves with a catlike grace and lightness not observed in any other canid. A male in captivity in Bogota ate raw meat, shoots of grass, and the common foods that people eat. This male and a female were brought to the Brookfield Zoo. They proved to be completely different in temperament: the male was exceedingly friendly and docile, but the female exhibited constant hostility. Although the male was shy in captivity before being sent to Brookfield, and growled and snarled when angry or frightened, it did not show any unfriendly actions at the Chicago zoo and, in fact, became very tame. It permitted itself to be hand fed and petted by persons it recognized, responding to petting by rolling over on its back and squealing. This male came to react to attention from familiar people by a weak but noticeable wagging of the back part of its tail. The female, on the other hand, when under direct observation, emitted a continuous growling sound, without opening its mouth or baring its teeth.

The odor from the anal glands was strong and musky in the male but scarcely detectable in the female. In both sexes the eyes glowed remarkably in dim light. The male, though smaller, was dominant in most activities. Although some snapping was observed between the two animals, no biting or fighting was noted, and they occupied a common sleeping box when not active. According to Jones (1982), one captive lived for 11 years.

According to Thornback and Jenkins (1982), *Atelocynus*

seems to be rare in some countries, though its status is difficult to assess because of its nocturnal, solitary habits. It may be at risk from indirect influence of human intrusion throughout its range. It is protected by law in Brazil and Peru.

CARNIVORA; CANIDAE; Genus SPEOTHOS
Lund, 1839

Bush Dog

The single living species, *S. venaticus*, occurs in Panama, Colombia, Venezuela, the Guianas, eastern Peru, Brazil, eastern Bolivia, Paraguay, and extreme northeastern Argentina (Cabrera 1957; Hall 1981; Thornback and Jenkins 1982). *Speothos* was first described from fossils collected in caves in Brazil. Although *Speothos* sometimes has been united with *Cuon* and *Lycaon* in the subfamily Simocyoninae, Berta (1984) showed that its true affinities lie with other South American canids, especially *Atelocynus*. She reported also that a second species, *S. pacivorus*, is known from late Pleistocene and early Recent deposits in Minas Gerais, Brazil.

Head and body length is 575–750 mm, tail length is 125–50 mm, and shoulder height is about 300 mm. Two males weighed 5 and 7 kg. The head and neck are ochraceous fawn or tawny, and this color merges into dark brown or black along the back and tail. The underparts are as dark as the back, though there may be a light patch behind the chin on the throat (Clutton-Brock, Corbet, and Hills 1976). The body is stocky, the muzzle is short and broad, and the legs are short. The tail is short and well haired but not bushy.

Speothos inhabits forests and wet savannahs, often near water. It seems to be mainly diurnal and to retire to a den at night, in either a burrow or a hollow tree trunk. It is report-

Bush dog *(Speothos venaticus)*, photo from San Diego Zoological Garden.

edly semiaquatic: one captive could dive and swim underwater with great facility. It preys mainly on relatively large rodents, such as *Agouti* and *Dasyprocta* (Husson 1978; Kleiman 1972; Langguth 1975*b*).

Speothos is a highly social canid, living and hunting cooperatively in packs of up to 10 individuals. Several captives of the same or opposite sex can stay confined together without fighting, though a dominance hierarchy may be established. There are a number of vocalizations, the most common of which is a high-pitched squeak that appears to help maintain contact as the group moves about in dense forest (Clutton-Brock, Corbet, and Hills 1976; Kleiman 1972).

Husson (1978) wrote that litters of 2–3 young are produced during the rainy season in the wild. Extended observations in captivity (Porton, Kleiman, and Rodden 1987), however, indicate that the reproductive pattern of *Speothos* is aseasonal and uninterrupted and is partly influenced by social factors. Young females typically do not experience estrus when living with their mother or sisters but quickly do so when paired with males. Estrus and births have been recorded throughout the year. Estrous periods average 4.1 days, and some females show polyestrous activity, which has not been recorded in any other canid. The interbirth interval averages 238 days, and gestation, 67 days. Litter size averages 3.8 young and ranges from 1 to 6. Earliest age of conception in a female was 10 months. According to Jones (1982), a captive lived for 10 years and 4 months.

Speothos is classified as vulnerable by the IUCN and is on appendix 1 of the CITES. It is still widespread but is scarce throughout its range, and it seems to disappear as settlement progresses and forests are cleared (Thornback and Jenkins 1982).

CARNIVORA; CANIDAE; Genus CANIS
Linnaeus, 1758

Dogs, Wolves, Coyotes, and Jackals

There are eight species (Clutton-Brock, Corbet, and Hills 1976; Coetzee, *in* Meester and Setzer 1977; Corbet 1978; Ferguson 1981; Hall 1981; Kingdon 1977; Nowak 1979; Thomas and Dibblee 1986; Vaughan 1983):

C. simensis (Simien jackal), mountains of central Ethiopia;

C. adustus (side-striped jackal), open country from Senegal to Somalia, and south to northern Namibia and eastern South Africa;

C. mesomelas (black-backed jackal), open country from Sudan to South Africa;

C. aureus (golden jackal), Balkan Peninsula to Thailand and Sri Lanka, Morocco to Egypt and northern Tanzania;

C. latrans (coyote), Alaska to Nova Scotia and Panama;

C. rufus (red wolf), central Texas to southern Pennsylvania and Florida;

C. lupus (gray wolf), Eurasia except tropical forests of southeastern corner, Egypt and Libya, Alaska, Canada, Greenland, conterminous United States except southeastern quarter and most of California, highlands of Mexico;

C. familiaris (domestic dog), worldwide in association with people, extensive feral populations in Australia and New Guinea.

Van Gelder (1977*b*, 1978) considered *Canis* to comprise *Vulpes* (including *Fennecus* and *Urocyon*), *Alopex, Lycalo-* *pex, Pseudalopex, Dusicyon, Cerdocyon,* and *Atelocynus* as subgenera. *C. simensis* sometimes has been placed in a separate genus or subgenus, *Simenia* Gray, 1868. *C. familiaris* sometimes is included in *C. lupus*. The feral populations of *C. familiaris* in Australia sometimes are considered to represent a distinct species, *C. dingo* or *C. antarcticus*. Lawrence and Bossert (1967, 1975) suggested that *C. rufus* is not more than subspecifically distinct from *C. lupus*. The presence of *C. lupus* in Egypt and Libya is based on Ferguson's (1981) remarkable determination that the population formerly known as *C. aureus lupaster* is actually a small wolf, rather than a large jackal.

Canis is characterized by a relatively high body, long legs, and a bushy, cylindrical tail. The pupils of the eyes generally appear round in strong light. Although most species have a scent gland near the base of the tail, it does not produce as strong an odor as that of *Vulpes*. The skull has large frontal sinuses and temporal ridges that are close together, often uniting to form a sagittal crest. The facial region of the skull, except in *C. simensis*, is relatively shorter than in *Vulpes* and *Pseudalopex* (Clutton-Brock, Corbet, and Hills 1976). Females have 8 or 10 mammae. Additional information is provided separately for each species.

Canis simensis (Simien jackal).

Head and body length averages about 990 mm, tail length about 250 mm, and shoulder height about 600 mm. The general coloration of the upper parts is tawny rufous with pale ginger underfur. The chin, insides of the ears, chest, and underparts are white. There is a distinctive white band around the ventral part of the neck. The tail is rather short, the facial region of the skull is elongated, and the teeth, especially the upper carnassials, are relatively small (Clutton-Brock, Corbet, and Hills 1976).

The present habitat is montane grassland and moorland at elevations of approximately 3,000–4,000 meters. *C. simensis* may have been found at lower elevations before becoming subject to severe human persecution (Yalden, Largen, and Kock 1980). The remainder of the account of this species is based largely on studies by Morris and Malcolm (1977) in the Bale Mountains of south-central Ethiopia. In this area the Simien jackal was found to be relatively numerous in grasslands supporting large rodent populations. It was uncommon in scrub and was not seen in forests at low altitudes. It was primarily diurnal, but there was some evidence of activity at night, especially in moonlight. Farther north, in the Simien Mountains, the few remaining jackals were reported to be almost entirely nocturnal, probably because of human disturbance. The animals made little effort to find shelter, and individuals slept in the open or in places with slightly longer grass than usual, even when temperatures were as low as −7° C.

The diet consisted mainly of diurnal rodents, especially *Tachyoryctes* and *Otomys*. Hunting was done mainly by walking slowly through areas of high rodent density, investigating holes and listening carefully, and then stealthily creeping toward the prey and capturing it with a final dash. Individuals generally hunted alone and did not come together during the day, though there was apparently considerable overlap in hunting ranges. Highest population density of *C. simensis* was perhaps 2/sq km. No direct evidence of territoriality was observed, but urine marking was frequent, and some animals seemed to avoid others. Around dawn and dusk, most jackals came together in groups of up to seven members. There was then much friendly social interaction and noise. Vocalizations varied from a howl to a staccato scream interspersed with barks and also included yips and whines. Pairs appeared to be forming in January, but births apparently did not occur until May or June.

Top, golden jackal *(Canis aureus)*. Bottom, dingos *(Canis familiaris)*. Photos by Lothar Schlawe.

The Simien jackal was reported from most Ethiopian provinces in the nineteenth century. It subsequently declined because of agricultural development in its range and environmental disruption that reduced prey populations. Also, it was incorrectly believed to be a predator of sheep and so was frequently shot. At present only about 40 individuals of the subspecies *C. s. simensis* remain in the Simien Mountains and nearby parts of north-central Ethiopia. There are estimated still to be 450–600 individuals of the subspecies *C. s.*

citerni in the Bale Mountains and a few others in neighboring highlands. There are national parks in the Simien and Bale mountains, but these areas are still being used for grazing. *C. simensis* is protected by law in Ethiopia and is classified as endangered by the IUCN (1978) and the USDI.

Canis adustus (side-striped jackal).

Head and body length is 650–810 mm, tail length is 300–410 mm, shoulder height is 410–500 mm, and weight is 6.5–14.0

kg (Kingdon 1977). Males are larger than females. The coat is long and soft. The upper parts are generally mottled gray; on each side of the body is a line of white hairs, followed below by a line of dark hairs; and the underparts and tip of the tail are white (Clutton-Brock, Corbet, and Hills 1976).

According to Kingdon (1977), the side-striped jackal is widespread in the moister parts of savannahs, thickets, forest edge, cultivated areas, and rough country up to 2,700 meters in elevation. Favored denning sites include old termite mounds, abandoned aardvark holes, and burrows dug into hillsides. The animal is strictly nocturnal in areas well settled by people. It is a more omnivorous scavenger than other kinds of jackal, taking a variety of invertebrates, small vertebrates, carrion, and plant material. Social groups are well spaced and usually consist of a mated pair and their young. In East Africa births occur mainly in June–July and September–October, gestation lasts 57–70 days, and litters consist of three to six young. Rosevear (1974) wrote that captives have lived up to 10 years.

Canis mesomelas (black-backed jackal).

Head and body length is 680–745 mm, tail length is 300–380 mm, shoulder height is 300–480 mm, and weight is 7.0–13.5 kg. Males average about 1 kg heavier than females. A dark saddle extends the length of the back to the black tip of the tail. The sides, head, limbs, and ears are rufous, and the underparts are pale ginger. The build is slender, and the ears are very large (Bekoff 1975; Clutton-Brock, Corbet, and Hills 1976; Kingdon 1977).

The black-backed jackal is found mainly in dry grassland, brushland, and open woodland. It generally dens in an old termite mound or aardvark hole. It is partly diurnal and crepuscular in undisturbed areas but becomes nocturnal in places intensively settled by people. The diet includes a substantial proportion of plant material, insects, and carrion, but *C. mesomelas* frequently hunts rodents and is capable of killing antelope up to the size of *Gazella thomsoni*. Some food may be cached (Bothma 1971b; Kingdon 1977; Smithers 1971).

In a study in southern Africa, average daily movement was found to be about 12 km, but one radio-collared individual covered 87 km in four nights. Average home range size in this area was about 11 sq km for adults, but some subadults wandered over much greater areas (Ferguson, Nel, and de Wet 1983). In the same region, Rowe-Rowe (1982) estimated a

population density of 1 jackal per 2.5–2.9 sq km and an average home range size of 18.2 sq km. According to these and other investigations (Bekoff 1975; Kingdon 1977; Moehlman 1978, 1983, 1987), all or a major part of the adult home range is a territory that is marked with urine and defended by both sexes. Each sex seems to repel mainly members of the same sex. Although 20–30 individuals may gather around a lion kill, the basic social unit is a mated pair and their young. Some young remain with the parents, do not breed, and assist in feeding, grooming, and guarding the next litter. This tendency seems to be stronger in *C. mesomelas* than in *C. aureus*, perhaps because the former maintain larger territories and, because of a dependency on smaller game, must make a greater effort to provide for the newborn. Final dispersal of the young may not occur until they are about 2 years old. Mated pairs are known to have maintained a monogamous relationship for at least 6 years.

Births occur mainly from July to October. The gestation period is about 60 days. Average litter size is four young, ranging from one to eight. The young emerge from the den after 3 weeks and may then be moved several times to new sites. They are weaned at 8–9 weeks, and the adults then regurgitate food for them. After about 3 months they no longer use the den, and by the age of 6 months they are hunting on their own. Sexual maturity may come at 10–11 months, though young that remain with the parents do not breed. Captives have lived up to 14 years (Bekoff 1975; Bothma 1971a; Kingdon 1977; Moehlman 1978, 1987).

Unlike *C. adustus*, the black-backed jackal is considered a serious predator of sheep and is therefore intensively hunted and poisoned (Clutton-Brock, Corbet, and Hills 1976). According to Bothma (1971b), *C. mesomelas* causes more problems than does any other animal in the sheep-farming areas of the Transvaal.

Canis aureus (golden jackal).

Head and body length is 600–1,060 mm, tail length is 200–300 mm, shoulder height is 380–500 mm, and weight is 7–15 kg (Kingdon 1977; Lekagul and McNeely 1977). The fur is generally rather coarse and not very long. The dorsal area is mottled black and gray; the head, ears, sides, and limbs are tawny or rufous; the underparts are pale ginger or nearly white; and the tip of the tail is black (Clutton-Brock, Corbet, and Hills 1976).

The golden jackal is found mainly in dry, open country. It

Black-backed jackal *(Canis mesomelas)*, photo by Ernest P. Walker.

Golden jackal *(Canis aureus)*, photo by Cyrille Barrette.

is strictly nocturnal in areas inhabited by people but may be partly diurnal elsewhere. Its opportunistic diet includes young gazelles, rodents, hares, ground birds and their eggs, reptiles, frogs, fish, insects, and fruit. It takes carrion on occasion but is a capable hunter (Kingdon 1977; Moehlman 1987; Van Lawick and Van Lawick–Goodall 1971).

The basic social unit is a mated pair and their young. Monogamy is the rule, though pair bonds do not seem to be as strong as in *C. mesomelas*. Sometimes the young of the previous year remain in the vicinity of the parents and even mate and bear their own litters. However, a more usual procedure is for one or two of the young to stay with the parents, without breeding, and help care for the next litter. The resulting associations are probably responsible for the reports of large packs hunting together. The usual hunting range of a family is about 2–3 sq km. A portion of this area is a territory marked with urine and defended against intruders. The young are born in a den within the territory. Births occur mainly in January–February in East Africa and in April–May in the Soviet Union but take place throughout the year in tropical Asia. The gestation period is 63 days. Litters contain one to nine, usually four to two, young. Their eyes open after about 10 days, and they begin to take some solid food at about 3 months. Both parents provide food and protection. Sexual maturity comes at 11 months. Captives have lived up to 16 years (Kingdon 1977; Lekagul and McNeely 1977; Moehlman 1983, 1987; Novikov 1962; Rosevear 1974; Van Lawick and Van Lawick–Goodall 1971).

As with other wild canids, the relationship of the golden jackal with people is controversial. In Bangladesh, for example, *C. aureus* plays an important scavenging role by eating garbage and animal carrion around towns and villages and may benefit agriculture by preventing increases in the numbers of rodents and lagomorphs. However, this jackal also raids such crops as corn, sugar cane, and watermelons. It may be involved in the spread of rabies, and in 1979 two young children were attacked and killed by jackals (Poché et al. 1987).

Canis latrans (coyote).

Head and body length is 750–1,000 mm and tail length is 300–400 mm. In a given population, males are generally larger than females. Bekoff (1977) listed weights of 8–20 kg for males and 7–18 kg for females. Northern animals are usually larger than those to the south. Gier (1975) gave average weight as 11.5 kg in the deserts of Mexico and 18 kg in Alaska. The largest coyotes of all are found in the northeastern United States, where the population apparently has been modified through hybridization with *C. lupus* (Nowak 1979). Pelage characters are variable, but usually the coat is long, the upper parts are buffy gray, and the underparts are paler. The legs and sides may be fulvous, and the tip of the tail is usually black (Clutton-Brock, Corbet, and Hills 1976). From *C. lupus* the coyote is distinguished by smaller size, narrower build, proportionally longer ears, and a much narrower snout.

The coyote is found in a variety of habitats, mainly open grasslands, brush country, and broken forests. The natal den is located in such places as brush-covered slopes, thickets, hollow logs, rocky ledges, and burrows made by either the parents themselves or other animals. Tunnels are about 1.5–7.5 meters long and lead to a chamber about 0.3 meter wide and 1 meter below the surface. Activity may take place at any time of day but is mainly nocturnal and crepuscular. The average distance covered in a night's hunting is 4 km. The coyote is one of the fastest terrestrial mammals in North America, sometimes running at speeds of up to 64 km/hr. There may be a migration to the high country in the summer and a return to the valleys in the fall. Some of the young disperse from the parental range in the fall and winter, generally covering a distance of 80–160 km (Banfield 1974; Bekoff 1977; Gier 1975). In a tagging study in Iowa, Andrews and Boggess (1978) found individuals to travel an average of about 31 km, but up to 323 km, from the point of capture.

About 90 percent of the diet is mammalian flesh, mostly jack rabbits, other lagomorphs, and rodents. Such small animals are usually taken by stalking and pouncing (Bekoff 1977; Gier 1975). Coyotes have been seen to enter shallow streams and snatch fish from the water (Springer 1980). Larger mammals, especially deer, are also eaten, often in the form of carrion but sometimes after a chase in which several coyotes work together. In northwestern Wisconsin, deer represent 21.3 percent of the diet (Niebauer and Rongstad 1977), and in northern Minnesota, deer, mainly as carrion, is the

Coyote *(Canis latrans)*, photo from U.S. Fish and Wildlife Service.

main food of the coyote (Berg and Chesness 1978). In Alberta, half of the food is carrion (Nellis and Keith 1976). Analyses of coyote predation on large mammals generally indicate that most of the individuals taken are immature, aged, or sick (Bekoff 1977). Livestock is killed by some coyotes, but the diet of the species seems to be largely neutral or beneficial with respect to human interests. *C. latrans* sometimes forms a "hunting partnership" with the badger *(Taxidea)*. The two move together, the coyote apparently using its keen sense of smell to locate burrowing rodents and the badger digging them up with its powerful claws. Both predators then share in the proceeds.

Population density is generally 0.2–0.4 per sq km but may be up to 2.0 per sq km under extremely favorable conditions (Bekoff 1977; Knowlton 1972). Reported home range size is 8–80 sq km. The ranges of males tend to be large and to overlap one another considerably, while the ranges of females are smaller and do not overlap (Bekoff 1977). In northern Minnesota, Berg and Chesness (1978) found home range size to average 68 sq km in males and 16 sq km in females. In Alberta, however, Bowen (1982) found home ranges to be shared by an adult male and female and to average about 14 sq km for both sexes. Laundré and Keller (1984) reviewed various reports of home range size of about 10–100 sq km and concluded that most were based on inadequate data.

Social structure and territoriality vary depending on habitat conditions and food supplies. *C. latrans* is found alone, in pairs, or in larger groups but generally is less social than *C. lupus* (Bekoff 1977). Several males may court a female during the mating season, but she eventually seems to select one for a lasting relationship. The pair then live and hunt together, sometimes for years. Gier (1975) suggested that pair territories tend to be bordered by natural features, cover about 1 sq km or less, and are defended only during the denning season. More recent work suggests that territories are larger and are

exclusive throughout the year, though the ranges of young or transient animals may be superimposed (Andelt 1985).

Some of the offspring of a pair disperse, but others may remain with the parents and thus form the basis of a pack. These helpers, whose main function apparently is to provide increased protection to the next litters, may not leave until they are 2–3 years old. Studies in northwestern Wyoming indicate that packs are larger and more stable if a major, clumped food resource, such as ungulate carrion, is readily available (Bekoff and Wells 1980, 1982). In this area, on the National Elk Refuge, Camenzind (1978) found that 61 percent of the coyotes composed resident packs of 3–7 members, 24 percent were resident mated pairs, and 15 percent were nomadic individuals. The latter ranged through the territories of the resident animals and were constantly chased away. The packs had well-defined social hierarchies and marked and defended territories. The sizes of two pack territories were 5 and 7.2 sq km. Sometimes more than one pair of a pack mated and bore young. Occasionally up to 22 individuals would come together around carrion, but such aggregations usually lasted less than an hour.

The coyote has a great variety of visual, auditory, olfactory, and tactile forms of communication (Lehner 1978). At least 11 different vocalizations have been identified; these are associated with alarm, threats, submission, greeting, and contact maintenance. The best-known sounds are the howls given by one or more individuals, apparently to announce location. Group howls may have a territorial and spacing function.

Mating usually occurs from January to March, and births take place in the spring. Females are monestrous. Estrus averages 10 (4–15) days, gestation averages 63 (58–65) days, and litter size averages about 6 (2–12) young. Larger numbers of offspring are sometimes found in one den, but these apparently represent the litters of two different females. The young weigh about 250 grams each at birth, open their eyes after 14 days, emerge from the den at 2–3 weeks, and are fully weaned at 5–6 weeks. The mother, and perhaps the father and older siblings, begins to regurgitate some solid food for the young when they are about 3 weeks old. Adult weight is attained at about 9 months. Some individuals mate in the breeding season following that in which they were born, but others wait another year. Maximum known longevity in the wild is 14.5 years, but few animals survive that long. In one unexploited population, 70 percent of the animals were under 3 years old (Bekoff 1977; Gier 1975; Kennelly 1978). A captive lived for 21 years and 10 months (Jones 1982).

The coyote long has been considered a serious predator of domestic animals, especially sheep. It is also sometimes alleged to spread rabies and to damage populations of game birds and mammals. Bounties against the coyote began in the United States in 1825 and are still paid in some areas but have never been effective in lasting prevention of depredations. In 1915 the U.S. government initiated a large-scale program of predator control, mainly in the West. The resulting known kill of C. latrans often exceeded 100,000 individuals per year. From 1971 to 1976 the annual take in 13 western states averaged 71,574 (Evans and Pearson 1980). Some state governments and private organizations also have control programs. Killing methods include trapping, poisoning, shooting from aircraft, and destroying the young in dens. In 1972 the federal government banned the general use of poison on its lands and in its programs.

Few wildlife issues have been as controversial as the question of coyote control (Bekoff 1977; Bekoff and Wells 1980; Hall 1946; Sterner and Shumake 1978; Wade 1978). No authorities deny that predation is a problem to some farmers and ranchers, but almost since the beginning of the federal program there have been serious doubts as to the extent of claimed livestock losses and degree of response. There was particular resentment against the widespread application of poison on western ranges, which was thought to kill not only far more coyotes than necessary but also many other carnivorous mammals. It has been argued that sheep raising is declining in the United States, regardless of the coyote and its control, and that damage by predators represents a relatively small part of overall losses to the industry. Surveys by trained biologists generally have indicated that losses are substantial in some areas but lower than often claimed. The stated intention of most current control operations is elimination of the specific individual coyotes that may be causing problems, and not the destruction of entire populations.

Information compiled by the U.S. Department of Agriculture (Gee et al. 1977) indicates that in 1974 losses attributed to the coyote in the western states amounted to 728,000 lambs (more than 8 percent of those born and one-third of the total lamb deaths) and 229,000 adult sheep (more than 2 percent of inventory and one-fourth of all adult deaths), valued at $27 million. In that same year the amount expended for coyote control in the West was $7 million (Gum, Arthur, and Magleby 1978). In 1974, 71,522 coyotes were killed by the federal control program. Data compiled by Pearson (1978) suggest that the total kill in the 17 western states that year (including the take by fur trappers and sport hunters) was around 300,000 individuals, approximately the same as in 1946. Data compiled by Terrill (1986) indicate that the rate of loss of sheep and lambs to coyote predation through 1985 generally remained about the same as given above, though there was a temporary decline in 1980, apparently caused by the spread of canine parovirus in the coyote population.

Coyote fur has varied in price. In 1974 skins taken in the United States sold for an average of $17.00 each. For the 1976/77 season 30 states reported a total harvest of 320,323 pelts, with an average value of $45.00; in 6 Canadian provinces the take was 65,819 pelts, with an average value of $59.76 (Deems and Pursley 1978). Coyote numbers evidently rebounded in the 1980s, but fur prices dropped. The total known kill in the United States during the 1983/84 season was 439,196. In Canada 68,975 pelts, with an average value of $15.90, were taken (Novak, Obbard, et al. 1987).

It is sometimes said that C. latrans, as a species, can easily sustain human-inflicted losses, because of its adaptability, cunning, and high reproductive potential. Such may not necessarily be the case. In the past, intensive control programs eliminated the species from central Texas, much of North Dakota, and certain large sheep-raising sections of Colorado, Wyoming, Montana, Utah, and Nevada (Gier 1975; Nowak 1979).

It does appear that C. latrans has substantially extended its range since the arrival of European colonists in North America (Brady and Campbell 1983; Gier 1975; Hill, Sumner, and Wooding 1987; Hilton 1978; Moore and Millar 1984; Nowak 1978, 1979; Thomas and Dibblee 1986; Vaughan 1983). The newly occupied regions may include Alaska, the Yukon, and Central America, though some evidence suggests that the species was already there, at least intermittently, in prehistoric times. It is known to have moved through Costa Rica and into western Panama since 1960. The main expansion was eastward, the two main factors apparently being (1) the creation of favorable habitat through the breaking up of the forests by settlement and (2) human elimination of the wolves (C. rufus and C. lupus) that might have competed with the coyote. In the late nineteenth and early twentieth centuries the coyote moved from its prairie homeland, through the Great Lakes region, and into southeastern Canada. From the 1930s to the 1960s the species established itself in New England and New York, and it now has pushed as far east as far as Nova Scotia and Prince Edward Island and spread down the Appalachians as far as southern Virginia. Other coyote populations pushed into southern Missouri and Arkansas in the 1920s, overran Louisiana in the 1950s, crossed into Mississippi by the 1960s, and subsequently became established as far as Florida and the Carolinas. The expanding coyote populations were modified by hybridization with the remnant pockets of wolves that they encountered, C. lupus in southeastern Canada and C. rufus in the south-central United States, with the result that the coyotes of eastern North America are larger and generally better adapted for forest life than are their western relatives.

Canis rufus (red wolf).

Head and body length is 1,000–1,300 mm, tail length is 300–420 mm, shoulder height is 660–790 mm, and weight is 20–40 kg. While the red element of the fur is sometimes pronounced, the upper parts are usually a mixture of cinnamon buff, cinnamon, or tawny, with gray or black; the dorsal area is generally heavily overlaid with black. The muzzle, ears, and outer surfaces of the limbs are usually tawny. The underparts are whitish to pinkish buff, and the tip of the tail is black. There was reported to be a locally common dark or fully black color variant in the forests of the Southeast. From C. lupus, the red wolf is distinguished by its narrower proportions of body and skull, shorter fur, and relatively longer legs and ears.

Habitats include upland and bottomland forests, swamps, and coastal prairies. Natal dens are located in the trunks of hollow trees, stream banks, and sand knolls. Dens, which are either self-excavated or taken over from some other animal, average about 2.4 meters in length and usually extend no further than 1 meter below the surface. The red wolf is primarily nocturnal but may increase its daytime activity during the winter. It hunts over a relatively small part of its home range for about 7–10 days and then shifts to another area.

Red wolf *(Canis rufus)*, photo from the *Washington Post.*

Reported foods include nutria, muskrats, other rodents, rabbits, deer, hogs, and carrion (Carley 1979; Nowak 1972; Riley and McBride 1975).

Home range in southeastern Texas has been reported: (1) to average 44 sq km for 7 individuals (Shaw and Jordan 1980); (2) to cover 65–130 sq km over 1–2 years (Riley and McBride 1975); and (3) to average 116.5 sq km for males and 77.7 sq km for females (Carley 1979). The basic social unit is apparently a mated, territorial pair. Groups of 2–3 individuals are most common, though larger packs have frequently been reported. Vocalizations are intermediate to those of *C. latrans* and *C. lupus* (Paradiso and Nowak 1972). Mating occurs from January to March, and offspring are produced in the spring. The gestation period is 60–63 days, and litters contain up to 12 young, usually about 4–7. Few individuals survive more than 4 years in the wild, but potential longevity in captivity is at least 14 years (Carley 1979).

Like most large carnivores, the red wolf was considered to be a threat to domestic livestock, if not to people themselves. It was therefore intensively hunted, trapped, and poisoned after the arrival of European settlers in North America. In addition, disruption of its habitat and reduction of its numbers allowed *C. latrans* to invade its range from the west and north and evidently stimulated interbreeding between the two species. This process led to the genetic swamping of the small pockets of red wolves that survived human persecution. By the 1960s the only pure populations of *C. rufus* were in the coastal prairies and swamps of southeastern Texas and southern Louisiana. Conservation efforts, especially by the U.S. Fish and Wildlife Service, were not successful. The hybridization process continued to spread, and by 1975 the pre-

vailing view was that the species could be saved only by securing and breeding some of the remaining animals that appeared to represent unmodified *C. rufus*. Many animals were captured and examined, and eventually 14 were selected for the breeding program. Since 1977 offspring have been produced from this stock, and by 1989 there were 83 living descendants (and one surviving wild-caught wolf). Most of these animals are still in captivity, the majority at the Point Defiance Zoo in Tacoma, but experimental reintroductions have been made on Bulls Island off South Carolina and Horn Island off Mississippi. Starting in 1987, pairs have been released as part of a large-scale effort to reestablish a permanent red wolf population at the Alligator River National Wildlife Refuge in eastern North Carolina; results thus far have been inconclusive (Carley 1979; McCarley and Carley 1979; Nowak 1972, 1974, 1979; Rees 1989). The red wolf is classified as endangered by the IUCN and the USDI.

Canis lupus (gray wolf).

This species includes the largest wild individuals in the family Canidae. Head and body length is generally about 1,000–1,600 mm and tail length is 350–560 mm. However, there are several small subspecies along the southern edge of the range of the gray wolf, especially *C. lupus arabs* of the Arabian Peninsula; an adult male of that subspecies had a head and body length of only 820 mm and a tail length of 320 mm (Harrison 1968). Ferguson (1981) listed the following data: *C. lupus arabs* of Israel and Arabia, head and body length 1,140–1,440 mm, tail length 230–360 mm, and weight 14–19 kg; *C. lupus lupaster* of Egypt and Libya, head and body length 1,058–1,300 mm, tail length 283–355 mm, and weight 10–

Gray wolf (Canis lupus), photo from Zoological Society of London.

16 kg. In North America the largest animals are found in Alaska and western Canada, and the smallest in Mexico. In a given population, males are larger on the average than females. Overall mean (and extreme) weights are about 40 (20–80) kg for males and 37 (18–55) kg for females (Mech 1970, 1974b). The pelage is long; the upper parts are usually light brown or gray, sprinkled with black, and the underparts and legs are yellow white. Entirely white individuals occur frequently in tundra regions and occasionally elsewhere. Black individuals are also common in some populations.

The gray wolf has the greatest natural range of any living terrestrial mammal other than *Homo sapiens*. It is found in all habitats of the Northern Hemisphere except tropical forests and arid deserts. Dens, used only for the rearing of young, may be in rock crevices, hollow logs, or overturned stumps but are usually in a burrow, either dug by the parents themselves or initially made by another animal and enlarged by the wolves. Several such burrows are sometimes excavated in the same season, perhaps as far apart as 16 km. A den may be used year after year. The animals prefer an elevated site near water. Tunnels are about 2–4 meters long and lead to an enlarged underground chamber with no bedding material. There may be several entrances, each marked by a large mound of excavated soil. After the young are about 8 weeks old, they are moved to a rendezvous site, an area of about 0.5 ha., usually near water and marked by trails, holes, and matted vegetation. Here the young romp and play, and the other members of the pack gather for daily rest during the summer. Such sites are frequently changed, but some may be used for one or two months (Mech 1970, 1974b; Peterson 1977).

Movements are extensive and usually take place at night, but diurnal activity may increase in cold weather. During the summer the pack usually sets out in the early evening and returns to the den or rendezvous site by morning. In winter the animals wander farther and do not necessarily return to a particular location. They tend to move in single file along regular pathways, roads, streams, and ice-covered lakes. The daily distance covered ranges from a few to 200 km (Mech 1970). In Finland, the mean daily movement was determined to be 23 km (Pulliainen 1975). On Isle Royale, in Lake Superior, Peterson (1977) found that packs averaged 11 km per day or 33 km per kill. When individuals permanently disperse from a pack, they move much farther than normal; one traveled 206 km in 2 months (Mech 1974b), and another covered a straight-line distance of 670 km in 81 days (Van Camp and Gluckie 1979). Recent record movements are of a single animal that dispersed 886 km from Minnesota to Saskatchewan (Fritts 1983) and a group of at least 2 and probably 4 wolves that moved at least 732 km across Alaska (Ballard, Farnell, and Stephenson 1983). Five dispersing wolves tracked by Fritts and Mech (1981) emigrated about 20–390 km, though others remained in the vicinity of the parental groups. Packs that depend on barren ground caribou make seasonal migrations with their prey and move as far as 360 km (Kuyt 1972; Mech 1970, 1974b).

The gray wolf usually moves at about 8 km/hr but has a running gait of 55–70 km/hr. It can cover up to 5 meters in a single bound and can maintain a rapid pursuit for at least 20 minutes. Prey is located by chance encounter, direct scenting, or following a fresh scent trail. Odors can be detected up to 2.4 km away. A careful stalk may be used to get as close as possible to the prey. If the objective is large and healthy and stands its ground, the pack usually does not risk an attack. Otherwise, a chase begins, usually covering 100–5,000 meters. If the wolves cannot quickly close with the intended victim, they generally give up. There seems to be a continuous process of testing the individual members of the prey population to find those that are easily captured. Most hunts are unsuccessful. On Isle Royale, for example, only about 8 percent of the moose tested are actually killed (Mech 1970, 1974b). In one exceptional case the chase of a deer went for 20.8 km (Mech and Korb 1978). Once the wolves overtake an ungulate, they strike mainly at its rump, flanks, and shoulders.

The gray wolf is primarily a predator of mammals larger than itself, such as deer, wapiti, moose, caribou, bison, muskox, and mountain sheep. The smallest consistent prey is beaver. Following a drastic decline of deer in central Ontario, beaver remains were found in most wolf droppings (Voight, Kolenosky, and Pimlott 1976). Kill rates vary from about one deer per individual wolf every 18 days to one moose per wolf every 45 days (Mech 1974b). The usual condition is that a pack of several wolves will make a kill, consume a large amount of food, and then make another kill some days later. An adult can eat about 9 kg of meat in one feeding. Studies in Minnesota indicate that an average daily consumption is about 2.5 kg of deer per wolf and that pack members generally remain in the vicinity of a kill for several days, eventually utilizing nearly the entire carcass, including much of the hair and bones (Mech et al. 1971). Most analyses of predation on wild species show that immature, aged, and otherwise inferior individuals constitute most of the prey taken by wolves (Mech 1974b).

There long has been controversy regarding the effects of the wolf on overall prey populations. It was once generally thought, even by some experienced zoologists, that the wolf could and did eliminate its prey. For example, Bailey (1930) wrote that "wolves and game animals can not be successfully maintained on the same range." Subsequently, a popular view developed that the wolf and the animals on which it depends are in precise balance, with the wolf taking just enough to prevent substantial population increases. Mech (1970) tentatively concluded that wolf predation is the major controlling factor where prey-predator ratios are about 11,000 kg of prey per wolf or less, as is the case on Isle Royale. More recently it became apparent that the wolf is not regulating the moose herd on Isle Royale but merely cropping part of the annual surplus production (Mech 1974a; Peterson 1977). On the other hand, studies by Mech and Karns (1977) indicate that wolf predation was a major contributing factor in a serious decline of deer in the Superior National Forest of Minnesota from 1968 to 1974. Primary causes in the decline there, and in other parts of the Great Lakes region, were maturation of the forests that had been cut around the turn of the century and a series of severe winters.

Wolf population density varies considerably, being as low as 1 per 520 sq km in parts of Canada. In Alberta, Fuller and Keith (1980) found densities of 1 per 73 sq km to 1 per 273 sq km. Several studies in the Great Lakes region found apparently stable densities of around 1 per 26 sq km and suggested that this level was the maximum allowed by social tolerance (Mech 1970; Pimlott, Shannon, and Kolenosky 1969). Subsequent investigations determined that the wolf could attain densities nearly twice as great in areas where prey concentrate for the winter (Kuyt 1972; Parker 1973; Van Ballenberghe, Erickson, and Byman 1975). On Isle Royale, an area of 544 sq km, the number of wolves remained relatively stable at a midwinter average of 22 from 1959 to 1973. There was an increase to 44 in 1976, evidently in response to rising moose numbers, thus indicating that food supply is the main factor in density level (Peterson 1977). The Isle Royale wolf population peaked at 50 in 1980 and then underwent a crash in

association with falling moose numbers (Peterson and Page 1988). By 1988 only 12 wolves survived there, even though the moose population again was on the rise, and there thus was concern about genetic viability.

Home range size depends on food availability, season, and number of wolves. The largest pack range, found in Alaska during winter, was 13,000 sq km, and the smallest, in southeastern Ontario during summer, was 18 sq km (Mech 1970; Pimlott, Shannon, and Kolenosky 1969). In Minnesota, ranges of 52–555 sq km have been reported (Fritts and Mech 1981; Harrington and Mech 1979; Van Ballenberghe, Erickson, and Byman 1975). One radio-collared pack in Minnesota used summer ranges of 117–32 sq km and winter ranges of 123–83 sq km (Mech 1977d). Alberta packs used ranges of 195–629 sq km in summer and 357–1,779 sq km in winter (Fuller and Keith 1980). Pack range in the Soviet Union varies from 30 sq km in areas of abundant food and good cover to over 1,000 sq km in deserts and tundra (Bibikov, Filimonov, and Kudaktin 1983).

The home range of a wolf pack usually corresponds to a defended territory. There is generally little or no overlap between the ranges of neighboring packs. Lone wolves that split off from a pack may disperse a considerable distance or are continuously pursued by the resident packs and forced to shift about (Mech 1974b; Peterson 1977; Rothman and Mech 1979). Territories are relatively stable, some being known to have been used at least 10 years (Mech 1979). Buffer zones, areas of little use, tend to develop between territories; lone wolves may sometimes center their activities there. In Minnesota, deer have been found to have higher densities in such border areas and to survive longer there than elsewhere during times of general population declines (Mech 1977b, 1979; Rogers et al. 1980). Under natural conditions, packs avoid areas in which they might encounter other packs. When food shortages induce stress, however, wolves move into the buffer zones to hunt and eventually trespass into the territories of other packs. Meetings between packs are agonistic and may result in chases, savage fighting, and mortality (Harrington and Mech 1979; Mech 1970, 1977c; Van Ballenberghe and Erickson 1973).

Although packs are hostile to one another, the gray wolf is among the most social of carnivores. Groups usually contain 5–8 individuals but have been reported to have as many as 36 (Mech 1974b). Such associations are probably essential for consistent success in the pursuit and overpowering of large prey. The number of wolves in a pack tends to increase with the size of the usual prey, being 7 or less where deer is the only important food, 6–14 where both deer and wapiti are eaten, and 15–20 on Isle Royale, where moose is the primary prey (Peterson 1977). In certain southern parts of its range, such as Mexico, Italy, and Arabia, C. lupus seems to be less gregarious than in the north (Harrison 1968; McBride 1980; Zimen 1981). This situation may be partly unnatural, resulting from intense human persecution and dependence on easily captured domestic animals and garbage.

A pack is essentially a family group, consisting of an adult pair, which may mate for life, and their offspring of one or more years (Mech 1970). The leader is usually a male, often referred to as the alpha male. He initiates activity, guides movements, and takes control at critical times, such as during a hunt. The males and females of a pack may have separate dominance hierarchies, reinforced by aggressive behavior and elaborate displays of greeting and submission by subordinate members. Generally, only the most dominant pair mate, and they inhibit sexual activity in the others. Social status is rather consistent, and a leader may retain its position for years, but roles can be reversed. Intragroup strife, perhaps resulting from increasing membership or declining food sup-

B

Gray wolf *(Canis lupus)*, photo from New York Zoological Society.

plies, can result in a division into two packs or the splitting off of individuals. The latter may maintain a loose association with the parental group, sometimes following at a distance and feeding on scraps left behind, or disperse to a new area to seek a mate and begin a new pack (Mech 1970, 1974*b*; Peterson 1977; Wolfe and Allen 1973; Zimen 1975).

Studies by Fritts and Mech (1981) have contributed to our understanding of new pack formation in an area of an expanding wolf population. A young individual evidently does not remain with its parental pack past breeding age (about 22 months) unless it becomes a breeder itself upon the death of an alpha animal of the same sex. As it approaches maturity, it may actively explore the fringes of the parental territory. After dispersal, it either joins another lone wolf to search for a new territory or establishes itself in an area and awaits the arrival of an animal of the opposite sex. A new pack territory is relatively small but may incorporate a portion of the parental territory. Therefore, as population size increases, average territory size decreases.

The gray wolf has a variety of visual, olfactory, and auditory means of communication. Vocalizations include growls, barks, and howls—continuous sounds usually lasting 3–11 seconds (Mech 1974*b*). Different individuals have distinctive howls (Peterson 1977). Humans can hear howls 16 km away on the open tundra, and wolves probably respond to howls at distances of 9.6–11.2 km. Howling functions to bring packs together and as an immediate, long-distance form of territorial expression (Harrington and Mech 1979).

Territories are also maintained by scent marking, via scratching, defecation, and especially urination. Scent marking differs from ordinary elimination in that there is a regular pattern of deposition at certain repeatedly used points. Peters and Mech (1975) determined that as a pack moves through its territory—visiting most parts at least every three weeks— signs are left at average intervals of 240 meters. Rothman and

Mech (1979) found that scent marking also is important in bringing new pairs together for breeding and in helping established pairs to achieve reproductive synchrony.

Threats and attacks by the dominant members of a pack probably prevent sexual synchronization in subordinates, and thus only the highest-ranking female normally bears a litter during the reproductive season (Zimen 1975). Some exceptions were reported by Van Ballenberghe (1983). Mating may occur anytime from January, in low latitudes, to April, in high latitudes. Births take place in the spring. Courtship may extend for days or months, estrus lasts 5–15 days, and the gestation period is usually 62–63 days. Mean litter size is 6 young, and the range is 1–11. The young weigh about 450 grams each at birth and are blind and deaf. Their eyes open after 11–15 days, they emerge from the den at 3 weeks, and they are weaned at around 5 weeks. The mother usually stays near the den for a period, while the father and other pack members hunt and bring food for both her and the pups. The young are commonly fed by regurgitation. At 8–10 weeks the young are shifted to the first in a series of rendezvous sites, each up to 8 km from the other. If in good condition, the young begin to travel with the pack in early autumn. Sexual maturity generally comes at around 22 months, but social restrictions often prevent mating at that time. A captive pair successfully bred when only 10 months old. Mortality is highest among the young. In times of sharply declining food supplies, all pups may be severely underweight or die from malnutrition, and reproduction may even cease. For adults in such a situation, the primary mortality factor has been found to be intraspecific strife. Annual survival of adults in a population not under nutritional stress or human exploitation has been calculated at 80 percent. Wild females are known to have given birth when 10 years old and to have lived to an age of 13 years and 8 months. Potential longevity is at least 16 years (Mech 1970, 1974*b*, 1977*c*, 1977*d*, 1989; Medjo and Mech

1976; Van Ballenberghe, Erickson, and Byman 1975; Van Ballenberghe and Mech 1975).

The wolf is often believed to be a direct threat to people. In Eurasia, attacks are unusual but evidently have occurred, sometimes resulting in death (Pulliainen 1980; Ricciuti 1978). In North America there appear to have been only four well-documented attacks by wild wolves (none resulting in death): (1) an injury inflicted by a female that may have been attracted to a camp by male dogs (Jenness 1985); (2) a prolonged assault involving a probably rabid individual (Mech 1970); (3) simply an aggressive leap that made contact with the person but caused no injury (Munthe and Hutchison 1978); and (4) several lunges by members of a pack at a group of three people (Scott, Bentley, and Warren 1985).

A far more substantive basis for the age-old warfare between people and the gray wolf is depredation by the latter on domestic animals, notably cattle, sheep, and reindeer. The wolf also has been persecuted, especially in the twentieth century, because of its alleged threat to populations of the wild ungulates that are desired by some persons for sport and subsistence hunting. The wolf was long taken by various kinds of traps and snares, as well as by pursuit with packs of specially trained dogs. In the nineteenth century, poison came into widespread use, and in the mid-twentieth century, hunting from aircraft became popular, especially on the open tundra.

The last wolves in the British Isles were exterminated in the 1700s. By the early twentieth century the species, except for occasional wandering individuals, had disappeared in most of western Europe and in Japan. Modest comebacks occurred in Europe during World Wars I and II, but currently the only substantial populations on the Continent west of Russia are in the Balkans. There are also very small remnant groups in Portugal, Spain, Italy, Czechoslovakia, Poland, and Scandinavia. *C. lupus* survives over much of its former range in southwestern and south-central Asia but is generally rare. Only 500–800 survive in India. There were estimated to be 150,000–200,000 wolves in the Soviet Union after World War II, but these became subject to an intensive government control program. The annual kill was 40,000–50,000 individuals from 1947 to 1962 and subsequently dropped to about 15,000. In the mid-1970s the estimated number of wolves in the Soviet Union was 50,000, about two-thirds of them in the Central Asian republics (Grzimek 1975; Pimlott 1975; Pulliainen 1980; Roberts 1977; Shahi 1982; Smit and Van Wijngaarden 1981; Zimen 1981). Recently, wolf numbers again increased in the Soviet Union, and there is now a bounty of up to 100 rubles (Bibikov 1980). There also is aerial hunting and poisoning by the government. In 1980 the number of wolf pelts taken there was 35,573 (Bibikov, Ovsyannikov, and Filimonov 1983).

The decline of the gray wolf was even more sweeping in the New World than in the Old. The species was largely eliminated along the east coast and in the Ohio Valley by the mid-nineteenth century. By 1914 the last gray wolves had been killed in Canada south of the St. Lawrence River and in the eastern half of the United States, except for northern parts of Minnesota and Wisconsin and the upper peninsula of Michigan. In the following year the federal government began a large-scale program to destroy predators. Partly as a result of this campaign, resident populations of the gray wolf apparently disappeared from the western half of the conterminous United States by the 1940s. The rate of decline subsequently slowed, mainly because the wolf had been eliminated from nearly all parts of the continent that were conducive to the raising of livestock. Conflict with agricultural interests does continue all along the lower edge of the current major range of the gray wolf, which extends across southern

Canada, from coast to coast, and dips down into northern Minnesota (Nowak 1975a). Regular government control operations are still carried out in parts of Canada and even have intensified recently in some areas in response to reported predation on livestock and big game (Gunson 1983; Stardom 1983; Tompa 1983).

The wolf still occurs over much of the northern part of Minnesota, primarily in boreal forest and bog country that is not suitable for agriculture but also in adjoining lands that are well settled. The species is protected by federal law, except that individuals preying on domestic animals may be taken by government agents. Many persons in Minnesota have argued that current protective regulations are overly restrictive and do not allow adequate control of depredations and that the wolf should be subject to sport hunting and commercial trapping. Recent studies in northwestern Minnesota, however, indicate that wolves infrequently attack domestic livestock, even when living nearby (Fritts 1982; Fritts and Mech 1981; Mech 1977a). In 1983 the U.S. Fish and Wildlife Service actually attempted to open the Minnesota wolf to sport hunting, but this effort was defeated in court.

In Alaska the wolf is legally open to sport hunting and fur trapping, but there is also controversy (Harbo and Dean 1983). Shooting from aircraft was a major hunting method for many years. In 1972 the practice was banned by federal law, except for government predator-control programs. Starting in the mid-1970s, the Alaska Department of Fish and Game authorized aerial hunting of wolves in an effort to increase the numbers of moose and caribou in certain parts of the state. A series of legal battles followed, involving opposing conservationist organizations, the state government, and the federal agencies that owned much of the land where the activity was to take place. Some of the hunts were halted, but others were allowed to proceed. The taking of the wolf in Alaska has been stimulated in part by rising fur prices. For the 1976/77 season the reported state harvest was 1,076 pelts with an average value of $200 (Deems and Pursley 1978). The fur take in Canada was 5,000–7,000 annually in the late 1970s but fell to about 4,000 in 1984 (Carbyn 1987).

The total number of gray wolves in North America can only be guessed at. The species was exterminated in Greenland by the 1930s, but individuals subsequently wandered back in from arctic Canada, and there now is a resident population of perhaps 50 animals (Dawes, Elander, and Ericson 1986). There are about 4,000–7,000 in Alaska, and there may be more than 30,000 in Canada (Carbyn 1983, 1987). Overall populations do not appear to have declined since the 1950s, but there is concern that excessive hunting, as well as oil and mineral exploitation, could adversely affect prey species, especially the caribou of tundra regions. In Minnesota there are about 1,200 wolves, probably as many as at any time in decades (Mech 1977a). Resident populations may have been eliminated in Wisconsin and Michigan (except on Isle Royale, as described above) in the 1950s, but a few individuals subsequently returned, and small packs are again present in Wisconsin (Jensen, Fuller, and Robinson 1986; Mech and Nowak 1981; Robinson and Smith 1977; Thiel 1985). In a 1974 experiment, four wolves were captured in northern Minnesota and released in the upper peninsula of Michigan, but within eight months all four had been killed by human agency (Weise et al. 1975). Since the 1940s there have been regular reports of wolves from the northwestern conterminous United States, especially along the Rocky Mountains between Glacier and Yellowstone national parks (Ream 1980; Weaver 1978). Such records probably represent individuals that have wandered from Canada, though there now is a small breeding population that utilizes land on both sides of the border in and around Glacier National Park. The occasional wolves that have appeared during the same period in the southwestern United States probably originated in Mexico. Mainly because of persecution by cattle ranchers, however, there may now be no more than 50 wolves left in Mexico itself (McBride 1980).

The IUCN classifies the species *C. lupus* as vulnerable. The USDI lists all populations of *C. lupus* in the conterminous United States and Mexico as endangered, except for that in Minnesota, which is designated as threatened. There have been repeated claims that the Minnesota population should be completely removed from the U.S. List of Endangered and Threatened Wildlife. However, it now is known that in addition to direct human persecution and long-term habitat disturbance, this population is jeopardized by parasitic heartworms and the deadly disease canine parovirus, both evidently spread by domestic dogs (Mech and Fritts 1987). *C. lupus* is on appendix 2 of the CITES, except for the populations of Pakistan, India, Nepal, and Bhutan, which are on appendix 1.

Canis familiaris (domestic dog).

There are approximately 400 breeds of domestic dog, the chihuahua being the smallest and the Irish wolfhound, the largest (Grzimek 1975). According to the National Geographic Society (1981), head and body length is 360–1,450 mm, tail length is 130–510 mm, shoulder height is 150–840 mm, and the normal weight range is 1–79 kg. The heaviest individual on record, a St. Bernard, weighs about 150 kg (*Washington Post*, 29 November 1987). In the wild subspecies of Australia, *C. f. dingo*, head and body length is 1,170–1,240 mm, tail length is 300–330 mm, shoulder height is about 500 mm, and weight is 10–20 kg. The dingo is usually tawny yellow in color, but some individuals are white, black, brown, rust, or other shades. The feet and tail tip are often white (Clutton-Brock, Corbet, and Hills 1976). From other forms of *C. familiaris* of comparable size and shape, the dingo can be distinguished by its longer muzzle, larger bullae, more massive molariform teeth, and longer, more slender canine teeth (Newsome, Corbett, and Carpenter 1980).

There have been various ideas regarding the origin of the domestic dog. The current consensus is that it was derived from one of the small south Eurasian subspecies of *C. lupus* and subsequently spread throughout the world in association with people (Nowak 1979). Some authorities, including Clutton-Brock, Corbet, and Hills (1976), think that the direct ancestor is probably the Indian wolf, *C. lupus pallipes*, but Olsen and Olsen (1977) argued in favor of the Chinese wolf, *C. l. chanco*. The wolf and dog still hybridize readily, but only when brought together in captivity or under very unusual natural conditions. The oldest well-documented remains of *C. familiaris*, dating from about 11,000 and 12,000 years ago, were found, respectively, in Idaho and Iraq. Beebe (1978) reported a specimen from the northern Yukon with a minimum age of 20,000 years, but Olsen (1985) considered that record doubtful.

At present, from the Balkans and North Africa to Southeast Asia dogs known as pariahs lead a semidomestic or even feral existence around villages (Bueler 1973; Fox 1978; Grzimek 1975). They may take food from people, scavenge, or actively hunt deer and other animals. They are socially flexible, being found either alone or in packs, but there seems to be a dominance hierarchy in any given area. These dogs, generally primitive in physical appearance, are probably closely related to the earliest dogs, as well as to the Australian dingo. The oldest definitely known fossils of the dingo date from about 3,500 years ago, but other remains may be as old as 8,600 years. People arrived in Australia at least 30,000 years ago. The dingo evidently was brought in long afterward but before true domestication had been achieved, and it was able to establish wild populations (Macintosh 1975). There are also wild dog populations in New Guinea and Timor, which are related to the primitive pariah-dingo group (Troughton 1971). It is possible that these dogs were spread by maritime peoples of south-central Asia rather than by migrating aboriginal peoples (Gollan 1984). There are many other populations of feral dogs, notably on islands and in Italy, but these animals are descendants of fully domesticated individuals (Lever 1985).

The dingo was once found throughout Australia, in forests, mountainous areas, and plains (Bueler 1973). Studies by Corbett and Newsome (1975) were carried out in an arid region of deserts and grasslands in central Australia, where the dingo depends in part on water holes made for cattle. Natal dens were found in caves, hollow logs, and modified rabbit warrens, usually within 2–3 km of water. Most activity in this region is nocturnal. In good seasons the diet consists mainly

Dingo *(Canis familiaris)*, photo from New York Zoological Society.

of small mammals, especially the introduced European rabbit (*Oryctolagus*). In times of drought the dingo takes kangaroos and cattle, mostly calves. Studies by Robertshaw and Harden (1985) in a forested area of northeastern New South Wales found the diet to consist mostly of larger native mammals, particularly wallabies.

According to Grzimek (1975), there may be regular seasonal migrations. Harden (1985) found daily movements to be about 10–20 km. Periods of movement usually were short and separated by even shorter periods of rest. On the average, individuals were active for 15.25 hours and at rest for 8.75 hours each day. Average home range for adults was 2,700 ha.

Corbett and Newsome (1975) learned that the dingo is basically solitary but that individuals in a given area form a loose, amicable association and sometimes come together. Fighting may develop between strangers. The dingo is not a particularly vocal canid but has a variety of sounds. Howls, which probably function to locate friends and repel strangers, are heard frequently in the single annual breeding season. Corbett (1988) reported that a captive group had separate male and female dominance hierarchies; all the females became pregnant, but the alpha female killed the newborn of the others. In the wild, pups are born in late winter and spring, following a gestation period of 63 days. Litter size is usually 4–5 young but ranges from 1 to 8 (Bueler 1973). Yearlings may assist an older pair to raise their pups. Independence is generally achieved by 3–4 months, but the young animals often then associate with a mature male (Corbett and Newsome 1975). Maximum known longevity is 14 years and 9 months (Grzimek 1975).

Although the dingo is said to be regularly captured and tamed by the natives and other people of Australia, Macintosh (1975) argued that it has never been successfully domesticated. It does have a close relationship with some groups of aborigines but evidently is not used in hunting nor inten-

tionally fed. Its main function may be to sleep in a huddle with persons and thereby provide protection from the cold.

Agriculturalists in Australia generally have a low opinion of the dingo, mainly because of its predation on sheep. Intensive persecution began in the nineteenth century and has continued to the present. During the 1960s about 30,000 dingos were killed for bounty annually in Queensland alone. Nearly 10,000 km of fencing has been constructed in eastern Australia in an effort to keep the wild dogs off the sheep ranges. Nonetheless, Macintosh (1975) suggested that depredations have been greatly exaggerated, and Grzimek (1975) pointed out that the dingo has actually aided agriculture by destroying introduced rabbits. Examination of stomach contents in Western Australia indicated that domestic stock was not a significant part of the dingo's diet, even though sheep and cattle were common in the study area (Whitehouse 1977). Human persecution has caused a decline in dingo populations, but another serious problem is hybridization with the domestic dog, a phenomenon that evidently is spreading with human settlement (Newsome and Corbett 1982, 1985).

Aside from the dingo, *C. familiaris* long was one of the least-known canids with respect to its behavior and ecology under noncaptive conditions, but there has been increasing study in recent years. Beck (1973, 1975) estimated that up to half of the 80,000–100,000 dogs in Baltimore, Maryland, are free-ranging, at least at times, giving an average density of about 230/sq km. They shelter in vacant buildings and garages and under parked cars and stairways. They are active mainly from 0500 to 0800 and from 1900 to 2200 hours in the summer, remaining out of sight during the midday heat. Their diet, consisting mostly of garbage but also including rats and ground-nesting birds, seems adequate to maintain weight and good health. Studies of unrestrained pets (Berman and Dunbar 1983; Rubin and Beck 1982) indicate that movement is concentrated in the early morning and late eve-

Domestic dogs *(Canis familiaris)*, photo from J. Nowak and E. Nowak.

ning but usually covers a very small area, probably because of the lack of a need to secure food. The distance a dog tends to move away from its home is positively correlated with the amount of time that it is allowed freedom.

Fully feral dogs sometimes occur in the countryside. Scott and Causey (1973) found Alabama packs to use moist floodplains in warm weather and dry uplands in cool weather and to cover distances of 0.5–8.2 km per day. Nesbitt (1975) determined that the females on a wildlife refuge in Illinois did not dig a den but gave birth in heavy cover. Activity in that area occurred anytime but was mainly nocturnal and crepuscular. The dogs traveled single file along roads, trails, and crop rows; if frightened, they took cover among trees and bushes. They fed on crippled waterfowl and deer, road kills and other carrion, small animals, some vegetation, and garbage.

In a study of free-ranging urban dogs in Newark, New Jersey, Daniels (1983a, 1983b) calculated a population density of about 150/sq km. Home range varied from 0.2 to 11.1 ha. in summer and from 0.1 to 5.7 ha. in winter. The relatively few groups that formed rarely contained more than two animals and often lasted only a few minutes; aggression was unusual, and there was no evidence of territoriality. However, presence of an estrous female in an area led to a congregation of males, usually about five or six, and to an increase in aggression and formation of a dominance hierarchy. Familiarity was observed to be an important basis for all social activity and for acceptance of a male by a female.

Home range was found to be 1.74 ha. for temporarily unrestrained suburban pets in Berkeley, California (Berman and Dunbar 1983); about 2.6 ha. each for 2 full-time free-ranging individuals in Baltimore (Beck 1975); 61 ha. for a group of 3 unowned dogs that lived in an abandoned building and used the streets and parks of St. Louis (Fox, Beck, and Blackman 1975); 444–1,050 ha. each for three feral packs of 2–5 dogs (Scott and Causey 1973); and 28,500 ha. for a feral pack of 5–6 (Nesbitt 1975). The last group was observed to have a dominance hierarchy and to be led by a female. Pets tend to be solitary and form social groups only randomly, while the three unowned urban animals maintained a long-term relationship, the single female member initiating movements. In the Baltimore study, half of the animals seen were solitary, 26 percent were in pairs, and the rest were in groups of up to 17 members. Spring and fall breeding peaks there were suggested by fluctuations in reports of unwanted dogs.

Female dogs enter estrus twice a year, usually in late winter or early spring and in the fall. Heat lasts about 12 days. At such times, males tend to leave their owner's home, mark territories, and fight rivals. The gestation period averages 63 days, litter size is usually 3–10 young, and nursing lasts about 6 weeks. Males often remain with the females and young. Sexual maturity comes after 10–24 months, and old age generally after 12 years, but a few individuals live for 20 years (Asdell 1964; Bueler 1973; Grzimek 1975).

There are estimated to be 50 million owned dogs in the United States, and many more lack owners. Although these animals have abundant uses and values, they may cause problems for some people. In Baltimore, for example, dogs have been implicated in the spread of several diseases (in addition to rabies), they may benefit rats by overturning garbage cans, and there are about 7,000 reported attacks on people each year (Beck 1973, 1975). There are about 1 million reported attacks annually for the whole United States. In 1986, 13 people, mostly children, were killed in the United States; in 7 of these cases the breed known as the pit bull was responsible (*Washington Post*, 1 September 1987, pp. B-1, B-5). *C. familiaris* also often is considered to be a serious predator of livestock and game animals, especially deer. Field studies, however, have indicated that feral dogs do not significantly affect deer

populations and may even have a sanitary function in eliminating carrion and crippled animals (Gipson and Sealander 1977; Nesbitt 1975; Scott and Causey 1973).

CARNIVORA; CANIDAE; Genus CHRYSOCYON
Hamilton-Smith, 1839

Maned Wolf

The single species, *C. brachyurus*, occurs in open country of central and eastern Brazil, eastern Bolivia, Paraguay, northern Argentina, and Uruguay (Cabrera 1957; Langguth 1975b). There may be a disjunct population on the llanos of Colombia (Dietz 1985).

Head and body length is 1,245–1,320 mm, tail length is 280–405 mm, and shoulder height is about 740 mm (Bueler 1973). Weight is 20–23 kg (Grzimek 1975). Overall it looks like a red fox (*Vulpes vulpes*) on stilts (Clutton-Brock, Corbet, and Hills 1976). The general coloration is yellow red. The hair along the nape of the neck and middle of the back is especially long and may be dark in color. The muzzle and lower parts of the legs are also dark, almost black. The throat and tail tuft may be white. The coat is fairly long, somewhat softer than that of *Canis*, and has an erectile mane on the back of the neck and top of the shoulders. The ears are large and erect, the skull is elongate, and the pupils of the eyes are round.

The maned wolf inhabits grasslands, savannahs, and swampy areas (Langguth 1975b). The natal nest is located in thick, secluded vegetation (Bueler 1973). Activity is mainly nocturnal and crepuscular (Dietz 1984). It has been suggested that the remarkably long legs are adaptations for fast running or for movement through swamps, but the actual function is probably to allow seeing above tall grass. *Chrysocyon* is not an especially swift canid, does not pursue prey for long distances, and generally stalks and pounces like a fox (Kleiman 1972). Its omnivorous diet includes rodents, other small mammals, birds, reptiles, insects, fruit, and other vegetable matter.

Chrysocyon is monogamous, with mated pairs sharing defended territories averaging 27 sq km. However, most activity is solitary, the male and female associating closely only during the breeding season (Dietz 1984). Captives sometimes can be kept together without apparent strife, though there is usually an initial period of fighting and then establishment of a dominance hierarchy (Brady and Ditton 1979). The father is known to regurgitate food for the young in captivity (Rasmussen and Tilson 1984) and may also have a significant parental role in the wild (Dietz 1984). The three main vocalizations are: a deep-throated single bark, heard mainly after dusk; a high-pitched whine; and a growl during agonistic behavior (Kleiman 1972).

Births in captivity have occurred in July and August in South America and in January and February in the Northern Hemisphere. Females are monestrous, heat lasts about 5 days, the gestation period is 62–66 days, and litter size is two to four young. The young weigh about 350 grams each at birth, open their eyes after 8 or 9 days, begin to take some regurgitated food at 4 weeks, and are weaned by 15 weeks (Brady and Ditton 1979; Da Silveira 1968; Faust and Scherpner 1967). One captive lived for 13 years (Jones 1982).

The maned wolf is not extensively hunted for its fur, but it is sometimes persecuted because of an unjustified belief that it kills domestic livestock (Grzimek 1975; Meritt 1973). It often is killed because of alleged depredations on chickens. It disappeared from Uruguay in the nineteenth century and is now threatened in other regions by the annual burning of its grassland habitat, as well as by hunting and live capture

Maned wolf *(Chrysocyon brachyurus)*, photo by Bernhard Grzimek.

(Thornback and Jenkins 1982). It also, however, recently extended its range into a deforested zone of central Brazil (Dietz 1985). It is classified as vulnerable by the IUCN and as endangered by the USDI and is on appendix 2 of the CITES.

CARNIVORA; CANIDAE; Genus OTOCYON
Müller, 1836

Bat-eared Fox

The single species, *O. megalotis*, is found from Ethiopia and southern Sudan to Tanzania and from southern Angola and Zimbabwe to South Africa (Coetzee, *in* Meester and Setzer 1977). According to Ansell (1978), it does not occur in Zambia. *Otocyon* sometimes has been placed in a separate subfamily, the Otocyoninae, on the basis of its unusual dentition. Clutton-Brock, Corbet, and Hills (1976), however, considered *Otocyon* to be simply an aberrant fox with systematic affinities to *Urocyon* (which they included in *Vulpes*) and some behavioral similarities to *Nyctereutes*.

Head and body length is 460–660 mm, tail length is 230–340 mm, shoulder height is 300–400 mm, and weight is 3.0–5.3 kg. The upper parts are generally yellow brown with gray agouti guard hairs. The throat, underparts, and insides of the ears are pale. The outsides of the ears, mask, lower legs, feet, and tail tip are black. In addition to coloration, distinguishing characters include the relatively short legs and the enormous ears (114–35 mm long).

Otocyon has more teeth than any other placental mammal that has a heterodont condition (the teeth being differentiated into several kinds). Whereas in all other canids there are no more than two upper and three lower molars, *Otocyon* has at least three upper and four lower molars. This condition is sometimes held to be primitive, but it more likely represents the results of a mutation that caused the appearance of the extra molar teeth in what had been a fox population with normal canid dentition (Clutton-Brock, Corbet, and Hills 1976).

The bat-eared fox is found in arid grasslands, savannahs, and brush country. It seems to prefer places with much bare ground or where the grass has been kept short by burning or grazing. When the grass again grows high, *Otocyon* may depart and wander about in search of a new place of residence. A capable digger, it either excavates its own den or enlarges the burrow of another animal. A family may have more than one den in its home range, each with multiple entrances and chambers and several meters of tunnels. In the Serengeti 85 percent of activity occurs at night, but in South Africa *Otocyon* is mainly diurnal in winter and nocturnal in summer. One female was observed to forage over about 12 km per night. The diet consists predominantly of insects, most notably termites, and also includes other arthropods, small rodents, the eggs and young of ground-nesting birds, and vegetable matter (Kingdon 1977; Lamprecht 1979; Nel 1978; Smithers 1971).

In the Masai Mara Reserve of Kenya, Malcolm (1986) found an overall population density of 0.8–0.9 individuals per sq km. Group home ranges there averaged 3.53 sq km and overlapped, in one case there being no area of exclusive use. Groups consisted of 2–5 members and generally had amicable relations with other groups. In South Africa, Nel (1978) also found home ranges to overlap extensively and observed no territorial defense or marking. Up to 15 individuals, representing four groups, were seen foraging within less than 0.5 sq km. In the Serengeti, however, Lamprecht (1979) found resident families to occupy largely exclusive home ranges of 0.25–1.5 sq km and to mark them with urine. Groups usually consisted of a mated adult pair, accompanied by their young

Bat-eared fox *(Otocyon megalotis)*, photo by Bernhard Grzimek.

of the year for a lengthy period. A few observations suggested that there may sometimes be two adult females with a male. Strangers of the same sex generally were hostile. Contact between members of a group was maintained by soft whistles. Nel and Bester (1983) reported that most communication involves visual signaling, mainly with the large ears and tail.

In both the Serengeti and Botswana, births occur mainly from September to November (Lamprecht 1979; Smithers 1971), but pups have been recorded in Uganda in March (Kingdon 1977), and reproduction may be year-round in some parts of East Africa (Malcolm 1986). Gestation is usually reported as 60–70 days, but Rosenberg (1971) calculated the period at 75 days for a birth in captivity. Litters contain two to six young. They are suckled for 15 weeks and then begin to forage with the parents. Regurgitation is evidently rare. The young are probably full grown by 5 or 6 months and separate from the parents prior to the breeding season. According to Jones (1982), a captive lived for 13 years and 9 months.

Otocyon has declined in settled parts of South Africa (Coetzee, *in* Meester and Setzer 1977). Nonetheless, it is apparently extending its range eastward into Mozambique and into previously unoccupied parts of Zimbabwe and Botswana (Pienaar 1970).

CARNIVORA; CANIDAE; Genus CUON
Hodgson, 1838

Dhole

The single species, *C. alpinus,* is found from southern Siberia and Soviet Central Asia to India and the Malay Peninsula and on the islands of Sumatra and Java but not Sri Lanka. Except as otherwise noted, the information for this account was taken from the review papers by Cohen (1977, 1978) and Davidar (1975).

Head and body length is 880–1,130 mm, tail length is 400–500 mm, and shoulder height is 420–550 mm. Males weigh 15–21 kg and females, 10–17 kg. The coloration is variable, but generally the upper parts are rusty red, the underparts are pale, and the tail is tipped with black. In the northern parts of the range the winter pelage is long, soft, dense, and bright red, and the summer coat is shorter, coarser, sparser, and less vivid in color. *Cuon* resembles *Canis* externally, but the skull has a relatively shorter and broader rostrum. Females have 12–16 mammae.

The dhole occupies many types of habitat but avoids deserts. In the Soviet Union it occurs mainly in alpine areas, and in India it is found in dense forest and thick scrub jungle. The preferred habitat in Thailand is dense montane forest at elevations of up to 3,000 meters (Lekagul and McNeely 1977). *Cuon* may excavate its own den, enlarge a burrow made by another animal, or use a rocky crevice. M. W. Fox (1984) found one earth den to have six entrances leading to a labyrinth of at least 30 meters of interconnected tunnels and four large chambers; many generations of dholes probably had developed this complex. *Cuon* may be active at any time, but mainly in early morning and early evening. Cohen et al. (1978) reported a major peak of activity at 0700–0800 hours and a lesser peak at 1700–1800 hours.

The dhole hunts in packs and is primarily a predator of mammals larger than itself. Prey is tracked by scent and then pursued, sometimes for a considerable distance. When the objective is overtaken, it is surrounded and attacked from different sides. Prey animals include deer, wild pigs, mountain sheep, gaur, and antelope. The chital *(Axis axis)* is probably the major prey in India, though Cohen et al. (1978) found remains of this deer to occur less frequently than those of *Lepus* in the droppings of *Cuon.* The diet also includes rodents, insects, and carrion. Reports of predation on tigers, leopards, and bears generally are not well documented, but those carnivores are sometimes driven from their kills by

Dhole *(Cuon alpinus)*, photo from New York Zoological Society.

packs of dholes. There are numerous records of leopards being treed.

In a study in southern India, Johnsingh (1982) found population densities of 0.35–0.90 individuals per sq km. A pack in this area contained an average of 8.3 adults and used a home range of 40 sq km. Other work suggests that there are usually 5–12 dholes in a pack, but up to 40 have been reported. This discrepancy was perhaps explained by M. W. Fox (1984), who pointed out that the larger groups are actually clans comprising several related packs. In parts of India the clans keep together during that part of the year when only juvenile and adult chital are available as prey but divide into smaller hunting packs when the chital have fawns. A pack apparently consists of a mated pair and their offspring. Although the social structure has not been closely studied, there seems to be a leader, a dominance hierarchy, and submissive behavior by lower-ranking animals. Intragroup fighting is rarely observed. More than one female sometimes den and rear litters together. Vocalizations include nearly all of those made by the domestic dog except loud and repeated barking; the most distinctive sound is a peculiar whistle that probably serves to keep the pack together during pursuit of prey.

In India, mating occurs from September to November, and births from November to March. In the Moscow Zoo, mating occurred in February, and gestation lasted 60–62 days (Sosnovskii 1967). Litters usually contain four to six young, but up to nine embryos have been recorded. In the wild, both mother and young are provided with regurgitated food by other pack members. The pups leave the den at 70–80 days and participate in kills at 7 months (Johnsingh 1982). Longevity in the Moscow Zoo is 15–16 years.

Although the dhole only rarely takes domestic livestock and was only once reported to attack a person, it has been intensively poisoned and hunted throughout its range. This situation seems based mainly on dislike by hunters, who see *Cuon* as a competitor for game and who are repulsed by its method of predation. Some of the village people of India actually welcome the dhole, following it in order to expropriate its kills. Because of direct persecution, elimination of natural prey, and destruction of its forest habitat, the dhole has declined seriously in range and numbers. It is classified as vulnerable by the IUCN and as endangered by the USDI and the Soviet Union and is on appendix 2 of the CITES.

CARNIVORA; CANIDAE; Genus LYCAON
Brookes, 1827

African Hunting Dog

The single species, *L. pictus,* occurs in most of Africa south of the Sahara Desert. It formerly occurred in suitable parts of

African hunting dog *(Lycaon pictus)*, photo by Bernhard Grzimek.

the Sahara and in Egypt (Coetzee, *in* Meester and Setzer 1977; Kingdon 1977).

Head and body length is 760–1,120 mm, tail length is 300–410 mm, shoulder height is 610–780 mm, and weight is 17–36 kg (Kingdon 1977). Males and females are about the same size (Frame et al. 1979). There is great variation in pelage, the mottled black, yellow, and white occurring in almost every conceivable arrangement and proportion. In most individuals, however, the head is dark and the tail has a white tip or brush. The fur is short and scant, sometimes so sparse that the blackish skin is plainly visible. The ears are long, rounded, and covered with short hairs. The legs are long and slender, and there are only four toes on each foot. The jaws are broad and powerful. *Lycaon* has a strong, musky odor. Females have 12–14 mammae (Van Lawick and Van Lawick–Goodall 1971).

Lycaon inhabits grassland, savannah, and open woodland (Coetzee, *in* Meester and Setzer 1977). Its den, usually an abandoned aardvark hole, is occupied only to bear young (Kingdon 1977). The pack does not wander very far from the den when pups are present (Frame et al. 1979). Otherwise, movements are generally correlated with hunting success: if prey is scarce, the entire home range may be traversed in two or three days. Hunts take place in the morning and early evening. Schaller (1972) recorded peaks of activity from 0700 to 0800 and from 1800 to 1900 hours. Prey is apparently located by sight, approached silently, and then pursued at speeds of up to 66 km/hr for 10–60 minutes (Kingdon 1977). Van Lawick and Van Lawick–Goodall (1971) observed a pack to maintain a speed of about 50 km/hr for 5.6 km. Their investigation indicated this distance to be about the maximum that *Lycaon* would usually follow before giving up. They observed 91 chases, 39 of which were successful. In all but 1 of the latter cases, the quarry was killed within five minutes of being caught. In his study, Schaller (1972) observed 70 percent of chases to be successful. Groups of *Lycaon* generally cooperate in hunting large mammals, but individuals sometimes pursue hares, rodents, or other small animals. The main prey seems to vary by area, being bushduiker and reed-

buck in the Kafue Valley of Zambia, impala in Kruger National Park of South Africa, and Thomson's gazelle and wildebeest in the Serengeti of Tanzania (Kingdon 1977). Certain packs in the Serengeti, however, specialize in the capture of zebra (Malcolm and Van Lawick 1975). Some food may be cached in holes (Malcolm 1980*b*).

From 1970 to 1977, population density in the Serengeti declined from one adult *Lycaon* per 35 sq km to one per 200 sq km. Pack home range in this area is generally 1,500–2,000 sq km (Frame et al. 1979). A pack in South Africa reportedly used a home range of about 3,900 sq km (Van Lawick and Van Lawick–Goodall 1971). Range contracts when there are small pups at a den; at such time one Serengeti pack used an area of only 160 sq km for 2.5 months (Schaller 1972). The home range of a pack overlaps by about 10–50 percent with those of several neighboring packs (Frame and Frame 1976). Territoriality does not seem to be well developed, and hundreds of individuals may once have gathered temporarily in response to migrations of the formerly vast herds of springbok in southern Africa (Kingdon 1977).

Studies on the Serengeti Plains of Tanzania have revealed that *Lycaon* has an intricate and unusual social structure (Frame and Frame 1976; Frame et al. 1979; Malcolm 1980*a*; Malcolm and Marten 1982; Schaller 1972; Van Lawick and Van Lawick–Goodall 1971). Groups were found to contain averages of 9.8 (1–26) individuals, 4.1 (0–10) adult males, and 2.1 (0–7) adult females. This sexual proportion is unlike the usual condition in social mammals. Some packs have as many as 8 adult males with only a single adult female. Moreover, in a reversal of the usual mammalian process, females emigrate from their natal group far more than do males. Commonly, several sibling females 18–24 months old leave their pack and join another that lacks sexually mature females. Following the transfer, one of the females achieves dominance, and its sisters may then depart. Whereas no female seems to stay in its natal pack past the age of 2.5 years, about half of the young males do remain; the other males emigrate, usually in sibling groups. The typical pack thus consists of several related males, often representing more than one generation, and one or more females that are genetically related to each other but not to the males. Some male lineages within a pack are known to have lasted at least 10 years.

There are separate dominance hierarchies for each sex. Only the highest-ranking male and female normally breed, and they inhibit reproduction by subordinates. There is intensive rivalry among the females for the breeding position. If a subordinate female does bear pups, the dominant one may steal them. Females sometime fight savagely, and the loser may leave the group and perish. Aside from this aspect of the social life of *Lycaon*, packs are remarkably amicable, with little overt strife. Food is shared, even by individuals that do not participate in the kill. An animal with a broken leg was allowed to feed, when it hobbled up after the others, throughout the time required for its leg to mend. Pups old enough to take solid food are given first priority at kills, eating even before the dominant pair. Subordinate animals, especially males, help feed and protect the pups. There are several vocalizations, the most striking of which is a series of wailing hoots that probably serves to keep the pack together during pursuit of prey.

Births may occur at any time of year but peak from March to June during the second half of the rainy season. The interval between births is normally 11–14 months but may be as short as 6 months if all the young perish. The gestation period is 79–80 days. Litter size averages about 10 and ranges from 6 to 16 young. The newborn weigh about 300 grams each and open their eyes after 13 days. At about 3 weeks the young emerge from the den and begin to take some solid food. Weaning is normally completed by the age of 11 weeks. All adult pack members regurgitate food to the young. Once, when a mother died, the males of the pack were able successfully to raise her 5-week-old pups. When the pack is hunting, 1 or 2 adults remain at the den to guard the pups. After 3 months the young begin to follow the pack, and at 9–11 months they can kill easy prey, but they are not proficient until they are about 12–14 months old. Social restrictions blur the actual time of sexual maturity. Five males were observed to first mate at 21, 33, 36, 36, and 60 months of age. The youngest female to give birth was 22 months old at the time. Maximum observed longevity is 11 years (Frame et al. 1979; Kingdon 1977; Malcolm 1980a; Schaller 1972; Van Heerden and Kuhn 1985; Van Lawick and Van Lawick–Goodall 1971).

Lycaon has been reported only rarely to attack people but is widely persecuted as a predator of domestic livestock and game. It has been wiped out in South Africa except in the

vicinity of Kruger National Park and has declined greatly in distribution in Namibia, Zimbabwe, Tanzania, and Kenya (Kingdon 1977; Lensing and Joubert 1977; Skinner, Fairall, and Bothma 1977). Although it still occurs over much of its original range, there are probably fewer than 7,000 individuals in existence (Malcolm 1980a). Its numbers in Zimbabwe have been estimated at 300–350, and it continues to be persecuted there, except in national parks (Childes 1988). *Lycaon* is classified as vulnerable by the IUCN and as endangered by the USDI.

CARNIVORA; Family URSIDAE

Bears

This family of three Recent genera and eight species occurred historically almost throughout Eurasia and North America, in the Atlas Mountains of North Africa, and in the Andes of South America. Hall (1981) recognized three living subfamilies: Tremarctinae, with the genus *Tremarctos*; Ursinae, with *Ursus*; and Ailuropodinae, with *Ailuropoda*. The sequence of genera presented here basically follows that of Simpson (1945), though he, like many other authorities, recognized additional genera.

Head and body length is 1,000 to approximately 2,800 mm, tail length is 65–210 mm, and weight is 27–780 kg. Males average about 20 percent larger than females. The coat is long and shaggy, and the fur is generally unicolored, usually brown, black, or white. Some genera have white or buffy crescents or semicircles on the chest. *Tremarctos*, the spectacled bear of South America, typically has a patch of white hairs encircling each eye. *Ailuropoda*, the giant panda, has a striking black and white color pattern.

Bears have a big head; a large, heavily built body; short, powerful limbs; a short tail; and small eyes. The ears are small, rounded, and erect. The soles are hairy in species that are mainly terrestrial but naked in species that climb considerably, such as *Ursus malayanus*. All limbs have five digits. The claws are strong, recurved, and used for tearing and digging. The lips are free from the gums.

The skull is massive, and the tympanic bullae are not inflated. In most genera the dental formula is: (i 3/3, c 1/1, pm 4/4, m 2/3) × 2 = 42. The species *Ursus ursinus*, however, has only 2 upper incisors and a total of 40 teeth. Ursid incisors are not specialized, the canines are elongate, the first 3 premolars are reduced or lost, and the molars have broad, flat, and tubercular crowns. The carnassials are not developed as such.

Habitats range from arctic ice floes to tropical forests. Those populations that occur in open areas often dig dens in hillsides. Others shelter in caves, hollow logs, or dense vegetation. Bears have a characteristic shuffling gait. They walk plantigrade, with the heel of the foot touching the ground. They are capable of walking on their hind legs for short distances. When need be, they are surprisingly agile and careful in their movements. Their eyesight and hearing are not particularly good, but their sense of smell is excellent. Bears are omnivorous, except that the polar bear *(Ursus maritimus)* feeds mainly on fish and seals.

During the autumn, in most parts of the range of the family, bears become fat. With the approach of cold weather they cease eating and go into a den that they have prepared in a protected location. Here they sleep through the winter, living mainly off stored fat reserves. With certain exceptions, especially pregnant females, the polar bear does not undergo winter sleep. Some authorities prefer not to call this process hibernation, as body temperature is not substantially reduced, body functions continue, and bears can usually be

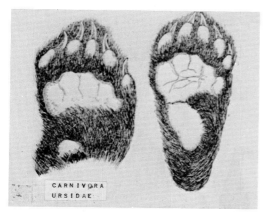

American black bear *(Ursus americanus)*, right forepaw and right hind foot, photo from *Proc. Zool. Soc. London*.

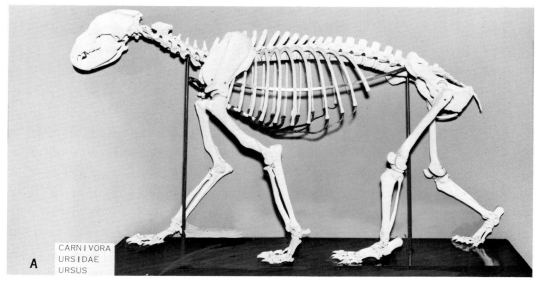

CARNIVORA
URSIDAE
URSUS

A. Brown bear *(Ursus arctos)*, skeleton from U.S. National Museum of Natural History. B. Photo of *U. arctos* by Leonard Lee Rue III. C. Giant panda *(Ailuropoda melanoleuca)*, photo by Elaine Anderson.

easily aroused. Sometimes they awaken on their own during periods of mild weather. Folk, Larson, and Folk (1976), however, found that the heart rate of a hibernating bear drops to less than half of normal and that other physiological changes occur. They concluded that bears do experience true mammalian hibernation.

Except for courting pairs and females with young, bears live alone. Litters are produced at intervals of 1–4 years. In most regions births occur from November to February, while the mother is hibernating. The period of pregnancy is commonly extended 6–9 months by delayed implantation of the fertilized egg. Litter size is one to four young. The young are relatively tiny at birth, ranging from 225–680 grams each. They remain with the mother at least through their first autumn. They become sexually mature at 2.5–6 years of age and normally live 15–30 years in the wild.

Bears are usually peaceful animals that try to avoid conflict. However, if they consider themselves, their young, or their food supply threatened, they can become formidable adversaries. Only a small proportion of the stories of unprovoked attacks by bears on people are true. When such cases are carefully investigated, it is usually found that there was provocation. Nonetheless, bears have been persecuted almost throughout their range because of alleged danger to humans and because they are sometimes considered to be serious predators of domestic livestock.

The geological range of the Ursidae is late Miocene to Recent in North America, late Pliocene to Recent in South America, late Eocene to Recent in Europe, early Miocene to Recent in Asia, Pleistocene to Recent in North Africa, and Pliocene in South Africa (Hendey 1977; Stains 1984).

CARNIVORA; URSIDAE; Genus URSUS
Gervais, 1855

Spectacled Bear

The single species, *T. ornatus*, is known to inhabit the mountainous regions of western Venezuela, Colombia, Ecuador, Peru, and western Bolivia (Cabrera 1957). It also has been

Spectacled bear *(Tremarctos ornatus)*, photo from San Diego Zoological Garden.

reported from eastern Panama and northern Argentina (Peyton 1986).

Head and body length is usually 1,200–1,800 mm, tail length is about 70 mm, and shoulder height is 700–800 mm. One male, 1,740 mm in length, weighed 140 kg. Peyton (1980) reported that a male 2,060 mm in total length weighed 175 kg. Grzimek (1975) gave the weight of females as 60–62 kg. The entire body is uniformly black or dark brown, except for large circles or semicircles of white around the eyes and a white semicircle on the lower side of the neck, from which lines of white extend onto the chest. The common name is derived from the white around the eyes. The head and chest markings are variable, however, and may be completely lacking in some individuals.

In the Andes of Peru, Peyton (1980) found *Tremarctos* to occupy a wide variety of habitats from 457 to 3,658 meters in elevation. The preferred habitats are humid forests between 1,900 and 2,350 meters and coastal thorn forests when water is available. High-altitude grasslands are also utilized. *Tremarctos* is apparently mainly nocturnal and crepuscular. During the day it beds down between or under large tree roots, on a tree trunk, or in a cave. It frequently climbs large trees to obtain fruit. While in a tree, it may assemble a large platform of broken branches, on which it positions itself to eat and to reach additional fruit. Peyton found one such platform at a height of 15 meters.

Tremarctos feeds extensively on fruit, moving about in response to seasonal ripening. It also depends on plants of the family Bromeliaceae, especially when ripe fruit is not available. It tears off the leaves of large bromeliads to feed on the white bases and obtains the edible hearts of small bromeliads by ripping the entire plant off the substrate. In addition, this bear climbs large cacti to get the fruits at the top, tears into the green stalks of young palms to eat the unopened inner leaves, and strips bark off trees to feed on the cortex. The diet also includes bamboo hearts, corn, rodents, and insects. Only about 4 percent of the food was found to be animal matter (Peyton 1980).

In Peru, Peyton (1980) received reports that a male often enters a cornfield with one or more females, and sometimes with yearling animals, during the months of March–July. According to Grzimek (1975), *Tremarctos* has a "striking shrill voice." In the Buenos Aires Zoo young were produced in July, while in European zoos births have occurred from late December to March. Pregnancy lasts 6.5–8.5 months and apparently involves delayed implantation. Litters contain one to three young, each weighing about 320 grams (Bloxam 1977; Gensch 1965; Grzimek 1975). One captive lived for 36 years and 5 months (Jones 1982).

Mittermeier et al. (1977) reported that the meat of *Tremarctos* is highly esteemed in northern Peru and that this bear also is killed by people for its skin and fat. Grimwood (1969) warned that *Tremarctos* had become rare and endangered in Peru through intensive hunting by sportsmen and landowners, who consider it to be a predator of domestic livestock. Peyton (1980) did not believe *Tremarctos* to be in immediate danger of extinction, because it is adapted to a diversity of habitats, some of which are largely inaccessible to people. He did note, however, that some bears become habituated to raiding cornfields and that these animals are frequently shot. Later, Peyton (1986) noted that vast parts of the original range have been replaced by agriculture and that surviving bear populations are fragmentary. Thornback and Jenkins (1982) stated that the spectacled bear is declining through much of its range because of habitat loss due to settlement, to human persecution that subsequently results from raids on crops and livestock, and to hunting for its meat and skin. *Tremarctos* is classified as vulnerable by the IUCN and is on appendix 1 of the CITES.

CARNIVORA; CANIDAE; **Genus URSUS**
Linnaeus, 1758

Black, Brown, Polar, Sun, and Sloth Bears

There are six species (Corbet 1978; Ellerman and Morrison-Scott 1966; Hall 1981; Kurten 1973; Laurie and Seidensticker 1977; Lay 1967; Lekagul and McNeely 1977; Ma 1983*a*; Simpson 1945):

U. thibetanus (Asiatic black bear), Afghanistan, southeastern Iran, Pakistan, Himalayan region, Burma, Thailand, Indochina, China, Manchuria, Korea, extreme southeastern Siberia, Japan, Taiwan, Hainan;

U. americanus (American black bear), Alaska, Canada, conterminous United States, northern Mexico;

U. arctos (brown or grizzly bear), western Europe and Palestine to eastern Siberia and Himalayan region, Atlas Mountains of northwestern Africa, Hokkaido, Alaska to Hudson Bay and northern Mexico;

U. maritimus (polar bear), primarily on arctic coasts, islands, and adjacent sea ice of Eurasia and North America;

U. malayanus (Malayan sun bear), Burma, Thailand, Indochina, Malay Peninsula, Sumatra, Borneo, Yunnan (southern China);

U. ursinus (sloth bear), India, Nepal, Bangladesh, Sri Lanka.

Each of these species has often been placed in its own genus or subgenus: *Selenarctos* Heude, 1901, for *U. thibetanus*; *Euarctos* Gray, 1864, for *U. americanus*; *Ursus* Linnaeus, 1758, for *U. arctos*; *Thalarctos* Gray, 1825, for *U. maritimus*; *Helarctos* Horsfield, 1825, for *U. malayanus*; and *Melursus* Meyer, 1793, for *U. ursinus*. Hall (1981), however, placed all of these names in the synonymy of *Ursus*. This arrangement is supported in part by the captive production of viable offspring through hybridization between several of the above species (Van Gelder 1977*b*).

The systematics of the brown or grizzly bear have caused considerable confusion. Old World populations long have been recognized to compose a single species, with the scientific name *U. arctos* and the general common name brown bear. In North America the name "grizzly" is applied over most of the range, while the term "big brown bear" is often used on the coast of southern Alaska and nearby islands, where the animals average much larger than those inland. Hall (1981) listed 77 Latin names that have been used in the specific sense for different populations of the brown or grizzly bear in North America. No one now thinks that there are actually so many species, but some authorities, such as Burt and Grossenheider (1976), have recognized the North American grizzly *(U. horribilis)* and the Alaskan big brown bear *(U. middendorffi)* as species distinct from *U. arctos* of the Old World. Other authorities (Erdbrink 1953; Kurten 1973; Rausch 1953, 1963), based on limited systematic work, have referred the North American brown and grizzly to *U. arctos*. This procedure is being used by most persons now studying or writing about bears and is followed here. Kurten (1973) distinguished three North American subspecies: *U. a. middendorfi*, on Kodiak and Afognak islands; *U. a. dalli*, on the south coast of Alaska and the west coast of British Columbia; and *U. a. horribilis*, in all other parts of the range of the species. Hall (1984), however, recognized nine North American subspecies of *U. arctos*.

From *Tremarctos*, *Ursus* is distinguished by its masseteric

fossa on the lower jaw not being divided by a bony septum into two fossae. From *Ailuropoda*, it is distinguished in having an alisphenoid canal (Hall 1981). Additional information is provided separately for each species.

Ursus thibetanus (Asiatic black bear).

Head and body length is 1,200–1,800 mm and tail length is 65–106 mm. Stroganov (1969) listed weight as 110–50 kg for males and 65–90 kg for females. Roberts (1977) stated that an exceptionally large male weighed 173 kg but that an adult female weighed only 47 kg. The coloration is usually black but is sometimes reddish brown or rich brown. There is some white on the chin and a white crescent or **V** on the chest.

The Asiatic black bear frequents moist deciduous forests and brushy areas, especially in the hills and mountains. It ascends to elevations as high as 3,600 meters in the summer and descends in the winter. It swims well. According to Lekagul and McNeely (1977), this bear is generally nocturnal, sleeping during the day in hollow trees, caves, or rock crevices. It is also seen abroad by day when favored fruits are ripening. It climbs expertly to reach fruit and beehives. It usually walks on all fours but often stands on its hind legs so that its forepaws can be used in fighting. The diet includes fruit, buds, invertebrates, small vertebrates, and carrion. Domestic livestock is sometimes taken, and animals as large as adult buffalo are killed by breaking the neck. Individuals become fat in late summer and early fall before hibernation, but some populations do not undergo winter sleep, or do so only for brief periods of severe weather. Roberts (1977) stated that in the Himalayas *U. thibetanus* hibernates, sometimes in a burrow of its own making, but that in southern Pakistan there is no evidence of hibernation. According to Stroganov (1969), hibernation in Siberia begins in November and lasts four or five months. Dens in that area are usually in tree holes. The bears are easily aroused during the first month but sleep more deeply from December to February.

In Siberia, individual home range is 500–600 ha., only one-third or one-fourth the size of that of *U. arctos*. Mating in Siberia occurs in June or July, and births take place from late December to late March, mostly in February (Stroganov 1969). In Pakistan, mating is thought to occur in October, and the young are born in February (Roberts 1977). According to Lekagul and McNeely (1977), pregnancy lasts 7–8 months, and usually two cubs are born in a cave or hollow tree in early winter. The eyes open when the cubs are about 1 week old, and shortly thereafter the young begin to follow the female as she forages. They are weaned at about 3.5 months but remain with the mother until they are 2–3 years old. Females have been seen with two sets of cubs. Sexual maturity comes at about 3 years, and longevity in captivity may be as much as 33 years.

The Asiatic black bear sometimes raids cornfields and attacks domestic livestock. It also has been occasionally reported to kill humans. Hayashi (1984), for example, reported that from 1970 to 1983 in Fukui Prefecture of Japan there were attacks on 16 people, one of whom was killed. For these reasons *U. thibetanus* is hunted by people, and it also has declined because of the destruction of its forest habitat (Cowan 1972). It is on appendix 1 of the CITES, and the subspecies *U. t. gedrosianus*, of southern Pakistan, is classified as endangered by the IUCN and USDI.

Ursus americanus (American black bear).

Head and body length is 1,500–1,800 mm, tail length is about 120 mm, and shoulder height is up to 910 mm. Banfield (1974) listed weights of 92–140 kg for females and 115–270 kg for males. The most common color phases are black, chocolate brown, and cinnamon brown. Different colors may occur in the same litter. A white phase is generally rare, and

Asiatic black bear *(Ursus thibetanus)*, photo from New York Zoological Society.

never in the majority, but seems to be most common on the Pacific coast of central British Columbia. A blue black phase is also generally rare but occurs frequently in the St. Elias Range of southeastern Alaska. Rounds (1987) reported that coloration varies geographically, with nearly all bears in eastern North America being black but most in some southwestern populations being non-black. Compared with *U. arctos*, *U. americanus* has a shorter and more uniform pelage, shorter claws, and shorter hind feet. Females have three pairs of mammae (Banfield 1974).

The American black bear occurs mainly in forested areas. It may originally have avoided open country because of the lack of trees in which to escape *U. arctos*. The latter species is known to be a competitor with, and sometimes a predator upon, *U. americanus* (Jonkel 1978). The black bear now appears to have extended its range northward onto the tundra, possibly in response to the decline of the barren ground grizzly (Jonkel and Miller 1970). Following the extermination of the grizzly in the mountains of southern California, *U. americanus* moved into the area (Hall 1981).

The usual locomotion is a lumbering walk, but *U. americanus* can be quick when the need arises. It swims and climbs well. It may move about at any hour but is most active at night (Banfield 1974). Like other bears that sleep through the winter, it becomes fat with the approach of cold weather, finally ceases eating, and goes into a den in a protected location. The shelter may be under a fallen tree, in a hollow tree or log, or in a burrow. In the Hudson Bay area, individuals may burrow into the snow. During hibernation body temperature drops from 38° C to 31–34° C, the respiration slows, and the metabolic rate is depressed (Banfield 1974). The winter sleep is interrupted by excursions outside during periods of relatively warm weather. Such emergences are more numerous at southern latitudes. Hibernation begins as early as October and may last until May. In Washington the average period is 126 days, while three Louisiana bears slept for 74–124 days each (Lindzey and Meslow 1976; Lowery 1974). At least 75 percent of the diet consists of vegetable matter, espe-

cially fruits, berries, nuts, acorns, grass, and roots. In some areas sapwood is important; to reach it the bear peels bark from trees, thereby causing forest damage (Poelker and Hartwell 1973). The diet also includes insects, fish, rodents, carrion, and occasionally large mammals.

Banfield (1974) suggested an overall population density of about one bear to every 14.5 sq km. Field studies in Alberta, Washington, and Montana, however, indicate a usual density of one per 2.6 sq km (Jonkel and Cowan 1971; Kemp 1976; Poelker and Hartwell 1973). Still higher densities have been reported: one per 1.3 sq km in southern California (Piekielek and Burton 1975) and one per 0.67 sq km on Long Island off southwestern Washington (Lindzey and Meslow 1977a). In the latter area, home range was found to average 505 ha. for adult males and 235 ha. for adult females (Lindzey and Meslow 1977b). Farther north in Washington, however, Poelker and Hartwell (1973) determined home range to average about 5,200 ha. for males and 520 ha. for females. The ranges of males did not overlap one another, but the ranges of females overlapped with those of males and occasionally with those of other females. In Idaho, Amstrup and Beecham (1976) found home ranges to vary from 1,660 to 13,030 ha., to remain stable from year to year, and to overlap extensively. Despite such overlap, individuals tend to avoid one another and to defend the space being used at a given time. A number of bears sometimes congregate at a large food source, such as a garbage dump, but they try to keep out of each other's way. More tolerance is shown to familiar individuals than to strangers (Banfield 1974; Jonkel 1978; Jonkel and Cowan 1971). There is a variety of vocalizations. When startled, the ordinary sound is a "woof." When cubs are lonely or frightened, they utter shrill howls.

The sexes come together briefly during the mating season, which generally peaks from June to mid-July. Females remain in estrus throughout the season until they mate. They usually give birth every other year but sometimes wait 3–4 years. Pregnancy generally lasts about 220 days, but there is delayed implantation. The fertilized eggs are not implanted in

A

American black bear (Ursus americanus), photo from San Diego Zoological Garden.

the uterus until the autumn, and embryonic development occurs only in the last 10 weeks of pregnancy. Births occur mainly in January and February, commonly while the female is hibernating. The number of young per litter ranges from one to five and is usually two or three. At birth the young weigh 225–330 grams each and are naked and blind. They usually are weaned at around 6–8 months but remain with the mother and den with her during their second winter of life. Upon emergence in the spring they usually depart in order to avoid the aggression of the adult males in the breeding season. Females reach sexual maturity at 4–5 years, and males about a year later. One female is known to have lived 26 years and to have been in estrus at that age (Banfield 1974; Jonkel 1978; Poelker and Hartwell 1973).

Except when wounded or attempting to protect its young, the black bear is generally harmless to people. In areas of total protection, such as national parks, the species has become accustomed to humans. It thus can be easily seen and is a popular attraction but is sometimes a nuisance, raiding campsites or begging for food along roads. Physical attacks are rare but occur with some regularity, often because the involved persons disregard safety regulations (Cole 1976; Jonkel 1978; Pelton, Scott, and Burghardt 1976). Black bears have killed people on occasion, most recently in Alberta in 1980 (J. R. Gunson, Alberta Fish and Wildlife Division, pers. comm., 1981).

People have intensively killed U. americanus because of fear, to prevent depredations on domestic animals and crops, for sport, and to obtain fur and meat. According to Lowery (1974), attacks on livestock are negligible, but the bear does serious damage to cornfields and honey production. The economic loss caused to beekeepers in the Peace River Valley of Alberta was estimated at $200,000 in 1973, and a government control program is directed against the bear in that area (Gilbert and Roy 1977). In most of the states and provinces

occupied by the black bear, it is treated as a game animal, subject to regulated hunting. An estimated 30,000 individuals are killed annually in North America (Jonkel 1978). Relatively few skins go to market now, as regulations sometimes forbid commerce and there is no great demand. The average price per pelt in the 1976/77 season was about $44 (Deems and Pursley 1978), but it fell to under $20 by 1983/84 (Novak, Obbard, et al. 1987).

The distribution of the black bear has declined substantially, but the species is still common in Alaska, Canada, the western conterminous United States, the upper Great Lakes region, northern New England and New York, and parts of the Appalachians. Small native populations also survive in coastal lowlands from the Dismal Swamp of Virginia to the Okefenokee of Georgia and in the bottom land forests of southern Alabama and southeastern Arkansas.

Data compiled by Cowan (1972) indicate the presence of about 170,000 black bears in the conterminous United States, and Raybourne (1987) stated that the estimate for all of North America is 400,000–500,000. Leopold (1959) considered the species still to be widespread in Mexico, but more recent information indicates that only a few hundred individuals survive there.

The subspecies U. a. floridanus, of Florida and adjacent areas, is considered to be threatened through habitat loss, fragmentation of remnant populations, and persecution by beekeepers (Brady and Maehr 1985; Layne 1978). The subspecies U. a. luteolus, formerly found from eastern Texas to Mississippi, was by the mid-twentieth century reduced to a few individuals along the Mississippi and lower Atchafalaya rivers in eastern Louisiana. It is now in imminent danger of extinction. During the 1960s the wildlife agencies of both Louisiana and Arkansas imported a number of bears from Minnesota (within the range of the subspecies U. a. americanus) to their respective states (Lowery 1974; Sealander

American black bear *(Ursus americanus)*, photo by J. Perley Fitzgerald.

1979), thus further jeopardizing the genetic viability of the native populations.

Ursus arctos (brown or grizzly bear).

Head and body length is 1,700–2,800 mm, tail length is 60–210 mm, and shoulder height is 900–1,500 mm. In any given population, adult males are larger, on the average, than adult females. The largest individuals—indeed, the largest of living carnivores—are found along the coast of southern Alaska and on nearby islands, such as Kodiak and Admiralty. In this area weight is as great as 780 kg. Size rapidly declines to the north and east. In southwestern Yukon, for example, Pearson (1975) found average weights of 139 kg for males and 95 kg for females. In the Yellowstone region, Knight, Blanchard, and Kendall (1981) found weights of full-grown animals to average 181 kg and to range from 102 to 324 kg. In Siberia and northern Europe weight is usually 150–250 kg. In parts of southern Europe average weight is only 70 kg (Grzimek 1975). Coloration is usually dark brown but varies from cream to almost black. In the Rocky Mountains the long hairs of the shoulders and back are often frosted with white, thus giving a grizzled appearance and the common name grizzly or silvertip. From *U. americanus*, *U. arctos* is distinguished in having a prominent hump on the shoulders, a snout that rises more abruptly into the forehead, longer pelage, and longer claws.

The brown bear has one of the greatest natural distributions of any mammal. It occupies a variety of habitats but in the New World seems to prefer open areas, such as tundra,

alpine meadows, and coastlines. It was apparently common on the Great Plains prior to the arrival of European settlers. In Siberia the brown bear occurs primarily in forests (Stroganov 1969). Surviving European populations are restricted mainly to mountain woodlands (Van Den Brink 1968). Even when living in generally open regions, *U. arctos* needs some areas with dense cover (Jonkel 1978). It shelters in such places by day, sometimes in a shallow excavation, and moves and feeds mainly during the cool of the evening and early morning. Egbert and Stokes (1976) noted that activity in coastal Alaska occurs throughout the day, but peaks from 1800 to 1900 hours. Seasonal movements are primarily toward major food sources, such as salmon streams and areas of high berry production (Jonkel 1978). In Siberia, individuals may travel hundreds of kilometers during the autumn to reach areas of favorable food supplies (Stroganov 1969).

According to Banfield (1974), the usual gait is a slow walk. *U. arctos* is capable of moving very quickly, however, and can easily catch a black bear. Its long foreclaws are not adapted for climbing trees. It has excellent senses of hearing and smell but relatively poor eyesight. The brown bear has great strength. Banfield saw one drag a carcass of a horse about 90 meters. In another case, a 360 kg grizzly killed and dragged a 450 kg bison.

Hibernation begins in October–December and ends in March–May. The exact period depends on the location, weather, and condition of the animal. In certain southerly areas hibernation is very brief or does not take place at all. In most cases the brown bear digs its own den and makes a bed of

Alaskan brown bear *(Ursus arctos)*, photo by Ernest P. Walker.

dry vegetation. The burrow is often located on a sheltered slope, either under a large stone or among the roots of a mature tree. The bed chamber has an average volume of around two cubic meters. A den is sometimes used repeatedly, year after year. During winter sleep there is a marked depression in heart rate and respiration but only a slight drop in body temperature. The animal can be aroused rather easily and can make a quick escape, if necessary (Craighead and Craighead 1972; Grzimek 1975; Stroganov 1969; Slobodyan 1976; Ustinov 1976).

The diet consists mainly of vegetation (Jonkel 1978). Early spring foods include grasses, sedges, roots, moss, and bulbs. In late spring, succulent, perennial forbs become important. During the summer and early autumn, berries are essential, and bulbs and tubers are also taken. Banfield (1974) wrote that *U. arctos* consumes insects, fungi, and roots at all times of the year and also digs mice, ground squirrels, and marmots out of their burrows. In the Canadian Rockies the grizzly is quite carnivorous, hunting moose, elk, mountain sheep and goats, and even black bears. In Mount McKinley National Park, Alaska, Murie (1981) found *U. arctos* to feed mostly on vegetation but also to eat carrion whenever available and occasionally to capture young calves of caribou and moose. During the summer, when salmon are moving upstream along the Pacific coasts of Canada, southern Alaska, and northeastern Siberia, brown bears gather to feed on the vulnerable fish (Banfield 1974; Egbert and Stokes 1976; Kistchinski 1972). Perhaps because of this abundant food supply, the bears of these areas are larger and are found at greater densities than anywhere else.

Some approximate reported population densities are: Carpathian Mountains, one bear per 20 sq km; Lake Baikal area, one per 60 sq km; coast of Sea of Okhotsk, one per 10 sq km; Kodiak Island, one per 1.5 sq km; Mount McKinley National Park, one per 30 sq km; northern parts of Alaska and Northwest Territories, one per 150 sq km; and Glacier National Park, Montana, one per 21 sq km (Dean 1976; Harding 1976;

Kistchinski 1972; Martinka 1974, 1976; Slobodyan 1976; Ustinov 1976). In the Yellowstone region of the western United States, overall average density is about one bear per 88 sq km. In summer, however, individuals have concentrated by night at feeding sites, so densities have reached about one per 0.05 sq km. Daytime dispersal has reduced density to about one per 0.36 sq km. In the Yellowstone ecosystem, individual home range averages about 80 km and varies from about 20 to 600 sq km, with respect to the area used in the course of a year. Lifetime individual range has been as great as 2,600 sq km (Craighead 1976; Knight, Blanchard, and Kendall 1981). The home ranges of males are generally substantially larger than those of females. In the northern Yukon, Pearson (1976) found averages of 414 sq km for males and 73 sq km for females.

Home ranges overlap extensively, and there is no evidence of territorial defense (Craighead 1976; Murie 1981). Although generally solitary, the grizzly is the most social of North American bears, occasionally gathering in large numbers at major food sources and often forming family foraging groups with more than one age class of young (Jonkel 1978). At a salmon stream in southern Alaska, Egbert and Stokes (1976) sometimes observed more than 30 bears at one time. Considerable intraspecific tolerance was demonstrated in such aggregations, but dominance hierarchies were enforced by aggression. The highest-ranking animals were the large adult males, which most other bears attempted to avoid. The most aggressive animals were females with young, and the least aggressive were adolescents. Overt fighting was usually brief, and no infliction of serious wounds was observed, but the researchers suspected that killing of young individuals by adult males was a factor in population regulation.

There are no lasting social bonds, except those between females and young. During the breeding season males may fight over females. Successful males attend one or two females for 1–3 weeks. Mating takes place from May to July, implantation of the fertilized eggs in the uterus is usually delayed until October or November, and births generally occur from January to March, while the mother is in hibernation. The total period of pregnancy may last 180–266 days. Females remain in estrus throughout the breeding season until mating and do not again enter estrus for at least 2, usually 3 or 4, years. The number of young in a litter averages about two and ranges from one to four. They weigh 340–680 grams each at birth and are naked and blind. They are weaned at about 5 months. They remain with the mother at least until their second spring of life, and usually until their third or fourth. Litter mates sometimes maintain an association for 2–3 years after leaving the mother. Puberty comes at around 4–6 years, but growth then continues. Males in southern Alaska may not reach full size until they are 10–11 years old. Females in the Yellowstone region are known to have lived 25 years and still be capable of reproduction (Craighead, Craighead, and Sumner 1976; Egbert and Stokes 1976; Glenn et al. 1976; Jonkel 1978; Murie 1981; Pearson 1976; Slobodyan 1976). Potential longevity in captivity may be as great as 50 years (Stroganov 1969).

The brown bear has the reputation of being the most dangerous animal in North America. If we disregard venomous insects, disease-spreading rodents, domestic animals, and people themselves, this may be true. Three persons were killed by grizzlies in Glacier National Park, Montana, in 1980 (*Washington Post*, 25 July 1980, p. A-14; 9 October 1980, p. A-54; 20 October 1980, p. A-5). Another was killed in Canada (J. R. Gunson, Alberta Fish and Wildlife Division, pers. comm., 1981). That was an unusually tragic year, but injuries and an occasional death have been reported from the western national parks since around 1900 (Cole 1976; Herrero 1970, 1976). During the nineteenth century, attacks on people apparently occurred with some regularity in California (Storer and Tevis 1955). Two men recently were killed by the grizzlies they were photographing, one in Yellowstone in October 1986 and one in Glacier Park in April 1987 (*Audubon*, July 1987, pp. 16–17; Mitchell 1987). According to Ustinov (1976), over 70 attacks and 17 deaths have been attributed to *U. arctos* in the Lake Baikal area of Siberia. Many such incidents have probably been provoked by an effort to shoot or harass the animal, as the brown bear normally tries to avoid humans. It is unpredictable, however, if startled at close quarters, especially when accompanied by young or engrossed in a search for food. Jonkel (1978) cautioned that there may be more difficulties as recreational and commercial activity increases in areas occupied by the grizzly. He suggested that problems could be reduced by improved education and planning and by not locating campsites, trails, and residential facilities in places regularly used by bears.

The brown bear long has been persecuted as a predator of domestic livestock, especially cattle and sheep. Those parts of North America from which it has been eliminated correspond closely to areas of intensive ranching and grazing. In the nineteenth and early twentieth centuries there were apparently some remarkably destructive bears (Hubbard and Harris 1960; Storer and Tevis 1955), and their activities earned the entire species the lasting enmity of cattle ranchers and sheepherders. The brown bear also has been widely sought as a big game trophy, and it is currently subject to regulated sport hunting in most of its range. During the 1983/84 season 1,441 brown bears were killed legally in the United States and Canada, and pelts sold in Canada for an average price of $162.05 (Novak et al. 1987).

The original eastern limits of *U. arctos* in North America are not certain. A skull found in a Labrador Eskimo midden dating from the late eighteenth century supports earlier stories that the grizzly used to occur to the east of Hudson Bay (Spiess 1976). The decline of the species on the Great Plains may have begun when the Indians of that region obtained the

horse and hence an improved hunting capability. A precipitous drop in grizzly numbers came in the nineteenth century as settlers and livestock filled the West, thereby setting up confrontations that usually ended to the detriment of the bear. This process was intensified by logging, mining, and road construction, which increased human presence in remote areas. The distinctive subspecies of California and northern Baja California, *U. a. californicus* (Hall 1984), evidently disappeared by the 1920s. In the early nineteenth century there may have been 100,000 grizzlies in the western conterminous United States. There are now probably fewer than 1,000. Of these, about 200 are in Glacier National Park (Martinka 1974), and several hundred more are in nearby parts of northwestern Montana, northern Idaho, and extreme northeastern Washington. The population in Yellowstone National Park and vicinity has been estimated at 136 (Craighead, Varney, and Craighead 1974) and 247 (Knight, Blanchard, and Kendall 1981). Some authorities have warned that the Yellowstone population may not be reproductively viable and that current management practices in and around the park are leading to conflicts between human interests and bears, often resulting in the death of the latter (Chase 1986). Such problems probably caused the recent disappearance of very small remnant groups in south-central Colorado and northwestern Mexico. The grizzly has been extirpated from the Great Plains of Canada, except for an isolated group in west-central Alberta (Banfield 1974). The species also has declined on the barrens of the Northwest Territories (Macpherson 1965). In the mountainous regions of western Canada and in Alaska, *U. arctos* is still relatively common, perhaps numbering about 50,000 individuals (Cowan 1972; Jonkel 1987; Schoen, Miller, and Reynolds 1987). In Eurasia there are an estimated 100,000 brown bears, about 70,000 of them in the Soviet Union (Vereschagin 1976). To the west of Russia, the only substantial surviving populations are in the Balkans and northern Scandinavia. There are small, isolated groups in Poland, Czechoslovakia, Austria, Italy, southern France, northern Spain, and southern Norway (Elgmork 1978; Smit and Van Wijngaarden 1981). *U. arctos* apparently disappeared from northwestern Africa around the middle of the nineteenth century (Harper 1945).

Appendix 2 of the CITES includes all North American populations of *U. arctos* except *U. a. nelsoni*, the Mexican grizzly bear (treated as a synonym of *U. a. arctos* by Hall 1984), and now also includes all European populations except that of the Soviet Union. Appendix 1 includes *U. a. nelsoni*, *U. a. pruinosus* of Tibet, and *U. a. isabellinus* of the mountains of Central Asia. The IUCN classifies *U. a. nelsoni* as extinct. The USDI lists *U. a. nelsoni*, *U. a. pruinosus*, and the Italian populations of *U. arctos* as endangered. Those populations of *U. arctos* in the conterminous United States are listed as threatened by the USDI.

Ursus maritimus (polar bear).

Head and body length is 2,000–2,500 mm, tail length is 76–127 mm, and shoulder height is up to 1,600 mm. DeMaster and Stirling (1981) gave the weight as 150–300 kg for females and 300–800 kg for males. Banfield (1974), however, wrote that males usually weigh 420–500 kg. The color is often pure white following the molt but may become yellowish in the summer, probably because of oxidation by the sun. The pelage also sometimes appears gray or almost brown, depending on season and light conditions. The neck is longer than that of other bears, and the head is relatively small and flat. The forefeet are well adapted for swimming, being large and oarlike. The soles are haired, probably for insulation from the cold and traction on the ice. Females have four functional mammae (DeMaster and Stirling 1981).

The polar bear is often considered to be a marine mammal.

Polar bears *(Ursus maritimus):* Top, photo from New York Zoological Society. Insets: A. Forefoot; B. Hind foot; photos from *Proc. Zool. Soc. London;* C. Young, 24 hours old, photo by Ernest P. Walker. Bottom, photo by Sue Ford, Washington Park Zoo, Portland, Oregon.

It is distributed mainly in arctic regions around the North Pole. The southern limits of its range are determined by distribution of the pack ice. It has been recorded from as far north as 88° N and from as far south as the Pribilof Islands in the Bering Sea, the island of Newfoundland, the southern tip of Greenland, and Iceland. There also are permanent populations in James Bay and the southern part of Hudson Bay. Although found generally in coastal areas or on ice hundreds of kilometers from shore, individuals have wandered up to 200 km inland (Stroganov 1969).

According to DeMaster and Stirling (1981), the preferred habitat is pack ice that is subject to periodic fracturing by wind and sea currents. The refreezing of such fractures provides places where hunting by the bear is most successful. Some animals spend both winter and summer along the lower edge of the pack ice, perhaps undergoing extensive north-south migrations as this edge shifts. Others move onto land for the summer and disperse across the ice as it forms along the coast and between islands during winter. The bears of the Labrador coast sometimes move north to Baffin Island, and some individuals have traveled as far as 1,050 km to the islands of northern Hudson Bay (Stirling and Kiliaan 1980). The population that summers along the southern shore of Hudson Bay spreads all across the partly ice-covered bay in November and returns to shore in July or August (Stirling et al. 1977). Despite such movements, the polar bear is not a true nomad. There are a number of discrete populations, each with its own consistently used areas for feeding and breeding (Stirling, Calvert, and Andriashek 1980).

The polar bear can outrun a reindeer for short distances on land and can attain a swimming speed of about 6.5 km/hr. It swims rather high, with head and shoulders above the water. If killed in the water, it will not immediately sink. According to DeMaster and Stirling (1981), it has been reported to swim for at least 65 km across open water. It is capable of diving under the ice and surfacing in holes utilized by seals. It seems to be most active during the first third of the day and least active in the final third. From July to December in the James Bay region, when a lack of ice prevents seal hunting, U. maritimus spends about 87 percent of its time resting, apparently living off of stored fat (Knudsen 1978). Depressions or complete earthen burrows are sometimes excavated on land during the summer in order to avoid the sun and keep cool (Jonkel et al. 1976).

Any individual bear may make a winter den for temporary shelter during severe weather, but only females, especially those that are pregnant, generally hibernate for lengthy periods. As with other bears, winter sleep involves a depressed respiratory rate and a slightly lowered body temperature but not deep torpor. Most pregnant females evidently do not spend the winter along the pack ice, but hibernate on land from October or November to March or April. Maternal dens are usually found within 8 km of the coast, but in the southern Hudson Bay region they are concentrated 30–60 km inland. They are excavated in the snow to depths of 1–3 meters, often on a steep slope. They usually consist of a tunnel several meters long that leads to an oval chamber of about 3 cubic meters. Some dens have several rooms and corridors (DeMaster and Stirling 1981; Harington 1968; Larsen 1975; Stirling, Calvert, and Andriashek 1980; Stirling et al. 1977; Uspenski and Belikov 1976).

The polar bear feeds primarily on the ringed seal (Phoca hispida) (DeMaster and Stirling 1981). The bear either remains still until a seal emerges from the water or stealthily stalks its prey on the ice (Stirling 1974). It may also dig out the subnivean dens of seals to obtain the young (Stirling, Calvert, and Andriashek 1980). During summer and autumn in the southern Hudson Bay region, U. maritimus often swims among sea birds and catches them as they sit on the water

(Russell 1975). The diet also includes the carcasses of stranded marine mammals, small land mammals, reindeer, fish, and vegetation. Berries become important for some individuals during summer and autumn (Jonkel 1978).

Reported population densities range from 1 bear per 37 sq km to 1 per 139 sq km (DeMaster and Stirling 1980). Home ranges are not well defined but are thought to vary from 150 to 300 km in diameter and to overlap extensively (Kolenosky 1987). Although U. maritimus is generally solitary, large aggregations may form around a major source of food (Jonkel 1978). Up to 40 individuals have been seen at one time in the vicinity of the Churchill garbage dump, on the southern shore of Hudson Bay (Stirling et al. 1977). Wintering females evidently tolerate one another well, as dens on Wrangel Island are sometimes found at densities of one per 50 sq meters (Uspenski and Belikov 1976). High concentrations of summer dens also have been reported (Jonkel et al. 1976). Adult females with young are not subordinate to any other age or sex class but tend to avoid interaction with adult males, presumably because the latter are potential predators of the cubs (DeMaster and Stirling 1981).

The sexes usually come together only briefly during the mating season, March–June. Delayed implantation apparently extends the period of pregnancy to 195–265 days. The young are born from November to January, while the mother is in her winter den. Females give birth every 2–4 years. The number of young per litter averages about two and ranges from one to four. They weigh about 600 grams each at birth and have some fur but are blind. Upon emergence from the den in March or April, the cubs weigh 10–15 kg each. They usually leave the mother at 24–28 months. The age of sexual maturity averages about 5–6 years. Adult weight is attained at about 5 years by females but not until 10–11 years by males. Wild females apparently have a reduced natality rate after the age of 20. Annual adult mortality in a population is about 8–16 percent. Potential longevity in the wild is estimated at 25–30 years (DeMaster and Stirling 1981; Ramsay and Stirling 1988; Stirling, Calvert, and Andriashek 1980; Uspenski and Belikov 1976). Several captives have lived for more than 30 years. A female at the Detroit Zoo gave birth at the age of 36 years and 11 months and was still living at 38 years and 2 months (Latinen 1987).

The polar bear often is considered to be dangerous to people, though usually the two species are not found in close proximity. An exception developed during the 1960s in the vicinity of the town of Churchill, on the southern shore of Hudson Bay (Stirling et al. 1977). Bears apparently increased in this area because of a decline in hunting. At the same time, more people moved in, and several large garbage dumps were established. A number of persons were attacked and one was killed. Many bears were shot or translocated by government personnel.

The native peoples of the Arctic have long hunted the polar bear for its fat and fur. Sport and commercial hunting increased in the twentieth century. The pelt of U. maritimus is the most valuable of any North American mammal's pelt that is now regularly marketed. During the 1976/77 season 530 skins from Canada were sold at an average price of $585.22 (Deems and Pursley 1978). Some individual prime pelts have brought over $3,000 each (Smith and Jonkel 1975).

The use of aircraft to locate polar bears and to land trophy hunters in their vicinity developed in Alaska in the late 1940s. The annual kill by such means increased to about 260 bears by 1972. In that year, however, the killing of U. maritimus, except for native subsistence, was prohibited by the United States Marine Mammal Protection Act. Canada and Denmark (for Greenland) also limit hunting to resident natives, and the Soviet Union and Norway (for Spitsbergen) provide complete protection. In 1973 these five nations drafted an

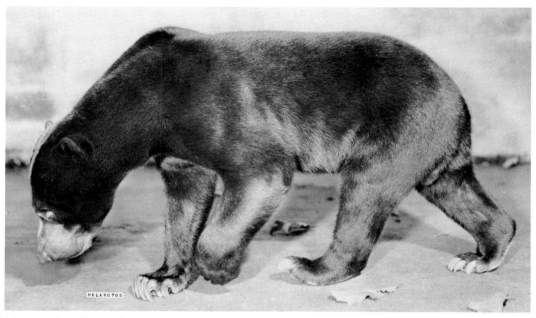

Malayan sun bear *(Ursus malayanus)*, photo by Ernest P. Walker.

agreement calling for the restriction of hunting, the protection of habitat, and the carrying out of cooperative research on polar bears. The agreement was ratified by the United States in 1976. The yearly worldwide kill is now estimated at around 1,000 animals. The total number of polar bears in the wild is perhaps 20,000, and populations are generally thought to be stable or increasing. *U. maritimus*, however, may be threatened by the exploitation of oil and gas reserves in the Arctic, especially with respect to development in the limited areas suitable for denning by pregnant females (DeMaster and Stirling 1981; Stirling and Kiliaan 1980; U.S. Fish and Wildlife Service 1980). There now also is concern that the influx of cash, as a result of oil and gas development, will stimulate increased hunting by native peoples and that such hunting, not being restricted to adult males, could damage the relatively small and vulnerable polar bear populations off northern Alaska and northwestern Canada (Amstrup, Stirling, and Lentfer 1986). The species is classified as vulnerable by the IUCN and is on appendix 2 of the CITES.

Ursus malayanus (Malayan sun bear).

This is the smallest bear. Head and body length is 1,000–1,400 mm, tail length is 30–70 mm, shoulder height is about 700 mm, and weight is 27–65 kg. The general coloration is black. There is a whitish or orange breast mark and a grayish or orange muzzle, and occasionally the feet are pale in color. The breast mark is often U-shaped but is variable and is sometimes wholly lacking. The body is stocky, the muzzle is short, the paws are large, and the claws are strongly curved and pointed. The soles are naked.

The sun bear inhabits dense forests at all elevations (Lekagul and McNeely 1977). It is active at night, usually sleeping and sunbathing by day in a tree, two to seven meters above the ground. Tree branches are broken or bent to form a nest and lookout post. *U. malayanus* has a curious gait in that all the legs are turned inward while walking. The species is usually shy and retiring and does not hibernate. An expert tree climber, it is cautious, wary, and intelligent. A young captive observed the way in which a cupboard containing a sugar pot was locked with a key. It then later opened the

cupboard by inserting a claw into the eye of the key and turning it. Another captive scattered rice from its feeding bowl in the vicinity of its cage, thus attracting chickens, which it then captured and ate.

The diet is omnivorous, and the front paws are used for most of the feeding activity. Trees are torn open in search of nests of wild bees and for insects and their larvae. The soft growing point of the coconut palm, known as palmite, is ripped apart and consumed. After digging up termite colonies, the animal places its forepaws alternately in the nest and licks the termites off. Jungle fowl, small rodents, and fruit juices also are included in the diet.

Births may occur at any time of the year. In the East Berlin Zoo a female produced one litter on 4 April 1961 and another on 30 August 1961. The gestation period for six births at that zoo was 95–96 days (Dathe 1970). At the Fort Worth Zoo, however, three pregnancies lasted 174, 228, and 240 days, evidently because of delayed fertilization or implantation (McCusker 1974). Litters usually contain one or two young, each weighing about 325 grams. They remain with the mother until nearly full-grown (Lekagul and McNeely 1977). A captive lived for 24 years and 9 months (Jones 1982).

Young individuals make interesting pets but become unruly within a few years. In the wild, *U. malayanus* is said to be one of the most dangerous animals within its range (Lekagul and McNeely 1977). It sometimes does great damage to coconut plantations. It is thought to be declining in some areas because of forest destruction (Cowan 1972). It is on appendix 1 of the CITES.

Ursus ursinus (sloth bear).

Head and body length is 1,400–1,800 mm, tail length is 100–125 mm, shoulder height is 610–915 mm, and weight is 55–145 kg. The shaggy black hairs are longest between the shoulders. The overall black coloration is often mixed with brown and gray, but cinnamon and red individuals also have been noted. The chest mark, typically shaped like a V or Y, varies from white or yellow to chestnut brown.

The sloth bear has a number of structural modifications associated with an unusual method of feeding. The lips are

Sloth bear *(Ursus ursinus)*: A. Photo by Hans-Jürg Kuhn; B. Photo from New York Zoological Society.

protrusible, mobile, and naked; the snout is mobile; the nostrils can be closed at will; the inner pair of upper incisors is absent, thus forming a gap in the front teeth; and the palate is hollowed. These features enable the bear to feed on termites (white ants) in the following manner: the nest is dug up, the dust and dirt blown off, and the occupants sucked up in a "vacuum cleaner" action. The resulting noises can be heard for over 185 meters and often lead to the bear's detection by hunters.

The sloth bear inhabits moist and dry forests, especially in areas of rocky outcrops. It may be active at any hour but is mainly nocturnal. During cool weather it spends the day in dense vegetation or shallow caves. The sense of smell is well developed, but sight and hearing are relatively poor. Hibernation is not known to occur. Termites are the most important food for most of the year, but the diet also includes other insects, grubs, honey, eggs, carrion, grass, flowers, and fruit.

In the Royal Chitawan National Park of Nepal, Sunquist (1982) found an adult male to move over a minimum area of 10 sq km during a period of about 2 years. In the same area, Laurie and Seidensticker (1977) found a minimum density of about 0.1 individual per sq km. Most observations were of lone bears or of females with cubs. Vocalizations, heard mainly in association with intraspecific agonistic encounters, included roars, howls, screams, and squeals. Births apparently occurred mostly from September to January.

Previous observations indicate that breeding takes place mainly in June in India and over most of the year in Sri Lanka. Pregnancy lasts about 6–7 months. The young, usually one or two and rarely three, are born in a ground shelter. They leave the den at 2–3 months and often ride on the mother's back. They remain with the mother until they are almost full grown, possibly 2 or 3 years. Captives have lived for 40 years.

The sloth bear normally is not aggressive but is held in great respect by some of the people that inhabit its range. Apparently because of its poor eyesight and hearing, it is sometimes closely approached by humans. It may then attack in what it considers to be self-defense and inflict severe wounds. Since it is thought to be dangerous, and since it

sometimes damages crops, it has been extensively hunted. It also seems not to tolerate regular human disturbance and is losing habitat to agriculture, logging, settlement, and hydroelectric projects. Fewer than 10,000 individuals are estimated to survive in India and Sri Lanka. The status of *U. ursinus* is considered to be indeterminate by the IUCN (1978).

CARNIVORA; URSIDAE; **Genus AILUROPODA**
Milne-Edwards, 1870

Giant Panda

The single species, *A. melanoleuca*, now is known from the central Chinese provinces of Gansu, Shaanxi, and Sichuan; there also are historical records from much of eastern China to the south of the Huang River (Schaller et al. 1985).

There is much controversy regarding the systematic position of this genus. It once was considered to be a close relative of *Ailurus* and was placed with that genus in the family Procyonidae, but most authorities eventually came to treat *Ailuropoda* as a bear. Hall (1981) put it in the ursid subfamily Ailuropodinae, but Chorn and Hoffmann (1978) referred it to the otherwise extinct ursid subfamily Agriotheriinae. While recognizing it as an offshoot of the Ursidae, Thenius (1979) suggested that the giant panda represents a distinct family, the Ailuropodidae. There has been a recent trend to again treat *Ailuropoda* and *Ailurus* as close relatives. Schaller et al. (1985) stated that on the basis of anatomical, biochemical, and paleontological evidence the giant panda's position remains equivocal but that certain characters of morphology, reproduction, and behavior indicate that *Ailuropoda* and *Ailurus* are related to both bears and raccoons, but more closely to the former, and that the two genera belong either in separate but closely related families of their own or together in their own family, the Ailuridae.

Head and body length is 1,200–1,500 mm, tail length is about 127 mm, and weight is 75–160 kg. The coat is thick and woolly. The eye patches, ears, legs, and band across the shoul-

Giant panda *(Ailuropoda melanoleuca)*, photo from New York Zoological Society. Insets: A. Right hind foot; B. Right forefoot; photos from *Proc. Zool. Soc. London*. C. Skull showing dentition, photo by P. F. Wright of specimen in U.S. National Museum of Natural History.

ders are black, sometimes with a brownish tinge. The remainder of the body is white but may become soiled with age. There are scent glands under the tail.

Ailuropoda resembles other bears in general appearance but is distinguished by its striking coloration and certain characters associated with its diet. The head is relatively massive because of the expanded zygomatic arches of the skull and the well-developed muscles of mastication. The second and third premolar teeth and the molars are relatively larger and broader than those of other bears. The forefoot has an unusual modification, thought to aid in the grasping of bamboo stems. The pad on the sole of each forepaw has an accessory lobe, and the pad of the first digit—and to a lesser extent the pad of the second digit—can be flexed onto the summit of this accessory lobe and its supporting bone.

The giant panda is found in montane forests with dense stands of bamboo. Its usual elevational range is 2,700–3,900 meters, but it may descend to as low as 800 meters in the winter. It does not make a permanent den, but takes shelter in hollow trees, rock crevices, and caves. It lives mainly on the ground but evidently can climb trees well. Activity is largely crepuscular and nocturnal. Schaller et al. (1985) found animals to be active about 14.2 hours per day. *Ailuropoda* does not hibernate, but descends to lower elevations in the winter and spring (Chorn and Hoffmann 1978). Average daily movement has been reported to cover 596 meters (Johnson, Schaller, and Hu 1988).

The diet consists mainly of bamboo shoots, up to 13 mm in diameter, and bamboo roots. *Ailuropoda* spends 10–12 hours a day feeding, usually in a sitting position with the forepaws free to manipulate the bamboo. Average daily consumption is

about 12.5 kg of bamboo (Johnson, Schaller, and Hu 1988). One radiotracked individual used an area of 3.8 sq km during a 9-month period, except for a 15-day foray during the spring bamboo shooting season, which increased the overall area to 6.8 sq km (Johnson, Schaller, and Hu 1988). Other plants, such as gentians, irises, crocuses, and tufted grasses, are also taken. *Ailuropoda* occasionally hunts for fish, pikas, and small rodents.

Schaller et al. (1985) studied wild individuals by radiotracking in the Wolong Natural Reserve of China. About 145 pandas occurred in this 2,000 sq km area. Within the 35 sq km study area there were 7 adult males, 5 or 6 adult females, 4 independent subadults, and 2 infants. Home range size varied from 3.9 to 6.4 sq km, with male ranges being only as large as or slightly larger than those of females. However, males were found to occupy greatly overlapping ranges lacking well-defined core areas. They shift frequently within their ranges, show no evidence of territorial behavior, and spend considerable time within the core areas of females and subadults. Females also may have overlapping ranges, but they spend most of their time within a discrete core area of only 30–40 ha. They evidently do not tolerate other females and subadults within their core areas. As the animals move about, they mark their routes by spraying urine, clawing tree trunks, and rubbing against objects. Captives are known to scent-mark with secretions from glands in the genital region (Kleiman 1983). The vocal repertoire consists of bleats, honks, squeals, growls, moans, barks, and chirps (Peters 1982).

Except as noted, the following information on reproduction and life history was taken from Schaller et al. (1985).

Ailuropoda is generally solitary, but during the breeding season several males may compete for access to a female. Mating generally occurs from March to May. Females have a single estrous period of 12–25 days, but peak receptivity lasts only 1–5 days. Births usually take place during August or September in a cave or hollow tree. The overall period of pregnancy lasts 97–163 days; the variation evidently results from a delay in implantation of 45–120 days. The number of young per litter is usually one or two, and occasionally three, but normally only a single cub is raised. The neonatal/maternal weight ratio may be the smallest among the eutherian mammals (Kleiman 1983). At birth the offspring weighs 90–130 grams, is covered with sparse white fur, and has a tail that is about one-third as long as the head and body; adult coloration is attained by the end of the first month, and the eyes open after 40–60 days (Chorn and Hoffmann 1978). The young begin to walk at 3–4 months and to eat bamboo at 5–6 months. They are fully weaned at 8–9 months, leave their mothers at about 18 months, and attain sexual maturity after 5.5–6.5 years. A captive specimen lived to an estimated age of 30 years.

The range of the giant panda began to decline in the late Pleistocene because of both climatic changes and the spread of people (Wang 1974). In the last 2,000 years the species disappeared from Henan, Hubei, Hunan, Guizhou, and Yunnan provinces (Schaller et al. 1985). At present about 1,000 individuals are thought to survive, and they are apparently divided into three isolated groups. In the mid-1970s about 100 pandas starved when an important food plant died over a large area. The species receives complete legal protection, and cooperative field investigations were recently begun by the Chinese government and the World Wildlife Fund (Schaller 1981). Nonetheless, poaching to obtain the valuable skin continues to be a major threat. The giant panda is classified as rare by the IUCN and as endangered by the USDI and is on appendix 1 of the CITES. It is among the most popular of zoo animals but has been extremely difficult to breed. The number of giant pandas in captivity is about 70 in China and 13 in other countries. A Chinese program of loaning pandas for exhibit in Western zoos became subject to intensive controversy in the late 1980s, with supporting parties arguing that the fees charged would be used for panda conservation and opponents claiming that the animals should be kept together in China for breeding purposes (Drew 1989).

CARNIVORA; **Family PROCYONIDAE**

Raccoons and Relatives

This family contains 7 Recent genera and 19 species. There are 2 subfamilies: Ailurinae, with the single genus *Ailurus* (lesser panda), found in the Himalayas and adjacent parts of eastern Asia; and Procyoninae, with the other 6 genera, which occur in temperate and tropical areas of the Western Hemisphere. *Ailurus* sometimes has been placed together with *Ailuropoda* (giant panda) in a separate family, the Ailuropodidae or Ailuridae.

Head and body length is 305–670 mm, tail length is 200–690 mm, and weight is about 0.8–12.0 kg. Males are about one-fifth larger and heavier than females. The pelage varies from gray to rich reddish brown. Facial markings are often present, and the tail is usually ringed with light and dark bands. The face is short and broad. The ears are short, furred, erect, and rounded or pointed. The tail is prehensile in the arboreal kinkajou *(Potos)* and is used as a balancing and semiprehensile organ in the coatis *(Nasua)*. Each limb bears five digits, the third being the longest. The claws are short, compressed, recurved, and, in some genera, semiretractile. The soles are haired in several genera. Males have a baculum.

The dental formula is usually: (i 3/3, c 1/1, pm 4/4, m 2/2) × 2 = 40. The premolars, however, number 3/3 in *Potos* and 3/4 in *Ailurus* (Stains 1984). The incisors are not specialized, the canines are elongate, the premolars are small and sharp, and the molars are broad and low-crowned. The carnassials are developed only in *Bassariscus*.

Procyonids walk on the sole of the foot, with the heel touching the ground, or partly on the sole and partly on the digits. The gait is usually bearlike. They are good climbers, and one genus *(Potos)* spends nearly its entire life in trees. Most procyonids shelter in hollow trees, on large branches, or in rock crevices. Most become active in the evening, but *Nasua* may be primarily diurnal. The diet is omnivorous, though *Potos* and *Bassaricyon* seem to depend largely on fruit, and *Ailurus* feeds mainly on bamboo. Most genera travel in pairs or family groups and give birth in the spring.

The geological range of the Procyonidae is early Oligocene to Recent in North America, late Miocene to Recent in South America, late Eocene to early Pleistocene in Europe, and early Miocene to Recent in Asia (Stains 1984).

CARNIVORA; PROCYONIDAE; **Genus AILURUS**
F. Cuvier, 1825

Lesser Panda

The single species, *A. fulgens*, is known to occur in Nepal, Sikkim, Bhutan, northern Burma, and the provinces of Yunnan and Sichuan in south-central China and probably also exists along the border of Tibet and Assam (Ellerman and Morrison-Scott 1966; Roberts and Gittleman 1984).

Head and body length is 510–635 mm, tail length is 280–485 mm, and weight is usually 3–6 kg. The coat is long and soft, and the tail is bushy. The upper parts are rusty to deep chestnut, being darkest along the middle of the back. The tail is inconspicuously ringed. Small, dark-colored eye patches are present, and the muzzle, lips, cheeks, and edges of the ears are white. The back of the ears, the limbs, and the underparts are dark reddish brown to black. The head is rather round, the ears are large and pointed, the feet have hairy soles, and the claws are semiretractile. The nonprehensile tail is about two-thirds as long as the head and body. There are glandular sacs in the anal region. Females have four mammae.

The lesser panda inhabits mountain forests and bamboo thickets at elevations of 1,800–4,000 meters. It seems to prefer colder temperatures than does the giant panda *(Ailuropoda)*. It is nocturnal and crepuscular, sleeping by day in a tree. When sleeping, it generally curls up like a cat or dog, with the tail over the head, but it may also sleep while sitting on top of a limb, with the head tucked under the chest and between the forelegs, as the American raccoon *(Procyon)* does at times. Although *Ailurus* is a capable climber, it seems to do most feeding on the ground. The diet consists mostly of bamboo sprouts, grasses, roots, fruits, and acorns. It also occasionally takes insects, eggs, young birds, and small rodents. A female radiotracked by Johnson, Schaller, and Hu (1988) for 9 months had an average daily movement of 481 meters and a home range of 3.4 sq km.

In the wild the lesser panda sometimes travels in pairs or small family groups. Such groups probably represent a consorting male and female, or a mother with cubs. The young seem to stay with the mother for about a year, or until the next litter is about to be born. The disposition of *Ailurus* is mild; when captured, it does not fight, tames readily, and is gentle, curious, and generally quiet. The usual call is a series

Lesser panda *(Ailurus fulgens)*, photo by Arthur Ellis, *Washington Post*.

of short whistles or squeaking notes; when provoked, it utters a sharp, spitting hiss or a series of snorts while standing on its hind legs. A musky odor is emitted from the anus when the animal is excited. According to Grzimek (1975), it scent-marks its territory by rubbing the anal region against objects. In the wild, births take place in the spring in a hollow tree or rock crevice.

Studies at the U.S. National Zoo in Washington, D.C. (Roberts 1975, 1980), indicate that adult males can be kept with females and young but that adult females are not tolerant of each other. Reproduction is most successful when a single adult male and female are placed together, though they will sleep and rest apart except during the breeding season. Females apparently have only one estrus annually and are then receptive for just 18–24 hours. They may begin to build a nest of sticks and leaves several weeks before giving birth.

Mating occurs from mid-January to early March, and births from mid-June to late July. Recorded gestation periods at the National Zoo are 114–45 days; however, there also is a record of only 90 days at the San Diego Zoo. It thus may be that there is delayed implantation in temperate zones but not in subtropical zones. Litters contain one to four, usually two, young. The cubs weigh about 200 grams at 1 week, open their eyes after 17–18 days, attain full adult coloration by 90 days, and take their first solid food at 125–35 days. The young are removed from the parents at 6–7 months and placed in small peer groups. Both sexes attain sexual maturity at about 18 months. According to Jones (1982), one individual lived for 13 years and 5 months in captivity.

The lesser panda is a very popular zoo animal and is frequently involved in the animal trade. It is on appendix 2 of the CITES.

Ringtail *(Bassariscus astutus)*, photo by Woodrow Goodpaster.

CARNIVORA; PROCYONIDAE; **Genus BASSARISCUS**
Coues, 1887

Ringtails, or Cacomistles

There are two species (Hall 1981):

B. astutus, southwestern Oregon and eastern Kansas to
   Baja California and southern Mexico;
B. sumichrasti, southern Mexico to western Panama.

The latter species formerly was often placed in a separate
genus, *Jentinkia* Trouessart, 1904.

In *B. astutus* head and body length is 305–420 mm, tail
length is 310–441 mm, and shoulder height is about 160 mm.
Armstrong, Jones, and Birney (1972) listed weights of 824–
1,338 grams. The upper parts are buffy, with a black or dark
brown wash, and the underparts are white or white washed
with buff. The eye is ringed by black or dark brown, and the
head has white to pinkish buff patches. The tail is bushy,
longer than the head and body, and banded with black and
white for its entire length. Females have four mammae.

In *B. sumichrasti* head and body length is 380–470 mm
and tail length is 390–530 mm. One individual weighed 900
grams. The color is usually buffy gray to brownish, and the
tail is ringed with buff and black. From *B. astutus, B.
sumichrasti* is distinguished in having pointed (rather than
rounded) ears, a longer tail, naked (rather than hairy) soles,
nonretractile (rather than semiretractile) claws, and low
(rather than high) ridges connecting the cusps of the molariform teeth.

B. astutus utilizes a varied habitat but seems to prefer
rocky, broken areas, often near water. It dens in rock crevices,
hollow trees, the ruins of old Indian dwellings, and the upper
parts of cabins. Activity occurs mainly at night. *B.
sumichrasti* is found in tropical forests and appears to be more
arboreal than *B. astutus.* The latter, however, is a good climber and travels quickly and agilely among cliffs and along
ledges. In a study of captive *B. astutus,* Trapp (1972) determined that the hind foot can rotate at least 180°, permitting a
rapid, headfirst descent and great dexterity. One individual
traveled upside down along a cord 5 mm in diameter. Ringtails sometimes climb in a crevice by pressing all four feet on
one wall and the back against the other. They also maneuver
by ricocheting off of smooth surfaces to gain momentum to
continue to an objective. The diet includes insects, rodents,
birds, fruit, and other vegetable matter.

Central American cacomistle *(Bassariscus sumichrasti),*
photo by I. Poglayen-Neuwall.

According to Grzimek (1975), recorded population densities are 1 per 8 sq km in California and 10 per 1.5 sq km on
the Edwards Plateau of Texas. Home range is about 3.2 sq km
at most but usually substantially smaller, and individuals
tend to stay in one area. *Bassariscus* is generally solitary,
except during the mating season. *B. astutus* scent-marks its
territory by regularly urinating at certain sites. *B. sumichrasti,* however, eliminates at random (Poglayen-Neuwall
1973). Both species have a variety of vocalizations. Adult *B.
astutus* may emit an explosive bark, a piercing scream, and a
long, plaintive, high-pitched call.

Female *B. sumichrasti* enter estrus in winter, spring, or
summer, but late winter appears to be the main breeding
season. *B. astutus* mates from February to May and gives
birth from April to July. Heat lasts only 24 hours, and the
gestation period is about 51–54 days. The number of young
per litter is one to five, usually two to four. The young are
born in a nest or den and may be given care by both parents.
They weigh about 28 grams each at birth, open their eyes at
31–34 days, begin taking some solid food at 4 weeks, begin to
forage with the adults at 2 months, are completely weaned at

Central American cacomistle *(Bassariscus sumichrasti),* immature, photo from Jorge A. Ibarra through Museo Nacional de
Historia Natural, Guatemala City.

4 months, and disperse in early winter. Both sexes attain sexual maturity at approximately 10 months (Grzimek 1975; Leopold 1959; Poglayen-Neuwall and Poglayen-Neuwall 1980). A captive *B. astutus* lived for 14 years and 3 months (Jones 1982).

Ringtails, especially females obtained when young, make charming pets. They were sometimes kept about the homes of early settlers as companions and to catch mice. The coat is not of particularly high quality. It is known commercially as "California mink" or "civet cat," but there is no scientific basis for either name. In the 1976/77 trapping season 88,329 pelts of *B. astutus* were reported taken in the United States and sold for an average price of $5.50 (Deems and Pursley 1978). The harvest peaked at about 135,000 in 1978/79 but subsequently declined, and prices fell to less than $3.00 (Kaufmann 1987). *B. astutus* apparently extended its range into Kansas, Arkansas, and Louisiana in the twentieth century and has even been reported from Alabama and Ohio (Hall 1981). These occurrences might result partly from the habit of boarding railroad cars (Sealander 1979).

CARNIVORA; PROCYONIADE; Genus PROCYON
Storr, 1780

Raccoons

Two subgenera and seven species are currently recognized (Cabrera 1957; Gardner 1976; Hall 1981):

subgenus *Procyon* Storr, 1780

P. lotor, southern Canada to Panama;
P. insularis, Tres Marias Islands off western Mexico;

P. maynardi, New Providence Island (Bahamas);
P. pygmaeus, Cozumel Island off northeastern Yucatan;
P. minor, Guadeloupe Island (Lesser Antilles);
P. gloveralleni, Barbados (Lesser Antilles);

subgenus *Euprocyon* Gray, 1865

P. cancrivorus (crab-eating raccoon), eastern Costa Rica to eastern Peru and Uruguay.

Lotze and Anderson (1979) suggested that several of the designated species of the subgenus *Procyon* might be conspecific with *P. lotor*. Olson and Pregill (1982) thought that *P. maynardi* probably is not a valid species and represents an introduced population of *P. lotor*. The latter species also now is established on Grand Bahama Island (Buden 1986).

Head and body length is 415–600 mm, tail length is 200–405 mm, shoulder height is 228–304 mm, and weight is usually 2–12 kg. Generally, males are larger than females, and northern animals are larger than southern ones. Five adult males in the Florida Keys averaged 2.4 kg. Mean weights in Alabama were 4.31 kg for males and 3.67 kg for females. Means in Missouri were 6.76 kg for males and 5.94 kg for females (Johnson 1970; Lotze and Anderson 1979). In Wisconsin the normal weight range is about 6–11 kg, but there is one record of a male weighing 28.3 kg (Jackson 1961).

The general coloration is gray to almost black, sometimes with a brown or red tinge. There are 5–10 black rings on the rather well-furred tail and a black "bandit" mask across the face. The head is broad posteriorly and has a pointed muzzle. The toes are not webbed, and the claws are not retractile. The front toes are rather long and can be widely spread. The footprints resemble those of people. Females have four pairs of mammae (Banfield 1974).

Raccoon *(Procyon lotor)*, photo from Zoological Society of Philadelphia.

Raccoons frequent timbered and brushy areas, usually near water. They are more nocturnal than diurnal and are good climbers and swimmers. The den is usually in a hollow tree, with an entrance more than 3 meters above the ground (Banfield 1974). The den may also be in a rock crevice, an overturned stump, a burrow made by another animal, or a human building. Urban (1970) found most raccoons in a marsh to den in muskrat houses. Except when sequestered during severe winter weather, or in cases of females with newborn, each den is usually occupied for only one or two days. The average distance between dens has been reported as 436 meters, and the general movements of *Procyon* are not extensive, but one individual was found to have traveled 266 km (Lotze and Anderson 1979).

Raccoons do not hibernate. In the southern parts of their range they are active throughout the year. In northern areas they may remain in a den for much of the winter but will emerge during intervals of relatively warm weather. While they are in winter sleep, their heartbeat does not decline, their body temperature stays above 35° C, and their metabolic rate remains high. They do, however, live mostly off of fat reserves accumulated the previous summer and fall and may lose up to 50 percent of their weight (Lotze and Anderson 1979).

Procyon has a well-developed sense of touch, especially in the nose and forepaws. The hands are regularly used almost as skillfully as monkeys use theirs. Food is generally picked up with the hands and then placed in the mouth. Although raccoons have sometimes been observed to dip food in water, especially under captive conditions, the legend that they actually wash their food is without foundation (Lowery 1974). The omnivorous diet consists mainly of crayfish, crabs, other arthropods, frogs, fish, nuts, seeds, acorns, and berries.

As many as 167 raccoons have been found in an area of 41 ha., but more typical population densities are 1 per 5–43 ha. (Lotze and Anderson 1979). Reported home range size varies from 0.2 to 4,946 ha. but seems to be typified by the situation found by Lotze (1979) on St. Catherine's Island, Georgia: he reported an annual average of 65 ha. for males and 39 ha. for females but indicated that there was much variation and that different study methods might give different results. About the smallest population density (0.5–1.0 per 100 ha.) and largest home range (means of 1,139 ha. for males and 806 ha. for females) were reported by Fritzell (1978) for the prairies of North Dakota. His study indicated that the ranges of adult males are largely exclusive of one another but do commonly overlap the ranges of 1–3 adult females and up to 4 yearlings. This and other studies (Lotze and Anderson 1979; Schneider, Mech, and Tester 1971) suggest that female ranges often are not exclusive and that territorial defense is not well developed in *Procyon* but that unrelated animals tend to avoid one another. Nonetheless, as many as 23 individuals have been found in the same winter den, and about the same number have congregated around artificial feeding sites (Lotze and Anderson 1979; Lowery 1974). Raccoons have a variety of vocalizations, most with little carrying power.

In the United States the reproductive season extends from December to August. Mating peaks in February and March, and births from April to June (Johnson 1970; Lotze and Anderson 1979). The breeding season of *P. cancrivorus* is July–September (Grzimek 1975). If a female *P. lotor* loses a newborn litter, she may ovulate a second time during the season (Sanderson and Nalbandov 1973). The gestation period averages 63 days and ranges from 60 to 73. The number of young per litter is one to seven, usually three or four. Captives in New York weighed 71 grams each at birth. The eyes open after about 3 weeks (Banfield 1974), and weaning takes place at from 7 weeks to 4 months (Lotze and Anderson 1979). In Minnesota, Schneider, Mech, and Tester (1971) found that

the young were kept in a den in a hollow tree until they were 7–9 weeks old and then were moved to one or a series of ground beds. At 10–11 weeks they were taken on short trips by the mother, and after another week the family began to move together. In November the members denned either together in one hollow tree or individually in nearby trees. The young usually separate from the mother at the end of winter. They may attain sexual maturity at about 1 year, but most do not mate until the following year. Few wild raccoons live more than 5 years, but some are estimated to have survived for 13–16 years (Lotze and Anderson 1979). One captive was still living after 20 years and 7 months (Jones 1982).

Raccoons sometimes damage corn and other crops but usually not to a serious extent (Jackson 1961). They make good pets and are interesting to observe in the wild; however, they carry pathogens known to cause such human diseases as leptospirosis, tularemia, and rabies. *P. lotor* is currently the most valuable wild fur bearer in the United States, though prices have fallen since the peak in the late 1970s. The harvest in 44 states during the 1976/77 season was 3,832,802 skins, which sold at an average price of $26.00 (Deems and Pursley 1978). In the 1983/84 season the take in the United States and Canada was 3,410,548 pelts, with an average value of $5.54 (Novak, Obbard, et al. 1987). Because of its commercial value, *P. lotor* was introduced in France, the Netherlands, Germany, and various parts of the Soviet Union, but now it is sometimes considered a nuisance in those areas (Corbet 1978; Grzimek 1975). *P. lotor* seems to have extended its range and increased in numbers in certain parts of North America since the nineteenth century. The various insular species are rare, however, and some may be extinct (Lotze and Anderson 1979).

CARNIVORA; PROCYONIDAE; **Genus NASUA**
Storr, 1780

Coatis, or Coatimundis

There are two species (Cabrera 1957; Hall 1981):

N. nasua, Arizona to Argentina;
N. nelsoni, Cozumel Island off northeastern Yucatan.

Head and body length is 410–670 mm, tail length is 320–690 mm, and shoulder height is up to 305 mm. Grzimek (1975) gave the weight as 3–6 kg. Males are usually larger than females. *N. nelsoni* has short, fairly soft, silky hair, but in *N. nasua* the fur is longer and somewhat harsh. In both species the general color is reddish brown to black above and yellowish to dark brown below. The muzzle, chin, and throat are usually whitish, and the feet blackish. Black and gray markings are present on the face, and the tail is banded. The muzzle is long and pointed, and the tip is very mobile. The forelegs are short, the hind legs are long, and the tapering tail is longer than the head and body.

Coatis are found mainly in wooded areas. They forage in trees, as well as on the ground, using the tail as a balancing and semiprehensile organ. While moving along the ground, the animals usually carry the tail erect, except for the curled tip. The long, highly mobile snout is well adapted for investigating crevices and holes. Adult males are often active at night, but coatis are primarily diurnal. They move about 1,500–2,000 meters a day in the search for food and usually retire to a roost tree at night (Kaufmann 1962). The diet includes both plant and animal matter. When fruit is abundant, coatis are almost exclusively frugivorous. At other times, females and young forage for invertebrates on the

Coatimundi *(Nasua nasua)*, photo from New York Zoological Society.

forest floor, and adult males tend to prey on large rodents (Smythe 1970).

Reported population densities are 26–42 per 100 ha. on Barro Colorado Island in the Panama Canal Zone and 1.2–2.0 per 100 ha. in Arizona (Lanning 1976). Home ranges of four solitary males in Arizona were 70–270 ha. each (Kaufmann, Lanning, and Poole 1976). On Barro Colorado Island, groups had overlapping, undefended home ranges of 35–45 ha. Each group, however, spent about 80 percent of its time in an exclusive core area within its home range (Kaufmann 1962).

The social system of *N. nasua,* as studied on Barro Colorado Island, is an interesting example of the interrelationship of ecology and behavior (Kaufmann 1962; Russell 1981; Smythe 1970). All females, and males up to two years old, are found in loosely organized bands usually comprising 4–20 individuals. Males over two years old become solitary except during the breeding season. They are usually excluded from membership in the bands by the collective aggression of the adult females, sometimes supported by the juveniles. In the breeding season an adult male is accepted into each group, but he is completely subordinate to the females. The breeding season corresponds with the period of maximum abundance of fruit. At this time there is thus a minimum of competition for food between the large males and the other animals. Moreover, at other times of the year, when the males become carnivorous, they may attempt to prey on young coatis and so would then threaten the survival of the group.

There is a single reproductive season, with births in Panama occurring from April to June, at the beginning of the rains. The gestation period lasts 10–11 weeks. Pregnant females separate from the group and construct a tree nest, where they give birth to a litter of two to seven young. When the young are 5 weeks old, they leave the nest, and they and the mother join the group. The young weigh 100–180 grams each at birth, open their eyes after 11 days, are weaned at 4 months, reach adult size at 15 months, and attain sexual maturity at 2 years (Grzimek 1975; Kaufmann 1962). A captive specimen was still living after 17 years and 8 months (Jones 1982).

Coatis seem to have extended their range northward in the twentieth century. Numbers reached a peak in Arizona in the late 1950s, crashed in the early 1960s, and then began a slow recovery (Kaufmann, Lanning, and Poole 1976). Coatis rarely damage crops and only infrequently take chickens. They are hunted for their meat by natives, who sometimes have dogs trained for this purpose. If cornered by a dog on the ground, a coati can inflict serious wounds with its large and sharp canine teeth. Coatis can be tamed and make interesting and inquisitive pets.

CARNIVORA; PROCYONIDAE; **Genus NASUELLA**
Hollister, 1915

Mountain Coati

The single species, *N. olivacea,* is known from the Andes of western Venezuela, Colombia, and Ecuador (Cabrera 1957). The species originally was placed in *Nasua,* and some au-

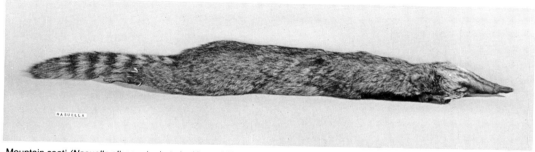

Mountain coati *(Nasuella olivacea)*, photo by Howard E. Uible of skin in U.S. National Museum of Natural History.

thorities, such as Grzimek (1975), still assign it to that genus.

Nasuella resembles *Nasua* but is usually smaller and has a shorter tail. A specimen of a male from Colombia in the U.S. National Museum of Natural History has a head and body length of 383 mm and a tail length of 242 mm. Respective measurements for a female are 394 and 201 mm (John Miles, U.S. National Museum of Natural History, pers. comm.). The general color is grayish sooty brown. The tail is ringed with alternating yellowish gray and dark brown bands. The skull of *Nasuella* is smaller and more slender than that of *Nasua*, the middle part of the facial portion is greatly constricted laterally, and the palate extends farther posteriorly.

Handley (1976) collected seven specimens in Venezuela at elevations of 2,000–3,020 meters. Of these, all were taken on the ground, four were taken at dry and three at moist sites, and four were taken in cloud forest and three in paramo. Like *Nasua*, *Nasuella* probably feeds on insects, small vertebrates, and fruit.

CARNIVORA; PROCYONIDAE; Genus POTOS
E. Geoffroy St.-Hilaire and G. Cuvier, 1795

Kinkajou

The single species, *P. flavus*, is found from southern Tamaulipas in eastern Mexico to the Mato Grosso of central Brazil (Cabrera 1957; Hall 1981). Confusion sometimes results from the application of the vernacular name "potto" to this species, because the same term is used for the African primate *Perodicticus*.

Head and body length is 405 to about 760 mm, tail length is 392–570 mm, shoulder height is up to 254 mm, and weight is 1.4–4.6 kg. Males are generally larger than females (Kortlucke 1973). The upper parts and upper surface of the tail are tawny olive, yellow tawny, or brownish; some individuals have a black middorsal line. The underparts and un-

Kinkajou *(Potos flavus)*, photo from New York Zoological Society.

dersurface of the tail are tawny yellow, buff, or brownish yellow, and the muzzle is dark brown to blackish. The hair is soft and woolly.

The kinkajou has a rounded head, a short face, a long, prehensile tail, and short, sharp claws. The hind feet are longer than the forefeet. The tongue is narrow and greatly extensible. *Potos* is similar to *Bassaricyon* but differs in its round, tapering, short-haired, prehensile tail; its stockier body form; and its face, which is not grayish. Females have two mammae (Grzimek 1975).

The kinkajou inhabits forests and is almost entirely arboreal. It spends the day in a hollow tree, sometimes emerging on hot, humid days to lie out on a limb or in a tangle of vines. By night it forages among the branches. Although it moves rapidly through a single tree, progress from one tree to another is made cautiously and relatively slowly. An individual probably returns to the same trees night after night. The long tongue of *Potos* is an adaptation for a frugivorous diet. It eats mainly fruit but also takes honey, insects, and small vertebrates (Husson 1978).

Population densities of 12.5–74.0 per sq km have been reported in various parts of the range of *Potos*, and an individual home range of 8 ha. was estimated in Veracruz ((Ford and Hoffmann 1988). *Potos* travels alone or in pairs. Small groups may form, but usually only temporarily in a fruit-bearing tree. Territories are not defended, though individuals do scent-mark, probably as a sexual signal (Grzimek 1975). *Potos* barks when disturbed and emits a variety of other vocalizations, but the usual call, given while feeding during the night, seems to be "a rather shrill, quavering scream that may be heard for nearly a mile" (Dalquest 1953:182).

According to Grzimek (1975), there is no particular breeding season, but Husson (1978) wrote that in Surinam births reportedly take place in April and May. The gestation period is 112–18 days. The number of young is usually one, rarely two. Born in a hollow tree, they weigh 150–200 grams at birth, open their eyes at 7–19 days, and begin to take solid food, and can hang by the tail, at 7 weeks. Sexual maturity comes at 1.5 years for males and 2.25 years for females. One pair lived together for 9 years and had their first litter at 12.5 years of age. One individual lived for 23 years and 7 months in the Amsterdam Zoological Gardens.

If captured when young and treated kindly, the kinkajou becomes a good pet. It is often sold under the name "honey bear." Its meat is said to be excellent (Husson 1978), and its pelt is used in making wallets and belts. It does little damage to cultivated fruit.

CARNIVORA; PROCYONIDAE; **Genus BASSARICYON**
J. A. Allen, 1876

Olingos

Five species are currently recognized (Cabrera 1957; Hall 1981; Handley 1976):

B. gabbii, Nicaragua to Ecuador and Venezuela;
B. pauli, known only from the type locality in western Panama;
B. lasius, known only from the type locality in central Costa Rica;
B. beddardi, Guyana and possibly adjacent parts of Venezuela and Brazil;
B. alleni, Ecuador, Peru, possibly Venezuela.

All of the above species may be no more than subspecies of *B. gabbii* (Cabrera 1957; Grimwood 1969; Hall 1981).

Head and body length is 350–475 mm and tail length is 400–480 mm. Grzimek (1975) gave the weight as 970–1,500 grams. The fur is thick and soft. The upper parts are pinkish buff to golden, mixed with black or grayish above, and the underparts are pale yellowish. The tail is somewhat flattened and more or less distinctly annulated along the median portion. The general body form is elongate, the head is flattened, the snout is pointed, and the ears are small and rounded. The limbs are short, the soles are partly furred, and the claws are sharply curved. The tail of *Bassaricyon*, unlike that of *Potos*, is long-haired and not prehensile. Females have a single pair of inguinal mammae.

The olingo is found in tropical forests from sea level to 2,000 meters. It is primarily arboreal and nocturnal. It is thought to spend the day in a nest of dry leaves in a hollow tree. According to Grzimek (1975), it is an excellent climber and jumper and can leap 3 meters, from limb to limb, without difficulty. It feeds mainly on fruit but also hunts for insects and warm-blooded animals. Poglayen-Neuwall (1966) found that captives required considerably more meat than did *Potos*.

In the wild, *Bassaricyon* lives alone or in pairs. It is sometimes found in association with *Potos*, as well as with opossums and douroucoulis *(Aotus)*. It scent-marks with urine, but the particular function is not known (Grzimek 1975). In studies at the Louisville Zoo, Poglayen-Neuwall (1966) found that *Bassaricyon* is less social than *Potos* and that two males could not be kept together. There was no definite breeding season, gestation lasted 73–74 days, and each of four births yielded a single offspring. The young weighed about 55 grams at birth, opened their eyes after 27 days, began taking solid food after 2 months, and attained sexual maturity at 21 months. According to Jones (1982), one captive was still living after 17 years and 1 month.

CARNIVORA; **Family MUSTELIDAE**

Weasels, Badgers, Skunks, and Otters

This family of 23 Recent genera and 65 species occurs in all land areas of the world except the West Indies, Madagascar, Sulawesi and islands to the east, most of the Philippines, New Guinea, Australia, New Zealand, Antarctica, and most oceanic islands. One genus, *Enhydra,* inhabits coastal waters of the North Pacific. The sequence of genera presented here follows basically that of Simpson (1945), who recognized five subfamilies: Mustelinae (weasels), with *Mustela, Vormela, Martes, Eira, Galictis, Lyncodon, Ictonyx, Poecilictis, Poecilogale,* and *Gulo;* Mellivorinae (honey badger), with *Mellivora;* Melinae (badgers), with *Meles, Arctonyx, Mydaus, Taxidea,* and *Melogale;* Mephitinae (skunks), with *Mephitis, Spilogale,* and *Conepatus;* and Lutrinae (otters), with *Lutra, Pteronura, Aonyx,* and *Enhydra.*

The smallest member of this family is the least weasel *(Mustela nivalis),* which has a head and body length of 114–260 mm and a weight as low as 25 grams. The largest members are the otters *Pteronura* and *Enhydra,* which have a head and body length of around 1 meter or more and a weight of 22–45 kg. Male mustelids are about one-fourth larger than females. The pelage is either uniformly colored, spotted, or striped. Some species of *Mustela* turn white in winter in the northern parts of their range. The body is long and slender in most genera but stocky in the wolverine *(Gulo)* and in badgers. The short ears are either rounded or pointed. The limbs are short and bear five digits each. The claws are compressed, curved, and nonretractile. The claws of badgers are

Olingo *(Bassaricyon gabbii)*, photo from New York Zoological Society. Inset: olingo *(B. pauli)*, photo by Hans-Jürg Kuhn.

large and heavy for burrowing. The digits of otters are usually webbed for swimming. Well-developed anal scent glands are present in most genera. Males have a baculum.

The skull is usually sturdy and has a short facial region. The dental formula is: (i 3/2–3, c 1/1, pm 2–4/2–4, m 1/1–2) × 2 = 28–38. *Enhydra* is the only genus with 2 lower incisors. The incisors of mustelids are not specialized, the canines are elongate, the premolars are small and sometimes reduced in number, and a constriction is usually present between the lateral and medial halves of the upper molar. The carnassials are developed. The second lower molar, if present, is reduced to a simple peg.

Mustelids are either nocturnal or diurnal and often shelter in crevices, burrows, and trees. Badgers usually dig elaborate burrows. Mustelids move about on their digits, or partly on their digits and partly on their soles. The smaller, slender forms usually travel by means of a scampering gait interspersed with a series of bounds. The larger, stocky forms proceed in a slow, rolling, bearlike shuffle. Mustelids often sit on their haunches to look around. Many genera are agile climbers, and otters and minks are skillful swimmers. Mustelids are mainly flesh eaters. They hunt by scent, though the senses of hearing and sight are also well developed. Some species occasionally feed on plant material, a few are omnivorous, and otters subsist mainly on aquatic animals. The storage of food has been reported for *Mustela, Poecilogale, Gulo,* and certain badgers.

Many mustelids use the secretions of their anal glands as a

defensive measure. Some genera have a contrasting pattern of body colors; for example, skunks have black and white stripes. This pattern is thought to be a form of warning coloration associated with the fetid anal gland secretion, and a reminder that the animal is better left alone. Some of the forms with contrasting body colors, such as the marbled polecat *(Vormela)* and certain skunks, expose and emphasize this contrasting pattern by means of bodily movements when they are alarmed.

Delayed implantation of the fertilized eggs in the uterus occurs in many genera. The actual period of developmental gestation is about 30–65 days, but with delayed implantation the total period of pregnancy is as long as 12.5 months in *Lutra canadensis*. There is usually a single litter per year. The offspring are usually tiny and blind at birth. The young of *Enhydra,* however, are born with their eyes open and in a more advanced stage than the young of other mustelids. The young of most genera can care for themselves at about 2 months and are sexually mature after a year or 2. The potential longevity in the wild is generally 5–20 years.

Mustelids sometimes kill poultry, but they also help to keep rodents in check. Many species are widely hunted for their fur.

The geological range of this family is early Oligocene to Recent in North America, Eocene to Recent in Europe, middle Oligocene to Recent in Asia; early Pliocene to Recent in Africa; and late Pliocene to Recent in South America (Stains 1984).

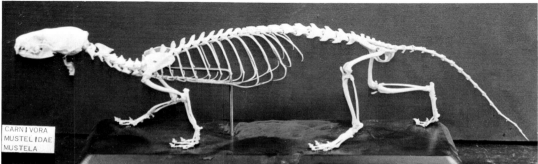

Giant otter *(Pteronura brasiliensis)*, photo by Bernhard Grzimek. Skeleton of mink *(Mustela vison)*, photo by P. F. Wright of specimen in U.S. National Museum of Natural History.

CARNIVORA; MUSTELIDAE; **Genus MUSTELA**
Linnaeus, 1758

Weasels, Ermines, Stoats, Minks, Ferrets, and Polecats

There are 5 subgenera and 17 species (Coetzee, *in* Meester and Setzer 1977; Corbet 1978; Ellerman and Morrison-Scott 1966; Hall 1951, 1981; Izor and de la Torre 1978; Izor and Peterson 1985; Lekagul and McNeely 1977; Schreiber et al. 1989; Van Bree and Boeadi 1978; Youngman 1982):

subgenus *Grammogale* Cabrera, 1940

M. felipei (Colombian weasel), known by four specimens from southwestern Colombia and northern Ecuador;
M. africana (tropical weasel), Amazon Basin of Brazil, eastern Ecuador, and northeastern Peru;

subgenus *Mustela* Linnaeus, 1758

M. erminea (ermine, or stoat), Scandinavia and Ireland to northeastern Siberia and the western Himalayan region, Japan, Alaska and northern Greenland to northern New Mexico and Maryland;
M. nivalis (least weasel), western Europe and Asia Minor to northeastern Siberia and Korea, parts of China and

possibly Indochina, Great Britain, several Mediterranean islands, Japan, northwestern Africa, Egypt, Alaska, Canada, north-central conterminous United States, Appalachian region;
M. frenata (long-tailed weasel), southern Canada to Guyana and Bolivia;
M. altaica (mountain weasel), southern Siberia to the Himalayan region and Korea;
M. kathiah (yellow-bellied weasel), Himalayan region to southern China;

subgenus *Lutreola* Wagner, 1841

M. lutreola (European mink), France to western Siberia and the Caucasus;
M. sibirica (Siberian weasel), eastern European Russia to eastern Siberia and Thailand, Japan, Taiwan;
M. lutreolina (Indonesian mountain weasel), southern Sumatra, Java;
M. nudipes (Malaysian weasel), Malay Peninsula, Sumatra, Borneo;
M. strigidorsa (back-striped weasel), Nepal to Thailand;

subgenus *Vison* Gray, 1843

M. vison (American mink), Alaska, Canada, conterminous United States except parts of southwest;
M. macrodon (sea mink), formerly on Atlantic coast from New Brunswick to Massachusetts;

Weasel (*Mustela* sp.), photo by Ernest P. Walker.

subgenus *Putorius* Cuvier, 1817

M. putorius (European polecat), western Europe to Ural
    Mountains;
M. eversmanni (steppe polecat), steppe zone from Austria
    to Manchuria and Tibet;
M. nigripes (black-footed ferret), plains region from
    Alberta and Saskatchewan to northeastern Arizona and
    Texas.

Grammogale was considered to be a full genus by Cabrera
(1957) and Stains (1967) but was given only subgeneric rank
by Hall (1951) and Izor and de la Torre (1978). Although Hall
(1981) also recognized *Lutreola* and *Putorius* as subgenera, he
noted that some members of these two taxa closely resemble
some members of the subgenus *Mustela* and that distinction
from the latter might not be warranted. Corbet (1978) did not
use any subgeneric designations. Youngman (1982) recognized the five subgenera with content as given above. *M.
macrodon* was considered a subspecies of *M. vison* by Manville (1966) but was given specific rank by Hall (1981). Several
authorities have suggested that *M. nigripes* of North America
is conspecific with *M. eversmanni* of the Old World (Anderson 1973; Corbet 1978), but recent electrophoretic evidence
indicates that the two are distinct species (Miller et al. 1988).
Both *M. erminea* and *M. putorius* have been reported, with
much doubt, from northwestern Africa (Coetzee, *in* Meester
and Setzer 1977).

There is considerable variation in size, but *Mustela*
includes the smallest species in the Mustelidae. The body is
usually long, lithe, and slender. The tail is shorter than the
head and body, often less than half as long. The legs are short,
and the ears are small and rounded. The skull has a relatively
shorter facial region than does that of *Martes*. The bullae are
long and inflated. Additional information is given separately
for each species.

Mustela felipei (Colombian weasel).

In the type specimen, an adult male, head and body length is
217 mm and tail length is 111 mm. The fur is relatively long,
soft, and dense. The upper parts and the entire tail are uniformly blackish brown, and the lower parts are light orange
buff. All four plantar surfaces are naked, and there is extensive interdigital webbing. These features are thought to be
adaptations for semiaquatic life. The known specimens were

collected in riparian areas at elevations of 1,750 and 2,700
meters (Izor and de la Torre 1978).

Mustela africana (tropical weasel).

Head and body length is 240–380 mm and tail length is about
160–210 mm. The upper parts are reddish to chocolate; the
underparts are pale and have a longitudinal median stripe of
the same color as the upper parts. All four plantar surfaces are
nearly naked, and there is extensive interdigital webbing. The
tropical weasel seems to be restricted to humid riparian forests (Izor and de la Torre 1978). It is reported to be a good
swimmer and climber.

Mustela erminea (ermine, or stoat).

Except as noted, the information for the account of this species was taken from Banfield (1974), Jackson (1961), C. M.
King (1983), Novikov (1962), and Stroganov (1969). Head and
body length is 170–325 mm, tail length is 42–120 mm, and
weight is 42–365 grams. Old World animals average larger
than those of North America, and males average larger than
females. Except in certain southern parts of its range, the
ermine changes color during three- to five-week molts in
April–May and October–November. In summer the back,
flanks, and outer sides of the limbs are rich chocolate brown,
the tip of the tail is black, and the underparts are white. In
winter the entire coat is white, except for the tip of the tail,
which remains black. The winter pelage is longer and denser
than that of the summer. Females have eight mammae.

The ermine is found in many habitats, from open tundra to
deep forest, but seems to prefer areas with vegetative or rocky
cover. It makes its den in a crevice, among tree roots, in a
hollow log, or in a burrow taken over from a rodent. It maintains several nests within its range, which are lined with dry
vegetation or the fur and feathers of its prey. It is primarily
terrestrial but climbs and swims well. It generally hunts in a
zigzag pattern, progressing by a series of leaps of up to 50 cm
each. It can easily run over the snow and, if pursued, may
move under the snow. It may travel 10–15 km in a night,
though the average hunt covers 1.3 km. Activity takes place
at any hour but is primarily nocturnal. The ermine is swift,
agile, and strong and has keen senses of smell and hearing. Its
slender body allows it to enter and move quickly through the
burrows of its prey. It generally kills by biting at the base of
the skull. It sometimes attacks animals considerably larger
than itself, such as adult hares. The diet consists mainly of
small rodents and also includes birds, eggs, frogs, and insects.

Ermine, or stoat *(Mustela erminea)* in winter coat, photo by Bernhard Grzimek.

Food may be stored underground for the winter.

Population density fluctuates with prey abundance. Under good conditions there may be an ermine for every 10 ha. Individual home range is up to about 200 ha., usually 10–40 ha., and is generally larger for males than for females. There are several vocalizations, including a loud and shrill squeaking. In a radiotracking study in southern Sweden, Erlinge (1977) found home range sizes of 2–3 ha. in females and 8–13 ha. in males. The ranges of the males included portions of those of the females. Resident animals of both sexes maintained exclusive territories. Boundaries were regularly patrolled and scent-marked, and neighbors usually avoided one another. Adult males were dominant over females and young. Females usually spent their lives in the vicinity of their birthplace, but juvenile males wandered extensively in the spring to find a territory.

Females are polyestrous but produce only one litter per year. Mating occurs in late spring or early summer, but implantation of the fertilized eggs in the uterus is delayed until around the following March, and birth takes place in April or May. Pregnancy thus lasts about 10 months, but embryonic development only a little over 1 month. Litter size is 3–18 young, averaging about 6 in the New World and 8–9 in the Old. The young weigh about 1.5–3.0 grams at birth, are blind and helpless, and are covered with fine white hair. They grow rapidly and by 8 weeks are able to hunt with the mother. Females attain sexual maturity at 2–3 months and sometimes can mate in their first summer of life. Males do not reach full size and sexual maturity until 1 year.

The ermine rarely molests poultry and is valuable to human interests because it destroys mice and rats. Its white winter fur has long been used in trimming coats and making stoles. The reported number of ermine pelts taken in eight Canadian provinces during the 1976/77 trapping season was 55,216, with an average price of $1.03 (Deems and Pursley 1978).

Mustela nivalis (least weasel).

Except as noted, the information for the account of this species was taken from Banfield (1974), Jackson (1961), and Stroganov (1969). The least weasel is the smallest carnivore. Head and body length is 114–260 mm, tail length is 17–78 mm, and weight is 25–250 grams. Old World animals average larger than those of North America, and males average larger than females. Except in certain southern parts of its range, the least weasel changes color during the spring and fall. In summer the upper parts are brown and the underparts are white. In winter, the entire coat is white, though there may be a few black hairs at the tip of the tail.

The least weasel is much like *M. erminea* in details of habitat, nest construction, and movements. It is, however, even more agile and does not travel over such large areas. It feeds almost entirely on small rodents and may store food for the winter. In an area of 27 ha. in England, King (1975) determined that not more than four adult males were resident at one time. Home range size was 7–15 ha. for males and 1–4 ha. for females.

Births apparently may occur at any time of the year. Females are polyestrous and are capable of bearing more than one litter annually. Delayed implantation does not occur, and the gestation period is 35–37 days. The number of young per litter averages 5 and ranges from 3 to 10. The offspring are weaned at 24 days and attain sexual maturity at 4 months.

The least weasel is rare and of little or no commercial value. It is not known to prey on domestic animals and is beneficial to people through its destruction of mice and rats. It evidently has been introduced by human agency on certain Mediterranean islands, on the Azores, and on Sao Tomes off west-central Africa (Corbet 1978).

Mustela frenata (long-tailed weasel).

In Canada and the United States, females have a head and body length of 203–28 mm, a tail length of 76–127 mm, and a weight of 85–198 grams; males have a head and body length of 228–60 mm, a tail length of 102–52 mm, and a weight of 198–340 grams (Burt and Grossenheider 1976). In Mexico, the head and body length is 250–300 mm, tail length is 140–205 mm, and weight of one male was 365 grams (Leopold 1959). Except as noted, the information for the remainder of this account was taken from Banfield (1974), Jack-

A. Least weasel *(Mustela nivalis)*. B. Least weasel changing into winter coat. Photos by Ernest P. Walker.

Immature long-tailed weasel *(Mustela frenata),* photo by Ernest P. Walker.

son (1961), and Lowery (1974). In Canada and the northern United States, a color change occurs in the course of 25- to 30-day molts from early October to early December and from late February to late April. During the summer in these regions, and throughout the year farther south, the upper parts are brown, the underparts are ochraceous or buff, and the tip of the tail is black. During the winter in these regions the entire coat is pure white, except for the terminal quarter of the tail, which is black. Females have eight mammae.

The long-tailed weasel occurs in a variety of habitats but shows preference for open, brushy or grassy areas near water. It dens in a hollow log or stump, among rocks, or in a burrow taken over from a rodent. Its den is lined with the fur of its victims. It is primarily nocturnal but is frequently active by day. It can climb and swim, but apparently not as well as *M. erminea*. Its long and slender shape allows it to follow a mouse to the end of a burrow or to enter a chicken coop through a knothole. This shape also prevents curling into a spherical resting posture to conserve heat, and thus the metabolic rate of *M. frenata* (and presumably of other weasels) is 50–100 percent higher than that of "normally" shaped mammals of the same weight (Brown and Lasiewski 1972). It thus has a voracious appetite but also speed, agility, and determination. *M. frenata* seizes its prey with its claws and teeth and usually kills by a bite to the back of the neck. It may kill animals larger than itself and has even been known to attack humans who get between it and its prey. The diet, however, consists mainly of rodents and other small mammals. Although weasels are sometimes said to suck blood, this behavior has not been scientifically documented.

Population density varies from about one per 2.6 ha. to one per 260 ha., and home range from 4 to 120 ha. Home ranges may overlap, but individuals seldom meet except during the reproductive season. The voice of *M. frenata* has been reported to consist of a "trill, screech, and squeal" (Svendsen 1976).

Females are monestrous. Mating occurs in July and August, but implantation of the fertilized eggs in the uterus is delayed until around the following March. Embryonic development then proceeds for approximately 27 days until birth in April or May. The total period of pregnancy is 205–337 days. The mean number of young per litter is six, and the range is three to nine. The young weigh about 3.1 grams each at birth, open their eyes after 35–37 days, and are weaned at 3.5 weeks. The father has been reported to assist in the care of the offspring. Females attain sexual maturity at 3–4 months, but males do not mate until the year following their birth.

The long-tailed weasel is more prone to raid henhouses than are other species of *Mustela* but is generally beneficial in the vicinity of poultry farms, because it destroys the rats that prey on young chickens. The white winter pelage, known collectively with that of *M. erminea* as "ermine," has varied in value, with prime pelts sometimes bringing up to $3.50 each. The reported number of pelts taken in the United States and Canada during the 1976/77 trapping season was 61,175, with an average price of about $1.00 (Deems and Pursley 1978).

Mustela altaica (mountain weasel).

The information for the account of this species was taken from Stroganov (1969). In males, head and body length is 224–87 mm, tail length is 108–45 mm, and weight is 217–350 grams. In females, head and body length is 217–49 mm, tail length is 90–117 mm, and weight is 122–220 grams. *M. altaica* resembles *M. sibirica* but is smaller and has shorter fur and a less luxuriant tail. There are spring and fall molts. The winter pelage is yellowish brown above and pale yellow below. In summer the coat is gray to grayish brown.

This weasel occurs in highland steppes and forests at eleva-

tions up to 3,500 meters. It nests in a rock crevice, among tree roots, or in an expropriated rodent burrow. It is quick, agile, and chiefly nocturnal or crepuscular. It feeds mainly on rodents, pikas, and small birds. In Kazakh, mating occurs in February and March. The gestation period is 40 days, litters contain two to eight young, and lactation lasts 2 months. Following independence, litter mates remain together until autumn. Of little importance in the fur trade, *M. altaica* is considered to be beneficial to agricultural interests.

Mustela kathiah (yellow-bellied weasel).

The information for the account of this species was taken from Mitchell (1977). Head and body length is 250–70 mm and tail length is 125–50 mm. The tail is more than half and sometimes nearly two-thirds as long as the head and body. The upper parts are dark brownish, and the underparts are deep yellow. This weasel inhabits pine forests and also occurs above the timber line. The elevational range is 1,800–4,000 meters. The diet includes birds, rodents, and other small mammals.

Mustela lutreola (European mink).

Except as noted, the information for the account of this species was taken from Novikov (1962) and Stroganov (1969). In males, head and body length is 280–430 mm, tail length is 124–90 mm, and weight is up to 739 grams. In females, head and body length is 320–400 mm, tail length is 120–80 mm, and weight is up to 440 grams. The general coloration is reddish brown to dark cinnamon. The underparts are somewhat paler than the back, and there may be some white on the chin, chest, and throat. The dense pelage is short, even in winter.

The European mink inhabits the densely vegetated banks of creeks, rivers, and lakes. It is rarely found more than 100 meters from fresh water. It may excavate its own burrow, take one from a water vole *(Arvicola)*, or den in a crevice, among tree roots, or in some other sheltered spot. It swims and dives well. Activity is mainly nocturnal and crepuscular. The summer is generally spent in an area of 15–20 ha., but there may be extensive fall and winter movements to locate swift, nonfrozen streams. The chief prey is the water vole. The diet also includes other small rodents, amphibians, mollusks, crabs, and insects. Food is often stored.

Mating occurs from February to March, and births in April and May. Pregnancy lasts 35–72 days, the variation probably resulting from delayed implantation in some females. The number of young per litter is two to seven, usually four or five. The young open their eyes after 4 weeks, are weaned at 10 weeks, disperse in the autumn, and attain sexual maturity the following year. Longevity is 7–10 years.

Although its fur is not as valuable as that of *M. vison*, the European mink has been widely trapped for commercial purposes. It has also been killed as a predator, has lost much habitat through hydroelectric developments and water pollution, and has suffered from competition with the introduced *M. vison*. In the Soviet Union its numbers have declined to 40,000–45,000 individuals (Tumanov and Zverev 1986), and to the west it now survives only in Finland, eastern Poland, parts of the Balkans, western France, and northern Spain. However, it may have spread through France as recently as the eighteenth and nineteenth centuries during a period of climatic amelioration, while at the same time it was disappearing from central Europe, and it appears to have entered Spain only about 1950 (Youngman 1982). It is classified as vulnerable by the IUCN.

Mustela sibirica (Siberian weasel).

Except as noted, the information for the account of this species was taken from Stroganov (1969). In males, head and body length is 280–390 mm, tail length is 155–210 mm, and weight is 650–820 grams. In females, head and body length is 250–305 mm, tail length is 133–64 mm, and weight is 360–430 grams. In winter the upper parts are bright ochre to straw yellow, and the flanks and underparts are somewhat paler. The summer pelage is darker, shorter, coarser, and sparser. Females have four pairs of mammae.

The Siberian weasel dwells mainly in forests, especially along streams, but sometimes enters towns and cities. It dens in tree hollows, under roots or logs, between stones, in a modified rodent burrow, or in a building. It lines its nest with fur, feathers, and dried vegetation. It is swift and agile, has good senses of smell and hearing, and can climb and swim well. It is mainly nocturnal and crepuscular and has been observed to cover a distance of 8 km in one night. In the fall it may move from upland areas to valleys. There are reports of mass migrations associated with food shortages. The diet consists mainly of small rodents but also includes pikas, birds, eggs, frogs, and fish. Food may be stored for winter use.

Several males may pursue and fight over a single female. Mating occurs in late winter and early spring, and births take place from April to June. The gestation period is 28–30 days, litter size is 2–12 young, the offspring open their eyes after 1 month, and lactation lasts 2 months. The young leave their mother by the end of August, but litter mates may travel together through the autumn. A captive lived for 8 years and 10 months (Jones 1982).

The Siberian weasel is important in the fur trade. It occasionally attacks domestic fowl but is generally considered beneficial because of its destruction of noxious rodents. It has been introduced on Sakhalin and Iriomote islands (Corbet 1978).

Mustela lutreolina (Indonesian mountain weasel).

The information for the account of this species was taken from Van Bree and Boeadi (1978). Based on six specimens of males, head and body length is 297–321 mm, tail length is 136–170 mm, and weight is 295–340 grams. The overall color is glossy dark russet and there is no mask or other facial markings. *M. lutreolina* has a striking resemblance to *M. lutreola* in size and color, but resembles *M. sibirica* in characters of the skull. The 11 known specimens have been collected at elevations of 1,000–2,200 meters in the mountains of Java and southern Sumatra. *M. lutreolina* is a close relative of *M. sibirica* and sometimes is considered conspecific with the latter. Its lifestyle probably is much the same. Apparently, during a glacial phase of the Pleistocene, a cooler climate and lower sea level allowed the ancestral stock of *M. lutreolina* to spread southward. Changing conditions subsequently isolated it in its present mountain habitat.

Mustela nudipes (Malaysian weasel).

The information for the account of this species was taken from Lekagul and McNeely (1977). Head and body length is 300–360 mm and tail length is 240–60 mm. The body colora-

tion varies from pale grayish white to reddish brown, and the head is much paler than the body. The soles of the feet are naked around the pads. Females have two pairs of mammae. This species is not well known, but its habits are thought to be like those of other weasels. A litter of four young has been recorded.

Mustela strigidorsa (back-striped weasel).

The information for the account of this species was taken from Lekagul and McNeely (1977) and Mitchell (1977). Head and body length is 250–325 mm and tail length is 130–205 mm. The general coloration is dark brown, but the upper lips, cheeks, chin, and throat are pale yellow. There is a narrow whitish stripe down the middle of the back and another along the venter. As in *M. nudipes*, the area around the foot pads is entirely naked. Females have two pairs of mammae.

The back-striped weasel inhabits evergreen forest at elevations of 1,200–2,200 meters. One was taken in a tree hole 3–4 meters above the ground. An individual was observed to attack a bandicoot rat three times its own size. According to Schreiber et al. (1989), this species is probably rare, though it now is known from at least 31 museum specimens.

Mustela vison (American mink).

Except as noted, the information for the account of this species was taken from Banfield (1974), Burt and Grossenheider (1976), Jackson (1961), and Lowery (1974). In males, head and body length is 330–430 mm, tail length is 158–230 mm, and weight is 681–2,310 grams. In females, head and body length is 300–400 mm, tail length is 128–200 mm, and weight is 790–1,089 grams. The pelage is soft and luxurious. Its general color varies from rich brown to almost black, but the ventral surface is paler and may have some white spotting. Captive breeding has produced a number of color variants. Anal scent glands emit a strong, musky odor, considered by some persons to be more obnoxious than that of skunks. Females have three pairs of mammae.

The mink is found along streams and lakes, as well as in swamps and marshes. It prefers densely vegetated areas. It dens under stones or the roots of trees, in an expropriated beaver or muskrat house, or in a self-excavated burrow. Such a burrow may be about 3 meters long and 1 meter beneath the surface and have one or more entrances just above water level. The mink is an excellent swimmer, can dive to depths of 5–6 meters, and can swim underwater for about 30 meters. It is primarily nocturnal and crepuscular but is sometimes active by day. It normally is not wide-ranging but may travel up to about 25 km in a night during times of food shortage. The most important dietary components are small mammals, fish, frogs, and crayfish; other foods include insects, worms, and birds.

Population densities of about 1–8 per sq km have been recorded. Females have home ranges of about 8–20 ha. The ranges of males are larger, sometimes up to 800 ha. Individu-

American mink *(Mustela vison)*, photo by Ernest P. Walker.

als are generally solitary and hostile to one another except when opposite sexes come together for breeding. Females are polyestrous but have only one litter per year. Mating occurs from February to April, and births in late April and early May. Because of a varying period of delay in implantation of the fertilized eggs, pregnancy may last from 39 to 78 days. Actual embryonic development takes 30–32 days. The number of young per litter averages 5 and ranges from 2 to 10. The young are born in a nest lined with fur, feathers, and dry vegetation. They are blind and naked at birth, open their eyes after 5 weeks, are weaned at 5–6 weeks, leave the nest and begin to hunt at 7–8 weeks, and separate from the mother in the autumn. Females reach adult weight at 4 months and sexual maturity at 12 months, while males reach adult weight at 9–11 months and sexual maturity at 18 months. Potential longevity is 10 years.

Most of the mink fur used in commerce is produced on farms. The preferred breeding stock results from crossing the large Alaskan and dark Labrador forms. Selective breeding has led to development of strains that regularly yield such colors as black, white, platinum, and blue (Grzimek 1975). In Canada during the 1971/72 season 72,674 wild and 1,155,020 farm-raised mink pelts were sold (Banfield 1974). During the 1976/77 season the reported number of wild mink taken and the average price per pelt were 116,537 and $19.67 in Canada, and 320,823 and $14.00 in the United States (Deems and Pursley 1978). In 1983/84 the total harvest of wild mink was 392,122, with an average value of $9.71 (Novak, Obbard, et al. 1987). *M. vison* has been introduced deliberately in many parts of the Soviet Union, and escaped animals have established populations in Iceland, Ireland, Great Britain, Scandinavia, and Germany (Corbet 1978; Smit and Van Wijngaarden 1981). The subspecies *M. v. evergladensis*, of southern Florida, appears to be rare and may be jeopardized by human water diversion projects (Smith and Cary 1982).

Mustela macrodon (sea mink).

The information for the account of this extinct species was taken from Allen (1942), Campbell (1988), and Manville (1966). The sea mink was said to resemble *M. vison* but to be much larger, to have a coarser and more reddish fur, and to have an entirely different odor. Head and body length has been estimated at 660 mm and tail length at 254 mm. No complete specimen is known to exist, and descriptions are based only on recorded observations and numerous bone fragments and teeth found at Indian middens along the New England coast.

The sea mink reportedly made its home among the rocks along the ocean. Its den had two entrances. It is thought to have been nocturnal and solitary. The diet consisted mainly of fish and probably also included mollusks. An adult and four young estimated to be three or four weeks old were seen along a beach in August. Because of the large size of *M. macrodon*, its pelt brought a higher price than that of *M. vison* and was persistently sought. Some persons pursued the species from island to island, using dogs to locate individuals on ledges and in rock crevices and then digging or smoking them out. The sea mink apparently had been exterminated by about 1880, and its range seems subsequently to have been occupied by *M. vison*.

Mustela putorius (European polecat).

Except as noted, the information for the account of this species was taken from Blandford (1987), Grzimek (1975), Novikov (1962), and Stroganov (1969). Males have a head and body length of 295–460 mm, a tail length of 105–90 mm, and a weight of 405–1,710 grams. Females have a head and body length of 205–385 mm, a tail length of 70–140 mm, and a weight of 205–915 grams. The general coloration is dark brown to black; the underfur is pale yellow and is clearly seen through the guard hairs; and the area between the eye and the ear is silvery white.

The European polecat is most common in open forests and meadows. It dens in such places as crevices, hollow logs, and burrows made by other animals. It sometimes enters settled areas and buildings occupied by people. It is nocturnal and terrestrial but is capable of climbing. The diet consists of small mammals, birds, frogs, fish, and invertebrates.

Population density has been calculated to be 1 per 1,000 ha., and home range about 100–150 ha. *M. putorius* is usually solitary and silent but has a variety of squeals, screams, and other sounds. Mating occurs from March to June, heat lasts 3–5 days, and the gestation period is about 42 days. The number of young per litter is 2–12, usually 3–7. The young weigh 9–10 grams at birth, open their eyes and are weaned after about 1 month, and become independent at around 3 months. Sexual maturity may come in the first or second year of life. Longevity in the wild is usually up to 5–6 years, but captives have lived as long as 14 years.

The domestic ferret, sometimes given the subspecific name *M. putorius furo*, is thought to be a descendant of the European polecat. It was bred in captivity as early as the fourth century B.C. It is usually tame and playful, and is used to control rodents and to drive rabbits from their burrows. It is now found in captivity in much of the world. An estimated 1 million are kept as pets in the United States, though there have been a few reported cases of rabid individuals and savage attacks on children (*Washington Post*, 5 April 1988). Unlike the wild polecat, it is generally white or pale yellow in color. According to Asdell (1964), females may have two or three

European polecat *(Mustela putorius)*, photo from Zoological Society of London.

Steppe polecat *(Mustela eversmanni)*, photo by Ernest P. Walker.

litters annually. Both the domestic ferret and the polecat apparently were introduced in New Zealand, and large feral populations are now established there. These animals are trapped for their fur in New Zealand, as well as in their original range. The pelt is sometimes referred to as "fitch." The wild polecat generally has been persecuted as a predator of small domestic mammals and birds. By the early twentieth century it had been eliminated throughout Great Britain, except in Wales, but there has been some recovery in recent years.

Mustela eversmanni (steppe polecat).

The information for the account of this species was taken from Stroganov (1969). Males have a head and body length of 370–562 mm, a tail length of 80–183 mm, and a weight of up to 2,050 grams. Females have a head and body length of 290–520 mm, a tail length of 70–180 mm, and a weight of up to 1,350 grams. There is much variation in color pattern, but generally the body is straw yellow or pale brown, being somewhat darker above than below. There is a dark mask across the face. The chest, limbs, groin area, and terminal third of the tail are dark brown to black. Some individuals bear a striking resemblance to the North American black-footed ferret *(M. nigripes)*.

The steppe polecat is found in open grassland and semidesert. It usually expropriates the burrow of a ground squirrel or some other animal and modifies the home for its own use. Some burrow systems, especially those of females, may be occupied for several years and become rather complex. *M. eversmanni* is quick and agile and has keen senses, especially of smell and hearing. It moves by leaps of up to 1 meter and constantly changes direction during hunts. It is nocturnal and has been known to cover up to 18 km in the course of a winter night. Local migrations may occur in response to extreme snow depth or food shortage. The diet consists mainly of pikas, voles, marmots, hamsters, and other rodents. Food is sometimes stored for later use.

Mating usually occurs from February to March, and births from April to May. If a litter is lost, however, the female may produce a second later in the year. The gestation period is 38–41 days. The average number of young per litter is about 8–10, and the range is 4–18. The young weigh 4–6 grams each at birth, open their eyes after 1 month, are weaned and start hunting with the mother at 1.5 months, disperse at 3 months, and attain sexual maturity at 9 months.

The steppe polecat is considered to be beneficial to agriculture because of its destruction of rodents. Its fur has commercial importance but is not as valuable as that of *M. putorius*.

Mustela nigripes (black-footed ferret).

Head and body length is 380–500 mm and tail length is 114–50 mm. The linear measurements of males are about 10 percent greater than those of females. Two males weighed 964 and 1,078 grams, and two females weighed 764 and 854 grams. The body is generally colored yellow buff and becomes palest on the underparts. The forehead, muzzle, and throat are nearly white. The top of the head and middle of the back are brown. The face mask, feet, and terminal fourth of the tail are black. Females have three pairs of mammae (Burt and Grossenheider 1976; Henderson, Springer, and Adrian 1969; Hillman and Clark 1980).

The black-footed ferret is found mainly on short and midgrass prairies. It is closely associated with prairie dogs *(Cynomys)* and utilizes their burrows for shelter and travel; it may modify such burrows for its own use. It is primarily nocturnal and is thought to have keen senses of hearing, smell, and sight. It depends largely on prairie dogs for food, but captives have readily accepted other small mammals. Studies of a recently discovered population near Meeteetse, Wyoming (Richardson et al. 1987), indicate that the ferret confines its activity primarily to prairie dog colonies and seldom moves from one colony to another. Average nightly movement during winter was 1,406 meters, and areas of activity for periods of 3–8 nights were 0.4–98.1 ha. Some food was stored in many small caches.

Studies in South Dakota (Hillman, Linder, and Dahlgren 1979) indicate that *M. nigripes* kills only enough to eat. A prairie dog town of 14 ha. was occupied by a single ferret for six months, but the prey population was not severely reduced. Females raising litters require relatively large prairie dog towns. Whereas overall average town size was found to be 8 ha., the average town size occupied by females with young was 36 (10–120) ha. The mean distance between a town and the nearest neighboring town was 2.4 km. The mean distance between two towns occupied by ferrets was 5.4 km. Investigation of the wild population near Meeteetse, Wyoming (Clark 1987b), found a density of about one ferret per 50 ha. of prairie dog colonies. This population consisted of 67 percent juveniles and 33 percent adults each August.

M. nigripes is solitary except during the breeding season, and males apparently do not assist in the rearing of young (Henderson, Springer, and Adrian 1969; Hillman and Clark 1980). Captives have mated in March and early April. The gestation periods of one female in two seasons were 42 and 45 days. The number of young per litter in the wild averages 3.3 and ranges from 1 to 6. The young emerge from the burrow in early July and separate from the mother in September or

Black-footed ferret *(Mustela nigripes)*, photo by Luther C. Goldman.

early October. Young males then disperse for a considerable distance, but young females often remain in the vicinity of their mother's territory. Sexual maturity is attained by 1 year (Forrest et al. 1988; Hillman and Clark 1980; Miller et al. 1988). Captives are estimated to have lived up to about 12 years (Hillman and Carpenter 1980).

The black-footed ferret may once have been common on the Great Plains, but its subterranean and nocturnal habits made it difficult to locate and observe. According to Clark (1976), there have been about 1,000 reports of the species since 1851, and there are about 100 specimens in museums. In 1920 ferret numbers were estimated at more than 500,000 (Clark 1987a). During the twentieth century, M. nigripes apparently declined in association with the extermination of prairie dogs by human agency. In Kansas, for example, the area occupied by prairie dog towns has been reduced by 98.6 percent since 1905. The ferret became so rare that some persons considered it to be extinct. Reports continued in several states, however, and in 1964 a population was discovered in South Dakota. This group was regularly observed and studied until 1974, when confirmed records ceased. Several captives were taken, but eventually all died without leaving surviving offspring. There again was fear that the species was extinct, though Clark (1978) gathered a number of reliable reports in Wyoming, and Boggess, Henderson, and Choate (1980) found the skull of a ferret that may have died as recently as 1977 in Kansas.

Then, on 26 September 1981 a ferret was found that had been killed by dogs on a ranch near Meeteetse, Wyoming. Agents of the U.S. Fish and Wildlife Service quickly live-captured, radio-collared, and released another individual in the same area and found evidence of several additional animals. It soon was realized that a substantial ferret population, indeed the largest ever scientifically observed, was present. Intensive field studies were initiated, and by July 1984 the population was estimated to contain 129 individuals. The following year, however, the number of ferrets began to fall, apparently as a result of a decline in their major prey, white-tailed prairie dogs, caused by plague. Amidst efforts to control the plague and live-capture a portion of the remaining ferrets, canine distemper somehow was introduced into the wild ferret population, which promptly collapsed (the first 6 ferrets to be caught also died of the deadly disease). From late 1985 to early 1987 all of the ferrets known to survive, a total of 18 individuals, were brought into captivity and used as the basis of a breeding program. By 1989 there were 58 ferrets in a

captive facility operated by the Wyoming Game and Fish Department and another 12 in the National Zoological Park at Front Royal, Virginia (Bender 1988; Clark 1987b; Collins 1989; Forrest et al. 1988; Miller et al. 1988; Solt 1981). There has been considerable controversy regarding the manner in which the ferret study and conservation program was handled, especially with respect to the length of time taken to establish a proper captive breeding program (Carr 1986; Clark 1987a, 1987b; May 1986; T. Williams 1986). There are plans to use the captive stock for eventual reintroduction projects. At present there are no known wild populations, though reports continue to come from various areas, and two skulls were found in Montana in 1984 that evidently were from animals that had died not more than 10 years previously (Clark, Anderson, et al. 1987). The species is thought to be present in Canada, beyond the range of prairie dogs but in areas of native grasslands with high densities of ground squirrels (Laing and Holroyd 1989). M. nigripes is classified as endangered by the IUCN and the USDI and is on appendix 1 of the CITES.

CARNIVORA; MUSTELIDAE; **Genus VORMELA**
Blasius, 1884

Marbled Polecat

The single species, *V. peregusna*, occurs in the steppe and subdesert zones from the Balkans and Palestine to Inner Mongolia and Pakistan (Corbet 1978).

Head and body length is 290–380 mm, tail length is 150–218 mm, and weight is 370–715 grams (Grzimek 1975; Stroganov 1969). *Vormela* resembles *Mustela putorius* but differs in its broken and mottled color pattern on the upper parts and in its long claws. The mottling on the back is reddish brown, and white or yellowish, and the tail is usually whitish with a dark tip. The underparts are dark brown or blackish, and the facial mask is dark brown. Females have five pairs of mammae (Roberts 1977).

Like most mustelids, *Vormela* possesses anal scent glands, from which a noxious-smelling substance is emitted. When this animal is threatened, it throws its head back, bares its teeth, erects its body hairs, and bristles and curls its tail over its back. This behavior results in the fullest display of the contrasting body colors, and the pattern thus exposed is

Marbled polecat *(Vormela peregusna)*, photo by Bernhard Grzimek.

thought to be a warning associated with the fetid anal gland secretion.

The marbled polecat seems to prefer steppes and foothills. With its strong paws and long claws, it excavates deep, roomy burrows. It may also shelter in the burrows of other animals. It is chiefly nocturnal and crepuscular but is sometimes active by day. It is a good climber but feeds mainly on the ground. It preys on rodents, birds, reptiles, and other animals.

Vormela is solitary except during the breeding season. Births occur from February to March, after a gestation period of about 9 weeks. Litter size is four to eight young (Stroganov 1969). Only the mother cares for the young, which are reared in a grass and leaf nest within a burrow. A captive specimen was still living after 8 years and 11 months (Jones 1982).

The fur of the marbled polecat has been sought at certain times but is not of major commercial importance (Grzimek 1975; Stroganov 1969). *Vormela* sometimes preys on poultry and has been eliminated in parts of its range. The subspecies *V. p. peregusna*, of Europe and Asia Minor, is classified as vulnerable by the IUCN and the Soviet Union. Schreiber et al. (1989) suggested that the major decline of this subspecies on the steppes of the Balkans and Ukraine, like the disappearance of the North American *Mustela nigripes*, has resulted from usurpation of grassland habitat by agriculture and the elimination of rodent prey.

CARNIVORA; MUSTELIDAE; Genus MARTES
Pinel, 1792

Martens, Fisher, and Sable

There are three subgenera and eight living species (Anderson 1970; Chasen 1940; Cholley 1982; Corbet 1978; Ellerman

and Morrison-Scott 1966; Hall 1981; Hsu and Wu 1981; Lekagul and McNeely 1977; Pilgrim 1980):

subgenus *Martes* Pinel, 1792

M. foina (beech marten, or stone marten), Denmark and Spain to Mongolia and the Himalayas, Crete, Rhodes, Corfu;
M. martes (European pine marten), western Europe to western Siberia and the Caucasus, Ireland, Great Britain, Corsica, Sardinia, Sicily;
M. zibellina (sable), originally the entire taiga zone from Scandinavia to eastern Siberia and North Korea, Sakhalin, Hokkaido;
M. melampus (Japanese marten), South Korea, Honshu, Kyushu, Shikoku, Tsushima;
M. americana (American pine marten), Alaska to Newfoundland, south in mountainous areas to central California and northern New Mexico, Great Lakes region, New England;

subgenus *Pekania* Gray, 1865

M. pennanti (fisher, or pekan), southern Yukon to Labrador, south in mountainous areas to central California and Utah, Great Lakes and Appalachian regions, New England;

subgenus *Charronia* Gray, 1865

M. flavigula (yellow-throated marten), southeastern Siberia to Malay Peninsula, Himalayan region, Hainan, Taiwan, Sumatra, Bangka Island, Java, Borneo;
M. gwatkinsi (Nilgiri marten), Nilgiri Hills of extreme southern India.

American pine marten *(Martes americana)*, photo by Howard E. Uible.

An additional species, *M. nobilis* (noble marten), is known from subfossil and fossil specimens from the Yukon, Idaho, Wyoming, Colorado, Nevada, and California and may have survived until less than 3,000 years ago. It was a large species, related to *M. americana* and evidently adapted to a wider variety of environments; the reason for its extinction is unknown (Grayson 1984). Anderson (1970) stated that additional study might show that *M. martes*, *M. zibellina*, *M. melampus*, and *M. americana* are conspecific. Corbet (1978) suggested that *M. gwatkinsi* is conspecific with *M. flavigula*, and Corbet (1984) cited information suggesting that *M. latinorum* of Sardinia is a species distinct from *M. martes*.

From *Mustela*, *Martes* is generally distinguished by a larger and heavier body, a longer and more pointed nose, larger ears, longer limbs, and a bushier tail. Additional information is provided separately for each species.

Martes foina (beech marten, or stone marten).

Except as noted, the information for the account of this species was taken from Anderson (1970), Grzimek (1975), Stroganov (1969), and Waechter (1975). Head and body length is about 400–540 mm, tail length is 220–300 mm, and weight is 1.1–2.3 kg. The general coloration is pale grayish brown to dark brown, and most specimens have a prominent white or pale yellow neck patch. The fur is coarser than that of *M. martes*, and the tail is relatively longer. The soles are covered with sparse hairs, through which the pads stand out markedly.

The stone marten is less dependent on forests than is *M. martes* and prefers rocky and open areas. It is found in mountains at elevations of up to 4,000 meters. It often enters towns and may occupy buildings. Natural nest sites include rocky crevices, stone heaps, abandoned burrows of other animals, and hollow trees. *M. foina* is a good climber but rarely goes high in trees. It is nocturnal and crepuscular. The diet consists of rodents, birds, eggs, and berries. Vegetable matter forms a major part of the summer food in some areas. The average home range in Alsace was found to be 80 ha.

Mating occurs in midsummer, but because of delayed implantation of the fertilized eggs in the uterus, births do not occur until the following spring. The total period of pregnancy is 230–75 days, though only about a month is true gestation. Litters usually contain three to four young but may have up to eight. Canivenc et al. (1981) reported that in southwestern France, where births take place in late March and early April, lactation lasts until mid-May. A captive was still living after 18 years and 1 month (Jones 1982).

Martes martes (European pine marten).

The information for the account of this species was taken from Grzimek (1975) and Stroganov (1969). Head and body length is 450–580 mm, tail length is 160–280 mm, and weight is 800–1,800 grams. The general coloration is chestnut to gray brown. There is a light yellow patch on the chest and lower neck. The winter pelage is luxuriant and silky, the summer pelage shorter and coarser. The paws are covered by dense hair. Females have four mammae.

The pine marten dwells in forests, both coniferous and deciduous. It is better adapted than *M. zibellina* for an arboreal life. An individual has several nests, located preferably in hollow trees. Activity is mainly nocturnal, and 20–30 km may be covered in a night's hunting. The diet consists of murids, sciurids, other small mammals, honey, fruit, and berries. There are several food storage sites within the home range, which may be 5 km in diameter.

Mating occurs in midsummer, but because of delayed implantation, births do not take place until March or April. Pregnancy lasts 230–75 days. The number of young per litter is two to eight, usually three to five. The young weigh about 30 grams each at birth, open their eyes after 32–38 days, are weaned at 6–7 weeks, separate from the mother in the autumn, and usually attain sexual maturity at 2 years old. Maximum known longevity is 17 years.

The fur of the pine marten is more valuable than that of *M. foina*. Wild populations have been excessively trapped and greatly reduced during the twentieth century, but the species is not in immediate danger of extinction. Efforts at captive breeding have had only limited success.

Martes zibellina (sable).

Except as noted, the information for the account of this species was taken from Grzimek (1975), Novikov (1962), and Stroganov (1969). Males have a head and body length of 380–560 mm, a tail length of 120–90 mm, and a weight of 880–1,800 grams. Females have a head and body length of 350–510 mm, a tail length of 115–72 mm, and a weight of 700–1,560 grams. The winter pelage is long, silky, and luxurious. Coloration varies but is generally pale gray brown to dark black brown. The summer pelage is shorter, coarser, duller, and darker. The soles are covered with extremely dense, stiff hairs. The body is very slender, long, and supple.

The sable dwells in both coniferous and deciduous forests, sometimes high in the mountains, and preferably near streams. It is mainly terrestrial but can climb. An individual may have several permanent and temporary dens, located in holes among or under rocks, logs, or roots. A burrow several meters long may lead to the enlarged nest chamber, which is lined with dry vegetation and fur. The sable hunts either by day or by night. It tends to remain in one part of its home range for several days and then move on. It sometimes stays in its nest for several days during severe winter weather. There may be migrations to higher country in summer and also large-scale movements associated with food shortage. The diet consists mostly of rodents but also includes pikas, birds, fish, honey, nuts, and berries.

Reported population densities vary from 1 sable per 1.5 sq km in some pine forests to 1 per 25 sq km in larch forests. Individual home range is usually several hundred hectares but may be as great as 3,000 ha. in more desolate parts of Siberia. At least part of the home range is defended against intruders, but a male may sometimes share its territory with a female. Mating occurs from June to August, and births usually in April or May. Actual embryonic development takes perhaps 25–40 days, but because of delayed implantation, the total period of pregnancy is 250–300 days. The number of young per litter ranges from one to five and is usually three or four. The young weigh 30–35 grams each at birth, open their eyes after 30–36 days, emerge from the den at 38 days, are weaned at about 7 weeks, and attain sexual maturity at 15–16 months. Maximum known longevity is 15 years.

The sable is one of the most valuable of fur bearers. Its pelt has been avidly sought since ancient times. Several hundred thousand skins were traded annually during the late eighteenth century in the west Siberian city of Irbit. Because of excessive trapping, this figure dropped to 20,000–25,000 by 1910–13, and the sable had then disappeared from much of its range. Subsequently, through programs of protection and reintroduction, the species increased in numbers and distribution. In the Soviet Union, according to Daniloff (1986), the skins of about 250,000 wild and 27,000 captive bred sables are harvested annually. About half are released to the world market, and these may sell for over $500 each.

Martes melampus (Japanese marten).
According to Anderson (1970), head and body length is 470–545 mm and tail length is 170–223 mm. The general coloration is yellowish brown to dark brown, and there is a whitish neck patch. Little is known about the habits of the Japanese marten. It is decreasing in numbers through excessive hunting for its fur and because of the harmful effects of agricultural insecticides. The subspecies *M. m. tsuensis*, of Tsushima Island, is classified as indeterminate by the IUCN. Schreiber et al. (1989) noted that this subspecies now is legally protected.

Martes americana (American pine marten).
Males have a head and body length of 360–450 mm, a tail length of 150–230 mm, and a weight of 470–1,300 grams. Females have a head and body length of 320–400 mm, a tail length of 135–200 mm, and a weight of 280–850 grams. The pelage is long and lustrous. The upper parts vary in color from dark brown to pale buff, the legs and tail are almost black, the head is pale gray, and the underparts are pale brown with irregular cream or orange spots. The body is slender, the ears and eyes are relatively large, and the claws are sharp and recurved. Females have eight mammae (Banfield 1974; Burt and Grossenheider 1976; Clark, Grensten, et al. 1987).

The American pine marten is found mainly in coniferous forest. It dens in hollow trees or logs, in rock crevices, or in burrows. The natal nest is lined with dry vegetation. The marten is primarily nocturnal and partly arboreal but spends considerable time on the ground. It can swim and dive well. It

is active all winter but may descend to lower elevations if living in a mountainous area. The diet consists mostly of rodents and other small mammals and also includes birds, insects, fruit, and carrion.

Population density varies from about 0.5 to 1.7 per sq km of good habitat (Banfield 1974). In Maine, Soutiere (1979) found densities of 1.2 resident martens per sq km in undisturbed and partly harvested forests but only 0.4 per sq km in commercially clear-cut forests. In a radiotracking study in northeastern Minnesota, Mech and Rogers (1977) determined home ranges to be 10.5, 16.6, and 19.9 sq km for 3 males and 4.3 sq km for 1 female. There was considerable overlap between the ranges of 2 of the males. The marten is primarily a solitary species, but Herman and Fuller (1974) regularly observed what seemed to be an adult male and female together, sometimes apparently with two of their offspring, even though it was not the breeding season.

Mating occurs from June to August. Implantation of the fertilized eggs in the uterus is delayed until February, and embryonic development then proceeds for about 28 days. The total period of pregnancy is 220–75 days. The average number of young per litter is 2.6, and the range is 1–5. The young weigh 28 grams each at birth, open their eyes after 39 days, are weaned at 6 weeks, reach adult size at 3.5 months, and attain sexual maturity at 15–24 months. Wild females have still been capable of reproduction at 12 years of age (Banfield 1974; Clark, Grensten, et al. 1987). Captives have lived up to 17 years.

The fur of the marten is valuable and is sometimes referred to as "American sable." In the 1940s, marten pelts were worth as much as $100 each. In the mid-nineteenth century the Hudson's Bay Company traded as many as 180,000 Canadian skins each year (Banfield 1974). By the early twentieth century excessive trapping had severely depleted *M. americana* in Alaska, Canada, and the western conterminous United States. Protective regulations subsequently allowed the species to make a comeback in some areas, but in the eastern United States the marten survives only in small parts of Minnesota, New York, and Maine (Blanchard 1974; Mech 1961; Mech and Rogers 1977; Yocom 1974). Reintroductions have been attempted in northern Michigan and Wisconsin (Knap 1975) and in New Hampshire (Strickland and Douglas 1987). During the 1976/77 trapping season 27,898 marten skins were taken in the United States, mostly in Alaska, and sold for an average price of $14.00. The respective figures for Canada were 102,632 skins and $19.92 (Deems and Pursley 1978). In 1983/84 the total kill was 188,647, and the average price was $18.61 (Novak, Obbard, et al. 1987).

Martes pennanti (fisher, or pekan).
Head and body length is 490–630 mm and tail length is 253–425 mm. Weight is 2.6–5.5 kg in males and 1.3–3.2 kg in females. The head, neck, shoulders, and upper back are dark brown to black. The underparts are brown, sometimes with small white spots. The thick pelage is coarser than that of *M. americana*. The body is slender but rather stocky for a weasel. Females have eight mammae.

Except as noted, the information for the remainder of this account was taken from Powell (1981, 1982). The fisher inhabits dense forests with an extensive overhead canopy and avoids open areas. It generally lacks a permanent den but seeks temporary shelter in hollow trees and logs, brush piles, abandoned beaver lodges, holes in the ground, and snow dens. All dens known to have been used for raising young were located high up in hollow trees. Activity may take place at any hour. The fisher is adapted for climbing but is primarily terrestrial. It is capable of traveling long distances; one individual covered 90 km in three days. Usual daily movement in New Hampshire, however, was found to be 1.5–3.0

Fisher *(Martes pennanti)*, photo from San Diego Zoological Garden.

km. The fisher generally forages in a zigzag pattern, constantly investigating places where prey might be concealed.

The diet consists mainly of small to medium-sized birds and mammals, and carrion. It has been calculated that a fisher requires either 1 porcupine every 10–35 days, 1 snowshoe hare every 2.5–8.0 days, 1 kg of deer carrion every 2.5–8.0 days, 1–2 squirrels per day, or 7–22 mice per day. Hares are killed with a quick rush and a bite to the back of the neck. Porcupines are taken only on the ground and are killed by repeatedly circling and biting at the face.

Population density in preferred habitat is one fisher per 2.6–7.5 sq km, but in other areas it may be as low as one per 200 sq km. Individual home range size is 15–35 sq km and is generally larger for males. There is little overlap between the ranges of animals of the same sex but extensive overlap between the ranges of opposite sexes. *M. pennanti* is solitary except during the breeding season. At this time, according to Leonard (1986), normal spacing mechanisms seem to break down, with males leaving their territories to seek as many females as possible and perhaps coming into physical conflict with one another.

Mating occurs from March to May, but implantation of the fertilized eggs in the uterus is delayed until the following January to early April, and births occur from February to May. Postimplantation embryonic development lasts about 30 days, but the total period of pregnancy is nearly a year. Females probably mate within 10 days of giving birth and thus are pregnant almost continually. Litters contain an average of about 3 young, but there may be as many as 6. The newborn are blind and only partly covered with fine hair. Their eyes do not open for about 7 weeks, and they do not walk until about 8–9 weeks. Weaning begins at 8–10 weeks, and separation from the mother occurs in the fifth month. Females reach adult weight after 6 months, and males after 1 year. Sexual maturity comes at 1–2 years. Longevities of about 10 years have been recorded for both wild and captive animals.

In the nineteenth and early twentieth centuries the fisher declined over most of its range because of excessive fur trapping and habitat destruction through logging. The species was almost totally eliminated in the United States and greatly reduced in eastern Canada. As a result, porcupines increased in numbers and began to do considerable forest damage. Widespread closed seasons and other protective regulations were initiated in the 1930s, and reintroductions were made in various areas in the 1950s and 1960s. To the south of Canada, populations of *M. pennanti* are now present in Washington, Oregon, northern California, Idaho, Montana, Minnesota, Wisconsin, Michigan, New York, Vermont, New Hampshire, Maine, Massachusetts, and West Virginia (Cottrell 1978; Coulter 1974; Handley 1980b; Mohler 1974; Mussehl and Howell 1971; Olterman and Verts 1972; Penrod 1976; Petersen, Martin, and Pils 1977; Schempf and White 1977; Yocom and McCollum 1973). In some of these states, during the past two decades the fisher has been taken legally for the fur market. The average price of a fisher pelt was $100 in the 1920s. The price subsequently declined but then again increased. During the 1976/77 trapping season 12,557 skins were taken in the United States and Canada, selling for an average of about $95 each (Deems and Pursley 1978). In 1983/84 the take was 20,248, and the average price was $49.45 (Novak, Obbard, et al. 1987), but in 1986 in Ontario the highest-quality pelts were selling for $450 (Douglas and Strickland 1987).

Martes flavigula (yellow-throated marten).

The information for the account of this species was taken from Lekagul and McNeely (1977) and Stroganov (1969). Head and body length is 450–650 mm, tail length is 370–450 mm, and weight is 2–3 kg. The fur is short, sparse, and coarse. There is much variation in color, but generally the top of the head and neck, the tail, the lower limbs, and parts of the back are dark brown to black. The rest of the body is pale brown, except for a bright yellow patch from the chin to the chest. Females have four mammae.

The yellow-throated marten is generally found in forests.

Yellow-throated marten *(Martes flavigula)*, photo from Zoological Society of London.

It climbs and maneuvers in trees with great agility but often comes to the ground to hunt. Activity is primarily diurnal. The diet includes rodents, pikas, eggs, frogs, insects, honey, and fruit. In the northern parts of its range, *M. flavigula* evidently preys heavily on the musk deer *(Moschus)* and the young of other ungulates. Individuals often hunt in pairs or family groups, and lifelong pair bonding has been suggested. The number of young per litter is usually two or three and may be up to five. Maximum known longevity is about 14 years. The pelt of this marten has little commercial value. The subspecies on Taiwan, *M. f. chrysospila,* has become very rare and is classified as endangered by the USDI and as indeterminate by the IUCN.

Martes gwatkinsi (Nilgiri marten).

According to Anderson (1970), no external measurements of this species are available. As in *M. flavigula,* the general coloration is dark brown and there is a yellowish patch from the chin to the chest. The dorsal profile of the skull of *M. gwatkinsi* is flat, not convex as in *M. flavigula.* The Nilgiri

marten is found only in hill forests, seldom occurring below 900 meters. It is classified as indeterminate by the IUCN.

CARNIVORA; MUSTELIDAE; **Genus EIRA**
Hamilton-Smith, 1842

Tayra

The single species, *E. barbara,* is found from southern Sinaloa (west-central Mexico) and southern Tamaulipas (east-central Mexico) to northern Argentina and on the island of Trinidad (Cabrera 1957; Goodwin and Greenhall 1961; Hall 1981).

Head and body length is 560–680 mm, tail length is 375–470 mm, and weight is 4–5 kg (Grzimek 1975; Leopold 1959). The short, coarse pelage is gray, brown, or black on the head and neck, has a yellow or white spot on the chest, and is black or dark brown on the body. There also is a rare, light-colored form, which is pale buffy and has a darker head. The

Tayra *(Eira barbara)*, photo by Ernest P. Walker.

tayra has a long and slender body, short limbs, and a long tail. The head is broad, the ears are short and rounded, and the neck is long. The soles are naked, and the strong claws are nonretractile.

The tayra is a forest dweller. It nests in a hollow tree or log, a burrow made by another animal, or tall grass. It can climb, run, and swim well. When pursued by dogs, it may run on the ground for some distance, then climb a tree, and then leap through the trees for about 100 meters before descending to the ground again. It thus gains time while the dogs are trying to pick up the trail. Leopold (1959) observed a group of four animals springing swiftly through the trees with incredible agility. *Eira* is active both at night and, especially when there is cloud cover, in the morning. The diet seems to consist mostly of rodents but also includes rabbits, birds, small deer (*Mazama*), honey, and fruit.

Eira is often seen alone, in pairs, or in small family groups. Hall and Dalquest (1963) wrote that the genus is not social to any extent, but Leopold (1959) referred to an old record of hunting troops made up of 15–20 individuals. He also cited a report that 2 young are born in February, that their eyes open after 2 weeks, and that they forage with their mother by the time they are 2 months old. Other reports suggest that the young may be born in any season. Poglayen-Neuwall (1975) and Poglayen-Neuwall and Poglayen-Neuwall (1976) determined that six gestation periods lasted 63–65 days, that newborn weigh about 74–92 grams, that their eyes do not open until they are 35–58 days old, and that they suckle for 2–3 months. According to Vaughn (1974), litters of 2 young each were produced in captivity on 4 March and 22 July, following gestation periods of about 67–70 days.

The tayra can live 18 years in captivity, loves to play, and can be tamed. It reportedly was used long ago by the Indians to control rodents (Grzimek 1975). It is of no particular importance as a fur bearer or predator of game. It may occasionally eat poultry but does no substantial damage (Leopold 1959). It also has been accused of raiding corn and sugar cane fields (Hall and Dalquest 1963). Schreiber et al. (1989) reported that the range of the tayra has been greatly reduced in Mexico because of the destruction of tropical forests and spread of agriculture. Remaining populations are small and threatened by habitat loss and hunting.

CARNIVORA; MUSTELIDAE; **Genus GALICTIS**
Bell, 1826

Grisóns

There are two species (Cabrera 1957; Hall 1981):

G. vittata (greater grisón), southern Mexico to central Peru and southeastern Brazil;

G. cuja (little grisón), central and southern South America.

Galictis has often been referred to as *Grison* Oken, 1816. Cabrera placed *G. cuja* in the subgenus *Grisonella* Thomas, 1912.

In *G. vittata* the head and body length is 475–550 mm, tail length is about 160 mm, and weight is 1.4–3.3 kg. In *G. cuja* the head and body length is 400–450 mm, tail length is 150–90 mm, and weight is about 1 kg. The color pattern is striking. In both species the black face, sides, and underparts, including the legs and feet, are sharply set off from the back. The back is smoky gray in *G. vittata* and yellow gray or brownish in *G. cuja*. A white stripe extends across the forehead and down the sides of the neck, separating the black of the face from the gray or brown of the back. Short legs and slender bodies give the animals of this genus an appearance somewhat like that of *Mustela*, but the color pattern immediately distinguishes them.

Grisóns are found in forests and open country, from sea level to 1,200 meters. They live under tree roots or rocks, in hollow logs, or in burrows made by other animals, such as the viscachas of South America. They are probably capable burrowers themselves. Quick and agile, they are are good climbers and swimmers and are active both by day and by night. The diet includes small mammals, birds and their eggs, cold-blooded vertebrates, invertebrates, and fruit.

Grisóns are sometimes seen in pairs or groups, and playing together. They have a number of vocalizations, including sharp, growling barks when threatened. Various reports indicate that offspring have been produced in March, August, September, and October (Grzimek 1975; Leopold 1959). Litter size is two to four young. A captive *G. vittata* was still living after 10 years and 6 months (Jones 1982).

Young grisóns tame readily and make affectionate pets. In early nineteenth-century Chile, grisóns reportedly were domesticated by natives and used in the same manner as ferrets to enter the crevices and holes of chinchillas to drive the latter out (Osgood 1943).

CARNIVORA; MUSTELIDAE; **Genus LYNCODON**
Gervais, 1844

Patagonian Weasel

The single species, *L. patagonicus*, is found in Argentina and southern Chile (Cabrera 1957).

Head and body length is usually 300–350 mm and tail length is 60–90 mm. The coloration on the back is grayish brown with a whitish tinge. The top of the head is creamy or white, and this color extends as a broad stripe on either side to each shoulder. The nape, throat, chest, and limbs are dark brown, and the rest of the lower surface is lighter brown varied with gray. The color pattern is characteristic and quite attractive. *Lyncodon* is somewhat similar externally to *Galic-*

Greater grisón *(Galictis vittata)*, photo by Ernest P. Walker.

Patagonian weasel *(Lyncodon patagonicus)*, photo by Tom Scott of mounted specimen in Royal Scottish Museum, Edinburgh.

tis but differs in such features as coloration and the shorter tail. Internally, *Lyncodon* has fewer teeth than *Galictis*, there being only two upper and two lower premolars, and one upper and one lower molar, on each side. Like most mustelids, *Lyncodon* has a slender body and short legs.

The Patagonian weasel inhabits the pampas. Its habits are little known, but Ewer (1973:177) wrote: "The reduced molars and cutting carnassials of *Lyncodon* strongly suggest that it is a highly carnivorous species." This weasel reportedly was sometimes kept in the houses of ranchers for the purpose of destroying rats. Miller et al. (1983) considered it to be rare in Chile.

CARNIVORA; MUSTELIDAE; Genus ICTONYX
Kaup, 1835

Zorilla, or Striped Polecat

The single species, *I. striatus*, occurs from Mauritania to Sudan and south to South Africa (Coetzee, *in* Meester and Setzer 1977).

Head and body length is 280–385 mm, tail length is 200–305 mm, and weight is 420–1,400 grams. Males are generally larger than females. The body is black with white dorsal stripes, the tail is more or less white, and the face has white markings. The appearance is somewhat like that of the spotted skunks *(Spilogale)* of North America, and early writers sometimes confused the two genera. *Ictonyx* also bears some resemblance to the other African genera *Poecilictis* and *Poecilogale* but may be recognized by its color pattern, long hair, and bushy tail. Females have two pairs of mammae (Rowe-Rowe 1978a).

The zorilla is found in a variety of habitats but avoids dense forest. It is mainly nocturnal, resting during the day in rock crevices or in burrows excavated by itself or some other animal. It occasionally shelters under buildings and in outhouses in farming areas. It is terrestrial but can climb and swim well. The usual pace is an easy trot, slower than that of a mongoose, with the back slightly hunched. The diet consists mainly of small rodents and large insects but also includes eggs, snakes, and other kinds of animals. The zorilla may take a chicken on occasion but is often useful in eliminating rodents from houses and stables.

At the sight of an enemy, such as a dog, a zorilla may erect its hair and tail and perhaps emit its anal gland secretion. Such behavior would seem to make the zorilla more formidable than it really is. When actually attacked, it usually emits fluid into the face of the enemy and then feigns death. The ejected fluid may vary in potency with the individual animal, and perhaps with age and the time of year. Some writers have remarked that the fluid is much less pungent than that of the American skunks, while others have stated that it is most repulsive, acrid, and persistent.

Ictonyx is solitary. Captive males are totally intolerant of one another, and even adults of opposite sexes are amicable only at the time of mating (Rowe-Rowe 1978b). There are several adult vocalizations associated with threat, defense, and greeting (Channing and Rowe-Rowe 1977). In a study of captives, Rowe-Rowe (1978a) determined the reproductive season to extend from early spring to late summer. All births occurred from September to December. There was usually a single annual litter, but if all young died at an early stage, the female sometimes mated and gave birth a second time. Gestation periods of 36 days were recorded. Litter size was one to three young. They weighed about 15 grams each at birth, began to take solid food after about 32 days, opened their eyes

Zorilla *(Ictonyx striatus)*, photo from Zoological Society of London.

at 40 days, were completely weaned at 18 weeks, and were almost full-grown at 20 weeks. A male first mated at 22 months of age, and a female bore its first litter at 10 months. According to Jones (1982), a captive zorilla lived for 13 years and 4 months.

CARNIVORA; MUSTELIDAE; **Genus POECILICTIS**
Thomas and Hinton, 1920

North African Striped Weasel

The single species, *P. libyca,* occurs from Morocco and Senegal to the Red Sea (Rosevear 1974). There is some question whether *Poecilictis* warrants generic distinction from *Ictonyx.* The two taxa are sometimes confused in northern Nigeria and Sudan, where their ranges overlap.

Head and body length is 200–285 mm and tail length is 100–180 mm. Three males weighed 200–250 grams each. The snout is black, the forehead is white, and the top of the head is black. The back is white with a variable pattern of black bands. The tail is white but becomes darker toward the tip. The underparts and limbs are black. From *Ictonyx, Poecilictis* differs in color pattern, in smaller size, in larger bullae, and in having hairy soles except for the pads. The well-developed anal glands are capable of ejecting a malodorous fluid (Rosevear 1974).

According to Rosevear (1974), this weasel is restricted to the edges of the Sahara and the contiguous arid zones. It is nocturnal, sheltering throughout the daylight hours in single subterranean burrows. These it digs for itself, either down into level surfaces or in the sides of dunes. Maternal dens consist of a single gallery ending in an unlined chamber. The diet apparently consists of rodents, young ground birds, eggs, lizards, and insects.

Births reportedly occur from January to March. Rosevear (1974) wrote that gestation may be as short as 37 or as long as 77 days. Litters usually contain two or three young. At birth they are blind and covered with short hair. Some captives have lived for 5 years or more. *Poecilictis* probably does not make a good pet, as it constantly has a disagreeable smell and is aggressive toward people.

CARNIVORA; MUSTELIDAE; **Genus POECILOGALE**
Thomas, 1883

African Striped Weasel

The single species, *P. albinucha,* is found from Zaire and Uganda to South Africa (Coetzee, *in* Meester and Setzer 1977).

Head and body length is 250–360 mm and tail length is 130–230 mm. Rowe-Rowe (1978*b*) listed weights of 283–380

North African striped weasel *(Poecilictis libyca)*, photo by Robert E. Kuntz.

African striped weasel *(Poecilogale albinucha):* Top, photo of mounted specimen from Field Museum of Natural History; Bottom, photo by D. T. Rowe-Rowe.

grams for males and 230–90 grams for females. From the white on the head and nape, four whitish to orange yellow stripes and three black stripes extend on the black back toward the tail, which is white. The legs and underparts are black. There is little variation in pattern.

Poecilogale is smaller and more slender than *Ictonyx* and has narrower back stripes. It resembles *Mustela* in its remarkably slender and elongate body and short legs. Like other weasels, it is able to enter any burrow that it can get its head into. Females have two pairs of mammae (Rowe-Rowe 1978a).

This weasel is found in a variety of habitats, including forest edge, grassland, and marsh. It is almost entirely nocturnal, usually spending the day in a burrow that is either self-excavated or taken over from another animal (Kingdon 1977). It can climb well but spends most of its time on the ground. It cannot run rapidly, and patiently trails prey by scent. The diet consists mainly of small mammals and birds but also includes snakes and insects.

Like most weasels, *Poecilogale* attacks by grabbing the throat or neck and hanging on and chewing until the victim is dead. Fair-sized prey, such as springhare, may run some distance before dropping from exhaustion, with the weasel retaining its hold. *Poecilogale* kills venomous snakes in much the same manner as the mongoose *Herpestes*. It repeatedly provokes the snake to strike until the reptile is tired and slower in recovery. Then, after awaiting an opportunity, it seizes the snake on the back of the head.

Poecilogale is generally solitary, but groups of two to four individuals—apparently family parties—have been observed on occasion. Adults of opposite sexes lived together quite amicably in captivity, but adult males fought each other at every encounter (Rowe-Rowe 1978b). The area around the den is marked by defecation (Kingdon 1977).

When attacked or under stress, *Poecilogale* emits a noxious odor from its anal glands. Although nauseating, this odor is not as strong or persistent as that of the American skunks or the African *Ictonyx*. Normally, *Poecilogale* is silent, but when alarmed it utters a loud sound, described as being between a growl and a shriek. Several adult vocalizations, associated with threat, defense, and greeting, were recorded by Channing and Rowe-Rowe (1977).

In studies of captives and wild-caught animals in South Africa, Rowe-Rowe (1978a) determined that births occurred from September to April. Females were polyestrous and would mate a second time in the season if their first litter was lost. Gestation periods of 31–33 days were recorded. Litters contained one to three young. They weighed four grams each at birth, started taking solid food after 35 days, opened their eyes at 51–54 days, were completely weaned at about 11 weeks, and were nearly full-grown at 20 weeks. A male first mated at 33 months, and a female had her first litter at 19 months. According to Smithers (1971), one individual lived for 5 years and 2 months after capture.

This weasel is considered rare, but not endangered, in South Africa (Meester 1976). Although accused of killing poultry on occasion, *Poecilogale* is generally beneficial to human interests because of its destruction of rats, mice, and springhares. It also eats quantities of locusts, when available, and digs their larvae out of the ground. Some African tribes use the skins of *Poecilogale* in ceremonial costumes or as ornaments, and it is said that some medicine men use parts of this animal in their rituals.

CARNIVORA; MUSTELIDAE; **Genus GULO**
Pallas, 1780

Wolverine

The single species, *G. gulo*, originally occurred from Scandinavia and Germany to northeastern Siberia, throughout Alaska and Canada, and as far south as central California,

Wolverine *(Gulo gulo)*, photo by James K. Drake.

southern Colorado, Indiana, and Pennsylvania (Corbet 1978; Hall 1981). Although Hall treated the North American populations as a separate species, *G. luscus*, he indicated that there was evidence supporting their recognition as a subspecies of *G. gulo*.

Head and body length is 650–1,050 mm, tail length is 170–260 mm, and weight is 7–32 kg. Females average 10 percent less than males in linear measurements and 30 percent less in weight (Hall 1981). The fur is long and dense. The general coloration is blackish brown. A light brown band extends along each side of the body, from shoulder to rump, and joins its opposite over and across the base of the tail. *Gulo* gives the impression of a giant marten, with a heavy build, a large head, relatively small and rounded ears, a short tail, and massive limbs (Stroganov 1969). Females have two pairs of mammae (Grzimek 1975).

The wolverine occurs in the tundra and taiga zones. It may be found in forests, mountains, or open plains. For shelter, it may construct a rough bed of grass or leaves in a cave or rock crevice, in a burrow made by another animal, or under a fallen tree (Banfield 1974; Stroganov 1969). Most maternal dens in Finland were found in holes under the snow (Pulliainen 1968). *Gulo* is mainly terrestrial, the usual gait being a sort of loping gallop, but it can climb trees with considerable speed and is an excellent swimmer. It has a keen sense of smell but apparently poor eyesight and indifferent hearing. It seems to be unexcelled in strength among mammals of its size and has been reported to drive bears and cougars from their kills. It is largely nocturnal but is occasionally active in daylight. In the far north, where there are extended times of light or darkness, the wolverine reportedly alternates 3- to 4-hour periods of activity and sleep. In winter it has been known to cover up to 45 km in a day. It can maintain a gallop for a lengthy period, sometimes moving 10–15 km without rest (Stroganov 1969). Juveniles normally disperse 30–100 km from their natal range, and an individual marked on 6 March 1981 was found 378 km away on 29 November 1982 (Gardner, Ballard, and Jessup 1986).

The diet includes carrion, the eggs of ground-nesting birds, lemmings, and berries. Large mammals, such as reindeer, roe deer, and wild sheep, are taken mainly in winter, when the snow cover allows *Gulo* to travel faster than its prey. Most large mammals, however, are obtained in the form of carrion. Caches of prey or carrion are covered with earth or snow or, sometimes, wedged in the forks of trees.

Gulo occurs at relatively low population densities (Van Zyll de Jong 1975). In Scandinavia the estimates vary from one individual per 200 sq km to one per 500 sq km. Each animal there has a home range that may cover as much as 2,000 sq km in winter. At least part of this range is a territory, within which individuals of the same sex are not tolerated. Territories are regularly marked, mainly by secretions from anal scent glands but also with urine (Grzimek 1975; Pulliainen and Ovaskainen 1975).

Somewhat different information was derived from studies in Montana (Hornocker 1983; Hornocker and Hash 1981; Koehler, Hornocker, and Hash 1980). A study area of 1,300 sq km there contained a minimum population of 20 wolverines. Average yearly home range was 422 sq km for males and 388 sq km for females; however, individuals sometimes left their ranges, and females occupied much reduced areas while they were nursing young. There was extensive overlap between the ranges of both the same and the opposite sexes, and no territorial defense was observed. There was extensive marking, but such was considered not to define a territory but rather to identify the area being utilized at a given time.

Investigations in arctic Alaska indicated that females generally exclude one another from their home ranges, while males tolerate the presence of females (Gipson 1985). In

south-central Alaska, Whitman, Ballard, and Gardner (1986) determined annual home range to average 535 sq km for males and 105 sq km for females with young.

The wolverine is solitary except during the breeding season. Females are monestrous and apparently give birth about every two years. Mating usually occurs from late April to July, but implantation of the fertilized eggs in the uterus is delayed until the following November to March, and births occur from January to April. The usual number of young per litter is two to four, and the range is one to five. Active gestation lasts 30–40 days. The young weigh 90–100 grams at birth, nurse for 8–10 weeks, separate from the mother in the autumn, and attain adult size after 1 year. Sexual maturity comes in the second or third year of life (Banci and Harestad 1988; Banfield 1974; Grzimek 1975; Pulliainen 1968; Rausch and Pearson 1972; Stroganov 1969). A captive female approximately 10 years of age gave birth after a pregnancy of 272 days (Mehrer 1976). Another captive wolverine lived for 17 years and 4 months (Jones 1982).

The pelt of the wolverine is not used widely in commerce but is valued for parkas by persons living in the Arctic, because it accumulates less frost than other kinds of fur. During the 1976/77 trapping season 1,922 skins were reported taken in Canada, Alaska, and Montana. The average selling price was about $182 (Deems and Pursley 1978). In 1983/84 the take was 1,377, with an average value of $77.16 (Novak et al. 1987). Fur trapping has contributed to a decline in the numbers and distribution of the wolverine, but a more important factor may be human consideration of the genus as a nuisance. In Scandinavia it was intensively hunted, often for bounty, because of alleged predation on domestic reindeer. Throughout its range the wolverine came into conflict with people by following traplines and devouring the fur bearers it found and by breaking into cabins and food caches and spraying the contents with its strong scent.

In Europe, *Gulo* is now found only in parts of Scandinavia and the northern Soviet Union (Smit and Van Wijngaarden 1981). The genus has disappeared over most of eastern and south-central Canada (Van Zyll de Jong 1975). It is classified as vulnerable by the IUCN and has been designated as endangered in eastern Canada. By the early twentieth century the wolverine also had been nearly eliminated in the conterminous United States. One population, however, held out and subsequently increased in the mountains of northern and eastern California, and another population reestablished itself in western Montana. Since 1960 there have been numerous reliable reports from Washington, Oregon, Idaho, Wyoming, and Colorado and a few questionable records from Minnesota, Iowa, and South Dakota (Birney 1974; Field and Feltner 1974; Johnson 1977; Nead, Halfpenny, and Bissell 1985; Nowak 1973; Schempf and White 1977). The subspecies *G. g. katschemakensis* of Alaska's Kenai Peninsula now numbers only about 50 individuals and seemingly continues to decline because of an excessively long hunting season (Schreiber et al. 1989).

CARNIVORA; MUSTELIDAE; Genus **MELLIVORA**
Storr, 1780

Honey Badger, or Ratel

The single species, *M. capensis*, originally occurred from Palestine and the Arabian Peninsula to Turkmen S.S.R. and eastern India and from Morocco and lower Egypt to South Africa (Coetzee, *in* Meester and Setzer 1977; Corbet 1978).

Head and body length is 600–770 mm, tail length is usually 200–300 mm, and shoulder height is usually 250–300

Honey badger, or ratel *(Mellivora capensis)*, photo by Bernhard Grzimek.

mm. Kingdon (1977) listed weight as 7–13 kg. The upper parts, from the top of the head to the base of the tail, vary from gray to pale yellow or whitish and contrast sharply with the dark brown or black of the underparts. Completely black individuals, however, have been found in Africa, particularly in the Ituri Forest of northern Zaire. The color pattern of the honey badger has been interpreted as being a warning coloration, because it makes the animal easily recognizable. Females have two pairs of mammae.

The body is heavily built, the legs and tail are relatively short, the ears are small, and the muzzle is blunt. The large forefeet are armed with very large and strong claws. The hair is coarse and quite scant on the underparts. The skin is exceedingly loose on the body and very tough. The skull is massive, and the teeth are robust. There are anal glands that secrete a vile-smelling liquid. This combination of characters provides an effective system of deterrence and defense.

Mellivora is very difficult to kill. The skin is so tough that a dog can make little impression, except on the belly. The ratel can twist about in its skin, so it can even bite an adversary that has seized it by the back of the neck. Porcupine quills and bee stings have little effect, and snake fangs are rarely able to penetrate. *Mellivora* seems to be devoid of fear, and it is doubtful that any animal of equivalent size can regularly kill it. It may rush out from its burrow and charge an intruder, especially in the breeding season. Horses, antelope, cattle, and even buffalo have been attacked and severely wounded in this manner.

The honey badger occupies a variety of habitats, mainly in dry areas but also in forests and wet grasslands. It lives among rocks, in hollow logs or trees, and in burrows. Its powerful limbs and large claws make it a capable and rapid digger. It is primarily terrestrial but can climb, especially when attracted by honey. It travels by a jog-trot but is tireless and trails its prey until the prey is run to the ground. The diet includes small mammals, the young of large mammals, birds, reptiles, arthropods, carrion, and vegetation. Honey and bees are important foods at certain times of the year.

A remarkable association has developed, at least in tropical Africa, between a bird—the honey guide *(Indicator indicator)*—and *Mellivora*. The association is mutually beneficial in the common exploitation of the nests of wild bees. In the presence of any mammal, even people, the honey guide has the unusual habit of uttering a series of characteristic calls. If a honey badger hears these calls, it follows the bird, which invariably leads it to the vicinity of a beehive. The badger

breaks open the nest, and the bird obtains enough of the honey and insects to pay for its work.

Mellivora may travel alone or in pairs. It is generally silent but may utter a very harsh, grating growl when annoyed. The sparse data on reproduction were summarized by Kingdon (1977), Rosevear (1974), and Smithers (1971). Mating has been noted in South Africa in February, June, and December. A lactating female was found in Botswana in November, and newborn were recorded in Zambia in December. Seasonal breeding has been reported in Turkmen S.S.R., with mating occurring in autumn and births in the spring. The gestation period is thought to last about 6 months. The number of young in a litter is commonly two, and the range is one to four. The young are born in a grass-lined chamber and evidently remain close to the burrow for a long time. *Mellivora* appears to thrive in captivity: one specimen lived for 26 years and 5 months (Jones 1982).

If captured before half-grown, the honey badger can become a satisfactory pet, as it is docile, affectionate, and active. It is, however, incredibly strong and energetic and can wreck cages and damage property in its explorations. Wild individuals sometimes prey on poultry, tearing through wire netting to effect entry (Smithers 1971). Destruction of commercial beehives in East Africa seems to be a significant problem, and *Mellivora* has thus been intensively poisoned, trapped, and hunted (Kingdon 1977). The overall range of the genus has declined, probably through human persecution. The ratel is classified as vulnerable in South Africa (Smithers 1986) and as rare by the Soviet Union.

CARNIVORA; MUSTELIDAE; **Genus MELES**
Storr, 1780

Old World Badger

The single species, *Meles meles*, originally occurred throughout Europe, including the British Isles and several Mediterranean islands, and in Asia as far east as Japan and as far south as Palestine, Iran, Tibet, and southern China (Corbet 1978; Long and Killingley 1983). The authority for the generic name *Meles* sometimes is given as Brisson, 1762.

Head and body length is 560–900 mm, tail length is 115–202 mm, and weight is usually 10–16 kg; Novikov (1962), however, wrote that old males attain weights of 30–34 kg in

Old World badger *(Meles meles)*, photo from Paignton Zoo, Great Britain.

the late fall. The upper parts are grayish, and the underparts and limbs are black. On each side of the face is a dark stripe that extends from the tip of the snout to the ear and encloses the eye; white stripes border the dark stripe. Like other badgers, *Meles* has a stocky body, short limbs, and a short tail. From *Arctonyx*, it is distinguished by its black throat and shorter tail, which is the same color as the back. Females have three pairs of mammae (Grzimek 1975).

The Old World badger is found mainly in forests and densely vegetated areas. It usually lives in a large communal burrow system that covers about 0.25 hectare. There are numerous entrances, passages, and chambers. Nests may be located 10 meters from an entrance and 2–3 meters below the surface of the ground and have a diameter of 1.5 meters. A burrow system may be used for decades or centuries, by one generation of badgers after another; it continually increases in complexity and may eventually cover several hectares (Grzimek 1975; Novikov 1962). The animals occupying a system may utilize one nest for several months and then suddenly move to another part of the burrow. The living quarters are kept quite clean. Bedding material, in the form of dry grass, brackens, moss, or leaves, is dragged backward into the den. Occasionally this bedding is brought up and strewn around the entrance to air for an hour or so in the early morning. Around the burrows are dung pits, sunning grounds, and areas for play (badgers play all sorts of games, including leapfrog). Well-defined foraging paths may extend outward for 2–3 km (Novikov 1962). Kruuk (1978a) distinguished two kinds of burrows: "main setts," with an average of 10.5 entrances; and small "outliers," usually with only a single entrance.

Meles usually does not emerge from its burrow until after sundown. During periods of very cold weather and high snow, it may spend days or weeks in the den. Such intervals of winter sleep extend to several months in northern Europe and up to seven months in Siberia. There is no substantial drop in body temperature, and the badger can be aroused easily, but the animal lives off of fat reserves accumulated in the summer and fall. The omnivorous diet includes almost any available food—small mammals, birds, reptiles, frogs, mollusks, insects, larvae of bees and wasps, carrion, nuts, acorns, berries, fruits, seeds, tubers, rhizomes, and mushrooms. Earthworms have been found to be of major dietary importance in some areas (Kruuk 1978b; Skoog 1970).

In a study in England, Kruuk (1978a) found badgers to be organized into clans of up to 12 individuals. The minimum distance between the main burrows of clans was 300 meters. Most clans used ranges of 50–150 ha., and there was little overlap. The ranges were marked by defecation and secretions from subcaudal glands. Several fights were observed along territorial boundaries. Most clans had more females than males, but one, which used a range of only 21 ha., consisted solely of males (it was suggested that in other parts of the range of *Meles* the bachelors may be nomadic). Individuals moved around alone within the clan range. Adult males always slept in the main setts, but females sometimes slept in the outliers, especially in the summer.

Mating occurs from late winter to midsummer. Development of the fertilized eggs stops at the blastocyst stage, and implantation in the uterus is delayed for about 10 months. The time of implantation seems to be controlled by conditions of light and temperature. Following implantation, embryonic development proceeds for 6–8 weeks, and births occur mainly from February to March. The total period of pregnancy may thus extend for about 1 year. Females may experience a postpartum estrus. The number of young in a litter is two to six, usually three or four. The young weigh 75 grams each at birth, open their eyes after about 1 month, nurse for 2.5 months, and usually separate from the female in the autumn. Both males and females apparently attain sexual maturity at the age of 1 year (Ahnlund 1980; Canivenc and Bonnin 1979; Grzimek 1975; Novikov 1962). One specimen lived in captivity for 16 years and 2 months (Jones 1982).

Meles sometimes damages ripening grapes, corn, and oats. Its hair is used to make various kinds of brushes, and its skin, at least formerly, has been used in northern China to make rugs. It has declined in much of its European range but apparently has recovered to some extent in Great Britain. It now occupies most of the island and has been estimated to number about 100,000–200,000 (Clements, Neal, and Yalden 1988).

Hog badger *(Arctonyx collaris)*, photo by Ernest P. Walker.

Hog Badger

The single species, *A. collaris,* occurs from Sikkim and north-eastern China to peninsular Thailand and on the island of Sumatra (Corbet 1978; Lekagul and McNeely 1977).

Head and body length is 550–700 mm, tail length is 120–70 mm, and weight is usually 7–14 kg. The back is yellowish, grayish, or blackish, and there is a pattern of white and black stripes on the head. The dark stripes run through the eyes and are bordered by white stripes that merge with the nape and with the white of the throat. The ears and tail also are white, and the feet and belly are black. The body form is stocky. This badger is distinguished from *Meles* by its white, rather than black, throat and by its long and mostly white tail, as distinct from the short tail, colored the same as the back, in *Meles.* Another external difference is that in *Arctonyx* the claws are pale in color, while in *Meles* they are dark. The common name of *Arctonyx* refers to the long, truncate, mobile, and naked snout, which is often compared to that of a pig *(Sus).*

The hog badger is usually found in forested areas at elevations of up to 3,500 meters (Lekagul and McNeely 1977). It is nocturnal, spending the day in natural shelters, such as rock crevices, or in self-excavated burrows. The snout is thought to be used in rooting for the various plants and small animals that compose the diet.

The general habits probably resemble those of *Meles.* As with *Meles* and *Mellivora,* the color pattern has been interpreted as a means of warning potential enemies that the animal so marked is best left alone. Like these other genera, *Arctonyx* is a savage and formidable antagonist. It has thick and loose skin, powerful jaws, fairly strong teeth, well-developed claws, and a potent anal gland secretion.

A female from northern China had four newborn young in April. Parker (1979) reported that a captive pair from China was received at the Toronto Zoo in July 1976. The female gave birth to two cubs in February 1977; one cub survived and reached approximate adult size at 7.5 months. In February 1978 the same female gave birth to four young. Matings had been observed from April to September 1977, but delayed implantation was suspected, and true gestation was postulated at less than 6 weeks. According to Jones (1982), a captive lived for 13 years and 11 months.

Stink Badgers

Long (1978) recognized two subgenera and two species:

subgenus *Mydaus* F. Cuvier, 1821

M. javanensis, Sumatra, Java, Borneo, Natuna Islands;

subgenus *Suillotaxus* Lawrence, 1939

M. marchei, Palawan and Calamian Islands (Philippines).

Suillotaxus often has been given generic rank.

In *M. javanensis* the head and body length is 375–510 mm, tail length is usually 50–75 mm, and weight is usually 1.4–3.6 kg. The coloration is blackish, except for a white crown and a complete or partial narrow white stripe down the back onto the tail. In *M. marchei* the head and body length is 320–460 mm, tail length is 15–45 mm, and weight is about 2.5 kg. The upper parts are brown to black, with a scattering of white or silvery hairs on the back and sometimes on the head, and the underparts are brown.

Both species have a pointed face, a somewhat elongate and mobile snout, short and stout limbs, and well-developed anal scent glands. Compared with *M. javanensis, M. marchei* has smaller ears, a shorter tail, and larger teeth.

M. javanensis reportedly is a montane species, often being found at elevations over 2,100 meters (Long and Killingley 1983). It is nocturnal, residing by day in holes in the ground, dug either by itself or by the porcupines with which it some-

Stink badger *(Mydaus javanensis)*, photo from Michael Riffel.

times lives. The burrows usually are not more than 60 cm deep. On Borneo this species reportedly inhabits caves. Captives have consumed worms, insects, and the entrails of chickens.

M. marchei inhabits grassland-thicket and cultivated areas (Long and Killingley 1983). Grimwood (1976) wrote that it is active both by day and by night. It is common, leaving its tracks and scent along roads and paths. It moves with a rather ponderous, fussy walk. One individual shammed death when first touched, and allowed itself to be carried, but finally squirted a jet of yellowish fluid from its anal glands into the lens of a camera about 1 meter away.

M. javanensis may growl and attempt to bite when handled. If molested or threatened, it raises the tail and ejects a pale greenish fluid. This vile-smelling secretion is reported by natives sometimes to asphyxiate dogs, or even blind them if they are struck in the eye. The old Javanese sultans used this fluid, in suitable dilution, in the manufacture of perfumes.

Some natives eat the flesh of *Mydaus,* removing the scent glands immediately after the animals are killed. Others mix shavings of the skin with water and drink the mixture as a cure for fever or rheumatism.

CARNIVORA; MUSTELIDAE; Genus TAXIDEA
Waterhouse, 1839

American Badger

The single species, *T. taxus,* is found from northern Alberta and southern British Columbia to Ohio, central Mexico, and Baja California (Hall 1981).

Head and body length is 420–720 mm, tail length is 100–155 mm, and weight is 4–12 kg. The upper parts are grayish to reddish, and a dorsal white stripe extends rearward from the nose. In the north this stripe usually reaches only to the neck or shoulders, but in the south the stripe usually extends to the rump (Long 1972c). Black patches are present on the face and cheeks; the chin, throat, and midventral region are whitish; the underparts are buffy; and the feet are dark

brown to black. The hairs are longest on the sides. *Taxidea* can be recognized by its flattened and stocky form, large foreclaws, distinctive black and white head pattern, long fur, and short, bushy tail. There are anal scent glands. Females have eight mammae (Jackson 1961).

The American badger is usually found in relatively dry, open country. It is a remarkable burrower and can quickly dig itself out of sight. The usual signs of its presence are the large holes that it digs when after rodents. For shelter, it either excavates a burrow or modifies one initially made by another animal. The burrow can be as long as 10 meters and can extend as much as 3 meters below the surface. A bulky nest of grass is located in an enlarged chamber, and the entrances are marked by mounds of earth (Banfield 1974).

Taxidea may be active at any hour but is mainly nocturnal. Its movements are restricted, especially in winter, and it shows a strong attachment to a home area. In the summer, however, the young animals disperse over a considerable distance; one traveled 110 km. The badger is active all year, but it may sleep in its den for several days or weeks during severe winter weather. One female was found to have emerged only once during a 72-day period (Messick and Hornocker 1981).

Most food is obtained by excavating the burrows of fossorial rodents, such as ground squirrels. Also eaten are other small mammals, birds, reptiles, and arthropods (Banfield 1974; Messick and Hornocker 1981). Food is sometimes buried and used later. If a sizable meal, such as a rabbit, is obtained, the badger may dig a hole, carry in the prey, and remain below ground with it for several days. There are reports that a badger sometimes forms a "hunting partnership" with a coyote (see account of *Canis latrans*).

On the basis of a radiotracking study in southwestern Idaho, Messick and Hornocker (1981) estimated a population density of up to 5/sq km and found average home ranges of 2.4 sq km for males and 1.6 sq km for females. The ranges overlapped, but individuals were solitary, except during the reproductive season. A female radiotracked in Minnesota used an area of 752 ha. during the summer. She had 50 dens within this area and was never found in the same den on two consecutive days. In the fall she shifted to an adjacent area of 52 ha. and often reused dens; in the winter she used a single

American badger *(Taxidea taxus)*, photo by E. P. Haddon from U.S. Fish and Wildlife Service.

den and traveled only infrequently within an area of 2 ha. (Sargeant and Warner 1972).

Mating occurs in summer and early autumn, but implantation of the fertilized eggs in the uterus is delayed until December–February, and births take place in March and early April (Long 1973b). The total period of pregnancy is thus about 7 months, but actual embryonic development lasts only about 6 weeks (Ewer 1973). The litter of one to five, usually two, young is born and raised in a nest of dry grass within a burrow. The young are weaned at about 6 weeks and disperse shortly thereafter. The study by Messick and Hornocker (1981) indicated that 30 percent of the young females mate in the first breeding season following birth, when they are about 4 months old, but that males wait until the following year. The oldest wild badger caught in this study had attained an age of 14 years. A captive badger lived for 26 years (Jones 1982).

Taxidea is generally beneficial to human interests, as it destroys many rodents and its burrows provide shelter for many kinds of wildlife, including cottontail rabbits. Its burrows and holes, however, may constitute a hazard to cattle, horses, and riders, and thus ranchers have often killed badgers. Many badgers also have been killed by poison put out for coyotes (Hall 1955). Although *Taxidea* has declined in numbers in some areas, it has extended its range eastward in the twentieth century. It now occupies most of Ohio and is invading southeastern Ontario (Hall 1981; Long 1978). There also are recent records from New York, New England, and southern Yukon, but it is not known whether these represent natural occurrences (Messick 1987).

For many years badger fur was used to make shaving brushes, but it has now been largely replaced by synthetic materials. The fur is still used for trimming garments. In the 1976/77 trapping season 49,807 pelts were reported taken in the United States and Canada, selling at an average price of about $38 (Deems and Pursley 1978). By 1983/84, the take had fallen to only about 20,000 and the average price to $10.00 (Messick 1987).

Ferret badger *(Melogale moschata)*, photo by Gwilym S. Jones.

CARNIVORA; MUSTELIDAE; Genus MELOGALE
I. Geoffroy St.-Hilaire, 1831

Ferret Badgers

There are three species (Long 1978; Zheng and Xu 1983):

M. moschata, Assam to central China and northern Indochina, Taiwan, Hainan;
M. personata, Nepal to Indochina, Java;
M. everetti, Borneo.

Long (1978) noted that *M. moschata sorella*, of southeastern China, and *M. personata orientalis*, of Java, might possibly be full species.

Head and body length is 330–430 mm, tail length is 145–230 mm, and weight is 1–3 kg. The general coloration of the upper parts is gray brown to brown black; the underparts are somewhat paler. A white or reddish dorsal stripe is usually present. *Melogale* is distinguished by the striking coloration of the head, which combines black with patches of white or yellow. The tail is bushy, the limbs are short, and the feet are broad and have long, strong claws for digging (Lekagul and McNeely 1977).

Ferret badgers are found in wooded country and grassland. They reside in burrows and natural shelters during the day and are active at dusk and during the night. They climb on occasion. *M. moschata* in Taiwan is reported to be a good climber and often to sleep on the branches of trees. Ferret badgers are savage and fearless when provoked or pressed and have an offensive odor. The conspicuous markings on the head have been interpreted as being a warning signal. The diet is omnivorous and is known to include small vertebrates, insects, earthworms, and fruit. A ferret badger is sometimes welcome to enter a native hut, because of its destruction of insect pests.

The young, usually one to three per litter, are born in a burrow in May and June. They apparently are dependent on the milk of the mother for some time, as two nearly full-grown suckling animals and their mother were once found in a burrow. According to Jones (1982), a specimen of *M. moschata* was still living after 10 years and 6 months in captivity.

CARNIVORA; MUSTELIDAE; Genus SPILOGALE
Gray, 1865

Spotted Skunks

There are two species (Hall 1981):

S. pygmaea, Pacific coast of Mexico from southern Sinaloa to Oaxaca;
S. putorius, southern British Columbia and Pennsylvania to Costa Rica and Florida.

Head and body length is 115–345 mm, tail length is 70–220 mm, and weight is usually 200–1,000 grams. Of the three genera of skunks, *Spilogale* has the finest fur. The hairs are longest on the tail and shortest on the face. The basic color pattern consists of six white stripes extending along the back

Spotted skunk *(Spilogale putorius)*, photo by Ernest P. Walker.

and sides, and these are broken into smaller stripes and spots on the rump. There is a triangular patch in the middle of the forehead, and the tail is usually tipped with white. The variations are infinite, as no two individuals have ever been found with exactly the same pattern. This genus may be distinguished from the other two genera by its small size, forehead patch, and pattern of stripes and spots, the white never being massed. There is a pair of scent glands under the base of the tail, from which a jet of strong-smelling fluid can be emitted through the anus. Females have 10 mammae (Lowery 1974).

Spotted skunks occur in a variety of brushy, rocky, and wooded habitats but avoid dense forests and wetlands. They generally remain under cover more than do striped skunks (*Mephitis*). They usually den underground but can climb well and sometimes shelter in trees. The dens are lined with dry vegetation. Spotted skunks are largely nocturnal and are active all year. The omnivorous diet consists mainly of vegetation and insects in the summer and of rodents and other small animals in the winter.

The white-plumed tail is used to warn other animals that spotted skunks should not be molested. If, however, the sudden erection of the tail is not sufficient deterrence, *Spilogale* may stand on its forefeet and sometimes even advance toward its adversary. Finally, the fluid from the anal glands is discharged at the enemy, usually after the skunk has returned its forefeet to the ground and assumed a horseshoe position (Lowery 1974).

According to Banfield (1974), population densities reach

5/sq km in good agricultural land, and winter home range is approximately 64 ha. In spring the males wander over an area of about 5–10 sq km, but females have smaller ranges. Spotted skunks are very playful with one another. As many as eight individuals sometimes share a den.

The reproductive pattern is not the same in all parts of the range (Foresman and Mead 1973; Mead 1968a, 1968b). Populations in South Dakota and Florida apparently mate mainly in March and April. Implantation of the fertilized eggs in the uterus occurs only 14–16 days later, and births take place in late May and June. Pregnancy is estimated to last 50–65 days. Populations farther to the west mate in September and October, but implantation is delayed until the following March or April, and births occur from April to June. The total period of pregnancy thus lasts 230–50 days, but actual embryonic development takes only 28–31 days. The number of young per litter is two to nine, usually three to six. The young weigh about 22.5 grams each at birth, have adult coloration after 21 days, open their eyes at 32 days, can spray musk at 46 days, are weaned at about 54 days, and attain adult size at about 15 weeks. A captive specimen lived for 9 years and 10 months (Egoscue, Bittmenn, and Petrovich 1970).

Spotted skunks have been reported to carry rabies and occasionally to take poultry and eggs but are generally beneficial to people through their destruction of rodents and insects. The pelts are very attractive and durable, but they generally sold for well under $1.00 each until about 1970 (Jackson 1961; Lowery 1974). In the 1976/77 trapping season the

Striped skunk *(Mephitis mephitis):* A. Facing danger that does not appear to be imminent; B. Aimed toward the enemy ready to spray its scent. Photos by Ernest P. Walker.

reported harvest in the United States was 41,952 skins, with an average selling price of $4.00 (Deems and Pursley 1978).

CARNIVORA; MUSTELIDAE; **Genus MEPHITIS**
E. Geoffroy St.-Hilaire and G. Cuvier, 1795

Striped Skunk and Hooded Skunk

There are two species (Hall 1981; Janzen and Hallwachs 1982):

M. mephitis (striped skunk), southern Canada to northern Mexico;
M. macroura (hooded skunk), Arizona and southwestern Texas to Costa Rica.

Head and body length is 280–380 mm, tail length is 185–435 mm, and weight is 700–2,500 grams. Both species have black and white color patterns, but with considerable variation. *M. mephitis* usually has white on top of the head and on the nape extended posteriorly and separated into stripes. In some individuals of this species the top and sides of the tail are white, while in others the white is limited to a small spot on the forehead. The white areas are composed entirely of white hairs, with no black hairs intermixed. *M. macroura* has a white-backed color phase and a black-headed color phase. In the former, there are some black hairs mixed with the white hairs of the back; in the latter, the two white stripes are widely separated and are situated on the sides of the animal, instead of being narrowly separated and situated on the back, as in *M. mephitis*. Female *Mephitis* have 10–14 mammae (Jackson 1961; Leopold 1959).

According to Lowery (1974), the well-known scent of *Mephitis* is expelled from two tiny nipples located just inside the anus, which mark the outlets of the two ducts leading from glands lying adjacent to the anus. This musk is discharged either as an atomized spray or as a short stream of rain-sized drops. The skunk usually employs this weapon only after much provocation. When confronted by an antagonist, it arches its back, elevates the tail, erects the hairs thereon, and sometimes stamps its feet on the ground. Finally, it turns the body in a U-shaped position, with the head and tail facing the intruder. The musk usually travels 2–3 meters, but the smell can be detected up to 2.5 km downwind. Lowery observed that one squirt is sufficient to send the most ferocious dog yelping in agony from burning eyes and nostrils and retching with nausea.

These skunks are found in a variety of habitats, including woods, grasslands, and deserts. They generally are active at dusk and through the night and spend the day in a burrow, under a building, or in any dry, sheltered spot. In Minnesota, Houseknecht and Tester (1978) found a general shift from underground, upland dens in winter to aboveground, lowland dens in summer. Although skunks tended to remain at a single den for a long time in winter, females with young able to travel changed dens every one or two days. Bjorge, Gunson, and Samuel (1981) reported that females moved a minimum daily distance of 1.5 km between dens. In summer, juveniles were found to disperse up to 22 km.

In northern parts of its range, *M. mephitis* stays in one den and sleeps through much of the winter. Males tend to sleep for shorter periods than females and to become active more readily during intervals of mild weather (Banfield 1974). The degree of lethargy achieved during the winter is not well understood. It does not appear to be deep torpor, but some skunks are known to have remained underground for over 100 consecutive days (Sunquist 1974). In Alberta, the overall period of female hibernation is 120–50 days, while in Illinois it is 62–87 days (Gunson and Bjorge 1979). A striped skunk may become very fat in the autumn before hibernation. The diet is omnivorous and includes rodents, other small vertebrates, insects, fruit, grains, and green vegetation.

Density estimates for striped skunk populations have ranged from 0.7 to 18.5 individuals per sq km, but most are 1.8–4.8 per sq km (Wade-Smith and Verts 1982). The home ranges of 6 females radiotracked for 45–105 days each in Alberta (Bjorge, Gunson, and Samuel 1981) averaged 208 ha. and varied from 110 to 370 ha. Males wandered over larger areas, most notably in the north. *Mephitis* is generally solitary, but there is a tendency for individuals to den together, especially in the north, as a means of optimizing winter survival and reproductive success. In Alberta, Gunson and Bjorge (1979) found that only males, both adults and juveniles, denned alone during the winter. Communal winter dens contained an average of 6.7 (2–19) individuals. Usually there was only a single adult male per den and an average of 5.8 females. A male apparently wanders in search of a group of females during the fall and then keeps other males away. *Mephitis* is usually silent but makes several sounds, such as low churrings, shrill screeches, and birdlike twitters (Lowery 1974).

Mating takes place from mid-February to mid-April, and births occur in May and early June. The period of pregnancy is 59–77 days, and delayed implantation may be involved. Females are usually monestrous but sometimes have a second estrus and parturition subsequent to the normal period if their first pregnancy is not successful. Litters contain 1–10 young, usually about 4–5. The young weigh about 30 grams at birth, open their eyes after 3 weeks, are weaned at 8–10 weeks, and separate from the mother by the fall. Females may bear their first litter at 1 year (Lowery 1974; Wade-Smith and Richmond 1975, 1978). The average longevity in captivity is about 6 years, but one individual was still living after 12 years and 11 months (Jones 1982).

Striped skunks are generally beneficial to human interests because of their destruction of rodents and insects. *Mephitis*, however, sometimes attacks poultry and is reportedly the principal carrier of rabies among North American wildlife (Wade-Smith and Richmond 1975). The fur is durable and of good texture, but demand and value have varied widely (Lowery 1974; Jackson 1961). During the 1976/77 season 175,884 skins were reported taken in the United States and Canada, selling at an average price of about $2.25 (Deems and Pursley 1978). In 1983/84 the take was about the same, but the price fell to under $1.00 (Rosatte 1987).

CARNIVORA; MUSTELIDAE; **Genus CONEPATUS**
Gray, 1837

Hog-nosed Skunks

Five species are now recognized (Cabrera 1957; Ewer 1973; Hall 1981; Kipp 1965; Manning, Jones, and Hollander 1986; Pine, Miller, and Schamberger 1979):

C. mesoleucus, southern Colorado and eastern Texas to Nicaragua;
C. leuconotus, southern Texas, eastern Mexico;
C. semistriatus, southern Mexico to northern Peru and eastern Brazil;
C. chinga, central and southern Peru, Bolivia, Chile, northwestern Argentina;

Hog-nosed skunk *(Conepatus mesoleucus)*, photo by Lloyd G. Ingles.

C. *humboldti*, Paraguay, southeastern Brazil, Uruguay, Argentina.

Statements by the various authorities cited above suggest that some, perhaps all, of the listed species are conspecific.

Head and body length is 300–490 mm, tail length is 160–410 mm, and weight is usually 2.3–4.5 kg. *Conepatus* has the coarsest fur of all skunks. There are two main color patterns, with variations. In one, the top of the head, the back, and the tail are white, and the remainder of the animal is black. This coloration occurs most commonly in areas where the ranges of *Conepatus* and *Mephitis* overlap. In the other pattern, the pelage is black except for two white stripes, beginning at the nape and extending on the hips, and a mostly white tail. This coloration resembles that of *Mephitis mephitis* and seems to be most common in areas where *Conepatus* is the only kind of skunk present. In all cases, hognosed skunks lack the thin white stripe down the center of the face that is present in *Mephitis*. *Conepatus* may be distinguished from the other two genera of skunks by its nose, which is bare, broad, and projecting. Females have three pairs of mammae (Leopold 1959).

Hog-nosed skunks are found in both open and wooded areas but avoid dense forests. They occur at all elevations up to at least 4,100 meters (Grimwood 1969). Dens are located in rocky places, hollow logs, or burrows made by other animals. Like other skunks, *Conepatus* is mainly nocturnal, is generally slow-moving, does not ordinarily climb, and defends itself by expelling musk from anal scent glands.

The diet may consist principally of insects and other invertebrates, though fruit and small vertebrates, including snakes, probably are also eaten. Hog-nosed skunks may turn over the soil in a considerable area with their bare snout and their claws when in search of food. Like *Mephitis*, they also pounce on insects. At least in the Andes, hog-nosed skunks are resistant to the venom of pit vipers. There is some evidence that the spotted skunks *(Spilogale)* also are resistant to rattlesnake venom. Since the musk of skunks produces an alarm reaction in rattlers (the same reaction that they exhibit in the presence of king snakes, which prey on them), it may be that skunks feed on rattlesnakes quite extensively.

In southern Chile, Fuller et al. (1987) found *C. humboldti* to be solitary but to occupy overlapping home ranges of 7–16 ha. According to Davis (1966b), *C. mesoleucus* is not as social as *Mephitis*, and usually only one individual lives in a den. The breeding season in Texas begins in February, most mature females are pregnant by March, and births occur in late April or early May. Gestation lasts approximately 2 months. Of six pregnant females on record, three contained three embryos each, and three had two each. By August most of the young are weaned and foraging for themselves. Available evidence indicates that in Mexico the young are also born in the spring (Hall and Dalquest 1963; Leopold 1959). The gestation period of one South American species is 42 days, and litter size is usually two to five young. A captive *C. chinga* lived for 6 years and 7 months (Jones 1982).

The pelt of *Conepatus* is inferior in quality to that of *Mephitis*, but large numbers have been marketed from Texas (Davis 1966b). Perhaps because of this factor, the subspecies *C. mesoleucus telmalestes*, isolated in the Big Thicket area of southeastern Texas, has declined to near the point of extinction (Schmidly 1983). It now is classified as indeterminate by the IUCN. *C. chinga rex* of northern Chile also is hunted for its pelt and now seems to be rare (Miller et al. 1983). Some natives use the skins for capes or blankets, and others consider the meat to have curative properties. *C. humboldti* is on appendix 2 of the CITES. Records of CITES indicate that during the 1970s about 155,000 skins of *Conepatus* were exported annually, each with a value of about U.S. $8.00; the trade may subsequently have declined (Broad, Luxmoore, and Jenkins 1988).

Canadian river otters *(Lutra canadensis)*, photos by Ernest P. Walker.

CARNIVORA; MUSTELIDAE; Genus LUTRA
Brünnich, 1771

River Otters

There are four subgenera and eight species (Chasen 1940; Coetzee, *in* Meester and Setzer 1977; Corbet 1978; Ellerman and Morrison-Scott 1966; Hall 1981; Lekagul and McNeely 1977; Rosevear 1974; Van Zyll de Jong 1972, 1987):

subgenus *Lutra* Brünnich, 1771

L. lutra, western Europe to northeastern Siberia and Korea, Asia Minor and certain other parts of southwestern Asia, Himalayan region, extreme southern India, China, Burma, Thailand, Indochina, northwestern Africa, British Isles, Sri Lanka, Sakhalin, Japan, Taiwan, Hainan, Sumatra, Java;

L. sumatrana, Indochina, Thailand, Malay Peninsula, Sumatra, Bangka, Java, Borneo;

subgenus *Hydrictis* Pocock, 1921

L. maculicollis, Sierra Leone to Ethiopia, and south to South Africa;

subgenus *Lutrogale* Gray, 1865

L. perspicillata, southern Iraq, Pakistan to Indochina and Malay Peninsula, Sumatra;

subgenus *Lontra* Gray, 1843

L. canadensis, Alaska, Canada, conterminous United States;

L. longicaudis, northwestern Mexico to Uruguay;

L. provocax, Chile, southern Argentina;

L. felina, Pacific coast from northern Peru to Tierra del Fuego.

Van Zyll de Jong (1972, 1987) considered *Lontra* and *Lutrogale* to be full genera, but this procedure was not followed by Corbet (1978), Corbet and Hill (1986), Hall (1981), or Jones et al. (1986). The basis on which various authors rejected generic status for *Lontra* was questioned by Kellnhauser (1983). Van Zyll de Jong also suggested a close relationship between *L. sumatrana*, which was referred to the subgenus *Lutra* by Ellerman and Morrison-Scott (1966), and *L. maculicollis* of the subgenus *Hydrictis*. Rosevear (1974), however, stated that there is a strong case for regarding *Hydrictis* as a full genus.

Head and body length is 460–820 mm, tail length is 300–

European river otter *(Lutra lutra)*, photo by Annelise Jensen.

500 mm, and weight is 3–14 kg. Males average larger than females (Van Zyll de Jong 1972). The upper parts are brownish, and the underparts are paler; the lower jaw and throat may be whitish. The fur is short and dense. The head is flattened and rounded; the neck is short and about as wide as the skull; the trunk is cylindrical; the tail is thick at the base, muscular, flexible, and tapering; the legs are short; and the digits are webbed. The small ears and the nostrils can be closed when the animal is in the water. Female *L. canadensis* have four mammae (Jackson 1961).

These aquatic mammals inhabit all types of inland waterways, as well as estuaries and marine coves. In southern Chile, *L. felina* is found largely or exclusively along the exposed rocky seashore, though farther north it may inhabit estuaries and fresh water (Estes 1986). Otters are excellent swimmers and divers and usually are found no more than a few hundred meters from water. They may shelter temporarily in shallow burrows or in piles of rocks or driftwood, but they also have at least one permanent burrow beside the water (Jackson 1961; Stroganov 1969). The main entrance may open underwater and then slope upward into the bank to a nest chamber that is above the high-water level. Erlinge (1967) found *L. lutra* to utilize the following types of facilities in southern Sweden: "dens," generally with several passages and a chamber lined with dry leaves and grass; "rolling places," bare spots near water, where the otters roll and groom themselves; "slides," either on slopes or in level places but most common on winter snow; "feeding places," including holes kept open through winter ice; "runways," well-defined paths on land that connect waterways and other facilities; and "sprainting spots and sign heaps," prominent points of land where the animals mark by scratching and elimination.

Otters swim by movements of the hind legs and tail and can remain underwater for six to eight minutes (Grzimek 1975). When traveling on ground, snow, or ice, they may use a combination of running and sliding. Although normally closely associated with water, river otters sometimes move

many kilometers overland to reach different river basins and to find ice-free water in winter (Stroganov 1969). When running on land, they can attain speeds of up to 29 km/hr (Banfield 1974). They may be either diurnal or nocturnal but are generally more active at night.

With the possible exception of the Old World badger *(Meles)*, river otters are the most playful of the Mustelidae. Some species engage in the year-round activity of sliding down mud and snow banks, and individuals of all ages participate. Sometimes they tunnel under snow to emerge some distance beyond.

The diet consists largely of fish, frogs, crayfish, crabs, and other aquatic invertebrates. Birds and land mammals, such as rodents and rabbits, are also taken. Studies indicate that the fish consumed are mainly nongame species. River otters capture their prey with the mouth, not the hands (Rowe-Rowe 1977).

Investigations by Erlinge (1967, 1968) in southern Sweden indicate that *L. lutra* occurs at population densities of 0.7–1.0 per sq km of water area, or 1 for every 2–3 km of lakeshore or 5 km of stream. The straight-line length of a home range, including land area, was found to average about 15 km for adult males and 7 km for females with young. The ranges of males constitute territories, which may overlap the ranges of 1 or more adult females and from which other males are excluded. Females also defend their ranges against individuals of the same sex. Territories are marked with scent, and fights occasionally take place. Apparently, males form a dominance hierarchy, with the highest-ranking animals occupying the most favorable ranges. Erlinge noted that the males generally are solitary and ignore the females and young. Studies in Scotland (Mason and Macdonald 1986) suggest a somewhat less rigid social system, with females occupying overlapping home ranges, though again the dominant males have relatively exclusive territories

In Idaho, Melquist and Hornocker (1983) found *L. canadensis* to have an overall population density of one per 3.9 km of waterway. Seasonal home range length ranged from 8 to 78 km, with males generally having larger ranges than females; however, there was extensive overlap between the ranges of both the same and opposite sexes. There was some mutual avoidance and defense of personal space but no strong territorial behavior. The basic social group consisted of an adult female and her juvenile offspring. Such families broke up before the female again gave birth, though yearlings occasionally associated with the group. Fully adult males were not observed to accompany family groups.

Observations in other areas, however, suggest that while the male is excluded from the vicinity of the female when the latter has small young, he joins the family when the cubs are about six months old (Banfield 1974; Jackson 1961). On the Malay Peninsula, L. perspicillata typically occurs in groups consisting of a mated adult pair and up to four young, which have a territory of 7–12 km of river (Lekagul and McNeely 1977). On Lake Victoria, L. maculicollis may undergo a regular cycle of aggregation and dispersal, with males and females each forming their own groups. The males' groups grow larger after the mating season, when males are not tolerated by the females with young. These groups may contain 8–20 individuals from January to May but then become smaller from June to August, when the older males leave to pair with the females (Kingdon 1977).

Within the wide geographic range of Lutra, there is considerable variation in reproductive pattern (Banfield 1974; Duplaix-Hall 1975; Ewer 1973; Kingdon 1977; Liers 1966; Mason and Macdonald 1986; Stroganov 1969). Female L. lutra are polyestrous, with the cycle lasting about 4–6 weeks and estrus about 2 weeks. In some areas (e.g., most of England), mating and birth may occur at any time of the year. In areas with a more severe climate (such as Sweden and Siberia), mating is in late winter or early spring, and births take place in April or May. The gestation period of this species is 60–63 days. The length of pregnancy is about the same in L. perspicillata, L. maculicollis, and L. felina. In L. canadensis, of North America, however, there is delayed implantation of the fertilized eggs in the uterus. Mating occurs in the winter or spring, and births take place the following year. The total period of pregnancy has been reported to vary from 245 to 380 days, though actual embryonic development is about 2 months, the same as in other kinds of river otters. Litter size in the genus is one to five young, usually two or three. The young weigh about 130 grams at birth, open their eyes after 1 month, emerge from the den and begin to swim at 2 months, nurse for 3–4 months, separate from the mother at about 1 year, and attain sexual maturity in the second or third year of life. A female L. canadensis lived in captivity for 23 years.

As a group, otters have suffered severely through habitat destruction, water pollution, misuse of pesticides, excessive fur trapping, and persecution as supposed predators of game and commercial fish. L. canadensis has disappeared or become rare throughout the conterminous United States, except in the northwest, the upper Great Lakes region, New York, New England, and the states along the Atlantic and Gulf coasts. The southwestern subspecies L. c. sonora has nearly disappeared, though there have been several recent reports in Arizona. Otters of another subspecies were released in Arizona in 1981. Since 1976 there also have been efforts to reintroduce otters in Colorado, Iowa, Kansas, Kentucky, Missouri, Nebraska, Oklahoma, Pennsylvania, and West Virginia (Melquist and Dronkert 1987).

The Old World L. lutra has declined drastically in such diverse places as Great Britain (Chanin and Jefferies 1978), Germany (Roben 1974), southeastern Siberia (Kucherenko 1976), and Japan (Mikuriya 1976). There have been extensive recent surveys of L. lutra in the Mediterranean region and southern Europe, with the species having been found to be still relatively common in parts of Spain, Portugal, Greece,

and North Africa but rare or absent in Italy, Austria, Switzerland, and most of France (Mason and Macdonald 1986).

The IUCN classifies the following as vulnerable: the subspecies L. lutra lutra of Europe and most of Asia; the subspecies L. longicaudis longicaudis of South America; and the species L. provocax and L. felina. The USDI lists L. longicaudis, L. provocax, and L. felina as endangered. L. lutra, L. longicaudis, L. provocax, and L. felina are on appendix 1 of the CITES, and the remaining species of Lutra are on appendix 2. There may now be fewer than 1,000 individuals of L. felina (Estes 1986); it has declined in Chile because of excessive hunting for its fur and in Peru because of persecution for alleged damage to prawn fisheries (Thornback and Jenkins 1982). L. maculicollis is now considered to be endangered in South Africa (Stuart 1985).

The beautiful and durable fur of river otters is used for coat collars and trimming. During the 1976/77 trapping season 32,846 pelts of L. canadensis were reported taken in the United States, and the average selling price was $53.00. Respective figures for Canada that season were 19,932 pelts and $69.04 (Deems and Pursley 1978). In 1983/84 the total take was 33,135, and the average value was $18.71 (Novak, Obbard, et al. 1987). L. longicaudis also was taken in large numbers for its valuable skin, with probably about 30,000 killed annually during the early 1970s in Colombia and Peru alone. Continued illegal hunting, along with habitat loss and water pollution, jeopardizes the survival of this species and L. provocax (Mason and Macdonald 1986; Thornback and Jenkins 1982).

CARNIVORA; MUSTELIDAE; Genus PTERONURA
Gray, 1867

Giant Otter

The single species, P. brasiliensis, is found in Colombia, Venezuela, the Guianas, eastern Ecuador and Peru, Brazil, Bolivia, Paraguay, Uruguay, and northeastern Argentina (Cabrera 1957; Thornback and Jenkins 1982).

The remainder of this account is based largely on Duplaix's (1980) study of Pteronura in Surinam. Head and body length is 864–1,400 mm and tail length is 330–1,000 mm. Males weigh 26–34 kg, and females, 22–26 kg. The short fur generally appears brown and velvetlike when dry and shiny black chocolate when wet. On the lips, chin, throat, and chest there are often creamy white to buff splotches, which may unite to form a large white "bib." The feet are large, and thick webbing extends to the ends of the five clawed digits. The tail is thick and muscular at the base but becomes dorsoventrally flattened with a noticeable bilateral flange. There are subcaudal anal glands for secretion of musk.

The giant otter is found mainly in slow-moving rivers and creeks within forests, swamps, and marshes. It prefers waterways with gently sloping banks that have good cover. At certain points along a stream, areas of about 50 sq meters are cleared and used for rest and grooming. Some of these sites have dens, which consist of one or more short tunnels leading to a chamber about 1.2–1.8 meters wide. Pteronura seems clumsy on land but may move a considerable distance between waterways. When swimming slowly or remaining stationary in the water, it paddles with all four feet. When swimming at top speed, it depends largely on undulations of the tail and uses the feet for steering. It is entirely diurnal. During the dry season, when cubs are being reared, activity is generally restricted to one portion of a waterway. In the wet season, movements are far more extensive, and spawning fish are followed into the flooded forest. Prey is caught with the

Giant otter *(Pteronura brasiliensis brasiliensis)*, photo from New York Zoological Society.

mouth and then may be held in the forepaws while being consumed. Small fish may be eaten in the water, but larger prey is taken to shore. The diet consists mainly of fish and crabs.

Group home range, including land area, measures about 12 km in both length and width. During the dry season, at least, several kilometers of stream form a defended territory. Both sexes regularly patrol and mark the area, but groups tend to avoid one another, and fighting is evidently rare. *Pteronura* is more social than *Lutra.* A population includes both resident groups and solitary transients. Up to 20 individuals have reportedly been seen together, but groups of 4–8 are usually observed. A group consists of a mated adult pair, 1 or more subadults, and 1 or more young of the year. There is a high degree of pair bonding and group cohesiveness. A male and female stay close together and share the same den, even when cubs are present. *Pteronura* is much noisier than *Lutra.* Nine vocalizations have been distinguished, including screams of excitement, often given while swimming with the forepart of the body steeply out of the water, and coos upon close intra-specific contact.

Although data are scanty, the young apparently are born from late August to early October, at the start of the dry season. If the first litter is lost, a second is sometimes produced from December to April. The gestation period is 65–70 days. The number of young per litter is one to five, usually one to three. The cubs weigh about 200 grams each at birth and are able to eat solid food by 3–4 months. The young remain with the parents at least until the birth of the next litter, and probably for some time afterward. A captive lived for 12 years and 10 months (Jones 1982).

The giant otter is classified as vulnerable by the IUCN and as endangered by the USDI and is on appendix 1 of the CITES. It has become very rare or has entirely disappeared over vast parts of its range. The main factor in its decline is excessive hunting by people for its large and valuable pelt. Because of its noise, diurnal habits, and tendency to approach intruders, it is relatively easy to locate and kill. In the early 1980s a skin could be sold by a hunter for the equivalent of U.S. $50, and on the market in Europe for $250. Such trade is now prohibited, but illegal hunting continues and is being facilitated by the opening of wilderness habitat (Thornback and Jenkins 1982).

CARNIVORA; MUSTELIDAE; **Genus AONYX**
Lesson, 1827

Clawless Otters

There are three subgenera and three species (Chasen 1940; Coetzee, *in* Meester and Setzer 1977; Ellerman and Morrison-Scott 1966; Lekagul and McNeely 1977; Rosevear 1974):

African clawless otter *(Aonyx capensis)*, photo from Zoological Society of London. Insets: A. Forefoot; B. Hind foot; photos by U. Rahm.

subgenus *Aonyx* Lesson, 1827

A. capensis, Senegal to Ethiopia, and south to South Africa;

subgenus *Paraonyx* Hinton, 1921

A. congica, southeastern Nigeria and Gabon to Uganda and Burundi;

subgenus *Amblonyx* Rafinesque, 1832

A. cinerea, northwestern India to southeastern China and Malay Peninsula, southern India, Hainan, Sumatra, Java, Borneo, Rhio Archipelago, Palawan.

Paraonyx and *Amblonyx* sometimes have been treated as full genera.

In the African species, *A. capensis* and *A. congica*, head and body length is 600–1,000 mm, tail length is 400–710 mm, and weight is 13–34 kg. In the smaller *A. cinerea*, of Asia, head and body length is 450–610 mm, tail length is 250–350 mm, and weight is 1–5 kg. The general coloration is brown, with paler underparts and sometimes white markings on the face, throat, and chest.

Aonyx differs from *Lutra* and *Pteronura* in having webbing that either does not extend to the ends of the digits or is entirely lacking, and in having much smaller claws. In *A. congica* all the toes bear small, blunt claws; in *A. cinerea* the claws of adults are only minute spikes that do not project beyond the ends of the digital pads; and in *A. capensis* there are no claws, except for tiny ones on the third and fourth toes of the hind feet. In association with these adaptations, *Aonyx* has developed very sensitive forepaws and considerable digital movement. *A. capensis* and *A. cinerea* have relatively large, broad cheek teeth, apparently for purposes of crushing the shells of crabs and mollusks. *A. congica* has lighter and sharper dentition, more adapted to cutting flesh.

The general habitat of *A. capensis* varies from dense rainforest to open coastal plain and semiarid country. The species is usually found near water, preferring quiet ponds and sluggish streams, but may sometimes wander a considerable distance overland. In coastal areas it has been seen to forage both in the sea and in adjoining freshwater streams and marshes (Verwoerd 1987). It is mainly nocturnal but may be active by day in areas remote from human disturbance. It dens under boulders or driftwood, in crannies under ledges, or in tangles of vegetation. It apparently does not dig its own burrow. *A. cinerea* occurs in rivers, creeks, estuaries, and coastal waters (Lekagul and McNeely 1977). *A. congica* is seemingly found only in small, torrential mountain streams, within heavy rainforest.

These otters use their sensitive and dexterous forepaws to locate prey in mud or under stones. Captive *A. capensis* usually take food with the forepaws and do not eat directly off the ground. In the wild this species also catches most of its food with the forefeet, not with the mouth as do *Lutra* and *Pteronura* (Rowe-Rowe 1978b). The diet of both *A. capensis* and *A. cinerea* seems to consist mainly of crabs, other crustaceans, mollusks, and frogs; fish are relatively unimportant. There is thus apparently little competition for food between *Aonyx* and the fish-eating *Lutra* where both genera occur together. Piles of cracked crab and mollusk shells are signs of the presence of *A. capensis*. Donnelly and Grobler (1976) observed this species to use hard objects as anvils on which to break open mussel shells.

Little has been recorded about the habits of *A. congica*. Because of its scanty hair, weakly developed facial vibrissae, digital structure, and dental features, there has been speculation that it is more terrestrial than other otters. It is thought to feed mainly on relatively soft matter, such as small land

Oriental small-clawed otter *(Aonyx cinerea)*, photo by Lim Boo Liat.

vertebrates, eggs, and frogs. If this supposition is correct, there would be little competition for food between *A. congica* and the other kinds of otters that may occur in the same region.

Arden-Clarke (1986) studied a population of *A. capensis* inhabiting a rugged, densely vegetated environment along the coast of South Africa. The mean population density there was 1 per 1.9 km of coast, and dens were spaced at intervals of 470 meters. A radiotracked adult male had a minimum home range of 19.5 km of coast, with a core area of 12.0 km, where it spent most of its time. An adult female had a 14.3 km home range with a 7.5 km core area. There apparently was a clan-type social organization, with groups of related animals defending joint territories. The home ranges of four adult males overlapped completely, and some of these animals were seen foraging together. In another coastal area, Verwoerd (1987) found that an adult male might maintain a loose association with a female and cubs. *A. capensis* emits powerful, high-pitched shrieks when disturbed or trying to attract attention. According to Timmis (1971), *A. cinerea* lives in loose family groups of about 12 individuals and has a vocabulary of 12 or more calls, not including basic instinctive cries.

Births of *A. capensis* have been recorded in July and August in Zambia, and young have been found in March and April in Uganda (Kingdon 1977). There is probably no set breeding season in West Africa (Rosevear 1974). Most births in a coastal area of South Africa occurred in December and January (Verwoerd 1987). This species has a gestation period of 63 days and a litter size of two young. The young remain with the parents for at least 1 year. Female *A. cinerea* have an estrous cycle of 24–30 days, with an estrus of 3 days. They may produce two litters annually. The gestation period is 60–64 days, and litters contain one to six young, usually one or two. Their eyes open after 40 days, they first swim at 9 weeks, and they take solid food after 80 days (Duplaix-Hall 1975; Lekagul and McNeely 1977; Leslie 1971; Timmis 1971). Little is known of the reproduction of *A. congica*, but it probably has a gestation period of about 2 months, gives birth to two or three young, and attains sexual maturity at about 1 year. One specimen of *A. capensis* was still living after 11 years in captivity (Jones 1982).

If captured when young, these otters make intelligent and charming pets. *A. cinerea* has been trained to catch fish by Malay fishermen. The fur of *Aonyx* is not as good as that of *Lutra;* nonetheless, the subspecies *A. congica microdon* of Nigeria and Cameroon has declined seriously through uncontrolled commercial hunting. It is classified as endangered by the IUCN (1976) and the USDI and is on appendix 1 of the CITES. The other species of *Aonyx* are on appendix 2 of the CITES.

CARNIVORA; MUSTELIDAE; **Genus ENHYDRA**
Fleming, 1822

Sea Otter

The single species, *E. lutris,* was originally found in coastal waters off Hokkaido, Sakhalin, Kamchatka, the Commander Islands, the Pribilof Islands, the Aleutians, southern Alaska, British Columbia, Washington, Oregon, California, and western Baja California (Estes 1980).

Head and body length is usually 1,000–1,200 mm and tail length is 250–370 mm. Males weigh 22–45 kg, and females, 15–32 kg (Estes 1980). The color varies from reddish brown to dark brown, almost black, except for the gray or creamy head, throat, and chest. Albinistic individuals are rare. The head is large and blunt, the neck is short and thick, and the legs and tail are short. The ears are short, thickened, pointed, and valvelike. The hind feet are webbed and are flattened into broad flippers; the forefeet are small and have retractile claws. *Enhydra* is the only carnivore with only four incisor teeth in the lower jaw. The molars are broad, flat, and well adapted to crushing the shells of such prey as crustaceans, snails, mussels, and sea urchins. Unlike most mustelids, the sea otter lacks anal scent glands. Females have two abdominal mammae (Estes 1980).

The sea otter differs from most marine mammals in that it lacks an insulating subcutaneous layer of fat. For protection against the cold water, it depends entirely on a layer of air trapped among its long, soft fibers of hair. If the hair becomes

Sea otters *(Enhydra lutris)*, A. Juvenile walking; B. *E. lutris* floating on its back eating the head of a large codfish; C. Mother floating on her back with newborn pup. Photos by Karl W. Kenyon.

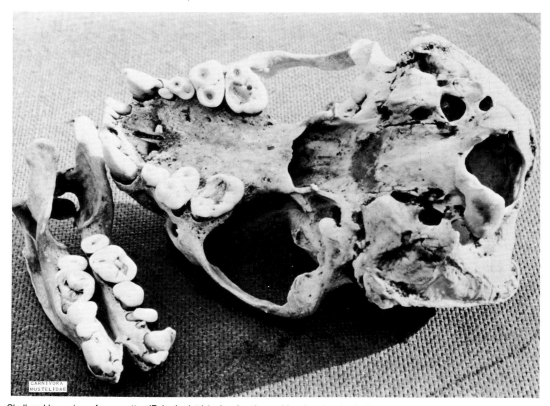

Skull and lower jaw of a sea otter *(Enhydra lutris),* showing the cavities that develop in the teeth of old animals because of the hard, rough materials that they eat, photo by H. Robert Krear.

soiled, as if by oil, the insulating qualities are lost and the otter may perish. The underfur, about 25 mm long, is the densest mammalian fur, averaging about 100,000 hairs per sq cm (Rotterman and Simon-Jackson 1988). It is protected by a scant coat of guard hairs.

Although the sea otter is a marine mammal, it rarely ventures more than 1 km from shore. According to Estes (1980), it forages in both rocky and soft-sediment communities, on or near the ocean floor. Off California, *Enhydra* seldom enters water of greater depth than 20 meters, but in the Aleutians it commonly forages at depths of 40 meters or more; the maximum confirmed depth of a dive was 97 meters. The usual period of submergence is 52–90 seconds, and the longest on record is 4 minutes and 25 seconds. The sea otter is capable of spending its entire life at sea but sometimes rests on rocks near the water. It walks awkwardly on land. When supine on the surface of the water, it moves by paddling with the hind limbs and sculling with the tail. For rapid swimming and diving, it uses dorsoventral undulations of the body. It can attain velocities of up to 1.5 km/hr on the surface and 9 km/hr underwater.

The sea otter is generally diurnal. It often spends the night in a kelp bed, lying under strands of kelp to avoid drifting while sleeping. It sometimes sleeps with a forepaw over the eyes. There may be local seasonal movements, but there are no extensive migrations.

The diet consists mainly of slow-moving fish and marine invertebrates, such as sea urchins, abalones, crabs, and mollusks (Estes 1980). Prey is usually captured with the forepaws, not the jaws. *Enhydra* floats on its back while eating and uses its chest as a "lunch counter." It is one of the few

mammals known to use a tool. While floating on its back, it places a rock on its chest and then employs the rock as an anvil for breaking the shells of mussels, clams, and large sea snails in order to obtain the soft internal parts. The sea otter requires a great deal of food: it must eat 20–25 percent of its body weight every day. It obtains about 23 percent of its water needs from drinking sea water and most of the rest from its food.

According to Estes (1980), *Enhydra* is basically solitary but sometimes rests in concentrations of up to 2,000 individuals. Animals of opposite sex usually come together only briefly for courtship and mating. At most times there is sexual segregation, with males and females occupying separate sections of coastline. Males usually occur at higher densities. Recent studies (Garshelis and Garshelis 1984; Garshelis, Johnson, and Garshelis 1984; Jameson 1989; Loughlin 1980; Rotterman and Simon-Jackson 1988) indicate that during the breeding season (which may be for most of the year) some males move into the areas occupied by females and establish territories. Such behavior has been documented both in Alaska and California, and males have been observed to return to the same place for up to seven years. The boundaries of these territories are vigorously patrolled, and intruding males are repulsed, but serious fighting is rare. The owner seeks to mate with any female that enters, though sometimes a pair bond is formed for a few days or weeks. Male territories are usually about 20–50 ha. and are smaller than female home ranges in the same area. Annual movements of both sexes frequently cover 50–100 km, considering foraging, breeding, and the passage of males between their territories and the all-male areas. Kenyon (1969) described nine vocalizations of *En-*

hydra, including screams of distress and coos of contentment.

Reproductive data have been summarized by Estes (1980). Breeding occurs throughout the year, but births peak in late May and June in the Aleutians and from December to February off California. Males may mate with more than one female during the season. Females are capable of giving birth every year but usually do so at greater intervals. If a litter does not survive, the female may experience a postpartum estrus. Females are known to adopt and nurse orphaned pups. Reports of the period of pregnancy range from 6.5 to 9 months, and delayed implantation is probably involved. Births probably occur most often in the water. There is normally a single offspring. About 2 percent of births are multiple, but only one young can be successfully reared. The pup weighs 1.4–2.3 kg at birth. While still small, it is carried, nursed, and groomed on the mother's chest as the mother swims on her back. The pup begins to dive in the second month of life. It may take some solid food shortly after birth but may nurse almost until it attains adult size. The period of dependency on the mother is thought to be about 6–8 months. Females become sexually mature at about 4 years. Males are capable of mating at 5–6 years but usually do not become active breeders until several more years have passed. Wendell, Ames, and Hardy (1984) concluded that the reproductive cycle of the California population is shorter than elsewhere, with females sometimes giving birth each year. It now is known that Alaskan females also are capable of annual reproduction (Garshelis, Johnson, and Garshelis 1984). According to Rotterman and Simon-Jackson (1988), a captive male fathered young when at least 19 years old, and maximum estimated longevity for wild females is 23 years.

The fur of the sea otter may be the most valuable of any mammal's. During the 1880s, prices on the London market ranged from $105 to $165 each. By 1903, when the species had become scarce, large, high-quality skins sold for up to $1,125. Pelts taken in Alaska in the late 1960s, during a brief reopening of commercial activity, sold for an average of $280 each (Kenyon 1969).

Intensive exploitation of the sea otter was begun by the Russians in 1741. Hunting was uncontrolled until 1799, when some conservation measures were established. Unregulated killing resumed in 1867, when Alaska was purchased by the United States. By 1911, when the sea otter was protected by a treaty among the United States, Russia, Japan, and Great Britain, probably only 1,000–2,000 of the animals survived worldwide (Kenyon 1969). Under protection of the treaty, state and national laws, and finally the United States Marine Mammal Protection Act of 1972, the sea otter has steadily increased in numbers and distribution. There are now probably 100,000–150,000 individuals in the major populations off southwestern and south-central Alaska and another 12,000 off Kamchatka and the Kuril and Commander islands in the Soviet Union. Alaskan populations are subject to limited killing for native subsistence purposes, may come into conflict with shellfisheries, and are potentially jeopardized by oil spills. Reintroduced populations (from Alaskan stock) apparently have been established off southeastern Alaska (now numbering 500–1,000 animals), Vancouver Island (about 350), and Washington (100). Reintroduced groups off Oregon and in the Pribilof Islands do not seem to have done well and have all but disappeared (Estes 1980; Jameson et al. 1982; Rotterman and Simon-Jackson 1988).

The southern sea otter (subspecies *E. l. nereis*), which originally ranged from Baja California to at least Washington and perhaps to south-central Alaska, was generally considered extinct by 1920. Apparently, however, a group of 50–100 individuals survived off central California in the vicinity of Monterey. In 1938 the presence of this population became generally known. By the 1970s it had grown to include about 1,800 animals, but subsequently numbers stabilized. The sea otter now regularly occurs along about 300 km of the central California coast, and there have been scattered reports of individuals from southern California and northern Baja California. As this population increased, there was concern that stocks of abalone and other shellfish were being depleted. Some parties with a commercial or recreational interest in these stocks have advocated control of the sea otter population, and there have been cases of illegal killing. Some otters also are being drowned accidentally in fishing nets. Another fear is that an oil spill, associated with either the extensive tanker traffic in the area or offshore drilling, could devastate the population (Armstrong 1979; Carey 1987; Estes and VanBlaricom 1985; Leatherwood, Harrington-Coulombe, and Hubbs 1978; U.S. Fish and Wildlife Service 1980). There has been concern that the genetic viability of *E. l. nereis* was severely reduced when it approached extinction earlier in the century, but Ralls, Ballou, and Brownell (1983) calculated that the existing population should theoretically retain about 77 percent of the original diversity and that transplanted colonies should also be viable. An effort to establish such a colony was started in 1987 when 63 otters taken from the main California population were released around San Nicolas Island (Brownell and Rathbun 1988). The southern sea otter is listed as threatened by the USDI and is on appendix 1 of the CITES. All other *E. lutris* are on appendix 2.

CARNIVORA; Family VIVERRIDAE

Civets, Genets, Linsangs, Mongooses, and Fossas

This family of 36 Recent genera and 71 species is found in southwestern Europe, southern Asia, the East Indies, Africa, and Madagascar. Certain genera have been introduced to areas in which the family does not naturally occur. The following sequence of genera, grouped in 7 subfamilies, is based on the classifications of Coetzee (*in* Meester and Setzer (1977), Ellerman and Morrison-Scott (1966), and Ewer (1973).

Subfamily Viverrinae (civets, genets, linsangs)

Viverra	*Genetta*	*Poiana*
Civettictis	*Osbornictis*	*Prionodon*
Viverricula		

Subfamily Paradoxurinae (palm civets)

Nandinia	*Paradoxurus*	*Macrogalidia*
Arctogalidia	*Paguma*	*Arctictis*

Subfamily Hemigalinae (palm and otter civets)

Hemigalus	*Chrotogale*	*Cynogale*

Subfamily Fossinae (Malagasy civets)

Fossa	*Eupleres*

Subfamily Galidiinae (Malagasy mongooses)

Galidia	*Mungotictis*	*Salanoia*
Galidictis		

Subfamily Herpestinae (mongooses)

Herpestes	*Liberiictis*	*Bdeogale*
Mungos	*Helogale*	*Rhynchogale*
Crossarchus	*Dologale*	*Ichneumia*

Subfamily Herpestinae (cont'd.)

Atilax   *Paracynictis*   *Suricata*
Cynictis

Subfamily Cryptoproctinae (fossa)

Cryptoprocta

Some authorities, including Corbet and Hill (1986) and Honacki, Kinman, and Koeppl (1982) now consider the mongoose subfamilies Herpestinae and Galidiinae to represent a separate family, the Herpestidae. However, such an arrangement was not followed by Meester et al. (1986), Stains (1984), or any of the other authorities cited in this account.

These are the characteristic small and medium-sized carnivores of the Old World. Head and body length is 170 to about 1,000 mm, tail length is 120–900 mm, and adult weight is 0.45 to approximately 14.0 kg. *Helogale* is the smallest genus and *Arctictis* is the largest. Viverrids have a variety of striped, spotted, and uniform color patterns. In some genera the tail is banded or ringed. The body is long and sinewy, with short legs and generally a long, bushy tail. One genus *(Arctictis)* has a truly prehensile tail. The head is elongate and the muzzle is pointed. Most genera have five toes on each foot, *Cynictis* has only four digits on the hind foot, and *Suricata* has only four digits on all feet (Stains 1967). The claws are semiretractile in some genera. Female viverrids usually have two or three pairs of abdominal mammae. Males have a baculum.

Most viverrids have scent glands in the anal region, which secrete a nauseous-smelling fluid as a defensive measure. In mongooses and some other members of the family, these glands open into a pouch or saclike depression, outside the anus proper, in which the secretion is stored. The conspicuous pattern of pelage in some genera has been interpreted as being a warning that the fetid anal gland secretion is present. Such a color pattern is also found in skunks and certain other members of the family Mustelidae. The secretion of the scent glands, when rubbed on various objects, is recognized by other individuals of the same species and is probably used to communicate various information.

The skull is usually long and flattened. The dental formula is: (i 3/3, c 1/1, pm 3–4/3–4, m 1–2/1–2) × 2 = 32–40. The second lower incisor is raised above the level of the first and third, the canines are elongate, and the carnassials are developed.

Viverrids are essentially forest inhabitants, but they also live in dense brush and thick grass. They are either diurnal or nocturnal and shelter in any convenient retreat, usually a hole in a tree, a tangle of vines, ground cover, a cave, a crevice, or a burrow. A few species dig their own burrows. Those species living near people sometimes seek refuge under rafters or in the drains of houses.

Those viverrids that walk on their digits (such as *Genetta*) have a gait described as "a waltzing trot," whereas the members of the family that walk on the sole, with the heel touching the ground (such as *Arctictis*), have a bearlike shuffle. Many genera are agile and extremely graceful in their movements. A number of species are skillful climbers; some apparently spend most of their lives in trees. Some genera take to water readily and swim well; three genera, *Osbornictis*, *Atilax*, and *Cynogale*, are semiaquatic. Sight, hearing, and smell are acute.

Viverrids may fight when cornered. They seek their prey in trees and on the ground, either by stalking it or by pouncing upon it from a hiding place. They eat small vertebrates and various invertebrates and occasionally consume vegetable matter, such as fruit, bulbs, and nuts. Carrion is taken by some species.

Viverrids are solitary or live in pairs or groups. Several genera, including *Cynictis* and *Suricata*, live in colonies in ground burrows. Some genera of mongooses associate in bands and take refuge as a group in any convenient shelter. Breeding may occur seasonally or throughout the year. A number of genera have two litters annually. The one to six offspring are born blind but haired. Most species probably have a potential longevity of 5–15 years.

The secretion of the scent glands, known as civet, is obtained from several genera (*Civettictis*, *Viverra*, and *Viverricula*) for both perfumery and medicinal purposes. Some viverrids are tamed and kept to extract the musk. They may also be kept as pets. Viverrids occasionally kill poultry but also prey on rodents. Mongooses, particularly of the genus *Herpestes*, have been introduced into several areas to check the numbers of rodents and venomous snakes. Such introductions, however, generally have not proven beneficial, as the mongooses quickly multiply and destroy many desirable forms of mammals and birds.

The geological range of this family is late Eocene to Recent in Europe, early Oligocene to Recent in Asia, early Miocene to Recent in Africa, and Pleistocene to Recent in Madagascar (Stains 1984).

CARNIVORA; VIVERRIDAE; Genus VIVERRA
Linnaeus, 1758

Oriental Civets

There are two subgenera and three species (Chasen 1940; Ellerman and Morrison-Scott 1966; Medway 1978; Taylor 1934):

subgenus *Viverra* Linnaeus, 1758

V. zibetha, Nepal and eastern India to southeastern China and Malay Peninsula;
V. tangalunga, Malay Peninsula, Sumatra, Bangka, Borneo, Rhio Archipelago, Philippines;

subgenus *Moschothera* Pocock, 1933

V. megaspila, peninsular India, Burma to Indochina and Malay Peninsula.

Civettictis sometimes is regarded as a third subgenus of *Viverra* (see Coetzee, *in* Meester and Setzer 1977).

Head and body length is 585–950 mm, tail length is 300–482 mm, and weight is 5–11 kg. The fur, especially in winter, is long and loose. It is usually elongated in the median line of the body, forming a low crest or mane. The color pattern of the body is composed of black spots on a grayish or tawny ground color. The sides of the neck and throat are marked with black and white stripes—usually three black and two white collars. The crest is marked by a black spinal stripe, which runs from the shoulders to the tail, and the tail is banded or ringed with black and white. The feet are black. In *V. zibetha* and *V. tangalunga* the third and fourth digits of the forefeet are provided with lobes of skin that act as protective sheaths for the retractile claws. *Viverra* is distinguished from *Viverricula* by its larger size, by the presence of a dorsal crest of erectile hairs, and by the insertion of the ears, the inner edges of which are set farther apart on the forehead.

Oriental civets occur in a wide variety of habitats in forest, brush, and grassland. They stay in dense cover by day and come out into the open at night. They are mainly terrestrial and often live in holes in the ground dug by other animals.

Oriental civet *(Viverra tanalunga)*, photo by Ernest P. Walker.

They apparently can climb readily but seldom do so. Like *Viverricula*, they are often found near villages and are common over most of their range. Like most civets, they are easily trapped. They are vigorous hunters, killing small mammals, birds, snakes, frogs, and insects, and taking eggs, fruit, and some roots. The species *V. zibetha* has been observed fishing in India, and the remains of crabs have been found in the stomachs of two individuals from China.

Viverra is generally solitary. *V. zibetha* is said to breed all year and to bear two litters annually (Lekagul and McNeely 1977). The number of young per litter is one to four, usually two or three. The young are born in a hole in the ground or in dense vegetation. The young of *V. zibetha* open their eyes after 10 days, and weaning begins at 1 month (Medway 1978). There are captive specimens of *V. zibetha* in the Ahmedabad Zoo in India that are more than 20 years old (Smielowski 1986).

Viverra is one of the sources of civet, a substance used commercially in producing perfume. Because of this function, *V. tangalunga* has been introduced to several islands of the East Indies, apparently including Sulawesi (Groves 1976; Laurie and Hill 1954). Civet is also obtained from the genera *Viverricula* and *Civettictis*. The subspecies *V. megaspila civettina* of peninsular India is classified as endangered by the IUCN (1972) and the USDI. It evidently has become very rare through persecution by people and loss of habitat to agriculture.

CARNIVORA; VIVERRIDAE; Genus CIVETTICTIS
Pocock, 1915

African Civet

The single species, *C. civetta*, is found from Senegal to Somalia and south to Namibia and eastern South Africa. Recognition of *Civettictis* as a distinct genus is in keeping with Ewer (1973), Kingdon (1977), and Rosevear (1974). Some other authorities, such as Coetzee (*in* Meester and Setzer 1977) and Rowe-Rowe (1978*b*), have included *C. civetta* in the genus *Viverra*.

Head and body length is 680–890 mm, tail length is 445–63 mm, and weight is 7–20 kg (Kingdon 1977). The color is black with white or yellowish spots, stripes, and bands. The long and coarse hair is thick on the tail. The perineal glands under the tail contain the oily scented matter used commercially in making perfume. All the feet have five claws, and the soles are hairy. From *Viverra*, *Civettictis* is distinguished by much larger molar teeth and a far broader lower carnassial (Rosevear 1974).

The African civet is widely distributed in both forests and savannahs, wherever long grass or thickets are sufficient to provide daytime cover (Ewer and Wemmer 1974). It seems to use a permanent burrow or nest only to bear young. It is nocturnal and almost completely terrestrial but takes to water readily and swims well. The omnivorous diet includes carrion, rodents, birds, eggs, reptiles, frogs, crabs, insects, fruits, and other vegetation. Poultry and young lambs are sometimes taken (Rosevear 1974).

Civettictis is generally solitary but has a variety of visual, olfactory, and auditory means of communication. Individuals may have defined and well-marked territories. The scent glands have a major social role, leaving scent along a path to convey information, such as whether a female is in estrus (Kingdon 1977). There are three agonistic vocalizations—the growl, cough-spit, and scream—but the most commonly heard sound is the "ha-ha-ha" used in making contact (Ewer and Wemmer 1974).

Available data on reproduction (Ewer and Wemmer 1974; Kingdon 1977; Mallinson 1973, 1974; Rosevear 1974) suggest that breeding occurs throughout the year. There may be two or even three litters annually. The gestation period is usually 60–72 days but is occasionally extended to as much as 81 days, perhaps because of delayed implantation. The number of young per litter is one to four, usually two or three. The young are born fully furred, open their eyes within a few days, cease suckling at 14–20 weeks, and attain sexual maturity at about 1 year. According to Jones (1982), a captive lived for 28 years.

In Ethiopia, and to a lesser extent in other parts of Africa, the natives keep civets in captivity and remove the musk from them several times a week. An average animal yields 3–4 grams weekly. The natives do not raise the civets, however,

African civet *(Civettictis civetta)*, photo from New York Zoological Society.

but merely capture wild ones. In 1934 Africa produced about 2,475 kg of musk with a value of U.S. $200,000. In that same year the United States imported 200 kg of musk. The production of civet musk is an old industry; King Solomon's supply came from East Africa. Rosevear (1974) reported that the trade in civet musk now has diminished considerably. However, Schreiber et al. (1989) indicated that in 1988 there still were over 2,700 captive civets in Ethiopia and that their musk, exported mainly to France, was selling for about $438 per kg.

CARNIVORA; VIVERRIDAE; **Genus VIVERRICULA**
Hodgson, 1838

Lesser Oriental Civet, or Rasse

The single species, *V. indica*, occurs from Pakistan to southeastern China and the Malay Peninsula, as well as on Sri Lanka, Taiwan, Hainan, Sumatra, Java, and Bali (Ellerman and Morrison-Scott 1966; Roberts 1977). The name *V. malaccensis* was used for this species by Medway (1978) and Lekagul and McNeely (1977).

Head and body length is 450–630 mm, tail length is 300–430 mm, and weight is usually 2–4 kg. The fur is harsh, rather coarse, and loose. The body color is buffy, brownish, or grayish, and the feet are black. Small spots are present on the forequarters, and larger spots, tending to run into longitudinal lines, are present on the flanks. There are six to eight dark stripes on the back, and the tail is ringed black and white by six to nine rings of each color.

Viverricula is distinguished from *Viverra* by smaller size, by the absence of a dorsal crest of erectile hairs, and by the insertion of the ears, the inner edges of which are set closer together on the forehead than those of *Viverra*. The muzzle is also shorter and more pointed. Internally, the two genera differ in a number of cranial and dental features.

The rasse inhabits grasslands or forests. It probably excavates its own burrow (Roberts 1977) but may also shelter in thick clumps of vegetation, buildings, or drains (Lekagul and McNeely 1977). It is generally nocturnal but may be seen hunting by day in areas not populated by humans. It is mainly terrestrial but is said to climb well. It usually tries to escape from dogs by dodging and twisting through the underbrush.

The diet consists of small vertebrates, carrion, insects and their grubs, fruits, and roots.

Viverricula is usually solitary but occasionally associates in pairs. It breeds throughout the year in Sri Lanka. The two to five young are born in a shelter on the ground. A captive lived for 10 years and 6 months (Jones 1982).

The rasse is kept in captivity by natives for the purpose of extracting the civet that is secreted and retained in sacs close to the genitals in both sexes. The removal of this secretion is accomplished by scraping the inside of the sac with a spoonlike implement. In India this secretion is used as a perfume to flavor the tobacco that is smoked by some natives. *Viverricula* was introduced by people to Socotra, the Comoro Islands, the Philippines, and Madagascar, probably for the production of civet. Its presence on the island of Sumbawa, to the east of Bali, also is thought to have resulted from introduction (Laurie and Hill 1954).

CARNIVORA; VIVERRIDAE; **Genus GENETTA**
Oken, 1816

Genets

There are 3 subgenera and 9 species (Ansell 1978; Coetzee, *in* Meester and Setzer 1977; Crawford-Cabral 1981; Lamotte and Tranier 1983; Meester et al. 1986; Schlawe 1980):

subgenus *Pseudogenetta* Dekeyser, 1949

G. thierryi, savannah zone from Senegal to area south of Lake Chad;
G. abyssinica, highlands of Ethiopia;

subgenus *Paragenetta* Kuhn, 1960

G. johnstoni, Liberia, Ivory Coast, Ghana;

subgenus *Genetta* Oken, 1816

G. servalina, southern Nigeria to western Kenya;
G. victoriae, northern and eastern Zaire, Uganda;
G. genetta, France, Spain, Portugal, Balearic Islands,

Lesser oriental civet *(Viverricula indica)*, photo by Ernest P. Walker.

southwestern Arabian Peninsula, northwestern Africa, savannah zones of Africa south of the Sahara;

G. angolensis, southern Zaire, central and northeastern Angola, western Zambia, northern Mozambique, probably southern Tanzania;

G. pardina, Gambia to Cameroon;

G. tigrina, Senegal to Somalia, and south to Namibia and South Africa.

Corbet (1978) suggested that the European populations of *G. genetta* are the result of introduction by human agency. Schlawe (1980) restricted the name *G. genetta* to Europe and northwestern Africa. He referred the populations to the south of the Sahara and in the southwestern Arabian Peninsula to a separate species, *G. felina*, and he showed that the reported presence of *Genetta* in Palestine is not correct. Later, Schlawe (1981) indicated that much work remains to be done before the systematics of this genus can be reasonably well understood. Corbet (1984) recognized the specific distinction of *G. felina*, but Meester et al. (1986) did not. Some of the authorities cited above restrict *G. tigrina* to South Africa and Lesotho and assign the populations in the remainder of the range given above to the separate species *G. rubiginosa*, but Meester et al. (1986) indicated that there was intergradation between the two.

Head and body length is usually 420–580 mm, tail length is 390–530 mm, and weight is 1–3 kg. Coloration is variable, but the body is generally grayish or yellowish, with brown or black spots and blotches on the sides, tending to be arranged in rows. A row of black erectile hairs is usually present along the middle of the back. The tail has black and white rings. Melanistic individuals seem to be fairly common. Genets have a long body, short legs, a pointed snout, prominent and rounded ears, short and curved retractile claws, and soft and dense hair. They have the ability to emit a musky-smelling fluid from their anal glands. Females have two pairs of abdominal mammae.

Genets inhabit forests, savannahs, and grasslands. They are active at night, usually spending the day in a rock crevice, in a burrow excavated by some other animal, in a hollow tree, or on a large branch. They seem to return daily to the same shelter. They climb trees to prey on nesting and roosting birds, but much of their food is taken on the ground. They are silent and stealthy hunters; when stalking prey, they crouch until the body and tail seem to glide along the ground. At the same time, the body seems to lengthen. Genets can go through any opening the head can enter, because of the slender and loosely jointed body. The diet consists of any small animals that can be captured, including rodents, birds, reptiles, and insects. Genets sometimes take game birds and poultry.

A young female radiotracked in Spain for about three months used a home range of 1.4 sq km, while an adult male wandered over 50 sq km during about five months (Palomares

Genets *(Genetta tigrina):* Top, photo by John Markham; Bottom, photo by John Visser.

and Delibes 1988). Genets travel alone or in pairs. They communicate with one another by a variety of vocal, olfactory, and visual signals. Breeding seems to correspond with the wet seasons in both West and East Africa. In Kenya, for example, pregnant and lactating females have been taken in May and from September to December. A pair of *G. genetta* in the National Zoo in Washington, D.C., regularly produced two litters per year, one in April–May and another in July–August. Gestation periods ranging from 56 to 77 days have been reported. The number of young per litter is one to four, usually two or three. The young weigh 61–82 grams each at birth, begin to take solid food at 2 months, and attain adult weight at 2 years. One female *G. genetta* became sexually mature at about 4 years and produced young regularly until

she died at the age of 13 years (Kingdon 1977; Rosevear 1974; Wemmer 1977). Another captive *G. genetta* lived for 21 years and 6 months (Jones 1982).

According to Smit and Van Wijngaarden (1981), the genet has declined in Europe because of persecution for alleged depredations on game birds and poultry. In addition, its winter pelt is highly esteemed. A recently described subspecies on Ibiza Island in the Balearics, *G. genetta isabelae*, is classified as rare by the IUCN.

CARNIVORA; VIVERRIDAE; Genus OSBORNICTIS
J. A. Allen, 1919

Aquatic Genet

The single species, *O. piscivora*, is known only by about 30 specimens taken in northeastern Zaire (Coetzee, *in* Meester and Setzer 1977; Hart and Timm 1978; Van Rompaey 1988).

An adult male had a head and body length of 445 mm, a tail length of 340 mm, and a weight of 1,430 grams; an adult female weighed 1,500 grams (Hart and Timm 1978). The body is chestnut red to dull red, and the tail is black. There is a pair of elongated white spots between the eyes. The front and sides of the muzzle and the sides of the head below the eyes are whitish. Black spots and bands are absent, and the tail is not ringed. The pelage is long and dense, especially on the tail. The palms and soles are bare, not furred as in *Genetta* and other related genera. The skull is long and lightly built, and the teeth are relatively small and weak.

Osbornictis is among the rarest genera of carnivores. All specimens probably originated in areas of dense forest at elevations of 500–1,500 meters (Hart and Timm 1978). The genus is generally thought to be semiaquatic, as several speci-

Aquatic genet *(Osbornictis piscivora),* photo of mounted specimen by M. Colyn.

mens have been taken in or near streams, and available evidence suggests that fish constitute a major part of the diet. Hart and Timm, for example, noted the following: the stomach of one specimen contained the remains of fish; natives of the area indicated that fish is the favored prey; the dentition of *Osbornictis* seems to be adapted to deal with slippery vertebrate prey, such as fish and frogs; and the bare palms may be an adaptation allowing the genet to feel for fish in muddy holes and then handle the prey. *Osbornictis* is apparently solitary. A pregnant female with a single embryo, 15 mm in length, was taken on 31 December.

CARNIVORA; VIVERRIDAE; Genus POIANA
Gray, 1864

African Linsang, or Oyan

The single species, *P. richardsoni,* occurs from Sierra Leone to northern Zaire and on the island of Bioko (Fernando Poo) (Coetzee, *in* Meester and Setzer 1977); the generic name reflects the occurrence on this island. Rosevear (1974) recognized the populations in the western part of the range of *Poiana* as a distinct species, *P. leightoni.*

The average head and body length is 384 mm, and the average tail length is 365 mm (Rosevear 1974). The general color effect is light brownish gray to rusty yellow; dark brown

to black spots and rings are present. Some individuals have alternating broad and narrow black bands on the tail, whereas others have only the broad bands. This genus differs from the Asiatic linsangs *(Prionodon)* in that the spots are smaller and show no tendency to run into bands or stripes, except in the region of the head and shoulder. It also differs from them, and resembles *Genetta,* in having a narrow bare line on the sole of each hind foot.

The oyan is a forest animal and is nocturnal. According to Dr. Hans-Jürg Kuhn (Anatomisches Institut der Universität Frankfurt am Mein, pers. comm.), *Poiana* builds a round nest of green material, in which several individuals sleep for a few days, and then moves on and builds a new nest. The nests are at least two meters above the ground, usually higher. Although it has been reported that the oyan sleeps in the abandoned nests of squirrels, reliable hunters say that the reverse is true: the squirrels sleep in abandoned nests of *Poiana.* The diet includes cola nuts, other plant material, insects, and young birds. In the Liberian hinterland, natives make medicine bags from the skins of *Poiana.*

Observations by Charles-Dominique (1978) in northeastern Gabon indicate a population density of 1/sq km. A lactating female has been noted in October. As in some other genera of viverrids, there may be two litters per year. The number of young per birth is two or three. A captive oyan lived for five years and four months (Jones 1982). The subspecies *P. r. liberiensis,* of Liberia, Ivory Coast, and Sierra Leone, is classified as indeterminate by the IUCN.

African linsang *(Poiana richardsoni)*, photo from *Proc. Zool. Soc. London.*

CARNIVORA; VIVERRIDAE; **Genus PRIONODON**
Horsfield, 1822

Oriental Linsangs

There are two subgenera and two species (Chasen 1940; Ellerman and Morrison-Scott 1966; Lekagul and McNeely 1977):

subgenus *Prionodon* Horsfield, 1822

P. linsang (banded linsang), western and southern
   Thailand, Malay Peninsula, Sumatra, Bangka, Java,
   Borneo;

subgenus *Pardictis* Thomas, 1925

P. pardicolor (spotted linsang), Nepal to Indochina.

Head and body length is 350–450 mm and tail length is 304–420 mm. Medway (1978) listed the weight of *P. linsang* as 598–798 grams; on the average, *P. pardicolor* is slightly smaller. In *P. linsang* the ground color varies from whitish gray to brownish gray and becomes creamy on the underparts. The dark pattern consists of four or five broad, transverse black or dark brown bands across the back; there is one large stripe on each side of the neck. The sides of the body and legs are marked with dark spots, and the tail is banded. Some individuals of *P. pardicolor* have a ground color of orange buff, whereas others are pale brown. Black spots on the upper parts are arranged more or less in longitudinal rows, and the tail has 8–10 dark rings.

These animals are extremely slender, graceful, and beautiful. The fur is short, dense, and soft; it has the appearance and feel of velvet. The claws are retractile; claw sheaths are present on the forepaws, and protective lobes of skin are present on the hind paws. The skull is long, low, and narrow, and the muzzle is narrow and elongate. Unlike many viverrids, *Prionodon* seems to be free from odor.

Oriental linsangs dwell mainly in forests. They are nocturnal and generally arboreal but frequently come to the ground in search of food (Lekagul and McNeely 1977). *P. linsang* constructs a nest of sticks and leaves; in one case a nest was located in a burrow at the base of a palm. This species is also said to live in tree hollows. The diet includes small mammals, birds, eggs, and insects.

The limited data on reproduction suggest that *P. linsang* has no clear breeding season (Lekagul and McNeely 1977). Two pregnant females, one with two embryos and the other with three, were collected in May, and two lactating females were found in April and October. *P. pardicolor* is said to breed in February and August and to have litters of two young. A captive *P. linsang* lived for 10 years and 8 months (Jones 1982).

The species *P. pardicolor* is listed as endangered by the USDI and is on appendix 1 of the CITES. *P. linsang* is on appendix 2 of the CITES. Schreiber et al. (1989) noted that the former seems to be very rare but the latter is still relatively numerous in certain areas.

CARNIVORA; VIVERRIDAE; **Genus NANDINIA**
Gray, 1843

African Palm Civet

The single species, *N. binotata*, occurs from Guinea-Bissau to southern Sudan and south to northern Angola and eastern Zimbabwe (Coetzee, *in* Meester and Setzer 1977).

Head and body length is 440–580 mm and tail length is 460–620 mm. Kingdon (1977) listed weight as 1.7–2.1 kg, but Charles-Dominique (1978) reported that males weighed as much as 5 kg. Coloration is quite variable but is usually grayish or brownish, tinged with buffy or chestnut. Often, two creamy spots are present between the shoulders, and obscure dark brown spots are present on the lower back and top of the tail. The tail, which is somewhat darker than the body, is the same color above and below and has a variable pattern of black rings. The throat tends to be grayish, and the underparts are grayish, tinged with yellow. The pelage is short and woolly but coarse-tipped. The ears are short and

Banded linsang *(Prionodon linsang)*, photo by Ernest P. Walker.

rounded, the tail is fairly thick, the legs are short, and the claws are sharp and curved. There are scent glands on the palms, between the toes, on the lower abdomen, and possibly on the chin (Kingdon 1977).

In a radiotracking study in Gabon, Charles-Dominique (1978) found *Nandinia* to be largely arboreal and to occur

African palm civet *(Nandinia binotata)*, photo by Ernest P. Walker.

mainly 10–30 meters above the ground in various types of forest. It was nocturnal, sleeping by day in a fork, on a large branch, or in a bundle of lianas. Stomach contents consisted of 80 percent fruit, on the average, but also included remains of rodents, bird eggs, large beetles, and caterpillars.

Charles-Dominique found a population density of five palm civets per sq km in his study area. Adult females established territories averaging 45 ha. They allowed immature females on these areas but did not tolerate trespassing by other adult females. Large, dominant adult males had territories averaging about 100 ha., which overlapped a number of female territories. The large males drove away other animals of the same size and sex but allowed smaller adult males to remain; however, the small adult males were not permitted access to the females. Territories were marked with scent. Fighting was severe, sometimes resulting in death. Loud calls were exchanged during courtship.

In West Africa, breeding apparently can occur during the wet or dry season (Rosevear 1974). Records from East Africa suggest that there are two birth peaks or seasons, May and October. The gestation period is 64 days. The number of young is usually two but up to four (Kingdon 1977). As soon as they are weaned, young males leave the territory of their mother. Sexual maturity comes in the third year of life (Charles-Dominique 1978). One individual was still alive after 15 years and 10 months in captivity (Jones 1982).

The African palm civet is easily tamed and will drink milk in captivity. It is said to be quite clean and to keep houses free of rats, mice, and cockroaches.

CARNIVORA; VIVERRIDAE; Genus ARCTOGALIDIA
Merriam, 1897

Small-toothed or Three-striped Palm Civet

The single species, *A. trivirgata*, is found from Assam to Indochina and the Malay Peninsula and on Sumatra, Bangka, Java, Borneo, and numerous small nearby islands of the East Indies (Chasen 1940; Ellerman and Morrison-Scott 1966).

Head and body length is 432–532 mm, tail length is 510–660 mm, and weight is usually 2.0–2.5 kg. The upper parts,

proximal part of the tail, and outside of the limbs are usually tawny, varying from dusky grayish tawny to bright orangish tawny. The head is usually darker and grayer, and the paws and distal part of the tail are brownish. There is a median white stripe on the muzzle, and three brown or black longitudinal stripes on the back. The median stripe is usually complete and distinct, whereas the laterals may be broken up into spots or may be almost absent. The undersides are grayish white or creamy buff, with a whitish patch on the chest.

Only the females of this genus possess the civet gland, which is located near the opening of the urinogenital tract. *Arctogalidia* closely resembles *Paradoxurus* in external form, as well as in the length of the legs and tail, but differs externally in characters of the feet. Internally, the skull differs from that of *Paradoxurus*, and the back teeth are smaller.

Arctogalidia inhabits dense forests. In some areas it frequents coconut plantations, though Lekagul and McNeely (1977) reported that it avoids human settlements. It is nocturnal, resting by day in the upper branches of tall trees (Medway 1978). It is arboreal, climbing actively and leaping from branch to branch with considerable agility. The omnivorous diet includes squirrels, birds, frogs, insects, and fruit.

Three animals, representing both sexes, occupied an empty nest of *Ratufa bicolor* about 20 meters above the ground in a tree. Mewing calls and light snarls, accompanied by playful leaps and chases, have been noted in a male and female at night in the wild. The young are reared in hollow trees. According to Lekagul and McNeely (1977), breeding probably continues throughout the year, there may be two litters annually, and litter size is two or three young. Batten and Batten (1966) reported that a female about 2 weeks old was captured in Borneo in August 1961. It entered estrus for the first time in December 1962 and then again at intervals of 6 months. In August 1964 it gave birth to its first litter (three young) after a gestation period of approximately 45 days. The young opened their eyes at 11 days and were suckled for over 2 months. The father was reintroduced to the family when the young were 2.5 months old and was soon accepted by the others. According to Jones (1982), one individual was still living after 15 years and 10 months in captivity.

The subspecies *A. t. trilineata*, of Java, is classified as indeterminate by the IUCN. Schreiber et al. (1989) indicated that it was already rare 50 years ago; the last confirmed record was in 1978.

Small-toothed palm civet *(Arctogalidia trivirgata)*, photo by Lim Boo Liat.

Palm civet (*Paradoxurus hermaphroditus*), photo by Ernest P. Walker.

PARADOXURUS

CARNIVORA; VIVERRIDAE; **Genus PARADOXURUS**
F. Cuvier, 1821

Palm Civets, Musangs, or Toddy Cats

There are three species (Chasen 1940; Ellerman and Morrison-Scott 1966; Laurie and Hill 1954):

P. hermaphroditus, Kashmir and peninsular India to southeastern China and Malay Peninsula, Sri Lanka, Hainan, Sumatra, Java, Borneo, Philippines, Sulawesi, many small islands of the East Indies;
P. zeylonensis, Sri Lanka;
P. jerdoni, southern India.

Taylor (1934) referred the populations of *Paradoxurus* in the Philippines to three species—*P. philippinensis, P. torvus,* and *P. minax*—but Lekagul and McNeely (1977) indicated that the only species in the Philippines is *P. hermaphroditus.*

Head and body length is 432–710 mm, tail length is 406–660 mm, and weight is 1.5–4.5 kg. The ground color is grayish to brownish but is often almost entirely masked by the black tips of the guard hairs. There is a definite pattern of dorsal stripes and lateral spots, at least in the new coat, but this is sometimes concealed by the long black hairs. The pattern is most plainly shown in the species *P. hermaphroditus,* where it consists of longitudinal stripes on the back and spots on the shoulders, sides, and thighs and sometimes on the base of the tail. A pattern of white patches and a white band across the forehead may be present on the head of this species.

The species *P. hermaphroditus* can always be distinguished from *P. zeylonensis* and *P. jerdoni* by the backward direction of the hairs on the neck. In the other species the hairs on the neck grow forward from the shoulders to the head.

Paradoxurus differs from *Arctogalidia* and *Paguma* in color or pattern and in characters of the skull and teeth. According to Lekagul and McNeely (1977), the teeth of *Paradoxurus* are less specialized for meat eating than those of most viverrids, having low, rounded cusps on rather square molars. There are well-developed anal scent glands in both sexes. Females have three pairs of mammae.

Musangs are nocturnal forest dwellers. They are expert climbers and spend most of their time in trees, where they utilize cavities or secluded nooks. They are often found about human habitations, probably because of the presence of rats and mice. Under such conditions, they shelter in thatched roofs and in dry drain tiles and pipes. They eat small vertebrates, insects, fruits, and seeds. They are fond of the palm juice, or "toddy," collected by the natives—thus one of the vernacular names.

A female with five kittens radiotracked for about a month in Nepal used a home range of 1.2 ha. (Dhungel and Edge 1985). Reproduction occurs throughout the year, though Lekagul and McNeely (1977) stated that the young of *P. hermaphroditus* seem to be seen most often from October to December. Litter size is two to five young. Sexual maturity is attained at 11–12 months in *P. hermaphroditus.* A captive of

Masked palm civet *(Paguma larvata)*, photo from New York Zoological Society.

this species lived for 22 years and 5 months (Medway 1978).

Groves (1976) noted that *P. hermaphroditus* has been carried about from island to island by people, for use as a rat catcher. Such activity is probably responsible for the presence of this species on Sulawesi, Timor, and various other islands. The species *P. jerdoni*, however, is classified as indeterminate by the IUCN.

CARNIVORA; VIVERRIDAE; Genus PAGUMA
Gray, 1831

Masked Palm Civet

The single species, *P. larvata*, occurs from Kashmir to Indochina and the Malay Peninsula, in much of eastern and southern China, and on the Andaman Islands, Taiwan, Hainan, Sumatra, and Borneo (Chasen 1940; Ellerman and Morrison-Scott 1966).

Head and body length is 508–762 mm, tail length is usually 508–636 mm, and weight is usually 3.6–5.0 kg. In the facial region there is generally a mask, which consists of a median white stripe from the top of the head to the nose, a white mark below each eye, and a white mark above each eye extending to the base of the ear and below. The general color is gray, gray tinged with buff, orange, or yellowish red. There are no stripes or spots on the body and no spots or bands on the tail. The distal part of the tail may be darker than the basal part, and the feet are blackish. This genus differs externally from *Paradoxurus* and *Arctogalidia* in the absence of the striping and spotting. Like *Paradoxurus*, *Paguma* has a potent anal gland secretion, which it uses to ward off predators. The conspicuously marked head has been interpreted as a warning signal of the presence of the secretion. Females have two pairs of mammae (Lekagul and McNeely 1977).

The masked palm civet frequents forests and brush country. It reportedly raises its young in tree holes, at least in Nepal. It is arboreal and nocturnal (Roberts 1977). The om-

nivorous diet includes small vertebrates, insects, and fruits.

Paguma is solitary, and apparently most young in the western parts of its range are born in spring and early summer (Roberts 1977). Births in Borneo have taken place in October (Banks 1978). Data on captives indicate that there may be two breeding seasons, in early spring and late autumn. Litters contain one to four young. They open their eyes at 9 days and are almost the size of adults by 3 months. Maximum known longevity is 15 years and 5 months (Lekagul and McNeely 1977; Medway 1978).

In Tenasserim, *Paguma* is reported to be a great ratter and not to destroy poultry. Medway (1978), however, wrote that it has been known to raid hen runs. The genus has been introduced on central Honshu, Japan (Corbet 1978).

CARNIVORA; VIVERRIDAE; Genus MACROGALIDIA
Schwarz, 1910

Sulawesian Palm Civet

The single species, *M. musschenbroeki*, occurs only on Sulawesi (Laurie and Hill 1954).

Wemmer et al. (1983) reported that an adult male had a head and body length of 715 mm, a tail length of 540 mm, and a weight of 6.1 kg, and that respective measurements for two adult females were 650 and 680 mm, 480 and 445 mm, and 3.8 and 4.5 kg. The upper parts are light brownish chestnut to dark brown. The underparts range from fulvous to whitish, with a reddish breast. The cheeks and a patch above the eye are usually buffy or grayish. Faint brown spots and bands are usually present on the sides and lower back, and the tail is ringed with dark and pale brown. The tail has more bands than that of *Arctogalidia* or *Paradoxurus*. Other distinguishing characters are the short, close fur and a whorl in the neck with the hairs directed forward. Both of the females noted above had two pairs of inguinal mammae.

According to Wemmer and Watling (1986), *Macrogalidia*

Sulawesian palm civet *(Macrogalidia musschenbroeki)*, photo by Christen Wemmer.

occurs in both lowland and montane forests up to about 2,600 meters. It seems to be dependent on primary forests but also moves onto adjacent grasslands and farms. It is a skillful climber and sometimes moves through trees but probably forages mainly on the ground. It feeds mainly on small mammals and fruits, especially palm fruit, and also raids domestic chickens and pigs. It is probably solitary, though a female and young may share a home range for a period after weaning. Although once thought to be extinct or restricted to the northern peninsula of Sulawesi, and currently classified as rare by the IUCN, recent observations indicate that *Macrogalidia* occurs in most parts of the island. It now seems to be neither abundant nor scarce but could be potentially jeopardized in some areas by logging and agricultural activity.

CARNIVORA; VIVERRIDAE; Genus ARCTICTIS
Temminck, 1824

Binturong

The single species, *A. binturong*, occurs from Burma, and possibly from Nepal, to Indochina and the Malay Peninsula and on Sumatra, Bangka, Java, the Rhio Archipelago, Borneo, and Palawan (Chasen 1940; Ellerman and Morrison-Scott 1966).

Head and body length is 610–965 mm, tail length is 560–890 mm, and weight is usually 9–14 kg. The fur is long and coarse, that on the tail being longer than that on the body. The hairs are black and lustrous, often with a gray, fulvous, or buff tip. The head is finely speckled with gray and buff, and the edges of the ears and the whiskers are white. The ears

have long hairs on the back that project beyond the tips and produce a fringed or tufted effect. The tail is particularly muscular at the base and is prehensile at the tip. The only other carnivore with a truly prehensile tail is the kinkajou *(Potos)*, which the binturong resembles in habits to some extent. Females have two pairs of mammae.

The binturong lives in dense forests and is nowhere abundant. It is mainly arboreal and nocturnal. When resting, it usually lies curled up, with the head tucked under the tail. It has never been observed to leap; rather, it progresses slowly but skillfully, using the tail as an extra hand. Its movements, at least during daylight hours, are rather slow and cautious, the tail slowly uncoiling from the last support as the animal moves carefully forward. According to Medway (1978), the binturong is reported to dive, swim, and catch fish. The diet also includes birds, carrion, fruit, leaves, and shoots.

Medway (1978) stated that *Arctictis* occurs either alone or in small groups of adults with immature offspring. Captives are very vocal, uttering high-pitched whines and howls, rasping growls, and, when excited, a variety of grunts and hisses. Lekagul and McNeely (1977) wrote that breeding seems to occur throughout the year. Numerous observations in captivity (Bulir 1972; Gensch 1963b; Grzimek 1975; Kuschinski 1974; Xanten, Kafka, and Olds 1976) indicate that females are nonseasonally polyestrous and may give birth to two litters annually. Wemmer and Murtaugh (1981) reported the following data for reproduction in captivity: breeding occurs year-round, but with a pronounced birth peak from January to March; the estrous cycle averages 81.8 days; the mean gestation period is 91.1 days, ranging from 84 to 99 days; the number of young per litter averages 1.98 and ranges from 1 to 6; the young weigh an average of 319 grams each at birth and begin to take solid food at 6–8 weeks; the mean age of first

Binturong *(Arctictis binturong)*, photo from New York Zoological Society.

mating is 30.4 months in females and 27.7 months in males; and both sexes can remain fertile until at least 15 years. According to Jones (1982), one binturong was still living after 22 years and 8 months in captivity.

Arctictis is sometimes kept as a pet. It is said to be easily domesticated, to become quite affectionate, and to follow its master like a dog.

CARNIVORA; VIVERRIDAE; **Genus HEMIGALUS**
Jourdan, 1837

Banded Palm Civets

There are two species (Chasen 1940; Ellerman and Morrison-Scott 1966; Lekagul and McNeely 1977; Medway 1977):

H. derbyanus, Tenasserim, Malay Peninsula, Sumatra and certain small islands to the west, Borneo;
H. hosei, Borneo.

The latter species sometimes has been placed in a separate genus, *Diplogale* Thomas, 1912.

In *H. derbyanus* head and body length is 410–510 mm, tail length is 255–383 mm, and weight is usually 1.75–3.0 kg. On the head there is a narrow, median dark streak extending from the nose to the nape, and on each side of this there is a broader dark stripe that encircles the eye and passes backward over the base of the ear. Two broad stripes, sometimes more or less broken into shorter stripes or spots, run backward from the neck and curve downward to the elbow. Behind these are two shorter stripes. The back behind the shoulders is marked with four or five broad transverse stripes separated by pale, usually narrower spaces, and there are two imperfect stripes at the base of the tail. The ground color is whitish to orange buff, usually lighter and more buffy below, and the tail is usually black. The hair on the back of the neck is reversed, in that the tips of the hairs point forward. The five-toed feet have strongly curved claws that are retractile like those of cats. Small scent glands are present.

In *H. hosei* head and body length is about 600 mm and tail

length is about 300 mm. Coloration is dark brown or black above and grayish, yellowish white, or slightly rufescent below. The ears are thinly haired and white inside. A buffy gray patch extends from above the eye to the cheek and terminates where it meets the white of the lips and throat. The inner side of the limb near the body is grayish, while the remainder of each limb is black. The tail is not banded but is dark throughout.

On the Malay Peninsula, *H. derbyanus* is restricted to tall forest and apparently is largely terrestrial (Medway 1978). This species, however, is at least partly arboreal and climbs well (Lekagul and McNeely 1977). *H. hosei*, of Borneo, is found mainly in montane forest and is largely terrestrial (Medway 1977). All specimens of *Hemigalus* taken by Davis (1962) in Borneo were collected on the ground. He observed that the animals were exclusively nocturnal and apparently foraged on the forest floor, picking up food from the surface. Orthopterans and worms made up 80 percent of the contents of 12 stomachs, and the rest of the food was mostly other invertebrates.

Gangloff (1975) reported that captive *H. derbyanus* were fond of fruit, did not construct nests, and marked with scent. A pregnant female *H. derbyanus* with one embryo was taken in Borneo in February. A captive female in the Wassenaar Zoo, Netherlands, had two young. They weighed 125 grams each at birth, opened their eyes after 8–12 days, and first took solid food at about 70 days (Ewer 1973). One specimen was still living after 12 years in captivity (Jones 1982). *H. derbyanus* is on appendix 2 of the CITES. *H. hosei* is known only from 15 museum specimens, the last of which was collected in 1955 (Schreiber et al. 1989).

CARNIVORA; VIVERRIDAE; **Genus CHROTOGALE**
Thomas, 1912

Owston's Palm Civet

The single species, *C. owstoni*, now is known from about 39 specimens taken in Yunnan in extreme southern China, Laos, and northern Viet Nam (Schreiber et al. 1989).

Banded palm civet *(Hemigalus derbyanus)*, photo from Duisburg Zoo.

Owston's palm civet *(Chrotogale owstoni)*, photo from Hanoi Zoo through Jorg Adler.

Otter civet *(Cynogale bennettii)*, photo from San Diego Zoological Society.

Head and body length is 508–635 mm and tail length is 381–482 mm. The body and base of the tail have alternating and sharply contrasting dark and light transverse bands, and longitudinal stripes are present on the neck. The pattern of stripes and bands resembles that of *Hemigalus derbyanus*, but it is supplemented by black spots on the sides of the neck, the forelimbs and thighs, and the flanks. Four seems to be the maximum number of dorsal bands. It has been suggested that this striking pattern serves as a warning signal, as *Chrotogale* is thought to possess a particularly foul-smelling anal gland secretion. The underparts are pale buffy, and a narrow orange midventral line runs from the chest to the inguinal region. The terminal two-thirds of the tail is completely black.

Form and markings are strikingly similar to those of *Hemigalus derbyanus*, but the hairs on the back of the neck of *Chrotogale* are not reversed in direction. *Chrotogale* is also distinguished by cranial and dental characters. The incisor teeth are remarkable in that they are broad, close-set, and arranged in practically a semicircle, a type unique among carnivores and only approached in certain of the marsupials. The other teeth and the skull are also peculiar, and indicate habits and a mode of life different from that of other genera in the family Viverridae. As yet, however, there has been no substantive investigation of the natural history of the genus. The stomachs of two specimens were found to contain earthworms.

Information cited by Schreiber et al. (1989) indicates that *Chrotogale* may be largely terrestrial and that it prefers habitat in the vicinity of rivers in primary and secondary forests. However, it can survive close to villages and even approaches houses in search of kitchen wastes. It is subject to considerable hunting pressure and is of much conservation concern.

CARNIVORA; VIVERRIDAE; Genus **CYNOGALE**
Gray, 1837

Otter Civet

The single species, *C. bennettii*, occurs in northern Viet Nam, the Malay Peninsula, Sumatra, and Borneo (Ellerman and

Morrison-Scott 1966; Lekagul and McNeely 1977). The one known specimen from Viet Nam was originally referred to a separate species, *C. lowei*. Schreiber et al. (1989), who continued to recognize *C. lowei*, reported likely records of it in Yunnan and northern Thailand.

Head and body length is 575–675 mm, tail length is 130–205 mm, and weight is usually 3–5 kg. The form is somewhat like that of the otters *(Lutra)*. The underfur is close, soft, and short. It is pale buff near the skin and shades to dark brown or almost black at the tip. The longer, coarser guard hairs are usually partially gray, which gives a frosted or speckled effect on the head and body. The lower side of the body is lighter brown and not speckled with gray. The whiskers are remarkably long and plentiful; those on the snout are fairly long, but those on a patch under the ear are the longest. The newly born young lack dorsal speckling; they have some gray on the forehead and ears and two longitudinal stripes down the sides of the neck extending under the throat.

Because of the deepening and expansion of the upper lip, the rhinarium occupies a horizontal position, with the nostrils opening upward on top of the muzzle. The nostrils can be closed by flaps, an adaptation for aquatic life. The ears can also be closed. Although the webbing between the feet does not extend farther toward the tips of the digits than in such genera as *Paradoxurus*, it is quite broad, and the fingers are capable of considerable flexion. A glandular area, merely three pores in the skin, is located near the genitals and secretes a mild scent material. The premolar teeth are elongate and sharp, adapted for capturing and holding prey, while the molars are broad and flat, for crushing. Females have four mammae.

The otter civet is usually found near streams and swampy areas. It can climb well and, when chased by dogs, often takes refuge in a tree. While walking, it usually carries its head and tail low and arches its back. Although it is partly adapted for an aquatic life, its tail is short and lacks special muscular power, and the webbing between the digits is only slightly developed. *Cynogale* is thus probably a slow swimmer and cannot turn quickly in the water. It probably captures aquatic animals only after they have taken shelter from the chase and catches some birds and mammals as they come to drink. It cannot be seen by its prey, because it is submerged with only

Otter civet *(Cynogale bennettii)*, photo from San Diego Zoological Society.

the tip of the nose exposed above the surface of the water. The diet includes crustaceans, perhaps mollusks, fish, birds, small mammals, and fruits. There are records of pregnant females with two and three embryos. Young, still with the mother, have been noted in May in Borneo. A captive specimen lived for five years (Jones 1982). *Cynogale* is on appendix 2 of the CITES. Schreiber et al. (1989) noted that it is rare and probably jeopardized by expanding human settlement and agriculture along rivers.

CARNIVORA; VIVERRIDAE; **Genus FOSSA**
Gray, 1864

Malagasy Civet

The single species, *F. fossa*, originally was found throughout the forested parts of Madagascar (IUCN 1972). Although the generic name is the same as the vernacular term for another Malagasy viverrid *(Cryptoprocta)*, the two animals are distinctly different.

Head and body length is 400–450 mm and tail length is 210–30 mm (Albignac 1972). Males weigh up to 2 kg, and females up to 1.5 kg (Coetzee, *in* Meester and Setzer 1977). The ground color is grayish, washed with reddish. There are four rows of black spots on each side of the back and a few black spots on the backs of the thighs. These spots may merge to form stripes, and the gray tail is banded with brown. The underparts are grayish or whitish and more or less obscurely spotted. The limbs are slender, perhaps being adapted for running. There are no anal scent glands, but there probably are marking glands on the cheeks and neck (Albignac 1972).

The Malagasy civet inhabits evergreen forests and shelters in hollow trees or crevices. It is nocturnal and may occur in trees or on the ground. The preferred foods are crustaceans, worms, small eels, and frogs. Other kinds of animal matter and fruit are also taken (Coetzee, *in* Meester and Setzer 1977).

Fossa lives in pairs, which share a territory. Vocalizations include cries, groans, and a characteristic "coq-coq," heard only in the presence of more than one individual. Births have been recorded from October to January. The gestation period is 3 months, and a single offspring is born. It weighs 65–70 grams at birth, is weaned after 2 months, and probably attains adult weight at 1 year (Albignac 1970a, 1972; Coetzee, *in* Meester and Setzer 1977). A captive specimen lived for 11 years (Jones 1982).

The Malagasy civet is classified as vulnerable by the IUCN

Malagasy civets *(Fossa fossa)*, photo from U.S. National Zoological Park.

(1972) and is on appendix 2 of the CITES. Loss of habitat and excessive hunting have restricted its range to the eastern and northwestern rainforests of Madagascar.

CARNIVORA; VIVERRIDAE; **Genus EUPLERES**
Doyère, 1835

Falanouc

The single species, *E. goudotii*, is found in the coastal forests of Madagascar (Coetzee, *in* Meester and Setzer 1977).

Head and body length is 450–650 mm, tail length is 220–50 mm, and weight is 2–4 kg (Albignac 1974). The subspecies *E. g. goudotii* is fawn-colored above and lighter below. In the subspecies *E. g. major* the males are brownish and the females are grayish. The pelage is woolly and soft, being made up of a dense underfur and longer guard hairs. The tail is covered by rather long hairs that give a bushy appearance.

Eupleres has a pointed muzzle, a narrow and elongate head, and short, conical teeth. It resembles the civets in some structural features and the mongooses in others. The small teeth are similar to each other and resemble those of insectivores, rather than carnivores. Indeed, *Eupleres* was classified as an insectivore before its somewhat obscure relationship with the mongooses was detected. The claws are relatively long and not retractile, or are only imperfectly so. The feet are peculiar in the comparatively large size and low position of the great toe and thumb.

The falanouc inhabits humid, lowland forests. It is crepuscular and nocturnal, resting by day in crevices and burrows. It is terrestrial, and if threatened, it may either run or remain motionless. In autumn, up to 800 grams of fat can be accumulated in the tail, and it has been suggested that *Eupleres* hibernates during winter. Observations of active indi-

viduals, however, have been made in winter. The diet consists mainly of earthworms and also includes other invertebrates and frogs, but apparently not reptiles, birds, rodents, or fruit (Albignac 1974; Coetzee, *in* Meester and Setzer 1977).

Eupleres may live alone or in small family groups. It has several vocalizations and other means of communication. Mating probably takes place in July or August (winter), and a birth was observed in November. Litters contain one or two young. Weight of the newborn is 150 grams, and weaning occurs at nine weeks (Albignac 1974; Coetzee, *in* Meester and Setzer 1977).

The falanouc is classified as vulnerable by the IUCN (1976) and is on appendix 2 of the CITES. It has declined in numbers and distribution because of excessive hunting, habitat destruction, and possibly competition from the introduced *Viverricula indica*.

CARNIVORA; VIVERRIDAE; **Genus GALIDIA**
I. Geoffroy St.-Hilaire, 1837

Malagasy Ring-tailed Mongoose

The single species, *G. elegans*, is found in eastern, west-central, and extreme northern Madagascar (Albignac 1969).

Head and body length is about 380 mm and tail length is about 305 mm. Weight is 700–900 grams (Coetzee, *in* Meester and Setzer 1977). The general coloration is dark chestnut brown, and the tail is ringed with dark brown and black. *Galidia* has some of the structural features of civets and some of mongooses. The feet differ from those of *Galidictis* in having shorter digits, fuller webbing, shorter claws, and more hairy soles. The lower canine teeth are smaller. From *Salanoia*, *Galidia* is distinguished by its ringed tail and very small second upper premolar. Dissection of two individuals

Falanoucs *(Eupleres goudotii)*, photos by R. Albignac.

Malagasy ring-tailed mongoose *(Galidia elegans)*, photo from U.S. National Zoological Park.

indicates the presence of a scent gland, closely associated with the external genitalia, in males but not in females.

Galidia occurs in humid forests. It shelters in burrows, which it digs very rapidly, and probably also in hollow trees. It is mainly diurnal but may also be active at night. More arboreal than most mongooses, it is able to climb and descend on vertical trunks only 4 cm in diameter. It can also swim. The diet consists mainly of small mammals, birds and their eggs, and frogs and also includes fruits, fish, reptiles, and invertebrates (Albignac 1972; Coetzee, *in* Meester and Setzer 1977).

Galidia is less social than most mongooses, being found alone or in pairs. Mating in Madagascar occurs from April to November, and births from July to February. The gestation period is 79–92 days, and a single young, weighing 50 grams, is produced. Physical maturity is reportedly attained at 1 year, and sexual maturity at 2 years (Coetzee, *in* Meester and Setzer 1977; Larkin and Roberts 1979). A captive animal was still living after 13 years and 2 months (Jones 1982).

CARNIVORA; VIVERRIDAE; **Genus GALIDICTIS**
I. Geoffroy St.-Hilaire, 1839

Malagasy Broad-striped Mongooses

There are two species (Coetzee, *in* Meester and Setzer 1977; Wozencraft 1986, 1987):

G. fasciata, eastern Madagascar;
G. grandidieri, southwestern Madagascar.

Head and body length is 320–40 mm and tail length is 280–300 mm (Albignac 1972). The general body color is pale

brown or grayish. In the subspecies *G. fasciata striata* there are usually 5 longitudinal black bands or stripes on the back and sides, and the tail is whitish. In *G. fasciata fasciata* there are usually 8–10 stripes, and the tail is bay-colored and somewhat bushy. *G. grandidieri* is larger than *G. fasciata* and has 8 dark brown longitudinal stripes on the back and sides that are narrower than the intervening spaces.

Galidictis differs from other viverrids in the color pattern and in cranial and dental characters. The feet differ from those of *Galidia* in having longer digits, less extensive webbing, and longer claws. A scent pouch is present in females.

The species *G. fasciata* occurs in forests and is nocturnal and crepuscular. The diet consists mainly of small vertebrates, especially rodents, and also includes invertebrates. The two known specimens of *G. grandidieri* were taken in an area of spiny desert vegetation. *Galidictis* is found in pairs or small social groups. It apparently produces one young per year, during the summer (Albignac 1972; Coetzee, *in* Meester and Setzer 1977). Young individuals are said to tame readily, to follow their master, and even to sleep in his lap. *G. fasciata* is classified as indeterminate by the IUCN. Schreiber et al. (1989) reported it to be locally common but threatened by habitat destruction.

CARNIVORA; VIVERRIDAE; **Genus MUNGOTICTIS**
Pocock, 1915

Malagasy Narrow-striped Mongoose

The single species, *M. decemlineata*, is found in western and southwestern Madagascar (Coetzee, *in* Meester and Setzer 1977).

Head and body length is 250–350 mm, tail length is 230–

Malagasy broad-striped mongoose *(Galidictis fasciata)*, photo from *Zoologie de Madagascar*, G. Grandidier and G. Petit.

Malagasy narrow-striped mongoose *(Mungotictis decemlineata)*. Inset: head *(M. decemlineata)*. Photos by Don Davis.

70 mm, and weight is 600–700 grams. The fur is rather dense and generally gray beige in color. There are usually 8–10 dark stripes on the back and flanks. The underparts are pale beige. The soles of the feet are naked, and the digits are partly webbed. There are glands on the side of the head and on the neck, which appear to be for marking with scent. Females have a single pair of inguinal mammae (Albignac 1971, 1972, 1976).

Mungotictis is found on sandy, open savannahs. It is diurnal and both arboreal and terrestrial. It spends the night in tree holes during the wet summer and in ground burrows during the dry winter. It can swim well. The diet consists mostly of insects but also includes small vertebrates, birds' eggs, and invertebrates. To break an egg or the shell of a snail, *Mungotictis* lies on one side, grasps the object with all four feet, and throws it abruptly until it breaks and the contents can be lapped up (Albignac 1971, 1972, 1976).

In a radiotracking study, Albignac (1976) determined that 22 individuals inhabited 300 ha. The animals were divided into two stable social units, which sometimes engaged in agonistic encounters where their ranges met. Each group contained several adults of both sexes, as well as juveniles and young of the year. Group cohesiveness and intrarelationships varied. Generally, adult males and females came together in the summer. In the winter there was division into small units, such as temporary pairs, maternal family parties, all-male groups, and solitary males.

The breeding season extends from December to April, peaking in February and March (summer). The gestation period is 90–105 days. There is usually a single offspring weighing 50 grams at birth. Weaning occurs at 2 months, and separation from the mother comes after 2 years (Albignac 1972, 1976).

This mongoose is now classified as vulnerable by the IUCN. Schreiber et al. (1989) reported that it appears subject to very little direct human persecution, but that its habitat is being burned and cleared at an alarming rate.

CARNIVORA; VIVERRIDAE; **Genus SALANOIA**
Gray, 1864

Malagasy Brown-tailed Mongoose, or Salano

The single species, *S. concolor,* is found in northeastern Madagascar (Coetzee, *in* Meester and Setzer 1977).

Head and body length is 250–300 mm and tail length is 200–250 mm (Albignac 1972). The general coloration is brown, and there are either dark or pale spots. The tail is the same color as the body and is not ringed. The claws are not strongly curved, the ears are broad and short, and the muzzle is pointed.

The salano occurs only in the evergreen forests on the northeastern part of the central plateau of Madagascar. It is typically diurnal, sheltering at night in tree trunks or burrows. It feeds mainly on frogs, small reptiles, and rodents. It occurs individually or in pairs, depending on the season. The young are born mainly during the summer (Albignac 1972; Coetzee, *in* Meester and Setzer 1977). A captive lived for 4 years and 9 months (Jones 1982).

CARNIVORA; VIVERRIDAE; **Genus HERPESTES**
Illiger, 1811

Mongooses

There are 3 subgenera and 14 species (Chasen 1940; Coetzee, *in* Meester and Setzer 1977; Corbet 1978; Ellerman and Morrison-Scott 1966; Lynch 1981; Medway 1977; Sanborn 1952; Wells and Francis 1988):

subgenus *Herpestes* Illiger, 1811

H. ichneumon, southern Spain and Portugal, Asia Minor, Palestine, Morocco to Tunisia, Egypt, possibly eastern Libya, most of Africa south of the Sahara;
H. javanicus, Thailand, Indochina, Malay Peninsula, Java;
H. auropunctatus, Iraq to Malay Peninsula, Hainan;
H. edwardsi, eastern Arabian Peninsula to Assam, Sri Lanka;
H. smithi, India, Sri Lanka;
H. fuscus, southern India, Sri Lanka;

Malagasy brown-tailed mongoose *(Salanoia concolor)*, photo of mounted specimen in Field Museum of Natural History.

A. African mongoose *(Herpestes ichneumon)*, photo by Ernest P. Walker. B. Crab-eating mongoose *(H. urva)*, photo by Robert E. Kuntz.

H. vitticollis, southern India, Sri Lanka;

H. urva, Nepal to southeastern China and peninsular Malaysia, Taiwan, Hainan;

H. brachyurus, Malay Peninsula, Sumatra, Borneo, Palawan;

H. hosei, known only by the type specimen from Sarawak in northern Borneo;

H. semitorquatus, Sumatra, Borneo;

subgenus *Xenogale* J. A. Allen, 1919

H. naso, southeastern Nigeria to eastern Zaire;

subgenus *Galerella* Gray, 1865

H. pulverulentus, southern Namibia, South Africa, Lesotho;

H. sanguineus, savannah and semiarid regions of Africa south of the Sahara.

An additional species, *H. palustris,* has been described from the vicinity of Calcutta in eastern India (Ghose 1965; Ghose and Chaturvedi 1972). Although said to be most closely related to *H. auropunctatus,* Ewer (1973) suggested that it may be a subspecies of *H. urva.* Lekagul and McNeely (1977) considered *H. auropunctatus* to be conspecific with *H. javanicus*. Some authorities, such as Rosevear (1974), treat *Xenogale* and *Galerella* as distinct genera. Watson and Dippenaar (1987) agreed that *Galerella* warrants generic rank and also recognized *G. nigrata*, of Namibia, as a full species.

Head and body length is 230–650 mm, tail length is 230–510 mm, and weight is 0.4–4.0 kg. Coloration varies considerably. Some forms are greenish gray, yellowish brown, or grayish brown. Others are finely speckled with white or buff. The underparts are generally lighter than the back and sides and are white in some species. The fur is short and soft in some species and rather long and coarse in others. The body is slender, the tail is long, there are five digits on each limb, the hind foot is naked to the heel, and the foreclaws are sharp and curved. Small scent glands are situated near the anus, and some species can eject a vile-smelling secretion. Females have four or six mammae.

Mongooses occupy a wide variety of habitats, ranging from densely forested hills to open, arid plains. They shelter in hollow logs or trees, holes in the ground, or rock crevices. They may be either diurnal or nocturnal. They are basically terrestrial but are very agile, and some species can climb skillfully (Grzimek 1975). During the morning, they frequently stretch out in an exposed area to sun themselves. The diet includes insects, crabs, fish, frogs, snakes, birds, small mammals, fruits, and other vegetable matter. Some species kill cobras and other venomous snakes. Contrary to popular belief, mongooses are not immune to the bites of these reptiles. Rather, they are so skillful and quick in their movements that they avoid being struck by the snake and almost invariably succeed in seizing it behind the head. The battle usually ends with the mammal's eating the snake.

Data cited by Ewer (1973) indicate that in Hawaii the individual home range diameter of *H. auropunctatus* is about 1.6 km for males and 0.8 km for females. Mongooses are found alone, in pairs, or in groups of up to 14 individuals. Ewer noted that family parties of *H. pulverulentus* den together but forage individually. Taylor (1975) wrote that *H. sanguineus* travels alone or in pairs, that its home range may be as small as 1 sq km but is much larger in desert areas, that it may be territorial, and that it is generally silent. Ben-Yaacov and Yom-Tov (1983) found that in Israel family groups of *H. ichneumon* occupied permanent home ranges and consisted of 1 adult male, 2–3 females, and young. The group produced a single annual litter in the spring, and the young remained with the family for a year or more.

Some species breed throughout the year, with females giving birth two or three times annually. One female *H. edwardsi* produced five litters in 18 months. Gestation periods range from about 42 days in *H. auropunctatus* to 84 days in *H. ichneumon*. Litters contain one to four young. The young of *H. auropunctatus* are weaned after 4–5 weeks. Captive *H. ichneumon* have lived over 20 years (Ewer 1973; Grzimek 1975; Kingdon 1977; Lekagul and McNeely 1977; Roberts 1977).

Mongooses have been widely introduced by people to kill rats and snakes. The presence of *H. ichneumon* in Spain and Portugal probably results from introduction in ancient times. That species also has been brought to Yugoslavia and Madagascar, and *H. edwardsi* has been brought to central Italy. Populations of *H. auropunctatus* have been established on most Caribbean islands, the northeastern coast of South America, Mafia Island off East Africa, and the Hawaiian and Fiji islands. Several individuals of this species have been taken on the mainland of North America. *H. edwardsi* has been introduced on the Malay Peninsula, Mauritius, and the Ryukyu Islands (Corbet 1978, 1984; Gorman 1976; Nellis et al. 1978; Van Gelder 1979). In addition to killing rats and snakes, mongooses have destroyed harmless birds and

mammals and have contributed to the extinction or endangerment of many desirable species of wildlife. They also have become pests by preying on poultry. The importation or possession of mongooses is therefore now forbidden by law in some countries.

CARNIVORA; VIVERRIDAE; **Genus MUNGOS**
E. Geoffroy St.-Hilaire and G. Cuvier, 1795

Banded and Gambian Mongooses

There are two species (Coetzee, *in* Meester and Setzer 1977):

M. gambianus (Gambian mongoose), savannah zone from Gambia to Nigeria;
M. mungo (banded mongoose), Gambia to northeastern Ethiopia and south to South Africa.

The name *Mungos* was formerly sometimes used for many other species of mongooses, including those now assigned to *Herpestes*.

Head and body length is 300–450 mm, tail length is 230–90 mm, and weight is 1.0–2.2 kg. *M. mungos* is brownish gray with dark brown and well-defined yellowish or whitish bands across the back. The banded pattern is produced by hair markings of the same type that produce the ground color of the remainder of the body. The hairs are alternately ringed with dark and light bands; the color rings of the individual hairs coincide with like colors on adjacent hairs. *M. gambianus* lacks the transverse bands but has a dark streak on the side of the neck (Coetzee, *in* Meester and Setzer 1977).

The pelage is coarse; compared with other mongooses, *Mungos* has little underfur. Although the tail is not bushy, it is covered with coarse hair and is tapered toward the tip. The foreclaws are elongate, the soles are naked to the wrist and heel, and there are five digits on each limb. There is no naked grooved line from the tip of the nose to the upper lip. Females have six mammae.

The information in the remainder of this account applies to *M. mungos* and was taken from Kingdon (1977), Neal (1970), Rood (1974, 1975), Rosevear (1974), and Simpson (1964). The banded mongoose is found in grassland, brushland, woodland, and rocky, broken country. It dens mainly in old termite mounds but also in such places as erosion gullies, abandoned aardvark holes, and hollow logs. Dens are communal and consist of one to nine entrance holes, a central sleeping chamber of about 1–2 cubic meters, and perhaps several smaller chambers. Most dens are used only for a few days, but some favorite sites may be occupied for as long as 2 months. *M. mungos* is terrestrial and diurnal and has excellent senses of vision, hearing, and smell. A group generally emerges from the den around 0700–0800 hours, forages for several hours, rests in a shady spot during the hottest part of the day, forages again, and returns to a den before sunset. More time is spent in the vicinity of the den if young are present therein. A group generally covers 2–3 km per day, moving in a zigzag pattern and searching among rocks and vegetation for food. The diet consists largely of invertebrates, especially beetles and millipedes, and also includes small vertebrates. To break an egg and obtain the contents, *M. mungos* grasps the item with its forefeet and propels it backward between its hind feet and against a hard object.

In the Ruwenzori National Park of Uganda, population density was found to be about 18/sq km, and group home range varied from about 38 to 130 ha. In the Serengeti, home

Banded mongoose *(Mungos mungo)*, photo from Zoological Garden Berlin-West through Ernst von Roy.

range may be over 400 ha. Ranges overlap, but intergroup encounters are generally noisy and hostile and sometimes involve chasing and fighting. Groups contain up to 40 individuals, usually about 10–20, including several adults of both sexes. Captive females have been seen to dominate males. The animals sleep in contact and forage in a fairly close, but not bunched, formation. Groups are cohesive, but some splitting off has been observed, and mating between members of different groups sometimes occurs. Individuals may mark one another with scent from anal glands. The most common vocalization is a continuous birdlike twittering that probably serves to keep the group together during foraging. There are also various agonistic growls and screams and an alarm chitter.

In East Africa, at least, reproduction continues throughout the year. Breeding is synchronized within a given group, with several females bearing litters at approximately the same time. Groups breed up to four times per year, though it cannot be said that any one female has that many litters annually. A period of mating often begins within 1–2 weeks after the birth of young. The gestation period is about 2 months. The number of young per litter seems usually to be about two or three but may be as high as six. The young weigh about 20 grams each at birth but grow rapidly. They are apparently kept together and raised commonly by the group. They suckle indiscriminantly from any lactating female and are usually guarded by one or two adult males while the rest of the group forages. They begin to travel with the others at

Banded mongooses *(Mungos mungo)*, photo by J. P. Rood.

Cusimanse *(Crossarchus obscurus)*, photo by Ernest P. Walker.

about 1 month. Females attain sexual maturity at 9–10 months. According to Van Rompaey (1978), a captive M. mungo lived to an age of approximately 12 years.

CARNIVORA; VIVERRIDAE; **Genus CROSSARCHUS**
F. Cuvier, 1825

Cusimanses

There are four species (Goldman 1984; Schreiber et al. 1989):

C. obscurus, Sierra Leone to Ghana;
C. platycephalus, southern Benin to southern Cameroon and possibly Congo;
C. alexandri, Zaire, Uganda, probably Congo and Central African Republic;
C. ansorgei, northwestern and north-central Zaire, one specimen from northwestern Angola.

Head and body length is 305–450 mm, tail length is 150–255 mm, and weight is 450–1,450 grams. The body is covered by relatively long, coarse hair that is a mixture of browns, grays, and yellows. The head is usually lighter-colored than the remainder of the body, while the feet and legs are usually the darkest. The legs are short, the tail is tapering, the ears are small, and the face is sharp.

Cusimanses live in forests and swampy areas. Various reports suggest that they may be active either by day or by night (Kingdon 1977). They travel about in groups, seldom remaining in any one locality longer than two days and taking temporary shelter in any convenient place. While seeking food, they scratch and dig in dead vegetation and in the soil. The diet consists principally of insects, larvae, small reptiles, crabs, tender fruits, and berries. It is said that cusimanses crack the shells of snails and eggs by hurling them with the forepaws back between the hind feet and against some hard object.

Groups contain 10–24 individuals. These probably represent 1–3 family units, each with a mated pair and the surviving members of 2–3 litters. There is no evidence of seasonal breeding in either East or West Africa. Observations in captivity indicate that there may be several litters annually (Kingdon 1977; Rosevear 1974). According to Goldman (1987), the gestation period averages 58 days, and litters con-

tain 2–4, usually 4, young; the young open their eyes after 12 days, take solid food at 3 weeks, and attain sexual maturity at about 9 months. Captives have been estimated to live for 9 years.

Cusimanses tame easily and make good pets. They are affectionate, playful, clean, and readily housebroken, but they sometimes mark objects with their anal scent glands (Rosevear 1974).

CARNIVORA; VIVERRIDAE; **Genus LIBERIICTIS**
Hayman, 1958

Liberian Mongoose

The single species, *L. kuhni,* now is known by 27 specimens from northeastern Liberia (Carnio 1989; Carnio and Taylor 1988).

Head and body length of an adult male is 423 mm, tail length is 197 mm, and weight is 2.3 kg. The predominant color of the pelage is dark brown. A dark stripe, bordered above and below by a pale stripe, is present on the neck. The throat is pale, the tail is slightly bicolored, and the legs are dark. From *Crossarchus, Liberiictis* is distinguished externally by the presence of neck stripes, a more robust body, and apparently longer ears (Schlitter 1974). The skull of *Liberiictis* is larger than that of *Crossarchus,* the rostrum and nasals are more elongate, the teeth are proportionally smaller and weaker, and there is an additional premolar in both the upper and lower jaws.

Schlitter (1974) noted that the long claws of the front feet, the long mobile snout, and the weak dentition of *Liberiictis* indicate that it is a terrestrial animal with a primarily insectivorous diet. Carnio and Taylor (1988) stated that the diet also includes worms, eggs, and small vertebrates. Schlitter's two specimens were taken in a densely forested area traversed by numerous streams. People native to the area said that *Liberiictis* is diurnal and found only on the ground. An adult male was taken in a snare on the ground. A juvenile female was excavated from a burrow associated with a termite mound on 29 July. An adult of unknown sex was also in the burrow but was not preserved. Schlitter stated that *Liberiictis* is eaten by human hunters. In a letter of 26 June 1963 to Ernest P. Walker, Dr. Hans-Jürg Kuhn, who had traveled in Liberia, wrote that native people say that *Liberiictis* is usually

Liberian mongoose *(Liberiictis kuhni)*, photo by F. Faigal, Metropolitan Toronto Zoo.

found in tree holes and lives in groups of 3–5 individuals. Carnio and Taylor (1988) reported that a group of 15 had been seen foraging. They added, however, that the genus seems to be very rare and may be jeopardized by human hunting and habitat destruction. In January 1989 a live specimen was obtained and placed in the Metro Toronto Zoo (Carnio 1989). The IUCN now classifies *Liberiictis* as endangered.

CARNIVORA; VIVERRIDAE; **Genus HELOGALE**
Gray, 1861

Dwarf Mongooses

There are two species (Coetzee, *in* Meester and Setzer 1977; Yalden, Largen, and Kock 1980):

H. parvula, Ethiopia to Angola and eastern South Africa;
H. hirtula, southern Ethiopia, southern Somalia, northern Kenya.

Head and body length is 180–260 mm, tail length is 120–200 mm, and weight is 230–680 grams (Kingdon 1977). Coloration is variable, but generally the upper parts are speckled brown to grayish. The lower parts are only slightly paler, and the tail and lower parts of the legs are dark. In some individuals there is a rufous patch on the throat and breast, and the basal portion of the lower side of the tail is reddish brown. Other individuals are entirely black.

Dwarf mongooses are found in savannahs, woodlands, brush country, and mountain scrub, from sea level to elevations of about 1,800 meters. They are mainly terrestrial and diurnal. They seek shelter at dusk in deserted or active termite mounds, among gnarled roots of trees, and in crevices. Their slender bodies enable them to squeeze into small openings. On occasion, they may excavate their own burrows. Dens are changed frequently, sometimes on a daily basis (Rood 1978). Most of the day is spent in an active and noisy search for food among brush, leaves, and rocks. The diet consists mainly of insects (Rasa 1977; Rood 1980) and also includes small vertebrates, eggs, and fruit.

Helogale is found in organized groups that are thought to

Liberian mongoose *(Liberiictis kuhni)*, photo of dead specimen by Lynn Robbins.

Dwarf mongoose (Helogale parvula), photo by Bernhard Grzimek.

use a definite home range, though there is conflicting information on group movements and relationships. According to Rasa (1977), a portion of the range is occupied for two or three months, and then there is a shift to another part, presumably because food supplies are depleted. Kingdon (1977) wrote that a group observed for seven years stayed mainly within an area of 2 ha.; another group used 2 ha. during the dry season but moved away with the coming of the rains. In the Serengeti, Rood (1978) found group home ranges to average 30 ha., to overlap by 5–40 percent with the ranges of one to four neighboring packs, and to contain 10–20 dens each. Rasa (1973, 1977) reported that captives have been observed to mark the vicinity of dens with secretions from cheek and anal glands and to show considerable intergroup aggression. Rasa (1986) stated that groups inhabit home ranges of 0.65–0.96 sq km, which show little overlap with one another and are traversed every 20–26 days, the same time taken for the decay of the marking secretions.

The social organization of *Helogale* is unique among mammals (Kingdon 1977; Rasa 1972, 1973, 1975, 1976, 1977, 1983; Rood 1978, 1980). There are as many as 40 individuals in a group, but usually about 10–12. The groups are matriarchal families, founded and led by an old female. She initiates movements and has priority to food. The second highest ranking member of the group is her mate, an old male. These two dominant animals are monogamous, only they usually produce offspring, and they suppress sexual activity in other group members. The latter form a hierarchy, with the *youngest* individuals ranking highest. This arrangement probably serves to allow the young animals to obtain sufficient food, without competition from older and stronger mongooses, during the unusually long period of growth in this genus. Within any age class in the group, females are dominant over males.

Despite the rigid class structure, or perhaps because of it, intragroup relations are generally harmonious, and severe fights are rare. Subordinate adults clean, carry, warm, and bring food to helpless young and take turns "baby-sitting" while the rest of the group forages. Females in addition to the mother sometimes nurse the young. The youngest mobile animals seem to have the role of watching for danger and alerting the others by means of visual signals or a shrill alarm call. Often, a single animal occupies an exposed position, where it serves as a group guard. One series of observations showed that when a low-ranking male became sick, it was allowed a higher than normal feeding priority and was also warmed by other group members. In another case, a group restricted its normal movements in order to provide care and

food for an injured member. Individuals communicate by depositing scent from the cheek and anal glands, and they also sometimes mark one another as an apparent sign of acceptance. As surviving subordinate animals grow older, they seem not to leave the group, even though they are not allowed to mate. If the dominant female dies, however, the group may split up.

In the Serengeti National Park of Tanzania, births occur mainly in the rainy season, from November to May, and the alpha female usually has three litters per year (Rood 1978, 1980). In a captive colony in Europe, the young are born regularly in spring and autumn (Rasa 1972). According to Rasa (1977), females there normally give birth twice a year, entering estrus 4–7 days after lactation ceases. If the newborn die, however, females may quickly remate, and they thus have the potential of producing five litters annually. The gestation period is 49–56 days. The number of young per litter averages about four and ranges from one to seven. Nursing lasts for at least 45 days, but group members begin to bring solid food to the young before weaning is complete. The young start to forage with the group by the time they are 6 months old. Females may reach physiological sexual maturity as early as the age of 107 days (Zannier 1965), but social restrictions normally delay breeding for several years. Apparently, full physical maturity is not attained until 3 years (Rasa 1972). One dominant pair was observed to mate first when about 3 years old and then to continue to breed for 7 years (Kingdon 1977).

CARNIVORA; VIVERRIDAE; **Genus DOLOGALE**
Thomas, 1926

African Tropical Savannah Mongoose

The single species, *D. dybowskii*, is found in the Central African Republic, northeastern Zaire, southern Sudan, and western Uganda (Coetzee, *in* Meester and Setzer 1977). This species was originally assigned to *Crossarchus* and has sometimes been referred to *Helogale*.

Head and body length is about 250–330 mm and tail length is 160–230 mm. Kingdon (1977) listed the weight as approximately 300–400 grams. Stripes are lacking. The head and neck are black, grizzled with grayish white. The back, tail, and limbs are lighter in color and have brownish spots. The

African tropical savannah mongoose *(Dologale dybowskii)*, photo by F. Petter of mounted specimen in Museum National d'Histoire Naturelle.

underparts are reddish gray. The fur is short, even, and fine, in contrast to the loose, coarse pelage of *Crossarchus*. The snout is not lengthened like that of *Crossarchus*. *Dologale* closely resembles *Helogale* but does not have a groove on the upper lip and has weaker teeth (Coetzee, *in* Meester and Setzer 1977).

Kingdon (1977) stated that the few known specimens of *Dologale* suggest adaptation to a variety of habitats—thick forest, savannah-forest, and montane forest grassland. A little evidence indicates that the genus is at least partly diurnal. It has robust claws, suggesting digging habits as in *Mungos*. Asdell (1964) wrote that a litter of four young had been noted in Zaire.

CARNIVORA; VIVERRIDAE; **Genus BDEOGALE**
Peters, 1850

Black-legged Mongooses

There are two subgenera and three species (Coetzee, *in* Meester and Setzer 1977):

subgenus *Bdeogale* Peters, 1850

B. crassicauda, southern Kenya to central Mozambique;

subgenus *Galeriscus* Thomas, 1894

B. nigripes, southeastern Nigeria to northern Zaire and northern Angola;
B. jacksoni, southeastern Uganda, central Kenya.

Rosevear (1974) considered *Galeriscus* to be a separate genus. Kingdon (1977) suggested that *B. jacksoni* is only a subspecies of *B. nigripes*.

Head and body length is 375 to at least 600 mm and tail length is 175–375 mm. Kingdon (1977) listed weight as 0.9–3.0 kg. There is considerable variation in color, both within and between species. The predominant general coloration is some shade of gray or brown, and the legs are usually black. The fur of adults is rather close, dense, and short, while that of the young is nearly twice as long and lighter in color.

Bdeogale resembles *Ichneumia* in having black feet, soft underfur, and long, coarse hair over the upper parts of the body. It differs in lacking the first or inner toe on each foot and in having larger premolar teeth. *Bdeogale* differs from *Rhynchogale* in having a naked groove from the nose to the upper lip. The foreparts of the feet of *Bdeogale* are naked, but the hind parts are well haired.

Taylor (1986, 1987) described *B. crassicauda* as rare, unspecialized, nocturnal, insectivorous, and solitary. According to Kingdon (1977), *B. crassicauda* inhabits woodland and moist savannah, while the other species live in tropical forest. *B. crassicauda* feeds almost entirely on insects, especially ants and termites, but may also take crabs and rodents. *B. nigripes* seems to prefer ants but also eats small vertebrates and carrion. These mongooses are not infrequently seen in pairs. A female and a quarter-grown young were taken in December on the coast of Kenya, a pregnant female with a large fetus was also taken in December, and a female with a newborn infant was found in southeastern Tanzania in late November. Information compiled by Rosevear (1974) suggests that adults are basically solitary in the wild but are not quarrelsome when kept together in captivity; that births in West Africa occur from November to January; and that litters nor-

Black-legged mongoose (*Bdeogale* sp.), photo by Don Davis.

Meller's mongoose *(Rhynchogale melleri)*, photo from *Proc. Zool. Soc. London.*

mally contain a single young. According to Jones (1982), a captive *B. nigripes* lived for 15 years and 10 months. The IUCN classifies the subspecies *B. crassicauda omnivora*, of coastal Kenya and Tanzania, as endangered, and *B. c. tenuis*, of Zanzibar, as indeterminate. Schreiber et al. (1989) noted that *B. c. omnivora* is very rare and that its restricted forest habitat is rapidly being cut over.

CARNIVORA; VIVERRIDAE Genus RHYNCHOGALE
Thomas, 1894

Meller's Mongoose

The single species, *R. melleri*, occurs from southern Zaire and Tanzania to eastern South Africa and possibly northeastern Angola (Coetzee, *in* Meester and Setzer 1977).

Head and body length is 440–85 mm and tail length is usually 300–400 mm. Kingdon (1977) listed weight as 1.7–3.0 kg. The general coloration is grayish or pale brown, the head and undersides are paler, and the feet are usually darker. *Rhynchogale* resembles *Ichneumia* in having coarse guard hairs protruding from the close underfur and the same dental formula. *Rhynchogale* differs from *Ichneumia* in the frequent reduction of the hallux and the lack of a naked crease from the nose to the upper lip. *Rhynchogale* has hind soles that are hairy to the roots of the toes. Females have two abdominal pairs of mammae.

According to Kingdon (1977), the habitat appears to be restricted to the woodland belt and possibly to moister and more heavily grassed or wooded areas, such as drainage lines and rock outcrops. Available information suggests that *Rhynchogale* is terrestrial, nocturnal, and solitary. The diet includes wild fruit, termites, and probably small vertebrates. A litter of two newborn with eyes still unopened was found in a small cave on a rocky hill in Zambia in December. A pregnant female containing two embryos was found in the same area, also in December. In Zimbabwe, births occur around November, and litters contain up to three young.

White-tailed mongoose *(Ichneumia albicauda)*, photo by Ernest P. Walker.

CARNIVORA; VIVERRIDAE; **Genus ICHNEUMIA**
I. Geoffroy St.-Hilaire, 1837

White-tailed Mongoose

The single species, *I. albicauda,* occurs in the southern part of the Arabian Peninsula and in most of Africa south of the Sahara (Coetzee, *in* Meester and Setzer 1977; Corbet 1978; Nader 1979).

Head and body length is 470–710 mm, tail length is 355–470 mm, and weight is 1.8–5.2 kg (Kingdon 1977; Taylor 1972). Long, coarse, black guard hairs protrude from a yellowish or whitish close, woolly underfur, producing a grayish general body color. The four extremities, from the elbows and knees, are black. The basal half of the tail is of the general body color. The terminal portion is usually white but occasionally black. *Ichneumia* is characterized by large size; its bushy, tapering tail; having the soles of the forelimbs naked to the wrist; and the division of the upper lip by a naked slit from the nose to the mouth. Females have four mammae.

The white-tailed mongoose is found mainly in savannahs and grassland. It prefers areas of thick cover, such as forest edge and bush-fringed streams. It is basically terrestrial and nocturnal, sheltering by day in porcupine or aardvark burrows, termite mounds, or cavities under roots or rocks. The diet consists mainly of insects and also includes snakes, other small vertebrates, and fruit. *Ichneumia* breaks eggs by hurling them back between its hind legs and against some hard object. It may defend itself by ejecting a particularly noxious secretion from its anal scent glands (Kingdon 1977).

In Kenya, Taylor (1972) found home range of one individual to be about 8 sq km. Detailed studies on the Serengeti of Tanzania (Waser and Waser 1985) revealed average home range to be 0.97 sq km for adult males and 0.64 sq km for females. The male ranges did not overlap, but there was complete overlap between the ranges of opposite sexes. Some female ranges were exclusive, but in other cases several females and their offspring used a common range, though they foraged separately. Such an arrangement was thought to represent a matrilineal clan of related individuals. Reported pairs and family groups probably reflect observations of, respectively, consorting individuals and mothers with young. *Ichneumia* is highly vocal, the most unusual sound being a doglike yap that may be associated with sexual behavior. Kingdon (1977) wrote that the two to four young are born in a burrow. All litters observed by Waser and Waser (1985) on the Serengeti contained one or two young, and such were seen most frequently from February to May. No young were seen during the August–November dry season. Independence apparently was attained there by 9 months of age. A captive was still living after 10 years (Jones 1982).

When the white-tailed mongoose dwells near a poultry raiser, it may prove to be a pest. In captivity it is the shyest of mongooses, but it is said to become a pleasing pet if captured young.

CARNIVORA; VIVERRIDAE; **Genus ATILAX**
F. Cuvier, 1826

Marsh Mongoose, or Water Mongoose

The single species, *A. paludinosus,* is found from Guinea-Bissau to Ethiopia, and south to South Africa (Rosevear 1974).

Head and body length is 460–620 mm, tail length is 320–530 mm, and weight is 2.5–4.1 kg (Kingdon 1977). The pelage is long, coarse, and generally brown in color. A sprinkling of black guard hairs often gives a dark effect. In some individuals, light rings on the hairs are prevalent and impart a grayish tinge. The head is usually lighter than the back, and the underparts are still paler.

Atilax is fairly heavily built. Although it is more aquatic than any other mongoose, it is the only one with toes that completely lack webbing. This feature may be associated with

Marsh mongoose, or water mongoose *(Atilax paludinosus),* photo by Ernest P. Walker.

the habit of feeling for aquatic prey in mud or under stones. There are five digits on each limb, the soles are naked, and the claws are short and blunt. The anal area is large and naked, and there is a narrow, naked slit between the nose and the upper lip. Females usually have two pairs of mammae.

Atilax is found in a variety of general habitat types, but its basic requirement seems to be permanent water bordered by dense vegetation (Rosevear 1974). Favored haunts are marshes, reed-grown streambeds, and tidal estuaries. Grassy patches and floating masses of vegetation often serve as feeding places and as dry resting spots. Like other mongooses, *Atilax* does little or no climbing but does run up leaning tree trunks or other inclines easy of access. It is an excellent swimmer and diver. When hard pressed, it submerges, leaving only the tip of the nose exposed for breathing. Normally, when it is swimming, the head and part of the back are exposed. Food is sought in the water and in travels on regular pathways along the borders of streams and marshes. Activity is usually said to be nocturnal and crepuscular, but Rowe-Rowe (1978b) referred to *Atilax* as diurnal and stated that it does much of its hunting while walking in shallow water.

The diet consists of almost any form of animal life that can be caught and killed. Regular foods include insects, mussels, crabs, fish, frogs, snakes, eggs, small rodents, and fruit (Kingdon 1977; Rosevear 1974). *Atilax* may throw such creatures as snails and crabs against hard surfaces to break the shell. A captive was seen to take a piece of beef rib between its forefeet, rear onto its hind feet with its forefeet held high, and then forcefully throw the bone to the floor of the cage in an effort to break it.

Kingdon (1977) wrote that *Atilax* is usually seen alone and that individuals are widely spaced and undoubtedly highly territorial. The young are said to be born in burrows in stream banks or on masses of vegetation gathered into heaps among reed beds. There is no evidence of a particular breeding season in West Africa (Rosevear 1974). Young have been found in South Africa in June, August, and October (Asdell 1964; Rowe-Rowe 1978b). The number of young per litter has been reported as one to three and usually seems to be two or three. The young weigh about 100 grams each at birth, open their eyes after 9–14 days, and are weaned at 30–46 days (Baker and Meester 1986). One water mongoose was still living after 17 years and 5 months in captivity (Jones 1982).

Rosevear (1974) observed that within the last 50 years *Atilax* has probably declined substantially in the drier parts of its range, because of human destruction of the available riverine habitat. *Atilax* is also widely hunted by people, as it is reputed to be a poultry thief.

CARNIVORA; VIVERRIDAE; Genus CYNICTIS
Ogilby, 1833

Yellow Mongoose

The single species, *C. penicillata*, is found in southern Angola, Namibia, Botswana, extreme southwestern Zimbabwe, and South Africa (Coetzee, *in* Meester and Setzer 1977).

Head and body length is 270–380 mm and tail length is 180–280 mm. Smithers (1971) listed weights of 440–797 grams. The hair is fairly long, especially on the tail, which is somewhat bushy. The general color is dark orange yellow to light yellow gray. The underfur is rich yellow, the chin is white, the underparts and limbs are lighter than the back, and the tail is tipped with white. The individual guard hairs are usually yellowish in the basal half, followed by a black band and a white tip. There is a seasonal color change in the coat:

Yellow mongoose *(Cynictis penicillata)*, photo by Hans-Jürg Kuhn.

the summer pelage, typical in January and February, is reddish, short, and thin; the winter coat, typical from June to August, is yellowish, long, and thick; and the transitional coat, typical in November and December, is pale yellow. The hands have five fingers, and the feet have four toes. The first, or inner, digit on the hand is small and above the level of the other four, so it does not touch the ground. Females have three pairs of mammae (Ewer 1973).

The yellow mongoose frequents open country, preferably with loose soil, but may take refuge among rocks or in brush along the banks of streams when disturbed. It lives in communal burrows that may cover 50 sq meters or more. It sometimes uses holes made by other animals, such as *Pedetes*, but usually excavates its own burrows. It is an energetic digger, constructing extensive underground tunnels and chambers with a number of entrance holes. Certain places within the burrow system are used for deposit of body wastes.

Cynictis is mainly diurnal but may be active at night when living near people. It basks in the sun and sits up on its haunches to obtain a better view of the surroundings. It is agile and capable of traveling at considerable speed. It seldom wanders more than 1 kilometer from the burrow. Apparently, pairs and perhaps entire colonies seek new homes when food becomes scarce in the vicinity of a burrow. The diet consists mainly of insects and other invertebrates (Herzig-Straschil 1977; Smithers 1971). Other reported foods of *Cynictis* include lizards, snakes, birds, the eggs of birds and turtles, small rodents, and even mammals as large as itself.

The number of individuals in a group has been reported to be as many as 50 or more, but Ewer (1973) wrote that such large colonies would be exceptional. In a study in South Africa, Earlé (1981) found the mean size of 5 colonies to be 8 individuals. The nucleus of each group consisted of an adult pair, their newest offspring, and 1 or 2 young or very old individuals. The other members had a loose, unclear association. The group hunting range of about 5–6 ha. corresponded to a territory, which was patrolled each day by the adult male and marked with urine and anal and cheek glands. Young animals from other groups were allowed to cross territorial boundaries if they showed submission by laying on their side and uttering high-pitched screams. Within the group, order of rank was: the adult male, the adult female, the youngest offspring, and then the other animals. The young showed reduced dominance by the age of 10 months. The mating season lasted from July to late September, and births occurred in October and November.

According to Ewer (1973), mating in South Africa occurs as late as December, births take place until January, estimates of gestation range from 45 to 57 days, litters contain 1–4 (usually 2) young, and weaning occurs at 6 weeks. Smithers (1971) stated that there may be sporadic breeding throughout the year in Botswana, with a peak at some interval from October to April, and that the number of embryos per pregnant female averages 3.2 and ranges from 2 to 5. One yellow mongoose lived in captivity for 15 years and 2 months (Jones 1982).

Gray Meerkat, or Selous's Mongoose

The single species, *P. selousi*, occurs in Angola, Zambia, Malawi, northern Namibia, Botswana, Zimbabwe, Mozambique, and eastern South Africa (Coetzee, *in* Meester and Setzer 1977). This species originally was referred to *Cynictis*.

Head and body length is 390–470 mm and tail length is 280–400 mm. Smithers (1971) listed weights of 1.4–2.2 kg. The upper parts are dull buff gray, the belly is buffy, the feet are black, and the tail is white-tipped. There is no rufous in the coloring, nor are there any spots or stripes. Unlike in *Cynictis*, there are only four digits on all four limbs. The

claws are long and slightly curved. These features are associated with a strong digging ability. *Paracynictis* can defend itself by expelling a strong-smelling secretion from its anal glands; its white-tipped tail, which makes the animal visible at night, may serve as a warning of this capability.

Paracynictis seems to prefer open scrub and woodland (Smithers 1971). It resides in labyrinthine burrows of its own construction. It is terrestrial and nocturnal but has been seen above ground by day. It has been described as shy and retiring. The diet consists of insects, other arthropods, frogs, lizards, and small rodents (Smithers 1971).

Apparently, each individual constructs its own burrow system, and there is less social activity than in *Cynictis*. On the basis of limited data, Smithers (1983) suggested that the births occur in the warm, wet months, probably from August to March, and that litter size is 2–4 young.

Suricate, or Slender-tailed Meerkat

The single species, *S. suricatta*, occurs in southwestern Angola, Namibia, Botswana, and South Africa (Coetzee, *in* Meester and Setzer 1977).

Head and body length is 250–350 mm and tail length is 175–250 mm. Smithers (1971) listed weights of 626–797 grams for males and 620–969 grams for females. The coloration is a light grizzled gray. The rear portion of the back is marked with black transverse bars, which result from the alternate light and black bands of individual hairs coinciding with similar markings of adjacent hairs. The head is almost white, the ears are black, and the tail is yellowish with a black tip. The coat is long and soft, and the underfur is dark rufous in color. The body is quite slender, though this feature is difficult to see because of the long fur. There are scent glands, peripheral to the anus, which open into a pouch that presumably stores the secretion (Ewer 1973). The forefeet

Selous's mongoose *(Paracynictis selousi)*, photo by Don Carter.

Suricates *(Suricata suricatta)*, photo by Bernhard Grzimek.

have very long and powerful claws. Females have six mammae (Grzimek 1975).

The suricate inhabits dry, open country, commonly with hard or stony ground (Smithers 1971). It is an efficient digger. Colonies on the plains may excavate their own burrows or share the holes of African ground squirrels *(Xerus)*. Colonies in stony areas live in crevices among the rocks. Outside activity is almost entirely diurnal. The suricate seems to enjoy basking in the sun, lying in various positions or sitting up on its haunches like a prairie dog *(Cynomys)*. If food supplies run low, a colony may establish a new den 1–2 km from the original site (Grzimek 1975). Individuals generally forage near the burrow, turning over stones and rooting in crevices. The diet is primarily insectivorous (Ewer 1973) but also includes small vertebrates, eggs, and vegetable matter.

According to Ewer (1973), *Suricata* is highly social. Groups usually have 2–3 family units and a total of 10–15 individuals. Each family contains a pair of adults and their young. The female may be larger than the male and may dominate him. At least 10 vocalizations have been identified, including a threatening growl and an alarm bark.

There is normally a single annual litter. Mating generally occurs in September and October, and births in November and December. The gestation period is 77 days, possibly less. The number of young per litter is two to five, usually four. The young weigh 25–36 grams each at birth, open their eyes after 10–14 days, and are weaned at 7–9 weeks (Ewer 1973). Sexual maturity is attained by 1 year (Grzimek 1975). One suricate was still living after 12 years and 6 months in captivity (Jones 1982).

Suricata tames readily, is affectionate, and enjoys the warmth of snuggling close to its master. It is sensitive to cold. It is often kept about homes in South Africa to kill mice and rats. It should not be confused with the yellow or thick-tailed meerkat *(Cynictis)*, with which it often associates. *Cynictis* is not so winning in its ways nor so pleasing as a pet.

CARNIVORA; VIVERRIDAE; **Genus CRYPTOPROCTA**
Bennett, 1833

Fossa

The single species, *C. ferox*, is found in Madagascar (Coetzee, *in* Meester and Setzer 1977). *Cryptoprocta* sometimes has been placed in the cat family (Felidae).

This is the largest carnivore of Madagascar. Head and body length is 610–800 mm, tail length is approximately the same, and shoulder height is about 370 mm. Weight is 7–12 kg (Albignac 1975). The fur is short, smooth, thick, soft, and usually reddish brown in color. Some black individuals have been captured. The mustache hairs are as long as the head. Like some other viverrids, *Cryptoprocta* has scent glands in the anal region which discharge a strong, disagreeable odor when the animal is irritated.

The general appearance is much like that of a large jaguarundi *(Felis yagouaroundi)* or small cougar *(Felis concolor)*. The curved claws are short, sharp, and retractile like those of a cat, but the head is relatively longer than in the Felidae. *Cryptoprocta* walks in a flat-footed manner on its soles, as do bears, rather than on its toes, like cats.

The fossa dwells in forests and woodland savannahs, from coastal lowlands to mountainous areas at elevations of 2,000 meters (Coetzee, *in* Meester and Setzer 1977). It is mainly nocturnal and crepuscular, is occasionally active by day, and often shelters in caves (Albignac 1972). A maternal den was located in an old termite mound and contained an unlined chamber about 70 cm deep, 100 cm wide, and 30 cm high (Albignac 1970b). The fossa is an excellent climber and pursues lemurs through the trees. Its diet consists mainly of small mammals and birds but also includes reptiles, frogs, and insects (Albignac 1972; Coetzee, *in* Meester and Setzer 1977).

Cryptoprocta is solitary except during the reproductive

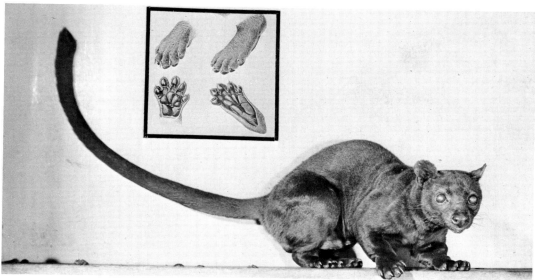

Fossa *(Cryptoprocta ferox)*, photo by Ernest P. Walker. Inset: forefeet and hind feet, top and bottom, photos from *Zoologie de Madagascar*, G. Grandidier and G. Petit.

season. Each individual has been estimated to require about 1 sq km. Mating occurs in September and October, and births take place in the austral summer. The gestation period lasts about 3 months. There are usually two young but sometimes three or four, each of which weighs about 100 grams at birth. The young leave the den after 4.5 months and are then weaned. Observations in captivity indicate that sexual maturity is not attained until 4 years. One specimen was still living after 20 years in captivity (Albignac 1972, 1975; Coetzee, *in* Meester and Setzer 1977; Köhncke and Leonhardt 1986).

The fossa is a powerful predator but has an exaggerated reputation for savagery and destructiveness. It normally flees at the sight of a human, though it may be dangerous if wounded. It sometimes preys on poultry, and some accounts claim that it attacks wild hogs and even oxen. It is widely hunted by people and is now depleted in numbers. The IUCN (1972) formerly classified it as vulnerable but now considers its status to be "insufficiently known." It also is on appendix 2 of the CITES.

CARNIVORA; **Family HYAENIDAE**

Aardwolf and Hyenas

This family of three Recent genera and four species is found in Africa and southwestern and south-central Asia. The sequence of genera presented here follows that of Simpson (1945), who recognized two subfamilies: Protelinae, with the single genus *Proteles* (aardwolf); and Hyaeninae, with *Crocuta* and *Hyaena*. Some authorities, such as Coetzee (*in* Meester and Setzer 1977) and Yalden, Largen, and Kock (1980), treat the Protelinae as a separate family.

Head and body length is 550–1,658 mm, tail length is 187–470 mm, and weight is 9–86 kg. The color pattern is striped in *Proteles* and *Hyaena hyaena*, spotted in *Crocuta*, and unmarked in *Hyaena brunnea*, except for its barred feet and pale head. A well-developed mane is present in *Proteles* and *Hyaena*. In all genera the guard hairs are coarse and the tail is bushy.

The head and forequarters are large, but the hindquarters are rather weak. The forelimbs, which are slender in *Proteles* but powerfully built in *Crocuta* and *Hyaena*, are longer than the hind limbs. *Proteles* has five digits on the forefoot and four on the hind foot, while *Crocuta* and *Hyaena* have four digits on each foot. The blunt claws are not retractile. Scent glands are present in the anal region. Males do not have a baculum, and females have one to three pairs of mammae.

In *Proteles* the skull and jaws are weak, the cheek teeth are reduced and widely spaced, and the carnassials are not developed, but the canines are sharp and fairly powerful. This genus may have as few as 24 teeth (Ewer 1973). In *Crocuta* and *Hyaena*, on the other hand, the skull and jaws are strong, and the teeth, including the carnassials, are powerfully developed for crushing bones. The dental formula of *Crocuta* and *Hyaena* is: (i 3/3, c 1/1, pm 4/4, m 1/1) × 2 = 34. In all genera of the family the incisors are not specialized and the canines are elongate.

Hyenas and the aardwolf generally inhabit grassland or bush country, but they may also occur in open forest. They live in caves, dense vegetation, or the abandoned burrows of other animals. They are mainly nocturnal. They move about on their digits and seem to trot tirelessly. The aardwolf feeds primarily on insects, while hyenas are efficient scavengers and predators of large mammals. Individuals may occur alone, in pairs, or in groups. Females normally bear a single litter per year.

The family is thought to have evolved from a branch of the Viverridae. The geological range of the Hyaenidae is Miocene to Recent in Asia and Africa, Pliocene to Pleistocene in Europe, and early Pleistocene in North America (Stains 1984).

CARNIVORA; HYAENIDAE; **Genus PROTELES**
I. Geoffroy St.-Hilaire, 1824

Aardwolf

The single species, *P. cristatus,* is found from the southern border of Egypt to central Tanzania and from southern An-

Aardwolf *(Proteles cristatus)*, photo by R. Pucholt.

gola and southern Zambia to the Cape of Good Hope (Coetzee, *in* Meester and Setzer 1977; Corbet 1978; Kingdon 1977).

Head and body length is 550–800 mm, tail length is usually 200–300 mm, and shoulder height is usually 450–500 mm. Kingdon (1977) listed weight as 9–14 kg. The underfur is long, loose, soft, and wavy and is interspersed with larger, coarser guard hairs. The body is yellow gray with black stripes. The legs are banded with black, and the part below the knee and hock is entirely black. The tail is bushy and tipped with black, and the hair along the back is long and crestlike, as evidenced in the Boer vernacular of "manhaarjakkal," or maned jackal. The jaws are weak, and the cheek teeth are vestigial and widely spaced, but the canine teeth are sharp and reasonably powerful.

The aardwolf is most common on open, sandy plains or in bush country. It dens in holes in the ground, usually abandoned burrows of the aardvark *(Orycteropus)*. It is primarily nocturnal and has been observed to cover just over 1 km/hr while foraging (Bothma and Nel 1980). The diet consists almost entirely of termites and insect larvae dug out of the ground. Although a captive juvenile killed a number of birds (Kingdon 1977), there are no substantive data on such behavior in the wild. *Proteles* has been accused of preying on chickens and lambs, but evidence to the contrary is overwhelming, and it is known that the aardwolf can scarcely be induced to eat meat unless it is finely ground or cooked. When seen near carrion, *Proteles* is usually there to pick up carrion beetles, maggots, and other insects.

Skinner and Van Aarde (1988) reported a population density of about one adult per sq km and also that an adult female used a home range of 3.8 sq km over a period of 16 months. *Proteles* is generally solitary, with individuals well spaced. Bothma and Nel (1980) usually found a single animal per den but noted that dens could be less than 500 meters apart. Pairs and family groups of five or six individuals are sometimes

observed. Several females have been found, together with their young, in a single den (Kingdon 1977). Both sexes mark their ranges with secretions from anal scent glands. The hairs of the mane can be erected to make the aardwolf look twice its normal size for purposes of intraspecific display or defense against predators. Surprisingly loud growls and roars can be produced under stress (Smithers 1971). If attacked by dogs, the aardwolf ejects musky fluid from its anal glands and may fight effectively with its formidable canine teeth.

The following data on reproduction are available: gestation is thought to last 90–110 days, an animal several months old was taken in Uganda in July, a birth in Kenya occurred in late May, the number of young per litter is usually two or three and ranges from one to five, and reports of larger litters may represent the utilization of one den by two mothers (Kingdon 1977); litters of two to four young are born in November and December in southern Africa (Ewer 1973); and in Botswana, lactating females have been taken in January and April, and pregnant females in July and October (Smithers 1971). A captive aardwolf was still living after 14 years and 3 months (Jones 1982).

Proteles is threatened in at least some areas by human hunting and habitat destruction. It was classified as rare in South Africa by Smithers (1986).

CARNIVORA; HYAENIDAE; Genus **CROCUTA**
Kaup, 1828

Spotted Hyena

In historical time, the single species, *C. crocuta*, was found throughout Africa south of the Sahara except in equatorial rainforests (Coetzee, *in* Meester and Setzer 1977). Through

Spotted hyena *(Crocuta crocuta)*, photo by E. L. Button.

the end of the Pleistocene the same species occurred in much of Europe and Asia (Kurten 1968).

Head and body length is 950–1,658 mm, tail length is 255–360 mm, shoulder height is 700–915 mm, and weight is 40–86 kg. On the average, females are 120 mm longer and 6.6 kg heavier than males (Kingdon 1977). The hair is coarse and woolly. The ground color is yellowish gray, and the round markings on the body are dark brown to black. There is no mane, or only a slight one. The jaws are probably the most powerful, in proportion to size, of any living mammal's. From *Hyaena, Crocuta* differs in its larger size, shorter ears, paler and spotted coat, larger and more swollen braincase, and greatly reduced upper molar tooth.

The external genitalia of the female so closely resemble those of the male that the two sexes are practically impossible to distinguish in the field (Kingdon 1977; Kruuk 1972): the clitoris looks like the penis, occupies the same position, and is capable of elongation and erection; in addition, the female has a pair of sacs, filled with nonfunctional fibrous tissue, that looks very much like the scrotum and is located in the same place. Both sexes have two anal scent glands that empty into the rectum. Females usually have a single pair of mammae.

The spotted hyena originally occupied nearly all the more open habitats of Africa south of the Sahara (Kingdon 1977). It is especially common in dry acacia bush, open plains, and rocky country. It is found at elevations of up to 4,000 meters. *Crocuta* is probably the most numerous of the large African predators, because of its ability quickly to eat and digest entire carcasses, including skin and bones, and because the plasticity of its behavior allows it to function effectively either as a solitary scavenger and predator of small animals or as a group-living hunter of ungulates. Dens vary greatly in size, sometimes accommodating entire communities of hyenas. They are usually in abandoned aardvark holes but may also be in natural caves. Most activity is nocturnal or crepuscular. Up to 80 km may be covered in a night's foraging. *Crocuta* has keen senses of sight, hearing, and smell.

Except as noted, the information for the remainder of this account was taken from Kruuk's (1972) report of his studies in the Serengeti National Park and the Ngorongoro Crater of northern Tanzania. In these areas the spotted hyena shelters in holes on the plains or in shady places with bushy vegetation on hillsides. Dens may have 12 or more entrances. Activity occurs mainly in the first half of the night, then declines, and then increases again toward dawn. The Ngorongoro population remains in one general area throughout the year, but the Serengeti hyenas follow the seasonal migrations of their prey.

The diet consists mainly of medium-sized ungulates, especially wildebeest *(Connochaetes)*, and mostly very young,

Spotted hyena (Crocuta crocuta), photo by E. L. Button.

very old, or otherwise inferior animals. One hyena commonly forces a herd of wildebeest to run, watches for a weak individual, and then begins a chase, which is soon joined by other hyenas. Zebras are hunted in a more organized manner by packs of 10–25 hyenas. Chases usually go for less than 2 km, with Crocuta averaging 40–50 km/hr. Maximum speed is about 60 km/hr. Roughly one-third of the hunts are successful. Far more animals are killed than are consumed as carrion. Although the spotted hyena is sometimes said to be a scavenger of the lion, most dead prey on which both hyenas and lions were seen feeding had been killed by the hyenas. The specialized teeth and digestive system of Crocuta allow it to crush and utilize much more bone than do other predators. Crocuta can consume 14.5 kg of food in one meal.

Population densities were estimated at 0.12 hyenas per sq km in the Serengeti and 0.16–0.24 per sq km in Ngorongoro. In the latter area, Crocuta is clearly organized into large communities, or "clans," each with up to 80 individuals and each occupying a territory of about 30 sq km. Clans are usually divided into smaller hunting packs and individuals, but all members evidently are recognizable to one another. The members mark the borders of their territory with secretions from their anal scent glands and defend the area from other clans. Groups that move about together commonly contain 2–10 individuals, but larger groups form to hunt zebra. In the Serengeti, the clan system is not so well developed, and average group size is smaller. Within any group of

Crocuta, females are dominant over males, and adults are dominant over young. The animals eat together and may compete by shoving one another, but there is generally no fighting. The spotted hyena is extremely vocal. The well-known laughing sound is emitted by an animal that is being attacked or chased. A whoop or howl is usually given spontaneously by a lone individual with the head held close to the ground; it begins low and deep and increases in volume as it runs to higher pitches.

Studies in Kenya by Frank (1986a, 1986b) have provided further understanding of the internal structure of spotted hyena clans. Each of these groups is a stable community of related females, among which unrelated males reside for varying periods. Whereas females remain in their natal clan for life, males disperse around the time of puberty and temporarily join together in a nomadic group. They subsequently settle within a new clan, sometimes remaining for several years, but then may depart again. Within a clan there is a separate dominance hierarchy for each sex. The highest-ranking female and her descendants are dominant over all other animals. Although all resident males were observed to court females, only the highest-ranking male actually was seen to mate.

There apparently is no permanent pair bonding. Breeding may occur at any time of the year, perhaps with a peak in the wet season. Females are polyestrous and have an estrous cycle of 14 days. Gestation lasts about 110 days. There are usually two young per birth, occasionally one or three. At birth they weigh about 1.5 kg and have their eyes open. Each clan has a single central denning site, where all females bear their cubs, but each mother suckles its own young. There is no cooperative rearing of young as found in *Hyaena brunnea* (Mills 1985). Food is not carried to the den or regurgitated for the young. Weaning comes after 12–16 months, when the young are nearly full-grown. Sexual maturity is attained at about 2 years by males and 3 years by females. A captive *Crocuta* lived for 41 years and 1 month (Jones 1982).

Human relationships with the spotted hyena have varied (Kingdon 1977; Kruuk 1972). Some native peoples protected it as a valuable scavenger, while others regarded it with superstitious dread. Certain tribes put their dead out for hyenas to consume. Fatal attacks on living humans have occurred. In the twentieth century, *Crocuta* has been considered to be a predator of domestic livestock and game and has been widely hunted, trapped, and poisoned. It has declined in numbers over much of its range and has been eliminated in parts of East and South Africa. Kurten (1968) suggested that the disappearance of *Crocuta* in Eurasia in the early postglacial period was associated with the development of agriculture.

CARNIVORA; HYAENIDAE; Genus HYAENA
Brünnich, 1771

Striped and Brown Hyenas

There are two species (Coetzee, *in* Meester and Setzer 1977; Corbet 1978):

H. hyaena (striped hyena), open country from Morocco and Senegal to Egypt and Tanzania, and from Asia Minor and the Arabian Peninsula to southern Soviet Central Asia and eastern India;
H. brunnea (brown hyena), Namibia, Botswana, western and southern Zimbabwe, southern Mozambique, South Africa.

Both species have pointed ears and erectile manes of long hairs, up to 305 mm in length, on the neck and back (in *Crocuta* the ears are rounded and there is little or no mane). In *Hyaena* the diameter of the upper molar tooth is at least twice that of the first upper premolar, and there is a small metaconid on the first lower molar (in *Crocuta* the upper molar is much smaller than the first premolar, and there is no metaconid on the first lower molar). The sexual organs of female *Hyaena*, unlike those of female *Crocuta*, do not closely resemble those of males. Like *Crocuta*, *Hyaena* has scent glands leading to an anal pouch that can be extruded for depositing secretions. *Hyaena* also has powerful jaws and teeth that can crush the largest bones of cattle, as well as a digestive system that can effectively utilize all parts of a carcass. Each front and hind foot has four digits. Additional information on each species is provided separately.

Hyaena hyaena (striped hyena).

Except as noted, the information for the account of this species was taken from the review papers by Rieger (1979, 1981). Head and body length is 1,036–1,190 mm, tail length is 265–470 mm, shoulder height is 600–942 mm, and weight is 25–55 kg. Males and females are about the same size. The general coloration is gray to pale brown, with dark brown to black stripes on the body and legs. The mane along the back is more distinct than in *H. brunnea*. The hairs of the mane are up to 200 mm long, while those of the rest of the body are about 70 mm long. Females have two or three pairs of mammae.

The striped hyena prefers open or rocky country and has an elevational range of up to 3,300 meters. It avoids true deserts and requires the presence of fresh water within 10 km. It is mainly crepuscular or nocturnal, resting by day in a temporary lair, usually under overhanging rocks (Kruuk 1976). Cubs are reared in natural caves, rocky crevices, or holes dug or enlarged by the parents. When searching for food, the striped hyena moves in a zigzag pattern at 2–4 km/hr (Kruuk 1976). The diet varies by season and area but seems to consist mainly of mammalian carrion. The larger, more northerly subspecies prey on sheep, goats, donkeys, and horses. Other important foods are small vertebrates, insects, and fruit. Some food is stored in dense vegetation for later use. Food is often brought to the den, and eventually a large number of bones may accumulate (Skinner, Davis, and Ilani 1980).

The striped hyena may have a small defended territory around its breeding den, surrounded by a larger home range. In a radiotracking study in the Serengeti, Kruuk (1976) determined home range to be about 44 sq km for a female and 72 sq km for a male. This species seems to be mainly solitary in East Africa, but there are indications of greater social activity farther north, where the animals are more prone to have a predatory life. In over half of his sightings of *H. hyaena* in Israel, Macdonald (1978) saw more than one animal. An adult pair and their offspring may sometimes forage together, and a family unit may persist for several years. Adult females are evidently intolerant of one another and are dominant over males. Scent marking with anal glands is an important means of communication. Aggressive displays involve erection of the long hairs of the mane and tail. *H. hyaena* is much less vocal than *Crocuta* but growls, whines, and makes several other sounds.

Females are polyestrous and breed throughout the year. The estrous cycle may be 40–50 days long. Estrus lasts 1 day and may follow birth by 20–21 days. The gestation period is 88–92 days. The number of young per litter is 1–5, averaging 2.4. The young weigh about 700 grams each at birth, open their eyes after 5–9 days, first take solid food at 30 days, and nurse for at least 4–5 months. Food is brought to the den for the young (Skinner and Ilani 1979). Sexual maturity usually comes at 2–3 years, and maximum longevity in captivity is 23–24 years.

The striped hyena has been known to attack and kill people, especially children. It can be tamed, however, and is said to become loyal and affectionate. Many of its parts are believed by some persons to have medicinal value. It is notorious in Israel for its destruction of melons, dates, grapes, apricots, peaches, and cucumbers (Kruuk 1976). The North African subspecies, *H. h. barbara*, is classified as endangered by the IUCN (1972) and the USDI. It has declined through habitat loss and persecution as a predator. Its range is now restricted to the highlands of Morocco, Algeria, and Tunisia.

Hyaena brunnea (brown hyena).

Head and body length is 1,100–1,356 mm, tail length is 187–265 mm, shoulder height is 640–880 mm, and weight is 37.0–47.5 kg (Skinner 1976; Smithers 1971). Males average slightly larger than females. The coat is long, coarse, and shaggy. The general coloration is dark brown. The head is gray, the neck and shoulders are tawny, and the lower legs and feet are gray with dark brown bars.

The brown hyena is found largely within arid habitat—open scrub, woodland savannah, grassland, and semidesert. It is mainly nocturnal and crepuscular, sheltering by day in a rocky lair, dense vegetation, or an underground burrow (Eaton 1976). It commonly uses an aardvark hole as a den but is capable of digging its own burrow. A favored maternal den site may be utilized for years (Owens and Owens 1979*a*).

Top, brown hyena *(Hyaena brunnea)* with a tumor near the left eye, photo from U.S. National Zoological Park. Bottom, striped hyena *(H. hyaena),* photo by Klaus Kussmann.

Young striped hyena *(Hyaena hyaena)*, photo by Reginald Bloom.

given area are organized into a "clan" and are recognized by one another. Foraging is usually done alone. In the dry season, individual home range averages about 40 sq km. Individuals do not maintain territories but use common hunting paths and frequently meet and exchange greetings. At times, especially during the wet season, up to six clan members join to exploit carrion. Each clan has a central breeding den and defends a surrounding territory of approximately 170 sq km.

A clan typically includes a dominant male, three or four subordinate males, four to six adult females, and associated young animals. A clan is not a strictly closed system, and emigration has been observed, especially among the young. There is a stable rank order, maintained by ritualized displays of aggression which commonly involve biting at the back of the neck, where the skin is tough. Individuals regularly deposit scent from their anal glands as they move about. This activity seems to be mainly for communication of information to other clan members, rather than demarcation of territory. Identified vocalizations include squeals, growls, yells, screams, and squeaks, all of which seem to be associated with conflict or submissive behavior.

Females are apparently seasonally polyestrous. Mating is thought to occur mainly from May to August, and births from August to November. The gestation period is about 3 months. The number of young per litter is two to five, usually three. The young open their eyes after 8 days and emerge from the den after 3 months (Eaton 1976; Skinner 1976). The young of several litters are raised together in a communal den, and each suckles from any lactating female. All clan females participate in bringing food to the den; there is no regurgitation. The cubs do not leave the vicinity of the den until they are about 14 months old (Owens and Owens 1979a, 1979b). Females usually produce their first litter late in their second year, interbirth interval is 12–41 months, natural mortality rates are relatively low until old age, and wild individuals are known to have lived at least 12 years (Mills 1982).

The brown hyena is classified as vulnerable by the IUCN and as endangered by the USDI and is on appendix 1 of the CITES. Although protected and still widely distributed in Botswana, it has declined drastically in range and numbers in Namibia and South Africa (Eaton 1976). The main problem is

Foraging is done in a zigzag pattern along regular pathways. Nightly movements cover 10–20 km in the wet season and 20–30 km in the dry season. Food is located mainly by scent, but hearing and night vision are also keen (Mills 1978; Owens and Owens 1978).

The brown hyena is primarily a scavenger of the remains of large mammals killed by other predators (Mills and Mills 1978; Owens and Owens 1978; Skinner 1976). Other important foods include rodents, insects, eggs, and fruit. At certain times of the year vegetation may constitute up to half of the diet of some individuals. *H. brunnea* is also known to frequent shorelines and to feed on dead crabs, fish, and seals. It often stores excess food in shrubs or holes and usually recovers it within 24 hours. Individuals have been seen to raid ostrich nests and to cache the eggs at scattered locations.

Eaton (1976) wrote that maximum population density is about one per 130 sq km, but Skinner (1976) found at least six adults and two cubs within 20 sq km. Social behavior evidently varies to some extent according to season (Mills 1978; Owens and Owens 1978, 1979a, 1979b). The animals in a

Brown hyenas *(Hyaena brunnea)*, photo by Bernhard Grzimek.

Leopard cat *(Felis bengalensis)* and young, photo from U.S. National Zoological Park.

killing by people, who consider it to be a predator of domestic animals. The damage that it does seems to have been greatly exaggerated; for example, on one large cattle ranch in the Transvaal where *H. brunnea* was studied no depredations were known to have occurred in 15 years (Skinner 1976).

CARNIVORA; **Family FELIDAE**

Cats

This family of 4 Recent genera and 37 species has a natural distribution that includes all land areas of the world except the West Indies, Madagascar, Japan, most of the Philippines, Sulawesi and islands to the east, New Guinea, Australia, New Zealand, Antarctica, and most arctic and oceanic islands. Recognition here of 4 genera—*Felis, Neofelis, Panthera,* and *Acinonyx*—is based on Cabrera (1957), Chasen (1940), Corbet (1978), Ellerman and Morrison-Scott (1966), Hall (1981), Lekagul and McNeely (1977), Medway (1977, 1978), Smithers (*in* Meester and Setzer 1977), Stains (1984), and Taylor (1934). There is, however, a great diversity of opinion as to how the cats should be classified. Some authorities, such as Romer (1968), divide the family into only 2 genera, *Felis* and *Acinonyx*. Others, including some who have recently done extensive research on the subject, recognize many more genera. The table on page 1185 shows the systematic arrangements of three of these authorities, plus a composite listing based on the sources cited above. The domestic cat *(Felis catus)* is not included in the table but is thought to be most closely related to *F. silvestris*.

The table (p. 1185) indicates that there is little disagreement at the specific level. The main controversy involves the division of what is here considered to be the genus *Felis* into genera and subgenera. Several authorities, including Chasen (1940), Hall (1981), and Smithers (*in* Meester and Setzer 1977), do not employ subgenera. All of the authorities cited here accept *Acinonyx* as a distinct genus; nearly all accept *Neofelis* (though with some disagreement as to content); and most do not accept *Uncia* as a full genus. *Panthera* is recognized as a separate genus by most but not by Hall (1981) or Van Gelder (1977b). Hall, like most American authorities,

Fishing cat *(Felis viverrina)*, photo from U.S. National Zoological Park.

Leyhausen 1979	Hemmer 1978	Ewer 1973	Sources Cited Above
Genus *Acinonyx*	Genus *Acinonyx*	Genus *Felis*	Genus *Felis*
A. jubatus	A. jubatus	F. silvestris	Subgenus *Felis*
Genus *Prionailurus*	Genus *Puma*	F. libyca (including	F. silvestris (including
P. bengalensis	P. concolor	ornata)	libyca and ornata)
P. rubiginosus	Genus *Herpailurus*	F. chaus	F. bieti
P. viverrinus	H. yagouaroundi	F. bieti	F. chaus
P. planiceps	Genus *Prionailurus*	F. margarita (including	F. margarita (including
P. iriomotensis	Subgenus *Prionailurus*	thinobia)	thinobia)
Genus *Leopardus*	P. bengalensis	F. nigripes	F. nigripes
L. pardalis	P. rubiginosus	Genus *Leptailurus*	Subgenus *Otocolobus*
L. wiedii	P. viverrinus	L. serval	F. manul
L. tigrinus	Subgenus *Mayailurus*	Genus *Prionailurus*	Subgenus *Lynx*
L. guigna	P. iriomotensis	P. bengalensis	F. lynx (including
L. geoffroyi	Subgenus *Ictailurus*	P. rubiginosus	canadensis)
Genus *Lynchailurus*	P. planiceps	P. viverrinus	F. pardina
L. pajeros (=Felis	Genus *Profelis*	Genus *Mayailurus*	F. rufus
colocolo)	P. aurata	M. iriomotensis	Subgenus *Caracal*
Genus *Oreailurus*	Genus *Catopuma*	Genus *Ictailurus*	F. caracal
O. jacobita	C. temmincki (including	I. planiceps	Subgenus *Leptailurus*
Genus *Herpailurus*	tristis)	Genus *Otocolobus*	F. serval
H. yagouaroundi	C. badia	O. manul	Subgenus *Pardofelis*
Genus *Felis*	Genus *Leopardus*	Genus *Pardofelis*	F. marmorata
F. silvestris	L. pardalis	P. marmorata	F. badia
F. libyca	L. wiedii	P. badia	Subgenus *Profelis*
F. ornata	Genus *Oncifelis*	Genus *Profelis*	F. temmincki (including
F. thinobia	Subgenus *Oncifelis*	P. temmincki (including	tristis)
F. margarita	O. tigrinus	tristis)	F. aurata
F. bieti	O. geoffroyi	P. aurata	Subgenus *Prionailurus*
F. nigripes	O. guigna	Genus *Caracal*	F. bengalensis
F. manul	Subgenus *Lynchailurus*	C. caracal	F. rubiginosa
F. chaus	O. pajeros (=Felis	Genus *Puma*	F. viverrina
Genus *Leptailurus*	colocolo)	P. concolor	F. planiceps
L. serval	Subgenus *Oreailurus*	Genus *Leopardus*	Subgenus *Mayailurus*
Genus *Lynx*	O. jacobita	L. pardalis	F. iriomotensis
L. lynx	Genus *Pardofelis*	L. tigrinus	Subgenus *Lynchailurus*
L. canadensis	P. marmorata	L. wiedii	F. colocolo
L. pardinus	Genus *Neofelis*	L. geoffroyi	Subgenus *Leopardus*
L. rufus	N. nebulosa	Genus *Oncifelis*	F. pardalis
Genus *Pardofelis*	Genus *Uncia*	O. guigna	F. wiedii
P. marmorata	U. uncia	Genus *Lynchailurus*	F. tigrina
P. badia	Genus *Panthera*	L. colocolo	F. geoffroyi
Genus *Profelis*	Subgenus *Tigris*	Genus *Oreailurus*	F. guigna
P. temmincki	P. tigris	O. jacobita	Subgenus *Oreailurus*
P. tristis	Subgenus *Panthera*	Genus *Herpailurus*	F. jacobita
P. aurata	P. onca	H. yagouaroundi	Subgenus *Herpailurus*
P. caracal	P. pardus	Genus *Lynx*	F. yagouaroundi
P. concolor	P. leo	L. lynx (including	Subgenus *Puma*
Genus *Uncia*	Genus *Leptailurus*	canadensis)	F. concolor
U. uncia	L. serval	L. pardinus	Genus *Neofelis*
Genus *Neofelis*	Genus *Caracal*	L. rufus	N. nebulosa
N. nebulosa	C. caracal	Genus *Panthera*	Genus *Panthera*
N. tigris	Genus *Felis*	P. leo	Subgenus *Uncia*
Genus *Panthera*	Subgenus *Lynx*	P. tigris	P. uncia
P. pardus	F. rufus	P. pardus	Subgenus *Tigris*
P. onca	F. canadensis	P. onca	P. tigris
P. leo	F. lynx	P. uncia	Subgenus *Panthera*
	F. pardinus	Genus *Neofelis*	P. pardus
	Subgenus *Otocolobus*	N. nebulosa	Subgenus *Jaguarius*
	F. manul	Genus *Acinonyx*	P. onca
	Subgenus *Felis*	A. jubatus	Subgenus *Leo*
	F. nigripes		P. leo
	F. margarita (including		Genus *Acinonyx*
	thinobia)		A. jubatus
	F. chaus		
	F. bieti		
	F. silvestris (including		
	libyca and ornata)		

treated *Lynx* as a distinct genus, and he included *Caracal* within *Lynx*. The preponderance of available information, however, suggests that if the genus *Felis* is accepted as comprising most species of the Felidae, *Lynx* should be included within *Felis* as no more than a subgenus, and *Caracal* should be a separate subgenus.

The living Felidae are sometimes divided into subfamilies. Stains (1984) recognized the Acinonychinae, for *Acinonyx* (cheetah); the Felinae, for *Felis;* and the Pantherinae, for *Neofelis* and *Panthera*. Leyhausen (1979) accepted the Acinonychinae, for *Acinonyx,* but put all other cats in the Felinae. The cheetah is usually placed last on lists of cats to express its aberrant characters. Actually, however, it seems in some respects to be little changed from the primitive stock that gave rise to all other cats. Locating it at the beginning of a systematic account, as was done by Leyhausen (1979) and Hemmer (1978), is thus fully appropriate. Moreover, Hemmer's placement of *Felis (Puma) concolor* next to *Acinonyx* may be fitting, as Adams (1979) has presented evidence that these two cats evolved from a common ancestor in North America.

In the family Felidae, head and body length is 337–2,800

A. Lion *(Panthera leo)*, photo by Leonard Lee Rue III. B. Clouded leopard *(Neofelis nebulosa)*, photo by Bernhard Grzimek. C. Leopard *(Panthera pardus)*, photo by Leonard Lee Rue III.

mm, tail length is 51–1,100 mm, and weight is 1.5–306.0 kg. The color varies from gray to reddish and yellowish brown, and there are often stripes, spots, or rosettes. The pelage is soft and woolly. Its beautiful, glossy appearance is maintained by frequent cleaning with the tongue and paws. The tail is well haired but not bushy, and the whiskers are well developed.

Cats have a lithe, muscular, compact, and deep-chested body. The limbs range from short to long and sinewy. The forefoot has five digits and the hind foot has four. In most species the claws are retractile (to prevent them from becoming blunted), large, compressed, sharp, and strongly curved (to aid in holding living prey). In the cheetah (Acinonyx), however, they are only semiretractile and relatively poorly developed. Except for the naked pads, the feet are well haired to assist in the silent stalking of prey. The baculum is vestigial or absent. Females have two to four pairs of mammae.

The head is rounded and shortened, the ears range from rounded to pointed, and the eyes have pupils that contract vertically. The tongue is suited for laceration and retaining food within the mouth, its surface being covered with sharp-pointed, recurved, horny papillae. The dental formula is: (i 3/3, c 1/1, pm 2–3/2, m 1/1) × 2 = 28 or 30. The incisors are small, unspecialized, and placed in a horizontal line. The canines are elongate, sharp, and slightly recurved. The carnassials, which cut the food, are large and well developed. The upper molar is small.

The dentition of the Felidae, with emphasis on the teeth used for seizing and cutting rather than on those used for grinding, reflects the highly predatory lifestyle of the family. Cats prey on almost any mammal or bird they can overpower and occasionally on fish and reptiles. They stalk their prey, or lie in wait, and then seize the quarry with a short rush. The cheetah (Acinonyx), however, is adapted for a more lengthy pursuit at high speeds. Cats walk or trot on their digits, often placing the hind foot in the track of the forefoot. They are agile climbers and good swimmers and have acute hearing and sight. Many are nocturnal, but some are active mainly during daylight. They shelter in trees, hollow logs, caves, crevices, abandoned burrows of other animals, or dense vegetation. They defend themselves with fang and claw, or flee, sometimes seeking refuge in trees.

Cats usually are solitary but sometimes are found in pairs or larger groups. The females of most species are polyestrous and give birth once a year, some can have two litters annually, and those of the larger species sometimes breed only every 2–3 years. Gestation periods of 55–119 days have been reported, and litter size is usually 1–6 young. At birth the kittens are usually blind and helpless but are haired and often spotted. They remain with their mother until they can hunt for themselves. Potential longevity is probably at least 15 years for most species, and some individuals have lived over 30 years.

Some of the larger species of cats have occasionally become a serious menace to human life in localized areas. Various felids are considered by some people to be a threat to domestic animals. Certain species, especially those with spotted or striped skins, are of considerable value in the fur trade. The big cats, and some of the smaller ones, are sought as trophies by hunters. For all of these reasons the Felidae have been extensively hunted and killed by people. As a result, many species and subspecies have become rare or endangered in at least parts of their ranges. The entire family has been placed on appendix 2 of the CITES, except for those species on appendix 1 and the domestic cat (Felis catus).

The geological range of the Felidae is late Eocene to Recent in North America and Eurasia, early Eocene to Recent in Africa, and late Pliocene to Recent in South America (Stains 1984). There are several extinct groups, including the saber-toothed cats (subfamily Machairodontinae), which persisted through the end of the Pleistocene.

CARNIVORA; FELIDAE; Genus FELIS
Linnaeus, 1758

Small Cats, Lynxes, and Cougar

There are 14 subgenera and 30 species (Cabrera 1957; Chasen 1940; Corbet 1978; Ellerman and Morrison-Scott 1966; Ewer 1973; Guggisberg 1975; Hall 1981; Lekagul and McNeely 1977; Medway 1977, 1978; Roberts 1977; Smithers, in Meester and Setzer 1977; Taylor 1934):

subgenus *Felis* Linnaeus, 1758

F. silvestris (wild cat), France and Spain to north-central China and central India, Great Britain, Balearic Islands, Sardinia, Corsica, Crete, woodland and savannah zones throughout Africa;
F. catus (domestic cat), worldwide in association with people;
F. bieti (Chinese desert cat), southern Mongolia, central China;
F. chaus (jungle cat), Volga River Delta and Egypt to Sinkiang and Indochina, Sri Lanka;
F. margarita (sand cat), desert zone from Morocco and northern Niger to Soviet Central Asia and Pakistan;
F. nigripes (black-footed cat), Namibia, Botswana, South Africa;

subgenus *Otocolobus* Brandt, 1841

F. manul (Pallas's cat), Caspian Sea and Iran to southeastern Siberia and Tibet;

subgenus *Lynx* Kerr, 1792

F. lynx (lynx), western mainland Europe to eastern Siberia and Tibet, possibly Sardinia, Sakhalin, Alaska, Canada, northern conterminous United States;
F. pardina (Spanish lynx), Spain, Portugal;
F. rufus (bobcat), southern Canada to Baja California and central Mexico;

subgenus *Caracal* Gray, 1843

F. caracal (caracal), Arabian Peninsula to Aral Sea and northwestern India, most of Africa;

subgenus *Leptailurus* Severtzow, 1858

F. serval (serval), Morocco, Algeria, most of Africa south of the Sahara;

subgenus *Pardofelis* Severtzow, 1858

F. marmorata (marbled cat), Nepal to Indochina and Malay Peninsula, Sumatra, Borneo;
F. badia (bay cat), Borneo;

subgenus *Profelis* Severtzow, 1858

F. temmincki (Asian golden cat), Tibet and Nepal to southeastern China and Malay Peninsula, Sumatra;
F. aurata (African golden cat), Senegal to Kenya and northern Angola;

A. Marbled cat *(Felis marmorata),* photo from *Jour. Bombay Nat. Hist. Soc.* B. Serval *(F. serval),* photo by Bernhard Grzimek. C. Golden cat *(F. temmincki),* photo by Ernest P. Walker. D. Pallas's cat *(F. manul),* photo by Howard E. Uible. E. Pumas *(F. concolor),* photo by Ernest P. Walker.

Top, black-footed cat *(Felis nigripes),* photo by John Visser. B. Little spotted cat *(F. tigrina).* C. Jungle cat *(F. chaus),* photo by Ernest P. Walker. D. European wild cat *(F. silvestris),* photo by Bernhard Grzimek. E. Jaguarundi *(F. yagouaroundi),* photo by Ernest P. Walker.

subgenus *Prionailurus* Severtzow, 1858

F. bengalensis (leopard cat), Ussuri region of southeastern
Siberia, Manchuria, Korea, Quelpart and Tsushima
islands (between Korea and Japan), eastern China,
Taiwan, Hainan, Pakistan to Indochina and Malay
Peninsula, Sumatra, Java, Bali, Borneo, several islands in
the western and central Philippines;
F. rubiginosa (rusty-spotted cat), southern India, Sri Lanka;
F. viverrina (fishing cat), Pakistan to Indochina, Sri Lanka,
Sumatra, Java;
F. planiceps (flat-headed cat), Malay Peninsula, Sumatra,
Borneo;

subgenus *Mayailurus* Imaizumi, 1967

F. iriomotensis (Iriomote cat), Iriomote Island (southern
Ryukyu Islands);

subgenus *Lynchailurus* Severtzow, 1858

F. colocolo (pampas cat), Ecuador and Mato Grosso region
of Brazil to central Chile and Patagonia;

subgenus *Leopardus* Gray, 1842

F. pardalis (ocelot), Arizona and Texas to northern
Argentina;
F. wiedii (margay), northern Mexico and possibly southern
Texas to northern Argentina and Uruguay;
F. tigrina (little spotted cat), Costa Rica to northern
Argentina;
F. geoffroyi (Geoffroy's cat), Bolivia and extreme southern
Brazil to Patagonia;
F. guigna (kodkod), central and southern Chile,
southwestern Argentina;

subgenus *Oreailurus* Cabrera, 1940

F. jacobita (mountain cat), the Andes of southern Peru,
southwestern Bolivia, northeastern Chile, and
northwestern Argentina;

subgenus *Herpailurus* Severtzow, 1858

F. yagouaroundi (jaguarundi), southern Arizona and
southern Texas to northern Argentina;

subgenus *Puma* Jardine, 1834

F. concolor (cougar, puma, panther, or mountain lion),
southern Yukon and Nova Scotia to southern Chile and
Patagonia.

All of the above subgenera are sometimes treated as full
genera, and still other genera have been used for certain
species here assigned to *Felis*. In contrast, some authorities
include all living cats except *Acinonyx* in the genus *Felis*, and
some do not employ subgenera (see account of the family
Felidae).

North American authorities, such as Hall (1981), commonly recognize *Lynx* as a distinct genus and the New World
populations of *Lynx lynx* as a separate species, *Lynx
canadensis*. Jones et al. (1986), however, used the name *Felis
lynx* instead of *Lynx canadensis*. There is also controversy
regarding the Spanish lynx, *F. pardina*: it is recognized as a
full species by most authorities, including Ewer (1973), Hemmer (1978), and Leyhausen (1979), but was considered to be
only a subspecies of *F. lynx* by Corbet (1978) and Ellerman

and Morrison-Scott (1966). In a recent evolutionary study,
Werdelin (1981) treated *Lynx* as a full genus and *F. canadensis* and *F. pardina* as species distinct from *F. lynx*.

The African and most Asian populations of *F. silvestris*
have been assigned to a separate species, *F. libyca*, by numerous authorities, including Smithers (*in* Meester and Setzer
1977). There is substantial evidence, however, that *F. silvestris* and *F. libyca* intergrade in the Middle East, and there is
also much doubt as to whether *F. bieti* of East Asia is distinct
from *F. silvestris* (Corbet 1978). Leyhausen (1979) listed *F.
silvestris* and *F. libyca* as separate species and considered the
populations found from Iran to India to represent still another
species, *F. ornata*. He united all three, however, in a single
superspecies. Leyhausen (1979) also recognized *F. thinobia*,
of Soviet Central Asia, and *F. tristis*, of Tibet, as species
distinct from, respectively, *F. margarita* and *F. temmincki*.

Except for *F. concolor* and some individuals of *F. lynx*, the
members of the genus *Felis* are smaller than those of the three
other genera of cats. Otherwise, the characters of *Felis* are the
same as those set forth for the family Felidae. From *Neofelis*,
Felis is distinguished by having relatively shorter canine teeth
and a smaller gap between the canines and the cheek teeth.
From *Panthera*, it is distinguished by having a completely
ossified hyoid apparatus without an elastic ligament. From
Acinonyx, it is distinguished by having a greater gap between
the canines and the cheek teeth, and usually by having shorter limbs and fully retractile claws. Additional information is
provided separately for each species. Except as noted, the
information for the following accounts was taken from Guggisberg (1975).

Felis silvestris (wild cat).

Head and body length is usually 500–750 mm, tail length is
210–350 mm, and weight is usually 3–8 kg. Males are larger,
on the average, than females. The fur is long and dense. In
European populations the ground color is yellowish gray, the
underparts are paler, and the throat is white. There are four or
five longitudinal stripes from the forehead to the nape, merging into a dorsal line that ends near the base of the tail. The
tail has several dark encircling marks and a blackish tip. The
legs are transversely striped. In African and Asian populations the general coloration varies from pale sandy to gray
brown and dark gray; there may be a pattern of distinct spots
or stripes. European animals are generally about one-third
larger than domestic cats and have longer legs, a broader
head, and a relatively shorter, more bluntly ending tail.
Females have eight mammae.

The wild cat occupies a variety of forested, open, and rocky
country. It is mainly nocturnal and crepuscular, spending the
day in a hollow tree, thicket, or rock crevice. It climbs with
great agility and seems to enjoy sunning itself on a branch. It
normally stays in one area, within which it has several dens
and a system of hunting paths. It usually stalks its prey,
attempting to approach to within a few bounds. The diet
consists mainly of rodents and other small mammals and also
includes birds, reptiles, and insects.

The wild cat is usually solitary, each individual having a
well-defined home range of about 50 ha. Males defend these
areas but may wander outside of their usual ranges during
times of food shortage or to locate estrous females. Mating
occurs from about January to March in Europe and Central
Asia. Females are polyestrous, with heat lasting 2–8 days.
Several males collect around a female in heat; there is considerable screeching and other vocalization, and sometimes
violent fighting. There is usually only a single litter per year,
though occasionally a second is produced in the summer.
Births in East Africa may occur at any time of year but seem
to peak there and in southern Africa during the wet season
(Kingdon 1977; Smithers 1971). The gestation period aver-

A. Leopard cat *(Felis bengalensis)*, photo by Lim Boo Liat. B. Margay *(F. wiedii)*, photo by Ernest P. Walker. C. Pampas cat *(F. colocolo)*, photo from San Diego Zoological Garden. D. Fishing cat *(F. viverrina)*, photo from New York Zoological Society. E. Geoffroy's cat *(F. geoffroyi)*, photo by Ernest P. Walker. F. African wild cat *(F. silvestris libyca)*, photo by Bernhard Grzimek.

Domestic cat *(Felis catus)*, photo by William J. Allen.

ages 66 days in Europe and about 1 week less in Africa. Litters usually contain two or three young in the wild. At the Berne Zoo, Meyer-Holzapfel (1968) observed that births occurred from March to August and that litter size averaged four and ranged from one to eight. The young weigh about 40 grams each at birth, open their eyes after about 10 days, nurse for about 30 days, emerge from the den at 4–5 weeks, begin to hunt with the mother at 12 weeks, probably separate from her at 5 months, and attain sexual maturity at around 1 year. According to Kingdon (1977), captives have lived up to 15 years.

The wild cat once occupied most of Europe but withdrew from Scandinavia and most of Russia by the Middle Ages, because of climatic deterioration. In modern times, especially during the nineteenth century, the species was intensively hunted by persons who considered it to be a threat to game and domestic animals and so was eliminated from much of western and central Europe. Diversion of human activity during World Wars I and II apparently stimulated recovery in such places as Scotland and West Germany. *F. silvestris* is now protected and encouraged in several nations (Smit and Van Wijngaarden 1981). It is classified as vulnerable by the Soviet Union.

Felis catus (domestic cat).

According to the National Geographic Society (1981), there are more than 30 different breeds of domestic cat, and the average measurements of several popular breeds are: head and body length, 460 mm, and tail length, 300 mm. E. Jones (1977) found that feral males on Macquarie Island, south of Australia, averaged 522 mm in head and body length, 269 mm in tail length, and 4.5 kg in weight, while females there averaged 478 mm in head and body length, 252 mm in tail length, and 3.3 kg in weight. On Macquarie, 90 percent of the cats were orange or tabby, and the remainder were black or tortoiseshell. Female *F. catus* have four pairs of mammae.

The domestic cat evidently is descended primarily from the wild cat of Africa and extreme southwestern Asia, *F. silvestris libyca*. The latter may have been present in towns in Palestine as long ago as 7,000 years, and actual domestication occurred in Egypt about 4,000 years ago. Introduction to Europe began around 2,000 years ago, and some interbreeding occurred there with the wild subspecies *F. silvestris silvestris*. Domestication seems originally to have had a re-

ligious basis (Grzimek 1975; Kingdon 1977). The cat was the object of a passionate cult in ancient Egypt, where a city, Bubastis, was dedicated to its worship. The followers of Bast, the goddess of pleasure, put bronze statues of cats in sanctuaries and carefully mummified the bodies of hundreds of thousands of the animals.

There have been relatively few detailed field studies of *F. catus*, but there is no reason to think that its behavior and ecology under noncaptive conditions differ greatly from what has been found for *F. silvestris*. On Macquarie Island, where the cat population has been feral since 1820, E. Jones (1977) obtained specimens in a variety of habitats by both day and night. The cats sheltered in rabbit burrows, thick vegetation, or piles of rocks. The diet consisted largely of rabbits (also introduced on the island) and also included rats, mice, birds, and carrion. Population density was estimated at two to seven cats per sq km.

In a rural area of southern Sweden, Liberg (1980) found a population density of 2.5–3.3 per sq km. About 10 percent of the cats were feral, and the rest, including all of the females, were associated with human households. Adult females lived alone or in groups of up to 8 usually closely related individuals. Each member of a group had a home range of 30–40 ha., which overlapped extensively with the ranges of other members of the same group but not with the ranges of the cats in other groups. Most females spent their life in the area in which they were born, seldom wandering more than 600 meters away. Nonferal males remained in their area of birth, along with females, until they were 1.5–3.5 years old but then left and tried to settle somewhere else. Males living in the same group had separate home ranges. There were 6–8 feral males in the study area; their home ranges were 2–4 km across, partly overlapped one another, and sometimes included the areas used by several groups of females. According to Haspel and Calhoon (1989), home ranges of unrestrained urban cats are much smaller than those in rural areas. In Brooklyn, range averaged 2.6 ha. for males and 1.7 ha. for females, the difference evidently being a function of body size.

According to Ewer (1973), the house cat is basically solitary, but individuals in a given area seem to have a social organization and hierarchy. A male newly introduced to an area normally must undergo a series of fights before its position is stabilized in relation to other males. Both males and

Jungle cat *(Felis chaus)*, photo from San Diego Zoological Society.

females sometimes gather within a few meters of each other without evident hostility. A male and female may form a bond that extends beyond the mating process. Females are polyestrous and normally produce two litters annually. They may mate with more than one male per season, and if a litter is lost, they soon enter estrus again. The gestation period averages 65 days. The number of young per litter averages four and ranges from one to eight. Kittens weigh 85–110 grams each at birth, open their eyes after 9–20 days, are weaned at 8 weeks, and attain independence at about 6 months. Hemmer (1976) listed age of sexual maturity in females as 7–12 months.

Although the cat has sometimes been venerated, it has also been associated with evil. Certain superstitions concerning it have persisted to modern times (Grzimek 1975). The species is now generally looked upon with favor by most cultures, but free-ranging cats often are considered to be among the greatest decimators of native wildlife, especially songbirds (Lowery 1974).

Felis bieti (Chinese desert cat).

Head and body length is 685–840 mm and tail length is 290–350 mm. The general coloration is yellowish gray, the back is somewhat darker, and there are practically no markings on the flanks. The tail is tipped with black and has three or four subterminal blackish rings. Despite the common name "desert cat," this species has been found only in steppe country and in mountains covered with brush and forest.

Felis chaus (jungle cat).

Head and body length is 500–750 mm, tail length is 250–90 mm, and weight is 4–16 kg (Lekagul and McNeely 1977; Novikov 1962). The general coloration varies from sandy or yellowish gray to grayish brown and tawny red, usually with no distinct markings on the body. The tail has several dark rings and a black tip. Lekagul and McNeely (1977) noted that the legs are proportionally the longest of any felid's in Thailand and are thus helpful in running down prey.

The jungle cat is found in either woodland or open country, from sea level to elevations of 2,400 meters. It dens in thick vegetation or in the abandoned burrow of a badger, fox, or porcupine. It is active either by day or by night. The diet consists mainly of hares and other small mammals and also includes birds, frogs, and snakes. Mating occurs in Soviet Central Asia in February and March, but 3-week-old kittens have been found in Assam in January and February. The gestation period is 66 days, and the usual litter size is three to five young. Sexual maturity comes at 18 months.

Felis margarita (sand cat).

Head and body length is 450–572 mm and tail length is 280–348 mm. The general coloration is pale sandy to gray straw ochre, the back is slightly darker, and the belly is white. A fulvous reddish streak runs across each cheek from the corner of the eye, and the tail has two or three subterminal rings and a black tip. The pelage is soft and dense. The soles of the feet are covered with dense hair. The limbs are short, and the ears are set low on the head.

The sand cat is adapted to extremely arid terrain, such as shifting dunes of sand. The padding on its soles facilitates progression over loose, sandy soil. Activity is mainly nocturnal and crepuscular, the day being spent in a shallow burrow. Prey consists of jerboas and other rodents and occasionally hares, birds, and reptiles. The sand cat apparently is able to subsist without drinking free water.

In Pakistan there may be two litters annually, as kittens have been found in both March–April and October (Roberts 1977). The gestation period is 59–63 days, litters contain two to four young, average weight at birth is 39 grams, and the eyes open after 12–16 days (Hemmer 1976).

The sand cat seems to be generally rare. The subspecies in Pakistan, *F. m. scheffeli*, was not discovered until 1966 and is now classified as endangered by the IUCN (1978) and the USDI. It apparently declined drastically through uncontrolled exploitation by commercial animal dealers from 1967 to 1972.

Sand cat *(Felis margarita)*, photo by Lothar Schlawe.

Felis nigripes (black-footed cat).

This is the smallest cat. Head and body length is 337–500 mm, tail length is 150–200 mm, and weight is 1.5–2.75 kg. The general coloration is dark ochre to pale ochre or sandy, being somewhat darker on the back and paler on the belly. There is a bold pattern of dark brown to black spots arranged in rows on the flanks, throat, chest, and belly. There are two streaks across each cheek, two transverse bars on the forelegs, and up to five transverse bars on the haunches. The tail has a black tip and two or three subterminal bands. The bottoms of the feet are black (since the animal walks on its toes, much of the black is usually visible).

The black-footed cat inhabits dry, open country. It shelters in old termite mounds and the abandoned burrows of other mammals. It is mainly nocturnal, but in captivity it has been found to be more active by day than most other small cats. The diet probably includes rodents, birds, and reptiles.

This felid seems to be highly unsocial. Even opposite sexes evidently come together only for 5–10 hours. Gestation lasts 59–68 days, and litters contain one to three young. The kittens leave the nest after 28–29 days and take their first solid food a few days later. Captive females have not initially entered heat until 15–21 months (Grzimek 1975).

F. nigripes is on appendix 1 of the CITES and is classified as rare in South Africa (Smithers 1986).

Felis manul (Pallas's cat).

Head and body length is 500–650 mm, tail length is 210–310 mm, and weight is 2.5–3.5 kg. The general coloration varies from light gray to yellowish buff and russet; the white tips of the hairs produce a frosted silvery appearance. There are two dark streaks across each side of the head and four rings on the dark-tipped tail. The coat is relatively longer and more dense than that of any other wild species of *Felis*. The fur is especially long near the end of the tail, and on the underparts of the body it is almost twice as long as on the back and sides. Such an arrangement provides good insulation for an animal that spends much time lying on frozen ground and snow. The body is massive, the legs are short and stout, the head is short and broad, and the ears are very short, bluntly rounded, and set low and wide apart.

Pallas's cat inhabits steppes, deserts, and rocky country up to elevations of over 4,000 meters. It dens in a cave, crevice, or burrow dug by another animal. It is usually nocturnal but is occasionally seen by day. It feeds on pikas and other small mammals. According to Stroganov (1969), the young are born in Siberia in late April and May, and litter size is five or six. Broad, Luxmoore, and Jenkins (1988) reported that at least 2,000 skins of *F. manul* are placed in international trade each year and that such activity may threaten the species.

Felis lynx (lynx).

In Eurasian populations, head and body length is 800–1,300 mm, tail length is 110–245 mm, shoulder height is 600–750 mm, and weight is 8–38 kg (Novikov 1962; Van Den Brink 1968). In North American populations, head and body length is about 800–1,000 mm, tail length is 51–138 mm, and weight is 5.1–17.2 kg (Banfield 1974; Burt and Grossenheider 1976). In any given area, males are generally larger than females. The coloration varies but is commonly yellowish brown; the upper parts may have a gray frosted appearance, the underparts are more buffy, and there is often a pattern of dark spots. The markedly short tail may have several dark rings and is tipped with black. The fur is long, lax, and thick. It is especially long on the lower cheeks in winter and gives the impression of a ruff around the neck. The

Pallas's cat *(Felis manul)*, photo by Bernhard Grzimek.

triangular ears are tipped by tufts of black hairs about 40 mm long. The legs are relatively long. The paws are large and densely furred, an adaptation for moving over winter snow. Females have four mammae (Banfield 1974).

The lynx is generally found in tall forests with dense undergrowth but may also enter open forest, rocky areas, or tundra. For shelter it constructs a rough bed under a rock ledge, fallen tree, or shrub (Banfield 1974). It is mainly nocturnal; reports of average nightly movement range from 5 to 19 km (Ewer 1973). The lynx usually keeps to one area but

may migrate under adverse conditions. The maximum known movement was by a female that was marked on 5 November 1974 in northern Minnesota and trapped on 20 January 1977 in Ontario, 483 km from the point of release (Mech 1977e). The lynx climbs well and is a good swimmer, sometimes crossing wide rivers. It hunts mainly by eye and also has well-developed hearing. It usually stalks its prey to within a few bounds, or it may wait in ambush for hours (Stroganov 1969). Reports from much of the range of the species indicate that leporids form a major part of the diet. The snowshoe rabbit *(Lepus americanus)* is of particular importance in North America. Deer and other ungulates are utilized heavily in certain areas, especially during the winter. Other foods include rodents, birds, and fish.

In Canada the numbers of lynx seem to fluctuate, together with those of the snowshoe rabbit, in a regular cycle. Numerical peaks occur at an average interval of 9.6 years, though not at the same time all over Canada. Several authorities have suggested that this phenomenon is actually an artifact, perhaps associated with the intensity of human trapping, but Finerty (1979) upheld the traditional view. Studies in central Alberta (Brand and Keith 1979; Brand, Keith, and Fischer 1976; Nellis, Wetmore, and Keith 1972) indicate that the main direct cause of the cyclic drop in lynx numbers is postpartum mortality of kittens through lack of food. There is also a reduced rate of pregnancy in females. Population density is this area varied from a low of 2.3 per 100 sq km in the winter of 1966/67 to a peak of 10 per 100 sq km in the winter of 1971/72. Maximum population density in eastern Europe and the Soviet Union is 5 per 100 sq km (Ewer 1973).

The reported size of individual home range varies from 11

Canada lynx *(Felis lynx)*, photo from San Diego Zoological Garden.

to 300 sq km (Brand, Keith, and Fischer 1976; Ewer 1973; Guggisberg 1975; Mech 1980). The lynx is probably territorial, but the ranges of females may overlap one another to some extent, and the range of a male may include that of a female and her young. Adults tend to avoid one another except during the breeding season. Females are monestrous and bear a single litter per year (Banfield 1974; Ewer 1973). Mating occurs mainly in February and March, gestation lasts 9–10 weeks, and litters contain one to five, usually two or three, young. Banfield (1974) gave birth weight as 197–211 grams. Lactation lasts for 5 months, but some meat is eaten by 1-month-old kittens. The young usually remain with their mother until the winter mating season, and siblings may stay together for a while afterward. The age of sexual maturity is 21 months in females and 33 months in males (Grzimek 1975). A captive individual lived for 26 years and 9 months (Rich 1983).

The lynx has been widely hunted by people because it is considered to be a predator of game and domestic animals and because its pelt has sometimes been of commercial value. It disappeared from much of Europe during the nineteenth century. Major populations there now occur only in the Soviet Union, Scandinavia, and the Carpathian region. Small, isolated groups are present in northeastern Poland, the southern Balkans, and southern France. Reintroductions have been carried out recently in parts of Germany, Austria, Switzerland, Italy, and Yugoslavia (Smit and Van Wijngaarden 1981). The lynx may have occurred in Palestine before that area was largely deforested; in the Arabian region the species is now found only in northern Iraq, where it is rare (Harrison 1968).

The range of *F. lynx* once extended at least as far south as Oregon, southern Colorado, Nebraska, southern Indiana, and Pennsylvania (Hall 1981). The only substantial population now remaining in the conterminous United States is that of western Montana and nearby parts of northern Idaho and northeastern Washington. On occasion, especially during periods of cyclically high populations in Canada, the species is found in the states of the northern plains and upper Great Lakes regions. There are small numbers in northern New England and Utah, and possibly in Oregon, Wyoming, and Colorado (Armstrong 1972; Cowan 1971; Deems and Pursley 1978; Gunderson 1978; Hunt 1974; Olterman and Verts 1972).

The lynx has been exterminated in parts of southeastern Canada but still occurs regularly over most of that country. Its numbers there seem to be affected more by the cyclical availability of food than by human hunting pressure. Populations evidently were relatively high until the early 1900s, declined until the mid-twentieth century, and then again increased (Cowan 1971). Subsequent highs and lows in the reported number of pelts taken in Canada are: 1949/50 trapping season, 3,734; 1954/55, 14,427; 1956/57, 8,748; 1962/63, 51,376; 1966/67, 13,038; 1971/72, 53,589; 1975/76, 13,162; and 1979/80, 34,366. The average price per pelt rose dramatically from $3.62 in 1953/54 to $30.52 in 1968/69 and to $336.36 in 1978/79 (Statistics Canada 1981). At fall 1984 sales in Ontario the average value of lynx pelts surged to nearly $600.00 (Quinn and Parker 1987).

Spanish lynx *(Felis pardina)*, photo by Lothar Schlawe.

Bobcat *(Felis rufus)*, photo by Lothar Schlawe.

Felis pardina (Spanish lynx).

Head and body length is 850–1,100 mm, tail length is 125–30 mm, and shoulder height is 600–700 mm. The upper parts are yellowish red and the underparts are white. There are round black spots on the body, tail, and limbs. The ears have tufts, and the face has a prominent fringe of whiskers (Smit and Van Wijngaarden 1981; Van Den Brink 1968).

The Spanish lynx inhabits open forests and thickets. Its general habits are much like those of *Felis lynx*. It feeds mainly on rabbits. Home range diameter is 4–10 km in summer and somewhat larger in winter. Mating is thought to occur mainly in January. According to the IUCN (1978), the gestation period is 63–73 days and litter size is two to three young.

The Spanish lynx is classified as endangered by the IUCN and the USDI. It formerly occurred throughout the Iberian Peninsula but now is restricted to scattered mountainous areas and the Guadalquivir Delta. The total number of animals is estimated at 1,000–1,500. There was a major decline during the 1950s and 1960s, when the disease myxomatosis hit the rabbit populations. The decline is continuing as suitable rabbit and lynx habitat is replaced by cereal cultivation and forest plantations.

Felis rufus (bobcat).

Head and body length is 650–1,050 mm, tail length is 110–90 mm, and shoulder height is 450–580 mm. Banfield (1974) gave the weight as 4.1–15.3 kg. There are various shades of buff and brown, spotted and lined with dark brown and black. The crown is streaked with black, and the backs of the ears are heavily marked with black. The short tail has a black tip, but only on the upper side. *F. rufus* resembles *F. lynx* but is usually smaller and has slenderer legs, smaller feet, shorter fur, and ears that are tufted less conspicuously or not at all. As in *F. lynx*, there is a ruff of fur extending from the ears to the jowls, giving the impression of sideburns. Females have four mammae (Lowery 1974).

The bobcat is more ubiquitous than the lynx, occurring in forests, mountainous areas, semideserts, and brushland. Its den is usually concealed in a thicket, hollow tree, or rocky crevice. It is mainly nocturnal and terrestrial but climbs with ease. Nightly movements of about 3–11 km have been reported. Prey is usually stalked with great stealth and patience, then seized after a swift leap. The diet consists mainly of small animals, such as rabbits, rodents, and birds. Larger prey, such as deer, is sometimes taken, especially in the winter. A study in Massachusetts found deer to be the most common winter food (McCord 1974).

Reported maximum population densities are one bobcat per 18.4 sq km in the western United States and one per 2.6 sq km in the southeast (Jachowski 1981). Average minimum home range in Louisiana was found to be about 5 sq km for males and 1 sq km for females (Hall and Newsom 1978). In a study in southeastern Idaho, Bailey (1974) found average (and extreme) home range size to be 42.1 (6.5–107.9) sq km for males and 19.3 (9.1–45.3) sq km for females. Female ranges were almost exclusive of one another, but the ranges of males overlapped each other, as well as those of females. There was a land tenure system, seemingly based on prior right: no neighboring resident or transient permanently settled in an area already occupied by a resident. All observed changes in resident home ranges were attributed to the death of a resident. Territoriality was pronounced, especially in females, and scent marking was accomplished by use of feces, urine, scrapes, and anal gland secretions. Individuals were solitary, avoiding each other even in areas of range overlap, except during the mating season. The bobcat is usually silent but may emit loud screams, hisses, and other sounds during courtship.

Females apparently are seasonally polyestrous; they usually produce a single annual litter during the spring, but there is evidence of a second birth peak in late summer and early autumn, perhaps involving younger females or ones that lost their first litters (Banfield 1974). According to Fritts and Sealander (1978), mating may occur as early as November or as late as August. The gestation period is 60–70 days (Hemmer 1976). The number of young per litter is one to six, commonly three. Weight at birth is 283–368 grams (Banfield 1974). The young open their eyes after 9–10 days, nurse for about 2 months, and begin to travel with the mother at 3–5 months. They separate from the mother in the winter (Jackson 1961), probably in association with the mating season. Females may reach sexual maturity by 1 year of age, but males do not mate until their second year of life. The oldest individuals captured in a Wyoming study were 12 years of age and were still sexually active (Crowe 1975). A captive bobcat lived for 32 years and 4 months (Jones 1982).

The bobcat occasionally preys on small domestic mammals and poultry and thus has been hunted and trapped by people. It has been exterminated in much of the Ohio Valley, upper Mississippi Valley, and southern Great Lakes region (Deems and Pursley 1978). The bobcat is uncommon in central Mexico (Leopold 1959); the subspecies there, *F. rufus escuinapae*, is listed as endangered by the USDI and is on appendix 1 of the CITES.

The value of bobcat fur has varied widely, depending on fashion and economic conditions. The average price per pelt rose from about $10 in the 1970/71 season (Deems and Pursley 1978) to $145 in 1978/79 (Jachowski 1981). There was a corresponding increase in the number of bobcats being harvested. The total known annual kill in the United States was approximately 92,000 in the late 1970s. This rate continued or declined slightly in the 1980s. About two-thirds of these animals were being taken primarily for their fur, and most of the pelts were being exported. There was concern that bobcat populations were being seriously reduced in some areas and that little was being done to manage the resource. Jachowski (1981), however, reported substantial improvement in management since *F. rufus* (except *F. r. escuinapae*) was placed on appendix 2 of the CITES in 1977. Prior to that time, few states had a closed season, and many had bounties. Presently, there are no bounties, 11 states provide complete protection, and the rest allow a regulated harvest during a limited season. Jachowski estimated that there are between 725,000 and 1,020,000 bobcats in the United States and suggested that current known harvest levels are not jeopardizing overall populations in the country. Export of skins was halted briefly by legal action in 1981 but was subsequently restored. In the 1982/83 season, about 75,000 skins were taken and sold for an average price of $142 (Rolley 1987).

Felis caracal (caracal).

Head and body length is 600–915 mm, tail length is 230–310 mm, shoulder height is 380–500 mm, and weight is 13–19 kg (Kingdon 1977). The pelage is dense but relatively short, and there are no side whiskers as in *F. lynx*. The general coloration is reddish brown. There is white on the chin, throat, and belly and a narrow black line from the eye to the nose. The ears are narrow, pointed, black on the outside, and adorned with black tufts up to 45 mm long. Smaller than *F. lynx*, *F. caracal* has a long and slender body and a tapering tail that is approximately one-third the length of the head and body.

The caracal is found mainly in dry country—woodland, savannah, and scrub—but avoids sandy deserts. Maternal dens are located in porcupine burrows, rocky crevices, or dense vegetation. This cat is largely nocturnal but sometimes is seen by day. It climbs and jumps well. It is mainly terrestrial and is apparently the fastest feline of its size. Prey is

Caracals *(Felis caracal)*, photo by Bernhard Grzimek.

stalked and then captured after a quick dash or leap. The diet includes birds, rodents, and small antelope.

The caracal has been reported to be territorial and to mark with urine. Vocalizations include miaows, growls, hisses, and coughing calls (Kingdon 1977). The species is usually seen alone, but Rowe-Rowe (1978b) reported a group of two adults and five young. Various observations suggest that the young may be born at any time of year. According to Kingdon (1977), the gestation period is 69–78 days, and the number of young per litter is one to six, usually three. The kittens open their eyes after 10 days, are weaned at 10–25 weeks, and attain sexual maturity between the ages of 6 and 24 months. Captives have lived up to 17 years.

The caracal is easily tamed and has been used to assist human hunters in Iran and India. It sometimes raids poultry, however, and thus has been killed by people. It apparently has become scarce in North Africa, South Africa, and parts of Asia. The subspecies in Soviet Central Asia, *F. c. michaelis*, is classified as rare by the IUCN (1978). All Asian populations are on appendix 1 of the CITES.

Felis serval (serval).

Head and body length is 670–1,000 mm, tail length is 240–450 mm, shoulder height is 540–620 mm, and weight is 8.7–18.0 kg. Males are generally larger than females (Kingdon 1977). The general coloration of the upper parts ranges from off-white to dark gold, and the underparts are paler, often white (Smithers 1978). The entire pelage is marked either with small, dark spots or with large spots that tend to merge into longitudinal stripes on the head and back. The tail has several rings and a black tip. The build is light, the legs and neck are long, and the ears are large and rounded.

According to Smithers (1978), the serval is generally a species of the savannah zone and is found in the vicinity of streams with densely vegetated banks. It is primarily nocturnal and may move 3–4 km per night. It is mainly terrestrial and can run or bound swiftly for short distances. Prey is apparently located both by sight and by hearing. Birds up to 3 meters above the ground may be captured by remarkable leaps, but the diet seems to consist mostly of murid rodents. Van Aarde and Skinner (1986b) reported home ranges of 2.1–2.7 sq km.

The serval is basically solitary. It has a shrill cry and also growls and purrs. There is no definite mating season, but Smithers (1978) reported that births in Zimbabwe occurred mainly in the warm months from September to April. Kingdon (1977) suggested that there are two birth peaks in East Africa, in March–April and September–November. Observations at the Basel Zoo (Wackernagel 1968) show that females can give birth twice a year, with a minimum normal interval of 184 days. Estrus usually lasted only one day, and the average gestation period was determined to be 74 days. The number of young in 20 litters averaged 2.35 and ranged from 1 to 4. Five newborn weighed 230–60 grams each, and one opened its eyes at 9 days. One female at the Basel Zoo gave birth to her last litter at 14 years and died at about 19 years and 9 months.

The serval is hunted for its skin in East Africa and now no longer occurs in areas heavily populated by people (Kingdon 1977). The species has been mercilessly hunted in farming areas of South Africa and is now considered to be rare in that country (Skinner, Fairall, and Bothma 1977).

Felis marmorata (marbled cat).

Head and body length is 450–530 mm, tail length is 475–550 mm, and weight is 2–5 kg (Lekagul and McNeely 1977). The ground color is brownish gray to bright yellow or rufous brown. The sides of the body are marked with large, irregular, dark blotches, each margined with black. There are solid black dots on the limbs and underparts. The tail is spotted, tipped with black, long, and bushy. The pelage is thick and soft, and the ears are short and rounded.

The marbled cat is a forest dweller, apparently nocturnal and partly arboreal in habit. It is thought to prey mostly on birds, to some extent on squirrels and rats, and possibly on lizards and frogs. Because of human disturbance and habitat destruction, it evidently has declined and become very rare in much of its range. It is classified as indeterminate by the IUCN (1978) and as endangered by the USDI and is on appendix 1 of the CITES.

Felis badia (bay cat).

Head and body length is 500–600 mm and tail length is 350–400 mm. The pelt is bright chestnut above and paler on the belly, with some obscure spots on the underparts and limbs. The long and tapering tail has a whitish median streak down the middle of its lower surface and becomes pure white at the tip. The bay cat has been reported to inhabit dense forests or areas of rocky limestone on the edge of the jungle. It apparently is known only by a few specimens and is classified as rare by the IUCN (1978).

Felis temmincki (Asian golden cat).

Head and body length is 730–1,050 mm and tail length is 430–560 mm. Lekagul and McNeely (1977) gave the weight as 12–15 kg. The pelage is of moderate length, dense, and rather harsh. The general coloration varies from golden red to dark brown and gray. Some specimens, especially from the northern parts of the range, have a pattern of spots on the body. The face is marked with white and black streaks, and the underside of the terminal third of the tail is white. The ears are short and rounded.

According to Lekagul and McNeely (1977), the Asian golden cat occurs in deciduous forests, tropical rainforests, and occasionally more open habitats. It is usually terrestrial but is capable of climbing. The diet includes hares, small deer, birds, lizards, and domestic livestock. This cat often hunts in pairs, and the male is said to play an active role in rearing the young. There is no confirmed breeding season. If a litter is lost, the female may produce another within 4 months. The gestation period is 95 days. Litters usually contain one or two young, which weigh about 250 grams each at birth. Individuals have lived nearly 18 years in captivity.

Because of habitat destruction and inability to adjust to the presence of human activity, *F. temmincki* has declined in much of its range. It is classified as indeterminate by the IUCN (1978) and as endangered by the USDI and is on appendix 1 of the CITES.

Felis aurata (African golden cat).

Head and body length is 616–1,016 mm, tail length is 160–460 mm, shoulder height is 380–510 mm, and weight is 5.3–16.0 kg. Males are generally larger than females (Kingdon 1977). The overall coloration varies from chestnut through fox red, fawn, gray brown, silver gray, and blue gray to dark slaty. The cheeks, chin, and underparts are white. Some specimens are marked all over with dark brown or dark gray dots, while in others the spots are restricted to the belly and insides of the limbs. Black specimens have been recorded. The legs are long, the head is small, and the paws are large.

According to Kingdon (1977), the African golden cat is found mainly in forests, sometimes in mountainous areas. It is said to be active both by day and by night. It is mainly terrestrial but climbs well. Prey is taken by stalking and rushing. The diet includes birds, small ungulates, and domestic animals. *F. aurata* is normally solitary. A pregnant female was taken in Uganda in September.

Asian golden cat *(Felis temmincki)*, photo from San Diego Zoological Society.

Felis bengalensis (leopard cat).

Head and body length is 445–1,070 mm, tail length is 230–440 mm, and weight is 3–7 kg (Lekagul and McNeely 1977; Stroganov 1969). There is much variation in color, but the upper parts are usually pale tawny and the underparts are white. The body and tail are covered with dark spots. There are usually four longitudinal black bands running from the forehead to behind the neck and breaking into short bands and rows of elongate spots on the shoulders. The tail is indistinctly ringed toward the tip. The head is small, the muzzle is short, and the ears are moderately long and rounded.

The leopard cat is found in many kinds of forested habitat at both high and low elevations. It dens in hollow trees or small caves or under overhangs or large roots. It is mainly nocturnal but often is seen by day. It is an excellent swimmer and has populated many offshore islands (Lekagul and McNeely 1977). It apparently hunts on the ground, as well as in trees, and feeds on hares, rodents, young deer, birds, reptiles, and fish.

Births have been reported to occur in May in both Siberia and India. According to Lekagul and McNeely (1977), however, breeding continues throughout the year in Southeast Asia. If one litter is lost, the female may mate and produce another within 4–5 months. The gestation period is 65–72 days. The number of young per litter is one to four, usually two or three. The father may participate in rearing the young. The latter open their eyes at about 10 days and reach sexual maturity at 18 months.

The subspecies *F. b. bengalensis*, of peninsular India and Southeast Asia, is listed as endangered by the USDI. It is, however, not particularly intolerant of human activity and is often found near villages, where it may raid poultry (Lekagul and McNeely 1977).

Felis rubiginosa (rusty-spotted cat).

Head and body length is 350–480 mm and tail length is 150–250 mm (Grzimek 1975). The upper parts are grizzled gray, with a rufous tinge of varying intensity, and marked with lines of brown, elongate blotches. The belly and insides of the limbs are white with large, dark spots. There are two dark streaks on the face, and four dark streaks run from the top of the head to the nape.

On the mainland of India, the rusty-spotted cat seems to prefer scrub, dry grassland, and open country. In Sri Lanka, however, it occurs in humid mountain forests. It is nocturnal and preys on birds and small mammals. Births occur in the

Leopard cat *(Felis bengalensis)*, photo from New York Zoological Society.

spring in India (Grzimek 1975). The Indian population is on appendix 1 of the CITES.

Felis viverrina (fishing cat).

Head and body length is 750–860 mm, tail length is 255–330 mm, shoulder height is 380–406 mm, and weight is 7.7–14.0 kg. The ground color is grizzled gray, sometimes tinged with brown, and there are elongate dark brown spots arranged in longitudinal rows. Six to eight dark lines run from the forehead, over the crown, and along the neck. The fur is short and rather coarse, the head is big and broad, and the tail is short and thick. The forefeet have moderately developed webbing between the digits. The claw sheaths are too small to allow the claws to retract completely.

The fishing cat is found in marshy thickets, mangrove swamps, and densely vegetated areas along creeks. It often wades in shallows and does not hesitate to swim in deep water. It catches prey fish by crouching on a rock or sand bank and using its paw as a scoop. The diet also includes crustaceans, mollusks, frogs, snakes, birds, and small mammals.

Kittens have been seen in the wild in April and June. At the Philadelphia Zoo, births occurred in March and August (Ulmer 1968). Observations there indicate that the gestation period is 63 days, birth weight is about 170 grams, the eyes of the young open by 16 days of age, the first meat is eaten at 53 days, and adult size is attained at 264 days. According to Grzimek (1975), the number of young per litter is one to four.

Felis planiceps (flat-headed cat).

Head and body length is 410–500 mm and tail length is 130–50 mm (Grzimek 1975). Two specimens from Malaysia weighed 1.6 and 2.1 kg (Muul and Lim 1970). The body is dark brown with a silvery tinge. The underparts are white, generally spotted and splashed with brown. The top of the head is reddish brown. The face below the eyes is light reddish, there are two narrow dark lines running across each cheek, and a yellow line runs from each eye to near the ear. The coat is thick, long, and soft. The pads of the feet are long and narrow, and the claws cannot be completely retracted. The skull is long, narrow, and flat; the nasals are short and narrow, and the orbits are placed well forward and close together (Lekagul and McNeely 1977).

The flat-headed cat is thought to be nocturnal and to hunt for frogs and fish along riverbanks. Observations of a captive kitten (Muul and Lim 1970) suggest that *F. planiceps* is a fishing cat. The kitten seemed to enjoy playing in water, took pieces of fish from the water, and captured live frogs but ignored live birds. Moreover, the long, narrow rostrum and the well-developed first upper premolar of the species would seem to be efficient for seizing slippery prey. *F. planiceps* is classified as indeterminate by the IUCN and as endangered by the USDI and is on appendix 1 of the CITES. It appears to be very rare, and some authorities consider it to be endangered in Thailand and Indonesia.

Felis iriomotensis (Iriomote cat).

Head and body length is 600 mm and tail length is 200 mm. The ground color is dark dusky brown, and spots arranged in longitudinal rows tend to merge into bands. Five to seven lines run from the back of the neck to the shoulders. The body is relatively elongate, the legs and tail are short, and the ears are rounded.

This cat dwells only on Iriomote, a Japanese-owned island of 292 sq km to the east of Taiwan. Its presence was not known to science until 1967. According to the IUCN (1978), *F. iriomotensis* appears to be restricted to lowland, subtropical rainforest. It is nocturnal and preys on small rodents, water birds, crabs, and mud skippers. It is territorial and has a home range not exceeding 2 sq km. Although totally protected by Japanese law, the Iriomote cat is rapidly losing habitat to

Flat-headed kitten *(Felis planiceps)*, photo from San Diego Zoological Society.

agricultural development. Its numbers have declined drastically since 1974, and probably only 40–80 individuals remain. The species is classified as endangered by the IUCN (1978) and the USDI.

Felis colocolo (pampas cat).

Head and body length is 567–700 mm, tail length is 295–322 mm, and shoulder height is 300–350 mm. The coloration ranges from yellowish white and grayish yellow to brown, gray brown, silvery gray, and light gray. Transverse bands of yellow or brown run obliquely from the back to the flanks. Two bars run from the eyes across the cheeks and meet beneath the throat. The coat is long, the tail is bushy, the face is broad, and the ears are pointed.

The pampas cat inhabits open grassland in some areas but also enters humid forests and mountainous regions. Grimwood (1969) reported it to occur throughout the Andes of Peru. It is mainly terrestrial but may climb a tree if pursued. It hunts at night, killing small mammals, especially guinea pigs *(Cavia)* and ground birds. Litters are said to contain one to three young.

Felis pardalis (ocelot).

Head and body length is 550–1,000 mm and tail length is 300–450 mm (Grzimek 1975; Leopold 1959). Weight is 11.3–15.8 kg. The ground color ranges from whitish or tawny yellow to reddish gray and gray. There are dark streaks and spots, which are arranged in small groups around areas that are darker than the ground color. There are two black stripes on each cheek, and one or two transverse bars on the insides of the legs. The tail is either ringed or marked with dark bars on the upper surface.

The ocelot occurs in a great variety of habitats, from humid tropical forests to fairly dry scrub country. It is generally nocturnal, sleeping by day in a hollow tree, in thick vegetation, or on a branch. It is mainly terrestrial but climbs, jumps, and swims well. The diet includes rodents, rabbits, young deer and peccaries, birds, snakes, and fish.

In a study in Peru, Emmons (1988*b*) found adult females to occupy exclusive home ranges of about 2 sq km. Male ranges were several times greater and also were exclusive of one another but did overlap a number of female ranges. Individuals generally moved about alone but appeared to make contact frequently and probably maintained a network of social ties. The ocelot communicates by mewing and, during courtship, by yowls not unlike those of *F. catus*. There is probably no seasonal breeding in the tropics (Grzimek 1975). In Mexico and Texas, births are reported to occur in fall and winter (Leopold 1959). The gestation period is 70 days. Usually there are two young, but there may be as many as four. A captive lived for 20 years and 3 months (Jones 1982).

The ocelot is classified as vulnerable by the IUCN and as endangered by the USDI. The subspecies *F. p. mearnsi*, of Central America, and *F. p. mitis*, of central and eastern Brazil, are on appendix 1 of the CITES. The subspecies *F. p. albescens* probably once ranged over most of Texas and at least as far east as Arkansas and Louisiana (Lowery 1974) but is currently restricted to the border region of extreme southern Texas and northeastern Mexico. Fewer than 1,000 individuals are thought to survive, and fewer than 100 of these are in Texas. The clearing of brush country for agricultural purposes is the main problem in this area. The ocelot still occurs over most of the remainder of its original range but is declining because of habitat loss and hunting for its fur. Prior to 1972, when importation into the United States was prohibited, enormous numbers of skins were brought into the country—133,069 in 1969 alone. As late as 1975 Great Britain imported 76,838 skins. In the early 1980s ocelot fur coats sold for up to U.S. $40,000 in West Germany. The ocelot is also reported to be in much demand for use as a pet, a live animal selling for $800. The species now is protected by the laws of most countries in which it lives, and its trade is partly regulated by the CITES, but enforcement is difficult (Thornback and Jenkins 1982).

Felis wiedii (margay).

Head and body length is 463–790 mm and tail length is 331–510 mm. The ground color is yellowish brown above and white below. There are longitudinal rows of dark brown spots, the centers of which are paler than the borders. The margay closely resembles the ocelot but is smaller, has a slenderer build, and has a relatively longer tail.

The margay is mainly, if not exclusively, a forest dweller. It is much more arboreal than the ocelot and is thought to forage in trees. The only specimen known from Texas was taken prior to 1852 and is thought to represent an individual that strayed far from the normal habitat (Leopold 1959). The arboreal acrobatics and effortless climbing of the margay are partly the result of limb structure (Grzimek 1975). The feet are broad and soft and have mobile metatarsals. The hind foot is much more flexible than that of other felids, being able to rotate 180°. There is probably no particular breeding season in the tropics, litters contain one or two young, and captives have lived over 13 years (Grzimek 1975).

The margay is classified as vulnerable by the IUCN and as endangered by the USDI. The subspecies *F. w. nicaraguae* and *F. w. salvinia*, of Central America, are on appendix 1 of the CITES. Leopold (1959) referred to the margay as "an exceedingly rare animal." Grimwood (1969) stated that exports of pelts from Peru were increasing. Paradiso (1972) indicated that at least 6,701 margays, including both live animals and skins, were imported into the United States in 1970. According to Broad, Luxmoore, and Jenkins (1988), the total number of skins known to be in international trade declined from at least 30,000 in 1977 to about 20,000 in 1980 and only 138 in 1985. Thornback and Jenkins (1982) noted that the margay now is protected in most of its range but is declining through extensive illegal hunting and habitat destruction. Because of its arboreal nature, it probably suffers more from deforestation than does the ocelot.

Felis tigrina (little spotted cat).

Head and body length is 400–550 mm and tail length is 250–400 mm (Grzimek 1975). A captive male weighed 2.75 kg and a captive female weighed 1.75 kg. The upper parts vary in color from light to rich ochre and have rows of large, dark spots. The underparts are paler and less spotted. The tail has 10 or 11 rings and a black tip. One-fifth of all specimens are melanistic.

The little spotted cat lives in forests, and its habits in the wild are not known. Captive females have an estrus of several days, a gestation period of 74–76 days, and litters of one or two young. The kittens develop slowly, opening their eyes at 17 days and starting to take solid food at 55 days.

Like the ocelot and margay, until recently the little spotted cat was subject to uncontrolled commerce. At least 3,170 individuals, including both live animals and skins, were imported into the United States in 1970 (Paradiso 1972). The species subsequently was listed as endangered by the USDI, and importation was banned. However, in the early 1980s *F. tigrina* became the leading spotted cat in the international fur trade, with the number of skins in commerce peaking at nearly 84,500 in 1983 (Broad 1987). The subspecies *F. t. oncilla*, of Central America, is on appendix 1 of the CITES. The entire species now is classified as vulnerable by the IUCN. The problems reported by Thornback and Jenkins (1982) are apparent natural rarity, uncontrolled hunting, and loss of habitat to agriculture and logging.

Felis geoffroyi (Geoffroy's cat).

Head and body length is 450–700 mm and tail length is 260–350 mm. The ground color varies widely, from brilliant ochre in the northern parts of the range to silvery gray in the south. The body and limbs are covered with small black spots. There may be several black streaks on the crown, two on each cheek, and one between the shoulders. The tail is ringed.

Geoffroy's cat inhabits scrubby woodland and open bush country. Ximenez (1975) reported the elevational range to be sea level to 3,300 meters and also that the species is nocturnal. According to Grzimek (1975), it is a good climber and swimmer; it sleeps in trees and readily enters water. It preys on small mammals, birds, and possibly fish. Hemmer (1976) listed a gestation period of 74–76 days. The single annual litter contains two or three young and in Uruguay is produced from December to May (Ximenez 1975).

F. geoffroyi is heavily exploited for the international fur trade; over 78,000 skins, mainly from Paraguay, were reported in commerce in 1983 (Broad 1987).

Felis guigna (kodkod).

Head and body length is 390–493 mm and tail length is 195–230 mm. The coat is buff or gray brown and is heavily marked with rounded, blackish spots on both the upper and lower parts. The tail has blackish rings. Melanistic individuals are not uncommon.

The kodkod is a nocturnal forest dweller. Some authorities have stated that it is an expert climber, living mainly in trees, while others have called it terrestrial. It probably preys mainly on small mammals but has been reported to raid henhouses.

Felis jacobita (mountain cat).

Head and body length is 600 mm and tail length is 350 mm. The coat is soft and fine. The upper parts are silvery gray and marked by irregular brown or orange yellow spots and transverse stripes. The underparts are whitish and have blackish spots. The tail is bushy, ringed with black or brown, and light-tipped.

The mountain cat is found in the arid and semiarid zone of the Andes. Elevational range is up to about 5,000 meters (Grzimek 1975). It preys on birds and small mammals, such as viscachas (Thornback and Jenkins 1982). It is classified as rare by the IUCN and as endangered by the USDI and is on appendix 1 of the CITES.

Felis yagouaroundi (jaguarundi).

Head and body length is 550–770 mm, tail length is 330–600 mm, and weight is 4.5–9.0 kg. There are two color phases:

Geoffroy's cat *(Felis geoffroyi)*, photo by K. Rudloff through East Berlin Zoo.

blackish to brownish gray, and fox red to chestnut. The body is slender and elongate, the head is small and flattened, the ears are short and rounded, the legs are short, and the tail is very long. This cat is sometimes said to resemble a weasel or otter in external appearance.

The jaguarundi inhabits lowland forests and thickets. It hunts in the morning and evening and is much less nocturnal than most cats. It forages mainly on the ground but is an agile climber. The diet includes birds and small mammals. *F. yagouaroundi* has been reported to live in pairs in Paraguay but to be solitary in Mexico. According to Grzimek (1975), there is no definite reproductive season in the tropics, but in Mexico young are produced around March and August. It is not known whether one female gives birth in both seasons. The gestation period is 63–70 days, and litters contain two to four young.

The pelt is of poor quality and of little value (Leopold 1959). The jaguarundi is widespread and not subject to commercial exploitation (Paradiso 1972). Nonetheless, four subspecies—*cacomitli, tolteca, fossata,* and *panamensis*—which range from southern Texas and Arizona to Panama, are listed as endangered by the USDI. The entire species is on appendix 1 of the CITES and is classified as indeterminate by the IUCN. Its status is not well understood, but it has declined in at least the northern portions of its range because of human persecution and habitat destruction (Thornback and Jenkins 1982). The jaguarundi is not now known to be resident in Arizona, and only a few individuals survive in Texas. During the late Pleistocene the species occurred as far as Florida, and a small population may have become established there recently through introduction by human agency. There have been sporadic reports of a cat resembling the jaguarundi from other southeastern states.

Felis concolor (cougar, puma, panther, or mountain lion).

The names cougar, puma, panther, and mountain lion are used interchangeably for this species, and various other vernacular terms are applied in certain areas. This is by far the largest species in the genus *Felis*, averaging about the same size as the leopard *(Panthera pardus)*. In males, head and body length is 1,050–1,959 mm, tail length is 660–784 mm, and weight is 67–103 kg. In females, head and body length is 966–1,517 mm, tail length is 534–815 mm, and weight is usually 36–60 kg. Shoulder height is 600–700 mm. Generally, the smallest animals are in the tropics, and the largest are in the far northern and southern parts of the range. There are two variable color phases: one ranges from buff, cinnamon, and tawny to cinnamon rufous and ferrugineous; the other ranges from silvery gray to bluish and slaty gray. The body is elongate, the head is small, the face is short, and the neck and tail are long. The limbs are powerfully built; the hind legs are larger than the forelegs. The ears are small, short, and rounded. Females have three pairs of mammae (Banfield 1974).

The cougar has the greatest natural distribution of any mammal in the Western Hemisphere except *Homo sapiens*. It can thrive in montane coniferous forests, lowland tropical forests, swamps, grassland, dry brush country, or any other area with adequate cover and prey. The elevational range extends from sea level to at least 3,350 meters in California and 4,500 meters in Ecuador. There is usually no fixed den, except as used by females to rear young. Temporary shelter is taken in such places as dense vegetation, rocky crevices, and caves. The cougar is agile and has great jumping power: it may leap from the ground to a height of up to 5.5 meters in a tree. It swims well but commonly prefers not to enter water. Sight is the most acute sense, and hearing is also good, but

Cougar *(Felis concolor)*, photo by Dom Deminick, Colorado Division of Wildlife.

smell is thought to be poorly developed. Activity may be either nocturnal or diurnal. The cougar hunts over a large area, sometimes taking a week to complete a circuit of its home range (Leopold 1959). In Idaho, Seidensticker et al. (1973) found residents to occupy fairly distinct, but usually contiguous, winter–spring and summer–fall home areas. The latter area was generally larger and at a higher elevation, reflecting the summer movements of ungulate herds.

The cougar carefully stalks its prey and may leap upon the victim's back or seize it after a swift dash. Throughout the range of the cougar, the most consistently important food is deer—*Odocoileus* in North America and *Blastoceros, Hippocamelus,* and *Mazama* in South America. Estimates of kill frequency vary from about one deer every three days for a female with large cubs to one every 16 days for a lone adult (Lindzey 1987). The diet also includes other ungulates, beaver, porcupines, and hares. The kill is usually dragged to a sheltered spot and then partly consumed. The remains are covered with leaves and debris, then visited for additional meals over the next several days.

Detailed studies in central Idaho (Hornocker 1969, 1970; Seidensticker et al. 1973) showed that the cougar population depends almost equally on mule deer *(Odocoileus hemionus)* and elk *(Cervus canadensis).* The study area contained 1 cougar for every 114 deer and 87 elk. About half of the animals killed were in poor condition. Deer and elk populations increased during a four-year study period, evidently being affected more by food availability than by cougar predation; nonetheless, predation was thought to moderate prey oscillations and to remove less fit individuals. During the same period, the cougar population remained stable, its numbers being regulated mainly by social factors rather than by food supply. Density was about 1 adult cougar per 35 sq km. The total area used by individuals varied from 31 to 243 sq km in winter–spring and from 106 to 293 sq km in summer–fall. There was little overlap in the areas occupied by resident adult

males, but the areas of resident females often overlapped one another completely and were overlapped by resident male areas. Young, transient individuals of both sexes moved through the areas used by residents. There was a land tenure system based on prior right; transients could not permanently settle in an occupied area unless the resident died. Dispersal and mortality of young individuals unable to establish an area for themselves seemed to limit the size of the cougar population.

According to Lindzey (1987), long-term population densities in other areas have been found to vary from about one cougar per 21 sq km to one per 200 sq km, and reported annual home range has been as great as 1,826 sq km for one animal in Texas. The cougar is essentially solitary, with individuals deliberately avoiding one another except during the brief period of courtship. Communication is mostly by visual and olfactory signals, and males regularly make scrapes in the soil or snow, sometimes depositing urine or feces therein. Vocalizations include growls, hisses, and birdlike whistles. A very loud scream has been reported on rare occasion, but its function is not known.

There is no specific breeding season, but most births in North America occur in late winter and early spring. Females are seasonally polyestrous and usually give birth every other year (Banfield 1974). Estrus lasts about 9 days, and the gestation period is 90–96 days. The number of young per litter is one to six, commonly three or four. The kittens weigh 226–453 grams at birth and are spotted until about 6 months old. They nurse for 3 months or more but begin to take some meat at 6 weeks. If born in the spring, they are able to accompany the mother by autumn and to make their own kills by the end of winter; nonetheless, they usually remain with their mother for several more months or even another year. Litter mates stay together for 2–3 months after leaving the mother. Sexual maturity is attained by females at about 2.5 years, but males may not mate until at least 3 years (Banfield 1974).

Clouded leopard *(Neofelis nebulosa)*, photo from San Diego Zoological Garden.

Regular reproductive activity does not begin until a young animal establishes itself on a permanent home area (Seidensticker et al. 1973). Females may remain reproductively active until at least 12 years of age, and males to at least 20 years; captives have lived for over 20 years (Currier 1983).

The remainder of the account of this species is based largely on Nowak (1976). The cougar is generally considered to be harmless to people, but attacks have occurred, and seven persons are known to have been killed in the United States and Canada since 1900. The cougar has long been viewed as a threat to domestic animals, such as horses and sheep, and is also sometimes thought to reduce populations of game. The species has thus been intensively hunted since the arrival of European colonists in the Western Hemisphere. Most successful hunting is done by using dogs to pursue the cat until it seeks refuge in a tree, where it can easily be shot.

By the early twentieth century the cougar appeared to have been eliminated everywhere to the north of Mexico except in the mountainous parts of the West, in southern Texas, and in Florida. The species still occupies the same regions, and about 16,000 individuals may be present. California provides almost complete protection to the cougar, Texas still allows it to be killed at any time, and all other western states and provinces permit regulated hunting for sport and predator control. Public antipathy seems to have moderated in the last few decades, and the general pattern of decline may have been halted, but loss of habitat and conflict with agricultural interests are still problems.

Since the late 1940s, evidence has accumulated indicating the presence of small surviving cougar populations in south-central Canada, New Brunswick, the southern Appalachians, and the Ozark region and adjoining forests of Arkansas, southern Missouri, eastern Oklahoma, and northern Louisiana. In southern Florida, numerous individuals have been killed or live-captured in the last two decades, and a federal-state program of study and conservation is under way (Belden 1986; Belden and Forrester 1980; Shapiro 1981). The subspecies *F. c. coryi*, which formerly occurred from eastern Texas to Florida, and the subspecies *F. c. couguar*, of the northeastern United States and southeastern Canada, are classified as endangered by the IUCN. These two subspecies, along with *F. c. costaricensis*, of Central America, are also listed as endangered by the USDI and are on appendix 1 of the CITES.

CARNIVORA; FELIDAE; **Genus NEOFELIS**
Gray, 1867

Clouded Leopard

The single species, *N. nebulosa*, occurs from Nepal to south-eastern China and the Malay Peninsula and on Taiwan, Hainan, Sumatra, and Borneo. Most authorities, including Ellerman and Morrison-Scott (1966), Ewer (1973), Guggisberg (1975), and Hemmer (1978), treat *Neofelis* as a distinct genus with one species. Simpson (1945) considered *Neofelis* to be a subgenus of *Panthera*. Leyhausen (1979) listed *Neofelis* as a full genus but included within it not only the clouded leopard but also the tiger (here referred to as *Panthera tigris*).

Head and body length is about 616–1,066 mm, tail length is 550–912 mm, and weight is usually 16–23 kg. The coat is grayish or yellowish, with dark markings ("clouds") in such forms as circles, ovals, and rosettes. The markings on the shoulders and back are darker on their posterior margins than on their front margins, thus suggesting that stripes can be evolved from blotches or spots. The forehead, legs, and base of the tail are spotted, and the remainder of the tail is banded. Melanistic specimens have been reported (Medway 1977).

The tail is long, the legs are stout, the paws are broad, and the pads are hard. The skull is long, low, and narrow (Guggisberg 1975). The upper canine teeth are relatively longer than those of any other living cat, having a length about three times greater than the basal width at the socket. The first upper premolar is greatly reduced or absent, leaving a wide gap between the canine and the cheek teeth. Unlike that of *Panthera*, the hyoid of *Neofelis* is ossified (Guggisberg 1975).

The clouded leopard inhabits various kinds of forest, perhaps to elevations of up to 2,500 meters. It is usually said to be highly arboreal, to hunt in trees, and to spring on ground prey from overhanging branches. Information compiled by Guggisberg (1975), however, suggests that *Neofelis* is more terrestrial and diurnal than is generally assumed. Rabinowitz, Andau, and Chai (1987) indicated that it travels mainly on the ground and uses trees primarily as resting sites. It has been reported to feed on birds, monkeys, pigs, cattle, young buffalo, goats, deer, and even porcupines.

Reproduction is known only through observations in captivity (Fellner 1965; Fontaine 1965; Guggisberg 1975; Murphy 1976). Births in Europe and Texas have occurred from

March to August. The gestation period seems normally to be 86–93 days. The number of young per litter ranges from one to five but is commonly two. The young weigh about 140–70 grams each at birth, open their eyes after 12 days, take some solid food at 10.5 weeks of age, and nurse for 5 months. The coat is initially black in the patterned areas, and full adult coloration is attained at about 6 months. Captives have lived over 17 years (Grzimek 1975). Many captives are gentle and playful and like to be petted by their custodians.

The clouded leopard is classified as vulnerable by the IUCN (1978) and as endangered by the USDI and is on appendix 1 of the CITES. The main problem is loss of forest habitat to agriculture. The species also has been excessively hunted in some areas for its beautiful pelt. It may already have disappeared from Hainan, and in Taiwan it is restricted to the wildest and most inaccessible parts of the central mountain range. After a survey on Taiwan, Rabinowitz (1988) reported that the most recent sighting was in 1983 but expressed hope that the species does survive.

CARNIVORA; FELIDAE; Genus PANTHERA
Oken, 1816

Big Cats

There are five subgenera and five species (Cabrera 1957; Corbet 1978; Ellerman and Morrison-Scott 1966; Guggisberg 1975; Hall 1981; Smithers, *in* Meester and Setzer 1977):

subgenus *Uncia* Gray, 1854

P. uncia (snow leopard), mountainous areas from Afghanistan to Lake Baikal and eastern Tibet;

subgenus *Tigris* Frisch, 1775

P. tigris (tiger), eastern Turkey to southeastern Siberia and Malay Peninsula, Sumatra, Java, Bali;

subgenus *Panthera* Oken, 1816

P. pardus (leopard), western Turkey and Arabian Peninsula to southeastern Siberia and Malay Peninsula, Sri Lanka, Java, Kangean Islands, most of Africa;

subgenus *Jaguarius* Severtzow, 1858

P. onca (jaguar), southern United States to northern Argentina;

subgenus *Leo* Brehm, 1829.

P. leo (lion), found in historical time from the Balkan and Arabian peninsulas to central India, and in almost all of Africa.

The use of the name *Panthera* for this genus is in keeping with Corbet (1978), Ellerman and Morrison-Scott (1966), Hemmer (1978), Leyhausen (1979), Mazak (1981), and most of the other authorities cited herein. A few authorities, such as Cabrera (1957) and Stains (1984), consider *Panthera* to be invalid for technical reasons of nomenclature and prefer to use the generic term *Leo*. Some authorities, including Hall (1981), do not consider *Panthera* to be generically distinct from *Felis*. *Uncia* was treated as a separate genus by Guggisberg (1975), Hemmer (1972, 1978), and Leyhausen (1979) but not by Corbet (1978), Ellerman and Morrison-Scott (1966), and Ewer (1973). Leyhausen placed *P. tigris* in the genus *Neofelis*. Both Hemmer (1978) and Mazak (1981) divided *Panthera* into only two subgenera: *Tigris*, with *P. tigris*; and *Panthera*, with *P. pardus*, *P. onca*, and *P. leo*.

The members of the genus *Panthera* have an incompletely ossified hyoid apparatus, with an elastic cartilaginous band replacing the bony structure found in other cats (Grzimek 1975). This elastic ligament is usually considered to be the anatomical feature that allows roaring but limits purring to times of exhaling. Other cats can purr both when exhaling and inhaling. The snow leopard *(P. uncia)* possesses an elastic ligament but does not roar. According to Paul Leyhausen (Max-Planck Institut für Verhaltensphysiologie, pers. comm.), certain small cats, such as *Felis nigripes*, produce a full roar but do not sound as formidable as the lion or tiger, simply because of the size difference. *Panthera* is also distinguished from *Felis* by having hair that extends to the front edge of the nose (Grzimek 1975). Additional information is provided separately for each species.

Panthera uncia (snow leopard).
Except as noted, the information for the account of this species was taken from Guggisberg (1975), Hemmer (1972), and Schaller (1977). Head and body length is 1,000–1,300 mm,

Snow leopard *(Panthera uncia)*, photo by Ernest P. Walker.

A. Lion *(Panthera leo)*, photo from P. D. Swanepoel. B. Lioness and cub *(P. leo)*, photo by Ernest P. Walker. C. Lion cubs *(P. leo)*, photo from San Diego Zoological Garden. D. Leopard *(P. pardus)*, photo from U.S. National Zoological Park.

tail length is 800–1,000 mm, shoulder height is about 600 mm, and weight is 25–75 kg. The ground color varies from pale gray to creamy smoke gray, and the underparts are whitish. On the head, neck, and lower limbs there are solid spots, and on the back, sides, and tail are large rings or rosettes, often enclosing some small spots. The coat is long and thick, and the head is relatively small.

The snow leopard is found in the high mountains of Central Asia. In summer it occurs commonly in alpine meadows and rocky areas at elevations of 2,700–6,000 meters. In the winter it may follow its prey down into the forests below 1,800 meters. It sometimes dens in a rocky cavern or crevice. It is often active by day, especially in the early morning and late afternoon. It is graceful and agile and has been reported to leap as far as 15 meters. Prey is either stalked or ambushed. The diet includes mountain goats and sheep, deer, boar, marmots, pikas, and domestic livestock.

In a study in Nepal, Jackson and Ahlborn (1988) found that an area of 100 sq km supported 5–10 snow leopards. The home ranges of five individuals in this region measured about 12–39 sq km; these ranges overlapped almost entirely, both between and within sexes, but the animals kept well apart. In general, *P. uncia* is thought to utilize a large home range that is traversed in the course of about a week. It is possible that a pair shares a range. Socially, the snow leopard is thought to be like the tiger, essentially solitary but not unsociable. The snow leopard does not roar but has several vocalizations, including a loud moaning associated with attraction of a mate. Births usually occur from April to June, both in the wild and in captivity, after a gestation period of 90–103 days. The young are born in a rocky shelter lined with the mother's fur. The number of young per litter is one to five, usually two or three. The cubs weigh about 450 grams each at birth, open their eyes after 7 days, eat their first solid food at 2 months, and follow their mother at 3 months. They hunt with the mother at least through their first winter of life and attain sexual maturity at about 2 years. Maximum longevity in captivity is 17–19 years.

The snow leopard is classified as endangered by the IUCN, the USDI, and the Soviet Union and is on appendix 1 of the CITES. It has declined in numbers through hunting by people, because it is considered to be a predator of domestic stock, it is valued as a trophy, and its fur is in demand by commerce. Although it is protected in China, it continues to be hunted there, and its skin is sold on the open market (Schaller et al. 1988). Contributing to the problem in Nepal, where perhaps 150–300 individuals remain, are a reduction in natural prey and increased use of alpine pastures by people and stock (Jackson 1979). The total number remaining in the wild has been estimated roughly as about 3,000–10,000 (Green 1988). There are another 378 in captivity (Blomqvist 1988).

Panthera tigris (tiger).

Head and body length is 1,400–2,800 mm and tail length is 600–950 mm (Grzimek 1975). The subspecies found in southeastern Siberia and Manchuria, *P. t. altaica*, is the largest living cat. The other mainland subspecies are also large, but those of the East Indies are much smaller. In *P. t. altaica* males weigh 180–306 kg, and females, 100–167 kg; in *P. t. tigris*, of India and adjoining countries, males weigh 180–258 kg, and females, 100–160 kg; in *P. t. sumatrae*, of Sumatra, males weigh 100–140 kg, and females, 75–110 kg; and in *P. t. balica*, of Bali, males weigh 90–100 kg, and females, 65–80 kg (Mazak 1981). The ground color of the upper parts ranges from reddish orange to reddish ochre, and the underparts are creamy or white. The head, body, tail, and limbs have a series of narrow black, gray, or brown stripes. On the flanks the stripes generally run in a vertical direction; and in some specimens the stripes are much reduced on the shoulders, forelegs, and anterior flanks (Guggisberg 1975). There also is a rare but much publicized variant with a chalky white coat, dark stripes, and icy blue eyes; 103 are now in captivity (Roychoudhury 1987).

Except as noted, the information for the remainder of this account was taken from Mazak (1981). The tiger is tolerant of a wide range of environmental conditions, its only requirements being adequate cover, water, and prey. It is found in such habitats as tropical rainforests, evergreen forests, mangrove swamps, grasslands, savannahs, and rocky country. An individual may have one or more favored dens within its territory, located in such places as caves, hollow trees, and dense vegetation. The tiger usually does not climb trees but is capable of doing so. It has been reported to cover up to 10 meters in a horizontal leap. It seems to like water and can swim well, easily crossing rivers 6–8 km wide and sometimes swimming up to 29 km. It is mainly nocturnal but may be active in daylight, especially in winter in the northern part of its range. Siberian animals have moved up to 60 km per day. In Nepal, Sunquist (1981) determined the usual daily movement to be 10–20 km.

To hunt, the tiger depends more on sight and hearing than on smell. It usually carefully stalks its prey, approaching from the side or rear and attempting to get as close as possible. It then leaps upon the quarry and tries simultaneously to throw it down and grab its throat. Killing is by strangulation or a bite to the back of the neck. The carcass is often dragged to an area within cover or near water. One individual dragged an adult gaur *(Bos gaurus)* 12 meters, and later 13 men tried to pull the carcass but could not move it (Grzimek 1975). After eating its fill, the tiger may cover the remains with grass or debris and then return for additional meals over the next several days. The diet consists mainly of large mammals, such as pigs, deer, antelope, buffalo, and gaur. Smaller mammals and birds occasionally are taken. A tiger can consume up to 40 kg of meat at one time, but individuals in zoos are given 5–6 kg per day. Although the tiger is an excellent hunter, it fails in at least 90 percent of its attempts to capture animals. The tiger thus cannot eliminate entire prey populations. Indeed, Sunquist (1981) found that prey numbers, with the exception of one species, were not even being limited by tiger predation.

In Kanha National Park, in central India, Schaller (1967) determined that 10–15 adult tigers were regularly resident in an area of about 320 sq km. In Royal Chitawan National Park, in Nepal, Sunquist (1981) found an overall population density of 1 adult per 36 sq km. Observations in these and other areas indicate much variation in home range size and social behavior, evidently depending on habitat conditions and prey availability. In India, individual home range seems usually to be 50–1,000 sq km. In Manchuria and southeastern Siberia the usual size is 500–4,000 sq km, and the maximum reported is 10,500 sq km. In Nepal, Smith, McDougal, and Sunquist (1987) found home range size to be 19–151 sq km for males and 10–51 sq km for females. These ranges essentially corresponded to defended territories, in that there was no overlap between those of adults of the same sex. A male range, however, overlapped the ranges of several females. Mothers evidently allowed their daughters to establish adjacent territories, but eventually the two generations may become agonistic. Schaller's studies indicate that the same kind of situation may exist in central India but also that females are not territorial there, adults of the same sex sometimes share a home range, and there are transient animals that lack an established range. All of these authorities suggested the presence of a land tenure system based on prior right, by which a resident animal is never replaced on its range until its death.

Top, Chinese tiger *(Panthera tigris amoyensis)*. Bottom, melanistic jaguar *(Panthera onca)*. Photos from East Berlin Zoo.

Territorial boundaries are not patrolled, but individuals do visit all parts of their ranges over a period of days or weeks. These areas are marked with urine and feces. Avoidance, rather than fighting, seems to be the rule for tigers; nonetheless, one individual transplanted from its normal range to a different area evidently was killed soon thereafter by another tiger (Seidensticker et al. 1976).

The tiger is essentially solitary, except for courting pairs and females with young. Even individuals that share a range usually keep 2–5 km apart (Sunquist 1981). The tiger is not unsociable, however, and the animals in a given area (probably close relatives) may know one another and have a generally amicable relationship (Schaller 1967). Several adults may come together briefly, especially to share a kill. Limited evidence suggests that a tiger roars to announce to its associates that it has made a kill. An additional function of roaring

Tiger *(Panthera tigris)*, photo from East Berlin Zoo.

seems to be the attraction of the opposite sex. There are a number of other vocalizations, such as purrs and grunts, and the tiger also communicates by marking with urine, feces, and scratches.

Mating may occur at any time but is most frequent from November to April. Females usually give birth every 2–2.5 years and occasionally wait 3–4 years; if all the newborn are lost, however, another litter can be produced within 5 months (Schaller 1967). Females enter estrus at intervals of 3–9 weeks, and receptivity lasts 3–6 days. The gestation period is usually 104–6 days but ranges from 93 to 111 days. Births occur in a cave, a rocky crevice, or dense vegetation. The number of young per litter is usually two or three and ranges from one to six. The cubs weigh 780–1,600 grams at birth, open their eyes after 6–14 days, nurse for 3–6 months, and begin to travel with the mother at 5–6 months. They are taught how to hunt prey and apparently are capable killers at 11 months (Schaller 1967). They usually separate from the mother at 2 years but may wait another year. Sexual maturity is attained at 3–4 years by females and at 4–5 years by males. About half of all cubs do not survive more than 2

years, but maximum known longevity is about 26 years both in the wild and in captivity.

The tiger probably has been responsible for more human deaths, through direct attack, than has any other wild mammal. In this regard, perhaps the most dangerous place in modern history was Singapore and nearby islands, where, following extensive settlement during the 1840s, over 1,000 persons were being killed annually (McDougal 1987). About 1,000 more people were reportedly killed each year in India during the early 1900s. Guggisberg (1975) questioned the accuracy of these statistics, but there seems to be little doubt that some tigers have preyed extensively, or almost exclusively, on people. One individual is said to have killed 430 persons in India. Although such man-eaters have declined with the general reduction in tiger numbers in the twentieth century, the problem does persist. In 1972, for example, India's production of honey and beeswax dropped by 50 percent when at least 29 persons who gathered these materials were devoured (Mainstone 1974).

Tigers currently seem to be especially dangerous in the Sundarbans mangrove forest, at the mouth of the Ganges

Top, Siberian tiger *(Panthera tigris altaica)*. Bottom, Sumatran tiger *(P. t. sumatrae)*. Photos from East Berlin Zoo.

River. Hendrichs (1975c) reported that 129 persons were killed in this area from 1969 to 1971 but noted that only 1 percent of the tigers there actually seem to seek out human prey. P. Jackson (1985) wrote that 429 persons had been killed in the same area during the previous 10 years, but Chakrabarti (1984) noted that unofficial estimates put the average annual toll at 100. A table compiled by Khan (1987) indicates a unique mammalian situation, in that the number of people killed by tigers in the Sundarbans during some recent years considerably exceeds the number of tigers killed by people.

Because it is considered to be a threat to human life and domestic livestock, and also because it is valued as a big game trophy, the tiger has been relentlessly hunted, trapped, and poisoned. Some European hunters and Indian maharajahs killed hundreds of tigers each. After World War II, hunting became even more widespread than previously (Guggisberg 1975). The commercial trade in tiger skins intensified in the 1960s, and by 1977 a pelt brought as much as U.S. $4,250 in Great Britain (IUCN 1978). Perhaps the greatest threat to the survival of the tiger is destruction of its habitat. With the

expansion of human populations, the logging of forests, the elimination of natural prey, and the spread of agriculture, there is continuous conflict between people and the tiger, and the latter species is almost always the loser. It is estimated that in 1920 there were about 100,000 tigers in the world. Estimates for wild populations ranged as low as 4,000 in the 1970s (Fisher 1978; Jackson 1978) but subsequently increased to about 6,000–8,000 (Foose 1987a; P. Jackson 1985; Luoma 1987). The increase has resulted largely from an intensive effort to protect the species and establish reserves in India (Karanth 1987; Panwar 1987), though there are doubts as to how long viable populations can be maintained there in the face of growing human encroachment (Ward 1987). The tiger is classified as endangered by the IUCN and the USDI and is on appendix 1 of the CITES.

Except as noted, the information for the following summaries of the status of the eight subspecies of *P. tigris* was taken from the IUCN (1978), Luoma (1987), and Mazak (1981):

P. t. virgata (Caspian tiger), formerly occurred from eastern Turkey and the Caucasus to the mountains of Soviet

Central Asia and Afghanistan, a few individuals still present in Turkey in the 1970s, now probably extinct;

P. t. tigris (Bengal tiger), originally found from Pakistan to western Burma, exterminated in Pakistan by 1906 (Roberts 1977), about 4,000 individuals now thought to survive in India, 400 in Bangladesh, 300 in Burma (Seal, Jackson, and Tilson 1987), at least 230 in Nepal (Mishra, Wemmer, and Smith 1987), and 200 in Bhutan (Dorji and Santiapillai 1989);

P. t. corbetti (Indochinese tiger), still found from eastern Burma to Viet Nam and the Malay Peninsula, about 2,000 individuals remain;

P. t. amoyensis (Chinese tiger), formerly occurred throughout eastern China, an estimated 4,000 individuals still survived in 1949 (Lu 1987), now confined largely to the Changjiang (Yangtze) Valley and apparently near extinction, with only 40–80 animals left in the wild, another 40 in captivity in China (Tan 1987; Xiang, Tan, and Jia 1987);

P. t. altaica (Siberian tiger), formerly found from Lake Baikal to the Pacific coast and Korea, apparently now very rare in Manchuria and Korea, protection in the Ussuri region of the Soviet Union seems to have resulted in an increase in numbers and distribution there—about 200–300 individuals are now present in the wild (Prynn 1980), and 567 in captivity (Foose 1987*a*);

P. t. sumatrae (Sumatran tiger), population declined rapidly from an estimated 1,000 in the 1970s to about 500–600 now in the wild, another 157 in captivity (Ballou and Seidensticker 1987);

P. t. sondaica (Javan tiger), almost all suitable habitat destroyed, only 4 or 5 individuals survived in the 1970s, now probably extinct;

P. t. balica (Bali tiger), probably extinct, last known specimen taken in 1937, though Van Den Brink (1980) held out some hope that it survives.

Panthera pardus (leopard).

Head and body length is 910–1,910 mm, tail length is 580–1,100 mm, and shoulder height is 450–780 mm. Males weigh 37–90 kg, and females, 28–60 kg. There is much variation in color and pattern. The ground color ranges from pale straw and gray buff to bright fulvous, deep ochre, and chestnut. The underparts are white. The shoulders, upper arms, back, flanks, and haunches have dark spots arranged in rosettes, which usually enclose an area darker than the ground color. The head, throat, and chest are marked with small black spots, and the belly with large black blotches. Melanistic leopards (black panthers) are common, especially in moist, dense forests (Guggisberg 1975; Kingdon 1977).

The leopard can adapt to almost any habitat that provides it with sufficient food and cover. It occupies lowland forests, mountains, grasslands, brush country, and deserts. A specimen was found at an elevation of 5,638 meters on Kilimanjaro. The leopard is usually nocturnal, resting by day on the branch of a tree, in dense vegetation, or among rocks. It may move 25 km in a night, or up to 75 km if disturbed. It generally progresses by a slow, silent walk but can briefly run at speeds of over 60 km/hr. It has been reported to leap over 6 meters horizontally and over 3 meters vertically. It climbs with great agility and can descend headfirst. It is a strong swimmer but is not as fond of water as is the tiger. Vision and hearing are acute, and the sense of smell seems to be better-developed than in the tiger (Guggisberg 1975).

Hunting is accomplished mainly by stalking and stealthily approaching as close as possible to the quarry. Larger animals are seized by the throat and killed by strangulation. Smaller prey may be dispatched by a bite to the back of the neck. The diet is varied but seems to consist mainly of whatever small or medium-sized ungulates are available, such as gazelles, impala, wildebeest, deer, wild goats and pigs, and domestic livestock. Monkeys and baboons also are commonly taken. If necessary, the leopard can switch to such prey as rodents, rabbits, birds, and even arthropods. Food is frequently stored in trees for later use. Such is the strength of the leopard that it can ascend a tree carrying a carcass larger than itself (Guggisberg 1975; Kingdon 1977).

Population density is usually about one leopard per 20–30 sq km, but under exceptionally favorable conditions it may be as high as one per sq km. Reported home range size is 8–63 sq km. Individuals apparently keep to a restricted area, which they usually defend against others of the same sex. Territories are marked with urine, and severe intraspecific fighting has been observed. The range of a male may include the range of one or two females. Several males sometimes follow and fight over a female. Apparently, the leopard is normally a solitary species, but there have been reports of males remaining with females after mating and even helping to rear the young. There are a variety of vocalizations, the most common being a coughing grunt and a rasping sound, which seem to function in communication (Grzimek 1975; Guggisberg 1975; Kingdon 1977; Muckenhirn and Eisenberg 1973; Myers 1976; Schaller 1972; R. M. Smith 1977).

Breeding occurs throughout the year in Africa and India, though there may be peaks in some areas. In Manchuria and southeastern Siberia the mating season apparently is January–February. A female may give birth every 1–2 years. The estrous cycle averages about 46 days, and heat lasts 6–7 days; the gestation period is 90–105 days. Births occur in a cave, crevice, hollow tree, or thicket. The number of young per litter is one to six, usually two or three. The cubs weigh 500–600 grams each at birth, open their eyes after 10 days, are weaned at 3 months, and usually separate from the mother at 18–24 months. Full size and sexual maturity are attained at around 3 years. Maximum longevity in captivity is over 23 years (Grzimek 1975; Guggisberg 1975; Kingdon 1977).

The leopard seems to be more adaptable to the presence and activities of people than is the tiger and still occurs over a greater portion of its original range. Nonetheless, it is confronted by the same problems—persecution as a predator, value as a trophy, commercial demand for its beautiful fur, and loss of habitat. Man-eating leopards, though representing only a tiny percentage of the species, have undeniably been a menace in some areas. One in India, for example, is said to have killed over 200 people (Guggisberg 1975). Intensification of agriculture, along with elimination of natural prey, sets up conflicts between herders and the leopard, usually to the detriment of the latter. The pelt of the leopard has been sought since ancient times, but in the 1960s there was a substantial increase in the worldwide market for the furs of spotted cats. Illegal hunting became rampant, and some leopard populations were decimated. As many as 50,000 leopard skins were marketed annually, of which nearly 10,000 were imported into the United States in some years (Myers 1976; Paradiso 1972). National laws and international agreements subsequently seem to have reduced this traffic.

The leopard is said to be still relatively common in parts of East and Central Africa. Martin and de Meulenaer (1988) estimated its total numbers in sub-Saharan Africa at about 700,000, including 226,000 in Zaire alone, but these and other such high figures have been questioned. Indeed, Stuart and Wilson (1988) estimated numbers at fewer than 15,000 for that part of Africa south of the Zambezi River, a region for which Martin and de Meulenaer had listed figures totaling about 75,000. Myers (1987*b*) indicated that the leopard populations of Ethiopia, Kenya, Namibia, and Zimbabwe were reduced by 90 percent during the 1970s, largely because of poisoning by livestock interests. Norton (1986) reported that the species has been eliminated from most of South Africa. It also has become rare in much of West and North Africa and in

Leopard *(Panthera pardus)*, photo from Zoological Society of London.

most of Asia. Santiapillai, Chambers, and Ishwaran (1982) noted that only 400–600 individuals survived in Sri Lanka and that the future of the leopard there looked bleak.

The IUCN now classifies *P. pardus* generally as threatened but also has indicated (1972, 1976) that the following subspecies are endangered: *P. p. orientalis*, of southeastern Siberia, Manchuria, and Korea; *P. p. tulliana*, of Asia Minor and adjacent areas; *P. p. nimr*, of the Arabian Peninsula, Jordan, and Israel; *P. p. jarvisi*, of the Sinai Peninsula; and *P. p. panthera*, of northwestern Africa. The USDI lists *P. pardus* as endangered, except in that part of Africa south of a line corresponding with the southern borders of Equatorial Guinea, Cameroon, Central African Republic, Sudan, Ethiopia, and Somalia. In the region so demarcated, the leopard is listed as threatened, and regulations allow the importation into the United States, under the provisions of the CITES, of trophy leopards taken in this region for purposes of sport hunting; importation for commercial purposes is prohibited. *P. pardus* is on appendix 1 of the CITES.

Panthera onca (jaguar).

Head and body length is 1,120–1,850 mm and tail length is 450–750 mm. Reported weights range from 36 to 158 kg. In Venezuela, males usually weigh 90–120 kg, and females, 60–90 kg. The ground color varies from pale yellow through reddish yellow to reddish brown and pales to white or light buff on the underparts. There are black spots on the head, neck, and limbs and large black blotches on the underparts. The shoulders, back, and flanks have spots, forming large rosettes that enclose one or more dots in a field darker than the ground color. Along the midline of the back is a row of elongate black spots that may merge into a solid line. Melanistic individuals are common, but the spots on such

animals can still be seen in oblique light. *P. onca* averages larger than *P. pardus* and has a relatively shorter tail, a more compact and more powerfully built body, and a larger and broader head (Grzimek 1975; Guggisberg 1975; Mondolfi and Hoogesteijn 1986).

The jaguar is commonly found in forests and savannahs but at the northern extremity of its range may enter scrub country and even deserts. It seems usually to require the presence of much fresh water and is an excellent swimmer. It may den in a cave, canyon, or ruin of a human building. It is not known to migrate, but individuals sometimes shift their range because of seasonal habitat changes, and males have been known to wander for hundreds of kilometers. Daily movement in Belize was found generally to be 2–5 km (Seymour 1989). The jaguar climbs well and is almost as arboreal as the leopard. Most hunting, however, is done on the ground and at night. Prey is stalked or ambushed, and carcasses may be dragged some distance to a sheltered spot. The most important foods are peccaries and capybaras. Tapirs, crocodilians, and fish are also taken (Grzimek 1975; Guggisberg 1975).

In a radiotracking study in southwestern Brazil, Schaller (1980a, 1980b) found a population density of about 1 jaguar per 25 sq km. There was a land tenure system much like that of the cougar and tiger. Females had home ranges of 25–38 sq km, which overlapped one another. Resident males used areas twice as large, which overlapped the ranges of several females. Studies in Belize indicate that 25–30 jaguars were present in about 250 sq km; male home ranges there averaged 33.4 sq km and overlapped extensively, while two females had nonoverlapping ranges of 10–11 sq km. Estimates of home range in other areas have been as high as 390 sq km (Seymour 1989). The jaguar seems to be basically solitary and ter-

Jaguar *(Panthera onca)*, photo from Zoological Society of London.

ritorial. It marks its territory with urine and tree scrapes and has a variety of vocalizations, including roars, grunts, and mews (Guggisberg 1975). When a female is in estrus, she may wander far from her regular home range and may sometimes be accompanied briefly by several males. Estrus lasts about 6–17 days (Mondolfi and Hoogesteijn 1986).

Births occur throughout the year in captivity and perhaps also in the wild, especially in tropical areas, but there is evidence for a breeding season in the more northerly and southerly parts of the range. Most births in Paraguay take place in November–December; in Brazil, in December–May; in Argentina, in March–July; in Mexico, in July–September; and in Belize, in June–August, the rainy season (Seymour 1989). The gestation period commonly is 93–105 days, and litters contain one to four, often two, young. The offspring weigh 700–900 grams each at birth, open their eyes after 3–13 days, suckle for 5–6 months, stay with the mother about 2 years, and attain full size and sexual maturity at 2–4 years. Captives have lived up to 22 years (Grzimek 1975; Guggisberg 1975; Mondolfi and Hoogesteijn 1986).

Until the end of the Pleistocene, *P. onca* occurred throughout the southern United States, and it seems to have been especially common in Florida. There is some evidence that the species still inhabited the southeastern United States in historical time (Daggett and Henning 1974; Nowak 1975*b*). Breeding may have continued in Arizona until about 1950 (Brown 1983). Otherwise, by the early twentieth century, resident populations had disappeared from the United States, though for many years wanderers continued to enter the country from Mexico, whereupon they were usually quickly shot. Such apparently occurred most recently in December 1986 in Arizona. The jaguar now has been exterminated in most of Mexico, much of Central America, and, at the other end of its range, in most of eastern Brazil, Uruguay, and all but the northernmost parts of Argentina (Swank and Teer 1989).

The jaguar declined because of the same factors that affected other large cats—persecution as a predator, habitat loss, and commercial fur hunting. The jaguar is thought to have killed people on rare occasion, but no individuals are known to have become systematic man-eaters. The species does have a reputation as a serious menace to domestic cattle; indeed, it may actually have increased in colonial times, when livestock was introduced to the savannahs of South America, a continent lacking vast natural herds of ungulates (Guggisberg 1975). As land was cleared and opened to ranching and other human activity, inevitable conflicts arose between the jaguar and people. As in the case of the leopard and other spotted cats, there was a great increase in the commercial demand for jaguar skins in the 1960s. An estimated 15,000 jaguars were then being killed annually in the Amazonian region of Brazil alone. The recorded number of pelts entering the United States reached 13,516 in 1968. Subsequent national and international conservation measures seem to have reduced the kill, but the problem continues (Thornback and Jenkins 1982). *P. onca* is classified as vulnerable by the IUCN and as endangered by the USDI and is on appendix 1 of the CITES.

Panthera leo (lion).

Except as noted, the information for the account of this species was taken from Grzimek (1975), Guggisberg (1975), Kingdon (1977), and Schaller (1972). In males head and body length is 1,700–2,500 mm, tail length is 900–1,050 mm, shoulder height is about 1,230 mm, and weight is 150–250 kg. In females head and body length is 1,400–1,750 mm, tail

African lions *(Panthera leo)*, photo by Bernhard Grzimek.

length is 700–1,000 mm, shoulder height is about 1,070 mm, and weight is 120–82 kg. The coloration varies widely, from light buff and silvery gray to yellowish red and dark ochraceous brown. The underparts and insides of the limbs are paler, and the tuft at the end of the tail is black. The male's mane, which apparently serves to protect the neck in intraspecific fighting, is usually yellow, brown, or reddish brown in younger animals but tends to darken with age and may be entirely black.

The preferred habitats of the lion are grassy plains, savannahs, open woodlands, and scrub country. It sometimes enters semideserts and forests and has been recorded in mountains at elevations of up to 5,000 meters. It normally walks at about 4 km/hr and can run for a short distance at 50–60 km/hr. Leaps of up to 12 meters have been reported. The lion readily enters trees by jumping but is not an adept climber. Senses of sight, hearing, and smell are all thought to be excellent. Activity may occur at any hour but is mainly nocturnal and crepuscular; in places where the lion is protected from human harassment it is commonly seen by day. The average period of inactivity is about 20–21 hours per day. Nightly movements in Nairobi National Park were found to cover 0.5–11.2 km (Rudnai 1973). In the Serengeti most lions remain throughout the year in one area, but about one-fifth of the animals are nomadic, following the migrations of ungulate herds.

The lion usually hunts by a slow stalk, alternately creeping and freezing, utilizing every available bit of cover; it then makes a final rush and leaps upon the objective. If the intended victim cannot be caught in a chase of 50–100 meters, the lion usually tires and gives up, but pursuits of up to 500 meters have been observed. Small prey may be dispatched by a swipe of the paw; large animals it seizes by the throat and strangles, or it suffocates them by clamping its jaws over the mouth and nostrils. Two lions sometimes approach prey from opposite directions; if one misses, the other tries to capture the victim as it flees by. An entire pride may fan out and then close in on the quarry from all sides. Groups have about twice the chance of lone individuals in capturing prey. Most hunts fail: of 61 stalks observed by Rudnai (1973), only 10 were successful. The lion eats anything it can catch and kill, but it depends mostly on animals weighing 50–300 kg. Important prey species are wildebeest, impala, other antelope, giraffe, buffalo, wild hogs, and zebra. Carrion is readily taken. Up to 40 kg of meat can be consumed by an adult male at one meal. After making a kill, a lion may rest in the vicinity of the

carcass for several days. About 10–20 large animals are killed per lion each year.

Population density in the whole Serengeti ecosystem of East Africa was determined to be one lion for every 10.0–12.7 sq km. Reported densities in other areas vary from one per 2.6 to one per 17.0 sq km. Nonmigratory lions in the Serengeti live in prides, each with a home range of 20–400 sq km. All or part of the range is a territory which is vigorously defended against other lions. Nomadic lions have ranges of up to 4,000 sq km; there is much overlap in these areas, and individuals behave amicably. Nomads are commonly found in groups of two to four animals, and membership changes freely.

The basis of a resident pride is a group of related females and their young. These associations may persist for many years, being generally closed to strange females. Daughters of group members are recruited into the pride, but young males depart as they approach maturity. Several adult males often come together. Such a group, or a single male, joins a pride of females and young for an indefinite period. The males cooperatively defend the pride against the approach of outside males. Some males associate with and defend several prides. Eventually, usually within three years, the pride males are driven off by another group of males (Bertram 1975). Studies by Hanby and Bygott (1987) indicate that the factor stimulating departure of subadult males from a pride is the arrival of a new group of adult males and that some subadult females also may depart at such time if they are not yet ready to mate.

In the Serengeti the average number of lions in a pride was found to be 15, the range was 4–37, and the number of adult males present was 2–4. Prides often were divided into widely scattered smaller groups with about 4 individuals each. In Nairobi National Park, Rudnai (1973) observed a single adult male to be associated with four prides of females. There was a rank order among the females, and a female led each group, even when the male was present, but the male was dominant with respect to access to food. Males living within a pride allow the females to do almost all of the hunting, but they arrive subsequent to a kill and sometimes drive the others away. Lions appear to behave asocially at a kill, there being much quarreling and snapping and little tolerance shown to subordinates and cubs.

The lion has at least nine distinct vocalizations, including a series of grunts that apparently serve to maintain contact as a pride moves about. The roar, which can be heard by people up

African lions *(Panthera leo)*, photo by Bernhard Grzimek.

to 9 km away, is usually given shortly after sundown for about an hour, and then again following a kill and after eating. It apparently has a territorial function. The lion also proclaims its territory by scent marking through urination, defecation, and rubbing its head in a bush.

Breeding occurs throughout the year in India and in Africa south of the Sahara. In any one pride, however, females tend to give birth at about the same time (Bertram 1975). Females are polyestrous, and heat lasts about 4 days. A female normally gives birth every 18–26 months, but if an entire litter is lost, she may mate again within a few days. The gestation period is 100–119 days, and litters contain one to six young, usually three or four. The newborn weigh about 1,300 grams each; their eyes may be open at birth or may take up to 2 weeks to open. Cubs follow their mother after 3 months, suckle from any lactating female in the pride, and usually are weaned by 6–7 months. They begin to participate in kills at about 11 months, are fully dependent on the adults for food until 16 months, and probably are not capable of surviving on their own until at least 30 months. Sexual maturity is attained at around 3–4 years, but growth continues to about the age of 6 years. The average longevity in zoos is about 13 years, but some captives have lived nearly 30 years.

With the exception of people, their domestic animals, and their commensals, the lion attained the greatest geographical distribution of any terrestrial mammal. Various populations, known from fossils by such names as *Panthera atrox* and *P. spelaea*, are now regarded as conspecific with *P. leo* (Hemmer 1974; Kurten 1985). About 10,000 years ago the lion apparently occurred in most of Africa, in all of Eurasia except probably the southeastern forests, throughout North America, and at least in northern South America. The lion is thought to have disappeared from most of Europe because of the development there of dense forests (Guggisberg 1975). It probably vanished from the Western Hemisphere when many of the large mammals on which it preyed were exterminated through the spread of advanced human hunters at the close of the Pleistocene. It was eliminated in the Balkan Peninsula, its last major stronghold in Europe, about 2,000 years ago, and in Palestine at the time of the Crusades.

The continued decline of *P. leo* in modern times has resulted primarily from the expansion of human activity and domestic livestock and the consequent persecution of the lion as a predator. A few lions became regular man-eaters—for example, a pair killed 124 people in Uganda in 1925—and thus gave a sinister reputation to the entire species. Hunting for sport was also a major factor in some areas—for example, one person killed over 300 lions in India in the mid-nineteenth century. At that time *P. leo* was still common from Asia Minor to central India and in northern Africa. By about 1940 the species had been eliminated throughout these regions except in the Gir Forest, Gujarat State, western India; the animals there have been under continuous pressure from livestock interests, but vigorous conservation efforts have been made, and the number of lions seems to have stabilized at around 180. There also are 196 in captivity (Olney, Ellis, and Sommerfelt 1988). The subspecies involved, *P. leo persica*, is classified as endangered by the IUCN and the USDI and is on appendix 1 of the CITES.

To the south of the Sahara, the lion has become rare in West Africa (Rosevear 1974) and has been exterminated in most of South Africa and much of East Africa. It is still present over a large region but is widely hunted and poisoned by persons owning livestock. Myers (1975a) wrote that since 1950 the number of lions in Africa may have been reduced by a half, to as few as 200,000 or fewer; later he (1984) set the estimate at only 50,000. He cautioned that the species was rapidly losing ground to agriculture and that by the end of the century it might number only a few thousand individuals and survive only in major parks and reserves. Stuart and Wilson (1988) suggested that in southern Africa there was little hope for the lion outside of these conservation areas; they estimated numbers south of the Zambezi River to be 6,100–9,100, of which about 4,200 already were in the protected reserves. Myers (1987b), warned, however, that such restriction and fragmentation of the lion population might result in inbreeding and loss of genetic viability.

CARNIVORA; FELIDAE; Genus ACINONYX
Brookes, 1828

Cheetah

The single species, *A. jubatus*, originally occurred from Palestine and the Arabian Peninsula to Tadzhik S.S.R. and central India, as well as throughout Africa, except in the tropical forest zone and the central Sahara (Ellerman and Morrison-Scott 1966; Guggisberg 1975; Kingdon 1977; Smithers, *in* Meester and Setzer 1977).

Except as noted, the information for the remainder of this account was taken from Eaton (1974), Grzimek (1975), Guggisberg (1975), and Kingdon (1977). Head and body length is 1,120–1,500 mm, tail length is 600–800 mm, shoulder height is 700–900 mm, and weight is 35–72 kg; on the average, males are larger than females. The ground color of the upper parts is tawny to pale buff or grayish white, and the underparts are paler, often white. The pelage is generally marked by round, black spots set closely together and not arranged in rosettes. A black stripe extends from the anterior corner of the eye to the mouth. The last third of the tail has a series of black rings. The coat is coarse. The hair is somewhat longer on the nape than elsewhere, forming a short mane; in young cubs the mane is much more pronounced and extends over the head, neck, and back. *Acinonyx* has a slim body, very long legs, a rounded head, and short ears. The pupil of the eye is round. The paws are very narrow compared with those of other cats and look something like those of dogs. The claws are blunt, only slightly curved, and only partly retractile.

An additional species, *A. rex* (king cheetah), was described in 1927. It was based on specimens that differed from other cheetahs in having longer and softer hair and partial replacement of the normal spots by dark bars. Only 13 skins have been recorded, all from Zimbabwe and adjacent areas. It is now generally accepted that these specimens represent merely a variety of *A. jubatus* (Hills and Smithers 1980). Indeed, individuals with the king cheetah markings now have been recorded within otherwise normal litters (Brand 1983; Van Aarde and Van Dyk 1986).

The habitat of the cheetah varies widely, from semidesert through open grassland to thick bush. Activity is mostly diurnal, and shelter is sought in dense vegetation. Recorded daily movements are about 3.7 km for a female with cubs and 7.1 km for adult males. The cheetah is capable of climbing and often plays about in trees. It is the fastest terrestrial mammal. Reported estimates of maximum speed range from 80 to 112 km/hr; such velocities, however, cannot be maintained for more than a few hundred meters. Unlike most cats, the cheetah does not usually ambush its prey or approach to within springing distance. It stalks an animal and then charges from about 70–100 meters away. It is seldom successful if it attacks from a point over 200 meters distant, and it can only continue a chase for about 500 meters. Most hunts fail.

Cheetah *(Acinonyx jubatus)*, photo by Bernhard Grzimek.

If an animal is overtaken, it is usually knocked down by the force of the cheetah's charge and then seized by the throat and strangled. The diet consists mainly of gazelles, impalas, other small and medium-sized ungulates, and the calves of large ungulates. A female with cubs may kill such an animal every day, while lone adults hunt every two to five days. Hares, other small mammals, and birds are sometimes taken. The cheetah seems to work harder for its living than do the other big cats of Africa and thus may be more vulnerable to environmental changes brought about by human disturbance.

Population density in good habitat varies from about one cheetah per 5 sq km to one per 100 sq km. In marginal habitat, density may be only one per 250 sq km or less (Myers 1975b). Reported home range size is about 50–130 sq km. The cheetah occurs alone or in small groups. The groups seem usually to be a female with cubs, or two to four related adult males. The male "coalitions" commonly defend a territory against other males, perhaps thus facilitating access to prey and mates (Caro and Collins 1987). Groups avoid one another and mark the area they are using at a given time. Marking is accomplished by regular urination on prominent objects. Such activity also serves to communicate sexual information. The cheetah is normally amicable toward others of its kind, but several males sometimes gather near and fight over an estrous female. There are a number of antagonistic vocalizations, purrs of contentment, a chirping sound made by a female to its cubs, and an explosive yelp that can be heard by people 2 km away.

Births are reported to occur from January to August in East Africa, November to January in Namibia, and November to March in Zambia. Wild females normally give birth at intervals of 17–20 months; however, if all young are lost, the mother may soon mate and bear another litter. Estrus lasts about 2 weeks, and the gestation period is 90–95 days. The number of young per litter is one to eight, usually three to five. The cubs weigh 150–300 grams each at birth, open their eyes after 4–11 days, and are weaned at 3–6 months. They begin to follow the mother after about 6 weeks, and the family may then shift its place of shelter on an almost daily basis. The cubs are taught by the mother to hunt, separate from her at 15–17 months, and attain sexual maturity at 21–22 months. Captives have lived up to 19 years.

People have tamed the cheetah and used it to run down game for at least 4,300 years. It was employed in ancient Egypt, Sumeria, and Assyria, and more recently by the royalty of Europe and India. It is usually hooded, like a falcon, when taken out for the chase, then freed when the game is in sight. If the hunt is successful, the cheetah is rewarded with a portion of the kill. If the cheetah should attempt to escape, it soon tires and can be easily caught by persons on horseback. Tame individuals are usually playful and affectionate.

The removal of live cheetahs from the wild has contributed to a decline in the species. Other factors are excessive hunting of both the cheetah and its prey, the spread of people and their livestock, and the fur market. New studies also have demonstrated that *Acinonyx* has an unusually low degree of genetic variation, perhaps reflecting a severe population contraction during recent evolutionary history, leaving the species especially vulnerable to environmental disruption (Cohn 1986; O'Brien 1983; O'Brien et al. 1985; Wayne, Modi, and O'Brien 1986). The cheetah seems to be much less adaptable than the leopard to the presence of people. It evidently has disappeared in Asia, except for Iran and possibly adjacent parts of Pakistan, Afghanistan, and Turkmen S.S.R. In the mid-1970s the population in Iran was estimated to include over 250 individuals and was considered to be well protected (IUCN 1976); this population has continued to maintain itself. In Africa the species is still widely distributed but has become very rare in the northern and western parts of the continent. It has been extirpated in most of South Africa and much of East Africa, though Hamilton (1986) suggested that it is holding its own in Kenya and that it has proved to be more resilient than anticipated. The total number of individuals remaining in Africa has been estimated at 10,000–15,000 (Myers 1987b), though Stuart and Wilson (1988) estimated 6,200–8,500 for just that part of the continent south of the Zambezi River. The IUCN classifies the cheetah generally as vulnerable and the Asiatic subspecies (*A. j. venaticus*) as endangered. The entire species is listed as endangered by the USDI and is on appendix 1 of the CITES.

Order Pinnipedia

Seals, Sea Lions, and Walrus

This order of aquatic mammals occurs along ice fronts and coastlines, mainly in polar and temperate parts of the oceans and adjoining seas of the world but also in some tropical areas and in certain inland bodies of water. The Pinnipedia traditionally are regarded as a full order, and some authorities, including Corbet (1978), Hall (1981), and Smithers (1983), continue to treat them as such. Other authorities, such as Simpson (1945), have considered the Pinnipedia to be only a suborder of the order Carnivora. Recently there was a growing consensus, based mainly on morphological evidence, that the pinnipeds belong within the arctoid division of the Carnivora and are biphyletic in origin, the families Otariidae (eared seals, sea lions) and Odobenidae (walrus) having arisen from bearlike ancestors and the family Phocidae (earless seals) being an early offshoot of the line leading to the otters (Corbet and Hill 1986; Honacki, Kinman, and Koeppl 1982; Jones et al. 1986; McLaren 1960; Rice 1977; Stains 1984; Tedford 1976). There also, however, were immunological and chromosomal data that support a monophyletic origin for the Pinnipedia (Arnason 1974; Sarich 1969). New studies of cranial and postcranial skeletal material, both fossil and Recent, now indicate that the pinnipeds are indeed monophyletic (Berta, Ray, and Wyss 1989; Wiig 1983; Wyss 1987, 1988a). There still is a question, however, whether this group warrants ordinal rank or would best be placed within the Carnivora. The former course has been followed here, with recognition of 3 Recent pinniped families (in the sequence suggested by Wyss 1987), 18 genera, and 34 species.

Pinnipeds are measured in a straight line from the tip of the nose to the tip of the tail. Total length varies from 120 cm to 600 cm; a short or vestigial tail between the hind limbs grows very little after birth. Adults weigh from about 35 kg to 3,700 kg, with *Phoca* containing some of the smallest species and *Mirounga* the largest. Pinnipeds have a streamlined, torpedo-shaped body, with all four limbs modified into flippers. The arm and leg bones are similar to those in the Carnivora, but the bases of the limbs, to or beyond the elbows and the knees, are deeply enclosed within the body. The hands and feet are long and flattened; hence the name Pinnipedia, which means "feather-footed." Each limb has five broadly webbed, oarlike digits, which form the flipper. In most species the head is flattened, and the face shortened, to aid in rapid propulsion through the water. External ears are small or entirely lacking, and the nostrils are slitlike; the ears and the nose can be tightly closed when the animal is underwater. The eyes, which are set in deep protective cushions of fat, are also adapted for underwater use. The cornea is flattened, and the pupil is capable of great enlargement to enable better sight in dark water. The neck in pinnipeds is generally thick and muscular, yet quite flexible. A reduction in the interlocking processes of the vertebrae enables these animals to bend backward to a greater degree than most other mammals. The overall design of the body is fluid, with great power and grace evident in the movements. This allows for absorbing the shock of the impact of ocean waves, hauling out on ice or rocky coasts, or executing agile maneuvers to capture prey at sea.

Pinnipeds are less modified for aquatic life than are the wholly aquatic cetaceans. Like the cetaceans, pinnipeds have a thick coat of subcutaneous blubber to provide energy, buoyancy, and insulation, but most also have a hairy coat to protect them from sand and rocks when ashore. All pinnipeds have whiskers and a hairy covering (though this is almost lacking in *Odobenus*), which is kept lubricated by secretions from sebaceous glands. Although all have a coarse coat of guard hairs, the fur seals (*Callorhinus* and *Arctocephalus*) also possess a dense layer of underfur. This underfur traps small bubbles and keeps the skin dry, while the stiffer guard hairs protect the body from abrasion. Most pinnipeds are born with a woolly coat called "lanugo," which is white in some species and jet black in others. Molting in pinnipeds usually occurs after the breeding season and is most spectacular in the elephant seals and monk seals, whose outer layer of skin is shed in patches along with the fur.

Pinnipeds are clumsy on land, but in the water they are skillful divers and swimmers. They swim by means of the flippers and by sinuous movements of the trunk. In the Otariidae and the Odobenidae, locomotion is accomplished mainly by use of the forelimbs; in the Phocidae, the hind limbs provide most of the thrust.

Expert diving by pinnipeds is dependent upon the animals' efficient use of oxygen, by means of which they remain submerged longer than terrestrial mammals without sustaining brain damage or the "bends." According to J. E. King (1983), just before diving the seal exhales; when breathing stops, the heartbeat slows, thus conserving oxygen. Adult seals can slow the heart rate from a normal speed of 55–120 beats per minute to 4–15 beats per minute. This phenomenon, known as bradycardia, develops more rapidly and lasts longer as the seal grows older. During the dive, the seal's peripheral blood vessels are constricted and circulation is reserved for the heart and brain, thus reducing oxygen consumption by one-third. In addition, pinnipeds have a high tolerance for carbon dioxide and lactic acid build-up in the blood. When the seal surfaces, its heart regains its normal beat within 5–10 minutes, and the blood is reoxygenated by the increased beat and a few deep breaths of air. This efficient use of oxygen enables pinnipeds to make long, deep dives. Maximum depth is known to be nearly 900 meters in some species, and submergence time may reach 73 minutes.

A. Guadalupe fur seal *(Arctocephalus philippii),* photo by Warren J. Houck. B. Leopard seal *(Hydrurga leptonyx),* photo by Michael C. T. Smith. C. A young northern elephant seal *(Mirounga angustirostris)* on a sandy beach, the remarkable modification of the hind limbs and body, which adapts it for an aquatic existence, being well illustrated; the front flippers are partially hidden by the loose sand; photo by Julio Berdegué.

Pinnipeds, unlike cetaceans, must keep some sort of link to the land, since they can mate and give birth to the young only on shore or on ice. The habitats in which they gather for mating and pupping vary from floe ice in the Arctic and Antarctic to ragged cliffs, sandy beaches, and lava caves elsewhere. The major necessity appears to be isolation from humans and other predators (Haley 1978).

Pinnipeds are carnivorous and consume a wide variety of animal matter, ranging from krill and other crustaceans to mollusks and fish. They usually eat the common seafood of an area, swallowing moderate-sized species whole and headfirst. Larger catches are shaken into bite-sized pieces (Haley 1978). Some pinnipeds seek food at night, and certain polar seals feed in total darkness for four months of the year. A 100-kg seal eats approximately 5–7 kg of food daily when not fasting.

Some species, such as the Ross seal *(Ommatophoca)*, live alone during the winter, but most pinnipeds are much more gregarious than land carnivores. A breeding colony of pinnipeds ranges from a few individuals to more than 1 million animals within a radius of 50 km. These mammals tend to frequent small, isolated breeding grounds and are polygamous (Otariidae and Odobenidae) or mostly monogamous (Phocidae). All give birth ashore, on land or ice, and mate once a year. The period of pregnancy is 8–15 months, with delayed implantation occurring in many species. Delayed implantation may represent an adaptation that allows the births to take place at approximately the same time of the year, an important feature for colonial and, in some cases, migratory species. Single births are the rule; twins are the exception. Newborn pinnipeds can swim, but the pups of some species do not have enough blubber to provide buoyancy and insulation until they are several weeks old. Growth during the nursing period is rapid, for the mother's milk is particularly rich, about 50 percent fat. The adult pelage is usually acquired near the end of the first summer. Pinnipeds are sexually mature at 2–5 years and may live to 40 years in the wild. Predators include large sharks, killer whales *(Orcinus)*, leopard seals *(Hydrurga)*, and polar bears *(Ursus)*.

Pinnipeds long have been greatly valued by humans because of their fur, oil, and ivory and for use as food or fertilizer. They generally have been easy targets, because of their tendency to congregate in large numbers in localized areas for breeding (Haley 1978). Pinnipeds have been hunted commercially for hundreds of years, and sealing expeditions have been responsible for the slaughter of millions of animals. With the decline of seal populations and the worldwide rise of conservation and humanitarian movements, commercial exploitation has been brought under control or stopped altogether. One of the last major organized hunts was the annual take of the valuable soft pelts of young harp seals ("white coats") and hooded seals ("blue coats") in the North Atlantic. Barzdo (1980) reported that 208,759 seals of both species were killed during the 1977 season. Public protests and official import bans subsequently led to a sharp curtailment of the harvest. The United States has prohibited the take of pinnipeds in its waters (except for certain specified purposes) since the passage of the Marine Mammal Protection Act of 1972.

The oldest known fossil pinniped, a nearly complete skeleton that shows many primitive characters that would be expected in a common ancestor of the entire group, recently was discovered in late Oligocene or early Miocene deposits of California (Berta, Ray, and Wyss 1989). Otherwise, the known geological range of the Pinnipedia is early Miocene to Recent in North America, Pliocene to Recent in South America and Europe, late Pliocene in Egypt, and Pleistocene to Recent in New Zealand, Australia, and Japan.

PINNIPEDIA; **Family OTARIIDAE**

Eared Seals, Fur Seals, and Sea Lions

This family of 7 Recent genera and 14 species occurs along the coasts of northeastern Asia, western North America, South America, southern Africa, southern Australia, New Zealand, and many, predominantly southern, oceanic islands. The fur seals are in the genera *Callorhinus* and *Arctocephalus*, and the sea lions are in *Eumetopias*, *Zalophus*, *Otaria*, *Neophoca*, and *Phocarctos*. According to Repenning and Tedford (1977), the lineage of *Callorhinus* evidently separated from that of *Arctocephalus* in the late Miocene, while the sea lions did not diverge from *Arctocephalus* until the late Pliocene or early Pleistocene. Warneke and Shaughnessy (1985) suggested that the species *Arctocephalus pusillus* is intermediate to the other fur seals and the sea lions. And Bonner (1984a) considered *Arctocephalus* to be more closely related to the sea lions than to *Callorhinus*, despite its resemblance to the latter in certain characters. These three viewpoints, along with J. E. King's (1983) arrangement of the sea lions, form the basis for the sequence of genera presented herein. Berta and Deméré (1986) supported the affinity of *Callorhinus* and *Arctocephalus* and placed both in the subfamily Arctocephalinae, while putting the sea lions in the subfamily Otariinae. They did indicate, however, that the latter is phylogenetically nearer to the sea lions than is the former. Some other authorities, including Stains (1984), use these same subfamilial names. Hall (1981) did not employ these terms but did regard the Odobenidae (walruses) as only a subfamily of the Otariidae and used the name Rosmarinae for this subfamily.

Total length in otariids is 120–350 cm and weight is about 27–1,100 kg, males always being much larger than females. Body form is slender and elongated, and the tail is small but always distinct. External ears are present, but they are small and entirely cartilaginous. The long, oarlike flippers bear rudimentary nails. The flippers, which are thick and cartilaginous, are thickest at the leading edge and have a smooth, leathery surface. In both the Otariidae and the Odobenidae, the hind flippers can be turned forward to help support the body, so that all four limbs can be used for traveling on land. Members of these two families walk or run on land in a somewhat doglike fashion. In the Phocidae, the hind flippers cannot be moved ahead, and the animals must wiggle and hunch to travel on land. The swimming mechanism of the Otariidae is centered near the forepart of the body, and locomotion in water is accomplished mainly by use of the forelimbs. Phocids swim primarily by strokes of the hind flippers.

Sea lions have a blunt snout and a coat of short, coarse guard hairs covering only a small amount of underfur. Fur seals have a more pointed snout and very thick underfur, which may be of considerable commercial value. The pelage of newborn otariids is silky, never woolly. Adult coloration varies from yellowish or red-brown to black; there generally are no stripes or sharp markings. Females usually have two pairs of mammae. In males the testes are scrotal and the baculum is well-developed.

The otariid skull is somewhat elongate and rounded, though rather bearlike in overall appearance. J. E. King (1983) pointed out a number of distinguishing characters. For example, the otariid skull has supraorbital processes and only slightly inflated tympanic bullae, the Odobenidae lack supraorbital processes and have moderately inflated bullae, and the Phocidae lack supraorbital processes and have well-inflated bullae. The normal dental formula of the Otariidae

New Zealand sea lion *(Phocarctos hookeri)*, photo by D. J. Griffiths.

is: (i 3/2, c 1/1, pm 4/4, m 1–2/1) × 2 = 34–36. The first and second upper incisors are small and divided by a deep groove into two cusps; the third (outer) upper incisor is caninelike, especially in the sea lions; the canine teeth are large, conical, pointed, and recurved; and the premolars and molars are similar, with one main cusp. The number of upper molars varies within and among genera.

Eared seals inhabit arctic, temperate, and subtropical waters. Their breeding habitat is exclusively marine, never freshwater. They shelter along sea coasts, in quiet bays, and on rocky, isolated islands. They may be active by both day and night. J. E. King (1983) explained that like all pinnipeds, the Otariidae have acute vision and good hearing underwater and apparently depend on olfaction to distinguish individuals. These seals protect themselves by tearing an adversary with their canine teeth, by hurling their weight against the adversary, or by diving and swimming away. They feed mainly on fish but also eat cephalopods and crustaceans. Dominant bulls generally fast during the breeding season.

Otariids are highly gregarious, especially during the reproductive season. The males arrive first on the breeding grounds, where dominant individuals establish territories. The females come later. The males are polygamous and may associate with a group of over 50 females. Mating occurs on land, soon after the females give birth to the young conceived during the previous season. The total time of pregnancy thus lasts nearly a year, but in some species it is known to include a period of delayed implantation. There normally is a single pup, which is cared for only by the mother. The young usually do not swim for at least two weeks; weaning occurs after 3–36 months.

The known geological range of this family is early Miocene to Recent in Pacific North America; Miocene to Recent in Europe and Asia; Pliocene to Recent in South America; Pleistocene to Recent in Africa, Australia, New Zealand, and Japan; and Recent in other parts of the current range (Stains 1984).

Northern Fur Seal

The single species, *C. ursinus,* occurs in a great arc across the North Pacific, from the Sea of Japan, through the Sea of Okhotsk and the Bering Sea, to the Channel Islands off southern California (Rice 1977; U.S. National Marine Fisheries Service 1981). The southerly limits of the winter–spring migration are at about 35° N on the Japanese side, and at San Diego, California, 33°10′ N, on the American side. On very rare occasions young individuals have been recorded from the Arctic coast, as far east as Letty Harbor, Northwest Territories (J. E. King 1983). One specimen was reported from the coast of northeastern China (Zhou 1986).

There is striking sexual dimorphism in size. Fully mature males have a length of about 213 cm and a weight of 181–272 kg; mature females measure 142 cm and weigh 43–50 kg (Baker, Wilke, and Baltzo 1970). Adult males have dark gray to brown upper parts, usually grayish shoulders and foreneck, a short mane, and reddish brown underparts and flippers. Adult females and immature males have grayish brown upper parts, reddish brown underparts, and a pale area on the chest.

Like *Arctocephalus, Callorhinus* is characterized by a sharper snout than that found in sea lions, as well as abundant underfur. Unlike that of *Arctocephalus,* the fur of the foreflipper of *Callorhinus* extends only to the wrist, where it terminates in a sharp, straight line. The facial angle of the skull is always less than 125° in *Callorhinus* but more than 125° in *Arctocephalus.* As a result, the rostrum of *Callorhinus* is shorter and down-curved. *Callorhinus* also is distinguished from other otariids by longer ear pinnae, longer hind flippers, and certain characters of the premaxillary bones, baculum, and pelage (Gentry 1981; J. E. King 1983; Repenning and Tedford 1977).

Except as noted, the remainder of this account is based largely on Gentry (1981) and J. E. King (1983). *Callorhinus* is more pelagic than are other otariids of the Northern Hemisphere, with only neonates spending more than 60–70 days per year on land. There appear to be no regular landing sites other than the breeding islands. Individuals concentrate in areas of upwelling over seamounts and along continental slopes and thus are rarely found close to shore. They are most common 48–100 km offshore in waters of 6°–11° C, probably because of the availability there of preferred prey. Dierauf (1984) reported the unusual occurrence of a northern fur seal approximately 144 km upstream from the Pacific in the Sacramento River.

The vast oceanic range of the northern fur seal contrasts with the relatively few and tiny islands where the genus concentrates for reproduction. Although such sites probably once extended along much of the Pacific coasts of North America and Asia, the only major breeding colonies today are on St. George and St. Paul islands of the Pribilofs in the eastern Bering Sea, Copper and Bering islands of the Commanders in the western Bering Sea, Robben Island in the Sea of Okhotsk, the central Kuril Islands, and San Miguel Island off southern California. Colonies at the last two sites have become reestablished only recently. Lloyd, McRoy, and Day (1981) reported the discovery of another small breeding group on Bogoslof Island in the eastern Bering Sea near the central Aleutians. Stein, Herder, and Miller (1986) reported the birth of one pup on the mainland coast of northern California in 1983 and cited a record of another birth on the Washington coast in 1959.

A northern fur seal (Callorhinus ursinus), member of the family Otariidae, which, like the Odobenidae, is able to move its foreflippers with great freedom and to turn its hind limbs forward and thus is able to assume an erect posture. Photo by Victor B. Scheffer through U.S. Fish and Wildlife Service.

Although the entire population of San Miguel Island probably remains in California waters all year, some *Callorhinus* make an annual round trip of more than 10,000 km, the most extensive migration of any pinniped's. Seasonal movements vary according to the sex, age, and breeding site of the individuals. Adult males leave the breeding grounds in early August and go to sea; most of those from the Pribilof Islands apparently winter south of the Aleutian Islands and eastward into the Gulf of Alaska. Adult females and juveniles of both sexes begin to leave the Pribilofs in October and are followed by pups of the year. By December most of the seals have left the islands, headed southeast through the Aleutian passes, and spread as far south as San Francisco. From January to April they may be found anywhere along the migration route from southeastern Alaska to the California-Mexico border. Generally, adult females deploy farther to the south than do the young seals. Northward migration begins in April, and by May large numbers of the seals are back in the Gulf of Alaska. From June to October most are back in the vicinity of the Pribilofs, though younger animals, not reproductively active, are still widely scattered.

In the western Pacific, most of the fur seals from Robben Island winter in the Sea of Japan, and most of those from the Commander Islands migrate toward Japan. The animals are first seen off Hokkaido in October, their numbers increasing there until December, and they then move south to Honshu. Northward migration begins later than on the American side, with the last seals still in Hokkaido waters in July. Tag-recapture studies cited by Lander and Kajimura (1982) indicate that thousands of seals born on the Pribilof Islands migrate into Japanese waters and intermingle with Asian seals and that a significant number return north to the Asian rookeries. Movement from the western to the eastern Pacific also occurs but is not as common.

While the seals are at sea, most activity occurs during the evening, night, and early morning. The animals commonly sleep in the middle of the day while floating on one side. On the island rookeries activity continues unabated day and night. *Callorhinus* travels well on land but must rest frequently. At sea, it is a skillful swimmer and expert diver. For short distances it can keep ahead of a ship moving at about 16–24 km/hr (Baker, Wilke, and Baltzo 1970).

Gentry, Kooyman, and Goebel (1986) carried out studies utilizing an instrument called the time-depth recorder, which they attached to fur seal mothers captured in the Pribilofs. These instrumented and released females were found to go to sea for 5.5–10.0 days at a time and to make 122–362 dives per trip. About 69 percent of dives occurred by night, and 80 percent of resting took place during daylight. Dives averaged 2.6 minutes, with some lasting 5–7 minutes. Mean depth was 68 meters, and maximum was 207 meters. Some individuals made only shallow dives, others made only deep dives, and still others made both. The most frequent depths attained were 50–60 and 175 meters. Female *Callorhinus* made relatively long trips and large numbers of dives compared with females of tropical species. Gentry et al. (1986) suggested that this contrast reflects the need of subpolar female seals to obtain a large amount of energy (through feeding) on each trip so that their young can be quickly brought to weaning weight.

Observations of noninstrumented females indicate that the first feeding trip is made about 7 days after giving birth and that the trips usually last 4–9 days, alternating with 2-day periods of suckling the young. Nonsuckling females make longer visits to shore but spend less total time on land and move to sea and back in an unpredictable pattern (Gentry and Holt 1986). Reproductively active males may remain on land for the entire breeding season, not feeding for up to two months and surviving on stored fat reserves. When *Callorhinus* does eat, it takes a wide variety of food, especially small, schooling fish. The diet consists mainly of juvenile pollock in continental shelf areas of the Bering Sea and of squid in oceanic areas. Other important prey species, among the 75 known, are anchovy, capelin, herring, and rockfish. *Callorhinus* sometimes is accused of being detrimental to commercial fisheries, but salmon occurred in only 239 of 9,580 seal stomachs containing food, collected from 1958 to 1966 in the northeastern Pacific (Baker, Wilke, and Baltzo 1970). Even off Japan, where the overlap between human and fur seal diet seems greatest, 70–90 percent of the volume of take by *Callorhinus* is lantern fish, which are not harvested by people.

Overall population density in favorable wintering waters can range up to 26/sq km. A concentration of food often will attract a group of 6–20 seals, and a loose assembly of about 100 has been observed (Baker, Wilke, and Baltzo 1970). Otherwise, *Callorhinus* usually is seen in smaller groups or alone while at sea. Despite the vast aggregations that form at rookeries, it is not a particularly sociable animal. There apparently are no lasting bonds other than those between mothers and suckling young.

The breeding season begins when adult males arrive at the rookeries, this time being early June in the Pribilof Islands. The males may return year after year to the site of their own birth to breed. They establish small territories on the beaches, with the animals often no more than a few meters apart. These areas are constantly defended through fights, threats, and roars. Immature and senescent males are forced away from the breeding grounds, often farther inland, where they form groups and maintain a size-related dominance hierarchy with respect to resting sites.

The adult females begin to arrive on the Pribilofs in mid-June. They are attracted to one another and form dense groups in favorable locations, often near the water. These groups comprise from 1 to over 100 females, the average being about 40 (Baker, Wilke, and Baltzo 1970). Although the females come together within the territory of a bull, the presence of the latter is incidental, and there is no actual harem formation. The males do not control the females and cannot prevent movement to the sea or across territorial boundaries. Adult females are aggressive toward each other, as well as toward the bulls, immature animals of both sexes, and neonates other than their own. Toward the end of the breeding season, the younger females are able to come ashore, and at least some of them mate with the bachelor males.

Adult females give birth about 2 days after they come ashore. They then mate again about 6 days later during an estrus of less than 48 hours. After fertilization the zygote develops to the blastocyst stage and enters the uterus. Implantation then is delayed for 3.5–4.0 months. The total gestation period is approximately 11.75 months.

Most pups are born from 20 June to 20 July. There normally is a single young. Twins are rare but have been recorded more often in *Callorhinus* than in any other otariid (Spotte 1982). At birth, males average 66 cm in length and 5.4 kg in

weight, and females average 63 cm and 4.5 kg. The newborn has coarse black hair, but after about 8 weeks this coat is shed for one that is steel gray dorsally and creamy white underneath. After giving birth, a female remains with her pup and is constantly attentive for about a week. She then begins extensive feeding trips at sea, returning to nurse the pup for only about 2 days a week. At other times the pups form peer groups of their own, wander about the rookery, and spend much time sleeping and playing. Although they are able to swim at birth, they usually do not enter the water until about 1 month old. They are not taught to swim by their mother, and they generally are ignored by their father. Females evidently use scent to distinguish their pups from among a large group. Weaning occurs after 3–4 months and is initiated by the young. Afterwards there evidently is no contact between mother and pup (Gentry and Holt 1986).

Females reach sexual maturity at 3–7 years and may then give birth once a year until age 23; the highest pregnancy rates occur in females 8–16 years old. Males become sexually fertile at 5–6 years but are unable to maintain a breeding territory until 10–12 years. Bulls breed until 15–20 years. Age determinations of up to 26 years have been made, and maximum life span is thought to be about 30 years (Baker, Wilke, and Baltzo 1970).

The thick underfur of *Callorhinus* makes its pelt the most valuable of any pinniped's. There are about 57,000 hairs per sq cm, which is half the density found in the sea otter. The search for the latter species brought Russian explorers to the waters of the North Pacific in the early 1700s and led to the discovery of the fur seal rookeries on the Commander Islands. As the sea otter became scarce, there was an increasing take of the fur seal. The Pribilof Island herd was found in 1786, and the Russians brought Aleut Indians there to hunt the seals (Bonner 1982a; Busch 1985). It is estimated that there originally were 1.5–2.0 million fur seals in the Commander Island rookeries and 2.0–2.5 million on the Pribilofs (Lander and Kajimura 1982). Many more occurred at the Kuril and Robben island rookeries and probably at numerous other sites along the coast of western North America. Some 130,000 skins taken by sealers in 1808 and 1809 at the Farallon Islands, off central California, probably represented *Callorhinus* (J. E. King 1983).

By the early nineteenth century a drastic depletion of the Pribilof herd was evident to the Russians. Controls were implemented, including a restriction of the harvest to immature males from 1835 to 1867. By the latter year, when the United States purchased Alaska, including the Pribilofs, the herd there seemed to have recovered to near its original size. An estimated 4 million northern fur seals had been taken through 1867, with the annual kill on the Pribilofs in the 1860s numbering 30,000–40,000. When the United States took over the Pribilofs, there was a failure to immediately reestablish hunting restrictions, and the kill there during 1868 and 1869 alone was about 329,000. Controls were then restored, but they were far more liberal than under the Russians, and the average kill of males on land averaged over 100,000 through 1889 (Bonner 1982a; Busch 1985; Lander and Kajimura 1982).

A new commercial lease and more stringent quotas reduced the average kill on the Pribilof rookeries to only about 17,000 over the next 20 years. By then, however, a severe new threat had developed in the form of pelagic sealing. Unlike the harvest on land, in which nonbreeding males could be separated easily from other animals, killing at sea involved shooting or harpooning any seals that were encountered. The individuals most affected by this new procedure were lactating females on their lengthy foraging trips. The death of one of these females also meant the loss of the unborn she carried and of her suckling young at the rookery. Over a million

Northern fur seals *(Callorhinus ursinus):* A. Harem bull with females and young; B. Courtship between fur seals. Photos by Victor B. Scheffer.

animals, of which 60–80 percent were females, were taken by pelagic sealers from 1868 to 1911, and at least as many were killed and not recovered. The resulting decline in the herds led to a series of diplomatic crises, generally pitting the major rookery owners, the United States and Russia, against the two main pelagic sealing nations, Canada and Japan. These conflicts never were fully resolved, but they eventually moderated simply because there were so few seals left. By the early 1900s the Kuril rookeries had been completely destroyed and those on the Commander and Robben islands nearly eliminated. Estimates of the number of seals then remaining in the Pribilof herd vary from about 130,000 to 300,000 (Baker, Wilke, and Baltzo 1970; Bonner 1982*a*; Busch 1985; Lander and Kajimura 1982).

In 1911 the United States, Russia, Japan, and Great Britain (representing Canada), through the North Pacific Fur Seal Convention, agreed to prohibit pelagic sealing, except by aboriginal peoples using primitive equipment. Under the agreement, Japan and Canada would each receive 15 percent of the seals taken on the Pribilofs and 15 percent of those taken on the Commander Islands. Canada, Russia, and the United States would each receive 10 percent of the skins taken on Robben Island (then Japanese-owned). From 1912 to 1917 all killing of the Pribilof seals, which had come under direct control of the U.S. government, was prohibited, except for local subsistence purposes. Subsequently, the authorized harvest there, consisting almost entirely of young males, climbed along with a dramatic increase of the seal herd. The annual kill averaged about 40,000 from 1918 to 1940. By the latter year the Pribilof herd was thought to be near its original size of 2.0–2.5 million animals (Bonner 1982a; Busch 1985; Lander and Kajimura 1982).

In October 1941 Japan terminated the agreement of 1911 and resumed pelagic sealing, on the grounds that the seals were damaging fisheries. There evidently was no substantial loss to the herds, however, and in 1957 a new convention was established by the United States, Canada, Japan, and the Soviet Union (now in control of Robben Island and the Kurils). Pelagic sealing again was prohibited, an international commission was established to coordinate management and research, and Canada and Japan were each provided 15 percent of the commercial harvest by the United States and the Soviet Union. There seemed to be every reason for optimism, and the story of the northern fur seal was held up as a model of conservation. The annual commercial harvest on the Pribilofs, still composed almost entirely of young males, averaged about 66,000 from 1941 to 1955, but the herds appeared to be thriving. The seals on the Soviet rookeries also were increasing under strict protection. By about 1970 there were an estimated 265,000 on the Commanders, 165,000 on Robben Island, and 33,000 on the Kurils, this last population having become reestablished only in the 1950s. A breeding group returned to San Miguel Island, off southern California, in 1968 (Baker, Wilke, and Baltzo 1970; Bonner 1982a; Busch 1985).

There were, however, some problems. By the mid-1950s mortality on the Pribilofs was considered to be too great. Apparently, high population densities were facilitating the spread of disease. A decision was made to reduce the size of the herd through a major commercial harvest of females. From 1956 to 1968 approximately 321,000 females were killed, along with an annual average of 52,000 males (Bonner 1982a; Lander and Kajimura 1982).

The expected improvement did not materialize, and a general deterioration of status continued on both sides of the North Pacific (Bonner 1982a; Busch 1985; Fowler 1985,

1987; Lander and Kajimura 1982; U.S. National Marine Fisheries Service 1984, 1985, 1986, 1987, 1989). The situation has been blamed partly on what now generally is considered to have been an excessive kill of females. Another critical factor seems to have been development of the pollock fishery in the North Pacific, which led to entanglement and drowning of fur seals in discarded nets and possibly also to reduced prey availability for the seals. The yearly kill on the Pribilofs, again restricted largely to males, averaged 34,000 from 1969 to 1975 and about 25,000 from 1976 to 1983. There has been no commercial harvest since 1973 on St. George and none since 1983 on St. Paul. The estimated number of pups born on St. Paul, where most of the rookeries are located, dropped from 450,000 per year in the mid-1950s to 168,000 in 1986. The total number of seals present on the Pribilofs was estimated in 1987 at 815,000, less than half that of the 1940s and early 1950s. There also were about 4,000 on San Miguel and 332,000 on the Soviet rookeries. The U.S. National Marine Fisheries Service has designated the northern fur seal as a depleted species pursuant to the Marine Mammal Protection Act of 1972 but has rejected a petition to add the species to the U.S. List of Endangered and Threatened Wildlife. Whatever classifications are applied, it is evident that *Callorhinus* is undergoing its most serious crisis since the era of pelagic sealing a century ago.

PINNIPEDIA; OTARIIDAE; Genus ARCTOCEPHALUS
E. Geoffroy St.-Hilaire and F. Cuvier, 1826

Southern Fur Seals

There are eight species (Bonner 1981a, 1984a; Carr, Carr, and David 1985; J. E. King 1983; Shaughnessy and Fletcher 1987; Smithers 1983):

A. townsendi, Guadalupe Island and occasionally other islands off southern California and Baja California;

Southern fur seal *(Arctocephalus forsteri),* photo by John Warham.

A. philippii, Juan Fernandez Islands off central Chile, San Felix and San Ambrosio islands off northern Chile;

A. galapagoensis, Galapagos Islands;

A. australis, coasts of South America from central Peru and southern Brazil to Tierra del Fuego, Falkland Islands;

A. tropicalis, primarily to the north of the Antarctic Convergence on Tristan, Inaccessible, Nightingale, Gough, Marion, Prince Edward, Amsterdam, and St. Paul islands, but wandering individuals also recorded from southern Brazil, Angola, South Africa, New Zealand, South Georgia, Macquarie Island (where breeding now is occurring), and the Juan Fernandez Islands;

A. gazella, primarily to the south of the Antarctic Convergence on South Shetland, South Orkney, South Sandwich, South Georgia, Kerguelen, Heard, and McDonald islands, but also breeds on Marion Island, and wandering individuals found on Prince Edward Island, Macquarie Island, and Tierra del Fuego;

A. forsteri, coastal waters of Western Australia, South Australia, and New Zealand, and many subantarctic islands east and south of New Zealand;

A. pusillus, coastal waters of Angola, Namibia, South Africa, southeastern Australia, and Tasmania, and a wandering individual recorded from Marion Island.

The species vary considerably in size, ranging from the comparatively small *A. galapagoensis* to the very large *A. pusillus*. Otherwise, all southern fur seals are rather alike in overall appearance and generally have grizzled dark gray-brown upper parts and slightly paler underparts. Like *Callorhinus*, but unlike the sea lions, *Arctocephalus* is characterized by dense underfur. Unlike that of *Callorhinus*, the fur of the foreflipper of *Arctocephalus* extends distally past the wrist and descends to a sinuous line over the metacarpals. The facial angle of the skull of *Arctocephalus* always is greater than 125°. The shape of the snout varies from very short in *A. galapagoensis* to very long in *A. philippii*, and the rhinarium may be smooth and inconspicuous, as in *A. gazella* and *A. galapagoensis*, or inflated and bulbous, as in *A. philippii* and *A. forsteri*. Based on the simple crowns of their cheek teeth, the island species appear to be the most primitive of the southern fur seals (Bonner 1981*a*; J. E. King 1983; Repenning and Tedford 1977). Additional information is provided separately for each species.

Arctocephalus townsendi.

Adult males are estimated to be 200 cm long and to weigh 140 kg; females are about 135 cm long and weigh 50 kg (U.S. National Marine Fisheries Service 1984). In color the males are dusky black, the head and shoulders appearing grayish because of the lighter tips of the guard hairs and the animals appearing a grizzled gray when dry. Both *A. townsendi* and *A. philippii* are distinguished from other species of *Arctocephalus* by their extremely long, pointed nose (J. E. King 1983).

Unlike many seals, *A. townsendi* seldom, if ever, lands on open sandy beaches but generally occurs on shores characterized by solid rock and large lava blocks, usually at the base of towering cliffs. Perhaps the most unusual feature of its behavior, however, is the tendency to frequent caves and rocky recesses while on land. It sometimes is found at least 25 meters back from the cave entrances. Based on these observations, Peterson et al. (1968) suggested that during the era of commercial sealing in the 1800s seals occurring in open rookeries were quickly slaughtered but that those with secretive behavior survived and passed on their traits to the current population.

Smoothly polished rock extending up to 30 meters above sea level indicates the former occurrence of great numbers of fur seals on much of Guadalupe Island, off Baja California, Mexico. *A. townsendi* also evidently once bred on the Channel Islands off southern California. Even today, individuals range widely at sea, sometimes appearing on San Miguel and San Nicolas in the Channel Islands and at Cedros Island, 300 km southeast of Guadalupe. Two specimens recently appeared on the coast of central California (Webber and Roletto 1987). In contrast to the seasonal landings of other pinnipeds, however, *A. townsendi* can be found on shore year-round and shows strong site tenacity. Males are territorial, defending a cave or recess and barking and puffing at other bulls. Associated with each territorial male is a group of females, usually only 2 or 3 but occasionally up to 10. Mating and the birth of young conceived the previous year take place from May to July (Peterson et al. 1968; Thornback and Jenkins 1982).

Estimates of the original population on Guadalupe alone range from 20,000 to 200,000 individuals, and there would have been many more on the Channel Islands and at other sites. Most of the seals were killed for their valuable skins around 1800–1820, and the last commercial catches were made from 1876 to 1894. *A. townsendi* was considered to be extinct from 1895 to 1926, but in 1928 two males were received by the San Diego Zoo after capture by fishermen who were aware of a small population on Guadalupe. Subsequently it was thought that the fishermen had killed off the population and that the species really was extinct. However, in 1949 a lone bull was seen on San Nicolas Island, and in 1954 a small breeding colony again was found on Guadalupe. Under complete protection by the laws of Mexico and the United States, the population appears to have increased substantially and currently contains about 1,600 seals (U.S. National Marine Fisheries Service 1987). *A. townsendi* is classified as vulnerable by the IUCN and as threatened by the USDI and is on appendix 1 of the CITES.

Arctocephalus philippii.

Adult males are estimated to be 150–200 cm long and to weigh about 140 kg; females are about 140 cm long and weigh 50 kg (Bonner 1981*a*). Males have a slim, pointed snout, a heavy mane of silver-tipped guard hairs, and shiny blackish brown fur on the posterior body parts and belly (Hubbs and Norris 1971).

This species, like *A. townsendi*, comes ashore mainly on solid lava rock at the base of cliffs, on ledges, and in caves and recesses. When it enters the sea, it tends to stay rather close to the rocks. In a characteristic behavior, also shared with *A. townsendi*, this seal often inverts itself, with the head straight down in the water, exposing and gently waving the spread rear flippers (Hubbs and Norris 1971). *A. philippii* may go to sea in the cold waters of the Humboldt Current in winter and spring. Breeding occurs in late spring and summer (November–January). The diet includes fish and cephalopods (Thornback and Jenkins 1982).

It is likely that there were over 4 million *A. philippii* in the late seventeenth century prior to the start of commercial exploitation (Torres N. 1987). An estimate of 2–3 million fur seals on Isla Alejandro Selkirk in the Juan Fernandez Archipelago was made in 1797. By that time, however, intensive slaughter by American and British sealers was under way, primarily to obtain skins for trade in China. An estimate made in 1798 put the number of surviving seals on the same island at 500,000–700,000. The take on this island from 1793 to 1807 may have exceeded 3.5 million skins. At times, crews from as many as 15 vessels were engaged simultaneously in killing the seals. Great numbers were taken in the same period on other of the Juan Fernandez Islands and farther north on San Felix and San Ambrosio. The seals seem to have been largely extirpated by 1824, though a few were killed in

Southern fur seal *(Arctocephalus galapagoensis)*, photo by H. Hoeck.

the late 1800s. Subsequently, *A. philippii* generally was considered to be extinct (Hubbs and Norris 1971).

In 1965 the species was discovered to still occur in small numbers on Isla Alejandro Selkirk, in 1968 it also was found on nearby Isla Robinson Crusoe, and in 1970 two individuals were recorded on San Ambrosio. A census in 1983–84 counted 6,300 animals throughout the Juan Fernandez Archipelago (Torres N. 1987). The species is completely protected by Chilean law, but there is some poaching and harassment by fishermen (Thornback and Jenkins 1982). *A. philippii* is classified as vulnerable by the IUCN.

Arctocephalus galapagoensis.

This is the smallest and least sexually dimorphic of the Otariidae. Length is about 154 cm in males and 120 cm in females, and weight is about 64 kg in males and 27 kg in females (Bonner 1984a). The upper parts are grizzled gray-brown and the underparts, muzzle, and ears are light tan (J. E. King 1983). The skull is very small and lightly constructed, and the snout is very short (Bonner 1981).

The Galapagos fur seal is found in an area unusually warm for an otariid. When ashore, it selects rugged terrain with caves or overhanging lava ledges where it can shelter from the sun. In contrast, *Zalophus californianus*, another otariid in the Galapagos Islands, prefers gently sloping, open beaches. All fur seal colonies have access to deep water (Bonner 1984b). There is no migration, but the animals move into the sea to take advantage of the cold and productive waters of the Humboldt Current. The diet consists of squid and fish (Trillmich 1987a).

In a study of lactating females instrumented with a time-depth recorder, Kooyman and Trillmich (1986) found trips to sea to average 16.4 hours. There was an average of 79 dives per trip, 95 percent of which were made at night. Most dives measured less than 30 meters, but maximum depth was 115 meters and maximum duration was 7.7 minutes.

On shore, 6–10 females may occupy an area of about 100 sq meters, a relatively low density compared with that of most other fur seals (Trillmich 1987a). Breeding males, in turn, establish comparatively large territories, of up to 200 sq meters, encompassing a number of females. Because of this size, as well as the broken terrain, a bull has difficulty defending his entire territory, and a rival male may sometimes invade and mate with a female. All territories have access to the sea, so that the bulls, overheated from threatening and fighting, can cool off at midday. However, during their entire territorial tenure, up to 51 days, they do not actually feed (Bonner 1984b).

Some unusual reproductive adaptations reflect the equatorial habitat of the Galapagos fur seal (Bonner 1984b; Trillmich 1986, 1987a). Breeding occurs during the cool season, August–November, when there is likely to be less heat stress and greater availability of prey. The peak of births comes in the first week of October, there being a single pup per female. The female enters estrus and mates about 8 days after giving birth. Because of delayed implantation, the total period of pregnancy lasts nearly a year. The mother initially remains with the newborn for 5–10 days and then establishes a routine of 1–3 days feeding at sea alternating with 1–2 days ashore with the pup. Foraging trips have been found to last 50–70 hours at the time of the new moon but only 10–20 hours at the time of the full moon. Lactation may last 2–3 years or even longer, a remarkable period for the Otariidae. While all females mate during the postpartum estrus, only 15 percent give birth the following year if they are still feeding a pup. Of females without dependent young, about 70 percent give birth. If a pup is born to a female that is feeding a year-old individual, the newborn usually starves or is occasionally killed by the yearling. If the older sibling is already 2 years old, the newborn has a 50 percent chance of survival. The extended lactation evidently is associated with diminishing food availability to the nonmigratory seals when the warm season begins in December. At that time the females must forage for 4–6 days and spend only 1 day ashore. The new pups, now weighing about 7 kg, do not have the strength to find their own food and are able to grow very little during the warm season. A year later they do some hunting on their own but still are not self-sufficient. A female has her first young at

Southern fur seals *(Arctocephalus tropicalis)*, photo by John Visser.

about 5 years and probably can raise only 5 offspring if she lives to age 15. Clark (1985) estimated the age of one female specimen to be 22 years.

A. galapagoensis was taken regularly by commercial sealers during the nineteenth century and was thought to be extinct by the early 1900s. A small colony was rediscovered in 1932–33, and numbers have increased substantially since then. A census in 1977–78 yielded a population estimate of about 30,000 (Trillmich 1987b). The species occurs on at least 15 islands of the Galapagos and is fully protected by Ecuadorean law, though some colonies may be jeopardized by feral dogs (Thornback and Jenkins 1982).

Arctocephalus australis.

Adult males are about 189 cm long, weigh about 150–200 kg, and are blackish gray in color with longer hairs on the neck and shoulders. Adult females are about 143 cm in length, 30–60 kg in weight, and usually grayish black above and lighter below. There is some evidence that animals on the mainland of South America are larger than those on the Falkland Islands (Bonner 1984a; Gentry and Kooyman 1986; J. E. King 1983).

In a study on the coast of Peru, Trillmich and Majluf (1981) found A. australis to inhabit rocky, vertically structured slopes that provided some areas of shade for most of the day. Some seals lived in a huge cave where they had to climb about 15 meters up and down. There was much local movement because of changing temperature. During the hot part of the day the animals concentrated near the sea or in tidepools. When at sea, A. australis may range widely but is not migratory (J. E. King 1983). Most hunting is done at night, and dives have been found to average 29 meters in depth and to reach a maximum of 170 meters (Trillmich et al. 1986). The

diet consists mostly of fish, cephalopods, and crustaceans (Bonner 1984a).

Trillmich and Majluf (1981) found an overall population density of 0.5–1.5 seals per sq meter. Bulls established breeding territories averaging about 50 sq meters, but males whose territories lacked access to water had to abandon these areas during midday, and some low-lying territories were lost each day when the tide came in. Nonterritorial males gathered on beaches where no females were present. Bulls drove other males away by threatening displays and a call that began with a growl and rose to a higher frequency. Males attempted to herd females, but the latter moved about freely in accordance with their own needs. The breeding season lasted from October to January, with the great majority of births in November–December. According to J. E. King (1983), pups weigh 3–5 kg and have soft black fur. Mating occurs 6–8 days after parturition, but implantation is delayed about 4 months, so that total gestation lasts about 11.75 months. Mothers and young locate one another through vocalizations and scent. During a period of poor feeding conditions on the coast of Peru, lactating females were found on the average to alternate 4.7 days of foraging with 1.3 days of suckling the young. As in A. galapagoensis, lactation is unusually long. Weaning occurs after 1–2 years, and a female may simultaneously suckle young of different ages (Trillmich and Majluf 1981; Trillmich et al. 1986). Sexual maturity is attained at 7 years by males and at 3 years by females (Bonner 1981a).

A. australis was used as a source of food and hides by prehistoric Indians. Commercial exploitation began on the coast of Uruguay shortly after 1515. For many years the kill was controlled, but by the 1940s a decline was evident. Subsequent management, restricting the take to young males, has allowed an increase. In recent years the kill in Uruguay has been around 12,000 seals per year from a total population of about 252,000. The Falkland Islands population of *A. australis*, which was greatly reduced by American and British sealers in the late 1700s, seems never to have fully recovered and now contains around 15,000 seals. There also are about 3,000 in Argentina, 40,000 in Chile, and 20,000 in Peru (Bonner 1981a, 1982a; Gentry and Kooyman 1986; J. E. King 1983; Majluf 1987).

Arctocephalus tropicalis.

Adult males are 150–80 cm long and weigh about 100–150 kg. Adult females are 120–45 cm long and weigh about 50 kg. This is the only fur seal with a clear color pattern. The upper parts are dark grayish brown, the belly is more ginger in color, and the chest, nose, and face are white to orange. There is a conspicuous crest on the top of the head of adult males, formed from longer guard hairs. The foreflippers are proportionately shorter in *A. tropicalis* than in *A. gazella* and *A. pusillus* (Bonner 1984a; J. E. King 1983; Smithers 1983).

This fur seal lives mainly on isolated islands to the north of the oceanic convergence where the cold currents of the Antarctic region sink beneath warmer waters. The convergence is not a total barrier, however, as wandering individuals occasionally cross to the south. Some evidently moved from Amsterdam Island, across the Antarctic Sea to New Zealand, a distance of about 5,000 km. Other wanderers have reached the coasts of Africa and South America. Aside from females that are feeding young, most individuals spend the winter and early spring (June–September) at sea (J. E. King 1983). This period corresponds with when most wandering seals have appeared on the coast of South Africa (Shaughnessy and Ross 1980).

There may be two main periods of activity on land, during summer for breeding and during autumn for molting. On Marion and Gough islands these numerical peaks ashore were found to occur, respectively, in December and March–April (Bester 1981; Kerley 1983a). On Amsterdam Island, however, there is a marked increase only of adult males ashore during the autumn peak (Roux and Hes 1984). *A. tropicalis* prefers to haul out in rugged, uneven terrain on the windward side of islands. The cooling effects of wind and spray are important in reducing heat stress, especially during the peak of breeding activity. Females may give birth in caves and rocky recesses (Smithers 1983). In a study on Gough Island, Bester (1982) found all habitat on the westward, windward side to be occupied by breeding colonies, provided that there was easy access and protection from high seas. Nonbreeding colonies occupied most parts of the leeward, eastern coast. The diet of *A. tropicalis* has been estimated to comprise 50 percent squid, 45 percent fish, and 5 percent krill but also has been reported to include penguins (Smithers 1983).

J. E. King (1983) wrote that the general breeding biology of *A. tropicalis* seems to be much the same throughout its range. The first adult males start to arrive on the breeding grounds in September, and the rest of the seals come in October and November. The pups are born from the end of November to February. Smithers (1983) provided additional details, especially with regard to Gough Island. Colonies at established breeding sites consist almost exclusively of territorial males, adult females, and newborn. Other colonies contain immature animals and adult males that have been unable to establish a territory. The latter have been found to spend 94 percent of their time in total inactivity.

There is a rapid increase in the number of breeding males on Gough Island during November. These animals engage in vicious fighting that may leave the contestants exhausted and badly injured. Territories are eventually established and are maintained by rushing to the boundaries with open-

mouthed, guttural threats and slashing briefly at the opponent's face and chest. Territories may vary from 5 to 63 sq meters in size and tend to be delineated by natural topographical features. Most of the adult females arrive about a week after the territorial males. They tend to space themselves out but can be highly aggressive toward one another and to pups other than their own. Soon after giving birth to the young conceived the previous season, they enter estrus and are then actively herded by the territorial males. On Amsterdam Island the average number of females associated with a territorial bull has been reported to be 6–8, with a maximum of 14 (Bonner 1968).

According to J. E. King (1983), the females give birth 5–6 days after coming ashore. Mating occurs 8–12 days after parturition. There is a period of delayed implantation, so that total gestation is approximately 11.75 months. The pup is about 60 cm long, weighs 4.5 kg, and has a black and chestnut coat, which is shed 8–12 weeks later for pelage more like that of the adult.

The peak pupping season on Gough Island occurs about mid-December. The female remains with her young for about a week and then begins to make trips into the water. Upon returning to land she gives a call, which may attract a number of young, and she then distinguishes her own by scent. Pups begin to associate with one another at about 2 weeks; most congregate at the back of the beach by the time they are 4–5 weeks. Bester (1981) reported that the young first enter the surf zone and rock pools at Gough Island at about 6 weeks. Weaning occurs at 8–11 months, and the young may not leave the breeding beaches until October. Normally, a single young is produced by a female each year, but Bester and Kerley (1983) reported that one mother on Marion Island apparently raised twins to weaning age.

All of the islands inhabited by *A. tropicalis* were regularly raided by commercial sealers during the late eighteenth and early nineteenth centuries. By the 1830s the breeding colonies had been greatly reduced, but a partial recovery led to renewed hunting in the late nineteenth and early twentieth centuries. Subsequent lessening of demand for seal skins and oil, plus legal restrictions, allowed increases in some areas. The most remarkable comeback has been that of the Gough Island population, which grew from around 300 individuals in 1892 to 13,000 in the 1950s and to 200,000 in 1978. There are about 75,000 *A. tropicalis* on the other islands of its range (Bester 1980, 1987; Bonner 1981a; Kerley 1987; J. E. King 1983; Roux 1987).

Arctocephalus gazella.

Except as noted, the information for the account of this species was taken from Bonner (1968, 1982b) and J. E. King (1983). Adult males are 172–97 cm long and weigh 126–60 kg. Adult females are 113–39 cm long and weigh 30–51 kg. The back and sides are gray to slightly brownish, and the neck and underparts are creamy. The male has a well-developed mane with many white hairs, which give it a grizzled appearance. Compared with *A. tropicalis*, with which it overlaps in range, *A. gazella* lacks the conspicuous yellow chest and has an apparently longer, less bulky body, a slenderer neck, relatively longer foreflippers, and smaller eyes.

A. gazella is found primarily on islands to the south of the Antarctic Convergence but also occurs to the north, especially on Marion and Prince Edward islands. These islands are shared by larger populations of *A. tropicalis*, and limited hybridization has been reported between the two species there. Southernmost limits are not well known; individuals have been recorded in winter on the pack ice southwest of Bouvet Island (Smithers 1983), and *A. gazella* may be capable of traveling long distances over ice, but the species does not seem to be well adapted to such habitat. There is a general movement away from the breeding islands and out to sea during winter (May–November), but where the animals go, and whether they have a directional migration or simply a dispersal, is not known. Some adult males and juveniles are found ashore or in the vicinity of the breeding islands throughout the year (Doidge, McCann, and Croxall 1986). In a survey of *A. gazella* on Marion Island, Kerley (1983*b*) found peak numbers during summer breeding in December and autumn molting in March. These two peaks were, respectively, earlier and later for *A. gazella* than for *A. tropicalis* on the same island.

Breeding colonies of *A. gazella* prefer rocky stretches of beach with some protection from the sea and with access to the interior. On South Georgia, where the largest colonies occur, there is a lush growth of tussock grass inland from the beaches. Numerous seals move into this area and frequently lie on top of the tussocks. The seals have a surprising agility in land travel, sometimes progressing over slippery rocks or through dense tussock considerably faster than a human. On a smooth surface they can gallop at around 20 km/hr. They probably can exceed this speed when swimming.

Kooyman, Davis, and Croxall (1986) captured 20 lactating female *A. gazella* at South Georgia Island, instrumented them with time-depth recorders and radio transmitters, and released them. The seals were found to forage at sea for 1–13 days at a time. Average trip length was 5.3 days for these individuals, compared with 4.3 days for noninstrumented seals. The first dive was made after an average journey of 57 km and 8 hours. Maximum range was about 150 km over 4–5 days. The mean number of dives per trip was 414, with 81 percent being made at night, when krill rose in the water. Dives averaged 1.9 minutes in length and 30 meters in depth, with respective maximums of 4.9 minutes and 101 meters. Krill *(Euphausia superba)*, comprising small shrimplike crustaceans, is the staple diet. Fish and squid are taken occasionally. Males do not feed while occupying a breeding territory ashore.

The adult males come ashore on South Georgia from late October to early December. They establish and maintain territories through fierce ritualized displays, high-pitched whimpering vocalizations, and fierce fighting. Opponents face each other and make ponderous slashes with open mouths. Most of the blows are received on the chest and sides of the neck, which are protected by a heavy mane, and thus cause no serious injury. Eventually a dominance hierarchy is established, with the most successful fighters winning territories on the beach, near the water's edge but well above the high-water mark. Next in status are males occupying areas higher up on the beach. Lower-ranking males are forced onto territories that are well inland or partly submerged, and some bulls become fully aquatic. Younger males move onto beaches not used for breeding or into areas of tussock vegetation behind the breeding beaches. Unlike the juveniles of some other otariids, they do not form groups but tend to avoid each other. Some young females also occupy these areas.

In a study at South Georgia, McCann (1980) found mean territory size to decline from about 60 sq meters in mid-November to 22 sq meters in December, as males continued to arrive and fight for breeding space. Territorial tenure lasted from less than 1 day to more than 53 days, averaging 34 days.

Adult females arrive on South Georgia mainly in late November and early December. They form groups in favorable areas that correspond to bull territories. The average number of females associated with each territorial male is about 15; McCann (1980) reported a range of 1–27. Bulls generally can prevent a single female from leaving a territory but have little control over a group. Studies by Doidge, Mc-

Cann, and Croxall (1986) show that within an average of 1.8 days of arrival females give birth to the young conceived the previous year. They enter estrus after another 6 days, mate, and then return to the sea about 6.9 days after parturition. The mean length of foraging trips was found to be 4.3 days, and the mean length of visits ashore to suckle the pup was 2.1 days.

Total gestation is about 11.75 months and probably includes a period of delayed implantation. Most pups are born in early December. They are about 65 cm long, weigh about 6 kg, and have a dark brown or black coat, which is shed for a more silvery yearling pelage in January or February. After the initial departure of the female, the pups begin to roam about and associate with one another. Upon the mother's return, she attracts the young by calling and then confirms identification by scent. As the season progresses, most mothers and young move inland, and suckling often occurs while the female lies on top of a tussock. By early January some pups already are venturing into the sea, but they do not swim competently until early March. Weaning, initiated by the pup, occurs at about 117 days of age (Doidge, McCann, and Croxall 1986). There normally is a single young, but Doidge (1987) reported that two sets of twins were raised to weaning on South Georgia. Sexual maturity is attained at 3–4 years, but males are unable to hold a territory until they are at least 8 years old.

The skin of *A. gazella*, with a density of about 40,000 hairs per sq cm (Bonner 1985), has a value comparable to that of *Callorhinus*. The colonies on South Georgia were discovered by Captain Cook in 1775, and commercial hunting for skins began there in the 1790s. At the peak of exploitation on South Georgia, in the season of 1800–1801, 17 American and British vessels took 112,000 skins. In 1819 the last great fur seal colonies were located on the South Shetland Islands, off the coast of the Antarctic Peninsula, and during the season of 1820–21, 250,000 skins were taken there. The smaller colonies at the South Orkney, the South Sandwich, and the Bouvet islands were subsequently devastated. By 1830 *A. gazella* was commercially extinct, but enough individuals survived to allow a partial recovery and a resumption of limited commercial sealing from the 1870s to the 1920s.

In the 1930s *A. gazella* again was at a low ebb, there being only about 100 individuals left in the vicinity of South Georgia. A remarkable recovery then began in that area, possibly in association with human extermination of baleen whales in the Southern Hemisphere during the twentieth century. These whales, like *A. gazella*, feed primarily on krill. The reduction of the whales to only about 16 percent of their original biomass may have resulted in a far greater availability of krill for the seals. Under legal protection, the seals began to increase at an annual rate of around 16.8 percent. By 1957 there were an estimated 15,000 at South Georgia, and in 1976 there were about 369,000. McCann and Doidge (1987) listed estimates of 1.2 million for South Georgia and about 15,000 in the remainder of the range. Wanderers from South Georgia may have reestablished colonies on some of the other islands formerly inhabited by *A. gazella*, but the current population on Bouvet (4,000) apparently represents the original stock. Following the spread of *A. gazella* to Marion Island, cases of hybridization between that species and *A. tropicalis*, which also occurs there, were reported (Kerley and Robinson 1987).

Given current conditions, and considering old records of abundance, the South Georgia population can be expected to peak at between 1 million and 2 million seals; large colonies may also grow on the South Shetlands and Kerguelen, and smaller groups (numbering in the tens of thousands) will develop on some of the other islands of the range. *A. gazella* is protected by the laws of all the nations in which it occurs, and also, south of 60° S, by the Antarctic Treaty and the Convention for the Conservation of Antarctic Seals. There is concern, however, that krill-harvesting technology, now being developed by several countries, may result in competition between human interests and the fur seals. Another problem is that the increasing seal population at South Georgia is leading to destruction of tussock vegetation, erosion, and jeopardy to certain bird species that depend on this habitat (Bonner 1985). Severe environmental damage also has been reported on Signy Island in the South Orkneys (Smith 1988).

Arctocephalus forsteri.

Except as noted, the information for this account was taken from Crawley (*in* Strahan 1983) and Crawley and Wilson (1976). Adult males are 150–250 cm long, weigh 120–85 kg, and have a massive neck and thick mane. Adult females are 130–50 cm long and weigh 40–70 kg. The upper parts are dark gray-brown and the underparts are paler.

This fur seal is found along rocky coasts and on islands. It generally tends to stay close to land and does not make long-distance migrations, but there is a seasonal population shift in New Zealand waters. Adult males and some subadults begin to leave the rookeries in January, move northward for the austral winter, and start back south in August. Nonbreeding colonies are established ashore as far north as Three Kings Island, off the northern tip of New Zealand. Three young individuals were found on New Caledonia in 1972–73 (King 1983). During the nonbreeding season, March–September, the rookery sites may be occupied by pups, yearlings, and small subadults of both sexes. Adult females divide their time during this season between the rookeries and the sea. Rookeries usually are on the exposed western coast of islands. They are characterized by rocky, irregular topography; reefs or outlying boulders that offer protection from the open sea; and areas above the splash zone where mothers and young pups can take shelter. *A. forsteri* feeds mainly on squid and octopus and also occasionally takes fish and penguins. Adult

males may fast for up to 10 weeks while defending breeding territories on land.

Adult males establish a size-based dominance hierarchy while ashore in nonbreeding colonies but are strictly territorial in the rookeries. They begin to arrive on the breeding grounds in October, their numbers increasing steadily to mid-November and then more slowly to late December. Agonistic behavior is ritualized and involves guttural and barking vocalizations and threatening displays that emphasize the size of the neck. Only about 30 percent of encounters need to be resolved by physical confrontations. In fighting, the opponents push with their chest, wave their neck from side to side, and attempt to inflict bites on the face, neck, or shoulder. Despite regular challenges and intrusions, the bulls spend about 75 percent of their time lying down. Territory size declines as the breeding season progresses and more bulls arrive; at the peak of the season few territories exceed 100 sq meters. Young males and other nonbreeding animals congregate on beaches near the breeding colonies and may move into the rookeries in January when the bulls begin to leave.

Adult females arrive at the rookeries in large numbers in late November and December. They come together in groups of about six in favorable locations near the water. They are highly agonistic and constantly threaten one another regarding the slightest disturbance. The nearest territorial bull attempts to control a group of females but has little success, and the latter move freely across territorial boundaries. Miller (1975a) found that after an average of 2.1 days of first coming ashore, females give birth to the young conceived the previous breeding season. They enter estrus after another 7.9 days and are receptive for up to 14 hours. Gestation lasts about 11.75 months (J. E. King 1983) and includes a 4-month period of delayed implantation.

Pups are born from late November to mid-January, mostly around mid-December. At birth they weigh 3.5 kg, are 55 cm long, and have black fur. The mother remains with her single pup for about 10 days, then goes to sea to forage for 3–5 days, and then returns to suckle the young for 2–4 days. As the pup grows, the mother's foraging trips become longer. At a few

Southern fur seals *(Arctocephalus forsteri)*, photo by Graham J. Wilson.

weeks of age the pups form small groups, or pods, of 4–5 individuals. Pups continue to suckle until August or September, when they and the females leave the rookeries. Females attain sexual maturity at 4–6 years of age, but males are not able to win breeding territories until they are 10–12 years old.

According to Warneke (1982), the pelt of *A. forsteri* was considered more valuable than that of the partly sympatric *A. pusillus*. Exploitation began about 1798 in the Bass Strait, between Australia and Tasmania, where *A. forsteri* formerly occurred. Sealers soon shifted east to New Zealand, and by 1825 all significant and accessible colonies had been destroyed or greatly reduced. Macquarie Island was discovered in 1810, 57,000 seal skins were taken in the season of 1810–11, and the seal population was thought to have been exterminated by 1815. Limited exploitation continued until the late nineteenth century, when both Australia and New Zealand began to apply regulations on sealing. There subsequently has been a general recovery in New Zealand, but not in Australia, and there has been no return of *A. forsteri* to the Bass Strait. A small population has returned to Macquarie Island, probably through immigration from other islands (J. E. King 1983). The estimated number of *A. forsteri* in Australia now is about 5,000 or fewer, and population trends are not clear (Ling 1987). The populations in New Zealand and nearby islands are estimated to number 50,000 individuals and are thought to be increasing (Mattlin 1987).

Arctocephalus pusillus.

Except as noted, the information for this account was taken from J. E. King (1983) and Warneke and Shaugnessy (1985). Adult males are 184–234 cm long and weigh 134–363 kg; adult females are 136–76 cm long and weigh 36–122 kg. In the African subspecies, *A. p. pusillus*, adult males have dark blackish gray upper parts and are lighter below, and females are brownish gray above and lighter brown ventrally. In the

Southern fur seal *(Arctocephalus pusillus)*, photo by John Visser.

Australian subspecies, *A. p. doriferus*, adult males are dark gray-brown all over, except for the paler mane, and females have silvery gray upper parts, a creamy yellow throat and chest, and a chocolate brown abdomen. Despite the great geographical difference between the two subspecies, they are almost identical in cranial characters.

Both subspecies are confined largely to waters of the continental shelf and their immediate vicinity. Maximum known range offshore is 220 km in Africa and 112 km in Australia; however, a wandering individual appeared on subantarctic Marion Island in 1982 (Kerley 1983c). *A. pusillus* generally remains near the breeding grounds throughout the year and does not make regular migrations. Some individuals following cold water currents along the West African coast have been recorded as far north as 11°19′ S. Preferred breeding beaches are characterized by bare rock, boulders, or ledges, but in Africa *A. pusillus* sometimes breeds on sandy beaches. The diet consists mostly of fish and also includes cephalopods and crustaceans. In a study off South Africa, Kooyman and Gentry (1986) found lactating females to make foraging trips of 5–6 days. Their dives averaged 2.1 minutes in length and 45 meters in depth, with respective maximums of 7.5 minutes and 204 meters.

Adult males begin to concentrate on shore and compete for territories in October. Territory size is 10–20 sq meters in Africa and 20–140 (averaging 62) sq meters in Australia. The larger size in the latter region may reflect an unnaturally low population density. *A. pusillus* differs from all other species of *Arctocephalus* in that the bulls tolerate the presence of some juveniles over a year old. Most young males, and older bulls that cannot hold a territory, congregate on beaches near the breeding sites.

Adult females form groups within the male territories. These groups contain averages of 28 individuals in Africa and 9 in Australia, though they may number as many as 66. In both Africa and Australia, females give birth from late October to late December, mostly around 1 December. They mate 5–7 days after parturition. Total gestation lasts 51 weeks and includes a period of delayed implantation of about 4 months. According to David and Rand (1986), females make a brief trip to sea between parturition and mating. The mean duration of foraging trips subsequently increases to about 4 days by the end of the third month after parturition.

Pups weigh 4.5–7.0 kg at birth and are 60–70 cm long. Lactation generally lasts 11–12 months but sometimes continues into the second or third year of life. By June or July the young are usually foraging effectively on their own. Sexual maturity is attained at about 4–5 months, but males usually are not capable of winning territories until they are about 11 years old. They seldom can hold territories for more than another 2 years. Maximum longevity is at least 18 years.

Commercial exploitation began in South Africa in 1610 and has continued to the present. Fur seal populations were reduced to very low levels by the end of the nineteenth century, when regulations were implemented. Numbers subsequently increased slowly until the 1930s and then began a more rapid growth. According to Gentry and Kooyman (1986), the African population of *A. pusillus* now numbers 1.1 million individuals, is growing at an annual rate of 3.9 percent, and sustains an annual commercial harvest of about 75,000.

Sealing began in the Bass Strait region of Australia in 1798. By 1825 about 200,000 *A. pusillus* had been killed, and populations were at a minimum. Less intensive exploitation continued until the late nineteenth century, when protective regulations were established. The populations slowly recovered to a size of about 20,000–25,000 individuals in the 1940s, but there has been no subsequent increase.

Steller sea lions *(Eumetopias jubatus):* A. Rookery, photo by Victor B. Scheffer; B & C. Male, female and pup, photos by Karl W. Kenyon.

PINNIPEDIA; OTARIIDAE; **Genus EUMETOPIAS**
Gill, 1866

Northern Sea Lion, or Steller Sea Lion

The single species, *E. jubatus*, occurs across the North Pacific from Hokkaido, through the Sea of Okhotsk and the Bering Sea, and down the west coast of North America to the Chan-

nel Islands off southern California. Zhou (1986) reported a specimen from the coast of Jiangsu Province in eastern China. Except as noted, the information for this account was taken from J. E. King (1983), Loughlin, Perez, and Merrick (1987), and Schusterman (1981).

Eumetopias is the largest otariid and exhibits pronounced sexual dimorphism. Adult males average about 300 cm in length and 1,000 kg in weight, with some reaching 325 cm and 1,120 kg. Adult females average about 240 cm in length and 270 kg in weight, with some attaining 350 kg. Both sexes are slightly variable in color, ranging from light buff to reddish brown, the chest and abdomen being a little darker. Adult males develop a massive neck with a heavy mane of long, coarse hairs. Females have 2–6, usually 4, retractable mammae. *Eumetopias* differs from *Zalophus* in being much larger in size and having a conspicuous diastema between the upper fourth and fifth postcanine teeth. The head of *Eumetopias* is bearlike, and its nose is short and straight, not upturned as in *Otaria*.

The Steller sea lion is not known to migrate but does disperse widely during the nonbreeding period. Loughlin, Rugh, and Fiscus (1984) listed 51 active breeding colonies. The largest are in the Aleutians and on the Gulf of Alaska. Others occur off Sakhalin, on the Kurils and islands in the Sea of Okhotsk, along the east coast of Kamchatka, on the Pribilof Islands, in southeastern Alaska, on the coast of British Columbia and Oregon, on Año Nuevo and Southeast Farallon Islands near San Francisco, and on San Miguel Island off southern California. The breeding season ends in August throughout the range of the genus, and Alaskan males may then move both southward and as far north as St. Lawrence Island. California males tend to move northward along the coast.

Rookeries, as well as nonbreeding haul-out sites, are located mainly on remote and rocky coasts and islands. They characteristically have access to the open sea and to abundant food resources. *Eumetopias* commonly is seen in waters about 200 meters deep and is known to dive to depths of 183 meters. It often feeds at night but may hunt schooling prey by day. The diet consists primarily of a wide variety of fish, mostly with no commercial value, and also includes substantial amounts of octopus and squid, bivalve mollusks, and crustaceans. Young fur seals, ringed seals, and sea otters sometimes are taken. When schooling prey is being sought, *Eumetopias* may hunt and dive in large groups. Aggregations of several hundred to several thousand sea lions sometimes depart from land and disperse into groups of fewer than 50 when 8–24 km offshore. Group feeding apparently helps to control the movement of schooling fish and squid.

Eumetopias is gregarious and polygynous. Adult males begin to return to the rookeries in large numbers in early May, reaching a maximum by early July. They are highly aggressive toward one another and compete for territories, sometimes inflicting severe wounds. Disputes involve roaring, hissing, and chest-to-chest confrontations with mouths open. Once territories have been established, boundaries can be maintained through ritualized displays with relatively little fighting. Boundaries follow topographical features, such as cracks and ridges in rocks. The most favored areas are semiaquatic, sloping down into the sea. Territories average about 225 sq meters in size and are held an average of 43 days, during which the bull remains ashore and does not feed. Males have been known to return to the same territory for 7 consecutive breeding seasons. Males that are too young or too weak to hold a territory may congregate on beaches near the rookeries, along with other nonbreeding animals.

Adult females assemble on the breeding grounds about 3 days before they are ready to give birth to the young conceived the previous year. They compete for the most favorable birth sites, areas of sloping rock just above high tide. They are highly aggressive and readily bite, push, and threaten one another and pups other than their own. Groups of 10–30 females form within the territory of each successful male (U.S. National Marine Fisheries Service 1978). Estrus occurs 10–14 days after parturition, and females actively solicit the male through vocalization and displays. The blastocyst implants in September or October after a delay of 3–4 months, and the total gestation period is about 11.5 months.

The young, normally one per female, are born from mid-May to mid-July, mostly about 5–16 June, in both Alaska and California. The pups weigh 18–22 kg and are dark brown to blackish in color. Their weight may double after 7 weeks. The mother remains with the pup for 5–13 days and then spends 9–40 hours foraging at sea. She continues to alternate feeding trips with visits to suckle the young. Pups are capable of swimming at birth. For several weeks the mother encourages the pup and stays with it in shallow water while it develops swimming skills. When 10–14 days old, the pups form groups of their own and play and sleep together while the mothers forage. Upon a female's return, contact is made with the young through mutual vocalizations and olfaction. Lactation continues at least until the female again gives birth, though the pups also do some foraging on their own. Occasionally, a female suckles a young individual into its second or even third year. Sexual maturity is attained between 3 and 8 years of age, but males generally cannot win breeding territories until they are 10 years old. Females may breed until they are in their early twenties and may live for 30 years.

Because of its massive size and relatively aggressive nature, *Eumetopias* rarely is seen in captivity and seldom is trained to perform tricks. Its skin formerly was used by the Aleutian natives for boat coverings, harness, clothing, and boats. It now is less in demand, though the very thick hide may be used for leather, and the meat may be used as food for ranched mink and fox. There were commercial harvests in the Aleutians as recently as 1972, and there still is a limited take by native peoples. Bigg (1988) stated that *Eumetopias* is well known throughout its range because of the damage it allegedly causes to commercial fish and fishing gear and that as a result there have been organized control programs and bounties.

The Pribilof Islands population, which originally numbered well over 15,000 individuals, was reduced to a few hundred animals by uncontrolled exploitation after the U.S. purchase of Alaska in 1867. Skins were sold and traded for sea otter pelts to Aleutian natives. Protection was established in 1914, and by 1960 the population was near 10,000 (Kenyon 1962).

Whether other populations underwent a comparative cycle of decline and recovery is not known, but the U.S. National Marine Fisheries Service (1978) indicated that *Eumetopias* had increased considerably throughout Alaska since the early 1900s and was at or near the carrying capacity of the ecosystem. The total world estimate in the late 1950s and early 1960s was 240,000–300,000 individuals, of which 200,000 were in Alaska. Loughlin, Rugh, and Fiscus (1984) obtained about the same estimate based on surveys from 1975 to 1980 but noted that there had been increases in some areas and serious declines in others. The Pribilof population, for example, had fallen again to about 2,000 animals. In British Columbia, numbers had dropped from 12,000 to 5,000, partly through commercial harvests and a deliberate control program. The California population, once numbering nearly 7,000 individuals, had declined to 3,000, and there were only 20 breeding animals left on San Miguel Island. It was suggested that numerical changes in some areas might have been caused by a shift of populations. Braham, Everitt, and Rugh (1980), in reporting a decline of over 50 percent in the population of the eastern Aleutians, also had suggested such explanations as disease, excessive commercial kill, and loss of prey as human fisheries greatly increased in the region.

Recent reports (Bigg 1988; Merrick, Loughlin, and Calkins 1987; U.S. National Marine Fisheries Service 1986, 1987) indicate that a general decline is continuing in the eastern Aleutians and spreading across most of the Alaskan region and into Canadian and Asian waters as well. New surveys show that there has been no population shift and that numbers are dropping sharply everywhere. Overall numbers in the Aleutians and southwestern Alaska fell from 140,000 in the late 1950s to 68,000 in 1965. The actual cause still is unknown, but factors under investigation include disease, loss of prey, entanglement and drowning of the sea lions in commercial fishing nets, and deliberate killing by fishermen. The total world estimate of *Eumetopias* is 95,000–122,000, of which 80,000–105,000 are in Alaska and 10,000 are in the rest of North America. The Asian population comprises only 5,000–7,000 individuals, compared with 20,000–30,000 a decade ago. On 10 April 1990 the USDI issued an emergency regulation classifying the Steller sea lion as a threatened species.

PINNIPEDIA; OTARIIDAE; **Genus ZALOPHUS**
Gill, 1866

California Sea Lion

The single species, *Z. californianus*, occurs as three isolated subspecies—*Z. c. japonicus*, on the coasts of Japan and Korea; *Z. c. californianus*, on the Pacific coast of North America from Vancouver Island to Nayarit (western Mexico); and *Z. c. wollebaeki*, in the Galapagos Islands. Except as noted, the information for this account was taken from Odell (1981) and Peterson and Bartholomew (1967).

Adult males are 200–250 cm long, weigh 200 to about 400

kg, and are generally brown in color; adult females are 150–200 cm long, weigh 50–110 kg, and are tan. J. E. King (1983) pointed out that adult males have a very noticeably raised forehead, because of the extremely high sagittal crest on the skull, and that the hair over the top of the crest may be lighter in coloration.

Zalophus is essentially a coastal animal, frequently hauling out on shore throughout the year. J. E. King (1983) noted that it is rarely found more than 16 km out to sea. Breeding in the subspecies *californianus* occurs from Mazatlan and the Tres Marias Islands, at the southern edge of the range, north through the Gulf of California and along the coast of Baja California, to the Channel Islands off southern California. A few pups are born on smaller islands farther north, including the Farallons, near San Francisco. When the breeding season ends, many of the adult and subadult males migrate north along the coast as far as British Columbia. One individual has been reported in the Gulf of Alaska. Another was seen recently in Acapulco Bay, far south of the normal breeding range (Gallo-R. and Ortega-O. 1986). Movements of the females and young are not fully understood, though at least some remain in the vicinity of the rookeries all year, and some may migrate southward.

On the Channel Islands off California, *Zalophus* breeds mainly on flat, open, sandy beaches but sometimes in rocky areas. Sites of rookeries are not permanently fixed, and the sea lions may shift location if disturbed. Eibl-Eibesfeldt

California sea lion (*Zalophus californianus*), photo by Victor B. Scheffer. Inset: photo by Daniel K. Odell.

(1984) reported that in the Galapagos *Zalophus* may use sandy beaches for resting but seldom for breeding. Trillmich (1986), however, stated that the Galapagos sea lion prefers flat beaches, either sandy or rocky, where there is easy access to relatively calm waters and where it can spend the hot hours around tidepools or in the shade of vegetation.

While on land, *Zalophus* may walk slowly, move at a rapid gallop, or stride over smooth surfaces using only the front flippers. When swimming, it frequently leaps from the water in a shallow arc and reenters headfirst (porpoising). Groups of 5–20 young sometimes swim in file, porpoising one after another. In California, activity on land and sea has been observed regularly both by day and by night. In the Galapagos, *Zalophus* is basically diurnal, with peaks of activity in the morning and late afternoon (Eibl-Eibesfeldt 1984). An investigation of lactating females in the Galapagos, utilizing the time-depth recorder, found 75 percent of dives to be made at night. Foraging trips averaged 15.7 hours, there were 85–198 dives per trip, average depth was 37 meters, and maximum was 186 meters (Kooyman and Trillmich 1986). Captive individuals have been trained to retrieve objects from depths as great as 250 meters. Several captives flown to and released on islands off California were able to find their way back to their pen in San Diego, 115–270 km away, in 2–7 days (Ridgway and Robinson 1985). The diet of *Zalophus* consists mainly of cephalopods and small fish.

During the nonbreeding season there is a conspicuous but not complete sexual segregation. When ashore at this time, individuals form temporary dominance hierarchies based on size, with smaller animals, for example, being forced from favorable resting sites. Nonetheless, *Zalophus* is highly gregarious, the animals packing themselves together tightly even when empty space is available. According to J. E. King (1983), about 13,000 *Zalophus* congregate during autumn and winter on Año Nuevo Island, an area of 6.5 ha.; on some of the more crowded beaches only about 0.6 sq meter is available for each animal.

On the California Channel Islands territorial behavior by adult males occurs from May to August and is especially intense in late June and early July. In contrast to some other otariids, male *Zalophus* do not establish territories until females, and some pups, are already present on the breeding beaches. Fighting for space occurs both on land and in water and includes chest-to-chest pushing and quick slashing bites. Once territories are established, physical combat is reduced, and boundaries are maintained by ritualized displays such as oblique stares, head shaking, and lunging without making contact. Bulls patrol their areas on land and swim back and forth along the water side, barking incessantly. Some territories are partly or mostly aquatic. In the Channel Islands they generally are arranged one deep along the beach, have access to the water for cooling, are not separated by topographical features, and are about 10–15 meters wide. In a rocky area, territories were determined to average 130 sq meters in size. In the Galapagos, Eibl-Eibesfeldt (1984) found territorial activity to occur throughout the year and diameter to be 40–100 meters. Male *Zalophus* hold their territories an average of only 27 days. Some territories are occupied by a succession of different males in the course of a breeding season, though sometimes the original owner may return to reclaim his space after an interval of feeding at sea. Males that are too young or too weak to win a territory form groups outside of the breeding zone, though occasionally one of these individuals attempts to slip into a territory and remain undetected.

Adult females average 16 per territorial male on the breeding beaches. Although occasionally blocked by the males, the females pay little attention to boundaries, and their movements may cause the males to follow suit. They are attracted to each other but defend their individual space with high-pitched barks and threatening gestures. Within a few days of coming ashore for the breeding season, they give birth to the young conceived during the previous season. They enter estrus about 3 weeks later (Odell 1984; Trillmich 1986) and actively solicit the male. Total gestation thus would be just over 11 months and probably includes about a 3-month period of delayed implantation.

In California births occur from mid-May to late June. Eibl-Eibesfeldt (1984) stated that births take place throughout the year in the Galapagos, except in April and May, and that there is commonly a peak in August–October. Trillmich (1986) noted that the Galapagos breeding season varies in onset and duration from year to year but usually lasts 16–40 weeks between June and December. Adult females there were found to spend an average of 6.8 days ashore after giving birth and then to return to sea before mating again. Subsequent visits to suckle the pup averaged 0.6 days, and foraging trips averaged 0.5 days. Females returned to their pups almost every night.

Each female usually produces a single pup per year, though twins have been reported in captivity (Spotte 1982). The pup is about 75 cm long, weighs about 6 kg, and is chestnut brown in color. The mother is very attentive for several days and then makes increasingly long trips to forage at sea. Adult males seem to take a greater interest in the young than do the males of other otariids, and they have been seen apparently joining in an effort to block the approach of sharks toward groups of swimming pups (Eibl-Eibesfeldt 1984). The young are capable of swimming awkwardly at birth and walk with coordination within 30 minutes. At 2–3 weeks they form pods of 5–200 individuals, which move and play together. When a female returns to land, she vocalizes, inspects any pup that gives a responding bleat, and confirms that it is her own through visual and olfactory inspection prior to allowing it to nurse. As time progresses, the mother and pup spend increasing periods of time together at sea. The young probably obtain some food for themselves before they are 5 months old, but complete weaning is a slow process (Trillmich 1986). Usually lactation terminates before 11–12 months, but occasionally a female suckles both a newborn pup and a yearling. Sexual maturity is reached at about 9 years by males and at 6–8 years by females.

Zalophus is the "seal" commonly seen in circuses and animal acts. According to J. E. King (1983), circus training is accomplished by constantly rewarding the animal with fish. Ball balancing may be taught by constantly throwing a ball at the sea lion until it is accidentally balanced, or by holding a ball on the animal's nose until the objective is understood. It may take a year of training before a trick is ready to be shown to the public, but *Zalophus* has a good memory and will be able to demonstrate it perfectly later, even after a complete rest of three months. The performing life of a sea lion may last 8–12 years. One captive individual lived for 30 years.

In addition to being captured for the circus trade, *Zalophus* once was exploited for its skin and oil. It still was common in the late nineteenth century, but subsequently it was intensively hunted for the so-called trimmings trade (Busch 1985). Certain of its internal parts were valuable in oriental medicine, and its whiskers were used as pipe cleaners. This trade led to the extermination of entire herds. The population on the Channel Islands was so reduced that only a single animal could be found there in 1908. A survey in 1938 indicated the presence of 2,020 individuals along the whole California coast. Protection then allowed partial recovery: by 1967 there were estimated to be 40,000 in California, primarily on the Channel Islands, and another 40,000 in Mexico; the respective figures for 1987 were 74,000 and 83,000. There are problems, however. As the breeding population increases to the south, so also does the fall and winter migration of males into the coastal waters off Washington, Oregon, and British Columbia. In this region, *Zalophus* is reported to be damaging the commercial salmon fishery, both by eating the fish and by tearing nets (U.S. National Marine Fisheries Service 1978, 1984, 1987). Poaching for meat and oil is occurring in the Gulf of California, though the population there has increased to about 20,000, 35 percent higher than in 1966 (Le Boeuf et al. 1983).

The subspecies in the Sea of Japan, *Z. c. japonicus*, may already be extinct because of persecution by fishermen. It evidently once occurred along the coasts of Honshu, Kyushu, Shikoku, and Korea, as well as on several small nearby islands. The last possible survivors may be on the South Korean island of Dokto. The subspecies is classified as endangered by the IUCN and the USDI.

The Galapagos population recovered from sealing activity at the turn of the century and now occurs throughout the archipelago. Estimates of 20,000–50,000 individuals have been made. However, numbers declined drastically after an epidemic in the 1970s and possibly again after adverse weather phenomena (El Niño) reduced food supplies in 1982–83 (Eibl-Eibesfeldt 1984; Gentry and Kooyman 1986).

PINNIPEDIA; OTARIIDAE; **Genus OTARIA**
Peron, 1816

Southern Sea Lion, or South American Sea Lion

The single species, *O. flavescens*, occurs in the coastal waters of South America, from northern Peru and southeastern Brazil, south to Tierra del Fuego and the Falkland Islands (Vaz-Ferreira 1981). Oliva (1988) presented a detailed argument supporting use of the name *O. byronia* for this species. In 1973 an individual was found dead in the Galapagos Islands (Wellington and De Vries 1976), and in 1977 one was killed on the Pacific coast of Panama (Méndez and Rodriguez 1984). Except as noted, the information for the remainder of this account was taken from J. E. King (1983) and Vaz-Ferreira (1981).

Adult males are 216–56 cm long and weigh 200–350 kg. They generally have dark brown upper parts, a slightly paler mane, and dark yellow underparts but are sometimes orange or pale gold in color. Adult females are about 180–200 cm long, weigh about 140 kg, are brownish orange to yellow in color, and often have paler markings on the face, neck, or back of the head. *Otaria* can be distinguished from other sea lions by its blunt, short, broad and high, slightly upturned, erectile muzzle and high, wide lower jaw. The widened neck and heavy mane reinforces the massive appearance of the head and foreparts of the male.

Otaria is not migratory, but males may go to sea for extensive periods after the breeding seasons. Some individuals move as far north as Rio de Janeiro on the Brazilian coast, about 1,600 km from the nearest breeding colonies in Uruguay. Some also may enter freshwater rivers. Breeding sites are usually on open, sand or pebble beaches or on flat zones of rock, but occasionally they are on steep beaches or in places with large boulders. The diet consists mainly of fish, squid, and crustaceans and also includes fur seals. When seeking schooling prey, *Otaria* sometimes feeds in groups. During the nonbreeding period, individuals of any age and sex may be found in the same area.

Studies on the Valdez Peninsula of Argentina (Campagna 1985; Campagna and Le Boeuf 1988) found that adults of both sexes began to assemble on the breeding beaches in the second week of December. Males reached a numerical peak

South American sea lions *(Otaria flavescens)*, male and female, photo by Lothar Schlawe.

during 15–21 January, and by the first week of February 90 percent of them had departed. Female numbers peaked at the end of January. Males compete for and maintain territories through vocalizations, ritual displays, and physical combat. As the season progresses, territorial emphasis shifts from defense of a specific area to defense of a group of females. Fighting, which is most common at the beginning of the breeding season, involves the two opponents' meeting face to face, moving their heads sideways, and attempting to bite and tear. The vocal challenge of the adult male consists of a roar followed by a series of grunts. Successful contenders may spend up to 2 months defending a territory, during which time they do not eat and get little sleep. Territories are about 3–10 meters in diameter and seem to be based on space and the presence of females rather than on topographical features. Defeated individuals and young males gather at the fringes of the breeding beaches.

There are up to 18 females in each male territory, though the average is only 2.8, considerably fewer than in most other otariids (Campagna and Le Boeuf 1988). Post-estrous individuals are free to move across boundaries or into the sea, but pre-estrous and estrous females frequently are blocked and even carried and tossed about by the males. This sequestering of females is much more strenuous than in other otariids. Also unlike the situation in other genera, the nonterritorial males of *Otaria* do not always remain outside of the breeding zones or limit themselves to individual, and largely ineffective, challenges. New studies (Campagna and Le Boeuf 1988; Conway 1986) indicate that the territorial bulls are constantly hard-pressed by the others and may not be able to hold more than 2 or 3 females at a time. The bachelor males form groups of about 25 animals that invade the territories, seize and mate with females, and even kidnap pups in an apparent effort to lure the mothers away. Some bachelor males are able to retain females in areas on the periphery of the main breeding area and to successfully defend them from other males. Bulls that lose all their females sometimes leave the breeding area. As a result, a relatively high proportion of male *Otaria* are able to participate in the mating process.

Within a few days of arriving on the breeding grounds, adult females give birth to the young conceived the previous year. Births occur from mid-December to early February, mostly in January. There is evidence that the period of pupping on the Pacific coast is earlier and longer than on the Atlantic side. Total gestation is 11.75 months and probably includes a period of delayed implantation. There normally is a single pup, which is 85 cm long and weighs 10–15 kg. The natal coat is black and fades to a dark chocolate color after a month. Following birth, the mothers remain with their young for a few days and then mate and go to sea to forage. They return at intervals to suckle the young, initially calling to the latter, receiving an answering bleat, and then confirming identification by olfaction. When left alone, the young form groups or pods of their own. Campagna (1985) reported that pups enter water for the first time at 3–4 weeks, in groups of 20–50 individuals. This activity occurs mainly during middle and low tides, when pools and channels are exposed. Lactation often continues until the mother gives birth the following year and sometimes lasts even longer. Sexual maturity probably is attained by males at 6 years and by females at 4 years. According to Jones (1982), a captive was estimated to have lived to an age of 24 years and 10 months.

Otaria was exploited heavily for its skin, meat, and oil until well into the twentieth century. About 250,000 individuals were killed in Argentina from 1917 to 1953, with a peak harvest of 15,000 in 1946 (Conway 1986). There still was a population of 400,000 animals in the Falkland Islands in 1937, but by 1966 only 30,000 survived. The reasons for this decline are not known; however, a reduction of a population on

Isle Verde, Uruguay, from 2,400 in 1953 to only 28 in 1977, was the result of deliberate shooting by fishermen, who blamed the sea lions for damaging nets (Pilleri and Gihr 1977c). The U.S. National Marine Fisheries Service (1987) estimated that there are 228,000 *Otaria* on the Pacific side of South America and 45,000 on the Atlantic side.

PINNIPEDIA; OTARIIDAE; **Genus NEOPHOCA**
Gray, 1866

Australian Sea Lion

The single species, *N. cinerea*, now occurs from Shark Bay, on the west-central coast of Western Australia, south and east to Bridgeport, at the southeastern tip of South Australia. Until about 1800 the species also was present farther east, in the area of Bass Strait, between Victoria and Tasmania. The information for this account was taken from J. E. King (1983) and Walker and Ling (1981b).

Adult males are 200–250 cm long, weigh up to 300 kg, are rich chocolate brown in general color, and have a white to yellowish mane of long, coarse hair. Adult females are 132–81 cm long, weigh 61–104 kg, and are silvery gray to fawn above and creamy below.

Neophoca is nonmigratory and may spend its entire life near the beach where it was born. Tagging studies of breeding colonies on Kangaroo Island have shown that some females and their 4-month-old pups move to other sites 20–40 km distant. The longest movement recorded was by a 24-month-old juvenile found at least 300 km away from the site of tagging. *Neophoca* evidently does not spend long periods at sea, and many individuals can be found ashore at any time of year. Resting sites include flat, sandy beaches, but females prefer rocky areas for giving birth. Most breeding colonies are found on small offshore islands. The current breeding range extends from Houtman Abrolhos, rocky islets off west-central Western Australia, to the Pages, a group of rocks just east of Kangaroo Island in South Australia. *Neophoca* is adept on land, can move at a surprisingly fast gallop, has a remarkable ability to climb steep cliffs, and has been found up to 9.7 km inland. It is a strong swimmer, makes series of leaps clear of the water (porpoising), and has been reported to dive to depths of about 40 meters. The diet probably consists mostly of fish and cephalopods but also has been reported to include crayfish, lobsters, and penguins.

Nonbreeding individuals form groups of up to 15, which may include all age and sex classes but seldom more than a single adult male. There is a dominance hierarchy, especially in adult males, which is manifested by older animals forcing younger ones from favorable rest sites. Breeding colonies characteristically include fewer than 100 individuals, the largest having 500–600. Male intolerance increases with age and the approach of the breeding season. The bulls establish and maintain territories through ritualized postures and violent combat. Agonistic encounters are accompanied by guttural threats, growls, and high-pitched barking. Unlike those of other otariids, male *Neophoca* do not remain on their territories for long periods without eating. A bull occasionally stays for over 2 weeks but usually leaves more frequently to forage at sea for 2–3 hours. Dominant individuals generally are able to regain their space upon returning to land. They are, however, harassed by groups of subadult males that move about the colony and attempt to mate with the females. The existence of a territory is based on the presence of females, and location may shift as the females move in response to environmental conditions.

Females are gregarious to some extent, but they also de-

Australian sea lions *(Neophoca cinerea):* A. Photo by Vincent Serventy; B. Photo by Eric Lindgren.

fend a small space with hisses, roars, and fighting. Up to 8 females are associated with each territorial male. They are actively herded, and if in or near estrus, they sometimes can be prevented from leaving a territory. The males are extremely rough in such activity and may even enter another territory to retrieve a female. In such behavior, *Neophoca* shows close resemblance to *Otaria* (Campagna and Le Bouef 1988). Females have been reported to be hostile to pups other than their own, but human investigators have seen some to exhibit "nanny" behavior. A single cow appeared to be in charge of a group of pups, including her own, and would defend the entire group against the humans. Later, she would be relieved by the mother of another pup in the group.

Breeding periodicity in *Neophoca* is not completely understood. Births may extend over several months or even throughout the year in a given area, and the pupping peak may vary from colony to colony. This flexibility in breeding may be associated with the mild climate that prevails throughout the range of *Neophoca*. Some evidence suggests that a period of 17–18 months elapses between the start of consecutive pupping seasons at several colonies in South Australia and thus that *Neophoca* has an 18-month breeding cycle rather than the usual annual pinniped cycle. One marked female, for example, gave birth on 29 September 1976, was seen afterwards on numerous occasions, and did not produce another pup until early February 1978. Direct observations, however, also have shown that females enter estrus 6–7 days after giving birth. Therefore, either there is subsequently a greatly extended period of delayed implantation prior to the normal pinniped active gestation of 8–9 months or there is a normal delay of about 3 months followed by a remarkably long active gestation. Observations made in 1979 suggest the validity of the former alternative—a delay in implantation of the blastocyst of up to 10–11 months.

Pregnant females come ashore about 3 days before giving birth. The single pup is 62–68 cm long, weighs 6–8 kg, and has a chocolate brown coat that is replaced after 2 months by pelage resembling that of the adult female. The mother remains with her young for about 14 days and then enters the sea to forage. She returns at intervals of about 2 days to nurse the pup. Upon coming ashore, she emits a "moo" that elicits a high-pitched call from the pup, and identification is confirmed by olfactory inspection. As the young grow older, they form small groups and begin swimming in rock pools. Mothers lead their pups into the sea, and by the time the latter are 3 months old they accompany their mothers continuously, both on land and in the water. Weaning normally occurs prior to the birth of the next pup, but females have been seen simultaneously suckling both a yearling and a newborn.

Neophoca is a relatively rare pinniped, there probably being no more than 3,000–5,000 individuals in existence. It may never have been very abundant, but limited evidence indicates that it was more numerous prior to European colonization of Australia and that early sealing activity eliminated the genus in the Bass Strait area. It now is completely protected, and populations seem stable.

PINNIPEDIA; OTARIIDAE; **Genus PHOCARCTOS**
Peters, 1866

Hooker's, New Zealand, or Auckland Sea Lion

The single species, *P. hookeri*, now occurs along the coast of South Island of New Zealand and on Stewart, Snares, Auckland, Campbell, and Macquarie islands to the south. In his-

torical time it also occurred and probably bred on North Island of New Zealand. This species formerly sometimes was placed in the genus *Neophoca* and even considered to be conspecific with *N. cinerea*, but based on morphological and behavioral distinctions, most authorities now treat *Phocarctos* as a full genus. The information for this account was taken from J. E. King (1983) and Walker and Ling (1981*a*).

Adult males are 200–250 cm long, are black or blackish brown in color, and have a dark mane of long, coarse hair. Adult females are 160–200 cm long and are silvery gray above and creamy below. Weight has not been reliably recorded but apparently is less than in *Neophoca*. The facial profile of *Phocarctos* is blunter and more rounded than that of *Neophoca*, and the muzzle appears to be shorter. The skull of *Phocarctos* can be distinguished by the posterior prolongation of the tympanic bullae and deeply concave palate. The postorbital process of the zygomatic arch is distinct in *Phocarctos* but almost entirely absent in *Neophoca*.

Phocarctos does not make lengthy migrations but may move far out to sea or well up the coast of New Zealand during the nonbreeding season. The current breeding range is restricted to the Auckland, Snares, and Campbell islands. Sandy beaches are preferred for breeding, but individuals also move up to 2 km inland to rest in the forest or in grass on top of high cliffs. *Phocarctos*, like *Neophoca*, moves about well on land and may porpoise while swimming. The diet consists of fish, cephalopods, crustaceans, and penguins.

Breeding has been reported to begin as early as October, but on Enderby Island in the Aucklands, where most scientific studies have been carried out, the adults of both sexes arrive to breed at the beginning of December. At the height of the season there are about 1,000 individuals present—200 males, 400 females, and 400 pups. Only 19 percent of these males, however, actually hold a territory with females. The defended area is no more than a 2-meter circle, from which the bull makes aggressive charges and which shifts location in relation to the movement of the females. Unlike those of *Neophoca*, male *Phocarctos* defend a territory even if no females are present. They hold these positions throughout the breeding season and do not leave to forage. Defense is largely ritualized, and serious fights are few. The males do not herd the females, which form compact groups on their own, and are much more gentle with them than are male *Neophoca*.

A few days after arrival the females give birth to the young conceived in the previous breeding season. Estrus and mating occur 6–7 days after parturition. Therefore total gestation lasts about 11.75 months and probably includes a period of delayed implantation. Most young are born on Enderby Island in December and early January. The single pup is about 75–80 cm long and chocolate brown in color. The young remain quietly with their mothers for 2–3 days but then become much more active than the pups of *Neophoca*. At about 1 month the young are brought to the vegetated area behind the beach, where they remain while the mothers forage at sea. Upon returning to land, mother-pup contact is made by mutual vocalization. When the females are away, the young form large pods and may enter streams and rock pools. When 2 months old, they can swim strongly and are able to visit neighboring islands with their mother. Lactation probably continues for up to a year, but the young can obtain some of their own food before being weaned.

Phocarctos, like *Neophoca*, is relatively uncommon, there now being only about 3,000–4,000 individuals in existence. Little is known of its original status, though scattered remains and historical records suggest that it formerly bred throughout New Zealand and that it was greatly reduced in numbers by human exploitation for its skin, meat, and oil. The Auckland Islands were discovered in 1806, and most of the sea lions

New Zealand sea lions *(Phocarctos hookeri):* Top, adult male; Bottom, adult female. Photos by Basil Marlow.

there had been eliminated by 1830. Some exploitation continued until New Zealand established legal protection in 1881. Partial recovery subsequently occurred, and the Auckland Islands, now uninhabited by people, have been made a reserve.

PINNIPEDIA; **Family ODOBENIDAE; Genus ODOBENUS**
Brisson, 1762

Walrus

The single living genus and species, *Odobenus rosmarus*, occurs primarily in coastal regions of the Arctic Ocean and adjoining seas. Within historical time the species has been present regularly as far south as Kamchatka, the Aleutians, southern Hudson Bay, Nova Scotia, southern Greenland, northern Norway, and the White Sea. There also are historical records of wandering individuals from Honshu (Japan), the Sea of Okhotsk, Yakutat Bay (southeastern Alaska), coastal Massachusetts, Iceland, Ireland, southern England, northern Spain, France, Belgium, the Netherlands, and coastal Germany (Allen 1942; Duguy 1986; Fay 1981, 1985; J. E. King 1983; Nores and Pérez 1988; Reeves 1978; Reijnders 1982). Hall (1981) considered the Odobenidae to be only a subfamily of the Otariidae, used the name Rosmarinae for this subfamily, and used the name *Rosmarus* Brünnich, 1772, in place of *Odobenus*. In contrast, Wyss (1987) pointed out that many characters indicate a close relationship of the Odobenidae and the Phocidae. The remainder of this account draws heavily from the extensive studies of Fay (1981, 1982, 1985).

Adult males are 270–356 cm long, are up to 150 cm in height, and weigh 800–1,700 kg. The smaller females are 225–312 cm long and weigh 400–1,250 kg. There is no external tail. The body is covered with short, coarse hair, most dense on young animals and least dense in old males. During their annual molt in June and July the males shed most of their hair and may appear to be naked. The older males often have numerous raised nodules on the neck and shoulders, which are almost entirely hairless. The overall color is cinnamon-brown, but there is variation. Old males have pale skin and hair and may appear almost white when immersed in cold water, or pinkish when lying in warm sun. All individuals have a conspicuous mustache consisting of about 450 thick bristles, the roots of which are richly supplied with blood vessels and nerves. The tough, wrinkled skin of the walrus is about 2–4 cm thick over most of the body, being thickest on the neck and shoulders of adult males. There is an underlying layer of blubber that is up to about 10 cm thick on the chest.

The walrus has a swollen body form, a rounded head and muzzle, and a short neck that is especially thick in old males. The eyes are small and piglike, and the external ears are represented only by a low wrinkle of skin. The foreflippers are nearly as wide as they are long, while the hind flippers are more triangular. All the flippers are thick and cartilaginous, being thickest on the forward or leading edge, and all have five digits. The palms and soles are bare, rough, and warty for traction on ice. As in the Otariidae, the hind limbs can be turned forward and used together with the forelimbs for maneuvering on land. The baculum of the male walrus, up to 63 cm in length, is the largest of any mammal's in both absolute and relative size. The male also has a pair of pouches that arise from the pharynx and pass backward along the insides of the neck. These sacs can be greatly expanded with air from the lungs and are thought to function principally as resonance chambers for production of a bell-like sound, usually made while underwater; they also are used as a flotation device. Females usually have four mammae, sometimes more.

The massive skull is approximately rectangular, and the halves of the lower jaw are solidly fused. The maximum number of teeth is 38, but more than half are rudimentary and occur with less than 50 percent frequency. The most common permanent dental formula is: (i 1/0, c 1/1, pm 3/3, m 0/0) $\times$ 2 = 18. The upper canines are an outstanding feature of the adult walrus. These teeth grow throughout life in both sexes, developing into tusks that may attain a length of over 100 cm in males and about 80 cm in females. About 80 percent of the tusk is exposed above the gumline. The tusk is peculiar in that it bears enamel only at the tip for a short time after eruption; the entire crown in the adult consists of dentin (ivory). A single tusk in an old male may weigh 5.35 kg. The tusks of females are more recurved in the middle and more slender than those of males. Skulls with three tusks are occasionally found. Tusks function mainly in intraspecific agonistic interaction and also are used for interspecific defense, cutting through ice, hooking over ice for stability while sleeping in water, and helping to pull the body out of the water. There is no evidence that the tusks are used for digging food from the ocean floor.

The walrus is principally an inhabitant of moving pack ice over shallow waters of the continental shelf. It depends for subsistence on benthic organisms that dwell at depths of not more than 80–100 meters. Most populations appear to be migratory, moving south with the arctic ice in the winter and then north as the ice recedes in the spring. There are currently four main populations. One, the Pacific population, is found mainly in the Bering Sea in winter (December–January), migrates north through Bering Strait in spring (April–June), spends the summer (July–September) along the edge of the permanent polar ice cap in the Chukchi Sea, and then returns to the Bering Sea again in autumn. Some individuals travel over 3,000 km in the course of the year. Several thousand males of the Pacific population do not participate in the northward migration, but remain in the Bering Sea during the summer. The other three populations are found from Hudson Bay to western Greenland, from eastern Greenland to Svalbard and the Kara Sea, and in the Laptev Sea off northern Siberia.

During winter, individuals tend to concentrate where the ice is relatively thin and dispersed and to avoid extensive areas of thick ice. In summer the males may utilize isolated coastal beaches and rocky islets. Females and young usually remain in the pack ice, especially on floes with a surface area of 100–200 sq meters. Ice seems to be preferred for resting, molting, and bearing young, but land is used if no ice is available. An individual can use its head to break through ice up to 20 cm thick and then maintain and enlarge the hole by abrading the perimeter with the tusks. Activity goes on both by day and by night. Terrestrial locomotion is quadrupedal and resembles that of the Otariidae but is accomplished with much less facility and speed. In the water the walrus propels itself by alternating strokes of the hind flippers, in a manner similar to that of the Phocidae. The forelimbs may be used as paddles at low speed but otherwise are held against the body or used only for steering. The normal swimming speed is about 7 km/hr, and the maximum is at least 35 km/hr.

Foraging dives usually last 2–10 minutes. Most feeding is done at depths of 10–50 meters; the deepest known dive reached about 80 meters. An animal locates food by swimming headfirst along the ocean bottom, the sensitive vibrissae on the snout constantly in motion and in contact with the floor. Organisms that reside deep in the sediments probably are unearthed by piglike rooting. Prey is examined and ma-

Walruses *(Odobenus rosmarus):* A. A herd of walruses, photo from U.S. Navy; B. Walrus bulls, photo by Leonard Lee Rue III; C. Young walrus, photo from Zoologisk Have, Copenhagen.

nipulated with the vibrissae, and soft-bodied organisms are swallowed whole. Organisms with shells are held in the lips, the fleshy parts removed by powerful suction and ingested, and the shells rejected. This remarkable process may account for thousands of small clams during a single meal. The diet consists mainly of bivalve mollusks (clams and mussels) but also includes a wide variety of other benthic invertebrates. Fish and seals are captured occasionally. Adult males evidently eat little during winter, when they accompany the females and young.

The walrus is gregarious, tending to haul out on land or ice in herds of up to several thousand individuals lying in close physical contact. In such groups, which are largely sexually segregated during the nonbreeding season, there are dominance hierarchies based on body and tusk size. In a study of an aggregation of males, Miller (1975b) found most social interactions to be agonistic and concluded that "walruses are

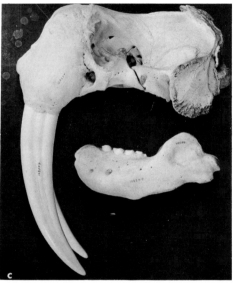

A. Mounted head of an adult male walrus from eastern Siberia *(Odobenus rosmarus divergens)*, photo by Grancel Fitz through New York Zoological Society. B. Yearling male walrus *(O. rosmarus rosmarus)*, photo by Victor B. Scheffer. C. Skull of Pacific walrus *(O. rosmarus divergens)*, photo from U.S. National Museum of Natural History. D. Walrus bulls *(O. rosmarus)*, photo by Leonard Lee Rue III.

bullies." When, for example, a large individual comes out of the water to seek a resting spot, it might throw back its head and point its tusks at a smaller animal. If this threatening display is not sufficient, and it often is not, there is striking with the tusks, and frequently bloodshed, but rarely serious injury.

During the breeding season the animals congregate in traditional favored areas of up to several hundred square kilometers. The females and young form groups of 20–50 individuals and are followed by a number of males. When the groups stop to rest on ice, the males compete for nearby locations in the water, about 7–10 meters apart. About 10 percent of the males are able to associate with the females, the younger and weaker males remaining nearby. The successful males display by making a variety of clicking and bell-like sounds underwater and then raising the head above the water and emitting a series of sharp clucks and whistles. When a female is attracted, she joins the male and mating takes place underwater. When nearly ready to give birth, a female often separates from her herd, but afterwards she joins a group of other mothers and young. A young female tends to remain in her mother's group. Young males drift away at about 2–3 years to join a male herd.

Mating occurs mainly during winter, especially in January and February. Implantation of the blastocyst is delayed for 4–5 months, mostly until June or July, and birth occurs 10–11 months thereafter, mainly from mid-April to mid-June; thus, the total gestation period is 15–16 months. At birth the single calf is about 113 cm long, about 63 kg in weight, grayish in color, and evidently capable of swimming. There is no external evidence of teeth, but the permanent tusks have formed internally and will erupt about a year later. The social bond between mother and calf is apparently stronger than in any other pinniped. The females are extremely solicitous and protective, and there are no lengthy periods of separation as in the Otariidae. There also is evidence of community care and even adoption of orphaned calves. Lactation usually continues for at least 2 years, though the young are capable of finding some of their own food long before final weaning. Females are sexually mature at 6–7 years and full-grown at 10–12 years. Males are fertile at 8–10 years but cannot successfully compete for mating privileges until they are at least 15 and fully grown. Fecundity is greatest in females at 9–11 years, when they can produce a calf every other year. The interval is longer in older animals. Maximum known longevity in the wild is just over 40 years.

For thousands of years the native peoples of the North regarded the walrus as having supernatural powers and human attributes, but at the same time they depended on the species as a major source of food, fuel, and materials to make tools, shelters, boats, sleds, and clothing. Such indigenous hunting probably had little effect on overall populations, but serious declines began with European commercial exploitation for ivory, oil, and hides. Such activity began in the Middle Ages and became especially intense on both sides of the North Atlantic in the sixteenth and seventeenth centuries. The original size of the populations in this region is not known, but given the magnitude of the kill, there must have been large numbers. Thousands of individuals, for example, were killed annually in the Gulf of St. Lawrence and off Nova Scotia during this period (Reeves 1978). Even in the late nineteenth century more than 1,000 were being killed each year at Svalbard, and in 1887 that same number was taken off Novaya Zemlya. Shortly thereafter the catch dropped to an insignificant level in the eastern North Atlantic, reflecting the near disappearance of the species. In the west, the walrus had been exterminated south of Labrador by the mid-nineteenth century and substantially depleted farther north. Ronald, Selley, and Healey (1982) indicated that

the last great populations in the Canadian Arctic, those around Baffin Island, were devastated from 1925 to 1931, when 175,000 individuals were killed. However, Richard and Campbell (1988) showed that this figure is grossly exaggerated, though they agreed that the former large populations in the region had been depleted. Protection and declining demand eventually prevented the total disappearance of the Atlantic subspecies (O. r. rosmarus), but there has been no substantial recovery. The continuing indigenous harvest is about equal to the annual recruitment rate of 6–10 percent. There are now about 25,000 individuals in the population that occurs from Hudson Bay to western Greenland and only 1,000–2,000 from eastern Greenland to the Kara Sea. The population in the Laptev Sea, sometimes considered a separate subspecies (O. r. laptevi), remained largely undisturbed by modern exploitation until the late nineteenth century. It, too, subsequently declined in the face of excessive hunting and now numbers about 4,000–5,000 individuals. The Soviet Union classifies O. r. rosmarus as vulnerable and O. r. laptevi as rare.

The North Pacific population (O. r. divergens) came under commercial pressure by the eighteenth century. More than 10,000 individuals were taken in some years through the late nineteenth century. In a two-year period about 16,000 kg of ivory was obtained on the Pribilof Islands alone. In 1909, when the population was at a low ebb, commercial harvesting was prohibited in Alaska, and a recovery began. From 1931 to 1957, however, the Soviet Union carried out intensive pelagic commercial hunting, and the population declined to only about 40,000–50,000 individuals. Subsequent Soviet and U.S. regulations ended commercial use and allowed another recovery. By the early 1980s the Pacific population was estimated to contain nearly 250,000 walruses and may have been close to its original size. However, Fay, Kelly, and Sease (1989) warned that the population again is being mismanaged through excessive subsistence harvests in the United States and the Soviet Union and by lack of coordination between the two countries. The total kill of 10,000–15,000 individuals annually, plus natural mortality, is leading to a decline that could halve the population during the 1990s. Recently, poaching has intensified as the demand for walrus ivory accelerated following the widespread destruction of Africa's elephant herds (see account of Loxodonta). There also is concern that expansion of the clam-dredging industry into the North Pacific, as has been proposed, would seriously reduce the food supply of the walrus (Ronald, Selley, and Healey 1982).

The walrus has a long geological history, and its family once was far more varied than it is today (Repenning and Tedford 1977). The primitive pinniped family Enaliarctidae evidently gave rise to the Odobenidae in the North Pacific during the early Miocene. In the late Miocene and early Pliocene there were at least six genera of odobenids in Pacific North America. The family then disappeared in the Pacific, but a branch evidently had entered the Atlantic in the late Miocene, and it is represented in the Pliocene, Pleistocene, and Recent of both Europe and Atlantic North America. The Odobenidae reentered the North Pacific in the late Pleistocene. There are late Pleistocene records as far south as San Francisco, Michigan, and North Carolina.

PINNIPEDIA; **Family PHOCIDAE**

True, Earless, or Hair Seals

This family of 10 Recent genera and 19 species occurs along ice fronts and coastlines, mainly in polar and temperate parts

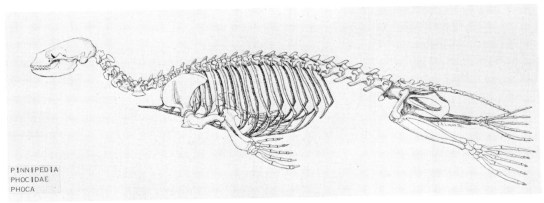

PINNIPEDIA
PHOCIDAE
PHOCA

Skeleton of harbor seal *(Phoca vitulina),* photo from American Museum of Natural History.

of the oceans and adjoining seas of the world but also in some tropical areas and in certain inland bodies of water. The sequence of genera presented here is based on a review of Burns and Fay (1970), J. E. King (1983), and McLaren (1984), which indicates recognition of two subfamilies: Monachinae, the southern phocids, with the genera *Monachus, Lobodon, Hydrurga, Leptonychotes, Ommatophoca,* and *Mirounga;* and Phocinae, the northern phocids, with *Erignathus, Cystophora, Halichoerus,* and *Phoca. Monachus* is considered to be the most primitive of all living phocids, and *Erignathus,* the most primitive member of the Phocinae. Based on a detailed morphological analysis, Wyss (1988b) doubted that the Monachinae comprise a monophyletic group. Some authorities, including Hall (1981) and Stains (1984), place *Cystophora* in a third subfamily, the Cystophorinae. Still other authorities, such as Corbet (1978) and Rice (1977), do not use subfamilial designations.

Total length in phocids is 120–600 cm and weight is 65–3,700 kg. In most genera males are about the same size as females, but in *Mirounga* and *Cystophora* the males are much larger. The body is streamlined, and the tail is small but obvious. External ears are represented only by a faint wrinkling of skin, and as in the Odobenidae, there is no supporting cartilage. The foreflippers, which are placed far forward, are smaller than the hind flippers and make up much less than one-fourth of the body length. The flexible flippers, of nearly uniform thickness, have five digits. In the subfamily Phocinae the metacarpals of the first and second digits are approximately the same size, and there are large claws on the fore- and hind flippers; in the Monachinae the first metacarpal is noticeably larger than the others, and the claws on the hind flippers are reduced (J. E. King 1983). The hind limbs cannot be turned forward, and thus phocids wriggle and hunch in order to travel on land. Such locomotion is rather laborious, so whenever possible the animals roll or slide; they can move fast when necessary. In water phocids propel themselves by moving the hind flippers in a vertical plane. The swimming mechanism is centered near the hind part of the body, not close to the forebody as in the Otariidae. Phocids frequently swim on their back, at least in captivity, and regularly stand upright in the water, maintaining their position by "treading" with their foreflippers.

The adult pelage is stiff and lacks appreciable amounts of underfur. Eyebrow vibrissae are well developed, and mustache vibrissae are often beaded. A number of species have spotted color patterns, and some of the species of *Phoca* are the only pinnipeds with a banded pattern and with sexual differences in this pattern. Some phocids have three distinct coats—newborn, subadult, and adult. The newborn are covered with a dense, soft, woolly, often white coat. There is an extensive amount of blubber. Females generally have four mammae. In males the testes are internal and the baculum is well developed.

According to J. E. King (1983), the phocid skull differs from that of the Otariidae in having well-inflated tympanic bullae, no supraorbital processes, and nasal bones that extend well posteriorly between the frontal bones. In the subfamily Phocinae the lateral swelling of the mastoid process forms an oblique ridge at an angle of about 60° from the long axis of the mastoid bone as a whole; in the Monachinae this ridge is absent. The dental formula is: (i 2–3/1–2, c 1/1, pm 4/4, m 0–2/0–2) × 2 = 26–36. The upper incisors have a simple pointed crown, the canines are elongate, and the postcanines usually have three or more distinct cusps.

Phocids shelter on rocky or sandy coasts, islands, and ice floes. Some stay on land or ice much of the time. Some migrate or disperse over a large area, and others make local movements corresponding to fluctuations of the ice. Activity is both diurnal and nocturnal. J. E. King (1983) indicated that phocids see and hear well underwater and that some phocids may have the most efficient hearing of pinnipeds in the air. Olfaction may be important in communicating intraspecific information. Phocids defend themselves by opening their mouth, uttering menacing cries, and advancing against the enemy, or they may flee to the water and dive. Most species eat fish, shellfish, and cephalopods. *Hydrurga* is the only pinniped that preys regularly on penguins and other seals. Some species can fast for long periods.

Phocids do not usually congregate in large rookeries as do otariids. Males may establish territories on land during the breeding season. Most species live in pairs during this period, but the males of some species, such as *Halichoerus* and *Mirounga,* may mate with a number of females. The total time of pregnancy is 270–350 days and usually includes a period of delayed implantation. At birth the single pup weighs 5–40 kg. In *Phoca* newborn can swim immediately, but in some other genera they often shun the water for several weeks. Lactation, lasting about 10–80 days, is generally much shorter than in the Otariidae.

The known geological range of the Phocidae is middle Miocene to Recent in North America and Europe, early Pliocene to Recent in South America, late Pliocene to Recent in Africa, and Recent in nearly all seas and oceans (J. E. King 1983).

A. Southern elephant seals *(Mirounga leonina)*, a nonbreeding group assembled on Macquarie Island. In such places they are helpless against human predators, and this group is composed of the descendants of a few that remained after the species was almost exterminated in the Antarctic up to 1918. Photo from Australian News and Information Bureau. B & C. Weddell seals *(Leptonychotes weddelli)*. About 14 days after birth of the pup, the mother tries to get it in the water by going in herself and calling it. While the pup gains courage, she saws a ramp in the ice with her teeth so that the pup can slide into the water. Photos from U.S. Navy.

Hawaiian monk seals *(Monachus schauinslandi):* A. Photo from San Diego Zoological Garden; B. Photo by Karl W. Kenyon.

PINNIPEDIA; PHOCIDAE; **Genus MONACHUS**
Fleming, 1822

Monk Seals

There are three species (Hall 1981; J. E. King 1983; Rice 1977):

M. schauinslandi (Hawaiian monk seal), Hawaiian Islands;
M. tropicalis (Caribbean monk seal), originally found throughout the West Indies and along the coasts of Florida, Yucatan, and eastern Central America;
M. monachus (Mediterranean monk seal), Mediterranean and Black seas, Atlantic coast of Morocco and Western Sahara, Madeira and Canary islands.

Wyss (1988*b*) stated that *M. monachus* and *M. tropicalis* seem to be more closely related to other phocids than to *M. schauinslandi* and that generic distinction for the latter seems justified. The more traditional view was set forth by J. E. King (1983), who observed that monachines were the dominant seals of the North Atlantic in the late Miocene and Pliocene and that certain populations presumably gave rise to *Monachus,* which then diverged into various species. However, it is agreed that *M. schauinslandi* has several skeletal features that are more primitive than those of the oldest known fossil monachine, the latter having lived about 14.5 million years ago. Therefore, *M. schauinslandi* apparently separated from the North Atlantic lineage of *Monachus* more than 15 million years ago, when an ancient seaway between

North and South America closed. The cranial characters of *M. schauinslandi* are more primitive than those of the other monk seals but are closer to those of *M. tropicalis* than to those of *M. monachus*. In general, *Monachus* is distinguished from other living monachines by having the nasal processes of the premaxillary broadly in contact with the nasals, rather than barely or not touching; having wide, heavy, crushing-type postcanines; having four, rather than two, mammae in females; having smooth, rather than faintly beaded, vibrissae; and having jet black, rather than white or grayish, neonatal pelage. The adult pelage of *Monachus* is never spotted as in some other monachines; the uniform brown or grayish dorsal color, supposedly resembling that of a monk's robe, gives the genus its common name. Additional information is provided separately for each species.

Monachus schauinslandi (Hawaiian monk seal).

Except as noted, the information for the account of this species was taken from Kenyon (1981) and J. E. King (1983). Females average larger than males and become much heavier when pregnant. One adult male was 214 cm long and weighed 173 kg, and a probably pregnant adult female was 234 cm long and weighed 272 kg. Coloration is slate gray dorsally and silvery gray ventrally. *M. schauinslandi* resembles *M. tropicalis* but has a more primitive bone structure of the inner ear. Another primitive and unique feature is that in *M. schauinslandi* the fibula is not fused to the proximal end of the tibia.

The Hawaiian monk seal is now largely restricted to the small, rocky Leeward Islands. Breeding now occurs regularly on five of these islets and atolls, and individuals tend to remain in the vicinity of their natal beaches. There is, however, much movement between islands and out to sea, and occasional wanderers reach the main Hawaiian Islands, to the southeast. Marked individuals have been found to move up to 1,165 km from the nearest breeding site. To give birth, females venture well up on sandy beaches, near shrubs and away from the tide line. Although *M. schauinslandi* is the only completely tropical phocid, it has no obvious physiological adaptations for warm habitat, and its blubber content is about the same as that of polar seals. It evidently avoids heat stress by remaining inactive during the day, lying in the shade or wet sand. It feeds at night, carrying out lengthy series of dives in shallow lagoons. During a period of 4.4 hours one individual made 25 dives, each lasting 5–14 minutes. Another individual was observed to remain underwater for 20 minutes. The diet consists primarily of reef- or floor-dwelling species of eels, other fish, cephalopods, and crustaceans.

Small, loose groups may assemble in environmentally favorable locations, but *M. schauinslandi* is basically solitary. During the spring and summer, adult males cruise constantly along favorite basking beaches in search of receptive females. Since the males outnumber the females by about three to one, the latter frequently are disturbed. Squabbles between rival males or males and upset females may be accompanied by bubbling and bellowing vocalizations, open-mouthed threats, and sometimes fighting. Mating takes place in the water. The young are born from late December to mid-August, mostly from mid-March to late May. The single pup weighs 16–18 kg, is about 100 cm long, and is covered with soft black hair. Lactation lasts about 35–40 days, during which time the pup reaches a weight of about 64 kg. The mother remains with the pup, fasting during this whole period, and loses about 90 kg of her own weight. The pup can swim weakly at birth and is taken into the water each day to practice under close maternal supervision. After weaning, the female returns to the sea, and the young must then fend entirely for itself. Its weight may fall to 45 kg by the time it is a year old. Weight then begins to increase again, but full size is not attained until at least 4 years (U.S. National Marine Fisheries Service 1978). Females can first bear young at about 5 years (McLaren 1984). Most give birth every other year, though some do so annually. Maximum known longevity in the wild is 30 years.

The Hawaiian monk seal evolved in an environment totally free of people and other sources of terrestrial predation and harassment. It thus unfortunately became both easily approachable and easily disturbed. It may once have bred throughout the Hawaiian Islands, but it disappeared from most beaches long ago. It probably survived in the Leewards because the Polynesians established no lasting settlements there. Remaining populations were largely eliminated by commercial sealing operations in the early nineteenth century, and the species was considered to be extinct by 1824. Some individuals survived, however, and in 1859 a single vessel took 1,500 skins. The species then remained generally undisturbed, though in very low numbers, until World War II. Subsequent establishment of military facilities and increased human presence led to new declines. The breeding colony on Midway disappeared by 1968. The main problems are that when pregnant females are disturbed by people or dogs, they may move to unfavorable areas to give birth, where the young cannot be properly sheltered; and that when lactating females are disturbed, they cannot adequately feed their young. There also is concern that increasing fishing activity in the Leeward Islands may lead to competition between the seals and human interests and to entanglement and drowning of the seals in fishing gear. The U.S. National Marine Fisheries Service (1987) estimated the total number of

M. schauinslandi at 500–1,500 individuals. The species is classified as endangered by the IUCN and USDI and is on appendix 1 of the CITES.

Monachus tropicalis (Caribbean monk seal).

According to J. E. King (1983), adults are about 200–240 cm long and grayish brown on the back, shading to yellowish white ventrally. Kenyon (1981) noted that very little is known about the biology of this species but that in many respects it appears to be similar to the other species of *Monachus*. An exception is that the peak of the pupping season was probably December.

European exploitation began during the voyages of Columbus and continued relentlessly as the seals were killed for their skins and oil. The species already was scarce in the mid-nineteenth century. More recently *M. tropicalis* has been persecuted extensively by fishermen, who view the animal as a competitor. An aerial survey in 1973 found fishing activity throughout the former habitat of *M. tropicalis* and led Kenyon (1981) to conclude that the species was extinct. This position was upheld by Le Boeuf, Kenyon, and Villa-Ramirez (1986) following a cruise through the Gulf of Mexico and around the Yucatan Peninsula. The last reliable records are of a small colony at Seranilla Bank, about midway between Jamaica and Honduras, in 1952. There have been a few more recent reports, however, and some authorities hope that *M. tropicalis* may yet survive (Mignucci Giannoni 1986). The species is classified as endangered by the USDI and is on appendix 1 of the CITES but is regarded as extinct by the IUCN.

Monachus monachus (Mediterranean monk seal).

Except as noted, the information for this account was taken from Kenyon (1981). Length was 238 cm and 262 cm in two adult males and 235 cm and 278 cm in two adult females. Usual adult weight is probably about 250–300 kg. Reported maximum weight is 400 kg. J. E. King (1983) wrote that coloration is generally dark brown to black above and a little lighter below and that a whitish gray patch occurs on the belly of some individuals.

Early reports indicate that *M. monachus* originally occurred on open sandy beaches (Bareham and Furreddu 1975). However, the species may always have depended mainly on beaches backed by extensive deserts or cliffs and thus inaccessible to terrestrial predators. Increasing human presence over the last few thousand years may have allowed this seal to survive primarily on small, rocky, waterless islands. Most births now occur in caves and grottoes, some of them with underwater entrances, but preferred breeding habitat may be beaches under cliffs. Foraging apparently is done mostly in waters less than 30 meters deep, though one individual was taken on fishing tackle set at 75 meters. The diet consists of fish and octopus.

Even though *M. monachus* has been in close contact with Western civilization far longer than has any other pinniped, surprisingly little is known of its ecology and behavior. A polygynous social structure is suggested by one observation of a large male visiting a group of five other seals and finding one receptive female. This mating occurred underwater off Deserta Grande Island, in the Madeiras, on 1 August 1976. Births have been recorded from May to November, with a peak in September and October. Presumably, gestation lasts about 11 months and includes a period of delayed implantation. The single pup weighs about 20 kg at birth, is about 84 cm long, and has a black woolly coat. Weaning occurs by about 6 weeks. Jones (1982) reported that a captive lived for 23 years and 8 months.

According to old reports and place names, *M. monachus* once was much more common than it is today. Hunting and disturbance by people resulted in major declines, though the species survived over much of its range into modern times. The rocky coasts of the Mediterranean, with numerous small islets and sea caves, offered an abundance of secluded breeding sites. However, recent increases in human presence, fishing activity, and motorized vessels have brought the seal under severe pressure. Although it is legally protected throughout its range, it is killed regularly by fishermen, who consider it to be a competitor and dislike it for damaging nets. The seals also may become entangled in nets and drown, and pregnant females may abort their young if disturbed by people.

Since the beginning of the twentieth century *M. monachus* has disappeared from the mainland coasts of Spain, southern France, the Crimea, Palestine, and Egypt. The total remaining number is estimated at 500–1,000 individuals and is declining. There are still breeding groups at the Deserta Islands and Rio de Oro on the Atlantic coast, along the North African coast from Morocco to Tunisia, and in the eastern Mediterranean and southern Black seas. The largest concentration is in the Aegean Sea, but Goedicke (1981) predicted that considering current trends, the species would disappear from Greek territory by the year 2000. *M. monachus* is classified as endangered by the IUCN and USDI and is on appendix 1 of the CITES.

PINNIPEDIA; PHOCIDAE; Genus LOBODON
Gray, 1844

Crabeater Seal

The single species, *L. carcinophagus,* is found primarily along the coasts and pack ice of Antarctica, but wandering individuals occasionally reach South Africa, Heard Island, southern Australia, Tasmania, New Zealand, and the Atlantic Coast of South America as far north as Rio de Janeiro. Except as noted, the information for this account was taken from J. E. King (1983), Kooyman (1981*b*), and Laws (1984*a*).

Adult males, which average slightly smaller than females, are usually 203–41 cm long, with a reported maximum of 257 cm. Adult females are usually 216–41 cm long, with a maximum of 262 cm. Few weights have been recorded, but the probable adult range is 200–300 kg. Following a summer molt in January–February, coloration is mainly dark brown dorsally, grading to blonde ventrally. On the back and sides are large chocolate brown markings, interrupted with lighter brown. The flippers are the darkest parts of all. The coat fades throughout the year, reaching an almost uniform blonde or creamy white color by summer. The body is relatively slim, and the snout is long. The cheek teeth are perhaps more complex than those of any other pinniped or carnivore. Spaces between the prominent tubercles intrude deeply into each tooth. The main cusps of the upper and lower teeth fit between each other, so that when the mouth is closed the only gaps are those between the tubercles.

The crabeater seal occurs largely in the vicinity of the Antarctic pack ice. During the maximum extent of the ice in the winter, the seal's range extends over about 22 million sq km. In summer it moves south and is confined to six regions of residual pack ice totaling 4 million sq km. Although concentrated along the edge of the drifting ice, the species sometimes moves onto the shore of Antarctica. One individual was found 113 km from open water, and another was found on a glacier 1,100 meters above sea level, perhaps an elevational record for pinnipeds.

Lobodon has remarkable agility on snow or ice and may be the fastest of pinnipeds there. Speeds of up to 25 km/hr have been reported. When it sprints, the forelimbs are thrust alternately against the snow, the pelvis is moved from side to side while also thrusting against the snow, and the hind flippers are held together and off the snow. Stirling and Kooyman (1971) reported that when *Lobodon* is caught, it rolls over many times. Such may be an instinctive reaction developed to escape the killer whale *(Orcinus)* or leopard seal *(Hydrurga).* Many adult *Lobodon* are scarred from attacks by these other two genera. The crabeater seal itself, despite its name, feeds almost exclusively on krill, small shrimplike animals of the family Euphausiidae. It catches these animals by swimming into a school of them with its mouth open, sucking them in, and then sieving the krill by forcing the water through the spaces between the tubercles of the postcanine teeth. Most feeding probably takes place at night. Other invertebrates and small fish are also eaten.

Population densities of around 1–7 per sq km have been reported. The animals usually are seen alone or in small groups. Siniff et al. (1979) reported that concentrations of 50–1,000, composed primarily of immature individuals, sometimes form on favorable areas of fast ice. Otherwise, *Lobodon* is found in small social units that apparently originate in the

Teeth of crabeater seal, postcanines 3, 4, and 5, right mandible, lingual aspect, photo by Victor B. Scheffer.

Crabeater seal *(Lobodon carcinophagus)*, photo by University of Minnesota Antarctic Seal Research Program.

spring breeding season when pregnant females move onto suitable ice floes to give birth. Just before and after parturition, each female is joined by a male. The latter protects the female and pup from other males, as well as from potential predators. An area of up to 50 meters from the female is so defended. The female, however, savagely keeps the male at bay until her pup is weaned and she enters estrus, about 4 weeks after giving birth. The male then becomes aggressive and forces the mother and pup apart. Mating takes place from October to December and occurs on the ice, rather than in the water as in most polar phocids. The male and female remain in close contact for a few days, then break up. As a result of fights with both males and females, the old males are heavily scarred about the head and neck. Agonistic interaction is accompanied by hissing and blowing sounds. *Lobodon* also emits a deep groaning sound underwater (Stirling and Siniff 1979).

Births occur from September to November, with a synchronized peak in early to mid-October. Total gestation lasts about 11 months and probably includes a period of delayed implantation. The single pup is about 150 cm long and has a soft, woolly, grayish brown coat. In the 4-week period of lactation it grows from 25 to 120 kg. The young are nearly full grown after 2 years, but some growth continues for years afterwards. Over 80 percent of females give birth each year. The life span may be up to 39 years.

Lobodon is the most numerous pinniped and may be the most abundant large mammal on earth. Population estimates range from 15 million to 40 million individuals. Numbers apparently increased in response to the greater availability of krill following the commercial extermination of the Antarctic baleen whales. It also has been reported that the age of sexual

maturity declined from 4 years in 1950 to only 2.5 years by about 1980. Bengtson and Siniff (1981), however, found that this trend may be changing in response to adjustments in Antarctic ecosystems and reported an average age at sexual maturity of 3.8 years in a sample of 94 females. They also cautioned that development of a major commercial krill fishery could further alter the ecosystems.

PINNIPEDIA; PHOCIDAE; **Genus HYDRURGA**
Gistel, 1848

Leopard Seal

The single species, *H. leptonyx*, occurs along the coasts and pack ice of Antarctica and also is regularly present on most subantarctic islands. Wandering individuals have reached South Africa, southern Australia, Tasmania, New Zealand, Lord Howe Island, the Cook Islands, Tierra del Fuego, and the Atlantic coast of South America nearly to Buenos Aires. The information for this account was taken from J. E. King (1983), Kooyman (1981c), and Siniff and Stone (1985).

Adult males, which average smaller than females, are about 300 cm long and weigh about 270 kg. Adult females are 290–380 cm long and weigh up to 500 kg. Coloration is dark gray to almost black dorsally, paler on the sides, and silvery below. There are gray spots on the throat, shoulders, and sides. The body is streamlined but massive, the head seems disproportionately large and reptilian, and the foreflippers are comparatively long. The skull and canine teeth are unusually long, and the postcanine teeth, having three prominent tu-

Leopard seals *(Hydrurga leptonyx):* A. Photo by U.S. Office of Territories through the National Archives; B. Photo by John Warham.

bercles with narrow clefts between them, are second in complexity to those of *Lobodon.*

Mature animals normally occur in the outer fringe of the pack ice. Young animals move throughout the subantarctic islands in the winter and early spring (June–October). Animals about 3–9 years old may reach the Antarctic coast in summer. There is some question whether these movements are regular migrations or occasional dispersals, but some islands are visited annually. There is a year-round presence on South Georgia and Heard Island.

The leopard seal has a variable diet and is the only pinniped that regularly preys on warm-blooded animals. The biomass taken has been estimated at 45 percent krill, 35 percent seals, 10 percent penguins, and 10 percent fish and cephalopods. Krill is consumed in much the same manner as described for *Lobodon,* with the organisms being filtered from the water by the complex cheek teeth. *Lobodon* itself is a major prey item, especially during the summer, when pups are available. Most crabeater seals bear scars from attacks by *Hydrurga.* Adult penguins are captured mostly in the water, while chicks are seized on the ice. Probably only a few individual leopard seals regularly take penguins.

A population density of about 0.12 animals per sq km of pack ice has been reported. *Hydrurga* seems to be solitary, and unlike in *Lobodon,* males have not been observed together with females and young. Bester and Roux (1986), however, reported groups of 2–5 adults, including both sexes, hauled out in the vicinity of abundant prey at Kerguelen Island. A long, deep, droning sound is produced underwater.

Mating occurs from November to February, apparently in the water. Gestation lasts about 11 months, and implantation evidently is delayed for about 2 months. Births take place from September to January, mostly in late October and November. The single pup is about 160 cm long and weighs about 30 kg. The natal coat is soft and thick, the dorsal color is dark gray with a darker central stripe, and the sides and ventral surface are almost white, irregularly spotted with black. Lactation may last about 4 weeks. Males apparently reach sexual maturity in their fourth year of life, and females, in their third year. Maximum known longevity in the wild is 26 years.

Population estimates for *Hydrurga* vary but seem to center around 400,000 individuals. The genus is highly adaptable and is not exploited, but it may face problems in the future. Young individuals evidently depend heavily on krill but are less efficient in foraging ability than are krill-feeding specialists such as *Lobodon.* Therefore, *Hydrurga* could be affected adversely by development of a commercial krill fishery.

PINNIPEDIA; PHOCIDAE; Genus LEPTONYCHOTES
Gill, 1872

Weddell Seal

The single species, *L. weddelli,* is found in areas of fast ice around Antarctica and some subantarctic islands, including the South Shetlands, the South Orkneys, and South Georgia. Wandering individuals have reached Heard, Kerguelen, Macquarie, and the Auckland, Juan Fernandez, and Falkland islands, as well as southeastern Australia, New Zealand, Patagonia, and Uruguay. Except as noted, the information for this account was taken from J. E. King (1983), Kooyman (1981*a,* 1981*d*), and Stirling (1971).

Adult males, which average slightly smaller than females, are about 250 cm long, with a reported maximum of 297 cm; adult females are about 260 cm long, with a reported maximum of 329 cm. In the spring both sexes commonly weigh around 400–450 kg. Coloration varies, but typically the upper parts are bluish black, with white streaks and splashes that increase toward the rear and bottom of the animal. The lower parts are gray with white streaks. During the summer the coat fades to rusty grayish brown, but it never bleaches to the whiteness often found in *Lobodon.* The head of *Lepto-*

Weddell seals *(Leptonychotes weddelli)*, photo from U.S. Navy.

nychotes is relatively small, the muzzle is very short, and the eyes and canine teeth are relatively large. The upper incisor teeth are markedly unequal, the outer being about four times longer than the inner. The cheek teeth are strong and fully functional but lack the complexity of those of *Hydrurga* and *Lobodon*.

The Weddell seal is perhaps the most southerly of mammals. It normally is found on fast ice in coastal areas within sight of land, and not on moving pack ice or isolated floes. There apparently are no major migrations, but there is a general northward movement before the onset of winter as the ice expands and a summer concentration to the south where fast ice remains. Some individuals remain in the most southerly parts of the range throughout the year, though they are less obvious in winter, when they presumably stay in the water where the temperature is constant. *Leptonychotes* depends heavily on holes through the ice for access to the water and for breathing. Holes are made initially by cutting with the incisor teeth and then are enlarged and maintained by abrading the ice with the canines. Cutting does not begin in areas where the ice is over 10 cm thick, but once a hole is made, it may be maintained even if the ice increases to a thickness of 100–200 cm. Holes generally are made near natural cracks in the ice, and *Leptonychotes*, while underwater, evidently uses the radiating pattern of cracks as a guide to find its way to the holes. Underwater vision is excellent and is thought to be the primary means of navigation. Dives are deeper and more frequent in daylight. During the long period of winter darkness, dives tend to be shallow and probably are in areas with which the animals had become familiar previously when light was available.

The underwater activity of the Weddell seal has been investigated by capturing some individuals, attaching an instrument known as the time-depth recorder, and then releasing them. Animals tend to make an initial series of shallow dives to familiarize themselves with an area and then begin to go farther and deeper. Horizontal swimming speed is around 8–12 km/hr, and average rate of descent is 35 meters/minute. Foraging usually involves a series of dives each lasting 8–15 minutes and alternating with periods of 2–4 minutes at the surface hole. The entire series may go on for 8–9 hours without stopping. It is no wonder that when *Leptonychotes* finally emerges to rest or sleep on the ice, it may give the impression of being lethargic and not easily aroused. It is, however, the supreme diver among pinnipeds and in this regard even outclasses some cetaceans. Its dives usually reach depths of 200–400 meters, but some have gone to 600 meters and lasted up to 73 minutes. Its foraging trips from an access hole may cover a round-trip distance of 10–12 km. The diet consists mostly of fish, both sluggish bottom dwellers and active midwater species. The Antarctic cod, which may weigh over 30 kg, frequently is taken. Cephalopods and crustaceans also are eaten.

Densities of 15–35 individuals per sq km have been recorded in favorable areas of the Weddell Sea. *Leptonychotes* is not gregarious, however, and adults keep apart when hauled out on the ice. Holes through the ice must be shared by a number of seals and sometimes are the scene of much aggressive activity and serious fighting. Younger animals may be forced to disperse into areas where the water is not completely covered by ice, and they sometimes form large aggregations. Dominant adult males establish territories below the ice, which vary in size from about 15 by 50 meters to 50 by 400 meters, and spend almost all their time from October to December defending these areas. During the same period the females haul out on the ice above to give birth to the young conceived during the previous breeding season. The females form loose groups at suitable pupping sites but keep well apart and may snap at each other. A sex ratio of 10 females to 1 male has been observed at breeding sites. After the season, subadults are admitted into these areas, and the distance between individuals decreases. *Leptonychotes* is among the noisiest of seals and produces a constant and varied stream of underwater sounds, including whistles, buzzes, tweets, and chirps. Some of these sounds probably are for territorial defense, and others are used to attract mates. Thomas and Stirling (1983) reported considerable variation in the vocalizations of different seal populations.

Mating takes place in December, evidently in the water. Implantation of the blastocyst in the uterus is delayed until the period from mid-January to mid-February. Births occur mostly in October in Antarctica but may be as early as late August at South Georgia. The single newborn is about 150 cm long, averages 29 kg in weight, and has a grayish coat that is replaced after 4 weeks by adultlike pelage. The mother remains with the pup almost constantly for about 2 weeks, then begins to spend about a third of her time foraging. Upon her return to the ice, she establishes vocal contact with her pup and confirms identification by sniffing. She may bring the pup into the water to practice swimming when it is only 8–10 days old. Lactation lasts for 6–7 weeks, after which the mother mates again and departs. By then the young averages 113 kg in weight and is able to dive to a depth of 92 meters. Females may reach sexual maturity as early as 2 years, but Croxall and Hiby (1983) reported that females in the South Orkney Islands first gave birth at an average age of 4–5 years. Up to 80 percent of adult females become pregnant each year. At McMurdo Sound, Antarctica, both sexes are recruited into the adult population at around 5 years, and annual reproductive rates are 46–79 percent (Testa 1987; Testa and Siniff 1987). Maximum known longevity is 25 years (McLaren 1984).

Unlike *Lobodon* and *Hydrurga*, the Weddell seal is easily approached and captured by people. Excessive hunting to supply Antarctic bases with meat for people and sled dogs has eliminated or reduced some breeding colonies. There is fear that if intensive sealing begins in Antarctica, *Leptonychotes* will be especially vulnerable because of its tendency to congregate on stable ice during the summer. Until now, however, there have been no major declines, and current numbers are estimated at 500,000 to 1 million.

PINNIPEDIA; PHOCIDAE; Genus OMMATOPHOCA
Gray, 1844

Ross Seal

The single species, *O. rossi*, is circumpolar in pack ice of the Antarctic. Wandering individuals have reached South Australia and Heard Island. The information for this account was taken from J. E. King (1983) and Ray (1981).

Adult males, which average smaller than females, are 168–208 cm long and weigh 129–216 kg; adult females are 190–250 cm long and weigh 159–204 kg. After the summer (January) molt, coloration is dark gray above and silvery white below, with streaks of pale gray on the sides of the head, shoulders, throat, and flanks. In the course of the year the coat fades to tan or brownish. The body hair and vibrissae are the shortest of any phocid's. The head is large and the neck thick, but the body is slenderer than that of *Leptonychotes*. The snout is very short, and the mouth appears to be relatively small. The eyes are proportionately larger than those of any other seal, though the slits are not unusually large. The front limbs, which are more specialized and flipperlike in shape than those of most phocids, have reduced claws and greatly elongated terminal phalanges. The hind flippers are proportionately longer than those of any other phocid. The skull is distinguished by the short rostrum and the enormous size of the eye sockets. The incisors and canine teeth are small compared with those of *Leptonychotes* but are sharp and recurved. The cheek teeth are relatively small and weak.

Ommatophoca is among the least-known of pinnipeds. It generally is considered to prefer the heavy consolidated pack ice, but some observations indicate that it also is found on smaller floes. It is slow and clumsy on the ice but is thought to be a swift, highly maneuverable swimmer and to pursue its prey under the ice, where its large eyes can perceive movements in the dimly lit waters. The diet consists mostly of cephalopods, including some large squid. Other invertebrates and fish also are taken.

Although the mouth is small externally, it can be opened remarkably wide and then closed forcefully. Swallowing of large prey is facilitated by an expanded trachea and powerfully developed muscles of the tongue and pharynx. The hard palate is short, but the soft palate is very long, extending far posteriorly to the level of the occipital condyles. The soft palate can be inflated with air, and both it and the expanded trachea seem to act as resonating chambers for the production of sound. When approached by people, *Ommatophoca* may lift its front end, open its mouth widely, inflate its soft palate to meet the back of the raised tongue, and emit a series of trills and thumping noises. It shows so little fear that it can be closely photographed and even touched during this display. A variety of birdlike chirps and other sounds are produced both in the air and underwater. The process occurs internally in the throat, and no air is expelled from the body until the end of the display.

The vocal capabilities of *Ommatophoca* suggest that it may establish underwater territories like those of *Leptonychotes* and maintain contacts over considerable distances. However, available evidence indicates that it is generally solitary and that individuals only occasionally come together in small groups. Highest recorded population density is 2.9/sq km. Some individuals are scarred on the neck and shoulders, but it is not known whether these marks result from intraspecific fighting or from attacks by *Orcinus* and *Hydrurga*. Females haul out on the ice in November to give birth to the young conceived during the previous breeding season. Mating apparently takes place in late December, after the young are weaned. Implantation of the blastocyst is delayed for 2–3 months, and the total gestation period is about 11 months. There seems to be a birth peak from 3 to 18 November. The single newborn is about 97 cm long, weighs 17 kg, has a coat similar to that of an adult, and is capable of swimming almost immediately. Weights of up to 75 kg are attained after 15 days, and lactation ceases after approximately 4 weeks. Most females become pregnant by the time they are 3 years old, and males may reach sexual maturity at 3–4 years. Maximum known longevity is 21 years.

Ommatophoca was not discovered until 1840, the latest of any pinniped, and fewer than 50 records accumulated over the next 100 years. Subsequent investigators have suggested that the genus is not rare but have provided an incomplete basis for an assessment of status. Overall population estimates of 100,000–650,000 have been made but are not considered reliable.

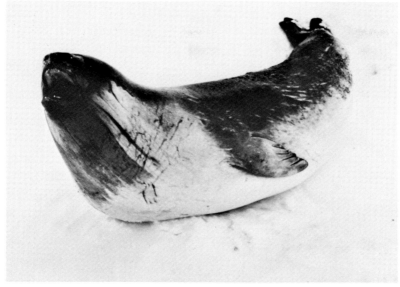

Ross seal *(Ommatophoca rossi)*, photos by University of Minnesota Antarctic Seal Research Program.

PINNIPEDIA; PHOCIDAE; Genus MIROUNGA
Gray, 1827

Elephant Seals

There are two species (J. E. King 1983; Ling and Bryden 1981; Rice 1977):

M. angustirostris (northern elephant seal), coastal waters from southeastern Alaska to Baja California;

M. leonina (southern elephant seal), most subantarctic islands between 40° S and 62° S, coast of southern Argentina, waters south to edge of antarctic ice at 78° S, occasionally on coasts of southern Africa and Australia as far north as St. Helena (16° S), Mauritius (20° S), and Lima (14° S).

According to J. E. King (1983), the ancestor of *Mirounga* probably descended from the same stock as that leading to *Monachus* and also passed through an ancient seaway from the Caribbean to the Pacific. Many characters of the skull,

Northern elephant seals *(Mirounga angustirostris):* A. Male, photo from Allen Hancock Foundation; B. Father and baby, photo by Julio Berdegué; C. Seal shedding its skin, photo from Colorado Museum of Natural History through Alfred M. Bailey and R. J. Niedrach; D. Adult male with proboscis turned downward into mouth, photo by Julio Berdegué; E. Group of females, animal on left shedding, photo by Rupert R. Bonner; F. South Atlantic elephant seal *(M. leonina),* photo from Zoological Garden Berlin-West.

dentition, and growth suggest that the northern elephant seal (*M. angustirostris*) is more primitive than *M. leonina*. Both species are characterized by large overall size and the trunk-like, inflatable proboscis of adult males. The proboscis reaches full size in animals about eight years old. Its tip then overhangs the mouth in front, so that the nostrils open downwards. An enlargement of the nasal cavity, internally it is divided into two parts by the nasal septum. Externally the proboscis is flattened and less obvious in the nonbreeding season. When breeding, *Mirounga* can erect this organ, partly by blood pressure, assisted by inflation, to form a high, bolster-shaped cushion on top of the snout. It may act as a resonating chamber to amplify the roar of the bull. The huge skull has an increase in the length of the narial basin and a high frontonasal area to provide space for attachment of the muscles that move the proboscis. The canine teeth of the adult male are relatively large and continue to grow for at least 12 years. The cheek teeth are much reduced. In color the males are dark gray, being a little lighter ventrally just after the molt, but through the year the coat fades to a rusty grayish brown. Fighting between bulls leads to intensive scarring of the neck region, and the scar tissue makes the skin of the chest extremely thick, tough, and cracked. The females are browner and generally darker than the males and have a light-colored yoke around the neck caused by the many small scars resulting from bites during mating. The hair is short, stiff, and harsh, and there is no underfur. Additional information is provided separately for each species.

Mirounga angustirostris (northern elephant seal).

Except as noted, the information for the account of this species was taken from J. E. King (1983) and McGinnis and Schusterman (1981). Adult males are 400–500 cm long and weigh 2,000–2,700 kg; the much smaller adult females are 200–300 cm long and weigh about 900 kg. From *M. leonina*, *M. angustirostris* differs in having a smaller and less flexible body, less sexual dimorphism, a narrower skull with a longer rostrum and a greater degree of suture closure with age, and a larger proboscis in the adult male. This proboscis is very long, hanging down over the mouth about 30 cm when relaxed and showing a very deep transverse groove that almost divides it in two when inflated. Inflation causes it to curve down between the jaws and directs the nostrils toward the pharynx, thus snorts are directed down into the open mouth and pharynx, which act as a resonating chamber.

The northern elephant seal currently breeds on numerous islands, from Cedros, off the west coast of central Baja California, north to the Farallons, near San Francisco. Preferred breeding sites are good sand or gravel beaches remote from human activity. According to Condit and Le Boeuf (1984), there now are six major and seven minor rookeries where the seals breed and molt. When not at these sites, the seals are feeding at sea, mainly in waters from northern Baja California to northern Vancouver Island. Juveniles move northward from their rookeries during the summer by an average of 900–1,000 km and return to haul out in the fall. During the winter, when adults are breeding, juveniles again go to sea. Adult males migrate northward during the spring, haul out in July and August to molt, move north again in the fall, and return to the rookeries to breed during the winter. Adult females are at sea for 10 weeks during the spring and again for about seven months during the summer and fall, available evidence suggesting that they do not move far from the rookeries where they breed. Individual seals occasionally appear off Alaska and have been found as far as the Aleutians, Midway Island, and the central Gulf of California.

M. angustirostris is capable of long, deep dives. Its diet consists mostly of deepwater and bottom-dwelling organisms, particularly squid and elasmobranch and teleost fish.

Using an instrument called the time-depth recorder, Le Boeuf et al. (1988) monitored movements of females during their first 14–27 days of foraging after lactation. They were found to dive almost continuously during this period. Mean dive duration was 19.2 minutes, the longest being 48 minutes. Modal depth was 350–650 meters, and the maximum, a record for the Pinnipedia, was 894 meters. Dives tended to be shallower at night. Evidence suggested that *M. angustirostris* feeds mostly in the deep scattering layer of the mesopelagic zone and preys on organisms that migrate vertically upwards at night. Most fish with such behavior are bioluminescent, as are several species of squid taken by the elephant seal. This factor may assist the seal to find its prey at great depths.

Adult males arrive on the breeding beaches in late November and early December and probably do not feed again until they depart in March. Adult females arrive in late December or January and also do not feed until their departure, about 34 days later. The females come together in compact groups, individuals usually not more than 1–2 meters apart. In newly established colonies, with relatively few animals, a dominant male may be able to defend a discrete group of females, but this pattern breaks down as more crowded conditions develop. Generally, a single male cannot control more than about 40–50 females. If a colony has a larger number of females, there will be several bulls present. The latter compete for access to the females and eventually form a dominance hierarchy. There is then no defense of a territorial space or a particular harem of females; rather, the highest-ranking males are able to mate the most, and subordinate males are forced to the periphery of the group. According to Reiter, Panken, and Le Boeuf (1981), the largest female group on Año Nuevo Island, southwest of San Francisco, contains 1,100 individuals and is associated with 10–25 males. A successful male in this colony mates with about 100 females during the season.

Males establish dominance through stares, gestures, vocalizations, and physical combat. A threat display may involve elevating the anterior part of the body to about 90°, inflating the proboscis, slamming the body to the ground, and emitting a series of snorts and low-pitched guttural sounds. If a fight develops, the opponents confront one another chest to chest and rock back and forth seeking an opportunity to deliver a downward blow with the canine teeth. Serious wounds occur frequently on the chest and proboscis but rarely lead to death. The roaring vocalization is rhythmic, resonant, and metallic and carries for nearly a kilometer. Distinctions have been detected between the sounds produced by the males of different islands.

Females have a loud and prolonged threat vocalization. After giving birth, they become aggressive toward each other and usually to pups other than their own, but some adoptions have been noted. The period from initial landing to the birth of the young conceived during the previous breeding season is 6–7 days. Toward the end of the lactation period, females enter estrus for 3–5 days. Most mating occurs in the third week of February. Gestation lasts for 11 months and includes a delayed implantation period of about 4 months. The peak of births occurs from 20 January to 1 February. The newborn pup is about 127 cm long, weighs 30–45 kg, and has grayish black pelage. At the time of weaning, 27 days later, it weighs about 158 kg, and the coat is silvery. Subsequently the young gather in small pods and practice swimming and diving but do not feed. When they leave the natal beaches in April or May, they are able to remain underwater for 15 minutes. Females are sexually mature in their third year of life, though most do not give birth until their fourth or fifth year. Nearly all adult females become pregnant each year and have the potential of producing 10–12 young during their lifetime, but individuals 6 or more years in age are much more likely to raise their pups to weaning (Reiter, Panken, and Le Bouef 1981). Males are fertile at 4 years but usually are not able to win mating privileges until they are 8–9 years old. Longevity in the wild is relatively short, not having been known to exceed 14 years. According to Jones (1982), a captive lived for 15 years.

A 4-meter-long elephant seal can yield about 325 liters of fine oil. Commercial exploitation of this resource began on the West Coast around 1818 and intensified during the California gold rush. By the 1860s, after perhaps 250,000 individuals had been killed (Busch 1985), there were almost too few individuals left to make hunting worthwhile. By 1892 M. angustirostris, which once had bred from Point Reyes, north of San Francisco, to Magdalena Bay, in southern Baja California, had been reduced to a single herd of about 100 animals on Guadalupe Island, off northern Baja. A subsequent recovery resulted from an end to commercial interest and the establishment of complete legal protection by Mexico in 1922. From Guadalupe, which still has the largest population, the species spread both south and north. Breeding colonies were reestablished on the California Channel Islands in the 1930s, in the Farallon Islands in 1972, and at Año Nuevo Point, on the California mainland, in 1975. Most of the original range now has been reoccupied. Cooper and Stewart (1983) reported that numbers were increasing at a rate of 14 percent annually and that about 25,000 pups were born in 1982, about half of them in the United States. The U.S. National Marine Fisheries Service (1989) estimated overall numbers at 100,000, the same figure that Busch (1985) had calculated for the period prior to exploitation. However, current elephant seal populations show a remarkable lack of genetic diversity, because of the great numerical reduction in the late nineteenth century, and may thus be vulnerable to changing conditions.

Mirounga leonina (southern elephant seal).

Except as noted, the information for the account of this species, the largest of pinnipeds, was taken from J. E. King (1983), Ling and Bryden (1981), and Smithers (1983). Adult males are usually 450–600 cm long and weigh up to 3,700 kg. Old records suggest that some males attained a length of 900 cm and a weight of 5,000 kg. The much smaller females are 200–300 cm long. Adult female weight seems to be the subject of some confusion, having been reported as 359 kg (Ling and Bryden 1981), 400 kg (McCann 1985), 680 kg (McLaren 1984), and 900 kg (Bonner 1982a; J. E. King 1983). M. leonina has a more flexible body than does M. angustirostris and can bend backwards and touch its tail with its head. The proboscis of the adult male is not so pendulous as in M. angustirostris, overhanging the mouth only about 10 cm. When erected, its transverse grooves are much less deep. It hangs down, so that the nostrils open in front of the mouth, but it does not actually go into the mouth. It functions as a resonator for sounds produced in the mouth.

The southern elephant seal breeds on most subantarctic islands and on the Valdez Peninsula, on the coast of southern Argentina. The three major populations are centered at South Georgia in the southwestern Atlantic, Kerguelen and Heard islands in the South Indian Ocean, and Macquarie Island south of New Zealand. Nonbreeding individuals regularly land at many points, including coastal Antarctica. M. D. Murray (1981) reported establishment of a small breeding colony at Peterson Island, just off the Indian Ocean side of Antarctica. Heimark and Heimark (1986) reported the same for Palmer Station on the Antarctic Peninsula. The annual life cycle involves lengthy stays at sea alternating with periods ashore for breeding, molting, and resting. The pelagic phases are not well known; the longest movement of a tagged animal was from South Georgia to South Africa, a distance of 4,800 km. Immature individuals, along with females that have not

recently given birth, come ashore to molt from November to January. Cows that gave birth during the previous breeding season molt in January and February, and adult males do so from February to May. Molting takes 3–40 days, and the animals do not feed during this period. They then go to sea again, returning to the breeding grounds in August or September. Preferred beaches are flat and smooth, with a substrate of sand, rounded stones, or pebbles. Some animals move into areas of tussock grass, and molting females often seek muddy wallows. Like M. angustirostris, M. leonina has flexible flippers and sometimes scoops sand over itself to lower body temperature. At the southern edge of its range, M. leonina hauls out on snow or ice. It can make access holes by butting though the ice with its head, and it may travel 20 km under the ice, utilizing air pockets along the way. It is capable of long, deep dives and has been known to remain underwater for over 30 minutes. Maximum swimming speed is about 25 km/hr. The diet consists almost entirely of cephalopods and fish.

Adult males arrive at the breeding beaches from early August to September. They remain there up to 9 weeks, during which time they do not feed. Adult females begin coming ashore a few weeks later and also do not feed until their departure, about 4–5 weeks after arrival. The females initially form small, peaceful groups, but eventually thousands of females may crowd onto a favorable beach. They become more aggressive and more widely spaced after giving birth. Social activity is much the same as in M. angustirostris. Bulls are not territorial but form a dominance hierarchy, with the highest-ranking individuals being able to mate with the most females. A single dominant male, or "beachmaster," may be able to control a small group of females, but when more than about 60 females are present in the cluster, a second bull gains some access. When the group grows to around 130 females, a third male wins a place, and so on. Some "harems" of this kind contain up to 1,350 females. Males hoping to enter the scene challenge the established males with a bubbling roar that may carry for several kilometers. If the call is answered, and the challenger is not scared off, a confrontation develops as described above for M. angustirostris. Serious fighting is not common; when it does occur, the opponents lift the anterior two-thirds of their bodies off the ground, move their heads from side to side, and lunge with the canine teeth. The loser deflates his proboscis and retreats to the water or to the periphery of the breeding area.

Within 2–7 days of arriving on land, females give birth to the young conceived in the previous breeding season. After another 17–22 days, they enter estrus and mate. Gestation lasts just over 11 months and begins with a delayed implantation period of about 4 months. The young are born in September or October, but there is a synchronous peak throughout the range of the species in mid-October. The newborn is 130 cm long, weighs 35–50 kg, and has black woolly hair. At the time of weaning, 20–25 days later, it weighs 140–80 kg, and its coat is silvery gray. The mother loses about a third of her body weight during lactation and then promptly returns to sea to forage, permanently abandoning her pup. The young then form groups of their own, learn to swim by themselves, and move out to sea when they are 8–10 weeks old. Females become sexually fertile at about 2 years, and nearly all females more than 4–6 years old become pregnant annually. Males are sexually mature at 6 years but do not begin to compete for mating privileges until they are 10 years old and usually cannot become beachmasters until 12–14 years old. In the formerly exploited population of South Georgia, where large bulls were killed regularly, males began to breed 2–3 years earlier than in other populations. Hindell and Little

(1988) reported that two females were observed suckling pups on Macquarie Island, within 1 km of where they had been born and marked 23 years earlier.

The southern elephant seal formerly bred in Tasmania but was eliminated there by aboriginal peoples. A colony on King Island, in Bass Strait, between Tasmania and Victoria, survived until the early nineteenth century. The species also once bred in the Juan Fernandez Islands and occurred on St. Helena and possibly the Seychelles. Commercial exploitation began around 1800 and intensified a few decades later as fur seals became scarce. Elephant sealing often was done in association with whaling, the main objective in each case being oil. A large bull *M. leonina* could produce about 750 liters of oil (Busch 1985). All of the major populations were greatly reduced, but most survived because some breeding beaches were not accessible to boats loading oil, as well as because the demand for whale oil declined. The last major sealing expeditions were undertaken in the 1870s, and the species subsequently recovered to some extent. In 1910 commercial exploitation resumed on South Georgia, but under regulations that limited the annual kill to a maximum of 6,000 bulls. Mistakes in management, including an inappropriate increase in the yearly quota, led to new declines by the 1940s. Improved controls were implemented in 1952, but hunting ceased in 1964, again in connection with a decline in whaling (Bonner 1982*a*). According to McCann (1985), southern elephant seal populations are no longer increasing, except for the group of 13,800 individuals in Argentina, and some are declining. Overall numbers are about 750,000, not including the approximately 216,000 pups born each year. South Georgia has about 350,000 individuals; Kerguelen, 157,500; Heard Island, 80,500; and Macquarie, 136,500.

PINNIPEDIA; PHOCIDAE; **Genus ERIGNATHUS**
Gill, 1866

Bearded Seal

The single species, *E. barbatus,* is found along coasts and ice floes in the Arctic Ocean and adjoining seas and regularly occurs as far south as the Sea of Okhotsk, Hokkaido, the Alaska Peninsula, Hudson Bay, and Labrador (Burns 1981; King 1983; Rice 1977). Individuals have reached eastern China, Tokyo Bay, the Gulf of St. Lawrence, Scotland, Normandy, and northern Spain (J. E. King 1983; McLaren 1984; Zhou 1986) but probably not Cape Cod Bay as reported previously (Ray and Spiess 1981). Except as noted, the information for the remainder of this account was taken from Burns (1981*b*) and J. E. King (1983).

Length is about 200–260 cm and weight is about 200–360 kg. The sexes are approximately the same size, though the heaviest individuals are pregnant females. Coloration is variable, most animals being light to dark gray, others being brownish, and all having a dark area on the back. The face and foreflippers are often rust or reddish brown. The hair is generally short and straight, but there is a great profusion of long, very sensitive, glistening white mustachial whiskers, which is the basis for the name of the species. The body is robust, but the head is disproportionately small. Norwegian sailors used the name "square flipper" for *Erignathus*, because the third digit on the forelimb is slightly longer than the others, resulting in a blunt, squared appearance to the appendage. The nails on the foreflippers are strong, while those on the hind flippers are slender and pointed. Females,

Bearded seal *(Erignathus barbatus)*, photo from Zoological Garden Berlin-West.

like those of *Monachus*, have four mammae (females of all other phocids have only two mammae).

The bearded seal generally is restricted to shallow waters over the continental shelf and near coastlines and often is associated with sea ice. It avoids regions of continuous, thick, shorefast or drifting ice but is common where ice is in constant motion and has many openings. Favorable ice conditions are especially prevalent in the Bering and Chukchi seas. There *Erignathus* moves as the ice seasonally advances and recedes. The population migrates south through the Bering Strait in late fall and early winter and spends the winter amidst the drifting ice of the Bering Sea. In the spring there is a northward movement into the Chukchi Sea. Certain other populations, such as that of the Sea of Okhotsk, have no ice available during the summer and so come ashore, preferably on gravel beaches, to rest or molt. *Erignathus* attempts to avoid the polar bear by resting very near the water and diving immediately when disturbed. It evidently has good vision and hearing and a fair sense of smell. It is able to make breathing holes through thin ice by ramming with its head. Dives to depths of 200 meters have been reported, but most feeding is done in shallower waters. The diet varies by area but consists almost entirely of bottom-dwelling animals. Important items are crabs, shrimps, gastropods, clams, whelks, arctic cod, and flounder. The whiskers may be used to help sort out smaller food organisms.

Erignathus tends to be solitary, though loose aggregations sometimes form at favorable hauling-out sites. It is highly vocal and produces a distinctive underwater song which Ray, Watkins, and Burns (1969) described as a long, oscillating warble that may last for more than one minute, followed by a short, low-frequency moan. This sound is known to be made only by mature males during the spring and may be a proclamation of territory or breeding condition. Mating occurs mainly in the first three weeks of May in Alaska and the Canadian Arctic, and from the end of March to late May in the Barents Sea. Implantation of the blastocyst is delayed about 2 months, until July or early August. The total period of pregnancy is approximately 11 months. The pups are born in the open on ice floes from mid-March to early May. Peak birth periods are late April in most regions and early to mid-April in the Sea of Okhotsk and the southern Bering Sea. Most females become pregnant each year and enter estrus just before or after stopping nursing of the young conceived in the previous breeding season. The single pup is about 130 cm long, weighs 30–40 kg, and has a short, woolly, grayish brown coat, with a light colored face and four broad, transverse, light bands on the crown and the back. It is weaned after 12–18 days, by which time it has molted and become adultlike in color and has grown to about 85 kg. It is promptly abandoned by its mother at this time, but is already actively swimming and feeding on its own. Sexual maturity usually is attained at 5–6 years of age by females and 6–7 years by males. Maximum known longevity is 31 years.

Erignathus formerly occurred in substantial numbers along the Norwegian coast as far south as Trondheim, but now it only rarely wanders into that area (Smit and Wijngaarden 1981). Otherwise, there appears to have been little reduction in distribution, and commercial exploitation has been limited because of the nongregarious habits of the genus. Its meat and skin are of importance to local subsistence economies. The strong, durable, and elastic hide is used to make boot soles, heavy ropes, dog harness, and kayak covering. Formerly, the intestines were used to make waterproof clothing, and the blubber was used to obtain oil for lamps. About 10,000–13,000 individuals are taken annually by subsistence and commercial hunters in Norwegian, Russian, and U.S. waters (U.S. National Marine Fisheries Service 1987). Population estimates are around 150,000 in the Sea of Ok-

hotsk, 300,000 in the Bering and Chukchi seas, and 300,000 along the rest of northern Eurasia. There are no recent estimates of numbers or harvests for the eastern Canadian Arctic, though the population there was thought to be about 185,000 in the late 1950s.

PINNIPEDIA; PHOCIDAE; **Genus CYSTOPHORA**
Nilsson, 1820

Hooded Seal

The single species, *C. cristata*, occurs mainly along ice floes in the North Atlantic region, from Baffin Island and Newfoundland, through the waters around Greenland, to Svalbard and Iceland. There also are records of wandering individuals from as far west as northern Alaska, as far north as eastern Ellesmere Island, as far east as the Kara Sea and the Yenisey River, and as far south as southwestern Hudson Bay, Montreal, Florida, the British Isles, and Portugal. Recently there were reports of the most southerly known occurrences on both sides of the Atlantic: a young individual at Fort Lauderdale, Florida (Fletemeyer 1985), and a pregnant female at the mouth of the Guadalquivir River, in southern Spain (Ibáñez et al. 1988). The information for this account was taken mostly from J. E. King (1983), Kovacs and Lavigne (1986), and Reeves and Ling (1981); major exceptions are noted.

Adult males, which average slightly larger than females, are about 250–70 cm long and weigh 200–400 kg; adult females are about 200–220 cm long and weigh 145–300 kg. The background color is silvery or bluish gray, and there are numerous black or brownish black patches of irregular size and shape, ranging from 50 to 80 sq cm on the back and sides to much smaller on the neck and abdomen. The face is black to behind the eyes. There are large claws on both the hind and foreflippers. The cranium is comparatively short, and the snout is long and wide. In association with the development of the prominent fleshy proboscis, the frontonasal area of the skull is elevated and the nasal openings are wider than in any other phocid.

The most striking feature of this seal is the enlargement of the nasal cavity of the adult male to form an inflatable crest, or hood, on top of the head. The hood begins to develop when a male is about 4 years old and continues to grow with age and increasing body size. It is divided into two by the nasal septum and is lined by a continuation of the mucous membrane of the nose. Its skin is very elastic, and there are muscle fibers, particularly around the nostrils. When not inflated, the hood, which starts just behind the level of the eyes, hangs down in front of the mouth. When blown up, however, it forms a high cushion on top of the head; it then is large enough to hold about 6.3 liters of water.

The presence of an inflatable proboscis in males of both *Cystophora* and *Mirounga* sometimes has been considered indicative of close phylogenetic relationship but now generally is thought to be the result of convergent evolution. In *Cystophora* the hood is inflated with air, and the nostrils are closed before inflation. In *Mirounga* the proboscis is erected partly by muscular action and blood pressure, assisted by inflation, and the nostrils remain open. In addition, *Cystophora* has the ability to blow a bright red balloon-shaped structure from one nostril (usually the left). This balloon, or bladder, is an inflation of the highly elastic mucous nasal septum and may reach at least the size of an ostrich egg. It is produced when an animal closes one nostril, fills the hood on that side with air, and then uses that air to force the nasal septum out of the other nostril. The hood and balloon may be used as a threat, as they have been observed during the mat-

Hooded seal *(Cystophora cristata):* A. Adult male, photo by Bernhard Grzimek; B. Bladder nose fully inflated while swimming; C. Adult male with hood inflated; D. Young in "blueback" stage. Photos by Erna Mohr.

ing season and when *Cystophora* is disturbed, but they also sometimes are inflated when an animal is lying quietly on the ice. Females also have a hood, but it is not well developed and not inflatable.

The hooded seal is found in the same general region as the harp seal *(Phoca groenlandica)* but tends to stay farther from shore and to occupy deeper water and thicker, drifting ice. It rarely hauls out on land or on shorefast ice, except in the Gulf of St. Lawrence. It follows a regular annual cycle of migration and dispersal. In February the adults concentrate in widely scattered groups on the thick ice of three areas: (1) off the east coast of Labrador and Newfoundland, and to a lesser extent in the Gulf of St. Lawrence; (2) in Davis Strait, between Canada and Greenland; and (3) around Jan Mayen Island in the Norwegian Sea east of Greenland. Following the breeding season in these areas, there is a period of dispersal for foraging, and then the adults from the first two concentrations move into the Denmark Strait, between Greenland and Iceland, for purposes of molting. The animals from the Jan Mayen area apparently molt at a point farther north along the east coast of Greenland. This molting phase occurs from June to early August, and the seals then disperse to little-known feeding waters until the next breeding season. Some individuals move to the south and then up the west coast of Greenland. Others spread to the vicinity of Svalbard and Bear Island. Available evidence indicates that there is much mixing of the breeding stocks, and Wiig and Lie (1984) found no significant differences between the skulls of the three groups.

Cystophora evidently does not feed while on the ice for breeding and molting. At other times it may move well out to sea to forage and is capable of diving to great depths. A one-month-old individual descended to 75 meters on its first re-

corded dive, and adults probably can reach depths greater than 180 meters. The diet consists mainly of cephalopods, shrimp, mussels, and fish such as halibut, cod, redfish, capelin, herring, and flounder.

Cystophora is solitary for most of the year, but during the breeding season females come together in favorable habitat to give birth, with individuals spaced about 50 meters apart on the ice. Based on aerial observations of sexual distribution at this time, Boness, Bowen, and Oftedal (1988) suggested a polygynous mating system, with one adult male usually associated with a cluster of two to five females. Prior to such association, however, up to seven males at a time compete fiercely for access to each female or cluster. Eventually, one large old male will chase the others away, either to the periphery of the ice floe or into the water. He then stays near the female while she nurses her newborn and is highly aggressive toward intruders, both people and other seals. Probably, however, he merely is defending his right to mate with the female when she enters estrus and is not primarily interested in protecting her or her newborn. After mating, he may depart to look for other available females. During this period the males frequently inflate their hood and nasal balloon, but known vocalizations are limited to a few low-frequency pulses.

Mating takes place in the water and after lactation ceases. Nearly all adult females become pregnant each year. Implantation of the blastocyst is delayed about 4 months, and the total gestation period is almost a full year. The young are born mainly from mid- to late March, occasionally in early April, and usually in the middle of a large ice floe. Births occur at about the same time in all three breeding areas. The newborn averages about 100 cm in length and 20 kg in weight and can swim and move about almost immediately. Its beautiful furry coat is silvery blue-gray above, darker on the face, and creamy white ventrally. The period of lactation usually has been reported as 5–12 days, but Bowen, Oftedal, and Boness (1985) found that pups were weaned after an average of only 4 days. This period, during which the pup grows to around 47 kg, is the shortest known for any mammal. Its significance may be related to a need to take rapid advantage of the interval between the time of heavy ice, when polar bears might have easy access to the young, and the final breakup of the ice. Bowen, Boness, and Oftedal (1987) found that females lose approximately 16 percent of their own weight during this period. They then abandon their young and return to the sea to forage. The pups remain on the ice until mid- to late April, not feeding, and lose about 13 kg. They then enter the water and are capable of fending for themselves. Most females reach sexual maturity at around 3 years of age. Males are sexually mature at 4–6 years but probably cannot compete successfully for mating until they are 13 years old. Maximum known longevity is 35 years.

Native people of the Greenland coast long have taken the hooded seal for its meat and skin. Commercial exploitation of the breeding aggregation around Jan Mayen Island started in the eighteenth century. Regular annual hunts on the ice off Newfoundland began at least 150 years ago and were carried out in conjunction with the killing of the more abundant harp seal. Harvests of the two species were not differentiated in the statistics until 1895, but the total take of harp and hooded seals was around 500,000 animals annually from 1820 to 1860, and subsequently declined. Before 1930 *Cystophora* was sought mainly to obtain oil and leather, and so both adults and pups were taken. Since then the beautiful pelt of the newborn has been the primary objective.

According to Wiig and Lie (1984), the annual kill of *Cystophora* averaged about 40,000 east of Greenland and 20,000 off Newfoundland around the turn of the century, and 24,000 at Jan Mayen and 12,000 off Newfoundland in the 1970s. Since 1971 there has been increasing regulation of the kill through international agreements and quotas. There has been concern about the hunt, both for humanitarian reasons and because of its possible adverse effects on overall popula-

tions. Numbers apparently declined during the first part of the twentieth century, partially recovered when hunting ceased during World War II, and then declined again in response to heavy kills in the 1950s and 1960s. J. E. King (1983) cited the following numerical estimates for, respectively, total population size and number of pups born annually: Newfoundland, 100,000 and 30,000; Davis Strait, 40,000 and 11,300; Jan Mayen, 250,000 and 50,000. Studies by Bowen, Myers, and Hay (1987) indicate larger population sizes, however, based on an estimate of annual pup production of 62,400 for Newfoundland and 19,000 for Davis Strait. They suggested that numbers may have increased recently and that the commercial kill dropped sharply because of poor markets for pelts.

Campbell (1987) provided total current population estimates of about 300,000 for Newfoundland and 90,000 for Davis Strait. He thought that these populations were at or above historic levels and that they would not be adversely affected by properly managed hunting. However, the market for seal pelts was largely eliminated in the late 1980s through an import ban by the European Community (the United States already had banned such trade through the Marine Mammal Protection Act of 1972). A kill of about 5,000 hooded seals annually, supposedly for subsistence purposes, continues off eastern Greenland.

PINNIPEDIA; PHOCIDAE; Genus HALICHOERUS
Nilsson, 1820

Gray Seal

The single species, *H. grypus*, occurs in temperate and subarctic waters of three regions of the North Atlantic: (1) along the coasts of Labrador, Newfoundland, Nova Scotia, and New England and in the Gulf of St. Lawrence and on associated islands as far south as Nantucket; (2) around Iceland, the British Isles and many associated islands, and Brittany and along the coasts of Norway and the Kola Peninsula; and (3) in the Baltic Sea, the Gulf of Bothnia, and the Gulf of Finland.

Gray seals *(Halichoerus grypus)*, male and female, photo by Lothar Schlawe.

Individuals occasionally reach New Jersey, Svalbard, the continental coast of the North Sea, and Portugal. Except as noted, the information for this account was taken from Bonner (1981b) and J. E. King (1983).

Halichoerus shows more sexual dimorphism than does any other phocid except *Mirounga*. Adult males, which may be up to three times the size of females, are 195–230 cm long and weigh 170–310 kg; adult females are 165–95 cm long and weigh 105–86 kg. Coloration varies widely, and all shades of gray, brown, black, and silver may be found. Both sexes are darker dorsally and lighter ventrally, with varying degrees of spotting, but in males the darker tone forms a continuous background upon which there may be irregular spotting, while in females the lighter tone forms the background and is marked with spots and patches of a darker color. Some males are almost completely black, and some females are predominantly creamy white with only a few scattered dark markings on the back.

Adult males have massive shoulders, and the skin there and on the chest is much scarred and thrown into heavy folds and wrinkles. They also have an elongated snout with a convex outline above the broad muzzle, which results in an equine appearance or "Roman nose." Adult females have a straight profile to the dorsal surface of the head. Miller and Boness (1979) suggested that the prominent snout of the males is used as a visual signal to communicate status; it is displayed prominently during short-range agonistic encounters in the breeding season. The massive skull of *Halichoerus* shows the elevated frontonasal area and large narial openings also found in *Cystophora*, but the cheek teeth are larger and stronger.

The gray seal generally breeds on rocky or cliff-backed coasts or on small offshore islands and is most often observed foraging in waters adjacent to such areas. It is also seen frequently in estuaries, particularly when salmon are present. Certain breeding colonies are present on open sandy beaches, most notably on Sable Island, a sand bar about 40 km long and 1.5 km wide east of Nova Scotia. Some breeding also occurs in sea caves. Bonner (1972) noted that North Rona Island, off northern Scotland, offers no beaches sheltered from oceanic swells, and so *Halichoerus* moved above the rocky shore to vegetated slopes for breeding purposes. In the Baltic Sea and the Gulf of St. Lawrence breeding usually takes place on drifting or shorefast ice.

There are no regular migrations, but young animals disperse widely. A pup tagged on Sable Island was found 25 days later on the coast of New Jersey, 1,280 km away. Pups from the Farne Islands sometimes appear in Norway, Denmark, and the Netherlands. Such individuals probably return to their natal beaches during the breeding season. Adults may forage in coastal waters after breeding but then come ashore again to molt, generally not at the same sites used for breeding. Molting occurs from around January to March in the eastern Atlantic, in April and May on the Baltic ice, and in May in the western Atlantic. The seals eat little or nothing during molting and generally fast for several weeks while ashore to breed. When they do seek prey, they forage at various depths down to at least 70 meters. The diet includes octopus, squid, and crustaceans but consists largely of whatever fish are most abundant in coastal waters. Salmon, cod, herring, skates, mackeral, and flounder are all of importance.

Aggregations of *Halichoerus* form ashore during the molting and breeding seasons. About a month before the young are born in the British Isles, large numbers of pregnant females and males of all ages assemble on or near the breeding beaches. The largest bulls tend to keep apart from the others. The birth of the first pups seems to signal the start of male territorial behavior. They compete for locations, generally inland, where there is access to the most females. Serious fighting is minimal, but through challenges and ritualized displays subordinate individuals are forced to the periphery of the breeding grounds or back into the water. If females are few in number and widely spaced, each male may defend only a single female. On crowded beaches, however, dominant bulls may be only 3–4 meters apart, and each may attempt to control 6–7 females. The area dominated varies from day to day; there is no fixed territory. The males remain ashore at least 18 days, often much longer, and attempt to approach all available estrous females. Mating occurs on land or in the water. A female usually mates with several males, and several times with each.

The timing of the breeding season varies widely between the three major populations and between individuals within the colonies. In the British Isles the young are born generally from September to December but occasionally from March to May. The birth peak is in September in Wales, in October in the Orkney Islands, in November in the Farne Islands, and from late December to early January in the Scroby Islands. In Iceland and Norway the peak also occurs in October, but in the Baltic breeding takes place on the ice in late February and early March. In the western Atlantic most births are from mid-January to mid-February.

Soon after coming ashore, females give birth to the young conceived during the previous breeding season. They enter estrus and mate within 3 weeks, at the end of the nursing period. The fertilized egg develops to the blastocyst stage after 8–10 days, implantation is then delayed 100 days, and active gestation lasts 240 days; the total period of pregnancy is thus about 11.5 months. There normally is a single pup, though Spotte (1982) recorded a number of twin births. The newborn averages 76 cm in length and 14 kg in weight and is covered with creamy white and rather silky fur. Lactation stops after 16–21 days, by which time the pup weighs around 50 kg and has acquired adultlike pelage. It is capable of swimming at birth but usually does not enter the water until its first molt is complete. The female may remain constantly with the pup until weaning or may go off to forage and then reestablish contact upon her return through vocalization and olfaction. She loses about a quarter of her own weight during lactation, then mates and quickly abandons her young. The latter stays ashore for another 2–4 weeks but then goes to sea and can provide for itself. Females usually have their first young at 4–5 years but continue to grow until they are nearly 15 years old. Males reach sexual maturity at about 6 years but do not usually appear on the breeding grounds until 8 years and continue to grow until age 11. Most breeding bulls are 12–18 years old. According to McLaren (1984), maximum known longevity is 46 years.

Halichoerus is of little commercial use but is persecuted extensively because of its alleged competition with valuable fisheries. There have been bounties and government control operations in Canada, but the population there, especially the colony on Sable Island, is increasing. According to Bonner (1982), gray seal numbers had been reduced greatly in the British Isles by 1900. Subsequently it received protection during part of the breeding season, and by the 1920s it had increased substantially in numbers and reportedly was causing problems for fishermen. Controversy has continued to the present, pitting economic interests, especially in Scotland, against seal preservationists. The closed season now has been extended generally to cover the entire breeding period, but provision has been made for some culling. Numbers more than doubled between the 1960s and the 1980s. There now are about 70,000 animals in the British Isles, most of them in island groups off northern Scotland. There are another 20,000 individuals throughout the remainder of the eastern Atlantic and about 30,000 in Canada. The only U.S. breeding colony, that at Nantucket and Muskeget islands, Massachusetts, con-

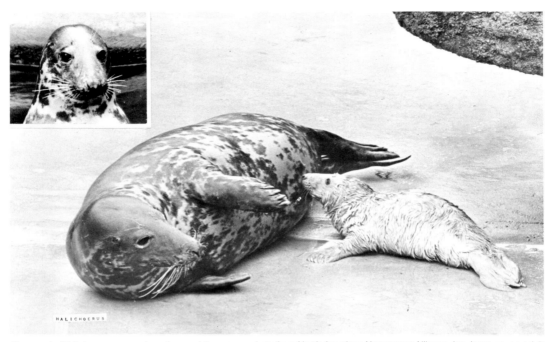

Gray seals *(Halichoerus grypus)*, mother suckling young, photo from North American Newspaper Alliance, Inc. Inset: young adult *(H. grypus)*, photo by Erna Mohr.

sists of about 20 seals. The Baltic subspecies, *H. g. macrorhynchus*, has been reduced drastically through bounties, water pollution, erosion of rookeries, and drowning of young in the wakes of ships. Numbers have fallen from over 100,000 a century ago to fewer than 1,500 (Wachtel 1986). This subspecies is classified as endangered by the Soviet Union.

PINNIPEDIA; PHOCIDAE; **Genus PHOCA**
Linnaeus, 1758

Harbor, Ringed, Harp, and Ribbon Seals

There are four subgenera and seven species (Bigg 1981; Burns and Fay 1970; G. S. Jones 1984; J. E. King 1983; Rice 1977; Shaughnessy and Fay 1977; Zhou 1986):

subgenus *Pagophilus* Gray, 1844

P. groenlandica (harp seal), primarily in coastal waters and pack ice from Hudson Bay and the Gulf of St. Lawrence across the Arctic to northwestern Siberia, occasional wanderers as far as the Mackenzie River Delta, Virginia, Iceland, the British Isles, and the Elbe River;

subgenus *Pusa* Scopoli, 1777

P. hispida (ringed seal), primarily in Arctic Ocean and adjoining seas, Sea of Japan, Sea of Okhotsk, Bering Sea, Hudson and James bays, Baltic Sea, Gulf of Bothnia, and several inland lakes in southern Finland and adjacent parts of the Soviet Union, occasional wanderers as far as northeastern China, the Pribilof Islands, Newfoundland, the British Isles, France, Germany, and Portugal;

P. sibirica (Baikal seal), Lake Baikal, a body of fresh water in south-central Siberia;

P. caspica (Caspian seal), Caspian Sea;

subgenus *Phoca* Linnaeus, 1758

P. largha (spotted seal, or larga seal), regularly in coastal waters and ice floes from northeastern Siberia and northern Yukon to northeastern China, occasional wanderers have reached southeastern China and possibly Taiwan;

P. vitulina (harbor seal, or common seal), coastal waters from Hokkaido around the North Pacific to Baja California, from northwestern Greenland and Hudson Bay to South Carolina, and from Iceland and northern Norway to the Baltic Sea and Portugal, and in several inland lakes near Hudson Bay;

subgenus *Histriophoca* Gill, 1873

P. fasciata (ribbon seal), mainly in pack ice from Hokkaido and the Sea of Okhotsk to northern Alaska and the Aleutian Islands.

Seals of the genus *Phoca* are the most abundant pinnipeds of the Northern Hemisphere. They are mostly small and show little sexual dimorphism in size but vary greatly in coloration. From *Halichoerus* they are distinguished by having a much smaller snout, smaller nasal openings, and more complex cheek teeth. From *Erignathus* they are distinguished by having a relatively longer and narrower jugal, having the first and second digits of the foreflipper longer (rather than shorter) than the third, and having only two mammae in the female (Hall 1981). Data compiled by Burns and Fay (1970) suggest that *Phoca* may still be in the process of differentiation and that the various subgenera—*Pagophilus, Pusa, Phoca,* and *Histriophoca*—may eventually reach generic level.

Some authorities have treated them as such (see Wyss 1988b), but Burns and Fay pointed out that there is no consistent cranial difference between them. *P. groenlandica* appears to be the most primitive species in the group, and *P. fasciata*, the most divergent from the ancestral stock. Additional information is provided separately for each species.

Phoca groenlandica (harp seal).

Except as noted, the information for the account of this species was taken from J. E. King (1983) and Ronald and Healey (1981). Adult males are 171–90 cm long and average 135 kg in weight; adult females are 168–83 cm long and average 120 kg. Coloration changes a number of times with age and also varies considerably in adults. Generally, adult males are silvery gray, with the head black to just behind the eyes and a black band running up each flank, the bands joining across the front of the back to form the shape of a horseshoe or harp. In adult females the facial and dorsal markings are paler and may be broken up into spots.

This species is divided into three major populations, one breeding in the Gulf of St. Lawrence and on the ice off eastern Newfoundland, one breeding to the north of Jan Mayen Island in the Greenland Sea, and one breeding in the White Sea. All three undergo an annual north-south migration of 6,000–8,000 km. The northwest Atlantic population spends the summer feeding in Hudson Bay and the waters off Baffin Island, northwestern Greenland, and northern Labrador. It starts south in September and by January has entered the breeding areas. After breeding, the adults molt in early May and then head back north. The Jan Mayen population summers between Svalbard and Greenland, moves south to the breeding area in late winter, molts on the ice north of Jan Mayen in April, and returns north by mid-May. The White Sea group summers mainly in the Kara and Barents seas close

to the ice edge, moves south in the fall, and rapidly enters the White Sea in January and early February. The adults molt from mid-April to late May and then migrate north out of the White Sea. The pups of this population are carried by drifting ice into the Barents Sea, where they begin to feed independently. The young of all populations generally do not move south until after the breeding season, and they join the adults to molt in the spring. Some young remain in the north throughout the year.

The harp seal is dependent on ice for breeding and molting and also may be found near the pack ice while foraging. It prefers rough, hummocky ice at least 0.25 meter thick and penetrates deep into large floes by following leads or channels. It maintains natural holes 60–90 cm in diameter for purposes of access to the water and breathing. It is capable of moving quickly on the ice and also is a powerful, high-speed swimmer. While migrating, it sometimes leaps from the water like a dolphin. Vision and hearing are thought to be acute, especially underwater, but not the sense of smell. Dives may reach a depth of 150–200 meters. The diet consists mainly of pelagic fish, especially capelin, and also pelagic and benthic crustaceans.

This species is gregarious, and a number of individuals may join in pursuing schooling fish. Tens of thousands come together in the spring molting aggregations, and up to 40 seals may share a breathing hole. The varied vocalizations include trills, clicks, and numerous birdlike sounds; they are heard largely during the breeding season. At that time the females form groups on the ice, spacing themselves about 1.5–2.5 meters apart. The males fight for access to the females, using teeth and flippers. Seasonal monogamy is thought to be the practice, and mating occurs on the ice.

A few days after hauling out on the ice, the females give birth to the young conceived the year before. They enter

Harp seal *(Phoca groenlandica)*: A. Adult female; B. Two-year-old young. Photos by A. Pedersen.

estrus and mate just before weaning the young. Implantation of the blastocyst is delayed about 4 months, and active gestation then lasts another 7.5 months. Births occur from late January to early April, mostly from late February to mid-March. The single pup is about 100 cm long at birth, weighs about 12 kg, and within 2–3 days develops a beautiful coat of silky white fur. It is weaned after 10–12 days and then is abandoned by the mother, which has fed little, if at all, during lactation. The pup's weight increases to about 33 kg at weaning but drops to 27 kg over the next 2 weeks as it remains on the ice and molts into a grayish pelage. It then enters the water and begins to forage, initially staying near the surface and taking small crustaceans. Both sexes become fertile at around 5.5 years, but females may not begin to produce young on an annual basis for several more years, and males are not able to actively compete for mating privileges until about 8 years. Both sexes remain sexually active through their twenties and may live until the early thirties.

Each of the three populations is thought originally to have numbered about 3 million individuals, and thus *P. groenlandica* would have been the second most abundant pinniped after *Lobodon carcinophagus*. Unlike the latter, however, the harp seal forms large seasonal concentrations in accessible areas and thus can be practically and intensively exploited by humans. It long has been taken by native peoples of the North as a source of food and fiber, and about 10,000 individuals per year still are killed for this purpose (Ronald, Selley, and Healey 1982). Commercial hunting was begun in the sixteenth century by Basque whalers off Newfoundland. The first recorded expedition specifically seeking the harp seal was to Jan Mayen in 1720. Large-scale hunts have been carried out on both sides of the Atlantic in most years since the early nineteenth century (Bonner 1982*a*; Busch 1985; Sergeant 1976). At first, seals of all ages were sought for skins and oil, but in recent years the main target has been the valuable white pelt of pups aged 2–10 days. Perhaps 50 million harp seals have been taken commercially in the nineteenth and twentieth centuries, far more than any other pinniped. The cost in human life also has been high, with about 1,000 men and 400 sailing vessels lost in the ice from 1800 to 1865, prior to the regular use of stronger, steam-powered craft to reach the seal concentrations. In 1857 nearly 400 vessels and 13,000 men participated in the hunt in the western Atlantic. The industry has been dominated by Norway since the 1880s, though many Canadians have been hired to do the hunting and processing. The Soviet Union began pelagic sealing with icebreakers in the White Sea in the 1920s and has been responsible for nearly the entire take there since the 1940s.

The Jan Mayen population was greatly reduced by hunting in the nineteenth century, and the kill there has been well under 50,000 in most years since 1890. As a result, the White Sea population came under increasing pressure, with the annual harvest there peaking at over 450,000 seals in 1924. The maximum recorded kill in the western Atlantic was around 687,000 in 1831. The annual take in each region was about 100,000–200,000 in the 1930s, but populations evidently then were declining. A cessation of hunting during World War II allowed some recovery, perhaps to near original numbers in the western Atlantic. After the war, sealing resumed, with the western kill alone reaching 456,000 in 1951 and averaging 316,000 per year from then until 1960. Subsequent evidence of declining populations, along with public protests in Europe and America on both conservation and humanitarian grounds, led to governmental restrictions on harp sealing. A regulatory and quota system was established in 1971 through an international agreement between Canada, Norway, and the Soviet Union. The annual kill quota has fallen steadily and was set between 170,000 and 186,000, including 100,000 pups, from 1977 to 1983. In response to public concern, the European Community in October 1983 placed a two-year ban on imports of products of harp and hooded seals (U.S. National Marine Fisheries Service 1984). This ban subsequently was extended. The United States, through the Marine Mammal Protection Act of 1972, already had prohibited importation. Therefore, commercial marketing and large-scale killing of the harp seal has ceased.

Recent numerical estimates place the Jan Mayen population at only 100,000–150,000 individuals, the White Sea population at 500,000–1 million, and the western Atlantic herds at 1–2 million. There is controversy regarding the future outlook. On the one hand, Roff and Bowen (1986) reported that the annual pup production in the western Atlantic, at around 480,000, is far in excess of what would be necessary to sustain the population, even if commercial kill quotas were being met, and thus a rapid increase is likely. On the other hand, Bonner (1982*a*) pointed out that recent intensification of the capelin fishery had depleted that important food source of the harp seal, with possibly ominous consequences.

Phoca hispida (ringed seal).

Except as noted, the information for the account of this species was taken from Frost and Lowry (1981) and J. E. King (1983). Adults in Alaskan waters average 115 cm in length and 49 kg in weight, with respective reported maximums of 168 cm and 113 kg. Seals of the Baltic population tend to be larger than those elsewhere, with males over 10 years old averaging 141 cm in length and weighing 71–128 kg, and females 137 cm and 61–142 kg (Helle 1980). Coloration is variable, but the upper parts are commonly gray, heavily spotted with black. The spots may merge to form large dark areas, but many of them are surrounded by a ring of lighter color.

The ringed seal is found in seasonally or permanently ice-covered waters and makes annual movements in relation to the advance and retreat of the ice. Highest densities of breeding adults are on and under stable landfast ice. Nonbreeders occur mainly in the flow zone and in association with moving pack ice. In the summer, all age and sex classes concentrate at the edge of the permanent arctic pack ice or on inshore ice remnants. Individuals, however, may be found wherever there is open water in the ice, even as far as the North Pole. In the Sea of Okhotsk, spring breeding occurs on the pack ice, but the seals must haul out on shore during the summer. Molting occurs from March to July, with a peak in June. Foraging in the water then slows but does not cease altogether. During the early spring, late summer, and fall the seals spend most of their time swimming among ice floes and feeding.

As the ice thickens during the late fall and winter, the seals maintain openings for breathing and access by abrading the ice with the claws of their foreflippers. Breathing holes, which can be kept open through ice over 2 meters thick, are cone-shaped, with the small end up. Rather than construct their own openings, the seals maintain and expand natural ones, which tend to occur along cracks. Freezing pressure may then force the ice up into ridges and heaps in the vicinity of the holes, and snow may accumulate around the uplifts and over the holes. The seals take advantage of the situation by coming up through the holes and hollowing out lairs in the snow. There may be several such dens in a single snowdrift. Some are used simply for resting by a single individual and usually have one oval chamber, about 30 cm high and 120 cm in diameter. Others are larger, have several chambers, and are used by a number of seals (Smith and Stirling 1975). Pregnant females construct more elaborate subnivean lairs, where they can give birth and nurse their young. The pup then sometimes digs additional tunnels into the surrounding snow. Since *P. hispida* has a longer lactation period than that

Ringed seal *(Phoca hispida)*, photo by Lothar Schlawe.

of most phocines, the snow-covered lair may provide necessary insulation while the pup builds up fat reserves. The lair also serves as a resting site for mother and young and may provide some protection against predators. On Lake Saimaa, Finland, birth lairs are constructed right along the shoreline and usually are partly over ground, rather than ice (Helle, Hyvarinen, and Sipila 1984). In the Sea of Okhotsk the young are born on shorefast ice, but not in lairs.

P. hispida will not use a breathing hole that has been at all disturbed and is very cautious when coming up onto open ice. It first uses its apparently excellent senses of sight, hearing, and smell to inspect the vicinity of the access hole. It evidently can detect a human at least 200 meters away by scent alone (Smith and Hammill 1981). While feeding in the water, *P. hispida* generally alternates periods of 12–130 seconds at the surface with dives of 18–720 seconds. However, it has been known to remain underwater for up to 43 minutes under natural conditions, and a forced dive in captivity lasted 68 minutes. It may reach depths as great as 91 meters while searching for food. The bulk of the diet is made up of fishes of the cod family (usually less than 20 cm long but occasionally up to 127 cm), pelagic amphipods, euphausiids (krill), shrimp, and other crustaceans.

P. hispida is abundant but not gregarious and is widely dispersed throughout the Arctic. Recorded population densities average around 1/sq km in favorable areas but are as high as 15/sq km in June on the fast ice off Baffin Island. During the nonbreeding season, seals are usually seen alone, occasionally in small groups, though herds of thousands once were reported in the Sea of Okhotsk. Several animals sometimes share a breathing or access hole and the associated ice surface or lair. A dominance hierarchy probably is established in such aggregations. According to Smith and Hammill (1981), the seals are then aggressive toward one another, slapping with their foreflippers, splashing water, and lunging and biting with the mouth. Overall social structure appears comparable to that of *Leptonychotes*. Adult males probably exhibit limited polygamy, establishing large underwater territories that comprise the areas and lairs used by several females. Four underwater vocalizations have been identified: high- and low-pitched barks, a yelp, and a chirp; they probably are involved in reproductive activity and in maintaining social structure around access holes (Calvert and Stirling 1985; Stirling 1973).

Mating takes place in the water, mostly in late April and early May, about a month after parturition. Implantation of the blastocyst is delayed for about 3.5 months, and the total gestation period is 11 months. Most births occur from mid-March to mid-April, but the peak is in early May on Lakes Saimaa and Ladoga. Twins have been reported, though normally there is a single pup. It averages 65 cm in length and 4.5 kg in weight and is covered with woolly white fur considerably longer and finer than that of young ribbon and spotted seals. It begins to shed this coat at 2–3 weeks and attains an adultlike pelage by 6–8 weeks. Nursing lasts 5–7 weeks in most areas but only 3 weeks in the Sea of Okhotsk, where births usually occur on drifting ice. The pups are abandoned after weaning, when they weigh 9–12 kg and the ice begins to break up. Growth continues for 8–10 years. Sexual maturity is reached at about 5–7 years, but the maximum annual pregnancy rate of around 78 percent is not achieved until females are over 10 years old. Maximum known longevity is 43 years. The major natural sources of mortality are predation on pups by the arctic fox, which evidently has little difficulty in locating and penetrating the subnivean lairs, and predation on older seals by the polar bear.

Many thousands of ringed seals are killed annually by native peoples for skins, furs, oil, and meat. Some commercial hunting takes place, but since *P. hispida* does not form large breeding aggregations, there have been no massive harvests and numerical declines comparable to those suffered by *P. groenlandica*. While there are no accurate population estimates, *P. hispida* is generally considered to be the most common northern pinniped, with perhaps 6–7 million individuals present in the Arctic Ocean and major adjoining bodies of water. The subspecies *P. hispida ladogensis*, in Lake Ladoga in the Soviet Union, now numbers about 5,000 individuals and is protected. However, the subspecies *P. h. saimensis*, in Lake Saimaa, Finland, has declined sharply because of the entanglement and drowning of young seals in fishing gear; only 100–120 individuals survive (Helle, Hyvarinen, and Sipila 1984). *P. h. saimensis* is classified as endangered by the IUCN.

The Baltic subspecies, *P. h. botnica*, formerly was com-

Baikal seal *(Phoca sibirica)*, photo by Eugene Maliniak.

mon, with up to 12,000 taken annually by hunters in the early 1900s. Excessive harvests and bounties subsequently led to greatly reduced numbers, and the total population now is only 10,000. The decline is continuing, evidently in connection with a reduction in the annual pregnancy rate to only 28 percent in mature females, which in turn is probably the result of water pollution (Helle 1980).

Phoca sibirica (Baikal seal).

The information for the account of this species was taken from J. E. King (1983) and Thomas et al. (1982). Adults are 110–42 cm long and weigh 50–130 kg. The color is dark brownish gray, shading to a lighter yellowish gray ventrally. The fur is dense and unspotted. The forelimbs and foreclaws are larger and stronger than those of *P. hispida* and *P. caspica*.

This seal is the only pinniped restricted to fresh water. It is confined almost entirely to Lake Baikal, though individuals are seen on occasion in connecting rivers. Seasonal movements and activities are determined primarily by ice conditions. During the winter the lake is completely covered by ice averaging 80–90 cm thick. At this time the seals may be found throughout the lake, mainly in the water and at breathing and access holes through the ice. They probably keep these holes open by abrading the ice with their strong foreclaws, but they may also use their head, teeth, and rear flippers. Immature seals generally utilize one hole, whereas adults have a primary hole and several auxiliary openings. In the late winter or early spring, pregnant females come onto the ice and make a subnivean lair for giving birth (see account of *P. hispida*). Around 1 April all seals begin to concentrate to feed along new natural openings in the ice, and in May they move toward the north end of the lake as the ice breaks up. The annual molt occurs from late May to early June. In the summer the seals concentrate to the southeast, using the shore and offshore rocks for hauling out. In the fall they again move to areas where the ice is forming.

Observations suggest that surface resting peaks in the winter between 1300 and 1700 hours, and during the summer at 1100 and 1800 hours, and that most foraging year-round is done at twilight and at night. Based on the seal's anatomy, the maximum natural diving time has been estimated at 20–25 minutes, but frightened captives have remained underwater for 2–3 times as long. The diet consists largely of noncommercially valuable fish, such as golomyanka, which reportedly move to depths of 20–180 meters at night, and also includes some invertebrates.

Although *P. sibirica* is basically solitary, a number of individuals may share access holes, and large aggregations form in areas of favorable habitat during certain times of the year. The spring feeding assemblies of 200–500 animals are begun by juveniles, which are then joined in turn by the adult males, pups of the year, and adult females. Other large groups form on shore during the summer. During the breeding season adult males are apparently polygynous. It is not known whether they are territorial, and available evidence suggests that they do not physically fight one another.

Mating is thought to occur underwater in May, at about the time of weaning. There apparently is a short period of delayed implantation, and total gestation lasts just over 9 months. The young are born on the ice from mid-February through March. There commonly is a single pup, but the twinning rate, about 4 percent, is unusually high for pinnipeds. Moreover, both twins often survive to weaning and then remain together for a while. The newborn is about 70 cm long and 3 kg in weight and is covered with a long, white, woolly coat. This pelage persists for about 6 weeks but then is replaced during the next 2 weeks with an adultlike coat. Lactation is unusually long, about 2.0–2.5 months. Males reach sexual maturity by 7 years. Most females are breeding by age 6 and can continue to age 30, with about 88 percent producing pups each spring. Maximum reported longevity is 56 years.

The Baikal seal is thought to have recovered from excessive

hunting that seriously reduced numbers in the 1930s. The total population now is estimated at about 70,000 individuals. There is a regulated annual commercial kill of 5,000–6,000 for pelts, oil, and meat.

Phoca caspica (Caspian seal).

Except as noted, the information for the account of this species was taken from J. E. King (1983). Adult males may reach a length of 150 cm and a weight of 86 kg. Females are slightly smaller, reaching about 140 cm. Both sexes are deep gray dorsally and grayish white ventrally. Males also generally have dark spots all over, while in females the spots are lighter and mainly on the back.

Annual movements are associated with the availability of ice in the Caspian Sea. In late fall the seals migrate into the northeastern part of the sea, where the water is shallower and freezes. The ice forms large floes that then smash against one another in the wind, creating upthrusts and rough hills. Here the seals maintain small breathing holes and larger exit holes; females give birth to their young on sheltered stretches of ice. Molting occurs in the spring, also in the north, but the ice then melts, the water becomes warm, and most of the seals migrate south to spend the summer in the deeper, cooler part of the sea. This shift also is associated with the abundance of prey, which includes mainly small fish and crustaceans.

P. caspica usually is found in pairs and may be monogamous. Mating occurs from late February to mid-March, about a month after the young conceived the previous year are born. The gestation period, which probably includes several months of delayed implantation, is about 11 months. Births take place from late January to early February. The newborn are about 64–79 cm long and 5 kg in weight and are covered with long, white fur that is replaced in about 3 weeks with short, dark gray hair. Lactation lasts about 1 month. Sexual maturity comes at 5–7 years. McLaren (1984) indicated that maximum known longevity is 50 years.

Total numbers are estimated at 500,000–600,000 animals and are considered reasonably stable. About 60,000 pups are taken each year for their skins.

Phoca largha (spotted seal, or larga seal).

The information for the account of this species was taken from Bigg (1981), J. E. King (1983), and McLaren (1984). Adult males are 150–70 cm long and weigh about 90 kg; females are 140–60 cm long and weigh about 80 kg. The coat is generally pale gray, darker dorsally, and sprinkled with small brown or black spots. Compared with that of *P. vitulina*, the skull of *P. largha* is small and fragile, and the teeth are set straight in the jaws.

This species is closely related to *P. hispida* but, unlike the latter, is strongly associated with ice. There is an annual migration in the fall and winter to the edge of the pack ice, where the seals haul out on floes. Breeding occurs during late winter and spring in eight areas of ice distributed from the Bering Sea to the northwestern Yellow Sea. After the breeding season, the seals remain on the ice to molt and then move back to the shoreline to spend the summer. They may then ascend rivers, possibly including the Yangtze. Adults are thought to be capable of diving to 300 meters and are known to take a wide variety of fish, cephalopods, and crustaceans. Newly weaned pups feed on small amphipods around ice floes.

Monogamy is evident during the breeding season. Pairs form about 10 days before the female gives birth to the young conceived the previous year and stay together for at least a month afterwards. Such family groups are separated by at least 0.25 km and may defend a territory around the natal ice floe. The reproductive cycle is thought to be similar to that of *P. vitulina*. Where the two species overlap in distribution, *P. largha* gives birth first, before the ice melts. The young are born as early as February in the southern parts of the range and from late March to mid-May in the Bering Sea. The pup

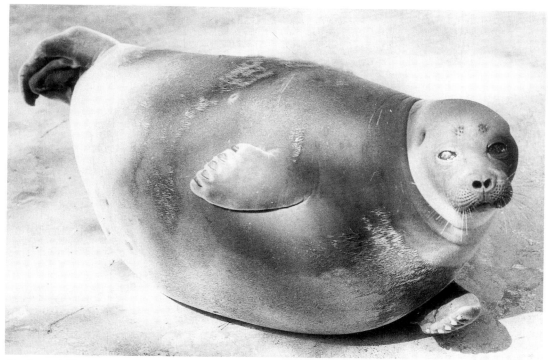

Caspian seal *(Phoca caspica)*, photo by Lothar Schlawe.

is about 85 cm long and has long white fur that is replaced by adultlike pelage after about a month. It cannot swim at birth, but by the time of weaning, a month later, its weight has tripled and it can dive to a depth of 80 meters. Sexual maturity is reached by 4–5 years of age, and maximum known longevity is 32 years.

P. largha has been reported to do some damage to commercial fisheries in the Sea of Okhotsk but apparently has not been extensively persecuted. Numbers are estimated at about 168,000 individuals in the Sea of Okhotsk and 200,000–250,000 in Alaskan waters.

Phoca vitulina (harbor seal, or common seal).

Except as noted, the information for the account of this species was taken from Bigg (1981) and J. E. King (1983). Adult males are about 150–200 cm long and weigh about 70–170 kg; females are about 120–50 cm long and weigh 50–150 kg. Coloration is variable, but there is generally a gray or brownish gray background, liberally sprinkled with small spots. In some populations the spots fuse, so that the entire dorsal region is dark, and in some cases there are pale rings around the spots. Compared with that of *P. groenlandica* and *P. his-*

Harbor seals, or common seals *(Phoca vitulina)*: A. Adult male, photo by Erna Mohr; B. Two females, photo from Miami Seaquarium; C. Immature seals, photo by A. Pedersen; D. Head, showing nostrils closed, photo by Erwin Kulzer.

pida, the skull of *P. vitulina* is more thickly boned and has a broader facial area.

The harbor seal is nonmigratory and makes only limited movements in association with foraging and breeding, though young have been known to disperse up to 300 km. This species, unlike most northern phocids, is not associated with ice. It lives mainly along shorelines and in estuaries, commonly resting on sandbanks, easily accessible beaches, reefs, and protected tidal rocks. It frequently ascends rivers for many kilometers and may remain year-round in freshwater lakes. Northerly populations tend to stay in areas where swift currents and tides maintain ice-free conditions in winter. *P. vitulina* is a strong swimmer and leaps completely out of the water (porpoising). Captive individuals have made dives as deep as 206 meters and lasting as long as 30 minutes, but the normal foraging time underwater is only 3 minutes. Newly weaned young feed mainly on shrimp and other small, bottom-dwelling crustaceans. Older animals take herring, anchovy, trout, smelt, cod, flounder, salmon, other fish, and octopus.

While this species usually is solitary, aggregations of several hundred may form ashore during the breeding season, and a number of individuals may also come together at a favorable hauling-out site. There then is much agonistic behavior involving biting, head butting, flipper waving, snorting, and growling. Sullivan (1982) reported that adults and juveniles form strong linear hierarchies based on size and sex, the adult males dominating all other animals. In a study of the largest breeding colony in eastern Canada, containing about 2,000 individuals, Godsell (1988) found no evidence of social cohesion other than that between mother and pup. Although *P. vitulina* generally is considered one of the least vocal of pinnipeds, Ralls, Fiorelli, and Gish (1985) found that an adult male made a variety of sounds and learned to mimic a number of human words and phrases.

Males are thought to be polygynous and may fight for mating privileges. Females enter estrus and mate shortly after weaning the young conceived during the previous breeding season. The fertilized egg develops to the blastocyst stage, and implantation then is delayed for 1.5–3.0 months. The peak of births varies widely by area, being in February in Baja California, March–April in California, May–June in Newfoundland and Nova Scotia, and June–July in Europe, the North Pacific, and the arctic North Atlantic. The single pup is 75–100 cm long, weighs 10–12 kg, usually has adultlike pel-

age, and is able to swim immediately. The mother is solicitous and will even take the pup in her mouth and dive if there is danger, but she promptly abandons it after weaning. Lactation lasts 3–6 weeks. Sexual maturity is attained at 3–5 years in females and about 6 years in males. About 90 percent of adult females bear a pup each year. Ohtaishi and Yoneda (1981) reported a 34-year-old individual.

P. vitulina is of little commercial value, though it is sometimes killed for its fur. It does come into conflict with commercial fisheries, especially in northwestern North America, the Sea of Okhotsk, and the Netherlands and therefore has been persecuted. As a result of excessive hunting, pollution, and disturbance, populations have declined in parts of Europe and along the coast of Asia. The subspecies in the latter region, *P. v. stejnegeri,* is classified as vulnerable by the IUCN. The Soviet Union regards *P. v. vitulina* of Europe as endangered. The U.S. National Marine Fisheries Service (1987) estimated numbers at 48,000–51,500 in Europe, 10,000–15,000 on the coast of eastern Asia, 260,000 in Alaska, 42,000 on the rest of the west coast of North America, and 30,000–45,000 in the western Atlantic. Heide-Jorgensen and Härkönen (1988) reported that populations in the waters between Norway, Sweden, and Denmark were recovering from past human persecution. Shortly thereafter, a severe epidemic, probably canine distemper, wiped out more than half of the seals in that region and adjacent areas and also spread to British waters (*Washington Post,* 30 August 1988, p. A-3).

Phoca fasciata (ribbon seal).

Except as noted, the information for the account of this species was taken from Burns (1981*a*) and J. E. King (1983). Adults of both sexes are about 150 cm long and weigh about 90 kg. The largest individual on record, a pregnant female, was 180 cm long and weighed 148 kg. Adult males are reddish brown before the molt and dark chocolate brown afterwards, with white or yellowish white bands encircling the neck, the hind end of the body, and each foreflipper. Females are paler with less distinct markings. Compared with other northern phocids, *P. fasciata* has a relatively longer and more flexible neck, a shorter rostrum, and more widely spaced teeth.

The ribbon seal is associated with sea ice during the late winter, spring, and early summer. At that time the species is concentrated in ice-covered parts of the Chukchi, Bering, and Okhotsk seas and adjacent straits and bays. It prefers to haul out on moderately thick, clean ice and depends on such for

Ribbon seal *(Phoca fasciata),* photo from San Diego Zoological Society.

giving birth and molting in the spring. It seldom hauls out on land and generally is found on ice floes well out to sea. The Bering Sea population evidently utilizes a remnant zone of ice in the middle of the sea that does not melt until mid-June. This population, like that of the Sea of Okhotsk, then apparently becomes completely pelagic for the summer and fall. Some individuals may wander a considerable distance, there being one record from central California and another from the central North Pacific south of the Aleutians (Stewart and Everett 1983). Possibly as an evolutionary consequence of its disassociation with the land, *P. fasciata* shows little caution when resting on ice and can easily be approached by humans or vessels. It can, however, move faster on ice than a person can sprint, using its foreflippers and body in a manner similar to that described for *Lobodon*. Its diet consists mainly of crustaceans, fish, and cephalopods.

The ribbon seal is usually solitary. Aggregations sometimes form on favorable ice floes, but individuals are well spaced. Males are apparently polygynous and do not stay long with any one female. Mating takes place at about the same time as weaning, and gestation, which probably includes a period of delayed implantation, lasts around 11 months. Births occur from 3 April to 10 May, mostly from 5 to 15 April. The single pup is about 86 cm long and 10.5 kg in weight and has a coat of long, white hair that is shed within 5 weeks. The mother is not especially solicitous, often goes off to forage, and promptly abandons her offspring after 3–4 weeks of lactation. The pup then is not a capable swimmer and remains mostly on the ice for a few more weeks, its weight falling from around 30 kg to 22 kg before it finally becomes proficient at foraging. Nearly all females reach sexual maturity and begin producing a pup annually by the time they are 4 years old. Males are sexually mature at 4–5 years. Potential longevity is probably about 30 years.

Because it is rarely found near land, *P. fasciata* is not regularly hunted by native peoples, but its lack of caution while on the ice makes it vulnerable to commercial pelagic sealers. In the 1960s the annual Soviet catch averaged about 13,000 animals, and overall populations began to fall. In 1969 the yearly quota was reduced to 3,000, and numbers rebounded. There now are estimated to be about 140,000 individuals in the Sea of Okhotsk and 100,000 in the Bering Sea.

Order Tubulidentata

TUBULIDENTATA; **Family ORYCTEROPODIDAE;**
Genus ORYCTEROPUS
E. Geoffroy St.-Hilaire, 1795

Aardvark, or Ant Bear

This order contains one family, Orycteropodidae, with the single living genus and species *Orycteropus afer,* which occurs throughout Africa, south of the Sahara, wherever suitable habitat is available (Meester, *in* Meester and Setzer 1977). Early Egyptian paintings suggest that this species also once ranged as far north as the Mediterranean (Kingdon 1971). The species *Plesiorycteropus madagascariensis,* which lived in Madagascar until about 1,000 years ago, formerly was considered to be an orycteropodid but now is thought to be an insectivore (R. D. E. Macphee, Duke University, pers. comm., 1988).

This animal resembles a medium-sized to large pig. Head and body length is 1,000–1,580 mm, tail length is 443–710 mm, and shoulder height is approximately 600–650 mm. Adults weigh from 40 to 100 kg, though most individuals weigh from about 50 to 70 kg. Males are slightly larger than females (Shoshani, Goldman, and Thewissen 1988). The thick skin is scantily covered with bristly hair that varies in color from dull brownish gray to dull yellowish gray. The hair on the legs is often darker than that on the body. Numerous vibrissae occur on the face around the muzzle and about the eyes. The dull pinkish gray skin is so tough that it sometimes saves the aardvark from the attacks of other animals.

The aardvark has a massive body, with a long head and snout. The round, blunt, piglike muzzle is pierced by circular nostrils, from which grow many curved whitish hairs 25–50 mm long. The tubular ears are 150–210 mm in length; they fold back to exclude dirt when the animal is burrowing and can be moved independently of one another. They are waxy and smooth, like scalded pig's skin. The tapering tongue often hangs out of the mouth, with the end coiled like a clock spring. The neck is short, the forequarters are low, and the back is arched. The strong and muscular tail is thick at the base and tapers to a point. The legs are short and stocky; the forefoot has four digits and the hind foot has five, the digits being webbed at the base. The long, straight, strong, blunt claws are suited to burrowing. The females have one pair of inguinal and one pair of abdominal mammae. In the males the penis has a fold of skin that covers scent glands at its base.

The teeth in the embryo are numerous and traversed by a number of parallel vertical pulp canals. The milk teeth do not break through the gums. In the adult the teeth are only in the posterior part of the jaw. The usual dental formula is: (i 0/0, c 0/0, pm 2/2, m 3/3) × 2 = 20; additional vestigial premolars

are sometimes present. The teeth do not grow simultaneously. Those nearest the front of the jaw develop first and fall out about the time the animal reaches maturity; they are succeeded by others farther back. The cheek teeth, which are covered externally by a layer of cement, each resemble a flat-crowned column and are composed of numerous hexagonal prisms of dentine surrounding tubular pulp cavities, hence the ordinal name Tubulidentata ("tubule-toothed"). The teeth of aardvarks grow continuously and lack enamel. The skull is elongate, and the lower jaw is straight, bladelike anteriorly, and swollen at the molars.

According to Smithers (1971), the aardvark occurs in a wide variety of habitats, including grassy plains, bush country, woodland, and savannah. In such areas it appears to prefer sandy soils. The presence of sufficient quantities of termites and ants is apparently the main factor in distribution. The word "aardvark" means "earth pig" in the Afrikaans language, an appropriate name, as *Orycteropus* looks somewhat like a pig and is an extraordinarily active burrower. If overtaken away from its den, it digs into the ground with amazing speed. It can dig faster in soft earth than can several persons with shovels, and even the hardest sun-baked ground is no obstacle to its powerful forefeet. When digging, it pushes the ground backward under its body while resting on its hind legs and tail. When a sufficient amount of soil has accumulated, the aardvark shoves it back or to one side with the hind feet, sometimes with the aid of the tail. *Orycteropus* digs shallow holes to search for food, burrows used for temporary shelter, and extensive tunnel systems in which the young are born (Smithers 1983). The temporary shelter is about 2–3 meters long and ends in a chamber large enough for the aardvark to turn in, as the animal generally enters and leaves its burrow headfirst. The more extensive tunnels may be 13 meters long, with numerous chambers and several entrances. A number of aardvarks may burrow in the same vicinity; there is one record of 60 entrances in an area approximately 300 by 100 meters. The burrows, when abandoned, are used by dozens of other kinds of animals. *Orycteropus* sometimes occupies a termite nest as a temporary shelter; its thick skin seems to be impervious to insect bites. It occasionally digs its holes in areas that seasonally flood and thus may have to evacuate at certain times of the year.

The aardvark tends to walk on its claws, and its tail often leaves a track on soft ground. It is an extremely powerful animal. In one case, a man with a firm grip on the tail of an aardvark in its den was slowly drawn into the burrow up to his waist and finally had to relinquish his hold, despite the additional leverage afforded by two other persons holding onto his legs. The aardvark's movements are awkward and slow, but the animal is said to escape with surprising rapidity if alarmed. Its hearing is acute, and at the least alarm it seeks a

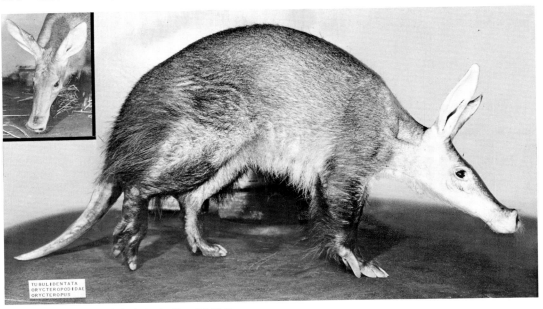

Aardvark *(Orycteropus afer)*, photos by Ernest P. Walker.

burrow. Its eyesight does not appear to be good, since the aardvark frequently crashes into bushes, tree trunks, and other obstructions when running (Smithers 1971). It avoids enemies by digging or running, but if cornered, it will fight by striking with the tail or shoulders, by rearing on the hind legs and slashing with the forefeet, or by rolling on the back and slashing with all feet.

The aardvark is mainly nocturnal but sometimes goes abroad by day and occasionally suns itself in the early morning at the burrow entrance. It usually sleeps during the day, curled up in a tight circle, with the snout protected by the hind limbs and tail. By night, it moves from one termite nest to another, following a regular route that is covered about once a week. When foraging, it travels in a zigzag path with its snout to the ground and inspecting a strip about 30 meters wide. It moves an average of about 10 km a night, with an estimated maximum of 30 km having been reported (Shoshani, Goldman, and Thewissen 1988). The diet consists principally of ants and termites, which are obtained by digging, tearing into nests, or seeking out the insects as they are on the march. A moving column of termites may contain tens of thousands of individuals and be about 40 meters long and may be detected by the aardvark by sound and smell. *Orycteropus* gathers ants and termites with its long, sticky tongue, which can be extended up to 300 mm. Other insects are taken occasionally, and Smithers (1971) reported apparent predation on the fat mouse *(Steatomys pratensis)*. Shoshani, Goldman, and Thewissen (1988) indicated that *Orycteropus* must have a source of water and sometimes digs up the fruits of a wild cucumber to obtain moisture. In captivity the aardvark accepts mealworms, boiled rice, meat, eggs, and milk and seems to thrive when some carbohydrate is included in the diet.

According to Dorst and Dandelot (1969), females are attached to a particular place, to which they come back regularly, whereas males are more vagabond. *Orycteropus* is generally solitary, but the young accompanies its mother for a long time. Births in Central Africa reportedly occur at the beginning of the second rainy season, in October or November. In southern Africa the young are produced during the cooler, drier months of the year, from about May to August (Smithers 1971). The gestation period is about 7 months (Dorst and Dandelot 1969). There is usually a single offspring, occasionally two. It weighs about 2 kg and is naked and flesh-colored. It remains in the burrow for about 2 weeks and then begins to accompany its mother on nightly excursions. For the next several months the mother and young occupy a series of burrows, moving from one to another. The young can dig for itself after about 6 months. Shoshani, Goldman, and Thewissen (1988) stated that sexual maturity is attained at about 2 years, that a female gave birth to 11 young during 16 years of captivity, and that a male sired 18 offspring by the time he was 24 years old.

The flesh of *Orycteropus* has the appearance of coarse beef and is prized by some persons, but others say that it is strong-smelling and as tough as leather. In some areas the hide is made into straps and bracelets, and the claws are worn as good-luck charms. The burrows sometimes cause serious damage to farming equipment and earthen dams and thus make the aardvark unpopular in agricultural areas. It has been greatly reduced in numbers and distribution and is now on appendix 2 of the CITES. In certain areas where the aardvark and other insectivorous animals have been exterminated, pasture and cereal crops have suffered enormous damage from termites (Shoshani, Goldman, and Thewissen 1988).

The geological range of the Tubulidentata is Eocene to Oligocene and Pliocene in Europe, late Miocene in Asia, and early Miocene and Recent in Africa (Shoshani, Goldman, and Thewissen 1988). Relationships with other orders, living and extinct, are not well understood (Thewissen 1985).

Order Proboscidea

Elephants

This order contains one living family, Elephantidae, with two living genera and species, *Elephas maximus*, of southern Asia, and *Loxodonta africana*, of Africa.

The most conspicuous external feature of living elephants is the trunk, which is flexible and muscular. The trunk is actually a great elongation of the nose, the nostrils being located at the tip. The fingerlike extremity of the trunk is used to pick up small objects, such as peanuts.

The head is huge, the ears are large (especially in *Loxodonta*) and fan-shaped, the neck is short, the body is long and massive, and the tail is of moderate length. Living members of the Proboscidea have a maximum height of nearly 400 cm and a weight of up to 7,500 kg (males). The limbs are long, massive, and columnar. All the limb bones are well developed and separate; lacking marrow cavities, they are filled with spongy bone. The feet are short and broad, columnar in shape. The weight of the animal rests on a pad of elastic tissue. There are five toes on each foot, but the outer pair may be vestigial, so some digits do not have hooves (nails). The Asiatic elephant has four hooves (occasionally five) on the hind foot and five on the forefoot, and the African elephant has three on the hind foot and five on the forefoot.

The skin of adult elephants is sparsely haired. The glands associated with the hair follicles in most mammals (sebaceous glands), which soften and lubricate the hair and skin, are not present in the Elephantidae. Females have two nipples just behind the front legs; the young nurse with the mouth, as do young of other mammals. Males retain the testes permanently within the abdomen.

The dental formula for Recent species of Elephantidae is: (i 1/0, c 0/0, pm 3/3, m 3/3) × 2 = 26. The single upper incisor grows throughout life into a large tusk (up to 330 cm long in the African elephant). The tusk, which is usually absent in females of the Asiatic elephant, has enamel only on the tip, where it is soon worn away (some extinct proboscideans had lower tusks, and others had longitudinal bands of enamel on their tusks). The grinding teeth are generally large and high-crowned, with a complex structure. Each tooth is composed of a large number of transverse plates of dentine covered with enamel; the spaces between the ridges of enamel are filled with cement. Ridges do not show on an unworn tooth, since it is covered with cement. The grinding teeth increase in size and in number of ridges from front to back. These teeth do not succeed one another vertically in the usual mammalian pattern, but come in successively from behind, the series thus moving obliquely forward. When the foremost tooth is so worn down as to be of no further use, it is pushed out, mostly in pieces. As these teeth are very large and the jaws are fairly short, on each side only one tooth above and one below are in use at the same time (part of a second tooth may also be in use).

The skull is huge and shortened. The premaxillary bones have been converted into sheaths for the tusks, and the nasal bones are extremely shortened. All the bones forming the braincase are greatly thickened and, at the same time, lightened by the development of an extensive system of communicating air cells and cavities. The brain chamber is hidden within the middle of the huge mass of the skull. The skeleton of a proboscidean is massive, comprising 12–15 percent of the body weight in Recent forms.

The living proboscideans occupy a variety of habitats but generally live in forests, savannahs, and river valleys. Vegetarians, they may consume more than 225 kg of forage a day. Modern elephants are gregarious herd animals, with a potential life span of about 80 years. The Asiatic elephant is commonly used as a beast of burden, and the African is sometimes used as a draft animal. Ivory is obtained from the tusks of both species, and this factor has led to the widespread disappearance of wild populations in modern times.

According to Dawson and Krishtalka (1984), elephants evidently originated in Africa and subsequently spread to Europe, Asia, and North America. The earliest definite proboscidean genus, *Palaeomastodon*, has been found in late Eocene deposits in Egypt and Libya and represents the extinct family Gomphotheriidae. From this group arose two main lineages, the living family Elephantidae and the extinct mastodons of the families Mastodontidae and Stegodontidae. The mastodontids appear to have arisen in the Oligocene in Africa; they survived into the Pleistocene in Africa, Europe, and Asia and until the very late Pleistocene or early Recent of North America. Like elephants, mastodons developed huge body size and a short, high skull. In the genus *Mastodon* there were never more than two molars in the jaw at the same time. These grinding teeth were low-crowned and fairly small and had three or four prominent transverse ridges of enamel. The tusks in the American species were directed nearly straight forward and were almost parallel with each other. Stegodontids evolved from the mastodontids and are known primarily from the Miocene to Pleistocene of Asia, as well as the Pliocene of Europe. They developed elephantlike molar teeth with transverse enamel ridges.

The first known members of the Elephantidae are from late Miocene or early Pliocene deposits in Africa. The genus *Primelephas* there may have given rise to the more advanced genera *Mammuthus*, *Elephas*, and *Loxodonta*. *Mammuthus*

Indian and African elephants (*Elephas maximus* and *Loxodonta africana*), photo by Elaine Anderson. The photo clearly shows the difference between the two species.

(mammoths) disappeared in Africa by the early Pleistocene but subsequently evolved into a number of species in Eurasia and North America; it became extinct in the late Pleistocene or early Recent, probably because of human predation. In the species *Mammuthus primigenius*, the woolly mammoth, the body was covered with a dense coat of woolly hair and long, coarse outer hair, as a protection against the cold. This is the species found in the frozen ground in Siberia, complete with hide and hair. Paleolithic humans left drawings of the woolly mammoth on the walls of caves. Agenbroad (1985) gave the date of the extinction of *M. primigenius* as about 8,000 B.P. The tusks in *Mammuthus*, though quite variable in form, had a tendency to spiral, first downward and outward, then upward and inward, and the grinding teeth had numerous ridges of enamel. Mammoths stood about 300 cm high at the shoulder. *Elephas* developed in Africa in the early Pliocene, spread through Eurasia in the middle and late Pliocene, and became restricted to Asia by the late Pleistocene. *Loxodonta* is known only from Africa, initially in the middle Pliocene.

In addition to the four families mentioned above, three more extinct families—Moeritheriidae, Barytheriidae, and Deinotheriidae—usually are associated with the Proboscidea. Moeritheres are known only from the early Eocene to early Oligocene of Pakistan and Egypt. They were about the size of a tapir and probably were amphibious. Barytheres are poorly known and have been recorded only from the late Eocene of Egypt and Libya. Deinotheres occurred from the Miocene to the Pleistocene of Europe, western Asia, and Africa. They lacked upper tusks but had a pair of lower tusks that curved downward and back. They were large, with a long and flattened skull, relatively small bilophodont and trilophodont cheek teeth, and probably a proboscis of some sort, and are thought to have been primarily terrestrial.

PROBOSCIDEA; ELEPHANTIDAE; **Genus ELEPHAS**
Linnaeus, 1758

Asiatic Elephant, or Indian Elephant

According to Olivier (1978), the single living species, *E. maximus*, occurred in historical time from Syria and Iraq east across Asia south of the Himalayas to Indochina and the Malay Peninsula, north in China to at least the Yangtze River, and on Sri Lanka, Sumatra, and possibly Java. In addition, the species apparently occurred on Borneo in the Pleistocene, and living populations have inhabited that island for at least several hundred years, but the latter may have resulted from introduction by human agency. There is also a small introduced population on the Andaman Islands. There have been suggestions that the elephants of ancient Syria actually represented *Loxodonta* or even *Mammuthus* (Corbet 1978; Sikes 1971). A pygmy species of *Elephas*, *E. falconeri*, occurred on certain Mediterranean and Aegean islands in the late Pleistocene and early Recent epochs. Radiocarbon dates as late as 4,390 B.P. have been reported for specimens from the Greek island of Tilos. This species had a shoulder height of only 90–140 cm (E. Anderson 1984; Martin 1984).

In *E. maximus*, head and body length is 550–640 cm, tail length is 120–50 cm, and shoulder height is 250–300 cm. Shoshani and Eisenberg (1982) reported that females weigh an average of 2,720 kg and that large bulls weigh 5,400 kg. One exceptionally huge circus animal reportedly weighs 6,700 kg (*Washington Post*, 31 March 1988, p. D-9). The hair covering is scant; the hairs are long, stiff, and bristly. There is a tuft of hair at the tip of the tail. The coloration of the skin is dark gray to brown, often mottled about the forehead, ears, base of the trunk, and chest with flesh-colored blotches that are perhaps caused by some skin disease. True coloration of the skin often is masked by the color of the soil on which the

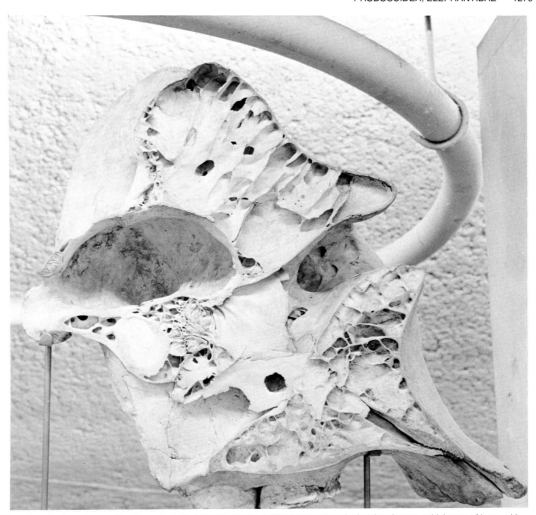

View of vertical median section through skull of Asiatic elephant *(Elephas maximus)*, showing the great thickness of bone with large cells or sinuses which almost completely surround the brain cavity. Photo by P. F. Wright of specimen in U.S. National Museum of Natural History.

animal lives, as it constantly throws dirt over its back and wallows in the mud. *Elephas* has a heart with a double apex that often is confused with a "double heart."

From *Loxodonta*, *Elephas* is distinguished by having considerably smaller ears, usually 4 hooves (nails) on the hind foot, 19 pairs of ribs, and 33 caudal vertebrae. Its forehead is flat, and the top of the head is the highest point of the animal. *Loxodonta* has large ears, generally 3 hooves on the hind foot, 21 pairs of ribs, and a maximum of 26 caudal vertebrae. The forehead is more convex, and the back more sloping, so the shoulders are the highest point. In addition, the trunk of the African elephant has 2 fingerlike processes at its tip in contrast to the single tip of the trunk of *Elephas*.

The Asiatic elephant originally occurred in a wide variety of habitats, from thick jungles to grassy plains. According to Shoshani and Eisenberg (1982), populations of the species in southern India and Sri Lanka now are restricted to single-monsoon, dry, thorn-scrub forest. Overall distribution is patchy and mostly limited to forest-grassland ecotone. Seasonal migrations formerly were extensive but now have been

reduced by human agricultural development. Seasonal movements of 30–40 km still occur in southern India and Sri Lanka. The animals usually do not feed for more than a few days in any one place. There are two major feeding peaks every 24 hours. The elephant feeds and moves about during the morning, evening, and night; it rests during the middle of the day. Drinking takes place at least once a day, the animals never being far from a source of fresh water. Shade is essential during much of the day, and excess heat is radiated through the ears, which are incessantly moving and flapping. The senses of hearing, vision, and olfaction are thought to be well developed. D. L. Johnson (1980) reported that the elephant is an excellent swimmer, covering up to 48 km at speeds of up to 2.7 km/hr.

McKay (1973) found that in Sri Lanka *Elephas* consumes a wide variety of grasses and also large amounts of bark, roots, leaves, and small stems. Cultivated crops, such as bananas, paddy, and sugar cane are favored foods, and as a result the elephant often becomes a pest in agricultural areas. An adult Asiatic elephant appears to have a daily food intake of approx-

Asiatic elephants *(Elephas maximus)*, photo by Lothar Schlawe.

imately 150 kg net weight. When feeding on long grasses, the animal uses its trunk as a "hand" to grasp a number of stems, which are pulled up and inserted directly into the mouth. When grasses are too short to be picked up in this way, the elephant scrapes the ground with its forefeet until a loose pile of grasses is formed, then sweeps the pile into its mouth with the "hand" of the trunk. When feeding on shrubs, the elephant uses the end of the trunk to break off small twigs and branches. Bark is removed from larger branches by inserting the branch into the mouth and then, with a turning motion of the trunk tip, rotating the branch against either the molars or a tusk, thus stripping the bark.

Shoshani and Eisenberg (1982) noted that overall population densities usually vary from 0.12/sq km to 1.00/sq km but can be greater than 7.00/sq km when temporary aggregations form during the wet season. Adult males in Sri Lanka have home ranges of 10–17 sq km. A herd of 23 females and young had a home range of 25 sq km in the wet season and 64 sq km in the dry season. There is no evidence of territoriality.

Elephas is gregarious, and although bulls sometimes live alone, cows always are found in herds. In Sri Lanka, bulls may live in temporary all-male groups of up to 7 animals, but cows and young are found in herds of sometimes great size (Kurt 1974). In the nineteenth century, these herds usually consisted of 30–50 animals, but much larger groups, with up to even 100 individuals, were not uncommon. More recently, female herd size has been reported as 15–40 (McKay 1973), 8–21 (Shoshani and Eisenberg 1982), and 3–11 (Santiapillai, Chambers, and Ishwaran 1984). A herd is a matriarchal family group consisting of mothers, daughters, and sisters, and not a group drawn together by accident or attachment. Movements are initiated by the oldest and usually largest female (Shoshani and Eisenberg 1982). In Sri Lanka, McKay (1973) found herds to contain both "nursing units," consisting of lactating females and their infants, and "juvenile care units,"

containing females with juveniles. There was a degree of flexibility within these units, in that the females might associate at a given time or divide into smaller subunits. All the unit members occupying a common range were considered to compose a herd. Such herds tended to remain distinct from all other herds occurring in the same general area. The units and subunits of each herd remained in close proximity but moved away from each other when foraging. Each herd had an overall range that included a rainy season range and a dry season range. Shoshani and Eisenberg (1982) listed eight basic vocalizations, including low-frequency, long-distance calls for maintaining contact between herd members; high-pitched, close-range calls that indicate mood; loud trumpeting for alarm; and low snorts to alert others to a change in environment.

McKay (1973) found that in Sri Lanka at least one adult male is present in 40 percent of the herds observed and that an adult male is associated with any one herd 25–30 percent of the time. Males stay with the herd when one or more females are in estrus. Males that are not associated with the female herd usually remain solitary and disperse over relatively small, widely overlapping home ranges. Sometimes they gather in small but temporary bull herds. There is a great amount of toleration between them, except possibly when cows are in estrus. If more than one male is in a herd, a dominant individual soon is recognized, and serious fighting is rare (Shoshani and Eisenberg 1982). Young males appear to leave the herds and become solitary at about the time of puberty, whereas young females tend to remain in their natal herd.

Shoshani and Eisenberg (1982) wrote that *Elephas* is polyestrous and that births can occur at any time of the year. Eisenberg and Lockhart (1972) and McKay (1973) found no evidence of reproductive seasonality in Sri Lanka. However, Santiapillai, Chambers, and Ishwaran (1984) reported that in

Asiatic elephants *(Elephas maximus)*, photo by Cyrille Barrette.

an area of relatively low rainfall in Sri Lanka mating seems to coincide with the dry season (June–September) and parturition with the rainy season (October–January). Eisenberg, McKay, and Jainudeen (1971) reported the estrous cycle in Sri Lanka to average 22 days, with the duration of estrus being 4 days. Dittrich (1966) reported that 32 births of male calves in European zoos and circuses showed gestation periods of 615–68 days, with an average of 644 days, and that 17 births of female calves showed gestations of 628–68 days, with an average of 648. There normally is a single calf, which at birth weighs 50–150 kg (the average is about 107 kg). Newborn baby elephants have a coat of widely spaced brown hairs that produces a halo effect as it stands out from the body. As the animal grows older, the hairy coat becomes less noticeable (adult elephants may appear to be naked, but throughout life they have a scattered coat of hair). According to Shoshani and Eisenberg (1982), an infant can stand shortly after birth and can follow its mother in her daily routine after a few days. It suckles with its mouth (not the trunk) from its mother or another lactating cow. After several months it begins to eat grass and foliage but may nurse occasionally for about 18 months. Parental supervision lasts for several years. Young males are capable of some independent movement at 4 years, and by 7–8 years they may form subgroups of their own or temporarily associate with older bulls. Growth slows in females at 10–12 years, in males at 15 years, and full size is attained at about 17 years. Both sexes may become sexually fertile at as early as 9 years, but males usually do not reach maturity until 14–15 years, and even then they are not capable of the social dominance that usually is necessary for successful reproductive activity. Females generally first give birth at 15–16 years. In favorable habitats they may produce a calf every 2.5–4.0 years, and in other areas, every 5–8 years. Although there have been some reports of elephants living over 100 years, the record documented age is of a captive that was estimated to be either 69 or 77 years old.

The Asiatic elephant has been domesticated for centuries; it is intelligent and docile when well treated. It is valued in Asia as a draft animal and also is used for transportation and hunting purposes. To capture it from the wild, a strong fence made of high posts set close together is usually built in a circular form, with wings leading outward from the front to make a V-shaped entrance to the enclosure. When a group of elephants is in the proper position, it is driven into the V and the corral. Then the entrance is closed, and the process of selecting and taming is started. Tame elephants that have been trained are of great assistance in this work. Wealthy Indians and royalty used to own elephants and ride on them, an elaborate howdah providing a comfortable seat for the riders. Tame elephants were used extensively in handling heavy items such as teak logs.

The Asiatic elephant is classified as endangered by the IUCN and the USDI and is on appendix 1 of the CITES. Like *Loxodonta* in Africa, it has been hunted ruthlessly for centuries because of its valuable ivory. Martin and Vigne (1989) indicated that the ivory carving industry of India dates back at least 4,000 years and that by the early nineteenth century this activity was being supported in large part by importation of tusks from Africa. Asiatic elephant populations also declined as a result of conflicts with agriculture and other activities stemming from increasing numbers of people. *Elephas* disappeared by about 2,000 years ago from southwestern Asia and most of China. In the latter country it was still widespread in the southeastern provinces in the fourth century A.D., and it survived until the seventeenth century in Guangsi, but today it is confined to the southern part of Yunnan (Gao 1981).

Olivier (1978) estimated that only 28,000–42,000 wild Asiatic elephants remained throughout the entire range of the genus. Of these, 9,950–15,050 occurred on the Indian subcontinent, 11,000–14,600 were in continental southeast Asia, and 7,330–12,330 occurred on the Malay Peninsula, Sri Lanka, Sumatra, Borneo, and the Andaman Islands. More recent estimates of numbers in India, 16,595–22,261 in the wild and about 3,000 in captivity, along with relatively few reports of poaching there (Martin and Vigne 1989), give some hope that populations may have stabilized for the time being. However, Santiapillai, Chambers, and Ishwaran (1984) reported that the number of elephants in Sri Lanka had declined from about 12,000 animals at the beginning of the nineteenth century to a present estimate of 1,800 and that half of those remaining might soon be jeopardized by agricultural development. Blower (1985) indicated that there are about 3,000 wild elephants in Burma but that their number and range are decreasing rapidly. About 120 are captured annually to maintain Burma's stock of 5,400 timber-working elephants. The number of captive elephants in Thailand fell from over 13,000 in 1950 to 4,800 in 1982 (Dobias 1987).

PROBOSCIDEA; ELEPHANTIDAE; Genus LOXODONTA
F. Cuvier, 1827

African Elephant

In historical time, the single species, *L. africana*, occurred throughout Africa, from the Mediterranean Sea to the Cape of Good Hope, except in parts of the Sahara and some other desert regions (Ansell, *in* Meester and Setzer 1977; Kingdon

1979; Laursen and Bekoff 1978). Sikes (1971) suggested that *Loxodonta* also occurred in southwestern Asia within the last 2,000 years. There have been reports that another species of elephant, *L. pumilio*, lives in the dense lowland jungles from Sierra Leone to Zaire (Meester et al. 1986). This so-called pygmy elephant is said to be very small and more solitary than *L. africana*. Most authorities consider such reports to reflect small individuals of *L. africana*. However, there is general acceptance that there are two kinds of *L. africana*, the smaller, darker forest elephant, now found in the tropical rainforest zone of West and Central Africa, and the larger, paler bush, or savannah, elephant of the remainder of the current range. Some authorities (including Kingdon 1979) regard these kinds as being equivalent to subspecies, *L. a. cyclotis* for the forest elephant and *L. a. africana* for the bush elephant. Others (including Ansell, *in* Meester and Setzer 1977; Laursen and Bekoff 1978; and Smithers 1983) recognize two subspecies of the forest elephant and four of the bush elephant. There have been suggestions that the forest elephant may warrant specific rank (Meester et al. 1986).

The African elephant is the largest living terrestrial mammal. Kingdon (1979) listed the following size ranges (means in parentheses): forest elephant, shoulder height 160–286 cm (250 cm) in males and 160–240 cm (210 cm) in females, weight 2,700–6,000 kg; bush elephant, shoulder height 300–400 cm (320 cm) in males and 240–340 cm (250 cm) in females, weight 4,000–6,300 kg (5,000 kg) in males and 2,400–3,500 kg (2,800 kg) in females. Other reports indicate an overall weight range of up to 7,500 kg. Head and body length (including the trunk, which is really an elongated

African elephant *(Loxodonta africana)*, photo by Lothar Schlawe.

nose) is 600–750 cm and tail length is 100–150 cm. Sparsely scattered with black bristly hairs, the skin generally is dull brownish gray in color. The flattened end of the tail has a tuft of coarse, crooked hairs 38–76 cm long. The elephant wallows in streams and pools and tosses dirt or mud onto its back; thus its coloration is usually similar to that of the soil it frequents.

In both sexes, one incisor tooth on each side of the upper jaw is greatly developed to form a tusk. The largest known tusk measures about 350 cm and weighs about 107 kg; the largest female tusk weighs about 18 kg (the average is about 7 kg). The tusks are used for fighting, digging, feeding, and marking (Laursen and Bekoff 1978). Interestingly, the largest tusks generally develop, not on the largest animals, but rather on those of mild temperament, which tend to avoid the combat and tree ramming that may break the teeth (Kingdon 1979). The tusks of males grow throughout life, both in length and bulk. The tusks of females also grow continuously, though only slowly past the age of 30 years, and after age 15 there is a considerable reduction in the rate of development of diameter in relation to length, so that the result is a rather slender tusk (Layser and Buss 1985).

The end of the long, tubular, muscular, and very sensitive trunk has a fingerlike projection both above and below. These projections are skillfully used to pick up food or other articles and to manipulate or examine objects. The trunk also is used for breathing (along with the mouth) and to suck up water, which then is squirted into the mouth for drinking or sprayed over the body for cooling.

Loxodonta has much larger ears than does *Elephas* and often has a hole through the lower portion of the earlobe caused by injury. The ears are sometimes nearly 200 cm in length from top to bottom. The undersides of the ears have an extensive supply of blood vessels. Ear flapping, characteristically seen when elephants stand in the shade on hot days, creates air currents over the blood vessels and promotes the radiation of excess body heat (Smithers 1983; Wright 1984).

Loxodonta lives in many kinds of habitats, including deep forests, open savannahs, wet marshes, thornbush, and semidesert scrub. It has been recorded from sea level to elevations of over 5,000 meters. According to Smithers (1983), its requirements are a supply of fresh water, plentiful food in the form of grass or browse, and some shade. To find these conditions, the elephant may have to make annual migrations of several hundred kilometers. If food, water, and shade all remain available, elephants or elephant units will not venture far; when these items are scarce, large-scale movements may occur. Rodgers and Elder (1977) stated that nearly all seasonal movements have the same general pattern, that is, a migration from permanent water sources at the start of the rainy season, followed by a movement back to permanent water when the rains end and water holes dry up. Often there is a concentration of animals in suitable areas during the dry season or in times of drought and a dispersal in the wet season. Human activity and agricultural development now have forced permanent, unnatural concentrations in many areas (Kingdon 1979). Shorter, nonseasonal movements of elephants may involve travel between water and feeding areas. Regular routes are followed, and well-worn paths may be created.

Activity is both diurnal and nocturnal but drops during the hottest hours of the day. Sleeping may take place either at noon or after midnight, with the animals lying down or standing and leaning against one another or a tree. The elephant usually seeks water once a day, or at least every few days, and then may bathe, wallow, or even submerge with only the nostrils at the tip of the trunk showing above the surface. In this manner an animal also may swim or walk underwater for several kilometers. A normal fast pace on land

proceeds at 10–16 km/hr, while a charge may reach 35–40 km/hr (Kingdon 1979). Daily average movements of 12 km have been reported (Merz 1986). *Loxodonta* may consume 200–300 kg of food and 160 liters of water per day (Kingdon 1979; Smithers 1983). The diet includes grass, tree foliage, bark, twigs, herbs, shrubs, roots, and fruit. There seems to be a pronounced seasonal variation in the elephant's diet, with grass intake increasing during the rainy season but falling to low levels during the dry season (Hanks 1979).

Because of its enormous size, the elephant can be destructive to vegetation, pushing over trees to obtain edible twigs and leaves and sometimes modifying the habitat over large areas. Bulls also seem to knock trees down in response to social pressures or other excitement (Kingdon 1979). In the past, when the species could roam more freely over the African continent, such environmental modification presented few problems and in fact may have been necessary for the maintenance of ecosystems that could support large and diversified populations of animals. As human populations increased in Africa, however, and as large areas were cleared for agricultural purposes, the range of the elephant became fragmented and more confined to restricted sites such as parks and reserves. Within these areas, elephants, being artificially concentrated, sometimes gave the impression of becoming very numerous. An elephant population can increase at an annual rate of 4–5 percent under favorable conditions, the maximum known being 7 percent (Cumming 1981; Douglas-Hamilton 1987). Unable to migrate or disperse naturally, a growing population can begin to damage its habitat. Therefore, some governments have carried out deliberate culling programs for the stated purposes of protecting habitat and preventing the elephants from destroying their own food supply.

Food quantity and quality and availability of water are the most important natural factors in determining the spatial distribution of the elephant. Differences in these factors have resulted in considerable variation in reported population densities, 0.26–5.00 individuals per sq km, and in home ranges, 14–3,120 sq km (Kingdon 1979). A relatively undisturbed population of about 500 elephants occupied an ecosystem of about 3,500 sq km in the Amboseli National Park of southern Kenya. Its overall density thus was 0.14 per sq km, increasing during the dry season to 0.4–0.9 per sq km and even up to 10 per sq km in certain habitats (Western and Lindsay 1984). In the Lake Manyara area of Tanzania, Douglas-Hamilton (1973) found home ranges of family units and of individual bulls to be 15–52 sq km in size and to overlap widely. In Tsavo National Park, Kenya, Leuthold and Sale (1973) recorded mean home ranges of 530 sq km in Tsavo West and 1,580 sq km in Tsavo East. Hanks (1979) found that on the whole, elephants in southern Africa had small home ranges, but some moved considerable distances for short periods of time. There is no evidence of territorial behavior in *Loxodonta* (Smithers 1983).

The African elephant is gregarious, sometimes being found in aggregations of hundreds or even over 1,000 individuals. These large groups, which have been observed mainly in East Africa, are thought to be temporary associations of organized herds and individuals that come together during drought, human interference, or other disruptions of the normal pattern of social life. In Amboseli groups seldom exceeded 20 individuals in the dry season, but aggregations of over 400 sometimes formed during the rains (Western and Lindsay 1984). Douglas-Hamilton (1973) suggested that this bunching could provide a collective defense against predators in time of social stress or disturbance. There also are temporary associations of males, the members of which join and leave at will. These bachelor herds usually are small but have been known to contain up to 144 animals (Kingdon 1979). Solitary males are common, and very old bulls often are found alone. Solitary females are extremely rare.

Other than described above, elephant society is essentially matriarchal and is organized around a stable family unit of cows and their calves (Hanks 1979). There may be only a single adult female with one or more of its offspring, but usually the units contain about 10 individuals (Kingdon 1979). Nearly all females are part of a family unit. Douglas-Hamilton (1973) found that sometimes two to four family units jointly numbering up to 50 animals may come together into slightly less stable associations, which he termed "kinship groups." Laursen and Bekoff (1978) referred to such groups as "clans," gave their membership as 6–70 individuals, and noted that they are led by a large female, the dominance of which is undisputed. Outsiders that may attempt to enter the group are rebuffed by the dominant female. She usually maintains her position until death and then is succeeded by her eldest daughter (Laursen and Bekoff 1978). At puberty, which occurs at 8–20 years of age, young males are driven out of the family group by the older females, and they then join or form bachelor groups. An adult bull joins a family group only when a female therein is in estrus. At any given time about half of the groups have a bull in attendance (Kingdon 1979).

Elephant society normally is peaceful, but females with young are unpredictable, and males compete with one another for dominance in their groups and for access to estrous females (Kingdon 1979; Laursen and Bekoff 1978; Smithers 1983). Aggressive tendencies are shown by raising the head and trunk, extending the ears perpendicular to the body, kicking dust, swaying the head, and making either a mock or serious charge. Most dominance struggles are resolved after some pushing and light tusking, but battles for mating privileges sometimes involve fatal use of the tusks. *Loxodonta* trumpets and screams loudly when upset. Members of a group maintain contact and greet one another by low rumblings or growls produced by the vocal chords.

Recent studies by Moss (1983) and Poole (1987a, 1987b, 1989a, 1989b) in Amboseli National Park, Kenya, have provided new insights into the mating system of *Loxodonta*. Males over about 25 years old annually enter a condition known as musth, characterized by copious secretions from the temporal gland behind the eye, continuous discharge of urine, a great increase in aggressive behavior, and the seeking of and association with female groups. This condition, perhaps resulting from high testosterone levels, initially lasts only a few days or weeks, but in males over 35 years old it continues for 2–5 months. Musth does not occur synchronously, but it does come at about the same time each year in any given male, and it is especially frequent during and just after the rainy season. A male in musth is dominant to other males and usually can defeat them in combat, even if the latter are larger and normally higher-ranking. Estrous females emit loud, very low-frequency calls that may attract potential mates up to several kilometers away but are mostly inaudible to humans. A female actively avoids most courting males, tending to choose a large male that is in musth, perhaps one with which she long has been familiar. This male then guards the female from other suitors. Tragically, this remarkable selective process may be disappearing more rapidly than is *Loxodonta* itself, because old males, which have the largest tusks, are the first victims of the ivory trade (see below).

Females are polyestrous and have an estrous cycle of about 2 months (Kingdon 1979). Estrus lasts 2–6 days (Moss 1983). Births may occur at any time of year, but there evidently is a calving peak just before the height of the rainy season, so that the young have a cool environment with an abundance of good cover. In Zambia nearly 88 percent of all conceptions occur from November to April, when almost all the rain falls (Hanks 1979). In Uganda breeding also occurs primarily during the rainy season (Smith and Buss 1973). It appears that a change in diet from browse and dry grass with a low crude protein content at the end of the dry season to fresh green grass with a high protein content at the height of the rains may stimulate ovulation and fertile mating (Hanks 1979). The average gestation period is 22 months, with a recorded range of 17–25 months. According to Laursen and Bekoff (1978), there is normally a single young, but twins occur in 1–2 percent of births. The newborn weighs 90–120 kg and can stand after a half-hour. The herd waits until the young have the strength to roam with them (usually about 2 days). Weaning takes place after 6–18 months, but the young occasionally nurse for 6 years or more (Kingdon 1979). Under optimum conditions sexual maturity is attained by males at about 10 years and by females at 11 years old but sometimes does not come until 20 years and 22 years, respectively. In any case, males cannot successfully compete for mating until they are over 20 years old. Females remain fertile until they are 55–60 years old, producing calves at intervals of 2.5–9 (usually about 5) years. Life expectancy is similar to that of humans, about 50–70 years (Kingdon 1979; Laursen and Bekoff 1978; Moss 1983; Smithers 1983).

The African elephant is intelligent and not difficult to tame, but it has not been employed by people for labor and transportation to the same extent as has the Asiatic elephant. Populations in North Africa were the source of war elephants for the ancient Carthaginians, Romans, and Ethiopians. *Loxodonta* disappeared from north of the Sahara by about the sixth century A.D. (Ansell, *in* Meester and Setzer 1977; Laursen and Bekoff 1978). The northern subspecies, *L. africana pharaohensis*, which is considered to be a relative of the living forest elephant, *L. a. cyclotis*, may have survived on the coast of Sudan and Eritrea until the mid-nineteenth century but now is extinct (Yalden, Largen, and Kock 1986). The distinctive subspecies *L. a. orleansi*, historically found in northern Somalia and adjacent Ethiopia, has been reduced to only 60–300 individuals in the vicinity of Harar in east-central Ethiopia (Largen and Yalden 1987). A combination of exploitation for ivory, sport hunting, and killing in the course of agricultural development also eliminated *Loxodonta* from almost all of South Africa by about 1900 and subsequently in much of West Africa. Kingdon (1979) reported that more than 500,000 elephants had been shot in East Africa during the previous 30 years, mostly along the frontiers of agricultural expansion.

Large populations continued to exist, but as noted above, they often became unnaturally restricted in their distribution and ability to migrate. In some cases, legal protection allowed substantial increases of formerly depleted elephant populations in national parks and other reserves but did not provide for the space, food, and water that the animals eventually would require (Cumming 1981; Laws 1981; Poché 1980). This process was especially prevalent in the savannah zone of eastern Africa, from Uganda to the Transvaal. On the grounds that constrained and often increasing herds were destroying their own habitat and that of other species, the governments of some countries have carried out large-scale programs to reduce elephant numbers. There has been considerable controversy about such killing, though most authorities have accepted its local necessity. Given the continued expansion of people and agriculture, this sort of management is likely to spread throughout the remaining range of the elephant. It therefore is unfortunate that even if some elephant populations are saved, the species may be lost with respect to its natural behavior and continent-wide ecological role. Recently, Western (1989b) pointed out that many field studies of elephants have contributed to a negative image of the ecology of the species, but only because the investigations dealt with artificially compressed populations that were

African elephants *(Loxodonta africana)*, photos by Bernhard Grzimek.

said to be damaging parks or commercial forests. He explained that the elephant has a key role throughout the savannah and tropical forest zones that it inhabits. When numbers and seasonal movements are natural, the elephant opens up dense woodland, prevents the spread of brush, and makes gaps through the jungle. It thereby allows the penetration of sunlight, promotes the growth of a great variety of plant species, and encourages a more abundant and more diverse fauna of smaller animals. If this role is applied across its entire original range, the elephant has been more important than any other species except people in shaping the ecology of Africa. Its loss will reduce biodiversity and increase extinction rates throughout the continent.

The most immediate factor jeopardizing the African elephant is the demand for its tusks by the international ivory trade. This problem is not new, as ivory has been an important commodity throughout recorded history. Its value in Europe was responsible for the killing of many elephants by colonists near the Cape of Good Hope in the 1600s (Smithers 1983). By the early nineteenth century tusks of *Loxodonta* were being exported to carvers in India to replace stocks from the depleted *Elephas* (Martin and Vigne 1989). From 1860 to 1930, 25,000–100,000 elephants were killed annually for their ivory, mostly to supply material for the manufacture of piano keys in Europe and the United States (Conniff 1987; Ricciuti 1980). For long after the Civil War, Americans unwittingly fostered a continuation of the slave trade; the commercial pursuit of ivory often was associated with the capture of slaves, who then were forced to carry the tusks before both were sold. This activity was concentrated along the east coast of Africa. The end of the slave trade and the development of wildlife protection regulations by the early 1900s may have allowed the partial recovery of remaining elephant populations in eastern and southern Africa. Indeed, Spinage (1973) reported that surviving elephants had entered an exponential phase of population increase.

A reintensification of the problem began in the 1970s, when the price of ivory rose along with that of gold and other precious materials in conjunction with the energy crisis. During that decade the price of raw ivory increased from about U.S. $5 to U.S. $50 per kilogram. As a result, illegal killing became widespread, and many elephant populations, especially in East Africa, were devastated. The estimated number of elephants in Kenya, for example, fell from 167,000 in 1973 to 70,000 in 1977 (Kingdon 1979). There also was renewed conservation interest during the decade, leading to such measures as the placing of *L. africana* on appendix 2 of the CITES and its classification as threatened by the USDI, along with issuance of regulations that limited, but did not prohibit, the importation of ivory. The species also was classified as vulnerable by the IUCN.

The conservation efforts failed, and by the late 1980s there was a general realization that the world's most spectacular land animal was in danger of extinction. Nearly all the major populations that had been monitored since the 1970s were found to be in rapid decline (Douglas-Hamilton 1987). The price of ivory continued to rise, reaching more than U.S. $100 per kilogram at several times in the 1980s, and poaching became more rampant than ever. Annual ivory exports from Africa, which had been about 200 metric tons in the early 1950s, were around 900 metric tons from 1979 to 1985 (Cobb and Western 1989). An estimated 100,000 elephants were being killed each year to supply this market, 80 percent of them illegally (Western 1989a). Most of the ivory went to the Far East, where it was carved into products for local use or export. After Japan, the United States was the largest importer of carved ivory. In 1986 the amount of ivory carvings brought into the United States represented 32,000 dead elephants. In that year the United States also imported the raw tusks of several hundred elephants and skins representing 11,000 elephants. The annual retail value of elephant products sold in the United States was estimated at $100 million (Thomsen 1988).

Such commerce proved disastrous, with the estimated number of African elephants falling from at least 1.3 million in 1979 to 625,000 in 1989 (Cobb and Western 1989). Again, the East African savannahs bore the brunt of the assault, with numerical declines during the period of 65,000 to 19,000 in Kenya, 134,000 to 40,000 in Sudan, 316,000 to 80,000 in Tanzania, and 150,000 to 41,000 in Zambia. Only about 16,000 elephants remained in all of West Africa—the region from Senegal to Nigeria (Western 1989b). Of particular concern was the devastation of what had been the last great, relatively undisturbed populations, those of the forests of Zaire. The estimated drop there, thought to have occurred mostly in the late 1980s, was from 378,000 to 85,000. Perhaps the only large elephant population left that has not been extensively exploited or forced into unnatural habitat is that of Gabon, numbering about 76,000 animals.

The problem involves not only a numerical decline but also the upsetting of the age and social structure of populations (Cobb and Western 1989; Ottichilo 1986; Poole and Thomsen 1989; Western 1989a). Since old bulls generally have the largest tusks, they are the main initial target of ivory hunters. In some areas males now compose only 5 percent of the adult population, and estrous females may not even be able to attract a mate. The next victims are the old females, which the family units depend upon for leadership and which are the most successful mothers. Present populations in some areas consist mostly of juveniles under 15 years old, but even these small-tusked animals are eventually taken. From 1979 to 1987 the average size of a tusk on the market fell from 9.8 kg to 4.7 kg, and thus twice as many elephants must be killed to obtain a metric ton of ivory.

Some authorities predict near total extinction of *Loxodonta* in the wild by about the year 2010 if current trends continue. Any annual rate of kill in excess of 5 percent, the normal rate at which an elephant population can increase under favorable conditions, will eventually lead to extinction, and yet at present the overall loss to the species is 10 percent, and some predict that it will rise exponentially (Poole and Thomsen 1989). Such concerns have led to intensified conservation efforts, including international programs to raise funds and carry out research by the International Union for Conservation of Nature, the World Wildlife Fund, the New York Zoological Society, and other private groups. In October 1989 the African elephant was transferred to appendix 1 of the CITES. The United States established the African Elephant Conservation Act to control commerce and help support conservation work in Africa, and in June 1989 it completely banned the importation of ivory. Bans also have been established by the European Community and, in part, by Japan. In 1990, after being petitioned to reclassify the African elephant from threatened to endangered, the U.S. Fish and Wildlife Service announced that it would propose such a measure but would exclude elephant populations in the countries of Zimbabwe, Botswana, and South Africa. There have been no substantial declines in these three countries during the last several decades, and management programs have been set up to cull excessive elephants, market the ivory, and use the profits for conservation.

Despite the deterioration of the elephant's status, there seems to be an official consensus that the ivory trade, if properly regulated, provides an economic value to the species that can encourage and fund its conservation. That this value is what is destroying the elephant herds of Africa does not seem to be of primary concern. There also may be insufficient appreciation of the historical trends that saw Africa's elephants eliminated initially in the extreme north and south, then along the east coast and in the west, then (in the 1970s and 1980s) on the eastern savannahs, and finally in the great central forests of Zaire. These trends suggest that the killing is moving from all directions toward the last undisturbed populations, in Gabon. Legal sales of ivory from countries in southern Africa might have some local and temporary conservation value, but they will not save the elephants of Gabon or anywhere else. Indeed, the total number of elephants in Zimbabwe, Botswana, and South Africa, about 100,000, is less than the number that was lost from 1979 to 1989 in the neighboring country of Zambia alone. The elephant populations of South Africa and Zimbabwe may be stable, but they are mostly compressed into reserves and should not exemplify conservation objectives. Populations in Botswana are expected to decline in the face of agriculture and other development (Melton 1985). The indefinite maintenance of remnant and ecologically unnatural herds should not provide a basis for a satisfactory appraisal of species status. Maintenance of such herds through the sale of culled ivory should not be done at the cost of losing most of the other elephants in Africa. And yet if society continues to give a substantial value to ivory, the next edition of this book is likely to record the collapse of the last major populations of *Loxodonta*.

Order Hyracoidea

HYRACOIDEA; **Family PROCAVIIDAE**

Hyraxes, or Dassies

This order contains the single Recent family Procaviidae, which comprises three Recent genera and seven species and occurs in southwestern Asia and most of Africa.

These animals are comparable in size and external appearance to rodents and lagomorphs. The length of the head and body is 300–600 mm; the tail is 10–30 mm in length or lacking. Adults sometimes weigh as much as 4.5 kg. The pelage consists of fine underhairs and coarser guard hairs. Scattered bristles, presumably tactile, are located mainly on the snout. A gland on the back is covered with hair of a different color from that on the rest of the body. The eye is unique in that a portion of the iris above the pupil bulges slightly into the aqueous humor, thus cutting off light from almost directly above the animal. Hyraxes have a short snout, a cleft upper lip, short ears, and short, sturdy legs. The vertebral column is convex from the neck to the tip of the tail. The forefoot has four digits, with flattened nails resembling hooves. The hind foot has three digits; the inner toe, the second digit, has a long curved claw, while the other digits have short, flattened, hooflike nails. The soles have special naked pads for traction that are kept continually moist by a glandular secretion. The pads have a muscle arrangement that retracts the middle of the sole; this forms a hollow that is a suction cup of considerable clinging power.

The dental formula of the deciduous teeth is: (i 2/2, c 1/1, pm 4/4) × 2 = 28; the formula for the permanent teeth is: (i 1/2, c 0/0, pm 4/4, m 3/3) × 2 = 34. The single pair of upper incisors grows continuously and is long and curved. The upper incisors are triangular in cross section and semicircular in form; the flattened back surfaces are without enamel and so produce pointed cutting edges. The lower incisors are chisel-shaped; the first pair has three cusps, the second pair only one. There is a wide space between the incisors and the cheek teeth. The milk canines are rarely persistent. From front to rear, the premolars become more like the molars in structure. The molars vary from low-crowned to high-crowned. Each has four roots, and the lower ones bear two crescents, as in the corresponding teeth of rhinos and horses. The skull is stout and the roof is flattened.

The terrestrial genera (*Procavia* and *Heterohyrax*) inhabit rocky areas, arid scrub, and open grassland, whereas tree hyraxes (*Dendrohyrax*) are usually arboreal and are found in forested areas, though in eastern Africa they inhabit lava flows. The elevational range of the family is sea level to 4,500 meters.

Although only *Dendrohyrax* is arboreal, all the genera apparently can climb well; *Heterohyrax*, for example, sometimes suns itself in a tree. Hyraxes move quickly and are extremely agile on rugged and steep surfaces, running and jumping with skill and gaining traction by means of specialized foot pads and probably the inner claw on the foot. This claw is apparently also used to groom the hair. Hyraxes travel on the sole of the foot, with the heel touching the ground, or partly on the digits. The terrestrial forms are similar in habits to the pikas, *Ochotona*, sheltering in colonies of 5 to about 50 individuals, usually among rocks. They are active mainly during the daylight hours and are fond of basking in the sun and rolling in the dust. Tree hyraxes shelter singly or in family groups, using tree hollows and dense foliage; they are active during the night. They are less gregarious than the terrestrial forms and are usually more tractable. The terrestrial hyraxes whistle, scream, and chatter; the tree hyraxes utter a series of loud cries.

Hyraxes have acute sight and hearing. They feed mainly on vegetation and are both grazers and browsers. Tree hyraxes feed on the ground as well as in trees. All the species habitually work their jaws in a manner reminiscent of cud chewing. Generally these animals are not limited in their distribution by a lack of water to drink. They sometimes travel more than 1.3 km for food. Hyraxes are preyed on mainly by rock pythons, eagles, and leopards.

Hyraxes, referred to in the Bible as conies, are in an order that originated in Africa and evolved for some time before the Oligocene. They once were more diverse than at present and ranged in size from that of a rabbit to that of a tapir, but they probably never spread beyond Africa and the Mediterranean region. There is disagreement as to whether their affinities lie with the Proboscidea and Sirenia or with the Perissodactyla. There is one extinct family, the Pliohyracidae. The geological range of the order and the family Procaviidae is early or middle Eocene to Recent in Africa, early Pliocene in Europe, and Recent in southwestern Asia (Dawson and Krishtalka 1984; C. Jones 1984).

Tree hyrax *(Dendrohyrax dorsalis),* photo by Jean-Luc Perret. The light-colored hairs indicate the location of glands that exist on the back of hyraxes.

A. Gray hyraxes (*Heterohyrax* sp.), photo by C. A. Spinage. B. Rock dassie (*Procavia* sp.), photo by Roy Pinney. C. Skull of rock dassie *(P. capensis)*, photo by P. F. Wright of specimen in U.S. National Museum of Natural History. D. Palmar surface of hand *(P. capensis);* E. Plantar surface of foot *(P. capensis);* photos by Ernest P. Walker.

HYRACOIDEA; PROCAVIIDAE; **Genus PROCAVIA**
Storr, 1780

Rock Dassie, or Hyrax

Most authorities now recognize only a single species, *P. capensis*, occurring in Syria, Lebanon, Israel, Jordan, Sinai, the southern Arabian Peninsula, and almost all of Africa southeast of a line from Senegal through southern Algeria and Libya to Egypt (Corbet 1979; Ellerman and Morrison-Scott 1966; Honacki, Kinman, and Koeppl 1982; Meester et al. 1986; Olds and Shoshani 1982; Roche 1972*a*; Yalden, Largen, and Kock 1986). Some other authorities, including Bothma (*in* Meester and Setzer 1977), consider *P. capensis* to be restricted to southern Africa and treat the following as distinct species: *P. welwitschii*, southwestern Angola, Namibia; *P. ruficeps*, southern Algeria and Senegal to Central African Republic; *P. johnstoni*, northeastern Zaire and central Kenya to Malawi; and *P. syriaca*, Egypt to Kenya, and the southwest Asian portion of the range of the genus. Corbet and Hill (1986) recognized three of those species but included *P. johnstoni* in *P. capensis* and *P. ruficeps* in *P. syriaca* and also regarded *P. habessinica*, of Ethiopia, as a full species.

Head and body length is 305–550 mm, an external tail is lacking, height at the shoulder is 202–305 mm, and average weight is about 4 kg in males and about 3.6 kg in females. The hair pelage is dense and rather coarse, with short, thick underfur and long, scattered guard hairs. The general coloration of the upper parts is brownish gray, the flanks are somewhat lighter, and the underparts are creamy; however, there is considerable variation in color and intensity. The black whiskers may be as long as 180 mm. In general appearance *Procavia* is similar to a large pika *(Ochotona)* or a tailless wood-

chuck *(Marmota)*. However, the similarities go no further than the superficial appearance. The soles are moist and rubberlike, which gives the animals traction on smooth surfaces and steep slopes. This genus is distinguished from *Heterohyrax* and *Dendrohyrax* by a black dorsal patch, which covers a gland that is exposed if the surrounding hairs are erected. Erection of these hairs is a demonstration of aggression (Smithers 1983). Also, in *Procavia* the premolar series of teeth is much shorter than the molar series; in *Heterohyrax* the two series are about equal in length, and in *Dendrohyrax* the premolar series is longer (Smithers 1983). Female *Procavia* have six mammae.

Procavia frequents rocky, scrub-covered habitat where there are suitable shelters in, between, or under rocks or where it can dig burrows of its own. In spite of its rather heavy build, *Procavia* is active and agile, running up steep, smooth rock surfaces with ease. Its senses are keen, and it becomes quite shy when persecuted. When alarmed, it quickly dashes into a rocky cranny or its own burrow. Although not a large animal, it can put up a vigorous fight in self-defense, biting savagely. Much of its lifestyle seems to be associated with its relatively poor thermoregulatory ability (Kingdon 1971; Olds and Shoshani 1982; Smithers 1983). Its body temperature fluctuates relative to ambient temperature, tending to increase during warmer parts of the day and after its extensive sessions of sunbasking. It is predominantly diurnal, emerging in cool weather well after sunrise and retiring to its rock shelter before sunset. In warm weather it may remain out if there is a half- to full moon. It is reluctant to emerge if the weather is overcast or cold and does not do so at all during rainy periods, but it will seek shade during very hot weather. Individuals regularly crowd together in holes to conserve heat.

Procavia may be found feeding at any time of the day,

Rock dassie *(Procavia capensis)*, photo from San Diego Zoological Garden.

provided it is warm and sunny, with peaks of activity in the early morning and late afternoon. However, the actual time spent feeding is remarkably short for a vegetarian, amounting to less than an hour per day. Animals tend to stay near shelter but may forage 50–100 meters away. The diet includes a great variety of grasses, shrubs, and forbs. *Procavia* is predominantly a browser in some areas and a grazer in others. It feeds on practically any plant, including some that are poisonous to most other animals, such as plants of the families Solanaceae and Euphorbiaceae. It is adept at climbing trees and going out on branches to obtain fresh leaves. It can secure sufficient moisture from its food but will drink regularly if water is available (Olds and Shoshani 1982; Smithers 1983).

According to Hoeck (1984), population density varies from 5/ha. to 40/ha., and group home range averages 4,250 sq meters. The dominant male of a group is territorial and repels other males. The females do not defend their range and may allow an outside female to join the group. The gregarious nature of *Procavia* partially reflects its needs with respect to heat conservation and shelter (Hoeck 1975; Olds and Shoshani 1982; Smithers 1983). Observations of captives indicate that as many as 25 animals will crowd together in a shelter to conserve warmth, either huddling on one level or heaping on top of each other. Depending on the amount of suitable habitat in a given location, especially the extent of shelter among rocks, a colony may be limited to a single family or may consist of hundreds of animals. The basic family unit consists of 1 adult territorial male, sometimes a subordinate male, and several adult females and young. Some males live alone after leaving their family group at 16–30 months. Group size is 2–26 individuals, usually about 4–10. The territorial male is the leader and most watchful member of the colony. Either he or one of the females stays on a high rock or branch while the others feed or sun themselves, and if danger threatens, this sentry gives an alarm call—a sharp bark—upon which the others take to cover. Fourie (1977) distinguished 21 adult vocalizations that apparently are important for information transfer.

Reproductive activity seems to vary geographically (Gombe 1983; Kingdon 1971; Mendelssohn 1965; Millar 1971; Smithers 1971, 1983). Mating has been reported in Israel from August to September, in Kenya from August to November, in South Africa in the second half of the summer (February and March), and in Tanzania at the end of the wet season, from April to June. Births peak in Israel from mid-March to early May, in Zimbabwe from March to April, and in Cape Province in September and October. The estrous cycle lasts about 13 days. The gestation period is around 202–45 days. Births usually occur in a crevice. The overall number of young per litter is 1–6, with reported averages of 3.2 in Israel, 1.9 in Zimbabwe, 2.2 in Transvaal, and 3.3 in Cape Province. The newborn weigh 170–240 grams each, are fully covered with hair, have their eyes open, and can move about with agility after a day. They begin taking solid food within 2 weeks but may continue to nurse for 1–5 months. Sexual maturity is reached at about 16–17 months. A captive rock dassie lived for 11 years (Jones 1982), and wild individuals are known to have lived at least 8.5 years (Hoeck 1989).

The chief natural enemies of *Procavia* are leopards and eagles, but it also is preyed upon by foxes, weasels, and mongooses. Its flesh is prized by some natives. Wanton killing for use as food and destruction of natural vegetation by people have led to a drastic reduction of the formerly abundant populations of *Procavia* in Egypt (Osborn and Helmy 1980). In some areas it takes up residence in culverts or stone walls and becomes a pest. There had been suggestions that it sometimes competes with sheep for forage in South Africa

and should be reduced in numbers, but Lensing (1983) showed that only about 1 percent of the potential agricultural yield of the involved area is lost each year to hyraxes. The young make interesting pets and will eat practically any vegetable matter.

HYRACOIDEA; PROCAVIIDAE; **Genus HETEROHYRAX**
Gray, 1868

Gray Hyraxes, or Yellow-spotted Hyraxes

There are three species (Bothma, *in* Meester and Setzer 1977):

H. antineae, Ahaggar Mountains of southern Algeria;
H. brucei, southeastern Egypt to central Angola and northeastern South Africa;
H. chapini, east-central Zaire.

Both Ellerman and Morrison-Scott (1966) and Roche (1972a) regarded *Heterohyrax* as only a subgenus of *Dendrohyrax*, but Hoeck (1978) showed that generic distinction is warranted. Both Roche (1972a) and Corbet and Hill (1986) considered *H. antineae* and *H. chapini* to be not more than subspecifically distinct from *H. brucei*, but Hoeck (1984) and Honacki, Kinman, and Koeppl (1982) listed all three taxa as full species.

Head and body length is 320–560 mm, an external tail is absent, height at the shoulder is about 305 mm, and weight is around 1.3–4.5 kg. The body is covered with thick, short, rather coarse hair. The upper parts are generally brownish or grayish, sometimes suffused with black, and the underparts are white. In general appearance it is much like a large guinea pig (*Cavia*). *Heterohyrax* is distinguished by a patch of yellowish, reddish, or whitish hair in the middle of the back. This patch covers a gland that is exposed when the surrounding hairs are erected in anger. Females usually have one pair of pectoral mammae and two pairs of inguinal mammae.

Sclater (1900) wrote: "The soles, which are naked, are covered by a very thick epithelium which is kept constantly moist by the secretion of the sudorific glands there present in extraordinary abundance; furthermore, a special arrangement of muscles enables the sole to be contracted so as to form a hollow air-tight cup which, when in contact with the rock, gives the animal great clinging power, so much so that even when shot dead it remains attached to almost perpendicular surfaces as if fixed there."

Smithers (1971) reported that in Botswana *H. brucei* is confined to areas of rocky kopjes, rocky hillsides, krantzes, and piles of loose boulders, particularly where there is a cover of trees and bushes on which it can feed. This species is more strictly confined to the larger areas of such a habitat and is less prone than is *Procavia* to colonize smaller outlying kopjes and piles of boulders. The altitudinal distribution of *Heterohyrax* is from sea level to at least 3,800 meters in mountains. *Heterohyrax* is sharp-sighted, keen of hearing, and quite aggressive, prepared to bite anything that attacks. Its presence often is detected by its calls, which resemble shrill screams. It is noted for its wariness, being ever alert and seeking shelter upon the slightest alarm, but is extremely curious and will soon show itself again. It enjoys basking in the sun and lying on rocky ledges during the afternoon, and individuals like to play and chase one another among the rocks.

Heterohyrax is primarily diurnal. Its habits and thermoregulatory needs are very much like those described above for

Gray hyrax (*Heterohyrax* sp.), photo by W. Leuthold.

Procavia, with which it may occur in close association. The two genera sometimes inhabit the same kopjes, they shelter in the same crevices and huddle together when outside, and their young play together. *Heterohyrax* is more of a browser than is *Procavia,* some 80 percent of its food being obtained in this manner, and it also is known to climb up in trees in order to feed. The diet consists of many kinds of bushes and trees, even those that are poisonous to most other mammals. Dobroruka (1973) reported that at the Numba Caves in Zambia *H. brucei* fed almost exclusively on the leaves of wild bitter yam *(Dioscorea dumetorum),* a plant often used by natives of the area to prepare poisoned arrows. According to Hoeck (1975), in Serengeti National Park *H. brucei* had feeding peaks in the morning and evening, and though primarily a browser, it grazed more in the wet season than in the dry season.

The following information on social life, demography, and reproduction was taken from the reports of Hoeck (1975, 1982, 1984, 1989) and Hoeck, Klein, and Hoeck (1982) on their studies in Serengeti National Park, Tanzania. In large areas of favorable habitat *Heterohyrax* lives in colonies of sometimes hundreds of animals, with population densities of 20–53 individuals per ha. Such aggregations are divided into cohesive, stable, polygynous family units of 5–34 animals each. On small kopjes (4,000 sq meters or less) there may be only a single family. One group observed for several years occupied an area of 3,600 sq meters and had an average of 16.3 members. Groups typically contain a single dominant, territorial, adult male; several peripheral males; 3–7 related adult females; and the young of one or two years. If several groups inhabit the same kopje, their ranges overlap, but the females of each family have their own core area, averaging about 2,100 sq meters, which is defended by the dominant male. He repulses and may fight all intruding adult males. He often stands guard on a high rock while the other animals feed or bask and is the first to give the shrill alarm call in the event of danger. He also emits a distinctive territorial call. The females do not defend the home range and sometimes allow an outside female to join their group. On a large kopje there may be a number of peripheral or young dispersing males that live alone and occupy ranges that overlap a number of group ranges. They may harass the territorial males, attempt to rush in and mate with their females, and await their chance to completely take over a family unit. On a small kopje a

young male must either quickly challenge the dominant male or depart the area.

The territorial male seeks to monopolize all females over 28 months old and shows little interest in younger ones. This behavior may serve to prevent inbreeding. In the Serengeti there are discrete mating seasons of about 7 weeks, during which adult females enter estrus several times for several days each. The two corresponding birth seasons are in May–July, following the long rainy season, and in December–January, after the short rains. Females of a given family group become receptive in only one of the annual breeding seasons. All the pregnant females of a group then give birth within 3 weeks of each other at one time of the year. Gestation lasts about 230 days. The average number of young is 1.6, and the range is 1–3. They are fully developed when born, are soon able to scamper about, and play together in their own nursery groups. They are weaned at 1–5 months and reach sexual maturity at 16–17 months. Females usually remain in their mother's family, but young males disperse at 16–30 months. Average longevity of a group of 17 females was 50.6 months, but one female is known to have lived at least 11 years. Males do not seem to live as long.

Native peoples eat hyrax flesh when other types of meat are scarce, but it is said to be tough, dry, and not very desirable. The chief foe of hyraxes is the rock python. They also are taken by leopards, birds of prey, and probably mongooses and other small carnivores.

HYRACOIDEA; PROCAVIIDAE; **Genus DENDROHYRAX**
Gray, 1868

Tree Hyraxes, or Bush Hyraxes

There are three species (Bothma, *in* Meester and Setzer 1977):

D. dorsalis, forest zone from Gambia to Uganda and extreme northwestern Angola, island of Bioko (Fernando Poo);
D. arboreus, eastern Zaire and central Kenya to eastern South Africa;

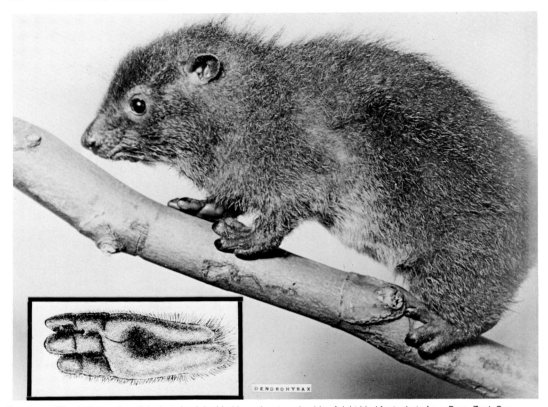

Tree hyrax *(Dendrohyrax dorsalis)*, photo by John Markham. Inset: underside of right hind foot, photo from *Proc. Zool. Soc. London.*

D. validus, eastern mainland of Tanzania and the islands of Zanzibar, Pemba, and Tumbatu.

Head and body length is 400–600 mm, tail length is 10–30 mm, and weight is usually 1,500–4,500 grams. The hair on the upper parts is brown, tipped with gray or yellow; black hairs are also present on the back. Some forms have two color phases, the darker coat and a yellow pelage. A white patch of hair, marking the location of a gland, is present on the back, and the ears are edged with white hairs. The underparts are usually brownish. The fur is longer and slightly more silky than that of other hyraxes.

According to Jones (1978), *D. dorsalis* occurs most frequently from sea level to about 1,500 meters' elevation but has been recorded at up to 3,000 meters. It inhabits upland and riverine areas within tropical rainforest or tropical closed forest and is more numerous in areas containing lianas than it is in other habitats. It is arboreal but frequently descends and moves about on the ground. This species is strictly herbivorous, feeding on leaves, fruits, twigs, and bark, usually from the upper canopy of the forest. It is nocturnal, with most activity taking place soon after dark and just before dawn.

Studies of captive *D. arboreus* by Rudnai (1984) indicate that the species is arboreal, sometimes nocturnal, and solitary. Smithers (1983) indicated that some populations of this species inhabit subalpine areas at altitudes of up to 4,500 meters but that most are found in forests at lower elevations, where they shelter in tree holes or dense vegetation. According to Kingdon (1971), *D. arboreus* is partly diurnal at high elevations, where it lives among rocks and is gregarious. It can be very noisy, and the calls often seem to follow a frequency pattern; most vocal activity seems to occur following inten-

sive feeding periods, between 2000 and 2300 hours and between 0300 and 0500 hours. The cry is made by both sexes but is much more powerful in the male, and it may have both sexual and territorial functions. Breeding seems to continue throughout the year.

Jones (1978) stated that *D. dorsalis* is usually solitary but that groups of two and occasionally three animals are to be found. Each animal uses a small area in the forest, usually centered around a single tree. In this species both mating and birth reportedly occur during the dry season. The young are born primarily during March and April in Gabon and Cameroon and from May to August in the western and southern parts of Zaire. In the eastern parts of the range of the species, young are born throughout the year. A single baby, occasionally two, is produced after a gestation period of about 8 months. The young are precocious; they reach adult length at about 120 days of age, and little growth occurs thereafter. Sexual maturity is attained at more than 1 year of age. One *D. dorsalis* lived in captivity for 5.5 years. According to Jones (1982), a specimen of *D. arboreus* was still living after 12 years and 3 months in captivity.

Predators of *D. dorsalis* include hawk eagles, leopards, golden cats, genets, servals, and pythons (Jones 1978). Tree hyraxes often assume a characteristic defensive position when threatened by a predator: the back and rump are turned toward the enemy, and the hairs around the dorsal gland are spread out and separated from each other, so that the naked glandular area is exposed. They are said to be less irritable than the other dassies and to tame readily. They are hunted by people for food and pelts throughout their range, and Hoeck (1984) suggested that they may be endangered in some areas through forest destruction.

Order Sirenia

Manatees, Dugong, and Sea Cow

This order of aquatic mammals contains two Recent families: Dugongidae for the genera *Dugong* (dugong, one species) and *Hydrodamalis* (Steller's sea cow, one species, extinct); and Trichechidae for the single genus *Trichechus* (manatees, three species). The dugong inhabits coastal regions in the tropical parts of the Old World, but some individuals go into the fresh water of estuaries and up rivers. Steller's sea cow occurred in the Bering Sea, being the only Recent member of this order adapted to cold waters. Manatees live along the coast and in coastal rivers in the southeastern United States, Central America, the West Indies, northern South America, and western Africa.

These massive, fusiform (spindle-shaped) animals have paddlelike forelimbs, no hind limbs or dorsal fin, and a tail in the form of a horizontally flattened fin. The adults of the living forms generally are 250–400 cm in length and weigh up to 908 kg. The skin is thick, tough, often wrinkled, and nearly hairless; stiff, thickened vibrissae (tactile hairs) are present around the lips. The head is rounded, the mouth is small, and the muzzle is abruptly cut off. The nostrils, which are valvular and separate, are located on the upper surface of the muzzle. The eyelids, though small, are capable of contraction, and a well-developed nictitating membrane is present. An external ear flap is lacking. The neck is short. The females have two pectoral mammae, one on each side in the axilla behind the flipper. The testes in males are abdominal (borne permanently within the abdomen).

The dense, heavy skeleton is characterized by an absence of pneumatic cavities (pachyostosis). The increased specific gravity is probably an adaptation to remaining submerged in shallow waters (Kaiser 1974). The skull is large in proportion to the size of the body; the nasal opening is located far back on the skull and directed posteriorly. The lower jaws are heavy and united for a considerable distance. The forelimb has a well-developed skeletal support, but there is no trace of the skeletal elements of the hind limb. In *T. senegalensis*, at least, there is no movement in the wrist joint (Kaiser 1974). The pelvis consists of one or two pairs of bones suspended in muscle. The vertebrae are separate and distinct throughout the spinal column; one genus *(Trichechus)* has six neck vertebrae (nearly all other mammals have seven).

The dentition is highly modified and often reduced (functional teeth are lacking in *Hydrodamalis*). Most of the incisor teeth are reduced or absent, and canines are present only in certain fossil species. When the incisors are present, there is a space between them. The cheek teeth, which are arranged in a continuous series, number from 3 to 10 in each half of each jaw. The anterior part of the palate and the corresponding surface of the lower jaw are covered with rough, horny plates, presumably used as an aid in chewing food; the small, fixed tongue is also supplied with rough plates.

The genera differ in a number of ways. In *Dugong* and *Hydrodamalis* the tail fin is deeply notched, but in *Trichechus* the tail fin is more or less evenly rounded. The upper lip is

 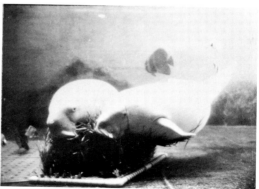

Left, Manatees *(Trichechus manatus)*, female and calf, photo by James A. Powell, Jr., U.S. Fish and Wildlife Service. Right, dugongs *(Dugong dugon)*, female on right and male on left, photo by T. Kataoka, Toba Aquarium, Japan.

more deeply cleft in manatees than in *Dugong* and *Hydro-damalis*. In *Dugong* there are 3/3 functional cheek teeth, and the males have one pair of tusklike incisors (usually un-erupted in females). In *Trichechus* functional incisors are not present, but the cheek teeth are numerous and indefinite in number (up to 10 in each half of each jaw). The cheek teeth of *Trichechus* are replaced consecutively from the rear, as in many proboscideans; an individual cheek tooth is worn down as it moves forward.

Sirenians are solitary, travel in pairs, or associate in groups of three to about six individuals. Generally slow and inoffen-sive, they spend all their life in the water. They are vege-tarians and feed on various water plants. They are the only mammals that have evolved to exploit plant life in the sea margin (Anderson 1979). The ordinal name Sirenia is related to the supposed mermaidlike nursing of dugongs (thought to be the origin of the myths of the sirens) and manatees. The only reliable observations of nursing in manatees, however, have revealed that the young suckle while the mother is un-derwater in a horizontal position, belly downward. P. K. Anderson (1984a) reported that suckling in the dugong is somewhat similar but that the calf usually is in an inverted position.

The Sirenia often are classified together with the Pro-boscidea and Hyracoidea in the mammalian superorder Paenungulata. The geological range of the order Sirenia is early Eocene to Recent. The earliest fossils are from Hungary and India. By the middle Eocene the order was present in southeastern North America, the West Indies, southern Eu-rope, northern and eastern Africa, and south-central Asia, and three distinct families—the Dugongidae, Prorastomidae, and Protosirenidae—had evolved (Dawson and Krishtalka 1984; Domning, Morgan, and Ray 1982). These aquatic mammals were apparently more abundant from the Oligo-cene to the Pliocene than they are now. Their comparative scarcity at the present time probably results from climatic changes in the Pliocene and Pleistocene and, more recently, exploitation by humans for food, hides, and oil. The number of individual sirenians remaining in the world, perhaps 60,000, is far smaller than that of any other mammalian order.

SIRENIA; Family DUGONGIDAE

Dugong and Sea Cow

This family contains two Recent subfamiles: Dugoninae, with the single Recent genus *Dugong* (dugong), found in coastal regions of the tropics of the Old World; and Hydro-damalinae, with the single Recent genus *Hydrodamalis* (Steller's sea cow), formerly found in the Bering Sea (Rath-bun 1984).

Dugong is generally 250–350 cm in length, whereas Steller's sea cow measured around 800 cm. The flippers in Steller's sea cow were curiously bent and were said to have been used to pull an individual along the bottom of the ocean as it foraged. Members of the family Dugongidae lack nails on their flippers. The deeply notched tail fin has two pointed, lateral lobes (the tail fin in manatees is more or less evenly rounded). The upper lip, which is more deeply cleft in the adult dugong than in the young, is not so deeply cleft as that of the manatees. The nostrils are located more dorsally than in other sirenians. In the dugong (and apparently all sire-nians) the eyelids contain a number of glands which produce an oily secretion to protect the eye against water.

The ribs of *Hydrodamalis* consisted entirely of dense bone; those of *Dugong* contain some porous bone. The skull

of a dugong has a downwardly bent rostrum (for holding the incisors), whereas the skull of a Steller's sea cow had a ros-trum that was only slightly inclined. *Hydrodamalis* lacked functional teeth but did possess functional, rough oral plates. Male *Dugong* have one pair of tusklike incisors (usually un-erupted in females), which are directed downward and for-ward and are partially covered with enamel. *Dugong* has 3/3 functional cheek teeth (up to 6/6 may be present in an indi-vidual's lifetime). The cheek teeth are columnar, are covered with cement, lack enamel, and have simple, open roots.

The geological range of this family, aside from the modern distribution, is early Eocene to Pliocene in Europe, middle Eocene and Pliocene in North Africa, Oligocene in Madagas-car, late Oligocene in southern Asia and Japan, Miocene in Sri Lanka, middle Eocene to Pleistocene in North America, and early Miocene to early Pliocene in South America (Domning and Ray 1986; De Muzion and Domning 1985; Rathbun 1984).

SIRENIA; DUGONGIDAE; Genus DUGONG
Lacépède, 1799

Dugong

The single species, *D. dugon*, originally occurred in the Red Sea and Gulf of Aqaba, along the eastern coast of Africa as far south as Mozambique and occasionally Natal, along the southern coast of Asia, around Madagascar and many of the islands in the Indian Ocean, throughout the East Indies, off eastern Asia as far north as Taiwan and the Ryukyu Islands, along the coast of Australia except in the south, and in the Pacific Ocean at least as far east as the Caroline Islands and the New Hebrides (Husar 1978a; Nishiwaki et al. 1979). Accord-ing to Nishiwaki and Marsh (1985), individuals occasionally are seen near Sydney, Australia, about 700 km south of the regularly occupied range. Kingdon (1971) indicated that this species also was present within historic times in the eastern Mediterranean Sea.

Adults usually have a total length of 240–70 cm and weigh 230–360 kg, but recorded maximums are 406 cm and 908 kg (Husar 1978a). There is little sexual dimorphism, though females may grow to a larger average size than do males (Marsh, Heinsohn, and Marsh 1984). The coloration is vari-able but is usually dull brownish gray above and somewhat lighter below. The skin is thick, tough, relatively smooth, and covered with widely scattered hairs. The front limbs are mod-ified into flippers about 35–45 cm in length; they are used for propulsion by the young, but adults propel themselves by means of the flukelike tail, using the flippers only for steer-ing. When the dugong grazes on the ocean floor, the flippers are used for "walking," and not to probe for food, but captives have been seen to use the flippers to convey food to the mouth. The dugong has a relatively small and simple stomach and also has unique adaptations for feeding on marine grasses. The upper lip protrudes considerably beyond the lower; it is deeply cleft, forming a large U-shaped, muscular pad that overhangs the small, downwardly opening mouth. On the sides of this facial pad are two ridges that bear short, sturdy, blunt bristles. The lower lip and the distal portion of the palate have rough, horny pads that are used to grasp sea grasses during feeding; the whole upper lip is extended and curved around the base of a plant, which is then grasped with the mouth pads and pulled up by the roots.

The dental formula is: (i 0/0, c 0/0, pm 0/0, m 2–3/2–3) × 2 = 10–14. The upper incisors are rootless and straight,

Dugong *(Dugong dugon)*, photos of a live female from the Surabaya Zoological Garden, temporarily stranded while her pool is being cleaned.

forming short, thick tusks in males over 12–15 years old. In females the incisors usually do not pierce the gum, which implies that these teeth have some sex-related function rather than a food-gathering function. Adult males invariably carry conspicuous scars that seem to have been made by competing males. The molars are circular in section, thick, rootless, and lacking in enamel. The last molar tooth is a double, rather than a single, cylinder (Lekagul and McNeely 1977). Young animals have up to six cheek teeth (including three premolars) on each side of each jaw; however, each row of cheek teeth migrates forward, worn teeth dropping out anteriorly, until only two molars remain in each quadrant of old animals (Rathbun 1984).

The dugong occurs in the shallow waters of coastal regions of tropical seas, where there is an abundance of vegetation. It is more strictly marine than are manatees and seldom is found in freshwater localities. Long-distance migration is unknown, but off some parts of Africa, southeastern Asia, and Australia seasonal changes in abundance are associated with the monsoons and may reflect movements in response to rough weather and availability of food; in some areas there also seem to be regular daily movements between feeding areas and deeper waters (Husar 1978a). The dugong may rest in deep water during the day and move toward the shore to feed at night, but Nishiwaki and Marsh (1985) noted that there is diurnal inshore feeding in some areas. Anderson (1986) determined that a population in Shark Bay, off Western Australia, has cyclic movements, abandoning feeding sites when temperatures fall below 19° C and seeking warmer waters for the winter. P. K. Anderson (1984b) noted that the distance to these winter waters is over 160 km. Average swimming speed is 10 km/hr, but animals can nearly double this rate if pressed (Husar 1978a). Dives seem to vary in length depending on such factors as water depth and forage species but have been reported normally to last around 1–3 minutes (P. K. Anderson 1982b; Nishiwaki and Marsh 1985). Feeding generally is at depths of 1–5 meters (P. K. Anderson 1984b).

Dugong is mostly herbivorous, and historic distribution broadly coincided with the tropical Indo-Pacific distribution of food plants, the phanerogamous sea grasses of the families Potomogetonaceae and Hydrocharitaceae. Such vegetation also is found in the Mediterranean Sea, where the dugong formerly may have occurred (Kingdon 1971). Although sea grasses are the primary food of the dugong, it has been reported to feed on brown algae following a cyclone that damaged the sea grass bed; green algae, marine algae, and some crabs also have been found in dugong stomachs (Husar 1978a).

According to Anderson (1982a), population densities in Shark Bay range from 0.12 to 12.8 individuals per kilometer of coastline. The dugong is essentially a gregarious animal, though it frequently is solitary. In the past, huge groups, sometimes containing thousands of individuals, were reported in various areas. Herds with hundreds of animals still can be found, but numbers usually are far smaller. Anderson suggested that such groups apparently are not simply aggregations in response to favorable feeding conditions but may involve deliberate assembly for protection against predators (especially sharks) and a process by which younger animals learn advantageous patterns of movement. Most dugongs in Shark Bay were found in the company of at least one other individual, and about 10–12 percent of the animals there were calves. Additional information, as summarized by Husar (1978a) and Lekagul and McNeely (1977), suggests that there may be lasting pair bonds between mated dugongs

Dugong *(Dugong dugon)*, photo by T. Preen / National Photographic Index of Australian Wildlife.

and that possibly family groups are formed within the larger groups. Calves may sometimes separate from the herd to form nursery subgroups of their own. Aggressive behavior evidently is rare. *Dugong* has been reported to make whistling sounds when frightened; calves make a bleating, lamblike cry. These vocalizations are thought to be used only for short-range communication.

Husar (1978*a*) wrote that breeding seems to occur throughout the year, with no well-defined season, though births apparently peak from July to September in Sri Lanka and parts of Australia. Marsh, Heinsohn, and Marsh (1984) reported that most births in northeastern Australia take place from September to December. Nishiwaki and Marsh (1985) stated that females are polyestrous and that the gestation period is 13–14 months. There normally is a single young; twins are rare (Husar 1978*a*). The newborn is 100–120 cm long and weighs 20–35 kg (Nishiwaki and Marsh 1985). It is born underwater and swims immediately to the surface for its first breath of air. The baby clings to its mother's back as she browses through the shoals of sea grass, submerging when the mother submerges and rising when she rises (Lekagul and McNeely 1977). The young have been seen to suckle while underwater in an inverted position behind the mother's axilla, and they sometimes are carried above the water when the mother rolls (P. K. Anderson 1984*a*). Lactation may last at least 18 months, though the calves begin to graze within 3 months of birth. Sexual maturity is attained by both males and females at as early as 9–10 years of age but may be delayed until 15 years. Females have a calving interval of 3–7 years, and the maximum rate of population increase is thought to be about 5 percent (Marsh, Heinsohn, and Marsh 1984). A pair of captives was maintained for 10 years (Husar 1978*a*), but examination of wild specimens indicates that maximum longevity is about 70 years (Nishiwaki and Marsh 1985).

Sharks are probably the main natural enemy of the dugong, but individuals have been seen to "gang up" on sharks in shallow water and drive them off by butting them with the head (Lekagul and McNeely 1977). Anderson and Prince (1985) related a report of a devastating attack by about 10 killer whales on a tightly bunched group of approximately 40 dugongs. People have had a far more serious long-term effect. Specialized human cultures based on dugong hunting have developed in several areas, such as the Torres Strait, between Australia and New Guinea, where an estimated 750 animals still are killed annually (Nishiwaki and Marsh 1985). The

dugong has been hunted for food throughout its range, its meat being likened to tender veal. The hide has been used to make a good grade of leather. The species also has been taken for its oil (24–56 liters for an average adult) and for its bones and teeth, which have been used to make ivory artifacts and a good grade of charcoal for sugar refining. Commercial dugong fisheries once operated from Sri Lanka, and several Asian cultures have prized dugong products for supposed medicinal and aphrodisiac properties. Large gill nets and harpooning are used to capture the dugong (Husar 1978*a*).

In many areas, the dugong has declined greatly in numbers, and fears have been expressed that it might be exterminated by continued hunting pressure. Kingdon (1971) wrote that it was known to the ancient Greeks, Egyptians, and Phoenicians of the Mediterranean; no populations have occurred in that region for centuries, though occasional individuals may have traversed the Suez Canal. According to Nishiwaki and Marsh (1985), populations also have disappeared from the Mascarene, Laccadive, and Maldive islands in the Indian Ocean. They have become very rare around Guam and Yap, in the Ryukyu Islands, and along much of the mainland coast of eastern Asia. There also has been a serious recent decline in the Torres Strait. Populations along the coasts of India, southwestern Asia, Africa, and Madagascar are thought to be in critical danger. The species is classified as vulnerable by the IUCN and is on appendix 1 of the CITES, except for the Australian population, which is on appendix 2. The USDI lists the dugong as endangered, but because of an uncorrected technicality this classification does not apply within territory of the United States, such as Guam. Nishiwaki et al. (1979) estimated that roughly 30,000 dugongs may still inhabit the whole Indo-Pacific region. There are still some thousands in Australian waters (Nishiwaki and Marsh 1985), including at least 1,000 that use Shark Bay (Anderson 1986).

SIRENIA; DUGONGIDAE; Genus **HYDRODAMALIS**
Retzius, 1794

Steller's Sea Cow, or Great Northern Sea Cow

The single Recent species, *H. gigas*, occurred in historical time around Bering and Copper islands in the Commander

Islands of the western Bering Sea (Rice 1977). Pleistocene remains of the same species have been found on Amchitka Island in the Aleutians and in Monterey Bay, off California. Pliocene remains of a related species, *H. cuestae*, have been found in Japan, California, and Baja California (Domning 1978).

This animal was the largest of the sirenians and possibly the largest noncetacean mammal to exist in historical times. Calculations by Scheffer (1972) indicate that the largest specimens were at least 800 cm long and weighed 10,000 kg. Observations by Steller (as reported by Stejneger 1936), the famous German naturalist who accompanied Bering (see below), provide an additional idea of the size and appearance of the animal in life. An adult female measured 752 cm from the tip of the nose to the point of the tail flipper. Its greatest circumference measured 620 cm, and its estimated weight was about 4,000 kg. The head was small in proportion to the body. The tail had two pointed lobes forming a caudal flipper. The forelimbs were very small, flipperlike, and somewhat truncated. There were no hind limbs externally, no trace even remaining of the pelvic elements. The skin was naked and was covered with a very thick, barklike, extremely uneven-appearing epidermis, from which the German name *Borkèntier* (bark animal) was derived. The skin was dark brown to gray brown in color, occasionally spotted or streaked with white. The forelimbs were covered with short, brushlike hairs. A large dried sample of this skin is at the Zoological Institute in Leningrad, but a reported specimen in the Hamburg Zoological Museum may actually represent a cetacean (Forsten and Youngman 1982). *Hydrodamalis* was parasitized by two or three species of small crustaceans that burrowed into the rough skin. Steller said that he found some of the holes made by the small crablike crustacean *Cyamus* to ooze a thin serum.

According to Forsten and Youngman (1982), the condylobasal length of the skull was 638–722 mm, the zygomatic width was 324–373 mm, the nasal basin on top extended past the eye opening, and the braincase was relatively small. *Hydrodamalis* had no functional teeth; mastication was accomplished by keratinized pads, one each in the rostral areas on the upper and lower jaws. The upper and lower pads had corresponding V-shaped crests and valleys, between which the food was ground.

Hydrodamalis was the only Recent sirenian adapted to cold waters. All that we know about its habits comes from Steller's account. It was quite numerous in the shallow bays and inlets around the coast of Bering Island when first discovered. It fed upon the kelp beds and the extensive growths of various marine algae that grew in the shallower waters. It was slow-moving, was utterly fearless of humans, and was said to be affectionate toward others of its kind, as individuals appar-

Steller's sea cow (*Hydrodamalis gigas*): A. Photo from *Extinct and Vanishing Mammals of the Western Hemisphere*, Glover M. Allen; Inset: Steller's sea cow's palate, from *Symbolae Sirenologicae*, drawing by J. F. Brandt; B. Outer surface of skin, photo by Erna Mohr.

ently tried to aid others that were either wounded or in distress.

The following are excerpts from Steller's own account (from Stejneger 1936):

Usually entire families keep together, the male with the female, one grown offspring and a little, tender one. To me they appear to be monogamous. They bring forth their young at all seasons, generally however in autumn, judging from the many new-born seen at that time; from the fact that I observed them to mate preferably in the early spring, I conclude that the fetus remains in the uterus more than a year. That they bear not more than one calf I conclude from the shortness of the uterine cornua and the dual number of mammae, nor have I ever seen more than one calf about each cow.

These gluttonous animals eat incessantly, and because of their enormous voracity keep their heads always under water with but slight concern for their life and security, so that one may pass in the very midst of them in a boat even unarmed and safely single out from the herd the one he wishes to hook. All they do while feeding is to lift the nostrils every four or five minutes out of the water, blowing out air and a little water with a noise like that of a horse snorting. While browsing they move slowly forward, one foot after the other, and in this manner half swim, half walk like cattle or sheep grazing. Half the body is always out of water. . . . In winter these animals become so emaciated that not only the ridge of the backbone but every rib shows.

Their capture was effected by a large iron hook, . . . the other end being fastened by means of an iron ring to a very long, stout rope, held by thirty men on shore. . . . The harpooner stood in the bow of the boat with the hook in his hand and struck as soon as he was near enough to do so, whereupon the men on shore grasping the other end of the rope pulled the desperately resisting animal laboriously towards them. Those in the boat, however, made the animal fast by means of another rope and wore it out with continual blows, until, tired and completely motionless, it was attacked with bayonets, knives and other weapons and pulled up on land. Immense slices were cut from the still living animal, but all it did was shake its tail furiously.

In the late Pleistocene, *Hydrodamalis* evidently occurred all around the rim of the North Pacific from Japan to California. Subsequent hunting by primitive peoples probably eventually restricted it to the vicinity of the Commander Islands (Domning 1978). This last population was discovered in 1741 when a Russian expedition led by Captain Vitus Bering was stranded on Bering Island. At that time the total number of sea cows probably did not exceed 1,000 to 2,000. Bering's crew, mostly sick with scurvy, and subsequent visitors to the area slaughtered the animals relentlessly. The meat of *Hydrodamalis* was used for food, and the hide for making skin boat covers and shoe leather. The genus probably became extinct by 1768, though there were later reports of its presence in various areas.

SIRENIA; Family TRICHECHIDAE; Genus TRICHECHUS
Linnaeus, 1758

Manatees

The single Recent genus, *Trichechus*, contains three species (Caldwell and Caldwell 1985; Domning and Hayek 1986; Husar 1977, 1978b, 1978c; Rice 1977):

T. inunguis (Amazonian manatee), throughout the Amazon Basin of northern South America;

T. manatus (West Indian manatee), coastal waters and some connecting rivers from Virginia around the Gulf of Mexico and the Caribbean Sea to eastern Brazil, Orinoco Basin, Greater and formerly Lesser Antilles, Bahamas;

T. senegalensis (West African manatee), coastal waters and connecting rivers from Senegal to Angola, occurs as far as 2,000 km up the Niger River and in tributaries of Lake Chad.

The range of *T. inunguis* sometimes is said to include the Orinoco Basin, but such reports apparently are based on misidentification of *T. manatus* in the early nineteenth century. It also is unlikely that *T. inunguis* reaches the waters that connect the Orinoco and Amazon basins. *T. inunguis* occurs at the mouth of the Amazon on the Atlantic coast, where it may be sympatric with *T. manatus*, and the latter species once was present as far south as the state of Espirito Santo at about 20° S (Domning 1981, 1982b; Thornback and Jenkins 1982).

T. manatus was introduced in the Panama Canal around 1960 to aid in weed control, and the population there subsequently increased and extended its range to the Pacific Ocean (Domning and Hayek 1986; Montgomery, Gale, and Murdoch 1982).

These animals have a rounded body, a small head, and a squarish snout. The upper lip is deeply split, and each half is capable of moving independently of the other. The nostrils are borne at the tip of the muzzle, the eyes are small, and there are no external ears. The flattened tail fin is more or less evenly rounded (not notched as in the dugong and sea cow) and is the sole means of propulsion in adults. The flexible flippers are used for aiding motion over the bottom, scratching, touching and even embracing other manatees, and moving food into and cleaning the mouth (Caldwell and Caldwell 1985). Vestigial nails may occur on the flippers. Stout bristles appear on the upper lip, and bristlelike short hairs are scattered singly over the body at intervals of about 1.25 cm. The skin is 5.1 cm thick. Females have two pectoral mammae, one on each side in the axilla behind the flipper.

The skeleton is of extremely dense (pachyostotic) bone, a condition that increases specific gravity and may contribute to neutral buoyancy (Caldwell and Caldwell 1985). The stout ribs consist wholly of dense bone. Manatees have only six neck vertebrae, whereas nearly all other mammals have seven. Nasal bones are present in the skull (these bones are absent or vestigial in the dugong and sea cow). Manatees have 2/2 incisors, which are concealed beneath horny plates and are lost before maturity. The cheek teeth number up to 10 in each half of each jaw (although more than 6/6 are rarely present at any one time). They are low-crowned, enameled, divided into cuspidate cross-crests, lack cement, and have closed roots. These teeth are replaced horizontally; they form at the back of the jaw and wear down as they move forward. This may be an adaptation to eating food mixed with sand or silt; a similar replacement of teeth occurs in many proboscideans and other mammals but not in the Dugongidae.

Rathbun (1984) wrote that the known geological range of the family Trichechidae is early Pliocene to Recent in Atlantic North America, early Miocene to Recent in South America, and Recent in West Africa. Domning (1982a) speculated that the family is descended from protosirenian stock that became isolated during the Eocene in South America. Until the late Miocene trichechids may have been restricted to the coastal rivers and estuaries of that continent. By the Pliocene a branch of the family leading to modern *T. inunguis* reached the interior of the Amazon Basin, where it adapted to feeding on the newly available floating meadows of nutrient-rich lakes. Another Pliocene branch, the forerunner of *T. manatus*, spread into the Caribbean, where its wear-resistant dentition proved superior to that of the dugongids then present in the region for dealing with a diet modified by the increased silt runoff of the period. A closely related branch dispersed across the Atlantic to West Africa during the Pliocene or Pleistocene and gave rise to *T. senegalensis*. Additional information on each of the three species is given separately below.

Trichechus inunguis (Amazonian manatee).

Husar (1977) wrote that this species is smaller and of more slender proportions than either *T. manatus* or *T. senegalensis*; the largest recorded specimen was a 280-cm male. Timm, Albuja V., and Clauson (1986) cited a commercial hunter in Ecuador as saying that the largest manatee he had taken weighed 480 kg. Domning and Hayek (1986) stated that the skin is smooth, slick, and rubberlike, not wrinkled as in the other two species. The general coloration is gray, and most individuals have a distinct white or bright pink patch on the breast. Fine hairs are sparsely distributed over the body, and there are thick bristles on both the upper and lower lips. The

West Indian manatees *(Trichechus manatus)*, mother and calf, photos by James A. Powell, Jr.

flippers, unlike those of the other two species, are elongate and usually lack nails. The skull is relatively long and narrow.

The Amazonian manatee occurs exclusively in fresh water (Thornback and Jenkins 1982). It has been found all around

Ilha de Marajó, at the mouth of the Amazon on the Atlantic coast, but even there salt water has little or no influence (Domning 1981). This species favors blackwater lakes, oxbows, and lagoons, and it has been maintained successfully in waters with temperatures from 22° to 30° C (Husar 1977). Its key requirements in the wild seem to be large blackwater lagoons with deep connections to large rivers and abundant macrophytic plants such as grasses, bladderworts, and water lilies (Timm, Albuja V., and Clauson 1986). It usually surfaces several times per minute to breathe, but record submergence is 14 minutes (Husar 1977). A captive juvenile released and radiotracked for 20 days was found to be equally active by day and night, to move about 2.6 km per day, and to spend most of its time where food, especially floating vegetation, was most abundant (Montgomery, Best, and Yamakoshi 1981). The diet of *T. inunguis* consists mostly of vascular aquatic vegetation (Caldwell and Caldwell 1985). Captive adults consume 9–15 kg of leafy vegetables per day (Husar 1977).

Best (1983) found that manatee populations of the central Amazon Basin make an annual movement in July–August, when water levels begin to fall. Some enter rivers, where the animals can continue to find food but must also spend energy to hold their position in the current. Others become restricted to deep parts of larger lakes during the dry season, September–March, where they do not have obvious food sources until water levels rise 1–2 meters. Available evidence suggests that the latter populations fast for nearly seven months, though they may take some vegetation. The manatee's large fat reserves and low metabolic rate, only 36 percent that usual for eutherians, allow the species to survive at this time. Best's study was done at Lago Amana, where 500–1,000 manatees congregate during the dry season.

There have been other reports of large aggregations of manatees in the middle reaches of the Amazon, but such assemblies appear rare today. Loose groups of 4–8 individuals may be present in feeding areas. A few short vocalizations have been noted (Husar 1977). Breeding has been reported to occur throughout the year, at least in some areas, but Timm, Albuja V., and Clauson (1986) received information that births occur mainly in January in one part of Amazonian

Ecuador and in June in another. Evidence gathered by Best (1982) indicates that in the central Amazon Basin breeding is seasonal, with nearly all births taking place from December to July, mainly from February to May, the period of rising river levels. According to Husar (1977), gestation lasts approximately 1 year, and there normally is a single young. The mother-calf bond seems to be long-lasting; females carry the young on their back or clasped to their side. Two individuals lived for 12.5 years in captivity.

Available records (Domning 1982b; Thornback and Jenkins 1982) indicate that in Brazil the Amazonian manatee was being exploited commercially for its meat as early as the 1600s and that even by the late eighteenth century there was some concern for its survival. Commercial hunting continued, however, with about 1,000–2,000 individuals being marketed each year from 1780 to 1925. Many other animals were taken for subsistence purposes. The main product during this period was *mixira*, fried meat packed in lard. The commercial kill rose sharply starting in 1935, when there was an increased demand for the manatee's hide to manufacture heavy-duty leather. This industry collapsed about 1954, when synthetic replacements became widely available, but there then was a renewed intensification of meat hunting. About 4,000–10,000 manatees were killed each year for commercial purposes in Brazil from 1935 to the early 1960s. The annual take then fell to about 1,000–2,000, probably reflecting a severe decline of the species. The manatee was protected legally in Brazil in 1973, but subsistence hunting and local commercial killing has continued.

The manatee also has declined in other parts of its range (Thornback and Jenkins 1982). The Siona Indians of Ecuador, who are traditional hunters of the manatee, now have a self-imposed ban on taking the species because of its low numbers. However, Timm, Albuja V., and Clauson (1986) thought that if meat hunting by settlers and the military continues at present levels, the manatee would be eliminated in Ecuador within 10–15 years. Other problems include accidental drowning in commercial fishing nets and degradation of food supplies by soil erosion resulting from deforestation. The large-scale destruction of Amazonian rainforests also would lead to a decline in rainfall, a drop in water levels, and increased mortality of the manatee (Best 1983). An overall minimum population estimate of 10,000 individuals was cited by Husar (1977). *T. inunguis* is classified as vulnerable by the IUCN and as endangered by the USDI and is on appendix 1 of the CITES.

Trichechus manatus (West Indian manatee).

According to Husar (1978a), total length of adults varies from 250 to more than 450 cm, and weight is 200 to more than 600 kg, but average adults are 300–400 cm long and weigh less than 500 kg. Sexual dimorphism in size has not been documented, though females seem bulkier. The finely wrinkled skin is slate gray to brown in color. Fine, colorless hairs, 30–45 mm long, are sparsely distributed over the body, and stiff, stout bristles are on the muscular, prehensile pads of the upper lip. Nails are present on the dorsal surface of the flippers. The skull is broad and has a relatively short and downturned snout.

Except as noted, the information for the remainder of this account was taken from Hartman (1979) and refers to *T. manatus* in Florida. This species inhabits shallow coastal waters, bays, estuaries, lagoons, and rivers. It utilizes both salt and fresh water, though there may be preference for the latter. It also seems to prefer water above about 20° C but can endure water as cold as 13.5° C. In temperate parts of its range, such as most of Florida, individuals may congregate during the winter at sources of warm water, such as natural hot springs and discharges from power plants. Drops in the

air temperature below 10° C seem to induce congregation in the warm headwaters of the Crystal and Homosassa rivers in central Florida. Some individuals winter in the tropical waters of southern Florida. When air temperatures rise above 10° C, Florida manatees may leave their refugia and wander all along the Gulf Coast and up the Atlantic coast as far as Virginia.

There is evidence of long-range, offshore migrations between population centers. Individuals have been caught up to 15 km off the coast of Guyana. They also have been reported in rivers 230 km from the sea in Florida and 800 km from the coast in South America (Husar 1978a). Most manatees appear to be nomadic and to move hundreds of kilometers, pausing for days, weeks, months, or seasons in those estuaries and rivers that supply their needs. They follow established travel routes, preferring channels that are 2 meters or more in depth and shunning those that are less than 1 meter in depth. Animals generally swim 1–3 meters below the surface of the water. They use their tails not only to propel themselves through the water but also as rudders by means of which they control roll, pitch, and yaw. The flippers are used in precise maneuvering and in minor corrective movements to stabilize, position, and orient the animal. Speed normally is 3–7 km/hr but can reach 25 km/hr when the animal is pressed. The manatee avoids fast currents, and animals migrating through the Intracoastal Canal were never seen swimming against currents that exceeded 6 km/hr. According to Husar (1978c), average submergence time is 259 seconds, but a dive of 980 seconds has been recorded.

T. manatus may be active both by day and by night. During a 24-hour period adults have been observed to feed 6–8 hours in sessions that usually lasted 1–2 hours, to rest 6–10 hours, and to move up to 12.5 km. The manatee rests by hanging suspended near the surface of the water or by lying prone on the bottom. In both positions, the animal lapses into a somnolent state, with eyes closed and body motionless. Sound appears to be the principal sensory mode, but in clear water the preferred method of environmental exploration is visual. In addition, the prevalence of mouthing in social interaction suggests that the species has a chemoreceptive sense by which it can recognize odors in the water. Comfort activities include stretching, using the flippers to scratch and clean the mouth, rubbing against logs and rocks, and rooting in the substrate. Rubbing also may involve deposition of scent, by which an estrous female signals receptivity to wandering males (Rathbun and O'Shea 1984).

The manatee is herbivorous, though some invertebrates are ingested together with vegetation and may provide an important amount of protein. Also, Powell (1978) reported that captives have deliberately eaten dead fish, and wild individuals off Jamaica have been seen to take fish entangled in nets. The diet consists mainly of a wide variety of submerged, vascular plants, but emergent and floating vegetation sometimes is eaten. Water hyacinths *(Hydrilla)* are a staple food in many rivers of Florida. Sea grasses are taken in salt-water areas. Captives have consumed up to a fourth of their body weight per day in wet greens. Feeding occurs from the surface to a depth of 4 meters. Individuals are drawn to sources of fresh water, which they appear to require for osmoregulation. However, such a need was questioned by Reynolds and Ferguson (1984), who observed two large manatees 61 km northeast of the Dry Tortugas Islands, which in turn are about 110 km west of Key West, Florida, and which lack fresh water.

In their search for receptive mates, males reportedly utilize a much larger home range than do females. Some new information also suggests that individuals come together deliberately in order to learn the location of favorable habitat and migration routes (Thornback and Jenkins 1982). Generally, however, the West Indian manatee is considered to be a weakly social, essentially solitary species. The only lasting associa-

West Indian manatees *(Trichechus manatus)*, mother and calf, U.S. Fish and Wildlife Service photo by Galen B. Rathbun.

tion seems to be that of a cow with its calf. Temporary, casual groups form in favorable areas for purposes of migration, feeding, resting, or cavorting. Such groups may be randomly made up of juveniles and adults of both sexes, though Reynolds (1981) suggested a tendency for subadult males to associate. There is no evidence of communal defense or mutual aid and little or no indication of a social hierarchy. Individuals indulge in what appears to be play. They exchange gentle nibbles, kisses, and embraces that are age- and sex-independent. They have exceptional acoustic sensitivity; sound is doubtless the major directional determinant in social interactions. *T. manatus* normally is silent but can emit high-pitched squeals, chirp-squeals, and screams under conditions of fear, aggravation, protest, internal conflict, male sexual arousal, and play. Its vocalizations are probably nonnavigational and lack ultrasonic signals, pulsed emissions, or directional sound fields. Bengtson and Fitzgerald (1985) suggested that certain calls help to relay warnings and to keep groups together during flight. However, the only predictable vocal exchange between manatees involves screams of alarm, by which a cow calls her calf in case of danger, and the responding squeals of the young.

A population wintering in the Crystal River of Florida one year contained 31 adults, 13 juveniles, and 6 calves and was divided about equally by sex. When a female is in estrus, she may be accompanied for a period of one week to more than a month by up to 17 males. The courtship of the latter is relentless, but the cow generally seeks to escape and to vigorously repulse their advances. Rathbun and O'Shea (1984) suggested that this process involves mate selection by the female. It also involves constant and sometimes violent pushing and shoving by the bulls as they compete for a position next to the estrous female; such activity seems to be the only display of aggression in the species. There appears to be no

definite breeding season in Florida, and there probably is none elsewhere (Caldwell and Caldwell 1985). Calves have been seen throughout the year in the Dominican Republic and Puerto Rico (Belitsky and Belitsky 1980; Powell, Belitsky, and Rathbun 1981). The interbirth interval probably is normally at least 2.5 years but can be less if an infant is lost. The gestation period is about 13 months. The cow seeks the shelter of a backwater to give birth, normally to a single young. The newborn is about 120–30 cm long, weighs 28–36 kg, and is dark in color (Caldwell and Caldwell 1985). Within half a day of birth it is capable of swimming and surfacing on its own, though it occasionally rides on the mother's back. The calf begins to take some vegetation at about 1–3 months, though it continues to suckle until leaving its mother at about 1–2 years. Males appear to reach sexual maturity at 9–10 years, females at 8–9 years (Odell, Forrester, and Asper 1978). A specimen of *T. manatus* was still living after 30 years in captivity (Jones 1982).

Commercial exploitation of the West Indian manatee began in the sixteenth century, and combined with extensive subsistence hunting, it has resulted in severely reduced populations in most areas. According to Thornback and Jenkins (1982), *T. manatus* probably has disappeared in the Lesser Antilles and is continuing to decline almost everywhere else despite legal protection. Problems include killing for meat, drainage of swamps, accidental drowning in fishing nets and flood control structures, and pollution. The largest remaining population, consisting of thousands of individuals, occurs in Guyana. Numbers may be in the hundreds in Belize, Surinam, French Guiana, and perhaps the other South American countries, but there are only around 100 or less in each of the other countries of Central America and the Greater Antilles. *T. manatus* has been extirpated in most of its former range in Brazil, though there evidently was a small population as far south as Bahia in the 1960s, and there still is an isolated group at the Rio Mearim in Maranhao (Domning 1981, 1982b; Whitehead 1977). The species is classified as vulnerable by the IUCN and as endangered by the USDI and is on appendix 1 of the CITES.

The population in Florida waters originally contained an estimated several thousand animals but was so greatly reduced by 1893 that it was then given legal protection by the state (Caldwell and Caldwell 1985). It now also is covered by the federal Marine Mammal Protection Act and the Endangered Species Act, but many animals are killed and injured by vandals, entrapment in gates of locks and dams, collision with barges, and the propellers of power boats (O'Shea et al. 1985). Indeed it is a sad fact that most manatees in Florida bear scars from propellers and that such marks are the chief means by which biologists identify the individuals they are studying. Baugh (1987) noted that there are an estimated 1,200 manatees remaining in Florida, about 600 on each coast. There are at least 125–30 deaths annually, at least 30 percent of which are caused by people, and there also are an estimated 120–30 births each year in Florida. During the spring and summer substantial numbers of manatees continue to enter Georgia waters, and some occasionally move farther north. A few individuals also recently have been found in Mississippi, Louisiana, and southern Texas (Powell and Rathbun 1984). Evidently the animals along most of the Gulf Coast belong to the subspecies T. manatus latirostris, which is centered in Florida and represented by wanderers elsewhere in the United States and in the Bahamas. Those in southern Texas come from Mexico and belong to T. m. manatus, which is found in the remainder of the range of the species (Domning and Hayek 1986). Occurrences in Texas have declined in association with a drastic reduction of the population in Mexico.

Trichechus senegalensis (West African manatee).

According to Husar (1978b), adults are 300–400 cm long and weigh less than 500 kg. The skin is finely wrinkled and grayish brown in color. Fine, colorless hairs are distributed over the body, and there are stout bristles on the upper and lower lip pads. The flippers have nails on the dorsal surface. The skull is broad and has a shortened, slightly deflected snout. Domning and Hayek (1986) found T. senegalensis and

T. manatus to be phenetically similar and noted that the reduced rostral deflection of the former probably is associated with its dependence on emergent or overhanging, rather than submerged, vegetation.

The West African manatee occurs in both shallow coastal waters and freshwater rivers but seems to prefer large, shallow estuaries and weedy swamps; its range apparently is limited to water with a temperature above 18° C (Husar 1978b). It seems to avoid salt water (Nishiwaki et al. 1982). The diet includes aquatic vegetation, of which a free-ranging adult might be expected to consume 8,000 kg in a year (Husar 1978b). Populations in some rivers depend heavily on overhanging bank growth, and those in estuarine areas feed exclusively on mangroves (Domning and Hayek 1986). In Sierra Leone manatees reportedly remove fish from nets and consume rice in such quantities that they are considered to be pests (Reeves, Tuboku-Metzger, and Kapindi 1988).

Roth and Waitkuwait (1986) reported that T. senegalensis lives singly or in family groups of 4, or rarely 6, animals. Husar (1978b) related a report of 15 individuals being seen together and indicated that social behavior probably is similar to that of T. manatus. The breeding period is uncertain and may last throughout the year. Parturition supposedly occurs in shallow lagoons, usually there is a single calf, and the newborn is about a meter long.

Husar (1978b) wrote that there had been no large-scale commercial exploitation of T. senegalensis comparable to that of other manatees but that populations have declined markedly in some areas because of local hunting for meat. Reeves, Tuboku-Metzger, and Kapindi (1988) also pointed out such problems as accidental capture in fishing nets and persecution for alleged depredations on netted fish and growing rice. Surveys in 1980–81 (Nishiwaki 1984; Nishiwaki et al. 1982) yielded a rough overall estimate of 9,000–15,000 individual T. senegalensis, exclusive of Angola and Congo. The largest numbers, perhaps several thousand in each area, are found in Gabon and the lower reaches of the Niger and Benue rivers. Numbers also are considerable in Cameroon, Ghana, Ivory Coast, and the upper reaches of the Niger but are fewer from Senegal to Liberia. The species is classified as vulnerable by the IUCN and as threatened by the USDI and is on appendix 1 of the CITES.

Order Perissodactyla

Odd-toed Ungulates
(Hoofed Mammals)

This order of 3 Recent families, 6 genera, and 17 species is native to eastern Europe, central and southern Asia, parts of the East Indies, Africa, and the region from southern Mexico to Argentina. Introduction by human agency has led to establishment of wild-living populations of two species (*Equus caballus* and *E. asinus*) in certain areas where the order does not naturally occur. Simpson (1945) divided the living Perissodactyla into two suborders: Hippomorpha, for the family Equidae (horses, asses, and zebras); and Ceratomorpha, for the families Tapiridae (tapirs) and Rhinocerotidae (rhinoceroses). A third suborder, the extinct Ancylopoda, contained the family Chalicotheriidae, which developed clawed feet and which survived until the Pleistocene.

These are medium-sized to large animals adapted to running (especially members of the family Equidae). All the Recent families are quite distinct, with tapirs and rhinos resembling one another more than either family resembles the horses. The main feature common to all is that the weight of the body is borne on the central digits, with the main axis of the foot passing through the third digit, which is the longest on all four feet. In the horses, only the third digit of each foot is functional; in tapirs, four digits are developed on the forefoot and three on the hind foot; in rhinos, three digits are present on all four feet. The first digit is not present in any Recent forms; it was vestigial in certain fossil species. The terminal digit bones are flattened and triangular, with evenly rounded free edges, and are encased by hooves (some members of the extinct family Chalicotheriidae had clawed digits). Perissodactyls progress on their hooves or on their digits, never on the sole of the foot with the heel touching the ground. The ulna and the fibula are reduced, so the movement of these bones is reduced or lacking. The ankle bone, or astragalus, has only a single, deeply grooved, pulleylike surface for the tibia, and its lower end is nearly flat; the calcaneum, or heel bone, which has a widened lower end, does not articulate with the fibula.

The skin usually is thickened, and sparsely to densely haired. The mammae are located in the region of the groin, and the males do not possess a baculum.

The dental formula for the order is as follows: (i 0–3/0–3, c 0–1/0–1, pm 2–4/2–4, m 3/3) × 2 = 20–44; for the Recent species it is: (i 0–3/0–3, c 0–1/0–1, pm 3–4/3–4, m 3/3) × 2 = 22–44. The canines, when present, are never tusklike in the Recent species. The cheek teeth are arranged in a continuous series; the premolars (at least the rear members of the series) are molarlike in the Recent species; the first cheek tooth is a persistent milk premolar. The grinding teeth are usually complex in structure, massive, and low-crowned to high-crowned; prominent transverse ridges are present in the cheek teeth of tapirs and rhinos, whereas the cheek teeth of horses, which are grazers rather than browsers, develop high crowns with four main columns and various infoldings. Some fossil species in this order had tubercles on the crowns of the grinding teeth. The skull is usually elongate with an abrupt slope in the back. The nasal bones are expanded posteriorly. Characteristic of the order is the arrangement of openings in the skull by which nerves and blood vessels enter and leave the braincase. The Recent species lack horns with true bony cores, though roughened cushions on the nasal bones of the skull bear horns in rhinos.

The development of the foot, in its highest form in the horses, is a specialization that enables the animals to be swift and strong runners. The foot is not developed to such an extent in the rhinos and tapirs. However, rhinos can run rapidly for short distances, and tapirs also can run well, though they usually inhabit a type of terrain that permits them to plunge into dense cover or water to escape their enemies.

Rhinos and horses usually live on grassy plains or in open scrub country, while tapirs are found in humid tropical forests. Members of this order are strict herbivores and can be classified as either browsers or grazers depending on how they feed. The structure of their lips and teeth facilitates the obtaining and chewing of coarse vegetable food. Adulthood in

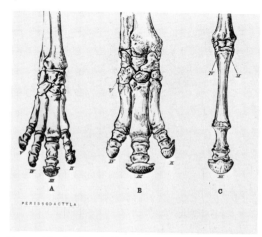

Bones of the forefeet: A. Tapir *(Tapirus indicus)*; B. Rhinoceros *(Dicerorhinus sumatrensis)*; C. Horse *(Equus caballus)*. Photos from *Mammalia*, F. E. Beddard.

these animals is attained in four to six years, and individuals may live five to seven times that long.

Perissodactyls flourished in early and middle Tertiary times. Carter (1984) listed 11 extinct families, gave the geological range of the order as early Eocene to Recent, and noted that it almost certainly had originated in the late Paleocene. Except for domestic horses and asses, however, perissodactyls are not now very numerous either in species or in individual animals. The living genera are mere remnants of a major evolutionary sequence that at one time was extremely diversified and widely distributed over the world. Even these genera each have one or more species that are in danger of extinction.

PERISSODACTYLA; Family EQUIDAE; Genus EQUUS
Linnaeus, 1758

Horses, Zebras, and Asses

The single Recent genus, *Equus*, contains six subgenera and eight species (Ansell, *in* Meester and Setzer 1977; Corbet 1978; Ellerman and Morrison-Scott 1966; Groves 1986b; Groves and Willoughby 1981):

subgenus *Asinus* Gray, 1822

E. asinus (African wild ass, donkey, or burro), probably once found in the wild from Morocco to Somalia and from Mesopotamia to Oman, now occurs in domestication throughout the world and feral populations established in some areas;

subgenus *Hemionus* Stehlin and Graziosi, 1935

E. hemionus (kulan and onager), originally found in the desert and dry steppe zone from the Ukraine and Palestine to Manchuria and western India;
E. kiang (kiang), Tibet and adjacent highland regions;

subgenus *Equus* Linnaeus, 1758

E. caballus (horse), probably once found in the wild throughout the steppe zone from Poland and Hungary to Mongolia, now occurs in domestication throughout the world and feral populations established in many areas;

subgenus *Dolichohippus* Heller, 1912

E. grevyi (Grevy's zebra), southern and eastern Ethiopia, Somalia, northern Kenya;

subgenus *Hippotigris* H. Smith, 1841

E. zebra (mountain zebra), southwestern Angola, Namibia, western and southern South Africa;

subgenus *Quagga* Shortridge, 1934

E. burchelli (Burchell's zebra), open country from southern Ethiopia to central Angola and eastern South Africa;
E. quagga (quagga), formerly found in South Africa.

Bennett (1980) concluded that there are only two subgenera: *Asinus* for the ass, kulan, and kiang; and *Equus* for the horse, quagga, and zebras. Dalquest (1988) recognized the two as full genera, based on his view that they had evolved sepa-

rately in the Pliocene, *Asinus* from the fossil genus *Astrohippus* and *Equus* from *Dinohippus*. Bennett also treated *E. onager* as a species distinct from *E. hemionus*, thereby restricting the range of the latter to regions east of Soviet Central Asia, and in this regard she was followed by Corbet and Hill (1986b) and Honacki, Kinman, and Koeppl (1982). The names *E. caballus* and *E. asinus* are based on domestic animals, and some authorities, such as Corbet (1978) and Groves (1986b), prefer to use the names *E. ferus* for the wild horse and *E. africanus* for the wild African ass. The diploid chromosome number is 64 in the domestic horse but 66 in the true wild horse, and some authorities thus treat the latter as a distinct species, *E. przewalskii*. Moreover, Ryder (1988) reported that analysis of mitochondrial DNA suggests that the ancestors of *E. caballus* and *E. przewalskii* diverged about 250,000 years ago. This methodology still is unproven, however, and Honacki, Kinman, and Koeppl (1982) regarded *caballus* and *przewalskii* as conspecific, noting that their fundamental chromosome numbers are the same and that matings between the two produce fertile offspring. Grubb (1981), R. E. Rau (1978), and Kingdon (1979) suggested that *E. quagga* and *E. burchelli* may be conspecific, but Meester et al. (1986) retained each as a species.

The general body form of *Equus* ranges from thickheaded, short-legged, and stocky to slender-headed and graceful. The size is variable, especially among the domesticated forms. The wild species measure 100–160 cm high at the shoulders. The tail is moderately long, with the hairs reaching at least to the middle of the leg when the tail is hanging down. Equids are heavily haired, but the length of the hair is variable. Most species have a mane on the neck and a lock of hair on the forepart of the head known as the forelock. The two mammae of the female are located in the groin region.

Coloration in most species is generally grayish or brownish above and white below, occasionally with stripes on the shoulders and legs. Striping is carried to an extreme in the zebras. Kingdon (1979, 1984) rejected the hypotheses that stripes serve as camouflage for zebras, that they visually confuse predators and pests, and that they assist in regulating body temperature through heat absorption; instead, he suggested that stripes facilitate group cohesion and socialization. It may be that stripes developed originally as foci for grooming behavior and that the animals then came to associate this attractive tactile stimulus with a visual pattern. There is evidence that zebras, even unrelated individuals, are drawn closely to one another by such a pattern, whereas other equids maintain greater distances. This behavior is important in tropical savannah habitat, where animals seasonally congregate at high densities to exploit resources. In colder regions equids do not assemble in such large and compact groups. In addition, a crisp pattern of stripes can not be maintained in a shaggy winter coat. It is likely that stripes are a primitive characteristic and that they initially appeared in the narrow form seen in *E. grevyi*. That species may be near the ancestral line leading to other zebras and the horse. The latter species, spreading into temperate, desert habitat, lost the need for an intense pattern of stripes. Likewise, such a pattern was greatly reduced in the most southerly of zebras, *E. quagga*, which was restricted to a temperate part of Africa.

The Recent species of equids have only one functional digit, the third. The terminal digit bone on each foot is widened and evenly rounded or spade-shaped; equids walk on the tips of their toes. The radius and ulna are united, although the ulna is greatly reduced in size, so all the weight is borne on the radius. In the hind leg, the enlarged tibia supports the weight, and the fibula is reduced and fused to the tibia.

The dental formula in the Recent forms is: (i 3/3, c 1/1, pm 3–4/3, m 3/3) × 2 = 40–42. The incisors are shaped like

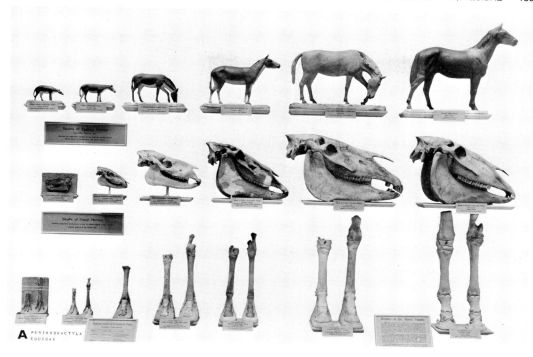

A. Exhibition in Chicago Natural History Museum showing the evolution of horses. B. A restoration of the primitive three-toed horse (*Mesohippus* sp.), which inhabited western America and other regions. Photos from Field Museum of Natural History.

chisels, the enamel on the tips folding inward to form a pit, or "mark," that is worn off in early life. The first permanent incisors appear at about 2.5–3.0 years in domestic horses. The canine teeth are vestigial or absent in female equids. The premolars are similar to the molars, and they are permanent, except for the first, which, when present, represents a persistent milk tooth. This first upper premolar is often reduced or entirely absent. The cheek teeth have a complex structure: they are high-crowned, with four main columns and various infoldings with much cement. The grinding teeth of certain fossil species had four tubercles and little cement. Age in horses is often estimated by the degree of wear of the surface pattern on the cheek teeth, but the rapidity of wear depends to a great extent on the abrasive character of the food. The skull is long, and the nasal bones are long and narrow, freely projecting anteriorly to points; they are always hornless. The eye socket is behind the teeth, thus avoiding pressure on the eye by the long molars.

Equids, which have been termed "admirable running machines," are active day and night but mainly during the evening. In the wild, they are alert for enemies and usually run from danger. In fighting among themselves or in attempting a defense, they kick with the hind feet, occasionally strike with the forefeet, and sometimes bite. Their teeth, though not adapted for lacerating or tearing (except in male zebras, which have pointed canines), can give quite hard pinches. Unfortunately, the curiosity of some of the species has led to their extermination or depletion. The members of this family are entirely vegetarian in habits, feeding mainly on grass, though some browsing is done. Most forms drink daily, but they can go without water for long periods of time. Female equids come into heat several times a year, or until they become pregnant. Some species give birth every two years; the number of young is usually one (twin births are generally abortive). The potential life span is 25–45 years.

The geological range of the family Equidae is early Eocene to late Pleistocene in North America, early Eocene to Recent in Europe, Miocene to Recent in Asia and Africa, and Pleistocene in South America. The Equidae have one of the best-documented fossil histories of any mammalian group and often are used to illustrate the course of evolution. Although this history has been studied extensively for more than a century, interpretations of the process have varied over the years, and new information continues to accumulate (MacFadden 1988).

The most primitive genus known is the early Eocene *Hyracotherium* (often referred to by its junior synonym *Eohippus*), which contained several species estimated to have weighed 20–35 kg. Based on dentition, they and other early Tertiary horses seem to have been generalized herbivores. Their grinding teeth bore conical cusps and were low-crowned, rooted, and free of cement. They still had four digits on the forefoot and three on the hind foot and were relatively plantigrade; that is, they ran on the bottoms of their feet rather than on the tips of elongated phalanges as in the digitigrade condition of later horses. During the remainder of the Eocene and Oligocene body size remained small—around 25–50 kg—but limbs underwent some elongation and reduction of the lateral digits. In most specimens of the Oligocene *Mesohippus*, for example, the forefoot contains only three digits. There was an increase in the chewing surface area of the dentition, with the premolars coming to look more like the molars and the development of crests or ridges that made the teeth more effective for browsing.

A major adaptive radiation of horses occurred during the Miocene, about 25 million to 8 million years ago. Body size diversified considerably, with average weight estimated at about 50 kg in some small species and 400 kg in some of the larger browsing forms. The browsing dentition was retained in one major division of the family, the anchithere horses, which spread across North America and the Old World before becoming extinct late in this period. The other division, which progressed through such North American forms as *Parahippus* and *Merychippus*, underwent a marked change in dentition, evidently to take advantage of the great expansion of savannah grasslands during the Miocene. Chewing surfaces became larger and more complex, cement was deposited outside the enamel, and tooth crowns became much higher (hypsodont) to deal with the abrasive grasses. Also in association with the prevalence of more open country were modifications that facilitated running, including a straightening of the previously arched back. There was an elongation of the limbs, resulting in a relatively longer stride, and the development of an expanded phalanx and dermal hoof at the end of the foot. The hipparion group of horses, which occurred in North America and the Old World from about the middle Miocene to Pliocene, retained three toes on each foot, though the lateral digits did not touch the ground in a standing gait (MacFadden 1984a). In the stock leading to modern equines the lateral digits became vestigial, leaving only one toe on each foot. The third (middle) digit grew and changed shape until it bore all the weight. The first digit was lost initially, then the fifth; the second and fourth digits were reduced first to dewclaws and finally to splints. In the evolution of the equids the lengthening of the legs was paralleled by the elongation of the head and neck, thus enabling the mouth to reach the ground.

During the late Miocene and early Pliocene there was a considerable decline in horse diversity. Most lines became extinct at this time, but one genus, *Dinohippus*, evidently gave rise to *Equus* by the middle Pliocene. The skull of *Dinohippus* resembles that of modern *Equus* in the loss or great reduction of facial fossae; the two genera also show affinity in their relatively large size, hypsodont dentition, elongated protocones, and highly developed, single-toed forefoot (MacFadden 1984b). There even are questions as to whether the two may be congeneric or whether *Dinohippus* itself may have split into asslike and zebralike stocks in the Pliocene, thus making *Equus* polyphyletic (Bennett 1980; Dalquest 1988; Groves 1986b; MacFadden 1988). In any case, the ancestral lines leading to the living species, as listed above, are identifiable in the fossil record of the Pleistocene. Additional information on each of those species is given separately below.

Equus asinus (African wild ass, donkey, or burro).

Approximate average size of wild individuals is: head and body length 200 cm, tail length 45 cm, height at the shoulder 125 cm, and weight 250 kg. Domestic breeds are as short as 80 cm and as tall as 150 cm at the shoulder. The upper parts generally are gray or brownish, may become more reddish in the summer, and shade into the white underparts. There often is a dark dorsal stripe, sometimes a transverse shoulder stripe, and sometimes there are bands on the legs. The mane is long, thin, and upright, and the tail is tufted. The hooves are long and the narrowest of any equid.

Groves (1986b) explained that the hoof of *E. asinus* is designed for surefootedness rather than speed. This condition reflects the habitat of the species, which is broken, undulating, stony desert country, in contrast to the broad, open plains inhabited by the related and swifter *E. hemionus*. Groves (1974b) noted, however, that *E. asinus* has been clocked at up to 50 km/hr. It generally grazes from dawn until late morning, rests during the hottest part of the day, and then grazes again in the late afternoon. In a study of a feral population in Arizona, Seegmiller and Ohmart (1981) found the diet to include 22 percent grasses and sedges, 33 percent forbs, and

Top, African wild asses *(Equus asinus)*, photo by Lothar Schlawe. Bottom, domestic asses *(E. asinus)*, photo from San Diego Zoological Garden.

40 percent browse. The animals there occurred in groups of variable composition, averaging 4.7 individuals, and had a mean annual home range of 19.2 sq km.

In the Danakil Desert of Ethiopia, Klingel (1977) found the social organization of E. asinus to be basically the same as that of E. hemionus and E. grevyi. Some individuals, almost always males, live alone, but most are found in unstable groups of variable composition. There is no indication of permanent bonds between adults; groups may form in the morning and break up in the evening. Herds comprise up to 49 animals. The smaller herds have individuals of only one sex or have a single stallion with a few females. The larger groups contain a number of adults of both sexes. Some males defend large territories, averaging 23 sq km, where they are recognized as dominant but within which they tolerate subordinate males. Woodward (1979) indicated that breeding in the wild was restricted to the wet season, though some feral populations in North America bred throughout the year. Groves (1986b) stated that both domestic and wild asses have a gestation period lasting a full year and sometimes more. Gooding (in Strahan 1983) noted that females reach sexual maturity in their second year and that in domesticity they may thereafter produce one young annually for some 20 years. Willoughby (1974) listed known longevities of 47 years for a domestic donkey and 43 years for a mule.

E. asinus was domesticated about 6,000 years ago, probably in either Egypt or Mesopotamia, and is mentioned frequently in the Bible. Its movements are much slower than those of other equids, but in portions of the world where horses do not thrive or are too expensive, it has been used extensively for riding and as a beast of burden. Because it is surefooted, persistent, and endurable, it is useful as a pack animal. It is capable of carrying well over 100 kg for days with little food. It can go for longer periods without water than can other species of equids, and it is remarkably capable of surviving on a minimum of food and of working under hot and difficult conditions. Another useful domestic animal, the mule, is the hybrid offspring of a male donkey and a female horse. The cross between a female donkey and a male horse is known as a hinny. Both mules and hinnies are usually sterile, though exceptions are known. Mules have been bred for over 3,000 years for heavy work or riding. They resemble donkeys but are usually larger, and they generally have greater strength and stamina than do horses of equivalent size. Mules bred in the United States average about 148 cm in height at the shoulder and 500 kg in weight (Willoughby 1974).

Feral populations descended from escaped or released domestic asses are now present in several places outside of the original range (Lever 1985). Perhaps the largest numbers, at least 1.5 million, are in Australia (Gooding, in Strahan 1983). In the southwestern United States, where it was brought by the Spanish in the sixteenth century and is known by the name burro, domestic E. asinus was used by prospectors to pack supplies into regions where access was difficult. Made largely obsolete by motor vehicles, many individuals were turned loose, and they sometimes formed substantial breeding populations. Most of these animals, about 5,500 in 1988, are in areas administered by the U.S. Bureau of Land Management. According to Mills (1982), however, the largest concentration in the United States, about 2,100 head, is in Death Valley National Monument, California. There, as in Grand Canyon National Park and some other areas, the burro reportedly has damaged the topsoil by trampling and caused a severe decline in the native bighorn sheep through competition for forage and water. Much controversy has centered on whether the burro should be eradicated as an undesirable alien or preserved for humane and aesthetic reasons. The U.S. Wild Free-roaming Horse and Burro Act of 1971 recog-

nized the burro as a valuable historic symbol and provided general protection to the species, but it also called for control measures. Some animals have been destroyed to protect natural habitat, but many others have been adopted by private parties, and a number of free-ranging herds are being maintained.

The original range of E. asinus is not completely known, and the issue is confused somewhat by the presence of feral asses in parts of the involved region. The species seems to have disappeared in the wild from most of its North African Range during Roman times and from southwestern Asia even earlier, though the latter region may have been the place of domestication (Groves 1986b). A continued drastic decline in modern times has resulted from such factors as excessive hunting, habitat deterioration, transmission of disease from livestock, and interbreeding with the domestic donkey. The wild African ass now apparently survives only in small parts of southern Sudan, Ethiopia, and Somalia and numbers perhaps 3,000 individuals. There were scattered reports from southeastern Egypt from the 1950s to the 1970s, though again these sightings may actually have been of feral animals (Osborn and Helmy 1980). A small herd recently was introduced on a reserve in Israel (Clark 1983a). E. asinus (not including domestic and feral animals) is classified as endangered by the IUCN and USDI and is on appendix 1 of the CITES.

Equus hemionus (kulan and onager).

Head and body length is 200–250 cm, tail length is 30–40 cm, height at the shoulder is 100–142 cm, and weight is 200–260 kg. Groves (1974b) indicated that average size slightly exceeds that of E. asinus. Coloration varies geographically and in some places by season. The upper parts may be reddish, tawny, yellowish, or gray. There usually is a broad, dark dorsal stripe, which usually has a white border, and there sometimes is a transverse shoulder stripe. The underparts, buttocks, and muzzle are white or buff. Compared with E. asinus, E. hemionus has broader hooves, shorter hairs on the mane and tail, and shorter ears.

According to Groves (1986b), E. hemionus lives in flat desert country, including salt flats and gravel plains. It is the swiftest equid, being capable of reaching 70 km/hr for brief periods and sustaining a pace of 50 km/hr. Groves (1974b) noted that it feeds on grass and low succulent plants. Its need for favorable feeding grounds in the winter and for watering sites in the hot summer may result in long-distance movements. Roberts (1977) indicated that E. h. khur, of India and Pakistan, grazes at night. It formerly congregated by the hundreds during the autumn after good rains. Smielowski and Raval (1988) observed a herd of more than 50 E. h. khur during the monsoon but noted that after the rains there is dispersal into small groups.

In Turkmenia, Klingel (1977) found that most individuals lived in unstable groups of variable composition, with no indication of permanent bonds between adults. The largest herd had 135 animals and included a number of adults of each sex. Another group, however, had only a single stallion, 77 mares, and 36 foals. Territorial behavior seemed comparable to that of E. asinus. Roberts (1977) wrote that the mares of E. h. khur enter estrus in August and September, and at that time the males fight each other viciously with teeth and hooves. Groves (1974b, 1986b) indicated that for the species as a whole breeding usually occurs from about late April to October, the gestation period is probably about 11 months, the interbirth interval is 2 years, and lactation lasts 1.0–1.5 years. Smielowski and Raval (1988) reported that female E. h. khur attain sexual maturity in their second year and males, a year later. According to Jones (1982), a captive specimen of E. hemionus lived for 35 years and 10 months.

Kulan *(Equus hemionus)*, photo by Lothar Schlawe.

There have been suggestions that *E. hemionus* was domesticated along with *E. asinus* during early historical times in southwestern Asia, but such now seems unlikely (Groves 1986*b*). By about 1800 the onager had been extirpated from the European part of Russia, but even in 1900 the species was distributed continuously from Palestine to the Gobi Desert. Subsequently its numbers declined sharply in response to excessive hunting for food, skins, and sport and as its habitat was taken over for agricultural and pastoral purposes. It currently occupies four limited areas in southern Mongolia (subspecies *E. h. hemionus*), Turkmenia *(E. h. kulan)*, northern Iran *(E. h. onager)*, and southwestern India *(E. h. khur)*. In the last area the population fell from about 4,000 individuals in the 1950s to only 362 in 1967, but subsequent conservation measures have allowed an increase to nearly 2,000 (Smielowski and Raval 1988). The IUCN (1976) cited estimates of 1,500 for *E. h. onager* and 1,000 for *E. h. kulan*. Mallon (1985) reported that while *E. h. hemionus* has been greatly reduced in numbers, it still is not rare. The subspecies that formerly occupied the region from Palestine to Iraq, the diminutive *E. h. hemippus*, became extinct when the last known captive died in the Vienna Zoo in 1927 (Groves 1974*b*), and the last wild animals disappeared at about the same time (Harrison 1968). However, several herds of *E. h. onager* from Iran recently were introduced into a reserve in Israel (Clark 1983*b*). The IUCN classifies *E. hemionus* generally as vulnerable and the subspecies *E. h. khur* as endangered. The USDI lists the entire species as endangered. *E. hemionus* is on appendix 2 of the CITES, except for the subspecies *E. h. khur* and *E. h. hemionus*, which are on appendix 1.

Equus kiang (kiang).

Except as noted, the information for the account of this species was taken from Groves (1974*b*). *E. kiang* is the largest wild ass; individuals in the eastern part of the Tibetan Plateau have a shoulder height of about 142 cm, females weigh 250–300 kg, and males often reach 400 kg. The upper parts are bright red in summer; the browner winter coat is very long

and thick. There is a well-marked black dorsal stripe. The underparts are pure white, and this color extends in wedges almost to the dorsal stripe. White also extends to the legs, throat, sides of the neck, and muzzle, and there are white rings around the eyes. Compared with *E. hemionus*, *E. kiang* has a larger head, a broader snout, relatively shorter ears, a longer mane, and broader hooves. The tail is tufted and has long hairs growing up either side.

The kiang inhabits high, undulating steppe at elevations of up to nearly 5,000 meters. Herds wander over wide moorlands, crossing the barren ridges by low passes and congregating where forage is available. In warm weather they often swim across rivers and seem to take pleasure in bathing. They feed on grass and low plants. Vegetation is abundant only in August and September, and at this time the animals may gain as much as 40–45 kg.

This species lives in herds of 5–400 individuals. The larger groups comprise adult females with their foals and younger animals of both sexes; each group is led by an old female. The herds are very cohesive, never become scattered, and move in file after the leader. Although the members run, turn, graze, and drink in unison, they differ from horses in that there is little mutual grooming or other physical contact. Adult males of seven to eight years tend to live alone for much of the year and combine into their own groups of up to 10 members during the winter. By about the end of July the stallions begin to follow the large herds, and by mid-August they are fighting one another, herding the females, and defending them against rivals. The mating season ends in mid-September, the gestation period lasts nearly a year, and births take place the following July or August. The interbirth interval is about two years. Pregnant females gather in groups of 2–5 and move to rocky places to give birth. The foals can walk and run within a few hours. The mothers and young rejoin the herd after a few weeks. The young achieve independence after about a year of life, though sexual maturity may be delayed for a considerable period.

Until recently the kiang was thought to be abundant and

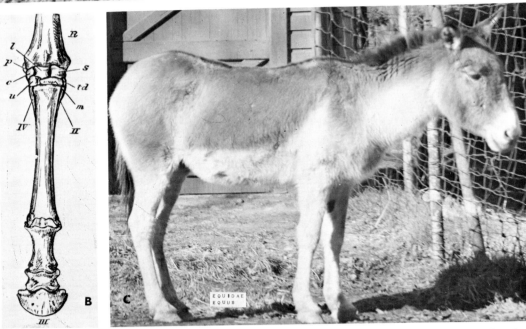

A. Przewalski's wild horse *(Equus caballus)*, photo by Bernhard Grzimek. B. Front foot of horse *(E. caballus)*: c = cuneiform; l = lunar; m = magnum; p = pisiform; r = radius; s = scaphoid; td = trapezoid; u = unciform; II & IV = rudimentary metacarpals; photo from *Mammalia*, F. E. Beddard. C. Kiang *(E. kiang)*, photo by Ernest P. Walker.

relatively secure in its remote habitat on the Tibetan Plateau. This situation may have begun to change after the Chinese occupation of Tibet in 1950. There has been a pronounced increase in human population, hunting activity, and agricultural and industrial development. Domestic livestock has multiplied in numbers and usurped the limited forage. The great herds of yak, Tibetan antelope, gazelle, and kiang that once characterized the region have all but disappeared (Schaller 1986; Van Dyk, Singh, and Rowell 1988). There are only 37 kiang in captivity (Olney, Ellis, and Sommerfelt 1988).

Equus caballus (horse).

The only wild population of this species that survived long enough to be studied scientifically was Przewalski's horse, which here is recognized as the subspecies *E. caballus przewalskii*. Available descriptions (Grzimek 1975; Willoughby 1974) indicate that compared with most domestic horse breeds it is small but heavily built and has a relatively large head, a short and erect mane, and a long tail. Head and body length is 220–80 cm, tail length (including hair) is 92–111 cm, height at the shoulder is 120–46 cm, and weight is 200–300 kg. The general color of the upper parts is yellow-reddish brown (known technically as dun or light bay). The flanks are paler, and the underparts are almost clear white. The mane and legs are dark brown or black. The winter pelage is paler, longer, and thicker than the summer coat. The tarpan (*E. c. gmelini*), a wild horse that survived until the late nineteenth or early twentieth century, stood about 125–35 cm high at the shoulder, was generally gray in color, and had a black tail and mane.

Domestic horses vary extensively in appearance (Willoughby 1974). Averages of typical size in 37 breeds are 155 cm in height at the shoulder and 530 kg in weight. The range is from 102 cm and 175 kg in the Shetland pony to 168 cm and 930 kg in the Shire draft horse. The smallest adult horse on record, a Falabella miniature pony, measured 51 cm in height at the shoulder and weighed 14 kg. The largest, a Belgian draft horse, measured 198 cm and weighed 1,453 kg. The term pony generally applies to various small breeds that stand less than 132 cm at the shoulder. Domestic horses and ponies display numerous colors and patterns, and Berger (1986) observed the same to be true for feral populations in the American West. Horses there include duns, bays, pintos, palaminos, sorrels, chestnuts, and others, and there apparently has been no selection for a uniform color.

Free-living horses usually are associated with grasslands, steppes, and semiarid shrub country, though some early reports indicate the former presence of true wild *E. caballus* in certain European forests. General requirements include a source of fresh water and some form of shelter, especially cover from wind during severe winter weather (Miller 1986). In the Great Basin Desert of Nevada, Berger (1986) found horses to shelter in ravines and juniper forests during snows. Most of the animals there migrated to higher elevations, about 2,300 meters, in the late spring and moved back to lower country, about 1,600 meters, in the fall. Daily movements there were mainly diurnal and covered around 1–3 km. Przewalski's horse occurred in both plains and hills up to an elevation of 2,500 meters and apparently made seasonal migrations in association with rainfall and vegetation growth. It reportedly spent the day in open desert and moved to grazing and watering areas after sundown (Grzimek 1975; Klimov 1988). Other observations indicate that feral horse activity may be either diurnal or nocturnal, that most of the day is spent seeking and consuming food, that feeding peaks may occur in the early morning and late afternoon, and that there are several daily periods of rest. Horses sleep in either standing or recumbent positions. They move with a variety of gaits, including a natural gallop of about 26–29 km/hr and a racing gallop of 64–69 km/hr (Waring 1983). Trained individuals have made record jumps up to 10 meters broad and 2.4 meters high (Willoughby 1974). Horses are primarily grazers, taking mainly grasses and low forage plants, but they may eat a variety of other vegetation.

The differing availability and concentration of food, water, and cover from area to area and season to season is reflected in considerable variation in reported population density, 0.1–11.0 per sq km, and home range size, 0.8–78.0 sq km (Waring 1983). In the Great Basin, Berger (1986) found range size to average 6.73 sq km for breeding bands and 8.97 sq km for bachelor bands during the summer at high altitude, and 25.12 sq km for breeding bands and 35.62 sq km for bachelor bands during fall, winter, and spring at low altitude. Most studies indicate that such areas overlap extensively and are not exclusive territories, in whole or in part. However, individuals and bands may defend the immediate space they occupy at a given

time and may fight one another for access to resources.

The basic social organization of free-ranging horses has been reviewed by Miller (1986) and Waring (1983). Although some individuals are solitary, most are found in discrete bands. These groups occasionally contain more than 20 but usually fewer than 10 members and are of several types. Breeding bands are relatively stable, typically having a single mature stallion and several mares with their offspring of two or three years. Most membership changes involve dispersal of the young of both sexes and the recruitment of young females into other bands. Sometimes more than a single adult male is present, though one is dominant and does most if not all of the breeding. Each male joins in herding and defending the females of the group. The dominant stallion frequently brings young mares into the group and may attempt to steal those of another band. Despite the controlling role of the males, group unity seems to depend more on affinity between adult females and on attachment of the young to their mothers. Bachelor bands, consisting entirely of males, are usually smaller and less stable, with members joining and leaving throughout the year. There also are bands of juveniles, about 2–3 years old, of both sexes.

A localized population of horses consisting of a number of bands of various types plus solitary animals sometimes is referred to as a herd. The member units may associate during migration, foraging, or other activity and may demonstrate an interband rank hierarchy that determines access to limited resources. A dominance hierarchy also is formed within all bands, there being separate orders for males and females within heterosexual groups. Rank is achieved through aggressive interaction, which may be limited to displays and threatening gestures or may involve shoving, kicking, and biting. While age, sex, size, and length of residence are involved in rank, it seems that generally the most persistent and aggressive animals attain a higher position than do passive individuals. A horse's rank influences its access to food and water and its ability to participate in reproduction. Dominant stallions maintain their position through highly ritualized displays involving stares, postures, olfaction, vocalization, and defecation. Disputes over mares or space sometimes lead to violent fights and severe injuries. Aside from such struggles, horses appear to seek and enjoy one another's company. They participate in play and mutual grooming (nibbling to remove parasites and foreign matter in the pelage) and communicate through a wide variety of visual, acoustical, and chemical signals. Common vocalizations include squeals of aggression and whinnies for facilitating social contact over distance.

There have been numerous studies of feral horse herds, with some minor variation in the results. In the Pryor Mountain Wild Horse Range in Wyoming and Montana, Feist and McCullough (1976) found the population organized into 44 breeding bands, each with a dominant stallion, 1–3 mares, and offspring. There also were 23 bachelor groups containing an average of 1.8 individuals and a maximum of 8. Breeding bands were long-lasting, and dominant stallions almost always were successful in retaining their mares when challenged. In the Granite Range of the Great Basin, Nevada, Berger (1986) determined that 88 percent of the breeding bands contained only a single stallion. Nearly half of all the females secured by the stallions there were won in combat. Young males frequently lost their harems. Bachelors sometimes deposed dominant males and sometimes formed their own groups from wandering females. Fathers evidently maintained a friendly relationship with their sons even after dispersal of the latter.

Although there never were formal field studies of Przewalski's horse in the wild, historical observations suggest a social structure much like that of feral horses. There were breeding

Domestic horse *(Equus caballus)*, photo by Lothar Schlawe.

bands of up to 15–20 individuals led by a stallion, and there also were groups of bachelors; these groups sometimes united in herds of 100 or more animals (Grzimek 1975; Klimov 1988). Recent investigations of captives have revealed no major qualitative behavioral differences between Przewalski's horse and domestic or feral horses (Houpt and Fraser 1988). One captive herd in a large reservation contained 40–43 individuals, including 3–4 adult stallions and 14–15 adult mares. One male was dominant for many years, showing concern for the members of the herd, attempting to protect them from danger, and coordinating their movements. There were subgroups of bachelors and juveniles (Klimov 1988). In another captive herd, consisting entirely of 8 males, the dominant stallion frequently associated with the youngest individuals and protected them from harassment by the other full-grown males (Tilson et al. 1988). According to Groves (1974b), the male Przewalski's horse reaches sexual maturity in his fifth year, the female in her fourth. Most young are born from April to June, gestation averages 335 days, birth weight is 25–30 kg, and weaning may take up to 2 years. Maximum recorded longevity of Przewalski's horse is 34 years.

The following additional information on reproduction was taken from Berger (1986), Waring (1983), and Willoughby (1974). Mares are seasonally polyestrous, have an estrous cycle of about 3 weeks, and are receptive for 3–7 days. They commonly first enter estrus in the early spring or within 11 days postpartum. Gestation averages about 332–42 days in different breeds and ranges from 315 to 387 days. There normally is a single foal, twins occurring only one out of several thousand times. Births occur throughout the year, but there is a peak from April to June in the American West. A newborn thoroughbred colt is about 100 cm high at the shoulder, weighs about 45 kg, is covered with hair, and can see and stand. It can follow its mother within a few hours and is weaned after 5–9 months. Large breeds are not full-grown until 5 years, but physiological sexual maturity is attained after only 1 year. Wild stallions may not be capable of maintaining a breeding band until they are 5–6 years old. Females are potentially capable of producing a foal annually from an age of 2 to at least 22 years, but the actual reproductive rate

depends on various environmental, physiological, and social factors. Few horses live past their twenties, though maximum known longevity is about 50 years.

The time and place of domestication of *E. caballus* is unknown. Volf (*in* Grzimek 1975) stated that it probably was attempted about 4000 B.C. in Mesopotamia and China and that by 2000 B.C. the domestic horse was familiar and widespread in China. Willoughby (1974) noted various possibilities, with a consensus being that domestication probably was first accomplished around 3100 B.C. by the Scythians on the steppes of southern Russia. However, he and Groves (1986b) suggested that part of the ancestry of the modern draft, or "cold-blooded," breeds might be traced to *E. caballus remagensis* (= *germanicus*), a large extinct subspecies of the late Pleistocene and possibly early postglacial times in central Europe. Today there are at least 180 extant breeds. With respect to numbers of individuals, the leading nations are the Soviet Union, the United States, China, and Brazil, each with 5–10 million horses. In various places where there are substantial numbers of domestic horses and suitable habitat, some animals have escaped captivity, taken up a feral existence, and established thriving breeding populations. At present the largest numbers, 128,000–205,000, are found in Australia (Lever 1985).

Although most of the evolutionary development of the Equidae occurred in North America, the family disappeared there at the end of the Pleistocene, probably in association with the first human invasion of the continent. Horses did not return until European colonies were established in the sixteenth century. According to Lever (1985), American Indian tribes probably initially obtained horses by raiding or trading with Spanish settlements in the southwest; many of these animals escaped or were freed. By the late eighteenth century domestic horses were a key element in the culture of many tribes, and several million feral horses, or mustangs, were present in the open country of the western United States and southwestern Canada. The spread of agriculture and fences eliminated feral horses from much of the Great Plains region in the late nineteenth century, but many persisted farther west. About 1 million were present in the 1920s

(Willoughby 1974), but a precipitous decline occurred from then through the 1950s, as feral horses were rounded up and slaughtered without limit, both because they were considered to be pests and competitors for forage with domestic stock and because their meat was in demand for use as food for pets, chickens, fur animals, and people. Subsequent controversy saw some parties arguing that feral horses should be eliminated to protect the habitat of native wildlife and others claiming that they should be saved for humane and historical reasons and even that they were filling a natural ecological niche that had been vacant since the extinction of Pleistocene horses. In 1971, when about 17,000 feral horses survived, the U.S. Wild Free-roaming Horse and Burro Act provided them general protection but also called for removal of animals that appeared to be in excess of what available habitat could support. Some populations increased, but many thousands of horses were rounded up and shot or given away to private parties. In 1988 there were an estimated 40,000 feral horses in the United States, the majority in Nevada, but there were questions as to how many of these animals represent recently escaped domestic stock and how many are descendants of the original mustangs.

The true wild horse has fared even worse than the feral herds. Of the three subspecies that lived in historical times, two are extinct and one apparently survives only in captivity. *E. c. sylvaticus*, the forest horse, may once have occupied parts of Central Europe as far as the Rhine but vanished from that region by the Middle Ages as its habitat was taken over by agriculture. It persisted in the wild in eastern Poland until about 1800, and it may be the predominant ancestor of the konik, a domestic breed of pony still used in that area (Groves 1974b; Volf, in Grzimek 1975). According to Groves (1986b), a captive forest horse may have survived as late as 1918, but there is some question whether the Polish and Rhineland animals represent the same subspecies.

The plains tarpan *(E. c. gmelini)* lived on the steppes of southern Russia about as far east as the Volga River. Its disappearance has been attributed partly to the persecution and interbreeding that resulted from a tendency of the stallions to run off with domestic mares. The last known wild individual died in the south-central Ukraine in 1879, and the last captive died in the Moscow Zoo a few years later (Volf, in Grzimek 1975). The tarpan is commonly thought to be the predominant ancestor of the domesticated horse. In the 1930s a group of dedicated horse lovers began a program to try to "breed back" to the tarpan from existing domestic horses. They selected breeds that most closely resembled descriptions of the original tarpan and by selective breeding produced an animal that probably gives a good idea of what the original wild European horse looked like. The most successful and legitimate efforts were made in Poland and centered on the konik, which is thought to retain a high percentage of tarpan genetic material (Chelminski 1978; Willoughby 1974). There now are about 200 tarpanlike horses at field stations in Poland, and others are established in zoos in Europe and the United States.

Przewalski's horse *(E. c. przewalskii)* formerly occurred in the steppes and deserts of Kazakhstan, Sinkiang, Mongolia, and probably the Transbaikal region of southern Siberia. It still occupied much of this region when it became known to science in 1879 but subsequently declined drastically because of excessive hunting by people and usurpation of grazing and watering sites by domestic animals. It was rare by the early twentieth century, and it was compressed into a small part of southwestern Mongolia and adjacent China by the 1950s. Although there still is a slight hope that it survives in the wild, the last reliable sightings were in 1968. A number of specimens were brought into captivity in the late nineteenth and early twentieth centuries. Twelve of those animals, plus one hybrid with a domestic horse, are the ancestors of a present intensively managed breeding pool of over 660 *E. c. przewalskii* in over 70 zoos and field stations. While there is concern that the long period in captivity has resulted in inbreeding and loss of genetic diversity, plans are being laid for reintroduction in the original habitat, probably in Mongolia (Groves 1974b; Ryder 1988; Ryder and Wedemeyer 1982; Sokolov and Orlov 1986; Volf, in Grzimek 1975; Willoughby 1974). Productivity of the captive population also is being increased through the transfer of early-stage embryos from Przewalski horse mares to the uteri of female domestic ponies, which then carry the embryos to term (Volf 1985). Przewalski's horse is classified as endangered by the IUCN and the USDI and is on appendix 1 of the CITES.

Equus grevyi (Grevy's zebra).

Head and body length is 250–300 cm, tail length is 38–60 cm, height at the shoulder is 140–60 cm, weight is 352–450 kg, and there is no size difference between the sexes (Kingdon 1979). Groves (1974b) referred to *E. grevyi* as the largest living wild equid and the most handsome of zebras. The stripes are very narrow and close together, cover most of the head and body, and extend down the legs to the hooves. Unlike in the pattern in *E. burchelli*, the stripes do not become predominantly horizontal on the hindquarters. The belly is white and unstriped, and there also is a narrow white zone on either side of the broad black dorsal stripe. *E. grevyi* has a relatively large head, a slender build, and long legs. The ears are very broad, and the mane is very tall and erect. The characters of *E. grevyi* are generally primitive and may indicate an ancestral form from which evolved the other zebras and perhaps the horse (see above discussion of stripes).

Grevy's zebra is found in a limited region of subdesert country between the more arid region to the north that is preferred by *E. asinus* and the better-watered country to the south favored by *E. burchelli* (Kingdon 1979; Yalden, Largen, and Kock 1986). In some localities there are seasonal migrations in search of forage and water. The diet includes tough, fibrous grasses that are inedible to cattle and some other ungulates.

The unusual social behavior of *E. grevyi* has been reviewed by Kingdon (1979) and Klingel (1974). Although thousands of animals formerly congregated in response to temporary environmental conditions, there are no lasting bonds between adults such as those of *E. burchelli* and *E. zebra*. The only strong association is that of a female and her offspring of one or more years. There are unstable groups of nursing mares and foals, other females, bachelor males, and various combinations thereof. There also are solitary stallions, some of which are territorial. The territories of these animals, together with those of *E. asinus*, are the largest known for any herbivore. Grevy's zebra stallions in Kenya were found to have territories with an average size of 5.75 sq km and a range of 2.70–10.50 sq km. The stallions may control their areas for years, leaving only briefly to find water. They tend to stay even under highly adverse conditions but sometimes join the rest of the population on seasonal migrations. The territories are only for purposes of mating, with the stallion attempting to herd and defend any estrous female passing through the area. He tolerates the presence of subordinate males within his territory, and they do not interfere when he is engaged in mating activity.

In areas where conditions force migrations, there evidently are seasonal peaks in mating when the territories are reestablished. These periods occur during the rains in July–August and October–November. However, some mating occurs throughout the year, even when territories are not occupied. At such times the males compete intensively for the estrous females, pursuing the latter and fighting one another

Grevy's zebra *(Equus grevyi)*, photo by Leonard Lee Rue III.

for lengthy periods. The estimated length of the gestation period is 390 days. The foal has a remarkable extension of the mane down the back to the tail. The newborn coloration is brown and black, with the adult pattern rapidly developing after 4 months of age. The young is relatively independent after 7 months but remains with its mother for up to 3 years. Both sexes are capable of breeding at 3 years, but wild stallions usually are not able to mate until about 6 years. Captives have lived up to 22 years.

Fossil remains indicate that the range of *E. grevyi* extended north to Egypt during the Neolithic period, about 6,000 years ago (Churcher 1986). In modern times its remnant distribution and numbers have been reduced severely in the face of hunting for its attractive skin and competition with domestic livestock for water. There has been an unconfirmed report of its presence in southern Sudan (IUCN 1978). It has been entirely exterminated in Somalia, has disappeared from most of its original range in Ethiopia, and is rare in much of Kenya (Kingdon 1979). In the 1970s numbers were estimated at about 1,500 in Ethiopia and 14,000 in Kenya, but subsequent military activity probably has contributed to a further decline in the former country, and drought and an upsurge in poaching may have cut populations in the latter by as much as 50 percent (Yalden, Largen, and Kock 1986). *E. grevyi* is classified as endangered by the IUCN and as threatened by the USDI and is on appendix 1 of the CITES.

Equus zebra (mountain zebra).

The account of this species is taken largely from Smithers (1983) but has been augmented by information from the other sources indicated. Head and body length is about 210–60 cm, tail length is 40–55 cm, height at the shoulder is 116–50 cm, and weight is 240–372 kg (Joubert 1974). The subspecies *E. z. zebra*, of South Africa, is usually smaller than *E. z. hartmannae*, of Angola and Namibia. The stripes on the body are narrower than those on the rump. The pattern of striping in *E. zebra* is intermediate to that of *E. burchelli* and *E. grevyi*. From *E. burchelli*, *E. zebra* can be distinguished by its generally narrower and more numerous stripes, having clearly much broader black stripes on the rump with no sign of shadow (faint) stripes between them (as sometimes found in *E. burchelli*), the presence of a gridiron pattern formed by the posterior portion of the dorsal stripe, having generally white underparts with a midventral black stripe on the chest and belly (in *E. burchelli* the body stripes continue around the underparts), in the presence of a dewlap, and in having ears over 200 mm long (less than 200 mm in *E. burchelli*).

As its common name implies, this zebra is found on slopes and plateaus in mountainous areas. *E. z. zebra* occurs at up to 2,000 meters but moves to lower elevations in winter. *E. z. hartmannae* sometimes wanders between mountains and sand flats. Both subspecies are predominantly diurnal, gener-

Hartmann's mountain zebra *(Equus zebra hartmannae)*, photo from Zoological Society of San Diego.

ally being most active early in the morning and from late afternoon to sunset. There usually is a daily dustbath (Penzhorn 1984). More than half of the daylight period is spent feeding. The diet consists mainly of grass but also includes some browse. Individuals usually drink once or twice a day. *E. z. hartmannae* may cover up to 100 km locally in search of forage. During winter, breeding bands of that subspecies have a grazing area of 6–20 sq km; summer areas are considerably smaller, and in Etosha National Park, Namibia, the two areas are separated by about 120 km (Penzhorn 1988). In Mountain Zebra National Park, South Africa, the home range of *E. z. zebra* was found to average 9.4 sq km and to vary from 3.1 to 16.0 sq km (Penzhorn 1982).

The mountain zebra occurs predominantly in small, cohesive, nonterritorial breeding bands with overlapping home ranges (Klingel 1974; Penzhorn 1979). The bands contain a single adult stallion and 1–5 mares with their young of 1 or more years (Penzhorn 1984). Dominant males may become aggressive toward one another when their groups approach, but sometimes the bands join to form temporary herds of up to 30 members. Stallions have been known to maintain control of a group for up to 15 years, but they may be ousted through combat with a younger male. Females usually remain in a group for life, but occasionally some will split off and may then join with a bachelor to form a new breeding band. The stallion is the dominant member of a group. If there is danger, he will alert the others and then take up a defensive position to the rear while movement away is led by a mare. The females have their own rank hierarchy maintained by threatening gestures. Young males, as well as some older stallions

that have lost their groups, commonly join together. Such groups have a rank hierarchy but are less stable than the breeding bands. Bachelor groups often attach themselves temporarily to a breeding band. Vocalizations include a high-pitched alarm call, by which a stallion alerts his group, and a drawn-out squeal of submission, uttered by a bachelor male when confronted by a dominant stallion (Penzhorn 1984).

Breeding occurs throughout the year, birth peaks having been noted in December–February (summer) for *E. z. zebra* and in November–April for *E. z. hartmannae*. The gestation period is approximately 1 year. Weight at birth is 25 kg (Penzhorn 1988). The foals can nibble grass when just a few days old and are weaned after about 10 months. The young of both sexes of *E. z. zebra* leave their natal group at 13–37 months (Penzhorn 1988). Mares of *E. z. hartmannae* have been seen trying to force juveniles 14–16 months old out of a group, but stallions of *E. z. zebra* may try to prevent young from leaving (Penzhorn 1984). Males are capable of holding a band at 5–6 years. Females produce their first foals at 3–6 years, have a usual interbirth interval of 1–3 years, and may remain reproductively active until they are 24 years old (Penzhorn 1985).

The subspecies *E. z. zebra* formerly occurred all across the mountains of southern and central Cape Province, South Africa, but by the 1930s had been brought almost to the point of extinction through excessive hunting (Harper 1945). A 1937 census showed just 45 known individuals, though apparently others were present in remote mountain ranges (Groves 1974b). Subsequent conservation efforts allowed a small recovery. Smithers (1986) reported numbers at 474, about half of them in the Mountain Zebra National Park, which was established for the protection of the subspecies. It is classified as endangered by the IUCN and USDI and is on appendix 1 of the CITES. *E. z. hartmannae* once ranged more or less continuously from southwestern Angola through the mountains of western Namibia to the border of South Africa. Even in the 1950s numbers were estimated at 50,000–75,000, but they subsequently declined through competition with domestic livestock and persecution by agricultural interests (Groves 1974*b*). Numbers recently were estimated to be 13,000 (Penzhorn 1988). The subspecies now is listed as vulnerable by the IUCN and as threatened by the USDI and is on appendix 2 of the CITES. Records of the CITES indicate that about 500–1,000 skins are still taken legally each year (Broad, Luxmoore, and Jenkins 1988).

Equus burchelli (Burchell's zebra).

Except as noted, the information for the account of this species was taken from Grubb (1981) and Kingdon (1979). Head and body length is 217–46 cm, tail length is 47–56 cm, height at the shoulder is 110–45 cm, and weight is 175–385 kg. Males average slightly larger than females. The body stripes are broad, those toward the rear of the flanks becoming especially broad and horizontal. In most populations the body stripes extend to the midventral line on the belly, in some the stripes extend down the legs to the hooves, and in some there are faint brown "shadow" stripes between the main flank stripes. There are various aberrations, including an almost entirely black coat and a reverse arrangement in which the ground color is dark and the stripes are white. There is much geographical variation in pattern, with a tendency in the south toward reduction and disappearance of the stripes on the legs, belly, and hindquarters. Since *E. quagga*, of South Africa, seems to represent a continuation of this trend (see below), there is increasing recognition that it is conspecific with *E. burchelli*, in which case the proper name for the resulting species would be, by nomenclatural priority, *E. quagga*.

Burchell's zebra occurs in a variety of habitats, including savannah, light woodland, open scrub, and grassland. It sometimes is found in broken, hilly country and on mountain slopes up to 4,400 meters. A general requirement is the regular availability of fresh water. There is a daily cycle of movement covering up to 13 kilometers and generally from the sleeping sites in higher, more open areas to the lower, lusher grazing grounds. The diet is 90 percent grass but also includes some browse. The differing habitats and the need to obtain forage and water have resulted in both sedentary and migratory populations. That of the Serengeti, in Tanzania, moves seasonally over a distance of 100–150 km. Home ranges there are 300–400 sq km in the rainy season and 400–600 sq km in the dry season. In Kruger National Park, of South Africa, where there are both sedentary and migratory populations, home range is 49–566 sq km and population density is 0.7–2.2 individuals per sq km. For the sedentary population of Ngorongoro Crater in Tanzania, the respective figures are 80–250 sq km and 19.2/sq km.

Although vast numbers may congregate during migration

Burchell's zebra *(Equus burchelli)*, photo by Lothar Schlawe.

Burchell's zebras *(Equus burchelli):* Top, southern specimen showing approach to *E. quagga* in reduction of striping, photo by Lothar Schlawe; Bottom, normally striped mother and young, photo by C. A. Spinage.

Quagga *(Equus quagga)*, photo from Zoological Society of London.

or in the vicinity of favorable resources, social organization is restricted to rather small family and stallion groups which are nonterritorial and occupy overlapping home ranges (Klingel 1969, 1974). The family units are stable, have a rank hierarchy, and consist of a dominant stallion, 1–6 adult mares, and the offspring of one or more years. Membership in different areas averages 4–8 individuals, with a maximum of 15. Mares normally stay in the same family group for their lifetime, even when they are very old and sick. Stallion groups have up to 16 members but usually just 2–3. These groups consist mostly of young bachelors but may also include very old males that have lost control of a family unit. In the latter situation, the sons may follow their father to form a stallion group. Although stallion groups generally are less stable than the family units, some males remain together for years. Upon entering her first estrus, a young female is courted by all the males of an area, both dominant stallions and bachelors, and she frequently is taken out of her natal group. Males fight fiercely to obtain or retain mares, circling one another and attempting to bite or kick the opponent. Dominant males also are highly protective of their units and take up a defensive position to the rear while the other animals flee. Intergroup relations vary, with the adult mares being antagonistic, the dominant stallions greeting one another in a ritualized ceremony, and the young sometimes playing together (Klingel 1969). *E. burchelli* communicates by a variety of gestures, facial expressions, and sounds. An explosive braying bark seems to identify stallions and maintain group cohesion. Affinity of animals also is shown through extensive bouts of mutual grooming.

Females are seasonally polyestrous (Smuts 1976). There may be a postpartum estrus after 8–10 days. Estrus lasts 2–19 days, and diestrus, 17–24 days. Estimates of the gestation period range from about 360 to 396 days. Births occur in all months but mostly during the wet season, October–March in East Africa. The single young weighs about 32 kg, can stand almost immediately, runs within an hour, and starts to eat grass within a week. However, weaning usually is completed after 7–11 months, and lactation sometimes continues for up to 16 months. The young normally leave their family after 1–3 years. Females reach puberty at 16–22 months, and males are able to compete for mares at 4 years. The interbirth interval usually falls between 1 and 3 years. Average longevity in the wild is about 9 years, reflecting an annual juvenile mortality that probably exceeds 50 percent, but captives have lived up to 40 years.

Although it is the only species of wild equid not at or near the point of extinction, *E. burchelli* has declined in numbers and distribution, mainly because of hunting by people for its skin and competition for habitat with domestic livestock. The most southerly subspecies, *E. b. burchelli*, like the neighboring quagga, is extinct. It once occurred in great numbers on the plains of the Orange Free State and some adjacent areas but was rare by the mid-nineteenth century, and the last known specimen died in the London Zoo in 1909. *E. burchelli* also has disappeared from the rest of South Africa, except the extreme east, and from much of Botswana, Zimbabwe, Zambia (Ansell 1978), Malawi (Ansell and Dowsett 1988), Tanzania, Kenya, Uganda, and Ethiopia (Yalden, Largen, and Kock 1986). The largest remaining population, numbering some 200,000 individuals (Sitwell 1983), occurs in the Serengeti of Tanzania.

Equus quagga (quagga).

Various reports indicate a total length of about 257 cm and a height at the shoulder of 125–35 (Smithers 1983). Despite its former abundance, this extinct species has been something of a mystery animal, with some observers considering it to be more of a wild horse than a zebra. Indeed, a recent analysis of mainly cranial characters suggested that *E. quagga* has closer phylogenetic affinity to *E. caballus* than to zebras (Bennett 1980). Based largely on color pattern, however, there is a growing consensus that *E. quagga* represents the southerly end of a cline seen in *E. burchelli* and that probably the two are conspecific (Groves, as cited by Meester et al. 1986; Grubb 1981; Kingdon 1979; R. E. Rau 1978). Although *E. quagga* sometimes is described as having much less pronounced striping than do other zebras, it actually demonstrates considerable variation. Some individuals are almost stripeless, but others have a pattern nearly the same as seen in the neighboring South African subspecies *E. burchelli burchelli,* with stripes covering most of the body except for the hindquarters, belly, and legs.

Some descriptions and photographs (perhaps including the one accompanying this account) suggest that the stripes of the quagga are light, rather than dark as in other zebras. Rau (1983), however, showed that this view is based on an illusion and that the quagga actually is a typical dark-striped zebra. The ground color of much of the head, neck, shoulders, and anterior portion of the back is creamy white and paler than the remainder of the upper parts, and it is this color, seen between the true stripes, that may give the impression of light stripes. The belly, legs, and tail also are white, and the ground color of the upper parts becomes reddish brown and progressively darker on the back and sides. The true stripes are darkest on the head and neck; they then become lighter, and the interspaces darker, until eventually they blend together and the stripes disappear at some point along the back.

Very little information was recorded about the quagga (Harper 1945; Smithers 1983). It had a restricted range in open, arid country and presumably was a grazer. Herds of 30–50 individuals were reported, and the animals sometimes traveled in single file. Early observations in South Africa often are not clear as to whether they refer to *E. quagga, E. burchelli,* or *E. zebra.* Nonetheless, the quagga evidently was found and hunted by European colonists near the Cape of Good Hope in the 1600s. It subsequently declined because of relentless killing for its meat and hide and loss of forage through overgrazing by domestic sheep. By the 1850s it was nearly gone south of the Orange River, and the last wild populations, those of the Orange Free State, evidently were extirpated in the late 1870s. A number of individuals were captured and sent to zoos in Europe. They were said to be much more tractable than *E. burchelli* and to quickly become docile and tamable. The last of them died in Berlin in 1875 and Amsterdam in 1883.

PERISSODACTYLA; **Family TAPIRIDAE; Genus TAPIRUS**
Brünnich, 1771

Tapirs

The single Recent genus, *Tapirus,* contains three subgenera and four species (Cabrera 1961; Eisenberg, Groves, and MacKinnon 1987; Ellerman and Morrison-Scott 1966; Hall 1981; Thornback and Jenkins 1982):

subgenus *Tapirus* Brünnich, 1771

T. terrestris (South American tapir), Colombia and
  Venezuela to northern Argentina and southern Brazil;
T. pinchaque (mountain tapir), the Andes from
  northwestern Venezuela to northwestern Peru;

subgenus *Tapirella* Palmer, 1903

T. bairdii (Baird's tapir), southern Mexico to Colombia and
  Ecuador west of the Andes;

subgenus *Acrocodia* Goldman, 1913

T. indicus (Malayan tapir), southern Burma and Thailand,
  Malay Peninsula, Sumatra, possibly Laos.

Acrocodia was treated as a full genus by Eisenberg, Groves, and MacKinnon (1987). According to Medway (1977), *T. indicus* was present on Borneo until within the last 8,000 years and possibly survived there into historical time. Honacki, Kinman, and Koeppl (1982) indicated that this species also formerly occurred in southern China and probably parts of Viet Nam, Laos, and Cambodia.

Head and body length is 180–250 cm, tail length is 5–13 cm, height at the shoulder is 73–120 cm, and weight is 180–320 kg. Short, bristly hairs are scattered on the body; they are thickest on *T. pinchaque.* A low, narrow mane, which is not always conspicuous, is present in *T. bairdii* and *T. terrestris.* The skin is quite thin in the mountain tapir but thick in the other species. *T. indicus* is readily distinguished by its color pattern; the front half of the body and the hind legs are black, and the rear half, above the legs, is white. This black and white pattern renders the animal practically invisible at night in the jungle, when moonlight on the vegetation assumes the same black and white pattern. *T. terrestris, T. pinchaque,* and *T. bairdii* are dark brown to reddish brown above and often paler below.

The general form of all tapirs is rounded in back and tapering in front, making them well suited for rapid movement through thick underbrush. The snout and upper lips are projected into a short, fleshy proboscis; the transverse nostrils are located at its tip. The proboscis is more elongated in the New World species than in the Malayan tapir. The eyes are small and flush with the side of the head. The ears are oval, erect, and not very mobile. The legs are rather short and slender. The radius and ulna are separate and about equally developed, and the fibula is complete. The forefoot has three main digits and a smaller one (the fifth), for a total of four; the small digit is functional only on soft ground. The hind foot has three digits. The tail is short and thick. Female tapirs have a single pair of mammae, located in the region of the groin.

Young Asiatic tapir *(Tapirus indicus)* in a position unusual among perissodactyls, photo by Ernest P. Walker.

A. Asiatic tapir *(Tapirus indicus)*, photo from New York Zoological Society. Inset: Asiatic tapir young *(T. indicus)*, photo from Zoological Society of London. B. Brazilian tapir *(T. terrestris)* and eight-day young, photo by Ernest P. Walker.

The dental formula in Recent species is: (i 3/3, c 1/1, pm 4/3–4, m 3/3) × 2 = 42–44. The incisors are shaped like chisels; the third upper incisor is shaped like a canine but is larger than the true canine, and the third lower incisor is reduced in size. The canines are conical and separated from the premolars by a space. In Recent species the posterior three premolars resemble the true molars in size and shape; in extinct species they were usually simpler than the molars. All of the cheek teeth lack cement and are low-crowned, with a series of transverse ridges and cusps. Horns are absent in all

Brazilian tapirs (*Tapirus terrestris*), photo by Bernhard Grzimek.

species. The skull is relatively short and laterally compressed, with a high braincase and a convex profile. The nasal bones are short, arched, and freely projecting.

The altitudinal range of tapirs is from sea level to about 4,500 meters. *T. pinchaque* has never been taken at altitudes of under 2,000 meters (Schauenberg 1969). Tapirs may live in nearly any wooded or grassy habitat where there is a permanent supply of water. They commonly shelter in forests and thickets by day, emerging at night to feed in bordering grassy or shrubby areas. Some species are partly diurnal (Eisenberg, Groves, and MacKinnon 1987). Tapirs are agile in closed or open habitat and in or under water. They are good hill climbers, runners, sliders, waders, divers, and swimmers. They generally walk with the snout close to the ground and are fond of splashing in water or wallowing in mud. Tapirs are generally shy and docile and seek refuge in water or crash off into the brush when threatened, but they can and will defend themselves by biting. They possess keen senses of hearing and scent.

These hoofed mammals wear paths to permanent bodies of water in areas where their populations are dense, and human engineers sometimes follow their trails up the sides of mountains in the construction of roads. Population density can reach 0.8/sq km in areas with lush vegetation (Eisenberg, Groves, and MacKinnon 1987). Tapirs consume aquatic vegetation and the leaves, buds, twigs, and fruits of low-growing terrestrial plants, but in any particular habitat they eat mainly the green shoots of the most common browsing plants. They also graze for food, and at times they have been known to damage young corn and other grains, especially in Mexico and Central America.

Except for females with young, tapirs are solitary animals. They communicate with shrill whistling sounds and scent-mark with urine. When two individuals meet in the wild, they generally behave aggressively toward one another (Eisenberg, Groves, and MacKinnon 1987). Breeding in *T. indicus* occurs in April or May, and a single young weighing about 6–7 kg is born after a gestation period of 390–95 days. The young stays with its mother for at least 6–8 months, by which time it is nearly of adult size. A female in her prime probably produces one calf every second year (Lekagul and McNeely 1977). Carter (1984) stated that captive female *T. terrestris* are polyestrous, with intervals of 50–80 days between estruses, and are reproductively aseasonal. However, he suggested that in the wild this species might mate just before the beginning of the rainy season and give birth early in the rainy season of the following year. The gestation period in *T. terrestris* is 385–412 days, the number of offspring is one or rarely two, and weight at birth is about 4–7 kg. The young initially remains in a sheltered spot, begins to follow its mother after a week, and remains with her for 10–11 months (Eisenberg, Groves, and MacKinnon 1987). Young tapirs of all species are dark reddish brown, with yellow and white stripes and spots. They usually lose this juvenile pattern at 5–8 months. They probably require 3–4 years to attain sexual maturity and full growth (Carter 1984). A captive *T. terrestris* lived for 35 years (Jones 1982).

In some areas, tapirs are hunted extensively for food and sport (for religious reasons, however, certain Indian tribes do not kill tapirs). In addition, all species have declined in recent years, mainly because of the clearing of the forests by humans for agricultural purposes and cattle grazing. Populations have almost or completely disappeared in several countries of Central and South America (Thornback and Jenkins 1982). The IUCN classifies *T. indicus* as endangered and *T. pinchaque* and *T. bairdii* as vulnerable. The USDI lists all four species as endangered. *T. pinchaque*, *T. bairdii*, and *T. indicus* are on appendix 1 of the CITES, and *T. terrestris* is on appendix 2.

The discontinuous distribution of modern tapirs suggests that they represent the remnants of a once widespread family. The fossil record shows that tapirs originated in the Northern Hemisphere and at various times occupied the land masses between where the present-day Asiatic and South American forms exist. The fact that tapirs are now living in both the American and Asiatic tropics supports the many pieces of evidence that the two continents were connected rather recently as measured by geological time and that during the period when the two were joined the climate was mild to warm in the northern portion of these continents, making conditions favorable for animals to move from one continent to the other. Subsequently, the continents were separated at the Bering Strait, or any other land bridge that might have existed, and the climate changed, so the animals were prevented from moving between the two continents by the strait and by colder climatic conditions.

Mountain tapir *(Tapirus pinchaque)*, photo by Lothar Schlawe.

The geological range of the Tapiridae is early Eocene to Recent in North America, early Oligocene to Pleistocene in Europe, Pleistocene to Recent in South America, and Miocene to Recent in Asia. The geological range of the genus *Tapirus* is late Miocene to Recent. *Megatapirus* is the only extinct tapir that has been found in Pleistocene deposits of the Old World; it is known from Sichuan Province of China. It was much larger than any Recent tapir and had a shorter and deeper skull.

PERISSODACTYLA; Family RHINOCEROTIDAE

Rhinoceroses

This family of four Recent genera and five species occurred in historical time in most of Africa south of the Sahara, perhaps in parts of North Africa, and in south-central and south-eastern Asia. Two or three subfamilies have been recognized by various authorities: Dicerorhinae, for *Dicerorhinus*; Rhinocerotinae, for *Rhinoceros* (and sometimes also for *Dicerorhinus*); and Dicerotinae, for *Diceros* and *Ceratotherium* (Groves 1967b, 1975b; Owen-Smith 1984). More recently, there has been recognition that the living genera should be placed in the same subfamily, the Rhinocerotinae (Groves 1983), or even the same tribe, the Rhinocerotini (Prothero, Manning, and Hanson 1986). Of the four, *Dicerorhinus* is thought to be the most primitive and to have affinity to *Rhinoceros*, while the other two genera are considered to be closely related to one another. There are additional subfamilies and families of fossil rhinoceroses.

Head and body length is 200–420 cm, tail length is 60–75

cm, height at the shoulder is 100–200 cm, and adult weight is around 1,000–3,500 kg. Females are smaller than males. The coloration is grayish to brownish, but the true color is often concealed by a coating of mud. The thick skin, which is scantily haired and wrinkled, is furrowed or pleated, having the appearance of riveted armor plate in some species. The tail bears stiff bristles.

Rhinoceroses have a massive body, a large head, one or two horns, a short neck, a broad chest, and short, stumpy legs. The radius and ulna, and the tibia and fibula, are only slightly movable but well developed and separate. The forefoot has three digits (four in some fossil forms), and the hind foot also has three; the hooves are distinct and separate for each digit. The upper lip is prehensile in two genera (*Rhinoceros* and *Diceros*). The small eyes are located on the side of the head, midway between the nostrils and the ears. The ears are fairly short but prominent and erect. The dental formula in the family is: (i 0–2/0–1, c 0/0–1, pm 3–4/3–4, m 3/3) $\times$ 2 = 24–34. The incisors and canines are vestigial. Except for the small first premolar, the premolars resemble the molars. The cheek teeth, which are high-crowned in *Ceratotherium* (the only species of Recent rhino that grazes rather than browses) and fairly low-crowned in the other Recent genera, are marked with transverse ridges of enamel. The skull, which is elongate and elevated posteriorly, has a small braincase. The nasal bones project freely beyond the skull. One or two median conical horns are present in rhinos, though they may be short or obscure in some forms (they were not present in many extinct species). If there is only one horn, it is borne on the nasal bones; if there are two horns, the posterior one is over the frontal bones of the skull. These horns are dermal in origin; although solid, they are composed of compressed keratin of a fibrous nature.

Rhinos generally inhabit savannahs, shrubby regions, and

Top, Sumatran rhinoceros *(Dicerorhinus sumatrensis)*. Bottom, Indian rhinoceros *(Rhinoceros unicornis)*. Photos by Lothar Schlawe.

dense forests in tropical and subtropical regions. The African species usually live in more open areas than do the Asiatic forms. Rhinos generally are restricted to areas where a daily trip to water is possible. Their paths between the watering and feeding places often pass through tunnels in the brush. They penetrate dense thorn thickets by sheer force. Rhinos are active mainly during the evening, through the night, and in the early morning, resting during the day in heavy cover

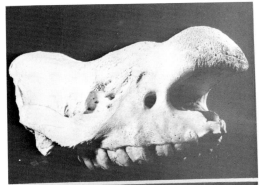

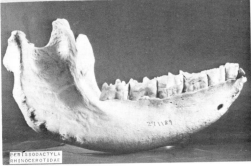

African black rhinoceros (Diceros bicornis), photos by P. F. Wright of skull in U.S. National Museum of Natural History.

extinct members of this group showed remarkable diversity in size and proportions; most lacked horns. Two genera of the Eurasian Pleistocene were *Coelodonta*, the woolly rhinoceros, and *Elasmotherium*, a huge animal with a single giant frontal horn. *Indricotherium* (= *Baluchitherium*), a hornless rhinoceros of the late Oligocene and early Miocene of Asia, was the largest known land mammal. It is estimated to have been 4–5 meters tall at the shoulder and to have weighed about 11,000 kg (Willoughby 1974).

Although modern rhinos are far more restricted in distribution and diversity than was the group in the geological past, it would be wrong to think that they are inevitably doomed to a natural extinction. Even in the nineteenth century they occurred in large numbers over much of Africa and Asia. The subsequent population crashes have been entirely the fault of relentless killing and habitat usurpation by people. Nearly all parts of rhinoceroses are used in folk medicine, but by far the greatest demand is for the horn, which in powdered form is reputed to cure numerous physical problems and which whole is used for artistic carving. Only about 10,000 rhinos survive throughout the world. There is much international interest in saving those that remain. Few other animals that occur over such large stretches of wilderness have been so well mapped and counted.

PERISSODACTYLA; RHINOCEROTIDAE; **Genus DICERORHINUS**
Gloger, 1841

Sumatran Rhinoceros, or Hairy Rhinoceros

The single species, *D. sumatrensis*, originally occurred from Assam and southeastern Bangladesh to the Malay Peninsula and possibly Viet Nam and on Sumatra and Borneo (Groves and Kurt 1972; Lekagul and McNeely 1977; Rookmaaker 1977, 1980; Van Strien 1975). Except as noted, the information for the remainder of this account was taken from Groves and Kurt (1972).

The smallest living rhinoceros, *D. sumatrensis* is relatively short-bodied. Head and body length is 236–318 cm, height at the shoulder is 112–45 cm, and two estimated weights were 800 kg and 2,000 kg; however, Van Strien (1986) stated that weight is up to about 1,000 kg. The skin is leathery, up to 16 mm thick, and dark gray-brown in color. The facial skin is characteristically wrinkled around the eye, but the muzzle is rounded and unwrinkled due to heavy keratinization. The body folds of the skin are less pronounced than in *Rhinoceros* but more so than in *Diceros* and *Ceratotherium*. There is more hair than in other rhinos. The pelage is fairly long and dense in calves, is still copious and is reddish brown in young adults, and becomes sparse, bristly, and almost black in older animals. There are two horns, but the one over the frontal bones often is inconspicuous. The nasal horn generally is short, the record well-authenticated specimen measuring 381 mm, but two horns that probably are referable to this species are 690 mm and 800 mm long. The dental formula is: (i 1/0, c 0/1, pm 3/3, m 3/3) × 2 = 28.

The Sumatran rhinoceros occurs mainly in hilly country near water. It inhabits both tropical rainforest and mountain moss forest but may be attracted to forest margin and secondary growth. Van Strien (1975) concluded that this species can live in a wide variety of habitats, from swamps at sea level to high in the mountains. It may make seasonal movements, keeping to the hills when the lowlands are flooded during the rains, descending when the weather becomes cool near the

that may be several kilometers from the waterholes. They sleep in both standing and recumbent positions and are fond of wallowing in muddy pools and sandy river beds. They run with a cumbersome motion, reaching their top speed at a canter, that is, at a gait resembling a gallop but with moderate and easy bounds or leaps. *Diceros* can attain speeds of up to 45 km/hr for short distances. Vision is poor, but smell and hearing apparently are acute. Rhinos eat a variety of vegetation, but succulent plants make up the bulk of the diet.

Rhinos often are accompanied by tick birds and egrets, which act as sentinels and feed on external parasites that often infest these mammals. Rhinos are usually timid but can be ferocious at bay. They sometimes charge an enemy, though their attack often is poorly directed. They may grunt or squeal when excited. Rhinos drop their dung in well-defined piles and often furrow the area around the piles with their horns; these piles may be scattered afterward. They are believed to act as "sign posts" or territory markers (urination spots and rubbing sticks also seem to serve this purpose).

During the breeding season, a pair of rhinos sometimes remains together for 4 months; females may give birth every 2 years. The gestation period in most species is about 420–570 days. The single offspring is active soon after birth and remains with the mother until the next youngster is born. The mother sometimes guides the baby with her horn. Rhinos have a potential lifespan of almost 50 years. The large cats prey on young rhinos, but the adults apparently have no enemies other than humans.

The geological range of the Rhinocerotidae is middle Eocene to Pleistocene in Europe, late Eocene to Recent in Asia, Miocene to Recent in Africa, and middle Eocene to Pliocene in North America (Carter 1984). However, this family is only part of a superfamily, the Rhinocerotoidea, which also contains two fossil families of rhinos and a total of over 50 known genera (Prothero, Manning, and Hanson 1986). The

Sumatran rhinoceros *(Dicerorhinus sumatrensis)*, photo by Erna Mohr.

end of the rains, and returning to high ground by March, possibly to escape the attacks of horse-flies, which abound at low elevations in the dry season. It can move up and down steep slopes with great agility, can swim well, and has been known to swim in the sea. It feeds before dawn and after sunset and moves mostly by night. Much of the day is spent in rainwater ponds or other wallows that are dug out or deepened by the animals themselves, usually located on a mountain top or a catchment area of a small stream. The surrounding area is cleared of vegetation for 10–35 meters and is used as a resting place. Wallowing is thought to be a cooling mechanism or to provide protection against insects. *Dicerorhinus* reportedly is regular in its movements, making well-defined trails to wallows and feeding sites, changing the latter every 10–15 days. Young saplings, which form a major food source, are bitten off, stepped on, or broken off with the horns. The diet includes fruit, leaves, twigs, and bark. Favored foods are wild mangoes, figs, bamboo, and all kinds of plants found in secondary growth. Cultivated crops sometimes are eaten. Average daily consumption may be over 50 kg (Van Strien 1986).

Dicerorhinus evidently depends on salt licks; during a study in Gunung Leuser National Park, in northern Sumatra, Van Strien (1986) found this factor to have considerable behavioral influence. Females with calves tended to remain at lower elevations and to visit the licks frequently. Average time between visits was 23 days for such pairs, 44 days for other adult females, 59 days for subadults, and 55 days for adult males. Population density in the study area was 13–14 individuals per sq km, probably considerably higher than in localities with fewer salt licks. Each individual rhino had a permanent, well-defined home range that included a salt lick. The home ranges of adult males averaged 30 sq km and overlapped extensively, but there appeared to be small, exclusive core areas. Female ranges were smaller but generally were separate from one another except in the vicinity of salt licks. Females were thought to be territorial and to avoid one another. When not involved in breeding, adult females tended to stay at higher elevations in an area of about 10 sq

km or less. When accompanied by a calf, the females moved to a lower area of about 10–15 sq km close to a salt lick. Following separation, the young rhinos remained for 2–3 years in their natal range. Adults of both sexes regularly marked their ranges with scrapes, bent or twisted saplings, feces, and sprayed urine.

Females commonly are found together with their offspring. Males usually are solitary but seem to visit the territories of females and possibly fight over the latter after the young are weaned. In his study area, Van Strien (1986) found males to sometimes frequent salt licks with the evident objective of meeting a female. Most births there (northern Sumatra) took place from October to May, the period of heaviest rainfall. A gestation period of 8 months has been reported but seems unlikely considering the 15–18 months recorded for other rhinos. One newborn was 914 mm long and weighed 23 kg. The coat is short, crisp, and black in the neonate; later it becomes long, shaggy, and almost fleecy. The calf separates from its mother at 16–17 months and possibly then associates with other young, but it later becomes solitary and probably does not begin to breed until at least 7–8 years; interbirth interval seems to be at least 3–4 years (Van Strien 1986). A captive specimen lived for 32 years and 8 months (Jones 1982).

The Sumatran rhino is classified as endangered by the IUCN and the USDI and is on appendix 1 of the CITES. It has disappeared from much of its original range, principally because of habitat destruction and overhunting for supposedly aphrodisiac and medicinal products made from the horn and other parts of its body by some peoples of Asia. Flynn and Abdullah (1984), reporting that extinction appeared imminent in peninsular Malaysia, noted that logging and clearance for agriculture not only reduces habitat and fragments populations but also facilitates access by poachers. Van Strien (1986) stated that the species is very sensitive to all forms of disturbance and is driven away from an area by logging operations, but he also suggested that numbers were not so small as once feared and that concerted conservation efforts might yet save the species. Khan (1989) reported that the total number of individuals remaining in the wild was 536–962 and that there were another 16 in captivity. Those known to be in the wild are at restricted sites, some of which are parks or reserves but others of which are totally unprotected. The subspecies *D. s. lasiotus*, formerly found in India, Bangladesh, and Burma, now may be represented only by 6–7 animals in Burma. There appear to be only 30–50 survivors of *D. s. harrisoni*, the subspecies of Borneo. All the remaining animals belong to *D. s. sumatrensis*, there being about 100 in peninsular Malaysia and 400–700 on Sumatra. A conservation program being coordinated by the IUCN includes monitoring and improved protection of wild populations and the transfer of animals from high-risk areas in the wild to breeding facilities in captivity.

PERISSODACTYLA; RHINOCEROTIDAE; **Genus RHINOCEROS**
Linnaeus, 1758

Asian One-horned Rhinoceroses

There are two species (Ellerman and Morrison-Scott 1966; Groves 1983; Khan 1989; Rookmaaker 1980):

R. sondaicus (Javan rhinoceros), originally found from Sikkim and eastern India to Viet Nam and apparently southern China, and on the Malay Peninsula, Sumatra, and Java;

R. unicornis (greater Indian rhinoceros), originally found in northern Pakistan, much of northern India, Nepal, northern Bangladesh, and Assam.

These rhinos are large, awkward-looking creatures with a large head, short, tubular legs, small eyes, and wide nostrils. They have a single horn on the nose, which is composed of agglutinated hairs and has no firm attachment to the bones of the skull. The dental formula is: (i 1/1, c 0/1, pm 3/3, m 3/3) × 2 = 30. These rhinos also may be distinguished from their African relatives by their skin, which has a number of loose folds, giving the animal the appearance of wearing armor; the African rhinos lack such folds. *R. unicornis* has a fold of skin that does not continue across the back of the neck; *R. sondaicus*, on the other hand, has a fold that continues across the

midline of the back. The skin is practically naked except for a fringe of stiff hairs around the ears and the tip of the tail. The skin of *R. unicornis* has large convex tubercles, whereas that of *R. sondaicus* is covered with small, polygonal, scalelike disks. Additional information for each species is provided separately.

Rhinoceros sondaicus (Javan rhinoceros).

Except as noted, the information for the account of this species was taken from Lekagul and McNeely (1977). Head and body length is 300–320 cm, tail length is about 70 cm, height at the shoulder is 160–75 cm, and weight is 1,500–2,000 kg. Analysis of cranial measurements by Groves (1982*b*) indicates that females are larger than males. This species is nearly as tall as *R. unicornis* and has the same dusky gray color but is less massive, has a much smaller head, and has less developed folds of skin on the neck. It has three folds of skin across the back, one in front of the shoulder, the second behind the shoulder, and the third over the rump. The single horn is short, record length being only 250 mm. Average length for males is closer to 150 mm, and females often lack a horn or have only a small bump. The upper lip is pointed and prehensile, being used for drawing browse toward the mouth.

The Javan rhinoceros inhabits dense rainforests with a good supply of water and plentiful mud wallows. It generally prefers low-lying areas, though some animals have been found above 1,000 meters. Individuals tend to have loosely defined centers of activity where they may spend several days at a time and to which they periodically return. Some animals may travel 15–20 km within 24 hours. In the course of feeding, branches up to 15–20 mm thick are torn off, stems at different heights above the ground are broken, and trees up to 150 mm in diameter are uprooted. The diet consists of shoots, twigs, young foliage, and fallen fruit.

This species is generally solitary except for mating pairs and mothers with young. Data cited by Laurie (1982) indicate that home ranges are small, population densities originally were greater than 0.30/sq km, there are a variety of vocalizations comparable to those of *R. unicornis*, there also is olfactory communication like that of *R. unicornis*, and courtship involves fighting between the sexes. Females probably are polyestrous and come into heat every 46–48 days. A single calf is born after a gestation period of about 16 months. It is suckled for at least 1 and perhaps as long as 2 years. Females reach sexual maturity at about 3–4 years, and males

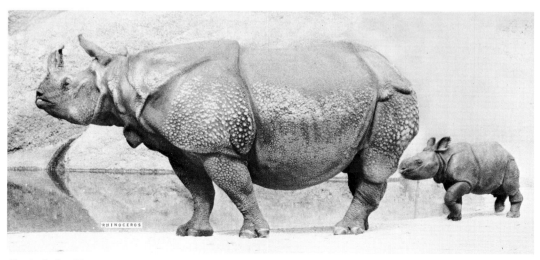

Greater Indian rhinoceroses *(Rhinoceros unicornis)*, photo by Dorothy Y. Mackenzie.

at about 6 years. Mature females probably do not breed more often than every fourth or fifth year. Record longevity in captivity is 21 years, though only 9 individuals are known ever to have been in captivity.

According to Khan (1989), the Javan rhinoceros is the rarest large mammal in the world. It once was widespread, and about 150 years ago the three recognized subspecies were distributed as follows: *R. s. inermis*, eastern India, Bangladesh, Assam, Burma; *R. s. annamiticus*, Viet Nam, Laos, Cambodia, eastern Thailand; *R. s. sondaicus*, Tenasserim, Malay Peninsula, Sumatra, western Java. Still earlier, perhaps until the sixteenth century, other populations ranged as far north as the Chinese provinces of Sichuan and Hunan (Rookmaaker 1980). Like other rhinoceroses, *R. sondaicus* declined because of habitat loss and persistent killing to obtain parts, especially the horn, for alleged medicinal purposes. Today the only substantial and relatively secure population consists of 50–54 individuals of *R. s. sondaicus* in Ujung Kulon National Park, at the extreme western tip of Java. Even that group is highly vulnerable to poaching, epidemics, and other problems. No animals currently are in captivity. The subspecies *inermis* is almost certainly extinct, and *annamiticus* is restricted to a few scattered groups in southern Laos, southern Viet Nam, and Cambodia. There long was doubt whether rhinos even survived in Indochina, and if so, whether they represented *R. sondaicus* or *Dicerorhinus sumatrensis*. Based on reports, Rookmaaker (1988) concluded that only *R. sondaicus* was present. Subsequently, Schaller et al. (1990) learned that a specimen of *R. sondaicus* had been taken illegally in November 1988 about 130 km northeast of Saigon, and they found evidence that perhaps 10–15 individuals still were present in this area. *R. sondaicus* is classified as endangered by the IUCN and the USDI and is on appendix 1 of the CITES.

Rhinoceros unicornis (greater Indian rhinoceros).

Except as noted, the information for the account of this species was taken from Laurie, Lang, and Groves (1983). Head and body length is 368–80 cm in males and 310–40 cm in females, tail length is 70–80 cm, height at the shoulder is 170–86 cm in males and 148–73 cm in females, and weight is about 2,200 kg in males and 1,600 kg in females (Owen-Smith 1984). A black nasal horn, reaching a maximum length of 529 mm, is present in both sexes. The hide is generally gray brown, becomes pinkish in the skin folds, and is covered with rivetlike knobs. Males show enormous development of the neck folds. Body hair may occasionally be apparent; eyelashes, ear fringes, and a tail brush are always present.

The greater Indian rhinoceros originally occurred mainly in alluvial plain grasslands, where the grass grew up to 8 meters tall. It also was found in adjacent swamps and forests. Its range now has been so restricted by human activity that it often must use cultivated areas, pastures, and modified woodlands. Activity takes place mostly at night, in the early morning, and in the late afternoon. The middle of the day commonly is spent resting in the shade or in mud wallows. Wallowing and bathing occur in lakes, rivers, and pools; this behavior is especially frequent during hot seasons and seems to be important for thermoregulation and to escape flies. The diet consists mainly of grass and also includes fruit, leaves, branches of trees and shrubs, and cultivated crops. When feeding on tall grasses, *R. unicornis* curls its prehensile upper lip around the grass stems, bends the stems over, and bites off and chews the tops, drawing the tips into the mouth from the side. Drinking takes place on a daily basis, and mineral licks are visited regularly.

Population densities of 0.4–2.0 individuals per sq km have been found in some areas, and densities of up to 4.85 per sq km have been reached in favored high-diversity habitat of the Chitawan Valley of Nepal (Laurie 1982). Apart from cow-calf pairs, groups are rare. Temporary associations of a few subadults or adult males sometimes form at wallows or on grazing grounds. Only the strongest bulls participate in breeding, and these animals have home ranges of at least 2 to more than 8 sq km. There is some degree of range exclusivity but no true territoriality; the ranges of dominant bulls overlap with each other and with the ranges of nonbreeding males. Individuals usually take to sudden flight away from a disturbance rather than attack, but on some occasions, especially when a cow with a young calf is disturbed at close quarters, they may charge with the head down. Rhinos, mainly cows, reportedly kill several people each year in India and Nepal. Encounters between two rhinos may result in agonistic displays, charges, chases, horn clashes, and lunges with the sharp-pointed lower tusks at the neck, flanks, and rump of the opponent. There are 10 distinct vocalizations, including a snort for initial contact and a honk, bleat, and roar heard during agonistic interaction. Olfactory communication is important and is carried out through urination, defecation, and pedal scent glands. Animals of all ages and both sexes defecate at a communal dung heap. Deposition of scent apparently aids males in determining the availability of receptive females.

Breeding occurs throughout the year. A male accompanies an estrous female intermittently for several days. Courtship often involves a lengthy chase of the female and severe fighting. Females are polyestrous, estrus takes place every 21–42 days, gestation lasts 462–91 days, and there normally is a single young. At birth head and body length is 96–122 cm, height at the shoulder is 56–67 cm, and weight is 40–81 kg. The calf suckles frequently up to 12 months of age and only rarely after 18 months. It is driven away at least one week before birth of the next calf, the interbirth interval being about 3 years. In the wild, females are full-grown at 6.5 years, and males at about 10 years. Females attain sexual maturity at 5–7 years. The record longevity in captivity is 47 years.

The rhinoceros remained common in northwestern India and Pakistan until about 1600 (Rookmaaker 1984). It disappeared from those regions shortly thereafter and declined sharply in the remainder of its range over the next 300 years. The main reason was the loss of alluvial plain grasslands to agricultural development, which destroyed the prime habitat of *R. unicornis*, led to conflicts with human interests, and made the rhino more accessible to hunters. Sport hunting of the species, by both Europeans and Asians, became very popular in the late nineteenth and early twentieth centuries. One maharajah killed 207 rhinos from 1871 to 1907 (Khan 1989). Surprisingly, even more were systematically slaughtered during this period for a government bounty established to protect tea plantations from the ravages of rhinos (Martin 1983; Martin, Martin, and Vigne 1987). By the first decade of the twentieth century the species was very near extinction; in India there were only a few scattered survivors, the main group comprising 12 individuals in the Kaziranga area of Assam, and in Nepal there were perhaps another 50. At that point there was a complete change in human treatment of the species: the bounty and sport hunting were halted, general legal protection was established, and Kaziranga was made a reserve (and eventually a national park).

Although there has been an encouraging overall recovery in the twentieth century, *R. unicornis* is jeopardized by loss of habitat to the expanding human population and illegal killing, especially in response to the astonishing rise in the value of the horn. The wholesale value of Asian rhino horn increased from U.S. $35 per kg in 1972 to $9,000 per kg in the mid-1980s. The retail price, after the horn has been shaved or powdered for sale, has at times and in certain East Asian markets reached $20,000–$30,000 per kg (Martin 1983; Martin, Martin, and Vigne 1987; Martin and Vigne 1987).

By contrast, in May 1990 pure gold was worth about $13,000 per kg. The processed horn is used extensively in oriental medicinal and pharmaceutical preparations as a pain reliever and fever suppressant and as a supposed cure for other problems. In India it is used as an aphrodisiac. The hide, internal organs, blood, and even the urine and dung of the rhino are also valuable. Strenuous efforts have been made by the governments of India and Nepal to control poaching and illicit trade in rhino products. The Chitawan National Park, which has Nepal's largest rhino population, now is protected by 700 armed troops and rangers (Martin 1985a). Projects are under way both in India and Nepal to reintroduce breeding populations in areas of former occurrence (Mishra and Dinerstein 1987; Sale and Singh 1987). The sources of the animals for these projects are Chitawan in Nepal, now with about 375 rhinos, and Kaziranga National Park in India, with 1,080. *R. unicornis* now numbers 1,724 in the wild and about 75 in captivity (Khan 1989). The species is classified as endangered by the IUCN and the USDI and is on appendix 1 of the CITES.

PERISSODACTYLA; RHINOCEROTIDAE; **Genus DICEROS**
Gray, 1821

Black Rhinoceros

The single species, *D. bicornis*, originally occurred throughout eastern and southern Africa and in the north ranged as far as northeastern Sudan and at least as far west as northeastern Nigeria (Ansell, *in* Meester and Setzer 1977). The extent of the former range in western Africa is not precisely known, though maps by Kingdon (1979) and Western and Vigne (1985) suggest that prior to 1900 *Diceros* was found in the savannah zone as far west as Guinea.

Head and body length is 300–375 cm, tail length is about 70 cm, height at the shoulder is 140–80 cm, and weight is 800–1,400 kg. The anterior horn is larger than the posterior one, averaging about 50 cm in length; sometimes the beginning of a third posterior horn is present. Both this rhino and the white rhino *(Ceratotherium)* are dark in color, but the black rhino is somewhat darker. Its coloration is dark yellow brown to dark brown or dark gray. An external feature more clearly distinguishing these genera is the upper lip: in *Diceros* it protrudes slightly in the middle and its tip is prehensile, whereas in *Ceratotherium* it is squared and nonprehensile. The dental formula of both genera is: (i 0/0, c 0/0, pm 3/3, m 3/3) × 2 = 24.

The black rhinoceros is found mainly in the transitional zone between grassland and forest, generally in thick thorn bush or acacia scrub but also in more open country (Schenkel and Schenkel-Hullinger 1969). It is not primarily a grassland animal but favors the edges of thickets and extensive areas of short woody growth and also is restricted to habitat within about 25 km of permanent water (Kingdon 1979). In Etosha National Park, Namibia, Joubert and Eloff (1971) reported the most important factor influencing distribution to be the presence of many natural, permanent water holes. According to Kingdon (1979), the black rhino frequents mud or water wallows to counteract heat and flies and commonly rests and sleeps therein. Well-worn paths lead to such areas. Normal movement is around 3–4 km/hr, but a charge can reach 50 km/hr. Sleeping usually occurs at midday, while the most intensive feeding takes place during the early morning and evening. *Diceros* is a browser, its main foods being the thin regenerating twigs of woody growth and legumes. A great variety of plant species are utilized, though acacias seem to be

a favorite. Twigs are gathered with the prehensile upper lip, drawn into the mouth, and snapped off with the premolars. Drinking occurs every day if water can be reached, and mineral licks are visited regularly.

Data cited by Kingdon (1979) suggest that in areas of favorable habitat the black rhinoceros can become remarkably plentiful and have a dominant ecological influence. Up to 23 individuals, including all ages and both sexes, have been known to reside in an area of less than 3 sq km in the Ngorongoro Crater, 17 of them permanently. Other reported natural densities have been around 0.1–1.0 per sq km. In the Ngorongoro Crater the home range of adults of both sexes averaged 15.5 sq km (2.6–44.0 sq km). In the more barren Olduvai Gorge the range averaged 25.0 sq km (3.6–90.0 sq km). Smithers (1983) noted that ranges overlap, there being no territoriality. However, in one South African reserve with an unusually dense population the breeding males do occupy mutually exclusive home ranges of about 4 sq km each (Owen-Smith 1984).

Diceros has a social system somewhat like that of *Ceratotherium*. Kingdon (1979) wrote that there seem to be clans of animals that are known to one another. Temporary aggregations of up to 13 such individuals have been observed at a wallow. Females usually are found together with a calf and sometimes an older daughter; those without young join a neighboring female. The young of both sexes also attach themselves to other animals. Only fully adult males become solitary, and even they may form temporary groups and move and feed together. Animals are usually tolerant of others that they know in adjacent ranges. Most conflicts involve strangers that move into the area used by a clan. Although at times several bulls may court a female simultaneously without apparent antagonism, Smithers (1983) noted that serious fights and frequent deaths result from conflicts over estrous females. There are a variety of vocalizations, including snorts for alarm, threat, and making contact. Olfactory communication also is important; males spray urine to mark the areas they utilize, and all animals utilize communal dung heaps, sometimes scraping their feet therein and thence leaving scent as they travel about. Such mechanisms may help individuals identify one another and facilitate contact between potential mates.

Breeding apparently occurs throughout the year, though Kingdon (1979) indicated that there may be mating peaks in Kenya during September–November and March–April, and Hitchins and Anderson (1983) indicated the same for Zululand during October–November and April–July. These and other reports suggest that births tend to take place in the rainy season. A premating bond develops between the bull and the cow, and the pair remain together during resting and feeding; they even sleep in contact with each other. There evidently is no serious fighting such as found in *Rhinoceros*. Females usually give birth every 2–5 years, the estrous cycle is 17–60 days, the gestation period is 419–78 days, and there is a single calf weighing about 40 kg at birth. Weaning occurs after about 2 years, independence at 2.5–3.5 years, and reproductive maturity at 4–6 years in females and 7–9 years in males (Grzimek 1975; Hitchins and Anderson 1983; Jarvis 1967; Kingdon 1979). One black rhino still was living after 45 years in captivity (Jones 1982).

Although its pugnacity has been greatly exaggerated, the black rhino is unpredictable and can be a dangerous animal, sometimes charging a disturbing sound or smell. It has tossed people in the air with the front horn and regularly charges vehicles and campfires. Catching the scent of humans, it usually crashes off through the brush and runs upwind, sometimes for several kilometers before stopping. Apparently the sense of smell is the primary method of detecting danger. Schenkel and Schenkel-Hullinger (1969) found that human

Black rhinoceroses *(Diceros bicornis):* Top, photo by Bernhard Grzimek; Bottom, photo from Zoological Society of London.

scent alone causes great alarm among black rhinos. On the other hand, if they detect no scent, rhinos will show no interest in a motionless person or car unless it is closer than 20–30 meters.

The black rhinoceros has been hunted by people since ancient times, but exploitation accelerated during the nineteenth and twentieth centuries. It was killed for sport, because it was considered dangerous, to obtain its durable hide,

and to secure its horn, which was carved into various ornamental objects or shaved or ground into powder for use as an alleged medicine or aphrodisiac. The largest subspecies, *D. bicornis bicornis,* of central and southern Namibia and most of South Africa, became extinct when the last known individual was shot in 1853 (Rookmaaker and Groves 1978; Smithers 1983). By about 1900 the black rhinoceros also had been eliminated in West Africa, though it still was distributed continuously from Cameroon to Ethiopia and south through East Africa to eastern South Africa (Cumming 1987). Numbers and distribution declined substantially in East Africa in the first half of the twentieth century, partly because of government-sponsored killing carried out on the grounds that the presence of the rhino was incompatible with human settlement (Kingdon 1979). Persecution and habitat destruction led to the disappearance of *Diceros* in most of Ethiopia and Somalia by the 1960s (Yalden, Largen, and Kock 1986). Nonetheless, in 1970 there were estimated still to be at least 65,000 black rhinos, with populations present in most countries of the original range. Over the next decade, however, there was increasing recognition that disaster was befalling the species as the value of its horn and consequent poaching increased. The black rhino was placed on appendix 1 of the CITES in 1975 and was listed as endangered by the USDI in 1980. It also was classified first as vulnerable and then as endangered by the IUCN.

The subsequent continued collapse of populations of *Diceros* represents perhaps the greatest single mammalian conservation failure of the late twentieth century. Total numbers in the wild fell to about 15,000 in 1980, to fewer than 9,000 in 1984, and to only 3,800 in 1986 (Cumming 1987; Western and Vigne 1985). The most recent estimates put the number in the wild at closer to 3,000, and there are approximately 200 in captivity. Populations of entire countries, such as that of the Central African Republic, which had about 3,000 rhinos in 1980, have been totally wiped out by poachers. There now are about 1,700 black rhinos in Zimbabwe, a few hundred each in Kenya, Namibia, South Africa, Tanzania, and Zambia, and a few dozen each in Cameroon, Malawi, and Rwanda. This situation has been brought about entirely by an irrational demand for the horn and to a large extent through a strictly ornamental utilization by a single class of persons in one small country. If such a narrow and needless desire has led to the near extinction of one of the world's most spectacular and popular animals at a time when wildlife conservation is receiving immense international interest and support, how can we ever hope to save the multitude of other creatures and ecosystems that are jeopardized by much more trenchant problems of human population growth and development or by far broader and more substantive commercial pressures?

The factor that triggered the recent collapse of rhino populations was a great increase in the demand for horn in the carving of ornamental handles for the traditional daggers (jambias) worn by many men in North Yemen (Martin 1979, 1985b; Martin and Martin 1987; Martin and Vigne 1987; Varisco 1989; Vigne and Martin 1987). Although this tradition dates back to the Middle Ages, many citizens of North Yemen recently went to work in nearby oil-producing regions, thus bringing an influx of wealth, with far more people being able to afford jambias made from rhino horn rather than from cheaper materials. From the early 1970s to 1984 about half of the entire supply of rhino horn on the world market went to North Yemen. Annual importation peaked in 1976 when over 8,300 kg entered the country. In 1982 the government banned importation. That measure had little initial effect, and, ironically, the source of most of the horn that continued to come in was Sudan, a member of the CITES. By 1987 intensified enforcement efforts, as well as the scarcity of rhinos, had reduced the trade. Illegal importation continues, however, with several hundred kilograms entering North Yemen each year. Moreover, there is an established investment value for rhino horn jambias, and market potential remains high.

The most substantial ongoing problem for the black rhino is the demand for horn for use in traditional oriental medicines (Martin 1989; Martin and Vigne 1987). The horn typically is sold in a pharmacy, shavings being made in front of the customer; they then are taken home, boiled, and given to the patient (Martin 1979). The horn may also be ground and then fabricated into pills or mixed in potions and tonics. It has a wide variety of uses, including fever suppression. While it evidently is not used as an aphrodisiac in oriental countries, it is in parts of India. China, the main producer of oriental medicines, apparently obtains most of its supply of raw horn from Hong Kong and other nearby countries. Hong Kong itself prohibited horn imports in 1979, but much illegal activity continues. Singapore was the largest importer until 1986, when it also implemented a ban, and now little horn goes there. The biggest importer for the last three years, and hence now the single greatest threat to the black rhino, is Taiwan (Vigne and Martin 1989). Although there has been a legal import ban since 1985, it is not enforced, and much horn enters, often by way of South Africa.

It has been estimated that 90 percent of all adult rhino deaths are caused by poaching to obtain the horn. About 200,000 kg of horn entered trade from 1970 to 1987; the average weight of a black rhino horn is 2.88 kg (Western 1989c). In the late 1960s the price of rhino horn was only about U.S. $20 per kg. It now fluctuates according to both time and place, but in 1987–88 in East Asian markets the retail value of black rhino horn was around U.S. $15,000–$20,000 per kg, more than the price of gold (Vigne and Martin 1989). So great is the demand for horn for medicinal purposes that antique carvings are being ground up and sold.

Several organizations are working to help the black rhino, and a number of individuals have made outstanding contributions, notably Esmond Bradley Martin, whose extensive travels and studies in Asia and Africa have yielded remarkably detailed accounts of the trade in rhino products. The IUCN, assisted by the World Wildlife Fund and other agencies, is developing a conservation plan for the black rhino. Strategy centers on a concentration of effort to save the most significant remaining wild populations while simultaneously working to halt the trade and utilization of rhino horn. There is concern regarding the genetic viability of many of the small, fragmented populations that remain, and there are questions whether attempts to consolidate such groups might upset natural systematic units (Du Toit 1989; Foose 1987b). Groves (1967c) recognized seven subspecies of *D. bicornis*, and Western and Vigne (1985) reported that all still survived but that several were approaching extinction (they mistakenly referred existing South African populations to the extinct *D. b. bicornis*). It now appears that *D. b. brucii*, of Sudan, Ethiopia, and Somalia, and *D. b. chobiensis*, of southeastern Angola, have indeed disappeared and that *D. b. longipes*, of Central Africa, is represented only by a few dozen individuals in Cameroon and possibly Chad. Recent studies suggest that taxonomic distinctions are less meaningful than once thought and thus that more flexibility in transfer and captive breeding programs may not be objectionable (Du Toit 1987).

PERISSODACTYLA; RHINOCEROTIDAE; **Genus
CERATOTHERIUM**
Gray, 1868

White Rhinoceros, or Square-lipped Rhinoceros

In the nineteenth century the single species, *C. simum*, inhabited two widely separated regions of Africa (Ansell, *in* Meester and Setzer 1977; Groves 1972b; Kingdon 1979). The subspecies *C. s. cottoni* occurred in southern Chad, the

White rhinoceroses, or square-lipped rhinoceroses *(Ceratotherium simum):* Top, photo from Société Royale de Zoologie d'Anvers through Walter Van den Bergh; Bottom, photo by K. Rudloff through East Berlin Zoo.

Central African Republic, southwestern Sudan, northeastern Zaire, and northwestern Uganda. The subspecies *C. s. simum* occurred in southeastern Angola, possibly southwestern Zambia, central and southern Mozambique, Zimbabwe, Botswana, eastern Namibia, and northern and eastern South Africa. About 2,000 years ago the range of *Ceratotherium* extended up the Nile Valley into southern Egypt and proba-

bly covered much of northwestern Africa. Rock paintings and skeletal remains show that rhinos once occurred as far as coastal Morocco and Algeria, but there are questions as to species identity and time of latest survival.

Except for *Elephas, Loxodonta,* and perhaps *Hippopotamus, Ceratotherium* is the largest living genus of land mammals. Head and body length is 335–420 cm, tail length is

50–70 cm, shoulder height is 150–85 cm, and weight is about 1,400–1,700 kg in females and 2,000–3,600 kg in males. Coloration is yellowish brown to slaty gray. This mammal is almost naked except for the ear fringes and tail bristles; there is copious but sparse body hair in *C. s. simum* (Groves 1972*b*). Additional hairs are present in the skin but do not protrude. The front horn averages about 60 cm in length but can reach more than 150 cm. From *Diceros*, *Ceratotherium* can be distinguished externally by its usually lighter coloration, squared upper lip with no trace of a proboscis, elongated and pointed ear conchae with a few bristly hairs at the tips (compared with rounded conchae with hairy edges in the black rhino), more sloping and less sharply defined forehead, shoulder hump, and less conspicuous skin folds on the body. The dental formula of both genera is: (i 0/0, c 0/0, pm 3/3, m 3/3) × 2 = 24.

In South Africa, the primary habitat of *Ceratotherium* is woodland interspersed with grassy openings. Its four main requirements seem to be relatively flat terrain, thick bush cover, short grass for eating, and water for drinking and wallowing (Smithers 1983). In East Africa this genus lives in open forest and nearby plains; it traverses but does not permanently inhabit steeply undulating country, and it may utilize swampy country along the Nile in the dry season and then move to higher ground 10 km away when the rains come (Groves 1972*b*). Daily movements of 4–15 km have been reported (Van Gyseghem 1984). Activity generally is in the early morning, late afternoon, and evening. The rhino wallows or rests in the shade during the middle of the day. Wallowing in the mud is especially important during hot weather for purposes of thermoregulation and for ridding the body of ectoparasites. As in other rhinos, vision is relatively poor, but senses of hearing and smell are acute. There is a graceful trot at about 24 km/hr and a gallop for short spurts at 40 km/hr. *Ceratotherium* differs from other rhinos in that it is entirely a grazer. It feeds largely on short grasses, using only the broad, flexible lips for cropping the stems (Groves 1972*b*; Kingdon 1979; Owen-Smith 1975).

Reported overall population densities vary from 0.03/sq km to 0.81/sq km (Groves 1972*b*), though local densities in favorable habitat may exceed 5.0/sq km (Owen-Smith 1981, 1984). Data obtained by investigations in Zululand (Kingdon 1979; Owen-Smith 1974; Smithers 1983) indicate that some adult males occupy territories of 0.75–2.60 sq km. They spend almost their entire life in these areas, unless water is unavailable, in which case they follow a narrow corridor to a drinking site every 3–4 days. Male territories are bordered by topographical features such as watercourses and ridges and overlap one another by only about 50–100 meters. Adult females in South Africa have home ranges of 6–8 sq km in good habitat and up to 20 sq km in less favorable areas. These ranges overlap one another extensively and are not defended; each may overlap up to seven male territories.

In a study of a small introduced population in Murchison Falls National Park, Uganda, Van Gyseghem (1984) found a somewhat different situation. The population, consisting of 15 individuals, occupied a total range of 130 sq km, of which 66 sq km was used in the rainy season and 74 sq km in the dry season. The single adult male was territorial, using 6 sq km during the rainy season and 24 sq km in the dry season. The other animals had overlapping home ranges of 30–97 sq km each.

Ceratotherium appears to have the most complex social structure among the rhinoceroses (Kingdon 1979; Owen-Smith 1974, 1984; Smithers 1983; Van Gyseghem 1984). Temporary associations of up to 14 individuals have been observed, and there are smaller, cohesive units. Territorial bulls are usually solitary; they mark and patrol the boundaries of their areas and challenge any intruding adult male.

Sometimes there are ritualized engagements involving repeated apposition of horns, but serious fighting is rare at such times, and usually one or both of the opponents retreats. More intensive conflicts, with head-on charges and the infliction of injuries by horning or ramming, may occur when males compete for estrous females. A dominant bull usually tolerates the presence of several subordinate males within his territory and also allows females and subadults to wander freely through the area. He attempts to prevent estrous females from leaving. Several females and their calves commonly form an association. Subadults, which are driven off by their mothers before the birth of the next calves, pair with one another; sometimes up to six young animals will join an adult female.

The white rhino has about 10 vocalizations, including a panting contact call, grunts and snorts associated with courtship, squeals of distress, and deep bellows or growls for threats. Dominant males spray urine to demarcate the boundaries of their territories; subordinate males and other animals do not spray urine. There are communal dung heaps, which facilitate olfactory identification between animals in an area. Territorial males have the habit of scattering their dung after defecation.

The following data on reproduction were taken from Groves (1972*b*), Kingdon (1979), Owen-Smith (1974, 1984), and Smithers (1983). A pair bond may last 5–20 days and involve some chasing and horn clashing. Breeding occurs throughout the year, but mating peaks have been observed in South Africa from October to December and in East Africa from February to June. The gestation period has been estimated to last as long as 18 months, though now it is thought to be closer to 16 months. The single calf weighs about 40–65 kg at birth and remains shaky for 2–3 days. When alarmed, it runs ahead of the cow, whereas the calf of *Diceros* tends to follow its mother. Weaning commences at 2 months of age, but nursing may continue for well over a year. Females commonly give birth every 2–3 years and drive off their previous calf just before parturition. Sexual maturity may come at 4–5 years, but females do not have their first calf until they are 6.5–7.0 years old, and males probably are 10–12 years old before they can claim a territory and mate. A wild, 36-year-old female was still reproductively active. Potential longevity is probably 40–50 years.

In contrast to *Diceros*, the white rhino is mild-tempered and interspecifically nonaggressive. It becomes tame and tractable in captivity and reportedly can be approached safely by people in the wild (Kingdon 1979; Owen-Smith 1984). Unfortunately these traits have contributed to its downfall, as it has been relentlessly hunted for the same reasons as those described above in the account of the black rhino. The subspecies *C. s. simum* was largely eliminated in the course of the settlement of southern Africa during the eighteenth and nineteenth centuries. It was considered to be totally extinct by 1893, but in the following year a small population was discovered in the Umfolozi area of Zululand in eastern South Africa. Some accounts indicate that fewer than 10 individuals survived at that time, but there probably were more than that, and the true low point may have come in the 1930s, when a drought reduced the population to under 100. Subsequent careful protection led to a growing population at Umfolozi, and adjoining reserves, that appeared to be in excess of available habitat. Numbers reached about 1,000 individuals by 1970 and 2,000 by 1980; a subsequent management program involved the live capture of many animals which then were used in massive reintroduction programs in other parts of South Africa as well as in other countries (Groves 1972*b*; Owen-Smith 1981; Smithers 1983). A recent count showed the existence of 4,404 southern white rhinos, including 4,062 in South Africa, 208 in Zimbabwe, 63 in Namibia, and 6 in

Zambia (Cumming and Du Toit 1989). Small groups introduced to Mozambique and Kenya (outside of the natural range) may have been eliminated by poachers. *C. s. simum* is on appendix 1 of the CITES.

The northern subspecies of the white rhino *(C. s. cottoni)* was not discovered by science until 1903, at which time it was still fairly numerous in the range given at the start of this account. Subsequently populations have fluctuated in response to the alternating prevalence of human exploitation and protection, but the overall trend has been disastrously downward (Hillman-Smith, Oyisenzoo, and Smith 1986; Kingdon 1979; Western and Vigne 1985). Numbers in Uganda fell to about 133 in 1928, increased to 500 in 1950, and then declined to 71 in 1963. In an effort to protect some of the last survivors from poachers, 15 individuals were captured and moved to Murchison Falls National Park. Their numbers grew to about 80, but all were killed in 1980. When Garamba National Park was established in Zaire in 1938 there were only about 100 white rhinos in the area. Numbers then grew to about 1,200 in 1963, fell sharply during a period of political turmoil in the 1960s, rebounded to about 500 after government control was restored in the 1970s, and finally collapsed as poaching for horn intensified in the early 1980s. Populations in Sudan and the Central African Republic also disappeared during this period. Groves (1972b) had stated that a few individuals might survive in Chad, but they evidently were lost by 1980. By the latest count, there now are known to be only 18 northern white rhinos in the wild, all at Garamba (Cumming and Du Toit 1989). There are another 13 in captivity. *C. s. cottoni* is classified as endangered by the IUCN and the USDI and is on appendix 1 of the CITES.

Order Artiodactyla

Even-toed Ungulates (Hoofed Mammals)

This order of 9 Recent families, 81 genera, and 211 species is native to all land areas of the world, except the West Indies, New Guinea and associated islands, Australia, New Zealand, Antarctica, and most oceanic islands. Introduction by human agency has led to the establishment of wild populations of some species in certain areas where the order does not naturally occur. Simpson (1945) divided the living Artiodactyla into the following groups:

Suborder Suiformes
  Infraorder Suina
      Family Suidae (pigs)
      Family Tayassuidae (peccaries)
  Infraorder Ancodonta
      Family Hippopotamidae (hippopotamuses)
Suborder Tylopoda
      Family Camelidae (llamas, camels)
Suborder Ruminantia
  Infraorder Tragulina
      Family Tragulidae (mouse deer)
  Infraorder Pecora
      Superfamily Cervoidea
        Family Cervidae (deer)
      Superfamily Giraffoidea
        Family Giraffidae (giraffes)
      Superfamily Bovoidea
        Family Antilocapridae (pronghorn)
        Family Bovidae (antelope, cattle, goats, sheep)

Although the Antilocapridae (see account thereof) are usually considered to be related to, if not confamilial with, the Bovidae, there is increasing evidence that they should be placed in the Cervoidea.

The smallest genus in the order is *Tragulus*, which has a head and body length of 400–750 mm and a weight of 0.7–8.0 kg. Maximum size is represented by *Hippopotamus*, which weighs up to 4,500 kg, and *Giraffa*, which attains a height of up to 5.8 meters. The principal distinguishing feature of the order is the foot, which has an even number of well-developed digits, except in the genus *Tayassu* (collared and white-lipped peccaries), in which the hind foot has three digits. A first digit (pollex of forelimb, hallux of hind limb) occurs only in certain fossil artiodactyls. The second and fifth (lateral) digits are slenderer than the third and fourth (medial) digits or are vestigial or absent. The main axis of the foot passes between the third and fourth digits, and the body

weight is borne thereon. In the two-toed, or "cloven-hoofed," genera of artiodactyls, the central wrist or ankle bones are absent. The humerus is usually shorter than the forearm, there being some few exceptions, and the radius and ulna are separated or fused. The ankle bone, or astragalus, has a rolling surface above the joint and a pulley surface below, giving free movement to the ankle. The fibula articulates with the heel bone, is usually slender or incomplete, and in some cases is fused with the tibia.

Many species have frontal appendages, commonly called horns or antlers. The nasal bones are not expanded posteriorly as they are in the order Perissodactyla. There is no alisphenoid canal. The upper incisor teeth are reduced or absent. The canines usually are reduced or lost, though in some species they are enlarged and tusklike. The molars, which are more complex than the premolars, are low-crowned with cusps in the suborder Suiformes and high-crowned with crescents in the suborders Tylopoda and Ruminantia.

In the infraorder Suina the stomach is two-chambered and nonruminating, in the infraorder Ancodonta it is three-chambered and nonruminating, in the suborder Tylopoda and the infraorder Tragulina it is three-chambered and ruminating, and in the infraorder Pecora it is four-chambered and ruminating. All the ruminants, or "cud chewers," crop or graze vegetable food, such as grasses and woody material, in which there is a relatively low amount of nutrients. They swallow the food rapidly, with little chewing, and then may retire to some secluded spot to digest it more thoroughly. In the ruminants, when the food is first swallowed, it enters the rumen, or paunch, and after undergoing a softening process there, it is regurgitated into the mouth, where it is chewed again and further mixed with salivary juices. The food then is swallowed a second time, entering the second compartment of the stomach (the reticulum, or honeycomb bag). It then progresses to the third stomach (the manyplies, omasum, or psalterium), and then to the fourth stomach, or digesting chamber (the reed, or abomasum), where the greatest digestive activity takes place. In this manner, cud-chewing animals can quickly consume a large quantity of low-grade food and, when no danger threatens, impart to it the thorough grinding and chemical treatment necessary to convert it to their use. Bacterial action also is involved in the breakdown of food by a ruminant.

Because most artiodactyls are massive and have large bones that resist decay and other destructive forces, there exists a good record of fossil forms going back to the early Eocene. The order Artiodactyla includes 18 extinct families (Simpson 1984). The Old World seems to have been the center of evolution of the order, whereas the perissodactyls developed mainly in North America. The earliest artiodactyls

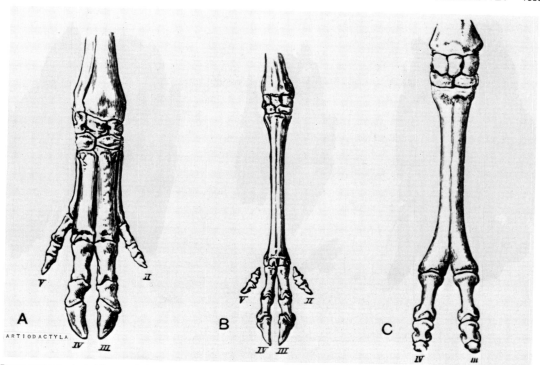

Bones of the forefeet of three artiodactyls: A. Hog *(Sus);* B. Deer *(Cervus);* C. Camel *(Camelus).* Photos from *Mammalia,* F. E. Beddard.

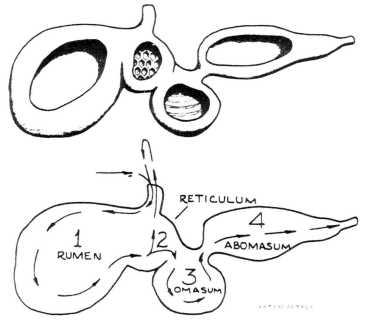

Diagrams of the complicated four-compartment stomachs of ruminants (the Cervoidea, Giraffoidea, and Bovoidea). The lower diagram shows the route traversed by the food from the time it is first swallowed until it is returned to the mouth for further mastication and then swallowed again to go into the second chamber and the remainder of the digestive tract. The Camelidae and Tragulidae also ruminate, or chew their cud, but their stomachs are slightly different and have only three compartments. Photo from *Detroit Zoo Guide Book.*

Wart hog *(Phacochoerus aethiopicus)*, photo by P. F. Wright of skull in U.S. National Museum of Natural History.

apparently had a full complement of teeth (a total of 44), four distinct digits on each foot, separate foot bones, no frontal appendages, and a simple, nonruminating stomach. Division into bunodont and selenodont dentition lines seems to have taken place at an early period, and there is evidence that intermediate forms existed.

ARTIODACTYLA; Family SUIDAE

Pigs, or Hogs

This family of five Recent genera and nine species originally occurred in Eurasia to the south of 48° N, on most associated continental islands as far to the southeast as the Philippines and Sulawesi, throughout Africa, and, probably through introduction, in Madagascar. Introduction by human agency also has led to the establishment of feral populations of one species, *Sus scrofa*, in parts of North America, in New Guinea and on many nearby islands, and in New Zealand. This species also has given rise to the domestic pig, which occurs throughout the world in association with people. Simpson (1984) recognized three subfamilies: Suinae, for *Sus, Potamochoerus*, and *Hylochoerus;* Phacochoerinae, for *Phacochoerus;* and Babyrousinae, for *Babyrousa*.

Head and body length is approximately 500–1,900 mm, tail length is about 35–450 mm, and adult weight is as much as 350 kg. The thick skin is usually sparsely covered with coarse bristles or bristly hairs, and some suids are almost naked. A mane occurs on some forms, and the tail has bristly hairs at its tip. The young are striped, except in *Babyrousa* and domestic *Sus scrofa*. The females of most genera have three or six pairs of mammae, but there is only one pair in *Babyrousa*. Suids have a two-chambered, simple, nonruminating stomach.

Suids are medium-sized mammals with a long and pointed head, a short neck, and a stocky, barrel-like body. The mobile snout is truncated terminally and has a disklike cartilage in the tip. This cartilaginous snout, used for turning up surface soil, is strengthened by an unusual bone, the prenasal, situated below the tip of the nasal bones of the skull. Grooving of the snout occurs in only one genus, *Babyrousa*. The nostrils are terminal, the eyes are small, and the ears are fairly long, often with a tassel of hairs near the tip. Some members of this family have warts or ridges on the face, which are skin growths without a bony support or core. The forelegs are half as long as the height at the shoulders. The foot bones are separate (not fused), and the feet are narrow. The first digit is absent, so each foot has four toes; the middle two (the third and fourth) are flattened and have hooves, whereas the other two (the second and fifth) are located higher up on the limb, do not reach to the ground in the ordinary walking position, and have smaller hooves than do the third and fourth digits.

The most striking feature of the skull in the family Suidae is the elevation and backward slope of the occipital crest, formed by the union of the supraoccipital and parietal bones. In *Potamochoerus, Sus*, and *Hylochoerus* the dental formula is: (i 3/3, c 1/1, pm 4/4, m 3/3) × 2 = 44. In *Babyrousa* it is: (i 2/3, c 1/1, pm 2/2, m 3/3) × 2 = 34. In *Phacochoerus* it is: (i 1/3, c 1/1, pm 3/2, m 3/3) × 2 = 34. The upper incisors decrease in size from the first to the third, and the lower incisors are long, narrow, set closely together, and almost horizontal in position. The incisors and the canines have sharp lateral edges. The large upper canines grow outward and backward, tending to form a complete circle. Usually, however, the canine teeth wear against each other, thus producing the sharp edges. These tusks, which are most prominent in males, reach their greatest development in *Babyrousa;* in this genus the upper canines are directed upward through the skin, never entering the mouth, and curve backward and downward, often touching the forehead. The cheek teeth are cuspidate, and the upper premolars are simpler in structure than are the molars. With age, the enamel wears away and all the teeth disappear except for the canines and back molars. The third molars of *Phacochoerus*, which often are the only teeth in that genus, are unlike those of any other mammal; they are composed of a number of closely set cylinders of dentine embedded in cement.

Pigs generally live in forests and woodlands. They shelter in tall grass or reedbeds and in burrows that are either self-excavated or abandoned by other animals. Pigs are surefooted and rapid runners and good swimmers, and they are fond of mud baths. When cornered or wounded, they frequently fight courageously. In such battles their tusks are deadly weapons. They are active mainly at night, especially in areas where they are molested by people. Some forms use their snout, and a few use their tusks, to dig for food. The omnivorous diet includes fungi, leaves, roots, bulbs, tubers, fruit, snails, earthworms, reptiles, young birds, eggs, small rodents, and carrion. Contrary to popular belief, a wild pig rarely will overeat.

The geological range of this family is early Oligocene to Recent in Europe, Oligocene to Recent in Asia, and early Miocene to Recent in Africa (Simpson 1984).

ARTIODACTYLA; SUIDAE; Genus SUS
Linnaeus, 1758

Pigs, Hogs, or Boars

There are two subgenera and five species (Ansell, *in* Meester and Setzer 1977; Chasen 1940; Ellerman and Morrison-Scott 1966; Groves 1981a; Grzimek 1975; Laurie and Hill 1954; Medway 1977, 1978; Oliver 1984; Sanborn 1952; Taylor 1934):

subgenus *Sus* Linnaeus, 1758

S. scrofa (wild boar, or pig), originally found from southern Scandinavia and Portugal to southeastern Siberia and the Malay Peninsula, from Western Sahara to Egypt, and on Britain, Ireland, Corsica, Sardinia, Sri Lanka, Japan, the Ryukyu Islands, Taiwan, Hainan, Sumatra, Java, and many small associated islands of the East Indies as far east as Komodo;

S. barbatus (bearded pig), Malay Peninsula, Rhio Archipelago, Sumatra, Bangka, Borneo, Philippines;

S. celebensis (Celebes wild boar), indigenous to Sulawesi and some nearby small islands;

S. verrucosus (Javan pig, or warty pig), Java and the nearby islands of Madura and Bawean;

subgenus *Porcula* Hodgson, 1847

S. salvanius (pygmy hog), Nepal, Sikkim, Bhutan, and adjacent parts of northeastern India.

A general view, as expressed by Honacki, Kinman, and Koeppl (1982), is that *S. verrucosus* includes *S. celebensis* and that the resulting species, and not *S. barbatus*, is found in the Philippines. Detailed studies by Groves (1981a), however, indicate a systematic and distributional situation as given above.

In *S. salvanius*, head and body length is 500–650 mm, tail length is about 30 mm, and shoulder height is 250–300 mm. In the other three species, head and body length is 900–1,800 mm, tail length is about 300 mm, shoulder height is 550–1,100 mm, and weight is 50–350 kg. Some domestic breeds of *S. scrofa* may attain a weight of 450 kg (Grzimek 1975). Males

European wild hog *(Sus scrofa)*, photo by Ernest P. Walker and William J. Schaldach, Jr.

usually are larger than females. Coloration of wild forms is dark gray to black or brown. The body is covered with stiff bristles and usually some finer fur, but the pelage often is quite scant, and the tail is only lightly covered with short hairs. Many individuals have side whiskers and a mane on the nape. The young are striped. Hogs have four continually growing tusks, two in each jaw. There are three pairs of mammae in female *S. salvanius* and six pairs in females of the other species.

Pigs live in many kinds of habitat but generally where there is some vegetation for cover. They are most plentiful in oak forests and in the reedbeds of Asia; a major limiting factor is thought to be snow depths greater than about 40–50 cm, which make travel and searching for food difficult (Groves 1981a). Individuals construct crude shelters by cutting grass and spreading it over a given area. They crawl under the grass and then raise themselves to lift the grass mat, which then attaches to uncut grass and forms a canopy. They wallow in mud and will do so for hours if the opportunity affords. They are swift runners and good, strong swimmers. Activity is mainly nocturnal and crepuscular. A great distance may be traveled during the night in search of food. The omnivorous diet includes fungi, tubers, bulbs, green vegetation, grains, nuts, cultivated crops, invertebrates, small vertebrates, and carrion.

In a study of feral *S. scrofa* in coastal South Carolina, Wood and Brenneman (1980) found a population density of 10–20 individuals per sq km and an average annual home range of 226 ha. for males and 181 ha. for females. In a nearby area, Kurz and Marchinton (1972) determined home range to average about 400 ha. and seldom observed groups of more than 3 individuals. Off the West Coast, on Santa Catalina, Baber and Coblentz (1986) found densities of 21–34 feral pigs per sq km and mean home ranges of around 200 ha. for males and 100 ha. for females. Singer (1981) listed densities of about 8–9 per sq km in Great Smoky Mountains National Park and around 1–30 per sq km in various parts of Europe.

In the Old World, *S. scrofa* has been seen in herds, or "sounders," of over 100, though average group size seems to be about 20 (Lekagul and McNeely 1977). According to Fradrich (1974), in both wild and feral *S. scrofa* the basic social unit is the female and her litter. After the young are weaned, two or more families may come together. This association remains stable until the beginning of the next mating season, when the previously solitary adult males join in fighting over the females. A male usually wins control of 1–3 females but sometimes obtains as many as 8. After mating, the males depart. Observations of wild boars by Dardaillon (1988) indicate that a sounder is basically a nursery unit; a near-term female separates from her group, but within a few weeks of birth there is a reassociation, first with other lactating mothers and then with young of the previous year. Barrette (1986) reported that conflicts between pigs begin with two animals walking parallel and edging closer and closer until their shoulders touch and shoving begins. They then sometimes rear up, lean against one another, and attempt to knock each other off balance. Finally, if neither yields, they turn and begin to thrust with mouths open and tusks bared.

Breeding occurs throughout the year in the tropics, but births peak shortly before or just after the rains. In temperate regions, the young are born in the spring. Females have an estrous cycle of about 21 days, are receptive for 2–3 days, and generally produce one litter annually. The gestation period is 100–140 days. The number of young per litter is 1–12, usually 4–8. Unlike the young of most ungulates, piglets are born in a nest and remain there following birth. They are weaned after about 3–4 months and may leave the mother prior to birth of the next litter, but young females often remain longer. Sexual maturity may be attained by as early

as 8–10 months, but females usually do not mate until around 18 months. Males generally are not able to compete successfully for mating privileges until they reach full size, at around 5 years (Fradrich 1974; Grzimek 1975; Henry 1968; Lekagul and McNeely 1977). Average longevity is about 10 years, but some pigs have lived up to 27 years.

Wild pigs have been extensively hunted by people for use as food, for sport, and because they sometimes are destructive to crops. Their size and sharp tusks make them potentially dangerous antagonists, but they usually do not attack unless molested. Wild *S. scrofa* was exterminated long ago in many parts of its original range, including the British Isles, Scandinavia, and Egypt, but has been reintroduced in Scandinavia (Lever 1985). In the period 1965–75 there was a dramatic increase in wild boar populations across Europe, with the numbers being taken by hunters approximately doubling and reaching 100,000 per year in Russia; subsequently populations have stabilized (Sáez-Royuela and Tellería 1986). Wild boars also have been deliberately released by human agency, mainly for purposes of sport hunting, in certain other parts of the world. There is sometimes difficulty, however, in determining whether the free-ranging pigs in these areas represent originally wild stock or are the descendants of domestic animals (Corbet 1978; Grzimek 1975; Laurie and Hill 1954; Wood and Barrett 1979).

Escaped domestic pigs certainly have formed large feral populations in many regions, including Central and South America, the West Indies, Australia, New Zealand, Indonesia, the Andaman Islands, New Guinea (but see below), Hawaii, the Galapagos, and many other oceanic islands (Lever 1985; Oliver 1984). They are generally considered detrimental in these areas, especially on islands, where they have been responsible for the destruction of many species of native animals and plants through either direct predation or habitat disruption. Pavlov (*in* Strahan 1983) referred to the feral pig as the most significant mammalian pest of agriculture in Australia and as a reservoir of many diseases.

According to Lekagul and McNeely (1977), domestication of *S. scrofa* took place in China around 4900 B.C. and may have occurred as early as 10,000 B.C. in Thailand. Many breeds have since been developed, especially in Europe. Pigs are valuable to agricultural economies, because they mature sooner than do other domestic ungulates, have larger litters, and can feed on human refuse. They thus have contributed to the spread of human populations to New Guinea and the Pacific islands. Several British breeds now are considered rare and are the subject of conservation efforts (Hall 1989).

The first pigs in the United States were those brought by the Polynesians to Hawaii around A.D. 1000 and those introduced by the Spanish to the Southeast in the early sixteenth century. Several valuable farm breeds of *S. scrofa* were developed in the United States, but large feral populations also became established. European wild boars were introduced in several places for purposes of sport hunting, and these interbred with the feral animals already present. Free-ranging pigs now occur from Texas to Florida and the Carolinas, throughout California, on eight of the major Hawaiian Islands, and on Puerto Rico and the Virgin Islands. They are valued as game animals in some areas, and over 100,000 are taken annually by sport hunters. Many persons, however, consider feral pigs to be detrimental to agriculture, forestry, and native wildlife (Singer 1981; Wood and Barrett 1979).

Groves (1981a) determined that some of the feral and domestic pigs of the East Indies are not *S. scrofa* but *S. celebensis*. He hypothesized that *S. celebensis* was domesticated initially on Sulawesi and then brought to the Moluccas, Flores, Timor, Halmahera, and some nearby small islands. In addition, the feral pig populations of New Guinea, Ceram, and some small islands in the Moluccas, which are to a large

A. European wild boars *(Sus scrofa)*, mother and young, photo by Eric Parbst through Zoologisk Have, Copenhagen. B. Domestic hog *(S. scrofa)*, photo by Roy Pinney.

extent genetically continuous with the domestic pig populations in the region, appear to have resulted from hybridization between introduced stocks of *S. celebensis* and *S. scrofa*.

The pygmy hog *(S. salvanius)* is classified as endangered by the IUCN and the USDI and is on appendix 1 of the CITES. Its range now appears to be restricted to northwestern Assam, where an estimated 100–150 individuals survive. It has declined mainly because of the modification and usurpation of its limited habitat by people. It is protected by law in India but is illegally hunted. The Visayan warty pig *(S. barbatus cebifrons)*, designated vulnerable by the IUCN and formerly found throughout the central Philippine archipelago, has become uncommon through human destruction of most of its rainforest habitat and excessive hunting (Cox 1987). The same problems were thought to have led to the extinction of the Javan pig *(S. verrucosus)*, which is restricted to a region having one of the world's densest human populations. A recent survey showed that it still survives in several parts of Java but that it is jeopardized not only by human persecution but also by hybridization with the more numerous *S. scrofa* (Blouch 1988). The IUCN now gives the vulnerable designation to *S. verrucosus* and also to *S. scrofa riukiuanus*, of the Ryukyu Islands, and *S. barbatus oi*, of the Malay Peninsula, Rhio Archipelago, Sumatra, and Bangka. Oliver (1984) indicated that *S. b. oi* is jeopardized through illegal importation of pork by Singapore, while the main threat to *S. scrofa riukiuanus* is excessive hunting for local consumption. He noted that the latter is a true endemic wild pig, not an introduced feral population as sometimes claimed.

ARTIODACTYLA; SUIDAE; Genus POTAMOCHOERUS
Gray, 1854

African Bush Pig, or Red River Hog

The single species, *P. porcus*, occurs through much of Africa south of the Sahara, though not in Namibia or in most of Botswana or South Africa (Ansell, *in* Meester and Setzer 1977; Smithers 1983). It also is found in Madagascar and on Mayotte Island in the Comoros, probably through introduction by human agency many years ago, but Kingdon (1979) suggested that these areas may have been colonized naturally, since *Potamochoerus* sometimes ventures into extensive papyrus beds which may detach and float out to sea.

Head and body length is 1,000–1,500 mm, tail length is 300–432 mm, shoulder height is 585–965 mm, and weight is 46–130 kg. Coloration varies from reddish brown to black, often with a heavy mixture of white or yellowish hairs. Some individuals are almost white or mottled white and black or brown. The young are longitudinally striped with pale yellow or buff on a dark brown ground color. The long, pointed ears often have long tufts or streamers of hairs at the tip, and there is a pronounced light-colored mane along the top of the neck and back.

The average upper tusk length is 76 mm, and the lower tusk measures 165–90 mm. The upper tusks point downward and wear against the lower ones. The male has warts in front of the eye, which, although they protrude 40 mm, frequently are not conspicuous, as they often are concealed by facial hair. This genus resembles *Phacochoerus* and *Sus* externally but differs from the former in having more hair and a greater number of teeth and from the latter in having ears that are more tufted. Females have three pairs of mammae.

The African bush pig is found in a variety of habitats. Kingdon (1979) wrote that it lives wherever there is sufficient moisture to support dense vegetation throughout the year and to keep the ground moderately soft and that it is numerous in all forests, in most riverine habitats, and in montane habitats with thick cover. It is most active at night and rests by day in a self-excavated burrow within an area of dense vegetation. It is a fast runner and a good swimmer. When cornered or wounded, it exhibits considerable courage and frequently attacks. It is wary of traps; when a trap is discovered, *Potamochoerus* will avoid the area for weeks. The omnivorous diet consists mainly of roots, berries, and fruit. Reptiles, eggs, and, occasionally, young birds also are eaten. The snout is used as a plough to "root" up subsoil vegetation, and in a short period of time a herd can do much damage to crops. Smithers (1983) noted that home ranges are very large in some areas, with an estimated distance of up to 4 km between feeding and resting places.

In Uganda, Laws, Parker, and Johnstone (1975) estimated

African bush pig *(Potamochoerus porcus)*: A. Photo by Bernhard Grzimek; B. Immature animal, photo from New York Zoological Society.

the average population density to be 1.29/sq km. *Potamochoerus* is gregarious and travels in groups of up to 20 individuals. In the Transvaal, Skinner, Breytenbach, and Maberly (1976) found most groups, or "sounders," to contain 4–6 pigs and to be led by a dominant male that defended the area being utilized at a given time. When two groups met, there were ritualized, threatening displays but rarely serious fighting. The tusks were used to mark trees along regularly traveled paths.

In the southern part of Africa, births seem to occur mainly from September to April, peaking during the warm, wet summer (November–February). The gestation period is about 4 months. The sow constructs a nest of grass about 3 meters across and 1 meter deep to protect the piglets. The number of young per litter is 1–8, usually about 3–4. They are given considerable care by the dominant boar. Females evidently attain sexual maturity at about 3 years (Skinner, Breytenbach, and Maberly 1976; Smithers 1971, 1983). A captive specimen lived for 20 years (Jones 1982).

The African bush pig has increased greatly in numbers because of the destruction of leopards by people and the creation of more favorable habitat by the spread of agriculture. *Potamochoerus* now damages many kinds of cultivated plants and has been known to wipe out entire peanut crops. It is widely hunted for this reason and also for its highly palatable flesh (Grzimek 1975; Smithers 1971). Kingdon (1979) suggested that it has potential for domestication.

ARTIODACTYLA; SUIDAE; Genus HYLOCHOERUS
Thomas, 1904

Giant Forest Hog

The single species, *H. meinertzhageni*, is found in the forest zone from Liberia to southwestern Ethiopia and northern Tanzania (Ansell, *in* Meester and Setzer 1977).

Head and body length is 1,300–2,100 mm, tail length is 300–450 mm, shoulder height is 762–1,100 mm, and weight is 130–275 kg. The pelage is long, coarse, and black and becomes sparse with age. The skin is blackish gray. The skin in front of each eye and on the upper part of the cheek below the eye is almost naked. Below and behind each eye there are two movable cutaneous thickenings or facial excrescences. One of these forms a wide ridge on each naked area of skin. A preorbital gland is present, marked externally by a slit on the naked area of the face in front of each eye. There are no facial protuberances as in *Phacochoerus*, and the upper canines are set horizontally, not at an angle as in *Phacochoerus*. The large skull of *Hylochoerus* contains a depression in the roof capable of holding nearly a cup of water.

Kingdon (1979) wrote that this species is scattered across tropical Africa in a series of rather localized populations in a variety of habitats from subalpine areas and bamboo groves through montane to lowland swamp forests, galleries, forest savannah mosaics, wooded savannahs, and postcultivation thickets. Peak activity is from dusk to nearly midnight, though daylight sightings increase as seasonal temperatures fall. According to D'Huart (1976), *Hylochoerus* is diurnal in areas where it is protected from human molestation and uses holes in the ground for nightly rest. It also makes tunnels and pathways, and it wallows in water and swamps. Kingdon (1979) noted that preferred sleeping sites are under fallen tree trunks and consist only of a shallow scoop of bare earth; a dense network of paths connects these sites with latrines, mineral licks, and grazing grounds. *Hylochoerus* feeds mainly on short grass, sedges, shrubs, and herbaceous growth and does not do much digging or grubbing for food.

Hylochoerus travels in groups, or "sounders," of up to 20 individuals. D'Huart (1976) reported that groups maintain a territory and are led by an old male. Kingdon (1979), however, wrote that group home ranges of 10 sq km have been suggested and that such areas evidently overlap extensively. The basic social unit, he noted, is the mother and her offspring of up to three generations. In one area 96 percent of

Giant forest hog (*Hylochoerus meinertzhageni*), photo by Bernhard Grzimek.

these families were accompanied by a male, and it was unusual for there to be more than one mature male in a sounder. Adult males also are often seen alone. On occasion several families and up to 40 individuals have associated, but sometimes when groups meet, one withdraws or the adult males push each other with their foreheads or repeatedly charge, ramming their heads together. Breeding is continuous, but mating peaks in East Africa have corresponded with the latter part of the rainy season. Litters of 2–11 young are born in the bedding-down place, following a gestation period of 125 days. Females probably reach sexual maturity at one year of age.

The giant forest hog was one of the last mammals of its size to become known to science. Its range still may not be completely known, but it apparently has disappeared from western Liberia (Ansell, *in* Meester and Setzer 1977). People seem to fear it more than they do *Potamochoerus*. Males often charge without warning or provocation, presumably to protect the sounder. *Hylochoerus* sometimes raids native shambas at the forest edge and causes much destruction. Certain tribes make war shields from the hide of this hog. Kingdon (1979) noted that it is tractable in captivity and easily tamed, that its meat is of excellent quality, and that this animal shows promise for domestication.

ARTIODACTYLA; SUIDAE; Genus PHACOCHOERUS
F. Cuvier, 1817

Wart Hog

The single species, *P. aethiopicus,* occurs in most of Africa south of the Sahara (Ansell, *in* Meester and Setzer 1977).

Head and body length is about 900–1,500 mm, tail length is 250–500 mm, shoulder height is 635–850 mm, and weight is 50–150 kg. The length of the upper tusks is 255–635 mm in males and 152–255 mm in females. A long, thin mane of coarse hair extends from the nape to the middle of the back, where it is broken by a bare space, and then continues on the rump. The remainder of the body is covered with bristles. The color of both the skin and hair is dark brown to blackish. The immature are reddish brown. A long, ridgelike fold on the cheek bears white hairs. The warts, which are prominent only on the males, are skin growths and have no bony support or core; they are located on the side of the head and in front of the eye. When an animal moves slowly, the tail hangs limply, but when it runs, the tail is carried in an upright position with the tufted tip hanging over.

The wart hog is usually found in savannah and lightly forested country. In contrast to most suids, it is usually diurnal. In places where it is molested by people, however, it may become almost nocturnal. To sleep, to rear young, and to find refuge from predators, *Phacochoerus* depends on holes, either natural ones or those made by the aardvark *(Orycteropus).* It usually backs into such holes, perhaps to confront a pursuer with its formidable tusks, but sometimes enters headfirst (Cumming 1975). It is usually inoffensive but will defend itself when cornered and can inflict severe wounds with its tusks. Fradrich (1974) wrote that the main weapons of defense are not the large upper tusks but their lower counterparts, which are smaller but sharper. Like other hogs, *Phacochoerus* enjoys mud baths. Its maximum running speed is 55 km/hr (Schaller 1972). Although its eyesight seems poor, its senses of hearing and smell are acute.

The wart hog feeds on grass, roots, berries, the bark of young trees, and occasionally carrion. While feeding, it drops on its padded wrists and frequently shuffles along in this position. In Zimbabwe, Cumming (1975) found *Phaco-*

choerus to be almost entirely graminivorous, being specialized both for grazing on short, seasonally succulent grasses and for digging grass rhizomes with its powerful rhinarium in hard, dry soils.

According to Cumming (1975), reported population densities vary from about 0.2/sq km to 20.0/sq km. In Zimbabwe, the maximum density of 10.6/sq km was found in grassland. Home range there varied from 64 to 374 ha. The population was divided into "clans," each consisting of several bands, or "sounders," and associated lone animals. Most bands contained 4–16 individuals, though groups of up to 40 have been reported. Most groups represent a temporary aggregation of young males or an association of adult females that may have lasting bonds and occupy an area for years. The different groups within a clan have overlapping ranges and share holes and other resources. Adult males are usually solitary but join the female groups briefly for mating. At such times, males engage in highly ritualized battles in which they push and strike with the head and the blunt upper tusks. The warts on the sides of the head serve to cushion the blows, and injuries are rare. There seems to be no territorial defense, but there sometimes is competition for resources, such as water holes, and the area being used at a given time is marked with saliva and secretions from glands around the eyes. Vocalizations include various grunts, growls, snorts, and squeals, which are used for greeting, contact maintenance, threats, warning, and submission.

There are clearly defined breeding seasons, with mating peaks occurring 4–5 months after the breaking of the rains and births mainly in the dry season; females are seasonally polyestrous, with estrus lasting about 72 hours at intervals of 6 weeks (Kingdon 1979). In Zimbabwe, mating occurs in May and June, and births in October and November. The gestation period is 171–75 days. The number of young per litter is 1–8, usually 2–3. The piglets begin to accompany their mother regularly at about 50 days of age and are completely weaned by 21 weeks. The young are temporarily driven away when the female is about to bear a new litter, but they may subsequently rejoin the family. Males separate from their mother by the age of 15 months. Females stay longer, perhaps in permanent association. Sexual maturity comes at 18–20 months, but males usually do not mate until around 4 years (Cumming 1975). A captive wart hog lived for 18 years and 9 months (Jones 1982).

Phacochoerus has been eliminated from most of South Africa but still occurs throughout the remainder of its original range (Ansell, *in* Meester and Setzer 1977). It is widely hunted by people for use as food. It is much less destructive to native crops than are the other wild pigs of Africa.

ARTIODACTYLA; SUIDAE; Genus BABYROUSA
Perry, 1811

Babirusa

The single species, *B. babyrussa,* occurs on Sulawesi, the nearby Togian and Sula islands, and Buru Island in the Moluccas (Laurie and Hill 1954). Its presence on Buru and possibly Sula is evidently the result of introduction by human agency (Groves 1980a).

Head and body length is usually 875–1,065 mm, tail length is 275–320 mm, shoulder height is 650–800 mm, and weight is up to 100 kg. The skin is either rough and brownish gray or smooth and sparsely covered with short whitish gray to yellowish hairs. The underside of the body and inner sides of the legs are sometimes lighter than the rest of the body.

Wart hog *(Phacochoerus aethiopicus):* Top, photo from New York Zoological Society; Bottom, photo by Karl H. Maslowski.

Frequently, this whitish color extends along the sides of the upper lip. The skin usually hangs in loose folds. Females have two pairs of mammae.

In most wild swine the tusks grow from the sides of the jaw. In *Babyrousa,* however, the upper tusks grow through the top of the muzzle and then curve backward toward the forehead. Although the tusks thus would appear to have little use as weapons, MacKinnon (1981) suggested that they are important to males for fighting. The upper tusks apparently have a general defensive function, and the daggerlike lower tusks are used offensively. Since the lowers are not honed by wearing against the uppers, however, the male actively sharp-

Babirusa *(Babyrousa babyrussa)*, photo from New York Zoological Society.

ens them on trees. The wear pattern of upper tusks from Sulawesi indicates that these teeth are used to interlock and hold the opponent's lower tusks. On Buru Island, though, the upper tusks evidently have lost this function and probably are used for butting. According to a native legend, this hog hangs itself by its tusks from a tree limb at night. The natives also say that the tusks are like the antlers of a deer; hence the name babirusa, which means "pig deer."

The preferred habitats of the babirusa are moist forests, canebrakes, and the shores of rivers and lakes. This pig is a swift runner and often swims in the sea to reach small islands. Its senses of hearing and smell are acute. Available evidence indicates that it is diurnal, with activity concentrated in the morning; captive individuals construct straw nests for resting and wallow in mud (Bowles 1986). *Babyrousa* seems not to root with its snout, as does *Sus,* and probably feeds on foliage and fallen fruit.

The babirusa travels about in small parties and reveals its presence by low grunting moans. Offspring are produced in the early months of the year and are not striped like the young of most pigs. The gestation period is about 160 days, females can produce two litters annually, and there are one or two young. The offspring are more precocial than those of other pigs and take some solid food after just 10 days of life. Captives have lived up to 24 years (Bowles 1986; Doherty 1989; Grzimek 1975).

The babirusa frequently is captured young and tamed by native people. The natives hunt it regularly for food, erecting an enclosure of poles and nets into which the animal is driven. Even though the babirusa long has been protected by law, it has declined substantially because of excessive hunting and habitat loss. The wild population has been estimated to number about 4,000 individuals (Bowles 1986). *Babyrousa* is classified as vulnerable by the IUCN (1978) and as endangered by the USDI and is on appendix 1 of the CITES.

ARTIODACTYLA; **Family TAYASSUIDAE**

Peccaries

This family of two Recent genera and three species occurs from the southwestern United States to central Argentina. The sequence of genera presented here is based on the suggestions by Wetzel (1977*a,* 1977*b,* 1981) that *Catagonus* is more primitive than *Tayassu.*

Head and body length is 750–1,112 mm and tail length is 15–102 mm. Peccaries have only 6–9 tail vertebrae, whereas suids have 20–23. The pelage is bristly, and there is a mane of long, stiff hairs on the middorsal line from the crown to the rump. The body form is piglike, but the legs are long and slim and the hooves are small. There are four digits on the forefoot, the two lateral ones being reduced and not touching the ground. There are two functional digits on the hind foot. There is a vestigial, median digit on the back of the hind foot in *Tayassu* but not in *Catagonus* (Wetzel 1977*b*). The third and fourth foot bones, which are completely separate in the Suidae, are united at their proximal ends in the Tayassuidae, as in the ruminants. The snout is the same as in the Suidae: elongate, mobile, and cartilaginous, with a nearly naked terminal surface, in which the nostrils are located. The ears are ovate and erect. Both genera have a scent gland about 75 mm in diameter and 125 mm thick on the rump in front of the tail. When a peccary is excited, the hairs on the neck and back bristle and the dorsal gland emits a musky secretion, the odor of which can be detected many meters away. The stomach is two-chambered and nonruminating but is more complex than that of the Suidae. Female *Tayassu* have two pairs of mammae, while female *Catagonus* have four pairs.

The dental formula is: (i 2/3, c 1/1, pm 3/3, m 3/3) × 2 = 38. The upper canines form tusks, but these are directed downward, not outward or upward as in the Suidae, and are smaller than those in pigs, the average length being about 40 mm. There is a space between the canines and the premolars. The premolars and molars form a continuous series of teeth, gradually increasing in size from the first to the last. The last premolar is nearly as complex as the molars. The molars have square crowns with four cusps.

The geological range of the Tayassuidae is early Oligocene to late Miocene in Europe, middle Miocene to early Pliocene in Asia, Miocene to Pliocene in Africa, early Oligocene to Recent in North America, and late Pliocene to Recent in South America (Simpson 1984). The late Pleistocene genus *Platygonus*, which was considerably larger than the living genera, evidently occurred throughout the conterminous United States about 12,000 years ago (Sowls 1984).

ARTIODACTYLA; TAYASSUIDAE; Genus CATAGONUS
Ameghino, 1904

Chacoan Peccary

The single species, *C. wagneri*, occurs in the Gran Chaco region of southeastern Bolivia, Paraguay, and northern Argentina (Eisentraut 1986; Wetzel 1977b). *Catagonus* long was known by fossil material from the Pleistocene, but the living population was not discovered scientifically until the early 1970s (Wetzel et al. 1975).

Except as noted, the information for the remainder of this account was taken from Mayer and Wetzel (1986) and Wetzel (1977b, 1981). Head and body length is about 900–1,112 mm, tail length is 24–102 mm, height at the shoulder is 520–690 mm, and weight is 29.5–40.0 kg. The general coloration is brownish gray; there is a faint collar of lighter hairs across the shoulders and a black middorsal stripe. From *Tayassu*, *Catagonus* differs externally in having larger overall size, longer and paler-colored hair on the ears and legs, a larger head, a longer snout, and longer ears, legs, and tail. There is no median dewclaw on the posterior side of the hind foot. The skull of *Catagonus* differs from that of *Tayassu* in having extreme development of the rostrum, nasal chamber, and sinuses and more posteriorly placed orbits. The canine teeth of *Catagonus* are more slender than those of *Tayassu*, and the molars are high-crowned (hypsodont) rather than low-crowned (bunodont).

The Chacoan peccary inhabits semiarid thorn forest and steppe. It is cursorial and largely diurnal, with peak activity in the late morning. Morphological characters indicate that it is superior to *Tayassu* in speed, distance vision, and olfaction but that it has less mental capacity and depends more on browsing for food. Legume seeds, roots, and cacti are thought to be important components of the diet.

In an area of favorable habitat where the peccary was protected, Mayer and Brandt (1982) calculated a population density of 9.24/sq km. *Catagonus* is highly social, being found in groups of 1–10, usually 4–5, individuals of all ages and both sexes. Like *Tayassu*, the animals scent-mark one

Chacoan peccary *(Catagonus wagneri)*, photo from West Berlin Zoo.

another, as well as the areas in which they live. The young are born from about July to early December, mostly in August–September (early spring). Mayer and Brandt (1978) found the number of young per litter to be 1–4, usually 2–3, and estimated minimum breeding age of females to be 3 years.

Although unknown to science as a living genus until the 1970s, *Catagonus* long was distinguished from other peccaries by the native Indians of the Chaco, who hunt it for its meat and hide. In this region it now has become the main source of meat for hunters, trappers, military personnel, and ranchers during the initial stage of land clearance. Its hide is exported for use as commercial leather and is sought by sportsmen and tourists in the Chaco. The main problem, however, is the replacement of native vegetation by grass to provide pasture for cattle. The Chacoan peccary is considered rare in Argentina and Bolivia and is rapidly declining in Paraguay (Thornback and Jenkins 1982). It is classified as vulnerable by the IUCN, and there is concern that this most recently discovered of large mammals could disappear by the end of the century.

ARTIODACTYLA; TAYASSUIDAE; Genus TAYASSU
Fischer, 1814

Collared and White-lipped Peccaries, or Javelinas

There are two species (Cabrera 1961; Hall 1981):

T. tajacu (collared peccary), Arizona and Texas to northern Argentina;

T. pecari (white-lipped peccary), southern Mexico to northeastern Argentina.

Hall (1981) placed *T. tajacu* in a separate genus, *Dicotyles* G. Cuvier, 1817, but Wetzel (1977*b*, 1981) considered *Dicotyles* to be congeneric with *Tayassu*.

Head and body length is 750 to about 1,000 mm, tail length is 15–55 mm, shoulder height is 440–575 mm, and weight is 14–30 kg. *T. pecari* is considerably larger, on average, than *T. tajacu*. Males and females are about the same size. *T. pecari* is dark reddish brown to black and has white on the sides of the jaws. *T. tajacu* is generally dark gray and has a whitish collar on the neck; its young are reddish and have a blackish stripe on the back. The name javelina is derived from the Spanish word for javelin or spear and refers to the sharp tusks of these animals.

Peccaries live in a great variety of habitats, including desert scrub, arid woodland, and rainforest. They usually shelter in a thicket or under a large boulder. Limestone caves often serve as winter quarters in some areas. Most activity is at night or in the cooler hours of the day. To escape the midday heat, a herd seeks the shade of rocks or vegetation, while in cold weather the animals may huddle in a self-excavated depression in the soil (Sowls 1984). Both species are known to use wallows in the mud or dust. *T. tajacu* is fairly sedentary and does not seem to travel far from its place of birth, but *T. pecari* wanders so far that it sometimes is considered to be nomadic or migratory (Sowls 1984). The speed, agility, and group defense of peccaries render them more than a match for dogs, coyotes, and even bobcats. They often clash their canine teeth together as a warning. When fleeing danger, they move with a fast running gait. Sowls (1984) listed speeds of up to 35 km/hr. They have poor vision and fair hearing. Their sense

Collared peccary and young *(Tayassu tajacu)*, photo from New York Zoological Society.

of smell is keen enough to locate a small covena bulb 5–8 cm underground before the new shoots are visible.

Peccaries, like pigs, are not "dirty" animals; on the contrary, they are quite clean. Their habit of pawing sand against the belly with the front feet is thought to be a cleansing action. They grub for food with their snout. They are mainly vegetarian, feeding on cactus fruit, berries, tubers, bulbs, and rhizomes. They also consume grubs and occasionally snakes and other small vertebrates; like some pigs, they do not seem to be harmed by rattlesnake bites. They frequent water holes or, in the tropics, stay near running streams. Populations of *T. pecari* reportedly wander over large areas of forest but seem always to stay near water; overall densities of 1.0/sq km and 1.6/sq km have been reported (Mayer and Wetzel 1987). In *T. tajacu* densities seem generally to be around 2/sq km to 8/sq km (Sowls 1984).

Peccaries are gregarious. In *T. tajacu*, group size ranges from 2 to 50 but is usually 5 to 15; both sexes and all ages are found in a group (Bigler 1974; Schweinsburg 1971; Sowls 1974, 1978, 1984). Groups of *T. tajacu* appear to remain permanently stable, but often subgroups and solitary animals split off for periods of hours or days (Oldenburg et al. 1985), and occasionally several groups join in temporary aggregations (Robinson and Eisenberg 1985). The groups evidently have a rank order, though there is disagreement among observers on this point, and while Sowls (1984) noted that females usually dominate males, Bissonette (1982) held that the dominant animal is always a male. Each group of *T. tajacu* has a home range of about 0.5–8.0 sq km. The central part of the range is an exclusive territory, while the peripheral area is shared with other herds. The dorsal scent gland is rubbed against tree trunks and other objects for territorial marking and also seems to identify group members and coordinate movements. *T. tajacu* has a number of vocalizations, including grunts and barks for establishing and maintaining contact between animals, growls of aggression, and repetitive "huffs" and "whoofs" to indicate alarm.

T. pecari sometimes is found in herds of several hundred individuals (Leopold 1959). Such groups contain animals of all ages and both sexes and have been estimated to have home ranges of 60–200 sq km (Mayer and Wetzel 1987). In Peru, Kiltie and Terborgh (1983) generally observed this species in bands of 100–200, while *T. tajacu* occurred in groups of fewer than 12. The difference was thought to be associated in part with the clumped distribution of the hard nuts and seeds favored by *T. pecari*. Exploitation of the feeding sites, together with defense against the predators that are drawn there, give advantages to large herd size. Like *T. tajacu*, *T. pecari* communicates through scent marking and a variety of vocalizations.

Females are polyestrous, with the estrous cycle of *T. tajacu* averaging 24 days and estrus about 4 days (Sowls 1984). Breeding of both species apparently can occur throughout the year. In Arizona, however, *T. tajacu* usually mates in February and March and gives birth during the summer. The gestation period in this species is around 145 days (Sowls 1984), while in *T. pecari* it is 156–62 days (Roots 1966). If a litter is lost, the female may quickly mate again and bear a second annual litter (Sowls 1978). The number of young per litter is 1–4, usually 2. The young, weighing about 500–900 grams each, are born in a thicket, hollow log, cave, or burrow dug by another animal. They can run in a few hours and accompany their mother a day or so after being born, when she rejoins the herd. In *T. tajacu*, lactation lasts 6–8 weeks; the milk of this species is lower in fat and total solids than that of domestic *Sus scrofa*. Young peccaries, at least of *T. tajacu*, reach the teats from the rear of the sow instead of standing parallel to the side. They remain with the mother for 2–3 months. Captives may breed in their first year of life (Sowls

1978). One captive *T. tajacu* lived for 24 years and 7 months (Jones 1982).

If unmolested, peccaries ordinarily do not bother humans, but if a member of a band is wounded or pursued, the entire herd may counterattack. Captive adults are sometimes unpredictable, but a pet kept at the U.S. National Zoo, in Washington, D.C., would come to the fence when it recognized a friend. It knew its name, would come promptly when called, and enjoyed being scratched. Wild peccaries are hunted for their meat and skin. There has been an extensive and wasteful trade in hides in parts of Latin America (Grimwood 1969; Leopold 1959). Millions of skins have been exported during the last few decades, with prices varying from about U.S. $0.20 to $5.00 each (Sowls 1984). In contrast, *T. tajacu* apparently has extended its range in the southwestern United States since the time of the early European explorers. The species is classified as a game animal there and is subject to regulated sport hunting (Sowls 1978, 1984). *T. pecari* depends on large tracts of wilderness, and such habitat is rapidly disappearing; the species now has disappeared or become rare in southern Mexico and northern Argentina (Mayer and Wetzel 1987). Both species of *Tayassu* were introduced to Cuba in 1930 (Hall 1981).

ARTIODACTYLA; **Family HIPPOPOTAMIDAE**

Hippopotamuses

This family of two Recent genera and three species originally occurred during historical time in suitable habitat throughout Africa south of the Sahara, all along the Nile River, and in Palestine and Madagascar.

The two genera, *Hippopotamus* and *Choeropsis*, differ greatly in size, but both are characterized by a broad snout, a large mouth, a short barrel-like body, and short, stocky legs. The belly is carried only a short distance above the ground. The nostrils are located on top of the snout and can be closed. The sparsely haired body contains special pores that secrete a pinkish substance known as "blood sweat." This material is thick, oily, and protective in nature, allowing the animal to remain in water or in a dry atmosphere on land for extended periods. The skin contains a layer of fat, which in *Hippopotamus* is 50 mm thick. The short tail is rather bristled. The foot bones are separate. All four toes on each foot support the body weight; the two lateral digits are nearly as well developed as the median digits. The terminal digital bones have nail-like hooves. The stomach is complex and three-chambered but nonruminating. Females have two mammae.

The large skull has an elongate facial region, but the braincase is relatively small. The dental formula is: (i 2–3/1–3, c 1/1, pm 4/4, m 3/3) × 2 = 38–42. There usually is only one pair of lower incisors in *Choeropsis* but two or three pairs in *Hippopotamus*. The incisors and canines are tusklike and grow continuously. The incisors are rounded, smooth, and widely separated. The upper incisors are quite short and project downward. The lower ones are longer, especially the inner pair (in *Hippopotamus*), and project forward and only slightly upward. The lower canines, which are the largest of the tusklike teeth, project upward and outward. The premolars are usually single-cusped, though this condition may vary even on opposite sides in the same animal. Two pairs of cusps are developed in the molars, except in the third molar, which has three pairs. These cusps wear down to various trefoil, figure-eight, or dumbbell-shaped enamel patterns.

Simpson (1984) listed the geological range of this family as late Miocene to late Pleistocene in Europe, late Miocene to Pleistocene in Asia (including Sri Lanka and Java), late

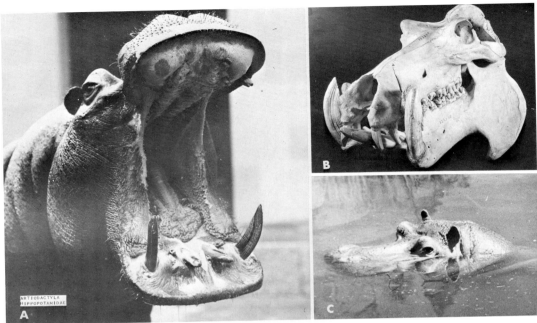

Hippopotamus (Hippopotamus amphibius): A. Photo from New York Zoological Society; B. Photo by P. F. Wright of specimen in U.S. National Museum of Natural History; C. Photo from Castle Films through U.S. National Zoological Park.

Miocene to Recent in Africa, and Pleistocene in Madagascar. However, this distribution disregards the known presence of *Hippopotamus* in early historical times in Palestine and Madagascar.

ARTIODACTYLA; HIPPOPOTAMIDAE; Genus HIPPOPOTAMUS
Linnaeus, 1758

Hippopotamus

The single living species, *H. amphibius*, originally occurred throughout Africa south of the Sahara, in areas with suitable waterways, and down the Nile River to its delta (Ansell, *in* Meester and Setzer 1977). It also was present in the Jordan River Valley of Palestine in early historical times (Por and Ortal 1985). Another species, *H. lemerlei*, was found in Madagascar but became extinct about 1,000 years ago, probably as a result of hunting and habitat destruction following the human invasion of that island; it was relatively small, being only about 200 cm long and 81 cm high at the shoulder (E. Anderson 1984; Mahe 1972).

The remainder of this account applies only to *H. amphibius*, in which head and body length is 290–505 cm, tail length is about 40–56 cm, shoulder height is 150–65 cm, and weight is 1,000–4,500 kg. Males are larger, on average, than females. The body is so scantily covered with short, fine hairs that it appears naked. The usual color of the skin is a slaty copper brown, which shades to dark brown above and purplish below. The skin is glandular and exudes droplets of moisture that contain red pigment; light reflected from the skin through these droplets appears red, thus giving rise to the belief that *Hippopotamus* "sweats blood." The eyes are protruding, and the ears (up to 100 mm long) are set high up and far back on the head. The upper canine teeth measure as much as 230 mm or more in circumference. The lower

canines may reach a total length of about 700 mm, of which 300 mm extends above the gum level (Smithers 1983), and they weigh as much as 3 kg. The jaws can open to 150°.

The preferred habitat of the hippopotamus is an area of deep, permanent water with adjacent reedbeds and grassland. Kingdon (1979) noted that the range may expand during the rainy season, when temporary pools are available. This amphibious mammal swims and dives well. When submerging, it closes the slitlike nostrils and ears. If it wants to see and breathe without exposing itself, it can keep the protruding eyes and nostrils out of the water and remain submerged. It normally stays underwater for 3–5 minutes but can remain longer, perhaps for 30 minutes. Its specific gravity is such that it can walk about on the bottom. It sometimes enters salt water adjacent to the mouth of a river. Apparently, hearing, sight, and smell are well developed (Grzimek 1975).

The hippopotamus spends practically the entire day sleeping and resting in or near water. If molested, it will lie in deep water in reedbeds. At night it may cover 33 km of water in search of food. It usually emerges to feed. Lock (1972) stated that the grazing range extends about 3.2 km from water, though some individuals may move farther. According to Kingdon (1979), an individual will walk up to 10 km to seek food on land; the diet consists mostly of grass, and vegetation is cropped entirely by action of the horny lips. The metabolic rate of this mammal is relatively low, and individuals have been known to survive for many weeks in a mud wallow with no food, water, or shade. Laws (1984b) noted that a hippo grazes 5–6 hours per night and rests in water for the remainder of its time; its daily consumption is only about 1–1.5 percent of its body weight, as compared with about 2.5 percent for most other hoofed mammals.

Hippopotamus sometimes attains very dense populations. In an 88-km stretch of the Nile River in Uganda, including 3 km of land on both sides of the river, the average density was 19.2/sq km; the recommended density for the habitat was 7.7/sq km (Laws, Parker, and Johnstone 1975). By another method of counting, an estimated 6,544 hippos occurred

Hippopotamus *(Hippopotamus amphibius)*, photo from New York Zoological Society.

along a 165-km stretch of the Luangwa River in Zambia (Tembo 1987). In favorable areas, hippos have been found to number about 7 per 100 meters of lake shore and 33 per 100 meters of river length; respective male territories in these habitats are 250–500 meters and 50–100 meters long. Such territories consist of a narrow strip of water and adjacent land, are held for up to 8 years, and are used either by just a solitary male or by a dominant bull plus a group of females and young and sometimes one or more subordinate males (Laws 1984b). A male that is alone in his territory when water levels are low may gain a group of females and young when levels rise, but particularly in riverine areas, he may have to abandon his territory altogether if levels fall excessively (Karstad and Hudson 1986).

Hippopotamus may occur alone or in aggregations of up to 150 individuals; usual group size is 10–15 (Laws 1984b). Apparently, the larger herds consist mostly of adult females and their young. Several females may separate from a herd and form a temporary nursery unit for the bearing of young (Grzimek 1975). Bachelor groups also have been reported,

Hippopotamuses *(Hippopotamus amphibius)*, photo from Amsterdam Zoo.

though Karstad and Hudson (1986) found that nonterritorial males preferred a solitary existence. According to Kingdon (1979), hippo groups are unstable and regularly split up. The only lasting relationship is that between a mother and her offspring; sometimes a female is accompanied by up to four young of different ages, though it is not certain that all were produced by her. The territories of dominant males are located within the most favorable areas occupied by an aggregation of hippos. There also are many solitary males on the periphery of such areas; these animals may be young or ones that have lost in competition for a territory.

Adult males may be highly aggressive toward all other individuals (Kingdon 1979). Their territories are held for purposes of mating, and they attempt to keep estrous females therein. At such times they may attack and kill any hippo in the proximity, even young animals. Adult females sometimes join together to protect their offspring; they also can be very aggressive, especially when they have young calves. Neighboring territorial males challenge one another with ritualized displays involving rearing the head out of the water, opening the jaws, scattering dung about with a whirring of the short tail, emitting a "wheeze-honk" sound, and rushing toward one another. Bulls also have a loud roar that can be heard over a great distance. They regularly mark certain land portions of their territory by defecation, eventually forming huge dung heaps. If neither opponent yields following initial displays during a confrontation, the bulls charge and slam their lower jaws together. Such fights are vicious and may last two hours. The chief weapons are the large lower canines. Hides of mature males invariably have numerous scars. Deaths are not infrequent.

Females are seasonally polyestrous, estrus lasts 3 days, and there seem to be seasonal peaks in breeding in at least some areas (Kingdon 1979). In Uganda, most mating is in February and August, and most births occur in October and April, months of maximum rainfall (Grzimek 1975). In South Africa, births also peak during the wet months, October–March (Smuts and Whyte 1981). The gestation period is 227–40 days. Normally a single calf is born; twins are rare. Births usually take place on land or in shallow water. The weight at birth is 25–55 kg. One individual weighed 250 kg at the end of its first year. The young can swim before they can walk, and they nurse underwater. The cow is a devoted mother; the calf may scramble onto her back and sun itself while she is floating at the surface. Such behavior may afford some protection against crocodiles. The interval between births is about 2 years (Kingdon 1979). Data compiled by Dittrich (1976) indicate that in captivity sexual maturity comes at 3–4 years but that in the wild mating does not occur until males are 6–13 years and females are 7–15 years. Upon reaching puberty, the males may be driven from a group by the territorial bull (Smithers 1983). Average longevity in a protected wild population is 41 years (Grzimek 1975). A captive hippopotamus lived for 54 years and 4 months (Jones 1982).

The hippopotamus has been hunted extensively by people for its highly prized flesh, its abundance of fat (about 90 kg per individual), the superior ivory of its teeth, and its hide. It also has been killed for sport, as well as because it sometimes enters cultivated fields and does extensive damage by both eating and trampling crops. It may be aggressive toward people and is said to be the most dangerous of wild artiodactyls. Smithers (1983) wrote that females with young are especially aggressive and that there have been many reports of small boats being overturned or damaged and the occupants bitten to death.

The hippopotamus became very rare in the Nile Delta during the eighteenth century, and the last in Egypt was killed about 1816 (Corbet 1978). In the Nile Valley, the hippopotamus now occurs only as far north as Khartoum. It also

has disappeared in most of West and South Africa, and it has become rare in much of the remainder of its range. There are still large populations in the upper Nile Valley and in certain other parts of eastern and southeastern Africa; numbers even have been considered to be excessive in some areas. Deliberate killing to prevent habitat degradation began in Uganda in 1933 (Kingdon 1979). Queen Elizabeth National Park, in the western part of that country, formerly supported the highest known biomass density of large mammals in the world, about 21,000 kg per sq km, including up to 31 hippos per sq km. A regular cropping program began there in 1957, and about 1,000 hippos were shot annually from 1962 to 1966 (Laws 1981, 1984b).

It must be remembered, however, that what seem to be excessive numbers may actually represent populations unnaturally compressed into remnant areas of suitable habitat, such as described in the account of *Loxodonta*. Like the elephant, the hippopotamus now occupies but a fraction of its overall original range (Kingdon 1979). Indeed, there is reason to expect that a crisis comparable to that now confronting *Loxodonta* eventually will strike *Hippopotamus*. Such is likely not only because of the conflicts between people and hippos for habitat but, remarkably, because hippo tusks (particularly the lower canines of males) are just as large as many elephant tusks now entering the market, closely resemble the latter when carved, and may be even more desirable, since they do not yellow with age. The disappearance of the elephant, and intensification of efforts to prevent trade in its ivory, could trigger the wholesale slaughter of the hippopotamus.

ARTIODACTYLA; HIPPOPOTAMIDAE; **Genus CHEOROPSIS**
Leidy, 1853

Pygmy Hippopotamus

The single species, *C. liberiensis*, has a discontinuous range in the lowland forest zone from Sierra Leone to Nigeria (Ansell, *in* Meester and Setzer 1977).

Head and body length is 1,500–1,750 mm, tail length is about 156 mm, shoulder height is 750–1,000 mm, and weight is 160–270 kg. The body is without hair except for a few bristles on the lips and tail. The color above is a slaty greenish black, the sides are more gray, and the underparts are grayish white to yellowish green. Superficially, this animal appears to be a miniature *Hippopotamus*, but there are notable structural differences. The head is rounder and not so broad or flat, the nostrils are large and almost circular, the eyes are set on the side of the head and do not protrude, and the toes are well separated and have sharp nails. In addition, *Choeropsis* usually has only a single pair of lower incisors, as compared with two or three pairs in *Hippopotamus*. The exudation from the pores of the skin of *Choeropsis* is a clear, viscous material that makes the animal sleek to the touch. In certain lights the reddish brown tone of the skin is reflected in the beads of secretion, which then look almost blood red. Erroneously, many people have believed the animal to be "sweating blood."

The pygmy hippopotamus is found along streams and in wet forests and swamps. It is less aquatic than *Hippopotamus*, but contrary to early reports, it does seek refuge in the water when there is potential danger (Grzimek 1975). It sometimes shelters in a tunnel or hollow, probably excavated by another animal, along the side of a stream (Robinson 1981). It sleeps by day and wanders in the forest at night, seeking tender shoots, leaves, and fallen fruit. *Choeropsis* is found alone, in pairs, or in triads of male, female, and calf (Laws 1984b). One

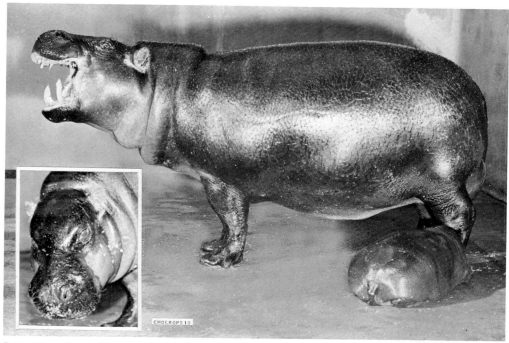

Pygmy hippopotamus *(Choeropsis liberiensis)* and 23-day-old young, photos by Ernest P. Walker.

young is born at a time. The gestation period ranges from 192 to 210 days, averaging 199 days (Grzimek 1975). Weight at birth is 3.4–6.4 kg (Stroman and Slaughter 1972). The estrous cycle is 28 days long, and estrus lasts 2–3 days (Partridge 1983). A captive was still living at an estimated age of 43 years and 10 months (Jones 1982).

The pygmy hippopotamus seems always to have been rare. A person may travel for days within its range and see no sign of it. *Choeropsis* is not an unduly vicious animal but can be dangerous when disturbed. It adjusts easily to captivity, and some individuals breed well. It is hunted for its flesh, which is said to taste like that of *Sus*. Its survival is jeopardized by uncontrolled hunting and destruction of habitat by logging. It is classified as vulnerable by the IUCN (1978) and is on appendix 2 of the CITES.

ARTIODACTYLA; Family CAMELIDAE

Camels, Guanaco, Llama, Alpaca, and Vicuña

This family of three Recent genera and six species apparently was found in the wild state during historical time from the Arabian Peninsula to Mongolia and in western and southern South America. Through human agency, there has been a drastic reduction in the range of wild camelids, but domesticated members of the family have spread over much of the world. The sequence of genera presented here follows that of Simpson (1945), though he did not give *Vicugna* generic rank.

There is a striking difference in size between the Old World genus *Camelus* and the New World *Lama* and *Vicugna*. All genera, however, are characterized by a long and thin neck, a small head, and a slender snout with a cleft upper lip. The hind part of the body is contracted. The stomach is three-chambered and ruminating. The Camelidae differ from all other mammals in the shape of their red blood corpuscles, which are oval instead of circular.

Each foot of Recent camelids has only two digits (the third and fourth). The proximal digital bones are expanded distally. The middle digital bones are wide, flattened, and embedded in a broad, cutaneous pad that forms the sole of the foot. The distal digital bones are small, not flattened on the inner surface, and not encased in hooves, bearing nails on the upper surface only. The digits are spread nearly flat on the ground. The feet are broad in *Camelus* and slender in *Lama* and *Vicugna*. The foot bones are united to form a cannon bone. In *Vicugna*, glands are associated with the cannon bone of the hind limb. In the forearm the ulna is reduced distally, and in the shank the fibula is reduced. The knee joint is low in position, because of the long femur and its vertical placement. All the limbs are long. The forelimbs have naked callosities in the guanaco, and prominent knee pads are present in camels.

The skull is low and elongate, and there are no horns or antlers. The usual dental formula in *Camelus* is: (i 1/3, c 1/1, pm 3/2, m 3/3) × 2 = 34. In *Lama* and *Vicugna* it is: (i 1/3, c 1/1, pm 2/1, m 3/3) × 2 = 30. However, there is some variation. The premaxillary bones of the skull bear the full number of upper incisors in the young, but only the outer incisor persists in the adult. The incisors, which are spatulate, are located in a forward, somewhat upward position. The lower incisors of *Vicugna* are unique among the Recent Artiodactyla—ever-growing, with the enamel on only one side. The canines, nearly erect and pointed, are sometimes absent in the lower jaw. The front premolars are simple and usually separated from the other cheek teeth. The molars have crescentic ridges of enamel on their crowns.

Wild camelids inhabit semiarid to arid plains, grasslands, and deserts. They are diurnal. *Camelus* and *Lama* run with a swinging stride, as the front and hind legs move in unison on

The head of a Bactrian camel *(Camelus bactrianus),* showing the closable, slitlike nostrils and the heavy eyebrows and eyelashes, which provide valuable protection in severe sandstorms. The slightly divided upper lip is also shown. Photo by Ernest P. Walker.

Left, Camel *(Camelus dromedarius),* photo by Bernhard Grzimek. Right, front foot of *C. bactrianus,* showing the hard ends of the toes and the broad foot with soft pads, which is well adapted to traversing soft sands, photo by Constance P. Warner.

each side of the body. Camelids lie down to rest and sleep. They are grazers, feeding on many kinds of grass, though camels, when hungry, will eat a wide variety of food. They have the habit of spitting the contents of their stomach at annoying objects, including incautious zoo visitors.

The Camelidae originated and evolved mainly in North America but disappeared there about 10,000 years ago (Franklin 1982). The overall geological range of the family is late Eocene to late Pleistocene in North America, middle Pliocene to Pleistocene in Europe, late Pliocene to Recent in North Africa, middle Pliocene to Recent in Asia, and late Pleistocene to Recent in South America (Simpson 1984).

ARTIODACTYLA; CAMELIDAE; **Genus LAMA**
G. Cuvier, 1800

Guanaco, Llama, and Alpaca

There are three species (Cabrera 1961):

L. guanicoe (guanaco), southern Peru to eastern Argentina and Tierra del Fuego;
L. glama (llama), found in domestication from southern Peru to northwestern Argentina;
L. pacos (alpaca), found in domestication in southern Peru and western Bolivia.

The llama and alpaca are evidently domesticated descendants of *L. guanicoe*, though there is some disagreement regarding their origin (Franklin 1982). Some authorities do not consider them to be more than subspecifically distinct from *L. guanicoe* (Grzimek 1975; Simpson 1984). They now have been bred for certain characters for such a long period that the question may never be resolved.

Head and body length is 1,200–2,250 mm, tail length is 150–250 mm, and height at the shoulder is 900–1,300 mm. Franklin listed weight as 100–120 kg in *L. guanicoe*, 130–55 kg in *L. glama*, and 55–65 kg in *L. pacos*. In *L. guanicoe* the upper parts are dark fawn brown, the underparts are white, and the face is blackish. The woolly coat is longest on the flanks, chest, and thighs. The limbs and neck are slender, and the body is trim. Females have four mammae. *L. glama* is brown to black, or is white, usually irregularly blotched with a darker color. The body has fairly long, dense, fine wool, and the hair on the head, neck, and limbs is shorter than it is elsewhere. *L. pacos* is somewhat smaller than *L. glama*; its hairs are up to 500 mm long on the body.

The guanaco inhabits dry, open country either in mountains or on plains. Populations are either sedentary or migratory. In the latter case there are either altitudinal or lateral shifts in range owing to snow cover or drought (Franklin 1982). The other two species are found mainly in the Andes at elevations of up to 4,800 meters. The guanaco can run at a speed of 56 km/hr. It enjoys standing and even lying in mountain streams and is said to be a good swimmer. When about to lie down, it first gets on the front knees (wrists), collapses the hindquarters, and then drops onto the chest, with the legs tucked under the body. Both the guanaco and the llama are graceful in their movements. The llama is capable of carrying a load of 96 kg at a rate of 26 km per day over rugged mountain terrain at an elevation of 5,000 meters. The hemoglobin of *Lama* has a much greater affinity for oxygen than does that of other mammals, and the blood contains more red corpuscles. These factors are partly responsible for the ability to function well at high altitudes. The guanaco is mainly a grazer, preferring grasses and forbs, but also takes considerable browse. The llama is likewise a grazer and browser, but the alpaca is strictly a grazer (Franklin 1982).

Franklin (1982) found that the social structure of *Lama guanicoe* resembles that of *Vicugna*. There are family groups with an average membership of 16 individuals, including a single adult male, several adult females, and their young less than 15 months old. These units are not so cohesive as those of the vicuña, and the members tend to spread out more. Sedentary family groups have a year-round feeding territory that is permanently occupied and defended by the adult male, though some of the females and young may leave and form their own unit during the winter. Territories on Tierra del Fuego averaged 29.5 ha. Young animals of both sexes are forcibly evicted by the adult male when they are 13–15 months old. There also are groups of young and nonterritorial

males, lone males, and aggregations of all ages and both sexes that form in migratory populations during the winter. Wilson and Franklin (1985) determined that young males spend 3–4 years in the bachelor groups, where they compete for dominance, practice fighting by neck wrestling and chest ramming, and learn the skills that may affect later reproductive success. They then disperse from these groups to establish or challenge for territories. It is said that if the dominant male of a family group is shot, the females will not flee but will nudge him in an effort to get him to his feet. When fighting a rival, the male emits a high-pitched scream, which changes to a low growl. Certain excretion sites are used by the guanaco and llama, as is indicated by dung heaps up to 2.4 meters in diameter and about 31 cm deep, composed of small, dry pellets.

Females give birth every other year. Mating occurs in August and September. The gestation period is 342–68 days, and the single young weighs 6–16 kg (Franklin 1982). Immediately following its birth, the young can run with surprising endurance. Lactation lasts 6–12 weeks. A captive guanaco lived for 28 years and 4 months (Jones 1982).

The guanaco is usually gentle and can become a good pet, though an adult can be aggressive. It has been exterminated by people in most of the eastern, lowland parts of its range. It is killed because it is considered to be a competitor with domestic sheep and because its skin is commercially valuable; during the 1970s, 400,000 guanaco skins were exported from Buenos Aires. There were an estimated 30–50 million guanacos when Europeans arrived in South America, but current populations are estimated to number 575,000 and are still declining (Franklin 1982). The guanaco is now on appendix 2 of the CITES.

The llama and alpaca evidently were first domesticated in Peru about 4,000–5,000 years ago and numbered in the tens of millions by the time of the Spanish conquest (Franklin 1982). The llama is used primarily as a beast of burden and is apparently the only animal with such a function to be domesticated by the native peoples of the New World. Franklin (1982) wrote that the Inca Empire's culture and economy revolved around the llama; the limits of the empire conformed closely to the range and ecological limitations of the llama. At the time of the Spanish conquest, 300,000 llamas were being used by the Incas at their silver mines. In addition, the meat of the llama is used for food, its fleece is woven into clothing, its hide is made into sandals, its fat is used in making candles, its hairs are braided into rope, and its dried excrement is used as a fuel. The alpaca is selectively bred for its wool, which is the finest of any animal's. It was formerly woven into the robes worn by the Inca royalty. While the llama (now numbering about 3 million individuals) is being replaced in many areas of South America by trucks and trains, the alpaca (3.5 million) is becoming more important because of the commercial value of its fleece (Franklin 1982, 1987).

ARTIODACTYLA; CAMELIDAE; **Genus VICUGNA**
Lesson, 1842

Vicuña

The single species, *V. vicugna*, is found in the Andes of southern Peru, western Bolivia, northwestern Argentina, and northern Chile (Cabrera 1961). There are reports that it once may have occurred as far north as Ecuador. Some authorities include *Vicugna* in *Lama*, but based on differences in the incisor teeth Franklin (1982) thought each to warrant full generic rank.

Head and body length is 1,250–1,900 mm, tail length is

A. Alpaca *(Lama pacos)*, photo by Ernest P. Walker. B. Llamas *(L. glama)*, photo from U.S. National Zoological Park. C. Leg of llama. D. Guanaco *(L. guanicoe)*.

Alpacas at livestock fair, Arequipa, Peru, 1952. This is the best wool breed of alpacas. Photo by Hilda H. Heller.

Guanacoes *(Lama guanicoe),* photo by Luiz Claudio Marigo.

Vicuña *(Vicugna vicugna)*, photo by C. B. Koford.

150–250 mm, and shoulder height is 700–1,100 mm (Grzimek 1975). Weight is 35–65 kg. The upper parts are tawny brown, the underparts are paler, and there is a white or yellowish red bib on the lower neck and chest. *Vicugna* resembles *Lama guanicoe* in general form but is about one-fourth smaller, is paler in color, lacks the dark face, and has no callosities on the inner sides of the forelimbs. The lower incisor teeth are unique among living artiodactyls, being rodentlike—ever-growing, with enamel on only one side.

The vicuña inhabits semiarid rolling grasslands and plains at elevations of 3,500–5,750 meters. It is capable of running 47 km/hr at an elevation of 4,500 meters. To dust its fleece and scratch its body, the vicuña paws and then vigorously rolls on the ground. Vision is the best-developed sense, hearing is only moderately acute, and olfaction is poor. *Vicugna* is extremely graceful in its movements, perhaps surpassing any other hoofed mammal in this respect. Juveniles often graze while lying prone, their legs tucked under the body, and both young and adults chew cud while resting. The vicuña is a grazer, the diet consisting almost entirely of low perennial grasses. Franklin (1982) reported that at the Pampas Galeras National Vicuña Reserve in southern Peru, population density increased from about 14/sq km in 1968 to 87/sq km in 1979.

In a study at Pampas Galeras, Franklin (1974) found the vicuña to be one of the few ungulates to defend a year-round feeding territory and a separate sleeping territory. The basic group consisted of a single dominant adult male, several adult females, and any associated young. The usual size of such a family was 5–10 individuals. Territories ranged from 7 ha. to 30 ha., the two main parts being connected by an undefended corridor. The male led the group, determined the extent of the territory, regulated group membership, and kept other males away. In addition to these groups, there were families that lacked permanent territories, as well as lone males and all-male groups, usually of 15–25 animals. In times of danger, a male with a family will warn the females with an alarm trill and interpose itself between the source of alarm and the other animals as they retreat. Members of a group urinate and defecate in a communal dung heap.

Mating in the wild occurs in March and April, and births take place in February and March. The gestation period is 330–50 days, and the single young weighs 4–6 kg at birth. The young can stand and walk about 15 minutes after being born. It sleeps at the side of its mother until at least 8 months and nurses several times a day until at least 10 months. The dominant male of the family drives juveniles of both sexes away, the young males at 4–9 months and the females at 10–11 months. The females usually then are accepted into another group. Most females mate at about 2 years, and some were still reproductively active at 19 years. Maximum known longevity in captivity is 24 years and 9 months (Franklin 1974, 1982; Schmidt 1975).

During the time of the Incas, vicuña were periodically rounded up, sheared of their wool, and then released. There were then an estimated 1 million to 1.5 million animals. Subsequent to the destruction of the Inca Empire, the vicuña began to be simply slaughtered in large numbers to obtain its wool and meat. In the 1950s there were still thought to be 400,000 left, but intensification of hunting pressure, commercial demand, and the spread of domestic livestock resulted in a drastic decline. By 1965 there were only about 6,000 vicuña. Conservation efforts have since allowed numbers to increase to 125,000 (Franklin 1982, 1987; Thornback and Jenkins 1982). About half are at the Pampas Galeras National Vicuña Reserve in Peru. Within the last decade much controversy has centered on the question whether expanding populations, especially that of Pampas Galeras, should be culled to protect the habitat, and their products marketed. Certain populations, including that of Pampas Galeras, now are on appendix 2 of the CITES, with the provision that only the cloth woven from the sheared wool of a live vicuña may be exported with respect to the treaty. Otherwise, the vicuña is on appendix 1 of the CITES, and the species is classified as vulnerable by the IUCN and as endangered by the USDI.

ARTIODACTYLA; CAMELIDAE; Genus CAMELUS
Linnaeus, 1758

Camels

There are two species (Corbet 1978; Ellerman and Morrison-Scott 1966):

C. bactrianus (Bactrian, or two-humped, camel), formerly found throughout the dry steppe and semidesert zone from Soviet Central Asia to Mongolia;

C. dromedarius (dromedary, or one-humped camel), probably once found as a wild animal throughout the Arabian region but known with certainty only in the domestic and feral state.

The name *C. ferus* often is used in place of *C. bactrianus*.

Head and body length is 225–345 cm, tail length is 35–55 cm, height at the shoulder is 180–230 cm, and weight is 300–690 kg. The color varies from deep brown to dusty gray. In *C. dromedarius*, which has only one hump on the back, the pelage is relatively short, soft, fine, and woolly. In *C. bactrianus*, which has two humps, the long hairs (255 mm) are thickest on the head, neck, humps, forelegs, and tip of the tail. Camels shed their winter coat so rapidly that it comes off in large masses, giving the animal the ragged appearance shown in the photograph of *C. bactrianus* that accompanies this account. The skin has almost no sweat glands.

Both species have a long head and neck and a relatively short tail. The eyes have heavy lashes, the ears are small and haired, and the upper lip is deeply divided. The slitlike nostrils can be closed to keep out dust and sand. There is a groove from each nostril to the cleft upper lip, so that any moisture from the nostrils can be caught in the mouth. The feet have two toes and undivided soles. The long, slender legs have prominent knee pads. *C. bactrianus* has shorter legs than *C. dromedarius* and therefore is not as tall; it also has shorter and harder feet and is more docile, slower, and easier to ride.

Camels live in arid areas. *C. bactrianus* is found on the Gobi Steppe along rivers but moves to the desert as soon as the snow melts. Camels can withstand extreme heat and cold and are said to be good swimmers. Their characteristic rolling gait is accomplished by simultaneously bringing up both legs on the same side. Speeds of over 65 km/hr have been recorded in a pursuit of *C. bactrianus* (Dash et al. 1978). Over a four-day period, a camel is able to carry 170–270 kg at a rate of 47 km/day and 4 km/hr. Sight is keen, and the sense of smell is extremely good.

Contrary to popular legend, there is no evidence that camels store water in the stomach. Although they are adapted for conservation of water, they will lose weight and strength if they go for long periods without drinking. According to Gauthier-Pilters (1974), domestic *C. dromedarius* in the Sahara can obtain sufficient water for months from desert vegetation and can survive a water loss exceeding 40 percent of its body weight. Camels can and will drink brackish and even salt water. By drinking as much as 57 liters of water, camels restore the normal amount of body fluid. They eat practically any vegetation that grows in the desert or semiarid regions. If forced by hunger, they will eat fish, flesh, bones, and skin. They thrive on salty plants that are wholly rejected by other grazing mammals. Camels are said to need halophytes in their diet and to lose weight if they are lacking. When camels are well fed, the hump is erect and plump, but when they do not have adequate food, the hump shrinks and often leans to one side.

Wild *C. bactrianus* is found alone or in groups, sometimes with over 30 individuals (Dash et al. 1978). A population density of about 5 individual *C. bactrianus* per 100 sq km has been calculated (Gu and Gao 1985). In the Sahara, domestic *C. dromedarius* commonly is left on its own for four or five months of the year, including the mating season (Gauthier-Pilters 1974). At that time, three kinds of herds form: bachelor males; adult females with their newborn; and up to 30 adult females, along with their one- and two-year-old offspring, all led by a single adult male. When young males are about 2 years old, they are driven away by the dominant male and then join a bachelor group. If a bachelor approaches a group of females with a male in attendance, he will be chased away. If two rival males approach, there is a series of intimidating displays involving urination, defecation, slapping the tail against the back, and spreading the hind legs. If neither retreats, the opponents fight by biting and thrashing with the forefeet (Gauthier-Pilters and Dagg 1981).

Schmidt (1973) stated that the peak birth season is March–April in *C. bactrianus* and February–May in *C. dromedarius*. According to Gauthier-Pilters and Dagg (1981), however, most females in the northwestern Sahara give birth in January–March; they are seasonally polyestrous, the estrous cycle lasting about 13–24 days and receptivity 3–4 days. A female camel usually gives birth every other year. The gestation period is 370–440 days. Litters contain a single offspring, rarely two. The calf weighs about 37 kg at birth, has soft fleece, utters a gentle "bah," and lacks the knee pads of the adult. By the end of the first day it can move about freely. The young generally are weaned by their human owners at around 1 year but suckle up to 1.5 years or more if left on their own. Full growth is attained at 5 years. Potential longevity is up to 50 years.

Camels are well known as beasts of burden. They are critical in maintenance of the still thriving nomadic cultures of the Sahara. They were employed in military combat in ancient times, made the Arab conquest of North Africa possible, and were used regularly for military purposes as recently as the 1960s (Gauthier-Pilters and Dagg 1981). In addition, they supply milk, meat, wool, hides, sinews, and bone. Chemicals can be extracted from their dung and urine. The milk is fermented to produce kumiss, a common intoxicating liquor. According to recent estimates, there are about 14 million domestic camels in Africa and Eurasia, the highest numbers being in Sudan, Somalia, and India. The great majority of domestic camels are *C. dromedarius*, but most of the 1.5 million or so in China and Mongolia are *C. bactrianus* (Gauthier-Pilters and Dagg 1981).

Camelus dromedarius may have persisted in Arabia as a wild animal until about 2,000 years ago (Corbet 1978). It was domesticated at least by 1800 b.c. and perhaps as early as 4000 b.c., probably in southern Arabia, and subsequently spread throughout southwestern Asia and North Africa. It had much to do with the development of nomadic culture and

A. Camel *(Camelus bactrianus)*, the unusually ragged appearance due to spring shedding of the winter coat, photo by Fritz Grogl.
B. Camel *(C. dromedarius)*, photo from U.S. National Zoological Park. Inset: *C. bactrianus,* photo by Ernest P. Walker.

trade routes in the Sahara. Interestingly, the spread of the camel there led to a consequent dying out of the use of wheeled vehicles (Gauthier-Pilters and Dagg 1981). *C. dromedarius* now occurs only in domestication, except for a feral population of about 25,000 individuals in Australia that is descended from camels brought there from 1840 to 1907 for purposes of desert transportation (Lever 1985; Newman, *in* Strahan 1983). In the middle of the nineteenth century, *C. dromedarius* was introduced into the southwestern United States as a potential source of efficient transportation. Enthusiastic advocates even predicted that because of their versatility, camels would supplant beef cattle in the Southwest. Due to various factors, such as the building of railroads, the experiment was a failure and some of the animals were released. Free-roaming camels survived in the Southwest until at least 1905, and a recaptured individual lived in a zoo until 1934 (Gauthier-Pilters and Dagg 1981). *C. dromedarius* also was introduced in the past to other countries, including Spain and Namibia, but feral populations did not become established (Lever 1985).

In Central Asia, *C. bactrianus* probably was domesticated before 2500 B.C. (Gauthier-Pilters and Dagg 1981). Through human agency, its range was extended from Asia Minor to northern China (Grzimek 1975). Wild populations also remained common until the 1920s but subsequently became restricted to relatively small areas of southwestern Mongolia and northwestern China. Recent surveys yielded estimates of about 200 individuals in China and 400–700 in Mongolia (Gu and Gao 1985; Mallon 1985). The wild camel is classified as vulnerable by the IUCN and endangered by the USDI.

ARTIODACTYLA; Family TRAGULIDAE

Chevrotains, or Mouse Deer

This family of two Recent genera and four species occurs in West and Central Africa and south-central and southeastern Asia.

These small, graceful animals bear some resemblance to deer (Cervidae) but also externally resemble agoutis (Rodentia: Dasyproctidae). The head is small and the snout is pointed; the narrow, slitlike nostrils are located in the naked nasal surfaces. There are no facial or foot glands. The legs are long, thin, and delicate, being about the size of a lead pencil.

In the genus *Tragulus* the pairs of metacarpal and metatarsal bones in the limbs are united into a single unit, known as a cannon bone. In *Hyemoschus* a cannon bone is not formed until old age. Each foot has four well-developed digits, though the lateral digits of *Hyemoschus* do not touch the ground when the animal stands. The stomach is three-chambered and ruminating. Females have four mammae.

There are no horns or antlers. The dental formula is: (i 0/3, c 1/1, pm 3/3, m 3/3) × 2 = 34. The upper canines are well developed (especially in the males), protrude below the lips, and are narrow, curved, and pointed. The lower canines look like incisors. All but the last premolars function in cutting the food; there are no premolars that resemble the canines. The molars bear crescentic ridges of enamel on their crowns. All the cheek teeth are arranged in a continuous series.

The geological range of this family is early Miocene to early Pliocene in Europe, early to middle Miocene and Pleistocene to Recent in Africa, and late Miocene to Recent in Asia (Simpson 1984).

ARTIODACTYLA; TRAGULIDAE; Genus HYEMOSCHUS Gray, 1845

Water Chevrotain

The single species, *H. aquaticus*, occurs in the lowland forest zone from Sierra Leone to western Uganda (Ansell, *in* Meester and Setzer 1977).

Head and body length is 600–850 mm, tail length is 75–150 mm, and height at the shoulder is 305–55 mm. Kingdon (1979) listed weight as 7–15 kg, noting that the average is 9.7 kg in males and 12.0 kg in females. Coloration is a rich brown, marked along the body with several longitudinal rows of white spots that become fused to form broken lines on the flanks. A white stripe runs along the edge of the jaw and the sides of the neck. The underside of the tail is white. The color and the arrangement of the spots are not constant. *Hyemoschus* has a small head, a pointed snout, a short tail, and a high body set on long, slender legs. Neither sex has antlers, but the upper canines of the male are developed into tusks similar to those of the musk deer *(Moschus)*.

Except as noted, the information for the remainder of this account was taken from Dubost's (1978) report of his study in

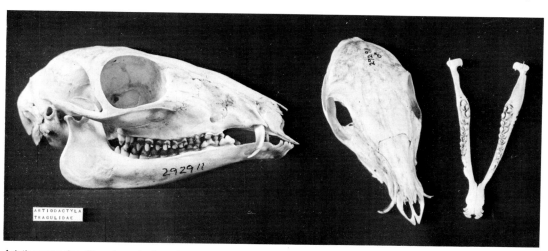

Asiatic mouse deer (*Tragulus* sp.), photos by P. F. Wright of specimen in U.S. National Museum of Natural History.

Water chevrotains *(Hyemoschus aquaticus)*, photo by Ernest P. Walker.

Gabon. *Hyemoschus* lives in evergreen forest, usually within 250 meters of water; however, it enters the water only when there is danger. It is nocturnal, resting by day in a hidden place, such as a pile of vegetation. The diet consists almost exclusively of fruit that has fallen to the ground.

Population density is 7.7–28.0 per sq km. Groups consist of an adult female and her young of one or two years. Such a group occupies a home range averaging about 13–14 ha. The young are usually found separately from the female, except during brief periods of suckling. Females are sedentary, and once a home range is established, they remain in an area all their life. Males are solitary and have home ranges of 20–30 ha., which overlap the ranges of two females. There seems to be a continuous replacement of males in any given area, none having been found in the same range for more than a year. There is no evident territoriality, but individuals are well spaced and seldom in contact.

Females reportedly bear a single young annually, though Kingdon (1979) noted that the four mammae of the female imply a capacity for larger litters. Gestation lasts 6–9 months, and lactation 3–6 months. The young reach maturity at 9–26 months, at which time they leave the home range of their mother. Some individuals live to an age of 11–13 years, which is considered old age. Hunting by people is evidently a serious cause of mortality, and overall numbers are decreasing as a result.

ARTIODACTYLA; TRAGULIDAE; **Genus TRAGULUS**
Brisson, 1762

Asiatic Chevrotains, or Asiatic Mouse Deer

There are two subgenera and three species (Chasen 1940; Ellerman and Morrison-Scott 1966; Lekagul and McNeely 1977; Medway 1977):

subgenus *Moschiola* Hodgson, 1843

T. meminna, peninsular India, Sri Lanka;

subgenus *Tragulus* Brisson, 1762

T. napu, southern Thailand and Indochina, Malay Peninsula, Sumatra, Borneo, many small islands of the East Indies, Balabac Island (extreme southwestern Philippines);

T. javanicus, Thailand, Indochina, Malay Peninsula, Sumatra, Java, Borneo, many small islands of the East Indies.

Groves and Grubb (1987) considered *Moschiola* to warrant full generic rank. Khan (1984) indicated that *T. javanicus* formerly occurred throughout Bangladesh but had been extirpated recently by human hunting.

These are the smallest artiodactyls, and *T. javanicus* is the smallest species. Head and body length is 400–750 mm, tail length is 25–125 mm, shoulder height is 200–350 mm, and weight is 0.7–8.0 kg. The hair, which is short, even, and close, is shortest on the head and underparts. All species are colored some shade of brown and have whitish underparts. The white expanse of the chin and throat is variously, though symmetrically, cut by patches of brown. Certain forms have longitudinal rows of spots or stripes of white or buff on the back and sides. There are no horns, but there are canine teeth in both jaws. The upper canines are especially well developed and are enlarged into tusks in males. Females have four mammae.

These mouse deer generally live among undergrowth on the edges of heavy lowland forests. They are seldom found far from water. They are present in large numbers but are seldom seen, as they are shy, retiring, and nocturnal. They customarily traverse tiny tunnel-like jungle trails. Lekagul and McNeely (1977) noted that *T. javanicus* may shelter in a rock crevice, a hollow tree, or dense vegetation. The diet consists chiefly of grass, the leaves of low bushes, and fallen jungle fruit and berries.

Tragulus has been reported to be solitary except during the breeding season. However, observations by Davison (1980) suggest that *T. javanicus* may live in monogamous pairs and that males defend territories, marking the areas and their females with an intermandibular scent gland located under the chin. They chase and fight one another for prolonged periods, slashing with the elongated canine teeth. Mating occurs throughout the year in some areas. Females may mate

Asiatic chevrotain: Top, *Tragulus meminna,* photo by Cyrille Barrette; Bottom, *T. napu,* photo from New York Zoological Society.

within 48 hours of giving birth and thus can bear young several times annually. Gestation periods of 140–77 days have been reported. There is usually a single offspring, occasionally two. The young is fully formed and active at birth and is able to stand in 30 minutes. Sexual and physical maturity comes by the age of 5 months (Lekagul and McNeely 1977; Medway 1978). A specimen of *T. napu* was still living after 14 years in captivity (Jones 1982).

These animals are preyed upon by a large number of carnivorous mammals and snakes and also are widely sought by native people for use as food. They tame readily and make good pets but are very delicate creatures. The mouse deer plays the same role in Malay folklore as the red fox does in European folklore.

ARTIODACTYLA; Family CERVIDAE

Deer

This family of 17 Recent genera and 45 species occurs nearly throughout North America, South America, and Eurasia, on most associated continental islands, and in northern Africa. Wild populations of deer have been established through introduction by people in Cuba, New Guinea, Australia, New Zealand, and certain other places where the family does not naturally occur. The sequence of genera presented here is based on the classifications of Simpson (1945) and Ellerman and Morrison-Scott (1966), in which five subfamilies are recognized: Moschinae, for the genus *Moschus*; Hydropotinae, for *Hydropotes*; Muntiacinae, for *Muntiacus* and *Elaphodus*; Cervinae, for *Dama, Axis, Cervus,* and *Elaphurus*; and Odocoileinae, for all other genera. However, there is a growing consensus that *Moschus* should be placed in an entirely separate family, the Moschidae (Corbet 1978; Corbet and Hill 1986; Groves and Grubb 1987; Scott and Janis 1987).

Head and body length is 720–2,900 mm, the tail is relatively short, shoulder height is 320–1,900 mm, and weight is 7–825 kg. *Pudu* is the smallest genus in the family and *Alces* is the largest. Females are generally slightly smaller and more delicately built than males; their neck in particular is not so full as that of males during the rutting season, and the hair on the neck is not so heavy. The general coloration of deer is brownish; the young of most species and the adults of some species are spotted. The winter coats are darker than the summer coats, at least in those species found in temperate latitudes.

These slim, long-legged artiodactyls are best characterized by the presence of antlers, which are lacking in only two genera, *Moschus* and *Hydropotes*. Antlers are borne only by the males, except in *Rangifer*, in which both sexes have antlers. Male deer are illustrated more frequently than females in the following pages, inasmuch as the antlers are often the principal external character by which the genera can be distinguished. Antlers are appendages of the skull, composed of a solid bony core and supported on permanent skin-covered pedicels. In the temperate zones the antlers begin to grow early in the summer, during which time they are well supplied with blood. They are soft and tender and are covered with a thin skin, which bears short fine hairs and has the appearance of velvet. By late summer the antlers have attained their maximum size. The blood then gradually recedes, and the thin skin with the velvety hair dries, loosens, and is rubbed off. Before the velvet is shed, all circulation of blood has ceased; thus, when shedding takes place, there is no bleeding and probably not even discomfort for the animal.

After the velvet is rubbed off, the antlers serve as sexual ornaments and weapons. In the Northern Hemisphere the antlers are shed each year from January to April, following the mating season. The species in temperate climates usually complete the shedding process in two to three weeks. In a single deer, both antlers are usually shed within several hours or days of each other. At least some of the tropical deer do not have a fixed breeding season, and these forms do not shed their antlers at any certain time. Deer usually grow their first set of antlers at one to two years of age, and these first-year horns are generally short, almost straight spikes. The antlers become larger and acquire more points in succeeding years until the animal is mature, when the antlers attain the shape typical of the species. Normal antler growth is dependent upon an adequate diet, for if certain minerals or vitamins are lacking, the antlers may be stunted or dwarfed.

Nearly all deer have facial glands, located in a depression in front of the eye. The pit is lined with a continuation of the skin of the face. Glands also occur on the limbs. There is a cannon bone that is formed by the fusion of the two main foot bones. The ulna and fibula are reduced. As in other artiodactyls, the first digit is absent from each foot. The third and fourth digits are well developed and bear the weight of the animal, and the second and fifth digits are small. The stomach is four-chambered and ruminating. A gall bladder is lacking, except in *Moschus*. Females usually have two pairs of mammae, but *Moschus* has only one pair.

The dental formula is: (i 0/3, c 0–1/1, pm 3/3, m 3/3) × 2 = 32–34. The upper canines are saber-shaped and enlarged in the antlerless species and absent in most other species. The lower canines resemble the incisors. The low cheek teeth bear crescentic ridges of enamel on their crowns.

The extremely varied habitats of deer include forests, swamps, brush country, deserts, and arctic tundra. Deer are usually good swimmers, some species being semiaquatic. Some undergo seasonal migrations. The diet is herbivorous and may include grass, bark, twigs, and young shoots. Most deer associate in groups. The stags often fight among themselves for possession of a harem. Females in tropical regions may enter estrus several times a year, but in temperate climates there is a definite mating season in the late fall or winter. The gestation period ranges from about 160 days in *Moschus* to 10 months in *Capreolus*, the latter being the only cervid genus known to have delayed implantation. The number of young is usually one or two, but three or four are sometimes born.

The geological range of this family is early Oligocene to Recent in Asia, late Oligocene to Recent in Europe, early Miocene to Recent in North America, and Pleistocene to Recent in South America. The extinct Pleistocene giant deer (*Megaloceros*), many complete skeletons of which have been found in peat bogs in Ireland and at other sites in Europe, had antlers that sometimes measured over 3.5 meters from tip to tip. These antlers are the largest known of any deer's, though the estimated body size of *Megaloceros* is exceeded by the largest of modern moose (*Alces*). The skeletal structure of *Megaloceros* suggests that this genus also was the most cursorial deer ever to evolve; it was highly specialized as a speedy and enduring runner. Large antlers are considered to be a means of attracting mates. They also are an excellent barometer of the owner's ability to find the best food resources and of his physiological potential to divert these resources away from mere body growth. The same potential is required by a female, who must give birth to well-developed offspring that will quickly achieve the same running speed as their parents. On the basis of these factors, Geist (1986) hypothesized that selection by female *Megaloceros* encouraged the evolution of the males' huge antlers.

A. Central American white-tailed deer *(Odocoileus virginianus)*; B. Père David's deer *(Elaphurus davidianus)*; photos by Ernest P. Walker. C. Red deer *(Cervus elaphus)*, photo by Erna Mohr. D. Père David's deer *(Elaphurus davidianus)*, showing deep preorbital lachrymal pits, photo by Ernest P. Walker.

Père David's deer *(Elaphurus davidianus)*, photo by D. W. Yalden.

ARTIODACTYLA; CERVIDAE; **Genus MOSCHUS**
Linnaeus, 1758

Musk Deer

The following five species now seem to be recognized (Groves 1975*a*, 1980*b*; Groves and Feng 1986; Groves and Grubb 1987; Grubb 1982*b*):

M. chrysogaster, Tibetan Plateau, Himalayan region from Afghanistan to south-central China;

M. leucogaster, Nepal, Sikkim, Bhutan, and adjacent parts of Tibet;

M. fuscus, Yunnan (south-central China), northern Burma;

M. berezovskii, eastern and southern China, northern Viet Nam;

M. moschiferus, Siberia, Mongolia, Manchuria, Korea, Sakhalin.

The systematics of this genus are confused. The name *M. sifanicus* sometimes has been applied to *M. chrysogaster. M. leucogaster* may be a subspecies of either *M. chrysogaster* or *M. berezovskii. Moschus* itself may warrant placement in a separate family, the Moschidae.

Head and body length is 700–1,000 mm, tail length is 18–60 mm, shoulder height is 500–610 mm, the rump is about 50 mm higher than the shoulder, and weight is 7–17 kg. The body is covered with long, thick, bristly hairs, which are usually white at the base; they are pithy and protect the animal from the severe weather encountered at high elevations. Coloration appears to be quite variable, possibly as a result of age and season, but generally it is a rich dark brown, mottled and speckled with light gray above, and paler beneath. The chin, inner borders of the ears, and insides of the thighs are whitish, and occasionally there is a spot of white on each side of the throat.

Although antlers are lacking, the upper canine teeth are developed as tusks in the male, about 75 mm in length; those

of the female are smaller. A musk gland in the abdomen of the male (three years of age or older) secretes a brownish waxlike substance. About 28 grams of this secretion can be obtained from a single animal. Unlike other Cervidae, *Moschus* has a gall bladder. Females have two mammae.

The preferred habitat of musk deer is forest and brushland at elevations of 2,600–3,600 meters. These deer have been reported to sleep in "forms" like those of hares *(Lepus)* during the day. In the Himalayas, Green (1987*b*) found that *M. chrysogaster* tends to remain in dense cover by day and to use open habitat at night, when it is usually more active. There was no evidence of any seasonal altitudinal movement in that region, though occasional migrations of 12–35 km have been reported for musk deer in Siberia. The diet consists of a variety of vegetation, such as grass, moss, and tender shoots. In winter, twigs, buds, and lichens are taken. Green (1987*b*) found a population density of 3–4 deer per sq km, which is lower than that reported in regions that experience less snow, thus suggesting that density varies with availability of food in winter.

Musk deer are shy, timid, and generally solitary. Almost never are more than two or three individuals found together. In a study of *M. chrysogaster*, Green (1987*b*) saw only solitary animals, except for two males that were fighting. Green (1987*a*) suggested that communication is chiefly by olfactory means, including urination, defecation, and, in the case of males, marking with the musk gland and also caudal and interdigital glands. During the fall and winter both sexes defecate at common latrines; such activity peaks in the Himalayas during the mating season in November and December. Green (1987*b*) reported that births in this region occur in June and July and that the gestation period is 196–98 days. The number of offspring is usually one in *M. chrysogaster* but usually two in *M. berezovskii* and *M. moschiferus*. The young are spotted, are nearly full-grown at 6 months, and usually reach sexual maturity by about 18 months. Captives in China have lived up to 20 years.

The substance obtained from the gland of the male is one of the musks used extensively in the manufacture of perfume, soap, and medicinal preparations. Because of the small quantity of musk obtained per animal, a high price is paid for this product, thus making *Moschus* much sought after by native and European hunters. Musk is reputed to be the most expensive animal product in the world, fetching up to U.S. $45,000 per kg on the international market (Green 1986). Also, the pouch in which the musk is stored is highly valued. Many females and young that do not produce musk are captured in the traps set for the adult males. Because of excessive hunting and also destruction of forest habitat, musk deer have declined seriously in numbers and have been extirpated in some areas.

The populations of the Himalayan region are classified as vulnerable by the IUCN and as endangered by the USDI and are on appendix 1 of the CITES; other populations are on appendix 2. The Soviet Union designates *M. moschiferus* as vulnerable. Green (1986) found that musk deer have disappeared from much of the region on the south slope of the Himalayas. He calculated that about 50,000 sq km of habitat is still available in this region, which is capable of supporting about 200,000 musk deer, but estimated that only about 30,000 of the animals survive and that at least 4,000 adult males are killed each year. He suggested development of a program to briefly live-capture wild males to extract their musk (which can be done safely). Such measures are thought to be more practical than breeding the deer in captivity, which has been attempted in China since 1958, though there is concern that either scheme would help to sustain the demand for musk and thus stimulate continued killing.

Musk deer (*Moschus moschiferus*), photo by K. Kutunidisz.

ARTIODACTYLA; CERVIDAE; **Genus HYDROPOTES**
Swinhoe, 1870

Chinese Water Deer

The single species, *H. inermis*, is found in the lower Yangtze Basin of east-central China and in Korea (Corbet 1978). The scientific name means "unarmed water drinker."

Head and body length is 775–1,000 mm, tail length is 60–75 mm, and shoulder height is 450–550 mm. Although males usually have been reported to weigh about 11–14 kg and to be heavier than females (Cockerill 1984; Whitehead 1972), Sheng and Lu (1985) stated that both sexes weigh about 30 kg. The hair is generally thick and coarse and is longest on the flanks and rump. The top of the face is grayish to reddish brown, the chin and upper throat are whitish, and the back and sides are usually a uniform yellowish brown, finely stippled with black. The underparts are white. Both sexes lack antlers, but the upper canine teeth, especially in the males, are enlarged, forming fairly long, slightly curved tusks. A small inguinal gland is present on each side in both sexes; this is the only known case of such glands in the Cervidae. Females have four mammae.

Hydropotes lives among tall reeds and rushes along rivers and also frequents tall grass on mountains and cultivated fields. When disturbed, it humps its back and travels by a series of leaps, much as do *Lepus* and *Sylvilagus*. It can swim for several kilometers and often moves back and forth between islets in accordance with availability of cover and food (Sheng and Lu 1985). The diet includes reeds, coarse grasses, and vegetables.

Hydropotes rarely congregates in herds. In China mating occurs from November to January and most young are born in late May and June (Sheng and Lu 1985). Estimates of the gestation period range from about 170 to 210 days (Sadleir 1987). Females are said sometimes to give birth to up to eight young at a time, more than are produced by any other kind of deer. In a survey of zoos, however, Dobroruka (1970a) found that there were usually only two offspring per birth, occasionally three. One litter contained five young, but two of these were born dead. Sheng and Lu (1985) found most wild pregnant females to have 2–3 fetuses. The young are marked with white spots and stripes. Males reach sexual maturity at 5–6 months, and females at 7–8 months (Sheng and Lu 1985). One specimen of *Hydropotes* was still living after 11 years and 5 months in captivity (Jones 1982).

The Chinese water deer has been bred extensively in captivity. Many individuals escaped from the Duke of Bedford's Woburn Park in England and established a wild-living population. *Hydropotes* also has been introduced in France (Corbet 1978; Lever 1985). In the wild this deer is heavily hunted because it is considered to be an agricultural pest and in order to obtain colostrum, which is sold commercially for use in folk medicine (Sheng and Lu 1985). Although it is not classified as an endangered species, there recently were estimated to be only 10,000 individuals remaining in the wild in China (Clutton-Brock 1987).

Chinese water deer *(Hydropotes inermis),* photo by Ernest P. Walker. Insets: head, photo by Erna Mohr; fawn, photo from New York Zoological Society; skull, photo from *Guide to the Great Game Animals,* British Museum (Natural History).

Muntjac *(Muntiacus muntjak),* photo from East Berlin Zoo.

ARTIODACTYLA; CERVIDAE; **Genus MUNTIACUS**
Rafinesque, 1815

Muntjacs, or Barking Deer

There are six species (Chasen 1940; Corbet 1978; Ellerman and Morrison-Scott 1966; Groves and Grubb 1982, 1987; Lekagul and McNeely 1977; Ma, Wang, and Xu 1986; Medway 1977; Sheng and Lu 1980):

M. reevesi, southern China, Taiwan;
M. rooseveltorum, Yunnan (south-central China), Laos;
M. crinifrons, east-central China;
M. feae, southern Tibet and China to Tenasserim (peninsular Burma) and adjacent parts of Thailand;
M. muntjak, India and Nepal to southern China and Malay Peninsula, Sri Lanka, Hainan, Sumatra, Bangka, Java, Bali, Kangean Islands, Rhio Archipelago, Borneo;
M. atherodes, Borneo.

Ma, Wang, and Xu (1986) considered *M. atherodes* to be a synonym of *M. muntjak,* but Groves and Grubb (1982, 1987) emphasized that the two are widely sympatric on Borneo and very distinctive. Wirth and Groves (1988) reported discovery of an unnamed species in the mountains of Yunnan and adjacent Tibet.

Head and body length is 640–1,320 mm, tail length is 86–229 mm, shoulder height is 406–780 mm, and weight is 14–28 kg. The body is covered with short, soft hairs, except for the ears, which are sparsely haired. Coloration varies from deep brown to yellowish or grayish brown with creamy or whitish markings. In *M. crinifrons* the head is decidedly lighter than the body. Antlers are carried only by males and are shed annually in most species. They rarely exceed 125–52 mm in length and usually are carried on long, bony, hair-

Muntjacs *(Muntiacus muntjak):* Top, photo from East Berlin Zoo; Bottom, photo by Cyrille Barrette.

covered pedicels. In *M. atherodes* the antlers are only 16.0–42.5 mm long, lack associated frontal tufts of hair, and normally are never shed (Groves and Grubb 1982, 1987). The upper canine teeth of the males are elongated into tusks, which curve strongly outward from the lips; they are thus capable of inflicting serious injuries to dogs and other animals. The females have small bony knobs and tufts of hair where the antlers occur in the males.

Muntjacs usually are found in forests and areas of dense vegetation, from sea level to medium elevations in hilly country. They generally do not occur far from water. Both diurnal and nocturnal activity have been reported in the wild, and observations in captivity indicate that *M. reevesi* is mainly crepuscular (Yahner 1980*a*). Barking deer are dainty little creatures; they are always on the alert and lift their feet high when walking. Their diet includes grasses, low-growing leaves, and tender shoots. In a study of *M. muntjak* in Sri Lanka, Barrette (1977) found fallen fruit to be a major food item. Population density in the study area was 2.5/sq km when the animals were concentrated during a drought but may have been 1.5/sq km when the animals were evenly distributed. Most sightings were of lone individuals, and no group of over 4 was seen. Miura (1984*a*) found that a group of 15 captive male *M. reevesi* interacted aggressively and established a dominance hierarchy; the 4 highest-ranking individuals partitioned the enclosure into nonoverlapping areas that were defended and scent-marked.

Muntjacs emit a deep, barklike sound; if they sense a predator in the area, they will "bark" for an hour or more. Yahner (1980*b*) found that barking occurs most often when visibility is reduced because of environmental conditions. His investigations indicate that barking is not, as sometimes claimed, a means of communication during the mating season but may serve to identify different individuals at any time of the year and also seems to be a mechanism that causes a predator to reveal itself or realize that it has been detected and thence move elsewhere.

Females are polyestrous, the estrous cycle lasts 14–21 days, and estrus is about 2 days long (Sadleir 1987). Medway (1978) stated that there was no evidence of a specific breeding season in the Malay Peninsula, but Lekagul and McNeely (1977) wrote that mating in Thailand occurs in December and January, and births in June or July. Sheng and Lu (1980) reported that newborn *M. crinifrons* are seen in April. Chaplin and Dangerfield (1973) reported that births of captive *M. reevesi* in England occurred mostly from March to August and that gestation was approximately 209–20 days. Reproduction in Sri Lanka occurred throughout the year, and the gestation period was about 210 days (Barrette 1977). Young muntjacs are usually born in dense jungle growth, where they remain hidden until they can move about with the mother. At birth the young weigh about 550–650 grams. Females reach sexual maturity within their first year of life (Sadleir 1987). A specimen of *M. muntjak* was still living after 17 years and 7 months in captivity (Jones 1982).

Barking deer are hunted for their meat and skins. They thrive in captivity and are found in many zoos. They are considered to be a nuisance in some areas, because they destroy trees by ripping off the bark. *M. feae*, restricted to a small area and subject to uncontrolled hunting, is classified as endangered by the IUCN and the USDI. *M. crinifrons*, long known by only three specimens, is classified as indeterminate by the IUCN but was found living in four provinces of east-central China in the 1970s (Sheng and Lu 1980). It generally is declining there because of heavy hunting and habitat destruction; numbers are estimated to be about 10,000 individuals in an overall range of 76,500 sq km (Lu and Sheng 1984). There also are thought to be about 140,000–150,000 *M. muntjak* and 650,000 *M. reevesi* in China (Clutton-Brock 1987). Introduction by human agency has resulted in the establishment of wild populations of *M. reevesi* throughout the southern half of England; wild populations of *M. muntjak* also once occurred there but evidently have died out (Chapman and Chapman 1982). *M. reevesi* also was introduced in France and is maintained there on some private estates but has not become established in the wild (Lever 1985).

Tufted Deer

The single species, *E. cephalophus*, occurs in central, eastern, and southern China and in northern Burma (Ellerman and Morrison-Scott 1966; Sheng and Lu 1982).

This dainty deer is somewhat larger than *Muntiacus*, having a head and body length of 1,100–1,600 mm, a tail length of 70–150 mm, a shoulder height of 500–700 mm, and a weight of 17–50 kg. The body is covered with coarse, almost spinelike hairs that give a shaggy appearance. The general color of the upper parts is deep chocolate brown, the underparts are white, and the head and neck are gray. In some cases, a pale streak extends forward from the pedicel and above the eye, and the tuft on the forehead is blackish brown. There are white markings on the back tips of the ears and on the underside of the tail.

Elaphodus is similar to *Muntiacus*. Although the canines are comparable in size, the antlers of *Elaphodus* are much smaller, frequently completely hidden by the tuft on the forehead. Other distinguishing characteristics are that the bony pedicels, on which the antlers grow, are shorter than those of *Muntiacus*, do not form heavy ridges on the forehead, and converge at the tips.

The tufted deer inhabits mountainous forests at elevations of 300–4,500 meters (Sheng and Lu 1982). It is said always to be found near water. When feeding, it carries its tail high; the tail flops with every bounce, displaying the white underside, much as does that of *Odocoileus virginianus*. The diet includes grasses and other vegetation.

Elaphodus is usually solitary but occasionally travels in pairs. Both sexes "bark" when suddenly alarmed and during the mating season. Mating occurs in late fall and early winter, and births in late spring and early summer (Sheng and Lu 1982). One or two young are born after a gestation period of about six months. The young are colored like their parents but have a row of spots along each side of the midline of the back. One specimen lived for seven years at the London Zoo.

There are estimated to be about 100,000 tufted deer in China (Clutton-Brock 1987).

Fallow Deer

The single species, *D. dama*, originally occurred in the Mediterranean region of southern Europe, from Asia Minor and Palestine to Iran, and probably in northern Africa (Ansell, *in* Meester and Setzer 1977; Chapman and Chapman 1975; Corbet 1978; Ellerman and Morrison-Scott 1966; Harrison 1968). Reports that *Dama* also lived in Ethiopia until about 1,000 years ago are based only on two questionable artistic representations (Yalden, Largen, and Kock 1984). The populations in Asia, to the east and south of Turkey, and in Africa sometimes have been referred to as a separate species, *D. mesopotamica*. Ferguson, Porath, and Paley (1985) reported discovery of the distinct remains of both *dama* and *mesopotamica* at the same 3,000-year-old archeological site in Palestine and thus regarded each as a valid species. However, Feldhamer, Farris-Renner, and Barker (1988) rejected this view and treated *mesopotamica* as a subspecies of *D. dama*. Opinion remains about evenly divided as to whether *Dama*

should be considered a distinct genus (Feldhamer, Farris-Renner, and Barker 1988; Groves and Grubb 1987) or included within *Cervus* (Corbet and Hill 1986; Honacki, Kinman, and Koeppl 1982).

Head and body length is 1,300–1,750 mm, tail length is 150–230 mm, shoulder height is 800–1,050 mm, and weight is 40–100 kg. Chapman and Chapman (1975) identified four main color varieties, more than in any other kind of deer: (1) common—upper parts rich brown with white spots and underparts whitish in summer, dark grayish brown with spots barely detectable in winter; (2) menil—pale fawn, heavily spotted with white all year; (3) white—not truly albino and probably restricted to parks; and (4) black—not jet black, but dark, especially in summer, with barely detectable gray brown spots. In the common phase, the summer coat is acquired in May and the winter coat in October.

Only the males have antlers; their peculiar shape is the basis for placing this deer in a genus distinct from *Cervus*. The antlers are flattened and palmate with numerous points. The males develop a single unbranched horn the second year, and each succeeding year the horns are larger and have more points, until the fifth or sixth year. The antlers are generally shed in April, and the new ones are full-grown and free of velvet in August. The front outer curve of the antlers is 635–940 mm, and the tip-to-tip measurement is 305–762 mm. There are no upper canine teeth.

There is a great variety of habitats, but generally some forest is required. Feeding occurs mainly in the early morning and from late afternoon to evening. The fallow deer is predominantly a grazer but also commonly browses trees and

Tufted deer *(Elaphodus cephalophus)*, photos from Zoological Society of San Diego.

Fallow deer *(Dama dama):* A. Male in spring pelage, photo from Zoological Society of London; B. Female in winter pelage, photo from New York Zoological Society.

shrubs (Chapman and Chapman 1975). Size of home range varies depending on availability of food and other factors (Feldhamer, Farris-Renner, and Barker 1988).

Reported population densities range from 8 to 43 individuals per 100 ha. Social behavior varies (Chapman and Chapman 1975; Feldhamer, Farris-Renner, and Barker 1988); in some areas *Dama* does not appear gregarious, but in others herds of up to 30 individuals are commonly seen throughout the year. There does not seem to be a dominance hierarchy, but a doe is usually the group leader. Adult males are usually solitary but may join in bachelor groups of fewer than 6 individuals during the summer. In the mating season the dominant males establish small territories, centered about 100 meters apart. Throughout this season the males perform a dancelike ritual and bellow in a deep voice, which presumably attract the females that gather around them. There is much fighting between rival males, but it mostly involves ritualized shoving with the antlers. Pemberton and Balmford (1987) referred to the mating system of *Dama* as "lekking," such as has been described for *Hypsignathus, Kobus,* and certain other mammals.

Mating takes place in September or October in the Northern Hemisphere, and births occur in the spring. Females are seasonally polyestrous, the estrous cycle lasts about 22–26 days, and the mean gestation period has been reported as 229 days in Germany and 237 days in New Zealand (Chapman and Chapman 1975; Feldhamer, Farris-Renner, and Barker 1988). There is usually a single fawn, which is slightly darker than commonly colored adults and spotted with white. Weight at birth is about 4–5 kg (Feldhamer, Farris-Renner, and Barker 1988). Females attain sexual maturity at about 16 months and subsequently breed each year. Males are physiologically capable of mating at about 17 months but generally do not breed until they are at least 4 years old. Full size is attained at 4–6 years by females and 5–9 years by males; captives have lived for 20 years or more (Feldhamer, Farris-Renner, and Barker 1988).

There is uncertainty regarding the natural distribution of this genus in Recent times. *Dama* occurred over much of Europe in the Pleistocene but by the end of that epoch was restricted to the Mediterranean region, or possibly only to Asia. It probably also occurred naturally from Morocco to Egypt. Within historical time, it was widely reintroduced in Europe by the Phoenicians and Romans and apparently was brought to Britain by the Normans. Wild populations now occur as far north as Sweden and western Russia. Free-living herds have also been established by people in the United States, Canada, the West Indies, Argentina, Chile, Peru, Uruguay, South Africa, Madagascar, Japan, Australia, Tasmania, New Zealand, and Fiji (Chapman and Chapman 1975, 1980; Grzimek 1975). Many fallow deer also are maintained in captivity for exhibit or for commercial production of meat and antler velvet (Feldhamer, Farris-Renner, and Barker 1988).

While the fallow deer was spreading into new areas, however, it was disappearing from its original range because of both excessive hunting and climatic changes. The genus apparently disappeared from Africa in the nineteenth century, from the mainland of Greece in the early 1900s, and from Sardinia in the 1950s. At the same time, it became very rare in the Asian parts of its range (Ansell, *in* Meester and Setzer 1977; Chapman and Chapman 1975, 1980).

The subspecies that formerly occurred from Palestine to Iran, *D. d. mesopotamica,* is classified as endangered by the IUCN and the USDI and is on appendix 1 of the CITES. This subspecies once was considered to be extinct, but in the 1950s a small population, probably containing fewer than 50 individuals, was found along several rivers in western Iran, near the border with Iraq. In the late 1970s, prior to recent disturbances in the region, this population reportedly was well protected and starting to increase in numbers; reintroduction and captive breeding projects were also under way (Chapman and Chapman 1980). There has been no recent news of this last wild population, though there is a captive breeding group at the East Berlin Zoo (Clutton-Brock 1987).

Axis deer *(Axis axis)*, photo from Zoological Society of London.

ARTIODACTYLA; CERVIDAE; **Genus AXIS**
Hamilton-Smith, 1827

Axis Deer

There are two subgenera and four species (Ellerman and Morrison-Scott 1966; Groves and Grubb 1987):

subgenus *Axis* Hamilton-Smith, 1827

A. axis (chital, or spotted deer), India, Nepal, Sikkim, Sri Lanka;

subgenus *Hyelaphus* Sundevall, 1846

A. porcinus (hog deer), northern India to Indochina;
A. kuhli, Bawean Island (Java Sea);
A. calamianensis, Calamian Islands (Philippines).

Corbet and Hill (1986), Honacki, Kinman, and Koeppl (1982), and various other authorities have placed *Axis* and *Hyelaphus* in the genus *Cervus* and have treated *A. kuhli* and *A. calamianensis* as subspecies of *A. porcinus*. Groves and Grubb (1987), however, followed the systematic arrangement given above.

Head and body length is 1,000–1,750 mm, tail length is 128–380 mm, shoulder height is 600–1,000 mm, and weight is 27–110 kg. The hair is coarse and longest on the flanks. There is no mane on the throat or neck. Coloration varies not only between the species but also from season to season. The general color is a bright rufous fawn or is yellowish brown to brownish. During part of the year the upper parts of *A. axis* are beautifully marked with small white spots. All species have a dark dorsal stripe, white underparts, and a whitish tail undersurface. Albino strains have been developed in captivity.

The species *A. axis* is slender, graceful, and of medium body build, whereas the other three species are stocky and

Axis deer *(Axis axis)*, photo from San Diego Zoological Garden.

have shorter legs. The antlers are carried on pedicels and are three-tined, the brow tine nearly forming a right angle with the beam. Upper canine teeth are usually absent, and the tail is usually relatively long and slender.

Axis deer frequent grasslands and open forest, seldom penetrating into heavy jungles. Normally they rest during the hotter part of the day and move about in the early morning and late afternoon. They may become nocturnal in the summer, or when molested by people. They take readily to water and are said to be good swimmers. Lekagul and McNeely (1977) noted that the term "hog deer" refers to the habit of *A. porcinus* of running through the underbrush with its head held low, in the manner of a hog, rather than leaping over obstacles, as do other deer. Both species are predominantly grazers, though they occasionally browse and are fond of various fallen flowers and fruits of forest trees. Blouch and Atmosoedirdjo (1987) found *A. kuhli* to feed largely on forbs and grasslike plants, to depend on dense cover, and to have a population density of up to 19.2/sq km in favorable secondary forest habitat.

In a study of *A. axis* in central India, Schaller (1967) found a population density of 23/sq km. Observations suggested that individuals spend their entire lives in a relatively small area. Home ranges during about half of the year were approximately 500 ha. for a male and 180 ha. for a female. The animals are gregarious, usually being found in herds of 5–10 individuals and sometimes forming aggregations of 100–200. Some groups seen by Schaller contained all ages and both sexes, while others had only adult males or only adult females and young. Group composition changed constantly, the only lasting association being that of a mother and her young. Herd movements were usually led by an adult female, but in social interaction the male was dominant. During the mating season, individual males roamed about in search of estrous females; they defended such females while courting but did not prevent other males from entering the herd, and no territorial behavior was evident. When a few males came together, a rank order was established based on displays and sparring with the antlers. Larger males, with the longest antlers, attained dominance and had a disproportionately large share of the females. Observations by Barrette (1987b) indicate that females sometimes reject the advances of one male and choose another to mate with.

Schaller (1967) called *A. porcinus* essentially solitary but stated that up to 40 individuals may assemble in certain feeding areas or during the mating season. At such congregation sites, each male seems to try to acquire an estrous female and then leaves with her. The mating season of *A. porcinus* peaks in September and October. Blouch and Atmosoedirdjo (1987) reported that *A. kuhli* also is solitary, the only associations being mother and young, courting pairs, and rival males fighting over females. There is much communication by short, sharp barks. *A. kuhli* is capable of breeding throughout the year, but in the wild most mating occurs from July to November, and most births from February to June. The gestation period is assumed to be about 7 months.

In *A. axis*, reproduction continues throughout the year, but mating peaks from March to June and most births occur in the cool season from January to May. If a fawn dies, the mother may mate again and give birth a second time in the course of a year. The gestation period is 8.0–8.5 months. A single offspring is the rule; twins are rare. The young remains hidden in dense cover, with the mother feeding nearby, until it has sufficient strength to roam with the herd. It may become independent after 12 months, but young females sometimes stay with the mother for nearly 2 years. Females often attain sexual maturity at 14–17 months (Schaller 1967). One specimen of *A. axis* lived for 20 years and 9 months in captivity (Jones 1982).

Human hunting and habitat destruction has caused axis deer to become rare in many parts of their natural range. People, however, have established wild-living populations of *A. axis* in Hawaii, California, Texas, Argentina, Brazil, Uruguay, Yugoslavia, parts of the Soviet Union, and the Andaman Islands; groups introduced in New Guinea and New Zealand appear to have died out (Lever 1985). *A. porcinus* occurs in Sri Lanka, and that island sometimes is given as part of its natural range, but the species probably was brought there by the Portuguese in the sixteenth century; another introduced population is in Victoria, Australia (Lever 1985). There is a possibility that *A. kuhli* and *A. calamianensis*, which are isolated on small islands far from their close relative *A. porcinus*, actually represent an ancient human introduction of the latter species, though Groves and Grubb (1987) suggested that they are survivors of widespread Pleistocene populations of hog deer. *A. kuhli* and *A. calamianensis*, as well as the subspecies *A. porcinus annamiticus*, of Thailand and Indochina, are listed as endangered by the USDI and are on appendix 1 of the CITES. The IUCN classifies *A. kuhli* as rare and *A. calamianensis* as vulnerable. Recently, Blouch and Atmosoedirdjo (1987) estimated that only about 300 individuals of *A. kuhli* survive in the wild but that there are at least another 70 in zoos.

ARTIODACTYLA; CERVIDAE; Genus CERVUS
Linnaeus, 1758

Red Deer, or Wapiti

There are 7 subgenera and 10 species (Chasen 1940; Corbet 1978; Ellerman and Morrison-Scott 1966; Groves 1982c; Groves and Grubb 1987; Grubb and Groves 1983; Grzimek 1975; Hall 1981; Laurie and Hill 1954; Lekagul and McNeely 1977; Medway 1977):

subgenus *Rusa* Hamilton-Smith, 1827

C. unicolor (sambar), India to southeastern China and Malay Peninsula, Sri Lanka, Taiwan, Hainan, Sumatra, Borneo, and many small nearby islands;
C. timorensis (Sunda sambar), Java, Bali, Sulawesi, Timor, Moluccas, and many small nearby islands;
C. mariannus (Philippine sambar), Luzon, Mindoro, Mindanao, and Basilan Islands in the Philippines;
C. alfredi (Visayan deer), Visayan Archipelago in the central Philippines;

subgenus *Rucervus* Hodgson, 1838

C. duvauceli (barasingha), India, Assam, Nepal;

subgenus *Thaocervus* Pocock, 1943

C. schomburgki (Schomburgk's deer), south-central Thailand;

subgenus *Panolia* Gray, 1843

C. eldi (thamin, or brow-antlered deer), Assam to Indochina, Hainan;

subgenus *Sika* Sclater, 1870

C. nippon (sika deer), Ussuri district of southeastern Siberia, Manchuria, Korea, eastern and southern China, northern Viet Nam, Japan, Ryukyu Islands, Taiwan;

subgenus *Przewalskium* Flerov, 1930

C. albirostris (Thorold's deer), Tibet, central China;

subgenus *Cervus* Linnaeus, 1758

C. elaphus (red deer, wapiti, or elk), Europe, Asia Minor, Caucasus, Central Asia, southern Siberia, Mongolia, Manchuria, Korea, northern and western China, Himalayan region, northwestern Africa, southern Canada, most of the conterminous United States, northern Mexico.

Groves and Grubb (1987) included *C. schomburgki* and *C. eldi* in the subgenus *Rucervus* and *C. nippon* in the subgenus *Cervus*. Some authorities consider the genus *Cervus* to include *Dama* and *Axis* (see accounts thereof). Taylor (1934) listed eight separate species of the subgenus *Rusa* in the Philippines. In contrast, Ellerman and Morrison-Scott (1966) suggested that the Philippine populations of *Rusa* are conspecific with *C. unicolor*. There now is a general consensus that the North American wapiti, or elk, once usually classified as the separate species *C. canadensis*, is at most subspecifically distinct from the Old World *C. elaphus* (Bryant and Maser 1982; Corbet and Hill 1986; Groves and Grubb 1987; Honacki, Kinman, and Koeppl 1982; Jones et al. 1986). Moreover, if *C. canadensis* and *C. elaphus* were to be regarded as distinct species, as they were by Cockerill (1984) and Whitehead (1972), the line between the two actually would lie along the Tien Shan Mountains and the Gobi Desert; the Asian populations to the north and east of that line have closer systematic affinity to the populations of North America than to those Eurasian populations to the south and west (Groves and Grubb 1987).

Groves and Grubb (1987) recognized a tribe, the Cervini, which includes the genera *Dama*, *Axis*, *Cervus*, and *Elaphurus*. This group is characterized by antlers that normally have at least three tines but that show much structural variation; therefore other characters are needed to assess phylogenetic relationships. *Dama* has a number of primitive features, including the presence of pedal glands on both the forefeet and hind feet, spotted pelage, no mane, a long tail, broad central incisor teeth, and brachyodont cheek teeth with almost no trace of basal pillars. *Axis* has more advanced, pouch-shaped pedal glands and only on the hind feet. The subgenus *Hyelaphus* of *Axis* is considered primitive because of various characters, including swollen auditory bullae. The subgenus *Axis* is more advanced in its reduced bullae, but it retains a full spotted adult pelage. *Cervus* and *Elaphurus* have no foot glands at all. They also are characterized generally by large size, coarse pelage, usually an unspotted adult pattern, a mane of long hairs in at least the male, a short tail, a prominent rump patch, small auditory bullae, all incisiform teeth with broad crowns, and hypsodont cheek teeth with basal pillars and prominent styles and columns on the molars. In addition, male *Cervus* and *Elaphurus* spray urine onto the belly and forequarters during the rut; this behavior is unknown in *Dama* and *Axis*. *Elaphurus* (see account thereof) has a number of distinctive morphological features that set it apart from the rest of the Cervini.

Not all species of *Cervus* have all of the above characters. *C. alfredi* and *C. nippon* have spotted adult pelage and narrow lateral incisiform teeth. *C. duvauceli*, *C. schomburgki*, *C. eldi*, and *C. nippon* retain swollen auditory bullae. Most species do not have an enlarged rump patch or a greatly shortened tail. Additional information is provided separately for each species.

Cervus unicolor (sambar).

Except as noted, the information for the account of this species was taken from Bentley (*in* Strahan 1983) and Lekagul and McNeely (1977). Head and body length is 1,620–2,460 mm, tail length is about 250–300 mm, height at the shoulder

A. North American wapiti *(Cervus elaphus)*, photo from U.S. Fish and Wildlife Service. B. European red deer *(C. elaphus)*, photo by Hans-Jürg Kuhn. C. Swamp deer, or barasingha *(C. duvauceli)*, photo from Zoological Society of London. D. Sika deer (spotted form) *(C. nippon)*, photo from A. Pedersen. E. Sika deer (unspotted form) *(C. nippon)*, photo by Leonard Lee Rue III.

Sambar *(Cervus unicolor)*, photo by Cyrille Barrette.

is 1,020–1,600 mm, and weight is 109–260 kg. Males are larger, on average, than females. The general coloration is dark brown. The rump and underside of the tail are white and serve as a signal when the tail is raised. The pelage is coarse and generally short; it becomes thicker and paler in the cool season. Males may have a long, dense mane on the neck and forequarters. The ears are large and rounded, and there is a round glandular area, devoid of hair, on the middle of the throat. The antlers are stouter and more rugose than those of most other *Cervus*. They have only three tines, but these become very prominent and grow to lengths of up to about 1,000 mm measured along the outside curve.

The sambar is found in a variety of habitats but seems to prefer wooded areas. It occurs regularly on forested mountain slopes up to elevations of 3,700 meters (Whitehead 1972). There may be movements from lower, more sheltered areas in winter to higher altitudes in the spring and summer. Most activity is crepuscular and nocturnal. The sambar is more of a browser than a grazer, generally feeding in areas of secondary growth and taking leaves, buds, grass, berries, and fallen fruit.

In a study in central India, Schaller (1967) found a population density of about 0.8/sq km. Groups there usually consisted of fewer than 6 animals, mainly females and young. There and in other areas adult males have been observed to roam about alone during the mating season and to then fight one another for small territories. The victors scent-mark these areas and mate with any females that enter. Up to 8 females may be gathered by a territorial male for various periods (Whitehead 1972). While breeding may occur during much of the year over the wide range of this species, the main mating season seems to be from about September to January. In Australia there is a birth peak in May and June. A single fawn is born after a gestation period of about 8 months. It stays with its mother for a year or more. Females reach sexual

maturity in their second year of life (Sadleir 1987).

The introduced population in Victoria, Australia, contains about 5,000 individuals. The sambar also has been established in the wild in New Zealand, California, Texas, and Florida (Lever 1985).

Cervus timorensis (Sunda sambar).

Except as noted, the information for the account of this species was taken from Bentley (*in* Strahan 1983) and Whitehead (1972). Head and body length is 1,420–1,850 mm, tail length is about 200 mm, height at the shoulder is 830–1,100 mm, and weight is 50–115 kg (averaging 73 kg in males and 53 kg in females). The pelage is grayish brown and often rough and coarse in appearance. This species resembles *C. unicolor* and also has broad ears, but it is smaller and less compact, its coat is even coarser (Groves and Grubb 1987), and its tail is not carried erect in alarm. The antlers are lyrelike and have three tines, the inner or posterior tine being longer; they usually are 600–750 mm long but have reached 1,115 mm.

The Sunda sambar occupies a variety of habitats, including forests, plantations, and open grassland. It is mainly nocturnal but sometimes grazes by day. The diet includes grass, shrubs, and herbs. Some animals have been reported to swim in the sea, to drink seawater, and apparently to relish certain seaweeds. Males form groups separate from the females and young except during the breeding season. Known vocalizations are an alarm bark and shrill roaring by males in the mating season. Breeding occurs throughout the year, with a mating peak from late June to August and a corresponding calving peak in March and April. The introduced population on the islands of the Torres Strait, off northeastern Australia, mates in September–October and gives birth in April–May. The calves are not spotted.

The natural distribution of *C. timorensis* is something of a mystery. Groves and Grubb (1987) stated that it is likely that

Sunda sambar *(Cervus timorensis)*, photo by Lothar Schlawe.

the species originally was confined to Java and Bali, though it also is known from prehistoric deposits in Sulawesi and Timor. It may have been moved to the latter two islands in association with early human migrations, and it certainly was spread through much of Indonesia by the seafarers of various nations. It is now known to be present on many of the Molucca Islands, the Obi Islands, Flores and numerous small islands between Java and Timor, and many small islands around Java and Sulawesi. Populations were established on Borneo in 1680, on New Guinea about 1900, and in various parts of Australia in the early twentieth century. There also are introduced populations on Mauritius, the Comoro Islands, New Zealand, New Britain, New Caledonia, and Horsburgh Island in the South Pacific.

Cervus mariannus (Philippine sambar).

Head and body length is 1,000–1,510 mm, tail length is 80–120 mm, shoulder height is 550–700 mm, and weight is 40–60 kg. In most populations the coloration is uniformly dark brown, sometimes being darker above and slightly paler below and on the legs. The underside of the tail is white. In one part of Mindanao deer were found with generally pale, sandy gray coats. *C. mariannus* resembles *C. unicolor* but differs in its smaller size, more compact build, very short (200–400 mm) antlers, relatively shorter tail with shorter hairs, and greater contrast between the relatively broad central incisors and narrow lateral incisiform teeth (Grubb and Groves 1983).

Whitehead (1972) indicated that *C. mariannus* occupies several habitats in the Philippines, including both lowlands and densely vegetated mountain slopes up to elevations of 2,900 meters. The species also was introduced by the Spanish to Guam and Rota islands in the Marianas (from where it was

first described, hence the scientific name) and to Bonin Island about 1850. These introduced populations now seem to have disappeared (Lever 1985; Whitehead 1972), but Groves and Grubb (1987) indicated that another established group is present on Ponape Island in the Carolines. It is likely that at least some of the natural populations are also in jeopardy because of the same sort of problems affecting *C. alfredi* (see below).

Cervus alfredi (Visayan deer).

This is one of the rarest, least known, and most narrowly distributed species of deer in the world. Until recently it commonly was treated as a subspecies of *C. mariannus* or *C. unicolor*, but Grubb and Groves (1983) pointed out that it is very distinctive. According to those authorities, it does resemble *C. mariannus* but differs from that and the other

Philippine sambar *(Cervus mariannus)*, photo from U.S. National Zoological Park.

species of the subgenus *Rusa* in the relative narrowness of its skull, in its remarkably fine and dense pelage, and in having a spotted coat. The ground color of the upper parts is dark brown, there is a dark dorsal band bordered with faint dull ochre spots, and on the flanks there are bigger and more scattered spots. The underparts and the underside of the tail are buffy. Antler length in the holotype is 244 mm. Height at the shoulder is 640 mm (Taylor 1934).

This deer formerly occurred throughout the dense tropical forests of the Visayan Archipelago, the group of islands in the central Philippines that includes Biliran, Bohol, Cebu, Guimaras, Leyte, Masbate, Negros, Panay, Samar, and Siquijor. Its numbers and range were greatly reduced by hunting and destruction of forest habitat by commercial logging and agricultural operations. Such activities were curtailed recently by termination of lumber concessions and depression of the sugar industry, but ironically these factors have led to an influx of unemployed people into the remaining forests in order to hunt for food and clear land for growing crops. The Visayan deer now is known to survive only in a few small scattered parts of Leyte, Negros, Panay, and Samar. If current rates of hunting and habitat loss continue, the species is unlikely to survive for longer than 10–15 years (Cox 1987). It now is classified as endangered by the IUCN and the USDI.

Cervus duvauceli (barasingha).

According to Whitehead (1972), height at the shoulder usually is about 1,190–1,240 mm and weight is about 172–81 kg. The general coloration is brown, males being darker than females. The underparts are paler, and the underside of the tail is white or light yellow. The overall coat becomes paler, yellowish brown in the summer, and some populations then show pale spots, comparable to those of *Axis axis*, on the back and sides. The antlers usually bear 10–15 points and are up to about 1,000 mm long. Groves and Grubb (1987) added that the antlers are smooth, not rugose as in *C. eldi*, and that both species have elongated and spreading hoofs as an adaptation for life in wetlands.

Two of the three remaining populations of the barasingha, those in Assam (subspecies *C. d. ranjitsinhi*) and along the Nepal-India border *(C. d. duvauceli)*, do dwell in marshy grassland. It is likely that the species originally was abundant all through the open floodplains of the great river systems of northern India. The third group, the so-called hard-ground barasingha, of Madhya Pradesh in central India *(C. d. branderi)*, lives in a habitat of dry meadows and woodland (Groves 1982c; Schaaf 1978). The species may be active either by day or by night and is primarily a grazer (Whitehead 1972).

On the floodplain of a tributary of the Ganges in Nepal, Schaaf (1978) determined that most of a group of about 1,000 barasingha spent the entire year in an area of about 32 sq km and formed large aggregations each year from February to April. In a study of the central Indian population of *C. duvauceli*, Schaller (1967) found a density of only about 0.2/sq km. Herds usually comprised 13–19 animals, and one had at least 500. Some groups consisted of only males or of only females and young, but in the mating season loose associations of both sexes formed. Males established a dominance hierarchy, with the highest-ranking individuals having priority access to estrous females. Schaller indicated that the

Swamp deer, or barasingha *(Cervus duvauceli)*, male and female, photo by Lothar Schlawe.

mating season extends from September to April. Whitehead (1972) also reported that the period was ill-defined but that it might peak in December and January and that the stags then fight for possession of a harem of up to 30 hinds. They frequently utter a roaring sound during the rut, and the species also has a shrill alarm call. According to Sadleir (1987), *C. duvauceli* is the only cervid in which females are definitely known to be monestrous; the gestation period is 240–50 days, and the females attain sexual maturity at more than 2 years of age.

The barasingha is classified as endangered by the IUCN and the USDI and is on appendix 1 of the CITES. Its once vast range from Assam to the Indus has been reduced to the three small areas noted above. The main problem is human usurpation or modification of grassland habitat. The status of the subspecies *C. d. branderi* has been best documented (Martin 1987). It numbered about 3,000 individuals in 1938 but declined to only about 100 in 1970. Since its range now is centered in Kanha National Park, efforts could be made to improve its habitat, and numbers now have climbed to around 400.

Cervus schomburgki (Schomburgk's deer).

The information for the account of this extinct species was taken from Lekagul and McNeely (1977). There are recorded measurements of 1,040 mm for height at the shoulder and 103 mm for tail length. The upper parts are uniform brown, the underparts are lighter, the ventral surface of the tail is white, the legs and the area between the antlers have a reddish tinge, and there is a mane of hair 50 mm long extending down the front of the foreleg. The antlers resemble those of *C. duvauceli* but are more complex and have a very short beam. There are at least 5 tines on each antler, and length along the outside curve is 320–830 mm.

This species may once have occurred as far north as Yunnan and Laos but is known with certainty only from south-

central Thailand, where it inhabited swampy plains with long grass, cane, and shrubs. It avoided densely vegetated areas. In the rainy season, when the plains were flooded, it sought refuge on small patches of raised ground. It usually spent the day resting in the shade and grazed during the evening and night. It typically stayed in family groups consisting of one buck, a few does, and a few fawns.

Large-scale commercial production of rice for foreign export began in the late nineteenth century in Thailand and by the early 1900s led to the loss of nearly all of the grasslands and swamps on which *C. schomburgki* depended. The species still was abundant in the Chao Phya River Valley at the turn of the century but came under intensive hunting pressure, especially when it was forced to crowd onto islands during the floods. A few survived in the wild until 1932, and the last known captive was killed in 1938.

Cervus eldi (thamin, or brow-antlered deer).

According to Lekagul and McNeely (1977), head and body length is 1,500–1,700 mm, tail length is 220–50 mm, height at the shoulder is 1,200–1,300 mm, and weight is 95–150 kg. The upper parts are brownish red in winter and somewhat lighter in summer, with does paler than stags all year. Some spots may be visible along the back. The hair is coarse and long in winter but shorter in summer; fully adult males have thick and long hair forming a mane on the neck. The antlers are unique in that the brow tine forms a continuous curve with the beam, so that in profile the antlers appear to be bow-shaped. Whitehead (1972) indicated that the antlers can reach about 1,000 mm in length excluding the brow tine. Groves and Grubb (1987) noted that *C. eldi* resembles *C. duvauceli* in having elongated and spreading hoofs but differs in having a deep lacrimal pit and a metatarsal tuft.

The natural history of *C. eldi* is comparable to that of *C. duvauceli* (Lekagul and McNeely 1977; Whitehead 1972). Populations in Assam and Burma live in low-lying swamps,

Schomburgk's deer *(Cervus schomburgki)*, photo from West Berlin Zoo through Lothar Schlawe.

Brow-antlered deer *(Cervus eldi)*, photo by Lothar Schlawe.

while those in Thailand and Cambodia usually are found in much drier areas. Stags are fond of wallowing in mud. *C. eldi* may enter the shady edge of forests during the hottest hours of the day but avoids dense vegetation. It is primarily a grazer but takes some browse and fruit. Aggregations of 50 or more animals have been reported to form during part of the year. Mature stags usually are solitary, but during the mating season they join the herds and fight for possession of the hinds. In the wild the rut reportedly falls between February and May. Studies of a captive population in Virginia (Wemmer and Grodinsky 1988) show that females become pregnant each year and are seasonally polyestrous during a period of 7.5 months, from February to September, the modal length of the estrous cycle is 17 days, estrus does not exceed 2 days, the gestation period averages 242 days, there normally is a single young, birth weight is about 4.7 kg, and weaning occurs after about 28 weeks. Females usually attain sexual maturity at 12–24 months (Sadleir 1987). Captive females have bred regularly until 15 years old and have lived as long as 19 years and 7 months (Prescott 1987).

The entire species *C. eldi* is listed as endangered by the USDI and is on appendix 1 of the CITES. The IUCN gives the endangered classification only to the subspecies *C. e. eldi*, of Assam, and *C. e. siamensis*, of Thailand and the region farther to the east. Salter and Sayer (1986) reported that *C. e. thamin*, of Burma, is still relatively widespread and abundant, there being an estimated 2,229 animals in a sanctuary measuring 268.2 sq km, but that the subspecies is declining nonetheless because of illegal hunting and loss of habitat to logging, livestock grazing, and agriculture. Similar problems led to the current restriction of *C. e. eldi* to one national park in eastern Assam and to the near extinction of *C. e. siamensis*. The status of the latter subspecies is not well known, though it may still survive in parts of Cambodia and there are about 140 individuals, 60 of which are in a semicaptive state, on Hainan (Clutton-Brock 1987). The only other captive *siamensis*, 11 animals, are in France (Prescott 1987).

Cervus nippon (sika deer).

Except as noted, the information for the account of this species was taken from Feldhamer (1980). Based on an introduced population in Maryland, head and body length is about 950–1,400 mm, tail length is 75–130 mm, and height at the shoulder is 640–811 mm. Males average 8.7 percent larger

than females. Average weights of dressed carcasses from the same population are 32.7 kg for males and 26.2 kg for females. However, Groves and Grubb (1987) referred to head and body lengths of up to 1,800 mm, and Whitehead (1972) listed shoulder heights of up to 1,090 mm and weights of up to 80 kg. The ground color of the pelage ranges from chestnut-brown to reddish olive. Numerous white spots, arranged in seven or eight rows, are present on the upper sides. The spots are more noticeable in the summer than in the winter. The chin, throat, and belly are off-white or gray. There is a large, white, erectile rump patch. Both sexes have a dark neck mane in the winter. The antlers are narrow and erect, have 2–5 points each, and measure about 300–660 mm.

The sika deer prefers forested areas with a dense understory but adapts well to a variety of other habitats, including marshes and grasslands. Seasonal altitudinal movements of about 700 meters have been reported in Japan, with summer ranges being generally higher and larger than winter ranges. Activity occurs primarily from dusk to dawn but may also be active at times during daylight. The species is highly adaptable in diet, being predominantly a grazer in certain situations and browsing on trees and shrubs in others.

This deer is not particularly gregarious, single individuals being seen about as often as small groups. Adult males are solitary for most of the year but sometimes band together when antlers are cast after the rut. During the calving season females and their young form groups of 2–3 that seem more stable than the male bands. Aggregations of 40–50 animals occasionally are seen, and Miura (1984b) reported that in Nara Park, on Honshu, 120 females and young gathered in one area each day to rest. During the summer, adult males begin to establish territories. Miura (1984b) found the size of these territories to average 4.76 ha., with a range of 2.69–7.70 ha. He also noted that nonterritorial males had an average home range of 11.74 ha. Only about a fifth of the males in Nara Park—those over five years old with antlers longer than 400 mm—exhibited territorial behavior. The boundaries of the territories are marked by urination and thrashing of the ground. In the mating season, each successful male gathers as many as 12 females on his territory. Other males are driven away, and there may be serious fighting with the antlers and hooves. *C. nippon* is highly vocal: 10 different sounds have been recorded, ranging from soft whistles between females to loud screams by the males.

Mating occurs in September and October, and births usually in May and June. Females are polyestrous. The gestation period is about 30 weeks. There is usually a single young weighing 4.5–7.0 kg. Nursing may continue for 8–10 months. Sexual maturity is attained at 16–18 months. Weight increases until the age of 4–6 years in females and 7–10 years in males.

The IUCN and the USDI classify the following subspecies as endangered: *C. nippon taiouanus* (Taiwan), *C. n. keramae* (Ryukyu Islands), *C. n. mandarinus* (northern China), *C. n. grassianus* (northern China), and *C. n. kopschi* (east-central China). The Soviet Union designates *C. n. hortulorum* (southeastern Siberia, Manchuria, Korea) as endangered. Problems generally involve uncontrolled hunting for food and commerce and usurpation of forest habitat by agriculture and the growing number of people in the involved regions. Except for *keramae*, which still survives on three islets not inhabited by people, these subspecies may have vanished in the wild, though some are present on farms (Cowan and Holloway 1978). Wild-living populations of sika deer are known to have been established by people in modern times in the British Isles, several countries of mainland Europe, Maryland, Texas, and New Zealand (Lever 1985). Such introduction, but in the distant past, might also explain the presence of the isolated subspecies *C. n. soloensis* on Jolo Island, south of

Thorold's deer *(Cervus albirostris)*, photo by J. F. Rock.

the Philippines, a locality far from the range of all other populations of sika deer (Groves and Grubb 1987).

Cervus albirostris (Thorold's deer).

According to Whitehead (1972), there are recorded measurements of 1,220 mm for height at the shoulder and 12–13 mm for tail length. The general summer color is brown, the underparts are creamier, and the nose, lips, and chin are white. In winter the color is slightly lighter and the hairs are long and wavy. There is a reversal of the hairs on the withers, which gives the appearance of a hump. The head has a somewhat flattened appearance, the ears are relatively long and narrow, and the antlers normally have five tines each. The hooves are high, short, and wide, different from those of all other deer and resembling those of cattle.

Groves and Grubb (1987) stated that the hooves and stout limbs of *C. albirostris* are adaptations for life at high altitudes. Cowan and Holloway (1978) noted that it is found in coniferous forest, rhododendron, and alpine grasslands and that it has been seriously depleted through hunting for the antler trade. Little other information has been recorded about this species; it is designated as indeterminate by the IUCN, but it may not be rare. It is thought to number over 100,000 individuals, and it is widely farmed in China (Clutton-Brock 1987).

Cervus elaphus (red deer, wapiti, or elk).

This is the largest and most phylogenetically advanced species of *Cervus* (Groves and Grubb 1987). Head and body length is 1,650–2,650 mm, tail length is 100–270 mm, height at the shoulder is 750–1,500 mm, and weight is 75–340 kg (Grzimek 1975). Animals in the populations of North America and northeastern Asia are usually larger than those of Europe and southern Asia, and within a given population males average larger than do females. The upper parts are usually some shade of brown, and the underparts are paler. There is a prominent pale-colored patch on the rump and buttocks. The pelage is coarse, and males have a long, dense mane. The well-developed antlers measure up to about 1,750 mm along the beam. There is a supernumerary tine (the bez) above the brow tine (Groves and Grubb 1987).

This deer utilizes a wide variety of habitats in both lowlands and mountains. In North America it originally was found in dense coniferous forests, open hardwood forests, chaparral, and grasslands (Skovlin 1982). It is usually active during the early morning and late afternoon. In some areas it

European red deer *(Cervus elaphus)*, photo from Zoological Society of London.

moves into higher country during the spring and returns to the lowlands in the fall. Populations in eastern North America did not migrate, but those in the western mountains frequently do, the autumn movement generally being to lower elevations to avoid snow cover. Summer ranges cover a much larger area than do winter ranges. In certain areas only a portion of a population migrates. About a fourth of the large herd wintering at the Jackson Hole Refuge in Wyoming remains in the same range all year, while the other animals move out for distances of up to 97 km (Adams 1982). On the other hand, a major segment of the herds in northern Yellowstone remain in high mountains even during severe winters (Houston 1982). The diet of *C. elaphus* varies and involves both grazing and browsing. In western North America up to 85 percent of spring forage is grass, there is a shift to forbs and woody plants in the summer, browse and dried grass is taken in the fall, and shrubs and conifers extending above the snow may be used in winter (Nelson and Leege 1982).

C. elaphus is highly gregarious. Discrete herds are formed, each usually occupying a definite area. For most of the year the sexes stay in separate herds. In South Dakota, Varland, Lovaas, and Dahlgren (1978) found three herds of cows and calves to be distributed as follows: 170 animals on 20.7 sq km, 90 animals on 25.9 sq km, and 40 animals on 11.7 sq km. Small groups of adult males also were present in the study area. In a study in Scotland, Clutton-Brock, Guinness, and Albon (1982) found population densities to vary from 1 per 10 ha. to 1 per 120 ha. Females there had home ranges of about 60 ha. in summer and 40 ha. in winter, whereas males had corresponding ranges of 40 ha. and 25 ha. Outside of the mating season the sexes remained apart in groups of usually 4–7 members (not counting the calves of females). Each group had a rank hierarchy maintained by threatening gestures, kicks, and chases. Females began to restrict their range in September and gathered in traditional rutting areas. The males also left their normal ranges and joined the females by early October. The stags then competed for the females through displays involving roars, thrashing, and spraying urine (see below). They fought frequently, and serious injuries were common. The younger and weaker individuals were driven away but sometimes remained in the vicinity and sought to split off females. Males under 4 years old rarely could hold a harem; maximum fighting success came at 7–10 years. Successful stags sometimes gained over 20 females, though seldom did a male actually father more than 4 young per season. Young females generally adopted a home range overlapping that of their mother, while young males left the vicinity at 2–3 years and joined a stag group.

Boyd (1978) described an annual progression in the social life of *C. elaphus* in North America. Following parturition in the late spring, the mother and newborn live alone for several weeks. Subsequently, the cows and their young, along with immature animals of both sexes, begin to congregate. By mid-July, herds of up to 400 individuals have formed and are led by older cows. Meanwhile, full-grown bulls live alone or in groups of up to 6 individuals. By September the bulls have shed their velvet and begun to attract the females. The older, more powerful bulls drive off the young males and gather harems of 15–20 cows each. No particular territories are defended. After mating is completed, by about mid-October, the males separate from the female groups. Later, aggregations of up to 1,000 animals may concentrate on the limited winter range.

Geist (1982) wrote that sexual competition during the rut is the most spectacular behavior of *C. elaphus*. Males attempt to attract a harem of females by advertising through displays and vocalizations. The mating call of stags of North American populations, which now generally are found in open country, is a high-pitched powerful bugle. The corresponding call of the European red deer, which often must carry through forests, is a deep roar. The stags also spray urine about as an olfactory signal, sometimes wallowing in the moistened earth, and thrash vegetation with their antlers. Once a group of females has been attracted, they are actively herded through the constant attention and threatening postures of the male. Probably, however, females are capable of escaping if they so desire. If a rival stag approaches and cannot be intimidated by display, there may be a fierce dominance fight. The opponents walk parallel for a while and then turn suddenly and lock antlers. They push and twist, attempting to throw one another off balance and seeking an opportunity to thrust with healed scars. The hides of old stags are covered with healed scars, and about 5 percent of males in a population may be expected to die annually from fighting.

In Europe and North America, mating occurs in the early fall, and births in the late spring. Females are seasonally polyestrous, the estrous cycle being about 18 days long and receptivity lasting 1–2 days (Sadleir 1987). The gestation period of European red deer is about 235 days, while that of the North American elk reportedly varies from 247 to 265 days (Clutton-Brock, Guinness, and Albon 1982; Taber, Raedeke, and McCaughran 1982). There is usually a single offspring, which weighs 13–18 kg at birth. The calf stands quickly, follows the cow after 3 days, grazes at 4 weeks of age, and loses its spots after 3 months. Lactation may last 4–7 months, sometimes longer. Females attain sexual maturity at about 28 months and are then capable of giving birth annually. Males are capable of mating in their second year of life but usually must wait considerably longer, because of competition from older bulls. More than half of all animals die before they are 1 year old, and few wild individuals survive more than about 12–15 years; males generally perish first, because of the intensity of their fighting. A captive *C. elaphus* lived for 26 years and 8 months (Jones 1982).

Few mammalian species have been so extensively affected by people. On the one hand, *C. elaphus* has been introduced to many areas beyond its natural range, and even within its natural range there has been much transplantation of populations and manipulation of herds in an effort to improve big game hunting. There now are major human-established populations in New Zealand, Australia, and Argentina (Lever 1985). On the other hand, excessive hunting and habitat modification have resulted in drastic declines in natural distribution and numbers.

There is some difficulty in assessing bioconservation status, because of the considerable disagreement regarding the division of *C. elaphus* into subspecies or even full species. The situation was reviewed by Groves and Grubb (1987), who recognized three general systematic groups. The first, the typical red deer, occurs in Europe, North Africa, and southwestern Asia as far east as the southern shores of the Caspian Sea. *C. elaphus* has disappeared from much of this region; existing populations in Europe generally are well protected, but some do not represent the original stock of the area in which they are found (Smit and Van Wijngaarden 1981). There now are about 1 million red deer in Europe, the largest populations being in Scotland, Germany, and Austria. The populations of West Germany, numbering about 85,000, are thought to be about as high as available habitat will allow and have sustained an annual legal sport hunting kill of about 30,000 for some years (Ueckermann 1987). The subspecies *C. e. corsicanus*, of Corsica and Sardinia, is classified as endangered by the IUCN and the USDI. Habitat degradation and illegal hunting reduced the number on Sardinia to about 200–400 individuals (Cowan and Holloway 1978; Smit and Van Wijngaarden 1981). The population on Corsica numbered around 1,000 in 1900 but was completely extirpated by the 1960s; there has been a recent effort to reintroduce the sub-

North American wapiti *(Cervus elaphus)*, photo by Leonard Lee Rue III.

species there, using animals from Sardinia *(World Wildlife Fund News,* no. 42 [1986]:3). *C. e. barbarus,* of North Africa, is designated vulnerable by the IUCN and endangered by the USDI. It evidently disappeared in Morocco during Roman times and later was reduced to very low numbers in Algeria and Tunisia. A modest recovery seems to have taken place recently, with numbers rising to more than 1,200 (Clutton-Brock 1987; Cowan and Holloway 1978). The subspecies of Asia Minor and the Caucasus, *C. e. maral,* was considered by Smit and Van Wijngaarden (1981) to be declining rapidly and in danger of extinction.

The second major systematic group of *C. elaphus* comprises the following six subspecies: *C. e. bactrianus,* of southern Soviet Central Asia and adjacent parts of Afghanistan; *C. e. hanglu,* of Kashmir; *C. e. wallichi* (shou), of southern Tibet and Bhutan; *C. e. yarkandensis,* of Sinkiang; *C. e. macneilli,* of the eastern Tibetan Plateau and adjacent Sinkiang; and *C. e. kansuensis,* of north-central China. This group is relatively primitive and appears to have given rise to both the red deer to the west and the wapiti to the north and east (Groves and Grubb 1987). It also is by far the most critically endangered component of *C. elaphus. Hanglu* is on appendix 1 of the CITES, and *bactrianus* is on appendix 2. The IUCN classifies *bactrianus, hanglu, wallichi,* and *yarkandensis* as endangered and *macneilli* as indeterminate; the USDI lists all five as endangered. Cowan and Holloway (1978) assessed the status of these five as follows: *bactrianus,* numbers 1,100 and increasing through conservation efforts in

the Soviet Union (having declined to only 300–400 in the 1960s because of uncontrolled hunting and usurpation of its range by agriculture and livestock [Bannikov 1978]); *hanglu,* greatly reduced through illegal killing and habitat disruption by people, known to survive only in one sanctuary, numbers estimated at 250; *wallichi,* probably extinct because of uncontrolled hunting; *yarkandensis,* possibly extinct through loss of habitat to agriculture; *macneilli,* populations seriously depleted by overhunting, principally for its antlers for use in the oriental medicine trade. The sixth subspecies in this group, *C. e. kansuensis,* does not have an official classification, but Whitehead (1972) noted that it was faced with extinction on account of the demand for antlers in velvet for alleged medicinal purposes. Recently, Dolan and Killmar (1988) discovered living specimens of *C. e. wallichi* in a park in Lhasa, Tibet, and indicated that there is evidence that the subspecies still survives in the wild in both Tibet and Bhutan.

The third group of *C. elaphus* occurs in the Tien Shan Mountains of eastern Kazakhstan and Sinkiang, Mongolia, Manchuria, Korea, southern Siberia, and, after a vast hiatus, North America. None of the various subspecies in these regions are now classified by the IUCN or the USDI, but most have been adversely affected by human activity, notably unregulated hunting during settlement in the nineteenth and early twentieth centuries. Indeed, the subspecies of eastern and central North America, *C. elaphus canadensis,* and of the southwestern United States and Mexico, *C. e. merriami,* became extinct in that period. The tule elk of California, *C. e.*

nannodes, was reduced to fewer than 100 individuals by the 1870s, but it was saved by private conservation efforts, and there now are several protected herds, with a total of about 900 animals. The other three subspecies of western North America also declined drastically in the nineteenth century. By 1900 only 41,000 individuals remained in the United States, and there were a few thousand more in Canada. Subsequent conservation measures allowed an increase to numbers estimated at about 500,000 (Bryant and Maser 1982) or nearly 1 million (Vogt 1978), and the wapiti is now generally subject to limited, legal sport hunting. About 100,000 are killed each year (Potter 1982). Recently, Schonewald-Cox, Bayless, and Schonewald (1985) cautioned that certain restricted populations on the West Coast appear to be as distinctive as some recognized subspecies and might warrant conservation attention.

ARTIODACTYLA; CERVIDAE; Genus ELAPHURUS
Milne-Edwards, 1866

Père David's Deer

The single species, *E. davidianus,* originally occurred in the lowlands of northeastern and east-central China. Fully wild animals disappeared from this region many years ago, but a herd was maintained in the large Imperial Hunting Park, south of Beijing, until about 1900. Captives descended from that herd are now present in many parts of the world (Cao 1978, 1985; Corbet 1978; Grzimek 1975). The date of final occurrence of *Elaphurus* in the wild usually has been given as about A.D. 200, but Xia (1986) indicated that it survived therein until the Ming Dynasty (A.D. 1368–1644). Also, Dobroruka (1970*b*) reported that two skins collected on Hainan in 1869 appear to represent *Elaphurus.*

According to Wemmer (1983*a,* 1983*b*), head and body length is about 1,830–2,160 mm, tail length is 220–355 mm, and height at the shoulder is 1,220–1,370 mm. Average weight is 214 kg for males and 159 kg for females. The summer pelage ranges from reddish buff to deep reddish brown with a black medial stripe on the shoulders. The throat mane of males and the tail tassel of both sexes tend to be

slightly darker. The winter coat is grayish brown with dark zones on the flanks and throat. The winter pelage is composed of a dense insulating woolly undercoat and coarser, longer guard hairs. *Elaphurus* probably is distinctive in having an abundance of long, somewhat wavy guard hairs most of the year. The legs are relatively heavy-boned, and the feet are broadly splayed, like those of *Rangifer,* with naked skin between the hooves. The tail is the longest of any cervid's.

The antlers are distinctive in having a main branched anterior segment, which may correspond to the simple brow tine of other cervids, and a relatively simple, unforked posterior segment, which extends backward and which may correspond to the main beam of the antlers of other deer. The antlers are shed in December and January, and a new set begins to develop almost immediately, reaching full size by May.

Père David's deer may originally have inhabited swampy, reed-covered marshlands. Although it supplements its grass diet with water plants in the summer, it is essentially a grazing animal. Wild herds with thousands of animals were present in ancient China (Xia 1986). At Woburn Park in England, where a large herd is now maintained, the sexes are together for half of the year. Adult males keep to themselves for about 2 months before and 2 months after the mating season, which begins in June. In that month, hinds begin to bunch together in several groups in definite areas. A stag joins each group of females and then fasts for several weeks, while he engages in mock combat and serious fights with rival males. In fighting, *Elaphurus* not only uses its antlers and teeth but also rises on its hind legs and boxes. In a process that continues to the end of the rut in August, males are successively ousted and replaced by other stags. After leaving a harem, a male begins feeding again and quickly regains weight. Females are seasonally polyestrous, the estrous cycle usually lasts about 20 days but sometimes is much longer, the average gestation period is 288 days, and the one or two spotted fawns usually are born in April or May (Curlewis, Loudon, and Coleman 1988; Jones and Manton 1983; Wemmer 1983*b*). Sexual maturity is usually attained at an age of 2 years and three months (Grzimek 1975). Maximum known longevity is 23 years and 3 months (Jones and Manton 1983).

This deer is named for Abbé Armand David, who procured two skins in 1865 by bribing guards at the Imperial Hunting Park. Subsequently, a number of live animals were sent from

Père David's deer *(Elaphurus davidianus),* photo from Cleveland Zoological Society.

China to various zoos in Europe. The imperial herd was largely destroyed by a flood in 1894 and by subsequent hunting during a famine and the Boxer Rebellion. The last member of this herd survived at the Beijing Zoo until 1922. Meanwhile, however, the captives from several European zoos had been pooled at the Duke of Bedford's estate, Woburn Abbey, in England. The herd thus established increased in numbers, and individual deer were later distributed to other areas. In 1956 two pairs were sent to Beijing, and in 1957 the first calf was born there (Bower 1979; Grzimek 1975). Jones and Manton (1983) reported the total number in captivity to be 873 in 82 collections. In November 1985 a herd was reintroduced in a 100-ha. area of original habitat near Beijing (Sitwell 1986).

ARTIODACTYLA; CERVIDAE; Genus ODOCOILEUS
Rafinesque, 1832

White-tailed Deer and Mule Deer

There are two species (Brokx 1984; Cabrera 1961; Hall 1981):

O. hemionus (mule deer), southern Yukon and Manitoba to Baja California and northern Mexico;
O. virginianus (white-tailed deer), southern Canada, conterminous United States except parts of the Southwest, Mexico to Bolivia and northeastern Brazil.

Hall (1981) used the name *Dama* Zimmermann, 1780, for this genus. Hybridization between the two species in the wild has been reported (Gavin and May 1988; Stubblefield, Warren, and Murphy 1986).

Head and body length is 850–2,100 mm, tail length is 100–350 mm, height at the shoulder is 550–1,100 mm, and weight is 18–215 kg. In a given population males average larger than females. Male *O. hemionus* commonly stand

about 1,000 mm at the shoulder and weigh 75–150 kg (Anderson and Wallmo 1984). *O. virginianus* shows much more geographic variation in size (R. H. Baker 1984; Hardin, Klimstra, and Silvy 1984; Whitehead 1972); adult males of some northern populations usually stand over 1,000 mm at the shoulder and weigh 100–150 kg, those in the south-central United States average closer to 900 mm and 50–100 kg, and those in parts of inland South America are about 800 mm tall and weigh 30–55 kg. The smallest deer of all are sometimes said to occur on the lower Florida Keys, where males average about 625 mm in height and 36 kg in weight. However, the Key deer show some overlap in size with populations in the nearby Everglades, and there are equally small deer in parts of Central America and coastal South America. In much of the range of *Odocoileus* the upper parts are brownish gray during winter and the underparts are then lighter. In summer the coat is reddish brown above and is spoken of as being "in the red." The hairs, especially those of the winter coat, are tubular and somewhat stiff and brittle. For this reason, the winter skins float in water and have at times been used as life preservers. In North America the antlers are shed from January to March, and the new ones begin to grow about April or May, losing their velvet in August or September. The antlers attain full size by the fourth or fifth year of life.

From *Cervus*, *Odocoileus* is distinguished by the absence of upper canine teeth. From the related South American genera *Blastocerus* and *Ozotoceros*, *Odocoileus* is distinguished by the presence of metatarsal glands. The two species of *Odocoileus* are much alike in appearance and behavior but usually can be separated by a number of morphological characters (Anderson and Wallmo 1984; R. H. Baker 1984). In *O. virginianus* the tail is brown above and white laterally and below, the antler has one main beam with minor branches, the sub-basal snag of the antler is long, the length of the ear is one-half the length of the head, the preorbital (lacrimal) pit is shallow, and the metatarsal gland is less than

Mule deer *(Odocoileus hemionus)*: A. Photo by Leonard Lee Rue IV; B. Photo by Leonard Lee Rue III.

42 mm long. In *O. hemionus* the tail is white or black above and tipped with black (and also is usually smaller in size), the antler branches into two nearly equal parts, the sub-basal snag is short, the length of the ear is two-thirds to three-fourths the length of the head, the preorbital pit is about 23 mm deep, and the metatarsal gland is more than 70 mm long.

These deer occur in a great variety of habitats. They prefer areas with enough vegetation for concealment but usually avoid dense forests. They walk about cautiously, flee from danger with a series of bounds, can run at speeds of up to 64 km/hr, and are excellent swimmers (Banfield 1974). They are generally most active at dawn and dusk. In some areas there is a fall migration to lower elevations or to the vicinity of favorable winter food supplies. Such migrations commonly cover 10–50 km (Marchinton and Hirth 1984). Young animals leaving their natal ranges may disperse across distances of 10–200 km (Anderson and Wallmo 1984). The diet includes grass, weeds, shrubs, twigs, mushrooms, nuts, and lichens. Both species browse for part of the year, but *O. hemionus* is predominantly a grazer in the summer. Hirth (1977) reported population densities of 25 *O. virginianus* per sq km in a study area in Michigan and 50 per sq km in an area in Texas. Anderson and Wallmo (1984) listed reported densities of about 5–50 per sq km for *O. hemionus*. Individual home ranges vary widely, but most reported figures, not including migratory routes, fall within the extremes of those found for *O. virginianus* in Texas: 24.3–137.6 ha. for females and 97.1–356.1 ha. for males (Halls 1978). Such areas usually have an elongated shape and include a good mix of food, water, and cover (Marchinton and Hirth 1984). The ranges of *O. hemionus* are comparable, though Anderson and Wallmo (1984) listed a maximum of 3,379 ha. for females on the Great Plains.

Unlike *Cervus* and some other kinds of deer, *Odocoileus* does not usually gather in large herds (Anderson and Wallmo 1984; Banfield 1974; Halls 1978; Hawkins and Klimstra 1970; Kucera 1978). A basic social unit comprises an adult doe, her yearling daughter, and the two fawns of the season. Several females may form a lasting association in a given area. Adult males occur either alone or in small groups for most of the year. Groups have a dominance hierarchy maintained by aggressive postures and stares, rearing and flailing with the forefeet, and chases. Vocalizations include grunts and snorts used in agonistic situations. In the mating season, the males rub their facial scent glands against vegetation and urinate into scrapes beneath marked bushes and trees; such activity may serve to communicate with females or to intimidate other males (Marchinton and Hirth 1984). The males compete with one another through ritualized combat for the right to associate temporarily with the females. If neither of two rivals will retreat after initial displays, the opponents may lower their antlers, charge, and then shove and twist until one is driven back; such engagements usually do not last more than 15–30 seconds (Marchinton and Hirth 1984). A male does not attempt to gather a group of females or to defend a territory but associates with one doe until he mates with her or is displaced by another male. During the winter, animals of both sexes and all ages may aggregate in a favorable area. Such groups appear to be generally larger and more organized in *O. hemionus*. This "yarding" behavior may have evolved originally as a means of reducing wolf predation during the period of maximum vulnerability in deep snow (Marchinton and Hirth 1984).

In parts of South America breeding may take place throughout the year, though on the llanos of Venezuela there is a mating peak from February to May and a birth peak from July to November (Brokx 1984). In Canada and the United States mating occurs from October to January, usually peaking in November, and births take place from April to Septem-

ber. Females are seasonally polyestrous, with an estrous cycle of about 28 days and an estrus of 24 hours. The gestation period usually is 195–212 days (Banfield 1974; Halls 1978). Females usually give birth to a single fawn in their first litter and subsequently to two, or occasionally three or four. The beautifully spotted offspring weigh about 1.5–3.5 kg at birth and are able to stand after a few hours. They are left hidden in dense vegetation for 1 month or so and are nursed about every 4 hours by the mother. They nibble on vegetation after a few days, can run at 3 weeks, and usually are completely weaned at about 4 months. Young females may follow their mother for 2 years, but the males usually leave after 1 year. In a study of *O. virginianus* in Iowa, Haugen (1975) determined that 65–74 percent of young females successfully bred within their first year of life. Most other reports indicate that neither sex of *Odocoileus* usually mates until the second year. Wild individuals seldom survive more than 10 years, or 4.5 years if in a regularly hunted herd (Halls 1978). Record longevities for *O. hemionus* are 20 years for a wild individual and 22 years for a semicaptive (Anderson and Wallmo 1984). A pregnant female *O. virginianus* is estimated to have been 19–23 years old (Tullar 1983).

It is thought that there were 23–40 million individuals of *O. virginianus* and 5–13 million of *O. hemionus* in North America prior to the arrival of Europeans. These deer long had been hunted for subsistence purposes by the native Indians, with no apparent adverse effects on overall populations. The venison is said to be tender, juicy, and of excellent flavor. Buckskin, a leather originally tanned by the Indians, is made from the skin of *Odocoileus*. Following the arrival of European settlers, hunting, including massive commercial procurement of hides and venison, became so intense that legal protection was established in some areas in colonial times. Such measures were largely unsuccessful, and by 1900 both species had been exterminated in most of their range. At that time the number of individuals remaining in each species is estimated to have been about 300,000–500,000. Subsequent regulations, management efforts, and environmental changes stimulated a great increase in numbers and distribution. In some places, such as the Great Lakes region, where logging created favorable habitat, white-tailed deer apparently became far more numerous than they had been prior to European colonization. However, because of human habitat disturbances in the West, mule deer evidently have declined substantially since reaching all-time highs in the 1960s. Currently in the United States there are at least 14 million *O. virginianus* and 5.5 million *O. hemionus*. Respective annual kills by sport hunters are 2 million and 1 million (Banfield 1974; Downing 1987; Halls 1978; McCabe and McCabe 1984; Mackie 1987; Vogt 1978; Wallmo 1978).

Although there has been a general recovery of *Odocoileus* in the United States and Canada, there has been much human manipulation and transfer of animals from one area to another. Many of the existing populations do not represent the original native stock, and the fate of several restricted subspecies is unknown. There have been introductions of *Odocoileus* to several places completely beyond the original range, the most successful being of *O. virginianus* to New Zealand (Lever 1985). Native white-tailed deer have disappeared from much of Mexico and Central America and are continuing to decline there because of human habitat destruction, illegal hunting, and expanding agricultural activity (Méndez 1984). In South America the illegal kill vastly exceeds the legal take, and there also are concerns about loss of habitat to the growing human population (Brokx 1984).

Three North American subspecies are listed as endangered by the USDI. The tiny key deer *(O. virginianus clavium)* inhabits the western Florida Keys. Because of excessive hunting and loss of habitat, only 26 individuals were estimated to

A. White-tailed deer *(Odocoileus virginianus),* male in velvet, photo by Earl W. Craven, U.S. Fish and Wildlife Service. B. White-tailed deer *(O. virginianus),* two-year-old doe with fawn, photo by Rex Gary Schmidt, U.S. Fish and Wildlife Service.

Marsh deer *(Blastocerus dichotomus)*: Top, photo of mounted group in Denver Museum of Natural History; Bottom, photo from West Berlin Zoo.

survive in 1945. Full legal protection and establishment of a refuge allowed numbers to increase to an estimated 350–400 in the early 1970s. This apparent recovery often was held up as a model conservation success story. However, continued loss of habitat outside of the refuge, killing by motor vehicles, and other mortality associated with the rapid development of the Keys has resulted in a new reduction of the population to only 250–300 animals, and there is concern for its long-term survival (Humphrey and Bell 1986). The Columbian white-tailed deer *(O. v. leucurus),* which now is classified as rare by the IUCN, occurred throughout western Washington and Oregon before that region was settled and developed for agriculture. It now exists only on a national wildlife refuge and some nearby islands and lowlands along the lower Columbia River, where there are about 400 individuals, and in the Umpqua Basin of southwestern Oregon, where numbers recently have grown to about 3,000 (Smith 1985). There has been concern that hybridization with *O. hemionus* is threatening the genetic integrity of *O. v. leucurus* (Gavin and May 1988; Thornback and Jenkins 1982). The Cedros Island mule deer *(O. hemionus cerrosensis),* also considered rare by the IUCN, is confined to a small island off Baja California, where it has

been subject to illegal hunting and repeated burning of its habitat; Povilitis and Ceballos (1986) estimated that about 300 individuals remained.

ARTIODACTYLA; CERVIDAE; **Genus BLASTOCERUS** *Wagner, 1844*

Marsh Deer

The single species, *B. dichotomus,* originally occurred in the southern half of Brazil, Bolivia, southeastern Peru, Paraguay, northeastern Argentina, and Uruguay (Cabrera 1961; Thornback and Jenkins 1982). Some authorities, such as Grzimek (1975), place this species in the genus *Odocoileus.*

This is the largest deer of South America. Head and body length is 1,800–1,950 mm, tail length is 100–150 mm, shoulder height is 1,100–1,200 mm, and weight is 100–150 kg (Grzimek 1975). The coat is long and coarse. In summer the coloration is bright rufous chestnut. In winter it is brownish red, becoming lighter on the flanks, neck, and chest. The muzzle and lower legs are black. The tail is yellowish rusty red above and black below. The antlers of adult males are usually doubly forked, that is, each of the two branches has a single fork, for a total of four points. The hooves can be spread widely and are bound by a strong membrane in the inner part of the divergence point. This represents an adaptation against sinking into soft ground. According to Groves and Grubb (1987), *Blastocerus* resembles *Odocoileus* in its very broad ears and broad incisiform teeth but differs from that genus, and resembles *Ozotoceros,* in its shorter and bushier tail and more developed face glands.

Blastocerus prefers marshes and wet savannahs with high grass and wooded islands, and damp forest edges. Although its general distribution overlaps to a large extent with that of *Ozotoceros,* Eisenberg (1987) pointed out that it formerly occupied the riverine habitats and that its range was concentrated along the Rio Parana and Rio Paraguay and their tributaries. It sometimes concentrates in dense cover in wetlands and then spreads out into surrounding grassland during annual flooding (Thornback and Jenkins 1982). It enters water frequently. It remains in seclusion most of the day and comes out into the clearings to feed in the evening and through the

Marsh deer *(Blastocerus dichotomus)*, photo by Lothar Schlawe.

night. The diet consists of grass, reeds, and numerous aquatic plants.

Schaller and Vasconcelos (1978) found population densities ranging from one deer per 3.8 sq km to one per 42.0 sq km in the Mato Grosso. According to Grzimek (1975), *Blastocerus* lives alone or in groups of up to six individuals, which usually include an older male as well as two females and their young. It has been reported that the males do not fight for possession of the females and that there is not a definite season in which the antlers are dropped. Newborn fawns have been noted at various times of the year. In the Mato Grosso the birth season extends at least from May to September (Schaller and Vasconcelos 1978). The gestation period is about nine months, and usually a single young is born.

The marsh deer is classified as vulnerable by the IUCN and endangered by the USDI and is on appendix 1 of the CITES. Numbers and distribution have declined substantially through loss of habitat to agriculture, marsh drainage, uncontrolled hunting, and transmission of disease from domestic livestock. During the wet season, when the grasslands are flooded, the deer are forced onto a limited amount of high ground with livestock. There is resultant competition and thus persecution by stockmen. *Blastocerus* evidently has disappeared in Uruguay and is rare in Peru, Bolivia, Paraguay, and Argentina. There still are about 7,000 individuals in the Patanal, a wet grassland in southern Brazil, but numbers are declining there because of transmission of a disease from cattle that causes spontaneous abortion (Thornback and Jenkins 1982).

ARTIODACTYLA; CERVIDAE; **Genus OZOTOCEROS**
Ameghino, 1891

Pampas Deer

The single species, *O. bezoarticus*, is found in Brazil to the south of the Amazon, eastern Bolivia, Paraguay, Uruguay, and northern and central Argentina (Cabrera 1961). Some authorities, such as Grzimek (1975), place this species in the genus *Odocoileus*. The generic name *Blastoceros* Fitzinger, 1860, which is disliked because it may be confused with

Pampas deer *(Ozotoceros bezoarticus):* Top, photo of mounted group in Denver Museum of Natural History; Bottom, photo from West Berlin Zoo.

Blastocerus Wagner, 1844, has sometimes been used in place of *Ozotoceros.*

Head and body length is 1,100–1,400 mm, tail length is 100–150 mm, shoulder height is 700–750 mm, and weight is 25–40 kg. The prevailing color of the upper parts and limbs is reddish brown or yellowish gray, the face is somewhat darker, the underparts are white, and the tail is dark brown above and white below. There is no marked difference between summer and winter pelage (Jackson and Langguth 1987). Fawns are spotted. The lower, or front, prong of the main fork of the antler is not divided, but the upper, or posterior, prong is. The usual number of tines is three. Males have glands on the posterior hooves; these glands emit a strong odor that can be detected at a distance of 1.5 km. Groves and Grubb (1987) indicated that the relatively small size, narrow ears, and absence of black muzzle markings of *Ozotoceros* distinguish it from the related *Blastocerus.*

Ozotoceros was one of the most abundant and most widely distributed deer in South America but also was one of the least known. A series of recent investigations in Brazil, Uruguay, and Argentina have substantially increased our knowledge of this genus (J. E. Jackson 1985, 1986, 1987; Jackson and Langguth 1987; Redford 1987). The pampas deer occupies a variety of open grassland habitats at low elevations. These habitats include areas temporarily inundated by fresh or estuarine waters, rolling hills, and areas with winter drought and no permanent surface water. In some places the grass is high enough to completely conceal a standing deer, but much of the original range has been modified by agriculture and other human activity. Individuals are largely sedentary, there being no regular daily or seasonal movements. Nocturnal activity has been reported, but if not disturbed, animals will feed at regular intervals throughout the daylight hours. If startled, an individual will freeze if protected by cover, or it will flee with a characteristic stiff-legged run or bound away with the tail curled over the back to reveal the white underside. The exact diet still is unknown, but the animals have been seen to selectively graze on new green growth and also to browse on shrubs and herbs. One population estimated to comprise 1,000 deer inhabited a 131,000-ha. park in Brazil, 60 percent of which was grassland.

The social behavior of *Ozotoceros* closely parallels that of *Odocoileus.* Even when it was abundant the pampas deer lived in small groups rarely exceeding 5–6 individuals, and it still does so. Most observations are of solitary animals. Groups are fluid in size and composition, the members, especially the males, freely moving from one to another. Males mix with females throughout the year. Aggregations of up to 50 or more deer sometimes form on common feeding grounds. During the mating season adult males compete with one another for the estrous females. They thrash vegetation with their antlers and rub the scent glands of their head and face on plants and other objects. Aggression is demonstrated by thrusting with the antlers or flailing with the forefeet; fighting is frequent. Sometimes several males may harry a single doe. There is no evidence of territoriality or of lasting pair or harem formation.

Antlers are normally cast near the end of the austral winter, in August or September. A new set then begins to grow and is fully developed by December. The rut is concentrated in the summer and autumn. While births have been recorded throughout the year, there is a major peak from September to November. The gestation period is just over 7 months. A female about to give birth separates from her group and then also keeps her offspring apart from others. There is nearly always a single young. It begins to follow the mother and take some solid food after about 6 weeks and loses its spots by the age of 2 months. It may remain with its mother at least a year. Sexual maturity also comes at about a year.

Ozotoceros originally occurred in great numbers, especially on the pampas of Argentina and Uruguay. Its role in the culture of the Indians of that region was comparable to that played by the bison in the lives of the plains Indians of North America. The Indians, expert horsemen who dominated the Argentine pampas until about 100 years ago, initially joined in the commercial hunting of the deer. Over 2 million skins of *Ozotoceros* are estimated to have been exported from Argentina in the years 1860–70, and many more were taken in other countries. The genus remained common, however, until the defeat of the Indians opened the way for settlement and agricultural expansion. Numbers subsequently declined rapidly through loss of habitat, transmission of disease from domestic livestock, and continued uncontrolled hunting. The species *O. bezoarticus* is listed as endangered by the USDI and is on appendix 1 of the CITES. The subspecies *O. b. celer,* formerly found in much of Argentina, is designated as endangered by the IUCN. It now is restricted to a few remnant but protected populations on the coast and another about 1,000 km inland, which together contain approximately 500 individuals. The subspecies *O. b. leucogaster* is still found across much of its original range in the Mato Grosso, Bolivia, Paraguay, and part of northern Argentina, but population densities are low and declining. The third subspecies, *O. b. bezoarticus,* occurs in reduced numbers in southeastern Brazil; in Uruguay only about 1,000 individuals survive at nine isolated sites.

ARTIODACTYLA; CERVIDAE; **Genus HIPPOCAMELUS**
Leuckart, 1816

Guemals, or Huemuls

There are two species (Cabrera 1961; Thornback and Jenkins 1982):

H. antisensis, the Andes of Peru, western Bolivia, northeastern Chile, and northwestern Argentina (contrary to one report, there is no evidence of occurrence in Ecuador);
H. bisulcus, the Andes of central and southern Chile and Argentina.

Head and body length is 1,400–1,650 mm, tail length is 115–30 mm, shoulder height is 775–900 mm, and weight is 45–65 kg. The coarse and brittle pelage is longest on the forehead and tail. Each individual hair has a whitish base. Generally, the speckled yellowish gray-brown coloration of *H. antisensis* is uniform during all seasons and in both sexes. This species has a dark brown tail with a white undersurface, and there is a black Y-shaped streak on the face. *H. bisulcus* is colored much the same but is paler beneath. This species has a brown spot on the rump, and its tail has a brown underside. The males of both species possess small, simple antlers, which usually branch only once, the front prong being the smaller. Both sexes of both species have canine teeth similar to those of *Moschus* and *Hydropotes,* but the tusks of *Hippocamelus* do not project beyond the lips. Metatarsal glands are absent. Groves and Grubb (1987) noted that the limbs are relatively short, the ears are rather narrow, and there are large unvaginated face glands. Povilitis (1985) suggested that the short legs and stocky build reflect adaptation to broken terrain.

Huemuls inhabit hills, rugged country, and steep mountain slopes, *H. antisensis* mainly at elevations of 2,500–5,200 meters and *H. bisulcus* at 1,300–1,700 meters. They tend to live at higher altitudes in the summer, move down the mountains in the fall, and spend the winter in valleys. *H. antisensis*

Huemul *(Hippocamelus antisensis):* A. With antlers in velvet; B. With fully developed antlers. Photos by Heinz-Georg Klos through Zoological Garden Berlin-West.

has been found in open alpine grassland, while *H. bisulcus* has been observed mainly in dense shrub or forest habitat (Merkt 1987; Povilitis 1983*b*). Although *Hippocamelus* sometimes is said to be nocturnal, Roe and Rees (1976) observed all activity during the day in southern Peru. The diet in that area consisted mainly of grasses and sedges. Groups contained up to 8 individuals, included adults and yearlings of both sexes, and appeared to be led by a female.

In another study in southern Peru, Merkt (1987) also observed *H. antisensis* exclusively during daylight. Adults had year-round home ranges, though there were marked seasonal shifts in the segments utilized. Social organization was found to be unique among the Cervidae. Both sexes commonly joined in cohesive groups each day while engaged in foraging and other activity. Although particular adult individuals did not continue to associate for more than a few days, the groups may have been components of more stable subpopulations that occupied certain restricted localities. On the average, these groups consisted of 2.4 adult males, 3.9 adult females, and, if present, 2.8 young; maximum size was 31. There also were smaller, all-male units. Solitary animals were uncommon. During the fawning season females about to give birth moved away from the males.

The social structure of *H. bisulcus* is somewhat different (Povilitis 1983*a*, 1985). Home ranges are 36–82 ha. and are largely shared by one adult male and one adult female. This pair evidently persists throughout much of the year. Solitary individuals are common, and occasionally a male is seen with several females, but there are no all-female groups, and all-male units are unusual. Although *H. bisulcus* thus is unlike *Odocoileus virginianus* with respect to grouping pattern, the actual social behavior of the two species is much the same. Courting adult male *H. bisulcus* thrash vegetation with their antlers and scent-mark it by rubbing with their facial glands. They block, threaten, and spar with other males that approach; part of the display involves emitting a rapid, laugh-like vocalization.

The breeding season of *H. antisensis* in southern Peru is strongly seasonal (Merkt 1987; Roe and Rees 1976). Mating there takes place mainly in June and extends until August, the dry winter period. Most small fawns are seen subse-

quently toward the end of the rainy season from February to April. Gestation lasts approximately 240 days. The young are not spotted and are kept hidden until well after birth. A captive specimen of *H. antisensis* lived for 10 years and 7 months (Jones 1982).

Both species of *Hippocamelus* are listed as endangered by the USDI and are on appendix 1 of the CITES. The IUCN classifies *H. antisensis* as vulnerable and *H. bisulcus* as endangered. The former is still widely but sparsely distributed at high altitudes in the northern Andes; its numbers are unknown but are declining, principally because of hunting by people for use as food, habitat destruction, and competition with domestic livestock (Thornback and Jenkins 1982). *H. bisulcus* has declined for the same reasons and also through competition with introduced *Cervus elaphus* and through exposure to attack by domestic dogs as a result of human destruction of vegetative cover. Formerly occurring all along the Andes south of 34° S in Chile and 40° S in Argentina, it now is restricted mainly to a few localities in the Aysen region of southern Chile. There are a few others in adjacent Argentina and also small herds at two protected sites farther north, one on each side of the border. The total number of individuals of *H. bisulcus* remaining has been estimated at 1,300 (Povilitis 1983*b*).

ARTIODACTYLA; CERVIDAE; Genus MAZAMA
Rafinesque, 1817

Brocket Deer

There are four species (Cabrera 1961; Hall 1981):

M. americana, eastern Mexico to northern Argentina;
M. gouazoubira, Isla San Jose off southern Panama, Colombia and Venezuela to northern Argentina and Uruguay;
M. rufina, the Andes from western Venezuela to Ecuador, southeastern Brazil, Paraguay, northeastern Argentina;
M. chunyi, the Andes of southern Peru and Bolivia.

Brocket deer (*Mazama* sp.): Top, photo of mounted group in Denver Museum of Natural History; Bottom, photo from Gladys Porter Zoo, Brownsville, Texas.

Head and body length is 720–1,350 mm, tail length is 50–200 mm, and shoulder height is 350–750 mm. Weight usually has been reported to be about 8–25 kg, but Branan and Marchinton (1987) found male *M. americana* occasionally to attain weights as great as 65 kg in Surinam. The hair on the face radiates in all directions from two whorls. The pelage of the adult is generally uniformly colored, with the species varying in color from light to dark brown. Usually, the body is bright reddish brown above and lighter below, and the underside of the tail is white. Some species are especially light-colored. The body is stout, the limbs are slender, and the back is arched. Although the antlers vary in size, they lack a brow tine and are usually simple spikes in all species. The upper canines may be present or absent. The metatarsal gland is absent. Females have four mammae.

Brocket deer are usually found in woodlands and forests

from sea level to elevations of 5,000 meters. They are relatively sedentary, sometimes staying in an area only a few hundred meters in circumference. Diurnal, nocturnal, and crepuscular activity has been reported. These small deer are seldom seen, however, because of their shyness, protective coloration, and habit of "freezing" when danger is sensed. If they are flushed, they frequently scamper a short distance and stop to look back at the pursuer. They do not seem to have the endurance of some other kinds of deer and can be overtaken and killed by a common dog. They enjoy water and are good swimmers. The diet includes many kinds of plants, some of the preferred items being grasses, vines, and tender green shoots. Stallings (1984) indicated that fruits and seeds are important for *Mazama* and found that in the Chaco of Paraguay *M. gouazoubira* feeds heavily on hard, dry fruits during the dry season and soft, fleshy fruits during the wet season. Branan, Werkhoven, and Marchinton (1985) reported fungi to be the most abundant item in the diet of *M. americana* in Surinam.

A population density of approximately one *M. americana* per sq km has been reported for the interior forests of Surinam (Branan and Marchinton 1987). Brocket deer are usually solitary, the sexes coming together only briefly during courtship (Grzimek 1975). According to Gardner (1971*b*), breeding of *M. americana* continues throughout the year in Peru, but mating peaks from July to September, and most young fawns are found from February to April. A female taken on 30 July in Peru was both lactating and pregnant, which suggests that a postpartum estrus had occurred. Continuous reproduction and a postpartum estrus also have been reported in other areas, including for *M. gouazoubira* (Stallings 1986). In Surinam, Branan and Marchinton (1987) found that *M. americana* mates mainly from April to July, during the longer of the two wet seasons there, and that fawning spanned a period of at least 8 months centered in December and January, in the shorter wet season. The gestation period is about 225 days in *M. americana* and 206 days in *M. gouazoubira* (Barrette 1987*a*). A single young is usually produced, and weight at birth is 510–67 grams (Grzimek 1975; Thomas 1975). The youngest pregnant female examined by Branan and Marchinton (1987) was estimated to be 13 months old. A captive *M. americana* lived for 13 years and 10 months (Jones 1982).

Brocket deer are intensively hunted by people for use as food and because they frequently damage bean and corn crops. Leopold (1959) noted, however, that *Mazama* is far superior to *Odocoileus* in its ability to avoid human hunters.

ARTIODACTYLA; CERVIDAE; Genus PUDU
Gray, 1852

Pudus

Except as noted, the information for this account was taken from Hershkovitz (1982). There are two species:

P. mephistophiles, the Andes of Colombia, Ecuador, and central Peru;
P. puda, southern Chile, southwestern Argentina.

P. mephistophiles sometimes is placed in a separate genus or subgenus, *Pudella* Thomas, 1913.

These are the smallest deer. Head and body length is 600–825 mm, tail length is 25–45 mm, shoulder height is 250–430 mm, and weight is 5.8–13.4 kg. The hairs of the coat are long, coarse, and brittle. *P. puda* is generally buffy agouti in color; the middle of the back is reddish brown agouti; the face is reddish buffy or brown; and the ears, narial patch, chin, underparts, and limbs are reddish. *P. mephistophiles* is generally buffy agouti; the middle of the back is dark brown; the underparts are buffy to rufous; the face, outer surface of the ears, chin, and feet are dark brown to blackish; and the sharply defined ears are whitish, grayish, or pale buff. The fawns of *P. puda* have white spots, whereas fawns of *P. mephistophiles* are not spotted.

The genus *Pudu* is characterized by a low-slung body, thick and short legs, slender hooves, comparatively small and rounded ears, small eyes, and a thick rhinarium. Both species possess short (less than 100 cm), spikelike antlers; tusks do not occur in the upper jaw; and an external tail is practically lacking. *P. mephistophiles* is characterized by relatively large size, a small preorbital gland, a shallow or dish-shaped lacrimal fossa, a somewhat bulbous rhinarium, an expanded supe-

rior border of the ascending ramus of the premaxillary bone, and comparatively long and narrow hooves. *P. puda* is smaller and has a large preorbital gland, a deep or bowl-shaped lacrimal fossa, a rhinarium that is not notably swollen, an ascending ramus of the premaxillary that is not expanded and that sometimes does not reach the nasal bone, and comparatively short hooves.

P. puda occurs in humid forests from sea level to an elevation of about 1,700 meters, and *P. mephistophiles* inhabits temperate-zone forests and fringing grasslands at 2,000–4,000 meters. Little is known about the latter species, and its current distribution in the high mountains may represent a last retreat rather than the original prime habitat. *P. puda* is considered to be a sedentary, cryptic species, its movements localized and likely limited to the narrowly circumscribed territory of the individual or family group. There is no evidence of seasonal migration. Although *P. puda* usually is reported to be nocturnal, Eldridge, MacNamara, and Pacheco (1987) found it to be active both by day and by night, mostly in the early morning, late afternoon, and evening. When pursued it runs in a zigzag pattern, it may climb trunks of trees that are inclined over streams or bluffs, and it is said to be very nimble among rocks. It is mainly a browser of many kinds of trees, shrubs, and vines but also eats bark, twigs, buds, blossoms, fruits, berries, nuts, acorns, and cultivated vegetables.

In the forests of Chile *P. puda* has been found to utilize home ranges of 16–26 ha. (Eldridge, MacNamara, and Pacheco 1987). Early accounts suggest that this species occurred in small herds, but modern reports indicate that it is solitary except in the breeding season. *P. mephistophiles* has been found only alone or in pairs and is said to have a whistling vocalization. Female *P. mephistophiles* with nearly full-term fetuses have been taken in Peru in April and November. Wild *P. puda* mates in the fall and gives birth in the spring (November–January). There is a postpartum estrus, the gestation period lasts 202–23 days, and there seems always to be a single fawn. The young weighs less than 1 kg at birth and is nearly full-grown after 3 months. Females reach sexual maturity at 6 months, but males may not do so until 18 months. One specimen of *P. puda* was still living after 12 years and 6 months in captivity (Jones 1982).

Human activity has caused an enormous shrinkage of the natural range of *Pudu* since the eighteenth century. The species *P. puda* is listed as endangered by the USDI and is on appendix 1 of the CITES, though Miller et al. (1983) reported that a substantial population survives on Chiloe Island. *P. mephistophiles* is classified as indeterminate by the IUCN and is on appendix 2 of the CITES. It has been relentlessly hunted throughout its range, and its habitat is being destroyed through burning and agricultural usurpation (Thornback and Jenkins 1982).

ARTIODACTYLA; CERVIDAE; Genus ALCES
Gray, 1821

Moose (North American term), or Elk (European term)

The single species, *A. alces*, originally occurred from northern Europe and the Caucasus to eastern Siberia and from Alaska to northern Colorado and the northeastern United States (Corbet 1978; Hall 1981; Pulliainen 1974).

This is the largest member of the deer family. Head and body length is 2,400–3,100 mm, tail length is 50–120 mm, shoulder height is 1,400–2,350 mm, and weight is 200–825

Pudu: Top, *Pudu puda,* photo by Lothar Schlawe; Bottom, *P. mephistophiles,* photo from West Berlin Zoo.

Alaskan moose *(Alces alces)*, photo by Bernhard Grzimek.

kg. The summer pelage is dark above, varying from black to dark brown, reddish brown, or grayish brown, and the underparts and lower legs are lighter. The winter pelage is grayer. The young are reddish brown, and unlike most immature members of this family, they are not spotted. *Alces* is easily identified by its broad overhanging muzzle, massive antlers, heavy mane, and characteristic pendulant flap of skin beneath the throat, known as the "bell." Franzmann (1981) noted that the maximum recorded antler spread is 2,048 mm.

The moose generally occurs in wooded areas that have a seasonal snow cover (Franzmann 1981). A favorite habitat is a moist area with abundant willows and poplars. Although vision is poorly developed, the senses of smell and hearing are acute. According to Banfield (1974), the moose normally walks carefully and quietly through the underbrush, can reach a speed of 56 km/hr, and is a strong swimmer. Activity occurs throughout the day, especially in the winter, but there are definite peaks at dawn and dusk. Some populations make regular seasonal movements between areas of favorable food supplies. Several distinct home ranges, separated by a considerable distance, are sometimes used in the course of a year. Migrations reportedly cover up to 179 km in North America

and 300 km in northeastern Europe (LeResche 1974; Pulliainen 1974).

The moose frequents marshy and timbered regions in search of the shrubs and trees on which it browses. It also feeds on water vegetation by wading into lakes and streams, often submerging entirely to obtain the roots and stems of plants on the bottom. By necessity, it eats twigs and the bark of trees and paws through the snow for small plants in the winter. Franzmann (1981) stated that *Alces* depends mainly on woody vegetation in an early developmental stage and requires 19.5 kg of food per day.

Normal population densities reportedly vary from about 0.1/sq km to 1.1/sq km, but local concentrations sometimes reach 100/sq km to 200/sq km (Bouchard and Moisan 1974; Filonov and Zykov 1974; Krefting 1974; Peek, LeResche, and Stevens 1974; Syroechkovskii and Rogacheva 1974). An individual seasonal home range generally covers 2.2–16.9 sq km, but much larger areas may be used in the course of a year or a lifetime (Franzmann 1981; LeResche 1974). Many moose may aggregate in a small, favorable area during late fall or winter (Peek, Urich, and Mackie 1976), but the genus is essentially solitary. In the breeding season, males compete for

North American moose *(Alces alces)*, juvenile, photo by Eugene Maliniak.

one female at a time, engaging in elaborate displays, shoving matches with the antlers, and occasionally serious fighting. The female also takes an active role at this time, moving about independently, attempting to attract males through sound and scent, and behaving aggressively toward others of her sex. The mating call of the female moose is a long moan, while that of the male is described as a "croak" (Franzmann 1981; Lent 1974).

Mating occurs in September and October, and births take place the following spring. Females are seasonally polyestrous, having an estrous cycle of 20–22 days and a true estrus of less than 24 hours. A litter is usually produced every year. The gestation period is 226–64 days. The number of young per litter is most often one, twins are common, and triplets are rare. The calf weighs 11–16 kg at birth, is able to browse and follow the mother after 2–3 weeks, and is completely weaned by the fifth month of life. It remains with the female at least 1 year, is driven away when the mother gives birth again, and may subsequently rejoin the family for some additional months. Both sexes are physiologically capable of mating in their second year of life. Females often do so at this time and continue to breed until they are 18 years old, but their maximum reproductive potential is reached between the ages of 4 and 12 (Franzmann 1978, 1981; Stringham 1974; Verme 1970). Maximum known longevity is 27 years (Peterson 1974b).

Because of uncontrolled hunting by people for the meat, leather, and bone of *Alces*, the distribution of the genus has been reduced substantially. By the thirteenth century A.D. it had disappeared from western Europe except for the Scandinavian Peninsula, by the early nineteenth century it was gone from the Caucasus, and by 1900 it had been all but wiped out in the conterminous United States. Subsequent measures for protection and management allowed the moose to rebuild its numbers in some regions in which it had been depleted. Small populations also have returned to New England, the upper Great Lakes region, and the central Rocky Mountains. Human habitat modifications, such as burning and logging, sometimes have benefited the moose by providing more abundant food and seem to have stimulated a northward range expansion in Alaska and parts of Canada. *Alces* was introduced by people on the island of Newfoundland, and a large population is now present there. Available numerical estimates suggest that there are about 900,000 moose in North America and 1 million in Eurasia (Banfield 1974; Coady 1980; Filonov and Zykov 1974; Hall 1973; Kelsall 1972, 1987; Krefting 1974; Markgren 1974; Mercer and Manuel 1974; Pulliainen 1974; Rausch, Somerville, and Bishop 1974; Ritcey 1974; Vogt 1978).

The moose is considered to be a game animal in most parts of its range and is subject to limited sport and subsistence hunting. The most productive populations in this regard are those of Sweden, where 152,000 individuals were killed in 1981. This harvest represented 14 percent of the total meat consumption of Sweden (Wilhelmson 1983). The annual kill in North America is about 72,000 (Kelsall 1987). In the Soviet Union, *Alces* has been domesticated for the production of meat and milk, as well as for use as a farm draft animal (Grzimek 1975; Knorre 1974). A small free-roaming population of moose was established through human agency in New Zealand in the early twentieth century but may now have died out (Lever 1985).

Caribou *(Rangifer tarandus)*, photo of mounted group in Field Museum of Natural History.

ARTIODACTYLA; CERVIDAE; **Genus RANGIFER**
Hamilton-Smith, 1827

Caribou (North American term), or Reindeer (European term)

In historical time, the single species, *R. tarandus,* occurred in Ireland, Scotland, Scandinavia, Germany, Poland, Svalbard, Russia north of 50° N, Mongolia, northeastern China, Sakhalin, Alaska, Canada, the extreme northern conterminous United States from Washington to Maine, and Greenland (Banfield 1961; Corbet 1978; Hall 1981; Smit and Van Wijngaarden 1981). In the late Pleistocene the species was found as far south as Nevada and Tennessee (Guilday, Hamilton, and Parmalee 1975). It apparently still was present in southern Idaho in the early nineteenth century (Meatte 1986).

Head and body length is 1,200–2,200 mm, tail length is 70–210 mm, shoulder height is 870–1,400 mm, and weight is 60–318 kg. Average weight in Canada is 110 kg for males and 81 kg for females (Banfield 1974). As protection against the rigorous climate in which it lives, the caribou has a heavy coat of woolly underfur and straight, stiff, tubular guard hairs. Coloration varies widely, but most individuals are predominantly brownish or grayish above and have white or pale underparts, inner legs, and buttocks. The winter pelage is somewhat paler. Some animals are dark brown or almost black, and some are quite pale. The subspecies *R. t. pearyi,* of northern Greenland, Ellesmere Island, and nearby islands, is almost pure white.

Rangifer is the only genus of the deer family in which both sexes are antlered. Although there is great diversity in the shape of the antlers, they are usually long, sweeping beams with forwardly projecting brow tines. Length is 520–1,300 mm in males and 230–500 mm in females (Banfield 1974). The broad, flat, and deeply cleft hooves aid in walking on soft

ground and snow. When the caribou walks, a clicking noise is produced by a tendon slipping over a bone in the foot.

Most caribou inhabit arctic tundra and surrounding boreal coniferous forest. Some populations are found in mountainous areas farther south. *Rangifer* is active primarily during the day and is almost constantly on the move. Its maximum running speed is 60–80 km/hr. Its eyesight is apparently poor, its hearing is adequate, and it depends in large part on the sense of smell to detect food and danger. Most populations have seasonal shifts in range. Those to the south may simply move to lower elevations for the winter or to areas of favorable food supplies and shelter. Northern populations, however, make extensive spring and fall migrations, sometimes traveling over 1,000 km between the summer range on the tundra and the wintering grounds in timbered areas. The rate of movement during migration is 19–55 km per day (Banfield 1974; Bergerud 1978). The populations on the arctic islands of Canada may make seasonal inter-island movements of up to 450 km (Miller, Russell, and Gunn 1977). Tracking of caribou in Alaska and the Yukon by satellite telemetry recently demonstrated that some individuals had minimal annual movements of up to 5,055 km, the longest documented for any terrestrial mammal (Fancy et al. 1989). In contrast, a nonmigratory population of woodland caribou in Manitoba had a home range that varied seasonally from about 100 to 200 sq km (Darby and Pruitt 1984). The diet of *Rangifer* comprises a wide variety of plants, including new growth in the spring, green leaves of deciduous shrubs in the summer, lichens and evergreen leaves in the fall, and fine twigs and greens dug through the snow in the winter (Bergerud 1978).

The original overall population density in North America is estimated to have been 0.4/sq km to 0.6/sq km. The maximum known density within the regular range of a particular herd is 1.5/sq km to 1.9/sq km. During migration, however, when a herd concentrates, there may be up to 19,000 animals per sq km. *Rangifer* is highly gregarious. The animals in a

Reindeer *(Rangifer tarandus)*: Top, photo by Bernhard Grzimek; Bottom, photo from Antwerp Zoo.

given region of the world are divided into populations, or herds, which move about as a unit, keep to the same general ranges and migration routes over the years, and do not mix extensively with other herds. About 100 discrete herds have been identified in North America. The smallest, that of northern Idaho and adjacent areas, contained only about 25 individuals prior to recent human manipulation. The 8 largest herds occur in a sequence all across the northern parts of Alaska and Canada. Although size fluctuates, in recent years

each of these herds has generally comprised 50,000–500,000 individuals. At most times the members of a large herd are divided into groups of 10–1,000 animals each, often with only one sex represented in the smaller groups. Typically, a large herd comes together for the spring migration, the females separate to give birth in late May and June, there is another concentration immediately after fawning, the animals disperse into small groups for the summer, the herd again masses for the fall migration, and there is dispersal for the winter (Banfield 1974; Bergerud 1978). Within the small groups there appears to be a dominance hierarchy, based mainly on antler size and aggressive interaction, that determines access to food resources (Barrette and Vandal 1986). Vocalizations include the snort of an alarmed adult, the bawl of a fawn, and the grunting roar of a rutting male.

In the mating season, males compete and fight for possession of females. Each male may either attempt to control a harem of 5–15 females or simply move about and try to mate with as many females as possible. Females are seasonally polyestrous, enter heat at intervals of 10–24 days, are receptive for about 50 hours, and usually give birth once a year. Most mating occurs in October, and most young are born in late May and early June. The gestation period averages about 227–29 days. There is almost always a single offspring, but twins have been reported. The young weighs 5–9 kg at birth, is not spotted, and is remarkably precocious. It is able to follow its mother after 1 hour of life and can outrun a human when 1 day old. It nurses for at least 1 month and sometimes until the winter. Sexual maturity is usually attained at between 17 and 41 months. In the wild, average longevity is 4.5 years and the maximum known is 15 years (Banfield 1974; Bergerud 1978; Sadleir 1987). A captive lived for 20 years and 2 months (Jones 1982).

Rangifer is hunted extensively by people for its meat, skin, antlers, and other parts. The genus survived in Germany until Roman times, in the British Isles until the Middle Ages, and in Poland until the sixteenth century (Banfield 1961).

Small populations still exist in Scandinavia (25,000) and the European part of the Soviet Union (20,000), but the range is shrinking (Smit and Van Wijngaarden 1981). The total number of wild reindeer in the Soviet Union is estimated to have been about 5 million originally but fell to only 300,000 by 1940. Under intensive management, there subsequently was a recovery to about 900,000 individuals, about half of which are in the Taimyr region of northwestern Siberia, but these animals are confronted by such problems as hostility from domestic reindeer herders and blockage of migratory routes by gas pipelines and icebreaker activity (Andreev 1978; Klein and Kuzyakin 1982; Syroechkovskii 1984). There are a few very small remnant herds in northern Mongolia and Manchuria (Ma 1983b; Mallon 1985).

In North America, uncontrolled hunting led to the disappearance of the caribou in much of the southern portion of its range by the early twentieth century. The subspecies *R. t. dawsoni*, of the Queen Charlotte Islands, British Columbia, was completely exterminated (Hall 1981). The woodland caribou of the mainland forests south of the tundra has been reduced to scattered herds now containing perhaps 20,000 animals. The near total loss of the herds in the northeastern United States and in Canada south of the St. Lawrence may be associated with human habitat modifications allowing a northward expansion of the white-tailed deer *(Odocoileus virginianus)*. The latter often carries a parasite that is harmless to the deer but that can be transmitted with lethal effect to the caribou (Burnett et al. 1989). The only population that still regularly occurs in the conterminous United States is the Selkirk herd, which wanders through the mountains of northern Idaho and adjacent parts of Washington and British Columbia and which had declined to about 25 members because of logging and other human disturbances in its habitat. This group was classified as endangered in 1983 by the USDI, which then, strangely, carried out large-scale transfers of caribou from other areas into the range of the Selkirk herd, thereby probably destroying the genetic integrity of the original endangered population. Mech and Nelson (1982) reported a few sightings of caribou in Minnesota in 1980–81.

The great barren ground herds of northern Canada declined drastically after 1900 as human activity increased in their ranges and the native people obtained modern firearms. The total number of caribou in North America is estimated to have been 3.5 million originally and 1.1 million in 1977 (Bergerud 1974, 1978). Surprisingly, there since has been a substantial increase, with total numbers reaching 2.3 million to 2.8 million; the largest herd, with some 600,000 members, is now in northern Quebec and Labrador (Williams and Heard 1986). There has been much concern, however, about how the remaining herds may be affected by oil and gas exploitation, blockage of migratory routes by transportation corridors, and other environmental modifications in the Arctic (Bergerud, Jakimchuk, and Carruthers 1984). Excessive and improperly regulated hunting by people also continues to be a problem (Miller 1987). Mostly for that reason, the largest herd in the United States, that of northwestern Alaska, declined from 242,000 individuals in 1970 to a minimum of 75,000 in 1976. Stricter limits on hunting contributed to an increase to about 113,000 by 1979 (Davis, Valkenburg, and Reynolds 1980) and to over 200,000 by 1984 (Williams and Heard 1986). The unique white, island-hopping subspecies, *R. t. pearyi*, has declined to only 3,300–3,600 individuals in the Canadian Arctic Archipelago and is considered to be threatened with extinction (Burnett et al. 1989; Gunn, Miller, and Thomas 1981). Caribou also have disappeared from eastern and northwestern Greenland, and the population in southwestern Greenland fell from around 100,000 in 1970 to only 8,000 by 1980 (Roby, Thing, and Brink 1984).

The domestication of reindeer may have begun in the Old World about 3,000 years ago and subsequently spread all across northern Eurasia. Domestic animals are generally allowed to live in herds and to move about freely, but selective breeding is practiced to some extent. Domestic reindeer have been introduced in Iceland, the Orkney Islands, Scotland, South Georgia, the Kerguelen Islands, Alaska, Canada, and Greenland (Banfield 1961; Smit and Van Wijngaarden 1981). The Alaskan animals, originally brought over in 1891, numbered around 600,000 in 1925 but subsequently declined to about 25,000 (Lever 1985). Currently, there are about 3 million domestic reindeer in the world, most of them in the Soviet Union. The annual production in that country is 32,000 tons of meat and 650,000 hides (Andreev 1978).

ARTIODACTYLA; CERVIDAE; **Genus CAPREOLUS**
Gray, 1821

Roe Deer

There are two species (Danilkin 1986; Sokolov et al. 1985):

C. capreolus, Great Britain and Spain to the Volga River and northern Iran;
C. pygarus, Volga River to southeastern Siberia, Korea, and central China.

C. pygarus usually has been considered a subspecies of *C. capreolus* (Corbet 1978; Corbet and Hill 1986; Honacki, Kinman, and Koeppl 1982), but Groves and Grubb (1987) recognized the specific distinction of the two.

Head and body length is 950–1,510 mm, tail length is about 20–40 mm, shoulder height is 650–1,000 mm, and weight is 15–50 kg. Coloration varies with the season. In summer the coat is reddish above, the ears are black, and the underparts are white. In winter the general color is buff or dark grayish brown, the throat and rump patch are white, and the face, chest, and legs are tawny. *Capreolus* is small and graceful. The antlers are slightly roughened at the base, only about 230–380 mm long, and seldom more than three-tined. The tail is inconspicuous.

Roe deer are found in forests, sparsely wooded valleys, open fields, and agricultural areas, usually at elevations not exceeding 2,400 meters. They are found in almost any area that furnishes a reasonable amount of cover and are able to live in parklike places among dense human populations. In the early morning and late evening, they emerge from cover and may seek open grassland, where they graze. Most populations are sedentary, remaining in relatively small areas throughout the year, but in the Soviet Union there are reported to be migrations of up to 300 km to traditional wintering sites with shallow snow cover (Stüwe and Hendrichs 1984). *Capreolus* is shy, but curious, and has well-developed senses. It is an excellent swimmer. The diet consists of grass, herbs, and cultivated crops.

Fruzinski, Labudzki, and Wlazelko (1983) found the most favorable habitat in Poland to be coniferous forest with dense thickets and adjacent productive fields; population density there averaged 16.1 individuals per 100 ha. Reported densities in other parts of Europe have been as high as 34 per 100 ha., though there generally is a decline during the winter. Also in Poland, Pielowski (1984) found annual home range to average 151 ha. for males and 141 ha. for females. Lifetime ranges were only slightly greater, 168 ha. for males and 219 ha. for females, thus indicating the sedentary nature of these animals.

Capreolus lives alone or in small groups, but aggregations of up to 100 animals frequently form at favorable feeding

Roe deer *(Capreolus capreolus)*, female with fawn born in the spring and male with winter coat and antlers, photos from Zoological Garden Berlin-West.

sites, especially during the fall and winter or in the course of migrations. Groups are unstable in membership, but the individuals in a given area have a dominance hierarchy (Bresinski 1982; Stüwe and Hendrichs 1984). During the spring and summer, when the animals live alone or in doe-fawn pairs, adults of both sexes establish defended territories. If two animals of the same sex meet at such times, they may fight. Reported size of territories is 7–25 ha. for males and 3–180 ha. for females; these areas compose only a portion of the total home range (Fruzinski, Labudzki, and Wlazelko 1983). Unlike most deer, male *Capreolus* usually escort only one female. They will savagely fight any intruding male. During the mating season, males often chase females in circles, leaving a track referred to as a "witch circle." Such chases also occur in other seasons, sometimes between individuals of the same sex. The roe deer utters a barking sound when disturbed. Females emit a screeching sound during the breeding season and to call fawns.

Mating usually occurs in July and August, and implantation of the fertilized egg in the uterus is subsequently delayed about 4 months. A few females, however, do not mate until November or December, do not experience delayed implantation, and have a total gestation of about 5.5 months (Grzimek 1975). Births occur in the spring. When about to bear its young, the doe chases away the offspring of the previous season and then retires to the forest, returning in about 10 days with the new family. The number of young per birth is usually two, occasionally one or three. Twins are born and suckled, not at the same spot, but usually 10–20 meters apart. The fawns have three longitudinal rows of white spots. Sexual maturity is attained at about 1 year and 4 months. *Capreolus* is delicate in captivity. The average life span of 11 specimens in the London Zoo was 40 months, and the maximum was 7 years. Normal longevity in the wild is 10–12 years, and the maximum may be 17 years.

Roe deer have adapted well to civilization. Their distribution and numbers seem to have increased in Europe; they occur almost throughout the continent, having recently extended their range northward in Scandinavia by several hundred kilometers (Danilkin 1986). Populations were reintroduced successfully in England after having been eliminated there by the eighteenth century (Lever 1985). About 500,000 individuals are killed annually in West Germany, but the population there remains stable (Grzimek 1975).

ARTIODACTYLA; Family GIRAFFIDAE

Okapi and Giraffe

This family of two Recent genera, each with a single species, was found over most of Africa in historical time. The sequence of genera presented here follows that of Simpson (1945), who recognized two subfamilies: Palaeotraginae, with the genus *Okapia*; and Giraffinae, with *Giraffa*. Geraads (1986) considered the Palaeotraginae to be part of the Giraffinae.

Giraffids have large eyes and ears, long and thin lips, and an extensible tongue. The neck, legs, and terminally tufted tail are long. The back inclines upward from the loins to the withers. The feet are large and heavy. There are two hoofed digits, the third and the fourth; the lateral digits are not developed. The stomach is four-chambered and ruminating, and there is no gall bladder. Females have two or four mammae.

The horns of the Giraffidae are unlike those of any other mammal. They are present at birth as cartilaginous knobs, which rapidly ossify and grow slowly throughout life. They consist of a bony core that at first is separate from, but later fuses with, the skull. In *Giraffa* they arise over the anterior part of the parietal bones behind the eyes, the forward base growing over the frontoparietal suture. In *Okapia* they arise on the frontals above the orbit. They are covered with skin and hair throughout life; the hair is worn away from the apex, but the skin is not, contrary to the common supposition. Horns occur in both sexes, but growth is less vigorous in the female. *Giraffa* also bears a median horn on the forepart of the frontal bones and the back of the nasal bones that develops to a greater or lesser extent and is identical in its mode of growth with the posterior horns.

The skull of *Giraffa* is characterized by large pneumatic sinuses above the nasal bones and cranium. These are developed to a much lesser extent in *Okapia* and do not extend back over the cranium. The dental formula is: (i 0/3, c 0/1, pm 3/3, m 3/3) × 2 = 32. The low-crowned molars are characteristically rugose in giraffids, being unlike those of all other mammals, in which the enamel is always smooth. The upper molars lack inner accessory columns.

The geological range of this family is middle Miocene to

Giraffe *(Giraffa camelopardalis)*, photo by James R. Anderson.

Pleistocene in Europe and Asia and early Miocene to Recent in Africa (Simpson 1984). *Sivatherium* was a large, heavily built member of this family known from the Pleistocene of southern Eurasia and Africa. The male often developed a variety of horns, frequently being four in number and greatly branched. This creature was more bovine in appearance than

the Recent giraffe, both the neck and legs being shorter than those of *Giraffa*.

ARTIODACTYLA; GIRAFFIDAE; **Genus OKAPIA**
Lankester, 1901

Okapi

The single species, *O. johnstoni*, is known to occur in the equatorial forests of northern, central, and eastern Zaire (Ansell, *in* Meester and Setzer 1977). There is evidence that the species also formerly entered western Uganda (Kingdon 1979).

Head and body length is about 1,970–2,150 mm, tail length is about 300–420 mm, shoulder height is 1,500–1,700 mm, and weight is about 200–250 kg. The general coloration is purplish, deep reddish, or almost black. The sides of the buttocks and upper portions of the limbs are transversely barred with black and white stripes of varying width. The shanks are white to maroon, and the facial markings are light. The hair is short and sleek, and the tail has a terminal tuft. The body is short and compact, the neck and legs are long, the ears are relatively large, and the eyes are large and dark. The tongue is so long that it can be used to clean the eyes. Males have small, hair-covered horns.

The preferred habitat of the okapi is dense, damp forest. It is diurnal and normally moves along well-trodden paths through the jungle. It is extremely wary and secretive, dashing through the forest at the least suspicion of danger. Hear-

Front foot of a giraffe *(Giraffa camelopardalis)*, from photo by Constance P. Warner.

Okapis *(Okapia johnstoni)*, photo by Bernhard Grzimek.

ing is the best-developed of its senses. It is a browser, feeding on the leaves, fruit, and seeds of many plants. Kingdon (1979) noted that the prehensile tongue is used to pluck leaves, buds, and even small branches. Hart and Hart (1988) reported that each individual has a home range of about 2.5 sq km and moves about 1 km per day as it forages.

Okapia is found alone, in pairs, or in small family parties but never in herds. Estimates of population density range from about 0.8/sq km to 2.3/sq km, communication seems to involve olfactory marking with pedal scent glands and urine, and males have a ritualized neck fight (Kingdon 1979). The young are born from August to October, the period of maximum rainfall. The single offspring is about 790 mm tall and weighs 16 kg at birth, and it starts to nurse within 6–12 hours. It is precocious, is suckled and defended by any postpartum female if penned with a group, and is weaned by 6 months (Kingdon 1979). According to Gijzen and Smet (1974), estimated gestation is 425–91 days. Females attain sexual maturity at 1 year and 7 months. One captive female gave birth to 12 calves, the last when she was 26 years old. One captive okapi is estimated to have lived 33 years (Jones 1982).

Although the okapi was long hunted by the pygmies of Zaire, it did not become known to science until 1900. The first live specimen reached Europe in 1918. Generally considered to be rare, the okapi has been protected in Zaire since 1933, but its densely forested habitat and elusive nature may conceal its true status (Grzimek 1975).

ARTIODACTYLA; GIRAFFIDAE; **Genus GIRAFFA**
Brünnich, 1772

Giraffe

In historical time, the single species, *G. camelopardalis*, occurred in most of the open country of Africa. For about the

last 1,400 years, the species has been restricted to areas south of the Sahara (Ansell, *in* Meester and Setzer 1977; Dagg 1971).

This is the tallest living terrestrial animal. Average height is 5,300 mm for males and 4,300 mm for females; the record is 5,880 mm. In seven males, head and body length was 3,810–4,724 mm and tail length was 787–1,041 mm (Dagg and Foster 1976). Shoulder height is 2,500–3,700 mm, and weight is 550–1,930 kg. Average adult weight is 800 kg (Dagg 1971). The color pattern varies but consists essentially of dark reddish to chestnut brown blotches of various shapes and sizes on a buff ground color. The underparts are generally light and unspotted. The coloration darkens with age. The long neck is maned with short hair, and the long tail is terminally tufted.

Like most other mammals, the giraffe has seven neck vertebrae, but they are greatly elongated. Both sexes possess two to four blunt, short, hornlike structures on top of the head. In certain forms there is another protuberance, sometimes only a knob, situated more or less between the eyes; such a giraffe is sometimes referred to as five-horned. The lower canine teeth are peculiarly flattened and are deeply grooved at right angles to the plane of flattening. The long, flexible tongue can be extended up to 456 mm and is used in plucking leaves from trees. The lips are prehensile and hairy. The nostrils can be closed at will. The eyes are large, dark brown, and shaded by long black lashes. The feet are large and heavy. The great height of *Giraffa* requires a series of valves to regulate the flow of blood to the head. Females have four mammae.

The giraffe dwells mainly on dry savannahs and in open woodland. It is usually associated with scattered acacia growth. In a study in Tsavo National Park, Kenya, Leuthold and Leuthold (1978) found animals to concentrate near rivers in the dry season and to disperse into deciduous woodland during the rains. Such seasonal movements generally covered 20–30 km. The giraffe is active mainly in the evening and early morning and rests during the heat of the day. It usually sleeps standing up but occasionally lies down. In real sleep, a giraffe rests its head on the lower part of one hind leg, its neck

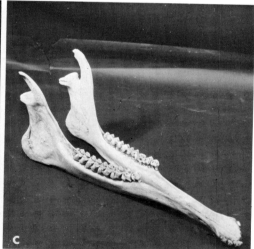

A. Giraffe *(Giraffa camelopardalis)*, with legs spread in a characteristic position for drinking, photo from New York Zoological Society. B & C. Skull of *G. camelopardalis*, photos by P. F. Wright of specimen in U.S. National Museum of Natural History.

forming an impressive arch. While dozing, which is more common, it rests on its withdrawn legs, but the neck remains outstretched, the eyes are half-closed, and the ears continue twitching. In order to drink or to pick food from the ground, the giraffe spreads its forelegs far apart and well to the front or bends at the knee until its head can reach the ground or water level.

The record running speed of *Giraffa* is 56 km/hr (Dagg and Foster 1976). Over moderate distances it can scarcely be overtaken by a good horse. If not crowded, a giraffe can lope for long distances without tiring. While the giraffe is running, the hind feet are swung forward of the forefeet, the head and neck swing widely, almost in a figure eight, and the tail is raised over the back. While it is walking, however, both feet

Giraffes (Giraffa camelopardalis), showing the considerable variation in pelage pattern, photos by Lothar Schlawe.

on a side are carried forward simultaneously. Although the giraffe has large feet, it is able to walk only on firm earth, since the long legs supporting such a heavy body would soon become bogged down in swampy terrain. As a result, large rivers are usually barriers.

The giraffe is shy, timid, and alert. It is especially vulnerable to lion predation when lying down, ground feeding, or drinking. A panicky movement indicates danger. When forced to defend itself, the giraffe kicks with its forefeet. In addition, it frequently uses the head to give blows, particularly to another giraffe. The senses of smell, hearing, and vision are acute. Among African big game species, this animal probably has the keenest sight, and its height gives it the greatest range of vision of any terrestrial creature's. The giraffe is a browser, feeding largely on leaves from acacia, mimosa, and wild apricot trees. It takes branches in its mouth and tears off the leaves by pulling its head away. It may chew its cud at any time of day. If water is available, it will take an occasional drink (about 7.5 liters a week); however, it is able to go without water for many weeks, if not months, at a time. Kingdon (1979) noted that feeding takes 16–20 hours per day and that the largest individuals consume up to 34 kg in that time.

Reported normal population density varies from about 0.1/sq km to 3.4/sq km. Mean individual home range in different areas, not including seasonal migration routes, varies from about 23 to 163 sq km. The average range size is about the same for both sexes; there is considerable overlap and no evidence of territoriality. *Giraffa* is usually observed alone or in small, loosely organized groups. When the genus was more common, herds of over 100 individuals were recorded, but most groups now contain 2–10 animals. Young males are commonly found in bachelor groups and tend to become solitary as they grow older. There is a dominance hierarchy among the males of each local population that is maintained by aggressive posturing and sparring. The latter involves two or more individuals' standing parallel, swinging their necks, and striking with their heads from the side. Sparring is usually ritualized and gentle, but on rare occasion, especially when several males are in the presence of an estrous female, it may evolve into serious fighting in which an animal may be knocked unconscious. Females with young sometimes aggregate, and one or two of the adults may remain with all of the calves while the rest of the adults go off to browse or drink. *Giraffa* has a variety of sounds but is rarely heard. It may grunt or snort when alarmed, a female may whistle to call its young, and calves bleat (Berry 1978; Dagg and Foster 1976; Kingdon 1979; Langman 1973, 1977; B. M. Leuthold 1979; Leuthold and Leuthold 1978; Pratt and Anderson 1985).

In at least some parts of Africa, such as Zambia, breeding may occur throughout the year (Berry 1973). However, there is a widespread tendency for the young to be born in the dry months, a peak having been reported for the Serengeti during May–August (Kingdon 1979). Females give birth at a mean interval of about 20–23 months (Hall-Martin and Skinner 1978; Leuthold and Leuthold 1978). The average gestation period is 457 days (Skinner and Hall-Martin 1975). The number of offspring is almost always one, though twins are known. At birth the young weigh 47–70 kg and are 1,700–2,000 mm tall. They are able to stand on their wobbly legs about 20 minutes after being born and begin to suckle within an hour. Contrary to early reports, there appears to be a strong bond between mother and calf. The calf may nurse for up to 13 months and then remain with the mother for another 2–5 months (Langman 1977; B. M. Leuthold 1979). Mothers initially keep their calves in relative isolation, thwarting the advances of other giraffes, but after a period of 8–30 days the young are allowed to associate with peers and older animals (Pratt and Anderson 1979). The age of sexual maturity usually is 3.5 years in females and 4.5 years in males. Full size is attained at 5 years by females and 7 years by males. Females are capable of reproduction until at least 20 years, and maximum known longevity in the wild is 26 years (Dagg and Foster 1976). A captive giraffe lived for 36 years and 2 months (Jones 1982).

The native peoples of Africa sometimes take giraffes in snares and pitfalls. They use the strong sinews for bowstrings and musical instruments and the thick hide as a covering for shields. The meat, although tough, has a good flavor. European settlers killed the animals in great numbers for their hides, which were used to make traces, long reins, whips, and other items. A combination of excessive hunting and climatic change has caused a great reduction in the distribution and numbers of *Giraffa* in historical time. The genus disappeared in Egypt about 2600 b.c. but may have survived in Morocco to a.d. 600. In the twentieth century, it has been wiped out in most of western and southern Africa. The only remaining large populations are in Tanzania and some adjacent areas (Dagg 1971; Dagg and Foster 1976). In 1976 a population of 10,750 giraffes inhabited the 10,984 sq km of woodlands in the Serengeti of Tanzania (Pellew 1983).

ARTIODACTYLA; Family ANTILOCAPRIDAE; Genus ANTILOCAPRA
Ord, 1818

Pronghorn

The single living genus and species, *Antilocapra americana*, originally occurred in open country from eastern Washington and southern Manitoba to Baja California and northeastern Mexico (Hall 1981). O'Gara and Matson (1975) suggested that the Antilocapridae be ranked only as a subfamily of the Bovidae, and in this regard they were followed by Honacki, Kinman, and Koeppl (1982). However, there now seems to be a growing consensus that the Antilocapridae should be retained as a separate family and placed together with the Cervidae in the superfamily Cervoidea (Baccus et al. 1983; Groves and Grubb 1987; Scott and Janis 1987; Solounias 1988).

Head and body length is 1,000–1,500 mm, tail length is 75–178 mm, shoulder height is 810–1,040 mm, and weight is 36–70 kg. Males average approximately 10 percent larger than females (Hall 1981). The woolly undercoat is overlaid with fairly long, straight, coarse, pithy, and brittle guard hairs. By flexing certain skin muscles, the pronghorn can maintain its pelage at different angles. Cold air is excluded when the hairs lie smooth and flat, but the hairs may be erected in the desert sun to allow air movement to cool the skin. The upper parts are reddish brown to tan, the neck has a black mane, and the underparts, the rump, and two bands across the neck are white. In the male, the face and a patch on the side of the neck are black, the horns are longer than the ears, and the nose is pointed downward slightly when it runs. In the female, the mask and patch are lacking, or nearly so; the horns, if present, seldom exceed the ears in length; and the nose is held more nearly horizontal when the animal is running. Females have four mammae.

Males, and most females, carry horns that consist of a permanent, laterally flattened bone core covered with a keratinous sheath that is shed annually after each breeding season. The new sheath grows upward under the old sheath. The shedding process is usually considered to be a major distinguishing feature between the Antilocapridae and the Bovidae, but O'Gara and Matson (1975) observed that a similar process occurs in at least one species of each of the five bovid subfamilies. In contrast, Solounias (1988) concluded that the horns of *Antilocapra* evolved independently from those of the Bovidae and actually are long frontal pedicels, homologous to structures found in certain fossil families related to modern deer. The horns of *Antilocapra* are erect, curving backward at the tips. Those of males are about 250 mm long and have a forward-directed prong arising from the upper half. Those of females, if present, are seldom over 120 mm long and seldom have prongs.

Pronghorn *(Antilocapra americana)*, photo by Bernhard Grzimek.

Antilocapra has large eyes (approximately 50 mm in diameter) and long, pointed ears. Only two digits are developed, the third and the fourth; the lateral toes are lacking. The hooves, especially those of the forefeet, are supplied with cartilaginous padding. The dental formula is: (i 0/3, c 0/1, pm 3/3, m 3/3) × 2 = 32. The high-crowned cheek teeth have crescentic ridges of enamel.

Habitats include grasslands and deserts. The elevational range is sea level to 3,353 meters (Yoakum 1978). The pronghorn is the swiftest terrestrial mammal in the New World. Kitchen (1974) clocked herds moving at 64–72 km/hr and observed a maximum speed of 86.5 km/hr. Such velocities can be attained only on hard ground and involve leaps of 3.5–6.0 meters. The cruising speed is approximately 48 km/hr. Fast runs of 5–6 km are common, but then exhaustion occurs rapidly. The pronghorn shakes the body after a fast run. The front feet carry most of the weight while the animal is running. The pronghorn is a good swimmer. It can see objects several kilometers away but apparently lacks visual acuity. A motionless person only 10–15 meters away may be ignored (Kitchen 1974). The pronghorn is a curious animal; it may approach a strange object if the object does not cause alarm by scent or sudden movement.

Activity is both diurnal and nocturnal, with slight peaks just after sunset and before sunrise (Kitchen 1974). Daily movement is usually 0.1–0.8 km in spring and summer and 3.2–9.7 km in fall and winter (Yoakum 1978). The range may

shift several times a year for purposes of obtaining food and water. The distance between the summer and winter range may be as much as 160 km (O'Gara 1978). The pronghorn both browses and grazes on a wide variety of shrubs, forbs, grasses, cacti, and other plants. It uses its front feet to dig food buried under the snow and also to scratch depressions for deposit of droppings. If water is available, it will drink freely, but it can derive sufficient moisture from plants if necessary.

According to O'Gara (1978), studies in Wyoming indicated that 48.6 ha. of desert could support 8 pronghorns year-round. Home range in Wyoming during the summer and early autumn was found to be 2.6–5.2 sq km. Yoakum (1978) stated that most herds have an overall range 8–16 km wide. During the fall and winter, *Antilocapra* forms large, loose aggregations of all ages and both sexes. In the northern part of the range, such groups contain up to 1,000 individuals (Yoakum 1978).

From late March to early October, groups are smaller and are segregated by sex (Kitchen 1974; O'Gara 1978). At this time, males over three years old compete with one another for possession of territories 0.23–4.34 sq km in size. A territory usually contains a permanent source of water, often has prominent physical borders, and may be separated by a no man's land of up to 0.8 km from other territories. An old male may return to the same territory each year after the herd breaks up and moves out of its winter range. He scent-marks the area with urine, feces, and secretions from subauricular glands. He constantly attempts to keep groups of females within his territory and to keep other males out. When a territorial male spots a rival, he first stares for a period. A stare is considered to be aggressive, and avoidance of a stare is considered submissive. If the other male holds his ground, there may be loud vocalization, an aggressive approach, a chase, and occasionally a fight. The sharp horns evidently often cause serious injury.

Old males that are unable to hold a territory may wander about alone. Younger males form bachelor herds with up to 36 members. These groups have a loose hierarchy and move about on the edges of the areas controlled by the territorial males. Female groups contain up to 23 members, have a permanent linear hierarchy, and move about freely on the territories of the dominant males. A female group seldom remains on a single territory through the entire period from March to October. Females are pursued by bachelor herds but try to avoid them. Individual does may separate from the group to give birth. Following the mating season, the horn sheaths are cast, and social distinctions become obscured, though the female associations may persist as subgroupings of the main herd.

The pronghorn has a variety of vocalizations. Mothers call calves with individually recognizable grunts, fawns bleat, and males sometimes roar during agonistic encounters. Mature animals of both sexes reveal anger or anxiety by forcefully expelling air through the nostrils. The hairs of the white rump patch are raised when danger is sensed, a conspicuous warning signal to other pronghorns. The white flash can be seen by a human for a distance of at least 4 km.

Mating occurs during a period of about 3 weeks between July and early October, and births take place in the spring. The gestation period averages 252 days (O'Gara 1978). A female generally produces a single young in her first pregnancy; thereafter she usually gives birth to twins, rarely triplets. At birth a fawn usually weighs 2–4 kg and has a beautiful, wavy, grayish pelage. The mother's milk is extremely rich in solids. At the age of 4 days the fawn can outrun a human, at 3 weeks it is nibbling at vegetation, and before 3 months it has acquired its first adultlike pelage. Apparently, the young may separate at least temporarily from the mother during the mating season and form small groups of their own, but there may subsequently be a reassociation with the mother until the following spring or summer (Autenrieth and Fichter 1975; O'Gara 1978). Females usually reach sexual maturity at 15–16 months. Males are capable of mating at this same age, but they generally do not do so until they are 3 years old. A captive specimen lived for 11 years and 10 months (Jones 1982).

There are estimated to have been about 35 million pronghorns in North America prior to the arrival of European explorers. Subsequent uncontrolled hunting for meat and sport, as well as human usurpation of habitat, resulted in a decline to fewer than 20,000 in the 1920s. Conservation efforts have since allowed numbers to expand to between 750,000 and 1 million in the United States and Canada (Cadieux 1987; Yates 1986). However, there is concern about the continuing spread of livestock fences that block natural routes, particularly those critical to winter range during times of deep snow. The pronghorn is now subject to limited sport hunting in most parts of its range, and the annual harvest is about 40,000. In Mexico, only about 1,200 individuals survive, and the number is declining through illegal hunting and habitat modification (O'Gara 1978; Yoakum 1978). The subspecies *A. a. sonoriensis*, of extreme southern Arizona and northwestern Mexico, and *A. a. peninsularis*, of Baja California, are classified as endangered by the IUCN and the USDI and are on appendix 1 of the CITES. *A. a. mexicana*, of Arizona, New Mexico, Texas, and northern Mexico, is on

appendix 2 of the CITES. A small pronghorn population has been introduced on Lanai in the Hawaiian Islands (Lever 1985).

The geological range of the family Antilocapridae is early Miocene to Recent in North America (Simpson 1984). Thirteen now-extinct genera were present during the Pliocene and Pleistocene (O'Gara 1978).

ARTIODACTYLA; Family BOVIDAE

Antelope, Cattle, Bison, Buffalo, Goats, and Sheep

This family of 47 Recent genera and 138 species has a natural distribution covering all of Africa, most of Eurasia and North America, and some islands of the Arctic and East Indies. The great majority of genera are native to Africa and to southern and Central Asia. Wild-living populations of some species have been introduced by human agency in New Guinea, New Zealand, Australia, and surrounding islands. Certain species have been domesticated and are now found throughout the world in association with people. The sequence of genera presented here follows basically that of G. G. Simpson (1945), who recognized five subfamilies:

Subfamily Bovinae

Tragelaphus	*Tetracerus*	*Bos*
Taurotragus	*Bubalus*	*Bison*
Boselaphus	*Syncerus*	

Subfamily Cephalophinae

Cephalophus	*Sylvicapra*

Subfamily Hippotraginae

Kobus	*Oryx*	*Alcelaphus*
Redunca	*Addax*	*Sigmoceros*
Pelea	*Damaliscus*	*Connochaetes*
Hippotragus		

Subfamily Antilopinae

Oreotragus	*Dorcatragus*	*Litocranius*
Ourebia	*Antilope*	*Gazella*
Raphicerus	*Aepyceros*	*Antidorcas*
Neotragus	*Ammodorcas*	*Procapra*
Madoqua		

Subfamily Caprinae

Pantholops	*Rupicapra*	*Capra*
Saiga	*Myotragus*	*Pseudois*
Capricornis	*Budorcas*	*Ammotragus*
Nemorhaedus	*Ovibos*	*Ovis*
Oreamnos	*Hemitragus*	

Simpson's list of genera has been modified extensively here, in accordance with various sources cited in the generic accounts that follow. There remains much controversy regarding the classification of the Bovidae. For example, C. D. Simpson (1984) recognized five additional subfamilies: Tragelaphinae, with the genera *Tragelaphus*, *Taurotragus*, and *Boselaphus*; Reduncinae, with *Redunca* and *Kobus*; Alcelaphinae, with *Damaliscus*, *Alcelaphus*, *Sigmoceros*, and *Connochaetes*; Aepycerotinae, with *Aepyceros*; and Neotraginae, with *Pelea*, *Oreotragus*, *Ourebia*, *Raphicerus*, *Neotragus*, *Madoqua*, and *Dorcatragus*.

Shoulder height ranges from as low as 255 mm in the pygmy antelope *(Neotragus)* to as great as 2,200 mm in some species of *Bos*. The pelage varies from smooth and sleek to rough and shaggy. The ulna and fibula are reduced, and the main foot bones are fused into a cannon bone. The front and hind feet are not the same length. On each foot, only two digits, the third and fourth, are well developed. The second and fifth digits are either absent or small and form the so-called dew hooves or dewclaws. The stomach is four-chambered and ruminating, and a gall bladder is usually present. Females have one or two pairs of functional mammae.

Horns are carried by all adult male bovids and by the females of most genera. The horns are composed of a bony core, attached to the frontal bones of the skull, and a hard sheath of horny material. *Tetracerus* is unique among bovids in having four horns; all other genera have only two. The dental formula is: (i 0/3, c 0/1, pm 3/2–3, m 3/3) × 2 = 30–32. The lower incisors project more or less forward, and the cheek teeth are low-crowned or high-crowned, with crescentic ridges of enamel on the crowns.

Most bovids inhabit grassland, scrubby country, or desert, but some live in forests, swamps, or arctic tundra. Goats and sheep generally occur in rocky or mountainous areas. Bovids graze or browse and are ruminants, that is, they chew the cud. The food is brought up from the first compartment of the stomach and chewed while the animal is at leisure, before being swallowed a second time for thorough digestion. Bovids feed by twisting grass, stems, or leaves around the tongue and cutting the vegetation off with the lower incisors.

According to Estes (1974), most species of Bovidae are gregarious and territorial. Exceptions include most species of the subfamily Bovinae, which are either gregarious and nonterritorial or, in the case of *Tragelaphus scriptus* and *T. spekei*, solitary and nonterritorial. The genera *Cephalophus, Redunca, Oreotragus, Ourebia, Raphicerus, Neotragus, Madoqua*, and *Dorcatragus* are generally solitary and territorial. Three social classes are found in all gregarious bovids: nursery herds of females and young, bachelor herds, and solitary adult males. If the species is also territorial, each adult male usually defends an exclusive area for all or part of the year. Glands on the hooves of the gregarious species release a substance onto the ground having a characteristic scent that an isolated animal can follow back to the herd.

Wild bovids have been hunted extensively by people for meat, hides, and sport. Many species and subspecies have become rare or threatened with extinction. Domestic cattle, sheep, and goats all have been derived from Eurasian species. The domestication of sheep and goats probably began in southwestern Asia 8,000–9,000 years ago.

The geological range of this family is early Miocene to Recent in Europe and Africa, middle Miocene to Recent in Asia, and Pleistocene to Recent in North America (Simpson 1984).

ARTIODACTYLA; BOVIDAE; Genus TRAGELAPHUS
De Blainville, 1816

Bongo, Sitatunga, Bushbuck, Nyalas, and Kudus

There are two subgenera and seven species (Ansell, *in* Meester and Setzer 1977; Büttiker 1982; Harrison 1972):

subgenus *Tragelaphus* De Blainville, 1816

T. buxtoni (mountain nyala), highlands of southern Ethiopia;

T. spekei (sitatunga), Gambia to southern Sudan, and south to northern Botswana;

T. angasi (nyala), southern Malawi, Mozambique, Zimbabwe, eastern South Africa;

T. scriptus (bushbuck), most of Africa south of the Sahara;

T. strepsiceros (greater kudu), southern Chad to Somalia, and south to South Africa;

T. imberbis (lesser kudu), Arabian Peninsula, southeastern Sudan, Ethiopia, Somalia, northeastern Uganda, Kenya, eastern Tanzania;

subgenus *Boocercus* Thomas, 1902

T. euryceros (bongo), forest zone from Sierra Leone to Kenya.

As explained by Ansell (*in* Meester and Setzer 1977) and Van Gelder (1977a, 1977b), there has been considerable disagreement regarding the classification of this genus. On the one hand, *Boocercus* often has been treated as a full genus, and some of the species listed above in the subgenus *Tragelaphus* sometimes have been separated into the genera or subgenera *Strepsiceros* Hamilton-Smith, 1827, and *Limnotragus* Pocock, 1900. On the other hand, *Taurotragus* occasionally has been considered to be only a subgenus of *Tragelaphus*.

According to Ansell (*in* Meester and Setzer 1977), both *Tragelaphus* and *Taurotragus* comprise medium-sized to large antelope with a variously developed face and body pattern of white spots and stripes. Males are distinctly larger than females. The tail is relatively short in the subgenus *Tragelaphus* but long and tufted in the subgenus *Boocercus* and the genus *Taurotragus*. The twisted horns are present in only the males of the subgenus *Tragelaphus* but in both sexes of *Boocercus* and *Taurotragus*. The skull has an ethmoid fissure and lacks a preorbital fossa. Females have two pairs of mammae. From *Taurotragus*, the genus *Tragelaphus* differs in having the horns in an open spiral, no dewlap, and feet that are non-oxlike and more elliptical. Additional information is given separately for each species.

Tragelaphus buxtoni (mountain nyala).

Head and body length is 1,900–2,600 mm, shoulder height is 900–1,350 mm, weight is 200–225 kg, and record horn length is 1,187 mm. According to Dorst and Dandelot (1969), the coat is rather shaggy, the general color is grayish chestnut, there is a white chevron between the eyes and two white spots on the cheek, there are poorly defined white stripes on the back and upper flanks, and there is a short brown mane on the neck.

Dorst and Dandelot (1969) wrote that the mountain nyala inhabits forests and heathland at elevations of 2,900–3,800 meters. It normally feeds in the evening and early morning and is mainly a browser. Old males are usually solitary, and other animals have been reported to occur alone or in groups of up to 15 individuals, but Hillman (1986a) indicated that aggregations of over 90 may form. *T. buxtoni* always has been restricted to a small highland area of Ethiopia and was not discovered until 1908. Since then it has declined substantially in numbers and distribution, mainly because of uncontrolled hunting and the clearing of forests and the burning of heaths for agricultural purposes (Yalden, Largen, and Kock 1984). Total numbers were estimated to be 7,000–8,000 in the 1960s, but the situation has continued to deteriorate. There now are about 3,000 in the wild, about a third of which are in the Bale Mountains National Park, and none in captivity (East 1988; Hillman 1986a).

Tragelaphus spekei (sitatunga).

Head and body length is 1,150–1,700 mm, tail length is about 200–260 mm, shoulder height is 750–1,250 mm, weight is

Sitatunga *(Tragelaphus spekei)*, photo by N. W. Labanoff through Bernhard Grzimek.

50–125 kg, and horn length in males is about 508–924 mm. Males average considerably larger than females. The general coloration is dull brown. There are two whitish areas on the throat, one near the head and one near the chest. There are also whitish marks on the face, ears, cheeks, body, legs, and feet. The females are slightly browner than the males, generally chestnut. The young are the color of the mother. The hair is much shaggier than in other antelope. The great elongation of the hooves and the peculiar flexibility of the joints at the feet (pasterns), which are bare and rest on the ground, represent striking structural adaptations for walking on boggy and marshy ground.

The sitatunga is semiaquatic, spending most of its life in dense beds of *Papyrus* and *Phragmites* within swamps (Smithers 1971). It is both diurnal and nocturnal and may move onto marshy land at night. It swims well and often submerges entirely when feeding in the water. If pursued, it may hide under the water with only the nostrils exposed. The sitatunga is a grazer of reeds, sedges, and grasses. Games (1983) estimated that 234 individuals occurred within an area of about 300 sq km in the Okavango Delta of Botswana. Observations there indicated that 60 percent of the animals occurred alone and the rest were in small, all-female groups. Data summarized by Kingdon (1982) also suggest that females are sociable, while males avoid one another. The latter have a loud barking vocalization that may serve to prevent direct confrontation. Breeding apparently occurs throughout the year, females produce a single young at an average interval of 11.6 months, the mean gestation period is 247 days, and sexual maturity is attained at approximately 1 year of age by females and 1.5 years by males (Densmore 1980). Potential longevity is about 20 years.

Tragelaphus angasi (nyala).

Head and body length is 1,350–1,550 mm, shoulder height is 800–1,150 mm, weight is 112–27 kg, and record horn length is 835 mm. *T. angasi* differs from other species of *Trag-*

elaphus in having a fringe of long pendant hairs that forms a line around the lower neck, lower shoulders, sides of the belly, lower thighs, and backs of the thighs. Adult males are charcoal gray in color; females and males under a year old have a red-brown coat (J. L. Anderson 1984).

The nyala is found on plains and mountains, usually in areas near water and with dense cover (Grzimek 1975). In a study in Zululand, Anderson (1980) found feeding activity mainly in the early morning and late afternoon. Average individual home range was 0.65 sq km in males and 0.83 sq km in females. There was extensive overlap both between and within sexes and no evidence of territoriality. Although herds of up to 30 nyala have been reported, Anderson observed an average of 2.38 individuals per sighting. Young males typically occurred in loose associations of about 3 animals, and females in groups of about 6, sometimes with their young. Older males tended to live alone for most of the year but attempted to join groups that contained females in estrus. Several males would compete for the females by postural displays and sparring with the horns; serious fights were apparently rare. The alarm call of the nyala is a staccato, doglike bark.

Anderson's (1984) studies also indicate that breeding continues throughout the year, with peaks of mating in the spring and autumn and resultant peaks of births the following autumn and spring. Females are polyestrous, having an estrous cycle of 10–34 days and an estrus of 2–3 days. There is a postpartum estrus 2–7 days following birth. The average interbirth interval is 297 days, but the shortest on record is 231 days, and hence the gestation period is estimated to be about 220 days. The single young weighs about 5.6 kg at birth and is nursed for at least 7 months. Males reach puberty at about 12 months but are not socially mature and able to mate until 5 years. Half of all females attain sexual maturity by 20 months, the rest before 36 months, and they are reproductively active until 14 years of age.

Tragelaphus scriptus (bushbuck).

Head and body length is 1,050–1,500 mm, tail length is about 200–275 mm, shoulder height is 650–1,100 mm, weight is 30–77 kg in males and 24–42 kg in females (Kingdon 1982), and horn length is 350–635 mm. The horns have pronounced keels in front and back and are twisted into spirals. The females are smaller than the males and are generally without horns. The color of the back and sides ranges from light tawny or reddish in females to dark brown or almost black in males. The underparts are usually slightly darker than the sides or back. There are a variety of white markings on the throat and lower neck, a line down the middle of the back, and stripes or rows of dots vertically on the sides. In some forms there is a pronounced mane the full length of the back.

The bushbuck inhabits the edges of swamps and other areas of dense vegetation near water. Kingdon (1982) noted that this species ranges up to 3,000 meters in the mountains of East Africa. It is adept at following small, overgrown tunnels through tangles of interwoven vines and shrubbery. It is either diurnal or nocturnal. It usually keeps to a restricted area, sometimes only a few hundred meters across, but may undertake seasonal movements in search of food. It is mainly a browser of the leaves and twigs of trees and shrubs, but it also grazes.

Odendaal and Bigalke (1979) reported that individual home range evidently decreases as population density increases. Studies in savannah areas revealed densities of over 25/sq km and ranges of 0.2 sq km or less. Studies in forests indicated densities of only 4/sq km but ranges of around 1 sq km. All available information (Allsopp 1978; Estes 1974; Jacobsen 1974; Odendaal and Bigalke 1979; Smithers 1971) indicates that *T. scriptus* is usually found alone. The only

A. Sitatunga *(Tragelaphus spekei),* photo by Ernest P. Walker. B. Bushbuck *(T. scriptus),* photo from U.S. National Zoological Park.

A. Nyala *(Tragelaphus angasi)*; B. Greater kudu *(T. strepsiceros)*. Photos from San Diego Zoological Garden.

Greater kudu *(Tragelaphus strepsiceros)*, photo by Lothar Schlawe.

regular exceptions are mothers with calves, and courting pairs. Males may compete vigorously for estrous females but are not territorial. Confrontations involve ritualized displays, charges, locking of the horns, powerful twists to throw the opponent off balance, and stabbing (Kingdon 1982). The bushbuck gives a call that is said to resemble closely the barking of a dog.

Breeding occurs throughout the year, though there are reproductive peaks in some areas. The gestation period is about 6 months, and the mean interval between births is about 8 months. The single young lies in concealment away from the mother for the first few weeks of life. Sexual maturity is attained after 11–12 months of life (Allsopp 1978; Dittrich 1972; Jacobsen 1974; Morris and Hanks 1974; Odendaal and Bigalke 1979; Von Ketelhodt 1976).

Tragelaphus strepsiceros (greater kudu).

Head and body length is 1,950–2,450 mm, tail length is 370–480 mm, and shoulder height is 1,000–1,500 mm. Weight is 190–315 kg in males and 120–215 kg in females (Kingdon 1982). The horns of males usually measure about 1,016 mm in a straight line and 1,320 mm along the curves; there is one record of 1,816 mm in a straight line. The hair of the neck is short and scant, and there is a crest down the midline of the back. The general coloration ranges from reddish

to pale slaty with blue-gray and white markings.

The greater kudu is found mainly in woodland and thickets and requires adequate cover for concealment (Smithers 1971). It may be active by day or by night. Its hearing is acute, and all observers agree that its extreme wariness makes it difficult to approach. It has the often fatal habit, however, of stopping after a short run to look back. Its ability to leap is marvelous; there is one report of its clearing bushes 2.5 meters high with ease. It feeds mainly by browsing (Smithers 1971).

According to Kingdon (1982), individuals disperse during the wet season, when food is available over a large area. Toward the end of the rains and in the dry season there is a concentration in favorable localities. Reported population densities are 1.9/sq km and 3.2/sq km. Two female groups had home ranges of 3.6 and 5.2 sq km, and two individual males had ranges of 11.1 and 11.2 sq km. Female ranges overlap each other, and during the mating period (May–August) the ranges of several solitary adult males encompass the ranges of two or three female groups. After the rut the males associate in unstable bachelor groups of 2–10 individuals and then again disperse with the coming of the rains. Female groups are more stable and usually have 5–6 members, including several adults and their young. Larger aggregations sometimes form for brief periods. During the rut the males compete with one another for estrous females,

Lesser kudu *(Tragelaphus imberbis)*, photo by Leonard Lee Rue III.

usually by displays and occasionally by serious combat. The spirals of the horns serve as a means for linking the opponents, which then shove and twist in an effort to throw one another off balance. Individual females separate from their groups to give birth and then rejoin as their calves grow. The young tend to be born during the early part of the wet season (January–March). Estimates of the gestation period range from 7 to nearly 9 months. The single offspring weighs about 16 kg (Smithers 1983). A captive kudu lived for 20 years and 9 months (Jones 1982).

The horns of the greater kudu are a prized trophy of sportsmen, and it also is killed for its excellent meat and because it allegedly damages crops. These factors, along with habitat destruction, have greatly reduced the range of *T. strepsiceros* (Kingdon 1982).

Tragelaphus imberbis (lesser kudu).

Head and body length is 1,100–1,750 mm, tail length is 260–300 mm, shoulder height is 900–1,050 mm, weight is about 60–100 kg, and horn length is about 600–900 mm. The general coloration of males is usually deep yellowish gray, but some individuals are darker. The general coloration of females is fawn. There are white patches on the throat, like those in *T. scriptus*, and 11–14 vertical stripes on the body. There is no throat mane like that of *T. strepsiceros*. The young are redder than the adult females and are strongly marked with white.

Except as noted, the remainder of this account is based on the studies by W. Leuthold (1974, 1979) in Tsavo National Park, Kenya. The lesser kudu lives mainly in relatively dry areas of dense vegetation, including bush country and woodland. It thus is especially susceptible to human clearing operations. It appears to be active mainly at night and in the early

morning and is seen to seek shade for resting only an hour after sunrise. Some individuals remain throughout the year in one area, while others make seasonal movements in search of food. There is a tendency to concentrate in riverine areas during the dry season and to disperse over a larger area in the wet season. Although there is a report of one individual making a horizontal leap of 9.2 meters, in which it went over the top of a bush 1.5 meters high and 1.8 meters in diameter, Leuthold saw no such spectacular leaps. The lesser kudu is almost exclusively a browser; the diet varies, but the bulk consists of leaves and shoots of trees and shrubs. Access to water is not critical.

Population density rarely exceeds 1/sq km, even in favorable habitat, and usually is much less. Individual home range varies from 0.4 to 6.7 sq km, averaging about 2.2 sq km for males and 1.8 sq km for females. Different parts of these areas are used in different seasons. There is much overlap, and territorial behavior is not apparent. Most groups contain 4 or fewer individuals, but aggregations of up to 24 occasionally form at favorable feeding sites. A group may be either unisexual or bisexual, but an adult male associates only temporarily with females. Typically, an original mother, several of her grown daughters, and their young form a unit that remains stable for years. Such units freely associate with one another but normally do not exchange members. A young male remains with its mother's group for 1.5–2 years, spends the next 2–3 years living alone or in a small, unstable, all-male group, and then becomes increasingly solitary. Consistent group leaders and dominance hierarchies are not apparent. Agonistic behavior consists mainly of displays, fighting is rare, and there is a sharp, barking alarm call.

Reproduction at Tsavo is nonseasonal; each female has her own cycle of estrus, pregnancy, and birth. There is a postpar-

Bongo *(Tragelaphus euryceros)*, female, photo from New York Zoological Society.

tum anestrous period of a few weeks, recorded gestation periods are 244–53 days, and the interbirth interval is about 9 months. A female temporarily separates from her group to bear her single calf. Birth weight is 4–7 kg, and the horns start to grow at 6–7 months of age. Females are sexually mature at about 15 months. Young males initially are colored like females, changing to the adult male coat during their second year of life. The thick neck, characteristic of the adult male, is not fully developed until the fifth year. While males reach puberty only a little later than do females, they are not socially mature until 4–5 years. A few wild individuals are known to have lived 8–10 years, and captives have survived for more than 15 years.

This species, as well as the entire genus *Tragelaphus*, long was known only in Africa. One of the most remarkable zoo-

geographic discoveries of modern times came in 1967 when a specimen was found in South Yemen (Harrison 1972). In 1981 the mounted horns of another individual, said to have been killed about 15 years earlier, were found in northwestern Saudi Arabia (Büttiker 1982). Unfortunately, efforts to locate the represented living population in the region have been unsuccessful, and it is not unlikely that it has been eliminated by human hunting. Populations across the Red Sea in Ethiopia, Somalia, Uganda, Kenya, and Tanzania also have been greatly reduced by poaching and habitat destruction (Kingdon 1982; Yalden, Largen, and Kock 1984).

Tragelaphus euryceros (bongo).

Head and body length is 1,700–2,500 mm, tail length is 450–650 mm, shoulder height is 1,100–1,400 mm, and weight is

150–220 kg. The bongo has short hair, an erect mane from the shoulders to the rump, and a long, tufted tail. It is among the most beautiful of bovids. The back and sides are bright chestnut red, and the belly is black. A pure white chevron crosses the forehead, and other white patches are located on the sides of the head. The breast bears a large white crescent. There are usually 11–12 narrow, vertical stripes on the sides of the body. Apparently, the number of stripes on each side is rarely the same. Females are usually brighter colored, and old males may become much darker, the chestnut red turning dark mahogany brown and the body stripes becoming buffy. There is a dark dorsal stripe, and the tail tuft is dark maroon or black. The outsides of the legs are dark or blackish, with a black chevron above white knees and a white patch above the hooves. The insides of the legs are white. The horns have yellow or buffy tips.

Both sexes have horns that spiral in one complete twist. The longest on record measured 1,002 mm along the front curve and were borne by a male from Kenya. The average horn length is about 835 mm, and smaller in females.

The bongo inhabits lowland forests in most of its range, but in Kenya it occurs in montane forests at elevations of 2,000–3,000 meters (Ralls 1978). It lives in the densest, most tangled parts of the forest. It has been reported to be diurnal, but Hillman (1986b) found most activity to be from dusk to early morning. It depends more on its sense of hearing than on sight or smell. It is very shy and swift and can quickly disappear when startled. It runs gracefully, and at full speed, through even the thickest tangles of lianas, laying its heavy, spiraled horns on its back, so that the brush cannot impede its flight. So common is this habit that most older animals have bare, rubbed patches on their back where the horn tips rest. Also, the front surfaces of the horns of older animals are often much worn and frayed from the friction caused by their rapid passage through the heavy undergrowth. The bongo prefers to go under or around obstacles rather than over them. It likes to wallow in mud puddles and then rub the mud against a tree, at the same time polishing its horns.

The diet is varied, but in general the bongo is a browser. It eats the tips, shoots, and trailers of many plants and, like the bushbuck and nyala, shows a marked preference for the tender bush herbage that grows around the bases of trees. Roots, bamboo leaves, cassava, and sweet potato leaves are also preferred items. The bongo sometimes raids coco yam farms for the tender leaves. It often uproots saplings with its horns in order to eat the roots. Like many other browsing animals, it can rear up on its hind legs, bracing the forelegs against a tree trunk. It can reach leaves and twigs as high as 2.5 meters above the ground. The bongo is said to eat earth at times and also to chew and swallow pieces of burned wood from lightning-killed forest trees, apparently to obtain salt.

In southern Sudan, Hillman (1986b) found population density to be about 1.2/sq km. Animals there occurred alone or in herds of up to 44, the average observation being of 9.1. The larger groups contained individuals of all ages and both sexes and sometimes more than one adult male. However, the larger herds seemed to be temporary aggregations of smaller, more stable units. Reports from certain other areas have suggested that groups are smaller and adult males are usually solitary. Births are said to occur in December or January in the wild and have taken place in December, April, and August in captivity. According to Ralls (1978) and Ralls et al. (1985), females have an estrous cycle of 21–22 days, estrus is about 3 days long, two interbirth intervals were 466 and 525 days, three gestation periods lasted 282–87 days, all 11 births in captivity yielded a single offspring, and average weight at birth is 19.5 kg. The young have the same color pattern as adults but are a rich tawny shade and much lighter. Ralls (1978) wrote that two captive-born females first con-

ceived at the ages of 27 and 31 months, respectively, and that a captive female lived for about 19 years and 5 months.

ARTIODACTYLA; BOVIDAE; **Genus TAUROTRAGUS**
Wagner, 1855

Elands

There are two species (Ansell, *in* Meester and Setzer 1977):

T. oryx (common eland), open country from Ethiopia and southern Zaire to South Africa;
T. derbianus (Derby eland, or giant eland), savannah zone from Senegal to southern Sudan.

Both Corbet and Hill (1986) and Honacki, Kinman, and Koeppl (1982) followed Van Gelder (1977b) in including *Taurotragus* within the genus *Tragelaphus*, but the two were regarded as generically distinct by Meester et al. (1986).

Head and body length is 1,800–3,450 mm, tail length is 500–900 mm, shoulder height is 1,000–1,800 mm, and weight is 400–1,000 kg. Males average larger than females. The color of *T. oryx* is a fairly uniform grayish fawn. The general coloration of *T. derbianus* is rich fawn, its head is lighter than its body, and its neck is black with a white band at the base. Both species have whitish or creamy vertical stripes on the upper parts. There is a short mane on the nape and longer hairs on the throat.

Taurotragus is characterized by oxlike massiveness, rounded hooves, a hump on the withers, a dewlap between the throat and chest, and heavy spiral horns that are carried by both sexes. The horns of *T. derbianus* are longer and more massive, the record length being about 1,200 mm.

The preferred habitat of *Taurotragus* is plains or moderately rolling country with brush and scattered trees. Elands are alert, keen of sense, and thus difficult to approach. They are sometimes said to be slow and easily caught, but Schaller (1972) reported maximum speed to be at least 70 km/hr. Despite their size, elands are good jumpers and can take a 1.5-meter fence with ease. They prefer to lie in some shelter during the heat of the day. In the morning and evening they seek more open areas, where they browse upon leaves and succulent fruits. Unlike many other antelope, elands do not disperse across the plains during the wet season; their ranges do increase substantially at this time, but they continue to form large groups, possibly because of an intense mutual attraction by calves (Kingdon 1982).

The home ranges of *T. oryx* vary considerably by sex and season. Hillman (1988) found that during the dry season adult males used an average area of 11.7 sq km out of their total year-round range of 41.1 sq km. In contrast, females and young had a dry season range of 26.1 sq km in brush country and then made extensive movements into open grassland during the wet season, so that their year-round range averaged 222.0 sq km. Leuthold (1977a) reported that male ranges are only 6–71 sq km in size and located mainly in wooded areas, while those of females are 34–360 sq km and generally on the open plains. There is much overlap of home range and evidently no exclusive use of space.

Schaller (1972) estimated that there were 7,000 *T. oryx* in the 25,000-sq-km Serengeti ecosystem (a more recent estimate is 18,000 [Sitwell 1983]). There they usually consisted of fewer than 25 individuals, including a number of cows and subadults and 1 or more large bulls. In Kenya, Hillman (1987) saw few lone animals, small groups with only adults, and large herds with up to 427 individuals. Apparently the larger groups, commonly with more than 100 members, were basi-

Eland *(Taurotragus oryx)*, photo by C. A. Spinage through East African Wildlife Society.

cally temporary aggregations of females and calves that formed during the wet season in open grassland. No enduring social bonds were observed other than between mothers and nursing young, and individuals moved freely in and out of groups. According to Leuthold (1977*a*), several adult males may be present in a female herd at a given time, but they have a strict dominance hierarchy that governs access to estrous females. Young elands have a strong tendency to associate in peer groups of their own.

There are distinct breeding seasons in some areas. In Zambia the calves are born in July and August (Wilson 1969). Gestation periods of 254 to 277 days have been recorded (Dittrich 1972). There is normally a single calf. It weighs 22–36 kg at birth (Kingdon 1982). Lactation may last 4–5 months (Hillman 1987). Sexual maturity is usually attained at about 3 years by females and 4 years by males. According to Treus

and Lobanov (1971), captive female *T. oryx* have given birth when as young as 22 months and as old as 19 years, and record longevity is 23 years and 6 months.

Elands have disappeared from large parts of their range, mainly because of excessive hunting by people. The western giant eland *(T. derbianus derbianus)*, originally found in Senegal, Gambia, Mali, Guinea, and Guinea-Bissau, is classified as endangered by the IUCN and the USDI. Elands yield a large quantity of tender meat, and the quality of the thick hide is excellent. These animals are docile and easily tamed, and efforts have been made in Africa and the Soviet Union to domesticate them for meat and milk production. Eland milk has about triple the fat content and twice the protein of the milk from a dairy cow (Treus and Lobanov 1971).

ARTIODACTYLA; BOVIDAE; **Genus BOSELAPHUS**
De Blainville, 1816

Nilgai, or Bluebuck

The single species, *B. tragocamelus*, is found in eastern Pakistan and India (Ellerman and Morrison-Scott 1966; Roberts 1977). It also evidently still occurs in much of Nepal (Shrestha 1989) but has been extirpated in Bangladesh (Khan 1984).

Head and body length is 1,800–2,100 mm, tail length is 456–535 mm, shoulder height is 1,200–1,500 mm, and weight is up to 300 kg. For an introduced population in Texas, Sheffield, Fall, and Brown (1983) reported average weights of 241 kg for males and 169 kg for females and noted that the length of the horns, which are carried only by males, was 150–240 mm. The hair on the body is short and wiry. Although in both sexes the neck is ornamented with a mane, only the bulls develop a tuft of hair on the throat. The upper parts of males are generally iron gray, but the lower surface of the tail, stripes inside the ears, rings on the fetlocks, and underparts are white. The head and limbs are tawny, and the throat tuft and tip of the tail are black. The females are more lightly colored. The forelegs are somewhat longer than the hind ones, and the head is long and pointed.

In its native range, the nilgai frequents forests, low jungles, and occasionally open plains. It is diurnal but also rests during the day. It is both a browser and a grazer and is fond of fruit and sugar cane. In a study in central India, Schaller (1967) found a population density of about 0.07/sq km. Males were observed to establish territories during the breeding season and to attempt to gather groups of 2–10 females thereon. The males fought one another, for possession of the territories and the females, by dropping to their knees and lunging with the horns. Nonterritorial males, in groups of up to 18 animals, were also present. Some breeding occurred throughout the year, but most calves were born from June to October.

The most detailed studies of the nilgai have been made on an introduced population in a mixed grassland-woodland habitat of southern Texas (Sheffield, Fall, and Brown 1983). Activity was predominantly diurnal, with peaks in the early morning and late afternoon. The eyesight and hearing of the nilgai appeared to be better than that of the white-tailed deer, but the sense of smell seemed to be less developed. The nilgai was found to be wary and swift, one attaining a speed of 48 km/hr when being chased by a truck. The preferred foods were grass, forbs, and browse, in that order. Population density was about 1 individual per 15–20 ha. Individuals tended to occupy an area for a few weeks and then move on; home ranges during these periods averaged 4.3 sq km. There was no evidence of territoriality. For most of the year males occurred in groups of about 2–5 individuals; younger males were in the larger groups, while old males frequently were solitary. Females spent most of the year in herds of 10–15 animals, tending to become less sociable when accompanied by newborn calves. All groups evidently represented temporary aggregations, with a membership that might change on a daily basis. During the mating season the social structure changed, with a single adult male joining a group of about 3 adult females and their young. Males competed with one another mostly by threatening postural displays and ritualized combat using only neck pressure; serious horn fighting was rare. There was a roaring vocalization, apparently to facilitate contact, but animals usually were silent.

Most mating in Texas occurred from December to March, most calving in September and October. Gestation periods of 243–47 days were considered to be normal. At least 58 percent of pregnancies resulted in twins, and a few in triplets. Young males remained with the female herds until they were 8–10 months old, then left to join a bachelor group, and probably did not start to participate in breeding until about 5 years old. Females reached sexual maturity in their third year of life. Few of the free-roaming animals in Texas survived for more than 10 years. However, a captive bluebuck lived for 21 years and 8 months (Jones 1982).

Boselaphus is rather docile for so large an animal, is said to tame easily, and thrives in captivity. It is regarded as a close relative of the sacred cow by the Hindu religion. Consequently, the nilgai traditionally enjoyed immunity from molestation and sometimes displayed remarkably little concern in the presence of people. It remained common in India and Pakistan until about 1900 but subsequently declined through habitat usurpation and intensified hunting. Sheffield, Fall, and Brown (1983) suggested that the free-roaming popula-

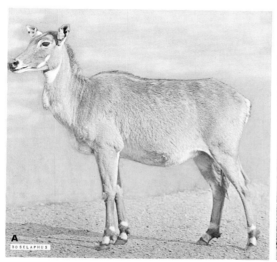

Nilgai *(Boselaphus tragocamelus)*: A. Female, photo from New York Zoological Society; B. Male, photo by Bernhard Grzimek.

tion in southern Texas, descended from animals released in the 1930s, may contain as many nilgai as currently survive in Asia. There now are an estimated 8,000–9,000 individuals in the 2,600-sq-km primary range in Kenedy and Willacy counties. Although the horns of the nilgai are not impressive and there is some question about the desirability of its meat, the Texas population is considered to be of potential interest to sportsmen because of the challenge of hunting this wary animal.

ARTIODACTYLA; BOVIDAE; **Genus TETRACERUS**
Leach, 1825

Four-horned Antelope, or Chousingha

The single species, *T. quadricornis*, occurs in India and Nepal (Ellerman and Morrison-Scott 1966; Mitchell 1977).

Head and body length is about 800–1,000 mm, tail length is about 126 mm, shoulder height is about 600 mm, and weight is 17–21 kg. The short, thin, coarse hair is a uniform brownish bay above, lighter on the lower sides, and white on the insides of the legs and middle of the belly. The muzzle, outer surface of the ears, and a line down the front of each leg

are blackish brown. The horns, borne only by males, are short, conical, smooth, and usually four in number. The posterior two horns are 80–100 mm long. The front two are often small, about 25–38 mm long, and are sometimes represented by only a slightly raised area of black hairless skin. In having four horns, *Tetracerus* is unique among the Bovidae. The small hooves are rounded in front.

The chousingha is found most frequently in open forests. It is shy and swift, dashing into dense cover at the first sign of danger. It is sometimes confused in the field with the hog deer *(Axis porcinus)* but can be distinguished by its peculiar jerky manner of walking or running. It is a grazer and drinks regularly, seldom being found far from water. Rice (1989) reported that it usually occurs at densities of less than 0.5/sq km. *Tetracerus* is not gregarious, and rarely are more than two individuals found together. Mating takes place during the rainy season, from July to September. The gestation period is 7.5–8.0 months (Grzimek 1975). There are one to three young per litter. A captive lived for 10 years (Jones 1982).

When captured young, the chousingha is easily tamed, but it is apparently quite delicate. Most persons agree that the meat is not as good as that of some of the other antelopes. Despite the small size of the horns, the presence of two pairs makes the chousingha much sought after by trophy hunters. According to Rice (1989), it is likely that populations are becoming increasingly isolated as human impacts intensify

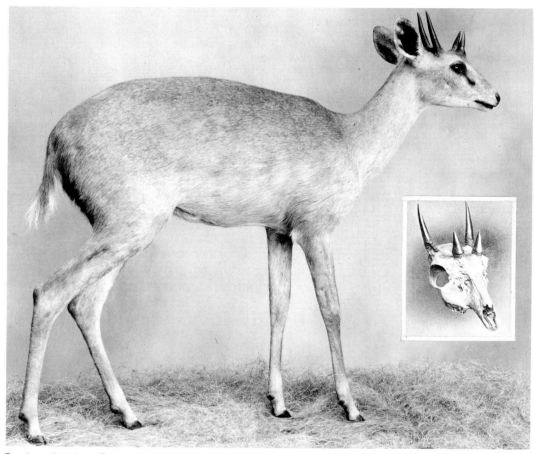

Four-horned antelope *(Tetracerus quadricornis)*, photo from Field Museum of Natural History. Inset: photo from *Guide to Great Game Animals*, British Museum (Natural History).

Four-horned antelope *(Tetracerus quadricornis)*, photo by N. Das.

and spread, but the species still is widely distributed and there is little immediate concern for its survival.

ARTIODACTYLA; BOVIDAE; Genus **BUBALUS**
Hamilton-Smith, 1827

Asian Water Buffalo, Tamaraw, and Anoas

There are two subgenera and four species (Groves 1969*a*; IUCN 1972, 1976):

subgenus *Bubalus* Hamilton-Smith, 1827

B. bubalis (Asian water buffalo), originally found from at least Nepal and India to Viet Nam and Malaysia;
B. mindorensis (tamaraw), Mindoro Island (Philippines);

subgenus *Anoa* Hamilton-Smith, 1827

B. depressicornis (lowland anoa), lowlands of Sulawesi;
B. quarlesi (mountain anoa), highlands of Sulawesi.

Anoa sometimes has been treated as a separate genus, and *B. mindorensis* sometimes has been placed in *Anoa*. *B. bubalis* has been widely domesticated and introduced by people, and

certain wild-living populations outside of the region indicated above are sometimes thought to represent the natural distribution of the species. Some authorities, including Groves (1969*a*), use the name *B. arnee* in place of *B. bubalis*. The tamaraw, or tamarao *(B. mindorensis)*, should not be confused with the carabao, a small domestic form of *B. bubalis* used in the Philippines.

In wild *B. bubalis*, head and body length is 2,400–3,000 mm, tail length is 600–1,000 mm, shoulder height is 1,500–1,900 mm, and weight is 700–1,200 kg. Domestic breeds are usually smaller, the weight range being about 250–550 kg (Popenoe 1981). The moderately long, coarse, and sparse hair is directed forward from the haunches to the head. There is a tuft on the forehead, and the tip of the tail is bushy. The general coloration is ash gray to black. Some domestic breeds are black and white, brown and white, or entirely white (Grzimek 1975). The face is long and narrow, the hooves are large and splayed, and the ears are comparatively small. The horns, carried by both sexes, are heavy at the base, normally curve backward and inward, are somewhat triangular in cross section, and are conspicuously marked with wrinkles. The spread of the horns, up to 1,200 mm along the outer edge, exceeds that of any other living bovid. Lekagul and McNeely (1977) stated that a wild *B. bubalis* is larger, quicker, and more aggressive, and has much more widely spreading horns, than a domestic buffalo.

Bubalus mindorensis is much smaller than *B. bubalis*, being only about 1,000 mm high at the shoulders and 300 kg in weight (Popenoe 1983). It has more hair on the body than

Asian water buffalo *(Bubalus bubalis)*, photo by R. Van Nostrand through San Diego Zoological Garden.

Tamaraw *(Bubalus mindorensis)*, photos by David Kuehn.

does *B. bubalis* and is dark brown to grayish black. Its horns are stout and short, only about 355–510 mm long.

In the subgenus *Anoa*, head and body length is about 1,600–1,720 mm, tail length is 180–310 mm, and shoulder height is 690–1,060 mm. Grzimek (1975) listed weight as 150–300 kg. Although the young are thickly covered with yellowish brown, woolly hair, the skin of old individuals is almost bare. The general color of adults varies from dark brown to blackish, and there are often blotches of white on the face, nape, throat, and lower limbs. The underparts are usually light brown. Males are generally darker than females. The hide is of exceptional thickness, the limbs are rather short, the body is plump, and the neck is thick. According to Groves (1969a), *B. depressicornis* has white forelegs, a long tail, and horns that are triangular in section, flattened, wrinkled, and 183–373 mm long; and *B. quarlesi* has legs that are generally the same color as the body, more hair than does *B. depressicornis*, a short tail, and horns that are rounded in section, nonwrinkled, and only 146–99 mm long.

The anoas and tamaraw are basically forest animals. The tamaraw requires dense vegetation for resting, water for drinking and wallowing, and open grazing land (IUCN 1976). Anoas feed in the morning and rest in the shade during the afternoon. Their gait is a trot, but at times they make clumsy leaps. Popenoe (1983) reported that they have a wide-ranging diet including grasses, ferns, saplings, palm, ginger, and fallen fruit.

The water buffalo is associated with wet grasslands, swamps, and densely vegetated river valleys. It is pestered by insects, and as a means of protection it wallows in the water and soil. It is thus often completely caked by a layer of mud through which insects cannot penetrate. It also escapes from insects by submerging in the water with only the nostrils exposed. The time of activity apparently varies, but animals of the introduced population in northern Australia leave the resting area at sunrise, graze for three to four hours, drink and wallow in the middle of the day, graze again in the late afternoon, and return to the resting area at sunset (Tulloch 1978). The favorite foods are lush grass and vegetation growing in or beside rivers and lakes.

Lowland anoas *(Bubalus depressicornis)*, photo from San Diego Zoological Garden.

Kuehn (1986) estimated that a minimum of 51 individual *B. mindorensis* occupied an area of 20 sq km on Mindoro. Adult bulls of that species were largely solitary and apparently aggressive toward each other. Adult females usually were seen alone, together with a bull, or accompanied by up to three young of different ages. Males up to 5 years old sometimes formed small groups. The social characteristics of the anoas may resemble those of the tamaraw; both are rare and wary animals.

The water buffalo is much more common, at least in the domestic form, and is gregarious. According to Tulloch (1978), about 100,000–200,000 feral buffalo inhabit 100,000 sq km of the Northern Territory of Australia. Adult females and their young form stable "clans" of up to 30 individuals. Each such clan has a home range of 170–1,000 ha., containing areas for resting, grazing, wallowing, and drinking. There is a dominance hierarchy, and the leader is always an old cow, even when bulls accompany the group. Young females remain in their mother's group, but young males are driven off at two to three years old. Several clans of females form a herd of 30–500 animals, which gathers at the same nightly resting area. There are also bachelor groups, usually containing about 10 males, which spend the dry season (May–September) apart from the females. Their home ranges are larger than those of the females and overlap one another. During the wet season (October–April) the bulls move into the areas used by the cows. Dominant males then mate with, but do not control, the females of a clan and are subsequently driven off. Very old males are usually solitary.

The young of *B. mindorensis* may be born throughout Mindoro's June–November rainy season and apparently separate from their mother between the ages of 1.5 and 4.5 years (Kuehn 1986). Anoas have no specific breeding season, a gestation period of 275–315 days, generally a single offspring, and a maximum known longevity in captivity of 28 years. In

B. bubalis, breeding is seasonal in some areas, the estrous cycle averages 21 days, estrus usually lasts about 24 hours, the interbirth interval is commonly about 2 years, the gestation period is 300–340 days, there is normally a single calf, weight at birth is 35–40 kg, nursing lasts 6–9 months, puberty is attained at about 18 months, and known longevity is up to 25 years in the wild and 29 years in captivity (Grzimek 1975; Popenoe 1981).

The species *B. mindorensis, B. depressicornis,* and *B. quarlesi* are classified as endangered by both the IUCN and the USDI and are on appendix 1 of the CITES. All have very restricted ranges and have declined through excessive hunting and loss of habitat. The tamaraw, one of the world's rarest mammals, attracted much attention earlier in the century, including the personal interest of Charles Lindbergh, the aviator. The Iglit-Baco National Park was established on Mindoro for the protection of the species. Nonetheless, poaching and habitat destruction resulted in a steady decline in numbers, from an estimated 10,000 in 1900 to 1,000 in 1949 and only 120 by 1975. A recent intensification of conservation measures, including establishment of a captive breeding center on Mindoro, seems to have allowed numbers to build back to a few hundred (Petocz 1989).

B. bubalis also now is designated as endangered by the IUCN. Wild members of this species have disappeared from most of the original range because of usurpation of habitat by agriculture, hunting by people, and competition from and diseases transmitted by domestic livestock. A few herds, thought to be descended from original native stock, are still scattered from India to Indochina (IUCN 1972). In India only about 1,000 individual wild buffalo survive, and only about 100 of those, located in the east-central state of Madhya Pradesh, are free of genetic influence from domestic or feral animals (Kane 1989).

In contrast, there now are at least 130 million domestic

Mountain anoa (Bubalus quarlesi), photo by Lothar Schlawe.

buffalo in the world, mostly in India, Southeast Asia, and the East Indies. Domestication of *B. bubalis* evidently began about 5,000 years ago in the Indus Valley. It was in use in China 4,000 years ago and was spread through the Middle East around A.D. 600. The species also has been brought to southern Europe, Asia Minor, northern and eastern Africa, Madagascar, Mauritius, Australia, Japan, Hawaii, and Central and South America. A few herds now are in use in Florida and Louisiana. The large feral population of northern Australia (see above) became established in the 1800s, and there are smaller feral groups in New Guinea, Tunisia, north-eastern Argentina, and certain other places. Domestic buffalo are especially suitable for tilling rice fields and also are used as beasts of burden. Their milk is richer in fat and protein than that of the dairy cow, they can yield tender meat that is difficult to distinguish from beef, and leather made from their skin is of superior quality. They are docile and tractable with persons whom they know and can even be controlled by young children (Grzimek 1975; Lever 1985; Popenoe 1981).

ARTIODACTYLA; BOVIDAE; Genus SYNCERUS
Hodgson, 1847

African Buffalo

The single species, *S. caffer*, originally occurred in most of Africa south of the Sahara (Ansell, *in* Meester and Setzer 1977).

Head and body length is 2,100–3,400 mm, tail length is 750–1,100 mm, shoulder height is 1,000–1,700 mm, and weight is 300–900 kg. There is much variation in size and other characters, and the subspecies *S. c. caffer*, of the eastern savannahs, may be twice as large as the subspecies *S. c. nana*, of the equatorial forests (Grubb 1977; Sinclair 1977a). In any particular region, males average considerably larger than females. There is a thick covering of hair on young individuals, a sparse covering on adults, and little hair on old animals. Coloration is brownish to black. Commonly, *S. c. caffer* is black and *S. c. nana* is red. The head and limbs are massive, the breadth of the chest is great, and the ears are relatively large, drooping, and fringed by soft hairs. The horns spread outward and downward and then upward in some animals, and out and back in others. In males the two horns are joined by a boss, a large shield covering the whole top of the head.

The African buffalo occurs in a great variety of habitats but prefers areas with grass, water, and some dense cover. It seems to enjoy splashing about in water and wallowing in mud. It is a powerful and deadly fighter and can run at speeds of up to 57 km/hr. Hearing is a much more important sense than is vision, to such an extent that blind buffalo that are otherwise fit have been found in herds (Kingdon 1982). Studies in East Africa by Mloszewski (1983) indicate that there is no migration but that groups occupy a particular home range for years and move through it on a circuitous route, perhaps 50–105 km long, repeatedly visiting favorable pastures, water sources, mineral licks, and resting sites. The animals are on the move for about 18 hours a day at an average rate of 5.4 km/hour. They simultaneously graze for about 8–10 hours, largely in the late afternoon and night. There are peaks of drinking in the morning and at dusk. Resting and ruminating

African buffalo *(Syncerus caffer):* Top, photo from San Diego Zoological Garden; Inset: photo from U.S. National Zoological Park; Bottom, photo from East Berlin Zoo.

may take place at any time of the day, but a favored period is 1200–1600 hours, usually the hottest part of the day. However, unless the heat is unusually intense, the animals prefer to rest in the open and not in deep shade. The African buffalo is mainly a grazer but occasionally browses on leaves.

Mloszewski (1983) calculated that in one region where the buffalo still occurred under fully natural conditions at least 3,000 animals occupied 5,625 sq km. He also reported that herds had home ranges of 126–1,075 sq km and that densitites therein were 0.17/sq km to 3.77/sq km. In another

study in East Africa, Sinclair (1977*a*) found herd ranges of about 10–300 sq km and population densities of about 3/sq km to 18/sq km. Generally, density was greatest, and range smallest, in areas with the highest rainfall. These two authorities differed to some extent in their interpretation of the social structure of *Syncerus* in the open country of East Africa, but a synthesis of their findings suggests that there are local populations or herds containing approximately 50–500 individuals. Larger gatherings, containing up to 3,000 buffalo, sometimes form but are temporary and lack social cohe-

Forest buffalo *(Syncerus caffer nana)*, photo by Lothar Schlawe.

sion. A true herd moves together through a home range that is largely separate from those of other herds. It comprises a number of permanent social groups consisting predominantly of adult females and young but including some adult males during at least the rainy part of the year. These groups in turn are composed of units consisting of a female and her young of the previous two birth seasons. A herd also contains several bachelor groups of up to 12 males each and a separate group of juvenile animals. Some males, especially very old ones, form small units that live apart from the rest of the herd. Dominance hierarchies exist within the various social units of a herd and are most evident in the bachelor groups. Rank is established by agonistic displays, especially head tossing, and occasional serious fighting that involves head-to-head charges. Although several mature males may be present in a mixed group, they compete for the estrous females. Males usually are dominant to females but have only a limited leadership role. Actual leadership of an entire herd seems to be shared by a number of ranking individuals of both sexes that alternate at the forefront during movements. Vocalizations resemble those of domestic cattle but tend to be lower in frequency. In the deep forests of West and Central Africa large herds are not evident, and groups seem to consist usually of 8–20 closely related individuals.

Reproduction occurs throughout the year in some areas, but there are seasonal peaks associated with rainfall. In the Serengeti, the heavy rains extend from about February to July. According to Sinclair (1977a), most conceptions there occur at the end of this season, and births take place mainly in the second half of the following wet season. Females have an estrous cycle of about 23 days, an estrus of 5–6 days, and a mean gestation period of 340 days. The single reddish or blackish brown calf has an average weight of 40 kg at birth. Young males leave their mother at about 2 years, but females remain until they produce their own young, or even longer. Sexual maturity seems to come mainly between 3.5 and 5

years. Wild individuals up to 18 years old were found in the Serengeti. A captive specimen lived for 29 years and 6 months (Jones 1982).

Syncerus often is considered to be the most dangerous big game animal of Africa. Old bulls are said to stalk human victims and attack without provocation. Sinclair (1977a) stated that the buffalo's reputation is unjustified and based mainly on the tales of hunters, especially those that tried to follow a wounded animal. The buffalo has been hunted heavily for sport and food. It has been exterminated in most of South Africa, and its range is becoming fragmentary in other regions, especially West Africa, as people occupy its range. It apparently became scarce in East Africa by the late nineteenth century, but numbers recovered from 1900 to 1920, when an epidemic of rinderpest prevented human populations from expanding into buffalo habitat. Schaller (1972) estimated that there were 50,000 buffalo in the Serengeti ecosystem.

ARTIODACTYLA; BOVIDAE; **Genus BOS**
Linnaeus, 1758

Oxen

There are four subgenera and five species (Corbet 1978; Ellerman and Morrison-Scott 1966; Grzimek 1975; Hoffmann 1986b; Lekagul and McNeely 1977; Medway 1977):

subgenus *Bos* Linnaeus, 1758

B. taurus (aurochs and domestic cattle), originally found throughout Europe, southern Asia, and northern Africa;
B. javanicus (banteng), Burma, Thailand, Indochina, Malay Peninsula, Java, Borneo;

Domestic bull, near aurochs *(Bos taurus)*, photo from West Berlin Zoo.

subgenus *Bibos* Hodgson, 1837

B. gaurus (gaur, or seladang), Nepal and India to Indochina and Malay Peninsula;

subgenus *Novibos* Coolidge, 1940

B. sauveli (kouprey), northern Cambodia and adjacent parts of Thailand and Laos, southern Viet Nam;

subgenus *Poephagus* Gray, 1843

B. grunniens (yak), Tibet and adjacent highland regions.

Each subgenus sometimes has been considered to be a full genus. The studies of Groves (1981b) suggest somewhat different relationships than those indicated above, namely that *B. sauveli* and not *B. javanicus* has affinity to *B. taurus* and that *B. grunniens* has affinity to *Bison*, the latter also belonging in the genus *Bos*. He and some other authorities use the names *B. primigenius* and *B. mutus*, respectively, in place of *B. taurus* and *B. grunniens*. The names *B. sondaicus* and *B. banteng* sometimes have been used in place of *B. javanicus*. The zebu, a domestic form of cattle from India, and the gayal, a domestic form of gaur from eastern India and Burma, sometimes have been given the respective scientific names *B. indicus* and *B. frontalis*. Based on an ornament found at an archeological site dating from the Han Dynasty (206 B.C.– A.D. 220), Hoffmann (1986b) suggested that the range of *B. sauveli* once extended as far north as Yunnan.

According to Lekagul and McNeely (1977), *Bos* is characterized by large size, a massive body, stout limbs, and a tail that is long and usually tufted at the tip. There are no suborbi-tal, inguinal, or interdigital glands. Horns, borne by both sexes, are larger in males, are inserted rather far apart on each extremity of the top of the skull, and are approximately oval in cross section. The occipital area of the skull forms an acute angle with the face, not a right angle as in *Bubalus*. Additional information is provided separately for each species.

Bos taurus (aurochs and domestic cattle).

According to Grzimek (1975), the wild aurochs, ancestor of domestic cattle, had a head and body length of up to 3,100 mm, a shoulder height of 1,750–1,850 mm in males, and a weight of 800–1,000 kg. Males were about one-fourth larger and heavier than females. The anterior part of the body was more massive than the hindquarters, the legs were rather long, the back was straight, and the head was slender. The pointed, sturdy horns were up to 800 mm long. The coat was short and smooth and was denser in winter. The general coloration was some shade of brown. The original habitat was open forest and meadows. Activity was probably mainly diurnal. The diet included grass, leaves, and acorns. Herds consisted of a bull, several females, and their young. During the mating season, in August and September, males had severe fights. The young were born in May and June, after a gestation period of about nine months. The aurochs apparently disappeared in southern Asia early in historical time but survived in western and central Europe until the Middle Ages. The decline was associated with habitat alteration and hunting by people. Efforts were made to protect the last known herd, which lived in Poland, but all the animals died by 1627. Harrison (1972) referred to a highly doubtful report that the aurochs may have survived in northern Iraq until the early twentieth century.

Top, Scotch cattle *(Bos taurus)*. Bottom, British park cattle *(Bos taurus)*. Photos by Ernest P. Walker.

The aurochs was domesticated by people about 8,000 years ago and gave rise to the numerous breeds of cattle now used for the production of milk, meat, leather, fertilizer, and other items. In Europe there are still some primitive breeds that resemble the aurochs in certain external characters. These forms include the Spanish fighting oxen, Corsican country cattle, Scottish highland cattle, and English park cattle (Grzimek 1975). Some of these kinds have been used experimentally in efforts to breed animals that look like the original aurochs (see accompanying photograph). The English park

A. Zebu *(Bos taurus)*, photo by R. Pucholt. B. Ankole cattle *(B. taurus)*, photo by C. A. Spinage.

cattle of the Chillingham and Chartley estates are generally white in color but are like the aurochs in build and behavior. Their precise origin is unknown. The Chillingham herd has been kept fenced, but in a semiwild state, since about A.D. 1220. The only change in this herd over the centuries, at least in external appearance, has been a slight decrease in size. The herd is an absolute monarchy, the "king" bull being the only male that mates with the females. He usually reigns for two or three years, or until he is deposed by another bull.

Modern domestic cattle usually have a shoulder height of 900–1,100 mm and a weight of 450–1,000 kg. The body is usually covered with short hair. Coloration depends on breed, ranges from white to black, and is often marked with spots and blotches. Cattle graze approximately 8 hours a day, during which time they consume about 70 kg of grass. The remaining time is spent resting or chewing the cud. Mating takes place throughout the year, and a single offspring (occasionally twins) is born after a gestation of 277–90 days. Females attain sexual maturity at 18 months and remain fertile for about 12 years. The life span may be more than 20 years.

The zebus, the sacred cattle of India, are forms of *Bos taurus*. Of the 30 or more breeds, each of which originated in a province of India, 4 main strains have been introduced to the United States. Zebus have a characteristic hump over the shoulder, drooping ears, and a large dewlap. The coloration is pale fawn, bay, gray, or black. In India, they are allowed to roam the streets and villages without molestation. They interbreed readily with other kinds of cattle, and both they and their hybrids are valued for their ability to resist heat, ticks,

and insects. A breed known as the Santa Gertrudis, recognized in 1940, was developed on the King Ranch of Texas by crossing zebu bulls with stock that originated from Texas longhorns, herefords, and shorthorns.

The Texas longhorn itself is one of a number of old breeds of cattle that now are attracting renewed attention from both a conservation and a commercial standpoint (Vietmeyer 1983). Introduced to the American West by the Spanish, it numbered in the millions during the nineteenth century and then declined to a few thousand individuals as it was replaced by more modern breeds. Now considered a valuable genetic resource that may lead to the breeding of hardier cattle, its numbers have increased to 62,000 registered individuals. There once were large feral herds of longhorns in the West, and there were vast numbers of free-living cattle in other parts of the world, notably on the pampas of Argentina (Jackson and Langguth 1987). Today there are scattered herds of feral cattle in both the western and eastern United States, as well as in Europe and on many islands, but the largest numbers, 58,000–129,000, are in Australia (Lever 1985).

Bos javanicus (banteng).

Head and body length is 1,800–2,250 mm, tail length is 650–700 mm, shoulder height is 1,200–1,900 mm, and weight is 400–900 kg. Coloration varies, the females usually being brown or reddish brown and the males blackish brown or blue-black. There are white stockings and a white rump patch. The horns of bulls are angular, turning out and then up, with inward pointing tips and a spread of 600–750 mm; the horns of cows are short and crescent-shaped (Grzimek

Banteng *(Bos javanicus)*, photo from West Berlin Zoo.

1975; Lekagul and McNeely 1977; Popenoe 1983).

The banteng is usually found in drier, more open areas than is *B. gaurus*. Nonetheless, it depends on dense thickets and forest for shelter. It may be active at any hour but has become nocturnal in areas where molested by people (Lekagul and McNeely 1977). It sometimes feeds continuously through the night, with many pauses for rest and chewing. It is extremely wary and shy. During the monsoon season, herds may leave the lowlands and drift up into the hill forests, where they feed on tender new herbage, including bamboo shoots. In the dry season, they return to the valleys and more open wooded districts, where they feed on grass. *B. javanicus* apparently does not have to drink as often as *B. gaurus*.

The banteng is usually found in groups of 2–40 animals. Generally, there is only a single fully adult bull per herd. Other males either are solitary or live in unisexual groups. Mating may occur at any time of year in captivity but takes place in May and June in the wild in Thailand. Females are capable of giving birth each year. The gestation period is 9.5–10.0 months, there are 1 or 2 offspring per birth, nursing lasts up to 9 months, and females attain sexual maturity at 2 years of age (Lekagul and McNeely 1977; Van Bemmel 1967).

The banteng is classified as vulnerable by the IUCN (1978) and as endangered by the USDI. It has disappeared in many areas through loss of habitat to the expanding human population, hunting pressure, and hybridization with domestic cattle. Several thousand individuals are still thought to survive from Burma to Borneo, but numbers are decreasing. In contrast, there now are an estimated 1.5 million domestic banteng, which are used as work animals and for the production of excellent meat. It is thought that domestication occurred in prehistoric times on Bali and Sumbawa and that the animals subsequently were brought to Sumatra, Java, Borneo, Sulawesi, Lombok, Timor, other small islands in the East Indies, and northern Australia. Feral herds now are established in Australia and some other areas. Mating between *B. javanicus*

and *B. taurus* occurs freely and yields fertile offspring; nearly all of the 575,000 domestic cattle on the island of Madura, off northern Java, are part of a population resulting from ancient hybridization between the banteng and the Indian zebu (see above; see also Lever 1985 and Popenoe 1983).

Bos gaurus (gaur, or seladang).

Head and body length is 2,500–3,300 mm, tail length is 700–1,050 mm, shoulder height is 1,650–2,200 mm, weight is 650–1,000 kg, and horn length is 600–1,150 mm. Males are approximately one-fourth larger and heavier than females (Grzimek 1975). The general coloration is dark reddish brown to almost blackish brown, and there are white stockings. In adult males there is a large hump over the shoulders.

The gaur inhabits forested hills and associated grassy clearings up to elevations of 1,800 meters. It requires water for drinking and bathing but seems not to wallow (Grzimek 1975). It is not excessively wary, but when startled, it crashes off through the jungle at high speed. In some areas it is basically diurnal, feeding in the morning and late afternoon and resting during the hot hours of the day. In other areas it has become largely nocturnal because of molestation by people (Lekagul and McNeely 1977). In a study in central India, Schaller (1967) found most activity at night; the animals filed from the meadows to the forest at dawn and were rarely seen in the open after 0800 hours. Most herds remained in a relatively small area until the beginning of the monsoon and then dispersed into the hills. The gaur was found to be both a grazer and a browser, preferring green grass when it was available but otherwise eating coarse, dry grasses and forbs and leaves.

In Schaller's study area, population density was about 0.6/sq km, and overall home range of a herd was around 78 sq km. Groups contained 2–40 individuals, usually about 8–11. There generally was not more than a single fully mature bull in a herd. Some males lived alone or in bachelor groups. At the peak of the mating season, the bulls wandered widely in

Gaur *(Bos gaurus)*, photo from West Berlin Zoo.

search of estrous females and did not spend more than a few days in any one herd. They competed and sparred with one another, but serious fighting was not observed. Dominance seemed to be based primarily on size. Schaller described the mating call of the bull as a pleasant song composed of clear resonant tones, each successively lower, that carries about 1.6 km. The gaur has also been reported to have a whistling snort as an alarm call and to bellow and "moo" like domestic cattle.

Reproduction takes place throughout the year, but in central India mating seems to occur mainly from December to June, and the peak calving season is in the cool months of December and January. The gestation period is 270–80 days. There is normally a single offspring which is nursed for 9 months. Females attain sexual maturity in their second or third year of life (Grzimek 1975; Medway 1978; Schaller 1967). A captive gaur lived for 26 years and 2 months (Jones 1982).

The gaur is classified as vulnerable by the IUCN (1976) and as endangered by the USDI and is on appendix 1 of the CITES. It still is found in scattered areas from India to Indochina and the Malay Peninsula but has declined drastically through hunting and habitat alteration by people and exposure to the diseases of domestic cattle. The gaur has been reported occasionally to ambush and kill persons that pursue it (Lekagul and McNeely 1977). The gayal, a domestic form of gaur, is smaller than the wild animal and has shorter legs and a less prominent dorsal hump. It may live in a semitame or feral state in eastern India and Burma, and it is used for work and meat production (Grzimek 1975). A gestation period of 293–303 days has been reported for the gayal (Scheurmann 1975).

Bos sauveli (kouprey).

Except as noted, the information for the account of this species was taken from Lekagul and McNeely (1977). Head and body length is 2,100–2,225 mm, tail length is 1,000–1,100 mm, shoulder height is 1,710–1,900 mm, and weight is 700–900 kg. In females and young animals the general coloration is gray, the underparts are lighter, and the chest, neck, and forelegs are darker. Old males are a very rich, dark brown. The lower legs are white or grayish. From the neck hangs a very long dewlap, which in old males almost reaches the ground. The tail is longer than in *B. gaurus* and *B. javanicus* and has a bushier tip. Suggestions that the kouprey is only a hybrid between *B. javanicus* and either *B. gaurus*, *B. taurus*, or *Bubalus bubalis* do not appear to be valid.

There is considerable difference between the horns of the sexes. In females the horns are lyre-shaped and corkscrew upward in a manner resembling those of *Tragelaphus imberbis*, of Africa. In males the horns are much larger, spread outward and downward from the base, and then curve forward and upward. When the bulls are about three years old, the horns split at the tip, and as the horns grow larger, the split pieces also continue to grow. This fraying of the horn tips is said to be caused largely by their being used for digging into the earth or thrusting into tree stumps.

The kouprey inhabits low, rolling hills covered by open country interrupted by patches of forest. It generally grazes in the open areas and enters the forest for shelter from the sun, for refuge from predators, and to seek food when the grasslands are dry. The kouprey moves considerably more than does *B. javanicus*, covering up to 15 km in a day's undisturbed feeding. Herds often contain more than 20 individuals, including several adult males. The herds are not particularly cohesive. Males are sometimes found alone or in groups of their own. Mating takes place in April, and births in December and January. A mother and young may stay away from the herd for about a month.

The kouprey is classified as endangered by the IUCN (1976) and the USDI. It did not become known to science

Kouprey *(Bos sauveli)*, young male with horns not fully developed, photo from Paris Zoo.

until 1937. Since its discovery, its limited range has been the scene of almost constant human warfare. Little information about the species has emerged, except that it is one of the world's most critically endangered mammals. Its total numbers were estimated at 1,000 in 1940, 500 in 1951, and 100 in 1969. At times it has been feared to be extinct, but surveys in 1986 obtained reliable reports indicating that small, fragmented populations still survive throughout most of the original known range of the species; development of an international conservation program recently was initiated by the World Wildlife Fund and other organizations (Stuart 1988; Thouless 1987).

Bos grunniens (yak).

In wild males, head and body length is up to 3,250 mm, shoulder height may be over 2,000 mm, and weight is about 1,000 kg; females weigh only one-third as much (Grzimek 1975). Long hairs that reach almost to the ground form a fringe around the lower part of the shoulders, the sides of the body, and the flanks and thighs. The tail also has long hairs. The general coloration is blackish brown. The large black horns curve upward and forward in the males. The domestic yak is much smaller than the wild form, has weaker horns, and has more varied coloration, being red, mottled, brown, or black.

The yak inhabits desolate steppes at elevations of up to 6,100 meters. In spite of its awkward appearance, it is an expert climber, sure-footed, and sturdy. According to Grzimek (1975), the yak stays in high areas with permanent snow during the relatively warm months of August and September and spends the rest of the year at lower elevations, feeding on grass, herbs, and lichens. Females and young congregate in large herds, formerly reported sometimes to contain thousands of individuals, and adult males spend most of the year alone or in groups of up to 12. The mating season begins in September, and for the next several weeks the bulls join the herds and fight one another for the females. At this time, the wild yak utters a strange grunting sound, but domestic animals emit this call throughout the year. In the wild, births occur in June, the gestation period is 9 months, females have a single calf every other year, independence is attained after 1 year of life, full size is reached at 6–8 years, and maximum longevity is 25 years. Domestic females have a less regular reproductive cycle and may give birth every year.

The wild yak is classified as endangered by the IUCN (1974) and the USDI. Although it is officially protected in China, numbers and distribution have declined drastically through uncontrolled hunting. The yak probably was domesticated in Tibet during the first millennium B.C., and it now occurs throughout the high plateaus and mountains of Central Asia, in association with people. This docile but powerful animal is the most useful of domestic mammals at elevations above 2,000 meters. It serves as a mount and beast of burden, is used for milk and meat production, and is sheared for its wool (Grzimek 1975). There now are as many as 12 million domestic yaks but probably no more than a few hundred wild yaks (Alexander 1987).

ARTIODACTYLA; BOVIDAE; Genus BISON
Hamilton-Smith, 1827

Bison

There are two species (Corbet 1978; Grzimek 1975; Hall 1981):

B. bonasus (European bison, wisent), found in historical time in much of Europe and possibly Asia;

B. bison (American bison), found in historical time in western Canada, most of the conterminous United States, northeastern Mexico, and possibly Alaska.

Yak *(Bos grunniens)*, photo from U.S. National Zoological Park.

European bison *(Bison bonasus)*, photo from East Berlin Zoo.

Some authorities, including Corbet (1978), consider *B. bonasus* to be conspecific with *B. bison*. Some authorities, including Van Gelder (1977b), consider *Bison* to be not more than subgenerically distinct from *Bos*. Groves (1981b), who adopted both of these views, suggested that bison are actually less distinct members of *Bos* than are *B. gaurus* and *B. javanicus*. The inclusion of *Bison* within *Bos* was not accepted by McDonald (1981), Meagher (1986), or Van Zyll de Jong (1986). Those three authorities also recognized the division of North American *B. bison* into two subspecies, *B. b. athabascae* to the north and *B. b. bison* to the south, with the dividing line between the two extending across west-central Canada and not reaching into the western United States. There is disagreement with respect to the fossil history of modern bison. McDonald (1981) concluded that *B. bison* is a remarkably young species, having evolved rather suddenly only 4,000–5,000 years ago from a surviving line of the larger Pleistocene *B. antiquus*. Van Zyll de Jong (1986) treated *B. antiquus* as a subspecies of *B. bison* and thought that it gave rise to *B. bison bison* about 5,000–7,000 years ago.

Head and body length is 2,100–3,500 mm, tail length is 300–600 mm, shoulder height is 1,500–2,000 mm, and weight is 350–1,000 kg. Males average larger than females. The pelage of the head, neck, shoulders, and forelegs is long, shaggy, and brownish black. There is usually a beard on the chin. The remainder of the body is covered with short hairs of a lighter color. Young calves are reddish brown. The forehead is short and broad, the head is heavy, the neck is short, and the shoulders have a high hump. The horns, borne by both sexes, are short, upcurving, and sharp. Females have a smaller hump, a thinner neck, and thinner horns than males.

Of the two species, *B. bison* has longer and more luxuriant hair on the neck, head, and forequarters, giving this species the appearance of larger size. Its body is lower, its pelvis smaller, and its hindquarters less powerful than in *B. bo-*

nasus, though its body on the whole is more massively built. The horns of *B. bison* are shorter and more curved, the front of its head is more convex, and its tail is shorter and less bushy.

The American bison is traditionally associated with the prairies, but it also occurred extensively in mountainous areas and open forests. The European bison inhabited both woodland and grassland. Bison frequently wallow in dust or mud and then rub against boulders, tree trunks, and other objects to rid themselves of parasites. They have acute senses of hearing and smell, can run at speeds of up to 60 km/hr, and regularly swim rivers about 1 km wide (Meagher 1986). Activity may occur at any hour. Banfield (1974) stated that *B. bison* is mostly diurnal and has an average daily movement of about 3 km. Populations of this species on the Great Plains formerly made migrations of several hundred kilometers, generally in a southward direction, to seek better feeding grounds for the winter. The existing population in northern Alberta still migrates up to 250 km each November and May, between the wooded hills and the Peace River Valley. The Yellowstone Park herd also shifts to the high country in the spring and to low valleys in the fall (Meagher 1973). *B. bison* feeds mostly on grass, while *B. bonasus* takes mainly the leaves, twigs, and bark of trees.

Population density of *B. bonasus* has been calculated at about 12 individuals per 1,000 ha. in Poland and 3–4 per 1,000 ha. in the Caucasus (Krasinski 1978). A small herd of *B. bison* has been estimated to use a range of about 30 sq km in summer and 100 sq km in winter (Banfield 1974). Both species seem to occur basically in small groups of probably related individuals, but sometimes, such as during migrations or on favorable feeding grounds, they associate in herds of hundreds of animals. Formerly, such aggregations of *B. bison* contained many thousands and perhaps millions of individuals. In a study of *B. bison* in Montana, Lott (1974) found adult

American bison *(Bison bison)*, photo from U.S. National Zoological Park.

females, and their young up to three years of age, to exist in groups with an average membership of 57 animals. Mature males moved about alone or in small groups for most of the year but joined the female groups during the mating season. At this time the males fought fiercely for the females, mainly by head-to-head ramming. A successful male would tend a female for several days and prevent other males from approaching within eight meters of her. However, Lott (1981) noted that a cow might break away and run past other bulls in an apparent attempt to stimulate competition and thereby provide herself with a mate of the highest rank. The distinctive roar or bellow of a rutting bull may carry nearly 5 km (Meagher 1986).

Mating occurs mainly from July to September in both North America and Europe, and births take place in the spring. There appears to be some degree of synchrony in local populations, however, with 90 percent of matings at the National Bison Range in Montana occurring in a two-week period in late July and early August. Females are seasonally polyestrous, have an estrous cycle of about 21 days and an estrus of 9–28 hours, and may produce young every 1 or 2 years. The gestation period averages 285 days in *B. bison* and is 260–70 days in *B. bonasus*. There is almost always a single calf. It weighs about 15–30 kg at birth, can run after 3 hours, and is weaned at about 7–12 months. The mother guards it closely and will charge intruders. Both sexes seem to attain sexual maturity at 2–4 years. Physical maturity may come at about 6 years for males and 3 years for females. Wild individuals are known to have lived about 20 years, and maximum potential longevity may be as much as 40 years (Banfield 1974; Grzimek 1975; Haugen 1974; Krasinski 1978; Lott 1981; Meagher 1973, 1986).

Mainly because of the spread of human settlements and agriculture, by the early twentieth century *B. bonasus* became restricted to the Caucasus and the Bialowieza Forest, on what is now the border between Poland and the Soviet Union. The population of the Caucasus became extinct by 1925. The Bialowieza population, which numbered 727 individuals and was well protected prior to World War I, was completely destroyed by 1919. Some animals had previously been distributed to zoos, however, and some of these bison were eventually reintroduced to Bialowieza and other areas. There are now about 400 *B. bonasus* in the Bialowieza Forest and about 2,700 in the world. A herd has been established in the Caucasus, but it contains a small admixture of genes from *B. bison* (Grzimek 1975; Krasinski 1978; Pucek 1984, 1986).

There were once an estimated 50 million *B. bison bison* in North America. Certain Indian tribes depended largely on the hunting of these animals. The decline of the great bison herds began almost as soon as European explorers arrived on the continent. The bison were hunted commercially and for subsistence in order to obtain both meat and skins. They also were shot to protect agricultural interests and to help subdue the Indians of the Great Plains. *B. bison bison* disappeared from east of the Mississippi early in the nineteenth century. By 1890 only a few hundred individuals survived. Subsequent private and governmental conservation efforts have allowed bison numbers to increase to over 100,000 (Thornback 1985). Many of these animals are captive or are descended from captive stock. Indeed, the only place where a wild herd of *B. bison bison* has been maintained continually is Yellowstone National Park (Meagher 1973). That herd, now containing about 1,500–2,000 animals, recently has been the subject of controversy, with many individuals wandering out of the park being shot for sport or to prevent the spread of disease to domestic cattle (Sabbag 1986). These problems give emphasis to the question whether today's bison herds represent a conservation triumph, as often is claimed. The meaning of a species involves not only its morphology but also ecological, behavioral, and historical factors. Maintenance of a series of living specimens that cannot be allowed to leave their expanded cages may be an inadequate representation of a creature that was rapidly evolving within the framework of an entire hemisphere and that was the dominant force in shaping the ecology of most of a continent.

The woodland subspecies, *B. bison athabascae*, was far less abundant than *B. b. bison*, and it may have disappeared from Alaska in prehistoric times. The original number in western Canada is estimated at 168,000. By 1891 no more than 300 survived, those being in northern Alberta and adjacent parts of the Northwest Territories. The Wood Buffalo National Park was established to protect them, but in the 1920s thousands of *B. b. bison* were released there, and consequently most of the population of *athabascae* was genetically swamped. In 1957 an isolated herd of about 200 wood bison was discovered in the northern part of the park. Although there is evidence of some former gene flow from the hybrid population, the newly found animals appear to be very close to original *athabascae*, and they have been subjected in part to a selective breeding program. The original herd now has grown to over 2,500 animals in six separate wild and captive groups (Burnett et al. 1989; Shoesmith 1989; Van Zyll de Jong 1986). *B. bison athabascae* is listed as endangered by the USDI and is on appendix 1 of the CITES.

ARTIODACTYLA; BOVIDAE; **Genus CEPHALOPHUS**
Hamilton-Smith, 1827

Duikers

There are 4 subgenera and 18 species (Ansell, *in* Meester and Setzer 1977; Groves and Grubb 1974; Kingdon 1982; Meester et al. 1986):

subgenus *Cephalophorus* Gray, 1842

C. adersi, east coast of Kenya, Zanzibar, Pemba Island;
C. natalensis, Zaire and southern Tanzania to South Africa;
C. harveyi, southern Somalia, Kenya, Tanzania, Malawi;
C. nigrifrons, southern Cameroon to western Kenya and northern Angola;
C. rufilatus, Senegal to southwestern Sudan;
C. rubidus, Ruwenzori Mountains of western Uganda;
C. leucogaster, southern Cameroon to Zaire;
C. ogilbyi, Sierra Leone to Gabon, island of Bioko (Fernando Poo);
C. callipygus, southern Cameroon, Gabon, Congo;
C. weynsi, Zaire, Uganda, Rwanda, western Kenya;
C. niger, Guinea to southwestern Nigeria;

subgenus *Cephalophus* Hamilton-Smith, 1827

C. spadix, northeastern and southern highlands of Tanzania;
C. sylvicultor, Gambia to Kenya, and south to northern Angola and Zambia;
C. jentinki, Liberia, probably Ivory Coast and Sierra Leone;
C. dorsalis, Guinea-Bissau or possibly Gambia to Zaire and northern Angola;

subgenus *Cephalophula* Knottnerus-Meyer, 1907

C. zebra, western Sierra Leone to central Ivory Coast;

subgenus *Philantomba* Blyth, 1840

C. monticola, southeastern Nigeria to Kenya, and south to South Africa;
C. maxwelli, Senegal to southwestern Nigeria.

A. Black-fronted duiker *(Cephalophus niger)*, photo by Ernest P. Walker. B. Striped-backed duikers *(C. zebra)*; C. Light-backed duiker *(C. sylvicultor)*; photos by Bernhard Grzimek.

Meester et al. (1986) recognized *Philantomba* (a synonym is *Guevei* Gray, 1852) as a full genus and *Cephalophorus*, *Cephalophus*, and *Cephalophula* as valid subgenera. That treatment, as well as Kingdon's (1982) assessment of specific relationships, placing *C. adersi* near the central evolutionary line of *Cephalophus*, is considered in the above arrangement. However, there is much controversy regarding the systematics of this genus. In contrast to most recent authorities (Ansell, *in* Meester and Setzer 1977; Corbet and Hill 1986; Honacki, Kinman, and Koeppl 1982), Kingdon did not regard *C. weynsi* as a species distinct from *C. callipygus* but did treat *C. rubidus* as a species separate from *C. nigrifrons*, and *C. harveyi* as a species separate from *C. natalensis*. He noted, however, probable hybridization between *C. nigrifrons* and *C. rubidus*. A species of *Cephalophus*, variously designated *C. natalensis*, *C. nigrifrons*, or *C. harveyi*, has been reported to occur in Ethiopia, but Yalden, Largen, and Kock (1984) stated that there were no satisfactory records.

Size varies considerably, ranging from that of a hare to that of a deer (Grzimek 1975). In *C. monticola,* one of the smallest species, head and body length is 600–650 mm, tail length is 90–100 mm, height at the shoulder is 350–410 mm, weight is 3.5–9.0 kg, and horn length is 30–90 mm. In *C. sylvicultor,* one of the largest species, head and body length is 1,500–1,900 mm, tail length is 180–200 mm, height at the shoulder is 700–870 mm, weight is 45–80 kg, and horn length is 110–210 mm (Kingdon 1982). The upper parts vary in color from almost buffy through shades of brown to almost black. Most forms have a stripe along the middle of the back that differs in color from the remainder of the back. One species, *C. sylvicultor,* is generally blackish brown in color but has a large yellow rump patch. Another species, *C. zebra,* has a bright orange coat marked with dark vertical stripes. The underparts range from white to almost as dark as the back. All species have relatively short legs, pointed hooves, and an arched back. The horns, possessed by both sexes, are short, project backward from the skull, and are frequently hidden by a long tuft of hair on the crown.

Kingdon saw *C. adersi* as a generalized species representing the central primitive stock of the genus that led to two main evolutionary branches: (1) the species with long, coarse neck fur—*C. natalensis, C. harveyi, C. nigrifrons, C. rufilatus,* and *C. rubidus;* and (2) the species with short, smooth neck fur—*C. leucogaster, C. ogilbyi, C. callipygus, C. weynsi, C. niger, C. spadix, C. sylvicultor, C. jentinki, C. dorsalis,* and *C. zebra.* A third branch, leading to the genus *Sylvicapra,* is represented by *C. monticola* and *C. maxwelli.* All three branches contain both small and large species.

Some species prefer open country with scattered trees and brush, whereas others are partial to the dense growth of the jungle. Some species are adapted to swamps, and others occur on mountain slopes at elevations up to 3,500 meters (Kingdon 1982). Although these antelope are quite numerous, they are seldom observed, owing to their shyness, largely nocturnal habits, and inaccessible habitats. When alarmed, they dart away with great speed into the protection of dense vegetation; hence the origin of the common name, duiker, which means "diving buck." They are sturdy and active and have acute vision and hearing. The horns serve several purposes, such as defense against enemies and combat with others of their own kind. They feed on grass and leaves, at times scrambling upon logs or climbing vine-entangled shrubs to obtain choice leaves and fruit. Recent investigations indicate that fruits make up the major part of the diet of some species (Dubost 1984; Lumpkin and Kranz 1984). Duikers also have been observed to eat insects and carrion. According to Farst et al. (1980), they kill and eat small animals, especially birds, and in captivity are given dog food in addition to vegetable matter.

Duikers are not gregarious and usually move about alone or in pairs. Radiotracking studies in Gabon showed that *C. monticola* occurs at a density of 70/sq km, pairs occupy exclusive permanent territories of 2.5–4.0 ha., and upon maturity the young sever all links with the parents (Kingdon 1982). According to Ralls (1973), *C. maxwelli* also is probably territorial and scent-marks an area with secretions from maxillary glands. Individuals are highly intolerant of others of the same sex, and fights are frequent. In the Ivory Coast, the young of this species are produced mainly during the two dry seasons, January–March and August–September. A female can give birth each year, the gestation period is 120 days, and there is a single offspring. It weighs 710–954 grams at birth, resembles the adults in color pattern, nibbles leaves after 14 days, and is completely weaned at 2 months. Observations of captives indicate that estrus occurs about once a month throughout the year, that gestation varies from 126 days in *C.*

niger to as long as 245 days in *C. rufilatus,* and that females of the latter species reach sexual maturity at around 9 months (Dittrich 1972; Farst et al. 1980). A captive *C. nigrifrons* lived for 19 years and 8 months (Jones 1982).

Duikers are hunted for food by native people but generally are not sought by sportsmen, because their head does not make a showy trophy. *C. jentinki,* restricted to a small and rapidly developing region of West Africa, is jeopardized by habitat destruction and probably numbers only a few hundred individuals; it is classified as endangered by the IUCN (1976) and the USDI. *C. zebra,* found in the same region, is classified as vulnerable by the IUCN, as are *C. adersi* and *C. spadix.* According to East (1988), the latter two species, along with *C. rubidus,* are endemic to small parts of East Africa and are threatened with extinction because of poaching and human encroachment on their limited habitat. *C. monticola* is on appendix 2 of the CITES; it declined sharply in South Africa during the 1940s but recovered to some extent by the 1980s following habitat protection and strict control of hunting (Crawford and Robinson 1984).

ARTIODACTYLA; BOVIDAE; Genus **SYLVICAPRA**
Ogilby, 1837

Gray Duiker, or Common Duiker

The single species, *S. grimmia,* occurs in the savannah zones and mountainous areas from Senegal to Ethiopia, and south to South Africa (Ansell, *in* Meester and Setzer 1977). Van Gelder (1977b) considered *Sylvicapra* to be a synonym of *Cephalophus.*

Head and body length is 700–1,150 mm, tail length is 75–195 mm, shoulder height is 450–700 mm, and weight is 12–25 kg. Females are usually larger than males. Coloration is grayish, yellowish, or reddish yellow. *Sylvicapra* is distinguished from *Cephalophus* in having ears that are pointed, rather than rounded, and horns that are sharply pointed, rise above the plane of the face, and are usually present only in males. Females occasionally carry small, stunted horns, and such females seem to be more common in some localities than in others.

The gray duiker avoids dense forests but usually is found in places with adequate vegetative cover. This hardy and adaptable antelope lives at higher elevations than any other hoofed mammal in Africa. It reportedly is common in alpine meadows in equatorial regions. It has great speed and stamina and is usually able to outdistance dogs. It rests during the day in scrub or grass, often in favorite hiding places, and comes out at night to feed. It is predominantly a browser, and the diet includes a great variety of leaves, stems, flowers, fruits, bulbs, tubers, and cultivated crops. It occasionally takes insects, frogs, birds, small mammals, and carrion (Kingdon 1982; Smithers 1971; Wilson 1966a).

Sylvicapra usually travels alone but occasionally is seen in pairs. Studies by Dunbar and Dunbar (1979) in Ethiopia indicate that there is little overlap between the home ranges of individuals of the same sex but more overlap between those of opposite sexes. Males are apparently territorial, scent-marking their areas with preorbital glands and fighting other males that intrude. Pregnant and lactating females have been taken throughout the year (Smithers 1971). A captive pair produced offspring at intervals of 232–98 days, and estimates of the actual gestation period range from 90 to about 210 days (Smithers 1983; Von Ketelhodt 1977). There is usually a single young per birth, occasionally two. Birth weight is 160–90 grams, adult size is attained at 6–7 months, and females

Common duikers *(Sylvicapra grimmia):* Top, photo from New York Zoological Society; Bottom, photo by John Visser.

reach sexual maturity at 8–9 months (Smithers 1983). A captive lived for 14 years and 4 months (Jones 1982).

ARTIODACTYLA; BOVIDAE; **Genus KOBUS**
A. Smith, 1840

Waterbuck, Lechwes, Kob, and Puku

There are five species (Ansell, *in* Meester and Setzer 1977):

K. ellipsiprymnus (waterbuck), savannah zones of Africa south of the Sahara;
K. megaceros (Nile lechwe), swamps of southern Sudan and western Ethiopia;

K. leche (lechwe), wetlands of southern Zaire, eastern Angola, the Caprivi Strip of Namibia, Zambia, and northern Botswana;
K. kob (kob), savannah zone from Senegal to western Kenya;
K. vardoni (puku), savannah zone of southern Zaire, eastern Angola, Zambia, southwestern Tanzania, Malawi, and northern Botswana.

The name *Adenota* Gray, 1872, sometimes has been used for a genus or subgenus comprising *K. kob* and *K. vardoni.* The name *Onotragus* Gray, 1872, sometimes has been used for a genus or subgenus comprising *K. megaceros* and *K. leche.* The northern and central populations of *K. ellipsiprymnus* often have been considered to be a distinct species, *K. defassa.*

Head and body length is 1,250–2,350 mm, tail length is 100–450 mm, shoulder height is 700–1,360 mm, and weight is 50–300 kg. Males average larger than females. The hair is long and coarse, being longest on the throat and tip of the tail. There is considerable color variation among the species; some are yellowish brown, and others range to almost black. Some species have white markings; *K. ellipsiprymnus,* for example, has a large white patch or an elliptical white ring on the buttocks.

The horns are long, usually measuring 510–1,020 mm along the front from base to tip. Normally carried only by males, they curve forward at the tip, lie at an angle slightly above the level of the face, and have transverse corrugations. The hooves vary in size and shape, but usually the false hooves (the second and fifth digits) are conspicuous. In *K. megaceros* and *K. leche* the backs of the pasterns are bare, an adaptation to wetland habitat.

These animals are seldom found far from water and often remain in swampy tracts, where they frequent reedbeds and shrubby growth. *K. megaceros* and *K. leche* are entirely restricted to floodplains and adjacent ground. *K. kob* and *K. vardoni* inhabit moist savannah, floodplains, and the margins of adjacent light woodlands. Despite the common name "waterbuck," *K. ellipsiprymnus* is less confined to wet areas than are the other species. Although it is always within reach of water, it ranges farther into woodlands than do *K. kob* and *K. vardoni.* Spinage (1982) noted that while it physiologically requires more water for drinking than do most bovids, it does

Lechwe *(Kobus leche)*, photo from New York Zoological Society.

not seem to actually like going into the water. All species are graceful in their movements and run swiftly. *K. megaceros* and *K. leche* are particularly adept at wading and swimming; they travel by a series of leaps in water that is too shallow for swimming. Like most antelopes, *Kobus* is active mainly in the morning and evening. There may be local seasonal movements in response to environmental changes. The population of *K. kob* in the Boma ecosystem of southeastern Sudan is found mainly on lightly wooded grasslands during the wet season, May–October, and then migrates 150–200 km to graze along watercourses in the dry season, November–April; at the latter time population density often exceeds 1,000/sq km (Fryxell and Sinclair 1988).

Detailed studies of *K. leche* on the Kafue Flats of Zambia

show that the species is highly specialized for life on floodplains (Sayer and Van Lavieren 1975; Schuster 1980). It often grazes in water up to shoulder height. Populations move back and forth in relation to the rising and falling of the annual floods, feeding on the grasses that are exposed at various times. Because of the abundance of forage produced by the floods, population density of the Kafue lechwe may be as great as 200/sq km. In contrast, densities of *K. kob* in two areas of Uganda were 8.6/sq km and 45/sq km (Modha and Eltringham 1976). Densities of the latter species actually vary widely over West, Central, and East Africa, the range being about 8/sq km to 124/sq km in suitable habitat (Mühlenberg and Roth 1985). Reported densities of *K. ellipsiprymnus* are about 0.15/sq km to 17.8/sq km, the overall mean in

Nile lechwe *(Kobus megaceros)*, photo from San Diego Zoological Society.

Rwenzori National Park, Uganda, being 2/sq km (Herbert 1972; Spinage 1982).

These antelope are gregarious, sometimes being found in aggregations of many thousands. Systematic observations, however, have revealed an intricate social structure that differs between species. In *K. ellipsiprymnus*, males leave their mother at about 8–9 months and join a bachelor group. Usually these groups have about 2–5 animals, but in some high-density populations there are up to 60. These groups have a dominance hierarchy based on size, strength, and horn length, and fighting is common. When a male reaches an age of about 6 years, he separates from his group and attempts to establish an individual territory. Such an area typically covers about 1–2 sq km, though there is much variation, and is associated with a water supply and other resources needed for the year-round, permanent support of the owner. Bachelor groups are tolerated and may wander over several territories, but other fully mature males are excluded through displays and combat. At any one time most males in a population are not actually holding territories. By the time a male is about 10 years old, he is ousted from his territory by a younger animal and may then become solitary or rejoin a bachelor group. In high-density populations territories are smaller, change ownership more frequently, and are won only by a small percentage of males, but the basic system remains the same, and there is no shift to lekking as found in *K. kob*. Females commonly wander in groups averaging about 5 individuals, occasionally up to 70, and have home ranges that cover several male territories. The female groups seem rather loose, with individuals often moving about separately; young females may be forced out of the home range (Herbert 1972; Spinage 1974, 1982; Wirtz 1982).

Perhaps because of the limited availability and seasonal nature of their habitat, the males of some populations of *K. kob* and *K. leche* do not maintain large, permanent territories. Instead, the adult males in given areas of high population density periodically gather on a cluster of tiny territories called a lek. In *K. kob* the leks are in an elevated area and seem to be used for many years. In *K. leche* leks may form in different spots, depending on the seasonal movements of the herd. Within a lek, 20–200 adult males defend areas ranging from about 15 to 200 meters in diameter. The smaller areas are in the middle of the lek, and there is such intensive competition for these central territories that they are seldom held by one individual for more than a day or two. Each group of lekking males is associated with a herd of up to about 1,000 females, most of which seek to mate with those males on the central territories. There is also an associated herd of young males, some of which may split off and attempt to gain a territory (Buechner 1974; Floody and Arnold 1975; E. Joubert 1972; Schuster 1980). In areas of lower population density, *K. kob* may not employ such an extreme lek system; the males are not so closely spaced during the breeding period and tend to hold their territories for longer periods, and the females and their young commonly form herds of about 5–6 individuals (Mühlenberg and Roth 1985). Some populations of *K. kob*, and some of *K. vardoni* as well, have a territorial system comparable to that of *Redunca* (Kingdon 1982).

Breeding may occur at any time of year, though there are peaks in some areas. In Uganda most births of *K. ellipsiprymnus* occur during the wet seasons, in August and November–December; females have an average interbirth interval of 325 days and enter estrus only once, for 12–24 hours, during this period. The Boma population of *K. kob* gives birth mainly at

Kob *(Kobus kob)*, photo by Leonard Lee Rue III.

Waterbucks *(Kobus ellipsiprymnus)*, photo by Peter Wirtz.

the end of the rains in November–December. Leks seem to form continuously in some populations of *K. kob* but form mainly from November to January in the population of *K. leche* on the Kafue Flats. In the latter area, most births occur from July to September, after a gestation period of about 230 days. Gestation is around 9 months in *K. ellipsiprymnus* and *K. kob*. There is usually a single offspring. Weaning occurs at about 6–9 months in *K. ellipsiprymnus*. Females begin to mate at about 1 year, but males normally must wait several more years (Buechner 1974; Dittrich 1972; Dobroruka 1987; Kingdon 1982; Sayer and Van Lavieren 1975; Schuster 1980; Spinage 1974, 1982). A captive *K. ellipsiprymnus* lived for 18 years and 8 months (Jones 1982), and a wild female of the same species was estimated to be 18.5 years old (Spinage 1982).

The lechwe *(K. leche)* is classified as vulnerable by the IUCN (1974) and as threatened by the USDI and is on appendix 2 of the CITES. Most populations have declined drastically through uncontrolled hunting and habitat loss. The greatest remaining concentration, that on the Kafue Flats of Zambia, may be jeopardized by the recent construction of two huge dams for hydroelectric power. It was feared that the dams would interfere with the natural fluctuation of water levels, thereby preventing maximum food production and perhaps disrupting the lek breeding system (Schuster 1980). Although the extensive disaster that some authorities predicted has not yet occurred, the population of lechwe fell from nearly 100,000 in 1970, before the dams began operation, to about 50,000 afterwards; the latter figure represents about half the total number of *K. leche* in Africa (East 1988; Jeffery, Bell, and Ansell 1989). The Nile lechwe *(K. megaceros)*, numbering 30,000–40,000 individuals and found only in a small swampy part of the upper Nile Valley, also may be jeopardized by water development projects. In the nearby Boma grassland ecosystem there are nearly 1 million *K. kob*, the largest antelope population of Africa except for that of *Connochaetes taurinus* in the Serengeti (East 1988; Hillman and Fryxell 1988). Despite this huge group and other major populations of *K. kob*, Kingdon (1982) noted that the range of the species is greatly diminished.

ARTIODACTYLA; BOVIDAE; **Genus REDUNCA**
Hamilton-Smith, 1827

Reedbucks

There are three species (Ansell, *in* Meester and Setzer 1977):

R. arundinum, savannah zone from Gabon and Tanzania to South Africa;

R. redunca, savannah zone from Senegal to Ethiopia and Tanzania;

R. fulvorufula, known from northern Cameroon, southeastern Sudan, southwestern Ethiopia,

Reedbuck (*Redunca* sp.), photo by E. L. Button.

Reedbuck *(Redunca fulvorufula)*, photo from New York Zoological Society.

northeastern Uganda, western Kenya, northeastern Tanzania, southern Botswana, southern Mozambique, and South Africa.

Head and body length is 1,100–1,600 mm, tail length is 150–445 mm, shoulder height is 600–1,050 mm, and weight is 19–95 kg (Grzimek 1975). The upper parts vary in color from bright fawn to grayish, the underparts are light, and the underside of the tail is white. The body hair is short and stiff, but the tail is bushy. The young have a woolly coat. The neck is thin, and the legs are long and slender. Horns, carried only by the males, are large at the base, prominently ringed, and usually 200–400 mm in length. They point backward, diverge in a graceful upward curve, and curve sharply forward at the tip. There is a conspicuous bare glandular patch below each ear. Females have four mammae.

Reedbucks frequent rolling grasslands, mountain plateaus, open forests, and reedbeds. They usually live near water but do not enter it freely. They are perhaps the least wary of riverside antelope, and once set to flight, they seldom go far before stopping to glance back. If suddenly alarmed, they crouch close to the ground. When fleeing, they hold the tail upright, exposing the white undersurface. They graze early in the morning and evening and frequently throughout the night, eating grass and tender shoots of reeds. Jungius (1971a, 1971b) reported that *R. arundinum* is largely nocturnal but is also active by day during the dry season and that grass makes up the bulk of its diet throughout the year.

In a study of *R. redunca* along the Seronera River in the Serengeti, Hendrichs (1975a) found a population density of 14/sq km to 21/sq km. Females lived alone with their fawns of one or two years in overlapping home ranges of 15–40 ha. Adult males defended territories that comprised the ranges of 1–5 females. *R. fulvorufula* also lives alone or in small groups, with males defending a territory of 15–48 ha. throughout the year (Irby 1979; Smithers 1971). Studies by Howard (1986a, 1986b) of *R. arundinum* on farmland in Natal, South Africa, indicated an average lifetime home range size following dispersal from the family group of 74 ha. for males and 123 ha. for females, with the areas showing considerable overlap. Adult males occurred alone or in bachelor groups for most of the year, and apparently because of limited suitable habitat, only the most dominant individuals were able to establish territories. In Kruger National Park, in South Africa, Jungius (1971b) found *R. arundinum* sometimes to form temporary aggregations of up to 20 animals around limited winter water sources. Otherwise, the animals lived in groups consisting of a mated pair and their young. The male defended a territory of 35–60 ha., with a core area containing favorite resting and grazing sites and a water supply. Other males were excluded, but fighting was not serious enough to cause injury. Vocalizations included clicks, pops, and whistles, the last being used for alarm, territorial expression, and contact between sexes.

According to Jungius (1970), *R. arundinum* breeds throughout the year, but most births occur from December to May in Kruger National Park. The female separates from the male for 3–4 months to give birth and rear the newborn. The calf remains hidden for about 2 months, during which time the mother visits it for 10–30 minutes a day. The maternal bond is broken shortly before the birth of the next calf. Irby (1979) stated that *R. fulvorufula* also is capable of reproduction throughout the year but that births peak from March to June in East Africa and in the wet summer in South Africa. Females have an interbirth interval of 9–14 months, an estrous cycle of 2–4 weeks, and a gestation period of about 8 months. Females reach sexual maturity at between 9 and 24 months of age. A captive *R. redunca* lived for 18 years (Jones 1982).

Most hunters agree that reedbucks are among the easiest of antelopes to approach and kill. *R. arundinum* was formerly common in South Africa but is now rare in that country, except in Natal and Transvaal (Jungius 1971b). In East Africa *R. fulvorufula* is restricted to isolated mountains and segments of broken country, its survival jeopardized by expanding agriculture and poaching (East 1988; Kingdon 1982).

ARTIODACTYLA; BOVIDAE; **Genus PELEA**
Gray, 1851

Rhebok

The single species, *P. capreolus*, is known to occur only in South Africa and Lesotho (Ansell, *in* Meester and Setzer 1977; Smithers 1983). There is controversy regarding the systematic affinity of *Pelea*, with various authorities placing it in its own subfamily, the Peleinae, or considering it to be a relative of *Redunca*, *Neotragus*, or even *Capra* (Meester et al. 1986).

Head and body length is 1,150–1,250 mm, tail length is 150–300 mm, shoulder height is 710–87 mm, and weight is 20–30 kg. The body is covered with hair that is woollier and curlier than that of other antelopes. The upper parts are brownish gray, the face and legs are yellowish, and the underparts of the body and tail are white. *Pelea* may be readily distinguished by its long, pointed, erect ears and by the absence of a bare patch below the ear. The horns, borne only by males, are straight, upright, and 200–250 mm in length. A naked area around the nostrils extends to the top of the nose and is swollen; it becomes studded with moisture when the animal is excited. Females have four mammae.

The rhebok lives among rocks and tangled growth on mountain sides and plateaus, but where protected, it will venture to grassy valleys. It probably frequented such valleys regularly before being driven out by human activity. It has a jerky gait resembling that of a rocking horse and is agile, fleet, and a good jumper. The tail is carried erect, so the white undersurface is conspicuous. Smithers (1983) wrote that the rhebok is active throughout the day, rests during the warmer

hours, and forages over a greater distance during the winter. It is a browser.

According to Smithers (1983), this antelope lives in family parties with up to about 12 individuals, including an adult male and several females and their young. There have been reports that a number of such groups will combine to form a temporary aggregation of up to 30 animals. The family units remain within a defined home range throughout the year. The lead male is territorial, defending a part of the home range against intruding males. Defense involves postural displays, snorting and stamping, and serious fighting. Males that are unable to win or maintain a territory are frequently solitary. Generally one of the group acts as a sentinel while the others feed and rest. If danger appears, the sentinel gives a warning grunt or cough and leads the herd to more rugged country. Males are extraordinarily aggressive, often killing others of their sex during the rutting season and even attacking and killing sheep and goats.

Mating takes place about April, and births occur about November or December, early in the warm, wet season. The gestation period is 261 days. There is a single calf, which is hidden by the mother for the first few days of life. The young males become sexually mature at about 18–21 months and then leave their group and begin to try to establish their own territory (Smithers 1983). Longevity in the wild has been reported as 8–10 years (Grzimek 1975). *Pelea* does not thrive in captivity and is seldom found in zoological parks.

ARTIODACTYLA; BOVIDAE; **Genus HIPPOTRAGUS**
Sundevall, 1846

Roan and Sable Antelope

There are three species (Ansell, *in* Meester and Setzer 1977):

H. leucophaeus (blue buck), formerly found in southwestern South Africa;
H. equinus (roan antelope), savannah zones from Senegal to western Ethiopia, and south to South Africa;
H. niger (sable antelope), savannah zone from southeastern Kenya to Angola and eastern South Africa.

Head and body length is 1,880–2,670 mm, tail length is 370–760 mm, shoulder height is 1,000–1,600 mm, and weight is 150–300 kg. Males average approximately one-fifth larger and heavier than females (Grzimek 1975). The general coloration is bluish gray in *H. leucophaeus*, pale reddish brown in *H. equinus*, and rich chestnut to black in *H. niger* (Ansell, *in* Meester and Setzer 1977). The underparts are white. *Hippotragus* is characterized by thick, tough skin; a well-developed and often upright mane on the nape; a short mane on the throat; a moderately long tail with a tufted tip; large, long, and pointed ears; and long white hairs below the eyes. The horns, borne by both sexes, are stout and heavily ringed and rise at an obtuse angle from the plane of the face. When fully developed, the horns are usually 510–1,020 mm long, and they rarely attain a length of 1,520 mm.

These antelope occur in a variety of habitats, but generally *H. equinus* is found in more open grassland, while *H. niger* requires some wooded country. Neither species moves more than 2–4 km from drinking water (Child and Wilson 1964; Estes and Estes 1974; Wilson and Hirst 1977). Where not persecuted, these antelope are not extremely wary, often running a short distance and then stopping to look back. When closely pursued, however, they can run as fast as 57 km/hr with great endurance. When wounded or cornered, they become savage, charging and using their horns with amazing

Rhebok (*Pelea capreolus*), photo by Herbert Lang through J. Meester.

speed and dexterity. The movements of *H. equinus* are fairly localized, but *H. niger* may change its range several times in the course of a year, sometimes covering 40 km or more. Both species feed primarily on grass, but the sable may also browse extensively in the dry season (Child and Wilson 1964; Estes 1974; Estes and Estes 1974).

Population density of *H. niger* has varied from 0.4/sq km to 9.2/sq km, but Wilson and Hirst (1977) calculated that maximum sustainable density is not over 4/sq km and that the minimum area needed to support a healthy population of 40–50 animals is 1,200–1,500 ha. There seems to be great variation in home range. In Zimbabwe, Grobler (1974) found herds of female and young *H. niger* to use ranges of 240–80 ha., while mature males had individual ranges of 25–40 ha. According to Estes and Estes (1974), however, home range of *H. niger* varies from about 1,000 to 32,000 ha. In a study of *H. equinus* in Kruger National Park, Joubert (1974) found herds

generally to use 6,400–10,400 ha. per year. The area regularly used at any given time was about 200–400 ha., and there was seldom overlap between the ranges of two herds.

Both species are gregarious, but herds of *H. equinus* commonly do not exceed 12–15 animals, while those of *H. niger* may be several times as large and occasionally contain over 100 antelope. The herds consist mostly of females and young and are dominated by a single adult male (Wilson and Hirst 1977). Joubert (1974) found herds of *H. equinus* to contain 6–12 individuals. The dominant male drove other males out when they reached 2.5 years of age. The latter then formed bachelor groups of 2–5 animals, which established a dominance hierarchy through much fighting. At about 5–6 years of age the males became solitary and sought to take over a herd of females. Victorious males defended an area extending outward about 300–500 meters from their herd. The mature females of a herd established a dominance hierarchy

A. Roan antelope *(Hippotragus equinus)*, photo from New York Zoological Society. B. Sable antelope and young *(H. niger)*, photo from San Diego Zoological Garden.

among themselves through extensive fighting, and the highest-ranking individual initiated most herd movements, even in the presence of the adult male.

Studies in Shimba Hills National Park, which is the last refuge of *H. niger* in Kenya and within which substantial seasonal migration is not feasible, indicate that about 200 sable are divided into six distinct female herds, each inhabiting a well-defined home range of 1,000–2,400 ha.; throughout the year the adult males maintain territories, one to six covering the range of each female herd (Sekulic 1981). Studies of the same species in Angola and Zimbabwe (Estes and Estes 1974; Grobler 1974) indicate that there is seasonal variation in the structure of herds of females and young. During the dry season the entire herd concentrates. When the rains begin, there is dispersal into small groups, and toward the end of the rainy season, when the calves are about to be born, there is still more fragmentation. The main herds form again after the calves leave concealment. Despite such fluctuations, Estes and Estes found herds to be stable from year to year and to have little interchange of members. The average number of individuals per herd was 25.6, and herds were generally spaced 5–10 km apart. In addition to the main herds of females and young, populations of *H. niger* may have bachelor groups of 2–12 males. Young males are driven from the main herds at about 3–4 years of age. At 5–6 years they become solitary and attempt to establish a territory. The size of such territories was reported by Grobler as 25–40 ha. but by Estes and Estes as 1,000–1,400 ha. Territories are marked by defecation and are vigorously defended against other fully mature males. A territory may be overlapped by the ranges of several female herds. The strongest males obtain the most centrally located and desirable territories. A male takes control of any female herd that enters his territory.

H. equinus seems to have no specific breeding season. Females of this species enter estrus 2–3 weeks after giving birth and have the potential of producing a calf every 10.0–10.5 months. *H. niger* is a seasonal breeder; the time and duration of calving varies across the range and seems to coincide with the height of the growing season. Births occur

mainly from January to March in the Transvaal, May to July in Angola, and June to September in Zambia. The gestation period has been variously reported as 268–80 days in *H. equinus* and 240–81 days in *H. niger.* The single calf weighs 13–18 kg at birth and remains hidden at least 10 days. Females attain sexual maturity at about 2 years (Dittrich 1972; Estes and Estes 1974; Grobler 1974; Wilson and Hirst 1977). A captive *H. niger* lived for 19 years and 9 months (Jones 1982).

The blue buck *(H. leucophaeus)* was one of the first mammals to become extinct in modern times. It apparently had been declining from natural causes since the Pleistocene, may also have been adversely affected by the introduction of domestic sheep around A.D. 400, and was restricted to the southern coastal portion of South Africa when that area was being settled by Europeans in the eighteenth century (Klein 1974; Smithers 1983). It was quickly eliminated by hunting, the last individual being killed about 1800. The other two species still occur in large areas but have declined drastically in recent decades through habitat deterioration, agricultural encroachment, illegal hunting, and deliberate slaughter for tsetse fly control (Wilson and Hirst 1977). *H. equinus* is on appendix 2 of the CITES. The subspecies *H. niger variani*, the giant sable antelope, is on appendix 1 and is classified as endangered by the IUCN and the USDI. This subspecies is restricted to a small part of northern Angola. An estimated 2,000–3,000 individuals were thought to survive in 1970; they occurred largely in government reserves and were totally protected by law. Subsequently numbers are thought to have declined as conservation efforts were disrupted by civil war and lawlessness in the area. The giant sable also is jeopardized by poaching and encroachment of the expanding human population (East 1989).

ARTIODACTYLA; BOVIDAE; **Genus ORYX**
De Blainville, 1816

Oryx and Gemsbok

There are three species (Ansell, *in* Meester and Setzer 1977; Corbet 1978):

O. dammah (scimitar oryx), originally found in the
  semidesert zones from Morocco and Senegal to Egypt
  and the Sudan;
O. leucoryx (Arabian oryx), originally found in Syria, Iraq,
  Israel, Jordan, Sinai, and the Arabian Peninsula;
O. gazella (gemsbok), arid country from Ethiopia and
  Somalia to Namibia and eastern South Africa.

Head and body length is 1,530–2,350 mm, tail length is 450–900 mm, shoulder height is 900–1,400 mm, and weight is 100–210 kg (Grzimek 1975). The general adult coloration varies from cream to grays and browns, and there are striking markings of black and brown. The young are brownish and have markings only on the tail and knees. A mane extends from the head to the shoulders, the tail is tufted, and males have a tuft of hair on the throat. The ears are fairly short, broad, and rounded at the tip. Both sexes have horns ranging from 600 to 1,500 mm in length. Those of females are usually longer and slenderer than those of males. The horns of *O. leucoryx* and *O. gazella* are fairly straight and directed backward from the eyes, but those of *O. dammah* curve back in a large arc.

These animals usually live on arid plains and deserts, but in some areas they also inhabit rocky hillsides and thick brush. According to Jungius (1978), the habitat of *O. leucoryx* consists of flat and undulating gravel plains, intersected by shallow wadis and depressions, and the dunes edging sand deserts, with a diverse vegetation of trees, shrubs, herbs, and grasses. It apparently digs shallow depressions in soft ground under trees and shrubs for resting. It is able to detect rainfall over great distances, and it moves accordingly in the direction

of fresh plant growth. Since rainfall is irregular, it must travel over hundreds of square kilometers in no fixed pattern. One individual is known to have moved over 90 km in 18 hours. In the Serengeti, Walther (1978) observed that *O. gazella* became active just after dawn, grazed until about 1000 hours, rested until 1400–1500 hours, grazed again, and began moving toward the sleeping area around sunset. A similar pattern was reported for an introduced herd of *O. leucoryx* in Jordan (Abu Jafer and Hays-Shahin 1988). Oryx have been described as alert, wary, and keen-sighted. They defend themselves by lowering the head, so that the sharp horns point forward. They feed on grasses and shrubs and go to streams and waterholes to drink. When free water is not available, they can obtain sufficient moisture for lengthy periods from such sources as melons and succulent bulbs.

An introduced herd of *O. leucoryx* in Oman uses an overall nomadic range of about 3,000 sq km comprising a series of separate suitable areas, mostly 100–300 sq km in size, which are occupied for 1–18 months at a time; population density is about 0.035/sq km (Price 1988). Female group home ranges of *O. gazella* in Africa are about 50–400 sq km, but solitary territorial males remain, sometimes for years, on areas of only 5–16 sq km; maximum population density is 1.4/sq km (Wacher 1988; Williamson and Williamson 1988).

Oryx are gregarious. *O. dammah* generally is found in groups of 20–40 individuals, but formerly, at certain times of the year in areas of fresh pasture or surface water after rainfall or during the wet season migrations, herds contained 1,000 animals or more (Newby 1978). The normal group size of *O. leucoryx* is 10 animals or fewer, but herds of up to 100 have been reported. Captive *O. leucoryx* have been maintained in herds consisting of a single dominant adult male and several adult females and young. Groups of bachelor males are kept separately, and these animals establish a dominance hierarchy through nonsevere fighting and chases (Jungius 1978). The males of an introduced herd in Jordan also have formed a hierarchy (Abu Jafer and Hays-Shahin 1988). Wacher (1988) reported that typical size of female groups of *O. gazella* in Kenya is 30–40 but that aggregations of several hundred may form. There also are solitary males that maintain territories

Gemsbok (*Oryx gazella*), photo by Karl H. Maslowski.

in which they seek to control and mate with the females that are present. In a study of *O. gazella* in the Serengeti, Walther (1978) found groups of up to 22 individuals, as well as lone males. Some herds contained only males, and others were harems controlled by a single dominant male. There was considerable competition between males, but it involved mainly ritualized, horn-to-horn sparring, with no serious injury.

In *O. gazella* there is no particular breeding season, females have a postpartum estrus and give birth at approximately 9-month intervals, the gestation period is about 8.5 months, and a single offspring is normal (Wacher 1988). In *O. leucoryx*, reproductive timing varies, but under favorable conditions a female can produce a calf once a year and during any month. Most births among introduced herds in Oman and Jordan occur from October to May. The gestation period in this species is about 240 days, the young are weaned by 4.5 months, and captive females initially give birth at 2.5–3.5 years (Abu Jafer and Hays-Shahin 1988; Jungius 1978). Female *O. dammah* have an estrous cycle of 21–22 days, an estrus of less than 24 hours, and a gestation period of 222–53 days; they attain sexual maturity at 11 months and give birth to a single calf weighing about 10 kg (Gill and Cave-Browne 1988). The potential longevity of *Oryx* appears to be about 20 years.

The Arabian oryx *(O. leucoryx)* is classified as endangered by the IUCN (1976) and the USDI and is on appendix 1 of the CITES. By the early twentieth century this species survived only in the Arabian Peninsula. It continued to decline as the expansion of the oil industry led to hunting from motor vehicles with modern firearms. Its meat is greatly esteemed, its hide is valued as leather, other parts have alleged medicinal uses, and the head makes a choice trophy. The last known individuals in the wild were killed in 1972, and there are unconfirmed reports from as late as 1979 (Abu-Zinada, Habibi, and Seitre 1988). However, in the 1950s efforts were made in several Arab countries to establish captive herds, and in 1962 some Arabian oryx were taken in the wild and brought to the United States (Grimwood 1988; Jones 1988). The latter animals served as the foundation of an international breeding effort and for reintroductions into the wild in Oman in 1982 and Jordan in 1983. There now are approximately 110 individuals in the wild, 300 in captivity in the Arabian Peninsula, and 370 held elsewhere, many at the Phoenix and San Diego zoos (Abu Jafer and Hays-Shahin 1988; Daly 1988; Dolan 1976; Jones 1988; Jungius 1978; Strassburger 1978).

The scimitar oryx *(O. dammah)* now is classified as endangered by the IUCN and is on appendix 1 of the CITES. It has undergone a long-term decline because of climatic changes, uncontrolled hunting by people, agricultural encroachment on its habitat, and excessive grazing of the limited vegetation by domestic livestock. It disappeared from Egypt and Senegal, the fringes of its range, in the 1850s. The deterioration of status was accelerated in the last few decades through increased hunting and a series of severe droughts. In the 1970s there still were an estimated 6,000 individuals distributed across several countries in the southern part of the Sahara. The only wild population known to exist at present is in Chad (Newby 1988). There also are 544 individuals in captivity (Olney, Ellis, and Sommerfelt 1988).

The gemsbok *(O. gazella)* is still common in some parts of Africa. The natives sometimes have used the sharp tips of its horns for spear points and its thick, tough skin for shield coverings. Although some populations have been extirpated

Scimitar oryx *(Oryx dammah),* photo from Paris Zoo.

by hunters (Kingdon 1982; Smithers 1983; Yalden, Largen, and Kock 1984), recent estimates (East 1988, 1989) set numbers of over 75,000 for the subspecies in eastern Africa (*O. g. beisa* and *O. g. callotis*) and about 200,000 for that of southern Africa *(O. g. gazella)*. There is a small introduced population of gemsbok in the White Sands area of southern New Mexico (Reid and Patrick 1983).

ARTIODACTYLA; BOVIDAE; Genus ADDAX
Rafinesque, 1815

Addax

The single species, *A. nasomaculatus*, originally occurred in desert and semidesert areas from Western Sahara and Mauritania to Egypt and Sudan. There are old, inconclusive reports from Palestine and the Arabian Peninsula (Ansell, *in* Meester and Setzer 1977).

Head and body length is 1,500–1,700 mm, tail length is 250–350 mm, shoulder height is 950–1,150 mm, and weight is 60–125 kg (Grzimek 1975). The body and neck are grayish brown in winter, but in summer the body coloration becomes sandy to almost white. The legs, hips, belly, ears, and facial markings are white, and the tuft on the forehead is black. Both sexes have horns bearing some resemblance to those of *Oryx*, but with a spiral twist of 1.5 to almost 3 turns. The horns measure 762–890 mm along the curves. The widely splayed hooves are adapted for traveling on desert sands.

The addax is the most desert-adapted of antelope (Corbet 1978). It lives most of its life without drinking, deriving sufficient moisture from the plants upon which it feeds. It travels great distances to search for the scant vegetation of the Sahara and, in spite of the seeming adversity of its habitat, is said always to appear in good condition. It seems to move about in herds of 5–20 individuals, led by an old male. A group of captive females were observed to establish a dominance hierarchy, the oldest animals ranking highest (Reason and Laird 1988). Wild populations generally give birth in the winter or early spring. Analysis of records of captive animals indicates that breeding may occur throughout the year but that there is a birth peak in winter and spring; reported gestation periods are 257 and 264 days, mean interbirth interval is 355 days, there almost always is a single young, a male reached sexual maturity at about 24 months, and females enter their first fertile estrus during their second or third summer (Densmore and Kraemer 1986; Dittrich 1972). A captive addax lived for 25 years and 4 months (Jones 1982).

The addax now is classified as endangered by the IUCN and is on appendix 1 of the CITES. This heavily built antelope is not capable of great speed and thus is easy prey to people with camels, horses, dogs, and modern weapons. Both the meat and the skin are prized by the natives, the latter being used for shoe and sandal soles. Ruthless hunting has eliminated resident populations in many parts of the original range. The species also has been adversely affected by recent severe droughts and by tourists who give chase in four-wheel-drive vehicles, exhausting the animals to the point of death. The decline of *Addax* has corresponded closely to that of *Oryx dammah* (see above). The only known viable herd of addax remaining in the wild, in northeastern Niger, contains only 50–200 animals (Newby and Magin 1989). Also, there are 409 individuals in captivity around the world (Olney, Ellis, and Sommerfelt 1988). During World War II, when the addax still was relatively common, American soldiers in

Addaxes *(Addax nasomaculatus)*, photo from New York Zoological Society.

North Africa observed the species and described it in letters to their families, who then contacted zoos to learn where such a white antelope lived, in order that they might know where the soldiers were.

ARTIODACTYLA; BOVIDAE; Genus DAMALISCUS
Sclater and Thomas, 1894

Sassabies

There are two subgenera and three species (Ansell, *in* Meester and Setzer 1977):

subgenus *Beatragus* Heller, 1912

D. hunteri (Hunter's hartebeest), eastern Kenya, extreme southwestern Somalia;

subgenus *Damaliscus* Sclater and Thomas, 1894

D. dorcas (bontebok and blesbok), South Africa;
D. lunatus (topi), open country from Senegal to Ethiopia, and south to South Africa.

Kingdon (1982) gave full generic rank to *Beatragus*. In contrast, Grzimek (1975) treated *D. hunteri* as a subspecies of *D. lunatus*. Yalden, Largen, and Kock (1984) regarded those populations of *D. lunatus* found from Senegal to Ethiopia and south as far as Tanzania to be a separate species, *D. korrigum*.

Van Gelder (1977b) considered *Damaliscus* to be a synonym of *Alcelaphus*.

Head and body length is 1,200–2,050 mm, tail length is 100–600 mm, shoulder height is 880–1,340 mm, and weight is 68–155 kg. The hair is generally soft and has an iridescent sheen, but some animals have a mixture of soft and coarse hairs. The general coloration ranges from grays and reds to rich brown and almost black. The underparts are lighter, and there may be various white markings on the face, legs, and hips. *D. dorcas* has a large white area on the face, and *D. hunteri* has a white line passing from one eye to the other across the forehead. *Damaliscus* is characterized by its slightly convex forehead and its somewhat elongated muzzle. The horns, which are well developed in both sexes, are angular and curved, are ringed for most of their length, lack pedicels, and are up to about 700 mm long.

These antelope inhabit grassland and sparsely timbered regions. Kingdon (1982) indicated that *D. lunatus* tends to concentrate in an area with abundant green forage and to then move on or perish in large numbers. If suddenly alarmed, sassabies may jump over each other's backs in their haste to escape. The normal gait is a lumbering canter, but maximum speed is at least 70 km/hr (Schaller 1972). Most activity occurs in the morning and early evening. There are seasonal migrations in some areas. The diet consists mainly of grass.

At one time, these antelope gathered in the thousands, at least in certain seasons. Recent studies show that populations have an intricate but highly variable structure (David 1973, 1975; DuPlessis 1972; Jewell 1972; S. C. J. Joubert 1972; Joubert and Bronkhorst 1977; Monfort-Braham 1975; Novellie 1979). In general, fully mature males are solitary and

Bontebok *(Damaliscus dorcas)*, photo by Karl H. Maslowski.

compete with one another, and with younger males, for possession of females. Most competition is through postural displays and ritualistic sparring with the horns rather than serious fighting. In Kruger National Park, a male *D. lunatus* establishes a territory of 200–400 ha., which he marks by urine, feces, secretions from preorbital glands, and horning of the soil. Within his territory, he permanently maintains a harem of about seven or eight adult females and associated offspring. Young males are ejected from the area before they are one year old and form bachelor groups of up to six individuals. When the young males are three or four years old, they begin to challenge the older bulls.

Males of *D. lunatus* in southern Rwanda also have large territories, up to 100 ha., and a permanent harem of up to 20 females. Males of *D. dorcas* in southern South Africa have territories that are small, 10–40 ha., but possibly occupied for life. Groups of adult females there, with an average of 3 individuals, are not permanently associated with any one male but wander from one territory to another. There are also large herds containing the young animals of both sexes. Some males of *D. dorcas* in northern South Africa, and of *D. lunatus* in Uganda and northern Rwanda, have small territories of about 1–3 ha., which are occupied only temporarily during the mating season. There are large associated herds of females, which spend only a brief time in any one territory. Other adult males of *D. dorcas* in northern South Africa have no territories, but move with and defend a herd of females.

D. hunteri apparently mates at the start of the long rains in March or April and gives birth mainly at the beginning of the short rainy season in October and November (Kingdon 1982). There also are generally well-defined seasons of birth in the other species: July–September for *D. lunatus* in Uganda, September–December for *D. lunatus* in Kruger National Park, September–October for *D. dorcas* in southern South Africa, and November–December for *D. dorcas* in northern South Africa. The gestation period is 7–8 months, and nor-

Hunter's hartebeests *(Damaliscus hunteri):* Top, photo from Gladys Porter Zoo, Brownsville, Texas; Left, photo from San Diego Zoological Society.

mally a single calf is born. Females attain sexual maturity in their second or third year of life. Males do not usually mate until they are old enough to compete successfully with their elders (David 1973, 1975; DuPlessis 1972; Jewell 1972; S. C. J. Joubert 1972). A captive *D. dorcas* lived for 21 years and 5 months (Jones 1982).

These antelope have been greatly reduced in numbers and distribution through excessive hunting and the encroachment of agriculture. The subspecies *D. dorcas dorcas* (bontebok), of southwestern Cape Province, was nearly exterminated by 1830, but several herds have been maintained on government and private preserves, and about 1,500 individuals now exist (East 1989). The subspecies is classified as vulnerable by the IUCN and as endangered by the USDI and is on appendix 1 of the CITES. The distribution of the related *D. dorcas phillipsi* (blesbok), of eastern South Africa, also has been severely curtailed; the subspecies now occurs mainly in fenced reserves, but there are still around 150,000 individuals (East 1989; Smithers 1983). *D. hunteri* is classified as rare by the IUCN. Estimates in the 1970s put its total numbers at around 14,000 (Bunderson 1977). Subsequently there has been a major decline because of competition with domestic livestock for forage and water, a severe drought, and poaching; current numbers are set at 5,000–10,000 (East 1988).

ARTIODACTYLA; BOVIDAE; Genus ALCELAPHUS
De Blainville, 1816

Hartebeest

The single species, *A. buselaphus*, originally was found in open country from Morocco to Egypt, from Senegal to Somalia and northeastern Tanzania, and from southern An-

gola and western Zimbabwe to South Africa (Ansell, *in* Meester and Setzer 1977). There are questionable reports of its former presence in Palestine (Harper 1945). *Sigmoceros* (see account thereof) sometimes is considered to be part of *Alcelaphus*.

Head and body length is 1,500–2,450 mm, tail length is 300–700 mm, shoulder height is 1,100–1,500 mm, and weight is 100–225 kg. The body hair is about 25 mm in length and is particularly fine textured. Coloration ranges from fawn through brownish gray to chestnut. *A. buselaphus* has prominent white patches on the hips and black on the forehead, muzzle, shoulders, and thighs. The genus is characterized by the rump being lower than the shoulders, a long head, slender legs, a tufted tail, and large, conspicuous glands below the eyes. The horns, borne by both sexes, unite or rise from a single pedicel, have heavy encircling rings, and are usually 450–700 mm long. Females have two mammae.

The hartebeest inhabits dry savannahs and grasslands. Owing to the difference in height between the forequarters and hindquarters, the gait may appear clumsy, but Schaller (1972) reported maximum speed to be 70–80 km/hr. The hartebeest is diurnal, feeding in the early morning and late afternoon and resting in the shade during the heat of the day. The diet consists mainly of grass.

Schaller (1972) estimated that there were 18,000 *A. buselaphus* in the 25,500-sq-km Serengeti ecosystem, and Laws, Parker, and Johnstone (1975) reported population density in Uganda to be 1.4/sq km. Hartebeests are gregarious, living in organized herds of up to 300 animals and sometimes uniting in aggregations of up to 10,000 (Smithers 1971). In Nairobi National Park, Gosling (1974) found *A. buselaphus* to have four social classes: territorial adult males, nonterritorial adult males, groups of up to 100 young males, and groups of females and young. Young males left the females at about 20 months and joined the male groups. At 3–4 years of age they

Hartebeests *(Alcelaphus buselaphus)*, photo by C. A. Spinage.

Lichtenstein's hartebeests *(Sigmoceros lichtensteinii)*, photo by Lorna Stanton, National Parks Board of South Africa, through Colleen Naudé.

began to attempt to take over a territory and hence control of the females thereon. Males generally lost their territory at 7–8 years. Animals holding a territory represented 38 percent of the adult male population and defended an average area of 31 ha. In territorial struggles, the males leapt forward in a kneeling position, with their horns lowered, but serious fights were rare.

In southern Ethiopia, according to Lewis and Wilson (1979), *A. buselaphus* occurs in groups of up to 180 animals but breaks into smaller units prior to the calving season. A herd of hartebeests may have a sentinel posted on a summit to warn the others of danger. When alarmed, the animals gallop off in a single file. These antelope usually are non-aggressive and silent, though Kingdon (1982) observed adult males during the rut to be intensely aggressive toward young males that were still with their mother.

Reproduction may be markedly seasonal. The young of *A. buselaphus* are born from October to November in South Africa and from mid-December to mid-February in southern Ethiopia (Lewis and Wilson 1979; Skinner, Van Zyl, and Van Heerden 1973). In Nairobi National Park, births of *A. buselaphus* occur throughout the year, with a major peak in February and March and a lesser peak in July (Gosling 1974). The gestation period is 214–42 days, and normally a single calf is produced. Potential longevity is 11–20 years.

The hartebeest, once abundant, has been greatly reduced in numbers through direct killing and habitat modification by people. Kingdon (1982) noted that it probably has suffered the greatest contraction in range of all African ruminants, primarily because it must compete for forage with domestic cattle and also because it is easy to hunt. The subspecies *A. buselaphus buselaphus*, of North Africa, still was widespread in 1900 but apparently became extinct in the late 1920s (Goodwin and Goodwin 1973; Harper 1945). The subspecies *A. b. tora* and *A. b. swaynei*, which formerly occurred from southern Egypt to Somalia, are classified as endangered by the

IUCN and the USDI. Much of the remaining range of *tora*, in the southern Sudan and northwestern Ethiopia, was devastated by a drought in the early 1980s, and only a few animals are thought to survive. *A. b. swaynei* has been exterminated in Somalia, but several relatively well protected herds, containing a total of 2,000–3,000 individuals, are present in central Ethiopia (East 1988). *A. buselaphus* also has disappeared from most of South Africa but is still locally common in parts of Botswana, Namibia, Tanzania, and Kenya.

ARTIODACTYLA; BOVIDAE; **Genus SIGMOCEROS**
Heller, 1912

Lichtenstein's Hartebeest

The single species, *S. lichtensteinii*, is found in the savannah zone of Tanzania, southeastern Zaire, northeastern Angola, Zambia, eastern Zimbabwe, Malawi, and Mozambique (Ansell, *in* Meester and Setzer 1977; Ansell and Dowsett 1988). *Sigmoceros* often has been treated as a synonym of *Alcelaphus*, and *S. lichtensteinii* as a second species of the latter genus. Kingdon (1982) regarded *S. lichtensteinii* only as a subspecies of *Alcelaphus buselaphus* or as an incipient species (one of Kingdon's maps, showing substantial overlap between the ranges of *lichtensteinii* and a neighboring subspecies of *A. buselaphus*, certainly suggests at least a specific distinction). Vrba (1979) thought that *Sigmoceros* should be restored to generic rank, and in this regard he has been followed by Ansell and Dowsett (1988), Corbet and Hill (1986), Honacki, Kinman, and Koeppl (1982), and Meester et al. (1986).

Except as noted, the information for the remainder of this account was taken from Kingdon (1982) and Smithers (1983). Head and body length is 1,600–2,030 mm, tail length is 400–510 mm, height at the shoulder is 1,190–1,360 mm, and weight is 157–204 kg in males and 125–81 kg in females. The general coloration is yellowish tawny, there is an indistinct saddle of a more rufous color from the shoulders to the base of the tail, the rump and base of the tail are white, and the chin, the fronts of the lower legs, and the tuft of hair at the end of the tail are black. Both sexes carry curved horns about 400–600 mm long. In *Sigmoceros* the bony horn pedicel is short and broad, the occiput is about level with the base of the horns, the horns curve inward toward each other before bending back, and the forehead is convex. In *Alcelaphus* the horn pedicel is long and much better developed, the occiput is in front of the base of the horns, the horns do not curve inward, and the forehead is flat.

Lichtenstein's hartebeest is a savannah species particularly associated with habitats that combine open woodland and floodplain grassland. There may be seasonal movements, with animals concentrating along low-lying, grassy drainages in the dry season and then moving into higher country during the rains. Most activity is in the early morning and late afternoon or evening, the animals lying down in the shade during the hottest hours of the day. *Sigmoceros* has good eyesight, but the sense of smell seems to be less well developed. It is almost exclusively a grazer.

Populations of *Sigmoceros* tend to fragment into smaller social units than are typical of *Alcelaphus*, perhaps because of the restricted extent of grassland within its habitat. Groups commonly contain about 3–10 females and young that are led for at least part of the year by a single adult male. He frequently takes up a position on elevated ground to watch for danger and stays at the rear of the group when it is in flight. He maintains a year-round territory of about 2.5 sq km in the most favorable habitat, which he marks by horning the ground and in which he will attempt to herd as many females as possible during the rut. At that time rival males have

frequent and long fights. Weaker and younger males may form small, unstable bachelor groups, which occupy poorer habitat. Booth (1985), however, noted that bachelors usually are found alone.

Breeding is seasonal, with young being born in the dry period from July to August in Zambia (Wilson 1966*b*) and about September in Zimbabwe and Mozambique. The gestation period is 240 days, there normally is a single young, weight at birth is about 15 kg, and females have their first calf at about 2 years. Young males are expelled from the family group at 10–12 months, and females leave at 15–18 months (Booth 1985).

Sigmoceros is declining wherever human settlement has invaded its habitat, particularly since the first areas to be cultivated are the fertile low-lying areas on which it depends for part of the year. However, its alertness and utilization of country in which hunting is difficult give it some advantage over *Alcelaphus*. Relatively large populations still occur in much of its original range (East 1988, 1989).

ARTIODACTYLA; BOVIDAE; **Genus CONNOCHAETES**
Lichtenstein, 1814

Wildebeests, or Gnus

There are two subgenera and two species (Ansell, *in* Meester and Setzer 1977):

subgenus *Connochaetes* Lichtenstein, 1814

C. gnou (black wildebeest, or white-tailed wildebeest), central and eastern South Africa;

Black wildebeest *(Connochaetes gnou)*, photo from East Berlin Zoo.

Blue wildebeest (Connochaetes taurinus), photo by Lothar Schlawe.

subgenus *Gorgon* Gray, 1850

C. taurinus (blue wildebeest, brindled gnu, or white-bearded wildebeest), southern Kenya and southern Angola to northern South Africa.

Head and body length is 1,500–2,400 mm, tail length is 350–560 mm (not including the tuft), shoulder height is 1,000–1,450 mm, and weight is 118–275 kg. In *C. gnou* the general coloration is buffy brown to black; tufts of long, black hair protrude from the muzzle, throat, and chest; the mane is upright; and the tail is black at the base but otherwise white. In *C. taurinus* the general coloration is grayish silver; there are brownish bands on the neck, shoulders, and forepart of the body; and the face, mane, beard, and tail are black, except in the subspecies *C. t. albojubatus* and *C. t. mearnsi*, of Kenya and Tanzania, in which the beard is white. Both sexes of *Connochaetes* possess horns, which arise separately and are usually heavy and recurved. The oxlike head and horns, the bristly facial hair, and the general body form give *Connochaetes* a ferocious appearance, which is misleading. Females have four mammae.

Wildebeest prefer open grassy plains with a nearby source of water. When disturbed, they begin curious antics, prancing about, pawing the ground, probing the earth with their horns, and thrashing their tails. If a person approaches to within 500 meters, they snort and dash off a short distance, wheel about to face the intruder, and then resume the ritual. They are active mainly in the early morning and late afternoon, resting in the hot hours of the day. Speeds of 80 km/hr have been clocked (Kingdon 1982). Some herds are migratory, constantly moving about in search of food, but others remain in one area throughout the year. Schaller (1972) estimated that there were 400,000 migratory and 10,000 nonmigratory *C. taurinus* in the 25,500-sq-km Serengeti eco-

system; both of these classes may now be several times as large. Wildebeest are primarily grazers, though they also feed on succulent plants and karroo bushes.

In some areas, during the dry season wildebeests form vast aggregations of all ages and both sexes. Usually, however, females and young are found in discrete herds of 10 to over 1,000 individuals. Such herds may be nomadic or may have a regular home range of perhaps 1 sq km. Males separate from these herds as yearlings and join a group of bachelors. When a male is about three or four years old, he becomes solitary and attempts to win a territory. Some males establish territories only temporarily, when the associated herd of females is not moving about. Other males may continuously occupy the same area for years. The distance between territorial males is usually about 100–400 meters but varies from as little as 9 meters in favorable areas to about 1,600 meters in poor habitat. Territorial competition between males involves ritualized displays, loud calls ("ge-nu"), and pushing with the horns but rarely severe fighting. Only males with a territory can mate, and they attempt to control all females that enter their area (Estes 1969; Schaller 1972; Von Richter 1972, 1974).

In a given area, reproduction is highly seasonal and births occur mainly within a period of 2–3 weeks at the beginning of the rains. This period falls from November to January in South Africa and in January or February in the Serengeti (Estes 1976; Skinner, Van Zyl, and Van Heerden 1973; Von Richter 1974). Sinclair (1977*b*) suggested that the lunar cycle is the external timer for synchronization of mating in the Serengeti. The gestation period is 8–9 months, and a single offspring is normal. It can stand within 15 minutes of birth, begins to follow its mother shortly thereafter, and is weaned by 9 months. Females attain sexual maturity in their second or third year of life and commonly give birth annually. A captive *C. taurinus* lived for 21 years and 5 months (Jones 1982).

The flesh of *Connochaetes* is coarse, dry, and hard; the skin makes good leather; and the silky tail is used for fly whisks called "chowries." Wildebeest populations have been reduced through hunting and human habitat modification and also are subject to decline periodically from drought. However, in contrast to the situation of large mammals in most of Africa, numbers of *C. taurinus* in the Serengeti area have increased dramatically in the last few decades. There now are 1 million to 1.5 million individuals in the Serengeti migratory population, the greatest concentration of wild grazing mammals left on earth, and 10,000 to 15,000 in each of several resident populations there. On the other hand, the Serengeti herds combined with the 100,000 additional *C. taurinus* in Tanzania represent over 80 percent of the entire species (Heminway 1987; Rodgers and Swai 1988). Populations in Botswana have fallen from the hundreds of thousands in the 1970s to only 38,700, partly because fences have cut access to water (Spinage, Williamson, and Williamson 1989). *C. gnou* was nearly exterminated by hunting during the settlement of South Africa in the nineteenth century but was maintained on a few government and private preserves. It has been widely reintroduced and currently numbers about 10,000 (East 1989).

ARTIODACTYLA; BOVIDAE; Genus OREOTRAGUS
A. Smith, 1834

Klipspringer

The single species, *O. oreotragus*, occurs in rocky hills and mountains in northern Nigeria, probably in the eastern Central African Republic, in northeastern and extreme southeastern Sudan, and from Ethiopia and Angola to South Africa (Ansell, *in* Meester and Setzer 1977).

Head and body length is 770–1,150 mm, tail length is 50–130 mm, shoulder height is 450–600 mm, and weight is 8–18 kg. The horns, ringed at the base and usually borne only by males, are 75–153 mm in length. The pelage has a yellowish

Klipspringer *(Oreotragus oreotragus)*, photo from San Diego Zoological Garden.

olive gloss and is speckled with yellow and brown or orange, shading to white below. The coat harmonizes with the background of rocks. The texture of the pelage has been variously described as mosslike and grasslike. The thick, pithy hair lies like a mat on the body to cushion the animal against bumps and bruises in its rocky environment. Females have four mammae (Grzimek 1975).

This little "cliff-springing" antelope might be likened to the chamois *(Rupicapra rupicapra)* or the mountain goat *(Oreamnos americanus)* in its mode of life among the rocks, where it makes rapid progress without apparent suitable footholds. It walks and stands on the tips of the relatively rounded hooves and can jump onto a rocky projection the size of a silver dollar, landing on it with all four feet. It is active mainly in the morning and late afternoon, sheltering at other times among rocks and under overhanging cliffs. It is mainly a browser but also eats some grass and apparently does not drink water regularly.

In three parts of Ethiopia, Dunbar and Dunbar (1974b) found population densities of 13.4, 19.1, and 46.7 individuals per ha. Groups had exclusive home ranges averaging 8.1 ha., which they defended against conspecifics and rarely left. Territories are larger, averaging up to 49 ha., in parts of South Africa with low rainfall and thus more sparsely distributed resources (Smithers 1983). In both of these countries and also in Namibia (Tilson 1980), groups were found to consist basically of a mated, monogamous pair and their offspring of one or two years. The adult female initiates most group movements, but the male seems to be mainly responsible for defense. Generally, one member of the group stands at a point above the others and constantly watches for danger. When alarmed, it gives a loud shrill whistle, and the animals ascend higher into the rocks. Observations by Tilson and Norton (1981) suggest that both members of a mated pair often simultaneously deliver an alarm call and that the function of this duet is to communicate the callers' alertness to a potential predator, thereby discouraging an attack. Both sexes scent-mark the territory with secretions from the preorbital gland and by defecation (Smithers 1983).

Reproduction seems to be nonseasonal in south-central Africa, but in Ethiopia mating apparently occurs in August and September (Dunbar and Dunbar 1974b; Smithers 1971, 1983; Wilson and Child 1965). The gestation period is about 214–25 days, a single young is produced, birth weight is about 1 kg, the offspring remain hidden for 2–3 months, and weaning occurs at about 4–5 months (Smithers 1983). Young males separate from the group before they are 1 year old, but females remain longer and may eventually mate with their father (Dunbar and Dunbar 1974b). A captive klipspringer lived for 12 years and 1 month (Jones 1982).

The skins of *Oreotragus* are used by the hill tribes to make bags for carrying bread. The meat is said to be excellent. Because of its elasticity and lightness, klipspringer hair was once in demand for stuffing saddlebags. The genus reportedly was once widespread in Nigeria but now is extremely rare in that country (Happold 1987).

ARTIODACTYLA; BOVIDAE; Genus OUREBIA
Laurillard, 1842

Oribi

The single species, *O. ourebi,* is found in the savannah zones of Africa south of the Sahara (Ansell, *in* Meester and Setzer 1977).

Head and body length is 920–1,400 mm, tail length is 60–

Oribi *(Ourebia ourebi),* photo by Lothar Schlawe.

150 mm, shoulder height is 500–700 mm, and weight is 14–21 kg. The hair is fine and silky, and the coloration of the upper parts is sandy rufous to tawny. The underparts and chin are white. The tail is black above and white below. Females usually have a dark crown patch. Both sexes have tufts of long hair on the knees. Beneath the ears is a bare glandular area, which is usually dark and conspicuous. The ears are large, the legs are slender, and the tail is short and bushy. The horns, borne only by males, are ringed at the base, set at an angle of about 45° to the plane of the face, and 75–125 mm long. Females have four mammae.

The oribi inhabits areas that are generally open but have a sufficient growth of vegetation for protection. It customarily shelters in tall grass and does not attempt to escape until an intruder is within a few meters. When flushed, it bounds along, leaping above the grass. When fleeing, it conspicuously displays the tail, which serves as a warning signal to conspecifics. It is capable of traveling at speeds of up to 40–50 km/hr (Kingdon 1982). It is active both day and night, lying in thickets during the heat of the day and emerging in the morning and evening to graze and browse.

Ourebia is commonly found in groups of up to five individuals. Such groups do not contain more than one adult male but often have more than one adult female. Groups are territorial and may be separated from each other by one to several kilometers. Both sexes defend the territory and mark it with urine, feces, and secretions from interdigital and antorbital glands (Leuthold 1977b; Monfort and Monfort 1974). A loud, shrill whistle may be uttered when an animal is alarmed. The single offspring generally is born from August to December, following a gestation period of about 210 days. Captives have lived up to 14 years (Kingdon 1982).

The flesh of *Ourebia* is said to be excellent, and it is extensively hunted. It is highly vulnerable to such hunting and to habitat alteration and has been greatly reduced in numbers and distribution (Ansell, *in* Meester and Setzer 1977; Skinner, Fairall, and Bothma 1977). The subspecies *O. o. keniae,* of the lower slopes of Mount Kenya, is probably extinct, and *O. o. haggardi,* of coastal Kenya and adjacent Somalia, is in jeopardy (East 1988). The oribi is relatively easy to raise and is said to make a delightful pet.

Steenbok *(Raphicerus campestris)*, photo by John Hanks.

ARTIODACTYLA; BOVIDAE; **Genus RAPHICERUS**
Hamilton-Smith, 1827

Steenbok and Grysboks

There are three species (Ansell, *in* Meester and Setzer 1977):

R. campestris (steenbok), Kenya and southern Angola to
   South Africa;
R. melanotis (grysbok), southern South Africa;
R. sharpei (Sharpe's grysbok), Tanzania to northeastern
   South Africa.

Kingdon (1982) suggested that *R. sharpei* might be more
closely related to *Neotragus* than to the other species of
Raphicerus and might warrant placement in a separate genus,
Nototragus Thomas and Schwann, 1906.

Head and body length is 610–950 mm, tail length is 40–80
mm, shoulder height is 450–600 mm, and weight is 7–16 kg.
The hair is rather coarse. Grysboks are tawny rufous to
chocolate red, speckled with white, and have dark markings
on the crown. The steenbok is reddish fawn to brown or gray.
The underside of the body and tail of all three species is white.
In contrast to *Ourebia*, *Raphicerus* has no bare patches below
the ears and no tufts on the knees, and the straight horns are
not conspicuously ridged but are rather smooth. The horns,
borne only by males, are 25–127 mm long. *R. campestris* and
R. sharpei do not have false hooves, but small ones are present
in *R. melanotis*. The ears are long and narrow, the tail is

short, and the legs are slender. Females have four mammae.

These antelope inhabit a variety of habitats but require
some dense cover. *R. campestris* is found mainly in lightly
wooded savannah grassland from sea level to 4,750 meters; it
is generally diurnal but in hot weather has peaks of activity in
the early morning and evening. *R. sharpei* favors woodland
savannah with thickets and secondary growth, often in
broken, rocky country; it is almost entirely nocturnal. *R.
melanotis* lives in thick scrub brush and is nocturnal (Kingdon 1982; Smithers 1983). Instead of running when danger
threatens, these little antelope often hide in the vegetation,
lying flat against the ground with the neck outstretched; they
do not run away until nearly trodden upon. Their gait is
unlike that of *Ourebia*, being a swift, less bounding run. It is
said that when hard pressed they will seek refuge in burrows
of the aardvark *(Orycteropus)*. They are both browsers and
grazers, also take some fruit, and kick up roots and tubers
with their hooves (Kingdon 1982; Smithers 1971, 1983). Although they will drink water, if available, they do not seem to
require free water.

Radiotracking studies indicate that male *R. melanotis*
occupy well-defined home ranges of about 2.5 ha. These areas
evidently are territories that are defended and scent-marked
with the preorbital glands, dung heaps, and urination. Males
of this species fight fiercely with their sharp horns and cannot
be kept together; females are more tolerant of one another
(Novellie, Manson, and Bigalke 1984). Limited investigation
of the other species (Kerr and Wilson 1967; Kingdon 1982;
Smithers 1971, 1983; Wilson and Kerr 1969) suggests that
they also are solitary and territorial but that loosely associ-

Steenbok (*Raphicerus campestris*), photo by Lothar Schlawe.

ated pairs may share and defend an area. Breeding in each species seems to occur throughout all or most of the year, though there may be peaks in southern Africa after the start of the spring rains (about November–December). Estimates of the gestation period vary from about 168 to 210 days, there is normally a single offspring, weaning occurs at 3 months, and sexual maturity is attained at 6–9 months.

The flesh of these antelope is palatable but somewhat dry. It has been claimed that they sometimes damage crops. They have disappeared or become rare in some parts of their original range. *R. melanotis* and *R. campestris* seem to be adaptable to the presence of people, but *R. sharpei* may be threatened by the spread of settlement in East Africa (East 1988).

ARTIODACTYLA; BOVIDAE; Genus NEOTRAGUS
Hamilton-Smith, 1827

Dwarf Antelopes

There are two subgenera and three species (Ansell, *in* Meester and Setzer 1977):

subgenus *Neotragus* Hamilton-Smith, 1827

N. pygmaeus (royal antelope), lowland forest zone from Sierra Leone to Ghana;

subgenus *Nesotragus* Von Deuben, 1846

N. batesi (Bates's dwarf antelope), lowland forest zone from southeastern Nigeria to western Uganda;
N. moschatus (suni), dry country from Kenya to eastern South Africa, Zanzibar, Mafia Island.

Nesotragus sometimes has been treated as a separate genus.

With the exception of some individuals of the genus *Tragulus*, *N. pygmaeus* is the smallest hoofed mammal. It has a head and body length of about 500 mm, a tail length of about 75 mm, and a shoulder height of 250–305 mm. In *N. batesi*, head and body length is 500–575 mm, tail length is 45–50 mm, height at the shoulder is 240–330 mm, and weight is 2–3 kg; females are slightly larger than males (Kingdon 1982). In *N. moschatus*, head and body length is 580–620 mm, tail length is 115–30 mm, shoulder height is about 300–410 mm, and weight is 4–9 kg. The general coloration of *N. pygmaeus* and *N. batesi* is cinnamon to russet above and white below. The tuft and underside of the tail are white in *N. pygmaeus* and usually dusky in *N. batesi*. *N. moschatus* has brownish gray to chestnut upper parts, rufous sides, and white underparts.

The genus *Neotragus* is characterized by a slender build, relatively high hindquarters, no tufts on the head and knees, and no false hooves. The horns, borne only by males, slope backward at about the same angle as that of the plane of the face. In *N. moschatus* the horns are 65–90 mm long and are heavily ringed for at least three-fourths of their length. In *N. pygmaeus* the horns are 12–25 mm long, smooth, and black in color. In *N. batesi* the horns are 38–50 mm long, smooth, and usually brown and/or fawn in color. Female *N. pygmaeus* have four mammae.

The species *N. pygmaeus* and *N. batesi* dwell in forests and clearings therein, whereas *N. moschatus* is found mainly in dry country with tangled underbrush. All species are shy and secretive. Their coloration harmonizes with their surroundings, and they generally are not seen until almost underfoot, when they bound away with considerable speed, dodging and twisting about the underbrush and then quickly disappearing. *N. pygmaeus* has been reported to make leaps of 2.8 meters. *N. moschatus* sleeps in the shade during the heat of the day; as evening approaches, it moves toward the glades, where it feeds. It is almost independent of free water, apparently deriving sufficient moisture from vegetation. Lawson (1989) found that its diet consists mainly of freshly fallen leaves and also includes fruit, flowers, and grass. *N. batesi* often eats the tops of peanut plants and is caught in snares set around peanut patches.

According to Kingdon (1982), studies in Gabon show that *N. batesi* rotates its feeding sites through an overall home range of 2–4 ha., spending one or two months in each part before moving on. Estimates of population density in favorable habitat there were 35/sq km to 75/sq km. Females were generally solitary, with an average separation of 150–90 meters, though some, perhaps related individuals, shared all or part of their range. Each male range, evidently a defended territory, overlapped at least two female ranges and was marked with secretions from the preorbital glands. Males seeking a female have a nasal call, and *N. batesi* also sometimes makes a short series of panting, raucous barks when fleeing. Mating in this species occurs throughout the year, with peaks in the late dry and early wet seasons, and gestation lasts about 6 months. Feer (1982) noted that sexual maturity in *N. batesi* comes at 8–18 months for males and about 16 months for females.

Kingdon (1982) reported that in suitable forest habitat male *N. moschatus* are spaced about 400 meters apart and

Top, suni *(Neotragus moschatus)*, photo from Los Angeles Zoo. Bottom, royal antelope *(N. pygmaeus)*, photo from *Proc. Zool. Soc. London.*

have defended and scent-marked territories of about 3 ha.; each male may be associated with 1–4 females. The young of *N. moschatus* are usually born in the wild from mid-November to mid-December and are somewhat darker than the adults. According to Izard and Umfleet (1971), births of

this species have occurred at the Dallas Zoo throughout the year, the gestation period is thought to be 180 days, and sexual maturity is attained at about 6 months. One captive *N. moschatus* lived for 10 years and 2 months (Jones 1982).

In the folklore of Liberia, *N. pygmaeus* has a position

comparable to that of our "Br'er Rabbit," being renowned for its speed and sagacity. Although the meat of *N. moschatus* is of mediocre quality and the species is protected in some preserves, it has declined drastically through illegal hunting and agricultural encroachment and is considered to be endangered (Skinner, Fairall, and Bothma 1977) or vulnerable (Smithers 1986) in South Africa. The subspecies *N. moschatus moschatus*, of Zanzibar, is classified as endangered by the IUCN and the USDI.

ARTIODACTYLA; BOVIDAE; Genus MADOQUA
Ogilby, 1837

Dik-diks

There are two subgenera and four species (Ansell, *in* Meester and Setzer 1977; Yalden 1978):

subgenus *Madoqua* Ogilby, 1837

M. saltiana, northeastern Sudan, northern and eastern Ethiopia, Somalia;
M. piacentinii, coast of southeastern Somalia;

subgenus *Rhynchotragus* Neumann, 1905

M. guentheri, extreme southeastern Sudan, northeastern Uganda, southern Ethiopia, northern Kenya, Somalia;
M. kirki, extreme southeastern Somalia, central and

southern Kenya, northern and central Tanzania, southwestern Angola, Namibia.

Kingdon (1982) thought the above subgeneric distinction to be unwarranted.

Head and body length is 520–720 mm, tail length is 35–56 mm, shoulder height is 305–430 mm, and weight is 3–7 kg. The pelage is soft and lax. The coloration varies from yellowish gray to reddish brown above and from grayish to white below. Horns, possessed only by the males, are ringed, stout at the base, somewhat longitudinally grooved, and occasionally partially concealed by a tuft of hair on the forehead. A characteristic feature of the genus is the elongated snout. Kingdon (1982) wrote that the specialized nose probably serves as a major thermoregulatory device, arterial blood being diverted to the membranes there and cooled through an evaporative process. In the subgenus *Rhynchotragus* the snout is especially long and can be turned in all directions (Grzimek 1975). The accessory hooves are small, and the tail is not conspicuous. Females have four mammae.

Dik-diks usually frequent areas that are rather dry but have considerable brush for concealment. When startled, they may dash off in a series of erratic, zigzag leaps, uttering a call resembling "zik-zik" or "dik-dik"—hence the origin of the common name. These shy, elusive animals usually stay in dense vegetation, but they utilize definite paths through their territories. They are active mainly in the morning and late afternoon but are also nocturnal to some extent (Kingdon 1982). Although at least one species regularly occurs along streambanks, others seem to live for months without moisture except that obtained from dew and the vegetation they

Dik-dik *(Madoqua saltiana)*, photo by J. Meester. Inset: *M. kirki*, photo by C. A. Spinage.

Dik-dik *(Madoqua kirki)*, photo from Zoological Society of San Diego.

Beira *(Dorcatragus megalotis)*, photo by Colin P. Groves.

consume. For food, dik-diks browse on such shrubs as acacias and also take some grass.

In a study of *M. kirki* in the Serengeti, Hendrichs (1975*b*) found a population density of 24/sq km. The animals lived in permanently mated pairs, together with their young, on territories of 5–20 ha. All members of the family marked the territory at regular points with urine and dung, but only the adult male actively defended the area. Young individuals were ejected at about 7–8 months.

Observations of *M. kirki* indicate that females are polyestrous and usually bear young twice a year. Births in the Serengeti area peak at the beginning of the rainy season and toward the end, November–December and April–May. The gestation period is 169–74 days, and there normally is a single offspring weighing about 600 grams at birth. Females produce their own first young at about 15–18 months. Wild individuals are known to have lived at least 10 years (Dittrich 1967, 1972; Hendrichs 1975*b*; Kellas 1955).

Some hunters dislike these little antelope, because they flush and warn the larger game. Dik-diks are hunted extensively in some areas, and their skins are used in the manufacture of gloves (Grzimek 1975).

ARTIODACTYLA; BOVIDAE; Genus DORCATRAGUS
Noack, 1894

Beira

The single species, *D. megalotis*, occurs only in a restricted arid region comprising northern Somalia, Djibouti, and a small adjacent part of eastern Ethiopia (Yalden, Largen, and Kock 1984).

Head and body length is 762–865 mm, tail length is 51–76 mm, shoulder height is 500–760 mm, and weight is 9–11 kg. The hair is thick and coarse. The upper parts are reddish gray, the underparts are white, the head is yellowish red, and the legs are fawn-colored. There are white areas around the eyes, and a distinct dark line occurs on each side from the lower shoulder to the flank. The horns, borne only by males, are upright, straight, spiked, spaced widely on top of the head, and 76–102 mm long. The ears, borne by both sexes, are large, about 152 mm long and 76 mm wide, and curiously marked inside with white hairs. *Dorcatragus* is also characterized by a relatively short body, long legs, a short and bushy tail, and short, padded hooves.

The beira inhabits stony hills and hot, dry plateaus. It appears to be goatlike in habit, leaping from rock to rock with great agility. Hunters report that it is extremely difficult to obtain, because of its protective coloration, extreme wariness, and great speed when alarmed. *Dorcatragus* feeds on coarse grass and the leaves of mimosas, and from these it seems to derive enough moisture to live without additional water (Dorst and Dandelot 1969). It associates in herds of four to seven individuals, which include one or two males. The beira apparently has always been uncommon, and its range is thought to have declined in recent decades. East (1988) regarded it to be threatened by uncontrolled hunting and habitat degradation caused by overgrazing and drought.

ARTIODACTYLA; BOVIDAE; Genus ANTILOPE
Pallas, 1766

Blackbuck

The single species, *A. cervicapra,* originally occurred in Pakistan and India, from the Punjab and Sind to Bengal and Cape Comorin, and in Nepal and Bangladesh (Ellerman and Morrison-Scott 1966; Lever 1985; Roberts 1977).

Head and body length is about 1,200 mm, tail length is about 178 mm, height at the shoulder is 737–838 mm, and weight is 32–43 kg. The blackbuck is one of the few antelope in which the male and female differ in coloration. The buck is rich dark brown above, on the sides, and on the outsides of the legs, whereas the doe is yellowish fawn on the head and back. In both sexes the underparts, insides of the legs, and an area encircling the eyes are white. The males gradually become darker with age. The build is graceful and slender. The horns, borne only by males, are 456–685 mm long, ringed at the base, and twisted spirally up to five turns. The narrow muzzle is sheeplike, the tail is short, and the hooves are delicate and sharply pointed.

The blackbuck lives on generally open plains, though according to Schaller (1967) it reaches greatest abundance in areas with thorn and dry deciduous forest. It is very fast and can usually outrun even a greyhound, but the cheetah (*Acinonyx*) has been trained to capture it. When danger approaches, the alert females are usually the first to warn the herd. When alarmed, a single animal will bound into the air to a surprising height and is soon followed by others until the whole herd is in motion. These bounds are continued for a few strides, after which the herd settles down to a regular gallop. Schaller (1967) reported *Antilope* to be diurnal during the cool season, with intermittent activity all day. In the hot season it is active in the very early morning and late afternoon, resting in the shade at other times. It is almost exclusively a grazer, feeding on short grass and various cultivated cereals.

Population densities of protected herds in restricted reserves have reached about 1 blackbuck per 2 ha. (Walther, Mungall, and Grau 1983). During periods of 6–22 months, 11 adult males used partially overlapping home ranges of 3.25–13.50 sq km (Prasad 1983). However, during part of the year dominant males establish exclusive defended territories that

Blackbucks *(Antilope cervicapra),* photo by Lothar Schlawe.

are much smaller. Observations by both Schaller (1967) and Nair (1976) indicate that the males are territorial at least during the breeding season and that the size of the defended areas apparently varies from about 25 ha. to 100 ha. Observations by Walther, Mungall, and Grau (1983) suggest that maximum size is under 20 ha. and that territories sometimes are occupied only for a few weeks or months at a time. In some situations the males hold small territories in an area of temporarily favorable habitat, and each group of females may have a range encompassing several male territories. A male attempts to hold a female group on his territory for as long as possible.

At one time, *Antilope* reportedly formed aggregations of thousands of individuals. Most recent observations are of herds of 5–50 animals, though in some areas several groups may join during December and January (Schaller 1967). In a study in southeastern India, Nair (1976) reported the number of individuals per herd to average 23 and vary from 2 to 129. Most herds were essentially harems, with a single adult male and a number of adult females and young. Some herds had 1 or 2 peripheral adult males, and there were also small separate groups of bachelors. The adult male drives younger males from the herd and competes with other adults for possession of territories and the females thereon. Dominance is achieved mainly by display of the horns and threatening gestures. The horns are potentially dangerous weapons, but serious fighting is rare.

Mating occurs throughout the year in central India, but the peak periods are March–April and August–October (Schaller 1967). The gestation period is 6 months, and there almost always seems to be a single young. Observations by Rajasingh (1984) indicate that three females lived at least 18 years in the wild and were still fertile at the end of that period.

The blackbuck is prized for its meat and as a sporting trophy. Probably once the most abundant hoofed mammal on the Indian subcontinent, it has been greatly reduced in numbers and distribution through excessive hunting and loss of habitat to agricultural development (Schaller 1967). It has disappeared in Bangladesh (Khan 1984) and now occurs in Pakistan only as a wanderer along the border (Roberts 1977). It also was thought to have been extirpated in Nepal, but two small populations were rediscovered in 1975; one subsequently was lost, but the other now contains 200 individuals (*Oryx* 22[1988]:233–34). Numbers in India fell from an estimated 4 million in the nineteenth century to 80,000 in 1947 and only 8,000 in 1964. There appears to have been subsequent stabilization, with most of the remnant herds in protected reserves. There also are large introduced populations in Texas and Argentina (Lever 1985; Roberts 1977; Walther, Mungall, and Grau 1983).

ARTIODACTYLA; BOVIDAE; Genus AEPYCEROS
Sundevall, 1847

Impala

The single species, *A. melampus*, occurs in the open country from Kenya and southern Angola to northern South Africa (Ansell, *in* Meester and Setzer 1977).

Head and body length is 1,100–1,500 mm, tail length is 250–400 mm, and shoulder height is 775–1,000 mm. The usual weight is 40–45 kg in females and 60–65 kg in males (Jarman and Jarman 1974). The hair is sleek and glossy. The coloration is dark fawn or reddish above and lighter on the thighs and legs. A distinct vertical black streak lies on each side of the hindquarters. The underside of the body and tail, the inside of the upper foreleg, the upper lip, and the chin are white. The horns, borne only by males, are lyrate, 500–750 mm long, and ridged only on the front surface. The hooves lack clefts, and there are glandular tufts of black hair on the hind feet.

The impala inhabits open woodlands, sandy bush country, and acacia savannahs. When alarmed, it makes prodigious leaps, seemingly without effort, the animals often jumping over bushes and over each other and even springing high into the air in places with no obstacle to clear. The speed of the impala is considerable; when running, it has been known to make successive leaps of eight, five, and nine meters. Although the gazelles (*Gazella*) and hartebeests (*Alcelaphus*), with which *Aepyceros* often grazes, run away on the open plains when disturbed, the impala seeks shelter in dense vegetation. It is active both day and night, alternately feeding and resting. It grazes on grass and browses on the leaves of bushes and trees and drinks at least once a day. Unlike certain other antelope, the impala apparently requires a source of free water.

Schaller (1972) estimated that there were 65,000 impala in the 25,500-sq-km Serengeti ecosystem. During the dry season, individuals of all ages and both sexes may form large aggregations, sometimes numbering in the hundreds, that move about in search of green vegetation. At other times of the year there is a more intricate social structure, as well as territorial activity (Fairall 1972; Jarman and Jarman 1973, 1974; Leuthold 1970). In parts of East Africa, where there are two wet seasons, territoriality may extend over most of the year, while in South Africa it may last only about three months. At such times the sexes segregate. Females and young form herds of usually 10–100 individuals, and males form groups of up to 60 bachelors. Such herds have regular home ranges of about 2–6 sq km. The male groups establish a dominance hierarchy through displays and fighting, and the highest-ranking animals attempt to take over a territory. At any given time, about one-third of the adult males in a particular population hold a territory. Males that lose their territory return to the bachelor herds. Territories are about 0.2–0.9 sq km in size and are marked by urination and defecation. The resident male attempts to control any herds of females and young that enter his area. From such groups, he ejects the males that have reached an age of about six to nine months, and the latter subsequently join bachelor herds. During the mating season, males frequently utter loud, hoarse grunts. Kingdon (1982) described tongue flashing, an extraordinary display of the male impala: upon approaching females or potential rivals, a dominant male opens his mouth widely and extrudes the tongue several times in rapid succession. The females tend to bunch at this signal, while the other males flee, unless they respond in kind, thereby indicating a challenge to combat.

Studies in Zimbabwe (M. G. Murray 1981, 1982) have yielded knowledge of still another element in the complex social behavior of the impala. Groups of females and young there evidently are associated in still larger units called clans. There are 30–150 animals in a clan. The home ranges of animals within one clan overlap by an average of 73 percent, whereas the ranges of animals in adjacent but different clans overlap by only 4 percent. There is no defense of clan boundaries and no overt hostility, but association is minimal, and young females usually remain in their natal clan.

In equatorial areas breeding is continuous, with two birth peaks in about March and November (Kingdon 1982). In South Africa there is a major peak in mating from April to June and a lesser peak in September and October (Anderson 1975). The gestation period is about 6–7 months, normally a single offspring is produced, weaning occurs at 5–7 months, males are physiologically capable of reproduction at 13 months, and individuals up to 13 years old have been found in

Impala *(Aepyceros melampus):* Top, female; Bottom, male. Photos from Paris Zoo.

the wild (Jarman and Jarman 1973; Kerr 1965). One impala was still living after 17 years and 5 months in captivity (Jones 1982).

Aepyceros is vulnerable to overhunting and has disappeared in much of southern Africa but has been introduced in some areas beyond the original range (Ansell, *in* Meester and Setzer 1977). The subspecies *A. m. petersi*, of Angola and Namibia, is classified as endangered by the IUCN and the USDI. Its numbers now are estimated at 1,000–2,000 individuals (East 1989).

ARTIODACTYLA; BOVIDAE; Genus AMMODORCAS
Thomas, 1891

Dibatag

The single species, *A. clarkei*, is endemic to the Ogaden region of eastern Ethiopia and adjacent parts of northern and central Somalia (Yalden, Largen, and Kock 1984).

Head and body length is about 1,170 mm, tail length is about 355 mm, shoulder height is 762–890 mm, and weight is 27–34 kg. The dark upper parts are purplish rufous, the underparts are white, and the buttocks are light and lack a dark band on the flank. The streak down the middle of the face is rich chestnut, and the tail is black. The peculiar purplish tint of the rufous coat blends so well with the surroundings that the dibatag is difficult to see.

Ammodorcas is slender and resembles *Gazella* in body form and *Redunca* in horn structure. The neck and tail are long, the skull is flat, and the hooves are small. The horns, borne only by males, are 150–250 mm long; the basal half is ringed and curved backward, and the terminal half is smooth and curved forward. The ears of *Ammodorcas* are rounded, whereas those of *Litocranius* are pointed.

Dibatags *(Ammodorcas clarkei)*, photos from Zoological Garden of Naples, Italy.

The dibatag frequents sandy areas with a scattered growth of thorn scrub and grass. It is not readily seen in its natural habitat, as it conceals its body behind vegetation and peers over the top. The neck is so slender, the head so pointed, and the coloration so much like that of the natural cover that the animal is practically invisible. It remains motionless until discovered. *Ammodorcas* bounds away with the head arched back and the tail thrown forward, whereas *Litocranius* runs with the head and tail outstretched in line with the body. The long neck of the dibatag, which allows it to reach a considerable height when browsing on shrubs, is reminiscent of the giraffe's. *Ammodorcas* often stands on its hind legs, with its forefeet in a tree, to reach as high as possible. The long upper lip also facilitates browsing. The dibatag probably can exist without free water.

Ammodorcas probably maintains defended territories that it marks by urination, defecation, and secretions from the preorbital glands (Walther, Mungall, and Grau 1983). It generally travels alone or in family parties of three to six individuals. Females are thought to give birth in October and November. A gestation period of 204 days has been recorded (Dittrich 1972).

The dibatag is classified as vulnerable by the IUCN and as endangered by the USDI. Its numbers appear to have dropped because of heavy poaching, the effects of drought, and competition from domestic livestock (East 1988).

ARTIODACTYLA; BOVIDAE; Genus LITOCRANIUS
Kohl, 1886

Gerenuk

The single species, *L. walleri*, is found in the arid parts of central and southern Ethiopia, Djibouti, Somalia, Kenya, and northeastern Tanzania. In early historical time this species also occurred in northeastern Sudan and eastern Egypt (Ansell, *in* Meester and Setzer 1977).

Head and body length is 1,400–1,600 mm, tail length is 220–350 mm, shoulder height is 800–1,050 mm, and weight is 28–52 kg (Kingdon 1982). The general coloration is reddish fawn. A broad, dark brown band runs down the back and along the upper third of the sides. The underparts and the front of the neck are white. The giraffelike neck is only 180–255 mm in circumference. The legs are long and slender, the muzzle elongate, the skull wide and flat, and the head wedge-shaped. The horns, developed only in males, are about 355 mm long and are comparatively massive. They curve back-

ward and upward and finally hook forward near the tips.

The gerenuk inhabits dry country with a light covering of brush and thorn scrub. Upon seeing a strange object, it usually stands motionless, hides behind a bush, and then looks over or around the cover by means of its long neck. When frightened, it usually leaves in a stealthy, crouched trot, with the neck and tail carried horizontally. It is not speedy compared with other antelope. It is active throughout the day.

Litocranius is rather sedentary but may make seasonal shifts within its home range. It is highly adapted for an arid environment and seems to be independent of free water, though individuals living near water may occasionally drink. It is exclusively a browser, taking the leaves and shoots of a wide variety of trees and shrubs (Leuthold 1978a). Its eating habits are somewhat like those of *Giraffa* in that it plucks acacia leaves with its long upper lip and long tongue. To reach the food, it often stands upright on its hind legs, with the body practically vertical. It then places its forelegs against a tree to reach for branches, from which it takes the highest and most succulent leaves.

In the Tsavo National Park of Kenya, Leuthold (1978a, 1978b) found population density to average 0.6/sq km and home range size to vary from 1.5 to 3.5 sq km. Ranges overlapped only slightly and seemed to constitute fairly stable territories. Males began to compete for territories when they matured, at around 3 years of age. Apparently, all adult males obtained an individual territory. Younger males wandered from place to place, alone or in small groups. Territories were marked with urine, feces, and secretions from antorbital glands. Adult females and their offspring occurred in groups of two to five individuals. Territorial males associated with the females and kept younger males away. Reproduction was nonseasonal. According to Leuthold, the gestation period is 6.5–7.0 months, a female may mate again within 1 month of giving birth, a single offspring is usual, young females attain sexual maturity at about 1 year, and maximum known longevity is about 8 years in the wild and well over 13 years in captivity.

Yalden, Largen, and Kock (1984) indicated that *Litocranius* was still common in Somalia and Ethiopia but was declining because of hunting for the skin trade. Simonetta (1988) re-

Gerenuks *(Litocranius walleri):* A. Photo by Hans-Jürg Kuhn; B. Photo by Leonard Lee Rue III.

Gerenuks *(Litocranius walleri)*: Left, male; Right, female and young. Photos by W. Leuthold.

ported that the genus now is rare in most of Somalia. East (1988) estimated its total numbers in Africa as 60,000–80,000.

ARTIODACTYLA; BOVIDAE; Genus GAZELLA
De Blainville, 1816

Gazelles

There are 3 subgenera and 16 species (Corbet 1978; Furley, Tichy, and Uerpmann 1988; Gentry, *in* Meester and Setzer 1977; Groves 1969*b*, 1985*b*, 1988; Groves and Lay 1985):

subgenus *Gazella* De Blainville, 1816

G. *dorcas*, desert and subdesert zones from Morocco and Senegal to Israel and Somalia;
G. *saudiya*, Saudi Arabia, Kuwait, southern Iraq;
G. *bennetti*, southern Iran, Pakistan, northern India;
G. *gazella*, Lebanon, Palestine, Sinai, Arabian Peninsula;
G. *bilkis*, North Yemen;
G. *arabica*, Farasan Island in the Red Sea;
G. *spekei*, eastern Ethiopia, Somalia;
G. *cuvieri*, Morocco, Algeria, Tunisia;
G. *rufifrons*, subdesert and savannah zones from Senegal to Ethiopia;

G. *thomsoni*, arid parts of southeastern Sudan, Kenya, and Tanzania;
G. *rufina*, known only by specimens purchased in markets in northern Algeria during the late nineteenth century;

subgenus *Trachelocele* Ellerman and Morrison-Scott, 1951

G. *subgutturosa*, desert and subdesert steppes from Palestine and the Arabian Peninsula to the Gobi Desert and northern China;
G. *leptoceros*, desert zone from central Algeria to Egypt;

subgenus *Nanger* Lataste, 1885

G. *dama*, southern Morocco and Senegal to Sudan;
G. *soemmerringi*, eastern Sudan, Ethiopia, Somalia;
G. *granti*, southeastern Sudan, southern Ethiopia, southwestern Somalia, northeastern Uganda, Kenya, northern Tanzania.

Groves (1985*b*, 1988) did not use subgenera but indicated that the species in *Nanger* might require placement in a separate genus, that the species in *Trachelocele* form a distinct group, that G. *cuvieri* is in a group by itself, that G. *rufifrons* and G. *rufina* form a separate group, and that the rest of the species in the subgenus *Gazella* also form a group. He treated G. *thomsoni* as a subspecies of G. *rufifrons*, but Corbet and Hill (1986) and Honacki, Kinman, and Koeppl (1982) regarded the two as specifically distinct. Those authorities also

Gazelles: Top, *Gazella thomsoni;* Bottom, *G. subgutturosa.* Photos from Paris Zoo.

Gazelles *(Gazella dorcas)*, photo by Lothar Schlawe.

did not accept his separation of *G. bennetti* and *G. saudiya* from *G. dorcas,* while Furley, Tichy, and Uerpmann (1988), on the other hand, considered *G. bennetti* to have affinity to *G. subgutturosa. G. bilkis* was described from old museum specimens in 1985 but subsequently was found to be represented by living animals in captivity in Qatar (*Oryx* 21 [1987]:53).

Head and body length is 850–1,700 mm, tail length is 150–300 mm, shoulder height is 500–1,100 mm, and weight is 12–85 kg (Grzimek 1975). The back and sides vary in color from rich brown through fawn and gray to white. In many species a dark band extends along the sides, immediately above the light area of the belly, and frequently there is a light band above the dark one. Most of the species have white areas around the base of the tail and on the back of the thighs.

The muzzle is normal and is neither expanded as in *Pantholops* nor elongated as in *Saiga.* The neck is not long as in *Litocranius,* and the back lacks the evertible folds present in *Antidorcas.* Both sexes of most species possess horns, which are smaller and more slender in females. One exception is *G. subgutturosa,* in which horns usually occur only in the males. Well-developed horns measure 150–760 mm but generally average 255–355 mm, and all are strongly ringed. Many gazelles have lyre-shaped horns, but there is considerable variation among the species.

Gazelles usually occur in dry open country or brushland. The elevational range of the genus is sea level to 5,750 meters. The name "gazelle" has come to suggest grace and beauty, for all species are dainty, alert, and graceful. Maximum speed of *G. thomsoni* is 80 km/hr (Schaller 1972). In some areas there are large-scale seasonal migrations to different elevations or new feeding grounds. The populations of *G. dama* to the south of the Sahara move northward into the desert during the rainy season. Most *G. thomsoni* in the Serengeti spend the rainy season in the grasslands and the dry season in bush country (Grzimek 1975). The diet varies; *G. thomsoni,* for example, is a grazer, whereas *G. granti* is a browser (Estes 1967).

Schaller (1972) estimated that there were 180,000 *G. thomsoni* in the 25,500-sq-km Serengeti ecosystem. This species, and some others, may form temporary aggregations of hundreds or thousands of individuals, but herd size is usually much smaller. Generally, groups contain about 10–

30 females and young. There are also small groups of males and lone adult males. The latter establish territories and temporarily win control of the females that enter. Female herds of *G. granti* wander over an annual home range of about 300 sq km, while males of this species have territories 500–2,000 meters in diameter. Defense against other territorial males and young males is mainly ritualistic and involves gestures and display of the horns. The territories of male *G. thomsoni* are smaller, the animals being spaced only about 200–300 meters apart, but defense is more vigorous (Estes 1967; Grzimek 1975; Walther 1972).

Baharav (1983) reported average annual densities of 13/sq km to 18/sq km for a population of *G. gazella* in the northern part of Israel. Reproductive activity was seasonal, the males establishing individual territories in December and January and attempting to mate with the groups of females that freely moved through. Resulting births occurred in May and June. Young females reached sexual maturity in their second breeding season and first gave birth at about 2 years. Another population, just 80 km to the south, was reproductively active all year, each female giving birth at intervals of 7 months. Females reached sexual maturity at an early age and initially produced their own young at only 1 year. The key difference between the two populations was the lack of free surface water in the north, which limited reproductive potential.

Reproduction occurs throughout the year in some other areas, though there may be peaks of breeding in the wet seasons. A female may experience a postpartum estrus and give birth more than once a year. *G. thomsoni,* for example, commonly calves in January–February, right after the rains, enters estrus within a month, and gives birth a second time in July. Reported gestation periods of *Gazella* vary from 156 to 224 days, that of *G. thomsoni* being about 160–80 days. There is normally a single offspring, but twins are frequent in *G. cuvieri.* Weight of newborn averages 1,250 grams for *G. spekei* and is 2,200–3,000 grams for *G. thomsoni.* Weaning in the latter species occurs after just 2 months, and the young are almost full-grown in a year. Females may attain sexual maturity in their first year of life, but reported age at first conception varies from 152 to 720 days (Dittrich 1972; Estes 1967; Furley 1986; Kingdon 1982; Read and Frueh 1980). A captive *G. dorcas* lived for 17 years and 1 month (Jones 1982).

Gazelles long have been hunted by people for use as food. The flesh of some species is considered to be excellent, while that of others is strong, coarse, and dry. Some hunters use falcons to harass a gazelle so that it can be readily overtaken by dogs. In Syria, Israel, Jordan, Sinai, and Saudi Arabia are the remains of large stone corrals into which gazelles were driven in prehistoric times and which were still in use in the early twentieth century (Legge and Rowley-Conwy 1987; Mendelssohn 1974). Excessive hunting by people, excessive grazing by domestic livestock, agricultural development, and other habitat modifications have adversely affected most populations of *Gazella* in the Middle East and North Africa. *G. rufina* apparently long has been extinct. The subspecies *G. leptoceros leptoceros,* known only from the Western Desert of Egypt, has been reduced to a tiny fraction of its original range and appears to be on the verge of extinction (Saleh 1987). The gazelles of Ethiopia, Somalia, and eastern Sudan also have experienced a serious decline (East 1988).

The IUCN classifies *G. cuvieri* and *G. subgutturosa marica* (Jordan, Arabian Peninsula) as endangered and *G. dama, G. dorcas* (including *G. bennetti, G. saudiya, G. bilkis,* and *G. arabica*), *G. gazella, G. leptoceros, G. rufifrons, G. soemmerringi,* and *G. spekei* as vulnerable. These IUCN designations cover all living members of the genus *Gazella* except *G. thomsoni* and *G. granti* of sub-Saharan Africa and the Central Asian subspecies of *G. subgutturosa.* However, the Soviet Union classifies the entire species *G. subgutturosa* as vulnerable. The USDI lists *G. subgutturosa marica*, *G. dorcas massaesyla* (Morocco to Tunisia), *G. dorcas pelzelni* (Somalia), *G. saudiya*, *G. gazella*, *G. leptoceros*, *G. cuvieri*, *G. dama lozanoi* (Western Sahara), and *G. dama mhorr* (Morocco) as endangered. *G. dama* is on appendix 1 of the CITES.

ARTIODACTYLA; BOVIDAE; Genus ANTIDORCAS
Sundevall, 1847

Springbuck, or Springbok

The single species, *A. marsupialis*, occurs in Angola, Namibia, Botswana, and South Africa (Ansell, *in* Meester and Setzer 1977).

Head and body length is 1,200–1,400 mm, tail length is 150–300 mm, shoulder height is 730–870 mm, and weight is about 30–48 kg. The springbok is cinnamon fawn above, with a dark reddish brown horizontal band extending from the upper foreleg to the edge of the hip, separating the upper color from the white underside. The inside of the legs, the back of the thighs, the tail, and a patch extending onto the rump are also white. Both sexes have black, ringed horns. False hooves are present, and there are no tufts of hair on the knees.

The general external appearance is very much like that of *Gazella*. *Antidorcas* was separated generically from *Gazella* because of its teeth; in the springbok there are five pairs of grinding teeth in the lower jaw, whereas there are six pairs in the gazelles. The one peculiar and striking external difference of *Antidorcas* is the fold of skin extending along the middle of the back to the base of the tail. This fold is covered with hair, much lighter in color than the rest of the back. When the animal becomes alarmed, it opens and raises this fold, so that the white hair shows as a conspicuous crest along the back; at the same time the white hairs on the rump are erected.

The springbok lives on open, dry savannahs and grassland. Its common name is based on its habit of leaping up to 3.5 meters into the air when startled or at play. In springing, the body is curved, the legs are held stiff and close together, and the head is lowered. As soon as the animal hits the ground, it rebounds with no apparent effort. Caro (1986) reviewed the possible functions of this behavior, known sometimes as stotting or pronking, and concluded that it most likely served to inform a potential predator that it had been detected and so

might as well give up any pursuit. The springbok is suspicious of roads and wagon paths and clears these obstacles at a bound. It can run at speeds of up to 88 km/hr (Smithers 1983). It may be active at any hour but sometimes rests during the hot part of the day (Smithers 1971). Drought occasionally forces the springbok to undertake large-scale migrations in search of new pastures. *Antidorcas* is both a grazer and a browser. It thrives on karroo shrubs and grass and is able to get along without water, though it will drink if water is available.

The springbok is highly gregarious; migratory herds formerly contained over 1 million individuals. The animals were so numerous that it would take days for the herd to pass a given point. Numbers have now been greatly reduced, but groups still occasionally have up to 1,500 individuals. The larger aggregations occur during the wet season, when the animals move toward areas of new vegetation. In the dry months, however, populations are divided into smaller groups. There then are herds of up to 100 females and young, each associated with a number of adult males. The latter establish individual territories of 10–40 ha., within which they attempt to retain the females that are moving about. There also are nonterritorial solitary males and separate groups consisting of up to 50 bachelors and sometimes a few young females. The territorial males make visual displays, mark their areas with urine and dung, and frequently chase and fight intruding males. When calving is about to occur, the females segregate and the young males leave to join the bachelor groups. Reproduction is not seasonal, but births are concentrated during the rains, which occur in the spring, summer, or winter, depending on location. The gestation period lasts about 24 weeks, there is normally a single offspring, and mean weight at birth is 3.82 kg. Females reach physiological sexual maturity at about 7 months and usually remain in their mother's group (Bigalke 1970; David 1978a, 1978b; Smithers 1971, 1983). A captive springbok lived for 19 years (Jones 1982).

Antidorcas is easily tamed and thrives in captivity. It has been extensively hunted by people for its excellent meat and because its mass migrations were ruinous to crops. As a result, it disappeared from most of South Africa, but there are reintroduced populations in numerous parks and reserves in the country (De Graaff and Penzhorn 1976; Smithers 1983). The springbok is the national emblem of the Republic of South Africa. To the north, it still occurs in most of its original range but is much less common than formerly (Ansell, *in*

Springbuck *(Antidorcas marsupialis):* Left, photo from San Diego Zoological Garden; Right, photo from Paris Zoo.

Meester and Setzer 1977; Smithers 1971). Total numbers recently were estimated at over 600,000 (East 1989).

ARTIODACTYLA; BOVIDAE; **Genus PROCAPRA**
Hodgson, 1846

Central Asian Gazelles

There are two subgenera and three species (Corbet 1978; Groves 1967*a*):

subgenus *Procapra* Hodgson, 1846

P. picticaudata (goa), Himalayan region, Tibet, highlands of central China;
P. przewalskii, subdesert steppes of north-central China;

subgenus *Prodorcas* Pocock, 1918

P. gutturosa (zeren), dry steppe and subdesert of Mongolia and adjacent parts of southern Siberia and northern China.

Head and body length is 950–1,480 mm, tail length is 20–120 mm, shoulder height is 540–840 mm, and weight is 20–40 kg

(Grzimek 1975). *P. gutturosa* is the largest species; it is orange buff above in the summer, with pinkish cinnamon sides, but is paler in the winter. The other species are generally brownish gray in summer and paler in winter. Both have a white rump patch, but that of *P. picticaudata* is continuous, whereas that of *P. przewalskii* is divided by a median line of darker color. All species have white underparts. The horns, present only in males, are about 200–250 mm long. The backward deflection of the horns is not as conspicuous in *P. gutturosa* as in the other species. The horns of *P. przewalskii* are curved in two places, with the points turned inward, while the horns of *P. picticaudata* are curved in only one plane.

These gazelles are found in dry grassland at elevations of up to 5,750 meters. *P. picticaudata* is a wary and swift animal that scrapes out a bedding place in the ground and feeds in areas of scant vegetation. Its mating season begins in December and lasts for about one month; the young are born the following May. A single offspring is normal.

The mating season of *P. gutturosa* is in the late fall or winter. At this time the males have a swollen throat. In the spring there is a northward migration, at which time herds of 6,000–8,000 individuals form. Upon reaching the summer pastures in June, the sexes separate and the young are born. Observations of captives indicate that females are seasonally polyestrous, the estrous cycle takes 29 days, estrus lasts less than 1 day, and the gestation period is 186 days (Miyashita

Zerens (Procapra gutturosa): Left, two views of male; Lower right, female and young; photos from Osaka Municipal Tennoji Zoological Garden. Upper right, *P.* sp., photo by Howard E. Uible of skull in U.S. National Museum of Natural History.

and Nagase 1981). There usually is a single young, but twins are not unusual (Grzimek 1975). Within a few days the young can keep up with the mother. A captive *P. gutturosa* lived for 7 years (Jones 1982).

These gazelles are hunted by people for their flesh and skin. Schaller (1989) reported that *P. picticaudata* once was probably the most abundant wild ungulate on the Tibetan Plateau but has been reduced drastically by hunting, especially in areas where human and livestock populations are large. Mallon (1985) noted that large numbers of *P. gutturosa* were being killed and that the species had disappeared from western and central Mongolia. The Soviet Union now classifies *P. gutturosa* as endangered.

ARTIODACTYLA; BOVIDAE; Genus PANTHOLOPS
Hodgson, 1834

Chiru, or Tibetan Antelope

The single species, *P. hodgsoni*, occurs in the highlands of Kashmir, Tibet, and adjacent parts of north-central China (Corbet 1978).

Head and body length is 1,300–1,400 mm, tail length is about 100 mm, shoulder height is 790–940 mm, and weight is 25–50 kg. The hair is short, dense, and woolly. The back and sides are pale fawn with a pinkish suffusion, the face and fronts of the legs are dark, and the underparts are white. The horns, carried only by males, are slender, black, ridged in front, almost vertical in relation to the head, and 510–710 mm long. The nose is swollen at the tip, the legs are slender, and the tail is relatively short.

The chiru inhabits plateau steppes at elevations of 3,700–5,500 meters. It is quite wary, and the gait is a trot that can be rapid enough to outdistance dogs and wolves. When a male is in motion, the horns are held high, greatly enhancing the appearance. When at rest, *Pantholops* often lies in a shallow depression that it has excavated to a depth of about 300 mm. Thus sheltered from the wind and partly concealed, it watches for enemies. In the morning and evening it grazes and perhaps browses along glacial streams.

According to Schaller and Ren (1988), there are early reports that 15,000–20,000 individuals were visible at one time, and even in 1985 at least 1,000 were observed in a day. In 1986 an estimated 3,500–4,000 animals occupied a general area of

20,000 sq km in Tibet, though they were locally concentrated at densities averaging 1.47/sq km. The average herd size in the area, not including lone individuals, generally adult males, was 16.9 in 1985 and 9.2 in 1986. Males and females are partially segregated in the summer, each sex sometimes concentrating in different areas.

During the mating season in late November and December, adult males eat little and are in a state of great excitement. Each attempts to form a harem of 10–20 females, which are jealously guarded. As soon as a male spots a rival, he lowers his saberlike horns and rushes to attack. The battles are fierce. The long, sharp horns can inflict terrible wounds, and sometimes both contestants die as a result. If one doe attempts to leave a harem, the male tries to drive her back. Meanwhile, the other does may use the opportunity to desert, as there is apparently no lasting bond between the sexes. The young are born in May or June.

Schaller (1986) and Schaller and Ren (1988) indicated that *Pantholops* has declined substantially in recent years because of intensified hunting and the effects of a disastrous snowstorm in 1985. The genus now survives primarily in the northwestern and north-central parts of the Tibetan Plateau.

ARTIODACTYLA; BOVIDAE; Genus SAIGA
Gray, 1843

Saiga

The single species, *S. tatarica*, was found in historical time in the steppe zone from western Ukraine to western Mongolia. In the Pleistocene this species evidently occurred from England to Alaska (Sokolov 1974).

Head and body length is 1,000–1,400 mm, tail length is 60–120 mm, shoulder height is 600–800 mm, and weight is 26–69 kg. The coat is heavy and wool-like, with a fringe of long hairs extending from the chin to the chest. In summer the upper parts are cinnamon buff, the nose and sides of the face are dark, the crown is grizzled, and the rump patch, underparts, and tail are white. In winter the coat is longer and thicker and uniformly whitish. The horns, possessed only by males, are 203–55 mm in length, irregularly lyrate, heavily ridged, and pale amber in color.

A remarkable character of *Saiga* is the inflated and proboscislike nose. The nostril openings point downward, and

Tibetan antelope, or chiru *(Pantholops hodgsoni)*, photo by George B. Schaller.

Saigas *(Saiga tatarica)*, photo from Dierenpark "Wassenaar," Wassenaar, Holland.

there are unusual internal structures. The bones of the nose are greatly developed and convoluted, and the nasal openings are lined with hairs, glands, and mucous tracts. In each nostril is a sac lined with mucous membranes, a feature found in no other mammal except whales. The inflated nose and associated structures may be adaptations for warming and moistening inhaled air. Kingdon (1982), however, suggested that blood is diverted to the membranes of the expanded proboscis and is there cooled through an evaporative process. The saiga's exceptionally keen sense of smell may also be related to the nasal development.

Except as noted, the information for the remainder of this account was taken from Bannikov et al. (1967) and Sokolov (1974). The saiga is found mainly on grassy plains, often in arid areas. It avoids broken country and dense vegetation but may enter the forest steppe zone in the summer. It can run at speeds of up to 80 km/hr. The saiga is active throughout the day for most of the year. During the summer it grazes in the early morning and evening and rests at midday. It usually has no fixed home range and commonly wanders several dozen kilometers per day. Some populations undertake extensive seasonal migrations. The animals on the west side of the Caspian Sea concentrate to the south in the winter, move northward in April and May, and return to the south in the fall. They may cover 80–120 km per day during such movements. The diet consists mainly of grass and also includes various herbs and shrubs. During the summer, under favorable conditions, the saiga visits water holes twice daily.

In 1958 an estimated 2 million saiga inhabited a total range of 2.5 million sq km, giving a mean density of 0.8/sq km. In local areas, however, density has reached 14/sq km during winter concentration and up to 40/sq km in concentrations around lakes during drought. On the wintering grounds to the west of the Caspian Sea, both sexes winter together. At the beginning of the spring migration the males form herds of

10–2,000 individuals and push out ahead of the females. The latter form vast aggregations and move off in search of a place to give birth. As soon as their young are able to travel, the females follow the males. In 1957 a continuous stream of females and young, estimated to contain 150,000–200,000 animals, was observed. In summer the saiga is generally found in groups of 30–40 individuals. Large migratory herds again form in the autumn. During the mating season in early winter, adult males become territorial, each attempting to gather a harem, generally consisting of 5–15 females. The male constantly herds the females and challenges other males that approach his area. There are fierce fights, often resulting in death. At the end of the rut the exhausted males perish in large numbers.

Mating extends from December to January, and births occur in late April and May. The gestation period is 139–52 days. About two-thirds of the females have twins, and the rest produce a single calf. The newborn weighs an average of 3.5 kg and can outrun a human by its second day of life. It begins to graze at 4–8 days of age but is not completely weaned for 4 months. Females usually attain sexual maturity before their first birthday and continue to grow until they are 20 months old. Males can mate at 19–20 months and grow until they reach 24 months. Known maximum longevity in the wild is 10–12 years.

The saiga was exterminated in the Crimea by the thirteenth century A.D. but survived elsewhere in the southern Ukraine until the eighteenth century. Subsequent uncontrolled hunting led to drastic declines in numbers and distribution, and by the early twentieth century fewer than 1,000 saiga were thought to survive. Many animals were killed merely to obtain the horns, which could be sold in the Orient for their alleged medicinal value. Total protection was established in Europe in 1919 and in Soviet Central Asia in 1923. The saiga made a remarkable comeback in the Soviet

Union: by 1958 an estimated 2 million individuals were present. Sport and commercial hunting is now allowed in some areas. Saiga products include muttonlike meat, hides, fat, and, still, horns for the pharmaceutical trade. The Mongolian subspecies, *S. t. mongolica*, which is listed as endangered by the USDI, has not shared in the general recovery of the species. Dash et al. (1978) estimated that about 200 individuals survived. Mallon (1985) indicated that numbers were increasing, but Sokolov (1989) stated that only a few dozen are now left.

ARTIODACTYLA; BOVIDAE; Genus CAPRICORNIS
Ogilby, 1837

Serows

There are three species (Corbet 1978; Ellerman and Morrison-Scott 1966; Groves and Grubb 1985; Lekagul and McNeely 1977):

C. sumatraensis, central and southern China, Himalayan region, Assam, Burma, Thailand, Indochina, Malay Peninsula, Sumatra;
C. swinhoei, Taiwan;
C. crispus, Honshu, Shikoku, Kyushu.

Capricornis was included in *Nemorhaedus* by Groves and Grubb (1985), but this view was not accepted by Mead (1989). *C. swinhoei* usually has been treated as a subspecies of *C. crispus* but was considered to be a distinct species by Groves and Grubb (1985).

Head and body length is 1,400–1,800 mm, tail length is 80–160 mm, shoulder height is 850–940 mm, and weight is 50–140 kg. The upper parts are generally gray or black, the mane ranges in color from white to black, and the underparts are whitish. *Capricornis* resembles *Nemorhaedus* (see account thereof) but has large preorbital glands and a straighter facial profile. The rhinarium is naked, and the ears are long, narrow, and pointed. The horns, carried by both sexes, are slightly curved, marked with narrow transverse ridges on the basal three-fourths, and 152–255 mm long. The hooves are short and the tail is moderately bushy.

Serows inhabit rugged mountains or ridges, covered with thick brush or forest, at elevations of up to 2,700 meters. Their gait is clumsy and not particularly rapid, but they are sure-footed in descending steep, rocky slopes. They are commonly hunted with dogs, and when brought to bay, they defend themselves with their horns in a deadly manner. According to Lekagul and McNeely (1977), they have acute senses of smell, vision, and hearing; they may have well-defined runways along steep rock faces; and they are known to swim between the small islands near the Malay Peninsula. They feed during the early morning and late evening, sheltering at other times in favorite resting places, often in caves or under overhanging rocks and cliffs. The diet consists of grass, shoots, and leaves.

Capricornis is usually solitary but sometimes occurs in groups of up to seven members. According to Schaller (1977), lone males, pairs, and small family groups of *C. crispus* tend to occupy small, discrete home ranges for most of the year. Home range size is 1.3–4.4 ha. for solitary individuals and 9.7–21.7 ha. for family units. In the summer and autumn, males roam beyond their home range, and grown young leave their mother's range. These home ranges may constitute exclusive territories, as males have been seen to chase other males away, and both sexes mark the areas with secretions

Formosa serow *(Capricornis crispus),* photo by Constance P. Warner.

Serow (*Capricornis sumatraensis*), photo by Wang Sung.

from their preorbital glands. Mating takes place in October and November, the gestation period lasts about 7 months, usually a single offspring is produced, weight at birth is about 3.5 kg, sexual maturity comes at about 30 months for females and at 30–36 months for males, and captives have lived for over 10 years (Lekagul and McNeely 1977; Sugimura et al. 1983; Tiba et al. 1988; Yamamoto 1967).

The meat of serows is of mediocre quality, but some people believe that various parts of the animals have medicinal value. *C. crispus* declined to the point that it was declared endangered in Japan in 1934. Subsequent protection led to a numerical increase, but simultaneous usurpation of habitat resulted in damage to new forestry plantations and to culling of the serow (Takatsuki and Suzuki 1984). The subspecies on Sumatra, *C. sumatraensis sumatraensis*, has been severely reduced in numbers and distribution through excessive hunting and habitat loss. It is classified as endangered by the IUCN (1972) and the USDI and is on appendix 1 of the CITES. *C. swinhoei* is designated as vulnerable by the IUCN.

ARTIODACTYLA; BOVIDAE; Genus **NEMORHAEDUS**
Hamilton-Smith, 1827

Gorals

There are three species (Groves and Grubb 1985; Mead 1989; Volf 1976):

N. baileyi, a small region where Tibet, Assam, and Burma come together;

N. caudatus, extreme southeastern Siberia to Burma and western Thailand;

N. goral, Himalayan region.

Groves and Grubb (1985) included *Capricornis* in *Nemorhaedus*. Mead (1989) included *N. baileyi* and *N. caudatus* in *N. goral*. Hayman (1961) regarded *N. cranbrooki*, of Tibet, Assam, and Burma, to be a distinct species, but Groves and Grubb (1985) thought it to be conspecific with *N. baileyi*, which occurs in the same region and which had been described previously.

Head and body length is 820–1,300 mm, tail length is 76–203 mm, shoulder height is 570–785 mm, and weight is 22–35 kg. The body is covered by a short, woolly undercoat, which is covered by long, coarse guard hairs. The male has a short, semierect mane. The general coloration of the upper parts varies from buffy gray to dark brown and foxy red. The underparts usually are paler. There is a black stripe on the foreleg, a white patch on the throat, and a dark stripe down the middle of the back. The conical horns, which are carried by both sexes, are 127–78 mm in length, curve backward, and are marked by small irregular ridges. The facial profile is concave, the back is somewhat arched, and the limbs are stout and long, well adapted to climbing and jumping. Females have four mammae.

Nemorhaedus resembles *Capricornis* in general appearance but is distinguished by smaller size, absence of facial glands, and shorter horns. Groves and Grubb (1985) suggested that gorals evolved from serows and are distinguished from the latter by having the facial part of the skull more flexed relative to the cranial part, more robust premaxillae, and nasal bones usually separate from the lacrimals and max-

Goral *(Nemorhaedus goral)*, photo by R. Pucholt.

illae, producing a deep narial notch. They noted also that the species *N. baileyi* has certain characters intermediate to those of serows and other gorals and that the species *Capricornis crispus* has several characters resembling those of gorals.

Gorals are found on rugged, wooded mountains, generally at elevations of 1,000–4,000 meters. They inhabit slopes even more precipitous than those used by *Capricornis*, seeming to prefer the most difficult terrain possible (Lekagul and McNeely 1977). They are most active during the early morning and late evening, but on cloudy days they roam throughout the day. After eating in the morning, they usually drink water, then retire to a sunny rock ledge to stretch out and rest until evening. They are quite difficult to recognize, even though they are in full view, for they lie motionless and their color blends with that of the rocks. The diet consists of twigs, low shrubs, grass, and nuts (Lekagul and McNeely 1977).

Nemorhaedus usually occurs in groups of 4–12, though old males commonly live alone most of the year. In the mating season males may occupy and mark a territory of 22–25 ha. When frightened, gorals emit a hissing or sneezing sound. Mating takes place from late September to November in Siberia and in November and December farther south. Estrus lasts 20–30 hours. Gestation periods of 6–8 months have been reported. There are 1 or 2 offspring per birth. Sexual maturity is attained in the third year of life (Dobroruka 1968; Lekagul and McNeely 1977; Mead 1989; Schaller 1977). A captive lived for 17 years and 7 months (Jones 1982).

Although the heads of gorals do not make showy trophies, they are frequently hunted for sport. *N. goral* (including *N. baileyi* and *N. caudatus*) is listed as endangered by the USDI and is on appendix 1 of the CITES. The IUCN classifies *N. baileyi* as vulnerable. The Soviet Union regards *N. caudatus* as endangered.

ARTIODACTYLA; BOVIDAE; **Genus OREAMNOS**
Rafinesque, 1817

Mountain Goat

The single species, *O. americanus*, originally occurred in the mountainous region from southeastern Alaska and southwestern Northwest Territories to north-central Oregon and western Montana (Hall 1981). *Oreamnos* is a member of the goat-antelope tribe (Rupicaprini) of the subfamily Caprinae and is not a true goat (tribe Caprini).

Head and body length is 1,200–1,600 mm, tail length is 100–200 mm, height at the shoulder is 900–1,200 mm, and weight is 46–140 kg. Males generally exceed females by 10–30 percent in linear dimensions. The pelage is white or yellowish white, and the underfur is thick and woolly. The hair is long and soft along the midline of the neck and shoulders, forming a ridge or hump. A beard is also present. The hooves have a hard, sharp rim, enclosing a soft inner pad, and are well suited to climbing over rocks and ice. Females have four mammae.

Except as noted, the information for the remainder of this account was taken from the review papers by Rideout (1978) and Rideout and Hoffmann (1975). The mountain goat is usually found among steep slopes and cliffs in alpine tundra or subalpine areas associated with low temperature and heavy snowfall. It is renowned for its ability to climb and jump easily through rugged terrain and has been known to gain 460 meters in elevation within 20 minutes. Activity generally occurs in the early morning and late afternoon and frequently goes on through the night. Daily movements generally cover several hundred meters. For bedding, shallow depressions may be excavated with the forefeet. In the fall there is a general downward movement to south- and west-facing slopes that often are free of snow. The distance between summer and winter centers of activity was found to range from

North American mountain goat *(Oreamnos americanus)*, photo by Leonard Lee Rue III.

1.7 to 11.1 km in Montana. The diet is broad, varies from place to place and from season to season, and may involve either browsing or grazing. Grass, mosses, lichens, woody plants, and herbs are all significant foods. Salt licks are important to the mountain goat in the spring and summer, and it may travel several miles to reach them.

Population densities of 0.03/sq km to 14.00/sq km have been reported, but a typical overall figure is thought to be 1/sq km or 2/sq km (J. L. Fox 1984). The average annual home range in Montana was found to be 21.5 sq km for adult males and 24 sq km for adult females. In winter, individual ranges are as small as 81 ha., and large groups of animals may form at that time. In other seasons, groups usually do not contain more than 4 animals, except around salt licks, and adult males are frequently seen alone. Social structure appears to be variable, with some reports indicating that males are dominant and others suggesting that males are subordinant to both females and juveniles. Except during the breeding season, the sexes seem indifferent to one another and sometimes clash over limited food resources. In agonistic interaction there is no butting of the heads as in *Ovis*, but the sharp horns may be thrust at the opponent's flanks and rump, sometimes causing serious injury.

Mating occurs from November to early January, and births take place in late May and early June. Females are seasonally polyestrous, the estrous cycle averages about 20 days, estrus lasts 48 hours or less, the gestation period is approximately 186 days, and weight at birth is about 3.2 kg (Hutchins et al. 1987). There is usually a single kid, but twins are not uncommon and triplets are rare. The young is able to follow its mother within a week and is completely weaned by

September. It is driven away by the aggression of the female, when she again gives birth. Sexual maturity is attained by both sexes at around 30 months. Maximum known longevity in the wild is 14 years for males and 18 years for females.

Because of its generally inaccessible habitat, the mountain goat has been affected less by human activity than has any other big game animal in North America. Nonetheless, some populations have declined seriously through excessive hunting, which has resulted in part from improved access by people on newly constructed roads. Introduced populations of *Oreamnos* have been successfully established in Colorado, central Montana, the Black Hills of South Dakota, northeastern Oregon, and Olympic National Park, and on the Alaskan islands of Kodiak, Baranof, and Chichagof. The mountain goat is subject to limited sport hunting in most parts of its range. Various population estimates suggest that the total number of individuals in North America is around 100,000 (Rideout 1978; Samuel and Macgregor 1977).

ARTIODACTYLA; BOVIDAE; **Genus RUPICAPRA**
De Blainville, 1816

Chamois

There are two species (Lovari 1987; Masini and Lovari 1988; Nascetti et al. 1985):

R. pyrenaica, mountains of northwestern Spain, Pyrenees, Apennines of central Italy;

R. rupicapra, the Alps and other mountains of south-central Europe, the Balkans, Asia Minor, and the Caucasus.

Fossil remains indicate that in the late Pleistocene and early Holocene *Rupicapra* occurred in most of Spain and Italy and in other parts of Europe where the genus is now absent (Masini 1985).

Head and body length is 900–1,300 mm, tail length is 30–40 mm, shoulder height is 760–810 mm, and weight is 24–50 kg. The pelage is stiff and coarse. The hairs of the summer coat are only about 40 mm long and are a tawny brown color in *R. rupicapra* and reddish in *R. pyrenaica.* The winter coat is composed of guard hairs 100–200 mm long and a thick, woolly underfur. In *R. rupicapra* the winter coat is blackish brown except for white markings on the head and throat and pale underparts. The winter coat of *R. pyrenaica* is similar but also has large white areas on the neck, shoulders, and flanks. The slender, black horns, borne by both sexes, are 152–203 mm long. They are set closely together, rise almost vertically, and then bend abruptly backward to form hooks. The pad of the hoof is slightly depressed and somewhat elastic, providing a sure foothold on uneven, slippery terrain.

During the warmer months of the year chamois generally stay above 1,800 meters in alpine meadows but never venture more than a few hundred meters from the cliffs where they can take refuge. In late fall and winter they move below 1,100 meters, sometimes entering forests, but stay on steep slopes where snow does not accumulate and block access to food (Lovari 1987). They return to alpine areas in the spring. These beautiful animals are nimble, agile, daring, and graceful. Their senses are acute. When alarmed, they flee to the most inaccessible places, often making prodigious leaps. They can jump almost 2 meters in height and at least 6 meters in length and can run at speeds of 50 km/hr on uneven ground (Lovari 1987). During the summer months the diet consists chiefly of herbs and flowers, but in winter the chamois eats lichens, mosses, and young pine shoots. It has been known to fast for two weeks and survive when the snow was so deep that food could not be secured.

Females and young commonly form herds of 15–30 individuals. Hamr (1985) reported that such groups occupied an average range of 74 ha. during summer–autumn and 60 ha. in winter and that the greatest distance between the two ranges was 2.7 ha. Lovari (1984) explained that there is seasonal variation in social structure, the largest group size being reached in late summer, dispersal coming when the animals move to the winter range, and the females isolating themselves to give birth in the spring. A herd is said to post a sentinel that warns the other animals of approaching danger by stamping its feet and uttering a sharp, high-pitched, whistling call. Fully adult males generally live alone for most of the year but begin to join the herds in the late summer. During the autumn rut, centered in November, the old males drive the younger males from the herds and occasionally kill them. After a gestation period of about 170 days, the kids are born during May and June in a shelter of grass and lichens. The usual number of young is one, but twins and triplets sometimes occur. Kids are able to follow their mother almost immediately after being born, and they rapidly improve their leaping ability within the first few days of life. If a mother is killed, other chamois take care of its young. Young males stay with their mother's group until they are 2–3 years old and then live in nomad fashion until they are fully mature at 8–9 years, when they become attached to a definite area (Lovari 1984). Potential longevity is 22 years.

The flesh of *Rupicapra* is prized as food by some people, the skin is made into "shammy" leather for cleaning glass and polishing automobiles, and the winter hair from the back is

Chamois *(Rupicapra rupicapra),* photos from Bernhard Grzimek.

used to make the "gamsbart," the brush of Tyrolean hats. Excessive hunting has greatly reduced the number of chamois in some areas. The subspecies *R. pyrenaica ornata*, of the Apennines, is now restricted to Abruzzo National Park in central Italy. Lovari (1977) reported that about 300–400 individuals survived, and Masini and Lovari (1988) indicated that numbers still are about the same. This subspecies is listed as vulnerable by the IUCN and as endangered by the USDI and is on appendix 1 of the CITES. The IUCN also classifies *R. rupicapra cartusiana*, of the Massif de la Chartreuse in southeastern France, as endangered and *R. r. tatrica*, of the Tatra Mountains on the border of Poland and Czechoslovakia, as rare. These two subspecies number, respectively, about 100 and 900 individuals (Masini and Lovari 1988). Otherwise, chamois now are generally increasing in number and have been introduced and reintroduced in various parts of Europe. Total numbers in Europe are about 523,000 *R. rupicapra* and 31,000 *R. pyrenaica*. There also are large introduced populations in New Zealand (Lovari 1987; Masini and Lovari 1988).

ARTIODACTYLA; BOVIDAE; Genus MYOTRAGUS
Bate, 1909

Cave Goat

The single species that lived in historic time, *M. balearicus*, is known only by skeletal remains found on Mallorca and Menorca in the Balearic Islands of the western Mediterranean Sea (E. Anderson 1984; Burleigh and Clutton-Brock 1980; Kurten 1968; Reumer and Sanders 1984; Vigne 1987). Its precise systematic affinities are uncertain, but probably the most closely related living genus is *Rupicapra*.

Myotragus was a relatively small animal, only about 500 mm tall at the shoulders and weighing about 14 kg. It probably had the appearance of a small, squat goat. Its most remarkable character was the development of the lower front teeth. Like all other bovids, *Myotragus* had no upper incisors, but it was unique in having only two lower incisors, what would be the middle pair in other genera, instead of six. These incisors grew into huge, chisel-like teeth with open roots; they grew constantly throughout life, as in rodents. The other incisors usually were missing, though a vestigial second pair was present in a few cases. The cheek teeth had very high crowns. Another peculiar feature was an unusual shortening of the limb bones. In the forelimb the humerus and metacarpal were very short compared with the radius and ulna, while in the hind limb the femur and metatarsal were very short compared with the tibia. Moreover, there was some fusion between the carpals and tarsals and the phalanges. The horns were short, straight, and sharply pointed.

Despite its common name, it is unlikely that *Myotragus* spent much time in caves. Most of its remains have been found in caves, but probably because the animals accidentally fell to their death therein. Over 500 individuals were recovered from one cave on Mallorca. The unusual features of the limbs probably were adaptations that enabled *Myotragus* to jump from crag to crag and to climb on the steep slopes of the Balearics. The huge incisor teeth resemble those of a beaver, and it is possible that *Myotragus* fed on the same kinds of woody plant tissue favored by *Castor*. It may have used its incisors to strip the bark off of trees. The high-crowned cheek teeth seem to have been an adaptation for dealing with coarse, grit-covered vegetation.

When neolithic people arrived on the Balearic Islands about 7,000 years ago, they encountered large populations of *Myotragus*. Remains of the latter have been found in association with pottery and the bones of humans and domestic

goats. *Myotragus* was hunted by people as a source of food, and it also may have been adversely affected through habitat modifications wrought by domestic animals. Nonetheless, *Myotragus* did not become extinct as rapidly as might be expected of a fairly large animal on a small island confronted with a human invasion. Some of its remains date from only about 3,800 years ago. It could have survived for a while by avoiding human hunters in remote, rugged habitat. There also have been suggestions, however, that it was domesticated for a certain period before being abandoned in favor of sheep and goats.

ARTIODACTYLA; BOVIDAE; Genus BUDORCAS
Hodgson, 1850

Takin

The single species, *B. taxicolor*, is found in eastern Tibet, Sikkim, Bhutan, northern Assam, northern Burma, and the central and southern Chinese provinces of Gansu, Shaanxi, Sichuan, and Yunnan; in early historic time it may also have occurred in northeastern China and Mongolia (Neas and Hoffmann 1987).

Head and body length is about 1,000–2,370 mm, tail length is 70–120 mm, height at the shoulder is 686–1,400 mm, and weight is 150–400 kg (Neas and Hoffmann 1987; Wu 1985). Coloration of the shaggy fur varies from yellowish white through straw brown to blackish brown. There is a dark stripe along the back. The build is heavy and oxlike, the front limbs are stout, the lateral hooves (dewclaws) are large, the profile is convex, and the muzzle is hairy. The horns, carried by both sexes, are fairly massive, transversely ribbed at the bases, and up to 635 mm long. They arise near the midline of the head, abruptly turn outward, and then sweep backward and upward.

The takin prefers dense thickets near the upper limit of tree growth at an elevation of around 1,000–4,250 meters. It makes narrow paths through this thick growth, which it uses regularly in passing to and from grazing areas and salt licks. It is slow and deliberate but can leap nimbly from rock to rock on rough slopes. There apparently are regular seasonal migrations between alpine areas in summer and forested valleys in winter (Neas and Hoffmann 1987). *Budorcas* sometimes spends much of the day in dense vegetation and emerges in the late afternoon and early morning to feed. On cloudy and foggy days, it may remain active throughout the day. Schaller et al. (1986) reported that the takin is a generalist herbivore, primarily a browser. In summer, it eats mostly forbs and deciduous leaves from shrubs and trees. The winter diet consists principally of twigs and evergreen leaves from woody species. The takin may push over or break saplings 80–100 mm in diameter to reach browse. It apparently requires a substantial amount of minerals and has been known to travel a great distance to reach salt licks, where individuals may congregate in large numbers and remain for several days (Neas and Hoffmann 1987).

Budorcas gathers in herds of considerable size near and above the tree line during the summer, but in winter there is dispersion into smaller bands for the movement to lower elevations. Neas and Hoffmann (1987) referred to records of summer herds of up to 300 individuals and of winter groups with 3–20 animals. Schaller et al. (1986) reported most herds to contain 10–35 takin and to include females, subadults, and some males. The old bulls are usually solitary, but during August and September the sexes are often seen together. When alarmed, an individual gives a warning cough to alert the others of its herd, and the animals then quickly dash for

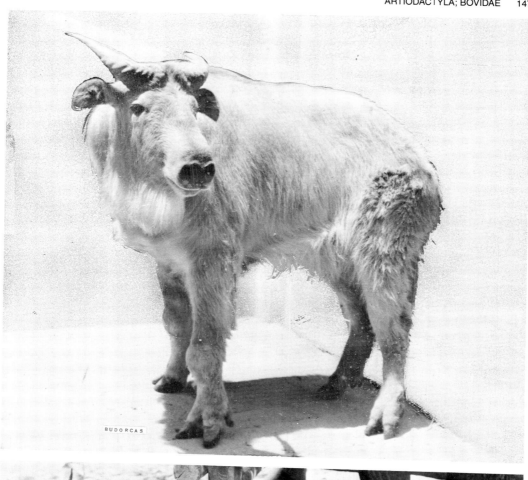

Takins *(Budorcas taxicolor):* Top, photo from New York Zoological Society; Bottom, photo by André Verstraete through Lothar
Schlawe.

the safety of dense underbrush. There also is a low bellow during the rut (Neas and Hoffmann 1987). In Sichuan mating occurs in July and August and births take place in March or April, but in Shaanxi the respective periods are said to be April and February–March (Neas and Hoffmann 1987). A gestation period of 200–220 days has been recorded in captivity (Aung 1968). Birth weights of about 5 kg and 7 kg have been reported (Penny 1989). There normally is a single kid, and within 3 days it is able to accompany its mother almost anywhere. One captive takin lived for 15 years and 10 months (Jones 1982).

Native peoples highly esteem the flesh of *Budorcas* and capture it by means of deadfalls, spear traps, and snares. Neas and Hoffmann (1987) considered the takin to be an endangered species because of overhunting and habitat destruction. The subspecies of central China, *B. t. tibetana* and *B. t. bedfordi*, are classified by the IUCN, respectively, as indeterminate and rare.

ARTIODACTYLA; BOVIDAE; Genus OVIBOS
De Blainville, 1816

Muskox

The single living species, *O. moschatus*, occurred in historical time from northern Alaska to Hudson Bay, on the northern and western islands of the Canadian Arctic, and in Greenland (Hall 1981). The same species, or a close relative, was present in northern Eurasia during the late Pleistocene and may have survived in Siberia until about 2,000 years ago (Corbet 1978; Tener 1965). Remains aged at several thousand years are commonly found on the Taimyr Peninsula, in northwestern Sibe-

ria, and it is possible that *Ovibos* existed there until only 100–200 years ago (Uspenski 1984).

Head and body length is 1,900–2,300 mm, tail length is 90–100 mm, shoulder height is 1,200–1,510 mm, and weight is generally 200–410 kg. Captive males have attained weights of up to 650 kg (Lent 1978). Males average larger than females. The dark brown, coarse guard hairs reach nearly to the ground, shed rain and snow, and take the wear. An inner coat of fine, soft, light brown hair is so dense that neither cold nor frost can penetrate it. The legs and middle of the back are pale. Gunn (1985) reported observations of wild cream-colored muskoxen. The general form of *Ovibos* is massive, there is a slight hump at the shoulders, and the neck, legs, and tail are short. The horns, borne by both sexes, are broad, curve down and outward, and nearly meet in the midline of the skull to form a large boss. The horns of the males are much more massive at the base than are those of the females. The common name refers to a characteristic odor that emanates from males during the rut. Females have four mammae.

The information for the remainder of this account was taken largely from Banfield (1974), Lent (1978, 1988), and Tener (1965); additional sources are noted. The muskox dwells exclusively on the arctic tundra. In the summer it prefers moist habitats, such as river valleys, lake shores, and seepage meadows. In the winter it may shift to hilltops, slopes, and plateaus, where prevailing winds keep snow levels to a minimum. Although the muskox appears to be clumsy and is usually slow and stolid, it moves with surprising agility and can run swiftly if necessary. It seems to have acute vision and hearing. A herd usually does not remain in one place for more than a day during the summer but may feed in a favorable spot for several days during the winter. Movements between winter and summer ranges do not cover more than

Four-year-old muskox bull *(Ovibos moschatus)*, photo from U.S. Fish and Wildlife Service.

Muskox *(Ovibos moschatus)*, photo by Lothar Schlawe.

80 km. The summer food includes mainly grasses and sedges, and the winter diet consists primarily of browse, such as crowberry, cowberry, and willow. Lassen (1984) reported population densities of 0.30/sq km, 0.37/sq km, and 0.44/sq km in northeastern Greenland.

Ovibos is gregarious, occurring in herds with up to 100 individuals. The usual number of animals per group is 15–20 in winter and about 10 in summer. When approached by an enemy, the members of a herd bunch together, often in the form of a circle or semicircle with the calves inside. Such a formation provides effective defense against wolves but has allowed entire herds to be easily killed by human hunters with modern firearms.

Some questions remain regarding social structure, but main herds generally seem to contain adults and young of both sexes. During the summer rut, however, a dominant bull, commonly 6–10 years of age, may drive the other males away and takes charge of the herd. Dominance is achieved by displays, threats, and serious fights. The latter involve repeated head-on charges at 40 km/hr, with the horns clashing together. The males that are forced out, mainly young animals and some that are very old, either become solitary or form small groups of their own. They usually rejoin the main herd in the autumn, but some may wander widely and attempt to acquire a herd of their own. Gray (1984) reported

that a dominance struggle involves visual displays and roars resembling those of an African lion, that fighting lasts up to 50 minutes, and that the loser seldom leaves the herd if he was already a member.

Mating takes place from July to September, and the young are born from mid-April to mid-June. A female gives birth every 1 or 2 years. The gestation period is about 8–9 months. A single offspring is normal, but twins have been reported on very rare occasion. The calf weighs 10–14 kg at birth and is highly precocious. It begins to nibble some vegetation within a week but may not be completely weaned for over a year. Mating under fully natural conditions does not usually occur until females are over 2 years and males are over 5 years, but captive animals have bred as yearlings. One free-ranging female produced 10 or 11 calves between 1968 and 1980. Wild individuals have been known to live up to 24 years (Alendal 1984).

The muskox has been exploited by people for its meat and hide, and the Eskimos have used the horns to fashion bows. As human activity increased in the Arctic during the nineteenth and early twentieth centuries, the muskox was exterminated in Alaska and nearly wiped out on the mainland of Canada. Subsequent conservation efforts allowed a limited recovery in some areas, and there are now an estimated 45,000 individuals in Canada and 21,000 in Greenland. There

Muskoxen *(Ovibos moschatus)* in defensive position, photo from U.S. Fish and Wildlife Service.

are small introduced populations in various parts of Alaska, the Soviet Union, Svalbard, Norway, and northern Quebec.

ARTIODACTYLA; BOVIDAE; Genus HEMITRAGUS
Hodgson, 1841

Tahrs

There are three species (Ellerman and Morrison-Scott 1966):

H. jemlahicus, the Himalayan region from Kashmir to Sikkim;
H. jayakari, Oman;
H. hylocrius, Nilgiri Hills and adjacent ranges in southern India.

These taxa sometimes have been considered to be only sub-specifically distinct, but Benirschke and Kumamoto (1980)

found a great difference in the chromosome number of *H. jemlahicus* and *H. hylocrius*, and most recent authorities have regarded all three taxa as full species.

Head and body length is about 900–1,400 mm, tail length is about 90–120 mm, shoulder height is 610–1,060 mm, and weight is about 50–100 kg. The pelage varies between species. *H. jemlahicus* is reddish to dark brown in color and has a shaggy mane around the neck and shoulders and extending to the knees. *H. jayakari* is covered with shaggy, brittle, short hairs that are grayish to tawny brown. *H. hylocrius* is dark yellowish to brownish and has a grizzled area across the back. Although *Hemitragus* bears a general resemblance to *Capra*, it differs in that the males lack the beard, the muzzle is naked, the feet have glands, and the horns are not twisted but are somewhat laterally flattened. Female *H. hylocrius* have two mammae, whereas female *H. jemlahicus* and *H. jayakari* have four.

All species occur on rugged hills and mountain slopes. *H. jayakari* occupies areas of sparse scrub vegetation, *H. jemlahicus* occurs mainly on wooded slopes, and *H. hylocrius*

Tahrs *(Hemitragus jemlahicus)*, photo from New York Zoological Society.

prefers grass-covered hills. These animals are wary and difficult to approach, especially from below. It is surprising with what ease and confidence they scamper about their uneven, rocky haunts. *H. hylocrius* is active intermittently from dawn to late evening and is mainly a grazer (Davidar 1978). *H. jayakari* is evidently crepuscular and nocturnal in the warm summer months and eats a variety of vegetation but seems to prefer such parts as seeds, fruits, and new growth (IUCN 1978).

In New Zealand, where *H. jemlahicus* has been introduced, population density was found to be 4.5/sq km to 6.8/sq km. In its natural habitat this species was observed to travel in herds of 2–23 animals. The average membership of all-female groups was 4.0, while that of herds with adults of both sexes was 10.2. When the sexes are together during the mating season the males compete for the females, sometimes locking horns and attempting to throw one another off balance, but the intensity of such struggles seems to be less than that observed in many other species of ungulates (Schaller 1977). In *H. hylocrius* there are herds consisting of 20–50 and occasionally over 100 females and young. Adult males associate in bachelor groups of 2–10 for much of the year, but during the rut they join the female herds. They compete for individual estrous females, not for the whole herd, and losers are not excluded from the group (Rice 1985). *H. jayakari* has been seen alone, in pairs, and in trios of a male, a female, and a kid. Males or pairs of this species are thought to be territorial (IUCN 1978).

Rice (1988*b*) reported that female *H. hylocrius* are polyestrous and capable of breeding throughout the year, though mating in the wild evidently is stimulated by the onset of the monsoon (July–August) and the main birth season lasts from January to mid-February. In the Himalayas, *H. jemlahicus* mates from mid-October to mid-January (Schaller 1977). A newborn *H. jayakari* was found in April (Harrison 1968). Gestation periods of 177 to 242 days have been reported. There is usually a single offspring, occasionally two. Female *H. hylocrius* attain sexual maturity at an average age of 19 months (Rice 1988*b*). A captive *H. jemlahicus* lived for 21 years and 9 months (Jones 1982).

The Arabian tahr (*H. jayakari*) is classified as endangered by the IUCN (1978) and the USDI. Although protected by law, it has declined through competition with domestic goats and illegal hunting. Fewer than 2,000 individuals are thought to survive in the limited available habitat. The Nilgiri tahr (*H. hylocrius*) is classified as vulnerable by the IUCN. Its range has declined drastically through poaching and habitat destruction, but under protection its total numbers appear to have stabilized at around 2,200 (Davidar 1978; Rice 1988*a*). *H. jemlahicus* has been introduced to New Zealand, where a population of 20,000–30,000 individuals is now established and is subject to sport and commercial hunting (Schaller 1977).

ARTIODACTYLA; BOVIDAE; Genus CAPRA
Linnaeus, 1758

Goats

There are eight species (Alados 1985*b*; Ansell, *in* Meester and Setzer 1977; Corbet 1978):

C. aegagrus (wild goat), mountains from Asia Minor to Afghanistan and Pakistan, Oman, Crete, Aegean Islands;

C. hircus (domestic goat), worldwide in association with people, feral in many areas;

C. ibex (ibex), European Alps, Palestine, Sinai, Arabian Peninsula, mountainous region from Afghanistan and northern India to Lake Baikal, Egypt and Sudan east of the Nile, northern Ethiopia;

C. walie (Walia ibex), Simien Mountains of north-central Ethiopia;

C. caucasica (west Caucasian tur), western Caucasus;

C. cylindricornis (east Caucasian tur), eastern Caucasus;

C. pyrenaica (Spanish ibex), Spain and formerly southern France and Portugal;

C. falconeri (markhor), mountains of southern Uzbek S.S.R., Tadzhik S.S.R., Afghanistan, northern and central Pakistan, and Kashmir.

There is much controversy regarding the systematics of this genus. Both Ansell (*in* Meester and Setzer 1977) and Corbet (1978) included *Ammotragus* in *Capra*. Van Gelder (1977*b*) considered *Ammotragus* and *Ovis* to be synonyms of *Capra*. Neither view has been widely accepted. Ellerman and Morrison-Scott (1966) placed *C. falconeri* in a separate subgenus, *Orthaegoceros* Trouessart, 1905, but Schaller (1977) did not think this designation warranted. Most recent authorities (Corbet 1978; Corbet and Hill 1986; Grzimek 1975; Honacki, Kinman, and Koeppl 1982; Nievergelt 1981; Schaller 1977) have treated *C. walie* as a subspecies of *C. ibex*, but Yalden, Largen, and Kock (1984) were convinced that the two are specifically distinct. Grzimek (1975) and Schaller (1977) included *C. caucasica* in *C. ibex*, and Grzimek (1975) also treated *C. cylindricornis* as a subspecies of *C. ibex*. Some authorities, such as Corbet (1978), prefer not to use the name *C. hircus*; others, such as Ellerman and Morrison-Scott (1966), include both *C. aegagrus* and the domestic goat in *C. hircus*; and others, such as Grzimek (1975), treat *C. hircus* as a subspecies of *C. aegagrus*. American authorities, including Schaller (1977) and Lowery (1974), have generally recognized *C. hircus* as a species distinct from *C. aegagrus*. Schaller (1977) indicated that *C. aegagrus* may have lived in the Balkan Peninsula until 1891.

Capra bears some resemblance to *Ovis* but differs in that males are odorous, there is a beard, the feet lack scent glands, and the forehead is convex, not concave. Females have two mammae. In wild species the horns of males are about 500–1,650 mm long, and those of females are 150–380 mm long; there is much structural variation between, and to some extent within, species (Grzimek 1975; Schaller 1977). In any given species or population the average size of males is substantially greater than that of females. Information on each species is given separately below.

Capra aegagrus (wild goat).

The information for the account of this species was compiled mainly from Grzimek (1975), Roberts (1977), and Schaller (1977); major exceptions are noted. Head and body length is 1,200–1,600 mm, tail length is 150–200 mm, height at the shoulder is 700–1,000 mm, and weight is 25–95 kg. In adult males the general coloration of the winter coat is almost silver-white; the chest, throat, and face are sooty gray; the belly, outside of the lower limbs, beard, and forepart of the face are black to deep chestnut-brown; there is a dark dorsal crest of longer hairs extending down the middle of the back; and there is a black stripe running from the withers down the front of the shoulders and merging with the dark chest. In females the general color is yellowish brown to reddish gray, there is a dark brown middorsal line, there may be dark brown markings on the face, and there is no beard. The summer coat is short and coarse, becoming more reddish buff in males.

The body is rather stocky, the limbs are strong, and the hooves are broad. The soles of the hooves are of a rubbery consistency, and the second and fifth digits consist only of

Cretian wild goat *(Capra aegagrus)*, photo by Ernest P. Walker.

rough, rounded horny knobs that are completely detached from the cannon bone. The horns of males are scimitar-shaped, 800–1,300 mm long, and have a generally sharp anterior keel, often with irregular knots or protuberances. The horns of females are slightly curved, 200–300 mm long, and relatively thin.

The wild goat occupies a variety of habitats on either gentle or very steep slopes and at elevations of up to 4,200 meters. It may be found in arid country with sparse vegetation, alpine meadows, or forests. Activity is mainly in the early morning and late afternoon, but in the hot season individuals sometimes lie in the shade for most of the day and forage at night. Food is taken by both grazing and browsing, depending on season and availability. Reported population densities are about 1/sq km to 4/sq km in Pakistan and 143/sq km on Theodoru Island in the Aegean.

Groups of over 100 individuals sometimes form, but average herd size commonly varies from about 5 to 25 depending on season and locality. Females spend most of the year in a group, separating only to give birth and then rejoining when their young are mobile. Males usually are found apart from the females in small bachelor groups of about 4–5. They form dominance hierarchies, and during the rut they associate with the female herds and compete with one another for mating privileges. Conflicts involve threats, visual displays with the horns, and then sometimes rearing up on the hind legs and coming down, the horns of the opponents clashing together.

Mating takes place from August to December in different areas, and births from January to May. The gestation period is 150–70 days. The kids number 1 or 2, occasionally 3, and are highly precocious. They can run within 24 hours of birth, suckle for 4–5 months, and remain with their mother until the following mating or birth season. The mother protects her kids by fighting with the horns or by decoying the intruder. Females initially give birth at about 3 years.

C. aegagrus has declined in many areas, largely through competition for forage with flocks of its domestic descendant, *C. hircus,* and excessive hunting by people for trophies and meat. It is classified as vulnerable by the Soviet Union. It survives in small numbers on Crete and has been introduced on Theodoru Island just to the north, but populations on other Aegean islands have hybridized with *C. hircus.* There are about 300 individuals on Crete and 100 on Theodoru (Husband, Davis, and Brown 1986; Smit and Van Wijngaarden 1981).

Capra hircus (domestic goat).

Present evidence indicates that the earliest domestication of *Capra* took place in southwestern Asia 8,000–9,000 years ago. *C. aegagrus* is generally considered to be directly ancestral to, if not conspecific with, *C. hircus.* There have been suggestions that *C. falconeri* also contributed to the ancestry of the domestic goat, but Schaller (1977) considered such a view unwarranted. There are numerous breeds of *C. hircus,*

Domestic goats *(Capra hircus),* photo by Lothar Schlawe.

with a multitude of forms and colors, some being completely black or white and some being hornless. When present, the horns may be either scimitar-shaped or spiraling, the former apparently being the original condition.

Domestic goats are used for the production of milk, meat, and wool (Grzimek 1975). Goats are very nimble and aggressive and can live and obtain food in places not generally accessible to other domestic mammals. Herds of goats have been highly destructive to natural vegetation, especially in the Mediterranean region and Middle East, thereby contributing to erosion, the spread of deserts, and the disappearance of native wildlife. Indeed, the domestic goat has been a major factor in the decline of its wild relatives by competing with them for available food. *C. hircus* has become feral on Hawaii, New Zealand, the Galapagos, and many other islands (Rudge 1984). Population density has reached 1,000/sq km on Macauley Island in New Zealand (Schaller 1977). Such island populations have been partly responsible for the endangerment and extinction of numerous species of forest birds. Feral goats also are established in many mainland areas and indeed are the most widespread feral ungulates in the United States (Lever 1985).

An assessment of reported information on feral goat populations (O'Brien 1988) indicates that most herds are relatively sedentary, occupying home ranges of about 1–5 sq km. Usual density is 33/sq km to 68/sq km. Modal group size is 2–4 individuals, and they tend to use a permanent night camp within the range, perhaps a holdover from domestic life. As in wild species of *Capra,* there usually is a separation of adult males from the females for most of the year. Female groups are relatively stable, though in a few cases there is dispersal during parturition. Reproduction is seasonal in some areas but not in others. In a study of an old feral population on Rhum Island, off northern Scotland, Dunbar (1986) found a

highly seasonal reproductive pattern. Permanent female matriarchal groups were joined by the adult males during the August rut. The latter fought one another by clashing horns, the largest individuals securing the most mates. Resulting births occurred in February, a time when the weather was severe but a time allowing maximum exploitation of the following spring growing season. Females probably produced a single kid annually over a reproductive life of 8 years.

Capra ibex (ibex).

The information for the account of this species was compiled mainly from Grzimek (1975), Roberts (1977), and Schaller (1977); major exceptions are noted. Head and body length is 1,150–1,700 mm, tail length is 100–200 mm, height at the shoulder is 650–1,050 mm, and weight is 35–150 kg. Measurements for the subspecies *C. ibex nubiana,* of the Arabian region and northeastern Africa, fall at the lower end of these ranges. Females average about a third smaller than males. Both sexes have a thick, woolly beard. In adult males the summer coat is rich chocolate brown with circular patches of yellowish white hair on the middle of the back and rump. The winter coat is thicker, with more white hairs and much variation in color. Females are generally reddish tan or almost golden in summer and grayer brown in winter. The body is relatively heavy, and the legs are short and sturdy. The horns of males are 700–1,400 mm long and are scimitar-shaped like those of *C. aegagrus,* but the anterior surface is relatively flat and is broken by prominent transverse ridges. The horns of females are 150–380 mm long and are relatively slender.

The ibex lives at elevations of up to 6,700 meters, generally at or above the tree line. It almost never enters dense forests. It tends to move upward in the summer and downward in the winter but descends the lowest in the spring to reach new grass. There also may be a daily descent to feed and a return

Ibex *(Capra ibex)*, photo by Lothar Schlawe.

to the highest, most precipitous crags in the evening. Most activity is in the early morning and late afternoon, the animals lying in the shade of rocks to avoid the midday heat. The ibex is predominantly a grazer in the spring and summer and may browse above the snow during winter. Population density varies widely, being around 1/sq km in Pakistan and 9/sq km in the Alps. Dzieciolowski et al. (1980) reported a general population density of 1.0/sq km in Mongolia, with maximum densities in favorable habitat of 3.3/sq km in winter and 4.8/sq km in summer.

Females and young live in fairly stable herds of about 10–20 individuals. Adult males spend most of the year in bachelor groups, where they establish a dominance hierarchy through displays and ritualized combat. Solitary males also occur. During the mating season the males join the herds, frequently fight among themselves, and seek to mate with estrous females. Combat is much the same as described above for *C. aegagrus*. The males may remain with the females through the winter and early spring but then depart. In Europe and Central Asia mating takes place in December and January, and births in May and June. The gestation period is 150–80 days. There may be 1, 2, or 3 young. They can jump on their first day of life and join in kid groups by their fourth week. Females may reach sexual maturity as early as their second year, but wild individuals usually do not give birth

until they are 3–6 years old. A captive ibex lived for 22 years and 3 months (Jones 1982).

Because of sport hunting, the European population of *C. ibex* survived only in the Gran Paradiso of northwestern Italy by the early nineteenth century. About 60 individuals were present there at that time. Those animals were carefully protected, and some of them were used later to reintroduce the species to other parts of the Alps. Smit and Van Wijngaarden (1981) reported that there were 3,000 in Gran Paradiso, now a national park, as well as 6,000 in Switzerland and some hundreds elsewhere. *C. ibex* also has been decimated by hunters in the Asian and African parts of its range and is continuing to decline through poaching, agricultural encroachment on its habitat, and competition with domestic goats (Grzimek 1975; Habibi 1986; Krausman and Shaw 1986; Schaller 1977).

Capra walie (Walia ibex).

Schaller (1977) listed height at the shoulder as 970 mm and average weight as 120 kg for males. According to Nievergelt (1981), *C. walie* resembles *C. ibex* but is less massive and more sleek than the subspecies *C. i. ibex*, of Europe, and stouter and heavier than *C. i. nubiana*, of Arabia and Africa. Males are larger than females. The upper parts are chestnut-brown, the underparts are whitish, there is a black stripe down the front of each limb, and old males have a black streak on the back. Males 4–7 years old have a small black beard; males over 7 years have a longer beard and a darker chest; and females have no beard. The horns of males have an average length of 1,160 mm and resemble those of *C. ibex* (Schaller 1977); however, *C. walie* is distinguished by a bony boss on the forehead (Ansell, *in* Meester and Setzer 1977).

The main habitat of the Walia ibex is steep, rocky slopes at elevations of 2,800–3,400 meters; 7 animals observed for 2–12 days used ranges of 0.5–3.0 sq km (Nievergelt 1981). Individuals seek stands of giant heath for browse and shelter; as many as 35 have been seen in one herd (Yalden, Largen, and Kock 1984). Rutting behavior occurs throughout the year, with a peak from March to May (Nievergelt 1974). As the rut approaches the previously separate adult male groups join the groups of females and young so that overall herd size increases (Dunbar and Dunbar 1981). However, the males then compete with one another, fighting is frequent, and the younger males will be forced away. There usually is a single young.

The Walia ibex is classified as endangered by the IUCN and the USDI. It remained relatively common until about 1930 but subsequently has been severely depleted by indiscriminate hunting and habitat usurpation as agriculture advanced up the mountain slopes to elevations of 3,000 meters. In 1963 an estimated 150–200 individuals survived, though there may since have been a modest increase in response to protection (Yalden, Largen, and Kock 1984).

Capra caucasica (west Caucasian tur).

According to Heptner, Nasimovich, and Bannikov (1988), males have a head and body length of 1,500–1,650 mm, a shoulder height of 950–1,090 mm, and a weight of 65–100 kg; respective figures for females are 1,200–1,400 mm, 780–900 mm, and 50–60 kg. The tail is very short and the beard of males is usually short and broad but sometimes long. In summer the upper parts are rusty gray to rusty chestnut, becoming lighter on the flanks, and the underparts are yel-

lowish gray or dirty white. In winter the upper parts are mainly gray brown. *C. caucasica* appears heavier and less slender than *C. ibex*. It has a very stout and long trunk, a thick and massive neck, and relatively short but strong legs. The horns of males are scimitar-shaped and resemble those of *C. ibex*, but are much more massive and thicker in relation to length. In cross section they are rounded or roughly in the form of a triangle with rounded angles. The horns of old males have an average length of about 740 mm; those of females are much smaller and thinner.

C. caucasica is confined to a narrow strip of montane habitat in the western Caucasus. Its total natural range, about 4,500 sq km, may be the smallest of any ungulate. Even this area has shrunk somewhat since the nineteenth century, because of human activity, and there has been some hybridization with the neighboring and closely related *C. cylindricornis*. Heptner, Nasimovich, and Bannikov (1988) noted that the biological features of *C. caucasica* are essentially identical to those of *C. cylindricornis* and thus discussed them together in the account of the latter species; the same is done here (see below).

Capra cylindricornis (east Caucasian tur).

The information for the account of this species was taken entirely from Heptner, Nasimovich, and Bannikov (1988). Males have a head and body length of 1,300–1,500 mm, a shoulder height of 790–980 mm, and a weight of 55–100 kg. Females are considerably smaller, being about 650–700 mm high at the shoulder and weighing 45–55 kg. The tail is very short and the beard of males usually is much shorter than that of *C. caucasica*. The winter coat is mostly dark brown-cinnamon or chestnut brown, with slightly paler underparts, and the much lighter summer coat is rusty brown. The body is massive and similar in structure to that of *C. caucasica*. However, the horns of male *C. cylindricornis* are larger and heavier and have a gentle spiral shape somewhat like that of *C. pyrenaica*. They curve up and out, then back and slightly

West Caucasian tur *(Capra caucasica)*, photo by Lothar Schlawe.

West Caucasian tur *(Capra caucasica)*, photo by K. Kutunidisz.

down and in, and finally in and up at the tips. In cross section they are circular or have the shape of a rounded triangle. The horns of adult males have a length of 800–1,030 mm; those of females reach only up to 300 mm.

The remainder of this account applies to both *C. cylindricornis* and *C. caucasica*. Tur live along steep rocky slopes at elevations of 800–4,200 meters. They may be found in alpine meadows, barren areas, or the forest zone. There are seasonal migrations covering a vertical distance of 1,500–2,000 meters, upward in May and downward in October–November. Adult males generally remain higher than do females and young. Steep slopes are ascended and descended with ease. In winter the animals may remain in open pastures throughout the day, alternately feeding and resting. In summer they commonly spend the warmest hours resting in sheltered places and then feed at intervals during the late afternoon, night, and morning. Daily movements may be as great as 15–20 km if the resting and feeding sites are well separated. The diet comprises a wide variety of plants, with grasses predominating in the summer and the leaves of trees and shrubs becoming more important in winter.

Recorded average population densities in protected areas are about 50–160 tur per 1,000 ha. Herds of up to 500 animals have been seen, though permanent social units probably contain only up to a few dozen individuals. Adult males usually live apart from the females and young, but mixed herds form during the breeding season. At that time the males fight vigorously for the estrous females. Mating takes place from late November to early January and births from mid-May to late June. The gestation period is 150–160 days. There usually is a single young, rarely two. Kids weigh 3.5–4.2 kg at birth, are difficult for a human to catch the next day, and start eating grass after a month, but continue to suckle until the end of summer. Females may attain sexual maturity in their second year and are capable of giving birth annually. Males usually are not able to participate in mating until their fourth or fifth year. Most wild individuals die before they are 10 years old but maximum longevity may be 22 years.

Both species of tur declined in the nineteenth and early twentieth centuries because of uncontrolled hunting and displacement by domestic livestock. Subsequent establishment of preserves and regulated hunting has allowed populations to increase in some areas. Attempts at introduction outside of the natural range have not been successful.

Capra pyrenaica (Spanish ibex).

Head and body length is 1,000–1,400 mm, tail length is 100–150 mm, height at the shoulder is 650–750 mm, and weight is 35–80 kg. The summer coat is light to red-brown (Grzimek 1975). The horns of the male, averaging about 750 mm in length, curve out and up, then back, inward, and up again, and they have a sharp posterior keel (Schaller 1977).

The Spanish ibex lives in rocky or mountainous areas and feeds on grass, herbs, and lichens (Grzimek 1975). There is seasonal variation in social structure (Alados 1985a). From November to January, the mating season, most of the animals live together in mixed herds of around 10 individuals. The adult males leave and form their own groups by February. In April there is a further segregation as the juveniles of both sexes form groups. The adult females are then left alone to give birth and rear their kids. The mixed groups form again in the fall.

East Caucasian tur *(Capra cylindricornis),* photo by Lothar Schlawe.

This species formerly was abundant in the Iberian Peninsula and adjacent areas but was drastically reduced by excessive hunting in modern times (Alados 1985b; Smit and Van Wijngaarden 1981). Of the four subspecies, *C. p. lusitanica,* of Portugal, became extinct by about 1890, and *C. p. victoriae,* of west-central Spain, declined to only a dozen animals in 1905. Conservation efforts subsequently allowed partial recovery of the latter subspecies, as well as of *C. p. hispanica,* of southern and eastern Spain, with the total ibex population of the country now estimated at about 28,000. However, the subspecies *C. p. pyrenaica,* of the Pyrenees, disappeared on the French side of the border in the mid-nineteenth century and today numbers only 25–30 in northern Spain. It is classified as endangered by the IUCN.

Capra falconeri (markhor).

The information for the account of this species was compiled mainly from Grzimek (1975), Roberts (1977), and Schaller (1977); major exceptions are noted. Head and body length is 1,400–1,800 mm, tail length is 80–140 mm, height at the shoulder is 650–1,040 mm, and weight is 80–110 kg in males and 32–40 kg in females. The coat is short and smooth in summer, longer in winter. Both sexes are generally reddish gray in color, there being more yellowish buff tones in summer and more gray in winter. There also is a dark brown stripe extending from the shoulders down the back to the base of the tail. Adult males have much white and gray in their winter pelage, a very extensive black beard, a shaggy mane of long dark hairs extending from the neck down the chest, and tufts of hair on the legs. Females sometimes have a thin beard. The limbs are relatively short and thick, and the

hooves are broad. The horns are sharp-keeled and begin close together but then spread apart, each twisting into a tight or open spiral. Those of males are up to 1,600 mm long, and those of females up to 250 mm.

The markhor lives at medium to high elevations around and above the tree line but generally below the habitat of the ibex. It sometimes is found in steep gorges, rocky areas, arid country, scrub forests, or grassy meadows. Seasonal altitudinal shifts may cover many hundreds of meters. Like other goats, it is active mainly in the early morning and late afternoon and is predominantly a grazer in the spring and summer and a browser in the winter.

Population densities in Pakistan vary from about 1/sq km to 9/sq km. Females and young live in herds with an average membership of about 9 individuals, though there sometimes are aggregations of 30–100. Adult males reportedly live alone for most of the year, joining the herds only during the rut, but in the population in northern Pakistan some males remain with the females throughout the year. Groups establish a dominance hierarchy through threats and fighting, and males become especially aggressive toward each other during the rut. Fights usually involve lunging and locking of the horns, the combatants pushing and twisting in an effort to throw each other off balance.

Mating takes place in winter, gestation lasts about 155 days, and births occur from late April to early June. There are one or two young. They remain with the mother until the following breeding season and reach sexual maturity at about 30 months. Probably few animals live for more than 11 or 12 years.

C. falconeri is generally classified as vulnerable by the

Spanish ibex *(Capra pyrenaica)*, photo by Lothar Schlawe.

IUCN (1976) and as endangered by the Soviet Union and is on appendix 2 of the CITES. The species has been sought as a trophy by sportsmen, killed for its meat and hide, and reduced through loss of habitat by agricultural expansion and competition with domestic livestock. Much of its limited range is in politically unstable areas and has been the scene of recent military activity. The subspecies *C. f. megaceros*, of central Pakistan and adjacent Afghanistan, perhaps numbering fewer than 2,000 individuals, is classified as endangered by the IUCN. The subspecies *C. falconeri jerdonensis* and *C. f. chialtanensis*, of Pakistan and Afghanistan, are officially listed as endangered by the USDI and are on appendix 1 of the CITES, but the former is probably a synonym of *C. f. megaceros* and the latter may be a form of *C. aegagrus* (Schaller 1977).

ARTIODACTYLA; BOVIDAE; Genus PSEUDOIS
Hodgson, 1846

Bharals, or Blue Sheep

There are two species (Corbet 1978; Groves 1978*b*; Wang and Hoffmann 1987):

P. nayaur, alpine zone from Himalayan region to Inner Mongolia;

P. schaeferi, upper Yangtze River Valley of central Sichuan.

Head and body length is 1,150–1,650 mm, tail length is 100–200 mm, shoulder height is 750–910 mm, and weight is 25–80 kg (Grzimek 1975; Schaller 1977). In *P. nayaur* the head and upper parts are brownish gray with a tinge of slaty blue, and the underparts and insides of the legs are white. This coloration blends well with the blue shale, rocks, and brown grasses of the open hillsides. *P. schaeferi* is generally smaller than *P. nayaur* and has a drabber coloration with a silvery sheen (Groves 1978*b*).

In structure and habits, *Pseudois* is intermediate to *Capra* and *Ovis* but is apparently more closely related to *Capra*. Schaller (1977) described bharals as "aberrant goats with sheep-like affinities." He noted that *Pseudois* resembles *Capra* in having a broad, flat tail with a bare central surface, large dewclaws, no inguinal glands, no preorbital glands, and usually no pedal glands. The rounded, smooth horns of *Pseudois* curve backward over the neck. They are borne by both sexes but are much larger in males, reaching a length of 820 mm.

Bharals occur on open slopes and plateaus, with abundant grass, at elevations of 2,500–6,500 meters. Because of their protective coloration, and in the absence of any brush in which to hide, these animals remain motionless when approached. When they discover that they have been seen, however, they take to the precipitous cliffs, ascending to the most difficult-to-reach and most inaccessible places. They feed and rest alternately throughout the day on the grassy slopes of mountains. The diet consists mainly of grasses and herbs during the summer and dried grass, browse, and lichens in the winter (Grzimek 1975; Wang and Hoffmann 1987).

Markhors *(Capra falconeri),* photo from San Diego Zoological Society.

Blue sheep *(Pseudois nayaur):* Left, male; Right, female and young. Photos by Lothar Schlawe.

Population density of *P. nayaur* in Nepal was found to be 0.9–2.7 individuals per sq km, increasing to 8.8–10.0 per sq km during winter concentrations in valleys (Schaller 1977; Wang and Hoffmann 1987). Social structure may vary seasonally and in relation to availability of resources. *P. nayaur* forms herds of up to 400 individuals, but the largest group seen by Schaller (1977) contained 61, and mean size in different study areas ranged from 4.8 to 18.4. Males tend to separate from the females after the mating season and either become solitary or form bachelor herds. A few males, however, associate with the females throughout the year. During the mating season, Schaller observed considerable aggression, usually directed against individuals of the same sex. In most fights, one combatant rears up and then lunges down at its opponent, which catches the blow between its horns. In another study of *P. nayaur*, Wegge (1979) determined average group size to be 11.1 individuals and observed no sexual segregation. *P. schaeferi* reportedly occurs in considerably smaller groups than does *P. nayaur* (Grzimek 1975).

Mating occurs from October to January, and the young are born from May to early July. The gestation period is 160 days, a single offspring is normal but twins reportedly are not unusual, lactation lasts 6 months, and sexual maturity is attained at 18 months (Grzimek 1975; Schaller 1977; Wang and Hoffmann 1987). Wegge (1979) observed that males do not reach trophy size until they are 7 years old, and it is unlikely that they are able to mate under natural conditions prior to that age. Blue sheep do well in captivity, and one specimen lived for 20 years and 3 months (Jones 1982).

ARTIODACTYLA; BOVIDAE; Genus AMMOTRAGUS
Blyth, 1840

Aoudad, or Barbary Sheep

The single species, *A. lervia*, occurs in highlands within the desert and subdesert zones from Morocco and Western Sahara to Egypt and Sudan. Ansell (*in* Meester and Setzer 1977), Corbet (1978), and Van Gelder (1977*b*) all included *Ammotragus* within the genus *Capra*, mainly on the basis of the production of fertile offspring from a cross between an aoudad and a domestic goat. In an assessment of the biochemical characters of *Ammotragus*, however, Manwell and Baker (1977) found 5 characters to be sheeplike, 5 to be goatlike, 7 to be common to both groups, and 10 to be unique. A recent review (Gray and Simpson 1980) maintained *Ammotragus* as a full genus, and that procedure was followed by Corbet and Hill (1986) and most other subsequent authorities.

Head and body length is 1,300–1,650 mm, tail length is 150–250 mm, shoulder height is 750–1,120 mm, and weight is 40–55 kg in females and 100–145 kg in males (Gray and Simpson 1980; Grzimek 1975). The general coloration is rufous tawny. The insides of the ears, the chin, a line on the underparts, and the insides of the legs are whitish. There is no beard, but there is a ventral mane of long, soft hairs on the throat, chest, and upper part of the forelegs. The horns of males sweep outward, backward, and then inward; they are rather heavy and wrinkled and measure up to 840 mm in length. Females also have prominent horns.

The aoudad is found mainly in rough, rocky, arid country. The almost total lack of vegetation tall enough to conceal this animal has seemingly resulted in its developing exceptional ability to hide by remaining motionless whenever danger threatens. Within its range, sources of water are few and far between, but it is usually able to obtain sufficient moisture from green vegetation and the dew that condenses on leaves during the cold desert nights. Nonetheless, populations decrease sharply during periods of drought. In captivity, the aoudad seems to like water and to enjoy taking a bath. The diet consists of grass, herbaceous plants, and stunted bushes.

Reported home range size for individuals introduced in western Texas is 1–5 sq km in winter and 13–31 sq km in summer (Gray and Simpson 1980). *Ammotragus* generally occurs alone or in small groups under natural conditions. A stable linear dominance hierarchy is established in captive herds, with adult males ranking highest but an adult female actually leading group movements (Gray and Simpson 1980). In the wild, adult males may fight savagely for control of a group of females (Grzimek 1975). Two males frequently stand about 10–15 meters apart, then walk rapidly toward each other, gradually gaining speed, and finally break into a run shortly before they collide. Just prior to impact the heads are lowered. A male will not attack, however, if the other is off balance or unprepared.

Schaller (1977) stated that mating takes place mainly in July and August in Niger. According to Gray and Simpson (1980), reproductive activity goes on throughout the year, but mating occurs largely from September to November, and births from March to May. Females are known to have given birth twice in 1 year. The gestation period is 154–61 days. A single offspring is usual, twins are common, and triplets are rare. The young weighs an average of 4.5 kg at birth and is able to negotiate moderately rugged terrain almost immediately. Sexual maturity is reached at around 18 months. A captive aoudad lived for 20 years and 11 months (Jones 1982).

The native peoples of the Sahara kill many *Ammotragus*, as this animal is an important source of meat, hides, hair, and sinews, all of which are valuable in the desert economy. The genus has thus declined alarmingly in much of its natural range and has disappeared entirely in some areas (Ansell, *in* Meester and Setzer 1977; Grzimek 1975). It now is classified as vulnerable by the IUCN. The subspecies *A. l. ornata*, of Egypt, may already be extinct (Gray 1985). The aoudad was introduced in the United States in the early to mid-twentieth century for purposes of sport hunting, however, and populations of some hundreds or thousands of animals now are established in California, New Mexico, and Texas. There is concern that *Ammotragus* will spread into the range of the native desert bighorn sheep *(Ovis canadensis)* and that the two animals will compete for limited resources, to the probable detriment of the bighorn. There also is an introduced population in Spain, but several groups released in Mexico have died out (Gray 1985; Lever 1985; Seegmiller and Simpson 1980).

ARTIODACTYLA; BOVIDAE; Genus OVIS
Linnaeus, 1758

Sheep

There are two subgenera and eight species (Ellerman and Morrison-Scott 1966; Hall 1981; Korobitsyna et al. 1974; Nadler et al. 1973*b*):

subgenus *Ovis* Linnaeus, 1758

O. vignei (urial), southern Soviet Central Asia, northeastern Iran, Afghanistan, Pakistan, Kashmir, northwestern India, Oman;

O. ammon (argali), southern Siberia, eastern Soviet Central Asia, Mongolia, northern China, Sinkiang, Tibet, Himalayan region;

O. orientalis (Asiatic mouflon), Asia Minor, western Iran;

Barbary sheep *(Ammotragus lervia):* Top, photo by Lothar Schlawe; Bottom, photo by Ernest P. Walker.

Argali *(Ovis ammon polii)*, photo by Lothar Schlawe.

O. musimon (mouflon), Corsica, Sardinia, Cyprus;
O. aries (domestic sheep), worldwide in association with
   people;

subgenus *Pachyceros* Gromova, 1936

O. canadensis (bighorn sheep), southwestern Canada,
   western conterminous United States, northern Mexico,
   Baja California;
O. dalli (Dall's sheep), Alaska, northwestern Canada;
O. nivicola (snow sheep), northeastern Siberia.

There is much controversy regarding the systematics of
sheep. Van Gelder (1977b) considered *Ovis* to be a synonym
of *Capra*, on the basis of the production of fertile offspring
through hybridization between *C. hircus* and *O. aries.* Some
authorities, including Corbet (1978) and Grzimek (1975), include all members of the subgenus *Ovis* in the species *O.
ammon,* and all members of *Pachyceros* in *O. canadensis.*
Schaller (1977) did not agree with the specific separation of *O.
vignei* from *O. ammon* or of *O. musimon* from *O. orientalis.*
Honacki, Kinman, and Koeppl (1982), while listing *O. musimon* as a species, noted that it probably represents early
primitive stocks of domestic *O. aries;* the latter species was
considered to include not only the domestic sheep but also
wild *O. orientalis.* Corbet (1984) stated that *O. musimon*
likely originated from the early introduction of partly domesticated *O. orientalis.* This view was followed by Corbet and
Hill (1986). Valdez, Nadler, and Bunch (1978) described a
zone of hybridization between *O. vignei* and *O. orientalis* in
northern Iran.

   Head and body length is 1,200–1,800 mm, tail length is
70–150 mm, shoulder height is 650–1,270 mm, and weight is
20–200 kg. There is much variation in size, the average

weight being about 32 kg for male *O. musimon* on Cyprus,
around 60 kg for male *O. vignei* in southern Central Asia, and
180 kg for male *O. ammon* in the Altai Mountains (Schaller
1977). In any given population, males average larger than
females. In *O. canadensis* of the northern Rocky Mountains,
males weigh 73–143 kg, and females, 53–91 kg. In *O.
canadensis* of the southwestern deserts, males weigh 58–86
kg, and females, 34–52 kg (Wishart 1978). Coloration varies
from creamy white to dark gray and brown. Some males
possess a fringe of long hair down the front of the neck, but
they do not have a beard such as is present in *Capra.* Also
unlike all goats, there is no gland at the base of the tail. *Ovis*
has a narrow nose and pointed ears. Males have massive spiral
horns that have an average length of 1,106 mm in *O. canadensis* of the northern Rockies (Wishart 1978) and a record
length of 1,690 mm in *O. ammon* (Schaller 1977). The horns
of females are much smaller and only slightly curved. Some
domestic breeds are hornless. Females have two mammae.

   Wild sheep are usually found in upland areas. Banfield
(1974) stated that *O. canadensis* inhabits alpine meadows,
grassy mountain slopes, and foothill country in proximity to
rugged, rocky cliffs and bluffs. This species is very alert, has
remarkable eyesight, is an excellent climber, and swims freely. When alarmed, it bounds away over jagged rocks with
surprising speed and agility. In contrast, Schaller (1977) reported that Eurasian wild sheep occur mainly in gentle to
steeply rolling but nonprecipitous terrain at elevations from
sea level to 5,000 meters. Geist (1971) noted that *O. ammon*
actually seems to be a poor jumper and avoids cliffs when
escaping. Sheep may feed intermittently throughout the day
or rest during the hottest hours, and they are sometimes
active at night. Most populations undergo seasonal movements, generally dispersing upward and over a larger area in
the summer and concentrating in sheltered valleys during the

Urial *(Ovis vignei)*, photo from Zoological Society of San Diego.

winter. Shackleton (1985) reported maximum migration distance of *O. canadensis* to approach 48 km. The diet consists largely of grasses, sedges, and forbs and also includes some browse.

Geist (1971) found that in the Canadian Rockies, males use up to 7 separate ranges, and females up to 4, in the course of a year, the different areas being about 1–32 km apart. The midwinter ranges average about 0.6 km across, while those of spring and fall have a mean diameter of 6 km. In a radiotracking study of the desert bighorn in Nevada, Leslie and Douglas (1979) found 10 individual home ranges to average 16.9 sq km. Some reported population densities are about 0.2/sq km for *O. dalli*, 2/sq km for *O. canadensis*, 11/sq km to 13/sq km for *O. vignei*, and 1.0/sq km to 1.2/sq km for *O. ammon* (Geist 1971; Schaller 1977).

Sheep are gregarious, sometimes gathering in herds of over 100 individuals. In some species, mature males stay apart from the females and young for most of the year. The sexes of feral *O. aries* usually remain apart but join during the rut, at which time the highest-ranking males are able to mate with the estrous females. In fully domestic flocks the sexes are usually kept apart, mature individuals being brought together for breeding, but even then there may be dominant rams that do most of the mating (Arnold and Dudzinski 1978). In Nevada, from January to May 1976, Leslie and Douglas (1979) found average group size of the desert bighorn to be 4.0 for

rams and 5.8 for ewes. In Yellowstone National Park, Woolf, O'Shea, and Gilbert (1970) obtained respective figures of 5.7 and 8.0. Observations in Central Asia indicate that groups are about the same size as in North America but that a few males may associate with the female herds throughout the year (Schaller 1977).

The most detailed studies of the social life of wild *Ovis* were made by Geist (1971) in the Canadian Rockies. He found that a young female remains in its mother's group but that a male departs at two to four years and joins a group of rams. Young are tolerated by the adults, seem to receive guidance from them, and eventually inherit their home range and migratory pattern. In male groups there is a strict dominance hierarchy based on age and size of the horns; the animal with the biggest horns is the leader. In the ritualized struggle for dominance, two individuals push and shove one another and finally back away some distance, rear up on their hind legs, and lunge forward and down, crashing their horns together. Young animals are the most aggressive, but the older rams easily neutralize the attacks with their massive horns, and thus injuries are rare. The most spectacular battles occur on the rutting grounds. A dominant ram is not territorial but courts one estrous female at a time and drives other males away. For their part, ewes prefer mates with large horns and thwart the advances of others. At other times of the year there are also fights for dominance within groups of males.

Argali *(Ovis ammon nigrimontana)*, photo by Lothar Schlawe.

Homosexual activity then may occur, with the dominant animal behaving like a courting male and the subordinate playing the role of an estrous female.

In both the Old and the New World, the rut occurs mainly in the autumn and early winter, and births take place in the spring. The desert bighorn of southwestern North America, however, has an unusually long mating season for a ruminant, lasting from July to December (Wishart 1978). Gestation periods of about 150–80 days have been reported, and litters contain one to four young. A single lamb seems to be the usual case in the subgenus *Pachyceros,* but multiple births are very common in the subgenus *Ovis,* especially in the domestic sheep (Geist 1971; Grzimek 1975). Female *O. canadensis* are seasonally polyestrous, have an average gestation period of 174.2 days, and normally produce one young (Shackleton 1985). The newborn weighs about 3–5 kg and is precocious. Within a few weeks of birth the lambs form bands of their own and seek their mothers only periodically to suckle. They are weaned by 4–6 months (Geist 1971). A captive female *O. canadensis* successfully mated when only 10–11 months (McCutchen 1977), but ewes generally do not breed until their second or third year of life. Because of social factors, males usually are not able to mate until they are 7 years old, but they will do so sooner if the dominant rams of their group are killed (Nichols 1978).

According to Geist (1971), longevity depends on population status. In a declining or stable population, with low recruitment, most sheep live over 10 years, with a maximum of 20 for rams and 20–24 for ewes. In an expanding population, with heavy reproduction, the average life span is only 6–7 years. The quality of an individual and its longevity are inversely related. Large, vigorous animals that reproduce well

quickly reach advanced physiological age and die young, but a social and reproductive failure, especially if in a group characterized by lethargic behavior and slow reproduction, may attain a venerable age.

Available chromosomal and archeological evidence indicates that the domestic sheep is descended from a mouflonlike animal and that domestication occurred about 10,000–11,000 years ago in the eastern Mediterranean region (Nadler et al. 1973*b*). No wild sheep has a woolly coat comparable to that of *O. aries. O. musimon* (which, as pointed out in the above systematic discussion, may actually represent primitive introduced populations of *O. aries*) has a woolly underfur in winter, but this is well hidden by the coarse heavy coat. It has been reported that when domestic sheep become feral, they gradually lose much of their woolly pelage and develop a coat of coarse hairs, approaching the kind found in the wild species. There are now more than 800 breeds of domestic sheep (Valdez, Nadler, and Bunch 1978) and over 800 million individuals (Grzimek 1975). These animals are valued for the production of wool and meat but also are sometimes considered to be detrimental to natural vegetation and wildlife. Feral populations of *O. aries* have become established in many parts of the world, notably islands, where they have contributed to the extinction and endangerment of numerous native species (Lever 1985; Rudge 1984).

The domestic sheep has adversely affected its wild relatives in many areas by competing with them for forage and spreading disease. The wild species have also declined through agricultural encroachment, other human habitat modifications, and indiscriminate hunting by people for meat and trophies. Several Central Asian subspecies have become very rare and may not be able to survive without increased protection (Schaller 1977). The Soviet Union considers a majority of the subspecies of *O. ammon* that occur within its territory to be rare, vulnerable, or endangered. *O. ammon hodgsoni,* of Tibet and the Himalayas, *O. vignei vignei,* of Kashmir, and *O. musimon ophion,* of Cyprus, are listed as endangered by the USDI and are on appendix 1 of the CITES. The other subspecies of *O. ammon* and the species *O. canadensis* are on appendix 2 of the CITES. Roberts (1985) and Valdez and DeForge (1985) reported that major declines have occurred throughout the ranges of *O. vignei* and *O. orientalis. O. v. vignei* now occupies a discontinuous range and numbers about 2,500 individuals. The subspecies *O. o. punjabiensis* seems to be in even worse condition, now occurring in just two small localities in the Punjab of Pakistan and numbering barely 1,000.

The IUCN considers the populations of *O. musimon* (mouflon) on Cyprus, Corsica, and Sardinia to be vulnerable. The mouflon was present on the mainland of Europe through the end of the Pleistocene but apparently has been restricted to the Mediterranean islands in historical time. Numbers in the natural habitat fell drastically after 1900, there being only a few hundred survivors in the 1960s. Subsequent protection has led to an increase, at least on Sardinia, where there now are about 1,100–1,600 (Cassola 1985). Since the mid-nineteenth century large introduced mouflon populations have become established in much of Europe, where there now are about 60,000 individuals, and in various other parts of the world (Lever 1985; Tomiczek 1985; Uloth and Prien 1985).

The IUCN also now classifies the entire species *O. ammon* (argali) as indeterminate. This, the largest of all wild sheep, and the one with the most magnificent horns, was eliminated in much of its range by sport hunting in the nineteenth and early twentieth centuries. More recent problems have been intensified subsistence hunting and competition for winter range with domestic livestock, especially as human activity expands into formerly remote regions. The argali has disap-

Mouflon *(Ovis musimon)*, photo by Lothar Schlawe.

Rocky Mountain bighorn sheep *(Ovis canadensis)*, photo by Lothar Schlawe.

Desert bighorn sheep *(Ovis canadensis)*, male and two females, photo from Zoological Society of San Diego.

peared in northeastern China, parts of Mongolia, southern Siberia, and much of Soviet Central Asia and has become rare in the Himalayan region and much of the Tibetan Plateau and Sinkiang (Fedosenko 1985; Harper 1945; Mallon 1985; Schaller 1977, 1986; Schaller et al. 1987, 1988).

In northwestern North America there are now about 110,000 Dall's sheep *(O. dalli)*. The largest numbers are in Alaska, where populations appear to have recovered from earlier declines (Elliott 1985; Heimer 1985; Hoefs and Bar-

ichello 1985; Nichols 1978; Poole and Graf 1985). To the south, the bighorn sheep *(O. canadensis)*, which originally may have numbered 1.5 million to 2 million, suffered severe losses during the nineteenth century, mainly through excessive hunting, competition from domestic livestock, and the introduction of disease from domestic sheep. The subspecies *O. c. auduboni*, of the Black Hills and adjacent areas, became extinct, and other subspecies underwent increasing fragmentation as settlement proceeded. Although hunting has been

Dall's sheep *(Ovis dalli)*, male and female, photo by Lothar Schlawe.

legally prohibited or controlled since the early 1900s, and while some bighorn populations appear to be doing well, there has been no general recovery comparable to that of certain other North American ungulates during the same period. There are now about 36,000 Rocky Mountain and California bighorns and 23,000 desert bighorns, roughly the same numbers as at the turn of the century (Boone and Crockett Club 1975; Buechner 1960; Hebert et al. 1985; Sandoval 1985; Thorne, Hickey, and Stewart 1985; Weaver 1985). There is particular concern for the desert subspecies of the southwestern United States and Mexico. Some authorities have suggested that the bighorn is not well adapted to the arid environment of that region. Others have pointed out that it is physiologically and behaviorally suited for life there and that its seemingly relictual distribution is a result of human disturbance, habitat usurpation, and introduction of domestic and feral competitors (Bailey 1980; Geist 1985; Hansen 1982; McCutchen 1981; Wehausen 1984).

Geist (1971, 1974) suggested that one reason for the failure of the bighorn to rebuild its numbers is the discontinuous nature of its habitat. Only certain terrain is suitable, and animals will not readily cross intervening areas to places formerly occupied by other populations. A bighorn establishes its home range not by venturing out on its own but by following and learning from the others of its group. A young animal depends on the knowledge and guidance of its elders, with respect to becoming familiar with the migratory paths and the rather complex system of ranges that are used in the course of a year. Moreover, the reproductive potential of a bighorn population seems to adjust itself to the limitations of traditionally available habitat, and there are no surplus animals to spread elsewhere. Human hunting of a restricted population may remove a disproportionately large number of dominant males, thereby upsetting the cohesiveness of groups and preventing the young from properly adapting to the intricacies of their social system and habitat. These problems demonstrate the value of behavioral and ecological information to people as they work to save their fellow mammals from extinction.

Dall's sheep *(Ovis dalli)*, photo by Leonard Lee Rue III.

Literature Cited

A

Abe, H. 1983. Variation and taxonomy of *Niviventer fulvescens* and notes on *Niviventer* group of rats in Thailand. J. Mamm. Soc. Japan 9:151–61.

———. 1985. Changing mole distributions in Japan. In Kawamichi (1985), pp. 108–12.

Ables, E. D. 1975. Ecology of the red fox in North America. In Fox (1975), pp. 216–36.

Abravaya, J. P., and J. O. Matson. 1975. Notes on a Brazilian mouse, *Blarinomys breviceps* (Winge). Los Angeles Co. Mus. Nat. Hist. Contrib. Sci., no. 270, 8 pp.

Abu Jafer, M. Z., and C. Hays-Shahin. 1988. Re-introduction of the Arabian oryx into Jordan. In Dixon and Jones (1988), pp. 35–40.

Abu-Zinada, A. H., K. Habibi, and R. Seitre. 1988. The Arabian oryx programme in Saudi Arabia. In Dixon and Jones (1988), pp. 41–46.

Acha, P. N., and A. M. Alba. 1988. Economic losses due to *Desmodus rotundus*. In Greenhall and Schmidt (1988), pp. 207–14.

Adams, A. W. 1982. Migration. In Thomas and Toweill (1982), pp. 301–21.

Adams, C. E. 1976. Measurements and characteristics of fox squirrel, *Sciurus niger rufiventer*, home ranges. Amer. Midl. Nat. 95:211–15.

Adams, D. B. 1979. The cheetah: native American. Science 205:1155–58.

Adams, M., P. R. Baverstock, C. H. S. Watts, and T. Reardon. 1987a. Electrophoretic resolution of species boundaries in Australian Microchiroptera. I. *Eptesicus* (Chiroptera: Vespertilionidae). Austral. J. Biol. Sci. 40:143–62.

———. 1987b. Electrophoretic resolution of species boundaries in Australian Microchiroptera. II. The *Pipistrellus* group (Chiroptera: Vespertilionidae). Austral. J. Biol. Sci. 40:163–70.

Adams, M., T. R. Reardon, P. R. Baverstock, and C. H. S. Watts. 1988. Electrophoretic resolution of species boundaries in Australian Microchiroptera. IV. The Molossidae (Chiroptera). Austral. J. Biol. Sci. 41:315–26.

Adams, W. 1958. A flying squirrel that really flew. Malayan Nat. J. 13:31.

Adler, H. E., and L. L. Adler. 1978. What can dolphins *(Tursiops truncatus)* learn by observation? Cetology, no. 30, 10 pp.

Aellen, V. 1973. Un *Rhinolophus* nouveau d'Afrique Centrale. Period. Biol. 75:101–5.

Agenbroad, L. D. 1985. The distribution and chronology of mammoth in the New World. Acta Zool. Fennica 170:221–24.

Aggundey, I. 1977. First record of *Potamogale velox* in Kenya. Mammalia 41:368.

Aggundey, I. R., and D. A. Schlitter. 1984. Annotated checklist of the mammals of Kenya. I. Chiroptera. Ann. Carnegie Mus. 53:119–61.

———. 1986. Annotated checklist of the mammals of Kenya. II. Insectivora and Macroscelidea. Ann. Carnegie Mus. 55:325–47.

Agrawal, V. C., and S. Chakraborty. 1969. Occurrence of the woolly flying squirrel, *Eupetaurus cinereus* Thomas (Mammalia: Rodentia: Sciuridae) in north Sikkim. J. Bombay Nat. Hist. Soc. 66:615–16.

———. 1971. Notes on a collection of small mammals from Nepal, with the description of a new mouse-hare (Lagomorpha:Ochotonidae). Proc. Zool. Soc. Calcutta 24:41–46.

Agrawal, V. C., and D. K. Ghosal. 1969. A new field-rat (Mammalia: Rodentia: Muridae) from Kerala, India. Proc. Zool. Soc. Calcutta 22:41–45.

Aguilar, A. 1986. A review of old Basque whaling and its effect on the right whales *(Eubalaena glacialis)* of the North Atlantic. In Brownell, Best, and Prescott (1986), pp. 191–99.

Aguilera M., M. 1985. Growth and reproduction in *Zygodontomys microtinus* (Rodentia: Cricetidae) from Venezuela in a laboratory colony. Mammalia 49:75–83.

Ahnlund, H. 1980. Sexual maturity and breeding season of the badger, *Meles meles*, in Sweden. J. Zool. 190:77–95.

Ahroon, J. K., and F. G. Fidura. 1976. The influence of the male on maternal behaviour in the Mongolian gerbil *(Meriones unguiculatus)*. Anim. Behav. 24:372–75.

Aiello, A. 1985. Sloth hair: unanswered questions. In Montgomery (1985a), pp. 213–18.

Aimi, M. 1980. A revised classification of the Japanese red-backed voles. Kyoto Univ. Ser. Biol. 8:35–84.

Aimi, M., and H. Inagaki. 1988. Grooved lower incisors in flying lemurs. J. Mamm. 69:138–40.

Aitken, P. F. 1971a. The distribution of the hairy-nosed wombat [*Lasiorhinus latifrons* (Owen)]. Part I: Yorke Peninsula, Eyre Peninsula, the Gawler Ranges and Lake Harris. S. Austral. Nat. 45:93–104.

———. 1971b. *Planigale tenuirostris* Troughton, the narrow-nosed planigale, an addition to the mammal fauna of South Australia. S. Austral. Nat. 46:18.

———. 1971c. Rediscovery of the large desert sminthopsis (*Sminthopsis psammophilus* Spencer) on Eyre Peninsula, South Australia. Victorian Nat. 88:103–11.

———. 1972. *Planigale gilesi* (Marsupialia, Dasyuridae); a new species from the interior of southeastern Australia. Rec. S. Austral. Mus. 16:1–14.

———. 1977a. The little pygmy possum [*Cercartetus lepidus* (Thomas)] found living on the Australian mainland. S. Austral. Nat. 51:63–66.

———. 1977b. Rediscovery of swamp antechinus in South Australia after 37 years. S. Austral. Nat. 52:28–30.

———. 1979. The status of endangered Australian wombats, bandicoots and the marsupial mole. In Tyler (1979), pp. 61–65.

Ajayi, S. S. 1975. Observations on the biology, domestication and reproductive performance of the African giant rat *Cricetomys gambianus* Waterhouse in Nigeria. Mammalia 39:344–64.

Akersten, W. A. 1973. Upper incisor grooves in the Geomyinae. J. Mamm. 54:349–55.

Alados, C. L. 1985a. Group size and composition of the Spanish ibex (*Capra pyrenaica* Schinz) in the Sierras of Cazorla and Segura. *In* Lovari (1985), pp. 134–47.

———. 1985b. Distribution and status of the Spanish ibex (*Capra pyrenaica* Schinz). *In* Lovari (1985), pp. 204–11.

Alberico, M. S. 1987. Notes on distribution of some bats from southwestern Colombia. *Fieldiana Zool.*, n.s., 39:133–35.

Albignac, R. 1969. Notes ethologiques sur quelques carnivores Malgaches: le *Galidia elegans* I. Geoffroy. Terre Vie 23:202–15.

———. 1970a. Notes ethologiques sur quelques carnivores Malgaches: le *Fossa fossa* (Schreber). Terre Vie 24:383–94.

———. 1970b. Notes ethologiques sur quelques carnivores Malgaches: le *Cryptoprocta ferox* (Bennett). Terre Vie 24:395–402.

———. 1971. Notes ethologiques sur quelques carnivores Malgaches: le *Mungotictis lineata* Pocock. Terre Vie 25:328–43.

———. 1972. The Carnivora of Madagascar. *In* Battistini and Richard-Vindard (1972), pp. 667–82.

———. 1974. Observations éco-ethologiques sur le genre *Eupleres*, viverridide de Madagascar. Terre Vie 28:321–51.

———. 1975. Breeding the fossa *Cryptoprocta ferox* at Montpelier Zoo. Internatl. Zoo Yearbook 15:147–50.

———. 1976. L'écologie de *Mungotictis decemlineata* dans les forêts décidues de l'ouest de Madagascar. Terre Vie 30:347–76.

———. 1987. Status of the aye-aye in Madagascar. Primate Conserv. 8:44–45.

Alendal, E. 1984. Muskoxen in captivity in Europe and Asia, April 1983. *In* Klein, White, and Keller (1984), pp. 9–11.

Alexander, B. 1987. A beast for all seasons. Internatl. Wildl. 17(3):44–51.

Alho, C. J. R. 1982. Brazilian rodents: their habitats and habits. Pymatuning Lab. Ecol. Spec. Publ. 6:143–66.

Alho, C. J. R., and M. J. de Souza. 1982. Home range and use of space in *Zygodontomys lasiurus* (Cricetidae, Rodentia) in the cerrado of central Brazil. Ann. Carnegie Mus. 51:127–32.

Ali, R. 1985. An overview of the status and distribution of the lion-tailed macaque. Monogr. Primatol. 7:13–25.

Al-Khalili, A. D., M. J. Delany, and I. A.

Nader. 1988. The ecology of the rock rat, *Praomys fumatus yemeni* Sanborn and Hoogstraal, in Saudi Arabia. Mammalia 52:35–43.

Allen, G. M. 1938. The mammals of China and Mongolia. Amer. Mus. Nat. Hist., New York, vol. 1.

———. 1939a. Bats. Harvard Univ. Press, Cambridge, 368 pp.

———. 1939b. A checklist of African mammals. Bull. Mus. Comp. Zool. 83:1–763.

———. 1940. The mammals of China and Mongolia. Amer. Mus. Nat. Hist., New York. Vol. 2.

———. 1942. Extinct and vanishing mammals of the Western Hemisphere with the marine species of all the oceans. Spec. Publ. Amer. Comm. Internatl. Wildl. Protection, no. 11, xv + 620 pp.

Allen, J. A. 1893. On a collection of mammals from the San Pedro Martir region of Lower California, with notes on other species, particularly of the genus *Sitomys*. Bull. Amer. Mus. Nat. Hist. 5:181–214.

———. 1905. Mammalia of southern Patagonia. Princeton Univ. Exped. Patagonia 3:1–210.

———. 1911. Mammals from Venezuela collected by Mr. M. A. Carriker, Jr., 1909–1911. Bull. Amer. Mus. Nat. Hist. 30:239–73.

———. 1912. Mammals from western Colombia. Bull. Amer. Mus. Nat. Hist. 31:71–95.

———. 1915. Review of the South American Sciuridae. Bull. Amer. Mus. Nat. Hist. 34:147–309.

———. 1917. The skeletal characters of *Scutisorex* Thomas. Bull. Amer. Mus. Nat. Hist. 37:769–84.

Allsopp, R. 1978. Social biology of bushbuck (*Tragelaphus scriptus* Pallas 1776) in the Nairobi National Park, Kenya. E. Afr. Wildl. J. 16:153–65.

Al-Robaae, K. 1977. Distribution of *Nesokia indica* (Gray and Hardwicke, 1830) in Basrah liwa, south Iraq; with some biological notes. Saugetierk. Mitt. 25:194–97.

Altmann, S. A., ed. 1967. Social communication among primates. Univ. Chicago Press, xiv + 392 pp.

Altmann, S. A., and J. Altmann. 1970. Baboon ecology. Univ. Chicago Press, vii + 220 pp.

Altuna, C. A., and E. P. Lessa. 1985. Penial morphology in Uruguayan species of *Ctenomys* (Rodentia: Octodontidae). J. Mamm. 66:483–88.

Alvarez del Toro, M. 1967. A note on the breeding of the Mexican tree porcupine. Internatl. Zoo Yearbook 7:118.

Ambrose, H. W., III. 1969. A comparison of *Microtus pennsylvanicus* home ranges as determined by isotope and live trap methods. Amer. Midl. Nat. 81:535–55.

Amstrup, S. C., and J. Beecham. 1976. Activity patterns of radio-collared black bears in ldaho. J. Wildl. Mgmt. 40:340–48.

Amstrup, S. C., I. Stirling, and J. W. Lentfer. 1986. Past and present status of polar bears in Alaska. Wildl. Soc. Bull. 14:241–54.

Anadu, P. A. 1979a. The occurrence of *Steatomys jacksoni* Hayman in south-western Nigeria. Acta Theriol. 24:513–17.

———. 1979b. Gestation period and early development in *Myomys daltoni* (Rodentia: Muridae). Terre Vie 33:59–69.

Anadu, P. A., and J. F. Oates. 1988. The olive colobus monkey in Nigeria. Nigerian Field 53:31–34.

Anand Kumar, T. C. 1965. Reproduction in the rat-tailed bat *Rhinopoma kinneari*. J. Zool. 147:147–55.

Andelt, W. F. 1985. Behavioral ecology of coyotes in south Texas. Wildl. Monogr., no. 94, 45 pp.

Andersen, D. C. 1978. Observations on reproduction, growth, and behavior of the northern pocket gopher (*Thomomys talpoides*). J. Mamm. 59:418–22.

Andersen, H. T., ed. 1969. The biology of marine mammals. Academic Press, New York, 511 pp.

Andersen, K. 1912. Catalogue of the Chiroptera in the collection of the British Museum. I. Megachiroptera. British Mus. (Nat. Hist.), London, ci + 854 pp.

Andersen, S. 1976. The taming and training of the harbor porpoise *Phocoena phocoena*. Cetology, no. 24, 9 pp.

Anderson, A. E., and O. C. Wallmo. 1984. *Odocoileus hemionus*. Mammalian Species, no. 219, 9 pp.

Anderson, E. 1970. Quaternary evolution of the genus *Martes* (Carnivora, Mustelidae). Acta Zool. Fennica, no. 130, 132 pp.

———. 1973. Ferret from the Pleistocene of central Alaska. J. Mamm. 54:778–79.

———. 1984. Who's who in the Pleistocene: a mammalian bestiary. *In* Martin and Klein (1984), pp. 40–89.

Anderson, J. L. 1975. The occurrence of a secondary breeding peak in the southern impala. E. Afr. Wildl. J. 13:149–51.

———. 1980. The social organisation and aspects of behaviour of the nyala *Tragelaphus angasi* Gray, 1849. Z. Saugetierk. 45:90–123.

———. 1984. Reproduction in the nyala

(Tragelaphus angasi) (Mammalia: Ungulata). J. Zool. 204:129–42.

Anderson, J. W., and W. A. Wimsatt. 1963. Placentation and fetal membranes of the Central American noctilionid bat, *Noctilio labialis minor*. Amer. J. Anat. 112:181–201.

Anderson, P. K. 1979. Dugong behavior: on being a marine mammalian grazer. Biologist 61:113–44.

———. 1982a. Studies of dugongs at Shark Bay, Western Australia. I. Analysis of population size, composition, dispersion and habitat use on the basis of aerial survey. Austral. Wildl. Res. 9:69–84.

———. 1982b. Studies of dugongs at Shark Bay, Western Australia. II. Surface and subsurface observations. Austral. Wildl. Res. 9:85–99.

———. 1984a. Suckling in *Dugong dugon*. J. Mamm. 65:510–11.

———. 1984b. Dugong. *In* Macdonald (1984), pp. 298–99.

———. 1986. Dugongs of Shark Bay, Australia—seasonal migration, water temperature, and forage. Natl. Geogr. Res. 2:473–90.

Anderson, P. K., and R. I. T. Prince. 1985. Predation on dugongs: attacks by killer whales. J. Mamm. 66:554–56.

Anderson, R. R., and K. N. Sinha. 1972. Number of mammary glands and litter size of the golden hamster. J. Mamm. 53:382–84.

Anderson, S. 1967. Introduction to the rodents. *In* Anderson and Jones (1967), pp. 206–9.

———. 1969. *Macrotus waterhousii*. Mammalian Species, no. 1, 4 pp.

———. 1982. *Monodelphis kunsi*. Mammalian Species, no. 190, 3 pp.

———. 1985. Taxonomy and systematics. *In* Tamarin, R. H., ed., Biology of New World *Microtus*, Amer. Soc. Mamm. Spec. Publ., no. 8, pp. 52–83.

Anderson, S., and J. K. Jones, Jr., eds. 1967. Recent mammals of the world. A synopsis of families. Ronald Press, New York, viii + 453 pp.

———. 1984. Orders and families of Recent mammals of the world. John Wiley & Sons, New York, xii + 686 pp.

Anderson, S., K. F. Koopman, and G. K. Creighton. 1982. Bats of Bolivia: an annotated checklist. Amer. Mus. Novit., no. 2750, 24 pp.

Anderson, S., and W. D. Webster. 1983. Notes on Bolivian mammals. 1. Additional records of bats. Amer. Mus. Novit., no. 2766, 3 pp.

Anderson, S., C. A. Woods, G. S. Morgan,

and W. L. R. Oliver. 1983. *Geocapromys brownii*. Mammalian Species, no. 201, 5 pp.

Anderson, S., T. L. Yates, and J. A. Cook. 1987. Notes on Bolivian mammals. 4. The genus *Ctenomys* (Rodentia, Ctenomyidae) in the eastern lowlands. Amer. Mus. Novit., no. 2891, 20 pp.

Anderson, T. J. C., A. J. Berry, J. N. Amos, and J. M. Cook. 1988. Spool-and-line tracking of the New Guinea bandicoot, *Echymipera kalubu* (Marsupialia, Peramelidae). J. Mamm. 69:114–20.

Ando, A., S. Shiraishi, and T. A. Uchida. 1987. Growth and development of the Smith's red-backed vole, *Eothenomys smithi*. J. Fac. Agr. Kyushu Univ. 31:309–20.

———. 1988. Reproduction in a laboratory colony of the Smith's red-backed vole, *Eothenomys smithi*. J. Mamm. Soc. Japan 13:11–20.

Andreev, V. N. 1978. State of world reindeer breeding and its classification. Soviet J. Ecol. 8:291–95.

Andrews, P. 1978. Taxonomy and relationships of fossil apes. *In* Chivers and Joysey (1978), pp. 43–56.

Andrews, P., and J. E. Cronin. 1982. The relationships of *Sivapithecus* and *Ramapithecus* and the evolution of the orang-utan. Nature 297:541–46.

Andrews, P., and C. P. Groves. 1976. Gibbons and brachiation. *In* Rumbaugh, D. M., ed., Gibbon and Siamang, IV, S. Karger, Basel, pp. 167–218.

Andrews, R. D., and E. K. Boggess. 1978. Ecology of coyotes in Iowa. *In* Bekoff (1978), pp. 249–65.

Angermann, R. 1983. The taxonomy of Old World *Lepus*. Acta Zool. Fennica 174:17–21.

Angst, W. 1975. Basic data and concepts on the social organization of *Macaca fascicularis*. Primate Behav. 4:325–88.

Ansell, W. F. H. 1960. Mammals of Northern Rhodesia. Government Printer, Lusaka, xxxi + 155 + 24 pp.

———. 1974. Some mammals from Zambia and adjacent countries. Occas. Pap. Natl. Parks and Wildl. Serv. Zambia, Suppl., no. 1, 48 pp.

———. 1978. The mammals of Zambia. Natl. Parks and Wildl. Serv., Chilanga, Zambia, ii + 126 pp.

———. 1982. The southeastern range limit of *Manis tricuspis* Rafinesque, and the type of *Manis tridentata* Focillon. Mammalia 46:559–60.

———. 1986. Some Chiroptera from south-central Africa. Mammalia 50:507–519.

Ansell, W. F. H., and P. D. H. Ansell. 1973.

Mammals of the north-eastern montane areas of Zambia. Puku 7:21–69.

Ansell, W. F. H., and R. J. Dowsett. 1988. Mammals of Malawi. An annotated check list and atlas. Trendrine Press, Cornwall, 170 pp.

Anthony, H. E. 1924. Preliminary report on Ecuadorean mammals. No. 6. Amer. Mus. Novit., no. 139, 9 pp.

———. 1929. Two new genera of rodents from South America. Amer. Mus. Novit., no. 383, 6 pp.

Antinucci, F., and E. Visalberghi. 1986. Tool use in *Cebus apella*: a case study. Internatl. J. Primatol. 7:351–63.

Aquino, R., and F. Encarnacion. 1986. Population structure of *Aotus nancymai* (Cebidae: Primates) in Peruvian Amazon lowland forest. Amer. J. Primatol. 11:1–7.

Arai, S., T. Mori, H. Yoshida, and S. Shiraishi. 1985. A note on the Japanese water shrew, *Chimarrogale himalayica platycephala* from Kyushu. J. Mamm. Soc. Japan 10:193–203.

Arata, A. A. 1967. Muroid, gliroid, and dipodoid rodents. *In* Anderson and Jones (1967), pp. 226–53.

Arata, A. A., and J. B. Vaughan. 1970. Analyses of the relative abundance and reproductive activity of bats in southwestern Colombia. Caldasia 10:517–28.

Archer, M. 1974a. New information about the Quaternary distribution of the thylacine (Marsupialia, Thylacinidae) in Australia. J. Roy. Soc. W. Austral. 57:43–50.

———. 1974b. Regurgitation or mercyism in the western native cat, *Dasyurus geoffroii*, and the red-tailed wambenger, *Phascogale calura* (Marsupialia, Dasyuridae). J. Mamm. 55:448–52.

———. 1975. *Ningaui*, a new genus of tiny dasyurids (Marsupialia) and two species, *N. timealeyi* and *N. ridei*, from arid western Australia. Mem. Queensland Mus. 17:237–49.

———. 1976. Revision of the marsupial genus *Planigale* Troughton (Dasyuridae). Mem. Queensland Mus. 17:341–65.

———. 1977. Revision of the dasyurid marsupial genus *Antechinomys* Krefft. Mem. Queensland Mus. 18:17–29.

———. 1979a. Two new species of *Sminthopsis* Thomas (Dasyuridae: Marsupialia) from northern Australia, *S. butleri* and *S. douglasi*. Austral. Zool. 20:327–45.

———. 1979b. The status of Australian dasyurids, thylacinids and myrmecobiids. *In* Tyler (1979), pp. 29–43.

———. 1981. Results of the Archbold Expeditions. No. 104. Systematic revision of the marsupial dasyurid genus *Sminthopsis*

Thomas. Bull. Amer. Mus. Nat. Hist. 168:61–224.

———, ed. 1982a. Carnivorous marsupials. Roy. Zool. Soc. New South Wales, Mosman, 2 vols.

———. 1982b. Review of the dasyurid (Marsupialia) fossil record, integration of data bearing on phylogenetic interpretation, and suprageneric classification. In Archer (1982a), pp. 397–443.

———. 1982c. A review of the Miocene thylacinids (Thylacinidae, Marsupialia), the phylogenetic position of the Thylacinidae and the problem of apiorisms in character analysis. In Archer (1982a), pp. 445–76.

———. 1984a. Origins and early radiations of marsupials. In Archer and Clayton (1984), pp. 585–625.

———. 1984b. The Australian marsupial radiation. In Archer and Clayton (1984), pp. 633–808.

———, ed. 1987. Possums and opossums. Surrey Beatty & Sons, Sydney.

Archer, M., and G. Clayton, eds. 1984. Vertebrate zoogeography and evolution in Australasia. Hesperian Press, Carlisle, Australia, xxiv + 1203 pp.

Archer, M., T. F. Flannery, A. Ritchie, and R. E. Molnar. 1985. First Mesozoic mammal from Australia—an early Cretaceous monotreme. Nature 318:363–66.

Archer, M., and J. A. W. Kirsch. 1977. The case for the Thylacomyidae and Myrmecobiidae Gill, 1872, or why are marsupial families so extended? Proc. Linnean Soc. New South Wales 102:18–25.

Archer, M., M. D. Plane, and N. S. Pledge. 1978. Additional evidence for interpreting the Miocene Obdurodon insignis Woodburne and Tedford, 1975, to be a fossil platypus (Ornithorhynchidae: Monotremata) and a reconsideration of the status of Ornithorhynchus agilis De Vis, 1885. Austral. Zool. 20:9–27.

Arden-Clarke, C. H. G. 1986. Population density, home range size and spatial organization of the Cape clawless otter, Aonyx capensis, in a marine habitat. J. Zool. 209:201–11.

Arita, H. T., and D. E. Wilson. 1987. Long-nosed bats and agaves: the tequila connection. Bats 5(4):3–5.

Armitage, K. B. 1973. Population changes and social behavior following colonization by the yellow-bellied marmot. J. Mamm. 54:842–54.

———. 1975. Social behavior and population dynamics of marmots. Oikos 26:341–54.

———. 1977. Social variety in the yellow-bellied marmot: a population-behavioural system. Anim. Behav. 25:585–93.

Armitage, K. B., and J. F. Downhower. 1974.

Demography of yellow-bellied marmot populations. Ecology 55:1233–45.

Armitage, K. B., and D. W. Johns. 1982. Kinship, reproductive strategies and social dynamics of yellow-bellied marmots. Behav. Ecol. Sociobiol. 11:55–63.

Armstrong, D. M. 1972. Distribution of mammals in Colorado. Monogr. Mus. Nat. Hist. Univ. Kansas, no. 3, x + 415 pp.

Armstrong, D. M., and J. K. Jones, Jr. 1971. Mammals from the Mexican state of Sinaloa. I. Marsupialia, Insectivora, Edentata, Lagomorpha. J. Mamm. 52:747–57.

———. 1972a. Megasorex gigas. Mammalian Species, no. 16, 2 pp.

———. 1972b. Notiosorex crawfordi. Mammalian Species, no. 17, 5 pp.

Armstrong, D. M., J. K. Jones, Jr., and E. C. Birney. 1972. Mammals from the Mexican state of Sinaloa. III. Carnivora and Artiodactyla. J. Mamm. 53:48–61.

Armstrong, J. J. 1979. The California sea otter: emerging conflicts in resource management. San Diego Law Rev. 16:249–85.

Arnason, U. 1974. Comparative chromosome studies in Pinnipedia. Hereditas 76:179–225.

Arnason, U., M. Höglund, and B. Widegren. 1984. Conservation of highly repetitive DNA in cetaceans. Chromosoma 89:238–42.

Arndt, R. G., F. C. Rohde, and J. A. Bosworth. 1978. Additional records of the rice rat, Oryzomys palustris (Harlan), from New Jersey and Delaware. Bull. New Jersey Acad. Sci. 23:65–72.

Arnold, G. W., and M. L. Dudzinski. 1978. Ethology of free-ranging domestic animals. Elsevier Scientific Publ. Co., Amsterdam, xi + 198 pp.

Arnold, P., H. Marsh, and G. Heinsohn. 1987. The occurrence of two forms of minke whale in east Australian waters with a description of external characters and skeleton of the diminutive or dwarf form. Sci. Rept. Whales Res. Inst. 38:1–46.

Arroyo-Cabrales, J., and J. K. Jones, Jr. 1988a. Balantiopteryx plicata. Mammalian Species, no. 301, 4 pp.

———. 1988b. Balantiopteryx io and Balantiopteryx infusca. Mammalian Species, no. 313, 3 pp.

Artois, M., and M.-J. Duchêne. 1982. Première identification du chien viverrin (Nyctereutes procyonoides Gray, 1834) en France. Mammalia 46:265–67.

Arundel, J. H., I. K. Barker, and I. Beveridge. 1977. Diseases of marsupials. In Stonehouse and Gilmore (1977), pp. 141–54.

Asdell, S. A. 1964. Patterns of mammalian reproduction. Cornell Univ. Press, Ithaca, viii + 670 pp.

Ashby, K. R. 1967. Studies on the ecology of field mice and voles (Apodemus sylvaticus, Clethrionomys glareolus and Microtus agrestis) in Houghall Wood, Durham. J. Zool. 152:389–513.

Asibey, E. O. A. 1974a. The grasscutter, Thryonomys swinderianus Temminck, in Ghana. Symp. Zool. Soc. London 34:161–70.

———. 1974b. Reproduction in the grasscutter (Thryonomys swinderianus Temminck) in Ghana. Symp. Zool. Soc. London 34:251–63.

Aslin, H. J. 1974. The behaviour of Dasyuroides byrnei (Marsupialia) in captivity. Z. Tierpsychol. 35:187–208.

———. 1975. Reproduction in Antechinus maculatus Gould (Dasyuridae). Austral. Wildl. Res. 2:77–80.

———. 1977. New records of Sminthopsis ooldea Troughton from South Australia. S. Austral. Nat. 52:9–12.

———. 1980. Biology of a laboratory colony of Dasyuroides byrnei (Marsupialia: Dasyuridae). Austral. Zool. 20:457–71.

———. 1983. Reproduction in Sminthopsis ooldea (Marsupialia: Dasyuridae). Austral. Mamm. 6:93–95.

Aslin, H. J., and C. H. S. Watts. 1980. Breeding of a captive colony of Notomys fuscus Wood Jones (Rodentia: Muridae). Austral. Wildl. Res. 7:379–83.

Asper, E. D., W. G. Young, and M. T. Walsh. 1988. Observations on the birth and development of a captive-born killer whale. Internatl. Zoo Yearbook 27:295–304.

Atallah, S. I. 1977. Mammals of the eastern Mediterranean region; their ecology, systematics and zoogeographical relationships. Saugetierk. Mitt. 25:241–320.

Atherton, R. G., and A. T. Haffenden. 1982. Observations on the reproduction and growth of the long-tailed pygmy possum, Cercartetus caudatus (Marsupialia: Burramyidae), in captivity. Austral. Mamm. 5:253–59.

Au, D. W. K., and W. L. Perryman. 1985. Dolphin habitats in the eastern tropical Pacific. Fishery Bull. 83:623–43.

Augee, M. L. 1978. Monotremes and the evolution of homeothermy. Austral. Zool. 20:111–19.

Augee, M. L., T. J. Bergin, and C. Morris. 1978. Observations on behaviour of echidnas at Taronga Zoo. Austral. Zool. 20:121–29.

Augee, M. L., E. H. M. Ealey, and I. P. Price. 1975. Movements of echidnas, Tachyglossus aculeatus, determined by marking-recapture and radio-tracking. Austral. Wildl. Res. 2:93–101.

Aulagnier, S., and R. Destre. 1985. Introduction a l'etude des chiropteres du Tafilalt (sudest marocain). Mammalia 49:329–37.

Aulagnier, S., and P. Mein. 1985. Note sur la presence d'*Otonycteris hemprichi* Peters, 1859 au Maroc. Mammalia 49:582–84.

Aung, H. 1968. A note on the birth of a Mishmi takin *Budorcas t. taxicolor* at Rangoon Zoo. Internatl. Zoo Yearbook 8:145.

Austin, T. A., and J. A. Kitchens. 1986. Expansion of *Baiomys taylori* into Hardeman County, Texas. Southwestern Nat. 31:547–48.

Australian National Parks and Wildlife Service. 1977, 1978. Australian endangered species. Mammals, nos. 1–23.

———. 1988. Kangaroos in Australia. Conservation status and management. Austral. Natl. Parks Wildl. Serv. Occas. Pap., no. 14, 45 pp.

Autenrieth, R. E., and E. Fichter. 1975. On the behavior and socialization of pronghorn fawns. Wildl. Monogr., no. 42, 111 pp.

Avila-Pires, F. D. de. 1960. Um novo gênero de roedor Sul-Americano (Rodentia—Cricetidae). Bol. Mus. Nac. Rio de Janeiro (Zool.), n.s., no. 220, 6 pp.

———. 1967. The type-locality of "*Chaetomys subspinosus*" (Olfers, 1818) (Rodentia, Caviomorpha). Rev. Brasil. Biol. 27:177–79.

———. 1972. A new subspecies of *Kunsia fronto* (Winge, 1888) from Brazil (Rodentia, Cricetidae). Rev. Brasil. Biol. 32:419–22.

Avila-Pires, F. D. de, and M. R. Caldatto Wutke. 1981. Taxonòmia e evolucao de *Clyomys* Thomas, 1916 (Rodentia, Echimyidae). Rev. Brasil. Biol. 41:529–34.

Ayala, S. C., and A. D'Alessandro. 1973. Insect feeding of some Colombian fruit-eating bats. J. Mamm. 54:266–67.

Ayres, J. M. 1985. On a new species of squirrel monkey, genus *Saimiri*, from Brazilian Amazona (Primates: Cebidae). Papeis Avulsos de Zoologia, Mus. Zool. Univ. Sao Paulo, 36:147–64.

Ayres, J. M., and A. D. Johns. 1987. Conservation of white uacaries in Amazonian várzea. Oryx 21:74–80.

Ayres, J. M., and J. L. Nessimian. 1982. Evidence for insectivory in *Chiropotes satanas*. Primates 23:458–59.

B

Baar, S. L., E. D. Fleharty, and M. F. Artman. 1975. Utilization of deep burrows and nests by cotton rats in west-central Kansas. Southwestern Nat. 19:440–44.

Baba, M., T. Doi, and Y. Ono. 1982. Home range utilization and nocturnal activity of the giant flying squirrel, *Petaurista leucogenys*. Jap. J. Ecol. 32:189–98.

Baber, D. W., and B. E. Coblentz. 1986. Density, home range, habitat use, and reproduction in feral pigs on Santa Catalina Island. J. Mamm. 67:512–25.

Babinska-Werka, J., J. Gliwicz, and J. Goszczynski. 1981. Demographic processes in an urban population of the striped field mouse. Acta Theriol. 26:275–83.

Baccus, R., N. Ryman, M. H. Smith, C. Reuterwall, and D. Cameron. 1983. Genetic variability and differentiation of large grazing mammals. J. Mamm. 64:109–20.

Badrian, A., and N. Badrian. 1984. Social organization of *Pan paniscus* in the Lomako Forest, Zaire. *In* Susman (1984a), pp. 325–46.

Badrian, N., and R. K. Malenky. 1984. Feeding ecology of *Pan paniscus* in the Lomako Forest, Zaire. *In* Susman (1984a), pp. 275–99.

Baeten, B., V. Van Cakenberghe, and F. De Vree. 1984. An annotated inventory of a collection of bats from Rwanda. Rev. Zool. Afr. 98:183–96.

Baharav, D. 1983. Observation on the ecology of the mountain gazelle in the upper Galilee, Israel. Mammalia 47:59–69.

Bailey, J. A. 1980. Desert bighorn, forage competition, and zoogeography. Wildl. Soc. Bull. 8:208–16.

Bailey, T. N. 1974. Social organization in a bobcat population. J. Wildl. Mgmt. 38:435–46.

Bailey, V. 1930. Animal life of Yellowstone National Park. Charles C. Thomas, Springfield, Illinois, 241 pp.

Bain, J. R. 1978. The breeding system of *Nycticeius humeralis* in Florida. Amer. Soc. Mamm., Abstr. Tech. Pap., 58th Ann. Mtg., p. 29.

Baker, A. N. 1978. The status of Hector's dolphin, *Cephalorhynchus hectori* (Van Beneden), in New Zealand waters. Rept. Internatl. Whaling Comm. 28:331–34.

———. 1985. Pygmy right whale—*Caperea marginata*. *In* Ridgway and Harrison (1985), pp. 345–54.

Baker, C. M., and J. Meester. 1986. Postnatal physical development of the water mongoose (*Atilax paludinosus*). Z. Saugetierk. 51:236–43.

Baker, C. S., and L. M. Herman. 1984. Aggressive behavior between humpback whales (*Megaptera novaeangliae*) wintering in Hawaiian waters. Can. J. Zool. 62:1922–37.

Baker, C. S., L. M. Herman, A. Perry, W. S. Lawton, J. M. Straley, A. A. Wolman, G. D. Kaufman, H. E. Winn, J. D. Hall, J. M. Reinke, and J. Östman. 1986. Migratory movement and population structure of humpback whales (*Megaptera novaeangliae*) in the central and eastern North Pacific. Mar. Ecol. Prog. Ser. 31:105–19.

Baker, H. G., and B. J. Harris. 1957. The pollination of *Parkia* by bats and its attendant evolutionary problems. Evolution 11:449–60.

Baker, R. C., F. Wilke, and C. H. Baltzo. 1970. The northern fur seal. U.S. Dept. Interior Circ., no. 336, iii + 19 pp.

Baker, R. H. 1968. Habitats and distribution. *In* King (1968), pp. 98–121.

———. 1974. Records of mammals from Ecuador. Michigan State Univ. Mus. Publ., Biol. Ser., 5:129–46.

———. 1977. Mammals of the Chihuahuan Desert region—future prospects. *In* Wauer, R. H., and D. H. Riskind, eds., Transactions of the symposium on the biological resources of the Chihuahuan Desert region, United States and Mexico, U.S. Natl. Park Serv. Trans. Proc. Ser., no. 3, pp. 221–25.

———. 1984. Origin, classification and distribution. *In* Halls (1984), pp. 1–18.

Baker, R. H., and J. K. Greer. 1962. Mammals of the Mexican state of Durango. Michigan State Univ. Mus. Publ., Biol. Ser., 2:25–54.

Baker, R. H., and M. K. Petersen. 1965. Notes on a climbing rat, *Tylomys*, from Oaxaca, Mexico. J. Mamm. 46:694–95.

Baker, R. J. 1984. A sympatric cryptic species of mammal: a new species of *Rhogeessa* (Chiroptera: Vespertilionidae). Syst. Zool. 33:178–83.

Baker, R. J., P. V. August, and A. A. Steuter. 1978. *Erophylla sezekorni*. Mammalian Species, no. 115, 5 pp.

Baker, R. J., J. W. Bickham, and M. L. Arnold. 1985. Chromosomal evolution in *Rhogeessa* (Chiroptera: Vespertilionidae): possible speciation by centric fusions. Evolution 39:233–43.

Baker, R. J., C. G. Dunn, and K. Nelson. 1988. Allozymic study of the relationships of *Phylloderma* and four species of *Phyllostomus*. Occas. Pap. Mus. Texas Tech Univ., no. 125, 14 pp.

Baker, R. J., H. H. Genoways, and J. C. Patton. 1978. Bats of Guadeloupe. Occas. Pap. Mus. Texas Tech Univ., no. 50, 16 pp.

Baker, R. J., H. H. Genoways, P. A. Seyfarth. 1981. Results of the Alcoa Foundation—Suriname Expeditions. VI. Additional chromosomal data for bats (Mammalia: Chiroptera) for Suriname. Ann. Carnegie Mus. 50:333–44.

Baker, R. J., R. L. Honeycutt, and R. A. Bass. 1988. Genetics. *In* Greenhall and Schmidt (1988), pp. 31–39.

Baker, R. J., and J. K. Jones, Jr. 1975. Additional records of bats from Nicaragua, with a revised checklist of Chiroptera. Occas. Pap. Mus. Texas Tech Univ., no. 32, 13 pp.

Baker, R. J., J. K. Jones, Jr., and D. C. Carter, eds. 1976. Biology of bats of the New World family Phyllostomatidae. Part I. Spec. Publ. Mus. Texas Tech Univ., no. 10, 218 pp.

———. 1977. Biology of bats of the New World family Phyllostomatidae. Part II. Spec. Publ. Mus. Texas Tech Univ., no. 13, 364 pp.

———. 1979. Biology of bats of the New World family Phyllostomatidae. Part III. Spec. Publ. Mus. Texas Tech Univ., no. 16, 441 pp.

Baker, R. J., B. F. Koop, and M. W. Haiduk. 1983. Resolving systematic relationships with G-bands: a study of five genera of South American cricetine rodents. Syst. Zool. 32:403–16.

Baker, R. J., J. C. Patton, H. H. Genoways, and J. W. Bickham. 1988. Genic studies of Lasiurus (Chiroptera: Vespertilionidae). Occas. Pap. Mus. Texas Tech Univ., no. 117, 15 pp.

Baker, R. J., and S. L. Williams. 1974. Geomys tropicalis. Mammalian Species, no. 35, 4 pp.

Balcomb, K. C., III, and M. A. Bigg. 1986. Population biology of the three resident killer whale pods in Puget Sound and off southern Vancouver Island. In Kirkevold and Lockard (1986), pp. 85–95.

Baldwin, J. D. 1970. Reproductive synchronization in squirrel monkeys (Saimiri). Primates 11:317–26.

Baldwin, J. D., and J. I. Baldwin. 1971. Squirrel monkeys (Saimiri) in natural habitats in Panama, Colombia, Brazil, and Peru. Primates 12:45–61.

———. 1972. Population density and use of space in howling monkeys (Alouatta villosa) in south-western Panama. Primates 13:371–79.

———. 1976. Primate populations in Chiriqui, Panama. In Thorington and Heltne (1976), pp. 20–31.

Ballard, W. B., R. Farnell, and R. O. Stephenson. 1983. Long distance movement by gray wolves, Canis lupus. Can. Field-Nat. 97:333.

Ballou, J. D., and J. Seidensticker. 1987. The genetic and demographic characteristics of the 1983 captive population of Sumatran tigers (Panthera tigris sumatrae). In Tilson and Seal (1987), pp. 329–47.

Banci, V., and A. Harestad. 1988. Reproduction and natality of wolverine (Gulo gulo) in Yukon. Ann. Zool. Fennici 25:265–70.

Banfield, A. W. F. 1961. A revision of the reindeer and caribou, genus Rangifer. Natl. Mus. Canada Bull., no. 177, vi + 137 pp.

———. 1974. The mammals of Canada. Univ. Toronto Press, xxv + 438 pp.

Banholzer, U. 1976. Water balance, metabo-

lism, and heart rate in the fennec. Naturwissenschaften 63:202.

Banks, E. 1978. Mammals from Borneo. Brunei Mus. J. 4:165–227.

Banks, E. M., R. J. Brooks, and J. Schnell. 1975. A radiotracking study of home range and activity of the brown lemming (Lemmus trimucronatus). J. Mamm. 56:888–901.

Banks, R. C., and R. L. Brownell. 1969. Taxonomy of the common dolphins of the eastern Pacific Ocean. J. Mamm. 50:262–71.

Bannikov, A. 1978. The present status of the Bactrian deer (Cervus elaphus bactrianus) in the USSR. In Threatened deer, Internatl. Union Conserv. Nat., Morges, Switzerland, pp. 159–72.

Bannikov, A. G., L. V. Zhirnov, L. S. Lebedeva, and A. A. Fandeev. 1967. Biology of the saiga. Israel Progr. Sci. Transl., Jerusalem, iv + 252 pp.

Barash, D. P. 1973a. Social variety in the yellow-bellied marmot (Marmota flaviventris). Anim. Behav. 21:570–84.

———. 1973b. The social biology of the Olympic marmot. Anim. Behav. Monogr. 6:173–44.

———. 1973c. Territorial and foraging behavior of pika (Ochotona princeps) in Montana. Amer. Midl. Nat. 89:202–7.

———. 1974a. The social behaviour of the hoary marmot (Marmota caligata). Anim. Behav. 22:256–61.

———. 1974b. Mother-infant relations in captive woodchucks (Marmota monax). Anim. Behav. 22:446–48.

———. 1976. Social behaviour and individual differences in free-living Alpine marmots (Marmota marmota). Anim. Behav. 24:27–35.

Barbehenn, K. R., J. P. Sumangil, and J. L. Libay. 1972–73. Rodents of the Philippine croplands. Philippine Agriculturist 56:217–42.

Barbour, D. B., and S. R. Humphrey. 1982. Status and habitat of the Key Largo woodrat and cotton mouse (Neotoma floridana smalli and Peromyscus gossypinus allapaticola). J. Mamm. 63:144–48.

Barbour, R. W., and W. H. Davis. 1969. Bats of America. University Press of Kentucky, Lexington, 286 pp.

———. 1974. Mammals of Kentucky. University Press of Kentucky, Lexington, xii + 322 pp.

Bareham, J. R., and A. Furreddu. 1975. Observations on the use of grottoes by Mediterranean monk seals. J. Zool. 175:291–98.

Barghoorn, S. F. 1977. New material of Vespertiliavus Schlosser (Mammalia, Chiroptera) and suggested relationships of emballonurid

bats based on cranial morphology. Amer. Mus. Novit., no. 2618, 29 pp.

Barkalow, F. S., Jr., and R. F. Soots, Jr. 1975. Life span and reproductive longevity of the gray squirrel, Sciurus c. carolinensis Gmelin. J. Mamm. 56:522–24.

Barker, I. K., I. Beveridge, A. J. Bradley, and A. K. Lee. 1978. Observations on spontaneous stress-related mortality among males of the dasyurid marsupial Antechinus stuartii Macleay. Austral. J. Zool. 26:435–47.

Barkley, L. J., and J. O. Whitaker, Jr. 1984. Confirmation of Caenolestes in Peru with information on diet. J. Mamm. 65:328–30.

Barlow, J. C. 1967. Edentates and pholidotes. In Anderson and Jones (1967), pp. 178–91.

———. 1969. Observations on the biology of rodents in Uruguay. Roy. Ontario Mus. Life Sci. Contrib., no. 75, 59 pp.

———. 1984. Xenarthrans and pholidotes. In Anderson and Jones (1984), pp. 219–39.

Barnes, L. G. 1985. Evolution, taxonomy and antitropical distributions of the porpoises (Phocoenidae, Mammalia). Mar. Mamm. Sci. 1:149–65.

Barnes, L. G., D. P. Domning, and C. E. Ray. 1985. Status of studies of fossil marine mammals. Mar. Mamm. Sci. 1:15–53.

Barnes, L. G., and S. A. McLeod. 1984. The fossil record and phyletic relationships of gray whales. In Jones, Swartz, and Leatherwood (1984), pp. 3–32.

Barnett, J. L., A. A. How, and W. F. Humphreys. 1977. Small mammal populations in pine and native forests in northeastern New South Wales. Austral. Wildl. Res. 4:233–40.

Barquez, R. M., and R. A. Ojeda. 1979. Nueva subespecie de Phylloderma (Chiroptera Phyllostomidae). Neotrópica 25:83–89.

Barquez, R. M., and C. C. Olrog. 1980. Tres nuevas especies de Vampyrops para Bolivia. Neotrópica 26:53–56.

Barrette, C. 1977. Some aspects of the behaviour of muntjacs in Wilpattu National Park. Mammalia 41:1–34.

———. 1986. Fighting behavior of wild Sus scrofa. J. Mamm. 67:177–79.

———. 1987a. The comparative behavior and ecology of chevrotains, musk deer, and morphologically conservative deer. In Wemmer (1987), pp. 200–213.

———. 1987b. Mating behaviour and mate choice by wild axis deer in Sri Lanka. J. Bombay Nat. Hist. Soc. 84:361–71.

Barrette, C., and D. Vandal. 1986. Social rank, dominance, antler size, and access to food in snow-bound wild woodland caribou. Behaviour 97:118–46.

Barritt, M. K. 1978. Two further specimens of the little pigmy possum [*Cercartetus lepidus* (Thomas)] from the Australian mainland. S. Austral. Nat. 53:12–13.

Barzdo, J. 1980. International trade in harp and hooded seals. Oryx 15:275–79.

Bateman, G. C., and T. A. Vaughan. 1974. Nightly activities of mormoopid bats. J. Mamm. 55:45–65.

Bates, P. J. 1985. Studies of gerbils of genus *Tatera*: the specific distinction of *Tatera robusta* (Cretzschmar, 1826), *Tatera nigricauda* (Peters, 1878) and *Tatera phillipsi* (de Winton, 1898). Mammalia 49:37–52.

———. 1988. Systematics and zoogeography of *Tatera* (Rodentia: Gerbillinae) of northeast Africa and Asia. Bonner Zool. Beitr. 39:265–303.

Bates, P. J. J., and D. L. Harrison. 1984. Significant new records of shrews (Soricidae) from the southern Arabian Peninsula, with remarks on the species occurring in the region. Mammalia 48:144–47.

Batten, P., and A. Batten. 1966. Notes on breeding the small-toothed palm civet *Arctogalidia trivirgata* at Santa Cruz Zoo. Internatl. Zoo Yearbook 6:172–73.

Battistini, R., and G. Richard-Vindard, eds. 1972. Biogeography and ecology in Madagascar. Dr. W. Junk B. V., The Hague, xv + 765 pp.

Baud. F. J. 1979. *Myotis aelleni*, nov. spec., chauve-souris nouvelle d'Argentine (Chiroptera: Vespertilionidae). Rev. Suisse Zool. 86:267–78.

———. 1982. Presence de *Rhinophylla alethina* (Mammalia, Chiroptera) en Equateur et repartition actuelle du genre en Amerique du Sud. Rev. Suisse Zool. 89:815–21.

Baugh, T. 1987. Man and manatee: planning for the future. Endangered Species Tech. Bull. 12(9):7.

Baumgardner, G. D., and D. J. Schmidly. 1981. Systematics of the southern races of two species of kangaroo rats (*Dipodomys compactus* and *D. ordii*). Occas. Pap. Mus. Texas Tech Univ., no. 73, 27 pp.

Baverstock, P. R. 1984. The molecular relationships of Australian possums and gliders. *In* Smith and Hume (1984), pp. 1–8.

Baverstock, P. R., M. Adams, and M. Archer. 1984. Electrophoretic resolution of species boundaries in the *Sminthopsis murina* complex (Dasyuridae). Austral. J. Zool. 32:823–32.

Baverstock, P. R., M. Adams, M. Archer, N. L. McKenzie, and R. How. 1983. An electrophoretic and chromosomal study of the dasyurid marsupial genus *Ningaui* Archer. Austral. J. Zool. 31:381–92.

Baverstock, P. R., M. Adams, T. Reardon, and C. H. S. Watts. 1987. Electrophoretic resolution of species boundaries in Australian Microchiroptera. III. The Nycticeiini—*Scotorepens* and *Scoteanax* (Chiroptera: Vespertilionidae). Austral. J. Biol. Sci. 40:417–33.

Baverstock, P. R., M. Archer, M. Adams, and B. J. Richardson. 1982. Genetic relationships among 32 species of Australian dasyurid marsupials. *In* Archer (1982a), pp. 641–50.

Baverstock, P. R., J. Birrell, and M. Krieg. 1987. Albumin immunologic relationships among Australian possums: a progress report. *In* Archer (1987), pp. 229–34.

Baverstock, P. R., C. H. S. Watts, M. Adams, and M. Gelder. 1980. Chromosomal and electrophoretic studies of Australian *Melomys* (Rodentia: Muridae). Austral. J. Zool. 28:553–74.

Baverstock. P. R., C. H. S. Watts, and S. R. Cole. 1977. Electrophoretic comparisons between allopatric populations of five Australian pseudomyine rodents (Muridae). Austral. J. Biol. Sci. 30:471–85.

Baverstock, P. R., C. H. S. Watts, J. T. Hogarth, A. C. Robinson, and J. F. Robinson. 1977. Chromosome evolution in Australian rodents. II. The *Rattus* group. Chromosoma 61:227–41.

Baxter, R. M., E. A. Goulden, and J. Meester. 1979. Activity patterns of *Myosorex varius* and *M. cafer* in captivity. S. Afr. J. Zool. 14:91–93.

Baxter, R. M., and C. N. V. Lloyd. 1980. Notes on the reproduction and postnatal development of the forest shrew. Acta Theriol. 25:31–38.

Baynes, A. 1982. Dasyurids (Marsupialia) in late Quaternary communities in southwestern Australia. *In* Archer (1982a), pp. 503–10.

———. 1984. Native mammal remains from Wilgie Mia aboriginal ochre mine: evidence on the pre-European fauna of the western arid zone. Rec. W. Austral. Mus. 11:297–310.

Baynes, A., A. Chapman, and A. J. Lynam. 1987. The rediscovery, after 56 years, of the heath rat *Pseudomys shortridgei* (Thomas, 1907) (Rodentia: Muridae) in Western Australia. Rec. W. Austral. Mus. 13:319–22.

Beamish, P. 1978. Evidence that a captive humpback whale (*Megaptera novaeangliae*) does not use sonar. Deep-Sea Res. 25:469–72.

Beatley, J. 1976. Environments of kangaroo rats (*Dipodomys*) and effects of environmental change on populations in southern Nevada. J. Mamm. 57:67–93.

Beck, A. J., and Lim Boo Liat. 1973. Reproductive biology of *Eonycteris spelaea*, Dobson (*Megachiroptera*) in West Malaysia. Acta Tropica 30:251–60.

Beck, A. M. 1973. The ecology of stray dogs. A study of free-ranging urban animals. York Press, Baltimore, xiv + 98 pp.

———. 1975. The ecology of "feral" and free-roving dogs in Baltimore. *In* Fox (1975), pp. 380–90.

Beck, B., T. G. Smith, and A. W. Mansfield. 1970. Occurrence of the harbour seal, *Phoca vitulina*, Linnaeus in the Thlewiaza River, N.W.T. Can. Field-Nat. 84:297–300.

Beck, B. B., D. Anderson, J. Ogden, B. Rettberg, C. Brejla, R. Scola, and M. Warneke. 1982. Breeding the Goeldi's monkey *Callimico goeldii* at Brookfield Zoo, Chicago. Internatl. Zoo Yearbook 22:106–14.

Beck, B. B., and C. Wemmer, eds. 1983. The biology and management of an extinct species Père David's deer. Noyes Publ., Park Ridge, New Jersey, xiv + 193 pp.

Beddington, J. R., R. J. H. Beverton, and D. M. Lavigne, eds. 1985. Marine mammals and fisheries. George Allen & Unwin Publ., London, 354 pp.

Bednarz, J. C., and J. A. Cook. 1984. Distribution and numbers of the white-sided jackrabbit (*Lepus callotis gaillardi*) in New Mexico. Southwestern Nat. 29:358–60.

Bee, J. W., and E. R. Hall. 1956. Mammals of northern Alaska on the Arctic Slope. Univ. Kansas Mus. Nat. Hist. Misc. Publ., no. 8, 309 pp.

Beebe, B. F. 1978. Two new Pleistocene mammal species from Beringia. Amer. Quaternary Assoc. Abstr., 5th Bien. Mtg., p. 159.

Beg, M. A., N. Adeeb, and S. A. Rana. 1981. Observations on reproduction in *Bandicota bengalensis* and *Nesokia indica*. Biologia 27:45–50.

Beg, M. A., and R. S. Hoffmann. 1977. Age determination and variation in the red-tailed chipmunk, *Eutamias ruficaudus*. Murrelet 58:26–36.

Begg, R. J. 1981. The small mammals of Little Nourlangie Rock, N.T. IV. Ecology of *Zyzomys woodwardi*, the large rock-rat, and *Z. argurus*, the common rock-rat (Rodentia: Muridae). Austral. Wildl. Res. 8:307–20.

Begg, R. J., and C. J. Dunlop. 1985. The diets of the large rock-rat, *Zyzomys woodwardi*, and the common rock-rat, *Z. argurus* (Rodentia: Muridae). Austral. Wildl. Res. 12:19–24.

Begg, R. J., B. Walsh, F. Woerle, and S. King. 1983. Ecology of *Melomys burtoni*, the grassland melomys (Rodentia: Muridae) at Cobourg Peninsula, N.T. Austral. Wildl. Res. 10:259–67.

Bekenov, A., and J. Mirzabekov. 1977. Reproduction of *Allactaga elater* in north Kyzylkum and Ustyurt. Zool. Zhur. 56:769–78.

Bekoff, M. 1975. Social behavior and ecology of the African Canidae: a review. *In* Fox (1975), pp. 120–42.

———. 1977. *Canis latrans.* Mammalian Species, no. 79, 9 pp.

———, ed. 1978. Coyotes. Biology, behavior, and management. Academic Press, New York, xx + 384 pp.

Bekoff, M., and M. C. Wells. 1980. The social ecology of coyotes. Sci. Amer. 242(4):130–48.

———. 1982. Behavioral ecology of coyotes: social organization, rearing patterns, space use, and resource defense. Z. Tierpsychol. 60:281–305.

Belden, R. C. 1986. Florida panther recovery plan implementation—a 1983 progress report. *In* Miller and Everett (1986), pp. 159–72.

Belden, R. C., and D. J. Forrester. 1980. A specimen of *Felis concolor coryi* from Florida. J. Mamm. 61:160–61.

Belitsky, D. W., and C. L. Belitsky. 1980. Distribution and abundance of manatees *Trichechus manatus* in the Dominican Republic. Biol. Conserv. 17:313–19.

Bell, D. J. 1986. A study of the hispid hare *Caprolagus hispidus* in Royal Suklaphanta Wildlife Reserve, western Nepal: a summary report. Dodo 23:24–31.

Bell, D. J., J. Hoth, A. Velazquez, F. J. Romero, L. Léon, and M. Aranda. 1985. A survey of the distribution of the volcano rabbit *Romerolagus diazi*: an endangered Mexican endemic. Dodo 22:42–48.

Bell, G. P. 1987. Evidence of a harem social system in *Hipposideros caffer* (Chiroptera: Hipposideridae) in Zimbabwe. J. Tropical Ecol. 3:87–90.

Bendel, P. R., and J. E. Gates. 1987. Home range and microhabitat partitioning of the southern flying squirrel *(Glaucomys volans).* J. Mamm. 68:243–55.

Bender, M. 1988. Regional news. Endangered Species Tech. Bull. 13(6–7):2–8.

Bengtson, J. L., and S. M. Fitzgerald. 1985. Potential role of vocalizations in West Indian manatees. J. Mamm. 66:816–19.

Bengtson, J. L., and D. B. Siniff. 1981. Reproductive aspects of female crabeater seals *(Lobodon carcinophagus)* along the Antarctic Peninsula. Can. J. Zool. 59:92–102.

Benirschke, K. 1986. Primates. The road to self-sustaining populations. Springer-Verlag, New York, xxiii + 1044 pp.

Benirschke, K., and A. T. Kumamoto. 1980. The chromosomes of the Nilgiri tahr *Hemitragus hylocrius.* Internatl. Zoo Yearbook 20:274–75.

Benjaminsen, T., and I. Christensen. 1979. The natural history of the bottlenose whale *Hyperoodon ampullatus* (Forster). *In* Winn and Olla (1979), pp. 143–64.

Bennett, D. K. 1980. Stripes do not a zebra make, part 1: a cladistic analysis of *Equus.* Syst. Zool. 29:272–87.

Bennett, E. L. 1987. Big noses of Borneo. Anim. Kingdom 90(2):9–15.

Bennett, E. L., and A. C. Sebastian. 1988. Social organization and ecology of proboscis monkeys *(Nasalis larvatus)* in mixed coastal forest in Sarawak. Internatl. J. Primatol. 9:233–55.

Bennett, N. C., and J. U. M. Jarvis. 1988a. The social structure and reproductive biology of colonies of the mole-rat, *Cryptomys damarensis* (Rodentia, Bathyergidae). J. Mamm. 69:293–302.

———. 1988b. The reproductive biology of the Cape mole-rat, *Georychus capensis* (Rodentia, Bathyergidae). J. Zool. 214:95–106.

Bennett, S., L. J. Alexander, R. H. Crozier, and A. G. Mackinlay. 1988. Are megabats flying primates? Contrary evidence from a mitochondrial DNA sequence. Austral. J. Biol. Sci. 41:327–32.

Benson, D. L., and F. R. Gehlbach. 1979. Ecological and taxonomic notes on the rice rat *(Oryzomys couesi)* in Texas. J. Mamm. 60:225–28.

Ben-Yaacov, R., and Y. Yom-Tov. 1983. On the biology of the Egyptian mongoose, *Herpestes ichneumon*, in Israel. Z. Saugetierk. 48:34–45.

Bere, R. M. 1962. The wild mammals of Uganda and neighbouring regions of East Africa. Longmans, Green, & Co., London, xii + 148 pp.

Berg, W. E., and R. A. Chesness. 1978. Ecology of coyotes in northern Minnesota. *In* Bekoff (1978), pp. 229–47.

Berger, J. 1986. Wild horses of the Great Basin. Univ. Chicago Press, xxi + 326 pp.

Bergerud, A. T. 1967. The distribution and abundance of arctic hares in Newfoundland. Can. Field-Nat. 81:242–48.

———. 1974. Decline of caribou in North America following settlement. J. Wildl. Mgmt. 38:757–70.

———. 1978. Caribou. *In* Schmidt and Gilbert (1978), pp. 83–101.

Bergerud, A. T., R. D. Jakimchuk, and D. R. Carruthers. 1984. The buffalo of the north: caribou *(Rangifer tarandus)* and human developments. Arctic 37:7–22.

Bergmans, W. 1973. New data on the rare African fruit bat *Scotonycteris ophiodon* Pohle, 1943. Z. Saugetierk. 38:285–89.

———. 1975a. A new species of *Dobsonia* Palmer, 1898 (Mammalia, Megachiroptera) from Waigeo, with notes on other members of the genus. Beaufortia 23:1–13.

———. 1975b. On the differences between sympatric *Epomops franqueti* (Tomes, 1860) and *Epomops buettikoferi* (Matschie, 1899), with additional notes on the latter species (Mammalia, Megachiroptera). Beaufortia 23:141–52.

———. 1976. A revision of the African genus *Myonycteris* Matschie, 1899 (Mammalia, Megachiroptera). Beaufortia 24:189–216.

———. 1977. Notes on new material of *Rousettus madagascariensis* Grandidier, 1929 (Mammalia, Megachiroptera). Mammalia 41:67–74.

———. 1978a. On *Dobsonia* Palmer 1898 from the Lesser Sunda Islands (Mammalia: Megachiroptera). Senckenberg. Biol. 59:1–18.

———. 1978b. Rediscovery of *Epomophorus pousarguesi* Trouessart, 1904 in the Central African Empire (Mammalia: Megachiroptera). J. Nat. Hist. 12:681–87.

———. 1979a. Taxonomy and zoogeography of the fruit bats of the People's Republic of Congo, with notes on their reproductive biology (Mammalia: Megachiroptera). Bijdragen Tot de Dierkunde 48:161–86.

———. 1979b. Taxonomy and zoogeography of *Dobsonia* Palmer, 1898, from the Louisiade Archipelago, the D'Entrecasteaux Group, Trobriand Island and Woodlark Island (Mammalia: Megachiroptera). Beaufortia 29:199–214.

———. 1979c. First records of *Epomops dobsonii* (Bocage, 1889) from Tanzania and Rwanda, with a note on its size range (Mammalia: Megachiroptera). Z. Saugetierk. 44:240–41.

———. 1980. A new fruit bat of the genus *Myonycteris* Matschie, 1899, from eastern Kenya and Tanzania (Mammalia, Megachiroptera). Zool. Meded. 55:171–81.

———. 1982. Noteworthy extensions of known ranges of three African fruit bat species (Mammalia, Megachiroptera). Bull. Zool. Mus. Univ. Amsterdam 8:157–63.

———. 1988. Taxonomy and biogeography of African fruit bats (Mammalia, Megachiroptera). I. General introduction; material and methods; results: the genus *Epomophorus* Bennett, 1836. Beaufortia 38:75–146.

Bergmans, W., L. Bellier, and J. Vissault. 1974. A taxonomical report on a collection of Megachiroptera (Mammalia) from the Ivory Coast. Rev. Zool. Afr. 88:18–48.

Bergmans, W., and J. E. Hill. 1980. On a new species of *Rousettus* Gray, 1821, from Sumatra and Borneo (Mammalia: Megachiroptera). Bull. British Mus. (Nat. Hist.) Zool. 38:95–104.

Bergmans, W., and H. Jachmann. 1983. Bat records from Malawi (Mammalia, Chiroptera). Bull. Zool. Mus. Univ. Amsterdam 9:117–22.

Bergmans, W., and F. G. Rozendaal. 1982. Notes on *Rhinolophus* Lacépède, 1799 from Sulawesi, Indonesia, with the description of a new species (Mammalia, Microchiroptera). Bijd. Dierkunde 52:169–74.

———. 1988. Notes on collections of fruit bats from Sulawesi and some off-lying islands (Mammalia, Megachiroptera). Zool. Verh., no. 248, 74 pp.

Bergmans, W., and S. Sarbini. 1985. Fruit bats of the genus *Dobsonia* Palmer, 1898 from the islands of Biak, Owii, Numfoor and Yapen, Irian Jaya (Mammalia, Megachiroptera). Beaufortia 34:181–89.

Bergmans, W., and P. J. H. Van Bree. 1972. The taxonomy of the African bat *Megaloglossus woermanni* Pagenstecher, 1885 (Megachiroptera: Macroglossinae). Biol. Gabonica 3–4:291–99.

———. 1986. On a collection of bats and rats from the Kangean Islands, Indonesia (Mammalia: Chiroptera and Rodentia). Z. Saugetierk. 51:329–44.

Berkes, F. 1977. Turkish dolphin fisheries. Oryx 14:163–67.

Berland, B. 1966. The hood and its extrusible balloon in the hooded seal—*Cystophora cristata* ERXL. Norsk Polarinst. Arbok, 1965, pp. 95–102.

Berman, M., and I. Dunbar. 1983. The social behaviour of free-ranging suburban dogs. Appl. Anim. Ethol. 10:5–17.

Bernard, R. T. F. 1982. Female reproductive cycle of *Nycteris thebaica* (Microchiroptera) from Natal, South Africa. Z. Saugetierk. 47:12–18.

Bernard, R. T. F., and J. A. J. Meester. 1982. Female reproduction and the female reproductive cycle of *Hipposideros caffer caffer* (Sundevall, 1846) in Natal, South Africa. Ann. Transvaal Mus. 33:131–44.

Bernstein, I. S., and T. P. Gordon. 1980. Mixed taxa introductions, hybrids and macaque systematics. *In* Lindburg (1980), pp. 125–47.

Berry, P. S. M. 1973. The Luangwa Valley giraffe. Puku 7:71–92.

———. 1978. Range movements of giraffe in the Luangwa Valley, Zambia. E. Afr. Wildl. J. 16:77–83.

Berry, R. J. 1970. The natural history of the house mouse. Field Studies 3:219–62.

———, ed. 1981. Biology of the house mouse. Symp. Zool. Soc. London, no. 47, xxx + 715 pp.

Berta, A. 1982. *Cerdocyon thous*. Mammalian Species, no. 186, 4 pp.

———. 1984. The Pleistocene bush dog *Speothos pacivorus* (Canidae) from the Lagoa Santa Caves, Brazil. J. Mamm. 65:549–59.

———. 1986. *Atelocynus microtis*. Mammalian Species, no. 256, 3 pp.

———. 1987. Origin, diversity, and zoogeography of the South American Canidae. Fieldiana Zool., n.s., 39:455–71.

Berta, A., and T. A. Deméré. 1986. *Callorhinus gilmorei* n. sp. (Carnivora: Otariidae) from the San Diego Formation (Blancan) and its implications for otariid phylogeny. Trans. San Diego Soc. Nat. Hist. 21:111–26.

Berta, A., C. E. Ray, and A. R. Wyss. 1989. Skeleton of the oldest known pinniped, *Enaliarctos mealsi*. Science 244:60–62.

Bertram, B. C. R. 1975. Social factors influencing reproduction in wild lions. J. Zool. 177:463–82.

Bertram, G. C. L. 1940. The biology of the Weddell and crabeater seal, with a study of the comparative behaviour of the Pinnipedia. British Graham Land Exped., 1934–1937, Sci. Rept., British Mus. (Nat. Hist.), 1:1–139.

Berzin, A. A. 1972. The sperm whale. Israel Progr. Sci. Transl., Jerusalem, v + 394 pp.

———. 1978. Whale distribution in tropical eastern Pacific waters. Rept. Internatl. Whaling Comm. 28:173–77.

Best, P. B. 1968. The sperm whale *(Physeter catodon)* off the west coast of South Africa. 2. Reproduction in the female. S. Afr. Div. Sea Fish. Investig. Rept., no. 66, 32 pp.

———. 1970a. The sperm whale *(Physeter catodon)* off the west coast of South Africa. 5. Age, growth and mortality. S. Afr. Div. Sea Fish. Investig. Rept., no. 79, 27 pp.

———. 1970b. Exploitation and recovery of right whales *Eubalaena australis* off the Cape Province. S. Afr. Div. Sea Fish. Investig. Rept., no. 80, 20 pp.

———. 1974. The biology of the sperm whale as it relates to stock management. *In* Schevill (1974), pp. 257–93.

———. 1979. Social organization in sperm whales, *Physeter macrocephalus*. *In* Winn and Olla (1979), pp. 227–89.

———. 1985. External characters of southern minke whales and the existence of a diminutive form. Sci. Rept. Whales Res. Inst. 36:1–33.

———. 1987. Estimates of the landed catch of right (and other whalebone) whales in the American fishery, 1805–1909. Fishery Bull. 85:403–18.

———. 1988. The external appearance of Heaviside's dolphin, *Cephalorhynchus heavisidii* (Gray, 1828). *In* Brownell and Donovan (1988), pp. 279–301.

Best, P. B., P. A. S. Canham, and N. Macleod. 1984. Patterns of reproduction in sperm whales, *Physeter macrocephalus*. *In* Perrin, Brownell, and DeMaster (1984), pp. 51–79.

Best, P. B., and P. D. Shaughnessy. 1981. First record of the melon-headed whale *Peponocephala electra* from South Africa. Ann. S. Afr. Mus. 83:33–47.

Best, R. C. 1981. The tusk of the narwhal (*Monodon monoceros* L.): interpretation of its function (Mammalia: Cetacea). Can. J. Zool. 59:2386–93.

———. 1982. Seasonal breeding in the Amazonian manatee, *Trichechus inunguis* (Mammalia: Sirenia). Biotrópica 14:76–78.

———. 1983. Apparent dry-season fasting in Amazonian manatees (Mammalia: Sirenia). Biotrópica 15:61–64.

Best, R. C., and V. M. F. Da Silva. 1984. Preliminary analysis of reproductive parameters of the boutu, *Inia geoffrensis*, and the tucuxi, *Sotalia fluviatilis*, in the Amazon River system. *In* Perrin, Brownell, and DeMaster (1984), pp. 361–69.

Best, R. C., and H. D. Fisher. 1974. Seasonal breeding of the narwhal (*Monodon monoceros* L.). Can. J. Zool. 52:429–31.

Best, T. L. 1972. Mound development by a pioneer population of the banner-tailed kangaroo rat, *Dipodomys spectabilis baileyi* Goldman in eastern New Mexico. Amer. Midl. Nat. 87:201–6.

———. 1978. Variation in kangaroo rats (genus *Dipodomys*) of the *heermanni* group in Baja California, Mexico. J. Mamm. 59:160–75.

———. 1988. Morphologic variation in the spotted bat *Euderma maculatum*. Amer. Midl. Nat. 119:244–52.

Best, T. L., and E. B. Hart. 1976. Swimming ability of pocket gophers (Geomyidae). Texas J. Sci. 27:361–66.

Best, T. L., and J. A. Lackey. 1985. *Dipodomys gravipes*. Mammalian Species, no. 236, 4 pp.

Bester, M. N. 1980. Population increase in the Amsterdam Island fur seal *Arctocephalus tropicalis* at Gough Island. S. Afr. J. Zool. 15:229–34.

———. 1981. Seasonal changes in the population composition of the fur seal *Arctocephalus tropicalis* at Gough Island. S. Afr. J. Wildl. Res. 11:49–55.

———. 1982. Distribution, habitat selection and colony types of the Amsterdam Island fur seal *Arctocephalus tropicalis* at Gough Island. J. Zool. 196:217–31.

———. 1987. Subantarctic fur seal, *Arctocephalus tropicalis*, at Gough Island (Tristan da Cunha group). *In* Croxall and Gentry (1987), pp. 57–60.

Bester, M. N., and G. I. H. Kerley. 1983. Rearing of twin pups to weaning by subantarctic fur seal *Arctocephalus tropicalis*. S. Afr. J. Wildl. Res. 13:86–87.

Bester, M. N., and J.-P. Roux. 1986. Summer presence of leopard seals *Hydrurga leptonyx* at the Courbet Peninsula, Iles Kerguelen. S. Afr. J. Antarctic Res. 16:29–32.

Betts, B. J. 1976. Behaviour in a population of Columbian ground squirrels, *Spermophilus columbianus columbianus*. Anim. Behav. 24:652–80.

Bhat, H. R., M. A. Sreenivasan, and P. G. Jacob. 1980. Breeding cycle of *Eonycteris spelaea* (Dobson, 1871) (Chiroptera, Pteropodidae, Macroglossinae) in India. Mammalia 44:341–47.

Bhat, S. K., and D. N. Mathew. 1984. Observations on the breeding biology of the Western Ghats squirrel, *Funambulus tristriatus* Waterhouse. Mammalia 48:573–84.

Bianchi, N. O., O. A. Reig, O. J. Molina, and F. N. Dulout. 1971. Cytogenetics of the South American akodont rodents (Cricetidae). I. A progress report of Argentinian and Venezuelan forms. Evolution 25:724–36.

Bibikov, D. I. 1980. Wolves in the USSR. Nat. Hist. 89(6):58–63.

Bibikov, D. I., A. N. Filimonov, and A. N. Kudaktin. 1983. Territoriality and migration of the wolf in the USSR. Acta Zool. Fennica 174:267–68.

Bibikov, D. I., N. G. Ovsyannikov, and A. N. Filimonov. 1983. The status and management of the wolf population in the USSR. Acta Zool. Fennica 174:269–71.

Bickham, J. W. 1987. Chromosomal variation among seven species of lasiurine bats (Chiroptera: Vespertilionidae). J. Mamm. 68:837–42.

Bigalke, R. C. 1970. Observations on springbok populations. Zool. Afr. 5:59–70.

Bigg, M. A. 1969. The harbour seal in British Columbia. Fish. Res. Bd. Can. Bull., no. 172, vii + 33 pp.

———. 1973. Census of California sea lions on southern Vancouver Island, British Columbia. J. Mamm. 54:285–87.

———. 1981. Harbour seal—*Phoca vitulina* and *P. largha*. *In* Ridgway and Harrison (1981b), pp. 1–27.

———. 1988. Status of the Steller sea lion, *Eumetopias jubatus*, in Canada. Can. Field-Nat. 102:315–36.

Bigg, M. A., and A. A. Wolman. 1975. Live-capture killer whale *(Orcinus orca)* fishery, British Columbia and Washington, 1962–73. J. Fish. Res. Bd. Can. 32:1213–21.

Bigler, W. J. 1974. Seasonal movements and activity patterns of the collared peccary. J. Mamm. 55:851–55.

Bindra, O. S., and P. Sagar. 1968. Breeding habits of the field rat *Millardia meltada*. J. Bombay Nat. Hist. Soc. 65:477–81.

Birkenholz, D. E. 1963. A study of the life history and ecology of the round-tailed muskrat *(Neofiber alleni* True) in north-central Florida. Ecol. Monogr. 33:255–80.

———. 1972. *Neofiber alleni.* Mammalian Species, no. 15, 4 pp.

Birkenholz, D. E., and W. O. Wirtz, II. 1965. Laboratory observations on the vesper rat. J. Mamm. 46:181–89.

Birkenstock, P. J., and J. A. J. Nel. 1977. Laboratory and field observations on *Zelotomys woosnami* (Rodentia: Muridae). Zool. Afr. 12:429–43.

Birney, E. C. 1973. Systematics of three species of woodrats (genus *Neotoma*) in central North America. Univ. Kansas Mus. Nat. Hist. Misc. Publ., no. 58, 173 pp.

———. 1974. Twentieth century records of wolverine in Minnesota. Loon 46:78–81.

Birney, E. C., J. B. Bowles, R. M. Timm, and S. L. Williams. 1974. Mammalian distributional records in Yucatan and Quintana Roo, with comments on reproduction, structure, and status of peninsular populations. Occas. Pap. Bell Mus. Nat. Hist., no. 13, 25 pp.

Bishop, I. R. 1979. Notes on *Praomys (Hylomyscus)* in eastern Africa. Mammalia 43:519–30.

Bissonette, J. A. 1982. Ecology and social behavior of the collared peccary in Big Bend National Park, Texas. U.S. Natl. Park Serv. Sci. Monogr., no. 16, xi + 85 pp.

Bjorge, R. R., J. R. Gunson, and W. M. Samuel. 1981. Population characteristics and movements of striped skunks *(Mephitis mephitis)* in central Alberta. Can. Field-Nat. 95:149–55.

Blanchard, H. M. 1974. The marten makes a comeback. Maine Fish and Game 16(4):21.

Bland, K. P. 1973. Reproduction in the female African tree rat *(Grammomys surdaster)*. J. Zool. 171:167–75.

Blandford, P. R. S. 1987. Biology of the polecat *Mustela putorius:* a literature review. Mamm. Rev. 17:155–98.

Bleich, V. C. 1977. *Dipodomys stephensi.* Mammalian Species, no. 73, 3 pp.

———. 1979. *Microtus californicus scirpensis* not extinct. J. Mamm. 60:851–52.

Bleich, V. C., and O. A. Schwartz. 1975. Observations on the home range of the desert woodrat, *Neotoma lepida intermedia.* J. Mamm. 56:518–19.

Blomqvist, L. 1988. The development of the captive snow leopard population between 1984 and 1985. Proc. 5th Internatl. Snow Leopard Symp., pp. 181–89.

Blood, B. R., and D. A. McFarlane. 1988. Notes on some bats from northern Thailand,

with comments on the subgeneric status of *Myotis altarium*. Z. Saugetierk. 53:276–80.

Blouch, R. A. 1988. Ecology and conservation of the Javan warty pig *Sus verrucosus* Müller, 1840. Biol. Conserv. 43:295–307.

Blouch, R. A., and S. Atmosoedirdjo. 1987. Biology of the Bawean deer and prospects for its management. *In* Wemmer (1987), pp. 320–27.

Blower, J. 1985. Conservation priorities in Burma. Oryx 19:79–85.

Bloxam, Q. 1977. Breeding the spectacled bear *Tremarctos ornatus* at Jersey Zoo. Internatl. Zoo Yearbook 17:158–61.

Blus, L. J. 1971. Reproduction and survival of short-tailed shrews *(Blarina brevicauda)* in captivity. Lab. Anim. Sci. 21:884–91.

Bockstoce, J. 1980a. Battle of the bowheads. Nat. Hist. 89(5):52–61.

———. 1980b. A preliminary estimate of the reduction of the western arctic bowhead whale population by the pelagic whaling industry: 1848–1915. Mar. Fish. Rev. 42:20–27.

Bockstoce, J., and D. B. Botkin. 1983. The historical status and reduction of the western arctic bowhead whale *(Balaena mysticetus)* population by the pelagic whaling industry, 1848–1914. *In* Tillman and Donovan (1983), pp. 107–41.

Bodini, R., and R. Pérez-Hernàndez. 1987. Distribution of the species and subspecies of cebids in Venezuela. Fieldiana Zool., n.s., 39:231–44.

Boeadi, B., and W. Bergmans. 1987. First record of *Dobsonia minor* (Dobson, 1879) from Sulawesi, Indonesia (Mammalia, Megachiroptera). Bull. Zool. Mus. Univ. Amsterdam 11:69–75.

Boeadi, B., and J. E. Hill. 1986. A new subspecies of *Aethalops alecto* (Thomas, 1923) (Chiroptera: Pteropodidae) from Java. Mammalia 50:263–66.

Bogan, M. A. 1972. Observations on parturition and development in the hoary bat, *Lasiurus cinereus*. J. Mamm. 53:611–14.

———. 1978. A new species of *Myotis* from the Islas Tres Marias, Nayarit, Mexico, with comments on variation in *Myotis nigricans*. J. Mamm. 59:519–30.

Bogart, M. H., R. W. Cooper, and K. Benirschke. 1977. Reproductive studies of black and ruffed lemurs *Lemur macaco macaco* and *L. variegatus*. Internatl. Zoo Yearbook 17:177–82.

Boggess, E. K., F. R. Henderson, and J. R. Choate. 1980. A black-footed ferret from Kansas. J. Mamm. 61:571.

Boher B., S., J. Naveua S., and L. Escobar M. 1988. First record of *Dinomys branickii* for Venezuela. J. Mamm. 69:433.

Bohlin, R. G., and E. G. Zimmerman. 1982. Genic differentiation of two chromosome races of the Geomys bursarius complex. J. Mamm. 63:218–28.

Boinski, S. 1985. Status of the squirrel monkey Saimiri oerstedi in Costa Rica. Primate Conserv. 6:15–16.

Boitani, L., and G. Reggiani. 1984. Movements and activity patterns of hedgehogs (Erinaceus europaeus) in Mediterranean coastal habitats. Z. Saugetierk. 49:193–206.

Bollinger, A., and T. C. Backhouse. 1960. Blood studies on the echidna Tachyglossus aculeatus. Proc. Zool. Soc. London 135:91–97.

Bolshakov, V. N., and O. N. Shubnikova. 1988. Common vole, Microtus arvalis (Rodentia, Muridae), on Spitzbergen Archipelago. Zool. Zhur. 67:308–10.

Bolton, B. L., and P. K. Latz. 1978. The western hare-wallaby, Lagorchestes hirsutus (Gould) (Macropodidae), in the Tanami Desert. Austral. Wildl. Res. 5:285–93.

Bonaccorso, F. J., and J. H. Brown. 1972. House construction of the desert woodrat, Neotoma lepida lepida. J. Mamm. 53:283–88.

Boness, D. J., W. D. Bowen, and O. T. Oftedal. 1988. Evidence of polygyny from spatial patterns of hooded seals (Cystophora cristata). Can. J. Zool. 66:703–6.

Bonhomme, F., J. Catalan, J. Britton-Davidian, V. M. Chapman, K. Moriwaki, E. Nevo, and L. Thaler. 1984. Biochemical diversity and evolution in the genus Mus. Biochem. Genetics 22:275–303.

Bonhomme, F., J. Fernandez, F. Palacios, J. Catalan, and A. Machordon. 1986. Caractérisation biochimique du complexe d'espèces du genre Lepus en Espagne. Mammalia 50:495–506.

Bonner, W. N. 1968. The fur seal of South Georgia. British Antarctic Surv. Sci. Rept., no. 56, 81 pp.

———. 1972. The grey seal and common seal in European waters. Oceanogr. Mar. Biol. Ann. Rev. 10:461–507.

———. 1981a. Southern fur seals—Arctocephalus. In Ridgway and Harrison (1981a), pp. 161–208.

———. 1981b. Grey seal—Halichoerus grypus. In Ridgway and Harrison (1981b), pp. 111–14.

———. 1982a. Seals and man. A study of interactions. Washington Sea Grant Publ., Seattle, xii + 170 pp.

———.1982b. The status of the Antarctic fur seal Arctocephalus gazella. Mammals in the Seas, FAO Fish. Ser. No. 5, 4:423–30.

———. 1984a. Eared seals. In Macdonald (1984), pp. 253–59.

———. 1984b. Antarctic renaissance. In Macdonald (1984), pp. 260–61.

———. 1985. Impact of fur seals on the terrestrial environment at South Georgia. In Siegfried, W. R., P. R. Condy, and R. M. Laws, eds., Antarctic nutrient cycles and food webs, Springer-Verlag, Berlin, pp. 641–46.

Boone and Crockett Club, ed. 1975. The wild sheep in modern North America. Winchester Press, New York, xv + 302 pp.

Boonstra, R., and C. J. Krebs. 1978. Pitfall trapping of Microtus townsendii. J. Mamm. 59:136–48.

Booth, V. R. 1985. Some notes on Lichtenstein's hartebeest, Alcelaphus lichtensteini (Peters). S. Afr. J. Zool. 20:57–60.

Borsboom, A. C. 1975. Pseudomys gracilicaudatus (Gould): range extension and notes on a little-known Australian murid rodent. Austral. Wildl. Res. 2:81–84.

Boskoff, K. J. 1977. Aspects of reproduction in ruffed lemurs (Lemur variegattus). Folia Primatol. 28:241–50.

Bothma, J. D. P. 1966. Food of the silver fox Vulpes chama. Zool. Afr. 2:205–10.

———. 1971a. Control and ecology of the black-backed jackal Canis mesomelas in the Transvaal. Zool. Afr. 6:187–93.

———. 1971b. Food of Canis mesomelas in South Africa. Zool. Afr. 6:195–203.

Bothma, J. D. P., and J. A. J. Nel. 1980. Winter food and foraging behaviour of the aardwolf Protetes cristatus in the Namib-Naukluft Park. Madoqua 12:141–49.

Bouchard, R., and G. Moisan. 1974. Chasse contrôlée á l'original dans les parcs et réserves du Quebec (1962–1972). Naturaliste Canadien 101:689–704.

Bouchardet da Fonseca, G. A., and T. E. Lacher, Jr. 1984. Exudate-feeding by Callithrix jacchus penicillata in semideciduous woodland (cerradao) in central Brazil. Primates 25:441–56.

Boulenge, E. L., and E. R. Fuentes. 1978. Preliminary population dynamics of Octodon degus. Abstr. 2nd Congr. Theriol. Internatl., Brno, p. 180.

Bourliere, F., C. Hunkeler, and M. Bertrand. 1970. Ecology and behavior of Lowe's guenon (Cercopithecus campbelli lowei) in the Ivory Coast. In Napier and Napier (1970), pp. 297–350.

Bowen, D. W., and R. J. Brooks. 1978. Social organization of confined male collared lemmings (Dicrostonyx groenlandicus Traill). Anim. Behav. 26:1126–35.

Bowen, W. D. 1982. Home range and spatial organization of coyotes in Jasper National Park, Alberta. J. Wildl. Mgmt. 46:201–16.

Bowen, W. D., D. J. Boness, and O. T. Ofte-

dal. 1987. Mass transfer from mother to pup and subsequent mass loss by the weaned pup in the hooded seal, Cystophora cristata. Can. J. Zool. 65:1–8.

Bowen, W. D., R. A. Myers, and K. Hay. 1987. Abundance estimation of a dispersed, dynamic population: hooded seals (Cystophora cristata) in the northwest Atlantic. Can. J. Fish. Aquat. Sci. 44:282–95.

Bowen, W. D., O. T. Oftedal, and D. J. Boness. 1985. Birth to weaning in 4 days: remarkable growth in the hooded seal, Cystophora cristata. Can. J. Zool. 63:2841–46.

Bowen, W. W. 1968. Variation and evolution of Gulf Coast populations of beach mice Peromyscus polionotus. Bull. Florida State Mus. 12:1–91.

Bower, J. N. 1979. Père David's deer: the trek from extinction. Natl. Parks and Conserv. Mag. 53(4):20–23.

Bowles, D. 1986. Social behaviour and breeding of babirusa Babyrousa babyrussa at the Jersey Wildlife Preservation Trust. Dodo 23:86–94.

Bowles, J. B. 1972. Notes on reproduction in four species of bats from Yucatan, Mexico. Trans. Kansas Acad. Sci. 75:271–72.

Bown, T. M., and E. L. Simons. 1984. First record of marsupials (Metatheria: Polyprotodonta) from the Oligocene in Africa. Nature 308:447–49.

Boyd, R. J. 1978. American elk. In Schmidt and Gilbert (1978), pp. 11–29.

Bozinovic, F., and M. Rosenmann. 1988. Daily torpor in Calomys musculinus, a South American rodent. J. Mamm. 69:150–52.

Brack, V., Jr., and J. C. Carter. 1985. Use of an underground burrow by Lasionycteris. Bat Research News 26:28–29.

Bradbury, J. W. 1977. Lek mating behavior in the hammer-headed bat. Z. Tierpsychol. 45:225–55.

Bradbury, J. W., and L. H. Emmons. 1974. Social organization of some Trinidad bats. I. Emballonuridae. Z. Tierpsychol. 36:137–83.

Bradbury, J. W., and S. L. Vehrencamp. 1977. Social organization and foraging in emballonurid bats. I. Field studies. Behav. Ecol. Sociobiol. 1:337–81.

Bradley, A. J., C. M. Kemper, D. J. Kitchener, W. F. Humphreys, R. A. How, and L. H. Schmitt. 1988. Population ecology and physiology of the common rock rat, Zyzomys argurus (Rodentia: Muridae) in tropical northwestern Australia. J. Mamm. 69:749–64.

Bradley, R. D., and D. J. Schmidly. 1987. The glans penes and bacula in Latin American taxa of the Peromyscus boylii group. J. Mamm. 68:595–616.

Bradley, W. G., and R. A. Mauer. 1971. Reproduction and food habits of Merriam's kangaroo rat, *Dipodomys merriami*. J. Mamm. 52:497–507.

Brady, C. A. 1978. Reproduction, growth and parental care in crab-eating foxes *Cerdocyon thous*. Internatl. Zoo Yearbook 18:130–34.

———. 1979. Observations on the behavior and ecology of the crab-eating fox *(Cerdocyon thous)*. In Eisenberg (1979), pp. 161–71.

———. 1981. The vocal repertoires of the bush dog *(Speothos venaticus)*, crab-eating fox *(Cerdocyon thous)*, and maned wolf *(Chrysocyon brachyurus)*. Anim. Behav. 29:649–69.

Brady, C. A., and M. K. Ditton. 1979. Management and breeding of maned wolves *Chrysocyon brachyurus* at the National Zoological Park, Washington. Internatl. Zoo Yearbook 19:171–76.

Brady, J. R., and H. W. Campbell. 1983. Distribution of coyotes in Florida. Florida Field Nat. 11:40–41.

Brady, J. R., and D. S. Maehr. 1985. Distribution of black bears in Florida. Florida Field Nat. 13:1–7.

Braham, H. W. 1983. Northern records of Risso's dolphin, *Grampus griseus*, in the northeast Pacific. Can. Field-Nat. 97:89–90.

———. 1984a. The bowhead whale, *Balaena mysticetus*. Mar. Fish. Rev. 46(4):45–53.

———. 1984b. Review of reproduction in the white whale, *Delphinapterus leucas*, narwhal, *Monodon monoceros*, and Irrawaddy dolphin, *Orcaella brevirostris*, with comments on stock assessment. In Perrin, Brownell, and DeMaster (1984), pp. 81–89.

Braham, H. W., F. E. Durham, G. H. Jarrell, and S. Leatherwood. 1980. Ingutuk: a morphological variant of the bowhead whale, *Balaena mysticetus*. Mar. Fish. Rev. 42:70–73.

Braham, H. W., R. D. Everitt, and D. J. Rugh. 1980. Northern sea lion population decline in the eastern Aleutian Islands. J. Wildl. Mgmt. 44:25–33.

Braham, H. W., B. D. Krogman, and G. M. Carroll. 1984. Bowhead and white whale migration, distribution, and abundance in the Bering, Chukchi, and Beaufort Seas, 1975–1978. U.S. Natl. Mar. Fish. Serv., NOAA Tech. Rept. NMFS SSRF-778, iv + 39 pp.

Braham, H. W., and D. W. Rice. 1984. The right whale, *Balaena glacialis*. Mar. Fish. Rev. 46(4):38–44.

Braithwaite, R. W. 1980. The ecology of *Rattus lutreolus*. III. The rise and fall of a commensal population. Austral. Wildl. Res. 7:199–215.

Bramblett, C. A., L. D. Pejaver, and D. J. Drickman. 1975. Reproduction in captive vervet and Sykes' monkeys. J. Mamm. 56:940–46.

Branan, W. V., and R. L. Marchinton. 1987.

Reproductive ecology of white-tailed and red brocket deer in Suriname. In Wemmer (1987), pp. 344–51.

Branan, W. V., M. C. M. Werkhoven, and R. L. Marchinton. 1985. Food habits of brocket and white-tailed deer in Suriname. J. Wildl. Mgmt. 49:972–76.

Brand, C. J., and L. B. Keith. 1979. Lynx demography during a snowshoe hare decline in Alberta. J. Wildl. Mgmt. 43:827–49.

Brand, C. J., L. B. Keith, and C. A. Fischer. 1976. Lynx responses to changing snowshoe hare densities in central Alberta. J. Wildl. Mgmt. 40:416–28.

Brand, C. J., R. H. Vowles, and L. B. Keith. 1975. Snowshoe hare mortality monitored by telemetry. J. Wildl. Mgmt. 9:741–47.

Brand, D. J. 1983. A "king cheetah" born at the cheetah breeding and research centre of the National Zoological Gardens of South Africa, Pretoria. Zool. Garten 53:366–68.

Brand, L. R. 1974. Tree nests of California chipmunks *(Eutamias)*. Amer. Midl. Nat. 91:489–91.

———. 1976. The vocal repertoire of chipmunks (genus *Eutamias*) in California. Anim. Behav. 24:319–35.

Brander, R. B. 1973. Life-history notes on the porcupine in a hardwood-hemlock forest in upper Michigan. Michigan Academician 5:425–33.

Brandon-Jones, D. 1984. Colobus and leaf monkeys. In Macdonald (1984), pp. 398–409.

Brattstrom, B. H. 1973. Social and maintenance behavior of the echidna, *Tachyglossus aculeatus*. J. Mamm. 54:50–70.

Breiwick, J. M., L. L. Eberhardt, and H. W. Braham. 1984. Population dynamics of western arctic bowhead whales *(Balaena mysticetus)*. Can. J. Fish. Aquatic Sci. 41:484–96.

Breiwick, J. M., and E. D. Mitchell. 1983. Estimated initial size of the Bering Sea stock of bowhead whales *(Balaena mysticetus)* from logbook and other catch data. In Tillman and Donovan (1983), pp. 147–51.

Breiwick, J. M., E. D. Mitchell, and D. G. Chapman. 1980. Estimated initial population size of the Bering Sea stock of bowhead whale, *Balaena mysticetus*: an iterative method. Fishery Bull. 78:843–53.

Brennan, E. J. 1985. De Brazza's monkeys *(Cercopithecus neglectus)* in Kenya: census, distribution, and conservation. Amer. J. Primatol. 8:269–77.

Bresinski, W. 1982. Grouping tendencies in roe deer under agrocenosis conditions. Acta Theriol. 27:427–47.

Broad, S. 1987. International trade in skins of Latin American spotted cats. Traffic Bull. 9:56–63.

Broad, S., R. Luxmoore, and M. Jenkins. 1988. Significant trade in wildlife: a review of selected species in CITES appendix II. Volume I: Mammals. Internatl. Union Conserv. Nat., Gland, Switzerland, xix + 183 pp.

Broadbooks, H. E. 1965. Ecology and distribution of the pikas of Washington and Alaska. Amer. Midl. Nat. 73:299–335.

———. 1970a. Home ranges and territorial behavior of the yellow-pine chipmunk, *Eutamias amoenus*. J. Mamm. 51:310–26.

———. 1970b. Populations of the yellow-pine chipmunk, *Eutamias amoenus*. Amer. Midl. Nat. 83:472–88.

———. 1974. Tree nests of chipmunks with comments on associated behavior and ecology. J. Mamm. 54:630–39.

Brockelman, W. Y., and D. J. Chivers. 1984. Gibbon conservation: looking to the future. In Preuschoft et al. (1984), pp. 3–12.

Brockelman, W. Y., and S. P. Gittins. 1984. Natural hybridization in the *Hylobates lar* species group: implications for speciation in gibbons. In Preuschoft et al. (1984), pp. 498–532.

Brockie, R. 1974. Self-annointing by wild hedgehogs, *Erinaceus europaeus*, in New Zealand. Anim. Behav. 24:68–71.

Brodie, E. D., E. D. Brodie, Jr., and J. A. Johnson. 1982. Breeding the African hedgehog *Atelerix pruneri* in captivity. Internatl. Zoo Yearbook 22:195–97.

Brodie, E. D., Jr. 1977. Hedgehogs use toad venom in their own defense. Nature 268:627–28.

Brodie, P. F. 1971. A reconsideration of aspects of growth, reproduction, and behavior of the white whale *(Delphinapterus leucas)*, with reference to the Cumberland Sound, Baffin Island, population. J. Fish. Res. Bd. Can. 28:1309–18.

Brokx, P. A. 1984. South America. In Halls (1984), pp. 525–46.

Bronner, G. N., and J. A. J. Meester. 1988. *Otomys angoniensis*. Mammalian Species, no. 306, 6 pp.

Bronson, F. H. 1979. The reproductive ecology of the house mouse. Quart. Rev. Biol. 54:265–99.

Brooke, A. P. 1987. Tent construction and social organization in *Vampyressa nymphaea* (Chiroptera: Phyllostomidae) in Costa Rica. J. Tropical Ecol. 3:171–75.

Brooks, P. M. 1972. Post-natal development of the African bush rat. Zool. Afr. 7:85–102.

———. 1982. Aspects of the reproduction, growth and development of the four-striped field mouse, *Rhabdomys pumilio* (Sparrman, 1784). Mammalia 46:53–63.

Brooks, R. J., and E. M. Banks. 1971. Radiotracking study of lemming home range. Commun. Behav. Biol. 6:1–5.

———. 1973. Behavioural biology of the collared lemming [*Dicrostonyx groenlandicus* (Traill)]: an analysis of acoustic communication. Anim. Behav. Monogr. 6:1–83.

Brosset, A. 1962. La reproduction des chiroptères de l'ouest et du centre de l'Inde. Mammalia 26:176–213.

———. 1976. Social organization in the African bat, *Myotis boccagei*. Z. Tierpsychol. 42:50–56.

———. 1984. Chiroptères d'altitude du Mont Nimba (Guinee). Description d'une espèce nouvelle, *Hipposideros lamottei*. Mammalia 48:545–55.

Brosset, A., L. Barbe, J.-C. Beaucournu, C. Faugier, H. Salvayre, and Y. Tupinier. 1988. La raréfaction du rhinolophe euryale (*Rhinolophus euryale* Blasius) en France. Recherche d'une explication. Mammalia 52:101–22.

Brosset, A., G. Dubost, and H. Heim de Balsac. 1965. Mammifères inédits récoltés au Gabon. Biol. Gabonica 1:147–74.

Brosset, A., and H. Saint Girons. 1980. Cycles de reproduction des microchiropteres troglophiles du nord-est du Gabon. Mammalia 44:225–32.

Brothwell, D. 1981. The Pleistocene and Holocene archaeology of the house mouse and related species. *In* Berry (1981), pp. 1–13.

Brown, D. E. 1983. On the status of the jaguar in the southwest. Southwestern Nat. 28:459–60.

Brown, D. H., D. K. Caldwell, and M. C. Caldwell. 1966. Observations on the behavior of wild and captive false killer whales, with notes on associated behavior of other genera of captive delphinids. Los Angeles Co. Mus. Nat. Hist. Contrib. Sci., no. 95, 32 pp.

Brown, J. C. 1964. Observations on the elephant shrews (Macroscelididae) of equatorial Africa. Proc. Zool. Soc. London 143:103–19.

Brown, J. H., and R. C. Lasiewski. 1972. Metabolism of weasels: the cost of being long and thin. Ecology 53:939–43.

Brown, L. N. 1970. Population dynamics of the western jumping mouse *(Zapus princeps)* during a four-year study. J. Mamm. 51:651–58.

———. 1975. Ecological relationships and breeding of the nutria *(Myocastor coypus)* in the Tampa, Florida area. J. Mamm. 56:928–30.

———. 1977. Litter size and notes on reproduction in the giant water vole *(Arvicola richardsoni)*. Southwestern Nat. 22:281–82.

Brown, M. K. 1979. Two old beavers from the Adirondacks. New York Fish and Game J. 26:92.

Brown, P. 1976. Vocal communication in the pallid bat, *Antrozous pallidus*. Z. Tierpsychol. 41:34–54.

Brown, P., T. W. Brown, and A. D. Grinnell. 1983. Echolocation, development, and vocal communication in the lesser bulldog bat, *Noctilio albiventris*. Behav. Ecol. Sociobiol. 13:287–98.

Brown, R. 1973. Has the thylacine really vanished? Animals 15:416–19.

Brown, R. E. 1978. First Iranian record of *Dipus sagitta* (Pallas, 1773). Mammalia 42:257.

Brown, S. G. 1986. Twentieth-century records of right whales *(Eubalaena glacialis)* in the northeast Atlantic Ocean. *In* Brownell, Best, and Prescott (1986), pp. 121–27.

Brownell, R. L., Jr. 1975a. Progress report on the biology of the Franciscana dolphin, *Pontoporia blainvillei*, in Uruguayan waters. J. Fish. Res. Bd. Can. 32: 1073–78.

———. 1975b. *Phocoena dioptrica*. Mammalian Species, no. 66, 3 pp.

———. 1983. *Phocoena sinus*. Mammalian Species, no. 198, 3 pp.

———. 1984. Review of reproduction in platanistid dolphins. *In* Perrin, Brownell, and DeMaster (1984), pp. 149–58.

———. 1986. Distribution of the vaquita, *Phocoena sinus*, in Mexican waters. Mar. Mamm. Sci. 2:299–305.

Brownell, R. L., Jr., A. Aguayo L., and D. Torres N. 1976. A Shepherd's beaked whale, *Tasmacetus shepherdi*, from the eastern South Pacific. Sci. Rept. Whales Res. Inst. 28:127–28.

Brownell, R. L., Jr., P. B. Best, and J. H. Prescott, eds. 1986. Right whales: past and present status. Rept. Internatl. Whaling Comm. Spec. Issue, no. 10, vi + 289 pp.

Brownell, R. L., Jr., and C. Chun. 1977. Probable existence of the Korean stock of the gray whale *(Eschrichtius robustus)*. J. Mamm. 58:237–39.

Brownell, R. L., Jr., and G. P. Donovan. 1988. Biology of the genus *Cephalorhynchus*. Rept. Internatl. Whaling Comm. Spec. Issue, no. 9, Cambridge, 344 pp.

Brownell, R. L., Jr., L. T. Findley, O. Vidal, A. Robles, and S. Manzanilla N. 1987. External morphology and pigmentation of the vaquita, *Phocoena sinus* (Cetacea: Mammalia). Mar. Mamm. Sci. 3:22–30.

Brownell, R. L., Jr., and E. S. Herald. 1972. *Lipotes vexillifer*. Mammalian Species, no. 10, 4 pp.

Brownell, R. L., Jr., J. E. Heyning, and W. P. Perrin. 1989. A porpoise, *Australophocaena dioptrica*, previously identified as *Phocoena spinipinnis*, from Heard Island. Mar. Mamm. Sci. 5:193–95.

Brownell, R. L., Jr., and R. Praderi. 1976. Records of the delphinid genus *Stenella* in western South Atlantic waters. Sci. Rept. Whales Res. Inst. 28:129–35.

———. 1984. *Phocoena spinipinnis*. Mammalian Species, no. 217, 4 pp.

———. 1985. Distribution of Commerson's dolphin, *Cephalorhynchus commersonii*, and the rediscovery of the type of *Lagenorhynchus floweri*. Sci. Rept. Whales Res. Inst. 36:153–64.

Brownell, R. L., Jr., K. Ralls, and W. F. Perrin. 1989. The plight of the 'forgotten' whales. Oceanus 32(1):5–11.

Brownell, R. L., Jr., and G. B. Rathbun. 1988. California sea otter translocation: a status report. Endangered Species Tech. Bull. 13(4):1, 6.

Bruch, D., and J. A. Chapman. 1983. Social behavior of the eastern cottontail, *Sylvilagus floridanus* (Lagomorpha: Leporidae), in a woodland habitat, with descriptions of new behaviors. Proc. Pennsylvania Acad. Sci. 57:74–78.

Brumback, R. A. 1974. A third species of the owl monkey *(Aotus)*. J. Hered. 65:321–23.

Bruner, P. L., and H. D. Pratt. 1979. Notes on the status and natural history of Micronesian bats. Elepaio 40:1–4.

Bryant, L. D., and C. Maser. 1982. Classification and distribution. *In* Thomas and Toweill (1982), pp. 1–59.

Bryden, M. M., R. J. Harrison, and R. J. Lear. 1977. Some aspects of the biology of *Peponocephala electra* (Cetacea: Delphinidae). I. General and reproductive biology. Austral. J. Mar. Freshwater Res. 28:703–15.

Bublitz, J. 1987. Untersuchungen zur Systematik der Rezenten Caenolestidae Trouessart, 1898. Bonner Zool. Monogr., no. 23, 96 pp.

Bucher, J. E., and H. I. Fritz. 1977. Behavior and maintenance of the woolly opposum *(Caluromys)* in captivity. Lab. Anim. Sci. 27:1007–12.

Buchler, E. R. 1976. The use of echolocation by the wandering shrew *(Sorex vagrans)*. Anim. Behav. 24:858–73.

Buchmann, O. L. K., and E. R. Guiler. 1974. Locomotion in the potoroo. J. Mamm. 55:203–6.

———. 1977. Behavior and ecology of the Tasmanian devil, *Sarcophilus harrisii*. *In* Stonehouse and Gilmore (1977), pp. 155–68.

Buckner, C. H. 1969. Some aspects of the population ecology of the common shrew, *Sorex araneus*, near Oxford, England. J. Mamm. 50:326–32.

Buden, D. W. 1976. A review of the endemic

West Indian genus *Erophylla*. Proc. Biol. Soc. Washington 89:1–16.

———. 1977. First records of bats of the genus *Brachyphylla* from the Caicos Islands with notes on geographic variation. J. Mamm. 58:221–25.

———. 1985. Additional records of bats from the Bahama Islands. Caribbean J. Sci. 21:19–25.

———. 1986. Distribution of mammals of the Bahamas. Florida Field Nat. 14:53–84.

Budnitz, N., and K. Dainis. 1975. *Lemur catta*: ecology and behavior. *In* Tattersall and Sussman (1975), pp. 219–35.

Buechner, H. K. 1960. The bighorn sheep in the United States, its past, present, and future. Wildl. Monogr., no. 4, 174 pp.

———. 1974. Implications of social behavior in the management of the Uganda Kob. *In* Geist and Walther (1974), pp. 853–70.

Bueler, L. E. 1973. Wild dogs of the world. Stein and Day, New York, 274 pp.

Buettner-Janusch, J. 1966. Origins of man; physical anthropology. John Wiley & Sons, New York, 674 pp.

Buettner-Janusch, J., and I. Tattersall. 1985. An annotated catalogue of Malagasy primates (families Lemuridae, Indriidae, Daubentoniidae, Megaladapidae, Cheirogaleidae) in the collections of the American Museum of Natural History. Amer. Mus. Novit., no. 2834, 45 pp.

Bulir, L. 1972. Breeding binturongs *Arctictis binturong* at Liberec Zoo. Internatl. Zoo Yearbook 12:11–12.

Bunderson, W. T. 1977. Hunter's antelope. Oryx 14:174–75.

Burleigh, R., and J. Clutton-Brock. 1980. The survival of *Myotragus balearicus* Bate, 1909 into the neolithic on Mallorca. J. Archaeol. Sci. 7:385–88.

Burnett, J. A., C. T. Dauphiné, Jr., S. H. Mc-Crindle, and T. Mosquin. 1989. On the brink. Endangered species in Canada. Western Producer Prairie Books, Saskatoon, Saskatchewan, viii + 192 pp.

Burns, J. A. 1980. The brown lemming, *Lemmus sibiricus* (Rodentia, Arvicolidae), in the late Pleistocene of Alberta and its postglacial dispersal. Can. J. Zool. 58:1507–11.

Burns, J. C., J. R. Choate, and E. G. Zimmerman. 1985. Systematic relationships of pocket gophers (genus *Geomys*) on the central Great Plains. J. Mamm. 66:102–18.

Burns, J. J. 1981a. Ribbon seal—*Phoca fasciata*. *In* Ridgway and Harrison (1981b), pp. 89–109.

———. 1981b. Bearded seal—*Erignathus barbatus*. *In* Ridgway and Harrison (1981b), pp. 145–70.

Burns, J. J., and F. H. Fay. 1970. Comparative morphology of the skull of the ribbon seal, *Histriophoca fasciata*, with remarks on systematics of the Phocidae. J. Zool. 161:363–94.

Burt, W. H., and R. P. Grossenheider. 1976. A field guide to the mammals. Houghton Mifflin, Boston, xxv + 289 pp.

Burton, M. 1955. Bulldog bats. III. London News 226:28.

Busch, B. C. 1985. The war against the seals. McGill–Queen's Univ. Press, Kingston, Ontario, xviii + 374 pp.

Butler, P. M. 1980. The tupaiid dentition. *In* Luckett (1980a), pp. 171–204.

Büttiker, W. 1982. Mammals of Saudi Arabia. The lesser kudu *(Tragelaphus imberbis)* (Blyth 1869). Fauna Saudi Arabia 4:483–87.

Butynski, T. M. 1979. Reproductive ecology of the springhaas *Pedetes capensis* in Botswana. J. Zool. 189:221–32.

———. 1984. Nocturnal ecology of the springhare, *Pedetes capensis*, in Botswana. Afr. J. Ecol. 22:7–22.

Butynski, T. M., and R. Mattingly. 1979. Burrow structure and fossorial ecology of the springhare *Pedetes capensis* in Botswana. Afr. J. Ecol. 17:205–15.

C

Cabrera, A. 1957, 1961. Catálogo de los mamiferos de América del Sur. Rev. Mus. Argentino Cien. Nat. "Bernardo Rivadavia," 4:1–732.

Cadieux, C. L. 1987. Pronghorn antelope: Great Plains rebound. *In* Kallman (1987), pp. 132–43.

Caire, W., J. E. Vaughan, and V. E. Diersing. 1978. First record of *Sorex arizonae* (Insectivora: Soricidae) from Mexico. Southwestern Nat. 23:532–33.

Calaby, J. H. 1966. Mammals of the upper Richmond and Clarence rivers, New South Wales. CSIRO Div. Wildl. Res. Tech. Pap., no. 10, 55 pp.

———. 1971. The current status of Australian Macropodidae. Austral. Zool. 16:17–31.

Calaby, J. H., L. K. Corbett, G. B. Sharman, and P. G. Johnston. 1974. The chromosomes and systematic position of the marsupial mole, *Notoryctes typhlops*. Austral. J. Biol. Sci. 27:529–32.

Calaby, J. H., H. Dimpel, and I. M. Cowan. 1971. The mountain pygmy possum, *Burramys parvus* Broom (Marsupialia) in the Kosciusko National Park, New South Wales. CSIRO Div. Wildl. Res. Tech. Pap., no. 23, 11 pp.

Calaby, J. H., and J. M. Taylor. 1980. Reevaluation of the holotype of *Mus ruber* Jentink, 1880 (Rodentia: Muridae) from western New Guinea (Irian Jaya). Zool. Meded. 55:215–19.

———. 1983. Breeding in wild populations of the Australian rock rats, *Zyzomys argurus* and *Z. woodwardi*. J. Mamm. 64:610–16.

Calaby, J. H., and C. White. 1967. The Tasmanian devil *(Sarcophilus harrisi)* in northern Australia in Recent times. Austral. J. Sci. 29:473–75.

Caldecott, J. O. 1986. Mating patterns, societies and the ecogeography of macaques. Anim. Behav. 34:208–20.

Calder, W. A. 1969. Temperature relations and underwater endurance of the smallest homeothermic diver, the water shrew. Comp. Biochem. Physiol. 30:1075–82.

Caldwell, D. K., and M. C. Caldwell. 1970. Echolocation-type signals by two dolphins, genus *Sotalia*. Quart. J. Florida Acad. Sci. 33:124–31.

———. 1971a. Underwater pulsed sounds produced by captive spotted dolphins, *Stenella plagiodon*. Cetology, no. 1, 7 pp.

———. 1971b. The pygmy killer whale, *Feresa attenuata*, in the western Atlantic, with a summary of world records. J. Mamm. 52:206–9.

———. 1977. Cetaceans. *In* Sebeok, T. A., ed., How animals communicate, Indiana Univ. Press, pp. 794–807.

———. 1985. Manatees—*Trichechus manatus*, *Trichechus senegalensis*, and *Trichechus inunguis*. *In* Ridgway and Harrison (1985), pp. 33–66.

Caldwell, D. K., M. C. Caldwell, and D. W. Rice. 1966. Behavior of the sperm whale, *Physeter catodon* L. *In* Norris (1966), pp. 678–717.

Caldwell, D. K., M. C. Caldwell, and R. V. Walker. 1976. First records for Fraser's dolphin *(Lagenodelphis hosei)* in the Atlantic and the melon-headed whale *(Peponocephala electra)* in the western Atlantic. Cetology, no. 25, 4 pp.

Caldwell, M. C., and D. K. Caldwell. 1969. The ugly dolphin. Sea Frontiers 15(5):1–7.

———. 1979. The whistle of the Atlantic bottlenosed dolphin *(Tursiops truncatus)*—ontogeny. *In* Winn and Olla (1979), pp. 369–401.

Caldwell, M. C., D. K. Caldwell, and W. E. Evans. 1966. Sounds and behavior of captive Amazon freshwater dolphins, *Inia geoffrensis*. Los Angeles Co. Mus. Nat. Hist. Contrib. Sci., no. 108, 24 pp.

Caldwell, M. C., D. K. Caldwell, and J. F. Miller. 1973. Statistical evidence for individual whistles in the spotted dolphin, *Stenella plagiodon*. Cetology, no. 16, 21 pp.

Calhoun, J. B. 1963. The ecology and sociology of the Norway rat. U.S. Pub. Health Serv., Baltimore, viii + 288 pp.

California Department of Fish and Game. 1978. At the Crossroads. Sacramento, 103 pp.

Callahan, J. R. 1980. Taxonomic status of *Eutamias bulleri*. Southwestern Nat. 25:1–8.

Callahan, J. R., and R. Davis. 1982. Reproductive tract and evolutionary relationships of the Chinese rock squirrel, *Sciurotamias davidianus*. J. Mamm. 63:42–47.

Calvert, W., and I. Stirling. 1985. Winter distribution of ringed seals *(Phoca hispida)* in the Barrow Strait area, Northwest Territories, determined by underwater vocalizations. Can. J. Fish. Aquat. Sci. 42:1238–43.

Camenzind, F. J. 1978. Behavioral ecology of coyotes on the National Elk Refuge, Jackson, Wyoming. *In* Bekoff (1978), pp. 267–94.

Cameron, G. N., and D. G. Rainey. 1972. Habitat utilization by *Neotoma lepida* in the Mohave Desert. J. Mamm. 53:251–66.

Cameron, G. N., and S. R. Spencer. 1981. *Sigmodon hispidus*. Mammalian Species, no. 158, 9 pp.

Campagna, C. 1985. The breeding cycle of the southern sea lion, *Otaria byronia*. Mar. Mamm. Sci. 1:210–18.

Campagna, C., and B. J. Le Boeuf. 1988. Reproductive behaviour of southern sea lions. Behaviour 104:233–61.

Campbell, C. B. G. 1974. On the phyletic relationships of the tree shrews. Mamm. Rev. 4:125–43.

Campbell, R. R. 1987. Status of the hooded seal, *Cystophora cristata*, in Canada. Can. Field-Nat. 101:253–65.

Campbell, R. R. 1988. Status of the sea mink, *Mustela macrodon*, in Canada. Can. Field-Nat. 102:304–6.

Campbell, T. M., III, T. W. Clark, and C. R. Groves. 1982. First record of pygmy rabbits *(Brachylagus idahoensis)* in Wyoming. Great Basin Nat. 42:100.

Canet, R. S., and V. B. Alvarez. 1984. Reproducción y ecologia de la jutia conga *(Capromys pilorides* Say). Poeyana 280:1–20.

Canivenc, R., and M. Bonnin. 1979. Delayed implantation is under environmental control in the badger *(Meles meles* L.). Nature 278:849–50.

Canivenc, R., C. Mauget, M. Bonnin, and R. J. Aitken. 1981. Delayed implantation in the beech marten *(Martes foiana)*. J. Zool. 193:325–32.

Cant, J. G. H. 1977. A census of the agouti *(Dasyprocta punctata)* in seasonally dry forest at Tikal, Guatemala, with some comments on strip censusing. J. Mamm. 58:688–90.

Cao Kequing. 1978. On the time of extinction of the wild mi-deer in China. Acta Zool. Sinica 24:289–91.

———. 1985. On the reasons of extinction of the wild mi-deer in China. Zool. Res. 6:111–15.

Carbyn, L. N., ed. 1983. Wolves in Canada and Alaska: their status, biology, and management. Can. Wildl. Serv. Rept. Ser., no. 45, 135 pp.

———. 1987. Gray wolf and red wolf. *In* Novak, Baker, et al. (1987), pp. 358–77.

———. 1989. Swift foxes in Canada. Recovery 1(1):8–9.

Carey, J. 1987. The sea otter's uncertain future. Natl. Wildl. 25(1):16–20.

Carl, E. A. 1971. Population control in arctic ground squirrels. Ecology 52:395–413.

Carleton, M. D. 1973. A survey of gross stomach morphology in the New World Cricetinae (Rodentia, Muroidea), with comments on functional interpretations. Misc. Publ. Mus. Zool. Univ. Michigan, no. 146, 43 pp.

———. 1977. Interrelationships of populations of the *Peromyscus boylii* species group (Rodentia, Muridae) in western Mexico. Occas. Pap. Mus. Zool. Univ. Michigan, no. 675, 47 pp.

———. 1979. Taxonomic status and relationships of *Peromyscus boylii* from El Salvador. J. Mamm. 60:280–96.

———. 1980. Phylogenetic relationships in neotomine-peromyscine rodents (Muroidea) and a reappraisal of the dichotomy within New World Cricetinae. Misc. Publ. Mus. Zool. Univ. Michigan, no. 157, vii + 146 pp.

———. 1984. Introduction to rodents. *In* Anderson and Jones (1984), pp. 255–65.

Carleton, M. D., and G. G. Musser. 1984. Muroid rodents. *In* Anderson and Jones (1984), pp. 289–379.

Carleton, M. D., and P. Myers. 1979. Karyotypes of some harvest mice, genus *Reithrodontomys*. J. Mamm. 60:307–13.

Carleton, M. D., and C. B. Robbins. 1985. On the status and affinities of *Hybomys planifrons* (Miller, 1900) (Rodentia:Muridae). Proc. Biol. Soc. Washington 98:956–1003.

Carleton, M. D., D. E. Wilson, A. L. Gardner, and M. A. Bogan. 1982. Distribution and systematics of *Peromyscus* (Mammalia: Rodentia) of Nayarit, Mexico. Smithson. Contrib. Zool., no. 352, iii + 46 pp.

Carley, C. J. 1979. Status summary: the red wolf *(Canis rufus)*. U.S. Fish and Wildl. Serv., Albuquerque, Endangered Species Rept., no. 7, iv + 36 pp.

Carman, M. 1979. The gestation and rearing

periods of the mandrill *Mandrillus sphinx*. Internatl. Zoo Yearbook 19:159–60.

Carnio, J. 1989. Liberian mongoose specimen found. Amer. Assoc. Zool. Parks Aquar. Newsl. 30(12):22.

Carnio, J., and M. Taylor. 1988. In search of the Liberian mongoose *(Liberiictis kuhni)*. Unpubl. ms. distributed at Conference on Breeding Endangered Species in Captivity, Cincinatti, October 1988.

Caro, T. M. 1976. Observations on the ranging behaviour and daily activity of lone silverback mountain gorillas *(Gorilla gorilla beringei)*. Anim. Behav. 24:889–97.

———. 1986. The functions of stotting: a review of the hypotheses. Anim. Behav. 34:649–62.

Caro, T. M., and D. A. Collins. 1987. Male cheetah social organization and territoriality. Ethology 74:52–64.

Caroll, J. B. 1984. The conservation and wild status of the Rodrigues fruit bat *Pteropus rodricensis*. Myotis 21–22:148–54.

Carpenter, C. R. 1965. The howlers of Barro Colorado Island. *In* De Vore (1965), pp. 250–91.

Carr, A., III. 1986. Introduction. *In* The black-footed ferret, Great Basin Nat. Mem., 8:1–7.

Carr, T., N. Carr, and J. H. M. David. 1985. A record of the sub-Antarctic fur seal *Arctocephalus tropicalis* in Angola. S. Afr. J. Zool. 20:77.

Carroll, L. E., and H. H. Genoways. 1980. *Lagurus curtatus*. Mammalian Species, no. 124, 6 pp.

Carter, C. H., and H. H. Genoways. 1978. *Liomys salvini*. Mammalian Species, no. 84, 5 pp.

Carter, C. H., H. H. Genoways, R. S. Loregnard, and R. J. Baker. 1981. Observations on bats from Trinidad, with a checklist of species occurring on the island. Occas. Pap. Mus. Texas Tech Univ., no. 72, 27 pp.

Carter, D. C. 1984. Perissodactyls. *In* Anderson and Jones (1984), pp. 549–62.

Carter, D. C., and J. K. Jones, Jr. 1978. Bats from the Mexican state of Hidalgo. Occas. Pap. Mus. Texas Tech Univ., no. 54, 12 pp.

Carter, T. S. 1983. The burrows of giant armadillos, *Priodontes maximus* (Edentata: Dasypodidae). Saugetierk. Mitt. 31:47–53.

———. 1985. Armadillos of Brazil. Natl. Geogr. Soc. Res. Rept. 20:101–7.

Carter, T. S., and C. D. Encarnaçao. 1983. Characteristics and use of burrows by four species of armadillos in Brazil. J. Mamm. 64:103–8.

Cary, J. R., and L. B. Keith. 1979. Reproductive change in the 10-year cycle of snowshoe hares. Can. J. Zool. 57:375–90.

Casimir, M. J. 1975. Some data on the systematic position of the eastern gorilla population of the Mt. Kahuzi region (République du Zaire). Z. Morph. Anthrop. 66:188–201.

Casimir, M. J., and E. Butenandt. 1973. Migration and core area shifting in relation to some ecological factors in a mountain gorilla group (Gorilla gorilla beringei) in the Mt. Kahuzi region (République du Zaire). Z. Tierpsychol. 33:514–22.

Cassola, F. 1985. Management and conservation of the Sardinian moufflon (Ovis musimon Schreber). An outline. In Lovari (1985), pp. 197–203.

Castro, R., and P. Soini. 1977. Field studies on Saguinus mystax and other callitrichids in Amazonian Peru. In Kleiman (1977a), pp. 73–78.

Catzeflis, F. 1983. Relations génétiques entre trois espèces du genre Crocidura (Soricidae, Mammalia) en Europe. Mammalia 47:229–36.

Catzeflis, F., T. Maddalena, S. Hellwing, and P. Vogel. 1985. Unexpected findings on the taxonomic status of east Mediterranean Crocidura russula auct. (Mammalia, Insectivora). Z. Saugetierk. 50:185–201.

Caubere, B., P. Gaucher, and J. F. Julien. 1984. Un record mondial de longévité in natura pour un chiroptère insectivore? Rev. Ecol. 39:351–53.

Caughley, G. 1984. The grey kangaroo overlap zone. Austral Wildl. Res. 11:1–10.

Caughley, G., R. G. Sinclair, and G. R. Wilson. 1977. Numbers, distribution and harvesting rate of kangaroos on the inland plains of New South Wales. Austral. Wildl. Res. 4:99–108.

Caughley, J. 1986. Distribution and abundance of the mountain pygmy-possum, Burramys parvus Broom, in Kosciusko National Park. Austral. Wildl. Res. 13:507–16.

Cawthorn, M. W. 1988. Recent observations of Hector's dolphin, Cephalorhynchus hectori, in New Zealand. In Brownell and Donovan (1988), pp. 303–14.

Ceballos, G., and R. A. Medellín L. 1988. Diclidurus albus. Mammalian Species, no. 316, 4 pp.

Ceballos-G., C., and D. E. Wilson. 1985. Cynomys mexicanus. Mammalian Species, no. 248, 3 pp.

Cerqueira, R. 1985. The distribution of Didelphis in South America (Polyprotodontia, Didelphidae). J. Biogeogr. 12:135–45.

Chakrabarti, K. 1984. The Sundarbans tiger. J. Bombay Nat. Hist. Soc. 81:459–60.

Chakraborty, S. 1985. Studies on the genus Callosciurus Gray (Rodentia: Sciuridae). Rec. Zool. Surv. India Misc. Publ. Occas. Pap., no. 63, 93 pp.

Chakraborty, S., and V. C. Agrawal. 1977. A melanistic example of woolly flying squirrel, Eupetaurus cinereus Thomas (Rodentia: Sciuridae). J. Bombay Nat. Hist. Soc. 74:346–47.

Chaline, J., P. Mein, and F. Petter. 1977. Les grandes pignes d'une classification évolutive des Muroidea. Mammalia 41:245–52.

Chaline, J., and J.-D. Graf. 1988. Phylogeny of the Arvicolidae (Rodentia): biochemical and paleontological evidence. J. Mamm. 69:22–33.

Chalmers, N. R. 1968a. Group composition, ecology, and daily activities of free living mangabeys in Uganda. Folia Primatol. 8:247–62.

———. 1968b. The social behaviour of free living mangabeys in Uganda. Folia Primatol. 8:263–81.

———. 1973. Differences in behavior between some arboreal and terrestrial species of African monkeys. In Michael and Crook (1973), pp. 69–100.

Chan, K. L., S. S. Dhaliwal, and Yong Hoi-Sen. 1978. Protein variation and systematics in Malayan rats of the subgenus Lenothrix (Rodentia: Muridae, genus Rattus Fischer). Comp. Biochem. Physiol. 59B:345–51.

Chanin, P. R. F., and D. J. Jefferies. 1978. The decline of the otter Lutra lutra L. in Britain: an analysis of hunting records and discussion of causes. Biol. J. Linnean Soc. 10:305–28.

Channing, A. 1984. Ecology of the namtap Graphiurus ocularis (Rodentia: Gliridae) in the Cedarberg, South Africa. S. Afr. J. Zool. 19:144–49.

Channing, A., and D. T. Rowe-Rowe. 1977. Vocalizations of South African mustelines. Z. Tierpsychol. 44:283–93.

Chaplin, R. E., and G. Dangerfield. 1973. Breeding records of muntjac deer (Muntiacus reevesi) in captivity. J. Zool. 170:150–51.

Chapman, B. R. 1972. Food habits of Loring's kangaroo rat, Dipodomys elator. J. Mamm. 53:877–80.

Chapman, B. R., and R. L. Packard. 1974. An ecological study of Merriam's pocket mouse in southeastern Texas. Southwestern Nat. 19:281–91.

Chapman, D., and N. Chapman. 1975. Fallow deer. Their history, distribution and biology. Terence Dalton, Ltd., Lavenham, United Kingdom, 271 pp.

———. 1982. The taxonomic status of feral muntjac deer (Muntiacus sp.) in England. J. Nat. Hist. 16:381–87.

Chapman, D. G. 1974. Status of Antarctic rorqual stocks. In Schevill (1974), pp. 218–38.

Chapman, J. A. 1974. Sylvilagus bachmani. Mammalian Species, no. 34, 4 pp.

———. 1975a. Sylvilagus transitionalis. Mammalian Species, no. 55, 4 pp.

———. 1975b. Sylvilagus nuttalii. Mammalian Species, no. 56, 3 pp.

———. 1984. Latitude and gestation period in New World rabbits (Leporidae: Sylvilagus and Romerolagus). Amer. Nat. 124:442–45.

Chapman, J. A., K. R. Dixon, W. Lopez-Forment, and D. E. Wilson. 1983. The New World jackrabbits and hares (genus Lepus).— 1. Taxonomic history and population status. Acta Zool. Fennica 174:49–51.

Chapman, J. A., and G. A. Feldhamer. 1981. Sylvilagus aquaticus. Mammalian Species, no. 151, 4 pp.

Chapman, J. A., A. L. Harman, and D. E. Samuel. 1977. Reproductive and physiologial cycles in the cottontail complex in western Maryland and nearby West Virginia. Wildl. Monogr., no. 56, 73 pp.

Chapman, J. A., J. G. Hockman, and M. M. Ojeda C. 1980. Sylvilagus floridanus. Mammalian Species, no. 136, 8 pp.

Chapman, J. A., and E. Schneider. 1984. Lagomorphs. In Macdonald (1984), pp. 712–21.

Chapman, J. A., and G. R. Willner. 1978. Sylvilagus audubonii. Mammalian Species, no. 106, 4 pp.

———. 1981. Sylvilagus palustris. Mammalian Species, no. 153, 3 pp.

Chapman, N. G., and D. I. Chapman. 1980. The distribution of fallow deer: a worldwide review. Mamm. Rev. 10:61–138.

Charles-Dominique, P. 1974. Aggression and territoriality in nocturnal prosimians. In Holloway (1974), pp. 31–48.

———. 1977. Ecology and behaviour of nocturnal primates. Prosimians of Equatorial West Africa. Duckworth, London, x + 277 pp.

———. 1978. Écologie et vie sociale de Nandinia binotata (Carnivores, Viverrides): comparaison avec les prosimiens sympatriques du Gabon. Terre Vie 32:477–528.

Charles-Dominique, P., and S. K. Bearder. 1979. Field studies of lorisid behavior: methodological aspects. In Doyle and Martin (1979), pp. 567–629.

Charles-Dominique, P., H. M. Cooper, A. Hladik, C. M. Hladik, E. Pages, G. F. Pariente, A. Petter-Rousseaux, and A. Schilling, eds. 1980. Nocturnal Malagasy primates. Academic Press, New York, xii + 215 pp.

Charles-Dominique, P., and J.-J. Petter. 1980. Ecology and social life of Phaner furcifer. In Charles-Dominique et al. (1980), pp. 75–95.

Chase, A. 1986. The grizzly and the juggernaut. Outside 11(1):28–34, 55–65.

Chasen, F. N. 1940. A handlist of Malaysian mammals. Bull. Raffles Mus., Singapore, no. 15, xx + 209 pp.

Chaudhry, M. A., and M. A. Beg. 1977. Reproductive cycle and population structure of the northern palm squirrel, Funambulus pennanti. Pakistan J. Zool. 9:183–89.

Cheatheam, L. K. 1977. Density and distribution of the black-tailed prairie dog in Texas. Texas J. Sci. 29:33–40.

Cheeseman, C. L., and R. B. Mitson, eds. 1982. Telemetric studies of vertebrates. Academic Press, London, 368 pp.

Cheke, A. S., and J. F. Dahl. 1981. The status of bats on western Indian Ocean islands, with special reference to Pteropus. Mammalia 45:205–38.

Chelminski, R. 1978. Polish forest, a time machine. Smithsonian 9(2):66–71.

Chen Fuguan, Min Zhilan, Luo Shiyou, and Xie Wenzhi. 1983. An observation on the behaviour and some ecological habits of the golden monkey (Rhinopithecus roxellanae) in Qing Mountains. Acta Theriol. Sinica 3:141–46.

Chen Peixun, Liu Peilin, Liu Renjun, Lin Kejie, and G. Pilleri. 1979. Distribution, ecology, behaviour and conservation of the dolphins of the middle reaches of Changjiang (Yangtze) River (Wuhan-Yueyang). Investig. Cetacea 10:87–103.

Chen Pei-Xun, Liu Renjun, and R. J. Harrison. 1982. Reproduction and reproductive organs in Neophocaena asiaeorientalis from the Yangtse River. Aquatic Mamm. 9:9–16.

Chesemore, D. L. 1972. History and economic importance of the white fox, Alopex, fur trade in northern Alaska 1798–1963. Can. Field-Nat. 86:259–67.

———. 1975. Ecology of the Arctic fox (Alopex lagopus) in North America—a review. In Fox (1975), pp. 143–63.

Chiarelli, A. B. 1972. Taxonomic atlas of living primates. Academic Press, London, vii + 363 pp.

Chiarelli, A. B., and R. S. Corruccini, eds. 1982. Advanced views in primate biology. Springer-Verlag, Berlin, viii + 266 pp.

Chidumayo, E. N. 1977. The ecology of the single striped grass mouse, Lemniscomys griselda, in Zambia. Mammalia 41:411–18.

Child, G., and V. J. Wilson. 1964. Observations on ecology and behaviour of roan and sable in three tsetse control areas. Arnoldia 1(16):1–8.

Childes, S. L. 1988. The past history, present status and distribution of the hunting dog Lycaon pictus in Zimbabwe. Biol. Conserv. 44:301–16.

Chimimba, C. T., and D. J. Kitchener. 1987. Breeding in the Australian yellow-bellied sheath-tailed bat, Saccolaimus flaviventris (Peters, 1867) (Chiroptera: Emballonuridae). Rec. W. Austral. Mus. 13:241–48.

Chivers, D. J. 1972. The siamang and the gibbon in the Malay Peninsula. In Rumbaugh (1972a), pp. 103–35.

———. 1973. An introduction to the socio-ecology of Malayan forest primates. In Michael and Crook (1973), pp. 101–46.

———. 1974. The siamang in Malaya. Contrib. Primatol. 4:i–xiii + 1–335.

———. 1977. The lesser apes. In Rainier III and Bourne (1977), pp. 539–98.

———. 1984. Feeding and ranging in gibbons: a summary. In Preuschoft et al. (1984), pp. 267–81.

———. 1986. Southeast Asian primates. In Benirschke (1986), pp. 127–51.

Chivers, D. J., and S. P. Gittins. 1978. Diagnostic features of gibbon species. Internatl. Zoo Yearbook 18:157–64.

Chivers, D. J., and J. Herbert, eds. 1978. Recent advances in primatology. I. Behaviour. Academic Press, London, xxiv + 980 pp.

Chivers, D. J., and K. A. Joysey, eds. 1978. Recent advances in primatology. III. Evolution. Academic Press, London, xvii + 509 pp.

Chivers, D. J., and W. Lane-Petter, eds. 1978. Recent advances in primatology. II. Conservation. Academic Press, London, xiii + 312 pp.

Choate, J. R. 1970. Systematics and zoogeography of Middle American shrews of the genus Cryptotis. Univ. Kansas Publ. Mus. Nat. Hist. 19:195–317.

Choate, J. R., and E. D. Fleharty. 1973. Habitat preference and spatial relations of shrews in a mixed grassland in Kansas. Southwestern Nat. 18:110–12.

———. 1974. Cryptotis goodwini. Mammalian Species, no. 44, 4 pp.

Choate, J. R., and J. K. Jones, Jr. 1970. Additional notes on reproduction in the Mexican vole, Microtus mexicanus. Southwestern Nat. 14:356–58.

Choate, T. S. 1972. Behavioural studies on some Rhodesian rodents. Zool. Afr. 7:103–18.

Choe, J. C., and R. M. Timm. 1985. Roosting site selection by Artibeus watsoni (Chiroptera: Phyllostomidae) on Anthurium ravenii (Araceae) in Costa Rica. J. Tropical Ecol. 1:241–47.

Cholley, B. 1982. Une martre (Martes martes L.) en Corse. Mammalia 46:267.

Chorazyna, H., and G. U. Kurup. 1975. Observations on the ecology and behaviour of Anathana ellioti in the wild. Proc. 5th Internatl. Congr. Primatol., pp. 342–44.

Chorn, J., and R. S. Hoffmann. 1978. Ailuropoda melanoleuca. Mammalian Species, no. 110, 6 pp.

Christensen, P. 1975. The breeding burrow of the banded ant-eater or numbat (Myrmecobius fasciatus). W. Austral. Nat. 13:32–34.

Christensen, P., K. Maisey, and D. H. Perry. 1984. Radiotracking the numbat, Myrmecobius fasciatus, in the Perup Forest of Western Australia. Austral. Wildl. Res. 11:275–88.

Christian, D. P. 1977. Diurnal activity of the four-striped mouse, Rhabdomys pumilio. Zool. Afr. 12:238–39.

Churcher, C. S. 1986. Equid remains from Neolithic horizons at Dakhleh Oasis, Western Desert of Egypt. In Meadow, R. H., and H.-P. Uerpmann, eds., Equids in the ancient world, Dr. Ludwing Reichert Verlag, Wiesbaden, pp. 413–21.

Churchfield, S. 1985. The feeding ecology of the European water shrew. Mamm. Rev. 15:13–21.

Churchill, S. K., L. S. Hall, and P. M. Helman. 1984. Observations on long-eared bats (Vespertilionidae: Nyctophilus) from northern Australia. Austral. Mamm. 7:17–28.

Churchill, S. K., P. M. Helman, and L. S. Hall. 1987. Distribution, populations and status of the orange horseshoe bat, Rhinonicteris aurantius (Chiroptera: Hipposideridae). Austral. Mamm. 11:27–33.

Ciochon, R. L. 1988. Gigantopithecus: the king of all apes. Anim. Kingdom 91(2):32–37.

Ciochon, R. L., and A. B. Chiarelli, eds. 1980. Evolutionary biology of the New World monkeys and continental drift. Plenum Press, New York, xvi + 528 pp.

Clark, B. 1983a. African wild ass. Oryx 17:28–31.

———. 1983b. Israel restores Asiatic wild ass. Oryx 17:113.

Clark, C. W. 1982. The acoustic repertoire of the southern right whale, a quantitative analysis. Anim. Behav. 30:1060–71.

———. 1985. Economic aspects of marine mammal–fishery interactions. In Beddington, Beverton, and Lavigne (1985), pp. 34–38.

Clark, C. W., and J. H. Johnson. 1984. The sounds of the bowhead whale, Balaena mysticetus, during the spring migrations of 1979 and 1980. Can. J. Zool. 62:1436–41.

Clark, D. A. 1984. Recent land mammals. In Perry (1984), pp. 225–45.

Clark, D. B. 1980. Population ecology of an endemic neotropical island rodent: Oryzomys *bauri* of Santa Fe Island, Galapagos, Ecuador. J. Anim. Ecol. 49:185–98.

Clark, M. J., and W. E. Poole. 1967. The reproductive system and embryonic diapause in the female grey kangaroo, *Macropus giganteus*. Austral. J. Zool. 15:441–59.

Clark, T. W. 1975. *Arctocephalus galapagoensis*. Mammalian Species, no. 64, 2 pp.

———. 1976. The black-footed ferret. Oryx 13:275–80.

———. 1977. Ecology and ethology of the white-tailed prairie dog *(Cynomys leucurus)*. Milwaukee Pub. Mus. Publ. Biol. Geol., no. 3, vi + 97 pp.

———. 1978. Current status of the black-footed ferret in Wyoming. J. Wildl. Mgmt. 42:128–34.

———. 1987a. Black-footed ferret recovery: a progress report. Conserv. Biol. 1:8–11.

———. 1987b. Restoring balance between the endangered black-footed ferret *(Mustela nigripes)* and human use of the Great Plains and intermountain west. J. Washington Acad. Sci. 77:168–73.

Clark, T. W., E. Anderson, C. Douglas, and M. Strickland. 1987. *Martes americana*. Mammalian Species, no. 289, 8 pp.

Clark, T. W., J. Grensten, M. Gorges, R. Crete, and J. Gill. 1987. Analysis of black-footed ferret translocation sites in Montana. Prairie Nat. 19:43–56.

Clarke, M. R. 1979. The head of the sperm whale. Sci. Amer. 240(1):128–41.

Clawson, R. L. 1987. Indiana bats: down for the count. Endangered Species Tech. Bull. 12(9):9–11.

Clements, E. D., E. G. Neal, and D. W. Yalden. 1988. The national badger sett survey. Mamm. Rev. 18:1–9.

Cleveland, A. G. 1970. The current geographic distribution of the armadillo in the United States. Texas J. Sci. 22:87–92.

———. 1986. First record of *Baiomys taylori* north of the Red River. Southwestern Nat. 31:547.

Clough, G. C. 1972. Biology of the Bahaman hutia, *Geocapromys ingrahami*. J. Mamm. 53:807–23.

———. 1973. A most peaceable rodent. Nat. Hist. 82(6):66–74.

———. 1974. Additional notes on the biology of the Bahamian hutia, *Geocapromys ingrahami*. J. Mamm. 55:670–72.

———. 1985. A rather remarkable rodent. Anim. Kingdom 88(3):41–45.

Clough, G. C., and J. J. Albright. 1987. Occurrence of the northern bog lemming, *Syn-*

aptomys borealis, in the northeastern United States. Can. Field-Nat. 101:611–13.

Clutton-Brock, J., G. B. Corbet, and M. Hills. 1976. A review of the family Canidae, with a classification by numerical methods. Bull. British Mus. (Nat. Hist.), Zool. 29:117–99.

Clutton-Brock, T. H., ed. 1977. Primate ecology: studies of feeding and ranging behaviour in lemurs, monkeys and apes. Academic Press, London, xxii + 631 pp.

———. 1987. Deer news. Internatl. Union Conserv. Nat. Species Survival Comm. Deer Specialist Group Newsletter 6:4–17.

Clutton-Brock, T. H., F. E. Guinness, and S. D. Albon. 1982. Red deer. Behavior and ecology of two sexes. Univ. Chicago Press, xxii + 378 pp.

Coady, J. W. 1980. History of moose in northern Alaska and adjacent regions. Can. Field-Nat. 94:61–68.

Coates-Estrada, R., and A. Estrada. 1985. Occurrence of the white bat, *Diclidurus virgo* (Chiroptera: Emballonuridae), in the region of "Los Tuxtlas," Veracruz. Southwestern Nat. 30:322–23.

Cobb, S., and D. Western. 1989. The ivory trade and the future of the African elephant. Pachyderm 12:32–37.

Cockburn, A. 1978. The distribution of *Pseudomys shortridgei* (Muridae: Rodentia) and its relevance to that of other heathland *Pseudomys*. Austral. Wildl. Res. 5:213–19.

Cockerill, R. A. 1984. Deer. *In* Macdonald (1984), pp. 520–29.

Cockrum, E. L. 1969. Migration in the guano bat, *Tadarida brasiliensis*. Univ. Kansas Mus. Nat. Hist. Misc. Publ., no. 51, pp. 303–36.

———. 1973. Additional longevity records for American bats. J. Arizona Acad. Sci. 8:108–10.

———. 1977. Status of the hairy footed gerbil, *Gerbillus latastei* Thomas and Trouessart. Mammalia 41:75–80.

Cockrum, E. L., T. C. Vaughan, and P. J. Vaughan. 1976a. A review of North African short-tailed gerbils *(Dipodillus)* with description of a new taxon from Tunisia. Mammalia 40:313–26.

———. 1976b. *Gerbillus andersoni* de Winton, a species new to Tunisia. Mammalia 40:467–73.

———. 1977. Status of the pale sand rat, *Psammomys vexillaris* Thomas, 1925. Mammalia 41:321–26.

Cohen, J. A. 1977. A review of the biology of the dhole or Asiatic wild dog *(Cuon alpinus* Pallas). Anim. Reg. Studies 1:141–58.

———. 1978. *Cuon alpinus*. Mammalian Species, no. 100, 3 pp.

Cohen, J. A., M. W. Fox, A. J. T. Johnsingh, and B. D. Barnett. 1978. Food habits of the dhole in south India. J. Wildl. Mgmt. 42:933–36.

Cohn, J. P. 1986. Surprising cheetah genetics. Bioscience 36:358–62.

Coimbra-Filho, A. F. 1972. Mamíferos ameaçados de extinçao no Brasil. *In* Acad. Brasil. Cien., Espécies da fauna Brasileira ameaçadas de extinçao, Rio de Janeiro, pp. 13–98.

Coimbra-Filho, A. F., and R. A. Mittermeier. 1973a. Distribution and ecology of the genus *Leontopithecus* Lesson, 1840 in Brazil. Primates 14:47–66.

———. 1973b. New data on the taxonomy of the Brazilian marmosets of the genus *Callithrix* Erxleben, 1777. Folia Primatol. 20:241–64.

———. 1977a. Tree-gouging, exudate eating and the "short-tusked" condition in *Callithrix* and *Cebuella*. *In* Kleiman (1977a), pp. 105–15.

———. 1977b. Conservation of the Brazilian lion tamarins *(Leontopithecus rosalia)*. *In* Rainier III and Bourne (1977), pp. 59–94.

Cole, G. F. 1976. Management involving grizzly and black bears in Yellowstone National Park 1970–75. U.S. Natl. Park Serv. Nat. Res. Rept., no. 9, 26 pp.

Cole, L. R. 1972. A comparison of *Malacomys longipes* and *Malacomys edwardsi* (Rodentia: Muridae) from a single locality in Ghana. J. Mamm. 53:616–19.

———. 1975. Foods and foraging places of rats (Rodentia: Muridae) in the lowland evergreen forest of Ghana. J. Zool. 175:453–71.

Collett, S. F. 1981. Population characteristics of *Agouti paca* (Rodentia) in Colombia. Michigan State Univ. Mus. Publ., Biol. Ser., 5:485–602.

Collier, G. D., and J. J. Spillett. 1973. The Utah prairie dog—decline of a legend. Utah Sci. 34:83–87.

Collins, L. 1989. Black-footed ferrets born at the National Zoo's Conservation and Research Center. Amer. Assoc. Zool. Parks Aquar. Newsl. 30(12):23.

Collins, L. R. 1973. Monotremes and marsupials. A reference for zoological institutions. Smithson. Inst. Press, Washington, D.C., v + 323 pp.

Collins, L. R., and J. F. Eisenberg. 1972. Notes on the behaviour and breeding of pacaranas *Dinomys branickii* in captivity. Internatl. Zoo Yearbook 12:108–14.

Committee for Whaling Statistics. 1980. International Whaling Statistics, Vols. 85 and 86. Sandefjord, Norway, 67 pp.

———. 1984. International Whaling Statistics. Vols. 93 and 94, 61 pp.

Condit, R., and B. J. Le Boeuf. 1984. Feeding habits and feeding grounds of the northern elephant seal. J. Mamm. 65:281–90.

Conner, D. A. 1985a. The function of the pika short call in individual recognition. Z. Tierpsychol. 67:131–43.

———. 1985b. Analysis of the vocal repertoire of adult pikas: ecological and evolutionary perspective. Anim. Behav. 33:124–34.

Conniff, R. 1987. When the music in our parlors brought death to darkest Africa. Audubon 89(4):76–93.

Constable, I. D., J. I. Pollock, J. Ratsirarson, and H. Simons. 1985. Sightings of aye-ayes and red ruffed lemurs on Nosy Mangabe and the Masoala Peninsula. Primate Conserv. 5:59–62.

Contreras, J. R. 1968. Akodon molinae una nueva especie de raton de campo del sur de la Provincia de Buenos Aires. Zool. Platense 1:9–12.

———. 1984a. Notas para servir de base a una revision del genero Ctenomys (Mammalia: Rodentia). II. Ctenomys saltarius Thomas, 1912. Hist. Nat. (Corrientes, Argentina) 3:249–52.

———. 1984b. Nota sobre Bibimys chacoensis (Shamel, 1931) (Rodentia: Cricetidae: Scapteromyinae). Hist. Nat. (Corrientes, Argentina) 4:280.

Contreras, J. R., and L. M. Berry. 1982a. Ctenomys bonettoi, una nueva especie de tucu-tucu procedente de la Provincia del Chaco, Republica Argentina (Rodentia, Octodontidae). Diagnosis preliminar. Hist. Nat. (Corrientes, Argentina) 2:123–24.

———. 1982b. Ctenomys argentinus, una nueva especie de tucu-tucu procedente de la Provincia del Chaco, Republica Argentina (Rodentia, Octodontidae). Hist. Nat. (Corrientes, Argentina) 2:165–73.

———. 1983. Notas acerca de los roedores del genero Oligoryzomys de la Provincia del Chaco, Republica Argentina. Hist. Nat. (Corrientes, Argentina) 3:145–48.

Contreras, J. R., and A. N. Ch. de Contreras. 1984. Diagnosis preliminar de una nueva especie de "anguya-tutu" (genero Ctenomys) para la Provincia de Corrientes, Argentina (Mammalia: Rodentia). Hist. Nat. (Corrientes, Argentina) 4:131–32.

Contreras, J. R., and E. R. Justo. 1974. Aportes a la mastozoologia pampeana. I. Nuevas localidades para roedores Cricetidae. Neotrópica 20:91–96.

Contreras, J. R., V. G. Roig, and C. M. Suzarte. 1977. Ctenomys validus, una nueva especie de "tunduque" de la Provincia de Mendoza (Rodentia: Octodontidae). Physis, sec. C, 36:159–62.

Contreras, J. R., and M. I. Rosi. 1980a. Una nueva subespecie del raton colilargo para la Provincia de Mendoza: Oligoryzomys flavescens occidentalis. Hist. Nat. (Mendoza, Argentina) 1:157–60.

———. 1980b. Acerca de la presencia en la Provincia de Mendoza del raton de campo Akodon molinae Contreras, 1968. Hist. Nat. (Mendoza, Argentina) 1:181–84.

Contreras, L. C., J. C. Torres-Mura, and J. L. Yáñez. 1987. Biogeography of octodontid rodents: an eco-evolutionary hypothesis. Fieldiana Zool., n.s., 39:401–11.

Conway, M. C., and C. G. Schmitt. 1978. Record of the Arizona shrew (Sorex arizonae) from New Mexico. J. Mamm. 59:631.

Conway, W. 1986. Sultan of South American shores. Anim. Kingdom 89(6):18–26.

Cooper, C. F., and B. S. Stewart. 1983. Demography of northern elephant seals, 1911–1982. Science 219:969–71.

Cooper, D. W., and P. A. Woolley. 1983. Confirmation of a new species of small dasyurid marsupial by electrophoretic analysis of enzymes and proteins. Austral. J. Zool. 31:743–51.

Cooper, J. E., S. S. Robinson, and J. B. Funderburg, eds. 1977. Endangered and threatened plants and animals of North Carolina. North Carolina State Mus. Nat. Hist., Raleigh, xvi + 444 pp.

Copley, P. B. 1983. Studies on the yellow-footed rock-wallaby, Petrogale xanthopus Gray (Marsupialia: Macropodidae). I. Distribution in South Australia. Austral. Wildl. Res. 10:47–61.

Corbet, G. B. 1978. The mammals of the Palaearctic Region: a taxonomic review. British Mus. (Nat. Hist.), London, 314 pp.

———. 1979. The taxonomy of Procavia capensis in Ethiopia, with special reference to the aberrant tusks of P. c. capillosa Brauer (Mammalia, Hyracoidea). Bull. British Mus. (Nat. Hist.) Zool. 36:251–59.

———. 1983. A review of classification in the family Leporidae. Acta Zool. Fennica 174:11–15.

———. 1984. The mammals of the Palaearctic Region: a taxonomic review. Supplement. British Mus. (Nat. Hist.) Publ., no. 944, vi + 45 pp.

———. 1986. Relationships and origins of the European lagomorphs. Mamm. Rev. 16:105–10.

———. 1988a. The family Erinaceidae: a synthesis of its taxonomy, phylogeny, ecology and zoogeography. Mamm. Rev. 18:117–72.

———. 1988b. Mus musculus domesticus Schwarz and Schwarz, 1943 (Mammalia, Rodentia): proposed conservation. Bull. Zool. Nomen. 45:214–15.

Corbet, G. B., and J. Hanks. 1968. A revision of the elephant-shrews, family Macroscelididae. Bull. British Mus. (Nat. Hist.) Zool. 16:1–111.

Corbet, G. B., and J. E. Hill. 1986. A world list of mammalian species. British Mus. (Nat. Hist.), London, 254 pp.

Corbet, G. B., and H. N. Southern. 1977. The handbook of British mammals. Blackwell Scientific Publications, London, xxxii + 520 pp.

Corbett, L. K. 1975. Geographical distribution and habitat of the marsupial mole, Notoryctes typhlops. Austral. Mamm. 1:375–78.

———. 1988. Social dynamics of a captive dingo pack: population regulation by dominant female infanticide. Ethology 78:177–98.

Corbett, L. K., and A. Newsome. 1975. Dingo society and its maintenance: a preliminary analysis. In Fox (1975), pp. 369–79.

Corn, P. S., and R. B. Bury. 1988. Distribution of the voles Arborimus longicaudus and Phenacomys intermedius in the central Oregon Cascades. J. Mamm. 69:427–29.

Corti, M., M. S. Merani, and G. de Villafañe. 1987. Multivariate morphometrics of vesper mice (Calomys): preliminary assessment of species, population and strain divergence. Z. Saugetierk. 52:236–42.

Costa, W. R., K. A. Nagy, and V. H. Shoemaker. 1976. Observations of the behavior of jackrabbits (Lepus californicus) in the Mojave Desert. J. Mamm. 57:399–402.

Cothran, E. G., and E. G. Zimmerman. 1985. Electrophoretic analysis of the contact zone between Geomys breviceps and Geomys bursarius. J. Mamm. 66:489–97.

Cottrell, W. 1978. The fisher (Martes pennanti) in Maryland. J. Mamm. 59:886.

Coulter, M. W. 1974. Maine's "black cat." Maine Fish and Game 16(3):23–26.

Cousins, D. 1976. The breeding of gorillas, Gorilla gorilla, in zoological collections. Zool. Garten 46:215–36.

Cousteau, J. Y., and P. Diole. 1972. Killer whales have fearsome teeth and a strange gentleness to man. Smithsonian 3(3):66–73.

Cowan, D. P., and D. J. Bell. 1986. Leporid social behaviour and social organization. Mamm. Rev. 16:169–79.

Cowan, I. M. 1971. Summary of the symposium on the native cats of North America. In Jorgenson, S. E., and L. D. Mech, eds., Proceedings of a symposium on the native cats of North America, U.S. Bur. Sport Fish. and Wildl., pp. 1–8.

———. 1972. The status and conservation of bears (Ursidae) of the world—1970. In Herrero (1972), pp. 343–67.

Cowan, I. M., and C. W. Holloway. 1978. Geographical location and current conservation status of the threatened deer of the world. *In* Threatened deer, Internatl. Union Conserv. Nat., Morges, Switzerland, pp. 11–22.

Cox, P. M. 1983. Observations on the natural history of Samoan bats. Mammalia 47:519–23.

Cox, R. 1987. The Philippine spotted deer and the Visayan warty pig. Oryx 21:37–42.

Craig, S. A. 1985. Social organization, reproduction and feeding behaviour of a population of yellow-bellied gliders, *Petaurus australis* (Marsupialia: Petauridae). Austral. Wildl. Res. 12:1–18.

———. 1986. A record of twins in the yellow-bellied glider (*Petaurus australis* Shaw) (Marsupialia: Petauridae) with notes on the litter size and reproductive strategy of the species. Victorian Nat. 103:72–75.

Craighead, F. C., Jr. 1976. Grizzly bear ranges and movement as determined by radiotracking. *In* Pelton, Lentfer, and Folk (1976), pp. 97–109.

Craighead, F. C., Jr., and J. J. Craighead. 1972. Grizzly bear prehibernation and denning activities as determined by radiotracking. Wildl. Monogr., no. 32, 35 pp.

Craighead, J. J., F. C. Craighead, Jr., and J. Sumner. 1976. Reproductive cycles and rates in the grizzly bear, *Ursus arctos horribilis*, of the Yellowstone ecosystem. *In* Pelton, Lentfer, and Folk (1976), pp. 337–56.

Craighead, J. J., J. R. Varney, and F. C. Craighead, Jr. 1974. A population analysis of the Yellowstone grizzly bears. Montana Cooperative Wildl. Res. Unit, 20 pp.

Cranbrook, Earl of. 1984. New and interesting records of mammals from Sarawak. Sarawak Mus. J. 33:137–44.

Crandall, L. S. 1964. The management of wild animals in captivity. Univ. Chicago Press, xv + 761 pp.

Cranford, J. A. 1977. Home range and habitat utilization by *Neotoma fuscipes* as determined by radiotelemetry. J. Mamm. 58:165–72.

Crawford, R. J. M., and G. A. Robinson. 1984. History of the blue duiker *Cephalophus monticola* population in the Tsitsikamma Forests, Republic of South Africa. Koedoe 27:61–71.

Crawford-Cabral, J. 1981. A new classification of genets. Afr. Small Mamm. Newsl., no. 6, pp. 8–10.

Crawley, M. C. 1969. Movements and home ranges of *Clethrionomys glareolus* Schreber and *Apodemus sylvaticus* L. in north-east England. Oikos 20:310–19.

———. 1973. A live-trapping study of Australian brush-tailed possums, *Trichosurus vulpecula* (Kerr), in the Orongorongo Valley, Wellington, New Zealand. Austral. J. Zool. 21:75–90.

Crawley, M. C., and G. J. Wilson. 1976. The natural history and behaviour of the New Zealand fur seal *(Arctocephalus forsteri)*. Tuatara 22:1–29.

Creighton, G. K. 1985. Phylogenetic inference, biogeographic interpretations, and the patterns of speciation in *Marmosa* (Marsupialia: Didelphidae). Acta Zool. Fennica 170:121–24.

Crespo, J. A. 1975. Ecology of the pampas gray fox and the large fox (culpeo). *In* Fox (1975), pp. 179–91.

Crichton, E. G. 1969. Reproduction in the pseudomyine rodent *Mesembriomys gouldii* (Gray) (Muridae). Austral. J. Zool. 17:785–97.

———. 1974. Aspects of reproduction in the genus *Notomys* (Muridae). Austral. J. Zool. 22:439–47.

Crichton, E. G., and P. H. Krutzsch. 1987. Reproductive biology of the female little mastiff bat, *Mormopterus planiceps* (Chiroptera: Molossidae) in southeast Australia. Amer. J. Anat. 178:369–86.

Crocker-Bedford, D. C., and J. J. Spillett. 1977. Home ranges of Utah prairie dogs. J. Mamm. 58:672–73.

Crockett, C. M., and J. F. Eisenberg. 1987. Howlers: variations in group size and demography. *In* Smuts, B. B., D. L. Cheyney, R. M. Seyfarth, R. W. Wrangham, and T. T. Struhsaker, eds., Primate societies, Univ. Chicago Press, pp. 54–68.

Crockett, C. M., and R. Rudran. 1987a. Red howler monkey birth data I: seasonal variation. Amer. J. Primatol. 13:347–68.

———. 1987b. Red howler monkey birth data II: interannual, habitat, and sex comparisons. Amer. J. Primatol. 13:369–84.

Crockett, C. M., and W. L. Wilson. 1980. The ecological separation of *Macaca nemestrina* and *M. fascicularis* in Sumatra. *In* Lindburg (1980), pp. 148–81.

Croft, D. B. 1980. Behaviour of red kangaroos, *Macropus rufus* (Desmarest, 1822) in northwestern New South Wales, Australia. Austral. Mamm. 4:5–58.

Cronin, J. E., R. Cann, and V. M. Sarich. 1980. Molecular evolution and systematics of the genus *Macaca*. *In* Lindburg (1980), pp. 31–51.

Crowcroft, P. 1977. Breeding of wombats *(Lasiorhinus latifrons)* in captivity. Zool. Garten 47:313–22.

Crowe, D. M. 1975. Aspects of ageing, growth, and reproduction of bobcats from Wyoming. J. Mamm. 56: 177–98.

Croxall, J. P., and R. L. Gentry, eds. 1987. Status, biology, and ecology of fur seals. Proceedings of an international symposium and workshop, Cambridge, England, 23–27 April 1984. Natl. Oceanic Atmos. Admin. Tech. Rept., NMFS 51.

Croxall, J. P., and L. Hiby. 1983. Fecundity, survival and site fidelity in Weddell seals, *Leptonychotes weddelli*. J. Appl. Ecol. 20:19–32.

Cumming, D. H. M. 1975. A field study of the ecology and behaviour of warthog. Trustees Natl. Museums and Monuments Rhodesia Mus. Mem., no. 7, 179 pp.

———. 1981. The management of elephant and other large mammals in Zimbabwe. *In* Jewell and Holt (1981), pp. 91–118.

———. 1987. Zimbawe and the conservation of black rhino. Zimbabwe Sci. News 21:59–62.

Cumming, D. H. M., and R. F. Du Toit. 1989. The African Elephant and Rhino Group Nyeri meeting. Pachyderm 11:4–6.

Cummings, W. C. 1985. Right whales—*Eubalaena glacialis* and *Eubalaena australis*. *In* Ridgway and Harrison (1985), pp. 275–304.

Cunha, F. L. D. S., and J. F. Cruz. 1979. Novo gênero de Cricetidae (Rodentia) de Castelo, Espírito Santo, Brasil. Bol. Mus. Biol. Prof. Mello Leitao, Santa Teresa, E. E. Santo, Brasil, ser. Zool., no. 96, 5 pp.

Curlewis, J. D., A. S. I. Loudon, and A. P. M. Coleman. 1988. Oestrous cycles and the breeding season of the Père David's deer hind *(Elaphurus davidianus)*. J. Reprod. Fert. 82:119–26.

Currier, M. J. P. 1983. *Felis concolor*. Mammalian Species, no. 200, 7 pp.

Cuttle, P. 1982. Life history of the dasyurid marsupial *Phascogale tapoatafa*. *In* Archer (1982a), pp. 13–22.

Czaplewski, N. J. 1983. *Idionycteris phyllotis*. Mammalian Species, no. 208, 4 pp.

Czekala, N. M., and K. Benirschke. 1974. Observations on a twin pregnancy in the African long-tongued fruit bat *(Megaloglossus woermanni)*. Bonner Zool. Beitr. 25:220–30.

D

Dagg, A. I. 1971. *Giraffa camelopardalis*. Mammalian Species, no. 5, 8 pp.

Dagg, A. I., and J. B. Foster. 1976. The giraffe. Its biology, behavior, and ecology. Van Nostrand Reinhold, New York, xiii + 210 pp.

Daggett, P. M., and D. R. Henning. 1974. The jaguar in North America. Amer. Antiquity 39:465–69.

Dahlheim, M. E., and F. Awbrey. 1982. A classification and comparison of vocalizations of captive killer whales *(Orcinus orca)*. J. Acoust. Soc. Amer. 72:661–70.

Dalby, P. L. 1975. Biology of pampa rodents.

Michigan State Univ. Mus. Publ., Biol. Ser., 5:149–272.

Dalby, P. L., and M. A. Mares. 1974. Notes on the distribution of the coney rat, *Reithrodon auritus*, in northwestern Argentina. Amer. Midl. Nat. 92:205–6.

Dalquest, W. W. 1953. Mammals of the Mexican state of San Luis Potosi. Louisiana State Univ. Studies, Biol. Ser., no. 1, 229 pp.

———. 1957. Observations on the sharpnosed bat, *Rhynchiscus naso* (Maximilian). Texas J. Sci. 9:219–26.

———. 1988. *Astrohippus* and the origin of Blancan and Pleistocene horses. Occas. Pap. Mus. Texas Tech Univ., no. 116, 23 pp.

Daly, M. 1979. Of Libyan jirds and fat sand rats. Nat. Hist. 88(2):64–71.

Daly, M., and S. Daly. 1973. On the feeding ecology of *Psammomys obesus* (Rodentia, Gerbillidae) in the Wadi Saoura, Algeria. Mammalia 37:545–61.

———. 1975a. Socio-ecology of Saharan gerbils, especially *Meriones libycus*. Mammalia 39:289–311.

———. 1975b. Behavior of *Psammomys obesus* (Rodentia: Gerbillinae) in the Algerian Sahara. Z. Tierpsychol. 37:289–321.

Daly, R. H. 1988. The early stages of reintroduction of the Arabian oryx in Oman. *In* Dixon and Jones (1988), pp. 14–17.

Danell, K. 1977. Dispersal and distribution of the muskrat *(Ondatra zibethica* (L.)) in Sweden. Viltrevy 10:1–26.

Daniel, M. J. 1975. First record of an Australian fruit bat (Megachiroptera: Pteropodidae) reaching New Zealand. New Zealand J. Zool. 2:227–31.

———. 1976. Feeding by the short-tailed bat *(Mystacina tuberculata)* on fruit and possibly nectar. New Zealand J. Zool. 3:391–98.

———. 1979. The New Zealand short-tailed bat, *Mystacina tuberculata*: a review of present knowledge. New Zealand J. Zool. 6:357–70.

Daniel, M. J., and G. R. Williams. 1983. Observations of a cave colony of the long-tailed bat *(Chalinolobus tuberculatus)* in North Island, New Zealand. Mammalia 47:71–80.

———. 1984. A survey of the distribution, seasonal activity and roost sites of New Zealand bats. New Zealand J. Ecol. 7:9–25.

Daniels, T. J. 1983a. The social organization of free-ranging urban dogs. I. Non-estrous social behavior. Appl. Anim. Ethol. 10:341–63.

———. 1983b. The social organization of free-ranging urban dogs. II. Estrous groups and the mating system. Appl. Anim. Ethol. 10:365–73.

Danilkin, A. A. 1986. Current geographic ranges of European (*Capreolus capreolus* L.) and Siberian (*C. pygargus* Pall) roe deer. Doklady Biol. Sci. 283:350–52.

Daniloff, R. 1986. Russian sables: "soft gold." Internatl. Wildl. 16(4):20–23.

Dapson, R. W. 1968. Reproduction and age structure in a population of short-tailed shrews, *Blarina brevicauda*. J. Mamm. 49:205–14.

Darby, W. R., and W. O. Pruitt, Jr. 1984. Habitat use, movements and grouping behaviour of woodland caribou, *Rangifer tarandus caribou*, in southeastern Manitoba. Can. Field-Nat. 98:184–90.

Dardaillon, M. 1988. Wild boar social groupings and their seasonal changes in the Camargue, southern France. Z. Saugetierk. 53:22–30.

Darling, J. D., and C. M. Jurasz. 1983. Migratory destinations of North Pacific humpback whales *(Megaptera novaeangliae)*. *In* Payne (1983), pp. 359–68.

Daschbach, N. J., M. W. Schein, and D. E. Haines. 1981. Vocalizations of the slow loris, *Nycticebus coucang* (Primates, Lorisidae). Internatl. J. Primatol. 2:71–80.

Dash, Y., A. Szaniawski, G. S. Child, and P. Hunkeler. 1978. Observations on some large mammals of the Transaltai, Djungarian and Shargin Gobi, Mongolia. Tigerpaper 5(2):5–10.

Da Silveira, E. K. P. 1968. Notes on the care and breeding of the maned wolf *Chrysocyon brachyurus* at Brasilia Zoo. Internatl. Zoo Yearbook 8:21–23.

Dathe, H. 1970. A second generation birth of captive sun bears *Helarctos malayanus* at East Berlin Zoo. Internatl. Zoo Yearbook 10:79.

David, J. H. M. 1973. The behaviour of the bontebok, *Damaliscus dorcas dorcas* (Pallas 1766), with special reference to territorial behaviour. Z. Tierpsychol. 33:38–107.

———. 1975. Observations on mating behaviour, parturition, suckling and the mother-young bond in the bontebok *(Damaliscus dorcas dorcas)*. J. Zool. 177:203–23.

———. 1978a. Observations on social organization of springbok, *Antidorcas marsupialis*, in the Bontebok National Park, Swellendam. Zool. Afr. 13:115–22.

———. 1978b. Observations on territorial behaviour of springbok, *Antidorcas marsupialis*, in the Bontebok National Park, Swellendam. Zool. Afr. 13:123–41.

David, J. H. M., and R. W. Rand. 1986. Attendance behavior of South African fur seals. *In* Gentry and Kooyman (1986), pp. 126–41.

Davidar, E. R. C. 1975. Ecology and behavior of the dhole or Indian wild dog *Cuon alpinus* (Pallas). *In* Fox (1975), 109–19.

———. 1978. Distribution and status of the Nilgiri tahr *(Hemitragus hylocrius)*—1975–1978. J. Bombay Nat. Hist. Soc. 75:815–44.

Davidge, C. 1978. Ecology of baboons *(Papio ursinus)* at Cape Point. Zool. Afr. 13:329–50.

Davis, B. L., and R. J. Baker. 1974. Morphometrics, evolution, and cytotaxonomy of mainland bats of the genus *Macrotus* (Chiroptera: Phyllostomatidae). Syst. Zool. 23:26–39.

Davis, D. D. 1962. Mammals of the lowland rain-forest of north Borneo. Bull. Singapore Natl. Mus., no. 31, 129 pp.

Davis, J. L., P. Valkenburg, and H. V. Reynolds. 1980. Population dynamics of Alaska's western Arctic caribou herd. *In* Reimers, E., E. Gaare, and S. Skjenneberg, eds., Proc. 2nd Internatl. Reindeer/caribou Symp., Direktoratet for vilt og ferskvannsfisk. Trondheim, Norway, pp. 595–604.

Davis, R. 1970. Carrying of young by flying female American bats. Amer. Midl. Nat. 83:186–96.

Davis, R., and C. Dunford. 1987. An example of contemporary colonization of montane islands by small, nonflying mammals in the American southwest. Amer. Nat. 129:398–406.

Davis, R. A., W. J. Richardson, S. R. Johnson, and W. E. Renaud. 1978. Status of Lancaster Sound narwhal population in 1976. Rept. Internatl. Whaling Comm. 28:209–15.

Davis, R. M. 1972. Behaviour of the vlei rat, *Otomys irroratus* (Brants, 1827). Zool. Afr. 7:119–40.

Davis, R. M., and J. Meester. 1981. Reproduction and postnatal development in the vlei rat, *Otomys irroratus*, on the Van Riebeeck Nature Reserve, Pretoria. Mammalia 45:99–116.

Davis, W. B. 1966a. Review of South American bats of the genus *Eptesicus*. Southwestern Nat. 11:245–74.

———. 1966b. The mammals of Texas. Texas Parks and Wildl. Dept. Bull., no. 41, 267 pp.

———. 1970a. *Tomopeas ravus* Miller (Chiroptera). J. Mamm. 51:244–47.

———. 1970b. A review of the small fruit bats (genus *Artibeus*) of Middle America. Part II. Southwestern Nat. 14:389–402.

———. 1975. Individual and sexual variation in *Vampyressa bidens*. J. Mamm. 56:262–65.

———. 1976a. Notes on the bats *Saccopteryx canescens* Thomas and *Micronycteris hirsuta* (Peters). J. Mamm. 57:604–7.

———. 1976b. Geographic variation in the lesser noctilio, *Noctilio albiventris* (Chiroptera). J. Mamm. 57:687–707.

———. 1980. New *Sturnira* (Chiroptera: Phyllostomidae) from Central and South

America, with key to currently recognized species. Occas. Pap. Mus. Texas Tech Univ., no. 70, 5 pp.

———. 1984. Review of the large fruit-eating bats of the *Artibeus "lituratus"* complex (Chiroptera: Phyllostomidae) in Middle America. Occas. Pap. Mus. Texas Tech Univ., no. 93, 16 pp.

Davis, W. B., and D. C. Carter. 1978. A review of the round-eared bats of the *Tonatia silvicola* complex, with descriptions of three new taxa. Occas. Pap. Mus. Texas Tech Univ., no. 53, 12 pp.

Davis, W. B., and J. R. Dixon. 1976. Activity of bats in a small village clearing near Iquitos, Peru. J. Mamm. 57:747–49.

Davis, W. B., and L. A. Follansbee. 1945. The Mexican volcano mouse, *Neotomodon*. J. Mamm. 26:401–11.

Davis, W. H., and J. W. Hardin. 1967. New records of mammals from Mesa Verde National Park, Colorado. J. Mamm. 48:322–23.

Davison, G. W. H. 1980. Territorial fighting by lesser mouse-deer. Malayan Nat. J. 34:1–6.

———. 1984. New records of peninsular Malaysian and Thai shrews. Malayan Nat. J. 36:211–15.

Dawbin, W. H. 1966. The seasonal migratory cycle of humpback whales. *In* Norris (1966), pp. 145–70.

Dawes, P. R., M. Elander, and M. Ericson. 1986. The wolf *(Canis lupus)* in Greenland: a historical review and present status. Arctic 39:119–32.

Dawson, G. A. 1977. Composition and stability of social groups of the tamarin, *Saguinus oedipus geoffroyi*, in Panama: ecological and behavioral considerations. *In* Kleiman (1977a), pp. 23–37.

Dawson, L. 1982a. Taxonomic status of fossil devils *(Sarcophilus*, Dasyuridae, Marsupialia) from late Quaternary eastern Australian localities. *In* Archer (1982a), pp. 517–25.

———. 1982b. Taxonomic status of fossil thylacines *(Thylacinus*, Thylacinidae, Marsupialia) from late Quaternary deposits in eastern Australia. *In* Archer (1982a), pp. 527–36.

Dawson, L., and T. Flannery. 1985. Taxonomic and phylogenetic status of living and fossil kangaroos and wallabies of the genus *Macropus* Shaw (Macropodidae: Marsupialia), with a new subgeneric name for the larger wallabies. Austral. J. Zool. 33:473–98.

Dawson, M. R., and L. Krishtalka. 1984. Fossil history of the families of Recent mammals. *In* Anderson and Jones (1984), pp. 11–58.

Dawson, S. M., and E. Slooten. 1988. Hector's dolphin. *Cephalorhynchus hectori*: distribution and abundance. *In* Brownell and Donovan (1988), pp. 315–24.

Dawson, T. J., D. Fanning, and T. J. Bergin. 1978. Metabolism and temperature regulation in the New Guinea monotreme *Zaglossus bruijni*. Austral. Zool. 20:99–103.

Deag, J. M. 1977. The status of the Barbary macaque *Macaca sylvanus* in captivity and factors influencing its distribution in the wild. *In* Rainier III and Bourne (1977), pp. 267–87.

Dean, F. C. 1976. Aspects of grizzly bear population ecology in Mount McKinley National Park. *In* Pelton, Lentfer, and Folk (1976), pp. 111–19.

Dean, W. R. J. 1978. Conservation of the white-tailed rat in South Africa. Biol. Conserv. 13:133–40.

Dearden, P. 1986. Status of the Vancouver Island marmot *(Marmota vancouverensis):* an update. Environ. Conserv. 13:168.

Dearden, P., and C. Hall. 1983. Non-consumptive recreation pressures and the case of the Vancouver Island marmot *(Marmota vancouverensis)*. Environ. Conserv. 10:63–66.

DeBlase, A. F. 1980. The bats of Iran: systematics, distribution, ecology. Fieldiana Zool., n.s., no. 4, xvii + 424 pp.

Deems, E. F., Jr., and D. Pursley, eds. 1978. North American furbearers. Internatl. Assoc. Fish and Wildl. Agencies, Univ. Maryland Press, College Park, x + 165 pp.

Deerson, D., R. Dunn, D. Spittall, and P. Williams. 1975. Mammals of the upper Lerderberg Valley. Victorian Nat. 92:28–43.

Defler, T. R. 1979a. On the ecology and behavior of *Cebus albifrons* in eastern Colombia: I. Ecology. Primates 20:475–90.

———. 1979b. On the ecology and behavior of *Cebus albifrons* in eastern Colombia: II. Behavior. Primates 20:491–502.

———. 1982. A comparison of intergroup behavior in *Cebus albifrons* and *C. apella*. Primates 23:385–92.

———. 1983. Associations of the giant river otter *(Pteronura brasiliensis)* with freshwater dolphins *(Inia geoffrensis)*. J. Mamm. 64:692.

DeFrees, S. L., and D. E. Wilson. 1988. *Eidolon helvum*. Mammalian Species, no. 312, 5 pp.

De Graaff, G. 1972. On the mole-rat *(Cryptomys hottentotus damarensis)* (Rodentia) in the Kalahari Gemsbok National Park. Koedoe 15:25–35.

De Graaff, G., and B. L. Penzhorn. 1976. The re-introduction of springbok *Antidorcas marsupialis* into South African national parks—a documentation. Koedoe 19:75–82.

De Jong, N., and W. Bergmans. 1981. A revision of the fruit bats of the genus *Dobsonia* Palmer, 1898 from Sulawesi and some nearby islands (Mammalia, Megachiroptera, Pteropodinae). Zool. Abhandl. (Dresden) 37:209–24.

Delany, M. J. 1969. The ecological distribution of small mammals on Bugala Island, Lake Victoria. Zool. Afr. 4:129–33.

———. 1971. The biology of small rodents in Mayanja Forest, Uganda. J. Zool. 165:85–129.

———. 1975. The rodents of Uganda. British Mus. (Nat. Hist.), London, vii + 165 pp.

———. 1986. Ecology of small rodents in Africa. Mamm. Rev. 16:1–41.

Delroy, L. B., J. Earl, I. Radbone, A. C. Robinson, and M. Hewett. 1986. The breeding and re-establishment of the brush-tailed bettong, *Bettongia penicillata*, in South Australia. Austral. Wildl. Res. 13:387–96.

Delson, E. 1980. Fossil macaques, phyletic relationships and a scenario of deployment. *In* Lindburg (1980), pp. 10–30.

DeMaster, D. P., and I. Stirling. 1981. *Ursus maritimus*. Mammalian Species, no. 145, 7 pp.

Demeter, A., and G. Topál. 1982. Ethiopian mammals in the Hungarian Natural History Museum. Ann. Hist.-Nat. Mus. Natl. Hung., Zool., 74:331–49.

De Muizon, C., and D. P. Domning. 1985. The first records of fossil sirenians in the southeastern Pacific Ocean. Bull. Mus. Natl. Hist. Paris, ser. 4 (Paleontol.-Geol.), 7:189–213.

Dene, H., M. Goodman, and W. Prychodko. 1978. An immunological examination of the systematics of Tupaioidea. J. Mamm. 59:697–706.

———. 1980. Immunodiffusion systematics of the Primates. IV: Lemuriformes. Mammalia 44:211–23.

Denham, W. W. 1982. History of green monkeys in the West Indies. Part II. Population dynamics of Barbadian monkeys. J. Barbados Mus. and Hist. Soc. 36:353–71.

Dennis, E., and J. I. Menzies. 1978. Systematics and chromosomes of New Guinea *Rattus*. Austral. J. Zool. 26:197–206.

———. 1979. A chromosomal and morphometric study of Papuan tree rats *Pogonomys* and *Chiruromys* (Rodentia, Muridae). J. Zool. 189:315–32.

Densmore, M. A. 1980. Reproduction of sitatunga *Tragelaphus spekei* in captivity. Internatl. Zoo Yearbook 20:227–29.

Densmore, M. A., and D. C. Kraemer. 1986. Analysis of reproductive data on the addax *Addax nasomaculatus*. Internatl. Zoo Yearbook 24/25:303–6.

De Santis, L. J. M., and E. R. Justo. 1980. *Akodon (Abrothrix) mansoensis* sp. nov. un nuevo "Raton Lanoso" de la provincia de Rio Negro, Argentina. Neotrópica 26:121–27.

Desy, E. A., and J. D. Druecker. 1979. The estrous cycle of the plains pocket gopher. *Geomys bursarius*, in the laboratory. J. Mamm. 60:235–36.

De Villafañe, G. 1981. Reproduccion y crecimiento de *Calomys musculinus murillus* (Thomas, 1916). Hist. Nat. (Mendoza, Argentina) 1:237–56.

De Vore, I., ed. 1965. Primate behavior. Field studies of monkeys and apes. Holt, Rinehart and Winston, New York, xiv + 654 pp.

De Vore, I., and K. R. L. Hall. 1965. Baboon ecology. *In* De Vore (1965), pp. 20–52.

De Vos, A., and A. Omar. 1971. Territories and movements of Sykes monkeys (*Cercopithecus mitis kolbi* Neuman) in Kenya. Folia Primatol. 16:196–205.

De Vree, F. 1972. Description of a new form of *Pipistrellus* from Ivory Coast. Rev. Zool. Bot. Afr. 85:412–16.

———. 1973. New data on *Scotophilus gigas* Dobson, 1875 (Microchiroptera—Vespertilionidae). Z. Saugetierk. 38:189–96.

Dewar, R. E. 1984. Extinctions in Madagascar. The loss of the subfossil fauna. *In* Martin and Klein (1984), pp. 574–93.

De Winton, W. E. 1898. On a small collection of mammals made by Mr. C. V. A. Peel in Somaliland. Ann. Mag. Nat. Hist., ser. 7, 1:247–51.

Dewsbury, D. A., B. Ferguson, and D. G. Webster. 1984. Aspects of reproduction, ovulation, and the estrous cycle in African four-striped grass mice (*Rhabdomys pumilio*). Mammalia 48:417–24.

D'Huart, J.-P. 1976. Actogramme journalier de l'hylochere (*Hylochoerus meinertzhageni* Thomas) au Parc National des Virunga, Zaire. Terre Vie 30:165–80.

Dhungel, S. K., and W. D. Edge. 1985. Notes on the natural history of *Paradoxurus hermaphroditus*. Mammalia 49:302–3.

Dierauf, L. A. 1984. A northern fur seal, *Callorhinus ursinus*, found in the Sacramento–San Joaquin Delta. California Fish and Game 70:189.

Diersing, V. E. 1980. Systematics and evolution of the pygmy shrews (subgenus *Microsorex*) of North America. J. Mamm. 61:76–101.

———. 1981. Systematic status of *Sylvilagus brasiliensis* and *S. insonus* from North America. J. Mamm. 62:539–56.

———. 1984. Lagomorphs. *In* Anderson and Jones (1984), pp. 241–54.

Diersing, V. E., and D. F. Hoffmeister. 1977. Revision of the shrews *Sorex merriami* and a

description of a new species of the subgenus *Sorex*. J. Mamm. 58:321–33.

Dieterlen, F. 1969. *Dendromus kahuziensis* (Dendromurinae; Cricetidae; Rodentia)— eine neue Art aus Zentralafrika. Z. Saugetierk. 34:348–53.

———. 1974. Bemerkungen zur Systematik der Gattung *Pelomys* (Muridae; Rodentia) in Athiopien. Z. Saugetierk. 39:229–31.

———. 1976. Die afrikanische Muridengattung *Lophuromys* Peters, 1874. Stuttgarter Beitr. Naturkunde, ser. A, no. 285, 96 pp.

———. 1983. Zur Systematik, Verbreitung und Okologie von *Colomys goslingi* Thomas and Wroughton, 1907 (Muridae; Rodentia). Bonner Zool. Beitr. 34:73–106.

Dieterlen, F., and G. Nikolaus. 1985. Zur Säugetierfauna des Sudan—weitre Erstnachweise und bemerkenswerte Funde. Saugetierk. Mitt. 32:205–9.

Dieterlen, F., and H. Rupp. 1976. Die Rotnasenratte *Oenomys hypoxanthus* (Pucheran, 1855) (Muriden, Rodentia)—Erstnachweis fur Athiopien und dritter Fund aus Tansania. Saugetierk. Mitt. 24:229–35.

———. 1978. *Megadendromus nikolausi*, gen. nov., sp. nov. (Dendromurinae; Rodentia), ein neuer Nager aus Athiopien. Z. Saugetierk. 43:129–43.

Dieterlen, F., and B. Statzner. 1981. The African rodent *Colomys goslingi* Thomas and Wroughton, 1907 (Rodentia: Muridae)—a predator in limnetic ecosystems. Z. Saugetierk. 46:369–83.

Dieterlen, F., and E. Van der Straeten. 1984. New specimens of *Malacomys verschureni* from eastern Zaire (Mammalia, Muridae). Rev. Zool. Afr. 98:861–68.

———. 1988. Deux nouveaux spécimens de *Lamottemys okuensis* Petter, 1986 du Cameroun (Muridae: Rodentia). Mammalia 52:379–85.

Dietz, J. D. 1983. Notes on the natural history of some small mammals in central Brazil. J. Mamm. 64:521–23.

Dietz, J. M. 1984. Ecology and social organization of the maned wolf (*Chrysocyon brachyurus*). Smithson. Contrib. Zool., no. 392, iv + 51 pp.

———. 1985. *Chrysocyon brachyurus*. Mammalian Species, no. 234, 4 pp.

Dimpel, H., and J. H. Calaby. 1972. Further observations on the mountain pigmy possum (*Burramys parvus*). Victorian Nat. 89:101–6.

Dinerstein, E. 1985. First records of *Lasiurus castaneus* and *Antrozous dubiaquercus* from Costa Rica. J. Mamm. 411–12.

Dippenaar, N. J. 1980. New species of *Crocidura* from Ethiopia and northern Tanzania

(Mammalia: Soricidae). Ann. Transvaal Mus. 32:125–54.

Dippenaar, N. J., and J. A. J. Meester. 1989. Revision of the *luna-fumosa* complex of Afrotropical *Crocidura* Wagler, 1832 (Mammalia: Soricidae). Ann. Transvaal Mus. 35:1–47.

Dippenaar, N. J., and I. L. Rautenbach. 1986. Morphometrics and karyology of the southern African species of the genus *Acomys* I. Geoffroy Saint-Hilaire, 1838 (Rodentia: Muridae). Ann. Transvaal Mus. 34:129–83.

Dittrich, L. 1966. Breeding Indian elephants *Elephas maximus* at Hanover Zoo. Internatl. Zoo Yearbook 6:193–96.

———. 1967. Breeding Kirk's dik-dik *Madoqua kirki thomasi* at Hanover Zoo. Internatl. Zoo Yearbook 7:171–73.

———. 1972. Gestation periods and age of sexual maturity of some African antelopes. Internatl. Zoo Yearbook 12:184–87.

———. 1976. Age of sexual maturity in the hippopotamus *Hippopotamus amphibius*. Internatl. Zoo Yearbook 16:171–73.

Dittus, W. 1977. The socioecological basis for the conservation of the toque monkey (*Macaca sinica*) of Sri Lanka (Ceylon). *In* Rainier III and Bourne (1977), pp. 237–65.

Dixon, A., and D. Jones. 1988. Conservation and biology of desert antelopes. Christopher Helm, London, xiv + 238 pp.

Dixon, J. M. 1971. *Burramys parvus* Broom (Marsupialia) from Falls Creek area of the Bogong high plains, Victoria. Victorian Nat. 88:133–38.

———. 1978. The first Victorian and other records of the little pigmy possum *Cercartetus lepidus* (Thomas). Victorian Nat. 95:4–7.

Dixon, K. R., J. A. Chapman, G. R. Willner, D. E. Wilson, and W. Lopez-Forment. 1983. The New World jackrabbits and hares (genus *Lepus*).—2. Numerical taxonomic analysis. Acta Zool. Fennica 174:53–56.

Dixson, A. F. 1977. Observations on the displays, menstrual cycles and sexual behaviour of the "black ape" of Celebes (*Macaca nigra*). J. Zool. 182:63–84.

———. 1981. The natural history of the gorilla. Columbia Univ. Press, New York, xviii + 202 pp.

Dixson, A. F., D. M. Scruton, and J. Herbert. 1975. Behaviour of the talapoin monkey (*Miopithecus talapoin*) studied in groups, in the laboratory. J. Zool. 176:177–210.

Dobias, R. J. 1987. Elephants in Thailand: an overview of their status and conservation. Tigerpaper 14(1):19–24.

Dobroruka, L. J. 1968. Breeding group of gorals *Nemorhaedus goral* at Prague Zoo. Internatl. Zoo Yearbook 8:143–45.

————. 1970a. Fecundity of the Chinese water deer, *Hydropotes inermis* Swinhoe, 1870. Mammalia 34:161–62.

————. 1970b. To the supposed former occurrence of the David's deer, *Elaphurus davidianus* Milne Edwards, 1866, in Hainan. Mammalia 34:162–64.

————. 1973. Yellow-spotted dassie *Heterohyrax brucei* (Gray, 1868) feeding on a poisonous plant. Saugetierk. Mitt. 21:365.

————. 1987. Breeding season of the lechwe *Kobus leche*. Internatl. Zoo Yearbook 26:294–96.

Doebl, J. H., and B. S. McGinnes. 1974. Home range and activity of a gray squirrel population. J. Wildl. Mgmt. 38:860–67.

Doerr, S. M., and F. W. Judd. 1984. Reproductive ecology of the pocket gopher, *Geomys personatus*, in south Texas. Texas J. Sci. 36:129–38.

Doherty, J. 1989. Twin babirusas born at the New York Zoological Park. Amer. Assoc. Zool. Parks Aquar. Newsl. 30(2):21.

Doidge, D. W. 1987. Rearing of twin offspring to weaning in Antarctic fur seals, *Arctocephalus gazella*. *In* Croxall and Gentry (1987), pp. 107–11.

Doidge, D. W., T. S. McCann, and J. P. Croxall. 1986. Attendance behavior of Antarctic fur seals. *In* Gentry and Kooyman (1986), pp. 102–14.

Dolan, J. M. 1976. The Arabian oryx *Oryx leucoryx*. Its destruction, captive history and propagation. Internatl. Zoo Yearbook 16:230–39.

Dolan, J. M., and L. E. Killmar. 1988. The shou, *Cervus elaphus wallichi* Cuvier, 1825, a rare and little-known cervid, with remarks on three additional Asiatic elaphines. Zool. Garten 58:84–96.

Dolan, P. G., and D. C. Carter. 1979. Distributional notes and records for Middle American Chiroptera. J. Mamm. 60:644–49.

Dolan, P. G., and T. L. Yates. 1981. Interspecific variation in *Apodemus* from the northern Adriatic islands of Yugoslavia. Z. Saugetierk. 46:151–61.

Dolgov, V. A., and R. S. Hoffmann. 1977. The Tibetan shrew, *Sorex thibetanus* Kastschenko, 1905 (Soricidae, Mammalia). Zool. Zhur. 56:1687–92.

Domning, D. P. 1978. Sirenian evolution in the North Pacific Ocean. Univ. California Publ. Geol. Sci. 118:1–176.

————. 1981. Distribution and status of manatees *Trichechus* spp. near the mouth of the Amazon River, Brazil. Biol. Conserv. 19:85–97.

————. 1982a. Evolution of manatees: a speculative history. J. Paleontol. 56:599–619.

————. 1982b. Commercial exploitation of manatees *Trichechus* in Brazil c. 1785–1973. Biol. Conserv. 22:101–26.

Domning, D. P., and L.-A. C. Hayek. 1986. Interspecific and intraspecific morphological variation in manatees (Sirenia: *Trichechus*). Mar. Mamm. Sci. 2:87–144.

Domning, D. P., G. S. Morgan, and C. E. Ray. 1982. North American Eocene sea cows (Mammalia: Sirenia). Smithson. Contrib. Paleobiol., no. 52, 69 pp.

Domning, D. P., and C. E. Ray. 1986. The earliest Sirenian (Mammalia: Dugongidae) from the eastern Pacific Ocean. Mar. Mamm. Sci. 2:263–76.

Donaldson, S. L., T. B, Wirtz, and A. E. Hite. 1975. The social behaviour of capybaras *Hydrochoerus hydrochaeris* at Evansville Zoo, Internatl. Zoo Yearbook 15:201–6.

Donnelly, B. G., and J. H. Grobler. 1976. Notes on food and anvil using behaviour by the Cape clawless otter, *Aonyx capensis*, in the Rhodes Matopos National Park, Rhodesia. Arnoldia 7(37):1–7.

Dorji, D. P., and C. Santiapillai. 1989. The status, distribution and conservation of the tiger *Panthera tigris* in Bhutan. Biol. Conserv. 48:311–19.

Dorsey, E. M. 1983. Exclusive adjoining ranges in individually identified minke whales *(Balaenoptera acutorostrata)* in Washington state. Can. J. Zool. 61:174–81.

Dorst, J., and P. Dandelot. 1969. A field guide to the larger mammals of Africa. Houghton Mifflin, Boston, 287 pp.

Douglas, A. M. 1967. The natural history of the ghost bat, *Macroderma gigas* (Microchiroptera, Megadermatidae), in Western Australia. W. Austral. Nat. 10:125–37.

Douglas, C. W., and M. A. Strickland. 1987. Fisher. *In* Novak, Baker, et al. (1987), pp. 510–29.

Douglas, M. E., G. D. Schnell, and D. J. Hough. 1984. Differentiation between inshore and offshore spotted dolphins in the eastern tropical Pacific Ocean. J. Mamm. 65:375–87.

Douglas-Hamilton, I. 1973. On the ecology and behavior of the Lake Manyara elephants. E. Afr. Wildl. J. 11:401–3.

————. 1987. African elephants: population trends and their causes. Oryx 21:11–24.

Douglass, R. J. 1977. Population dynamics, home ranges, and habitat associations of the yellow-cheeked vole, *Microtus xanthognathus*, in the Northwest Territories. Can. Field-Nat. 91:237–47.

Dowler, R, C., and H. H. Genoways. 1978. *Liomys irroratus*. Mammalian Species, no. 82, 6 pp.

Downing, R. L. 1987. Success story: white-tailed deer. *In* Kallman (1987), pp. 44–57.

Doyle, G. A. 1979. Development of behavior in prosimians with special reference to the lesser bushbaby, *Galago senegalensis moholi*. *In* Doyle and Martin (1979), pp. 157–206.

Doyle, G. A., and S. K. Bearder. 1977. The galagines of South Africa. *In* Rainier III and Bourne (1977), pp. 1–35.

Doyle, G. A., and R. D. Martin, eds. 1979. The study of Prosimian behavior. Academic Press, New York, xvii + 696 pp.

Drabek, C. M. 1973. Home range and daily activity of the round-tailed ground squirrel, *Spermophilus tereticaudus neglectus*. Amer. Midl. Nat. 89:287–93.

Drew, L. 1989. Are we loving the panda to death? Natl. Wildl. 27(1):14–17.

Drickamer, L. C., J. G. Vandenbergh, and D. R. Colby. 1973. Predictors of dominance in the male golden hamster *(Mesocricetus auratus)*. Anim. Behav. 21:557–63.

Drickamer, L. C., and B. M. Vestal. 1973. Patterns of reproduction in a laboratory colony of *Peromyscus*. J. Mamm. 54:523–28.

Dubost, G. 1978. Un aperçu sur l'écologie du chevrotain africain *Hyemoschus aquaticus* Ogilby, Artiodactyle Tragulide. Mammalia 42:1–62.

————. 1984. Comparison of the diets of frugivorous forest ruminants of Gabon. J. Mamm. 65:298–316.

————. 1988. Ecology and social life of the red acouchy, *Myoprocta exilis*; comparison with the orange-rumped agouti, *Dasyprocta leporina*. J. Zool. 214:107–23.

Dubost, G., and H. Genest. 1974. Le comportement social d'une colonie de maras *Dolichotis patagonum* Z. dans le Parc de Branféré. Z. Tierpsychol, 35:225–302.

Dubost, G., and F. Petter. 1978. Une espèce nouvelle de "rat-pecheur" de Guyane française: *Daptomys oyapocki* sp. nov. (Rongeurs, Cricetidae). Mammalia 42:435–39.

Duguy, R. 1986. Observation d'un morse *(Odobenus rosmarus)* sur la côte de Gironde, France. Mammalia 50:563–64.

Du Mond, F. V. 1968. The squirrel monkey in a seminatural environment. *In* Rosenblum and Cooper (1968), pp. 87–145.

Dunaway, P. B. 1968. Life history and population aspects of the eastern harvest mouse. Amer. Midl. Nat. 79:48–67.

Dunbar, R. 1986. Rhum deal for goats. Nat. Hist. 95(11):40–47.

Dunbar, R. I. M. 1974. Observations on the ecology and social organization of the green monkey, *Cercopithecus sabaeas*, in Senegal. Primates 15:341–50.

————. 1977a. The gelada baboon: status and conservation, In Rainier III and Bourne (1977), pp. 363–83.

————. 1977b. Feeding ecology of gelada baboons: a preliminary report. In Clutton-Brock (1977), pp. 251–73.

————. 1983a. Structure of gelada baboon reproductive units. II. Social relationships between reproductive females. Anim. Behav. 31:556–64.

————. 1983b. Structure of gelada baboon reproductive units. III. The male's relationship with his females. Anim. Behav. 31:565–75.

Dunbar, R. I. M., and E. P. Dunbar. 1981. The grouping behaviour of male Walia ibex with special reference to the rut. Afr. J. Ecol. 19:251–63.

Dunbar, R. I. M., and P. Dunbar. 1974a. On hybridization between Theropithecus gelada and Papio anubis in the wild. J. Human Evol. 3:187–92.

————. 1974b. Social organization and ecology of the klipspringer (Oreotragus oreotragus) in Ethiopia. Z. Tierpsychol. 35:481–93.

————. 1975. Social dynamics of gelada baboons. Contrib. Primatol. 6:i–vii + 1–157.

————. 1979. Observations on the social organization of common duiker in Ethiopia. Afr. J. Ecol. 17:249–52.

Duncan, P., and R. W. Wrangham. 1970. On the ecology and distribution of subterranean insectivores in Kenya. J. Zool. 164:149–63.

Dunford, C. 1972. Summer activity of eastern chipmunks. J. Mamm. 53:176–80.

————. 1974. Annual cycle of cliff chipmunks in the Santa Catalina Mountains, Arizona. J. Mamm. 55:401–16.

————. 1977. Social system of round-tailed ground squirrels. Anim. Behav. 25:885–906.

Dunlop, J. N., and I. R. Pound. 1981. Observations on the pebble-mound mouse Pseudomys chapmani Kitchener, 1980. Rec. W. Austral. Mus. 9:1–5.

Duplaix, N. 1980. Observations on the ecology and behavior of the giant river otter Pteronura brasiliensis in Suriname. Rev. Ecol. (Terre Vie) 34:496–620.

Duplaix-Hall, N. 1975. River otters in captivity: a review. In Martin (1975b), pp. 315–27.

DuPlessis, S. S. 1972. Ecology of blesbok with special reference to productivity. Wildl. Monogr., no. 30, 70 pp.

Duran, J. C., P. E. Cattan, and J. L. Yáñez. 1985. The grey fox Canis griseus (Gray) in Chilean Patagonia (southern Chile). Biol. Conserv. 34:141–48.

Durham, N. M. 1971. Effects of altitude differences on group organization of wild black spider monkeys (Ateles paniscus). Proc. 3rd Internatl. Congr. Primatol. 3:32–40.

Durrell, G., and J. Mallinson. 1970. The volcano rabbit Romerolagus diazi in the wild and at Jersey Zoo. Internatl. Zoo Yearbook 10:118–22.

Dusi, J. L. 1976. Mammals. In Boschung, H., ed., Endangered and threatened plants and animals of Alabama. Bull. Alabama Mus. Nat. Hist., no. 2, pp. 88–92.

Duthie, A. 1987. The endangered riverine rabbit—a research update. Afr. Wildl. 41(4):168–71.

Duthie, A. G., J. D. Skinner, and T. J. Robinson. 1989. The distribution and status of the riverine rabbit, Bunolagus monticularis, South Africa. Biol. Conserv. 47:195–202.

Du Toit, R. 1987. The existing basis for subspecies classifications of black and white rhinos. Pachyderm 9:3–5.

————. 1989. Suggested procedure for priority ranking of black rhino populations. Pachyderm 11:7–10.

Dutrillaux, B., A.-M. Dutrillaux, M. Lombard, J.-P. Gautier, R. Cooper, F. Moysan, and J. M. Lernould. 1988. The karyotype of Cercopithecus solatus Harrison 1988, a new species belonging to C. lhoesti, and its phylogenetic relationships with other guenons. J. Zool. 215:611–17.

Dwyer, P. D. 1966. Observations on Chalinolobus dwyeri (Chiroptera: Vespertilionidae) in Australia. J. Mamm. 47:716–18.

————. 1968. The biology, origin, and adaption of Miniopterus australis (Chiroptera) in New South Wales. Austral. J. Zool. 16:49–68.

————. 1970. Latitude and breeding season in a polyestrous species of Myotis. J. Mamm. 51:405–10.

————. 1975a. Notes on Dobsonia moluccensis (Chiroptera) in the New Guinea highlands. Mammalia 39:113–18.

————. 1975b. Observations on the breeding biology of some New Guinea murid rodents. Austral. Wildl. Res. 2:33–45.

————. 1977. Notes on Antechinus and Cercartetus (Marsupialia) in the New Guinea highlands. Proc. Roy. Soc. Queensland 88:69–73.

————. 1983. An annotated list of mammals from Mt. Erimbari, Eastern Highlands Province, Papua New Guinea. Sci. New Guinea 10:28–38.

Dzieciolowski, R., J. Krupka, Bajandelger, and R. Dziedzic. 1980. Argali and Siberian ibex populations in the Khuhsyrh Reserve in Mongolian Altai. Acta Theriol. 25:213–19.

Dzneladze, A. M. 1974. Distribution of the long-clawed mole-vole (Prometheomys schaposchnikovi Satun. 1901) in the Georgian SSR. Vestn. Zool. 1974(6):80–82.

E

Earlé, R. A. 1981. Aspects of the social and feeding behaviour of the yellow mongoose Cynictis penicillata (G. Cuvier). Mammalia 45:143–55.

Earle, R. D., and K. R. Kramm. 1980. Techniques for age determination in the Canadian porcupine. J. Wildl. Mgmt. 44:413–19.

East, R. 1988. Antelopes. Global survey and regional action plans. Part 1. East and northeast Africa. Internatl. Union Conserv. Nat., Gland, Switzerland, iv + 96 pp.

————. 1989. Antelopes. Global survey and regional action plans. Part 2. Southern and south-central Africa. Internatl. Union Conserv. Nat., Gland, Switzerland, iv + 96 pp.

Easterla, D. A. 1973. Ecology of the 18 species of Chiroptera at Big Bend National Park, Texas. Part II. Northwest Missouri State Univ. Studies 34:54–165.

————. 1976. Notes on the second and third newborn of the spotted bat, Euderma maculatum, and comments on the species in Texas. Amer. Midl. Nat. 96:499–501.

Easterla, D. A., and L. C. Watkins. 1970. Breeding of Lasionycteris noctivagans and Nycticeius humeralis in southwestern Iowa. Amer. Midl. Nat. 84:254–55.

Eaton, R. L. 1974. The cheetah. Van Nostrand Reinhold, New York, xii + 178 pp.

————. 1976. The brown hyena: a review of biology, status and conservation. Mammalia 40:377–99.

Eberhard, I. H. 1978. Ecology of the koala, Phascolarctos cinereus (Goldfuss) Marsupialia: Phascolarctidae, in Australia. In Montgomery (1978), pp. 315–27.

Eberhardt, L. E., and W. C. Hanson. 1978. Long-distance movements of arctic foxes tagged in northern Alaska. Can. Field-Nat. 92:386–89.

Edmunds, R. M., J. W. Goertz, and G. Linscombe. 1978. Age ratios, weights, and reproduction of the Virginia opossum in northern Louisiana. J. Mamm. 59:884–85.

Edwards, G. P., and E. H. M. Ealey. 1975. Aspects of the ecology of the swamp wallaby Wallabia bicolor (Marsupialia: Macropodidae). Austral Mamm. 1:307–17.

Egbert, A. L., and A. W. Stokes. 1976. The social behavior of brown bears on an Alaskan salmon stream. In Pelton, Lentfer, and Folk (1976), pp. 41–56.

Eger, J. L. 1977. Systematics of the genus Eumops (Chiroptera: Molossidae). Roy. Ontario Mus. Life Sci. Contrib., no. 110, 69 pp.

Eger, J. L., and R. L. Peterson. 1979. Distribution and systematic relationship of *Tadarida bivittata* and *Tadarida ansorgei* (Chiroptera: Molossidae). Can. J. Zool. 57:1887–95.

Egoscue, H. J. 1970. A laboratory colony of the Polynesian rat, *Rattus exulans*. J. Mamm. 51:261–68.

———. 1972. Breeding the long-tailed pouched rat, *Beamys hindei*, in captivity. J. Mamm. 53:296–302.

———. 1979. *Vulpes velox*. Mammalian Species, no. 122, 5 pp.

———. 1981. Additional records of the dark kangaroo mouse *(Microdipodops megacephalus nasutus)*, with a new maximum altitude. Great Basin Nat. 41:333–34.

Egoscue, H. J., J. G. Bittmenn, and J. A. Petrovich. 1970. Some fecundity and longevity records for captive small mammals. J. Mamm. 51:622–23.

Eibl-Eibesfeldt, I. 1984. The Galapagos seals. Part 1. Natural history of the Galapagos sea lion (*Zalophus californianus wollebaeki*, Sivertsen). *In* Perry (1984), pp. 207–14.

Eisenberg, J. F. 1963. The behavior of heteromyid rodents. Univ. California Publ. Zool. 69:i–iv + 1–100.

———. 1974. The function and motivational basis of hystricomorph vocalizations. Symp. Zool. Soc. London 34:211–47.

———. 1975. Tenrecs and solenodons in captivity. Internatl. Zoo Yearbook 15:6–12.

———. 1977. Comparative ecology and reproduction of New World monkeys. *In* Kleiman (1977a), pp. 13–22.

———, ed. 1979. Vertebrate ecology in the northern neotropics. Smithson. Inst. Press, Washington, D. C., 271 pp.

———. 1981. The mammalian radiations. Univ. Chicago Press, xx + 610 pp.

———. 1984. Mouse-like rodents. *In* Macdonald (1984), pp. 636–49.

———. 1987. The evolutionary history of the Cervidae with special reference to the South American radiation. *In* Wemmer (1987), pp. 60–64.

Eisenberg, J. F., and N. Gonzalez Gotera. 1985. Observations on the natural history of *Solenodon cubanus*. Acta Zool. Fennica 173:275–77.

Eisenberg, J. F., and E. Gould. 1966. The behavior of *Solenodon paradoxus* in captivity with comments on the behavior of other Insectivora. Zoologica 51:49–58.

———. 1967. The maintenance of tenrecoid insectivores in captivity. Internatl. Zoo Yearbook 7:194–96.

———. 1970. The tenrecs: a study in mammalian behavior and evolution. Smithson. Contrib. Zool., no. 27, v + 138 pp.

———. 1984. The insectivores. *In* Jolly, A., et al., eds., Madagascar, Pergamon Press, London, pp. 155–64.

Eisenberg, J. F., C. P. Groves, and K. MacKinnon. 1987. Tapire. *In* Keienburg, W., ed., Grzimeks Enzyklopädie, Kindler Verlag, Munich, Säugetiere, 4:598–608.

Eisenberg, J. F., and M. C. Lockhart. 1972. An ecological reconnaissance of Wilpattu National Park, Ceylon. Smithson. Contrib. Zool., no. 101, 118 pp.

Eisenberg, J. F., and G. M. McKay. 1970. An annotated checklist of the Recent mammals of Ceylon. Ceylon J. Sci., Biol. Sci. 8:69–99.

Eisenberg, J. F., G. M. McKay, and M. R. Jainudeen. 1971. Reproductive behavior of the Asiatic elephant (*Elephas maximus maximus* L.). Behaviour 38:193–225.

Eisenberg, J. F., and E. Maliniak. 1973. Breeding and captive maintenance of the lesser bamboo rat *Cannomys badius*. Internatl. Zoo Yearbook 13:204–7.

———. 1974. The reproduction of the genus *Microgale* in captivity. Internatl. Zoo Yearbook 14:108–10.

———. 1978. Reproduction by the two-toed sloth, *Choloepus hoffmanni*, in captivity. Amer. Soc. Mamm., Abstr. Tech. Pap., 58th Ann. Mtg., pp. 41–42.

Eisentraut, M. 1970. Beitrag zur Fortpflanzungsbiologie der Zwergbeutelratte *Marmosa murina* (Didelphidae, Marsupialia). Z. Saugetierk 35:159–72.

———. 1986. Über das Vorkommen des Chaco-Pekari, *Catagonus wagneri*, in Bolivien. Bonner Zool. Beitr. 37:43–47.

Elbl, A., U. H. Rahm, and G. Mathys. 1966. Les mammifères et leurs tiques dannս la forêt du Rueggege (République Rwandaise). Acta Tropica 23:223–63.

Elder, F. F. B., and M. R. Lee. 1985. The chromosomes of *Sigmodon ochrognathus* and *S. fulviventer* suggest a realignment of *Sigmodon* species groups. J. Mamm. 66:511–18.

Eldridge, W. D., M. MacNamara, and N. Pacheco. 1987. Activity patterns and habitat utilization of pudus *(Pudu puda)* in south-central Chile. *In* Wemmer (1987), pp. 352–70.

Elgmork, K. 1978. Human impact on a brown bear population (*Ursus arctos* L.). Biol. Conserv. 13:81–103.

El Hilali, M., and J.-P. Veillat. 1975. *Jaculus orientalis:* a true hibernator. Mammalia 39:401–4.

Ellefson, J. O. 1974. A natural history of white-handed gibbons in the Malayan peninsula. *In* Rumbaugh, D. M., ed., Gibbon and siamang, III, S. Karger, Basel, pp. 1–136.

Ellerman, J. R. 1941. The families and genera of living rodents. II. Family Muridae. British Mus. (Nat. Hist.), London, xii + 690 pp.

———. 1946. Further notes on two little-known Indian murine genera, and preliminary diagnosis of a new species of *Rattus* (subgenus *Cremnomys*) from the eastern Ghats. Ann. Mag. Nat. Hist., ser. 11, 13:204–08.

Ellerman, J. R., and T. C. S. Morrison-Scott. 1966. Checklist of Palaearctic and Indian mammals. British Mus. (Nat. Hist.), London, 810 pp.

Elliott, J. P. 1985. The status of thinhorn sheep *(Ovis dalli)* in British Columbia. *In* Hoefs (1985), pp. 43–47.

Elliott, L. 1978. Social behavior and foraging ecology of the eastern chipmunk *(Tamias striatus)* in the Adirondack Mountains. Smithson. Contrib. Zool., no. 265, vi + 107 pp.

Ellis, L. S., V. E. Diersing, and D. F. Hoffmeister. 1978. Taxonomic status of short-tailed shrews *(Blarina)* in Illinois. J. Mamm. 59:305–11.

Ellis, L. S., and L. R. Maxson. 1979. Evolution of the chipmunk genera *Eutamias* and *Tamias*. J. Mamm. 60:331–34.

Ellis, R. 1980. The book of whales. Alfred A. Knopf, New York, xvii + 202 pp.

El-Rayah, M. A. 1981. A new species of bat of the genus *Tadarida* (family Molossidae) from West Africa. Roy. Ontario Mus. Life Sci. Occas. Pap., no. 36, 10 pp.

Elwood, R. W. 1975. Paternal and maternal behavior in the Mongolian gerbil. Anim. Behav. 23:766–72.

Emison, W. B., J. W. Porter, K. C. Norris, and G. J. Apps. 1975. Ecological distribution of the vertebrate animals of the volcanic plains–Otway Range area of Victoria. Victoria Fish. and Wild. Pap., no. 6, 93 pp.

———. 1978. Survey of the vertebrate fauna in the Grampians-Edenhope area of southwestern Victoria. Mem. Natl. Mus. Victoria 39:281–363.

Emmons, L. H. 1978. Sound communication among African rainforest squirrels. Z. Tierpsychol. 47:1–49.

———. 1979. A note on the forefoot of *Myosciurus pumilio*. J. Mamm. 60:431–32.

———. 1980. Ecology and resource partitioning among nine species of African rain forest squirrels. Ecol. Monogr. 50:31–54.

———. 1981. Morphological, ecological, and behavioral adaptations for arboreal browsing in *Dactylomys dactylinus* (Rodentia, Echimyidae). J. Mamm. 62:183–89.

———. 1982. Ecology of *Proechimys* (Rodentia, Echimyidae) in southeastern Peru. Tropical Ecol. 23:280–90.

————. 1983. A field study of the African brush-tailed porcupine, *Atherurus africanus*, by radiotelemetry. Mammalia 47:183–94.

————. 1988a. Replacement name for a genus of South American rodent (Echimyidae). J. Mamm. 69:421.

————. 1988b. A field study of ocelots *(Felis pardalis)* in Peru. Rev. Ecol. (Terre Vie) 43:133–57.

Enders, R. K. 1966. Attachment, nursing and survival of young in some didelphids. Symp. Zool. Soc. London 15:195–203.

————. 1980. Observations on *Syntheosciurus:* taxonomy and behavior. J. Mamm. 61:725–27.

Engelmann, G. F. 1985. The phylogeny of the Xenarthra. *In* Montgomery (1985a), pp. 51–64.

Engstrom, M. D., and J. W. Bickham. 1983. Karyotype of *Nelsonia neotomodon*, with notes on the primitive karyotype of peromyscine rodents. J. Mamm. 64:685–88.

Engstrom, M. D., T. E. Lee, and D. E. Wilson. 1987. *Bauerus dubiaquercus.* Mammalian Species, no. 282, 3 pp.

Engstrom, M. D., and D. E. Wilson. 1981. Systematics of *Antrozous dubiaquercus* (Chiroptera: Vespertilionidae), with comments on the status of *Bauerus* Van Gelder. Ann. Carnegie Mus. 50:371–83.

Erdbrink, D. P. 1953. A review of fossil and Recent bears of the Old World with remarks on their phylogeny based upon their dentition. Deventer, Netherlands, 597 pp.

Erlinge, S. 1967. Home range of the otter *Lutra lutra* L. in southern Sweden. Oikos 18:186–209.

————. 1968. Territoriality of the otter *Lutra lutra* L. Oikos 19:81–98.

————. 1977. Spacing strategy in stoat *Mustela erminea*. Oikos 28:32–42.

Ernest, K. A. 1986. *Nectomys squamipes.* Mammalian Species, no. 265, 5 pp.

Ernest, K. A., and M. A. Mares. 1986. Ecology of *Nectomys squamipes*, the neotropical water rat, in central Brazil: home range, habitat selection, reproduction and behaviour. J. Zool. 210:599–612.

Escherich, P. C. 1981. Social biology of the bushy-tailed woodrat. Univ. California Publ. Zool. 110:i–xiv + 1–132.

Eshelkin, I. I. 1976. On the reproduction of *Alticola strelzovi* in the south-east Altai. Zool. Zhur. 55:437–42.

Eshelman, B. D., and G. N. Cameron. 1987. *Baiomys taylori.* Mammalian Species, no. 285, 7 pp.

Esher, R. J., J. L. Wolfe, and J. N. Layne. 1978. Swimming behavior of rice rats *(Oryzomys palustris)* and cotton rats *(Sigmodon hispidus)*. J. Mamm. 59:551–58.

Estes, J. A. 1980. *Enhydra lutris.* Mammalian Species, no. 133, 8 pp.

————. 1986. Marine otters and their environment. Ambio 15:181–83.

Estes, J. A., and G. R. VanBlaricom. 1985. Sea otters and shellfisheries. *In* Beddington, Beverton, and Levigne (1985), pp. 187–235.

Estes, R. D. 1967. The comparative behavior of Grant's and Thomson's gazelles. J. Mamm. 48:189–209.

————. 1969. Territorial behavior of the wildebeest (*Connochaetes taurinus* Burchell, 1823). Z. Tierpsychol. 26:284–370.

————. 1974. Social organization of the African Bovidae. *In* Geist and Walther (1974), pp. 166–205.

————. 1976. The significance of breeding synchrony in the wildebeest. E. Afr. Wildl. J. 14:135–52.

Estes, R. D., and R. K. Estes. 1974. The biology and conservation of the giant sable antelope, *Hippotragus niger variani* Thomas, 1916. Proc. Acad. Nat. Sci. Philadelphia 126:73–104.

Estrada, A., and R. Coates-Estrada. 1984. Some observations on the present distribution and conservation of *Alouatta* and *Ateles* in southern Mexico. Amer. J. Primatol. 7:133–37.

————. 1988. Tropical rain forest conversion and perspectives in the conservation of wild primates (*Alouatta* and *Ateles*) in Mexico. Amer. J. Primatol. 14:315–27.

Eudey, A. A. 1987. Action plan for Asian primate conservation: 1987–1991. Internatl. Union Conserv. Nat., Gland, Switzerland, 65 pp.

Evans, C. D., and L. S. Underwood. 1978. How many bowheads? Oceanus 21(2):17–23.

Evans, G. D., and E. W. Pearson. 1980. Federal coyote control methods used in the western United States, 1971–77. Wildl. Soc. Bull. 8:34–39.

Everard, C. O. R., and E. S. Tikasingh. 1973. Ecology of the rodents, *Proechimys guyannensis trinitatis* and *Oryzomys capito velutinus*, on Trinidad. J. Mamm. 54:875–86.

Ewer, R. F. 1967. The behaviour of the African giant rat (*Cricetomys gambianus* Waterhouse). Z. Tierpsychol. 24:6–79.

————. 1968. A preliminary survey of the behaviour in captivity of the dasyurid marsupial, *Sminthopsis crassicaudata* (Gould). Z. Tierpsychol. 25:319–65.

————. 1971. The biology and behaviour of a free-living population of black rats (*Rattus rattus*). Anim. Behav. Monogr. 4:127–74.

————. 1973. The carnivores. Cornell Univ. Press, Ithaca, xv + 494 pp.

Ewer, R. F., and C. Wemmer. 1974. The behaviour in captivity of the African civet, *Civettictis civetta* (Schreber). Z. Tierpsychol. 34:359–94.

F

Fa, J. E. 1984. The Barbary macaque. A case study in conservation. Plenum Press, New York, xvii + 369 pp.

Fa, J. E., D. M. Taub, N. Menard, and P. J. Stewart. 1984. The distribution and current status of the Barbary macaque in North Africa. *In* Fa (1984), pp. 79–111.

Fadem, B. H., and R. S. Rayve. 1985. Characteristics of the oestrous cycle and influence of social factors in grey short-tailed opossums *(Monodelphis domestica)*. J. Reprod. Fert. 73:337–42.

Fairall, N. 1972. Behavioural aspects of the reproductive physiology of the impala, *Aepyceros melampus* (Licht.). Zool. Afr. 7:167–74.

Fairley, J. S. 1971. The present distribution of the bank vole *Clethrionomys glareolus* Schreber in Ireland. Proc. Roy. Irish Acad., sec. B, 71:183–88.

Fairley, J. S., and J. M. Jones. 1976. A woodland population of small rodents (*Apodemus sylvaticus* (L.) and *Clethrionomys glareolus* Schreber) at Adare, Co. Limerick. Proc. Roy. Irish Acad., sec. B, 76:323–36.

Fairon, J. 1980. Deux nouvelles espèces de chiroptères pour la faune du Massif de l'Air (Niger): *Otonycteris hemprichi* Peters, 1859 et *Pipistrellus nanus* (Peters, 1852). Bull. Inst. Roy. Soc. Nat. Belg. 52(17):1–7.

Fan Naichang and Shi Yinzhu. 1982. A revision of the zokors of subgenus *Eospalax*. Acta Theriol. Sinica 2:183–200.

Fancy, S. G., L. F. Pank, K. R. Whitten, and W. L. Regelin. 1989. Seasonal movements of caribou in arctic Alaska as determined by satellite. Can. J. Zool. 67:644–50.

Fanning, F. D. 1982. Reproduction, growth and development in *Ningaui* sp. (Dasyuridae, Marsupialia) from the Northern Territory. *In* Archer (1982a), pp. 23–37.

Farentinos, R. C. 1972. Observations on the ecology of the tassel-eared squirrel. J. Wildl. Mgmt. 36:1234–39.

Farst, D. D., D. P. Thompson, G. A. Stones, P. M. Burchfield, and M. L. Hughes. 1980. Maintenance and breeding of duikers *Cephalophus* spp. at Gladys Porter Zoo, Brownsville. Internatl. Zoo Yearbook 20:93–99.

Faust, R., and C. Scherpner. 1967. A note on the breeding of the maned wolf. Internatl. Zoo Yearbook 7:119.

Fay, F. H. 1978. Belukha whale. *In* Haley (1978), pp. 132–37.

————. 1981. Walrus—*Odobenus rosmarus*. *In* Ridgway and Harrison (1981*a*), pp. 1–23.

————. 1982. Ecology and biology of the Pacific walrus, *Odobenus rosmarus divergens* Illiger. N. Amer. Fauna, no. 74, vi + 279 pp.

————. 1985. *Odobenus rosmarus*. Mammalian Species, no. 238, 7 pp.

Fay, F. H., B. P. Kelly, and J. L. Sease. 1989. Managing the exploitation of Pacific walruses: a tragedy of delayed response and poor communication. Mar. Mamm. Sci. 5:1–16.

Fayenuwo, J. O., and L. B. Halstead. 1974. Breeding cycle of straw-colored fruit bat, *Eidolon helvum*, Ile-Ife, Nigeria. J. Mamm. 55:453–54.

Fedosenko, A. K. 1985. Present status of argali sheep populations in the U.S.S.R. *In* Hoefs (1985), pp. 200–210.

Feer, F. 1982. Maturité sexuelle et cycle annuel de reproduction de *Neotragus batesi* de Winton, 1903 (Bovidé forestier africain). Mammalia 46:65–74.

Feiler, A. 1978*a*. Bemerkungen uber *Phalanger* der "orientalis-Gruppe" nach Tate (1945). Zool. Abhandl. (Dresden) 34:385–95.

————. 1978*b*. Über artliche Abgrenzung und innerartliche Ausformung bei *Phalanger maculatus* (Mammalia, Marsupialia, Phalangeridae). Zool. Abhandl. (Dresden) 35:1–30.

————. 1980. *Taphozous saccolaimus* Temminck, 1841 auf Sulawesi (Celebes) (Mammalia, Chiroptera, Emballonuridae). Zool. Abhandl. (Dresden) 36:225–28.

Feist, J. D., and D. R. McCullough. 1976. Behavior patterns and communication in feral horses. Z. Tierpsychol. 41:337–71.

Feldhamer, G. A. 1980. *Cervus nippon*. Mammalian Species, no. 128, 7 pp.

Feldhamer, G. A., K. C. Farris-Renner, and C. M. Barker. 1988. *Dama dama*. Mammalian Species, no. 317, 8 pp.

Feldhamer, G. A., and J. E. Gates. 1980. A black rat population in western Maryland. Proc. Pennsylvania Acad. Sci. 54:191–92.

Fellner, K. 1965. Natural rearing of clouded leopards *Neofelis nebulosa* at Frankfurt Zoo. Internatl. Zoo Yearbook 5:111–13.

Felten, H. 1964. Flughunde der Gattung *Pteropus* von Neukaledonien und den Loyalty-Inseln (Mammalia, Chiroptera). Senckenberg. Biol. 45:671–83.

Felten, H., and D. Kock. 1972. Weitere Flughunde der Gattung *Pteropus* von den Neuen Hebriden, sowie den Banks- und Torres- Inseln, Pazifischer Ozean. Senckenberg. Biol. 53:179–88.

Feng Tso-Chien. 1973. A new species of *Ochotona* (Ochotonidae, Mammalia) from Mount Jolmo-Lungma area. Acta Zool. Sinica 19:69–75.

Fenton, M. B. 1975. Observations on the biology of some Rhodesian bats, including a key to the Chiroptera of Rhodesia. Roy. Ontario Mus. Life Sci. Contrib., no. 104, 27 pp.

Fenton, M. B., R. M. Brigham, A. M. Mills, and I. L. Rautenbach. 1985. The roosting and foraging areas of *Epomophorus wahlbergi* (Pteropodidae) and *Scotophilus viridis* (Vespertilionidae) in Kruger National Park, South Africa. J. Mamm. 66:461–68.

Fenton, M. B., C. L. Gaudet, and M. L. Leonard. 1983. Feeding behaviour of the bats *Nycteris grandis* and *Nycteris thebaica* (Nycteridae) in captivity. J. Zool. 200:347–54.

Fenton, M. B., and T. H. Kunz. 1977. Movements and behavior. *In* Baker, Jones, and Carter (1977), pp. 351–64.

Fenton, M. B., and R. L. Peterson. 1972. Further notes on *Tadarida aloysiisabaudiae* and *Tadarida russata* (Chiroptera: Molossidae—Africa). Can. J. Zool. 50:19–24.

Fenton, M. B., D. C. Tennant, and J. Wyszecki. 1987. Using echolocation calls to measure the distribution of bats: the case of *Euderma maculatum*. J. Mamm. 68:142–44.

Fenton, M. B., D. W. Thomas, and R. Sasseen. 1981. *Nycteris grandis* (Nycteridae): an African carnivorous bat. J. Zool. 194:461–65.

Ferguson, J. W. H., J. A. J. Nel, and M. J. de Wet. 1983. Social organization and movement patterns of black-backed jackals *Canis mesomelas* in South Africa. J. Zool. 199:487–502.

Ferguson, W. W. 1981. The systematic position of *Canis aureus lupaster* (Carnivora: Canidae) and the occurrence of *Canis lupus* in North Africa, Egypt and Sinai. Mammalia 45:459–65.

Ferguson, W. W., Y. Porath, and S. Paley. 1985. Late Bronze period yields first osteological evidence of *Dama dama* (Artiodactyla: Cervidae) from Israel and Arabia. Mammalia 49:209–14.

Ferron, J. 1976. Cycle annuel d'activité de l'écureuil roux *(Tamiasciurus hudsonicus)*, adultes et jeunes en semi-liberté au Quebec. Naturaliste Canadien 103:1–10.

————. 1977. Le comportement de marquage chez le spermophile à mante dorée *(Spermophilus lateralis)*. Naturaliste Canadien 104:407–18.

Ferron, J., and J. Prescott. 1977. Gestation, litter size, and number of litters of the red squirrel *(Tamiasciurus hudsonicus)* in Quebec. Can. Field-Nat. 91:83–85.

Festa-Bianchet, M., and D. A. Boag. 1982. Territoriality in adult female Columbian ground squirrels. Can. J. Zool. 60:1060–66.

Field, R. J., and G. Feltner. 1974. Wolverine. Colorado Outdoors 23(2):1–6.

Figala, J., K. Hoffmann, and G. Goldau. 1973. Zur Jahresperiodik beim Dsungarischen Zwerghamster *Phodopus sungorus* Pallas. Oecologia 12:89–118.

Filewood, L. W. 1983. The possible occurrence in New Guinea of the ghost bat (*Macroderma gigas;* Chiroptera, Megadermatidae). Austral. Mamm. 6:35–36.

Filippucci, M. G., G. Nascetti, E. Capanna, and L. Bullini. 1987. Allozyme variation and systematics of European moles of the genus *Talpa* (Mammalia, Insectivora). J. Mamm. 68:487–99.

Filonov, C. P., and C. D. Zykov. 1974. Dynamics of moose populations in the forest zone of the European part of the USSR and in the Urals. Naturaliste Canadien 101:605–13.

Findley, J. S. 1972. Phenetic relationships among bats of the genus *Myotis*. Syst. Zool. 21:31–52.

Findley, J. S., and D. E. Wilson. 1974. Observations on the neotropical disk-winged bat, *Thyroptera tricolor* Spix. J. Mamm. 55:562–71.

Finerty, J. P. 1979. Cycles in Canadian lynx. Amer. Nat. 114:453–55.

Finley, K. J., J. P. Hickie, and R. A. Davis. 1987. Status of the beluga, *Delphinapterus leucas*, in the Beaufort Sea. Can. Field-Nat. 101:271–78.

Finley, R. B., Jr., and J. Creasy. 1982. First specimen of the spotted bat *(Euderma maculatum)* from Colorado. Great Basin Nat. 42:360.

Fish, J. F., J. L. Sumich, and G. L. Lingle. 1974. Sounds produced by the gray whale, *Eschrichtius robustus*. Mar. Fish. Rev. 36(4):38–45.

Fisher, J. 1978. Tiger! Tiger! Internatl. Wildl. 8(3):4–12.

Fisher, J., N. Simon, and J. Vincent. 1969. Wildlife in danger. Viking Press, New York, 368 pp.

Fisler, G. F. 1971. Age structure and sex ratio in populations of *Reithrodontomys*. J. Mamm. 52:653–62.

————. 1977. Potential behavioral dominance by the Mongolian gerbil. Amer. Midl. Nat. 97:33–41.

Fitch, H. S., P. Goodrum, and C. Newman. 1952. The armadillo in the southeastern United States. J. Mamm. 33:21–37.

Fitch, J. E., and R. L. Brownell, Jr. 1968. Fish otoliths in cetacean stomachs and their importance in interpreting feeding habits. J. Fish. Res. Bd. Can. 25:2561–74.

————. 1971. Food habits of the franciscana *Pontoporia blainvillei* (Cetacea: Platanistidae) from South America. Bull. Mar. Sci. 21:626–36.

Fitch, J. H., and K. A. Shump, Jr. 1979. *Myotis keenii*. Mammalian Species, no. 121, 3 pp.

Fittinghoff, N. A., Jr., and D. G. Lindburg. 1980. Riverine refuging in east Bornean *Macaca fascicularis*. *In* Lindburg (1980), pp. 182–214.

Fitzgerald, A. E. 1976. Diet of the opossum, *Trichosurus vulpecula* (Kerr) in the Orongorongo Valley, Wellington, New Zealand, in relation to food-plant availability. New Zealand J. Zool. 3:399–419.

———. 1978. Aspects of the food and nutrition of the brush-tailed opossum, *Trichosurus vulpecula* (Kerr, 1792), Marsupialia: Phalangeridae, in New Zealand. *In* Montgomery (1978), pp. 289–303.

Fitzgerald, J. P., and R. R. Lechleitner. 1974. Observations on the biology of Gunnison's prairie dog in central Colorado. Amer. Midl. Nat. 92:146–63.

Flannery, T. F. 1983. Revision in the macropodid subfamily Sthenurinae (Marsupialia: Macropodoidea) and the relationships of the species of *Troposodon* and *Lagostrophus*. Austral. Mamm. 6:15–28.

———. 1985. Notes on the distribution, abundance, diet and habitat of the New Guinea murid (Rodentia) *Xenuromys barbatus* (Milne-Edwards, 1900). Austral. Mamm. 8:111–15.

———. 1987a. A new species of *Phalanger* (Phalangeridae: Marsupialia) from montane western Papua New Guinea. Rec. Austral. Mus. 39:183–93.

———. 1987b. An historic record of the New Zealand greater short-tailed bat, *Mystacina robusta* (Microchiroptera: Mystacinidae) from the South Island, New Zealand. Austral. Mamm. 10:45–46.

———. 1988. *Pogonomys championi* n. sp., a new murid (Rodentia) from montane western Papua New Guinea. Rec. Austral. Mus. 40:333–41.

Flannery, T. F., K. Aplin, C. P. Groves, and M. Adams. 1989. Revision of the New Guinean genus *Mallomys* (Muridae: Rodentia), with descriptions of two new species from subalpine habitats. Rec. Austral. Mus. 41:83–105.

Flannery, T. F., M. Archer, and G. Maynes. 1987. The phylogenetic relationships of living phalangerids (Phalangeroidea: Marsupialia) with a suggested new taxonomy. *In* Archer (1987), pp. 477–506.

Flannery, T. F., and J. H. Calaby. 1987. Notes on the species of *Spilocuscus* (Marsupialia: Phalangeridae) from northern New Guinea and the Admiralty and St. Matthias Island groups. *In* Archer (1987), pp. 547–58.

Flannery, T. F., P. V. Kirch, J. Specht, and M. Spriggs. 1988. Holocene mammal faunas from archaeological sites in island Melanesia. Archaeol. Oceania 23:89–94.

Flannery, T. F., and F. Szalay. 1982. *Bohra paulae*, a new giant fossil tree kangaroo (Marsupialia: Macropodidae) from New South Wales, Australia. Austral. Mamm. 5:83–94.

Fleagle, J. G. 1984. Are there any fossil gibbons? *In* Preuschoft et al. (1984), pp. 431–47.

Fleharty, E. D., and M. A. Mares. 1973. Habitat preference and spatial relations of *Sigmodon hispidus* on a remnant prairie in west-central Kansas. Southwestern Nat. 18:21–29.

Fleming, K., and N. R. Holler. 1989. Endangered beach mice repopulate Florida beaches. Endangered Species Tech. Bull. 14(1–2):9, 11.

Fleming, M. R. 1980. Thermoregulation and torpor in the sugar glider, *Petaurus breviceps* (Marsupialia: Petauridae). Austral. J. Zool. 28:521–34.

———. 1985. The thermal physiology of the mountain pygmy-possum *Burramys parvus* (Marsupialia: Burramyidae). Austral. Mamm. 8:79–90.

Fleming, M. R., and A. Cockburn. 1979. *Ningaui*: a new genus of dasyurid for Victoria. Victorian Nat. 96:142–45.

Fleming, M. R., and H. Frey. 1984. Aspects of the natural history of feathertail gliders *(Acrobates pygmaeus)* in Victoria. *In* Smith and Hume (1984), pp. 403–8.

Fleming, T. H. 1971. Population ecology of three species of neotropical rodents. Misc. Publ. Mus. Zool. Univ. Michigan, no. 143, 77 pp.

———. 1972. Aspects of the population dynamics of three species of opossums in the Panama Canal Zone. J. Mamm. 53:619–23.

———. 1973. The reproductive cycles of three species of opossums and other mammals in the Panama Canal Zone. J. Mamm. 54:439–55.

———. 1974. The population ecology of two species of Costa Rican heteromyid rodents. Ecology 55:493–510.

———. 1977. Growth and development of two species of tropical heteromyid rodents. Amer. Midl. Nat. 98:109–23.

———. 1988. The short-tailed fruit bat. A study in plant-animal interactions. Univ. Chicago Press, xvi + 365 pp.

Fleming, T. H., E. T. Hooper, and D. E. Wilson. 1972. Three Central American bat communities: structure, reproductive cycles, and movement patterns. Ecology 53:555–69.

Fletcher, T. P. 1985. Aspects of reproduction in the male eastern quoll, *Dasyurus viverrinus* (Shaw) (Marsupialia: Dasyuridae), with notes on polyoestry in the female. Austral. J. Zool. 33:101–10.

Fletemeyer, J. 1985. A hooded seal. Florida Nat. 58(2):9.

Floody, O. R., and A. P. Arnold. 1975. Uganda kob *(Adenota kob thomasi)*: territoriality and the spatial distribution of sexual and agnostic behaviors at a territorial ground. Z. Tierpsychol. 37:192–212.

Floyd, B. L., and M. R. Stromberg. 1981. New records of the swift fox *(Vulpes velox)* in Wyoming. J. Mamm. 62:650–51.

Flux, J. E. C. 1970. Life history of the mountain hare *(Lepus timidus scoticus)* in northeast Scotland. J. Zool. 161:75–123.

———. 1983. Introduction to taxonomic problems in hares. Acta Zool. Fennica 174:7–10.

Flux, J. E. C., and P. J. Fullagar. 1983. World distribution of the rabbit *(Oryctolagus cuniculus)*. Acta Zool. Fennica 174:75–77.

Flynn, R. W., and M. T. Abdullah. 1984. Distribution and status of the Sumatran rhinoceros in peninsular Malaysia. Biol. Conserv. 28:253–73.

Foerg, R. 1982. Reproductive behavior in *Varecia variegata*. Folia Primatol. 38:108–21.

Fogden, M. P. L. 1974. A preliminary field study of the western tarsier, *Tarsius bancanus* Horsfield. *In* Martin, Doyle, and Walker (1974), pp. 151–65.

Folk, G. E., Jr., A. Larson, and M. A. Folk. 1976. Physiology of hibernating bears. *In* Pelton, Lentfer, and Folk (1976), pp. 373–80.

Foltz, D. W. 1981. Genetic evidence for long-term monogamy in a small rodent, *Peromyscus polionotus*. Amer. Nat. 117:665–75.

Foltz, D. W., and J. L. Hoogland. 1981. Analysis of the mating system in the black-tailed prairie dog *(Cynomys ludovicianus)* by likelihood of paternity. J. Mamm. 62:706–12.

Fons, R. 1974. Le répertoire comportemental de la pachyure Etrusque, *Suncus etruscus* (Savi, 1822). Terre Vie 28:131–57.

Fontaine, P. A. 1965. Breeding clouded leopards *Neofelis nebulosa* at Dallas Zoo. Internatl. Zoo Yearbook 5:113–14.

Fontaine, R., and F. V. Du Mond. 1977. The red ouakari in a seminatural environment: potentials for propagation and study. *In* Rainier III and Bourne (1977), pp. 167–236.

Fooden, J. 1963. A revision of the woolly monkeys (genus *Lagothrix*). J. Mamm. 44:213–47.

———. 1969. Taxonomy and evolution of the monkeys of Celebes (Primates: Cercopithecidae). Bibl. Primatol. 10:1–148.

———. 1971. Report on primates collected in western Thailand January–April, 1967. Fieldiana Zool. 59:1–62.

———. 1975. Taxonomy and evolution of liontail and pigtail macaques (Primates: Cercopithecidae). Fieldiana Zool. 67:1–169.

———. 1976a. Primates obtained in peninsular Thailand June–July, 1973, with notes on the distribution of continental Southeast Asian leaf-monkeys *(Presbytis)*. Primates 17:95–118.

———. 1976b. Provisional classification and key to living species of macaques (Primates: *Macaca*). Folia Primatol. 25:225–36.

———. 1979. Taxonomy and evolution of the *sinica* group of macaques: I. Species and subspecies accounts of *Macaca sinica*. Primates 10:109–40.

———. 1982. Ecogeographic segregation of macaque species. Primates 23:574–79.

———. 1987. Gibbon distribution in China. Acta Theriol. Sinica 7:161–67.

———. 1988. Taxonomy and evolution of the *sinica* group of macaques: 6. Interspecific comparisons and synthesis. Fieldiana Zool., no. 45, vi + 44 pp.

Fooden, J., and R. J. Izor. 1983. Growth curves, dental emergence norms, and supplementary morphological observations in known-age captive orangutans. Amer. J. Primatol. 5:285–301.

Fooden, J., A. Mahabal, and S. S. Saha. 1981. Redefinition of rhesus macaque—bonnet macaque boundary in peninsular India (Primates: *Macaca mulatta, M. radiata*). J. Bombay Nat. Hist. Soc. 78:463–74.

Fooden, J., Quan Guoqiang, and Luo Yining. 1987. Gibbon distribution in China. Acta Theriol. Sinica 7:161–67.

Fooden, J., Quan Guoqiang, Wang Zongren, and Wang Yingxiang. 1985. The stumptail macaques of China. Amer. J. Primatol. 8:11–30.

Foose, T. 1987a. Species survival plans and overall management strategies. *In* Tilson and Seal (1987), pp. 304–16.

———. 1987b. Long-term management of small rhino populations. Pachyderm 9:13–15.

Ford, J. K. B., and H. D. Fisher. 1978. Underwater acoustic signals of the narwhal *(Monodon monoceros)*. Can. J. Zool. 56:552–60.

———. 1983. Group-specific dialects of killer whales *(Orcinus orca)* in British Columbia. *In* Payne (1983), pp. 129–61.

Ford, L. S., and R. S. Hoffmann. 1988. *Potos flavus*. Mammalian Species, no. 321, 9 pp.

Ford, S. D. 1977. Range, distribution and habitat of the western harvest mouse, *Reithrodontomys megalotis*, in Indiana. Amer. Midl. Nat. 98:422–32.

Ford, S. M. 1986a. Systematics of the New World monkeys. *In* Swindler and Erwin (1986), pp. 73–135.

———. 1986b. Subfossil platyrrhine tibia (Primates: Callitrichidae) from Hispaniola: a possible further example of island gigantism. Amer. J. Phys. Anthropol. 70:47–62.

Fordyce, R. E., R. H. Mattlin, and J. M. Dixon. 1984. Second record of spectacled porpoise from subantarctic southwest Pacific. Sci. Rept. Whales Res. Inst. 35:159–64.

Foresman, K. R., and R. A. Mead. 1973. Duration of post-implantation in a western subspecies of the spotted skunk *(Spilogale putorius)*. J. Mamm. 54:521–23.

Forrest, S. C., D. E. Biggins, L. Richardson, T. W. Clark, T. M. Campbell, III, K. A. Fagerstone, and E. T. Thorne. 1988. Population attributes for the black-footed ferret *(Mustela nigripes)* at Meeteetse, Wyoming, 1981–1985. J. Mamm. 69:261–73.

Forsten, A., and P. M. Youngman. 1982. *Hydrodamalis gigas*. Mammalian Species, no. 165, 3 pp.

Fossey, D. 1972. Vocalizations of the mountain gorilla *(Gorilla gorilla beringei)*. Anim. Behav. 20:36–53.

———. 1974. Observations on the home range of one group of mountain gorillas *(Gorilla gorilla beringei)*. Anim. Behav. 22:568–81.

Fossey, D., and A. H. Harcourt. 1977. Feeding ecology of free-ranging mountain gorilla *(Gorilla gorilla beringei)*. *In* Clutton-Brock (1977), pp. 415–47.

Foster, M. S., and R. M. Timm. 1976. Tentmaking by *Artibeus jamaicensis* (Chiroptera: Phyllostomatidae) with comments on plants used by bats for tents. Biotrópica 8:265–69.

Fourie, P. B. 1977. Acoustic communication in the rock hyrax. Z. Tierpsychol. 44:194–219.

Fowler, C. W. 1985. An evaluation of the role of entanglement in the population dynamics of northern fur seals on the Pribilof Islands. *In* Shomura, R. S., and H. O. Yoshida, eds., Proceedings of the workshop on the fate and impact of marine debris, Natl. Oceanic Atmos. Admin. Mem., NOAA-TM-NMFS-SWFC-54, pp. 291–307.

———. 1987. Marine debris and northern fur seals: a case study. Marine Pollution Bull. 18:326–35.

Fox, B. J., and D. A. Briscoe. 1980. *Pseudomys pilligaensis*, a new species of murid rodent from the Pilliga Scrub, northern New South Wales. Austral. Mamm. 3:109–26.

Fox, J. L. 1984. Population density of mountain goats in southeast Alaska. Proc. Bien. Symp. N. Wild Sheep and Goat Council, 4:51–60.

Fox, M. W. 1975. The wild canids. Van Nostrand Reinhold, New York, xvi + 508 pp.

———. 1978. The dog. Its domestication and behavior. Garland STPM Press, New York, vii + 296 pp.

———. 1984. The whistling hunters. Field studies of the Asiatic wild dog *(Cuon alpinus)*. State Univ. New York Press, Albany, viii + 150 pp.

Fox, M. W., A. M. Beck, and E. Blackman. 1975. Behavior and ecology of a small group of urban dogs *(Canis familiaris)*. Appl. Anim. Ethol. 1:119–37.

Fradrich, H. 1974. A comparison of behaviour in the Suidae. *In* Geist and Walther (1974), pp. 133–43.

Frame, L. H., and G. W. Frame. 1976. Female African wild dogs emigrate. Nature 263:227–29.

Frame, L. H., J. R. Malcolm, G. W. Frame, and H. Van Lawick. 1979. Social organization of African wild dogs *(Lycaon pictus)* on the Serengeti Plains, Tanzania (1967–1978). Z. Tierpsychol. 50:225–49.

Francis, C. M., and J. E. Hill. 1986. A review of the Bornean *Pipistrellus* (Mammalia: Chiroptera). Mammalia 50:43–55.

Frank, L. G. 1986a. Social organization of the spotted hyaena *(Crocuta crocuta)*. I. Demography. Anim. Behav. 34:1500–1509.

———. 1986b. Social organization of the spotted hyaena *Crocuta crocuta*. II. Dominance and reproduction. Anim. Behav. 34:1510–27.

Franklin, W. L. 1974. The social behavior of the vicuna. *In* Geist and Walther (1974), pp. 477–87.

———. 1982. Biology, ecology, and relationship to man of the South American camelids. Pymatuning Lab. Ecol. Spec. Publ. 6:457–89.

———. 1987. My two decades with America's camels. Internatl. Wildl. 17(5):34–43.

Franzmann, A. W. 1978. Moose. *In* Schmidt and Gilbert (1978), pp. 67–81.

———. 1981. *Alces alces*. Mammalian Species, no. 154, 7 pp.

Fraser, F. C. 1970. An early 17th century record of the Californian grey whale in Icelandic waters. Investig. Cetacea 2:13–20.

Frazer, J. F. D. 1973. Specific foetal growth rates of cetaceans. J. Zool. 169:111–26.

Freeman, P. W. 1981. A multivariate study of the family Molossidae (Mammalia, Chiroptera): morphology, ecology, and evolution. Fieldiana Zool., n.s., no. 7, vii + 173 pp.

Freese, C. 1976. Censusing *Alouatta palliata, Ateles geoffroyi*, and *Cebus capucinus* in the Costa Rican dry forest. *In* Thorington and Heltne (1976), pp. 4–19.

Freese, C. H., M. A. Freese, and N. Castro R. 1977. The status of callitrichids in Peru. *In* Kleiman (1977a), pp. 121–30.

Freitas, T. R. O., and E. P. Lessa. 1984. Cytogenetics and morphology of *Ctenomys tor-*

quatus (Rodentia: Octodontidae). J. Mamm. 65:637–42.

French, N. R., D. M. Stoddart, and B. Bobek. 1975. Patterns of demography in small mammal populations. *In* Golley, F. B., K. Petrusewicz, and L. Ryszkowski, eds., Small mammals: their productivity and population dynamics. Cambridge Univ. Press, London, pp. 73–102.

French, T. W. 1981. Notes on the distribution and taxonomy of short-tailed shrews (genus *Blarina*) in the southeast. Brimleyana 6:101–10.

Fritts, S. H. 1982. Wolf depredations on livestock in Minnesota. U.S. Fish and Wildl. Serv. Res. Publ., no. 145, 11 pp.

———. 1983. Record dispersal by a wolf from Minnesota. J. Mamm. 64:166–67.

Fritts, S. H., and L. D. Mech. 1981. Dynamics, movements, and feeding ecology of a newly protected wolf population in northwestern Minnesota. Wildl. Monogr., no. 80, 79 pp.

Fritts, S. H., and J. A. Sealander. 1978. Reproductive biology and population characteristics of bobcats *(Lynx rufus)* in Arkansas. J. Mamm. 59:347–53.

Fritzell, E. K. 1978. Aspects of raccoon *(Procyon lotor)* social organization. Can. J. Zool. 56:260–71.

Froehlich, J. W., and P. H. Froehlich. 1987. The status of Panama's endemic howling monkeys. Primate Conserv. 8:58–62.

Froehlich, J. W., R. W. Thorington, Jr., and J. S. Otis. 1981. The demography of howler monkeys *(Alouatta palliata)* on Barro Colorado Island, Panama. Internatl. J. Primatol. 2:207–36.

Frost, K. J., and L. F. Lowry. 1981. Ringed, Baikal and Caspian seals—*Phoca hispida*, *Phoca sibirica* and *Phoca caspica*. *In* Ridgway and Harrison (1981*b*), pp. 29–53.

Fruzinski, B., L. Labudzki, and M. Wlazelko. 1983. Habitat, density and spatial structure of the forest roe deer population. Acta Theriol. 28:243–58.

Fry, E. 1971. The scaly-tailed possum *Wyulda squamicaudata* in captivity. Internatl. Zoo Yearbook 11:44–45.

Fryxell, J. M., and A. R. E. Sinclair. 1988. Seasonal migration by white-eared kob in relation to resources. Afr. J. Ecol. 26:17–31.

Fujita, M. S., and T. H. Kunz. 1984. *Pipistrellus subflavus*. Mammalian Species, no. 228, 6 pp.

Fulk, G. W. 1975. Population ecology of rodents in the semiarid shrublands of Chile. Occas. Pap. Mus. Texas Tech Univ., no. 33, 40 pp.

———. 1976. Notes on the activity, reproduction, and social behavior of *Octodon degus*. J. Mamm. 57:495–505.

Fulk, G. W., and A. R. Khokhar. 1980. Observations on the natural history of pika *(Ochotona rufescens)* from Pakistan. Mammalia 44:51–58.

Fulk, G. W., A. C. Smiet, and A. R. Khokhar. 1979. Movements of *Bandicota bengalensis* (Gray 1873) and *Tatera indica* (Hardwicke 1807) as revealed by radio telemetry. J. Bombay Nat. Hist. Soc. 76:457–62.

Fuller, P. J., and A. A. Burbidge. 1987. Discovery of the dibbler, *Parantechinus apicalis*, on islands at Jurien Bay. W. Austral. Nat. 16:177–81.

Fuller, T. K. 1978. Variable home-range sizes of female gray foxes. J. Mamm. 59:446–49.

Fuller, T. K., W. E. Johnson, W. L. Franklin, and K. A. Johnson. 1987. Notes on the Patagonian hog-nosed skunk *(Conepatus humboldti)* in southern Chile. J. Mamm. 68:864–67.

Fuller, T. K., and L. B. Keith. 1980. Wolf population dynamics and prey relationships in northwestern Alberta. J. Wildl. Mgmt. 44:583–601.

Funaioli, U. 1971. Guida breve dei mammiferi della Somalia. Istituto Agronomico per l'Oltremare, Firenze, 232 pp.

Funmilayo, O. 1977. Distribution and abundance of moles *(Talpa europaea* L.) in relation to physical habitat and food supply. Oecologia 30:277–83.

Furley, C. W. 1986. Reproductive parameters of African gazelles: gestation, first fertile matings, first parturition and twinning. Afr. J. Ecol. 24:121–28.

Furley, C. W., H. Tichy, and H.-P. Uerpmann. 1988. Systematics and chromosomes of the Indian gazelle, *Gazella bennetti* (Sykes, 1831). Z. Saugetierk. 53:48–54.

Furuta, M. 1984. Note on a gray whale found in the Ise Bay on the Pacific coast of Japan. Sci. Rept. Whales Res. Inst. 35:195–97.

G

Gabow, S. A. 1975. Behavioral stabilization of a baboon hybrid zone. Amer. Nat. 109:701–12.

Gabuniya, L. K., N. S. Shevyreva, and V. D. Gabuniya. 1985. On the first find of fossil Marsupialia in Asia. Doklady Akad. Nauk SSSR 281:684–85.

Gade, D. W. 1967. The guinea pig in Andean folk culture. Geogr. Rev. 57:213–24.

Gaines, M. S., and R. K. Rose. 1976. Population dynamics of *Microtus ochrogaster* in eastern Kansas. Ecology 57:1145–61.

Gaisler, J. 1983. Nouvelles données sur les chiroptères du nord Algerien. Mammalia 47:359–69.

Gaisler, J., V. Holas, and M. Homolka. 1976. Ecology and reproduction of Gliridae (Mammalia) in northern Moravia. Folia Zool. 26:213–28.

Gaisler, J., G. Madkour, and J. Pelikan. 1972. On the bats (Chiroptera) of Egypt. Acta Sci. Nat. Brno 6:1–40.

Galat, G., and A. Galat-Luong. 1976. La colonisation de la mangrove par *Cercopithecus aethiops sabaeus* au Senegal. Terre Vie 30:3–30.

———. 1977. Démographie et régime alimentaire d'une troupe de *Cercopithecus aethiops sabaeus* en habitat marginal au nord Senegal. Terre Vie 31:557–77.

———. 1978. Diet of green monkeys in Senegal. *In* Chivers and Herbert (1978), pp. 257–58.

Galat-Luong, A. 1975. Notes préliminaires sur l'écologie de *Cercopithecus ayscanius schmidti* dans les environs de Bangui (R. C. A.). Terre Vie 29:288–97.

Galdikas, B. M. F. 1981. Orangutan reproduction in the wild. *In* Graham, C. E., ed., Reproductive biology of the great apes, Academic Press, New York, pp. 281–300.

———. 1982. Orang-utan tool-use at Tanjung Puting Reserve, central Indonesian Borneo (Kalimantan Tengah). J. Human Evol. 10:19–33.

———. 1985*a*. Sub-adult male orangutan sociality and reproductive behavior at Tanjung Puting. Amer. J. Primatol. 8:87–99.

———. 1985*b*. Orangutan sociality at Tanjung Puting. Amer. J. Primatol. 9:101–19.

———. 1988. Orangutan diet, range, and activity at Tanjung Puting, central Borneo. Internatl. J. Primatol. 9:1–35.

Gallagher, M. D., and D. L. Harrison. 1977. Report on the bats (Chiroptera) obtained by the Zaire River Expedition. Bonner Zool. Beitr. 28:19–32.

Gallo-R., J. P., and A. Ortega-O. 1986. The first report of *Zalophus californianus* in Acapulco, Mexico. Mar. Mamm. Sci. 2:158.

Gambell, R. 1985. Fin whale—*Balaenoptera physalus*. *In* Ridgway and Harrison (1985), pp. 171–92.

Games, I. 1983. Observations on the sitatunga *Tragelaphus spekei selousi* in the Okavango Delta of Botswana. Biol. Conserv. 27:157–70.

Gangloff, B. 1975. Beitrag zur Ethologie der Schleichkatzen (Banderlinsang, *Prionodon linsang* [Hardw.] und Banderpalmenroller, *Hemigalus derbyanus* [Gray]). Zool. Garten 45:329–76.

Gannon, W. L. 1988. *Zapus trinotatus*. Mammalian Species, no. 315, 5 pp.

Ganslosser, U. 1984. On the occurrence of female coalitions in tree kangaroos (Marsupialia, Macropodidae, *Dendrolagus*). Austral. Mamm. 7:219–21.

Ganzhorn, J. U., J. P. Abraham, and M. Razanahoera-Rakotomalala. 1985. Some aspects of the natural history and food selection of *Avahi laniger*. Primates 26:452–63.

Ganzhorn, J. U., and J. Rabesoa. 1986. The aye-aye *(Daubentonia madagascariensis)* found in the eastern rainforest of Madagascar. Folia Primatol. 46:125–26.

Gao Yaoting. 1981. On the present status, historical distribution and conservation of wild elephant in China. Acta Theriol. Sinica 1:19–26.

Garber, P. A., L. Moya, and C. Malaga. 1984. A preliminary field study of moustached tamarin monkey *(Saguinus mystax)* in northeastern Peru: questions concerned with the evolution of a communal breeding system. Folia Primatol. 42:17–32.

Garber, P. A., and M. F. Teaford. 1986. Body weights in mixed species troops of *Saguinus mystax mystax* and *Saguinus fuscicollis nigrifrons* in Amazonian Peru. Amer. J. Phys. Anthropol. 71:331–36.

Gardner, A. L. 1971a. Karyotypes of two rodents from Peru, with a description of the highest diploid number recorded from a mammal. Experientia 27:1088–89.

———. 1971b. Postpartum estrus in a red brocket deer, *Mazama americana*, from Peru. J. Mamm. 52:623–24.

———. 1973. The systematics of the genus *Didelphis* (Marsupialia: Didelphidae) in North and Middle America. Spec. Publ. Mus. Texas Tech Univ., no. 4, 81 pp.

———. 1976. The distributional status of some Peruvian mammals. Occas. Pap. Mus. Zool. Louisiana State Univ., no. 48, 18 pp.

———. 1977. Feeding habits. *In* Baker, Jones, and Carter (1977), pp. 293–350.

———. 1981. Review of *The mammals of Suriname*, by A. M. Husson. J. Mamm. 62:445–48.

———. 1986. The taxonomic status of *Glossophaga morenoi* Martinez and Villa, 1938 (Mammalia: Chiroptera: Phyllostomidae). Proc. Biol. Soc. Washington 99:489–92.

Gardner, A. L., and L. H. Emmons. 1984. Species groups in *Proechimys* (Rodentia, Echimyidae) as indicated by karyology and bullar morphology. J. Mamm. 65:10–25.

Gardner, A. L., R. K. LaVal, and D. E. Wilson. 1970. The distributional status of some Costa Rican bats. J. Mamm. 51:712–29.

Gardner, A. L., and J. P. O'Neill. 1969. The taxonomic status of *Sturnira bidens* (Chirop-

tera: Phyllostomatidae) with notes on its karyotype and life history. Occas. Pap. Mus. Zool. Louisiana State Univ., no. 38, 8 pp.

———. 1971. A new species of *Sturnira* (Chiroptera: Phyllostomatidae) from Peru. Occas. Pap. Mus. Zool. Louisiana State Univ., no. 42, 7 pp.

Gardner, A. L., and J. L. Patton. 1972. New species of *Philander* (Marsupialia: Didelphidae) and *Mimon* (Chiroptera: Phyllostomatidae) from Peru. Occas. Pap. Mus. Zool. Louisiana State Univ., no. 43, 12 pp.

———. 1976. Karyotypic variation in oryzomyine rodents (Cricetinae) with comments on chromosomal evolution in the neotropical cricetine complex. Occas. Pap. Mus. Zool. Louisiana State Univ., no. 49, 48 pp.

Gardner, C. L., W. B. Ballard, and H. Jessup. 1986. Long distance movement by an adult wolverine. J. Mamm. 67:603.

Garrott, R. A., L. E. Eberhardt, and W. C. Hanson. 1983. Arctic fox den identification and characteristics in northern Alaska. Can. J. Zool. 61:423–26.

Garrott, R. A., and D. A. Jenni. 1978. Arboreal behavior of yellow-bellied marmots. J. Mamm. 59:433–34.

Garshelis, D. L., and J. A. Garshelis. 1984. Movements and management of sea otters in Alaska. J. Wildl. Mgmt. 48:665–78.

Garshelis, D. L., A. M. Johnson, and J. A. Garshelis. 1984. Social organization of sea otters in Prince William Sound, Alaska. Can. J. Zool. 62:2648–58.

Gartlan, J. S. 1970. Preliminary notes on the ecology and behavior of the drill, *Mandrillus leucophaeus* Ritgen, 1824. *In* Napier and Napier (1970), pp. 445–80.

Gartlan, J. S., and C. K. Brain. 1968. Ecology and social variability in *Cercopithecus aethiops* and *C. mitis*. *In* Jay (1968), pp. 253–92.

Gartlan, J. S., and T. T. Struhsaker. 1972. Polyspecific associations and niche separation of rain-forest anthropoids in Cameroon, West Africa. J. Zool. 168:221–66.

Gashwiler, J. S. 1972. Life history notes on the Oregon vole, *Microtus oregoni*. J. Mamm. 53:558–69.

———. 1976. Notes on the reproduction of Trowbridge shrews in western Oregon. Murrelet 57:58–62.

Gaski, A. L. 1988. 'Roo update: a short history of U.S. kangaroo skin and product imports, 1984–1987. Traffic (U.S.A.) 8(2):1–5.

Gaskin, D. E. 1968. The New Zealand Cetacea. New Zealand Mar. Dept. Fish. Res. Bull., no. 1, n.s., 92 pp.

———. 1970. Composition of schools of sperm whales *Physeter catodon* Linn. east of New Zealand. New Zealand J. Mar. Freshwater Res. 4:456–71.

———. 1976. The evolution, zoogeography and ecology of Cetacea. Oceanogr. Mar. Biol. Ann. Rev. 14:247–346.

———. 1977. Harbour porpoise *Phocoena phocoena* (L.) in the western approaches to the Bay of Fundy 1969–75. Rept. Internatl. Whaling Comm. 27:487–92.

———. 1982. The ecology of whales and dolphins. Heinemann, London, xii + 459 pp.

———. 1984. The harbour porpoise *Phocoena phocoena* (L.): regional populations, status, and information on direct and indirect catches. Rept. Internatl. Whaling Comm. 34:569–86.

———. 1987. Updated status of the right whale, *Eubalaena glacialis*, in Canada. Can. Field-Nat. 101:295–309.

Gaskin, D. E., P. W. Arnold, and B. A. Blair. 1974. *Phocoena phocoena*. Mammalian Species, no. 42, 8 pp.

Gaskin, D. E., and B. A. Blair. 1977. Age determination of harbour porpoise, *Phocoena phocoena* (L.), in the western North Atlantic. Can. J. Zool. 55:18–30.

Gaskin, D. E., G. J. D. Smith, A. P. Watson, W. Y. Yasui, and D. B. Yurick. 1984. Reproduction in the porpoises (Phocoenidae): implications for management. *In* Perrin, Brownell, and DeMaster (1984), pp. 135–48.

Gaskin, D. E., and A. P. Watson. 1984. The harbor porpoise, *Phocoena phocoena*, in Fish Harbour, New Brunswick, Canada: occupancy, distribution, and movements. Fishery Bull. 83:427–42.

Gasperetti, J., D. L. Harrison, and W. Büttiker. 1985. The Carnivora of Arabia. Fauna Saudi Arabia 7:397–461.

Gates, G. R. 1978. Vision in the monotreme echidna *(Tachyglossus aculeatus)*. Austral. Zool. 20:147–69.

Gauckler, A., and M. Kraus. 1970. Kennzeichen und Verbreitung von *Myotis brandti* (Eversman, 1845). Z. Saugetierk. 35:113–24.

Gauthier-Pilters, H. 1974. The behaviour and ecology of camels in the Sahara, with special reference to nomadism and water management. *In* Geist and Walther (1974), pp. 542–51.

Gauthier-Pilters, H., and A. I. Dagg. 1981. The camel. Univ. Chicago Press, xii + 208 pp.

Gautier, J.-P. 1985. Quelques caractéristiques écologiques du singe des marais: *Allenopithecus nigroviridis* Lang 1923. Rev. Ecol. 40:331–42.

Gautier-Hion, A. 1971. L'écologie du talapoin du Gabon. Terre Vie 25:427–90.

———. 1973. Social and ecological features of talapoin monkey—comparisons with sympatric cercopithecines. *In* Michael and Crook (1973), pp. 147–70.

Gautun, J. 1975. Périodicité de la reproduction de quelques rongeurs d'une savane préforestiere du centre de la Côte-D'Ivoire. Terre Vie 29:265–87.

Gautun, J.-C., I. Sankhon, and M. Tranier. 1986. Nouvelle contribution à la connaissance des rongeurs du massif guinéen des monts Nimba (Afrique occidentale). Systématique et aperçu quantitatif. Mammalia 50:205–17.

Gautun, J.-C., M. Tranier, and B. Sicard. 1985. Liste préliminaire des rongeurs du Burkina Faso (ex Haute-Volta). Mammalia 49:537–42.

Gavin, T. A., and B. May. 1988. Taxonomic status and genetic purity of Columbian white-tailed deer. J. Wildl. Mgmt. 52:1–10.

Gebo, D. L., and D. T. Rasmussen. 1985. The earliest fossil pangolin (Pholidota: Manidae) from Africa. J. Mamm. 66:538–41.

Gee, C. K., R. Magleby, W. R. Bailey, R. L. Gum, and L. M. Arthur. 1977. Sheep and lamb losses to predators and other causes in the western United States. U.S. Dept. Agric., Agric. Econ. Rept., no. 369, 41 pp.

Geiser, F. 1986. Thermoregulation and torpor in the kultarr, Antechinomys laniger (Marsupialia: Dasyuridae). J. Comp. Physiol., ser. B, 156:751–57.

Geiser, F., M. L. Augee, H. C. K. McCarron, and J. Raison. 1984. Correlates of torpor in the insectivorous marsupial Sminthopsis murina. Austral. Mamm. 7:185–91.

Geiser, F., L. Matwiejczyk, and R. V. Baudinette. 1986. From ectothermy to heterothermy: the energetics of the kowari, Dasyuroides byrnei (Marsupialia: Dasyuridae). Physiol. Zool. 59:220–29.

Geist, V. 1971. Mountain sheep. A study in behavior and evolution. Univ. Chicago Press, xv + 383 pp.

———. 1974. Bound by tradition. Anim. Kingdom 77(6):2–9.

———. 1982. Adaptive behavioral strategies. In Thomas and Toweill (1982), pp. 219–77.

———. 1984. Goat antelopes. In Macdonald (1984), pp. 584–89.

———. 1985. On Pleistocene bighorn sheep: some problems of adaptation, and relevance to today's American megafauna. Wildl. Soc. Bull. 13:351–59.

———. 1986. The paradox of the great Irish stags. Nat. Hist. 95(3):54–65.

Geist, V., and F. Walther. 1974. The behaviour of ungulates and its relation to management. Internatl. Union Conserv. Nat. Publ., n.s., no. 24, 2 vols.

Geluso, K. N., J. S. Altenbach, and D. E. Wilson. 1976. Bat mortality: pesticide poisoning and migratory stress. Science 194:184–86.

———. 1981. Organochlorine residues in young Mexican free-tailed bats from several roosts. Amer. Midl. Nat. 105:249–57.

Gelvin, B. R. 1980. Morphometric affinities of Gigantopithecus. Amer. J. Phys. Anthropol. 53:541–68.

Genest, H., and G. Dubost. 1974. Pair-living in the mara (Dolichotis patagonum Z.). Mammalia 38:155–62.

Genest-Villard, H. 1978a. Radio-tracking of a small rodent, Hybomys univittatus, in an African equatorial forest. Bull. Carnegie Mus. Nat. Hist., no. 6, pp. 92–96.

———. 1978b. Révision systématique du genre Graphiurus (Rongeurs, Gliridae). Mammalia 42:391–426.

———. 1979. Écologie de Steatomys opimus Pousargues, 1894 (Rongeurs, Dendromuridés) en Afrique centrale. Mammalia 43:275–94.

———. 1980. Régime alimentaire des rongeurs myomorphes de forêt équatoriale (région de M'Baiki, République Centrafricaine). Mammalia 44:423–84.

Gengozian, N., J. S. Batson, and T. A. Smith. 1977. Breeding of tamarins (Saguinus spp.) in the laboratory. In Kleiman (1977a), pp. 207–13.

Genoways, H. H., and R. J. Baker. 1972. Stenoderma rufum. Mammalian Species, no. 18, 4 pp.

———. 1988. Lasiurus blossevillii (Chiroptera: Vespertilionidae) in Texas. Texas J. Sci. 40:111–13.

Genoways, H. H., and E. C. Birney. 1974. Neotoma alleni. Mammalian Species, no. 41, 4 pp.

Genoways, H. H., and J. R. Choate. 1972. A multivariate analysis of systematic relationships among populations of the short-tailed shrew (genus Blarina) in Nebraska. Syst. Zool. 21:106–16.

Genoways, H. H., and J. K. Jones, Jr. 1972. Variation and ecology in a local population of the vesper mouse (Nyctomys sumichrasti). Occas. Pap. Mus. Texas Tech Univ., no. 3, 22 pp.

Genoways, H. H., and S. L. Williams. 1979a. Notes on bats (Mammalia: Chiroptera) from Bonaire and Curacao, Dutch West Indies. Ann. Carnegie Mus. 48:311–21.

———. 1979b. Records of bats (Mammalia: Chiroptera) from Suriname. Ann. Carnegie Mus. 48:323–35.

———. 1980. Results of the Alcoa Foundation–Suriname Expeditions. I. A new species of bat of the genus Tonatia (Mammalia: Phyllostomatidae). Ann. Carnegie Mus. 49:203–11.

———. 1984. Results of the Alcoa Foundation–Suriname Expeditions. IX. Bats of the genus Tonatia (Mammalia: Chiroptera) in Suriname. Ann. Carnegie Mus. 53:327–46.

———. 1986. Results of the Alcoa Foundation–Suriname Expeditions. XI. Bats of the genus Micronycteris (Mammalia: Chiroptera) in Suriname. Ann. Carnegie Mus. 55:303–24.

Genoways, H. H., S. L. Williams, and J. A. Groen. 1981. Results of the Alcoa Foundation–Suriname Expeditions. V. Noteworthy records of Surinamese mammals. Ann. Carnegie Mus. 50:319–32.

Gensch, W. 1963a. Breeding tupaias. Internatl. Zoo Yearbook 4:75–76.

———. 1963b. Successful rearing of the binturong Arctictis binturong. Internatl. Zoo Yearbook 4:79–80.

———. 1965. Birth and rearing of a spectacled bear Tremarctos ornatus at Dresden Zoo. Internatl. Zoo Yearbook 5:11.

Gentry, R. L. 1981. Northern fur seal—Callorhinus ursinus. In Ridgway and Harrison (1981a), pp. 143–60.

Gentry, R. L., D. P. Costa, J. P. Croxall, J. H. M. David, R. W. Davis, G. L. Kooyman, P. Majluf, T. S. McCann, and F. Trillmich. 1986. Synthesis and conclusions. In Gentry and Kooyman (1986), pp. 220–64.

Gentry, R. L., and J. R. Holt. 1986. Attendance behavior of northern fur seals. In Gentry and Kooyman (1986), pp. 41–60.

Gentry, R. L., and G. L. Kooyman, eds. 1986. Fur seals. Maternal strategies on land and at sea. Princeton Univ. Press, xviii + 291 pp.

Gentry, R. L., G. L. Kooyman, and M. E. Goebel. 1986. Feeding and diving behavior of northern fur seals. In Gentry and Kooyman (1986), pp. 61–78.

George, G. G. 1979. The status of endangered Papua New Guinea mammals. In Tyler (1979), pp. 93–100.

George, G. G., and U. Schürer. 1978. Some notes on macropods commonly misidentified in zoos. Internatl. Zoo Yearbook 18:152–56.

George, S. B. 1986. Evolution and historical biogeography of soricine shrews. Syst. Zool. 35:153–62.

George, S. B., J. R. Choate, and H. H. Genoways. 1981. Distribution and taxonomic status of Blarina hylophaga Elliot (Insectivora: Soricidae). Ann. Carnegie Mus. 50:493–513.

———. 1986. Blarina brevicauda. Mammalian Species, no. 261, 9 pp.

George, S. B., H. H. Genoways, J. R. Choate, and R. J. Baker. 1982. Karyotypic relationships within the short-tailed shrews, genus Blarina. J. Mamm. 63:639–45.

George, W. 1974. Notes on the ecology of

gundis (F. Ctenodactylidae). Symp. Zool. Soc. London 34:143–60.

———. 1978. Reproduction in female gundis (Rodentia: Ctenodactylidae). J. Zool. 185:57–71.

———. 1979. The chromosomes of the hystricomorphous family Ctenodactylidae (Rodentia: Sciuromorpha) and their bearing on the relationships of the four living genera. Zool. J. Linnean Soc. 65:261–80.

———. 1982. Ctenodactylus (Ctenodactylidae, Rodentia): one species or two? Mammalia 46:375–80.

———. 1985. Cluster analysis and phylogenetics of five species of Ctenodactylidae (Rodentia). Mammalia 49:53–63.

———. 1988. Massoutiera mzabi (Rodentia, Ctenodactylidae) in a climatological trap. Mammalia 52:331–38.

Geraads, D. 1986. Remarques sur la systématique et la phylogénie des Giraffidae (Artiodactyla, Mammalia). Geobios 19:465–77.

Gerell, R., and K. Lundberg. 1985. Social organization in the bat Pipistrellus pipistrellus. Behav. Ecol. Sociobiol. 16:177–84.

Germain, M., and F. Petter. 1973. Addition à la liste des rongeurs myomorphes de la République Centrafricaine: Zelotomys instans Thomas, 1915. Mammalia 37:683.

Gerson, H. B., and J. P. Hickie. 1985. Head scarring on male narwhals (Monodon monoceros): evidence for aggressive tusk use. Can. J. Zool. 63:2083–87.

Gettinger, R. D. 1984. A field study of activity patterns of Thomomys bottae. J. Mamm. 65:76–84.

Getz, L. L. 1972. Social structure and aggressive behavior in a population of Microtus pennsylvanicus. J. Mamm. 53:310–17.

Gewalt, W. 1979a. Eine "neue" Delphinart in Delphinarien—der Karib-Delphin Sotalia guianensis (Van Beneden, 1864) (?). Saugetierk. Mitt. 27:288–91.

———. 1979b. The Commerson's dolphin (Cephalorhynchus commersonii)—capture and first experiences. Aquatic Mamm. 7:37–40.

Ghobrial, L. I., and A. S. K. Hodieb. 1982. Seasonal variations in the breeding of the Nile rat (Arvicanthis niloticus). Mammalia 46:319–33.

Ghose, R. K. 1965. A new species of mongoose (Mammalia: Carnivora: Viverridae) from West Bengal, India. Proc. Zool. Soc. Calcutta 18:173–78.

———. 1978. Observations on the ecology and status of the hispid hare in Rajagarh Forest, Darrang District, Assam, in 1975 and 1976. J. Bombay Nat. Hist. Soc. 75:206–9.

Ghose, R. K., and T. K. Chakraborty. 1983. A note on the status of the flying squirrels of Darjeeling and Sikkim. J. Bombay Nat. Hist. Soc. 80:411.

Ghose, R. K., and U. Chaturvedi. 1972. Extension of range of the mongoose Herpestes palustris Ghose (Mammalia: Carnivora: Viverridae), with a note on its endoparasitic nematode. J. Bombay Nat. Hist. Soc. 69:412–13.

Ghose, R. K., and D. K. Ghosal. 1984. Record of the fulvous fruit bat, Rousettus leschenaulti (Desmarest, 1820) from Sikkim, with notes on its interesting feeding habit and status. J. Bombay Nat. Hist. Soc. 81:178–79.

Ghose, R. K., and S. S. Saha. 1981. Taxonomic review of Hodgson's giant flying squirrel, Petaurista magnificus (Hodgson) [Sciuridae: Rodentia], with description of a new subspecies from Darjeeling District, West Bengal, India. J. Bombay Nat. Hist. Soc. 78:93–102.

Giacalone, J., N. Wells, and G. Willis. 1987. Observations on Syntheosciurus brochus (Sciuridae) in Volcán Poás National Park, Costa Rica. J. Mamm. 68:145–47.

Giagia, E., I. Savic, and B. Soldatovic. 1982. Chromosomal forms of the mole rat Microspalax from Greece and Turkey. Z. Saugetierk. 47:231–36.

Gier, H. T. 1975. Ecology and behavior of the coyote (Canis latrans). In Fox (1975), pp. 247–62.

Giger, R. D. 1973. Movements and homing in Townsend's mole near Tillamook, Oregon. J. Mamm. 54:648–59.

Gijzen, A., and S. Smet. 1974. Seventy years okapi. Acta Zool. Pathol., no. 59, 111 pp.

Gilbert, B. K., and L. D. Roy. 1977. Prevention of black bear damage to beeyards using aversive conditioning. In Phillips and Jonkel (1977), pp. 93–102.

Gill, J. P., and A. Cave-Browne. 1988. Scimitar-horned oryx (Oryx dammah) at Edinburgh Zoo. In Dixon and Jones (1988), pp. 119–35.

Gilmore, D. P. 1977. The success of marsupials as introduced species. In Stonehouse and Gilmore (1977), pp. 169–78.

Gilmore, R. M. 1978. Right whale. In Haley (1978), pp. 62–69.

Gingerich, P. D. 1984. Paleobiology of tarsiiform primates. In Niemitz (1984a), pp. 33–44.

Gipson, P. S. 1985. Ecology of wolverines in an arctic ecosystem. Natl. Geogr. Soc. Res. Rept. 20:253–63.

Gipson, P. S., and J. A. Sealander. 1977. Ecological relationships of white-tailed deer and dogs in Arkansas. In Phillips and Jonkel (1977), pp. 3–16.

Glander, K. E. 1978. Howling monkey feeding behavior and plant secondary compounds: a study of strategies. In Montgomery (1978), pp. 561–73.

Glanz, W. E., R. W. Thorington, Jr., J. Giacalone-Madden, and L. W. Heaney. 1982. Seasonal food use and demographic trends in Sciurus granatensis. In Leigh, E. G., Jr., A. S. Rand, and D. M. Windsor, eds., The ecology of a tropical forest, Smithson. Inst. Press, Washington, D.C., pp. 239–52.

Glass, B. P. 1985. History of classification and nomenclature in Xenarthra (Edentata). In Montgomery (1985a), pp. 1–3.

Glenn, L. P., J. W. Lentfer, J. B. Faro, and L. H. Miller. 1976. Reproductive biology of female brown bears (Ursus arctos), McNeil River, Alaska. In Pelton, Lentfer, and Folk (1976), pp. 381–90.

Gliwicz, J. 1973. Characteristics of a population of Proechimys semispinosus (Tome 1860)—a rodent species of the tropical rain forest. Bull. Acad. Polonaise Sci., Ser. Sci. Biol. 21:413–18.

Glockner-Ferrari, D. A., and M. J. Ferrari. 1984. Reproduction in the humpback whales, Megaptera novaeangliae, in Hawaiian waters. In Perrin, Brownell, and DeMaster (1984), pp. 237–42.

Godfrey, G. K. 1969. Reproduction in a laboratory colony of the marsupial mouse Sminthopsis larapinta (Marsupialia: Dasyuridae). Austral. J. Zool. 17:637–54.

———. 1975. A study of oestrus and fecundity in a laboratory colony of mouse opossums (Marmosa robinsoni). J. Zool. 175:541–55.

———. 1979. Gestation period in the common shrew, Sorex coronatus (araneus) fretalis. J. Zool. 189:548–55.

Godfrey, G. K., and P. Crowcroft. 1971. Breeding the fat-tailed marsupial mouse Sminthopsis crassicaudata in captivity. Internatl. Zoo Yearbook 11:33–38.

Godfrey, G. K., and W. L. R. Oliver. 1978. The reproduction and development of the pigmy hedgehog. Dodo 15:38–52.

Godfrey, L. R., and A. J. Petto. 1981. Clinal size variation of Archaeolemur spp. on Madagascar. In Chiarelli, A. B., and R. S. Corruccini, eds., Primate evolutionary biology, Springer-Verlag, Berlin, pp. 14–34.

Godin, A. J. 1977. Wild mammals of New England. Johns Hopkins Univ. Press, Baltimore, xii + 304 pp.

Godsell, J. 1988. Herd formation and haul-out behaviour in harbour seals (Phoca vitulina). J. Zool. 215:83–98.

Goedicke, T. R. 1981. Life expectancy of monk seal colonies in Greece. Biol Conserv. 20:173–81.

Goehring, H. H. 1972. Twenty-year study of

Eptesicus fuscus in Minnesota. J. Mamm 53:201–7.

Goertz, J. W. 1970. An ecological study of *Neotoma floridana* in Oklahoma. J. Mamm. 51:91–104.

———. 1971. An ecological study of *Microtus pinetorum* in Oklahoma. Amer. Midl. Nat. 86:1–12.

Goertz, J. W., R. M. Dawson, and E. E Mowbray. 1975. Response to nest boxes and reproduction by *Glaucomys volans* in northern Louisiana. J. Mamm. 56:933–39.

Goldman, C. A. 1984. Systematic revision of the African mongoose genus *Crossarchus* (Mammalia: Viverridae). Can. J. Zool. 62:1618–30.

———. 1987. *Crossarchus obscurus*. Mammalian Species, no. 290, 5 pp.

Goldman, E. A. 1920. Mammals of Panama. Smithson. Misc. Coll. 69:1–309.

Gollan, K. 1984. The Australian dingo: in the shadow of man. *In* Archer and Clayton (1984), pp. 921–27.

Gombe, S. 1983. Reproductive cycle of the rock hyrax *(Procavia capensis)*. Afr. J. Ecol. 21:129–33.

Goodall, A. G. 1977. Feeding and ranging behaviour of a mountain gorilla group *(Gorilla gorilla beringei)* in the Tshibinda-Khuzi region (Zaire). *In* Clutton-Brock (1977), pp. 449–79.

Goodall, A. G., and C. P. Groves. 1977. The conservation of eastern gorillas. *In* Rainier III and Bourne (1977), pp. 599–637.

Goodall, J. 1977. Infant killing and cannibalism in free-living chimpanzees. Folia Primatol. 28:259–82.

———. 1986. The chimpanzees of Gombe. Patterns of behavior. Harvard Univ. Press, Belknap Press, Cambridge, 673 pp.

Goodall, R. N. P. 1978. Report on the small cetaceans stranded on the coasts of Tierra del Fuego. Sci. Rept. Whales Res. Inst. 30:197–230.

Goodall, R. N. P., A. R. Galeazzi, S. Leatherwood, K. W. Miller, I. S. Cameron, R. K. Kastelein, and A. P. Sobral. 1988. Studies of Commerson's dolphins, *Cephalorhynchus commersonii*, off Tierra del Fuego, 1976–1984, with a review of information on the species in the South Atlantic. *In* Brownell and Donovan (1988), pp. 3–70.

Goodall, R. N. P., K. S. Norris, A. R. Galeazzi, J. A. Oporto, and I. S. Cameron. 1988. On the Chilean dolphin, *Cephalorhynchus eutropia* (Gray, 1846). *In* Brownell and Donovan (1988), pp. 197–257.

Goodman, M. 1975. Protein sequence and immunological specificity. Their role in phylogenetic studies of primates. *In* Luckett and Szalay (1975), pp. 219–48.

Goodwin, G. G., and A. M. Greenhall. 1961. A review of the bats of Trinidad and Tobago. Bull. Amer. Mus. Nat. Hist. 122:187–302.

Goodwin, H. A., and J. M. Goodwin. 1973. List of mammals which have become extinct or are possibly extinct since 1600. Internatl. Union Conserv. Nat. Occas. Pap., no. 8, 20 pp.

Goodwin, M. K. 1979. Notes on caravan and play behavior in young captive *Sorex cinereus*. J. Mamm. 60:411–13.

Goodwin, R. E. 1970. The ecology of Jamaican bats. J. Mamm. 51:571–79.

———. 1979. The bats of Timor: systematics and ecology. Bull. Amer. Mus. Nat. Hist. 163:75–122.

Goodyear, N. C. 1987. Distribution and habitat of the silver rice rat, *Oryzomys argentatus*. J. Mamm. 68:692–95.

Goodyear, N. C., and J. D. Lazell, Jr. 1986. Relationships of the silver rice rat *Oryzomys argentatus* (Rodentia: Muridae). Postilla, no. 198, 7 pp.

Gopalakrishna, A. 1955. Observations on the breeding habits and ovarian cycle in the Indian sheath-tailed bat, *Taphozous longimanus* (Hardwicke). Proc. Natl. Inst. Sci. India 21B:29–41.

Gopalakrishna, A., and P. N. Choudhari. 1977. Breeding habits and associated phenomena in some Indian bats, Part 1—*Rousettus leschenaulti* (Desmarest)—Megachiroptera. J. Bombay Nat. Hist. Soc. 74:1–16.

Gopalakrishna, A., and K. B. Karim. 1980. Female genital anatomy and the morphogenesis of foetal membranes of Chiroptera and their bearing on the phylogenetic relationships of the group. Natl. Acad. Sci. India Golden Jubilee Commemoration Vol., pp. 379–428.

Gopalakrishna, A., R. S. Thakur, and A. Madhavan. 1975. Breeding biology of the southern dwarf pipistrelle, *Pipistrellus mimus mimus* (Wroughton) from Maharashtra, India. Dr. B. S. Chauhan Commemoration Vol., pp. 225–40.

Gordon, D. H. 1986. Extensive chromosomal variation in the pouched mouse, *Saccostomus campestris* (Rodentia, Cricetidae) from southern Africa: a preliminary investigation of evolutionary status. Cimbebasia, ser. A, 8:37–47.

———. 1987. Discovery of another species of tree rat. Transvaal Mus. Bull. 22:30–32.

Gordon, G., and P. M. Johnson. 1973. Rock rat in north Queensland. Queensland Div. Plant Industry Adv. Leaf., no. 1192, 3 pp.

Gordon, G., and B. C. Lawrie. 1977. The rufescent bandicoot, *Echymipera rufescens* (Peters and Doria) on Cape York Peninsula. Austral. Wildl. Res. 5:41–45.

———. 1980. The rediscovery of the bridled

nail-tailed wallaby, *Onychogalea fraenata* (Gould) (Marsupialia: Macropodidae) in Queensland. Austral. Wildl. Res. 7:339–45.

Gordon, G., D. G. McGreevy, and B. C. Lawrie. 1978. The yellow-footed rock-wallaby, *Petrogale xanthopus* Gray (Macropodidae), in Queensland. Austral. Wildl. Res. 5:295–97.

Gordon, G. H. 1978. Distribution of sibling species of the *Praomys (Mastomys) natalensis* group in Rhodesia (Mammalia: Rodentia). J. Zool. 186:397–401.

Gorman, M. L. 1976. Seasonal changes in the reproductive pattern of feral *Herpestes auropunctatus* (Carnivora: Viverridae) in the Fijian Islands. J. Zool. 178:237–46.

Gosling, L. M. 1974. The social behavior of Coke's hartebeest *(Alcelaphus buselaphus cokei)*. *In* Geist and Walther (1974), pp. 488–511.

Gosling, L. M., and S. J. Baker. 1981. Coypu *(Myocastor coypus)* potential longevity. J. Zool. 197:285–312.

Gosnell, M. 1977. Carlsbad's famous bats are dying off. Natl. Wildl. 15(4):28–33.

Goss, C. M., L. T. Popejoy, II, J. L. Fusiler, and T. M. Smith. 1968. Observations on the relationship between embryological development, time of conception, and gestation. *In* Rosenblum and Cooper (1968), pp. 171–91.

Gouat, P., J. Gouat, and J. Coulon. 1984. Répartition et habitat de *Massoutiera mzabi* (Rongeur Cténodactylidé) en Algerié. Mammalia 48:351–62.

Gould, E. 1965. Evidence for echolocation in the Tenrecidae of Madagascar. Proc. Amer. Philos. Soc. 109:352–60.

———. 1969. Communication in three genera of shrews (Soricidae): *Suncus, Blarina,* and *Cryptotis*. *In* Communications in Behavioral Biology, Academic Press, New York, pt. A, 3:11–31.

———. 1977. Foraging behavior of *Pteropus vampyrus* on the flowers of *Durio zibethinus*. Malayan Nat. J. 30:53–57.

———. 1978a. Rediscovery of *Hipposideros ridleyi* and seasonal reproduction in Malaysian bats. Biotrópica 10:30–32.

———. 1978b. The behavior of the moonrat, *Echinosorex gymnurus* (Erinaceidae) and the pentail shrew, *Ptilocercus lowi* (Tupaiidae) with comments on the behavior of other insectivora. Z. Tierpsychol. 48:1–27.

Gould, E., and J. F. Eisenberg. 1966. Notes on the biology of the Tenrecidae. J. Mamm. 47:660–86.

Gould, E., N. C. Negus, and A. Novick. 1964. Evidence for echolocation in shrews. J. Exp. Zool. 156:19–38.

Goulden, E. A., and J. Meester. 1978. Notes on the behaviour of *Crocidura* and *Myosorex*

(Mammalia: Soricidae) in captivity. Mammalia 42:197–207.

Grabert, H. 1984. Migration and speciation of the South American Iniidae (Cetacea, Mammalia). Z. Saugetierk. 49:334–41.

Graf, R. P. 1985. Social organization of snowshoe hares. Can. J. Zool. 63:468–74.

Graham, G. L. 1987. Seasonality of reproduction in Peruvian bats. Fieldiana Zool., n.s., 39:173–86.

Graham, G. L., and L. J. Barkley. 1984. Noteworthy records of bats from Peru. J. Mamm. 65:709–11.

Grand, T. I., and J. F. Eisenberg. 1982. On the affinities of the Dinomyidae. Saugetierk. Mitt. 30:151–57.

Grant, T. R. 1973. Dominance and association among members of a captive and free-ranging group of grey kangaroos *(Macropus giganteus)*. Anim. Behav. 21:449–56.

———. 1983. Behavioral ecology of monotremes. Amer. Soc. Mamm. Spec. Publ. 7:360–94.

Grant, T. R., and F. N. Carrick. 1978. Some aspects of the ecology of the platypus, *Ornithorhynchus anatinus*, in the upper Shoalhaven River, New South Wales. Austral. Zool. 20:181–99.

Gratz, N. G. 1984. The global public health importance of rodents. *In* Dubock, A. C., ed., Proceedings of a conference on the organisation and practice of vertebrate pest control, World Health Organization, Geneva, pp. 413–35.

Gray, D. R. 1984. Dominance fighting in muskoxen of different social status. *In* Klein, White, and Keller (1984), pp. 118–22.

Gray, G. G. 1985. Status and distribution of *Ammotragus lervia*: a worldwide review. *In* Hoefs (1985), pp. 95–126.

Gray, G. G., and C. D. Simpson. 1980. *Ammotragus lervia*. Mammalian Species, no. 144, 7 pp.

Grayson, D. K. 1981. A mid-Holocene record for the heather vole, *Phenacomys* cf. *intermedius*, in the central Great Basin and its biogeographic significance. J. Mamm. 62:115–21.

———. 1984. Time of extinction and nature of adaptation of the noble marten, *Martes nobilis*. Carnegie Mus. Nat. Hist. Spec. Publ. 8:233–40.

———. 1987. The biogeographic history of small mammals in the Great Basin: observations on the last 20,000 years. J. Mamm. 68:359–75.

Greegor, D. H., Jr. 1980a. Diet of the little hairy armadillo, *Chaetophractus vellerosus*, of northwestern Argentina. J. Mamm. 61:331–34.

———. 1980b. Preliminary study of movements and home range of the armadillo, *Chaetophractus vellerosus*. J. Mamm. 61:334–35.

———. 1985. Ecology of the little hairy armadillo *Chaetophractus vellerosus*. *In* Montgomery (1985a), pp. 397–405.

Green, C. A., H. Keogh, D. H. Gordon, M. Pinto, and E. K. Hartwig. 1980. The distribution, identification, and naming of the *Mastomys natalensis* species complex in southern Africa (Rodentia: Muridae). J. Zool. 192:17–23.

Green, J. S., and J. T. Flinders. 1980. *Brachylagus idahoensis*. Mammalian Species, no. 125, 4 pp.

Green, M. J. B. 1986. The distribution, status and conservation of the Himalayan musk deer *Moschus chrysogaster*. Biol. Conserv. 35:347–75.

———. 1987a. Some ecological aspects of a Himalayan population of musk deer. *In* Wemmer (1987), pp. 307–19.

———. 1987b. Scent-marking in the Himalayan musk deer *(Moschus chrysogaster)*. J. Zool., ser. B, 1:721–37.

———. 1988. Protected areas and snow leopards: their distribution and status. Proc. 5th Internatl. Snow Leopard Symp., pp. 3–19.

Green, R. H. 1967. Notes on the devil *(Sarcophilus harrisi)* and the quoll *(Dasyurus viverrinus)* in north-eastern Tasmania. Rec. Queen Victoria Mus., no. 27, 13 pp.

———. 1968. The murids and small dasyurids in Tasmania. Parts 3 and 4. Rec. Queen Victoria Mus., no. 32, 19 pp.

———. 1972. The murids and small dasyurids in Tasmania. Parts 5, 6 and 7. Rec. Queen Victoria Mus., no. 46, 34 pp.

Green, R. H., and J. L. Rainbird. 1984. The bat genus *Eptesicus* Gray in Tasmania. Tasmanian Nat. 76:1–5.

———. 1987. The common wombat *Vombatus ursinus* (Shaw, 1800) in northern Tasmania—Part 1. Breeding, growth and development. Rec. Queen Victoria Mus., no. 91, 20 pp.

Green, S., and K. Minkowski. 1977. The lion-tailed monkey and its South Indian forest habitat. *In* Rainier III and Bourne (1977), pp. 289–337.

Greenbaum, I. F., R. J. Baker, and D. E. Wilson. 1975. Evolutionary implications of the karyotypes of the stenodermine genera *Ardops*, *Ariteus*, *Phyllops*, and *Ectophylla*. Bull. S. California Acad. Sci. 74:156–59.

Greenbaum, I. F., and J. K. Jones. Jr. 1978. Noteworthy records of bats from El Salvador, Honduras, and Nicaragua. Occas. Pap. Mus. Texas Tech Univ., no. 55, 7 pp.

Greenhall, A. M. 1968. Notes on the behavior of the false vampire bat. J. Mamm. 49:337–40.

———. 1972. The biting and feeding habits of the vampire bat, *Desmodus rotundus*. J. Zool. 168:451–61.

Greenhall, A. M., G. Joermann, U. Schmidt, and M. R. Seidel. 1983. *Desmodus rotundus*. Mammalian Species, no. 202, 6 pp.

Greenhall, A. M., and U. Schmidt, eds. 1988. Natural history of vampire bats. CRC Press, Boca Raton, Florida, 272 pp.

Greenhall, A. M., U. Schmidt, and G. Joermann. 1984. *Diphylla ecaudata*. Mammalian Species, no. 227, 3 pp.

Greer, J. K. 1965. Mammals of Malleco Province Chile. Michigan State Univ. Mus. Publ., Biol. Ser., 3:49–152.

Griffiths, M. 1968. Echidnas. Pergamon Press, Oxford, ix + 282 pp.

———. 1978. The biology of monotremes. Academic Press, New York, 367 pp.

Griffiths, T. A. 1982. Systematics of the New World nectar-feeding bats (Mammalia, Phyllostomidae), based on the morphology of the hyoid and lingual regions. Amer. Mus. Novit., no. 2742, 45 pp.

———. 1985. Molar cusp patterns in the bat genus *Brachyphylla*: some functional and systematic observations. J. Mamm. 66:544–49.

Grigg, G. C., L. A. Beard, G. Caughley, D. Grice, J. A. Caughley, N. Sheppard, M. Fletcher, and C. Southwell. 1985. The Australian kangaroo populations, 1984. Search 16:9–12.

Grimwood, I. 1988. Operation oryx: the start of it all. *In* Dixon and Jones (1988), pp. 1–3.

Grimwood, I. R. 1969. Notes on the distribution and status of some Peruvian mammals. Spec. Publ. Amer. Comm. Internatl. Wildl. Protection, no. 21, v + 86 pp.

———. 1976. The Palawan stink badger. Oryx 13:297.

Griner, L. A. 1980. East Bornean proboscis monkey, *Nasalis larvatus orientalis*. Zool. Garten 50:73–81.

Grobler, J. H. 1974. Aspects of the biology, population ecology and behaviour of the sable *Hippotragus niger niger* (Harris, 1838) in the Rhodes Matopos National Park, Rhodesia. Arnoldia 7(6):1–36.

Gross, J. E., L. C. Stoddart, and F. H. Wagner. 1974. Demographic analysis of a northern Utah jackrabbit population. Wildl. Monogr., no. 40, 68 pp.

Groves, C. P. 1967a. On the gazelles of the genus *Procapra* Hodgson, 1864. Z. Saugetierk. 32:144–49.

———. 1967b. On the rhinoceroses of southeast Asia. Saugetierk. Mitt. 15:221–37.

————. 1967c. Geographic variation in the black rhinoceros *Diceros bicornis* (L., 1758). Z. Saugetierk. 32:267–76.

————. 1969a. Systematics of the anoa (Mammalia, Bovidae). Beaufortia 17:1–12.

————. 1969b. On the smaller gazelles of the genus *Gazella* de Blainville, 1916. Z. Saugetierk. 34:38–60.

————. 1970a. The forgotten leaf-eaters, and the phylogeny of the Colobinae. *In* Napier and Napier (1970), pp. 555–87.

————. 1970b. Population systematics of the gorilla. J. Zool. 161:287–300.

————. 1971a. Systematics of the genus *Nycticebus*. Proc. 3rd Internatl. Congr. Primatol. 1:44–53.

————. 1971b. Distribution and place of origin of the gorilla. Man 6:44–51.

————. 1972a. Systematics and phylogeny of gibbons. *In* Rumbaugh (1972), pp. 1–89.

————. 1972b. *Ceratotherium simum*. Mammalian Species, no. 8, 6 pp.

————. 1974a. Taxonomy and phylogeny of prosimians. *In* Martin, Doyle, and Walker (1974), pp. 449–73.

————. 1974b. Horses, asses and zebras in the wild. Ralph Curtis Books, Hollywood, Florida, 192 pp.

————. 1975a. The taxonomy of *Moschus* (Mammalia, Artiodactyla), with particular reference to the Indian region. J. Bombay Nat. Hist. Soc. 72:662–76.

————. 1975b. Taxonomic notes on the white rhinoceros *Ceratotherium simum* (Burchell, 1817). Saugetierk. Mitt. 23:200–212.

————. 1976. The origin of the mammalian fauna of Sulawesi (Celebes). Z. Saugetierk. 41:201–16.

————. 1978a. Phylogenetic and population systematics of the mangabeys (Primates: Cercopithecoidea). Primates 19:1–34.

————. 1978b. The taxonomic status of the dwarf blue sheep (Artiodactyla: Bovidae). Saugetierk. Mitt. 26:177–83.

————. 1980a. Notes on the systematics of *Babyrousa* (Artiodactyla, Suidae). Zool. Meded. 55:29–46.

————. 1980b. A further note on *Moschus*. J. Bombay Nat. Hist. Soc. 77:130–32.

————. 1980c. Speciation in *Macaca*: the view from Sulawesi. *In* Lindburg (1980), pp. 84–124.

————. 1981a. Ancestors for the pigs: taxonomy and phylogeny of the genus *Sus*. Australian Natl. Univ. Dept. Prehistory Tech. Bull., no. 3, vii + 96 pp.

————. 1981b. Systematic relationships in the Bovini (Artiodactyla, Bovidae). Z. Saugetierk. 19:264–78.

————. 1982a. The systematics of tree kangaroos (*Dendrolagus*; Marsupialia, Macropodidae). Austral. Mamm. 5:157–86.

————. 1982b. The skulls of Asian rhinoceroses: wild and captive. Zoo Biol. 1:251–61.

————. 1982c. Geographic variation in the barasingha or swamp deer *(Cervus duvauceli)*. J. Bombay Nat. Hist. Soc. 79:620–29.

————. 1983. Phylogeny of the living species of rhinoceroses. Z. Zool. Syst. Evol. 21:293–313.

————. 1984. A new look at the taxonomy and phylogeny of gibbons. *In* Preuschoft et al. (1984), pp. 542–61.

————. 1985a. The case of the pygmy gorilla: a cautionary tale for cryptozoology. Cryptozoology 4:37–44.

————. 1985b. An introduction to the gazelles. Chinkara 1(1):4–10.

————. 1986a. Systematics of the great apes. *In* Swindler and Erwin (1986), pp. 187–217.

————. 1986b. The taxonomy, distribution, and adaptations of Recent equids. *In* Meadow, R. H., and H.-P. Uerpmann, eds., Equids in the ancient world, Dr. Ludwig Reichert Verlag, Wiesbaden, pp. 11–65.

————. 1987a. On the highland cuscuses (Marsupialia: Phalangeridae) of New Guinea. *In* Archer (1987), pp. 559–67.

————. 1987b. On the cuscuses (Marsupialia: Phalangeridae) of the *Phalanger orientalis* group from Indonesian territory. *In* Archer (1987), pp. 569–79.

————. 1988. A catalogue of the genus *Gazella*. *In* Dixon and Jones (1988), pp. 193–98.

Groves, C. P., and R. H. Eaglen. 1988. Systematics of the Lemuridae (Primates, Strepsirhini). J. Human Evol. 17:513–38.

Groves, C. P., and Feng Zuojian. 1986. The status of musk-deer from Anhui Province, China. Acta Theriol. Sinica 6:103–6.

Groves, C. P., and P. Grubb. 1974. A new duiker frorn Rwanda. Rev. Zool. Afr. 88:189–96.

————. 1982. The species of muntjac (genus *Muntiacus*) in Borneo: unrecognized sympatry in tropical deer. Zool. Meded. 56:203–16.

————. 1985. Reclassification of the serows and gorals (*Nemorhaedus*: Bovidae). *In* Lovari (1985), pp. 45–50.

————. 1987. Relationships of living deer. *In* Wemmer (1987), pp. 21–59.

Groves, C. P., and L. B. Holthius. 1985. The nomenclature of the orangutan. Zool. Meded. 59:411–17.

Groves, C. P., and F. Kurt. 1972. *Dicerorhinus sumatrensis*. Mammalian Species, no. 21, 6 pp.

Groves, C. P., and D. M. Lay. 1985. A new species of the genus *Gazella* (Mammalia: Artiodactyla: Bovidae) from the Arabian Peninsula. Mammalia 49:27–36.

Groves, C. P., and D. P. Willoughby. 1981. Studies on the taxonomy and phylogeny of the genus *Equus*. I. Subgeneric classification of the Recent species. Mammalia 45:321–54.

Grubb, P. 1973. Distribution, divergence and speciation of the drill and mandrill. Folia Primatol. 20:161–77.

————. 1977. Variation and incipient speciation in the African buffalo. Z. Saugetierk. 37:121–44.

————. 1981. *Equus burchelli*. Mammalian Species, no. 157, 9 pp.

————. 1982a. Systematics of sun-squirrels *(Heliosciurus)* in eastern Africa. Bonner Zool. Beitr. 33:191–204.

————. 1982b. The systematics of Sino-Himalayan musk deer *(Moschus)*, with particular reference to the species described by B. H. Hodgson. Saugetierk. Mitt. 30:127–35.

Grubb, P., and C. P. Groves. 1983. Notes on the taxonomy of the deer (Mammalia, Cervidae) of the Philippines. Zool. Anz., Jena, 210:119–44.

Grzimek, B., ed. 1975. Grzimek's animal life encyclopedia. Mammals, I–IV. Van Nostrand Reinhold, New York, vols. 10–13.

Gu Jing-he and Gao Xing-yi. 1985. The wild two-humped camel in the Lop-Nor region. *In* Kawamichi (1985), pp. 117–20.

Gucwinska, H. 1971. Development of six-banded armadillos *Euphractus sexcinctus* at Wroclaw Zoo. Internatl. Zoo Yearbook 11:88–89.

Gudynas, E. 1989. The conservation status of South American rodents: many questions but few answers. *In* Lidicker (1989), pp. 20–25.

Guggisberg, C. A. W. 1975. Wild cats of the world. Taplinger, New York, 328 pp.

Guilday, J. E., H. W. Hamilton, and P. W. Parmalee. 1975. Caribou (*Rangifer tarandus* L.) from the Pleistocene of Tennessee. J. Tennessee Acad. Sci. 50:109–12.

Guiler, E. R. 1961. Breeding season of the thylacine. J. Mamm. 42:396–97.

————. 1970a. Observations on the Tasmanian devil, *Sarcophilus harrisii* (Marsupialia: Dasyuridae). I. Numbers, home range, movements, and food in two populations. Austral. J. Zool. 18:49–62.

————. 1970b. Observations on the Tasmanian devil, *Sarcophilus harrisii* (Marsupialia: Dasyuridae). II. Reproduction, breeding, and growth of pouch young. Austral. J. Zool. 18:63–70.

———. 1971*a*. Food of the Potoroo (Marsupialia, Macropodidae). J. Mamm. 52:232–34.

———. 1971*b*. The husbandry of the potoroo *Potorous tridactylus*. Internatl. Zoo Yearbook 11:21–22.

———. 1971*c*. The Tasmanian devil *Sarcophilus harrisii* in captivity. Internatl. Zoo Yearbook 11:32–33.

———. 1982. Temporal and spatial distribution of the Tasmanian devil, *Sarcophilus harrisii* (Dasyuridae: Marsupialia). Pap. Proc. Roy. Soc. Tasmania 116:153–63.

Guiler, E. R., H. R. Burton, and N. J. Gales. 1987. On three odontocete skulls from Heard Island. Sci. Rept. Whales Res. Inst. 38:117–24.

Guiler, E. R., and D. A. Kitchener. 1967. Further observations on longevity in the wild potoroo, *Potorous tridactylus*. Austral. J. Sci. 30:105–6.

Guillotin, M., and F. Petter. 1984. Un *Rhipidomys* nouveau de Guyane française, *R. leucodactylus aratayas* ssp. nov. (Rongeurs, Cricétidés). Mammalia 48:541–44.

Gulland, J. A. 1974. Distribution and abundance of whales in relation to basic productivity. *In* Schevill (1974), pp. 27–52.

———. 1976. Antarctic baleen whales: history and prospects. Polar Record 18:5–13.

Gulotta, E. F. 1971. *Meriones unguiculatus*. Mammalian Species, no. 3, 5 pp.

Gum, R. L., L. M. Arthur, and R. S. Magleby. 1978. Coyote control: a simulation evaluation of alternative strategies. U.S. Dept. Agric., Agric. Econ. Rept., no. 408, 49 pp.

Gunderson, H. L. 1978. A mid-continent irruption of Canada lynx, 1962–1963. Prairie Nat. 10:71–80.

Gunn, A. 1985. Observations of cream-colored muskoxen in the Queen Maud Gulf area of Northwest Territories. J. Mamm. 66:803–4.

Gunn, A., F. L. Miller, and D. C. Thomas. 1981. The current status and future of Peary Caribou *Rangifer tarandus pearyi* on the Arctic islands of Canada. Biol. Conserv. 19:283–96.

Gunn, S. J., and I. F. Greenbaum. 1986. Systematic implications of karyotypic and morphologic variation in mainland *Peromyscus* from the Pacific Northwest. J. Mamm. 67:294–304.

Gunson, J. R. 1983. Status and management of wolves in Alberta. *In* Carbyn (1983), pp. 25–29.

Gunson, J. R., and R. R. Bjorge. 1979. Winter denning of the striped skunk in Alberta. Can. Field-Nat. 93:252–58.

H

Habibi, K. 1986. Arabian ungulates—their status and future protection. Oryx 20:100–103.

Hadidian, J., and I. S. Bernstein. 1979. Female reproductive cycles and birth data from an Old World monkey colony. Primates 20:429–42.

Hafner, D. J., and K. N. Geluso. 1983. Systematic relationships and historical zoogeography of the desert pocket gopher, *Geomys arenarius*. J. Mamm. 64:405–13.

Hafner, D. J., J. C. Hafner, and M. S. Hafner. 1979. Systematic status of kangaroo mice, genus *Microdipodops*: morphometric, chromosomal, and protein analyses. J. Mamm. 60:1–10.

Hafner, D. J., K. E. Petersen, and T. L. Yates. 1981. Evolutionary relationships of jumping mice (genus *Zapus*) of the southwestern United States. J. Mamm. 62:501–12.

Hafner, J. C., D. J. Hafner, J. L. Patton, and M. F. Smith. 1983. Contact zones and the genetics of differentiation in the pocket gopher *Thomomys bottae* (Rodentia: Geomyidae). Syst. Zool. 32:1–20.

Hafner, J. C., and M. S. Hafner. 1983. Evolutionary relationships of heteromyid rodents. Great Basin Nat. Mem. 7:3–29.

Hafner, M. S., and L. J. Barkley. 1984. Genetics and natural history of a relictual pocket gopher, *Zygogeomys* (Rodentia: Geomyidae). J. Mamm. 65:474–79.

Hafner, M. S., and D. J. Hafner. 1979. Vocalizations of grasshopper mice (genus *Onychomys*). J. Mamm. 60:85–94.

Hafner, M. S., and J. C. Hafner. 1982. Structure of surface mounds of *Zygogeomys* (Rodentia: Geomyidae). J. Mamm. 63:536–38.

Haiduk, M. W., and R. J. Baker. 1982. Cladistical analysis of g-banded chromosomes of nectar feeding bats (Glossophaginae: Phyllostomidae). Syst. Zool. 31:252–65.

Haiduk, M. W., J. W. Bickham, and D. J. Schmidly. 1979. Karyotypes of six species of *Oryzomys* from Mexico and Central America. J. Mamm. 60:610–15.

Haimoff, E. H. 1984. The organization of song in the Hainan black gibbon *(Hylobates concolor hainanus)*. Primates 25:225–35.

Haimoff, E. H., D. J. Chivers, S. P. Gittins, and T. Whitten. 1982. A phylogeny of gibbons (*Hylobates* spp.) based on morphological and behavioural characters. Folia Primatol. 39:213–37.

Hain, J. H. W., and S. Leatherwood. 1982. Two sightings of white pilot whales, *Globi-*

cephala melaena, and summarized records of anomalously white cetaceans. J. Mamm. 63:338–43.

Haley, D., ed. 1978. Marine mammals of eastern North Pacific and Arctic waters. Pacific Search Press, Seattle, 256 pp.

Hall, E. R. 1946. Mammals of Nevada. Univ. California Press, Berkeley, xi + 710 pp.

———. 1951. American weasels. Univ. Kansas Mus. Nat. Hist. Publ. 4:1–466.

———. 1955. Handbook of mammals of Kansas. Univ. Kansas Mus. Nat. Hist. Misc. Publ., no. 7, 303 pp.

———. 1981. The mammals of North America. John Wiley & Sons, New York, 2 vols.

———. 1984. Geographic variation among brown and grizzly bears *(Ursus arctos)* in North America. Univ. Kansas Mus. Nat. Hist. Spec. Publ., no. 13, 16 pp.

Hall, E. R., and W. W. Dalquest. 1963. The mammals of Veracruz. Univ. Kansas Publ. Mus. Nat. Hist. 14:165–362.

Hall, E. R., and B. Villa R. 1949. An annotated checklist of the mammals of Michoacan, Mexico. Univ. Kansas Mus. Nat. Hist. Publ. 1:431–72.

Hall, E. S., Jr. 1973. Archaeological and recent evidence for expansion of moose range in northern Alaska. J. Mamm. 54:294–95.

Hall, H. T., and J. D. Newsom. 1978. Summer home ranges and movements of bobcats in bottomland hardwoods of southern Louisiana. Proc. Ann. Conf. S.E. Assoc. Fish and Wildl. Agencies 30:427–36.

Hall, J. G. 1981. A field study of the Kaibab squirrel in Grand Canyon National Park. Wildl. Monogr., no. 75, 54 pp.

Hall, K. R. L. 1967. Social interactions of the adult male and adult females of a patas monkey group. *In* Altmann (1967), pp. 261–80.

———. 1968. Behaviour and ecology of the wild patas monkey, *Erythrocebus patas*, in Uganda. *In* Jay (1968), pp. 32–119 (reprinted from J. Zool. 148:15–87).

Hall, K. R. L., and I. De Vore. 1965. Baboon social behavior. *In* De Vore (1965), pp. 53–110.

Hall, L. S. 1987. Identification, distribution and taxonomy of Australian flying-foxes (Chiroptera: Pteropodidae). Austral. Mamm. 10:75–79.

Hall, S. J. G. 1989. Breed structures of rare pigs: implications for conservation of the Berkshire, Tamworth, middle white, large black, Gloucester old spot, British saddleback, and British lop. Conserv. Biol. 3:30–38.

Hallett, A. F., and J. Meester. 1971. Early postnatal development of the South African hamster *Mystromys albicaudatus*. Zool. Afr. 6:221–28.

Hallett, J. G. 1978. *Parascalops breweri*. Mammalian Species, no. 98, 4 pp.

Hall-Martin, A. J., and J. D. Skinner. 1978. Observations on puberty and pregnancy in female giraffe *(Giraffa camelopardalis)*. S. Afr. J. Wildl. Res. 8:91–94.

Halls, L. K. 1978. White-tailed deer. *In* Schmidt and Gilbert (1978), pp. 43–65.

———, ed. 1984. White-tailed deer. Ecology and management. Stackpole Books, Harrisburg, Pennsylvania, xxiii + 870 pp.

Haltenorth, T., and H. H. Roth. 1968. Short review of the biology and ecology of the red fox. Saugetierk. Mitt. 16:339–52.

Hamilton, P. H. 1986. Status of the cheetah in Kenya, with reference to sub-Saharan Africa. *In* Miller and Everett (1986), pp. 65–76.

Hamilton, R. B., and D. T. Stalling. 1972. *Lasiurus borealis* with five young. J. Mamm. 53:190.

Hamilton, W. J., III. 1962. Reproductive adaptations of the red tree mouse. J. Mamm. 43:486–504.

Hamilton, W. J., III, R. E. Buskirk, and W. H. Buskirk. 1978. Omnivory and utilization of food resources by chacma baboons, *Papio ursinus*. Amer. Nat. 112:911–24.

Hamilton-Smith, E. 1979. Endangered and threatened Chiroptera of Australia and the Pacific region. *In* Tyler (1979), pp. 85–91.

———. 1980. The status of Australian Chiroptera. Proc. 5th Internatl. Bat Res. Conf., pp. 199–205.

Hamr, J. 1985. Seasonal home range size and utilisation by female chamois *(Rupicapra rupicapra* L.) in northern Tyrol. *In* Lovari (1985), pp. 106–16.

Hanák, V., and A. Elgadi. 1984. On the bat fauna (Chiroptera) of Libya. Vest. Cesk. Spol. Zool. 48:165–87.

Hanák, V., and J. Gaisler. 1983. *Nyctalus leisleri* (Kuhl, 1818), une espèce nouvelle pour le continent africain. Mammalia 47:585–87.

Hanby, J. P., and J. D. Bygott. 1987. Emigration of subadult lions. Anim. Behav. 35:161–69.

Hand, S. J. 1985. New Miocene megadermatids (Chiroptera: Megadermatidae) from Australia with comments on megadermatid phylogenetics. Austral. Mamm. 8:5–43.

Handley, C. O., Jr. 1959. A review of the genus *Hoplomys* (thick-spined rats), with description of a new form from Isla Escudo de Veraguas, Panama. Smithson. Misc. Coll. 139:1–10.

———. 1976. Mammals of the Smithsonian Venezuelan Project. Brigham Young Univ. Sci. Bull., Biol. Ser., 20(5):1–89.

———. 1980a. Inconsistencies in formation of family-group and subfamily-group names in Chiroptera. Proc. 5th Internatl. Bat Res. Conf., pp. 9–13.

———. 1980b. Mammals. *In* Linzey, D. W., ed., Endangered and threatened plants and animals of Virginia, Virginia Polytechnic Inst., Blacksburg, pp. 483–621.

———. 1984. New species of mammals from northern South America: a long-tongued bat, genus *Anoura* Gray. Proc. Biol. Soc. Washington 97:513–21.

———. 1987. New species of mammals from northern South America: fruit-eating bats, genus *Artibeus* Leach. Fieldiana Zool., n.s., 39:163–72.

Handley, C. O., Jr., and L. K. Gordon. 1979. New species of mammals from northern South America: mouse possums, genus *Marmosa* Gray. *In* Eisenberg (1979), pp. 65–72.

Handley, C. O., Jr., and E. Mondolfi. 1963. A new species of fish-eating rat, *Ichthyomys*, from Venezuela (Rodentia, Cricetidae). Acta Biol. Venezuelica 3:417–19.

Hanks, J. 1979. The struggle for survival. The elephant problem. Mayflower Books, New York, 176 pp.

Hannah, A. C., and W. C. McGrew. 1987. Chimpanzees using stones to crack open oil palm nuts in Liberia. Primates 28:31–46.

Hanselka, C. W., J. M. Inglis, and H. G. Applegate. 1971. Reproduction in the blacktailed jackrabbit in southwestern Texas. Southwestern Nat. 16:214–17.

Hansen, C., and E. D. Fleharty. 1974. Structural ecological parameters of a population of *Peromyscus maniculatus* in west-central Kansas. Southwestern Nat. 19:293–303.

Hansen, M. C. 1982. Desert bighorn sheep: another view. Wildl. Soc. Bull. 10:133–40.

Hansson, L. 1968. Population densities of small rodents in open field habitats in south Sweden in 1964–1967. Oikos 19:53–60.

Happel, R., and T. Cheek. 1986. Evolutionary biology and ecology of *Rhinopithecus*. *In* Taub, D. M., and F. A. King, eds., Current perspectives in primate social dynamics, Van Nostrand Reinhold, New York, pp. 305–24.

Happel, R. E. 1982. Ecology of *Pithecia hirsuta* in Peru. J. Human Evol. 11:581–90.

Happold, D. C. D. 1977. A population study on small rodents in the tropical rain forest of Nigeria. Terre Vie 31:385–457.

———. 1987. The mammals of Nigeria. Clarendon Press, Oxford, xvii + 402 pp.

Happold, D. C. D., and M. Happold. 1978a. The fruit bats of western Nigeria. Nigerian Field 43:30–37.

———. 1978b. The fruit bats of western Nigeria. Part 2. Nigerian Field 43:72–77.

Happold, M. 1976a. Social behavior of the conilurine rodents (Muridae) of Australia. Z. Tierpsychol. 40:113–82.

———. 1976b. Reproductive biology and developments in the conilurine rodents (Muridae) of Australia. Austral. J. Zool. 24:19–26.

Harako, R. 1981. The cultural ecology of hunting behaviour among Mbuti pygmies in the Ituri Forest, Zaire. *In* Harding and Teleki (1981), pp. 499–555.

Harbo, S. J., Jr., and F. C. Dean. 1983. Historical and current perspectives on wolf management in Alaska. *In* Carbyn (1983), pp. 51–64.

Harcourt, A. H. 1978. Strategies of emigration and transfer by primates, with particular reference to gorillas. Z. Tierpsychol. 48:401–20.

———. 1979. Social relationships between adult male and female mountain gorillas in the wild. Anim. Behav. 27:325–42.

———. 1981. Can Uganda's gorillas survive?—A survey of the Bwindi Forest Reserve. Biol. Conserv. 19:269–82.

———. 1986. Gorilla conservation: anatomy of a campaign. *In* Benirschke (1986), pp. 31–46.

Harcourt, A. H., and Kai Curry-Lindhal. 1978. The FPS mountain gorilla project—a report from Rwanda. Oryx 14:316–24.

Harcourt, A. H., and D. Fossey. 1981. The Virunga gorillas: decline of an 'island' population. Afr. J. Ecol. 19:83–97.

Harcourt, A. H., D. Fossey, and J. Sabater Pi. 1981. Demography of *Gorilla gorilla*. J. Zool. 195:215–33.

Harcourt, A. H., and S. A. Harcourt. 1984. Insectivory by gorillas. Folia Primatol. 43:229–33.

Harcourt, A. H., and K. J. Stewart. 1984. Gorillas' time feeding: aspects of methodology, body size, competition and diet. Afr. J. Ecol. 22:207–15.

Harcourt, A. H., K. J. Stewart, and D. Fossey. 1976. Male emigration and female transfer in wild mountain gorilla. Nature 263:226–27.

———. 1981. Gorilla reproduction in the wild. *In* Graham, C. E., ed., Reproductive biology of the great apes, Academic Press, New York, pp. 265–79.

Harcourt, C. 1981. An examination of the function of urine washing in *Galago senegalensis*. Z. Tierpsychol. 55:119–28.

———. 1986. *Galago zanzibaricus*: birth seasonality, litter size and perinatal behaviour of females. J. Zool. 210:451–57.

Harcourt, C. S., and L. T. Nash. 1986. Social

organization of galagos in Kenyan coastal forests: I. *Galago zanzibaricus*. Amer. J. Primatol. 10:339–55.

Harden, R. H. 1985. The ecology of the dingo in north-eastern New South Wales. I. Movements and home range. Austral. Wildl. Res. 12:25–37.

Hardin, J. W., R. W. Barbour, and W. H. Davis. 1970. Observations on the home range of a yellow-nosed cotton rat, *Sigmodon ochrognathus*. Southwestern Nat. 14:353–55.

Hardin, J. W., W. D. Klimstra, and N. J. Silvy. 1984. Florida Keys. *In* Halls (1984), pp. 381–90.

Harding, H. R., F. N. Carrick, and C. D. Shorey. 1987. The affinities of the koala *Phascolarctos cinereus* (Marsupialia: Phascolarctidae) on the basis of sperm ultrastructure and development. *In* Archer (1987), pp. 353–64.

Harding, L. E. 1976. Den-site characteristics of arctic coastal grizzly bears (*Ursus arctos* L.) on Richards Island, Northwest Territories, Canada. Can. J. Zool. 54:1357–63.

Harding, R. S. O., and D. K. Olson. 1986. Patterns of mating among male patas monkeys *(Erythrocebus patas)*. Amer. J. Primatol. 11:343–58.

Harding, R. S. O., and G. Teleki, eds. 1981. Omnivorous primates. Columbia Univ. Press, New York, vi + 673 pp.

Harington, C. R. 1968. Denning habits of the polar bear (*Ursus maritimus* Phipps). Can. Wildl. Serv. Rept. Ser., no. 5, 30 pp.

Harland, R. M., P. J. Blancher, and J. S. Millar. 1979. Demography of a population of *Peromyscus leucopus*. Can. J. Zool. 57:323–28.

Harper, F. 1945. Extinct and vanishing mammals of the Old World. Spec. Publ. Amer. Comm. Internatl. Wildl. Protection, no. 12, xv + 850 pp.

Harper, R. J. 1977. "Caravanning" in *Sorex* species. J. Zool. 183:541.

Harrington, F. H., and L. D. Mech. 1979. Wolf howling and its role in territory maintenance. Behaviour 68:207–49.

Harrington, J. E. 1975. Field observations of social behavior of *Lemur fulvus fulvus* E. Geoffroy 1812. *In* Tattersall and Sussman (1975), pp. 259–79.

———. 1978. Diurnal behavior of *Lemur mongoz* at Ampijora, Madagascar. Folia Primatol. 29:291–302.

Harris, S., and G. C. Smith. 1987. Demography of two urban fox *(Vulpes vulpes)* populations. J. Appl. Ecol. 24:75–86.

Harrison, C. S. 1979. Sighting of Cuvier's beaked whale *(Ziphius cavirostris)* in the Gulf of Alaska. Murrelet 60:35–36.

Harrison, C. S., and J. D. Hall. 1978. Alaskan distribution of the beluga whale, *Delphinapterus leucas*. Can. Field-Nat. 92:235–41.

Harrison, D. L. 1964, 1968, 1972. The mammals of Arabia. Ernest Benn, London, 3 vols.

———. 1975a. *Macrophyllum macrophyllum*. Mammalian Species, no. 62, 3 pp.

———. 1975b. A new species of African free-tailed bat (Chiroptera: Molossidae) obtained by the Zaire River Expedition. Mammalia 39:313–18.

———. 1978. A critical examination of alleged sibling species in the lesser three-toed jerboas (subgenus *Jaculus*) of the North African and Arabian deserts. Bull. Carnegie Mus. Nat. Hist., no. 6, pp. 77–80.

———. 1979. A new species of pipistrelle bat (*Pipistrellus*: Vespertilionidae) from Oman, Arabia. Mammalia 43:573–76.

Harrison, D. L., and P. J. J. Bates. 1984. Mammals of Saudi Arabia. On the occurrence of the European free-tailed bat, *Tadarida teniotis* Rafinesque, 1814 (Chiroptera: Molossidae) in Saudi Arabia with a zoogeographical review of the molossids of the Kingdom. Fauna Saudi Arabia 6:551–56.

———. 1986. New records of a rare African shrew, *Crocidura butleri percivali* Dollman, 1915 (Insectivora: Soricidae). Ann. Naturhist. Mus. Wien 88/89:205–11.

Harrison, D. L., and M. C. Jennings. 1980. Occurrence of the noctule, *Nyctalus noctula* Schreber, 1774 (Chiroptera: Vespertilionidae) in Oman, Arabia. Mammalia 44:409–10.

Harrison, M. J. S. 1988. A new species of guenon (genus *Cercopithecus*) from Gabon. J. Zool. 215:561–75.

Harrison, R. J. 1969. Reproduction and reproductive organs. *In* Andersen (1969), pp. 296–322.

Harrison, R. J., and R. L. Brownell, Jr. 1971. The gonads of the South American dolphins, *Inia geoffrensis*, *Pontoporia blainvillei*, and *Sotalia fluviatilis*. J. Mamm. 52:413–19.

Harrison, R. J., M. M. Bryden, and D. A. McBrearty. 1981. The ovaries and reproduction in *Pontoporia blainvillei* (Cetacea: Platanistidae). J. Zool. 193:563–80.

Hart, J. A., and R. M. Timm. 1978. Observations on the aquatic genet in Zaire. Carnivore 1:130–31.

Hart, R. P., and D. J. Kitchener. 1986. First record of *Sminthopsis psammophila* (Marsupialia: Dasyuridae) from Western Australia. Rec. W. Austral. Mus. 13:139–44.

Hart, T., and J. A. Hart. 1988. Tracking the rainforest giraffe. Anim. Kingdom 91(1):26–32.

Hartenberger, J.-L. 1985. The order Rodentia: major questions on their evolutionary origin, relationships and suprafamilial systemat-ics. *In* Luckett and Hartenberger (1985), pp. 1–33.

Hartman, D. S. 1979. Ecology and behavior of the manatee *(Trichechus manatus)* in Florida. Amer. Soc. Mamm. Spec. Publ., no. 5, viii + 153 pp.

Hartman, G. D., and T. L. Yates. 1985. *Scapanus orarius*. Mammalian Species, no. 253, 5 pp.

Harvey, M. J. 1976. Home range, movements, and diel activity of the eastern mole, *Scalopus aquaticus*. Amer. Midl. Nat. 95:436–45.

Harvie, A. E. 1973. Diet of the opossum *(Trichosurus vulpecula* Kerr) on farmland northeast of Waverly, New Zealand. Proc. New Zealand Ecol. Soc. 20:48–52.

Hasler, J. F. 1975. A review of reproduction and sexual maturation in the microtine rodents. Biologist 57:52–86.

Hasler, J. F., and E. M. Banks. 1975. Reproductive performance and growth in captive collared lemmings *(Dicrostonyx groenlandicus)*. Can. J. Zool. 53:777–87.

Hasler, J. F., and M. W. Sorenson. 1974. Behavior of the tree shrew, *Tupaia chinensis*, in captivity. Amer. Midl. Nat. 91:294–314.

Hasler, M. J., J. F. Hasler, and A. V. Nalbandov. 1977. Comparative breeding biology of musk shrews *(Suncus murinus)* from Guam and Madagascar. J. Mamm. 58:285–90.

Haspel, C., and R. E. Calhoon. 1989. Home ranges of free-ranging cats *(Felis catus)* in Brooklyn, New York. Can. J. Zool. 67:178–81.

Hassinger, J. D. 1973. A survey of the mammals of Afghanistan resulting from the 1965 Street Expedition (excluding bats). Fieldiana Zool. 60:i–xi + 1–195.

Hatcher, R. T. 1982. Distribution and status of red foxes (Canidae) in Oklahoma. Southwestern Nat. 27:183–86.

Hatt, R. T. 1959. The mammals of Iraq. Misc. Publ. Mus. Zool. Univ. Michigan, no. 106, 113 pp.

Hatton, J., N. Smart, and K. Thomson. 1984. In urgent need of protection—habitat for the woolly spider monkey. Oryx 18:24–29.

Haugen, A. O. 1974. Reproduction in the plains bison. Iowa State J. Res. 49:1–8.

———. 1975. Reproductive performance of white-tailed deer in Iowa. J. Mamm. 56:151–59.

Hauser, J., F. Catzeflis, A. Meylan, and P. Vogel. 1985. Speciation in the *Sorex araneus* complex (Mammalia: Insectivora). Acta Zool. Fennica 170:125–30.

Hausfater, G. 1974. History of three little-known New World populations of macaques. Lab. Primate Newsl. 13(1):16–18.

Hawes, M. L. 1977. Home range, territoriality, and ecological separation in sympatric shrews, *Sorex vagrans* and *Sorex obscurus*. J. Mamm. 58:354–67.

Hawkins, R. E., and W. D. Klimstra. 1970. A preliminary study of the social organization of white-tailed deer. J. Wildl. Mgmt. 34:407–19.

Hayashi, T. 1984. The people attacked by the Japanese black bear—instances in Fukui Prefecture from 1970 to 1983. J. Mamm. Soc. Japan 10:55–62.

Hayden, B. 1981. Subsistence and ecological adaptations of modern hunter/gatherers. *In* Harding and Teleki (1981), pp. 344–421.

Hayden, P., and R. G. Lindberg. 1976. Survival of laboratory-reared pocket mice. *Perognathus longimembris*. J. Mamm. 57:266–72.

Hayes, S. R. 1977. Home range of *Marmota monax* (Sciuridae) in Arkansas. Southwestern Nat. 22:547–50.

Hayman, R. W. 1946. A new genus of fruit-bat and a new squirrel, from Celebes. Ann. Mag. Nat. Hist., ser. 11, 12:766–75.

———. 1961. The red goral of the northeast frontier region. Proc. Zool. Soc. London 136:317–24.

Hayward, B. J., and E. L. Cockrum. 1971. The natural history of the western long-nosed bat *Leptonycteris sanborni*. Western New Mexico State Univ. Res. Sci. 1:75–123.

Heaney, L. R. 1979. A new species of tree squirrel (*Sundasciurus*) from Palawan Island, Philippines (Mammalia: Sciuridae). Proc. Biol. Soc. Washington 92:280–86.

———. 1985. Systematics of Oriental pygmy squirrels of the genera *Exilisciurus* and *Nannosciurus* (Mammalia: Sciuridae). Misc. Publ. Mus. Zool. Univ. Michigan, no. 170, iv + 58 pp.

Heaney, L. R., and A. C. Alcala. 1986. Flatheaded bats (Mammalia, *Tylonycteris*) from the Philippine Islands. Silliman J. 33:117–23.

Heaney, L. R., and E. C. Birney. 1977. Distribution and natural history notes on some mammals from Puebla, Mexico. Southwestern Nat. 21:543–59.

Heaney, L. R., and P. D. Heideman. 1987. Philippine fruit bats: endangered and extinct. Bats 5(1):3–5.

Heaney, L. R., P. D. Heideman, and K. M. Mudar. 1981. Ecological notes on mammals in the Lake Balinsasayao region, Negros Oriental, Philippines. Silliman J. 28:122–31.

Heaney, L. R., and R. S. Hoffmann. 1978. A second specimen of the neotropical montane squirrel, *Syntheosciurus poasensis*. J. Mamm. 59:854–55.

Heaney, L. R., and G. S. Morgan. 1982. A new species of gymnure, *Podogymnura* (Mammalia: Erinaceidae) from Dinagat Island, Philippines. Proc. Biol. Soc. Washington 95:13–26.

Heaney, L. R., and R. L. Peterson. 1984. A new species of tube-nosed fruit bat (*Nyctimene*) from Negros Island, Philippines (Mammalia: Pteropodidae). Occas. Pap. Mus. Zool. Univ. Michigan, no. 708, 16 pp.

Heaney, L. R., and D. S. Rabor. 1982. Mammals of Dinagat and Siargao Islands, Philippines. Occas. Pap. Mus. Zool. Univ. Michigan, no. 699, 30 pp.

Heaney, L. R., and R. W. Thorington. 1978. Ecology of neotropical red-tailed squirrels, *Sciurus granatensis*, in the Panama Canal Zone. J. Mamm. 59:846–51.

Heaney, L. R., and R. M. Timm. 1983a. Systematics and distribution of shrews of the genus *Crocidura* (Mammalia: Insectivora) in Vietnam. Proc. Biol. Soc. Washington 96:115–20.

———. 1983b. Relationships of pocket gophers of the genus *Geomys* from the central and northern Great Plains. Univ. Kansas Mus. Nat. Hist. Misc. Publ., no. 74, 59 pp.

———. 1985. Morphology, genetics, and ecology of pocket gophers (genus *Geomys*) in a narrow hybrid zone. Biol. J. Linnean Soc. 25:301–17.

Hearn, B. J., L. B. Keith, and O. J. Rongstad. 1987. Demography and ecology of the arctic hare (*Lepus arcticus*) in southwestern Newfoundland. Can. J. Zool. 65:852–61.

Hebert, D., W. Wishart, J. Jorgenson, and M. Festa-Bianchet. 1985. Bighorn sheep population status in Alberta and British Columbia. *In* Hoefs (1985), pp. 48–55.

Heide-Jorgensen, M.-P., and T. J. Härkönen. 1988. Rebuilding seal stocks in the Kattegat-Skagerrak. Mar. Mamm. Sci. 4:231–46.

Heideman, P. D. 1988. The timing of reproduction in the fruit bat *Haplonycteris fischeri* (Pteropodidae): geographic variation and delayed development. J. Zool. 215:577–95.

Heimark, R. J., and G. M. Heimark. 1986. Southern elephant seal pupping at Palmer Station, Antarctica. J. Mamm. 67:189–90.

Heim de Balsac, H. 1972. Insectivores. *In* Battistini and Richard-Vindard (1972), pp. 629–60.

Heim de Balsac, H., and R. Hutterer. 1982. Les soricides (Mammifères Insectivores) des îles du Golfe de Guinée: faits nouveaux et problèmes biogéographiques. Bonner Zool. Beitr. 33:133–50.

Heimer, W. E. 1985. Population status and management of Dall sheep. *In* Hoefs (1985), pp. 1–15.

Heimlich-Boran, J. R. 1986. Cohesive relationships among Puget Sound killer whales. *In* Kirkevold and Lockard (1986), pp. 261–84.

Heithaus, E. R., and T. H. Fleming. 1978. Foraging movements of a frugivorous bat, *Carollia perspicillata* (Phyllostomatidae). Ecol. Monogr. 48:127–43.

Heldmaier, G., and S. Steinlechner. 1981. Seasonal pattern and energetics of short daily torpor in the Djungarian hamster, *Phodopus sungorus*. Oecologia 48:265–70.

Helle, E. 1980. Reproduction, size and structure of the Baltic ringed seal population of the Bothnian Bay. Acta Univ. Ouluensis, ser. A, no. 106, 47 pp.

Helle, E., H. Hyvärinen, and T. Sipilä. 1984. Breeding habitat and lair structure of the Saimaa ringed seal *Phoca hispida saimensis* Nordq. in Finland. Acta Zool. Fennica 172:125–27.

Heller, K.-G., and M. Volleth. 1984. Taxonomic position of "*Pipistrellus societatis*' Hill, 1972 and the karyological characteristics of the genus *Eptesicus* (Chiroptera: Vespertilionidae). Z. Zool. Syst. Evol. 22:65–77.

Hellwing, S. 1970. Reproduction in the white-toothed shrew *Crocidura russula monacha* Thomas in captivity. Israel J. Zool. 19:177–78.

———. 1973. Husbandry and breeding of white-toothed shrews (Crocidurinae) in the research zoo of the Tel-Aviv University. Internatl. Zoo Yearbook 13:127–34.

Helm, J. D., III. 1975. Reproductive biology of *Ototylomys* (Cricetidae). J. Mamm. 56:575–90.

Heltne, P. G., D. C. Turner, and N. J. Scott, Jr. 1976. Comparison of census data on *Alouatta palliata* from Costa Rica and Panama. *In* Thorington and Heltne (1976), pp. 10–19.

Heminway, J. 1987. Wildebeests: the awkward survivors. Smithsonian 17(11):50–61.

Hemmer, H. 1972. *Uncia uncia*. Mammalian Species, no. 20, 5 pp.

———. 1974. Untersuchungen zur Stammesgeschichte der Pantherkatzen (Pantherinae). III. Zur Artgeschichte des Löwen *Panthera (Panthera) leo* (Linnaeus 1758). Veröff. Zool. Staatssaml. München 17:167–280.

———. 1976. Gestation period and postnatal development in felids. *In* Eaton, R. L., ed., Proc. 3rd Internatl. Symp. World's Cats, vol. 3, Carnivore Res. Inst., Univ. Washington, pp. 143–65.

———. 1978. The evolutionary systematics of living Felidae: present status and current problems. Carnivore 1:71–79.

Henderson, F. R., P. F. Springer, and R. Adrian. 1969. The black-footed ferret in South Dakota. South Dakota Dept. Game, Fish, and Parks Tech. Bull., no. 4, 37 pp.

Henderson, M. T., G. Merriam, and J. Wegner. 1985. Patchy environments and species survival: chipmunks in an agricultural mosaic. Biol. Conserv. 31:95–105.

Hendey, Q. H. 1977. Fossil bear from South Africa. S. Afr. J. Sci. 73:112–16.

Hendrichs, H. 1975a. Observations on a population of Bohor reedbuck, *Redunca redunca* (Pallas 1767). Z. Tierpsychol. 38:44–54.

———. 1975b. Changes in a population of dikdik, *Madoqua (Rhynchotragus) kirki* (Günther 1880). Z. Tierpsychol. 38:55–69.

———. 1975c. The status of the tiger *Panthera tigris* (Linne, 1758) in the Sundarbans Mangrove Forest (Bay of Bengal). Saugetierk. Mitt. 23:161–99.

Hennings, D., and R. S. Hoffmann. 1977. A review of the taxonomy of the *Sorex vagrans* species complex from western North America. Occas. Pap. Mus. Nat. Hist. Univ. Kansas, no. 68, 35 pp.

Henry, S. 1984. Social organisation of the greater glider in Victoria. *In* Smith and Hume (1984), pp. 221–28.

Henry, S., and S. A. Craig. 1984. Diet, ranging behaviour and social organization of the yellow-bellied glider (*Petaurus australis* Shaw) in Victoria. *In* Smith and Hume (1984), pp. 331–41.

Henry, V. G. 1968. Length of estrous cycle and gestation in European wild hogs. J. Wildl. Mgmt. 32:406–8.

Hensley, A. P., and K. T. Wilkins. 1988. *Leptonycteris nivalis*. Mammalian Species, no. 307, 4 pp.

Henson, O. W., Jr., and A. Novick. 1966. An additional record of the bat *Phyllonycteris aphylla*. J. Mamm. 47:351–52.

Heptner, V. G. 1975. Uber einige Besonderheiter der Formbildung und der geographischen Verbreitung der Rennmaus, *Meriones (Pallasiomys) meridianus* Pallas, 1773, in den Wusten Mittelasiens. Z. Saugetierk. 40:261–69.

Heptner, V. G., A. A. Nasimovich, and A. G. Bannikov. 1988. Mammals of the Soviet Union. I. Artiodactyla and Perissodactyla. Smithsonian Inst. Libraries and Natl. Sci. Foundation, Washington, D. C., xxvii + 1147 pp.

Herald, E. S., R. L. Brownell, Jr., F. L. Frye, E. J. Morris, W. E. Evans, and A. B. Scott. 1969. Blind river dolphin: first side-swimming cetacean. Science 166:1408–10.

Herbert, H. 1985. Echoortungsverhalten des Flughundes *Rousettus aegyptiacus* (Megachiroptera). Z. Saugetierk. 50:141–52.

Herbert, H. J. 1972. The population dynamics of the waterbuck *Kobus ellipsiprymnus* (Ogilby, 1833) in the Sabi-Sand Wildtuin. Mammalia Depicta, Verlag Paul Parey, Hamburg, 68 pp.

Herman, L. M. 1980. Humpback whales in Hawaiian waters: a study in historical ecology. Pacific Sci. 33:1–15.

Herman, L. M., and R. C. Antinoja. 1977. Humpback whales in the Hawaiian breeding waters: population and pod characteristics. Sci. Rept. Whales Res. Inst. 29:59–85.

Herman, L. M., C. S. Baker, P. H. Forestell, and R. C. Antinoja. 1980. Right whale *Balaena glacialis* sightings near Hawaii: a clue to the wintering grounds? Mar. Ecol. Prog. Ser. 2:271–75.

Herman, T., and K. Fuller. 1974. Observations of the marten, *Martes americana*, in the Mackenzie District, Northwest Territories. Can. Field-Nat. 88:501–3.

Hermanson, J. W., and T. J. O'Shea. 1983. *Antrozous pallidus*. Mammalian Species, no. 213, 8 pp.

Hernandez, C. S., C. B. C. Tapia, A. N. Garduno, E. C. Corona, and M. A. G. Hidalgo. 1985. Notes on distribution and reproduction of bats from coastal regions of Michoacan, Mexico. J. Mamm. 66:549–53.

Hernandez-Camacho, J. 1960. Primitiae mastozoologicae Colombianae—I. Status taxonómico de *Sciurus pucheranii santanderensis*. Caldasia 8:359–68.

Hernandez-Camacho, J., and A. Cadena. 1978. Notas para la revisión del genero *Lonchorhina* (Chiroptera, Phyllostomidae). Caldasia 12:199–251.

Hernandez-Camacho, J., and R. W. Cooper. 1976. The nonhuman primates of Colombia. *In* Thorington and Heltne (1976), pp. 35–69.

Herrero, S. 1970. Human injury inflicted by grizzly bears. Science 170:593–98.

———, ed. 1972. Bears—their biology and management. Internatl. Union Conserv. Nat. Publ., n.s., no. 23, 371 pp.

———. 1976. Conflicts between man and grizzly bears in the national parks of North America. *In* Pelton, Lentfer, and Folk (1976), pp. 121–45.

Herrero, S., C. Schroeder, and M. Scott-Brown. 1986. Are Canadian foxes swift enough? Biol. Conserv. 36:159–67.

Herselman, J. C., and P. M. Norton. 1985. The distribution and status of bats (Mammalia: Chiroptera) in the Cape Province. Ann. Cape Prov. Mus. (Nat. Hist.) 16:73–126.

Hersh, S. L., and D. K. Odell. 1986. Mass stranding of Fraser's dolphin, *Lagenodelphis hosei*, in the western North Atlantic. Mar. Mamm. Sci. 2:73–76.

Hershkovitz, P. 1959a. Nomenclature and taxonomy of the neotropical mammals described by Olfers, 1818. J. Mamm. 40:337–53.

———. 1959b. Two new genera of South American rodents (Cricetinae). Proc. Biol. Soc. Washington 72:5–10.

———. 1960. Mammals of northern Colombia, preliminary report no. 8: arboreal rice rats, a systematic revision of the subgenus *Oecomys*, genus *Oryzomys*. Proc. U.S. Natl. Mus. 110:513–68.

———. 1962. Evolution of neotropical cricetine rodents (Muridae) with special reference to the phyllotine group. Fieldiana Zool. 46:1–524.

———. 1963. A systematic and zoogeographic account of the monkeys of the genus *Callicebus* (Cebidae) of the Amazonas and Orinoco river basins. Mammalia 27:1–79.

———. 1966a. South American swamp and fossorial rats of the scapteromyine group (Cricetinae, Muridae) with comments on the glans penis in murid taxonomy. Z. Saugetierk. 31:81–149.

———. 1966b. Mice, land bridges and Latin American faunal exchange. *In* Wenzel, R. L., and V. J. Tipton, eds., Ectoparasites of Panama, Field Mus. Nat. Hist., Chicago, pp. 725–51.

———. 1970. Supplementary notes on neotropical *Oryzomys dimidiatus* and *Oryzomys hammondi* (Cricetinae). J. Mamm. 51:789–94.

———. 1971. A new rice rat of the *Oryzomys palustris* group (Cricetinae, Murinae) from northwestern Colombia, with remarks on distribution. J. Mamm. 52:700–709.

———. 1972. Notes on New World monkeys. Internatl. Zoo Yearbook 12:3–12.

———. 1975a. The scientific name of the lesser *Noctilio* (Chiroptera), with notes on the chauve-souris de la Vallee d'Ylo (Peru). J. Mamm. 56:242–47.

———. 1975b. Comments on the taxonomy of Brazilian marmosets (Callithrix, Callithricidae). Folia Primatol. 24:137–72.

———. 1976. Comments on generic names of four-eyed opossums (family Didelphidae). Proc. Biol. Soc. Washington 89:295–304.

———. 1977. Living New World monkeys (Platyrrhini). Volume I. Univ. Chicago Press, xiv + 1117 pp.

———. 1979. The species of sakis, genus *Pithecia* (Cebidae, Primates), with notes on sexual dichromatism. Folia Primatol. 31:1–22.

———. 1981. *Philander* and four-eyed opossums once again. Proc. Biol. Soc. Washington 93:943–46.

———. 1982. Neotropical deer (Cervidae). Part I. Pudus, genus *Pudu* Gray. Fieldiana Zool., n.s., no. 11, 86 pp.

———. 1983. Two new species of night monkeys, genus *Aotus* (Cebidae, Platyrrhini): a preliminary report on *Aotus* taxonomy. Amer. J. Primatol. 4:209–43.

———. 1984a. On the validity of the family-

Hodgdon, H. E., and R. A. Lancia. 1983. Behavior of the North American beaver, *Castor canadensis*. Acta Zool. Fennica 174:99–103.

Hoeck, H. N. 1975. Differential feeding behaviour of the sympatric hyrax *Procavia johnstoni* and *Heterohyrax brucei*. Oecologia 22:15–47.

———. 1978. Systematics of the Hyracoidea: toward a clarification. Bull. Carnegie Mus. Nat. Hist., no. 6, pp. 146–51.

———. 1982. Population dynamics, dispersal and genetic isolation in two species of hyrax (*Heterohyrax brucei* and *Procavia johnstoni*) on habitat islands in the Serengeti. Z. Tierpsychol. 59:177–210.

———. 1984. Hyraxes. *In* Macdonald (1984), pp. 462–65.

———. 1989. Demography and competition in hyrax. Oecologia 79:353–60.

Hoeck, H. N., H. Klein, and P. Hoeck. 1982. Flexible social organization in hyrax. Z. Tierpsychol. 59:265–98.

Hoefs, M., ed. 1985. Wild sheep. Distribution, abundance, management and conservation of the sheep of the world and closely related mountain ungulates. Northern Wild Sheep and Goat Council, Whitehorse, Yukon, 218 pp.

Hoefs, M., and N. Barichello. 1985. Distribution, abundance and management of wild sheep in Yukon. *In* Hoefs (1985), pp. 16–34.

Hoese, H. D. 1971. Dolphin feeding out of water in a salt marsh. J. Mamm. 52:222–23.

Hoffmann, R. S. 1984. A review of the shrewmoles (genus *Uropsilus*) of China and Burma. J. Mamm. Soc. Japan 10:69–80.

———. 1985. The correct name for the Palearctic brown, or flat-skulled, shrew is *Sorex roboratus*. Proc. Biol. Soc. Washington 98:17–28.

———. 1986*a*. A review of the genus *Soriculus* (Mammalia: Insectivora). J. Bombay Nat. Hist. Soc. 82:459–81.

———. 1986*b*. A new locality record for the kouprey from Viet-Nam, and an archaeological record from China. Mammalia 50:391–96.

———. 1987. A review of the systematics and distribution of Chinese red-toothed shrews (Mammalia: Soricinae). Acta Theriol. Sinica 7:100–139.

Hoffmann, R. S., and R. D. Fisher. 1978. Additional distributional records of Preble's shrew (*Sorex preblei*). J. Mamm. 59:893–94.

Hoffmann, R. S., J. K. Jones, Jr., and J. A. Campbell. 1987. First record of *Myotis auricularis* from Guatemala. Southwestern Nat. 32:391.

Hoffmann, R. S., J. W. Koeppl, and C. F. Nadler. 1979. The relationships of the amphi-

beringian marmots (Mammalia: Sciuridae). Occas. Pap. Mus. Nat. Hist. Univ. Kansas, no. 83, 56 pp.

Hoffmeister, D. F. 1986. Mammals of Arizona. Univ. Arizona Press, xix + 602 pp.

Hoffmeister, D. F., and V. E. Diersing. 1978. Review of the tassel-eared squirrels of the subgenus *Otosciurus*. J. Mamm. 59:402–13.

Hoffmeister, D. F., and W. W. Goodpaster. 1962. Life history of the desert shrew *Notiosorex crawfordi*. Southwestern Nat. 7:236–52.

Holisova, V., J. Pelikan, and J. Zejoa. 1962. Ecology and population dynamics in *Apodemus microps* Krat. and Ros. (Mamm.: Muridae). Acta. Acad. Sci. Cechoslovenciae, Brno 34:493–540.

Hollander, R. R., and J. K. Jones, Jr. 1988. Northernmost record of the tropical brown bat, *Eptesicus furinalis*. Southwestern Nat. 33:100.

Holley, A. J. F. 1986. A hierarchy of hares: dominance status and access to oestrous does. Mamm. Rev. 16:181–86.

Hollien, H., P. Hollien, D. K. Caldwell, and M. C. Caldwell. 1976. Sound production by the Atlantic bottlenosed dolphin *Tursiops truncatus*. Cetology, no. 26, 8 pp.

Holliman, D. C. 1983. Status and habitat of Alabama Gulf Coast beach mice *Peromyscus polionotus ammobates* and *P. p. trissyllepsis*. Northeast Gulf Sci. 6:121–29.

Holloway, R. L., ed. 1974. Primate aggression, territoriality, and xenophobia: a comparative perspective. Academic Press, New York, xiv + 513 pp.

Holthuis, L. B. 1987. The scientific name of the sperm whale. Mar. Mamm. Sci. 3:87–90.

Homan, J. A., and J. K. Jones, Jr. 1975*a*. *Monophyllus redmani*, Mammalian Species, no. 57, 3 pp.

———. 1975*b*. *Monophyllus plethodon*, Mammalian Species, no. 58. 2 pp.

Homewood, K. 1975. Can the Tana mangabey survive? Oryx 13:53–59.

Homewood, K., and W. A. Rodgers. 1981. A previously undescribed mangabey from southern Tanzania. Internatl. J. Primatol. 2:47–55.

Honacki, J. H., K. E. Kinman, and J. W. Koeppl. 1982. Mammal species of the world: a taxonomic and geographic reference. Allen Press, Lawrence, Kansas, ix + 694 pp.

Honeycutt, R. L., R. J. Baker, and H. H. Genoways. 1980. Results of the Alcoa Foundation–Suriname Expeditions. III. Chromosomal data for bats (Mammalia: Chiroptera) from Suriname. Ann. Carnegie Mus. 49:237–50.

Honeycutt, R. L., and V. M. Sarich. 1987. Albumin evolution and subfamilial relation-

ships among New World leaf-nosed bats (family Phyllostomidae). J. Mamm. 68:508–17.

Honeycutt, R. L., and S. L. Williams. 1982. Genic differentiation in pocket gophers of the genus *Pappogeomys*, with comments on intergeneric relationships in the subfamily Geomyinae. J. Mamm. 63:208–17.

Hood, C. S., and J. K. Jones, Jr. 1984. *Noctilio leporinus*. Mammalian Species, no. 216, 7 pp.

Hood, C. S., and J. Pitocchelli. 1983. *Noctilio albiventris*. Mammalian Species, no. 197, 5 pp.

Hood, C. S., L. W. Robbins, R. J. Baker, and H. S. Shellhammer. 1984. Chromosomal studies and evolutionary relationships of an endangered species, *Reithrodontomys raviventris*. J. Mamm. 65:655–67.

Hoogland, J. L. 1981. The evolution of coloniality in white-tailed and black-tailed prairie dogs (Sciuridae: *Cynomys leucurus* and *C. ludovicianus*). Ecology 62:252–72.

———. 1982. Prairie dogs avoid extreme inbreeding. Science 215:1639–41.

———. 1983. Black-tailed prairie dog coteries are cooperatively breeding units. Amer. Nat. 121:275–80.

———. 1985. Infanticide in prairie dogs: lactating females kill offspring of close kin. Science 230:1037–40.

Hoogstraal, H. 1962. A brief review of contemporary land mammals of Egypt (including Sinai). 1. Insectivora and Chiroptera. J. Egyptian Pub. Health Assoc. 37:143–62.

Hooijer, D. A. 1965. Notes on *Coryphomys bühleri* Schaub, a gigantic murine rodent from Timor. Israel J. Zool. 14:128–33.

Hooper, E. T. 1968. Classification. *In* King (1968), pp. 27–74.

———. 1972. A synopsis of the rodent genus *Scotinomys*. Occas. Pap. Mus. Zool. Univ. Michigan, no. 665, 32 pp.

Hooper, E. T., and M. D. Carleton. 1976. Reproduction, growth and development in two contiguously allopatric rodent species, genus *Scotinomys*. Misc. Publ. Mus. Zool. Univ. Michigan, no. 151, 52 pp.

Hope, J. H. 1981. A new species of *Thylogale* (Marsupialia: Macropodidae) from Mapala Rock Shelter, Jaya (Carstensz) Mountains, Irian Jaya (western New Guinea), Indonesia. Rec. Austral. Mus. 33:369–87.

Horacek, I. 1975. Notes on the ecology of the bats of the genus *Plecotus* Geoffroy, 1818 (Mammalia: Chiroptera). Vest. Cesk. Spol. Zool. 39:195–210.

Horacek, I., and V. Hanák. 1984. Comments on the systematics and phylogeny of *Myotis nattereri* (Kuhl, 1818). Myotis 21–22:20–29.

———. 1986. Generic status of *Pipistrellus*

savii and comments on classification of the genus *Pipistrellus* (Chiroptera, Vespertilionidae). Myotis 23–24:9–16.

Horn, A. D. 1979. The taxonomic status of the bonobo chimpanzee. Amer. J. Phys. Anthropol. 51:273–82.

———. 1987a. The socioecology of the black mangabey *(Cercocebus aterrimus)* near Lake Tumba, Zaire. Amer. J. Primatol. 12:165–80.

———. 1987b. Taxonomic assessment of the allopatric gray-cheeked mangabey *(Cercocebus albigena)* and black mangabey *(C. aterrimus)*: comparative socioecological data and the species concept. Amer. J. Primatol. 12:181–87.

Hornocker, M. G. 1969. Winter territoriality in mountain lions. J. Wildl. Mgmt. 33:457–64.

———. 1970. An analysis of mountain lion predation upon mule deer and elk in the Idaho Primitive Area. Wildl. Monogr., no. 21, 39 pp.

———. 1983. Tracking the truth about wolverines. Natl. Wildl. 21(5):32–39.

Hornocker, M. G., and H. S. Hash. 1981. Ecology of the wolverine in northwest Montana. Can. J. Zool. 59:1286–1301.

Horr, D. A. 1975. The Borneo orang-utan: population structure and dynamics in relationship to ecology and reproductive strategy. Primate Behav. 4:307–23.

Horrocks, J. A. 1986. Life-history characteristics of a wild population of vervets *(Cercopithecus aethiops sabaeus)* in Barbados, West Indies. Internatl. J. Primatol. 7:31–47.

Horwich, R. H. 1983. Species status of the black howler monkey, *Alouatta pigra*, of Belize. Primates 24:288–89.

Horwich, R. H., and K. Gebhard. 1983. Roaring rhythms in black howler monkeys *(Alouatta pigra)* of Belize. Primates 24:290–96.

Horwich, R. H., and E. D. Johnson. 1986. Geographical distribution of the black howler *(Alouatta pigra)* in Central America. Primates 27:53–62.

Horwood, J. W. 1986. The distribution of the southern blue whale in relation to recent estimates of abundance. Sci. Rept. Whales Res. Inst. 37:155–65.

———. 1987. The sei whale: population biology, ecology and management. Croom Helm, London, 375 pp.

Hoshino, J. 1985. Feeding ecology of mandrills *(Mandrillus sphinx)* in Campo Animal Reserve, Cameroon. Primates 26:248–73.

Hoshino, J., A. Mori, H. Kudo, and M. Kawai. 1984. Preliminary report on the grouping of mandrills *(Mandrillus sphinx)* in Cameroon. Primates 25:295–307.

Hoth, J., A. Velazquez, F. J. Romero, L. Leon, M. Aranda, and D. J. Bell. 1987. The volcano rabbit—a shrinking distribution and a threatened habitat. Oryx 21:85–91.

Houpt, K. A., and A. F. Fraser. 1988. Przewalski horses. Appl. Anim. Behav. Sci. 21:1–3.

Houseal, T. W., I. F. Greenbaum, D. J. Schmidly, S. A. Smith, and K. M. Davis. 1987. Karyotypic variation in *Peromyscus boylii* from Mexico. J. Mamm. 68:281–96.

Houseknecht, C. R., and J. R. Tester. 1978. Denning habits of striped skunks *(Mephitis mephitis)*. Amer. Midl. Nat. 100:424–30.

Houston, D. B. 1982. The northern Yellowstone elk. Macmillan, New York, xix + 474 pp.

How, R. A. 1976. Reproduction, growth and survival of young in the mountain possum, *Trichosurus caninus* (Marsupialia). Austral. J. Zool. 24:189–99.

———. 1978. Population strategies of four species of Australian "possums." *In* Montgomery (1978), pp. 305–13.

How, R. A., J. L. Barnett, A. J. Bradley, W. J. Humphreys, and R. W. Martin. 1984. The population biology of *Pseudocheirus peregrinus* in a *Leptospermum laevigatum* thicket. *In* Smith and How (1984), pp. 261–68.

Howard, P. C. 1986a. Social organization of common reedbuck and its role in population regulation. Afr. J. Ecol. 24:143–54.

———. 1986b. Spatial organization of common reedbuck with special reference to the role of juvenile dispersal in population regulation. Afr. J. Ecol. 24:155–71.

Howard, W. E., and J. N. Amaya. 1975. European rabbit invades western Argentina. J. Wildl. Mgmt. 39:757–61.

Howe, D. 1975. Observations on a captive marsupial mole, *Notoryctes typhlops*. Austral. Mamm. 1:361–65.

Howe, H. F. 1974. Additional records of *Phyllonycteris aphylla* and *Ariteus flavescens* from Jamaica. J. Mamm. 55:662–63.

Howe, R. J. 1974. Marking behaviour of the Bahaman hutia *(Geocapromys ingrahami)*. Anim. Behav. 22:645–49.

———. 1976. Social behavior of Bahamian hutias in captivity. Florida Sci. 39:8–14.

Howell, K. M., and P. D. Jenkins. 1984. Records of shrews (Insectivora, Soricidae) from Tanzania. Afr. J. Ecol. 22:67–68.

Hoyt, E. 1977. *Orcinus orca*. Separating facts from fantasies. Oceans 10(4):22–26.

Hsu Lunghui and Wu Jiayan. 1981. A new subspecies of *M. flavigula* from Hainan Island. Acta Theriol. Sinica 1:145–48.

Hubbard, C. A. 1970. A new species of *Tatera* from Tanzania with a description of its life history and habits studied in captivity. Zool. Afr. 5:237–47.

———. 1972. Observations on the life histories and behaviour of some small rodents from Tanzania. Zool. Afr. 7:419–49.

Hubbard, W. P., and S. Harris. 1960. Notorious grizzly bears. Sage Books, Denver, 205 pp.

Hubbs, C. L., and K. S. Norris. 1971. Original teeming abundance, supposed extinction, and survival of the Juan Fernandez fur seal. *In* Burt, W. H., ed., Antarctic Pinnipedia, Amer. Geophys. Union, Washington, D.C., pp. 35–51.

Hubbs, C. L., W. F. Perrin, and K. C. Balcomb. 1973. *Stenella coeruleoalba* in the eastern and central tropical Pacific. J. Mamm. 54:549–52.

Hubert, B. 1978a. Revision of the genus *Saccostomus* (Rodentia, Cricetomyinae), with new morphological and chromosomal data from specimens from the lower Omo Valley, Ethiopia. Bull. Carnegie Mus. Nat. Hist., no. 6, pp. 48–52.

———. 1978b. Modern rodent fauna of the lower Omo Valley, Ethiopia. Bull. Carnegie Mus. Nat. Hist., no. 6, pp. 109–12.

Hubert, B., F. Adam, and A. Poulet. 1973. Liste préliminaire des rongeurs du Sénégal. Mammalia 37:76–87.

Huckaby, D. G. 1980. Species limits in the *Peromyscus mexicanus* group (Mammalia: Rodentia: Muroidea). Los Angeles Co. Mus. Nat. Hist. Contrib. Sci., no. 326, 24 pp.

Hudson, J. W. 1974. The estrous cycle, reproduction, growth, and development of temperature regulation in the pygmy mouse, *Baiomys taylori*. J. Mamm. 55:572–88.

Hudson, W. S., and D. E. Wilson. 1986. *Macroderma gigas*. Mammalian Species, no. 260, 4 pp.

Hulbert, A. J., G. Gordon, and T. J. Dawson. 1971. Rediscovery of the marsupial *Echymipera rufescens* in Australia. Nature 231:330–31.

Hult, R. W. 1982. Another function of echolocation for bottlenosed dolphins *(Tursiops truncatus)*. Cetology, no. 47, 7 pp.

Humphrey, S. R. 1975. Nursery roosts and community diversity of Nearctic bats. J. Mamm. 56:321–46.

———. 1978. Status, winter habitat, and management of the endangered Indiana bat, *Myotis sodalis*. Florida Sci. 41:65–76.

———. 1981. Goff's pocket gopher *(Geomys pinetis goffi)* is extinct. Florida Sci. 44:250–52.

Humphrey, S. R., and D. B. Barbour. 1981. Status and habitat of three subspecies of *Peromyscus polionotus* in Florida. J. Mamm. 62:840–44.

Humphrey, S. R., and B. Bell. 1986. The Key deer population is declining. Wildl. Soc. Bull. 14:261–65.

Humphrey, S. R., and L. N. Brown. 1986. Report of a new bat (Chiroptera: *Artibeus jamaicensis*) in the United States is erroneous. Florida Sci. 49:262–63.

Humphrey, S. R., and T. H. Kunz. 1976. Ecology of a Pleistocene relict, the western big-eared bat *(Plecotus townsendii)*, in the southern Great Plains. J. Mamm. 57:470–94.

Humphrey, S. R., A. R. Richter, and J. B. Cope. 1977. Summer habitat and ecology of the endangered Indiana bat, *Myotis sodalis*. J. Mamm. 58:334–46.

Humphreys, W. F., R. A. How, A. J. Bradley, C. M. Kemper, and D. J. Kitchener. 1984. The biology of *Wyulda squamicaudata* Alexander 1919. *In* Smith and Hume (1984), pp. 162–69.

Hunsaker, D., II, ed. 1977a. The biology of marsupials. Academic Press, New York, xv + 537 pp.

———. 1977b. Ecology of New World marsupials. *In* Hunsaker (1977a), pp. 95–156.

Hunsaker, D., II, and D. Shupe. 1977. Behavior of New World marsupials. *In* Hunsaker (1977a), pp. 279–347.

Hunt, J. H. 1974. The little-known lynx. Maine Fish and Game 16(2):28.

Huntly, N. J., A. T. Smith, and B. L. Ivins. 1986. Foraging behavior of the pika *(Ochotona princeps)*, with comparisons of grazing versus haying. J. Mamm. 67:139–48.

Husar, S. L. 1977. *Trichechus inunguis*. Mammalian Species, no. 72, 4 pp.

———. 1978a. *Dugong dugon*. Mammalian Species, no. 88, 7 pp.

———. 1978b. *Trichechus senegalensis*. Mammalian Species, no. 89, 3 pp.

———. 1978c. *Trichechus manatus*. Mammalian Species, no. 93, 5 pp.

Husband, T. P., P. B. Davis, and J. H. Brown, Jr. 1986. Population measurements of the Cretan agrimi. J. Mamm. 67:757–59.

Husson, A. M. 1978. The mammals of Suriname. E. J. Brill, Leiden, xxxiv + 569 pp.

Husson, A. M., and L. B. Holthuis. 1974. *Physeter macrocephalus* Linnaeus, 1758, the valid name for the sperm whale. Zool. Meded. 48:205–17.

Hutchins, M., G. Thompson, B. Sleeper, and J. W. Foster. 1987. Management and breeding of the Rocky Mountain goat *Oreamnos americanus* at Woodland Park Zoo. Internatl. Zoo Yearbook 26:297–308.

Hutterer, R. 1980. A record of Goodwin's shrew, *Cryptotis goodwini*, from Mexico. Mammalia 44:413.

———. 1981a. Nachweis der Spitzmaus *Crocidura roosevelti* für Tanzania. Stuttgarter Beitr. Naturkunde, ser. A, no. 342, 9 pp.

———. 1981b. Der Status von *Crocidura ariadne* Pieper, 1979 (Mammalia: Soricidae). Bonner Zool. Beitr. 32:3–12.

———. 1981c. *Crocidura manengubae* n. sp. (Mammalia: Soricidae) eine neue Spitzmaus aus Kamerun. Bonner Zool. Beitr. 32:241–48.

———. 1981d. Range extension of *Crocidura smithii*, with description of a new subspecies from Senegal. Mammalia 45:388–91.

———. 1983a. Taxonomy and distribution of *Crocidura fuscomurina*. Mammalia 47:221–27.

———. 1983b. *Crocidura grandiceps*, eine neue Spitzmaus aus Westafrika. Rev. Suisse Zool. 90:699–707.

———. 1983c. Status of some African *Crocidura* by Isidore Geoffroy Saint-Hilaire, Carl J. Sundevall and Theodor von Heuglin. Ann. Mus. Roy. Afr. Cent., Sec. Zool., 237:207–17.

———. 1986a. Diagnosen neuer Spitzmäuse aus Tansania (Mammalia: Soricidae). Bonner Zool. Beitr. 37:23–33.

———. 1986b. African shrews allied to *Crocidura fischeri*: taxonomy, distribution and relationships. Cimbebasia, ser. A, 8:24–35.

———. 1986c. The species of *Crocidura* in Morocco. Mammalia 50:521–34.

———. 1986d. Synopsis der Gattung *Paracrocidura* (Mammalia: Soricidae), mit Beschreibung einer neuen Art. Bonner Zool. Beitr. 37:73–90.

———. 1986e. Eine neue Soricidengattung aus Zentralafrika (Mammalia: Soricidae). Z. Saugetierk. 51:257–66.

Hutterer, R., and F. Dieterlen. 1984. Zwei neue Arten der Gattung *Grammomys* aus Äthiopien und Kenia (Mammalia: Muridae). Stuttgarter Beitr. Naturkunde, ser. A, 18:1–18.

Hutterer, R., and N. J. Dippenaar. 1987a. A new species of *Crocidura* Wagler, 1832 (Soricidae) from Zambia. Bonner Zool. Beitr. 38:1–7.

———. 1987b. *Crocidura ansellorum*, emended species name for a recently described African shrew. Bonner Zool. Beitr. 38:269.

Hutterer, R., and D. C. D. Happold. 1983. The shrews of Nigeria (Mammalia: Soricidae). Bonner Zool. Monogr., no. 18, 79 pp.

Hutterer, R., and D. L. Harrison. 1988. A new look at the shrews (Soricidae) of Arabia. Bonner Zool. Beitr. 39:59–72.

Hutterer, R., and U. Hirsch. 1979. Ein neuer *Nesoryzomys* von der Insel Fernandina, Galapagos (Mammalia, Rodentia). Bonner Zool. Beitr. 30:276–83.

Hutterer, R., and P. D. Jenkins. 1980. A new species of *Crocidura* from Nigeria (Mammalia: Insectivora). Bull. British Mus. (Nat. Hist.) Zool. 39:305–10.

———. 1983. Species-limits of *Crocidura somalica* Thomas, 1895 and *Crocidura yankariensis* Hutterer and Jenkins, 1980 (Insectivora: Soricidae). Z. Saugetierk. 48:193–201.

Hutterer, R., and U. Joger. 1982. Kleinsäuger aus dem Hochland von Adamaoua, Kamerun. Bonner Zool. Beitr. 33:119–32.

Hutterer, R., and D. Kock. 1983. Spitzmäuse aus den Nuba-Bergen Kordofans, Sudan. Senckenberg. Biol. 63:17–26.

Hutterer, R., L. F. Lopez-Jurado, and P. Vogel. 1987. The shrews of the eastern Canary Islands: a new species (Mammalia: Soricidae). J. Nat. Hist. 21:1347–57.

Hutterer, R., N. Lopez-Martínez, and J. Michaux. 1988. A new rodent from Quaternary deposits of the Canary Islands and its relationships with Neogene and Recent murids of Europe and Africa. Palaeovertebrata 18:241–62.

Hutterer, R., and W. Verheyen. 1985. A new species of shrew, genus *Sylvisorex*, from Rwanda and Zaire (Insectivora: Soricidae). Z. Saugetierk. 50:266–71.

Hyndman, D., and J. I. Menzies. 1980. *Aproteles bulmerae* (Chiroptera: Pteropodidae) of New Guinea is not extinct. J. Mamm. 61:159–60.

I

Ibáñez, C. 1980. Descripcion de un nuevo genero de quiroptero neotropical de la familia Molossidae. Doñana Acta Vert. 7:104–11.

———. 1985. Notes on *Amorphochilus schnablii* Peters (Chiroptera, Furipteridae). Mammalia 49:584–87.

———. 1988. Notes on bats from Morocco. Mammalia 52:278–81.

Ibáñez, C., M. Delibes, J. Castroviejo, R. Martín, J. F. Beltrán, and S. Moreno. 1988. An unusual record of hooded seal *(Cystophora cristata)* in SW Spain. Z. Saugetierk. 53:189–90.

Ibáñez, C., and R. Fernandez. 1985a. Murcielagos (Mammalia, Chiroptera) de las Canarias. Doñana Acta Vert. 12:307–15.

———. 1985b. Systematic status of the long-eared bat *Plecotus teneriffae* Barret-Hamilton, 1907 (Chiroptera; Vespertilionidae). Saugetierk. Mitt. 32:143–49.

Ibáñez, C., and J. Ochoa G. 1985. Distribucion y taxonomia de *Molossops temminckii* (Chiroptera, Molossidae) en Venezuela. Doñana Acta Vert. 12:141–50.

Ibáñez, C., and J. A. Valverde. 1985. Taxonomic status of *Eptesicus platyops* (Thomas, 1901) (Chiroptera, Vespertilionidae). Z. Saugetierk. 50:241–42.

Ichihara, T. 1966. The pygmy blue whale *Balaenoptera musculus brevicauda*, a new subspecies from the Antarctic. *In* Norris (1966), pp. 79–113.

Ikeda, H. 1986. Old dogs, new treks. Nat. Hist. 95(8):38–45.

Ikeda, H., K. Eguchi, and Y. Ono. 1979. Home range utilization of a raccoon dog, *Nyctereutes procyonides viverrinus*, Temminck, in a small islet in western Kyushu. Japanese J. Ecol. 29:35–48.

Ilmen, M., and S. Lahti. 1968. Reproduction, growth and behaviour in the captive wood lemming, *Myopus schisticolor* (Lilljeb). Ann. Zool. Fennici 5:207–19.

Ingles, J. M., P. N. Newton, M. R. W. Rands, and G. R. Bowden. 1980. The first record of a rare murine rodent *Diomys* and further records of three shrew species from Nepal. Bull. British Mus. (Nat. Hist.) Zool. 39:205–11.

Ingles, L. G. 1961. Home range and habitats of the wandering shrew. J. Mamm. 42:455–62.

Innes, D. G. L. 1978. A reexamination of litter size in some North American microtines. Can. J. Zool. 56:1488–96.

Insley, H. 1977. An estimate of the population density of the red fox *(Vulpes vulpes)* in the New Forest, Hampshire. J. Zool. 183:549–53.

Irby, L. R. 1979. Reproduction in mountain reedbuck *(Redunca fulvorufula)*. Mammalia 43:191–213.

Irvine, A. B., M. D. Scott, R. S. Wells, and J. H. Kaufmann. 1981. Movements and activities of the Atlantic bottlenose dolphin, *Tursiops truncatus*, near Sarasota, Florida. Fishery Bull. 79:671–88.

Ishii, N. 1982. Reproductive activity of the Japanese shrew-mole, *Urotrichus talpoides* Temminck. J. Mamm. Soc. Japan 9:25–36.

Iskandar, D., J.-M. Duplantier, F. Bonhomme, F. Petter, and L. Thaler. 1988. Mise en évidence de deux espèces jumelles sympatriques du genre *Hylomyscus* dans le nord-est du Gabon. Mammalia 52:126–30.

IUCN (International Union for Conservation of Nature and Natural Resources). 1972–78. Red data book. I. Mammalia. Morges, Switzerland.

Ivanter, E. V. 1972. Contribution to the ecology of *Sicista betulina* Pall. Aquilo, Ser. Zool., 13:103–8.

Iverson, S. L., and B. N. Turner. 1972. Natural history of a Manitoba population of Franklin's ground squirrels. Can. Field-Nat. 86:145–49.

Iwano, T. 1975. Distribution of Japanese monkey *(Macaca fuscatta)*. Proc. 5th Internatl. Congr. Primatol., pp. 389–91.

Izard, J., and K. Umfleet. 1971. Notes on the care and breeding of the suni *Nesotragus moschatus* at Dallas Zoo. Internatl. Zoo Yearbook 11:129.

Izard, M. K. 1987. Lactation length in three species of *Galago*. Amer. J. Primatol. 13:73–76.

Izard, M. K., and D. T. Rasmussen. 1985. Reproduction in the slender loris *(Loris tardigradus malabaricus)*. Amer. J. Primatol. 8:153–65.

Izard, M. K., P. C. Wright, and E. L. Simons. 1985. Gestation length in *Tarsius bancanus*. Amer. J. Primatol. 9:327–31.

Izawa, K. 1970. Unit groups of chimpanzees and their nomadism in the savanna woodland. Primates 11:1–46.

———. 1976. Group sizes and compositions of monkeys in the upper Amazon basin. Primates 17:367–99.

———. 1978. A field study of the ecology and behavior of the black-mantle tamarin *(Saguinus nigricollis)*. Primates 19:241–74.

———. 1980. Social behavior of the wild black-capped capuchin *(Cebus apella)*. Primates 21:443–67.

Izor, R. J. 1979. Winter range of the silver-haired bat. J. Mamm. 60:641–43.

Izor, R. J., and L. de la Torre. 1978. A new species of weasel *(Mustela)* from the highlands of Colombia, with comments on the evolution and distribution of South American weasels. J. Mamm. 59:92–102.

Izor, R. J., and T. J. McCarthy. 1984. *Heteromys gaumeri* (Rodentia: Heteromyidae) in the northern plain of Belize. Mammalia 48:465–67.

Izor, R. J., and N. E. Peterson. 1985. Notes on South American weasels. J. Mamm. 66:788–90.

Izor, R. J., and R. H. Pine. 1987. Notes on the black-shouldered opossum, *Caluromysiops irrupta*. Fieldiana Zool., n.s., 39:117–24.

J

Jabir, H. A., D. Bajomi, and A. Demeter. 1985. New record of the black rat *(Rattus rattus* L.) from Hungary, and a review of its distribution in Central Europe (Mammalia). Ann. Hist.-Nat. Mus. Natl. Hung., Zool, 77:263–67.

Jachowski, R. L. 1981. Proposal to remove the bobcat from Appendix II of the Convention on International Trade in Endangered Species of Wild Fauna and Flora. Federal Register 46:45652–56.

Jackson, H. H. T. 1961. Mammals of Wisconsin. Univ. Wisconsin Press, Madison, xii + 504 pp.

Jackson, J. E. 1985. Behavioural observations on the Argentine pampas deer *(Ozotoceros bezoarticus celer* Cabrera, 1943). Z. Saugetierk. 50:107–16.

———. 1986. Antler cycle in pampas deer *(Ozotoceros bezoarticus)* from San Luis, Argentina. J. Mamm. 67:175–76.

———. 1987. *Ozotoceros bezoarticus*. Mammalian Species, no. 295, 5 pp.

Jackson, J. E., and A. Langguth. 1987. Ecology and status of the pampas deer *(Ozotoceros bezoarticus)* in the Argentinian pampas and Uruguay. *In* Wemmer (1987), pp. 402–9.

Jackson, P. 1985. Man-eaters. Internatl. Wildl. 15(6):4–11.

Jackson, P. F. R. 1978. Scientists hunt the Bengal tiger—but only in order to trace and save it. Smithsonian 9(5):28–37.

Jackson, R. 1979. Snow leopards in Nepal. Oryx 15:191–95.

Jackson, R., and G. G. Ahlborn. 1988. Observations on the ecology of snow leopard in west Nepal. Proc. 5th Internatl. Snow Lepard Symp., pp. 65–87.

Jacobs, L. L. 1980. Siwalik fossil tree shrews. *In* Luckett (1980a), pp. 205–16.

Jacobsen, N. H. G. 1974. Distribution, home ranges, and behaviour patterns of bushbuck in the Lutope and Sengwa valleys, Rhodesia. J. S. Afr. Wildl. Mgmt. Assoc. 4:75–93.

Jacobsen, N. H. G., and E. Du Plessis. 1976. Observations on the ecology and biology of the Cape fruit bat *Rousettus aegyptiacus leachi* in the eastern Transvaal. S. Afr. J. Sci. 72:270–73.

Jaeger, J.-J., C. Denys, and B. Coiffait. 1985. New Phiomorpha and Anomaluridae from the Late Eocene of north-west Africa: phylogenetic implications. *In* Luckett and Hartenberger (1985), pp. 567–87.

Jameson, E. W., Jr., amd G. S. Jones. 1978. Thc Soricidae of Taiwan. Proc. Biol. Soc. Washington 90:459–82.

Jameson, R. J. 1989. Movements, home range, and territories of male sea otters off central California. Mar. Mamm. Sci. 5:159–72.

Jameson, R. J., K. W. Kenyon, A. M. Johnson, and H. M. Wight. 1982. History and status of translocated sea otter populations in North America. Wildl. Soc. Bull. 10:100–107.

Jannett, F. J., Jr. 1984. Reproduction of the montane vole, *Microtus montanus*, in subnivean populations. Carnegie Mus. Nat. Hist. Spec. Publ. 10:215–24.

Jannett, F. J., and J. Z. Jannett. 1974. Drum-marking by *Arvicola richardsoni* and its taxonomic significance. Amer. Midl. Nat. 92:230–34.

Janzen, D. H., and W. Hallwachs. 1982. The hooded skunk, *Mephitis macroura*, in lowland northwestern Costa Rica. Brenesia 19/20:549–52.

Jarman, P. J., and M. V. Jarman. 1973. Social behaviour, population structure and reproductive potential in impala. E. Afr. Wildl. J. 11:329–38.

———. 1974. Impala behaviour and its relevance to management. *In* Geist and Walther (1974), pp. 871–81.

Jarvis, C. 1967. Tabulated data on the breeding biology of the black rhinoceros *Diceros bicornis* compiled from reports in the yearbook. Internatl. Zoo Yearbook 7:166.

Jarvis, J. U. M. 1969. The breeding season and litter size of African mole-rats. J. Reprod. Fert., Suppl. 6:237–48.

———. 1973a. Activity patterns in the mole-rats *Tachyoryctes splendens* and *Heliophobius argenteocinereus*. Zool. Afr. 8:101–19.

———. 1973b. The structure of a population of mole-rats, *Tachyoryctes splendens* (Rodentia: Rhizomyidae). J. Zool. 171:1–14.

———. 1978. Energetics of survival in *Heterocephalus glaber* (Ruppell), the naked mole-rat (Rodentia: Bathyergidae). Bull. Carnegie Mus. Nat. Hist., no. 6, pp. 81–87.

———. 1985. Ecological studies on *Heterocephalus glaber*, the naked mole-rat, in Kenya. Natl. Geogr. Soc. Res. Rept. 20:429–37.

Jarvis, J. U. M., and J. B. Sale. 1971. Burrowing and burrow patterns of East African mole-rats *Tachyoryctes*, *Heliophobius* and *Heterocephalus*. J. Zool. 163:451–79.

Jay, P. C., ed. 1968. Primates: studies in adaptation and variability. Holt, Rinehart, and Winston, New York, xii + 529 pp.

Jeanne, R. L. 1970. Note on a bat *(Phylloderma stenops)* preying upon the brood of a social wasp. J. Mamm. 51:624–25.

Jefferson, T. A. 1988. *Phocoenoides dalli*. Mammalian Species, no. 319, 7 pp.

———. 1989. Calving seasonality of Dall's porpoise in the eastern North Pacific. Mar. Mamm. Sci. 5:196–200.

Jeffery, R. C. V., R. H. V. Bell, and W. F. H. Ansell. 1989. Zambia. *In* East (1989), pp. 11–19.

Jenkins, P. D. 1976. Variation in Eurasian shrews of the genus *Crocidura* (Insectivora: Soricidae). Bull. British Mus. (Nat. Hist.) Zool. 30:269–309.

———. 1982. A discussion of Malayan and Indonesian shrews of the genus *Crocidura* (Insectivora: Soricidae). Zool. Meded. 56:267–79.

———. 1984. Description of a new species of *Sylvisorex* (Insectivora: Soricidae) from Tanzania. Bull. British Mus. (Nat. Hist.) Zool. 47:65–76.

———. 1987. Catalogue of primates in the British Museum (Natural History) and elsewhere in the British Isles, Part IV: Suborder Strepsirhini, including the subfossil Madagascan lemurs and family Tarsiidae. British Mus. (Nat. Hist.), London, x + 189 pp.

———. 1988. A new species of *Microgale* (Insectivora: Tenrecidae) from northeastern Madagascar. Amer. Mus. Novit., no. 2910, 7 pp.

Jenkins, P. D., and J. E. Hill. 1981. The status of *Hipposideros galeritus* Cantor, 1846 and *Hipposideros cervinus* (Gould, 1854) (Chiroptera: Hipposideridae). Bull. British Mus. (Nat. Hist.) Zool. 41:279–94.

———. 1982. Mammals from Siberut, Mentawei Islands. Mammalia 46:219–24.

Jenkins, S. H., and P. E. Busher. 1979. *Castor canadensis*. Mammalian Species, no. 120, 8 pp.

Jenness, S. E. 1985. Arctic wolf attacks scientist—a unique Canadian incident. Arctic 38:129–32.

Jennings, T. J. 1975. Notes on the burrow systems of woodmice *(Apodemus sylvaticus)*. J. Zool. 177:500–504.

Jensen, I. M. 1983. Metabolic rates of the hairy-tailed mole, *Parascalops breweri* (Bachman, 1842). J. Mamm. 64:453–62.

Jensen, W. F., T. K. Fuller, and W. L. Robinson. 1986. Wolf, *Canis lupus*, distribution on the Ontario-Michigan border near Sault Ste. Marie. Can. Field-Nat. 100:363–66.

Jerison, H. J. 1986. The perceptual worlds of dolphins. *In* Schusterman, Thomas, and Wood (1986), pp. 141–66.

Jewell, P. A. 1972. Social organisation and movements of topi *(Damaliscus korrigum)* during the rut, at Ishasha, Queen Elizabeth Park, Uganda. Zool. Afr. 7:233–55.

Jewell, P. A., and S. Holt. 1981. Problems in management of locally abundant wild mammals. Academic Press, New York, xiv + 361 pp.

Jiménez M., P., and J. Pefaur. 1982. Aspectos sistemáticos y ecologicos de *Platalina genovensium* (Chiroptera: Mammalia). Actas Congr. Latinoamer. Zool. 8:707–18.

Jing Xianying, Xiao Youfu, and Jing Rongcai. 1981. Acoustic signals and acoustic behaviour of Chinese river dolphin *(Lipotes vexillifer)*. Sci. Sinica 24:407–15.

Johanson, D. C., and T. D. White. 1979. A systematic assessment of early African hominids. Science 203:321–30.

Johns, A. 1985. Current status of the southern

bearded saki *(Chiropotes satanas satanas)*. Primate Conserv. 5:28.

———. 1986. Notes on the ecology and current status of the buffy saki, *Pithecia albicans*. Primate Conserv. 7:26–29.

Johnsingh, A. J. T. 1982. Reproductive and social behaviour of the dhole, *Cuon alpinus* (Canidae). J. Zool. 198:443–63.

Johnson, A. S. 1970. Biology of the raccoon (*Procyon lotor varius* Nelson and Goldman) in Alabama. Auburn Univ. Agric. Exp. Sta. Bull., no. 402, vi + 148 pp.

Johnson, C. N., and K. A. Johnson. 1983. Behaviour of the bilby, *Macrotis lagotis* (Reid) (Marsupialia: Thylacomyidae) in captivity. Austral. Wildl. Res. 10:77–87.

Johnson, D. H. 1962. Two new murine rodents. Proc. Biol. Soc. Washington 75:317–19.

Johnson, D. L. 1980. Problems in the land vertebrate zoogeography of certain islands and the swimming powers of elephants. J. Biogeogr. 7:383–98.

Johnson, G. L., and R. L. Packard. 1974. Electrophoretic analysis of *Peromyscus comanche* Blair, with comments on its systematic status. Occas. Pap. Mus. Texas Tech Univ., no. 24, 16 pp.

Johnson, J. H., and A. A. Wolman. 1984. The humpback whale, *Megaptera novaeangliae*. Mar. Fish. Rev. 46(4):30–37.

Johnson, K. 1981. Social organization in a colony of rock squirrels (*Spermophilus variegatus*, Sciuridae). Southwestern Nat. 26:237–42.

Johnson, K. A. 1980. Spatial and temporal use of habitat by the red-necked pademelon, *Thylogale thetis* (Marsupialia: Macropodidae). Austral. Wildl. Res. 7:157–66.

Johnson, K. G., G. B. Schaller, and Hu Jinchu. 1988. Comparative behaviour of red and giant pandas in the Wolong Reserve, China. J. Mamm. 69:552–64.

Johnson, M. L. 1973. Characters of the heather vole, *Phenacomys*, and the red tree vole, *Arborimus*. J. Mamm. 54:239–44.

Johnson, M. L., and C. Maser. 1982. Generic relationships of *Phenacomys albipes*. Northwest Sci. 56:17–19.

Johnson, M. L., R. H. Taylor, and N. W. Winnick. 1975. The breeding and exhibition of capromyid rodents at Tacoma Zoo. Internatl. Zoo Yearbook 15:53–56.

Johnson, P. M. 1978. Husbandry of the rufous rat-kangaroo *Aeprymnus rufescens* and brush-tailed rock wallaby *Petrogale penicillata* in captivity. Internatl. Zoo Yearbook 18:156–57.

———. 1979. Reproduction in the plain rock-wallaby, *Petrogale penicillata inornata* Gould, in captivity, with age estimation of the pouch young. Austral. Wildl. Res. 6:1–4.

―――. 1980. Observations of the behaviour of the rufous rat-kangaroo, *Aeprymnus rufescens* (Gray), in captivity. Austral. Wildl. Res. 7:347–57.

Johnson, P. M., and R. Strahan. 1982. A further description of the musky rat-kangaroo, *Hypsiprymnodon moschatus* Ramsay, 1876 (Marsupialia, Potoroidae), with notes on its biology. Austral. Zool. 21:27–46.

Johnson, R. E. 1977. An historical analysis of wolverine abundance and distribution in Washington. Murrelet 58:13–16.

Johnson, R. H. 1982. Food-sharing behavior in captive Amazon river dolphins *(Inia geoffrensis)*. Cetology, no. 43, 2 pp.

Johnson-Murray, J. L. 1987. The comparative myology of the gliding membranes of *Acrobates, Petauroides,* and *Petaurus* contrasted with the cutaneous myology of *Hemibelideus* and *Pseudocheirus* (Marsupialia: Phalangeridae) and with selected gliding Rodentia (Sciuridae and Anomaluridae). Austral. Wildl. Res. 35:101–13.

Johnston, P. G., and G. B. Sharman. 1976. Studies on populations of *Potorous* Desmarest (Marsupialia). I. Morphological variation. Austral. J. Zool. 24:573–88.

―――. 1977. Studies on populations of *Potorous* Desmarest (Marsupialia). II. Electrophoretic, chromosomal and breeding studies. Austral. J. Zool. 25:733–47.

Jolly, A. 1966. Lemur behavior: a Madagascar field study. Univ. Chicago Press, xiv + 187 pp.

―――. 1972. Troop continuity and troop spacing in *Propithecus verreauxi* and *Lemur catta* at Berenty (Madagascar). Folia Primatol. 17:335–62.

Jones, C. 1971a. The bats of Rio Muni, West Africa. J. Mamm. 52:121–40.

―――. 1971b. Notes on the anomalurids of Rio Muni and adjacent areas. J. Mamm. 52:568–72.

―――. 1972. Comparative ecology of three pteropid bats in Rio Muni, West Africa. J. Zool. 167:353–70.

―――. 1977. *Plecotus rafinesquii.* Mammalian Species, no. 69, 4 pp.

―――. 1978. *Dendrohyrax dorsalis.* Mammalian Species, no. 113, 4 pp.

―――. 1984. Tubulidentates, proboscideans, and hyracoideans. *In* Anderson and Jones (1984), pp. 523–36.

Jones, C., and S. Anderson. 1978. *Callicebus moloch.* Mammalian Species, no. 112, 5 pp.

Jones, C., M. A. Bogan, and L. M. Mount. 1988. Status of the Texas kangaroo rat *(Dipodomys elator)*. Texas J. Sci. 40:249–58.

Jones, C., and J. Sabater Pi. 1968. Comparative ecology of *Cercocebus albigena* (Gray) and *Cercocebus torquatus* (Kerr) in Rio Muni, West Africa. Folia Primatol. 9:99–113.

―――. 1971. Comparative ecology of *Gorilla gorilla* (Savage and Wyman) and *Pan troglodytes* (Blumenbach) in Rio Muni, West Africa. Bibl. Primatol., no. 13, 96 pp.

Jones, C., and H. W. Setzer. 1970. Comments on *Myosciurus pumilio.* J. Mamm. 51:813–14.

―――. 1971. The designation of a holotype of the West African pygmy squirrel, *Myosciurus pumilio* (LeConte, 1857) (Mammalia: Rodentia). Proc. Biol. Soc. Washington 84:59–64.

Jones, C. A., J. R. Choate, and H. H. Genoways. 1984. Phylogeny and paleobiogeography of short-tailed shrews (genus *Blarina*). Carnegie Mus. Nat. Hist. Spec. Publ. 8:56–148.

Jones, D. M. 1988. The Arabian oryx in captivity with particular reference to the herds in Saudi Arabia. *In* Dixon and Jones (1988), pp. 47–57.

Jones, E. 1977. Ecology of the feral cat, *Felis catus* (L.) (Carnivora: Felidae), on Macquarie Island. Austral. Wildl. Res. 4:249–62.

Jones, F. W. 1923–25. The mammals of South Australia. A. B. Jones, Government Printer, Adelaide, 458 pp.

Jones, G. S. 1983. Ecological and distributional notes on mammals from Vietnam, including the first record of *Nyctalus.* Mammalia 47:339–44.

―――. 1984. *Phoca* sp. (Mammalia: Carnivora; Pinnipedia) on Taiwan: extra-limital records. J. Taiwan Mus. 37:75–76.

Jones, G. S., and R. E. Mumford. 1971. *Chimarrogale* from Taiwan. J. Mamm. 52:228–32.

Jones, J. K., Jr. 1964. Distribution and taxonomy of mammals of Nebraska. Univ. Kansas Mus. Nat. Hist. Publ. 16:1–356.

―――. 1966. Bats from Guatemala. Univ. Kansas Publ. Mus. Nat. Hist. 16:439–72.

―――. 1977. *Rhogeessa gracilis.* Mammalian Species, no. 76, 2 pp.

Jones, J. K., Jr., J. Arroyo-Cabrales, and R. D. Owen. 1988. Revised checklist of bats (Chiroptera) of Mexico and Central America. Occas. Pap. Mus. Texas Tech Univ., no. 120, 34 pp.

Jones, J. K., Jr., and R. J. Baker. 1979. Notes on a collection of bats from Montserrat, Lesser Antilles. Occas. Pap. Mus. Texas Tech Univ., no. 60, 6 pp.

Jones, J. K., Jr., and D. C. Carter. 1976. Annotated checklist, with keys to subfamilies and genera. *In* Baker, Jones, and Carter (1976), pp. 7–38.

―――. 1979. Systematic and distributional notes. *In* Baker, Jones, and Carter (1979), pp. 7–11.

Jones, J. K., Jr., D. C. Carter, H. H. Genoways, R. S. Hoffmann, D. W. Rice, and C. Jones. 1986. Revised checklist of North American mammals north of Mexico, 1986. Occas. Pap. Mus. Texas Tech Univ., no. 107, 22 pp.

Jones, J. K., Jr., J. R. Choate, and A. Cadena. 1972. Mammals from the Mexican State of Sinaloa. II. Chiroptera. Occas. Pap. Mus. Nat. Hist. Univ. Kansas, no. 6, 29 pp.

Jones, J. K., Jr., and M. D. Engstrom. 1986a. The black-eared rice rat, *Oryzomys melanotis*, in Nicaragua. Southwestern Nat. 31:137.

―――. 1986b. Synopsis of the rice rats (genus *Oryzomys*) of Nicaragua. Occas. Pap. Mus. Texas Tech Univ., no. 103, 23 pp.

Jones, J. K., Jr., and H. H. Genoways. 1970. Harvest mice (genus *Reithrodontomys*) of Nicaragua. Occas. Pap. Western Foundation Vert. Zool., no. 2, 16 pp.

―――. 1971. Notes on the biology of the Central American squirrel, *Sciurus richmondi.* Amer. Midl. Nat. 86:242–47.

―――. 1973. *Ardops nichollsi.* Mammalian Species, no. 24, 2 pp.

Jones, J. K., Jr., H. H. Genoways, and R. J. Baker. 1971. Morphological variation in *Stenoderma rufum.* J. Mamm. 52:244–47.

Jones, J. K., Jr., H. H. Genoways, and T. E. Lawlor. 1974. Annotated checklist of mammals of the Yucatan Peninsula, Mexico. II. Rodentia. Occas. Pap. Mus. Texas Tech Univ., no. 22, 24 pp.

Jones, J. K., Jr., H. H. Genoways, and J. D. Smith. 1974. Annotated checklist of mammals of the Yucatan Peninsula, Mexico. III. Marsupialia, Insectivora, Primates, Edentata, Lagomorpha. Occas. Pap. Mus. Texas Tech Univ., no. 23, 12 pp.

Jones, J. K., Jr., and J. A. Homan. 1974. *Hylonycteris underwoodi.* Mammalian Species, no. 32, 2 pp.

Jones, J. K., Jr., J. D. Smith, and H. H. Genoways. 1973. Annotated checklist of mammals of the Yucatan Peninsula, Mexico. I. Chiroptera. Occas. Pap. Mus. Texas Tech Univ., no. 13, 31 pp.

Jones, J. K., Jr., J. D. Smith, and R. W. Turner. 1971. Noteworthy records of bats from Nicaragua, with a checklist of the chiropteran fauna of the country. Occas. Pap. Mus. Nat. Hist. Univ. Kansas, no. 2, 35 pp.

Jones, J. K., Jr., P. Swanepoel, and D. C. Carter. 1977. Annotated checklist of the bats of Mexico and Central America. Occas. Pap. Mus. Texas Tech Univ., no. 47, 35 pp.

Jones, J. K., Jr., and T. L. Yates. 1983. Review of the white-footed mice, genus *Peromyscus,* of Nicaragua. Occas. Pap. Mus. Texas Tech Univ., no. 82, 15 pp.

Jones, M. J., and J. B. Theberge. 1982. Summer home range and habitat utilisation of the red fox (Vulpes vulpes) in a tundra habitat, northwest British Columbia. Can. J. Zool. 60:807–12.

Jones, M. L. 1982. Longevity of captive mammals. Zool. Garten 52:113–28.

Jones, M. L., and V. J. A. Manton. 1983. History in captivity. In Beck and Wemmer (1983), pp. 1–14.

Jones, M. L., S. L. Swartz, and S. Leatherwood, eds. 1984. The gray whale Eschrichtius robustus. Academic Press, Orlando, xxiv + 600 pp.

Jonkel, C. J. 1978. Black, brown (grizzly), and polar bears. In Schmidt and Gilbert (1978), pp. 227–48.

———. 1987. Brown bear. In Novak, Baker, et al. (1987), pp. 456–73.

Jonkel, C. J., and I. M. Cowan. 1971. The black bear in the spruce-fir forest. Wildl. Monogr., no. 27, 57 pp.

Jonkel, C. J., and F. L. Miller. 1970. Recent records of black bears (Ursus americanus) on the barren grounds of Canada. J. Mamm. 51:826–28.

Jonkel, C. J., P. Smith, I. Stirling, and G. B. Kolenosky. 1976. The present status of the polar bear in the James Bay and Belcher Islands area. Can. Wildl. Serv. Occas. Pap., no. 26, 42 pp.

Jordan, C. 1982. Object manipulation and tool-use in captive pygmy chimpanzee (Pan paniscus). J. Human Evol. 11:35–39.

Jorge, W., D. A. Meritt, Jr., and K. Benirschke. 1977. Chromosome studies in Edentata. Cytobios 18:157–72.

Joubert, E. 1972. A note on the challenge rituals of territorial male lechwe. Madoqua, ser. 1, no. 5, pp. 63–67.

Joubert, E., and F. C. Eloff. 1971. Notes on the ecology and behaviour of the black rhinoceros Diceros bicornis Linn. 1758 in South West Africa. Madoqua, ser. 1, no. 3, pp. 5–53.

Joubert, S. C. J. 1972. Territorial behaviour of the tsessebe (Damaliscus lunatus lunatus Burchell) in the Kruger National Park. Zool. Afr. 7:141–56.

———. 1974. The social organization of the roan antelope Hippotragus equinus and its influence on the special distribution of herds in the Kruger National Park. In Geist and Walther (1974), pp. 661–75.

Joubert, S. C. J., and P. J. L. Bronkhorst. 1977. Some aspects of the history and population ecology of the tsessebe Damaliscus lunatus lunatus in the Kruger National Park. Koedoe 20:125–45.

Jouventin, P. 1975a. Les roles des colorations du mandrill (Mandrillus sphinx). Z. Tierpsychol. 39:455–62.

———. 1975b. Observations sur la socio-écologie du mandrill. Terre Vie 29:493–532.

Juarez G., J., T. Jimenez A., and D. Navarro L. 1988. Additional records of Bauerus dubiaquercus (Chiroptera: Vespertilionidae) in Mexico. Southwestern Nat. 33:385.

Junge, J. A., and R. S. Hoffmann. 1981. An annotated key to the long-tailed shrews (genus Sorex) of the United States and Canada, with notes on Middle American Sorex. Occas. Pap. Mus. Nat. Hist. Univ. Kansas, no. 94, 48 pp.

Junge, J. A., R. S. Hoffmann, and R. W. De Bry. 1983. Relationships within the Holarctic Sorex arcticus–Sorex tundrensis species complex. Acta Theriol. 21:339–50.

Jungers, W. L., and R. L. Susman. 1984. Body size and skeletal allometry in African apes. In Susman (1984a), pp. 131–77.

Jungius, H. 1970. Studies on the breeding biology of the reedbuck (Redunca arundinum Boddaert, 1785) in the Kruger National Park. Z. Saugetierk. 35:129–46.

———. 1971a. Studies on the food and feeding behavior of the reedbuck Redunca arundinum Boddaert, 1785 in the Kruger National Park. Koedoe 14:65–97.

———. 1971b. The biology and behaviour of the reedbuck (Redunca arundinum Boddaert 1785) in the Kruger National Park. Mammalia Depicta, Paul Parey, Hamburg, 106 pp.

———. 1978. Plan to restore Arabian oryx in Oman. Oryx 14:328–36.

K

Kahlke, H. D. 1973. A review of the Pleistocene history of the orang-utan (Pongo Lacépède 1799). Asian Perspectives 15:514.

Kaiser, H. E. 1974. Morphology of the Sirenia. S. Karger, Basel, 76 pp.

Kale, H. W., II. 1972. A high concentration of Cryptotis parva in a forest in Florida. J. Mamm. 53:216–18.

Kallman, H., ed. 1987. Restoring America's wildlife. U.S. Fish and Wildl. Serv., xiii + 394 pp.

Kamiya, T., and F. Yamasaki. 1974. Organ weights of Pontoporia blainvillei and Platanista gangetica. Sci. Rept. Whales Res. Inst. 26:265–70.

Kane, R. 1989. The wild buffalo. Internatl. Union Conserv. Nat. Species Survival Comm. Asian Wild Cattle Specialist Group Newsl., no. 2.

Kano, T., and M. Mulavwa. 1984. Feeding ecology of the pygmy chimpanzees (Pan paniscus) of Wamba. In Susman (1984a), pp. 233–74.

Kaplan, H., and S. O. Hyland. 1972. Behavioural development in the Mongolian gerbil (Meriones unguiculatus). Anim. Behav. 20:147–54.

Kappeler, M. 1984. The gibbon in Java. In Preuschoft et al. (1984), pp. 19–31.

Karanth, K. U. 1987. Tigers in India: a critical review of field censuses. In Tilson and Seal (1987), pp. 118–32.

Karstad, E. L., and R. J. Hudson. 1986. Social organization and communication of riverine hippopotami in southwestern Kenya. Mammalia 50:153–64.

Kasuya, T. 1972a. Growth and reproduction of Stenella coeruleoalba based on the age determination by means of dentinal growth layers. Sci. Rept. Whales Res. Inst. 24:57–79.

———. 1972b. Some information on the growth of the Ganges dolphin with a comment on the Indus dolphin. Sci. Rept. Whales Res. Inst. 24:87–108.

———. 1973. Systematic consideration of the Recent toothed whales based on the morphology of tympano-periotic bone. Sci. Rept. Whales Res. Inst. 25:1–103.

———. 1975. Past occurrence of Globicephala melaena in the western North Pacific. Sci. Rept. Whales Res. Inst. 27:95–110.

———. 1976. Reconsideration of life history parameters of the spotted and striped dolphins based on cemental layers. Sci. Rept. Whales Res. Inst. 28:73–106.

———. 1977. Age determination and growth of the Baird's beaked whale with a comment on the fetal growth rate. Sci. Rept. Whales Res. Inst. 29:1–20.

———. 1978. The life history of Dall's porpoise with special reference to the stock off the Pacific coast of Japan. Sci. Rept. Whales Res. Inst. 30:1–63.

———. 1985. Effect of exploitation on reproductive parameters of the spotted and striped dolphins off the Pacific coast of Japan. Sci. Rept. Whales Res. Inst. 36:107–38.

———. 1986. Distribution and behavior of Baird's beaked whales off the Pacific coast of Japan. Sci. Rept. Whales Res. Inst. 37:61–83.

Kasuya, T., and R. L., Brownell, Jr. 1979. Age determination, reproduction, and growth of Franciscana dolphin Pontoporia blainvillei. Sci. Rept. Whales Res. Inst. 31:45–67.

Kasuya, T., and A. K. M. Aminul Haque. 1972. Some information on distribution and seasonal movement of the Ganges dolphin. Sci. Rept. Whales Res. Inst. 24:109–15.

Kasuya, T., and K. Kureha. 1979. The population of finless porpoise in the Inland Sea of Japan. Sci. Rept. Whales Res. Inst. 31:1–44.

Kasuya, T., and H. Marsh. 1984. Life history

and reproductive biology of the short-finned pilot whale, *Globicephala macrorhynchus*, off the Pacific coast of Japan. *In* Perrin, Brownell, and DeMaster (1984), pp. 259–310.

Kasuya, T., N. Miyazaki, and W. H. Dawbin. 1974. Growth and reproduction of *Stenella attenuata* in the Pacific Coast of Japan. Sci. Rept. Whales Res. Inst. 26:157–226.

Kasuya, T., and M. Nishiwaki. 1975. Recent status of the population of Indus dolphin. Sci. Rept. Whales Res. Inst. 27:81–94.

Kaufmann, J. H. 1962. Ecology and social behavior of the coati, *Nasua narica*, on Barro Colorado Island, Panama. Univ. California Publ. Zool. 60:95–222.

———. 1974. Social ethology of the whiptail wallaby, *Macropus parryi*, in northeastern New South Wales. Anim. Behav. 22:281–369.

———. 1975. Field observations of the social behaviour of the eastern grey kangaroo, *Macropus giganteus*. Anim. Behav. 23:214–21.

———. 1987. Ringtail and coati. *In* Novak, Baker, et al. (1987), pp. 500–509.

Kaufmann, J. H., D. V. Lanning, and S. E. Poole. 1976. Current status and distribution of the coati in the United States. J. Mamm. 57:621–37.

Kavanagh, M., A. A. Eudey, and D. Mack. 1987. The effects of live trapping and trade on primate populations. *In* Marsh and Mittermeier (1987), pp. 147–77.

Kavanagh, M., and E. Laursen. 1984. Breeding seasonality among long-tailed macaques, *Macaca fascicularis*, in peninsular Malaysia. Internatl. J. Primatol. 5:17–29.

Kawabe, M., and T. Mano. 1972. Ecology and behavior of the wild proboscis monkey, *Nasalis larvatus* (Wurmb), in Sabah, Malaysia. Primates 13:213–28.

Kawai, M., ed. 1979. Ecological and sociological studies of gelada baboons. Contrib. Primatol. 16:i–xxiv + 1–344.

Kawai, M., R. Dunbar, H. Ohsawa, and U. Mori. 1983. Social organization of gelada baboons: social units and definitions. Primates 24:13–24.

Kawamichi, T. 1976. Hay territory and dominance rank of pikas (*Ochotona princeps*). J. Mamm. 57:133–48.

———, ed. 1985. Contemporary mammalogy in China and Japan. Mamm. Soc. Japan, Osaka, 194 pp.

Kawamichi, T., and M. Kawamichi. 1979. Spatial organization and territory of tree shrews (*Tupaia glis*). Anim. Behav. 27:381–93.

———. 1982. Social system and independence of offspring in tree shrews. Primates 23:189–205.

Kawamura, A. 1973. Food and feeding of sei whale caught in the waters south of 40°N in the North Pacific. Sci. Rept. Whales Res. Inst. 25:219–36.

———. 1974. Food and feeding ecology in the southern sei whale. Sci. Rept. Whales Res. Inst. 26:25–144.

Keen, R., and H. B. Hitchcock. 1980. Survival and longevity of the little brown bat *(Myotis lucifugus)* in southeastern Ontario. J. Mamm. 61:1–7.

Keim, M. H., and I. J. Stout. 1987. Longevity record for the Florida mouse, *Peromyscus floridanus*. Florida Sci. 50:41.

Keith, K., and J. H. Calaby. 1968. The New Holland mouse, *Pseudomys novaehollandiae* (Waterhouse), in the Port Stephens district, New South Wales. CSIRO Wildl. Res. 13:45–58.

Keith, L. B., J. R. Cary, O. J. Rongstad, and M. C. Brittingham. 1984. Demography and ecology of a declining snowshoe hare population. Wildl. Monogr., no. 90, 43 pp.

Keith, L. B., and L. A. Windberg. 1978. A demographic analysis of the snowshoe hare cycle. Wildl. Monogr., no. 58, 70 pp.

Kellas, L. M. 1955. Observations on the reproductive activities, measurements, and growth rate of the dikdik (*Rhynchotragus kirkii thomasi* Neumann). Proc. Zool. Soc. London 124:751–84.

Kelley, J., and D. Pilbeam. 1986. The dryopithecines: taxonomy, comparative anatomy, and phylogeny of Miocene large hominoids. *In* Swindler and Erwin (1986), pp. 361–411.

Kellnhauser, J. T. 1983. The acceptance of *Lontra* Gray for the New World river otters. Can. J. Zool. 61:278–79.

Kellogg, R. 1940. Whales, giants of the sea. Natl. Geogr. 77:35–90.

Kelsall, J. P. 1972. The northern limits of moose *(Alces alces)* in western Canada. J. Mamm. 53:129–38.

———. 1987. The distribution and status of moose *(Alces alces)* in North America. Swedish Wildl. Res. Suppl. 1:1–10.

Kelt, D. A., and D. R. Martínez. 1989. Notes on the distribution and ecology of two marsupials endemic to the Valdivian forests of southern South America. J. Mamm. 70:220–24.

Kemp, G. A. 1976. The dynamics and regulation of black bear *Ursus americanus* populations in northern Alberta. *In* Pelton, Lentfer, and Folk (1976), pp. 191–97.

Kemp, G. A., and L. B. Keith. 1970. Dynamics and regulation of red squirrel (*Tamiasciurus hudsonicus*) populations. Ecology 51:763–79.

Kemp, T. S. 1982. Mammal-like reptiles and the origin of mammals. Academic Press, London, xiv + 363 pp.

———. 1983. The relationships of mammals. Zool. J. Linnean Soc. 77:353–84.

Kemper, C. M. 1976a. Growth and development of the Australian murid *Pseudomys novaehollandiae*. Austral. J. Zool. 24:27–37.

———. 1976b. Reproduction of *Pseudomys novaehollandiae* (Muridae) in the laboratory. Austral. J. Zool. 24:159–67.

Kenagy, G. J. 1976. Field observations of male fighting, drumming, and copulation in the Great Basin kangaroo rat, *Dipodomys microps*. J. Mamm. 57:781–85.

Kennelly, J. J. 1978. Coyote reproduction. *In* Bekoff (1978), pp. 73–93.

Kenyon, K. W. 1962. History of the Steller sea lion at the Pribilof Islands, Alaska. J. Mamm. 43:68–75.

———. 1969. The sea otter in the eastern Pacific Ocean. N. Amer. Fauna, no. 68, ix + 352 pp.

———. 1981. Monk seals—*Monachus*. *In* Ridgway and Harrison (1981b), pp. 195–220.

Keogh, H. J. 1973. Behaviour and breeding in captivity of the Namaqua gerbil *Desmodillus auricularis* (Cricetidae: Gerbillinae). Zool. Afr. 8:231–40.

Kerle, J. A. 1984a. The behaviour of *Burramys parvus* Broom (Marsupialia) in captivity. Mammalia 48:317–25.

———. 1984b. Growth and development of *Burramys parvus* in captivity. *In* Smith and Hume (1984), pp. 409–12.

Kerle, J. C., and A. Borsboom. 1984. Home range, den tree use and activity patterns in the greater glider, *Petauroides volans*. *In* Smith and Hume (1984), pp. 229–36.

Kerley, G. I. H. 1983a. Record for the Cape fur seal *Arctocephalus pusillus pusillus* from subantarctic Marion Island. S. Afr. J. Zool. 18:139–40.

———. 1983b. Comparison haul-out patterns of fur seals *Arctocephalus tropicalis* and *A. gazella* on subantarctic Marion Island. S. Afr. J. Wildl. Res. 13:71–77.

———. 1983c. Relative population sizes and trends, and hybridization of fur seals *Arctocephalus tropicalis* and *A. gazella* at the Prince Edward Islands, Southern Ocean. S. Afr. J. Zool. 18:388–92.

———. 1987. *Arctocephalus tropicalis* on the Prince Edward Islands. *In* Croxall and Gentry (1987), pp. 61–64.

Kerley, G. I. H., and T. J. Robinson. 1987. Skull morphometrics of male antarctic and subantarctic fur seals, *Arctocephalus gazella* and *A. tropicalis*, and their interspecific hybrids. *In* Croxall and Gentry (1987), pp. 121–31.

Kern, J. A. 1964. Observations on the habits of the proboscis monkey, *Nasalis larvatus* (Wurmb), made in the Brunei Bay area, Borneo. Zoologica 49:183–92.

Kerr, M. A. 1965. The age at sexual maturity in male impala. Arnoldia 1(24):1–6.

Kerr, M. A., and V. J Wilson. 1967. Notes on reproduction in Sharpe's grysbok. Arnoldia 3(17):1–4.

Kerridge, D. C., and R. J. Baker. 1978. *Natalus micropus*. Mammalian Species, no. 114, 3 pp.

Khajuria, H. 1981. A new bandicoot rat, *Erythronesokia bunnii* gen. et sp. nov. (Rodentia: Muridae), from Iraq. Bull. Nat. Hist. Res. Centre, Baghdad, 7:157–64.

Khan, M. A. R. 1984. Endangered mammals of Bangladesh. Oryx 18:152–56.

———. 1987. The problem tiger of Bangladesh. *In* Tilson and Seal (1987), pp. 92–96.

Khan, Mohd Khan Bin Momin. 1978. Man's impact on the primates of peninsular Malaysia. *In* Chivers and Lane-Petter (1978), pp. 41–46.

———. 1989. Asian rhinos: an action plan for their conservation. Internatl. Union Conserv. Nat. Species Survival Comm., Asian Rhino Specialist Group, Gland, Switzerland, 23 pp.

Kilgore, D. L., Jr. 1969. An ecological study of the swift fox *(Vulpes velox)* in the Oklahoma Panhandle. Amer. Midl. Nat. 81:512–34.

Kiltie, R. A., and J. Terborgh. 1983. Observations on the behavior of rain forest peccaries in Peru: why do white-lipped peccaries form herds? Z. Tierpsychol. 62:241–55.

King, C. 1975. The home range of the weasel *(Mustela nivalis)* in an English woodland. J. Anim. Ecol. 44:639–69.

King, C. M. 1983. *Mustela erminea*. Mammalian Species, no. 195, 8 pp.

King, J. A., ed. 1968. Biology of *Peromyscus* (Rodentia). Amer. Soc. Mamm. Spec. Publ., no. 2, xiii + 593 pp.

King, J. E. 1978. On the specific name of the southern sea lion (Pinnipedia, Otariidae). J. Mamm. 59:861–63.

———. 1983. Seals of the world. British Mus. (Nat. Hist.), London, 240 pp.

Kingdon, J. 1971. East African mammals. An atlas of evolution in Africa. I. Academic Press, London, ix + 446 pp.

———. 1974a. East African mammals. An atlas of evolution in Africa. II(A). Insectivores and bats. Academic Press, London, xi + 341 + l pp.

———. 1974b. East African mammals. An atlas of evolution in Africa. II(B). Hares and rodents. Academic Press, London, ix + 362 + lvii pp.

———. 1977. East African mammals. An atlas of evolution in Africa. III(A). Carnivores. Academic Press, London, viii + 475 pp.

———. 1979. East African mammals. An atlas of evolution in Africa. III(B). Large mammals. Academic Press, London, v + 436 pp.

———. 1982. East African mammals. An atlas of evolution in Africa. III(C and D). Bovids. Academic Press, London, x + 746 pp.

———. 1984. The zebra's stripes: an aid to group cohesion? *In* Macdonald (1984), pp. 486–87.

Kinsey, K. P. 1976. Social behaviour in confined populations of the Allegheny woodrat, *Neotoma floridana magister*. Anim. Behav. 24:181–87.

Kinzey, W. G. 1977. Diet and feeding behaviour of *Callicebus torquatus*. *In* Clutton-Brock (1977), pp. 127–51.

Kinzey, W. G., A. L. Rosenberger, P. S. Heisler, D. L. Prowse, and J. S. Trilling. 1977. A preliminary field investigation of the yellow-handed titi monkey, *Callicebus torquatus torquatus*, in northern Peru. Primates 18:159–81.

Kipp, H. 1965. Beitrag zur Kenntnis der Gattung *Conepatus* Molina, 1782. Z. Saugetierk. 30:193–232.

Kirkby, R. J. 1977. Learning and problem-solving in marsupials. *In* Stonehouse and Gilmore (1977), pp. 193–208.

Kirkevold, B. C., and J. S. Lockard, eds. 1986. Behavioral biology of killer whales. Zoo Biol. Monogr., vol. 1, Alan R. Liss, New York, 474 pp.

Kirkland, G. L., Jr., and J. M. Levengood. 1987. First record of the Maryland shrew *(Sorex fontinalis)* from West Virginia. Proc. Pennsylvania Acad. Sci. 61:35–37.

Kirkland, G. L., Jr., D. F. Schmidt, and C. J. Kirkland. 1979. First record of the long-tailed shrew *(Sorex dispar)* in New Brunswick. Can. Field-Nat. 93:195–98.

Kirkland, G. L., Jr., and H. M. Van Deusen. 1979. The shrews of the *Sorex dispar* group: *Sorex dispar* Batchelder and *Sorex gaspensis* Anthony and Goodwin. Amer. Mus. Novit., no. 2675, 21 pp.

Kirkpatrick, R. L., and G. L. Valentine. 1970. Reproduction in captive pine voles, *Microtus pinetorum*. J. Mamm. 51:779–85.

Kirkpatrick, T. H. 1965. Studies of Macropodidae in Queensland. 2. Age estimation in the grey kangaroo, the red kangaroo, the eastern wallaroo and the red-necked wallaby, with notes on dental abnormalities. Queensland J. Agric. Anim. Sci. 22:301–17.

———. 1967. The grey kangaroo in Queensland. Queensland Agric. J. 93:550–52.

———. 1968. Studies on the wallaroo. Queensland Agric. J. 94:362–65.

———. 1970a. The agile wallaby in Queensland. Queensland Agric. J. 96:169–70.

———. 1970b. The swamp wallaby in Queensland. Queensland Agric. J. 96:335–36.

Kirsch, J. A. W. 1968. Burrowing by the quenda. *Isoodon obesulus*. W. Austral. Nat. 10:178–80.

———. 1977a. The six-percent solution: second thoughts on the adaptedness of the Marsupialia. Amer. Sci. 65:276–88.

———. 1977b. The classification of marsupials. *In* Hunsaker (1977a), pp. 1–50.

———. 1977c. The comparative serology of Marsupialia, and a classification of marsupials. Austral. J. Zool., Suppl. Ser., no. 52, 152 pp.

Kirsch, J. A. W., and J. H. Calaby. 1977. The species of living marsupials: an annotated list. *In* Stonehouse and Gilmore (1977), pp. 9–26.

Kirsch, J. A. W., and W. E. Poole. 1972. Taxonomy and distribution of the grey kangaroos, *Macropus giganteus* Shaw and *Macropus fuliginosus* (Desmarest), and their subspecies (Marsupialia: Macropodidae). Austral. J. Zool. 20:315–39.

Kirsch, J. A. W., and P. F. Waller. 1979. Notes on the trapping and behavior of the Caenolestidae (Marsupialia). J. Mamm. 60:390–95.

Kistchinski, A. A. 1972. Life history of the brown bear *(Ursus arctos* L.) in north-east Siberia. *In* Herrero (1972), pp. 67–73.

Kitchen, D. W. 1974. Social behavior and ecology of the pronghorn. Wildl. Monogr., no. 38, 96 pp.

Kitchener, D. J. 1972. The importance of shelter to the quokka, *Setonix brachyurus* (Marsupialia), on Rottnest Island. Austral. J. Zool. 20:281–99.

———. 1973. Reproduction in the common sheath-tailed bat, *Taphozous georgianus* (Thomas) (Microchiroptera: Emballonuridae), in Western Australia. Austral. J. Zool. 21:375–89.

———. 1975. Reproduction in female Gould's wattled bat, *Chalinolobus gouldii* (Gray) (Vespertilionidae), in Western Australia. Austral. J. Zool. 23:29–42.

———. 1976. Further observations on reproduction in the common sheath-tailed bat, *Taphozous georgianus* Thomas, 1915 in Western Australia, with notes on the gular pouch. Rec. W. Austral. Mus. 4:335–47.

———. 1980a. *Taphozous hilli* sp. nov. (Chiroptera: Emballonuridae), a new sheath-tailed bat from Western Australia and Northern Territory. Rec. W. Austral. Mus. 8:161–69.

———. 1980*b*. A new species of *Pseudomys* (Rodentia: Muridae) from Western Australia. Rec. W. Austral. Mus. 8:405–14.

———. 1985. Description of a new species of *Pseudomys* (Rodentia: Muridae) from Northern Territory. Rec. W. Austral. Mus. 12:207–21.

Kitchener, D. J., M. Adams, and P. Baverstock. 1984. Redescription of *Pseudomys bolami* Troughton, 1932 (Rodentia: Muridae). Austral. Mamm. 7:149–59.

Kitchener, D. J., and P. Coster. 1981. Reproduction in female *Chalinolobus morio* (Gray) (Vespertilionidae) in south-western Australia. Austral. J. Zool. 29:305–20.

Kitchener, D. J., and N. Caputi. 1985. Systematic revision of Australian *Scoteanax* and *Scotorepens* (Chiroptera: Vespertilionidae), with remarks on relationships to other Nycticeiini. Rec. W. Austral. Mus. 12:85–146.

Kitchener, D. J., N. Caputi, and B. Jones. 1986. Revision of Australo-Papuan *Pipistrellus* and of *Falsistrellus* (Microchiroptera: Vespertilionidae). Rec. W. Austral. Mus. 12:435–95.

Kitchener, D. J., and S. Foley. 1985. Notes on a collection of bats (Mammalia: Chiroptera) from Bali I., Indonesia. Rec. W. Austral. Mus. 12:223–32.

Kitchener, D. J., and W. F. Humphreys. 1985. Description of a new species of *Pseudomys* (Rodentia: Muridae) from the Kimberley region, Western Australia. Rec. W. Austral. Mus. 12:419–34.

Kitchener, D. J., and W. F. Humphreys. 1986. Description of a new species of *Pseudomys* (Rodentia: Muridae) from the Kimberley region, Western Australia. Rec. W. Austral. Mus. 12:419–34.

———. 1987. Description of a new subspecies of *Pseudomys* (Rodentia: Muridae) from Northern Territory. Rec. W. Austral. Mus. 13:285–95.

Kitchener, D. J., and G. Sanson. 1978. *Petrogale burbidgei* (Marsupialia, Macropodidae), a new rock wallaby from Kimberley, Western Australia. Rec. W. Austral. Mus. 6:269–85.

Kitchener, D. J., J. Stoddart, and J. Henry. 1983. A taxonomic appraisal of the genus *Ningaui* Archer (Marsupialia: Dasyuridae), including description of a new species. Austral. J. Zool. 31:361–79.

———. 1984. A taxonomic revision of the *Sminthopsis murina* complex (Marsupialia: Dasyuridae) in Australia, including descriptions of four new species. Rec. W. Austral. Mus. 11:201–48.

Kivanc, E. 1986. *Microtus (Pitymys) majori* Thomas, 1906 in der europaischen Turkei. Bonner Zool. Beitr. 37:39–42.

———. 1988. Geographic variations of Turkish *Spalax* species (Spalacidae, Rodentia,

Mammalia). Ankara Univ., Ankara, Turkey, iv + 88 pp.

Kleiman, D. G. 1970. Reproduction in the female green acouchi, *Myoprocta pratti* Pocock. J. Reprod. Fert. 23:55–65.

———. 1971. The courtship and copulatory behaviour of the green acouchi *(Myoprocta pratti)*. Z. Tierpsychol. 29:259–78.

———. 1972. Social behavior of the maned wolf *(Chrysocyon brachyurus)* and bush dog *(Speothos venaticus)*: a study in contrast. J. Mamm. 53:791–806.

———, ed. 1977*a*. The biology and conservation of the Callitrichidae. Smithson. Inst. Press, Washington, D.C., 354 pp.

———. 1977*b*. Characteristics of reproduction and sociosexual interactions in pairs of lion tamarins *(Leontopithecus rosalia)* during the reproductive cycle. *In* Kleiman (1977*a*), pp. 181–92.

———. 1981. *Leontopithecus rosalia*. Mammalian Species, no. 148, 7 pp.

———. 1983. Ethology and reproduction of captive giant pandas *(Ailuropoda melanoleuca)*. Z. Tierpsychol. 62:1–46.

Kleiman, D. G., B. B. Beck, J. M. Dietz, L. A. Dietz, J. D. Ballou, and A. F. Coimbra-Filho. 1986. Conservation program for the golden lion tamarin: captive research management, ecological studies, educational strategies, and reintroduction. *In* Benirschke (1986), pp. 959–79.

Kleiman, D. G., J. F. Eisenberg, and E. Maliniak. 1979. Reproductive parameters and productivity of caviomorph rodents. *In* Eisenberg (1979), pp. 173–83.

Kleiman, D. G., and P. A. Racey. 1969. Observations on noctule bats *(Nyctalus noctula)* breeding in captivity. Lynx 10:65–77.

Klein, D. R., and V. Kuzyakin. 1982. Distribution and status of wild reindeer in the Soviet Union. J. Wildl. Mgmt. 46:728–33.

Klein, D. R., R. G. White, and S. Keller. 1984. Proceedings of the First International Muskox Symposium. Biol. Pap. Univ. Alaska Spec. Rept., no. 4, x + 218 pp.

Klein, L. L., and D. J. Klein. 1976. Neotropical primates: aspects of habitat usage, population density, and regional distribution in La Macarena, Colombia. *In* Thorington and Heltne (1976), pp. 70–78.

———. 1977. Feeding behaviour of the Colombian spider monkey. *In* Clutton-Brock (1977), pp. 153–81.

Klein, R. G. 1974. On the taxonomic status, distribution and ecology of the blue antelope, *Hippotragus leucophaeus* (Pallas, 1766). Ann. S. Afr. Mus. 65:99–143.

Kleinenberg, S. E., A. V. Yablokov, B. M. Bel'Kovich, and M. N. Tarasevich. 1969. Beluga *(Delphinapterus leucas)*: investigation

of the species. Israel Progr. Sci. Transl., Jerusalem, vi + 376 pp.

Klimchenko, I. Z., et al. 1975. Change of the population structure of Vinogradov's gerbil and red-tailed Libyan jird in Azerbaidzhan after their extermination by the bait method. Soviet J. Ecol. 6:436–39.

Klimov, V. V. 1988. Spatial-ethological organization of the herd of Przewalski horses *(Equus przewalski)* in Askania-Nova. Appl. Anim. Behav. Sci. 21:99–115.

Klingel, H. 1969. The social organisation and population ecology of the plains zebra *(Equus quagga)*. Zool. Afr. 4:249–63.

———. 1974. A comparison of the social behaviour of the Equidae. *In* Geist and Walther (1974), pp. 124–32.

———. 1977. Observations on social organization and behaviour of African and Asiatic wild asses *(Equus africanus* and *E. hemionus)*. Z. Tierpsychol. 44:323–31.

Klingener, D. 1968. Anatomy. *In* King (1968), pp. 127–47.

———. 1984. Gliroid and dipodoid rodents. *In* Anderson and Jones (1984), pp. 381–88.

Klingener, D., and G. K. Creighton. 1984. On small bats of the genus *Pteropus* from the Philippines. Proc. Biol. Soc. Washington 97:395–403.

Klingener, D., H. H. Genoways, and R. J. Baker. 1978. Bats from southern Haiti. Ann. Carnegie Mus. 47:81–97.

Klippel, W. E., and P. W. Parmalee. 1984. Armadillos in North American late Pleistocene contexts. Carnegie Mus. Nat. Hist. Spec. Publ. 8:149–60.

Klopfer, P. H., and K. J. Boskoff. 1979. Maternal behavior in prosimians. *In* Doyle and Martin (1979), pp. 123–56.

Knap, J. J. 1975. Martens on the move. Internatl. Wildl. 5(5):32–35.

Knight, R. R., B. M. Blanchard, and K. C. Kendall. 1981. Yellowstone grizzly bear investigations: report of the Interagency Study Team. U.S. Natl. Park Serv., v + 55 pp.

Knopf, F. L., and D. F. Balph. 1977. Annual periodicity of Uinta ground squirrels. Southwestern Nat. 22:213–24.

Knorre, E. P. 1974. Changes in the behavior of moose with age and during the process of domestication. Naturaliste Canadien 101:371–77.

Knowlton, F. F. 1972. Preliminary interpretations of coyote population mechanics with some management implications. J. Wildl. Mgmt. 36:369–82.

Knox, E. 1978. A note on the identification of *Melomys* species (Rodentia: Muridae) in Australia. J. Zool. 185:276–77.

Knudsen, B. 1978. Time budgets of polar bears (Ursus maritimus) on North Twin Island, James Bay, during summer. Can. J. Zool. 56:1627–28.

Koch-Weser, S. 1984. Fledermause aus Obervolta, W-Afrika (Mammalia: Chiroptera). Senckenberg. Biol. 64:255–311.

Kock, D. 1969a. Dyacopterus spadiceus (Thomas 1890) auf den Philippinen (Mammalia, Chiroptera). Senckenberg. Biol. 50:1–7.

———. 1969b. Eine neue Gattung und Art cynopteriner Flughunde von Mindanao, Philippinen (Mammalia, Chiroptera). Senckenberg. Biol. 50:319–27.

———. 1969c. Eine bemerkenswerte neue Gattung und Art Flughunde von Luzon, Philippinen. Senckenberg. Biol. 50:329–38.

———. 1974a. Eine neue Suncus-Art von Flores, Kleine Sunda-Inseln (Mammalia: Insectivora). Senckenberg. Biol. 55:197–203.

———. 1974b. Egyptian tomb bat Taphozous perforatus E. Geoffroy 1818: first record from Uganda. E. Afr. Nat. Hist. Soc. Bull., July, p. 130.

———. 1975. Ein originalexemplar von Nyctinomus ventralis Heuglin 1861 (Mammalia: Chiroptera: Molossidae). Stuttgarter Beitr. Naturkunde, ser. A, no. 272, 9 pp.

———. 1978a. A new fruit bat of the genus Rousettus Gray 1821, from the Comoro Islands, western Indian Ocean (Mammalia: Chiroptera). Proc. 4th Internatl. Bat Res. Conf., Nairobi, pp. 205–16.

———. 1978b. The identity of Gerbillus bottai Lataste, 1882 (Mammalia: Rodentia), from Sennar, Sudan. Bull. Carnegie Mus. Nat. Hist., no. 6, pp. 31–37.

———. 1981a. Philetor brachypterus auf Neu-Britannien und den Philippinen (Mammalia: Chiroptera: Vespertilionidae). Senckenberg. Biol. 61:313–19.

———. 1981b. Zur Chiropteren-Fauna von Burundi. Senckenberg. Biol. 61:329–36.

———. 1987. Micropterus intermedius Hayman 1963 und andere Fledermause vom unteren Zaire. Senckenberg. Biol. 67:219–24.

Kock, D., and H. Felten. 1980. Zwei Fledermause neu fur Pakistan. Senckenberg. Biol. 61:1–9.

Kock, D., and H. Posamentier. 1983. Cannomys badius (Hodgson, 1842) in Bangladesh (Rodentia: Rhizomyidae). Z. Saugetierk. 48:314–16.

Koehler, G. M., M. G. Hornocker, and H. S. Hash. 1980. Wolverine marking behavior. Can. Field-Nat. 94:339–41.

Koenig, L. 1970. Zur Fortpflanzung und Jugendentwicklung des Wüstenfuchses (Fennecus zerda Zimm. 1780). Z. Tierpsychol. 27:205–46.

Koenigswald, W. V., and L. D. Martin. 1984. Revision of the fossil and Recent Lemminae (Rodentia, Mammalia). Carnegie Mus. Nat. Hist. Spec. Publ. 9:122–37.

Koepcke, J. 1984. "Blattzelte" als Schafplatze der Fledermaus Ectophylla macconnelli (Thomas, 1901) (Phyllostomidae) im tropischen Regenwald von Peru. Saugetierk. Mitt. 31:123–26.

Koepcke, J., and R. Kraft. 1984. Cranial and external characters of the larger fruit bats of the genus Artibeus from Amazonian Peru. Spixiana 7:75–84.

Koffler, B. R. 1972. Meriones crassus. Mammalian Species, no. 9, 4 pp.

Köhncke, M., and K. Leonhardt. 1986. Cryptoprocta ferox. Mammalian Species, no. 254, 5 pp.

Kolenosky, G. B. 1987. Polar bear. In Novak, Baker, et al. (1987), pp. 474–87.

Koontz, F. W., and N. J. Roeper. 1983. Elephantulus rufescens. Mammalian Species, no. 204, 5 pp.

Koop, B. F., and R. J. Baker. 1983. Electrophoretic studies of relationships of six species of Artibeus (Chiroptera: Phyllostomidae). Occas. Pap. Mus. Texas Tech Univ., no. 83, 12 pp.

Koop, B. F., R. J. Baker, and H. H. Genoways. 1983. Numerous chromosomal polymorphisms in a natural population of rice rats (Oryzomys, Cricetidae). Cytogenet. Cell Genet. 35:131–35.

Koop, B. F., R. J. Baker, and J. T. Mascarello. 1985. Cladistical analysis of chromosomal evolution within the genus Neotoma. Occas. Pap. Mus. Texas Tech Univ., no. 96, 9 pp.

Koopman, K. F. 1971. Taxonomic notes on Chalinolobus and Glauconycteris (Chiroptera, Vespertilionidae). Amer. Mus. Novit., no. 2451, 10 pp.

———. 1972. Eudiscopus denticulus. Mammalian Species, no. 19, 2 pp.

———. 1973. Systematics of Indo-Australian Pipistrellus. Period. Biol. 75:113–16.

———. 1975. Bats of the Sudan. Bull. Amer. Mus. Nat. Hist. 154:353–444.

———. 1978a. Zoogeography of Peruvian bats with special emphasis on the role of the Andes. Amer. Mus. Novit., no. 2651, 33 pp.

———. 1978b. The genus Nycticeius (Vespertilionidae), with special reference to tropical Australia. Proc. 4th Internatl. Bat Res. Conf., Nairobi, pp. 165–71.

———. 1979. Zoogeography of mammals from islands off the northeastern coast of New Guinea. Amer. Mus. Novit., no. 2690, 17 pp.

———. 1982a. Results of the Archbold Expeditions. No. 109. Bats from eastern Papua and the East Papuan Islands. Amer. Mus. Novit., no. 2747, 34 pp.

———. 1982b. Biogeography of the bats of South America. Pymatuning Lab. Ecol. Spec. Publ. 6:273–302.

———. 1983. A significant range extension for Philetor (Chiroptera, Vespertilionidae) with remarks on geographical variation. J. Mamm. 64:525–26.

———. 1984a. Taxonomic and distributional notes of tropical Australian bats. Amer. Mus. Novit., no. 2778, 48 pp.

———. 1984b. A synopsis of the familes of bats. Part VII. Bat Research News 25:25–27.

———. 1984c. Bats. In Anderson and Jones (1984), pp. 145–86.

———. 1986. Sudan bats revisited: an update of "Bats of the Sudan." Cimbebasia, ser. A, 8:9–13.

———. 1988. Systematics and distribution. In Greenhall and Schmidt (1988), pp. 8–17.

Koopman, K. F., and F. Gudmundsson. 1966. Bats in Iceland. Amer. Mus. Novit., no. 2262, 6 pp.

Koopman, K. F., and J. K. Jones, Jr. 1970. Classification of bats. In Slaughter, B. H., and D. W. Walton, eds., About bats, S. Methodist Univ. Press, Dallas, pp. 22–28.

Koopman, K. F., R. E. Mumford, and J. F. Heisterberg. 1978. Bat records from Upper Volta, West Africa. Amer. Mus. Novit., no. 2643, 6 pp.

Kooyman, G. L. 1981a. Weddell seal. Consummate diver. Cambridge Univ. Press, viii + 135 pp.

———. 1981b. Crabeater seal—Lobodon carcinophagus. In Ridgway and Harrison (1981b), pp. 221–35.

———. 1981c. Leopard seal—Hydrurga leptonyx. In Ridgway and Harrison (1981b), pp. 261–74.

———. 1981d. Weddell seal—Leptonychotes weddelli. In Ridgway and Harrison (1981b), pp. 275–96.

Kooyman, G. L., R. W. Davis, and J. P. Croxall. 1986. Diving behaviour of Antarctic fur seals. In Gentry and Kooyman (1986), pp. 115–25.

Kooyman, G. L., and R. L. Gentry. 1986. Diving behavior of South African fur seals. In Gentry and Kooyman (1986), pp. 145–52.

Kooyman, G. L., and F. Trillmich. 1986. Diving behaviour of Galapagos sea lions. In Gentry and Kooyman (1986), pp. 209–19.

Korobitsyna, K. V., C. F. Nadler, N. N. Vorontsov, and R. S. Hoffmann. 1974. Chromosomes of the Siberian snow sheep, Ovis nivicola, and implications concerning the origin of amphiberingian wild sheep (subgenus *Pachyceros*). Quaternary Res. 4:235–45.

Kortlandt, A., and E. Holzhaus. 1987. New data on the use of stone tools by chimpanzees in Guinea and Liberia. Primates 28:473–96.

Kortlucke, S. M. 1973. Morphological variation in the kinkajou, *Potos flavus* (Mammalia: Procyonidae), in Middle America. Occas. Pap. Mus. Nat. Hist. Univ. Kansas, no. 17, 36 pp.

Kovacs, K. M., and D. M. Lavigne. 1986. *Cystophora cristata*. Mammalian Species, no. 258, 9 pp.

Kovalskaya, J. M., and V. E. Sokolov. 1980. *Microtus evoronensis* sp. n. (Rodentia, Cricetidae) from the Lower Amur territory. Zool. Zhur. 59:1409–16.

Kowalski, K. 1955. Our bats and their protection. Polish Acad. Sci. Nat. Protection Res. Cent. Publ., no. 11, 111 pp.

Koyama, N., K. Norikoshi, and T. Mano. 1975. Population dynamics of Japanese monkeys at Arashiyama. Proc. 5th Internatl. Congr. Primatol., pp. 411–17.

Krasinski, Z. A. 1978. Dynamics and structure of the European bison population in the Bialowieza primeval Forest. Acta Theriol. 23:3–48.

Kratochvil, J., L. Rodriguez, and V. Barus. 1978. Capromyinae (Rodentia) of Cuba. I. Prirod. Prace Ustavu Cesk. Akad. Ved v Brne, Acta Sci. Nat. Brno 12(11):1–60.

Kraus, S. D., J. H. Prescott, A. R. Knowlton, amd G. S. Stone. 1986. Migration and calving of right whales *(Eubalaena glacialis)* in the western North Atlantic. *In* Brownell, Best, and Prescott (1986), pp. 139–44.

Krausman, P. R., and W. W. Shaw. 1986. Nubian ibex in the Eastern Desert, Egypt. Oryx 20:176–77.

Krefting, L. W. 1974. Moose distribution and habitat selection in north central North America. Naturaliste Canadien 101:81–100.

Krishna, A., and C. J. Dominic. 1981. Reproduction in the vespertilionid bat, *Scotophilus heathi* Horsefield. Arch. Biol. (Bruxelles) 92:247–58.

Kristiansson, H. 1981. Distribution of the European hedgehog (*Erinaceus europaeus* L.) in Sweden and Finland. Ann. Zool. Fennici 18:115–19.

Krupnik, I. I. 1984. Gray whales and the aborigines of the Pacific Northwest: the history of aboriginal whaling. *In* Jones, Swartz, and Leatherwood (1984), pp. 103–20.

Krutzsch, P. H., and E. G. Crichton. 1985. Observations on the reproductive cycle of female *Molossus fortis* (Chiroptera: Molossidae) in Puerto Rico. J. Zool. 207:137–50.

Kruuk, H. 1972. The spotted hyena: a study of predation and social behavior. Univ. Chicago Press, xvi + 335 pp.

———. 1976. Feeding and social behaviour of the striped hyaena (*Hyaena vulgaris* Desmarest). E. Afr. Wildl. J. 14:91–111.

———. 1978a. Spatial organization and territorial behaviour of the European badger, *Meles meles*. J. Zool. 184:1–19.

———. 1978b. Foraging and spatial organization of the European badger, *Meles meles* L. Behav. Ecol. Sociobiol. 4:75–89.

Krzanowski, A. 1977. Contribution to the history of bats on Iceland. Acta Theriol. 22:272–73.

Kubiak, H. 1965. The appearance of a raccoon dog, *Nyctereutes procyonides* (Gray, 1834) in Cracow District (Poland). Prezelgl. Zool. 9:417–22.

Kucera, T. E. 1978. Social behavior and breeding system of the desert mule deer. J. Mamm. 59:463–76.

Kucherenko, S. P. 1976. The common otter *(Lutra lutra)* in the Amur-Ussury district. Zool. Zhur. 55:904–11.

Kucherenko, S. P., and V. G. Yudin. 1973. Distribution, population density and economical importance of the raccoon-dog *(Nyctereutes procyonides)* in the Amur-Ussury district. Zool. Zhur. 52:1039–45.

Kudo, H., and M. Mitani. 1985. New record of predatory behavior by the mandrill in Cameroon. Primates 26:161–67.

Kuehn, D. W. 1986. Population and social characteristics of the tamarao *(Bubalus mindorensis)*. Biotrópica 18:263–66.

Kuhn, H. 1971. An adult female *Micropotamogale lamottei*. J. Mamm. 52:477–78.

Kummer, H. 1968. Social organization of hamadryas baboons: a field study. Univ. Chicago Press, vi + 189 pp.

Kummer, H., W. Goetz, and W. Angst. 1970. Cross-species modifications of social behavior in baboons. *In* Napier and Napier (1970), pp. 351–63.

Kuntz, R. E., and B. J. Myers. 1969. A checklist of parasites and commensals reported for the Taiwan macaque. Primates 10:71–80.

Kunz, T. H. 1973a. Population studies of the cave bat *(Myotis velifer)*: reproduction, growth, and development. Occas. Pap. Mus. Nat. Hist. Univ. Kansas, no. 15, 43 pp.

———. 1973b. Resource utilizations: temporal and spatial components of bat activity in central Iowa. J. Mamm. 54:14–32.

———. 1982. *Lasionycteris noctivagans*. Mammalian Species, no. 172, 5 pp.

Kunz, T. H., P. V. August, and C. D. Burnett. 1983. Harem social organization in cave roosting *Artibeus jamaicensis* (Chiroptera: Phyllostomidae). Biotrópica 15:133–38.

Kurland, J. A. 1973. A natural history of Kra

macaques (*Macaca fascicularis* Raffles, 1821) at the Kutai Reserve, Kalimantan Timur, Indonesia. Primates 14:245–62.

———. 1977. Kin selection in the Japanese monkey. Contrib. Primatol. 12:i–x + 1–145 pp.

Kuroda, S., T. Kano, and K. Muhindo. 1985. Further information on the new monkey species *Cercopithecus salongo* Thys van den Audenaerde, 1977. Primates 26:325–33.

Kurt, F. 1974. Remarks on the social structure and ecology of the Ceylon elephant in the Yala National Park. *In* Geist and Walther (1974), pp. 618–34.

Kurten, B. 1968. Pleistocene mammals of Europe. Aldine, Chicago, viii + 317 pp.

———. 1973. Transberingean relationships of *Ursus arctos* Linne (brown and grizzly bears). Commentat. Biol. 65:1–10.

———. 1985. The Pleistocene lion of Beringia. Ann. Zool. Fennici 22:117–21.

Kurz, J. C., and R. L. Marchinton. 1972. Radiotelemetry studies of feral hogs in South Carolina. J. Wildl. Mgmt. 36:1240–48.

Kuschinski, L. 1974. Breeding binturongs *Arctictis binturong* at Glasgow Zoo. Internatl. Zoo Yearbook 14:124–26.

Kuyper, M. A. 1985. The ecology of the golden mole *Amblysomus hottentotus*. Mamm. Rev. 15:3–11.

Kuyt, E. 1972. Food habits and ecology of wolves on barren-ground caribou range in the Northwest Territories. Can. Wildl. Serv. Rept. Ser., no. 21, 36 pp.

L

Lacher, T. E., Jr. 1979. Rates of growth in *Kerodon rupestris* and an assessment of its potential as a domesticated food source. Papéis Avulsos de Zoologia, Mus. Zool. Univ. Sao Paulo, 33:67–76.

———. 1981. The comparative social behavior of *Kerodon rupestris* and *Galea spixii* and the evolution of behavior in the Caviidae. Bull. Carnegie Mus. Nat. Hist., no. 17, 71 pp.

Lacher, T. E., Jr., and C. J. R. Alho. 1989. Microhabitat use among small mammals in the Brazilian Pantanal. J. Mamm. 70:396–401.

Lackey, J. A. 1976. Reproduction, growth, and development in the Yucatan deer mouse, *Peromyscus yucatanicus*. J. Mamm. 57:638–55.

Laerm, J. 1981. Systematic status of the Cumberland Island pocket gopher, *Geomys cumberlandius*. Brimleyana 6:141–51.

———. 1982. Genetic determination of the status of an endangered species of pocket gopher in Georgia. J. Wildl. Mgmt. 46:513–18.

Laing, R. I., and G. L. Holroyd. 1989. The

status of the black-footed ferret in Canada. Blue Jay 47:121–25.

Lair, H. 1985. Length of gestation in the red squirrel, *Tamiasciurus hudsonicus*. J. Mamm. 66:809–10.

Lamotte, M., and F. Petter. 1981. Une taupe doree nouvelle do Cameroun (Mt Oku, 6°15′N, 10°26′E): *Chrysochloris stuhlmanni balsaci* ssp. nov. Mammalia 45:43–48.

Lamotte, M., and M. Tranier. 1983. Un spécimen de *Genetta (Paragenetta) johnstoni* collecté dans la région du Nimba (Côte d'Ivoire). Mammalia 47:430–32.

Lamprecht, J. 1979. Field observations on the behaviour and social system of the bat-eared fox *Otocyon megalotis* Desmarest. Z. Tierpsychol. 49:260–84.

Lander, R. H., and H. Kajimura. 1982. Status of northern fur seals. Mammals in the Seas, FAO Fish. Ser. No. 5, 4:319–45.

Lang, H., and J. P. Chapin. 1917. Notes on the distribution and ecology of Central African Chiroptera. Bull. Amer. Mus. Nat. Hist. 37:479–563.

Langguth, A. 1975*a*. La identidad de *Mus lasiotis* Lund y el status del gênero *Thalpomys* Thomas (Mammalia, Cricetidae). Papeis Avulsos de Zoologia, Mus. Zool. Univ. Sao Paulo, 29:45–54.

———. 1975*b*. Ecology and evolution in the South American canids. *In* Fox (1975), pp. 192–206.

Langham, N. P. E. 1982. The ecology of the common tree shrew, *Tupaia glis*, in peninsular Malaysia. J. Zool. 197:323–44.

Langley, W. 1981. The effect of prey defenses on the attack behavior of the southern grasshopper mouse *(Onychomys torridus)*. Z. Tierpsychol. 56:115–27.

Langman, V. A. 1973. Radio-tracking giraffe for ecological studies. J. S. Afr. Wildl. Mgmt. Assoc. 3:75–78.

———. 1977. Cow-calf relationships in giraffe *(Giraffa camelopardalis giraffa)*. Z. Tierpsychol. 43:264–86.

Lanning, D. V. 1976. Density and movements of the coati in Arizona. J. Mamm. 57:609–11.

Largen, M. J., D. Kock, and D. W. Yalden. 1974. Catalogue of the mammals of Ethiopia. 1. Chiroptera. Italian J. Zool., Suppl., n.s., 5:221–98.

Largen, M. J., and D. W. Yalden. 1987. The decline of elephant and black rhinoceros in Ethiopia. Oryx 21:103–6.

Larkin, P., and M. Roberts. 1979. Reproduction in the ring-tailed mongoose *Galidia elegans* at the National Zoological Park, Washington. Internatl. Zoo Yearbook 19:189–93.

Larsen, T. 1975. Polar bear den surveys in

Svalbard in 1973. Norsk Polarinst. Arbok, 1973, pp. 101–12.

Larson, J. S., and J. R. Gunson. 1983. Status of the beaver in North America. Acta Zool. Fennica 174:91–93.

Larsson, T., L. Hansson, and E. Nyholm. 1973. Winter reproduction in small rodents in Sweden. Oikos 24:475–76.

Lassen, P. 1984. Muskox distribution and population structure in Jameson Land, northeast Greenland, 1981–1983. *In* Klein, White, and Keller (1984), pp. 19–24.

Latinen, K. 1987. Longevity and fertility of the polar bear, *Ursus maritimus* Phipps, in captivity. Zool. Garten 57:197–99.

Laundré, J. W., and B. L. Keller. 1984. Home-range size of coyotes: a critical review. J. Wildl. Mgmt. 48:127–39.

Laurie, A. 1982. Behavioural ecology of the greater one-horned rhinoceros *(Rhinoceros unicornis)*. J. Zool. 196:307–41.

Laurie, A., and J. Seidensticker. 1977. Behavioural ecology of the sloth bear *(Melursus ursinus)*. J. Zool. 182:187–204.

Laurie, E. M. O., and J. E. Hill. 1954. List of land mammals of New Guinea, Celebes and adjacent islands, 1758–1952. British Mus. (Nat. Hist.), London, 175 pp.

Laurie, W. A., E. M. Lang, and C. P. Groves. 1983. *Rhinoceros unicornis*. Mammalian Species, no. 211, 6 pp.

Laursen, L., and M. Bekoff. 1978. *Loxodonta africana*. Mammalian Species, no. 92, 8 pp.

LaVal, R. K. 1973*a*. A revision of the neotropical bats of the genus *Myotis*. Los Angeles Co. Nat. Hist. Mus. Sci. Bull., no. 15, 54 pp.

———. 1973*b*. Systematics of the genus *Rhogeessa* (Chiroptera: Vespertilionidae). Occas. Pap. Mus. Nat. Hist. Univ. Kansas, no. 19, 47 pp.

———. 1973*c*. Observations on the biology of *Tadarida brasiliensis cynocephala* in southeastern Louisiana. Amer. Midl. Nat. 89:112–20.

———. 1976. Voice and habitat of *Dactylomys dactylinus* (Rodentia: Echimyidae) in Ecuador. J. Mamm. 57:402–4.

———. 1977. Notes on some Costa Rican bats. Brenesia 10/11:77–83.

LaVal, R. K., and H. S. Fitch. 1977. Structure, movements and reproduction in three Costa Rican bat communities. Occas. Pap. Mus. Nat. Hist. Univ. Kansas, no. 69, 28 pp.

LaVal, R. K., and M. L. LaVal. 1977. Reproduction and behavior of the African banana bat, *Pipistrellus nanus*. J. Mamm. 58:403–10.

Lavrov, L. S. 1983. Evolutionary development of the genus *Castor* and taxonomy of

the contemporary beavers of Eurasia. Acta Zool. Fennica 174:87–90.

Lawlor, T. E. 1969. A systematic study of the rodent genus *Ototylomys*. J. Mamm. 50:28–42.

———. 1982. *Ototylomys phyllotis*. Mammalian Species, no. 181, 3 pp.

Lawrence, B. 1939. Collections from the Philippine Islands. Mammals. Bull. Mus. Comp. Zool. 86:28–73.

Lawrence, B., and W. H. Bossert. 1967. Multiple character analysis of *Canis lupus, latrans,* and *familiaris*, with a discussion of the relationships of *Canis niger*. Amer. Zool. 7:223–32.

———. 1975. Relationships of North American *Canis* shown by a multiple character analysis of selected populations. *In* Fox (1975), pp. 73–86.

Lawrence, M. A. 1982. Western Chinese arvicolines (Rodentia) collected by the Sage Expedition. Amer. Mus. Novit., no. 2745, 19 pp.

Laws, R. M. 1981. Large mammal feeding strategies and related overabundance problems. *In* Jewell and Holt (1981), pp. 217–32.

———. 1984*a*. The krill-eating crabeater. *In* Macdonald (1984), pp. 280–81.

———. 1984*b*. Hippopotamuses. *In* Macdonald (1984), pp. 506–11.

———. 1985. Animal conservation in the Antarctic. *In* Hearn, J. P., and J. K. Hodges, eds., Advances in animal conservation, Symp. Zool. Soc. London 54:3–23.

Laws, R. M., I. S. C. Parker, and R. C. B. Johnstone. 1975. Elephants and their habitats: the ecology of elephants in north Bunyoro, Uganda. Oxford Univ. Press, London, xii + 376 pp.

Lawson, D. 1989. The food habits of suni antelopes *(Neotragus moschatus)* (Mammalia: Artiodactyla). J. Zool. 217:441–48.

Lay, D. M. 1967. A study of the mammals of Iran. Fieldiana Zool. 54:1–282.

———. 1975. Notes on rodents of the genus *Gerbillus* (Mammalia: Muridae: Gerbillinae) from Morocco. Fieldiana Zool. 65:89–101.

———. 1978. Observations on reproduction in a population of pocket gophers, *Thomomys bottae*, from Nevada. Southwestern Nat. 23:375–80.

———. 1983. Taxonomy of the genus *Gerbillus* (Rodentia: Gerbillinae) with comments on the applications of generic and subgeneric names and an annotated list of species. Z. Saugetierk. 48:329–54.

Lay, D. M., and C. F. Nadler. 1975. A study of *Gerbillus* (Rodentia: Muridae) east of the Euphrates River. Mammalia 39:423–45.

Layne, J. N. 1967. Lagomorphs. *In* Anderson and Jones (1967), pp. 192–205.

———. 1968. Ontogeny. *In* King (1968), pp. 148–253.

———. 1972. Tail autotomy in the Florida mouse, *Peromyscus floridanus*. J. Mamm. 53:62–71.

———, ed. 1978. Rare and endangered biota of Florida. I. Mammals. University Presses of Florida, Gainesville, xx + 52 pp.

Layne, J. N., and D. Glover. 1977. Home range of the armadillo in Florida. J. Mamm. 58:411–13.

———. 1985. Activity patterns of the common long-nosed armadillo *Dasypus novemcinctus* in south-central Florida. *In* Montgomery (1985*a*), pp. 407–17.

Layne, J. N., and A. M. Waggener, Jr. 1984. Above-ground nests of the nine-banded armadillo in Florida. Florida Field Nat. 12(3): 58–61.

Layser, T. R., and I. O. Buss. 1985. Observations on morphological characteristics of elephant tusks. Mammalia 49:407–14.

Lazell, J. D., Jr. 1984. A new marsh rabbit *(Sylvilagus palustris)* from Florida's lower keys. J. Mamm. 65:26–33.

Lazell, J. D., Jr., and K. F. Koopman. 1985. Notes on bats of Florida's lower keys. Florida Sci. 48:37–42.

Lazell, J. D., Jr., T. W. Sutterfield, and W. D. Giezentanner. 1984. The population of rock wallabies (genus *Petrogale*) on Oahu, Hawaii. Biol. Conserv. 30:99–108.

Lazenby-Cohen, K. A., and A. Cockburn. 1988. Lek promiscuity in a semelparous mammal, *Antechinus stuartii* (Marsupialia: Dasyuridae)? Behav. Ecol. Sociobiol. 22:195–202.

Leakey, L. S. B., P. V. Tobias, and J. R. Napier. 1964. A new species of the genus *Homo* from Olduvai Gorge. Nature 202:7–9.

Leatherwood, J. S. 1974. Aerial observations of migrating gray whales, *Eschrichtius robustus*, off southern California, 1969–72. Mar. Fish. Rev. 36(4):45–49.

Leatherwood, S., D. K. Caldwell, and H. E. Winn. 1976. Whales, dolphins, and porpoises of the western North Atlantic: a guide to their identification. U.S. Natl. Mar. Fish. Serv., NOAA Tech. Rept. NMFS Circ. 396, iv + 176 pp.

Leatherwood, S., L. J. Harrington-Coulombe, and C. L. Hubbs. 1978. Relict survival of the sea otter in central California and evidence of its recent dispersal south of Point Conception. Bull. S. California Acad. Sci. 77:109–15.

Leatherwood, S., R. A. Kastelein, and K. W. Miller. 1988. Observations of Commerson's dolphin and other cetaceans in southern Chile,

January–February 1984. *In* Brownell and Donovan (1988), pp. 71–83.

Leatherwood, S., and R. R. Reeves. 1978. Porpoises and dolphins. *In* Haley (1978), pp. 97–111.

———. 1983. The Sierra Club handbook of whales and dolphins. Sierra Club Books, San Francisco, xviii + 302 pp.

Leatherwood, S., R. R. Reeves, W. F. Perrin, and W. E. Evans. 1982. Whales, dolphins, and porpoises of the eastern North Pacific and adjacent arctic waters. U.S. Natl. Mar. Fish. Serv., NOAA Tech. Rept. NMFS Circ. 444, v + 245 pp.

Leatherwood, S., and W. A. Walker. 1979. The northern right whale dolphin *Lissodelphis borealis* Peale in the eastern North Pacific. *In* Winn and Olla (1979), pp. 85–141.

Le Boeuf, B. J., D. Aurioles, R. Condit, C. Fox, R. Gisiner, R. Romero, and F. Sinsel. 1983. Size and distribution of the California sea lion population in Mexico. Proc. California Acad. Sci. 43:77–85.

Le Boeuf, B. J., D. P. Costa, A. C. Huntley, and S. D. Feldkamp. 1988. Continuous deep diving in female northern elephant seals, *Mirounga angustirostris*. Can. J. Zool. 66: 446–58.

Le Boeuf, B. J., K. W. Kenyon, and B. Villa-Ramirez. 1986. The Caribbean monk seal is extinct. Mar. Mamm. Sci. 2:70–72.

Lee, A. K., A. J. Bradley, and R. W. Braithwaite. 1977. Corticosteroid levels and male mortality in *Antechinus stuartii*. *In* Stonehouse and Gilmore (1977), pp. 209–22.

Lee, H. K., and R. J. Baker. 1987. Cladistical analysis of chromosomal evolution in pocket gophers of the *Cratogeomys castanops* complex (Rodentia: Geomyidae). Occas. Pap. Mus. Texas Tech Univ., no. 114, 15 pp.

Lee, M. R., and D. J. Schmidly. 1977. A new species of *Peromyscus* (Rodentia: Muridae) from Coahuila, Mexico. J. Mamm. 58:263–68.

Lee, M. R., D. J. Schmidly, and C. C. Huheey. 1972. Chromosomal variation in certain populations of *Peromyscus boylii* and its systematic implications. J. Mamm. 53: 697–707.

Lee, P. C., J. Thornback, and E. L. Bennett. 1988. Threatened primates of Africa. The IUCN Red Data Book. Internatl. Union Conserv. Nat., Gland, Switzerland, xx + 155 pp.

Legendre, S. 1984. Etude odontologique des representants actuels du groupe *Tadarida* (Chiroptera, Molossidae). Implications phylogeniques, systematiques et zoogeographiques. Rev. Suisse Zool. 91:399–442.

Legge, A. J., and P. A. Rowley-Conwy. 1987. Gazelle killing in Stone Age Syria. Sci. Amer. 257(2):76–83.

Le Gros Clark, W. E. 1955. The fossil evidence

for human evolution. Univ. Chicago Press, 180 pp.

Lehner, P. N. 1978. Coyote communication. *In* Bekoff (1978), pp. 128–62.

Lekagul, B., and J. A. McNeely. 1977. Mammals of Thailand. Sahakarnbhat, Bangkok, li + 758 pp.

Lemke, T. O. 1984. Foraging ecology of the long-nosed bat, *Glossophaga soricina*, with respect to resource availability. Ecology 65: 538–48.

———. 1986. Distribution and status of the sheath-tailed bat *(Emballonura semicaudata)* in the Mariana Islands. J. Mamm. 67:743–46.

Lemke, T. O., A. Cadena, R. H. Pine, and J. Hernandez-Camacho. 1982. Notes on opossums, bats, and rodents new to the fauna of Colombia. Mammalia 46:225–34.

Lemke, T. O., and J. R. Tamsitt. 1979. *Anoura cultrata* (Chiroptera: Phyllostomatidae) from Colombia. Mammalia 43:579–81.

Lenglet, G., and G. Coppois. 1979. Description du crâne et de quelques ossements d'un genre nouveau éteint de Cricetidae (Mammalia—Rodentia) géant des Galapagos: *Megaoryzomys* (gen. nov.). Bull. Acad. Roy. Belgique, Classe des Sciences, 5:632–48.

Lensing, J. E. 1983. Feeding strategy of the rock hyrax and its relation to the rock hyrax problem in southern South West Africa. Madoqua 13:177–96.

Lensing, J. E., and E. Joubert. 1977. Intensity distribution patterns for five species of problem animals in South West Africa. Madoqua 10:131–41.

Lent, P. C. 1974. A review of rutting behavior in moose. Naturaliste Canadien 101:307–23.

———. 1978. Musk-ox. *In* Schmidt and Gilbert (1978), pp. 135–47.

———. 1988. *Ovibos moschatus*. Mammalian Species, no. 302, 9 pp.

Leonard, M. L., and M. B. Fenton. 1983. Habitat use by spotted bats (*Euderma maculatum*, Chiroptera: Vespertilionidae): roosting and foraging behavior. Can. J. Zool. 61:1487–91.

Leonard, R. D. 1986. Aspects of reproduction of the fisher, *Martes pennanti*, in Manitoba. Can. Field-Nat. 100:32–44.

Leopold, A. S. 1959. Wildlife of Mexico. Univ. California Press, Berkeley, xiii + 568.

LeResche, R. E. 1974. Moose migrations in North America. Naturaliste Canadien 101: 393–415.

Leslie, D. M., Jr., and C. L. Douglas. 1979. Desert bighorn sheep of the River Mountains, Nevada. Wildl. Monogr., no. 66, 56 pp.

Leslie, G. 1971. Further observations on the

oriental short-clawed otter *Amblonyx cinerea* at Aberdeen Zoo. Internatl. Zoo Yearbook 11:112–13.

Leuthold, B. M. 1979. Social organization and behaviour of giraffe in Tsavo East National Park. Afr. J. Ecol. 17:19–34.

Leuthold, B. M., and W. Leuthold. 1978. Ecology of the giraffe in Tsavo East National Park, Kenya. E. Afr. Wildl. J. 16:1–20.

Leuthold, W. 1970. Observations on the social organization of impala *(Aepyceros melampus)*. Z. Tierpsychol. 27:693–721.

———. 1974. Observations on home range and social organization of lesser kudu, *Tragelaphus imberbis* (Blyth, 1869). *In* Geist and Walther (1974), pp. 206–34.

———. 1977a. African ungulates. Springer-Verlag, Berlin, xiii + 307 pp.

———. 1977b. A note on group size and composition in the oribi *Ourebia ourebi* (Zimmerman, 1783) (Bovidae). Saugetierk. Mitt. 25:233–35.

———. 1978a. On the ecology of the gerenuk *Litocranius walleri*. J. Anim. Ecol. 47:561–80.

———. 1978b. On social organization and behaviour of the gerenuk *Litocranius walleri* (Brooke 1878). Z. Tierpsychol. 47:194–216.

———. 1979. The lesser kudu, *Tragelaphus imberbis* (Blyth, 1869). Ecology and behaviour of an African antelope. Saugetierk. Mitt. 27:1–75.

Leuthold, W., and J. B. Sale. 1973. Movements and patterns of habitat utilization of elephants in Tsavo National Park, Kenya. E. Afr. Wildl. J. 11:369–84.

Levenson, H., and R. S. Hoffmann. 1984. Systematic relationships among taxa in the Townsend chipmunk group. Southwestern Nat. 29:157–68.

Levenson, H., R. S. Hoffmann, C. F. Nadler, L. Deutsch, and S. D. Freeman. 1985. Systematics of the Holarctic chipmunks *(Tamias)*. J. Mamm. 66:219–42.

Lever, C. 1985. Naturalized mammals of the world. Longman, London, xvii + 487 pp.

Lewis, J. G., and R. T. Wilson. 1979. The ecology of Swayne's hartebeest. Biol. Conserv. 15:1–12.

Leyhausen, P. 1979. Cat behavior. Garland STPM Press, New York, 340 pp.

Li Weidong, and Ma Yong. 1986. A new species of *Ochotona*, Ochotonidae, Lagomorpha. Acta Zool. Sinica 32:375–79.

Li Zhixiang, Ma Shilai, Hua Chenghui, and Wang Yingxiang. 1982. The distribution and habits of the Yunnan golden monkey, *Rhinopithecus bieti*. J. Human Evol. 11:633–38.

Liberg, O. 1980. Spacing patterns in a population of rural free roaming domestic cats. Oikos 35:336–49.

Lichter, A. A. 1986. Records of beaked whales (Ziphiidae) from the western South Atlantic. Sci. Rept. Whales Res. Inst. 37:109–27.

Lidicker, W. Z., Jr. 1976. Social behaviour and density regulation in house mice living in large enclosures. J. Anim. Ecol. 45:677–97.

———. 1988. Solving the enigma of microtine "cycles." J. Mamm. 69:225–35.

———. 1989. Rodents: a world survey of species of conservation concern. Occas. Pap. Internatl. Union Conserv. Nat. Species Survival Comm., no. 4, iv + 60 pp.

Lidicker, W. Z., Jr., and P. V. Brylski. 1987. The conilurine rodent radiation of Australia, analyzed on the basis of phallic morphology. J. Mamm. 68:617–41.

Lidicker, W. Z., Jr., and B. J. Marlow. 1970. A review of the dasyurid marsupial genus *Antechinomys* Krefft. Mammalia 34:212–27.

Lidicker, W. Z., Jr., and A. C. Ziegler. 1968. Report on a collection of mammals from eastern New Guinea including species keys for fourteen genera. Univ. California Publ. Zool. 87:i–v + 1–64.

Liers, E. E. 1966. Notes on breeding the Canadian otter *Lutra canadensis* in captivity and longevity records of beavers *Castor canadensis*. Internatl. Zoo Yearbook 6:171–72.

Lilly, J. C. 1977. The cetacean brain. Oceans 10(4):4–7.

Lim Boo Liat. 1967. Note on the food habits of *Ptilocercus lowii* Gray (pentail tree-shrew) and *Echinosorex gymnurus* (Raffles) (moonrat) in Malaya with remarks on "ecological labelling" by parasite patterns. J. Zool. 152:375–79.

———. 1970. Distribution, relative abundance, food habits, and parasite patterns of giant rats *(Rattus)* in West Malaysia. J. Mamm. 51:730–40.

Lim Boo Liat, and I. Muul. 1975. Notes on a rare species of arboreal rat, *Pithecheir parvus* Kloss. Malayan Nat. J. 28:181–85.

Lim Boo Liat, Chai Koh Shin, and I. Muul. 1972. Notes on the food habit of bats from the fourth division, Sarawak, with special reference to a new record of Bornean bat. Sarawak Mus. J. 20:351–57.

Limpus, C. J. 1983. *Melomys rubicola*, an endangered murid rodent endemic to the Great Barrier Reef of Queensland. Austral. Mamm. 6:77–79.

Linares, O. J. 1969. Notas acerca de la captura de una rata acuática *(Nectomys squamipes)* en la Cueva del Agua (AN-1), Anzoátegui, Venezuela. Bol. Soc. Venezolana Espeleol. 2:31–34.

———. 1973. Présence de l'oreillard d'Amérique du Sud dans les Andes Vénézuéliennes (Chiroptères, Vespertilionidae). Mammalia 37:433–38.

Lindburg, D. G. 1971. The rhesus monkey in north India: an ecological and behavioral study. Primate Behav. 2:1–106.

———, ed. 1980. The macaques: studies in ecology, behavior and evolution. Van Nostrand Reinhold, New York, xii + 384 pp.

Linder, R. L., and C. N. Hillman, eds. 1973. Proceedings of the black-footed ferret and prairie dog workshop. Dept. Wildl. and Fish. Sci., S. Dakota State Univ., v + 208 pp.

Lindsay, I. M., and D. W. Macdonald. 1986. Behaviour and ecology of the Ruppell's fox, *Vulpes ruppelli*, in Oman. Mammalia 50:461–74.

Lindsay, S. L. 1981. Taxonomic and biogeographic relationships of Baja California chickarees *(Tamiasciurus)*. J. Mamm. 62:673–82.

Lindstedt, S. L. 1980. Energetics and water economy of the smallest desert mammal. Physiol. Zool. 53:82–97.

Lindzey, F. 1987. Mountain lion. *In* Novak, Baker, et al. (1987), pp. 656–69.

Lindzey, F. G., and E. C. Meslow. 1976. Winter dormancy in black bears in southwestern Washington. J. Wildl. Mgmt. 40:408–15.

———. 1977a. Population characteristics of black bears on an island in Washington. J. Wildl. Mgmt. 41:408–12.

———. 1977b. Home range and habitat use by black bears in southwestern Washington. J. Wildl. Mgmt. 41:413–25.

Ling, J. K. 1987. New Zealand fur seal, *Arctocephalus forsteri*, in South Australia. *In* Croxall and Gentry (1987), pp. 53–55.

Ling, J. K., and M. M. Bryden. 1981. Southern elephant seal—*Mirounga leonina*. *In* Ridgway and Harrison (1981b), pp. 297–327.

———, eds. 1985. Studies of sea mammals in south latitudes. S. Austral. Mus., Sydney, vii + 132 pp.

Linzey, A. V. 1983. *Synaptomys cooperi*. Mammalian Species, no. 210, 5 pp.

Linzey, D. W., and R. L. Packard. 1977. *Ochrotomys nuttalli*. Mammalian Species, no. 75, 6 pp.

Lippold, L. K. 1977. The duoc langur: a time for conservation. *In* Rainier III and Bourne (1977), pp. 513–38.

Lloyd, D. S., C. P. McRoy, and R. H. Day. 1981. Discovery of northern fur seals *(Callorhinus ursinus)* breeding on Bogoslof Island, southeastern Bering Sea. Arctic 34:318–20.

Lloyd, H. G. 1975. The red fox in Britain. *In* Fox (1975), pp. 207–15.

———. 1983. Past and present distribution of red and gray squirrels. Mamm. Rev. 13:69–80.

Lloyd, J. A. 1975. Social structure and reproduction in two freely-growing populations of house mice (*Mus musculus* L.). Anim. Behav. 23:413–24.

Lo, N. C. H., and T. D. Smith. 1986. Incidental mortality of dolphins in the eastern tropical Pacific, 1959–72. Fishery Bull. 84:27–33.

Lock, J. M. 1972. The effects of hippopotamus grazing on grasslands. J. Ecol. 60:445–67.

Lockard, R. B. 1978. Seasonal change in the activity pattern of *Dipodomys spectabilis*. J. Mamm. 59:563–68.

Lockard, R. B., and D. H. Owings. 1974. Moon-related surface activity of bannertail (*Dipodomys spectabilis*) and Fresno (*D. nitratoides*) kangaroo rats. Anim. Behav. 22:262–73.

Lockyer, C. 1984. Review of baleen whale (Mysticeti) reproduction and implications for management. *In* Perrin, Brownell, and DeMaster (1984), pp. 27–50.

Lockyer, C., R. N. P. Goodall, and A. R. Galeazzi. 1988. Age and body-length characteristics of *Cephalorhynchus commersonii* from incidentally-caught specimens off Tierra del Fuego. *In* Brownell and Donovan (1988), pp. 103–18.

Long, C. A. 1972*a*. Taxonomic revision of the mammalian genus *Microsorex* Coues. Trans. Kansas Acad. Sci. 74:181–96.

———. 1972*b*. Notes on habitat preference and reproduction in pigmy shrews, *Microsorex*. Can. Field-Nat. 86:155–60.

———. 1972*c*. Taxonomic revision of the North American badger, *Taxidea taxus*. J. Mamm. 53:725–59.

———. 1973*a*. Reproduction in the white-footed mouse at the northern limits of its geographical range. Southwestern Nat. 18:11–20.

———. 1973*b*. *Taxidea taxus*. Mammalian Species, no. 26, 4 pp.

———. 1974. *Microsorex hoyi* and *Microsorex thompsoni*. Mammalian Species, no. 33, 4 pp.

———. 1978. A listing of Recent badgers of the world, with remarks on taxonomic problems in *Mydaus* and *Melogale*. Rept. Fauna and Flora Wisconsin, Univ. Wisconsin Mus. Nat. Hist., 14:1–6.

Long, C. A., and C. A. Killingley. 1983. The badgers of the world. Charles C. Thomas, Springfield, Ill., xxiv + 404 pp.

Lopez, J. C., and D. Lopez. 1985. Killer whales (*Orcinus orca*) of Patagonia, and their behavior of intentional stranding while hunting near shore. J. Mamm. 66:181–83.

Lopez-Forment, W. 1980. Longevity of wild *Desmodus rotundus* in Mexico. Proc. 5th Internatl. Bat Res. Conf., pp. 143–44.

Lorenz, R., C. O. Anderson, and W. A. Mason. 1973. Notes on reproduction in captive squirrel monkeys (*Saimiri sciureus*). Folia Primatol. 19:286–92.

Lott, D. F. 1974. Sexual and aggressive behaviour of adult male American bison (*Bison bison*). *In* Geist and Walther (1974), pp. 382–94.

———. 1981. Sexual behavior and intersexual strategies in American bison. Z. Tierpsychol. 56:97–114.

Lotze, J. 1979. The raccoon (*Procyon lotor*) on St. Catherines Island, Georgia. 4. Comparisons of home ranges determined by live-trapping and radiotracking. Amer. Mus. Novit., no. 2664, 25 pp.

Lotze, J., and S. Anderson. 1979. *Procyon lotor*. Mammalian Species, no. 119, 8 pp.

Loughlin, T. R. 1980. Home range and territoriality of sea otters near Monterey, California. J. Wildl. Mgmt. 44:576–82.

Loughlin, T. R., C. H. Fiscus, A. M. Johnson, and D. J. Rugh. 1982. Observations of *Mesoplodon stejnegeri* (Ziphiidae). J. Mamm. 63:697–700.

Loughlin, T. R., and M. A. Perez. 1985. *Mesoplodon stejnegeri*. Mammalian Species, no. 250, 6 pp.

Loughlin, T. R., M. A. Perez, and R. L. Merrick. 1987. *Eumetopias jubatus*. Mammalian Species, no. 283, 7 pp.

Loughlin, T. R., D. J. Rugh, and C. H. Fiscus. 1984. Northern sea lion distribution and abundance: 1956–80. J. Wildl. Mgmt. 48:729–40.

Loukashkin, A. S. 1940. On the pikas of north Manchuria. J. Mamm. 21:402–5.

Louwman, J. W. W. 1973. Breeding the tailless tenrec *Tenrec ecaudatus* at Wassenaar Zoo. Internatl. Zoo Yearbook 11:125–26.

Lovari, S. 1977. The Abruzzo chamois. Oryx 14:47–50.

———. 1984. The nimble chamois. *In* Macdonald (1984), pp. 590–91.

———, ed. 1985. The biology and management of mountain ungulates. Croom Helm, London, xii + 271 pp.

———. 1987. Evolutionary aspects of the biology of chamois, *Rupicapra* spp. (Bovidae, Caprinae). *In* Soma, H., ed., The biology and management of *Capricornis* and related mountain antelopes, Croom Helm, London, pp. 51–61.

Lovejoy, B. P., and H. C. Black. 1974. Growth and weight of the mountain beaver, *Aplodontia rufa pacifica*. J. Mamm. 55:364–69.

———. 1979*a*. Movements and home range of the Pacific mountain beaver *Aplodontia rufa pacifica*. Amer. Midl. Nat. 101:393–402.

———. 1979*b*. Population analysis of the mountain beaver, *Aplodontia rufa pacifica*, in western Oregon. Northwest Sci. 53:82–89.

Lovejoy, B. P., H. C. Black, and E. F. Hooven. 1978. Reproduction, growth, and development of the mountain beaver (*Aplodontia rufa pacifica*). Northwest Sci. 52:323–28.

Lowery, G. H., Jr. 1974. The mammals of Louisiana and its adjacent waters. Louisiana State Univ. Press, xxiii + 565 pp.

Loy, J. 1981. The reproductive and heterosexual behaviours of adult patas monkeys in captivity. Anim. Behav. 29:714–26.

Lu Ho-Gee, and Sheng He-Lin. 1984. Status of the black muntjac, *Muntiacus crinifrons*, in eastern China. Mamm. Rev. 14:29–36.

Lu Houji. 1987. Habitat availability and prospects for tigers in China. *In* Tilson and Seal (1987), pp. 71–74.

Luckett, W. P., ed. 1980*a*. Comparative biology and evolutionary relationships of tree shrews. Plenum Press, New York, xv + 314 pp.

———. 1980*b*. The suggested evolutionary relationships and classification of tree shrews. *In* Luckett (1980*a*), pp. 3–31.

Luckett, W. P., and J.-L. Hartenberger, eds. 1985. Evolutionary relationships among rodents: a multidisciplinary analysis. Plenum Press, New York, xiii + 721 pp.

Luckett, W. P., and F. S. Szalay, eds. 1975. Phylogeny of the primates: a multidisciplinary approach. Plenum Press, New York, xiv + 483 pp.

Ludwig, D. R. 1984. *Microtus richardsoni*. Mammalian Species, no. 223, 6 pp.

Lumpkin, S., and K. R. Kranz. 1984. *Cephalophus sylvicultor*. Mammalian Species, no. 225, 7 pp.

Lumsden, L. F., A. F. Bennett, and P. Robertson. 1988. First record of the paucident planigale, *Planigale gilesi* (Marsupialia: Dasyuridae), for Victoria. Victorian Nat. 105:81–87.

Lunney, D., and J. Barker. 1986. The occurrence of *Phoniscus papuensis* (Dobson) (Chiroptera: Vespertilionidae) on the south coast of New South Wales. Austral. Mamm. 9:57–58.

Lunney, D., J. Barker, and D. Priddel. 1985. Movements and day roosts of the chocolate wattled bat *Chalinolobus morio* (Gray) (Microchiroptera: Vespertilionidae) in a logged forest. Austral. Mamm. 8:313–17.

Luoma, J. R. 1987. The state of the tiger. Audubon 89(4):61–63.

———. 1989. The chimp connection. Anim. Kingdom 92(1):38–51.

Lutton, L. M. 1975. On the territorial behavior and response to predators of the pika, *Ochotona princeps.* J. Mamm. 56:231–34.

Lyman, C. P., and R. C. O'Brien. 1977. Laboratory study of the Turkish hamster *Mesocricetus brandti.* Breviora, no. 442, 27 pp.

Lyman, C. P., R. C. O'Brien, G. C. Greene, and E. D. Papafrangos. 1981. Hibernation and longevity in the Turkish hamster *Mesocricetus brandti.* Science 212:668–70.

Lynch, C. D. 1981. The status of the Cape grey mongoose, *Herpestes pulverulentus* Wagner, 1839 (Mammalia: Viverridae). Navors. Nas. Mus., Bloemfontein 4:121–68.

———. 1986. The ecology of the lesser dwarf shrew, *Suncus varilla*, with reference to the use of termite mounds of *Trinervitermes trinervoides.* Navors. Nas. Mus., Bloemfontein 5:277–97.

Lynch, G. R., F. D. Vogt, and H. R. Smith. 1978. Seasonal study of spontaneous daily torpor in the white-footed mouse, *Peromyscus leucopus.* Physiol. Zool. 51:289–99.

Lyne, A. G. 1974. Gestation period and birth in the marsupial *Isoodon macrourus.* Austral. J. Zool. 22:303–9.

———. 1976. Observations on oestrus and the oestrous cycle in the marsupials *Isoodon macrourus* and *Perameles nasuta.* Austral. J. Zool. 24:513–21.

Lyne, A. G., and P. A. Mort. 1981. Comparison of skull morphology in the marsupial bandicoot genus *Isoodon*: its taxonomic implications and notes on a new species, *Isoodon arnhemensis.* Austral. Mamm. 4:107–33.

M

Ma Shilai and Wang Yingxiang. 1986. The taxonomy and distribution of the gibbons in southern China and its adjacent region—with description of three new subspecies. Zool. Res. 7:393–410.

Ma Shilai, Wang Yingxiang, and Xu Longhui. 1986. Taxonomic and phylogenetic studies on the genus *Muntiacus.* Acta Theriol. Sinica 6:191–209.

Ma Yi-ching. 1983a. The status of bears in China. Acta Zool. Fennica 174:165–66.

———. 1983b. Status of reindeer in China. Acta Zool. Fennica 175:157–58.

McAllister, J. A., and R. S. Hoffmann. 1988. *Phenacomys intermedius.* Mammalian Species, no. 305, 8 pp.

MacArthur, R. A. 1978. Winter movements and home range of the muskrat. Can. Field-Nat. 92:345–49.

McBee, K., D. A. Schlitter, and R. L. Robbins. 1987. Systematics of the African bats of the genus *Eptesicus* (Mammalia: Vespertilionidae). 2. Karyotypes of African species and their generic relationships. Ann. Carnegie Mus. 56:213–22.

McBride, R. T. 1980. The Mexican wolf *(Canis lupus baileyi)*: a historical review and observations on its status and distribution. U.S. Fish and Wildl. Serv., Albuquerque, Endangered Species Rept., no. 8, vi + 38 pp.

McCabe, R. E., and T. R. McCabe. 1984. Of slings and arrows: an historical retrospection. *In* Halls (1984), pp. 19–72.

McCann, C. 1975. A study of the genus *Berardius* Duvernoy. Sci. Rept. Whales Res. Inst. 27:111–37.

McCann, T. S. 1980. Territoriality and breeding behaviour of adult male Antarctic fur seal, *Arctocephalus gazella.* J. Zool. 192:295–310.

———. 1985. Size, status and demography of southern elephant seal *(Mirounga leonina)* populations. *In* Ling and Bryden (1985), pp. 1–17.

McCann, T. S., and D. W. Doidge. 1987. Antarctic fur seal, *Arctocephalus gazella. In* Croxall and Gentry (1987), pp. 5–8.

McCarley, H., and C. J. Carley. 1979. Recent changes in distribution and status of wild red wolves *(Canis rufus).* U.S. Fish and Wildl. Serv., Albuquerque, Endangered Species Rept., no. 4, v + 38 pp.

McCarthy, T. J. 1982a. *Chironectes, Cyclopes, Cabassous*, and probably *Cebus* in southern Belize. Mammalia 46:397–400.

———. 1982b. Bat records from the Caribbean lowlands of El Peten, Guatemala. J. Mamm. 63:683–85.

———. 1987. Distributional records of bats from the Caribbean lowlands of Belize and adjacent Guatemala and Mexico. Fieldiana Zool., n.s., 39:137–62.

McCarthy, T. J., and N. A. Bitar. 1983. New bat records *(Enchisthenes* and *Myotis)* from the Guatemalan central highlands. J. Mamm. 64:526–27.

McCarthy, T. J., and M. Blake. 1987. Noteworthy bat records from the Maya Mountains Forest Reserve, Belize. Mammalia 51:161–64.

McCarthy, T. J., A. Cadena G., and T. O. Lemke. 1983. Comments on the first *Tonatia carrikeri* (Chiroptera: Phyllostomatidae) from Colombia. Lozania (Acta Zool. Colombiana), no. 40, 6 pp.

McCarthy, T. J., and C. O. Handley, Jr. 1987. Records of *Tonatia carrikeri* (Chiroptera: Phyllostomidae) from the Brazilian Amazon and *Tonatia schulzi* in Guyana. Bat Research News 28:20–23.

McCarthy, T. J., P. Robertson, and J. Mitchell. 1988. The occurrence of *Tonatia schulzi* (Chiroptera: Phyllostomidae) in French Guiana with comments on the female genitalia. Mammalia 52:583–84.

McCarty, R. 1975. *Onychomys torridus.* Mammalian Species, no. 59, 5 pp.

———. 1978. *Onychomys leucogaster.* Mammalian Species, no. 87, 6 pp.

McClenaghan, L. R., Jr. 1983. Notes on the population ecology of *Perognathus fallax* (Rodentia: Heteromyidae) in southern California. Southwestern Nat. 28:429–36.

———. 1984. Comparative ecology of sympatric *Dipodomys agilis* and *Dipodomys merriami* populations in southern California. Acta Theriol. 29:175–85.

McClenaghan, L. R., Jr., and M. S. Gaines. 1978. Reproduction in marginal populations of the hispid cotton rat *(Sigmodon hispidus)* in northwestern Kansas. Occas. Pap. Mus Nat. Hist. Univ. Kansas, no. 74, 16 pp.

McCord, C. 1974. Selection of winter habitat by bobcats *(Lynx rufus)* on the Quabbin Reservoir, Massachusetts. J. Mamm. 55:428–37.

McCracken, G. F. 1984. Communal nursing in Mexican free-tailed bat maternity colonies. Science 223:1090–91.

McCracken, G. F., and J. W. Bradbury. 1981. Social organization and kinship in the polygynous bat *Phyllostomus hastatus.* Behav. Ecol. Sociobiol. 8:11–34.

McCracken, G. F., and M. K. Gustin. 1987. Batmom's daily nightmare. Nat. Hist. 96 (10):66–73.

McCracken, H. E. 1986. Observations on the oestrous cycle and gestation period of the greater bilby, *Macrotis lagotis* (Reid) (Marsupialia: Thylacomyidae). Austral. Mamm. 9:5–16.

McCrane, M. P. 1966. Birth, behaviour and development of a hand-reared two-toed sloth. Internatl. Zoo Yearbook 6:153–63.

McCully, H. 1967. The broad-handed mole, *Scapanus latimanus*, in a marine littoral environment. J. Mamm. 48:480–82.

McCusker, J. S. 1974. Breeding Malayan sun bears *Helarctos malayanus* at Fort Worth Zoo. Internatl. Zoo Yearbook 14:118–19.

McCutchen, H. E. 1977. A minimum breeding age for a desert bighorn ewe. Southwestern Nat. 22:153.

———. 1981. Desert bighorn zoogeography and adaptation in relation to historic land use. Wildl. Soc. Bull. 9:171–79.

Macdonald, D., ed. 1984. The encyclopedia of mammals. Facts on File Publ., New York, xlviii + 895 pp.

Macdonald, D. W. 1978. Observations on the behaviour and ecology of the striped hyaena, *Hyaena hyaena*, in Israel. Israel J. Zool. 27:189–98.

———. 1982. Notes on the size and composition of groups of proboscis monkey, *Nasalis larvatus.* Folia Primatol. 37:95–98.

McDonald, J. N. 1981. North American

bison: their classification and evolution. Univ. California Press, Berkeley, 316 pp.

McDougal, C. 1987. The man-eating tiger in geographical and historical perspective. *In* Tilson and Seal (1987), pp. 435–48.

McDowell, S. B., Jr. 1958. The Greater Antillean insectivores. Bull. Amer. Mus. Nat. Hist. 115:113–214.

Macêdo, R. H., and M. A. Mares. 1987. Geographic variation in the South American cricetine rodent *Bolomys lasiurus*. J. Mamm. 68:578–94.

McEvoy, J. S. 1970. Red-necked wallaby in Queensland. Queensland Div. Plant Industry Adv. Leaf., no. 1050, 4 pp.

MacFadden, B. J. 1980. Rafting mammals or drifting islands? Biogeography of the Greater Antillean insectivores *Nesophontes* and *Solenodon*. J. Biogeogr. 7:11–22.

———. 1984*a*. Systematics and phylogeny of *Hipparion, Neohipparion, Nannippus,* and *Cormohipparion* (Mammalia, Equidae) from the Miocene and Pliocene of the New World. Bull. Amer. Mus. Nat. Hist. 179:1–196.

———. 1984*b*. *Astrohippus* and *Dinohippus* from the Yepómera Local Fauna (Hemphillian, Mexico) and implications for the phylogeny of one-toed horses. J. Vert. Paleontol. 4:273–83.

———. 1988. Horses, the fossil record, and evolution. Evol. Biol. 22:131–58.

McFarlane, D. A. 1986. Cave bats in Jamaica. Oryx 20:27–30.

McFarlane, D. A., and B. R. Blood. 1986. Taxonomic notes on a collection of Rhinolophidae (Chiroptera) from northern Thailand, with a description of a new subspecies of *Rhinolophus robinsoni*. Z. Saugetierk. 51:218–23.

McGhee, M. E., and H. H. Genoways. 1978. *Liomys pictus*. Mammalian Species, no. 83, 5 pp.

McGinnis, S. M., and R. J. Schusterman. 1981. Northern elephant seal—*Mirounga angustirostris*. *In* Ridgway and Harrison (1981*b*), pp. 329–49.

McGrew, J. C. 1979. *Vulpes macrotis*. Mammalian Species, no. 123, 6 pp.

McGrew, W. C., and D. A. Collins. 1985. Tool use by wild chimpanzees *(Pan troglodytes)* to obtain termites *(Macrotermes hersus)* in the Mahale Mountains, Tanzania. Amer. J. Primatol. 9:47–62.

McGuire, M. T., et al. 1974. The St. Kitts vervet. Contrib. Primatol. l:i–x + 1–199.

McHugh, J. L. 1974. The role and history of the International Whaling Commission. *In* Schevill (1974), pp. 305–35.

Macintosh, N. W. G. 1975. The origin of the dingo: an enigma. *In* Fox (1975), pp. 87–106.

McKay, G. M. 1973. Behavior and ecology of the Asiatic elephant in southeastern Ceylon. Smithson. Contrib. Zool., no. 125, v + 113 pp.

———. 1982. Nomenclature of the gliding possum genera *Petaurus* and *Petauroides* (Marsupialia: Petauridae). Austral. Mamm. 5:37–39.

McKean, J. L. 1972. Notes on some collections of bats (order Chiroptera) from Papua New Guinea and Bougainville Island. CSIRO Div. Wildl. Res. Tech. Pap., no. 26, 5 pp.

———. 1975. The bats of Lord Howe Island with the description of a new nyctophiline bat. Austral. Mamm. 1:329–32.

McKean, J. L., and J. H. Calaby. 1968. A new genus and two new species of bats from New Guinea. Mammalia 32:372–78.

McKean, J. L., and G. R. Friend. 1979. *Taphozous kapalgensis,* a new species of sheath-tailed bat from the Northern Territory, Australia. Victorian Nat. 96:239–41.

McKean, J. L., and L. S. Hall. 1964. Notes on microchiropteran bats. Victorian Nat. 81:36–37.

McKean, J. L., and W. J. Price. 1967. Notes on some Chiroptera from Queensland, Australia. Mammalia 31:101–19.

McKean, J. L., G. C. Richards, and W. J. Price. 1978. A taxonomic appraisal of *Eptesicus* (Chiroptera: Mammalia) in Australia. Austral. J. Zool. 26:529–37.

McKenna, M. C. 1962. *Eupetaurus* and the living petauristine sciurids. Amer. Mus. Novit., no. 2104, 38 pp.

———. 1975. Toward a phylogenetic classification of the Mammalia. *In* Luckett and Szalay (1975), pp. 21–46.

McKenzie, N. L., and M. Archer. 1982. *Sminthopsis youngsoni* (Marsupialia: Dasyuridae) the lesser hairy-footed dunnart, a new species from arid Australia. Austral. Mamm. 5:267–79.

McKey, D., and P. G. Waterman. 1982. Ranging behaviour of a group of black colobus *(Colobus satanas)* in the Doula-Edea Reserve, Cameroon. Folia Primatol. 39:264–304.

Mackie, R. J. 1987. Mule deer. *In* Kallman (1987), pp. 265–71.

MacKinnon, J. 1971. The orang-utan in Sabah today. Oryx 11:141–91.

———. 1973. Orang-utans in Sumatra. Oryx 12:234–42.

———. 1974. The behaviour and ecology of wild orang-utans *(Pongo Pygmaeus)*. Anim. Behav. 22:3–74.

———. 1981. The structure and function of the tusks of babirusa. Mamm. Rev. 11:37–40.

———. 1984. The distribution and status of gibbons in Indonesia. *In* Preuschoft et al. (1984), pp. 16–18.

MacKinnon, J., and K. MacKinnon. 1980. The behavior of wild spectral tarsiers. Internatl. J. Primatol. 1:361–79.

———. 1987. Conservation status of the primates of the Indo-Chinese subregion. Primate Conserv. 8:187–95.

MacKinnon, K. 1984. Flying lemurs. *In* Macdonald (1984), pp. 446–47.

———. 1986. The conservation status of nonhuman primates in Indonesia. *In* Benirschke (1986), pp. 99–126.

Mackintosh, J. H. 1973. Factors affecting the recognition of territory boundaries by mice *(Mus musculus)*. Anim. Behav. 21:464–70.

Mackintosh, N. A. 1966. The distribution of southern blue and fin whales. *In* Norris (1966), pp. 125–44.

Mackowski, C. M. 1986. Distribution, habitat and status of the yellow-bellied glider, *Petaurus australis* Shaw (Marsupialia: Petauridae) in northeastern New South Wales. Austral. Mamm. 9:141–44.

McLanahan, E. B., and K. M. Green. 1977. The vocal repertoire and an analysis of the contexts of vocalizations in *Leontopithecus rosalia*. *In* Kleiman (1977*a*), pp. 251–69.

McLaren, I. A. 1960. Are the Pinnipedia biphyletic? Syst. Zool. 9:18–28.

———. 1984. True seals. *In* Macdonald (1984), pp. 270–79.

McLaughlin, C. A. 1967. Aplodontoid, Sciuroid, Geomyoid, Castoroid, and Anomaluroid rodents. *In* Anderson and Jones (1967), pp. 210–25.

———. 1984. Protogomorph, sciuromorph, castorimorph, myomorph (Geomyoid, anomaluroid, pedetoid, and ctenodactyloid) rodents. *In* Anderson and Jones (1984), pp. 267–88.

McLean, I. G. 1982. The association of female kin in the arctic ground squirrel *Spermophilus parryii*. Behav. Ecol. Sociobiol. 10:91–99.

McLellan, L. J. 1986. Notes on bats of Sudan. Amer. Mus. Novit., no. 2839, 12 pp.

McMahon, E. E., C. C. Oakley, and S. P. Cross. 1981. First record of the spotted bat *(Euderma maculatum)* from Oregon. Great Basin Nat. 41:270.

McManus, J. J. 1970. Behavior of captive opossums, *Didelphis marsupialis virginiana*. Amer. Midl. Nat. 84:144–69.

———. 1974. *Didelphis virginiana*. Mammalian Species, no. 40, 6 pp.

McNally, R. 1977. Echolocation. Cetaceans' sixth sense. Oceans 10(4):27–33.

McNeely, J. A. 1981. Conservation needs of *Nesolagus netscheri*. *In* Myers, K., and C. D. MacInnes, eds., Proceedings of the World Lagomorph Conference, Guelph Univ. Press, pp. 929–32.

MacPhee, R. D. E. 1984. Quaternary mammal localities and heptaxodontid rodents of Jamaica. Amer. Mus. Novit., no. 2803, 34 pp.

———. 1987a. The shrew tenrecs of Madagascar: systematic revision and Holocene distribution of *Microgale* (Tenrecidae, Insectivora). Amer. Mus. Novit., no. 2889, 45 pp.

———. 1987b. Systematic status of *Dasogale fontoynonti* (Tenrecidae, Insectivora). J. Mamm. 68:133–35.

MacPhee, R. D. E., and C. A. Woods. 1982. A new fossil cebine from Hispaniola. Amer. J. Phys. Anthropol. 58:419–36.

MacPhee, R. D. E., C. A. Woods, and G. S. Morgan. 1983. The Pleistocene rodent *Alterodon major* and the mammalian biogeography of Jamaica. Palaeontology 26:831–37.

McPhee, E. C. 1988. Ecology and diet of some rodents from the lower montane region of Papua New Guinea. Austral. Wildl. Res. 15:91–102.

Macpherson, A. H. 1965. The barren-ground grizzly bear and its survival in northern Canada. Can. Audubon 27(1):2–8.

———. 1969. The dynamics of Canadian arctic fox populations. Can. Wildl. Serv. Rept. Ser., no. 8, 52 pp.

McQuade, L. R. 1984. Taxonomic relationship of the greater glider *Petauroides volans* and lemur-like possum *Hemibelideus lemuroides*. *In* Smith and Hume (1984), pp. 303–10.

McQuaid, C. D. 1986. Post-1980 sightings of bowhead whales *(Balaena mysticetus)* from the Spitsbergen stock. Mar. Mamm. Sci. 2:316–18.

McWhinnie, M., and C. J. Denys. 1980. The high importance of the lowly krill. Nat. Hist. 89(3):66–73.

McWilliam, A. N. 1987a. The reproductive and social biology of *Coleura afra* in a seasonal environment. *In* Fenton, M. B., P. Racey, and J. M. V. Rayner, eds., Recent advances in the study of bats, Cambridge Univ. Press, London, pp. 324–50.

———. 1987b. Territorial and pair behavior of the African false vampire bat, *Cardioderma cor* (Chiroptera: Megadermatidae), in coastal Kenya. J. Zool. 213:243–52.

———. 1988. Social organization of the bat *Tadarida (Chaerephon) pumila* (Chiroptera: Molossidae) in Ghana, West Africa. Ethology 77:115–24.

Madden, J. R. 1974. Female territoriality in a Suffolk County, Long Island, population of *Glaucomys volans*. J. Mamm. 55:647–52.

Maddock, A. H. 1986. *Chrysospalax trevelyani*: an unknown and rare mammal endemic to southern Africa. Cimbebasia, ser. A, 8:88–90.

Maddock, T. H., and A. McLeod. 1974. Polyoestry in the little brown bat, *Eptesicus pumilus*, in central Australia. S. Austral. Nat. 48:50, 63.

Madhavan, A. 1971. Breeding habits in the Indian vespertilionid bat, *Pipistrellus ceylonicus chrysothrix* (Wroughton). Mammalia 25:283–306.

Madhavan, A., D. R. Patil, and A. Gopalakrishna. 1978. Breeding habits and associated phenomena in some Indian bats. Part IV—*Hipposideros fulvus fulvus* (Gray)—Hipposideridae. J. Bombay Nat. Hist. Soc. 75:96–103.

Madison, D. M. 1977. Movements and habitat use among interacting *Peromyscus leucopus* as revealed by radiotelemetry. Can. Field-Nat. 91:273–81.

Madureira, M. L. 1981. Discriminant analysis in Portuguese pine voles: *Pitymys lusitanicus* Gerbe and *Pitymys duodecimcostatus* de Selys-Longchamps (Mammalia: Rodentia). Arquivos Mus. Bocage (Lisbon) 1:111–22.

Maeda, K. 1980. Review on the classification of little tube-nosed bats, *Murina aurata* group. Mammalia 44:531–51.

———. 1982. Studies on the classification of *Miniopterus* in Eurasia, Australia and Melanesia. Honyurui Kagaku (Mammalian Science), Suppl., no. 1, 176 pp.

———. 1983. Classificatory study of the Japanese large noctule, *Nyctalus lasiopterus aviator* Thomas, 1911. Zool. Mag. 92:21–36.

Magnusson, W. E., G. J. W. Webb, and J. A. Taylor. 1976. Two new locality records, a new habitat and a nest description for *Xeromys myoides* Thomas (Rodentia: Muridae). Austral. Wildl. Res. 3:153–57.

Mahe, J. 1972. The Malagasy subfossils. *In* Battistini and Richard-Vindard (1972), pp. 339–65.

———. 1976. Crâniométrie des lémuriéns analyses multivariables-phylogénie. Mem. Mus. Natl. Hist. Nat., ser. C, 32:1–342.

Maheshwari, U. K. 1982. Some observations on reproduction of the Indian hedgehog (*Hemiechinus auritus collaris* Gray, 1830). Saugetierk. Mitt. 30:184–89.

Mahoney, J. A. 1968. *Baiyankamys* Hinton 1943 (Muridae, Hydromyinae), a New Guinea rodent genus named for an incorrectly associated skin and skull (Hydromyinae, *Hydromys*) and mandible (Murinae, *Rattus*). Mammalia 32:64–71.

———. 1975. *Notomys macrotis* Thomas, 1921, a poorly known Australian hopping mouse (Rodentia: Muridae). Austral. Mamm. 1:367–74.

———. 1977. Skull characters and relationships of *Notomys mordax* Thomas (Rodentia: Muridae), a poorly known Queensland hopping-mouse. Austral. J. Zool. 25:749–54.

———. 1981. The specific name of the honey possum (Marsupialia: Tarsipedidae: *Tarsipes rostratus* Gervais and Verreaux, 1842). Austral. Mamm. 4:135–38.

Mahoney, J. A., and B. J. Marlow. 1968. The rediscovery of the New Holland mouse. Austral. J. Sci. 31:221–23.

Mahoney, J. A., and H. Posamentier. 1975. The occurrence of the native rodent, *Pseudomys gracilicaudatus* (Gould, 1845) (Rodentia: Muridae), in New South Wales. Austral. Mamm. 1:333–46.

Main, A. R., and M. Yadav. 1971. Conservation of macropods in reserves in Western Australia. Biol. Conserv. 3:123–33.

Mainstone, B. J. 1974. Tigers reduce honey production. Malayan Nat. J. 28:36.

Majluf, O. 1987. South American fur seal, *Arctocephalus australis*, in Peru. *In* Croxall and Gentry (1987), pp. 33–35.

Malagnoux, M., and J. C. Gautun. 1976. Un ennemi des plantations d'araucaria en Côte d'Ivoire. Bois et Forêts Tropiques 165:35–38.

Malcolm, J. R. 1980a. African wild dogs play every game by their own rules. Smithsonian 11(8):62–71.

———. 1980b. Food caching by African wild dogs (*Lycaon pictus*). J. Mamm. 61:743–44.

———. 1986. Socio-ecology of bat-eared foxes (*Otocyon megalotis*). J. Zool. 208:457–67.

Malcolm, J. R., and K. Marten. 1982. Natural selection and the communal rearing of pups in African wild dogs (*Lycaon pictus*). Behav. Ecol. Sociobiol. 10:1–13.

Malcolm, J. R., and H. Van Lawick. 1975. Notes on wild dogs (*Lycaon pictus*) hunting zebras. Mammalia 39:231–40.

Maliniak, E., and J. F. Eisenberg. 1971. Breeding spiny rats *Proechimys semispinosus* in captivity. Internatl. Zoo Yearbook 11:93–98.

Mallinson, J. J. C. 1973. The reproduction of the African civet *Viverra civetta* at Jersey Zoo. Internatl. Zoo Yearbook 13:147–50.

———. 1974. Establishing mammal gestation at the Jersey Zoological Park. Internatl. Zoo Yearbook 14:184–87.

———. 1977. Maintenance of marmosets and tamarins at Jersey Zoological Park with special reference to the design of the new marmoset complex. *In* Kleiman (1977a), pp. 323–29.

———. 1987. International efforts to secure a viable population of the golden-headed lion tamarin. Primate Conserv. 8:124–25.

Mallon, D. P. 1985. The mammals of the Mongolian People's Republic. Mamm. Rev. 15:71–102.

Mallory, F. F., and R. J. Brooks. 1978. Infanticide and other reproductive strategies in the collared lemming, *Dicrostonyx groenlandicus*. Nature 273:144–46.

Malygin, V. M., and V. N. Yatsenko. 1986. Taxonomic nomenclature of sibling species of the common vole (Rodentia, Cricetidae). Zool. Zhur. 65:579–91.

Manning, R. W., C. Jones, R. R. Hollander, and J. K. Jones, Jr. 1987. An unusual number of fetuses in the pallid bat. Prairie Nat. 19(4):261.

Manning, R. W., J. K. Jones, Jr., and R. R. Hollander. 1986. Northern limits of distribution of the hog-nosed skunk, *Conepatus mesoleucus*, in Texas. Texas J. Sci. 38:289–91.

Mansergh, I. 1984a. The mountain pygmy-possum *(Burramys parvus)* (Broom): a review. *In* Smith and Hume (1984), pp. 413–16.

———. 1984b. Ecological studies and conservation of *Burramys parvus*. *In* Smith and Hume (1984), pp. 545–52.

———. 1984c. The status, distribution and abundance of *Dasyurus maculatus* (tiger quoll) in Australia, with particular reference to Victoria. Austral. Zool. 21:109–22.

Mansfield, A. W., T. G. Smith, and B. Beck. 1975. The narwhal, *Monodon monoceros*, in eastern Canadian waters. J. Fish. Res. Bd. Can. 32:1041–46.

Manville, R. H. 1966. The extinct sea mink, with taxonomic notes. Proc. U.S. Natl. Mus. 122:1–12.

Manwell, C., and C. M. A. Baker. 1977. *Ammotragus lervia*: Barbary sheep or Barbary goat. Comp. Biochem. Physiol. 58:267–71.

Maples, W. R. 1972. Systematic reconsideration and a revision of the nomenclature of Kenya baboons. Amer. J. Phys. Anthropol. 36:9–19.

Marchinton, R. L., and D. H. Hirth. 1984. Behavior. *In* Halls (1984), pp. 129–68.

Marcuzzi, G. 1986. Man-beaver relations. *In* Pilleri, G., ed., Investigations on beavers, vol. V, Brain Anatomy Inst., Berne, pp. 16–72.

Mares, M. A. 1977. Water balance and other ecological observations on three species of *Phyllotis* in northwestern Argentina. J. Mamm. 58:514–20.

———. 1988. Reproduction, growth, and development in Argentine gerbil mice, *Eligmodontia typus*. J. Mamm. 69:852–54.

Mares, M. A., and R. A. Ojeda. 1982. Patterns of diversity and adaptation in South American hystricognath rodents. Pymatuning Lab. Ecol. Spec. Publ. 6:393–432.

Mares, M. A., M. D. Watson, and T. E. Lacher, Jr. 1976. Home range perturbations in *Tamias striatus*. Oecologia 25:1–12.

Mares, M. A., M. R. Willig, K. E. Streilen, and T. E. Lacher, Jr. 1981. The mammals of northeastern Brazil: a preliminary assessment. Ann. Carnegie Mus. 50:81–137.

Marinina, L. S. 1971. Vital activity of the *Rhombomys opimus* population in the central Karakum after the winter of 1968/69. Soviet J. Ecol. 2:77–80.

Marinkelle, C. J., and A. Cadena. 1972. Notes on bats new to the fauna of Colombia. Mammalia 36:50–58.

Markgren, G. 1974. The moose in Fennoscandia. Naturaliste Canadien 101:185–94.

Markham, D. D., and F. W. Whicker. 1973. Seasonal data on reproduction and body weights of pikas *(Ochotona princeps)*. J. Mamm. 54:496–98.

Marlow, B. J. 1961. Reproductive behavior of the marsupial mouse, *Antechinus flavipes* (Waterhouse) (Marsupialia), and the development of pouch young. Austral. J. Zool. 9:203–18.

Marples, T. G. 1973. Studies on the marsupial glider, *Schoinobates volans* (Kerr). IV. Feeding biology. Austral. J. Zool. 21:213–16.

Marques, S. A. 1986. Activity cycle, feeding and reproduction of *Molossus ater* (Chiroptera: Molossidae) in Brazil. Bol. Mus. Paraense Emilio Goeldi, Ser. Zool., 2:159–79.

Marques, S. A., and D. C. Oren. 1987. First Brazilian record for *Tonatia schulzi* and *Sturnira bidens* (Chiroptera: Phyllostomidae). Bol. Mus. Par. Emilio Goeldi, Ser. Zool., 3:159–60.

Marquette, W. M. 1978. Bowhead whale. *In* Haley (1978), pp. 70–81.

———. 1979. The 1977 catch of bowhead whales *(Balaena mysticetus)* by Alaskan Eskimos. Rept. Internatl. Whaling Comm. 29:281–89.

Marsh, C. W., and R. A. Mittermeier, eds. 1987. Primate conservation in the tropical rain forest. Alan R. Liss, New York, xviii + 365 pp.

Marsh, H., G. E. Heinsohn, and L. M. Marsh. 1984. Breeding cycle, life history and population dynamics of the dugong, *Dugong dugon* (Sirenia: Dugongidae). Austral. J. Zool. 32:767–88.

Marsh, R. E. 1984. Evidence does not support the existence of a Bajan mouse. J. Barbados Mus. and Hist. Soc. 37:159–60.

———. 1985. More about the Bajan mouse. J. Barbados Mus. Hist. Soc. 37:310.

Marshall, A. G., and A. N. McWilliam. 1982. Ecological observations on epomorphorine fruit-bats (Megachiroptera) in West African savanna woodland. J. Zool. 198:53–67.

Marshall, J., and J. Sugardjito. 1986. Gibbon systematics. *In* Swindler and Erwin (1986), pp. 137–85.

Marshall, J. T., Jr. 1977. A synopsis of Asian species of *Mus* (Rodentia, Muridae). Bull. Amer. Mus. Nat. Hist. 158:173–220.

———. 1979. Classification of genus *Mus*. *In* Altman, P. C., and D. D. Katz, eds., Inbred and genetically defined strains of laboratory animals, part I, Mouse and rat, Fed. Amer. Soc. Exp. Biol., Bethesda, Md., pp. 212–20.

———. 1981a. The agile gibbon in south Thailand. Nat. Hist. Bull. Siam Soc. 29:129–36.

———. 1981b. Taxonomy. *In* Foster, H. L., J. D. Small, and J. G. Fox, eds., The mouse in biomedical research, vol. I, History, genetics, and wild mice, Academic Press, New York, pp. 17–26.

———. 1986. Systematics of the genus *Mus*. Current Topics Microbiol. Immunol. 127:12–18.

Marshall, J. T., Jr., and R. O. Sage. 1981. Taxonomy of the house mouse. *In* Berry (1981), pp. 15–25.

Marshall, L. G. 1977. *Lestodelphys halli*. Mammalian Species, no. 81, 3 pp.

———. 1978a. *Lutreolina crassicaudata*. Mammalian Species, no. 91, 4 pp.

———. 1978b. *Dromiciops australis*. Mammalian Species, no. 99, 5 pp.

———. 1978c. *Glironia venusta*. Mammalian Species, no. 107, 3 pp.

———. 1978d. *Chironectes minimus*. Mammalian Species, no. 109, 6 pp.

———. 1980. Systematics of the South American marsupial family Caenolestidae. Fieldiana Zool., n.s., no. 5, 145 pp.

———. 1981. The families and genera of Marsupialia. Fieldiana Geol., n.s., no. 8, 65 pp.

———. 1982. Systematics of the South American marsupial family Microbiotheriidae. Fieldiana Geol., n.s., no. 10, 75 pp.

———. 1984. Monotremes and marsupials. *In* Anderson and Jones (1984), pp. 59–116.

Martín, A. R., R. Hutterer, and G. B. Corbet. 1984. On the presence of shrews (Soricidae) in the Canary Islands. Bonner Zool. Beitr. 35:5–14.

Martín, A. R., P. Reynolds, and M. G. Richardson. 1987. Aspects of the biology of pilot whales *(Globicephala melaena)* in recent mass strandings on the British coast. J. Zool. 211:11–23.

Martin, C. 1987. Interspecific relationships between barasingha *(Cervus duvauceli)* and axis deer *(Axis axis)* in Kanha National Park,

India, and relevance to management. *In* Wemmer (1987), pp. 299–306.

Martin, C. O., and D. J. Schmidly. 1982. Taxonomic review of the pallid bat, *Antrozous pallidus* (Le Conte). Spec. Publ. Mus. Texas Tech Univ., no. 18, 48 pp.

Martin, E. B. 1979. The international trade in rhinoceros products. World Wildlife Fund, Gland, Switzerland, 82 pp.

———. 1983. Rhino exploitation: the trade in rhino products in India, Indonesia, Malaysia, Burma, Japan, and South Korea. World Wildlife Fund, Hong Kong, 122 pp.

———. 1985a. Religion, royalty and rhino conservation in Nepal. Oryx 19:11–16.

———. 1985b. Rhinos and daggers: a major conservation problem. Oryx 19:198–201.

———. 1989. Report on the trade in rhino products in eastern Asia and India. Pachyderm 12:13–22.

Martin, E. B., and C. B. Martin. 1987. Combatting the illegal trade in rhinoceros products. Oryx 21:143–48.

Martin, E. B., C. B. Martin, and L. Vigne. 1987. Conservation crisis—the rhinoceros in India. Oryx 21:212–18.

Martin, E. B., and L. Vigne. 1987. Recent developments in the rhino horn trade. Traffic Bull. 9(2–3):49–53.

———. 1989. The decline and fall of India's ivory industry. Pachyderm 12:4–21.

Martin, I. G. 1981. Venom of the short-tailed shrew (*Blarina brevicauda*) as an insect immobilizing agent. J. Mamm. 62:189–92.

Martin, P. 1971. Movements and activities of the mountain beaver (*Aplodontia rufa*). J. Mamm. 52:717–23.

Martin, P. G. 1977. Marsupial biogeography and plate tectonics. *In* Stonehouse and Gilmore (1977), pp. 97–115.

Martin, P. S. 1984. Prehistoric overkill: the global model. *In* Martin and Klein (1984), pp. 354–403.

Martin, P. S., and R. G. Klein. 1984. Quaternary extinctions: a prehistoric revolution. Univ. Arizona Press, Tucson, x + 892 pp.

Martin, P. S., and H. E. Wright, eds. 1967. Pleistocene extinctions. Yale Univ. Press, New Haven, x + 453 pp.

Martin. R. A. 1972. Synopsis of late Pliocene and Pleistocene bats of North America and the Antilles. Amer. Midl. Nat. 87:326–35.

Martin, R. B., and T. de Meulenaer. 1988. Survey of the status of the leopard (*Panthera pardus*) in sub-Saharan Africa. Secretariat of the Convention on International Trade in Endangered Species of Wild Fauna and Flora, Lausanne, Switzerland, xx + 106 pp.

Martin, R. D. 1968. Reproduction and ontogeny in tree-shrews (*Tupaia belangeri*), with reference to their general behaviour and taxonomic relationships. Z. Tierpsychol. 25:409–95, 505–32.

———. 1973. A review of the behaviour and ecology of the lesser mouse lemur (*Microcebus murinus* J. F. Miller 1777). *In* Michael and Crook (1973), pp. 1–68.

———. 1975a. The bearing of reproductive behavior and ontogeny on strepsirhine phylogeny. *In* Luckett and Szalay (1975), pp. 265–97.

———, ed. 1975b. Breeding endangered species in captivity. Academic Press, London, xxv + 420 pp.

———. 1984. Tree shrews. *In* Macdonald (1984), pp. 441–45.

Martin, R. D., G. A. Doyle, and A. C. Walker, eds. 1974. Prosimian biology. Duckworth, London, xxi + 983 pp.

Martin, R. E. 1970. Cranial and bacular variation in populations of spiny rats of the genus *Proechimys* (Rodentia: Echimyidae) from South America. Smithson. Contrib. Zool., no. 35, 19 pp.

Martin, R. E., and K. G. Matocha. 1972. Distributional status of the kangaroo rat, *Dipodomys elator*. J. Mamm. 53:873–77.

Martin, R. W., and A. Lee. 1984. The koala, *Phascolarctos cinereus*, the largest marsupial folivore. *In* Smith and Hume (1984), pp. 463–67.

Martinka, C. J. 1974. Population characteristics of grizzly bears in Glacier National Park, Montana. J. Mamm. 55:21–29.

———. 1976. Ecological role and management of grizzly bears in Glacier National Park, Montana. *In* Pelton, Lentfer, and Folk (1976), pp. 147–56.

Masaki, Y. 1976. Biological studies on the North Pacific sei whale. Bull. Far Seas Fish. Res. Lab. 14:1–104.

———. 1978. Yearly change in the biological parameters of the Antarctic sei whale. Rept. Internatl. Whaling Comm. 28:421–29.

Masataka, N. 1981a. A field study of the social behavior of Goeldi's monkeys (*Callimico goeldii*) in north Bolivia. I. Group composition, breeding cycle, and infant development. *In* Kyoto Univ. Overseas Res. Rept. of New World Monkeys, pp. 23–32.

———. 1981b. A field study of the social behavior of Goeldi's monkeys (*Callimico goeldii*) in north Bolivia. II. Grouping pattern and intragroup relationship. *In* Kyoto Univ. Overseas Res. Rept. of New World Monkeys, pp. 33–41.

———. 1982. A field study on the vocalizations of Goeldi's monkeys (*Callimico goeldii*). Primates 23:206–19.

Maser, C., ed. 1974. The sage vole, *Lagurus curtatus* (Cope, 1868), in the Crooked River National Grassland, Jefferson County, Oregon: a contribution to its life history and ecology. Saugetierk. Mitt. 22:193–222.

Masini, F. 1985. Würmian and Holocene chamois of Italy. *In* Lovari (1985), pp. 31–44.

Masini, F., and S. Lovari. 1988. Systematics, phylogenetic relationships and dispersal of the chamois (*Rupicapra* spp.). Quaternary Res. 30:339–49.

Mason, C. F., and S. M. Macdonald. 1986. Otters: ecology and conservation. Cambridge Univ. Press, London, vii + 236 pp.

Mason, W. A. 1968. Use of space by *Callicebus* groups. *In* Jay (1968), pp. 200–216.

———. 1971. Field and laboratory studies of social organization in *Saimiri* and *Callicebus*. Primate Behav. 2:107–37.

Massoia, E. 1963. *Oxymycterus iheringi* (Rodentia—Cricetidae) nueva especie para la Argentina. Physis, sec. C, 24:129–36.

———. 1973. Descripción de *Oryzomys fornesi*, nueva especie y nuevos datos sobre algunas especies y subespecies argentinas del subgénero *Oryzomys* (*Oligoryzomys*) (Mammalia—Rodentia—Cricetidae). Rev. Investig. Agropec. Buenos Aires, ser. 1, 10:21–37.

———. 1977. Sobre la identidad del holotipo de *Graomys hypogaeus* Cabrera, 1934 (Mammalia—Rodentia—Cricetidae). Rev. Investig. Agropec. Buenos Aires, ser. 5, 13:15–20.

———. 1979. Descripción de un género y especie nuevos: *Bibimys torresi* (Mammalia — Rodentia — Cricetidae — Sigmodontinae —Scapteromyini). Physis, sec. C, 38:1–7.

———. 1980. El estado sistemático de cuatro especies de cricetidos Sudamericanos y comentarios sobre otras especies congenéricas. Ameghiniana, Rev. Asoc. Paleontol. Argentina 17:280–87.

———. 1981. El estado sistemático y zoogeografía de *Mus brasiliensis* Desmarest y *Holochilus sciureus* Wagner. Physis, sec. C, 39:31–34.

———. 1982. Diagnosis previa de *Cabreramys temchuki*, nueva especie (Rodentia, Cricetidae). Hist. Nat. (Corrientes, Argentina) 2:91–92.

Massoia, E., and J. Foerster. 1974. Un mamífero nuevo para la Republica Argentina: *Caluromys lanatus lanatus* (Illiger), (Mammalia—Marsupialia—Didelphidae). IDIA, January–February 1974, pp. 5–7.

Massoia, E., and A. Fornes. 1967. El estado sistemático, distribución geográfica y datos etoecológicos de algunos mamíferos neotropicales (Marsupialia y Rodentia) con la descripción de *Cabreramys* género nuevo (Cricetidae). Acta Zool. Lilloana 23:407–30.

Masters, J. 1986. Geographic distributions of karyotypes and morphotypes within the greater galagines. Folia Primatol. 46:127–41.

———. 1988. Speciation in the greater galagos (Prosimii: Galaginae): review and synthesis. Biol. J. Linnean Soc. 34:149–74.

Masui, K., Y. Sugiyama, A. Nishimura, and H. Ohsawa. 1975. The life table of Japanese monkeys at Takasakiyama: a preliminary report. Proc. 5th Internatl. Congr. Primatol., pp. 401–6.

Matocha, K. G. 1977. The vocal repertoire of *Spermophilus tridecemlineatus*. Amer. Midl. Nat. 98:482–87.

Matson, J. O., and J. P. Abravaya. 1977. *Blarinomys breviceps*. Mammalian Species, no. 74, 3 pp.

Mattlin, R. H. 1987. New Zealand fur seal, *Arctocephalus forsteri*, within the New Zealand region. *In* Croxall and Gentry (1987), pp. 49–51.

May, R. M. 1986. The cautionary tale of the black-footed ferret. Nature 320:13–14.

Mayer, J. J., and P. N. Brandt. 1978. Pelage description and reproductive data of the Chacoan peccary, *Catagonus wagneri*. American Soc. Mamm., Abstr. Tech. Pap., 58th Ann. Mtg., p. 44.

———. 1982. Identity, distribution, and natural history of the peccaries, Tayassuidae. Pymatuning Lab. Ecol. Spec. Publ. 6:433–56.

Mayer, J. J., and R. M. Wetzel. 1986. *Catagonus wagneri*. Mammalian Species, no. 259, 5 pp.

———. 1987. *Tayassu pecari*. Mammalian Species, no. 293, 7 pp.

Maynes, G. M. 1973. Reproduction in the parma wallaby, *Macropus parma* Waterhouse. Austral. J. Zool. 21:331–51.

———. 1974. Occurrence and field recognition of *Macropus parma*. Austral. Zool. 18:72–87.

———. 1977a. Distribution and aspects of the biology of the parma wallaby, *Macropus parma*, in New South Wales. Austral. Wildl. Res. 4:109–25.

———. 1977b. Breeding and age structure of the population of *Macropus parma* on Kawau Island, New Zealand. Austral. J. Ecol. 2:207–14.

———. 1982. A new species of rock wallaby, *Petrogale persephone* (Marsupialia: Macropodidae), from Proserpine, central Queensland. Austral. Mamm. 5:47–58.

Mayr, E. 1951. Taxonomic categories in fossil hominids. Cold Spring Harbor Symp. Quant. Biol. 15:109–18.

———. 1963. The taxonomic evaluation of fossil hominids. *In* Washburn, S. L., ed., Classification and human evolution, Aldine, Chicago, pp. 332–45.

Maza, B. G., N. R. French, and A. P. Aschwanden. 1973. Home range dynamics in a population of heteromyid rodents. J. Mamm. 54:405–25.

Mazak, V. 1981. *Panthera tigris*. Mammalian Species, no. 152, 8 pp.

Mazin, V. N. 1975. Data on reproduction of jerboas of the genus *Allactaga* in southeast Kazakhstan. Soviet J. Ecol. 6:334–39.

Mead, J. G. 1975. Preliminary report on the former net fisheries for *Tursiops truncatus* in the western North Atlantic. J. Fish. Res. Bd. Can. 32:1155–62.

———. 1981. First records of *Mesoplodon hectori* (Ziphiidae) from the Northern Hemisphere and a description of the adult male. J. Mamm. 62:430–32.

———. 1984. Survey of reproductive data for the beaked whales (Ziphiidae). *In* Perrin, Brownell, and DeMaster (1984), pp. 91–96.

Mead, J. G., and A. N. Baker. 1987. Notes on the rare beaked whale, *Mesoplodon hectori* (Gray). J. Roy. Soc. New Zealand 17:303–12.

Mead, J. G., and E. D. Mitchell. 1984. Atlantic gray whales. *In* Jones, Swartz, and Leatherwood (1984), pp. 33–53.

Mead, J. G., D. K. Odell, R. S. Wells, and M. D. Scott. 1980. Observations on a mass stranding of spinner dolphin, *Stenella longirostris*, from the west coast of Florida. Fishery Bull. 78:353–60.

Mead, J. G., and R. S. Payne. 1975. A specimen of the Tasman beaked whale, *Tasmacetus shepherdi*, from Argentina. J. Mamm. 56: 213–18.

Mead, J. G., W. A. Walker, and W. J. Houck. 1982. Biological observations on *Mesoplodon carlhubbsi* (Cetacea: Ziphiidae). Smithson. Contrib. Zool., no. 344, iii + 25 pp.

Mead, J. I. 1989. *Nemorhaedus goral*. Mammalian Species, no. 335, 5 pp.

Mead, R. A. 1968a. Reproduction in eastern forms of the spotted skunk (genus *Spilogale*). J. Zool. 156:119–36.

———. 1968b. Reproduction in western forms of the spotted skunk (genus *Spilogale*). J. Mamm. 49:373–90.

Meagher, M. M. 1973. The bison of Yellowstone National Park. U. S. Natl. Park Serv. Sci. Monogr., no. 1, xv + 161 pp.

———. 1986. *Bison bison*. Mammalian Species, no. 266, 8 pp.

Meaney, C. A., S. J. Bissell, and J. S. Slater. 1987. A nine-banded armadillo, *Dasypus novemcinctus* (Dasypodidae), in Colorado. Southwestern Nat. 32:507–8.

Meatte, D. S. 1986. Evidence for the historic occurrence and aboriginal utilization of caribou in southern Idaho. Paper Presented at 13th Annual Conference of Idaho Archae-

ological Society, Twin Falls, 27 September 1986.

Mech, L. D. 1961. The marten: symbol of wilderness. Anim. Kingdom 44(5):133–37.

———. 1970. The wolf: the ecology and behavior of an endangered species. Natural History Press, Garden City, N.Y., xx + 384 pp.

———. 1974a. A new profile for the wolf. Nat. Hist. 83(4):26–31.

———. 1974b. *Canis lupus*. Mammalian Species, no. 37, 6 pp.

———. 1977a. A recovery plan for the eastern timber wolf. Natl. Parks and Conserv. Mag. 50(1):17–21.

———. 1977b. Wolf-pack buffer zones as prey reservoirs. Science 198:320–21.

———. 1977c. Productivity, mortality, and population trends of wolves in northeastern Minnesota. J. Mamm. 58:559–74.

———. 1977d. Population trend and winter deer consumption in a Minnesota wolf pack. *In* Phillips and Jonkel (1977), pp. 55–83.

———. 1977e. Record movement of a Canadian lynx. J. Mamm. 58:676–77.

———. 1979. Why some deer are safe from wolves. Nat. Hist. 88(1):70–77.

———. 1980. Age, sex, reproduction, and spatial organization of lynxes colonizing northeastern Minnesota. J. Mamm. 61:261–67.

———. 1989. Wolf longevity in the wild. Endangered Species Tech. Bull. 14(5):8.

Mech, L. D., L. D. Frenzel, Jr., R. R. Ream, and J. W. Winship. 1971. Movements, behavior, and ecology of timber wolves in northeastern Minnesota. *In* Mech, L. D., and L. D. Frenzel, Jr., eds., Ecological studies of the timber wolf in northeastern Minnesota, U.S. Forest Serv. Res. Pap., no. NC-52, pp. 1–35.

Mech, L. D., and S. H. Fritts. 1987. Parovirus and heartworm found in Minnesota wolves. Endangered Species Tech. Bull. 12(5–6):5–6.

Mech, L. D., and P. D. Karns. 1977. Role of the wolf in a deer decline in the Superior National Forest. U. S. Forest Serv. Res. Pap., no. NC-148, 23 pp.

Mech, L. D., and M. Korb. 1978. An unusually long pursuit of a deer by a wolf. J. Mamm. 59:860–61.

Mech, L. D., and M. E. Nelson. 1982. Reoccurrence of caribou in Minnesota. Amer. Midl. Nat. 108:206–8.

Mech, L. D., and R. M. Nowak. 1981. Return of the gray wolf to Wisconsin. Amer. Midl. Nat. 105:408–9.

Mech, L. D., and L. L. Rogers. 1977. Status, distribution, and movements of martens in northeastern Minnesota. U. S. Forest Serv. Res. Pap., no. NC-143, 7 pp.

Medellín L., R. A. 1983. *Tonatia bidens* and *Mimon crenulatum* in Chiapas, Mexico. J. Mamm. 64:150.

———. 1989. *Chrotopterus auritus*. Mammalian Species, no. 343, 5 pp.

Medellín L., R. A., D. Navarro L., W. B. Davis, and V. J. Romero. 1983. Notes on the biology of *Micronycteris brachyotis* (Dobson) (Chiroptera), in southern Veracruz, Mexico. Brenesia 21:7–11.

Medjo, D. C., and L. D. Mech. 1976. Reproductive activity in nine- and ten-month-old wolves. J. Mamm. 57:406–8.

Medway, Lord. 1964. The marmoset rat, *Hapalomys longicaudatus* Blyth. Malayan Nat. J. 18:104–10.

———. 1970. The monkeys of Sundaland. *In* Napier and Napier (1970), pp. 513–53.

———. 1972. Reproductive cycles of the flat-headed bats *Tylonycteris pachypus* and *T. robustula* (Chiroptera: Vespertilioninae) in a humid equatorial environment. Zool. J. Linnean Soc. 51:33–61.

———. 1973. The taxonomic status of *Tylonycteris malayana* Chasen 1940 (Chiroptera). J. Nat. Hist. 7:125–31.

———. 1977. Mammals of Borneo. Monogr. Malaysian Branch Roy. Asiatic Soc., no. 7, xii + 172 pp.

———. 1978. The wild mammals of Malaya (peninsular Malaysia) and Singapore. Oxford Univ. Press, Kuala Lumpur, xxii + 128 pp.

Medway, Lord, and A. G. Marshall. 1972. Roosting associations of flat-headed bats, *Tylonycteris* species (Chiroptera: Vespertilionidae) in Malaysia. J. Zool. 168:463–82.

Medway, Lord, and Yong Hoi-Sen. 1976. Problems in the systematics of the rats (Muridae) of peninsular Malaysia. Malaysian J. Sci. 4:43–53.

Meester, J. 1972. A new golden mole from the Transvaal (Mammalia: Chrysochloridae). Ann. Transvaal Mus. 28:37–46.

———. 1973. Mammals collected during the Bernard Carp Expedition to the Western Province of Zambia. Puku 7:137–49.

———. 1976. South African red data book—small mammals. S. Afr. Natl. Sci. Programmes Rept., no. 11, vi + 59 pp.

Meester, J., and N. J. Dippenaar. 1978. A new species of *Myosorex* from Knysna, South Africa (Mammalia: Soricidae). Ann. Transvaal Mus. 31:29–42.

Meester, J., and A. F. Hallett. 1970. Notes on early postnatal development in certain southern African Muridae and Cricetidae. J. Mamm. 51:703–11.

Meester, J., and H. W. Setzer. 1977. The mammals of Africa: an identification manual. Smithson. Inst. Press, Washington, D.C.

Meester, J. A. J., I. L. Rautenbach, N. J. Dippenaar, and C. M. Baker. 1986. Classification of southern African mammals. Transvaal Mus. Monogr., no. 5, x + 359 pp.

Mehrer, C. F. 1976. Gestation period in the wolverine, *Gulo gulo*. J. Mamm. 57:570.

Meier, B., R. Albignac, A. Peyrieras, Y. Rumpler, and P. Wright. 1987. A new species of *Hapalemur* from south east Madagascar. Folia Primatol. 48:211–15.

Meier, B., and Y. Rumpler. 1987. Preliminary survey of *Hapalemur simus* and of a new species of *Hapalemur* in eastern Betsileo, Madagascar. Primate Conserv. 8:40–43.

Meier, M. N., and V. N. Yatsenko. 1980. On taxonomic status and distribution of the common (*Microtus arvalis* Pallas 1778) and Kirghiz (*M. kirgisorum* Ognev 1950) voles in south-east Kazakhstan. Zool. Zhur. 59:283–88.

Mein, P., and Y. Tupinier. 1977. Formule dentaire et position systématique du Minoptere (Mammalia, Chiroptera). Mammalia 41:207–11.

Meirte, D. 1983. New data on *Casinycteris argynnis* Thomas 1910 (Megachiroptera, Pteropodidae). Ann. Mus. Roy. Afr. Cent., Sec. Zool., 237:9–17.

Melchior, H. R. 1971. Characteristics of arctic ground squirrel alarm calls. Oecologia 7:184–90.

Melquist, W. E., and A. E. Dronkert. 1987. River otter. *In* Novak, Baker, et al. (1987), pp. 626–41.

Melquist, W. E., and M. G. Hornocker. 1983. Ecology of river otters in west central Idaho. Wildl. Monogr., no. 83, 60 pp.

Melton, D. A. 1985. The status of elephants in northern Botswana. Biol. Conserv. 31:317–33.

Mendel, F. C. 1981. Use of hands and feet of two-toed sloths (*Choloepus hoffmanni*) during climbing and terrestrial locomotion. J. Mamm. 62:413–21.

———. 1985. Use of hands and feet of three-toed sloths (*Bradypus variegatus*) during climbing and terrestrial locomotion. J. Mamm. 66:359–66.

Mendelssohn, H. 1965. Breeding the Syrian hyrax *Procavia capensis siriaca* Schreber, 1784. Internatl. Zoo Yearbook 5:116–25.

———. 1974. The development of the populations of gazelles in Israel and their behavioural adaptions. *In* Geist and Walther (1974), pp. 722–43.

Mendelssohn, H., Y. Yom-Tov, G. Ilany, and D. Meninger. 1987. On the occurrence of

Blanford's fox, *Vulpes cana* Blanford, 1877, in Israel and Sinai. Mammalia 51:459–67.

Méndez, E. 1984. Mexico and Central America. *In* Halls (1984), pp. 513–24.

Méndez, E., and B. Rodriguez. 1984. A southern sea lion *Otaria flavescens* (Shaw) found in Panama. Caribbean J. Sci. 20:105–8.

Menkhorst, P. W., and J. H. Seebeck. 1981. The distribution, habitat and status of *Pseudomys fumeus* Brazenor (Rodentia: Muridae). Austral. Wildl. Res. 8:87–96.

Menu, H. 1984. Revision du statut de *Pipistrellus subflavus* (F. Cuvier, 1832). Proposition d'un taxon generique nouveau: *Perimyotis* nov. gen. Mammalia 48:409–16.

Menzies, J. I. 1971. The lobe-lipped bat (*Chalinolobus nigrogriseus* Gould) in New Guinea. Rec. Papua and New Guinea Mus. 1:6–8.

———. 1973. A study of leaf-nosed bats (*Hipposideros caffer* and *Rhinolophus landeri*) in a cave in northern Nigeria. J. Mamm. 54:930–45.

———. 1977. Fossil and subfossil fruit bats from the mountains of New Guinea. Austral. J. Zool. 25:329–36.

Menzies, J. I., and E. Dennis. 1979. Handbook of New Guinea rodents. Wau Ecology Inst., Wau, Papua New Guinea, v + 68 pp.

———. 1984. Systematics and chromosomes of New Guinea *Rattus*: a correction. Austral. J. Zool. 32:603.

Menzies, J. I., and J. C. Pernetta. 1986. A taxonomic revision of cuscuses allied to *Phalanger orientalis* (Marsupialia: Phalangeridae). J. Zool., ser. B, 1:551–618.

Mercer, W. E., and F. Manuel. 1974. Some aspects of moose management in Newfoundland. Naturaliste Canadien 101:657–71.

Merchant, J. C. 1976. Breeding biology of the agile wallaby, *Macropus agilis* (Gould) (Marsupialia: Macropodidae), in captivity. Austral. Wildl. Res. 3:93–103.

Meritt, D. A., Jr. 1973. Some observations on the maned wolf, *Chrysocyon brachyurus*, in Paraguay. Zoologica 58:53.

———. 1975. The lesser anteater *Tamandua tetradactyla* in captivity. Internatl. Zoo Yearbook 15:41–44.

———. 1976. The La Plata three-banded armadillo *Tolypeutes matacus* in captivity. Internatl. Zoo Yearbook 16:153–55.

———. 1977. Second-generation owl monkey birth. Amer. Assoc. Zool. Parks Aquar. Newsl. 18(3):12.

———. 1980. Captive reproduction and husbandry of the douroucouli *Aotus trivirgatus* and the titi monkey *Callicebus* spp. Internatl. Zoo Yearbook 20:52–59.

———. 1985a. The two-toed Hoffman's sloth, *Choloepus hoffmanni* Peters. *In* Montgomery (1985a), pp. 333–41.

———. 1985b. Naked-tailed armadillos *Cabassous* sp. *In* Montgomery (1985a), pp. 389–91.

Meritt, D. A., Jr., and G. F. Meritt. 1976. Sex ratios of Hoffmann's sloth, *Choloepus hoffmanni* Peters and three-toed sloth, *Bradypus infuscatus* Wagler in Panama. Amer. Midl. Nat. 96:472–73.

Merkt, J. R. 1987. Reproductive seasonality and grouping patterns of the North Andean deer or taruca *(Hippocamelus antisensis)* in southern Peru. *In* Wemmer (1987), pp. 388–401.

Merrett, P. K. 1983. Edentates. Zoological Trust of Guernsey, iii + 88 pp.

Merrick, R. L., T. R. Loughlin, and D. G. Calkins. 1987. Decline in abundance of the northern sea lion, *Eumetopias jubatus*, in Alaska, 1956–86. Fishery Bull. 85:351–65.

Merritt, J. F. 1978. *Peromyscus californicus.* Mammalian Species, no. 85, 6 pp.

Merritt, J. F., and J. M. Merritt. 1978. Population ecology and energy of *Clethrionomys gapperi* in a Colorado subalpine forest. J. Mamm. 59:576–98.

Merz, G. 1986. Movement patterns and group size of the African forest elephant *Loxodonta africana cyclotis* in the Tai National Park, Ivory Coast. Afr. J. Ecol. 24:133–36.

Meserve, P. L. 1971. Population ecology of the prairie vole, *Microtus ochrogaster*, in the western mixed prairie of Nebraska. Amer. Midl. Nat. 86:417–33.

Meserve, P. L., B. K. Lang, and B. D. Patterson. 1988. Trophic relationships of small mammals in a Chilean temperate rainforest. J. Mamm. 69:721–30.

Meserve, P. L., R. Murúa, O. Lopetegui N., and J. R. Rau. 1982. Observations on the small mammal fauna of a primary temperate rain forest in southern Chile. J. Mamm. 63:315–17.

Meserve, P. L., J. Rodriguez-M., and R. E. Martin. 1978. Demography and community ecology of the caviomorph rodent, *Octodon degus*, in central and northern Chile. Amer. Soc. Mamm., Abstr. Tech. Pap., 58th Ann. Mtg., p. 69.

Messick, J. P. 1987. North American badger. *In* Novak, Baker, et al. (1987), pp. 586–97.

Messick, J. P., and M. G. Hornocker. 1981. Ecology of the badger in southwestern Idaho. Wildl. Monogr., no. 76, 53 pp.

Meyer-Holzapfel, M. 1968. Breeding the European wild cat *Felis s. silvestris* at Berne Zoo. Internatl. Zoo Yearbook 8:31–38.

Meylan, A. 1977. Fossorial forms of the water vole, *Arvicola terrestris* (L.), in Europe. EPPO Bull. 7:209–21.

Michael, R. P., and J. H. Crook, eds. 1973. Comparative ecology and behaviour of primates. Academic Press, London, 847 pp.

Michalak, I. 1983. Reproduction, maternal and social behaviour of the European water shrew under laboratory conditions. Acta Theriol. 28:3–24.

Michener, G. R., and J. W. Koeppl. 1985. *Spermophilus richardsonii.* Mammalian Species, no. 243, 8 pp.

Michener, G. R., and J. O. Murie. 1983. Black-tailed prairie dog coteries: are they cooperatively breeding units? Amer. Nat. 121:266–74.

Mignucci Giannoni, A. A. 1986. Caribbean monk seals in Puerto Rico. J. Colorado-Wyoming Acad. Sci. 18:54.

Mihok, S. 1976. Behaviour of subarctic red-backed voles *(Clethrionomys gapperi athabascae)*. Can. J. Zool. 54:1932–45.

Mikkola, H. 1974. The raccoon dog spreads to western Europe. Wildlife 16:344–45.

Mikuriya, M. 1976. Notes on the Japanese otter, *Lutra lutra whiteleyi* (Gray). J. Mamm. Soc. Japan 6:214–17.

Millar, J. S. 1972. Timing of breeding of pikas in southwestern Alberta. Can. J. Zool. 50:665–69.

———. 1973. Evolution of litter-size in the pika, *Ochotona princeps* (Richardson). Evolution 27:134–43.

———. 1974. Success of reproduction in pikas, *Ochotona princeps* (Richardson). J. Mamm. 55:527–42.

Millar, J. S., and F. C. Zwickel. 1972a. Determination of age, age structure, and mortality of the pika, *Ochotona princeps* (Richardson). Can. J. Zool. 50:229–32.

———. 1972b. Characteristics and ecological significance of hay piles of pikas. Mammalia 36:657–67.

Millar, R. P. 1971. Reproduction in the rock hyrax *(Procavia capensis)*. Zool. Afr. 6:243–61.

Miller, B. J., S. H. Anderson, M. W. DonCarlos, and E. T. Thorne. 1988. Biology of the endangered black-footed ferret and the role of captive propagation in its conservation. Can. J. Zool. 66:765–73.

Miller, D. H., and L. L. Getz. 1977. Comparisons of population dynamics of *Peromyscus* and *Clethrionomys* in New England. J. Mamm. 58:1–16.

Miller, E. H. 1975a. Body and organ measurements of fur seals, *Arctocephalus forsteri* (Lesson), from New Zealand. J. Mamm. 56:511–13.

———. 1975b. Walrus ethology. I. The social role of tusks and applications of multidimensional scaling. Can. J. Zool. 53:590–613.

Miller, E. H., and D. J. Boness. 1979. Remarks on display functions of the snout of the grey seal, *Halichoerus grypus* (Fab.) with comparative notes. Can. J. Zool. 57:140–48.

Miller, F. L. 1987. Management of barren-ground caribou *(Rangifer tarandus groenlandicus)* in Canada. *In* Wemmer (1987), pp. 523–34.

Miller, F. L., R. H. Russell, and A. Gunn. 1977. Interisland movements of Peary caribou *(Rangifer tarandus pearyi)* on western Queen Elizabeth Islands, arctic Canada. Can. J. Zool. 55:1029–37.

Miller, G. S., Jr. 1907. The families and genera of bats. Bull. U.S. Natl. Mus. 57:i–xvii + 1–282.

Miller, G. S., and N. Hollister. 1922. A new phalanger from Celebes. Proc. Biol. Soc. Washington 35:115–16.

Miller, R. 1986. A review of feral horse research pertinent to the reintroduction of Przewalski's horses. *In* The Przewalski horse and restoration to its natural habitat in Mongolia, FAO, Rome, pp. 135–41.

Miller, S. D., and D. D. Everett, eds. 1986. Cats of the world: biology, conservation, and management. Natl. Wildl. Fed., Washington, D.C., xi + 501 pp.

Miller, S. D., J. Rottmann, K. J. Raedeke, and R. D. Taber. 1983. Endangered mammals of Chile: status and conservation. Biol. Conserv. 25:335–52.

Mills, M. G. L. 1978. Foraging behaviour of the brown hyaena *(Hyaena brunnea* Thunberg, 1820) in the southern Kalahari. Z. Tierpsychol. 48:113–41.

———. 1982. *Hyaena brunnea.* Mammalian Species, no. 194, 5 pp.

———. 1985. Related spotted hyaenas forage together but do not cooperate in rearing young. Nature 316:61–62.

Mills, M. G. L., and M. E. J. Mills. 1978. The diet of the brown hyaena *Hyaena brunnea* in the southern Kalahari. Koedoe 21:125–49.

Mills, R. S., G. W. Barrett, and M. P. Farrell. 1975. Population dynamics of the big brown bat *(Eptesicus fuscus)* in southwestern Ohio. J. Mamm. 56:591–604.

Mills, S. 1982. The burros in Death Valley. Oryx 16:411–14.

Milton, K. 1984. Habitat, diet, and activity patterns of free-ranging woolly spider monkeys *(Brachyteles arachnoides* E. Geoffroy 1806). Internatl. J. Primatol. 5:491–514.

———. 1985. Mating patterns of woolly spider monkeys, *Brachyteles arachnoides:* implications for female choice. Behav. Ecol. Sociobiol. 17:53–59.

———. 1986. Ecological background and conservation priorities for woolly spider monkeys *(Brachyteles arachnoides).* *In* Benirschke (1986), pp. 241–50.

Mineau, P., and D. Madison. 1977. Radio-tracking of *Peromyscus leucopus*. Can. J. Zool. 55:465–68.

Mishra, A. C., and V. Dhanda. 1975. Review of the genus *Millardia* (Rodentia: Muridae), with description of a new species. J. Mamm. 56:76–80.

Mishra, H. R., and E. Dinerstein. 1987. Relocating rhinos in Nepal. Smithsonian 18(6):66–73.

Mishra, H. R., C. Wemmer, and J. L. D. Smith. 1987. Tigers in Nepal: management conflicts with human interests. *In* Tilson and Seal (1987), pp. 449–63.

Misonne, X. 1969. African and Indo-Australian Muridae: evolutionary trends. Ann. Mus. Roy. Afr. Cent. (Tervuren, Belgium), ser. IN-8°, 172:1–219.

————. 1974. Une nouvelle gerbille de Libye (Mammalia, Rodentia). Bull. Inst. Roy. Sci. Nat. Belg. 50(6):1–6.

————. 1979. Muridae collected in Irian Jaya, Indonesia. Bull. Inst. Roy. Sci. Nat. Belg. 51(8):1–16.

Misonne, X., and J. Verschuren. 1976. Les rongeurs du Nimba Liberien. Acta Zool. Pathol. 66:199–220.

Mitchell, E. D. 1975a. Report of the meeting on smaller Cetaceans, Montreal, April 1–11, 1974. J. Fish. Res. Bd. Can. 32:889–983.

————. 1975b. Porpoise, dolphin and small whale fisheries of the world. Internatl. Union Conserv. Nat. Monogr., no. 3, 129 pp.

————. 1978. Finner whales. *In* Haley (1978), pp. 36–45.

Mitchell, E. D., and V. M. Kozicki. 1975. Supplementary information on minke whale *(Balaenoptera acutorostrata)* from Newfoundland fishery. J. Fish. Res. Bd. Can. 32:985–94.

Mitchell, E. D., and J. G. Mead. 1977. History of the gray whale in the Atlantic Ocean (abstr.). Proc. 2nd Conf. Biol. Mar. Mamm., San Diego.

Mitchell, E. D., and R. R. Reeves. 1982. Factors affecting abundance of bowhead whales *Balaena mysticetus* in the eastern Arctic of North America, 1915–1980. Biol. Conserv. 22:59–78.

————. 1983. Catch history, abundance, and present status of northwest Atlantic humpback whales. *In* Tillman and Donovan (1983), pp. 153–212.

Mitchell, J. G. 1987. The photographer who got too close. Audubon 89(2):28–34.

Mitchell, R. M. 1975. A checklist of Nepalese mammals (excluding bats). Saugetierk. Mitt. 23:152–57.

————. 1977. Accounts of Nepalese mammals and analysis of the host-ectoparasite data

by computer techniques. Ph.D. diss., Iowa State Univ., 557 pp.

————. 1979. The sciurid rodents (Rodentia: Sciuridae) of Nepal. J. Asian Ecol. 1:21–28.

————. 1980. New records of bats (Chiroptera) from Nepal. Mammalia 44:339–42.

Mitchell, R. M., and F. Punzo. 1975. *Ochotona lama* sp. n. (Lagomorpha: Ochotonidae), a new pika from the Tibetan highlands of Nepal. Mammalia 39:419–22.

Mittermeier, R. A. 1973. Group activity and population dynamics of the howler monkey on Barro Colorado Island. Primates 14:1–19.

————. 1987. Rescuing Brazil's muriqui: monkey in peril. Natl. Geogr. 171:386–95.

Mittermeier, R. A., and A. F. Coimbra-Filho. 1977. Primate conservation in Brazilian Amazonia. *In* Rainier III and Bourne (1977), pp. 117–66.

Mittermeier, R. A., W. R. Konstant, H. Ginsberg, M. G. M. Van Roosmalen, and E. C. Da Silva, Jr. 1983. Further evidence of insect consumption in the bearded saki monkey, *Chiropotes satanas chiropotes*. Primates 24:602–5.

Mittermeier, R. A., H. de Macedo-Ruiz, B. A. Luscombe, and J. Cassidy. 1977. Rediscovery and conservation of the Peruvian yellow-tailed woolly monkey *(Lagothrix flavicauda)*. *In* Rainier III and Bourne (1977), pp. 95–115.

Mittermeier, R. A., J. F. Oates, A. E. Eudey, and J. Thornback. 1986. Primate conservation. *In* Mitchell, G., and J. Erwin, eds., Comparative primate biology, vol. 2, Behavior, conservation, and ecology, John Wiley & Sons, New York, pp. 3–72.

Mittermeier, R. A., C. V. Padua, C. Valle, and A. F. Coimbra-Filho. 1985. Major program underway to save the black lion tamarin in Sao Paulo, Brazil. Primate Conserv. 6:19–21.

Miura, S. 1984a. Dominance hierarchy and space use pattern in male captive muntjacs, *Muntiacus reevesi*. J. Ethol. 2:69–75.

————. 1984b. Social behavior and territoriality in male sika deer *(Cervus nippon* Temminck 1838) during the rut. Z. Tierpsychol. 64:33–73.

Miyashita, M., and K. Nagase. 1981. Breeding of the Mongolian gazelle *Procapra gutturosa* at Osaka Zoo. Internatl. Zoo Yearbook 21:158–62.

Miyazaki, N. 1984. Further analyses of reproduction in the striped dolphin, *Stenella coeruleoalba*, off the Pacific coast of Japan. *In* Perrin, Brownell, and DeMaster (1984), pp. 343–54.

Miyazaki, N., L. L. Jones, and R. Beach. 1984. Some observations on the schools of *dalli-* and *truei-*type Dall's porpoises in the northwestern Pacific. Sci. Rept. Whales Res. Inst. 35:93–105.

Mizroch, S. A., D. W. Rice, and J. M. Breiwick. 1984. The blue whale, *Balaenoptera musculus*. Mar. Fish. Rev. 46(4):15–19.

Mizue, K., M. Nishiwaki, and A. Takemura. 1971. The underwater sounds of Ganges River dolphins *(Platanista gangetica)*. Sci. Rept. Whales Res. Inst. 23:123–28.

Mloszewski, M. J. 1983. The behavior and ecology of the African buffalo. Cambridge Univ. Press, London, ix + 256 pp.

Mock, O. B., and C. H. Conaway. 1975. Reproduction of the least shrew *(Cryptotis parva)* in captivity. *In* Antikatzides, T., S. Erichsen, and A. Spiegel, eds., The laboratory animal in the study of reproduction, Gustav Fischer Verlag, Stuttgart, pp. 59–74.

Modha, K. L., and S. K. Eltringham. 1976. Population ecology of the Uganda kob *(Adenota kob thomasi* (Neumann)) in relation to the territorial system in the Rwenzori National Park, Uganda. J. Appl. Ecol. 13:453–73.

Modi, W. S., and M. R. Lee. 1984. Systematic implications of chromosomal banding analyses of populations of *Peromyscus truei* (Rodentia: Muridae). Proc. Biol. Soc. Washington 97:716–23.

Moehlman, P. D. 1978. Jackals of the Serengeti. Wildl. News, African Wildl. Leadership Foundation, 13(3):2–6.

————. 1983. Socioecology of silverbacked and golden jackals *(Canis mesomelas* and *Canis aureus)*. Amer. Soc. Mamm. Spec. Publ. 7:423–53.

————. 1987. Social organization in jackals. Amer. Sci. 75:366–75.

Mohler, L. L. 1974. Threatened wildlife of Idaho. Idaho Wildl. Rev. 26(5):3–5.

Mohnot, S. M. 1978. The conservation of non-human Primates in India. *In* Chivers and Lane-Petter (1978), pp. 47–53.

Mohr, C. E. 1972. The status of threatened species of cave-dwelling bats. Bull. Natl. Speleol. Soc. 34:33–47.

Molinari, J., and P. J. Soriano. 1987. *Sturnira bidens*. Mammalian Species, no. 276, 4 pp.

Molnar, R. E., L. S. Hall, and J. H. Mahoney. 1984. New fossil localities for *Macroderma* Miller, 1906 (Chiroptera: Megadermatidae) in New South Wales and its past and present distribution in Australia. Austral. Mamm. 7:63–73.

Moncrief, N. D., J. R. Choate, and H. H. Genoways. 1982. Morphometric and geographic relationships of short-tailed shrews (genus *Blarina*) in Kansas, Iowa, and Missouri. Ann. Carnegie Mus. 51:157–80.

Mondolfi, E., and R. Hoogesteijn. 1986. Notes on the biology and status of the jaguar in Venezuela. *In* Miller and Everett (1986), pp. 85–123.

Mones, A. 1973. Estudios sobre la familia Hydrochoeridae (Rodentia). I.—Introduccion e historia taxonómica. Rev. Brasil. Biol. 33: 277–83.

———. 1975. Estudios sobre la familia Hydrochoeridae (Rodentia). VI. Catálogo anotado de los ejemplares tipo. Comun. Palentol. Mus. Hist. Nat. Montevideo 1:99–128.

Mones, A., and J. Ojasti. 1986. *Hydrochoerus hydrochaeris*. Mammalian Species, no. 264, 7 pp.

Monfort, A., and N. Monfort. 1974. Notes sur l'écologie et le comportement des oribis (*Ourebia ourebi*, Zimmerman, 1783). Terre Vie 28:169–208.

Monfort-Braham, N. 1975. Variations dans la structure sociale du topi, *Damaliscus korrigum* Ogilby, au Parc National de l'Akagera, Rwanda. Z. Tierpsychol. 39:332–64.

Montgomery, G. G., ed. 1978. The ecology of arboreal folivores. Smithson. Inst. Press, Washington, D.C., 573 pp.

———. 1985a. The evolution and ecology of armadillos, sloths, and vermilinguas. Smithson. Inst. Press, Washington, D.C., 451 pp.

———. 1985b. Impact of vermilinguas (*Cyclopes, Tamandua*: Xenarthra = Edentata) on arboreal ant populations. *In* Montgomery (1985a), pp. 351–63.

———. 1985c. Movements, foraging and food habits of the four extant species of Neotropical vermilinguas (Mammalia; Myrmecophagidae). *In* Montgomery (1985a), pp. 365–77.

Montgomery, G. G., R. C. Best, and M. Yamakoshi. 1981. A radio-tracking study of the Amazonian manatee *Trichechus inunguis*. Biotrópica 13:81–85.

Montgomery, G. G., N. B. Gale, and W. P. Murdoch, Jr. 1982. Have manatee entered the eastern Pacific Ocean? Mammalia 46:257–58.

Montgomery, G. G., and Y. D. Lubin. 1977. Prey influences on movements of neotropical anteaters. *In* Phillips and Jonkel (1977), pp. 103–31.

———. 1978. Movements of *Coendou prehensilis* in the Venezuelan llanos. J. Mamm. 59:887–88.

Montgomery, G. G., and M. E. Sunquist. 1978. Habitat selection and use by two-toed and three-toed sloths. *In* Montgomery (1978), pp. 329–59.

Montgomery, W. I. 1976. On the relationship between yellow-necked mouse (*Apodemus flavicollis*) and woodmouse (*A. sylvaticus*) in a Cotswold valley. J. Zool. 179:229–33.

Moore, G. C., and J. S. Millar. 1984. A comparative study of colonizing and longer established eastern coyote populations. J. Wildl. Mgmt. 48:691–99.

Moore, J. C. 1958a. A new species and a re-

definition of the squirrel genus *Prosciurillus* of Celebes. Amer. Mus. Novit., no. 1890, 5 pp.

———. 1958b. New genera of East Indian squirrels. Amer. Mus. Novit., no. 1914, 5 pp.

———. 1959. Relationships among living squirrels of the Sciurinae. Bull. Amer. Mus. Nat. Hist. 118:153–206.

———. 1968. Relationships among the living genera of beaked whales with classifications, diagnoses and keys. Fieldiana Zool. 53:209–98.

———. 1972. More skull characters of the beaked whale *Indopacetus pacificus* and comparative measurements of austral relatives. Fieldiana Zool. 62:1–19.

Moore, J. C., and G. H. H. Tate. 1965. A study of the diurnal squirrels, Sciurinae, of the Indian and Indochinese subregions. Fieldiana Zool. 48:1–351.

Moore, N. W. 1975. The diurnal flight of the Azorean bat *(Nyctalus azoreum)* and the avifauna of the Azores. J. Zool. 177:483–86.

Moore, R. E., and N. S. Martin. 1980. A recent record of the swift fox *(Vulpes velox)* in Montana. J. Mamm. 61:161.

Moors, P. J. 1975. The urogenital system and notes on the reproductive biology of the female rufous rat-kangaroo, *Aepyprymnus rufescens* (Gray) (Macropodidae). Austral. J. Zool. 23:355–61.

Morcombe, M. K. 1967. The rediscovery after 83 years of the dibbler *Antechinus apicalis* (Marsupialia, Dasyuridae). W. Austral. Nat. 10:103—11.

Morejohn, G. V. 1979. The natural history of Dall's porpoise in the North Pacific Ocean. *In* Winn and Olla (1979), pp. 45–83.

Moreno, S. 1988. Reproduction of garden dormouse *Eliomys quercinus lusitanicus* in southwest Spain. Mammalia 52:401–7.

Moreno, S., and M. Delibes. 1982. Notes on the garden dormouse (*Eliomys*; Rodentia, Gliridae) of northern Morocco. Saugetierk. Mitt. 30:212–15.

Morgan, G. S., C. E. Ray, and O. Arredondo. 1980. A giant extinct insectivore from Cuba (Mammalia: Insectivora: Solenodontidae). Proc. Biol. Soc. Washington 93:597–608.

Morgan, G. S., and C. A. Woods. 1986. Extinction and the zoogeography of West Indian mammals. Biol. J. Linnean Soc. 28:167–203.

Morlok, W. F. 1978. Nagetiere aus der Turkei. Senckenberg. Biol. 59:155–62.

Morrell, S. 1972. Life history of the San Joaquin kit fox. California Fish and Game 53:162–74.

Morris, N. E., and J. Hanks. 1974. Reproduction in the bushbuck *Tragelaphus scriptus ornatus*. Arnoldia 1(7):1–8.

Morris, P. A., and J. R. Malcolm. 1977. The Simien fox in the Bale Mountains. Oryx 14:151–60.

Morrison, D. W. 1978. Influence of habitat on the foraging distance of the fruit bat, *Artibeus jamaicensis*. J. Mamm. 59:622–24.

———. 1979. Apparent male defense of tree hollows in the fruit bat, *Artibeus jamaicensis*. J. Mamm. 60:11–15.

———. 1980. Foraging and day-roosting dynamics of canopy fruit bats in Panama. J. Mamm. 61:20–29.

Morton, S. R. 1974. First record of Forrest's mouse *Leggadina forresti* (Thomas, 1906) in N.S.W. Victorian Nat. 91:91–94.

———. 1978a. An ecological study of *Sminthopsis crassicaudata* (Marsupialia: Dasyuridae). I. Distribution, study areas and methods. Austral. Wildl. Res. 5:151–62.

———. 1978b. An ecological study of *Sminthopsis crassicaudata* (Marsupialia: Dasyuridae). II. Behaviour and social organization. Austral. Wildl. Res. 5:163–82.

———. 1978c. An ecological study of *Sminthopsis crassicaudata* (Marsupialia: Dasyuridae). III. Reproduction and life history. Austral. Wildl. Res. 5:183–211.

———. 1978d. Torpor and nest-sharing in free-living *Sminthopsis crassicaudata* (Marsupialia) and *Mus musculus* (Rodentia). J. Mamm. 59:569–75.

Morton, S. R., and T. C. Burton. 1973. Observations on the behaviour of the macropodid marsupial *Thylogale billardieri* (Desmarest) in captivity. Austral. Zool. 18:1–14.

Morton, S. R., and A. K. Lee. 1978. Thermoregulation and metabolism in *Planigale maculata* (Marsupialia: Dasyuridae). J. Thermal Biol. 3:117–20.

Morton, S. R., J. W. Wainer, and T. P. Thwaites. 1980. Distributions and habitats of *Sminthopsis leucopus* and *S. murina* (Marsupialia: Dasyuridae) in southeastern Australia. Austral. Mamm. 3:19–30.

Moss, C. J. 1983. Oestrous behaviour and female choice in the African elephant. Behaviour 86:167–96.

Moyer, C. A., G. H. Adler, and R. H. Tamarin. 1988. Systematics of New England *Microtus*, with emphasis on *Microtus breweri*. J. Mamm. 69:782–94.

Moynihan, M. 1976. Notes on the ecology and behavior of the pygmy marmoset (*Cebuella pygmaea*) in Amazonian Colombia. *In* Thorington and Heltne (1976), pp. 79–84.

Mu Wenwei and Yang Dehua. 1982. A primary observation on the group figures, moving lines and food of *Rhinopithecus bieti* at the east side of Baima-Snow Mountain. Acta Theriol. Sinica 2:125–31.

Muckenhirn, N. A., and J. F. Eisenberg. 1973. Home ranges and predation of the Ceylon leopard. *In* Eaton, R. L., ed., The world's cats, vol. I, World Wildlife Safari, Winston, Oregon, pp. 142–75.

Mudar, K. M., and M. S. Allen. 1986. A list of bats from northeastern Luzon, Philippines. Mammalia 50:219–25.

Mühlenberg, M., and H. H. Roth. 1985. Comparative investigations into the ecology of the kob antelope (*Kobus kob kob* (Erxleben 1777)) in the Comoe National Park, Ivory Coast. S. Afr. J. Wildl. Res. 15:25–31.

Muir, B. G. 1985. The dibbler *(Parantechinus apicalis)* found in Fitzgerald River National Park, Western Australia. W. Austral. Nat. 16:48–51.

Mukherjee, R. P. 1982. Phayre's leaf monkey (*Presbytis phayrei* Blyth, 1847) of Tripura. J. Bombay Nat. Hist. Soc. 79:47–56.

Mullen, D. A. 1977. The striped dolphin, *Stenella coeruleoalba*, in the Gulf of California. Bull. S. California Acad. Sci. 76:131–32.

Muller, J. P. 1977. Populationsökologie von *Arvicanthis abyssinicus* in der Grassteppe des Semien Mountains National Park (Äthiopien). Z. Saugetierk. 42:145–72.

Mullican, T. R., and B. L. Keller. 1986. Ecology of the sagebrush vole *(Lemmiscus curtatus)* in southeastern Idaho. Can. J. Zool. 64:1218–23.

———. 1987. Burrows of the sagebrush vole *(Lemmiscus curtatus)* in southeastern Idaho. Great Basin Nat. 47:276–79.

Mumford, R. E. 1973. Natural history of the red bat *(Lasiurus borealis)* in Indiana. Period. Biol. 75:155–58.

Mumford, R. E., and D. M. Knudson. 1978. Ecology of bats at Vicosa, Brazil. Proc. 4th Internatl. Bat Res. Conf., Nairobi, pp. 287–95.

Munthe, K., and J. H. Hutchison. 1978. A wolf-human encounter on Ellesmere Island, Canada. J. Mamm. 59:876–78.

Murie, A. 1981. The grizzlies of Mount McKinley. U.S. Natl. Park Serv. Sci. Monogr., no. 14, xvi + 251 pp.

Murie, J. O. 1973. Population characteristics and phenology of a Franklin ground squirrel *(Spermophilus franklinii)* colony in central Alberta. Amer. Midl. Nat. 90:334–40.

Murie, J. O., and M. A. Harris. 1978. Territoriality and dominance in male Columbian ground squirrels *(Spermophilus columbianus).* Can. J. Zool. 56:2402–12.

Murie, O. J. 1959. Fauna of the Aleutian Islands and Alaska Peninsula. N. Amer. Fauna, no. 61, xiv + 406 pp.

Murphy, E. T. 1976. Breeding the clouded leopard *Neofelis nebulosa* at Dublin Zoo. Internatl. Zoo Yearbook 16:122–24.

Murphy, M. R. 1985. History of the capture and domestication of the Syrian golden hamster (*Mesocricetus auratus* Waterhouse). *In* Siegel, H. I., ed., The hamster, Plenum Press, New York, pp. 3–20.

Murray, M. D. 1981. The breeding of the southern elephant seal, *Mirounga leonina*, on the Antarctic continent. Polar Record 20:370–71.

Murray, M. G. 1981. Structure of association in impala, *Aepyceros melampus*. Behav. Ecol. Sociobiol. 9:23–33.

———. 1982. Home range, dispersal and the clan system of impala. Afr. J. Ecol. 20:253–69.

Muskin, A. 1984. Field notes and geographic distribution of *Callithrix aurita* in eastern Brazil. Amer. J. Primatol. 7:377–80.

Mussehl, T. W., and F. W. Howell. 1971. Game management in Montana. Montana Fish and Game Dept., Helena, xi + 238 pp.

Musser, G. G. 1964. Notes on the geographic distribution, habitat, and taxonomy of some Mexican mammals. Occas. Pap. Mus. Zool. Univ. Michigan, no. 636, 22 pp.

———. 1969a. Results of the Archbold Expeditions. No. 91. A new genus and species of murid rodent from Celebes, with a discussion of its relationships. Amer. Mus. Novit., no. 2384, 41 pp.

———. 1969b. Results of the Archbold Expeditions. No. 92. Taxonomic notes on *Rattus dollmani* and *Rattus hellwaldi* (Rodentia, Muridae) of Celebes. Amer. Mus. Novit., no. 2386, 24 pp.

———. 1970a. Species-limits of *Rattus brahma*, a murid rodent of northeastern India and northern Burma. Amer. Mus. Novit., no. 2406, 27 pp.

———. 1970b. *Rattus masaretes:* a synonym of *Rattus rattus moluccarius*. J. Mamm. 51:606–9.

———. 1970c. Results of the Archbold Expeditions. No. 93. Reidentification and reallocation of *Mus callitrichus* and allocations of *Rattus maculipilis*, *R. m. jentinki*, and *R. microbullatus* (Rodentia, Muridae). Amer. Mus. Novit., no. 2440, 35 pp.

———. 1971a. The taxonomic association of *Mus faberi* Jentink with *Rattus xanthurus* (Gray), a species known only from Celebes (Rodentia: Muridae). Zool. Meded. 45:107–18.

———. 1971b. The identities and allocations of *Taeromys paraxanthus* and *T. tatei*, two taxa based on composite holotypes (Rodentia, Muridae). Zool. Meded. 45:127–38.

———. 1971c. Results of the Archbold Expeditions. No. 94. Taxonomic status of *Rattus tatei* and *Rattus frosti*, two taxa of murid rodents known from middle Celebes. Amer. Mus. Novit., no. 2454, 19 pp.

———. 1971d. The taxonomic status of *Rattus tondanus* Sody and notes on the holotypes of *R. beccarii* (Jentink) and *R. thysanurus* Sody (Rodentia: Muridae). Zool. Meded. 45:147–57.

———. 1971e. The taxonomic status of *Rattus dammermani* Thomas and *Rattus toxi* (Rodentia, Muridae) of Celebes. Beaufortia 18:205–16.

———. 1972a. The species of *Hapalomys* (Rodentia, Muridae). Amer. Mus. Novit., no. 2503, 27 pp.

———. 1972b. Identities of taxa associated with *Rattus rattus* (Rodentia, Muridae) of Sumba Island, Indonesia. J. Mamm. 53:861–65.

———. 1973a. Zoogeographical significance of the ricefield rat, *Rattus argentiventer*, on Celebes and New Guinea and the identity of *Rattus pesticulus*. Amer. Mus. Novit., no. 2511, 30 pp.

———. 1973b. Notes on additional specimens of *Rattus brahma*. J. Mamm. 54:267–70.

———. 1973c. Species-limits of *Rattus cremoriventer* and *Rattus langibanis*, murid rodents of Southeast Asia and the Greater Sunda Islands. Amer. Mus. Novit., no. 2525, 65 pp.

———. 1977a. *Epimys benguetensis*, a composite, and one zoogeographical view of rat and mouse faunas in the Philippines and Celebes. Amer. Mus. Novit., no. 2624, 15 pp.

———. 1977b. Results of the Archbold Expeditions. No. 100. Notes on the Philippine rat, *Limnomys*, and the identity of *Limnomys picinus*, a composite. Amer. Mus. Novit., no. 2636, 14 pp.

———. 1979. Results of the Archbold Expeditions. No. 102. The species of *Chiropodomys*, arboreal mice of Indochina and the Malay Archipelago. Bull. Amer. Mus. Nat. Hist. 162:377–445.

———. 1981a. Results of the Archbold Expeditions. No. 105. Notes on systematics of Indo-Malayan murid rodents, and descriptions of new genera and species from Ceylon, Sulawesi, and the Philippines. Bull. Amer. Mus. Nat. Hist. 168:225–334.

———. 1981b. The giant rat of Flores and its relatives east of Borneo and Bali. Bull. Amer. Mus. Nat. Hist. 169:67–176.

———. 1981c. A new genus of arboreal rat from West Java, Indonesia. Zool. Verh., no. 189, 35 pp.

———. 1982a. Results of the Archbold Expeditions. No. 108. The definition of *Apomys*, a native rat of the Philippine Islands. Amer. Mus. Novit., no. 2746, 43 pp.

———. 1982b. Results of the Archbold Expeditions. No. 110. *Crunomys* and the small-bodied shrew rats native to the Philippine Islands and Sulawesi (Celebes). Bull. Amer. Mus. Nat. Hist. 174:1–95.

———. 1982c. Results of the Archbold Expeditions. No. 107. A new genus of arboreal rat from Luzon Island in the Philippines. Amer. Mus. Novit., no. 2730, 23 pp.

———. 1982d. The Trinil rats. Modern Quaternary Res. Southeast Asia 7:65–85.

———. 1984. Identities of subfossil rats from caves in southwestern Sulawesi. Modern Quaternary Res. Southeast Asia 8:61–94.

———. 1986. Sundaic Rattus: definitions of Rattus baluensis and Rattus korinchi. Amer. Mus. Novit., no. 2862, 24 pp.

Musser, G. G., and B. Boeadi. 1980. A new genus of murid rodent from the Komodo Islands in Nusatenggara, Indonesia. J. Mamm. 61:395–413.

Musser, G. G., and D. Califia. 1982. Results of the Archbold Expeditions. No. 106. Identities of rats from Palau Maratua and other islands off East Borneo. Amer. Mus. Novit., no. 2726, 30 pp.

Musser, G. G., and S. Chiu. 1979. Notes on taxonomy of Rattus andersoni and R. excelsior, murids endemic to western China. J. Mamm. 60:581–92.

Musser, G. G., and M. Dagosto. 1987. The identity of Tarsius pumilus, a pygmy species endemic to the montane mossy forests of central Sulawesi. Amer. Mus. Novit., no. 2867, 53 pp.

Musser, G. G., and P. W. Freeman. 1981. A new species of Rhynchomys (Muridae) from the Philippines. J. Mamm. 62:154–59.

Musser, G. G., and A. L. Gardner. 1974. A new species of the Ichthyomine Daptomys from Peru. Amer. Mus. Novit., no. 2537, 23 pp.

Musser, G. G., and L. K. Gordon. 1981. A new species of Crateromys (Muridae) from the Philippines. J. Mamm. 62:513–25.

Musser, G. G., L. K. Gordon, and H. Sommer. 1982. Species-limits in the Philippine murid, Chrotomys. J. Mamm. 63:514–21.

Musser, G. G., and L. R. Heaney. 1985. Philippine Rattus: a new species from the Sulu Archipelago. Amer. Mus. Novit., no. 2818, 32 pp.

Musser, G. G., L. R. Heaney, and D. S. Rabor. 1985. Philippine rats: a new species of Crateromys from Dinagat Island. Amer. Mus. Novit., no. 2821, 25 pp.

Musser, G. G., K. F. Koopman, and D. Califia. 1982. The Sulawesian Pteropus arquatus and P. argentatus are Acerodon celebensis; the Philippine P. leucotis is an Acerodon. J. Mamm. 63:319–28.

Musser, G. G., J. T. Marshall, Jr., and B. Boeadi. 1979. Definition and contents of the Sundaic genus Maxomys (Rodentia, Muridae). J. Mamm. 60:592–606.

Musser, G. G., and C. Newcomb. 1983. Ma-

laysian murids and the giant rat of Sumatra. Bull. Amer. Mus. Nat. Hist. 174:327–598.

———. 1985. Definitions of Rattus losea and a new species from Vietnam. Amer. Mus. Novit., no 2814, 32 pp.

Musser, G. G., and E. Piik. 1982. A new species of Hydromys (Muridae) from western New Guinea (Irian Jaya). Zool. Meded. 56:153–67.

Musser, G. G., A. Van de Weerd, and E. Strasser. 1986. Paulamys, a replacement name for Floresomys Musser, 1981 (Muridae), and new material of that taxon from Flores, Indonesia. Amer. Mus. Novit., no. 2850, 10 pp.

Musser, G. G., and M. M. Williams. 1985. Systematic studies of oryzomyine rodents (Muridae): definitions of Oryzomys villosus and Oryzomys talamancae. Amer. Mus. Novit., no. 2810, 22 pp.

Mutere, F. A. 1967. The breeding biology of equatorial vertebrates: reproduction in the fruit bat, Eidolon helvum, at latitude 0°20′ N. J. Zool. 153:153–61.

———. 1970. The breeding biology of equatorial vertebrates: reproduction in the insectivorous bat, Hipposideros caffer, living at 0°27′ N. Bijdragen Tot de Dierkunde 40:56–58.

———. 1973a. Reproduction in two species of equatorial free-tailed bats (Molossidae). E. Afr. Wildl. J. 11:271–80.

———. 1973b. A comparative study of reproduction in two populations of the insectivorous bats, Otomops martiensseni, at latitudes 1°5′ S and 2°30′ S. J. Zool. 171:79–92.

Muul, I. 1989. Rodents of conservation concern in the Southeast Asian region. In Lidicker (1989), pp. 51–52.

Muul, I., and Lim Boo Liat. 1970. Ecological and morphological observations of Felis planiceps. J. Mamm. 51:806–8.

———. 1971. New locality records for some mammals of West Malaysia. J. Mamm. 52:430–37.

———. 1974. Reproductive frequency in Malaysian flying squirrels, Hylopetes and Pteromyscus. J. Mamm. 55:393–400.

———. 1978. Comparative morphology, food habits, and ecology of some Malaysian arboreal rodents. In Montgomery (1978), pp. 361–68.

Muul, I., and K. Thonglongya. 1971. Taxonomic status of Petinomys morrisi (Carter) and its relationship to Petinomys setosus (Temminck and Schlegel). J. Mamm. 52:362–69.

Myers, L. G. 1969. Home range and longevity in Zapus princeps in Colorado. Amer. Midl. Nat. 82:628–29.

Myers, N. 1975a. The silent savannahs. Internatl. Wildl. 5(5):411.

———. 1975b. The cheetah Acinonyx jubatus in Africa. Internatl. Union Conserv. Nat. Monogr., no. 4, 88 pp.

———. 1976. The leopard Panthera pardus in Africa. Internatl. Union Conserv. Nat. Monogr., no. 5, 79 pp.

———. 1984. Cats in crisis. Internatl. Wildl. 14(6):42–48.

———. 1987a. Trends in the destruction of rain forests. In Marsh and Mittermeier (1987), pp. 3–22.

———. 1987b. Africa. In Kingdom of cats, Natl. Wildl. Fed., pp. 124–77.

Myers, P. 1977a. Patterns of reproduction of four species of vespertilionid bats in Paraguay. Univ. California Publ. Zool. 107:1–41.

———. 1977b. A new phyllotine rodent (genus Graomys) from Paraguay. Occas. Pap. Mus. Zool. Univ. Michigan, no. 676, 5 pp.

Myers, P., and M. D. Carleton. 1981. The species of Oryzomys (Oligoryzomys) in Paraguay and the identity of Azara's "Rat sixième ou Rat à Tarse noir." Misc. Publ. Mus. Zool. Univ. Michigan, no. 161, iii + 41 pp.

Myers, P., and R. M. Wetzel. 1979. New records of mammals from Paraguay. J. Mamm. 60:638–41.

———. 1983. Systematics and zoogeography pf the bats of the Chaco Boreal. Misc. Publ. Mus. Zool. Univ. Michigan, no. 165, iv + 59 pp.

Myers, P., R. White, and J. Stallings. 1983. Additional records of bats from Paraguay. J. Mamm. 64:143–45.

Myllymäki, A. 1977a. Demographic mechanisms in the fluctuating populations of the field vole Microtus agrestis. Oikos 29:468–93.

———. 1977b. Intraspecific competition and home range dynamics in the field vole Microtus agrestis. Oikos 29:553–69.

Myrick, A. C., Jr., A. A. Hohn, J. Barlow, and P. A. Sloan. 1986. Reproductive biology of female spotted dolphins, Stenella attenuata, from the eastern tropical Pacific. Fishery Bull. 84:247–59.

Mysterud, I., J. Viitala, and S. Lahti. 1972. On winter breeding of the wood lemming (Myopus schisticolor). Norwegian J. Zool. 20:91–92.

N

Nadachowski, A. 1984. On a collection of small mammals from the People's Republic of Korea. Acta Zool. Cracoviensia 27:47–60.

Nader, I. A. 1974. A new record, bushy-tailed jird, Sekeetamys calurus calurus (Thomas, 1892) from Saudi Arabia. Mammalia 38:347–49.

———. 1975. On the bats (Chiroptera) of the

Kingdom of Saudi Arabia. J. Zool. 176:331–40.

———. 1979. The present status of the viverrids of the Arabian Peninsula (Mammalia: Carnivora: Viverridae). Senckenberg. Biol. 59:311–16.

———. 1982. New distributional records of bats from the Kingdom of Saudi Arabia (Mammalia: Chiroptera). J. Zool. 198:69–82.

———. 1989. The status of rodents in the western Asian region. In Lidicker (1989), pp. 45–47.

Nader, I. A., and D. Kock. 1980. First record of Tadarida nigeriae (Thomas 1913) from the Arabian Peninsula. Senckenberg. Biol. 60:131–35.

———. 1983a. A new slit-faced bat from central Saudi Arabia (Mammalia: Chiroptera: Nycteridae). Senckenberg. Biol. 63:9–15.

———. 1983b. Rhinopoma microphyllum asirensis n. subsp. from southwestern Saudi Arabia (Mammalia: Chiroptera: Rhinopomatidae). Senckenberg. Biol. 63:147–52.

Nader, I. A., D. Kock, and A. D. Al-Khalili. 1983. Eliomys melanurus (Wagner 1839) and Praomys fumatus (Peters 1878) from the Kingdom of Saudi Arabia. Senckenberg. Biol. 63:313–24.

Nadler, C. F., N. N. Vorontsov, R. S. Hoffmann, I. I. Formichova, and C. F. Nadler, Jr. 1973a. Zoogeography of transferins in arctic and long-tailed ground squirrel populations. Comp. Biochem. Physiol. 44B:33–40.

———. 1973b. Cytogenetic differentiation, geographic distribution, and domestication in Palearctic sheep (Ovis). Z. Saugetierk. 38:109–25.

———. 1974. Evolution in ground squirrels—I. Transferins in holarctic populations of Spermophilus. Comp. Biochem. Physiol. 47A:663–81.

———. 1975. Chromosomal evolution in holarctic ground squirrels (Spermophilus). I. Giemsa-band homologies in Spermophilus columbianus and S. undulatus. Z. Saugetierk. 40:1–7.

———. 1977. Chromosomal evolution in chipmunks, with special emphasis on A and B karyotypes of the subgenus Neotamias. Amer. Midl. Nat. 98:343–53.

———. 1978. Biochemical relationships of the holarctic vole genera [Clethrionomys, Microtus, and Arvicola (Rodentia: Arvicolinae)]. Can. J. Zool. 56:1564–75.

———. 1982. Evolution in ground squirrels. II. Biochemical comparisons in holarctic populations of Spermophilus. Z. Saugetierk. 47:198–215.

Nagel, U. 1971. Social organization in a baboon hybrid zone. Proc. 3rd Internatl. Congr. Primatol. 3:48–57.

———. 1973. A comparison of anubis baboons, hamadryas baboons and their hybrids at a species border in Ethiopia. Folia Primatol. 19:104–65.

Nagorsen, D. 1985. Kogia simus. Mammalian Species, no. 239, 6 pp.

Nagorsen, D., and J. R. Tamsitt. 1981. Systematics of Anoura cultrata, A. brevirostrum, and A. werckleae. J. Mamm. 62:82–100.

Nagorsen, D. W. 1983. Winter pelage colour in snowshoe hares (Lepus americanus) from the Pacific Northwest. Can. J. Zool. 61:2313–18.

———. 1987. Summer and winter food caches of the heather vole, Phenacomys intermedius, in Quetico Provincial Park, Ontario. Can. Field-Nat. 101:82–85.

Nagorsen, D. W., and D. M. Jones. 1981. First records of the tundra shrew (Sorex tundrensis) in British Columbia. Can. Field-Nat. 95:93–94.

Nagy, K. A., R. S. Seymour, A. K. Lee, and R. Braithwaite. 1978. Energy and water budgets in free-living Antechinus stuartii (Marsupialia: Dasyuridae). J. Mamm. 59:60–68.

Naiman, R. J., C. A. Johnston, and J. C. Kelley. 1988. Alteration of North American streams by beaver. Bioscience 38:753–62.

Nair, S. S. 1976. A population survey and observations on the behaviour of the blackbuck in the Point Calimere Sanctuary, Tamil Nadu. J. Bombay Nat. Hist. Soc. 73:304–10.

Napier, J. R., and P. H. Napier. 1967. A handbook of living primates. Academic Press, New York, xiv + 456 pp.

———, eds. 1970. Old World monkeys: evolution, systematics, and behavior. Academic Press, New York, xiv + 660 pp.

———. 1985. The natural history of primates. MIT Press, Cambridge, 200 pp.

Nascetti, G., S. Lovari, P. Lanfranchi, C. Berducou, S. Mattiucci, L. Rossi, and L. Bullini. 1985. Revision of Rupicapra genus. III. Electrophoretic studies demonstrating species distinction of chamois populations of the Alps from those of the Apennines and Pyrenees. In Lovari (1985), pp. 55–62.

Nash, L. T., and C. S. Harcourt. 1986. Social organization of galagos in Kenyan coastal forests: II. Galago garnettii. Amer. J. Primatol. 10:357–69.

Nass, R. D. 1977. Movements and home ranges of Polynesian rats in Hawaiian sugarcane. Pacific Sci. 31:135–42.

National Geographic Society. 1981. Book of mammals. Spec. Publ. Div., Natl. Geogr. Soc., Washington, D.C., 2 vols.

Natori, M. 1988. A cladistic analysis of interspecific relationships of Saguinus. Primates 29:263–76.

Natori, M., and T. Hanihara. 1988. An analysis of interspecific relationships of Saguinus based on cranial measurements. Primates 29:255–62.

Naumov, N. P., and V. S. Lobachev. 1975. Ecology of desert rodents of the U.S.S.R. In Prakash and Ghosh (1975), pp. 465–598.

Nead, D. M., J. C. Halfpenny, and S. Bissell. 1985. The status of wolverines in Colorado. Northwest Sci. 8:286–89.

Neal, B. R. 1977a. Reproduction of the punctated grass-mouse, Lemniscomys striatus in the Ruwenzori National Park, Uganda (Rodentia: Muridae). Zool. Afr. 12:419–28.

———. 1977b. Reproduction of the multimammate rat, Praomys (Mastomys) natalensis (Smith), in Uganda. Z. Saugetierk. 42:221–31.

———. 1983. The breeding pattern of two species of spiny mice, Acomys percivali and A. wilsoni (Muridae: Rodentia), in central Kenya. Mammalia 47:311–21.

Neal, E. 1970. The banded mongoose, Mungos mungo Gmelin. E. Afr. Wildl. J. 8:63–71.

Neas, J. F., and R. S. Hoffmann. 1987. Budorcas taxicolor. Mammalian Species, no. 277, 7 pp.

Nel, J. A. J. 1978. Notes on the food and foraging behavior of the bat-eared fox, Otocyon megalotis. Bull. Carnegie Mus. Nat. Hist., no. 6, pp. 132–37.

Nel, J. A. J., and M. H. Bester. 1983. Communication in the southern bat-eared fox Otocyon m. megalotis (Desmarest, 1822). Z. Saugetierk. 48:277–90.

Nel, J. A. J., and C. J. Stutterheim. 1973. Notes on early post-natal development of the Namaqua gerbil Desmodillus auricularis. Koedoe 16:117–25.

Nellis, C. H., and L. B. Keith. 1976. Population dynamics of coyotes in central Alberta, 1964–1968. J. Wildl. Mgmt. 40:389–99.

Nellis, C. H., S. P. Wetmore, and L. B. Keith. 1972. Lynx-prey interactions in central Alberta. J. Wildl. Mgmt. 36:320–29.

Nellis, D. W., and C. P. Ehle. 1977. Observations on the behavior of Brachyphylla cavernarum (Chiroptera) in Virgin Islands. Mammalia 41:403–9.

Nellis, D. W., N. F. Eichholz, T. W. Regan, and C. Feinstein. 1978. Mongoose in Florida. Wildl. Soc. Bull. 6:249–50.

Nelson, J. E., and A. Goldstone. 1986. Reproduction in Peradorcas concinna (Marsupialia: Macropodidae). Austral. Wildl. Res. 13:501–5.

Nelson, J. E., and E. Hamilton-Smith. 1982. Some observations on Notopteris macdonaldi (Chiroptera: Pteropodidae). Austral. Mamm. 5:247–52.

Nelson, J. R., and T. A. Leege. 1982. Nutritional requirements and food habits. *In* Thomas and Toweill (1982), pp. 323–67.

Nelson, K., R. J. Baker, H. S. Shellhammer, and R. K. Chesser. 1984. Test of alternative hypotheses concerning the origin of *Reithrodontomys raviventris:* genetic analysis. J. Mamm. 65:668–73.

Nerini, M. K., H. W. Braham, W. M. Marquette, and D. J. Rugh. 1984. Life history of the bowhead whale, *Balaena mysticetus* (Mammalia: Cetacea). J. Zool. 204:443–68.

Nesbitt, W. H. 1975. Ecology of a feral dog pack on a wildlife refuge. *In* Fox (1975), pp. 391–95.

Neugebauer, W. 1980. The status and management of the pygmy chimpanzee *Pan paniscus* in European zoos. Internatl. Zoo Yearbook 20:64–70.

Neuhauser, H. N., and A. F. DeBlase. 1974. Notes on bats (Chiroptera: Vespertilionidae) new to the faunal lists of Afghanistan and Iran. Fieldiana Zool. 62:85–96.

Neuweiler, G. 1969. Verhaltensbeobachtungen an einer indischen Flughundkolonie (*Pteropus g. giganteus* Brünn). Z. Tierpsychol. 26:166–99.

Neville, M., N. Castro, A. Marmol, and J. Revilla. 1976. Censusing primate populations in the reserved area of the Pacaya and Samiria rivers, Department Loreto, Peru. Primates 17:151–81.

Nevo, E. 1961. Observations of Israeli populations of the mole-rat *Spalax ehrenbergi* Nehring 1898. Mammalia 25:127–44.

———. 1982. Genetic structure and differentiation during speciation in fossorial gerbil rodents. Mammalia 46:523–30.

Nevo, E., and H. Bar-El. 1976. Hybridization and speciation in fossorial mole rats. Evolution 30:831–40.

Nevo, E., R. Ben-Shlomo, A. Beiles, J. U. M. Jarvis, and G. C. Hickman. 1987. Allozyme differentiation and systematics of the endemic subterranean mole rats of South Africa. Biochem. Syst. Ecol. 15:489–502.

Nevo, E., M. Corti, G. Heth, A. Beiles, and S. Simson. 1988. Chromosomal polymorphisms in subterranean mole rats: origins and evolutionary significance. Biol. J. Linnean Soc. 33:309–22.

Nevo, E., G. Heth, and A. Beiles. 1982. Population structure and evolution in subterranean mole rats. Evolution 36:1283–89.

Newby, J. E. 1978. Scimitar-horned oryx—the end of the line? Oryx 14:219–21.

———. 1988. Aridland wildlife in decline: the case of the scimitar-horned oryx. *In* Dixon and Jones (1988), pp. 146–66.

Newby, J. E., and C. Magin. 1989. Addax in Niger: distribution, status, and conservation

options. Internatl. Union Conserv. Nat. Species Survival Comm. Captive Breeding Specialist Group Conf., San Antonio, 16–17 September 1989, 11 pp.

Newman, M. A. 1978. Narwhal. *In* Haley (1978), pp. 138–44.

Newsome, A. E. 1971a. The ecology of red kangaroos. Austral. Zool. 16:32–50.

———. 1971b. Competition between wildlife and domestic livestock. Austral. Vet. J. 47:577–86.

———. 1975. An ecological comparison of the two arid-zone kangaroos of Australia, and their anomalous prosperity since the introduction of ruminant stock to their environment. Quart. Rev. Biol. 50:389–424.

Newsome, A. E., and L. K. Corbett. 1982. The identity of the dingo. II. Hybridization with domestic dogs in captivity and in the wild. Austral. J. Zool. 30:365–74.

———. 1985. The identity of the dingo. III. The incidence of dingoes, dogs and hybrids and their coat colours in remote and settled regions of Australia. Austral. J. Zool. 33:363–75.

Newsome, A. E., L. K. Corbett, and S. M. Carpenter. 1980. The identity of the dingo. I. Morphological discriminants of dingo and dog skulls. Austral. J. Zool. 28:615–25.

Neyman, P. F. 1977. Aspects of the ecology and social organization of free-ranging cotton-top tamarins *(Saguinus oedipus)* and the conservation status of the species. *In* Kleiman (1977a), pp. 39–71.

Nicholls, D. G. 1971. Daily and seasonal movements of the quokka, *Setonix brachyurus* (Marsupialia), on Rottnest Island. Austral. J. Zool. 19:215–26.

Nichols, L., Jr. 1978. Dall's sheep. *In* Schmidt and Gilbert (1978), pp. 173–89.

Nicoll, M. E. 1985. The biology of the giant otter shrew *Potamogale velox*. Natl. Geogr. Soc. Res. Rept. 21:331–37.

Nicoll, M. E., and P. A. Racey. 1981. The Seychelles fruit bat, *Pteropus seychellensis seychellensis*. Afr. J. Ecol. 19:361–64.

Nicoll, M. E., and J. M. Suttie. 1982. The sheath-tailed bat, *Coleura seychellensis* (Chiroptera: Emballonuridae) in the Seychelles Islands. J. Zool. 197:421–26.

Niebauer, T. J., and O. J. Rongstad. 1977. Coyote food habits in northwestern Wisconsin. *In* Phillips and Jonkel (1977), pp. 237–51.

Niemitz, C. 1984a. Biology of tarsiers. Gustav Fischer Verlag, Stuttgart, ix + 357 pp.

———. 1984b. Taxonomy and distribution of the genus *Tarsius. In* Niemitz (1984a), pp. 1–15.

———. 1984c. An investigation and review of the territorial behaviour and social

organisation of the genus *Tarsius. In* Niemitz (1984a), pp. 117–27.

Niethammer, G. 1970. Beobachtungen am Pyrenaen-Desman, *Galemys pyrenaica*. Bonner Zool. Beitr. 21:157–82.

Nieuwenhuijsen, K., Ad J. J. C. Lammers, K. J. de Neef, and A. K. Slob. 1985. Reproduction and social rank in female stumptail macaques *(Macaca arctoides)*. Internatl. J. Primatol. 6:77–99.

Nievergelt, B. 1974. A comparison of rutting behaviour and grouping in the Ethiopian and Alpine ibex. *In* Geist and Walther (1974), pp. 324–40.

———. 1981. Ibexes in an African environment. Springer-Verlag, Berlin, 230 pp.

Nishida, T. 1972a. A note on the ecology of the red-colobus monkeys *(Colobus badius tephrosceles)* living in the Mahali Mountains. Primates 13:57–64.

———. 1972b. Preliminary information of the pygmy chimpanzees *(Pan paniscus)* of the Congo Basin. Primates 13:415–25.

Nishida, T., S. Uehara, and R. Nyundo. 1979. Predatory behavior among wild chimpanzees of the Mahale Mountains. Primates 20:1–20.

Nishimura, A., and K. Izawa. 1975. The group characteristics of woolly monkeys *(Lagothrix lagothrica)* in the upper Amazonian Basin. Proc. 5th Internatl. Congr. Primatol., pp. 351–57.

Nishiwaki, M. 1966. A discussion of rarities among the smaller cetaceans caught in Japanese waters. *In* Norris (1966), pp. 192–204.

———. 1975. Ecological aspects of smaller cetaceans, with emphasis on the striped dolphin (*Stenella coeruleoalba*). J. Fish. Res. Bd. Can. 32:1069–72.

———. 1984. Current status of the African manatee. Acta Zool. Fennica 172:135–36.

Nishiwaki, M., and T. Kasuya. 1970. A Greenland right whale caught at Osaka Bay. Sci. Rept. Whales Res. Inst. 22:45–62.

Nishiwaki, M., T. Kasuya, N. Miyazoki, T. Tobayama, and T. Kataoka. 1979. Present distribution of the dugong in the world. Sci. Rept. Whales Res. Inst. 31:133–41.

Nishiwaki, M., and H. Marsh. 1985. Dugong—*Dugong dugon. In* Ridgway and Harrison (1985), pp. 1–31.

Nishiwaki, M., and K. S. Norris. 1966. A new genus, *Peponocephala*, for the odontocete cetacean species *Electra electra*. Sci. Rept. Whales Res. Inst. 20:95–100.

Nishiwaki, M., and N. Oguro. 1971. Baird's beaked whales caught on the coast of Japan in recent 10 years. Sci. Rept. Whales Res. Inst. 23:111–22.

Nishiwaki, M., M. Yamaguchi, S. Shokita, S. Uchida, and T. Kataoka. 1982. Recent survey on the distribution of the African manatee. Sci. Rept. Whales Res. Inst. 34:137–47.

Nitikman, L. Z., and M. A. Mares. 1987. Ecology of small mammals in a gallery forest of central Brazil. Ann. Carnegie Mus. 56:75–95.

Norberg, U. M., and M. B. Fenton. 1988. Carnivorous bats? Biol. J. Linnean Soc. 33:383–94.

Nores, C., and C. Pérez. 1988. The occurrence of walrus (Odobenus rosmarus) in southern Europe. J. Zool. 216:593–96.

Norris, K. S. 1966. Whales, dolphins, and porpoises. Univ. California Press, Berkeley, xv + 789 pp.

Norris, K. S., and G. W. Harvey. 1972. A theory for the function of the spermaceti organ of the sperm whale (Physeter catodon L.). In Galler, S. R., K. Schmidt-Koenig, G. J. Jacobs, and R. E. Belleville, eds., Animal orientation and navigation, U.S. Natl. Aeronautics and Space Admin., Washington, D.C., pp. 397–417.

Norris, M. L., and C. E. Adams. 1972. The growth of the Mongolian gerbil, Meriones unguiculatus, from birth to maturity. J. Zool. 166:277–82.

Norton, P. M. 1986. Historical changes in the distribution of leopards in the Cape Province, South Africa. Bontebok 5:1–9.

Nosek, J., O. Kozuch, and J. Chmela. 1972. Contribution to the knowledge of home range in common shrew Sorex araneus L. Oecologia 9:59–63.

Novacek, M. J. 1985. Evidence for echolocation in the oldest known bats. Nature 315:140–41.

———. 1986. The skull of leptictid insectivorans and the higher-level classification of eutherian mammals. Bull. Amer. Mus. Nat. Hist. 183:1–112.

Novak, M., J. A. Baker, M. E. Obbard, and B. Malloch. 1987. Wild furbearer management and conservation in North America. Ontario Ministry Nat. Res., xviii + 1150 pp.

Novak, M., M. E. Obbard, J. G. Jones, R. Newman, A. Booth, A. J. Satterthwaite, and G. Linscombe. 1987. Furbearer harvests in North America, 1600–1984. Ontario Ministry Nat. Res., xvi + 270 pp.

Novellie, P. A. 1979. Courtship behaviour of the blesbok (Damaliscus dorcas phillipsi). Mammalia 43:263–74.

Novellie, P. A., J. Manson, and R. C. Bigalke. 1984. Behavioural ecology and communication in the Cape grysbok. S. Afr. J. Zool. 19:22–30.

Novick, A., and B. A. Dale. 1971. Foraging behavior in fishing bats and their insectivorous relatives. J. Mamm. 52:817–18.

Novikov, G. A. 1962. Carnivorous mammals of the fauna of the U.S.S.R. Israel Progr. Sci. Transl., Jerusalem, 284 pp.

Nowak, E. 1984. Verbreitungs- und Bestandsentwicklung des Marderhundes, Nyctereutes procyonides (Gray, 1834) in Europa. Z. Jagdwiss 30:137–54.

Nowak, R. M. 1972. The mysterious wolf of the south. Nat. Hist. 81(1):50–53, 74–77.

———. 1973. Return of the wolverine. Natl. Parks and Conserv. Mag. 47(2):20–23.

———. 1974. Red wolf: our most endangered mammal. Natl. Parks and Conserv. Mag. 48(8):9–12.

———. 1975a. The cosmopolitan wolf. Natl. Rifle Assoc. Conserv. Yearbook, pp. 76–82.

———. 1975b. Retreat of the jaguar. Natl. Parks and Conserv. Mag. 49(12):10–13.

———. 1976. The cougar in the United States and Canada. Unpubl. Rept. to U.S. Fish and Wildl. Serv., 190 pp.

———. 1978. Evolution and taxonomy of coyotes and related Canis. In Bekoff (1978), pp. 3–16.

———. 1979. North American Quaternary Canis. Monogr. Mus. Nat. Hist. Univ. Kansas, no. 6, 154 pp.

O

Oates, J. F. 1977a. The guereza and man. In Rainier III and Bourne (1977), pp. 419–67.

———. 1977b. The guereza and its food. In Clutton-Brock (1977), pp. 276–321.

———. 1977c. The social life of a black-and-white colobus monkey, Colobus guereza. Z. Tierpsychol. 45:1–60.

———. 1982. In search of rare forest primates in Nigeria. Oryx 16:431–36.

———. 1984. The niche of the potto, Perodicticus potto. Internatl. J. Primatol. 5:51–61.

———. 1985. Action plan for African primate conservation: 1986–1990. Internatl. Union Conserv. Nat., Gland, Switzerland, 41 pp.

Oates, J. F., and T. F. Trocco. 1983. Taxonomy and phylogeny of black-and-white colobus monkeys: inferences from an analysis of loud call variation. Folia Primatol. 40:83–113.

Obidina, V. A. 1972. Winter breeding of Alticola argentatus under natural conditions. Soviet J. Ecol. 3:567–68.

O'Brien, P. H. 1988. Feral goat social organization: a review and comparative analysis. Appl. Anim. Behav. Sci. 21:209–21.

O'Brien, S. J. 1983. The cheetah is depauperate in genetic variation. Science 221:459–62.

O'Brien, S. J., M. E. Roelke, L. Marker, A. Newman, C. A. Winkler, D. Meltzer, L. Col-ly, J. F. Evermann, M. Bush, and D. E. Wildt. 1985. Genetic basis for species vulnerability in the cheetah. Science 227:1428–34.

Ochoa G., J. 1984. Presencia de Nyctinomops aurispinosa en Venezuela (Chiroptera: Molossidae). Acta Cien. Venezolana 35:147–50.

Ochoa G., J., H. Castellanos, and C. Ibáñez. 1988. Records of bats and rodents from Venezuela. Mammalia 52:175–80.

Ochoa G., J., and C. Ibáñez. 1982. Nuevo murcielago del genero Lonchorhina (Chiroptera: Phyllostomidae). Mem. Soc. Ciencias Nat. La Salle 42:145–59.

———. 1985. Distributional status of some bats from Venezuela. Mammalia 49:65–73.

O'Connell, M. A. 1982. Population biology of North and South American grassland rodents: a comparative review. Pymatuning Lab. Ecol. Spec. Publ. 6:167–85.

O'Connor, T. P. 1986. The garden dormouse Eliomys quercinus from Roman York. J. Zool. 210:620–22.

Odell, D. K. 1975. Status and aspects of the life history of the bottlenose dolphin, Tursiops truncatus, in Florida. J. Fish. Res. Bd. Can. 32:1055–58.

———. 1981. California sea lion—Zalophus californianus. In Ridgway and Harrison (1981a), pp. 67–97.

———. 1984. The fight to mate. In Macdonald (1984), pp. 262–63.

Odell, D. K., D. Forrester, and E. Asper. 1978. Growth and sexual maturation in the West Indian manatee. Amer. Soc. Mamm., Abstr. Tech. Pap., 58th Ann. Mtg., pp. 7–8.

Odendaal, P. B., and R. C. Bigalke. 1979. Home range and groupings of bushbuck in the southern Cape. S. Afr. J. Wildl. Res. 9:96–101.

O'Farrell, M. J. 1978. Home range dynamics of rodents in a sagebrush community. J. Mamm. 59:657–68.

O'Farrell, M. J., and A. R. Blaustein. 1974a. Microdipodops megacephalus. Mammalian Species, no. 46, 3 pp.

———. 1974b. Microdipodops pallidus. Mammalian Species, no. 47, 2 pp.

O'Farrell, M. J., and E. H. Studier. 1973. Reproduction, growth, and development in Myotis thysanodes and M. lucifugus (Chiroptera: Vespertilionidae). Ecology 54:18–30.

O'Farrell, T. P. 1987. Kit fox. In Novak, Baker, et al. (1987), pp. 422–31.

O'Farrell, T. P., R. J. Olson, R. O. Gilbert, and J. D. Hedlund. 1975. A population of Great Basin pocket mice, Perognathus parvus, in the shrub-steppe of south-central Washington. Ecol. Monogr. 45:1–28.

O'Gara, B. W. 1978. Antilocapra americana. Mammalian Species, no. 90, 7 pp.

O'Gara, B. W., and G. Matson. 1975. Growth and casting of horns by pronghorns and exfoliation of horns by bovids. J. Mamm. 56:829–46.

Ognev, S. I. 1962–64. Mammals of eastern Europe and northern Asia. Israel Progr. Sci. Transl., Jerusalem, 8 vols.

Ohsawa, H., and M. Kawai. 1975. Social structure of gelada baboons: studies of the gelada society (I). Proc. 5th Internatl. Congr. Primatol., pp. 464–69.

Ohsumi, S. 1965. Reproduction of the sperm whale in the north-west Pacific. Sci. Rept. Whales Res. Inst. 19:1–35.

———. 1966. Sexual segregation of the sperm whale in the North Pacific. Sci. Rept. Whales Res. Inst. 20:1–16.

———. 1971. Some investigations on the school structure of sperm whale. Sci. Rept. Whales Res. Inst. 23:1–25.

———. 1979. Population assessment of the Antarctic minke whale. Rept. Internatl. Whaling Comm. 29:407–20.

Ohsumi, S., and Y. Masaki. 1975. Biological parameters of the Antarctic minke whale at the virginal population level. J. Fish. Res. Bd. Can. 32:995–1004.

Ohsumi, S., Y. Masaki, and A. Kawamura. 1970. Stock of the Antarctic minke whale. Sci. Rept. Whales Res. Inst. 22:75–110.

Ohtaishi, N., and M. Yoneda. 1981. A thirty four years old male Kuril seal from Shiretoko Pen., Hokkaido. Sci. Rept. Whales Res. Inst. 33:131–35.

Ojasti, J. 1972. Revisión preliminar de los picures o aguties de Venezuela (Rodentia, Dasyproctidae). Mem. Soc. Ciencias Nat. La Salle 32:159–204.

Ojasti, J., and O. J. Linares. 1971. Adiciones a la fauna de murciélagos de Venezuela con notas sobre las especies del género Diclidurus (Chiroptera). Acta Biol. Venezuelica 7:421–41.

Ojasti, J., and C. J. Naranjo. 1974. First record of Tonatia nicaraguae in Venezuela. J. Mamm. 55:248–49.

Ojasti, J., and G. M. Padilla. 1972. The management of capybara in Venezuela. Trans. N. Amer. Wildl. Conf. 33:268–77.

Ojeda, R. A., V. G. Roig, E. P. De Cristaldo, and C. N. De Moyano. 1989. A new record of Tympanoctomys (Octodontidae) from Mendoza Province, Argentina. Texas J. Sci. 41:333–36.

Okia, N. O. 1974a. Breeding in Franquet's bat, Epomops franqueti (Tomes), in Uganda. J. Mamm. 55:462–65.

———. 1974b. The breeding pattern of the eastern epauletted bat, Epomophorus anurus Heuglin, in Uganda. J. Reprod. Fert. 37:27–31.

———. 1976. The biology of the bush rat, Aethomys hindei Thomas in southern Uganda. J. Zool. 180:41–56.

Oldenberg, P. W., P. J. Ettestad, W. E. Grant, and E. Davis. 1985. Structure of collared peccary herds in south Texas: spatial and temporal dispersion of herd members. J. Mamm. 66:764–70.

Olds, N., and S. Anderson. 1987. Notes on Bolivian mammals. 2. Taxonomy and distribution of rice rats of the subgenus Oligoryzomys. Fieldiana Zool., n.s., 39:261–81.

Olds, N., S. Anderson, and T. L. Yates. 1987. Notes on Bolivian mammals. 3. A revised diagnosis of Andalgalomys (Rodentia, Muridae) and the description of a new subspecies. Amer. Mus. Novit., no. 2890, 17 pp.

Olds, N., and J. Shoshani. 1982. Procavia capensis. Mammalian Species, no. 171, 7 pp.

Olds, T. J., and L. R. Collins. 1973. Breeding Matschie's tree kangaroo Dendrolagus matschiei in captivity. Internatl. Zoo Yearbook 13:123–25.

Olert, J., F. Dieterlen, and H. Rupp. 1978. Eine neue Muriden—Art aus Südäthiopien. Z. Zool. Syst. Evol. 16:297–308.

Oliva, D. 1988. Otaria byronia (de Blainville, 1820), the valid scientific name for the southern sea lion (Carnivora: Otariidae). J. Nat. Hist. 22:767–72.

Oliveira, J. M. S., M. G. Lima, C. Bonvincino, J. M. Ayres, and J. G. Fleagle. 1985. Preliminary notes on the ecology and behavior of the Guianan saki. Acta Amazonica 15:249–63.

Oliver, W. L. R. 1977. The hutias of the West Indies. Internatl. Zoo Yearbook 17:14–20.

———. 1978. The doubtful future of the pigmy hog and the hispid hare. J. Bombay Nat. Hist. Soc. 75:341–72.

———. 1982. The coney and the yellow snake: the distribution and status of the Jamaican hutia Geocapromys brownii and the Jamaican boa Epicrates subflavus. Dodo 19:6–33.

———. 1984. Introduced feral pigs. Papers Presented at the Workshop on Feral Mammals, Internatl. Union Conserv. Nat. Species Survival Comm., Caprinae Specialist Group, Helsinki, pp. 85–126.

Oliver, W. L. R., L. Wilkins, R. H. Kerr, and D. L. Kelly. 1986. The Jamaican hutia Geocapromys brownii captive breeding and reintroduction programme—history and progress. Dodo 23:32–58.

Olivera, J., J. Ramirez-Pulido, and S. L. Williams. 1986. Reproduccion de Peromyscus (Neotomodon) alstoni (Mammalia: Muridae) en condiciones de laboratorio. Acta Zool. Mexicana, n.s., no. 16, 27 pp.

Olivier, R. 1978. Distribution and status of the Asian elephant. Oryx 14:380–424.

Olney, P. J. S. 1988. Studbooks and world registers for rare species of wild animals in captivity. Internatl. Zoo Yearbook 27:482–90.

Olney, P. J. S., P. Ellis, and B. Sommerfelt. 1988. Census of rare animals in captivity. Internatl. Zoo Yearbook 27:436–81.

Olsen, P. D. 1982. Reproductive biology and development of the water rat, Hydromys chrysogaster, in captivity. Austral. Wildl. Res. 9:39–53.

Olsen, S. J. 1985. Origins of the domestic dog: the fossil record. Univ. Arizona Press, Tucson, xiv + 118 pp.

Olsen, S. J., and J. W. Olsen. 1977. The Chinese wolf, ancestor of New World dogs. Science 197:533–35.

Olson, S. L., and G. K. Pregill. 1982. Introduction to the paleontology of Bahaman vertebrates. In Olson, S. L., ed., Fossil vertebrates from the Bahamas, Smithson. Contrib. Paleobiol., no. 48, pp. 1–7.

Olterman, J. H., and B. J. Verts. 1972. Endangered plants and animals of Oregon. IV. Mammals. Oregon State Univ. Agric. Exp. Sta. Spec. Rept., no. 364, 47 pp.

Omar, A., and A. DeVos. 1971. The annual reproductive cycle of an African monkey (Cercopithecus mitis kolbi Neuman). Folia Primatol. 16:206–15.

Omura, H. 1974. Possible migration route of the gray whale on the coast of Japan. Sci. Rept. Whales Res. Inst. 26:1–14.

———. 1984. History of gray whales in Japan. In Jones, Swartz, and Leatherwood (1984), pp. 57–77.

———. 1986. History of right whale catches in the waters around Japan. In Brownell, Best, and Prescott (1986), pp. 35–41.

Oppenheimer, J. R. 1969. Changes in forehead patterns and group composition of the white-faced monkey (Cebus capucinus). Proc. 2nd Internatl. Congr. Primatol. 1:36–42.

———. 1977. Presbytis entellus, the hanuman langur. In Rainier III and Bourne (1977), pp. 469–512.

Oppenheimer, J. R., and E. C. Oppenheimer. 1973. Preliminary observations of Cebus nigrivittatus (Primates: Cebidae) on the Venezuelan llanos. Folia Primatol. 19:409–36.

Orlov, V. N., and Yu. M. Kovalskaya. 1978. Microtus mujanensis sp. n. from the Vitim River Basin. Zool. Zhur. 57:1224–32.

Orlov, V. N., and V. M. Malygin. 1988. A new species of hamster—Cricetulus sokolovi sp. n. (Rodentia, Cricetidae)—from People's Republic of Mongolia. Zool. Zhur. 67:304–8.

Orsini, Ph., F. Bonhomme, J. Britton-Davidian, H. Croset, S. Gerasimov, and L. Thaler. 1983. Le complex d'espèces du genre Mus en Europe Centrale et Orientale. II. Criteres d'identification, repartition et caracteristiques ecologiques. Z. Saugetierk. 48:86–95.

Osborn, D. J., and I. Helmy. 1980. The contemporary land mammals of Egypt (including Sinai). Fieldiana Zool., n.s., no. 5, xix + 579 pp.

Osgood, W. H. 1932. Mammals of the Kelley-Roosevelts and Delacour Asiatic Expeditions. Field Mus. Nat. Hist. Publ., Zool. Ser., 18:193–339.

———. 1943. The mammals of Chile. Field Mus. Nat. Hist. Zool. Ser., 30:1–268.

———. 1947. Cricetine rodents allied to *Phyllotis*. J. Mamm. 28:165–74.

O'Shea, T. J. 1976. Home range, social behavior, and dominance relationships in the African unstriped ground squirrel, *Xerus rutilus*. J. Mamm. 57:450–60.

O'Shea, T. J., C. A. Beck, R. K. Bonde, H. I. Kochman, and D. K. Odell. 1985. An analysis of manatee mortality patterns in Florida, 1976–81. J. Wildl. Mgmt. 49:1–11.

O'Shea, T. J., and T. A. Vaughan. 1977. Nocturnal and seasonal activities of the pallid bat, *Antrozous pallidus*. J. Mamm. 58:269–84.

Ostermeyer, M. C., and R. W. Elwood. 1984. Helpers(?) at the nest in the Mongolian gerbil, *Meriones unguiculatus*. Behaviour 91:61–77.

Ottenwalder, J. A., and H. H. Genoways. 1982. Systematic review of the Antillean bats of the *Natalus micropus*–complex (Chiroptera: Natalidae). Ann. Carnegie Mus. 51:17–38.

Ottichilo, W. K. 1986. Age structure of elephants in Tsavo National Park, Kenya. Afr. J. Ecol. 24:69–75.

Owen, R. C. 1984. The Americas: the case against an Ice-Age human population. *In* Smith, F. H., and F. Spencer, eds., The origins of modern humans: A world survey of the fossil evidence, Alan R. Liss, New York, pp. 517–63.

Owen, R. D. 1986. Second record of *Cryptotis parva* (Soricidae: Insectivora) in New Mexico, with review of its status on the Llano Estacado. Southwestern Nat. 31:403–5.

———. 1987. Phylogenetic analyses of the bat subfamily Stenodermatinae (Mammalia: Chiroptera). Spec. Publ. Mus. Texas Tech Univ., no. 26, 65 pp.

Owen, R. D., and W. D. Webster. 1983. Morphological variation in the Ipanema bat, *Pygoderma bilabiatum*, with description of a new subspecies. J. Mamm. 64:146–49.

Owens, D. D., and M. J. Owens. 1979a. Communal denning and clan associations in brown hyenas (*Hyaena brunnea*, Thunberg) of the central Kalahari Desert. Afr. J. Ecol. 17:35–44.

———. 1979b. Notes on social organization and behavior in brown hyenas (*Hyaena brunnea*). J. Mamm. 60:405–8.

Owens, M. J., and D. D. Owens. 1978. Feeding ecology and its influence on social organization in brown hyenas (*Hyaena brunnea*, Thunberg) of the central Kalahari Desert. E. Afr. Wildl. J. 16:113–35.

Owen-Smith, N. 1974. The social system of the white rhinoceros. *In* Geist and Walther (1974), pp. 341–51.

———. 1975. The social ethology of the white rhinoceros *Ceratotherium simum* (Burchell 1817). Z. Tierpsychol. 38:337–84.

———. 1981. The white rhino overpopulation problem and a proposed solution. *In* Jewell and Holt (1981), pp. 129–50.

———. 1984. Rhinoceroses. *In* Macdonald (1984), pp. 490–97.

Owings, D. H., M. Borchert, and R. A. Virginia. 1977. The behaviour of California ground squirrels. Anim. Behav. 25:221–30.

Owings, D. H., and R. A. Virginia. 1978. Alarm calls of California squirrels (*Spermophilus beecheyi*). Z. Tierpsychol. 46:58–70.

Oxnard, C. E. 1981. The uniqueness of *Daubentonia*. Amer. J. Phys. Anthropol. 54:1–21.

———. 1985. Humans, apes and Chinese fossils. New implication for human evolution. Hong Kong Univ. Press Occas. Pap. Ser., no. 4, 46 pp.

P

Packard, R. L. 1967. Octodontoid, bathyergoid, and ctenodactyloid rodents. *In* Anderson and Jones (1967), pp. 273–90.

———. 1968. An ecological study of the fulvous harvest mouse in eastern Texas. Amer. Midl. Nat. 79:68–88.

Packard, R. L., and J. B. Montgomery, Jr. 1978. *Baiomys musculus*. Mammalian Species, no. 102, 3 pp.

Pagels, J. F., and C. Jones. 1974. Growth and development of the free-tailed bat, *Tadarida brasiliensis cynocephala* (Le Conte). Southwestern Nat. 19:267–76.

Pages, E. 1978. Home range, behaviour and tactile communication in a nocturnal Malagasy lemur *Microcebus coquereli*. *In* Chivers and Joysey (1978), pp. 171–77.

———. 1980. Ethoecology of *Microcebus coquereli* during the dry season. *In* Charles-Dominique et al. (1980), pp. 97–116.

Pagnoni, G., and S. Saba. 1989. New record of the spectacle porpoise. Mar. Mamm. Sci. 5:201–2.

Palacios, F. 1976. Descripción de una nueva especie de liebre (*Lepus castroviejoi*), en-démica de la Cordillera Cantábrica. Doñana Acta Vert. 3:205–23.

———. 1983. On the taxonomic status of the genus *Lepus* in Spain. Acta Zool. Fennica 174:27–130.

Palmeirim, J. M. 1982. On the presence of *Nyctalus lasiopterus* in North Africa (Mammalia: Chiroptera). Mammalia 46:401–3.

Palmeirim, J. M., and R. S. Hoffmann. 1983. *Galemys pyrenaicus*. Mammalian Species, no. 207, 5 pp.

Palomares, F., and M. Delibes. 1988. Time and space use by two common genets (*Genetta genetta*) in the Doñana National Park, Spain. J. Mamm. 69:635–37.

Panwar, H. S. 1987. Project tiger: the reserves, the tigers and their future. *In* Tilson and Seal (1987), pp. 110–17.

Paradiso, J. L. 1971. A new subspecies of *Cynopterus sphinx* (Chiroptera: Pteropodidae) from Serasan (South Natuna) Island, Indonesia. Proc. Biol. Soc. Washington 84:293–300.

———. 1972. Status report on cats (Felidae) of the world, 1971. U.S. Bur. Sport Fish. and Wildl. Spec. Sci. Rept.—Wildl., no. 157, iv + 43 pp.

Paradiso, J. L., and A. M. Greenhall. 1967. Longevity records for American bats. Amer. Midl. Nat. 78:251–52.

Paradiso, J. L., and R. M. Nowak. 1972. *Canis rufus*. Mammalian Species, no. 22, 4 pp.

Parker, C. 1979. Birth, care and development of Chinese hog badgers *Arctonyx collaris albogularis* at Metro Toronto Zoo. Internatl. Zoo Yearbook 19:182–85.

Parker, G. R. 1973. Distribution and densities of wolves within barren-ground caribou range in northern mainland Canada. J. Mamm. 54:341–48.

———. 1977. Morphology, reproduction, diet, and behavior of the arctic hare (*Lepus arcticus monstrabilis*) on Axel Heiberg Island, Northwest Territories. Can. Field-Nat. 91:8–18.

Parker, P. 1977. An ecological comparison of marsupial and placental patterns of reproduction. *In* Stonehouse and Gilmore (1977), pp. 273–86.

Parker, S. A. 1971. Notes on the small black wallaroo *Macropus bernardus* (Rothschild, 1904) of Arnhem Land. Victorian Nat. 88:41–43.

Parsons, D. A. Smith, and R. F. Whittam. 1986. Maternity colonies of silver-haired bats, *Lasionycteris noctivagans*, in Ontario and Saskatchewan. J. Mamm. 67:598–600.

Partridge, J. 1967. A 3,300 year old thylacine (Marsupialia: Thylacinidae) from the Nullarbor Plain, Western Australia. J. Roy. Soc. W. Austral. 50:57–59.

———. 1983. The management of the pygmy hippopotamus *(Choeropsis liberiensis)* at Bristol Zoo. Internatl. Zoo News 30(3):28–41.

Patten, D. R., and L. T. Findley. 1970. Observations and records of *Myotis (Pizonyx) vivesi* Menegaux (Chiroptera: Vespertilionidae). Los Angeles Co. Mus. Nat. Hist. Contrib. Sci., no. 183, 9 pp.

Patterson, B. D. 1980. A new subspecies of *Eutamias quadrivittatus* (Rodentia: Sciuridae) from the Organ Mountains, New Mexico. J. Mamm. 61:455–64.

———. 1984. Geographic variation and taxonomy of Colorado and Hopi chipmunks (genus *Eutamias*). J. Mamm. 65:442–56.

Patterson, B. D., and M. H. Gallardo. 1987. *Rhyncholestes raphanurus*. Mammalian Species, no. 286, 5 pp.

Patterson, B. D., M. H. Gallardo, and K. E. Freas. 1984. Systematics of mice of the subgenus *Akodon* (Rodentia: Cricetidae) in southern South America, with the description of a new species. Fieldiana Zool., n.s., no. 23, 16 pp.

Patterson, F. 1986. The mind of the gorilla: conversation and conservation. *In* Benirschke (1986), pp. 933–47.

Pattie, D. 1973. *Sorex bendirii*. Mammalian Species, no. 27, 2 pp.

Patton, J. L. 1973. An analysis of natural hybridization between the pocket gophers, *Thomomys bottae* and *Thomomys umbrinus*, in Arizona. J. Mamm. 54:561–84.

———. 1984. Systematic status of the large squirrels (subgenus *Urosciurus*) of the western Amazon Basin. Studies on Neotropical Fauna and Environment 19:53–72.

———. 1987. Species groups of spiny rats, genus *Proechimys* (Rodentia: Echimyidae). Fieldiana Zool., n.s., 39:305–45.

Patton, J. L., and L. H. Emmons. 1985. A review of the genus *Isothrix* (Rodentia, Echimyidae). Amer. Mus. Novit., no. 2817, 14 pp.

Patton, J. L., and A. L. Gardner. 1972. Notes on the systematics of *Proechimys* (Rodentia: Echimyidae), with emphasis on Peruvian forms. Occas. Pap. Mus. Zool. Louisiana State Univ., no. 44, 30 pp.

Patton, J. L., and M. S. Hafner. 1983. Biosystematics of the native rodents of the Galapagos Archipelago, Ecuador. *In* Bowman, R. I., M. Benson, and A. E. Leviton, eds., Patterns of evolution in Galapagos organisms, Pacific Division Amer. Assoc. Adv. Sci., pp. 539–68.

Patton, J. L., H. MacArthur, and S. Y. Yang. 1976. Systematic relationships of the fourtoed populations of *Dipodomys heermanni*. J. Mamm. 57:159–63.

Patton, J. L., S. W. Sherwood, and S. Y. Yang. 1981. Biochemical systematics of chaetodipine pocket mice, genus *Perognathus*. J. Mamm. 62:477–92.

Patton, J. L., and M. F. Smith. 1981. Molecular evolution in *Thomomys*: phyletic systematics, paraphyly, and rates of evolution. J. Mamm. 62:493–500.

Patton, J. L., M. F. Smith, R. D. Price, and R. A. Hellenthal. 1984. Genetics of hybridization between the pocket gophers *Thomomys bottae* and *Thomomys townsendii* in northeastern California. Great Basin Nat. 44:431–40.

Paul, J. R. 1968. Risso's dolphin, *Grampus griseus*, in the Gulf of Mexico. J. Mamm. 49:746–48.

Paula Couto, C. 1979. Tratado de Paleomastozoologia. Acad. Brasil. Cien., Rio de Janeiro, 590 pp.

Paulson, D. D. 1988a. *Chaetodipus baileyi*. Mammalian Species, no. 297, 5 pp.

———. 1988b. *Chaetodipus hispidus*. Mammalian Species, no. 320, 4 pp.

Pavlenko, T. A., and A. G. Daveletshina. 1971. Nutrition of lesser jerboa in the Fergana Valley. Soviet J. Ecol. 2:69–71.

Pavlinov, I. Ya. 1980a. Superspecies groupings Cardiocraniinae Satunin (Mammalia, Dipodidae). Vestn. Zool. 1980(2):47–50.

———. 1980b. Taxonomic status of *Calomyscus* Thomas (Rodentia, Cricetidae) on the basis of structure of auditory ossicles. Zool. Zhur. 59:312–16.

———. 1982a. Phylogeny and classification of the subfamily Gerbillinae. Byull. Mosk. Obshch. Ispyt. Prirody, Otd. Biol., 87:19–31.

———. 1982b. Species names of jirds of the group *libycus-erythrourus-shawi-caudatus* (Rodentia, Gerbillinae, *Meriones*). Zool. Zhur. 61:1766–68.

Payne, J., C. M. Francis, and K. Phillipps. 1985. A field guide to the mammals of Borneo. Sabah Society with World Wildlife Fund Malaysia, 332 pp.

Payne, K., and R. Payne. 1985. Large scale changes over 19 years in songs of humpback whales in Bermuda. Z. Tierpsychol. 68:89–114.

Payne, K., P. Tyack, and R. Payne. 1983. Progressive changes in the songs of humpback whales *(Megaptera novaeangliae)*: a detailed analysis of two seasons in Hawaii. *In* Payne (1983), pp. 9–57.

Payne, M. R. 1978. Population size and age determination in the Antarctic fur seal *Arctocephalus gazella*. Mamm. Rev. 8:67–73.

Payne, N. 1986. The trade in Pacific fruit bats. Traffic Bull. 8(2):25–27.

Payne, R. 1976. At home with right whales. Natl. Geogr. 149:322–39.

———, ed. 1983. Communication and behavior of whales. Westview Press, Boulder, Colorado, 643 pp.

———. 1986. Long term behavioral studies of the southern right whale *(Eubalaena australis)*. *In* Brownell, Best, and Prescott (1986), pp. 161–67.

Payne, R., and E. M. Dorsey. 1983. Sexual dimorphism and aggressive use of callosities in right whales *(Eubalaena australis)*. *In* Payne (1983), pp. 295–329.

Payne, R., and L. N. Guinee. 1983. Humpback whale *(Megaptera novaeangliae)* songs as an indicator of "stocks." *In* Payne (1983), pp. 333–58.

Pearson, A. M. 1975. The northern interior grizzly bear *Ursus arctos* L. Can. Wildl. Serv. Rept. Ser., no. 34, 86 pp.

———. 1976. Population characteristics of the arctic mountain grizzly bear. *In* Pelton, Lentfer, and Folk (1976), pp. 247–60.

Pearson, E. W. 1978. A 1974 coyote harvest estimate for 17 western states. Wildl. Soc. Bull. 6:25–32.

Pearson, O. P. 1948. Life history of mountain viscachas in Peru. J. Mamm. 29:345–74.

———. 1972. New information on ranges and relationships within the rodent genus *Phyllotis* in Peru and Ecuador. J. Mamm. 53:677–86.

———. 1983. Characteristics of a mammalian fauna from forests in Patagonia, southern Argentina. J. Mamm. 64:476–92.

———. 1984. Taxonomy and natural history of some fossorial rodents of Patagonia, southern Argentina. J. Zool. 202:225–37.

Pearson, O. P., S. Martin, and J. Bellati. 1987. Demography and reproduction of the silky desert mouse *(Eligmodontia)* in Argentina. Fieldiana Zool., n.s., 39:433–46.

Pearson, O. P., and J. L. Patton. 1976. Relationships among South American phyllotine rodents based on chromosome analysis. J. Mamm. 57:339–50.

Pearson, O. P., et al. 1968. Estructura social, distribución espacial y composición por edades de una población de tuco-tucos *(Ctenomys talarum)*. Inv. Zool. Chilenas 13:47–80.

Peek, J. M., R. E. LeResche, and D. R. Stevens. 1974. Dynamics of moose aggregations in Alaska, Minnesota, and Montana. J. Mamm. 55:126–37.

Peek, J. M., M. R. Pelton, H. D. Picton, J. W. Schoen, and P. Zager. 1987. Grizzly bear conservation and management: a review. Wildl. Soc. Bull. 15:160–69.

Peek, J. M., D. L. Urich, and R. J. Mackie. 1976. Moose habitat selection and relationships to forest management in northeastern Minnesota. Wildl. Monogr., no. 48, 65 pp.

Pefaur, J., W. Hermosilla, F. Di Castri, R. Gonzalez, and F. Salinas. 1968. Estudio preliminar de mamiferos silvestres Chilenos: su distribución, valor económico e importancia zoonótica. Rev. Soc. Med. Vet. (Chile) 18(1–4):3–15.

Pelikan, J., and V. Holisova. 1969. Movements and home ranges of *Arvicola terrestris* on a brook. Zool. Listy 18:207–24.

Pellew, R. A. 1983. The giraffe and its food resource in the Serengeti. II. Response of the giraffe population to changes in the food supply. Afr. J. Ecol. 21:269–83.

Pelton, M. R., J. W. Lentfer, and G. E. Folk. 1976. Bears—their biology and management. Internatl. Union Conserv. Nat. Publ., n.s., no. 40, 467 pp.

Pelton, M. R., C. D. Scott, and G. M. Burghardt. 1976. Attitudes and opinions of persons experiencing property damage and/or injury by black bears in the Great Smoky Mountains National Park. *In* Pelton, Lentfer, and Folk (1976), pp. 157–67.

Pemberton, J. M., and A. P. Balmford. 1987. Lekking in fallow deer. J. Zool. 213:762–65.

Pembleton, E. F., and S. L. Williams. 1978. *Geomys pinetis*. Mammalian Species, no. 86, 3 pp.

Peng Hungshou and Wang Yingxiang. 1981. New mammals from the Gaoligong Mountains. Acta Theriol. Sinica 1:167–76.

Penny, C. 1989. Sichuan takin calves born at the San Diego Zoo. Amer. Assoc. Zool. Parks Aquar. Newsl. 30(10):17.

Penrod, B. 1976. Fisher in New York. Conservationist 31(2):20.

Penzhorn, B. L. 1979. Social organisation of the Cape mountain zebra *Equus z. zebra* in the Mountain Zebra National Park. Koedoe 22:115–56.

———. 1982. Home range sizes of Cape mountain zebras *Equus zebra zebra* in the Mountain Zebra National Park. Koedoe 25:103–8.

———. 1984. A long-term study of social organisation and behaviour of Cape mountain zebras *Equus zebra zebra*. Z. Tierpsychol. 64:97–146.

———. 1985. Reproductive characteristics of a free-ranging population of Cape mountain zebra *(Equus zebra zebra)*. J. Reprod. Fert. 73:51–57.

———. 1988. *Equus zebra*. Mammalian Species, no. 314, 7 pp.

Peracchi, A. L. 1968. Sobre os habitos de "*Histiotus velatus*" (Geoffroy, 1824) (Chiroptera, Vespertilionidae). Rev. Brasil. Biol. 28:469–73.

Perez, G. S. A. 1972. Observations of Guam bats. Micronesia 8:141–49.

———. 1973. Notes on the ecology and life history of Pteropodidae on Guam. Period. Biol. 75:163–68.

Pernetta, J. C. 1977. Population ecology of British shrews in grassland. Acta Theriol. 22:279–96.

Perrers, C. 1965. Notes on a pigmy possum, *Cercartetus nanus* Desmarest. Austral. Zool. 13:126.

Perrin, W. F. 1975a. Distribution and differentiation of populations of dolphins of the genus *Stenella* in the eastern tropical Pacific. J. Fish. Res. Bd. Can. 32:1059–67.

———. 1975b. Variation of spotted and spinner porpoise (genus *Stenella*) in the eastern tropical Pacific and Hawaii. Bull. Scripps Inst. Oceanogr. Univ. California 21:1–206.

———. 1976. First record of the melon-headed whale, *Peponocephala electra*, in the eastern Pacific, with a summary of world distribution. Fishery Bull. 74:457–58.

Perrin, W. F., P. B. Best, W. H. Dawbin, K. C. Balcomb, R. Gambell, and G. J. B. Ross. 1973. Rediscovery of Fraser's dolphin *Lagenodelphis hosei*. Nature 241:345–50.

Perrin, W. F., R. L. Brownell, Jr., and D. P. DeMaster, eds. 1984. Reproduction in whales, dolphins and porpoises. Rept. Internatl. Whaling Comm., Spec. Issue, no. 6, xii + 495 pp.

Perrin, W. F., J. M. Coe, and J. R. Zweifel. 1976. Growth and reproduction of the spotted porpoise, *Stenella attenuata*, in the offshore eastern tropical Pacific. Fishery Bull. 74:229–69.

Perrin, W. F., and J. R. Henderson. 1984. Growth and reproductive rates in two populations of spinner dolphins, *Stenella longirostris*, with different histories of exploitation. *In* Perrin, Brownell, and DeMaster (1984), pp. 417–30.

Perrin, W. F., D. B. Holts, and R. B. Miller. 1977. Growth and reproduction of the eastern spinner dolphin, a geographical form of *Stenella longirostris* in the eastern tropical Pacific. Fishery Bull. 75:725–50.

Perrin, W. F., R. B. Miller, and P. A. Sloan. 1977. Reproductive parameters of the offshore spotted dolphin, a form of *Stenella attenuata*, in the eastern tropical Pacific, 1973–75. Fishery Bull. 75:629–33.

Perrin, W. F., E. D. Mitchell, J. G. Mead, D. K. Caldwell, M. C. Caldwell, P. J. H. Van Bree, and W. H. Dawbin. 1987. Revision of the spotted dolphins, *Stenella* spp. Mar. Mamm. Sci. 3:99–170.

Perrin, W. F., E. D. Mitchell, J. G. Mead, D. K. Caldwell, and P. J. H. Van Bree. 1981. *Stenella clymene*, a rediscovered tropical dolphin of the Atlantic. J. Mamm. 62:583–98.

Perrin, W. F., and S. B. Reilly. 1984. Reproductive parameters of dolphins and small

whales of the family Delphinidae. *In* Perrin, Brownell, and DeMaster (1984), pp. 97–133.

Perrin, W. F., M. D. Scott, G. J. Walker, and V. L. Cass. 1985. Review of geographical stocks of tropical dolphins (*Stenella* spp. and *Delphinus delphis*) in the eastern Pacific. U.S. Natl. Mar. Fish. Serv., NOAA Tech. Rept. NMFS 28, iv + 28 pp.

Perrin, W. F., and W. A. Walker. 1975. The rough-toothed porpoise, *Steno bredanensis*, in the eastern tropical Pacific. J. Mamm. 56:905–9.

Perry, R. 1984. Key environments: Galapagos. Pergamon Press, Oxford, x + 321 pp.

Peshev, D. 1983. New karyotype forms of the mole rat, *Nannospalax leucodon* Nordmann (Spalacidae, Rodentia), in Bulgaria. Zool. Anz., Jena, 211:65–72.

Peters, G. 1982. A note on the vocal behaviour of the giant panda, *Ailuropoda melanoleuca* (David, 1869). Z. Saugetierk. 47:236–46.

Peters, R. P., and L. D. Mech. 1975. Scent-marking in wolves. Amer. Sci. 63:628–37.

Petersen, L. R., M. A. Martin, and C. M. Pils. 1977. Status of fishers in Wisconsin, 1975. Wisconsin Dept. Nat. Resources Res. Rept., no. 92, 9 pp.

Peterson, R. L. 1965a. A review of the flat-headed bats of the family Molossidae from South America and Africa. Roy. Ontario Mus. Life Sci. Contrib., no. 64, 32 pp.

———. 1965b. A review of the bats of the genus *Ametrida*, family Phyllostomidae. Roy. Ontario Mus. Life Sci. Contrib., no. 65, 13 pp.

———. 1969. Notes on the Malaysian fruit bats of the genus *Dyacopterus*. Roy. Ontario Mus. Life Sci. Occas. Pap., no. 13, 4 pp.

———. 1971a. The African molossid bat *Tadarida russata*. Can. J. Zool. 49:297–301.

———. 1971b. Notes on the African long-eared bats of the genus *Laephotis* (family Vespertilionidae). Can. J. Zool. 49:885–88.

———. 1971c. The systematic status of the African molossid bats *Tadarida bemmeleni* and *Tadarida cistura*. Can. J. Zool. 49:1347–54.

———. 1972. Systematic status of the African molossid bats *Tadarida congica*, *T. niangarae* and *T. trevori*. Roy. Ontario Mus. Life Sci. Contrib., no. 85, 32 pp.

———. 1973. The first known female of the African long-eared bat *Laephotis wintoni* (Vespertilionidae: Chiroptera). Can. J. Zool. 51:601–3.

———. 1974a. Variation in the African bat, *Tadarida lobata*, with notes on habitat and habits (Chiroptera: Molossidae). Roy. Ontario Mus. Life Sci. Occas. Pap., no. 24, 8 pp.

———. 1974b. A review of the general life

history of the moose. Naturaliste Canadien 101:9–21.

———. 1981. Systematic variation in the *tristis* group of the bent-winged bats of the genus *Miniopterus* (Chiroptera: Vespertilionidae). Can. J. Zool. 59:828–43.

———. 1982. A new species of *Glauconycteris* from the east coast of Kenya (Chiroptera: Vespertilionidae). Can. J. Zool. 60:2521–25.

———. 1985. A systematic review of the molossid bats allied with the genus *Mormopterus* (Chiroptera: Molossidae). Acta Zool. Fennica 170:205–8.

Peterson, R. L., and M. B. Fenton. 1970. Variation in the bats of the genus *Harpyionycteris*, with the description of a new race. Roy. Ontario Mus. Life Sci. Occas. Pap., no. 17, 15 pp.

Peterson, R. L., and D. L. Harrison. 1970. The second and third known specimens of the African molossid bat, *Tadarida lobata*. Roy. Ontario Mus. Life Sci. Occas. Pap., no. 16, 6 pp.

Peterson, R. L., and D. A. Smith. 1973. A new species of *Glauconycteris* (Vespertilionidae, Chiroptera). Roy. Ontario Mus. Life Sci. Occas. Pap., no. 22, 9 pp.

Peterson, R. O. 1977. Wolf ecology and prey relationships on Isle Royale. U.S. Natl. Park Serv. Sci. Monogr., no. 11, xx + 210 pp.

Peterson, R. O., and R. E. Page. 1988. The rise and fall of Isle Royale wolves, 1975–1986. J. Mamm. 69:89–99.

Peterson, R. S., and G. A. Bartholomew. 1967. The natural history and behavior of the California sea lion. Amer. Soc. Mamm. Spec. Publ., no. 1, xi + 79 pp.

Peterson, R. S., C. L. Hubbs, R. L. Gentry, and R. L. DeLong. 1968. The Guadalupe fur seal: habitat, behavior, population size, and field identification. J. Mamm. 49:665–75.

Petocz, R. 1989. The tamaraw. Internatl. Union Conserv. Nat. Species Survival Comm., Asian Wild Cattle Specialist Group Newsl., no. 2.

Petrov, B., and M. Todorović. 1982. *Dinaromys bogdanovi* (V. et E. Martino, 1922)—Bergmaus. Handbuch Säugetiere Europas 2(1):195–208.

Petter, F. 1972. The rodents of Madagascar: the seven genera of Malagasy rodents. *In* Battistini and Richard-Vindard (1972), pp. 661–65.

———. 1973. Les noms de genre *Cercomys*, *Trichomys* et *Proechimys* (Rongeurs, Echimyides). Mammalia 37:422–26.

———. 1975. Les *Praomys* de République Centrafricaine (Rongeurs, Muridés). Mammalia 39:51–56.

———. 1977. Les rats à mamelles multiples d'Afrique occidentale et centrale: *Mastomys*

erythroleucus (Temminck, 1853) et *M. huberti* (Wroughton, 1908). Mammalia 41:441–44.

———. 1978a. Une souris nouvelle du sud de l'Afrique: *Mus setzeri* sp. nov. Mammalia 42:377–79.

———. 1978b. Epidémiologie de la leishmaniose en Guyane française, en relation avec l'existence d'une espèce nouvelle de rongeurs echimyides, *Proechimys cuvieri* sp. n. Compt. Rend. Acad. Sci. Paris, ser. D, 287:261–64.

———. 1979. Une nouvelle espèce de rat d'eau de Guyane française, *Nectomys parvipes* sp. nov. (Rongeurs, Cricetidae). Mammalia 43:507–10.

———. 1981a. Les souris africaines du groupe *sorella* (Rongeurs, Muridés). Mammalia 45:313–20.

———. 1981b. Remarques sur la systematique des chrysochlorides. Mammalia 45:49–53.

———. 1982. Les parentes des *Otomys* du Mont Oku (Cameroun) et des autres formes rapportees a *O. irroratus* (Brants, 1827) (Rodentia, Muridae). Bonner Zool. Beitr. 33:215–22.

———. 1983a. *Paratatera*, sous-genre nouveau de *Gerbillurus* Shortridge, 1942 (Rongeurs, Gerbillides). Mammalia 47:265–66.

———. 1983b. Elements d'une revision des *Acomys* Africains. Un sous-genre nouveau, *Peracomys* Petter and Roche, 1981 (Rongeurs, Muridés). Ann. Mus. Roy. Afr. Cent., Sec. Zool., 237:109–19.

———. 1986. Un rongeur nouveau du Mont Oku (Cameroun) *Lamottemys okuensis*, gen nov, sp nov (Rodentia, Muridae). Cimbebasia, ser. A, 8:97–105.

Petter, F., and F. J. Baud. 1981. *Calomys laucha* en Colombie (Rongeurs, Cricetinae, Phyllotini). Mammalia 45:513–14.

Petter, F., and H. Cuenca Aguirre. 1982. Un *Isothrix* nouveau de Bolivie (Rongeurs, Echimyidés): La denture des *Isothrix*. Mammalia 46:191–203.

Petter, F., and J. Roche. 1981. Remarques preliminaires sur la systematique des *Acomys* (Rongeurs, Muridae). *Peracomys*, sous-genre nouveau. Mammalia 45:381–83.

Petter, J.-J. 1965. The lemurs of Madagascar. *In* De Vore (1965), pp. 292–319.

———. 1975. Breeding of Malagasy lemurs in captivity. *In* Martin (1975b), pp. 187–202.

———. 1977. The aye-aye. *In* Rainier III and Bourne (1977), pp. 37–57.

———. 1978. Ecological and physiological adaptations of five sympatric nocturnal lemurs to seasonal variations in food production. *In* Chivers and Herbert (1978), pp. 211–23.

Petter, J.-J., and P. Charles-Dominique. 1979.

Vocal communication in prosimians. *In* Doyle and Martin (1979), pp. 247–305.

Petter, J.-J., and A. Petter. 1967. The aye-aye of Madagascar. *In* Altmann (1967), pp. 195–205.

Petter, J.-J., and A. Petter-Rousseaux. 1979. Classification of the prosimians. *In* Doyle and Martin (1979), pp. 1–44.

Petter, J.-J., and A. Peyrieras. 1975. Preliminary notes on the behavior and ecology of *Hapalemur griseus*. *In* Tattersall and Sussman (1975), pp. 281–86.

Petter, J.-J., A. Schilling, and G. Pariente. 1975. Observations on behavior and ecology of *Phaner furcifer*. *In* Tattersall and Sussman (1975), pp. 209–18.

Petter-Rousseaux, A., and J.-J. Petter. 1967. Contribution á la systématique des Cheirogaleinae (lémuriens malagaches). *Allocebus*, gen. nov. pour *Cheirogaleus trichotis* Günther 1875. Mammalia 31:574–82.

Pettifer, H. L., and J. A. J. Nel. 1977. Hoarding in four southern African rodent species. Zool. Afr. 12:409–18.

Pettigrew, J. D. 1986. Flying primates? Megabats have the advanced pathway from eye to midbrain. Science 231:1304–6.

———. 1987. Are flying foxes (Chiroptera: Pteropodidae) really primates? Austral. Mamm. 10:119–24.

Peyton, B. 1980. Ecology, distribution, and food habits of spectacled bears, *Tremarctos ornatus*, in Peru. J. Mamm. 61:639–52.

———. 1986. Spectacled bear news. Species 6:15–16.

Phillips, C. J. 1966. A new species of bat of the genus *Melonycteris* from the Solomon Islands. J. Mamm. 47:23–27.

———. 1967. A collection of bats from Laos. J. Mamm. 48:633–36.

———. 1968. Systematics of megachiropteran bats in the Solomon Islands. Univ. Kansas Mus. Nat. Hist. Publ. 16:777–837.

Phillips, C. J., and E. C. Birney. 1968. Taxonomic status of the vespertilionid genus *Anamygdon* (Mammalia; Chiroptera). Proc. Biol. Soc. Washington 81:491–98.

Phillips, C. J., and J. K. Jones, Jr. 1968. Additional comments on reproduction in the woolly opossum *(Caluromys derbianus)* in Nicaragua. J. Mamm. 49:320–21.

———. 1969. Notes on reproduction and development in the four-eyed opossum, *Philander opossum*, in Nicaragua. J. Mamm. 50:345–48.

———. 1971. A new subspecies of the long-nosed bat, *Hylonycteris underwoodi*, from Mexico. J. Mamm. 52:77–80.

Phillips, R. L., and C. Jonkel, eds. 1977. Proceedings of the 1975 predator symposium. Montana Forest and Conserv. Exp. Sta., Univ. Montana, Missoula, ix + 268 pp.

Phillips, W. R., and S. J. Inwards. 1985. The annual activity and breeding cycles of Gould's long-eared bat, *Nyctophilus gouldi* (Microchiroptera: Vespertilionidae). Austral. J. Zool. 33:111–26.

Pidduck, E. R., and J. B. Falls. 1973. Reproduction and emergence of juveniles in *Tamias striatus* (Rodentia: Sciuridae) at two localities in Ontario, Canada. J. Mamm. 54:693–707.

Pidoplichko, I. G. 1973. On time of *Lagurus lagurus* Pall. extinction on the Ukraine right bank area. Vestn. Zool. 1973(5):35–41.

Piechocki, R. 1966. Uber die Nachweise der Langohr-Fledermause, *Plecotus auritus* L. und *Plecotus austriacus* Fischer in mitteldeutschen Raum. Hercynia 3:407–15.

Piekielek, W., and T. S. Burton. 1975. A black bear population study in northern California. California Fish and Game 61:4–25.

Pielowski, Z. 1972. Home range and degree of residence of the European hare. Acta Theriol. 17:93–103.

———. 1976. On the present state and perspectives of the European hare breeding in Poland. *In* Ecology and management of European hare populations, Proc. Internatl. Symp., 23–24 December 1974, Poznan, Poland, pp. 25–27.

———. 1984. Some aspects of population structure and longevity of field roe deer. Acta Theriol. 29:17–33.

Pienaar, U. D. V. 1970. A note on the occurrence of bat-eared fox *Otocyon megalotis megalotis* (Desmarest) in the Kruger National Park. Koedoe 13:23–27.

———. 1972. A new bat record for the Kruger National Park. Koedoe 15:91–93.

Pieper, H. 1978. Eine neue *Crocidura*-Art (Mammalia: Soricidae) von der Insel Kreta. Bonner Zool. Beitr. 4:281–86.

Pierson, E. D., V. M. Sarich, J. M. Lowenstein, M. J. Daniel, and W. E. Rainey. 1986. A molecular link between the bats of New Zealand and South America. Nature 323:60–63.

Pilbeam, D. 1984. The descent of hominoids and hominids. Sci. Amer. 250(3):84–96.

Pilgrim, W. 1980. Fisher, *Martes pennanti* (Carnivora: Mustelidae) in Labrador. Can. Field-Nat. 94:468.

Pilleri, G. 1971. On the La Plata dolphin *Pontoporia blainvillei* off the Uruguayan coast. Investig. Cetacea 3:59–67.

———. 1979. Observations on the ecology of *Inia geoffrensis* from the Rio Apure, Venezuela. Investig. Cetacea 10:137–43.

Pilleri, G., and Chen Peixun. 1979. How the finless porpoise *(Neophocaena asiaeorien-*

talis) carries its calves on its back, and the function of the denticulated area of skin, as observed in the Changjiang River, China. Investig. Cetacea 10:105–8.

Pilleri, G., and M. Gihr. 1971. Zur Systematik der Gattung *Platanista* (Cetacea). Rev. Suisse Zool. 78:746–59.

———. 1975. On the taxonomy and ecology of the finless black porpoise, *Neophocaena* (Cetacea, Delphinidae). Mammalia 39:657–73.

———. 1977a. Observations on the Bolivian (*Inia boliviensis* d'Orbigny, 1834) and the Amazonian bufeo (*Inia geoffrensis* de Blainville, 1817) with description of a new subspecies *(Inia geoffrensis humboldtiana)*. Investig. Cetacea 8:11–76.

———. 1977b. Neotype for *Platanista indi* Blyth, 1859. Investig. Cetacea 8:77–81.

———. 1977c. Radical extermination of the South American sea lion *Otaria byronia* (Pinnipedia, Otariidae) from Isla Verde, Uruguay. Brain Anat. Inst., Univ. Berne, Switzerland, 15 pp.

———. 1981a. Additional considerations on the taxonomy of the genus *Inia*. Investig. Cetacea 11:15–24.

———. 1981b. Checklist of the cetacean genera *Platanista*, *Inia*, *Lipotes*, *Pontoporia*, *Sousa*, and *Neophocaena*. Investig. Cetacea 11:33–36.

Pilleri, G., M. Gihr, P. E. Purves, K. Zbinden, and C. Kraus. 1976. On the behaviour, bioacoustics and functional morphology of the Indus River dolphin (*Platanista indi* Blyth, 1859). Investig. Cetacea 6:1–151.

Pilleri, G., and O. Pilleri. 1979a. Precarious situation of the dolphin population (*Platanista indi* Blyth, 1859) in the Punjab, upstream from the Taunsa Barrage, Indus River. Investig. Cetacea 10:121–27.

———. 1979b. Observations on the dolphins in the Indus Delta (*Sousa plumbea* and *Neophocaena phocaenoides*) in winter 1978–1979. Investig. Cetacea 10:129–35.

Pilleri, G., K. Zbinden, and M. Gihr. 1976. The "black finless porpoise" (*Neophocaena phocaenoides* Cuvier, 1829) is not black. Investig. Cetacea 7:161–64.

Pils, C. M., and M. A. Martin. 1978. Population dynamics, predator-prey relationships and management of the red fox in Wisconsin. Wisconsin Dept. Nat. Res. Tech. Bull., no. 105, 56 pp.

Pimlott, D. H., ed. 1975. Wolves. Internatl. Union Conserv. Nat. Suppl. Pap., n. s., no. 43, 144 pp.

Pimlott, D. H., J. A. Shannon, and G. B. Kolenosky. 1969. The ecology of the timber wolf in Algonquin Provincial Park. Ontario Dept. Lands and Forests, 92 pp.

Pine, R. H. 1971. A review of the long-

whiskered rice rat, *Oryzomys bombycinus* Goldman. J. Mamm. 52:590–96.

———. 1972a. A new subgenus and species of murine opossum (genus *Marmosa*) from Peru. J. Mamm. 53:279–82.

———. 1972b. The bats of the genus *Carollia*. Texas Agric. Exp. Sta. Tech. Monogr., no. 8, 125 pp.

———. 1973a. Anatomical and nomenclatural notes on opossums. Proc. Biol. Soc. Washington 86:391–402.

———. 1973b. Una nueva especie de *Akodon* (Mammalia: Rodentia: Muridae) de la Isla de Wellington, Magallanes, Chile. An. Inst. Patagonia 4:423–26.

———. 1975. A new species of *Monodelphis* (Mammalia: Didelphidae) from Bolivia. Mammalia 39:320–22.

———. 1976a. A new species of *Akodon* (Mammalia: Rodentia: Muridae: Cricetinae) from Isla de los Estados, Argentina. Mammalia 40:63–68.

———. 1976b. *Monodelphis umbristriata* (A. de Miranda-Ribeiro) is a distinct species of opossum. J. Mamm. 57:785–87.

———. 1977. *Monodelphis iheringi* (Thomas) is a recognizable species of Brazilian opossum (Mammalia: Marsupialia: Didelphidae). Mammalia 41:235–37.

———. 1979. Taxonomic notes on "*Monodelphis dimidiata itatiayae* (Miranda-Ribeiro)," *Monodelphis domestica* (Wagner) and *Monodelphis maraxina* Thomas (Mammalia: Marsupialia: Didelphidae). Mammalia 43:495–99.

———. 1980. Notes on rodents of the genera *Wiedomys* and *Thomasomys* (including *Wilfredomys*). Mammalia 44:195–202.

———. 1981. Reviews of the mouse opossums *Marmosa parvidens* Tate and *Marmosa invicta* Goldman (Mammalia: Marsupialia: Didelphidae) with description of a new species. Mammalia 45:56–70.

Pine, R. H., and J. P. Abravaya. 1978. Notes on the Brazilian opossum *Monodelphis scalops* (Thomas) (Mammalia: Marsupialia: Didelphidae). Mammalia 42:379–82.

Pine, R. H., D. C. Carter, and R. K. LaVal. 1971. Status of *Bauerus* Van Gelder and its relationships to other nyctophiline bats. J. Mamm. 52:663–69.

Pine, R. H., P. L. Dalby, Jr., and J. O. Matson. 1985. Ecology, postnatal development, morphometrics, and taxonomic status of the short-tailed opossum, *Monodelphis dimidiata*, an apparently semelparous annual marsupial. Ann. Carnegie Mus. 54:195–231.

Pine, R. H., and C. O. Handley, Jr. 1984. A review of the Amazonian short-tailed opossum *Monodelphis emiliae*. Mammalia 48:239–45.

Pine, R. H., S. D. Miller, and M. L. Schamberger. 1979. Contributions to the mammalogy of Chile. Mammalia 43:339–76.

Pine, R. H., and A. Ruschi. 1976. Concerning certain bats described and recorded from Espirito Santo, Brazil. An. Inst. Biol. Univ. Nal. Autón. Mexico, Ser. Zool., 47:183–96.

Pine, R. H., and R. M. Wetzel. 1975. A new subspecies of Pseudoryzomys wavrini (Mammalia: Rodentia: Muridae: Cricetinae) from Bolivia. Mammalia 39:649–55.

Pinedo, M. C. 1987. First record of a dwarf sperm whale from southwest Atlantic, with reference to osteology, food habits and reproduction. Sci. Rept. Whales Res. Inst. 38:171–86.

Pinter, A. J. 1970. Reproduction and growth for two species of grasshopper mice (Onychomys) in the laboratory. J. Mamm. 51:236–43.

Pinto da Silveira, E. K. 1968. Notas sôbre a historia natural do tamanduá mirim (Tamandua tetradactyla chiriquensis J. A. Allen 1904, Myrmecophagidae), com referências a fauna do Istmo do Panamá. Vellozia, Rio de Janeiro, no. 6, pp. 9–31.

———. 1969. História natural do tamanduá-bandeira Myrmecophaga tridactyla Linn. 1758, Myrmecophagidae. Vellozia, Rio de Janeiro, no. 7, 20 pp.

Pitman, R. L., A. Aguayo L., and J. Urbán R. 1987. Observations of an unidentified beaked whale (Mesoplodon sp.) in the eastern tropical Pacific. Mar. Mamm. Sci. 3:345–52.

Pitts, R. M. 1978. Carnivorous behavior in pygmy mice (Baiomys taylori). Bios 49:107–8.

Pitts, R. M., and M. J. Smolen. 1988. Record of Cryptotis parva from Brown County, Texas. Texas J. Sci. 40:113–14.

Pizzimenti, J. J. 1975. Evolution of the prairie dog genus Cynomys. Occas. Pap. Mus. Nat. Hist. Univ. Kansas, no. 39, 73 pp.

Pizzimenti, J. J., and L. R. McClenaghan, Jr. 1974. Reproduction, growth and development, and behavior in the Mexican prairie dog Cynomys mexicanus Merriam. Amer. Midl. Nat. 92:130–45.

Platt, W. J. 1976. The social organization and territoriality of short-tailed shrew (Blarina brevicauda) populations in old-field habitats. Anim. Behav. 24:305–18.

Poché, R. M. 1975. Notes on reproduction in Funisciurus anerythrus from Niger, Africa. J. Mamm. 56:700–701.

———. 1980. Elephant management in Africa. Wildl. Soc. Bull. 8:199–207.

Poché, R. M., and G. L. Bailie. 1974. Notes on the spotted bat (Euderma maculatum) from southwest Utah. Great Basin Nat. 34:254–56.

Poché, R. M., S. J. Evans, P. Sultana, M. E.

Hague, R. Sterner, and M. A. Siddique. 1987. Notes on the golden jackal (Canis aureus) in Bangladesh. Mammalia 51:259–70.

Poché, R. M., P. Sultana, S. J. Evans, M. E. Hague, M. A. Karim, and M. A. Siddique. 1982. Nesokia indica from Bangladesh. Mammalia 46:547–49.

Poduschka, W. 1980. Notes on the giant golden mole Chrysospalax trevelyani Gunther, 1875 (Mammalia: Insectivora) and its survival chances. Z. Saugetierk. 45:193–206.

Poduschka, W., and B. Richard. 1986. The Pyrenean desman—an endangered insectivore. Oryx 20:230–32.

Poelker, R. J., and H. D. Hartwell. 1973. Black bear of Washington. Washington State Game Dept. Biol. Bull., no. 14, viii + 180 pp.

Poglayen-Neuwall, I. 1966. Notes on care, display and breeding of olingos Bassaricyon. Internatl. Zoo Yearbook 6:169–71.

———. 1973. Preliminary notes on maintenance and behaviour of the Central American cacomistle Bassariscus sumichrasti. Internatl. Zoo Yearbook 13:207–11.

———. 1975. Copulatory behavior, gestation and parturition of the tayra (Eira barbara L., 1758). Z. Saugetierk. 40:176–89.

Poglayen-Neuwall, I., and I. Poglayen-Neuwall. 1976. Postnatal development of tayras (Carnivora: Eira barbara L., 1758). Zool. Beitr. 22:345–405.

———. 1980. Gestation period and parturition of the ringtail Bassariscus astutus (Liechtenstein, 1830). Z. Saugetierk. 45:73–81.

Poirier, F. E. 1970. The Nilgiri langur (Presbytis johnii) of south India. Primate Behav. 1:254–383.

———. 1972. The St. Kitts green monkey (Cercopithecus aethiops sabaeus): ecology, population dynamics, and selected behavior traits. Folia Primatol. 17:20–55.

———. 1982. Taiwan macaques: ecology and conservation needs. In Chiarelli and Corruccini (1982), pp. 138–42.

Pola, Y., and C. T. Snowdon. 1975. The vocalizations of pygmy marmosets (Cebuella pygmaea). Anim. Behav. 23:826–42.

Polaco, O. J. 1987. First record of Noctilio albiventris (Chiroptera: Noctilionidae) in Mexico. Southwestern Nat. 32:508–9.

Pollock, J. I. 1975. Field observations on Indri indri: a preliminary report. In Tattersall and Sussman (1975), pp. 287–311.

———. 1977. The ecology and sociobiology of feeding in Indri indri. In Clutton-Brock (1977), pp. 37–69.

———. 1979. Spatial distribution and ranging behavior in lemurs. In Doyle and Martin (1979), pp. 359–409.

———. 1986. The song of the indris (Indri indri; Primates: Lemuroidea): natural history, form, and function. Internatl. J. Primatol. 7:225–64.

Pollock, J. I., I. D. Constable, R. A. Mittermeier, J. Ratsirarson, and H. Simons. 1985. A note on the diet and feeding behavior of the aye-aye Daubentonia madagascariensis. Internatl. J. Primatol. 6:435–47.

Pook, A. G., and G. Pook. 1981. A field study of the socio-ecology of the Goeldi's monkey (Callimico goeldii) in northern Bolivia. Folia Primatol. 35:288–312.

Poole, J. H. 1987a. Rutting behaviour in African elephants: the phenomenon of musth. Behaviour 102:283–316.

———. 1987b. Elephants in musth, lust. Nat. Hist. 96(11):46–55.

———. 1989a. Announcing intent: the aggressive state of musth in African elephants. Anim. Behav. 37:140–52.

———. 1989b. Mate guarding, reproductive success and female choice in African elephants. Anim. Behav. 37:842–49.

Poole, J. H., and J. B. Thomsen. 1989. Elephants are not beetles: implications of the ivory trade for the survival of the African elephant. Oryx 23:188–98.

Poole, K. G., and R. P. Graf. 1985. Status of Dall's sheep in the Northwest Territories, Canada. In Hoefs (1985), pp. 35–42.

Poole, T. B., and H. D. R. Morgan. 1973. Differences in aggressive behaviour between male mice (Mus musculus L.) in colonies of different sizes. Anim. Behav. 21:788–95.

———. 1976. Social and territorial behaviour of laboratory mice (Mus musculus L.) in small complex areas. Anim. Behav. 24:476–80.

Poole, W. E. 1973. A study of breeding in grey kangaroos, Macropus giganteus Shaw and M. fuliginosus (Desmarest), in central New South Wales. Austral. J. Zool. 21:183–212.

———. 1975. Reproduction in the two species of grey kangaroos, Macropus giganteus Shaw and M. fuliginosus (Desmarest). II. Gestation, parturition and pouch life. Austral. J. Zool. 23:333–53.

———. 1976. Breeding biology and current status of the grey kangaroo, Macropus fuliginosus fuliginosus, of Kangaroo Island, South Australia. Austral. J. Zool. 24:169–87.

———. 1977. The eastern grey kangaroo, Macropus giganteus, in south-east South Australia: its limited distribution and need of conservation. CSIRO Div. Wildl. Res. Tech. Pap., no. 31, 15 pp.

———. 1978. Management of kangaroo harvesting in Australia. Austral. Natl. Parks Wildl. Serv. Occas. Pap., no. 2, 28 pp.

———. 1979. The status of the Australian Macropodidae. In Tyler (1979), pp. 13–27.

———. 1984. Management of kangaroo harvesting in Australia (1984). Austral. Natl. Parks Wildl. Serv. Occas. Pap., no. 9, 25 pp.

Poole, W. E., and P. C. Catling. 1974. Reproduction in the two species of grey kangaroos, *Macropus giganteus* Shaw and *M. fuliginosus* (Desmarest). I. Sexual maturity and oestrus. Austral. J. Zool. 22:277–302.

Popenoe, H. 1981. The water buffalo: new prospects for an underutilized animal. National Academy Press, Washington, D.C., vii + 118 pp.

———. 1983. Little-known Asian animals with a promising economic future. National Academy Press, Washington, D.C., xiv + 131 pp.

Por, F. D., and R. Ortal. 1985. River Jordan—the survival and future of a very special river. Environ. Conserv. 12:264–68.

Porter, F. L. 1978. Roosting patterns and social behavior in captive *Carollia perspicillata*. J. Mamm. 59:627–30.

———. 1979a. Social behavior in the leaf-nosed bat, *Carollia perspicillata*. I. Social organization. Z. Tierpsychol. 49:406–17.

———. 1979b. Social behavior in the leaf-nosed bat, *Carollia perspicillata*. II. Social communication. Z. Tierpsychol. 50:1–8.

Porter, F. L., and G. F. McCracken. 1983. Social behavior and allozyme variation in a captive colony of *Carollia perspicillata*. J. Mamm. 64:295–98.

Porton, I. J., D. G. Kleiman, and M. Rodden. 1987. Aseasonality of bush dog reproduction and the influence of social factors on the estrous cycle. J. Mamm. 68:867–71.

Posamentier, H., and H. F. Recher. 1974. The status of *Pseudomys novaehollandiae* (the New Holland mouse). Austral. Zool. 18:66–71.

Potter, D. R. 1982. Recreational use of elk. *In* Thomas and Toweill (1982), pp. 509–59.

Poulet, A. R. 1972. Recherches écologiques sur une savane Sahélienne du Ferlo Septentrional Sénégal: les mammifères. Terre Vie 26:440–72.

———. 1978. Evolution of the rodent population of a dry bush savanna in the Senegalese Sahel from 1969 to 1977. Bull. Carnegie Mus. Nat. Hist., no. 6, pp. 113–17.

Poulet, A. R., and H. Poupon. 1978. L'invasion d'*Arvicanthis niloticus* dans le Sahél Sénégalais en 1975–1976 et ses conséquences pour la strate ligneuse. Terre Vie 32:161–94.

Povilitis, A. 1983a. Social organization and mating strategy of the huemul *(Hippocamelus bisulcus)*. J. Mamm. 64:156–58.

———. 1983b. The huemul in Chile: national symbol in jeopardy? Oryx 17:34–40.

———. 1985. Social behavior of the huemul *(Hippocamelus bisulcus)* during the breeding season. Z. Tierpsychol. 68:261–86.

Povilitis, A., and E. Ceballos. 1986. The rare deer of Cedros Island, Mexico. Oryx 20:111–14.

Powell, J. A., Jr. 1978. Evidence of carnivory in manatees *(Trichechus manatus)*. J. Mamm. 59:442.

Powell, J. A., Jr., D. W. Belitsky, and G. B. Rathbun. 1981. Status of the West Indian manatee *(Trichechus manatus)* in Puerto Rico. J. Mamm. 62:642–46.

Powell, J. A., Jr., and G. B. Rathbun. 1984. Distribution and abundance of manatees along the northern coast of the Gulf of Mexico. Northeast Gulf Sci. 7:1–28.

Powell, R. A. 1981. *Martes pennanti*. Mammalian Species, no. 156, 6 pp.

———. 1982. The fisher. Life history, ecology, and behavior. Univ. Minnesota Press, Minneapolis, xvi + 217 pp.

Powell, R. A., and R. B. Brander. 1977. Adaptations of fishers and porcupines to their predator prey system. *In* Phillips and Jonkel (1977), pp. 45–53.

Prakash, I. 1971. Breeding season and litter size of Indian desert rodents. Z. Angewandte Zool. 58:441–54.

———. 1975. The population ecology of the rodents of the Rajasthan Desert, India. *In* Prakash and Ghosh (1975), pp. 75–116.

———. 1989. Rodent conservation in the Indian subcontinent. *In* Lidicker (1989), pp. 48–50.

Prakash, I., and P. K. Ghosh, eds. 1975. Rodents in desert environments. W. Junk, The Hague, xv + 624 pp.

Prakash, I., A. P. Jain, and B. D. Rana. 1975. A study of field populations of rodents in the Indian Desert. IV. Ruderal habitat. Z. Angewandte Zool. 62:339–48.

Prakash, I., and B. D. Rana. 1972. A study of field populations of rodents in the Indian Desert. II. Rocky and piedmont zones. Z. Angewandte Zool. 59:129–39.

Prasad, N. L. N. S. 1983. Home range size of blackbuck, *Antilope cervicapra*, at Mudmal. Z. Saugetierk. 48:109–17.

Pratt, D. M., and V. H. Anderson. 1979. Giraffe cow-calf relationships and social development of the calf in the Serengeti. Z. Tierpsychol. 51:233–51.

———. 1985. Giraffe social behaviour. J. Nat. Hist. 19:771–81.

Pratt, H. D., B. F. Bjornson, and K. S. Littig. 1977. Control of domestic rats and mice. U.S. Pub. Health Serv., Center for Disease Control, Atlanta, 47 pp.

Prescott, J. 1987. The status of the Thailand brow-antlered deer *(Cervus eldi siamensis)* in captivity. Mammalia 51:571–77.

Prescott, J., and J. Ferron. 1978. Breeding and behaviour development of the American red squirrel *Tamiasciurus hudsonicus*. Internatl. Zoo Yearbook 18:125–30.

Preston, E. M. 1975. Home range defense in the red fox *Vulpes vulpes* L. J. Mamm. 56:645–52.

Preuschoft, H. 1971. Mode of locomotion in subfossil giant lemuroids from Madagascar. Proc. 3rd Internatl. Congr. Primatol. 1:79–90.

Preuschoft, H., D. J. Chivers, W. Y. Brockelman, and N. Creel. 1984. The lesser apes: evolutionary and behavioural biology. Edinburgh Univ. Press, xiii + 709 pp.

Price, M. R. S. 1988. Field operations and research in Oman. *In* Dixon and Jones (1988), pp. 18–34.

Pritchard, J. S. 1989. Ilin Island cloud rat extinct? Oryx 23:126.

Prociv, P. 1983. Seasonal behaviour of *Pteropus scapulatus* (Chiroptera: Pteropodidae). Austral. Mamm. 6:45–46.

Prothero, D. R., E. Manning, and C. B. Hanson. 1986. The phylogeny of the Rhinocerotoidea (Mammalia, Perissodactyla). Zool. J. Linnean Soc. 87:341–66.

Prouty, L. A., P. D. Buchanan, W. S. Politzer, and A. R. Mootnick. 1983. *Bunopithecus*: a genus-level taxon for the hoolock gibbon *(Hylobates hoolock)*. Amer. J. Primatol. 5:83–87.

Prynn, D. 1980. Tigers and leopards in Russia's Far East. Oryx 15:496–503.

Pucek, Z. 1984. What to do with the European bison, now saved from extinction? Acta Zool. Fennica 172:187–90.

———. 1986. European bison. Species 7:9–10.

———. 1989. A preliminary report on threatened rodents in Europe. *In* Lidicker (1989), pp. 26–32.

Pulliainen, E. 1965. On the distribution and migrations of the arctic fox (*Alopex lagopus* L.) in Finland. Aquilo, Ser. Zool., 2:25–40.

———. 1968. Breeding biology of the wolverine (*Gulo gulo* L.) in Finland. Ann. Zool. Fennici 5:338–44.

———. 1974. Seasonal movements of moose in Europe. Naturaliste Canadien 101:379–92.

———. 1975. Wolf ecology in northern Europe. *In* Fox (1975), pp. 292–99.

———. 1980. The status, structure and behaviour of populations of the wolf (*Canis l. lupus* L.) along the Fenno-Soviet border. Ann. Zool. Fennici 17:107–12.

Pulliainen, E., and P. Ovaskainen. 1975. Territory marking by a wolverine *(Gulo gulo)* in northeastern Lapland. Ann. Zool. Fennici 12:268–70.

Purves, P. E., and G. Pilleri. 1974–75. Observations on the ear, nose, throat and eye of *Platanista indi.* Investig. Cetacea 5:13–57.

———. 1978. The functional anatomy and general biology of *Pseudorca crassidens* (Owen) with a review of the hydrodynamics and acoustics in Cetacea. Investig. Cetacea 9:68–227.

———. 1983. Echolocation in whales and dolphins. Academic Press, London, xiv + 261 pp.

Pye, T., and W. N. Bonner. 1980. Feral brown rats, *Rattus norvegicus,* in South Georgia (South Atlantic Ocean). J. Zool. 192:237–55.

Q

Quadagno, D. M., J. T. Allin, R. J. Brooks, R. D. St. John, and E. M. Banks. 1970. Some aspects of the reproductive biology of *Baiomys taylori ater.* Amer. Midl. Nat. 84:550–51.

Quinn, N. W. S., and G. Parker. 1987. Lynx. *In* Novak, Baker, et al. (1987), pp. 682–94.

Qumsiyeh, M. B. 1980. New records of bats from Jordan. Saugetierk. Mitt. 40:36–39.

———. 1983. Occurrence and zoogeographical implications of *Myotis blythi* (Tomes, 1857) in Libya. Mammalia 47:429–30.

———. 1985. The bats of Egypt. Spec. Publ. Mus. Texas Tech Univ., no. 23, 102 pp.

———. 1986. Phylogenetic studies of the rodent family Gerbillidae: I. Chromosomal evolution in the southern African complex. J. Mamm. 67:680–92.

Qumsiyeh, M. B., and J. K. Jones, Jr. 1986. *Rhinopoma hardwickii* and *Rhinopoma muscatellum.* Mammalian Species, no. 263, 5 pp.

Qumsiyeh, M. B., and D. A. Schlitter. 1982. The bat fauna of Jabal Al Akhdar, northeast Libya. Ann. Carnegie Mus. 51:377–89.

Quris, R. 1975. Ecologie et organisation sociale de *Cercocebus galeritus agilis* dans le nord-est du Gabon. Terre Vie 20:337–98.

R

Rabinowitz, A. 1988. The clouded leopard in Taiwan. Oryx 22:46–47.

Rabinowitz, A., P. Andau, and P. P. K. Chai. 1987. The clouded leopard in Malaysian Borneo. Oryx 21:107–11.

Rabor, D. S. 1939. *Sciuropterus mindanensis* sp. nov., a new species of flying squirrel from Mindanao. Philippine J. Sci. 69:389–93.

———. 1952. Two new mammals from Negros Island, Philippines. Nat. Hist. Misc. (Chicago Acad. Sci.), no. 96, 7 pp.

Racey, P. A. 1973. The time of onset of hibernation in pipistrelle bats, *Pipistrellus pipistrellus.* J. Zool. 171:465–67.

Rado, R., N. Levi, H. Hauser, J. Witcher, N. Adler, N. Intrator, Z. Wollberg, and J. Terkel. 1987. Seismic signalling as a means of communication in a subterranean mammal. Anim. Behav. 35:1249–66.

Raemaekers, J. 1984. Large versus small gibbons: relative roles of bioenergetics and competition in their ecological segregation in sympatry. *In* Preuschoft et al. (1984), pp. 209–18.

Rageot, R. 1978. Observaciones sobre el monito del monte. Chile Min. Agric. Corp. Nac. For. Dept. Tec. IX—Reg. Interp.—V. Silvestre, 16 pp.

Rahm, U. 1962. L'élevage et la reproduction en captivité de l'*Atherurus africanus* (Rongeurs, Hystricidae). Mammalia 26:1–9.

———. 1966. Les mammifères de la forêt équatoriale de l'Est du Congo. Ann. Mus. Roy. Afr. Cent. (Tervuren, Belgium), ser. IN-8°, 149:39–121.

———. 1969a. Dokumente über *Anomalurus* und *Idiurus* des ostlichen Kongo. Z. Saugetierk. 34:75–84.

———. 1969b. Zur Fortpflanzungsbiologie von *Tachyoryctes ruandae* (Rodentia, Rhizomyidae). Rev. Suisse Zool. 76:695–702.

———. 1970a. Note sur la reproduction des sciurides et murides dans la forêt équatoriale au Congo. Rev. Suisse Zool. 77:635–46.

———. 1970b. Ecology, zoogeography, and systematics of some African forest monkeys. *In* Napier and Napier (1970), pp. 589–626.

———. 1972. Zur Oekologie der Muriden in Regenwaldgebiet des ostlichen Kongo (Zaire). Rev. Suisse Zool. 79:1121–30.

Rainier III, Prince of Monaco, and G. H. Bourne. 1977. Primate Conservation. Academic Press, New York, xviii + 658 pp.

Rajagopalan, P. K. 1968. Notes on the Malabar spiny dormouse, *Platacanthomys lasiurus* Blyth, 1859, with new distribution record. J. Bombay Nat. Hist. Soc. 65:214–15.

Rajasingh, G. J. 1984. A note on the longevity and fertility of the blackbuck, *Antilope cervicapra* (Linnaeus). J. Bombay Nat. Hist. Soc. 80:632–34.

Ralls, K. 1973. *Cephalophus maxwelli.* Mammalian Species, no. 31, 4 pp.

———. 1976. Mammals in which females are larger than males. Quart. Rev. Biol. 51:245–76.

———. 1978. *Tragelaphus eurycerus.* Mammalian Species, no. 111, 4 pp.

Ralls, K., J. Ballou, and R. L. Brownell, Jr. 1983. Diversity in California sea otters: theoretical considerations and management implications. Biol. Conserv. 25:209–32.

Ralls, K., H. K. Buechner, R. Kiltie, and K. Kranz. 1985. Behavior and reproduction of captive bongo, *Tragelaphus eurycerus.* Zool. Garten 55:41–67.

Ralls, K., P. Fiorelli, and S. Gish. 1985. Vocalizations and vocal mimicry in captive harbor seals, *Phoca vitulina.* Can. J. Zool. 63:1050–56.

Ramirez, M. F., C. H. Freese, and J. Revilla C. 1977. Feeding ecology of the pygmy marmoset, *Cebuella pygmaea,* in northeastern Peru. *In* Kleiman (1977a), pp. 91–104.

Ramirez-Pulido, J., and W. Lopez-Forment. 1979. Additional records of some Mexican bats. Southwestern Nat. 24:541–44.

Ramsay, M. A., and I. Stirling. 1988. Reproductive biology and ecology of female polar bears *(Ursus maritimus).* J. Zool. 214:601–34.

Ranck, G. L. 1968. The rodents of Libya. Bull. U.S. Natl. Mus. 275:1–264.

Randall, J. A. 1984. Territorial defense and advertisement by footdrumming in bannertail kangaroo rats *(Dipodomys spectabilis)* at high and low population densities. Behav. Ecol. Sociobiol. 16:11–20.

———. 1987. Sandbathing as a territorial scent-mark in the bannertail kangaroo rat, *Dipodomys spectabilis.* Anim. Behav. 35:426–34.

Randall, J. A., and C. M. Stevens. 1987. Footdrumming and other anti-predator responses in the bannertail kangaroo rat *(Dipodomys spectabilis).* Behav. Ecol. Sociobiol. 20:187–94.

Raphael, M. G. 1984. Late fall breeding of the northern flying squirrel, *Glaucomys sabrinus.* J. Mamm. 65:138–39.

Rasa, O. A. E. 1972. Aspects of social organization in captive dwarf mongooses. J. Mamm. 53:181–85.

———. 1973. Marking behaviour and its social significance in the African dwarf mongoose, *Helogale undulata rufula.* Z. Tierpsychol. 32:293–318.

———. 1975. Mongoose sociology and behaviour as related to zoo exhibition. Internatl. Zoo Yearbook 15:65–73.

———. 1976. Invalid care in the dwarf mongoose *(Helogale undulata rufula).* Z. Tierpsychol. 42:337–42.

———. 1977. The ethology and sociology of the dwarf mongoose *(Helogale undulata rufula).* Z. Tierpsychol. 43:337–406.

———. 1983. A case of invalid care in wild dwarf mongooses. Z. Tierpsychol. 62:235–40.

———. 1986. Ecological factors and their relationship to group size, mortality and behaviour in the dwarf mongoose *Helogale undulata* (Peters, 1852). Cimbebasia, ser. A, 8:15–21.

Rasmussen, J. L., and R. L. Tilson. 1984. Food provisioning by adult maned wolves *(Chrysocyon brachyurus)*. Z. Tierpsychol. 65:346–52.

Rasweiler, J. J., IV. 1973. Care and management of the long-tongued bat, *Glossophaga soricina* (Chiroptera: Phyllostomatidae), in the laboratory, with observations on estivation induced by food deprivation. J. Mamm. 54:391–404.

Rathbun, G. B. 1973. Territoriality in the golden-rumped elephant shrew. E. Afr. Wildl. J. 11:405.

———. 1979. *Rhynchocyon chrysopygus*. Mammalian Species, no. 117, 4 pp.

———. 1984. Sirenians. *In* Anderson and Jones (1984), pp. 537–48.

Rathbun, G. B., and M. Gache. 1980. Ecological survey of the night monkey, *Aotus trivirgatus*, in Formosa Province, Argentina. Primates 21:211–19.

Rathbun, G. B., and T. J. O'Shea. 1984. The manatee's simple social life. *In* Macdonald (1984), pp. 300–301.

Ratomponirina, C., J. Andrianivo, and Y. Rumpler. 1982. Spermatogenesis in several intra- and interspecific hybrids of the lemur *(Lemur)*. J. Reprod. Fert. 66:717–21.

Rau, R. 1978. The end of the hunt.? Natl. Wildl. 16(3):4–11.

Rau, R. E. 1978. Additions to the revised fist of preserved material of the extinct Cape Colony quagga and notes on the relationship and distribution of southern plains zebras. Ann. S. Afr. Mus. 77:27–45.

———. 1983. The colouration of the extinct Cape Colony quagga. Afr. Wildl. 37(4):136–41.

Rausch, R. A., and A. M. Pearson. 1972. Notes on the wolverine in Alaska and the Yukon Territory. J. Wildl. Mgmt. 36:249–68.

Rausch, R. A., R. J. Somerville, and R. H. Bishop. 1974. Moose management in Alaska. Naturaliste Canadien 101:705–21.

Rausch, R. L. 1953. On the status of some arctic mammals. Arctic 6:91–148.

———. 1963. Geographic variation in size in North American brown bears, *Ursus arctos* L., as indicated by condylobasal length. Can. J. Zool. 41:33–45.

Rausch, R. L., and V. R. Rausch. 1975. Taxonomy and zoogeography of *Lemmus* sp. (Rodentia: Arvicolinae), with notes on laboratory-reared lemmings. Z. Saugetierk. 40:8–34.

Rautenbach, I. L., and I. W. Espie. 1982. First records of occurrence for two species of bats in the Kruger National Park. Koedoe 25:111–12.

Rautenbach, I. L., M. B. Fenton, and L. E. O. Braack. 1985. First records of five species of insectivorous bats from the Kruger National Park. Koedoe 28:73–80.

Rautenbach, I. L., and J. A. J. Nel. 1975. Further records of smaller mammals from the Kalahari Gemsbok National Park. Koedoe 18:195–98.

Rautenbach, I. L., and D. A. Schlitter. 1978. Revision of genus *Malacomys* of Africa (Mammalia: Muridae). Ann. Carnegie Mus. 47:385–422.

Rautenbach, I. L., D. A. Schlitter, and L. E. O. Braack. 1984. New distributional records of bats for the Republic of South Africa, with special reference to the Kruger National Park. Koedoe 27:131–35.

Ray, C. E. 1964. The taxonomic status of *Heptaxodon* and dental ontogeny in *Elasmodontomys* and *Amblyrhiza*. Bull. Mus. Comp. Zool. 131:107–27.

Ray, C. E., O. J. Linares, and G. S. Morgan. 1988. Paleontology. *In* Greenhall and Schmidt (1988), pp. 19–30.

Ray, C. E., and A. E. Spiess. 1981. The bearded seal, *Erignathus barbatus*, in the Pleistocene of Maine. J. Mamm. 62:423–27.

Ray, G. C. 1981. Ross seal—*Ommatophoca rossi*. *In* Ridgway and Harrison (1981b), pp. 237–60.

Ray, G. C., and W. E. Schevill. 1974. Feeding of a captive gray whale, *Eschrichtius robustus*. Mar. Fish. Rev. 36(4):31–38.

Ray, G. C., W. A. Watkins, and J. J. Burns. 1969. The underwater song of *Erignathus* (bearded seal). Zoologica 54:79–83.

Raybourne, J. W. 1987. Black bear: home in the highlands. *In* Kallman (1987), pp. 105–17.

Read, A. J., and D. E. Gaskin. 1988. Incidental catch of harbor porpoises by gill nets. J. Wildl. Mgmt. 52:517–23.

Read, A. J., K. Van Waerebeek, J. C. Reyes, J. S. McKinnon, and L. C. Lehman. 1988. The exploitation of small cetaceans in coastal Peru. Biol. Conserv. 46:53–70.

Read, B., and R. J. Frueh. 1980. Management and breeding of Speke's gazelle *Gazella spekei* at the St. Louis Zoo, with a note on artificial insemination. Internatl. Zoo Yearbook 20:99–104.

Read, D. G. 1984. Reproduction and breeding season of *Planigale gilesi* and *P. tenuirostris* (Marsupialia: Dasyuridae). Austral. Mamm. 7:161–73.

Ream, R. R. 1980. Wolf ecology project: annual report. Wilderness Inst., Univ. Montana, 58 pp.

Reason, R. C., and E. W. Laird. 1988. Determinants of dominance in captive female addax *(Addax nasomaculatus)*. J. Mamm. 69:375–77.

Redding, R. W., and D. M. Lay. 1978. Description of a new species of shrew of the genus *Crocidura* (Mammalia: Insectivora: Soricidae) from southwestern Iran. Z. Saugetierk. 43:306–10.

Redford, K. H. 1987. The pampas deer *(Ozotoceros bezoarticus)* in central Brazil. *In* Wemmer (1987), pp. 410–14.

Redford, K. H., and R. M. Wetzel. 1985. *Euphractus sexcinctus*. Mammalian Species, no. 252, 4 pp.

Redhead, T. D., and J. L. McKean. 1975. A new record of the false water-rat, *Xeromys myoides*, from the Northern Territory of Australia. Austral. Mamm. 1:347–54.

Reduker, D. W., T. L. Yates, and I. F. Greenbaum. 1983. Evolutionary affinities among southwestern long-eared *Myotis* (Chiroptera: Vespertilionidae). J. Mamm. 64:666–77.

Rees, M. D. 1989. Red wolf recovery effort intensifies. Endangered Species Tech. Bull. 14(1–2):3.

Reeve, N. J. 1982. The home range of the hedgehog as revealed by a radio tracking study. *In* Cheeseman and Mitson (1982), pp. 207–30.

Reeves, R. R. 1977. Hunt for the narwhal. Oceans 10(4):50–57.

———. 1978. Atlantic walrus *(Odobenus rosmarus rosmarus)*: a literature survey and status report. U.S. Fish and Wildl. Serv. Wildl. Res. Rept., no. 10, 41 pp.

Reeves, R. R., and S. Leatherwood. 1985. Bowhead whale—*Balaena mysticetus*. *In* Ridgway and Harrison (1985), pp. 305–44.

Reeves, R. R., and J. K. Ling. 1981. Hooded seal—*Cystophora cristata*. *In* Ridgway and Harrison (1981b), pp. 171–94.

Reeves, R. R., J. G. Mead, and S. Katona. 1978. The right whale, *Eubalaena glacialis*, in the western North Atlantic. Rept. Internatl. Whaling Comm. 28:303–12.

Reeves, R. R., and E. Mitchell. 1981. The whale behind the tusk. Nat. Hist. 90(8):50–57.

———. 1984. Catch history and initial population of white whales *(Delphinapterus leucas)* in the River and Gulf of St. Lawrence, eastern Canada. Naturaliste Canadien 111:63–121.

———. 1987a. Catch history, former abundance, and distribution of white whales in Hudson Strait and Ungava Bay. Naturaliste Canadien 114:1–65.

———. 1987b. History of white whale *(Delphinapterus leucas)* exploitation in eastern Hudson Bay and James Bay. Can. Spec. Publ. Fish. Aquat. Sci., no. 95, vi + 45 pp.

———. 1987c. Distribution and migration, exploitation, and former abundance of white whales *(Delphinapterus leucas)* in Baffin Bay and adjacent waters. Can. Spec. Publ. Fish. Aquat. Sci., no. 99, vi + 34 pp.

———. 1988. Current status of the gray whale, *Eschrichtius robustus*. Can. Field-Nat. 102:369–90.

Reeves, R. R., E. Mitchell, A. Mansfield, and M. McLaughlin. 1983. Distribution and migration of the bowhead whale, *Balaena mysticetus*, in the eastern North American Arctic. Arctic 36:5–64.

Reeves, R. R., and S. Tracey. 1980. *Monodon monoceros*. Mammalian Species, no. 127, 7 pp.

Reeves, R. R., D. Tuboku-Metzger, and R. A. Kapindi. 1988. Distribution and exploitation of manatees in Sierra Leone. Oryx 22:75–84.

Reichman, O. J. 1975. Relation of desert rodent diets to available resources. J. Mamm. 56:731–51.

Reichman, O. J., and R. J. Baker. 1972. Distribution and movements of two species of pocket gophers (Geomyidae) in an area of sympatry in the Davis Mountains, Texas. J. Mamm. 53:21–33.

Reid, W. H., and G. R. Patrick. 1983. Gemsbok *(Oryx gazella)* in White Sands National Monument. Southwestern Nat. 28:97–99.

Reig, O. A. 1970. Ecological notes on the fossorial octodont rodent *Spalacopus cyanus* (Molina). J. Mamm. 51:592–601.

———. 1978. Roedores cricetidos del Plioceno Superior de la Provincia de Buenos Aires (Argentina). Publ. Mus. Munic. Cien. Nat. Mar del Plata "Lorenzo Scaglia," 2:164–90.

———. 1980. A new fossil genus of South American cricetid rodents allied to *Wiedomys*, with an assessment of the Sigmodontinae. J. Zool. 192:257–81.

———. 1987. An assessment of the systematics and evolution of the Akodontini, with the description of new fossil species of *Akodon* (Cricetidae: Sigmodontinae). Fieldiana Zool., n.s., 39:347–99.

Reijnders, P. J. H. 1982. Observation, catch and release of a walrus, *Odobenus rosmarus*, on the Isle of Ameland, Dutch Wadden Sea. Zool. Anz., Jena, 209:88–90.

Reilly, S. B. 1978. Pilot Whale. *In* Haley (1978), pp. 112–19.

Reiner, F. 1986. First record of Sowerby's beaked whale from Azores. Sci. Rept. Whales Res. Inst. 37:103–7.

Reinhardt, J. 1852. Description of *Carterodon sulcidens*. Lund. Ann. Mag. Nat. Hist., ser. 2, 10:420.

Reiter, J., K. J. Panken, and B. J. Le Boeuf. 1981. Female competition and reproductive success in northern elephant seals. Anim. Behav. 29:670–87.

Renfree, M. B. 1980. Embryonic diapause in the honey possum *Tarsipes spencerae*. Search 11:81.

———. 1983. Marsupial reproduction: the choice between placentation and lactation. Oxford Res. Reprod. Biol. 5:1–29.

Renfree, M. B., E. M. Russell, and R. D. Wooller. 1984. Reproduction and life history of the honey possum, *Tarsipes rostratus*. *In* Smith and Hume (1984), pp. 427–37.

Rennert, P. D., and C. W. Kilpatrick. 1986. Biochemical systematics of populations of *Peromyscus boylii*. I. Populations from east-central Mexico with low fundamental numbers. J. Mamm. 67:481–88.

———. 1987. Biochemical systematics of *Peromyscus boylii*. II. Chromosomally variable populations from eastern and southern Mexico. J. Mamm. 68:799–811.

Repenning, C. A. 1967. Subfamilies and genera of the Soricidae. U.S. Geol. Surv. Prof. Pap., no. 565, iv + 74 pp.

———. 1983. *Pitymys meadensis* Hibbard from the Valley of Mexico and the classification of North American species of *Pitymys* (Rodentia: Cricetidae). J. Vert. Paleontol. 2:471–82.

Repenning, C. A., and R. H. Tedford. 1977. Otarioid seals of the Neogene. U.S. Geol. Surv. Prof. Pap., no. 992, vi + 93 pp.

Repenning, R. W., and S. R. Humphrey. 1986. The Chadwick Beach cotton mouse (Rodentia: *Peromyscus gossypinus restrictus*) may be extinct. Florida Sci. 49:259–62.

Reumer, J. W. F., and S. Payne. 1986. Notes on the Soricidae (Insectivora, Mammalia) from Crete. II. The shrew remains from Minoan and classical Kommos. Bonner Zool. Beitr. 37:173–82.

Reumer, J. W. F., and E. A. C. Sanders. 1984. Changes in the vertebrate fauna of Menorca in prehistoric and classical times. Z. Saugetierk. 49:321–25.

Reynolds, H. G., and F. Turkowski. 1972. Reproductive variations in the round-tailed ground squirrel as related to wlnter rainfall. J. Mamm. 53:893–98.

Reynolds, J. C. 1985. Details of the geographic replacement of the red squirrel *(Sciurus vulgaris)* by the grey squirrel *(Sciurus carolinensis)* in eastern England. J. Anim. Ecol. 54:149–62.

Reynolds, J. E., III. 1981. Aspects of the social behaviour and herd structure of a semi-isolated colony of West Indian manatees, *Trichechus manatus*. Mammalia 45:431–51.

Reynolds, J. E., III, and J. C. Ferguson. 1984. Implications of the presence of manatees *(Trichechus manatus)* near the Dry Tortugas Islands. Florida Sci. 47:187–89.

Reynolds, V., and F. Reynolds. 1965. Chimpanzees of the Budongo Forest. *In* De Vore (1965), pp. 368–424.

Ricciuti, E. R. 1978. Dogs of war. Internatl. Wildl. 8(5):36–40.

———. 1980. The ivory wars. Anim. Kingdom 83(1):6–58.

Rice, C. 1989. Four-horned antelope. Gnusletter 8(1):7.

Rice, C. G. 1985. Courting 'round the mountain. Anim. Kingdom 88(6):22–31.

———. 1988a. Habitat, population dynamics, and conservation of the Nilgiri tahr, *Hemitragus hylocrius*. Biol. Conserv. 44:137–56.

———. 1988b. Reproductive biology of Nilgiri tahr, *Hemitragus hylocrius* (Mammalia: Bovidae). J. Zool. 214:269–84.

Rice, D. W. 1967. Cetaceans. *In* Anderson and Jones (1967), pp. 291–324.

———. 1977. A list of the marine mammals of the world (third edition). U.S. Natl. Mar. Fish. Serv., NOAA Tech. Rept. NMFS SSRF-711, iii + 15 pp.

———. 1978a. Blue whale. *In* Haley (1978), pp. 30–35.

———. 1978b. Gray whale. *In* Haley (1978), pp. 54–61.

———. 1978c. Sperm whales. *In* Haley (1978), pp. 82–87.

———. 1978d. Beaked whales. *In* Haley (1978), pp. 88–95.

———. 1984. Cetaceans. *In* Anderson and Jones (1984), pp. 447–90.

Rice, D. W., and A. A. Wolman. 1971. The life history and ecology of the gray whale *(Eschrichtius robustus)*. Amer. Soc. Mamm. Spec. Publ., no. 3, viii + 142 pp.

Rice, D. W., A. A. Wolman, and H. W. Braham. 1984. The gray whale, *Eschrichtius robustus*. Mar. Fish. Rev. 46(4):7–14.

Rice, E. K. 1988. The chinchilla: endangered whistler of the Andes. Anim. Kingdom 91(1):6–7.

Rich, M. S. 1983. The longevity record for *Lynx canadensis* Kerr, 1792. Zool. Garten 53:365.

Richard, A. 1974. Patterns of mating in *Propithecus verreauxi verreauxi*. *In* Martin, Doyle, and Walker (1974), pp. 49–74.

———. 1977. The feeding behaviour of *Propithecus verreauxi*. *In* Clutton-Brock (1977), pp. 72–96.

———. 1978. Variability in the feeding behavior of a Malagasy prosimian, *Propithecus verreauxi*: Lemuriformes. *In* Montgomery (1978), pp. 519–33.

———. 1985. Social boundaries in a Malagasy prosimian, the sifaka *(Propithecus verreauxi)*. Internatl. J. Primatol. 6:553–68.

Richard, A. F., and R. W. Sussman. 1975. Future of the Malagasy lemurs: conservation or extinction? *In* Tattersall and Sussman (1975), pp. 335–50.

Richard, P. B. 1973. Capture, transport and husbandry of the Pyrenean desman *Galemys pyrenaicus*. Internatl. Zoo Yearbook 13:175–77.

———. 1976. Détermination de l'age et de la longévité chez le desman des Pyrénées *(Galemys pyrenaicus)*. Terre Vie 30:181–92.

Richard, P. B., and A. V. Viallard. 1969. Le desman des Pyrénées *(Galemys pyrenaicus)*: premières notes sur sa biologie. Terre Vie 23:225–45.

Richard, P. R., and R. R. Campbell. 1988. Status of the Atlantic walrus, *Odobenus rosmarus rosmarus*, in Canada. Can. Field-Nat. 102:337–50.

Richards, R. L. 1980. Rice rat (*Oryzomys* cf. *palustris*) remains from southern Indiana caves. Proc. Indiana Acad. Sci. 89:425–31.

Richardson, B. J., and G. B. Sharman. 1976. Biochemical and morphological observations on the wallaroos (Macropodidae: Marsupialia) with a suggested new taxonomy. J. Zool. 179:499–513.

Richardson, E. G. 1977. The biology and evolution of the reproductive cycle of *Miniopterus schreibersii* and *M. australis* (Chiroptera: Vespertilionidae). J. Zool. 183:353–75.

Richardson, L., T. W. Clark, S. C. Forrest, and T. M. Campbell, III. 1987. Winter ecology of black-footed ferrets *(Mustela nigripes)* at Meeteetse, Wyoming. Amer. Midl. Nat. 117:225–39.

Rick, A. M. 1968. Notes on bats from Tikal, Guatemala. J. Mamm. 49:516–20.

Rickart, E. A. 1977. Reproduction, growth and development in two species of cloud forest *Peromyscus* from southern Mexico. Occas. Pap. Mus. Nat. Hist. Univ. Kansas, no. 67, 22 pp.

Riddle, B. R., and J. R. Choate. 1986. Systematics and biogeography of *Onychomys leucogaster* in western North America. J. Mamm. 67:233–55.

Ride, W. D. L. 1964. A review of Australian fossil marsupials. J. Roy. Soc. W. Austral. 47:97–131.

———. 1970. A guide to the native mammals of Australia. Oxford Univ. Press, Melbourne, xiv + 249 pp.

Rideout, C. B. 1978. Mountain goat. *In* Schmidt and Gilbert (1978), pp. 149–59.

Rideout, C. B., and R. S. Hoffmann. 1975. *Oreamnos americanus*. Mammalian Species, no. 63, 6 pp.

Ridgway, S. H. 1986a. Physiological observations on dolphin brains. *In* Schusterman, Thomas, and Wood (1986), pp. 31–59.

———. 1986b. Diving by cetaceans. *In* Brubakk, A. O., J. W. Kanwisher, and G. Sundness, eds., Diving in animals and man, Roy. Norwegian Soc. Sci. Letters, Trondheim, pp. 33–62.

Ridgway, S. H., and R. Harrison, eds. 1981a. Handbook of marine mammals. Volume 1. The walrus, sea lions, fur seals and sea otter. Academic Press, London, xiv + 235 pp.

———. 1981b. Handbook of marine mammals. Volume 2. Seals. Academic Press, London, xv + 359 pp.

———. 1985. Handbook of marine mammals. Volume 3. The sirenians and baleen whales. Academic Press, London, xviii + 362 pp.

Ridgway, S. H., and C. C. Robinson. 1985. Homing by released captive California sea lions, *Zalophus californianus*, following release on distant islands. Can. J. Zool. 63:2162–64.

Rieger, I. 1979. A review of the biology of striped hyaenas, *Hyaena hyaena* (Linne, 1758). Saugetierk. Mitt. 27:81–95.

———. 1981. *Hyaena hyaena*. Mammalian Species, no. 150, 5 pp.

Rigby, R. G. 1972. A study of the behaviour of caged *Antechinus stuartii*. Z. Tierpsychol. 31:15–25.

Rijksen, H. D. 1978. A field study on Sumatran orang utans (*Pongo pygmaeus abelii* Lesson 1827). Meded. Landbouwhogeschool, Wageningen, Netherlands, 420 pp.

———. 1986. Conservation of orangutans: a status report, 1985. *In* Benirschke (1986), pp. 153–59.

Riley, G. A., and R. T. McBride. 1975. A survey of the red wolf *(Canis rufus)*. *In* Fox (1975), pp. 263–77.

Ritcey, R. W. 1974. Moose harvesting programs in Canada. Naturaliste Canadien 101:631–42.

Robbins, C. B. 1973. Nongeographic variation in *Taterillus gracilis* (Thomas) (Rodentia: Cricetidae). J. Mamm. 54:222–38.

———. 1974. Comments on the taxonomy of the West African *Taterillus* (Rodentia: Cricetidae) with the description of a new species. Proc. Biol. Soc. Washington 87:395–404.

———. 1977. A review of the taxonomy of the African gerbils, *Taterillus* (Rodentia: Cricetidae). *In* Sokolov, V. E., ed., Advances in modern theriology, Publishing House "Nauka," Moscow, pp. 178–94.

———. 1978. Taxonomic identification and history of *Scotophilus nigrita* (Schreber) (Chiroptera: Vespertilionidae). J. Mamm. 59:212–13.

———. 1980. Small mammals of Togo and Benin. I. Chiroptera. Mammalia 44:83–88.

———. 1984. A new high forest species in the African bat genus *Scotophilus* (Vespertilionidae). Ann. Mus. Roy. Afr. Cent., Sec. Zool., 237:19–24.

Robbins, C. B., F. De Vree, and V. Van Cakenberghe. 1985. A systematic revision of the African bat genus *Scotophilus* (Vespertilionidae). Zool. Wetenschappen-Ann. 246:53–84.

Robbins, C. B., J. W. Krebs, Jr., and K. M. Johnson. 1983. *Mastomys* (Rodentia: Muridae) species distinguished by hemoglobin pattern differences. Amer. J. Tropical Med. Hyg. 32:624–30.

Robbins, C. B., and H. W. Setzer. 1985. Morphometrics and distinctness of the hedgehog genera (Insectivora: Erinaceidae). Proc. Biol. Soc. Washington 98:112–20.

Robbins, L. W., J. R. Choate, and R. L. Robbins. 1980. Nongeographic and interspecific variation in four species of *Hylomyscus* (Rodentia: Muridae) in southern Cameroon. Ann. Carnegie Mus. 49:31–48.

Robbins, L. W., and V. M. Sarich. 1988. Evolutionary relationships in the family Emballonuridae (Chiroptera). J. Mamm. 69:1–13.

Robbins, L. W., and D. A. Schlitter. 1981. Systematic status of dormice (Rodentia: Gliridae) from southern Cameroon, Africa. Ann. Carnegie Mus. 50:271–88.

Robbins, L. W., and H. W. Setzer. 1979. Additional records of *Hylomyscus baeri* Heim de Balsac and Aellen (Rodentia, Muridae) from western Africa. J. Mamm. 60:649–50.

Roben, P. 1974. Zum vorkommen des Otters, *Lutra lutra* (Linne, 1758) in der Bündesrepublik Deutschland. Saugetierk. Mitt. 22:29–36.

———. 1975. Zur ausbreitung des Waschbaren, *Procyon lotor* (Linne, 1758), und der Marderhundes, *Nyctereutes procyonides* (Gray, 1834), in der Bündesrepublik Deutschland. Saugetierk. Mitt. 23:93–101.

Roberts, A. 1951. Mammals of South Africa. Central News Agency, Johannesburg, xlviii + 700 pp.

Roberts, M., S. Brand, and E. Maliniak. 1985. The biology of captive prehensile-tailed porcupines, *Coendou prehensilis*. J. Mamm. 66:476–82.

Roberts, M., F. Koontz, L. Phillips, and E. Maliniak. 1987. Management and biology of the prehensile-tailed porcupine *Coendou prehensilis*. Internatl. Zoo Yearbook 26:265–75.

Roberts, M., E. Maliniak, and M. Deal. 1984. The reproductive biology of the rock cavy, *Kerodon rupestris*, in captivity: a study of reproductive adaptation in a trophic specialist. Mammalia 48:253–66.

Roberts, M. S. 1975. Growth and development of mother-reared red pandas *Ailurus fulgens*. Internatl. Zoo Yearbook 15:57–63.

———. 1980. Breeding the red panda (Ailurus fulgens) at the National Zoological Park. Zool. Garten 50:253–63.

Roberts, M. S., and J. L. Gittleman. 1984. Ailurus fulgens. Mammalian Species, no. 222, 8 pp.

Roberts, M. S., K. V. Thompson, and J. A. Cranford. 1988. Reproduction and growth in captive punare (Thrichomys apereoides Rodentia: Echimyidae) of the Brazilian Caatinga with reference to the reproductive strategies of the Echimyidae. J. Mamm. 69:542–51.

Roberts, T. J. 1977. The mammals of Pakistan. Ernest Benn, London, xxvi + 361 pp.

———. 1985. Distribution and present status of wild sheep in Pakistan. In Hoefs (1985), pp. 159–63.

Robertshaw, J. D., and R. H. Harden. 1985. The ecology of the dingo in north-eastern New South Wales. II. Diet. Austral. Wildl. Res. 12:39–50.

Robinson, A. C. 1975. The sticknest rat, Leporillus conditor, on Franklin Island, Nuyts Archipelago, South Australia. Austral. Mamm. 1:319–27.

Robinson, F. J., A. C. Robinson, C. H. S. Watts, and P. R. Baverstock. 1978. Notes on rodents and marsupials and their ectoparasites collected in Australia in 1974–75. Trans. Roy. Soc. S. Austral. 102:59–70.

Robinson, J. G., and J. F. Eisenberg. 1985. Group size and foraging habits of the collared peccary Tayassu tajacu. J. Mamm. 66:153–55.

Robinson, J. W., and R. S. Hoffmann. 1975. Geographical and interspecific cranial variation in big-eared ground squirrels (Spermophilus): a multivariate study. Syst. Zool. 24:79–88.

Robinson, P. T. 1981. The reported use of denning structures by the pygmy hippopotamus Choeropsis liberiensis. Mammalia 45:506–8.

Robinson, T. J., and N. J. Dippenaar. 1983a. Morphometrics of the South African Leporidae. I: Genus Prolagus Lyon, 1904. Ann. Mus. Roy. Afr. Cent., Sec. Zool., 237:43–61.

———. 1983b. The status of Lepus saxatilis, L. whytei, and L. crawshayi in southern Africa. Acta Zool. Fennica 174:35–39.

Robinson, T. J., F. F. B. Elder, and J. A. Chapman. 1984. Evolution of chromosomal variation in cottontails, genus Sylvilagus (Mammalia: Lagomorpha). II. Sylvilagus aubonii, S. idahoensis, S. nuttallii, and S. palustris. Cytogenet. Cell Genet. 38:282–89.

Robinson, T. J., and J. D. Skinner. 1983. Karyology of the riverine rabbit, Bunolagus monticularis, and its taxonomic implications. J. Mamm. 64:678–81.

Robinson, W. L., and G. J. Smith. 1977. Observations on recently killed wolves in upper Michigan. Wildl. Soc. Bull. 5:25–26.

Robson, S. K. 1984. Myotis adversus (Chiroptera: Vespertilionidae): Australia's fish-eating bat. Austral. Mamm. 7:51–52.

Roby, D. D., H. Thing, and K. L. Brink. 1984. History, status, and taxonomic identity of caribou (Rangifer tarandus) in northwest Greenland. Arctic 37:23–30.

Roche, J. 1972a. Systématique du genre Procavia et des damans en général. Mammalia 36:22–49.

———. 1972b. Capture de Zenkerella insignis (Rongeurs, Anomalurides) en République Centralafricaine. Mammalia 36:305–6.

Roche, J., E. Capanna, M. V. Civitelli, and A. Ceraso. 1984. Caryotypes des rongeurs de Somalie. 4. Premiere capture de rongeurs arboricoles du sous-genre Grammomys (genre Thamnomys, Murides) en Republique de Somalie. Italian J. Zool., Suppl., n.s., 19:259–77.

Rodgers, D. H., and W. H. Elder. 1977. Movements of elephants in Luangwa Valley, Zambia. J. Wildl. Mgmt. 41:56–62.

Rodgers, W. A., and I. Swai. 1988. Tanzania. In East (1988), pp. 53–65.

Rodman, P. S. 1973. Population composition and adaptive organisation among orangutans of the Kutai Reserve. In Michael and Crook (1973), pp. 171–209.

———. 1977. Feeding behaviour of orangutans of the Kutai Nature Reserve, East Kalimantan. In Clutton-Brock (1977), pp. 384–413.

———. 1978. Diets, densities, and distributions of Bornean primates. In Montgomery (1978), pp. 465–78.

Roe, N. A., and W. E. Rees. 1976. Preliminary observations of the taruca (Hippocamelus antisensis: Cervidae) in southern Peru. J. Mamm. 57:722–30.

Roff, D., and W. D. Bowen. 1986. Further analysis of population trends in the northwest Atlantic harp seal (Phoca groenlandica) from 1967 to 1985. Can. J. Fish. Aquat. Sci. 43:553–64.

Rogers, D. S. 1983. Phylogenetic affinities of Peromyscus (Megadontomys) thomasi: evidence from differentially stained chromosomes. J. Mamm. 64:617–23.

Rogers, D. S., I. F. Greenbaum, S. J. Gunn, and M. D. Engstrom. 1984. Cytosystematic value of chromosomal inversion data in the genus Peromyscus (Rodentia: Cricetidae). J. Mamm. 65:457–65.

Rogers, D. S., and D. J. Schmidly. 1982. Systematics of spiny pocket mice (Heteromys) of the desmarestianus species group from Mexico and northern Central America. J. Mamm. 63:375–86.

Rogers, L. L., L. D. Mech, D. K. Dawson, J. M. Peek, and M. Korb. 1980. Deer distribution in relation to wolf pack territory edges. J. Wildl. Mgmt. 44:253–58.

Rogovin, K. A. 1981. Utilization of territory by little earth hares (Alactagulus acontion) and structure of their groups by materials of marking. Zool. Zhur. 60:568–78.

Roguin, L. D. 1986. Les mammifères du Paraguay dans les collections du Muséum de Genève. Rev. Suisse Zool. 93:1009–22.

———. 1988. Notes sur quelques mammifères du Baluchistan iranien. Rev. Suisse Zool. 95:595–606.

Rohwer, S. A., and D. L. Kilgore, Jr. 1973. Interbreeding in the arid-land foxes, Vulpes velox and V. macrotis. Syst. Zool. 22:157–65.

Roig, V. G., and O. A. Reig. 1969. Precipitin test relationships among Argentinian species of the genus Ctenomys (Rodentia, Octodontidae). Comp. Biochem. Physiol. 30:665–72.

Rolley, R. E. 1987. Bobcat. In Novak, Baker, et al. (1987), pp. 670–81.

Romer, A. S. 1968. Notes and comments on vertebrate paleontology. Univ. Chicago Press, viii + 304 pp.

Romer, J. D. 1974. Notes on the care and breeding of tree squirrels Callosciurus spp. Internatl. Zoo Yearbook 14:115–16.

Ronald, K., and P. J. Healey. 1981. Harp seal—Phoca groenlandica. In Ridgway and Harrison (1981b), pp. 55–87.

Ronald, K., J. Selley, and P. Healey. 1982. Seals. In Chapman, J. A., and G. A. Feldhamer, eds., Wild mammals of North America, Johns Hopkins Univ. Press, Baltimore, pp. 769–827.

Rongstad, O. J., and J. R. Tester. 1971. Behavior and maternal relations of young snowshoe hares. J. Wildl. Mgmt. 35:338–46.

Rood, J. P. 1970a. Notes on the behavior of the pygmy armadillo. J. Mamm. 51:179.

———. 1970b. Ecology and social behavior of the desert cavy (Microcavia australis). Amer. Midl. Nat. 83:415–54.

———. 1972. Ecological and behavioural comparisons of three genera of Argentine cavies. Anim. Behav. Monogr. 5:1–83.

———. 1974. Banded mongoose males guard young. Nature 248:176.

———. 1975. Population dynamics and food habits of the banded mongoose. E. Afr. Wildl. J. 13:89–111.

———. 1978. Dwarf mongoose helpers at the den. Z. Tierpsychol. 48:277–87.

———. 1980. Mating relationships and breeding suppression in the dwarf mongoose. Anim. Behav. 28:143–50.

Rood, J. P., and F. H. Test. 1968. Ecology of the spiny rat, *Heteromys anomalus*, at Rancho Grande, Venezuela. Amer. Midl. Nat. 79:89–102.

Rookmaaker, L. C. 1977. The distribution and status of the rhinoceros, *Dicerorhinus sumatrensis*, in Borneo—a review. Bijdragen Tot de Dierkunde 47:197–204.

———. 1980. The distribution of the rhinoceros in eastern India, Bangladesh, China, and the Indo-Chinese region. Zool. Anz., Jena, 205:253–68.

———. 1984. The former distribution of the Indian rhinoceros *(Rhinoceros unicornis)* in India and Pakistan. J. Bombay Nat. Hist. Soc. 80:555–63.

———. 1988. The Indo-Chinese rhinoceros. Oryx 22:241.

Rookmaaker, L. C., and W. Bergmans. 1981. Taxonomy and geography of *Rousettus amplexicaudatus* (Geoffroy, 1810) with comparative notes on sympatric congeners (Mammalia, Megachiroptera). Beaufortia 31:1–29.

Rookmaaker, L. C., and C. P. Groves. 1978. The extinct Cape rhinoceros, *Diceros bicornis bicornis* (Linnaeus, 1758). Saugetierk. Mitt. 26:117–26.

Roonwal, M. L. 1949. Systematics, ecology and bionomics of mammals studied in connection with tsutsugamushi disease (scrub typhus) in the Assam-Burma war theatre during 1945. Trans. Natl. Inst. Sci. India 3:67–122.

Roonwal, M. L., and S. M. Mohnot. 1977. Primates of south Asia. Harvard Univ. Press, Cambridge, xviii + 421 pp.

Roots, C. G. 1966. Notes on the breeding of white-lipped peccaries *Tayassu albirostris* at Dudley Zoo. Internatl. Zoo Yearbook 6:198–99.

Rosatte, R. C. 1987. Striped, spotted, hooded, and hog-nosed skunk. *In* Novak, Baker, et al. (1987), pp. 598–613.

———. 1988a. Bat rabies in Canada: history, epidemiology and prevention. Can. Vet. J. 28:754–56.

———. 1988b. Rabies in Canada: history, epidemiology and control. Can. Vet. J. 29:362–65.

Rose, K. D. 1975. *Elpidophorus*, the earliest dermopteran (Dermoptera, Plagiomenidae). J. Mamm. 56:675–79.

Rose, R. K. 1981. *Synaptomys* not extinct in the Dismal Swamp. J. Mamm. 62:844–45.

Rose, R. K., R. K. Everton, and T. M. Padgett. 1987. Distribution and current status of the threatened Dismal Swamp southeastern shrew, *Sorex longirostris fisheri*. Virginia J. Sci. 38:358–63.

Rose, R. K., and M. S. Gaines. 1976. Levels of aggression in fluctuating populations of the prairie vole, *Microtus ochrogaster*, in eastern Kansas. J. Mamm. 57:43–57.

———. 1978. The reproductive cycle of *Microtus ochrogaster* in eastern Kansas. Ecol. Monogr. 48:21–42.

Rose, R. W. 1978. Reproduction and evolution in female Macropodidae. Austral. Mamm. 2:65–72.

———. 1986. The habitat, distribution and conservation status of the Tasmanian bettong, *Bettongia gaimardi* (Desmarest). Austral. Wildl. Res. 13:1–6.

———. 1987. Reproductive biology of the Tasmanian bettong (*Bettongia gaimardi*: Macropodidae). J. Zool. 212:59–67.

Rose, R. W., and D. J. McCartney. 1982. Reproduction of the red-bellied pademelon, *Thylogale billardierii* (Marsupialia). Austral. Wildl. Res. 9:27–32.

Rosenberg, H. 1971. Breeding the bat-eared fox *Otocyon megalotis* at Utica Zoo. Internatl. Zoo Yearbook 11:101–2.

Rosenberger, A. L. 1977. *Xenothrix* and ceboid phylogeny. J. Human Evol. 6:461–81.

Rosenberger, A. L., and A. F. Coimbra-Filho. 1984. Morphology, taxonomic status and affinities of the lion tamarins, *Leontopithecus* (Callitrichinae, Cebidae). Folia Primatol. 42:149–79.

Rosenblum, L. A. 1968. Some aspects of female reproductive physiology in the squirrel monkey. *In* Rosenblum and Cooper (1968), pp. 147–69.

Rosenblum, L. A., and R. W. Cooper, eds. 1968. The squirrel monkey. Academic Press, New York, xii + 451 pp.

Rosenthal, M. A. 1975a. Observations on the water opossum or yapok *Chironectes minimus* in captivity. Internatl. Zoo Yearbook 15:4–6.

———. 1975b. The management, behavior and reproduction of the short-eared elephant shrew *Macroscelides proboscideus* (Shaw). M.A. thesis, Northeastern Illinois Univ., 61 pp.

Rosenthal, M. A., and D. A. Meritt, Jr. 1973. Hand-rearing springhaas *Pedetes capensis* at Lincoln Park Zoo, Chicago. Internatl. Zoo Yearbook 13:135–37.

Rosevear, D. R. 1965. The bats of West Africa. British Mus. (Nat. Hist.), London, xvii + 418 pp.

———. 1969. The rodents of West Africa. British Mus. (Nat. Hist.), London, xii + 604 pp.

———. 1974. The carnivores of West Africa. British Mus. (Nat. Hist.), London, xii + 548 pp.

Ross, G. J. B. 1977. The taxonomy of bottlenosed dolphins *Tursiops* species in South African waters, with notes on their biology. Ann. Cape Prov. Mus. (Nat. Hist.) 11:135–94.

———. 1979. Records of pygmy and dwarf sperm whales, genus *Kogia*, from southern Africa, with some biological notes and some comparisons. Ann. Cape Prov. Mus. (Nat. Hist.) 11:259–327.

———. 1984. The smaller cetaceans of the south east coast of southern Africa. Ann. Cape Prov. Mus. (Nat. Hist.) 15:173–410.

Ross, G. J. B., P. B. Best, and B. G. Donnelly. 1975. New records of the pygmy right whale *(Caperea marginata)* from South Africa, with comments on distribution, migration, appearance, and behavior. J. Fish. Res. Bd. Can. 32:1005–17.

Ross, W. G. 1974. Distribution, migration, and depletion of bowhead whales in Hudson Bay, 1860 to 1915. Arctic and Alpine Res. 6:85–98.

———. 1979. The annual catch of Greenland (bowhead) whales in waters north of Canada, 1719–1915: a preliminary compilation. Arctic 32:91–121.

Rossolimo, O. L. 1976a. Taxonomic status of the mouse-like dormouse *Myomys* (Mammalia, Myoxidae) from Bulgaria. Zool. Zhur. 55:1515–24.

———. 1976b. *Myomimus setzeri* (Mammalia, Myoxidae), a new species of mouselike dormouse from Iran. Vestn. Zool. 1976(4):51–53.

Roth, H. H., and E. Waitkuwait. 1986. Répartition et statut des grandes espèces de mammifères en Côte-d'Ivoire. III. Lamantins. Mammalia 50:227–42.

Rothman, R. J., and L. D. Mech. 1979. Scent-marking in lone wolves and newly formed pairs. Anim. Behav. 27:750–60.

Rotterman, L. M., and T. Simon-Jackson. 1988. Sea otter. *Enhydra lutris. In* Lentfer, J. W., ed., Selected marine mammals of Alaska, Mar. Mamm. Comm., Washington, D.C., pp. 239–75.

Rounds, R. C. 1987. Distribution and analysis of colourmorphs of the black bear *(Ursus americanus)*. J. Biogeogr. 14:521–38.

Rounsevell, D. E., and S. J. Smith. 1982. Recent alleged sightings of the thylacine (Marsupialia, Thylacinidae) in Tasmania. *In* Archer (1982a), pp. 233–36.

Roux, J.-P. 1987. Recolonization process in the subantarctic fur seal, *Arctocephalus tropicalis*, on Amsterdam Island. *In* Croxall and Gentry (1987), pp. 189–94.

Roux, J.-P., and A. D. Hes. 1984. The seasonal haul-out cycle of the fur seal *Arctocephalus tropicalis* (Gray, 1872) on Amsterdam Island. Mammalia 48:377–89.

Rowell, T. E. 1971. Organization of caged groups of *Cercopithecus* monkeys. Anim. Behav. 19:625–45.

————. 1977a. Variation in age at puberty in monkeys. Folia Primatol. 27:284–96.

————. 1977b. Reproductive cycles of the talapoin monkey (Miopithecus talapoin). Folia Primatol. 28:188–202.

Rowe-Rowe, D. T. 1977. Prey capture and feeding behaviour of South African otters. Lammergeyer, no. 23, pp. 13–21.

————. 1978a. Reproduction and post-natal development of South African mustelines (Carnivora: Mustelidae). Zool. Afr. 13:103–14.

————. 1978b. The small carnivores of Natal. Lammergeyer, no. 25, 48 pp.

————. 1982. Home range and movements of black-backed jackals in an African montane region. S. Afr. J. Wildl. Res. 12:79–84.

Rowlands, I. W. 1974. Mountain viscacha. Symp. Zool. Soc. London 34:131–41.

Roychoudhury, A. K. 1987. White tigers and their conservation. In Tilson and Seal (1987), pp. 380–88.

Rozendaal, F. G. 1984. Notes on macroglossine bats from Sulawesi and the Moluccas, Indonesia, with the description of a new species of Syconycteris Matschie, 1899 from Halmahera (Mammalia: Megachiroptera). Zool. Meded. 58:187–212.

Rubin, H. D., and A. M. Beck. 1982. Ecological behavior of free-ranging urban dogs. Appl. Anim. Ethol. 8:161–68.

Rudd, R. L. 1980. Population density and movements of the lesser gymnure, Hylomys suillus. Malayan Nat. J. 34:111–12.

Rudge, M. R. 1984. The occurrence and status of populations of feral goats and sheep throughout the world. Papers Presented at the Workshop on Feral Mammals, Internatl. Union Conserv. Nat. Species Survival Comm., Caprinae Specialist Group, Helsinki, pp. 55–84.

Rudnai, J. A. 1973. The social life of the lion. Washington Square East, Publishers, Wallingford, Pennsylvania, 122 pp.

————. 1984. Activity cycle and space utilization in captive Dendrohyrax arboreus. S. Afr. J. Zool. 19:124–28.

Rudran, R. 1973. The reproductive cycles of two subspecies of purple-faced langurs (Presbytis senex) with relation to environmental factors. Folia Primatol. 19:41–60.

————. 1978. Socioecology of the blue monkeys (Cercopithecus mitis stuhlmani) of the Kibale Forest, Uganda. Smithson. Contrib. Zool., no. 249, iv + 88 pp.

Ruhiyat, Y. 1983. Socio-ecological study of Presbytis aygula in west Java. Primates 24:344–59.

Rumbaugh, D. M. 1970. Learning skills of anthropoids. Primate Behav. 1:1–70.

————, ed. 1972. Gibbon and siamang. I. Evolution, ecology, behavior, and captive maintenance. S. Karger, Basel, x + 263 pp.

Rumpler, Y. 1974. Cytogenetic contributions to a new classification of lemurs. In Martin, Doyle, and Walker (1974), pp. 865–69.

————. 1975. The significance of chromosomal studies in the systematics of the Malagasy lemurs. In Tattersall and Sussman (1975), pp. 25–40.

Rumpler, Y., and R. Albignac. 1975. Intraspecific chromosome variability in a lemur from the north of Madagascar: Lepilemur septentrionalis, species nova. Amer. J. Phys. Anthropol. 42:425–30.

————. 1977. Chromosome studies of the Lepilemur, an endemic Malagasy genus of lemurs: contribution of the cytogenetics to their taxonomy. J. Human Evol. 7:191–96.

Rumpler, Y., and B. Dutrillaux. 1978. Chromosomal evolution in Malagasy lemurs. III. Chromosome banding studies in the genus Hapalemur and the species Lemur catta. Cytogenet. Cell Genet. 21:201–11.

Rumpler, Y., S. Warter, J.-J. Petter, R. Albignac, and B. Dutrillaux. 1988. Chromosomal evolution of Malagasy lemurs. XI. Phylogenetic position of Daubentonia madagascariensis. Folia Primatol. 50:124–29.

Rusch, D. A., and W. G. Reeder. 1978. Population ecology of Alberta red squirrels. Ecology 59:400–420.

Russell, E. M. 1974a. The biology of kangaroos (Marsupialia-Macropodidae). Mamm. Rev. 4:1–59.

————. 1974b. Recent ecological studies on Australian marsupials. Austral. Mamm. 1:189–211.

————. 1979. The size and composition of groups in the red kangaroo, Macropus rufus. Austral. Wildl. Res. 6:237–44.

Russell, J. K. 1981. Exclusion of adult male coatis from social groups: protection from predation. J. Mamm. 62:206–8.

Russell, R. H. 1975. The food habits of polar bears of James Bay and southwest Hudson Bay in summer and autumn. Arctic 28:117–29.

Russell, R. J. 1968. Evolution and classification of the pocket gophers of the subfamily Geomyinae. Univ. Kansas Mus. Nat. Hist. Publ. 16:473–579.

Ryan, M. J., and M. D. Tuttle. 1983. The ability of the frog-eating bat to discriminate among novel and potentially poisonous frog species using acoustic cues. Anim. Behav. 31:827–33.

Ryan, M. J., M. D. Tuttle, and R. M. R. Barclay. 1983. Behavioral responses of the frog-eating bat, Trachops cirrhosus, to sonic frequencies. J. Comp. Physiol., ser. A, 150:413–18.

Ryan, R. M. 1966. A new and some imperfectly known Australian Chalinolobus and the taxonomic status of African Glauconycteris. J. Mamm. 47:86–91.

Ryberg, O. 1947. Studies on bats and bat parasites. Univ. Lund and Zool. Lab. Agric., Dairy, and Hort. Inst. Alnarp, Stockholm, 330 pp.

Rydell, J. 1986. Feeding territoriality in female northern bats, Eptesicus nilssoni. Ethology 72:329–37.

Ryder, O. A. 1988. Przewalski's horse—putting the wild horse back in the wild. Oryx 22:154–57.

Ryder, O. A., and E. A. Wedemeyer. 1982. A cooperative breeding program for the Mongolian wild horse, Equus przewalskii, in the United States. Biol. Conserv. 22:259–71.

Rzebik-Kowalska, B., B. W. Woloszyn, and A. Nadachowski. 1978. A new bat, Myotis nattereri (Kuhl, 1818) (Vespertilionidae), in the fauna of Iraq. Acta Theriol. 23:541–50.

S

Saayman, G. S., and C. K. Tayler. 1973. Social organisation of inshore dolphins (Tursiops aduncus and Sousa) in the Indian Ocean. J. Mamm. 54:993–96.

————. 1979. The socioecology of humpback dolphins (Sousa sp.). In Winn and Olla (1979), pp. 165–226.

Sabater Pi, J. 1972. Contribution to the ecology of Mandrillus sphinx Linnaeus 1758 of Rio Muni (Republic of Equatorial Guinea). Folia Primatol. 17:304–19.

————. 1973. Contribution to the ecology of Colobus polykomos satanas (Waterhouse 1838) of Rio Muni, Republic of Equatorial Guinea. Folia Primatol. 19:193–207.

Sabbag, G. E. 1986. The wild buffalo of Yellowstone. Mainstream 17(3):16–18.

Sade, D. S. 1967. Determinants of dominance in a group of free-ranging rhesus monkeys. In Altmann (1967), pp. 99–114.

Sadleir, R. M. F. S. 1970. The establishment of a dominance rank order in male Peromyscus maniculatus and its stability with time. Anim. Behav. 18:55–59.

————. 1987. Reproduction of female cervids. In Wemmer (1987), pp. 123–44.

Sáez-Royuela, C., and J. L. Tellería. 1986. The increased population of the wild boar (Sus scrofa L.) in Europe. Mamm. Rev. 16:97–101.

Sage, R. D., J. R. Contreras, V. G. Roig, and J. L. Patton. 1986. Genetic variation in the South American burrowing rodents of the genus Ctenomys (Rodentia: Ctenomyidae). Z. Saugetierk. 51:158–72.

Saha, S. S. 1981. A new genus and a new species of flying squirrel (Mammalia: Rodentia: Sciuridae) from northeastern India. Bull. Zool. Surv. India 4:331–36.

Sahu, A., and B. R. Maiti. 1978. Estrous cycle of the bandicoot rat—a rodent pest. Zool. J. Linnean Soc. 63:309–14.

Sailler, H., and U. Schmidt. 1978. Die sozialen Laute der Gemeinen vampirfledermaus Desmodus rotundus bei Konfrontation am Futterplatz unter experimentellen Bedingungen. Z. Saugetierk. 43:249–61.

Sale, J. B., and S. Singh. 1987. Reintroduction of greater Indian rhinoceros into Dudhwa National Park. Oryx 21:81–84.

Saleh, M. A. 1987. The decline of gazelles in Egypt. Biol. Conserv. 39:83–95.

Salter, R. E., N. A. MacKenzie, N. Nightingale, K. M. Aken, and P. Chai P. K. 1985. Habitat use, ranging behaviour, and food habits of the proboscis monkey, Nasalis larvatus (van Wurmb), in Sarawak. Primates 26:436–51.

Salter, R. E., and J. A. Sayer. 1986. The brow-antlered deer in Burma—its distribution and status. Oryx 20:241–45.

Samaras, W. F. 1974. Reproductive behavior of the gray whale Eschrichtius robustus in Baja California. Bull. S. California Acad. Sci. 73:57–64.

Samuel, W., and W. G. Macgregor, eds. 1977. Proceedings of the First International Mountain Goat Symposium. British Columbia Fish and Wildlife Branch, iii + 243 pp.

Sanborn, C. C. 1931. Bats from Polynesia, Melanesia, and Micronesia. Field Mus. Nat. Hist. Publ., Zool. Ser., 18:7–29.

———. 1950. New Philippine fruit bats. Proc. Biol. Soc. Washington 63:189–90.

———. 1952. Philippine Zoological Expedition 1946–1947: mammals. Fieldiana Zool. 33:89–158.

———. 1953. Mammals from Mindanao, Philippine Islands collected by the Danish Philippine Expedition 1951–1952. Vidensk. Medd. fra Dansk Naturhist. Foren. 115:283–88.

Sanborn, C. C., and A. J. Nicholson. 1950. Bats from New Caledonia, the Solomon Islands, and New Hebrides. Fieldiana Zool. 31:313–38.

Sanderson, G. C., and A. V. Nalbandov. 1973. The reproductive cycle of the raccoon in Illinois. Illinois Nat. Hist. Surv. Bull. 31:29–85.

Sandhu, S. 1984. Breeding biology of the Indian fruit bat, Cynopterus sphinx (Vahl) in central India. J. Bombay Nat. Hist. Soc. 81:600–611.

Sandoval, A. V. 1985. Status of bighorn sheep in the Republic of Mexico. In Hoefs (1985), pp. 86–94.

Santiapillai, C., M. R. Chambers, and N. Ishwaran. 1982. The leopard Panthera pardus fusca (Meyer 1794) in the Ruhuna National Park, Sri Lanka, and observations relevant to its conservation. Biol. Conserv. 23:5–14.

———. 1984. Aspects of the ecology of the Asian elephant Elephas maximus L. in the Ruhuna National Park, Sri Lanka. Biol. Conserv. 29:47–61.

Santos, I. B., W. L. R. Oliver, and A. B. Rylands. 1987. Distribution and status of two species of tree porcupines, Chaetomys subspinosus and Sphiggurus insidiosus, in southeast Brazil. Dodo 24:43–60.

Sapargeldyev, M. S. 1984. A contribution to the ecology of the mouse-like hamster Calomyscus mystax (Rodentia, Cricetidae) in Turkmenistan. Zool. Zhur. 63:1388–95.

Sargeant, A. B., and D. W. Warner. 1972. Movements and denning habits of a badger. J. Mamm. 53:207–10.

Sarich, V., J. M. Lowenstein, and B. J. Richardson. 1982. Phylogenetic relationships of Thylacinus cynocephalus (Marsupialia) as reflected in comparative serology. In Archer (1982a), pp. 703–9.

Sarich, V. M. 1969. Pinniped phylogeny. Syst. Zool. 18:416–22.

———. 1985. Xenarthran systematics: albumin immunological evidence. In Montgomery (1985a), pp. 77–81.

Sauer, E. G. F. 1973. Zum Sozialverhalten der kurzohrigen Elefantenspitzmaus, Macroscelides proboscidens. Z. Saugetierk. 38:65–97.

Savic, I. 1973. Ecology of the species Spalax leucodon Nordm. in Yugoslavia. Zbornik za prir. nauke Matice srpske 44:66–70.

Savic, I., and B. Soldatovic. 1979. Distribution range and evolution of chromosomal forms in the Spalacidae of the Balkan Peninsula and bordering regions. J. Biogeogr. 6:363–74.

Sawrey, D. K., D. J. Baumgardner, M. J. Campa, B. Ferguson, A. W. Hodges, and D. A. Dewsbury. 1984. Behavioral patterns of Djungarian hamsters: an adaptive profile. Anim. Learning and Behav. 12:297–306.

Sayer, J. A., and L. P. Van Lavieren. 1975. The ecology of the Kafue lechwe population of Zambia before the operation of hydroelectric dams on the Kafue River. E. Afr. Wildl. J. 13:9–37.

Sazima, I. 1976. Observations on the feeding habits of phyllostomatid bats (Carollia, Anoura, and Vampyrops) in southeastern Brazil. J. Mamm. 57:381–82.

———. 1978. Vertebrates as food items of the woolly false vampire, Chrotopterus auritus. J. Mamm. 59:617–18.

Sazima, I., and W. Uieda. 1977. O morcego Promops nasutus no sudeste Brasileiro

(Chiroptera, Molossidae). Ciencia e Cultura 29:312–14.

Sazima, I., L. D. Vizotto, and V. A. Taddei. 1978. Uma nova especie de Lonchophylla da serra do Cipo, Minas Gerais, Brasil (Mammalia, Chiroptera, Phyllostomatidae). Rev. Brasil. Biol. 38:81–89.

Scarff, J. E. 1986. Historic and present distribution of the right whale (Eubalaena glacialis) in the eastern North Pacific south of 50° N and east of 180° W. In Brownell, Best, and Prescott (1986), pp. 43–63.

Scarlett, N. 1969. The bilby, Thylacomys lagotis in Victoria. Victorian Nat. 86:292–94.

Schaaf, D. 1978. Some aspects of the ecology of the swamp deer or barasingha (Cervus d. duvauceli) in Nepal. In Threatened deer, Internatl. Union Conserv. Nat., Morges, Switzerland, pp. 65–85.

Schaller, G. B. 1963. The mountain gorilla. Univ. Chicago Press, xvii + 431 pp.

———. 1965. The behavior of the mountain gorilla. In De Vore (1965), pp. 324–67.

———. 1967. The deer and the tiger. Univ. Chicago Press, 370 pp.

———. 1972. The Serengeti lion: a study of predator-prey relations. Univ. Chicago Press, xiii + 480 pp.

———. 1976. The mouse that barks. Internatl. Wildl. 6(5):12–16.

———. 1977. Mountain monarchs: wild sheep and goats of the Himalaya. Univ. Chicago Press, xviii + 425 pp.

———. 1980a. Movement patterns of jaguar. Biotrópica 12:161–68.

———. 1980b. Epitaph for a jaguar. Anim. Kingdom 83(2):4–11.

———. 1981. Pandas in the wild. Natl. Geogr. 160:735–49.

———. 1986. A Tibetan tragedy. Anim. Kingdom 89(4):22–33.

———. 1989. Notes on the distribution of the Tibetan antelope. Gnusletter 8(3):14–15.

Schaller, G. B., and P. G. Cranshaw, Jr. 1981. Social organization in a capybara population. Saugetierk. Mitt. 29:3–16.

Schaller, G. B., N. X. Dang, L. D. Thuy, and V. T. Son. 1990. Javan rhinoceros in Vietnam. Oryx 24:77–80.

Schaller, G. B., Hu Jinchu, Pan Wenshi, and Zhu Jing. 1985. The giant pandas of Wolong. Univ. Chicago Press, xix + 298 pp.

Schaller, G. B., Li Hong, Talipu, Lu Hua, Ren Junrang, Qiu Mingjiang, and Wang Haibin. 1987. Status of large mammals in the Taxkorgan Reserve, Xinjiang, China. Biol. Conserv. 42:53–71.

Schaller, G. B., Li Hong, Talipu, Ren Junrang, and Qiu Mingjiang. 1988. The snow leopard in Xinjiang, China. Oryx 22:197–204.

Schaller, G. B., and Ren Junrang. 1988. Effects of a snowstorm on Tibetan antelope. J. Mamm. 69:631–34.

Schaller, G. B., Teng Qitao, Pan Wenshi, Qin Zisheng, Wang Xiaoming, Hu Jinchu, and Shen Heming. 1986. Feeding behavior of Sichuan takin *(Budorcas taxicolor)*. Mammalia 50:311–22.

Schaller, G. B., and J. M. C. Vasconcelos. 1978. A marsh deer census in Brazil. Oryx 14:345–51.

Schauenberg, P. 1969. Contribution à l'étude du tapir pinchaque *Tapirus pinchaque* Roulin 1829. Rev. Suisse Zool. 76:211–56.

———. 1978. Note sur le rat de Cuming *Phloeomys cumingi* Waterhouse 1839 (Rodentia, Phloeomyidae). Rev. Suisse Zool. 85:341–47.

Scheffer, V. B. 1972. The weight of the Steller sea cow. J. Mamm. 53:912–14.

———. 1978a. Killer whale. *In* Haley (1978), pp. 120–27.

———. 1978b. False killer whale. *In* Haley (1978), pp. 128–31.

Scheich, H., G. Langner, C. Tidemann, R. B. Coles, and A. Guppy. 1986. Electroreception and electrolocation in platypus. Nature 319:401–2.

Schempf, P. F., and M. White. 1977. Status of six furbearer populations in the mountains of northern California. U.S. Forest Serv., iv + 51 pp.

Schenkel, R., and L. Schenkel-Hullinger. 1969. Ecology and behaviour of the black rhinoceros *(Diceros bicornis* L.). Mammalia Depicta, Paul Parey, Hamburg, 100 pp.

Scheurmann, E. 1975. Beobachtungen zur Fortflanzung des Gayal, *Bibos frontalis* Lambert, 1837. Z. Saugetierk. 40:113–27.

Schevill, W. E. 1974. The whale problem: a status report. Harvard Univ. Press, Cambridge, x + 419 pp.

———. 1986. The International Code of Zoological Nomenclature and a paradigm: the name *Physeter catodon* Linnaeus 1758. Mar. Mamm. Sci. 2:153–57.

Schlawe, L. 1980. Zur geographischen Verbreitung der Ginsterkatzen, Gattung *Genetta* Cuvier, 1816. Faun. Abh. Mus. Tierk. Dresden 7:147–61.

———. 1981. Material, Fundorte, Text- und Bildquellan als Grundlagen für eine Artenliste zur Revision der Gattung *Genetta* G. Cuvier, 1816 (Mammalia, Carnivora, Viverridae). Zool. Abhandl. 37:85–182.

Schlieman, H. 1982. A first record of wild cavies for Ecuador and description of *Cavia*

aperea patzelti subsp. nov. Z. Saugetierk. 47:79–90.

Schliemann, H., and B. Maas. 1978. *Myzopoda aurita*. Mammalian Species, no. 116, 2 pp.

Schlitter, D. A. 1973. A new species of gerbil from South West Africa with remarks on *Gerbillus tytonis* Bauer and Niethammer, 1959 (Rodentia: Gerbillinae). Bull. S. California Acad. Sci. 72:13–18.

———. 1974. Notes on the Liberian mongoose, *Liberiictis kuhni* Hayman, 1958. J. Mamm. 55:438–42.

———. 1976. Taxonomy, zoogeography and evolutionary relationships of hairy-footed gerbils, *Gerbillus (Gerbillus)*, of Africa and Asia (Mammalia: Rodentia). Ph.D. Diss., Univ. Maryland, xvi + 558 pp.

———. 1989. African rodents of special concern: a preliminary assessment. *In* Lidicker (1989), pp. 33–39.

Schlitter, D. A., and I. R. Aggundey. 1986. Systematics of African bats of the genus *Eptesicus* (Mammalia: Vespertilionidae). 1. Taxonomic status of the large serotines of eastern and southern Africa. Cimbebasia, ser. A, 8:167–74.

Schlitter, D. A., I. R. Aggundey, M. B. Qumsiyeh, K. Nelson, and R. L. Honeycutt. 1986. Taxonomic and distributional notes on bats from Kenya. Ann. Carnegie Mus. 55:297–302.

Schlitter, D. A., and S. B. McLaren. 1981. An additional record of *Myonycteris relicta* Bergmans, 1980, from Tanzania (Mammalia: Chiroptera). Ann. Carnegie Mus. 50:385–89.

Schlitter, D. A., J. Phillips, and G. E. Kemp. 1973. The distribution of the white-collared mangabey, *Cercocebus torquatus* in Nigeria. Folia Primatol. 19:380–83.

Schlitter, D. A., I. L. Rautenbach, and C. G. Coetzee. 1984. Karyotypes of southern African gerbils, genus *Gerbillurus* Shortridge, 1942 (Rodentia: Cricetidae). Ann. Carnegie Mus. 53:549–57.

Schlitter, D. A., L. W. Robbins, and S. A. Buchanan. 1982. Bats of the Central African Republic (Mammalia: Chiroptera). Ann. Carnegie Mus. 51:133–55.

Schlitter, D. A., L. W. Robbins, and S. L. Williams. 1985. Taxonomic status of dormice (genus *Graphiurus*) from West and Central Africa. Ann. Carnegie Mus. 54:1–9.

Schlitter, D. A., and H. W. Setzer. 1972. A new species of short-tailed gerbil *(Dipodillus)* from Morocco. Proc. Biol. Soc. Washington 84:385–92.

Schlitter, D. A., and K. Thonglongya. 1971. *Rattus turkestanicus* (Satunin, 1903), the valid name for *Rattus ratoides* Hodgson, 1845 (Mammalia: Rodentia). Proc. Biol Soc. Washington 84:171–74.

Schlitter, D. A., S. L. Williams, and J. E. Hill. 1983. Taxonomic review of Temminck's trident bat, *Aselliscus tricuspidatus* (Temminck, 1834) (Mammalia: Hipposideridae). Ann. Carnegie Mus. 52:337–58.

Schmidly, D. J. 1973. The systematic status of *Peromyscus comanche*. Southwestern Nat. 18:269–78.

———. 1974a. *Peromyscus attwateri*. Mammalian Species, no. 48, 3 pp.

———. 1974b. *Peromyscus pectoralis*. Mammalian Species, no. 49, 3 pp.

———. 1977. The mammals of trans-Pecos Texas. Texas A & M Univ. Press, College Station, xii + 225 pp.

———. 1983. Texas mammals east of the Balcones Fault Zone. Texas A & M Univ. Press, College Station, xviii + 400 pp.

Schmidly, D. J., M. H. Beleau, and H. Hildebran. 1972. First record of Cuvier's dolphin from the Gulf of Mexico with comments on the taxonomic status of *Stenella frontalis*. J. Mamm. 53:625–28.

Schmidly, D. J., R. D. Bradley, and P. S. Cato. 1988. Morphometric differentiation and taxonomy of three chromosomally characterized groups of *Peromyscus boylii* from east-central Mexico. J. Mamm. 69:462–80.

Schmidly, D. J., and W. A. Brown. 1979. Systematics of short-tailed shrews (genus *Blarina*) in Texas. Southwestern Nat. 24:39–48.

Schmidly, D. J., and F. S. Hendricks. 1976. Systematics of the southern races of Ord's kangaroo rat, *Dipodomys ordii*. Bull. S. California Acad. Sci. 75:225–37.

Schmidly, D. J., M. R. Lee, W. S. Modi, and E. G. Zimmerman. 1985. Systematics and notes on the biology of *Peromyscus hooperi*. Occas. Pap. Mus. Texas Tech Univ., no. 97, 40 pp.

Schmidt, C. R. 1973. Breeding season and notes on some aspects of reproduction in captive camelids. Internatl. Zoo Yearbook 13:387–90.

———. 1975. Captive breeding of the vicuna. *In* Martin (1975b), pp. 271–83.

Schmidt, J. L., and D. L. Gilbert, eds. 1978. Big game of North America. Stackpole Books, Harrisburg, Pennsylvania, xv + 494 pp.

Schmidt, U. 1988. Reproduction. *In* Greenhall and Schmidt (1988), pp. 99–109.

Schmidt, U., C. Schmidt, W. Lopez-Forment, and R. F. Crespo. 1978. Rückfunde beringter Vampirfledermause *Desmodus rotundus* in Mexiko. Z. Saugetierk. 43:65–70.

Schneider, D. G., L. D. Mech, and J. R. Tester. 1971. Movements of female raccoons and their young as determined by radio-tracking. Anim. Behav. Monogr. 4:1–43.

Schneider, E., and M. Leipoldt. 1983. DNA relationship within the genus *Lepus* in S.W. Europe. Acta Zool. Fennica 174:31–33.

Schnell, G. D., M. E. Douglas, and D. J. Hough. 1985. Sexual dimorphism in spotted dolphins *(Stenella attenuata)* in the eastern tropical Pacific Ocean. Mar. Mamm. Sci. 1:1–14.

———. 1986. Geographic patterns of variation in offshore spotted dolphins *(Stenella attenuata)* of the eastern tropical Pacific Ocean. Mar. Mamm. Sci. 2:186–213.

Schoen, J. W., S. D. Miller, and H. V. Reynolds, III. 1987. Last stronghold of the grizzly. Nat. Hist. 96(1):50–61.

Schoenfeld, M., and Y. Yom-tov. 1985. The biology of two species of hedgehogs, *Erinaceus europaeus concolor* and *Hemiechinus auritus aegypticus*, in Israel. Mammalia 49:339–55.

Schonewald-Cox, C. M., J. W. Bayless, and J. Schonewald. 1985. Cranial morphometry of Pacific coast elk *(Cervus elaphus)*. J. Mamm. 66:63–74.

Schowalter, D. B., J. R. Gunson, and L. D. Harder. 1979. Life history characteristics of little brown bats *(Myotis lucifugus)* in Alberta. Can. Field-Nat. 93:243–51.

Schowalter, D. B., L. D. Harder, and B. H. Treichel. 1978. Age composition of some vespertilionid bats as determined by dental annuli. Can. J. Zool. 56:355–58.

Schreiber, A., R. Wirth, M. Riffel, and H. Van Rompaey. 1989. Weasels, civets, mongooses, and their relatives. An action plan for the conservation of mustelids and viverrids. Internatl. Union Conserv. Nat., Gland, Switzerland, iv + 99 pp.

Schroder, G. D. 1979. Foraging behavior of the bannertail kangaroo rat *(Dipodomys spectabilis)*. Ecology 60:657–65.

Schultz, A. H. 1942. Growth and development of the proboscis monkey. Bull. Mus. Comp. Zool. 89:279–314.

Schuster, R. 1980. Will the Kafue lechwe survive the Kafue dams? Oryx 15:476–89.

Schusterman, R. J. 1981. Steller sea lion— *Eumetopias jubatus*. *In* Ridgway and Harrison (1981a), pp. 119–41.

Schusterman, R. J., J. A. Thomas, and F. G. Wood. 1986. Dolphin cognition and behavior: a comparative approach. Lawrence Erlbaum Associates, Hillsdale, New Jersey, xv + 393 pp.

Schwartz, C. W., and E. R. Schwartz. 1959. The wild mammals of Missouri. Univ. Missouri Press, vi + 341 pp.

Schwartz, J. H. 1986. Primate systematics and a classification of the order. *In* Swindler and Erwin (1986), pp. 1–41.

———. 1987. The red ape: orang-utans and human origins. Houghton Mifflin Co., Boston, 352 pp..

Schwartz, J. H., and I. Tattersall. 1985. Evolutionary relationships of living lemurs and lorises (Mammalia, Primates) and their potential affinities with European Eocene Adapidae. Anthropol. Pap. Amer. Mus. Nat. Hist. 60:1–100.

Schweinsburg, R. E. 1971. Home range, movements, and herd integrity of the collared peccary. J. Wildl. Mgmt. 35:455–60.

Sclater, W. L. 1900. The mammals of South Africa. R. H. Porter, London, 2 vols.

Scott, F. W. 1987. First record of the long-tailed shrew, *Sorex dispar*, for Nova Scotia. Can. Field-Nat. 101:404–7.

Scott, K. M., and C. M. Janis. 1987. Phylogenetic relationships of the Cervidae, and the case for a superfamily "Cervoidea." *In* Wemmer (1987), pp. 3–20.

Scott, M. D., and K. Causey. 1973. Ecology of feral dogs in Alabama. J. Wildl. Mgmt. 37:253–65.

Scott, M. P. 1987. The effect of mating and agonistic experience on adrenal function and mortality of male *Antechinus stuartii* (Marsupialia). J. Mamm. 68:479–86.

Scott, P. A., C. V. Bentley, and J. J. Warren. 1985. Aggressive behavior by wolves toward humans. J. Mamm. 66:807–9.

Seal, U. S., P. Jackson, and R. L. Tilson. 1987. A global tiger conservation plan. *In* Tilson and Seal (1987), pp. 487–98.

Sealander, J. A. 1979. A guide to Arkansas mammals. River Road Press, Conway, Arkansas, x + 313 pp.

Seebeck, J. H. 1981. *Potorous tridactylus* (Kerr) (Marsupialia: Macropodidae): its distribution, status and habitat preferences in Victoria. Austral. Wildl. Res. 8:285–306.

Seebeck, J. H., and P. G. Johnston. 1980. *Potorous longipes* (Marsupialia: Macropodidae); a new species from eastern Victoria. Austral. J. Zool. 28:119–34.

Seegmiller, R. F., and R. D. Ohmart. 1981. Ecological relationships of feral burros and desert bighorn sheep. Wildl. Monogr., no. 78, 58 pp.

Seegmiller, R. F., and C. D. Simpson. 1980. The Barbary sheep: some conceptual implications of competition with desert bighorn. Desert Bighorn Council Trans., 1979, pp. 47–49.

Seidensticker, J., R. K. Lahiri, K. C. Das, and A. Wright. 1976. Problem tiger in the Sundarbans. Oryx 13:267–73.

Seidensticker, J. C., IV, M. G. Hornocker, W. V. Wiles, and J. P. Messick. 1973. Mountain lion social organization in the Idaho Primitive Area. Wildl. Monogr., no. 35, 60 pp.

Sekulic, R. 1981. Conservation of the sable *Hippotragus niger roosevelti* in the Shimba Hills, Kenya. Afr. J. Ecol. 19:153–65.

———. 1982a. The function of howling in red howler monkeys *(Alouatta seniculus)*. Behaviour 81:38–54.

———. 1982b. Daily and seasonal patterns of roaring and spacing in four red howler *Alouatta seniculus* troops. Folia Primatol. 39:22–48.

Sergeant, D. E. 1973. Biology of white whales *(Delphinapterus leucas)* in western Hudson Bay. J. Fish. Res. Bd. Can. 30:1065–90.

———. 1976. History and present status of populations of harp and hooded seals. Biol. Conserv. 10:96–117.

Sergeant, D. E., and P. F. Brodie. 1975. Identity, abundance, and present status of populations of white whales, *Delphinapterus leucas*, in North America. J. Fish. Res. Bd. Can. 32:1047–54.

Sergeant, D. E., D. K. Caldwell, and M. C. Caldwell. 1973. Age, growth, and maturity of bottlenosed dolphin *(Tursiops truncatus)* from northeast Florida. J. Fish. Res. Bd. Can. 30:1009–11.

Sergeant, D. E., and W. Hoek. 1988. An update of the status of white whales *Delphinapterus leucas* in the Saint Lawrence Estuary, Canada. Biol. Conserv. 45:287–302.

Sergeant, D. E., D. J. St. Aubin, and J. R. Geraci. 1980. Life history and northwest Atlantic status of the Atlantic white-sided dolphin, *Lagenorhynchus acutus*. Cetology, no. 37, 12 pp.

Setoguchi, T., and A. L. Rosenberger. 1985. Miocene marmosets: first fossil evidence. Internatl. J. Primatol. 6:615–25.

Setzer, H. W. 1971. New bats of the genus *Laephotis* from Africa (Mammalia: Chiroptera). Proc. Biol. Soc. Washington 84:259–64.

Seymour, C., and R. W. Dickerman. 1982. Observations on the long-legged bat, *Macrophyllum macrophyllum*, in Guatemala. J. Mamm. 63:530–32.

Seymour, K. 1989. *Panthera onca*. Mammalian Species, no. 340, 9 pp.

Shackleton, D. M. 1985. *Ovis canadensis*. Mammalian Species, no. 230, 9 pp.

Shahi, S. P. 1982. Status of the grey wolf *(Canis lupus pallipes* Sykes) in India—a preliminary survey. J. Bombay Nat. Hist. Soc. 79:493–502.

Shaimardanov, R. T. 1982. *Otonycteris hemprichi* and *Barbastella leucomelas* (Chiroptera) in Kazakhstan. Zool. Zhur. 61:1765.

Shane, S. H., R. S. Wells, and B. Würsig. 1986. Ecology, behavior and social organization of the bottlenose dolphin: a review. Mar. Mamm. Sci. 2:34–63.

Shapiro, A. E. 1981. Florida panther population studies. Endangered Species Tech. Bull. 6(7):1, 3.

Sharman, G. B. 1970. Reproductive physiology of marsupials. Science 167:1221–228.

Sharman, G. B., C. E. Murtagh, P. M. Johnson, and C. M. Weaver. 1980. The chromosomes of a rat-kangaroo attributable to *Bettongia tropica* (Marsupialia: Macropodidae). Austral. J. Zool. 28:59–63.

Sharman, G. B., and P. E. Pilton. 1964. The life history and reproduction of the red kangaroo *(Megaleia rufa)*. Proc. Zool. Soc. London 142:29–48.

Shaughnessy, P. D., and F. H. Fay. 1977. A review of the taxonomy and nomenclature of North Pacific harbour seals. J. Zool. 182:385–419.

Shaughnessy, P. D., and L. Fletcher. 1987. Fur seals, *Arctocephalus* spp., at Macquarie Island. *In* Croxall and Gentry (1987), pp. 177–88.

Shaughnessy, P. D., and G. J. B. Ross. 1980. Records of the subantarctic fur seal *(Arctocephalus tropicalis)* from South Africa with notes on its biology and some observations of captive animals. Ann. S. Afr. Mus. 82:71–89.

Shaw, C. A., and H. G. McDonald. 1987. First record of giant anteater (Xenarthra, Myrmecophagidae) in North America. Science 236:186–88.

Shaw, J. H., and P. A. Jordan. 1977. The wolf that lost its genes. Nat. Hist. 86(10):80–88.

Shaw, J. S., T. S. Carter, and J. C. Machado-Neto. 1985. Ecology of the giant anteater *Myrmecophaga tridactyla* in Serra da Canastra, Minas Gerais, Brazil: a pilot study. *In* Montgomery (1985a), pp. 379–84.

Shaw, J. S., J. C. Machado-Neto, and T. S. Carter. 1987. Behavior of free-living giant anteaters *(Myrmecophaga tridactyla)*. Biotrópica 19:255–59.

Shea, B. T. 1984. Between the gorilla and the chimpanzee: a history of debate concerning the existence of the *kooloo-kamba* or gorilla-like chimpanzee. J. Ethnobiol. 4:1–13.

Sheets, R. G., R. L. Linder, and R. B. Dahlgren. 1971. Burrow systems of prairie dogs in South Dakota. J. Mamm. 52:451–53.

Sheffield, W. J., B. A. Fall, and B. A. Brown. 1983. The nilgai antelope in Texas. Kleburg Studies Nat. Res., Texas A & M Univ., 100 pp.

Shellhammer, H. S. 1989. Salt marsh harvest mice, urban development, and rising sea levels. Conserv. Biol. 3:59–65.

Sheng Helin and Lu Hogee. 1980. Current studies on the rare Chinese black muntjac. J. Nat. Hist. 14:803–7.

Sheng Helin and Lu Houji. 1982. Distribution, habits and resource status of the tufted

deer *(Elaphodus cephalophus)*. Acta Zool. Sinica 28:307–11.

———. 1985. A preliminary study on the Chinese river deer population of Zhoushan Island and adjacent islets. *In* Kawamichi (1985), pp. 6–9.

Sheppe, W. A. 1972. The annual cycle of small mammal populations in a Zambian floodplain. J. Mamm. 53:445–60.

———. 1973. Notes on Zambian rodents and shrews. Puku 7:167–90.

Sherman, P. W., and M. L. Morton. 1984. Demography of Belding's ground squirrels. Ecology 65:1617–28.

Shield, J. 1968. Reproduction of the quokka, *Setonix brachyurus*, in captivity. J. Zool. 155:427–44.

Shoemaker, A. H. 1982a. Fecundity in the captive howler monkey, *Alouatta caraya*. Zoo Biol. 1:149–56.

———. 1982b. Notes on the reproductive biology of the white-faced saki *Pithecia pithecia* in captivity. Internatl. Zoo Yearbook 22:124–27.

Shoesmith, M. 1989. Bison—an endangered species in Canada? Recovery 1(1):5–6.

Short, H. L. 1961. Age at sexual maturity of Mexican free-tailed bats. J. Mamm. 42:533–36.

Shoshani, J., and J. F. Eisenberg. 1982. *Elephas maximus*. Mammalian Species, no. 182, 8 pp.

Shoshani, J., C. A. Goldman, and J. G. M. Thewissen. 1988. *Orycteropus afer*. Mammalian Species, no. 300, 8 pp.

Shrestha, S. P. 1989. Nepal's three species of antelopes. Gnusletter 8(2):13.

Shryer, J., and D. L. Flath. 1980. First record of the pallid bat *(Antrozous pallidus)* from Montana. Great Basin Nat. 40:115.

Shubin, I. G. 1972. Reproduction and numbers of steppe lemming in northern Balkhash area. Soviet J. Ecol. 3:450–52.

———. 1974. Ecology of *Lagurus luteus* in the Zaisan Hollow. Zool. Zhur. 53:272–77.

Shubin, I. G., and N. G. Suchkova. 1975. The biology of the Altai birch mouse *(Sicista napaea)*. Zool. Zhur. 54:475–79.

Sicard, B., M. Tranier, and J.-C. Gautun. 1988. Un rongeur nouveau du Burkina Faso (ex Haute-Volta): *Taterillus petteri*, sp. nov. (Rodentia, Gerbillidae). Mammalia 52:187–98.

Sigg, H., A. Stolba, J.-J. Abegglen, and V. Dasser. 1982. Life history of hamadryas baboons: physical development, infant mortality, reproductive parameters and family relationships. Primates 23:473–87.

Sikes, S. K. 1971. The natural history of the African elephant. American Elsevier, New York, xxv + 397 pp.

Silber, G. K. 1988. Recent sightings of the Gulf of California harbor porpoise, *Phocoena sinus*. J. Mamm. 69:430–33.

Silva Taboada, G., and R. H. Pine. 1969. Morphological and behavioral evidence for the relationship between the bat genus *Brachyphylla* and the Phyllonycterinae. Biotrópica 1:10–19.

Silverman, H. B., and M. J. Dunbar. 1980. Aggressive tusk use by the narwhal (*Monodon monoceros* L.). Nature 284:57–58.

Simonds, P. E. 1965. The bonnet macaque in south India. *In* De Vore (1965), pp. 175–96.

Simonetta, A. M. 1968. A new golden mole from Somalia with an appendix on the taxonomy of the family Chrysochloridae (Mammalia, Insectivora). Italian J. Zool., Suppl. n.s., 2:27–55.

———. 1979. First record of *Caluromysiops* from Colombia. Mammalia 43:247–48.

———. 1988. Somalia. *In* East (1988), pp. 27–33.

Simonetti, J., and A. Spotorno. 1980. Posición taxonómica de *Phyllotis micropus* (Rodentia: Cricetidae). An. Mus. Hist. Nat. Valparaiso 13:285–97.

Simonetti, J. A. 1983. Occurrence of the black rat *(Rattus rattus)* in central Chile. Mammalia 47:131–32.

Simons, E. L. 1988. A new species of *Propithecus* (Primates) from northeast Madagascar. Folia Primatol. 50:143–51.

———. 1989. Human origins. Science 245:1343–50.

Simons, E. L., and T. M. Bown. 1984. A new species of *Peratherium* (Didelphidae: Polyprotodonta): the first African marsupial. J. Mamm. 65:539–48.

———. 1985. *Afrotarsius chatrathi*, first tarsiform primate (? Tarsiidae) from Africa. Nature 313:475–77.

Simons, E. L., and Y. Rumpler. 1988. *Eulemur*: new generic name for species of *Lemur* other than *Lemur catta*. C. R. Acad. Sci. Paris, ser. III, 307:547–51.

Simons, L. S. 1984. Seasonality of reproduction and dentinal structures in the harbor porpoise *(Phocoena phocoena)* of the North Pacific. J. Mamm. 65:491–95.

Simpson, C. D. 1964. Notes on the banded mongoose, *Mungos mungo* (Gmelin). Arnoldia 1(19):1–8.

———. 1984. Artiodactyls. *In* Anderson and Jones (1984), pp. 563–88.

Simpson, G. G. 1945. The principles of classification and a classification of the mammals.

Bull. Amer. Mus. Nat. Hist. 85:i–xvi + 1–350.

Sinclair, A. R. E. 1977a. The African buffalo. Univ. Chicago Press, xii + 355 pp.

———. 1977b. Lunar cycle and timing of mating season in Serengeti wildebeest. Nature 267:832–33.

Sineo, L., R. Stanyon, and B. Chiarelli. 1986. Chromosomes of the *Cercopithecus aethiops* species group: *C. aethiops* (Linnaeus, 1758), *C. cynosurus* (Scopoli, 1786), *C. pygerythrus* (Cuvier, 1821), and *C. sabaeus* (Linnaeus, 1766). Internatl. J. Primatol. 7:569–82.

Singer, F. J. 1981. Wild pig populations in the national parks. Environ. Mgmt. 5:263–70.

Sinha, Y. P., and S. Chakraborty. 1971. Taxonomic status of the vespertilionid bat, *Nycticeius emarginatus* Dobson. Proc. Zool. Soc. Calcutta 24:53–59.

Siniff, D. B., I. Stirling, J. L. Bengtson, and R. A. Reichle. 1979. Social and reproductive behavior of crabeater seals *(Lobodon carcinophagus)* during the austral spring. Can. J. Zool. 57:2243–55.

Siniff, D. B., and S. Stone. 1985. The role of the leopard seal in the tropho-dynamics of the Antarctic marine ecosystem. *In* Siegfried, W. R., P. R. Condy, and R. M. Laws, eds. Antarctic nutrient cycles and food webs, Springer-Verlag, Berlin, 555–60.

Sisk, T., and C. Vaughan. 1984. Notes on some aspects of the natural history of the giant pocket gopher (*Orthogeomys* Merriam) in Costa Rica. Brenesia 22:233–47.

Sitwell, N. 1983. The great herds of the Serengeti. Wildlife (London) 25:174–79.

———. 1986. Back home at last, China's *milu* get a new lease on life. Smithsonian 17(3):114–19.

Skinner, J. D. 1976. Ecology of the brown hyaena *Hyaena brunnea* in the Transvaal with a distribution map for southern Africa. S. Afr. J. Sci. 72:262–69.

Skinner, J. D., G. J. Breytenbach, and C. T. A. Maberly. 1976. Observations on the ecology and biology of the bushpig *Potamochoerus porcus* Linn. in the northern Transvaal. S. Afr. J. Wildl. Res. 6:123–28.

Skinner, J. D., S. Davis, and G. Ilani. 1980. Bone collecting by striped hyaenas, *Hyaena hyaena*, in Israel. Paleontol. Afr. 23:99–104.

Skinner, J. D., N. Fairall, and J. D. P. Bothma. 1977. South African red data book—large mammals. S. Afr. Natl. Sci. Programmes Rept., no. 18, v + 29 pp.

Skinner, J. D., and A. J. Hall-Martin. 1975. A note on foetal growth and development of the giraffe *Giraffa camelopardalis giraffa*. J. Zool. 177:73–79.

Skinner, J. D., and G. Ilani. 1979. The striped hyaena *Hyaena hyaena* of the Judean and Negev deserts and a comparison with the brown hyaena *H. brunnea*. Israel J. Zool. 28:229–32.

Skinner, J. D., and R. J. Van Aarde. 1988. The use of space by the aardwolf *Proteles cristatus*. J. Zool. 214:299–301.

Skinner, J. D., J. H. M. Van Zyl, and J. A. H. Van Heerden. 1973. The effect of season on reproduction in the black wildebeest and red hartebeest in South Africa. J. Reprod. Fert., Suppl., 19:101–10.

Skirrow, M. H., and M. Rysan. 1976. Observations on the social behaviour of the Chinese hamster, *Cricetulus griseus*. Can. J. Zool. 54:361–68.

Skoog, P. 1970. The food of the Swedish badger, *Meles meles* L. Viltrevy 7:1–120.

Skovlin, J. M. 1982. Habitat requirements and evaluations. *In* Thomas and Toweill (1982), pp. 369–413.

Slade, N. A., and D. F. Balph. 1974. Population ecology of Uinta ground squirrels. Ecology 55:989–1003.

Slaughter, B. H., and J. E. Ubelaker. 1984. Relationship of South American cricetines to rodents of North America and the Old World. J. Vert. Paleontol. 4:255–64.

Slobodyan, A. A. 1976. The European brown bear in the Carpathians. *In* Pelton, Lentfer, and Folk (1976), pp. 313–19.

Slooten, E., and S. M. Dawson. 1988. Studies on Hector's dolphin, *Cephalorhynchus hectori*: a progress report. *In* Brownell and Donovan (1988), pp. 325–38.

Sly, G. R. 1975. Second record of the bronzed tube-nosed bat *(Murina aenea)* in peninsular Malaysia. Malayan Nat. J. 28:217.

Smales, L. R. 1984. A survey of *Hydromys chrysogaster*. Victorian Nat. 101:115–18.

Small, G. L. 1971. The blue whale. Columbia Univ. Press, New York, xiii + 248 pp.

Smeenk, C. 1987. The harbour porpoise, *Phocoena phocoena* (L., 1758) in the Netherlands: stranding records and decline. Lutra 30:77–90.

Smielowski, J. 1985. Longevity of carnivores from South America. Zool. Garten 55:177.

———. 1986. Longevity of large Indian civet, *Viverra zibetha* L., 1758. Zool. Garten 56:436–37.

Smielowski, J. M., and P. P. Raval. 1988. The Indian wild ass—wild and captive populations. Oryx 22:85–88.

Smit, C. J., and A. Van Wijngaarden. 1981. Threatened mammals in Europe. Akademische Verlagsgesellschaft, Wiesbaden, 259 pp.

Smith, A. 1982. Is the striped possum (*Dactylopsila trivirgata*; Marsupialia, Petauridae) an arboreal anteater? Austral. Mamm. 5:229–34.

———. 1984a. Diet of Leadbeater's possum, *Gymnobelideus leadbeateri* (Marsupialia). Austral. Wildl. Res. 11:265–73.

———. 1984b. Demographic consequences of reproduction, dispersal and social interaction in a population of Leadbeater's possum. *In* Smith and Hume (1984), pp. 359–73.

Smith, A., and I. Hume, eds. 1984. Possums and gliders. Surrey Beatty & Sons, Norton, New South Wales, xv + 598 pp.

Smith, A. T. 1981a. Territoriality and social behavior of *Ochotona princeps*. *In* Myers, K., and C. D. MacInnes, eds., Proceedings of the World Lagomorph Conference, Guelph, Univ. Press, pp. 310–23.

———. 1981b. Population dynamics of pikas (genus *Ochotona*). *In* Myers, K., and C. D. MacInnes, eds., Proceedings of the World Lagomorph Conference, Guelph, Univ. Press, pp. 572–86..

Smith, A. T., and D. M. Cary. 1982. Distribution of Everglades mink. Florida Sci. 45:106–12.

Smith, A. T., and B. L. Ivins. 1983. Colonization in a pika population: dispersal vs philopatry. Behav. Ecol. Sociobiol. 13:37–47.

Smith, A. T., H. J. Smith, Wang Xu Gao, Yin Xiangchu, and Liang Junxiun. 1986. Social behavior of the steppe-dwelling black-lipped pika. Natl. Geogr. Res. 2:57–74.

Smith, C. C. 1977. Feeding behaviour and social organization in howling monkeys. *In* Clutton-Brock (1977), pp. 96–126.

———. 1978. Structure and function of the vocalizations of tree squirrels *(Tamiasciurus)*. J. Mamm. 59:793–808.

Smith, D. A., and L. C. Smith. 1975. Oestrus, copulation, and related aspects of reproduction in female eastern chipmunks, *Tamias striatus* (Rodentia: Sciuridae). Can. J. Zool. 53:756–67.

Smith, G. W. 1977. Population characteristics of the porcupine in northeastern Oregon. J. Mamm. 58:674–76.

Smith, H. C., and E. J. Edmonds. 1985. The brown lemming, *Lemmus sibiricus*, in Alberta. Can. Field-Nat. 99:99–100.

Smith, H. D., and C. D. Jorgensen. 1975. Reproductive biology of North American desert rodents. *In* Prakash and Ghosh (1975), pp. 305–30.

Smith, J. D. 1972. Systematics of the chiropteran family Mormoopidae. Univ. Kansas Mus. Nat. Hist. Misc. Publ., no. 56, 32 pp.

———. 1977. On the nomenclatorial status of *Chilonycteris gymnonotus* Natterer, 1843. J. Mamm. 58:245–46.

Smith, J. D., and H. H. Genoways. 1974. Bats

of Margarita Island, Venezuela, with zoogeographic comments. Bull. S. California Acad. Sci. 73:64–79.

Smith, J. D., and J. E. Hill. 1981. A new species and subspecies of bat of the *Hipposideros bicolor* group from Papua New Guinea, and the systematic status of *Hipposideros calcaratus* and *Hipposideros cupidus* (Mammalia: Chiroptera: Hipposideridae). Los Angeles Co. Mus. Nat. Hist. Contrib. Sci., no. 331, 19 pp.

Smith, J. D., and C. S. Hood. 1981. Preliminary notes on bats from the Bismarck Archipelago (Mammalia: Chiroptera). Sci. New Guinea 8:81–121.

———. 1983. A new species of tube-nosed fruit bat *(Nyctimene)* from the Bismarck Archipelago, Papua New Guinea. Occas. Pap. Mus. Texas Tech Univ., no. 81, 14 pp.

———. 1984. Genealogy of the New World nectar-feeding bats reexamined: a reply to Griffiths. Syst. Zool. 33:435–60.

Smith, J. L. D., C. W. McDougal, and M. E. Sunquist. 1987. Female land tenure system in tigers. *In* Tilson and Seal (1987), pp. 97–109.

Smith, J. R., C. H. S. Watts, and E. G. Crichton. 1972. Reproduction in the Australian desert rodents *Notomys alexis* and *Pseudomys australis* (Muridae). Austral. Mamm. 1:1–7.

Smith, L. C., and D. A. Smith. 1972. Reproductive biology, breeding seasons, and growth of eastern chipmunks, *Tamias striatus* (Rodentia: Sciuridae) in Canada. Can. J. Zool. 50:1069–85.

Smith, M. J. 1971. Breeding the sugar-glider *Petaurus breviceps* in captivity; and growth of pouch young. Internatl. Zoo Yearbook 11:26–28.

———. 1973. *Petaurus breviceps*. Mammalian Species, no. 30, 5 pp.

Smith, M. J., B. K. Brown, and H. J. Frith. 1969. Breeding of the brush-tailed possum, *Trichosurus vulpecula* (Kerr), in New South Wales. CSIRO Wildl. Res. 14:181–93.

Smith, M. J., and R. A. How. 1973. Reproduction in the mountain possum, *Trichosurus caninus* (Ogilby), in captivity. Austral. J. Zool. 21:321–29.

Smith, M. J., and G. C. Medlin. 1982. Dasyurids of the northern Flinders Ranges before pastoral development. *In* Archer (1982a), pp. 563–72.

Smith, N. 1974. Agouti and babassu. Oryx 12:581–82.

Smith, N. S., and I. O. Buss. 1973. Reproductive ecology of the female African elephant. J. Wildl. Mgmt. 37:524–34.

Smith, P. A., and C. J. Jonkel. 1975. Résumé of the trade in polar bear hides in Canada, 1973–74. Can. Wildl. Serv. Progress Notes, no. 48, 5 pp.

Smith, R. F. C. 1969. Studies on the marsupial glider, *Schoinobates volans* (Kerr). Austral. J. Zool. 17:625–36.

Smith, R. I. L. 1988. Destruction of Antarctic terrestrial ecosystems by a rapidly increasing fur seal population. Biol. Conserv. 45:55–72.

Smith, R. M. 1977. Movement patterns and feeding behaviour of leopard in the Rhodes Matopos National Park, Rhodesia. Arnoldia 8(13):1–16.

Smith, T. G. 1976. Predation of ringed seal pups *(Phoca hispida)* by the arctic fox *(Alopex lagopus)*. Can. J. Zool. 54:1610–16.

Smith, T. G., and M. O. Hammill. 1981. Ecology of the ringed seal, *Phoca hispida*, in its fast ice breeding habitat. Can J. Zool. 59:966–81.

———. 1986. Population estimates of white whale, *Delphinapterus leucas*, in James Bay, eastern Hudson Bay, and Ungava Bay. Can. J. Fish. Aquat. Sci. 43:1982–87.

Smith, T. G., M. O. Hammill, D. J. Burrage, and G. A. Sleno. 1985. Distribution and abundance of belugas, *Delphinapterus leucas*, and narwhals, *Monodon monoceros*, in the Canadian high Arctic. Can. J. Fish. Aquat. Sci. 42:676–84.

Smith, T. G., and I. Stirling. 1975. The breeding habitat of the ringed seal *(Phoca hispida)*: the birth lair and associated structures. Can. J. Zool. 53:1297–1305.

Smith, W. J., S. L. Smith, E. C. Oppenheimer, and J. G. Devilla. 1977. Vocalizations of the black-tailed prairie dog, *Cynomys ludovicianus*. Anim. Behav. 25:152–64.

Smith, W. P. 1985. Current geographic distribution and abundance of Columbian white-tailed deer, *Odocoileus virginianus leucurus* (Douglas). Northwest Sci. 59:243–51.

Smith Canet, R., and V. Berovides Alvarez. 1984. Reproducción y ecología de la jutía conga *(Capromys pilorides* Say). Poeyana 280:1–20.

Smithers, R. H. N. 1971. The mammals of Botswana. Trustees Natl. Museums and Monuments Rhodesia Mus. Mem., no. 4, 340 pp.

———. 1978. The serval *Felis serval* Schreber, 1776. S. Afr. J. Wildl. Res. 8:29–37.

———. 1983. The mammals of the southern African subregion. Univ. Pretoria, xxii + 736 pp.

———. 1986. South African red data book—terrestrial mammals. S. Afr. Natl. Sci. Programmes Rept., no. 125, ix + 216 pp.

Smolen, M. J., H. H. Genoways, and R. J. Baker. 1980. Demographic and reproductive parameters of the yellow-cheeked pocket gopher *(Pappogeomys castanops)*. J. Mamm. 61:224–36.

Smuts, G. L. 1976. Reproduction in the zebra mare *Equus burchelli antiquorum* from the Kruger National Park. Koedoe 19:89–132.

Smuts, G. L., and I. J. Whyte. 1981. Relationships between reproduction and environment in the hippopotamus *Hippopotamus amphibius* in the Kruger National Park. Koedoe 24:169–85.

Smythe, N. 1970. The adaptive value of the social organization of the coati *(Nasua narica)*. J. Mamm. 51:818–20.

———. 1978. The natural history of the Central American agouti *(Dasyprocta punctata)*. Smithson. Contrib. Zool., no. 257, iii + 52 pp.

———. 1987. The paca *(Cuniculus paca)* as a domestic source of protein for the neotropical, humid lowlands. Appl. Anim. Behav. Sci. 17:155–70.

Snyder, P. A. 1974. Behavior of *Leontopithecus rosalia* (golden-lion marmoset) and related species: a review. J. Human Evol. 3:109–22.

Snyder, R. L., and S. C. Moore. 1968. Longevity of captive mammals in Philadelphia Zoo. Internatl. Zoo Yearbook 8:175–83.

Soini, P. 1982. Ecology and population dynamics of the pygmy marmoset, *Cebuella pygmaea*. Folia Primatol. 39:1–21.

Sokolov, V. E. 1974. *Saiga tatarica*. Mammalian Species, no. 38, 4 pp.

———. 1981. A new species of five-toed jerboa *Allactaga nataliae* sp. n. (Rodentia, Dipodidae) from Mongolia. Zool. Zhur. 60:793–95.

———. 1989. Antelopes of Kazakh and Mongolian Republic. Gnusletter 8(3):14.

Sokolov, V. E., M. I. Baskevich, and Yu. M. Kovalskaya. 1981. Revision of birch mice of the Caucasus: sibling species *Sicista caucasica* Vinogradov, 1925 and *S. kluchorica* sp. n. (Rodentia, Dipodidae). Zool. Zhur. 60:1386–93.

———. 1986a. The karyotype variability in the southern birch mouse *(Sicista subtilis* Pallas) and substantiation of the species validity for *S. severtzovi* Ognev. Zool. Zhur. 65:1684–92.

———. 1986b. *Sicista kazbegica* sp. n. (Rodentia, Dipodidae) from the basin of the Terek River upper reaches. Zool. Zhur. 65:949–52.

Sokolov, V. E., Yu. M. Kovalskaya, and M. I. Baskevich. 1982. Taxonomy and comparative cytogenetics of some species of the genus *Sicista* (Rodentia, Dipodidae). Zool. Zhur. 61:102–8.

Sokolov, V. E., G. G. Markov, A. A. Danilkin, Kh. M. Nikolov, and S. Gerasimov. 1985. Species status of the European *(Capreolus capreolus)* and Siberian *(C. pygargus)* roe deer (craniometric investigations). Doklady Biol. Sci. 280:90–94.

Sokolov, V. E., and V. N. Orlov. 1986. Introduction of Przewalski horses into the wild. *In* The Przewalski horse and restoration to its natural habitat in Mongolia, FAO, Rome, pp. 77–88.

Sokolov, V. E., and G. I. Shenbrot. 1987. A new species of thick-tailed three-toed jerboa, *Stylodipus sungorus* sp. n. (Rodentia, Dipodidae), from western Mongolia. Zool. Zhur. 66:579–87.

Soldatovic, B., and I. Savic. 1974. The karyotype forms of the genus *Spalax* Guld. in Yugoslavia and their areas of distribution. *In* Kratochvil, J., and R. Obrtel, eds., Proceedings of the International Symposium on Species and Zoogeography of European Mammals, Academia Publishing House, Prague, pp. 125–30.

Soliman, S. 1983. Changes in population densities of two sympatric species of genus *Gerbillus* (Rodentia: Cricetidae) from the western desert of Egypt. Bull. Zool. Soc. Egypt 33:27–36.

Solounias, N. 1988. Evidence from horn morphology on the phylogenetic relationships of the pronghorn *(Antilocapra americana)*. J. Mamm. 69:140–43.

Solt, V. 1981. Black-footed ferret discovered on Wyoming ranch. Fish and Wildl. News, Oct.–Nov., p. 3.

Soota, T. D., and Y. Chaturvedi. 1980. New locality record of *Pipistrellus camortae* Miller from Car Nicobar and its systematic status. Rec. Zool. Surv. India 77:83–87.

Sorenson, M. W. 1970. Behavior of tree shrews. Primate Behav. 1:141–94.

———. 1974. A review of aggressive behavior in the tree shrews. *In* Holloway (1974), pp. 13–30.

Sorenson, M. W., and C. H. Conway. 1968. The social and reproductive behavior of *Tupaia montana* in captivity. J. Mamm. 49:502–12.

Soriano, P. J. 1987. On the presence of the short-tailed opossum *Monodelphis adusta* (Thomas) in Venezuela. Mammalia 51:321–24.

Soriano, P. J., and J. Molinari. 1987. *Sturnira aratathomasi*. Mammalian Species, no. 284, 4 pp.

Sosnovskii, I. P. 1967. Breeding the red dog *Cuon alpinus* at Moscow Zoo. Internatl. Zoo Yearbook 7:120–22.

Southwick, C. H., M. A. Beg, and M. F. Siddiqi. 1965. Rhesus monkeys in north India. *In* De Vore (1965), pp. 111–59.

Southwick, C. H., and D. G. Lindburg. 1986. The primates of India: status, trends, and conservation. *In* Benirschke (1986), pp. 171–87.

Southwick, C. H., and M. F. Siddiqi. 1977. Population dynamics of rhesus monkeys in northern India. *In* Rainier III and Bourne (1977), pp. 339–62.

———. 1988. Partial recovery and a new population estimate of rhesus populations in India. Amer. J. Primatol. 16:187–97.

Southwick, C. H., M. F. Siddiqi, J. A. Cohen, J. R. Oppenheimer, J. Khan, and S. W. Ashraf. 1982. Further declines in rhesus populations of India. *In* Chiarelli and Corruccini (1982), pp. 128–37.

Southwick, C. H., M. F. Siddiqi, M. Y. Farooqui, and B. C. Pal. 1974. Xenophobia among free-ranging rhesus groups in India. *In* Holloway (1974), pp. 185–209.

Southwick, C. H., M. F. Siddiqi, and J. R. Oppenheimer. 1983. Twenty-year changes in rhesus monkey populations in agricultural areas of northern India. Ecology 64:434–39.

Soutiere, E. C. 1979. Effects of timber harvesting on marten in Maine. J. Wildl. Mgmt. 43:850–60.

Sowls, L. K. 1974. Social behaviour of the collared peccary *Dicotyles tajacu* (L.). *In* Geist and Walther (1974), pp. 144–65.

———. 1978. Collared peccary. *In* Schmidt and Gilbert (1978), pp. 191–205.

———. 1984. The peccaries. Univ. Arizona Press, Tucson, xvi + 251 pp.

Spiess, A. 1976. Labrador grizzly *(Ursus arctos* L.): first skeletal evidence. J. Mamm. 57:787–90.

Spinage, C. A. 1973. A review of ivory exploitation and elephant population trends in Africa. E. Afr. Wildl. J. 11:281–89.

———. 1974. Territoriality and population regulation in the Uganda Defassa waterbuck. *In* Geist and Walther (1974), pp. 635–43.

———. 1982. A territorial antelope: the Uganda waterbuck. Academic Press, London, xvi + 334 pp.

Spinage, C. A., D. T. Williamson, and J. E. Williamson. 1989. Botswana. *In* East (1989), pp. 41–49.

Spitzenberger, F. 1978. Die Stachelmaus von Kleinasien, *Acomys cilicicus* n. sp. Ann. Naturhist. Mus. Wien 81:443–46.

Spitzer, N. C., and J. D. Lazell, Jr. 1978. A new rice rat (genus *Oryzomys*) from Florida's lower keys. J. Mamm. 59:787–92.

Spotte, S. 1982. The incidence of twins in pinnipeds. Can. J. Zool. 60:2226–33.

Spotte, S., C. W. Radcliffe, and J. L. Dunn. 1979. Notes on Commerson's dolphin *(Cephalorhynchus commersonii)* in captivity. Cetology, no. 35, 9 pp.

Springer, J. T. 1980. Fishing behavior of coyotes on the Columbia River, southcentral Washington. J. Mamm. 61:373–74.

Sreenivasan, M. A., H. R. Bhat, and G. Geevarghese. 1973. Breeding cycle of *Rhinolophus rouxi* Temminck, 1835 (Chiroptera: Rhinolophidae), in India. J. Mamm. 54:1013–17.

———. 1974. Observations on the reproductive cycle of *Cynopterus sphinx sphinx* Vahl, 1797 (Chiroptera: Pteropodidae). J. Mamm. 55:200–202.

Srikosamatara, S. 1984. Ecology of pileated gibbons in south-east Thailand. *In* Preuschoft et al. (1984), pp. 242–57.

Stains, H. J. 1967. Carnivores and pinnipeds. *In* Anderson and Jones (1967), pp. 325–54.

———. 1984. Carnivores. *In* Anderson and Jones (1984), pp. 491–522.

Stallings, J. R. 1984. Notes on feeding habits of *Mazama gouazoubira* in the Chaco Boreal of Paraguay. Biotrópica 16:155–57.

———. 1986. Notes on the reproductive biology of the grey brocket deer *(Mazama gouazoubira)* in Paraguay. J. Mamm. 67:172–75.

Stangl, F. B., Jr., and W. W. Dalquest. 1986. Two noteworthy records of Oklahoma mammals. Southwestern Nat. 31:123–24.

Stangl, F. B., B. F. Koop, and C. S. Hood. 1983. Occurrence of *Baiomys taylori* (Rodentia: Cricetidae) on the Texas High Plains. Occas. Pap. Mus. Texas Tech Univ., no. 85, 4 pp.

Stanley, W. C. 1963. Habits of the red fox in northeastern Kansas. Univ. Kansas Mus. Nat. Hist. Misc. Publ., no. 34, 31 pp.

Stardom, R. R. P. 1983. Status and management of wolves in Manitoba. *In* Carbyn (1983), pp. 30–34.

Starin, E. D. 1981. Monkey moves. Nat. Hist. 90(9):36–43.

Starrett, A. 1967. Hystricoid, erethizontoid, cavioid, and chinchilloid rodents. *In* Anderson and Jones (1967), pp. 254–72.

———. 1972. *Cyttarops alecto.* Mammalian Species, no. 13, 2 pp.

Starrett, A., and R. S. Casebeer. 1968. Records of bats from Costa Rica. Los Angeles Co. Mus. Nat. Hist. Contrib. Sci., no. 148, 21 pp.

Starrett, A., and G. F. Fisler. 1970. Aquatic adaptations of the water mouse, *Rheomys underwoodi.* Los Angeles Co. Mus. Nat. Hist. Contrib. Sci., no. 182, 14 pp.

Start, A. N. 1972*a*. Notes on *Dyacopterus spadiceus* from Sarawak. Sarawak Mus. J. 20:367–69.

———. 1972*b*. Some bats of Bako National Park, Sarawak. Sarawak Mus. J. 20:371–76.

———. 1975. Another specimen of *Dyacopterus spadiceus* from Sarawak. Sarawak Mus. J. 23:267.

Statistics Canada. 1981. Fur production. Can. Min. Supply and Services, Ottawa.

Steadman, D. W., and C. E. Ray. 1982. The relationships of *Megaoryzomys curioi*, an extinct cricetine rodent (Muroidea: Muridae) from the Galapagos Islands, Ecuador. Smithson. Contrib. Paleobiol., no. 51, 23 pp.

Stebbings, R. E. 1982. Radio tracking greater horseshoe bats with preliminary observations on flight patterns. *In* Cheeseman and Mitson (1982), pp. 161–73.

Stebbings, R. E., and F. Griffith. 1986. Distribution and status of bats in Europe. Inst. Terr. Ecol., Nat. Environ. Res. Council, 142 pp.

Steck, F., and A. Wandeler. 1980. The epidemiology of fox rabies in Europe. Epidemiol. Rev. 2:71–95.

Stein, B. R. 1987. Phylogenetic relationships among four arvicolid genera. Z. Saugetierk. 52:140–56.

Stein, J. L., M. Herder, and K. Miller. 1986. Birth of a northern fur seal on the mainland California coast. California Fish and Game 72:179–81.

Stejneger, L. 1936. Georg Wilhelm Steller. Harvard Univ. Press, Cambridge, 623 pp.

Steltner, H., S. Steltner, and D. E. Sergeant. 1984. Killer whales, *Orcinus orca*, prey on narwhals, *Monodon monoceros*: an eyewitness account. Can. Field-Nat. 98:458–62.

Stenseth, N. C. 1985. Geographic distribution of *Clethrionomys* species. Ann. Zool. Fennici 22:215–19.

Sterner, R. T., and S. A. Shumake. 1978. Coyote damage-control research: a review and analysis. *In* Bekoff (1978), pp. 297–325.

Stevenson, M. F. 1973a. Notes on pregnancy in the sooty mangabey *Cercocebus atys*. Internatl. Zoo Yearbook 13:134–35.

———. 1973b. Observations of maternal behaviour and infant development in the DeBrazza monkey *Cercopithecus neglectus* in captivity. Internatl. Zoo Yearbook 13:179–84.

Stewart, B. S., and W. T. Everett. 1983. Incidental catch of a ribbon seal (*Phoca fasciata*) in the central North Pacific. Arctic 36:369.

Stickel, L. F. 1968. Home range and travels. *In* King (1968), pp. 373–411.

Stiemie, S., and J. A. J. Nel. 1973. Nest-building behaviour in *Aethomys chrysophilus*, *Praomys* (*Mastomys*) *natalensis* and *Rhabdomys pumilio*. Zool. Afr. 8:91–100.

Stinson, N., Jr. 1977. Home range of the western jumping mouse, *Zapus princeps*, in the Colorado Rocky Mountains. Great Basin Nat. 37:87–90.

Stirling, I. 1971. *Leptonychotes weddelli*. Mammalian Species, no. 6, 5 pp.

———. 1973. Vocalization in the ringed seal (*Phoca hispida*). J. Fish. Res. Bd. Can. 30:1592–94.

———. 1974. Midsummer observations on the behavior of wild polar bears (*Ursus maritimus*). Can. J. Zool. 52:1191–98.

Stirling, I., W. Calvert, and D. Andriashek. 1980. Population ecology studies of the polar bear in the area of southeastern Baffin Island. Can. Wildl. Serv. Occas. Pap., no. 44, 33 pp.

Stirling, I., C. Jonkel, P. Smith, R. Robertson, and D. Cross. 1977. The ecology of the polar bear (*Ursus maritimus*) along the western coast of Hudson Bay. Can. Wildl. Serv. Occas. Pap., no. 33, 64 pp.

Stirling, I., and H. P. L. Kiliaan. 1980. Population ecology studies of the polar bear in northern Labrador. Can. Wildl. Serv. Occas. Pap., no. 42, 21 pp.

Stirling, I., and G. L. Kooyman. 1971. The crabeater seal (*Lobodon carcinophagus*) in McMurdo Sound, Antarctica, and the origin of mummified seals. J. Mamm. 52:175–80.

Stirling, I., and D. B. Siniff. 1979. Underwater vocalizations of leopard seals (*Hydrurga leptonyx*) and crabeater seals (*Lobodon carcinophagus*) near the South Shetland Islands, Antarctica. Can. J. Zool. 57:1244–48.

Stock, A. D. 1972. Swimming ability in kangaroo rats. Southwestern Nat. 17:98–99.

Stodart, E. 1977. Breeding and behaviour of Australian bandicoots. *In* Stonehouse and Gilmore (1977), pp. 179–91.

Stogov, I. I. 1985. On two little studied species of white-toothed shrews (Insectivora, Soricidae, *Crocidura*) from the mountain regions in the south of the USSR. Zool. Zhur. 64:264–68.

Stone, R. D. 1987. The social ecology of the Pyrenean desman (*Galemys pyrenaicus*) (Insectivora: Talpidae), as revealed by radio-telemetry. J. Zool. 212:117–29.

Stone, R. D., and M. L. Gorman. 1985. Social organization of the European mole (*Talpa europaea*) and the Pyrenean desman (*Galemys pyrenaicus*). Mamm. Rev. 15:35–42.

Stonehouse, B., and D. Gilmore, eds. 1977. The biology of marsupials. University Park Press, Baltimore, viii + 486 pp.

Stones, R. C., and C. L. Hayward. 1968. Natural history of the desert woodrat, *Neotoma lepida*. Amer. Midl. Nat. 80:458–76.

Storer, T. I., and L. P. Tevis, Jr. 1955. California grizzly. Univ. California Press, Berkeley, vii + 335 pp.

Storm, G. L., R. D. Andrews, R. L. Phillips, R. A. Bishop, D. B. Siniff, and J. R. Tester. 1976. Morphology, reproduction, dispersal, and mortality of midwestern red fox populations. Wildl. Monogr., no. 49, 81 pp.

Storm, G. L., and G. G. Montgomery. 1975. Dispersal and social contact among red foxes: results from telemetry and computer simulation. *In* Fox (1975), pp. 237–46.

Storro-Patterson, R. 1977. Gray whale pro-

tection. How well is it working? Oceans 10(4):44–49.

Storrs, E. E., and H. P. Burchfield. 1985. Leprosy in wild common long-nosed armadillos *Dasypus novemcinctus*. *In* Montgomery (1985a), pp. 265–68.

Strahan, R. 1975. Status and husbandry of Australian monotremes and marsupials. *In* Martin (1975b), pp. 171–82.

———, ed. 1983. The Australian Museum complete book of Australian mammals. Angus & Robertson Publ., London, xxi + 530 pp.

Strassburger, S. 1978. The oryx express. Zoonooz 51(6):4–8.

Streilen, K. E. 1982a. Ecology of small mammals in the semiarid Brazilian Caatinga. I. Climate and faunal composition. Ann. Carnegie Mus. 51:79–106.

———. 1982b. Ecology of small mammals in the semiarid Brazilian Caatinga. II. Water relations. Ann. Carnegie Mus. 51:109–26.

———. 1982c. The ecology of small mammals in the semiarid Brazilian Caatinga. III. Reproductive biology and population ecology. Ann. Carnegie Mus. 51:251–69.

———. 1982d. The ecology of small mammals in the semiarid Brazilian Caatinga. V. Agonistic behavior and overview. Ann. Carnegie Mus. 51:345–69.

———. 1982e. Behavior, ecology, and distribution of the South American marsupials. Pymatuning Lab. Ecol. Spec. Publ. 6:231–50.

Strelkov, P. P. 1986. The Gobi bat (*Eptesicus gobiensis* Bobrinskoy, 1926), a new species of Chiroptera of Palaearctic fauna. Zool. Zhur. 65:1103–8.

Strickland, M. A., and C. W. Douglas. 1987. Marten. *In* Novak, Baker, et al. (1987), pp. 530–47.

Stringham, S. F. 1974. Mother-infant relations in moose. Naturaliste Canadien 101:325–69.

Stroganov, S. U. 1969. Carnivorous mammals of Siberia. Israel Progr. Sci. Transl., Jerusalem, x + 522 pp.

Stroman, H. R., and L. M. Slaughter. 1972. The care and breeding of the pygmy hippopotamus *Choeropsis liberiensis* in captivity. Internatl. Zoo Yearbook 12:126–31.

Stromberg, M. R., and M. S. Boyce. 1986. Systematics and conservation of the swift fox, *Vulpes velox*, in North America. Biol. Conserv. 35:97–110.

Strong, J. T. 1988. Status of the narwhal, *Monodon monoceros*, in Canada. Can. Field-Nat. 102:391–98.

Struhsaker, T. T. 1967a. Ecology of vervet monkeys (*Cercopithecus aethiops*) in the Masai-Amboseli Game Reserve, Kenya. Ecology 48:891–904.

————. 1967b. Auditory communication among vervet monkeys *(Cercopithecus aethiops)*. *In* Altmann (1967), pp. 281–324.

————. 1971. Notes on *Cercocebus a. atys* in Senegal, West Africa. Mammalia 35:343–44.

————. 1975. The red colobus monkey. Univ. Chicago Press. xiv + 311 pp.

————. 1981. Vocalizations, phylogeny and palaeogeography of red colobus monkeys *(Colobus badius)*. Afr. J. Ecol. 19:265–83.

Struhsaker, T. T., and J. S. Gartlan. 1970. Observations on the behaviour and ecology of the patas monkey *(Erythrocebus patas)* in the Waza Reserve. Cameroon. J. Zool. 161:49–63.

Stuart, C. T. 1980. The distribution and status of *Manis temmincki* Smuts, 1832 (Pholidota: Manidae). Saugetierk. Mitt. 28:123–29.

————. 1985. The status of two endangered carnivores occurring in the Cape Province, South Africa, *Felis serval* and *Lutra maculicollis*. Biol. Conserv. 32:375–82.

Stuart, C. T., and V. J. Wilson. 1988. The cats of southern Africa. Internatl. Union Conserv. Nat. Species Survival Comm. Cat Specialist Group and Chipangali Wildl. Trust, Bulawayo, Zimbabwe, 32 pp.

Stuart, S. N. 1988. New hope for the kouprey. Species 10:17.

Stubblefield, S. S., R. J. Warren, and B. R. Murphy. 1986. Hybridization of free-ranging white-tailed and mule deer in Texas. J. Wildl. Mgmt. 50:688–90.

Stüwe, M., and H. Hendrichs. 1984. Organization of roe deer *(Capreolus capreolus)* in an open field habitat. Z. Saugetierk. 49:359–67.

Subbaraj, R., and M. K. Chandrashekaran. 1977. 'Rigid' internal timing in the circadian rhythm of flight activity in a tropical bat. Oecologia 29:341–48.

Suckling, G. C. 1984. Population ecology of the sugar glider, *Petaurus breviceps*, in a system of fragmented habitats. Austral. Wildl. Res. 11:49–75.

Sudman, P. D., J. C. Burns, and J. R. Choate. 1986. Gestation and postnatal development of the plains pocket gopher. Texas J. Sci. 38:91–94.

Sudman, P. D., J. R. Choate, and E. G. Zimmerman. 1987. Taxonomy of chromosomal races of *Geomys bursarius lutescens* Merriam. J. Mamm. 68:526–43.

Sugardjito, J., and N. Nurhuda. 1981. Meat-eating behaviour in wild orang utans, *Pongo pygmaeus*. Primates 22:414–16.

Sugimura, M., Y. Suzuki, I. Kita, Y. Ide, S. Kodera, and M. Yoshizawa. 1983. Prenatal development of Japanese serows, *Capricornis crispus*, and reproduction in females. J. Mamm. 64:302–4.

Sugiyama, Y. 1973. The social structure of wild chimpanzees: a review of field studies. *In* Michael and Crook (1973), pp. 375–41.

Sullivan, R. M. 1982. Agonistic behavior and dominance relationships in the harbor seal, *Phoca vitulina*. J. Mamm. 63:554–69.

Sullivan, R. M., D. J. Hafner, and T. L. Yates. 1986. Genetics of a contact zone between three chromosomal forms of the grasshopper mouse (genus *Onychomys*): a reassessment. J. Mamm. 67:640–59.

Sung, C. V. 1976. New data on morphology and biology of some rare small mammals from North Vietnam. Zool. Zhur. 55:1880–85.

————. 1984. Inventaire des rongeurs du Vietnam. Mammalia 48:391–95.

Sunquist, M. E. 1974. Winter activity of striped skunks *(Mephitis mephitis)* in east-central Minnesota. Amer. Midl. Nat. 92:434–46.

————. 1981. The social organization of tigers *(Panthera tigris)* in Royal Chitawan National Park, Nepal. Smithson. Contrib. Zool., no. 336, vi + 98 pp.

————. 1982. Movements and habitat use of a sloth bear. Mammalia 46:545–47.

Sunquist, M. E., S. N. Austad, and F. Sunquist. 1987. Movement patterns and home range in the common opossum *(Didelphis marsupialis)*. J. Mamm. 68:173–76.

Sunquist, M. E., and G. G. Montgomery. 1973. Activity patterns and rates of movement of two-toed and three-toed sloths *(Choloepus hoffmanni* and *Bradypus infuscatus)*. J. Mamm. 54:946–54.

Susman, R. L., ed. 1984a. The pygmy chimpanzee: evolutionary biology and behavior. Plenum Press, New York, xxviii + 435 pp.

————. 1984b. The locomotor behavior of *Pan paniscus* in the Lomako Forest. *In* Susman (1984a), pp. 369–93.

Sussman, R. W. 1975. A preliminary study of the behavior and ecology of *Lemur fulvus rufus* Audebert 1800. *In* Tattersall and Sussman (1975), pp. 237–58.

————. 1977. Feeding behaviour of *Lemur catta* and *Lemur fulvus*. *In* Clutton-Brock (1977), pp. 1–36.

Sussman, R. W., and W. G. Kinzey. 1984. The ecological role of the Callitrichidae: a review. Amer. J. Phys. Anthropol. 64:419–49.

Sussman, R. W., and A. Richard. 1974. The role of aggression among diurnal prosimians. *In* Holloway (1974), pp. 49–76.

Sussman, R. W., and I. Tattersall. 1986. Distribution, abundance, and putative ecological strategy of *Macaca fascicularis* on the island of Mauritius, southwestern Indian Ocean. Folia Primatol. 46:28–43.

Suthers, R. A., and J. M. Fattu. 1973. Fishing behaviour and acoustic orientation by the bat *(Noctilio labialis)*. Anim. Behav. 21:61–66.

Suzuki, A. 1971. Carnivority and cannibalism observed among forest-living chimpanzees. J. Anthropol. Soc. Nippon 79:30–48.

Svendsen, G. E. 1970. Notes on the ecology of the harvest mouse, *Reithrodontomys megalotis*, in southwestern Wisconsin. Trans. Wisconsin Acad. Sci., Arts, and Letters 58:163–66.

————. 1974. Behavioral and environmental factors in the spatial distribution and population dynamics of a yellow-bellied marmot population. Ecology 55:760–71.

————. 1976. Vocalizations of the long-tailed weasel *(Mustela frenata)*. J. Mamm. 57:398–99.

————. 1978. Castor and anal glands of the beaver *(Castor canadensis)*. J. Mamm. 59:618–20.

————. 1979. Territoriality and behavior in a population of pikas *(Ochotona princeps)*. J. Mamm. 60:324–30.

Swanepoel, P. 1975. Small mammals of the Addo Elephant National Park. Koedoe 18:103–30.

Swanepoel, P., and H. H. Genoways. 1978. Revision of the Antillean bats of the genus *Brachyphylla* (Mammalia: Phyllostomatidae). Bull. Carnegie Mus. Nat. Hist., no. 12, 53 pp.

————. 1983a. *Brachyphylla cavernarum*. Mammalian Species, no. 205, 6 pp.

————. 1983b. *Brachyphylla nana*. Mammalian Species, no. 206, 3 pp.

Swanepoel, P., and D. A. Schlitter. 1978. Taxonomic review of the fat mice (genus *Steatomys*) of West Africa (Mammalia: Rodentia). Bull. Carnegie Mus. Nat. Hist., no. 6, pp. 53–76.

Swanepoel, P., R. H. N. Smithers, and I. L. Rautenbach. 1980. A checklist and numbering system of the extant mammals of the Southern African subregion. Ann. Transvaal Mus. 32:155–96.

Swank, W. G., and J. G. Teer. 1989. Status of the jaguar—1987. Oryx 23:14–21.

Swanson, H. H. 1974. Sex differences in behaviour of the Mongolian gerbil *(Meriones unguiculatus)* in encounters between pairs of same or opposite sex. Anim. Behav. 22:638–44.

Swindler, D. R., and J. Erwin, eds. 1986. Comparative primate biology. I. Systematics, evolution, and anatomy. Alan R. Liss, New York, xvi + 820 pp.

Sylvestre, J.-P. 1983. Review of *Kogia* specimens (Physeteridae, Kogiinae) kept alive in captivity. Investig. Cetacea 15:201–19.

Sylvestre, J.-P., and S. Tasaka. 1985. On the intergeneric hybrids in cetaceans. Aquatic Mamm. 11:101–8.

Symington, M. M. 1988. Demography, ranging patterns, and activity budgets of black spider monkeys *(Ateles paniscus chamek)* in the Manu National Park, Peru. Amer. J. Primatol. 15:45–67.

Syroechkovskii, E. E., ed. 1984. Wild reindeer of the Soviet Union. Amerind Publ. Co., New Delhi, xii + 309 pp.

Syroechkovskii, E. E., and E. V. Rogacheva. 1974. Moose of the Asiatic part of the U.S.S.R. Naturaliste Canadien 101:595–604.

Szalay, F. S. 1982. A new appraisal of marsupial phylogeny and classification. *In* Archer (1982a), pp. 621–40.

———. 1985. Rodent and lagomorph morphotype adaptations, origins, and relationships: some postcranial attributes analyzed. *In* Luckett and Hartenberger (1985), pp. 83–129.

Szalay, F. S., A. L. Rosenberger, and M. Dagosto. 1987. Diagnosis and differentiation of the order Primates. Yearbook Phys. Anthropol. 30:75–105.

T

Taber, R. D., K. Raedeke, and D. A. McCaughran. 1982. Population characteristics. *In* Thomas and Toweill (1982), pp. 279–98.

Taber, R. D., A. N. Sheri, and M. S. Ahmad. 1967. Mammals of the Lyallpur region, West Pakistan. J. Mamm. 48:392–407.

Taddei, V. A. 1976. The reproduction of some Phyllostomatidae (Chiroptera) from the northwestern region of the state of Sao Paulo. Bol. Zool. Univ. Sao Paulo 1:313–30.

———. 1979. Phyllostomidae (Chiroptera) do norte-ocidental do estado de Sao Paulo. III—Stenodermatinae. Ciencia e Cultura 31: 900–914.

Taddei, V. A., and V. Garutti. 1981. The southernmost record of the free-tailed bat, *Tadarida aurispinosa*. J. Mamm. 62:851–52.

Taddei, V. A., L. D. Vizotto, and S. M. Martins. 1976. Notas taxónomicas e biológicas sobre *Molossops brachymeles cerastes* (Thomas, 1901) (Chiroptera—Molossidae). Naturalia 2:61–69.

Taddei, V. A., L. D. Vizotto, and I. Sazima. 1983. Uma nova especie de *Lonchophylla* do Brasil e chave para identificacao das especies do genero (Chiroptera, Phyllostomidae). Ciencia e Cultura 35:625–30.

Takatsuki, S., and K. Suzuki. 1984. Status and food habits of Japanese serow. Proc. Bien. Symp. N. Wild Sheep and Goat Council 4:231–40.

Tamarin, R. H., and S. R. Malecha. 1972. Reproductive parameters in *Rattus rattus* and *R. exulans* of Hawaii, 1968 to 1970. J. Mamm. 53:513–28.

Tamsitt, J. R., A. Cadena, and E. Villarraga. 1986. Records of bats (*Sturnira magna* and *Sturnira aratathomasi*) from Colombia. J. Mamm. 67:754–57.

Tamsitt, J. R., and C. Hauser. 1985. *Sturnira magna*. Mammalian Species, no. 240, 4 pp.

Tamsitt, J. R., and D. Nagorsen. 1982. *Anoura cultrata*. Mammalian Species, no. 179, 5 pp.

Tamura, N., F. Hayashi, and K. Miyashita. 1988. Dominance hierarchy and mating behavior of the Formosan squirrel, *Callosciurus erythraeus thaiwanensis*. J. Mamm. 69:320–31.

Tan Bangjie. 1985. The status of primates in China. Primate Conserv. 5:63–81.

———. 1987. Status and problems of captive tigers in China. *In* Tilson and Seal (1987), pp. 134–48.

Taruski, A. G. 1979. The whistle repertoire of the North Atlantic pilot whale *(Globicephala melaena)* and its relationship to behavior and environment. *In* Winn and Olla (1979), pp. 345–68.

Tate, C. M., J. F. Pagels, and C. O. Handley, Jr. 1980. Distribution and systematic relationship of two kinds of short-tailed shrews (Soricidae: *Blarina*) in south-central Virginia. Proc. Biol. Soc. Washington 93:50–60.

Tate, G. H. H. 1931. Random observations on habits of South American mammals. J. Mamm. 12:248–56.

———. 1932. Distribution of the South American shrews. J. Mamm. 13:223–28.

———. 1933. A systematic revision of the marsupial genus *Marmosa*, with a discussion of the adaptative radiation of the murine opossums *(Marmosa)*. Bull. Amer. Mus. Nat. Hist. 66:1–250.

———. 1934. Bats from the Pacific Islands, including a new fruit bat from Guam. Amer. Mus. Novit., no. 713, 3 pp.

———. 1940. The mammals of the Guiana region. Bull. Amer. Mus. Nat. Hist. 76:151–229.

———. 1942. Results of the Archbold Expeditions. No. 47. Review of the vespertilionine bats, with special attention to genera and species of the Archbold Collections. Bull. Amer. Mus. Nat. Hist. 80:221–97.

———. 1945. Results of the Archbold Expeditions. No. 52. The marsupial genus *Phalanger*. Amer. Mus. Novit., no. 1283, 30 pp.

———. 1951a. *Harpyionycteris*, a genus of rare fruit bats. Amer. Mus. Novit., no. 1522, 9 pp.

———. 1951b. Results of the Archbold Expeditions. No. 65. The rodents of Australia and New Guinea. Bull. Amer. Mus. Nat. Hist. 97:183–430.

Tate, G. H. H., and R. Archbold. 1939. A revision of the genus *Emballonura* (Chiroptera). Amer. Mus. Novit., no. 1035, 14 pp.

Tattersall, I. 1971. Revision of the subfossil Indriinae. Folia Primatol. 16:257–69.

———. 1977a. Ecology and behavior of *Lemur fulvus mayottensis* (Primates, Lemuriformes). Amer. Mus. Nat. Hist. Anthropol. Pap. 54:421–82.

———. 1977b. The lemurs of the Comoro Islands. Oryx 13:445–48.

———. 1978a. Behavioural variation in *Lemur mongoz* (=*L. m. mongoz*). *In* Chivers and Joysey (1978), pp. 127–32.

———. 1978b. Functional cranial anatomy of the subfossil Malagasy lemurs. Natl. Geogr. Soc. Res. Rept., 1969 Proj., pp. 559–68.

———. 1982. The primates of Madagascar. Columbia Univ. Press, New York, xiv + 382 pp.

———. 1986. Systematics of the Malagasy strepsirhine primates. *In* Swindler and Erwin (1986), pp. 43–72.

Tattersall, I., and J. H. Schwartz. 1975. Relationships among the Malagasy lemurs: the craniodental evidence. *In* Luckett and Szalay (1975), pp. 299–312.

Tattersall, I., and R. W. Sussman. 1975. Lemur biology. Plenum Press, New York, xiii + 365 pp.

Taub, D. M. 1977. Geographic distribution and habitat diversity of the Barbary macaque *Macaca sylvanus* L. Folia Primatol. 27:108–33.

———. 1984. A brief historical account of the recent decline in geographic distribution of the Barbary macaque in North Africa. *In* Fa (1984), pp. 71–78.

Taylor, E. H. 1934. Philippine land mammals. Philippine Bur. Sci. Monogr., no. 30, 548 pp.

Taylor, J. M., J. H. Calaby, and T. D. Redhead. 1982. Breeding in wild populations of the marsupial-mouse *Planigale maculata sinualis* (Dasyuridae, Marsupialia). *In* Archer (1982a), pp. 83–87.

Taylor, J. M., J. H. Calaby, and S. C. Smith. 1983. Native *Rattus*, land bridges, and the Australian region. J. Mamm. 64:463–75.

Taylor, J. M., J. H. Calaby, and H. M. Van Deusen. 1982. A revision of the genus *Rattus* (Rodentia, Muridae) in the New Guinean region. Bull. Amer. Mus. Nat. Hist. 173:177–336.

Taylor, J. M., and B. E. Horner. 1970a. Gonadal activity in the marsupial mouse, *Antechinus bellus*, with notes on other species of the genus (Marsupialia: Dasyuridae). J. Mamm. 51:659–68.

———. 1970b. Reproduction in the mosaic-tailed rat, *Melomys cervinipes* (Rodentia: Muridae). Austral. J. Zool. 18:171–84.

————. 1971. Reproduction in the Australian tree-rat *Conilurus penicillatus* (Rodentia: Muridae). CSIRO Wildl. Res. 16:1–9.

————. 1972. Observations on the reproductive biology of *Pseudomys* (Rodentia: Muridae). J. Mamm. 53:318–28.

————. 1973*a*. Results of the Archbold Expeditions. No. 98. Systematics of native Australian *Rattus* (Rodentia, Muridae). Bull. Amer. Mus. Nat. Hist. 150:1–130.

————. 1973*b*. Reproductive characteristics of wild native Australian *Rattus* (Rodentia: Muridae). Austral. J. Zool. 21:437–75.

Taylor, M. E. 1972. *Ichneumia albicauda*. Mammalian Species, no. 12, 4 pp.

————. 1975. *Herpestes sanguineus*. Mammalian Species, no. 65, 5 pp.

————. 1986. Aspects on the biology of the four-toed mongoose, *Bdeogale crassicauda*. Cimbebasia, ser. A, 8:187–93.

————. 1987. *Bdeogale crassicauda*. Mammalian Species, no. 294, 4 pp.

Taylor, P. J., J. U. M. Jarvis, T. M. Crowe, and K. C. Davies. 1985. Age determination in the Cape molerat *Georychus capensis*. S. Afr. J. Zool. 20:261–67.

Tedford, R. H. 1976. Relationships of pinnipeds to other carnivores (Mammalia). Syst. Zool. 25:363–74.

Tembo, A. 1987. Population status of the hippopotamus on the Luangwa River, Zambia. Afr. J. Ecol. 25:71–77.

Temme, M. 1974. Neue Belege der Philippinischen Streifenratte *Chrotomys whiteheadi* Thomas, 1895. Z. Saugetierk. 39:342–45.

Tenaza, R. R. 1975. Pangolins rolling away from predation risks. J. Mamm. 56:257.

Tener, J. S. 1965. Muskoxen in Canada. Can. Wildl. Serv., Ottawa, 166 pp.

Terrill, C. E. 1986. Trends of predator losses of sheep and lambs from 1940 through 1985. Proc. 12th Vert. Pest Conf., Univ. California, Davis, pp. 347–51.

Terry, R. P. 1986. The behaviour and trainability of *Sotalia fluviatilis guianensis* in captivity: a survey. Aquatic Mamm. 12:71–79.

Tesh, R. B. 1970*a*. Observations on the natural history of *Diplomys darlingi*. J. Mamm. 51:197–99.

————. 1970*b*. Notes on the reproduction, growth, and development of echimyid rodents in Panama. J. Mamm. 51:199–202.

Testa, J. W. 1987. Long-term reproductive patterns and sighting bias in Weddell seals *(Leptonychotes weddelli)*. Can. J. Zool. 65:1091–99.

Testa, J. W., and D. B. Siniff. 1987. Population dynamics of Weddell seals *(Leptonychotes weddelli)* in McMurdo Sound, Antarctica. Ecol. Monogr. 57:149–65.

Testaverde, S. A., and J. G. Mead. 1980. Southern distribution of the Atlantic whitesided dolphin, *Lagenorhynchus acutus*, in the western North Atlantic. Fishery Bull. 78:167–69.

Thaeler, C. S., Jr. 1968. An analysis of three hybrid populations of pocket gophers (genus *Thomomys*). Evolution 22:543–55.

————. 1972. Taxonomic status of the pocket gophers, *Thomomys idahoensis* and *Thomomys pygmaeus* (Rodentia, Geomyidae). J. Mamm. 53:417–28.

————. 1980. Chromosome numbers and systematic relations in the genus *Thomomys* (Rodentia: Geomyidae). J. Mamm. 61:414–22.

Thaeler, C. S., Jr., and L. L. Hinesley. 1979. *Thomomys clusius*, a rediscovered species of pocket gopher. J. Mamm. 60:480–88.

Thein, U. T. 1977. The Burmese freshwater dolphin. Mammalia 41:233–34.

Thenius, E. 1979. Zur systematischen und phylogenetischen Stellung des Bambusbäaren: *Ailuropoda melanoleuca* David (Carnivora, Mammalia). Z. Saugetierk. 44:286–305.

Thewissen, J. G. M. 1985. Cephalic evidence for the affinities of Tubulidentata. Mammalia 49:257–84.

Thiel, R. P. 1985. Relationship between road densities and wolf habitat suitability in Wisconsin. Amer. Midl. Nat. 113:404–7.

Thomas, D. W. 1983. The annual migrations of three species of West African fruit bats (Chiroptera: Pteropodidae). Can. J. Zool. 61:2266–72.

Thomas, D. W., and A. G. Marshall. 1984. Reproduction and growth in three species of West African fruit bats. J. Zool. 202:265–81.

Thomas, H. H., and R. L. Dibblee. 1986. A coyote, *Canis latrans*, on Prince Edward Island. Can. Field-Nat. 100:565–67.

Thomas, J., V. Pastukhov, R. Elsner, and E. Petrov. 1982. *Phoca sibirica*. Mammalian Species, no. 188, 6 pp.

Thomas, J. A., and I. Stirling. 1983. Geographic variation in the underwater vocalizations of Weddell seals *(Leptonychotes weddelli)* from Palmer Peninsula and McMurdo Sound, Antarctica. Can. J. Zool. 61:2203–12.

Thomas, J. W., and D. E. Toweill. 1982. Elk of North America: ecology and management. Stackpole Books, Harrisburg, Pennsylvania, xx + 698 pp.

Thomas, K. R. 1974. Burrow systems of the eastern chipmunk (*Tamias striatus pipilans* Lowery) in Louisiana. J. Mamm. 55:454–59.

Thomas, M. E., and D. N. McMurray. 1974. Observations on *Sturnira aratathomasi* from Colombia. J. Mamm. 55:834–36.

Thomas, O. 1881. Description of a new species of *Mus* from southern India. Ann. Mag. Nat. Hist., ser. 5, 7:24.

————. 1894. Descriptions of some new neotropical murids. Ann. Mag. Nat. Hist., ser. 6, 14:346–66.

————. 1895. Descriptions of four small mammals from South America, including one belonging to the peculiar marsupial genus "Hyracodon," Tomes. Ann. Mag. Nat. Hist., ser. 6, 16:367–70.

————. 1896. On new small mammals from the neotropical region. Ann. Mag. Nat. Hist., ser. 6, 18:301–14.

————. 1898*a*. On seven new small mammals from Ecuador and Venezuela. Ann. Mag. Nat. Hist., ser. 7, 1:451–57.

————. 1898*b*. Descriptions of new mammals from South America. Ann. Mag. Nat. Hist., ser. 7, 2:265–75.

————. 1899. Descriptions of new rodents from the Orinoco and Ecuador. Ann. Mag. Nat. Hist., ser. 7, 4:378–83.

————. 1900. New Peruvian species of *Conepatus*, *Phyllotis*, and *Akodon*. Ann. Mag. Nat. Hist., ser. 7, 6:466–70.

————. 1901. New mammals from Peru and Bolivia, with a list of those recorded from the Inambari River, upper Madre de Dios. Ann. Mag. Nat. Hist., ser. 7, 7:178–90.

————. 1906*a*. On a second species of *Lenothrix* from the Liu Kiu Islands. Ann. Mag. Nat. Hist., ser. 7, 17:88–89.

————. 1906*b*. Notes on South American rodents. Ann. Mag. Nat. Hist., ser. 7, 18:442–45.

————. 1909*a*. New species of *Oecomys* and *Marmosa* from Amazonia. Ann. Mag. Nat. Hist., ser. 8, 3:378–81.

————. 1909*b*. Notes on some South American mammals, with descriptions of new species. Ann. Mag. Nat. Hist., ser. 8, 4:230–42.

————. 1910*a*. Further new African mammals. Ann. Mag. Nat. Hist., ser. 8, 5:191–202.

————. 1910*b*. A new genus of fruit-bats and two new shrews from Africa. Ann. Mag. Nat. Hist., ser. 8, 6:111–14.

————. 1910*c*. Mammals from the River Supinaam, Demerara, presented by Mr. F. V. McConnell to the British Museum. Ann. Mag. Nat. Hist., ser. 8, 6:184–89.

————. 1915. On bats of the genera *Nyctalus*, *Tylonycteris*, and *Pipistrellus*. Ann. Mag. Nat. Hist., ser. 8, 15:225–32.

————. 1916. On Muridae from Darjiling

and the Chin Hills. J. Bombay Nat. Hist. Soc. 24:404–15.

———. 1918. On small mammals from Salta and Jujuy collected by Mr. E. Budin. Ann. Mag. Nat. Hist., ser. 9, 1:186–93.

———. 1921. A new genus of opossum from southern Patagonia. Ann. Mag. Nat. Hist., ser. 9, 8:136–39.

———. 1924. New *Callicebus, Conepatus,* and *Oecomys* from Peru. Ann. Mag. Nat. Hist., ser. 9, 14:286–88.

Thomas, W. D. 1975. Observations on captive brockets *Mazama americana* and *M. gouazoubira.* Internatl. Zoo Yearbook 15:77–78.

Thompson, B. G. 1982. Records of *Eptesicus vulturnus* (Thomas) (Vespertilionidae: Chiroptera) from the Alice Springs area, Northern Territory. Austral. Mamm. 5:69–70.

Thompson, D. C. 1977. Diurnal and seasonal activity of the grey squirrel *(Sciurus carolinensis).* Can. J. Zool. 55:1185–89.

Thompson, T. J., H. E. Winn, and P. J. Perkins. 1979. Mysticete sounds. *In* Winn and Olla (1979), pp. 403–31.

Thompson-Handler, N., R. K. Malenky, and N. Badrian. 1984. Sexual behavior of *Pan paniscus* under natural conditions in the Lomako Forest, Equateur, Zaire. *In* Susman (1984*a*), pp. 347–68.

Thomsen, J. B. 1988. Recent U.S. imports of certain products from the African elephant. Pachyderm 10:1–5, 21.

Thonglongya, K. 1973. First record of *Rhinolophus paradoxolophus* (Bourret, 1951) from Thailand, with the description of a new species of the *Rhinolophus philippinensis* group (Chiroptera, Rhinolophidae). Mammalia 37:587–97.

Thorington, R. W., Jr. 1968. Observations of squirrel monkeys in a Colombian forest. *In* Rosenblum and Cooper (1968), pp. 69–85.

———. 1985. The taxonomy and distribution of squirrel monkeys *(Saimiri). In* Rosenblum, L. A., and C. L. Coe, eds., Handbook of squirrel monkey research, Plenum Publ. Corp., pp. 1–33.

———. 1988. Taxonomic status of *Saguinus tripartitus* (Milne-Edwards, 1878). Amer. J. Primatol. 15:367–71.

Thorington, R. W., Jr., and S. Anderson. 1984. Primates. *In* Anderson and Jones (1984), pp. 187–217.

Thorington, R. W., Jr., and C. P. Groves. 1970. An annotated classification of the Cercopithecoidea. *In* Napier and Napier (1970), pp. 629–47.

Thorington, R. W., Jr., and P. G. Heltne, eds. 1976. Neotropical primates: field studies and conservation. Natl. Acad. Sci., Washington, D.C., v + 135 pp.

Thorington, R. W., Jr., N. A. Muckenhirn, and G. G. Montgomery. 1976. Movements of a wild night monkey *(Aotus trivirgatus). In* Thorington and Heltne (1976), pp. 32–34.

Thorington, R. W., Jr., J. C. Ruiz, and J. F. Eisenberg. 1984. A study of a black howling monkey *(Alouatta caraya)* population in northern Argentina. Amer. J. Primatol. 6:357–66.

Thorington, R. W., Jr., and R. E. Vorek. 1976. Observations on the geographic variation and skeletal development of *Aotus.* Lab. Anim. Sci. 26:1006–21.

Thornback, J. 1985. Wild cattle: a genetic heritage under threat. IUCN Bull. 16(1–3):22.

Thornback, J., and M. Jenkins. 1982. The IUCN mammal red data book. Part 1: Threatened mammalian taxa of the Americas and the Australasian zoogeographic region (excluding Cetacea). Internatl. Union Conserv. Nat., Gland, Switzerland, xl + 516 pp.

Thorne, E. T., W. O. Hickey, and S. T. Stewart. 1985. Status of California and Rocky Mountain bighorn sheep in the United States. *In* Hoefs (1985), pp. 56–81.

Thornton, W. A., and G. C. Creel. 1975. The taxonomic status of kit foxes. Texas J. Sci. 26:127–36.

Thouless, C. 1987. Kampuchean wildlife—survival against the odds. Oryx 21:223–28.

Tiba, T., M. Sato, T. Hirano, I. Kita, M. Sugimura, and Y. Suzuki. 1988. An annual rhythm in reproductive activities and sexual maturation in male Japanese serows *(Capricornis crispus).* Z. Saugetierk. 53:178–87.

Tidemann, C. R. 1986. Morphological variation in Australian and island populations of Gould's wattled bat, *Chalinolobus gouldii* (Gray) (Chiroptera: Vespertilionidae). Austral. J. Zool. 34:503–14.

———. 1987*a*. Notes on the flying-fox, *Pteropus melanotus* (Chiroptera: Pteropodidae), on Christmas Island, Indian Ocean. Austral. Mamm. 10:89–91.

———. 1987*b*. Flying-foxes (Chiroptera: Pteropodidae) and bananas: some interactions. Austral. Mamm. 10:133–35.

Tidemann, C. R., D. M. Priddel, J. E. Nelson, and J. D. Pettigrew. 1985. Foraging behaviour of the Australian ghost bat, *Macroderma gigas* (Microchiroptera: Megadermatidae). Austral. J. Zool. 33:705–13.

Tien, D. V. 1960. Recherches zoologiques dans la region de Vinh-Linh (Province de Quang-tri, centre Vietnam). Zool. Anz., Jena, 164:222–39.

———. 1966. Notes sur une collection de petits mammiferes des regions de Thanh-hoa, Nghe-an, Ha-tinh et Quang-binh (centre Vietnam). Zool. Anz., Jena, 176:428–37.

———. 1972. Données écologiques sur l'écureuil géant de McClelland *(Ratufa bicolor gigantea)* (Rodentia, Sciuridae) au Vietnam. Zool. Garten 41:240–43.

Tien, D. V., and C. V. Sung. 1971. Données écologiques sur les rats de bambou (*Rhizomys pruinosus* Blyth et *Rhizomys sumatrensis cinereus* McClelland) au Vietnam. Zool. Garten 40:227–31.

Tillman, M. F., and G. P. Donovan, eds. 1983. Historical whaling records. Rept. Internatl. Whaling Comm. Spec. Issue, no. 5.

Tilson, R. L. 1977. Social organization of simakobu monkeys *(Nasalis concolor)* in Siberut Island, Indonesia. J. Mamm. 58:202–12.

———. 1980. Klipspringer *(Oreotragus oreotragus)* social structure and predator avoidance in a desert canyon. Madoqua 11:303–14.

———. 1981. Family formation strategies of Kloss's gibbons. Folia Primatol. 35:259–87.

Tilson, R. L., and P. M. Norton. 1981. Alarm duetting and pursuit deterrence in an African antelope. Amer. Nat. 118:455–62.

Tilson, R. L., and U. S. Seal, eds. 1987. Tigers of the world. Noyes Publ., Park Ridge, New Jersey, xxx + 510 pp.

Tilson, R. L., K. A. Sweeny, G. A. Binczik, and N. J. Reindl. 1988. Buddies and bullies: social structure of a bachelor group of Przewalski horses. Appl. Anim. Behav. Sci. 21:169–85.

Timm, R. M. 1982. *Ectophylla alba.* Mammalian Species, no. 166, 4 pp.

———. 1984. Tent construction by *Vampyressa* in Costa Rica. J. Mamm. 65:166–67.

———. 1985. *Artibeus phaeotis.* Mammalian Species, no. 235, 6 pp.

———. 1987. Tent construction by bats of the genera *Artibeus* and *Uroderma.* Fieldiana Zool., n.s., 39:187–212.

Timm, R. M., L. Albuja V., and B. L. Clauson. 1986. Ecology, distribution, harvest, and conservation of the Amazonian manatee *Trichechus inunguis* in Ecuador. Biotrópica 18:150–56.

Timm, R. M., L. R. Heaney, and D. D. Baird. 1977. Natural history of rock voles *(Microtus chrotorrhinus)* in Minnesota. Can. Field-Nat. 91:177–81.

Timm, R. M., and J. Mortimer. 1976. Selection of roost sites by Honduran white bats, *Ectophylla alba* (Chiroptera: Phyllostomatidae). Ecology 57:385–89.

Timmis, W. H. 1971. Observations on breeding the oriental short-clawed otter *Amblonyx cinerea* at Chester Zoo. Internatl. Zoo Yearbook 11:109–11.

Tiwari, K. K., R. K. Ghose, and S. Chakraborty. 1971. Notes on a collection of small mammals from Western Ghats, with remarks on the status of *Rattus rufescens* (Gray) and *Bandicota indica malabarica* (Shaw). J. Bombay Nat. Hist. Soc. 68:378–84.

Tobias, P. V. 1978. The place of *Australopithecus africanus* in hominid evolution. *In* Chivers and Joysey (1978), pp. 373–94.

Tomasi, T. E., and R. S. Hoffmann. 1984. *Sorex preblei* in Utah and Wyoming. J. Mamm. 65:708.

Tomiczek, H. 1985. The mouflon (*Ovis ammon musimon* Schreber, 1782) in the southern and western countries of Europe. *In* Hoefs (1985), pp. 127–32.

Tompa, F. S. 1983. Problem wolf management in British Columbia: conflict and program evaluation. *In* Carbyn (1983), pp. 112–19.

Toop, J. 1985. Habitat requirements, survival strategies and ecology of the ghost bat *Macroderma gigas* Dobson (Microchiroptera, Megadermatidae) in central coastal Queensland. Macroderma 1:37–41.

Topachevskii, V. A. 1976. Fauna of the USSR: Mammals. Mole rats, Spalacidae. Amerind Publ. Co., New Delhi, x + 308 pp.

Topal, G. 1970a. The first record of *Ia io* Thomas, 1902 in Vietnam and India, and some remarks on the taxonomic position of *Parascotomanes beaulieui* Bourret, 1942, *Ia longimana* Pen, 1962, and the genus *Ia* Thomas, 1902 (Chiroptera: Vespertilionidae). Opusc. Zool. Budapest 10:341–47.

———. 1970b. On the systematic status of *Pipistrellus annectans* Dobson, 1871 and *Myotis primula* Thomas, 1920 (Mammalia). Ann. Hist.-Nat. Mus. Natl. Hung., Zool., 62:373–79.

———. 1974. Field observations on Oriental bats. Vert. Hung. 15:83–94.

Torres de Assumpcao, C. 1983. Ecological and behavioural information on *Brachyteles arachnoides*. Primates 24:584–93.

Torres-Mura, J. C., M. L. Lemus, and L. C. Contreras. 1989. Herbivorous specialization of the South American desert rodent *Tympanoctomys* Barrerae. J. Mamm. 70:646–48.

Torres N., D. 1987. Juan Fernandez fur seal, *Arctocephalus philippii*. *In* Croxall and Gentry (1987), pp. 37–41.

Trajano, E. 1982. New records of bats from southeastern Brazil. J. Mamm. 63:529.

Tranier, M. M. 1976. Nouvelles données sur l'évolution non parallèle du caryotype et de la morphologie chez les phyllotines (Rongeurs, Cricetides). Compt. Rend. Acad. Sci. Paris, ser. D, 283:1201–3.

———. 1985. Un *Glyphotes canalvus* Moore 1959 dans les collections du Muséum National d'Histoire Naturelle (Rodentia, Sciuridae). Mammalia 49:294–96.

Trapp, G. R. 1972. Some anatomical and behavioral adaptations of ringtails *Bassariscus astutus*. J. Mamm. 53:549–57.

Trapp, G. R., and D. L. Hallberg. 1975. Ecology of the gray fox (*Urocyon cinereoargenteus*): a review. *In* Fox (1975), pp. 164–78.

Travi, V. H. 1981. Nota prévia sobre nova espécie do genero *Ctenomys* Blainville, 1826 (Rodentia, Ctenomyidae). Iheringia, Ser. Zool., 60:123–24.

Trebbau, P. 1975. Measurements and some observations on the freshwater dolphin, *Inia geoffrensis*, in the Apure River, Venezuela. Zool. Garten 45:153–67.

Trebbau, P., and P. J. H. Van Bree. 1974. Notes concerning the freshwater dolphin *Inia geoffrensis* (de Blainville, 1817) in Venezuela. Z. Saugetierk. 39:50–57.

Trent, T. T., and O. J. Rongstad. 1974. Home range and survival of cottontail rabbits in southwestern Wisconsin. J. Wildl. Mgmt. 38:459–72.

Treus, V. D., and N. V. Lobanov. 1971. Acclimatisation and domestication of the eland *Taurotragus oryx* at Askanya-Nova Zoo. Internatl. Zoo Yearbook 11:147–56.

Trillmich, F. 1986. Attendance behavior of Galapagos sea lions. *In* Gentry and Kooyman (1986), pp. 196–208.

———. 1987a. Seals under the sun. Nat. Hist. 96(10):42–49.

———. 1987b. Galapagos fur seal, *Arctocephalus galapagoensis*. *In* Croxall and Gentry (1987), pp. 23–27.

Trillmich, F., G. L. Kooyman, P. Majluf, and M. Sanchez-Griñan. 1986. Attendance and diving behavior of South American fur seals during El Niño in 1983. *In* Gentry and Kooyman (1986), pp. 153–67.

Trillmich, F., and P. Majluf. 1981. First observations on colony structure, behavior, and vocal repertoire of the South American fur seal (*Arctocephalus australis* Zimmermann, 1783) in Peru. Z. Saugetierk. 46:310–22.

Troughton, E. Le G. 1931. Three new bats of the genera *Pteropus, Nyctimene,* and *Chaerephon* from Melanesia. Proc. Linnaean Soc. New South Wales 56:204–9.

———. 1936. The mammalian fauna of Bougainville Island, Solomons Group. Rec. Austral. Mus. 19:341–53.

———. 1971. The early history and relationships of the New Guinea highland dog (*Canis hallstromi*). Proc. Linnean Soc. New South Wales 96:93–98.

Trout, R. C. 1978a. A review of studies on populations of wild harvest mice [*Micromys minutus* (Pallas)]. Mamm. Rev. 8:143–58.

———. 1978b. A review of studies on captive harvest mice [*Micromys minutus* (Pallas)]. Mamm. Rev. 8:159–75.

Tsingalia, H. M., and T. E. Rowell. 1984. The behaviour of adult male blue monkeys. Z. Tierpsychol. 64:253–68.

Tucker, P. K., and D. J. Schmidly. 1981. Studies of a contact zone among three chromosomal races of *Geomys bursarius* in east Texas. J. Mamm. 62:258–72.

Tullar, B. F., Jr. 1983. A long-lived white-tailed deer. New York Fish and Game J. 30:119.

Tulloch, D. G. 1978. The water buffalo, *Bubalus bubalis*, in Australia: grouping and home range. Austral. Wildl. Res. 5:327–54.

Tumanov, I. L., and E. L. Zverev. 1986. Present distribution and numbers of the European mink (*Mustela lutreola*) in the USSR. Zool. Zhur. 65:426–35.

Tupinier, Y. 1977. Description d'une chauve-souris nouvelle: *Myotis nathalinae* nov. sp. (Chiroptera—Vespertilionidae). Mammalia 41:327–40.

Turnbull, C. M. 1976. Man in Africa. Anchor Press/Doubleday, Garden City, New York, xx + 313 pp.

Turner, D. C. 1975. The vampire bat: a field study in behavior and ecology, Johns Hopkins Univ. Press, Baltimore, viii + 145 pp.

Turner, K. 1970. Breeding Tasmanian devils *Sarcophilus harrisii* at Westbury Zoo. Internatl. Zoo. Yearbook 10:65.

Tutin, C. E. G., and M. Fernandez. 1983. Gorillas feeding on termites in Gabon, West Africa. J. Mamm. 64:530–31.

———. 1984. Nationwide census of gorilla (*Gorilla g. gorilla*) and chimpanzee (*Pan t. troglodytes*) populations in Gabon. Amer. J. Primatol. 6:313–36.

Tutin, C. E. G., and P. R. McGinnis. 1981. Chimpanzee reproduction in the wild. *In* Graham, C. E., ed., Reproductive biology of the great apes, Academic Press, New York, pp. 239–64.

Tutin, C. E. G., W. C. McGrew, and P. J. Baldwin. 1983. Social organization of savanna-dwelling chimpanzees, *Pan troglodytes verus*, at Mt. Assirik, Senegal. Primates 24:154–73.

Tuttle, M. D. 1970. Distribution and zoogeography of Peruvian bats, with comments on natural history. Univ. Kansas Sci. Bull. 49:45–86.

———. 1976a. Population ecology of the gray bat (*Myotis grisescens*): philopatry, timing and patterns of movement, weight loss during migration, and seasonal adaptive strategies. Occas. Pap. Mus. Nat. Hist. Univ. Kansas, no. 54, 38 pp.

———. 1976b. Population ecology of the gray bat (*Myotis grisescens*): factors influencing growth and survival of newly volant young. Ecology 57:587–95.

———. 1979. Status, causes of decline, and management of endangered gray bats. J. Wildl. Mgmt. 43:1–17.

————. 1987. Endangered gray bat benefits from protection. Endangered Species Tech. Bull. 12(3):4–5.

Tuttle, M. D., and S. J. Kern. 1981. Bats and public health. Milwaukee Pub. Mus. Publ. Biol. Geol., no. 48, 11 pp.

Tuttle, R. H. 1986. Apes of the world. Noyes Publ., Park Ridge, New Jersey, xix + 421 pp.

Twigg, G. I. 1965. Studies on *Holochilus sciureus berbicensis*, a cricetine rodent from the coastal region of British Guiana. Proc. Zool. Soc. London 145:263–83.

Tyack, P., and H. Whitehead. 1983. Male competition in large groups of wintering humpback whales. Behaviour 83:132–54.

Tyler, M. J., ed. 1979. The status of endangered Australian wildlife. Proc. Cent. Symp. Roy. Zool. Soc. S. Australia, Adelaide, ix + 210 pp.

Tyndale-Biscoe, C. H. 1968. Reproduction and post-natal development in the marsupial *Bettongia lesueur* (Quoy and Gaimard). Austral. J. Zool. 16:577–602.

————. 1973. Life of marsupials. American Elsevier, New York, viii + 254 pp.

Tyndale-Biscoe, C. H., and R. B. MacKenzie. 1976. Reproduction in *Didelphis marsupialis* and *D. albiventris* in Colombia. J. Mamm. 57:249–65.

Tyndale-Biscoe, C. H., and R. F. C. Smith. 1969. Studies on the marsupial glider, *Schoinobates volans* (Kerr). II. Population structure and regulatory mechanisms. J. Anim. Ecol. 38:637–49.

U

Ueckermann, E. 1987. Managing German red deer (*Cervus elaphus* L.) populations. *In* Wemmer (1987), pp. 505–16.

Uieda, W., I. Sazima, and A. Storti Filho. 1980. Aspectos da biologia do morcego *Furipterus horrens* (Mammalia, Chiroptera, Furipteridae). Rev. Brasil. Biol. 40:59–66.

Ulmer, F. A., Jr. 1968. Breeding fishing cats *Felis vierrina* at Philadelphia Zoo. Internatl. Zoo Yearbook 8:49–55.

Uloth, W., and S. Prien. 1985. The history of introductions of mouflon sheep (*Ovis ammon musimon* Schreber 1782) in central and eastern Europe, and the development and management of these wild sheep populations. *In* Hoefs (1985), pp. 133–37.

Union of Soviet Socialist Republics, Ministry of Agriculture. 1978. Red data book of USSR. Lesnaya Promyshlennost Publ., Moscow, 460 pp.

United States Department of Health, Education and Welfare. 1975. Restrictions on importation of nonhuman primates. Federal Register 40:33659.

USDI (United States Department of the Inte-

rior). 1976. Proposal to list 27 species of primates as endangered or threatened species. Federal Register 41:1646–49.

U.S. (United States) Fish and Wildlife Service. 1980. Administration of the Marine Mammal Protection Act of 1972, April 1, 1979 to March 31, 1980. Washington, D.C., v + 86 pp.

U.S. (United States) National Marine Fisheries Service. 1978. The Marine Mammal Protection Act of 1972. Annual Report, 1977–78. Washington, D.C., v + 183 pp.

————. 1981. Marine Mammal Protection Act of 1972. Annual Report 1980/81. Washington, D.C., v + 143 pp.

————. 1984. Marine Mammal Protection Act of 1972. Annual Report 1983/84. Washington, D.C., 146 pp.

————. 1985. Marine Mammal Protection Act of 1972. Annual Report 1984/85. Washington, D.C., 50 pp.

————. 1986. Marine Mammal Protection Act of 1972. Annual Report 1985/86. Washington, D.C., 57 pp.

————. 1987. Marine Mammal Protection Act of 1972. Annual Report 1986/87. Washington, D.C., 47 pp.

————. 1989. Marine Mammal Protection Act of 1972. Annual Report 1987/88. Washington, D.C., 68 pp.

Urban, D. 1970. Raccoon populations, movement patterns, and predation on a managed waterfowl marsh. J. Wildl. Mgmt. 34:372–82.

Uspenski, S. M. 1984. Muskoxen in the USSR: some results of and perspectives on their introduction. *In* Klein, White, and Keller (1984), pp. 12–14.

Uspenski, S. M., and S. E. Belikov. 1976. Research on the polar bear in the USSR. *In* Pelton, Lentfer, and Folk (1976), pp. 321–23.

Ustinov, S. K. 1976. The brown bear on Baikal: a few features of vital activity. *In* Pelton, Lentfer, and Folk (1976), pp. 325–26.

V

Valdez, R., and J. Deforge. 1985. Status of moufloniform (urial) sheep in Asia. *In* Hoefs (1985), pp. 146–50.

Valdez, R., and R. K. LaVal. 1971. Records of bats from Honduras and Nicaragua. J. Mamm. 52:247–50.

Valdez, R., C. F. Nadler, and T. D. Bunch. 1978. Evolution of wild sheep in Iran. Evolution 32:56–72.

Valtonen, M. H., E. J. Rajakoski, and J. I. Mäkelä. 1977. Reproductive features in the female raccoon dog. J. Reprod. Fert. 51:517–18.

Van Aarde, R. J. 1985. Reproduction in captive female Cape porcupines (*Hystrix africaeaustralis*). J. Reprod. Fert. 75:577–82.

————. 1987a. Pre- and postnatal growth of the Cape porcupine *Hystrix africaeaustralis*. J. Zool. 211:25–33.

————. 1987b. Demography of a Cape porcupine, *Hystrix africaeaustralis*, population. J. Zool. 213:205–12.

Van Aarde, R. J., and J. D. Skinner. 1986a. Reproductive biology of the male Cape porcupine, *Hystrix africaeaustralis*. J. Reprod. Fert. 76:545–52.

————. 1986b. Pattern of space use by relocated servals *Felis serval*. Afr. J. Ecol. 24:97–101.

Van Aarde, R. J., and A. Van Dyk. 1986. Inheritance of the king coat colour pattern in cheetahs *Acinonyx jubatus*. J. Zool. 209:573–78.

Van Ballenberghe, V. 1983. Two litters raised in one year by a wolf pack. J. Mamm. 64:171–72.

Van Ballenberghe, V., and A. W. Erickson. 1973. A wolf pack kills another wolf. Amer. Midl. Nat. 90:490–93.

Van Ballenberghe, V., A. W. Erickson, and D. Byman. 1975. Ecology of the timber wolf in northeastern Minnesota. Wildl. Monogr., no. 43, 43 pp.

Van Ballenberghe, V., and L. D. Mech. 1975. Weights, growth, and survival of timber wolf pups in Minnesota. J. Mamm. 56:44–63.

Van Bemmel, A. C. V. 1967. The banteng *Bos javanicus* in captivity. Internatl. Zoo Yearbook 7:222–23.

Van Bree, P. J. H. 1971. On *Globicephala sieboldii* Gray, 1846, and other species of pilot whales (Notes on Cetacea, Delphinoidea III). Beaufortia 19:79–87.

————. 1976. On the correct Latin name of the Indus susu (Cetacea, Platanistoidea). Bull. Zool. Mus. Univ. Amsterdam 5:139–40.

Van Bree, P. J. H., and M. Sc. Boeadi. 1978. Notes on the Indonesian mountain weasel, *Mustela lutreolina* Robinson and Thomas, 1917. Z. Saugetierk. 43:166–71.

Van Bree, P. J. H., and M. D. Gallagher. 1978. On the taxonomic status of *Delphinus tropicalis* Van Bree, 1971 (Notes on Cetacea Delphinoidea IX). Beaufortia 28:1–8.

Van Bree, P. J. H., and P. E. Purves. 1972. Remarks on the validity of *Delphinus bairdii* (Cetacea, Delphinidae). J. Mamm. 53:372–74.

Van Cakenberghe, V., and F. De Vree. 1985. Systematics of African *Nycteris* (Mammalia: Chiroptera). Proc. Internatl. Symp. Afr. Vert., Bonn, pp. 53–90.

Van Camp, J., and R. Gluckie. 1979. A record long-distance movement by a wolf (*Canis lupus*). J. Mamm. 60:236–37.

Van Den Brink, F. H. 1968. A field guide

to the mammals of Britain and Europe. Houghton Mifflin, Boston, 221 pp.

Van Den Brink, J. 1980. The former distribution of the Bali tiger, *Panthera tigris balica* (Schwarz, 1912). Saugetierk. Mitt. 28:286–89.

Van der Horst, G. 1972. Seasonal effects on the anatomy and histology of the reproductive tract of the male rodent mole *Bathyergus suillus suillus* (Schreber). Zool. Afr. 7:491–520.

Van der Merwe, M. 1975. Preliminary study on the annual movements of the Natal clinging bat. S. Afr. J. Sci. 71:237–41.

———. 1978. Postnatal development and mother-infant relationships in the Natal clinging bat *Miniopterus schreibersi natalensis* (A. Smith 1834). Proc. 4th Internatl. Bat Res. Conf., Nairobi, pp. 309–22.

Van der Merwe, M., and I. L. Rautenbach. 1986. Multiple births in Schlieffen's bat, *Nycticeius schlieffenii* (Peters, 1859) (Chiroptera: Vespertilionidae) from the southern African subregion. S. Afr. J. Zool. 21:48–50.

Van der Merwe, M., I. L. Rautenbach, and W. J. Van der Colf. 1986. Reproduction in females of the little free-tailed bat, *Tadarida (Chaerephon) pumila*, in the eastern Transvaal, South Africa. J. Reprod. Fert. 77:355–64.

Van der Merwe, M., J. D. Skinner, and R. P. Millar. 1980. Annual reproductive pattern in the springhaas, *Pedetes capensis*. J. Reprod. Fert. 58:259–66.

Van der Straeten, E. 1975. *Lemniscomys bellieri*, a new species of Muridae from the Ivory Coast. Rev. Zool. Afr. 89:906–8.

———. 1976. *Lemniscomys striatus dieterleni*, a new subspecies of Muridae from Zaire. Rev. Zool. Afr. 90:431–34.

———. 1980a. Etude biometrique de *Lemniscomys linulus* (Afrique Occidentale) (Mammalia, Muridae). Rev. Zool. Afr. 94:185–201.

———. 1980b. A new species of *Lemniscomys* (Muridae) from Zambia. Ann. Cape Prov. Mus. (Nat. Hist.) 13:55–62.

———. 1981. Some biometrical data for *Stenocephalemys albocaudata* and *Stenocephalemys griseicauda*. Afr. Small Mamm. Newsl., no. 6. pp. 11–14.

———. 1984. Etude biometrique des genres *Dephomys* et *Stochomys* avec quelques notes taxonomiques (Mammalia, Muridae). Rev. Zool. Afr. 98:771–98.

———. 1985. Note sur *Hybomys basilii* Eisentraut, 1965. Bonner Zool. Beitr. 36:1–8.

Van der Straeten, E., and F. Dieterlen. 1983. Description de *Praomys ruppi*, une nouvelle espece de Muridae d'Ethiopie. Ann. Mus. Roy. Afr. Cent., Sec. Zool., 237:121–27.

———. 1987. *Praomys misonnei*, a new species of Muridae from eastern Zaire (Mammalia). Stuttgarter Beitr. Naturkunde, ser. A, 11:1–11.

Van der Straeten, E., and R. Hutterer. 1986. *Hybomys eisentrauti*, une nouvelle espece de Muridae du Cameroun (Mammalia, Rodentia). Mammalia 50:35–42.

Van der Straeten, E., and W. N. Verheyen. 1978a. Karyological and morphological comparisons of *Lemniscomys striatus* (Linnaeus, 1758) and *Lemniscomys bellieri* Van der Straeten, 1975, from Ivory Coast (Mammalia: Muridae). Bull. Carnegie Mus. Nat. Hist., no. 6, pp. 41–47.

———. 1978b. Taxonomical notes on the West-African *Myomys* with the description of *Myomys derooi* (Mammalia—Muridae). Z. Saugetierk. 43:31–41.

———. 1979a. Notes taxonomiques sur les *Malacomys* de l'Ouest africain avec redescription du patron chromosomique de *Malacomys edwardsi* (Mammalia, Muridae). Rev. Zool. Afr. 93:10–35.

———. 1979b. Note sur la position systematique de *Lemniscomys macculus* (Thomas et Wroughton, 1910) (Mammalia, Muridae). Mammalia 43:377–89.

———. 1980. Relations biométriques dans le groupe spécifique *Lemniscomys striatus* (Mammalia, Muridae). Mammalia 44:73–82.

———. 1981. Etude biometrique du genre *Praomys* en Cote d'Ivoire. Bonner Zool. Beitr. 32:249–64.

———. 1983. Nouvelles captures de *Lophuromys rahmi* et *Delanymys brooksi* en Republique Rwandaise. Mammalia 47:426–29.

Van der Straeten, E., W. N. Verheyen, and B. Harrie. 1986. The taxonomic status of *Hybomys univittatus lunaris* Thomas, 1906 (Mammalia: Muridae). Cimbebasia, ser. A, 8:209–18.

Van Deusen, H. M. 1968. Carnivorous habits of *Hypsignathus monstrosus*. J. Mamm. 49:335–36.

———. 1969. Results of the 1958–1959 Gilliard New Britain Expedition. 5. A new species of *Pteropus* (Mammalia, Pteropodidae) from New Britain, Bismarck Archipelago. Amer. Mus. Novit., no. 2371, 16 pp.

Van Deusen, H. M., and J. K. Jones, Jr. 1967. Marsupials. *In* Anderson and Jones (1967), pp. 61–86.

Van Deusen, H. M., and K. Keith. 1966. Range and habitat of the bandicoot, *Echymipera clara*, in New Guinea. J. Mamm. 47:721–23.

Van Deusen, H. M., and K. F. Koopman. 1971. Results of the Archbold Expeditions. No. 95. The genus *Chalinolobus* (Chiroptera, Vespertilionidae). Taxonomic review of *Chalinolobus picatus*, *C. nigrogriseus*, and *C. rogersi*. Amer. Mus. Novit., no. 2468, 30 pp.

Van Dyck, S. 1980. The cinnamon antechinus, *Antechinus leo* (Marsupialia: Dasyuridae), a new species from the vineforests of Cape York Peninsula. Austral. Mamm. 3:5–17.

———. 1982a. The relationships of *Antechinus stuartii* and *A. flavipes* (Dasyuridae, Marsupialia) with special reference to Queensland. *In* Archer (1982a), pp. 723–66.

———. 1982b. The status and relationships of the Atherton antechinus, *Antechinus godmani* (Marsupialia: Dasyuridae). Austral. Mamm. 5:195–210.

———. 1985. *Sminthopsis leucopus* (Marsupialia: Dasyuridae) in north Queensland rainforest. Austral. Mamm. 8:53–60.

———. 1986. The chestnut dunnart, *Sminthopsis archeri* (Marsupialia: Dasyuridae), a new species from the savannahs of Papua New Guinea and Cape York Peninsula, Australia. Austral. Mamm. 9:111–24.

———. 1987. The bronze quoll, *Dasyurus spartacus* (Marsupialia: Dasyuridae), a new species from the savannahs of Papua New Guinea. Austral. Mamm. 11:145–56.

Van Dyk, J., R. Singh, and G. Rowell. 1988. Long journey of the Brahmaputra. Natl. Geogr. 174:672–711.

Van Ee, C. A. 1966. A note on breeding the Cape pangolin *Manis temmincki* at Bloemfontein Zoo. Internatl. Zoo Yearbook 6:163–64.

Van Gelder, R. G. 1977a. An eland x kudu hybrid, and the content of the genus *Tragelaphus*. Lammergeyer, no. 23, 6 pp.

———. 1977b. Mammalian hybrids and generic limits. Amer. Mus. Novit., no. 2635, 25 pp.

———. 1978. A review of canid classification. Amer. Mus. Novit., no. 2646, 10 pp.

———. 1979. Mongooses on mainland North America. Wildl. Soc. Bull. 7:197–98.

Van Gyseghem, R. 1984. Observations on the ecology and behaviour of the northern white rhinoceros *(Ceratotherium simum cottoni)*. Z. Saugetierk. 49:348–58.

Van Heerden, J., and F. Kuhn. 1985. Reproduction in captive hunting dogs *Lycaon pictus*. S. Afr. J. Wildl. Res. 15:80–84.

Van Horn, R. N., and G. G. Eaton. 1979. Reproductive physiology and behavior in prosimians. *In* Doyle and Martin (1979), pp. 79–122.

Van Lawick, H., and J. Van Lawick–Goodall. 1971. Innocent killers. Houghton Mifflin, Boston, 222 pp.

Van Lawick–Goodall, J. 1968. The behaviour of free-living chimpanzees in the Gombe Stream Reserve. Anim. Behav. Monogr. 1:165–311.

———. 1973. Cultural elements in a chimpanzee community. Symp. 4th Internatl. Congr. Primatol. 1:144–84.

Van Peenen, P. F. D., R. H. Light, F. J. Duncan, R. See, J. Sulianti Saroso, B. Boeadi, and W. P. Carney. 1974. Observations on *Rattus bartelsii* (Rodentia: Muridae). Treubia 28:83–117.

Van Peenen, P. F. D., P. F. Ryan, and R. H. Light. 1969. Preliminary identification manual for mammals of South Vietnam. Smithson. Inst., Washington, D.C., vi + 310 pp.

Van Rompaey, H. 1978. Longevity of a banded mongoose *(Mungos mungo gmelin)* in captivity. Internatl. Zoo News 25(3):32–33.

———. 1988. *Osbornictis piscivora*. Mammalian Species, no. 309, 4 pp.

Van Strien, N. J. 1975. *Dicerorhinus sumatrensis* (Fischer) the Sumatran or two-horned Asiatic rhinoceros: a study of literature. Netherlands Comm. Internatl. Nat. Protection Meded., no. 22, 82 pp.

———. 1986. The Sumatran rhinoceros *Dicerorhinus sumatrensis* (Fischer, 1814) in the Gunung Leuser National Park Sumatra, Indonesia. Mammalia Depicta, Verlag Paul Parey, Hamburg, vii + 200 pp.

Van Valen, L. 1967. New Paleocene insectivores and insectivore classification. Bull. Amer. Mus. Nat. Hist. 135:217–84.

Van Weers, D. J. 1976. Notes on Southeast Asian porcupines (Hystricidae, Rodentia). I. On the taxonomy of the genus *Trichys* Günther, 1877. Beaufortia 25:15–31.

———. 1977. Notes on Southeast Asian porcupines (Hystricidae, Rodentia). II. On the taxonomy of the genus *Atherurus* F. Cuvier, 1829. Beaufortia 26:205–30.

———. 1978. Notes on Southeast Asian porcupines (Hystricidae, Rodentia). III. On the taxonomy of the subgenus *Thecurus* Lyon, 1907 (genus *Hystrix* Linnaeus, 1758). Beaufortia 28:17–33.

———. 1979. Notes on Southeast Asian porcupines (Hystricidae, Rodentia). IV. On the taxonomy of the subgenus *Acanthion* F. Cuvier, 1823 with notes on the other taxa of the family. Beaufortia 29:215–72.

Van Zyll de Jong, C. G. 1972. A systematic review of the nearctic and neotropical river otters (genus *Lutra*, Mustelidae, Carnivora). Roy. Ontario Mus. Life Sci. Contrib., no. 80, 104 pp.

———. 1975. The distribution and abundance of the wolverine *(Gulo gulo)* in Canada. Can. Field-Nat. 89:431–37.

———. 1976. Are there two species of pygmy shrews *(Microsorex)?* Can. Field-Nat. 90:485–87.

———. 1982. Relationships of amphiberingian shrews of the *Sorex cinereus* group. Can. J. Zool. 60:1580–87.

———. 1983. A morphometric analysis of North American shrews of the *Sorex arcticus* group, with special consideration of the taxonomic status of *S. a. maritimensis*. Nat. Can. 110:373–78.

———. 1984. Taxonomic relationships of nearctic small-footed bats of the *Myotis leibii* group (Chiroptera: Vespertilionidae). Can. J. Zool. 62:2519–26.

———. 1986. A systematic study of Recent bison, with particular consideration of the wood bison *(Bison bison athabascae* Rhoads 1898). Natl. Mus. Can. Publ. Nat. Sci., no. 6, viii + 69 pp.

———. 1987. A phylogenetic study of the Lutrinae (Carnivora; Mustelidae) using morphological data. Can J. Zool. 65:2536–44.

Varisco, D. M. 1989. Beyond rhino horn—wildlife conservation for North Yemen. Oryx 23:215–19.

Varland, K. L., A. L. Lovaas, and R. B. Dahlgren. 1978. Herd organization and movements of elk in Wind Cave National Park, South Dakota. U.S. Natl. Park Serv. Nat. Res. Rept., no. 13, 28 pp.

Varona, L. S. 1979. Subgenero y especie nuevos de *Capromys* (Rodentia: Caviomorpha) para Cuba. Poeyana, no. 194, 33 pp.

Varona, L. S., and O. Arredondo. 1979. Nuevos taxones fosiles de Capromyidae (Rodentia: Caviomorpha). Poeyana, no. 195, 51 pp.

Varty, N., and J. E. Hill. 1988. Notes on a collection of bats from the riverine forests of the Jubba Valley, southern Somalia, including two species new to Somalia. Mammalia 52:533–92.

Vasey, D. E. 1979. Capybara ranching for Amazonia? Oryx 15:47–49.

Vaughan, C. 1983. Coyote range expansion in Costa Rica and Panama. Brenesia 21:27–32.

Vaughan, T. A. 1976. Nocturnal behavior of the African false vampire bat *(Cardioderma cor)*. J. Mamm. 57:227–48.

Vaughan, T. A., and G. C. Bateman. 1970. Functional morphology of the forelimb of mormoopid bats. J. Mamm. 51:217–35.

Vaughan, T. A., and T. J. O'Shea. 1976. Roosting ecology of the pallid bat, *Antrozous pallidus*. J. Mamm. 57:19–41.

Vaughan, T. A., and R. P. Vaughan. 1986. Seasonality and the behavior of the African yellow-winged bat. J. Mamm. 67:91–102.

———. 1987. Parental behavior in the African yellow-winged bat. J. Mamm. 68:217–23.

Vaughn, R. 1974. Breeding the tayra *Eira barbara* at Antelope Zoo, Lincoln. Internatl. Zoo Yearbook 14:120–22.

Vaz-Ferreira, R. 1981. South American sea lion—*Otaria flavescens. In* Ridgway and Harrison (1981a), pp. 39–65.

Veal, R., and W. Caire. 1979. *Peromyscus eremicus*. Mammalian Species, no. 118, 6 pp.

Vedder, A. 1987. Report from the Gorilla Advisory Committee on the status of *Gorilla gorilla*. Primate Conserv. 8:75–81.

Vehrencamp, S. L., F. G. Stiles, and J. W. Bradbury. 1977. Observations on the foraging behavior and avian prey of the neotropical carnivorous bat, *Vampyrum spectrum*. J. Mamm. 58:469–78.

Velte, F. F. 1978. Hand-rearing springhaas *Pedetes capensis* at Rochester Zoo. Internatl. Zoo Yearbook 18:206–8.

Vereschagin, N. K. 1976. The brown bear in Eurasia, particularly the Soviet Union. *In* Pelton, Lentfer, and Folk (1976), pp. 327–35.

Verheyen, W. N., and E. Van der Straeten. 1977. Description of *Malacomys verschureni*, a new murid-species from Central Africa (Mammalia-Muridae). Rev. Zool. Afr. 91:737–44.

———. 1985. Karyological comparison of three different species of *Hybomys* (Mammalia: Muridae). Proc. Internatl. Symp. Afr. Vert., Bonn, pp. 29–34.

Verme, L. J. 1970. Some characteristics of captive Michigan moose. J. Mamm. 51:403–5.

Vermeiren, L. J. P., and W. N. Verheyen. 1980. Notes sur les *Leggada* de Lamto, Côte d'Ivoire, avec la description de *Leggada baoulei* sp. n. Rev. Zool. Afr. 94:570–90.

Verschuren, J. 1980. Note sur les chiropteres du Burundi. Bull. Inst. Roy. Soc. Nat. Belg. 52(19):1–9.

———. 1985. Note sur les mammiferes des Seychelles: un facteur de mortalite de la rousette endemique, *Pteropus seychellensis*. Mammalia 49:424–26.

Verts, B. J., D. Gehman, and K. J. Hundertmark. 1984. *Sylvilagus nuttallii*: a semiarboreal lagomorph. J. Mamm. 65:131–35.

Verwoerd, D. J. 1987. Observations on the food and status of the Cape clawless otter *Aonyx capensis* at Betty's Bay, South Africa. S. Afr. J. Zool. 22:33–39.

Veselovsky, Z. 1966. A contribution to the knowledge of the reproduction and growth of the two-toed sloth *Choloepus didactylus* at Prague Zoo. Internatl. Zoo Yearbook 6:147–53.

Vesmanis, I. 1976. Beitrag zur Kenntnis der Crociduren-Fauna Siziliens (Mammalia: Insectivora). Z. Saugetierk. 41:257–73.

———. 1977. Eine neue *Crocidura*-Art aus der Cyrenaica, Libyen: *Crocidura aleksandrisi* n. sp. (Mammalia: Insectivora: *Crocidura).* Bonner Zool. Beitr. 28:3–12.

——. 1986. Ein weiteres Exemplar von *Crocidura lucina* Dippenaar, 1980 aus Äthiopien. Zool. Abhandl. 41:195–202.

Vessey, S. H., B. K. Mortenson, and N. Muckenhirn. 1978. Size and characteristics of primate groups in Guyana. *In* Chivers and Herbert (1978), pp. 187–88.

Vestjens, W. J. M., and L. S. Hall. 1977. Stomach contents of forty-two species of bats from the Australian region. Austral. Wildl. Res. 4:25–35.

Vietmeyer, N. 1983. Not all endangered animals are wild. Smithsonian 14(5):46–55.

Vigne, J.-D. 1982. Nouvelles données sur le peuplement de la Corse par les rongeurs subactuels et actuels. Mammalia 46:261–65.

——. 1987. L'extinction holocène du fond de peuplement mammalien indigène des îles de Méditerranée occidentale. Mem. Soc. Geol. France, n.s., 150:167–77.

Vigne, L., and E. B. Martin. 1987. The North Yemen government and the rhino horn trade. Swara 10(4):25–27.

——. 1989. Taiwan: the greatest threat to the survival of Africa's rhinos. Pachyderm 11:23–28.

Viljoen, S. 1977. Behaviour of the bush squirrel, *Paraxerus cepapi cepapi* (A. Smith, 1836). Mammalia 41:119–66.

——. 1978. Notes on the western striped squirrel *Funisciurus congicus* (Kuhl 1820). Madoqua 11:119–28.

——. 1986. Use of space in southern African tree squirrels. Mammalia 50:293–309.

Viljoen, S., and S. H. C. Du Toit. 1985. Postnatal development and growth of southern African tree squirrels in the genera *Funisciurus* and *Paraxerus*. J. Mamm. 66:119–27.

Villa, B. 1976. Report on the status of *Phocoena sinus*, Norris and McFarland 1958, in the Gulf of California. An. Inst. Biol. Univ. Nal. Auton. México, Ser. Zool. 47:203–08.

Visser, D. S., and T. J. Robinson. 1986. Cytosystematics of the South African *Aethomys* (Rodentia: Muridae). S. Afr. J. Zool. 21:264–68.

Vitullo, A. D., E. R. S. Roldan, and M. S. Merani. 1988. On the morphology of spermatozoa of tuco-tucos, *Ctenomys* (Rodentia: Ctenomyidae): new data and its implications for the evolution of the genus. J. Zool. 215:675–83.

Vivas, A. M. 1986. Population ecology of *Sigmodon alstoni* (Rodentia: Cricetidae) in the Venezuelan llanos. Rev. Chilena Hist. Nat. 59:179–91.

Vizotto, L. D., and V. A. Taddei. 1976. Notas sobre *Molossops temminckii temminckii* e *Molossops planirostris* (Chiroptera—Molossidae). Naturalia 2:47–59.

Vogel, P. 1970. Biologische Beobachtunge an Etruckerspitzmausen (*Suncus etruscus* Savi, 1832). Z. Saugetierk. 35:173–85.

——. 1988. Taxonomical and biogeographical problems in Mediterranean shrews of the genus *Crocidura* (Mammalia, Insectivora) with reference to a new karyotype from Sicily (Italy). Bull. Soc. Vaud. Sc. Nat. 79:39–48.

Vogel, P., T. Maddalena, and F. Catzeflis. 1986. A contribution to the taxonomy and ecology of shrews (*Crocidura zimmermanni* and *C. suaveolens*) from Crete and Turkey. Acta Theriol. 31:537–45.

Vogt, B. 1978. The big ones are back! Natl. Wildl. 16(6):4–13.

Voigt, D. R. 1987. Red fox. *In* Novak, Baker, et al. (1987), pp. 378–93.

Voigt, D. R., G. B. Kolenosky, and D. H. Pimlott. 1976. Changes in summer foods of wolves in central Ontario. J. Wildl. Mgmt. 40:663–68.

Volf, J. 1976. Some remarks on the taxonomy of the genus *Nemorhaedus* H. Smith 1827. Vest. Cesk. Spol. Zool. 40:75–80.

——. 1985. World's first Przewalski wild horse born to surrogate pony. *In* Volf, J., ed., Pedigree Book of the Przewalski Horse, Zool. Garden Prague, pp. 920–21.

Von Ketelhodt, H. F. 1976. Observations of the lambing interval of the Cape bushbuck *Tragelaphus scriptus sylvaticus*. Zool. Afr. 11:221–25.

——. 1977. Observations on the lambing interval of the grey duiker, *Sylvicapra grimmia grimmia*. Zool. Afr. 12:232–33.

Von Richter, W. 1972. Territorial behaviour of the black wildebeest *Connochaetes gnou*. Zool. Afr. 7:207–31.

——. 1974. *Connochaetes gnou*. Mammalian Species, no. 50, 6 pp.

Vorontsov, N. N., I. V. Kartavtseva, and E. G. Potapova. 1979. Systematics of the genus *Calomyscus* (Cricetinae). Zool. Zhur. 59:283–88.

Vorontsov, N. N., and G. I. Shenbrot. 1984. A systematic review of the genus *Salpingotus* (Rodentia, Dipodidae), with a description of *Salpingotus pallidus* sp. n. from Kazakhstan. Zool. Zhur. 63:731–44.

Vose, H. M. 1973. Feeding habits of the western Australian honey possum, *Tarsipes spenserae*. J. Mamm. 54:245–47.

Voss, R. S. 1988. Systematics and ecology of ichthyomyine rodents (Muroidea): patterns of morphological evolution in a small adaptive radiation. Bull. Amer. Mus. Nat. Hist. 188:259–493.

Voss, R. S., and A. V. Linzey. 1981. Comparative gross morphology of male accessory glands among neotropical Muridae (Mammalia: Rodentia) with comments on systematic implications. Misc. Publ. Mus. Zool. Univ. Michigan, no. 159, 41 pp.

Voss, R. S., J. L. Silva L., and J. A. Valdes L. 1982. Feeding behavior and diets of neotropical water rats, genus *Ichthyomys* Thomas, 1893. Z. Saugetierk. 47:364–69.

Voth, E. H., C. Maser, and M. L. Johnson. 1983. Food habits of *Arborimus albipes*, the white-footed vole, in Oregon. Northwest Sci. 57:1–7.

Vrba, E. S. 1979. Phylogenetic analysis and classification of fossil and Recent Alcelaphini—Mammalia: Bovidae. Biol. J. Linnean Soc. 11:207–28.

Vuillaume-Randriamanantena, M., L. R. Godfrey, and M. R. Sutherland. 1985. Revision of *Hapalemur (Prohapalemur) gallieni* (Standing 1905). Folia Primatol. 45:89–116.

W

Wace, N. M. 1986. The rat problem on oceanic islands—research is needed. Oryx 20:79–86.

Wacher, T. 1988. Social organisation and ranging behaviour in the Hippotraginae. *In* Dixon and Jones (1988), pp. 102–13.

Wachtel, P. S. 1986. Silja saves the grey Baltic seal. World Wildl. Fund News, no. 43, p. 7.

Wackernagel, H. 1968. A note on breeding the serval cat *Felis serval* at Basle Zoo. Internatl. Zoo Yearbook 8:46–47.

Wade, D. A. 1978. Coyote damage: a survey of its nature and scope, control measures and their application. *In* Bekoff (1978), pp. 347–68.

Wade-Smith, J., and M. E. Richmond. 1975. Care, management, and biology of captive striped skunks (*Mephitis mephitis*). Lab. Anim. Sci. 25:575–84.

——. 1978. Reproduction in captive striped skunks (*Mephitis mephitis*). Amer. Midl. Nat. 100:452–55.

Wade-Smith, J., and B. J. Verts. 1982. *Mephitis mephitis*. Mammalian Species, no. 173, 7 pp.

Wadsworth, C. E. 1972. Observations of the Colorado chipmunk in southeastern Utah. Southwestern Nat. 16:451–54.

Waechter, A. 1975. Ecologie de la fouine en Alsace. Terre Vie 29:399–457.

Wainer, J. W. 1976. Studies of an island population of *Antechinus minimus* (Marsupialia, Dasyuridae). Austral. Zool. 19:1–7.

Waithman, J. 1979. A report on a collection of mammals from southwest Papua, 1972–1973. Austral. Zool. 20:313–26.

Wakefield, N. A. 1970a. Notes on Australian pigmy-possums (*Cercartetus*, Phalangeridae, Marsupialia). Victorian Nat. 87:11–18.

———. 1970b. Notes on the glider-possum, *Petaurus australis* (Phalangeridae, Marsupialia). Victorian Nat. 87:221–36.

———. 1971. The brush-tailed rock-wallaby *(Petrogale penicillata)* in western Victoria. Victorian Nat. 88:92–102.

———. 1972. Studies in Australian Muridae: review of *Mastacomys fuscus*, and description of a new subspecies of *Pseudomys higginsi*. Mem. Natl. Mus. Victoria 33:15–31.

Walker, A. 1970. Nuchal adaptations in *Perodicticus potto*. Primates 11:134–44.

———. 1979. Prosimian locomotor behavior. *In* Doyle and Martin (1979), pp. 543–65.

Walker, G. E., and J. K. Ling. 1981a. New Zealand sea lion—*Phocarctos hookeri*. *In* Ridgway and Harrison (1981a), pp. 25–38.

———. 1981b. Australian sea lion—*Neophoca cinerea*. *In* Ridgway and Harrison (1981a), pp. 99–118.

Walker, P. L. 1980. Archaeological evidence for the recent extinction of three terrestrial mammals on San Miguel Island. *In* Power, D. M., ed., The California Islands: proceedings of a multidisciplinary symposium, Santa Barbara Mus. Nat. Hist., pp. 703–17.

Waller, G. N. H. 1983. Is the blind river dolphin sightless? Aquatic Mamm. 10:106–8.

Walley, H. D., and W. L. Jarvis. 1971. Longevity record for *Pipistrellus subflavus*. Trans. Illinois Acad. Sci. 64:305.

Wallin, L. 1969. The Japanese bat fauna. Zool. Bidrag Uppsala 37:223–440.

Wallmo, O. C. 1978. Mule and black-tailed deer. *In* Schmidt and Gilbert (1978), pp. 31–41.

Walther, F. R. 1972. Social grouping in Grant's gazelle (*Gazella granti* Brooke 1827) in the Serengeti National Park. Z. Tierpsychol. 31:348–403.

———. 1978. Behavioral observations on oryx antelope *(Oryx beisa)* invading Serengeti National Park, Tanzania. J. Mamm. 59:243–60.

Walther, F. R., E. C. Mungall, and G. A. Grau. 1983. Gazelles and their relatives: a study in territorial behavior. Noyes Publ., Park Ridge, New Jersey, xiii + 239 pp.

Walton, D. W., J. E. Brooks, U. M. M. Tun, and U. H. Naing. 1978. Observations on reproductive activity among female *Bandicota bengalensis* in Rangoon. Acta Theriol. 23:489–501.

Walton, G. M., and D. W. Walton. 1973. Notes on hedgehogs of the lower Indus Valley. Korean J. Zool. 16:161–70.

Walton, R., and B. J. Trowbridge. 1983. The use of radio-tracking in studying the foraging

behaviour of the Indian flying fox *(Pteropus giganteus)*. J. Zool. 201:575–79.

Wang Fulin. 1985. Preliminary study of the complex-toothed flying squirrel. *In* Kawamichi (1985), pp. 67–69.

Wang Peilie. 1984. Distribution of the gray whale *(Eschrichtius gibbosus)* off the coast of China. Acta Theriol. Sinica 4:21–26.

Wang Sung and Quan Guoqiang. 1986. Primate status and conservation in China. *In* Benirschke (1986), pp. 213–20.

Wang Sung, Zheng Changlin, and Tsuneaki Kobayashi. 1989. A tentative list of threatened rodents in China and Japan with notes on their distribution, habitat, and status. *In* Lidicker (1989), pp. 42–44.

Wang Tsiang-Ke. 1974. On the taxonomic status of species, geological distribution and evolutionary history of *Ailuropoda*. Acta Zool. Sinica 20:201.

Wang Xiaoming, and R. S. Hoffmann. 1987. *Pseudois nayaur* and *Pseudois schaeferi*. Mammalian Species, no. 278, 6 pp.

Wang Yingxiang. 1982. New subspecies of the pipistrels (Chiroptera, Mammalia) from Yunnan, China. Zool. Res., Suppl., 3:343–48.

Wang Yingxiang, Gong Zhengda, and Duan Xingde. 1988. A new species of *Ochotona* (Ochotonidae, Lagomorpha) from Mt. Gaoligong, Northwest Yunnan. Zool. Res. 9:201–7.

Wang Yingxiang, Luo Zexun, and Feng Zuojian. 1985. Taxonomic revision of Yunnan hare, *Lepus comus* G. Allen, with description of two new subspecies. Zool. Res. 6:101–9.

Wang Youzhi. 1985. A new genus and species of Gliridae—*Chaetocauda sichuanensis* gen. et sp. nov. Acta Theriol. Sinica 5:67–75.

Wang Youzhi, Hu Jinchu, and Chen Ke. 1980. A new species of Murinae—*Vernaya foramena* sp. nov. Acta Zool. Sinica 26:393–97.

Ward, D. W., and J. A. Randall. 1987. Territorial defense in the bannertail kangaroo rat *(Dipodomys spectabilis)*: footdrumming and visual threats. Behav. Ecol. Sociobiol. 20:323–28.

Ward, G. C. 1987. India's intensifying dilemma: can tigers and people coexist? Smithsonian 18(8):52–65.

Ward, G. D. 1978. Habitat use and home range of radio-tagged opossums *Trichosurus vulpecula* (Kerr) in New Zealand lowland forest. *In* Montgomery (1978), pp. 267–87.

Waring, G. H. 1970. Sound communications of black-tailed, white-tailed, and Gunnison's prairie dogs. Amer. Midl. Nat. 83:167–85.

———. 1983. Horse behavior. Noyes Publ., Park Ridge, New Jersey, xii + 292 pp.

Warneke, R. M. 1982. The distribution and

abundance of seals in the Australasian region, with summaries of biology and current research. Mammals in the Seas, FAO Fish. Ser. No. 5, 4:431–75.

Warneke, R. M., and P. D. Shaughnessy. 1985. *Arctocephalus pusillus*, the South African and Australian fur seal: taxonomy, evolution, biogeography, and life history. *In* Ling and Bryden (1985), pp. 53–77.

Waser, P. M. 1977. Feeding, ranging and group size in the mangabey *Cercocebus albigena*. *In* Clutton-Brock (1977), pp. 183–222.

Waser, P. M., and O. Floody. 1974. Ranging patterns of the mangabey, *Cercocebus albigena*, in the Kibale Forest, Uganda. Z. Tierpsychol. 35:85–101.

Waser, P. M., and M. S. Waser. 1985. *Ichneumia* and the evolution of viverrid gregariousness. Z. Tierpsychol. 68:137–51.

Wassmer, D. A., and J. L. Wolfe. 1983. New Florida localities for the round-tailed muskrat. Northeast Gulf Sci. 6:197–99.

Watkins, L. C. 1972. *Nycticeius humeralis*. Mammalian Species, no. 23, 4 pp.

———. 1977. *Euderma maculatum*. Mammalian Species, no. 77, 4 pp.

Watkins, L. C., J. K. Jones, Jr., and H. H. Genoways. 1972. Bats of Jalisco, Mexico. Spec. Publ. Mus. Texas Tech Univ., no. 1, 44 pp.

Watkins, W. A. 1976. A probable sighting of a live *Tasmacetus shepherdi* in New Zealand waters. J. Mamm. 57:415.

Watkins, W. A., P. Tyack, K. E. Moore, and G. Notarbartolo-Di-Sciara. 1987. *Steno bredanensis* in the Mediterranean Sea. Mar. Mamm. Sci. 3:78–82.

Watkins, W. A., W. E. Schevill, and P. B. Best. 1977. Underwater sounds of *Cephalorhynchus heavisidii* (Mammalia: Cetacea). J. Mamm. 58:316–18.

Watson, J. P., and N. J. Dippenaar. 1987. The species limits of *Galerella sanguinea* (Rüppell, 1836), *G. pulverulenta* (Wagner, 1839) and *G. nigrita* (Thomas, 1928) in southern Africa (Carnivora: Viverridae). Navors. Nas. Mus. Bloemfontein 5:355–414.

Watts, C. H. S. 1974. The native rodents of Australia: a personal view. Austral. Mamm. 1:109–16.

———. 1975a. Vocalizations of Australian hopping-mice (Rodentia: *Notomys*). J. Zool. 177:247–63.

———. 1975b. The neck and chest glands of Australian hopping-mice, *Notomys*. Austral. J. Zool. 23:151–57.

———. 1976a. Vocalizations of the plains rat *Pseudomys australis* Gray (Rodentia: Muridae). Austral. J. Zool. 24:95–103.

————. 1976b. *Leggadina lakedownensis*, a new species of murid rodent from north Queensland. Trans. Roy. Soc. S. Austral. 100:105–8.

————. 1977. The foods eaten by some Australian rodents (Muridae). Austral. Wildl. Res. 4:151–57.

————. 1979. The status of endangered Australian rodents. *In* Tyler (1979), pp. 75–83.

————. 1980. Vocalizations of nine species of rat (*Rattus*; Muridae). J. Zool. 191:531–55.

Wayne, R. K., W. S. Modi, and S. J. O'Brien. 1986. Morphological variability and asymmetry in the cheetah *(Acinonyx jubatus)*, a genetically uniform species. Evolution 40:78–85.

Weaver, J. 1978. The wolves of Yellowstone. U.S. Natl. Park Serv. Nat. Res. Rept., no. 14, 38 pp.

Weaver, R. A. 1985. The status of desert bighorn in the United States. *In* Hoefs (1985), pp. 82–85.

Webb, S. D. 1985. The interrelationships of tree sloths and ground sloths. *In* Montgomery (1985a), pp. 105–12.

Webber, M. A., and J. Roletto. 1987. Two recent occurrences of the Guadalupe fur seal *Arctocephalus townsendi* in central California. Bull. S. California Acad. Sci. 86:159–63.

Weber, A. W., and A. Vedder. 1983. Population dynamics of the Virunga gorillas: 1959–1978. Biol. Conserv. 26:341–66.

Webster, W. D., and C. O. Handley. 1986. Systematics of Miller's long-tongued bat, *Glossophaga longirostris*, with description of two new subspecies. Occas. Pap. Mus. Texas Tech Univ., no. 100, 22 pp.

Webster, W. D., and J. K. Jones, Jr. 1980a. Noteworthy records of bats from Bolivia. Occas. Pap. Mus. Texas Tech Univ., no. 68, 6 pp.

————. 1980b. Taxonomic and nomenclatorial notes on bats of the genus *Glossophaga* in North America, with description of a new species. Occas. Pap. Mus. Texas Tech Univ., no. 71, 12 pp.

————. 1982. A new subspecies of *Glossophaga commissarisi* (Chiroptera: Phyllostomidae) from western Mexico. Occas. Pap. Mus. Texas Tech Univ., no. 76, 6 pp.

————. 1983. First record of *Glossophaga commissarisi* (Chiroptera: Phyllostomidae) from South America. J. Mamm. 64:150.

————. 1984a. Notes on a collection of bats from Amazonian Ecuador. Mammalia 48:247–52.

————. 1984b. *Glossophaga leachii*. Mammalian Species, no. 226, 3 pp.

————. 1982. A new subspecies of *Glossophaga mexicana* (Chiroptera: Phyllostomi-

dae) from southern Mexico. Occas. Pap. Mus. Texas Tech Univ., no. 91, 5 pp.

————. 1985. *Glossophaga mexicana*. Mammalian Species, no. 245, 2 pp.

————. 1987. A new subspecies of *Glossophaga commissarisi* (Chiroptera: Phyllostomidae) from South America. Occas. Pap. Mus. Texas Tech Univ., no. 109, 6 pp.

Webster, W. D., J. K. Jones, Jr., and R. J. Baker. 1980. *Lasiurus intermedius*. Mammalian Species, no. 132, 3 pp.

Webster, W. D., and W. B. McGillivray. 1984. Additional records of bats from French Guiana. Mammalia 48:463–65.

Webster, W. D., and R. D. Owen. 1984. *Pygoderma bilabiatum*. Mammalian Species, no. 220, 3 pp.

Webster, W. D., L. W. Robbins, R. L. Robbins, and R. J. Baker. 1982. Comments on the status of *Musonycteris harrisoni* (Chiroptera: Phyllostomidae). Occas. Pap. Mus. Texas Tech Univ., no. 78, 5 pp.

Wegge, P. 1979. Aspects of the population ecology of blue sheep in Nepal. J. Asian Ecol. 1:10–20.

Wehausen, J. D. 1984. Comment on desert bighorn as relicts: further considerations. Wildl. Soc. Bull. 12:82–85.

Weir, B. J. 1974a. The tuco-tuco and plains viscacha. Symp. Zool. Soc. London 34:113–30.

————. 1974b. Reproductive characteristics of hystricomorph rodents. Symp. Zool. Soc. London 34:265–301.

————. 1974c. Notes on the origin of the domestic guinea-pig. Symp. Zool. Soc. London 34:437–46.

Weise, T. F., W. L. Robinson, R. A. Hook, and L. D. Mech. 1975. An experimental translocation of the eastern timber wolf. Audubon Conserv. Rept., no. 5, 28 pp.

Weitzel, V., and C. P. Groves. 1985. The nomenclature and taxonomy of the colobine monkeys of Java. Internatl. J. Primatol. 6:399–409.

Wellington, G. M., and Tj. De Vries. 1976. The South American sea lion, *Otaria byronia*, in the Galapagos Islands. J. Mamm. 57:166–67.

Wells, D. R., and C. M. Francis. 1988. Crab-eating mongoose *Herpestes urva*, a mammal new to peninsular Malaysia. Malayan Nat. J. 42:37–41.

Wells, N. M., and J. Giacalone. 1985. *Syntheosciurus brochus*. Mammalian Species, no. 249, 3 pp.

Wells, R. T. 1978. Field observations of the hairy-nosed wombat, *Lasiorhinus latifrons* (Owen). Austral. Wildl. Res. 5:299–303.

Wells-Gosling, N., and L. R. Heaney. 1984. *Glaucomys sabrinus*. Mammalian Species, no. 229, 8 pp.

Wemmer, C. M. 1977. Comparative ethology of the large-spotted genet *Genetta tigrina* and some related viverrids. Smithson. Contrib. Zool., no. 239, iii + 93 pp.

————. 1983a. Systematic position and anatomical traits. *In* Beck and Wemmer (1983), pp. 15–25.

————. 1983b. Sociology and management. *In* Beck and Wemmer (1983), pp. 126–32.

————, ed. 1987. Biology and management of the Cervidae. Smithson. Inst. Press, Washington, D.C., xiii + 577 pp.

Wemmer, C. M., and C. Grodinsky. 1988. Reproduction in captive female brow-antlered deer *(Cervus eldi thamin)*. J. Mamm. 69:389–93.

Wemmer, C. M., and J. Murtaugh. 1981. Copulatory behavior and reproduction in the binturong, *Arctictis binturong*. J. Mamm. 62:342–52.

Wemmer, C. M., and D. Watling. 1986. Ecology and status of the Sulawesi palm civet *Macrogalidia musschenbroekii* Schlegel. Biol. Conserv. 35:1–17.

Wemmer, C. M., J. West, D. Watling, L. Collins, and K. Lang. 1983. External characters of the Sulawesi palm civet, *Macrogalidia musschenbroekii* Schlegel, 1879. J. Mamm. 64:133–36.

Wendell, F. E., J. A. Ames, and R. A. Hardy. 1984. Pup dependency period and length of reproductive cycle: estimates from observations of tagged sea otters, *Enhydra lutris*, in California. California Fish and Game 70:89–100.

Wenstrup, J. J., and R. A. Suthers. 1984. Echolocation of moving targets by the fish-catching bat, *Noctilio leporinus*. J. Comp. Physiol., ser. A, 155:75–89.

Werbitsky, D., and C. W. Kilpatrick. 1987. Genetic variation and genetic differentiation among allopatric populations of *Megadontomys*. J. Mamm. 68:305–12.

Werdelin, L. 1981. The evolution of lynxes. Ann. Zool. Fennici 18:37–71.

West, S. D. 1977. Midwinter aggregation in the northern red-backed vole, *Clethrionomys rutilus*. Can. J. Zool. 55:1404–9.

Westerman, M., A. H. Sinclair, and P. A. Woolley. 1984. Cytology of the feathertail possum *Distoechurus pennatus*. *In* Smith and Hume (1984), pp. 423–25.

Western, D. 1989a. The undetected trade in rhino horn. Pachyderm 11:26–28.

————. 1989b. Ivory trade under scrutiny. Pachyderm 12:2–3.

————. 1989c. The ecological role of elephants in Africa. Pachyderm 12:42–45.

Western, D., and W. K. Lindsay. 1984. Seasonal herd dynamics of a savanna elephant population. Afr. J. Ecol. 22:229–44.

Western, D., and L. Vigne. 1985. The deteriorating status of African rhinos. Oryx 19:215–20.

Wetzel, R. M. 1975. The species of *Tamandua* Gray (Edentata, Myrmecophagidae). Proc. Biol. Soc. Washington 88:95–112.

———. 1977a. The extinction of peccaries and a new case of survival. Ann. New York Acad. Sci. 288:538–44.

———. 1977b. The Chacoan peccary *Catagonus wagneri* (Rusconi). Bull. Carnegie Mus. Nat. Hist., no. 3, 36 pp.

———. 1980. Revision of the naked-tailed armadillos, genus *Cabassous* McMurtie. Ann. Carnegie Mus. 49:323–57.

———. 1981. The hidden Chacoan peccary. Carnegie Mag. 55(2):24–32.

———. 1982. Systematics, distribution, ecology, and conservation of South American edentates. Pymatuning Lab. Ecol. Spec. Publ. 6:345–75.

———. 1985a. The identification and distribution of Recent Xenarthra (= Edentata). *In* Montgomery (1985a), pp. 5–21.

———. 1985b. Taxonomy and distribution of armadillos, Dasypodidae. *In* Montgomery (1985a), pp. 23–46.

Wetzel, R. M., and F. D. de Avila-Pires. 1980. Identification and distribution of the Recent sloths of Brazil (Edentata). Rev. Brasil. Biol. 40:831–36.

Wetzel, R. M., R. E. Dubos, R. L. Martin, and P. Myers. 1975. *Catagonus*, an "extinct" peccary, alive in Paraguay. Science 189:379–81.

Wetzel, R. M., and D. Kock. 1973. The identity of *Bradypus variegatus* Schinz (Mammalia, Edentata). Proc. Biol. Soc. Washington 86:25–34.

Wetzel, R. M., and J. W. Lovett. 1974. A collection of mammals from the Chaco of Paraguay. Univ. Connecticut Occas. Pap., Biol. Sci. Ser., 2:203–16.

Wetzel, R. M., and E. Mondolfi. 1979. The subgenera and species of long-nosed armadillos, genus *Dasypus* L. *In* Eisenberg (1979), pp. 43–63.

Wheatley, B. P. 1980. Feeding and ranging of east Bornean *Macaca fascicularis*. *In* Lindburg (1980), pp. 215–46.

Wheeler, M. E., and C. F. Aguon. 1978. The current status and distribution of the Marianas fruit bat on Guam. Guam Aquatic and Wildl. Res. Div. Tech. Rept., no. 1, 29 pp.

Whitaker, J. O., Jr. 1972. *Zapus hudsonius*. Mammalian Species, no. 11, 7 pp.

Whitaker, J. O., Jr., and B. Abrell. 1986. The swamp rabbit, *Sylvilagus aquaticus*, in Indiana, 1984–1985. Indiana Acad. Sci. 95:563–70.

Whitaker, J. O., Jr., and H. Black. 1976. Food habits of cave bats from Zambia. Africa. J. Mamm. 57:199–204.

Whitaker, J. O., Jr., and R. E. Mumford. 1972. Ecological studies on *Reithrodontomys megalotis* in Indiana. J. Mamm. 53:850–60.

Whitaker, J. O., Jr., and R. E. Wrigley. 1972. *Napaeozapus insignis*. Mammalian Species, no. 14, 6 pp.

Whitehead, G. K. 1972. Deer of the world. Viking Press, New York, xii + 194 pp.

Whitehead, H., and T. Arnbom. 1987. Social organization of sperm whales off the Galapagos Islands, February–April 1985. Can. J. Zool. 65:913–19.

Whitehead, H., R. Payne, and M. Payne. 1986. Population estimate for the right whales off Peninsula Valdes, Argentina, 1971–1976. *In* Brownell, Best, and Prescott (1986), pp. 169–71.

Whitehead, P. J. P. 1977. The former southern distribution of New World manatees (*Trichechus* spp.). Biol. J. Linnean Soc. 9:165–89.

Whitehouse, S. J. O. 1977. The diet of the dingo in Western Australia. Austral. Wildl. Res. 4:145–50.

Whitman, D. W., M. S. Blum, and C. G. Jones. 1986. Prey-specific attack behaviour in the southern grasshopper mouse *(Onychomys torridus)* (Coues). Anim. Behav. 34:295–97.

Whitman, J. S., W. B. Ballard, and C. L. Gardner. 1986. Home range and habitat use by wolverines in southcentral Alaska. J. Wildl. Mgmt. 50:460–63.

Whitten, A. J. 1982. Home range use by Kloss gibbons *(Hylobates klossii)* on Siberut Island, Indonesia. Anim. Behav. 30:182–98.

Whittow, G. C., E. Gould, and D. Rand. 1977. Body temperature, oxygen consumption, and evaporative water loss in a primitive insectivore, the moon rat, *Echinosorex gymnurus*. J. Mamm. 58:233–35.

Whitworth, M. R. 1984. Maternal care and behavioural development in pikas *Ochotona princeps*. Anim. Behav. 32:743–52.

Wickler, W., and U. Seibt. 1976. Field studies of the African fruit bat *Epomophorus wahlbergi* (Sundevall), with special reference to male calling. Z. Tierpsychol. 40:345–76.

Widholzer, F. L., and W. A. Voss. 1978. Breeding the giant anteater *Myrmecophaga tridactyla* at Sao Leopoldo Zoo. Internatl. Zoo Yearbook 18:122–23.

Wiegl, P. D. 1974. Study of the northern flying squirrel, *Glaucomys sabrinus*, by temperature telemetry. Amer. Midl. Nat. 92:482–86.

Wiig, O. 1983. On the relationship of pinnipeds to other carnivores. Zool. Scripta 12:225–27.

Wiig, O., and R. W. Lie. 1984. An analysis of the morphological relationships between the hooded seals *(Cystophora cristata)* of Newfoundland, the Denmark Strait, and Jan Mayen. J. Zool. 203:227–40.

Wiles, G. J., and N. H. Payne. 1986. The trade in fruit bats *Pteropus* spp. on Guam and other Pacific islands. Biol. Conserv. 38:143–61.

Wiley, R. W. 1980. *Neotoma floridana*. Mammalian Species, no. 139, 7 pp.

Wilhelmson, M. 1983. Moose as a source of meat in Sweden. Spam 26(1):1–4.

Wilkins, K. T. 1987. *Lasiurus seminolus*. Mammalian Species, no. 280, 5 pp.

Wilkinson, G. S. 1985a. The social organization of the common vampire bat. I. Pattern and cause of association. Behav. Ecol. Sociobiol. 17:111–21.

———. 1985b. The social organization of the common vampire bat. II. Mating system, genetic structure, and relatedness. Behav. Ecol. Sociobiol. 17:123–34.

———. 1988. Social organization and behavior. *In* Greenhall and Schmidt (1988), pp. 85–97.

Williams, C. F. 1986. Social organization of the bat, *Carollia perspicillata* (Chiroptera: Phyllostomidae). Ethology 71:265–82.

Williams, D. F. 1975. Distributional records for shrew-moles, *Neurotrichus gibbsii*. Murrelet 56:2–3.

———. 1978a. Karyological affinities of the species groups of silky pocket mice (Rodentia, Heteromyidae). J. Mamm. 59:599–612.

———. 1978b. Taxonomic and karyologic comments on small brown bats, genus *Eptesicus*, from South America. Ann. Carnegie Mus. 47:361–83.

———. 1986. Mammalian species of special concern in California. California Dept. Fish and Game, 112 pp.

Williams, D. F., J. D. Druecker, and H. L. Black. 1970. The karyotype of *Euderma maculatum* and comments on the evolution of the plecotine bats. J. Mamm. 51:602–6.

Williams, D. F., and M. A. Mares. 1978. A new genus and species of phyllotine rodent (Mammalia: Muridae) from northwestern Argentina. Ann. Carnegie Mus. 47:193–221.

Williams, E. E., and K. F. Koopman. 1952. West Indian fossil monkeys. Amer. Mus. Novit., no. 1546, 16 pp.

Williams, S. L. 1982. Phalli of Recent genera

and species of the family Geomyidae (Mammalia: Rodentia). Bull. Carnegie Mus. Nat. Hist., no. 20, 62 pp.

Williams, S. L., and R. J. Baker. 1974. *Geomys arenarius*. Mammalian Species, no. 36, 3 pp.

Williams, S. L., and H. H. Genoways. 1980a. Results of the Alcoa Foundation–Suriname Expeditions. II. Additional records of bats (Mammalia: Chiroptera) from Suriname. Ann. Carnegie Mus. 49:213–36.

———. 1980b. Results of the Alcoa Foundation–Suriname Expeditions. IV. A new species of bat of the genus *Molossops* (Mammalia: Molossidae). Ann. Carnegie Mus. 49:487–98.

———. 1980c. Morphological variation in the southeastern pocket gopher, *Geomys pinetis* (Mammalia: Rodentia). Ann. Carnegie Mus. 49:405–53.

Williams, S. L., H. H. Genoways, and J. A. Groen. 1983. Results of the Alcoa Foundation–Suriname Expeditions. VII. Records of mammals from central and southern Suriname. Ann. Carnegie Mus. 52:329–36.

Williams, S. L., and J. Ramirez-Pulido. 1984. Morphometric variation in the volcano mouse, *Peromyscus (Neotomodon) alstoni* (Mammalia: Cricetidae). Ann. Carnegie Mus. 53:163–83.

Williams, S. L., J. Ramirez-Pulido, and R. J. Baker. 1985. *Peromyscus alstoni*. Mammalian Species, no. 242, 4 pp.

Williams, T. 1986. The final ferret fiasco. Audubon 88(3):111–19.

Williams, T. C., L. C. Ireland, and J. M. Williams. 1973. High altitude flights of the free-tailed bat, *Tadarida brasiliensis*, observed with radar. J. Mamm. 54:807–21.

Williams, T. M., and D. C. Heard. 1986. World status of wild *Rangifer* populations. Rangifer, Spec. Issue, no. 1, pp. 19–28.

Williamson, D., and J. Williamson. 1988. Habitat use and ranging behaviour of Kalahari gemsbok. In Dixon and Jones (1988), pp. 114–18.

Willig, M. R. 1983. Composition, microgeographic variation, and sexual dimorphism in caatingas and cerrado bat communities from northeast Brazil. Bull. Carnegie Mus. Nat. Hist., no. 23, 131 pp.

———. 1985a. Ecology, reproductive biology, and systematics of *Neoplatymops mattogrossensis* (Chiroptera: Molossidae). J. Mamm. 66:618–28.

———. 1985b. Reproductive patterns of bats from caatingas and cerrado biomes in northeast Brazil. J. Mamm. 66:668–81.

Willig, M. R., and R. R. Hollander. 1987. *Vampyrops lineatus*. Mammalian Species, no. 275, 4 pp.

Willig, M. R., and J. K. Jones, Jr. 1985. *Neoplatymops mattogrossensis*. Mammalian Species, no. 244, 3 pp.

Willner, G. R., J. A. Chapman, and J. R. Goldsberry. 1975. A study and review of muskrat food habits with special reference to Maryland. Maryland Dept. Nat. Res. Publ. Wildl. Ecol., no. 1, 25 pp.

Willner, G. R., J. A. Chapman, and D. Pursley. 1979. Reproduction, physiological responses, food habits, and abundance of nutria on Maryland marshes. Wildl. Monogr., no. 65, 43 pp.

Willner, G. R., K. R. Dixon, and J. A. Chapman. 1983. Age determination and mortality of the nutria *(Myocastor coypus)* in Maryland, U.S.A. Z. Saugetierk. 48:19–34.

Willner, G. R., G. A. Feldhamer, E. E. Zucker, and J. A. Chapman. 1980. *Ondatra zibethicus*. Mammalian Species, no. 141, 8 pp.

Willoughby, D. P. 1974. The empire of *Equus*. Thomas Yoseloff, London, 475 pp.

Wilson, B. A., and A. R. Bourne. 1984. Reproduction in the male dasyurid *Antechinus minimus maritimus* (Marsupialia: Dasyuridae). Austral. J. Zool. 32:311–18.

Wilson, D(avid). E., and S. M. Hirst. 1977. Ecology and factors limiting roan and sable antelope populations in South Africa. Wildl. Monogr., no. 54, 111 pp.

Wilson, D(on). E. 1971. Ecology of *Myotis nigricans* (Mammalia: Chiroptera) on Barro Colorado Island, Panama Canal Zone. J. Zool. 163:1–13.

———. 1976. The subspecies of *Thyroptera discifera* (Lichtenstein and Peters). Proc. Biol. Soc. Washington 89:305–12.

———. 1979. Reproductive patterns. In Baker, Jones, and Carter (1979), pp. 317–78.

Wilson, D(on). E., and E. L. Tyson. 1970. Longevity records for *Artibeus jamaicensis* and *Myotis nigricans*. J. Mamm. 51:203.

Wilson, M., M. Daly, and P. Behrends. 1985. The estrous cycle of two species of kangaroo rats (*Dipodomys microps* and *D. merriami*). J. Mamm. 66:726–32.

Wilson, P. 1985. Maeda's *Miniopterus* taxonomy. Macroderma 1:29–36.

Wilson, P., and W. L. Franklin. 1985. Male group dynamics and inter-male aggression of guanacos in southern Chile. Z. Tierpsychol. 69:305–28.

Wilson, V. J. 1966a. Notes on the food and feeding habits of the common duiker, *Sylvicapra grimmia* in eastern Zambia. Arnoldia 2(14):1–19.

———. 1966b. Observations on Lichtenstein's hartebeest, *Alcelaphus lichtensteini* over a three-year period, and their response to various tsetse control measures in eastern Zambia. Arnoldia 2(15):1–14.

———. 1969. Eland, *Taurotragus oryx*, in eastern Zambia. Arnoldia 4(12):1–9.

———. 1975. Mammals of the Wankie National Park. Trustees Natl. Museums and Monuments Rhodesia Mus. Mem., no. 5, i + 147 pp.

Wilson, V. J., and G. Child. 1965. Notes on klipspringer from tsetse fly control areas in eastern Zambia. Arnoldia 1(35):1–9.

Wilson, V. J., and M. A. Kerr. 1969. Brief notes on reproduction in Steenbok, *Raphicerus campestris*, Thunberg. Arnoldia 4(23):1–5.

Wilsson, L. 1971. Observations and experiments on the ethology of the European beaver (*Castor fiber* L.). Viltrevy 8:115–266.

Winchester, B. H., R. S. Delotelle, J. R. Newman, and J. T. McClave. 1978. Ecology and management of the colonial pocket gopher: a progress report. In Odom, R. R., and L. Landers, eds., Proceedings of the rare and endangered wildlife symposium, Georgia Dept. Nat. Resources, pp. 173–84.

Winkel, K., and I. Humphrey-Smith. 1987. Diet of the marsupial mole, *Notoryctes typhlops* (Stirling 1889) (Marsupialia: Notoryctidae). Austral. Mamm. 11:159–61.

Winkler, P., H. Loch, and C. Vogel. 1984. Life history of hanuman langurs *(Presbytis entellus)*. Folia Primatol. 43:1–23.

Winn, H. E., R. K. Edel, and A. G. Taruski. 1975. Population estimate of the humpback whale *(Megaptera novaeangliae)* in the West Indies by visual and acoustic techniques. J. Fish. Res. Bd. Can. 32:499–506.

Winn, H. E., and B. L. Olla, eds. 1979. Behavior of marine animals. Volume 3: cetaceans. Plenum Press, New York, xix + 438 pp.

Winn, H. E., P. J. Perkins, and L. Winn. 1970. Sounds and behavior of the northern bottle-nosed whale. Proc. Ann. Conf. Biol. Sonar and Diving Mammals 7:53–59.

Winn, H. E., C. A. Price, and P. W. Sorensen. 1986. The distributional biology of the right whale *(Eubalaena glacialis)* in the western North Atlantic. In Brownell, Best, and Prescott (1986), pp. 129–37.

Winn, H. E., and N. E. Reichley. 1985. Humpback whale—*Megaptera novaeangliae*. In Ridgway and Harrison (1985), pp. 241–73.

Winn, H. E., T. J. Thompson, W. C. Cummings, J. Hain, J. Hudnall, H. Hays, and W. W. Steiner. 1981. Song of the humpback whale—population comparisons. Behav. Ecol. Sociobiol. 8:41–46.

Winn, H. E., and L. K. Winn. 1978. The song of the humpback whale *(Megaptera novaeangliae)* in the West Indies. Mar. Biol. 47:97–114.

Winter, J. W. 1979. The status of endangered Australian Phalangeridae, Petauridae, Burramyidae, Tarsipedidae and the Koala. *In* Tyler (1979), pp. 45–59.

———. 1984a. The Thornton Peak melomys, *Melomys hadrourus* (Rodentia: Muridae): a new rainforest species from northeastern Queensland, Australia. Mem. Queensland Mus. 21:519–39.

———. 1984b. Conservation studies of tropical rainforest possums. *In* Smith and Hume (1984), pp. 469–81.

Winter, P. 1968. Social communication in the squirrel monkey. *In* Rosenblum and Cooper (1968), pp. 235–53.

Wirth, R., and C. Groves. 1988. A new deer in China. Species 11:15.

Wirtz, P. 1982. Territory holders, satellite males and bachelor males in a high density population of waterbuck *(Kobus ellipsiprymnus)* and their associations with conspecifics. Z. Tierpsychol. 58:277–300.

Wirtz, W. O., II. 1972. Population ecology of the Polynesian rat, *Rattus exulans*, on Kure Atoll, Hawaii. Pacific Sci. 26:433–64.

———. 1973. Growth and development of *Rattus exulans*. J. Mamm. 54:189–202.

Wishart, W. 1978. Bighorn sheep. *In* Schmidt and Gilbert (1978), pp. 161–71.

Wistrand, H. 1974. Individual, social, and seasonal behavior of the thirteen-lined ground squirrel *(Spermophilus tridecemlineatus)*. J. Mamm. 55:329–47.

Withers, P. C. 1983. Seasonal reproduction by small mammals of the Namib Desert. Mammalia 47:195–204.

Witte, G. R. 1971. Jungentransport in den Backentaschen beim Syrischen Goldhamster *(Mesocricetus auratus* Waterhouse, 1839). Z. Saugetierk. 36:216–19.

Wodzicki, K., and H. Felten. 1975. The peka, or fruit bat *(Pteropus tonganus tonganus)* (Mammalia, Chiroptera), of Niue Island, South Pacific. Pacific Sci. 29:131–38.

———. 1981. Fruit bats of the genus *Pteropus* from the islands Rarotonga and Mangaia, Cook Islands, Pacific Ocean. Senckenberg. Biol. 61:143–51.

Wodzicki, K., and J. E. C. Flux. 1971. The parma wallaby and its future. Oryx 11:40–47.

Wolfe, J. L. 1982. *Oryzomys palustris.* Mammalian Species, no. 176, 5 pp.

———. 1985. Population ecology of the rice rat *(Oryzomys palustris)* in a coastal marsh. J. Zool. 205:235–44.

Wolfe, J. L., and A. V. Linzey. 1977. *Peromyscus gossypinus.* Mammalian Species, no. 70, 5 pp.

Wolfe, L. D., and E. H. Peters. 1987. History of the free-ranging rhesus monkeys *(Macaca mulatta)* of Silver Springs. Florida Sci. 50:234–45.

Wolfe, M. L., and D. L. Allen. 1973. Continued studies of the status, socialization, and relationships of Isle Royale wolves, 1967 to 1970. J. Mamm. 54:611–33.

Wolff, J. O., and G. C. Bateman. 1978. Effects of food availability and ambient temperature on torpor cycles of *Perognathus flavus* (Heteromyidae). J. Mamm. 59:707–16.

Wolfheim, J. H. 1983. Primates of the world. Distribution, abundance, and conservation. Univ. Washington Press, Seattle, xxiii + 831 pp.

Wolman, A. A. 1978. Humpback whale. *In* Haley (1978), pp. 46–53.

Wolman, A. A., and D. W. Rice. 1979. Current status of the gray whale. Rept. Internatl. Whaling Comm. 29:275–79.

Wolton, R. J., P. A. Arak, H. C. J. Godfray, and R. P. Wilson. 1982. Ecological and behavioural studies of the Megachiroptera at Mount Nimba, Liberia, with notes on Microchiroptera. Mammalia 46:419–48.

Womochel, D. R. 1978. A new species of *Allactaga* (Rodentia: Dipodidae) from Iran. Fieldiana Zool. 72:65–73.

Won, Pyong-Oh. 1968. Notes on the first propagating record of the Palaearctic flying squirrel, *Pteromys volans aluco* (Thomas) from Korea. J. Mamm. Soc. Japan 4:40–43.

Wood, A. E. 1977. The evolution of the rodent family Ctenodactylidae. J. Paleontol. Soc. India 20:120–37.

———. 1985. The relationships, origin and dispersal of the hystricognathous rodents. *In* Luckett and Hartenberger (1985), pp. 475–514.

Wood, D. H. 1970. An ecological study of *Antechinus stuartii* (Marsupialia) in a southeast Queensland rain forest. Austral. J. Zool. 18:185–207.

———. 1971. The ecology of *Rattus fuscipes* and *Melomys cervinipes* (Rodentia: Muridae) in a south-east Queensland rain forest. Austral. J. Zool. 19:371–92.

Wood, G. W., and R. H. Barrett. 1979. Status of wild pigs in the United States. Wildl. Soc. Bull. 7:237–46.

Wood, G. W., and R. E. Brenneman. 1980. Feral hog movements and habitat use in coastal South Carolina. J. Wildl. Mgmt. 44:420–27.

Woodburne, M. O., and W. J. Zinsmeister. 1982. Fossil land mammal from Antarctica. Science 218:284–86.

———. 1984. The first land mammal from Antarctica and its biogeographic implications. J. Paleontol. 58:913–48.

Woodman, N. 1988. Subfossil remains of *Per-*

omyscus stirtoni (Mammalia: Rodentia) from Costa Rica. Rev. Biol. Tropical 36:247–53.

Woods, C. A. 1973. *Erethizon dorsatum.* Mammalian Species, no. 29, 6 pp.

———. 1981. Last endemic mammals in Hispaniola. Oryx 16:146–52.

———. 1984. Hystricognath rodents. *In* Anderson and Jones (1984), pp. 389–446.

———. 1985. Evolution and systematics of endemic Antillean rodents. Acta Zool. Fennica 170:199–200.

———. 1989. Endemic rodents of the West Indies: the end of a splendid isolation. *In* Lidicker (1989), pp. 11–19.

Woods, C. A., and D. K. Boraker. 1975. *Octodon degus.* Mammalian Species, no. 67, 5 pp.

Woods, C. A., J. A. Ottenwalder, and W. L. R. Oliver. 1985. Lost mammals of the Greater Antilles: the summarised findings of a ten weeks field survey in the Dominican Republic, Haiti and Puerto Rico. Dodo 22:23–42.

Woods, C. A., W. Post, and C. W. Kilpatrick. 1982. *Microtus pennsylvanicus* (Rodentia: Muridae) in Florida: a Pleistocene relict in a coastal saltmarsh. Bull. Florida State Mus., Biol. Sci., 28:25–52.

Woodsworth, G. C., G. P. Bell, and M. B. Fenton. 1981. Observations of the echolocation, feeding behaviour, and habitat use of *Euderma maculatum* (Chiroptera: Vespertilionidae) in southcentral British Columbia. Can. J. Zool. 59:1099–1102.

Woodward, S. L. 1979. The social system of feral asses *(Equus asinus).* Z. Tierpsychol. 49:304–16.

Woolard, P., W. J. M. Vestjens, and L. MacLean. 1978. The ecology of the eastern water rat *Hydromys chrysogaster* at Griffith, N.S.W.: food and feeding habits. Austral. Wildl. Res. 5:59–73.

Woolf, A., T. O'Shea, and D. L. Gilbert. 1970. Movements and behavior of bighorn sheep on summer ranges in Yellowstone National Park. J. Wildl. Mgmt. 34:446–50.

Woolley, P. 1971a. Observations on the reproductive biology of the dibbler, *Antechinus apicalis* (Marsupialia: Dasyuridae). J. Roy. Soc. W. Austral. 54:99–102.

———. 1971b. Maintenance and breeding of laboratory colonies of *Dasyuroides byrnei* and *Dasycercus cristicauda.* Internatl. Zoo Yearbook 11:351–54.

———. 1977. In search of the dibbler, *Antechinus apicalis* (Marsupialia: Dasyuridae). J. Roy. Soc. W. Austral. 59:111–17.

———. 1980. Further searches for the dibbler, *Antechinus apicalis* (Marsupialia: Dasyuridae). J. Roy. Soc. W. Austral. 63:47–52.

———. 1982a. Phallic morphology of the

Australian species of *Antechinus* (Dasyuridae, Marsupialia): a new taxonomic tool? *In* Archer (1982a), pp. 767–81.

————. 1982b. Observations on the feeding and reproductive status of captive feather-tailed possums, *Distoechurus pennatus* (Marsupialia: Burramyidae). Austral. Mamm. 5:285–87.

————. 1984. Reproduction in *Antechinomys laniger* ("spenceri" form) (Marsupialia: Dasyuridae): field and laboratory investigations. Austral. Wildl. Res. 11:481–89.

Woolley, P., and A. Valente. 1982. The dibbler, *Parantechinus apicalis* (Marsupialia: Dasyuridae): failure to locate populations in four regions in the south of Western Australia. Austral. Mamm. 5:241–45.

Wozencraft, W. C. 1986. A new species of striped mongoose from Madagascar. J. Mamm. 67:561–71.

————. 1987. Emendation of species name. J. Mamm. 68:198.

Wrangham, R. W. 1977. Feeding behavior of chimpanzees in Gombe National Park, Tanzania. *In* Clutton-Brock (1977), pp. 503–37.

Wright, P. C., D. Haring, E. L. Simons, and P. Andau. 1987. Tarsiers: a conservation perspective. Primate Conserv. 8:51–54.

Wright, P. C., M. K. Izard, and E. L. Simons. 1986. Reproductive cycles in *Tarsius bancanus*. Amer. J. Primatol. 11:207–15.

Wright, P. G. 1984. Why do elephants flap their ears? S. Afr. J. Zool. 19:266–69.

Wrigley, R. E. 1972. Systematics and biology of the woodland jumping mouse, *Napaeozapus insignis*. Illinois Biol. Monogr., no. 47, 117 pp.

Wrigley, R. E., J. E. Dubois, and H. W. R. Copland. 1979. Habitat, abundance, and distribution of six species of shrews in Manitoba. J. Mamm. 60:505–20.

Wrigley, R. E., and D. R. M. Hatch. 1976. Arctic fox migrations in Manitoba. Arctic 29:147–58.

Wroughton, R. C. 1912. Some new Indian rodents. J. Bombay Nat. Hist. Soc. 21:338–42.

Wu Deling and Deng Xiangfu. 1984. A new species of tree mice from Yunnan, China. Acta Theriol. Sinica 4:206–12.

Wu Deling and Wang Guanghuan. 1984. A new subspecies of *Typhlomys cinereus* Milne-Edwards from Yunnan, China. Acta Theriol. Sinica 4:213–15.

Wu Jia-yan. 1985. Systematics and distribution of Chinese takin. *In* Kawamichi (1985), pp. 95–100.

Würsig, B. 1982. Radio tracking dusky porpoises in the South Atlantic. Mammals in the Seas, FAO Fish. Ser. No. 5, 4:145–60.

————. 1986. Delphinid foraging strategies. *In* Schusterman, Thomas, and Wood (1986), pp. 347–59.

Würsig, B., and R. Bastida. 1986. Long-range movement and individual associations of two dusky dolphins *(Lagenorhynchus obscurus)* off Argentina. J. Mamm. 67:773–74.

Würsig, B., E. M. Dorsey, M. A. Fraker, R. S. Payne, and W. J. Richardson. 1985. Behavior of bowhead whales, *Balaena mysticetus*, summering in the Beaufort Sea: a description. Fishery Bull. 83:357–77.

Würsig, B., and M. Würsig. 1980. Behavior and ecology of the dusky dolphin, *Lagenorhynchus obscurus*, in the South Atlantic. Fishery Bull. 77:871–90.

Wyss, A. R. 1987. The walrus auditory region and the monophyly of pinnipeds. Amer. Mus. Novit., no. 2871, 31 pp.

————. 1988a. Evidence from flipper structure for a single origin of pinnipeds. Nature 334:427–28.

————. 1988b. On "retrogression" in the evolution of the Phocinae and phylogenetic affinities of the monk seals. Amer. Mus. Novit., no. 2924, 38 pp.

X

Xanten, W. A., H. Kafka, and E. Olds. 1976. Breeding the binturong *Arctictis binturong* at the National Zoological Park, Washington. Internatl. Zoo Yearbook 16:117–19.

Xia Jingshi. 1986. Some records of mi-lu *(Elaphurus davidianus)* in Chinese ancient books. Acta Theriol. Sinica 6:267–72.

Xia Wuping. 1985. A study on Chinese *Apodemus* and its relation to Japanese species. *In* Kawamichi (1985), pp. 76–79.

Xiang Peilon, Tan Bangjie, and Jia Xianggang. 1987. South China tiger recovery program. *In* Tilson and Seal (1987), pp. 323–28.

Ximenez, A. 1975. *Felis geoffroyi*. Mammalian Species, no. 54, 4 pp.

————. 1980. Notas sobre el genéro *Cavia* Pallas con la descripcion de *Cavia magna* sp. n. (Mammalia—Caviidae). Rev. Nordest. Biol. 3 (Spec. Issue):145–79.

Ximenez, A., and A. Langguth. 1970. *Akodon cursor montensis* en el Uruguay (Mammalia—Cricetinae). Comun. Zool. Mus. Hist. Nat. Montevideo 10(128):1–7.

Ximenez, A., A. Langguth, and R. Praderi. 1972. Lista sistemática de los mamíferos del Uruguay. An. Mus. Nac. Hist. Nat. Montevideo 2(5):1–49.

Y

Yablokov, A. V., and L. S. Bogoslovskaya. 1984. A review of Russian research on the biology and commercial whaling of the gray whale. *In* Jones, Swartz, and Leatherwood (1984), pp. 465–85.

Yager, R. H., and C. B. Frank. 1972. The nine-banded armadillo for medical research. Inst. Lab. Anim. Res. News 15(2):4–5.

Yahner, R. H. 1978. Burrow system and home range use by eastern chipmunks, *Tamias striatus*: ecological and behavioral considerations. J. Mamm. 59:324–29.

————. 1980a. Activity patterns of captive Reeve's muntjacs *(Muntiacus reevesi)*. J. Mamm. 61:368–71.

————. 1980b. Barking in a primitive ungulate, *Muntiacus reevesi*: function and adaptiveness. Amer. Nat. 116:157–77.

Yahr, P. 1977. Social subordination and scent-marking in male Mongolian gerbils *(Meriones unguiculatus)*. Anim. Behav. 25:292–97.

Yalden, D. W. 1975. Some observations on the giant mole-rat *Tachyoryctes macrocephalus* (Ruppell, 1842) (Mammalia: Rhizomyidae) of Ethiopia. Italian J. Zool., Suppl., n.s., 15:275–303.

————. 1978. A revision of the dik-diks of the subgenus *Madoqua (Madoqua)*. Italian J. Zool., Suppl., n.s., 11:245–64.

Yalden, D. W., M. J. Largen, and D. Kock. 1976. Catalogue of the mammals of Ethiopia. 2. Insectivora and Rodentia. Italian J. Zool., Suppl., n.s., 8:1–118.

————. 1977. Catalogue of the mammals of Ethiopia. 3. Primates. Italian J. Zool., Suppl., n.s., 9:1–52.

————. 1980. Catalogue of the mammals of Ethiopia. 4. Carnivora. Italian J. Zool., Suppl., n.s., 13:169–272.

————. 1984. Catalogue of the mammals of Ethiopia. 5. Artiodactyla. Italian J. Zool., Suppl., n.s., 19:67–221.

————. 1986. Catalogue of the mammals of Ethiopia. 6. Perissodactyla, Proboscidea, Hyracoidea, Lagomorpha, Tubulidentata, Sirenia and Cetacea. Italian J. Zool., Suppl., n.s., 21:31–103.

Yamamoto, S. 1967. Breeding Japanese serows *Capricornis crispus* in captivity. Internatl. Zoo Yearbook 7:174–75.

Yáñez, J. L., J. C. Torres-Mura, J. R. Rau, and L. C. Contreras. 1987. New records and current status of *Euneomys* (Cricetidae) in southern South America. Fieldiana Zool., n.s., 39:283–87.

Yang Guangrong and Wang Yingxiang. 1987. A new subspecies of *Hadromys humei* (Muridae, Mammalia) from Yunnan, China. Acta Theriol. Sinica 7:46–50.

Yao-ting, G. 1983. Current studies on the Chinese Yarkand hare. Acta Zool. Fennica 174:23–25.

Yates, S. 1986. A pronghorn needs freedom to feel at home on the range. Smithsonian 17(9):86–95.

Yates, T. L. 1984. Insectivores, elephant shrews, tree shrews, and dermopterans. In Anderson and Jones (1984), pp. 117–44.

Yates, T. L., R. J. Baker, and R. K. Barnett. 1979. Phylogenetic analysis of karyological variation in three genera of peromyscine rodents. Syst. Zool. 28:40–48.

Yates, T. L., and I. F. Greenbaum. 1982. Biochemical systematics of North American moles (Insectivora: Talpidae). J. Mamm. 63:368–74.

Yates, T. L., and D. J. Schmidly. 1977. Systematics of Scalopus aquaticus (Linnaeus) in Texas and adjacent states. Occas. Pap. Mus. Texas Tech Univ., no. 45, 36 pp.

———. 1978. Scalopus aquaticus. Mammalian Species, no. 105, 4 pp.

Yates, T. L., D. J. Schmidly, and K. L. Culbertson. 1976. Silver-haired bat in Mexico. J. Mamm. 57:205.

Yeaton, R. I. 1972. Social behavior and social organization in Richardson's ground squirrel (Spermophilus richardsonii) in Saskatchewan. J. Mamm. 53:139–47.

Yenbutra, S., and H. Felten. 1983. A new species of the fruit bat genus Megaerops from SE-Asia (Mammalia: Chiroptera: Pteropodidae). Senckenberg. Biol. 64:1–11.

Yensen, E., and W. H. Clark. 1986. Records of desert shrews (Notiosorex crawfordi) from Baja California, Mexico. Southwestern Nat. 31:529–30.

Yoakum, J. D. 1978. Pronghorn. In Schmidt and Gilbert (1978), pp. 103–21.

Yochem, P. K., and S. Leatherwood. 1985. Blue whale—Balaenoptera musculus. In Ridgway and Harrison (1985), pp. 193–240.

Yocom, C. F. 1974. Status of marten in northern California, Oregon and Washington. California Fish and Game 60:54–57.

Yocom, C. F., and M. T. McCollum. 1973. Status of the fisher in northern California, Oregon and Washington. California Fish and Game 59:305–9.

Yoder, A. D. 1989. Survey of cheirogaleid primates in Madagascar. Unpubl. Rept. to World Wildl. Fund, 7 pp.

Yong Hoi-Sen. 1970. A Malayan view of Rattus edwardsi and R. sabanus (Rodentia: Muridae). Zool. J. Linnean Soc. 49:359–69.

Yoshiyuki, M. 1979. A new species of the genus Ptenochirus (Chiroptera, Pteropodidae) from the Philippine Islands. Bull. Natl. Sci. Mus. (Tokyo), ser. A, 5:75–81.

———. 1983. A new species of Murina from Japan (Chiroptera, Vespertilionidae). Bull. Natl. Sci. Mus. (Tokyo), ser. A, 9:141–50.

———. 1984. A new species of Myotis (Chiroptera, Vespertilionidae) from Hokkaido, Japan. Bull. Natl. Sci. Mus. (Tokyo), ser. A, 10:153–58.

Youngman, P. M. 1975. Mammals of the Yukon Territory. Natl. Mus. Can. Publ. Zool., no. 10, 192 pp.

———. 1982. Distribution and systematics of the European mink Mustela lutreola Linnaeus 1761. Acta Zool. Fennica 166:1–48.

Yurick, D. B., and D. E. Gaskin. 1988. Asymmetry in the skull of the harbour porpoise Phocoena phocoena (L.) and its relationship to sound production and echolocation. Can. J. Zool. 66:399–402.

Z

Zaitsev, M. V. 1984. A contribution to the taxonomy and diagnostics of the subgenus Erinaceus (Mammalia, Erinaceinae) of the fauna of the USSR. Zool. Zhur. 63:720–30.

Zannier, F. 1965. Verhaltensuntersuchungen an der Zwergmanguste Helogale undulata rufula in Zoologischen Garten Frankfurt am Main. Z. Tierpsychol. 22:672–95.

Zara, J. L. 1973. Breeding and husbandry of the capybara Hydrochoerus hydrochaeris at Evansville Zoo. Internatl. Zoo Yearbook 13:137–39.

Zheng Yonglie and Xu Longhui. 1983. Subspecific study on the ferret badger (Melogale moschata) in China, with description of a new subspecies. Acta Theriol. Sinica 3:165–71.

Zhou Kaiya. 1982. Classification and phylogeny of the superfamily Platanistoidea, with notes on evidence of the monophyly of the Cetacea. Sci. Rept. Whales Res. Inst. 34:93–108.

———. 1986. The ringed seal and other pinnipeds wandering off the coast of China. Acta Theriol. Sinica 6:107–13.

Zhou Kaiya, G. Pilleri, and Li Yuemin. 1979. Observations on the baiji (Lipotes vexillifer) and the finless porpoise (Neophocaena asiaeorientalis) in the Changjiang (Yangtze) River between Nanjing and Taiyangzhou, with remarks on some physiological adaptations of the baiji to its environment. Investig. Cetacea 10:109–20.

———. 1980. Observations on baiji (Lipotes vexillifer) and finless porpoise (Neophocaena asiaeorientalis) in the lower reaches of the Chang Jiang. Sci. Sinica 23:785–94.

Zhou Kaiya and Qian Weijuan. 1985. Distribution of the dolphins of the genus Tursiops in the China Seas. Aquatic Mamm. 11:16–19.

Zhou Kaiya, Qian Weijuan, and Li Yuemin. 1977. Studies on the distribution of baiji, Lipotes vexillifer Miller. Acta Zool. Sinica 23:72–79.

———. 1979. The osteology and the systematic position of the baiji, Lipotes vexillifer. Acta Zool. Sinica 25:58–74.

Ziegler, A. C. 1972. Additional specimens of Planigale novaeguineae (Dasyuridae: Marsupialia) from Territory of Papua. Austral. Mamm. 1:43–45.

———. 1977. Evolution of New Guinea's marsupial fauna in response to a forested environment. In Stonehouse and Gilmore (1977), pp. 117–38.

———. 1981. Petaurus abidi, a new species of glider (Marsupialia: Petauridae) from Papua New Guinea. Austral. Mamm. 4:81–88.

———. 1982a. The Australo-Papuan genus Syconycteris (Chiroptera: Pteropodidae) with the description of a new Papua New Guinea species. Occas. Pap. Bishop Mus. 25(5):1–22.

———. 1982b. An ecological check-list of New Guinea Recent mammals. Monogr. Biol. 42:863–94.

———. 1984. A Papua New Guinea specimen of Hydromys hussoni Musser and Piik, 1982 (Rodentia: Muridae). Austral. Mamm. 7:101–5.

Zihlman, A. 1984. Body build and tissue composition in Pan paniscus and Pan troglodytes, with comparisons to other hominoids. In Susman (1984a), pp. 179–200.

Zihlman, A., and J. M. Lowenstein. 1979. False start of the human parade. Nat. Hist. 88(7):86–91.

Zimen, E. 1975. Social dynamics of the wolf pack. In Fox (1975), pp. 336–62.

———. 1981. Italian wolves. Nat. Hist. 90(2):66–81.

Zimina, R. P., and I. P. Gerasimov. 1973. The periglacial expansion of marmots (Marmota) in middle Europe during late Pleistocene. J. Mamm. 54:327–40.

Zimina, R. P., and E. V. Yasny. 1977. Observations on the ecology of Prometheomys schaposchnikovi Sat. Bull. Moscow Soc. Nat., Biol. Ser., 82:24–30.

Zimmermann, E. 1985. Vocalizations and associated behaviours in adult slow loris (Nycticebus coucang). Folia Primatol. 44:52–64.

Zubaid, A. 1988. The second record of a trident horseshoe bat, Aselliscus stoliczkanus (Hipposiderinae) from peninsular Malaysia. Malayan Nat. J. 42:29–30.

Zumbaugh, D. M., and J. R. Choate. 1985. Historical biogeography of foxes in Kansas. Trans. Kansas Acad. Sci. 88:1–13.

Zyadi, F. 1988. Répartition de Gerbillus hoogstrali Lay, 1975 (Rongeurs, Gerbillidés) au sud du Maroc. Mammalia 52:132–33.

Index

The scientific names of orders, families, and genera, which have titled accounts in the text, are in boldface type. The page numbers on which such accounts begin are also in boldface type. Other scientific names, and vernacular names, appear in ordinary type. If the scientific and common names of a mammal are identical, they usually are indexed separately (for example, **Addax** and Addax). If, however, the scientific name does not qualify for boldface type, and the common name is used in the singular, the two names are indexed as one (for example, Anoa). Names in legends of illustrations usually are not indexed independently unless an illustration appears separately from the main text on the subject; usually in such a case only the scientific name is indexed. If no illustration of a genus appears on the page indexed for that genus, the reader should look one or two pages either before or after the account. However, not every account has an accompanying illustration (see preface).

Designed by Glen Burris
Set in Aldus text and Helvetica display by The Composing Room of Michigan, Inc.
Printed on 70-lb. S. D. Warren Lustro Offset Enamel and bound in Holliston
Aqualite cloth by The Maple Press Company

GEOLOGICAL TIME

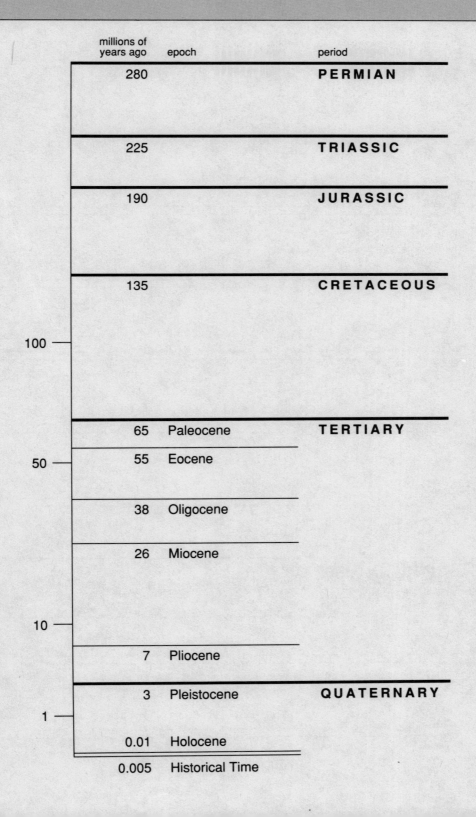

millions of years ago	epoch	period
280		**PERMIAN**
225		**TRIASSIC**
190		**JURASSIC**
135		**CRETACEOUS**
100		
65	Paleocene	**TERTIARY**
55	Eocene	
50		
38	Oligocene	
26	Miocene	
10		
7	Pliocene	
3	Pleistocene	**QUATERNARY**
1		
0.01	Holocene	
0.005	Historical Time	

WEIGHT
scales for comparison of metric and U.S. units of measurement

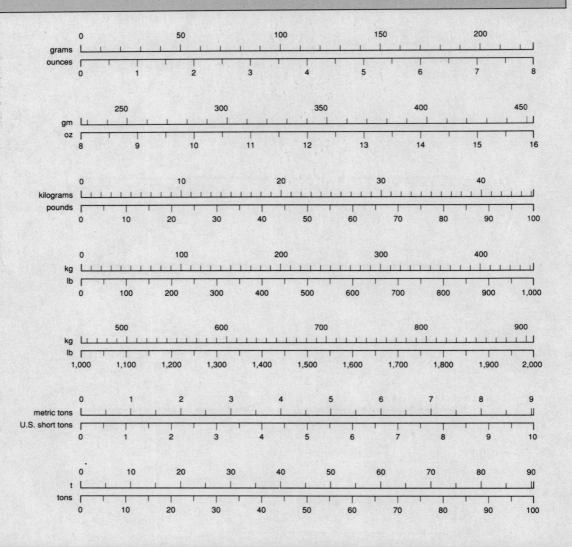

TEMPERATURE
scales for comparison of metric and U.S. units of measurement

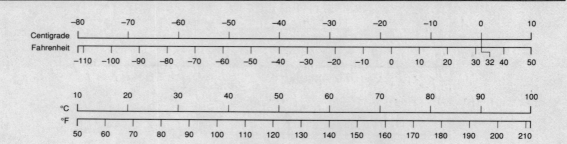